UHLIG'S CORROSION HANDBOOK

THE ELECTROCHEMICAL SOCIETY SERIES

THE ELECTROCHEMICAL SOCIETY, INC.

ECS-The Electrochemical Society
65 South Main Street
Pennington, NJ 08534-2839
http://www.electrochem.org

UHLIG'S CORROSION HANDBOOK

THIRD EDITION

Edited by

R. WINSTON REVIE
CANMET Materials Technology Laboratory
Ottawa, Ontario, Canada

THE ELECTROCHEMICAL SOCIETY, INC.

WILEY

A JOHN WILEY & SONS, INC., PUBLICATION

Published by John Wiley & Sons, Inc., Hoboken, New Jersey
Published simultaneously in Canada

For general information on our other products and services or for technical support, please contact our Customer Care Department within the United States at (800) 762-2974, outside the United States at (317) 572-3993 or fax (317) 572-4002.

Wiley also publishes its books in a variety of electronic formats. Some content that appears in print may not be available in electronic formats. For more information about Wiley products, visit our web site at www.wiley.com.

Library of Congress Cataloging-in-Publication Data:

Uhlig's corrosion handbook / edited by R. Winston Revie–3rd ed.
        p. cm. –(The ECS series of texts and monographs)
    Includes index.
    ISBN 978-0-470-08032-0

Printed in the United States of America

eBook ISBN: 978-0-470-87285-7
oBook ISBN: 978-0-470-87286-4

10 9 8 7 6 5 4

Herbert H. Uhlig
March 3, 1907–July 3, 1993

This Handbook is dedicated to the memory of Herbert H. Uhlig.

Herbert Uhlig began his career at MIT in 1936 where, with the exception of the interruption caused by World War II, he remained until his retirement nearly 40 years later, bringing the MIT Corrosion Laboratory to a level of international prominence that it retains to this day as a major center of excellence. He helped to establish the Corrosion Division of The Electrochemical Society in 1942 and served as President of the Society in 1955–1956. His characteristics as an uncompromising innovator and meticulous scientist who insisted on reliable data and on achieving results led to the success of his many endeavors as educator and mentor, including the Corrosion Handbook, published in 1948, that he conceived, organized, and edited.

EDITORIAL ADVISORY BOARD

CONTENTS

PART IV   CORROSION PROTECTION

PART VI   CORROSION MONITORING

FOREWORD

In the roughly 10 years since the appearance of the second edition of the *Corrosion Handbook,* new technologies and new engineering systems have found their way into the global marketplace at an increasing rate. It is no wonder now that a third edition would be timely and appropriate. It is also no surprise that the third edition would expand the scope of the previous editions to include chapters on composites such as are used in airframes, shape memory alloys which find application in medical devices, and electrodeposited nanocrystals as well as application-specific chapters which address the materials of construction of the engineered barriers for nuclear waste containment, ethanol-induced stress corrosion cracking of carbon steels, and other such topics. An entirely new section on corrosion monitoring also appears in the third edition.

Corrosion is ubiquitous: All engineering systems are subject to environmental degradation in service environments, whether these systems are used to meet the energy needs of the inhabitants of this planet; to provide clean air; to treat and transport water, food, and other products typical of our commercial world; to both save and improve the quality of our lives; and to ensure the readiness of those engineering systems that are of importance in terms of national defense and homeland security as well as many others. From heart stents to nuclear electric generating stations, corrosion is part of our world.

The *Corrosion Handbook* continues today, as it has since its first appearance over 60 years ago, to serve as a trusted resource to generations of corrosion engineers. There are many reasons to believe that its presence in the libraries of engineering practitioners of all kinds is greater now than ever before. First, it appears that much of the expertise in this area of technology, which resided for decades in the staff and laboratories of metal producers, has retired and is not being replaced as many of the metal producers have responded to the global economy of the past decade and more. Second, the interest of young people in engineering education, including corrosion engineering, is also in decline. Third, as the global economy recovers from the meltdown of the recent past, nations with a strong manufacturing base that creates products of value to the market will respond most quickly. But this will require an educated and informed engineering workforce. It is a concern to me that industrialized nations all over the world are on the brink of losing this technological infrastructure through retirement, the decline of traditional manufacturing industries, and declining student interest. Without a means of capturing this expertise in a useful form the next generation of engineers are going to find a gap in their knowledge base. I am confident that this volume will be of value in that context. Every industrialized nation must have the capacity and intellectual strength necessary to design, manufacture, and maintain either contemporary engineering systems or emerging engineering systems that may find their way into the marketplace of the future. The *Corrosion Handbook* remains an invaluable resource in that regard, and once again Winston Revie has assembled a world-class group of authors in producing a comprehensive volume covering the entire field of contemporary corrosion engineering.

R. M. LATANISION
February 2010

Director (Emeritus)
The H. H. Uhlig Corrosion Laboratory, MIT, and
Corporate Vice President
Exponent–Failure Analysis Associates, Inc.
Natick, Massachusetts

FOREWORD TO THE SECOND EDITION

The first and, prior to the current volume, only edition of the *Corrosion Handbook* was published in 1948. It represented a heroic effort by Professor Herbert Uhlig and the leadership of the relatively newly established Corrosion Division of the Electrochemical Society. It was intended, as Professor Uhlig recorded in the Preface to the 1948 edition, to serve as "... a convenient reference volume covering the entire field of corrosion, to bring together, in effect, much of the information scattered broadly throughout the scientific and engineering literature." Its success was equally heroic: the *Corrosion Handbook* has served generation after generation of corrosion engineer and today, more than a one-half of a century since its first appearance, the volume remains a trusted resource in the personal libraries of many of those who populate the world's engineering community.

Over the years that I knew Professor Uhlig personally, he often mentioned to me his concern for the need to produce a revised edition of the Handbook. I am confident that he would have been very pleased that one of his doctoral students at MIT, Winston Revie, had taken up this challenge. Winston, just as his mentor, is a meticulous and innovative corrosion scientist. This truly monumental revision of the *Corrosion Handbook* is certain to serve the engineering community well as we enter the new millennium. Much has happened in corrosion science and engineering since 1948, and the contributors to this volume, an assembly of the international leaders in the field, have captured these changes wonderfully well. The breadth of corrosion and corrosion control is made clear by the inclusion of ceramics, polymers, glass, concrete and other materials as well as of metals, the focus of the first edition. The introduction of standards into corrosion science and engineering is emphasized as is life prediction, and economic and risk analyses associated with environmental degradation of materials.

While the introduction of new technologies has dramatically changed virtually every aspect of life on the Earth in the fifty years since the appearance of the *Corrosion Handbook*, what remains a persistent reality in the engineering enterprise is that engineering systems are built of materials. Whether an airframe, integrated circuit, bridge, prosthetic device or, perhaps as we shall see in the not too distant future, implantable drug delivery systems—the chemical stability of the materials of construction of such systems continues to be a key element in determining their useful life. This new edition of the *Corrosion Handbook* will serve, among others, designers, inspectors, owners and operators of engineering systems of all kinds, many of which are unknown today, for generations to come. Dr. Revie has succeeded, just as did his mentor in 1948, in producing a convenient reference volume covering the entire field of corrosion.

R. M. Latanision

H. H. Uhlig Corrosion Laboratory
Massachusetts Institute of Technology
Cambridge, Massachusetts

PREFACE

The objective in preparing this third edition of *Uhlig's Corrosion Handbook* has been to provide an updated book—one affordable volume—in which the current state of knowledge on corrosion is summarized. The fundamental scientific aspects and engineering applications of new and traditional materials and corrosion control methods are discussed, along with indications of future trends. The book is intended to meet the needs of scientists, engineers, technologists, students, and all those who require an up-to-date source of corrosion knowledge. This new edition contains a total of 88 chapters divided among six parts:

 I. Basics of Corrosion Science and Engineering
 II. Nonmetals
 III. Metals
 IV. Corrosion Protection
 V. Testing for Corrosion Resistance
 VI. Corrosion Monitoring

Topics discussed in chapters that are new in this edition include failure analysis (Chapter 1), principles of accelerated corrosion testing (Chapter 73), metal–matrix composites (Chapter 35), nanocrystals (Chapter 37), ethanol stress corrosion cracking (Chapter 50), computation of Pourbaix diagrams at elevated temperature (Chapter 9), high-temperature oxidation (Chapters 20 and 74), dealloying (Chapter 11), and diagnosing, measuring, and monitoring microbiologically

influenced corrosion (MIC) (Chapter 88). Dr. Tomomi Murata has provided some very insightful introductory notes on the effects of climate change, life-cycle design, and corrosion of steel under changing atmospheric conditions.

Throughout the book, extensive reference lists are included to help readers identify sources of information beyond what could be included in this one-volume handbook.

It is a pleasure to acknowledge the authors who wrote the chapters of this edition as well as the reviewers, who, in anonymity, carried out their work in the spirit of continuous improvement. I would also like to acknowledge the members of the Editorial Advisory Committee, who made many constructive suggestions to help define, focus, and clarify the discussions in this new edition. I would like to acknowledge Mary Yess and her staff at The Electrochemical Society Headquarters in Pennington, New Jersey, for their support during the preparation of this book. I greatly appreciate the encouragement and support of Bob Esposito and his staff at John Wiley & Sons, Inc. in Hoboken, New Jersey.

Finally, I would like to thank my many friends and colleagues at the CANMET Materials Technology Laboratory, where it has been my privilege to work for the past 32 years.

R. WINSTON REVIE

Ottawa, Ontario, Canada

CONTRIBUTORS

*Agarwal, D. C., DNV Columbus, Inc., Dublin, Ohio, USA

Bale, C. W., Département de génie physique et de génie des matériaux, Ecole Polytechnique, Montréal, Québec, Canada

Beavers, J, A., DNV Columbus, Inc., Dublin, Ohio, USA

Been, J., Alberta Innovates Technology Futures, Calgary, Alberta, Canada

Böhni, H., Institute of Materials Chemistry and Corrosion, Swiss Federal Institute of Technology, Zürich, Switzerland (Retired)

Broomfield, J. P., Corrosion Consultant, London, UK

Campion, R. P., MERL Ltd., Wilbury Way, Hitchin, UK

Cox, B., Centre for Nuclear Engineering, University of Toronto, Toronto, Ontario, Canada (Retired)

Crook, P., Haynes International, Kokomo, Indiana, USA (Retired)

*Eden, D. A., Honeywell Process Solutions, Houston, Texas, USA

Eiselstein, L. E., Exponent-Failure Analysis Associates, Inc., Menlo Park, California, USA

Efird, K. D., Efird Corrosion International, Inc., The Woodlands, Texas, USA

Elboujdaini, M., CANMET Materials Technology Laboratory, Ottawa, Ontario, Canada

Erb, U., Department of Materials Science and Engineering, University of Toronto, Toronto, Ontario, Canada

Falkland, M. L., Outokumpu Stainless AB, Avesta, Sweden

Fitzgerald III, J. H., Grosse Pointe Park, Michigan, USA

Ford, T. E., University of New England, Biddeford, Maine USA

Frankenthal, R. P., Bell Laboratories, Lucent Technologies, Murray Hill, New Jersey, USA (Retired)

Garfias-Mesias, L. F., DNV Columbus, Inc., Dublin, Ohio, USA

Ghali, E., Department of Mining, Metallurgy and Materials, Laval University, Québec, Canada

Glaes, M., Outokumpu Stainless AB, Avesta, Sweden

Goodwin, F. E., International Lead Zinc Research Organization, Inc., Research Triangle Park, North Carolina, USA

Grambow, B., La Chantrerie, Laboratoire SUBATECH (UMR 6457), Ecole des Mines de Nantes, Nantes Cedex 3, France

Grauman, J. S., TIMET, Henderson, Nevada, USA

Grubb, J. F., Technical & Commercial Center, ATI Allegheny Ludlum Corp., Brackenridge, Pennsylvania, USA

Gu, J.-D., School of Biological Science, The University of Hong Kong, Hong Kong, China

Gui, F., DNV Columbus, Inc., Dublin, Ohio, USA

Hare, C. H., Coating System Design Inc., Lakeville, Massachusetts, USA (Retired)

Hashimoto, K., Tohoku Institute of Technology, Sendai, Japan

*Deceased.

Heidersbach, R., Dr. Rust, Inc., Cape Canaveral, Florida, USA

Hihara, L. H., Department of Mechanical Engineering, University of Hawaii at Manoa, Honolulu, Hawaii, USA

Huet, R., Exponent-Failure Analysis Associates, Inc., Menlo Park, California, USA

Jack, T. R., University of Calgary, Calgary, Alberta, Canada

John, R. C., Shell International E&P, Inc., Houston, Texas, USA

Jordan, D. L., Ford Motor Company, Dearborn, Michigan, USA

Kane, R. D., iCorrosion LLC, Houston, Texas, USA

Kaye, M. H., Faculty of Energy Systems and Nuclear Science, University of Ontario Institute of Technology, Oshawa, Ontario, Canada

Khaladkar, P. R., E. I. DuPont de Nemours & Co, Inc., Wilmington, Delaware, USA

King, G. A., CSIRO Building, Construction and Engineering, Highett, Victoria, Australia (Retired)

Kruger, J., Department of Materials Science and Engineering, Johns Hopkins University, Baltimore, Maryland, USA (Retired)

Latanision, R. M., Exponent-Failure Analysis Associates, Inc., Natick, Massachusetts, USA

Lee, J. S., Naval Research Laboratory, Stennis Space Center, Mississippi, USA

Lewis, B. J., Department of Chemistry and Chemical Engineering, Royal Military College of Canada, Kingston, Ontario, Canada

Liljas, M., Outokumpu Stainless AB, Avesta, Sweden

Little, B. J., Naval Research Laboratory, Stennis Space Center, Mississippi, USA

Malhotra, V. M., Consultant, Ottawa, Ontario, Canada

*****Matsushima, I.**, Maebashi Institute of Technology, Maebashi, Japan

Mendez, M., Honeywell Corrosion Solutions, Houston, Texas, USA

Meng, Q. J., Honeywell Corrosion Solutions, Houston, Texas, USA

Mitchell, R., Laboratory of Microbial Ecology, Harvard School of Engineering and Applied Sciences, Harvard University, Cambridge, Massachusetts, USA

Mitton, D. B., Gold Standard Corrosion Science Group, LLC, Boston, Massachusetts, USA

Morris, P. I., FPInnovations, Vancouver, BC, Canada

Murata, T., Office of Technology Transfer Innovation Headquarters, Japan Science and Technology Agency, Tokyo, Japan

*****Murphy, T. P.**, Campion Hall, University of Oxford, Oxford, UK

Nešić, S., Institute for Corrosion and Multiphase Flow Technology, Ohio University, Athens, Ohio, USA

Nessim, M., C-FER Technologies Inc., Edmonton, Alberta, Canada

Norsworthy, R., Lone Star Corrosion Services, Lancaster, Texas, USA

Papavinasam, S., CANMET Materials Technology Laboratory, Hamilton, Ontario, Canada

*****Parkins, R. N.**, University of Newcastle upon Tyne, Newcastle upon Tyne, UK

Pelton, A. D., Département de génie physique et de génie des matériaux, Ecole Polytechnique, Montréal, Québec, Canada

Piro, M. H., Department of Chemistry and Chemical Engineering, Royal Military College of Canada, Kingston, Ontario, Canada

Postlethwaite, J., Department of Chemical Engineering, University of Saskatchewan, Saskatoon, Saskatchewan, Canada (Retired)

Ray, R. I., Naval Research Laboratory, Stennis Space Center, Mississippi, USA

Rebak, R. B., GE Global Research, Niskayuna, New York, USA

Reid, M., Stokes Research Institute, University of Limerick, Limerick, Ireland

Revie, R. W., CANMET Materials Technology Laboratory, Ottawa, Ontario, Canada

Rigaud, M., Département de génie physique et de génie des matériaux, Ecole Polytechnique, Montréal, Québec, Canada

Roberge, P. R., Department of Chemistry and Chemical Engineering, Royal Military College of Canada, Kingston, Ontario, Canada

Ruddick, J. N. R., Department of Wood Science, Forest Sciences Centre, University of British Columbia, Vancouver, B.C., Canada

*Deceased.

Sequeira, C. A. C., Instituto Superior Técnico, Lisboa, Portugal

Shibata, T., Department of Materials Science and Processing, Graduate School of Engineering, Osaka University, Japan (Retired)

Silence, W. L., Consultant, Fairfield Glade, Tennessee, USA

Silverman, D. C., Argentum Solutions, Inc., Chesterfield, Missouri, USA

Sridhar, N., DNV Columbus, Inc., Dublin, Ohio, USA

Srinivasan, S., Advanced Solutions–Americas, Honeywell International, Inc., Houston, Texas, USA

Staehle, R. W., University of Minnesota, Minneapolis and Industrial Consultant, North Oaks, Minnesota, USA

*****Streicher, M. A.**, E. I. DuPont de Nemours & Co., and the University of Delaware, Newark, Delaware

Thompson, W. T., Centre for Research in Computational Thermochemistry, Royal Military College of Canada, Kingston, Ontario, Canada

Thomson, B., MERL Ltd., Wilbury Way, Hitchin, UK

Verink, Jr., E. D., Department of Materials Science and Engineering, University of Florida, Gainesville, Florida, USA (Retired)

Wang, Y.-Z., Canadian Nuclear Safety Commission, Ottawa, Ontario, Canada

Wilmott, M., Wasco Coatings Ltd., Kuala Lumpur, Malaysia

Young, A. L., Humberside Solutions Ltd., Toronto, Ontario, Canada

Yunovich, M., Honeywell Corrosion Solutions, Houston, Texas, USA

Zhang, X. G., Teck Metals Ltd., Mississauga, Ontario, Canada

*Deceased.

INTRODUCTORY NOTES ON CLIMATE CHANGE, LIFE-CYCLE DESIGN, AND CORROSION OF STEEL

T. MURATA

Japan Science & Technology Agency, Saitama, Japan

A. CLIMATE CHANGE

Climate change is attributed mainly to increased CO_2 in our atmosphere because of anthropogenic activities and is expected to increase as much as 50% by 2030 compared to the concentration in 2005, that is, 359 ppm [1, 2]. Such a change will affect the corrosion of carbon steel through acidification due to increased concentration of HCO_3^- and Ca^{2+} in waters at temperatures a few degrees Celsius higher than those in 1990. In addition, other influential factors that will arise from climate change include the following:

1. Increase in precipitation
2. Formation of aerosols with CO_2 emission
3. Increased SO_x emissions caused by the use of sulfur-bearing coal due to oil shortages
4. Enhanced biological growth in waters

For these reasons, the corrosivity of environments in the future will be complex, and a simple acidification model will not be adequate. To predict the effects of climate change on corrosion, computational analyses and systematic corrosion studies are required to develop models based on projected climate change.

"Time of wetness" is universally considered to be a key corrosion index for atmospheric corrosion. In recent years, weather instability has led to changes in global rainfall distribution, changes that could lead to new and different predictive indices for atmospheric corrosion. For corrosion in waters, microbiological factors are expected to increase in importance with the changing climate. In contrast to the environmental factors that pertain to corrosion in air and water, the heterogeneous distribution of chemicals in contaminated soils in industrialized areas results in nonuniform soil corrosivity. Dynamic corrosion models are required with on-site monitoring systems.

B. LIFE-CYCLE DESIGN

To minimize the environmental burden and to attain a sustainable society, life-cycle design of steel structures is required to ensure safety, reliability, durability, and the best use of materials and energy throughout the life cycle. The life-cycle concept will be required for future design and construction of social as well as industrial infrastructure. For example, in developing a life-cycle design for weathering steels, discussed in Chapter 48, reliable corrosion data for long-term service and a systematic approach to minimize both corrosion damage and social costs are necessary.

In general, corrosion is studied using a set of parameters under simplified or fixed conditions. In the real world, in response to constantly changing environmental parameters, corrosion behavior also changes. For this reason, an understanding of corrosion dynamics is required, and the corrosion protection models that are implemented must have a capacity to reflect dynamic environmental conditions that are subject to constant change.

REFERENCES

1. Intergovernmental Panel on Climate Change (IPCC), "Climate and Water," Technical Paper VI, Geneva, June 2008.
2. Intergovernmental Panel on Climate Change (IPCC), "Implications of Proposed CO_2 Emissions Limitations," Technical Paper IV, Geneva, Oct. 1997, Figure 6, p. 16.

PART I

BASICS OF CORROSION SCIENCE AND ENGINEERING

1

CORROSION FAILURE ANALYSIS WITH CASE HISTORIES

L. E. Eiselstein and R. Huet

Exponent-Failure Analysis Associates, Inc., Menlo Park, California

A. INTRODUCTION

Arc failure analyses useful? The answer is an emphatic *Yes*. There are many reasons to perform a failure analysis; the most common one is to help prevent future failures. To ensure that corrective actions will be effective, it is necessary to understand why failures have occurred in the first place. Otherwise, any design or manufacturing changes that are implemented may not be effective or simpler ways to prevent future failures may be overlooked. Another common reason for performing a failure analysis is to establish responsibilities for the mishap. For instance, an insurance company may want to determine if an event is covered by the policy or not or financial responsibilities must be established for the resolution of a lawsuit.

There is always value in performing some level of failure analysis, even if it seems that it would be better to try something new rather than finding out exactly what went wrong. A tremendous amount of information can be gleaned from understanding how things fail, and this knowledge is invaluable in making things (equipment, machines, and processes) work better in the future. There is a natural tendency to move past the setback of a failure, maybe to avoid dwelling on unpleasant facts or assuming that nothing can be learned from something that did not work out. However, much can be learned from understanding what went wrong, and a good learning opportunity should not be thrown away with the failed parts.

The end point of a failure analysis depends on the specific circumstances and the type of answers needed. In some cases, it may be enough to rule out a specific failure mode, rather than establishing exactly what happened. In an industrial setting, the goal may be to understand the failure enough to be able to identify corrective measures or to determine if a product recall is required. Finally, in serious accidents an exhaustive failure analysis may be necessary for insurance, legal, or safety reasons.

B. FAILURE ANALYSIS PROCEDURES

The analysis of corrosion failures is not fundamentally different from any other failure analysis. Although some

Uhlig's Corrosion Handbook, Third-Edition, Edited by R. Winston Revie
Copyright © 2011 John Wiley & Sons, Inc.

3

unique techniques may be used and the failure modes are specifically related to corrosion, the methodology used is much the same for every type of failure analysis. Failure analysis follows the scientific method. Typically there is a question to be answered, such as "How or why did this failure occur?" Hypotheses are proposed to answer the question or questions. The hypotheses are checked against facts, experiments, and analyses. In the end, some hypotheses are ruled out and others are confirmed.

In practice, there may be many hypotheses put forward to explain a failure, some likely and others very unlikely. Conducting an efficient failure analysis means that most effort is spent proving the hypothesis that ultimately turns out to be the correct explanation while dismissing early (but with good reason) those hypotheses that turn out to be unfounded. Thus it is important to identify the relevant information early on, even though one cannot be sure of what information is ultimately going to be relevant. Performing a failure analysis results in an interplay between hunches, developing likely hypotheses, testing them rigorously, dismissing other hypotheses for good cause, and keeping an open mind for other possible scenarios if the ones that appeared likely at first turn out to not fit all of the facts.

C. GENERAL APPROACH FOR CONDUCTING A FAILURE ANALYSIS

There is no firm set of rules to conduct a correct failure analysis, but the following approach will help. Gather some general information, formulate hypotheses, and then use these hypotheses to gather more targeted information. Use this targeted information from observations, testing, and analyses to validate or rule out the hypotheses.

C1. Gather General Information

The first step of a failure analysis should be to understand the role of the failed component and its environment. Is the failed part available for examination? Does a cursory examination provide some clues as to why it failed? By definition, the failure was not desired, so were any steps taken in the design or operation to prevent it? Was the part or equipment that failed a recent design, had it been modified recently, or did it have a long history of good service? Does the manufacturer, designer, operator, eyewitnesses, or end user have any hypotheses about the cause of this failure? All these questions will help orient the investigator at the start of a failure analysis.

C2. Formulate Hypotheses

It is important to formulate hypotheses early in a failure investigation, because they will guide the collection of further information. Without some hypotheses, relevant information may be overlooked, or to the contrary too much information will be gathered in an effort to be inclusive, which may also impair getting to the truly important facts. One should also think about hypotheses that have to be considered even if they may be ruled out in the end. For instance, if the corrosion failure involves dissimilar metals, galvanic corrosion should be investigated because it is an obvious possibility, even if other aspects of the situation make it unlikely.

C3. Gather Further Information

The hypotheses will help the investigator gather relevant information. It is important to collect facts that may tend to disprove a given hypothesis as well as those that may support it. One of the traps to be avoided is to bias the information collected toward proving one particular scenario.

The Royal Society's motto *Nullius in verba*, roughly translated as "Take nobody's word for it," is a valuable principle to follow when gathering information. Reported observations, hearsay ("Joe told me that. . ."), and sweeping generalizations ("This has never happened before") should be noted but not considered reliable until they have been checked for accuracy. Often the simple act of verifying information will separate fact from fiction and considerably clarify a picture that may have appeared confused at first.

C4. Validate or Reject Hypotheses

It is very important that hypotheses be tested or validated in some fashion; without this step they are nothing more than speculation. The validation may be very simple in some cases or it may require extensive analysis in other cases, but this step should never be overlooked. The validation process must be based on physical and engineering principles, not merely on a process of elimination based on commonalities and differences. Sometimes a failure analysis proceeds by listing common factors and differences between failures and instances of successful operation. This method may be useful as a guide to formulate hypotheses (although not as useful as trying to understand the physical factors affecting a failure), but it should not be used as the exclusive means to validate or reject any hypothesis.

The validation of hypotheses must not be biased in favor of a specific scenario. It is not always easy to recognize that a hypothesis that looked promising at first should actually be rejected or modified, but one should remain alert to this possibility if the validation does not turn out as expected.

D. TECHNIQUES TYPICALLY USED TO INVESTIGATE CORROSION FAILURES

Corrosion failures often involve the use of some specific information or techniques. Several guidance documents suggest checklists or procedures that are specific to corrosion failures, for instance the American Society for Testing and Materials (ASTM) has issued a *Standard Guide for Corrosion-Related Failure Analysis* [1]. These guides supplement the general failure analysis process and adapt it to corrosion failures. A few topics that recur regularly in corrosion failures are discussed here, but the reader may wish to consult some of the extensive literature on failure analysis [1–14].

D1. Sampling and Collecting Corrosion Evidence

Sampling and collection of corrosion evidence are often key steps in corrosion failure analysis. Sampling should be done carefully because corrosion products and deposits often contain valuable information that can be easily damaged or contaminated. For example, if microbial activity is suspected, samples should be collected in sterile containers under appropriate conditions to avoid contamination by other microorganisms. These live samples should be analyzed promptly. If the pH or dissolved oxygen level of the aqueous environment may help to explain the corrosion, field measurements should be made of these parameters as they can change rapidly during storage. Samples should be protected from contamination by other debris and stored such that they will be protected from further corrosion damage.

In addition, samples must be representative. In many cases a few well-chosen samples will be enough. The investigator may easily choose a few "typical" samples from the affected and nonaffected areas. However, in cases where one must assess the condition of a large number of items (e.g., to assess the extent of damage), a statistically valid sampling method must be used. There is no sampling method that always produces a "statistically valid" sample, so the sampling must be defined for each case with the help of a competent statistician.

D2. Determining Corrosion Rates

A question often arises about the timing of some corrosion event. For instance, one may want to know for how long some corrosive conditions have been present or how long before some corrosion will result in a leak or vessel rupture. Unfortunately, corrosion rates are notoriously variable, and in some cases it is nearly impossible to make precise predictions. However, corrosion rates have been published for many combinations of materials and environment and they can be useful if their limitations are understood. These rates are typically averages over many samples or observations; although the performance of any single sample may deviate significantly from the average, in aggregate these rates can be useful. In general, corrosion rates tend to slow down with time, so it is usually important to know the time period over which the rate was measured because linear extrapolation cannot be used in most cases.

Direct measurements of corrosion rates in the laboratory usually take a long time, which may not be practical in the context of the failure analysis. Where this is not possible, accelerated corrosion tests are sometimes performed; however, it is generally quite difficult to determine the acceleration factor over the actual environment. When possible, long-term monitoring of the corrosion process in service is the best way to obtain relevant rates, and this monitoring should be started as soon as possible after the equipment, product, or process is placed in service.

D3. Characterizing the Form of Corrosion

Various types of corrosion have been defined, based generally on morphology or mechanism. Authors differ on this classification, but generally some variation of the following eight categories of corrosion are used: (1) uniform, (2) galvanic, (3) crevice, (4) pitting or localized, (5) intergranular, (6) dealloying, (7) erosion–corrosion, (8) environmentally assisted cracking (EAC), or stress corrosion cracking (SCC) [15]. Other named forms of corrosion such as microbiological-induced corrosion (MIC), filiform corrosion, and liquid-metal embrittlement are typically covered as subcategories of these types. In this chapter we discuss corrosion failures of various types and illustrate them with case histories where applicable.

D3.1. Uniform Corrosion. Uniform corrosion, also known as general corrosion, is a very common corrosion type where the metal is corroding more or less uniformly. Most often, this type of corrosion is easily investigated, since it occurs whenever a susceptible metal is in contact with an aggressive environment: Leave a nail in water and it will corrode. Questions that are more difficult to answer may include the rate at which the corrosion damage occurred, for instance, why the corrosion was particularly rapid in a specific case. In other cases, the key may be to find out how the environment came into contact with the susceptible metal. Finally, questions may arise regarding the effect of corrosion on the strength or other properties of the corroding material.

D3.1.1. Corrosion Rates. Although corrosion rates are extremely variable, they are useful to indicate the average behavior of many samples. They are also useful as a broad indicator of the intensity of corrosion: In cases where the observed rates are far different from the published ones, there should be some explanation of the difference.

Case Study: Corrosion Rates Indicate That Product Is Not Suitable for Intended Use. A company decided to manufacture a copper-covered stainless sheet to make roof panels, gutters, and flashing products. In this process the stainless steel was electroplated with copper on both sides: 10 μm on the side expected to be exposed to the weather and

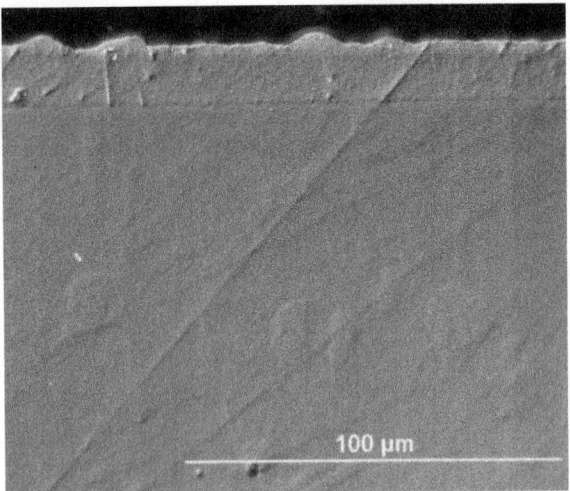

FIGURE 1.1. Copper-coated stainless steel (thick side).

FIGURE 1.2. Rapid corrosion of copper coating from areas underneath wood shake. Arrows indicate bare stainless steel exposed after the copper layer has corroded away.

3 μm on the other (see Fig. 1.1). Atmospheric exposure tests performed by the manufacturer and others indicated a corrosion rate of the order of 1 μm per year for coastal marine atmospheric environments. At this rate, the copper coating would have been consumed in about 10 years (or 3 years if the wrong side of the sheet is exposed to atmospheric corrosion), well short of the 30-year life that was contemplated for this product and somewhat shorter than the time to form a patina. Once the copper layer is removed, the roof looks like stainless steel rather than the intended copper patina, and furthermore the now-exposed stainless steel may pit. Although this material may perform well in dry and noncoastal marine climates, it was clearly not suitable for general use under all outdoor atmospheric corrosion conditions.

The field experience indicated that after less than one year of service there were complaints of excessive corrosion. The copper layer was completely removed from areas of severe exposure, such as in chimney flashings exposed to the acidic flue gases or from the water runoff from wood shakes (see Fig. 1.2). Even though most of the installations had not failed after a few years, the occurrence of several early failures corroborated the reported corrosion rates and indicated that the coatings were too thin for this application. By comparison, a similar product with 50 μm of copper roll bonded to both sides of stainless steel sheet has demonstrated good performance for more than 20 years of service.

Case Study: Extraordinarily High Corrosion Rates. Dilute nitric acid can be extremely corrosive to carbon steel, but concentrated nitric acid passivates carbon steel. This passivation is temporary and can be reversed. A well-known experiment illustrates this behavior: A nail is placed in a test tube and concentrated nitric acid is added to cover about half the nail. Nothing happens because the nail is

passivated by the acid. Water is added slowly so that it forms a separate layer on top of the nitric acid. Nothing happens at first, but after a minute or so, corrosion starts at the interface where water dilutes the nitric acid. The passivation breaks down and the corrosion reaction becomes extremely violent.

This experiment was repeated unwittingly inside a nitrogen tetroxide (N_2O_4) tank car through a series of errors [16]. The carbon steel tank car was used to carry N_2O_4 that was being used in a paper plant. The N_2O_4 will react with water to form nitric acid. This N_2O_4 tank car had been involved in an earlier incident in which a significant amount of water had entered it undetected. Sometime later, the presence of water was detected and it was decided to drain the car of the nitrogen tetroxide and nitric acid that had formed. After draining some material, the car was erroneously thought to be nearly empty and water was added to dilute what was thought to be a small "heel" of liquid but was in fact a significant amount of nitric acid and nitrogen tetroxide. The operation was repeated twice in the following days, every time with the same result. Each time, a complicated process of mixing and diluting was taking place inside the car: The inflow of water may have stirred the car's content, but water, concentrated nitric acid, and nitrogen tetroxide tended to separate in layers because of their different densities. The result was unpredictable and at some point the carbon steel passivation broke down. Very rapid corrosion occurred, a massive release of nitrogen oxides overwhelmed the venting capacity of the pressure relief valve, and the tank car ruptured. There were three distinct corrosion bands on the inside of the car where severe loss of material had occurred in a matter of hours or days.

D3.1.2. Cosmetic Corrosion Failure. Sometimes, uniform corrosion does not affect the structural properties of the corroding part and the only effects are cosmetic. Whether or not this represents a failure depends on the circumstances.

If the appearance of the part is of no concern, a slight corrosion may be acceptable; otherwise corrosion may be deemed a serious concern well before any structural failure may occur.

Case Study: Atmospheric Corrosion on the Underside of Roofing Panels. The roof of a California building was made of steel panels with a thick coating on the external side. The underside, which was exposed to an attic space, was left with a thin shop primer coating because there were no concerns about its appearance. After a few years in service, the structural performance of the panels was called into question, in part because the underside had visibly corroded. Some atmospheric corrosion had appeared in spite of the shop primer. Cross sections of the corroded areas demonstrated that the depth of corrosion was minimal and that, at the observed rate, perforation of the roofing panel would not occur during the expected lifetime of the building. Consequently, this was an instance where corrosion was not a failure.

Case Study: Atmospheric Corrosion of Terne-Coated Roof. In two buildings, one located in Alaska and the other in Louisiana, roofs made from lead–tin (terne)–coated stainless steel became severely discolored instead of developing the dull gray appearance typically associated with lead roofs. Metals such as copper and lead used for roofs develop a patina or surface film on exposure to the atmosphere; the patina for terne is usually dull gray. However, in these two instances, the roofs developed irregular patterns of reddish-yellowish corrosion patterns (Fig. 1.3). Our investigation showed that the discoloration of the terne-coated roofs was not caused by rusting of the stainless steel substrate. Rather, it resulted from the normal patination of the terne coating being disrupted due to adverse environmental conditions. The sequence of patina formation is orthorhombic lead oxide (PbO, yellow) $\rightarrow$ basic lead carbonate $\rightarrow$ normal lead carbonate $\rightarrow$ lead sulfite $\rightarrow$ lead sulfate. If access to the air is restricted, there may not be enough carbon dioxide to form the basic lead carbonate, so the yellow lead oxides remain. There was no attack of the stainless steel substrate and no risk of leaks from the roof corrosion, so this was purely a cosmetic failure. However, this is an example where the appearance of the roof was important—the terne coating had been chosen for its color—so this was indeed a corrosion failure.

D3.2. Galvanic Corrosion. Galvanic corrosion is a common failure mode. It occurs where a less noble metal is in electrical contact with a more noble metal in an electrolyte. ASTM standard G71 provides a test procedure to evaluate the potential for galvanic corrosion [17].

A related failure mode, but not strictly galvanic corrosion, occurs when the source of potential difference results from

FIGURE 1.3. Terne-coated roof showing severe reddish-yellowish discoloration instead of the expected gray patina.

a difference in the electrolyte composition between two zones. For instance, differences in oxygen concentration can accelerate corrosion of the area depleted in oxygen. Yet another source of potential differences may be electrical currents generated by some external cause, for instance, stray ground currents from large electrical equipment.

Typically, galvanic corrosion results in fairly rapid attack. But potential differences may be used to protect a piece of metal by forcing the corrosion to occur on a sacrificial anode. This, of course, is the principle of cathodic protection. However, the cathodic protection may not perform as anticipated, leading to corrosion failures.

Case History: Ineffective Cathodic Protection. In some small ships, the propeller shafts are enclosed in stern tubes that are part of the hull. The inside of the stern tubes is exposed to the seawater whereas the outside is dry, being in the hold of the ship. In a specific model of ships, the propeller shafts are made of stainless steel while the stern tubes and the hull are made of low-carbon steel. The shafts are supported by three bearings in the tubes: one aft, one midlength of the tubes, and one that is part of the forward end of the tubes. The hull is painted and there are zinc anodes on the hull and propeller for cathodic protection. However, severe corrosion was observed after less than one year in service on the inside of the stern tubes, particularly at their forward end, which is the farthest from the aft opening of the tubes (Fig. 1.4).

The corrosion pattern was typical of painted low-carbon steel exposed to aerated seawater without cathodic protection. Potential measurements on a boat in service confirmed that the steel was not cathodically protected in the forward region of the stern tube. This is not surprising: There were no zinc anodes inside the tubes and the "throwing power" of the anodes on the outside of the hull was simply not enough to reach the inner surface at the forward end of the tubes, especially given the shaft bearing supports that are in the way. Further, the stainless steel shaft, sealing flange, and

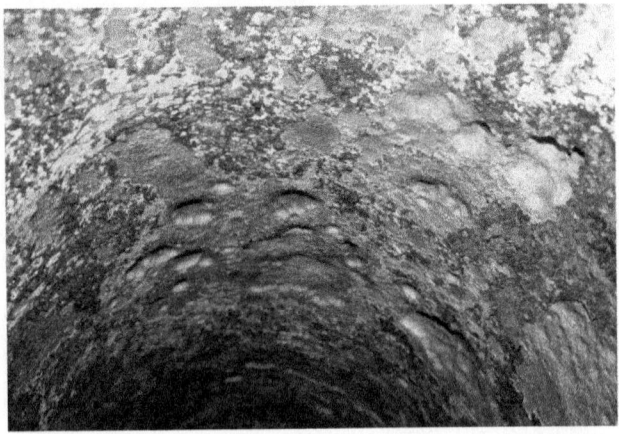

FIGURE 1.4. Severe corrosion of the inside of a stern tube. The pattern of general corrosion with some deeper pits is typical of low-carbon steel in seawater.

rotating seal were likely to have anodically polarized the steel tubes, accelerating the corrosion. Under these conditions, any small defect or holiday in the paint coat (and the inside surfaces of the tubes are difficult to paint) would lead to the observed corrosion. It was recommended that zinc anodes should be placed on the inside of the tubes for effective cathodic protection.

Case History: Thermogalvanic Corrosion. A relatively uncommon source of potential difference between two electrodes is a difference in temperature. The electrode potential of a metal piece in an electrolyte depends in part on the temperature, but temperature effects are usually small and negligible. However, in some cases these temperature differences lead to significant failures. A common occurrence is the potential difference between the hot and cold water lines in homes when the copper pipes are buried under a concrete slab. In several residential developments in the western United States, the hot and cold water lines were buried under the home cement slab foundation. The temperature difference between the hot and cold copper water lines provides a potential difference, which in some environments results in the hot line being anodic to the cold. The soil and soil moisture provide the electrolyte. The result is pitting corrosion on the outside of the hot water lines. This failure mode happens more often on homes with hot recirculation pumps. These pumps keep hot water circulating constantly in the lines, so that there is almost instantly hot water at the faucets when they are open. The result is that the lines are hot 24 hours a day, whereas without recirculation pumps the lines would be cold most of the time.

In this situation, statistical methods can be used to predict future leaks based on prior experience. The Weibull distribution is commonly used to model failures. In a particular development, the number of leaks in homes with recirculation pumps could be modeled accurately by a three-parameter Weibull distribution (Fig. 1.5) whereas the homes without recirculation could not be easily characterized without accounting for home location or hot water usage.

D3.3. Crevice Corrosion. Crevice corrosion is a type of localized corrosion at an area that is shielded from full exposure to the environment. This type of attack is usually associated with small volumes of stagnant solution caused by holes, gasket surfaces, lap joints, surface deposits, and crevices under bolt and rivet heads. This form of corrosion is sometimes also called deposit or gasket corrosion [15]. ASTM provides a guide for evaluation of the crevice corrosion resistance of stainless steel and nickel-based corrosion-resistant alloys in chloride-containing environments [18].

Crevices are formed in a variety of design situations. For instance, aircraft skin panels are joined by lap splice joints, where moisture and corrosive atmospheric gases can become trapped between the two panels, resulting in crevice corrosion [19, 20]. This mechanism was blamed for the 1988 Aloha Airline accident in which a 20-year-old Boeing 737 lost a major portion of the upper fuselage at 25,000 ft [19, 21, 22]. In this particular instance, crevice corrosion resulted in the formation of voluminous hydrated aluminum oxides which acted to separate the two skin panels, stressing the rivets and resulting in fatigue failures.

D3.4. Pitting. Alloys that maintain their resistance to corrosion through the formation of a protective passive layer, such as stainless steels, aluminum alloys, and titanium alloys, generally do not suffer from uniform corrosion; rather they will usually corrode as a result of the localized breakdown of a small region of the passive film. Corrosion occurs rapidly at this defect compared to the surrounding material covered with the passive film, resulting in the formation of a pit. ASTM has a standard that helps to characterize the nature of pitting [23].

Case History: Pitting Corrosion of Aluminum Due to Copper-Containing Fungicides. Irrigation pipes used in several California farms suffered from rapid pitting corrosion starting on the inside of the pipe. The pipes are used to irrigate crops and spray fungicide and other chemicals. The pipes that experienced severe pitting were used to spray copper-based fungicides.

A literature review indicates that dissolved copper (in the form of copper hydroxide, the active ingredient of the fungicides) is not enough to promote severe pitting of aluminum alloys; there must be some chlorine and bicarbonate ions as well. The pitting mechanism has not been identified in detail, but it probably involves penetration of the aluminum oxide layer by chloride ions, plating of copper on the exposed aluminum surface, and starting the formation of corrosion products nodules over the corroding areas. Once

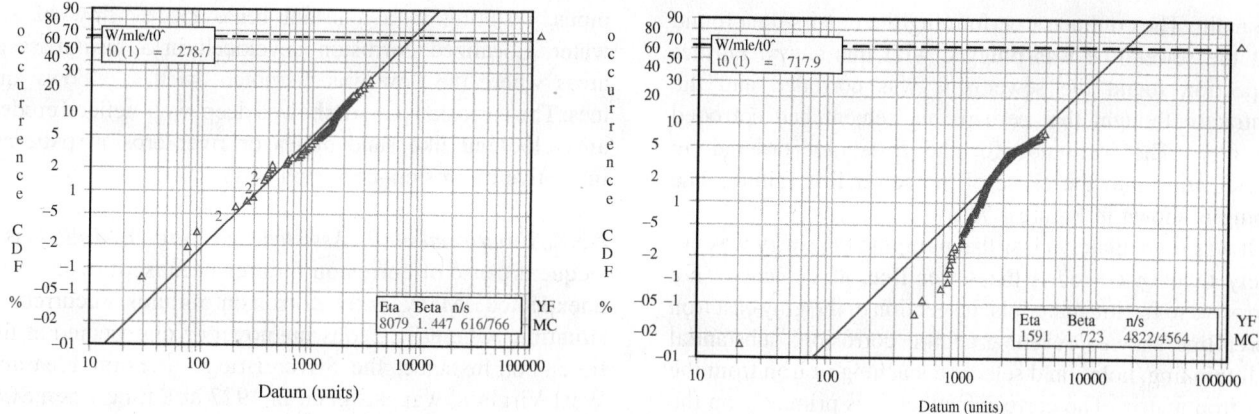

FIGURE 1.5. Three-parameter Weibull distribution fits very well the cumlative failure distribution of homes with recirculation pumps (left), but homes without pumps cannot be characterized with a two- or three-parameter Weibull distribution without accounting for geographic distribution or other effects such as hot water usage (right).

the nodules are formed, they set up concentration cells and severe pits grow under the nodules.

Experiments confirmed that all three ions were necessary for severe pitting. Aluminum pipes exposed to fungicide dissolved in deionized water were slightly attacked, with few shallow pits forming. Similarly, pipes exposed to farm water (containing bicarbonate hardness as well as about 9 ppm chlorides) without fungicide experienced mild pitting. Only the combination of farm water and fungicides caused severe pitting, comparable to the damage observed in the irrigation pipes on the farms (Fig. 1.6).

D3.5. Dealloying.
Dealloying (also known as selective dissolution) includes two commonly occurring phenomena known as graphitic corrosion of cast irons and dezincification of brasses. These are two related corrosion modes, in which

one of the components of an alloy leaches out selectively, leaving behind a spongy and weak matrix. In dezincification, zinc leaches out of the brass, leaving behind a copper matrix. In graphitic corrosion of cast irons, the iron corrodes away, leaving behind a porous and weak graphite matrix.

This dealloying corrosion may not be easily recognized by a simple visual examination; frequently the matrix does not look appreciably different from the intact material. A metallographic cross section may be required to identify this type of corrosion and determine its extent.

Case History: Graphitic Corrosion of a Sewer. A fire protection water main at a restaurant failed. As part of the repairs, slurry comprised of cement and pea gravel was pumped under the restaurant foundation to fill the void that was created from water gushing out of the ruptured

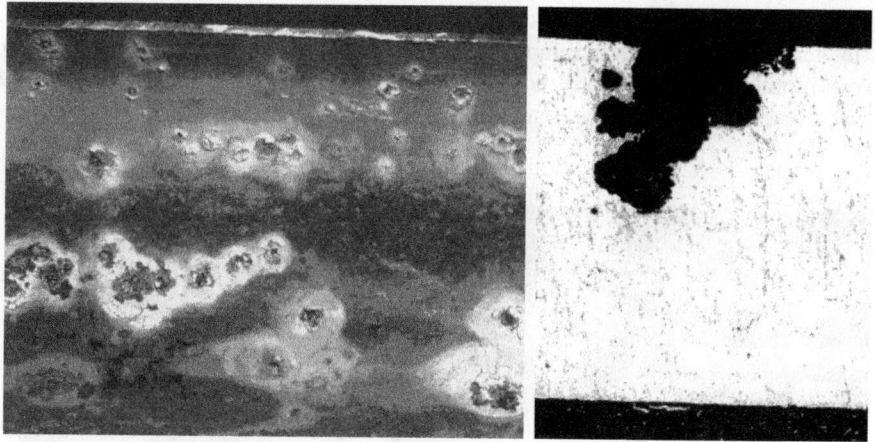

FIGURE 1.6. Pitting observed on the inside surface of an aluminum irrigation pipe (left) and cross-section of a pit in a laboratory sample of aluminum pipe exposed to copper-containing fungicide (right).

fire main. The contractor performing the repairs then found that the sewer line was plugged with this slurry. Further inspection found the sewer line was corroded and the contractor thought that perhaps his cement had corroded the pipe. The insurance for the restaurant ordered an investigation into the cause of the sewer line failure. The piping is shown in Figure 1.7.

It was immediately clear that pumping the slurry was not likely to have corroded the sewer line, as cement is not corrosive to ferritic materials. Inspection of the gray cast iron sewer line showed severe graphitic corrosion: substantial wall thinning, holes, and selective leaching of iron from the cast iron matrix. The corrosive attack was primarily on the sewer pipe inner surface, with little or no corrosive attack from the outside. The graphitic corrosion was extensive, indicating that the sewer may have been leaking for a long time, probably years. The most severe corrosion, including holes in the sewer line, was near the foundation on the south side of the restaurant where the fire protection water main entered the building. Examination of the fire main showed that it had failed due to weakening of the pipe wall from external corrosion. Thus, it is likely that the sewer line leak was the source of moisture that caused the external corrosion damage to the fire protection water main.

D3.6. Erosion–Corrosion.
Some metals corrode more rapidly when exposed to flowing water. A good example is copper: Corrosion is much accelerated when exposed to water flowing at more than about 1 m/s past the surface. These conditions may be found in home plumbing systems with hot recirculation lines, as shown below.

Case History: Erosion–Corrosion of Copper Hot Recirculation Lines. Some homes have hot recirculation lines; that is, the hot water lines form a loop in which a small pump keeps a constant flow of water. This allows hot water to flow almost instantly from any tap in the home. However, the constant flow and high temperature could lead to severe corrosion. In a recent residential development where these loops were installed in fairly small homes, the available pumps were oversized with respect to the

pipes, resulting in a constant high-velocity flow of hot water. Localized corrosion developed in a few years at areas where the flow was disrupted, such as elbows and tees. The corrosion pattern showed deep, well-defined eroded areas looking like sand dunes or river erosion patterns, indicative of erosion–corrosion (Fig. 1.8).

D3.7. Environmentally Assisted Cracking.
EAC/SCC is a frequent cause of corrosion failures because it tends to be unexpected. Many early corrosion failures occurred in situations where SCC was unknown or unexpected at the time. For instance, the Silver Bridge at Point Pleasant, West Virginia, was designed in 1927 at a time when SCC was not known to occur under rural atmospheric conditions in the classes of bridge steels used for construction [24–26]. Yet SCC developed in an eyebar suspension link and led to the catastrophic collapse of the bridge after 40 years in service. Similarly, the Flixborough explosion of 1974, which killed 28 people in Great Britain when 50 tons of cyclohexane was released in a chemical plant, was (in part) a result of nitrate SCC and liquid metal embrittlement [27, 28].

Generally, SCC requires a specific combination of stress, material, and environment. Since the fundamental mechanisms for SCC are not always well understood, new combinations that can cause SCC are unfortunately found by accident.

Case History: Room Temperature Transgranular SCC of Austenitic Stainless Steel. Austenitic stainless steel, such as type 304 or 316, is known to be susceptible to transgranular SCC (TGSCC), but it was thought that this required exposure to concentrated chloride environments above 60°C. However, there have been several recent reports of such SCC at room temperature. For instance, SCC has been observed in permanent anchors used for rock climbing in Thailand (Fig. 1.9), in limestone formations at sea level [29].

After a few years, some of the anchors developed extensive cracking, to the point that they broke under normal climbing loads. Examination of the broken pieces showed

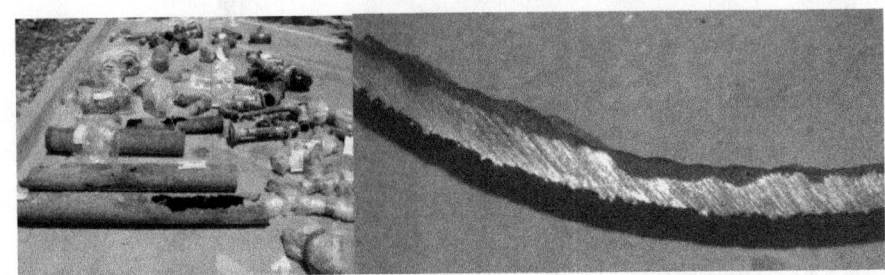

FIGURE 1.7. Fire main and sewer piping from restaurant (left) and graphitic corrosion on inner surface of sewer piping (right).

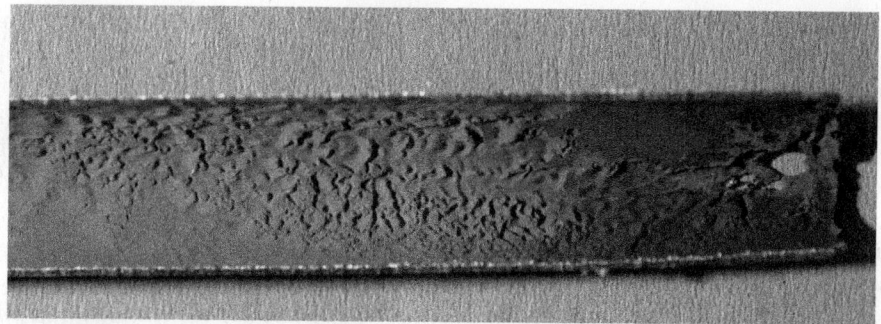

FIGURE 1.8. Copper hot recirculation line showing signs of erosion–corrosion.

pervasive intergranular SCC and the fracture surfaces contained a high concentration of magnesium, rather than the sodium that would be expected in a marine environment. Review of the known cases of room temperature transgranular SCC showed that the climbing anchors are in an environment that combines known promoters of TGSCC. Essentially, this type of SCC requires that the steel be exposed to very high concentrations of chlorides, which are promoted by salts such as magnesium or calcium chlorides, and by low relative humidity (a high relative humidity tends to dilute any salt water film on the metal). The climbing cliffs are located on tower karsts, which get their characteristic steep sides because of very active dissolution and redeposition of the limestone. Thus the climbing anchors can be exposed to calcium and magnesium salts as well as sodium chloride from the ocean. Although the environment is usually very humid, there are times where the relative humidity is fairly low, concentrating any solution that has formed on the stainless steel. Thus the climbing anchors are exposed to an environment that is extremely severe for transgranular chloride SCC.

D4. Complex failure analysis

Sometimes the incident under investigation is particularly complex or involves significant costs, either human or financial. In these instances, it is more important than ever to perform the failure analysis in a careful manner, separating various issues that may be involved and thoroughly validating hypotheses before coming to final conclusions. Such an example is described here.

Case Study: Chlorine Release at a Manufacturing Facility. A massive amount of chlorine gas was released to the atmosphere at a chemical plant making chlorine from calcium chloride [30, 31]. In the plant, the chlorine gas is liquefied before transport by railcar. In the liquefaction process, the chlorine gas is first compressed, then cooled down in a shell-and-tube heat exchanger, in which the chlorine flows inside tubes while chilled calcium chloride brine at about $-23°C$ ($-10°F$) circulates on the outside of the tubes. The liquefied chlorine is sent to a storage tank via a long transfer pipe containing several tees and elbows. The release

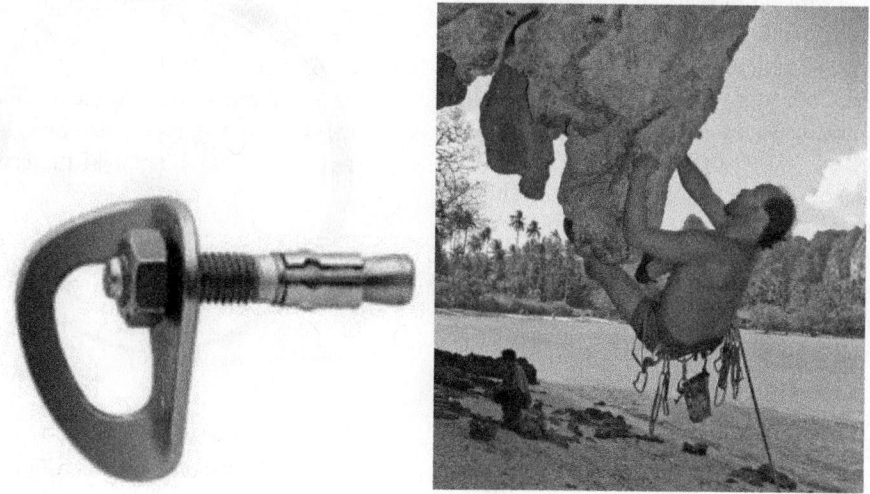

FIGURE 1.9. Typical 316L rock climbing bolt (left) installed permanently on climbing routes in cliff formations located on Thai beaches (right).

occurred through a large corrosion hole in an elbow in the transfer line between the liquefier and the storage tank.

Early indications showed that the hole in the elbow was due to severe general corrosion on the inside surface. Dry chlorine is not corrosive to the steel elbow, but addition of water to the chlorine could make the mixture extremely corrosive. Potential sources of water included the chlorine liquefier: If there was a leak in the tubes, brine solution could enter the chlorine stream. The liquefier was pressure tested and several tubes were found to be leaking.

The liquefier was cut open to expose the leaking tubes. To the general surprise, an old rag was found stuck in the shell of the liquefier, right at the brine inlet piping. It appeared that the rag had been there for a long time, most probably from the time of installation of the liquefier, some 25 years earlier. The rag partially blocked the brine flow path, resulting in accelerated flow in the areas that remained unobstructed. The leaking tubes were found in the area of accelerated brine flow.

With these early findings, the following scenario was hypothesized: The rag had been in the liquefier since installation, leading to increased brine flow rates over some tubes. This eroded the tubes, and, after about 25 years, one or more tubes were perforated. The brine flowed into the chlorine, creating a very corrosive mixture that corroded through the elbow in the transfer line within a few days, before the water contamination of the chlorine could be detected. Each step in the scenario was validated with tests and analyses.

1. *The rag increased the flow rate over some tubes in the bundle.* A fluid flow finite-element analysis of the liquefier inlet section was performed. With the cloth, the brine velocity in the area of the holes was about 3.94 m/s, whereas without the cloth it would have been only 1.27 m/s, or about one-third.

2. *This increased flow rate led to through erosion of the tube in about 25 years.* The observed corrosion rate of the tubes in areas where the brine flow was not accelerated by the cloth was very slow, about 10 μm per year. At this rate, it would take over 200 years to puncture a tube, so the design and operation of the liquefier were not the cause of the tube leak. The corrosion rate of the tube material exposed to brine flowing at high velocities was measured in a test bed in which chilled brine of various pH values was flowed over dummy tubes at various velocities. For a brine solution at its natural pH (no chemical additions), flowing at about 4 m/s, the interpolated corrosion rate would lead to through-wall erosion in about 22 years, very close to the actual service life of 25 years.

3. *Once the tube was perforated, brine flowed into the chlorine stream.* This conclusion was actually not immediate, because both the brine and chlorine systems were pressurized to roughly the same value. A careful pressure drop analysis of both the chlorine and the brine systems, coupled with review of plant data and some actual measurements made on a mock-up of the brine system, confirmed that the brine pressure was likely higher than that of the chlorine, so that brine would be entrained into the chlorine stream.

4. *The transfer line elbow corroded mostly from the inside out.* This was not immediately obvious, as both the inside and the outside of the elbow were severely corroded. The outside surface may have suffered from underinsulation corrosion during its life, and perhaps more rapid corrosion had occurred during the incident, when liquid chlorine was released from the leak site, but the extent of damage on the inside and outside needed to be compared accurately. A cross section of the failed elbow was traced over the outline of an undamaged elbow; this illustrated clearly that most of the corrosion had occurred on the inside surface (Fig. 1.10).

5. *The corrosion rate at the elbow was high enough to lead to a leak in a few days.* It is likely that the brine leak in the chlorine liquefier occurred relatively shortly before the incident, because it is likely that small amounts of water in the chlorine would have been detected prior to distribution. The best estimate of the start of the brine leak was a few days before the incident. The elbow was about 5.3 mm thick originally, so the corrosion rate would have to be extraordinarily high for through-wall attack in a few days. A related observation that required explanation was that the corrosion at the leak site was much more severe than

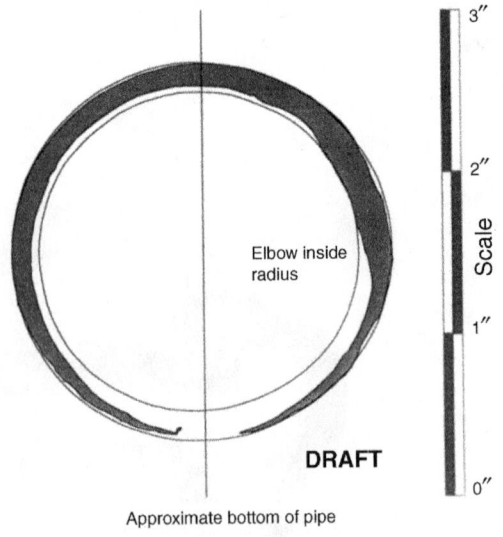

FIGURE 1.10. Erosion–corrosion wall loss of liquefied chlorine run down elbow compared to exemplar elbow.

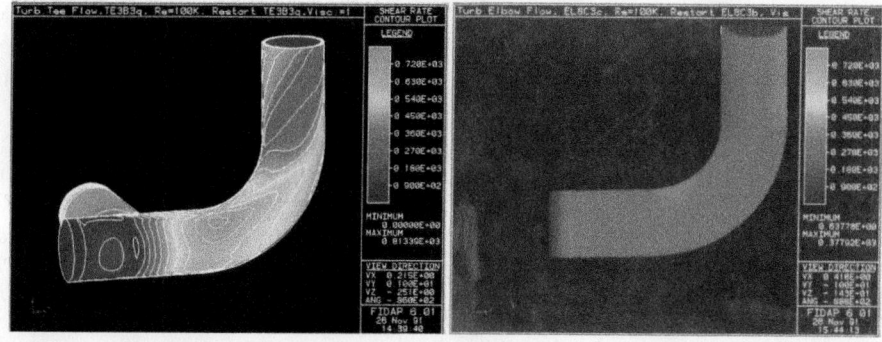

FIGURE 1.11. CFD calculated shear rate of (a) failed elbow-tee configuration and (b) elbow closest to storage tank without tee.

elsewhere in the transfer line, even though the line contained at least one other elbow. This was investigated by a combination of tests and analysis. The failed elbow happened to be located immediately after a T fitting, so that the flow went through two changes of direction in rapid succession. A computer fluid dynamics (CFD) model showed that the flow velocity and the shear rate in this elbow were much higher than in the other elbow in the transfer line, explaining why the corrosion was so much more severe at the leak point (Fig. 1.11). Actual corrosion rate measurements with several mixtures of brine and chlorine, both static and flowing, showed very high corrosion rates, but not quite high enough to achieve penetration in less than seven days. However, given the variability of the measured rates and the uncertainty about the flow conditions and the composition of the corroding mixture, it was concluded that the elbow most likely corroded in a few days after the tubes had started leaking.

The hypothesized scenario was thus validated step by step. This was a very serious incident, in which the stakes were high enough that a thorough failure analysis had to be performed, and the resources were available. Most failure analyses cannot be this detailed, but the basic steps must be the same: gather initial information, formulate hypotheses, and, most importantly, validate these hypotheses with the level of detail appropriate for the conclusions to be reached.

REFERENCES

1. American Society for Testing and Materials (ASTM), ASTM G161-00(2006): Standard Guide for Corrosion-Related Failure Analysis, ASTM, West Conshohocken, PA, 2006.

2. American Society for Testing and Materials (ASTM), ASTM E2332-04: Standard Practice for Investigation and Analysis of Physical Component Failures, ASTM, West Conshohocken, PA, 2004.

3. American Society for Testing and Materials (ASTM), ASTM E1492-05: Standard Practice for Receiving, Documenting, Storing, and Retrieving Evidence in a Forensic Science Laboratory, ASTM, West Conshohocken, PA, 2005.

4. American Society for Testing and Materials (ASTM), ASTM E1459-92(2005): Standard Guide for Physical Evidence Labeling and Related Documentation, ASTM, West Conshohocken, PA, 2005.

5. Journal of Engineering Failure Analysis, Elsevier Ltd., Oxford, UK, 2009.

6. Journal of Failure Analysis and Prevention, ASM International, Materials Park, OH, 2009.

7. P. de Castro and A. Fernandes, "Methodologies for Failure Analysis: A Critical Survey," Mater. Design, **25**(2), 117–123 (2004).

8. H. Herro and R. Port, The Nalco Guide to Cooling Water System Failure Analysis, McGraw-Hill Professional, New York, 1992.

9. E. D.D. During, Corrosion Atlas: A Collection of Illustrated Case Histories, Third Edition, Elsevier, Amsterdam, 1997.

10. American Society for Testing and Materials (ASTM), ASTM E860-07: Standard Practice for Examining and Preparing Items That Are or May Become Involved in Criminal or Civil Litigation, ASTM, West Conshohocken, PA, 2007.

11. American Society for Testing and Materials (ASTM), ASTM E1188-05: Standard Practice for Collection and Preservation of Information and Physical Items by a Technical Investigator, ASTM, West Conshohocken, PA, 2005.

12. K. Esaklul, Handbook of Case Histories in Failure Analysis, Vols. I and II, ASM International, Materials Park, OH, 1993.

13. ASM, Metals Handbook, Vol. 11, Failure Analysis and Prevention, 11th ed., ASM International, Materials Park, OH, 2002.

14. C. Brooks and A. Choudhury, Failure Analysis of Engineering Materials, McGraw-Hill Professional, New York, 2001.

15. M. G. Fontana and N. D. Greene, Corrosion Engineering, McGraw-Hill Book Company, New York, 1978.

16. National Transportation Safety Board (NTSB), Tank Car Failure and Release of Poisonous and Corrosive Vapors, Gaylord Chemical Corporation, Bogalusa, Louisiana, Oct. 23, 1995NTSB Report No. HZB-98-01, Washington, D.C. 1998.

17. American Society for Testing and Materials (ASTM), ASTM G71-81 (2009): Standard Guide for Conducting and Evaluating Galvanic Corrosion Tests in Electrolytes, ASTM, West Conshohocken, PA, 2007.

18. American Society for Testing and Materials (ASTM), ASTM G78-01(2007): Standard Guide for Crevice Corrosion Testing of Iron-Base and Nickel-Base Stainless Alloys in Seawater and Other Chloride-Containing Aqueous Environments, ASTM, West Conshohocken, PA, 2007.

19. R. G., Kelly, "Crevice Corrosion," in ASM Handbook: Corrosion Fundamentals, Testing, and Protection, ASM International, Metals Park, OH, 2003, pp. 242–247.

20. W. Wallace, D. Hoeppner, and P. Kandachar, AGARD Corrosion Handbook, Vol. 1, Aircraft Corrosion: Causes and Case Histories, Advisory Group for Aerospace Research and Development, Neuilly sur Seine, France, 1985.

21. National Transportation Safety Board (NTSB), Aircraft Accident Report Aloha Airlines, flight 243, Boeing 737-200, N73711, near Maui, Hawaii April 28, 1988, NTSB No. AAR-89/03, Washington, D.C. 1989.

22. P. R. Roberge, Handbook of Corrosion Engineering, McGraw-Hill, New York, 2000.

23. American Society for Testing and Materials (ASTM), ASTM G46-94(2005) Standard Guide for Examination and Evaluation of Pitting Corrosion, ASTM, West Conshohocken, PA, 2005.

24. J. Bennett and H. Mindlin, "Metallurgical Aspects of the Failure of the Point Pleasant Bridge," J. Test. Eval. JTEVA, **1**(2), 152–161 (1973).

25. C. LeRose, "The Collapse of the Silver Bridge," West Virginia Historical Soc. Quart., Charleston, WV, **15**(4) (2001).

26. National Transportation Safety Board, (NTSB), Collapse of U.S. 35 Highway Bridge Point Pleasant, West Virginia December 15, 1967, NTSB No. HAR-71/01, Washington, D.C. 1970.

27. S. Mannan and, F., Lees, "Flixborough," in Lees' Loss Prevention in the Process Industries: Hazard Identification, Assessment, and Control, Vol. 3, Elsevier, Amsterdam, 2005.

28. A. Cottrell and P. Swann, "Technical Lessons of Flixborough—A Metallurgical Examination of the 8-Inch Line," Chem. Eng., **308**, 266–274 (1976).

29. A. Sjong and L. E. Eiselstein, "Marine Atmospheric SCC of Unsensitized Stainless Steel Rock Climbing Protection," J. Failure Anal. Prevention, **8**(5) (2008).

30. S. P. Andrew et al. "Evaluation of a Failure in a Chlorine Production Facility," in Proceedings of IMEC2001: 2001 ASME International Mechanical Engineering Congress and Exposition, ASME, New York, 2001.

31. J. G. Routley, "Massive Leak of Liquified Chlorine Gas Henderson, Nevada (May 6, 1991)," Report 052, FEMA—United States Fire Administration, Washington, DC, 1991.

2

COST OF METALLIC CORROSION

J. KRUGER[†]

Department of Materials Science and Engineering, Johns Hopkins University, Baltimore, Maryland

A. INTRODUCTION

Metallic corrosion seriously affects many sectors of a nation's economy or, on a vastly smaller scale, the design choices made by an engineer. This finding is so because corrosion and protective measures to control corrosion result in the utilization of materials, energy, labor, and technical expertise that would otherwise be available for alternative uses. Corrosion causes users of metal products to incur added expenses such as more costly corrosion-resistant materials, painting and other corrosion-protective measures, earlier replacement of capital goods, increased spare-parts inventories, and increased maintenance. Some of these costs are avoidable and could be lowered by applying the economically best available corrosion prevention technology. However, decreasing the presently remaining unavoidable costs would require advances in technology.

The cost of the corrosion of metals will be considered from two standpoints: the cost to the economy of a nation and the cost of selected corrosion control measures. The main basis for the discussion of the cost to a nation will be the study published in 1978 [1] that the U.S. Congress directed the National Bureau of Standards (NBS; currently National Institute of Standards and Technology, NIST) to undertake. This study was, and still remains, probably the most comprehensive investigation of the full extent of corrosion on the economy of a nation. The analysis required was contracted out to Battelle Columbus Laboratories (BCL). The NBS–BCL study, unlike previous ones (B3), was based upon a solid technical–economic method that attempted to evaluate, in a rigorous way, all costs of corrosion, direct and indirect, over the entire economy and to evaluate the uncertainties in these cost estimates. "A significant feature of the study was that the method employed—input/output analysis—provides a methodological framework that permits comprehensive treatment of all elements of the costs of corrosion: production costs, capital costs, and changes in useful lives, for example. The input/output model allows analysis of interindustry relationships in the national economy and attribution of relative costs to specific segments of the economy" [1a, p. 30].

To carry out the second aspect of this discussion, a National Association of Corrosion Engineers (NACE) Recommended Practice [2] and a more recent NACE update [3] will be the source of the discussion of the economic appraisal of a selected corrosion control method.

[†]Retired.

B. ECONOMIC EFFECTS OF METALLIC CORROSION IN THE UNITED STATES

The NBS–BCL study [1] of metallic corrosion costs in the United States in 1975 employed a modified version of the BCL National Input/Output (I/O) model. A series of articles giving a detailed discussion of the NBS–BCL study [4] was published in the more available corrosion literature.

B1. Approach

B1.1. Elements of Corrosion Costs Used in the BCL I/O Model. The BCL I/O model using 130 sectors of the U.S. economy included the following as costs associated with corrosion: (a) material and labor expenditures associated with the protective measures of painting, applying cathodic protection, coatings, and use of inhibitors; (b) the expenses arising from the extra material and labor for prevention; (c) the partial corrosion losses that result in replacement costs and lost production; and (d) the expenses incurred by using information, technology transfer, research, development, and demonstration of methods to cope with the destructive effects of metallic corrosion.

A basic factor to be considered in measuring macrocorrosion costs in a sector of an economy was the lifetime and replacement value of a given component that could suffer from corrosion damage. Table 2.1 gives some of the elements of the costs of corrosion used in the BCL I/O model.

Corrosion costs used in the I/O approach can be either direct or indirect. The NBS report [1a, p. 11] describes these in the following way: Direct costs include all reductions in the requirements for inputs for production which would become possible if there were no corrosion. These include flow inputs (e.g., pig iron into steel), capital inputs for expansion and replacement of capacity (e.g., blast furnaces for steel), and

TABLE 2.1. Some Elements of the Costs of Corrosion[a]

Capital costs
   Replacement of equipment and buildings
   Excess capacity
   Redundant equipment
Control costs
   Maintenance and repair
   Corrosion control
Design costs
   Materials of construction
   Corrosion allowance
   Special processing
Associated costs
   Loss of product
   Technical support
   Insurance
   Parts and equipment inventory

[a]See [1a, p. 10].

value added. The direct flow effects include reduced maintenance costs and the use of less expensive materials for embodiment in outputs. Among the direct capital effects are the reduced need for equipment due to less time down for maintenance and the lower replacement cost because of increased equipment life. The value-added effects include reduced costs of labor and lower depreciation allowances for the smaller capital requirements. The indirect effects include two elements: (1) In addition to the reduced input requirements, the inputs also cost less, because of savings in their own and earlier production processes: (2) the general interactive effects of reductions in production levels on one another.

Value added is the additional value accruing to a product's ingredients as they are fashioned into the product itself. It includes wages, salaries, rents, profits, interest, taxes, and depreciation. It can also be defined as the value of the productive factors contributed by the industry itself, rather than those purchased from other industries.

B1.2. Scenarios for the I/O Model. Three scenarios were developed and used in the I/O model to quantify corrosion costs: Three "worlds" were formulated: *world I*—the real-world economy in 1975, *world II*—an imaginary corrosion-free world, and *world III*—a hypothetical world in which everyone applies the best economically practical corrosion control measures. The scenarios used in the model involved the gross national product (GNP—a more recently used term is gross domestic product, GDP). Three sets of these worlds were (a) world II minus world I is *the total national cost of corrosion* and represents resources that are wasted because of corrosion; (b) world III minus world I is *the total national avoidable costs* of corrosion and represents resources that would be available if economically best preventive practices were used throughout the economy; and (c) world III minus world II measures *presently unavoidable costs*.

B1.3. The I/O Model. To describe the BCL I/O model, it is best to quote directly from the NBS report [1a, p. 12]: "A number of characteristics make input–output analysis, pioneered by W. W. Leontief, and the modified Battelle model well suited for use in estimating the total direct and indirect costs of corrosion. The model is quite detailed. In this study, it has 130 economic sectors and each is represented by a production function consisting of the respective inputs from that sector plus value added. As a result, relatively detailed industry corrosion cost data may be incorporated into the model for simulation purposes. The complex structure serves as a guide for the precise analysis of corrosion costs and a means for integrating the results."

"The model is comprehensive. It has sufficient components to allow all the contributions to corrosion costs (production expenses, capital cost, reductions in replacement, and excess capital capacity, for example) to be considered in the analysis. Because of the model's structure, all of these aspects, and their interactions, may be evaluated in a coordinated and systematic manner."

"The model is simultaneous and, therefore, able to account for both direct and indirect effects of certain changes in the economy. This is critical to estimating the total costs of corrosion to society."

"Because the model simultaneously determines equilibrium values, comparative static analysis (i.e., comparison of alternate growth scenarios at the same moment in time) is an obvious application. For example, the costs of corrosion in the existing world (World I) are compared to those in each of the two hypothetical worlds mentioned previously—World II in which no corrosion exists and, thus, the costs are zero, and World III in which 'best practice' corrosion control methods are employed."

B2. Results

If we use the approach outlined above, the NBS–BCL study separates the total costs of corrosion into two costs: (a) *avoidable costs*, which are costs that can be reduced "by the most economically effective use of presently available corrosion technology," and (b) *unavoidable costs*, which result from "presently unavoidable losses." For the study's base year 1975, total costs of metallic corrosion (materials, labor, energy, and technical capabilities) were estimated by the BCL I/O model used in the study to be $82 billion, 4.9% of the $1677 billion GNP. Approximately 40% of this ($33 billion, 2.0% of the GNP) was estimated to be avoidable. Combining the BCL results and the NBS detailed analysis of the uncertainty, the total national yearly cost of metallic corrosion was reported to be about $70 billion (4.2% of GNP), with an uncertainty of ≈±30%. The NBS report found the 1975 avoidable cost of corrosion was roughly 15% of the total, but it estimated that it could have a range of 10–45%. An analysis of the errors, especially the estimates of the avoidable costs, led to the conclusion in the NBS report that the values cited above were reasonable.

Table 8 of the NBS report [1a] lists the total costs that were allocated to the 130 economic sectors of the United States that were produced by the BCL I/O model. This table gives corrosion losses on a dollar basis and as cost per unit of sales for the total and avoidable costs of both direct and direct plus indirect costs.

In addition to the 130 sectors of the economy provided by Table 8 of this report, the following special area costs were covered in more detail in the NBS report:

(a) U.S. federal government—total costs attributable to corrosion were estimated to be 2% of the federal budget with 20% of this total being judged to be avoidable.

(b) Personally owned automobiles—total expenses of corrosion in the ownership of an automobile were

found to be $6–14 billion, 1975 dollars, with avoidable costs being $2–8 billion.

(c) Electric power—total direct costs arising from the effects of corrosion on the operation of power generation plants were estimated to be $4.1 billion with avoidable costs of $120 million (3% of total costs).

(d) Fossil fuel energy and materials—total costs arising from the additional energy and materials losses resulting from corrosion were estimated to be $1.4 billion, with avoidable costs of $248.5 million for energy, and $1.705 billion, with avoidable costs of $212 million, for nonrenewable raw material sectors.

B3. 1995 Update of the NBS–BCL Study

The Specialty Steel Industry of North America engaged Battelle to produce a report [5] updating the NBS–BCL study [1, 3] and reflecting the changes resulting from economic growth, inflation, and 20 years of scientific research and technological advances. In Table 2.2, the revised estimates for 1995 are compared to the 1975 BCL values—not taking the NBS uncertainty analysis into consideration. In part, the Battelle panel that produced the report attributed the 1995 reductions in the percent of GNP to the following factors: (a) the anticorrosion technology of the motor vehicle industry (the most significant factor); (b) increased use of stainless steels, coated metals, and more protective coatings; (c) substitution of material to reduce weight; and (d) reclassification of unavoidable costs as avoidable.

In a more recent study, the annual direct cost of corrosion to industry and to governments in the United States was estimated to be approximately $276 billion, or 3.1% of the GDP [5a].

TABLE 2.2. Metallic Corrosion in the United States[a]

	1975 (billions of current dollars)	1995
All Industries		
Total	82.0	296.0
Avoidable	33.0	104.0
Motor Vehicles		
Total	31.4	94.0
Avoidable	23.1	65.0
Aircraft		
Total	3.0	13.0
Avoidable	6	3.0
Other Industries		
Total	47.6	189.0
Avoidable	9.3	36.0

[a]See [5].

C. CORROSION COSTS IN VARIOUS COUNTRIES

The fact that many countries have attempted to assess their national corrosion costs points up the worldwide awareness that corrosion can be a serious economic concern. It is useful to compare the United States results to those obtained in various industrial countries where such assessments of national corrosion costs were undertaken (including other past U.S. studies). These other studies of corrosion losses either involved major data gathering and interpretation efforts, which were the results of the analysis by a single authority of the country's corrosion costs in terms of the knowledge of the best presently available corrosion control measures and industrial practices, or, simply, assertions by an author with no reference as to how the costs were computed. This is the rationale for using the NBS–BCL study as the major focus of this discussion, because it looks at the entire economy of a nation and seeks to estimate the uncertainties in the numbers reported. Because of this more rigorous approach, the results of the NBS–BCL analysis gave higher total costs of corrosion but, surprisingly, agreed qualitatively with those found in previous studies as discussed in the NBS report. In 1986, a more recent examination of corrosion costs in various countries was the topic of an NACE symposium entitled "International Approaches to Reducing Corrosion Costs" [6]. The NBS study considered the following countries.

C1. United Kingdom

In 1969, a major data-gathering and interpretation endeavor, which was the precursor to the NBS–BCL U.S. study and unlike the less meticulous efforts of the countries described below, was initiated in the United Kingdom with the appointment by the Minister of Technology of a 25-member committee headed by one of the leading corrosionists in the United Kingdom, T. P. Hoar, to determine the cost of corrosion [7].

The committee contacted 800 industries in the country, all government departments, corrosion protection companies, and corrosion consultants. They were to gather from these sources information on the effects of corrosion, including the amount of shutdowns, rejection of product losses, structural failures, and the loss to industries from these. The committee added to these losses the costs of items replaced because of corrosion, expenditures on corrosion protection, and information services, research, and development in the various industries. Using these collected data, the Hoar committee arrived at an industrywide estimate of the cost of corrosion [7].

The Hoar report reported losses to the United Kingdom (Table 2.3) of £1.365 billion ($3.2 billion, 1969 U.S. dollars) for 1969–1970. This amounts to ≈3.5% of the GNP of the United Kingdom for that period. In addition, the committee

TABLE 2.3.   National Cost of Corrosion and Corrosion in the United Kingdom[a]

Industry or Agency	Estimated Cost (£M)
Building and construction	250
General engineering	11
Marine	280
Metal refining and semifabrication	15
Oil and chemical	180
Power	60
Transport	350
Water	25
Total	1365

[a]See [7].

found that some £310 million, or 23% of this total figure, was potentially avoidable. The estimated potential savings were £310 million, or 22.7% of the GNP. They suggested approaches toward achieving these savings, such as improved materials selection, specification and control of protective measures, improved awareness of corrosion, especially in design, and greater use of cathodic protection. An informal conference was held in 1971 as a supplement to the Hoar report to discuss the findings. Six sessions were held, one for each section of the report [8].

C2.   German Federal Republic

Behrens [9] estimated that total losses for the period 1968–1969 were 19 billion DM ($6 billion, 1969 dollars), with avoidable costs of 4.3 billion DM ($1.5 billion, 1969 dollars). No details were given as to what these figures include or how they were computed. Total costs were reported to be about 3% of the West German GNP for 1969, and avoidable losses were roughly 25% of total costs. These figures, with respect to GNP and percentage of avoidable cost, are in good agreement with figures found for other nations.

C3.   Sweden

A partial study of corrosion costs in Sweden by Trädgåidh [10] in which painting expenditures to combat corrosion were analyzed for the year 1964 found these costs to be 300–400 million crowns ($58–77 million, 1964 dollars) with between 25 and 35% being avoidable.

C4.   Finland

Costs to Finland for the year 1965 have been estimated by Vläsaari [11] to be 150–200 million markaa ($47–62 million, 1965 dollars). Linderborg [12], referring to these losses, described the factors that must be taken into account in assessing corrosion costs to the Finnish nation. He recognized that an important factor was the variable lifetimes for a variety of items using the specific example of the automobile.

C5. Union of Soviet Socialist Republics (Now Russia)

Kolotyrkin [13] reported in 1969 that corrosion costs were ≈2% of the GNP, or 6 billion rubles ($6.7 billion, 1969 U.S. dollars), giving no indication as to what this figure includes or how it was computed.

C6. Australia

The direct costs of corrosion in 1973 were estimated by Revie and Uhlig [14] to be A$470 million ($550 million, 1973 dollars). The authors decided that these costs are "probably too low" considering the factors they used to develop this figure. Some additional direct costs—mostly labor—for the mining, transportation, and communications industries were unavailable. Only muffler corrosion was considered as contributing to automobile losses. Lifetimes were not taken into account quantitatively. The amount of $470 million was 1.5% of Australia's GNP for 1973. However, since indirect costs may equal or exceed this figure, total corrosion costs to Australia were estimated to be ~3% of GNP. No quantitative effort was made to assess uncertainties or to separate these costs into avoidable and unavoidable components.

C7. India

For the period 1960–1961 Rajagopalan [15] estimated the cost of corrosion to India was 1.54 billion rupees ($320 million, 1961 dollars). He calculated the expenditures of certain measures to prevent or control corrosion, including direct material and labor expenses for protection, additional costs for increased corrosion resistance or redundancy, costs of information transfer, and funds spent on research and development. No quantitative estimate of uncertainty was attempted nor were avoidable and unavoidable costs broken down.

C8. Japan

A survey [16] conducted from 1976 to 1977 in Japan found that the annual direct cost of corrosion was 2500 billion yen ($9.2 billion, 1974 dollars), which amounts to 1.8% of the Japanese GNP. If the indirect costs were included, the total would increase severalfold.

C9. Previous U.S. Studies

Probably the first itemized measure of the costs of corrosion in the United States was carried out by Uhlig [17], who arrived at a value for the total direct corrosion losses of $5.5 billion for the late 1940s. A more recent study was carried out by NACE [18]. This study, based on replies of 1006 persons to a questionnaire, estimated the cost to NACE members of direct expenditures of corrosion control measures to be $9.67 billion for 1975.

D. COST OF SELECTED CORROSION CONTROL MEASURES

In order for the corrosion engineer to select the economically optimum corrosion control measure for a given specific problem, it is necessary that the corrosion engineer recognize the pertinent economic factors that bear on the choice of appropriate corrosion technology. To address this need, NACE Technical Unit Committee T-3C on Economics of Corrosion produced and issued in 1972 the Recommended Practice NACE Standard RP-02-72, "Direct Calculation of Economic Appraisals of Corrosion Control Measures" [2]. In 1994, the NACE Task Group T-3C-1 issued a technical report [3], "Economics of Corrosion," to replace, the simpler and less rigorous from an accounting practice standpoint, RP-02-72. The objectives of the 1994 report were(1) present the economic techniques in a form that can be readily understood and used by engineers as a decision-making tool; (2) facilitate the communication of decisions between the corrosion technologist and management; and (3) justify investments in anticorrosion methods that have long-term benefits.

The 1994 report used more advanced and standardized accounting notation and terminology that was based on American National Standards Institute (ANSI) standard Z94.5 entitled "Engineering Economy" [19]. Another more accessible source of the calculational techniques that enable an economic appraisal of corrosion control measures has been published by Verink [20]. A useful feature of the report (also present in the 1972 document) was a section devoted to worked examples and applications using the calculational techniques in the report to select process equipment and the best alternate cathodic protection proposal.

REFERENCES

1. L. H. Bennett, J. Kruger, R. I. Parker, E. Passaglia, C. Reimann, A. W. Ruff, H. Yakowitz, and E. B. Berman, "Economic Effects of Metallic Corrosion in the United States—A Three Part Study for Congress." (a) Part I, NBS Special Publication 511-1, SD Stock No. 003-003-01926-7. (b) Part II, NBS Special Publication 511-2 Appendix B. A Report to NBS by Battelle Columbus Laboratories, SD Stock No. 003-003-01927-5, U.S. Government Printing Office, Washington, DC, 1978. (c) Part III Appendix C, Battelle Columbus Input/Output Tables, NBS GCR78-122, PB-279 430, National Technical Information Service, Springfield, VA, 1978.

2. NACE Standard RP-02-72, "Recommended Practice Direct Calculation of Economic Appraisals of Corrosion Control Measures," National Association of Corrosion Engineers, Houston, TX, 1972.

3. NACE Technical Committee Report, Task Group T-3C-1 on Industrial Economic Calculational Techniques, "Economics of Corrosion," National Association of Corrosion Engineers, Houston, TX, 1994.

4. J. H. Payer, W. K. Boyd, D. G. Dippold, and W. H. Fisher, Mater. Perform., **19**(1–7) (1980).

5. Report by Battelle to Specialty Steel Industry of North America, "Economic Effects of Metallic Corrosion in the United States—A 1995 Update," Apr. 1995. (a) G. H. Koch, M. P. H. Brongers, N. G. Thompson, Y. Paul Virmani, and J. H. Payer, Corrosion Costs and Preventive Strategies in the United States, Supplement to *Materials Performance,* July 2002, Report No. FHWA-RD-01-156, Federal Highway Administration, McLean, VA, 2002. See also www.corrosioncost.com.

6. NACE Symposium, "International Approaches to Reducing Corrosion Costs," R. N. Parkins (Ed.), NACE, Houston, TX, March 1986.

7. T. P. Hoar, "Report of the Committee on Corrosion and Protection," Department of Trade and Industry, H.M.S.O., London, U.K. 1971.

8. T. P. Hoar, Information Conference "Corrosion and Protection," presented at the Instn. Mech. Engrs., April 20–21, 1971, to discuss the Hoar Committee.

9. D. Behrens, Br. Corros. J., **10**(3), 122 (1975).

10. K. F. Trädgåidh, Tekn. Tedskrift (Sweden), **95**(43), 1191 (1965). (quoted by Linderborg [12]).

11. V. Vläsaari, Talouselämä' (Economy), No. 14/15, **351**(1965) (quoted by Payer et al. [4]).

12. S. Linderborg, Kemian Teollusius (Finland), **24**(3), 234 (1967).

13. Y. Kolotyrkin, quoted in Sov. Life, **9**, 168 (1970).

14. R. W. Revie and H. H. Uhlig, J. Inst. Engr. Aust., **46**(3–4), 3 (1974).

15. K. S. Rajagopalan, Report on Metallic Corrosion and Its Prevention in India, CSIR. Summary published in "The Hindu," English language newspaper (Madras), Nov. 12, 1973.

16. Committee on Corrosion and Protection, Boshoku Gijutsu (Corrosion Eng.), **26**(7), 401 (1977).

17. H. H. Uhlig, Corrosion, **6**(1), 29 (1950).

18. NACE Committee Survey Report, Corrosion, 31 (10) 1975.

19. ANSI Standard Z94.5 (latest revision), "Engineering Economy," American National Standards Institute, New York.

20. E. D. Verink, "Corrosion Economic Calculations," in ASM Handbook, Vol. 13, 9th ed., Corrosion, ASM International, Materials Park, OH, 1987, pp. 369–374.

3

ECONOMICS OF CORROSION[*]

E. D. VERINK, JR.[†]

Distinguished Service Professor Emeritus, Department of Materials Science and Engineering, University of Florida, Gainesville, Florida

A. INTRODUCTION

Over $220 billion is lost to corrosion in the United States each year, according to government and industry studies. This corrosion cost is equivalent to 3 or 4% of the gross national product (GNP). The real tragedy of this annual corrosion cost is that ~ 15% or more could be saved by the application of existing technology to prevent and control corrosion. The existing technology includes the following methodologies to prevent and control corrosion: proper design, selection of materials, coatings and linings, cathodic protection, and inhibitors. Standards, reports, books, and thousands of technical articles attest to the successful use of the existing technology to prevent and control corrosion and thus to reduce the annual losses to corrosion.

B. CORROSION—AN ECONOMIC PROBLEM

Corrosion is essentially an economic problem. It is vitally important that engineers and engineering managers be aware of the economic impact their decisions have on the ability of a business to meet its corporate goals. A discipline that assists in the measurement of the economic impact of such decisions is called "engineering economy."

Engineering economy is concerned with money, both as a resource and as the price of other resources. Business success is dependent on the prudent and efficient use of all resources, including money. The principles of engineering economy permit direct comparisons of potential alternatives in monetary terms. In this way, they encourage efficient use of resources.

The economics of corrosion evaluation involves the assessment of the technical validity and economic justification of each alternative. Corrosion technologists generally spend much effort exploring the technical validity of materials or processes. However, each corrosion problem may have more than one material or process that could satisfactorily solve the technical problem. Each candidate material or process probably has a unique stream of investment and operating

---

[*] Adapted from "Economics of Corrosion," Publication 3C194, Technical Committee Report, copyright © 1994 by NACE International. Reprinted with permission. Further reproduction is prohibited. NACE technical committee reports are reviewed every 10 years; users should contact NACE at P.O. Box 218340, Houston, Texas 77218-8340.
[†] Retired.

Uhlig's Corrosion Handbook, Third Edition, Edited by R. Winston Revie
Copyright © 2011 John Wiley & Sons, Inc.

or maintenance costs while providing equivalent technical benefits. Indeed, once the technically viable alternatives are chosen and their respective costs and performance characteristics are isolated, the technologist's decision becomes one of financial analysis because almost all of the factors in the decision process can be reduced to the magnitude and timing of cash flows.

The purposes of this chapter are:

1. To present the economic techniques in a form that can be readily understood and used by engineers as a decision-making tool
2. To facilitate the communication of decisions between the corrosion technologist and management
3. To justify investment in anticorrosion methods that have long-term benefits

B1. Basics—Money and Time

Consider the effects of time and earning power on $20.00. If $20.00 is placed in the bank and earns interest at a rate of 5% per year, it grows to $21.00 when the interest, $1.00, is paid at the end of the year. Thus, $20.00 today at 5% is equivalent to $21.00 a year from now. Stated another way, in order to have $21.00 one year from now, only $20.00 has to be deposited today if the interest rate is 5%. The $20.00 is called the "discounted present value" of the $21.00 needed one year hence.

The initial deposit as well as the earned interest left in the account have earning power because the interest is compounded, which means that it is computed on both the principal and the accrued interest.

Suppose the $20.00 were used to pay four equal annual installments of $5.00 each. Without interest, the $20.00 would be exhausted after making the last payment. If the $20.00 is deposited in the bank at 5% interest, it will be worth $21.00 at the end of the first year. Paying out $5.00 would leave $16.00 to be held at 5% interest for the second year. The $16.00 invested at 5% will earn $0.80 in one year. Subtracting $5.00 from $16.80 leaves $11.80 to be held at 5% interest for the third year. The $11.80 will earn $0.59 by the end of the third year at 5%. Another annual payment of $5.00 leaves $7.39 to earn interest during the fourth year. Adding the $0.37 interest earned during the fourth year and subtracting the final $5.00 annual payment leaves a balance of accrued interest of $2.76.

The earning power of money permits another strategy. If the initial deposit were reduced to $17.73 at 5% interest, $5.00 could be paid out each year for four years and nothing would be left. This example illustrates the distinction between the terms "equivalent" and "equal." The $20.00 is equal to four payments of $5.00 each. It also would be equivalent to four $5.00 annual payments (only) if the interest rate were zero. The $17.75 is not equal to the sum of four payments of $5.00 each. However, when $17.73 is invested at 5%, it is equivalent to four annual payments of $5.00.

The term "equivalent" implies that the concept of the time value of money is applied at some specific interest rate. Therefore, for an amount of money to have a precise meaning, it must be fixed both in time and amount. Mathematical formulas and tables are available to translate an amount of money at any particular time into an equivalent amount at another date.

Many kinds of translations are possible. For example, a single amount of money can be translated into an equivalent amount at either a later or an earlier date. This is accomplished by calculating the present worth (PW) or the future worth (FW) as of the present date. Single amounts of money can be translated into equivalent annuities (A) involving a series of uniform amounts occurring each year. Conversely, annuities can be translated into equivalent single amounts at either an earlier or a later date. The present worth of an annuity (P/A) is the single amount of money equivalent to a future annuity. The single amount equivalent to a past annuity is referred to as the future worth of an annuity (F/A).

It is also possible to calculate the amount of money that would be equivalent to a nonuniform series of cash flows. Two types of nonlinear series that find application are an arithmetic progression, in which the series changes by a constant amount, and a geometric progression, in which the series changes by a constant rate. The arithmetic progression is considered to be representative of variable costs, such as maintenance costs, which may increase as equipment ages. The geometric progression is used to represent the effects of inflation or deflation.

C. NOTATION AND TERMINOLOGY

The American National Standards Institute (ANSI) standard Z94.5 titled "Engineering Economy" consists of a compilation of the symbology and terminology of the field so that the improved communication benefits of standardization are available to practitioners [1]. With the development and publication of this standard and its adoption by the Institute of Industrial Engineers and the Engineering Economy Division of the American Society of Engineering Education, it is expected that future books and articles will utilize these symbols common to engineering economy because they represent the consensus choice of the prominent modern authors and educators in this field. This would avoid one of the significant previous deterrents to the use of these methods in the past.

The reader is referred to ANSI standard Z94.5 for further details, including functional forms and uses of compound interest factors, and formulas involving annual compounding and others involving continuous compounding [1].

The basic form of the notation used for all time value factors consists of a ratio of two letters representing two amounts of money (e.g., P/A or F/P) plus an interest rate ($i\%$) and a number of periods (n). The customary manner of writing these is

$$(P/A, i\%, n)$$

or

$$(F/P, i\%, n)$$

The present worth (P) of a known annuity (A) can also be expressed as follows:

$$P = A(P/A) \qquad (3.1)$$

or the future worth (F) of a known present amount (P) is

$$F = P(F/P) \qquad (3.2)$$

Other forms also are common, such as PW($\cdot$) or FW($\cdot$), which are called "operators" because they represent some computational operation, such as the PW or the FW of whatever is inside the parentheses.

It should be evident that (F/P) is the reciprocal of (P/F) and that (F/A) and (A/F), and so on, also are reciprocals. This observation is useful because it means that only three time value factors need to be tabulated in order to conduct six operations.

Another algebraic relationship shows that if two time value factors are multiplied, the product is a third time value factor. For example,

$$(A/F) \times (P/A) = (P/F) \qquad (3.3)$$

D. METHODS OF ECONOMIC ANALYSIS

Economic analysis methods that are concerned with the entire service life sometimes are called "life-cycle cost" methods. Those that lead to single-measure numbers include:

(a) Internal rate of return (IROR)
(b) Discounted payback (DPB)
(c) Present worth (PW) method, also referred to as the net present value (NPV)
(d) Present worth of future revenue requirements (PWRR)
(e) Benefit–cost (BCR) ratios

All five methods employ the concept of PW. While each method has certain advantages, the individual methods vary considerably with respect to their application and complexity.

The IROR method compares the initial capital investment with the PW of a series of net revenues or savings over the anticipated service life. Expenses include all operation, maintenance, taxes, insurance, and overheads but do not include return on (or of) the invested capital. From an economic standpoint, IROR consists essentially of the interest cost on borrowed capital plus any existing (positive or negative) profit margin. The disadvantage of this method is that it ignores benefits extending beyond the assumed life of the equipment and thereby may omit a substantial part of the actual service life. This may lead to unnecessarily pessimistic measures of long-range economy.

The PWRR method is particularly applicable to regulated public utilities, at which the rates of return are more or less fixed by regulation. It is particularly applicable when it has already been determined by IROR analysis that a project is economically viable, and the engineer wishes to determine which is the most economic alternative under circumstances wherein several alternatives produce the same revenue but some of them create less expense (requirement for revenue) and consequently a greater profit margin (or lower losses) than others. The principal objection to the PWRR method is that it is inadequate when alternatives are competing for a limited amount of capital because it does not identify the alternative that produces the greatest return on invested capital.

The DPB method is somewhat more complicated than the PWRR method. The BCR method is similar to the IROR method because both methods involve assessment of alternatives, not only for economic measures compared with a "do-nothing" scenario, but also for incremental measures associated with incremental capital investments.

The PW method, also referred to as the NPV method, is considered the easiest and most direct of the five methods and has the broadest application to engineering economy problems. Many industries refer to this method as the "discounted cash flow" method of analysis. This method often is used as the "referee" method to test the results of other methods of analysis. Under the circumstances, it is not surprising that there is a preference for this method and that primary attention is given to this method in this chapter. Those interested in exploring the other methods are referred to standard texts on the subject.

D1. Annual Versus Continuous Compounding

Because actual cash flow (both inward and outward) is continuous, it appears that continuous compounding is the more accurate assumption for engineering economy studies. However, although the overall cash flows tend to be continuous, the cash flow data seldom are sufficiently precise in economy studies to take full advantage of continuous compounding. It is also true that the normal purpose of the economy study is to analyze some specific event that will occur at a specific time, so a procedure that is readily associated with a specific time is usually the most appropriate. For these reasons, and because annual compounding is conceptually easier to understand and apply, annual compounding is the method primarily used in this chapter. The PW method is a form of discounted cash flow (DCF), wherein cash flow data, which include dates of receipts and disbursements, are discounted to PW. Before applying these methods, a management judgment is made as to the desired life (usually expressed as a number of years, n) and the minimum acceptable rate of return (ROR) on invested capital for a project (expressed in terms of the effective interest rate, i, or the nominal rate of return, r). Rates of return (before taxes) vary among industries, ranging from 10–15%, where obsolescence is not high, to 25–40% (or perhaps higher) for dynamic industries. Obviously, it is convenient if the minimum acceptable ROR is known before making engineering economy analyses. However, it is not always easy to learn what is considered an acceptable ROR. Under such circumstances, it has been helpful to prepare a series of economic alternatives in which the ROR is varied so that management can make a choice.

An example illustrates the features of the method. In this case, the "longhand" method is used. In practice, there are several shortcuts.

Should an expenditure of $15,000 be made to reduce labor and maintenance costs from $8200 to $5100 per year? Money is worth 10% and the life of the project is 10 years. For simplicity, the effects of taxes and depreciation have been neglected in this example.

Table 3.1 shows the projected pattern of cash flow for the "defender" (the present method) and the "challenger"

TABLE 3.1.   Tabulation of Cash Flow

Period (year)	Plan A (dollars) (Defender)	Plan B (dollars) (Challenger)	B − A (dollars)
0	—	− 15,000	− 15,000
1	− 8,200	− 5,100	+ 3,100
2	− 8,200	− 5,100	+ 3,100
3	− 8,200	− 5,100	+ 3,100
4	− 8,200	− 5,100	+ 3,100
5	− 8,200	− 5,100	+ 3,100
6	− 8,200	− 5,100	+ 3,100
7	− 8,200	− 5,100	+ 3,100
8	− 8,200	− 5,100	+ 3,100
9	− 8,200	− 5,100	+ 3,100
10	− 8,200	− 5,100	+ 3,100
Totals	− 82,000	− 66,000	+ 16,000

Source: E. Verink, "Corrosion Economic Calculations," ASM Handbook, Vol. 13, 9th ed., Corrosion, ASM International, Materials Park, OH, 1987, p. 372 (Table 4).

(the proposed method) over the life of the project. A minus sign means that money leaves the "bank," whereas a plus sign means that the "bank balance" increases in size. The net cash flow for each year appears in the right-hand column under the heading "B − A."

It is apparent that selection of the challenger (plan B) results in a net positive cash flow. That is, the net amount of money in the bank is increased over the life of the project when plan B is selected. Before reaching a conclusion regarding implementation of plan B, these cash flows are reduced to a common basis for comparison and to determine whether the objective of a 10% ROR has been achieved. The PW (or present value) of these cash flows provides such a basis. The ROR for plan B is calculated by iteration using interest tables and interpolating between values.

D1.1. First Iteration. Assume a rate of return of 10% and refer to Table 3.2:

$$(PW) = -15,500 + 3100(P/A, 10\%, \ 10 \text{ years}) \quad (3.4)$$
$$= -15,500 + 3100(6.145) = +\$4047.95 \quad (3.5)$$

The ROR for which the discounted cash flow is equal to zero (i.e., the first term on the right of the above expression is balanced by the second term) is the actual rate of return. From the first iteration, it already is evident that plan B returns more than 10% because the net cash flow is positive. Thus, plan B, the challenger, is the more economical. The numerical value of the actual rate of return can be determined by additional iterations. Such an exercise reveals that the ROR in this case is 16.1%.

In this example, plans A and B could represent alternative materials of construction having different corrosion rates, with the annual dollar difference being related to the consequences of corrosion on maintenance costs. This presupposes the availability of corrosion data that can be used to estimate expected life, maintenance costs, and so on. Other examples illustrate how to account for the effects of salvage value, taxes, and depreciation.

E. DEPRECIATION

Depreciation has been defined as the lessening in value of an asset with the passage of time. All physical assets (with the possible exception of land) depreciate with time. There are several types of depreciation. Two of the more common types are physical depreciation and functional depreciation. Accidents can also cause loss of value, but this cause is often accommodated in other ways (i.e., insurance or reserves) and is not considered here.

Physical depreciation includes such phenomena as deterioration resulting from corrosion, rotting of wood, bacterial

TABLE 3.2. 10% Interest Factors for Annual Compounding

	Single Payment		Equal-Payment Series				
	Compound-Amount Factor (To find F Given P F/P, i, n)	Present-Worth Factor (To find P Given F P/F, i, n)	Compound-Amount Factor (To find F Given A F/A, i, n)	Sinking-Fund Factor (To find A Given F A/F, i, n)	Present-Worth Factor (To find P Given A P/A, i, n)	Capital Recovery Factor (To find A Given P A/G, i, n)	Uniform Gradient-Series Factor (To find A Given G A/G, i, n)
n							
1	1.100	0.9091	1.000	1.0000	0.9091	1.1000	0.0000
2	1.210	0.8265	2.100	0.4762	1.7355	0.5762	0.4762
3	1.331	0.7513	3.310	0.3021	2.4869	0.4021	0.9366
4	1.464	0.6830	4.641	0.2155	3.1699	0.3155	1.3812
5	1.611	0.6209	6.105	0.1638	3.7908	0.2638	1.8101
6	1.772	0.5645	7.716	0.1296	4.3553	0.2296	2.2236
7	1.949	0.5132	9.487	0.1054	4.8684	0.2054	2.6216
8	2.144	0.4665	11.436	0.0875	5.3349	0.1875	3.0045
9	2.358	0.4241	13.579	0.0738	5.7590	0.1737	3.3724
10	2.594	0.3856	15.937	0.0628	6.1446	0.1628	3.7255
11	2.853	0.3505	18.531	0.0540	6.4951	0.1540	4.0641
12	3.138	0.3186	21.384	0.0468	6.8137	0.1468	4.3884
13	3.452	0.2897	24.523	0.0408	7.1034	0.1408	4.6988
14	3.798	0.2633	27.975	0.0358	7.3667	0.1358	4.9955
15	4.177	0.2394	31.772	0.0315	7.6061	0.1315	5.2789
16	4.595	0.2176	35.950	0.0278	7.8237	0.1278	5.5493
17	5.054	0.1979	40.545	0.0247	8.0216	0.1247	5.8071
18	5.560	0.1799	45.599	0.0219	8.2014	0.1219	6.0526
19	6.116	0.1635	51.159	0.0196	8.3649	0.1196	6.2861
20	6.728	0.1487	57.275	0.0175	8.5136	0.1175	6.5081
21	7.400	0.1351	64.003	0.0156	8.6487	0.1156	6.7189
22	8.140	0.1229	71.403	0.0140	8.7716	0.1140	6.9189
23	8.954	0.1117	79.543	0.0126	8.8832	0.1126	7.1085
24	9.850	0.1015	86.497	0.0113	8.9848	0.1113	7.2881
25	10.835	0.0923	96.347	0.0102	9.0771	0.1102	7.4580
26	11.918	0.0839	109.182	0.0092	9.1610	0.1092	7.6187
27	13.110	0.0763	121.100	0.0083	9.2372	0.1083	7.7704
28	14.421	0.0694	134.210	0.0075	9.3066	0.1075	7.9137
29	15.863	0.0630	148.631	0.0067	9.3696	0.1067	8.0489
30	17.449	0.0573	164.494	0.0061	9.4269	0.1061	8.1762
31	19.194	0.0521	181.943	0.0055	9.4790	0.1055	8.2962
32	21.114	0.0474	201.138	0.0050	9.5264	0.1050	8.4091
33	23.225	0.0431	222.252	0.0045	9.5694	0.1045	8.5152
34	25.548	0.0392	245.477	0.0041	9.6086	0.1041	8.6149
35	28.102	0.0356	271.024	0.0037	9.6442	0.1037	8.7086
40	45.259	0.0221	442.593	0.0023	9.7791	0.1023	9.0962
45	72.890	0.0137	718.905	0.0014	9.6628	0.1014	9.3741
50	117.391	0.0085	1,163.909	0.0009	9.9148	0.1009	9.5704
55	189.059	0.0053	1,880.591	0.0005	9.9471	0.1005	9.7075
60	304.482	0.0033	3,034.816	0.0003	9.9672	0.1003	9.8023
65	490.371	0.0020	4,893.707	0.0002	9.9796	0.1002	9.8672
70	789.747	0.0013	7,887.470	0.0001	9.9873	0.1001	9.9113
75	1,271.695	0.0008	12,708.954	0.0001	9.9921	0.1001	9.9410
80	2,048.400	0.0005	20,474.002	0.0001	9.9951	0.1001	9.9609
85	3,298.969	0.0003	32,979.690	0.0000	9.9970	0.1000	9.9742
90	5,313.023	0.0002	53,120.226	0.0000	9.9981	0.1000	9.9831
95	8,556.676	0.0001	85,556.760	0.0000	9.9988	0.1000	9.9889
100	13,780.612	0.0001	137,796.123	0.0000	9.9993	0.1000	9.9928

Source: G. J. Thuesen and W. J. Fabrycky, Engineering Economy, 6th ed., Prentice-Hall, Englewood Cliffs, NJ, 1984, p. 574.

action, chemical decomposition, wear and tear, and so on, which can reduce the ability of an asset to render its intended service.

Functional depreciation results not from the inability of an asset to be available to serve its intended purpose, but rather from the fact that some other asset is available that can perform the desired function more economically. Thus, obsolescence and/or inadequacy, or inability to meet the demands placed on the asset, lead to functional depreciation. Technological advances produce improvements that often result in obsolescence of existing assets.

The manner in which depreciation is accounted for is largely a tax question. The language of the tax laws specifies the procedures that are permissible. Tax laws change from time to time, so procedures that are attractive under a given set of circumstances may become unattractive (or even forbidden) under other circumstances. Some of the more common methods of depreciation include:

1. Straight line
2. Declining balance
3. Declining balance switching to straight line
4. Sum of the year's digits
5. Accelerated cost recovery system

E1. Straight-Line Method

The straight-line depreciation method assumes that the value of an asset declines at a constant rate. If the asset originally cost $6000 and had a salvage value of $1000 after five years' life, the annual depreciation would be ($6000 − 1000)/5 = $1000/year. Symbolically, this can be expressed

$$\text{Annual depreciation,} \quad D = \frac{P - S}{n} \qquad (3.6)$$

The "book value" of the asset would decrease at the end of each year by the amount of the annual depreciation until, at the end of the fifth year, the book value would be the same as the salvage value.

E2. Declining-Balance Method

The declining-balance method assumes that the asset depreciates more rapidly during early years than in later years. A certain percent of depreciation is applied each year to the remaining book value of the asset. Under these circumstances, the size of the depreciation declines each successive year until the asset is fully depreciated. When the declining-balance method is used, the maximum rate that has been permissible for tax purposes is double the straight-line rate. This accounts for the term "double-declining balance."

E3. Declining Balance Switching to Straight Line

Prior to 1981 in the United States, it was allowable to depreciate an asset using declining-balance depreciation for the early years and then switch to straight-line depreciation when the allowable depreciation (using declining balance) falls below the amount permissible under straight-line depreciation. Switching to straight-line depreciation permits the book value to go to zero eventually. The declining-balance method switching to straight line has been incorporated in part of the 1986 version of the ACRS depreciation method discussed below.

E4. Sum-of-the-Year's-Digits Method

The sum-of-the-year's-digits method assumes that the value of an asset decreases at a decreasing rate. Assume an asset has a five-year life. The sum of the year's digits equals $1 + 2 + 3 + 4 + 5 = 15$. For the $6000 asset mentioned above, which has a salvage value after five years of $1000, the first year's depreciation would be ($6000 − 1000)(5/15) = $1666.67. The second year's depreciation would be ($6000 − 1000) (4/15) = $1333.33, and so on, until the asset is fully depreciated. This method is no longer permitted in the United States for new assets.

Most of the property acquired before 1981 is still being depreciated by one of the methods mentioned above. Each of these (pre-1981) methods involves the taxpayer estimating a "useful life" either on the basis of experience or based on a guideline from the U.S. Internal Revenue Service in Washington, DC. The guidelines are presented as ranges of values in the class life asset depreciation range (ADR) system. The midpoint of the range is referred to as the "ADR life."

E5. Accelerated Cost Recovery System Method

For property placed in service after 1981, the accelerated cost recovery system (ACRS), which is a part of the Economic Recovery Act of 1981, prescribes a different method for recovering the cost of depreciable property. The Tax Reform Act of 1986 revamped the ACRS enacted in 1981. The 1986 act generally is less generous than the prior law, which combined ACRS with investment tax credit (ITC). While ACRS was originally designed primarily as an incentive for investment, the new rules are intended to provide a more even match between the class lives of particular assets and their useful lives. Nonetheless, deductions are more accelerated than under pre-1981 law.

The principal differences between the prior version of ACRS and the new law are in the class lives of assets and the methods of recovering their costs. As in the prior law, salvage continues to be disregarded. The new system is generally effective for assets placed in service after 1986, but by special election, property placed in service after July 1, 1986, also may qualify under the new rules.

TABLE 3.3. Recovery Property

ACRS Class and Method	ADR Midpoint	Special Rules
3-year, 200% DB	4 years or less	Excludes cars, light trucks
5-year, 200% DB	> 4 to <10 years	Includes cars, light trucks, semiconductor manufacturing equipment, qualified technical equipment, R & D property, and so on
7-year, 200% DB	10 to <16 years	Includes agricultural, horticultural structures, and RR track. Property with no ADR midpoint
10-year, 200% DB	16 to <20 years	None
15-year, 150% DB–SL	20 to <25 years	Includes sewage treatment plants, telephone, data, and voice communication
20-year, 150% DB–SL	25 years or more	Excludes real property with ADR midpoint 27.5 years or more. Includes municipal sewers
27.5-year, SL	N/A	Residential rental property
31.5-year, SL	N/A	Nonresidential real property

Note: DB—declining balance; SL—straight line; N/A—not applicable.

Assets are assigned to one of six classes of depreciable property or to one of two classes of real property. The ACRS classes are tabulated in Table 3.3. Costs in the more short-lived classes (e.g., 3-, 5-, and 10-year classes) are recovered using double-declining-balance depreciation. It is permissible to optimize deductions by switching to straight-line depreciation when this becomes advantageous. The 15- and 20-year classes recover using the 150% declining-balance method, switching to the straight line when the depreciation based on the straight-line method (for the remaining years of depreciable life) is greater than the declining-balance amount. All real estate is depreciated by the straight-line method.

The new law does not provide statutory cost recovery allowances; therefore, averaging conventions are used. Generally, a half-year convention applies for the first-year and last-year allowances for depreciable property, and the midmonth convention applies to real property. A special rule provides that when more than 40% of asset additions in any year are placed into service in the last quarter of the taxable year, a midquarter convention is used to compute the allowances for all additions during the year. Under prior ACRS rules, a half-year convention was built into the statutory allowance for the first year of service, but the cost of three-year property, for example, still was "recovered in three years. Under the new statute, the half-year convention for personal property effectively adds another year to the cost recovery period because for the same three-year asset the half-year convention applies to the first year; a full allowance applies to years 2 and 3, and the remainder is recovered in the fourth year. Thus, the deductions are spread over four years although the recovery period is considered to be three years.

E6. Alternate Cost Recovery System

The 1981 statute provided a variety of alternate cost recovery options. By contrast, the 1986 statute unifies into one alternative cost recovery system several variations of ACRS enacted for various purposes. The alternative system provides longer lives for assets (i.e., slower rates of depreciation) and utilizes straight-line depreciation. The alternative system life is its ADR midpoint life. For real property, the ADR midpoint life is 40 years. If there is no ADR midpoint for personal property, a 12-year life is assigned.

There also have been changes in ACRS class for particular assets. Under prior law, cars and light trucks were 3-year property. Now they are classified as 5-year property, as listed in Table 3.3. Similarly, most equipment that was not a public utility property was in the 5-year class. Now many types of equipment are classified as 7- or 10-year equipment.

Notwithstanding these general rules, there are several assets for which there are special rules. These are listed in Table 3.4.

Taxpayers may elect the alternative system of determining the applicable depreciation allowance, or they may elect to use the straight-line method over the ACRS life. Either election is made on a class-by-class basis and, once made, is irrevocable. For example, for property put into service in a given year, a taxpayer may choose the regular system for the three- and five-year property, while choosing the alternative system for the seven-year property put into service that year. Property put into service the subsequent year is subject to a new election (on a class-by-class basis).

TABLE 3.4. Special Alternative ACRS Lives

Type of Property	Alternative Class Life (years)
Semiconductor property	5
Telephone central switching equipment	9.5
Railroad track	10
Single-purpose agricultural structures	15
Municipal wastewater treatment plant	24
Telephone distribution plant and equipment	24
Municipal sewers	50

The 1986 act repeals the ITC. Assets placed into service after December 31, 1985, are not eligible for ITC. A transition rule is provided to permit credit for assets for which binding contracts existed on this date. The transition rule requires that the asset must be placed into service by a specific date based on the property's depreciation recovery class. The ITC carryovers that had not expired as of December 31, 1985, are preserved and may be used in 1986 (and thereafter); however, they are reduced by 35%.

For detailed information, readers are referred to the act itself or to explanatory treatises [2].

F. GENERALIZED EQUATIONS

Generalized equations that simplify the solution of a large percentage of engineering economy problems have been set up. These equations take into account the influence of taxes, depreciation, operating expenses, and salvage value in the calculation of present worth and annual cost. Using these equations, an individual problem can be solved merely by entering data into the equations with the assistance of the compound interest tables and solving for the unknown value. The tax rate, t, is expressed as a decimal. Operating expense is represented by X. The letter S represents salvage value.

For straight-line depreciation

$$(PW) = -P + \frac{t(P-S)}{n}(P/A, \, i\%, n)$$
$$- (1-t)(X)(P/A, \, i\%, n) + S(P/F, \, i\%, n)$$
$$(3.7)$$

In this expression for the present worth, the first term, P, is the cost at "time zero" of the initial investment. The value at time zero is a definition of PW (at time zero). Hence, it is not necessary to translate this amount to another date because evaluation is made at time zero. Because the investment involves flow of money out of the bank, the sign of this term is minus.

The second term is concerned with depreciation. The depreciation method influences the mathematical formulation of this term. The annual amount of tax credit permitted by this method of depreciation, in this case, straight-line depreciation, is expressed by $[t(P-S)/n]$. These equal annual amounts are translated back to zero time by converting them to present worth using $(P/A, \, i\%, n)$. The ACRS depreciation methods allow specific straight-line depreciation options. When one of these options is chosen, the formulation of this term remains the same as shown above with the exception that salvage is ignored. If the straight-line option is not chosen, this term in the generalized equation becomes a series of terms each having the tax rate multiplied by the appropriate depreciation percentage of the initial cost multiplied by the appropriate single-payment present worth factor (P/F) for the applicable year and interest rate. The sum of these terms represents the effect of depreciation on PW.

The third term in the general equation actually consists of two terms. One is $[(X)(P/A, \, i\%, n)]$, and it represents the cost of items properly chargeable as expense. Examples include such things as cost of maintenance or insurance, cost of inhibitors, and so on. Because this term involves expenditure of money from the bank, it carries a minus sign. The term $[t(X)(P/A, \, i\%, n)]$ accounts for the tax credit for this business expense. Because it represents a saving, the sign is plus.

The fourth term translates the anticipated (future) value of salvage to present value. This is a one-time event rather than a uniform series; therefore, it involves the single-payment present-worth factor. As noted above, the ACRS depreciation method ignores the salvage value of an asset, so when using the ACRS method, the fourth term is zero.

Present worth can be converted to equivalent annual cost by use of the following equation:

$$A = (PW)(A/P, \, i\%, n) \qquad (3.8)$$

Alternative materials having the same life can be compared from an engineering economy standpoint merely by comparing the magnitude of their present values. However, if the candidate materials have different life expectancies, it is necessary to convert present worth to equivalent annual cost, A, to compare the materials from an engineering economy standpoint. The material with the lowest annual cost is the material of choice.

G. WORKED EXAMPLES AND APPLICATIONS

Example 3.1. Process Equipment. A new heat exchanger is required in conjunction with rearrangement of existing facilities. Because of corrosion, the expected life of a carbon steel exchanger is 5 years. The installed cost is $9500. It is proposed to substitute a Unified Numbering System (UNS) S31600 [American Iron and Steel Institute (AISI) 316 stainless steel] exchanger having an installed cost of $26,500, and an estimated life of 15 years, written off in 11 years. Which is the more economical choice based on annual cost? The minimum acceptable rate of return is 10%, the tax rate is 48%, and the depreciation method is straight line.

Because the lives of the alternatives are unequal, the economic choice cannot be based merely on the discounted cash flow over a single life of each alternative. Instead, comparison is made on the basis of equivalent uniform annual costs, commonly referred to as "annual cost," as mentioned above. Referring to the generalized equation (3.7), data in this example are available for only the first two terms of the equation. The third term, which involves maintenance expense, and the fourth term, involving salvage value, are both assumed to be zero.

The first step is to compute the discounted cash flow over one life span for each alternative. This involves the calculation of the PW for each material:

$$(PW)_{steel} = -\$9500 + \frac{0.48(9500 - 0)}{5}(3.791) = -\$6043 \tag{3.9}$$

$$(Pw)_{316} = -\$26,500 + \frac{0.48(26,500 - 0)}{11} \tag{3.10}$$

To compare the two alternatives, the discounted cash flows (PWs) each are converted to annual costs:

$$A_{steel} = -\$6043(0.2638) = -\$1594 \tag{3.11}$$

$$A_{316} = -\$18,989(0.1540) = -\$2924 \tag{3.12}$$

Therefore, the carbon steel heat exchanger, which has the lower annual cost, is the more economical alternative under these conditions.

Example 3.1a. Process Equipment. If the carbon steel exchanger were to require $3000 yearly maintenance (e.g., painting, use of inhibitors, and cathodic protection), would carbon steel still be the preferred alternative?

Maintenance costs are treated as expense items with one-year life (end-of-year costs). This particular exercise involves the third term in the generalized equation shown above. In Example 3.1, the annual cost without the yearly maintenance cost has already been computed, so in this case the annual cost including maintenance (stated to be $3000 per year) is obtained simply by subtracting the after-tax maintenance costs from the result given in Example 3.1:

$$A_{steel} = -\$1594 - (1 - 0.48)(\$3000) = -\$3154 \tag{3.13}$$

Because this is an after-tax cash flow, the quantity $(1 - 0.48)$ ($3000) is the after-tax maintenance cost where 0.48 equals the tax rate expressed as a decimal. The $3000 is the present value of the first year's maintenance, so no further discounting is necessary in this case.

By comparing this result with the annual cost of UNS S31600 in Example 3.1, it is evident that if $3000 is required each year to keep the carbon steel unit operative, the UNS S31600 unit would be more economical.

Example 3.1b. Process Equipment. Under the conditions described in Example 3.1, if it is not certain that a five-year life will be attained for unprotected carbon steel, at what life (for carbon steel) is it economically equivalent to the UNS S31600 heat exchanger?

The carbon steel heat exchanger will be economically equivalent to the UNS S31600 heat exchanger when their annual costs are equal. Therefore, the problem can be solved by determining how many years of life for steel is equivalent to an annual cost of $2924. Trial-and-error methods are useful for such problems.

Trying $n = 3$ years

$$(PW)_{steel} = -\$9500 + 0.48\frac{9500 - 0}{3}(2487) = -\$5719 \tag{3.14}$$

$$A_{steel} = -\$5719(0.40211) = -\$2300 \tag{3.15}$$

Trying $n = 2$ years

$$(PW)_{steel} = -\$9500 + 0.48\frac{9500 - 0}{2}(1.736) = -\$5542 \tag{3.16}$$

$$A_{steel} = -\$5542(0.57619) = -\$3319 \tag{3.17}$$

Thus, a carbon steel heat exchanger must last more than two years but is economically favored in less than three years under the conditions given.

Example 3.1c. Process Equipment. Under the described conditions, how much product loss, X, could be tolerated after two of the five years of anticipated life (e.g., from roll leaks or a few tube failures) before the selection of UNS S31600 is justified?

Equating

$$A_{316} = A_{steel} + A_{product\ loss} \tag{3.18}$$

$$-2924 = -1594 + [(1 - 0.48)(X)(0.8264)][0.2638] \tag{3.19}$$

Solving for X

$$-1430 = 0.1134(X) \tag{3.20}$$

$$X = -\$12,610 \tag{3.21}$$

The term 1–0.48 is recognized from term 3 of the general equation (3.7) as $1 - t$, wherein t is the tax rate. The quantity 0.8264 is the single-payment PW factor (P/F, 10%, 2 years) and translates the product loss, X, to its present worth at time zero. The quantity 0.2638 is the uniform series capital recovery factor (A/P, 10%, 5 years) and translates the present worth of the product loss to annual cost so that it can be added to the annual cost of steel for comparison with the annual cost of UNS S31600. If production losses exceed $12,610 in year 2 (no losses in other years), the UNS S31600 heat exchanger is justified.

Example 3.2. Cathodic Protection. It is proposed to cathodically protect an underground pipeline installation, assuming the following: $i = 10\%$, $t = 34\%$, straight-line depreciation. Three proposals have been made:

1. Anodes, 4250 at $100,000, installed. Useful life, 10 years.
2. Thirty (30) rectifiers with groundbeds at $100,000, installed. Expected life, 20 years with annual maintenance cost of $5900.
3. A "mixed" system involving rectifiers ($18,000 installed plus $1200/year maintenance for 20-year life) plus anodes ($82,000 installed for 10-year life).

For proposal 1,

$$(\text{PW})_1 = -100,000 + (0.34)\frac{100,000 - 0}{10}(6.1446)$$

$$= -\$79,108 \tag{3.22}$$

$$A_1 = -79,108(0.1628) = -\$12,879 \tag{3.23}$$

For proposal 2,

$$(\text{PW})_2 = -100,000 + (0.34)\frac{100,000 - 0}{20}(8.5136)$$

$$- (1 - 0.34)(5900)(8.5136) = -118,679 \tag{3.24}$$

$$A_2 = -118,679(0.1175) = -\$13,945 \tag{3.25}$$

For proposal 3, for the anode portion,

$$(\text{PW})_3 = -82,000 + (0.34)\frac{82,000 - 0}{10}(6.1446)$$

$$= -\$64,869 \tag{3.26}$$

$$A_3 = -118,679(0.1628) = -\$10,561 \tag{3.27}$$

For the rectifier-driven portion,

$$(\text{PW})_3 = -18,000 + (0.34)\frac{18,000 - 0}{20}(8.5136)$$

$$- (1 - 0.34)(1200)(8.5136) = -\$22,138 \tag{3.28}$$

$$A_3 = -22,138(0.1175) = -\$2601 \tag{3.29}$$

$$A_T = A_3(\text{anodes}) + A_3(\text{rectifiers}) = -10,561 - 2610$$

$$= -\$13,162 \tag{3.30}$$

Under these particular conditions, proposal 1 is calculated to be the least expensive alternative on the basis of lowest annual cost, followed by proposal 3.

REFERENCES

1. "Engineering Economy," ANSI Standard Z94.5, American National Standards Institute, 11 West 42nd Street, New York, 10036.
2. M. Bender, Tax Reform 1986, Analysis and Planning, Arthur Andersen and Co., Times Mirror Books, New York, 1986.

BIBLIOGRAPHY

American Telephone and Telegraph Co. Engineering Economy, 3rd. ed., McGraw-Hill, New York, 1977.

E. L. Grant, W. G. Ireson, and R. S. Leavenworth, Principles of Engineering Economy, 7th ed., Wiley, New York, 1982.

F. C. Jelen, Cost and Optimization Engineering, McGraw-Hill, New York, 1983.

G. J. Thuesen and W. J. Fabrycky, Engineering Economy, 6th ed., Prentice-Hall, Englewood Cliffs, NJ, 1984.

E. D. Verink, JEMMSE **3**(2), 239 (1981).

E. D. Verink, "Corrosion Economic Calculations," ASM Handbook, Vol. 13, 9th ed., Corrosion, ASM, Materials Park, OH, 1987, pp. 369–374.

4

LIFETIME PREDICTION OF MATERIALS IN ENVIRONMENTS

R. W. Staehle

University of Minnesota, Minneapolis, and Industrial Consultant, North Oaks, Minnesota

A. BACKGROUND

Predicting the performance of materials in operating environments can be approached by an orderly stepwise procedure; this procedure is described in this chapter and examples are given. Before describing this procedure, it should be understood that all engineering materials are reactive chemicals. As table salt dissolves in water, so any engineering material can dissolve rapidly and completely in some environments. In view of this inherent reactivity, the surprise is not that materials fail; the surprise is that they work.

For metals in aqueous solutions, the barrier that prevents inherently reactive materials from dissolving is a thin insoluble compound on the surface between the reactive metal and the chemical environment. Such a barrier or film is called a "passivating film" because it passivates the inherent reactivity and because the term "film" implies thinness and fragility. Thus, predicting and assuring reliable performance need to be undertaken with great care and with full understanding of both the inherent reactivity of metals and the fragility of the thin barriers that provide protection. Similar patterns of reactivity and self-protection apply to all solids. Figure 4.1 shows the relative thinness of protective films on iron alloys in neutral aqueous solutions.

Much of this discussion is oriented toward large systems since these are the most complex, and it is often thought that predicting their corrosion-affected lifetimes either is not possible or must necessarily be guessed. This is not true, and a straightforward approach to predicting the corrosion behavior of complex systems is described in this chapter. Simpler systems can be approached using the same methods, but shortened.

In this chapter, while predicting performance is the nominal topic, "assuring performance" is implicit within the same framework. In this discussion, "predicting" carries with it the responsibility of "assuring." For simplicity in this chapter "predicting" is used, but it should be understood that "assuring" is implied.

Uhlig's Corrosion Handbook, Third Edition, Edited by R. Winston Revie
Copyright © 2011 John Wiley & Sons, Inc.

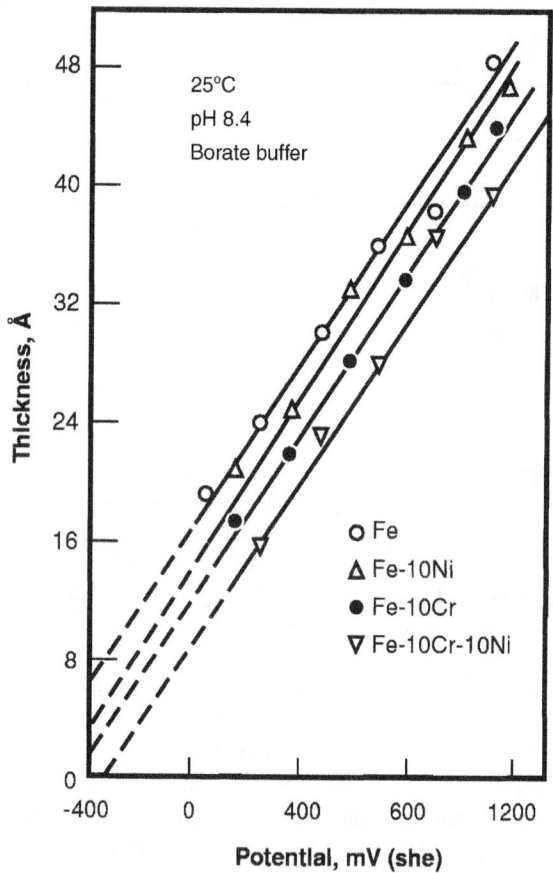

FIGURE 4.1. Film thickness versus electrochemical potential for Fe, Fe–10 Ni, Fe–10 Cr, and Fe–10 Ni–10 Cr alloys after 1 h of polarization at 25°C measured in pH 8.4 in borate buffer solution. (After Goswami and Staehle [9].)

performance of large systems is not included here. Such a broader subject embraces prediction and assurance procedures such as HAZOP (hazard and operability) and HAZAN (hazard analysis). The former is concerned with "what if" kinds of questions that are traditionally more related to system design and arrangement of components; the latter is concerned with mathematical analyses and is often called probabilistic risk assessment (PRA) or quantitative risk assessment (QRA). These approaches and similar ones are dealt with comprehensively in a series of texts published by the American Institute of Chemical Engineers (AIChE) and others, including Kletz [3–8].

It is not the province of this chapter to consider whether pipes in a complex chemicals plant are connected properly or whether a given check valve is inserted in the wrong direction; such subjects are the province of other methodologies such as HAZOP. Operator error is not a part of the present discussion except as it may relate to factors that affect degradation. On the other hand, many failures that lead to the shutdowns of plants result from degradation that is not traditionally included (but should be) in such methodologies as HAZOP and HAZAN.

The degradation of materials generally occurs via three well-known avenues: corrosion, fracture, and wear. Corrosion is traditionally related to chemical processes that break chemical bonds; fracture is related to mechanical processes that break bonds; and wear is related to relative motion that breaks bonds. These are, to some extent, separate considerations, but they are also interconnected. Chemical environments accelerate fracture; chemical environments accelerate wear and vice versa as wear products produce deposits that accelerate corrosion; and fracture processes can permit one component to contaminate another. It is not possible in this chapter to deal with the three traditional subjects of degradation or with their interactions except for stress corrosion cracking (SCC) and corrosion fatigue (CF), both of which are traditionally considered to be part of corrosion. The subject generally understood as corrosion is the focal point of this chapter. However, the procedures developed herein are generally applicable to the three general modes of degradation.

There are some nomenclature problems here. "Corrosion" is often considered to be what happens to metals, whereas the subject of "durability" is often considered to embrace the chemical degradation of concrete. Degradation of polymers is often considered as "bond breaking." No effort is made here to deal broadly with the degradation of all types of materials and with their respective historical terminology. However, the chemical-, mechanical-, and wear-type degradations occur in all solids, and the treatment of all of these can be readily embraced in the framework of this chapter. However, most of the examples here are based on the behavior of metallic materials in aqueous solutions, and many of the examples are taken from the nuclear industry.

The approach of this chapter follows the corrosion-based design approach (CBDA) [1, 2]. This is based on the idea that the subject of corrosion is a design science; the specific study and subject of corrosion would not be important except as it relates to assuring the reliable performance of equipment. The CBDA is not directly related to designing components for their functions; it is the purpose of the CBDA to assure that components work reliably in their intended applications and for their intended lives.

This chapter should be as useful to designers as to specialists in corrosion. This chapter is organized mainly around the design process and associated decision making. Lists and charts are included for identifying issues that need to be dealt with and for measuring progress toward achieving goals. This chapter also recognizes that much of design is a feedback process. As new information is acquired, it can be introduced into the design and operations.

While predicting performance and assuring reliability, as discussed here, follow an underlying theme that is related to degradation of materials, dealing with predicting the

Regardless of whether corrosion, fracture, or wear is considered individually or as their interactions (i.e., corrosion-accelerated wear) the generic subject covered here is "degradation." However, the term "corrosion" is more general, as it is derived from the Latin word that means "to gnaw," and this term is used.

This chapter takes a bottom up approach. Here, predictions start with individual materials in individual components as these materials sustain different environments on their surfaces. Such an approach is characteristic of corrosion problems; they must be dealt with specifically at the interfaces between materials and environments. In contrast, other approaches to reliability (e.g., HAZOP) ask the questions about the performance of components as they interact to produce a reliable system. Unfortunately, many of these approaches ignore questions of the chemical and mechanical stability of materials, and the conclusions of these exercises are often incomplete since the degradation of materials was not considered.

Predicting the performance of materials in components usually involves more than one possible mode of failure in more than one location. Figure 4.2 illustrates different sub-modes* [9] of corrosion that have occurred at different locations on an example U-tube in a recirculating steam generator of a pressurized water nuclear power system. Here, at seven different locations, eight modes and submodes of corrosion have occurred. For a prediction of performance in the application typified in the case of Figure 4.2, the seven different locations have to be identified explicitly together with the possible modes and submodes which can occur at each of the locations. A complete consideration of possible failures in the case of Figure 4.2 would require a matrix of 56 different possible interactions; approaching such apparent complexity is dealt with in this chapter. Each combination of locations and modes which is analyzed for corrosion in this chapter is called a "mode–location" case.

Generally, in predicting corrosion performance, the best guidance is obtained from applicable and prior engineering experience. Such experience provides bases for determining what mode–location cases should be considered and where additional research is necessary. Engineering experience may also be misleading in the sense of providing unwarranted confidence; small changes of environments in new applications often produce enormous differences in corrosive conditions, especially where concentration processes occur. Also, the experience obtained from several units operating for relatively short times is often mistakenly used to assure long-time experience. In developing predictions, the CBDA method includes both engineering experience and potentially important effects that may not have been obvious in previous applications.

Predicting the performance of materials as a part of a larger system of components is not accomplished with a single model for a single process. Predicting performance is more generally a "bookkeeping" problem as described here. Predicting performance requires, above all, both accounting for numerous possible circumstances that can produce failure and developing explicit responses. Some of these responses that describe specific mode–location cases can be intensely mathematical but as parts of this larger framework. The bookkeeping approach described here requires that explicit decisions be made either to include or to neglect considerations of certain possible failure conditions. Such decisions can become part of the design assumptions and part of the written record of the design.

The approach taken here considers "design" to embrace the full set of factors that affect reliable performance. Thus, considerations of corrosion, as they affect reliability, are as important as the product meeting its initial commitments to functions. Design, then, should include the elements of the CBDA.

The CBDA is based on the premise that materials have properties that depend on their environments. For example, the strength of a stainless steel listed in a reputable handbook may be useless in the presence of chloride ions since the useful strength is no longer the handbook strength but rather is the threshold stress for the occurrence of stress corrosion cracking, SCC[†]; this threshold strength may be 10% of the yield stress. The same is true for nominally high strength materials; in the presence of small defects, the actual strength of these materials decreases with increasing strength when exposed to many common environments. This modified strength is called "situation-dependent strength" and has been described in detail [11].

Predicting performance should also be considered in the context of the "cause" and "mode" of failure. The cause of failure is what has to be fixed and is generally in the realm of institutional and economic decisions. The mode of failure refers to the process by which failure occurs. For example, SCC occurs via the mode of a chloride- and oxygen-containing aqueous environment being exposed to austenitic stainless steel. How this process occurred, the sources of chloride, oxygen, temperature, and stress relate to the mode. The cause of this failure may be related to the choice of stainless steel for reasons of false economics, lack of instrumentation (or maloperation or lack of redundancy) to detect the conductivity that results from chloride leaking from a condenser, or lack of training of operators. The emphasis in the present

---

*The concept of "modes" and "submodes" is explained in Section E. The mode is a generic term (e.g., SCC), and submodes of SCC have, respectively, different dependencies.

[†] The abbreviation SCC is used throughout for "stress corrosion cracking." Also, the abbreviation. IGC and CF are used, respectively, for "intergranular corrosion" and "corrosion fatigue." The terms IGSCC and TGSCC refer similarly to "intergranular SCC" and "transgranular SCC."

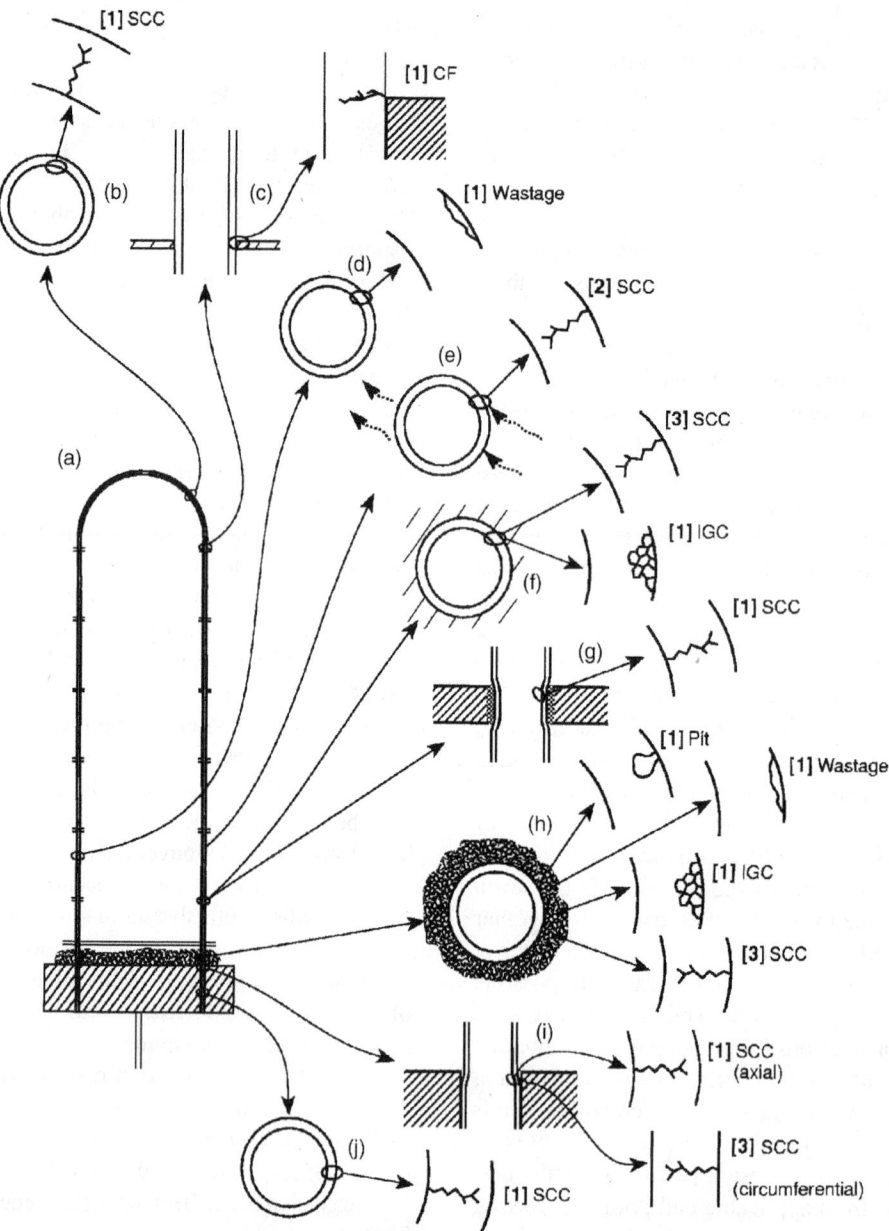

FIGURE 4.2. Schematic relationship of location to possible mode–location cases of corrosion for a U-tube steam generator used in pressurized water nuclear reactor systems. Tube sheet shown at bottom. Full-width tube support shown above tube sheet; other periodic locations of tube sheets noted. (a) Single tube of U-tube in steam generator of pressurized water reactor (PWR) systems. (Numbers in brackets indicate similar submodes although they may occur at different stresses or temperatures. Curved vignettes indicate that the view is perpendicular to the axis of the tube; straight vignettes indicate that the view is parallel to the axis of the tube.) (b) High stress from bending and forming in U bend produces axial SCC, [1], from primary side. (c) Immediately above the tube support with circumferential corrosion fatigue, [1], from secondary side. (d) Free span from secondary side with, [1], wastage on cold leg. (e) Free span from secondary side with axial SCC on hot leg side, [1]. (f) Tube at tube support with axial SCC, [3], and IGC, [2]. (g) Denting at intersection of tube support and tube gives stresses in tube which produce axial SCC, [1], from the primary side. (h) Tube inside sludge pile above the tube sheet with corrosion on secondary side: pitting [1]; wastage, [2]; IGC, [1]; axial SCC, [1]. (i) Roll transition with circumferential SCC, [3], on secondary side and axial SCC, [1], on the primary side. (j) Expanded tube in tube sheet with axial SCC, [1], from primary side. (After Staehle [10].)

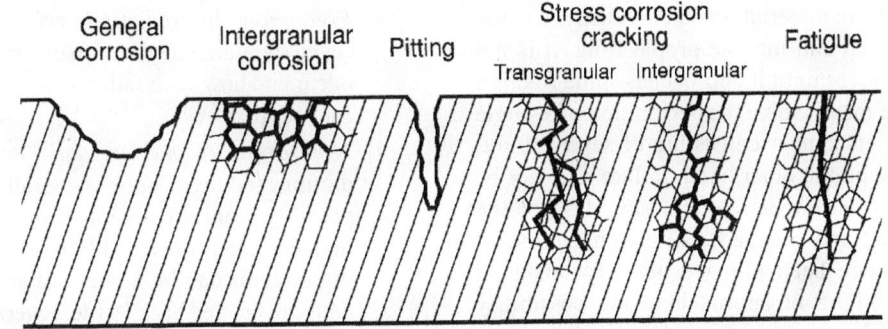

FIGURE 4.3. Schematic views of five intrinsic modes of corrosion penetration: general (including wear, erosion, and fretting), intergranular, pitting, SCC, and fatigue cracking.

approach to prediction is upon using a format that requires explicit considerations of all the possible modes by which failure can occur. Avoiding serious consideration of a mode–location case for reasons of economics or haste is a "cause" of failure.

There is an approach to analyze failures used in some industries called "root-cause" analysis. This is a misnomer since these exercises usually deal with mode and not cause. Furthermore, these exercises almost never get to a root of anything. In many cases, identifying real cause would embarrass the responsible organization.

The comprehensive approach to predicting and assuring lifetime used here is based on the CBDA. This CBDA is implemented according to the following steps[*]:

[1] *Environmental Definition.* Defining the environment involves specifying the conditions in which a material is required to perform reliably. The most broadly acknowledged "environments" are usually chemical, but environment should be considered more broadly to include, for example, thermal conditions and conditions of stressing as described in Sections C3 and C4.

[2] *Material Definition.* Defining a material involves more than specifying, for example, a stainless steel of some well-known composition and properties; the reactivity of even well-known materials is affected by seemingly subtle processes such as the preferential concentration of impurities at grain boundaries. Defining a material means to define it with respect to its reactivity as such reactivity may be exacerbated, for example, by the composition of grain boundaries, second phases, cold work, or particular combinations of impurities.

[3] *Mode Definition.* Defining the mode of corrosion means to define the morphology of the degradation,

or the avenues followed by the degradation and their dependence on the "principal variables" (for metals in aqueous environments the principal variables are potential, pH, species, temperature, stress, alloy composition, and alloy structure). When there is more than one occurrence of a mode, and the dependencies of such occurrences (e.g., alkaline and acidic SCC) on the principal variables are different such different occurrences are called "submodes." Figure 4.3 shows schematically the morphologies of the intrinsic[†] modes of degradation.

[4] *Superposition.* Superposition involves comparing, for a specific material, the various possible modes of degradation and their dependencies with the prevailing environments identified in the "environmental definition." If the modes of degradation can occur within these environments for a defined material, then further work may be required to analyze performance, or the conditions of the overlap should be avoided and the design should be changed. At this point, it is common to add inhibitors to the environment or change the material, change the operating conditions (e.g., lower the temperature), lower the mean stress, minimize residual stress, change aspects of the design, eliminate crevices, or eliminate stress intensifications.

[5] *Failure Definition.* Defining failure means to define the objectives of the prediction. Failure implies quite different outcomes in different industries as well as within even a single component, depending on requirements for reliable performance and/or safety. Failure in the food or the architecture industries may involve the appearance of rust on steel parts, although

---

[*] In this text, boldface numbers in brackets are used to indicate an action or an event that requires checking or action as opposed to a numerical list of aspects or features.

[†] The term "intrinsic" refers to modes thai are sustained by the material regardless of the geometric features of die environment. For example, "crevice corrosion" is not a mode of corrosion but rather is general corrosion that occurs inside a crevice just as the same general corrosion could occur outside the crevice. Similarly, "microbiological corrosion" is not a mode of corrosion but an environment that can produce any of the intrinsic modes.

there is little loss of material; in other industries, while a crack may exist and may be propagating, it is not considered a failure until it approaches some fraction of K_{Ic}.* Failure may involve the perforation of a single tube in a heat exchanger consisting of several thousands of tubes; on the other hand, failure may not be defined until some larger fraction, say 10%, of the tubes are failed and plugged. (Although individual tubes would have been plugged at early outages.) Furthermore, failure or approaching the possibility of failure implies certain responses and actions; developing or defining such action levels is also a part of defining failure.

[6] *Developing a Statistical Framework.* All degradation phenomena are inherently statistical, even under ideal conditions, owing to the multiple pathways by which a single corrosion mode can proceed. Furthermore, even when the same nominal materials are used (but with different vendors and different times of manufacturing), subtle and not so subtle differences occur among them (e.g., heat-to-heat variations). Finally, the properties of environments vary even, again, under ideal conditions. Thus, to account for these intrinsic and extrinsic variabilities, it is necessary to treat the properties of materials and environments in a statistical framework. It should be recognized that failure does not usually occur when the mean or median of failure is reached (e.g., failure of half the tubes in a condenser); by this point the equipment is generally decimated. Rather, failure occurs when possibly 0.01 or 0.1% of subcomponents (e.g., tubes, welds, and area) fail. Thus, it is usually necessary to predict the very earliest failures; such predictions are necessarily based on statistical methods.

[7] *Accelerated Testing.* Since most equipment is intended to perform for many years, whereas the tune available for prior testing is short, some accelerated testing may be necessary. Accelerated tests are invoked for obtaining quantitative predictions or at least some indication of the capacity of materials to endure extended circumstances. Results from such testing provide credibility to predictions. Acceleration is achieved through the use of various "stressors" (higher temperatures, higher stress, lower pH, higher chloride concentration, and cold work of the material) that can be used to intensify the exposure; using several different stressors helps to confirm that correlations being used are credible. The essential requirement of accelerated testing is that the acceleration produce the same mode of failure that is expected under the nonaccelerated nominal and longer term conditions.

* See Section C5 for a discussion of K_{Ic}.

[8] *Prediction.* In predicting performance, the foregoing seven steps are integrated to determine when failure can occur and how such failures would relate to the criteria for failure.

[9] *Monitoring, Inspection, and Feedback.* Perfect predictions could require generally large efforts and possibly prohibitively long times; further, some eventualities cannot be readily predicted. Most engineering systems can be monitored and inspected, and the data can be "fed back" to designers and operators. This feedback process is implicit in prediction although there are some engineering circumstances (storage of radioactive waste) where feedback from inspection and monitoring is not practical over the life of the project.

[10] *Modification in Design, Materials, Manufacturing, Operation, Monitoring, and Inspection.* Data taken from operation, including failures, can be compared with predictions and design life objectives to determine what changes need to be made in the design, materials, manufacturing, operation, monitoring, and inspection.

This chapter describes the implications and implementation of these 10 steps of the CBDA. Certain of these steps (e.g., environmental definition) is emphasized in proportion to their importance to reliable performance.

B. SCOPE OF PREDICTION

B1. Chronological Stage

It is often assumed that prediction applies mainly to steady-state operation of equipment Actually, failure can occur at any time in the life of a component starting from the earliest stages of manufacturing. The stages at which degradation can occur, and therefore needs to be considered, are identified in Figure 4.4; examples of failures at these various stages are well known. The stages in Figure 4.4 apply to complex equipment. Less complex equipment and products may involve fewer stages, but analyzing the successive stages, as in Figure 4.4, is still incumbent upon designers.

B2. Hierarchical Levels

Often, a single failure in a component causes an entire system to be shut down. On the other hand, the failure in this single component may result from ingesting chemicals produced in another part (although connected) of the system. Furthermore, any system contains many subsystems, components, and materials all of which are vulnerable to failure one way or another. Thus, in general, predicting failure means considering a larger system than a single component and a single material. While such a comprehensive approach may not

Stages for Analysis
(Chronological)

Before Startup

[1]. Manufacture

[2]. Testing (e.g., hydrotest)

[3]. Storage at Manufacture

[4]. Shipping

[5]. Site Storage

[6]. Installation

[7]. Prestart Testing

After Startup

[8]. Startup

[9]. Steady State Operation

[10]. Shutdown (planned and forced)

[11]. Maintenance During Shutdown

[12]. Cleaning During Shutdown

[13]. Inspection During Shutdown

[14]. Startup During Shutdown

[15]. Long Time Operation

FIGURE 4.4. Schematic illustration of the 15 stages in the chronology of equipment. Stages separated in two categories according to whether they relate to conditions before startup or after. At each of these stages the possibility of aggressive degradation has to be considered.

always be necessary, it needs to be considered, and the relevant parts of connected systems need to be identified. Such an integrated consideration involves eight hierarchical levels:

[1] *Stage [SG_j].* * The chronological or sequential stages which need to be considered are enumerated in Figure 4.4. Analysis needs to be carried out at each of the chronological stages with the extent appropriate to the stage. Failures caused at early stages in the chronology of Figure 4.4 are sometimes helpful in refining the manufacturing process or in reacting to fullscale testing. More important is the possible initiation of failure that develops during the early stages and that produces eventual failures during steady-state operations (e.g., improper processing, machining lubricants that contain MoS_2, or cleaning operations that contain HF where subsequent rinsing has been inadequate).

[2] *Site [ST_j].* An overall site may include several systems. A power station contains the transmission equipment, electrical transformers, power plant, fuel handling, and others.

[3] *System [SY_j].* A single system may have a higher priority for analysis. This is often the power plant in a power station. The overall analysis of a system is usually the province of HAZOP- and HAZAN-type analyses.

[4] *Subsystem [SS_j].* Within a power plant there are numerous subsystems. These would include, with their connected parts, turbine, primary water system, secondary cooling water system, and tertiary cooling system.

[5] *Component [CM_j].* Within a subsystem [SS_j] there are numerous components. For example, a secondary system of a pressurized water nuclear power plant includes a steam generator, condenser, feedwater heaters, and turbine.

[6] *Subcomponent [SC_j].* In a particular component there are subcomponents. For example, in a nuclear steam generator there are tubes, vessel walls, steam separators, baffles, tube supports, and tubesheets.

[7] *Locations for Analysis [LA_i].* At the LA_i the analysis begins. An LA_i is a location where failure is likely and should receive careful considerations. Such locations are usually identified by some combination of corrosion-prone materials, locally high stresses, stress intensifications, high heat transfer rates, high flows, and/or the occurrences of crevices and deposits. At an LA_i it is possible to be specific and consider the surface of a material with respect to predicting its performance. However, as described in Section C.3, the possible failure at any LA_i may be influenced by factors such as corrosion products from other components within the proximate system or from other systems (copper from condensers, leaks through turbine seals, insoluble iron products from piping). Further, there may be multiple LA_i on a single component as noted in the example of Figure 4.2.

[8] *Modes and Submodes of Degradation [MD_j, SD_j].* The MD_j or SD_j are illustrated in Figure 4.3. The full set of MD_j, or SD_j need to be identified as they apply to each of the LA_i. At this point, the analysis begins and may include either approximate correlations or detailed

* In each of the hierarchical levels, the subscripts i and j arc used since there may be multiple elements. The subscript j is used when the elements have a horizontal character; i is used when the elements have a vertical character.

mathematical models as they are required to describe a given mode–location case.

B3. Mode–Location Approach: Explicit Considerations

The central activity of the CBDA involves identifying and analyzing specific combinations of locations and modes called "mode–location" cases. The details of how locations and modes are identified is discussed in Sections C and E. Figure 4.2 shows examples of 14 mode–location cases on a single subcomponent. The specific steps in implementing the mode–location approach follow:

[1] *Identify reference subcomponent for analysis.*

[2] *On the reference subcomponent, identify the LA_i.* At this step, it is necessary to recognize that there may be multiple LA_i on a single component, as shown in Figure 4.2. While they may vary depending on how failure is defined (see Section G) and with other considerations in the overall system, the LA_i are selected by identifying where failure, or aggressive conditions, are reasonably most likely. LA_i include locations, for example, where:

  a. Chemical environments are the most aggressive, most concentrated, at the extremes of pH, include geometries that produce deposits, involve electrochemical cells, or have crevices.

  b. Stresses are high due to combinations of applied and residual contributions, and stress intensifications are present.

  c. Thermal conditions provide the highest stresses, highest surface temperatures, and highest heat fluxes.

  d. Flows are highest, flows change direction, flows are stagnant, or flows are erosive.

  e. Other aggressive conditions.

  Identifying such locations, generally, is based upon the judgment of experienced engineers but needs to include considerations of disciplines familiar with special inputs such as stress analysis, fabrication, material properties, and corrosion. Initially, it may be useful to select more locations rather than fewer since predicting failure is the inverse of justice: guilty until proven innocent.

[3] *Identify the set of modes, MD_j, and submodes, SD_j, that may reasonably occur over the range of LA_i.* Disciplines that are knowledgeable in such modes should be involved.

[4] *Develop a matrix where the relevant LA_i and MD_j/SD_j can be related.* Such a matrix together with examples of LA_i and MD_j/SD_j is shown in Figure 4.5(a and b). Here, for the steam generator component that connects the primary (inside of tubes) with the secondary (outside of tubes) system, the significant locations for analysis are shown in Figure 4.5(a). For each of these locations, the possible modes and submodes of degradation are also shown in Figure 4.5(a). Figure 4.2 provides perspective for Figure 4.5.

  In Figure 4.5(b), a table is shown where the locations for analysis, LA_i, are listed in the vertical column and possible modes, MD_j, and submodes, SD_j, are shown along the horizontal axis. Also, in the second and third columns, the locations on the inside (primary) and outside (secondary) of the tubes where the LA_i, are relevant are noted. The table in Figure 4.5(b) is the focal point of analysis in the CBDA.

[5] *Define specific actions that might be taken at each cell of Figure 4.5b.* Figure 4.6(a) shows the explicit actions that can be considered for each of these matrix elements defined by LA_i and MD_j/SM_j. For example, in the $j = 13$, $i = 2$ cell, no action is necessary because wear at this location is not reasonable; also fatigue in the sludge, $j = 12$, $i = 5$, can be rejected. On the other hand, wastage in the sludge pile, $j = 10$, $i = 5$, can occur and needs to be considered as a possible failure mode at that location.

  Continuing to consider how the chart of Figure 4.5(b) might be used, after substituting a new alloy such as alloy 690 for the older alloy 600 in the steam generator shown in Figure 4.5, the mode of LPSCC, $j = 1$ can be omitted from consideration at all locations on the primary side, $i = 1, 3, 6, 10$; and pitting, $j = 11$, might be eliminated from the primary and in some instances from the secondary side. The justification for eliminating other cells is based on likely risks. Figure 4.6 shows how the mode–location chart might appear for two different alloys in Figure 4.6(b).

[6] *Develop quantitative models for each of the mode-location cells.* Many of the cells in Figure 4.5(b) can be modeled by well-known relationships. While there is always an academic interest in modeling phenomena such as those in the cells of Figure 4.5(b) using relationships developed from first principles, such relationships are rarely better or more general than good correlations. Further, the development of such fundamental models after requires more than the available time.

  For most of the SCC submodes, the mean-value data might be modeled with a simple equation of a power law-exponential type as shown in Eq. (4.1):

$$x = A[\mathrm{H^+}]^n [\chi]^p \sigma^m e^{E - E_0/\beta} e^{Q/RT} t^q \qquad (4.1)$$

Pitting might be modeled as Eq. (4.2):

$$x = B[\mathrm{H^+}]^n [\chi]^p e^{E - E_0/\beta} e^{Q/RT} t^{1/3} \qquad (4.2)$$

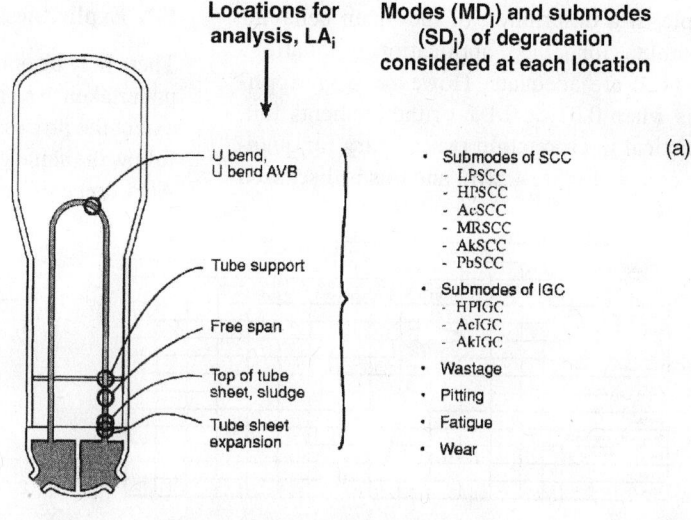

Locations for analysis, LA$_i$

↓

Modes (MD$_j$) and submodes (SD$_j$) of degradation considered at each location

(a)

- U bend, U bend AVB
- Tube support
- Free span
- Top of tube sheet, sludge
- Tube sheet expansion

- Submodes of SCC
  - LPSCC
  - HPSCC
  - AcSCC
  - MRSCC
  - AkSCC
  - PbSCC
- Submodes of IGC
  - HPIGC
  - AcIGC
  - AkIGC
- Wastage
- Pitting
- Fatigue
- Wear

(b)

Locations for Analysis, LA$_i$	Tube Side		Submodes of SCC						Submodes of IGC			Wastage (j=10)	Pitting (j=11)	Fatigue (j=12)	Wear (j=13)
	ID	OD	LPSCC (j=1)	HPSCC (j=2)	AcSCC (j=3)	MRSCC (j=4)	AkSCC (j=5)	PbSCC (j=6)	HPIGC (j=7)	AcIGC (j=8)	AkIGC (j=9)				
Tubesheet Expansion (i=1)	x														
Tubesheet Expansion (i=2)		x													
Top of tubesheet (i=3)	x														
Top of tubesheet (i=4)		x													
Sludge (i=5)		x													
Free span (i=6)	x														
Free span (i=7)		x													
Tube support (hot leg) (i=8)		x													
Tube support (cold leg) (i=9)		x													
U-bend (i=10)	x														
U-bend AVB (i=11)		x													

FIGURE 4.5. Bases for mode–location analysis. (a) Schematic view of steam generator with different locations for analysis, LA$_i$, and possible modes, MD$_i$, and submodes SD$_i$ noted for a single tube (out of several thousands), (b) Chart for aggregating mode–location cells as applied to a steam generator in a pressurized water nuclear power plant. The prefixes LP, HP, Ac, MR, Ak, and Pb for SCC and ICG refer to "low potential," "high potential," "acidic," "mid-range pH," "alkaline," or "lead" for "stress corrosion cracking" and "intergranular corrosion."

Linear processes such as wastage might be modeled as Eq. (4.3):

$$x = C[\mathrm{H}^+]^n[\chi]^p e^{E-E_0/\beta} e^{-Q/RT} t \qquad (4.3)$$

where

A, B, C, n, p, m, β, q = constants unique to specific correlations
x = depth of penetration
H = hydrogen ion concentration
χ = concentration of relevant species (e.g., chloride)
σ = stress
E = electrochemical potential
E_0 = equilibrium or corrosion potential
Q = apparent activation energy
R = gas constant
T = absolute temperature
t = time

The constants in Eqs. (4.1)–(4.3) can be determined often from previous engineering experience or from laboratory experiments; these values might also be determined by extrapolations from accelerated tests as described in Section I. Depending on what constitutes failure (Section G), various statistical functions (Section F) can be inserted into these cells as described in Section H.

[7] *Develop statistical expressions.* Developing statistical expressions for describing failure data is described in Section H. However, how such expressions are developed depends on what constitutes failure.

For example, if a description of the mean behavior is acceptable for the application, equations like (4.1)–(4.3) are adequate. However, if an application fails when 0.01 or 0.1% of the elements fail, then a statistical interpretation is necessary for quantifying the cells in Figure 4.5(b), and this is discussed in Section **H**.

B4. Explicitness and Simplification

There may be practical limits to the extent of work that can be undertaken in responding to the cells in Figure 4.5(b). To some extent the process can be simplified since many of the modes follow the same general patterns as shown in Eqs. (4.1)–(4.3). Also, some of the cells can be neglected, as in Figure 4.6, as

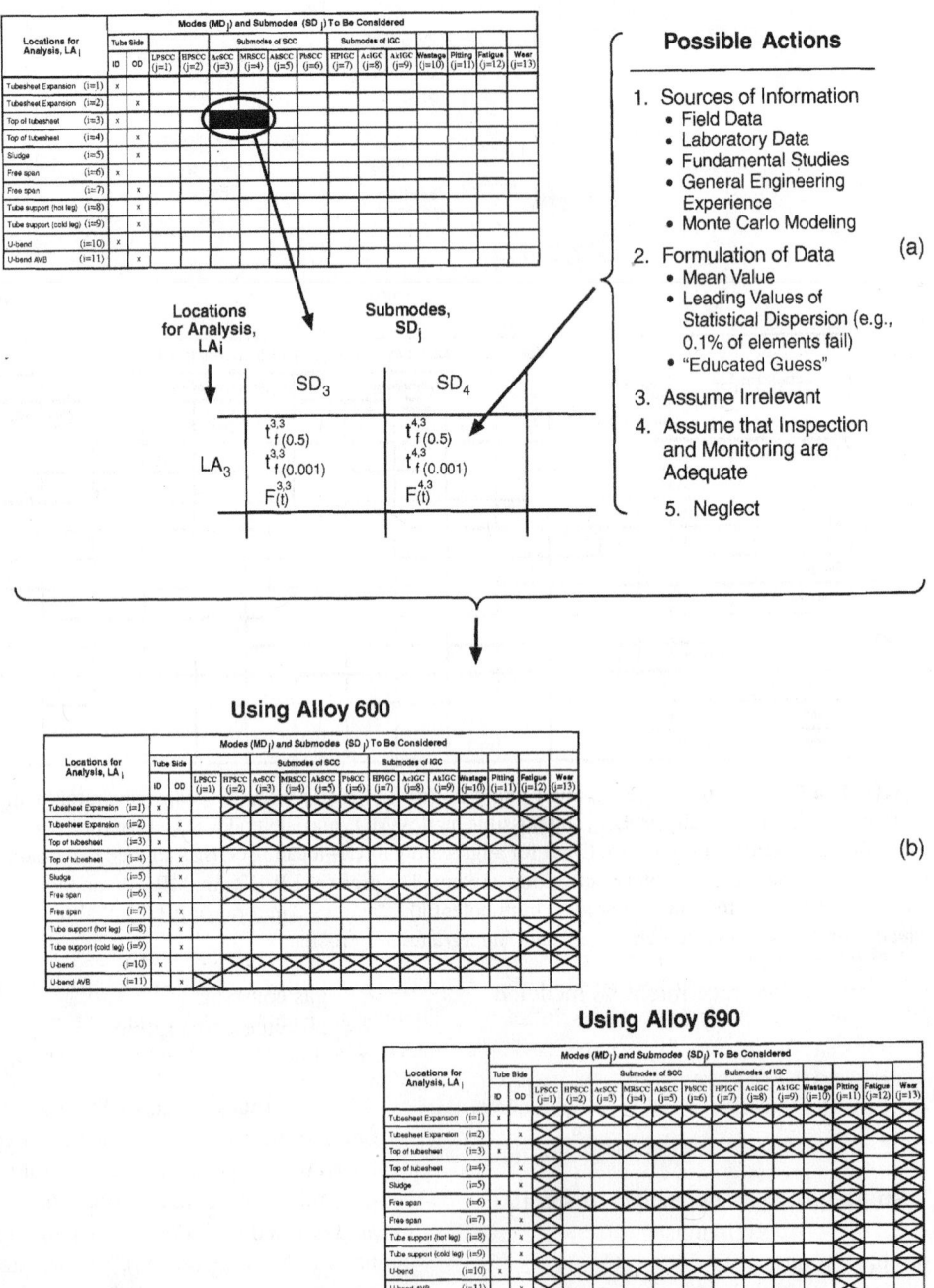

FIGURE 4.6. Possible actions and results for preparing mode–location charts. (a) Possible actions for each cell. The use of statistical results is discussed in Section H and Figure 4.30. Here the time to fail at a cumulative fraction failed of 0.001 is noted. Also, the cumulative expression $F(t)$ is inserted. (b) Examples of possible charts where two different alloys, alloy 600 (76 Ni–16 Cr–7 Fe) and alloy 690 (60 Ni–30 Cr–10 Fe), are applied to the pressurized water nuclear steam generator application shown in Figure 4.5. Where a cell can be neglected, it is noted as crossed out.

irrelevant to the specific application, that is, mechanical wear may not occur at the free span.

The mode–location matrix shown in Figure 4.5(b) provides a simple but rigorous basis for prediction in that it requires explicit considerations of each cell. Thus, explicit action has to be taken for each mode–location cell, and possible explicit actions are described in Figure 4.6(a). These actions can include developing a specific experimental program to develop the necessary correlation, or they may include deciding to not consider or include the cell. Such decisions could then become a matter of record, the wisdom of which can be reviewed over time.

Once the array of necessary information is defined from explicit consideration of the cells in Figure 4.5(b), a program of experiments and/or review of engineering experience can be organized to obtain the information in some efficient way, that is, some of the cells may depend on the same or similar variables or exhibit similar dependencies.

Responses to completing the information for each cell depend on the application. For example, failure (as defined in Section G) may occur when 10% of tubes in a heat exchanger fail. Thus, the information needed in a cell is the time to fail 10% of tubes. In other cells, the time to penetrate 50% of the wall is the appropriate information. In the case of welded pipes, failure can occur when one weld out of 1000 fails (0.1%). Thus, the information to be gathered for each of the cells could be different possibly even in the same subcomponent and would depend on the definition of failure from Section G.

C. CBDA STEP [1]: ENVIRONMENTAL DEFINITION

C1. Perspective

The environments to which materials are exposed dominate considerations in predicting and assuring their reliable performance. Further, defining environments is often the most poorly treated part of design. Environments, at best, are difficult to define, and their broad and uncertain variability increases the difficulty of prediction. Environments include a broad array of chemical and nonchemical considerations as noted in Section C2. In addition, environments that constitute aggressive conditions are not always obvious; for example, many corrosion reactions that proceed by complexing in the liquid or gas state are not commonly appreciated (e.g., the SCC of zirconium alloys in dilute iodine gas and the SCC of titanium in alcohols). Sometimes, pure environments (e.g., pure water) are very corrosive, as for the case where pure water produces SCC of alloy 600 in the range of 250–400°C. For these reasons, special emphasis is placed on step 1 of the CBDA, environmental definition.

Environments of concern are local environments on the surface, and their characters are often quite different from bulk environments. Such differences are produced, for example, by heat transfer, evaporation, flow, and electrochemical cells. Differences between bulk environments and the surface environments, where corrosion actually occurs, are illustrated in Figure 4.7(a).

Environments act in corrosion when the necessary conditions are simultaneously present in time and are congruent is location. An environment that promotes SCC is significant only when the necessary stress is also present at the same time (the principle of "simultaneity"). An environment that promotes SCC in sensitized stainless steel may be significant only when it is exposed to welds where the heat-affected zones have been sensitized, and the stresses are high (the principle of "congruence"). These conditions are illustrated in Figures 4.2 and 4.7.

As shown in Figures 4.2 and 4.5(a), multiple and different environments can occur on the same subcomponent (e.g., "multiplicity" of environments). Thus, to have evaluated the environment at one crevice location, for example, a tube sheet crevice noted in Figure 4.5(b), may have nothing to do with the sludge crevice owing to the different chemistries in the two cases.

The discussion in Section B1 concerning chronological stages indicates that length of exposure and the circumstances associated with each of the stages are important in predicting and assuring performance. Also, time spent within a stage affects environments on surfaces owing to possible gradual buildup or changes in the identity of chemicals.

In addition to the stages shown in Figure 4.4 having different conditions, the compositions of environments, even in the steady state, change with time. Such changes involve, for example, accumulations of chemicals and the buildup of deposits, as shown schematically in Figure 4.8. For example, during steady state, the depth of a sludge pile shown in Figure 4.5(a) increases with time, as shown in the three stages of Figure 4.8. In tube sheet crevices, the concentration of chemicals increases with time as more solution enters the crevice and is evaporated by the high heat flux. In the case of reinforced concrete, the original pH is in the range where iron is maximally corrosion resistant (pH of 12–13); however, with time, CO_2 enters the concrete and causes the pH to be lowered through a process called "carbonation" [12]. This lower pH surrounding the steel accelerates its corrosion and the consequent degradation of the reinforced concrete.

Environments are also changed, and made possibly more aggressive, as inhibitors are added to mitigate corrosion. For example, chemicals such as ammonia or phosphate are often added to control the pH; hydrazine or sulfite may be added to minimize the oxygen concentration. These species together with their possibly aggressive byproducts need to be considered in predicting corrosion performance. Ammonia from hydrazine may accelerate the corrosion of copper alloys;

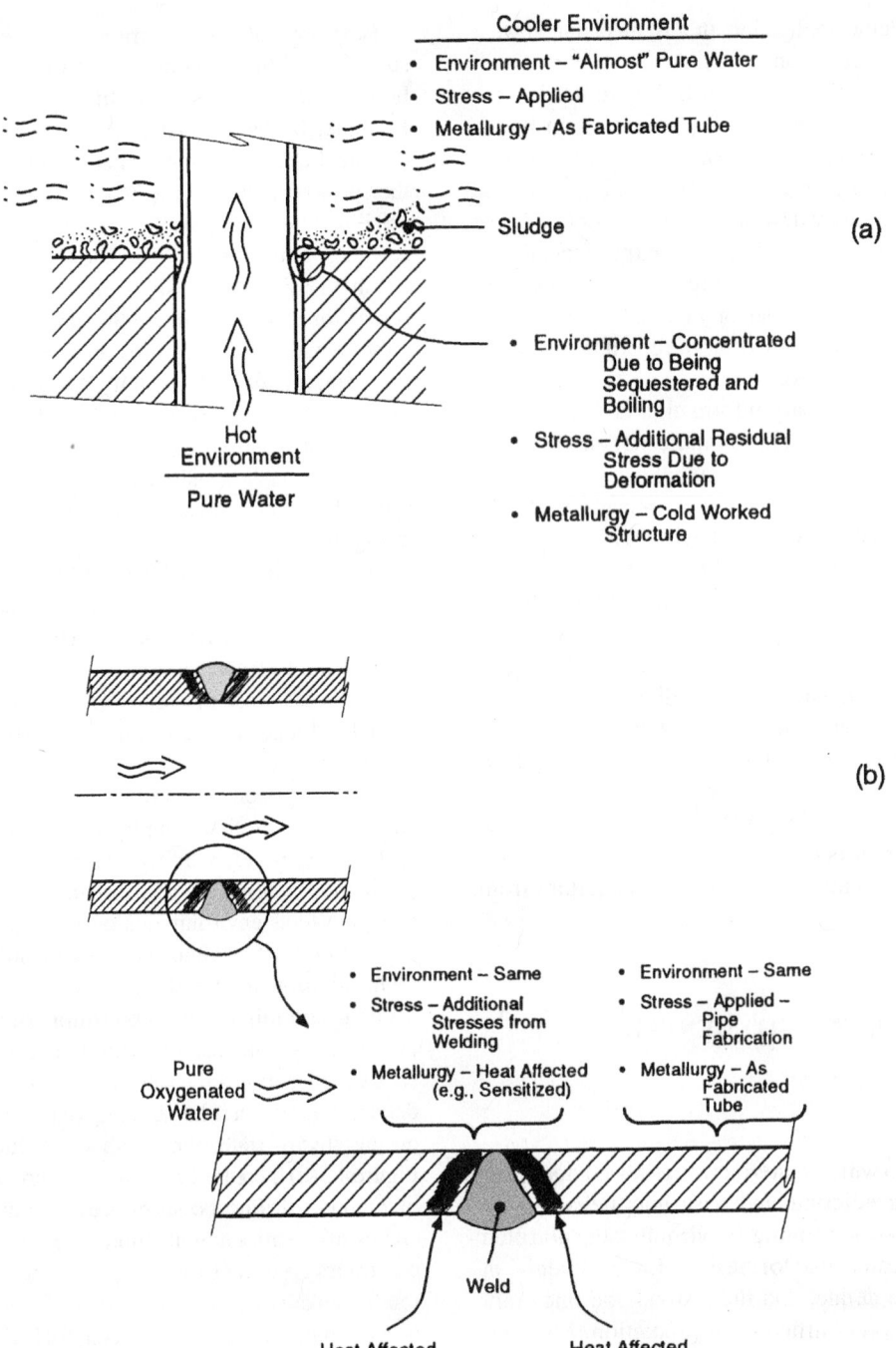

FIGURE 4.7. Schematic views showing that, for corrosion to occur, (1) the local environment needs to be specifically considered and (2) the chemical environment, the stress, and the corrosion-prone metallurgy must occur at the same locations at the same time. (a) Tube expanded into a tube sheet with surrounding sludge. (b) A welded pipe with heat-affected zone and welding stresses at the same location but with an environment the same on both welded and nonwelded regions.

sulfite may be reduced to sulfide, which accelerates the corrosion of iron-based alloys; precipitation of phosphate compounds above the retrograde temperature produces phosphoric acid.

C2. Components of Environment

As conceived by those who think about the subject of corrosion, "the environment" is often thought to be the

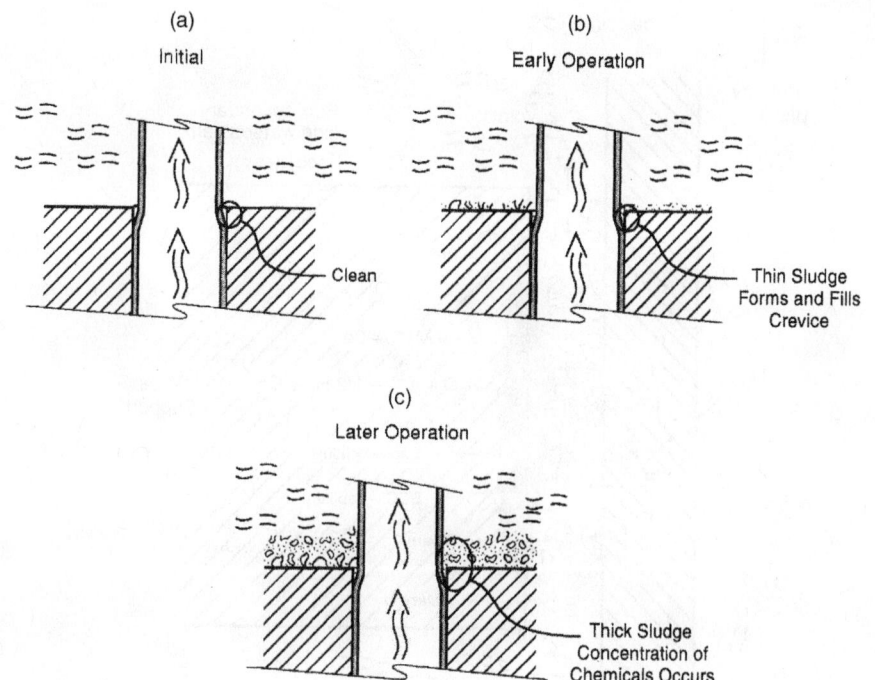

FIGURE 4.8. Schematic view showing how local environments, such as depths of deposits, change with time during steady-state operation: (a) initial condition; (b) after short-time operation; (c) after longer time operation.

chemical environment; this is an inadequate conception. Rather, considerations and components of the environment that affect degradation and that need to be included in predicting are the following:

[1] *Environmental type* refers to whether the environment is aqueous, organic, molten salt, liquid metal, hot gases of various types (oxygen, chlorine, sulfur species), electrolytes, and nonelectrolytes. The environments of principal practical interest are usually aqueous or high-temperature air, and the examples in this discussion are taken from such environments, mainly aqueous. However, the framework provided by the CBDA applies to predictions in all environments.

[2] *Chemical composition* is the chemistry of the environment on the surface at a specific LA_i, as shown in Figure 4.7. It is not the chemical composition of the bulk, although this is useful in assessing eventual chemical composition at an LA_i. The chemical composition includes the major and minor species in their dissolved, gaseous, or solid forms.

Figure 4.9 shows a schematic view of the chemical composition in a tube-to-tubesheet crevice of a steam generator of the type shown in Figure 4.5(a) where the bulk environment was virtually pure. These chemicals have concentrated as a result of the high temperature in the crevice within a lower temperature enclave on the

secondary side. This high temperature and lack of flow in the secondary crevice concentrate even the most dilute chemicals from the bulk environment. Figure 4.9 emphasizes that the environment, which is relevant to corrosion, at the location of tube support crevices is not the very dilute bulk environment but the very concentrated environment in the heat transfer crevice. Figure 4.9 also shows the chemical and physical complexity of this heat transfer crevice. Numerous chemicals are present and can transform to complex compounds and different soluble chemicals. Furthermore, four local gradients cause movement and reactions: electrochemical potential, temperature, concentration, and density.

[3] *Electrochemical composition* is both the chemistry of those species and the geometric conditions that affect the electrochemical parameters mainly of potential and pH or equivalent properties. In aqueous environments, this includes oxygen as it increases the potential or hydrazine as it decreases the potential. The electrochemical potential produces a powerful effect on chemical reactions and is the product not only of environmental chemistry but also of separated electrochemical cells, especially where the relative areas of anodes and cathodes are different.

[4] *Microbiological species* [13] include organisms that metabolize the surrounding chemistry to produce

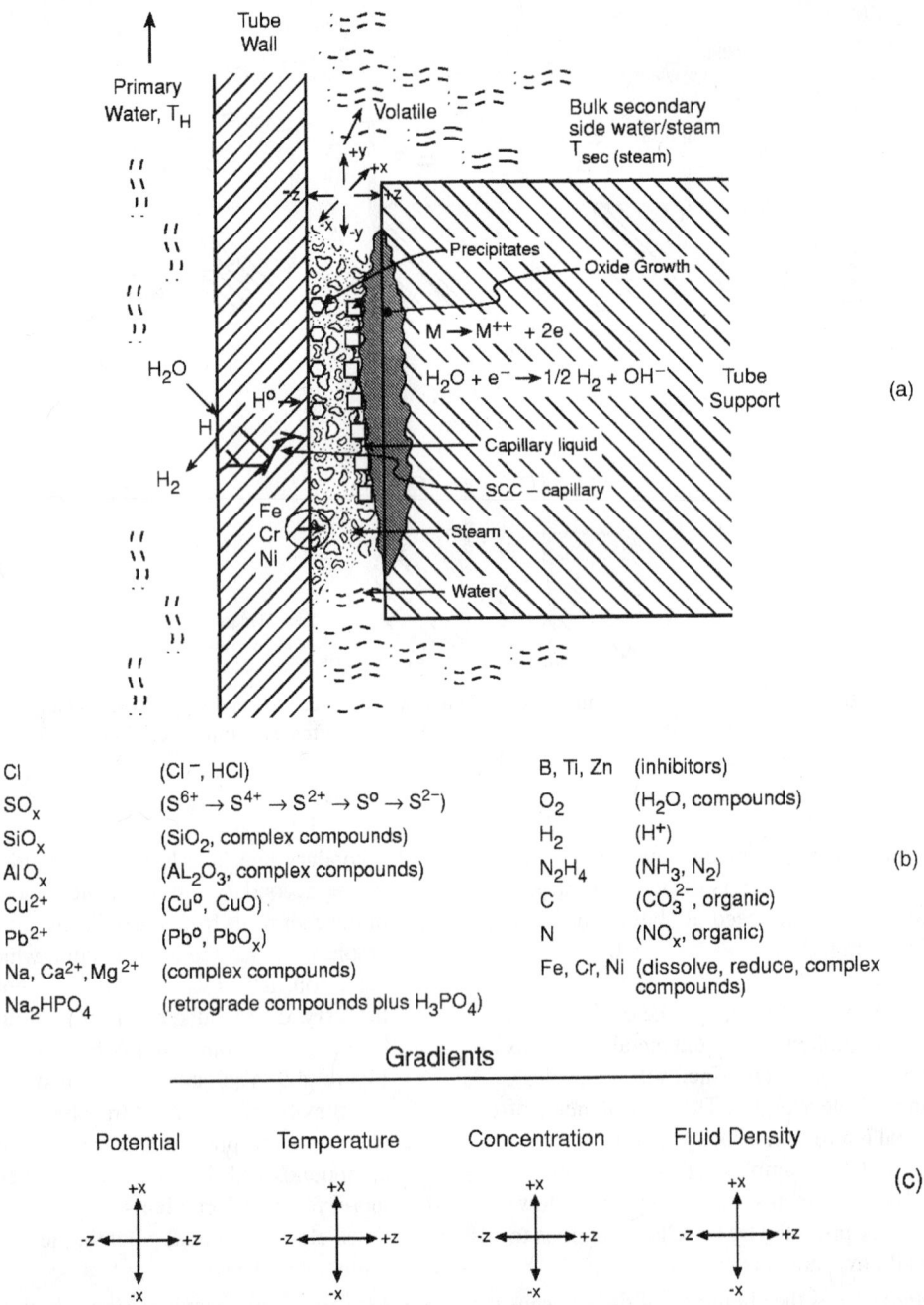

Cl	(Cl$^-$, HCl)	B, Ti, Zn	(inhibitors)
SO$_x$	(S^{6+} → S^{4+} → S^{2+} → S^o → S^{2-})	O$_2$	(H$_2$O, compounds)
SiO$_x$	(SiO$_2$, complex compounds)	H$_2$	(H$^+$)
AlO$_x$	(AL$_2$O$_3$, complex compounds)	N$_2$H$_4$	(NH$_3$, N$_2$)
Cu^{2+}	(Cu^o, CuO)	C	(CO$_3^{2-}$, organic)
Pb^{2+}	(Pb^o, PbO$_x$)	N	(NO$_x$, organic)
Na, Ca^{2+},Mg^{2+}	(complex compounds)	Fe, Cr, Ni	(dissolve, reduce, complex compounds)
Na$_2$HPO$_4$	(retrograde compounds plus H$_3$PO$_4$)		

FIGURE 4.9. Schematic illustration of chemical processes that can occur in heat transfer crevices at the intersection of tubes and tube supports on the secondary side of steam generators used in pressurized water nuclear reactors, (a) The crevice geometry occurs between a tube containing hotter water than the secondary side and a tube support that is cooler, (b) Chemicals concentrate and react in the sequestered geometry. Typical species of products are shown. (c) Inside the crevice, gradients in potential, temperature, concentration, and density occur; these cause movements, further concentrations, and electrochemical cells.

different and possibly corrosive species. Some species, for example, metabolize SO$_4^{2-}$ to produce sulfide; others metabolize oils and greases to produce organic acids. Microbial colonies, owing to their geometry alone, can produce sequestrations that accelerate localized corrosion. These microbiologically induced transformations lead to local environments that are often very corrosive, producing all the modes of corrosion identified in Figure 4.3. Effects of microbial species on corrosion are more pervasive than are commonly recognized.

[5] *Surface alterations* involve exterior coatings of organic and inorganic materials as well as changes in the substrate produced by carburizing/decarburizing, dealloying, surface machining, and adsorbed contaminants (e.g., fluorine remaining from pickling processes). Each of these surface alterations involves its own chemistry and sequestration (e.g., some organic coatings permit filiform corrosion).

[6] *Flow (chemical and erosive effects)* [14–16] affects the rates of reduction and oxidation reactions as it affects the boundary layer thickness; mixing inside pits and SCC depending on the axis of flow relative to the surface axis especially of SCC; cavitation and associated mechanical damage; deposits of suspended solids that result when flow is reduced; erosion due only to the fluid velocity when it is locally very high (e.g., wire drawing); gradients in electrochemical potential produced by flow-induced streaming potentials; and erosion due to the presence of abrasive particles. Examples of effects of flow on electrochemical parameters, flow-assisted corrosion (FAC) and sequestered geometries, are shown in Figure 4.10.

[7] *Phase* of the environment refers to there being gas, liquid, or solid but also to two phase circumstances such as at liquid–gas interfaces (water lines), solid–liquid–gas interfaces (such as at deposits), nucleate boiling, and film blanketing.

[8] *Geometrically induced chemical environments* are those due to crevices, deposits, heat transfer crevices, galvanic juxtapositions, differences in pH of nearby areas, relative area ratios, and gravitational effects (accumulation of deposits). This category includes what, in the past, have been called crevice corrosion and galvanic corrosion.

[9] *Temperature effects (affect chemical environments)* are those related to reaction rate, surface temperatures, heat flux and associated surface concentrations, and temperature gradient chemical transfer.

[10] *Stress environments* are those related to applied and residual stresses "together with variations in time" that affect cyclic stresses, R ratio, frequency, randomness, and wave shape. Here, the importance of residual stresses is always and greatly underestimated; over 50% of all SCC is related to residual stresses [18]. Considerations of differences due to plane stress and plane strain are included here, although their distinction is usually small relative to environmental effects.

[11] *Stress environments affected* by (a) fluid flow that produces stresses owing to differences in velocity and produces vibrations in certain critical ranges;

(b) temperature gradients that produce stresses and very rapid rates of temperature change that produce thermal shock; (c) geometric effects that intensify stresses at sharp corners, crack tips, irregular shapes such as threads, and corrosion pits.

[12] '*Relative motion of adjacent surfaces* produces wear, fretting, galling, and seizing. Wear and fretting may be greatly influenced by the chemical environment.

[13] *Neutron flux* (mainly relevant to nuclear reactors) produces several different effects on materials: First, neutrons, depending on their energies, produce transmutations or changes in chemical identities of many isotopes. Such transmutations may be chemically significant. For example, iodine produced in the fission of uranium produces SCC of zirconium alloy fuel elements in nuclear reactor cores. Second, neutrons, especially at higher energies, displace atoms from their lattice sites and produce hardening of the material. This hardening increases the strength and, thereby, increases the proneness to SCC. Third, the often large concentrations of vacancies that are produced in the displacement of atoms lead to vacancy condensation with consequent increases in volume; this increase in volume leads to the inevitable stresses and dimensional incompatibilities in adjoining subcomponents [19]. Fourth, neutrons produce the well-known fissioning of uranium, thorium, and plutonium with the consequent fission fragments of greatly different chemistries and increase in volume.

[14] *Electromagnetic radiations* such as gamma rays, X rays, and solar rays produce some important effects: First, gamma rays produce internal heating depending on their energy, intensity, absorption rate, and geometry of the subcomponent. Second, solar rays produce degradation of some polymers such as crazing and forms of SCC. Third, electromagnetic radiations catalyze some chemical reactions.

[15] *Charged particles* such as alpha particles (helium nuclei) and beta particles (electrons) can displace atoms in materials and can produce transmutations.

Thus, when environments are defined at any LA_i in the chart of Figure 4.5(b), items [1]–[15] need to be considered. In subsequent sections, only a few of the 15 environments are considered. However, the list of 15 provides a more general view of environments than the only chemical ones that are often considered These environments affect the modes and submodes differently. For example, the stresses noted in Section C5 and the accompanying Figure 4.13 are relevant not usually to general corrosion but mainly to SCC and CF. Neutrons, energetic electromagnetic radiation (gamma rays), and other energetic particles are not relevant except in applications involving nuclear reactors.

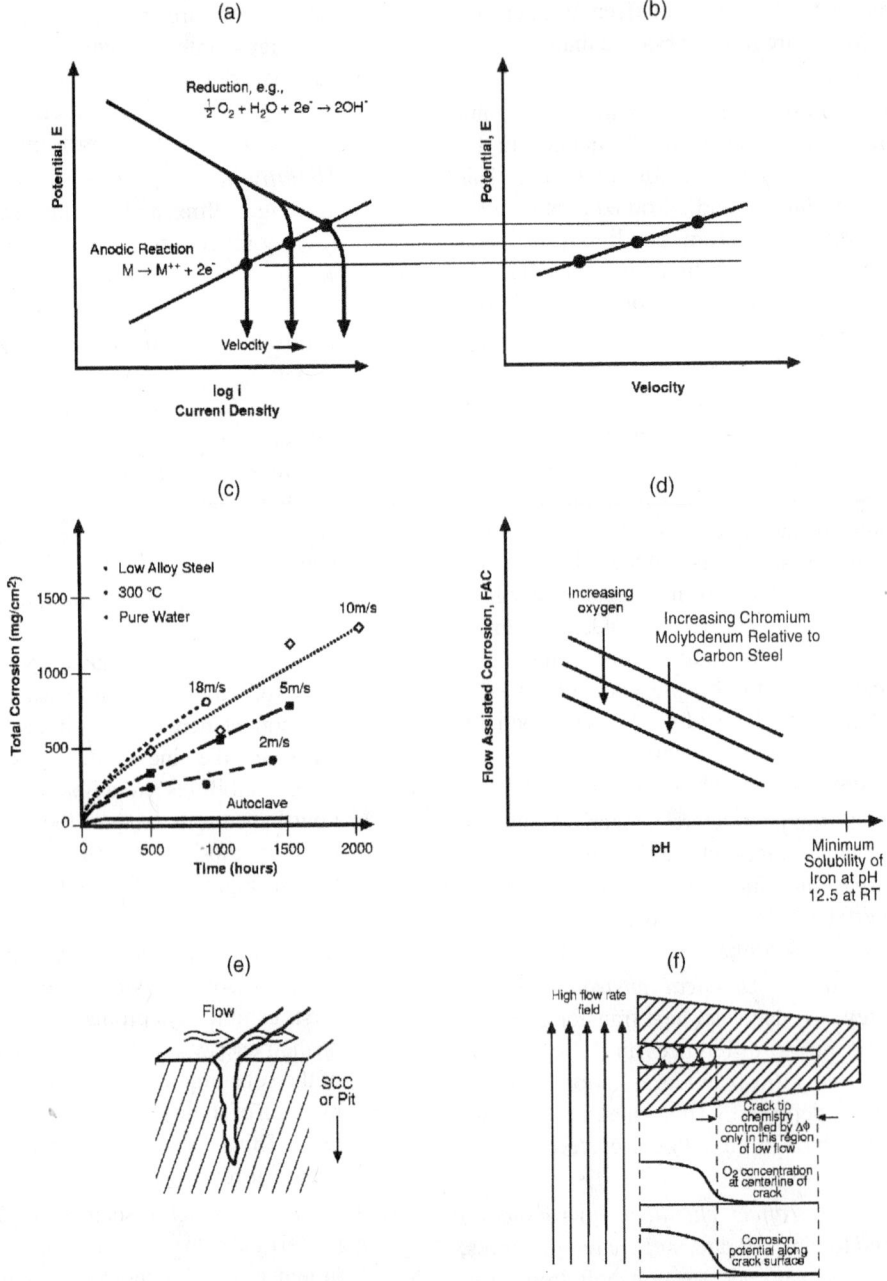

FIGURE 4.10. Three different effects of flow on corrosion. (a) Increasing flow accelerates the rate of the reduction of reducible species such as oxygen. Schematic polarization curves representing increasing velocities of flow are shown, (b) Corrosion potential versus velocity is shown corresponding to the effects of velocity on the polarization behavior in (a). (c) Total corrosion versus time for low-alloy steel exposed in 300°C pure water exposed to increasing velocities. (After Berge [17].) (d) Schematic relationship between FAC and pH, increasing oxygen concentration, and increasing chromium and/or molybdenum added to low-alloy steels. (e) Schematic view of advancing SCC (could be pitting) exposed to flow. (f) Schematic view of effects of flow on convection inside an advancing SCC. Here, the effective "electrochemical mouth" of the advancing SCC is moved inward as a result of mixing induced by external flow. (After Andresen and Young [14].)

C3. Chemical Environment

The "chemical environment" as described here is that identified in item **[2]** of Section C2. Figure 4.11 illustrates schematically how to approach defining the chemical environment. The objective noted in Figure 4.11 is to develop the chemical conditions at the LA_i (i.e., the i column of Figure 4.5(b) and **[8]** in Figure 4.11). Defining the chemical environment starts, **[1]**, with defining the major and minor species in "nominal chemistry" as they occur at the specific surface of an LA_i. Major species include water molecules and whatever is intentionally dissolved; minor species may include impurities, such as those in tap water. Considering the prior chemistry history, **[2]** supplies information on possible contaminants that may have formed on surfaces that would affect the LA_i either directly or as these species may dissolve and find their way to the LA_i. **[2]** refers generally to contaminants that may preexist in the system from construction (e.g., lead hammers left in the system), contaminants from cleaning operations, or defects produced by prior corrosion.

System sources, **[3]** in Figure 4.11, refer, for example, to copper that is released from copper alloy tubed condensers and then deposits on other surfaces in a secondary system to accelerate localized corrosion such as pitting. Species enter turbines from steam sources that are contaminated by river water that leaks into condensers. Oxygen enters through leaks in turbine seals. Chloride may be leached from platinum catalysts that have been prepared with chloroplatinate. System sources, **[3]**, then are those that are not normally expected to be produced or related to the component but may be transported from other parts of the system. These can be readily identified and in most cases predicted.

Physical features, **[4]**, refer to crevices, deposits, flow, phases, and local electrochemical cells that influence the chemistry of an LA_i. An LA_i with a crevice involves a quite different situation from the nominal chemistry of the bulk, as indicated in Figures 4.7, 4.8 and 4.9. An LA_i involving a waterline involves an electrochemical cell that changes the local oxidizing conditions.

Transformations, **[5]**, involve changes in chemical identity: for example, the process of microbes changing a hydrocarbon to an organic acid. Another transformation might involve the reduction of a sulfate (SO_4^{2-}) to a lower valence such as thiosulfate or bisulfide in a low-potential environment such as one containing hydrogen or hydrazine. These lower valences of sulfur are usually more chemically aggressive and, in some cases, accelerate hydrogen entry leading to SCC; sulfate by itself is not so aggressive.

Concentration, **[6]**, involves the concentration of species usually from the bulk; thus, whereas the bulk may contain chloride at a few parts per million (ppm) or even a few parts per billion (ppb), heat transfer crevices such as shown in Figure 4.9 or evaporation or a wicking process may concentrate the environment at an LA_i. Thus, the environmental species can attain concentrations much greater than the bulk; often such local concentrations are fully saturated.

Inhibition, **[7]**, involves chemical species added to minimize corrosion usually by affecting the pH, lowering the oxygen, or preferentially blocking anodic sites. Depending on the nature of the environments, as noted in Section C2, there are many possible approaches to inhibition. Adding inhibitors needs to be undertaken with the recognition that inhibitors may produce undesirable and possibly aggressive byproducts. For example, phosphate that is added to buffer and control pH can undergo a precipitation, as temperatures are increased, through a retrograde insolubilization process that produces local phosphoric acid. Using hydrazine for lowering the concentration of oxygen sometimes accelerates

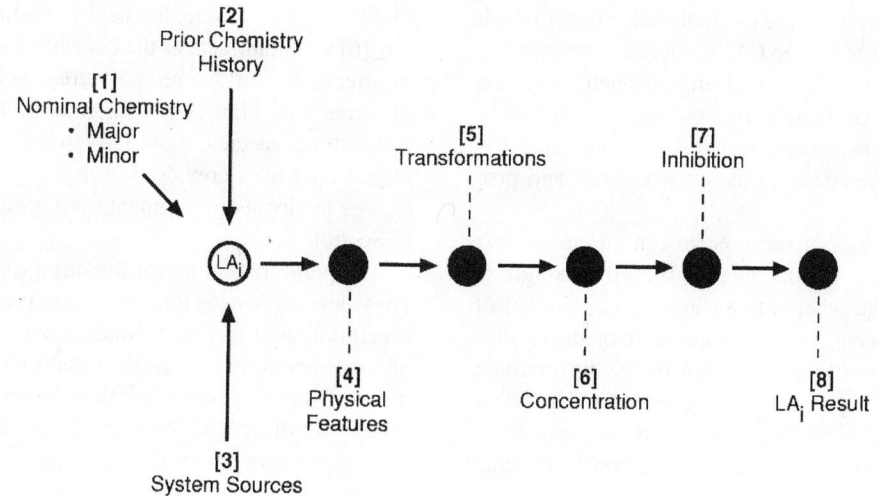

FIGURE 4.11. Analysis sequence for determining chemical environment at a location for analysis, LA_i.

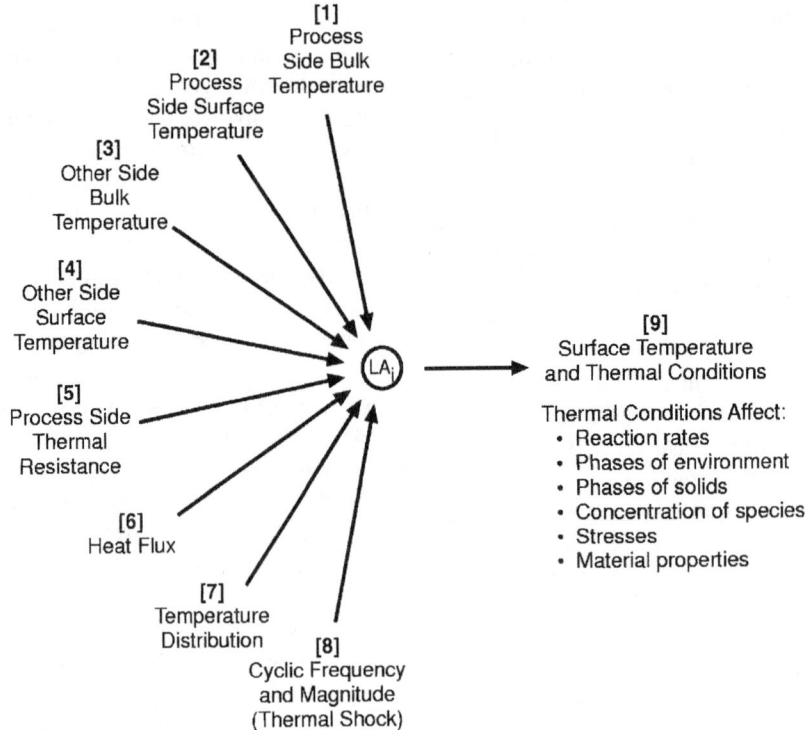

FIGURE 4.12. Analysis sequence for assessing effects of thermal environment on a location for analysis, LA_i.

the corrosion of copper-based alloys through the complexation of the copper with the ammonia produced in the decomposition of hydrazine.

Consideration of items [1]–[7] together provides the bases for defining the chemical environment at a single LA_i at [8] and in Figure 4.5(b).

C4. Thermal Environment

Figure 4.12 shows factors of the thermal environment which affect the conditions for a given LA_i. Temperature enters this discussion first because of its direct proportionality to some function of the rate of most corrosion reactions noted in Eqs. (4.1)–(4.3). Temperature gradients affect corrosion reactions as heat fluxes concentrate environments and produce thermal stresses.

For this illustrative discussion, corrosion on the process side is discussed. The temperature of the process side is usually lower than the other side as noted in [3] and [4] of Figure 4.12.* This means that the temperature on the surface of the process side is higher than that of the surrounding bulk environment. The process side is chosen for discussion since it is usually the side where chemicals concentrate on surfaces or in heat transfer crevices; furthermore, the thermal

stresses on the process side are generally tensile, providing that the process side is cooler.

The process side bulk temperature, [1] in Figure 4.12, is cooler than the surface, [2]; however, corrosion occurs at the temperature of the surface and not at the bulk so that degradation is occurring at [2]. This difference is exacerbated when the bulk environment is a gas where the resistance to heat transfer is higher. Regardless, the temperature of main interest for corrosion is at the surface and not the bulk.

At the process side, the next important factor is the heat flux [6] which influences the concentration of species as well as affects the surface temperature. Such a concentration is illustrated in Figure 4.9. Increasing heat fluxes tend to concentrate species, especially when the surface may be sequestered by a crevice or a deposit. Such concentrations change the local environments and need to be accounted for in Section C3.

Temperatures influence the onset of retrograde solubilities where increasing temperature may cause some phases to precipitate and produce simultaneously new compositions of supernatant liquids. In the case of the retrograde precipitation from phosphate solutions, phosphoric acid may be produced adjacent to the precipitated compounds.

Temperature enters the chemical aspect of corrosion at LA_i as temperature gradients. For example, thermal gradients produce tensile stresses on the process side (if cooler); thermal gradients in exothermic hydrogen-occluding

---

* Not all "process sides" are at the lower temperature side of a tube.

metals cause hydrogen to redistribute to the colder locations, thereby increasing brittleness. Thermal cycling produces fatigue effects as the associated stresses are cycled, and a very high rate of temperature change produces thermal shock.

C5. Stress Environment

The stress environment is usually considered to be the province of mechanical engineering and mechanics as they analyze stresses for the purpose of sizing components. These disciplines are also interested in cyclic stressing as it affects dry fatigue and design life. However, the stress analysis and corrosion communities have common interests in stresses as related to SCC and CF. Analyses by both communities are affected by residual stresses which, while difficult to analyze and predict, cause a large fraction of all failures by SCC [18].

Figure 4.13 shows how factors of the stress environment combine to produce conditions at an LA_i.

Generally, the applied stresses, [1], are well defined and are produced by the usually obvious pressures, bending, tensile loads, flow, and temperature gradients. Applied stresses include those due to cyclic stressing, including various R ratios, cyclic frequencies, and wave shapes. Less well defined but often the critical contribution to modes of corrosion such as SCC are residual stresses [2]. Whereas the applied stresses

are usually required to be something less than half the yield stress or lower, residual stresses are usually in the range of the yield stress. Residual stresses result from fabrication processes such as welding, cold deformation, and surface machining. Quantifying such residual stresses is often omitted in design with the erroneous conclusion that such stresses are irrelevant to design and performance.

"Other stresses" [3] result from sources such as corrosion products forming in restricted geometries where the specific volume of the corrosion product is greater than that of the metal that is corroded [20]. Such stresses can cause cracks to initiate and grow, as shown in Figure 4.14(a). Here, the stresses to cause SCC have been produced in a crevice between an insert and a notched specimen. These cracks then continue after the insert is removed, thereby indicating that the stresses to propagate SCC are totally due to the action of expanding corrosion products inside the advancing cracks.

Stresses from expanding corrosion products can readily cause adjacent metals to flow plastically as occurs in nuclear steam generators; this process, called "denting" [21], is illustrated schematically in Figure 4.14(b). Here, corrosion of the adjacent tube support produces corrosion products that expand and begin to strangle the tube. The chemistry that produces such corrosion is described also in Figure 4.9. This process is called "denting" since, when seen from the inside

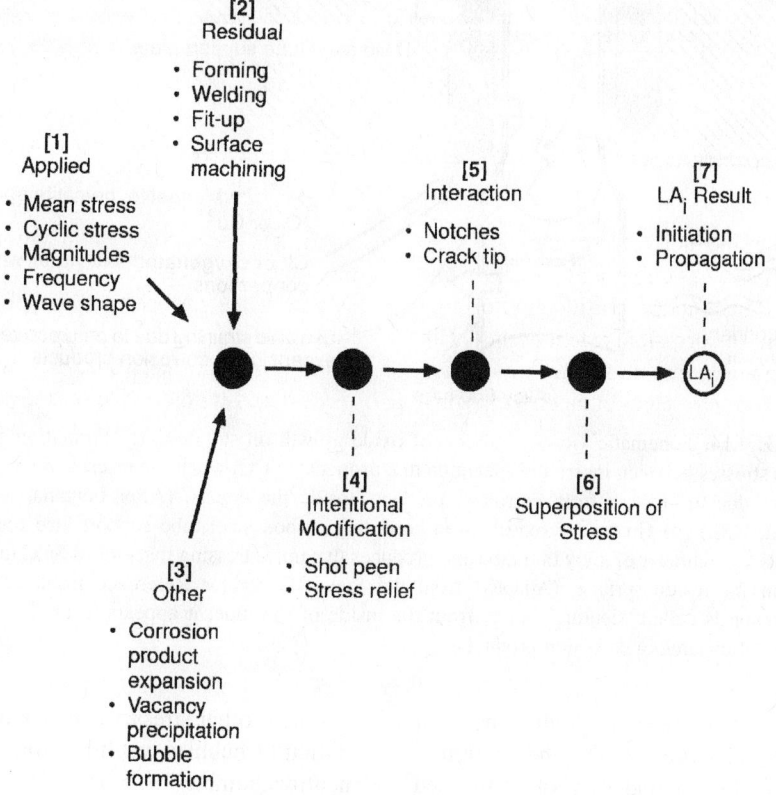

FIGURE 4.13. Analysis sequence for determining stress environment at a location for analysis, LA_i.

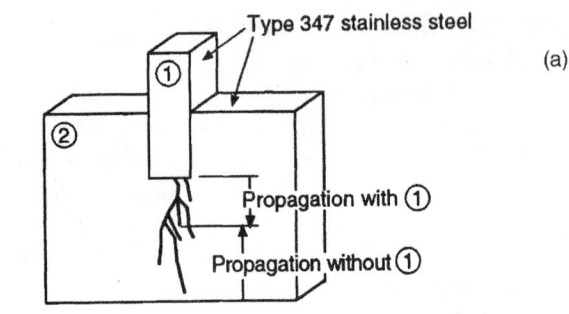

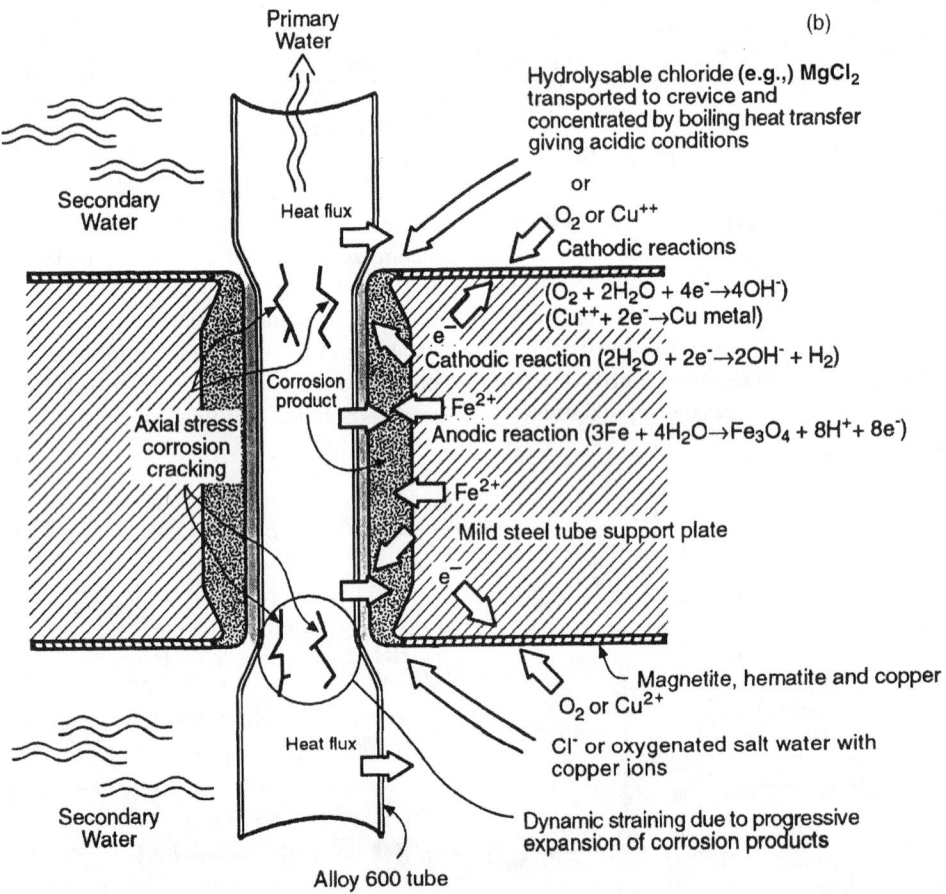

FIGURE 4.14. Schematic views of effects of oxide growth on stresses. (a) Formation of oxides produce stresses between insert and specimen that propagate SCC; when the insert is removed, SCC continues due to stresses from corrosion products inside the cracks. (After Fontana, Beck, and Pickering [23].) (b) Growth of oxides from corroding carbon steel tube support [see Fig. 4.5(a)] constricts the diameter of alloy 600 tube and produces dynamic stressing that causes SCC in the pure water on the inside surface. (Adapted from the Steam Generator Reference Book [21].) This phenomenon is called "denting" since, from the inside of the tube, it appears to be dented at the locations of expanding corrosion products.

of the tubes, these deformations seem to produce dents at the tubesheet locations. Similar stresses from the buildup of corrosion products cause the degradation of reinforced concrete [22].

These other stresses can also be produced by the precipitation of bubbles resulting from the lattice disarray due to neutron bombardment [19], from the formation of helium from the nuclear degradation of B^{10}, which produces an alpha particle and Li, or from the formation of hydrogen or methane bubbles that embrittle grain boundaries or produce blisters, usually in steel.

Stresses can be modified or minimized, **[4]** in Figure 4.13, by thermal stress relief, shot peening, or surface treatments such as carburizing. Appropriately placed welds can produce intentional contractions in locations where material is prone to SCC as a result of high local tensile stresses.

Stresses interact, **[5]** in Figure 4.13, with geometries and sharp cracks to intensify the local stresses. Such interactions have been studied in great detail as part of fracture mechanics [24]. A principal interest in fracture mechanics is to determine circumstances under which the intensified stresses that result from sharp geometries reach a critical intensity for catastrophic fracture. Catastrophic or fast fracture occurs when the value of a critical stress intensity, K_{I_c} (I corresponds to the opening mode of the crack), is exceeded. This criterion is given by Eq. (4.4):

$$K_I = \sigma_m \sqrt{C\pi a} \qquad (4.4a)$$

$$K_{I_c} = \sigma_m \sqrt{C\pi a} \qquad (4.4b)$$

where

σ_m = mean stress
a = depth of defect
C = geometric constant
K_I = stress intensity for any combination of mean stress and defect depth less than K_{I_c}
K_{I_c} = critical stress intensity at which catastrophic fracture occurs

Stresses are additive, **[6]** in Figure 4.13, and the total stressing condition needs to be applied to the LA_i**[7]** in Figure 4.13. The stress environment needs to be considered together with the chemical environment of Figure 4.11 in assessing the location conditions in the chart of Figure 4.5 (b). At the point where stresses are being calculated in detail and in the presence of discontinuities, careful attention needs to be given to stresses at LA_i. This is usually undertaken with finite-element stress analysis together with detailed considerations of geometries of defects.

The interactions between mechanical and corrosion considerations are illustrated in Figure 4.15. In Figure 4.15(a) the continuum of stress intensity [following Eq. (4.4a)] is plotted schematically to emphasize the relative contributions of corrosion and purely mechanical fracture. First, there is a relatively low critical value, usually called $K_{I_{scc}}$, K_{I_H}, or $K_{I_{CF}}$ as these symbols relate to SCC, hydrogen effects, or corrosion fatigue. When these lower thresholds are exceeded, cracks start to grow, but slowly. This slow growth may occur at constant stress as for SCC or may be

more related to a small increment of crack growth, $\Delta a/\Delta n$, for each cycle of stressing, ΔK, in fatigue. The fatigue crack growth rate may be increased by environmental contributions as noted in Figure 4.15(b). This increase due to environments in Figure 4.15(d) often corresponds to the threshold stress intensity for the onset of SCC (see Fig. 4.15c) although some environments accelerate CF while they do not produce SCC.

For a given material (assuming that the material properties are constant) at a definite temperature, the value of K_{I_c} is more or less invariant; however, the onset of slow crack growth, especially for SCC, is not. The value of $K_{I_{scc}}$ for a given alloy depends greatly upon the environment on the surface and may easily vary by factors of 2 or 3. Thus, precision in K_{I_c} is not nearly so critical to predicting performance as precision in $K_{I_{scc}}$ (as well as in defining the local environment), which is more difficult to determine owing to its dependence on subtle combinations of environments and materials.

Once a crack starts to grow, reaching K_{I_c} is inevitable unless there is some intervention to change materials, change the environment, lower the mean stress, or blunt the crack. Thus, $K_{I_{scc}}$ and similar onset criteria are possibly more critical to long-term performance than K_{I_c}. In addition, it is desirable to know the growth rate, either for SCC or CF, reasonably well so that intervention strategies can be developed: That is, if the crack growth is sufficiently slow and the instantaneous depth produces a K_I significantly less than K_{I_c}, a crack may be allowed to grow until it reaches a value below K_{I_c} that is regarded as a safe upper limit.

This discussion has not considered the region of K_I below the thresholds, for example, $K_{I_{scc}}$. In the subthreshold region, SCC and other forms of localized corrosion produce magnitudes of K that are less than $K_{I_{scc}}$. These subthreshold modes are usually corrosion related, as noted in Figure 4.15b, unless they relate to defects in the material, design or accidental nicks, and notches. Thus, the first stage of degradation in the subthreshold corrosion may be pitting or SCC. When these reach some critical depth and $K_{I_{scc}}$ is reached, then the crack growth follows a pattern shown in Figure 4.15(c or d). Thus, there are three sequential stages for the growth of defects:

1. Subthreshold (mainly chemically-related growth): corrosion (localized corrosion as SCC, IGC, or pitting); fatigue (dry and wet). Generally, the chemical environment dominates here.

2. Above chemical growth threshold (slow growth): chemical–mechanical (static and cyclic) interaction. In the absence of necessary environments, the static load cracks stop, and the magnitude of $\Delta a/\Delta n$ decreases.

3. Above K_{I_c}: catastrophic fracture. Totally mechanical.

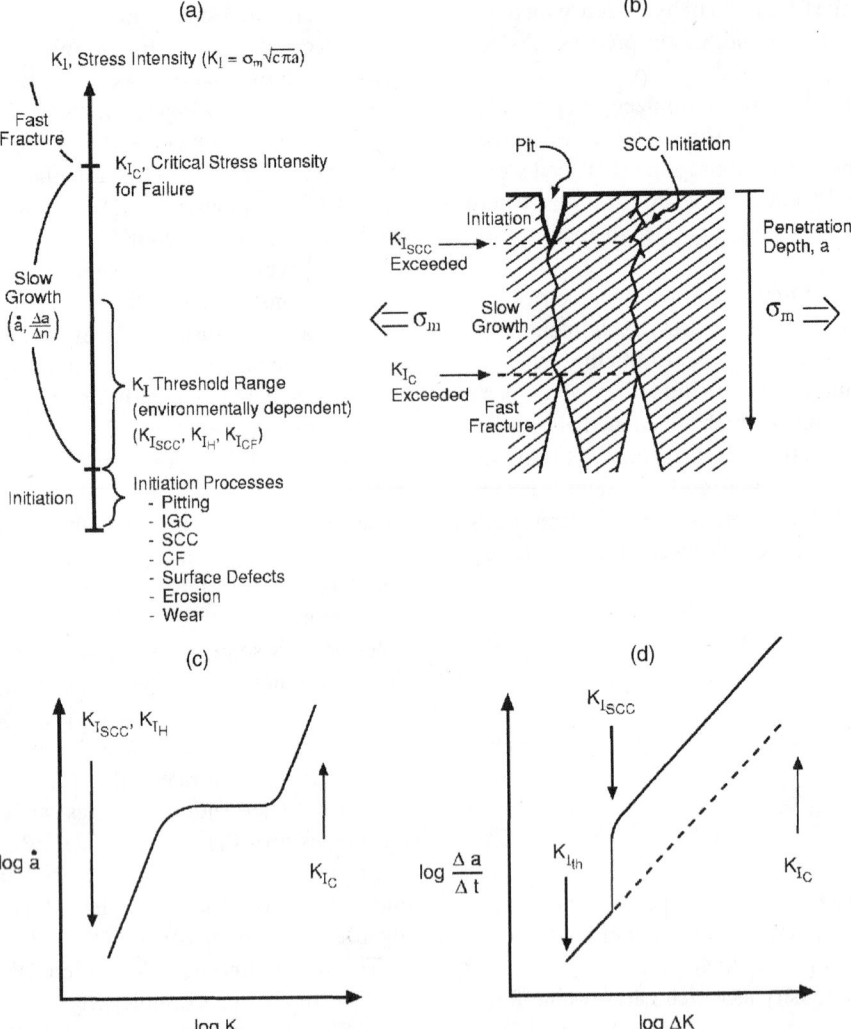

FIGURE 4.15. (a) Schematic view of relative magnitudes of K_I shown with transitions and principal dependencies. The K_I parameter for threshold range is shown to be variable. (b) Morphological regimes. (c) The da/dt versus K for SCC and similar environmentally dependent slow growth (e.g., hydrogen), (d) $\Delta a/\Delta n$ versus ΔK for fatigue crack growth showing possible effect of SCC in increasing the slow growth.

In the subthreshold regime, the stress at which SCC initiates depends greatly on the combinations of environments and materials. These interactions are discussed in Section E. The useful stress below which SCC cannot initiate depends on the local situations and is called "situation-dependent strength" [11].

C6. Other Environments

In Section C2, 15 classes of environments are identified, and Figure 4.11 shows how these enter considerations in the LA_i of Figure 4.5(b). All of these environments deserve detailed considerations; however, here, we emphasize the broad and illustrative views of environments; the details of how they are envisioned and incorporated depends on specific applications.

D. CBDA STEP [2]: DEFINING MATERIALS

It would seem that a material such as type 304 stainless steel is well defined and should respond reproducibly when purchased to historically well-established specifications. Unfortunately, this is not always so, and such well-known and extensively used materials sometimes exhibit unexpected behavior—mainly because important details (surface cold work or erroneous heat treatments) of the material are not identified. Sometimes the term "space age material" applied to materials such as titanium is also misleading. Despite the often excellent corrosion resistance of titanium, for example, and partly due to its very high reactivity, it may corrode rapidly and sometimes unexpectedly.

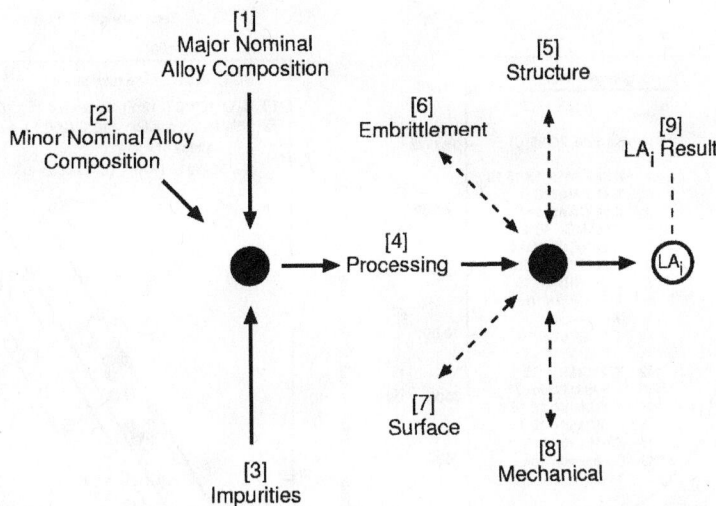

FIGURE 4.16. Analysis sequence for defining materials at a location for analysis, LA$_i$.

Usually, in environments that can be reasonably well defined, unexpected corrosion results from lack of definition of the material itself. Defining a material by its specification is often, especially for new designs, if not usually, quite inadequate.

As shown in Figure 4.16, defining a material starts with defining its major alloy composition, **[1]**. This step is usually straightforward and is not the reason for unexpected behavior. The next step, **[2]**, involves defining the "minor alloy composition"; this would include carbon, manganese, and sulfur in stainless steel. Such species often increase the intensity of corrosion reactions, especially SCC. Impurities, **[3]**, identify the importance of defining the impurities in the material; some materials, like titanium, are prone to absorbing (e.g., in welding) high concentrations of species such as oxygen, nitrogen, and carbon owing to the high affinity of titanium for these species. Sometimes, the aggregation of even minor elements all on the high side accelerates degradation reactions. Impurities are sometimes absorbed during fabrication operations such as welding and heat treating; this is especially a concern in reactive materials such as titanium.

Probably the most important consideration in defining materials is defining the compositions of their grain boundaries, since many degradation reactions either follow grain boundaries or are initiated at grain boundaries. In addition to corrosion reactions that follow compositionally altered grain boundaries, embrittlement processes often follow the same compositionally altered boundaries, producing, for example, the phenomenon of "temper embrittlement."

Figure 4.17 shows schematically the composition of grain boundaries depending on whether adsorption to grain boundaries occurs as in Figure 4.17(a) or a compound is formed (e.g., a carbide), as in Figure 4.17(b). In the case of

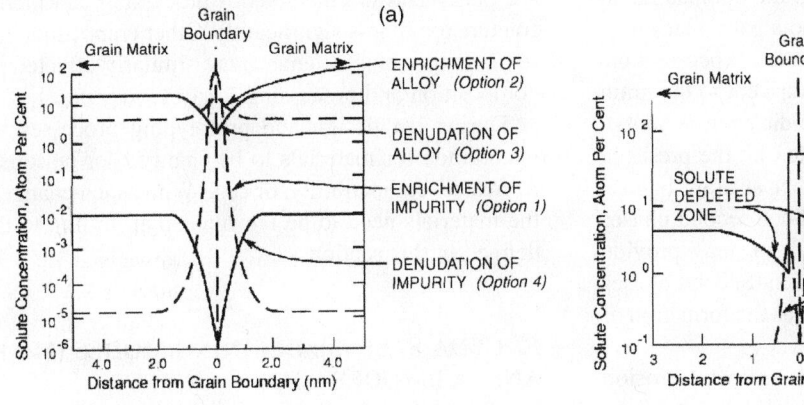

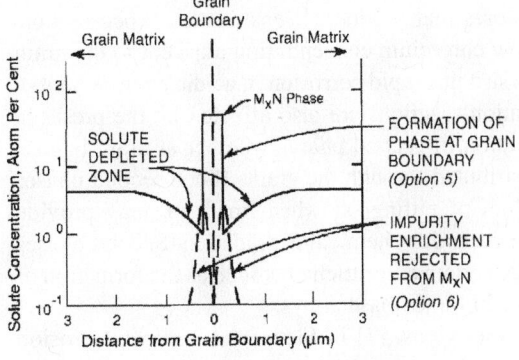

FIGURE 4.17. Schematic view of concentration versus distance from grain boundaries for circumstances where atomic species are adsorbed to the grain boundaries or where precipitates are formed. (a) Four options for concentration versus distance where atoms are adsorbed or rejected at grain boundaries. (b) Principal features of composition versus distance for the case where precipitates are formed at grain boundaries. (After Staehle [11].)

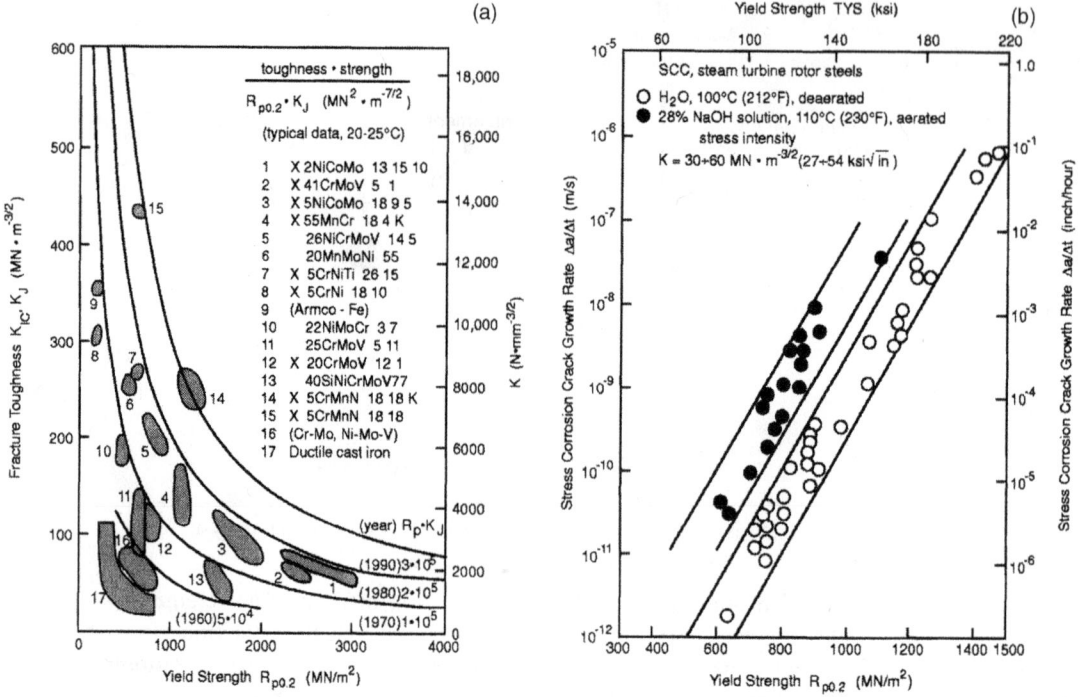

FIGURE 4.18. Effects of strength of iron-based alloys on their toughness (a) and SCC velocity (b). (Adapted from Speidel and Attens [25].)

adsorption, impurities may be concentrated by factors as much as 10^5; alloying species are concentrated less. In the case of compound formation, its composition may be enriched in either a passivating or nonpassivating species. Immediately adjacent to this compound there may be a narrow region of enriched species that are rejected as the compound forms. Finally, there are the "solute-depleted regions" the deficiency of which has supplied the necessary species for forming the compound. A good example of such a region is the depletion of chromium associated with the formation of chromium carbides at grain boundaries in stainless steels when they are exposed to a particular range of temperatures that produces "sensitization" (because the resulting low chromium concentration adjacent to the grain boundaries sustains rapid corrosion in acidic environments).

Degradation reactions are also affected by the presence and distribution of second phases whether at grain boundaries or distributed through the grains. Such second phases are often loci of pitting or, when lined up, may provide preferential paths for chemical degradation. Second phases also produce certain embrittlements such as the formation of sigma phase in Fe–Cr base alloys.

Surface alterations, **[7]** in Figure 4.16, affect corrosion behavior through altered compositions or changed mechanical properties due to surface finishing.

Figure 4.16 shows that **[4]** (processing) affects **[5]** (structure), **[6]** (embrittlement), **[7]** (surface), and **[8]** (mechanical) properties. "Processing" generally includes cycles of

temperature, time, and extent of deformation. The fact that unexpected and undesirable behaviors occur involving **[5]**–**[8]** usually results mostly from processing, the important details of which are often obscure. Therefore, it is necessary to determine the aspects of **[5]**–**[8]** by direct measurements (e.g., of grain boundary composition or cold work) as appropriate to their relevance to degradation.

Processing and alloy chemistry also affect the fracture toughness as shown in Figure 4.18. Here, while increasing the strength generally lowers the fracture toughness, improving the purity and the methods of processing can increase the fracture toughness significantly. Other embrittling processes (e.g., temper embrittlement) are similarly affected by alloy composition and processing.

During the design and prototyping processes it is not uncommon for materials to be changed for reasons of performance, compatibility, or cost. When such changes occur, the materials need to be reevaluated according to the steps defined in this section.

E. CBDA STEP [3]: DEFINING MODES (MD$_j$) AND SUBMODES (SD$_j$)

Figure 4.3 identifies five modes by which corrosion-related degradation proceeds through solids. These modes occur in every type of engineering solid with some, usually minor, differences that account for the structure and composition of

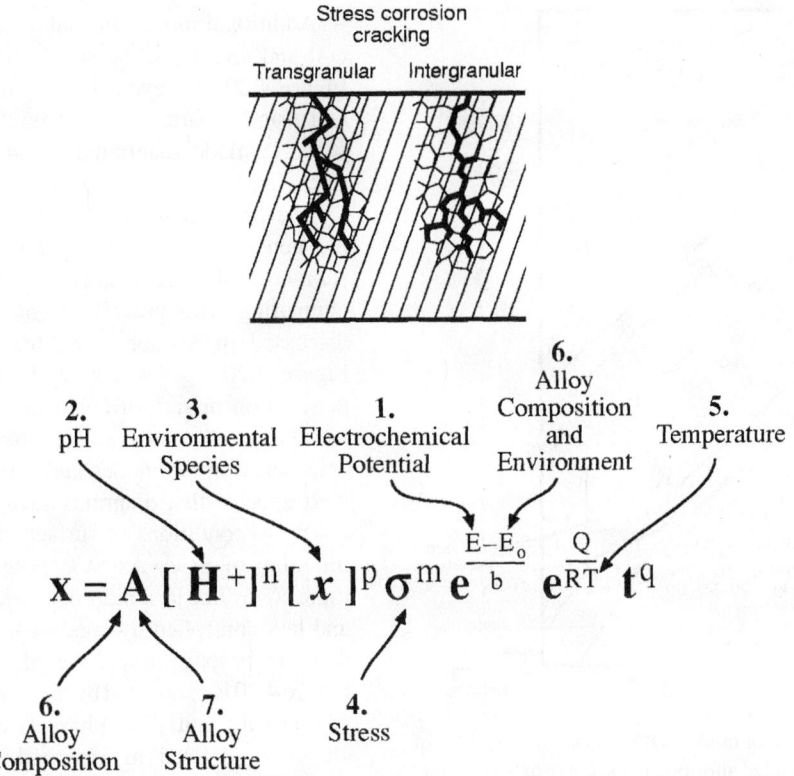

FIGURE 4.19. Schematic view of principal variables that control corrosion of metals in aqueous solutions illustrated here for SCC [as in Eq. (4.1)].

the solid. Similarly, the various submodes of these principal modes also follow the same morphologies except that their dependencies upon principal variables are different

Defining modes and submodes means to define (a) the morphologies as shown in Figure 4.3 and (b) the dependencies of the modes and submodes upon principal variables. These modes and submodes are inserted into the chart of Figure 4.5(b).

Not extensively discussed here are the totally mechanical modes of fast fracture, wear, and wear-related processes (e.g., fretting) [26] and dry fatigue. Fast fracture is not discussed as a mode since its morphology, while possibly important to analyzing failure, is of little importance to prediction. The dependence of K_{Ic} on relevant principal variables and the theory of K_{Ic} are not discussed here since this chapter is concerned primarily with environmental effects. Much of wear produces morphologies similar to general corrosion and pitting as depicted in Figure 4.3; however, wear in either its dry or environmental conditions is not considered here, although such a topic in the context of environmental effects as discussed in Section C would be appropriate.

Defining possible modes of degradation is important because they vary so much in their penetration velocity and their extent At a single LA$_i$ it is possible that all modes could occur or only a single one. Such possibilities are illustrated in Figure 4.2.

Defining modes means specifically to define their dependencies upon the "primary variables," as illustrated in Figure 4.19. The primary variables, as defined for metals in aqueous solutions, are potential, pH, species (e.g., chloride as opposed to species that affect potential or pH), temperature, stress, alloy composition, and alloy structure. The dependent variable here would be either penetration or penetration rate depending on which is useful. When a given mode (e.g., SCC) exhibits several submodes (as for alkaline and acidic SCC), they are distinguished by their different dependencies on these principal variables.

The various modes of corrosion (Fig. 4.3) differ greatly in their response to these principal variables depending on whether the initiation or propagation stages are important.

The array of possible modes and submodes can be understood more readily using a diagram such as that in Figure 4.20, which shows submodes of SCC for alloy 600 in aqueous solutions in the range of 300–350°C. These submodes are shown with reference to a potential–pH diagram containing pertinent lines for both iron and nickel (two alloy constituents) in water. The development of this diagram and similar ones is discussed in separate publications [27].

In Figure 4.20, four submodes of SCC are shown for alloy 600. One occurs in the alkaline region, AkSCC, and is bounded at higher and lower potentials to occur in a range

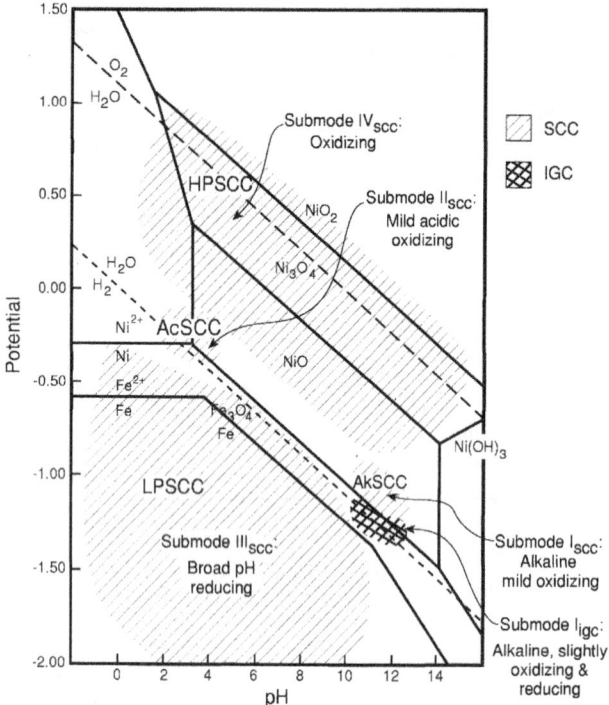

FIGURE 4.20. Occurrence of modes, MD_i, and submodes, SD_i, of SCC and IGC for mill-annealed alloy 600 in the range of 300–350°C in the framework of the electrochemical potential and pH. Selected lines, taken at 300°C, from the Fe–H$_2$O and Ni–H$_2$O equilibria shown for reference; also domains of selected species and compounds shown.

of potential of about 300 mV with the onset being about 100 mV above the standard hydrogen equilibrium. This submode is common to many alloys in the Fe–Cr–Ni system, including low-alloy steel. A second submode, AcSCC, occurs in the acidic region and seems to include about the same extent of potential and pH as the alkaline submode; however, the data describing this submode are not so extensive, and the boundaries are not so well defined, although they are better defined for low-alloy steels in the same submode. A third submode occurs generally below a line corresponding to 0.1 atm of hydrogen pressure. This is called low-potential SCC (LPSCC); historically this submode was called "primary water SCC" since it was often observed in the primary side of steam generators in PWR-type nuclear plants [28]. This latter nomenclature is not useful since it would compel this process to occur only in one location and to lack generality. Finally, a fourth submode, high-potential SCC (HPSCC), occurs at higher potentials and generally in a range of about 50–300 mV above the standard hydrogen line. The pH dependence of this submode is not well defined.

Figure 4.20 is called a "mode digagram." A similar mode diagram for various modes of mechanical deformation has been developed by Ashby and Tomkins [29].

Additional modes and submodes of SCC occur for alloy 600 and could also be shown on the same coordinates in Figure 4.20; however, the information becomes less clear, and using separate mode diagrams is more convenient. A separate mode diagram for minor environmental species (e.g., Cu, Pb, S, Si, Al) is not shown here.

The mode diagram of Figure 4.20 is useful for identifying the general conditions for the occurrence of modes and submodes of corrosion. This presentation is useful also for comparing with prevailing environmental chemistries, as discussed in Section F on the subject of superposition. Figure 4.20 does not give the detailed dependencies of penetration or rate of corrosion as a function of principal variables; rather, this figure provides perspectives for the occurrence of the modes and submodes.

It appears that diagrams such as Figure 4.20 relate primarily to conditions of surface chemistry that relate to the initiation of SCC. As SCC progresses, it is progressively more controlled by conditions at the tip of the advancing SCC and less controlled by conditions at the surface. It is likely that the propagation phase of each of the submodes in Figure 4.20 is controlled by the same crack tip chemistry [30], whereas the initiation phase is controlled by the electrochemistry defined in Figure 4.20. Evidence for the pH independence of crack propagation in the same chemical system is shown in Figure 4.21. Thus, while the AkSCC and AcSCC submodes are well defined and are limited, respectively, by a lower and higher pH, the propagation of SCC as shown in Figure 4.21 shows no such dependence. Whether the propagation of SCC can be properly considered as a

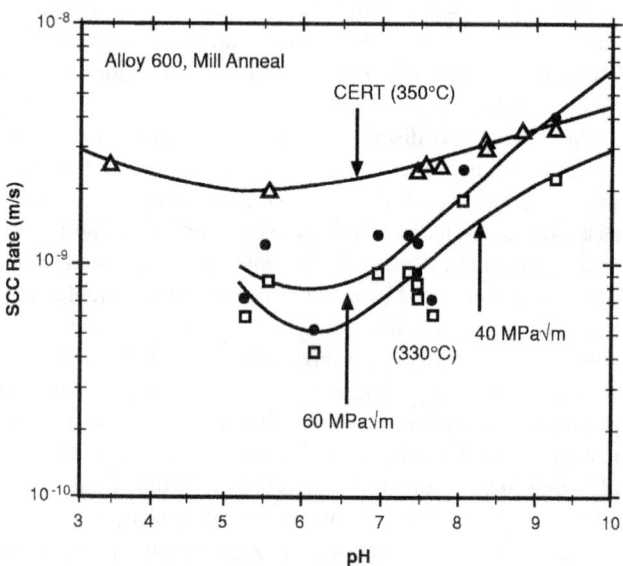

FIGURE 4.21. Effect of pH on the velocity of SCC for alloy 600. Constant Extension Rate Test (CERT) results obtained at 350°C and constant-load test results obtained at 330°C, (After Szklarska-Smialowska et al. [31].)

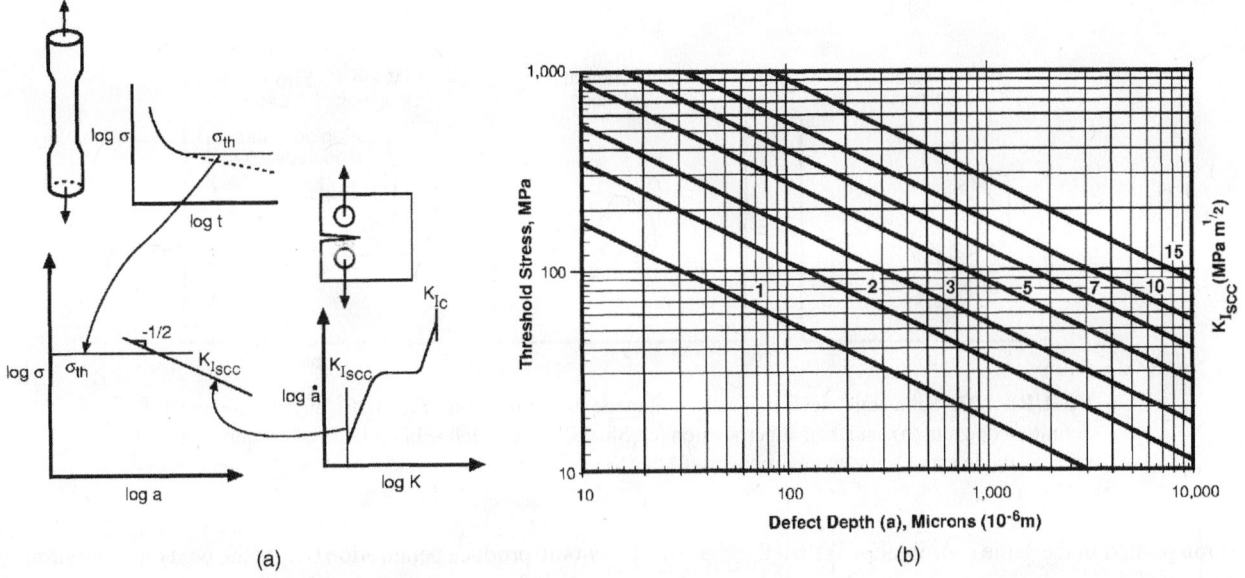

FIGURE 4.22. Procedure for distinguishing between initiation and propagation. (a) Schematic view of log stress versus log of defect depth for the cases of initiation stress as determined from smooth surface and SCC growth as determined from a precracked specimen. (After Staehle [1].) (b) Log–log plot of parametric values of threshold stress (horizontal lines) and $K_{I_{scc}}$ ($-\frac{1}{2}$ slope).

single chemical process independent of the surface environment is not clear.

In developing the dependencies of SCC in the initiation and propagation stages, it is useful to understand the depth of the transition between initiation and propagation. Figure 4.22 provides such bases. Figure 4.22(a) shows a schematic plot of log stress, σ, on the ordinate and log defect depth, a. The horizontal line, σ_{th}, corresponds to a threshold stress for SCC based on testing smooth specimens; a line with a slope of $-\frac{1}{2}$ [cf. with Eq. 4.4(a)] is shown in Figure 4.22(a) corresponding to $K_{I_{scc}}$ (see Fig. 4.15) obtained from SCC testing with precracked specimens. The intersection between the horizontal and sloping lines defines the transition between initiation and propagation.

A parametric plot is shown in Figure 4.22(b) where specific values of horizontal and $-\frac{1}{2}$ sloping lines are plotted. Suppose, for example, using the plot in Figure 4.22(b), that the threshold stress for SCC is 100 MPa and the $K_{I_{scc}}$ is 5 MPam$^{1/2}$; these lines intersect at a defect depth a of 800 μm or 32 mils, which is most of the wall thickness of the tubes shown in Figures 4.2 and 4.5(a). This depth is the nominal transition between initiation and propagation for a specific material in a specific environment; the numbers of this example are typical for AkSCC of alloy 600 at 300°C.

In addition to there being multiple submodes of SCC as shown in Figure 4.20, multiple submodes often occur for general corrosion, pitting, and intergranular corrosion as these submodes depend differently on principal variables. Other submodes than for SCC are noted in the chart of Figure 4.5(b).

These different modes of corrosion and degradation, in the past, have been described as "forms" of corrosion. However, in the formulation of the historic so-called forms of corrosion, differences between the intrinsic response of materials and the prevailing geometric conditions were never distinguished. For example, in addition to the forms of corrosion being identified as general corrosion, pitting, intergranular corrosion, SCC, and corrosion fatigue (as in Fig. 4.3), such geometry-dependent forms of corrosion were identified as "crevice corrosion" and "galvanic corrosion." The latter are not intrinsic to the materials but are rather consequences of how local geometries affect chemical environments on local surfaces. For example, intergranular corrosion can initiate from a free surface or within a crevice or in a galvanic cell. Similarly, all modes of corrosion can initiate within a microbiological enclave.

This chapter defines modes of corrosion in Figure 4.3 as the intrinsic response of materials regardless of the geometry and environment. Geometric effects as they affect local chemistries are discussed in Sections C2–C6.

F. CBDA STEP [4]: SUPERPOSITION

At the CBDA step of superposition the domains of modes and submodes are compared with the domains of the environments as they are known at this point in the analysis. These domains can be compared as shown in the simple schematic illustration of Figure 4.23. Figure 4.23 compares a schematic view of an environment in potential–pH space with a mode of

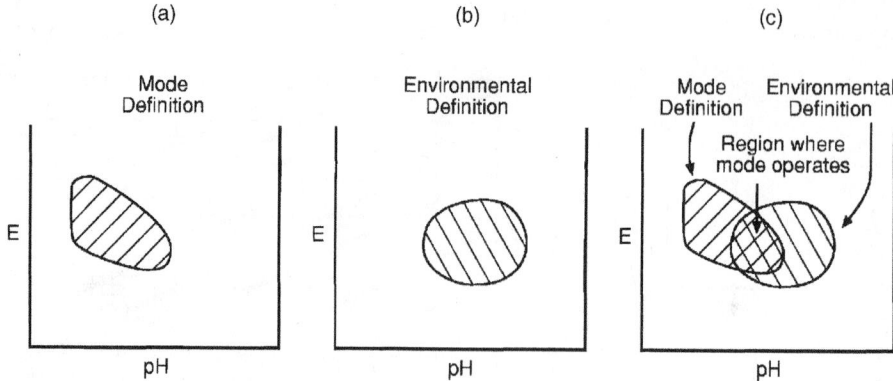

FIGURE 4.23. Schematic views of a corrosion mode diagram (a) (see Fig. 4.20), an environmental definition diagram (b), and their superposition (c). Shaded regions define boundaries of existence of a specific mode and a relevant environment. (After Staehle [27].)

corrosion plotted in the same coordinates. Where the shaded regions overlap is where significant degradation can be expected. In this region of overlap, it is desirable to determine how rapidly degradation can advance within whatever design life is expected. This depends on evaluating the principal variables that describe both the mode and the environmental conditions that apply to a specific LA_i.

If the overlap determined at this step of superposition involves significant corrosion, then changes may be required Such changes might involve changing alloys, revising the environment, changing the temperature, lowering stresses, or changing the design. Inhibitive environmental species can be identified at this point.

The step of superposition is, generally, the first where the need for changes is identified. The need for changes also becomes apparent during accelerated testing (Section I), prediction (Section J), and operation (Section K). There is, naturally, an obvious incentive to make necessary changes early.

G. CBDA STEP [5]: DEFINE FAILURE AND ACTION LEVELS

Failure needs to be defined because it provides the objective for analyses and ultimately for prediction. For example, in the food or architecture industries, a barely visible rusting may be a failure since it is unsightly. In much industrial equipment, a corrosion rate of 0.001 in./year for steel is quite acceptable, whereas such amounts of rust from the point of view of the food industry would be unacceptable. Each of these applications would be analyzed differently.

It is not uncommon for growing cracks to be considered acceptable (see Fig 4.15) so long as they are substantially less than a critical crack size [Eq. (4.4b) and Fig. 4.15(a)] and can be repaired at the next outage or shutdown. Until an inspection identifies a rate of crack growth or general corrosion that

would produce penetration before the next outage (with some safety factor), remedial actions are usually not taken.

In the development of sites for storing radioactive waste, failure is taken as a minimum amount of radioactivity appearing in the groundwater after some time in the range of $10,000-10^6$ years. Thus, the overall concept of failure here may have nothing to do with the integrity of the storage container but everything to do with the transport of radioactive species in the surroundings. This particular problem is challenging since it is not so easy to monitor the performance of containers of the radioactive waste owing to the long times and their relative inaccessibility associated with radioactivity.

Defining failure also involves statistical implications. Failure may not relate to failure at the mean lifetime (i.e., 50% failure) but to the occurrence of failure of 1%, 0.01%, or 0.001% of the tubes, welds, area, or units. Thus, failure might be defined for an entire heat exchanger, for example, as perforation of the first 10% of tubes; failure might be defined for other applications as failure of 0.1 or 0.01% loss of the elements. If the analysis and testing has not considered such a statistical objective, which occurs at times substantially less than the mean time to failure, the analysis is deficient, The application of statistical methods is discussed in Section H.

In designing components, it is common to define a "design life" [design life objective (DLO)] within which a component is expected to perform reliably. Sometimes a conservative additional life ["design life conservative objective" (DLCO)] is required so that there is no question about reliable performance within the desired design life. This design life or the conservative version becomes the target for prediction, and failure is defined when perforation occurs earlier than either DLO or the DLCO.

In addition to the initial DLO and DLCO concepts, certain equipment is built initially or modified later for life extension. Such an objective carries with it the need for even longer life. Thus, a life extension objective (LEO) and a life extension conservative objective (LECO) can be defined. At each increment in these objectives longer life is needed from the components.

In addition to defining failure, it is necessary to define the extent of response in the event that failures occur. Such responses take the form of "action levels." Presumably, following the steps in the CBDA would minimize or even prevent the occurrence of failures. However, when failures occur and, almost by definition, are unexpected despite the attention given during a responsible CBDA, a clear idea of response is necessary. Such responses can be calibrated in terms of the DLO, DLCO, LEO, and LECO objectives. Thus, when failure occurs at a time less than that DLO, a substantial response is required. Such a response would be defined as "action level number one" (AL-1). Action levels are discussed in connection with Figure 4.36 in Section J. Thus, five action levels can be defined in terms of life objectives (where t_f is time to failure).

Action Level	Corresponding Time to Failure
AL-1	$t_f < DLO$
AL-2	$t_f < DLCO$
AL-3	$t_f < LEO$
AL-4	$t_f < LECO$
AL-5	$t_f > LECO$

The implementation of these action levels is discussed in Section J on prediction. Such targets as DLO, DLCO, LEO, and LECO provide objectives for the durations over which no failures should occur. Choosing which of these is a reasonable objective for design is a management problem.

In defining failure there are certain implications that need to be considered as follows:

[1] *Small Cumulative Fractions.* Most experimental work and resulting correlation equations are based on obtaining mean values, although such a condition is rarely stated (i.e., failure of 50% of the population). Often, the time required to fail 50% of anything is significantly beyond a practically acceptable fraction failed. More likely, failure occurs when a small fraction of elements, welds, or areas fail; and such small fractions failed may be taken as 0.001, 0.01, or 0.1%. Such smaller fractions of failure often occur several orders of magnitude earlier in time than the mean failure time depending greatly on the slope of the statistical dispersion curve. This situation is discussed in connection with Figures 4.27 and 4.33 in Section J.

Thus, if a small fraction failed is, in fact, an operating definition of failure, then serious considerations need to be given to how such a failure is actually predicted since there is no prevailing theory for the physical bases of dispersion, and there are few data available that describe dispersion (dispersion relates to the distribution of data and is discussed in Section J) as applied to corrosion

data. This subject is considered in Section J, where step [6] of the CBDA, establish statistical framework, is discussed.

[2] *Location of Initial Perforation Relative to a Catastrophic Path.* If perforation were to occur in the middle of a relatively tough material, its significance might be relatively minor except for the leak that might ensue. On the other hand, if the same perforation occurs in the path of a low-toughness element (e.g., a weld), the perforation is more important, especially since one perforation rarely occurs by itself. Owing to a possibly much lower K_{I_c} or lower $K_{I_{scc}}$ in the weld region compared with the surroundings, an initially minor penetration could become critical at a relatively small size. Such possibilities are illustrated in Figure 4.24. Figure 4.24(a) shows a circumferential weld, Figure 4.24(b) shows a longitudinal weld, and in Figure 4.24(c) a multiple welded vessel is shown. In Figure 4.24, a relatively small defect is magnified in its influence by the extensive low-toughness paths available for propagation.

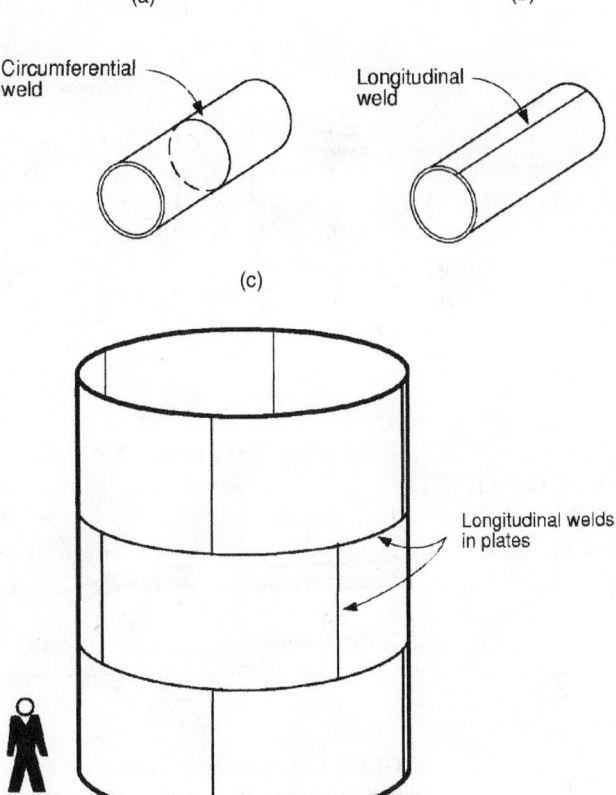

FIGURE 4.24. Schematic illustrations of continuous weld paths in piping and fabricated vessels. (a) Circumferential weld in pipe. (b) Longitudinal weld in pipe. (c) Longitudinal and peripheral welds in a weld-fabricated vessel.

[3] *Simultaneous Penetration Due to Multiple Similar Conditions.* In the case of a cylindrically welded pipe, there are three factors that act to promote simultaneous penetration around the periphery: (a) similar weld heating conditions that might, for example, produce a corrosion-prone microstructure such as sensitization; (b) similar stresses owing especially to automatic welding; and (c) a uniform crevice, for example, from a peripheral thermal shield (similarly uniform crevices are often produced in heat transfer crevices and in sludge piles). These are illustrated in Figure 4.25. Figure 4.25 shows three cases of uniform conditions around a circular geometry where combinations of uniform environment, uniform metallurgy, and/or uniform stresses can be simultaneously present ("congruent" condition). Figure 4.25(a) shows the case of a tube in a tubesheet [similar to Figs. 4.2 and 4.5(a)]; Figure 4.25 (b) shows the case for a circumferential weld of a pipe; Figure 4.25(c) shows the case for a thermal sleeve in a nozzle attached to a pressure vessel. The occurrence of a failure related to the uniform conditions illustrated in Figure 4.25(c) is shown in Figure 4.26, where the SCC

penetration extended 270° of the periphery and the penetration was approximately uniform. All of the penetration noted in this figure occurred before leaking was detected.

While Figure 4.25 suggests that all three factors of metallurgical structure, stress, and crevice chemistry can produce extensive penetration before leaking is observed; such relatively uniform penetration can occur when only one of these factors exists. Such a circumstance is favored in some cases where the velocity of cracks is independent of stress intensity in the "plateau region" in the plot of crack velocity versus stress intensity shown in Figure 4.15(c). The main factor that affects uniform penetration is uniformity in any or several of the three main influences.

[4] *Direction of Defect Relative to "Double-Ended Rupture."* When a corrosion defect occurs and is parallel to the axis of a tube or pipe (an axial defect), its perforation produces a leak. However, when a similar defect occurs and is circumferential, then the leak may lead to the rupture of the total cross section of the pipe. Thus, the direction of perforation relative to

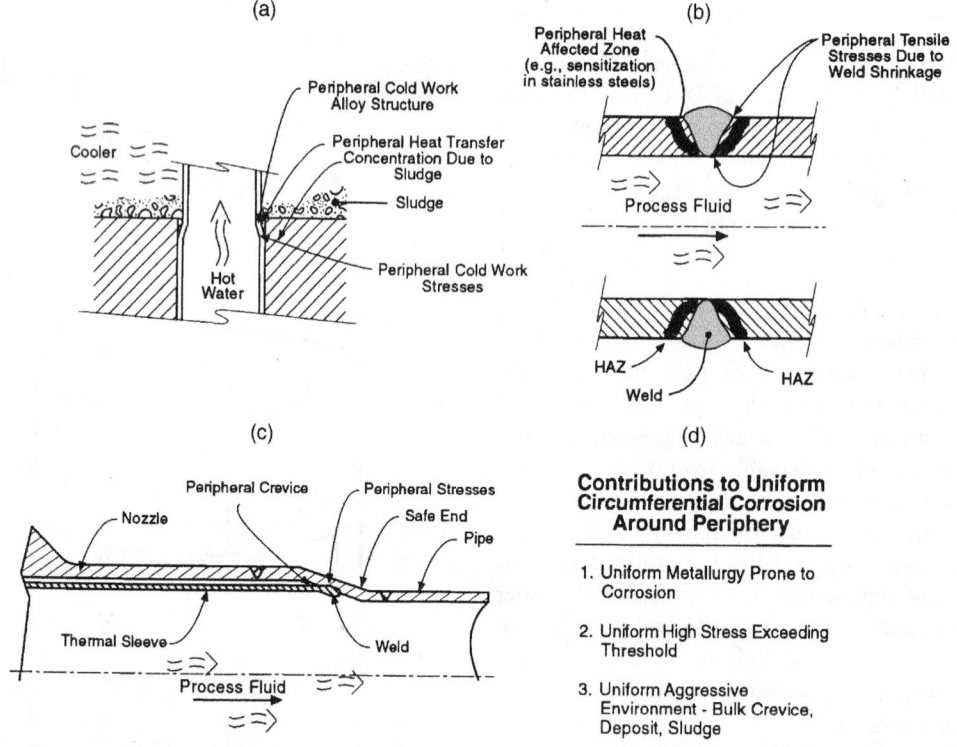

FIGURE 4.25. Schematic views of circumstances where relatively uniform conditions occur around periphery of components with circular cross sections. The major uniform peripheral conditions, when they occur, are usually, metallurgy, stress, and aggressive environments. (a) Sludge pile adjacent to cold-expanded tube in tubesheet at the bottom of a heat exchanger. (b) Weld around pipe with process fluid inside, uniform metallurgy, and uniform stresses. (c) Thermal sleeve in nozzle of pressure vessel with uniform crevice, uniform weld metallurgy, and uniform weld stresses. (d) Summary of uniform conditions leading to uniform SCC or other corrosion.

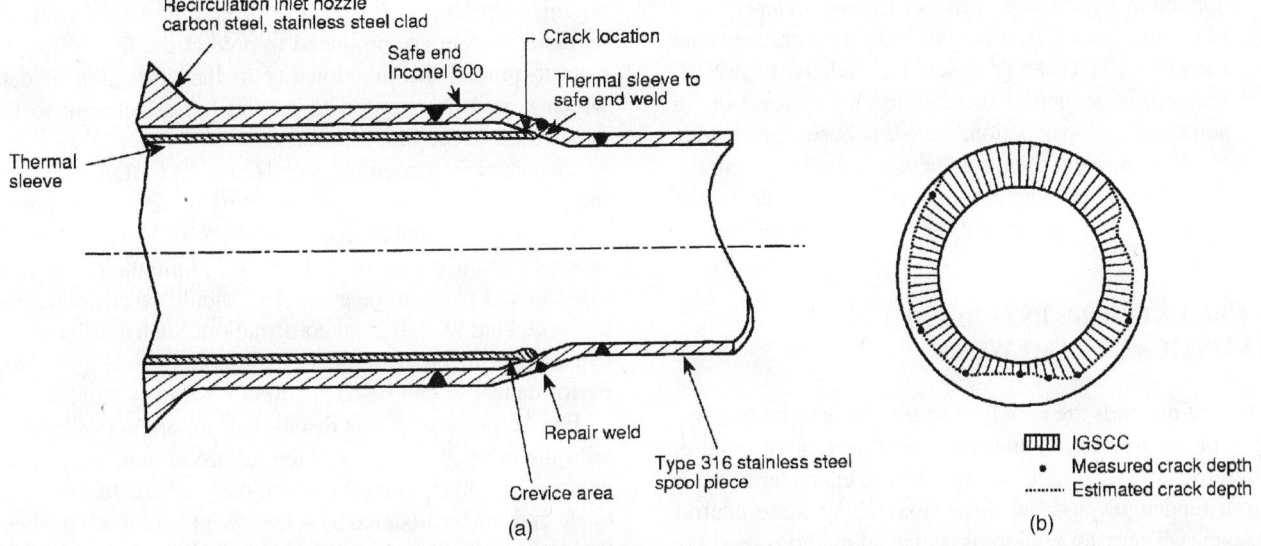

FIGURE 4.26. Crosssection of a recirculation inlet nozzle safe end from the Duane Arnold boiling water nuclear reactor. (a) Crosssection of safe end. (Pressure vessel to the left.) (b) Plan view of crack depth and peripheral distribution looking in the direction of the axis of the pipe. (After [32] modified by Staehle [1].)

the axis of a tube determines whether the tube will sustain only leaks or a rupture of the total cross section. The latter is called a double-ended rupture.

[5] *Redundant and Nonredundant Elements.* An example of redundant and nonredundant elements is found in bridge suspensions. The failure of the Silver Bridge in Ohio involved the complete collapse of a suspension bridge [33]. The suspension was provided by a series of bars connected by dowel joints. Failure occurred when a SCC defect propagated, eventually exceeded K_{I_c}, and catastrophic fracture occurred.

The type of nonredundant suspension used in the Silver Bridge is no longer used. Instead multiple cables are used which, in turn, have multiple wires; such cables are also substantially overdesigned. If a wire breaks, it is a minor problem, and its failure can usually be detected.

The cable-type suspension is a redundant structure whereas the Silver Bridge was a nonredundant structure. In a nonredundant structure, more attention has to be given to corrosion protection and to fracture toughness (a high K_{I_c}) as well as to more frequent and more careful inspections.

[6] *Sequential Failures.* Figure 4.15 identifies the sequence of events that lead to exceeding K_{I_c}. The earliest step may involve simply a penetration due to localized corrosion. When this defect exceeds either a SCC or CF threshold (e.g., $K_{I_{scc}}$), slow growth is initiated. After an SCC crack grows to a critical depth for an existing mean stress, K_{I_c} is exceeded, and fast fracture occurs.

Figure 4.15 suggests that when localized corrosion is observed, the eventual occurrence of catastrophic failure

can be foreseen. This predictable sequence of events suggests that the occurrence of the initial defects should be prevented and that, should they occur, they or their cause should be removed.

[7] *LBB and BBL.* It is common to assume that corrosion-related failures will produce detectable leaks before any serious failure occurs. Such a condition is called "leak before break" (LBB). Such an assumption is supported by the fact that many industrial alloys have relatively high fracture toughnesses so that any penetration by corrosion would be less extensive than could be reasonably expected before K_{I_c} would be exceeded. Thus, a leak due to corrosion would occur before K_{I_c} is exceeded. However, the considerations associated with the discussion of Figures 4.25 and 4.26 indicate that the LBB criterion for design may not be justified in some cases and especially where the simultaneous conditions described in Figure 4.25d obtain.

Had a serious mechanical transient occurred before the leak associated with Figure 4.26, a "break before leak" (BBL) would have occurred. Assuming that the BBL will not occur is not justified, in general, although such an occurrence is much less frequent than LBB.

[8] *Fraction Penetrated as Determined by Nondestuctive Examination (NDE).* A common means of defining failure or of defining when an element should be removed or repaired is related to a fraction of the crosssection penetrated by corrosion. These penetrations arc usually determined by some NDE method. However, these NDE methods are often quite inaccurate unless they have been carefully calibrated and are applied by experienced personnel. Even under the best of circumstances, results from NDE measurements are inaccurate. Thus, if 50% penetration (relative to an NDE method) is defined as the criterion for removal or for some kind of repair action, it needs to be recognized that the NDE measurement can be substantially inaccurate; thus, a measured penetration (by NDE) of less than some target extent could exceed the criterion for failure.

H. CBDA STEP [6]: ESTABLISH A STATISTICAL FRAMEWORK

Statistical methods are widely used for characterizing properties of environments, materials, and modes of failure. Generally, statistical methods are used to characterize both central tendencies and the dispersions about these central tendencies. There are numerous statistical methods used for such applications, and these are described in numerous textbooks [34, 35]. Distributions such as the normal, log-normal, Weibull, and extreme value are often used in characterizing failure phenomena. In this discussion only the Weibull distribution is discussed since it usually provides the best and most flexible characterization for corrosion and failure data.

Corrosion phenomena are particularly suited to statistical characterization owing to the inherently variable behavior of environments, materials, and modes even under the best of circumstances. While statistical methods have potentially great utility for characterizing and interpreting corrosion phenomena, such methods have not been used extensively. A useful overview of applying statistical methods specifically to corrosion has been prepared by Shibata [36]. Also, statistical methods have been applied to characterizing SCC by Akashi and Nakayama [37], where they have characterized several sequential steps with different statistical expressions and then combined the results in a single mathematical expression.

In addition to using statistical methods for characterizing the central tendencies and the dispersions of data, these methods can be used for predicting and analyzing the very early stages of failure where relatively small fractions of a system have failed (e.g., failure of a few tubes from thousands in a heat exchanger). The application of statistical methods is essential for predicting both central tendencies and the very early stages of failure; some of the important approaches are discussed here.

In many applications, it is necessary to predict when a small fraction or a single element of a system (e.g., one of many welds) would fail; that is, when does the first tube in a heat exchanger or the first weld of many in a long pipe fail? When does the first canister of radioactive waste fail? Unfortunately, achieving such predictions is not often an objective of experimental work. Most experimental work is organized to determine when the mean failure occurs, and no methods or data are produced to predict the first failure or even to predict the functionality of the dispersion of data which would be required to predict trends relevant to the occurrence of the first failure.

This section concerning step [6] of the CBDA considers the statistical framework for predicting failure in the context of the mode–location chart of Figure 4.5(b). Thus, each of the mode–location cases, or a given j–i combination, if it is relevant and needs to be evaluated, should be characterized by some kind of statistical construction. Such constructions can be estimated, can be based upon prior engineering performance, or can be based upon laboratory data.

For the purpose of this discussion, we are not concerned with the more gross cases of the failure of automobiles, jet engines, or other complex machinery where failures can occur and can be modeled by statistical procedures but where the details of failure may be related to multiple mechanical and human, but not necessarily degradation, factors.

There are numerous texts that describe the application of statistical methods to problems such as the mode–locations of concern here. Such texts from Abernethy [38] and Nelson [39, 40] should be reviewed.

There are three main sources of variability in considering the statistical predictions necessary for evaluating the j–i cells of the mode–location matrix in Figure 4.5(b). The first source is the mode or submode itself. While sentiment has been expressed for the "deterministic" prediction of modes and submodes of corrosion (or any degradation), the occurrence of these modes under the best of circumstances is inherently variable. For example, the location of initiation sites is inherently variable. Local stresses vary by a factor of about 3, and propagation of stress-related modes depends generally on a power law function. Any intergranular path contains variable chemistries. As the mode propagates, the region in which propagation has occurred is filled with reaction products of nondefined porosity and chemistry. Thus, the initiation and propagation of any mode is inherently and intrinsically variable.

The second source of variability arises from the heat-to-heat, weld-to-weld, or fabrication and processing variability. These sources are extrinsic to the inherent variability. The third source of variability arises from the environments (Section C2)], which are both variable and often difficult to specify.

The most widely used statistical correlation for characterizing failures is the Weibull distribution. Its probability density function is given by Eq. (4.5) and its cumulative distribution (probability of failure) is given by Eq. (4.6):

$$f(t) = \left[\frac{b}{\theta - t_0} \left(\frac{t - t_0}{\theta - t_0} \right)^{b-1} \right] \left\{ \exp\left[-\left(\frac{t - t_0}{\theta - t_0} \right)^b \right] \right\}$$

$$(4.5)$$

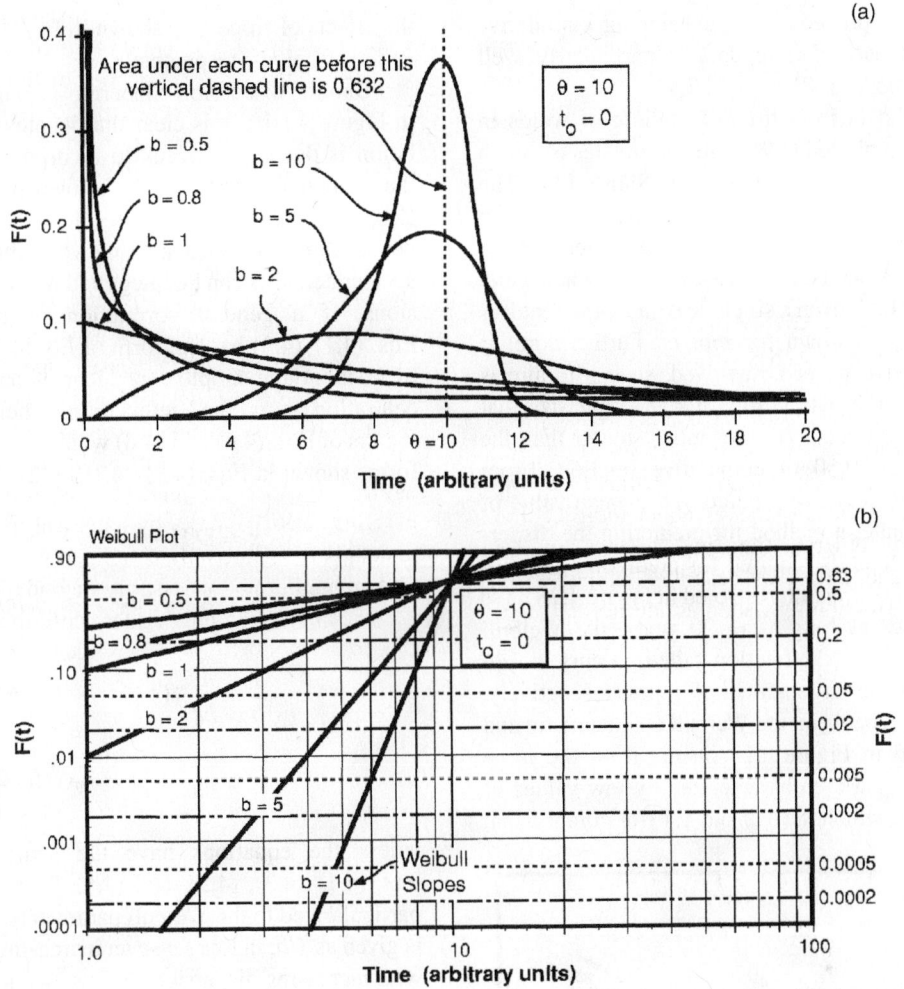

FIGURE 4.27. Weibull plots for constant characteristic, $\theta = 10$, and various slopes, $b = 0.5, 0.8, 1.0,$ 2, 5, 10. The parameter t_0 is taken as zero. Time in arbitrary units. (a) Probability density function, f(t), versus time. (b) Cumulative fraction failed, F(t), versus time. (After Staehle et al. [42].)

$$F(t) = 1 - \exp\left[-\left(\frac{t - t_0}{\theta - t_0}\right)^b\right] \qquad (4.6)$$

where

$t =$ time

$f(t) =$ probability density function

$F(t) =$ cumulative distribution [integral of $f(t)$]

$\theta =$ Weibull characteristic value

$t_0 =$ initiation time (a statistical constant)

$b =$ Weibull slope (higher values indicate less dispersion)

Figure 4.27 shows examples of the probability density function and cumulative distribution for a constant value of the Weibull characteristic, $\theta = 10$, and the $t_0 = 0$. Of particular interest in prediction is Figure 4.27(b), which shows $F(t)$ versus time; typical slopes for materials-related failures vary from $b = 0.5$ to $b = 10$, although some failure processes occur outside this range. Taking, for example, the lower slopes of $b = 1$ or 2, the time of occurrence of failure at 0.01 or 0.001 (1 or 0.1%) in Figure 4.27(b) is seen to be substantially less than the Weibull characteristic θ (close to the mean), which is taken as 10 in this schematic plot. Thus, having determined the mean behavior is useless for predicting early failures without a knowledge of the dispersion or slope of the probability plot of Figure 4.27(b).

Some clear quantification of the Weibull slope b (dispersion) is desirable for predicting performance, although, in practice, such information is not often available. In fact, the slope of the dispersion, the Weibull b, is often developed initially by plotting the fractions failed versus time for early field failures; such fractions would be in the range of 0.001–0.1% for large numbers of elements (e.g., tubes in a heat exchanger). Such fractions are shown on the ordinate of

Figure 4.27(b). The procedure for developing cumulative distributions from early failure data is particularly well described by Abernethy [38].

An example of data from the SCC failure of welds in stainless steel piping plotted in Weibull coordinates is shown in Figure 4.28 from the work of Eason and Shusto [41]. The data here are taken over a large population of welds in stainless steel pipes in numerous boiling water reactor (BWR)–type nuclear plants. In the case of leaking at a weld, as noted in Figure 4.24, even a single leak in a pipe requires that the plant be shut down for repairs. Furthermore, if leaking of radioactive water is involved, such a failure is even more serious. The data in Figure 4.28 emphasize that important failures can occur at times much shorter than the mean value, that is, 0.50 of cumulative fraction. From Figure 4.28, it is evident that predicting the mean value of failure is not adequate; a method for predicting the dispersion, the Weibull slope, is required for predicting the very earliest failures. While the failures noted in Figure 4.28 have occurred over times as long as $\sim$ 20 years, the Weibull characteristic time θ (0.632 fraction failed) occurs at 144 and 1675 years for the large and small pipes, respectively. This large difference between the Weibull characteristic and the times observed in Figure 4.28 results from the large dispersions of data that are characterized by low values of the Weibull slope b, which are 1.2 and 1.4 [for comparison,

the effects of slopes are shown in Fig. 4.27(b)] for the large and small pipes, respectively.

From the discussion concerning the mode–location cases in Figure 4.5(b), it is clear that whatever statistical formulation is developed needs to incorporate dependencies on the principal variables, as shown in Figure 4.19 and Eqs. (4.1)–(4.3). Furthermore, these dependencies need to be developed in such a framework that predictions from accelerated tests can be integrated into the statistical expressions. If a general correlation of data, starting with Eqs. (4.1)–(4.3) has the form of Eq. (4.7), for example, for SCC, involving simple dependencies on hydrogen ion concentration, stress, and temperature, then the Weibull parameters of Eqs. (4.5) and (4.6) would be expected to have the forms shown in Eqs. (4.8)–(4.10) [42]:

$$x = D[\mathrm{H}^+]^n \sigma^m e^{-Q/RT} t^q \qquad (4.7)$$

$$\theta = \theta_0 [\mathrm{H}^+]^{n_\theta} \sigma^{-m_\theta} e^{Q_\theta/RT} \qquad (4.8)$$

$$\frac{1}{b} = \frac{1}{b_0}[\mathrm{H}^+]^{-n_b} \sigma^{-m_b} e^{Q_b/RT} \qquad (4.9)$$

$$t_0 = t_{0,0}[\mathrm{H}^+]^{-n_{t_0}} \sigma^{-m_{t_0}} e^{Q_{t_0}/RT} \qquad (4.10)$$

where the equations have the same meanings as for Eqs. (4.1)–(4.3), (4.5), and (4.6). The constants would be particularized to the Weibull parameters. Note that Eq. (4.8) is given as $1/b$; in this sense an increasing value corresponds with increasing dispersion. In Eqs. (4.7)–(4.10), some terms have been omitted for simplicity; more extensive expressions are given in Eqs. (4.1)–(4.3).

An example of relating Weibull parameters to a physical variable has been investigated by Shimada and Nagai [43] where they determined the effect of stress as shown in Figure 4.29; here, they determined the cumulative distributions for the SCC failure of Zircaloy-2 in iodine gas and from these data obtained correlations for the Weibull parameters. The correlation equations for the Weibull parameters were calculated by Fang and Staehle [44].

If data for use in Figure 4.5(b) are obtained from statistical methods, it is likely that the information to be entered into the i–j cells would be failure times for some cumulative failure fraction. For example, time to failure of 0.1% might be taken as the appropriate data. For comparison, it might also be useful to enter the mean time to failure (i.e., time for 50% failure). The relationship between statistically defined data and j–i cells of Figure 4.5(b) is shown schematically in Figure 4.30.

As a final note, despite the great importance of the dispersion as well as the initiation time to predicting early failures, there is no theory or physical foundation for their character. Most physical theory in corrosion is relevant only

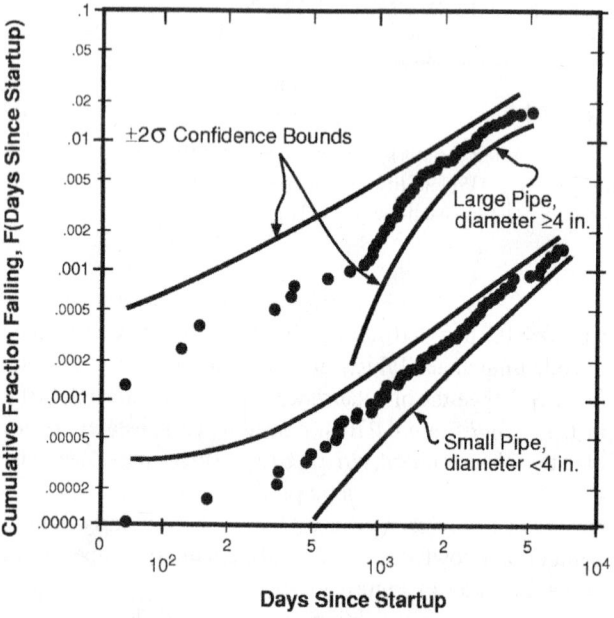

FIGURE 4.28. Cumulative fraction of welds failed in terms of repair rates in stainless steel pipes designated as "large diameter, $\geq$4 in." or "small pipe diameter, <4 in." from BWR-type nuclear plants. The larger diameter pipes include only IGSCC, and the smaller pipes include all failure modes. The Weibull characteristic θ and the Weibull slope b for the larger diameter data are approximately 144 years and 1.2 and for the smaller diameter are approximately 1675 years and 1.4, respectively. (After Eason and Shusto [41].)

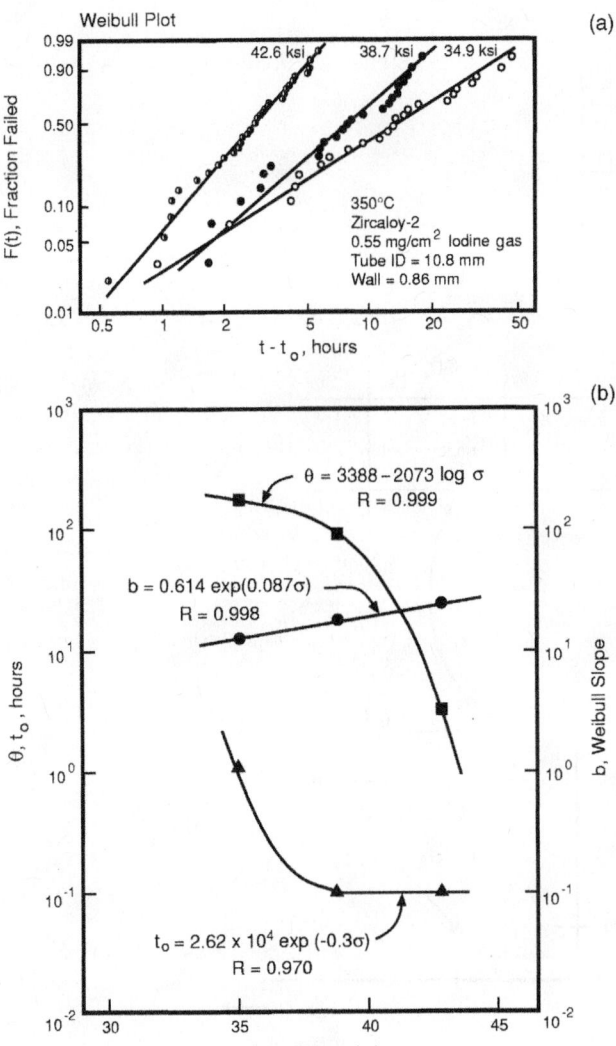

FIGURE 4.29. Effect of stress on the cumulative failure rate of Zircaloy-2 (1.5 Sn–0.15 Fe–0.10 Cr–0.05 Ni) in iodine gas. (a) Cumulative distributions for three stresses. (b) Effect of stress on Weibull characteristic θ, Weibull slope b, and Weibull initiation time t_0. (Adapted from Shimada and Nagai [43].) Correlation equations in (b) calculated by Fang and Staehle [44]. The coefficient of correlation for these curves with the data are given as *R*.

to the mean value or the Weibull characteristic, and even such theories require adjustable parameters.

I. CBDA STEP [7]: ACCELERATED TESTING

The purpose of accelerated testing is to confirm or improve the credibility of predictions that support achieving an objective such as design life (DLO). Accelerated testing is usually carried out by conducting experiments at more intense conditions using specifically intensified variables, "stressors," and, then, extrapolating the results to the conditions of interest. Such stressors might include temperature,

stress, pH, increased concentrations of species, number of fatigue cycles per unit time, and similarly more intense conditions. For example, the "salt spray test" is often used to qualify materials and components for use on salted highways or for ocean and seaside use. In this case, the stressor involves more salt and more frequent applications together with continued high humidity relative to conditions expected for automobiles on winter highways. Incidentally, this test does not always provide reliable results.

The most important consideration in accelerated testing is that the mode of failure in the accelerated test must be the same as the mode of failure expected in the operating equipment. A well-known problem of erroneous accelerated testing involves corrosion fatigue. In much of the fatigue testing over the years, it has been found that the effects of stress cycling expected of a component over time can be assessed by conducting experiments where the number of expected stress cycles is accelerated by using higher cyclic frequencies in order to obtain data in a relatively short time.

However, when fatigue is affected by environments, the incremental propagation of fatigue cracks has been shown to depend on cyclic frequency with increasing propagation per fatigue cycle occurring at lower cyclic frequencies in the range of 1 Hz and lower with little effect of environments observed above 10 Hz. Thus, to have conducted experiments in excess of 1 Hz, when assessing the effects of environments on fatigue, would give a nonconservative result [45]. Premature failures, especially of ship propellers, have occurred because of such errors.

Figure 4.31 illustrates three stressors (temperature, stress, and pH) that could be used relative to an application involving relatively less intense conditions (i.e., target performance conditions). Along the coordinates for each stressor is noted the amount of acceleration based on a key parameter such as the activation energy (temperature), the stress exponent (stress), and the hydrogen ion exponent (concentration of hydrogen ions). In practice, these accelerations have to be determined.

Following the suggestion from Figure 4.29 [43], Figure 4.32 suggests that the effect of temperature could be studied at more intense conditions within a statistical framework and then extrapolated to lower temperatures. This plot suggests that both the mean value and the distributions can be extrapolated.

Statistical interpretations of data provide insights into apparent inconsistencies that result from accelerated testing. Figure 4.33 shows two hypothetical lines on a cumulative distribution plot similar to Figure 4.27(b). Line N-1 corresponds, schematically, to a typical result for the mean and dispersion for the nominal condition. It is similar to the low Weibull slopes of Figure 4.28. The schematic line A-1 corresponds to a cumulative distribution for a hypothetical accelerated test, and *b* is taken as 5.0, which is typical of such

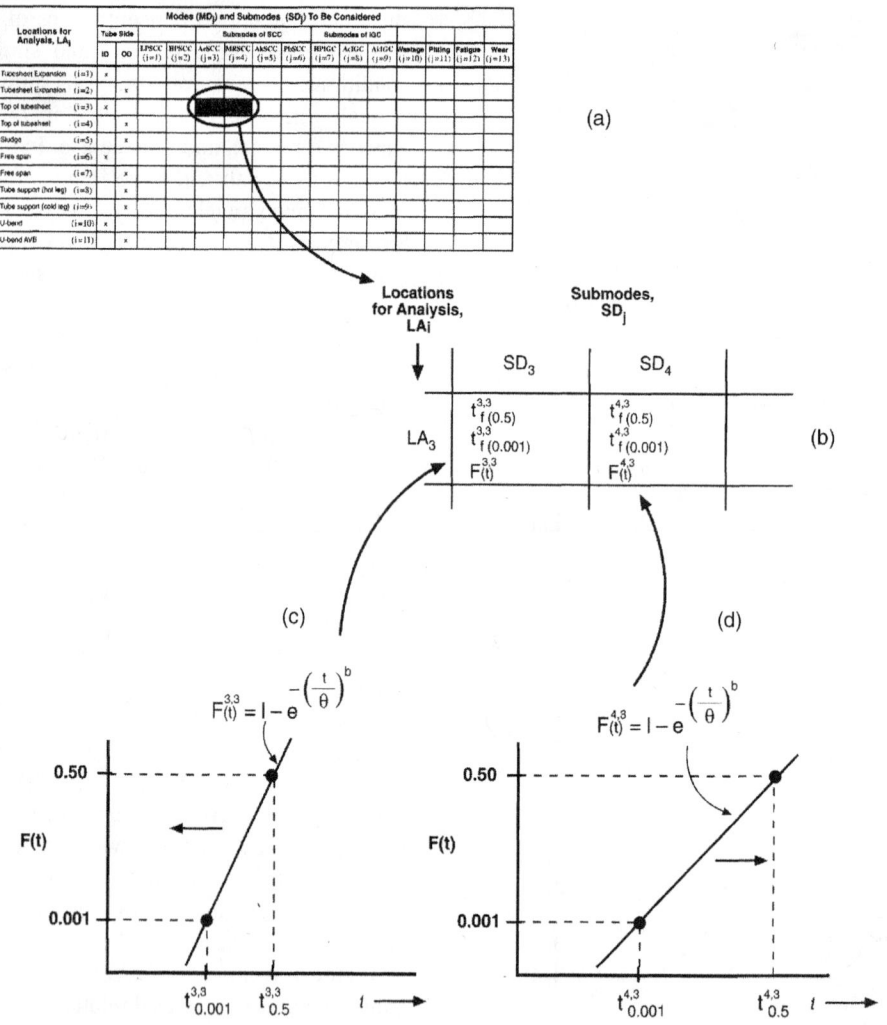

FIGURE 4.30. Including statistically calculated results in j–i cells of the mode–location matrix of Figure 4.5(b). From the mode–location chart of (a), sample cells are shown in (b), where values from two schematic data plots are shown in (c) and (d). Included in the cells are the times to failure for cumulative fractions failed of 0.001 and 0.50 (mean) together with the expressions for $F(t)$ from (c) and (d).

accelerated tests. Line N-l represents a real result from field performance after the equipment was designed, built, and operated. For the accelerated test, which might have been conducted ahead of time, the mean value of line A-1 is determined to be 0.4 year. Thus, for the accelerated test the mean value occurs at 0.4 year, which is 100-fold less than the expected mean based on extrapolating early failure points on line A-l. A 100-fold acceleration is a common objective for accelerated tests, although such a factor of acceleration is often considered to be at the edge of credibility.

Comparing the hypothetical field result, N-l, with the hypothetical accelerated tests, A-1, shows that the ratio of the mean of the field results to the mean of the accelerated test is 100; however, comparing these results at a cumulative fraction failed of 0.01 gives a ratio of 3.4 rather than 100

between the two. At a cumulative fraction failed of 0.001, the ratio is <1 and is ~0.7. This ratio means that the failure of the field elements would occur earlier than the failure of the accelerated test at a cumulative fraction failed of 0.001. The ratio would be even less at 0.0001, which is in the early range of the field data of Figure 4.28.

The implication of the comparison in Figure 4.33 is that the mean values from accelerated tests are not generally good predictors for failure. Thus, a mean value from an accelerated test, following the pattern of stressors in Figure 4.31, would be both inadequate and misleading. This dilemma can be rationalized by referring to Eqs. (4.7)–(4.10). The comparison of mean values essentially follows Eq. (4.8). However, the dispersion follows Eq. (4.9); the dispersions provide the primary differences in lines N-l and A-l at the lower cumulative fractions.

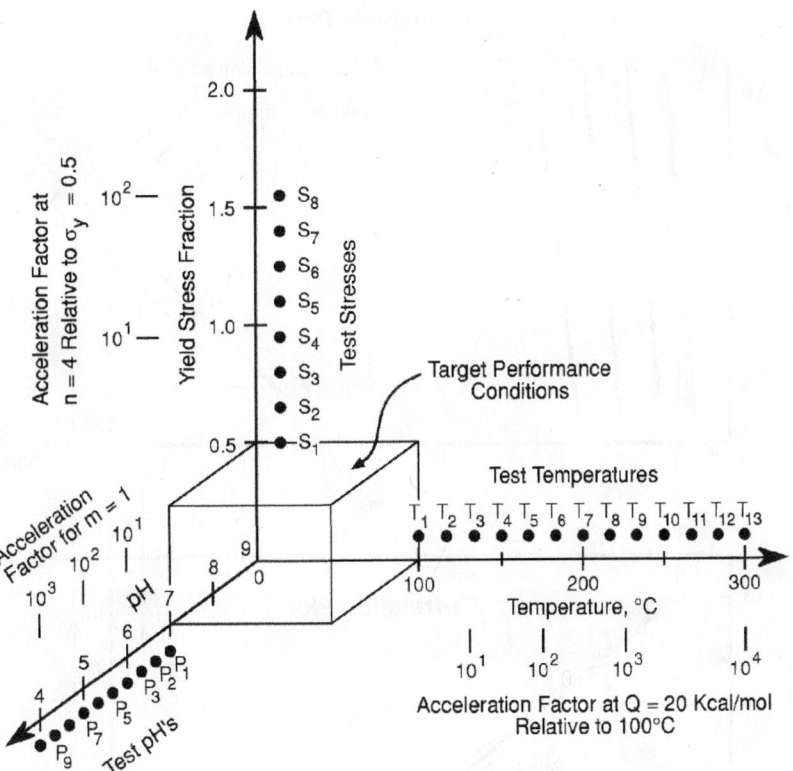

FIGURE 4.31. Stressors of temperature, stress, and pH relative to target performance conditions. Possible test conditions shown along each stressor with the magnitude of acceleration shown for selected values of stressor constants. (After Staehle [1].)

Equation (4.9) suggests that increased values of the stressors decrease the dispersion. While this is assumed in the formulation of Eq. (4.9), the pattern is well known. Thus, the steeper slope of A-l in Figure 4.33 compared with N-l results from the more narrow dispersion (greater slope of the Weibull b) produced by the accelerating stressors.

Line A-2 in Figure 4.33 shows how a predicted failure rate would appear if the Weibull slope did not change from the accelerated test in A-1 for a 40-year life. However, even with such a prediction the cumulative failure would be ∼10 years for the 0.001 fraction. This does not provide the 40-year life if the 0.001 cumulative fraction is taken as the 40-year objective. In fact, if the Weibull slope remains 5.0, then a cumulative rate of the type in A-3 would be required with a mean time of ∼170 years. Line N-2 shows the characteristic failure rate that would be required if the criterion for failure is a cumulative failure fraction of 0.001 and the Weibull slope is 1.0.

Comparing Figure 4.33 with Figure 4.32 suggests an inconsistency. However, had several cumulative distributions been determined as suggested in Figure 4.32, the result of Figure 4.33 could have been predicted by extrapolation.

Validating accelerated tests is a continuing concern. On the other hand, were it not for accelerated tests, performance

of equipment would have to be guessed, or substantial delays would be required until credible data could be accumulated. To develop useful accelerated tests, it is necessary to quantify and predict both the mean and the dispersion.

J. CBDA STEP [8]: PREDICTION

Predicting failure (as defined in step 5 of the CBDA, Section G) and assuring performance are the goals of steps [1]–[7]. Step [8], prediction, then involves synthesizing the results of steps [1]–[7] of the CBDA. The format for organizing the necessary data to support predictions is that in Figure 4.5(b), and the steps are identified in Section B3. Possible approaches for responding to each of the j–i cells are shown in Figure 4.6(a), and the application of statistical analyses to answering the questions of each cell is illustrated in Figure 4.30.

At this point, it is necessary to decide what specific information should be placed in each of the j–i cells and how the information in all the cells for a given subcomponent should be aggregated. [Actually, such a step would have been resolved in connection with CBDA steps [5] and [6]—define failure and action levels and establish a statistical framework

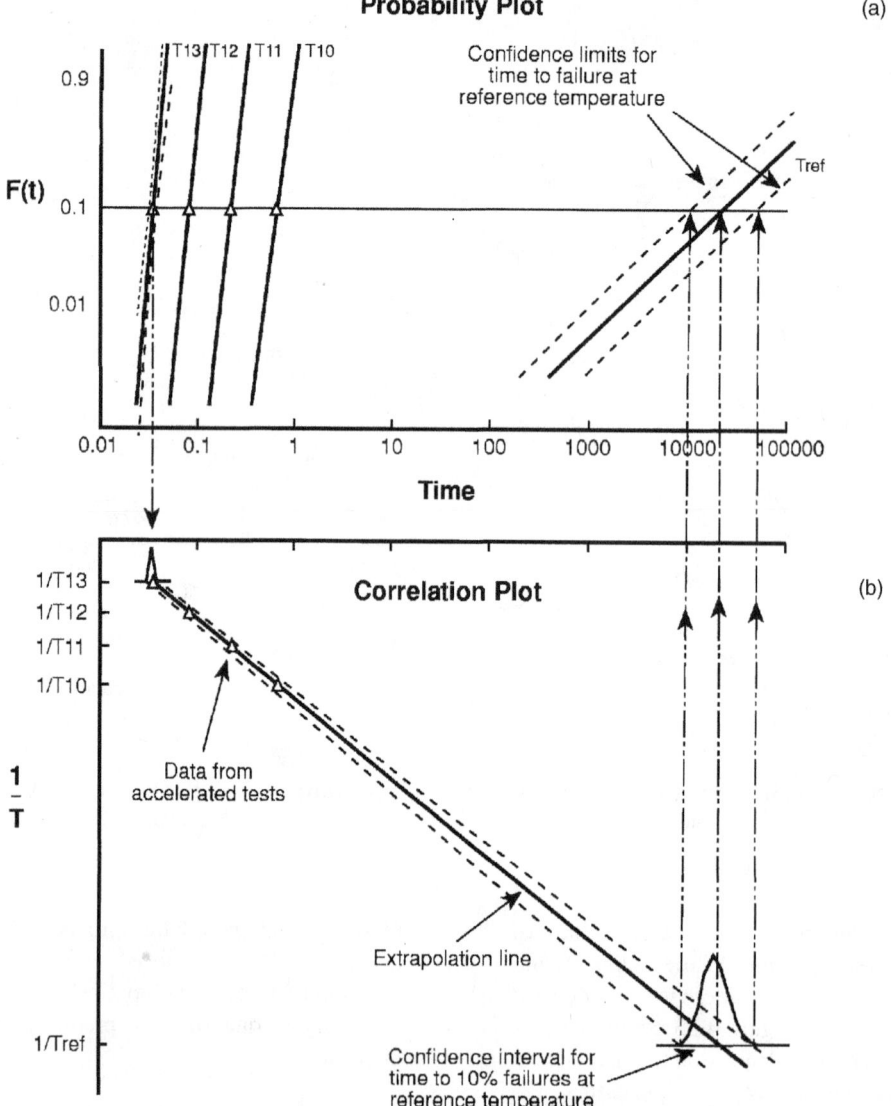

FIGURE 4.32. (a) Schematic cumulative failure plots showing curves for high temperatures associated with acceleration tests and for a reference temperature associated with the expected application. Confidence limits are also noted. (b) Plot of $1/T$ versus time showing probability density functions corresponding to cumulative probability plots. (After Staehle et al. [42].)

(Sections G and H)]. A simple approach could involve determining the time for (e.g., 0.1%) failure in each of the cells assuming that something is known about the relationship between accelerated testing and field performance as described in Section I. Such an approach is illustrated in Figure 4.30. If the necessary statistical distributions are available, then this or any percent failure as a criterion can be calculated and used in the j–i cells. Other approaches can be developed, especially when the statistical distributions can be developed from either laboratory data or field experience.

A specific approach to evaluating the contribution of multiple mode–location cases is illustrated in Figure 4.34

for the case of a steam generator considered in Figures 4.2 and 4.5 [46], The mode–location cases of Figure 4.34 include many of those that are illustrated in Figure 4.2. The data for cumulative failures (dots) shown in Figure 4.34 are based on information taken by NDE methods when examining tubes during outages.

At each outage, the total probability of failure is calculated by the relationship in Eq. (4.11). Equation (4.11) is based on the relationship that the total probability of success is $1 - F_T(t)$, that is, 1 minus the total probability of failure, $F_T(t)$. The total probability of success is then the product of the individual probabilities of success, that is, $1 - F_{j-i}(t)$, where $F_{j-i}(t)$ is the cumulative probability of failure of each

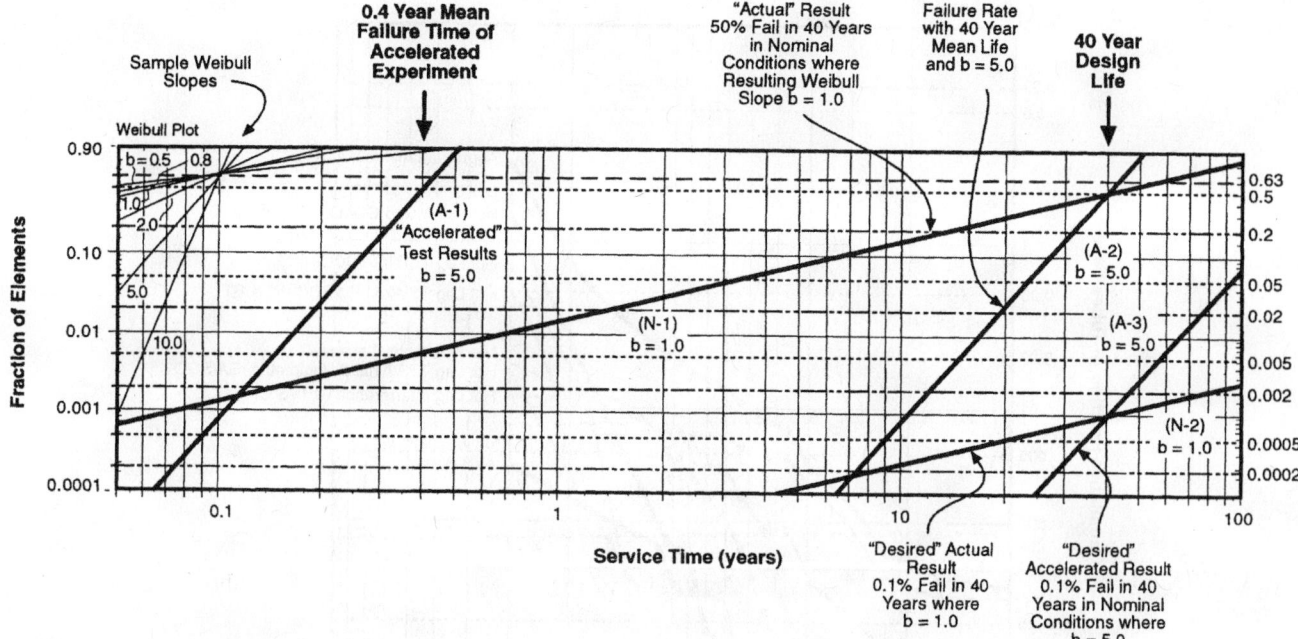

FIGURE 4.33. Schematic comparison of hypothetical actual field results with hypothetical accelerated testing using cumulative distributions versus time. The line N-1 corresponds to the nominal field failure rate with $b = 1$. Line A-I with $b = 5.0$ corresponds to an accelerated test. Line N-2 is the desired field failure rate that does not exceed 0.001 cumulative failures in 40 years. Line A-2 has the same slope as A-I for a mean life of 40 years; line A-3 has the same slope but with 0.001 failure in 40 years.

mode–location case. For simplicity, the F_{j-i} (mode–location cases) are given as $F_1, F_2,\ldots, F_n$:

$$F_T(t) = 1 - [1 - F_1(t)][1 - F_2(t)] \cdots [1 - F_n(t)] \quad (4.11)$$

After the last measurements shown in Figure 4.34, given by the last dot on the respective plots of F_{j-i}, the curves are extrapolated using expected functionalities. The individual cumulative probabilities can then be summed as in Eq. (4.11) to give the overall aggregate of all mechanisms. In the case of steam generators failure occurs at some point where the number of tubes that have been "plugged" (i.e., taken out of service by plugging to prevent access of the primary to secondary side because perforation failures were identified by NDE) exceeds a designated fraction of the total number of tubes that corresponds to excessive loss of power. This point of overall failure might be, for example, 10% of all tubes or a 0.10 cumulative fraction failed of "the aggregate of all mechanisms." A prediction such as that in Figure 4.34 results from an orderly aggregation of all failures as identified in Figure 4.5(b).

Once specific versions of Figure 4.5(b) are organized for various subcomponents, then it is possible to predict overall performance. Another layer of analysis may be required to deal with the collected functionalities of the data in Figure 4.5(b). Such an approach is followed in applying Eq. (4.11) for the separate mode–location cases in

Figure 4.34. However, the most appropriate approaches for aggregating data from the set of mode–location cases depend on both the forms of the results in the cells of Figure 4.5(b) for the various components and the criteria for failure of the engineering project.

The process of engaging in the CBDA and developing predictions is dynamic. As the $j-i$ cells are evaluated, the design is changed to minimize the intensity of degradation in any cell. The overall performance will then be based upon a set of modified and rejected $j-i$ cells. Once the conditions of individual cells are stabilized and agreed upon, it is possible to predict the overall performance, recognizing, of course, that certain cells have been assumed to be irrelevant or that certain approximations to their functional character can be developed. However, even such decisions would be (could be) a matter of record for later review and possible modification.

Once the $j-i$ matrices are completed for various subcomponents, they can be aggregated as shown in the first step of Figure 4.35 for components. The results from the various components can then be aggregated as shown in the successive steps of Figure 4.35.

The predicted times to failure or similar results, as summarized in the appropriate versions of Figure 4.5(b), need to be compared with objectives for DLO, DLCO, LEO, and LECO as described in Section G, where step [5] of the CBDA is described. Providing that the predicted times to failure exceed the goals for DLO, DLCO, LEO, or LECO as required

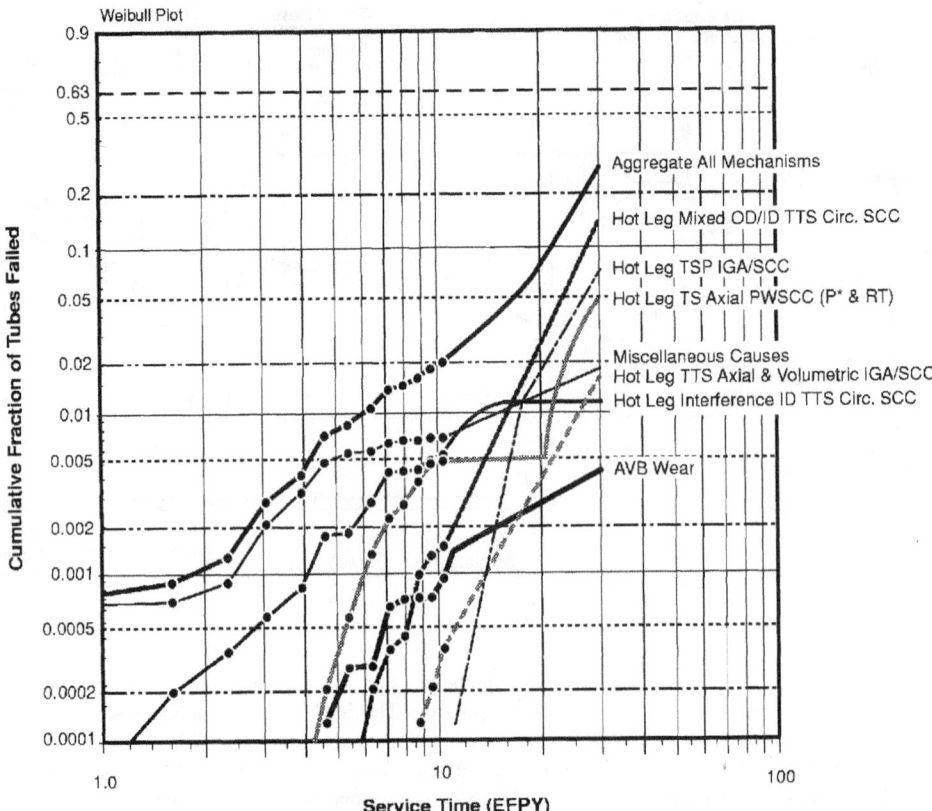

FIGURE 4.34. Cumulative fraction of tubes failed versus service time in equivalent full power years (EFPY) for seven mode–location cases from the set of steam generators in the Ringhals 4 pressurized water nuclear plant: TTS refers to "top of tube sheet," TS to "tubesheet," Circ. SCC to "circumferential SCC," P* to a special location where SCC may not be serious, RT to SCC at the "roll transition" location, and the AVB to "antivibration bars" (at top of steam generator to stabilize the tops of the U–bends), (Unpublished data provided by L. Bjornkvist of Vattenfall and J. Gorman of Dominion Engineering.)

by the application, no further work of prediction is required. Such conclusions assume that the *j–i* cells, which have been judged irrelevant or where correlations have been assumed based on marginal data, do not pose excessive risks.

One of the major benefits of the CBDA approach implicit in Figure 4.5(b) is the necessity for explicit action on every cell. Presumably, the design group would require that every cell be accounted for by some formal procedure. Whether the cell is evaluated (as in Figs. 4.6 and 4.30) by laboratory data, prior history, engineering judgment, assumption of irrelevance, or assumption of no problem, these actions would become part of the written record of the design.

The times specified for DLO, DLCO, LEO, and LECO then become action levels for operators; if failures occur in times less than these objectives, then certain actions are required. Figure 4.36 indicates hypothetically possible responses if failures in the field occur earlier than predicted. For example, if action level 1 (AL 1) is taken as failures occurring at times less than DLO, then numerous and

serious actions might be required. These are illustrated by the black dots in Figure 4.35. If failures occur at longer times, the action levels require less extensive responses; progressively lower priority actions are noted by shaded and then nonshaded dots.

Figure 4.35 becomes the agenda for a major design review with respect to the CBDA. By the time it is possible to prepare a complete version of Figure 4.35, sufficient information should be available for a comprehensive assessment of the design. Often, owing to the necessities of schedules and costs, it is sometimes thought necessary to move ahead on construction before all the detailed designs are completed. The CBDA, through the chart of Figure 4.5(b), provides a continuing framework for assessing what decisions are necessary as detailed designs are completed.

In some designs, there are circumstances where inspection and design are not possible, although such circumstances should be avoided. For example, in the storage of radioactive waste where times of possibly 1 million years of nonfailure (i.e., nonrelease of more than a minimum amount

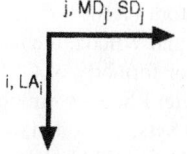

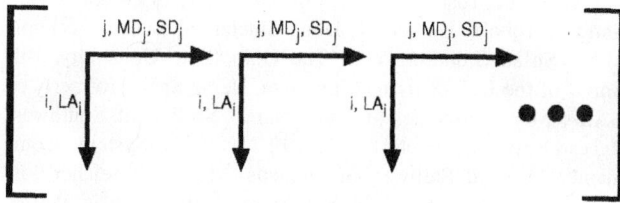

Component with Subcomponents (e.g., steam generator)

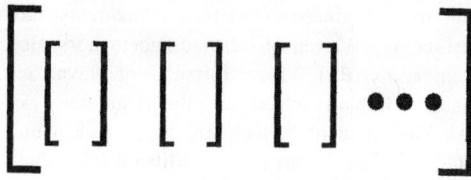

Subsystem (e.g., secondary side)

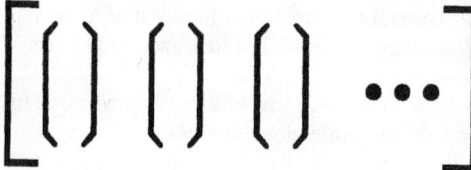

System (e.g., nuclear steam supply system)

FIGURE 4.35. Mode–location, MD_j/SD_j–LA_i, charts aggregated for successively higher levels of system hierarchy.

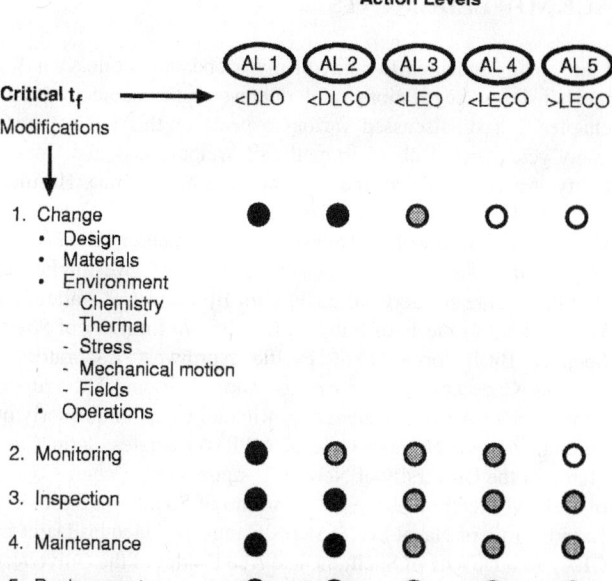

FIGURE 4.36. Schematic view of relationships among hypothetical times to failure, targets for design life (DLO, DLCO, LEO, and LECO), action levels (AL 1, AL 2, AL 3, AL 4, and AL 5), and responses appropriate for each action level. Black dots indicate high-priority actions; shaded dots indicate lower priority actions; open circles indicate no action.

of radioactivity) may be desired, a greater emphasis needs to be placed upon reliability than cases where much shorter design lives are specified. Other cases where explosions, release of radioactivity, or failure of implantable medical devices could occur, higher levels of assurance are required and need to be acknowledged in preparing Figure 4.5(b). Finally, in some relatively ordinary lifetimes certain subcomponents are not inspectable or cannot be monitored. Here, again, higher priorities need to be given to assuring performance in preparing Figure 4.5(b).

K. CBDA STEP [9]: MONITORING, INSPECTION, AND FEEDBACK

Monitoring and inspection provide feedback to designers and operators for comparing with predictions and for defining

action levels. These responses vary greatly with applications and industries.

A discussion of monitoring, inspection, and feedback is a large subject and is not elaborated upon here. However, the choice of methods for monitoring and inspection should be based upon cues provided by the CBDA steps [1]–[5] (Sections C to G). Furthermore, the frequency of monitoring and inspection should consider the statistical character, step [6] (Section H), of the mode–location combinations.

Certain components and subcomponents are not readily inspectable, which implies that certain features of failure must be assumed, and higher levels of certainty of reliable performance should be required for such applications.

L. CBDA STEP [10]: MODIFICATION

As failures occur, certain inadequacies in the assumptions of high-priority mode–location cases and in what i–j cells can be neglected become apparent The occurrence of premature failures guides appropriate modifications and reconsideration of these assumptions. Once indications of performance and failures are available, then designers and operators can act to modify the design and/or operation.

As modifications are implemented, steps in the CBDA need to be repeated since modifications change the system and its response to various stressors.

ACKNOWLEDGMENTS

I am greatly indebted to many of my friends and colleagues for their insights, suggestions, and interest in the subject of this chapter. I have discussed various aspects of this subject over many years with: Bob Abernmethy of Weibull Risk and Uncertainty Analysis, Masatsune Akashi of Ishikawajima Harima, Peter Andresen of General Electric, Mick Apted of QuantiSci, Hans Arup formerly of the Danish Corrosion Center, Allen Baum of Westinghouse Bettis, Richard Begley of Westinghouse, Philippe Berge formerly of EDF, Lars Bjornkvist of Vattenfall, Walter Boyd formerly of Battelle, Ron Brown formerly of Shell, Spencer Bush formerly of Pacific Northwest Laboratories, Dwayne Chesnut of Lawrence Livermore National Laboratory, Kim Clark of Aptech Engineering, Richard Claassen formerly of Sandia, the late Norman Cole of MPR Associates, John Congleton of the University of Newcastle-upon-Tyne, Robert Cowan of General Electric, Gustavo Cragnolino of Southwest Research, David Curtis of Naval Sea Systems Command, Jacques Daret of CEA, Tom DeWitz of Shell, the late Abe Dukler of the University of Houston, David Duquette of Rensselaer Polytechnic Institute, Ernie Eason of Modeling and Computing Services, Geof Egan of Aptech Engineering, Jan Engstrom of Vattenfall, Zhi Fang of the University of Minnesota, the late Mars Fontana of The Ohio State University, Peter Ford of General Electric, Rick Gangloff of the University of Virginia, Frank Garner of Pacific Northwest Laboratories, Dick Garnsey of NNC Limited, Tracy Gendron of Atomic Energy Canada Limited, Bill Gerberich of the University of Minnesota, Gerry Gordon formerly of General Electric, Jeff Gorman of Dominion Engineering, Mac Hall of Westinghouse Bettis, Carolyn Hansson of the University of Waterloo, Bob Hermer of Westinghouse Bettis, Dick Horvath of Shell, Keizo Hosoya of JGC, the late Bob Jaffee of the Electric Power Research Institute, Lee James formerly of Westinghouse Bettis, Randy John of Shell, Joung Soo Kim of Korea Atomic Energy Research Institute, Uh Chul Kim of Korea Atomic Energy Research Institute, Yong Soo Kim of Yonsei University, Tatsuo Kondo formerly of Japan Atomic Energy Research Institute, Jerry Kruger formerly of Johns Hopkins University, Il Hiun Kuk of Korea Atomic Energy Research Institute, Jiro Kuniya of Hitachi, Ron Latanision of the Massachusetts Institute of Technology, Phil Lindsay of Aptech Engineering, Bill Lindsay formerly of Westinghouse, Brenda Little of the Stennis Space Center, Jesse Lumsden of Rockwell International, Digby Macdonald of Stanford Research Institute, Dan McCright of Lawrence Livermore National Laboratory, Art McEvily of the University of Connecticut, Al McIlree of the Electric Power Research Institute, Harry Mandil formerly of Naval Sea Systems Commands, Tom Marciniak of Engineering Systems, Bruce Miglin of Shell, Peter Millett of the Electric Power Research Institute, Anders Molander of Studsvik Energy, Charles Morin of Engineering Systems, Tomomi Murata of Nippon Steel, Joe Muscara of the Nuclear Regulatory Commission, Hiroo Nagano formerly of Sumitomo, Paul Nelson of Shell, Roger Newman of the University of Manchester, Suat Odar of Siemens, Hideya Okada formerly of Nippon Steel, Takatsugu Okada of Toshiba, Bob O'Shea of Engineering Systems, Redvers Parkins of the University of Newcastle-upon-Tyne, Tom Passell of the Electric Power Research Institute, Joe Payer of Case Western Reserve University, Tom Pigford of U.C. Berkeley, Fred Pocock formerly of Babcock and Wilcox, the late Marcel Pourbaix of CEBELCOR, Mike Pryor formerly of Olin, Robert Rapp of The Ohio State University, Peter Rhodes formerly of Shell, Ted Rockwell formerly of Naval Sea Systems Command, Dan van Rooyen formerly of the Brookhaven National Laboratory, Bo Rosborg of Studsvik Energy, Norio Sato of Hokkaido University, Bill Schneider of Babcock and Wilcox, Peter Scott of Framatome, Toshio Shibata of the University of Osaka, Tetsuo Shoji of Tohoku University, Doug Sinclair of Lucent Technologies, Susan Smialowska of The Ohio State University, Bill Smyrl of the University of Minnesota, Heinz Spähn formerly of BASF, Markus Speidel of ETH, Narasi Sridhar of Southwest Research, Robert Steele formerly of Naval Sea Systems Command, Bernhard Stellwag of Siemens, Michael Streicher formerly of Dupont, Shunichi Suzuki of Tokyo Electric Power, Hiroshi Takamatsu of Kansai Electric Power, Bob Tapping of Atomic Energy Canada Limited, George Theus of Engineering Systems, Shigeo Tsujikawa of the University of Tokyo, Art Turner of Dominion Engineering, Hisao Wakamatsu of Ishikawajima Harima, Jay Warren of Cardiac Pacemakers, Bob Wei of Lehigh University, Bill Wilson formerly of Naval Sea Systems Command, Graham Wood of the University of Manchester, Brian Woodman of Aptech Engineering, Katsumi Yamamoto of JGC, Toshio Yonezawa of Mitsubishi, Edwin Zebroski formerly of the Electric Power Research Institute, Zhang Sen Ru of the Nuclear Power Design Research Institute of China, Zhang Weiguo of the Chinese Institute for Atomic Energy, Zhu Xu Hui of the China National Nuclear Corporation.

I am also indebted to several who supplied figures including Peter Andresen of General Electric, Lars Bjornkvist of Vattenfall, Ken Burrill of Atomic Energy Canada Limited, and Jeff Gorman of Dominion Engineering.

Appreciation is also expressed to my office and support staff for their enthusiastic help and professionalism: Nancy Clasen, Julie Daugherty, John Ilg, Mary Ilg, Barbara Lea, and Tim Springfield.

Finally, I am indebted to Winston Revie, who invited me to prepare this chapter and who encouraged me in its preparation.

REFERENCES

1. R.W. Staehle, "Combining Design and Corrosion for Predicting Life," in Life Prediction of Corrodible Structures, R.N. Parkins (Ed.), Vol. 1. NACE International, Houston, TX, 1994, pp. 138–291.
2. R. W. Staehle, "Environmental Definition," in Materials Performance Maintenance, R. W. Revie, V. S. Sastri, M. Elboujdaini, E. Ghali, D. L. Piron, P. R. Roberge, and P. Mayer (Eds.), Pergamon, New York, 1991, pp. 3–43.
3. Guidelines for Chemical Process Quantitative Risk Analysis, Center for Chemical Process Safety, American Institute of Chemical Engineers, New York, 1989.
4. Tools for Making Acute Risk Decisions with Chemical Process Safety Applications, Center for Chemical Process Safety, American Institute of Chemical Engineers, New York, 1995.

5. Guidelines for Investigating Chemical Process Incidents, Center for Chemical Process Safety. American Institute of Chemical Engineers, New York, 1992.

6. Guidelines for Chemical Process Quantitative Risk Analysis, Center for Chemical Process Safety, American Institute of Chemical Engineers, New York, 1989.

7. T. A. Kletz, Hazop and Hazan: Identifying and Assessing Process Industry Hazards, 3rd ed., Institution of Chemical Engineers, Rugby, Warwickshire, 1992.

8. T. A. Kletz, Improving Chemical Engineering Practices: A New Look at Old Myths of the Chemical Industry, 2nd ed., Hemisphere Publishing Corporation, New York, 1990.

9. F. N. Goswami and R.W. Staehle, Electrochim. Acta, **16**, 1895 (1971).

10. R. W. Staehle, "Combining Design and Corrosion for Predicting Life," in Life Prediction of Corrodible Structures, R. N. Parkins (Ed.), Vol. **1**, NACE International, Houston, TX, 1994, pp. 138–291.

11. R. W. Staehle, "Understanding 'Situation-Dependent Strength': A Fundamental Objective in Assessing the History of Stress Corrosion Cracking," in Environment Induced Cracking of Metals, R. P. Gangloff and M. B. Ives (Eds.), NACE-10, NACE International, Houston, TX, 1990, pp. 561–612.

12. J. P. Broomfield, Corrosion of Steel in Concrete: Understanding, Investigation and Repair, E & FN Spon, London, 1997.

13. B. Little, P. Wagner, and F. Mansfeld, Int. Mater. Rev., **36**(6), 253 (1991).

14. P. D. Andresen and L. M. Young, "Characterization of the Roles of Electrochemistry, Convection and Crack Chemistry in Stress Corrosion Cracking," in Seventh International Symposium on Environmental Degradation of Materials in Nuclear Power Systems—Water Reactors, G. Airey et al. (Eds.), NACE International, Houston, TX, 1995, pp. 579–596.

15. B. Chexal, J, Horowitz, R. Jones, B, Dooley, C. Wood, M. Bouchacourt, F. Remey, F. Nordmann, and P. St. Paul, Flow-Acoelerated Corrosion in Power Plants. TR-106611, Electric Power Research Institute, Palo Alto, CA, 1996.

16. G. J. Bignold, C. H. De Whalley, K. Garbett, and I. S. Woolsey, "Mechanistic Aspects of Erosion-Corrosion Under Boiler Feedwater Conditions," in Water Chemistry of Nuclear Reactor Systems 3, Vol. 1, British Nuclear Energy Society, London, UK 1983, pp. 219–226.

17. Ph. Beige, Proceedings of the Conference ADRP on Water Chemistry and Corrosion in the Steam-Water Loops of Nuclear Power Stations, Ermenonville, France, 1972.

18. H. Spähn, G. H. Wagner, and U. Steinhoff, "Stress Corrosion Cracking and Catbodic Hydrogen Embrittlement in the Chemical Industry," in Stress Corrosion Cracking and Hydrogen Embrittlement of Iron Base Alloys, R. W. Staehle, J. Hochmann, R. D. McCright, and J. E. Slater (Eds.), NACE-5, NACE International, Houston, TX, 1977, pp. 80–110.

19. F. A. Garner, "Irradiation Performance of Cladding and Structural Steels in Liquid Metal Reactors," in Nuclear Materials, B. R. T. Frost (Ed.), VCH, Weinheim, Germany, 1994, pp. 419–543.

20. N. B. Pilling and R. E. Bedworth, J. Inst. Metals, **29**, 534 (1923).

21. Steam Generator Reference Book, Steam Generator Owners Group, Electric Power Research Institute, Palo Alto, CA, 1985.

22. E. B. Rosa, B, McCollum, and O.S. Peters, "Electrolysis in Concrete," 2nd ed., Technologic Papers of the Bureau of Standards **18**, Department of Commerce, Government Printing Office, Washington, DC, 1919.

23. H. W. Pickering, F. H. Beck, and M. G. Fontana, Corrosion, **18**, 230 (1962).

24. T. L. Anderson, Fracture Mechanics: Fundamentals and Applications, 2nd ed., CRC Press, Boca Raton, FL, 1995.

25. M. O. Speidel and A. Atrens (Eds.), Corrosion in Power Generating Equipment, Plenum, New York, 1984, pp. 88, 343.

26. R. B. Waterhouse, Fretting Corrosion, Pergamon, Oxford, UK, 1972.

27. R. W. Staehle, "Development and Application of Corrosion Mode Diagrams," in Parkins Symposium on Stress Corrosion Cracking, S. M. Bruemmer, E. I. Meletis, R. H. Jones, W. W. Gerberich, F. P. Ford, and R.W. Staehle (Eds.), TMS, Warrendale, PA, 1992, pp. 447–491.

28. R. W. Staehle, "General Patterns and Mechanistic Alternatives for the Stress Corrosion Cracking of Inconel 600 at Low Potentials," in Proceedings: Specialist Meeting on Environmental Degradation of Alloy 600, R. G. Ballinger, A. R. McIlree, and J. P. N. Paine (Eds.), TR-I04898 Electric Power Research Institute, Palo Alto, CA, 1996, pp. 20/1-20/98.

29. M. F. Ashby and B. Tomkins, "Micromechanisms of Fracture and Elevated Temperature Fracture Mechanics," in Mechanical Behavior of Materials, Vol. 1, K. J. Miller and R. F. Smith (Eds.), Pergamon, Oxford, UK, 1980, pp. 47–89.

30. R. W. Staehle, "Occurrence of Modes and Submodes of IGC and SCC," in Control of Corrosion on the Secondary Side of Steam Generators, R. W. Staehle, J. A. Gorman, and A. R. McIlree (Eds.), NACE International, Houston, TX, 1996, pp. 135–208.

31. Z. Szklarska-Smialowska, Z. Xia, and R. R. Valbuena, Corrosion, **50**(9), 676 (1994).

32. NRC Pipe Crack Study Group, Investigation and Evaluation of Stress Corrosion Cracking in Piping of Light Water Reactor Plants, NUREG-0531, U.S. Nuclear Regulatory Commission, Washington, Jan., 1979.

33. B. F. Brown, Stress Corrosion Cracking Control Measures, NACE International, Houston, TX, 1981, p. 50.

34. M. G. Natrella, Experimental Statistics, National Bureau of Standards Handbook 91, U.S. Government Printing Office, Washington, DC, 1966.

35. R. V. Hogg and J. Ledolter, Engineering Statistics, Macmillan, New York, 1987.

36. T. Shibata, Corrosion, **52**(11), 813 (1996).

37. M. Akashi and G. Nakayama, "A Process Model for the Initiation of Stress Corrosion Crack Growth in BWR Plant Materials," in Effects of the Environment on the Initiation of Crack Growth, W. A. Van Der Sluys, R. S. Piascik, and R.

Zawierucha (Eds.), ASTM STP 1298, ASTM, West Conshohocken, PA, 1997, pp. 150–165.

38. R. B. Abernethy, The New Weibull Handbook, 2nd ed, R. B. Abernethy, North Palm Beach, Florida, July 1996.

39. W. Nelson, Applied Life Data Analysis, Wiley, New York, 1982.

40. W. Nelson, Accelerated Testing, Wiley, New York, 1990.

41. E. D. Eason and L. M. Shusto, Analysis of Cracking in Small-Diameter BWR Piping, NP-4394, Electric Power Research Institute, Palo Alto, CA, 1986.

42. R. W. Staehle, J. A. Gonnan, K. D. Stavropoulos, and C. S. Welty, Jr., "Application of Statistical Distributions to Characterizing and Predicting Corrosion of Tubing in Steam Generators of Pressurized Water Reactors," in Life Prediction of Corrodible Structures, Vol. 2, R.N. Parkins (Ed.), NACE International, Houston, TX, 1994, pp. 1374–1399.

43. S. Shimada and M. Nagai, Reliability Eng., **9**, 19 (1984).

44. Z. Fang and R. W. Staehle, Calculations with KaleidaGraph software, version 3.0, Synergy Software, Reading, PA.

45. R. P. Wei, "Fatigue Crack Growth in Aqueous and Gaseous Environments," in Environmental Degradation of Engineering Materials in Aggressive Environments, M. R. Louthan, Jr., R. P. McNitt, and R. D. Sisson, Jr. (Eds.), Virginia Polytechnic Institute, Blacksburg, VA, 1981, pp. 73–81.

46. L. Bjornkvist of Vattenfall and J. Gorman of Dominion Engineering provided these nonpublished data for use in this chapter.

5

ESTIMATING THE RISK OF PIPELINE FAILURE DUE TO CORROSION

M. Nessim

C-FER Technologies (1999) Inc., Edmonton, Alberta, Canada

A. INTRODUCTION

Metal loss corrosion is one of the major causes of pipeline failures. For example, failure statistics collected by the U.S. Department of Transportation indicate that 17% of reportable gas pipeline failures and 27% of reportable liquid pipeline failures are caused by corrosion [1]. In Alberta, where there is a large percentage of gathering pipelines and consequently more significant internal corrosion problems [2], corrosion accounts for approximately 40% of all pipeline failures.

As the pipeline network ages, it is important to understand the risk of failure due to corrosion and take appropriate measures to keep it at tolerable levels. Risk is defined in the Concise Oxford Dictionary as "the chance of loss," which captures the two main components of risk, namely an uncertain event (chance) that can cause adverse consequences (loss). In the context of pipeline corrosion, risk R can be defined as the probability of a corrosion-caused failure P multiplied by a measure of the failure consequences C (i.e., $R = PC$).

There is a significant body of literature dealing with pipeline failure risks due to corrosion [3–10]. Existing approaches can be classified as either qualitative or quantitative methods. Qualitative methods provide approximate rankings of pipeline segments with respect to the risk of failure. These rankings are based on combining subjective scores assigned to each attribute that affects the risk of corrosion (e.g., age, operating temperature, cathodic protection, and soil corrosivity). Quantitative risk assessment methods on the other hand attempt to provide actual estimates of the probability and consequences of failure based on historical data and analytical models. As such, they require a more significant effort to implement; however, they have the potential to provide more objective and reliable results.

The objective of this chapter is to describe an approach for estimating the risk of pipeline failure due to corrosion and for making efficient maintenance decisions to control corrosion-related problems. Although the focus is on pipeline corrosion, the overall framework and many of the concepts are applicable to other systems such as downhole casing, boilers, and pressure vessels.

B. CHARACTERIZING PIPELINE CORROSION

Figure 5.1 shows a sketch of a typical external corrosion defect, which usually consists of a number of individual pits

Uhlig's Corrosion Handbook, Third Edition, Edited by R. Winston Revie
Copyright © 2011 John Wiley & Sons, Inc.

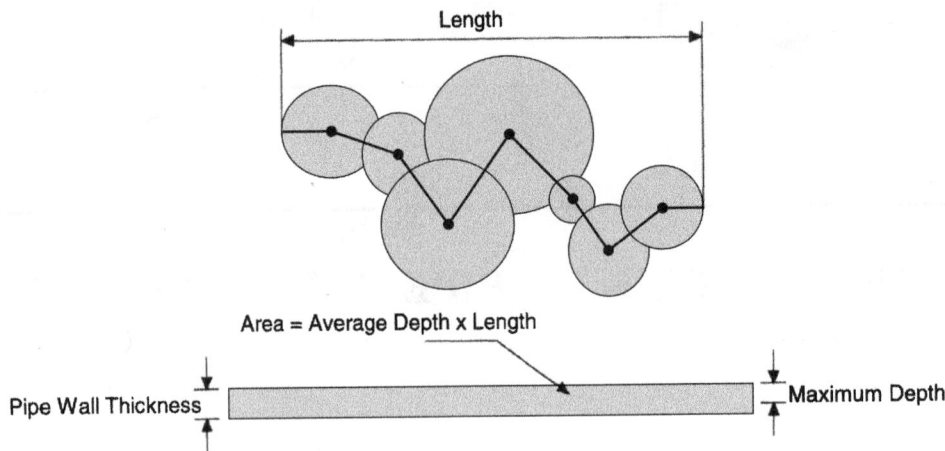

FIGURE 5.1. Geometry of a corrosion defect.

that grow until they join together into a single feature. The parameters used to characterize a corrosion feature are shown on the figure. They may include defect length (defined as the maximum length of the corrosion feature along the pipe axis) and defect depth (which can be characterized by either the maximum depth or the average depth along the deepest route through the feature).

There are several inspection technologies that can be used to provide direct or indirect information on pipeline corrosion defects. These include the following:

Coating damage survey methods such as Pearson surveys, current attenuation surveys, close interval potential surveys (CIPSs), and the direct-current (dc) voltage gradient (DCVG) survey technique. These methods detect coating damage and provide an indication of the severity (or size) of damage. Differences between these methods relate to the physical principles used and the type of information and interpretation required. Some survey methods produce additional information (beyond coating damage indications) that is relevant to corrosion. The CIPSs, for example, determine the effectiveness of cathodic protection, whereas DCVG surveys provide an indication of whether or not corrosion is active at a given location. Harvey [11] gives a summary of the capabilities and limitations of different coating damage survey methods.

Low-resolution or high-resolution in-line inspection tools. Low-resolution tools generally provide the location of metal loss corrosion defects and a coarse estimate of the maximum defect depth (e.g., < 30%, 30–50%, and > 50% of the pipe wall thickness). High-resolution inspection tools provide information on the number, location, and geometry of defects (including estimates of defect length, width, and average or maximum defect depth). Some inspection vendors offer an incremental approach in which low-resolution data are provided, with the option of an information upgrade that provides high-resolution inspection data.

Regardless of how sophisticated a certain inspection method is, there are always accuracy limitations associated with the information provided.

C. PRESSURE RESISTANCE OF CORRODED PIPELINES

The degree of reduction in the pressure capacity of a pipeline at a corrosion defect depends on the defect size (specifically depth and length). The pipe resistance R can be calculated as a function of defect geometry, pipe geometry, and material yield strength using the following relationship [12]:

$$R(\tau) = \frac{2.3T}{D} S \left[\frac{1 - H(\tau)/T}{1 - H(\tau)/M(\tau)T} \right] + C \qquad (5.1a)$$

$$M(\tau) = \sqrt{1 + 0.6275 \frac{[L(\tau)]^2}{DT} - 0.003375 \frac{[L(\tau)]^4}{D^2 T^2}} \qquad (5.1b)$$

where T is the wall thickness, D is the pipe diameter, S is the yield strength, $M(\tau)$ is a geometric factor (called the Folias factor) that accounts for bulging of the pipe before failure, L is the defect length, C is a model uncertainty factor (see below), and τ is time. It is noted that the resistance is a function of time since both defect depth and length are treated as functions of time to account for defect growth:

$$H(\tau) = H(0) + \tau G_h \qquad (5.2a)$$

$$L(\tau) = L(0) + \tau G_l \qquad (5.2b)$$

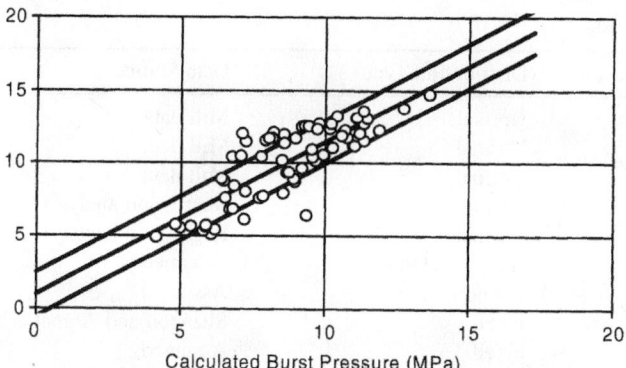

FIGURE 5.2. Burst test versus corrosion model results.

where $H(0)$ and $L(0)$ represent the defect depth and length at present, and G_h and G_l are the defect depth and length growth rates.

Equation (5.1) is based on a semiempirical relationship that was developed in the early 1970s [13, 14]. Figure 5.2 shows the results of this model plotted against the results of burst tests carried out on corroded pipe taken out of service [15]. This figure indicates that there is some uncertainty associated with the results of the model—hence the model uncertainty factor C in Eq. (5.1). Regression analysis of the data in Figure 5.2 shows that C has an average value of 1.38 MPa.

Example 5.1. Consider a 914-mm outside diameter (OD) X60 pipeline with a wall thickness of 8.74 mm and an operating pressure of 5.7 MPa. Assume that the line has a corrosion feature with a length of 75 mm and an average depth of 1.5 mm. Also assume that defect depth and length have growth rates of 0.35 and 1.0 mm/year, respectively. Figure 5.3 shows the resistance of this corrosion defect as a function of time as calculated from Eqs. (5.1) and (5.3). Figure 5.3 shows that, if not repaired, this feature will fail after 17 years.

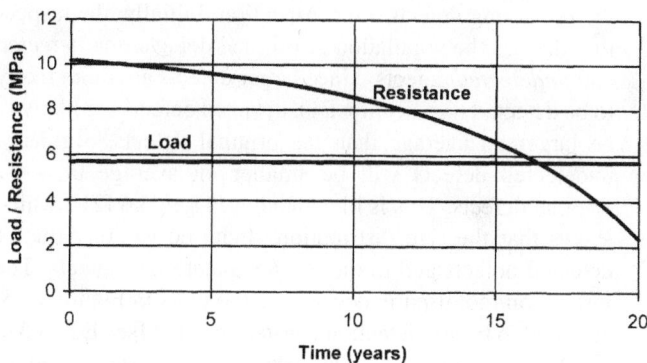

FIGURE 5.3. Resistance at a single corrosion defect as a function of time (Example 5.1).

D. PROBABILITY OF FAILURE DUE TO CORROSION

As demonstrated in Section C, a corrosion failure occurs when the pressure resistance at the defect drops below the maximum operating pressure (see Figure. 5.3). Because of the uncertainties associated with defect sizes, growth rates, pipe material yield strength, and model error, there is some uncertainty regarding the calculated value of the time to failure for a specific corrosion defect. The time to failure is therefore best characterized in probabilistic terms. The probability of failure on or before time τ, $p_f(\tau)$, is equal to the probability that the resistance, $R(\tau)$, will drop below the applied pressure, A. This can be expressed as follows:

$$p_f(\tau) = p[R(\tau) < A] = p[R(\tau) - A < 0] \qquad (5.3)$$

Substituting the value of $R(\tau)$ from Eqs. (5.1a) and (5.2a) gives the probability of failure as the probability of occurrence of a particular combination of pipeline and defect attributes (viz. diameter wall thickness, yield strength, defect dimensions, and growth rates) that lead to a lower resistance than the applied pressure. There are a number of standard methods that can be used to calculate this probability from the probability distributions of the basic pipeline and defect attributes [16, 17].

It is also possible to calculate the probability of failure during a given time interval (τ_1 to τ_2) given that the pipeline survives to the beginning of this interval from Eg. (5.5) [17, p. 287]:

$$p(\tau_1 < \tau < \tau_2) = \frac{p(\tau < \tau_2) - p(\tau < \tau_1)}{1 - p(\tau < \tau_1)} \qquad (5.4)$$

If the time interval is taken as one year, Eq. (5.4) gives the annual probability of failure as a function of time.

Finally, if it is assumed that failures at individual corrosion features are independent events, the probability of failure per kilometer of pipe can be calculated by multiplying the probability of failure per defect by the number of defects per kilometer.

Example 5.2. Assume that the pipeline in Example 5.1 has an average of 2.5 corrosion defects per kilometer and that the pipeline and defect attributes used in Eqs. (5.1) and (5.2) are as given in Table 5.1. Figure 5.4 gives the annual probability of failure as a function of time [as calculated from Eqs. (5.3) and (5.4)] for this pipeline.

E. IMPACT OF MAINTENANCE ON RELIABILITY

E1. Characterization of Inspection Accuracy

E1.1. Detection Power. An inspection method is not guaranteed to detect all defects and therefore there is a

TABLE 5.1.   Probability Distributions of Input Parameters (Example 5.2)

Parameter	Mean Value	COV (%)	Distribution Type	Data Source
Yield strength	461 MPa	3.5	Normal	Mill data
Pipe wall thickness	8.74 mm	1.0	Normal	Mill data
Pipe diameter	914 mm	0.06	Normal	Mill data
Model error A	1.0	0	Fixed	Regression analysis
Model error B	1.38	110	Normal	Regression analysis
Defect depth	1.8 mm	45	Lognormal	Assumed
Defect length	120 mm	40	Lognormal	Assumed
Depth growth rate	0.1 mm/year	0	Fixed	Shannon and Argent [18]
Length growth rate	5 mm/year	0	Fixed	Assumed

Note: COV = coefficient of variation.

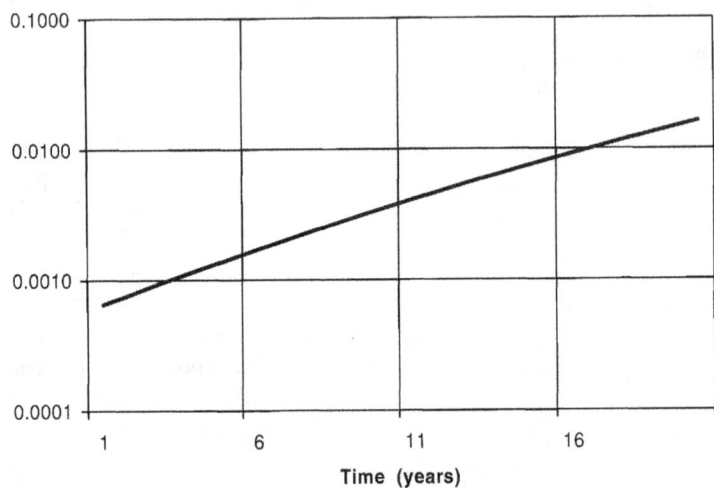

FIGURE 5.4. Probability of failure as a function of time for pipeline in Example 5.2.

chance that an existing defect will be missed. The probability of detecting a specific defect can be characterized as a function of defect size; the larger the defect, the higher the probability that it will be detected. Rodriguez and Provan [19] suggested the following characterization of the probability of detection:

$$p_{d/h} = 1 - e^{-qh} \qquad (5.5)$$

where $p_{d/h}$ is the probability of detection for a defect with size h, and q is a constant that determines the overall power of the detection method.

E1.2. Sizing Accuracy. There is also a random measurement error E associated with defect sizes estimated from in-line inspection. Measurement errors are usually modeled by a normal distribution. The mean value of E is equal to the systematic bias of the measurement (or zero if there is no bias) and the standard deviation represents the random error component. These parameters can be obtained from vendor specifications or from verification excavation data [20].

E2. Effect of Inspection and Repair on Defect Size

The process of inspection and repair improves the pipeline condition by modifying the probability distributions of defect size and defect frequency. The degree of modification depends on the accuracy of the inspection method and the threshold used for repair.

Figure 5.5 illustrates the steps involved in characterizing the remaining defects after inspection. Initially, the inspection divides the population of original defects into *detected* and *undetected* defects. Since larger defects are more likely to be detected (see Section E1), detected defects are likely to be larger on average than the original defects. Similarly, undetected defects will be smaller on average than the original defects. This is illustrated in Figure 5.6 [21], which shows that the size distribution is shifted to the right for detected defects and to the left for undetected defects. The inspection tool used in developing the plots in Figure 5.6 is assumed to have a detection constant $q = 0.4$ [see Eq. (5.5)].

The inspection tool will provide a *measured* size of detected defects, which will be somewhat different from the *actual* size because of tool accuracy limitations. A repair

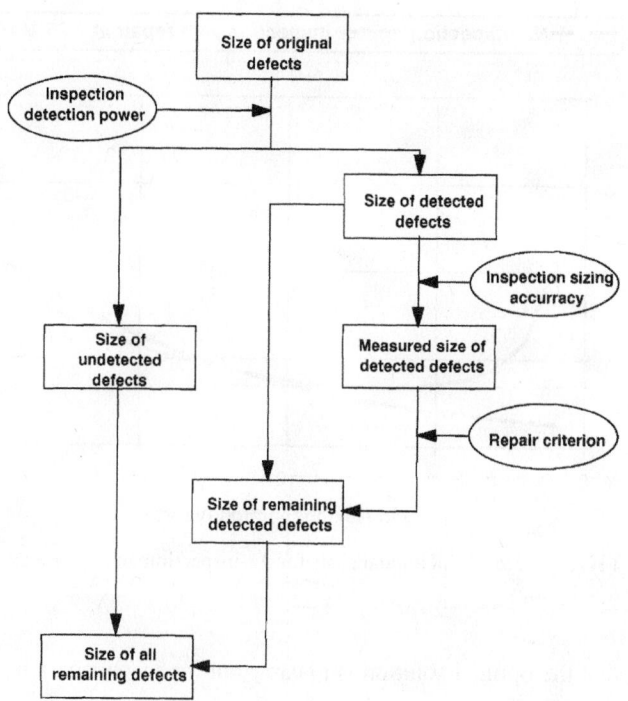

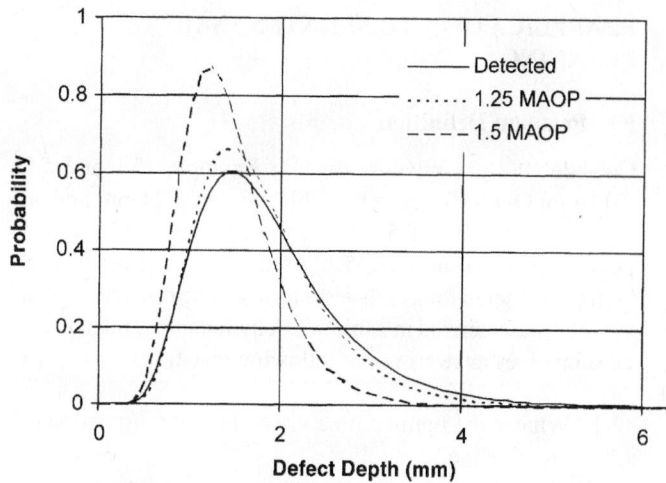

FIGURE 5.7. Probability distributions of defect size before and after repair.

FIGURE 5.5. Modeling the impact of inspection and repair on corrosion defect population.

criterion will typically be applied to each detected defect to identify ones that require repair. The repair criterion may be based on the measured defect depth or the calculated failure pressure (from the measured defect depth and length) at the defect location. The population of *remaining* defects will then consist of defects that did not meet the repair criterion. Because the repair decision is based on the measured (rather

than the actual) defect size, some critical defects may have been undersized by the inspection tool and will therefore be left not repaired.

The population of *remaining defects* can be obtained by combining undetected defects with defects that were detected but not repaired. Figure 5.7 shows a comparison between the size distributions of an original defect population and the corresponding population of defects remaining after inspection and repair of all defects with a failure pressure < 1.25 and 1.5 of the maximum allowable operating pressure (MAOP). The measurement error parameters for the inspection tool used are assumed to be as shown in Table 5.2. The figure shows that the inspection and repair shifts the defect size to the left (lower values). The average number of defects per kilometer remaining after repair can be obtained by subtracting the number of defects that are detected and repaired from the original average number of defects per kilometer.

Details of the calculations used to produce the results in this section are given by Nessim and Pandey [21].

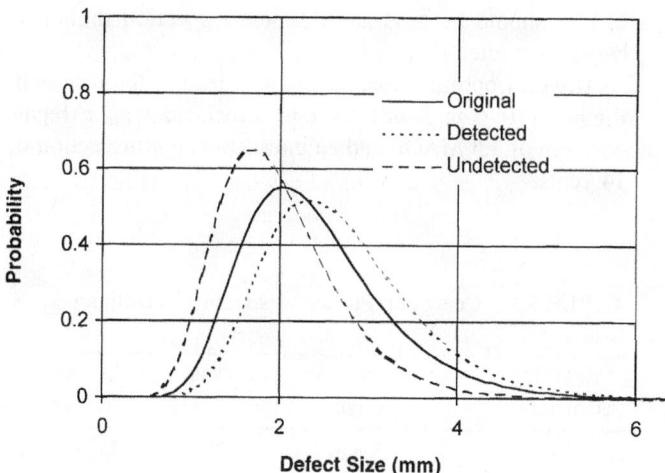

FIGURE 5.6. Size distributions of original, detected, and undetected defects.

TABLE 5.2. Measurement Error Parameters for a Typical High-Resolution In-line Inspection Tool

Parameter	Average (mm)	Standard Deviation (mm)	Distribution Type
Defect depth measurement error	0	0.68	Normal
Defect length measurement error	15	27	Normal

F. APPLICATION TO MAINTENANCE PLANNING

F1. Problem Definition

Consider the gas pipeline used in Examples 5.1 and 5.2 (914-mm OD X60 with a wall thickness of 8.74 mm and an operating pressure of 5.7 MPa) with the corrosion defect population given in Table 5.1. Assume that the pipeline is being considered for a high-resolution in-line inspection with the tool characterized in Table 5.2. A maintenance plan in this case involves answering the following questions:

1. What is the optimal time interval to first inspection of the pipeline?
2. Once the first inspection is performed, what are the optimal repair criterion and interval to next inspection?

F2. Time to First Inspection

The annual failure probability for this pipeline in its initial state is plotted in Figure 5.4. The time to first inspection can be estimated from this figure by defining a maximum allowable probability of failure and finding the time at which this probability will be exceeded. Figure 5.4 shows that if the maximum allowable probability were 10^{-3}/km·year, the inspection would be required after 3 years.

It is also possible to base the decision on a cost optimization analysis. The expected total annual cost c_t for each choice can be calculated from

$$c_t = p_f c_f + (c_i + n_r c_r) \left[\frac{u}{1 - (1+u)^{-\tau_i}} \right] \quad (5.6)$$

in which the first term is the expected annual failure cost calculated as the annual probability of failure, p_f, multiplied by the cost of failure, c_f, in present-day currency, and the second term is the total maintenance cost per kilometer amortized over the time to next inspection. The total maintenance cost is calculated as the sum of the inspection cost per kilometer, c_i, and the average number of repairs per kilometer, n_r, multiplied by the cost per repair, c_r. This cost is amortized over the interval to next inspection, τ_i, based on a real interest rate of u.

If we assume that the cost of failure is $1 million, the inspection cost is $4000/km, the cost of repair is $5000 per defect, and the real interest rate is 5%, the total expected cost will be as shown in Figure 5.8. This figure plots the results for the status quo and for an inspection followed by repair of all defects with a failure pressure less than 1.25 MAOP. The minimum cost associated with the inspection option is $1.8 million, corresponding to an inspection interval of ~7 years. The cost associated with the status quo is less than $1.8 million for the first 9 years. This means

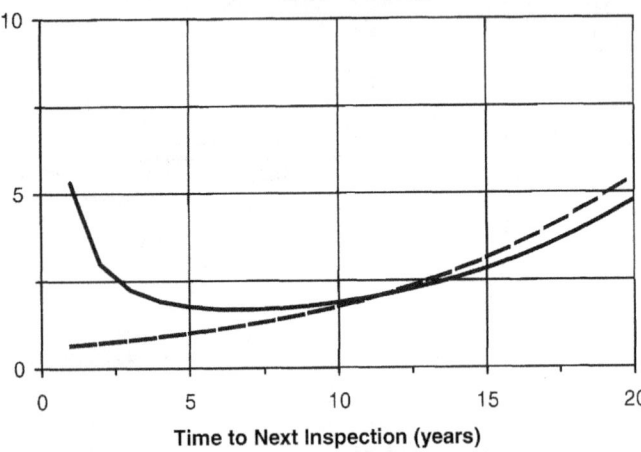

FIGURE 5.8. Total annual costs for the inspection and no-inspection options.

that the optimal solution is to carry out the inspection after 9 years.

F3. Inspection Interval and Repair Criterion

It is now assumed that the inspection has been carried out after 9 years and that an average number of 1.3 defects per kilometer is found, with the depth and length distributions given in Table 5.3. The maintenance planner wishes to choose a repair criterion and an interval to next inspection. Figure 5.9 can be used to make this decision on the basis of the maximum allowable failure probability. It shows that a repair criterion of 1.25 MAOP is inadequate to meet a target annual failure probability of 10^{-3}. A repair criterion of 1.5 MAOP would lead to an acceptable probability of failure for the next 15 years. Based on this, a repair criterion of 1.5 MAOP should be used and the next inspection should be carried out after 15 years.

The cost optimization calculations lead to the results in Figure 5.10. The lowest cost is associated with a repair criterion of 1.5 MAOP and an interval to next inspection of 14 years.

TABLE 5.3. Corrosion Parameters Obtained from an Inspection

Parameter	Average	COV (%)	Distribution Type
Defect depth	2.5 mm	60	Lognormal
Defect length	120 mm	50	Lognormal
Number of detected flaws	1.3 per km	0	Fixed

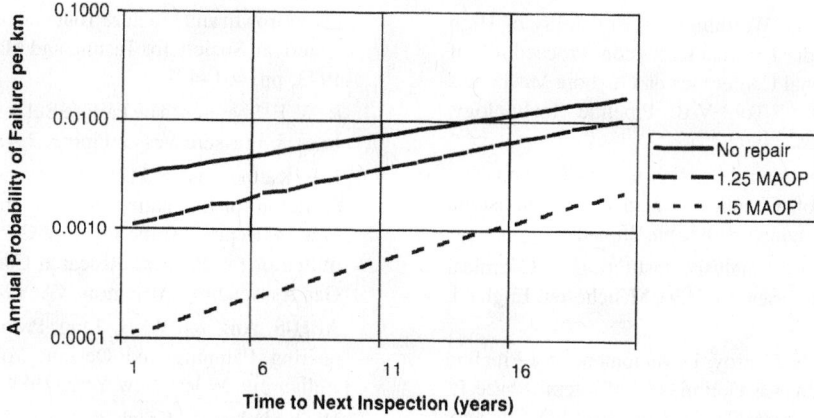

FIGURE 5.9. Annual probability of failure for three different repair criteria.

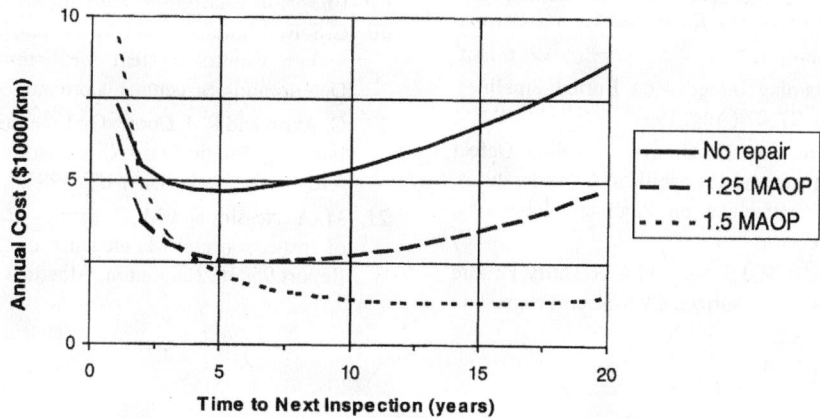

FIGURE 5.10. Total annual cost for three different repair criteria.

G. SUMMARY

This chapter described a framework for risk-based assessment of pipeline corrosion risk and demonstrated maintenance planning with a risk-based approach. The examples developed serve to demonstrate the benefits associated with risk-based planning:

1. *Consistent Safety Levels.* Minimum safety targets can be set and actions taken to ensure that they are met across a whole system.
2. *Optimal Balance between Maintenance Costs and Failure Risks.* Maintenance plans that achieve safety goals at a minimum possible cost can be defined. This can result in significant savings in overall operating costs.
3. *Documentation of Rationale Behind Decisions.* The analysis process provides the reasoning and documentation needed to communicate prudent risk management

REFERENCES

1. AGA, An Analysis of Reportable Incidents for Natural Gas Transmission and Gathering Lines—June 1994–1990, NG-1S Report No. 200, prepared by Battelle Memorial Research Institute for the Line Pipe Research Supervisory Committee of the American Gas Association, Arlington, VA, 1992.
2. Alberta Energy and Utilities Board, Pipeline Performance in Alberta 1980–1997, EUB Report 98-G, Calgary, Alberta, Canada, 1998, p.12.
3. M. A. Nessim and M. J. Stephens, Optimization of Pipeline Integrity Maintenance Based on Quantitative Risk Analysis, Proceedings of the Pipeline Reliability Conference, Houston, TX, 1995.
4. W. K. Muhlbauer, Pipeline Risk Management Manual, Gulf Publishing Company, Houston, TX, 1992.
5. M. Urednicek, R. I. Coote, and R. Coutts, Optimizing Rehabilitation Process with Risk Assessment and Inspection. Proceedings of the CANMET International Conference on Pipeline Reliability, Calgary, June II-12-1 to II-12-14, Gulf Publishing, Houston, TX, 1992.

6. T. B. Morrison and R. G. Worthingham, Reliability of High Pressure Line Pipe Under External Corrosion, Proceedings of the Eleventh International Conference on Offshore Mechanics and Arctic Engineering, Vol. V-B, Pipeline Technology, Calgary, Alberta, Canada, 1992.

7. G. D. Fernehough,The Control of Risk in Gas Transmission Pipeline, Institution of Chemical Engineers, Symposium, No. 93, Manchester, England, 1985, pp. 25–44.

8. R. T. Hill, Pipeline Risk Analysis, Institution of Chemical Engineers Symposium Series, No. 130, Manchester, England, 1992, pp. 637–670.

9. R. B. Kulkarni and J. E. Conroy, Development of a Pipeline Inspection and Maintenance Optimization System (Phase I), Contract No. 5091–271-2086, Gas Research Institute, Chicago, IL, 1991.

10. J. F. Kiefner, P. H. Vieth, J. E. Orban, and P. I. Feder, Methods for Prioritizing Pipeline Maintenance and Rehabilitation, Final Report on Project PR3-919 to the American Gas Association, Arlington, VA, 1990.

11. D. W. Harvey, Maintaining Integrity on Buried Pipelines, Mater. Perform., **33**(8), 22–27 (Aug. 1994).

12. M. Brown, M. Nessim, and H. Greaves, Pipeline Defect Assessment: Deterministic and Probabilistic Considerations, Second International Conference on Pipeline Technology, Ostend, Belgium, Sept. 1995.

13. J. F. Kiefner, W. A. Maxey, R. J. Eiber, and A. R. Duffy, Failure Stress Levels of Flaws in Pressurized Cylinders, Progress in Flaw Growth and Fracture Touchness Testing, ASTM STP 536, American Society for Testing and Materials, Philadelphia, PA, 1973, pp. 461–481.

14. R. W. E. Shannon, The Failure Behaviour of Line Pipe Defects, Inter. J. Pressure Vessel Piping, **2**(4), 243–255 (1974).

15. J. F. Kiefner and P. H. Vieth, Project PR 3-805: A Modified Criterion for Evaluating the Remaining Strength of Corroded Pipe, A Report for the Pipeline Corrosion Supervisory Committee of the Pipeline Research Committee of the American Gas Association, Arlington, VA, 1989.

16. A. H-S. Ang and W. H. Tang, Probability Concepts in Engineering Planning and Design, Vol. II:Decision, Risk and Reliability, Wiley, New York, 1984.

17. H. O. Madsen, S. Krenk, and N. C. Lind, Method of Structural Safety, Prentice-Hall, Englewood Cliffs, NJ, 1986.

18. R. W. E. Shannon and C. J. Argent, Maintenance Strategy Set by Cost Effectiveness, Oil and Gas J., **87**(6), 41–44 (1989).

19. E. S. Rodriguez III and J. W. Provan, Development of a General Failure Control System for Estimating the Reliability of Deteriorating Structures, Corrosion, **45**(3), 193–206 (1989).

20. G. Avrin and R. I. Coote, On Line Inspection and Analysis for Integrity, Pacific Coast Gas Association, Transmission Conference, Salt Lake City, UT, 1987.

21. M. A. Nessim and M. D. Pandey, Reliability Based Planning of Inspection and Maintenance of Pipeline Integrity, C-FER Report 95036, Edmonton, Alberta, Canada, 1996.

6

DESIGNING TO PREVENT CORROSION

E. D. VERINK, JR.[†]

Distinguished Service Professor Emeritus, Department of Materials Science and Engineering, University of Florida, Gainesville, Florida

A. Introduction
B. Design-related causes of corrosion
    B1. Dissimilar metals
    B2. Drainage
    B3. Sealants
    B4. Cathodic protection
    B5. Grounding
    B6. Flowing systems
    B7. Liquid–vapor systems
    B8. Redundant systems
    B9. Welded joints
    B10. Nonmetallics
Bibliography

A. INTRODUCTION

The choice (and detailed specification) of a material known to be chemically resistant to a given environment is one step in the selection process. Another step, of equal importance, is the choice of proper design configurations and principles to permit the material to perform in the desired manner. Listed below are a number of design-related causes of corrosion in metallic systems:

1. Dissimilar metals
2. Improper drainage
3. Joints between metals and nonmetals
4. Crevices
5. Stray currents
6. Complex cells

[†]Retired.

7. Relative motion between two interacting parts or between a part and its environment
8. Selective loss of one or more ingredients of the alloy
9. Inability to clean the surface properly

There is a third step in assuring good performance, which unfortunately is often given inadequate attention. That is careful inspection and verification that design specifications and recommendations actually are carried out in the installation.

The detailed methods by which corrosion may be mitigated by design are listed below and are illustrated in the accompanying figures:

1. Where dissimilar metals are involved, select materials that have a minimum difference in electrode potential under the conditions of temperature and electrolyte composition encountered.
2. Where feasible, design structures so that butt joints rather than lap joints are employed, use drip skirts to avoid moisture collecting under structures.
3. Support tanks on stanchions rather than pads if possible. If pads are required for tanks, provide sufficient "crown" on the pads to assure drainage and to avoid "oil can" effects. For large field-erected tanks use domed, compacted, oiled sand where appropriate as a support for tank bottoms.
4. Employ resilient sealants to exclude moisture from potential crevices.
5. Employ cathodic protection where appropriate. This includes the use of galvanized steel or alclad aluminum products or sprayed metals to provide cathodic protection in crevices.

Uhlig's Corrosion Handbook, Third Edition, Edited by R. Winston Revie
Copyright © 2011 John Wiley & Sons, Inc.

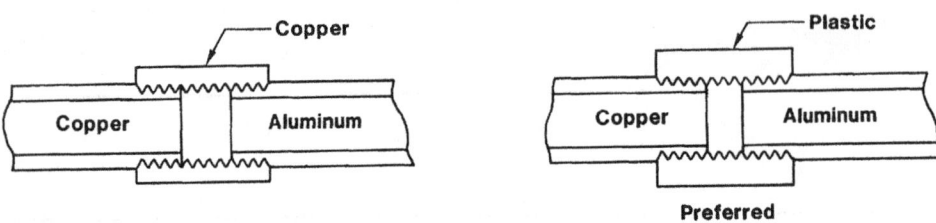

FIGURE 6.1. Insulating material between dissimilar metals, such as copper and aluminum, prevents galvanic corrosion.

6. Equalize the electrode potential between electrically interfering structures that are exposed in the same electrolyte (e.g., underground or in large tanks). Cross-bonding, use of cathodic protection, and careful grounding are important methods for accomplishing this.

7. Use special care to avoid turbulence. This involves study of flow patterns, avoidance of constrictions, or sharp changes in direction and attention to the relationship between pressure and amplitude of motion between parts.

8. Select materials known to be compatible with the environment and with each other. For important structures this implies pretesting.

9. Provide redundant systems for critical applications. This includes spare heat exchanger bundles, replaceable spools in pipe systems, or scavengers for removal of dissolved heavy metals.

10. Use nonmetallic materials when required.

B. DESIGN-RELATED CAUSES OF CORROSION

B1. Dissimilar Metals

The electromotive force (emf) series can be of qualitative guidance in predicting which material is likely to be sacrificed in a dissimilar metal couple. Unfortunately, the emf series (which lists pure metals in solutions of their own ions of unit activity) bears little relationship to actual applications. Consequently, the predictions of the emf series are likely to be reliable only for those circumstances where there is a very large difference in electrode potential. An example might be aluminum versus copper (Fig. 6.1). In most circumstances, aluminum would sacrifice itself to protect copper in a conductive electrolyte. The actual electrode potential of a given metal will be strongly influenced by the composition of the electrolyte and the temperature. In practical cases, pure metals seldom are used as materials of construction. More commonly alloys are employed. The addition of alloy ingredients changes the electrode potential from that of the predominant pure metal ingredient. Thus it is important to

know the electrochemical relationships between dissimilar metal alloys under the real conditions of exposure in order to assess the likelihood of dissimilar metal action. Figure 6.2 shows a galvanic series for alloys exposed in seawater.

The intensity of dissimilar metal action will be strongly influenced by the relative areas of the dissimilar metals exposed in the electrolyte that is common to them both. Of

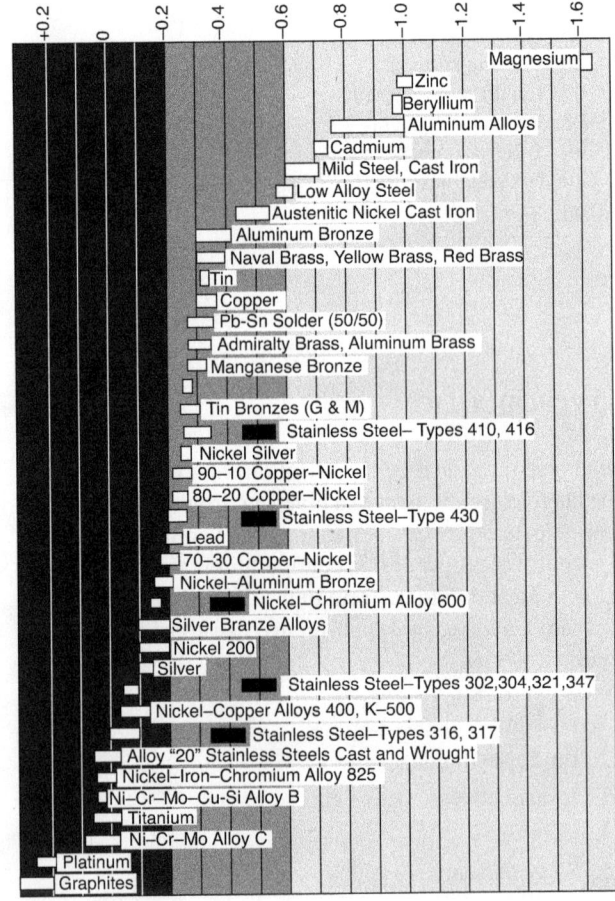

Alloys are listed in the order of the potential they exhibit in flowing sea water. Certain alloys indicated by the symbol: ▬ in low–velocity or poorly aerated water, and at shielded areas, may become active and exhibit a potential near –0.5 volts.

FIGURE 6.2. The galvanic series in seawater.

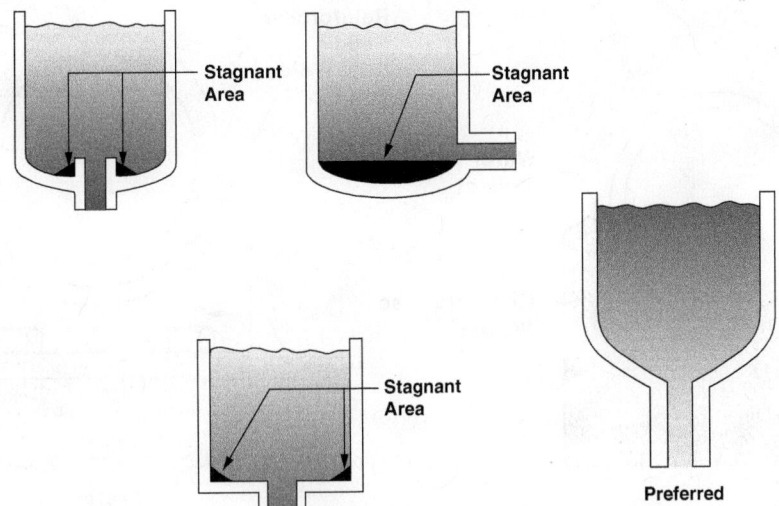

FIGURE 6.3. Design for drainage.

particular interest is the anode current density. In most metal systems exposed to aqueous environments, the corrosion process is under "cathodic control." This means that the size of the cathode will control the anode current density. The larger the area of the cathode as compared to the anode area, the greater will be the anode current density. Thus the design should maximize the anode area and minimize the cathode area where possible. It is good practice to avoid using rivets that are anodic to metal parts being joined since such a practice would lead to failure of the joint by corrosion of the rivets.

B2. Drainage

Drains should always be located at the lowest point in a tank. The joint between the tank and the drain pipe should be designed to permit free drainage (Fig. 6.3). Usually, this will mean the avoidance of weld beads projecting into the tank that could trap residues or moisture. The use of sumps in large tanks is a useful procedure to mitigate problems related to drainage; however, special care must be made in the design of sumps to be sure that they are also free draining or can be inspected and cleaned easily. Inspection flanges are convenient for such purposes. Large flat-bottom tanks present a particular problem from the standpoint of drainage because of the difficulty of assembling such tank bottoms in a manner that will avoid areas that are difficult to drain. The design should incorporate some positive means of providing drainage, perhaps by sloping the bottom. Rain water or other moisture on the exterior of tanks can be drawn under flat-bottom tanks by capillary action. Use of "drip skirts" around the edges of such tanks will direct moisture to the ground rather than allowing it to cling to the tanks and be drawn under the tanks and will reduce danger of trapping moisture (Fig. 6.4).

Particularly for pipes and small- to moderate-size tanks, it is good practice to support them on stanchions rather than on pads. This provides free access of air around the bottoms of the tanks and facilitates periodic inspection, Metal cradles welded to the tanks for support above grade provide another option (Fig. 6.5).

Flat cement pads should be avoided as a supporting means for flat-bottom tanks. Instead, the pads should have a crown from center to edge of $\sim$1 mm/12 mm radius (1 in./foot of radius). It is difficult to manufacture a flat-bottom tank with a tank bottom that is actually flat. Consequently, when the tank is empty, the tendency is for the tank bottom to stand away from the base irregularly. When the tank is filled, the tank settles onto the pad. However, each time the contents of the tank are emptied, there is a tendency for portions of the bottom to rise away from the pad. Thus the bottom can snap

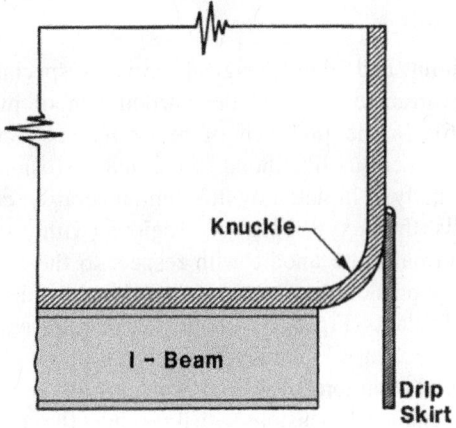

FIGURE 6.4. A drip skirt reduces moisture collection under flat-bottom tanks.

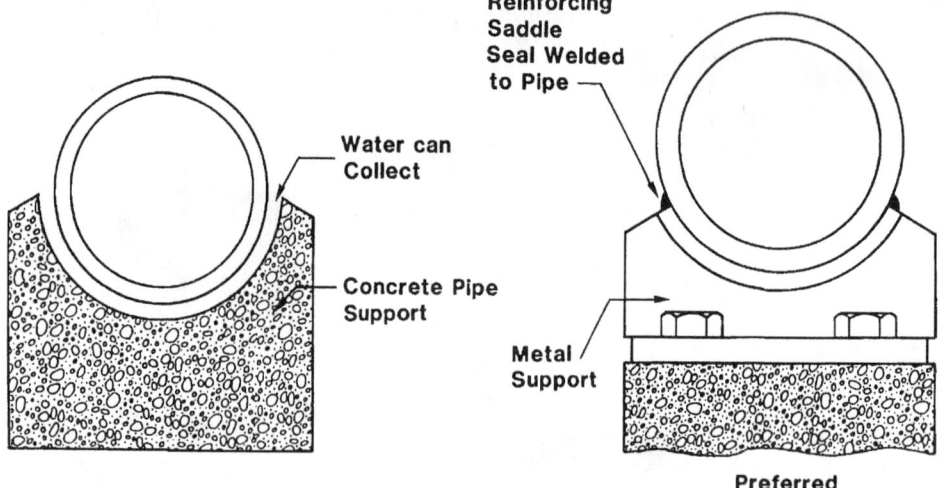

FIGURE 6.5. Sealing and a metal cradle avoid accumulation of water under a pipe in outdoor construction.

back and forth much in the manner of the bottom of an oil can and in so doing "pump" moisture under the tank. The use of a crowned pad will prevent this pumping action. The diameter of the supporting pad should be *smaller* than the tank being supported. The use of smaller size pads in conjuction with drip skirts is effective in reducing the amount of moisture that collects under a tank. For very large tanks set on the ground (e.g., large field storage tanks for oil), it is customary to use a cement ring to support the edge of the tank and support the bottom on compacted, crowned, oiled sand. The use of oiled sand serves to provide corrosion protection for the underside of the tank bottom.

The use of butt welds rather than lap welds in piping systems tends to reduce the danger of entrained solids collecting in the bottoms of pipes. It may be desirable in critical applications to employ special procedures (e.g., grinding) to reduce the tendency for weld beads to trap solids.

B3. Sealants

For lap joints and other "designed" crevices, special attention is warranted to avoid the introduction of moisture (Fig. 6.6). In the presence of moisture, such crevices represent a serious likelihood of crevice corrosion. Such action usually is initiated by differential aeration cells. In such cells, the "oxygen-starved" regions (within the crevices) normally are anodic with respect to the "oxygen-rich" areas outside the crevice and tend to corrode preferentially. Sealants (Fig. 6.7) used must be selected with care. Ideally, a sealant will remain resilient during its expected life. Special attention should be given to avoid sealants that harden or change dimensions with time under the conditions of exposure. Where sealants cannot be employed the use of galvanized steel mating parts or the use of alclad aluminum mating parts (e.g., in a riveted structure) can provide

cathodic protection within the crevice, thereby mitigating the corrosion problem.

B4. Cathodic Protection

The use of cathodic protection either by use of galvanic anodes or by driven electrodes can be effective in preserving the integrity of a designed structure. Attention must be given to adjacent structures that may inadvertently either provide cathodic protection to the structure under consideration or receive cathodic protection from the structure. Examples are buried pipelines that may be in close proximity. It is not uncommon under these circumstances for current to flow from one structure to the other. When this occurs, special corrosion is observed where current flows from the structure into the electrolyte and protection occurs where the current flows from the electrolyte into the structure. A common design procedure is to connect the neighboring structures

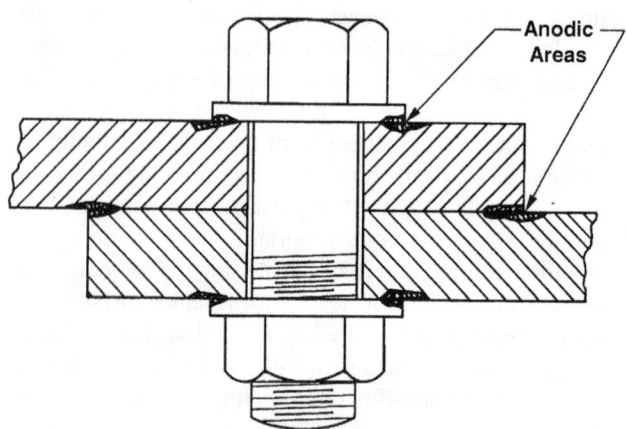

FIGURE 6.6. Crevices lead to concentration cell corrosion.

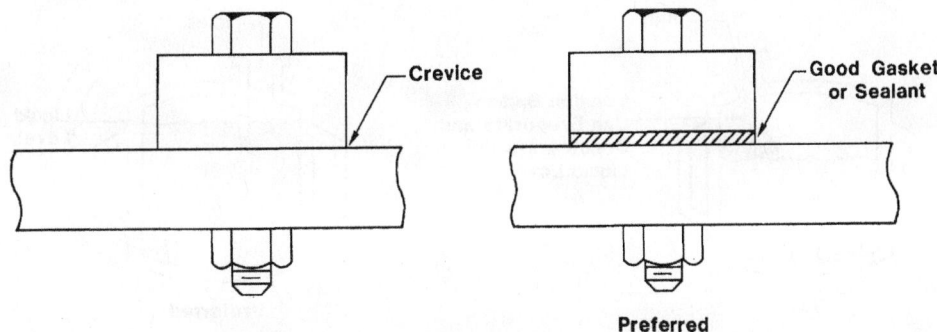

FIGURE 6.7. Gasket or sealant to avoid crevices.

electrically with a suitable bond so that their potentials will become the same. Often such cross-bonding procedures are incorporated with installation of cathodic protection.

B5. Grounding

Leakage currents have been discovered on a number of metallic structures because of inadequate grounding. Particularly in environments where there is a considerable fluctuation in the water table, buried structures have suffered serious deterioration from leakage currents. The grounding of electrical services to buried water lines is a common practice. Such structures should then be connected to ground rods that are deep enough to be in permanent soil moisture. If the water table falls below the level of the grounding system, there is increased danger that leakage currents will result in wastage of the water piping.

B6. Flowing Systems

In flowing systems, the relative motion between the environment and the parts operating in the environment can lead to erosion corrosion, cavitation, or abrasive wear. All of these types of damage are related exponentially to the flow rate

(relative motion). Thus it is good practice to reduce the flow rate if feasible and to avoid localized constrictions or sharp changes in direction in flowing systems. The entrance end of heat exchanger tubes is an example of a location where special damage may occur. One way to mitigate inlet end corrosion (or erosion) of heat exchanger tubes is by the use of metallic or nonmetallic ferrules inserted in the ends of the tubes. Damage ultimately requires the replacement of such ferrules; however, the integrity of the system is maintained. Where ferrules are cemented in place, special care must be exercised to avoid extruding sealant beyond the ends of the ferrules since "stalactites" or collections of sealant can serve as local sources of turbulence in the flowing system. Entrained trash, sand, or marine organisms such as crustaceans may collect in flowing systems setting up not only local turbulence but also the possibility of crevice corrosion from differential aeration cells. Structures should be designed so that they may be cleaned periodically by use of "pipe pigs," brushes, or sponge balls that can be pumped through the system. Such procedures can result in dislodging of collected debris and can restore heat transfer in heat exchangers. At sharp (unavoidable) changes in direction, localized impingement damage can be mitigated by incorporation of wear-resistant devices (Fig. 6.8).

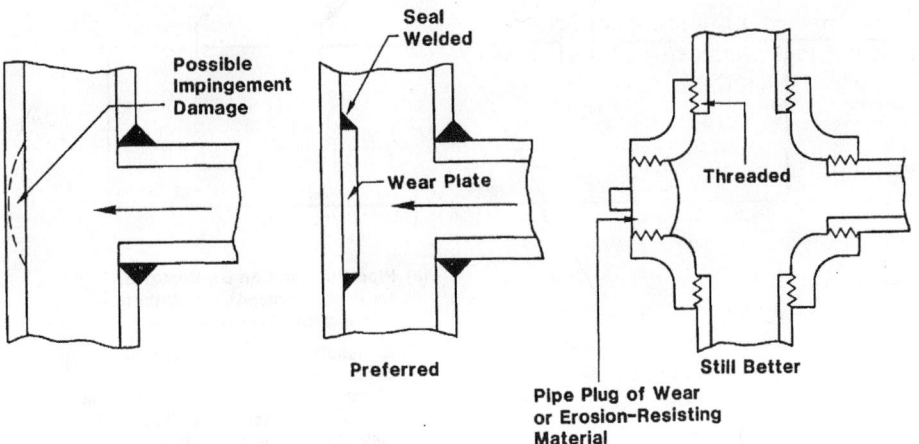

FIGURE 6.8. Wear-resisting designs to minimize corrosive wear.

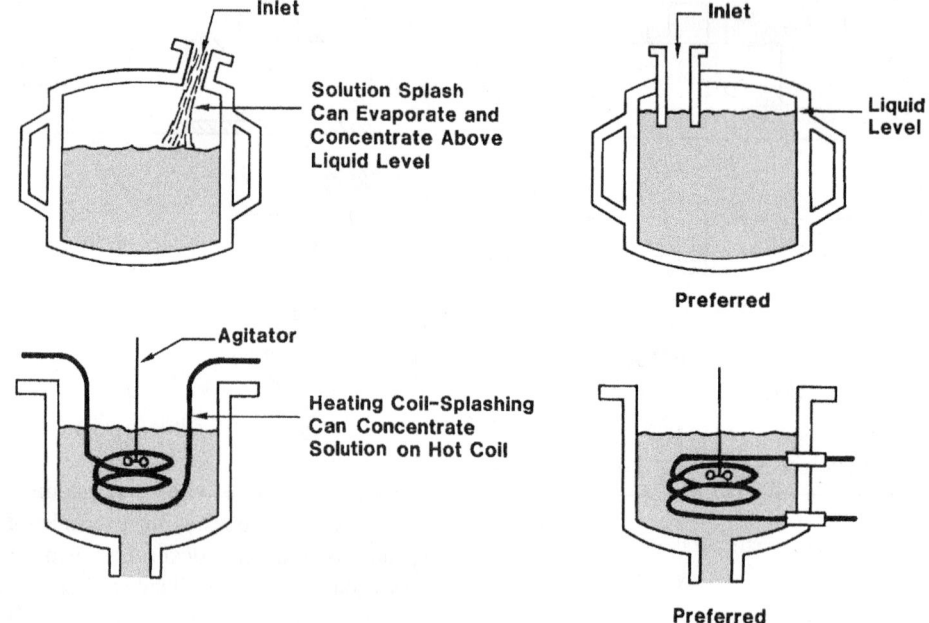

FIGURE 6.9. Designs to minimize evaporation and concentration of solutions.

B7. Liquid–Vapor Systems

Conditions that permit evaporation of splashed liquids or localized condensation of vapors, which then can evaporate and become more concentrated, often can lead to special corrosion. Splashing can be avoided by designs such as those shown in Fig. 6.9. Localized condensation usually can be avoided by thermal insulation or by preheating liquids so that surfaces of piping or equipment are above the ambient temperature (and dew point).

B8. Redundant Systems

In critical process applications, such as heat exchangers, it is common practice to design the system so that several extra heat exchanger bundles are installed on a manifold to permit periodic inspection and/or repair or replacement of systems without shutting down the entire plant operation. In piping systems, it also is common practice to install heavy wall "spool" sections to bear the brunt of special corrosion problems. An example is a piping system involving steel and aluminum alloy pipes (Fig. 6.10). To mitigate the effects of this potential dissimilar metal cell, a short length of heavy wall (Schedule 80, e.g.) alclad aluminum pipe may be inserted between the steel pipe and the aluminum pipe. The alclad coating will provide cathodic protection to both the adjacent steel and aluminum pipes and also will scavenge dissolved copper ions that may be present in the flowing aqueous system, thereby sparing the aluminum

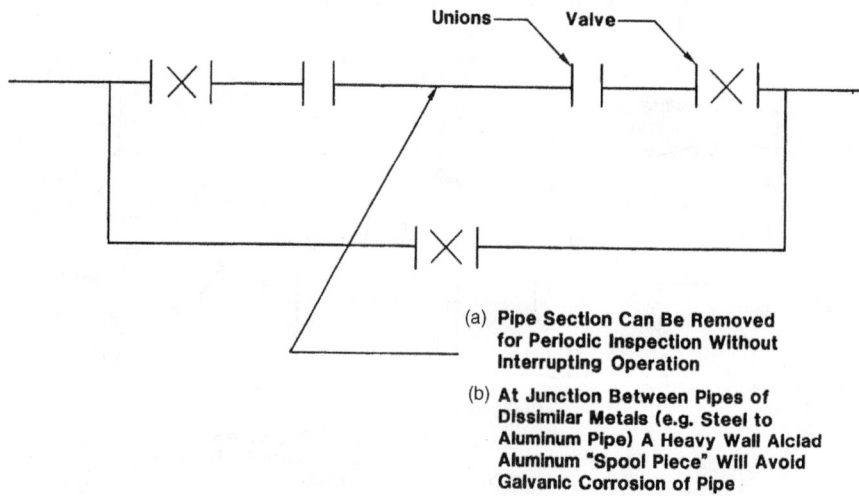

(a) **Pipe Section Can Be Removed for Periodic Inspection Without Interrupting Operation**

(b) **At Junction Between Pipes of Dissimilar Metals (e.g. Steel to Aluminum Pipe) A Heavy Wall Alclad Aluminum "Spool Piece" Will Avoid Galvanic Corrosion of Pipe**

FIGURE 6.10. Design that allows inspection to be carried out without interrupting operation.

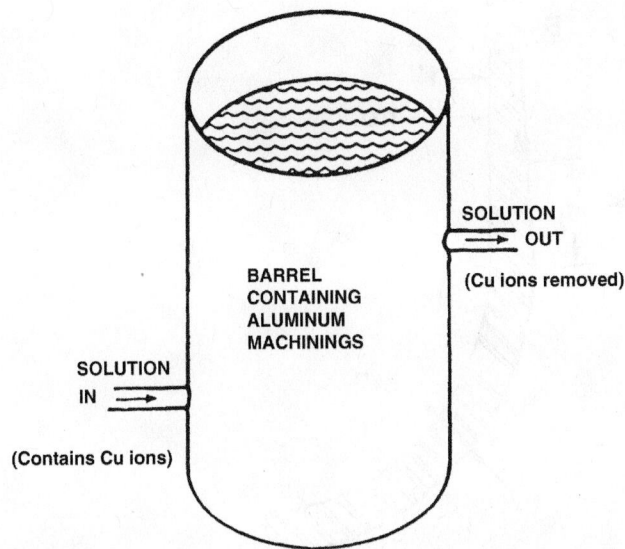

FIGURE 6.11. Heavy metals trap to remove heavy metals from a process stream.

pipe downstream from dissimilar metal action. Periodically, the system is shut off and the spool is examined and/or replaced if necessary. It is possible to put in two such spool sections in parallel with suitable valving so that the system

need not be taken out of service during inspection and/or replacement.

Some flowing systems are significantly contaminated by heavy metal ions such as copper, lead, and so on. These ions tend to be plated out on steel or aluminum piping downstream and can result in corrosion of the downstream piping. This can be avoided readily by the inclusion of a "heavy metals trap" upstream from the steel or aluminum piping (Fig. 6.11). A typical trap might be a tank or drum filled with aluminum machinings. Periodically, it would be necessary to dispose of exhausted machinings by replenishing with new machinings.

B9. Welded Joints

In design of equipment involving use of carbon steel to stainless steel welded joints, there is the possibility of alloy dilution in the welded joints. This presents the potential problem of reduced corrosion resistance at welded joints. This problem may be mitigated by designs such as that shown in Figure 6.12. Joints in piping systems provide opportunities for the inadvertent incorporation of crevices. Vigilance to avoid such crevices is urged (see Fig. 6.13). When attaching dished heads to the sidewall of pressure vessels it may be necessary to machine the surface of the dished head to

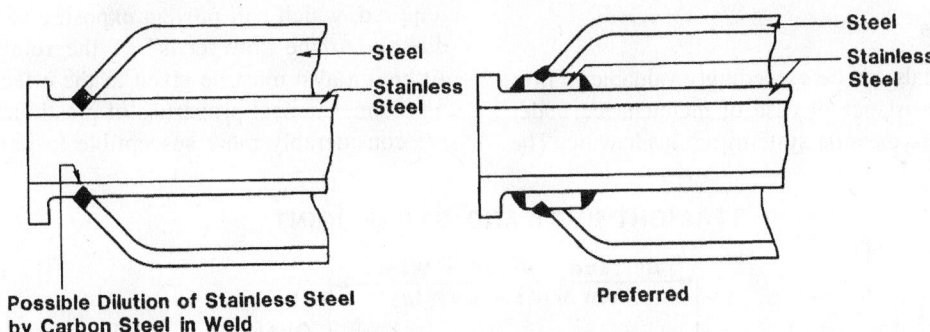

FIGURE 6.12. Design to prevent base metal dilution.

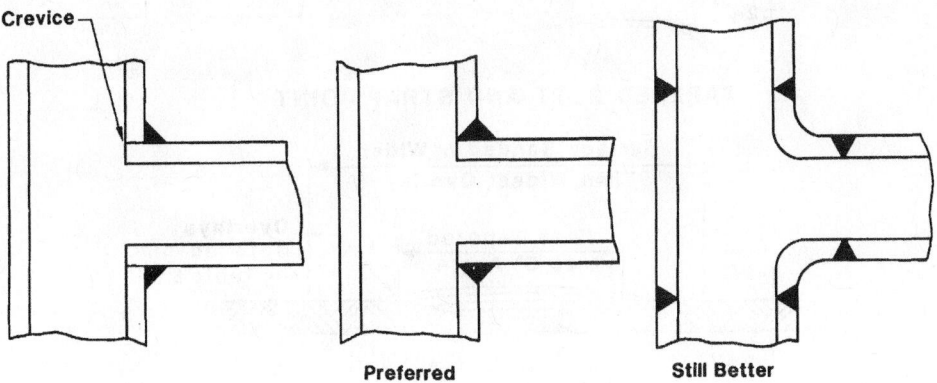

FIGURE 6.13. Designs to avoid crevices caused by incomplete weld penetration.

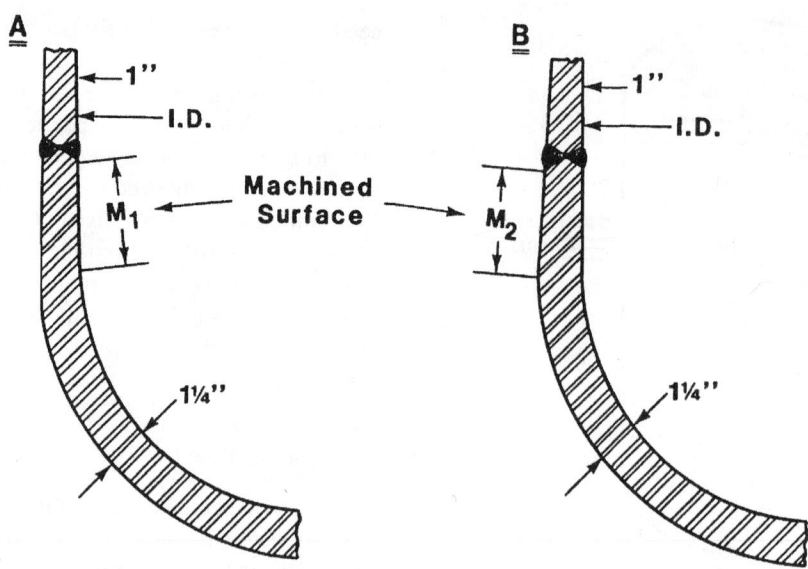

FIGURE 6.14. Design of attachments to the sidewall of pressure vessels so that the final machining operation is carried out on the exterior of the vessel.

accommodate the thickness transition. It is recommended that the dished heads be sized such that this final machining operation be performed on the *exterior* of the vessel (Fig. 6.14).

B10. Nonmetallics

Nonmetallic materials can be exceedingly valuable in providing corrosion resistance or ease of maintenance under circumstances where metallic systems are inadequate. The use of nonmetallic materials such as fiber-reinforced plastics requires special design attention to avoid premature failure. In the case of the fiber-reinforced products, special attention must be given to avoid wicking of moisture up the fibers. This normally means that a finish coat is required, which will prevent exposure of the environment directly to the "raw ends" of the reinforcing material. Attention also must be given to the effect of temperature on the mechanical properties of plastic materials. Plastics are considerably more susceptible to fatigue failure than

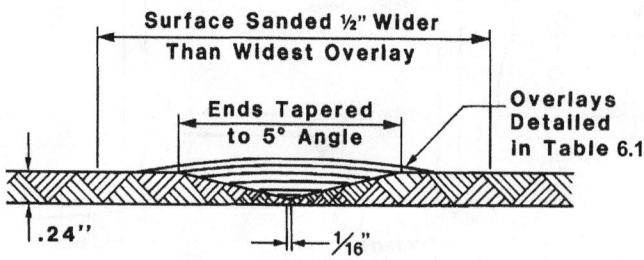

FIGURE 6.15. Joint designs.

TABLE 6.1. Major Types of Failures in Reinforced Plastics and Their Causes

System	Type of Failure	Cause
Piping	Cemented joint	Failure to follow recommended practice
	Chemical attack	Poor resin selection or change in environmental conditions
	Impact damage	Various (generally handling damage)
Tanks and process vessels	Chemical attack	Poor resin selection or environmental data wrong
	Internal pressure	Operational errors and poor process design
	External pressure (vacuum)	Poor equipment design
	Secondary bond failure	Faulty fabrication
	Impact	Handling damage
Scrubbers and absorbers	Chemical attack	Service conditions wrong and change in process specs
	Vacuum damage	Poor equipment design
	Impact	Handling damage
	Fire	Lack of maintenance, lack of adequate safeguards, lack of interlocks, possible lack of employee training, or no scrubbing liquid
Ducts, fans, and stacks	Joint failures	Generally—glue line
	Chemical attack	Poor resin selection, wrong or changed service conditions
	Fire	80% of fires originate from an internal or process source, while 20% of fires originate externally. Lack of sprinkler protection is by far the largest single reason for large losses occurring.

are metals. They must also be protected from degradation by ultraviolet light. Thermal expansion of plastic materials is considerably greater than metals, a particularly important consideration when considering piping systems or composite structures such as plastic-lined pipe. Special attention must also be given to the possibility of fire damage. A number of suggested joint details are given (Fig. 6.15). Table 6.1 lists a number of types of failures that have occurred in plastic systems. Many of these could have been avoided by careful design, fabrication, and installation.

There are numerous applications of ceramic or glass materials in the chemical and pharmaceutical industries.

Successful use of this class of materials requires that they be protected from impact loadings or flexure.

BIBLIOGRAPHY

R. B. Mears and R. H. Brown, Corrosion, **3**(3), 97 (1947).
W. H. Burton, Mater. Protect., **22** (Feb. 1967).
J. T. N. Atkinson and H. Van Droffelaar, Corrosion and Its Control: An Introduction to the Subject, NACE, Houston, TX, 1982, pp. 92–98.
A. de S. Brasunas (Ed.), NACE Basic Corrosion Course, NACE, Houston, TX, 1970, pp. 1–17, 21–27.

7

SIMPLIFIED PROCEDURE FOR CONSTRUCTING POURBAIX DIAGRAMS*

E. D. VERINK, JR.[†]

Distinguished Service Professor Emeritus, Department of Materials Science and Engineering, University of Florida, Gainesville, Florida

A. INTRODUCTION

For the behavior of metals in aqueous solutions, one of the most important contributions to the corrosion literature has been the work of Pourbaix [1, 2] and his associates in the development of thermodynamic equilibrium diagrams (E vs. pH), called Pourbaix diagrams. A vast amount of data may be presented simply and concisely in Pourbaix diagrams. When the advantages and limitations of such diagrams are understood, valuable inferences may be made regarding corrosion phenomena. The selection of conditions for cathodic and anodic protection is simplified. Candidates for consideration as inhibitor species may be selected with greater efficiency. Critical corrosion experiments may be designed with equal efficiency.

Corrosion processes involve both chemical and electrochemical phenomena. In 1923, Evans [3] observed that, if two samples of iron connected by a galvanometer are immersed in two solutions of potassium chloride separated

by a porous membrane and if a stream of air is bubbled through one of these solutions, an electric current circulates between the aerated sample, which becomes the cathode, and the nonaerated sample, which becomes the anode and corrodes. On the other hand, if a sample of iron and a sample of another metal (copper, zinc, or magnesium) are connected as above, a passage of electric current is also observed. Under these circumstances, iron becomes the anode and corrodes when connected to copper, whereas zinc or magnesium become the anode and corrode providing protection to iron. Thus, it is necessary to consider not only chemical thermodynamics but also electrochemical thermodynamics when considering corrosion reactions.

Chemical equilibria are defined as those that do not involve oxidation–reduction processes but do involve the law of mass action and the law of solubility product (involving partial pressures or fugacities and concentrations or activities). By contrast, electrochemical reactions are defined as those in which free electric charges, or electrons, participate.

B. THERMODYNAMIC BACKGROUND

The procedure for calculating Pourbaix diagrams is straightforward and is amenable to computer calculation. On the other hand, certain assumptions are made that must be borne in mind when applying the information available from Pourbaix diagrams in "real" situations. First is the assumption of equilibrium. Since Pourbaix diagrams are equilibrium diagrams, they give no information on the kinetics of the reactions considered. Kinetic information may be obtained experimentally by methods described elsewhere [4]. It also is assumed that the reaction products are known and that the

*Adapted with permission from J. of Educ. Modules Mater. Sci. Eng. **1**, 535 (1979); current title, J. Mater. Educ., published by the Materials Education Council. Copyright © Pennsylvania State University.
[†] Retired.

Uhlig's Corrosion Handbook, Third Edition, Edited by R. Winston Revie
Copyright © 2011 John Wiley & Sons, Inc.

free energy of formation of each solid and ionic species is known for the conditions of temperature and pressure of interest. The pH of the solution is assumed to be known and constant in the bulk as well as at the metal or reaction product surface. Temperature and pressure are considered to be constant and are usually assumed to be 298 K (25°C) and 1 atm, respectively. Pourbaix diagrams may be calculated for other temperatures if thermodynamic data are available or may be estimated [5]. Generally, the features of Pourbaix diagrams are not significantly altered by increased pressures since thermodynamic properties are relatively insensitive to pressure (as compared with temperature).

The simple graphical methods described herein greatly facilitate the practical consideration of the various equilibrium reactions involved. For chemical reactions, it is convenient to make use of the Van't Hoff equation, which involves the equilibrium constant:

$$\Delta G^\circ = -RT \ln K = -2.303 \times 1.987 \times 298 \log K$$

for electrochemical reactions, the procedures involve manipulation of the Nernst equation, which can be written as

$$\phi = \phi^\circ + \frac{0.0591}{n} \log \frac{(a_P)^p (a_{H^+})^h}{(a_R)^r (a_{H_2O})^w} \qquad (7.1)$$

where

ϕ = reduction potential
$\phi^\circ = \Delta G^\circ / n\mathcal{F}$
ΔG° = standard free energy change
$\mathcal{F}$ = Faraday's constant
n = number of free electrons

for the general reaction,

$$rR + wH_2O \rightarrow pP + hH^+ + ne^-$$

Note that electrochemical reactions are written as oxidation reactions; that is, electrons are on the right.

Taking

$$a_{H_2O} = 1$$

and

$$pH = -\log a_{H^+}$$

then

$$\phi = \frac{\Delta G^\circ}{n\mathcal{F}} + \frac{0.0591}{n} \log \frac{(a_P)^p}{(a_R)^r} + \frac{[-0.0591h]}{n} pH \quad (7.2)$$

The standard free-energy change, ΔG°, for the reaction can be obtained readily from tabulated thermodynamic data.

Therefore, the first term to the right of the equality sign in Eq. (7.2) is a constant. The second term also becomes a constant when values of a_P and a_R are chosen in the normal manner. In constructing Pourbaix diagrams, the concentration of the ionic species at the boundary between a solid substance and a dissolved substance is usually taken as a very low value, such as $10^{-6}M$. The sum of the first two terms in Eq. (7.2) gives a constant equal to the value of the potential, ϕ, at pH 0. The resulting expression is the equation of a straight line of slope equal to the coefficient of the pH term, $-0.0591h/n$, and intercept equal to ϕ at pH 0.

Pourbaix diagrams are constructed from the three, and only three, types of straight-line relationships, which result from the analysis of the possible chemical and electrochemical equilibria in the system under consideration. Depending on the reactants and products of the assumed reactions, these straight lines will be either horizontal, vertical, or sloping:

1. A reaction involving a solid substance, a dissolved substance, and hydrogen ion in water without free electrons gives a vertical straight line, that is, independent of potential (when $n = 0$, the slope of the line equals ∞).
2. A reaction involving a solid substance and a dissolved substance in water plus free electrons but without hydrogen ion gives a horizontal straight line, that is, independent of pH (when $h = 0$, the slope = 0).
3. A reaction involving a solid substance, a dissolved substance, free electrons, and hydrogen ion will give a straight line with a slope equal to $-0.0591h/n$.

After plotting the straight lines on potential versus pH coordinates, the domain of the thermodynamic stability for each individual species is determined by requiring that all equations involving that species be satisfied simultaneously.

Example 7.1. One solid substance, one dissolved substance, and hydrogen ion in water without free elections:

$$2Fe^{3+} + 3H_2O = Fe_2O_3 + 6H^+ \qquad (7.3)$$

Assuming the activities of H_2O and Fe_2O_3 to be unity, the equilibrium constant may be expressed as

$$K = \frac{(a_{H^+})^6}{(a_{Fe^{3+}})^2} \qquad (7.4)$$

$$\begin{aligned} \log K &= 6 \log(a_{H^+}) - 2 \log(a_{Fe^{3+}}) \\ &= -6\,pH - 2 \log(a_{Fe^{3+}}) \end{aligned} \qquad (7.5)$$

Referring to tabulated thermodynamic data and substituting in the Van't Hoff equation, $\log K$ may be calculated:

$$\Delta G^\circ = -RT \ln K = -2.303 \times 1.987 \times 298 \log K$$

$$\log K = \frac{-\Delta G^\circ}{2.303 \times 1.987 \times 298} \qquad (7.6)$$

$$= \frac{-(-1970)}{1373} \qquad (7.7)$$

$$= 1.43$$

Substituting in Eq. (7.5) and rearranging, we obtain a generalized Pourbaix equation,

$$\log(a_{Fe^{3+}}) = -0.72 - 3pH \qquad (7.8)$$

In the case where

$$(a_{Fe^{3+}}) = 10^{-6}$$

then $pH = 1.76$.

This gives a vertical line on the Pourbaix diagram.

Example 7.2. One solid substance, one dissolved substance, and electrons, but without H^+ as a reactant or product:

$$Fe \rightarrow Fe^{2+} + 2e^- \qquad (7.9)$$

$$K = a_{Fe^{2+}} \qquad (7.10)$$

$$\log K = \log(a_{Fe^{2+}}) \qquad (7.11)$$

Using tabulated thermodynamic data, substitution in Eq. (7.2) gives

$$\phi = \frac{-20,300}{2 \times 23,060} + \frac{0.0591}{2} \log(a_{Fe^{2+}}) \qquad (7.12)$$

$$\phi = -0.440 + 0.0295 \log(a_{Fe^{2+}})$$

which is a generalized Pourbaix equation.

In the case where

$$a_{Fe^{2+}} = 10^{-6}$$

then

$$\phi = -0.617\,V$$

This gives a horizontal line on the Pourbaix diagram.

Example 7.3. One solid substance, one dissolved substance in water, plus free electrons, and hydrogen ion:

$$2Fe^{2+} + 3H_2O = Fe_2O_3 + 6H^+ + 2e^- \qquad (7.13)$$

The equilibrium constant for this reaction may be expressed as

$$K = \frac{(a_{H^+})^6}{(a_{Fe^{2+}})^2} \qquad (7.14)$$

and

$$\log K = 6\log(a_{H^+}) - 2\log(a_{Fe^{2+}})$$
$$= -6\,pH - 2\log(a_{Fe^{2+}}) \qquad (7.15)$$

Substitution into Eq. (7.2) gives

$$\phi = \frac{33,570}{2 \times 23,060} + \frac{0.0591}{2}[-6\,pH - 2\log(a_{Fe^{2+}})]$$

$$= 0.728 - 0.1773\,pH - 0.0591\log(a_{Fe^{2+}}) \qquad (7.16)$$

a generalized Pourbaix equation.

In the case where

$$a_{Fe^{2+}} = 10^{-6}$$

then

$$\phi = 1.0826 - 0.1773\,pH$$

This gives a sloping line on the Pourbaix diagram.

C. CONSTRUCTION OF DIAGRAMS

Table 7.1 lists the data for the iron–water diagram together with the various reactions and equilibrium formulas [1]. Figure 7.1 shows a resulting Pourbaix diagram considering that the only solid species are iron, Fe_3O_4, and Fe_2O_3. Naturally, a number of other assumptions could have been made. However, this serves to illustrate the construction of a diagram. Lines ⓐ and ⓑ designate the limits of thermodynamic stability of water at 298 K and 1 atm pressure. Above line ⓑ, water is unstable with regard to the evolution of oxygen, and below line ⓐ, water is unstable with respect to the evolution of hydrogen. The other dashed lines on the diagram comprise an "ionic species diagram." For ionic species (dashed lines), the coexistence lines represent the condition wherein the thermodynamic activity of the species on each side of that line is the same. For example, on line $6'$, the activities of Fe^{2+} and $Fe(OH)_2^+$ are equal. The triple point involving lines $1'$, $6'$, and $7'$ is an invariant point at which the activities of Fe^{2+}, $Fe(OH)_2^+$, and $HFeO_2^-$ are the same.

A line on the diagram represents a univariant system, whereas a family of lines, each of which is related to a value of a parameter, represents a divariant or trivariant system depending on whether the parameter contains one component (concentration) or two components (a term containing two concentrations). Heavy solid lines are used to separate solid species, whereas lighter weight solid lines are used to delineate the boundaries between a solid species and an ionic species.

TABLE 7.1.   Information, Reactions, and Equilibrium Formulas for the Iron–Water System at 298 K (25°C) and 1 atm

A. Substances Considered and Substances not Considered

	Oxidation Number (Z)	Considered	Not Considered	$\mu°$(cal)	Name, Color Crystalline System
Solid	0	Fe	—	0	α-Iron, light gray, face-centered-cubic (fcc)
substances	+2	FeO hydr.	—	$-58{,}880$	Ferrous hydroxide $Fe(OH)_2$, white, rhomb.
	+2	—	FeO anh.	—	Ferrous oxide, black, cubic
	+2.67	Fe_3O_4 anh.	—	$-242{,}400$	Magnetite, black, cubic
	+2.67	—	$Fe_3O_4 \cdot xH_2O$	—	Hydrated magnetite, green-black
	+3	Fe_2O_3 anh.	—	(a) $-177{,}100$	Haematite, red-brown, rhomb. or cubic
	+3	Fe_2O_3 hydr.	—	(b) $-161{,}930$	Ferric hydroxide $Fe(OH)_3$, red-brown, fcc
Dissolved	+2	Fe^{2+}	—	$-20{,}300$	Ferrous ion, green
substances	+2	$HFeO_2^-$	—	$-90{,}627$	Dihypoferrite ion, green
	+2	—	FeO_2^{2-}	—	Hypoferrite ion
	+3	Fe^{3+}	—	$-2{,}530$	Ferric ion, colorless
	+3	$FeOH^{2+}$	—	$-55{,}910$	Ferric ion, colorless
	+3	$Fe(OH)_2^+$	—	$-106{,}200$	Ferric ion, colorless
	+3	—	FeO_2^-	—	Ferrite ion
	+4	—	FeO^{2+}	—	Ferryl ion
	+4	—	FeO_3^{2-}	—	Perferrite ion
	+5	—	FeO_2^+	—	Perferryl ion
	+6	FeO_4^{2-}	—	$-111{,}685$	Ferrate ion, violet

B. Reactions and Equilibrium Formulas
Two Dissolved Substances
Relative Stability of the Dissolved Substances

$Z = +2$

1. $Fe^{2+} + 2H_2O = HFeO_2^- + 3H^+$ $\log \dfrac{(HFeO_2^-)}{(Fe^{2+})} = -31.58 + 3\,pH$

$Z = +3$

2. $Fe^{3+} + H_2O = FeOH^{2+} + H^+$ $\log \dfrac{(FeOH^{2+})}{(Fe^{3+})} = -2.43 + pH$

3. $FeOH^{2+} + H_2O = Fe(OH)_2^+ + H^+$ $\log \dfrac{[Fe(OH)_2^+]}{(FeOH^{2+})} = -4.69 + pH$

$+2 \rightarrow +3$

4. $Fe^{2+} = Fe^{3+} + e^-$ $E_0 = 0.771 + 0.0591 \log \dfrac{(Fe^{3+})}{Fe^{2+}}$

5. $Fe^{2+} + H_2O = FeOH^{2+} + H^+ + e^-$ $E_0 = 0.914 - 0.0591\,pH + 0.0591 \log \dfrac{(FeOH^{2+})}{(Fe^{2+})}$

6. $Fe^{2+} + 2H_2O = Fe(OH)_2^+ + 2H^+ + e^-$ $E_0 = 1.191 - 0.1182\,pH + 0.0591 \log \dfrac{[Fe(OH)_2^+]}{(Fe^{2+})}$

7. $HFeO_2^- + H^+ = Fe(OH)_2^+ + e^-$ $E_0 = -0.675 + 0.0591\,pH + 0.0591 \log \dfrac{[Fe(OH)_2^+]}{(HFeO_2^-)}$

$+2 \rightarrow +6$

8. $HFeO_2^- + 2H_2O = FeO_4^{2-} + 5H^+ + 4e^-$ $E_0 = 1.001 - 0.0738\,pH + 0.0148 \log \dfrac{(FeO_4^{2-})}{(HFeO_2^-)}$

$+3 \rightarrow +6$

9. $Fe^{3+} + 4H_2O = FeO_4^{2-} + 8H^+ + 3e^-$ $E_0 = 1.700 - 0.1580\,pH + 0.0197 \log \dfrac{(FeO_4^{2-})}{(Fe^{3+})}$

10. $FeOH^{2+} + 3H_2O = FeO_4^{2-} + 7H^+ + 3e^-$ $E_0 = 1.652 - 0.1379\,pH + 0.0197 \log \dfrac{(FeO_4^{2-})}{(FeOH^{2+})}$

11. $Fe(OH)_2^+ + 2H_2O = FeO_4^{2-} + 6H^+ + 3e^-$ $E_0 = 1.559 - 0.1182\,pH + 0.0197 \log \dfrac{(FeO_4^{2-})}{[Fe(OH)_2^+]}$

TABLE 7.1 (*Continued*)

Limits of the Domains of Relative Predominance of the Dissolved Substances

1′.	$Fe^{2+}/HFeO_2^{-}$	$pH = 10.53$
2′.	$Fe^{3+}/FeOH^{2+}$	$pH = 2.43$
3′.	$FeOH^{2+}/Fe(OH)_2^{+}$	$pH = 4.69$
4′.	Fe^{2+}/Fe^{3+}	$E_0 = 0.771$
5′.	$Fe^{2+}/FeOH^{2+}$	$E_0 = 0.914 - 0.0591\,pH$
6′.	$Fe^{2+}/Fe(OH)_2^{+}$	$E_0 = 1.191 - 0.1182\,pH$
7′.	$HFeO_2^{-}/Fe(OH)_2^{+}$	$E_0 = -0.675 + 0.0591\,pH$
8′.	$HFeO_2^{-}/FeO_4^{2-}$	$E_0 = 1.001 - 0.0738\,pH$
9′.	Fe^{3+}/FeO_4^{2-}	$E_0 = 1.700 - 0.1580\,pH$
10′.	$FeOH^{2+}/FeO_4^{2-}$	$E_0 = 1.652 - 0.1379\,pH$
11′.	$Fe(OH)_2^{+}/FeO_4^{2-}$	$E_0 = 1.599 - 0.1182\,pH$

Two Solid Substances
Limits of the Domains of Relative Stability of Iron and Its Oxides and Hydroxides

12.	$0 \rightarrow +2$	$Fe + H_2O = FeO + 2H^{+} + 2e^{-}$	$E_0 = -0.047 - 0.0591\,pH$
13.	$0 \rightarrow +2.67$	$3Fe + 4H_2O = Fe_3O_4 + 8H^{+} + 8e^{-}$	$E_0 = -0.085 - 0.0591\,pH$
14.	$0 \rightarrow +3$	$2Fe + 3H_2O = Fe_2O_3 + 6H^{+} + 6e^{-}$	(a) $E_0 = -0.051 - 0.0591\,pH$
			(b) $E_0 = -0.059 - 0.0591\,pH$
15.	$+2 \rightarrow 2.67$	$3FeO + H_2O = Fe_3O_4 + 2H^{+} + 2e^{-}$	$E_0 = -0.197 - 0.0591\,pH$
16.	$+2 \rightarrow +3$	$2FeO + H_2O = Fe_2O_3 + 2H^{+} + 2e^{-}$	(a) $E_0 = -0.057 - 0.0591\,pH$
			(b) $E_0 = 0.271 - 0.0591\,pH$
17.	$+2.67 \rightarrow +3$	$2Fe_3O_4 + H_2O = 3Fe_2O_3 + 2H^{+} + 2e^{-}$	(a) $E_0 = 0.221 - 0.0591\,pH$
			(b) $E_0 = 1.208 - 0.0591\,pH$

One Solid Substance and One Dissolved Substance
Solubility of Iron and Its Oxides and Hydroxides

	$Z = +2$		
18.		$Fe^{2+} + H_2O = FeO + 2H^{+}$	$\log(Fe^{2+}) = 13.29 - 2pH$
19.		$FeO + H_2O = HFeO_2^{-} + H^{+}$	$\log(HFeO_2^{-}) = -18.30 + pH$
	$Z = +3$		
20.		$2Fe^{3+} + 3H_2O = Fe_2O_3 + 6H^{+}$	(a) $\log(Fe^{3+}) = -0.72 - 3pH$
			(b) $\log(Fe^{3+}) = 4.84 - 3pH$
21.		$2FeOH^{2+} + H_2O = Fe_2O_3 + 4H^{+}$	(a) $\log(FeOH^{2+}) = -3.15 - 2pH$
			(b) $\log(FeOH^{2+}) = 2.41 - 2pH$
22.		$2Fe(OH)_2^{+} = Fe_2O_3 + H_2O + 2H^{+}$	(a) $\log[Fe(OH)_{2+}] = -7.84 - pH$
			(b) $\log[Fe(OH)_{2+}] = -2.28 - pH$
23.	$0 \rightarrow +2$	$Fe = Fe^{2+} + 2e^{-}$	$E_0 = -0.440 + 0.0295\,\log(Fe^{2+})$
24.		$Fe + 2H_2O = HFeO_2^{-} + 3H^{+} + 2e^{-}$	$E_0 = 0.493 - 0.0886pH + 0.0295\,\log(HFeO_2^{-})$
25.	$0 \rightarrow +3$	$Fe = Fe^{3+} + 3e^{-}$	$E_0 = -0.037 + 0.0197\,\log(Fe^{3+})$
26.	$+2 \rightarrow +2.67$	$3Fe^{2+} + 4H_2O = Fe_3O_4 + 8H^{+} + 2e^{-}$	$E_0 = 0.980 - 0.2364pH - 0.0886\,\log(Fe^{2+})$
27.		$3HFeO_2^{-} + H^{+} = Fe_3O_4 + 2H_2O + 2e^{-}$	$E_0 = -1.819 + 0.0295\,pH - 0.0886\,\log(HFeO_2^{-})$
28.	$+2 \rightarrow +3$	$2Fe^{2+} + 3H_2O = Fe_2O_3 + 6H^{+} + 2e^{-}$	(a) $E_0 = 0.728 - 0.1773\,pH - 0.0591\,\log(Fe^{2+})$
			(b) $E_0 = 1.057 - 0.1773pH - 0.0591\,\log(Fe^{2+})$
29.		$2HFeO_2^{-} = Fe_2O_3 + H_2O + 2e^{-}$	(a) $E_0 = -1.139 - 0.0591\,\log(HFeO_2^{-})$
			(b) $E_0 = -0.810 - 0.0591\,\log(HFeO_2^{-})$

For reactions involving Fe_2O_3, (a) indicates anhydrous Fe_2O_3, whereas (b) indicates $Fe(OH)_3$.
Source: Excerpted from [1].

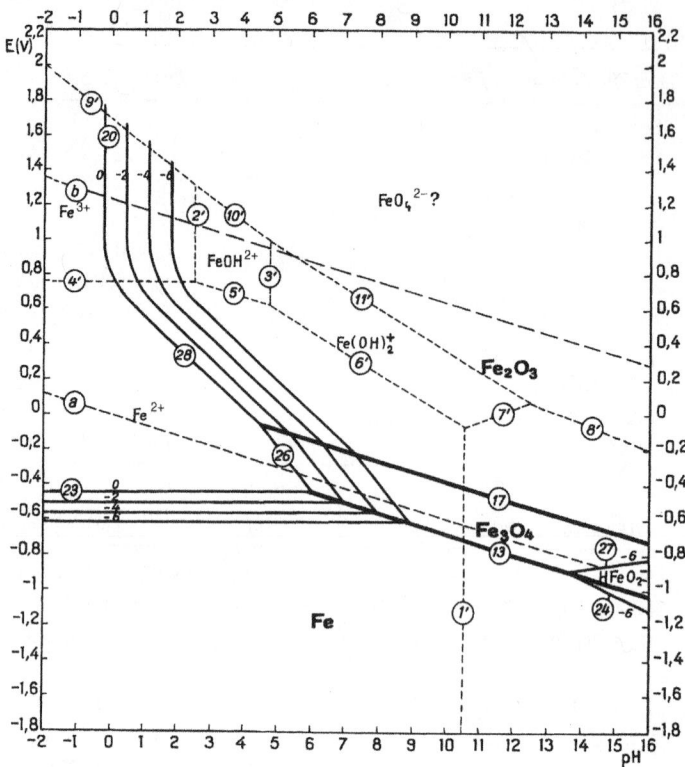

FIGURE 7.1. Potential–pH equilibrium diagram for the system iron–water at 25°C (considering as solid substances only Fe, Fe_3O_4. and Fe_2O_3) [1], Reproduced with permission from [1]. Copyright © Marcel Pourbaix.

Example Problem. It is suggested that the reader construct the Pourbaix diagram for the Fe–H_2O system (considering the solid species Fe, Fe_3O_4, and Fe_2O_3) at 298 K and 1 atm pressure. Plot the diagram on graph paper using pH ranging from -2 to $+16$ and the electrode potential from -1.8 to $+2.2\,V_{SHE}$. The equations for each coexistence already have been calculated and are listed in Table 7.1. It will be necessary to define what ionic activity will be considered to represent "significant corrosion." The establishment of what is considered to be "corrosion" evolves into a determination as to the amount of metal dissolution which is permissible "for all practical purposes." It often has been found convenient to consider the solubility of 10^{-6} g atoms of soluble ion per liter as representing "no corrosion for all practical purposes." On Figure 7.1, this assumption is represented by the lines marked -6 (the logarithm of the activity).

The domain of the metal is found at the bottom of the diagram. For simplicity, it is suggested that the equilibrium involving the metal and its least highly oxidized ionic form is an easy equilibrium to start with. In this case, we are considering Eq. (7.23) in Table 7.1 involving iron and ferrous ion. In as much as no hydrogen ion is involved, it is apparent that this equilibrium coexistence involves a family of horizontal lines on potential versus pH coordinates the position of which depends on the assumed value of thermodynamic

activity chosen for the ferrous ion. If you choose 10^{-6} for the ionic activity, you will obtain only one line. The question arises as to whether the domain of iron is above or below the line. Referring to the electrochemical equation,

$$Fe \rightarrow Fe^{2+} + 2e^-$$

a "thought experiment" is helpful in answering this question. If electrons were added, the reaction would be driven in the direction of iron; thus, the addition of electrons (and hence more negative potentials) favors iron rather than Fe^{2+}. Therefore, Fe is below the line and Fe^{2+} above the line.

The next question is "How far along the pH axis does the horizontal Fe/Fe^{2+} equilibrium coexistence extend?" To the left, the boundary is usually chosen arbitrarily by selecting the range of pH of interest. The limit for this line in the direction of higher pH depends on the activities of the ionic species and the restrictions imposed by other equilibria, for example, line 13 in Fig. 7.1, which is the coexistence between iron and Fe_3O_4.

$$3Fe + 4H_2O \rightarrow Fe_3O_4 + 8H^+ + 8e^-$$

This is a logical choice for next consideration, since Fe_3O_4 is the solid species that involves oxide with the lowest level of oxidation of iron. Reaction 13 in Table 7.1 involves both electrons and hydrogen ion and will be a sloping line on the diagram. Another thought experiment will reveal on which side of line 13 iron is stable and on which side Fe_3O_4 is stable. Since this is a sloping line, there are two ways in which this decision can be made. Adding electrons will favor Fe; hence, Fe will be on the more negative side of the line (i.e., below the line). Addition of hydrogen ion (lower pH) also stabilizes the species iron. Thus, iron should be to the left of (or below) the line and Fe_3O_4 to the right (or above). Simultaneous solution of the equations for lines 23 and 13 will reveal the point of intersection and consequently the termination of dominance of the Fe/Fe^{2+} coexistence.

A similar calculation for reaction 26 in Table 7.1 between Fe^{2+} and Fe_3O_4 yields another line with a slope different from reaction 13:

$$3Fe^{2+} + 4H_2O \rightarrow Fe_3O_4 + 8H^+ + 2e^-$$

Adding either H^+ or electrons to the right side pushes the reaction to the left to restore equilibrium. This favors the

species on the left (Fe^{2+}). By using the same procedures, it is observed that lines 23, 26, and 13 in Figure 7.1 intersect at an invariant point for a given activity of ferrous ion.

Line 20, the equilibrium between Fe^{3+} and Fe_2O_3, represents a reaction in which there is no electron transfer. Thus, by inspection of the equation, this coexistence will appear as a vertical line on the Pourbaix diagram. The decision of which species is on which side of the line can easily be made by assessing the effect of adding hydrogen ions and observing that this addition favors Fe^{3+}. Thus, Fe^{3+} is on the low-pH side of the line.

It is suggested that the equilibrium coexistences be taken one at a time progressing from the elemental state through the various oxidized states to establish the limits of each of the lines. It usually becomes complicated to draw all the lines and to remove those lines and portions of lines that are redundant or improper.

When the diagram is complete, it is possible to test the predictions of the Pourbaix diagram you have just drawn. Table 7.2 reports a number of experiments conducted by

TABLE 7.2.   Corrosion and Noncorrosion of Iron at 298 K^{a,b}

Experiment	Sample No.	Solution		pH	$E_H(V)$	State of Metal^c	Gas
a	1	H_2O distilled		8.1	−0.486	Y	—
	2	NaCl	1 g/L	6.9	−0.445	Y	—
	3	H_2SO_4	1 g/L	2.3	−0.351	Y	H_2
	4	$NaHSO_4$	1 g/L	6.4	−0.372	Y	—
	5	NaOH	1 g/L	11.2	+0.026	N	—
	6	K_2CrO_4	1 g/L	8.5	+0.235	N	—
	7	K_2CrO_4 + NaCl	1 g/L	8.6	−0.200	Y/N	—
	8	$KMnO_4$	0.2 g/L	6.7	−0.460	Y	—
	9	$KMnO_4$	1 g/L	7.1	+0.900	N	—
	10	H_2O_2	0.3 g/L	5.7	−0.200	Y	—
	11	H_2O_2	3.0 g/L	3.4	+0.720	N	O_2
	12	Brussels city water		7.0	−0.450	Y	—
b	13	NaOH	40 g/L degassed	13.7	−0.810	Y	H_2
c	14	City water–iron–copper		7.5	−0.445	Y	—
	15	City water–iron–zinc		7.5	−0.690	N	H_2
	16	City water–iron–magnesium		7.5	−0.910	N	H_2
	17	City water–iron–platinum		7.5	−0.444	Y	—
c′	14′	City water–iron–copper		7.8	−0.385	Y	—
	15′	City water–iron–zinc		7.7	−0.690	N	H_2
	16′	City water–iron–magnesium		8.7	−0.495	N	H_2
	17′	City water–iron–platinum		—	—	Y	—
d	18	$NaHCO_3$ 0.1 M Pole −		8.4	−0.860	N	H_2
	19	$NaHCO_3$ 0.1 M Pole +		8.4	−0.350	Y	—
	20	$NaHCO_3$ 0.1 M Pole −		8.4	−0.885	N	H_2
	21	$NaHCO_3$ 0.1 M Pole −		8.4	+1.380	N	O_2
e	22	$NaHCO_3$ 0.1 M Pole −		8.4	−0.500	Y	—
	23	$NaHCO_3$ 0.1 M Pole +		8.4	+1.550	N	O_2
	24	$NaHCO_3$ 0.1 M Pole −		8.4	−1.000	N	H_2
	25	$NaHCO_3$ 0.1 M Pole +		8.4	+1.550	N	O_2

[a]Experimental conditions used to test the predictions of the Pourbaix diagram for the iron–water system.
[b]Excerpted from M. Pourbaix. Lectures on Electrochemical Corrosion, NACE, Houston, 1995 p. 18.
[c]Abbreviations: Y = corrosion; N = no corrosion.

Pourbaix in which iron electrodes were immersed in various solutions. The pH and the electrode potential were measured in each case, and the specimens were allowed to stand in beakers containing each of the indicated chemical environments. Now plot on the diagram the data from the experiments numbered 1–25 in Table 7.2. For convenience, the sample numbers should be written beside each data point. This will make it possible to compare the experimental results of Pourbaix with the predictions based on the Pourbaix diagram. Remember, the term "immunity" is reserved for noncorrosion and represents the case in which corrosion cannot occur for thermodynamic reasons. The term "corrosion" is reserved for areas of the diagram where an ionic species is the stable species thermodynamically.

"Passivation" describes the portion of the diagram where a solid reaction product is formed. Presumably, if the solid reaction product is protective, corrosion will stop. Thus, the term "passivation" might be said to apply to a region in the diagram where (thermodynamically) corrosion is possible, but it does not occur, because of the formation of a barrier coating. The diagram is not sufficient to decide whether a solid reaction product is also protective. This information can only be gained by performing an experiment, Comparison between the data points on your diagram with the information contained in Table 7.2 should reveal that in each case shown the Pourbaix diagram would have predicted correctly whether or not corrosion would occur merely by knowing the electrode potential and the pH of the solution.

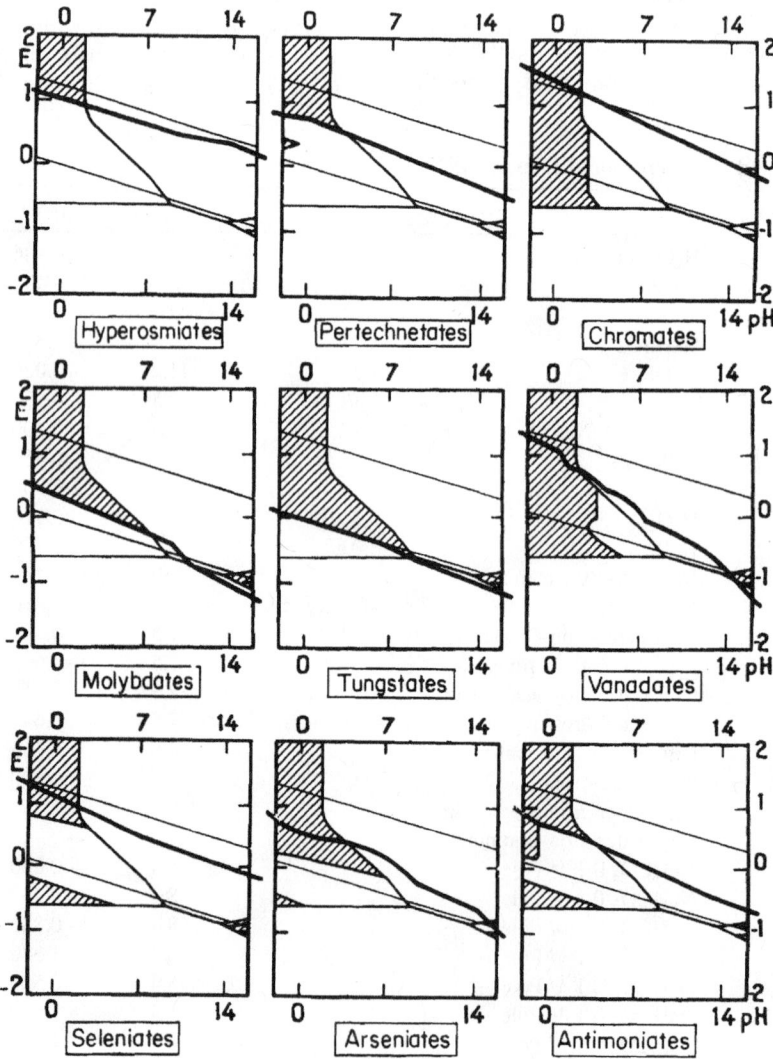

FIGURE 7.2. Oxidizing corrosion inhibitors. The hatched regions indicate theoretical corrosion domains in the presence of 0.01 M solutions of inhibitor. Reproduced with permission from [1]. Copyright © Marcel Pourbaix.

D. APPLICATIONS OF POURBAIX DIAGRAMS

It is possible to predict conditions under which corrosion, noncorrosion, and passivation are possible. It is also possible to make a number of other useful predictions. For example, the electrode potential for cathodic protection is represented by the equilibrium coexistence line between ferrous ion and iron in Figure 7.1. The domain of potential and pH in which anodic protection may be considered is represented by the passive region (either Fe_3O_4 or Fe_2O_3), but care should be exercised to avoid the domains where Fe^{2+}, Fe^{3+}, or $HFeO_2^-$ are stable.

If the electrode potential falls in a corrosion regime (e.g., in the region where ferrous ion is stable), it is possible to stop corrosion by adding an oxidant that would bring the electrode potential into the region of stability for Fe_2O_3 by raising the electrode potential, or by changing the pH in the alkaline direction so as to move horizontally into the passive region, or by cathodic protection that has the effect of lowering the potential into the immunity region. It should be emphasized that the predictions made by using the Pourbaix diagram should be tested prior to actual use, since the formation of a reaction product film does not necessarily mean that this film is protective. In addition, cathodic protection may result in hydrogen evolution at the cathode, which could have an adverse effect on protective coatings

or might under some circumstances induce hydrogen embrittlement of certain metals.

It is also possible to predict the types of ions that have promise as oxidizing corrosion inhibitors. Superposition of the chromium–water diagram over the iron–water diagram, for example, shows that the region of stability for Cr_2O_3 coincides with a portion of the iron diagram wherein ferrous ion is the stable species.

Consequently, in the absence of an inhibitor, corrosion of iron would be anticipated in this domain of potential and pH. The effect of adding chromates is to provide a means of forming a protective Cr_2O_3 film that inhibits corrosion. Figure 7.2 suggests the influence of various oxidizing inhibitors on the corrosion of iron [1]. These predictions should be tested before actual use.

REFERENCES

1. M. Pourbaix, Atlas of Electrochemical Equilibria in Aqueous Solutions, National Association of Corrosion Engineers (NACE), Houston, TX, and Centre Belge d'Etude de la Corrosion (CEBELCOR), Brussels, 1974.
2. M. Pourbaix, Lectures on Electrochemical Corrosion, Plenum, New York, 1973.
3. U. R, Evans, J. Inst. Metals, **30**, 263, 267 (1923).
4. E. D. Verink and M. Pourbaix, Corrosion, **27**(12), 495 (1971).
5. C. M. Criss and J. W. Cobble, J. Am. Chem. Soc., **86**, 5394 (1964).

8

POURBAIX DIAGRAMS FOR MULTIELEMENT SYSTEMS

W. T. Thompson

Centre for Research in Computational Thermochemistry, Royal Military College of Canada, Kingston, Ontario, Canada

M. H. Kaye

Faculty of Energy Systems and Nuclear Science, University of Ontario Institute of Technology, Oshawa, Ontario, Canada

C. W. Bale and A. D. Pelton

Départment de Génie Physique et de Genie des Materiaux, Ecole Polytechnique, Montréal, Québec, Canada

A. Introduction
B. Computation using gibbs energy minimization
C. Additional element in the aqueous phase
D. Additional element in the metal phase
E. Conclusions
References

A. INTRODUCTION

The thermodynamics of aqueous corrosion is conveniently summarized for practical purposes in the diagrams devised by Pourbaix [1]. These diagrams, assembled from the Gibbs energy of formation of ionic species and phases, draw attention to the combined importance of pH (abscissa) and redox potential (ordinate) in distinguishing conditions of active corrosion, passivity, and immunity. In their original form, these isothermal diagrams indicate the regions or domains of stability of one particular metal (or element) in several possible chemical forms, including complex ions and phases, limited, however, to those chemical forms of the metal containing only hydrogen and/or oxygen. Typical diagrams of this type are shown later for Ag, Au, Cu, Fe, and Ni in Figures 8.1, 8.3, 8.5, 8.7, and 8.8. A previous chapter reviews the principles involved in constructing these classical three-element (M–H–O) dia-

grams from the underlying thermodynamic principles related to the equilibrium between phases and species. This chapter considers the issues that arise when *additional elements* are introduced either as species in the aqueous phase or as alloying in the metal [2].

It is important at the outset to realize that in general it is not possible to progress in the understanding of an alloy corroding in a complex aqueous medium by superimposing a series of conventional Pourbaix diagrams. The corrosion of pure silver or gold in chloride media cannot be understood with the conventional Pourbaix diagrams because important (quite stable) compounds or complex ions involving chloride are not represented in those diagrams. Likewise, the corrosion of iron–nickel alloys cannot be understood by examining conventional Pourbaix diagrams for Fe and Ni; this approach overlooks the importance of $NiFe_2O_4$ that is not found in either diagram [3]. The use of conventional Pourbaix diagrams not surprisingly, therefore, often leads to false conclusions or, worse, the belief that thermodynamic considerations may be of little value generally in matters of aqueous corrosion.

The inclusion of necessarily many more species and phases in the development of multielement diagrams leads to obstacles in diagram construction not encountered in systems of the M–H–O type. The first of these obstacles is the lack of a priori knowledge for a complex system of which species and phases form an equilibrium associated with a domain boundary on the diagram; a formal procedure is required to exclude equilibria such as between Ag_2O_2 and Ag (see Fig. 8.1), which in the traditional diagram development

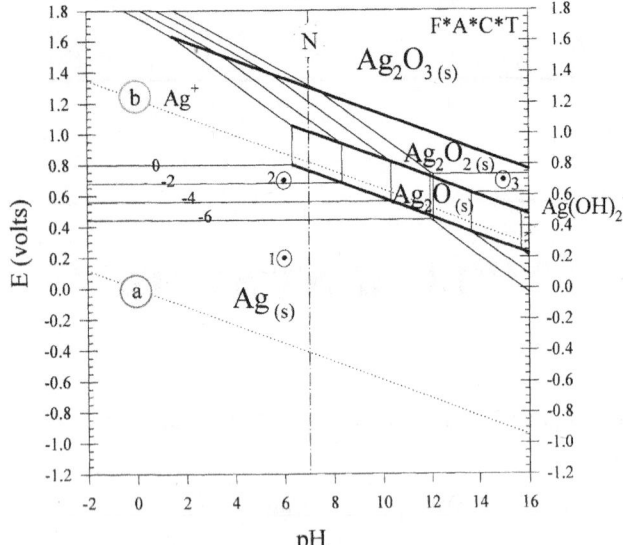

FIGURE 8.1. The Ag Pourbaix diagram at 298 K (25°C). Concentration of aqueous species range from 1 to $10^{-6}m$. Points 1, 2, and 3 should be compared with Gibbs energy changes in Table 8.2.

for Ag may be excluded because of independent knowledge that these phases cannot coexist.

A second practical obstacle in dealing with systems of several elements is that a comprehensive set of diagrams leading to a compilation such as Pourbaix's Atlas is not possible; the enormous number of diagrams resulting from the combinatorial possibilities precludes publication. An easy-to-apply systematic approach not based on any a priori assumptions and therefore amenable to computing is essential.

The third obstacle is more fundamental and subtle in nature and requires decisions about how additional elements are to be incorporated into the diagram development. To oversimplify this matter for the present, it will suffice to say that it is necessary, at the outset of diagram construction, to know whether the additional elements are associated with the corrosive media or with the alloy. The former places constraint(s) on chemical potential(s) or concentrations; the latter places constraints on mass balances. The topology of the multielement Pourbaix diagram therefore depends on how additional elements are treated.

B. COMPUTATION USING GIBBS ENERGY MINIMIZATION

The circumvention of the matters raised in Section A is to be found in the process of Gibbs energy minimization. It is helpful in understanding this process as a basis for its extension into multielement systems by first considering the treatment of a simple M–H–O system. To be more

specific, consider the case of silver and the development of the diagram in Figure 8.1. The reactions (involving only H and O) whereby silver may become an ionic species, a compound, or remain in its metallic form (as a degenerate case) are

$$Ag \rightarrow \mathbf{Ag} \tag{8.1}$$

$$Ag \rightarrow \mathbf{Ag}^+ + e^- \tag{8.2}$$

$$Ag + \tfrac{1}{2}H_2O \rightarrow \tfrac{1}{2}\mathbf{Ag_2O} + H^+ + e^- \tag{8.3}$$

$$Ag + H_2O \rightarrow \tfrac{1}{2}\mathbf{Ag_2O_2} + 2H^+ + 2e^- \tag{8.4}$$

$$Ag + \tfrac{3}{2}H_2O \rightarrow \tfrac{1}{2}\mathbf{Ag_2O_3} + 3H^+ + 3e^- \tag{8.5}$$

$$Ag + 2H_2O \rightarrow \mathbf{Ag(OH)_2^-} + 2H^+ + e^- \tag{8.6}$$

The Gibbs energy change for each reaction depends not only on the standard Gibbs energy of formation [4] of the reactants and products (Table 8.1) but also on the activity or to a good approximation the molal concentration m of the aqueous silver–containing species [add $RT \ln(m)$ to the standard Gibbs energy of formation] as well as the pH and redox potential. The Gibbs energy of the H^+ is given by

$$\Delta G_{H^+} = -2.303RT(pH) \tag{8.7}$$

and the effective Gibbs energy of the electron (which represents an unspecified redox reaction) is given by

$$\Delta G_{e^-} = -\Im(E_H) \tag{8.8}$$

At any particular condition of E_H and pH there will be, in general, one reaction that gives rise to the largest negative

TABLE 8.1. Thermodynamic Data

Species	$\Delta G°$ (kJ/mol)
H^+	0
H_2O	-237.2
Ag	0
Ag^+	77.0
Ag_2O	-10.8
Ag_2O_2	27.4
Ag_2O_3	121.1
$Ag(OH)_2^-$	-260.2
Cl^-	-131.1
$AgCl$	-109.5
$AgCl_2^-$	-215.5
$AgClO_2$	75.7

Gibbs energy change per mole of Ag; this is the species or compound associated with the domain in which that point resides. If all of the Gibbs energy changes are positive, the E_H–pH coordinate must reside in the domain of silver immunity. As a convenience in understanding, the Gibbs energy changes in Table 8.2 for reactions (8.1)–(8.6) should be compared with arbitrarily selected points 1, 2, and 3 in Figure 8.1. The domains for all species can be located as precisely as required by repetition of the foregoing for a suitably fine matrix of E_H and pH points. The number of computations can be considerably reduced when it is realized that a domain for a particular phase/species must be contiguous. With any personal computer, this approach, which was completely impractical in the not too distant past, leads to an almost instant construction of the diagram for a particular concentration of the aqueous species. The diagram in Figure 8.1 is actually four such figures superimposed for concentrations of Ag^+ and $Ag(OH)_2^-$ of 1, 10^{-2}, 10^{-4}, and $10^{-6} m$ (strictly speaking activity) developed using data in Table 8.1. The dotted lines *@* and *ⓑ* correspond to redox potentials determined by hydrogen and oxygen saturation at one standard atmosphere, respectively.

C. ADDITIONAL ELEMENT IN THE AQUEOUS PHASE

The flexibility of the Gibbs energy minimization approach is particularly striking when, in addition to H and O, an element such as chlorine (as Cl^-) is considered in the aqueous phase. To continue with silver as the basis for discussion, the reactions to consider in addition to those above are

$$Ag + Cl^- \rightarrow \mathbf{AgCl} + e^- \qquad (8.9)$$

$$Ag + 2Cl^- \rightarrow \mathbf{AgCl_2^-} + e^- \qquad (8.10)$$

$$Ag + 2H_2O + Cl^- \rightarrow \mathbf{AgClO_2} + 4H^+ + 5e^- \qquad (8.11)$$

By using the data in Table 8.1, the Gibbs energy change for each reaction can be found as indicated in Section B. Some of the Gibbs energy changes are shown in Table 8.2. In examining Table 8.2 and Figure 8.2, it will be clear than the Ag-containing phase/species associated with the lowest Gibbs energy change is the one labeling the domain at the corresponding E_H–pH coordinate in Figure 8.2.

With the mechanics of construction summarized, it is now appropriate to comment on the role of Cl^-. Chloride ion at the 1 m concentration used in developing Figure 8.2 results in a complete eclipsing of the Ag^+ and Ag_2O fields in Figure 8.1 with AgCl and $AgCl_2^-$ has become the most important Ag-containing ion in all but the most concentrated alkaline solutions. Clearly, the corrosion of silver in aqueous chloride media could not be properly understood

TABLE 8.2. Gibbs Energy Minimization at 1 m Concentration for Ag-Containing Aqueous Species (and 1 m Cl^- concentration for the case of Figure 8.2)[a]

Species	Reaction	Point 1 $E_H = 0.2$ V pH = 6.0 (kJ)	Point 2 $E_H = 0.7$ V pH = 6.0 (kJ)	Point 3 $E_H = 0.7$V pH = 15.0 (kJ)
Ag	1	0	0	0
Ag^+	2	57.7	9.4	9.4
Ag_2O	4	59.7	11.4	−39.9
Ag_2O_2	3	143.8	47.3	−55.4
Ag_2O_3	5	255.7	111.0	−43.1
$Ag(OH)_2^-$	6	126.4	78.2	−24.5
AgCl	9	2.4	−45.8	−45.8
$AgCl_2^-$	10	27.7	−20.6	−20.6
$AgClO_2$	11	448.0	206.7	1.3

[a]$AgClO_3$ and $AgClO_4$ have been excluded as possible phases.

using the classical Pourbaix diagram in Figure 8.1, as Pourbaix himself recognized.

Figures 8.3 and 8.4 show the influence of chloride on the gold Pourbaix diagram. Note the solubility of Au as tetrachloraurate ion in acidic oxidizing media. This accounts for both the well-known attack of gold by aqua regia and the use of an acidic chloride electrolyte in the Wohlwill gold electrorefining process [5].

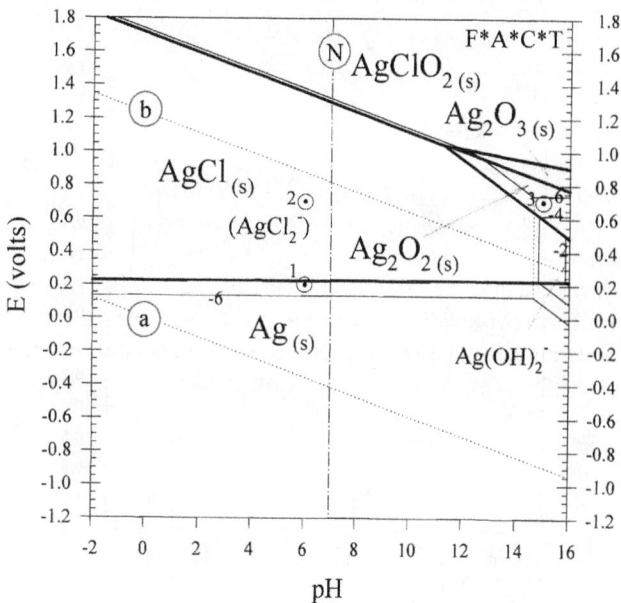

FIGURE 8.2. The Ag Pourbaix diagram at 298 K (25°C) in chloride solution at 1 m Cl^- concentration. The concentration of the Ag-containing species $AgCl_2^-$ and $Ag(OH)_2^-$ is shown at 10^{-6}, 10^{-4}, and $10^{-2} m$. Points 1, 2, and 3 should be compared with Gibbs energy changes in Table 8.2. Note that AgCl completely eclipses the Ag^+ and Ag_2O fields on Figure 8.1 for this concentration of Cl^-.

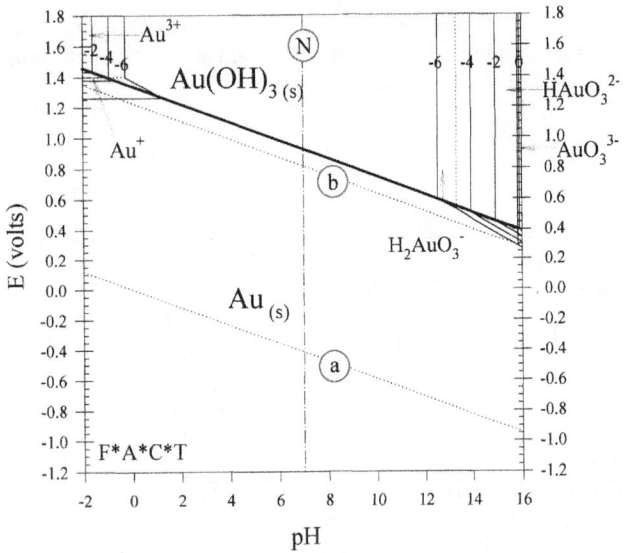

FIGURE 8.3. The Au Pourbaix diagram at 298 K (25°C). Concentration of aqueous species range from 1 to $10^{-6}m$. Gold immunity extends into the strong acid region even under oxygen-saturated conditions (line b).

The diagrams in Figures 8.5 and 8.6 pertain to the effect of ammonia on copper corrosion. When deaeration additives such as hydrazine, N_2H_4, are added in low concentration to water, the slow chemical breakdown that consumes dissolved oxygen [6] and lowers the redox potential to slightly above

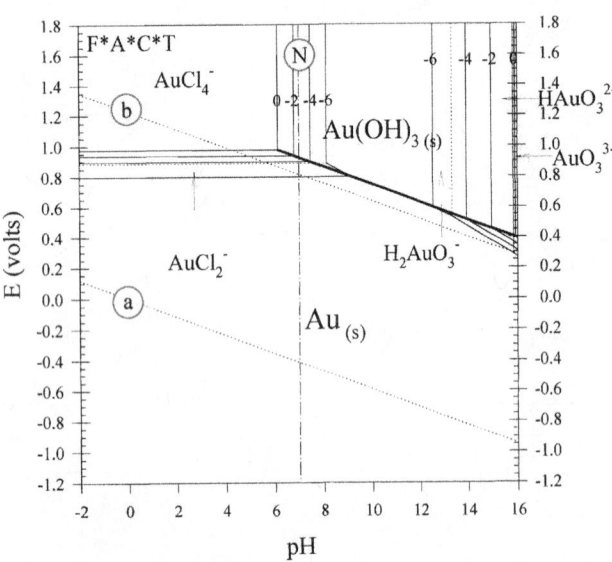

FIGURE 8.4. The Au Pourbaix diagram at 298 K (25°C) in chloride solution at 1 m Cl$^-$ concentration. The concentration of the Au-containing species is shown at 10^{-6}, 10^{-4}, 10^{-2}, and 1 m. The domain of gold immunity does not extend into strong acid solutions for oxygen saturated conditions (line b) because of the stability of $AuCl_4^-$.

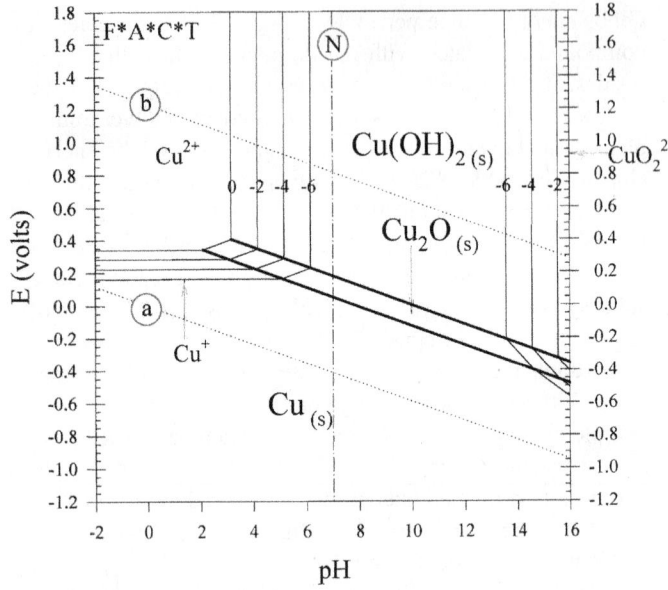

FIGURE 8.5. The Cu Pourbaix diagram at 298 K (25°C). Concentrations of aqueous species range from 1 to $10^{-6}m$. Compare with Figure 8.6 to gauge the effect of a low NH_3 concentration.

line a also contributes a low concentration ($\sim 10^{-3}m$) of residual ammonia, which tends to raise the pH above neutral. Furthermore, the residual ammonia forms very stable complex ions with copper. Fortunately, as the diagrams show, the

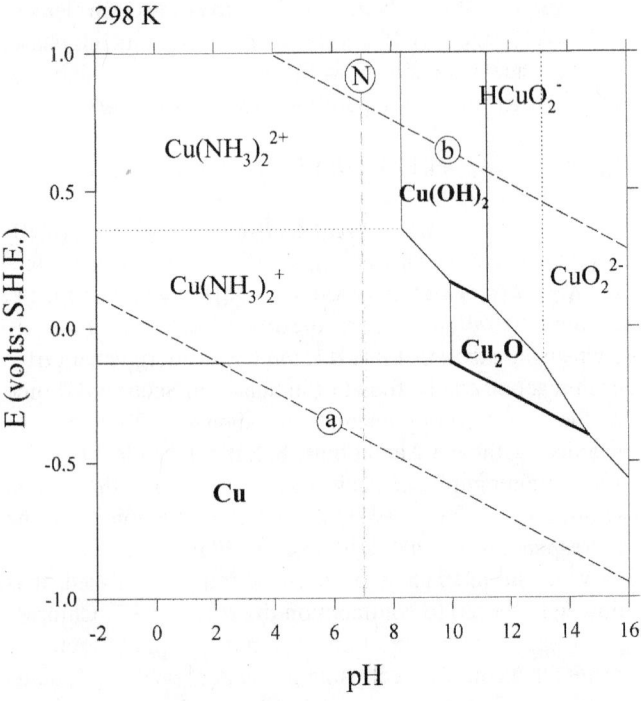

FIGURE 8.6. The Cu Pourbaix diagram at 298 K (25°C) in ammonia-containing solutions at $10^{-3}m$ NH_3. The concentration of copper-containing aqueous species is shown only at $10^{-6}m$.

stability of copper amines is not so great as to contribute to corrosion of copper. Of course, if the ammonia concentration were allowed to increase drastically, it would be necessary to recompute Figure 8.6 to examine the extent to which the copper amine fields enlarge.

The computation of Figure 8.6 is slightly different than Figures 8.2 and 8.4 because the additional species involve more than one element. The formation reactions are not based on ammonia as in

$$Cu + 2\,NH_3 \rightarrow Cu(NH_3)_2^+ + e^- \qquad (8.12)$$

but rather on the elements

$$Cu + N_2 + 6H^+ + 5e^- \rightarrow Cu(NH_3)_2^+ \qquad (8.13)$$

At each E_H–pH coordinate, the effective hydrogen partial pressure can be calculated using

$$E_H = -\frac{RT}{2\mathfrak{J}}\ln(P_{H_2}) - \frac{2.303RT}{\mathfrak{J}}(pH) \qquad (8.14)$$

The effective N_2 partial pressure can be calculated from the equilibrium constant for

$$N_2 + 3H_2 \rightleftarrows 2\,NH_3 \qquad (8.15)$$

$$K_{eq} = \frac{(m_{NH_3})^2}{(P_{N_2})(P_{H_2})^3} \qquad (8.16)$$

The Gibbs energy changes for formation reactions such as (8.13) can now be calculated since corrections can be made for the effective partial pressure of N_2 by adding $RT\ln(P_{N_2})$.

D. ADDITIONAL ELEMENT IN THE METAL PHASE

When the metal phase is an alloy, the Gibbs energy minimization differs from that described above. In the previous sections, the voluminous nature of the aqueous phase in comparison with the limited quantity of metal corroding permits the concentration of the additional species in the aqueous phase to be considered as constant. In the case of an alloy, it is necessary to specify (and preserve in the diagram development) the molar ratio of the alloy elements that are regarded as being insignificant in total molar amount to other elements in the voluminous aqueous phase.

The general form of the corrosion reactions for a binary alloy, M_1–M_2, is modified from that described above:

$$\{rM_1 + (1-r)M_2\}_{alloy} + \{aH_2O + bH^+ + cX^-\}_{aqueous}$$

$$\rightarrow m(M_1, M_2, H, O, X)^\alpha + n(M_1, M_2, H, O, X)^\beta$$
$$(8.17)$$

where the two product compounds, α and β, contain M_1 and M_2 *in different proportions* and may or may not contain H, O, and additional element(s) in the aqueous phase X and where m and n are nonnegative mole numbers for the value of r that is selected for the alloy. This may be called the mass balance constraint. To establish at each coordinate which pair of compounds is most stable, it is first necessary to find from among all possible compounds the pairs that satisfy the mass balance constraint. Thereafter, the Gibbs energy change for all of these reactions (based on 1 mol of alloy) is computed. Finally, the reaction with the most negative Gibbs energy change is found, thereby identifying the most stable pair of compounds at that particular E_H and pH. The resulting diagram has domains that are doubly labeled for the case of a two-component alloy. As a special case, when there are no product compounds that contain both M_1 and M_2, the resultant diagram is the superimposition of the two conventional Pourbaix diagrams for M_1 and M_2. This, unfortunately, is indeed a special case that is the exception rather than the rule.

Figures 8.7 and 8.8 show the conventional Pourbaix diagrams for Fe and Ni. These may be compared with Figures 8.9–8.12, which show the computed Fe–Ni composite diagram at four different levels of aqueous species concentration ranging from 1 to $10^{-6}m$. The complexity of the diagram is such that it is virtually essential to provide a series of diagrams for particular concentrations of aqueous species. The compound that gives rise to the complex appearance, namely, the $NiFe_2O_4$ phase, is boldly outlined. The inability to comprehend the thermodynamics of Fe–Ni alloy using

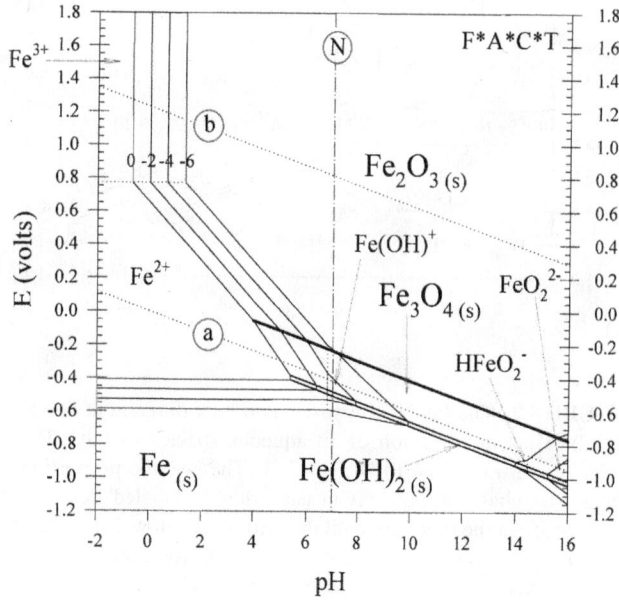

FIGURE 8.7. The Fe Pourbaix diagram at 298 K (25°C). Concentrations of aqueous species range from 1 to $10^{-6}m$.

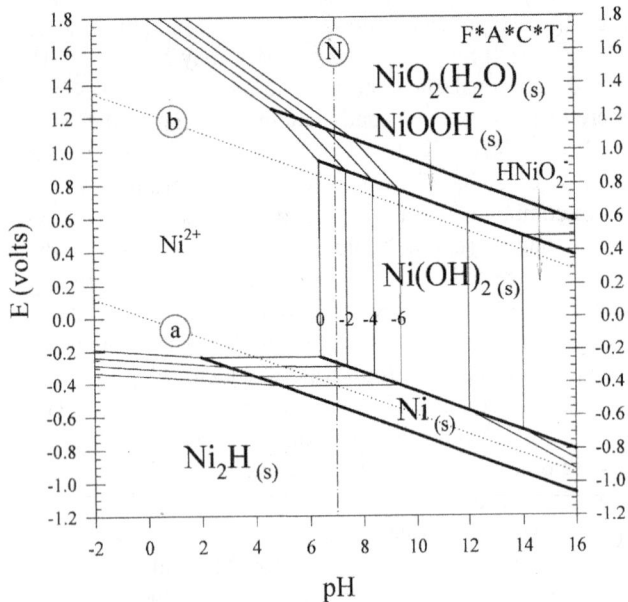

FIGURE 8.8. The Ni Pourbaix diagram at 298 K (25°C). Concentrations of aqueous species range from 1 to $10^{-6}m$.

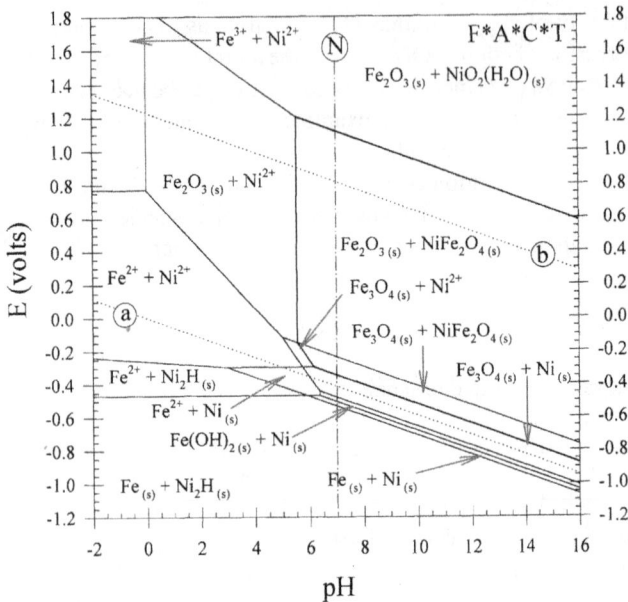

FIGURE 8.10. The Fe–Ni composite Pourbaix diagram at 298 K (25°C). The concentration of all aqueous species is $10^{-2}m$. The molar proportion of Fe to Ni is $>2:1$. Note the placement of $NiFe_2O_4$, solid spinel, which is outlined in bold.

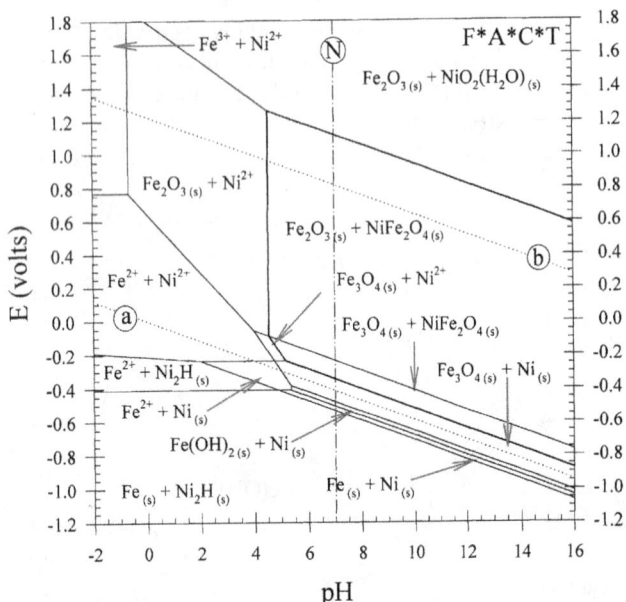

FIGURE 8.9. The Fe–Ni composite Pourbaix diagram at 298 K (25°C). The concentration of all aqueous species is 1 m. The molar proportion of Fe to Ni is $>2:1$. The specific proportion affects the phase proportions in each doubly labeled field but does not affect the topology until the ratio falls below 2:1 (when, e.g., $NiFe_2O_4$ could not coexist with Fe_2O_3 for mass balance reasons). Note the placement of $NiFe_2O_4$ solid spinel, which is outlined in bold.

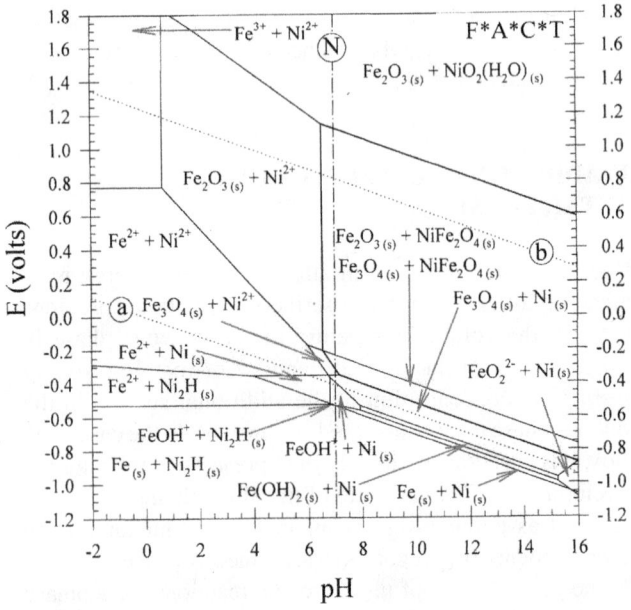

FIGURE 8.11. The Fe–Ni composite Pourbaix diagram at 298 K (25°C). The concentration of all aqueous species is $10^{-4}m$. The molar proportion of Fe to Ni is $>2:1$. Note the placement of $NiFe_2O_4$ solid spinel, which is outlined in bold.

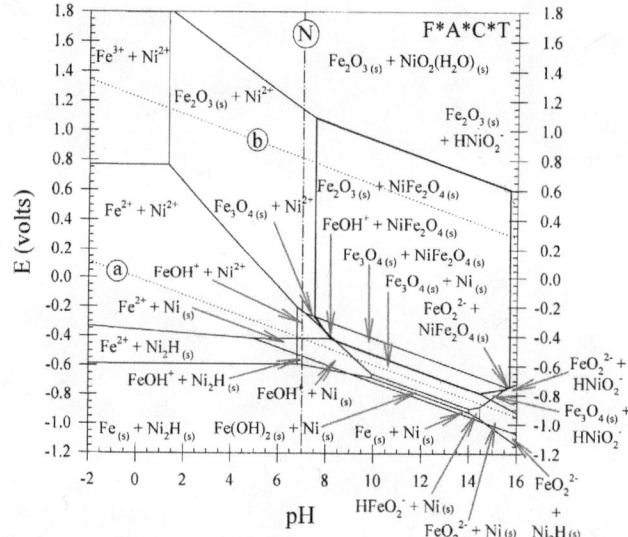

FIGURE 8.12. The Fe–Ni composite Pourbaix diagram at 298 K (25°C). The concentration of all aqueous species is $10^{-6}m$. The molar proportion of Fe to Ni is $>2:1$. Note the placement of $NiFe_2O_4$ solid spinel, which is outlined in bold.

as suggested in this chapter. The technique, therefore, is well suited to computational methods and is especially convenient to use when the programming is linked to a database of thermodynamic properties. In this regard, all of the figures shown were prepared using the Facility for the Analysis of Chemical Thermodynamics (F*A*C*T) [7], although certain enhancements were applied for the purposes of this chapter.

ACKNOWLEDGMENT

The authors thank Dr. V. F. Baston, Principal Engineer, GE Nuclear Energy, San Jose, CA, for helpful discussions and data in connection with the Fe–Ni system.

only the conventional Fe and Ni diagrams does not require further comment [3].

E. CONCLUSIONS

Pourbaix diagrams are invaluable tools in the understanding of the thermodynamics of aqueous corrosion. However, it is usually the case that these must be specially constructed to incorporate all of the elements, whether found in the alloy or aqueous phase, that can lead to the formation of significant phases or species. The methodology of Gibbs energy minimization is particularly effective for this purpose, in that no suppositions about coexistence of phases or species is necessary and it can be extended to handle many elements

REFERENCES

1. M. Pourbaix, Atlas of Electrochemical Equilibria in Aqueous Solutions, NACE, Houston, TX, 1974.

2. W. T. Thompson, C. W. Bale, and A, D. Pelton, Ernest Peters International Symposium on Hydrometallurgy Theory and Practice, Metallurgical Society of CIM, Vancouver, British Columbia, 1992.

3. M. Ullberg and M. Tanse Larsson, Forsmarks Kraftgrupp AB, Sweden, VIIth International Conference of Water Chemistry of Reactor Systems, Bournemouth, UK, Oct. 1996.

4.   A. D. Pelton, W. T. Thompson, C. W. Bale, and G. Eriksson, "F*A*C*T Thermochemical Databases for Calculations in Materials Chemistry at High Temperature," High Temp. Sci., **26**, 231 (1990).

5. C. L. Mantel, Electrochemical Engineering, McGraw Hill, New York, 1960.

6. H. H. Uhlig and R. W. Revie, Corrosion and Corrosion Control, 3rd ed, Wiley, New York, 1985.

7. C. W. Bale, A. D. Pelton, and W. T. Thompson, Facility for the Analysis of Chemical Thermodynamics—User Manual 2.1, Ecole Polytechnique de Montreal/McGill University, 1996.

9

COMPUTATION OF POURBAIX DIAGRAMS AT ELEVATED TEMPERATURE

M. H. KAYE

Faculty of Energy Systems and Nuclear Science, University of Ontario Institute of Technology, Oshawa, Ontario, Canada

W. T. THOMPSON

Center of Research in Computational Thermochemistry, Royal Military College of Canada, Kingston, Ontario, Canada

A. INTRODUCTION

As other chapters in this handbook indicate, redox potential–pH (or Pourbaix) diagrams contribute to a better understanding of aqueous corrosion phenomena, in particular, the depiction of the relative stability of the metallic elements in various possible phases or water-soluble species. However, to achieve this objective, it may be necessary to custom construct diagrams showing the influence of water impurities and alloying additions as discussed in Chapter 8. There is sometimes the effect of temperature to consider as well. This matter, with some attention given to the activity effect on the metallic elements caused by alloying, is the principal subject of this chapter.

The Zircaloy system provides a good basis to discuss the effect of temperature on Pourbaix diagrams. This alloy, important in nuclear energy technologies, is essentially a binary combination of Zr and Sn (0.25–2.5 wt %) [1] and is used principally to isolate the natural or isotopically enriched UO_2 fuel from direct contact with pressurized ($\sim$100 atm) heavy water in the primary heat transport system. The high hydrostatic pressure to suppress boiling creates a situation where an aqueous phase contacts the alloy at temperatures in the range 250–300°C, nearly the maximum possible (critical condition for water is 374°C at 218 atm [2]). Although heavy water, D_2O, actually contacts the Zircaloy in some designs (e.g., CANDU reactors), the very small chemical distinction between D_2O and H_2O does not detectably affect the construction of Pourbaix diagrams. Accordingly, the thermodynamic properties for H_2O may substitute for D_2O for computational purposes in the construction of the Sn and Zr diagrams.

B. THERMODYNAMIC DATA

The construction of Pourbaix diagrams at elevated temperatures follows the practice recommended in Chapter 8 except for the matter of handling the Gibbs energies for the various possible species. In Chapter 8, the standard Gibbs energies *of formation* were supplied as constants at 298 K. In the present case, a temperature-dependent formulation for the standard Gibbs energy of formation, $\Delta G°$, is required. This formulation is generally based on experimental or estimated data for the enthalpy of formation, $\Delta H°_{298\ K}$, the absolute entropy, $\Delta S°_{298\ K}$, and the heat capacity, c_p. The heat capacity may be a function of temperature and is typically represented by a series such as

$$c_p = a + bT + cT^2 + dT^{-2} + \cdots \qquad (9.1)$$

The formulation for the standard Gibbs energy (pure phase or aqueous species at unit activity, that is, approximately 1 m) at any temperature and pressure is

$$G^\circ_{T,P} = \Delta H^\circ_{298\,K} + \int c_p\, dT - T\left(S^\circ_{298\,K} + \int \frac{c_p}{T}\, dT\right) + \int V\, dP \qquad (9.2)$$

This standard Gibbs energy (no Δ) is sometimes called the "absolute" Gibbs energy [3] in reference to the use of absolute entropy in place of the entropy of formation. However, in truth, the Gibbs energy calculated in this way remains a relative term intended only as an initial step in finding Gibbs energy *differences* for isothermal, isobaric chemical reactions.

For example, for the process

$$Sn^{4+} + 2H_2O \rightarrow SnO_2 + 4H^+ \qquad (9.3)$$

the correct standard Gibbs energy *change* can be found by summing the products of the number of moles and "absolute" standard (molar) Gibbs energies for the species on the right and deducting from it a similar summation from the left. Conventions used in recording the properties in Table 9.1 result in the Gibbs energy of formation, ΔG°, for H^+ being zero at all temperatures [4]. The last term in Eq. (9.2) involving the small molar volume is typically (for nongaseous species) well within the uncertainty in the other terms when

the hydrostatic pressure does not greatly exceed 100 atm. Accordingly, the $V\, dP$ term may be dropped or equivalently the hydrostatic pressure may be ignored as a practically significant variable in Pourbaix diagram construction.

A large body of experimental data have been gathered for the Gibbs energy of formation of aqueous species at 298 K or its equivalent expressed in electrochemical terms (tables of standard electrode potential). Less experimental data are available for partial molar enthalpy of formation and partial molar heat capacity of aqueous species. Partial molar properties for solutes in water are to be understood as the disturbing effect on the heat capacity of an aqueous solution property by the small addition of those solutes, typically ions. Thus, the partial molar heat capacity for aqueous species may be positive or negative. Fortunately, the effect of heat capacity on ΔG° calculated using differences in "absolute" G° is very small, so a lack of heat capacity information for an aqueous species has no major consequence on the computation of ΔG° at the modest elevated temperatures associated with Pourbaix diagram construction. The entropy (that may also be positive or negative for an aqueous species) is the main factor conferring temperature dependence to ΔG°, or equivalently in assessing the contribution of ΔH° to ΔG°. It is sometimes necessary to estimate the standard entropy based on chemical similarities and trends with other aqueous species for which measurements have been reported. The choice of the estimated entropy and its influence on diagram construction are pursued below.

C. Sn POURBAIX DIAGRAM

Data for all species of Sn involving H and O are shown in Table 9.1 [5–8]. Data in italics were estimated in a way that is

TABLE 9.1. Data Used to Calculate Sn Pourbaix Diagram

Species	$\Delta H^\circ_{298\,K}$ (J/mol)	$S^\circ_{298\,K}$ (J/mol/K)	$c_p = a + bT + cT^2 + dT^{-2}$ (J/mol/K)				Additional Terms
			a	b	c	d	
$Sn_{(s)}$	0	51.195	21.5936	0.01810	0	0	—
$Sn_{(l)}$	6,263	62.893	21.5392	0.00615	0	1,288,254	—
$SnO_{(s)}$	−285,767	56.484	39.9572	0.01464	0	0	—
$SnO_{2(s)}$	−580,739	52.300	73.8894	0.01004	0	−2,158,944	—
$SnOH^+_{(aq)}$	−285,767	50.208	−94.0103	0.22864	0	7,137,904	—
$Sn^{4+}_{(aq)}$	−56,856	−410.000	−54.8690	0.09394	0	6,755,486	—
$Sn^{2+}_{(aq)}$	−10,042	−24.686	−54.8690	0.09394	0	6,755,486	—
$HSnO_{2(aq)}^-$	−523,834	10.000	0	0	0	0	—
$SnO_{2(aq)}^-$	−724,228	10.000	0	0	0	0	—
$H^+_{(aq)}$	0	0	0	0	0	0	—
$OH^-_{(aq)}$	−229,987	−10.878	506.377	−1.18134	0	−24,602,300	—
$H_2O_{(l)}$	−285,830	69.950	−203.1190	1.52070	−0.0032	3,848,758	$2.471 \times 10^{-6} T^3$
$H_{2(gas)}$	0	130.571	19.8256	0.00308	1.43×10^{-6}	−295,180	$194.861 T^{-\frac{1}{2}}$
$O_{2(gas)}$	0	205.038	26.9241	0.01698	-6.77×10^{-6}	229,329	$-79.162 T^{-\frac{1}{2}}$

Note: Data in italics were estimated.

intended to be representative of practices that have been applied to many other high-temperature Pourbaix diagram computations [9]. In the case of Sn^{4+}, the partial molar heat capacity expression was taken to be the same as Sn^{2+} [7]. For the other aqueous ions, where there was no basis for estimation, the partial molar heat capacity was set to zero. The entropy of Sn^{4+} was estimated based on chemical similarities with other $4+$ ions (e.g., U^{4+}). The enthalpy change for Sn^{4+} was then calculated to give a Gibbs energy difference in relation to Sn^{2+} so that the redox potential [$+0.151$ V standard hydrogen electrode (SHE) at 298 K] would be respected for

$$Sn^{4+} + 2e^- \rightarrow Sn^{2+} \tag{9.4}$$

For $SnOH^+$, SnO_3^{2-}, and $HSnO_2^-$ the partial molar heat capacity was set to zero, the entropies were estimated, and the enthalpies of formation were calculated to duplicate features of the Sn diagram in the Pourbaix Atlas [8] at 298 K. The hydrogen–oxygen species [4, 7, 10] are included at the bottom of the table to justify the conventional placement of the a and b lines associated with redox potential conferred by 1 atm partial pressure of H_2 or 1 atm partial pressure of O_2, respectively.

The data in Table 9.1 generate the "absolute" Gibbs energies in Table 9.2 at 298 K, 373 K, and 550 K. As support for the relative significance of hydrostatic pressure, the Gibbs energy for water is given at both 1 and 100 atm; the difference is 0.2 kJ/mol. The heat capacity is also given.

A plot of $G°$ versus T for Sn^{4+} in Figure 9.1 gives evidence of the small effect of c_p in causing departure from linear variation for aqueous species. This is particularly so when the entropy is large.

C1. Electron in Aqueous Thermodynamic Computations

The use of the electron symbol, e^-, in aqueous thermodynamics is in reference to electrons understood to be exchanged with a SHE for which the potential is defined to be zero *at all temperatures* [4] when hydrogen gas at 1 atm partial pressure coexists with hydrogen ions at pH zero. Thus, the electrochemical reaction

$$Sn^{2+} + 2e^- \rightarrow Sn \tag{9.5}$$

actually refers to the reaction

$$Sn^{2+} + H_2 \rightarrow Sn + 2H^+ \tag{9.6}$$

and, in particular, the Gibbs energy change for both processes are the same.

Equivalently, one can say that the standard Gibbs energy change for the following process [understood to be combined with Eq. (9.5) to give Eq. (9.6)]

$$H^+ + e^- \rightarrow \tfrac{1}{2}H_2 \tag{9.7}$$

is zero at all temperatures. This change may be expressed as a difference in "absolute" Gibbs energies

$$\Delta G° = \tfrac{1}{2}G°_{H_2} - G°_{H^+} - G°_{e^-} = 0 \tag{9.8}$$

But, since $G°_{H^+}$ is also defined to be zero at all temperatures (Table 9.2), then

TABLE 9.2. Absolute Gibbs Energies and Heat Capacities at Selected Temperatures

Species	$G°_{298\,K}$ (J/mol)	$G°_{373\,K}$ (J/mol)	$G°_{550\,K}$ (J/mol)	$c_{p(298\,K)}$ (J/mol/K)	$c_{p(373\,K)}$ (J/mol/K)	$c_{p(550\,K)}$ (J/mol/K)
$Sn_{(s)}$	$-152,634$	$-19,334$	$-30,565$	27.0	28.3	30.7
$Sn_{(l)}$	$-12,479$	$-17,509$	$-31,188$	37.9	33.1	29.2
$SnO_{(s)}$	$-302,608$	$-307,224$	$-320,691$	44.3	45.4	48.0
$SnO_{2(s)}$	$-596,332$	$-600,735$	$-614,697$	52.6	62.1	72.3
$SnOH^+_{(aq)}$	$-300,737$	$-304,921$	$-317,279$	54.5	42.6	55.3
$Sn^{4+}_{(aq)}$	$65,386$	$96,244$	$171,901$	-6.9	-41.4	-84.2
$Sn^{2+}_{(aq)}$	$2,682$	-664	$6,792$	-6.9	-41.4	-84.2
$HSnO_2^-{}_{(aq)}$	$-526,815$	$-527,564$	$-529,334$	0.0	0.0	0.0
$SnO_3^{2-}{}_{(aq)}$	$-727,209$	$-727,958$	$-729,728$	0.0	0.0	0.0
$H^+_{(aq)}$	0	0	0	0.0	0.0	0.0
$OH^-_{(aq)}$	$-226,744$	$-224,945$	$-213,585$	-122.6	-111.1	-224.7
$H_2O_{(l,1\,atm)}$	$-306,686$	$-312,577$	$-330,816$	75.4	76.0	83.7
$H_2O_{(l,100\,atm)}$	$-306,505$	$-312,396$	$-330,635$	75.4	76.0	83.7
$H_{2\,(gas)}$	$-38,930$	$-48,955$	$-74,284$	28.8	29.1	29.3
$O_{2\,(gas)}$	$-61,132$	$-76,736$	$-115,312$	29.4	29.9	31.6

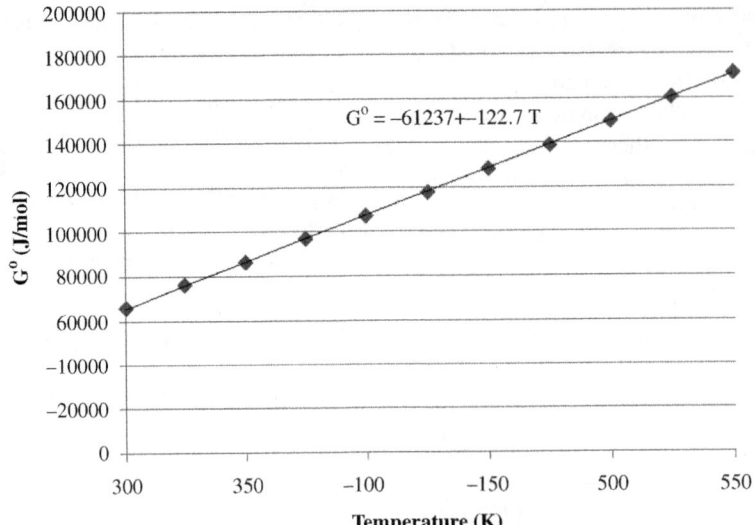

FIGURE 9.1. Plot of absolute $G°$ [Eq. (9.2)] as a function of temperature for Sn^{4+} showing the small effect of c_p on departure from otherwise linear behavior.

$$G°_{e^-} = \left\{ \tfrac{1}{2} G°_{H_2} \right\} \tag{9.9}$$

Thus, for example, in computing the standard reduction potential for

$$Sn^{2+} + 2e^- \rightarrow Sn \tag{9.10}$$

from thermodynamic data, such as given in Table 1, the standard Gibbs energy difference is first calculated using

$$\Delta G° = G°_{Sn} - G°_{Sn^{2+}} - 2 \left\{ \tfrac{1}{2} G°_{H_2} \right\} \tag{9.11}$$

where the term in braces accounts for each electron in Eq. (9.10). The standard electrode potential $E°_h$ (-0.137 V SHE) is then found using

$$\Delta G° = -2F\left(E°_h\right) \tag{9.12}$$

where F is the faraday (96,485 C/mol).

C2. Development of the Pourbaix Diagram

The construction of the Pourbaix diagram from the absolute Gibbs energies in Table 9.2 is as described in Chapter 8 in this handbook except that in this case the Gibbs energy *changes* for the reactions (calculated using "absolute" Gibbs energies) must take into account the formalities in dealing with the electron as described above. The computed Pourbaix diagram for tin at 298 K is shown in Figure 9.2 for the conventional concentration for aqueous ions of 10^{-6} molar deemed to be significant in most corrosion applications [8].

As an affirmation of the methodologies involved in this construction, consider the placement of triple points, 1 and 2, so labeled in Figure 9.2. Table 9.3 provides the Gibbs energy changes for the reactions (9.13a)–(9.17a). These involve Sn^{2+}, Sn^{4+}, SnO_2, $SnOH^+$, and SnO. Details for determining the Gibbs energy change at 298 K and $10^{-6}m$ for aqueous Sn ions are shown below each reaction (9.13b)–(9.17b); note the inclusion of the term in braces for the "absolute" Gibbs energy of e^- and the addition of $-F(E°_h)$ for each e^- when the redox potential is not zero [as described in Eq. (8.8) in Chapter 8]:

$$Sn \rightarrow Sn^{2+} + 2e^- \tag{9.13a}$$

$$\Delta G = \left[G°_{Sn^{2+}} + RT \ln(10^{-6}) \right] + 2\left[\left\{ \tfrac{1}{2} G°_{H_2} \right\} - F(E_h) \right] - G°_{Sn} \tag{9.13b}$$

$$Sn \rightarrow Sn^{4+} + 4e^- \tag{9.14a}$$

$$\Delta G = \left[G°_{Sn^{4+}} + RT \ln(10^{-6}) \right] + 4\left[\left\{ \tfrac{1}{2} G°_{H_2} \right\} - F(E_h) \right] - G°_{Sn} \tag{9.14b}$$

$$Sn + 2H_2O \rightarrow SnO_2 + 4H^+ + 4e^- \tag{9.15a}$$

$$\Delta G = G°_{SnO_2} - 4(2.303)RT\, pH + 4\left[\left\{ \tfrac{1}{2} G°_{H_2} \right\} - F(E_h) \right] - G°_{Sn} - 2G°_{H_2O} \tag{9.15b}$$

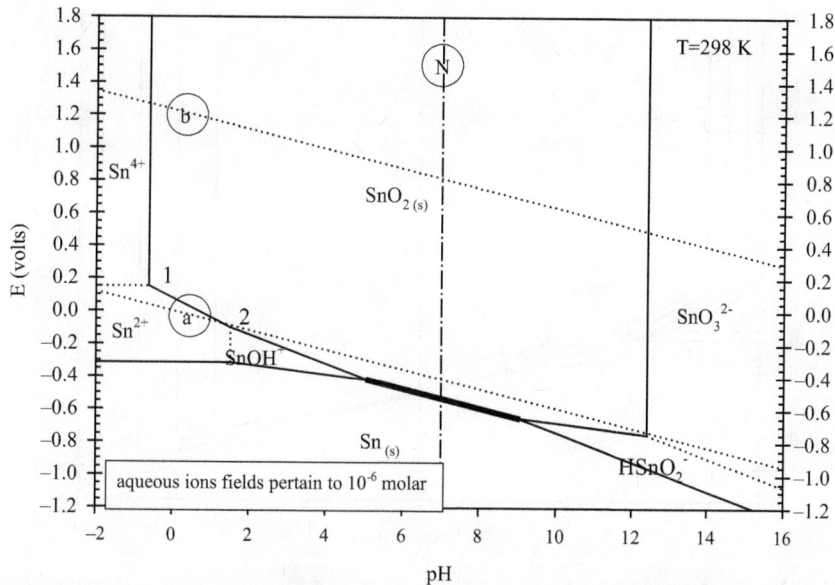

FIGURE 9.2. Pourbaix diagram for Sn at 298 K. The heaviest line represents the phase boundary between coexisting solid phases. The lighter solid lines represent conditions of a condensed phase coexisting with an aqueous solution of the adjacent ion at a concentration of $10^{-6}m$. The broken lines represent conditions where the concentrations of ions flanking those lines are equal. The downward-sloping dotted lines [(a) and (b)] correspond to H_2 and O_2 saturation at 1 atm partial pressure. The pH of neutrality is indicated with a vertical line labeled (N).

$$Sn + H_2O \rightarrow SnOH^+ + H^+ + 2e^- \qquad (9.16a)$$

$$\Delta G = \left[G^\circ_{SnOH^+} + RT \ln(10^{-6}) \right] - 2.303RT \, pH \\ + 2\left[\left\{ \tfrac{1}{2} G^\circ_{H_2} \right\} - F(E_h) \right] - G^\circ_{Sn} - G^\circ_{H_2O} \qquad (9.16b)$$

$$Sn + H_2O \rightarrow SnO + 2H^+ + 2e^- \qquad (9.17a)$$

$$\Delta G = G^\circ_{SnO} - 2(2.303)RT \, pH \\ + 2\left[\left\{ \tfrac{1}{2} G^\circ_{H_2} \right\} - F(E_h) \right] - G^\circ_{Sn} - G^\circ_{H_2O} \qquad (9.17b)$$

An examination of Table 9.3 shows that Sn^{2+}, Sn^{4+}, and SnO_2 are equally stable at triple point 1 (same ΔG) and that no other species shown has a more negative ΔG. Similarly, Sn^{2+}, $SnOH^+$, and SnO_2 coexist at triple point 2. SnO is

included in the Table 9.3 to give evidence that SnO_2 is the more stable oxide of tin at both triple points (i.e., ΔG for SnO_2 is lower than SnO).

The size of the fields for the aqueous species of tin is sensitive to the concentration to which they refer. Figure 9.3 shows the complete development of the Pourbaix diagram for the conventional series of concentrations (taken to be the same as activity following Pourbaix [8]): 1, 10^{-2}, 10^{-4}, and $10^{-6}m$.

Using the same methodology, the diagram for tin at elevated temperatures (550 K) may be constructed as shown in Figure 9.3. Notable points of comparison between the diagram at 298 and 550 K are the following:

- Lines a and b, corresponding to redox potentials resultant from saturation with H_2 and O_2, respectively, at 1 atm, are steeper at higher temperature (proportional to absolute temperature); the lines also move closer together.

TABLE 9.3. Gibbs Energy Change for Reactions Evaluated at Location of Triple Points 1 and 2 in Figure 9.2

Reaction	Point 1 ($E_h = 0.151$ V SHE, pH = -0.6175)	Point 2 ($E_h = -0.101$ V SHE, pH = 1.512)
13 (Sn^{2+})	$-89{,}732$	$-41{,}103$
14 (Sn^{4+})	$-89{,}732$	$7{,}524$
15 (SnO_2)	$-89{,}732$	$-41{,}103$
16 ($SnOH^+$)	$-77{,}576$	$-41{,}103$
17 (SnO)	$-41{,}676$	$-17{,}361$

TABLE 9.4. Data Used for SnO_3^{2-} in Construction of Figure 9.5

	Species	$\Delta H^\circ_{298\,K}$	$\Delta S^\circ_{298\,K}$	$\Delta G^\circ_{298\,K}$	$\Delta G^\circ_{550\,K}$
A	SnO_3^{2-}	$-697{,}394$	100.000	$-727{,}209$	$-752{,}394$
B	SnO_3^{2-}	$-724{,}228$	10.000	$-727{,}209$	$-729{,}728$
C	SnO_3^{2-}	$-757{,}024$	-100.000	$-727{,}209$	$-702{,}024$

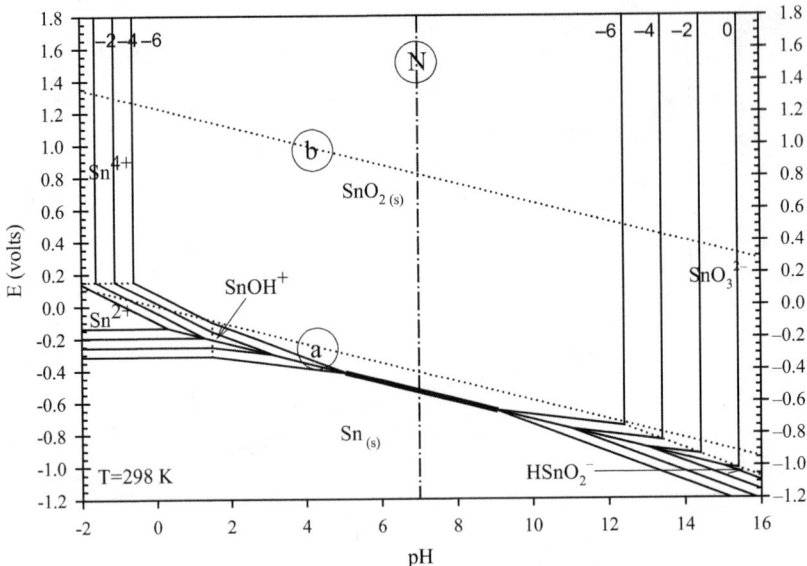

FIGURE 9.3. Pourbaix diagram for Sn at 298 K showing aqueous concentrations (activity) for the Sn species at 1, 10^{-2}, 10^{-4}, and $10^{-6}m$.

- The fields on the diagram for the aqueous species tend to shift to the left at higher temperature.
- The solubility of SnO_2 tends to increase with temperature in alkaline solution.
- Pure metallic tin is more stable as a liquid at 550 K.
- The pH associated with neutrality shifts to lower values with increasing temperature, as the equilibrium constant for the dissociation of H_2O increases with increasing temperature.

With regard to the last point, for

$$H_2O \rightarrow H^+ + OH^- \qquad (9.18)$$

$\Delta G° = +117{,}050\,J/mol$ at 550 K as calculated from the data shown in Table 9.2. This corresponds to an equilibrium constant of 7.36×10^{-12} (instead of very nearly 10^{-14} at 298 K), implying that neutrality at 550 K occurs at a pH of 5.56 instead of 7. It is apparent that care must be taken in reference

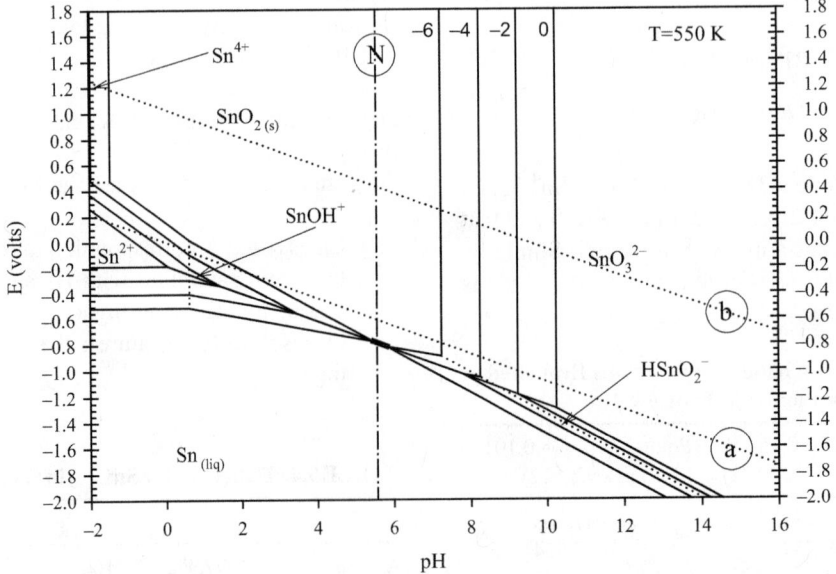

FIGURE 9.4. Pourbaix diagram for Sn at 550 K. The pressure is understood to be sufficient to suppress boiling but otherwise has a negligible effect on the placement of domain boundaries. Note that pure Sn is more stable as a liquid at this temperature.

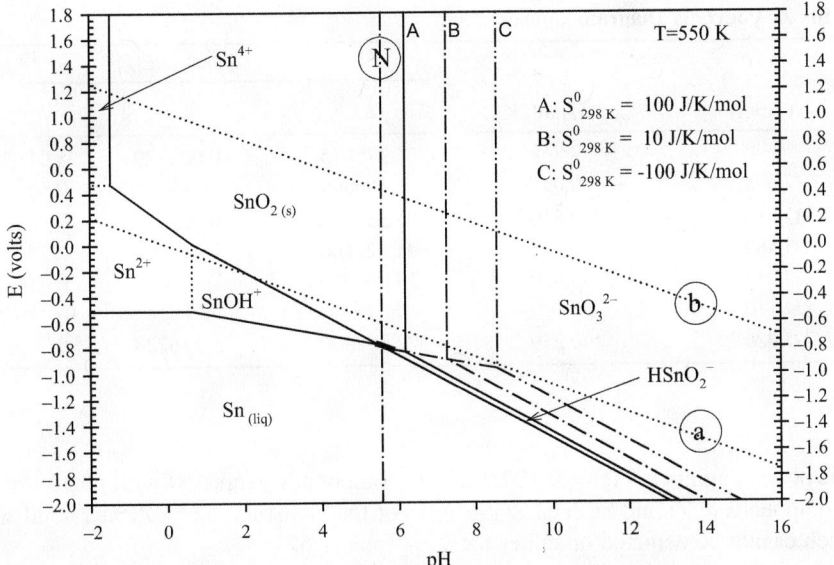

FIGURE 9.5. Pourbaix diagram for Sn at 550 K showing the effect of varying S° for SnO_3^{2-}. Pure metallic tin is more stable as a liquid at this temperature. The concentration of all aqueous species is $10^{-6}m$.

to discussions concerning pH when the temperature is substantially different from 298 K. In particular, it is important to distinguish the temperature of the water when sampled from the temperature of that sample when the pH is measured.

C3. Effect of Estimated Gibbs Energies on Diagram Development

As implied above, it is generally necessary to estimate the thermodynamic properties for at least some of the aqueous species at elevated temperatures.

Table 9.4 shows the effect on diagram construction resulting from different ways of extrapolating data for SnO_3^{2-} from 298 to 550 K. In this case, $S_{298\,K}^\circ$ was arbitrarily changed to a substantial degree from case B (used in the development of Fig. 9.4) to cases A and C, which represent a large variation. The values of $\Delta H_{298\,K}^\circ$ were correspondingly changed to preserve the desired value of $\Delta G_{298\,K}^\circ$. With the consequent different ways of projecting the Gibbs energy to temperatures above 298 K, the Pourbaix diagram was redeveloped as shown in Fig. 9.5 at 550 K, but just for the case, for purposes of clarity, where the concentration of the aqueous ions was $10^{-6}m$. The area of the SnO_3^{2-} field is reduced in case C as the Gibbs energy estimate becomes less negative as a consequence of the $-TS_{298\,K}^\circ$ term in Eq. (9.2). However, in spite of this rather extreme variation in the estimated $-G_{550\,K}^\circ$, the topology of the diagram is not affected. Therefore, in the absence of experimental data, if there is any basis to make a reasonable judgment of the way in which the Gibbs energy may project from 298 K to higher temperatures, it is still worthwhile to undertake Pourbaix

diagram development to assist in the understanding of high-temperature aqueous corrosion behavior.

D. Zr POURBAIX DIAGRAMS

The Pourbaix diagrams for Zr, Figures 9.6 and 9.7, were developed using the thermodynamic data [7, 8, 10, 11] in Table 9.5 and following the same methodology as described above for Sn. Data in italics were estimated in a way that is intended to be representative of practices that have been applied to many other high-temperature Pourbaix diagram computations [9].

Zirconium hydride is an extremely stable metal hydride. As a consequence, this field eclipses the field for metallic zirconium. Nevertheless, following the practice of Pourbaix, it is customary to show the metal field in relation to the other compounds of zirconium. These diagrams are shown in Figures 9.8 and 9.9 and were obtained by simply withdrawing the data for ZrH_2 during the computational process. There is the possibility that ZrO_2 may form a monohydrate instead of the dihydrate at 550 K, but the monohydrate was not considered as a possible species effecting the construction at 550 K for lack of reliable data.

E. Zr–Sn PHASE DIAGRAM AND INTERMETALLIC COMPOUNDS

In using the calculated high-temperature Pourbaix diagrams for Zr and Sn in relation to the corrosion of Zircaloy, it is

TABLE 9.5. Data Used for Zr Pourbaix Diagram Construction

Species	$\Delta H^\circ_{298\,K}$ (J/mol)	$\Delta S^\circ_{298\,K}$ (J/mol/K)	$c_p = a + bT + cT^2 + dT^{-2}$ (J/mol/K)			
			a	b	c	d
$\alpha Zr_{(s)}$	0	38.869	25.275675	0.000229	5.64×10^{-6}	−56,334
$Zr^{4+}_{(aq)}$	−647,898	−410.000	95.700000	−0.620000	0	2,842,000
$ZrO^{2+}_{(aq)}$	−933,450	−300.830	155.331000	0.023652	0	−4,678,967
$ZrO_{2\,(s)}$	−1,097,463	50.359	94.621160[a]	0	0	0
$ZrH_{2\,(s)}$	−169,034	35.020	30.961600	0	0	0
$HZrO^-_{3(aq)}$	−1,334,110	30.000	0	0	0	0
$ZrO_2 \cdot 2H_2O_{(s)}$	−1,701,426	190.259	172.102600	0.116728	0	0

Note: Data in italic were estimated.
[a] $-584.476T^{-1/2} - 120410834T^{-3}$.

necessary to consider the phase diagram in Figure 9.10 [12]. This calls attention to compounds of Zr and Sn (i.e., Zr_4Sn, Zr_5Sn_3, and $ZrSn_2$) which cannot be depicted on either the Sn or Zr Pourbaix diagram. For zirconium–tin alloys involving low concentrations of tin, nearly pure α-Zr coexists with Zr_4Sn at temperatures below 550 K. Therefore, the matter of significance is mainly the activity of tin. The phase diagram development shown in Figure 9.10 fortunately was undertaken in such a way [12, 13] as to provide the data for these compounds as given in Table 9.6. It is therefore possible to calculate the activity of Sn from the equilibrium constant for:

$$4Zr + Sn = Zr_4Sn \qquad (9.19)$$

at any temperature. The activity of Sn with respect to both solid white tin and liquid at 298 K and 550 K is given in Table 9.7. The Gibbs energies of formation of the Zr–Sn

compounds expressed with respect to the most stable form of the elements (alpha Zr and solid white Sn) are given in Table 9.6.

To account for the chemical behaviour of Sn in Zircaloy, the activities for Sn in Table 9.7 (referred to either solid or liquid) can be applied to the Gibbs energies for pure Sn at the top of Table 9.2 by the additions shown in the equation.

$$G_{Sn} = G^\circ_{Sn,solid} + RT \ln a_{Sn,solid} = G^\circ_{Sn,liquid} + RT \ln a_{Sn,liquid}$$
$$(9.20)$$

Since the effect of the very low activities of Sn in Zircaloy is to substantially lower the partial molar Gibbs energy for Sn, the effect on the diagram construction is to greatly enlarge the field for Sn. This field must now be

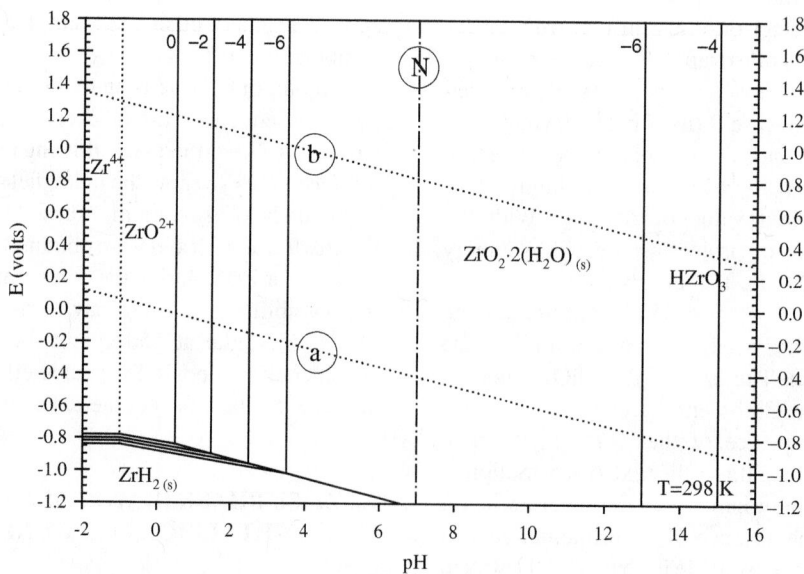

FIGURE 9.6. Pourbaix diagram for Zr at 298 K showing aqueous concentrations for the Zr species at 1, 10^{-2}, 10^{-4}, and $10^{-6}m$.

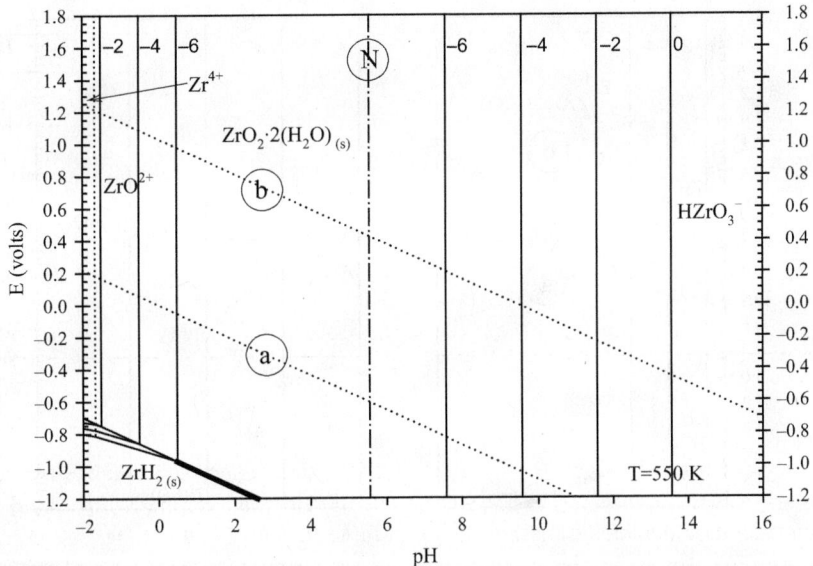

FIGURE 9.7. Pourbaix diagram for Zr at 550 K showing aqueous concentrations for the Zr species at 1, 10^{-2}, 10^{-4}, and $10^{-6}m$. The pressure is understood to be sufficient to suppress boiling but otherwise has a negligible effect on the placement of domain boundaries.

understood to refer to "Sn" that is chemically combined with Zr in Zr_4Sn coexisting with nearly pure alpha Zr. The Pourbaix diagrams are shown in Figures 9.11 and 9.12 only for concentrations of aqueous ions of $10^{-6}m$. The notable feature is that, in nearly hydrogen saturated conditions characteristic of the use of Zircaloy as nuclear fuel sheathing, Sn is only very sparingly soluble in water typically made slightly alkaline with LiOH additions (pH ~ 10.5 at room temperature).

F. CONCLUSION

Since the implications of thermodynamics tend to be better respected as temperature increases, Pourbaix diagrams may have greater value in corrosion interpretation at elevated temperature than at the ambient conditions where they are much more often used. This chapter has focused on computational matters leading to generalizations in diagram appearance associated with increasing temperature as well as

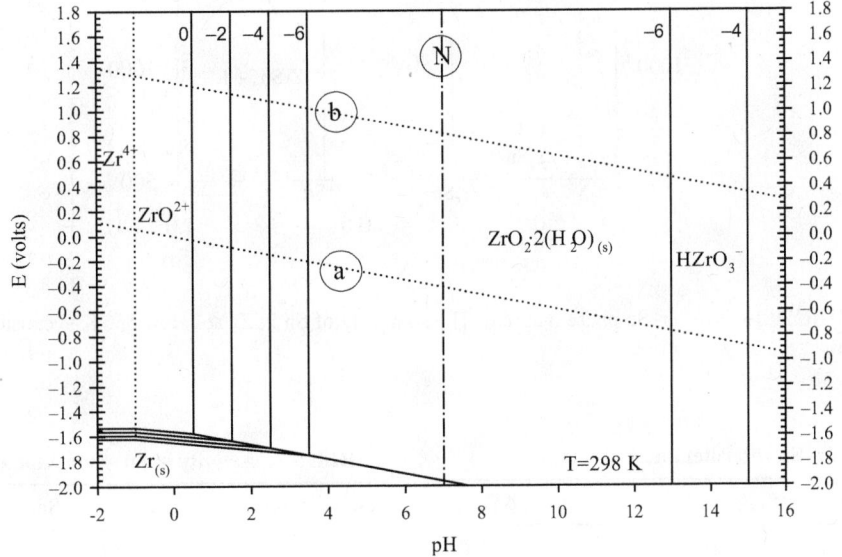

FIGURE 9.8. Pourbaix diagram for Zr at 298 K when ZrH_2 is withdrawn from the calculation. Aqueous concentrations for the Zr species are at 1, 10^{-2}, 10^{-4}, and $10^{-6}m$.

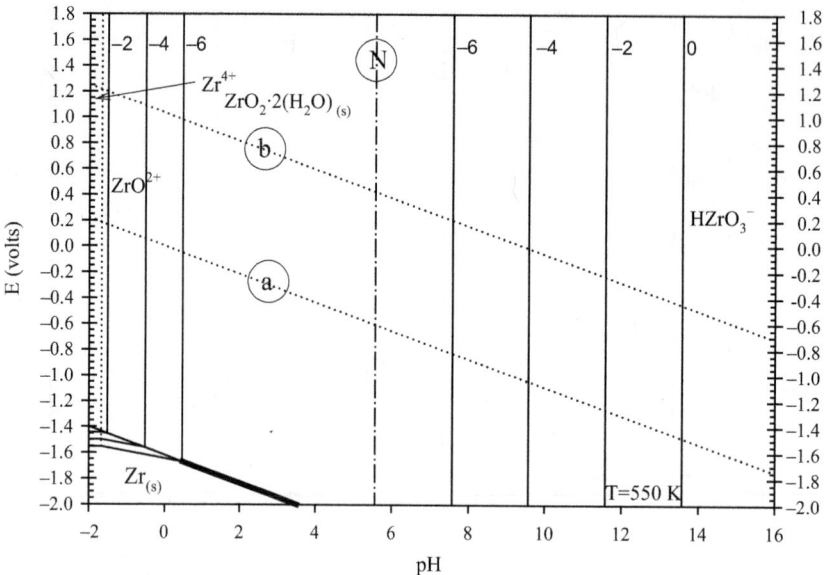

FIGURE 9.9. Pourbaix diagram for Zr at 550 K when ZrH_2 is withdrawn from the calculation. Aqueous concentrations for the Zr species at 1, 10^{-2}, 10^{-4}, and $10^{-6}m$. The pressure is understood to be sufficient to suppress boiling but otherwise has a negligible effect on the placement of domain boundaries.

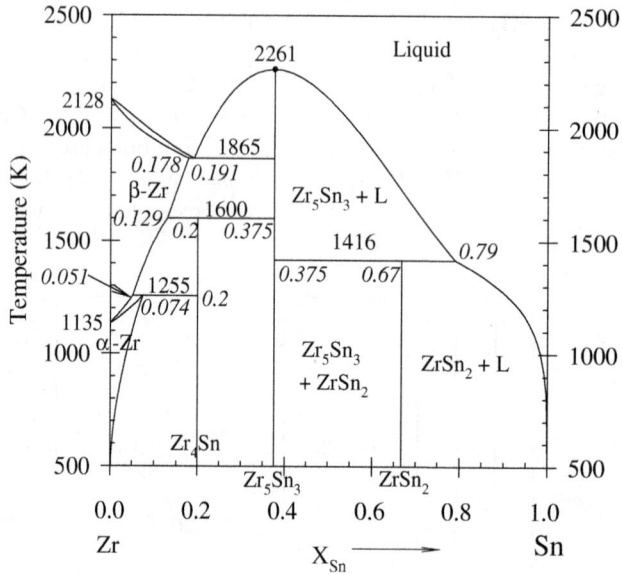

FIGURE 9.10. The Zr–Sn phase diagram. The solubility of Sn in Zr at below 550 K is considered negligible.

TABLE 9.6. Data for Zr–Sn Intermetallic Compounds

Species	$\Delta H^\circ_{298\ K}$	$\Delta S^\circ_{298\ K}$
Zr_4Sn	$-133,820$	202.79
Zr_5Sn_3	$-348,085$	346.67
$ZrSn_2$	$-71,924$	152.38

TABLE 9.7. Activity of Sn when Alpha Zr Coexists with Zr_4Sn

Temperature	Activity w.r.t. $Sn_{(s)}$	Activity w.r.t. $Sn_{(l)}$
298 K	5.7×10^{-24}	1.9×10^{-24}
550 K	2.8×10^{-13}	3.2×10^{-13}

Note: w.r.t = With respect to.

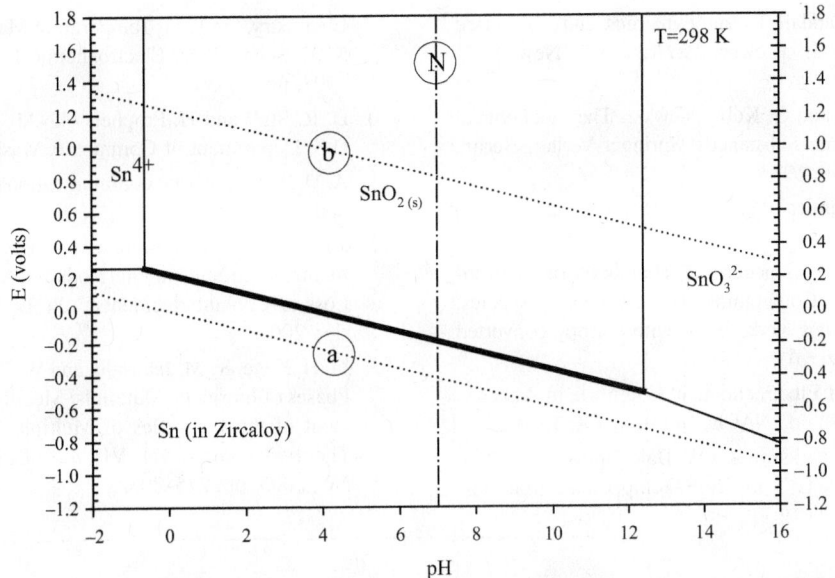

FIGURE 9.11. Sn (reduced activity) at 298 K. The concentration associated with the aqueous fields is $10^{-6}m$.

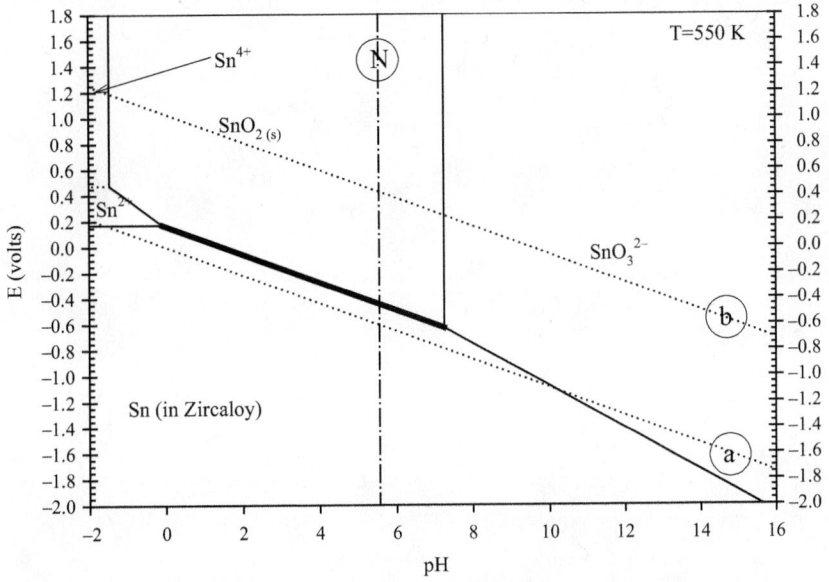

FIGURE 9.12. Sn (reduced activity) at 550 K. The concentration associated with the aqueous fields is $10^{-6}m$.

the sensitivity of field placement to judgments about the way to project Gibbs energies above 298 K when experimental data may be absent. It also shows the effect of alloying and a way by which this factor may be introduced into custom-constructed Pourbaix diagrams to depict the behaviour of a solute element.

ACKNOWLEDGMENTS

The authors thank Markus H. Piro for helpful discussions and recommendations.

REFERENCES

1. B. Cox, "Zirconium and Zirconium Alloys," in Uhlig's Corrosion Handbook, 3rd ed. R. W. Revie (Ed.), The Electrochemical Society, Pennington, NJ, 2011.
2. D.R. Lide (Ed.), CRC Handbook of Chemistry and Physics, CRC Press, Boca Raton, LA, 2000.
3. G. Eriksson and W. T. Thompson, "A Procedure to Estimate Equilibrium Concentrations in Multicomponent Systems and Related Applications," CALPHAD, **13**(4), 389–400 (1989).

4. G. N. Lewis and M. Randall, Thermodynamics, 2nd ed., revised by K. S. Pitzer and L. Brewer, McGraw-Hill, New York, 1961.

5. I. Barin, O. Knacke, and O. Kubaschewski, Thermochemical Properties of Inorganic Substances, Springer-Verlag, Berlin, 1973, and Supplement, 1977.

6. I. Barin, Thermochemical Data of Pure Substances, VCH, Weinheim, 1989.

7. H. E. Barner and R. V. Scheuerman, Handbook of Thermochemical Data for Compounds and Aqueous Species, Wiley-Interscience, New York, 1978 (with entropy converted to conventional H^+ zero).

8. M. Pourbaix, Atlas of Electrochemical Equilibria in Aqueous Solutions, 2nd English ed., NACE, Houston, TX, 1974.

9. W. T. Thompson, G. Eriksson, C. W. Bale, and A.D. Pelton, "Applications of F*A*C*T in High Temperature Materials Chemistry," in High Temperature Materials Chemistry, Vol 9, K. E. Spear (Ed.), Electrochemical Society, Pennington, NJ, 1997, pp. 16–30.

10. D. R. Stull and H. Prophet, JANAF Thermochemical Tables, U.S. Department of Commerce, Washington, DC, 1985.

11. A. D. Pelton, private communication (based on phase diagram optimzations).

12. M. H. Kaye, "A Thermodynamic Model for Noble Metal Alloy Inclusions in Nuclear Fuel Rods and Application to the Study of Loss-of-Coolant Accidents," Ph. D. Thesis, Queen's University, 2001.

13. M. H. Kaye, K. M. Jaansalu, and W. T. Thompson, "Condensed Phases of Inorganic Materials: Metallic Systems," in Measurement of the Properties of Multiple Phases — Experimental Thermodynamics, Vol. VII, IUPAC, Research Triangle Park, NC, 2005, pp. 275–305.

10

GALVANIC CORROSION

X. G. ZHANG

Teck Metals Ltd., Mississauga, Ontario, Canada

A. INTRODUCTION

Galvanic corrosion, resulting from a metal contacting another conducting material in a corrosive medium, is one of the most common types of corrosion. It may be found at the junction of a water main, where a copper pipe meets a steel pipe, or in a microelectronic device, where different metals and semiconductors are placed together, or in a metal matrix composite material in which reinforcing materials, such as graphite, are dispersed in a metal, or on a ship, where the various components immersed in water are made of different metal alloys. In many cases, galvanic corrosion may result in quick deterioration of the metals but, in other cases, the galvanic corrosion of one metal may result in the corrosion protection of an attached metal, which is the basis of cathodic protection by sacrificial anodes.

Galvanic corrosion is an extensively investigated subject, as shown in Table 10.1, and is qualitatively well understood but, due to its highly complex nature, it has been difficult to deal with in a quantitative way until recently. The widespread use of computers and the development of software have made great advances in understanding and predicting galvanic corrosion.

B. DEFINITION

When two dissimilar conducting materials in electrical contact with each other are exposed to an electrolyte, a current, called the galvanic current, flows from one to the other. Galvanic corrosion is that part of the corrosion that occurs at the anodic member of such a couple and is directly related to the galvanic current by Faraday's law.

Under a coupling condition, the simultaneous additional corrosion taking place on the anode of the couple is called the local corrosion. The local corrosion may or may not equal the corrosion, called the normal corrosion, taking place when the two metals are not electrically connected. The difference between the local and the normal corrosion is called the difference effect, which may be positive or negative. A galvanic current generally causes a reduction in the total corrosion rate of the cathodic member of the couple. In this case, the cathodic member is cathodically protected.

Uhlig's Corrosion Handbook, Third Edition, Edited by R. Winston Revie
Copyright © 2011 John Wiley & Sons, Inc.

TABLE 10.1. Studies on Galvanic Actions of Miscellaneous Alloys In Various Environments

Alloy 1	Alloy 2	Measurements[a]	Focus	References
		Atmosphere		
Steel, S. steel	Al	Weight loss	Automotive parts	1
Al, Cu, Pb, Sn, Mg, Ni, Zn, steel, S. steel	Miscellaneous	Weight loss	Corrosion rate	2
Pt	Zn	I_g	Humidity sensor	3
Al, Cu alloys, Ni, Pb, Zn, steel, S. steels	Miscellaneous	Weight loss	Tropical data	4
Cu	Steel	I_g, weight loss	Corrosion probe	5
Al alloys	S. steel	E_{corr}, I_g	Inhibitors	6
Al, Cu alloys, Ni, Pb, S. steels	Miscellaneous	Weight loss	Clad metals	7
		Fresh Water (pure, river, lake, and underground)		
Co alloys	Carbon	$E-I$ curves	Magnetic disc	8
Cu	Ag	E_{corr}^d	Electrical contact	9
Steel	Zn	E_g, I_g	Polarity reversal	10, 11
Steel	Zn	E_g, I_g	Polarity reversal	12, 13
Al, Cu alloys, Ni, Pb, Zn, steel, S. steels	Miscellaneous	Weight loss	Damage data	14
Al	Steel	E_{corr}, E_g	Polarity reversal	15
		Seawater		
S. steel	Steel	Thickness loss	Metallic joints	16
S. steel, Ti	Brass, bronze	I_g	Corrosion rate	17
S. steel	Cu	$E-I$ curves, E_{corr}	Localized corrosion	18
Steel	Zn	E_g	Transient $E-t$	19
Cu alloys	Cu alloys	Weight loss	Effect of sulfide	20
Miscellaneous	Miscellaneous	E_{corr}, $E-I$ curves	Review	21
Al, Cu alloys, Ni, Pb, Zn, steel, S. steels	Miscellaneous	Weight loss	Damage data	14
Steels, S. steels, Cu alloys	Ti alloys, S. steel	I_g, E_c,	Power plant condenser	22
Cu–Ni, Ti	Bronze, Zn	E distribution	Cathodic protection	23
Al alloys	S. steel	E_{corr}, I_g	Inhibitors	6
S. steel	Ni alloys, graphite	Miscellaneous	Review	24
Bronze, Ti,	Cu–Ni, Zn, bronze	$E-I$ curves	Time effect	25
Cu alloy, Fe, Zn, S. steel	Miscellaneous	E_g, I_g, E_{corr}	Materials interaction	26
		Soils		
Brass, steel	Brass, Pb, Cu, Zn	Weight loss	Corr. and protection	27
S. steel	Zn	I_g	Soil resistance	28
Steel, Zn, Pb	Pb, Cu, steel	IR drop	$E_c - E_a$	29
		Acids		
S. steels, Ti	Ni alloys	I_g, E_g, $E-I$, weight loss	Prediction	30
Ni alloys	Ni alloys	Thickness loss	Welds	31
Fe–Cr–Ni alloys	Austenite/ferrite	Weight loss, morphology	Phase interaction	32
Fe–Ni, Ti, S. steel, Ni alloys	Graphite, Ni alloys	I_g, E_g, $E-I$ curves	Polarization effects	33
Fe–Cr alloys	Fe–Cr alloys	Weight loss	Phase interaction	34
Ag, Au, Al, Ti, Pt, Fe, Cu, Zn	Minerals	E_{corr}, $E-I$ curves	Processing equipment	35
		Salt Solutions		
Al, Cd films	4340 steel	I_g, E_g, stress	Hydrogen embritt.	36
SiC	Mg alloys	I_g, i_{corr}	Composite material	37
Steel	Al, Zn	Weight loss, $E-I$ curves	Mechanism	38
Steel	Al	E_g, I_g, weight loss	pH, dissov. oxygen	39
Steel	G. steel	I_g	Corrosion products	40
Steel	G. steel	E_a, E_c, I_g, $E-I$ curves	Zn coating composition	41
Steel	Zn coating	I_g	Effect of paint	42
Steel	G. steel	I_g	Polarity reversal	43

TABLE 10.1. (*Continued*)

Alloy 1	Alloy 2	Measurements[a]	Focus	References
Steel	Zn alloys	Weight loss	Solution composition	44
Steel	Cd, Zn, Al	Cracking	SCC protection	45
Al, Ti alloys, Pt, Cu	Cu, Zn, Fe	I_g	Current measurement	46
S. steel	Graphite-epoxy	E_g, I_g, E_{corr}	Area effect	47
Al alloys, S. steels	Graphite-epoxy	I_g	Composite materials	48, 49
Al alloys	Cu, Cd, Zn, Ti, steels	I_g	Area effect	50
Al alloys	Ag, Cu, Ni, Sn, Zn	E_g, I_g	Galvanic series	51
Al	Graphite, TiB, SiC	E_{corr}, E_g, I_g	Composite materials	52, 53
Cu	Zn	E, I distributions	Solution resistance	54
Cu	Zn	E distribution	Modeling	55
Cu, Brass	Zn	E distribution, I_g	Corrosion rate	56
Al, Ti, Fe, Ni, Cu, S. steel	Oxides	I_g, E–I curves	Corrosion products	57
Cu	Zn	E distribution	Geometry analysis	25
Sn, Cd, Zn, Steel	S. Steel, Ni, Cu, Ti, Sn, steel, Zn	E_g, I_g, weight loss	Corrosion rate	58
G. steel, S. steel	Polyethylene	I_g	Telephone cable	59
Al, Au, Ag, Pt, Si, Mg, Cu		E_{corr}	Galvanic series	60
Other Environments				
S. steel[b]	Ti, Nb, Ta	E_g, I_g	Biocompatibility	61
Cu[c]	Zn	E distribution	Kelvin probe	62
Steel[d]	Zn	E_c, I_g	Cathodic protection	63
Steel[e]	Zn	Morphology	Paint adhesion	64
Steel[f]	Zn	Morphology	Humidity, time	65
S, steel[g]	S. steel	E_{corr}, I_g, E–I curves	Abrasion–corrosion	66
Miscellaneous	Miscellaneous	E_{corr}, E–I curves	Review	21
Steel, S. steels[h]	Steel, S. steels	Weight loss	Database	67
Steel[d]	Zinc	Weight loss	Galvanic protection	68
Zinc	Miscellaneous	Miscellaneous	Review	69

[a]I_g galvanic current; E_g, potential of couple; E_c, potential of cathode; E_{corr}, corrosion potential; E_a potential of anode; S. steel, stainless steel; G. steel, galvanized steel; i_{corr} corrosion current.
[b]Ringer's solution.
[c]Humid gas.
[d]Concrete.
[e]Painted.
[f]Cyclic test.
[g]Wet minerals.
[h]Oil and gas.

C. FACTORS IN GALVANIC CORROSION

Many factors play a role in galvanic corrosion in addition to the potential difference between the two coupled metals. Depending on the circumstances, some or all of the factors illustrated in Figure 10.1 may be involved. Generally, for a given couple, the factors in categories (a)–(c) vary less from one situation to another than the factors in categories (d)–(g). Effects of geometric factors on galvanic actions can, in many cases, be mathematically analyzed. On the other hand, effects of electrode surface conditions on reaction kinetics in real situations can be very difficult to determine. Compared to normal corrosion, galvanic corrosion is generally more complex because, in addition to material and environmental factors, it involves geometrical factors.

D. MATERIAL FACTORS

D1. Effects of Coupled Materials

As listed in Figure 10.1, all the factors affecting the electrode properties, such as those under categories (a)–(g), have an influence on galvanic action between any two metals. The reversible electrode potentials of the two coupled metals determine the intrinsic polarity of a galvanic couple, whereas the reactions, metallurgical factors, and surface conditions determine the actual polarity under a given situation because the actual potential (the corrosion potential) of a metal in an electrolyte is usually very different from its thermodynamic equilibrium value due to kinetic processes. For example, titanium has a very negative reversible electrode potential

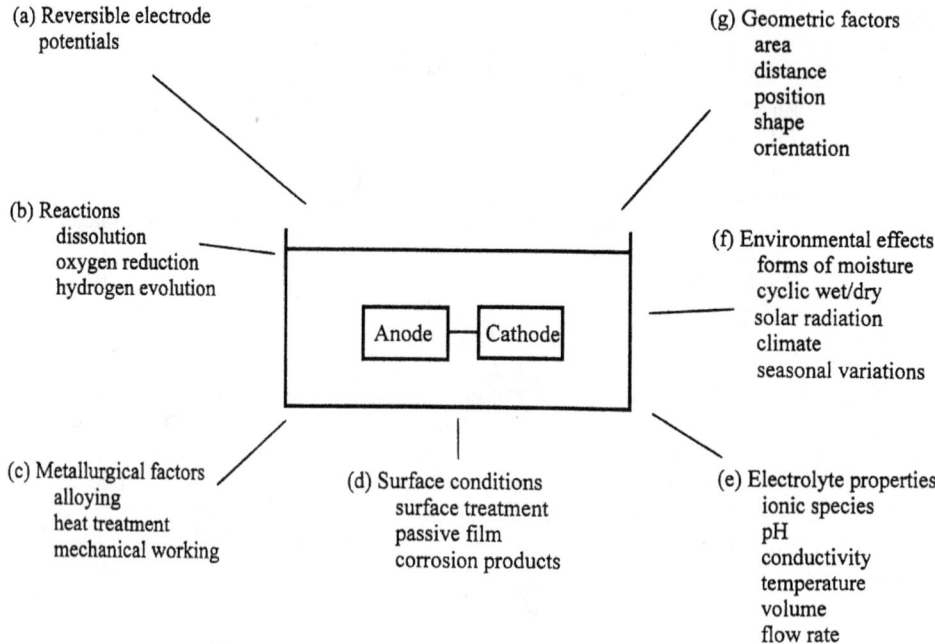

FIGURE 10.1. Factors involved in galvanic corrosion of bimetallic couple.

and has an active position in the emf series. However, titanium occupies a noble position in the galvanic series in many practical environments due to passivation of the surface.

The extent of galvanic activity is not always related to the difference in the corrosion potentials of two metals. Table 10.2 shows that, for steel, the galvanic corrosion is much higher when coupled to nickel and copper than when coupled to 304 stainless steel and Ti–6Al–4V, for which the potential differences were larger. The galvanic corrosion of zinc is the highest when coupled to steel, although the potential difference between zinc and steel is much less than between zinc and most other alloys.

Similar results have been reported on galvanic corrosion in atmospheres [2] where, in addition to the potential differences between the two metals, other factors, such as reaction kinetics and formation of corrosion products, are important in determining the galvanic corrosion rate. When the cathodic reaction is oxygen reduction and diffusion limited, different galvanic corrosion rates of an anode, coupled to different cathode materials, can be explained by the different diffusion rates of oxygen through the oxide films. When diffusion is not the limiting process, differences in galvanic corrosion rates can result from differences in cathodic efficiency of oxygen reduction in the oxide scale on the cathode surface [58], which may not depend on the corrosion potential. The difference in corrosion potentials of uncoupled metals is, thus, not a reliable indicator of the rate of galvanic corrosion.

The extent of galvanic corrosion can be ranked with actual corrosion loss data (i.e., the increase in corrosion rate relative to uncoupled conditions) [51, 58]. There is a difference between the corrosion loss determined by weight loss, which

includes the loss due to local corrosion, and due to galvanic current, which measures the true loss due to galvanic action. As noted in Table 10.2, the weight loss of zinc, when galvanically coupled to other metal alloys, can be much larger than the sum of the galvanic corrosion calculated from

TABLE 10.2. Galvanic Corrosion Rate of Steel and Zinc Coupled to Various Metal Alloys Tested in 3.5% NaCl Solution[a,b]

Coupled Alloy	r_g^c (μm/year)	r_{wl}^d (μm/year)	ΔV^e (mV)
4130 Steel, $r_0 = 90$			
SS 304	119	625	−439
Ti–6Al–4V	79	589	−338
Cu	343	1260	−316
Ni	341	1050	−299
Sn	122	581	−69
Cd		38	+221
Zn		14	+483
Zn, $r_0 = 101$			
SS 304	244	705	−905
Ni	990	1390	−817
Cu	1065	1450	−811
Ti–6Al–4V	315	815	−729
Sn	320	810	−435
4130 steel	1060	1550	−483
Cd	600	660	−258

[a]Tested for 24 h, equal size surface area of 20 cm².
[b]See [58].
[c]Measured as galvanic current.
[d]Measured as weight loss.
[e]Potential difference between the coupled metals before testing.

the Faradaic current plus the normal corrosion measured in an uncoupled condition. This indicates that the local corrosion of zinc is increased by galvanic coupling to another alloy. Some of the factors that determine the relationship of galvanic current and weight loss have been discussed in the literature [51].

In general, addition of small amounts of alloying elements does not change the reversible potential of a metal to a large extent, but may change significantly the kinetics of the electrochemical processes and, thus, behavior in galvanic action. For example, significant differences have been found in the corrosion behavior of different aluminum alloys in galvanic couples [51].

For alloys with a microstructure of more than two phases, there can be significant galvanic action among the different phases. Microscale galvanic action has been studied for the active dissolution of duplex stainless steel in acidic solutions [34] and for the interaction between martensite and ferrite in grinding media [66]. Potential and current distributions on the surface of a metal, consisting of two randomly distributed phases, have been mathematically modeled by Morris and Smyrl [70].

Increased corrosion of the cathodic member in a galvanic couple may also occur (e.g., the zinc–aluminum couple). Although aluminum is cathodic to zinc in 3.5% solution, the rate of aluminum corrosion is greater when coupled to zinc than in the uncoupled condition [51]. The higher corrosion rate of the coupled aluminum is attributed to die increased alkalinity near the surface due to the cathodic reaction, since aluminum is not stable in a solution of high alkalinity. Similar effects have been reported for tin–zinc and cadmium–zinc couples, where the corrosion of tin and cadmium, being the cathodic members, in 3.5% NaCl solution increased compared to the uncoupled condition [58].

Historically, galvanic corrosion has been reported to occur mostly in bimetallic couples. With the ever-increasing use of nonmetallic materials, galvanic corrosion is now being identified in many situations where a metal is in contact with a nonmetallic material (e.g., galvanic corrosion of metals occurs in metal-reinforced polymer matrix composites and graphite metal matrix composites [49, 53], in processing of semiconducting minerals [35], in contact with conducting polymers [59], with semiconducting metal oxides [57], and with conducting inorganic compounds [8]). It has been found that minerals, in general, exhibit potentials more noble than most metals and, therefore, may cause galvanic corrosion of metals used in processing equipment [35].

D2. Effect of Area

The effect of anode and cathode areas on galvanic corrosion depends on the type of control in the system, as illustrated later in Figure 10.10. If the galvanic system is under cathodic control, variation in the anode area has little effect on the total

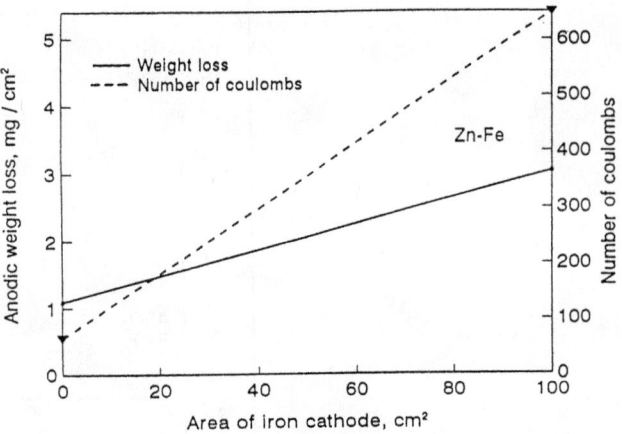

FIGURE 10.2. Effect of area of mild steel cathode on weight loss of Zn anode (area of $100 \, cm^2$) and on number of coulombs flowing between Zn–steel couple over a 96-h period in 1 N NaCl solution at 25°C [38].

rate of corrosion, but variation of the cathode area has a significant effect. The opposite is true if the system is under anodic control.

Galvanic currents in many situations are proportional to the surface area of the cathode (e.g., Figure 10.2 shows that the galvanic corrosion of zinc increases with increasing iron cathode area). On the other hand, the galvanic corrosion of zinc changes only very slightly with increasing zinc anode area. These results indicate that the galvanic corrosion of zinc in the system is mainly cathodically controlled. Similar results were found for aluminum alloys, coupled to copper, stainless steels, or Ti–6A1–4V, where the total galvanic current is independent of the surface area of the anode but is proportional to the cathode area [50].

D3. Effect of Surface Condition

The surface of metals in contact with an electrolyte is generally not "bare" but is covered with a surface layer, at least an adsorption layer, but often a solid surface film. This is the most important factor that causes the difference between the intrinsic polarity and apparent polarity and between the difference in potentials and the extent of galvanic corrosion. Formation of a surface film, whether a salt film or an oxide film, may significantly change the electrochemical properties of the metal surfaces, resulting in very different galvanic action.

A corrosion product film may serve as a physical barrier between the metal surface and the environment. It may also be directly involved in the electrochemical reactions if it conducts electrical current, either as a conductor or a semiconductor. Most metal oxides, common corrosion products, are conductive materials, mainly as semiconductors [71]. Depending on the electronic structure, oxide films exhibit potentials that are generally very different from the base metals. In many situations, these oxides, rather than the

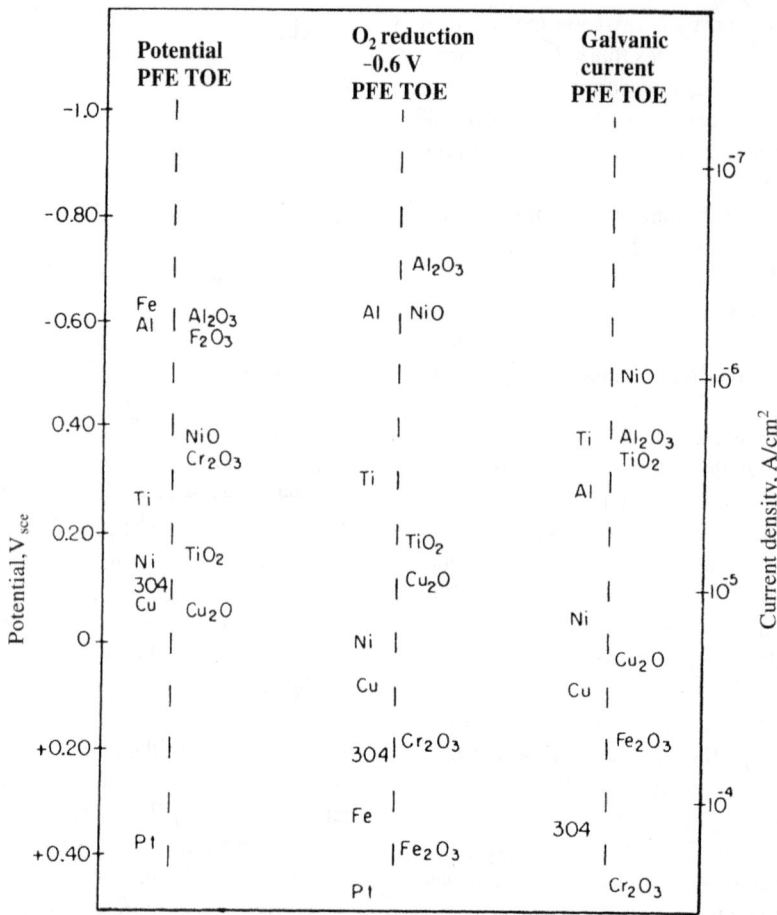

FIGURE 10.3. Corrosion potentials and oxygen reduction rates of metal oxides. PFE: passive film electrode; TOE: thermal oxide electrode [57], (Copyright ASTM. Reprinted with permission.)

metals themselves, determine the electrode potential and the position in a galvanic series. Figure 10.3 shows a galvanic series and the cathodic efficiency for O_2 reduction on a number of metal oxides [57]. The highest current densities for O_2 reduction are observed for n-type semiconductor oxides (Fe_2O_3) and metal-like oxides (Cr_2O_3). Insulators (Al_2O_3) and p-type oxides (NiO) are inefficient cathodes. The oxides, having a high cathodic efficiency and exhibiting a more noble potential value in a galvanic couple, result in a larger galvanic corrosion rate of the coupled metal.

According to Stratman and Müller [72], oxygen reduction of an iron electrode is greatly increased due to the formation of rust because oxygen can be reduced in the iron oxide scale, which is generally porous and has a large effective surface area. The corroded steel surface is, thus, a highly effective cathode when coupled to metals that have more negative potentials, such as zinc, aluminum, and magnesium [40, 73].

Surface passivation is important in the galvanic action of a bimetallic couple (e.g., aluminum is normally passivated in neutral aqueous solutions), but the extent of passivity is relatively low in solutions containing species such as chloride

ions and may break down under certain conditions. When aluminum is coupled to steel, it acts as an anode in chloride solutions, whereas it acts as a cathode in tap water and distilled water [37].

When a surface film does not fully cover the entire surface, part of the metal surface is passivated and acts as the cathode, forming a local galvanic cell, increasing the corrosion rate of the nonpassivated part of the surface, and possibly causing pitting corrosion [57]. For example, galvanic current develops between a passivated zinc sample and a partially passivated zinc sample in a cell of two compartments, containing $0.1\,M$ K_2CrO_4 in one and $0.1\,M$ K_2CrO_4 and NaCl in the other, respectively [74], Pitting occurred on the sample placed in the compartment containing NaCl [74].

When considering surface condition, the effects of time should also be included. With the passage of time, two basic changes invariably occur in a corrosion system: (1) a change of the physical structure and chemical composition of the corroding metal surface and (2) a change in the composition of the solution, particularly in the vicinity of the surface [75]. Specific changes may occur in surface roughness and area, adsorption of species, formation of passive films, saturation of dissolution products, precipitation of a solid layer, and exhaustion of reactants. Mechanistically, these changes may lead to alterations in the equilibrium potentials, the type of reactions involved, the rate-controlling process, and so on. As a result, the corrosion potential may vary greatly depending on the nature and extent of these changes.

The steady-state corrosion potential of a metal electrode depends on whether the surface is active or passive, and the time required for reaching a steady-state value varies with the conditions. The rate of galvanic corrosion may change with time as a result of changes in polarity and in potential difference between the metals in the couple. It has been reported that the potentials of various bimetallic couples, including iron, stainless steel, copper, bronze, and zinc, exposed in flowing seawater are highly variable, and the galvanic currents are reduced by about one order of magnitude within the first 120 days [26].

E. ENVIRONMENTAL FACTORS

A corrosive environment is characterized by its physical and chemical nature, which may affect the electrochemical properties. Given that the electrochemical properties of each metal are distinctive in a given electrolyte, galvanic corrosion is essentially unique for each metal couple in each environment. The combination of metal couples and environmental conditions is, thus, limitless, as can be appreciated from Table 10.1.

E1. Effects of Solution

As discussed in Section D, galvanic action of a bimetallic couple depends on the surface condition of the metals, which, in turn, is determined by environmental conditions. A metal surface exhibits different potentials in different electrolytes, as shown in Table 10.3, which lists the corrosion potentials of a number of metals in four different electrolytes of similar ionic strength. A galvanic series provides information on the polarity of a bimetallic couple but is environment-specific because the relative position of each metal changes with solution.

The extent of galvanic corrosion also varies with solution composition. The corrosion rates of zinc and steel in coupled and uncoupled conditions in several solutions [44] can be seen in Table 10.4. In all the solutions, galvanic action results in protection of the steel, but the amount of zinc corrosion

TABLE 10.3. Corrosion Potentials (mV_{sce}) of Metals after 24-h Immersion in Four Different Solutions, Compared with the emf Series[a]

emf	0.1 M HCl	0.1 M NaCl	0.1 M Na$_2$SO$_4$	0.1 M NaOH
Ag +799	Ag +48	Ag -60	Ag +147	Ag -64
Cu +324	Ni -135	Ni -142	Cu -43	S. steel -96
Pb -126	Cu -139	Zr -150	Cr -45	Ni -171
Sn -138	Ta -213	Cu -189	Ti -66	Cu -231
Ni -257	Ti -221	Cr -270	Ni -70	Cr -303
In -338	Zr -297	Ti -272	Ta -154	Fe -389
Fe -447	Cr -347	Ta -295	Zr -218	Ta -500
	S. steel -473	S. steel -320	S. steel -348	Zn -555
Cr -744	Pb -487	Pb -565	Sn -421	Ti -591
Ta -750	Sn -497	Sn -565	Al -505	In -600
Zn -762	Fe -557	In -646	Pb -545	Zr -631
Zr -1553	In -680	Fe -710	In -651	Pb -757
Ti -1630	Al -731	Al -712	Fe -720	Mg -809
Al -1660	Zn -989	Zn -1019	Zn -1049	Sn -1096
Mg -2370	Mg -1894	Mg -1548	Mg -1588	Al -1351

[a]Sample surface area about 1 cm^2; polished with 600-grade emery paper; solution open at air at room temperature.

TABLE 10.4.   Corrosion Rate of a Zinc–Steel Couple in Various Solutions (μm/year)^{a,b}

	Uncoupled		Coupled	
Solution	Zinc	Steel	Zinc	Steel
0.05 M MgSO$_4$	+	66	86.4	+
0.05 M Na$_2$SO$_4$	285	254	838	+
0.05 M NaCl	254	254	762	+
0.005 M NaCl	112	178	218	+
Carbonic acid	10.2	73.7	38.1	+
Calcium carbonate	+	150	+	+
Tap water	+	71.1	+	+

^{a}See [44].
^{b}Specimen of equal surface area partially immersed for 39 days. Plus signs indicate specimens gained weight.

varies with solution composition. The difference in the corrosion rates in magnesium sulfate and sodium sulfate solutions indicates the significant effect of cations on the reaction kinetics.

The conductivity of the electrolyte is a very important factor because it determines the distribution of galvanic corrosion across the anode surface. When conductivity is high, as in seawater, the galvanic corrosion of the anodic metal is distributed uniformly across the surface. As the conductivity decreases, galvanic corrosion becomes concentrated in a narrow region near the junction, as illustrated in Figure 10.4. Usually, the total galvanic corrosion is less in a poorly conducting electrolyte than in a highly conducting one.

Ions of noble elements in solution may cause galvanic corrosion of a less noble metal immersed in the solution because precipitation of the noble element can cause small galvanic cells to form [76].

If there is only a limited amount of electrolyte, the composition of the electrolyte may significantly change as a result of electrochemical reactions. Massinon et al. [77, 78] found an increase of pH in a confined electrolyte after a certain time of galvanic action for a zinc–steel couple. Pryor and Keir [79] pointed out that, when the distance between the

anode and cathode is small compared to the dimension of the electrodes, the galvanic corrosion is small due to the limitation in the mass transport of the reactants and reaction products.

One important solution factor is the thickness of thin-layer electrolytes, which is encountered in atmospheric environments. The thickness of an electrolyte affects corrosion processes in several different ways. First, it affects the lateral resistance of the electrolyte and, thus, affects the potential and current distribution across the surface of the coupled metals. Second, it affects the transport rate of oxygen across the electrolyte layer and, thus, the rate of cathodic reaction. Third, it changes the volume and the solvation capacity of the electrolyte and, thus, affects the formation of corrosion products.

As shown later in Figures 10.11 and 10.12, changes in potential and galvanic current across a metal surface are greater for thinner electrolytes due to the larger electrical resistance involved. Under a thin-layer electrolyte, the galvanic corrosion is the most intense at the anode area near the anode–cathode boundary, while there is very little galvanic corrosion away from the boundary, as illustrated in Figure 10.4(a). Variation of electrolyte thickness also determines the rate-controlling process for a given cell dimension. Figure 10.5 shows that, for a thinner electrolyte, the galvanic current is larger when the anode and the cathode are close, but the reverse applies when the two electrodes are far apart [80]. Since the oxygen diffusion rate under thin-layer electrolytes changes with electrolyte thickness, the change of the relative galvanic current values for small and large distances, shown in Figure 10.5, is due to the change of the rate-limiting process from oxygen diffusion at a close distance to ohmic conduction in the electrolyte at a large distance [80].

The physical position of a galvanic couple in solution can also affect the galvanic action between coupled metals. Shams El Din et al. [74] found a large potential variation near the solution surface between a zinc anode and a copper cathode that were half-immersed in solution, due to the higher oxygen concentration near the surface than in the bulk solution.

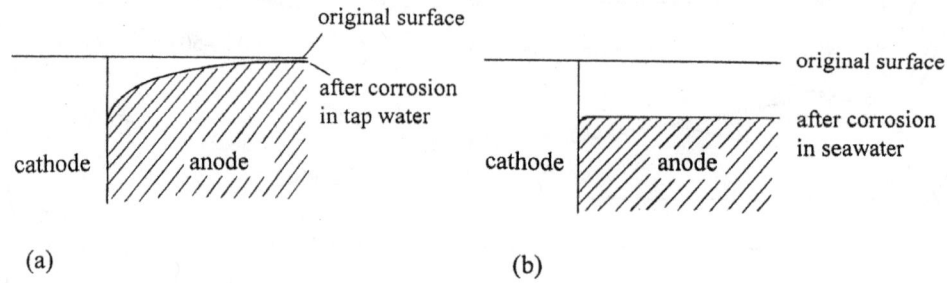

FIGURE 10.4. Effect of solution conductivity on distribution of galvanic corrosion: (a) low conductivity and (b) high conductivity.

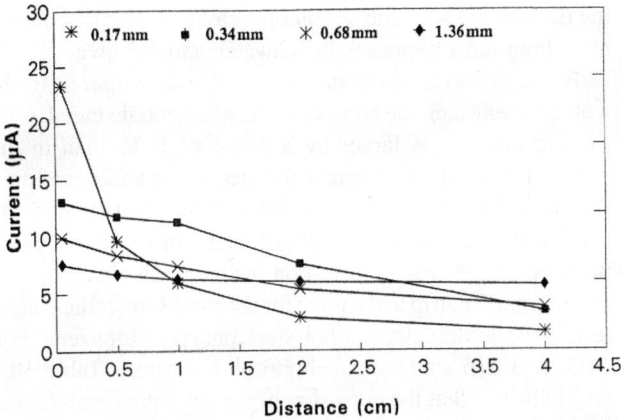

FIGURE 10.5. Galvanic current as function of distance between zinc and steel in 0.001 M Na_2SO_4 solutions of different electrolyte thickness [80]. [Reprinted from *Corrosion Science*, **34**, X. G. Zhang and E. M. Valeriote, "Galvanic Protection of Steel and Galvanic Corrosion of Zinc under Thin Layer Electrolytes," p. 1957 (1993), with permission from Elsevier Science.]

E2. Atmospheric Environments

Galvanic corrosion occurs commonly in atmospheric environments as different combinations of materials are used in buildings and structures exposed to indoor and outdoor atmospheres. A test program of galvanic corrosion in atmospheres was started as early as 1931 by the American Society for Testing and Materials (ASTM) [2]. Since then, a number of extensive exposure programs have been carried out all over the world [81–84]. The various aspects of atmospheric galvanic corrosion have been discussed in a comprehensive review by Kucera and Mattsson [2].

Galvanic corrosion under atmospheric environments is most often evaluated by weight loss measurement. In other environments, potentials and/or galvanic currents of coupled metals can be measured, but it is very difficult to measure in situ potentials of metals under atmospheric conditions due to the thin layer of the electrolyte.

Data in Table 10.5, reported by Kucera and Mattsson [2], show the galvanic corrosion rates of a number of metal alloys,

in the form of wire coupled to bolts of various metals. Depending on the bolt metal and the type of atmosphere, the galvanic corrosion rate of the wire can be many times that of the normal corrosion rate. For example, the galvanic corrosion of zinc is as much as five times the normal corrosion in a rural atmosphere and three times that in a marine atmosphere [85]. The amount of corrosion shown in Table 10.5 does not appear to relate to differences between the reversible potentials of the coupled metals. Steel and copper, as cathodic members of a couple, cause the most galvanic corrosion of the coupled anodic material.

Galvanic corrosion in atmospheres is usually restricted to a narrow region of the anode metal near the bimetallic junction because of the high resistance of thin-layer electrolytes formed by rain and water condensation [80, 86, 87]. Even for the most incompatible metals, direct galvanic action will not extend more than a few millimeters from the junction. Because of the very narrow range of galvanic action, geometrical factors of the coupled metals, such as shape and size, generally do not have a strong effect on galvanic corrosion in atmospheric environments.

Galvanic action is most significant in marine atmospheres because of the high conductivity of seawater. Compared to other types of moisture formed under atmospheric conditions, rain is particularly effective in causing galvanic corrosion. The galvanic corrosion rate is several times that of the normal corrosion rate in an open exposure, whereas they are similar when under a rain shelter because the electrolyte layer formed by rain is thicker and has a smaller lateral electric resistance than the moisture formed by condensation [2].

E3. Natural Waters

Natural waters are commonly classified as seawater and freshwater, such as river, lake, and underground waters. A distinct difference between seawater and freshwater is that seawater has a high conductivity due to its high salt content, whereas freshwaters generally have low conductivities. In comparison to other environments, waters as corrosion

TABLE 10.5. Corrosion Rate (μm/year) of Wire Specimens Coupled to Bolts of Other Materials Exposed for 1 Year in an Urban Environment[a]

Bolt	Nylon	Steel	S. Steel	Cu	Pb	Zn	Ni	Al	Sn	Cr	Mg
Wire											
Steel	25.7		31	32	23	1.2	29	22	32	27	0.6
S. steel				0.2			0.2			0.02	
Cu		0.3	1.0		0.6		1.0		0.7	0.4	0.1
Zn	1.2	3.3	1.8	2.0	2.4		1,9	1.1	2.6	1.4	0.04
Ni			1.3	0.1						1.0	
Al	0.2	1.8	0.6	5.3	0.6	0.0	0.6		0.6	0.3	0.0
Sn		0.4	1.5	3.5		0.0	1.5	0.4		0.3	0.0
Mg		18	10	10	13	9.0	20	5.3	8.1	9.2	

[a]See [2].

TABLE 10.6. Galvanic Series of Some Commercial Metals and Alloys in Seawater[a]

↑	Platinum
Noble or	Gold
cathodic	Graphite
	Titanium
	Silver
	Chlorimet 3 (62 Ni–18 Cr–18 Mo)
	Hastelloy C (62 Ni–17 Ct–15 Mo)
	18-8 Mo stainless steel (passive)
	18-8 stainless steel (passive)
	Chromium stainless steel 11–30% Cr (passive)
	Inconel (passive) (80 Ni–13 Cr–7 Fe)
	Nickel (passive)
	Silver solder
	Monel (70 Ni–30 Cu)
	Cupronickels (60–90 Cu–40–10 Ni)
	Bronzes (Cu–Sn)
	Copper
	Brasses (Cu–Zn)
	Chlorimet 2 (66 Ni–32 Mo–1 Fe)
	Hastelloy B (60 Ni–30 Mo–6 Fe–1 Mn)
	Inconel (active)
	Nickel (active)
	Tin
	Lead
	Lead–tin solders
	18-8 Mo stainless steel (active)
	18-8 stainless steel (active)
	Ni-Resist (high Ni cast iron)
	Chromium stainless steel, 13% Cr (active)
	Cast iron
	Steel or iron
	2024 aluminum (4.5 Cu, 1.5 Mg, 0.6 Mn)
Active or	Cadmium
anodic	Commercially pure aluminum (1100)
↓	Zinc
	Magnesium and magnesium alloys

[a]See [88]. Reprinted from M. G. Fontana and N. D. Greene, *Corrosion Engineering*, 2nd ed., 1978, McGraw–Hill, with permission of The McGraw–Hill Companies.

environments are homogeneous (e.g., seawater is almost constant with respect to time and geographic location). The galvanic action in seawater, due to its high conductivity and uniformity, is long range and spreads uniformly across the entire surface area of a metallic structure. The galvanic effect in freshwaters is generally much less than in seawater because of the lower conductivity [14].

Table 10.6 presents the galvanic series of some commercial metals and alloys obtained in seawater [88]. As discussed previously, such a galvanic series differs from the emf series and is specific to seawater. A galvanic series of a number of common alloy couples in flowing seawater has also been reported [26, 89, 90].

Galvanic corrosion in seawater has been extensively investigated, as indicated in Table 10.1. Table 10.7 shows

the data on the galvanic action of various bimetallic couples after long-term exposure in seawater and freshwater [14]. Galvanic action is much stronger in seawater than in freshwater. In seawater, the corrosion rate of an anodic metal, such as zinc or steel, is larger by a factor of 5–12 than in the uncoupled condition, whereas the increase is a factor of only 2–5 in freshwater. The data also indicate the great effect of the relative sizes of the anode and cathode; a factor of 6–7 times more corrosion was observed on the anode by changing the anode from the strip to the plate for the same bimetallic couple (e.g., 316 stainless steel/carbon steel, phosphor bronze/carbon steel, and 316 stainless steel/phosphor bronze). Table 10.7 also indicates that the corrosion of the cathodic member was generally decreased, in varying degrees, as a result of the galvanic corrosion of the anodic member.

F. POLARITY REVERSAL

The normal polarity of some galvanic couples under certain conditions may reverse with the passage of time. This phenomenon was first reported by Schikorr in 1939 on a zinc–steel couple in hot supply water with iron becoming anodic to zinc, which has been a serious problem for galvanized steel hot water tanks [91]. It has subsequently been extensively investigated [10, 12, 43].

Polarity reversal is invariably caused by the change of surface condition of at least one of the coupled metals, such as formation of a passive film. The degree of passivity, the nature of the redox couples in the solution, and the stability of the system determine the polarity and its variation with time. For a zinc–steel couple, the change in the zinc electrode potential is chiefly responsible for the reversal of polarity since the potential of the steel remains relatively unchanged with time in hot water [12]. It has generally been found that polarity reversal does not occur in distilled water up to 65°C and without the presence of oxygen [12, 43, 92].

Depending on the conditions, it may occur rather quickly, taking several minutes, or rather slowly, taking many days. In addition to temperature, other factors, such as dissolved ions, pH, and time of immersion, affect the polarity of a zinc–steel couple. The necessary conditions for polarity reversal of a zinc–steel couple are passivation of the zinc surface and sufficient reducing species, such as dissolved oxygen, in the water to provide cathodic depolarization.

Polarity reversal of an aluminum–steel couple has also been found to occur in natural environments where aluminum alloys are used as anodes for cathodic protection of steel. The general mechanism is similar to that occurring with a zinc–steel couple. However, unlike zinc, aluminum is normally passivated by a thin oxide film in most natural environments. The potential of aluminum depends on the degree of passivity, which is sensitive to the ionic species

TABLE 10.7.   Galvanic Corrosion of Various Metal Alloys in Seawater and Freshwaters After 16 Years Exposure[a,b]

Strip/Plate[c]	Seawater	Freshwater
316 S. steel/carbon steel	0.1/49.5(2.0, 48.1)	0.0/32.4 (0.0, 26.0)
316 S. steel/naval brass	0.0/16.2 (2.0, 12,4)	0.0/2.5 (0.0,1.9)
316 S. steel/phosphor bronze	0.0/9.4 (2.0, 5.5)	0.0/0.7 (0.0, 0.7)
Phosphor bronze/carbon steel	0.4/55.1 (5.5, 48.1)	0.1/28.9 (0.7, 26.0)
Phosphor bronze/aluminum	0.7/6.5 (5.5, 1.0)	0.1/12.4 (0.7, 5.0)
Phosphor bronze/2% Ni steel	0.5/60.8 (5.5, 52)	0.2/24.6 (0.7, 21.1)
Phosphor bronze/cast steel	0.2/51.6 (5.5, 43.3)	0.1/30.1 (0.7, 26.3)
Phosphor bronze/302 S. steel	21.9/6.3 (5.5, 9.3)	3.8/0.0 (0.7, 0.0)
Phosphor bronze/316 S. steel	41.3/0.2 (5.5, 2.0)	1.8/0.0 (0.7, 0.0)
Phosphor bronze/70 Cu–30 Ni	4.6/3.8 (5.5, 2.3)	0.5/1.3 (0.7, 1.3)
Phosphor bronze/monel	71.8/7.0 (5.5, 8.7)	14.1/0.8 (0.7, 0.6)
Carbon steel/aluminum	1.6/7.8 (48.1, 0.9)	17/9.4 (26.0, 1.5)
Carbon steel/2%Ni steel	63.1/48.2 (48.1, 52)	33.9/19.1 (26.0, 22.9)
Carbon steel/70 Cu–30 Ni	310[d]/1.6 (48.1, 2.3)	68.2/0.3 (26.0, 1.3)
Carbon steel/nickel	320[d]/4.8 (48.1, 19)	71.3/0.2 (26.0, 0.0)
Carbon steel/copper	350[d]/2.9 (48.1, 6)	77.7/0.2 (26.0, 1.0)
Carbon steel/phosphor bronze	318[d]/1.9 (48.1, 5.5)	65.5/0.3 (26.0, 0.7)
Carbon steel/302 S. steel	298[d]/0.8 (48.1, 9.5)	52.7/0.0 (26.0, 0.0)
Carbon steel/316 S. steel	260[d]/0.0 (48.1, 2.0)	44.4/0.0 (26.0, 0.0)
Carbon steel/20%Zn brass	281[d]/2.4 (48.1, 3.7)	63.4/0.3 (26.0, 1.7)
Zinc/carbon steel	5/15.5 (14.9, 48.1)	43.1/19.9 (7.9. 26.0)
Zinc/2%Ni steel	187[d]/14.9 (14.9, 52)	36.6/17.4 (7.9, 22.8)
Zinc/cast steel	198/11.4 (14.9, 43.3)	45.6/21.9 (7.9, 26.3)
Zinc/18% Ni cast iron	167[d]/0.1 (14.9, 22.8)	23.7/7.7 (7.9, 8.0)

[a]See [97].

[b]Average penetration m mils (1 mil = 25.4 μm), the values in parentheses are the corrosion loss of the allays in uncoupled conditions.

[c]Strip area = 141 cm^2, plate area = 972 cm^2, carbon steel (0.24%C), 316 S. steel (18Cr–1.3Mo), 302 S. steel (18 Cr–8 Ni), phosphor bronze (4 Sn–0.25 P), low brass (20%Zn), aluminum (99%), zinc (99.5%), lead (99.5%), aluminum bronze (5%Al), Monel (70 Ni–30 Cu).

[d]Estimated according to the data at 8 years.

in the environment. For example, carbonate and bicarbonate ions promote passivity and, thus, produce more noble potential values, whereas ions like chloride give the opposite effect [15]. In practice, sacrificial aluminum anodes are alloyed with various elements to prevent reversal of polarity.

The consequence of polarity reversal can be serious. In the zinc–steel and the aluminum–steel couples, zinc and aluminum serve as sacrificial anodes for protecting the steel. Polarity reversal results in the loss of cathodic protection of steel, causes galvanic corrosion of the steel, and shortens the life of the steel structure.

G. PREVENTIVE MEASURES

The essential condition for galvanic corrosion to occur is two dissimilar metals that are both electrically and electrolytically connected. Theoretically, prevention of galvanic corrosion can be achieved by avoiding the use of dissimilar metals in an assembly, by electrically separating the dissimilar metals with an insulating material or by physically insulating the environment from the metal surface with a coating impermeable

to water. In reality, however, complete prevention is often not practical, as dissimilar metals need often to be used in direct contact and exposed to a corrosive environment and there is no absolutely impermeable coating. Thus, measures to minimize the possibility and extent of galvanic corrosion must be implemented All the factors listed in Figure 10.1 can be considered and controlled in order to reduce galvanic corrosion. Some practical approaches are as follows:

(a) Avoid combinations of dissimilar metals that are far apart in the galvanic series applicable to the environment.

(b) Avoid situations with small anodes and large cathodes.

(c) Isolate the coupled metals from the environment.

(d) Reduce the aggressiveness of the environment by adding inhibitors.

(e) Use cathodic protection of the bimetallic couple with a rectifier or a sacrificial anode.

(f) Increase the length of solution path between the two metals. This method is beneficial only in electrolytes of low conductivity, such as freshwaters, because strong galvanic action exists several meters away in highly conductive media, such as seawater.

The use of these approaches must meet the specific requirements of each application [93, 94]. Sometimes one is sufficient, but a combination of two or more may be required in other situations. It must be emphasized that the most effective and efficient way to prevent or minimize galvanic corrosion is to consider the problem and take measures early in the design stage.

H. BENEFICIAL EFFECTS OF GALVANIC CORROSION

As a result of galvanic corrosion of the anodic metal, the corrosion of the cathodic, coupled metal or alloy is generally reduced (i.e., cathodically protected). This effect has been well utilized in the application of sacrificial anodes, coatings, and paints for corrosion protection of many metal components and structures in various environments.

Sacrificial anodes, mainly made of zinc, aluminum, and magnesium and their alloys, are widely used in corrosion prevention underwater and underground for structures such as pipelines, tanks, bridges, and ships. Each alloy possesses a unique set of electrochemical and engineering properties and has its own characteristic advantages as an anode for galvanic protection of a more noble alloy, mostly steel, in a given situation [95, 96]. Anodes can be designed for composition, shape, and size according to specific applications [95].

Galvanized (i.e., zinc coated) steel is a typical example of a metallic coating that provides a barrier layer to protect the steel and also sacrificially protects the locations where discontinuities occur in the coating [39, 97]. The combination of barrier and galvanic protection by the zinc coating results in very effective corrosion protection of steels. Table 10.8 shows that galvanic corrosion resulted in a reduction of the corrosion of steel by 3 times in rural, 40 times in industrial, and 300 times in seacoast industrial atmospheres. On the

other hand, the galvanic corrosion of zinc, an increase of corrosion by a factor of 1.6–3 compared to uncoupled conditions, is very little compared to the reduction of steel corrosion. Galvanic protection of the steel is more effective in industrial and marine atmospheres than in rural ones, suggesting that the pollutants in the atmospheres are beneficial to the galvanic protection of steel, although they are very harmful to the normal corrosion of the uncoupled steel.

The protection distance of steel by a zinc coating in atmospheric environments is limited to a region only a few millimeters from the zinc coating because of the high resistance of thin-layer electrolytes formed in the atmosphere [87]. The protection distance, as a function of electrolyte thickness and surface area of steel, is shown in Figure 10.6. Figure 10.7 shows the protection distance as a function of separation distance and width of steel determined in an atmospheric environment [86]. The data indicate that the largest protection distance is ~1 mm, implying that the width of a scratch on a zinc-coated steel, which is fully protected is ~2 mm in the atmosphere. However, the actual protected area, which also includes the areas under partial protection, is considerably larger [86].

I. FUNDAMENTAL CONSIDERATIONS

I1. Electrode Potential and Kirchhoff's Law

The direction of galvanic current flow between two connected bare metals is determined by the actual electrode potentials (i.e., corrosion potentials of the metals in a corrosion environment). The metal which has a higher (i.e., more positive, more noble, or more cathodic) electrode potential is the cathode in the galvanic couple, and the other is the anode.

The polarity of galvanic couples in real situations may be different from that predicted by the thermodynamic reversible potential in the emf series, because the corrosion potentials are determined by the reaction kinetics at the metal– electrolyte interface. Thus, the actual position of each metal or alloy in a specific environment forms a galvanic

TABLE 10.8.   Corrosion of Galvanic Couples in Different Atmospheres after 7 Years Exposure[a]

Couple	Industrial		Rural		Industrial, Marine	
	W^b	R^c	W	R	W	R
Zn/Zn	187		27		195	
Zn/Fe	332	1.8	81	3.0	349	1.8
Fe/Fe	1825		470		1534	
Fe/Zn	43	1/40	147	1/3	5	1/300

[a] Weight loss of the first metal in a couple (e.g., Zn in Zn/Fe). Samples consisted of two 1.5-in. diameter disks 1/16 in. in thickness, clamped together with 1-in. diameter Bakelite washers, giving an exposed area of 1/16 in. all round the edge of the disk, and an annular area 1/4 in. deep = 1.275 in.^{2}.
[b] See [94]. Weight loss in milligrams.
[c] Corrosion ratio of galvanic couple to nongalvanic couple.

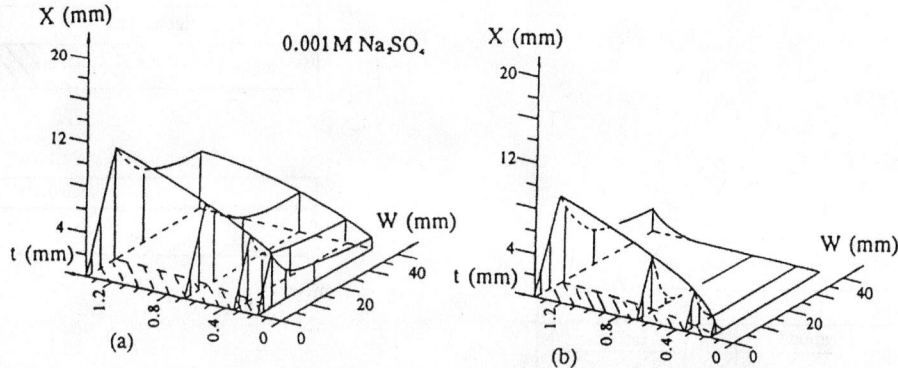

FIGURE 10.6. Protection distance X as function of electrolyte thickness (t), steel width (W), and distance between zinc and steel (D): (a) D = 0; (b) D = 5 mm [87]. (Copyright ASTM. Reprinted with permission.)

series that generally differs from the emf series. Also, as has been discussed earlier, the relative positions of two coupled metals in a galvanic series indicate only the polarity or the flow direction of the galvanic current, but not the magnitude of the current or the rate of corrosion, which is also determined by many other factors. The fundamental relationship in galvanic corrosion is described by Kirchhoff's second law:

$$E_c - E_a = IR_e + IR_m \tag{10.1}$$

where R_e is the resistance of the electrolytic portion of the galvanic circuit, R_m the resistance of the metallic portion, E_c the effective (polarized) potential of the cathodic member of the couple, and E_a the effective (polarized) potential of the anodic member. Generally, R_m is very small and can be

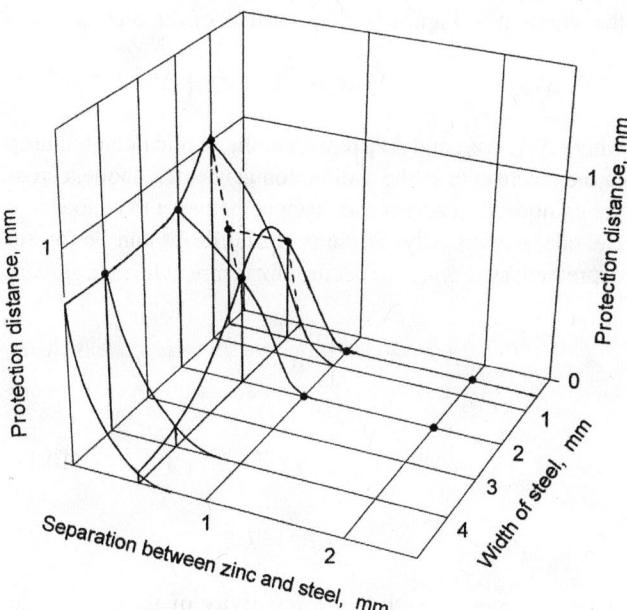

FIGURE 10.7. Protection distance of zinc–steel couple as function of steel width and separation distance under natural atmospheric exposure.

neglected. Both E_a and E_c are functions of the galvanic current I; hence, the potential difference between the two metals, when there is a current flow through the electrolyte, does not equal the open-circuit cell potential.

I2. Analysis

Although the mathematical description of galvanic corrosion can be very complex because of the many factors involved, particularly geometric factors, it can be simplified for certain situations. Following is an analysis of coplanar, coupled metals, as illustrated in Figure 10.8(a). Such a geometry applies to a wide range of situations. The distance between anode and cathode (d) may equal zero (e.g., metal joints or a coated metal with the coating partially removed), as shown in Figure 10.8(b). On the other hand, when one metal is used as a sacrificial anode to cathodically protect another metal, the distance between the anode and cathode may be very large, as shown in Figure 10.8(c), that is, $d \gg (x_{ae} - d)$ and $d \gg x_{ce}$ (where $x_{ae} - d$ and x_{ce} are the lengths of anode and cathode, respectively).

The basic current and potential relationships for the geometrical arrangement shown in Figure 10.8(a) can be expressed as follows:

$$I_a = I_c \tag{10.2}$$

and

$$E_{c,corr} - E_{a,corr} = \eta_a(x^a) - \eta_c(x^c) + \Delta V_R(x^a, x^c) \quad \begin{matrix} x^a \leq 0 \\ x^c \geq 0 \end{matrix} \tag{10.3}$$

where $E_{a,corr}$ and $E_{c,corr}$ are the uncoupled corrosion potentials of the anode and cathode, respectively; η_a and η_c, the overpotentials of the anode and cathode, respectively, in the couple; and ΔV_R, the ohmic potential drop across the electrolyte between x^a on the anodic surface and x^c on the

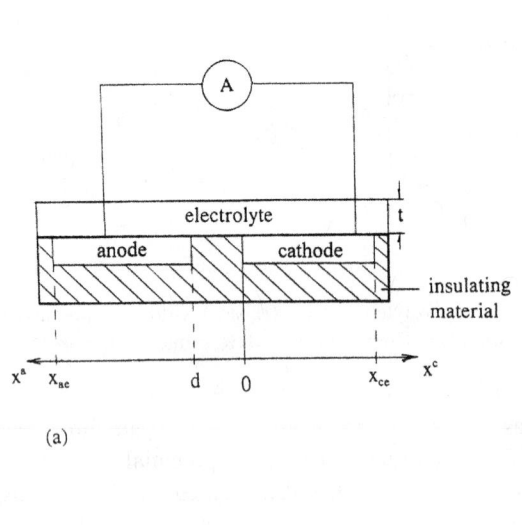

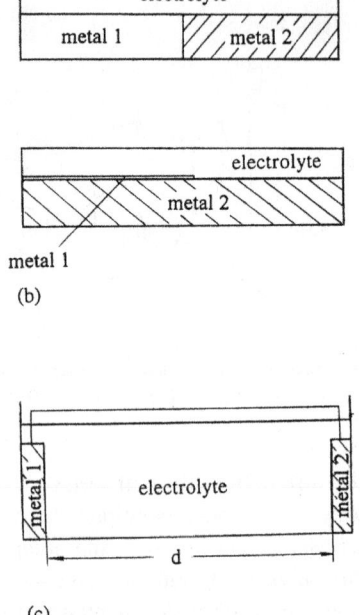

FIGURE 10.8. (a) General geometry of bimetallic couple; (b) bimetallic joint and a metal partially coated with metallic coating; and (c) anode coupled to distant cathode.

cathodic surface; I_a, the total anodic current; and I_c, the total cathodic current. Then,

$$I_a = \int_d^{x_{ae}} i_a(x^a) l\, dx^a \qquad (10.4)$$

$$I_c = \int_0^{x_{ce}} i_c(x^c) l\, dx^c \qquad (10.5)$$

where l is the width of the electrodes, and $i_a(x^a)$ and $i_c(x^c)$ are the anodic and cathodic current densities, respectively. When both the anodic and cathodic reactions are activation controlled, they can be expressed by the Butler–Volmer equation:

$$I_a = i_{0a}\,\theta_a\{\exp[\beta_{aa}\eta_a(x^a)] - \exp[-\beta_{ac}\eta_a(x^a)]\} \quad (10.6)$$

$$I_c = i_{0c}\,\theta_c\{\exp[\beta_{ac}\eta_c(x^c)] - \exp[-\beta_{cc}\eta_c(x^c)]\} \quad (10.7)$$

where i_{0a} and i_{0c} are the exchange currents for the anodic and cathodic reactions, respectively; β_{aa}, β_{ac}, β_{ca}, and β_{cc}, the kinetic constants; and θ_a and θ_c, the area factors, varying between 0 and 1. Here $\theta = 1$ when the whole surface is fully active and θ is close to zero if the surface is fully passivated. When the cathodic reaction is limited by oxygen diffusion in the electrolyte, Eq. (10.7) is replaced by

$$i_c = 4\mathcal{F}D_O C_{O2}/\delta \qquad (10.8)$$

where $\mathcal{F}$ is the Faraday constant; D_O, the diffusion coefficient of oxygen in the electrolyte; C_{O2}, the oxygen concentration in the bulk electrolyte; and δ, the thickness of the diffusion layer.

The total ohmic potential drop in the electrolyte between any two points on the surface of the anode and the cathode for the situation in Figure 10.8(a) consists of three parts:

$$\Delta V_R(x^a, x^c) = \Delta V_a(x^a) + \Delta V_c(x^c) + \Delta V_d \qquad (10.9)$$

where ΔV_a, ΔV_c, and ΔV_d represent the ohmic potential drop in the electrolyte in the x direction across the anode, across the cathode, and across the distance between the anode and cathode, respectively. These potential drops can be further expressed by

$$\Delta V_a(x^a) = \int_d^{x^a} j_a(x^a)\, dR(x^a) \qquad (10.10)$$

$$\Delta V_c(x^c) = \int_0^{x^c} j_c(x^c)\, dR(x^c) \qquad (10.11)$$

$$\Delta V_d = I_a R_d = I_c R_d \qquad (10.12)$$

where $R_d = \rho d/tl$, with ρ the resistivity of the electrolyte; t the electrolyte thickness; d the distance between the anode and cathode; l the width of the electrodes; and j_a and j_c the sums of the current from x^a to x_{ae} on the anode and from x^c to

x_{ce} on the cathode, respectively, given by the following Eqs. (10.13) and (10.14):

$$j_a(x^a) = \int_{x^a}^{x_{ae}} i_a(x^a) l \, dx^a \qquad (10.13)$$

$$j_c(x^c) = \int_{x^c}^{x_{ce}} i_c(x^c) l \, dx^c \qquad (10.14)$$

The factors listed under categories (a)–(f) in Figure 10.1 contribute to galvanic action through the electrochemical reaction kinetics given by Eqs. (10.6) and (10.7). For example, changing the pH of the solution may cause a change of the kinetic parameters, $i_{0a}, i_{0c}, \beta_a,$ or β_c On the other hand, the geometric factors under category (g) affect galvanic corrosion through the parameters in all the equations from (10.4) to (10.14).

Equations (10.4)–(10.14) describe a general situation. It can be simplified for specific applications and geometry. For example, for Figure 10.8(b), representing the galvanic action of a metal joint or a partially coated metal, the term ΔV_d in Eq. (10.9) becomes zero. For the geometry in Figure 10.8(c), representing galvanic action of two metals separated by a large distance [i.e., $d \gg (x_{ae} - d)$ and $d \gg x_{ce}$], I_a and I_c in Eqs. (10.4) and (10.5) become $i_a A_a$, and $i_c A_c$ with $A_a = l(x_{ae} - d)$ and $A_c = lx_{ce}$, the areas for the anode and the cathode, respectively. In addition, ΔV_a and ΔV_c in Eq. (10.9) can be taken as zero because they are very small compared to ΔV_d. In such a case, the geometry in the galvanic cell (i.e., shape and orientation of electrodes, and size of the electrode) becomes insignificant in the galvanic action of the couple, and the galvanic corrosion of the anode, as well as the galvanic protection of the cathode surface, become uniform. Thus, the galvanic action can be fully described by the polarization characteristics of the anode and the electrolyte resistance without consideration of geometric factors.

I3. Polarization and Resistance

In a galvanic couple, it is important to know the relative contributions from the polarization of the coupled metals and the electrolyte resistance, as described by

$$E_{c,corr} - E_{a,corr} = \Delta V_c + \Delta V_a + IR \qquad (10.15)$$

which is essentially Eq. (10.3) simplified when geometric factors are not considered.

Equation (10.15) can be graphically illustrated by the anodic and cathodic polarization curves shown in Figure (10.9). When the solution resistance, R, is infinite, no current flows, and $E_c - E_a$ equals the difference in corrosion potentials of the separated (not coupled) metals (i.e., $E_{c,corr} - E_{a,corr}$). As R decreases, I increases and $E_c - E_a$ becomes smaller because of polarization. When R

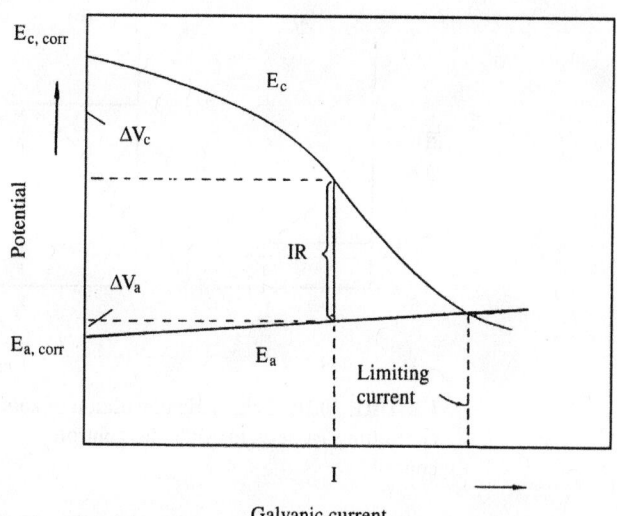

FIGURE 10.9. Graphic estimation of galvanic current.

is zero, $E_c - E_a$, becomes zero and the galvanic current reaches the maximum, known as the "limiting galvanic current," which is at the intersection of the polarization curves of the anode and cathode. The exact shapes of the anodic and cathodic polarization curves depend on the electrochemical reaction kinetics of each metal in the electrolyte and, thus, are functions of pH, temperature, solution concentration, diffusion, formation of passive films, and so on. Often, the anodic dissolution of a nonpassivated metal is activation controlled with a relatively small Tafel slope, while the cathodic reactions on the other metal surface, on the other hand, can either be activation or diffusion controlled depending on the conditions, particularly solution pH and aeration conditions.

The controlling mechanisms in a galvanic corrosion system depend on the relative extent of the anodic and cathodic polarization, on the potential drop in the solution, and on the total potential difference between the coupled metals. If the anode does not polarize and the cathode does, then, in solutions of low resistivity, the current flow is controlled entirely by the cathode. Such a situation is considered to be under cathodic control [Fig. 10.10(a)]. If the anode polarizes and the cathode does not, the status is reversed and the system is said to be under anodic control [Fig. 10.10(b)]. If neither electrode polarizes and the current flow is controlled by the resistivity of the path, mostly in the electrolyte, then the system is said to be under resistance control [Fig. 10.10(c)]. In most situations, a galvanic system is under mixed control, by anodic and cathodic polarization and electrolyte resistance [Fig. 10.10(d)].

The relative magnitude of polarization resistance and solution resistance determines the effective dimension of a galvanic cell, which can be estimated using the polarization parameter, L_i:

$$L_i = 1/\rho |d\eta_i/dI_i| \qquad (10.16)$$

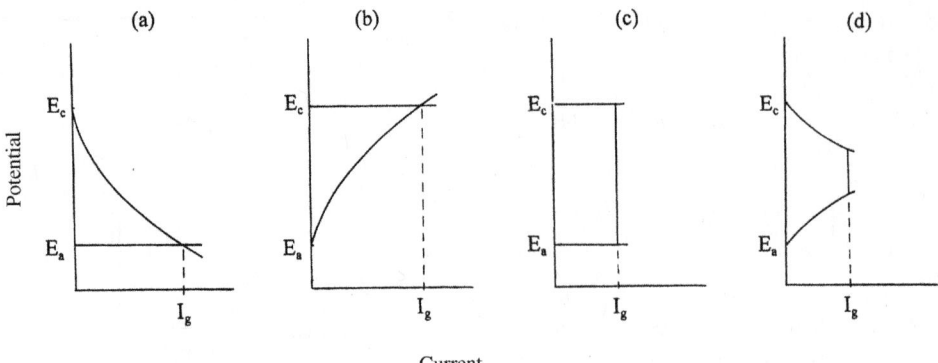

FIGURE 10.10. Schematic illustration of anodic and cathodic polarization carves for four different controlling modes: (a) cathodic control, (b) anodic control, (c) resistance control, and (d) mixed control.

where ρ is the specific resistivity of the electrolyte; I_i is the current density, and η_i is the overpotential of the anode or the cathode. The polarization parameter, defined by Wagner [98], has the dimension of length and provides an electrochemical yardstick for classifying electrochemical systems. It has been widely used to describe the behavior of galvanic corrosion cells [99–102]. Whether the anode and cathode behave "microscopically" or "macroscopically" is determined by the ratio of the dimension of either electrode C_i, divided by the polarization parameter L_i [100]. Mathematical modeling has indicated that, when the ratio, C_i/L_i, is small, the variation of current density across an electrode is small (i.e., the electrode behaves microscopically), On the other hand, when the characterizing ratio is large (i.e., when the electrode dimension is much larger than L_i), the electrode process can be regarded as macroscopic, and the variation of current density across the electrode surface is large.

I4. Potential and Current Distributions

The galvanic action between two metals is governed essentially by the potential distribution across the surface of each electrode. The galvanic current distribution can be determined from the potential distribution when the potential–current relationships for the electrodes are known. Potential distribution can be calculated theoretically or determined experimentally.

Theoretically, a complete description of the potential distribution on the surfaces of a galvanic couple can be obtained by solving Laplace's equation:

$$\nabla^2 E(x, y, z) = 0 \qquad (10.17)$$

This equation is derived from Ohm's law, which states that, at any point in the electrolyte, the current density is proportional to the potential gradient

$$I = \sigma \nabla E \qquad (10.18)$$

and from the electroneutrality law, which states that, at any point in the electrolyte, the net current under the steady state must be zero

$$\nabla I = 0 \qquad (10.19)$$

Many numerical models, with varying mathematical methods and in geometrical and polarization boundary conditions, have been developed for different galvanic systems, as listed in Table 10.9.

These numerical models provide many useful insights to galvanic corrosion. As an example, McCafferty [111] modeled the potential distribution of a concentric circular galvanic corrosion cell, assuming a linear polarization for both the anodic and the cathodic reactions. Figures 10.11 and 10.12 show the results of the potential distribution and current distribution, respectively, as a function of electrolyte thickness. In the bulk electrolyte, the potential variation across the electrodes is small, but both the anode and the cathode are strongly polarized; thus, the actual electrode potentials are far away from E_a^0 and E_c^0. Under a thin-layer electrolyte, the potential variation is large from the anode to the cathode, but both the anode and cathode are only slightly polarized, except for the areas near the boundary between the anode and the cathode. The galvanic current increases with increasing electrolyte thickness. Also, the current is distributed on the electrode surface more uniformly in bulk solutions than in thin-layer solutions where the current is more concentrated near the contact line in the thin electrolyte. According to the calculations of Doig and Flewitt [55], the potential distribution is uniform in the thickness direction under a thin layer of electrolyte (e.g., 1 mm), whereas it is nonuniform under a thick layer of electrolyte. Similar results were reported by Morris and Smyrl [114] for a galvanic cell with coplanar electrodes. The potential distribution under more general geometrical conditions has also been modeled [99, 105].

The results of numerical modeling can be used to predict the galvanic action for the entire surface area of coupled

TABLE 10.9. Studies of Mathematical Modeling for Various Galvanic Systems

Galvanic Couple	Solution	Geometry	Focus	Reference
Cu, Ti, Ni alloys/Zn	Seawater	Cylindrical	Seawater systems	103
Steel/I-600	EDTA solution[a]	Various	Numerical models	104
Fe/Zn	Seawater	General	Modeling	99
Steel/zinc	Seawater	General	Modeling	105
Fe/Cu	0.6 M NaCl	Circular disk	Local current	106
S. steel/steel	Water	Coplanar and tubular	Simulation	107
Miscellaneous	Seawater	Tube/sheet	Heat exchanger	23
Carbon/Co	Generic	Sandwich structure	Defects in films	108
Miscellaneous	Seawater	Cylindrical	Heat exchanger	109
Generic	Generic	Random distribution	Heterogeneous surface	70
Generic	Generic	General	Numeric method	110
Generic	Generic	Coplanar strips	Size effects	101
Generic	Generic	Annular electrodes	Thin-layer electrolyte	29
Pt/Fe	0.05 M NaCl	Circular cells	Polarization parameters	111
Cu/Zn	0.01 M HCl	Coplanar electrodes	Finite differ. analysis	112
Fe/Zn	Seawater	General geometries	Boundary conditions	113

[a]Ethylenediaminetetraacetic acid = EDTA.

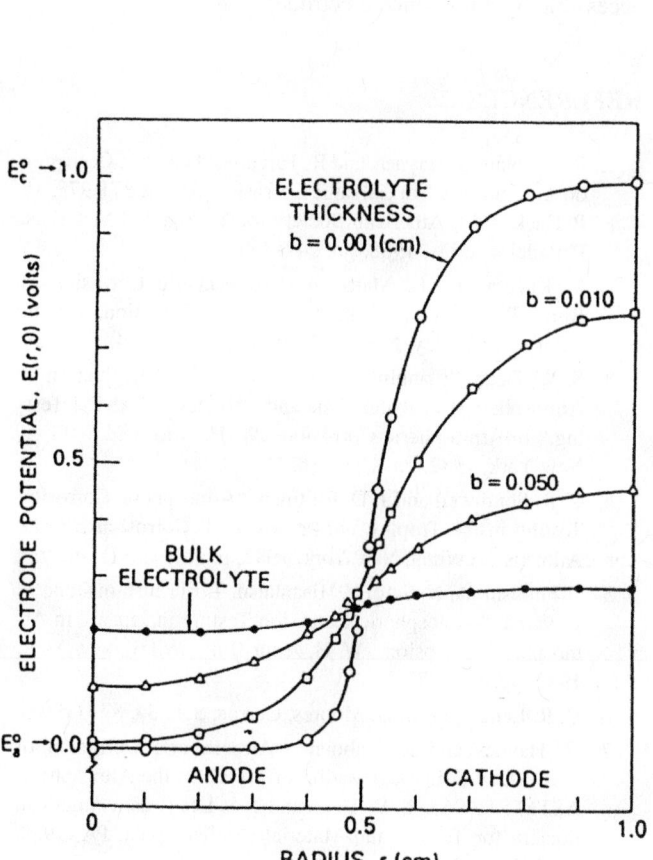

FIGURE 10.11. Distribution of electrode potential for length of anode $L_a = 1$ cm and length of cathode $L_c = 10$ cm for different electrolyte thicknesses. (Anode radius $a = 0.5$ cm, cathode radius $c = 1.0$ cm; $E_0^a = 0$V, $E_0^c = 1$V) [111].

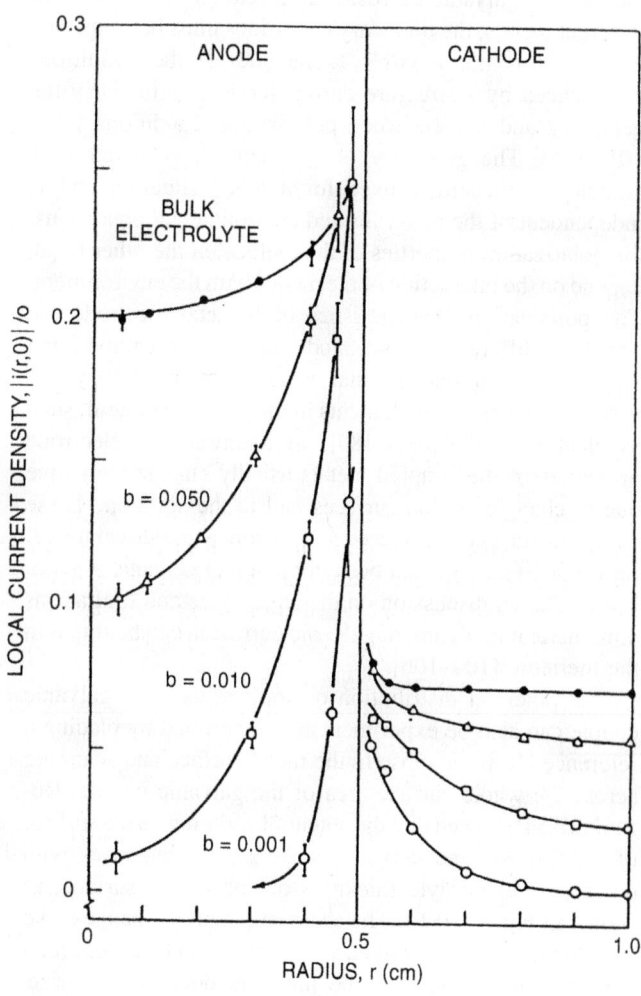

FIGURE 10.12. Current distribution for different electrolyte thicknesses under same conditions as in Figure 10.11 [111].

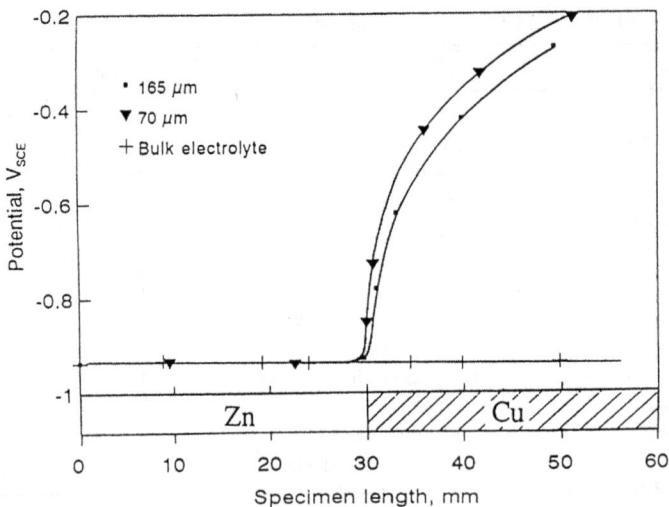

FIGURE 10.13. Distribution of potentials on electrode surface of galvanic couple Cu–Zn in 0.1 N NaCl solution as function of electrolyte thickness [54].

metals. For galvanic corrosion in a real structure made of different metals, the boundary conditions must be simplified because it is not possible to include all the conditions experienced by a structure during service, particularly the geometry and the electrode polarization conditions [104, 105, 115]. The geometry of a structure, no matter how complex, is generally fixed for a given situation and is independent of the materials and environmental conditions; the polarization properties of the metals, on the other hand, depend on the interaction of the metals with the environment. The polarization characteristics of a metal electrode are generally different for the anode and for the cathode and they vary in different potential ranges. Sometimes, they also vary with the physical elements in the galvanic system, such as electrolyte thickness [80]. In addition, the electrode properties of the coupled metals usually change with time due to changes on the surfaces and in the solution. These elements must be considered when using a numerical model for predicting long-term behavior is a real galvanic system. More detailed discussion on the advantages and limitations of numerical modeling of galvanic corrosion can be found in the literature [104–106].

The potential distribution on the surface of a galvanic couple can also be experimentally determined by placing a reference electrode close to the metal surface and scanning across the whole surface area of the galvanic couple. Rozenfeld [54] showed that the potential variation on the surface of a coplanar zinc–copper couple greatly increases with decreasing electrolyte thickness on top of the surface, as shown in Figure 10.13. The sharpest potential changes take place on the copper cathode, whereas the anode, except for a very narrow region near the junction, does not polarize. Using this experimental approach, data on potential distribution and galvanic action of the system can be obtained, but this method can be impractical in many situations (e.g., when the structure is so complex that not all the surface area is accessible by a reference electrode).

REFERENCES

1. R. Baboian, G. Haynes, and R. Turcotte, "Galvanic Corrosion on Automobiles," in Galvanic Corrosion, ASTM STP 978, H. P. Hack (Ed.), American Society for Testing and Materials, Philadelphia, PA, 1988, pp. 249–259.

2. V. Kucera and E. Mattsson, "Atmospheric Corrosion of Bimetallic Structures," in Atmospheric Corrosion, W. H. Ailor (Ed.), Wiley, New York, 1982, pp. 561–574.

3. S. W. Dean, "Planning, Instrumentation, and Evaluation of Atmospheric Corrosion Tests and a Review of ASTM Testing," in Atmospheric Corrosion, W. H. Ailor (Ed.), Wiley, New York, 1982, pp. 195–216.

4. C. R. Southwell and J. D. Bultman, "Atmospheric Corrosion Testing in the Tropics," in Atmospheric Corrosion, W. H. Ailor (Ed.), Wiley, New York, 1982, p. 967.

5. T. Fukushima, N. Sato, Y. Hisamatsu, T. Matsushima, and Y. Aoyama, "Atmospheric Corrosion Testing in Japan," in Atmospheric Corrosion, W. H. Ailor (Ed.), Wiley, New York, 1982, pp. 841–872.

6. D. R. Lenard and J. G. Moores, Corros. Sci., **34**, 871 (1993).

7. G. Haynes and R. Baboian, "Atmospheric Corrosion of Clad Metals," in Degradation of Metals in the Atmosphere, ASTM STP 965, S. W. Dean and T. S. Lee (Eds.), American Society for Testing and Materials, Philadelphia, PA, 1988, pp. 145–190.

8. V. Brusic, M. Russak, R. Schad, G. Frankel, A. Selius, D. DiMilia, and D. Edmonson, "Corrosion of Thin Film Magnetic Disc: Galvanic Effects of the Carbon Overcoat," J. Electrochem. Soc., **136**, 42 (1989).

9. A. M. Shams El Din and L. Wang, Br. Corros. J., **28**(4), 271 (1994).

10. G. K. Glass and V. Ashworth, Corros. Sci., **25**(11), 971 (1985).

11. R. B. Hoxeng and C. F. Prutton, Corrosion, **5**(10), 330 (1949).

12. P. T. Gilbert, "An Investigation into the Corrosion of Zinc and Zinc-Coated Steel in Hot Waters," Sheet Metal Industries, Oct.–Dec., 1948.

13. H. L. Shuldener and L. Lehrmen, Corrosion, **14**(12), 17 (1958).

14. C. R. Southwell, J. D. Bultman, and A.L. Alexander, Mater. Perform., **15**(7), 9 (1976).

15. D. R. Gabe and A. M. El Hassan, Br. Corros. J., **21**(3), 185 (1986).

16. S. G. Al Zaharani, B. Todd, and J. W. Oldfield, "Bimetallic Joints in Multistage Flash Desalination Plants," in Galvanic Corrosion, ASTM STP 978, H. P. Hack, (Ed.), American Society for Testing and Materials, Philadelphia, PA, 1988, pp. 323–335.

17. G. A. Gehring, Jr., "Galvanic Corrosion in Power Plant Condensers," in Galvanic Corrosion, ASTM STP 978, H. P. Hack (Ed.), American Society for Testing and Materials, Philadelphia, PA, 1988, pp. 301–309.

18. R. Baboian and G. Haynes, "Galvanic Corrosion of Ferritic Stainless Steels in Seawater," in Corrosion in Natural Environments, ASTM STP 558, W. H. Ailor, S. W. Dean, and F. H. Haynie (Eds.), American Society for Testing and Materials, Philadelphia, PA, 1974, pp. 171–184.

19. S. Schuldiner and R. E. White, J. Electrochem. Soc., **97**(12), 433 (1950).

20. H. P. Hack, "Galvanic Corrosion of Piping and Fitting Alloys in Sulfide-Modified Seawater," in Galvanic Corrosion, ASTM STP 978, H. P, Hack (Ed.), American Society for Testing and Materials, Philadelphia, PA, 1988, pp. 339–351.

21. K. D. Efird, "Galvanic Corrosion in Oil and Gas Production," in Galvanic Corrosion, ASTM STP 978, H. P. Hack (Ed.). American Society for Testing and Materials, Philadelphia, PA, 1988, pp. 260–282.

22. L. S. Redmerski, J. J. Eckenrod, K. E. Pinnow, and W. Kovach, "Experience with Cathodic Protection of Power Plant Condensers Operating with High Performance Ferritic Stainless Steel Tubing," in Galvanic Corrosion, ASTM STP 978, H. P. Hack (Ed.), American Society for Testing and Materials, Philadelphia, PA, 1988, pp. 310–322.

23. J. R. Scully and H. P. Hack, "Prediction of Tube-Tubesheet Galvanic Corrosion Using Finite Element and Wagner Number Analyses," in Galvanic Corrosion, ASTM STP 978, Philadelphia, PA, 1988, pp. 136–157.

24. R. Francis, Br. Corros. J., **29**(1), 53 (1994).

25. H. P. Hack and J. R. Scully, Corrosion, **42**, 79 (1986).

26. R. Foster, H. Hack, and K. Lucas, "Long-Term Current and Potential Data for Selected Galvanic Couples," Paper No. 517, CORROSION/96, NACE, Houston, TX, 1996.

27. M. Romanoff, "Underground Corrosion," Circular 579, U. S. National Bureau of Standards, Washington, DC, 1957.

28. E. Escalante, "The Effect of Soil Resistivity and Soil Temperature on the Corrosion of Galvanically Coupled Metals in Soil," in Galvanic Corrosion, ASTM STP 978, Philadelphia, PA, 1988, pp. 193–202.

29. K. G. Compton, Corrosion, **16**, 87 (1960).

30. N. Sridhar and J. Kolts, "Evaluation and Prediction of Galvanic Corrosion in Oxidizing Solutions," in Galvanic Corrosion, ASTM STP 978, H. P. Hack (Ed.), American Society for Testing and Materials, Philadelphia, PA, 1988, pp. 203–219.

31. R. A. Corbett, W. S. Morrison, and R. Snyder, "Galvanic Corrosion Resistance of Weld Dissimilar Nickel-Base Alloys," in Galvanic Corrosion, ASTM STP 978, H. P. Hack (Ed.), American Society for Testing and Materials, Philadelphia, PA, 1988, pp. 235–245.

32. Y. H. Yau and M. A. Streicher, "Galvanic Corrosion of Duplex Fe–Cr–10%Ni Alloys in Reducing Acids," in Galvanic Corrosion, ASTM STP 978, H. P. Hack (Ed.), American Society for Testing and Materials, Philadelphia, PA, 1988, pp. 220–234.

33. G. O. Davis, J. Kolts, and N. Sridhar, Corrosion, **42**, 329 (1986).

34. E. Symniotis, Corrosion, **46**, 2 (1990).

35. D. A. Jones and A. J. P. Paul, Corrosion, **50**, 516 (1994).

36. W. J. Polock and B. R. Hinton, "Hydrogen Embrittlement of Plated High-Strength 4340 Steel by Galvanic Corrosion," in Galvanic Corrosion, ASTM STP 978, H. P. Hack (Ed.), American Society for Testing and Materials, Philadelphia, PA, 1988, pp. 35–50.

37. L. H. Hihara and P. K. Kondepudi, Corros. Sci., **34**, 1761 (1993).

38. M. J. Pryor and D. S. Keir, J. Electrochem. Soc., **104**(5), 269 (1957).

39. L. M. Wing, J. Commander, J. O'Grady, and T. Koga, "Specifying Zinc Alloy Coatings for Improved Galvanic Corrosion Performance," Paper 971004, SAE, Warrendale, PA, 1997.

40. D. L. Jordan, "Influence of Iron Corrosion Products on the Underfilm Corrosion of Painted Steel and Galvanized Steel," in Zinc-Based Steel Coating Systems: Metallurgy and Performance, G. Krauss and D. K. Matlock (Eds.), TMS, Warrendale, PA, 1990, pp. 195–205.

41. C. Cabrillac and A. Exertier, "The Effect of Coating Composition on the Properties of a Galvanized Coating," in Proceedings of the 7th International Conference on Hot Dip Galvanizing, Paris, Pergamon. New York, June 1964, pp. 289–313.

42. S. Kurokawa, K. Yamato, and T. Ichida, "A Study on Cosmetic and Perforation Corrosion Test Procedures for Automotive Steel Sheets," Paper No. 396, NACE CORROSION'91 Conference, Cincinnati, OH, Mar. 11–15,1991, NACE, Houston, TX, 1991.

43. J. A. von Fraunhofer and A. T. Lubinski, Corros. Sci., **14**, 225 (1974).

44. E. A. Anderson, in Corrosion Resistance of Metals and Alloys, 2nd ed., F. L. LaQue and H. R. Copson (Eds.), Reinhold Publishing, New York, 1963, pp. 223–247.

45. A. Asphahani and H. H. Uhlig, J, Electrochem. Soc., **122**(2), 174 (1975).

46. G. Lauer and F. Mansfeld, Corrosion, **26**(11), 504 (1970).

47. F. Bellucci, Corrosion, **48**(4), 281 (1992).

48. P. Bellucci and G. Capobianco, Br. Corros. J., **24**(3), 219 (1989).

49. F. Bellucci, Corrosion, **47**, 808 (1991).

50. F. Mansfeld and J. V. Kenkel, Corros. Sci., **15**, 239 (1975).

51. F. Mansfeld, D. H. Hengstenberg, and J. V. Kenkel, Corrosion, **30**(10), 343 (1974).

52. L. H. Hihara and R. M. Latanision, Corros. Sci., **34**, 655 (1993).

53. L. H. Hihara and R. M. Latanision, Corrosion, **48**, 546 (1992).

54. I. L. Rozenfeld, Atmospheric Corrosion of Metals, NACE, Houston, TX, 1972.

55. P. Doig and P. E. J. Flewitt, J. Electrochem. Soc., **126**(12), 2057 (1979).

56. A. M. Shams El Din, J. M. Abd El Kader, and M. M. Badran, Br. Corros. J., **16**(1), 32 (1981).

57. S. M. Wilhelm, "Galvanic Corrosion Caused by Corrosion Products," in Galvanic Corrosion, ASTM STP 978, H. P. Hack (Ed), American Society for Testing and Materials, Philadelphia, PA, 1988, pp. 23–34.

58. F. Mansfeld and J.V. Kenkel. Corrosion, **31**, 298 (1974).

59. G. Schick, "Avoiding Galvanic Corrosion Problems in the Telephone Cable Plant," in Galvanic Corrosion, ASTM STP 978, Philadelphia, PA, 1988, pp. 283–290.

60. A. J. Griffin, Jr., S. E. Henandez, F. R. Brotzen, and C. F. Dunn, J. Electrochem. Soc., **141**, 807 (1994).

61. J. Gluszek and J. Masalski, Br. Corros. J., **27**(2), 135 (1992).

62. S. Huang and R. A. Oriani, "The Corrosion Potential of Galvanically Coupled Copper and Zinc Under Humid Gases," Abstract No. 91, Electrochemical Society Extended Abstracts, Fall Meeting, Oct. 10–15, 1993, New Orleans, LA, Electrochemical Society, Pennington, NJ, Vol. 2–93, p. 156, 1993.

63. D. Whiting, D. Stark, and W. Schutt, "Galvanic Anode Cathodic Protection System for Bridge Decks—Updated Results," Paper 41, CORROSION/81, International Corrosion Forum Sponsored by the National Association of Corrosion Engineers, Toronto, ON, Apr. 6–10, 1981.

64. A. Al-Hashem and D. Thomas, "The Effect of Corona Discharge Treatment on the Corrosion Behaviour of Metallic Zinc Spots Contained in Polymeric Coatings," in Proceedings of the International Conference on Zinc and Zinc Alloy Coated Steel Sheet, GALVATECH'89, Tokyo, Japan, Sept. 5–7, 1989, pp. 611–618.

65. X. Sun and S. Tsujikawa, Corros. Eng. (Jpn), **41**, 741 (1992).

66. J. W. Jang, I. Iwasaki, and J. J. Moore, Corrosion, **45**, 402 (1989).

67. S. M. Wilhelm, Corrosion, **48**, 691 (1992).

68. X. G. Zhang and J. Hwang, Mater. Perform., **36**(2), 22 (Feb. 1997).

69. X. G. Zhang, J. Electrochem. Soc., **143**, 1472 (1996).

70. R. Morris and W. Smyrl, J. Electrochem. Soc., **136**, 3237 (1989).

71. S. R. Morrison, Electrochemistry at Semiconductor and Oxidized Metal Electrodes, Plenum, New York, 1980.

72. M. Stratmann and J. Müller, Corros. Sci., **36**(2), 327 (1994).

73. D. L. Jordan, "Galvanic Interactions between Corrosion Products and Their Bare Metal Precursors: A Contribution to the Theory of Underfilm Corrosion," in Proceedings of the Symposium on Advances in Corrosion Protection by Organic Coatings, Vol. B9-3, Electrochemical Society, Pennington, NJ, 1989, pp. 30–43.

74. A. M. Shams El Din, J. M. Abd El Kader, and A. T. Kuhn, Br. Corros. J., **15**(4), 208 (1980).

75. X. G. Zhang, Corrosion and Electrochemistry of Zinc, Plenum, New York, 1996.

76. L. Kenworthy, J. Inst. Metals, **69**, 67 (1943).

77. D. Massinon and D. Thierry, "Rate Controlling Factors in the Cosmetic Corrosion of Coated Steels," Paper No. 574, NACE CORROSION'91 Conference, Mar. 11–15, NACE, Houston, TX, 1991.

78. D. Massinon and D. Dauchelle, "Recent Progress Towards the Understanding of Underfilm Corrosion of Coaled Steels Used in the Automotive Industry," in Proceedings of the International Conference on Zinc and Zinc Alloy Coated Steel Sheet, GALVATECH'89, Tokyo, Japan, Sept. 5–7 1989, pp. 585–595.

79. M. J. Pryor and D. S. Keir, J. Electrochem, Soc., **105**(11), 629 (1958).

80. X. G. Zhang and E. M. Valeriote, Corros. Sci., **34**, 1957 (1993).

81. I. L. Rosenfeld, "Atmospheric Corrosion of Metals. Some Questions of Theory," The 1st International Congress on Metallic Corrosion, London, UK, Apr. 1961, pp. 243–253.

82. D. P. Doyle and T. E. Wright, "Quantitative Assessment of Atmospheric Galvanic Corrosion," in Galvanic Corrosion, ASTM STP 978, American Society for Testing and Materials, Philadelphia, PA, 1988, pp. 161–171.

83. K. G. Compton and A. Mendizza, "Galvanic Couple Corrosion Studies by Means of the Threaded Bolt and Wire Test," ASTM 58th Annual Meeting, Symposium on Atmospheric Corrosion of Non-Ferrous Metals, American Society for Testing and Materials, ASTM STP 175, Philadelphia, PA, 1955, pp. 116–125.

84. A. K. Dey, A. K. Sinha Mahapatra, D. K. Khan, A. N. Mukherjee, R. Narain, K. P. Mukherjee, and T. Banerjee, NML Tech. J. India, **8**(4), 11 (1966).

85. M. E. Warwick and W. B. Hampshire, "Atmospheric Corrosion of Tin and Tin Alloys," in Atmospheric Corrosion, W. H. Ailor (Ed.), Wiley, New York, 1982, pp. 509–527.

86. X. G. Zhang, "Galvanic Protection Distance of Zinc Coated Steels Under Various Environmental Conditions," Paper No. 747, CORROSION/98, NACE, Houston, TX. 1998.

87. X. G. Zhang and E. M. Valeriote, "Galvanic Protection of Steel by Zinc Under Thin Layer Electrolytes," in Atmospheric Corrosion, ASTM STP 1239, W. W. Kirk and H. H. Lawson (Eds.), American Society for Testing and Materials, Philadelphia, PA, 1995, pp. 230–239.

88. M. G. Fontana and N. D. Greene, Corrosion Engineering, 2nd ed., McGraw-Hill, New York, 1978, p. 32.

89. F. Mansfeld and J. V. Kenkel, Corrosion, **33**(7), 236 (1977).

90. F. Mansfeld and J. V. Kenkel, Corrosion, **31**(8), 298 (1975).

91. G. Schikorr, Trans. Electrochem. Soc., **76**, 247 (1939).

92. R. B. Hoxeng, Corrosion, **6**(9), 308 (1950).

93. A. G. S. Morton, "Galvanic Corrosion in Navy Ships," in Galvanic Corrosion, ASTM STP 978, H. P. Hack (Ed.), American Society for Testing and Materials, Philadelphia, PA, 1988, pp. 291–300.

94. H. P. Hack, Galvanic Corrosion Test Methods, NACE International, Houston, TX, 1993.

95. L. Sherwood, "Sacrificial Anodes," in Corrosion, vol. 2, L. L. Shreier, R. A. Jarman, and G. T. Burstein (Eds.), Butterworth Heinemann, 1995, pp. 1029–1054.

96. T. J. Lennox, Jr., "Electrochemical Properties of Mg, Zn and Al Galvanic Anodes in Sea Water," Proceedings of the Third International Congress on Marine Corrosion and Fouling, Gaithersburg, MD, Oct. 2–6, 1972, National Technical Information Service (NTIS), Alexandria, VA, 1974, pp. 176–190.

97. D. Massinon, D. Dauchelle, and J. C. Charbonnier, Mater. Sci. Forum, **44–45**, 461 (1989).

98. C. Wagner, J. Electrochem. Soc., **99**(1), 1 (1952).

99. R. S. Munn and O. F. Devereux, Corrosion, **47**(8), 618 (1991).

100. S. M. Abd El Haleem, Br. Corros. J., **11**(4), 215 (1976).

101. J. T. Waber, "Analysis of Size Effects in Corrosion Processes," in Localized Corrosion, NACE, Houston, TX, 1974, pp. 221–237.

102. J. T. Waber et al., J. Electrochem. Soc., **101**(6), 271 (1954); **102**(6), 344 (1955); **102**(7), 420 (1955); **103**(1), 64 (1956); **103**(2), 138 (1956); **103**(10), 567 (1956).

103. D. J. Astlry, "Use of the Microcomputer for Calculation of the Distribution of Galvanic Corrosion and Cathodic Protection in Seawater Systems," in Galvanic Corrosion, ASTM STP 978, H. P. Hack (Ed.), American Society for Testing and Materials, Philadelphia, PA, 1988, pp. 53–78.

104. J. W. Fu, "Galvanic Corrosion Prediction and Experiments Assisted by Numerical Analysis," in Galvanic Corrosion, ASTM STP 978, H. P. Hack (Ed.), American Society for Testing and Materials, Philadelphia, PA, 1988, pp. 79–85.

105. R. A. Adey and S. M. Niku, "Computer Modelling of Galvanic Corrosion," in Galvanic Corrosion, ASTM STP 978, H. P. Hack (Ed.), American Society for Testing and Materials, Philadelphia, PA, 1988, pp. 96–117.

106. R. G. Kasper and C. R. Crowe, "Comparison of Localized Ionic Currents as Measured From 1-D and 3-D Vibrating Probes with Finite-Element Predictions for an Iron-Copper Galvanic Couple," in Galvanic Corrosion, ASTM STP 978, H. P. Hack (Ed.), American Society for Testing and Materials, Philadelphia, PA, 1988, pp. 118–135.

107. E. Bardal, R. Johnsen, and P. O. Gartland, Corrosion, **40**, 628 (1984).

108. A. Kassimati and W. H. Smyrl, J. Electrochem. Soc., **136**, 2158 (1989).

109. D. J. Astley and J. C. Bowlands, Br. Corros. J., **20**(2), 90 (1985).

110. R. B. Morris, J. Electrochem. Soc., **137**, 3039 (1990).

111. E. McCafferty, J. Electrochem. Soc., **124**(12), 1869 (1977).

112. P. Doig and P. E. J. Flewitt, Br. Corros. J., **13**(3), 118 (1978).

113. R. S. Munn and O. F. Devereux. Corrosion, **47**(8), 612 (1991).

114. R. Morris and W. Smyrl, J. Electrochem. Soc., **136**(11), 3229 (1989).

115. F. LaQue, Corrosion, **39**, 36 (1983).

11

DEALLOYING

R. Heidersbach

Dr. Rust, Inc. Cape Canaveral, Florida

A. DEFINITION

Dealloying is a corrosion process whereby one constituent of an alloy is removed leaving an altered residual structure. It was first reported in 1886 on copper–zinc alloys (brasses) and has since been reported on virtually all copper alloys as well as on cast irons and many other alloy systems [1, 2]. Other terms for dealloying include parting, selective leaching, and selective attack. Terms such as dezincification, dealuminification, cation, denickelificaton, and the like indicate the loss of one constituent of the alloy, but the general term dealloying has gained wider use in recent years.

B. HISTORY

During World War I the dezincification of brass condenser tubes was a major problem that kept more British ships out of action than the efforts of the German navy. This problem lessened with the development of more corrosion-resistant alloys, and dealloying is no longer a major problem for most marine operators.

Cast iron gas mains also dealloyed in soil and produced weakened pipes that could not withstand increased pressures. These pipes leaked as demands for natural gas increased and gas company operators increased pressures on their lines. The continued use of cast iron water piping results in occasional reports of dealloying caused by corrosive waters or corrosive soils.

In recent years research has concentrated on using dealloying as a means of producing high-surface-area substrates for catalysis and other purposes [3–8]. The subject of deliberate dealloying as a manufacturing process has been reviewed in the American Society for Metals (ASM) International *Metals Handbook* [9] and will not be discussed further in this chapter.

C. MECHANISM

Dealloying has been shown to occur by at least two different mechanisms. Sometimes the entire alloy dissolves, and one constituent redeposits on the corroded metal surface. In other circumstances diffusion removes only the more corrosion-susceptible constituent, leaving an altered porous matrix. Both mechanisms have been shown to occur simultaneously on the same metal surface [10].

Figure 11.1 shows copper deposits on the surface of dezincified alpha brass. Similar deposits have been reported on cupronickels and Monel [11, 12].

Figure 11.2 is a more typical picture of dealloying showing a porous discolored structure surrounded by unattacked metal. This picture is from a corroded scuba tank valve, and the porous chrome plating allowed seawater to penetrate the protective chrome plating and corrode the underlying brass.

Dealloying is also a problem with cast irons. While both diffusion and noble metal deposition are discussed as

Uhlig's Corrosion Handbook, Third Edition, Edited by R. Winston Revie
Copyright © 2011 John Wiley & Sons, Inc.

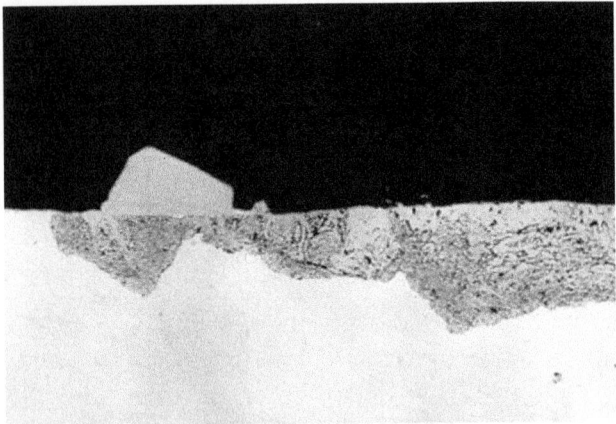

FIGURE 11.1. Copper deposit on surface of dezincified alpha brass.

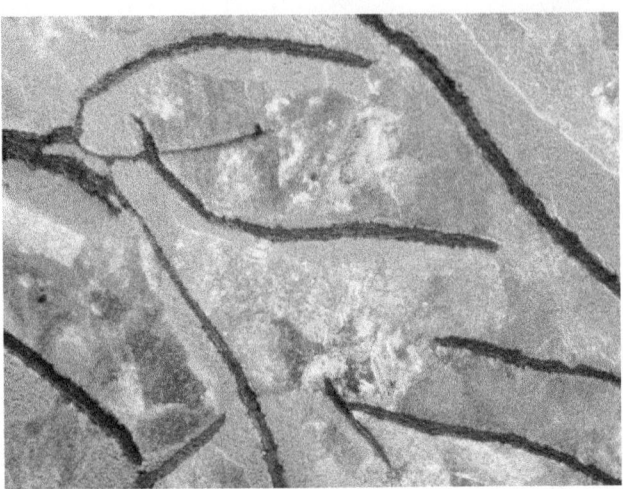

FIGURE 11.4. Microstructure of gray cast iron.

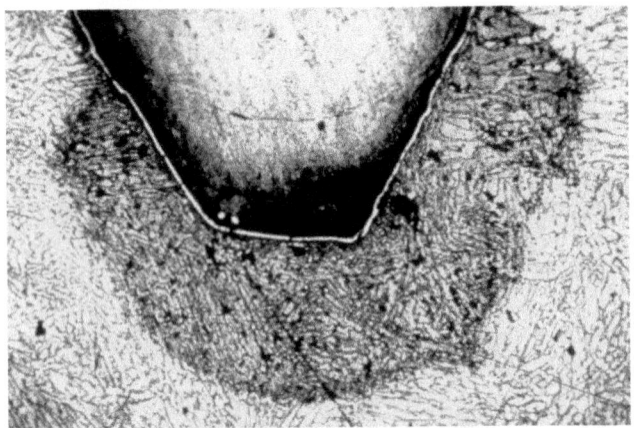

FIGURE 11.2. Dezincified brass scuba tank valve [12]. (Reproduced with permission. © NACE International 1982.)

mechanisms associated with dealloying in copper-based alloys, the mechanism of dealloying in cast irons involves the dissolution of the iron-rich phases, leaving a porous matrix of graphite and iron corrosion products. Figure 11.3 shows a porous graphite plug in a cast iron water pipe.

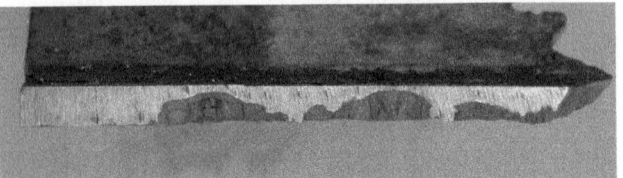

FIGURE 11.3. Dark, graphitic corrosion on the exterior of a cast iron water main. Arrow points to graphitic corrosion (dealloyed) region. (Photo courtesy Testlabs International Ltd., Winnipeg, Manitoba, sole owner of the photograph.)

Gray cast iron, which is considered obsolete for most corrosion-related applications, is most susceptible to this problem. Figure 11.4 shows the microstructure of gray cast iron. The dark regions are graphite flakes, essentially pure carbon, while the lighter phases are essentially pure iron. While carbon is only 2–4% by weight of the alloy, the volume percent is much higher, and the almost continuous graphite matrix maintains the original profile of the cast iron, but with much reduced mechanical integrity. This is a form of selective phase attack, which is discussed further in the section on copper alloys.

At one time it was assumed that only certain alloys would undergo dealloying, and the concept of a "parting limit" was popular. The idea was that some alloys, for example, the 60Cu–40Zn duplex alpha-plus-beta-phased brasses, were more susceptible to this corrosion than the single-phased alpha brasses containing 30% zinc or less. Many authorities claimed that "red brasses" having zinc contents of 15% or less would not dezincify and suggested a parting limit somewhere between 15 and 30% zinc. In recent decades dealloying has been reported in brasses down to 2 wt %, and the concept of parting limits has been generally discarded, although it is still believed in some circles [9, 13].

Parrish and Verink [14] proposed that the potential differences between the equilibrium potentials for copper and nickel could be used to explain the dealloying of copper–nickel alloys. They superimposed the potential–pH (Pourbaix) diagrams for these two pure metals and hypothesized that the potential region between the two equilibrium potentials would be a region where dealloying was likely to occur in copper–nickel alloys [14]. This concept has been applied to other systems [15] and has gained widespread acceptance as a means of explaining why dealloying can occur [16]. The shaded area in Figure 11.5 shows the potential–pH region where dealloying is thought to occur

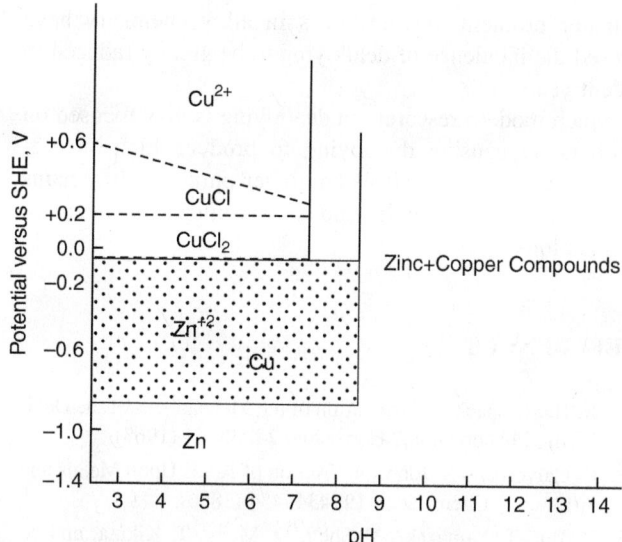

FIGURE 11.5. Superimposed potential–pH diagrams for copper and zinc. Shaded region indicates dezincification susceptibility.

for copper–zinc (brass) alloys. Pure copper is immune from corrosion in this region, and pure zinc will corrode producing zinc ions at low pHs. The result is a copper-rich dealloyed surface.

The concepts described in the preceding paragraph are also used in varying ways by the researchers working on producing high-surface-area dealloyed structures for manufacturing processes [5].

D. CONTROL OF DEALLOYING CORROSION

Almost any method that will lower corrosivity of an environment will reduce the rate of dealloying, but most dealloying is controlled by proper alloy selection.

D1. Copper Alloys

The three most common alloy systems used in corrosion service are copper–zinc brasses, bronzes (historically copper–tin alloys, but now more often aluminum bronzes), and cupronickels.

Single-phase alpha brasses (30% zinc or less) and duplex alpha-plus-beta brasses (typically 40% zinc) are the only forms of copper–zinc alloys that have engineering significance. The problems with dezincification of marine service brass condenser tubes led to extensive research in the early part of the twentieth century. Single-phase admiralty and duplex naval brasses, containing tin additions of approximately 1%, were found to reduce most forms of corrosion, to include dezincification. These became standard tubing alloys in the 1920s and 1930s.

Dealloying of brasses is normally controlled by the use of alloys with small additions of arsenic, antimony, or phosphorus. Bengough and May reported, in 1922, that small additions of arsenic would prevent dezincification of alpha brasses (usually about 30% zinc) but would not protect the beta phases of duplex alloys (60Cu–40Zn alloys) [17]. The reasons for arsenic protecting alpha brass but not the beta phase of duplex alloys have remained controversial but may be related to differential solubilities of arsenic in the alpha and beta phases [18]. The toxicity of arsenic has led to research into alternatives for arsenic [19].

In freshwater systems aluminum brass (approximately 70Cu–29Zn–1Al) is a common tubing material, for example, for condensers using freshwater for cooling. The aluminum is primarily useful for erosion resistance and, like tin in admiralty and naval brasses, lowers most corrosion rates but does not specifically affect dealloying.

Duplex brasses were at one time used in high-pressure condensers on some marine vessels, but cupronickels have largely replaced brasses for condenser tubes. Brasses and bronzes are, however, still used for pumps and fittings on these systems, and dealloying remains a problem.

Cupronickels have largely replaced brasses for marine piping applications. Both cupronickels (90–10 and 70–30 Cu–Ni) and Monels (approximately 70Ni–30Cu) have been reported to dealloy by a redeposition process, usually in stagnant conditions; but this is rare [1, 10, 11].

Pumps and fittings in copper-based piping systems are typically made of bronzes, multiphased alloys with high erosion resistance. At one time most bronze casting alloys had tin as the primary alloying addition, but the high cost of tin has led to the development and widespread use of alternative aluminum bronzes, which are cheaper and have most of the corrosion resistance and other properties of copper–tin bronzes. Common alloys include copper–manganese–aluminum and nickel–aluminum bronzes. Unfortunately, both of these alloy systems can form corrosion-susceptible phases that can lead to selective phase attack, a form of dealloying. This is a foundry control problem, but it can also occur when welding is necessary for repairs [20, 21]. The extent to which this attack occurs depends on the electrochemical potential difference between the phases. Since these alloys are usually used in castings, which are relatively thick compared to wrought products like tubing, the consequences of corrosion may be relatively minor, but loss of sealing surfaces and loss of strength can become problems. Figure 11.6 shows selective phase attack on a nickel–aluminum bronze casting in seawater service.

D2. Cast Irons

Gray cast iron has a long history of dealloying. At one time this was a common material used for buried pipelines and similar applications.

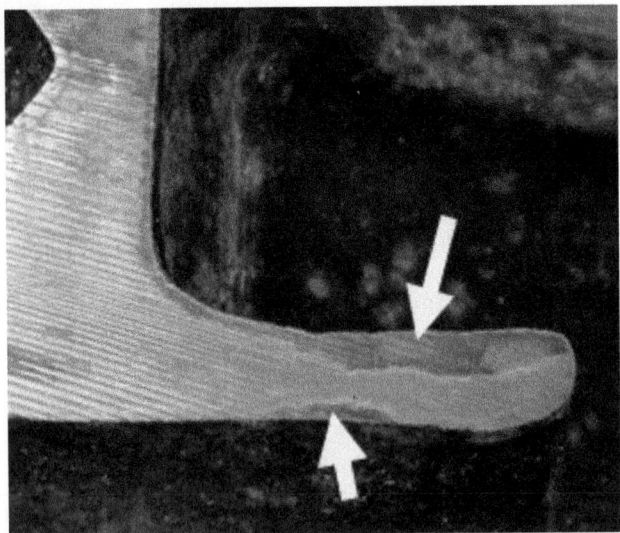

FIGURE 11.6. Selective phase attack of nickel–aluminum bronze [22]. (Photo courtesy Defence R&D Canada – Atlantic.)

The development of malleable and nodular (marketed in North America as "ductile cast iron") cast irons has caused gray cast iron to become obsolete for corrosive environment applications such as water piping and buried pipelines. The rounded graphite second phases in malleable or nodular cast iron present less grain boundary area to act as cathodes for galvanic corrosion of the high-iron phases in these metals. This improves their corrosion resistance somewhat. Buried water lines made from nodular (ductile) cast iron piping have been in use for several decades, and many utilities are reporting dealloying of this metal, which was once considered to be immune to dealloying.

Dealloying of cast irons is called "graphitic corrosion" because the previous term for this corrosion, "graphitization," is also applied to foundry heat treatment processes.

E. INSPECTION

Inspection for dealloying has been very difficult because this form of corrosion usually leaves a surface profile similar to uncorroded metal. The only way of finding dealloying has been by visual inspection for color changes or by mechanical probing to identify a loss of integrity.

Several researchers have reported on attempts to develop methods of in situ nondestructive inspection for dealloying, but they have not achieved widespread acceptance [20, 22].

F. SUMMARY

Dealloying has been known to occur in copper and iron-based alloys since the nineteenth century. At one time this was a major problem, but advances in alloy chemistry have caused the incidence of dealloying to be greatly reduced in recent years.

Much modern research on dealloying is now focused on deliberately causing dealloying to produce high-surface-area structures for catalysis and other purposes. It remains to be seen if this will lead to significant commercial applications.

REFERENCES

1. R. Heidersbach, "Clarification of the Mechanisms of the Dealloying Phenomenon," Corrosion, **24**, 38–44 (1968).

2. C. Calvert and R. Johnson, "Action of Acids Upon Metals and Alloys," J. Chem. Soc., **19**, 434–454 (1886).

3. Z. Liu, T. Yamazaki, Y. Shen, D. Meng, T. Kikuta, and N. Nakatani, "Fabrication of WO3 Nanoflakes by a Dealloying-based Approach," Chem. Lett., **37**(3), 296–297 (2008).

4. J. Hayes, A. Hodge, J. Biener, A. Hamza, and K. Sieradzki, "Monolithic Nanoporous Copper by Dealloying Mn-Cu," J. Mater. Res., **21**(10), 2611–2616 (Oct. 2006).

5. J. Erlebacher and K. Sieradzki, "Pattern Formation During Dealloying," Scripta Materialia, **49**(10), 991–996 (2003).

6. J. Y, Y. Ding, C. Xu, and A. Inoue, "Nanoporous Metals by Dealloying Multicomponent Metallic Glasses," Chem. Mater., **20**(14), 4548–4550 (2008).

7. J. Ehlebacher, M. Aziz, A. Dimitrov, and K. Sieradzki, "Evolution of Nanoporosity in Dealloying," Nature, **410**, 450–453 (2001).

8. B. Ateya, G. Geh, A. Carim, and H. Pickering, "Characterization of the Product Layer after Selective Dissolution below the Critical Potential and Back Alloying in Copper-Gold Alloys," J. Electrochem. Soc., **149**, B27–B33 (2002).

9. S. Corcoran, "Effects of Metallurgical Variables on Dealloying Corrosion," in Metals Handbook, Vol. 13A, Corrosion: Fundamentals, Testing and Protection, ASM International, Metals Park, OH, 2003, pp. 287–293.

10. E. Verink and R. Heidersbach, "The Dezincification of Alpha and Beta Brasses," Corrosion, **28**, 397–418 (1972).

11. D. R. Lenard, J. Martin, and R. Heidersbach, "Dealloying of Cupro-Nickels in Stagnant Seawater," CORROSION/99 paper no. 99314 (Houston, TX: NACE International, 1999).

12. R. Heidersbach, "Dealloying Corrosion," in NACE Handbook, Vol. 1, Forms of Corrosion—Recognition and Prevention, C. P. Dillon (ed.), NACE, Houston, TX, 1982, Chapter 7, pp. 99–104.

13. R. W. Revie and H. H. Uhlig, Corrosion and Corrosion Control, 4th ed., Wiley-Interscience, Hoboken, NJ, 2008, pp. **17**, 334, 374.

14. E. Verink and P. Parrish, "Use of Pourbaix Diagrams in Predicting Susceptibility to Dealloying Phenomena," Corrosion, **26**, 214 (1970).

15. E. Verink and R. Heidersbach, "Evaluation of the Tendency for Dealloying in Metal Systems," in Localized Corrosion—Cause

of Metal Failure, ASTM STP **516**, American Society for Testing and Materials, West Conshocken, PA, 1972, pp. 303–322.

16. R. Oliphant and M. Schock, "Copper Alloys and Solders," in Internal Corrosion of Water Distribution Systems, 2nd ed. American Water Works Association and AWWA Research Foundation, Denver, CO, 1996, Chapter 6.

17. R. B. Abrams, Trans. Am. Electrochem. Soc., **42**, 39 (1922), with discussion by G. D. Bengough and R. May.

18. M. J. Pryor and K-K Giam, "The Effect of Arsenic on the Dealloying of α-Brass," J. Electrochem. Soc., **129**, 2157–2163 (1982).

19. J. Chen, Z. Li, and Y. Zhao, "Corrosion Characteristic of Ce Al Brass in Comparison with As Al Brass," Mater. Design, **30**, 1743–1747 (2009).

20. E. Culpan and A. Foley, "The Detection of Selective Phase Corrosion of Nickel Aluminum Bronze by Acoustic Emission Techniques," J. Mater. Sci., **17**, 953–964 (1982).

21. Z. Han, Y. He, and C. Lin, "Dealloying Characterizations of Cu-Al Alloy in Marine Environment," J. Mater. Sci. Lett., **19**, 393–395 (2000).

22. D. R. Lenard, C. J. Bayley, and B. A. Noren, "Electrochemical Monitoring of Selective Phase Corrosion of Nickel Aluminum Bronze in Seawater," Corrosion, **64**, 764–772 (2008).

12

PASSIVITY

J. Kruger[†]

Department of Materials Science and Engineering, Johns Hopkins University, Baltimore, Maryland

A. INTRODUCTION

All metals and alloys (it is commonly held that gold is an exception) have a thin protective corrosion product film present on their surface resulting from reaction with the environment. If such a film did not exist on metallic materials exposed to the environment, they would revert back to the thermodynamically stable condition of their origin—the ores used to produce them. Some of these films—the passive films—on some, but not all, metals and alloys have special characteristics that enable them to provide superior corrosion-resistant metal surfaces. These protective "passive" films are responsible for the phenomenon of passivity.

B. IMPORTANCE OF PASSIVITY TO CORROSION CONTROL TECHNOLOGY

As mentioned in the introduction, if the passive film did not exist, most of the technologies that depend on the use of

[†]Retired

metals in any society could not exist because the phenomenon of passivity is a critical element controlling corrosion processes. One of these corrosion processes is the destruction of passivity (breakdown) that leads to a large part of the corrosion failures of metal and alloy structures—localized attack such as pitting, crevice corrosion, stress corrosion, corrosion fatigue, and others. The passivity of metallic materials, however, is not only of predominant significance to corrosion science and engineering. For example, semiconductor device technology depends on the superior passive films that form on silicon [1].

The development of the stainless steels in the third decade of the twentieth century is regarded as a major application of the phenomenon of passivity. The consequence of this has significantly contributed to modern technology by providing the design engineer with engineering materials such as the large number of iron and nickel base alloys as well as other alloy systems that exhibit superior corrosion resistance—this effort continues to the present day.

C. DEFINITIONS

Two types of passivity have been defined by Revie and Uhlig [2]:

Type 1—"A metal is passive if it substantially resists corrosion in a given environment resulting from marked anodic polarization" (low corrosion rate, noble potential).

Type 2—"A metal is passive if it substantially resists corrosion in a given environment despite a marked thermodynamic tendency to react" (low corrosion rate, active potential).

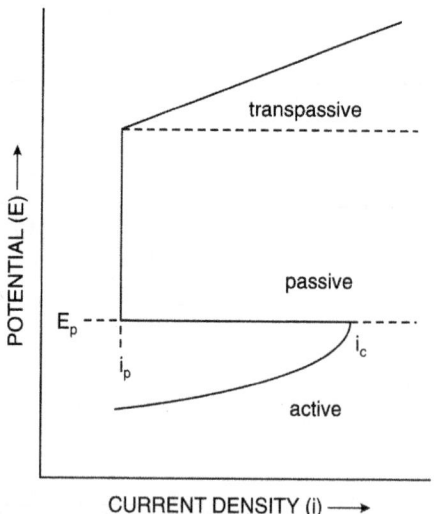

FIGURE 12.1. Schematic and idealized anodic polarization curve for passive metal. Three different potential regions are identified: the active, the passive, and the transpassive. When the current density reaches the critical current density for passivation i_c, at the passivation potential E_p, the potential above which the system becomes passive and exhibits the very low passive current density i_p.

Only type 1 passivity will be covered here. Metals or alloys exhibiting type 1 passivity are nickel, chromium, titanium, iron in oxidizing environments (e.g., chemical passivating solutions such as chromate), stainless steels, and numerous others. Type 2 passivity examples are lead in sulfuric acid and iron in an inhibited pickling acid.

A polarization curve [current density (rate) vs. potential (driving force)], of the sort shown in Figure 12.1 distinguishes a type 1 passive system. The curve in Figure 12.1 epitomizes the definition of type 1 passivity proposed by Wagner [3] who extended the type 1 definition by stating that a metal becomes passive when its potential is increased in the positive or anodic (oxidizing) direction to a potential where the current density (rate of anodic dissolution) decreases (in many cases by orders of magnitude) to a value less than that observed at a less anodic potential. This decrease occurs even though the driving force for anodic dissolution is brought to a higher value because of the formation of the passive film.

Another definition of passivity has been provided by the NACE/ASTM Committee J01, Joint Committee on Corrosion "passive—the state of a metal surface characterized by low corrosion rates in a potential region that is strongly oxidizing for the metal" [4].

D. CORROSION SCIENCE AND ENGINEERING

Although virtually all of the progress in the development of alloys that exhibit superior corrosion behavior has resulted

from empirical rather than fundamental research, this section will briefly go into those fundamental aspects that have a bearing on corrosion engineering. It will concentrate on the basic aspects that are concerned with the film responsible for passivity, its formation and breakdown and its nature. Detailed discussions of the fundamental aspects of the passive film and the electrochemistry of passivation can be found in a number of sources [5–10].

D1. Nature and Properties of the Passive Film

Because it is generally accepted that a film is responsible for the phenomenon of passivity, the nature and properties of the passive film that make it protective are significant issues. For the purposes of this chapter this important, and still controversial, subject cannot be covered in any detail. Instead, some of the large number of publications [5–10] that review and discuss the concepts and research that go extensively into the nature (e.g., composition and thickness) and properties (e.g., structure, electronic properties, and mechanical properties) of the passive film should be referred to by those who want to go into this aspect of passivity.

D2. Thermodynamics

Passivity has been defined by the fact that it lowers the rate of corrosion, sometimes drastically—a kinetic consideration. However, before addressing the kinetics of passivation, the thermodynamic aspects must be discussed. This is needed when considering type 2 passivity, which involves any kind of corrosion product film rather than the special type 1 "passive film." This is so because thermodynamics provides a guide to the conditions under which passivation by any kind of corrosion product film, type 1 or 2, becomes possible. A most worthwhile and practical guide produced by corrosion science is the potential–pH diagram, the Pourbaix diagram. Pourbaix's *Atlas of Electrochemical Equilibria in Aqueous Solutions* [11] describes applications of these potential–pH equilibrium diagrams to corrosion science and engineering for most of the metals. An application that is especially valuable for the field of corrosion is the establishment in the Pourbaix diagrams of the theoretical conditions or domains of corrosion, immunity, and passivation. A simplified diagram for the iron–water system is given in Figure 12.2. Thermodynamic data enable the determination of the potential–pH conditions for *immunity* where no corrosion is possible, for *passivation* where a corrosion product film forms that may confer protection against corrosion, and for *corrosion* where corrosion or activation is expected. Whether a given system's film is protective (passive) or not is a kinetic consideration and not a thermodynamic one that depends on the nature of the passive film. Pourbaix's book presents a set of simplified diagrams like the one shown in Figure 12.2 for the majority

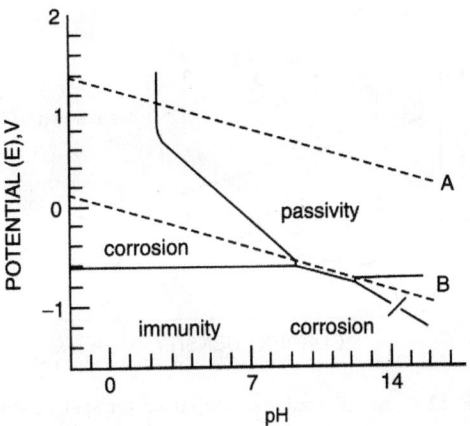

FIGURE 12.2. Simplified potential–pH equilibrium diagram (Pourbaix diagram) for iron–water.

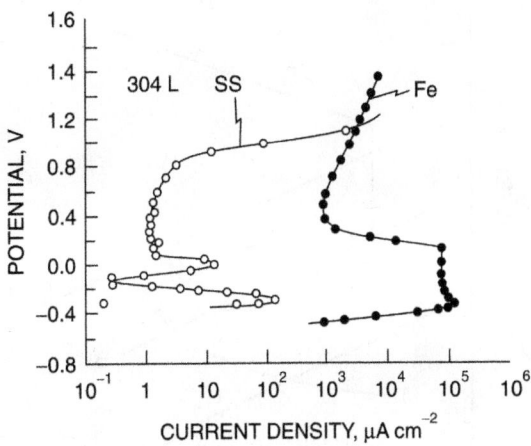

FIGURE 12.3. Anodic polarization curves for iron and 304L stainless steel in $1N\,H_2SO_4$. (Adapted from Fontana and Greene [13].)

of metals that enable one to identify the conditions for corrosion, passivation, and immunity. These equilibrium diagrams could be called "road maps of the thermodynamically possible conditions for passivity."

Consequently, to identify the active, passive, and transpassive regions of an active–passive polarization curve (Fig. 12.1), one need only examine the potential–pH diagram for a given system. For example, the transpassive region lies at potentials above the oxygen evolution potential–pH line (line *A* in Fig. 12.2). Another valuable use of the diagrams in passivity considerations is their use to explain the reasons for loss of the protective nature of the passive film in the transpassive region. For example, the protective layer on stainless steels contains chromium as Cr(III); when the potential is raised above the transpassive potential, Cr(III) is oxidized to Cr(VI) and the protective Cr(III) in the passive film is removed because it becomes the soluble chromate ion, as the potential–pH diagram for chromium shows. Usually, the passive regions of the polarization curves exist at potentials in the equilibrium diagrams where passive films are stable. It should finally be mentioned that the thermodynamically based Pourbaix diagrams sometimes fail to predict a passive region in a polarization curve. Hoar [12] sought to explain how iron can passivate in sulfuric acid solutions under conditions where the diagrams would predict an active condition, by proposing the existence of metastable passive regions.

D3. Kinetics

Passivity affects the kinetics of the corrosion process because it produces a protective (passive) film that acts as a barrier to attack of the metal surface by the environment. The passive film substantially lowers the rate of corrosion, even though from thermodynamic (corrosion tendency) considerations active corrosion should be expected. The

hypothetical anodic polarization curve (Fig. 12.1) that is the basis for the definition of type 1 passivity describes in a general way the kinetics of passivity. Anodic polarization curves for iron and 304L stainless steel in H_2SO_4 (Fig. 12.3) show the three general features of the schematic curve for a passive system (Fig. 12.1): (1) The current initially increases with an increase in potential. (2) When the potential reaches the value of the passivating potential E_p, the critical current density for passivation i_c is reached. (3) The current density (corrosion rate) decreases markedly (more for the corrosion-resistant stainless steel than for the iron) to the passive current density i_p, signaling the onset of passivity. As Figure 12.1 shows, the potential cannot be increased indefinitely because at sufficiently high values the current density starts to increase, resulting in either the initiation of pitting or entry into the transpassive region. In the transpassive region oxygen evolution and possibly increased corrosion are observed.

Since corrosion is the result of the interaction of the anodic and cathodic reactions, the corrosion potential of a metal surface is determined by the intersection of the anodic and cathodic polarization curves, the intersection occurring where the anodic and cathodic reaction rates are equal to the corrosion rate. Therefore, the corrosion rate of a metal capable of exhibiting passivity, will be determined by the location of the intersection of the cathodic polarization curve and the passive metal anodic polarization curve of the type shown in Figure 12.1. Three possible cases can arise from the intersection of different cathodic curves with the passive anodic polarization curve (shown schematically in Fig. 12.4): (1) A cathodic reaction arising from oxidizing conditions, polarization curve *A*, sets the corrosion potential in the passive region (low corrosion rate). (2) A cathodic reaction, the result of reducing conditions, curve *C*, places the corrosion potential in the active region (high corrosion rate). (3) The cathodic reaction, polarization curve *B*, locates the

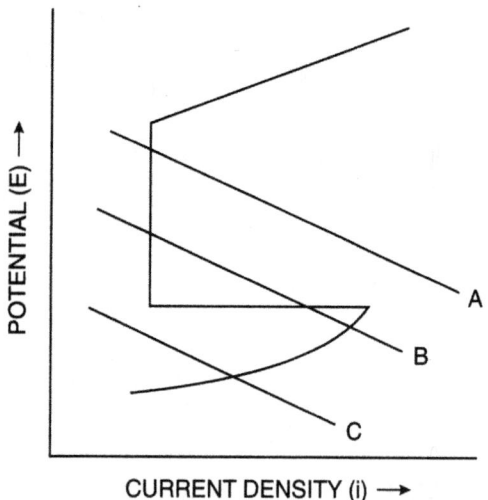

FIGURE 12.4. Intersection of three possible cathodic polarization curves (straight lines A, B, and C) with anodic polarization curve for system capable of exhibiting passivity. The corrosion rate depends on the current density at the intersection. Curve A produces a passive system, curve C an active system, and curve B an unstable system that can exhibit both low and high corrosion rates.

corrosion potential in both the passive and active regions, where the surface will oscillate between active and passive states, creating unstable conditions.

D4. Breakdown of Passivity

Another possible anodic–cathodic intersection site exists when conditions exist that lead to breakdown of passivity, localized attack, and then to pitting. The part of the passive anodic polarization curve shown in Figure 12.5 at potentials above the potential for breakdown, E_{bd} (called by some the critical breakdown or pitting potential E_{pit}), is the potential below the transpassive region at which the current density increases above i_{pass}. The intersection of cathodic curve 1 with the anodic curve of such a system results in breakdown and pitting while no breakdown occurs when the anodic polarization curve is intersected by cathodic curve 2 at a potential below E_{bd}.

The breakdown of passivity (the breaching of the protective barrier provided by the passive film) initiates the most damaging kinds of corrosion, the localized forms of corrosion, pitting, crevice corrosion, intergranular attack, and stress corrosion. A 1982 review by Kruger and Rhyne [14] extensively reviewed the corrosion science and engineering of the breakdown processes that are brought about by the chemical alteration of the passive film or the environment so that the film becomes unable to effectively prevent destructive local attack. Other reviews of breakdown have been given by Kruger [15], Galvele [16], and by Janik-Czachor [17].

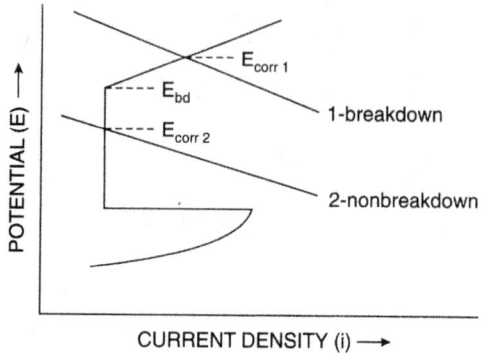

FIGURE 12.5. Anodic polarization curve for system capable of exhibiting passivity but is subject to breakdown at potentials above breakdown potential E_{bd} where pitting is initiated. The intersection of cathodic curve 1 with the anodic curve of a system exhibiting passive results in the breakdown of passivity that leads to pitting, while the intersection of cathodic curve 2 at a potential below E_{bd} results in no breakdown.

Hoar [18] stated that four conditions (that are usually but not always) considered to be required for the breakdown that initiates localized attack: (1) *Critical Potential*—a certain critical potential E_{bd} must be exceeded. (2) *Damaging Species*—damaging species (examples are chloride or the higher atomic weight halides) are needed in the environment to initiate breakdown and propagate localized corrosion processes like pitting. (3) *Induction Time*—an induction time exists, which starts with the initiation of the breakdown process by the introduction of breakdown conditions and ends when the localized corrosion density begins to rise. (4) *Local Sites*—the presence of highly localized sites where breakdown takes place. Various models for initiation have been developed that satisfy these four conditions [15–17].

E. USING PASSIVITY TO CONTROL CORROSION

Using the phenomenon of passivity to control corrosion entails employing methods that bring the potential of the surface to be protected to a value in the passive region. Some of the tactics that can be used to do this are:

1. A current can be applied by means of a device called a potentiostat, which can set and control the potential at a value greater than the passivating potential E_p. This method of producing passivity is called anodic protection.

2. For environments containing the damaging species, such as chloride ions that cause local corrosion, the potentiostat or other devices that control the potential can be used as in tactic 1 to set the potential to a value

in the passive region below the critical potential for pitting E_{pit}.

3. Alloys or metals that spontaneously form a passive film, for example, stainless steels, nickel, or titanium alloys, can be used in applications that require resistance to corrosion. Usually a pretreatment such as that described in tactic 4 is desirable.

4. A surface pretreatment can be carried out on an alloy capable of being passivated. The use of such a pretreatment has been standard practice for stainless steels for many years. The passivating procedure involves immersion of thoroughly degreased stainless steel parts in a nitric acid solution followed by a thorough rinsing in clean hot water. The most popular solution and conditions of operation for passivating stainless steel is a 30-min immersion in a 20 vol % nitric acid solution operating at 120°F (49°C). However, other solutions and treatments may be used, depending on the type of stainless being treated [19].

5. The environment can be modified to produce a passive surface. Oxidizing agents such as chromate and concentrated nitric acid are examples of passivating solutions that maintain a passive state on some metals and alloys.

REFERENCES

1. A. G. Revesz and J. Kruger, in Passivity of Metals, R. P. Frankenthal and J. Kruger (Eds.), Electrochemical Society, Pennington, NJ, 1978, pp. 138–143.
2. R. Revie and H. H. Uhlig, Corrosion and Corrosion Control, 4th ed., Wiley, New York, 2008, p. 84.
3. C. Wagner, Corros. Sci., **5**, 751 (1965).
4. NACE/ASTM Standard G193-10b, 2010, "Standard Terminology and Acronyms Relating to Corrosion," ASTM International, West Conshohocken, PA, and NACE International, Houston, TX, 2010, p. 11, DOI: 10.1520/G0193-10B, www.nace.org and www.astm.org.
5. R. P. Frankenthal and J. Kruger (Eds.), Passivity of Metals, Electrochemical Society, Pennington, NJ, 1978.
6. N. Sato and G. Okamoto, in Comprehensive Treatise of Electrochemistry, Vol. 4, Bockris et al. (Eds.), Plenum, New York, 1981, pp. 193–306.
7. J. Kruger, Int. Mater. Rev., **33**, 113 (1988).
8. M. Froment (Ed.), Passivity of Metals and Semiconductors, Elsevier, Amsterdam, The Netherlands, 1983.
9. (a) N. Sato and K. Hashimoto (Eds.), Passivity of Metals and Semiconductors, Part I, Pergamon, Oxford, UK, 1990; (b) Corros. Sci., **31**(1990).
10. K. E. Heusler (Ed.), Passivity of Metals and Semiconductors, Materials Science Forum, **185–188**, Trans. Tech. Publs., Aedermann, Switzerland, 1995.
11. M. Pourbaix, Atlas of Electrochemical Equilibria in Aqueous Solutions, 2nd ed., National Association of Corrosion Engineers, Houston, TX, 1974.
12. T. P. Hoar, Corros. Sci., **7**, 341 (1967).
13. M. G. Fontana and N. D. Greene. Corrosion Engineering, McGraw-Hill, New York, 1967, pp. 336–337.
14. J. Kruger and K. Rhyne, Nucl. Chem. Waste Management, **3**, 205 (1982).
15. J. Kruger, in Passivity and Its Breakdown on Iron and Iron Base Alloys, R. W. Staehle and H. Okada (Eds.), National Association of Corrosion Engineers, Houston, TX, 1976, pp. 91–98.
16. J. R. Galvele, in Passivity of Metals, R. P. Frankenthal and J. Kruger (Eds.), Electrochemical Society, Pennington, NJ, 1978, pp. 285–327.
17. M. Janik-Czachor, J. Electrochem. Soc., **128**, 513C (1981).
18. T. P. Hoar, Corros. Sci., **7**, 355 (1967).
19. D. M. Blott, in Handbook of Stainless Steels, D. Peckner and I. M. Bernstein, (Eds.), McGraw-Hill, New York, 1977, p. 24–24.

13

LOCALIZED CORROSION OF PASSIVE METALS

H. Böhni[†]

Institute of Materials Chemistry and Corrosion, Swiss Federal Institute of Technology, Zurich, Switzerland

A. INTRODUCTION

Localized corrosion of passive metals includes various types of corrosion phenomena, such as pitting, crevice corrosion, intergranular attack as well as stress corrosion cracking. The latter form of corrosion is discussed in Chapter 14 and will not be addressed here, but it has to be pointed out that crack initiation quite often starts at sites where localized corrosion processes occur. Therefore localized corrosion and stress corrosion cracking are quite often correlated.

[†]Retired

Since Part I of this book is devoted primarily to the fundamental aspects of corrosion, the scope of this chapter will also be limited to the fundamentals of localized corrosion of passive metals, where several detailed mechanisms for the different stages of localized corrosion have been extensively studied. Pitting and crevice corrosion are the two major forms of localized corrosion. They are often treated separately, especially when dealing with more practical aspects of localized corrosion. The different morphological appearance of the two types of corrosion suggest different mechanisms, a point of view that does not hold when looking at the electrochemical fundamentals of these processes. Therefore this chapter is focused mainly on pitting corrosion phenomena. With respect to intergranular corrosion, the structural aspects of the various alloys are of primary importance. Since this form of localized corrosion depends strongly on the alloy under consideration, the readers are referred to Part III of this book where the various metals and alloys are treated separately.

B. PHENOMENOLOGICAL ASPECTS

During pitting corrosion of passive metals and alloys, local metal dissolution occurs leading to the formation of cavities within a passivated surface area. In practice, pitting corrosion of passive metals is commonly observed in the presence of chlorides or other halides. Therefore the question arises whether the presence of specific, so-called localized, corrosion-inducing anions, such as halides, are always required for pitting to take place. Some results indicate that pitting may also occur in pure water as in the case of carbon steel in high-purity water at elevated temperature or aluminum in nitrate solutions at high potentials. In all these forms of localized corrosion, active and passive surface states are simultaneously stable on the same metal surface over an extended period of time, so that local pits can grow to macroscopic size.

Therefore, from an electrochemical point of view, more appropriate questions would be: What are the necessary requirements to initiate pits and what is required to stabilize a heterogeneous electrode state consisting of simultaneously active and passive surface areas, exhibiting significantly different dissolution rates?

B1. Electrochemical Potential

Electrochemical studies of pitting corrosion usually indicate that stable pitting occurs only within or above a critical potential or potential range. Therefore the susceptibility of passive metals to pitting corrosion is often investigated by electrochemical methods. Most commonly, potential current curves are measured either by applying potentials stepwise (potentiostatically) or by applying a constant potential sweep rate (potentiodynamically) and recording the resulting current. Typical potential current curves exhibit an active–passive transition and subsequently a sudden current increase within the passive potential range, as shown schematically in Figure 13.1. The following values are often determined and used to characterize metals and alloys with respect to pitting and crevice corrosion: (1) the critical current density i_{crit}, characterizing the active–passive transition, (2) the pitting potential E_p where stable pits start to grow, and (3) the repassivation or protection potential E_{rep} or E_{pp} (after reversal of the potential sweep direction) below which the already growing pits are repassivated and the growth is stopped.

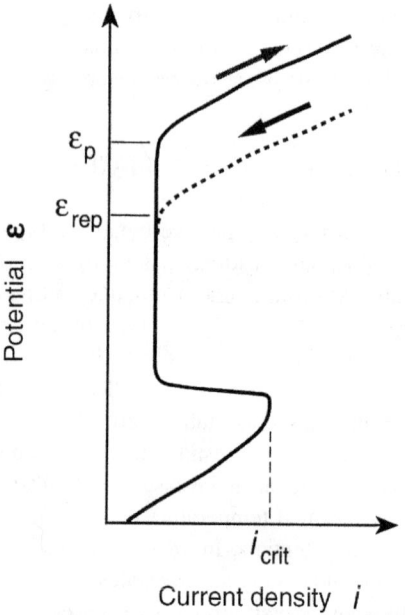

FIGURE 13.1. Schematic polarization curve for metal showing active–passive transition as well as pitting in the passive potential range: $E > E_p$, pitting will occur; $E < E_{rep}$, growing pits will repassivate.

More detailed reviews of the various electrochemical methods used to measure the susceptibility to pitting corrosion are given by Szlarska-Smialowska [1] and Sedriks [2]. Obviously, these quantities mentioned above can be measured as functions of the alloy composition or of the composition of the environment. Without discrediting the practical value of such investigations, especially in comparing alloys as well as environments with respect to their pitting-susceptibility or pitting-promoting tendency, they give neither any direct insight into the mechanisms of localized corrosion processes nor can the values be used as true limiting potential values to prevent localized corrosion processes in engineering applications.

B2. Effects of Alloying and Microstructure

The alloy composition as well as the microstructure can have a strong influence on the pitting resistance of an alloy, as shown, for instance, by Horvath and Uhlig [3], who demonstrated the beneficial effect of chromium and molybdenum in stainless steels. The pitting potential was found to increase dramatically with chromium contents $\gtrsim 20$ wt %, whereas molybdenum is effective at minor concentrations of 2–6 wt %, but only in the presence of chromium. Similar effects have also been reported for small amounts of alloyed nitrogen or tungsten. Various explanations have been given to explain the strong influence of molybdenum on the pitting behavior of stainless steels, which is also well confirmed in engineering practice. It was suggested that molybdenum is adsorbed on the surface as molybdate or acts by blocking active surface sites, inhibiting active metal dissolution and finally favoring repassivation [4–6]. Other models suggest that molybdenum, as well as other elements, improves the cation-selective properties of the passive film, hindering the migration of aggressive anions, such as chlorides, to the metal surface [7, 8] or reduce the flux of cation vacancies in the passive film [9]. Though the exact mechanism is not clear yet, most of the research done so far favors an effect on the growth stage rather than on the initiation stage of localized corrosion. Recent results, applying microelectrochemical techniques, confirmed that even in the superaustenitic stainless steels molybdenum strongly improves the repassivation behavior but has no influence on pit initiation [10, 11].

Similar effects can also be observed on aluminum, where small additions of alloying elements may increase the pitting potential as long as the structure is single phase. Aluminum–copper alloys, widely used in the aircraft industry, are well known and intensively studied examples with respect to their pitting behavior [12]. The pitting potential increases with increasing copper concentration as long as copper is in solid solution. Since aluminum usually exhibits rather low solubility limits, alloying elements tend to form second-phase intermetallic compounds, such as Al_2Cu. In the presence of

Al$_2$Cu the pitting resistance decreases to values of aluminum with small copper concentrations. The corresponding decrease in pitting potential is explained by the existence of a copper-depleted zone around these particles, where pits would initiate first.

Localized corrosion of passive metals almost always initiates at local heterogeneities, such as inclusions and second-phase precipitates as well as grain boundaries, dislocations, flaws, or sites of mechanical damage. In the case of stainless steel surfaces, pit initiation occurs almost exclusively at sites of MnS inclusions, which are found in commercial as well as in high-purity alloys. A more detailed discussion of this subject will follow. To prevent inclusions and precipitates, nonequilibrium single-phase conditions can be attained by special preparation techniques, such as rapidly quenching or physical vapor deposition. The resulting microstructure is either nanocrystalline or amorphous. It was recently shown that sputter-deposited aluminum alloys containing only a few atomic percent of metal solute such as Cr, Ta, Nb, W, Mo, or Ti exhibit a strong increase of E_p of 0.2–1V [13–17]. Similar results were also obtained with sputter-deposited stainless steels, where nonequilibrium single-phase structures with molybdenum concentrations up to 14 wt % could be obtained [18, 19]. In both cases, the increase in pitting resistance was explained by the reduced pit initiation tendency as well as by a more protective passive film, favoring rapid repassivation.

B3. Effect of Temperature

Increasing temperature usually also increases the pitting tendency of metals and alloys. At low temperature, high pitting potentials are observed. In the case of stainless steels, a strong decrease of pitting potentials in the temperature range between 0 and 70°C of $\sim$ 0.5 V can be observed as shown in Figure 13.2 [20, 21]. This strong dependence on temperature has led to experimental techniques which allow stainless steels to be ranked according to their pitting susceptibility. A critical pitting temperature (CPT) has been defined, below which a steel in an aggressive Cl$^-$-containing solution, usually a FeCl$_3$ solution, would not pit regardless of potential and exposure time [20, 22].

Furthermore, the CPT used to characterize the pitting resistance of alloys is often correlated with the composition of stainless steels, especially with Cr, Mo, and N [1, 2, 20]. The effect of temperature on the pitting behavior of other metals and alloys such as Ni, Al, Ti, and its alloys has also been studied but to a lesser extent. Important earlier results are summarized in [20]. Nevertheless most of the investigations on temperature effects of localized corrosion deal with the significance of temperature to evaluate and compare pitting susceptibilities of different metals and alloys rather than with the more fundamental aspects of temperature effects on pit initiation and pit growth mechanisms.

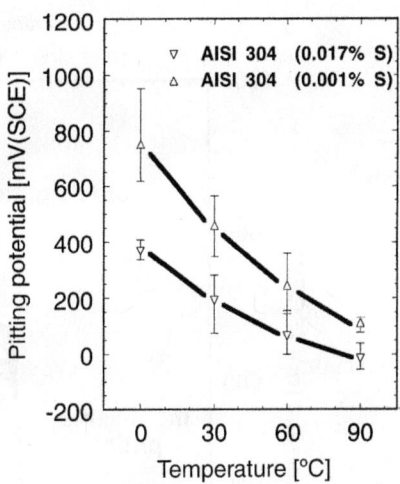

FIGURE 13.2. Effect of temperature on pitting potential E_p of 304 stainless steel (SS) in 0.1M NaCl with different sulfur contents [21].

B4. Stochastic Aspects

Corrosion processes are based mainly upon deterministic approaches, such as the electrochemical theory of corrosion. Localized corrosion events, however, due to their unpredictable occurrence, cannot be explained without using statistical methods to evaluate the experimental data. Stochastic aspects of pitting corrosion were studied in the late 1970s, especially by authors in Japan [23–26]. Recently, a convincing review of statistical and stochastic approaches to localized corrosion was published by Shibata [27]. He evaluated large numbers of pitting potential values using a Gaussian distribution, whereas the Poisson distribution was found to be a better approach for pit generation. The results indicate that different pit generation rates can be observed as a function of time. He proposed two groups of models considering either pit generation events alone or assuming pit generation and subsequent repassivation processes. The latter model could be fitted more satisfactorily to the various cases studied experimentally. More detailed information will be given in Chapter 27.

C. STAGES OF LOCALIZED CORROSION

The different stages of localized corrosion can best be explained and discussed in connection with its potential dependence. In Figure 13.3, a typical potential current curve of a passive metal, such as stainless steel, measured in chloride solutions, shows the different stages of pitting corrosion.

At lower potentials, pit initiation is followed by rapid repassivation. This stage is usually referred to as metastable pitting. The resulting current transients differ widely with

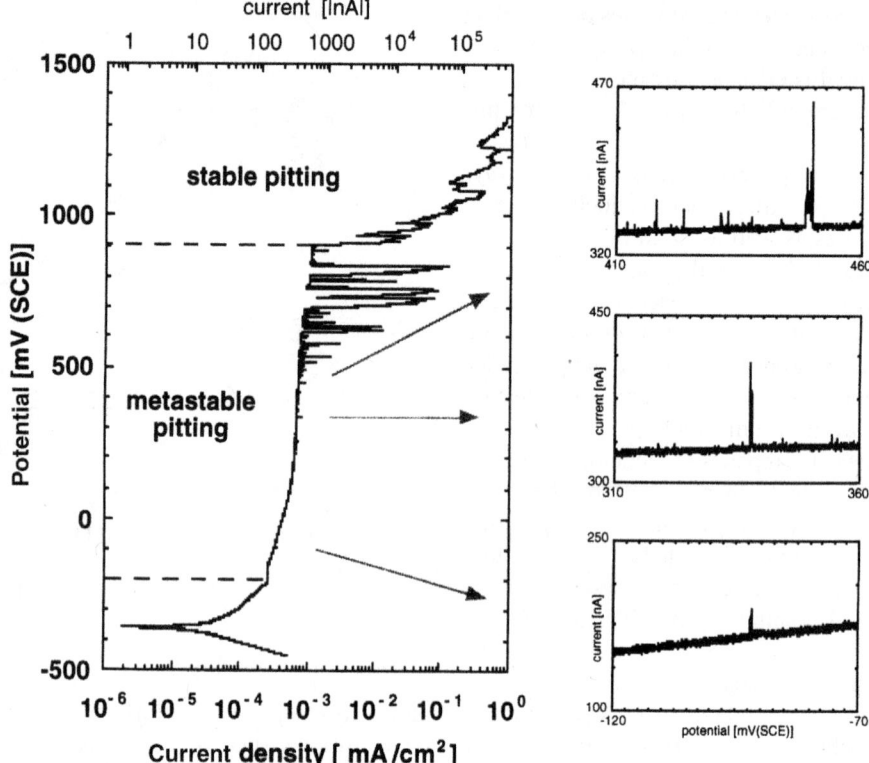

FIGURE 13.3. Typical potential current curve of stainless steel in chloride solution showing different stages of localized corrosion (see text).

respect to the peak current height as well as the lifetime. Small current transients in the femtoampere (fA) and pico-ampere (pA) range, corresponding to pits of nanometer and micrometer size can be detected only by applying micro- and nanoelectrochemical methods, as will be discussed later. Increasing the potential generally leads to larger current transients with higher peak currents and longer lifetimes, indicating an extended pit growth period. Above a certain potential or potential range, a transition to stable pit growth occurs. Even above the pitting potential, repassivation may still occur, showing the stochastic character of localized corrosion processes. Whether well-defined pitting potentials can be determined depends on a variety of factors, such as the type of metal, the chloride concentration, and the temperature. The higher the chloride concentration or the temperature, the more precise the resulting pitting potentials usually are. Furthermore the use of crevice-free experimental techniques may also be very decisive in obtaining reproducible results.

Metastable pitting is usually not considered as a real corrosion risk from an engineering point of view. Nevertheless, studies of metastable pitting as a precursor to stable pitting may provide valuable insights into fundamental aspects of pitting corrosion, since pit initiation as well as the trans formation of metastable into stable pits are key factors in localized corrosion processes. Investigations of metastable pitting also allow a statistical evaluation of corrosion data necessary to study stochastic pitting models.

With respect to engineering application, studies of metastable pitting may also substantially improve the evaluation of metal–environment systems. The occurrence of metastable pitting below the pitting potential indicates a potential corrosion risk, especially when crevice conditions cannot be completely excluded. Furthermore the repassivation behavior of metals and alloys can easily be studied, which is very important for developing highly corrosion-resistant alloys.

D. METASTABLE PITTING: PIT INITIATION AND REPASSIVATION

A number of models have been proposed to describe the initiation of localized corrosion of passive metals based either on the breakdown processes of the passive film itself or on structural defects or heterogeneities of the underlying metal or alloy such as dislocations, grain boundaries, second-phase precipitates, or nonmetallic inclusions. Certainly a strict differentiation of the two approaches is not always possible, since film breakdown and structural parameters of the underlying metal may be correlated.

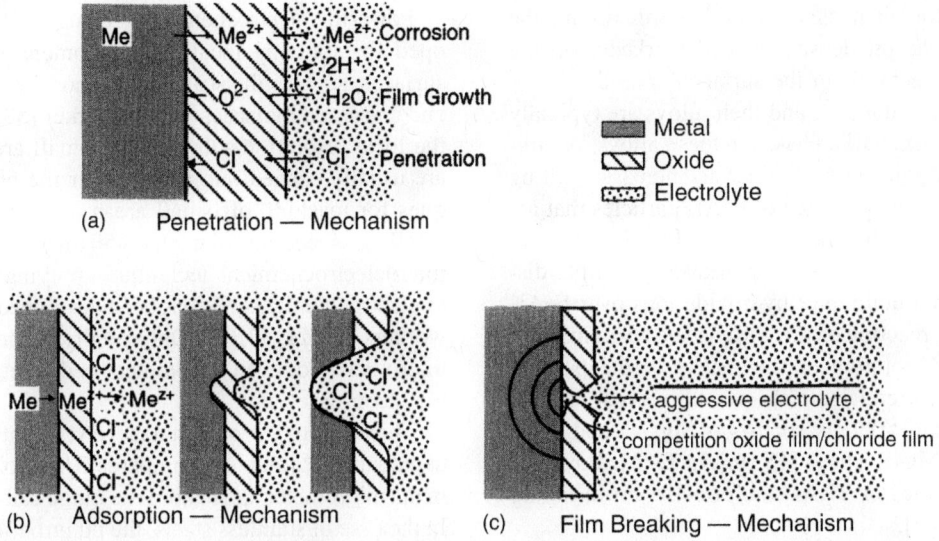

FIGURE 13.4. Models for pit initiation leading to passive film breakdown. [After Strehblow and co-workers [28–30]. Reproduced with permission from H. Kaesche, *Die Korrosion der Metalle*, 3rd edition. Springer, Berlin, Germany, 1990, Fig. 12.12a–c, p. 308. Copyright © Springer.]

D1. Passive Film Breakdown

Theoretical models that describe the initiation process leading to passive film breakdown may be grouped into three classes: (1) adsorption and adsorption-induced mechanisms, where the adsorption of aggressive ions like Cl^- is of major importance, (2) ion migration and penetration models, and (3) mechanical film breakdown theories, as shown in Figure 13.4 [28–30].

In the case of the cluster adsorption model, originally proposed by Heusler and Fischer [31] several years ago for iron, localized adsorption of chlorides leads to an enhanced oxide dissolution at these sites with subsequent thinning of the oxide film until finally a complete removal is achieved and active dissolution starts. During the latter years further refinements of this model were obtained evaluating the measured induction times statistically [32]. Further evidence for the stochastic nature of these initiation processes were also obtained on stainless steels [10]. For the mechanical breakdown of passive films models have been discussed, where the breakdown is either the principal step, giving the environment direct access to the metal surface, as shown by Sato [33], or an additional step combined with other processes as in the case of the defect model as pointed out by Macdonald and co-workers [34, 35]. A critical review of these mechanisms is given in [36].

Most of the theoretical models for pit initiation have not been sufficiently verified experimentally. Since the relationships used to test the different models are very general in nature, such as the correlation between pitting potential and chloride concentration or the quantitative prediction of induction times, they do not really cover specific model-sensitive aspects. Furthermore, the pitting potentials proposed in these models are actually critical pit initiation potentials, which often do not correspond with the experimentally determined pitting potentials, at which the transition to stable pit growth takes place. Pit initiation processes, on the other hand, often occur at much lower potentials, as was recently shown on stainless steels by applying a new microcapillary technique to measure current transients in the picoampere and femtoampere range [10, 37, 38].

Recently, new results on the initiation and formation of porous semiconductors were obtained [39]. In case of *n*- and *p*-type GaAs, local dissolution can be electrochemically triggered in chloride-containing solutions resulting in a porous structure. In contrast to pitting corrosion of passive metals, local dissolution can be achieved not only by the presence of an oxide film (*p*-type GaAs) but also by depletion conditions in the semiconductor space-charge layer (*n*-type GaAs). In the latter case, the presence or absence of an oxide film is not significant for the pitting process. This is of particular significance since the previously discussed theoretical approaches ascribe a key role in the localized nature of pitting to the properties of a surface oxide film.

D2. Structural Parameters

Numerous investigations during recent years have shown that the sites of pit initiation on passive metal surfaces may generally be related to defect structures of the underlying metals. Detailed summaries are given in [1, 2], In case of Ni single crystals, for instance, it has been clearly demonstrated that the emerging points of screw dislocations are especially susceptible sites for pit nucleation [40]. Furthermore, the shape as well as the pit density depend markedly on the crystallographic orientation of the surface exposed.

Pits in aluminum, titanium, and their alloys are typically associated with intermetallic phases in these alloys. Aluminum contaminated with iron exhibits an increased pitting susceptibility due to the presence of $FeAl_3$ particles that act as local cathodes on the metal surface [41–43]. Recent investigations using microsensors to measure local pH distributions revealed a buildup of hydroxide ions over $FeAl_3$ due to the cathodic reaction on the particle surface. This may lead to alkaline dissolution of the matrix at the particle interface [44]. Preferential attack can also occur on the surface of intermetallic phases, such as on Mg_2Al_3 and MgSi in Al–Mg and Al–Mg–Si alloys [45, 46], In Ti–Al alloys it is assumed that particles of Ti_3Al_2 are effective in initiating pitting corrosion [47].

On the other hand, nonmetallic inclusions may also act as potential nucleation sites [48–50]. Sulfide inclusions in stainless steels are particularly susceptible. Figure 13.5 shows the initial stage of localized corrosion on stainless steel at the site of a MnS inclusion [11]. This is observed not only in austenitic [51–53] but also in ferritic [54, 55] stainless steels. These inclusions are often manganese sulfides or manganese-containing sulfide compounds [53–56]. It has been suggested that, at low manganese levels in steel, CrS is the thermodynamically stable sulfide, while above some level of manganese the stable sulfide is an iron–manganese spinel, which appears to be a better initiation site for pitting than CrS. In a recent survey, Srivastava and Ives [57] summarized the different types of attack on nonmetallic inclusions in stainless steels from a phenomenological point of view.

Powerful electrochemical techniques have been developed for studying localized phenomena on passive metal surfaces, such as the scanning methods, extensively applied and discussed by Isaacs and co-worker [58, 59]. To improve the local resolution substantially, small area measurements are usually carried out, using either the photoresist techniques for masking off small areas or embedded wires with small cross sections (diameter ∼50 μm) [60]. Recently, a new microelectrochemical technique applying microcapillaries as electrochemical cells has been developed by Suter and co-workers [10, 11, 37, 38]. Only small surface areas with a few micrometers or even nanometers in diameter are exposed to the electrolyte.

This leads to a strongly enhanced current resolution, down to picoamperes and femtoamperes. Therefore local processes in the micrometer and nanometer range can easily be studied. In the case of stainless steels, the pit initiation process due to the oxidative dissolution of active MnS inclusions as a precursor of pitting corrosion can be investigated directly as shown in Figure 13.6.

Additionally, the results clearly indicate that the dissolution of inclusions takes place even in chloride-free solutions,

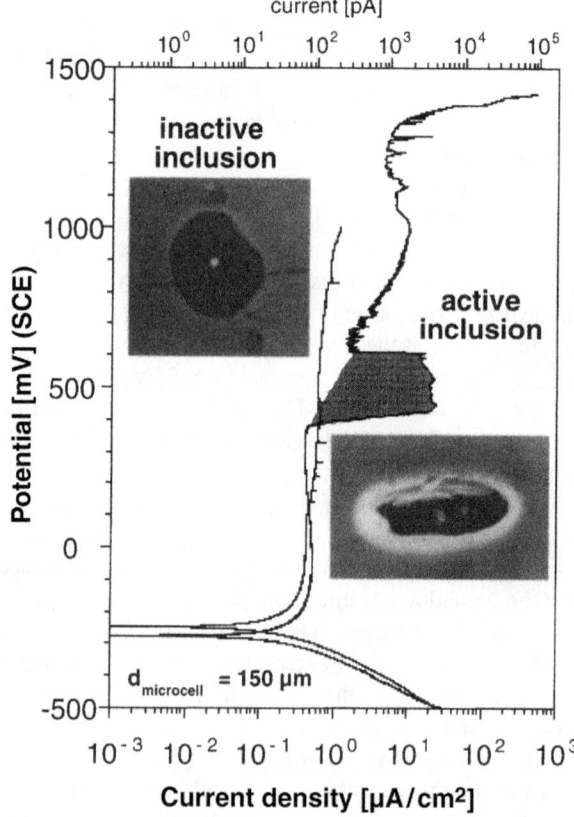

FIGURE 13.6. Local potentiodynamic polarization curves at active and at inactive MnS inclusion site with corresponding scanning electron microscopy (SEM) photographs taken after polarization measurements [38].

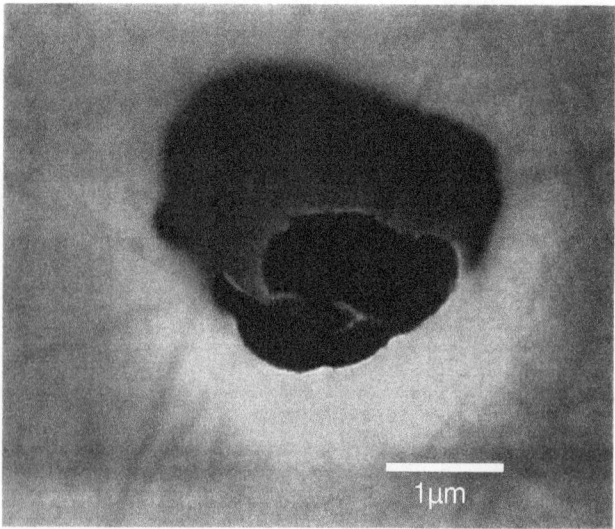

FIGURE 13.5. Pit initiation and early pit growth at MnS inclusion in 304 SS [11].

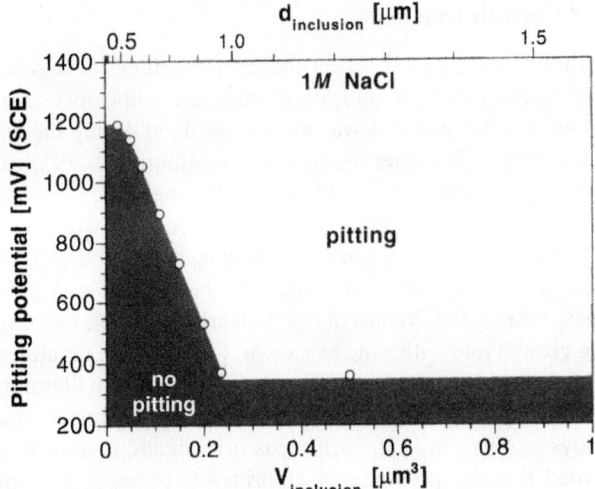

FIGURE 13.7. Correlation between pitting potential and inclusion size [10].

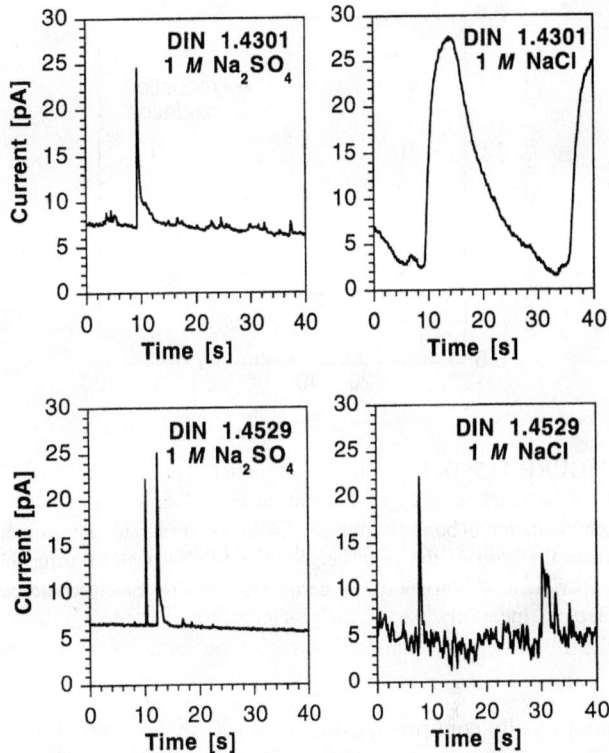

FIGURE 13.8. Microtransients of DIN 1.4301 (0% Mo) and DIN 1.4529 (6.4% Mo) stainless steels in chloride-free and chloride-containing solutions [10].

whereas chlorides are required for metal dissolution and stable pit growth processes.

The results in Figure 13.6 show that the microcapillary technique also makes it possible to distinguish between active and inactive inclusions with respect to pit initiation. An increase of the sulfur content drastically increases the number of large current transients due to an increasing number of large MnS inclusions, verified by microstructural investigations [37, 38]. Similar effects have also been observed on iron in contact with Cl^--containing borate solutions [61].

Noise analysis obtained from microelectrochemical investigations of stainless steels under potentiostatic conditions revealed that the current noise, expressed as standard deviation σ_i of the passive current, increases linearly with the size of the exposed area, whereas the pitting potential decreases. Computer simulations showed [10, 38] that the current noise, largely caused by the dissolution of small inclusions, can be correlated to the size of the inclusions. Therefore the pitting potential is also related to the size of active inclusions [62], as shown in Figure 13.7. Specifically the size of the inclusions in stainless steels has to be kept below ~1 μm to improve substantially the pitting resistance of stainless steel. This effect was already observed much earlier by simple immersion tests [63].

The effect of molybdenum on the pitting behavior of stainless steels can also be studied by microelectrochemical techniques and then compared to molybdenum free alloys having approximately the same impurity level, as shown in Figure 13.8 [10]. Molybdenum in superaustenitic stainless steels has only a minor effect on the initiation process. The superior corrosion resistance of these high-molybdenum-

containing alloys rather has to be attributed to a considerably improved repassivation behavior.

D3. Stability of Passive Films

The stability of passive films plays an important role with respect to the pitting behavior of passive alloys. Fast and effective repassivation of locally activated metal surfaces only occurs if a stable passive film is formed. Since pit initiation can hardly be neglected on commercial alloys in real environments, rapid repassivation in which a passive film of high stability is formed, is very important to obtain highly corrosion-resistant alloys. In order to understand the importance of the chemical and electrochemical stability of passive films with respect to localized corrosion, the behavior of naturally grown as well as synthetically prepared thin oxide films has to be studied [64, 65]. In several contributions Schmuki and co-workers [66, 67] showed that the semiconductive properties and the chemical stability of the thin oxide films of iron and chromium correlate quite well with those of naturally grown films. The results indicate that the presence of Fe(II), acting as doping species in iron oxide, strongly effects its stability. The good correlation between the semiconductive properties and the stability of passive films was

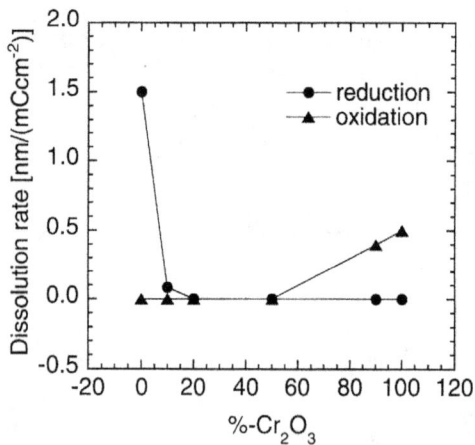

FIGURE 13.9 Dissolution rate of sputter-deposited Fe/Cr oxide films as function of Cr_2O_3 content during galvanostatic reduction and oxidation in borate buffer, pH 8.4. Dissolution rates determined from the drop of the edge height of XANES spectra during the experiment. (Reproduced by permission of The Electrochemical Society, Inc. [70]).

additionally confirmed, using in situ X-ray absorption near-edge spectroscopy (XANES) techniques [68, 69]. The experiments showed that the mixed oxides are far more stable than the pure oxides as shown in Figure 13.9.

The presence of sufficient amounts of chromium oxide protects iron oxide against reductive dissolution, whereas iron oxide protects chromium oxide from oxidative dissolution. During anodic polarization of iron oxide films in acidic solutions, a deleterious effect of chloride anions compared with sulfates was found. In HCl solutions of increasing concentration, not only the increased acidity but also the increased chloride concentration accelerates the dissolution markedly [70].

E. PIT GROWTH

When the active state within pits and crevices is maintained over an extended period of time, rapid metal dissolution usually occurs. The resulting pit and crevice geometries as well as the surface state within the pits vary markedly from open and polished hemispherical pits on free surfaces to etched crack-like shapes within crevices, depending largely on the type of rate-controlling reactions during the growth stage.

The kinetics of pit growth is not only of scientific interest but is also of great engineering importance for commercial corrosion-resistant materials, since the possibility of local breakdown should not be ignored under practical conditions. Sufficient knowledge of the mechanisms of growth and stability is therefore a necessary requirement in order to predict the corrosion behavior of passive metals correctly as well as for developing new corrosion-resistant materials.

E1. Growth Kinetics

Although pit growth is experimentally much easier to study and quantify than pit nucleation, substantive information on pit growth mechanisms was mainly obtained during the last two decades. In earlier research on aluminum [71–79], iron [80], and stainless steels [81, 82] under open-circuit conditions, studies were limited to the description of growth rate by a simple power law $d_p = at^b$, whereby d_p is the pit depth, t is time, and a and b are constants, with the latter averaging in many cases $\sim$0.5. Values of $b < 1$ clearly indicate a decreasing growth rate with time as shown, for example, on aluminum in various tap waters [79]. A comprehensive literature survey, summarizing the rate laws of various metals and alloys is given in [83]. Numerous investigations have confirmed that the presence of chlorides is necessary for pit growth and that the growth rate increases with increasing chloride concentration [84–90]; other anions, however, behave differently.

To overcome the problems associated with accurate pit current density measurements, special techniques have been developed to determine pit growth kinetics. Hunkeler and co-workers [83, 91, 92] used the time for pits to penetrate metal foils of different thickness to determine the growth rate of the fastest growing pits. Using this simple method pit growth rates in aluminum as well as stainless steels were measured and several parameters, such as the potential, the composition of the electrolyte, and the temperature, were varied. Another elegant approach to study the growth kinetics involved the investigations of two-dimensional pits in thin metal films as demonstrated by Frankel et at. [93–96]. The measurements of lateral pit growth rates from analysis of images of the growing two-dimensional pits provides a very simple and direct way via Faraday's law, with no need for any further assumptions. Investigations were carried out on different materials such as nickel and aluminum alloys. The study of single pits, formed in different ways have also been performed either by masking off a small area, implanting an activating species at a small spot, or using single-pit electrodes, such as embedded wires [60, 97–99].

E1.1. Diffusion Control. Several detailed studies on the kinetics of growing pits, performed in the 1970s, recognized the presence as well as the importance of salt layers during pit growth [100–104]. In Figure 13.10, the pit and crevice growth of 304 SS in a chloride-containing solution as a function of the potential, using the foil technique as mentioned above, is shown [105, 106]. At high potentials, where pits grow on open surfaces as well as crevices, a potential independent growth rate is observed, suggesting diffusion-controlled growth mechanism.

When lower potentials are applied, the mode of localized corrosion changes to etch-type crevice corrosion with a strong potential dependence. Below a critical growth rate,

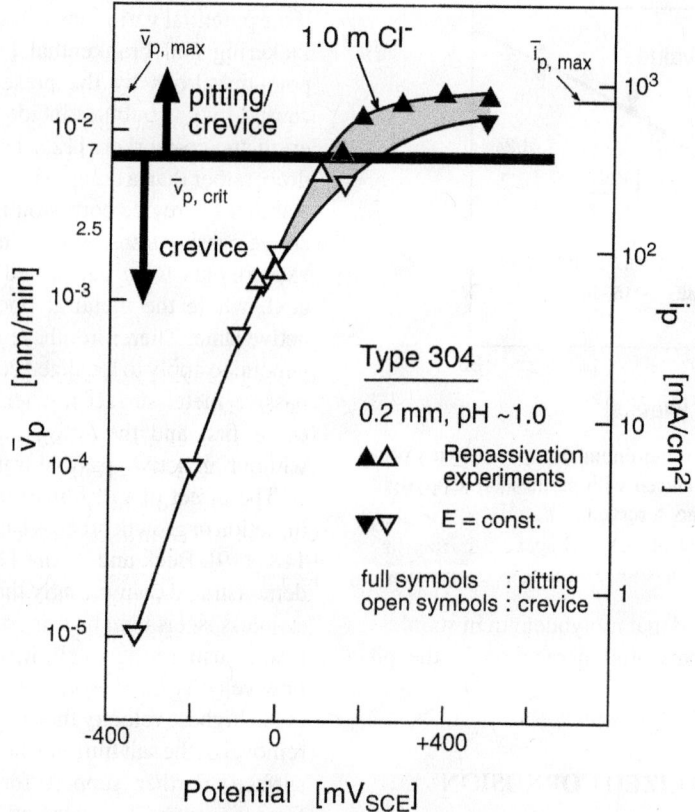

FIGURE 13.10. Mean pit and crevice growth rate $\bar{v}_p$ vs. potential of type 304 SS [105, 106, 112]. (Reproduced with permission from *Advances in Localized Corrosion*, p. 70, Copyright © NACE International. All rights reserved.)

depending on the chloride concentration, etch-type crevice corrosion is the only stable form of localized corrosion, since saturation conditions and salt film formation are not attained. Concerning the practical significance of diffusion-controlled pit growth, it must be pointed out that this type of pit growth is not often observed on stainless steels since it usually requires high potentials and/or high chloride concentration.

In the case of unidirectional, single-pit growth using wire electrodes, the diffusion-controlled pit growth can easily be calculated using Fick's first law. Excellent experimental proof was obtained on stainless steel and nickel, for which a parabolic rate law was observed [107]. Additionally, it was found that if the bulk electrolyte contains the corresponding diffusing metal ion in solution, the pit growth is much faster in less concentrated solutions, as expected from theoretical considerations [108].

E1.2. Ohmic/Charge Transfer Control. As pointed out by Beck [109], pitting on titanium and aluminum occurs at high ohmic limited current densities. Generation of large amounts of hydrogen bubbles within the pit strongly increases the mass transport rate. Therefore the fluid flow of the bulk electrolyte has little effect on pit growth under such conditions [110]. Further and more detailed support for

ohmic-controlled pit growth on aluminum was obtained by Hunkeler [83]. For small Tafel constants, as in the case of aluminum, and sufficiently large pits (>10 μm), contributions from charge transfer as well as ohmic transport outside the pit may be neglected, and a simple parabolic rate law [92] can be derived in which the preexponential factor depends directly on the electrolytic conductivity of the bulk electrolyte. Due to the generation of hydrogen bubbles during pitting of aluminum, no significant change in the composition of the electrolyte within the pit takes place, in contrast to situations in which diffusion processes control pit growth. These findings are in excellent agreement with the evaluation of long-term pit growth measurement under open-circuit conditions on aluminum in tap water of known conductivity, as shown in Figure 13.11 [111, 112]. The experimental values fit very nicely into the parabolic rate law. The resulting potential difference of ∼ 15 mV is in good agreement with the open-circuit conditions in the absence of strong oxidants.

For materials such as stainless steels or nickel, the charge transfer reaction cannot be neglected, as shown in Figure 13.10. A semilogarithmic relationship between current density and potential is observed at low potentials, indicating that mixed ohmic/charge transfer control is probably effective. Similar results obtained by Newman [6] using

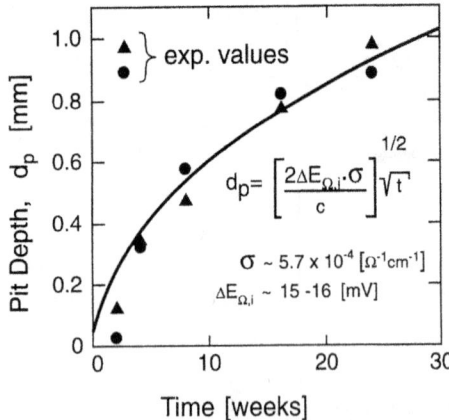

FIGURE 13.11. Pit growth on aluminum in tap water at open-circuit conditions [111]. (Reproduced with permission. Copyright © NACE International. All rights reserved.)

artificial pit electrodes showed that molybdenum in stainless steels ennobles the anodic dissolution reaction in the pit environment.

F. STABILITY OF LOCALIZED CORROSION

The concepts of stability of pitting and crevice corrosion are based mainly on either compositional changes of the electrolyte within the pit, salt film formation, or sufficiently large ohmic potential drops within pits. Various authors have suggested that critical concentrations of ionic species must be exceeded for stable pit growth to occur. Galvele [113–115] considered localized acidification as the main reason for stable pit growth. Sato [116, 117] pointed out that two models of metal dissolution have to be distinguished: (1) active or etching dissolution, which occurs at lower potentials, and (2) transpassive or brightening dissolution, which occurs at relatively noble potentials. For etch pitting, a critical hydrogen ion concentration should be reached, whereas for brightening pits, the necessity of a critical concentration of aggressive anions, such as chlorides, was proposed. On the other hand Vetter and Strehblow [118, 119] concluded from theoretical considerations that during the early stage of growth of an open hemispherical pit, the metal chloride concentration within the pit increases, but not sufficiently for precipitation to occur, which they considered as a possible requirement for the stability of pits. Instead, they proposed that an ion-conducting salt layer on the metal surface results from adsorption of aggressive anions, such as chlorides.

The significance of an ohmic potential drop within pits was also questioned by the same authors [118, 119]. For iron, they estimated potential drops of ~18 mV, which again cannot in any case explain the stability of growing pits. However, several groups at about this time also measured

large potential variations within crevices and pits [120–124], Pickering and Frankenthal [121–123] explained the large potential drops by the presence of high resistance paths caused by gas bubbles inside the pits. More recently, Pickering and co-workers [125, 126] showed that the ohmic (IR) drop, rather than a composition variation, was responsible for stabilizing crevice corrosion in some systems which exhibit active/passive transitions in the bulk solution. Most of these experiments were performed on iron or nickel in sulfuric acid, where the metal at open-circuit conditions is in the active state. Therefore, these results may not be sufficiently general to apply to localized corrosion processes of originally passive metal surfaces, where an activation process has to occur first and the active state has to be stabilized even without an active–passive transition under bulk conditions.

The effect of salt films formed within the pit during the initiation or growth process has been discussed frequently [84, 118, 119]. Beck and Alkire [103] and Beck and Chan [127] demonstrated convincingly that the formation of salt films on stainless steels may be important for the stability of growing pits. In using a flow cell, it was shown that with increasing flow velocity the dissolution current increases at first, before at the highest velocity the current drops drastically due to the removal of the salt film and the subsequent repassivation of the surface. Further support for the importance of salt film formation was also obtained from evaluation of metastable pitting on stainless steels [128]. Transition from metastable to stable pit growth occurs only when salt-film precipitation takes place.

Additional results showing the significance of a salt film were obtained from electrochemical impedance spectroscopy. The investigation of the anodic dissolution of artificial pit electrodes indicated that, in the case of stainless steel and nickel, significant ohmic potential drops are present within the salt film during diffusion controlled pit growth. The thin, conducting salt film stabilizes the active metal surface at the pit bottom and prevents the repassivation process [108]. Furthermore, a critical potential for salt-film formation E_{sf} can be determined, below which no salt film can form. The critical potential for salt-film formation amounts to $-0.19\,V_{sce}$ and $-0.08\,V_{sce}$ for 302 stainless steel and nickel, respectively, and does not depend significantly on the bulk concentration. For 304 SS, the lowest potential, E_{pi}, where pit initiation occurs and current transients (metastable pitting) have been found by microelectrochemical measurements [11] coincides with the critical potential for salt-film formation, E_{sf}, as well as with the photoelectrochemically determined flatband potential E_{fb} [129], below which no chloride adsorption takes place. Further investigations are necessary to clarify whether these values coincide accidentally or a possible correlation exists.

The above discussion of pitting corrosion leads to the following concept of localized corrosion. At potentials even below the pitting potential but above E_{pi} a nonzero probability of pit initiation exists, depending largely on the type, size, amount, and chemical properties of these local heterogeneities. Once a pit begins to grow in the presence of aggressive anions, such as chlorides a thin salt film may be formed after a transition time. As long as the salt film is present, stable dissolution occurs, and the growth rate is diffusion controlled. If the potential is then lowered, pits may continue to grow if the composition of the pit electrolyte, the current density, and the potential at the pit bottom are such that repassivation is not possible. The growth process in this case is then mixed ohmic/charge transfer controlled [98, 105, 107]. Under these conditions, the ohmic potential drop inside and outside of the pit may play an important role.

G. SUMMARY

Localized corrosion is an important but complex problem responsible for many corrosion failures in engineering applications. The local breakdown of passivity of commercially available engineering materials, such as stainless steels, nickel, or aluminum, occurs preferentially at sites of local heterogeneities, such as inclusions, second-phase precipitates, or even dislocations. The size, shape, distribution, as well as the chemical or electrochemical dissolution behavior (active or inactive) of these heterogeneities in a given environment, determine to a large extent whether pit initiation is followed either by repassivation (metastable pitting) or stable pit growth, Microelectrochemical techniques, combined with statistical evaluation of the experimental results allow to gain more insight into the mechanisms of these processes.

In addition to the local activation or pit initiation process, the stability of the passive film is decisive for the corrosion resistance of passive metals and alloys. Fast and effective repassivation, necessary for highly corrosion-resistant alloys, may only occur if highly stable films are formed during repassivation. Therefore, further investigations should be focused not only on the initiation of localized corrosion, but also on the stability of passive films. The stability of passive films is often reflected by the semiconductive properties of these films. Therefore electrochemical impedance spectroscopy, photoelectrochemical methods as well as in situ analytical techniques are very valuable tools to study the chemical and electrochemical behavior of these passivating oxide films.

REFERENCES

1. Z. Szlarska-Smialowska, Pitting Corrosion of Metals, NACE, Houston, TX, 1986.
2. A. J. Sedriks, Corrosion of Stainless Steels, Corrosion Monograph Series, Wiley, New York, 1996.
3. H. Horvath and H. H. Uhlig, J. Electrochem. Soc., **115**, 791 (1968).
4. H. Ogawa, H. Omata, I. Itoh, and H. Okada, Corrosion, **34**, 52 (1978).
5. K. Hashimoto, K. Asami, and K. Teramoto, Corros. Sri., **19**, 3 (1979).
6. R. C. Newman, Corros. Sci., **25**, 341 (1985).
7. M. Sakashita and N. Sato, Corros. Sci., **17**, 473 (1977).
8. S. Virtanen and H. Böhni, Corros. Sci., **31**, 333 (1990).
9. M. Urquidi and D. D. Macdonald, J. Electrochem. Soc., **132**, 555 (1985).
10. T. Suter and H. Böhni, Electrochim. Acta, **43**, 2843 (1998).
11. T. Suter, Microelectrochemical Studies of Austenitic Stainless Steels, Ph.D. Thesis, ETH Zürich Nr. 11962, 1997.
12. L. Muller and J. R. Galvele, Corros. Sci., **17**, 1201 (1977).
13. G. S. Frankel, M. S. Russak, C. V. Jahnes, and V. A. Brusic, J. Electrochem. Soc., **136**, 1243 (1989).
14. W. C. Moshier, G. D. Davis, J. S. Ahearn, and H. F. Hough, J. Electrochem. Soc., **133**, 1063 (1986).
15. W. C. Moshier, G. D. Davis, J. S. Ahearn, and H. F. Hough, J. Electrochem. Soc., 2677 (1987).
16. W. C. Moshier, G. D. Davis, and G. O. Cote, J. Electrochem. Soc., **136**, 356 (1989).
17. W. C, Moshier, G. D. Davis, T. L. Fritz, and G. O. Cote, J. Electrochem. Soc., **137**, 1317 (1990).
18. M. Kraak, H. Böhni, and W. Muster, Mater. Sci. Forum, **192–194**, 165 (1995).
19. H. Böhni, T. Suter, M. Büchler, P. Schmuki, and S. Virtanen, Metall. Foundry Eng., **23**, 139 (1997).
20. R. J. Brigham and E. W. Tozer, Corrosion, **29**, 33 (1973).
21. S. Matsch, T. Suter, and H. Böhni, Mater. Sci. Forum, **289–292**, 1127 (1998).
22. M. Renner, U. Heubner, M. B. Rockel, and E Wallis, Werkst Korros., **8**, 182 (1986).
23. G. Okamoto, T. Sugita, S. Nishiyama, and T. Tachibana, Boshoku Gijutsu, **23**, 439 (1974).
24. N. Sato, J. Electrochem. Soc., **123**, 1 (1976).
25. T. Shibata and T. Takeyama, Nature (London), **260**, 315 (1976).
26. T. Shibata and T. Takeyama, Corrosion, **33**, 243 (1977).
27. T. Shibata, Corrosion, **52**, 813 (1996).
28. H. H. Strehblow, Werkst. Korros., **27**, 793 (1976).
29. B. P. Löchel and H. H. Strehblow, Werkst. Korros., **31**, 353 (1980).
30. H. H. Strehblow, in Proc. Ninth International Congress on Metallic Corrosion, Toronto, Publ. NRC, Ottawa, Canada, Vol. 2, 1984, p. 99.
31. K. E. Heusler and L. Fischer, Werkst. Korros., **27**, 550 (1976).
32. R. Dölling and K. E. Heusler, in Proc. Ninth International Congress on Metallic Corrosion, Toronto, Publ. NRC, Ottawa, Canada, Vol. 2, 1984, p. 129.
33. N. Sato, Electrochim. Acta, **16**, 1683 (1971).
34. C. Y. Chao, L. F. Lin, and D. D. Macdonald, J. Electrochem. Soc., **128**, 1187 (1981).

35. L. F. Lin, C. Y. Chao, and D. D. Macdonald, J. Electrochem. Soc., **128**, 1194 (1981).

36. H. H. Strehblow, in Corrosion Mechanisms in Theory and Practice, P. Marcus and J. Oudar, (Eds.), Marcel Dekker, New York, 1995, p. 201.

37. T. Suter, T. Peter, and H. Böhni, Mater. Sci. Forum, **192–194**, 25 (1995).

38. H. Böhni, T. Suter, and A. Schreyer, Electrochim. Acta, **40**, 1361 (1995).

39. P. Schmuki, J. Fraser, C. M. Vitus, M. J. Graham, and H. S. Isaacs, J. Electrochem. Soc., **143**, 3316 (1996).

40. I. Graz and H. Worch, Corros. Sci., **9**, 71 (1969).

41. W. Hubner and G. Wranglen, in Current Corrosion Research in Scandinavia, J. Larin-kari (Ed.), Publ. Sanoma Osakeyhtio, Helsinki, Finland, 1964, p. 6,069.

42. G. A. W. Murray, H. J. Lamb, and H. P. Godard, Br. Corros. J., **2**, 216 (1967).

43. P. M. Aziz and H. P. Godard, Corrosion, **10**, 269 (1954).

44. J. O. Park, C. H. Paik, and R. C. Alkire, in Critical Factors in Localized Corrosion II, P. M. Natishan, R. J. Kelly, G. S. Frankel, and R.C.Newman (Eds.), Electrochemical Society, Pennington, NJ, 1995, p. 218.

45. V. P. Batrakov, in 3rd International Congress on Metallic Corrosion, Moscow, Y. M. Kolotyrkin (Ed.), Vol. 1, Publ. MIR, Moscow,1969, p. 313.

46. M. N. Ronzhin. V. G. Pedanova, A. I. Golubiev, and V. Koshechkin, Dokl. Akad. Nauk SSSR, **180**, 1161 (1968).

47. N. D. Tomashov and L. N. Volkov, in Korrozi. Met. Mater. Primienienye, Moscow, 1974, p. 159.

48. B. E. Wilde and J. S. Armijo, Corrosion, **23**, 208 (1967).

49. G. S. Eklund, J. Electrochem. Soc., **123**, 170 (1976).

50. G. S. Eklund, Scand. J. Metall., **1**, 331 (1972).

51. P. E. Manning, D. J. Duquette and N. T. Savage, Corrosion, **36**, 313 (1980).

52. J. Degerbeck, Werkst. Korros., **29**, 179 (1978).

53. M. Henthorne, Corrosion, **26**, 511 (1970).

54. Z. Szklarska-Smialowska, Br. Corros. J., **5**, 159 (1970).

55. B. R. T. Anderson and B. Solly, Scand. J. Metall., **4**, 85 (1975).

56. G. Wranglen, Corros. Sci., **14**, 331 (1974).

57. S. C. Srivastava and M. B. Ives, Corrosion, **45**, 488 (1989).

58. H. S. Isaacs and G. Kissel, J. Electrochem. Soc., **119**, 1628 (1972).

59. H. S. Isaacs, Corros. Sci., **29**, 313 (1989).

60. G. T. Burstein and S. P. Mattin, in Critical Factors in Localized Corrosion II, P. M. Natishan, R. G. Kelly, G. S. Frankel and R. C. Newman (Eds.), Electrochemical Society, Pennington, NJ, 1995, p. 1.

61. S. Virtanen, Y. Kobayashi, and H. Böhni, in Critical Factors in Localized Corrosion III, R. G. Kelly, G. S. Frankel P. M. Natishan, and R. C. Newman,(Eds.) Electrochemical Society, Pennington, NJ, 1998, p. 281.

62. T. Suter and H. Böhni, Electrochim. Acta, **42**, 3275 (1997).

63. K. Osozawa and N. Okato, in First Soviet-Japanese Seminar on Corrosion and Protection of Metals, Moscow, Y. Kolotyr-kin (Ed.), NAUKA, 1979, p. 229.

64. K. Sugimoto, M. Seto, S. Tanaka, and S. Hara, J. Electrochem. Soc., **140**, 1586 (1993).

65. S. Tanaka, N. Hara, and K. Sugimoto, Mater. Sci. Eng., **A198**, 63 (1995).

66. P. Schmuki, M. Büchler, S. Virtanen, H. Böhni, H. Müller, and L. J. Gauckler, J. Electrochem. Soc., **142**, 3336 (1995).

67. S. Virtanen, P, Schmuki, H. Böhni, P. Vuoristo, and T. Mäntylä, J. Electrochem. Soc., **142**, 3067 (1995).

68. S. Virtanen, P. Schmuki, A. J. Davenport, and C. M Vitus, J. Electrochem. Soc., **144**, 198 (1997).

69. S. Virtanen, P. Schmuki, M. Büchler, and H. Böhni, Analusis, **25**, M 22 (1997).

70. S. Virtanen, P. Schmuki, H. S. Isaacs, M. P. Ryan, L. Oblonsky, and H. Böhni, in Passivity and Its Breakdown, P. M. Natishan, H. S. Isaacs, M. Janik-Czachor, V. A. Macagno, P. Marcus, and M. Seo (Eds.), Electrochemical Society, Pennington, NJ, 1998, p. 171.

71. E. Otero, R. Lizarbe, and S. Feliu, Br. Corros. J., **13**, 82 (1978).

72. W. H. Ailor, Br. Corros, J., **1**, 237 (1966).

73. H. Godard, W. P. Jepson, M. R. Bothwell, and R. L. Kane, The Corrosion of Light Metals, Wiley, New York, 1967.

74. T. E. Wright, H. P. Godard, and I. H. Jenks, Corrosion, **13**, 481t (1957).

75. H. S. Campbell, J. Inst. Met., **93**, 97 (1964).

76. D. O. Sprowls and M. E. Carlisle, Corrosion, **17**, 125t (1966).

77. P. M. Aziz and H. P. Godard, Ind. Eng. Chem., **44**, 1791 (1951).

78. F. C. Porter and S. E. Hadden, J. Appl. Chem., **3**, 385 (1953).

79. W. A. Bell and H. S. Campbell, Br. Corros. J., **1**, 72 (1966).

80. G. Butler, P. Stretton, and J. G. Beyon, Br. Corros. J., **7**, 168 (1972).

81. N. D. Tomashov, G. P. Chernova, and O. N. Markova, Prot. Met., **7**, 85 (1971).

82. I. L. Rosenfeld and I. S. Danilov, Corros. Sci., **7**, 129 (1967).

83. F. Hunkeler, On the Pitting Mechanism of Aluminum with Special Emphasis on Pit Growth Kinetics (in German), Ph.D. Thesis, ETH Zürich Nr. 6663, 1980.

84. H. Kaesche, Z. Phys. Chem., **NF 34**, 87 (1962).

85. H.-J. Engell and N. D. Stolica, Arch. Eisenhüttenw., **30**, 239 (1959).

86. J. Tousek, Corros. Sci., **15**, 147 (1975).

87. J. Tousek, Werkst. Korros., **25**, 496 (1974).

88. G. Herbsleb and H.-J. Engell, Z. Phys. Chem., **215**, 167 (1960).

89. G. Herbsleb and H.-J. Engell, Werkst. Korros., **17**, 365 (1966).

90. J. Wenners, Dissolution Mechanisms within Pits on Iron and Nickel (in German), Ph.D. Thesis, Freie Universität Berlin, 1977.

91. F. Hunkeler and H. Böhni, Werkst. Korros., **32**, 129 (1981).

92. F. Hunkeler and H. Böhni, Corrosion, **37**, 645 (1981).

93. G. S. Frankel, R. C. Newman, C. V Jahnes, and M. A. Russak, J. Electrochem. Soc., **140**, 2192 (1993).

94. G. S. Frankel, Corros. Sci., **30**, 1203 (1990).

95. G. S. Frankel, J. O. Dukovic, B. M. Rush, V, Brusic, and C. V. Jahnes, J. Electrochem. Soc., **139**, 2196 (1992).

96. G. S. Frankel, J. R. Scully, and C. V. Jahnes, J. Electrochem. Soc., **143**, 1834 (1996).

97. R. C. Alkire and K. P. Wong, Corros. Sci., **28**, 411 (1988).

98. R. C. Newman and E. M. Franz, Corrosion, **40**, 325 (1984).

99. K. P. Wong and R. C. Alkire, J. Electrochem. Soc., **137**, 3010 (1990).

100. H.-J. Engell, Electrochim. Acta, **22**, 987 (1977).

101. H. H. Strehblow, K. J. Vetter, and A. Willigallis, Ber, Bunsenges. Phys. Chem., **75**, 822 (1971).

102. I. L. Rosenfeld, I. S. Danilov, and R. N. Oranskays, J. Electrochem. Soc., **125**, 1729 (1978).

103. T. R. Beck and R. C. Alkire, J. Electrochem. Soc., **126**, 1662 (1979).

104. R. Alkire, D. Ernsberger, and T. R. Beck, J. Electrochem. Soc., **125**, 1382 (1978).

105. F. Hunkeler and H. Böhni, in Passivity of Metals and Semiconductors, M. Froment (Ed.), Elsevier, Amsterdam, The Netherlands, 1983, p. 655.

106. H. Böhni, in Corrosion in Power Generating Equipment, M. O. Speidel and A. Atrens (Eds.), Plenum, New York, 1984, p. 29.

107. F. Hunkeler and H. Böhni, in Corrosion Chemistry within Pits Crevices and Cracks, A. Turnbull (Ed.), HMSO Publications Centre, London, 1987, p. 27.

108. F. Hunkeler, A. Krolikowski, and H. Böhni, Electrochim. Acta, **32**, 615 (1987).

109. T. R. Beck, Corrosion, **33**, 9 (1977).

110. C. Edeleanu, J, Inst. Met, **89**, 90 (1960/1961).

111. F. Hunkeler and H. Böhni, Corrosion, **40**, 10 (1984).

112. H. Böhni and F. Hunkeler, in Advances in Localized Corrosion, H. Isaacs, U. Bertocci, J. Kruger, and S. Smialowska (Eds.), NACE-9, NACE, Houston, TX, 1990, p. 69.

113. J. R. Galvele, in Passivity of Metals, R. P. Frankenthal and J. Kruger (Eds.), Electrochemical Society, Pennington, 1978, p. 249.

114. J. R. Galvele, J. Electrochem. Soc., **123**, 464 (1976).

115. J. R. Galvele, Corros. Sci., **21**, 551 (1981).

116. N. Sato, in Corrosion and Corrosion Protection, R. P. Frankenthal and F. Mansfeld (Eds.), Electrochemical Society, Pennington, NJ, 1981, p. 101.

117. N. Sato, J. Electrochem. Soc., **129**, 260 (1982).

118. K. J. Vetter and H. H. Strehblow, Ber. Bunsenges. Phys. Chem., **74/75**, 1024/449 (1970).

119. K. J. Vetter and H. H. Strehblow, in Localized Corrosion, R. Staehle, B. Brown, J. Kruger, and A.K. Agrawal (Eds.), NACE-3, NACE, Houston, TX, 1974, p. 240.

120. G. Herbsleb and H. J. Engell, Z. Elektrochem., **65**, 881 (1961).

121. H. W. Pickering and R. P. Frankenthal, J. Electrochem. Soc., **119**, 1297 (1972).

122. R. P. Frankenthal and H. W. Pickering, J. Electrochem. Soc., **119**, 1304 (1972).

123. H. W. Pickering and R. P. Frankenthal, in Localized Corrosion, R. Staehle, B. Brown, J. Kruger, and A. K. Agrawal (Eds.), NACE-3, NACE, Houston, TX, 1974, p. 261.

124. C. M. Chen, F. H. Beck, and M. G. Fontana, Corrosion, **24**, 234 (1971).

125. M. I. Abdulsalam and H. W. Pickering, Corros. Sci., **41**, 351 (1999).

126. M. I. Abdulsalam and H. W. Pickering, J. Electrochem. Soc., **145**, 2276 (1998).

127. T. R. Beck and S. G. Chan, Corrosion, **37**, 665 (1981).

128. G. S. Frankel, L. Stockert, F. Hunkeler, and H. Böhni, Corrosion, **43**, 429 (1987).

129. P. Schmuki and H. Böhni, J. Electrochem. Soc., **139**, 1908 (1992).

14

STRESS CORROSION CRACKING

R. N. Parkins*

University of Newcastle upon Tyne, Newcastle upon Tyne, UK

A. INTRODUCTION

Metals and alloys subjected to tensile stresses and exposed to certain environmental conditions may develop cracks that would not occur in the absence of either of those controlling parameters. Not all environments that are corrosive to a particular metal promote stress corrosion cracking (SCC), but even some apparently innocuous substances, such as water, may induce cracking in some materials [1], the composition and structure of which can play a critical role in the incidence or otherwise of SCC. The manifestation of the cracks may create the impression of brittleness in the metal because the cracks often propagate with little attendant deformation, although almost invariably the properties of the metal conform to ductility specifications.

The incidence of SCC appears to have increased over the last few decades, possibly because as the problem of general corrosion has been overcome, by control of environmental factors and the use of inherently more corrosion-resistant materials, the probability of more localized forms of corrosion has increased. Those trends have been accompanied by moves toward higher operating stresses, deriving from the more efficient use of materials, and the more extensive use of welding as a method of fabrication. The result gives rise to residual stresses, as do other methods of fabrication involving inhomogeneous deformation, and unless such stresses are relieved, they can promote SCC, the incidence of which is probably greater from the presence of residual than from operating stresses alone. An example of SCC resulting from residual welding stresses is shown in Figure 14.1. Here liquor seeping from the cracks reveals the positions of the latter, which are associated with a longitudinal weld where the principal tensile residual stresses would be parallel to the weld and of the order of the yield stress in this nonstress relieved structure [2]. The low rate of growth of those cracks, the facts that they are visible, and because the operating pressure is relatively low, would allow appropriate action to be taken. But invariably that is not the case since the cracks may grow undetected. This was the case in the failure shown in Figure 14.2, which is from a buried high-pressure gas pipeline that failed in service and where the main fracture is apparent from the separation of the fracture surfaces toward the top of the photograph. But secondary cracks, emanating from the soil side, are also apparent [3]. These cracks are not associated with a weld and the operating stresses undoubtedly played a major role in the failure. Both Figures 14.1 and 14.2 show the presence of multiple cracks, which would be seen to be even more prevalent if examined at higher magnifications. Multiple cracking frequently also accompanies SCC failures.

Stress corrosion cracking may be associated with intergranular or transgranular paths through the metal and, in some cases, with a mixture of those modes. Examination of metallographic sections from the steels involved with Figures 14.1

*Deceased.

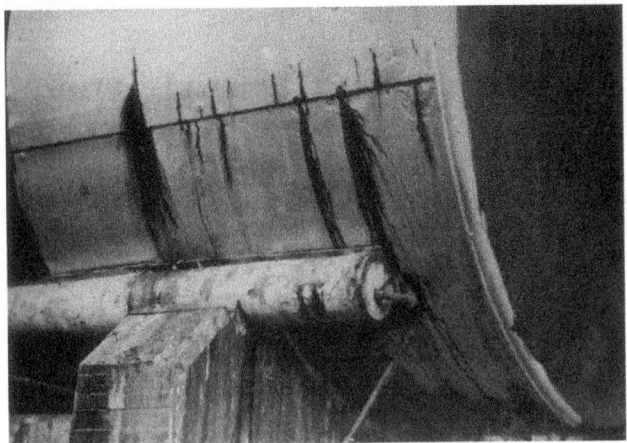

FIGURE 14.1. Stress corrosion cracks in shell of coal gas liquor recirculating tank [2].

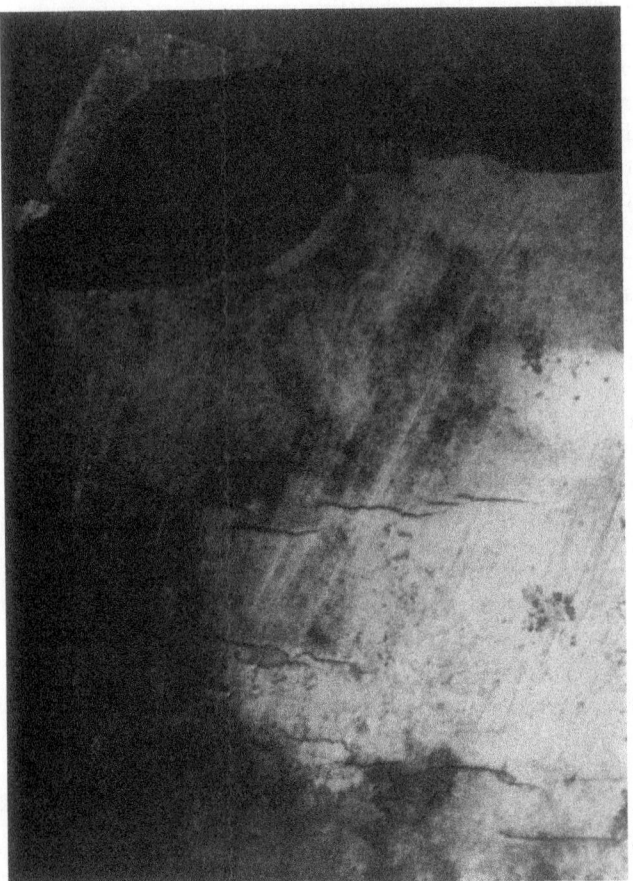

FIGURE 14.2. Secondary SCC on outer surface of high-pressure gas transmission pipeline near origin of SCC failure [3].

and 14.2 revealed intergranular cracks, with a typical example in Figure 14.3. An example of an essentially transgranular crack, in a different pipeline steel from that involved with Figure 14.3 and exposed to a different environment [4], is shown in Figure 14.4. Figure 14.5 shows an example of

mixed-mode cracking in an arsenical α brass exposed to water containing a small amount of sulfur dioxide [5]. In the same system, cracks may initiate in the intergranular mode, then change to transgranular as the stress concentration increases with increasing crack depth, and even revert to intergranular if the crack bifurcates and reduces the stress concentration. Since the most common mode of failure of metals is transgranular (e.g., by overload or fatigue), but not high-temperature creep, intergranular SCC is sometimes regarded as unusual and has often been associated with the collection of segregates or precipitates at grain boundaries. That is undoubtedly so in some cases, of which the propensity for intergranular SCC in sensitized austenitic stainless steels, due to the precipitation of chromium carbides in boundaries and the impoverishment of the surrounding material in chromium is probably the best-known instance [6]. However, as the examples quoted above involving pipeline steels indicate, a change in environment may result in a change of crack path, while the examples involving brass indicate that a change in the stressing conditions may also result in a change in cracking mode. Thus, while electrochemical heterogeneity due to segregates or precipitates at grain boundaries may play a role in some instances of intergranular SCC, the incidence of the latter or otherwise is dependent on the environmental and stressing conditions.

The interactions between environmental and stressing conditions upon the paths of cracks indicate the possible problems in deducing some mechanistic hypothesis for SCC based upon limited studies of a particular system. Mechanisms of SCC have been, and continue to be, widely discussed. Most suggested mechanisms invoke either a process of localized embrittlement of the metal in the vicinity of the crack tip or of localized dissolution in that region [7]. There are variations on both of these themes, so that, for example, while hydrogen ingress may result in the embrittlement of some metals, and the mechanism of such embrittlement is a matter of debate, the formation of films with certain properties may induce cleavage which, once initiated, continues to advance into the underlying metal before arresting. Such film-induced cleavage may involve dealloyed layers in some alloys, but those layers may play the same role as oxide films, in preventing lateral dissolution on the crack sides and concentrating dissolution in the crack tip region, when bare metal is exposed by straining in the adjacent metal. Such strains may be associated with transgranular cracking where bare metal slip steps emerge through films in the crack tip region, or they may induce intergranular cracking, especially if electrochemical heterogeneity causes preferential dissolution at the grain boundaries. It is possible that with some environments, both dissolution and hydrogen ingress may facilitate crack growth. It is not surprising that mechanistic aspects of SCC are the subject of almost annual conferences, but for our present purposes the emphasis is on assisting the practicing engineer predict those combinations of metal and environment that promote SCC and how they may be

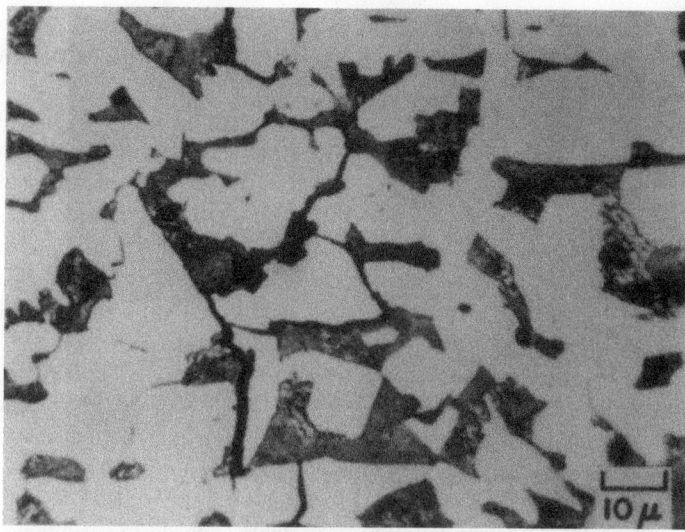

FIGURE 14.3. Intergranular cracks in pipeline steel [3].

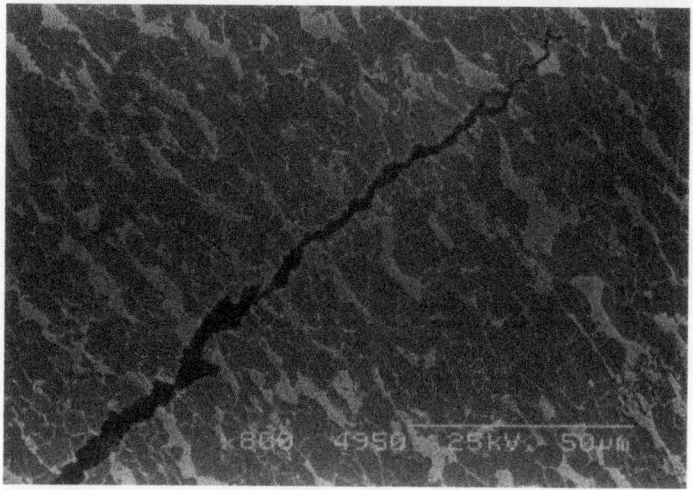

FIGURE 14.4. Transgranular cracking in different pipeline steel from that of Figure 14.3 produced by different environment [4].

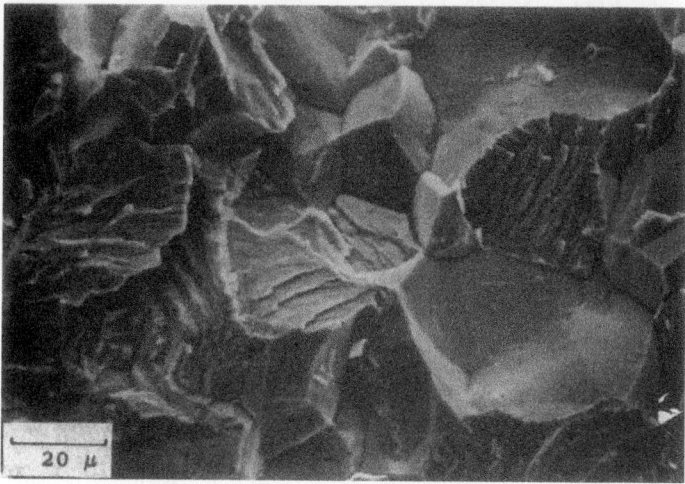

FIGURE 14.5. Mixed-mode SCC in α brass exposed to $H_2O + SO_2$ [5].

controlled or avoided in particular engineering structures. Complete mechanistic understanding of SCC would assist in achieving such goals, but in the absence of such understanding it is still possible to make some advances in relation to practical control.

B. ENVIRONMENTAL CONDITIONS FOR SCC

Until about four decades ago, it was thought that SCC occurred only in particular alloys exposed to a few very specific environments (e.g., ferritic steels exposed to hydroxides or nitrates, brasses exposed to ammoniacal environments, and austenitic stainless steels or aluminum alloys to chlorides). While the concept of solution specificity remains, since not all environments corrosive toward a particular alloy promote SCC, the number of environments that will promote that mode of failure has increased considerably in recent times. It is impossible to list here all of the environments that have been shown to promote SCC in the commonly used alloys, but, in terms of predicting whether or not a particular metal–environment combination is likely to suffer SCC, certain circumstances need to obtain. Where crack growth is by a localized dissolution process, potent solutions will need to promote a critical balance between activity and passivity since a highly active condition will result in general corrosion, while a completely passive condition cannot lead

to SCC. For most engineering alloys, inactivity at exposed surfaces is due to the presence of adherent oxide films on those surfaces. Then, it is not surprising that the alloys of high inherent corrosion resistance (e.g., austenitic stainless steels that readily develop protective films) require an aggressive ion, such as a halide, to promote SCC. Alternatively, metals of low inherent corrosion resistance (e.g., C steels) require the presence of an environment that is itself partially passivating for SCC. Such steels can fail in solutions of anodic inhibitors (e.g., nitrates, hydroxides, carbonates, or phosphates).

The SCC is influenced not only by the presence of particular ions and their concentration but also by the electrode potential, so that cracking occurs only within certain potential ranges for particular metal–environment combinations. Figure 14.6 shows the cracking domains for ferritic steels exposed to various environments (also involving different temperatures, another parameter that can influence SCC), and pH and potential dependent cracking domains have been shown to exist for other alloys. The boundaries of such domains would be expected to relate to particular reactions involving film formation. For each of the systems shown in Figure 14.6, the upper boundaries correspond to the formation of Fe_2O_3, with ductile failure in slow strain rate tests at potentials above the cracking domain. There are two exceptions in that at potentials high enough to form Fe_2O_3, nitrates and high-temperature water promote cracking from within pits. The initiation of cracks from pits has been

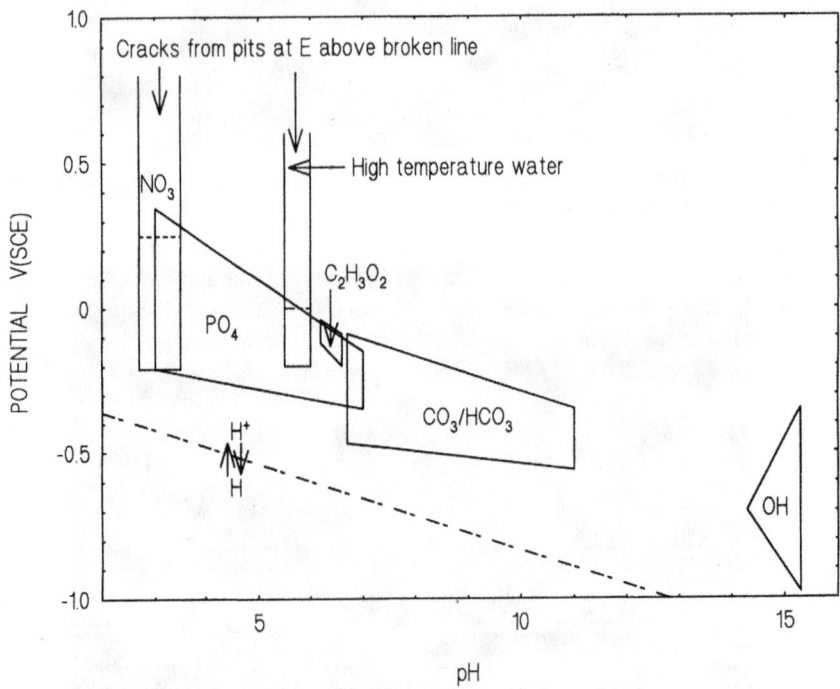

FIGURE 14.6. SCC potential–pH domains for ferritic steels in various environments, the latter at a variety of temperatures, together with equilibrium potentials for discharge of hydrogen at 25°C.

observed in many systems and while that has sometimes been related to stress intensification at the base of pits, it is probably at least as likely to relate to changes in the composition of the solution and potential within the pits. With some bulk solutions, lowering of the pH and potential will occur in pits, so that the SCC domain for the bulk solution may be irrelevant and indeed the mechanism of cracking may change. Thus, Figure 14.6 shows that the cracking domains are above the equilibrium potentials for hydrogen discharge from the solutions of various pH values, but even the buffered solutions (OH, CO_3/HCO_3, $C_2H_3O_2$, and PO_4), which do not cause pitting at the concentrations involved with Figure 14.6, can promote hydrogen-related cracking if the potential is reduced sufficiently for hydrogen discharge. However, where acidification of the localized environment within a pit occurs, hydrogen discharge is likely to be facilitated.

While it is possible to predict cracking domains from thermodynamic data for some of the systems to which Figure 14.6 refers, there are difficulties with such predictions when the bulk environment is changed within a pit or crack enclave. Moreover, even successful predictions have followed from experimental determination of the cracking domain boundaries and, at present, there is no theoretical approach to determining whether or not a particular metal–environment combination, for which there is no prior experience, will promote SCC. However, there are some relatively rapid experimental approaches to determining the potency of systems for SCC, where the bulk environment is itself the potent solution. The latter may derive from the bulk solution, which itself may be incapable of promoting SCC, by concentration in a crevice or at a heat transfer surface. In these circumstances predictability may be more difficult, unless the possibility of such concentration is recognized. The instances of SCC in pipelines mentioned in the context of Figures 14.2–14.4 illustrate the point. The transgranular cracking (Fig 14.4) is due to the presence of a dilute groundwater containing CO_2 (pH $\approx$ 6.5), while the intergranular cracking (Figs. 14.2 and 14.3) is due to the generation of a relatively concentrated carbonate–bicarbonate solution derived from groundwater but concentrated due to the flow of cathodic current at the pipe surface and ion transport, as well as heat transfer in the crevice between the pipe and a disbonded coating. Where crack advance is by dissolution, the crack tip must be active, but as the crack advances the crack sides must become relatively inactive, otherwise the sides will extend laterally and the geometry will be changed to that of a pit. Figure 14.4 shows evidence of corrosion on the crack sides increasing in moving away from the tip, so that the amount of activity acceptable on the crack sides will depend on the rate of crack growth. Transitions from electrochemically active to relatively inactive behavior may be expected to be reflected in the current response of the bare metal exposed to the appropriate environment. Thus, dissolution will be associated with the passage of relatively high anodic current densities, but

with the passage of time this current will decay if filming occurs. Very rapid rates of decay are not likely to permit much dissolution and so are not likely to be indicative of conditions conducive to SCC. Very slow rates of decay are more likely to be indicative of insufficient development of inactivity to retain crack geometry. There are various techniques for measuring these features, including scratching or rapidly straining electrodes previously filmed at particular potentials. A convenient way of anticipating the range of potentials in which SCC may occur is through potentiodynamic polarization curves determined at different sweep rates [8]. If the potential of an initially film-free surface is rapidly changed ($\approx$ 1 V/min) over an appropriate range, then the currents passed at the surface will indicate ranges of potential in which relatively high anodic activity is likely. The rapid sweep of the potential range has the object of minimizing film formation. If the experiment is repeated at a slow rate ($\approx$ 10 mV/min) of potential change allowing filming, comparison of the two curves will indicate ranges of potential within which high anodic activity in the film-free condition reduces to insignificant activity at the slow sweep rate, thereby identifying the range of potentials in which SCC is likely. The method correctly anticipates SCC of ferritic steels in a number of very different environments, but is only applicable where air-formed oxide films can be reduced. Thus for metals with very stable films one resorts to scraping or rapid straining as a means of creating bare metal. For systems where cracks may be initiated from pits, measurement of the pitting potential can give an indication of the minimum potential for cracking, while for systems where hydrogen-induced cracking is possible calculation of the equilibrium potential for hydrogen discharge, from the pH of the solution, will give an indication of the highest potential for such cracking. However, in the application of any of these approaches to estimating potentials for cracking it is well to remember that the surfaces in plant may differ appreciably in terms of potentials from those measured in the laboratory, the latter usually on carefully polished surfaces, in the same environment. Consequently, a laboratory SCC test at open circuit potential may not give an adequate indication of the propensity for cracking in a particular system. A few additional tests at controlled potentials, defined by some appropriate electrochemical measurements, are therefore necessary. This is especially so if cracking does not occur at open circuit.

In relation to the stressing of specimens for assessing the propensity for SCC in any system, slow strain rate tests (SSRTs) offer a rapid method of arriving at a result [9, 10]. These, which may be conducted on initially plain or precracked specimens, are simply tensile tests conducted at relatively slow strain rates, typically of the order of $\sim$$10^{-6}$/s for steels and Cu or Ni alloys, and 10^{-5}/s for Ti or Mg alloys. Failure will usually occur in 1 or 2 days and various parameters may be employed for quantifying the results, particularly those related to ductility, although fractographic or metallographic examination of the failed samples should be conducted. Obviously, the stressing and/or straining conditions in SSRTs are beyond those likely to be experienced in service. Strain rate effects in the latter are considered later, but where the objective of the laboratory test is to give a "go–no go" result quickly, then SSRTs are particularly useful.

C. ROLE OF STRESS IN SCC

It is usual to consider the role of stress, or the stress intensity factor in the case of precracked specimens, in terms of its influence upon the time to failure in a given system. Figure 14.7 shows some typical results [11], which indicates that there is a stress, often referred to as the threshold stress, below which failure does not occur in an extended test time. The threshold stress is not only a function of alloy composition and structure, but also of the environmental conditions, including solution composition, potential, and temperature, so that it is not a unique property of a material in the sense of a yield or tensile strength. Moreover, while the threshold stress is sometimes defined as the stress below which cracking does not occur, this is not necessarily so and it is better defined as the stress above which total failure occurs, since for some systems cracks have been shown to initiate below die threshold but to cease to propagate after some growth. It is difficult to explain why cracks should cease to propagate on any stress-based argument, since stress concentration or intensification would be expected to increase with crack growth under constant load. A

feasible explanation is that it is not stress per se but the strain rate it engenders that is the controlling factor, and that cracks cease to propagate when the crack tip strain rate falls below some critical value related to the rate of film growth. Such an explanation is consistent with the influence of the relative times at which the stress and the environmental conditions for cracking are established, creep at constant load prior to the establishment of the environmental conditions delaying or preventing cracking in laboratory tests. Obviously, it is also consistent with the demonstration of the influence of applied strain rate upon cracking for a wide variety of combinations of metal and environment, for some of which it has been shown that sufficiently slow rates of straining can result in ductile failure without impairment of the tensile strength of the alloy despite its exposure to a potent cracking environment.

There is another important consequence of the significance of strain rate in facilitating crack growth and it relates to cyclic, as opposed to static, loading. Hysteresis effects are well known to accompany cyclic loading and the cyclic stress–strain curves of materials often fall appreciably below their monotonic loading counterparts, reflecting the fact that cyclic loading facilitates microplastic deformation [12]. It follows from such observations that load cycling may produce SCC at significantly lower stresses than those needed with static loading. Figure 14.8 shows this to be so for a pipeline steel exposed to a carbonate–bicarbonate solution by comparison of the data for static and cyclic loading conditions [13]. In the absence of cyclic loading ($\Delta K = 0$), the threshold stress intensity factor, $K_{I_{scc}}$, is ~ 21 MN/m$^{3/2}$, and cyclic loading at the frequency of 11 Hz did not alter that

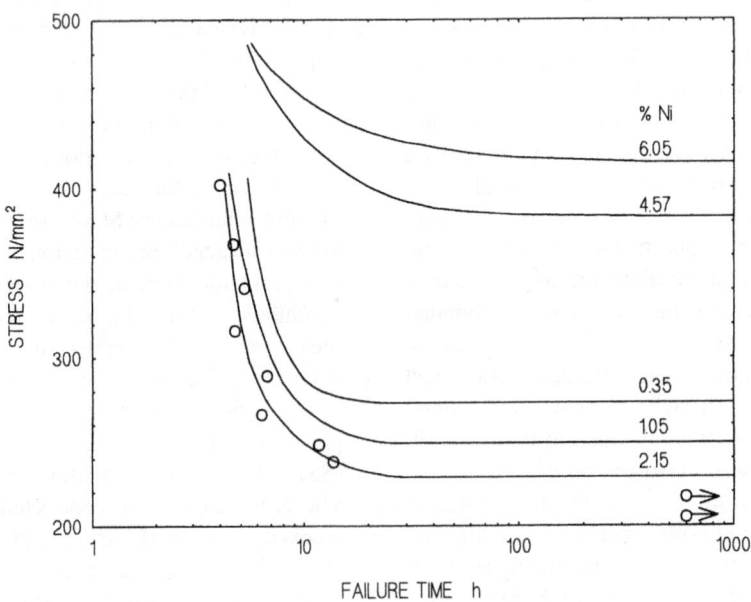

FIGURE 14.7. Initial stress–time to failure curves for ferritic steels with different Ni contents in boiling $4N$ NH$_4$NO$_3$. Data points for only one steel are shown but those give an indication of the scatter for each steel [11].

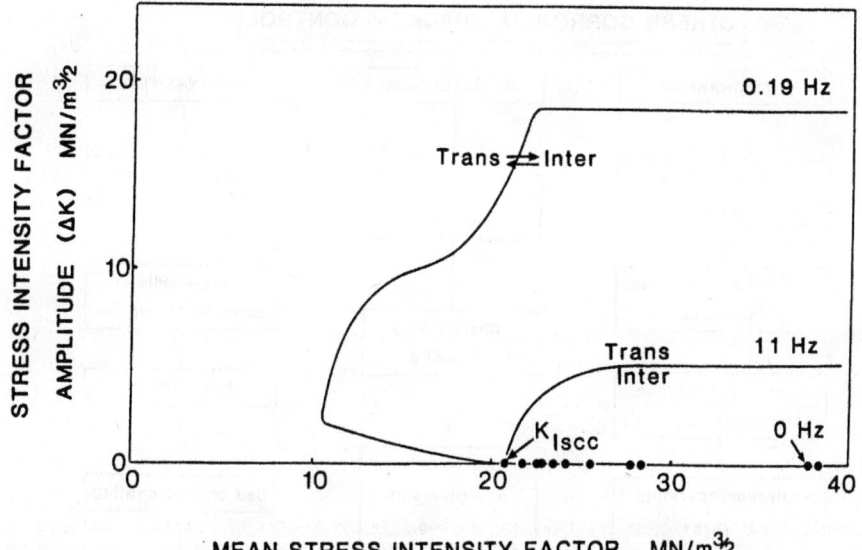

FIGURE 14.8. Modified Goodman diagram indicating loading parameters for which intergranular or transgranular cracking were observed in pipeline steel exposed to carbonate–bicarbonate solution at 75°C and −0.65 V(SCE) [13].

threshold, although intergranular cracking extended to ΔK values of ~ 5 MN/m$^{3/2}$ before a transition to transgranular cracking. With reduction of the frequency of load cycling to 0.19 Hz, not only is the purely intergranular cracking observed at higher ΔK values, but the threshold mean stress intensity factor is reduced to ~ 10 MN/m$^{3/2}$ at small values of ΔK. Similar large reductions in threshold stresses have been observed with initially plain specimens when subjected to cyclic loading at low frequencies and stress amplitudes of $\sim 20\%$ of the mean stress.

Of course, it may be argued that with cyclic loading what is being studied is corrosion fatigue, but fractographically the cyclically loaded specimens are indistinguishable from statically loaded samples, providing that the ΔK values are maintained below those levels that promote transgranular cracking. Indeed it may be argued that the latter mode of cracking is more typical of what may be expected in corrosion fatigue in such a system. The distinction here then is not between static and cyclic loading, or SCC and corrosion fatigue, but rather between relatively small stress amplitudes at low frequencies, sometimes referred to as ripple loading, and the much higher stress amplitudes usually involved with fatigue. Such distinctions are inevitably rather arbitrary, but there is another point to be remembered in this context, which relates to environmental influences. The environments that will promote corrosion fatigue in, say, ferritic steels are much more extensive than those that will promote SCC. A possible reason is that, as already mentioned, for SCC the environment needs to have characteristics that assist in the retention of crack geometry by filming of crack sides, but with large amplitude stresses mechanical crack sharpening occurs, so

that there are less stringent requirements for the properties of the environment.

D. PREVENTION AND CONTROL OF SCC

Since the incidence of SCC requires a susceptible alloy to be exposed to a specific environment at stresses above some limiting value, it follows that control of the problem may be through manipulation of any or all of these three parameters. Ideally, approaches to prevention should begin with the selection of a resistant alloy, which is the most usual approach, followed by consideration of possible modification of the stress or environment, with variations within these three themes as outlined in Figure 14.9 [14]. However, it sometimes happens that SCC occurs in an existing plant when such had not been anticipated at the design stage or, indeed, that a susceptible material had to be used because of other considerations, in which cases the approaches to prevention are restricted. Whatever approach is pursued, it is likely to depend on experience or laboratory test data, and it is important to realize that alloy susceptibility is not simply a function of alloy composition or structure but also of the environmental conditions. Thus, Figure 14.7 shows the beneficial effects of sufficient Ni additions to a ferritic steel upon resistance to cracking, as measured by the threshold stress, in a nitrate environment. However, if the same steels are exposed to a boiling MgCl$_2$ solution, then the Ni additions have the effect of increasing cracking susceptibility, although much higher ($\approx 50\%$) additions may promote immunity, based upon data for austenitic steels containing some 18–20% Cr [15].

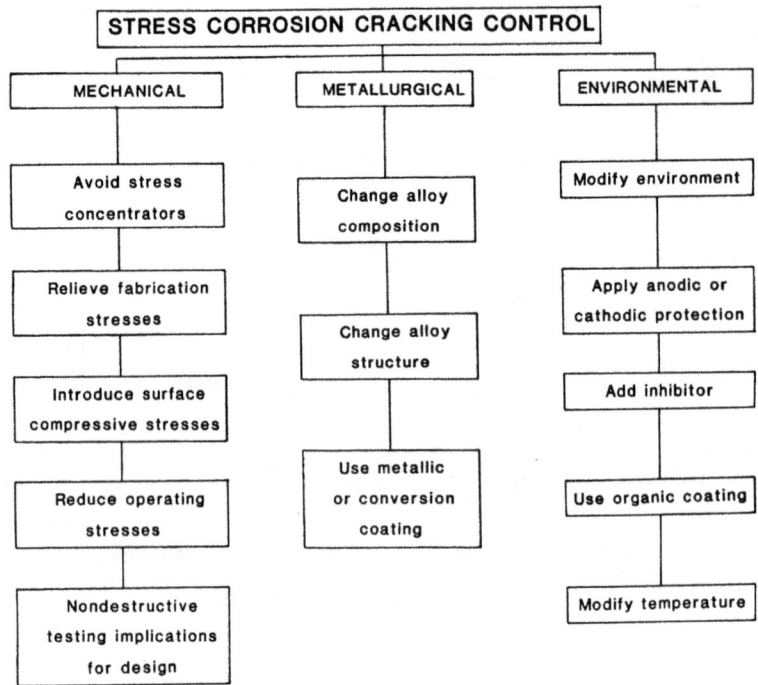

FIGURE 14.9. Approaches to SCC control.

D1. Metallurgical Approaches

There are many published studies of the manipulation of the composition or structure of alloys to control SCC [16]. Where intergranular cracking is due to segregation at the grain boundaries, control of the segregant affords a means of controlling SCC. Thus, restricting the carbon content of stainless steels to a maximum of 0.02%, whereby weld sensitization is avoided, can markedly improve cracking resistance [17]. There are other examples of relatively small changes in alloy composition markedly influencing cracking resistance (e.g., in aluminum alloys [18]), but in many instances, relatively large additions of alloying elements are necessary to achieve significant improvements in resistance. The point may be illustrated by data for additions to ferritic steels in relation to their SCC susceptibility in NO$_3$ or OH solutions [11]. It has already been mentioned that cracking is a function of potential for a given environment, and determination of that relationship in SSRTs affords a means of comparing the effects of alloying additions. The susceptibility in an SSRT may be expressed in terms of the ratio of the time to failure in the potent solution to that in an inert environment (oil) at the same temperature, a ratio of 1 indicating no susceptibility and increasing departure from 1 indicating increasing susceptibility. A plot of that ratio against potential bounds an area within which SCC occurs, that area being a measure of the stress corrosion index (SCI), and reductions or increases in that area indicate beneficial or deleterious effects, respectively, for alloying additions.

Multiple regression analysis of the data resulted in the following equations for the relative effects of various alloying additions upon SCI in NO$_3$ and OH solutions:

$$SCI_{(NO_3)} = 1777 - 996\% \, C - 390\% \, Ti - 343\%$$
$$Al(-132\% Mn) - 111\% \, Cr - 90\% \, Mo$$
$$- 62\% \, Ni + 292\% Si$$

$$SCI_{(OH)} = 105 - 45\% \, C - 40\% \, Mn - 13.7\% \, Ni$$
$$- 12.3\% \, Cr - 11\% \, Ti + 2.5\% \, Al$$
$$+ 87\% \, Si + 413\% \, Mo$$

The first constant on the right-hand side of those expressions reflects the greater propensity for cracking and the wider potential range involved with NO$_3$ than OH solutions, while negative coefficients for the various alloying elements indicate beneficial effects and positive coefficients deleterious influences. (Where a coefficient is bracketed, the t ratio, the coefficient/standard error of the coefficient, was <2 and only the remaining elements should be regarded as having significant effects.) The data reflected in these equations underline a point mentioned earlier, that alloying additions that are beneficial in relation to cracking in one environment may not have a similar influence in a different solution. Thus, while Mo additions are beneficial in relation to the cracking of ferritic steels in NO$_3$, they are markedly deleterious in relation to OH-induced cracking, largely because they extend the range of potentials for cracking in OH solutions. It is apparent from the coefficients involved that relatively large alloying additions are necessary for large decreases in susceptibility, varying with the environment and element involved, but averaging some 8–9%, neglecting carbon where other considerations usually inhibit its involvement in the prevention of SCC. By the same token, some possible compositions that could be deduced from those equations as rendering a ferritic steel immune, would not produce usable materials.

The structure of an alloy is often determined by strength or ductility considerations rather than SCC resistance, although the latter has been shown to be influenced by structure in many materials. In general, large grain sizes are often associated with relatively low threshold stresses for SCC, reflecting the influence of grain size upon yield strength, but increasing the latter by quenching and tempering treatments, where feasible, or heat-affected zones associated with welds, may result in higher crack growth rates. In relation to intergranular SCC, heat treatments may be effective in redistributing those features at grain boundaries that promote electrochemical heterogeneity. Thus, solution heat treatment of stainless steels to redissolve carbides and eliminate Cr depletion at previously sensitized grain boundaries can be effective in avoiding SCC [17]. There are other examples involving materials as diverse as Al and Ni alloys where cracking can be influenced by appropriate heat treatments. Structural modification associated with cold work may also influence SCC resistance, but it will rarely be the case that this can be used in plant in view of its implications in other directions.

A review [19] of information on the use of metallic coatings applied to Al, Mg, and Ti alloys, together with low alloy and stainless steels, concluded that sacrificial coatings are beneficial on all those materials except low alloy steels. Medium-to-higher strength steels are particularly prone to hydrogen-related cracking, so metallic coatings, especially Zn, that promote low electrode potentials must be used with care, with galvanized high-strength steel bolts being capable of delayed failure when subjected to atmospheric corrosion. Where SCC occurs in potential ranges above those that lead to hydrogen discharge, the use of metallic coatings based on Al, Cr, or Ni can be beneficial, depending on their influence on potential in relation to the range of the latter in which SCC occurs. The same review [19] indicates that conversion coatings applied to Al, Mg, or Ti alloys give little protection against SCC when applied alone, or may even be deleterious, but in conjunction with high-grade paint schemes, they give fair protection.

D2. Environmental Approaches

In relation to control through environmental factors, Figure 14.9 indicates a number of different approaches.

Where the offending chemical species can be removed from the environment, there is an obvious action, as has been practiced in cases where small amounts of chloride have caused cracking of stainless steel. Of course, where the offending substance is a reactant or product in a process, that approach is not available and recourse must be to some alternative, of which anodic or cathodic protection looks attractive in light of the restricted ranges of potential in which SCC occurs. Cathodic protection could be applied to preventing SCC in the systems to which Figure 14.6 refers, although as already mentioned, each of those environments can promote hydrogen-related cracking in steels at sufficiently low potentials. Consequently, the potential needs to be carefully controlled below the lowest value for dissolution-related cracking but above the highest potential for hydrogen-induced failure. Anodic protection could be an effective means of control for some of the systems involved with Figure 14.6, although not where cracking is associated with pitting, since the primary requirement for such protection is an environment that allows the formation of stable passivating films at potentials above the cracking domain. For either of these approaches to SCC control it is vital that the current required to change the potential reaches all of the surfaces exposed to the environment This can present some difficulties where narrow crevices are present in the structure or where intermittent wetting or discontinuous liquid films are involved.

Inhibitive additions to environments to control SCC have long been practiced, especially in relation to the caustic cracking of boilers, although other systems may be similarly treated [20, 21]. Oxidizing substances that simply raise the potential above the range for SCC can be effective providing they are maintained at an appropriate concentration. However, some such inhibitors may be unsafe in that they are ineffective if some other agency operates to cause the potential to remain within the cracking range. Thus, while raising the pH of NO_3 solutions to about 7 inhibits intergranular cracking at the open-circuit potential, steels tested at even higher pH values can suffer SCC if the potential is held within the cracking range. Clearly the ideal, or safe, inhibitor is one that prevents cracking even when the potential is within the range that promotes SCC in the absence of the inhibitor. Safe inhibitors have been identified for particular systems and can be effective in preventing the growth of existing cracks as well as avoiding initiation, while substances capable of preventing the ingress of hydrogen to metals are also known. Some of those inhibitors are suitable for incorporation into paints or other organic coatings, preferably within the priming coat as leach primers, in view of the well-established facility of coatings to allow the passage of water and other molecules.

Temperature is an important parameter in most instances of SCC, the conditions for cracking in terms of potential range and crack velocities being typical thermally activated processes. While temperatures may be dictated by other considerations than SCC control (e.g., in process plant), lower temperatures are less likely to result in cracking in many situations or will be associated with lower crack growth rates.

D3. Stress Control

Many practical instances of SCC result from the presence of residual stresses, so thermal stress relief may be expected to be beneficial. The most usual problems with applying full stress relieving heat treatments to structures is that they are either too large for available furnace capacity or they distort at the relatively high temperatures involved ($\approx 650°C$ for ferritic and $\approx 800°C$ for austenitic steels). However, partial stress relief by heating to lower temperatures can be adequate where the total residual and operating stresses are reduced below the threshold stress. Moreover, these lower temperatures can be achieved with less distortion in furnace annealing and can be achieved with locally applied heating on large structures. The application of local heating to obtain reduction of the peak tensile residual stresses needs to be practiced with care, if the thermal stresses involved in the relieving process are not to simply move the high stresses to other areas. A variation on local heating for redistributing stresses has been applied to the stainless steel cracking problem associated with boiling water reactors. Induction heating of the outside of a pipe to 500–550°C, while maintaining the inside surface at 100°C with cooling water, results in compressive stresses on the inside surface where SCC may otherwise occur [22]. Similar changes in the residual stress distribution may be achieved with last pass heat sink welding, involving the application of a high weld heat input during the last pass while maintaining the inside surface of the pipe at $\sim 100°C$ with flowing water [22]. Compressive stresses at surfaces where SCC would otherwise initiate can be induced by other means, shot peening or grit blasting having been shown to be effective in that respect [14].

While residual stresses probably account for most SCC failures in service, operational stresses may be becoming an increasing contributor to such failures, possibly deriving from the more efficient use of materials by employing higher operating stresses. Operating conditions are largely dictated by considerations other than control of SCC, but it is important to remember that stress cycles in particular may lower the threshold stress for cracking below that associated with static loading. In some service situations, the most damaging stress fluctuations are likely to be associated with start-up and shutdown, a matter that has been considered in some detail in relation to the damage to boiling water reactor components [23]. Otherwise relatively little consideration has been given to whether or not start-up or shutdown procedures for plant may be identified that would minimize SCC risks while remaining realistic from other operations viewpoints. The latter considerations will most often override the former, but there are likely to be some circumstances where operational conditions can be manipulated to minimize the risk associated with stress cycles.

While ideally structures should be designed and fabricated so that SCC is avoided, in practice it is sometimes necessary to live with the problem. This implies an ability to detect and measure the size of cracks before they reach the critical size that may result in catastrophic failure. Such inspection has important implications for plant design, which obviously should allow inspection at appropriate locations. Failures have occurred at locations where inspection was not possible, but could have been avoided by inspection had the problem been anticipated. The most likely locations for cracking, with their implications for inspection, are regions of high residual stress (welded, bolted, or riveted joints) and regions where stress or environment concentration can occur (notches and crevices). A variety of nondestructive test (NDT) techniques are available for monitoring crack growth and provision for such is now being made at the design stage in some cases so that early detection can be achieved. There is one NDT technique that may have widespread use, even though it is not one for monitoring crack growth so much as detection, and which is worthy of special mention in the present context. It involves the use of liquids, such as dye penetrants, which can themselves induce cracking in some materials, especially high-strength steels prone to hydrogen-induced cracking [14].

REFERENCES

1. P. L. Andresen and F. P. Ford, Research Topical Symposia, Proc. CORROSION 96, NACE International, Houston, TX, 1996, p. 51.
2. R. N. Parkins and R. Usher, J. Appl. Chem., **9**, 445 (1959).
3. R. N. Parkins, "Overview of Intergranular Stress Corrosion Cracking Research Activities," Report No. PR-232–9401, American Gas Association, Arlington, VA, 1994.
4. R. N. Parkins, W. K. Blanchard, Jr., and B. S. Delanty, Corrosion, **50**, 394 (1994).
5. R. N. Parkins, C. M. Rangel, and J. Yu, Met. Trans. A, **16A**, 1671 (1985).
6. A. J. Sedriks, Corrosion of Stainless Steels, Electrochemical Society, Princeton, NJ, 1979.
7. R. P. Gangloff and M. B. Ives (Eds.), Environment-Induced Cracking of Metals, NACE, Houston, TX, 1990.
8. R. N. Parkins, Corros. Sci., **20**, 147 (1980).
9. G. M. Ugiansky and J. H. Payer (Eds.), Stress Corrosion Cracking—The Slow Strain Rate Technique, ASTM Special Technical Publication 665, American Society for Testing and Materials, Philadelphia, PA, 1979.

10. R. D. Kane (Ed.), Slow Strain Rate Testing for the Evaluation of Environmentally Induced Cracking, ASTM Special Technical Publication 1210, American Society for Testing and Materials, Philadelphia, PA, 1993.

11. R. N. Parkins, P. W. Slattery, and B. S. Poulson, Corrosion, **37**, 650 (1981).

12. R. N. Parkins, E. Belhimer, and W. K. Blanchard Jr., Corrosion, **49**, 951 (1993).

13. R. N. Parkins and B. S. Greenwell, Met. Sci., **11**, 405 (1977).

14. R. N. Parkins, Mater. Perform., **24**(8), 9 (1985).

15. H. R. Copson, Physical Metallurgy of Stress Corrosion Fracture, T. N. Rhodin (Ed.), Interscience, New York, 1959, p. 247.

16. R. W. Staehle, J. Hochmann, R. D. McCright, and J. E. Slater (Eds.), Stress Corrosion Cracking and Hydrogen Embrittlement of Iron Base alloys, NACE, Houston, TX, 1977.

17. J. C. Danko, "Boiling Water Reactors Group Research Program; 4 Years Later," CORROSION/84 Paper No. 162, NACE, Houston, TX, 1984.

18. M. O. Speidel, Stress Corrosion Research, NATO Advanced Study Institute Series, Series E, Applied Science, No. 30, H. Arup and R. N. Parkins, (Eds.), Sijthoff & Noordhoff, Alphen aan den Rijn, Denmark, 1979, p. 117.

19. H. G. Cole, AGARD Conf. Proc., Paper No. 53 NATO, 13, 1970.

20. C. S. O'Dell and B. F. Brown, in Corrosion Control by Coatings, H. Leidheiser (Ed.), Science Press, Princeton, N.J., 1979, p. 339.

21. R. N. Parkins, in Embrittlement by the Localized Crack Environment, R. P. Gangloff (Ed.), Metallurgical Society of AIME, Warrendale, PA, 1984, p. 385.

22. S. Iwasaki, in Predictive Methods for Assessing Corrosion Damage to BWR Piping and PWR Steam Generators, H. Okada and R. W. Staehle (Eds.), NACE, Houston, TX, 1982, p. 144.

23. Y. S. Garud and T. L. Gerber, in ASME Conference on Advances in Life Prediction Methods, ASME, New York, 1983.

15

HYDROGEN-INDUCED CRACKING AND SULFIDE STRESS CRACKING

M. Elboujdaini

CANMET Materials Technology Laboratory, Ottawa, Ontario, Canada

A. INTRODUCTION

The four main environmental cracking phenomena of concern [1–5] when steels are exposed to wet H_2S environments are

1. Hydrogen blistering: blister formation, originating at nonmetallic inclusions
2. Hydrogen-induced cracking (HIC): also known as stepwise cracking (SWC)
3. Stress-oriented hydrogen-induced cracking (SOHIC)
4. Sulfide stress cracking (SSC): cracking in steels of high strength or high hardness

These four types of hydrogen damage are illustrated in Figure 15.1. For any of these to occur, there must be a corrosive environment and a material that is susceptible to

Uhlig's Corrosion Handbook, Third Edition, Edited by R. Winston Revie
Copyright © 2011 John Wiley & Sons, Inc.

cracking. The stress may be applied (or residual) stress, as in SOHIC and SSC, or internal stress resulting from the internal pressure of hydrogen, as in HIC and hydrogen blistering.

A variation of HIC, called SOHIC, has also caused failures; for example, SOHIC in a hard weld microstructure caused a major explosion of a pressure vessel [1, 6]. Evidence suggested that cracks propagated to the critical size by a HIC mechanism. Hydrogen evolved at the reacting surface can be absorbed by the steel as atomic hydrogen, as shown in Figure 15.2, which can diffuse through the steel and enter hydrogen traps, such as voids around nonmetallic inclusions. There, hydrogen atoms can combine to form molecular hydrogen, which can cause the internal pressure to increase to a level at which cracks initiate and propagate, as shown in Figure 15.3. In SOHIC, the laminations are arranged in parallel arrays perpendicular to the surface. Numerous catastrophic failures have been reported, many of them of steels of high yield strength for tubing and casing, although accidents have also occurred with low alloy steels used in pipelines and pressure vessels [7].

B. HYDROGEN-INDUCED CRACKING

Hydrogen-induced cracking, also known as SWC, can lead to failures of pipelines, tubulars, and pressure vessels exposed to sour environments, such as sour gas, sour crude oil, and other H_2S-containing environments [8–10]. Hydrogen-induced cracking in pipeline steel in the United States was first reported in 1954 [11]. In 1972, an undersea pipeline made of Grade 448 (X-65) steel in the Persian Gulf began to leak after a few weeks of service [12, 13]. The pipe had been made using a low-temperature (690–750°C) controlled

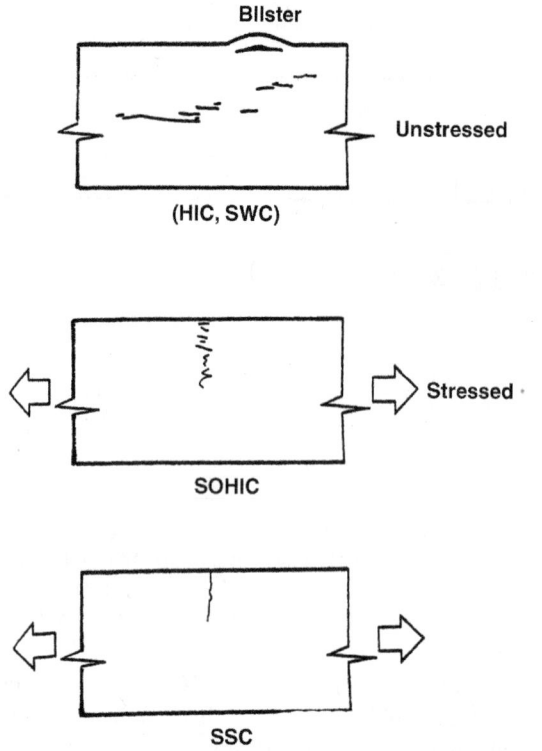

FIGURE 15.1. Schematic presentation of various hydrogen-induced cracks in steel.

rolling process, which was suspected as contributing to the cause of failure. Another major failure occurred in 1974 with a sour-gas pipeline in Saudi Arabia, which occurred within a few weeks of startup and affected a length of about 10 km [14].

Hydrogen-induced cracking occurs in three steps [15]:

1. Formation of hydrogen atoms at the steel surface and adsorption on the surface
2. Diffusion of adsorbed hydrogen atoms into the steel substrate
3. Accumulation of hydrogen atoms at hydrogen traps, such as voids around inclusions in the steel matrix, leading to increased internal pressure and crack initiation and propagation

Cracking requires the production of nascent hydrogen atoms (H^0) at the steel surface, usually by a corrosion reaction in an H_2S-containing, aqueous solution:

$$H_2S + Fe^{2+} \xrightarrow[H_2O]{} FeS + 2H^0$$

The hydrogen atoms produced at the steel surface may combine to form innocuous hydrogen gas molecules (H_2); however, in the presence of sulfide or cyanide, the hydrogen recombination reaction is poisoned, so that the nascent hydrogen atoms (H^0) diffuse into the steel rather than recombining on the metal surface to form hydrogen gas.

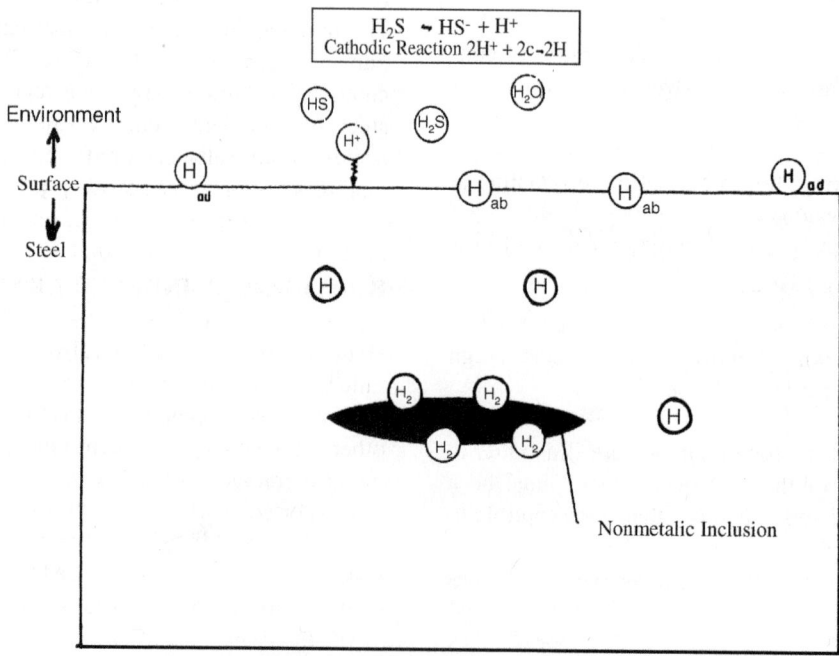

FIGURE 15.2. Mechanism of hydrogen entry.

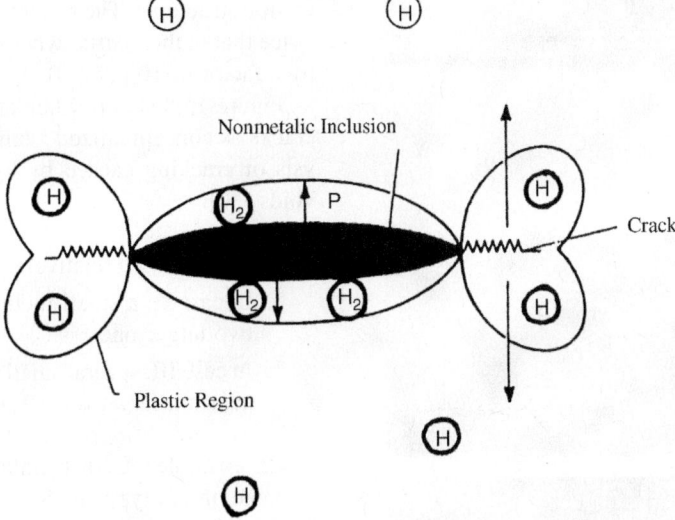

FIGURE 15.3. Schematic illustration of cracking mechanism.

Hydrogen atoms that enter the metallic lattice and permeate through the metal can cause embrittlement and failure of structures in service environments. It is generally observed that, if large amounts of hydrogen are absorbed, there may be a general loss in ductility. Internal blisters may occur if large amounts of hydrogen collect in localized areas [16, 17]. Small amounts of dissolved hydrogen may also react with microstructural features of alloys to produce failures at applied stress far below the yield strength. All these phenomena are referred to as hydrogen embrittlement [18–23].

Chemical species that have been reported to accelerate hydrogen damage include hydrogen sulfide (H_2S), carbon dioxide (CO_2), chloride (Cl^-), cyanide (CN^-), and ammonium ion (NH_4^+).

Some of these species help to produce severe hydrogen charging of steel equipment and may lead to HIC and SOHIC, either of which can cause failure. It is essential to characterize the cracking severity of environments, so that either aggressive environments can be modified and/or materials can be selected with adequate resistance to cracking.

The performance of steels is generally considered to be affected by

Material condition (composition, processing history, microstructure, and mechanical properties)

Fabrication (welding and joining)

Total stress (applied stress plus residual stress)

Environmental effects (corrosion, HIC, SSC, etc.)

Specific chemical agents in service environments may result in degradation, and, additionally, degradation is a function of time, temperature, and other environmental factors. In certain processes, these effects may be complicated since some environmental variables cannot be easily characterized because of the dynamic nature of engineering systems; for example, welded construction in refinery service involves local variations in chemical composition, inhomogeneous microstructure, and residual stresses. These effects are superimposed on influences of environmental factors, which add to the complexity of the situation. Careful assessment of service experience, evaluation of failures in plant tests, and laboratory research are all important and can provide useful information on parameters that are critical to serviceability.

B1. Hydrogen Blistering and HIC

Hydrogen blistering occurs as a result of nascent hydrogen (H^0) atoms diffusing through the steel and accumulating at hydrogen traps, typically voids around inclusions. When hydrogen atoms meet in a trap and combine, they form hydrogen gas (H_2) molecules in the trap. As more gas molecules form, the pressure increases, causing HIC and blister formation. Blisters occur primarily in low-strength steels (<80 ksi or 535 MPa yield strength) and are formed preferentially along elongated nonmetallic inclusions or laminations in the steels used for linepipe [24, 25].

Two types of HIC cracks, centerline cracks and blister cracks, are shown in Figure 15.4. Blister cracks, Figure 15.4(b), are HIC that form near the surface, where hydrogen pressure forms blisters. The formation of blister cracks is related to the type and distribution of nonmetallic inclusions in the steel. Elongated, type II MnS inclusions as well as planar arrays of other inclusions are the predominant initiation sites for cracking. Since inclusions are elongated and/or aligned in the longitudinal (rolling) direction, the cracks propagate in the longitudinal direction.

The steels most susceptible to this form of attack are those with high concentrations of sulfur and manganese, which

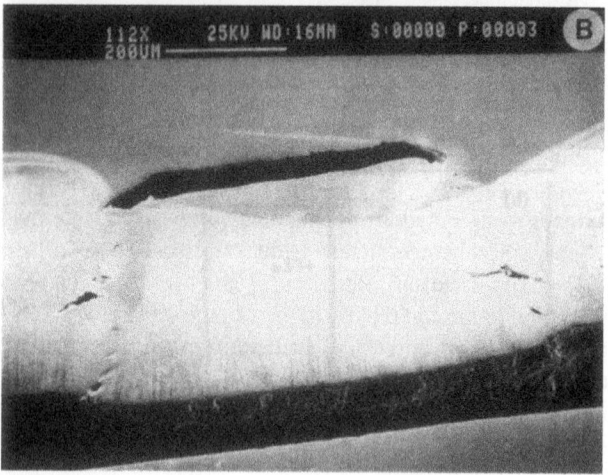

FIGURE 15.4. Two types of HIC. (a) centerline cracks and (b) blister cracks.

combine during melting to form MnS inclusions. Rolling tends to elongate these inclusions (Figure 15.5), increasing the surface area of hydrogen traps. Nevertheless, low-S steels are not necessarily HIC resistant since alloying for inclusion shape control, reduced centerline segregation, and reduction of nitrides and oxides also is necessary [26]. In addition, high-S steels are not necessarily susceptible to HIC. Microsegregation and inclusion shape are more important than bulk sulfur content [26].

Hydrogen-induced cracking is a form of blistering in which lamination-type fissures parallel to the steel surface link in the through-thickness direction, as shown schematically in Figure 15.6. This type of damage (Figure 15.7), caused by linkage of individual defects, can result in rapid through-wall penetration.

Hydrogen-induced cracking may propagate in a straight or stepwise manner. Straight growth occurs in steels having ferrite-pearlite structures if there are high levels of Mn and P segregation or if there are martensitic or bainitic transfor-

mation structures. The Mn level around linear cracks may be twice that in the matrix, whereas the P level may be elevated by a factor of 10 [27–29].

Figure 15.8 shows schematically the stepwise growth of cracks as conceptualized from models based on stress analysis of cracking caused by hydrogen accumulation within voids [30].

1. In case I, two relatively large inclusions are connected.
2. In case II, a small inclusion bridges the gap between two larger ones.
3. In case III, several small inclusions are connected to an adjacent larger one.

An example of crack linkage is shown in Figure 15.9. Linkage of this type has been found to occur if the spacing between inclusions is <0.3 mm.

Like HIC, SOHIC is caused by atomic hydrogen dissolved in steel combining irreversibly to form molecular hydrogen. The molecular hydrogen collects at defects in the metal lattice, as in HIC. However, due to either applied or residual stresses, the trapped molecular hydrogen produces microfissures that align and interconnect in the through-wall direction, as shown in Figure 15.10(a). Although SOHIC can propagate from blisters caused by HIC and SSC, and from prior weld defects [31, 32], neither HIC nor SSC are preconditions for SOHIC [31]. As with HIC, the primary cause of SOHIC is probably atomic hydrogen produced at the steel surface by wet acid gas corrosion [9, 33–35].

Stress-oriented hydrogen-induced cracking tends to occur in the base metal adjacent to hard weldments in pipe and plate steels where cracks may initiate by SSC. The SOHIC is characterized by interlinking microscopic cracks oriented both in the direction perpendicular to the stress and in the plane defined by nonmetallic inclusions, as shown in Figure 15.10(b). The SOHIC is a process by which SSC can propagate in relatively low-strength steels [hardness <22 HRC (Rockwell C hardness)], which would otherwise be considered resistant to SSC.

B2. Factors Influencing HIC

B2.1. Environmental Factors.
Figure 15.11 shows the hydrogen uptake in steel as a function of pH. Hydrogen-induced cracking can occur if the hydrogen content exceeds C_{th} (threshold hydrogen concentration, about 0.8 mL/100 g), so it can occur for plain carbon steel at pH < 5.8, whereas with Cu-bearing steel, it would occur only at pH < 4. Copper alloying improves HIC resistance, particularly in media containing H_2S at pH 4–6. Copper does not improve resistance at pH < 4, and thus, effects of Cu are detected in the British Petroleum (BP) test, but not in NACE TM0177 solution, as shown in Figure 15.11.

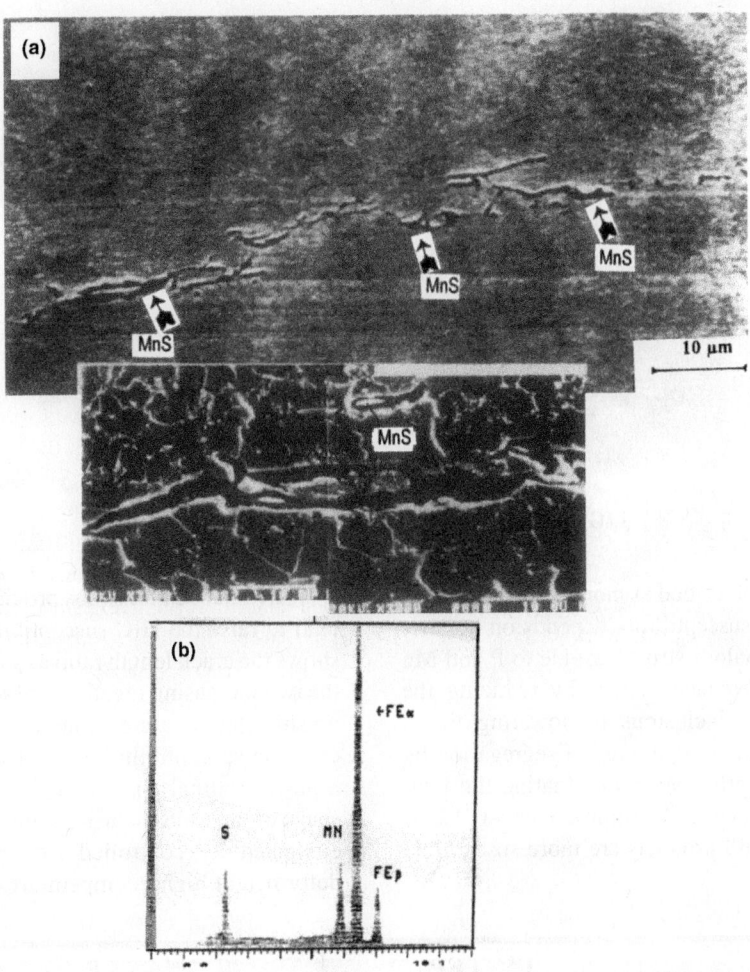

FIGURE 15.5. Scanning electron micrograph of coupons tested in sour environment showing, (a) Massive and elongated nonmetallic inclusion (MnS) and (b) energy-dispersive X-ray (EDX) analysis of the above cracking surface.

Hydrogen penetration decreases as the pH increases and becomes minimal at pH 7.5; at higher pH, hydrogen penetration increases [36]. Hydrogen-induced cracking in alkaline media has been reported [37]; in such cases, NH_3 and CN may accompany the H_2S.

Hydrogen-induced cracking is accelerated when Cl^- accompanies H_2S; mill scale inhibits HIC in media containing H_2S but lacking Cl^-, whereas there is no such effect when the two occur together [38]. Carbon dioxide often

accompanies H_2S in a sour well; the combination of H_2S and CO_2 tends to increase the occurrence of HIC and to increase the concentration of hydrogen at the surface, C_0.

Water is almost always present and is essential to HIC since hydrogen from the cathodic corrosion reaction penetrates the steel. When there is no water, corrosion does not occur, so there is no hydrogen evolved to penetrate, and no HIC should occur. At elevated H_2S contents, the medium is generally more corrosive [39].

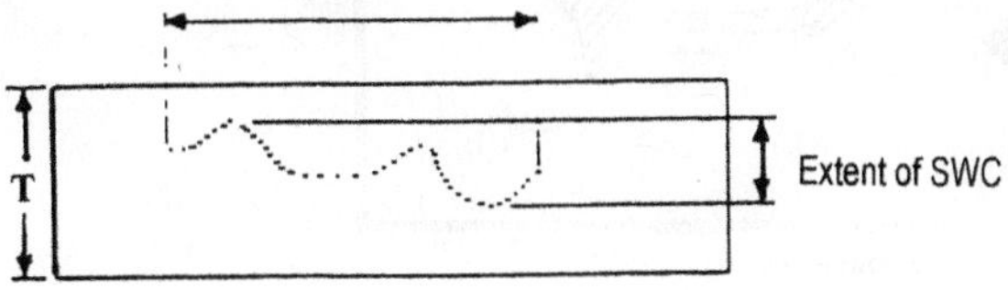

B2.2. Material Factors. The HIC-resistant pipeline steels have been developed [40–44]. Figure 15.12 [45] compares susceptibility of various commercial steels based on ingot cores (where susceptibility is highest). The larger the ingot the higher the HIC susceptibility. The susceptibility of the ingot core is ascribed to the high-volume proportion of segregating materials there, whereas the low susceptibility at the rim is caused by the low-volume fraction of segregating materials, as the cooling rate at the surface is high. The upper

Length of SWC

FIGURE 15.6. Schematic illustrating how length and extent of SWC are defined. "T" indicates the through-thickness direction in plate or pipe sample.

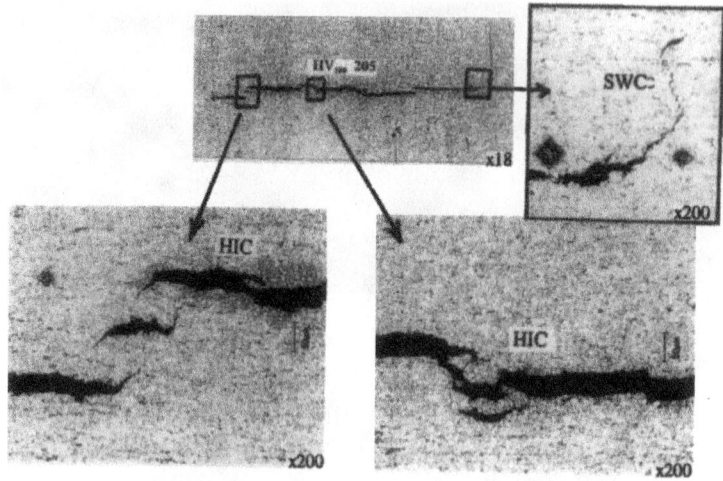

FIGURE 15.7. Typical example of stepwise cracking.

part of the ingot has been described as more susceptible than the bottom part. Because susceptibility depends on nonmetallic inclusions and anomalous structures due to P and Mn segregation, HIC can be reduced either by reducing the proportion of nonmetallic inclusions by lowering the S content or by modifying the morphology of segregation by adding Ca. Tempering is effective in eliminating the low-temperature anomalous structure. Addition of Cu ($> 0.2\%$) is also effective. Hot-strip mill products are more susceptible than plate mill products.

Plate-shaped inclusions provide hydrogen traps and, thus, tend to raise the HIC susceptibility [46]. Figure 15.13 [29] shows the crack length ratio as a function of inclusion length, the two increasing together and not greatly influenced by the finishing temperature. It has also been found [47, 48] that the cracking susceptibility in commercial steel is usually larger at lower rolling temperatures, which may be related to the shape of the MnS inclusions; that is, those inclusions may be elongated by controlled rolling at $\sim 750°C$ but are not deformed at higher temperatures.

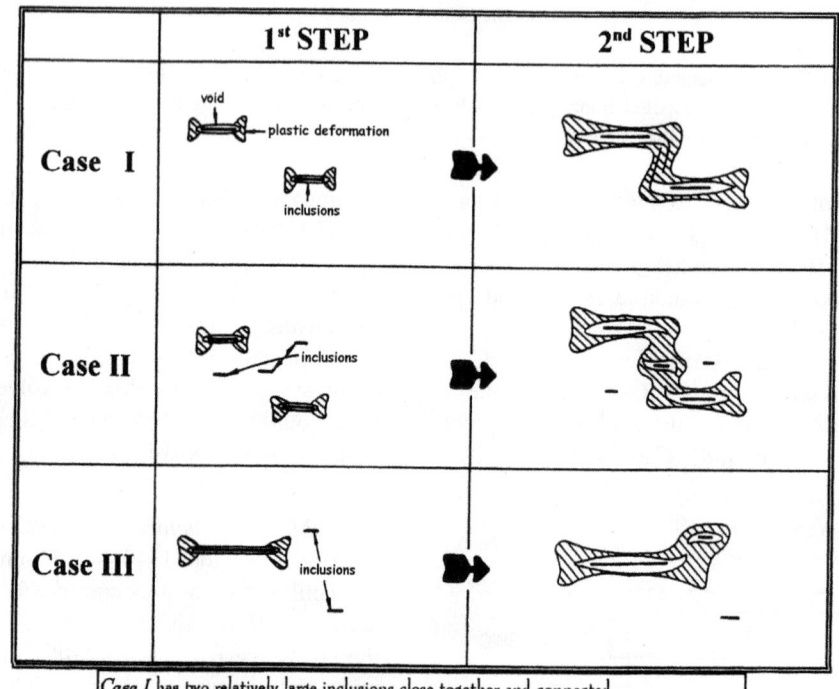

FIGURE 15.8. Stepwise crack growth [30].

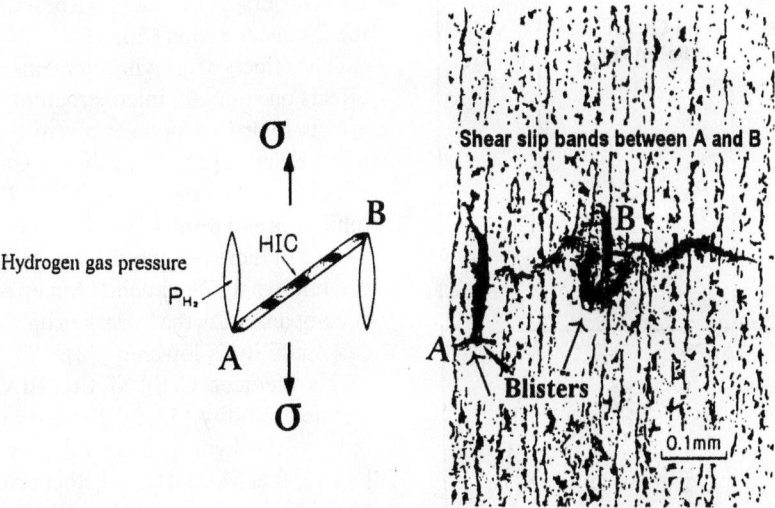

FIGURE 15.9. Link-up of hydrogen blister arrays formed under stress. [R. T. Hill and M. Iino, in *Current Solutions to Hydrogen Problems in Steels* (1982), ASM International, Materials Park, OH 44073-0002 (formerly the American Society for Metals, Metals Park, OH 44073), p. 198 (Fig. 6).]

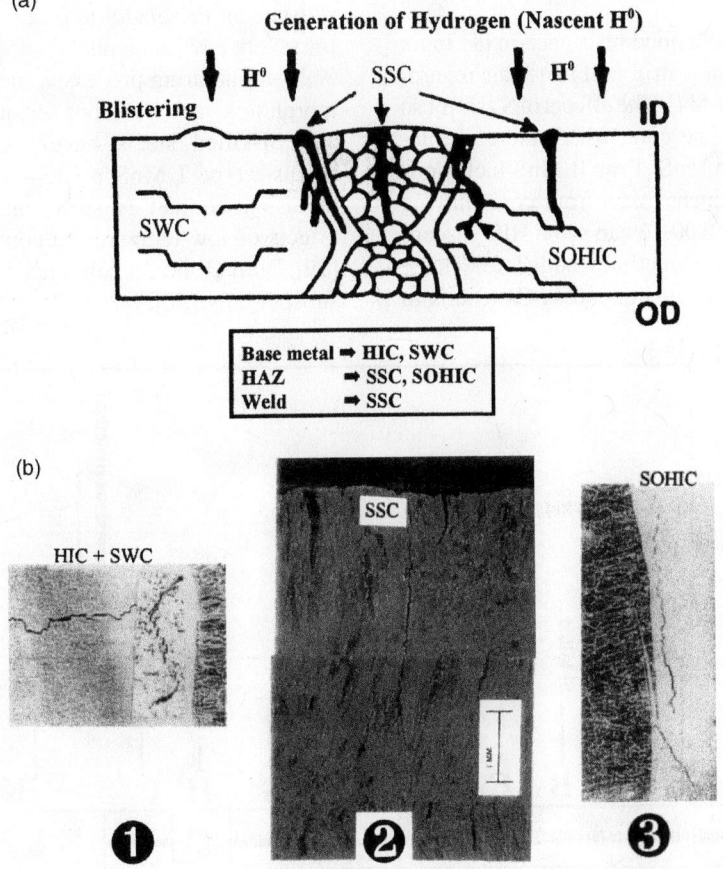

FIGURE 15.10. (a) Forms of hydrogen damage in H$_2$S service and (b) metallographic sections of weld coupons of pipe showing: (1) SWC in the base metal, (2) apparent SSC in weld zone, and (3) SOHIC and SSC in heat-affected zone (HAZ).

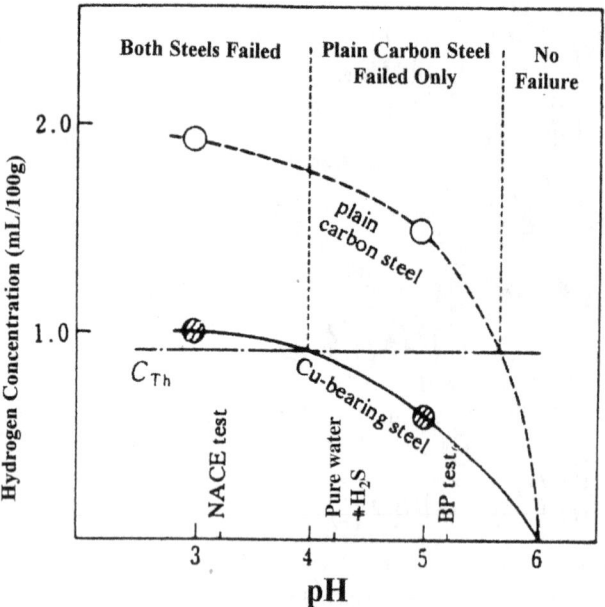

FIGURE 15.11. Relationship between pH and hydrogen uptake for plain and Cu-bearing steels [30].

CaS clusters, so that there is a best Ca/S range that depends on the S concentration [50, 51].

The effects of alloying elements on HIC occur through the effects on strength, microstructure, and segregation. Copper affects hydrogen uptake; above 0.2% Cu, susceptibility is much reduced [29, 52]. Copper reduces C_0, but does not alter C_{th} [41], so that the NACE TM0177 solution shows no HIC inhibition as a result of Cu addition to the steel [53–55], Over 1.0% Mn increases the susceptibility, but Q and T treatment can remove the detrimental Mn effect [40, 42, 56]. When Co accompanies Cu, the hydrogen uptake is reduced and the HIC susceptibility is lowered [53].

The elements C, Si, Ni, Cr, and V have only minor effects on susceptibility [53, 56], although it is stated [53] that Ni reduces the hydrogen uptake, and it has also been found [53, 57] that Mo and Cu together result in the beneficial effect of Cu being reduced. It has also been reported [55, 58–59] that Cu + W has only slight effects on the sensitivity. It is claimed [59] that Bi, Pd, or Pt improves the resistance by forming surface films, but the use of these expensive metals may not be a practical solution in most applications.

Calcium and rare earth metals, such as La and Ce, spheroidize nonmetallic inclusions, raising C_{th}, (i.e., the resistance); these additions are often used in making steels for severe environments [41–45].

Manufacturing processes and treatments affect the MnS morphology and influence sensitivity; for example, rimmed and Si-killed steels have relatively low susceptibility because type I MnS predominates [60, 61]. If the MnS inclusion content is sufficiently low, even the adverse effects of low-temperature controlled rolling are reduced [29]. Both Q and T treatments can reduce the susceptibility substantially [50].

When S < 0.007, it produces good resistance in the ingot rim of controlled-rolled and hot-strip mill materials regardless of the finishing procedure [44]. The effects of S are not so evident for specimens from the core, where there are numerous inclusions other than MnS. Type II MnS inclusions occur in the core, and specimens taken from cores having average S contents as low as 0.001% can show HIC susceptibility [47, 49]. Calcium is sometimes added for shape control of type II MnS inclusions, but excess Ca results in

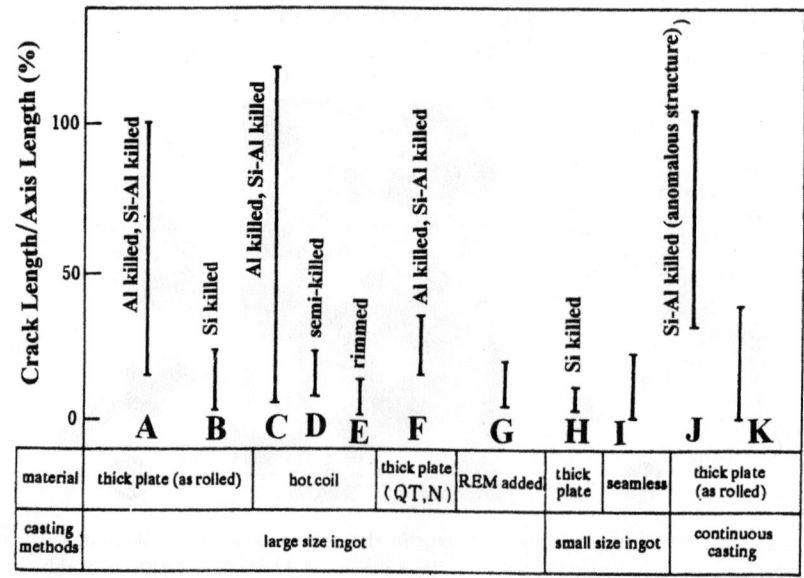

FIGURE 15.12. Hydrogen-induced cracking susceptibility for commercial steels [28].

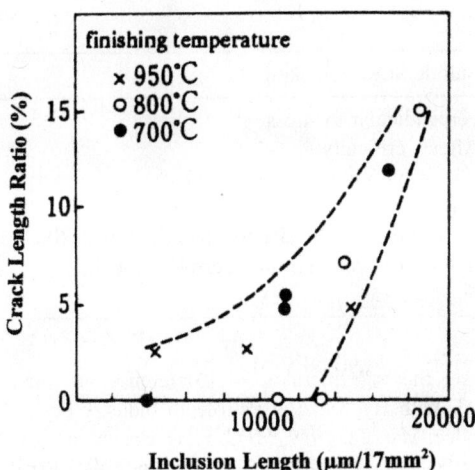

FIGURE 15.13. Relationship between total inclusion length and crack length ratio [29]. (Copyright 1976 by The American Society of Mechanical Engineers. Reprinted with permission.)

B3. Prevention and Control

There are two ways of inhibiting HIC: One is to prevent hydrogen from penetrating the steel and another is to reduce the concentration of hydrogen traps. The HIC can be inhibited by using over 0.2% Cu. In addition, Co, Cu + Co, Cu + W, and Ni are also stated to be effective. Alternatively, an inhibitor can be added to the solution to produce a protective film [60–63], or the pH can be raised. Using a paint or lining to isolate the steel may also inhibit hydrogen penetration and prevent HIC [64–66].

If S is reduced to 0.002–0.005%, the number of non-metallic inclusions, such as MnS, may be reduced and HIC inhibited. It is claimed that Ca and rare earth metals inhibit HIC susceptibility by modifying the morphology of inclusions [67]. Tempering is also effective in eliminating localized Mn and P segregation if the Mn level is more

than 1%; tempering reduces hardness around inclusions, and thus, the HIC susceptibility. Calcium added to steels low in S and Mn for use in sour-gas pipelines improves HIC resistance.

C. SULFIDE STRESS CRACKING

Sulfide stress cracking was first recognized as a serious problem in the early 1950s, when the oil industry experienced failures of tubular steels and well-head equipment [11] constructed from steels with hardness values greater than HRC 22. The NACE International recommendations to prevent SSC were based on an approach that has been reasonably successful in practice; that is, heat treat steels to hardness levels less than HRC 22. Although the base metal of steel pipe normally has hardness levels well below this value, service failures have occurred in regions of high hardness in the weld heat-affected zone. It is, therefore, common to apply the HRC 22 limit (equivalent to Vickers hardness "HV 248") to the weld/HAZ areas in pipeline steels. Test procedures for SSC are reviewed in Chapter 78.

Sulfide stress cracking is a form of hydrogen embrittlement that occurs in high-strength steels [3] and in localized hard zones in weldment [39] of susceptible materials. The graph in Figure 15.14 shows the synergistic effects of high steel strength and high H_2S concentration in the environment to cause SSC [68]. In the heat-affected zones adjacent to welds, there are often very narrow hard zones combined with regions of high residual tensile stress that may become embrittled to such an extent by dissolved atomic hydrogen that they crack, as shown in Figure 15.10(b). Sulfide stress cracking is directly related to the amount of atomic hydrogen dissolved in the metal lattice and usually occurs at temperatures below 90°C (194°F) [32]. Sulfide stress cracking also depends on the composition, microstructure, strength, and

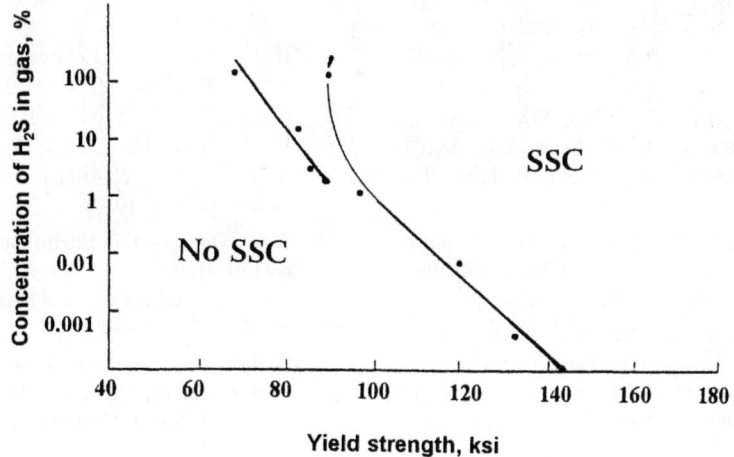

FIGURE 15.14. The maximum H_2S concentration limit for SSC-free behavior at 100% of yield strength [68]. (Copyright Materials Properties Council. Reprinted with permission.)

TABLE 15.1. Comparison of Features of HIC and SSC

	Hydrogen-Induced Cracking	Sulfide Stress Cracking
Crack direction	Dependent on microstructure	Perpendicular to stress
Applied stress	No effect	Affects critically
Material strength	Primarily in low-strength steel	Primarily in high-strength steel
Location	Ingot core	Anywhere
Microstructure	Cleanliness and nonmetallic inclusions are critical	Critical effect, Q and T treatment enhances SSC resistance
Environment	Highly corrosive conditions, appreciable hydrogen uptake	Can occur even in mildly corrosive media

the total stress (residual stress plus applied stress) levels of the steel [31].

C1. Comparison of HIC and SSC

Features of HIC and SSC are compared in Table 15.1; the orientation of HIC depends on microstructure and morphology of nonmetallic inclusions, whereas SSC is perpendicular to the stress. Sulfide stress cracking occurs only under certain stress conditions, whereas HIC can occur without external stress. Sulfide stress cracking occurs in high-strength steel, whereas HIC occurs in low-strength steel. The occurrence of HIC depends on nonmetallic inclusions, so that the steel-making process and the location within the steel ingot are very important. This resistance can be increased by quench and temper heat treatments, although SSC of high-strength steel can occur in mild environments where the steel absorbs only small amounts of hydrogen. On the other hand, HIC occurs in low-strength steel in severe environments, in which appreciable amounts of hydrogen are formed by cathodic reduction on the steel surface and absorbed by the steel.

REFERENCES

1. R. D. Merrick, Mater. Perform., **27**(6), 30 (1988).
2. API Survey "Extent of Equipment Cracking in Wet Hydrogen Sulfide Environments," by T. McLaury at the American Petroleum Institute mid-year refining meeting, Chicago, IL, Apr. 1989.
3. "Standard Materials Requirements-Sulnde Stress Cracking Resistant Metallic Materials for Oilfield Equipment," NACE Standard MR0175-99, NACE International, Houston, TX, 1999.
4. R. D. Merrick and M. L. Bullen, "Prevention of Cracking in Wet H_2S Environments," Paper No. 269, CORROSION/89, NACE international, Houston, TX, 1989.
5. W. Bruckhoff, O. Geier, K. Hofbauer, O. Schmitt, and D. Steinmetz, "Rupture of a Sour Line Pipe due to SOHIC: Failure Analysis, Experimental Results and Corrosion Prevention," Paper No. 389, CORROSION/85, NACE International, Houston, TX, 1985.
6. M. J. Humphries, J. E. McLaughlin, and R. Pargeter, "Toughness Characteristics of Hydrogen Charged Pressure Vessel Steels," International Conference on Interaction of Steels with Hydrogen in Petroleum Industry Pressure Vessel Service, Mar. 28, 1989, MFC, Paris, France, p. 243.
7. R. S. Treseder, "Historical Review of SSC of Alloy Steels," International Conference on "Sour Service in Oil, Gas and Petrochemical Industries," London, UK, May 11–12, 1987.
8. G. J. Biefer, "Stepwise Cracking of Linepipe Steels in Sour Environment," Mater. Perform., **21**(6), 19 (1982).
9. M.G. Hay and D.W. Rider, "Integrity Management of a HIC Damaged Pipeline and Refinery Pressure Vessel through Hydrogen Permeation Measurements," Paper No. 395, CORROSION/98, NACE International, Houston, TX, 1998.
10. R. W. Revie and M, Fichera, CANMET MTL Report: ERP/ PMRL 81-39 (TR), Ottawa, Canada, Oct. 1981.
11. F. Parades and W. W. Mize, Oil Gas J., **52**, 99 (1954).
12. H. P. Cotton, Sixth Symposium on Line Pipe Research, Houston, TX, Oct. 29 – Nov. 1, 1979, American Gas Association, Arlington, VA, 1979.
13. R. R. Irving, Iron Age, **24**(6), 8 (1974).
14. E. M. Moore and J. J. Warga, Mater. Perform., **15**, 17 (1976).
15. H. I. McHenry, T. R. Shives, D. T. Read, J. D. McColskey, C. H, Brady, and P. T. Portscher, "Examination of a Pressure Vessel That Ruptured at the Chicago Refinery of the Union Oil Company on July 23, 1984," Report No. NBSIR 86-3049, National Bureau of Standards, Boulder, CO, Mar. 1986.
16. H. I. McHenry, D. T. Read, and T. R. Shives, Mater. Perform., **26**(8), 18 (1987).
17. H. K. Birnbaum, in Environment-Sensitive Fracture of Engineering Materials, Z. A. Foroulis (Ed.), American Institute of Mining, Metallurgical, and Petroleum Engineers, Warrendale, PA, 1979, pp. 326–357.
18. A. S. Tetelman, The Fracture of Engineering Metals, Wiley, New York, July 1967.
19. A. W. Thompson, in Environment-Sensitive Fracture of Engineering Materials, Z. A. Foroulis (Ed.), American Institute of Mining, Metallurgical, and Petroleum Engineers, Warrendale, PA, 1979, pp. 379–410.
20. W. Beck, E. J. Jankowsky, and P. Fischer, "Hydrogen Stress Cracking of High Strength Steel," Technical Report NADC-MA-7140, Naval Air Development Centre, Warminster, PA, Dec. 1971.
21. J. J. DeLuccia and D. A. Berman, "An Electrochemical Technique to Measure Diffusible Hydrogen in Metals (Barnacle

Electrode)," in Electrochemical Corrosion Testing, ASTM STP 727, F. Mansfeld and U. Bertocci (Eds.), American Society for Testing and Materials, West Conshohoken, PA, 1969, pp. 259–273.

22. C. F. Barth, E. A. Steigerwald, and A. R. Troiano, Corrosion, **25**, 9, 353 (1969).

23. T. P. Groeneveld, E. E. Fletcher, and A. R. Elsea, "A Study of Hydrogen Embrittlement of Various Alloys," Battelle Columbus Laboratories, Columbus, OH, June 23, 1966.

24. M. Elboujdaini, M. T. Shehata, W. Revie, and R. R. Ramsingh, "Hydrogen Induced Cracking and Effect of Nonmetallic Inclusions in Linepipe Steels," Paper No. 748, CORROSION/98, NACE International, Houston, TX, 1998.

25. M. H. Bartz and C. E. Rawlings, Corrosion. **4**, 187 (1948).

26. E. M. Moore and D. R. McIntyre, Mater. Perform., **37**(10), 77 (1998).

27. T. Taira, Y. Kobayachi, N. Seki, K. Tsukada, M. Tanimura, and H. Inagaki, Technol. Rep. NKK (Nippon Kokan), No. 67, 421 (1980).

28. F. Terazaki, A. Ikeda, S. Okamoto, and M. Takeyam, Sumitomo Met., **30**, 40 (1978).

29. E. Miyoshi, T. Tanaka, F. Terazaki, and A. Ikeda, "Hydrogen-Induced Cracking of Steels Under Wet Hydrogen Sulfide Environment", J. Eng. Ind., **98**, 1221 (1976).

30. A. Ikeda, Y. Morita, T. Tanaka, and M. Takeyama, Proceedings of Second International Conference on Hydrogen in Metals, 4A7, Paris, France, 1977.

31. G. M. Buchheim, "Ways to Deal with Wet H$_2$S Cracking Revealed by Study," Oil Gas J., **9**, 92 (1990).

32. J. Gutzeit, "Cracking of Carbon Steel Components in Amine Service," Mater. Perform., **29**(9), 54 (1990).

33. R. D. Merrick, "Refinery Experiences with Cracking in Wet H$_2$S Environment," Paper No. 190, CORROSION/87, NACE International, Houston, TX, 1987.

34. A. Ikeda, T. Kaneko. T. Hashimoto, M. Takeyama, Y. Sumitomo, and T. Yamura,"Development of HIC Resistant Steels and HIC Test Methods for H$_2$S Service," Symposium on the Effects of Hydrogen Sulfide on Steel, 22nd Annual Conference of Metallurgists, CIM, Edmonton, Alberta, Canada, Aug. 1983.

35. R. Kane, M. Cayard, and M. Prager, "Evaluation of advanced steels for resistance to HIC and SOHIC in wet H$_2$S Environments," in 2nd Int. Conf. on Interaction of Steels with Hydrogen in Petroleum Industry Pressure Vessel and Pipeline Service, Vienna, Austria, M. Prager (Ed.), Vol. **1**, Pub. Materials Properties Council, New York, 1994.

36. W. A. Bonner, H. D. Burnham, I. J. Conradi, and T. Skei, Proc. API, **33**, 255 (1953).

37. H. E. Knowlton, J. W. Coombs, and E. R. Allen, Oil Gas J., **78**, 50 (1980).

38. A. Ikeda, F. Terazaki, and M. Kowaka, J. Iron Steel Jpn., **63**, 433 (1978).

39. R. D. Kane and J. B. Greer, J. Pet. Techn., 29, 1483 (1977).

40. O. Nakasugi, H. Matsuda, S. Sugimura, and T. Murata, Steelmaking Res., **297**(12), 878 (1979).

41. F. Terazaki, A. Ikeda, M. Nakanishi, T. Kaneko, Y. Sumitomo, and M. Takeyama, Sumitomo Met., **32**, 147 (1980).

42. M. Watkins and R. Ayer, "Microstrucmre—The Critical Variable Controlling the SSC Resistance of Low-Alloy Steels," Paper No. 50, CORROSION/95, NACE International, Houston, TX, 1995.

43. M. Ueda and T. Kudo, Effect of Alloying Elements on Corrosion Resistance of Ni-Base Alloy in Sulfur Containing Sour Environments, Paper No. 69, CORROSION/94, NACE International, Houston, TX, 1994.

44. F. Terazaki, A. Ikeda, S. Okamoto, and M. Takeyama, The Sumitomo Search, **19**, 103 (1978).

45. R. D. Kane and M. S. Cayard, "Roles of H$_2$ and H$_2$S in Behaviour of Engineering Alloys in Petroleum Applications," in Proceedings, Materials for Resource Recovery and Transport, L. Collins (Ed.), The Metallurgical Society of CIM, Montreal, Canada, Aug. 1998, pp. 3–49.

46. M. Elboujdaini, M. T. Shenata, and W. Revie, "Performance of Pipeline Steels in Sour Service," in Proceedings, Materials for Resource Recovery and Transport, L. Collins (Ed.), The Metallurgical Society of CIM, Montreal, Canada, Aug. 1998, pp. 109–127.

47. Y. Nakai, H. Kurahashi, T. Emi, and O. Haida, Trans. ISIJ, **19** (7),401 (1979).

48. G. Biefer and M. J. Fichera, "Factors Affecting the SWC of Line-Pipe Steels in Sour Environments," CANMET Report 84-13 (TR), Ottawa, Canada, 1984.

49. R. D. Kane and S. Srinivasan, "Serviceability of Petroleum Process and Power Equipment," D. Bagnoli, M. Prager, and D. M. Schlader (Eds.), ASME, PVP Vol. 239/MPC Vol. 33, 170, ASME, New York, 1992.

50. T. Taira, Y. Kobayashi, K. Matsumoto, and T. Tsukada, Corrosion **40**(9), 478 (1984).

51. O. Haida, T. Emi, K. Sanbongi, T. Shiraishi, and A. Fujiwara, "Optimizing Sulfide Shape Control in Large HSLA Steel Ingots by Treating the Melt with Calcium or Rare Earths," Tetsu-to-Hagane (J. Iron Steel Inst. Jpn.) **64**, 1538 (1978).

52. M. F. Galis and D. Petelot,"Scale Formation and Growth on Steels in Contact with Various H$_2$S Environments," in Ref. 6, p. 109.

53. A. Ikeda, T. Kaneko, and F. Terasaki, "Influence of Environmental Conditions and Metallurgical Factors in Hydrogen-Induced Cracking of Line Pipe Steel," Paper No. 8, CORROSION/80, NACE International, Houston, TX, 1980.

54. A. Ikeda, F. Terasaki, M. Takeyama, I. Takeuchi, and Y. Nara, "Hydrogen Induced Cracking Susceptibility of Various Steel Line Pipes in the Wet H$_2$S Environment," Paper No. 43 CORROSION/78, NACE International, Houston, TX, 1978.

55. H. Inagaki, M. Tanimura, I. Matsushima, and T. Nishimura, Trans. ISIJ, **18**(3), 149 (1978).

56. J. Charles, L. Coudreuse, R. Blondeau, and L. Cadiou, "Clean Steel to Resist Hydrogen Embrittlement," Paper No. 202, CORROSION/90, NACE International Houston, TX, 1990.

57. T. Nishimura, H. Inagaki, and M. Tanimura, "Hydrogen Cracking in Sour Gas Pipeline Steels," Paper No. 3E9,

Proceedings of 2nd International Conference on Hydrogen in Metals, Paris, France, 1977, Pergamon, Oxford, UK, 1978.

58. C. J. Bennett and A. Brown, "A Round Robin Laboratory Assessment of Hydrogen Induced Cracking and Sulfide Stress Corrosion Cracking of Line Pipe Steels—Results Variation and Proposals for a Modified Test Standard," Paper No. 4, CORRDSION/91, NACE International, Houston, TX 1991.

59. M. Iino, T. Nomura, H. Takezawa, and H. Gondo, J. Iron Steel Inst. Jpn., **65**, A65 (1979).

60. E. M. Moore and J. J. Warga, Mater. Perform., **15**, 17 (1976).

61. R. L. Schuyler, Mater. Perform., **18**, 9 (1979).

62. J. R. Perumareddi, M. Elboujdaini, and V. S. Sastri, "Inhibition of Hydrogen Entry into Steel," in Proceedings Materials for Resource Recovery and Transport, L. Collins (Ed.), The Metallurgical Society of CTM, Calgary, Aug. 1998, pp. 117–187.

63. V. S. Sastri, Corrosion Inhibitors—Principles and Applications, Wiley, Chichester, UK, 1998.

64. A. Miyasaka, Y. Yamaguchi, T. Miyagawa, and A. Nakamura, "Selection of Metallic and Nanmetallic Materials Suitable for I.D. Surface Coating of Production Tubing for Sour Service," Paper No. 83, CORROSION/94, NACE International, Houston, TX, 1994.

65. K. Masamura, Y. Takeuchi, K. Tamaki, T. Miyagawa, and A. Nakamura, "Corrosion resistance of Newly Developed Inside Coated Material in a Sour Environment," Paper No. 84, CORROSION/94, NACE International, Houston, TX, 1994.

66. K. Tamaki, A. Nakamura, T. Miyagawa, T. Tamaki, and M. Ogasawara, "Integrity of the Internally Coated Premium Connections," Paper No. 85, CORROSION/94, NACE International, Houston, TX, 1994.

67. T. E. Pérez, H. Quintanilla, and E. Rey, "Effect of Ca/S Ratio on HIC Resistance of Seamless Line Pipes," Paper No. 121, CORROSION/98, NACE International, Houston, TX, 1998.

68. R. D. Kane, S. M. Wilhelm, and J. W. Oldfield, "Review of Hydrogen Induced Cracking of Steels in Wet H_2S Refinery Service," in Proceedings of International Conference, Interaction of Steels with Hydrogen in Petroleum Industry Pressure Vessel Service, Paris, France, 1989, M. Prager (Ed.), Materials Properties Council, New York, 1993, pp. 7–15.

16

CORROSION FATIGUE

Y.-Z. Wang*

CANMET Materials Technology Laboratory, Ottawa, Ontario, Canada

A. INTRODUCTION

Corrosion fatigue is caused by crack development under the simultaneous action of corrosion and fluctuating, or cyclic, stress. Many instances of environment-assisted cracking are caused by corrosion fatigue because loads on most engineering structures do vary to some extent Corrosion fatigue is a subject of international conferences [1–3], major review papers [4–6], and books [7].

Metal subjected to a fluctuating stress will fail at a stress much lower than is required to cause failure under constant load. The extent of stress fluctuation is defined by the stress ratio, R = minimum stress/maximum stress. The number of cycles to failure, the fatigue life, increases as the maximum stress during cycling decreases until the endurance limit, or fatigue limit, is reached; at or below this stress, the material undergoes an infinite number of cycles without failure. True fatigue limits exist for only a limited number of materials; for the majority of engineering alloys, the fatigue limit refers to the stress level below which failure does not occur within a specified number of cycles, usually 10^7 or 10^8 cycles.

Fatigue crack growth rate, the increment of crack size per load cycle, is important for risk assessment and for predicting remaining life, and is often described by a relationship with stress intensity factor, K, which includes stress and crack sizes.

Both the fatigue life and the fatigue limit can be markedly reduced in the presence of a corrosive environment, and, in many cases, the endurance limit is no longer observed. In addition, corrosive environments can accelerate crack growth. The damage due to corrosion fatigue is almost always much greater than the sum of the damage by corrosion and fatigue acting separately. Figure 16.1 shows an example of the reduction of fatigue life and the elimination of the fatigue limit of high-strength steel in a sodium chloride solution [8]. This figure also shows that cathodic polarization restores the fatigue properties of the steel.

In general, a corrosive environment can decrease the fatigue properties of any engineering alloy, meaning that corrosion fatigue is not material–environment specific. Although fatigue cracks are typically transgranular, corrosion fatigue cracks can be transgranular, intergranular, or a combination of both, depending on the mechanical loading and environmental conditions. Localized corrosion, such as pitting, often produces favorable sites for corrosion fatigue crack initiation, but pits are not the only initiation sites, and pitting is not a necessary precursor to failure. Although multiple cracks can initiate, fatigue failure often results from the propagation of a single crack; whereas crack interaction and coalescence are important in the corrosion fatigue failure process.

* Current address: Canadian Nuclear Safety Commission, Ottawa, Ontario, Canada.

Uhlig's Corrosion Handbook, Third Edition, Edited by R. Winston Revie
Copyright © 2011 John Wiley & Sons, Inc.

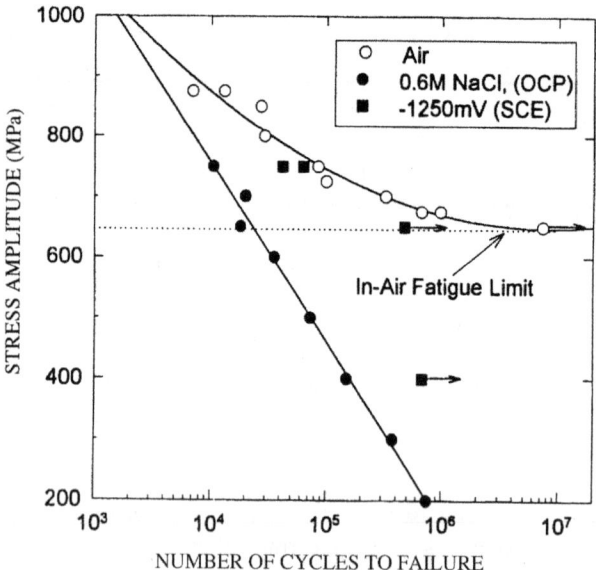

FIGURE 16.1. Fatigue life data, *S–N* curves, for a high-strength steel under different environmental conditions. Stress ratio $R = -1$. Loading frequency 1 Hz for tests in 0.6 *M* NaCl solution. Horizontal arrows indicate failure condition not attained. OCP = open-circuit potential. [Reproduced with permission from Y.-Z. Wang, R. Akid, and K. J. Miller, "The Effect of Cathodic Polarization on Corrosion Fatigue of a High Strength Steel in Salt Water," Fatigue Fract. Engng. Mater. Struct. 18(3), 295 (1995). Blackwell Science Ltd., Oxford, UK.]

B. MECHANISTIC ASPECTS OF CORROSION FATIGUE

Like stress corrosion cracking, corrosion fatigue depends on interactions among the material, environmental, chemical, and electrochemical parameters and mechanical loading conditions.

Cracking phenomena for ductile alloys involve plastic deformation, and it is the localization of plastic deformation, due to cyclic loading, that causes fatigue failure at a stress level far below the yield stress of the material. There are two main processes associated with corrosion damage, anodic metal dissolution and cathodic reactions (often hydrogen reduction). The reduction of material resistance to fatigue under the influence of a corrosive medium can be regarded as a result of the synergistic enhancements of these processes.

There are two main categories for the mechanisms of corrosion fatigue: anodic slip dissolution and hydrogen embrittlement, as schematically summarized in Figure 16.2 [9]. As shown in Figure 16.2(a), cracks grow by slip dissolution that results from diffusion of the active species (e.g., water molecules or halide anions) to the crack tip; rupture of the protective oxide film at a slip step or in the immediate wake of a crack tip by strain concentration or fretting contact between the crack faces; dissolution of the exposed surface; nucleation and growth of oxide on the bare surface.

For the alternative mechanism of hydrogen embrittlement in aqueous media, the critical steps [Fig. 16.2(b)] involve: diffusion of water molecules or hydrogen ions to the crack tip; reduction to hydrogen atoms adsorbed at the crack tip; surface diffusion of adsorbed atoms to preferential surface locations; absorption and diffusion to critical locations (e.g., grain boundaries, the region of high triaxiality ahead of a crack tip, or a void).

Under cyclic loading, fretting contact between the mating crack faces, pumping of the aqueous environments to the crack tip by the crack walls, and continual blunting and resharpening of the crack tip by the reversing load influence the rate of dissolution. Consequently, both cyclic frequency and stress waveform strongly influence crack development by corrosion fatigue, whereas for fatigue alone these factors are usually less significant.

Fatigue damage can be divided into the following four stages:

1. *Precrack Cyclic Deformation.* Repetitive mechanical damage is accumulated in some local regions; dislocation and other substructures may develop; and persistent slip bands (PSBs, slip bands that develop on the sample surface during cyclic deformation and that reappear at the same locations during further cyclic deformation after polishing the surface), extrusions, and intrusions form.

2. *Crack Initiation and Stage I Growth.* Cracks initiate as a result of deepening of the intrusions; crack growth in this stage is within the planes of high shear stress.

3. *Stage II Crack Propagation.* Well-defined cracks propagate on the planes of high tensile stress in the direction normal to the maximum tensile stress.

4. *Ductile Fracture.* When the crack reaches sufficient length so that the remaining cross section cannot support the applied load, ductile fracture occurs.

The relative proportion of the total cycles to failure that are involved in each stage depends on mechanical loading conditions and on the material. There is considerable ambiguity in deciding when a cracklike surface feature should be called a crack. In general, a larger proportion of the total cycles to failure are involved in propagation of stage II cracks in low-cycle fatigue than in high-cycle fatigue, whereas initiation and stage I crack growth comprise the largest segment for low-stress, high-cycle fatigue. The surface conditions of the material also influence the proportion of each stage in the total fatigue lifetime. Surface discontinuities, such as sharp notches and nonmetallic inclusions, can significantly reduce the number of cycles required for crack initiation and early stages of propagation.

A corrosive environment can influence all the stages of crack development except the last one, in which ductile

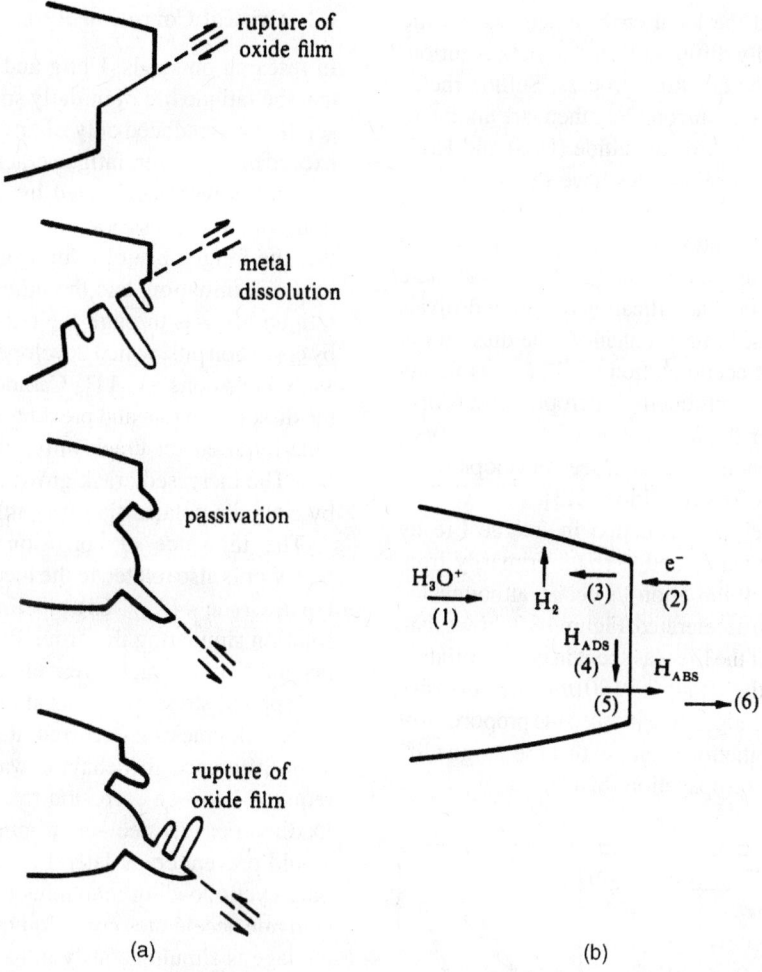

FIGURE 16.2. Schematic of (a) slip dissolution and (b) hydrogen embrittlement in aqueous media; (1) liquid diffusion, (2) discharge and reduction, (3) hydrogen adatom recombination, (4) adatom surface diffusion, (5) hydrogen absorption in metal, and (6) diffusion of absorbed hydrogen. (Reproduced with permission from Subra Suresh, Fatigue of Materials, Cambridge University Press, Cambridge, UK, 1991, p. 363.)

fracture occurs, and can also influence the relative proportion of the total cycles to failure that take place in each stage.

C. CORROSION FATIGUE CRACK INITIATION

C1. Role of Nonmetallic Inclusions

For low-stress, high-cycle fatigue, crack initiation consumes a large portion of the total lifetime. Fatigue crack initiation in commercial alloys occurs on the surface or the subsurface and is usually associated with surface defects, especially nonmetallic inclusions. For an inclusion to be a potential source of fatigue failure, two main criteria must be fulfilled: The inclusion should have a critical size and the inclusion should have a low deformability, related to the expansion coefficient at the temperature during fatigue [10]. For steels, the "dangerous" inclusions include single-phase alumina

(Al_2O_3), spinels, and calcium-aluminates $> 10 \mu m$ in size. The most common elongated sulfide inclusions (MnS) appear to be the least harmful. Surface discontinuities can act as stress raisers causing local stress concentration, but it is the enhanced localization of plastic deformation around an inclusion that reduces the fatigue resistance of a material.

Preferential attack by an environment at specific surface locations may provide the most favorable sites for crack initiation when a cyclically loaded engineering component or structure is exposed to a corrosive medium during service. For a high-strength steel exposed to a sodium chloride solution, it was found that sulfide inclusions contribute sites for corrosion pits and subsequent fatigue crack initiation, whereas the angular-shaped calcium-aluminates, which are responsible for fatigue crack initiation in air, did not affect corrosion fatigue [8, 11]. The relative chemical and electrochemical activity of an inclusion determines whether it is preferentially dissolved. Both the stress concentration associated with a pit and the local environment within the pit, which can be markedly different from the bulk solution, can significantly affect the cracking process. Sulfide inclusions and the immediate area surrounding them are anodic to the steel matrix [12, 13]. Hydrogen sulfide (H_2S) and HS^- ions formed by dissolution of sulfides have the most deleterious effects on development of corrosion pits. The H_2S and HS^- ions produced in solution can catalyze the anodic dissolution of iron from the matrix and poison the cathodic discharge of hydrogen. Local acidification due to hydrolysis of ferrous and ferric ions, in turn, enhances the dissolution of sulfide inclusions, and accumulation of HS^- ions favors continued localized attack, producing micropits. It has also been reported that, under the application of a cyclic stress, corrosion at the inclusion–matrix interfaces develops more rapidly than under stress-free conditions [13].

For high-strength steel, the reduction in fatigue life by sodium chloride solution was found to result primarily from the shorter time for crack initiation to occur, although the crack growth rate was also accelerated. Figure 16.3 shows that, for fatigue in air, >80% of the life was spent in crack initiation and propagation below the length of 100 μm; the corrosive environment reduced the fatigue life and also the proportion of time spent in the crack initiation stage, so that the fatigue life was predominantly crack propagation [8, 11].

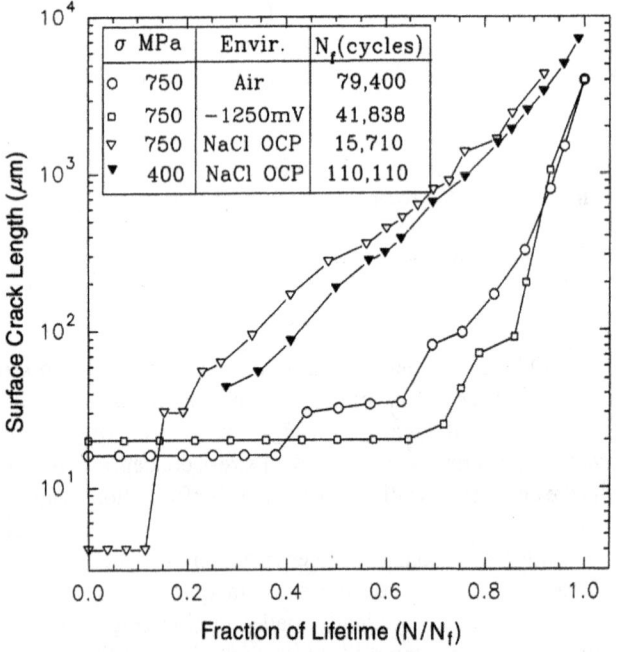

σ MPa	Envir.	N_f(cycles)
○ 750	Air	79,400
□ 750	−1250mV	41,838
▽ 750	NaCl OCP	15,710
▼ 400	NaCl OCP	110,110

FIGURE 16.3. Surface crack length versus fraction of lifetime for high-strength steel at different conditions. The fatigue life data are shown in Figure 16.1. Number of cycles to failure N_f. [Reproduced with permission from Y.-Z. Wang, R. Akid, and K. J. Miller, "The Effect of Cathodic Polarization on Corrosion Fatigue of a High Strength Steel in Salt Water," Fatigue Fract. Engng. Mater. Struct. 18(3), 295 (1995). Blackwell Science Ltd., Oxford, UK. and Reproduced with permission from Y.-Z. Wang and U. Akid, Corrosion, 52, 92 (1996)]

C2. Critical Corrosion Rate

In research on steels, Uhlig and co-workers [14, 15] found that the fatigue life of initially smooth specimens in aqueous solution was reduced only when a critical dissolution rate was exceeded. However, fatigue crack propagation in precracked specimens was accelerated by applying cathodic polarization, which suppresses anodic dissolution. Later, it was found that the fatigue life of a high-strength steel was dominated by crack initiation, and the influence of the environment on fatigue life was through the reduction of the initiation time by corrosion pits, which developed by selective dissolution of MnS inclusions [8, 11]. Cathodic polarization suppresses the dissolution rate and prevents pit formation, extending the time required for crack initiation and restoring the fatigue life. The increased crack growth rate of well-defined cracks by cathodic polarization was attributed to hydrogen effects.

The influence of corrosion rate on corrosion fatigue behavior is also related to the mechanical loading conditions. For a Grade 448 (X-65) pipeline steel exposed to a dilute solution simulating the groundwater environment, the number and sizes of cracks were larger at pH 5.6 than at 6.9 when the applied stress ratio was at or below 0.6; at higher stress ratio, 0.8, cracking occurred at pH 6.9, but not at 5.6 [16]. This difference in behavior was attributed to the balance required between corrosion rate and severity of mechanical loading. For this steel–environment system, passivation that would prevent crack lateral dissolution does not occur, and only cyclic loading maintains crack sharpness. High corrosion rate accelerates corrosion fatigue when sufficient cyclic damage is simultaneously induced.

Corrosion fatigue cracks tend to initiate at surface discontinuities, such as notches and pits. Nevertheless, crack initiation is a competitive process, occurring first at the most favorable sites. Eliminating one type of initiation site can extend the fatigue life but may not prevent failure by corrosion fatigue. The beneficial effect of shot peening is attributed to the compressive stress generated at the surface layer that delays corrosion fatigue crack initiation.

C3. Three Corrosion Situations

Duquette [6] divided material and corrosive environments into three groups on the basis of surface conditions:

1. Active dissolution conditions
2. Electrochemically passive conditions
3. Bulk surface films, such as three-dimensional oxides

In the first group, emerging persistent slip bands (PSBs) are preferentially attacked by dissolution. This preferential attack leads to mechanical instability of the free surface and the generation of new and larger PSBs, which localize corrosion attack and lead to crack initiation. Under passive conditions, the relative rates of periodic rupture and reformation of the passive film control the extent to which corrosion reduces fatigue resistance. When bulk oxide films are present on a surface, rupture of the films by PSBs leads to preferential dissolution of the fresh metal that is produced.

D. CORROSION FATIGUE CRACK PROPAGATION

D1. Fracture Mechanics Characterization

As in stress corrosion cracking (SCC), the propagation behavior of well-defined corrosion fatigue cracks is often described using fracture mechanics, where the average crack extension per cycle (da/dN) is described as a function of the applied stress intensity factor range ($\Delta K = K_{max} - K_{min}$). Under cyclic loading, corrosion fatigue cracks can propagate when the maximum stress intensity factor is considerably below the threshold stress intensity factor for stress corrosion

cracking; and in the presence of a corrosive environment the crack growth rate can be markedly increased.

It is convenient to characterize the effects of environment on the rates of fatigue crack growth by considering different combinations of crack growth rates measured under purely mechanical fatigue and under SCC conditions. Figure 16.4(a) schematically illustrates the sigmoidal variation of fatigue crack growth as a function of stress intensity factor range (on a log–log scale) under purely mechanical cyclic loading conditions. The typical variation in crack velocity, da/dt, as a function of the applied stress intensity factor, K, is plotted (on a log–log scale) in Figure 16.4 (b) for growth of cracks in metallic materials under sustained loading in the presence of an environment. As shown in this figure, the environment has no effect on fracture behavior below a static stress intensity factor, K_{Iscc}, the threshold stress intensity factor for the growth of stress corrosion cracks in mode I (tensile opening mode). Above K_{Iscc} the crack velocity increases precipitously with increasing

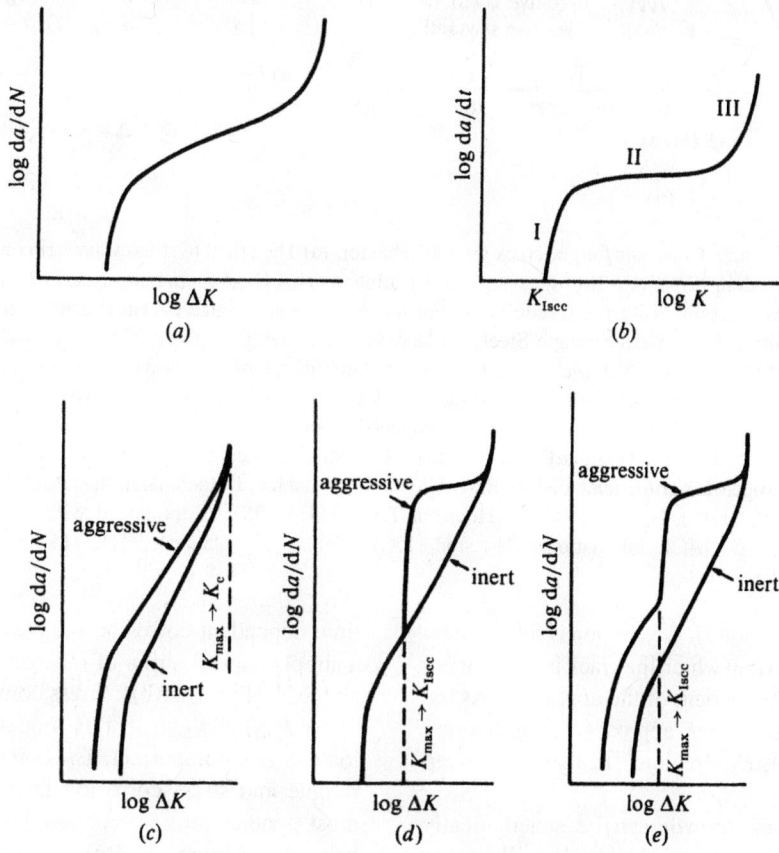

FIGURE 16.4. Schematic representations of the combinations of mechanical fatigue and environmentally assisted crack growth; (a) fatigue crack growth behavior in inert environments, (b) stress corrosion crack growth under sustained loads, (c) type A corrosion fatigue crack growth, true corrosion fatigue, arising from synergistic effects of cyclic loads and aggressive environment, (d) stress corrosion–fatigue behavior obtained from a superposition of mechanical fatigue (a) and stress corrosion cracking (b), and (e) mixed corrosion fatigue behavior obtained from a combination of (c) and (d). Reproduced with permission from A. J. McEvily and R. P. Wei, "Fracture Mechanics and Corrosion Fatigue," in Corrosion Fatigue: Chemistry, Mechanics and Microstructure, O. Devereux, A. J. McEvily, and R. W. Staehle (Eds.), NACE-2, NACE, Houston, TX, 1972, p. 381.

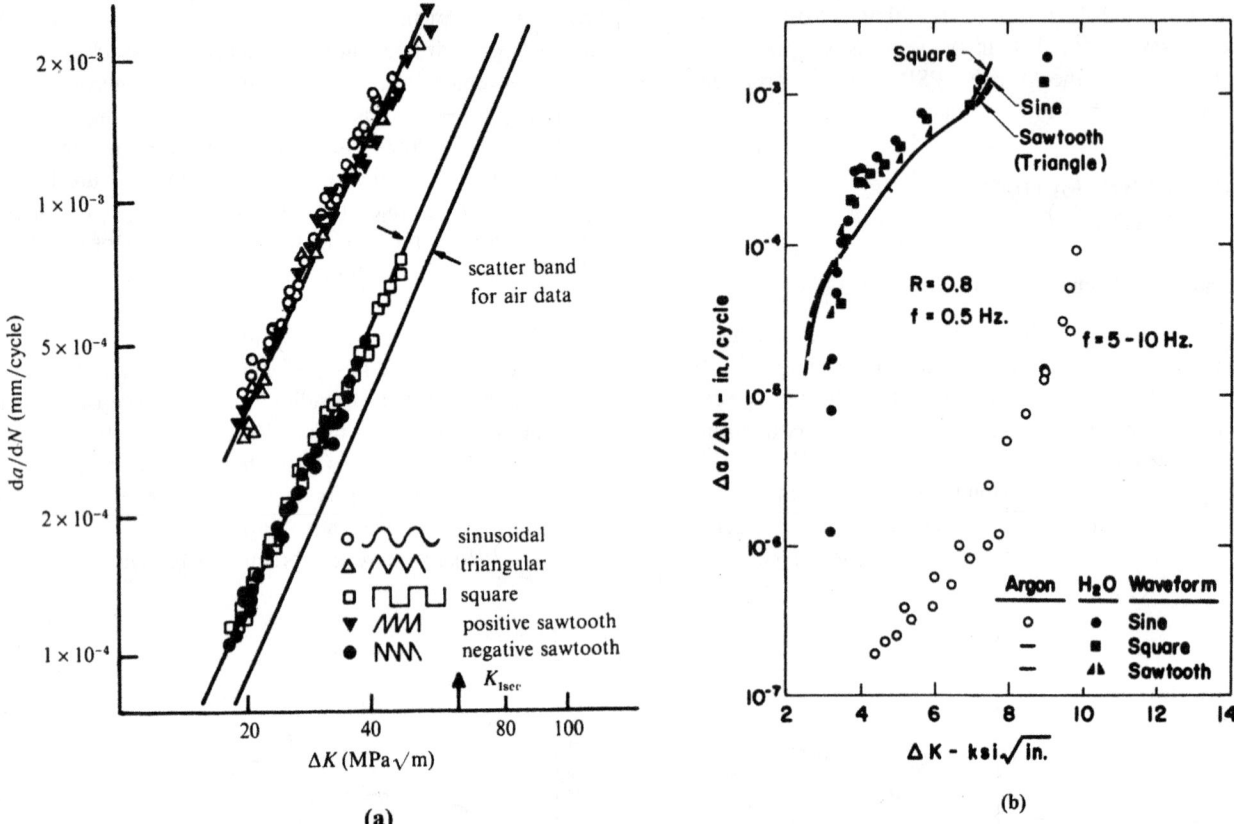

FIGURE 16.5. Corrosion fatigue crack growth behavior, (a) The effect of stress waveform on fatigue crack growth in 12Ni-%Cr-3Mo steel is 3% NaCl solution at 0.1 Hz at room temperature. Reproduced from J. M. Barsom, "Effect of Cyclic Stress Form on Corrosion Fatigue Crack Propagation Below KISCC in a High-Yield-Strength Steel," in Corrosion Fatigue: Chemistry, Mechanics and Microstructure, O. Devereux, A. J. McEvily, and R. W. Staehle (Eds.), NACE-2, NACE, Houston. TX, 1972, p. 426. (b) Time-dependent corrosion fatigue above K_{Iscc} for high-strength type 4340 steel in water vapor, modeled by linear superposition. Reproduced from R. P. Wei and G. W. Simmons, "Environment Enhanced Fatigue Crack Growth in High-Strength Steels," in Stress Corrosion Cracking and Hydrogen Embrittlement of Iron Base Alloys, R. W. Staehle, J. Hochmann, R. D. McCright, and J. E. Slater (Eds.), NACE-5, NACE, Houston, TX, 1973, p. 751. (Reproduced with permission. Copyright © NACE International, Houston, TX.) (1 in./cycle = 25.4 mm/cycle; 1 ksi · in.$^{1/2}$ = 1.098 MPa · $\sqrt{m}$).

stress intensity factor, K (region I). This region is followed by a region of growth (region II) in which the crack increment per unit time is essentially independent of the applied K. As the maximum stress intensity factor approaches the fracture toughness of the material, K_{Ic} (region III), there is a steep increase in crack velocity.

Corrosion fatigue crack growth can be schematically represented in three ways [17]. Figure 16.4(c) illustrates type A true corrosion fatigue growth behavior in which the synergistic interaction between cyclic plastic deformation and environment produces cycle- and time-dependent crack growth rates. True corrosion fatigue influences cyclic fracture even at maximum stress intensity factor K_{max} in fatigue $< K_{Iscc}$. The cyclic load form is important. Figure 16.4(d) shows the stress corrosion fatigue process, type B, purely

time-dependent corrosion fatigue crack propagation that is a simple superposition of mechanical fatigue [Fig. 16.4(a)] and SCC [Fig. 16.4(b)]. Stress corrosion fatigue occurs only when $K_{max} > K_{Iscc}$. In this model, the cyclic character of loading is not important. The combination of true corrosion fatigue and stress corrosion fatigue results in type C, the most general form of corrosion fatigue crack propagation behavior, Figure 16.4(e), which involves cyclic time-dependent acceleration in da/dN below K_{Iscc}, combined with time-dependent cracking (SCC) above the threshold.

Figure 16.5 shows examples of corrosion fatigue crack propagation behavior. Figure 16.5(a) illustrates the behavior of maraging steel exposed to 3% NaCl [18], representing Type A growth, and Figure 16.5(b) shows the behavior of high-strength type 4340 [Unified Numbering System (UNS)

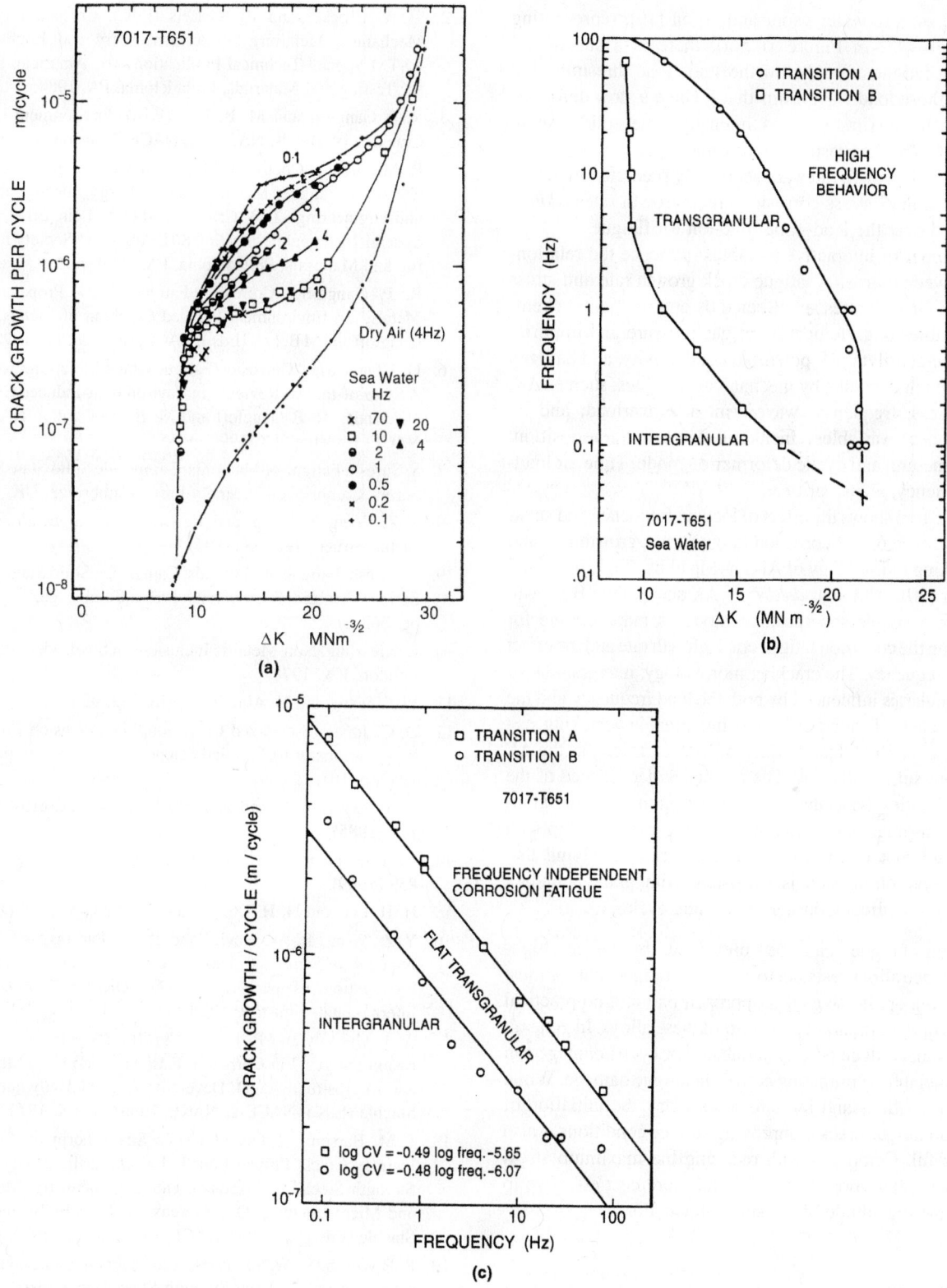

FIGURE 16.6. Corrosion fatigue of Al–Zn–Mg alloy, 7017 in natural seawater. (a) Crack growth rate as a function of ΔK for a range of cyclic loading frequencies, (b) The dependence of corrosion fatigue fracture morphologies in terms of cyclic loading frequency and ΔK. (c) Fracture morphologies in terms of crack growth rate and cyclic loading frequency. [Reprinted from N. J. H. Holroyd and D. Hardie, "Factors Controlling Crack Velocity in 7000 Series Aluminum Alloys during Fatigue in an Aggressive Environment," Corrosion Science, 23, pp. 529, 533, 535 (1983), with permission from Elsevier Science.]

G43400] steel in water vapor and argon [19], representing type B growth. In Figure 16.5(a), there is a substantial corrosion fatigue effect below the static load threshold, but only for those load waveforms that include a slow deformation rate to maximum stress intensity. The solid line in Figure 16.5(b) demonstrates that time-dependent corrosion fatigue crack growth rates are accurately predicted by linear superposition of stress corrosion crack growth rates (da/dt) integrated over the load–time function for fatigue.

A number of interactive variables influence the relationship between corrosion fatigue crack growth rate and stress intensity. Growth rates are affected by environmental chemical variables (e.g., temperature; gas pressure and impurity content; electrolyte pH, potential, conductivity, and halogen or sulfide ion content); by mechanical variables, such as ΔK, mean stress, frequency, waveform, and overload; and by metallurgical variables, including impurity composition, microstructure, and cyclic deformation mode. Time, or loading frequency, is also critical.

Figure 16.6 shows the effect of loading frequency and stress intensity range on the corrosion fatigue crack growth rate and the cracking morphology of Al–Zn–Mg alloy, 7017, in natural seawater [20]. Although $da/dN - \Delta K$ shows type B growth behavior, a simple superposition model is inappropriate for describing the corrosion fatigue crack growth rate and the effect of load frequency. The cracking morphology, intergranular or transgranular, is influenced by both the load frequency and the stress intensity factor range, and intergranular cracking can occur at a very high load frequency (70 Hz) as long as the ΔK values are sufficiently low. The frequency dependence of the crack velocities associated with the transition from intergranular to transgranular cracking shows a linear relationship with the square root of the loading cycle period, implying that the rate-controlling step is consistent with grain boundary diffusion of hydrogen during the loading cycle.

Corrosion fatigue can be prevented by using high-performance alloys resistant to corrosion fatigue; but for most engineering applications this approach may not be practical because of the availability and cost of these alloys. In general, methods that reduce corrosion rate and/or cyclic damage can be beneficial for eliminating corrosion fatigue damage. While effective coatings and inhibitors can delay the initiation of corrosion fatigue cracks, improving surface conditions is also very useful. Compared with reducing the maximum stress level, it is often more beneficial and more cost effective to reduce the magnitude of the stress fluctuation.

REFERENCES

1. O. Devereux, A. J. McEvily, and R. W. Staehle (Eds.), Corrosion Fatigue: Chemistry, Mechanics and Microstructure, NACE-2, NACE, Houston, TX, 1972.

2. T. W. Crocker and B. N. Leis (Eds.), Corrosion Fatigue: Mechanics, Metallurgy, Electrochemistry and Engineering, ASTM Special Technical Publication 801, American Society for Testing and Materials, Philadelphia, PA, 1984.

3. R. P. Gangloff and M. B. Ives (Eds.), Environment-Induced Cracking of Metals, NACE-10, NACE, Houston, TX, 1990.

4. P. M. Scott, "Chemical Effects in Corrosion Fatigue," in Corrosion Fatigue: Mechanics, Metallurgy, Electrochemistry and Engineering, T. W. Crocker and B. N. Leis (Eds.), ASTM Special Technical Publication 801, American Society for Testing and Materials, Philadelphia, PA, 1984, p. 319.

5. R. P. Gangloff, "Corrosion Fatigue Crack Propagation in Metals," in Environment-Induced Cracking of Metals, R. P. Gangloff and M B. Ives (Eds.), NACE, Houston, TX, 1990, p. 45.

6. D. J. Duquette, "Corrosion Fatigue Crack Initiation Processes: A State-of-the Art Review," in Environment-Induced Cracking of Metals, R. P. Gangloff and M. B. Ives (Eds.), NACE-10, NACE, Houston, TX, 1990, p. 45.

7. S. Suresh, Fatigue of Materials, Cambridge Solid State Science Series, Cambridge University Press, Cambridge, UK, 1991.

8. Y.-Z. Wang, R. Akid, and K. J. Miller, Fatigue Fract. Eng. Mater. Struct., **18**, 293 (1995).

9. S. Suresh, Fatigue of Materials, Cambridge Solid State Science Series, Cambridge University Press, Cambridge, UK, 1991, pp. 363–368.

10. R. Kiessling, Non-Metallic Inclusions in Steel, Metals Society, London, UK, 1978.

11. Y.-Z. Wang and U. Akid, Corrosion, **52**, 92 (1996).

12. D. C. Jones, "Localized Corrosion," in Corrosion Processes, R. N. Parkins (Ed.), Applied Science Publishers, London, UK, 1982, p. 161.

13. G. P. Ray, R. A. Jaman, and J. G. N. Thomas, Corros. Sci., **25**, 171 (1985).

14. D. J. Duquette and H. H. Uhlig, Trans. Am. Soc. Metals, **62**, 839 (1969).

15. H. H. Lee and H. H. Uhlig, Metall. Trans., **3**, 2949 (1971).

16. Y.-Z. Wing, R. W. Revie, and R. N. Parkins, "Mechanistic Aspects of Stress Corrosion Crack Initiation and Early Propagation," Paper No. 99143, CORROSION/99, NACE International, Houston, TX, 1999.

17. A. J. McEvily and R. P. Wei, "Fracture Mechanics and Corrosion Fatigue," in Corrosion Fatigue: Chemistry, Mechanics and Microstructure, O. Devereux, A. J. McEvily, and R. W. Staehle (Eds.), NACE-2, NACE, Houston, TX, 1972, p. 381.

18. J. M. Barsom, "Effect of Cyclic Stress Form on Corrosion Fatigue Crack Propagation Below $K_{I_{SCC}}$ in a High-Yield-Strength Steel," in Corrosion Fatigue: Chemistry, Mechanics and Microstructure, O. Devereux, A. J. McEvily, and R. W. Staehle (Eds.), NACE-2, NACE, Houston. TX, 1972, p. 426.

19. R. P. Wei and G. W. Simmons, "Environment Enhanced Fatigue Crack Growth in High-Strength Steels," in Stress Corrosion Cracking and Hydrogen Embrittlement of Iron Base Alloys, R. W. Staehle, J. Hochmann, R. D. McCright, and J. E. Slater (Eds.), NACE-5, NACE, Houston, TX, 1973, p. 751.

20. N. J. H. Holroyd and D. Hardie, Corros. Sci., **23**, 527 (1983).

17

FLOW EFFECTS ON CORROSION

K. D. EFIRD

Efird Corrosion International, Inc., The Woodlands, Texas

A. INTRODUCTION

The effect of flow on corrosion is complex and varied and is dependent on both the chemistry and physics of a system. The effect of chemistry is typical of most corrosion mechanisms except that flow has a significant effect on the kinetics of the corrosion reactions and the phases participating in the corrosion reaction. These effects are related to the influence of flow on the movement, distribution, and mixing of fluids in the flowing system.

The key variable defining the effect of flow on corrosion is turbulence. High turbulence can result in flow-induced corrosion, erosion–corrosion, or cavitation. Low turbulence can result in corrosion in a separated water phase and allows the occurrence of corrosion under deposits and/or in separated liquid water. The emphasis is on flow in pipes since pipe flow is the location of the vast majority of flow-induced corrosion problems.

"Flow-induced corrosion" is the term used to describe the increase in corrosion resulting from high fluid turbulence due to the flow of a fluid over a surface in a flowing single or multiphase system. "Underdeposit corrosion" is the term used to describe the increased corrosion occurring in a separated water phase beneath deposits of nonmetallic solids on a metal surface resulting from low-flow turbulence.

A clear understanding of the difference between flow-induced corrosion and "erosion–corrosion" aids in the discussion. The two are not the same, as indicated by the definitions of these terms in the Glossary. Flow-induced corrosion is the increase in corrosion resulting from increased fluid turbulence intensity and mass transfer as a result of the flow of a fluid over a surface. Erosion–corrosion is the increased corrosion due to the physical impact on a surface causing mechanical damage. The impact can be from solid particles entrained in a liquid or gas phase or liquid droplets entrained in a gas phase. The effect of erosion on corrosion is covered in Chapter 18, Erosion–Corrosion in Single- and Multiphase Flow.

B. FLOW BASICS RELATED TO CORROSION

Corrosion is a surface phenomenon, and, as such, what goes on at the metal surface has a profound effect on corrosion. Many aspects of fluid dynamics relate to and define the interactions of a fluid with the surface that are important to corrosion.

When a fluid flows over a solid surface, the flow is characterized as either laminar or turbulent. In most situations where the effect of fluid flow on corrosion is important, the flow is turbulent; so the physical structure of turbulent flow is a primary consideration. A number of possibly unfamiliar terms introduced in this chapter are defined in the Glossary. The definitions of symbols used in equations are given in the Nomenclature at the end of this chapter.

B1. Turbulent Boundary Layer

Fully developed turbulent flow consists of a turbulent core where the mean velocity is essentially constant and a boundary layer of varying fluid velocity near the solid–fluid interface. The majority of the changes in fluid stress characteristics, turbulence, mass transfer, and fluid interaction with the wall occur within this boundary layer.

All mean and turbulent velocity components of the flow must go to zero at the wall and must be very small in the immediate vicinity of the wall. Therefore, the components of all turbulent stresses must also go to zero at the wall, leaving only the viscous stresses of laminar flow to act on the wall. This layer near the wall within the boundary layer, where viscous forces dominate over turbulent forces and flow is laminar-like, is termed the viscous sublayer. There is a transfer of turbulent energy from the outer layer to the viscous sublayer and a turbulent diffusion of energy from the laminar sublayer to the outer layer at approximately equal rates for established flow. A diagram of these layers and the transfer processes is shown in Figure 17.1 [1].

The calculated boundary layer thickness generally referred to in the corrosion literature is related to and representative of the physical boundary layer but is actually a defined thickness that is a linear approximation [2]. This approximation for the diffusion boundary layer ($\delta_{\mathbf{d}}$) and the hydrodynamic boundary layer (δ_{ν}) is shown graphically in Figure 17.2 [3]. The approximation assumes that the flowing fluid can be broken down into two parts: (1) a thin boundary layer near the solid surface and (2) the bulk solution, and that

the transition occurs abruptly, defining the boundary layer thickness. The basic assumption is that momentum and mass transport occur by viscous flow and diffusion within the boundary layer, as measured by the mass transfer coefficient (k) and the shear stress (τ), and by turbulent flow and convection outside the boundary layer [2].

The turbulent boundary layer is further resolved into three regions: an outer "logarithmic" region ($30 < y_+ < 100$), an intermediate "buffer" region ($5 < y_+ < 30$), and, very close to the solid wall, a "viscous" region ($y_+ < 5$), where y_+ is the dimensionless viscous length perpendicular to the wall, defined as $\boldsymbol{y_+} = \boldsymbol{yu_\tau}/\nu$ [4]. A diagram of these layers along with an idealized plot of the normalized turbulent velocity variation is shown in Figure 17.3 [4]. Experimental details of the variation in boundary layer turbulence with distance from the wall are given by Kline et al. in their extensive study into the structure of the turbulent boundary layer [3].

The nature of the turbulent boundary layer dictates that this is where the processes that control corrosion and film formation will occur. The majority of the movement of corroding species to, and of corrosion products from, the wall and chemical reactions with the wall must occur in this region. Therefore, any disturbance in the turbulent boundary layer must be a primary factor affecting the corrosion process.

B2. Boundary Layer Disruption

Disruption of the boundary layer in turbulent flow occurs primarily by the formation of turbulent bursts and sweeps. The turbulent burst is an ejection of fluid from the wall, which also causes fluid to impinge on the wall by the simultaneous formation of sweeps, or movement of fluid toward the wall. Turbulent bursts and sweeps occur through the formation of vortices and the lift-up of wall streaks [5].

One process of vortex formation and evolution into a turbulent burst proposed by Praturi and Brodkey [6] is illustrated in Figure 17.4. The vortex moves with the flow and is increasingly tilted by the mean shear in the viscous

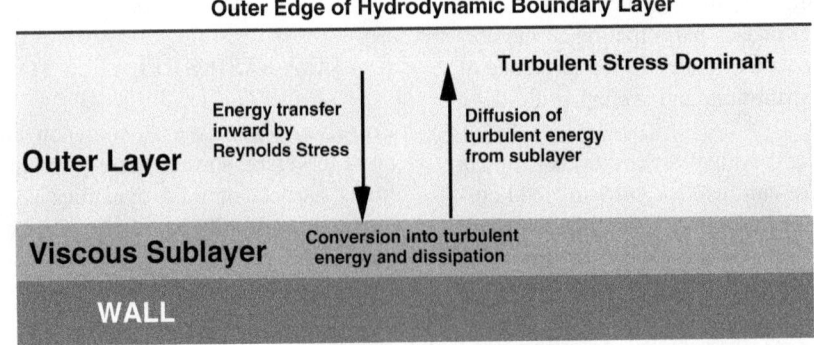

FIGURE 17.1. Basic structure of turbulent boundary layer and associated mechanical energy flow within the boundary layer [1].

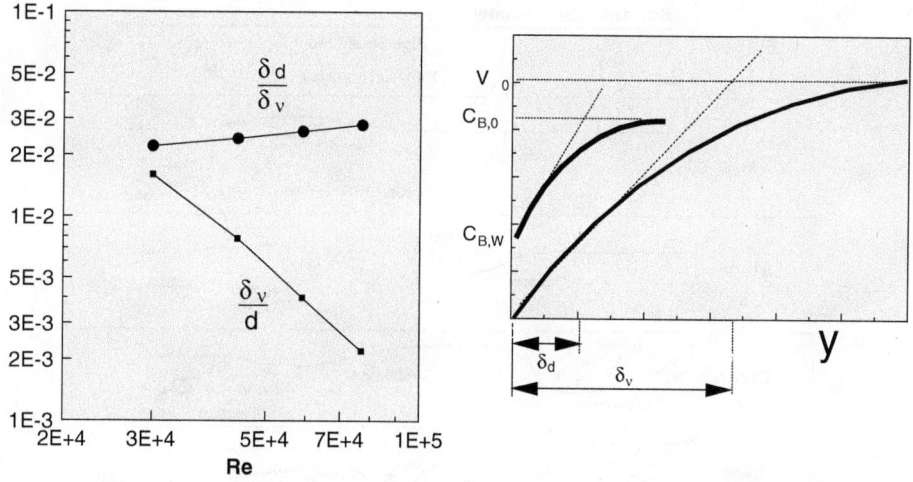

FIGURE 17.2. Relative boundary layer thickness (δ) for concentration (c) and velocity (v) profiles in a turbulent flowing fluid (solid line = true profile; dotted line = linear approximation) [3].

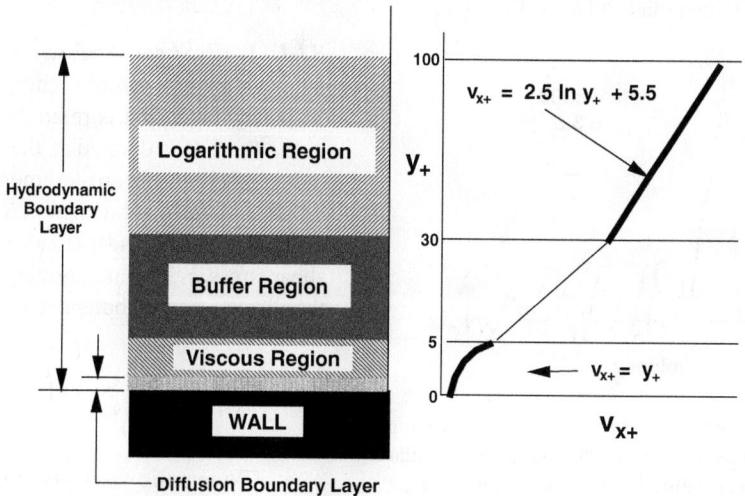

FIGURE 17.3. Structure of the turbulent boundary layer with an idealized plot of the normalized turbulent velocity variation with distance from the wall, expressed as dimensionless viscous length [4].

region (stages 1–4). At some point, the vortex becomes unstable and a strong ejection occurs along with a rapid sweep of fluid into the viscous region (stage 5). It is evident that severe pressure fluctuations must occur with the formation of a turbulent burst and during the rapid ejection of fluid.

A slightly different model of the bursting process is given by Often and Kline, where a description of the complete burst cycle is given [7]. In this model, bursting is associated with wall streaks and stretched and lifted vortices. All of the current theories of wall turbulence agree in the basic structure of the turbulence, that is, the existence and interaction of turbulent bursts, ejections, sweeps, and wall streaks. The differences lie in the mechanisms of formation of these structures and the details of their interaction.

An excellent review of the present understanding of the turbulent boundary layer structure is given by Robinson [8]. As a result of this review, Robinson concluded that coherent motions exist in the viscous region, that they consist of elongated, unsteady regions of high- and low-speed streaks, and that sweep motions dominate the viscous region. The majority of the coherent flow structures occur in the buffer region of the turbulent boundary layer, with only sweeps and ejections penetrating into the viscous layer [8]. Laser Doppler measurements have clearly demonstrated the presence of turbulence in the viscous region near the wall as shown in Figure 17.5 [4].

The relationship of turbulent structures in the boundary layer, specifically near wall bursting activity, and fluctuations

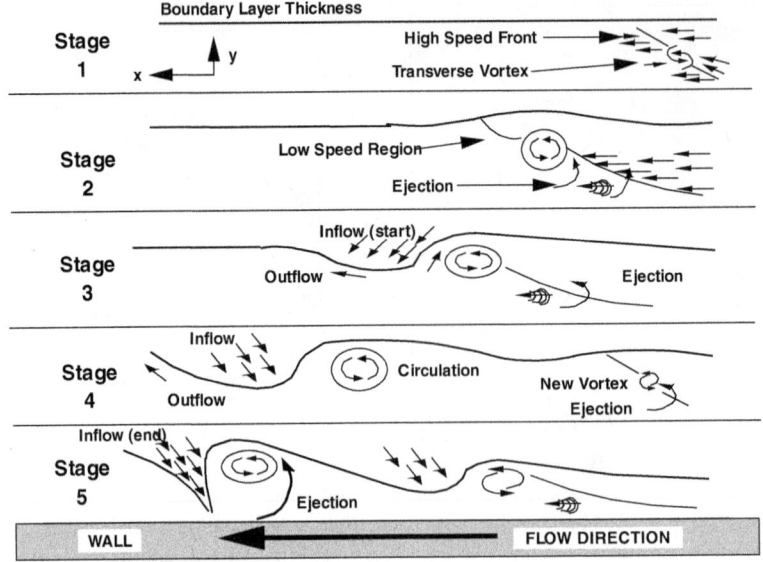

FIGURE 17.4. Progression of near-wall flow illustrating the formation of shear layer vortices and evolution into turbulent bursts [6].

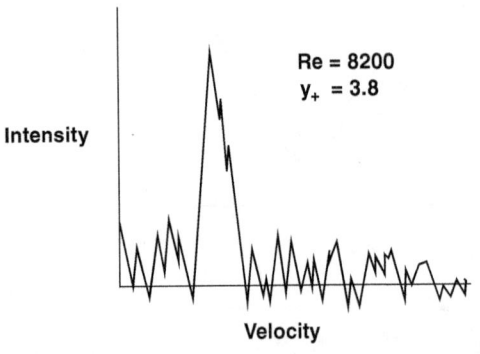

FIGURE 17.5. Laser Doppler measurements at $y_+ = 3.8$, showing turbulence inside viscous region for water flowing in rectangular pipe. The large peak is the main velocity component, and the small peaks represent instantaneous velocity fluctuations [4].

in the wall shear stress and wall pressure were measured by Thomas and Bull [9]. A diagram of their results is given in Figure 17.6. This work provides evidence of significant variations in wall shear stress related to the boundary layer turbulence generation process that could have a major effect on corrosion processes.

B3. Flow Parameters

Fluid flow must be expressed in terms broadly related to flow parameters that are common to all hydrodynamic systems and that effectively define the interaction of the fluid with the metal surface. The parameters that best fit this requirement are wall shear stress (τw) and mass transfer coefficient (k). These parameters are calculated from empirical equations developed to characterize fluid flow.

B3.1. Wall Shear Stress. Wall shear stress is a direct measure of the viscous energy loss within the turbulent boundary layer and is related to the intensity of turbulence in the fluid. It is defined as the isothermal pressure loss in a moving fluid within an incremental length due to fluid friction as a result of contact with a stationary wall. The mathematical definition of wall shear stress, τ_w, is as follows. [2] The total shear stress, τ, in a fluid moving past a fixed wall is the sum of the viscous and turbulent stresses, expressed as

$$\tau = \nu\left(\frac{\partial \bar{U}}{\partial y}\right) - u_x u_y \qquad (17.1)$$

where

ν = kinematic viscosity (m²/s)
U = mean velocity (m/s)
y = direction $\perp$ to the surface (m)
u_x, and u_y = fluctuating velocity component (m/s) in the x and y direction

The Reynolds stresses ($u_x u_y$) go to zero at the wall, leaving only the viscous stress in the fluid. The wall shear stress, τ_w, is defined as this viscous shear stress at the wall ($y = 0$), expressed as

$$\tau_{\mathbf{w}} = \nu\left(\frac{\partial U}{\partial y}\right)_{y=0} \qquad (17.2)$$

B3.2. Mass Transfer Coefficient. Mass transfer defines the chemical and electrochemical effects of fluid flow. It relates to the transport of reactive chemical species from the bulk solution to and from the solid surface through the boundary layer. It defines the movement of corrosive species to the metal surface and the movement of corrosion products away

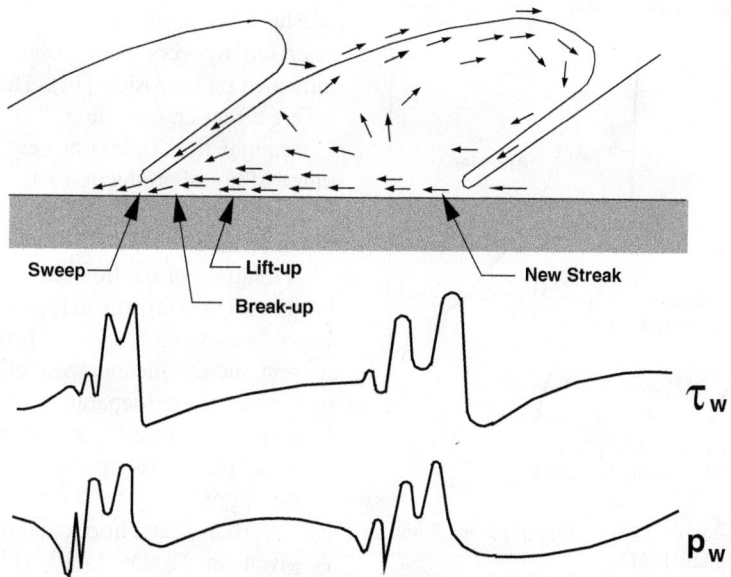

FIGURE 17.6. Wall shear stress and wall pressure fluctuations in relation to location of near-wall bursting activity in turbulent boundary layer [9].

from the metal surface. The diffusion layer, in which the mass transfer occurs, is roughly defined by the diffusion mass flux density ($N_{b,y}$) as expressed in the Nernst diffusion model [10]:

$$N_{j,y} = -D_j \left(\frac{dC_j}{dy}\right)_y = -D_j \frac{C_{j,0} - C_{j,y}}{\delta_d} \qquad (17.3)$$

where
$N_{j,y}$ = mass flux density of species j in the y direction (mol/m^2 s)
D_j = diffusion coefficient for species j (m^2/s)
C_j = concentration of species j (mol/L)
δ_d = diffusion boundary layer thickness (m)

The mass transfer coefficient, k_d, is defined as the proportionality between the mass flux density and the concentration gradient, expressed as

$$k_d = \frac{N_{j,y}}{C_{j,0} - C_{j,y}} \qquad (17.4)$$

Substituting the definition of the flux density from Eq. (17.3), the mass transfer coefficient is defined as

$$k_d = \frac{D_j}{\delta_d} \qquad (17.5)$$

B3.3. Interrelationship of Mass Transfer Coefficient and Wall Shear Stress. Mass transfer is related to wall shear stress in that changes in flow parameters that affect one result in changes in the other. The diffusion coefficient can be related to the wall shear stress by the Chilton–Colburn analogy:

$$k_d \approx \left(\frac{\tau_w}{\rho}\right)^{0.5} \qquad (17.6)$$

where ρ is fluid density (kg/m^3). This relationship is more precisely defined as

$$k_d = 17.24 \left(\frac{\tau_w}{\rho}\right)^{0.5} Sc^{2/3} \qquad \text{for Sc} > 100 \qquad (17.7)$$

where Sc is the Schmidt number.

C. FLOW REGIME AND FLOW CORROSION

C1. Single-Phase Flow

Single-phase flow consists of only one phase, gas or liquid. In single-phase aqueous systems there is a flow pattern that consists of a turbulent core and a boundary layer near the solid–fluid interface as described earlier. All of the mass and momentum transfer to the boundary layer occurs through this turbulent core. The equations and relationships discussed in the sections on wall shear stress and mass transfer are directly applicable to the single-phase flow situation.

C2. Multiphase Flow and Flow Regime

For multiphase flow the volume flow rates of the various phases affect the resulting flow pattern, called the "flow regime." The flow regime is a primary consideration in the determination of flow-induced corrosion in multiphase flow. The basic criterion for flow-induced corrosion in multiphase

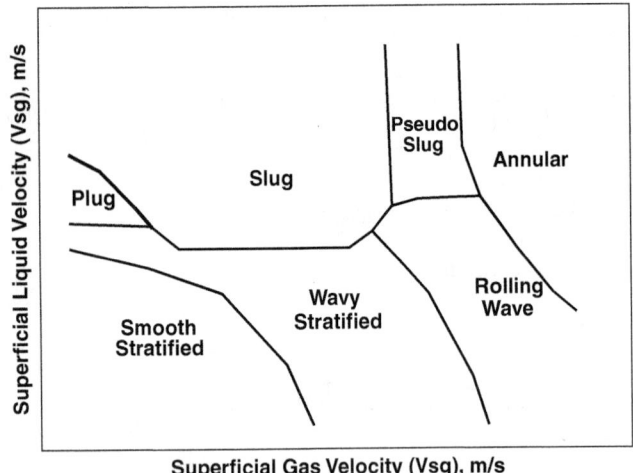

FIGURE 17.7. Typical gas–liquid flow regime map showing boundaries between various regions [11].

flow is that a liquid corrosive phase be present and be in contact with the metal surface. The flow regime affects the turbulence in the liquid phase, the location of the liquid phase in the flowing stream, and whether or not solids can settle out of the flowing stream. An example of a flow regime map showing the flow regimes encountered in multiphase gas–liquid horizontal flow is given in Figure 17.7 [11].

The bubbly flow regime occurs in liquid systems with low gas and high liquid rates. Stratified flow occurs at low liquid rates or when a liquid phase condenses. The liquid in stratified flow is generally stagnant or very low flow, with low turbulence.

Slug flow occurs at higher liquid flow rates and is characterized by very high turbulence, with a corresponding influence on corrosion [10]. This flow regime has the most severe effect on corrosion.

Annular flow occurs at very high gas rates over a wide range of liquid production rates. Turbulence in annular flow with respect to the effects on corrosion is not well characterized.

A diagram of the flow regimes encountered in gas–water horizontal flow is given in Figure 17.8 [11] and hydrocarbon–water flow in Figure 17.9 [12]. When a hydrocarbon phase is present, the significant aspect of the flow regime with respect to corrosion is the separation of the hydrocarbon and water–liquid phases, or the level of turbulence required to eliminate the water phase as a separate entity contacting the metal wall.

A diagram of the flow regimes encountered in gas–hydrocarbon–water horizontal flow and the flow regime map is given in Figure 17.10 [13]. Gas–hydrocarbon–water flow combines the characteristics of the gas–water and hydrocarbon–water flow regimes. The gas phase adds a high degree of turbulence, but the primary corrosion concern is the contact of water with the metal wall.

The flow regime maps shown are for horizontal flow and are significantly altered for inclined flow [14]. The effect of inclination is particularly important for the occurrence and severity of slug flow with upward inclination, where both the frequency and intensity of the slug increases [14]. This is particularly significant due to the effect of slug flow on turbulence and corrosion [13].

A generalized summary of the types of corrosion expected for the various flow regimes, along with the location of free

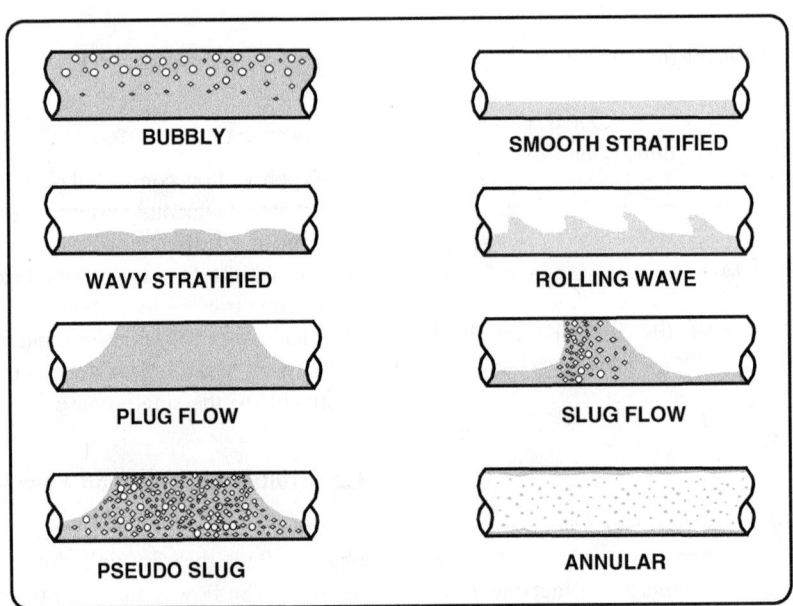

FIGURE 17.8. Gas–water two-phase flow regimes in horizontal pipes [11].

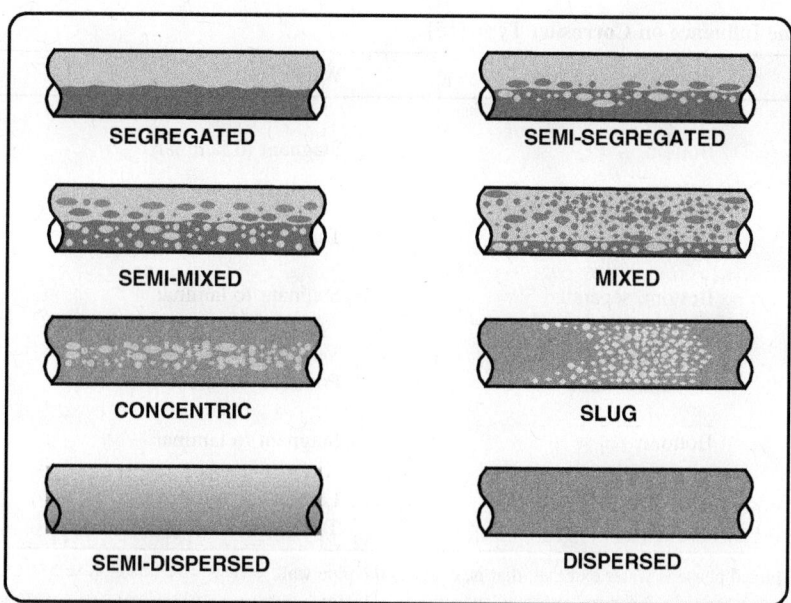

FIGURE 17.9. Hydrocarbon–water flow regimes in horizontal pipes [12].

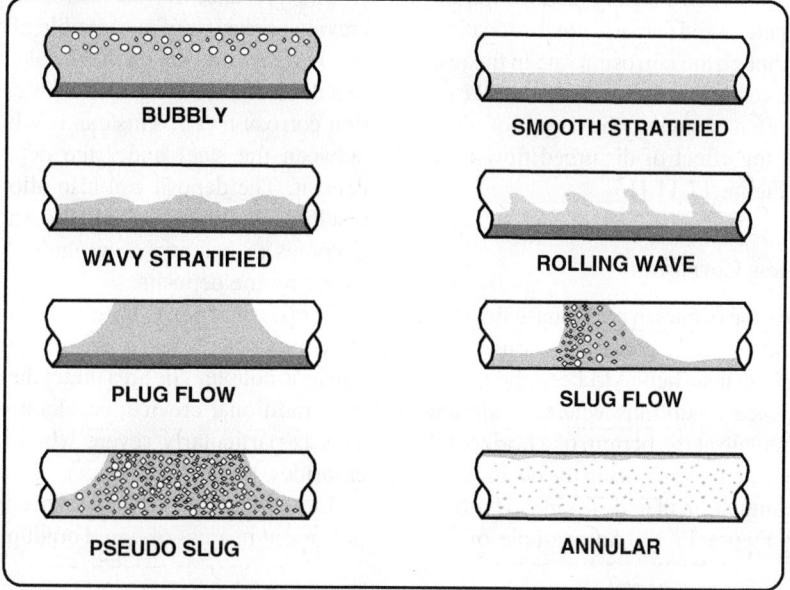

FIGURE 17.10. Gas–oil–water three-phase flow regimes in horizontal pipes [13].

water and the water turbulence, is outlined in Table 17.1 [15]. The flow regimes where either underdeposit corrosion or flow-induced corrosion is expected are given in this table.

C3. Effect of Disturbed Flow

Flow-induced corrosion failures in many environments occur in areas where fully developed flow patterns are disrupted, termed regions of disturbed flow [15–18]. Examples of these disturbed flow locations are:

- Downstream of weld beads, at pipe joints and upsets
- Downstream of pipe fittings, at preexisting pits
- Downstream of valves, at bends and elbows in piping
- Downstream of orifice plates, at heat exchanger tube inlets

The flow disturbance destroys the fully developed hydrodynamic boundary layer and the diffusion boundary layer. The effect of this boundary layer disruption is the production

TABLE 17.1. Flow Regime Influence on Corrosion Type [15]

Flow Regime	Free Water Location	Water Turbulence	Corrosion Type
Gas/water			
Stratified flow	Bottom	Stagnant to laminar	Underdeposit corrosion Pitting of stainless steels
Slug flow	Mostly bottom, mixed	Very turbulent	Flow-induced corrosion
Annular flow	Circumferential	Turbulent	Flow-induced corrosion
Gas–liquid–water			
Stratified flow	Bottom, separated	Stagnant to laminar	Underdeposit corrosion Pitting of stainless steels
Slug flow	Mostly bottom, mixed	Very turbulent	Flow-induced corrosion
Annular flow	Circumferential	Possibly turbulent	Flow-induced corrosion
Liquid–water			
Segregated flow	Bottom	Stagnant to laminar	Underdeposit corrosion Pitting of stainless steels
Mixed flow	Mostly bottom, mixed	Laminar to turbulent	Underdeposit corrosion
Dispersed flow	Mixed	Turbulent	Flow-induced corrosion[a]

[a]Only if the hydrocarbon water mixed phase is water external, that is, contacts the pipe wall.

of a steady-state condition as opposed to an equilibrium condition. In this case, the normal equilibrium corrosion reactions cannot be maintained, and a steady-state condition is established. The result is increased corrosion at the location of the flow disturbance, although the corrosion rate in the rest of the pipe wall is low. The corrosion pattern appears as large pits or corroded areas, often showing the signs of flow direction. A schematic of the effect of disturbed flow over a weld bead is shown in Figure 17.11 [15].

C4. Low-Turbulence-Flow Corrosion

Low-turbulence-flow corrosion occurs in multiphase flow for a stratified flow regime and in single-phase flow containing solids where the turbulence is insufficient to keep the solids suspended. In low-turbulence conditions where solids are present deposits can accumulate at the bottom of a horizontal or slightly inclined line. A schematic showing the occurrence of corrosion under the accumulation of deposits in the bottom of a pipeline is shown in Figure 17.12. An example of the

severe corrosion that can occur in the bottom of a pipeline under deposits is shown in Figure 17.13.

The presence of the solids deposit can also result in crevice corrosion of susceptible alloys, where the deposited solids form a crevice on the metal underneath. This can result in a differential aeration or differential metal ion concentration corrosion cell. This can result in a galvanic interaction between the steel under the deposit and that outside the deposit. The deposit can also allow bacteria to proliferate, resulting in microbiologically induced corrosion (MIC). Colonies of bacteria grow under the deposits and are protected by the deposits.

For passive metals such as stainless steels, the area under the deposit can lose passivity, resulting in an active–passive galvanic couple. The area under the deposit can acidify just as in a traditional crevice, accelerating corrosion in that area. This is particularly severe when the water phase contains chlorides.

Liquid and possibly solids accumulation in gas lines can be present in areas of liquid holdup. This occurs primarily on

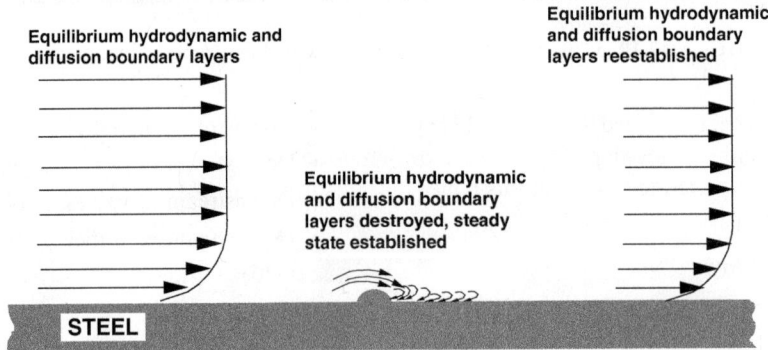

FIGURE 17.11. Effect of disrupted flow over weld bead on boundary layer [15].

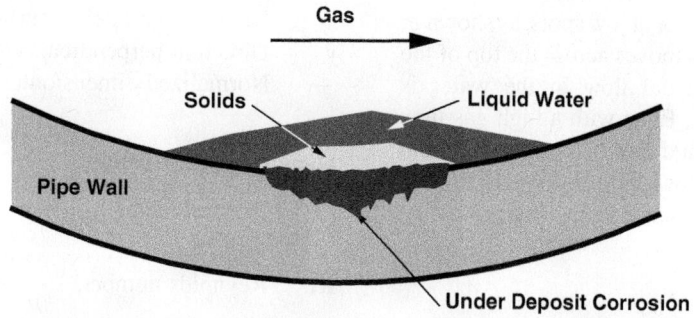

FIGURE 17.12. Occurrence of corrosion under water–solids deposits in bottom of pipeline at locations where liquid holdup occurs.

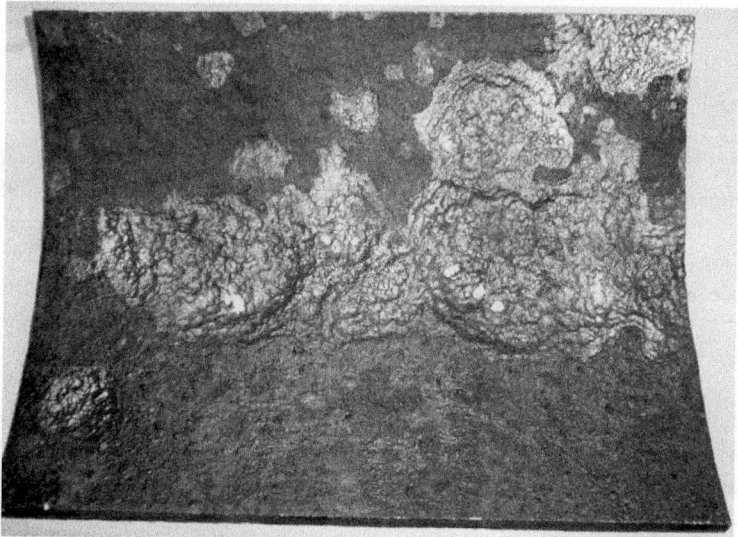

FIGURE 17.13. Corrosion in bottom of pipeline resulting from solids deposits under low-flow conditions.

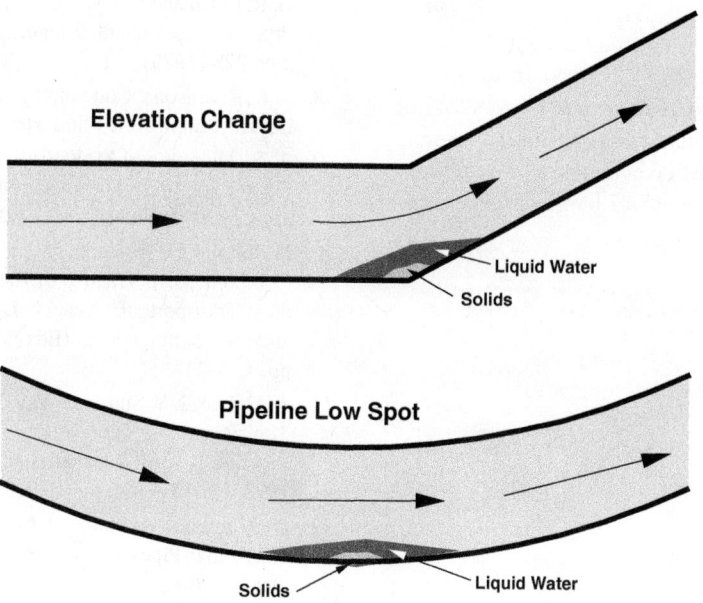

FIGURE 17.14. Locations where liquid holdup and solids deposition can occur.

areas of upward elevation change or at low spots, as shown in Figure 17.14. In these areas, gas moves across the top of the accumulated water, but the actual flow in the water is generally very slow to stagnant. Even with a high gas flow rate through the pipeline, the liquid flow in the areas of liquid holdup is sufficiently low to allow entrained solids to drop out from the liquid stream and settle to the bottom of the pipeline.

NOMENCLATURE

C Concentration (mol/L)
D Diffusion coefficient (m²/s)
d Diameter, m
e Roughness factor, m
i Current density, A/m²
K mass transfer coefficient (m/s)
L Characteristic length, m
N Mass flux density (mol/m²s)
P Pressure, Pa, kg/m s²
R_{cor} Corrosion rate, mm/y, mpy
r Radius or radial distance, m
T Temperature, °C
t Time, s
U Mean velocity, m/s
u' Local velocity fluctuation, m/s
x Direction parallel to surface, m
y Direction perpendicular to the surface, m
y_+ Viscous length $= yu\tau/\nu$, dimensionless
z Direction parallel to the surface but perpendicular to the flow direction, m

Greek Symbols

Δ Gradient of a property
δ Boundary layer thickness (m)
ρ Density (kg/m³)
μ Dynamic viscosity (kg/m s)
ν Kinematic viscosity (m²/s)
τ Shear stress (N/m²)

Subscripts

b Boundary layer
cor   Corrosion
d Diffusion
f Final or ending
i Initial or beginning
j Species "j"
lim   Limiting
0   Standard or primary
t Value at time t
w Wall or electrode surface

x Direction parallel to surface
y Direction perpendicular to surface
$+$ Normalized dimensionless form

Dimensionless Groups

f Friction factor, $\dfrac{(2\tau)}{(\rho U^2)}$

Re   Reynolds number, $\dfrac{(\rho U d)}{\mu}$

Sc   Schmidt number, $\dfrac{\mu}{(\rho d)}$

Sh   Sherwood number, $\dfrac{(kd)}{D}$

REFERENCES

1. A. A. Townsend, The Structure of Turbulent Flow, Cambridge University Press, Cambridge, 1956, pp. 232–237.
2. H. Schlichting, Boundary-Layer Theory, McGraw-Hill, New York, 1979, p. 28.
3. S. J. Kline, W. C. Reynolds, F. A. Schraub, and P. W. Runstadler, "The Structure of Turbulent Boundary Layers," J. Fluid Mech., **30**, 741–773 (1967).
4. J. T. Davies, "Eddy Transfer Near Solid Surfaces," in Turbulence Phenomena, Academic, New York, 1972, pp. 121–143.
5. M. T. Landahl and E. Mollo-Cristensen, Turbulence and Random Processes in Fluid Mechanics, Cambridge University Press, Cambridge, 1987, pp. 111–120.
6. A. K. Praturi and R. S. Brodkey, "A Stereoscopic Visual Study of Coherent Structures in Turbulent Shear Flow," J. Fluid Mech., **89**(Pt 2), 251–272 (1978).
7. G. R. Often and S. J. Kline, "A Proposed Model of the Bursting Process in Turbulent Boundary Layers," J. Fluid Mech., **70**, 209–228 (1975).
8. S. K. Robinson, "Coherent Motions in the Turbulent Boundary Layer," Annu. Rev. Fluid Mech., **23**, 601–639 (1991).
9. A. S. Thomas and M. K. Bull, "On the Role of Wall Pressure Fluctuations in Deterministic Motions in the Turbulent Boundary Layer," J. Fluid Mech., **128**, 283–322 (1983).
10. N. Ibl and O. Dossenbach, "Convective Mass Transport," in Comprehensive Treatise of Electrochemistry, Vol. 6, Electrodics: Transport, E. Yeager, J. O'M. Bockris, B. E. Conway, and S. Sarangapani (Eds.), Plenum, New York, 1983, pp. 133–237.
11. A. H. Lee, J. Y. Sun, and W. P. Jepson, "Study of Flow Regime Transitions of Oil-Water-Gas Mixtures in Horizontal Pipelines," 3rd International Conference ISOPE, Vol. II, 1993, pp. 159–164.
12. A. Maholtra,"A Study of Oil/Water Flow Characteristics in Horizontal Pipes," MS Thesis, Ohio University, Athens, OH, Oct., 1996.
13. X. Zhou and W. P. Jepson, "Corrosion in Three-Phase Oil/Water/Gas Slug Flow in Horizontal Pipes," Paper No.

94026, CORROSION/94, NACE International, New Orleans, LA, Mar., 1994.

14. C. Kang, R. Wilkins, and W. P. Jepson, "The Effect of Slug Frequency on Corrosion in High Pressure Inclined Pipelines," Paper No. 96020, CORROSION/96, NACE International, Denver, CO, Mar. 1996.

15. K. D. Efird, "Disturbed Flow and Flow Accelerated Corrosion in Oil and Gas Production," Proceedings: ASME Energy Resources Technology Conference, Houston, TX, Feb. 1998.

16. J. Postlethwaite, S. Nesic, G. Adamopoulos, and D. J. Bergstrom, "Predictive Modeling for Erosion Corrosion under Disturbed Flow Conditions," Corros. Sci., **35**, 627–633 (1993).

17. S. Nesic and J. Postlethwaite, "Relationship between the Structure of Disturbed Flow and Erosion Corrosion," Corrosion, **46**, 874–880 (1990).

18. G. Schmitt and T. Gudde, "Local Mass Transport Coefficients and Wall Shear Stresses at Flow Disturbances," Paper No. 95102, CORROSION/95, Orlando, FL, Mar. 1995.

18

EROSION–CORROSION IN SINGLE- AND MULTIPHASE FLOW

J. Postlethwaite*

Department of Chemical Engineering, University of Saskatchewan, Saskatoon, Saskatchewan, Canada

S. Nešić

Institute for Corrosion and Multiphase Flow Technology, Ohio University, Athens, Ohio

A. INTRODUCTION

Erosion–corrosion encompasses a wide range of flow-induced corrosion processes. Flowing fluids can damage protective films on metals resulting in greatly accelerated corrosion. Damage to the films may be the result of mechanical forces or flow-enhanced dissolution and the accelerated corrosion may be accompanied by erosion of the underlying metal. This conjoint action of erosion and corrosion is known as erosion–corrosion [1]. Impingement attack by liquid droplets and solid particles and cavitation attack are also included here under this broad definition of erosion–corrosion. In practice, the relative contributions of *accelerated corrosion* and erosion to the total metal loss vary widely with the type of erosion–corrosion and the hydrodynamic intensity of the flow.

B. FLOW CONDITIONS

Erosion–corrosion normally occurs under *turbulent-flow* conditions. The flowing fluid may be single phase (Fig. 18.1) as in the erosion–corrosion of copper tubing by potable water (Fig. 18.2). Multiphase flows [2–4] (Fig. 18.3) with various combinations of gas, water, oil, and sand can cause severe erosion–corrosion of oil/gas production systems [5–7] as shown in Figures 18.4–18.6.

The most severe erosion–corrosion problems occur under conditions of *disturbed* turbulent flow [8, 9] at sudden changes in the flow system geometry, such as bends, heat exchanger tube inlets, orifice plates, valves, fittings, and turbomachinery, including pumps, compressors, turbines, and propellers. Surface defects in the form of small protrusions or depressions such as corrosion pits, deposits, and weld beads can give rise to disturbed flow on a smaller scale but sufficient to initiate erosion–corrosion [10, 11]. The presence of suspended solid particles, gas bubbles, or vapor bubbles in aqueous flow and liquid droplets in high-speed gas flow can be especially damaging.

* Retired.

Uhlig's Corrosion Handbook, Third Edition, Edited by R. Winston Revie
Copyright © 2011 John Wiley & Sons, Inc.

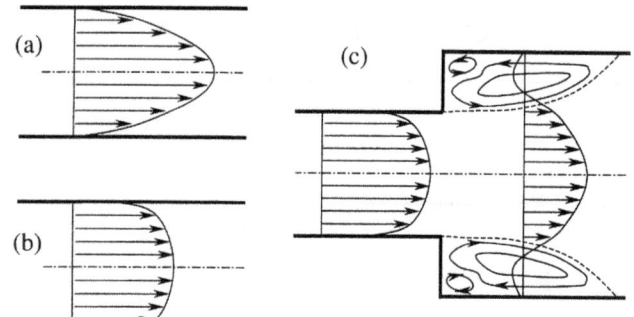

FIGURE 18.1. Single-phase pipe flow. (a) Developed laminar flow, showing parabolic velocity profile. (b) Developed turbulent flow, showing logarithmic velocity profile with large gradients near the wall (nondisturbed flow). (c) Disturbed turbulent flow with separation, recirculation, and reattachment, showing complex velocity field. (Adapted from Heitz [18].)

C. PROTECTIVE FILMS

Most metals and alloys used in industry owe their corrosion resistance to the formation and retention of a protective film. Protective films fall into two categories:

- The relatively thick porous *diffusion barriers* such as those formed on carbon steel (red rust) and copper (cuprous oxide)
- The thin invisible *passive films* such as those on stainless steels, nickel alloys, and other passive metals such as titanium

Diffusion barriers are typically formed by anodic dissolution followed by precipitation whereas passive films are formed by direct oxidation without the metal ions entering the solution. The softer thicker diffusion barriers are more easily damaged and more slowly repaired than the passive films, and passive alloys can withstand much more severe service conditions.

D. EROSION–CORROSION RATE

The effect of film destruction on the corrosion rate is illustrated by the following example. Carbon steel pipe carrying water is usually protected by a film of rust, which slows down the rate of mass transfer of dissolved oxygen to the pipe wall. The resulting corrosion rates are typically <1 mm/a. The removal of the film by a flowing sand slurry gives corrosion rates of the order of 10 mm/a in addition to any erosion of the underlying metal [12]. When corrosion is controlled by the rate of dissolved oxygen mass transfer, the corrosion rate can be calculated by the application of well-established mass transfer correlations of dimensionless groups. In general,

$$\text{Sh} = \alpha \, \text{Re}^{\beta} \, \text{Sc}^{\gamma}$$

where

$\text{Re} = l u_b \rho / \mu$, Reynolds number, ratio of inertial forces to viscous forces

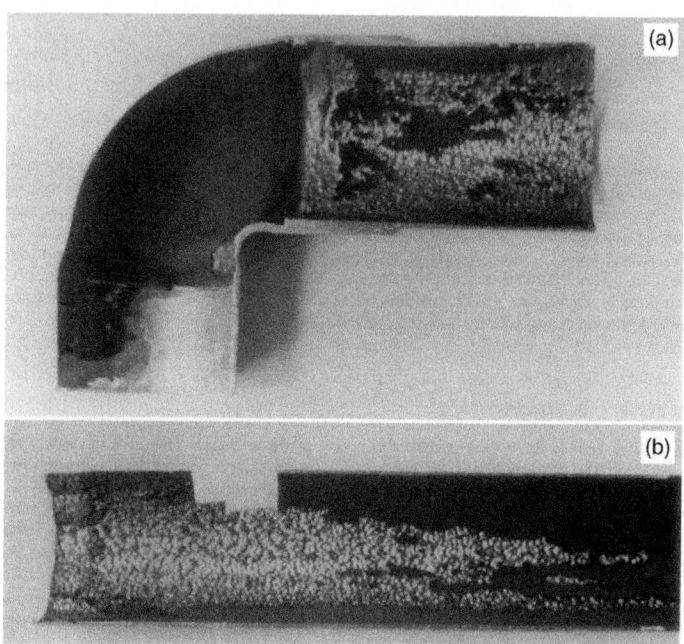

FIGURE 18.2. Erosion–corrosion of 53-mm copper tubing by potable water. (a) The attack started at the step whore the tubing fitted into the elbow. (b) Once started, the attack progressed along the tube as a result of additional disturbed flow created by the erosion–corrosion roughened surface.

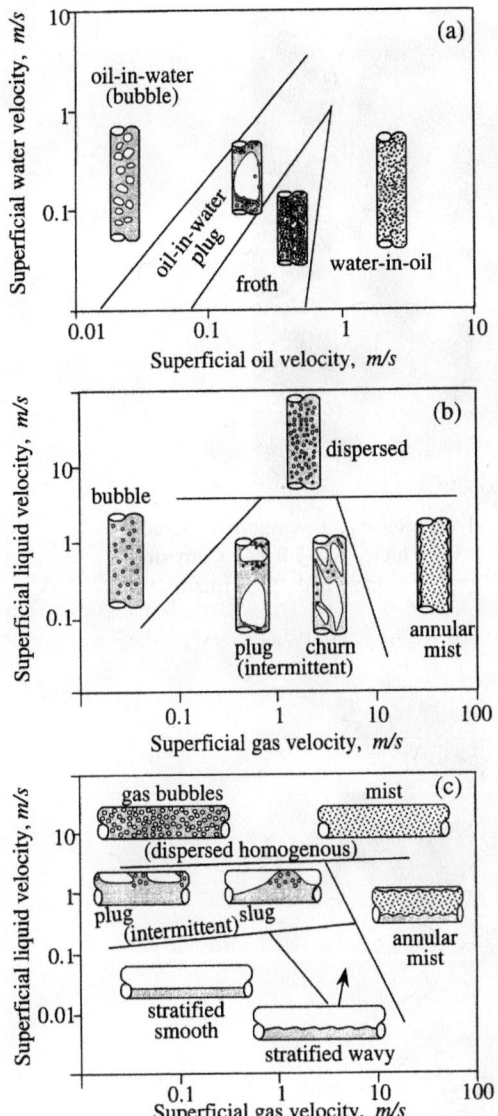

FIGURE 18.3. Qualitative two-phase flow structures in pipes: (a) oil–water in vertical pipe; (b) liquid–gas in vertical pipe; (c) liquid–gas in horizontal pipe. (Adapted from Lotz [2], Weisman [3], and Govier and Aziz [4].)

Sc = $\mu/\rho D$, Schmidt number, ratio of momentum diffusivity to mass diffusivity

Sh = kl/D, Sherwood number, ratio of convective mass transport to diffusive transport

k = mass transfer coefficient, m/s

l = characteristic dimension (e.g. pipe diameter), m

D = diffusion coefficient, m^2/s

u_b = bulk flow velocity, m/s

μ = viscosity, Pa·s

ρ = density, kg/m^3

α, β, γ = experimental constants

The Berger–Hau [13] correlation, one of the most widely accepted mass transfer correlations for fully developed turbulent flow in smooth pipes, gives $\alpha = 0.0165$, $\beta = 0.86$, and $\gamma = 0.33$. Based on the above correlation, a corrosion rate (CR in mm/a) calculated using the rate of mass transfer of dissolved oxygen to a film-free carbon steel pipe wall is given by

$$CR = 4923 C_b \left(\frac{D_{O_2}}{d} \right) Re^{0.86} Sc^{0.33}$$

where

C_b = bulk oxygen concentration, mol/m^3

D_{O_2} = diffusion coefficient for oxygen in water, m^2/s

d = pipe diameter, m

This calculation assumes that two-thirds of the oxygen reaching the wall is used in oxidizing the iron to ferrous ions and that one-third is used in the oxidation of the ferrous ions to ferric ions close to the wall [12]. On this basis, a 100-mm-diameter pipe carrying an aerated aqueous solution with $C_b = 8$ ppm $\equiv 0.25$ mol/m^3 of dissolved oxygen, $D_{O_2} = 2 \times 10^{-9}$ m^2/s flowing at 2 m/s would corrode at 7 mm/a.

The above calculation demonstrates the aggressive nature of erosion–corrosion and gives an indication of the high and unacceptable rates of corrosion whenever a protective film is damaged by erosion. Uniform erosion–corrosion and fully developed mass transfer is assumed. In practice erosion–corrosion often occurs under disturbed flow conditions at localized areas. Consider, for example, iron, where the rust film formed at ambient temperatures is a poor cathode for oxygen reduction. The rate of oxygen mass transfer (and corrosion) at the small surface area where the protective film is removed will be substantially enhanced by:

- The disturbed flow effect [9, 14, 15]
- The mass transfer entrance length effect [16]
- The surface roughness effect [17] shown in Figure 18.7

In addition, the situation may be complicated by the formation of flow-induced macroscopic corrosion cells following localized film disruption with differing behavior observed for copper and iron alloys [18].

E. RELATIVE ROLES OF EROSION AND CORROSION

The *accelerated corrosion*, following damage to the protective film, may be accompanied by *erosion* of the underlying metal. In some cases the erosion of the base metal is not a factor. In other cases erosion is the predominant factor and a wide spectrum of behaviors is observed [19]. The relative roles of corrosion and erosion following damage to the

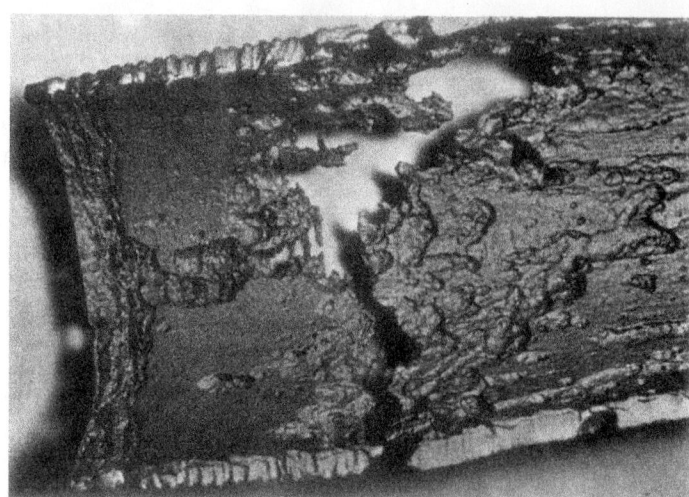

FIGURE 18.4. Erosion–corrosion of 115-mm API L-80 oil well tubing. Environment: Crude oil/CH_4/CO_2 and 1% H_2O in tight emulsion; temperature 200°C; velocity 6.4–7.9 m/s. Corrosion rate >10 mm/y. Remedy: Continuous inhibitor injection. (From Houghton and Westermark [6]. Reprinted by permission of NACE International.)

FIGURE 18.5. Impingement-type erosion–corrosion of AISI 4140 (UNS 941400) 115-mm flow coupling and subsurface safety valve in natural gas condensate production. Minor species: CO_2 and H_2O. Temperature: 79°C. Exit velocity from valve 91 m/s. Remedy: replacement with 13 Cr martensitic ss. (From Houghton and Westermark [6]. Reprinted by permission of NACE International.)

protective film can be estimated (Table 18.1) on the basis of the exponent y in the relationship between the metal loss rate and the bulk flow velocity, u_b [2]:

$$\text{erosion} + \text{corrosion} \propto u_b^y$$

The value of y will depend on the relative contributions of corrosion and erosion to the total metal loss. In disturbed flow it is the flow characteristics in the direct vicinity of the wall rather than the bulk flow velocity that are important. In practice, however, the superficial flow velocity is the flow parameter that is readily measured and controlled [2].

F. EROSION–CORROSION MECHANISMS

The sources of the various *mechanical* forces involved in the erosion of protective films and/or the underlying metal at flow system walls (Fig. 18.8) are:

- Turbulent flow, fluctuating shear stresses, and pressure impacts
- Impact of suspended solid particles
- Impact of suspended liquid droplets in high-speed gas flow

FIGURE 18.6. Erosion of tungsten carbide choke beans inside a steel holder from an oil well with sand production. Note the highly polished, streamlined appearance of the erosion pattern. Subcritical flow occurred in this well and these assemblies lasted between 6 and 9 months. (From Smart [7]. Reprinted by permission of NACE International.)

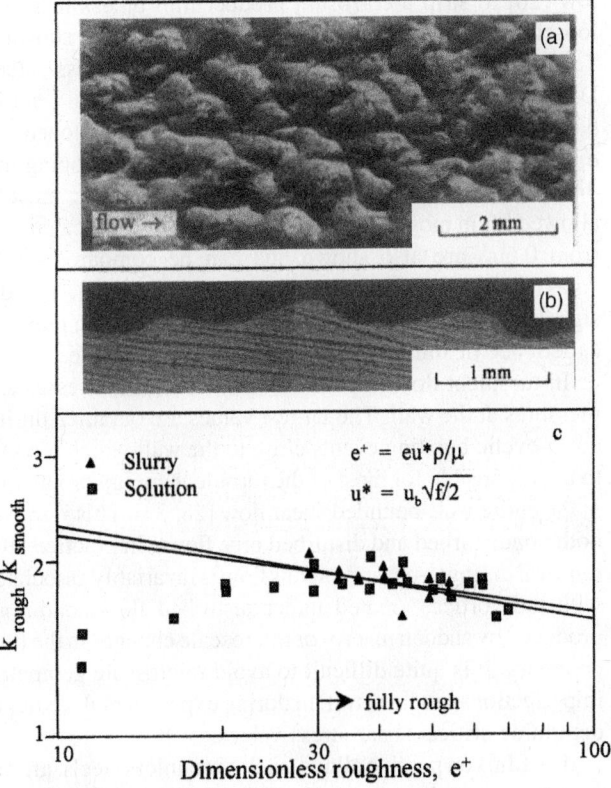

FIGURE 18.7. Effect of pipe wall roughness on mass transfer rates. (a) Rough surface, preroughened by erosion–corrosion prior to mass transfer study.(b) Roughness profile details: roughness height e ≈ 0.2 mm, pitch/height in the range 5–10. (c) The ratio of measured mass transfer coefficients for preroughened and smooth surfaces at pipe wall. (Adapted from Postlethwaite and Lotz [17].)

TABLE 18.1. Flow Velocity as Diagnostic Tool for Erosion–Corrosion Rates Following Damage to Protective Film

Mechanism of Metal Loss	Velocity Exponent, y
Corrosion	
Liquid-phase mass transfer control	0.8–1
Charge transfer (activation) control	0
Mixed (charge/mass transfer) control	0–1
Activation/repassivation (passive films)	1
Erosion	
Solid-particle impingement	2–3
Liquid-droplet impingement in high-speed gas flow	5–8
Cavitation attack	5–8

- Impact of suspended gas bubbles in aqueous flow
- Violent collapse of vapor bubbles following cavitation

Flow-enhanced film dissolution and thinning are "chemical" forms of protective film erosion leading to accelerated corrosion of the underlying metal.

F1. Turbulent Flow

Erosion–corrosion occurs in single-phase turbulent pipe flow. The exact mechanism of protective film damage is still in doubt. There is uncertainty regarding the roles of mechanical forces and mass transfer in film disruption [20–22] since both are directly related to the turbulence intensity.

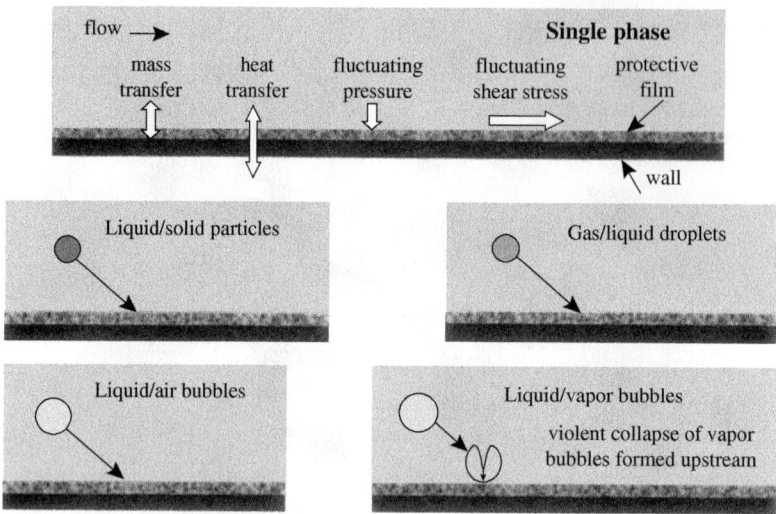

FIGURE 18.8. Interaction of flowing fluid with wall of the flow system leading to erosion–corrosion. (Adapted from Heitz [18].)

Distinct "breakaway" velocities above which damage to normally protective films occurs in copper alloy tubes are observed [20] and this gave rise to the concept of a *critical shear stress* for film disruption.

The wall shear stress τ_w for flow in pipes is given by

$$\tau_w = f(\tfrac{1}{2}\rho \, u_b^2)$$

where f is the Fanning friction factor [23]. In general [24]

$$f = f\left(\text{Re}, \frac{\varepsilon}{d}, \frac{\varepsilon'}{d}, m\right)$$

where ε is a measure of the size of the roughness projections, ε' is a measure of the arrangement of the roughness elements, and m is a form factor that is dependent on the shape of roughness elements.

For flow in smooth pipes

$$f = 0.046 \, \text{Re}^{-0.2}$$

is valid to higher Reynolds numbers than the Blasius equation [25]. Values of the Fanning friction factor for both smooth and rough commercial pipes and tubes can be obtained from a Moody* chart [23]. For fully rough† pipe flow, f is independent of Re, and $\tau_w \propto u_b^2$.

Early attempts [26] to correlate film damage with the wall shear stresses in pipe flow based on the bulk flow velocity resulted in shear stress values (Table 18.2) that seem too low [20] to strip a corrosion product film unless it is very loosely adhering. The "breakaway" velocities u_{cr} shown in Table 18.2 were calculated from critical shear stress values. Design velocities for heat exchanger tubing should not be based on critical shear stress values for fully developed *non disturbed* tube flow since the flow is both developing and disturbed at the tube inlets. In fact, disturbed flow must be allowed for in most industrial systems. The values calculated from $0.5u_{cr}$ are also shown and can be compared to the "maximum recommended design velocities" [27]. The design velocity must clearly be chosen with care with previous experience in similar systems being the best guide.

In turbulent flow there are fluctuating shear stresses and pressures at the wall. The largest values are obtained during quasi-cyclic bursting events close to the wall which are said to be responsible for most of the turbulent energy production in the entire wall-bounded shear flow [28, 29]. This is true of both nondisturbed and disturbed pipe flows. In practice, film removal in single-phase aqueous flow is invariably associated with the vortices created under disturbed flow conditions produced by sudden macro- or microscale changes in the flow geometry. It is quite difficult to avoid microscale geometric imperfections in a flow system during experimental studies to determine critical shear stress values.

The films on passive alloys, such as stainless steels, are not usually damaged by single-phase aqueous flow [18, 30]. Rust films on carbon steel at ambient temperature are more mechanically stable than the protective films formed on copper. The compactness and protectiveness of rust films (FeOOH) increase with velocity [31]. The magnetite films formed in high-temperature water and sulfide films formed

---

* The Fanning friction factor should not be confused with the Darcy friction factor, which is four times greater and is used on some Moody charts.
† In flow through rough pipes, the roughness elements penetrate the boundary layer into the main fluid stream at high Re resulting in form drag, and, under fully rough conditions, the viscous forces become negligible.

TABLE 18.2.   Critical Flow Parameters for Copper Alloy Tubing in Seawater

Alloy	Critical Shear Stress[a] (N/m^2)	Critical Velocity 25-mm tube[b] (m/s)	"Design" Velocity[c] Based on 50% $\tau_{w,crit}$ (m/s)	Accepted Maximum Design Velocity[d] (m/s)
Cupro nickel with Cr	297	12.6	8.6	9
70–30 Cupro nickel	48	4.6	3.1	4.5–4.6
90–10 Cupro nickel	43	4.3	2.9	3–3.6
Aluminum bronzes	—	—	—	2.7
Arsenical Al brass	19	2.7	1.9	2.4
Inhib. Admiralty	—	—	—	1.2–1.8
Low Si bronze	—	—	—	0.9
P deoxidized copper	9.6	1.9	1.3	0.6–0.9

[a]From Efird [26], rectangular duct.
[b]Critical velocities, calculated using Efird shear stress values.
[c]Calculated values based on 50% of critical shear stress.
[d]"Accepted" maximum tubular design velocities [27].

on carbon steel have a high mechanical stability in single-phase aqueous flow [18]. Thus the velocity limit, around 1 m/s, suggested for copper tubing [32, 33] is lower than the velocities tolerated by carbon steel and low-alloy piping at ambient temperatures and much lower than the allowable velocities at elevated temperatures.

In *slug flow* [Fig. 18.3(c)] large but short-lived shear stress fluctuations are observed [2]. The slugs of liquid travel at much higher velocities than is encountered in single-phase liquid flows and have been suspected to be the cause of severe erosion–corrosion problems in oil/gas production systems.

F2. Solid-Particle Impingement

Solid-particle impacts can damage both types of protective films (thick diffusion barriers and thin passive films) leading to erosion–corrosion. The particles may also erode the underlying metal, adding to the overall metal loss.

As might be expected, the erosion rate ER is a function of the kinetic energy of the particles (proportional to u_p^2) and the frequency of impacts (proportional to u_p), and to a first approximation [2],

$$ER \propto u_p^3$$

The kinetic energy of the particle normal to the wall will be determined by the *impact velocity* and *impact angle* along with the particle *density* and *size*. The impacts of small particles in aqueous slurries may be damped by the presence of a thick boundary layer at the wall [36]. In practice, hard abrasive slurries are often transported through mild-steel pipes with the particles finely ground. The erosion is reduced because of the smaller particle size and the lower flow velocity required to maintain the solids in suspension [37].

Both the impact velocity and the impact angle will be strongly affected by disturbed flow and this is where the most severe erosion is found in practice. Impact angles in non-disturbed turbulent pipe flow are <5° whereas a wide range of angles will be encountered in disturbed flow [38]. Thus carbon–steel pipe can transport sand slurries with an *erosion rate* of < 1 mm/a under nondisturbed flow conditions [12, 36] (Fig. 18.9). Similar sand slurry flows caused the erosive failure of a reducer located immediately downstream of a T-junction and pinch valve in eight days, illustrating [Fig. 18.9(c)] the very severe erosion that can occur under disturbed flow conditions [36].

In dilute slurries <5 vol % solids, the impact frequency and erosion rate are proportional to the *solids concentration* [8]. At higher concentrations particle–particle interference reduces this dependency [37, 39].

The *shape* of the particle [34] and the *microroughness* of the particle surface [40] influence the effective forces generated at the wall by the impact. Thus rounded sand particles, which were apparently smooth, resulted in erosion rates [39] two orders of magnitude greater than the erosion by glass beads in the same system. The difference in behavior was related to the micro-roughness of the sand particles, which could be seen at high magnification under an electron microscope.

The damage done will be strongly related to the *relative hardness* (Fig. 18.10) of the impinging particles and the flow system walls. Erosion drops dramatically [35, 37, 41] when the impacting particles are softer than the wall. For example, the Mohs hardness of coal is ~2, iron ore 6, and sand 7. Carbon steel pipe API 5LX or similar used for long-distance slurry transportation [42] has a Brinell hardness of ~120 (Mohs hardness ~3). Thus it might be expected that whereas coal slurries could possibly damage rust films and cause accelerated corrosion, there would be no erosion of the underlying metal. Sand, iron ore, and other hard minerals

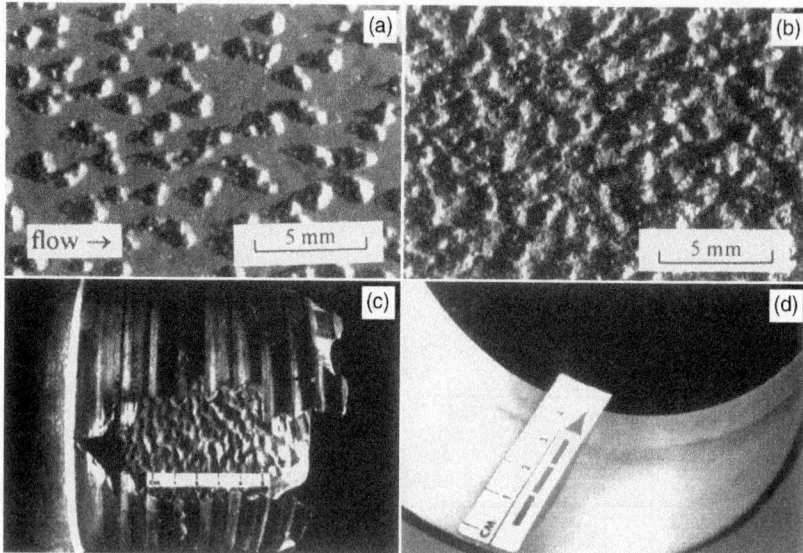

FIGURE 18.9. Variations in erosion–corrosion wear patterns in a 100-mm-diameter pipe carrying an abrasive sand slurry. (a) Developing erosion–corrosion wear pattern in nondisturbed flow showing individual pits. (b) Fully developed erosion–corrosion wear pattern with roughened surface. (c) Effect of disturbed flow, with failure of reducer placed downstream of T-junction and pinch valve in 8 days. (d) Erosion pattern with the corrosion component suppressed by application of cathodic protection. *Note:* Cathodic protection not suitable for the inside of pipelines. (From Postlethwaite et al. [36]. Reprinted by permission of NACE International.)

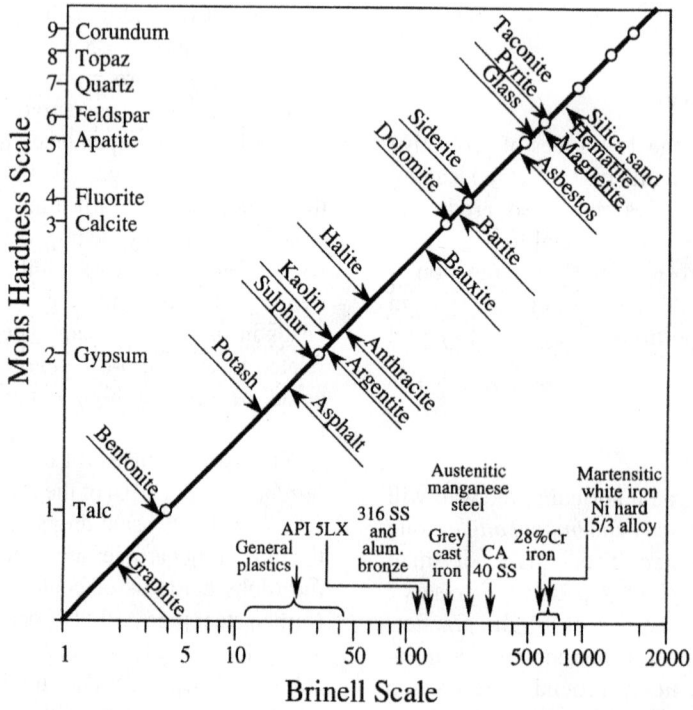

FIGURE 18.10. Approximate comparison of hardness values of various common minerals and metals. (Adapted from Wilson [55].)

could cause both accelerated corrosion and erosion. A complicating factor sometimes found with coal and other slurries is the presence of small amounts of silica and other hard minerals.

The ability of impacting particles to damage or interfere with the development of protective films is difficult to quantify in the absence of information regarding the mechanical properties of the in situ films [18] and in the case of passive films the damage and healing characteristics. In general, impacting particles will often damage protective films even under circumstances where they cannot erode the base metal. Lotz [2] has proposed that solid-particle damage to passive films with subsequent rapid repassivation will result in a corrosion rate directly proportional to the particle concentration and flow velocity. Any erosion of the underlying metal will increase the exponent on the velocity. Metal loss measurements [43] with dilute slurries in fully developed turbulent pipe flow resulted in values of the velocity exponent up to 3 for corrosion-resistant materials and values down to 1.4 for less corrosion resistant materials. Measurements [12] with sand slurries in carbon steel pipes gave an exponent close to 1 with erosion playing a minor role at the velocities involved. These two sets of results indicate the utility of using the velocity as a diagnostic tool to determine the relative roles of accelerated corrosion and erosion of the underlying metal discussed above.

Sand particles can play a major role in the erosion–corrosion of oil/gas production systems [5]. At low velocities accelerated corrosion is the major factor, whereas at the high velocities found in chokes erosion can lead to the rapid destruction of the component [7], as shown in Figure 18.6.

F3. Liquid-Droplet Impingement

Erosion by impinging liquid droplets carried in high-speed vapor or gas flows is usually called liquid impingement attack. The attack involves the exposure of the solid to repeated discrete impacts by liquid droplets which generate impulsive and destructive contact pressures far higher than those produced by steady flows [44]. Impingement erosion has been a problem with low-pressure steam turbine blades operating with wet steam and rain erosion of aircraft and helicopter rotors. Liquid-droplet impingement attack has

much in common with cavitation attack [45] in terms of both the mechanism, which is largely one of deformation erosion, and the morphology of the attack, with continued deepening of discrete sharp-edged pits, which coalesce to a honeycomb-like appearance [46] (Fig. 18.11). Corrosion, as with cavitation attack, usually plays a secondary role, which varies with the hydrodynamic intensity.

F4. Air Bubble Impingement

The erosion–corrosion of copper alloy heat exchanger tube inlets has been linked [10] to the local damage of protective films by impinging air bubbles which have a substantial radial velocity component immediately downstream from the tubesheet inlets. Nitrogen bubbles [20] have a similar effect, and it seems clear that the damage to the film is hydromechanical. The attack, which initially takes the form of isolated "horse-shoes," spreads to a general roughening characteristic of erosion–corrosion. Similar horseshoe patterns were produced by small-scale turbulence at grain boundaries on dissolving solid benzoic acid [10] illustrating the role of hydrodynamics in the erosion–corrosion pattern produced on copper tubing.

F5. Cavitation

Cavitation attack takes the form of steep-sided pits, as shown in Figure 18.12 [33], which sometimes coalesce to form a honeycomb-like structure [45]. The attack is the result of the violent collapse of *vapor cavities*, or bubbles, <1 mm diameter [47, 48] that are formed in a flowing liquid when the hydrostatic pressure drops below the vapor pressure of the fluid. The cavities are carried downstream to higher pressure regions where collapse occurs. The *dynamic forces* associated with the collapse of the cavities at the walls of the system lead to high-frequency fatigue-stress damage. The shock waves from spherical collapse and microjets from asymmetrical collapse have both been suggested as the cause of the damage. The inconsistencies with previous theories are now considered to be explained by the concept of the collapse of clusters of cavities in concert [47]. Cavitation attack often involves an induction period, especially for ductile materials, following which the erosion rate is nonlinear [49]. The time

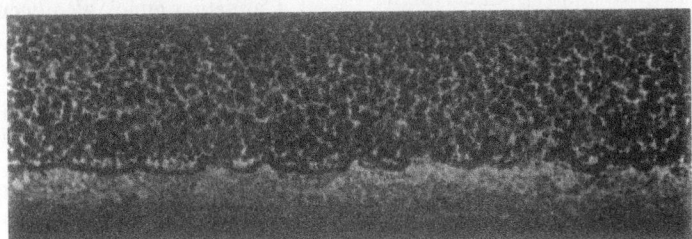

FIGURE 18.11. Impingement attack of outside of cupro nickel tube by wet steam. (From Century Brass [46].)

FIGURE 18.12. Gavitation attack of a stainless steel (AISI 316) pump impeller. Environment: skimmed milk at 70°C. Remedy: Replace impeller by one made of Teflon PFA (perfluoroalkoxy). (From [33]. Reprinted from E. D. D. During, Corrosion Atlas: A Collection of Illustrated Case Histories, Vol. 2: Stainless Steels and Nonferrous Materials, Copyright © 1988, with permission from Elsevier Science.)

dependency is similar to that observed in the related phenomenon of liquid-droplet impingement [44].

The cavitation resistance of metals and alloys spans a range of at least two orders of magnitude [44, 48], far greater than the conventional strength properties. The velocity exponent varies widely with values usually given in the range 5–8 [48]. With such a high-velocity exponent, a small change in operating conditions could lead to serious problems. The effects of velocity should not be considered independently of pressure, since the cavitation number σ, which is a measure of the intensity of cavitation, is a function of both parameters [49]:

$$\sigma = \frac{p - p_v}{0.5 \, \rho \, u^2}$$

where p and p_v are the absolute static pressure and the liquid vapor pressure, respectively, and u the free-stream velocity. When $\sigma = 0$, the pressure is reduced to the vapor pressure and cavitation will occur.

The role of corrosion, which varies with the hydrodynamic intensity, is difficult to predict [49]. Under some circumstances it is claimed that hydrogen bubbles liberated by the application of cathodic protection provide a mechanical cushion reducing the cavitation damage [45] in addition to the normal cathodic protection effect. In other cases corrosion may contribute to the overall attack directly, following damage to protective films. In the latter case, the velocity exponent for a given system might be expected to be reduced as the corrosion component increases, as is the case with erosion–corrosion by solid particles discussed above.

Cavitation is a problem with ship propellers, hydraulic pumps and turbines, valves, orifice plates, and all places where the static pressure varies very abruptly on the basis of the Bernoulli[*] principle:

$$gz + \frac{u^2}{2} + \frac{p}{\rho} = \text{const}$$

where z is the elevation. Strictly speaking, the Bernoulli equation applies to flow along a streamline[†]; however, in turbulent disturbed flow it can be used qualitatively. Thus the increase in the velocity as the liquid is accelerated through an orifice plate or over the curved impeller of a centrifugal pump results in a fall in the static pressure. As the liquid slows down, after it passes the vena contracta or approaches the volute in the pump, the pressure rises again, leading to the collapse of the cavities described above. Friction losses and the resulting drop in pressure in pump suction lines can lead to serious cavitation damage at the pump inlet. Objectionable noise, vibration, and loss of equipment performance often accompany cavitation attack.

Cavitation attack also occurs at vibrating solid surfaces such as the cooling-water side of diesel engine cylinder liners [49] and on the shell side of heat exchanger tubing [50].

---

[*] The engineering form of the Bernoulli equation includes terms for shaft work and friction losses [23].

[†] The use of the term streamline in turbulent flow is confusing to some who might think the term applies only to laminar (viscous) flow. A streamline is a continuous line drawn through a fluid so that it has the direction of the velocity vector at every point. There can be no flow across a streamline. When very small solid particles are used to study the flow pattern, by the use of laser Doppler anemometry, the particles follow the streamlines [24].

FIGURE 18.13. Wire drawing in a bronze cam ring for a hydraulic pump. (From Glaeser [51]. Reprinted, by permission of NACE International.)

F6. Wire Drawing

Wire drawing is a form of erosion–corrosion found in very high velocity fluid flow through small gaps. This type of attack, which is difficult to classify, takes the form of single or multiple clean smooth grooves, as shown in Figure 18.13 [51]. The attack could involve liquid-droplet impingement attack when it occurs at valve seats with escaping steam and cavitation attack when it occurs with a single-phase fluid system [44]. The fluid may be aqueous or nonaqueous.

F7. Flow-Enhanced Film Dissolution

Flow-enhanced film dissolution (with thinning) is a "chemical" form of protective film erosion leading to accelerated corrosion. This type of erosion–corrosion has caused some serious problems in carbon steel lines (protected by magnetite films) carrying hot (100–250°C) deoxygenated water or wet steam in fossil-fueled and nuclear power plants.* The problem occurs at locations with high mass transfer rates, corresponding to highly turbulent flow conditions found at bends and downstream of pumps, valves, and orifice plates. At these locations the film is dissolved more rapidly, and a higher anodic current is needed to replenish the protective film. The resulting steady state involves a thinner and less protective film and a much higher corrosion rate, leading to rapid localized thinning of the equipment walls.

---

*The term flow-accelerated corrosion (FAC) has been used to describe this specific problem [54]. However, the term erosion–corrosion is normally used in the literature and is the preferred term. The term FAC could relate to a whole range of flow-dependent corrosion processes in addition to its application to flow-enhanced film thinning.

At high velocities in wet steam, convective dissolution of magnetite films is said to be supplemented by oxide fatigue by liquid-droplet impacts. At the outside of bends, liquid droplets under the influence of secondary flow and centrifugal acceleration impinge on the magnetite, fatiguing it and causing it to erode away, exposing new metal to the steam. On the inside of the bend, where an inward flowing stagnation point occurs with a very high mass transfer coefficient, convective dissolution is the major mode of oxide damage [52]. Giralt and Trass [53] have suggested that the contribution of dissolution can be additive to film erosion and sometimes acts in a synergistic way by loosening the crystal grains of the film, making them more prone to erosion.

The effect of velocity on flow-enhanced film dissolution is complex and, although much has been learned with respect to the identification and control of this problem, much more work is required on the fundamentals of the processes involved [54].

REFERENCES

1. American Society for Testing and Materials (ASTM), "Standard Terminology Relating to Corrosion and Corrosion Testing," in Annual Book of ASTM Standards, Vol. 03.02, Wear and Erosion: Metal Corrosion, ASTM G 15-97, ASTM, West Conshohocken, PA, 1997, pp. 65–68.
2. U. Lotz, "Velocity Effects in Flow Induced Corrosion," in Proceedings of Symposium on Flow-Induced Corrosion; Fundamental Studies and Industry Experience, K. H. Kennelley, R. H. Hausler, and D. C. Silverman (Eds.), NACE, Houston, TX, 1991, pp. 8:1–8:22.
3. J. Weisman, "Two-Phase Flow Patterns," in Handbook of Fluids in Motion, N. P. Cheremisinoff and R. Gupta (Eds.), Ann Arbor Science Publishers, Ann Arbor, MI, 1983, pp. 409–424.

4. G. W. Govier and K. Aziz, The Flow of Complex Mixtures in Pipes, Van Nostrand Reinhold Company, New York, 1972, pp. 324, 326, 506.

5. American Petroleum Institute (API), Corrosion of Oil and Gas-Well Equipment, Book 2 of the Vocational Training Series, 2nd ed., API, Washington, DC, 1990, p. 8.

6. C. J. Houghton and R. V. Westermark, "Downhole Corrosion Mitigation in Ekofisk (North Sea) Field," Mater. Perform. 22(1), 16–22 (1983).

7. J. S. Smart III, "A Review of Erosion Corrosion in Oil and Gas Production," in Proceedings of Symposium on Flow-Induced Corrosion; Fundamental Studies and Industry Experience, K. H. Kennelley, R. H. Hausler, and D. C. Silverman (Eds.), NACE, Houston, TX, 1991, pp. 18:1–18:18.

8. W. Blatt, T. Kohley, U. Lotz, and E. Heitz, "The Influence of Hydrodynamics on Erosion-Corrosion in Two-Phase Liquid-Particle Flow," Corrosion, 45, 793–804 (1989).

9. U. Lotz and J. Postlethwaite, "Erosion-Corrosion in Disturbed Two Phase Liquid/Particle Flow," Corros. Sci., 30, 95–106 (1990).

10. G. Bianchi, G. Fiori, P. Longhi, and F. Mazza, "Horse Shoe Corrosion of Copper Alloys in Flowing Sea Water: Mechanism, and Possibility of Cathodic Protection of Condenser Tubes in Power Stations," Corrosion, 34, 396–406 (1978).

11. G. Schmitt, W. Bücken, and R. Fanebust, "Modeling Micro-Turbulences at Surface Imperfections as Related to Flow-Induced Localized Corrosion and Its Prevention," Paper No. 465, Corrosion'91, NACE, Houston TX, 1991.

12. J. Postlethwaite, M. H. Dobbin, and K. Bergevin, "The Role of Oxygen Mass Transfer in the Erosion-Corrosion of Slurry Pipelines," Corrosion, 42, 514–521 (1986).

13. F. P. Berger and K.-F., F.-L. Hau, "Mass Transfer in Turbulent Flow Measured by the Electrochemical Method," Int. J. Heat Mass Transfer, 20, 1185–1194 (1977).

14. T. Sydberger and U. Lotz, "Relation between Mass Transfer and Corrosion in a Turbulent Pipe Flow," J. Electrochem. Soc., 129, 276–283 (1982).

15. S. Nesic, G. Adamopoulos, J. Postlethwaite, and D. J. Bergstrom, "Modelling of Turbulent Flow and Mass Transfer with Wall Function and Low-Reynolds Number Closures," Can. J. Chem. Eng., 71, 28–34 (1993).

16. Y. Wang and J. Postlethwaite, "The Application of Low Reynolds Number k–ε Turbulence Model to Corrosion Modelling in the Mass Transfer Entrance Region," Corros. Sci., 39, 1265–1283 (1997).

17. J. Postlethwaite and U. Lotz, "Mass Transfer at Erosion-Corrosion Roughened Surfaces," Can. J. Chem. Eng., 66, 75–78 (1988).

18. E. Heitz, "Chemo-Mechanical Effects of Flow on Corrosion," in Proceedings of Symposium on Flow-Induced Corrosion; Fundamental Studies and Industry Experience, K. H. Kennelley, R. H. Hausler, and D. C. Silverman (Eds.), NACE, Houston, TX, 1991, pp. 1:1–1:29.

19. B. S. Poulson, "Erosion Corrosion," in Corrosion, 3rd ed., L. L. Shreir, R. A. Jarman, and G. T. Burstein (Eds.), Butterworth-Heinemann, Oxford, England, 1994, pp. 1:293–1:303.

20. B. C. Syrett, "Erosion-Corrosion of Copper-Nickel Alloys in Sea Water and Other Aqueous Environments—A Literature Review," Corrosion, 32, 242–252 (1976).

21. E. F. C. Somerscales and H. Sanatgar, " Hydrodynamic Removal of Corrosion Products from a Surface," Br. Corros. J., 27, 36–55 (1992).

22. J. Postlethwaite, Y. Wang, G. Adamopoulos, and S. Nesic, "Relationship between Modelled Turbulence Parameters and Corrosion Product Film Stability in Disturbed Single-Phase Aqueous Flow," in Modelling Aqueous Corrosion, K. R. Trethaway and P. R. Roberge (Eds.), Kluwer Academic, 1994, pp. 297–316.

23. R. H. Perry and D. Green (Eds.), Perry's Chemical Engineers' Handbook, 7th ed., McGraw Hill, New York, 1997, p. 6:10.

24. V. L. Streeter, E. B. Wylie, and K. W. Bedford, Fluid Mechanics, 9th ed. McGraw-Hill, New York, 1998, pp. 105, 291.

25. C. O. Bennett and J. E. Myers, Momentum Heat and Mass Transfer, McGraw-Hill, New York, 1982, p. 168.

26. K. D. Efird, "Effect of Fluid Dynamics on the Corrosion of Copper-Base Alloys in Sea Water," Corrosion, 33, 3–8 (1977).

27. ASM, "Corrosion of Copper and Copper Alloys," in ASM Metals Handbook, Vol. 13, Corrosion, ASM, Metals Park, OH, 1987, p. 624.

28. J. Kim, "Turbulence Structures Associated with the Bursting Event," Phys. Fluids, 28, 52–58 (1985).

29. J. L. Dawson and C. C. Shih, "Corrosion Under Flowing Conditions—An Overview and Model," in Proceedings of Symposium on Flow-Induced Corrosion; Fundamental Studies and Industry Experience, K. H. Kennelley, R. H. Hausler, and D. C. Silverman (Eds.), NACE, Houston, 1991, pp. 2:1–2:12.

30. G. J. Danek, Jr., "The Effect of Sea-Water Velocity on the Corrosion Behavior of Metals," Naval Eng. J., 78, 763–769 (1966).

31. G. Butler and E. G. Stroud, "The Influence of Movement and Temperature on the Corrosion of Mild Steel: II. High Purity Water," J. Appl. Chem., 15, 325–338 (1965).

32. A. Cohen, "Corrosion by Potable Waters in Building Systems," Mater. Perform., 32, 56–61 (1993).

33. E. D. D. During, Corrosion Atlas: A Collection of Illustrated Case Histories, Vol. 1, Carbon Steels; Vol. 2, Stainless Steels and Non-Ferrous Materials, "Erosion-Corrosion of Copper Tubing," 06.05.34.01; "Valve Erosion," 04.01.32.01; "Pump Cavitation," 04.11.33.01; Elsevier, Amsterdam, 1988.

34. J. G. A. Bitter, "A Study of Erosion Phenomena," Parts I and II, Wear, 6, 5–21, 169–190 (1963).

35. T. H. Kosel, "Solid Particle Erosion," in ASM Metals Handbook, Vol. 18, Friction, Wear and Lubrication Technology, ASM, Metals Park, OH, 1992, p. 199.

36. J. Postlethwaite, B. J. Brady, M. W. Hawrylak, and E. B. Tinker, "Effects of Corrosion on the Wear Patterns in Horizontal Slurry Pipelines," Corrosion, 34, 245–250 (1978).

37. E. J. Wasp, J. P. Kenny, and R. L. Gandhi, "Solid-Liquid Flow: Slurry Pipeline Transportation," in Series on Bulk Materials Handling, Vol. 1(1975/77), No. 4, Trans. Tech. Publications, Clausthal, Germany, 1977, p. 144.

38. S. Nesic, "Erosion-Corrosion in Disturbed Flow," Ph.D. Thesis, University of Saskatchewan, Canada, 1991.

39. J. Postlethwaite and S. Nesic, "Erosion in Disturbed Liquid/ Particle Pipe Flow: Effects of Flow Geometry and Particle Surface Roughness," Corrosion, **49**, 850–857 (1993).

40. W. Madsen and R. Blickensderfer, "A New Flow-Through Slurry Erosion Wear Test," in J. E. Miller and F. E. Schmidt, Jr. (Eds.), Slurry Erosion: Uses Applications and Test Methods, ASTM STP 946, ASTM, West Conshohocken, PA, 1987, pp. 169–184.

41. I. Finnie, "Erosion of Surfaces by Solid Particles," Wear, **3**, 87–103 (1960).

42. P. J. Baker and B. E. A. Jacobs, A Guide to Slurry Pipeline Systems, BHRA Fluid Engineering, Cranfield, England, 1979, p. 15.

43. E. Heitz, S. Weber, and R. Liebe, "Erosion Corrosion and Erosion of Various Materials in High Velocity Flows Containing Particles," in Proceedings of Symposium on Flow-Induced Corrosion; Fundamental Studies and Industry Experience, K. H. Kennelley, R. H. Hausler, and D. C. Silverman (Eds.), NACE, Houston, TX, 1991, pp. 5:1–5:15.

44. F. J. Heymann, "Liquid Impingement Erosion," in ASM Metals Handbook, Vol. 18, Friction, Wear and Lubrication Technology, ASM, Metals Park, OH, 1992, pp. 221–232.

45. F. J. Heymann, "Erosion by Liquids," Machine Design, Dec. **10**, 118–124 (1970).

46. Century Brass, The Century Heat Exchanger Tube Manual, Century Brass Products, Waterbury, CT, 1977.

47. C. M. Hansson and L. H. Hansson, "Cavitation Corrosion," in ASM Metals Handbook, Vol. 18, Friction, Wear and Lubrication Technology, ASM, Metals Park, OH, 1992, pp. 214–220.

48. B. Angell, "Cavitation Damage," in Corrosion, 3rd ed., L. L. Shreir, R. A. Jarman, and G. T. Burstein (Eds.), Butterworth-Heinemann, Oxford, England 1994, pp. 8:197–8:207.

49. C. M. Preece, "Cavitation Erosion," in Treatise on Materials Science and Technology, Vol. 16, Erosion, C. M. Preece (Ed.), Academic, London, 1979, pp. 253, 259, 261, 294.

50. D. McIntyre (Ed.), Forms of Corrosion, Recognition and Prevention, NACE Handbook 1, Vol. 2, NACE, Houston, TX, 1997, pp. 89, 93.

51. W. Glaeser, "Erosion-Corrosion Cavitation and Fretting," in NACE Handbook 1, Forms of Corrosion, Recognition and Prevention, C. P. Dillon (Ed.), NACE, Houston, TX, 1982, p. 72.

52. P. Griffith, "Multiphase Flow in Pipes," J. Petrol. Technol., **Mar.** 361–367 (1984).

53. F. Giralt and O. Trass, "Mass Transfer from Crystalline Surfaces in a Turbulent Impinging Jet, Part II: Erosion and Diffusional Transfer," Can. J. Chem. Eng. **54**, 148 (1976).

54. Electric Power Research Institute, (EPRI), Flow Accelerated Corrosion in Power Plants, TR-106611, EPRI, Pleasant Hill, CA, 1996, pp. 4:2, 6:25.

55. G. Wilson, "The Design Aspects of Centrifugal Pumps for Abrasive Slurries," in Proceedings of the 2nd International Conference on Hydraulic Transport of Solids in Pipes, BHRA Fluid Engineering, Cranfield, UK, 1972, pp. H2:25–H2:52.

19

CARBON DIOXIDE CORROSION OF MILD STEEL

S. Nešić

Institute for Corrosion and Multiphase Flow Technology, Ohio University, Athens, Ohio

A. INTRODUCTION

Aqueous carbon dioxide (CO_2) corrosion is usually associated with mild steel. It becomes a serious issue when, in oil and gas production and transportation, significant amounts of CO_2 and water are present. CO_2 is very soluble in water, where it forms carbonic acid (H_2CO_3), which attacks mild steel. This has been known for many decades, yet aqueous CO_2 corrosion of mild steel remains a significant problem for the oil and gas industry [1] (see Fig. 19.1). Despite thousands of papers published on CO_2 corrosion in the open literature, one cannot easily piece together a picture of CO_2 corrosion, since much of the useful information is scattered. The text below is a compilation of the current understanding we have on this subject, with a focus on the basic processes and mechanisms.

It should be noted at the outset that corrosion-resistant alloys (CRAs) exist which are able to withstand CO_2 corrosion, such as many stainless steels and other iron- or nickel-based alloys. Nevertheless, due to cost, mild steel is almost exclusively used as a construction material for large facilities such as pipelines. The cost of pipeline failure due to internal CO_2 corrosion can be enormous, including repair/replacement costs, lost production, environmental impact, and losses in the downstream industries such as refineries.

B. THEORY OF UNIFORM CO_2 CORROSION OF MILD STEEL IN AQUEOUS SOLUTIONS

For the case of mild steel, one can write the overall CO_2 corrosion reaction as

$$Fe + CO_2 + H_2O \Rightarrow Fe^{2+} + CO_3^{2-} + H_2 \quad (19.1)$$

As iron (Fe) from the steel is oxidized to ferrous ions (Fe^{2+}) in the presence of CO_2 and water (H_2O), hydrogen gas (H_2) is evolved due to reduction of hydrogen ions present in water. It is also common to show the overall corrosion reaction in a

Uhlig's Corrosion Handbook, Third Edition, Edited by R. Winston Revie
Copyright © 2011 John Wiley & Sons, Inc.

FIGURE 19.1. Section of corroded pipe showing localized CO_2 attack. (With permission from ConocoPhillips.)

more compact form:

$$Fe + H_2CO_3 \Rightarrow FeCO_3 + H_2 \qquad (19.2)$$

To understand aqueous CO_2 corrosion of mild steel, one first needs to define the species present in a CO_2-saturated aqueous solution before looking into the reduction/oxidation processes at the steel surface, which are behind corrosion. The former is best described by *homogeneous aqueous chemistry*, while the latter is the subject of *heterogeneous (electro)chemical analysis*. The state of the solution at the metal surface and that in the bulk are related, as species move back and forth due to concentration gradients; this can be accounted for by proper *mass transfer analysis*. Therefore, the text below is structured to first cover homogeneous chemical reactions in a CO_2 saturated aqueous solution, followed by the description of the heterogeneous (electro) chemical processes occurring at the steel surface and finally a mass transfer analysis which links the two. Means of quantifying the various physicochemical processes involved are provided whenever they are known.

B1. Homogeneous Reactions in CO_2-Saturated Aqueous Solutions

The CO_2 gas is rather soluble in water. In terms of volumetric molar concentrations, almost as much CO_2 can be found in the aqueous phase as in the gas phase. Only a rather small fraction (0.2%) of the dissolved CO_2 molecules is hydrated to form a "weak" carbonic acid, H_2CO_3. The term "weak" refers to the fact that H_2CO_3 does not fully dissociate. This reaction and all the other important homogeneous chemical reactions occurring in an aqueous CO_2 solution are listed in Table 19.1, along with the corresponding equilibrium expressions. The equilibrium "constants" typically depend on temperature and in some cases pressure and ionic strength of the solution, as listed in Table 19.2.

One can use the equilibrium equations listed in Tables 19.1 and 19.2 to calculate the bulk concentrations of all the species in an aqueous CO_2 solution, including the concentration of protons, c_{H^+}, which yields the bulk pH. To achieve this, one

TABLE 19.1. Key Chemical Reactions Occurring in Aqueous CO_2 Solution and Corresponding Equilibrium Expressions

Name	Reaction	Equilibrium Expression	
Dissolution of carbon dioxide	$CO_{2(g)} \overset{K_{sol}}{\Leftrightarrow} CO_2$	$H_{sol(CO_2)} = \dfrac{1}{K_{sol}} = \dfrac{p_{CO_2}}{c_{CO_2}}$	(19.3)
Carbon dioxide hydration	$CO_2 + H_2O \overset{K_{hyd}}{\Leftrightarrow} H_2CO_3$	$K_{hyd} = \dfrac{c_{H_2CO_3}}{c_{H_2O}\, c_{CO_2}}$	(19.4)
Carbonic acid dissociation	$H_2CO_3 \overset{K_{ca}}{\Leftrightarrow} H^+ + HCO_3^-$	$K_{ca} = \dfrac{c_{H^+}\, c_{HCO_3^-}}{c_{H_2CO_3}}$	(19.5)
Bicarbonate anion dissociation	$HCO_3^- \overset{K_{bi}}{\Leftrightarrow} H^+ + CO_3^{2-}$	$K_{bi} = \dfrac{c_{H^+}\, c_{CO_3^{2-}}}{c_{HCO_3^-}}$	(19.6)
Water dissociation	$H_2O \overset{K_{wa}}{\Leftrightarrow} H^+ + OH^-$	$K_{wa} = \dfrac{c_{H^+}\, c_{OH^-}}{c_{H_2O}}$	(19.7)

Note: Means for calculating equilibrium constants are given in Table 19.2 below.

TABLE 19.2.   Equilibrium Constant for Homogeneous Reactions Listed in Table 19.1 as Function of Temperature, Pressure, and Ionic Strength

Equilibrium Constant For	Source	
Dissolution of carbon dioxide:	Oddo and Tomson [2]	(19.8)
$$K_{sol} = 14.463 \times 10^{-(2.27 + 5.65\times10^{-3}\times T_f - 8.06\times10^{-6}\times T_f^2 + 0.075\times I)}$$		
Carbon dioxide hydration:	Palmer and van Eldik [3]	(19.9)
$$K_{hy} = 2.58 \times 10^{-3}$$		
Carbonic acid dissociation:	Oddo and Tomson [2]	(19.10)
$$K_{ca} = 387.6 \times 10^{-(6.41 - 1.594\times10^{-3}\times T_f + 8.52\times10^{-6}\times T_f^2 - 3.07\times10^{-5}\times p - 0.4772\times I^{0.5} + 0.118\times I)}$$		
Bicarbonate anion dissociation:	Oddo and Tomson [2]	(19.11)
$$K_{bi} = 10^{-(10.61 - 4.97\times10^{-3}\times T_f + 1.331\times10^{-5}\times T_f^2 - 2.624\times10^{-5}\times p - 1.166\times I^{0.5} + 0.3466\times I)}$$		
Water dissociation:	Kharaka et al. [4]	(19.12)
$$K_{wa} = 10^{-(29.3868 - 0.0737549\times T_k + 7.47881\times10^{-5}\times T_k^2)}$$		

Note: T_f is temperature in degrees Fahrenheit, T_k is absolute temperature in Kelvin, $I = \left(c_1 z_1^2 + c_2 z_2^2 + \cdots\right)/2$ is ionic strength in molar, and p is the pressure in psi.

more equation is needed to find a unique mathematical solution: a constraint that describes charge conservation, that is, electroneutrality of the solution.

Clearly, chemical reactions shown in Table 19.1, which involve ions, always remain balanced with respect to charge and therefore one can write

$$c_{H^+} = c_{HCO_3^-} + 2c_{CO_3^{2-}} + c_{OH^-} \qquad (19.13)$$

This simple method of solution speciation is valid only for a pure ideal CO_2-saturated aqueous solution. If the aqueous solution contains other ions, such as, for example, Fe^{2+} produced by corrosion of steel or Na^+ or Cl^- etc., then Eq. (19.13) needs to be extended to include these species. Likewise, if other homogeneous reactions occur (in addition to those listed in Table 19.1), they need to be included in the analysis and the corresponding equilibrium expressions accounted for by adding them to the list in Table 19.2.

An example of calculated carbonic species distribution in a CO_2-saturated aqueous solution as a function of pH is given in Figure 19.2 for a system where the partial pressure of the CO_2 gas, $p_{CO_2} = 1$ bar, and temperature, $T = 25°C$, are kept constant. The conditions chosen for this plot are not uncommon; for example, the p_{CO_2} of mild steel pipelines typically varies from a fraction of a bar to a few bars. The temperature range is anywhere from 1 to 100°C, and the corresponding pH is slightly below pH 4 for pure (condensed) water and up to pH 7 for buffered (formation) water—often referred to as "brine." The concentration of various salts can be rather high, typically a few weight percent, even if for some "heavy" brines it may exceed 20 wt %.

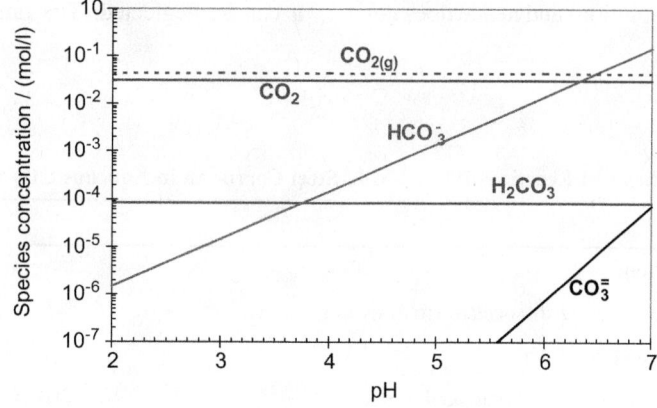

FIGURE 19.2. Calculated carbonic species concentrations as function of pH for a CO_2-saturated aqueous solution; $p_{CO_2} = 1$ bar, 25°C, 1 wt % NaCl.

It should be noted that in Figure 19.2, due to constant p_{CO_2}, the concentration of carbonic acid, $c_{H_2CO_3}$, is also constant across the whole pH range, while the concentrations of the bicarbonate ion, $c_{HCO_3^-}$, and the carbonate ion, $c_{CO_3^{2-}}$, increase with pH.

Some reactions listed in Table 19.1, such as, for example, the dissociation of carbonic acid, are very fast compared to all other processes occurring simultaneously; thus it is reasonable to assume that they are in chemical equilibrium. Other reactions, such as the CO_2 dissolution in water and the hydration of dissolved CO_2, are much slower. In that case, other faster processes occurring simultaneously can lead to local nonequilibrium in the solution. More importantly, when being a part of the overall corrosion process, slow reactions can become the rate-determining step and limit the corrosion rate, as explained further below.

B2. Heterogeneous Reactions in Corrosion of Mild Steel in CO_2-Saturated Aqueous Solutions

The most important reactions which are "behind" the CO_2 corrosion of mild steel are the heterogeneous reduction/oxidation reactions occurring at the steel surface, that is, the electrochemical reactions. The key ones are listed in Table 19.3.

B2.1. Anodic Reaction.
The electrochemical dissolution of iron from the steel, reaction 19.14 in Table 19.3, is the main reaction behind aqueous CO_2 corrosion of mild steel. Note that this reversible electrochemical reaction is written as if it occurs only in one direction, that is, as an oxidation or anodic step. The reverse-reduction reaction—the cathodic iron deposition step—has been ignored here. This is because in a corrosion situation reaction (19.14) is not in equilibrium, and there is a net dissolution (oxidation) of iron. In other words, at the corrosion potential, the reverse cathodic partial reaction (iron deposition) can be assumed to proceed much slower than the anodic partial (iron dissolution) and hence does not

need to be taken into account. For this to be a sustainable situation, other reduction (cathodic) reaction(s) must be occurring simultaneously to balance the overall corrosion process electrically, as discussed below.

In reality, the anodic iron dissolution reaction is more complicated than it appears in reaction (19.14). For example, the two electrons are not "released" in one step; rather a sequence of steps occurs, which all add up to the overall reaction. These and many other details on the mechanism of the iron dissolution reaction can be found in the open literature, both for general acidic corrosion [5, 6] and more specifically for CO_2 corrosion [7].

The rate of the active dissolution of iron, reaction (19.14) in Table 19.3, is independent of flow and mass transfer and is not a strong function of pH, or of p_{CO_2}, but does depend on temperature. Electrochemically, there can be some variation in the behavior from one type of mild steel to another; however, the corresponding corrosion rate does not vary much, as it is typically controlled by the cathodic reaction, described below. At more positive (anodic) potentials and at high pH often seen in the presence of ferrous carbonate layers, reversible (pseudo)passivation of mild steel may occur, leading to very low corrosion rates.

B2.2. Cathodic Reactions.
Mild steel does not corrode significantly in slightly acidic or near-neutral water ($4 < pH < 6$) but does so vigorously when CO_2 is present. The presence of CO_2 increases the corrosion rate primarily by making it easier for hydrogen to evolve from water. How this happens is explained in the paragraphs below.

It is well known that in *strong* mineral acids, which are almost fully dissociated, hydrogen evolution occurs by reduction of the dissociated (free) hydrogen ions, H^+, according to the overall reaction (19.15), given in Table 19.3. Note that the reverse partial reaction (oxidation of dissolved hydrogen gas, H_2) is ignored here for the same reasons as listed above for iron dissolution; at the corrosion potential the oxidation of dissolved hydrogen gas proceeds so slowly that it can be neglected. The rate of hydrogen evolution reac-

TABLE 19.3. Possible Electrochemical Reactions Behind Mild Steel Corrosion in Aqueous CO_2 Solutions

Name	Reaction	
Anodic dissolution (oxidation) of iron	$Fe \rightarrow Fe^{2+} + 2e^-$	(19.14)
Cathodic hydrogen evolution by reduction of dissociated (free) hydrogen ions	$2H^+ + 2e^- \rightarrow H_2$	(19.15)
Cathodic hydrogen evolution by reduction of water	$2H_2O + 2e^- \rightarrow H_2 + 2OH^-$	(19.16)
Cathodic hydrogen evolution by reduction of carbonic acid	$2H_2CO_3 + 2e^- \rightarrow H_2 + 2HCO_3^-$	(19.17)
Cathodic hydrogen evolution by reduction of bicarbonate ion	$2HCO_3^- + 2e^- \rightarrow H_2 + 2CO_3^{2-}$	(19.18)

tion (19.15) depends on the concentration of H$^+$ ions, making it a strong function of pH. However, in the case of mild steel corrosion in strong acids, the rate of this electrochemical reaction would be extremely high if it were not limited by the rate at which the main reactant, H$^+$ ions, can be transported from the bulk solution to the steel surface by mass transfer (often referred to as *diffusion*, hence the term *diffusion limitation*). This makes the reaction rate "flow sensitive," which really refers to the fact that turbulent flow can increase the mass transfer rate of H$^+$ ions and indirectly the rate of the hydrogen evolution reaction and corrosion.

For the pH range seen in typical CO$_2$-saturated aqueous solutions (4 < pH < 6), this limiting diffusion rate is rather small due to a low concentration of H$^+$ ions in the bulk. The corresponding corrosion rate should be small as well; however, this is not the case. Therefore, other cathodic reactions must be involved.

The so-called direct reduction of water is one obvious possibility, amounting to the reduction of H$^+$ ions from the water molecules adsorbed on the steel surface [reaction (19.16) in Table 19.3]. This reaction is thermodynamically equivalent to the hydrogen evolution reaction (19.15), which can be easily shown by combining it with the water dissociation reaction given in Table 19.1. However, the "direct reduction of water" pathway for hydrogen evolution is comparatively slow (kinetically hindered) and cannot be used to explain the high corrosion rates seen in CO$_2$-saturated aqueous solutions [8, 9]. Therefore it need not be considered in further analysis.

It is the presence of the *weak* H$_2$CO$_3$ which enables hydrogen evolution and corrosion at a much higher rate in CO$_2$-saturated aqueous solutions. Actually, the presence of CO$_2$ leads to a much higher corrosion rate than would be found in an aqueous solution of a *strong* acid at the same pH.

This apparent paradox can be readily explained by considering that in an aqueous solution the homogeneous dissociation of H$_2$CO$_3$ (see Table 19.1) provides additional H$^+$ ions, which can readily be reduced at the steel surface according to reaction (19.15) [1]. A different pathway is also possible where the H$^+$ ions are "directly" reduced from the H$_2$CO$_3$ molecules adsorbed on the steel surface, much in the same way as happens for adsorbed water molecules; see reaction (19.17) in Table 19.3. This is often referred to as "direct reduction of H$_2$CO$_3$" [10–12]. Clearly, this hydrogen evolution reaction also amounts to reduction of H$^+$ ions, this time from the adsorbed H$_2$CO$_3$ molecules, and is thermodynamically equivalent to the other two hydrogen evolution reactions (19.15) and (19.16). This point can be easily proven by combining the various hydrogen evolution reactions [(19.15)–(19.17)] with the homogeneous chemical reactions listed in Table 19.1. For the three cases the distinction is only in the pathway and consequently in the kinetics.

The rate of hydrogen evolution from H$_2$CO$_3$ [reaction (19.17)] is much faster than that from water [reaction (19.16)]. Due to the abundance of water and dissolved CO$_2$, there is a sufficiently large "reservoir" of H$_2$CO$_3$ and ultimately H$^+$ ions in a CO$_2$-saturated aqueous solution. This would lead to some very high corrosion rates, were it not for the slow CO$_2$ hydration step (seen in Table 19.1), which gives rise to chemical reaction limiting rates for this cathodic reaction [12, 13], as described below.

In CO$_2$-saturated aqueous solutions, when the pH increases, so does the concentration of the bicarbonate ions, $c_{HCO_3^-}$ (see Fig. 19.2). Since, HCO$_3^-$ is another weak acid (it only partially dissociates to give CO$_3^{2-}$; see Table 19.1), then by analogy with H$_2$CO$_3$ one can assume that "direct reduction of HCO$_3^-$" is yet another pathway for hydrogen evolution [14], according to reaction (19.18) in Table 19.3. Again this reaction is thermodynamically equivalent to the previous three pathways for hydrogen evolution, and the distinction is in the kinetics.

Experimental evidence suggests that this pathway for hydrogen evolution is slower than direct reduction of H$_2$CO$_3$. However, in neutral and alkaline conditions, the concentration of HCO$_3^-$ is much higher than that of H$_2$CO$_3$ (see Fig. 19.2), and it is likely that an accelerated rate of reaction (19.18) can make it the dominant cathodic reaction in CO$_2$-saturated aqueous solutions. However, in the practical range of interest considered here (4 < pH < 6), the two concentrations are in the similar range, and due to kinetic hindrance, the direct reduction of HCO$_3^-$ can be ignored in further analysis.

This leaves only two significant possibilities for hydrogen evolution in CO$_2$-saturated aqueous solutions: reduction of the dissociated (free) H$^+$ ions [reaction (19.15)] and direct reduction of H$_2$CO$_3$ molecules adsorbed on the steel surface [reaction (19.17)]. The former is a strong function of pH, while the latter is a function of H$_2$CO$_3$ concentration, which is a linear function of CO$_2$ partial pressure in the gas phase; see Table 19.1. Both are kinetically limited, reaction (19.15) by diffusion of H$^+$ ions from the bulk to the steel surface and reaction (19.17) by the slow CO$_2$ hydration step (see Table 19.1), as described in more detail below. The overall CO$_2$ corrosion process outlined above is shown in the simplified schematic given in Figure 19.3.

B2.3. Calculating the Charge Transfer Rate. A corrosion process proceeds so that the rate of the main anodic process (iron dissolution) is electrically balanced by the rate of the cathodic (hydrogen evolution) reaction(s), as there can be no charge (electron) accumulation in the steel. Means of calculating the rates of the various processes involved are presented below.

A rate $\Re$ [in kmol/(m^2 s)], for any of the cathodic or anodic reactions described above can be readily expressed in terms of current density i (in A/m^2) since the two are directly related

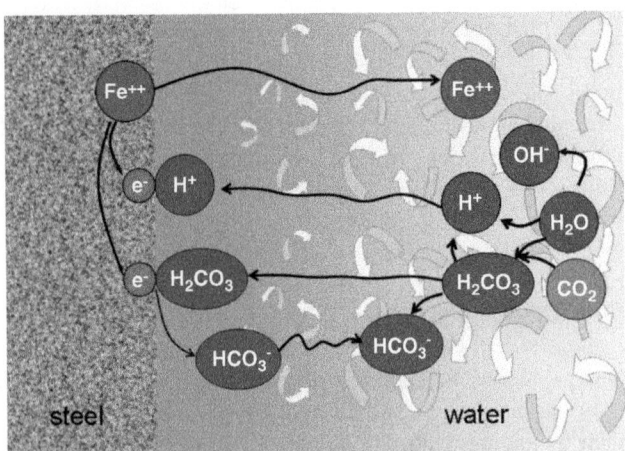

FIGURE 19.3. Simplified schematic representing key reactions occurring during CO_2 corrosion of mild steel.

via Faraday's law:

$$\Re = \frac{i}{nF} \qquad (19.19)$$

When the rate of the overall corrosion process is limited by the rate of an electrochemical step in the overall sequence, this is often referred to as "charge transfer"–controlled corrosion rate. Generally, the rate of any electrochemical reaction at the steel surface depends on the concentration of all the reacting species involved in that reaction and temperature. Since electrons are one of the key "reactants" in the electrochemical reactions (19.14)–(19.18), their influence is expressed via the electrical potential at the steel surface using a simplified version of the Butler–Volmer equation:

$$i_\alpha = \pm i_0 \times 10^{\pm(E - E_{rev})/b} \qquad (19.20)$$

In this formalism, the effect of all the other aqueous reactants on the kinetics is captured via the function i_0 as explained below. The plus sign in Eq. (19.20) applies for the anodic (oxidation) reactions, in this case the dissolution-of-iron reaction (19.14), while the minus sign applies for the cathodic (reduction) reactions, such as the various pathways for hydrogen evolution [reactions (19.15)–(19.18)].

It should be noted that in Eq. (19.20), according to standard textbook practice, i_0 is the exchange current density, E_{rev} is the reversible potential, while b is the Tafel slope, all characteristic for a particular reversible electrochemical reaction. However, in order to keep calculations simple, we need to avoid accounting for the reverse partials of the electrochemical reactions (19.14), (19.15), and (19.17), which are deemed negligible. To achieve this, it is easiest not to treat i_0 and E_{rev} in a strict theoretical sense. It is sufficient to quantify i_0, E_{rev}, and b for one important partial of the reversible electrochemical reaction and define the influence of various parameters (temperature, pH, etc.) on that partial without worrying what happens to the reverse partial of the same reaction.

The i_0 for all the important partial electrochemical reactions [(19.14), (19.15), and (19.17)] is a function of temperature, which is best expressed in terms of an Arrhenius-type equation:

$$i_0 = i_0^{ref} e^{-\Delta H/R\left(1/T_k - 1/T_{k,ref}\right)} \qquad (19.21)$$

The parameters for this function are defined in Table 19.4 below for the three key electrochemical reactions.

For cathodic reactions, the i_0 is also a function of the surface concentration of the dissolved species. In the case of the H^+ ion reduction reaction (19.15), $i_{0(H^+)}$ is a function of pH:

$$\frac{\partial \log i_{0(H^+)}}{\partial pH} = -0.5 \qquad (19.22)$$

while for the H_2CO_3 reduction reaction (19.17), the $i_{0(H_2CO_3)}$ depends on H_2CO_3 concentration:

$$\frac{\partial \log i_{0(H_2CO_3)}}{\partial \log c_{H_2CO_3}} = 1 \qquad (19.23)$$

The Tafel slope in Eq. (19.20) is defined as

$$b = \frac{2.303 R T_k}{\alpha F} \qquad (19.24)$$

TABLE 19.4. Parameters for Key Electrochemical Reactions in Aqueous CO_2 Corrosion of Mild Steel with Reference to Equations (19.21) and (19.24)

Parameter	Iron Oxidation, $Fe \rightarrow Fe^{2+} + 2e^-$	Hydrogen Ion Reduction $2H^+ + 2e^- \rightarrow H_2$	Carbonic Acid Reduction $2H_2CO_3 + 2e^- \rightarrow H_2 + 2HCO_3^-$	Unit
i_0^{ref}	1	0.03	0.06	A/m^2
$T_{k,ref}$	300	300	300	K
ΔH	50	30	57.5	kJ/mol
α	1.5	0.5	0.5	

The theoretical value of the apparent transfer coefficient factor α is given in Table 19.4 for the three key electrochemical reactions [(19.14), (19.15), and (19.17)], resulting in Tafel slopes in the range of 40–50 mV for the iron dissolution reaction (19.14) and approximately 120–140 mV for the hydrogen evolution reactions (19.15) and (19.17).

B2.4. Calculating the Limiting Rate. Frequently, due to a high rate of the electrochemical reactions, the cathodic reactants at the steel surface are consumed rapidly and their concentration at the steel surface can be much lower than that in the bulk. In that case, the rate of the overall reaction is given by the rate at which these reactants can be replenished. This leads to the concept of limiting current densities, i_{lim}, which are independent from the electrical potential. The overall current density can be written for that case as

$$i = \frac{1}{1/i_\alpha + 1/i_{lim}} \quad (19.25)$$

When the limiting current/rate is very large (such as in the case of iron dissolution from steel when we can assume the supply of Fe is "unlimited"), we can write $i_{lim} \gg i_\alpha$ and the overall rate of the reaction is limited by the rate of the charge transfer process; that is from (19.25) we obtain $i = i_\alpha$. When the limiting current is very small, as in the case the hydrogen evolution reactions (19.15) and (19.17), we can assume that $i_{lim} \ll i_\alpha$, and from (19.25) we obtain $i = i_{lim}$; that is, the rate of the overall reaction is limited by how fast the cathodic reactants can be replenished at the steel surface.

For the case of the H$^+$ ion reduction reaction (19.15), the limiting current density is related to the rate of transport (diffusion) of H$^+$ ions from the bulk of the solution through the boundary layer to the steel surface:

$$i_{lim(H^+)}^d = k_{m(H^+)} F c_{H^+} \quad (19.26)$$

In turbulent flow, the mass transfer coefficient $k_{m(H^+)}$ can be calculated from a correlation of the Sherwood, Reynolds, and Schmidt numbers, such as the straight pipe correlation of Berger and Hau [15]:

$$Sh_p = \frac{k_m d_p}{D} = 0.0165 \times Re^{0.86} \times Sc^{0.33} \quad (19.27)$$

or the rotating cylinder correlation of Eisenberg et al. [16]:

$$Sh_r = \frac{k_m d_c}{D} = 0.0791 \times Re^{0.7} \times Sc^{0.356} \quad (19.28)$$

This type of correlation is limited to fully developed, steady-state, single-phase flow situations and is valid only for a specific flow geometry such as the two listed above, which are typical for corrosion. For other flow geometries, transient situations, developing and/or multiphase flow, and so on, different correlations must be found which apply to the given condition.

In the case of H$_2$CO$_3$ reduction reaction (19.17), the limiting current is due to a slow CO$_2$ hydration step preceding it [12, 13] (see Table 19.1). The limiting current density can be calculated from the CO$_2$ hydration rate constant [17] k_{hyd}^f as

$$i_{lim(H_2CO_3)}^r = F c_{CO_2} f_{H_2CO_3} \sqrt{D_{H_2CO_3} K_{hyd} k_{hyd}^f} \quad (19.29)$$

where k_{hyd}^f is a function of temperature:

$$k_{hyd}^f = 10^{169.2 - 53.0 \times \log(T_k) - 11715/T_k} \quad (19.30)$$

The equilibrium constant for this reaction (listed in Tables 19.1 and 19.2) is by definition $K_{hyd} = k_{hyd}^f / k_{hyd}^b$.

The flow factor $f_{H_2CO_3}$ captures the small effect that turbulent mixing has on this limiting current density [34]:

$$f_{H_2CO_3} = \coth \left(\zeta_{H_2CO_3} \right) \quad (19.31)$$

where

$$\zeta_{H_2CO_3} = \frac{\delta_{m(H_2CO_3)}}{\delta_{r(H_2CO_3)}} \quad (19.32)$$

and

$$\delta_{m(H_2CO_3)} = \frac{D_{H_2CO_3}}{k_{m(H_2CO_3)}} \quad (19.33)$$

$$\delta_{r(H_2CO_3)} = \sqrt{\frac{D_{H_2CO_3}}{k_{hyd}^b}} \quad (19.34)$$

The carbonic acid mass transfer coefficient $k_{m(H_2CO_3)}$ is calculated in the same way as the one described above for H$^+$ ions, via the Sherwood number correlations.

The diffusion coefficient of either species is a function of temperature according to the Einstein's relation:

$$D = D_{ref} \left(\frac{T_k}{T_{k,ref}} \right) \left(\frac{\mu_{H_2O,ref}}{\mu_{H_2O}} \right) \quad (19.35)$$

B2.5. Calculating the Balance. The rate of mild steel corrosion can now be calculated. The unknown corrosion potential E_{corr} in Eq. (19.20) can be found from the electrical current (charge) balance equation at the steel surface:

$$i_{(H^+)} + i_{(H_2CO_3)} = i_{(Fe)} \qquad (19.36)$$

When the calculated value of E_{corr} is returned to Eq. (19.20) written out for iron dissolution, the corrosion current can be calculated:

$$i_{corr} = i_{(Fe)} \qquad (19.37)$$

and the CO_2 corrosion rate is obtained by using Faraday's law:

$$CR = \frac{i_{corr} M_{Fe}}{\rho_{Fe} 2F} \qquad (19.38)$$

It happens that, for iron in steel, when i_{corr} is expressed in amperes per meters squared, CR in millimeters per year takes on almost the same numerical value, more precisely, $CR = 1.155 \times i_{corr}$.

B2.6. Ferrous Carbonate Precipitation.

Under certain conditions, a ferrous carbonate ($FeCO_3$), layer can form in the CO_2 corrosion of mild steel by precipitation and can reduce the corrosion rate [17]. The causes for the precipitation of $FeCO_3$ are the high local concentrations of Fe^{2+} and CO_3^{2-} species in the aqueous solution. Most of the Fe^{2+} ions are provided by corrosion [see reaction (19.14) in Table 19.3], while the CO_3^{2-} ions come from dissolved CO_2 [see reaction (19.11) in Table 19.1]. When the product of their concentrations exceeds the so-called solubility limit, they form solid ferrous carbonate according to

$$Fe^{2+} + CO_3^{2-} \overset{K_{sp(FeCO_3)}}{\Longleftrightarrow} FeCO_{3(s)} \qquad (19.39)$$

where the solubility product constant is a function of temperature and ionic strength [18]:

$$K_{sp(FeCO_3)} =$$
$$10^{(-59.3498 - 0.041377 \times T_k - 2.1963/T_k + 24.5724 \times \log T_k + 2.518 \times I^{0.5} - 0.657 \times I)} \qquad (19.40)$$

Due to corrosion, the product of concentrations of Fe^{2+} and CO_3^{2-} ions in the aqueous solution is frequently much higher than the solubility limit. This is the case particularly at the steel surface where the Fe^{2+} ions are generated and the concentration can be very high. The solubility limit is also easily exceeded at high pH, when the concentration

CO_3^{2-} ions is high (see Fig. 19.2). In those cases, reaction (19.39) is not in equilibrium and there is a net rate from left to right amounting to a formation of solid $FeCO_{3(s)}$. This nonequilibrium situation is termed *supersaturation*, defined as

$$SS_{(FeCO_3)} = \frac{c_{Fe^{2+}} c_{CO_3^{2-}}}{K_{sp(FeCO_3)}} \qquad (19.41)$$

The higher the supersaturation is, the faster precipitation happens, since the imbalance in reaction (19.39) is larger, driving the process. In simple words, one can say that the $FeCO_3$ precipitation process is the "return" of the solution toward equilibrium. The higher the temperature, the faster this process will happen.

The $FeCO_3$ precipitation starts by heterogeneous nucleation, a process which happens relatively fast due to the many imperfections on the surface of the steel, which serve as good nucleation sites. Nucleation is followed by crystalline $FeCO_3$ layer growth. The rate of precipitation ($\Re_{FeCO_3}$) on the steel surface is therefore limited by the rate of crystal growth, which can be expressed as a function of supersaturation, surface area, and temperature:

$$\Re_{FeCO_3} = k_{r(FeCO_3)} \frac{A}{V} K_{sp(FeCO_3)} (SS_{(FeCO_3)} - 1) \qquad (19.42)$$

where $k_{r(FeCO_3)}$ is a kinetic constant which is a function of temperature, best expressed using an Arrhenius-type equation [19]:

$$k_{r(FeCO_3)} = e^{A_{(FeCO_3)} - \frac{B_{(FeCO_3)}}{RT_k}} \qquad (19.43)$$

where $A_{(FeCO_3)} = 28.2$ and $B_{(FeCO_3)} = 64{,}851$ J/mol. Rapid formation of $FeCO_3$ layers happens only at higher temperature (typically above 50°C) and at high pH (typically at pH > 5). Rapid $FeCO_3$ layer formation is a precondition for its protectiveness, as discussed below.

Scanning electron microscopy (SEM) images of a crystalline ferrous carbonate layer formed on a mild steel substrate are shown in Figures 19.4 and 19.5. When dense, the $FeCO_3$ layer can slow down the corrosion process by being a transport barrier for the corrosive species. Protection from corrosion is also achieved by the so-called coverage of the surface; that is, in places where the $FeCO_3$ crystals adhere to the steel, there is little or no corrosion. This is clearly shown in the TEM image in Figure 19.5, where the attack on the steel seems to proceed only in the area "between" the $FeCO_3$ crystals. In the presence of dense $FeCO_3$ layers, the conditions at the steel surface are different compared to those in the bulk. The pH is much higher, which may lead to formation of other solids such as oxides and hydroxides. Small amounts of magnetite (Fe_3O_4) were detected underneath the $FeCO_3$

FIGURE 19.4. SEM image showing top view of ferrous carbonate layer formed on mild steel: 80°C, pH 6.6, $p_{CO_2} = 0.5$ bar, stagnant conditions.

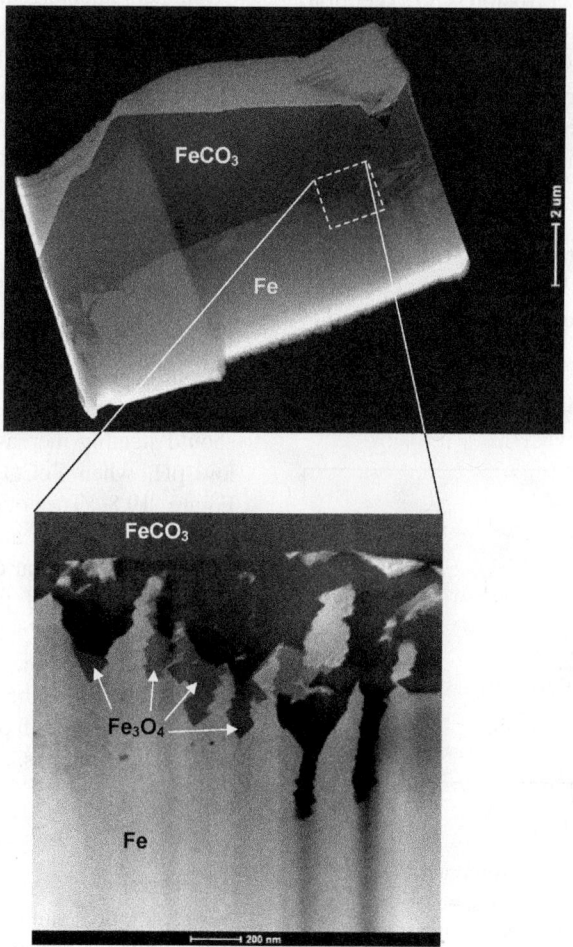

FIGURE 19.5. TEM of ferrous carbonate crystal formed on mild steel: 80°C, pH 6.6, $p_{CO2} = 0.5$ bar, stagnant conditions.

layer, as shown in Figure 19.5, which led to an even better protection from CO_2 corrosion by passivation.

C. FACTORS AFFECTING AQUEOUS CO_2 CORROSION OF MILD STEEL

The primary factors which affect the aqueous CO_2 corrosion of mild steel are discussed below. In many instances it is impossible to fully separate one effect from the other, and there is no perfect way of presenting the overall situation. In the discussion below, the effects of primary factors on CO_2 corrosion are put into the context of the theory described above. The most common range of conditions is considered for each primary parameter, and when it is justifiable, the key secondary parameter is introduced on the same graph.

C1. Effect of pH

The pH is probably one of the most influential parameters in acidic corrosion of steel and CO_2 corrosion is no exception. Even in the limited range of pH ($4 < \text{pH} < 6$), lowering the pH leads to higher corrosion rates and vice versa, as shown in Figure 19.6. If we recall that reduction of free H^+ ions [reaction (19.15)], is one of the two important cathodic reactions in CO_2 corrosion, this effect of pH is to be expected. However, the effect of pH seems to be exaggerated at higher temperature, as shown in Figure 19.6. This is because pH promotes the formation of $FeCO_3$ layers, and the kinetics of this process is highly temperature dependent. There are other more complicated effects of pH, some of which are discussed below.

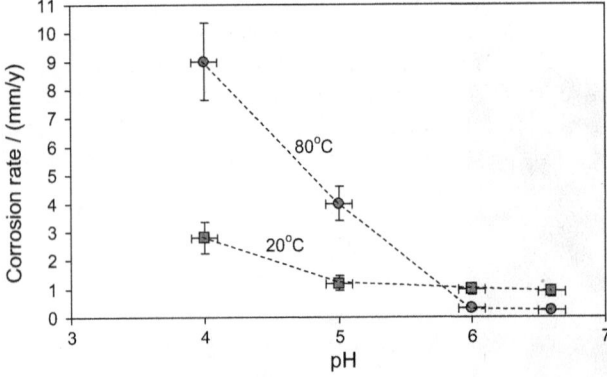

FIGURE 19.6. Effect of pH on CO_2 corrosion rate of mild steel measured at 20°C ($p_{CO_2} = 1$ bar) and 80°C ($p_{CO_2} = 0.5$ bar), 3 wt % NaCl, using rotating cylinder flow with outer diameter (OD) of 10 mm at 1000 rpm. The error bars represent typical variations seen in the experiments. The dotted lines are added to indicate trends.

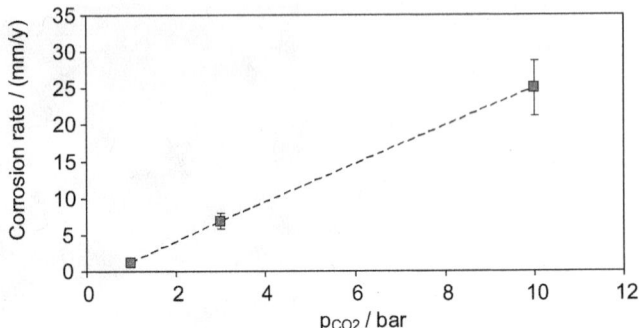

FIGURE 19.7. Effect of p_{CO_2} on CO_2 corrosion rate of mild steel measured at 60°C, pH 5, in single-phase pipe flow, internal diameter (ID) 100 mm, at 1 m/s, 1 wt % NaCl. The error bars represent typical variations seen in the experiments. The dotted line is added to indicate the trend.

C2. Effect of CO_2 Partial Pressure

It has been known for a while that p_{CO_2} leads to an increase in the corrosion rate, even at constant pH and with all other parameters unchanged. Based on the theory laid out above, one can see that with increasing p_{CO_2}, the concentration of H_2CO_3 increases, accelerating the other significant cathodic reaction (19.17) and ultimately the corrosion rate. This effect of p_{CO_2} at a constant pH is illustrated in Figure 19.7. As partial pressures of CO_2 exceed 10 bars, the corrosion rate does not increase as much. It is thought that this is related to formation of protective $FeCO_3$ layers, which is aided at very high p_{CO_2} [20].

C3. Effect of Temperature

Increased temperature accelerates all the processes involved in corrosion. It would then appear that the corrosion rate should steadily increase with temperature. This happens at low pH, when $FeCO_3$ layers do not form, as shown in Figure 19.8. The opposite happens when solubility of $FeCO_3$ is exceeded and a protective layer forms, typically when pH > 5. In that case, increasing temperature accelerates the kinetics of precipitation of $FeCO_3$ and leads to formation of a more protective layer, as shown in Figure 19.8. It should be noted that at low temperature the two curves do not differ much as the corrosion rate is not a function of pH and even at very high pH an unprotective $FeCO_3$ layer forms due to slow kinetics.

C4. Effect of Flow

Theoretically there are two main ways in which flow may affect CO_2 corrosion of mild steel: through *mass transfer* or via *mechanical means*. Turbulent flow enhances mass transport of species to and away from the steel surface by affecting transport through the boundary layer [21]. On the other hand,

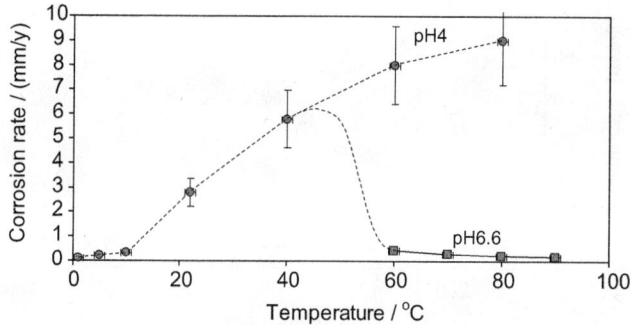

FIGURE 19.8. The effect of temperature on CO_2 corrosion rate of mild steel measured at pH 4 and pH 6.6, 1 wt % NaCl, using rotating cylinder flow with an OD of 10 mm at 1000 rpm. *Note*: In these atmospheric experiments p_{CO_2} decreased with temperature; e.g., at 20°C it was almost 1 bar while at 80°C it was approximately 0.5 bar. The error bars represent typical variations seen in the experiments. The dotted lines are added to indicate trends.

intense flow could theoretically lead to mechanical damage of protective $FeCO_3$ layers or inhibitor films. Both these mechanisms are aggravated by flow disturbances such as valves, constrictions, expansions, and bends, where a local increase of near-wall turbulence is seen. The same is true for gas–liquid flow and particularly in a slug flow regime, where very high short-lived fluctuations of both mass transfer and wall shear stress are measured.

However, very little hard evidence exists that either of these two flow-affected mechanisms is of great significance in CO_2 corrosion. Mass transfer plays a secondary role, as the CO_2 corrosion rate is usually controlled by charge transfer and/or a slow chemical reaction (see Section B above). Wall shear stress varies from a few Pascals seen in single-phase liquid flow to a few hundred or even thousands of pascals for high peaks in slug flow, yet it seems stresses of the order of megapascals are needed to detach a protective $FeCO_3$ layer or an inhibitor film. Therefore it seems both flow effects play only a secondary role in CO_2 corrosion of mild steel; that is, they become important only when the conditions are "at the margin." For example, if water chemistry is such that a protective $FeCO_3$ layer may dissolve, flow may accelerate the process locally and even enhance it by mechanical means. Similarly, flow may interfere locally with protective $FeCO_3$ layer buildup when the driving force for precipitation is not very strong (e.g., low supersaturation and/or low temperature), which can lead to localized corrosion.

Somewhat counterintuitively, flow seems to create more problems in CO_2 corrosion of mild steel when the flow rate is low rather than high. For example, in oil/water flow in pipelines the two liquids can flow stratified or mixed, usually with the oil phase being continuous and the water phase dispersed. In the stratified case, which is typical for low flow rates, corrosion of the bottom of the line is seen, while in the dispersed case, characteristic for high flow rates, no corrosion

occurs. Another example is the problem of settling of solids at low flow rates, which leads to the so-called underdeposit attack. The steel under the deposit is hard to protect with conventional corrosion inhibitors, while in some cases galvanic cells or even bacterial attack can aggravate the problem.

C5. Effect of Hydrogen Sulfide

In many field cases, in addition to CO_2, varying amounts of hydrogen sulfide (H_2S) gas are present, from a fraction of a millibar to tens of bars partial pressure. Corrosion issues related to H_2S still represent a significant problem for the oil and gas industry [22–28]. Mild steel corrosion in the presence of H_2S almost inevitably leads to formation of ferrous sulfide layers, which are a dominant factor governing the overall corrosion rate. This brief section will only cover cases where very small amounts of H_2S are present in CO_2 corrosion yet H_2S produces a big impact on the corrosion rate.

Dissolved H_2S is another weak acid, so it can be seen as an additional reservoir of H^+ ions and we must allow possibility of "direct H_2S reduction," in a similar way as was done above for H_2CO_3, both pathways causing an increase in the rate of the hydrogen evolution reaction. The overall corrosion reaction due to H_2S can be written as

$$Fe_{(s)} + H_2S \rightarrow FeS_{(s)} + H_2 \qquad (19.44)$$

However, what is even more important is that this reaction almost always results in the formation of a solid ferrous sulfide ($FeS_{(s)}$), layer, usually in the form of a nonstoichiometric sulfur-deficient mackinawite ($Fe_{1+x}S$), which can be found on the corroding steel surface even in the presence of very small amounts of H_2S and even below its thermodynamic solubility limit.

The tightly adherent mackinawite film is very thin ($\ll 1$ μm) but apparently rather dense, since it acts as an effective solid-state diffusion barrier for all the species involved in the corrosion reaction. Therefore this thin mackinawite film is one of the most important factors governing the corrosion rate in CO_2/H_2S corrosion. Experiments show that CO_2 corrosion rates are affected even if very small amounts of H_2S are present in the gas phase (as little as 10^{-5} bar). In CO_2 corrosion of mild steel in the presence of significant amounts H_2S, there is usually an outer carbonate/sulfide layer which thickens over time (typically $\gg 1$ μm) and forms an additional diffusion barrier. However, this outer layer is more porous and rather loosely attached to the steel surface, so that it rather easily cracks, peels, and spalls, which are accelerated by turbulent flow.

Mackinawite will transform over time into other forms of less soluble and more stable ferrous sulfide: pyrrhotite and troilite. At very high H_2S concentrations, pyrite and elemental sulfur appear. However, there is no clearly defined relationship between the type of the sulfide layer and the nature or

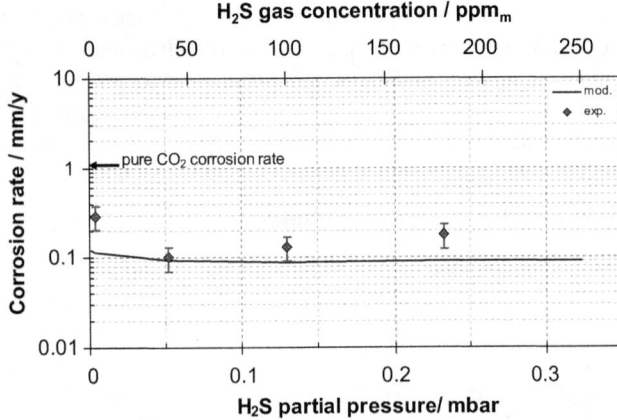

FIGURE 19.9. Effect of H₂S on CO₂ corrosion rate of mild steel measured at $p_{H_2S} = 0.0013$–0.32 mbar, $p_{CO_2} = 1$ bar, 20°C, pH 5, 1 wt % NaCl, using rotating cylinder flow with an OD of 10 mm at 1000 rpm. The error bars represent typical variations seen in the experiments. The dotted lines are added to indicate trends.

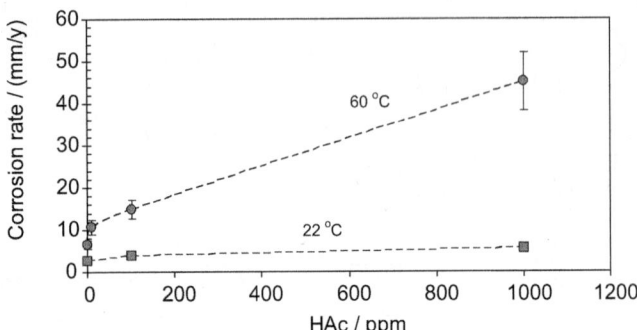

FIGURE 19.10. Effect of organic acid concentration on CO₂ corrosion rate of mild steel at different temperature measured at pH 4, 3 wt % NaCl, using rotating cylinder flow with an OD of 10 mm at 1000 rpm. The error bars represent typical variations seen in the experiments. The dotted lines are added to indicate trends.

intensity of the corrosion process. It is generally thought that all types of ferrous sulfide layers offer some degree of corrosion protection for mild steel.

An example of the effect of very small amounts of H₂S on CO₂ corrosion is seen in Figure 19.9, where, for $p_{CO_2} = 1$ bar, the p_{H_2S} ranges from 0.0013 to 0.32 mbar, corresponding to 1–250 ppm_m in the gas phase; that is, H₂S is present in "traces." Clearly this is a CO₂-dominated corrosion scenario (p_{CO_2}/p_{H_2S} ratio is in the range 10^3–10^6); however, the presence of H₂S and the thin mackinawite layer greatly affects the corrosion rate. Even when present in such small amounts, H₂S reduces the pure CO₂ (H₂S-free) corrosion rate by a factor of 3–10.

C6. Effect of Organic Acids

In many fields, organic acids are found in the fluid stream. The low-molecular-weight acids are primarily soluble in water while the high-molecular-weight acids are oil soluble and are termed *naphthenic acids*. The former can lead to corrosion of upstream mild steel facilities (tubing, pipelines, separators, etc.) while the latter are a problem in refineries at high temperatures. From the low-molecular-weight organic acid found in brines, the acetic acid CH₃COOH (shorthand: HAc) is most prevalent, with varying amounts of other acids, such as propionic, formic, present. Their corrosiveness is very similar to HAc when present in the same concentration. Therefore the HAc is often used to represent all the organic acids found in the brine.

HAc is another weak acid, and many of the electrochemical arguments made above for H₂CO₃ hold for HAc. The presence of HAc can lead to a significant acceleration of the hydrogen evolution reaction, particularly at high temperature (>50°C) and lower pH (<5) when much of the acid is present

in undissociated form [29], as seen in Figure 19.10. The overall corrosion reaction due to HAc can be written as

$$Fe_{(s)} + 2HAc \rightarrow Fe^{2+} + 2Ac^- + H_2 \qquad (19.45)$$

Since iron acetate solubility is much higher than that of ferrous carbonate, all corrosion products are soluble. There are some indications that the presence of organic acids impairs the protectiveness of ferrous carbonate layers; however, the mechanism is still not clear.

C7. Effect of Crude Oil

Crude oil affects internal CO₂ corrosion of mild steel pipelines in one of two ways:

- Inhibitive effect—by providing naturally occurring chemicals which act as corrosion inhibitors
- Wettability effect—by sweeping the water away from the internal pipe walls and entraining it in the stream

Adding corrosion inhibitors to fluid stream in the form of large organic surface-active organic compounds is one of the main methods of CO₂ corrosion mitigation. However, in many cases the crude oil contains compounds which can adsorb onto the steel surface by either direct wetting or by first partitioning into the water phase. The most common surface-active organic compounds found in crude oil which have mitigating effects are those containing oxygen, sulfur, and nitrogen in their molecular structure. Furthermore, both asphaltenes and waxes have shown some positive inhibitive effects on corrosion. However, our present understanding of these phenomena is at best qualitative, making it virtually impossible to make any reliable predictions.

Wettability was already briefly discussed in the context of flow effects on corrosion. Low velocities lead to stratification of water and oil and continuous water wetting of the pipe bottom, resulting in corrosion. If the amount of water is relatively small (<10%), a fast-moving oil phase can entrain the water, which may lead to intermittent oil/water wetting or continuous oil wetting of the pipe walls. In the former case the corrosion rate is much reduced while in the latter case there is no corrosion. Various crude oils have widely varying capacities to entrain water. Typically it takes much higher flow rates for light oils to entrain water ($v > 1.5$ m/s) due to their lower density and viscosity. Some heavier oils are able to do the same at velocities as low as 0.5 m/s. However, chemical properties of crude oils and particularly the content of surface-active substances are just as important for wettability as are their physical properties. Again our present understanding of these effects is fairly qualitative and predictions are difficult.

C8. Effect of Glycol/Methanol

Glycol and methanol are often injected into pipeline streams to prevent hydrates from forming at low temperatures. It is not unusual to have them comprise more than 50% of the total liquid phase. It appears that some retardation of corrosion seen in the field is primarily related to dilution of the water phase by glycol/methanol; that is, it is due to the decreased activity of water. However, our current understanding of this effect is rudimentary.

C9. Effect of Condensation in Wet-Gas Flow

In long pipelines, cooling of a wet-gas stream leads to condensation of water vapor and light hydrocarbons on the internal pipe walls. The condensed water is very pure and contains dissolved CO$_2$, making it corrosive. This leads to the so-called *top-of-the-line corrosion* (TLC) scenario. Condensation rate is an important factor in TLC as high condensation rates (e.g., 1 mL/m^2/s or higher) are related to very high corrosion rates, and vice versa; low condensation rates (<0.1 mL/m^2/s) lead to formation of protective ferrous carbonate layers and low corrosion rates. Incidents of localized attack in TLC were reported, causes of which are presently not well understood [30]. Other important factors are co-condensation of hydrocarbons, temperature, and CO$_2$ content in the gas. TLC is difficult to inhibit as it is not easy for corrosion inhibitors (which are typically high-molecular-weight liquids) to reach the upper portions of the internal pipe wall. TLC is complicated to predict because, besides water chemistry and surface electrochemistry, considerations must include hydrodynamics as well as heat and mass transfer, which all play a very important role. A full description of TLC exceeds the scope of this text, and the reader is directed to some of the more recent articles on this topic [30, 31].

C10. Effect of Nonideal Solutions and Gases

The infinite dilution theory which is at the core of the chemical expressions used above does not often hold and corrections are needed. A simple way to do this is to utilize the concept of ionic strength, which was already introduced into the various equilibrium expressions in Table 19.2. A more accurate way is to use activity coefficients [32]; however, this path is significantly more complicated. The effect of nonideal solutions on electrochemical reactions behind CO$_2$ corrosion (given in Table 19.3) is not known and cannot be accounted for presently. Some simple studies have indicated that corrosion is mitigated in very concentrated brines (>10 wt % salt); however, causes of this effect are not well understood.

At very high pressures, the simple assumptions about the ideality of the gas phase as well as its solubility in water, break down. One can make corrections based on the fugacity coefficient [33] and use more accurate equations of state; however, these are complicated and their description exceeds the scope of this text. Cases in which the critical point for CO$_2$ is exceeded are even more different and much of what was discussed above needs to be reevaluated for liquid and supercritical CO$_2$ solutions. These topics are beyond the aims of this review.

D. LOCALIZED CO$_2$ CORROSION OF MILD STEEL IN AQUEOUS SOLUTIONS

When compared to uniform CO$_2$ corrosion of mild steel, not nearly as much is known about localized CO$_2$ corrosion. The reason is probably that there are many rather different primary causes of localized CO$_2$ corrosion, and for each of them there is usually more than one secondary factor which complicates the situation. In addition, the term *localized corrosion* is very broadly defined and used because it only implies that corrosion attack occurs in certain locations while it does not happen elsewhere. Given this definition, one can encounter very different scales of localized corrosion. One example of localized corrosion on a macroscopic scale is shown in Figure 19.1, where the severely attacked area is of the order of centimeters or even meters and is still fairly "localized" when compared to the overall length of the pipeline, which typically measures in kilometers. The specific morphology seen in Figure 19.1 with large, flat receded areas free of corrosion products which have corroded severely, sharply divided from surrounding protected areas covered with corrosion products, is termed *mesa attack*, a name borrowed from geological literature. On the other extreme is the very small scale localized attack measured on the micrometer or millimeter scale, which is often called *pitting*. An example of localized corrosion on this microscopic scale is shown in Figure 19.11, which

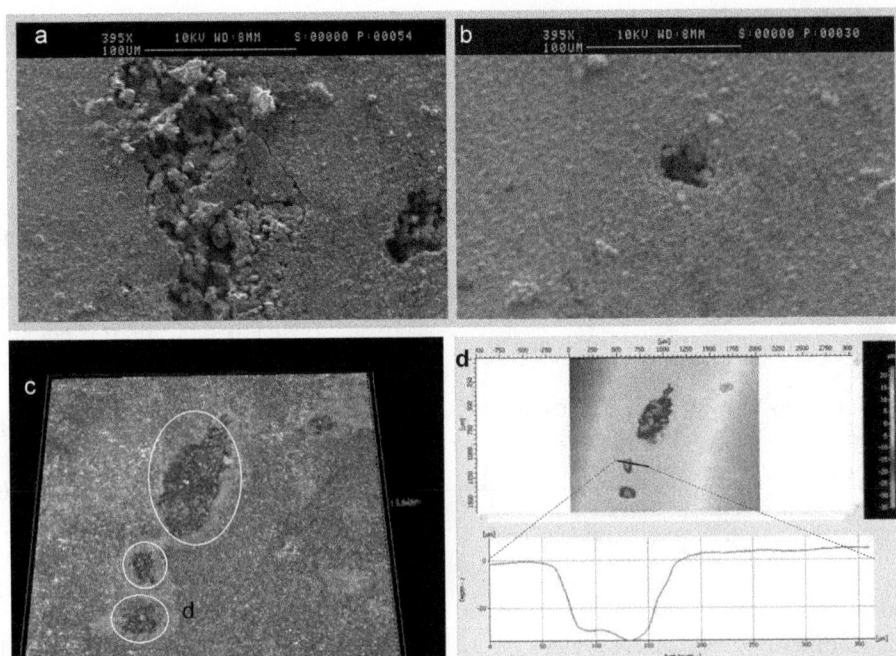

FIGURE 19.11 Isolated pitting on a mild steel surface following partial failure by dissolution of a protective ferrous carbonate layer. Conditions: $80°C$, $p_{CO_2} = 0.5$ bar, 10 wt % NaCl, 10 days exposure, pH 6.3 (during 4 days of layer formation), pH 5.0 (during 6 days of layer dissolution). (a and b) SEM images of ferrous carbonate layer surface covering the steel. (c) Corresponding IFM image approximately 1.5×1.5 mm in size of bare steel surface after ferrous carbonate layer removal. (d) Corresponding bare steel surface elevation map and a depth profile across a characteristic pit.

shows localized attack following a partial failure of a $FeCO_3$ layer by chemical dissolution. This type of attack can be isolated or so widespread over a steel surface that it becomes the dominant mode of corrosion (see Fig. 19.12). In some cases the numerous pits merge to give a mesalike morphology.

The causes of localized CO_2 corrosion defined in this broad sense are many, and they can be divided into those that lead to localized corrosion on a *macroscopic* scale and those that lead to a localized corrosion on a *microscopic* scale. The first group includes:

- Poor or incomplete inhibition
- Water and/or solids separation in multiphase flow
- Flow disturbances and extreme multiphase flow conditions
- Localized condensation in wet-gas flow
- Preferential corrosion of welds
- Ingress of oxygen
- Presence of organic acids
- Presence of H_2S and elemental sulfur

While some of these factors were already discussed above, a more detailed explanation of all the different mechanisms exceeds the scope of the present review. It should be noted that many of them appear simultaneously; for example, precipitation of solids at low flow rates may lead to poor inhibition and a galvanic effect which can be complicated by

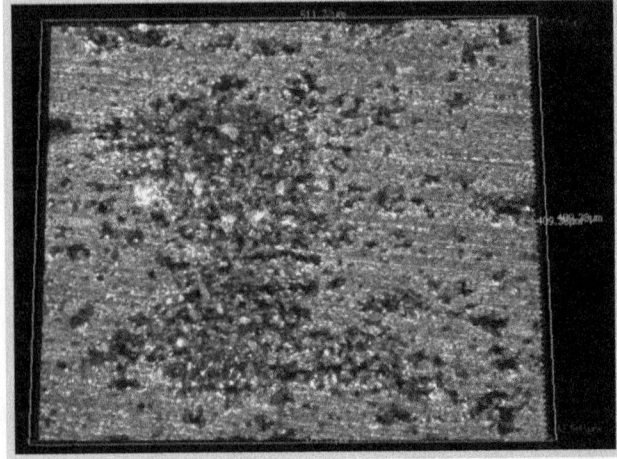

FIGURE 19.12. Image of a mild steel surface approximately 0.5×0.5 mm in size showing widespread pitting following partial dissolution of a protective ferrous carbonate layer, $80°C$, $p_{CO_2} = 0.5$ bar, pH 6.3 (during layer formation), pH 5.0 (during layer dissolution), 20 wt % NaCl, 10 days exposure.

an underdeposit bacterial attack. Another example is a flow disturbance which can lead to localized separation of water in oil/water flow or localized condensation of water in wet-gas flow, resulting in partial removal of a protective scale and aggravated galvanic attack.

The localized CO_2 corrosion rate due any combination of these macroscopic factors can be as high as the corresponding uniform CO_2 corrosion rate of unprotected ("bare") steel, which can be anywhere from high to catastrophic, depending on the conditions (e.g., see Fig. 19.6–19.10). However, in some instances the localized CO_2 corrosion rate is even much higher than the corresponding unmitigated uniform corrosion rate. The reason for this is a different mechanism of CO_2 corrosion which evolves on a microscopic scale. In many instances localized CO_2 corrosion on a microscopic scale is caused by one of these two factors:

- Galvanic attack
- Bacteria

Both effects, which are rather complicated and still not properly understood, are the subject of ongoing investigations.

NOMENCLATURE

A	Surface area of steel, m^2
A/V	Surface-to-volume ratio, m^{-1}
$A_{(FeCO_3)}$	Constant in Arrhenius-type equation for $k_{r(FeCO_3)}$
$B_{(FeCO_3)}$	Constant in Arrhenius-type equation for $k_{r(FeCO_3)}$, kJ/kmol
c_{CO_2}	Bulk aqueous concentration of CO_2, $kmol/m^3$
$c_{CO_3^{2-}}$	Bulk aqueous concentration of CO_3^{2-} ions, $kmol/m^3$
$c_{Fe^{2+}}$	Bulk aqueous concentration of Fe^{2+} ions, $kmol/m^3$
c_{H^+}	Bulk aqueous concentration of H^+ ions, $kmol/m^3$
$c_{s(H^+)}$	"Near-zero" concentration of H^+ underneath mackinawite film at steel surface, set to 1.0×10^{-7} $kmol/m^3$
$c_{HCO_3^-}$	Bulk aqueous concentration of HCO_3^- ions, $kmol/m^3$
$c_{H_2CO_3}$	Bulk aqueous concentration of H_2CO_3, $kmol/m^3$
c_i	Bulk aqueous concentration of given aqueous species, $kmol/m^3$
$c_{s(CO_2)}$	Aqueous concentration of CO_2 underneath mackinawite film at steel surface
CR	Corrosion rate, mm/y
d	Characteristic dimension for given flow geometry, m
d_p	Diameter of pipe, m
d_c	Diameter of rotating cylinder, m
D	Diffusion coefficient of given species, m^2/s
$D_{H_2CO_3}$	Aqueous diffusion coefficient of H_2CO_3, m^2/s
$D_{ref(H_2CO_3)}$	Reference aqueous diffusion coefficient of H_2CO_3, $D_{ref,H_2CO_3} = 1.3 \times 10^{-9}$ m^2/s at 25°C
D_{H^+}	Aqueous diffusion coefficient for H^+
$D_{ref(H^+)}$	Reference aqueous diffusion coefficient for H^+, $D_{ref(H^+)} = 2.80 \times 10^{-8}$ m^2/s at 25°C
D_{CO_2}	Aqueous diffusion coefficient for dissolved CO_2, $D_{CO_2} = 1.96 \times 10^{-9}$ m^2/s
E	Potential, V
E_{corr}	Corrosion (open circuit) potential, V
E_{rev}	Reversible potential
$f_{H_2CO_3}$	Flow factor for chemical reaction boundary layer
F	Faraday constant, $F = 96,485$ C/mol
$H_{sol(CO_2)}$	Henry's constant for dissolution of CO_2, bar/$(kmol/m^3)$
ΔH_{Fe}	Activation enthalpy for Feoxidation, $\Delta H_{Fe} = 50$ kJ/mol
$\Delta H_{(H^+)}$	Activation enthalpy for H^+ ion reduction, $\Delta H_{(H^+)} = 30$ kJ/mol
$\Delta H_{(H_2CO_3)}$	Activation enthalpy for H_2CO_3 reduction, $\Delta H_{(H_2CO_3)} = 57.5$ kJ/mol
i	Current density, A/m^2
i_{corr}	Corrosion current density, A/m^2
i_a	Anodic current density, A/m^2
i_c	Cathodic current density, A/m^2
$i_{c(H^+)}$	Cathodic current density for H^+ ion reduction, A/m^2
$i_{c(H_2CO_3)}$	Cathodic current density for H_2CO_3 reduction, A/m^2
$i_{lim(H^+)}^d$	Mass transfer (diffusion) limiting current density for H^+ ion reduction, A/m^2
$i_{lim(H_2CO_3)}^r$	Chemical reaction limiting current density for H_2CO_3 reduction, A/m^2
$i_{0(Fe)}$	Exchange current density of iron oxidation, A/m^2
$i_{0(H^+)}$	Exchange current density for H^+ ion reduction, A/m^2
$i_{0(H_2CO_3)}$	Exchange current density for H_2CO_3 reduction, A/m^2
$i_{0(Fe)}^{ref}$	Reference exchange current density of Feoxidation, $i_{0(Fe)}^{ref} = 1$ A/m^2
$i_{0(H^+)}^{ref}$	Reference exchange current density of H^+ oxidation, $i_{0(H^+)}^{ref} = 0.03$ A/m^2 at $T_{c,ref} = 25$°C and pH 4
$i_{0(H_2CO_3)}^{ref}$	Reference exchange current density for H_2CO_3 reduction, $i_{0(H_2CO_3)}^{ref} = 0.06$ A/m^2 at $T_{c,ref} = 25$ °C, pH 5, and $c_{H_2CO_3,ref} = 10^{-4}$ $kmol/m^3$
$i_{\alpha(H^+)}$	Charge transfer current density for H^+ ion reduction, A/m^2
$i_{\alpha(H_2CO_3)}$	Charge transfer current density for H_2CO_3 reduction, A/m^2

I Ionic strength, $kmol/m^3$

k_{hyd}^b Backward reaction rate of H_2CO_3 dehydration reaction, s^{-1}, $k_{hyd}^b = k_{hyd}^f / K_{hyd}$

k_{hyd}^f Forward reaction rate for CO_2 hydration reaction, s^{-1}

$k_{m(H^+)}$ Aqueous mass transfer coefficient for H^+, m/s

$k_{m(H_2CO_3)}$ Aqueous mass transfer coefficient for H_2CO_3, m/s

$k_{m(CO_2)}$ Aqueous mass transfer coefficient for CO_2, m/s

$k_{r(FeCO_3)}$ Kinetic constant in ferrous carbonate precipitation rate equation $mol^{-1} s^{-1}$

K_{hyd} Equilibrium hydration constant for CO_2, $K_{hyd} = k_{hyd}^f / k_{hyd}^b = 2.58 \times 10^{-3}$

K_{bi} Equilibrium constant for dissociation of HCO_3^-, $kmol/m^3$

K_{bs} Equilibrium constant for dissociation HS^-, $kmol/m^3$

K_{ca} Equilibrium constant for dissociation of H_2CO_3, $kmol/m^3$

$K_{sol(CO_2)}$ Solubility constant for dissolution of CO_2, $kmol/m^3/bar$

$K_{sp(FeCO_3)}$ Solubility product constant for ferrous carbonate, $(kmol/m^3)^2$

M_{Fe} Molecular mass of iron, kg/kmol

n Number of electrons used in reducing or oxidizing given species, kmol/kmol

p_{CO_2} Partial pressure of CO_2, bars

p_{H_2S} Partial pressure of H_2S, bars

$\mathfrak{R}$ Electrochemical reaction rate, $kmol/(m^2 s)$

$\mathfrak{R}_{FeCO_3}$ Precipitation rate for iron carbonate, $kmol/(m^3 s)$

R Universal gas constant, $R = 8.314 \, J/(mol \, K)$

Re Reynolds number, $Re = v\rho_{H_2O}d/\mu_{H_2O}$

Sc Schmidt number of a given species, $Sc = \mu_{H_2O}/(\rho_{H_2O}D)$

Sh_p Sherwood number of a given species for a straight pipe flow geometry, $Sh_p = k_m d_p/D$

Sh_r Sherwood number of a given species for a rotating cylinder flow geometry, $Sh_r = k_m d_c/D$

$SS_{(FeCO_3)}$ Supersaturation of iron carbonate

T_c Temperature, °C

$T_{c,ref}$ Reference temperature, $T_{c,ref} = 25$°C

T_f Temperature, °F

T_k Temperature, K

v Water characteristic velocity, m/s

z_i Species charge of various aqueous species

Greek Characters

$\delta_{m(H_2CO_3)}$ Thickness of mass transfer layer for H_2CO_3, m

$\delta_{r(H_2CO_3)}$ Thickness of chemical reaction layer for H_2CO_3, m

μ_{H_2O} Water dynamic viscosity, Pa · s

$\mu_{H_2O,ref}$ Reference water dynamic viscosity, at reference temperature, Pa · s, $\mu_{H_2O,ref} = 1.002 \times 10^{-4}$ Pa · s at 20°C

$\zeta_{H_2CO_3}$ Ratio of mass transfer layer and chemical reaction thicknesses for H_2CO_3

ρ_{H_2O} Density of water, kg/m^3

ρ_{Fe} Density of iron, kg/m^3

ACKNOWLEDGMENTS

The author is indebted to the many individuals who contributed to the evolution of this review. The list includes fellow scientists and researchers, corrosion engineers from the industry, and graduate students. Many are already mentioned in the list of references. Others, who are too numerous to be listed by name here, are mainly past and present staff and students at the Institute for Corrosion and Multiphase Technology, Ohio University. Also, my gratitude goes to the many companies who have been supporting research efforts in CO_2 corrosion for a sustained period of time by providing continued guidance and funding.

REFERENCES

1. M. R. Bonis and J. L. Crolet, "Basics of the Prediction of the Risks of CO_2 Corrosion in Oil and Gas Wells," Paper No. 466, CORROSION/1989, NACE International, Houston, TX, 1989.

2. J. Oddo and M. Tomson, "Simplified Calculation of $CaCO_3$ Saturation at High Temperatures and Pressures in Brine Solutions," Society of Petroleum Engineers, Richardson, TX, 1982, pp. 1583–1590.

3. D. A. Palmer and R. van Eldik, Chem. Rev., **83**, 651 (1983).

4. Y. K. Kharaka et al., "SOLMINEQ 88: A Computer Program code for Geochemical Modelling of Water-Rock Interactions," U. S. Geological Survey Water Resources Investigations Report 88-4227, Menlo Park, CA, 1988.

5. D. M. Drazic, "Iron and Its Electrochemistry in an Active State," Aspects of Electrochem., **19**, 79 (1989).

6. W. Lorenz and K. Heusler, "Anodic Dissolution of Iron Group Metals," in Corrosion Mechanisms, F. Mansfeld (Ed.), Marcel Dekker, New York, 1987.

7. S. Nešić, N. Thevenot, and J. L. Crolet, "Electrochemical Properties of Iron Dissolution in CO_2 Solutions—Basics Revisited," Paper No. 3, CORROSION/1996, NACE International, Houston, TX, 1996.

8. P. Delahay, J. Am. Chem. Soc., **74**, 3497 (1952).

9. S. Nešić, J. Postlethwaite, and S. Olsen, "An Electrochemical Model for Prediction of the Corrosion of Mild Steel in Aqueous CO_2 Solutions," J. Corrosion, **52**, 280 (1996).

10. C. de Waard and D. E. Milliams, Corrosion, **31**, 131 (1975).

11. L. G. S. Gray, B. G. Anderson, M. J. Danysh, and P. G. Tremaine, "Mechanism of Carbon Steel Corrosion in Brines Containing Dissolved Carbon Dioxide at pH 4," Paper No. 464, CORROSION/1989, NACE International, Houston, TX, 1989.

12. E. Eriksrud and T. Søntvedt, "Effect of Flow on CO_2 Corrosion Rates in Real and Synthetic Formation Waters," in Advances in CO_2 Corrosion, Vol. 1, Proceedings of the Corrosion/83 Symposium on CO_2 Corrosion in the Oil and Gas Industry, R. H. Hausler and H. P. Goddard (Eds.), NACE, 1984, Houston, TX, p. 20.

13. G. Schmitt and B. Rothman, Werkstoffe und Korrosion, **28**, 816 (1977).

14. L. G. S. Gray, B. G. Anderson, M. J. Danysh, and P. R. Tremaine, "Effect of pH and Temperature on the Mechanism of Carbon Steel Corrosion by Aqueous Carbon Dioxide," Paper No. 40, CORROSION/1990, NACE International, Houston, TX, 1990.

15. F. P. Berger and K.-F. F.-L. Hau, Int. J. Heat Mass Transfer, **20**, 1185 (1977).

16. M. Eisenberg, C. W. Tobias, and C. R. Wilke, J. Electrochem. Soc., **101**, 306 (1954).

17. K. Chokshi, W. Sun, and S. Nešić, "An Investigation of Corrosion Inhibitor—Iron Carbonate Scale Interaction in Carbon Dioxide Corrosion," J. Corros. Sci. Eng., **10**, preprint 24 (2008).

18. W. Sun, S. Nešić, and R. C. Woollam, "The Effect of Temperature and Ionic Strength on Iron Carbonate ($FeCO_3$) Solubility Limit," Corros. Sci. **51**, 1273 (2009).

19. W. Sun and S. Nešić, "Kinetics of Corrosion Layer Formation, Part 1. Iron Carbonate Layers in Carbon Dioxide Corrosion," Corrosion, **64**, 334 (2008).

20. Y. Sun and S. Nešić, "A Parametric Study and Modeling on Localized CO_2 Corrosion in Horizontal Wet Gas Flow," Paper No. 380, CORROSION/2004, NACE International, Houston, TX, 2004.

21. S. Nešić, G.T. Solvi, and J. Enerhaug, Corrosion, **51**, 773 (1995).

22. D. W. Shoesmith, P. Taylor, M. G. Bailey, and D. G. Owen, "The Formation of Ferrous Monosulfide Polymorphs During the Corrosion of Iron by Aqueous Hydrogen Sulfide at 21°C," J. Electrochem. Soc., **125**, 1007–1015 (1980).

23. D. W. Shoesmith, "Formation, Transformation and Dissolution of Phases Formed on Surfaces," Lash Miller Award Address, Electrochemical Society Meeting, Ottawa, Nov. 27, 1981.

24. S. N. Smith, "A Proposed Mechanism for Corrosion in Slightly Sour Oil and Gas Production," Paper no. 385, Twelfth International Corrosion Congress, Houston, TX, Sept. 19–24 1993.

25. S. N. Smith and E. J. Wright, "Prediction of Minimum H_2S Levels Required for Slightly Sour Corrosion," Paper No. 11, CORROSION/1994, NACE International, Houston, TX, 1994.

26. S. N. Smith and J. L. Pacheco, "Prediction of Corrosion in Slightly Sour Environments," Paper No. 02241, CORROSION/2002, NACE International, Houston, TX, 2002.

27. S. N. Smith and M. Joosten, "Corrosion of Carbon Steel by H_2S in CO_2 Containing Oilfield Environments," Paper No. 06115, CORROSION/2006, NACE International, Houston, TX, 2006.

28. M. Bonis, M. Girgis, K. Goerz, and R. MacDonald, "Weight Loss Corrosion with H_2S: Using Past Operations for Designing Future Facilities," Paper No. 06122, CORROSION/2006, NACE International, Houston, TX, 2006.

29. K. George and S. Nešić, "Investigation of Carbon Dioxide Corrosion of Mild Steel in the Presence of Acetic Acid, Part I—Basic Mechanisms," J. Corros., **63**, 178 (2007).

30. Y. M. Gunaltun, D. Larrey, "Correlation of Cases of Top of Line Corrosion with Calculated Water Condensation Rates," Paper No. 71, CORROSION/2000, NACE International, Houston, TX, 2000.

31. Z. Zhang, D. Hinkson, M. Singer, H. Wang, and S. Nesic, "A Mechanistic Model of Top of the Line Corrosion," J. Corros., **63**, 1051 (2007).

32. A. Anderko and R. Young, "Simulation of CO_2/H_2S Corrosion Using Thermodynamic and Electrochemical Models," Paper No. 31, CORROSION/1999, NACE International, Houston, TX, 1999.

33. C. de Waard and U. Lotz, "Prediction of CO_2 Corrosion of Carbon Steel," Paper No. 69, CORROSION/1993, NACE International, Houston, TX, 1993.

34. S. Nešić, B. F. M. Pots, J. Postlethwaite, and N. Thevenot, "Superposition of Diffusion and Chemical Reaction Limiting Currents—Application to CO_2 Corrosion," J. Corros. Sci. Eng., **1**, Paper No. 3, available: http://www.cp.umist.ac.uk/JCSE/Vol1/PAPER3/V1_p3int.htm, The Corrosion Information Server, The Corrosion & Protection Centre at UMIST, Manchester, UK, 1995.

20

HIGH-TEMPERATURE OXIDATION[*]

C. A. C. Sequeira

Instituto Superior Técnico, Lisboa, Portugal

[*] Editor's Note: In his forthcoming book, *High Temperature Corrosion: Fundamentals and Engineering*, in the Wiley Series in Corrosion, C. A. C. Sequeira discusses the subject of high-temperature oxidation in more detail than is possible in this overview chapter.

A. INTRODUCTION

A1. Historical Perspective

Oxidation is an important high-temperature corrosion phenomenon. Metals or alloys are oxidized when heated to elevated temperatures in air or in highly oxidizing environments, such as combustion atmospheres with excess air or oxygen. Many metallic components are subject to oxidation in engineering applications.

The first paper that expressly addressed high-temperature oxidation was written by Gustav Tammann in 1920. He articulated the "parabolic law," that is, the rate of oxidation of metal decreases as oxide layer thickness increases. In 1922 he established the logarithmic law of oxidation of metals. However, the first paper that laid out the basics of the problem as we know it was that by N. B. Pilling and R. E. Bedworth in 1923. They defined "high temperature" as that at which the transport of the reactive components through the protective layer was the principal determinant of the reaction rate (as opposed to the situation in aqueous corrosion processes at close to ambient temperatures). They showed that under these circumstances it could be expected that the rate of reaction would diminish as the protective scale thickened, leading to a "parabolic rate law." They also highlighted the problems associated with forming an adherent crack-free protective oxide layer on the oxidizing surface because of the volume changes associated with the oxidation process.

The discussion to this seminal paper shows that others were thinking along similar lines at the time. Six years later, Leonard B. Pfeil introduced the concept of movement of metal outward rather than oxygen inward into the oxide layer, and in 1934 Portevin, Prétet, and Jolivet carried out extensive studies on the oxidation of iron and its alloys. At the same time, the discovery that oxides contained lattice defects and that the transport processes within them are determined by the motion of these defects allowed a more quantitative approach. This was recognized by Carl Wagner, who produced an important body of work over the course of the next 20 years largely defining how we now look at the basic theory of the bulk transport processes in oxides. Of significance was the derivation of Wagner's equation by Hoar and Price in 1938. In recent years, however, the recognition that in many cases the transport processes involve short-circuit paths, such as grain boundaries, has introduced further complications into this elegant picture.

The problem of the integrity of the protective oxide was not part of Wagner's contribution, and its practical solution for the high-temperature alloys that were developed for applications mentioned above was discovered essentially by accident in the early 1940s. It was found that very small amounts of what are now called "reactive elements" added to the alloys introduced a remarkable improvement in the apparent integrity of the protective oxide, particularly in its resistance to thermal cycling. The reasons for this effect are still a matter of considerable debate.

Other important contributions in the 1939–1948 period were those of Cabrera and Mott postulating that oxide film growth is controlled by ions jumping from site to site over intervening energy barriers. Mott's theory was then highly criticized by Karl Hauffe, who studied the oxidation of alloys. During the 1920–1940 period, other relevant studies on oxidation at high temperature deserve reference: the interference method of obtaining thickness of oxide films (Tammann, 1920–1926), the spectroscopic method to obtain thickness of oxide film (Constable, 1927), and the X-ray and electron diffraction methods to study oxide films (Finch Quarrell, 1933).

The earliest treatments of oxidation problems considered simple systems, with a single oxidant (usually oxygen) and a pure metal, although in practice high-temperature-resistant materials were always alloys. More recently, approaches such as in multicomponent diffusion theory have been applied, and the growth of oxides on polyphase materials have been analyzed. These approaches have been greatly assisted by the development of modern characterization techniques.

A2. Purpose of This Chapter

The oxidation of metals is usually a reaction between a gas and a solid which produces a solid reaction product. At a first glance, this would seem to be a very simple process but,

actually, it is considerably more complex. The metal is usually not pure but contains, in addition to metallic impurities, O, N, H, C, S, and so on. The gas atmosphere is also usually complex, containing (in addition to O_2), N_2, H_2, CO_2, H_2O, and so on. One would think that the reaction product, that is, the scale that formed on the metal, acts as a physical barrier between the reactants, and thus the reaction should cease after the barrier is established. We know that this is not the case, because transport of matter through the scale causes the reaction to continue. We also know that the scale may not be dense and adherent to the substrate, but it may be cracked, partially spalled, partially detached (wrinkled), or even very porous. In some extreme cases, the scale may be a liquid which simply drips from the surface or it may volatilize at very high temperatures. Indeed, the reaction between a gas and a metal is very complicated.

Our interest in this multidisciplinary field of physical chemistry, solid-state chemistry, metallurgy, materials science, and engineering arises from the fact that the chemical activity of a metal in various environments is an important factor in the winning, processing, and use of the metal. Demands are pressing from aerospace/gas turbine, chemical processing, refining and petrochemical, fossil-fired power generation, coal gasification, waste-to-energy industry, pulp and paper, heat treating, mineral and metallurgical processing, nuclear power, space exploration, molecular electronics, and other sides for better metals and alloys for high-temperature service in special reactive atmospheres and for metals with special physical properties.

The science of gas–solid reactions, oxidation referred to in the generic sense, can involve reactions with, for example, sulfur, nitrogen, carbon dioxide, and water vapor and has greatly evolved in the past 50 years. Numerous symposia and colloquia [1–7] have been held on the subject and have been widely attended by researchers from all over the world. One of the many factors enabling advancement of our understanding of the field is the creation and evolution of new, sophisticated instruments and techniques which allow a much better analysis of scale compositions and structures. Details regarding this subject have led to thousands of publications and the writing of several books. Of the more recent books on oxidation, those indicated in references [8–17] have been very useful.

The present chapter summarizes the main factors for determining the nature and extent of gas–metal reactions, which are of paramount importance to understand the subject.

B. THERMODYNAMIC CONSIDERATIONS

An important tool in the analysis of oxidation problems is equilibrium thermodynamics which, although not predictive of kinetics, allows one to ascertain which reaction products are possible, whether or not significant evaporation or condensation of a given species is possible, the conditions under which a given reaction product can react with a condensed deposit, and so on. The complexity of the oxidation phenomena usually dictates that the thermodynamic analysis be represented in graphical form. The types of thermodynamic diagrams most often used in oxidation research are:

1. Gibbs free energy versus composition diagrams and activity versus composition diagrams, which are used for describing thermodynamics of solutions [18–22]
2. Standard free energy of formation versus temperature diagrams which allow the thermodynamic data for a given class of compounds, oxides, sulfides, carbides, and so on, to be represented in a compact form [23–29]
3. Vapor species diagrams, which allow the vapor pressures of compounds to be presented as a function of convenient variables such as partial pressure of a gaseous component [30–33]
4. Two-dimensional, isothermal stability diagrams, which map the stable phases in systems involving one metallic and two reactive, nonmetallic components [34–37]
5. Two-dimensional, isothermal stability diagrams, which map the stable phases in systems involving two metallic components and one reactive, nonmetallic component [38, 39]
6. Two-dimensional, isothermal stability diagrams, which map the stable phases in systems involving two metallic and two reactive, nonmetallic components [40]

The basic concepts pertinent to the construction and analysis of those thermodynamic diagrams are partial Gibbs free energy, chemical potential, free energy of mixing, Gibbs–Duhem equation, activity, Raoult's law, Henry's law, equilibrium state, variance (number of degrees of freedom of a system), pressure dependence of activity in the case of equilibrium between a condensed solution and a gas phase, and types of solution (ideal, dilute, concentrated). These concepts are described in numerous thermodynamics books [19, 41–43].

Determination of the conditions under which a given corrosion product is likely to form is often required, for example, in selective oxidation of alloys. In this regard, Ellingham diagrams, that is, plots of the standard free energy of formation ($\Delta G°$) versus temperature for the compounds of a type (e.g., oxides, sulfides, carbides) are useful in that they allow comparison of the relative stabilities of each compound. In the next section these free-energy/temperature diagrams will be discussed, and then our considerations on the thermodynamics of high-temperature oxidation will finish with a brief reference to the volatility of oxides.

B1. Ellingham Diagrams

The Ellingham plot is a graphical representation of the standard free energies of oxidation of pure metals versus temperature. The most useful form of representation is to express the quantities in terms of 1 mol of the oxidising species:

$$M + O_2 = MO_2 \qquad (20.1)$$

At a given temperature T, the standard free energy $\Delta G_T°$ of the reaction is

$$\Delta G_T° = \Delta H_T° - T \, \Delta S_T° \qquad (20.2)$$

where $\Delta H_T°$ and $\Delta S_T°$ are, respectively, the standard enthalpy and entropy of the reaction. At equilibrium, we can write

$$\Delta G_T° = -RT \ln K = -RT \ln \left(\frac{a_{HO_2}}{a_M} \cdot p_{O_2}^{-1} \right) \qquad (20.3)$$

Assuming that the oxide MO_2 is stoichiometric and that there is no miscibility between metal and oxide, the activities of M and MO_2 are taken as unity. Thus

$$\Delta G_T° = RT \ln p_{O_2} \qquad (20.4)$$

The oxygen partial pressure at which the metal and oxide coexist is the dissociation pressure of the oxide. In Ellingham diagrams, $\Delta G_T°$ (or $RT \ln p_{O_2}$) is plotted versus temperature for the compounds of a given type (e.g., oxides, sulfides, chlorides, carbides). The standard enthalpy and entropy of formation of the compounds are also considered to be independent of temperature over large temperature ranges (the "Ellingham approximation"). For these conditions

$$\Delta G_T° = \Delta H_{298}° - T \, \Delta S_{298}° \qquad (20.5)$$

Accordingly, the $\Delta G_T°$ values fall on straight lines in the diagram although changes in the slope of the lines occur in the diagram corresponding to phase changes of the metal or the compound. Figure 20.1 is such a plot for many simple oxides. The values of $\Delta G_T°$ are expressed as kilojoules per mole of O_2 so the stabilities of various oxides may be compared directly; that is, the lower the position of the line on the diagram, the more stable the oxide.

The values of p_{O_2} may be obtained directly from the oxygen monograph on the diagram by drawing a straight line from the origin marked O through the free energy line at the temperature of interest and reading the oxygen partial pressure from its intersection with the scale at the right side labeled p_{O_2}. Values for the pressure ratio H_2/H_2O for equilibrium between a given metal and oxide may be obtained by drawing a similar line from the point marked H to the scale

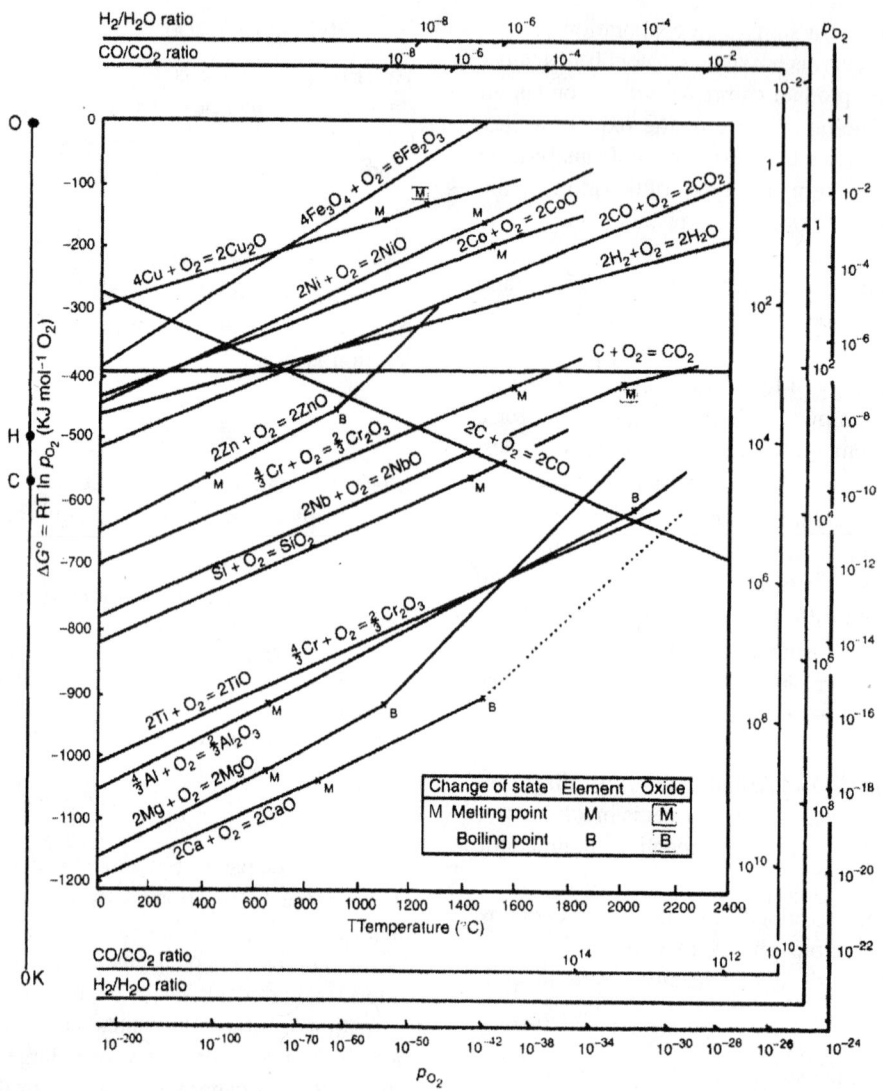

FIGURE 20.1. Standard free energy of formation of selected oxides as function of temperature [12].

labelled H_2/H_2O ratio and values for the equilibrium CO/CO_2 ratio may be obtained by drawing a line from point C to the scale CO/CO_2 ratio. Thus, it is possible to obtain the oxygen potential of the environment in terms of $p_{O_2}, p_{H_2}/p_{H_2O}, p_{CO}/p_{CO_2}$ and the oxygen partial pressure in equilibrium with the oxide of interest from Figure 20.1 to determine whether or not the oxide is likely to form thermodynamically.

Figure 20.1 also illustrates the relative stability of various oxides. The most stable oxides have the largest negative values of ΔG_T°, or the lowest values of p_{O_2}, or the highest values of p_{H_2}/p_{H_2O} and p_{CO}/p_{CO_2}.

It is clear from Figure 20.1 that oxides of iron, nickel, and cobalt, which are the alloy bases for the majority of engineering alloys, are significantly less stable than the oxides of some solutes (e.g., chromium, aluminum, silicon) in engineering alloys. When one of these solute elements is added

to iron, nickel, or cobalt, internal oxidation of the solute is expected to occur if the concentration of the solute is relatively low. As the solute concentration increases to a sufficiently high level, oxidation of the solute will be changed from internal oxidation to external oxidation, resulting in an oxide scale that protects the alloy from rapid oxidation. This process is known as "selective oxidation." The majority of iron-, nickel-, and cobalt-based alloys rely on selective oxidation of chromium to form a Cr_2O_3 scale for oxidation resistance. Some high-temperature alloys use aluminum to form an Al_2O_3 scale for oxidation resistance.

Most oxides exhibit high melting points and remain in a solid state for the temperature range in which the alloys are used. Oxides of molybdenum (MoO_3) and vanadium (V_2O_5), however, exhibit very low melting points. Vanadium (V), which is a strong carbide former, is often used in alloy steels for increasing the strength of the material. However, the amount used typically is quite small and is not likely to form V_2O_5. Molybdenum (Mo) is also a strong carbide former and is used in a small amount to strengthen low-alloy steels (e.g., Cr–Mo steels). It is unlikely that these steels will be affected by MoO_3 oxidation problems.

Industrial gaseous atmospheres may, in addition to air, consist of complex gases (CO_2, H_2O, SO_2, SO_3, etc.) or mixtures of several gases or contain the gaseous products of corrosion or of the thermal decomposition of corrosion products [44–46]. Despite the complexity of corrosion phenomena, the problem can be solved, in some cases, using simple thermodynamic calculations from simplifying hypotheses. In other situations involving numerous components, it is necessary to use software that minimizes the total free energy of the system [47–53].

B2. Volatility of Oxides

Some oxides exhibit high vapor pressures at very high temperatures (e.g., above 1000°C). Oxide scales become less protective when their vapor pressures are high. Chromium, molybdenum, tungsten, vanadium, platinum, rhodium, and silicon are metals for which volatile species are important at high temperature. Vanadium is typically used in small quantities as a carbide former in alloy steels. Thus, the volatility of VO_2 is generally of no concern in oxidation of alloys.

The oxidation of Pt and Pt group metals at high temperatures is influenced by oxide volatility in that the only stable oxides are volatile. This results in a continuous mass loss. Alcock and Hooper [54] studied the mass loss of Pt and Rh at 1400°C as a function of oxygen pressure. The gaseous species were identified as PtO_2 and RhO_2. These results have an extra significance because Pt and Pt–Rh wires are often used to support specimens during high-temperature oxidation experiments. If these experiments involve mass change measurements, it must be recognized that there will be a mass loss associated with volatilization of oxides from the support wires.

The oxidation of pure Cr is, in principle, a simple process since a single oxide, Cr_2O_3, is observed to form. However, under uncertain exposure conditions, several complications arise which are important both for the oxidation of pure Cr and for many important engineering alloys which rely in a protective Cr_2O_3 layer for oxidation protection. The two most important features are scale thinning by CrO_3 evaporation and scale buckling as a result of compressive stress development [55–58].

The formation of CrO_3 by the reaction

$$Cr_2O_3 + \tfrac{3}{2}O_2 = 2CrO_3 \qquad (20.6)$$

becomes significant at high temperatures and high oxygen partial pressures. The evaporation of CrO_3, shown schemat-

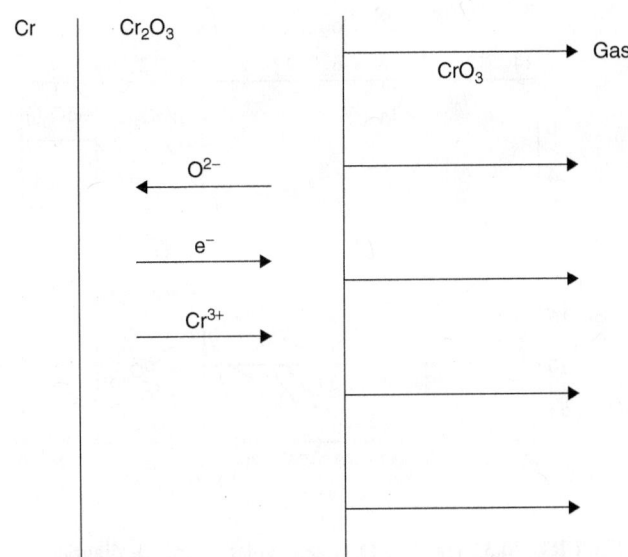

FIGURE 20.2. Schematic diagram of combined scale growth and oxide volatilization from Cr [12].

ically in Figure 20.2, results in the continuous thinning of the protective Cr_2O_3 scale so the diffusive transport through it is rapid. The effect of the volatilization on the oxidation kinetics has been analysed by Tedmon [59]. Caplan and Cohen [57] also observed that resistance promoted volatilization of Cr_2O_3. Asteman et al. [58] indicated that high vapor pressure of $CrO_2(OH)_2$ can form by reacting Cr_2O_3 with H_2O in O_2-containing environments.

The volatilization of oxides is particularly important in the oxidation of Mo and W at high temperatures and high oxygen pressures. Unlike Cr, which develops a limiting scale thickness, complete oxide volatilization can occur in these systems. The condensed and vapor species for the Mo–O and W–O systems have been reviewed by Gulbransen and Meier [60] and the vapor species diagrams for a temperature of 1250 K are presented in Figures 20.3 and 20.4. The effects of oxide volatility on the oxidation of Mo have been observed by Gulbransen and Wysong [61] at temperatures as low as 475°C and the rate of oxide evaporation above 725°C was such that gas-phase diffusion became the rate-controlling process [62]. Naturally, under these conditions, the rate of oxidation is catastrophic. Similar behavior is observed for the oxidation of tungsten, but at higher temperatures because of the lower vapor pressures of the tungsten oxides. The oxidation behavior of tungsten has been reviewed in detail by Kofstad [8].

The formation of SiO_2 on silicon-containing alloys and Si-based ceramics results in very low oxidation rates. However, this system is also one that can be influenced markedly by oxide vapor species. Whereas the oxidation of Cr is influenced by such species at high oxygen pressures, the effects for Si are important at low oxygen partial pressures.

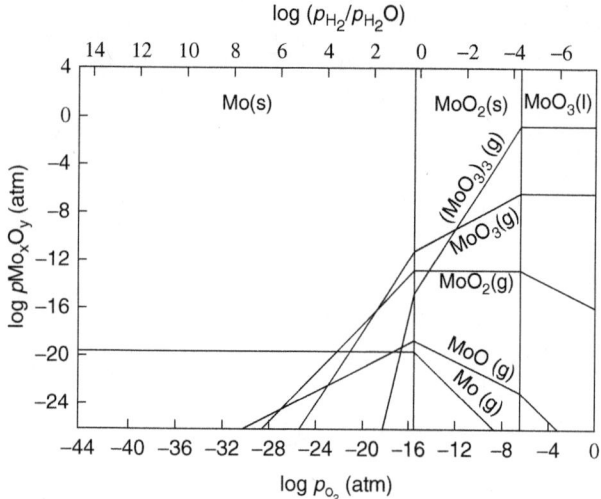

FIGURE 20.3. The Mo–O system volatile-species diagram for 1250 K [12].

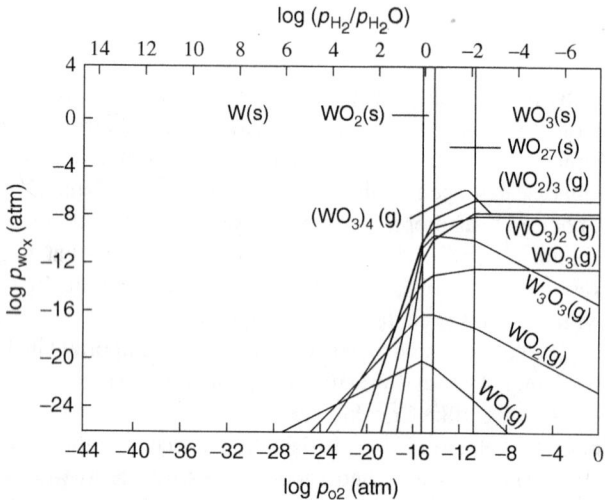

FIGURE 20.4. The W–O system volatile-species diagram for 1250 K [12].

The reason for this may be seen from the volatile-species diagram for the Si–O system [45]. A significant pressure of SiO is seen to be in equilibrium with $SiO_2(s)$ and $Si(s)$ at oxygen pressures near the dissociation pressure of SiO_2. This can result in a rapid flux of SiO away from the specimen surface and the subsequent formation of a nonprotective SiO_2 smoke. This formation of the SiO_2 as a smoke, rather than as a continuous layer, allows continued rapid reaction [63, 64].

C. KINETIC CONSIDERATIONS

Consider the oxidation reaction

$$M + \tfrac{1}{2}O_2 = MO \tag{20.7}$$

The progress, W, of this reaction can be characterized using several definitions of the reaction rate. It can, for example, be defined as the rate of oxygen pickup, $dn_o/A\, dt$, where dn_o corresponds to the number of moles of oxygen consumed during time dt and A the sample area. From an experimental point of view, though, it is easier to use the sample weight gain Δm. The reaction rate can then be expressed as $d\,\Delta m/A\, dt$, where $d\,\Delta m$ is the weight change occurring during time dt. These expressions are linked by the equation

$$\frac{dn_o}{dt} = \frac{1}{16}\frac{d\,\Delta m}{dt} \tag{20.8}$$

where Δm is expressed in grams. Integration of the rate equation leads to the rate law corresponding to the corrosion process and defines the progress, W, of the reaction with time. We obtain either an implicit form,

$$f(W) = kt \tag{20.9}$$

where k is the rate constant for the reaction process, or an explicit form,

$$W = g(t) \tag{20.10}$$

In any fundamental study of the oxidation mechanism of a metal or an alloy, one of the main factors that needs to be determined is the variation of the oxidation rate with temperature and with the pressure of the oxidizing gas.

This type of investigation has sometimes been neglected because of difficulties controlling precisely the gas pressure in thermobalances. However, for some years, it has been possible to couple such thermogravimetric equipment to devices capable of controlling and monitoring the oxygen partial pressure in a gas (e.g., an electrochemical oxygen pump, oxygen sensor). The oxygen pressure then can be precisely controlled between 1 and 10^{-25} bar in gas mixtures, for example, inert gas–oxygen, CO–CO_2, or H_2–H_2O mixtures [65].

The rate constants of the kinetic laws often obey, under constant pressure, an Arrhenius-type equation:

$$k = k^0 \exp\left(-\frac{E_a}{RT}\right) \quad \text{(at given } P\text{)} \tag{20.11}$$

where E_a is the apparent activation energy of the process, R the gas constant, and T the absolute temperature. The apparent activation energy can be easily determined by plotting k as a function of $1/T$. The slope of the straight line obtained is equal to $-E_a/2.303R$. A change in the activation energy could indicate a corresponding change in the limiting process for the corrosion reaction.

The main kinetic laws are of linear, parabolic, logarithmic, or cubic types, but it should be noted that these are limiting cases and deviations from them are often encountered. In some cases, it is difficult, or even impossible, to obtain such simple kinetic laws from the experimental results [11, 66–74].

At high temperatures, the oxidation kinetics of numerous metals obey a parabolic law:

$$W^2 = k_p t \qquad (20.12)$$

where k_p is the parabolic constant. Such a law corresponds, as will be shown later, to a corrosion rate limited by diffusion through the compact scale that is formed.

The reaction rate constant may be expressed in different units depending on the actual parameter used to define the progress of the reaction. For example, if the extent of reaction is characterized by the mass gain per unit surface area of the metal during the exposure period t, the kinetic law is given by $(\Delta m/A)^2 = k_p t$ and the rate constant is expressed in $kg^2/m^4/s$. If the reaction rate is defined by the increase in thickness, y, of the scale, the kinetic law has the form $y^2 = k'_p t$ and the parabolic rate constant k'_p is expressed in m^2/s. On the other hand, if the reaction rate is defined by the number of moles of the compound MX formed per unit area during the exposure period t, the kinetic law has the form $(\Delta n_{MX}/A)^2 = k''_p t$. In this case, the rate constant k''_p is expressed in $mol^2/m^4/s$. There is a simple relationship between the rate constants k_p, k'_p, and k''_p:

$$k'_p = \Omega^2 k''_p = \left(\frac{\Omega}{M_x}\right)^2 k_p \qquad (20.13)$$

where Ω is the molar volume of compound MX and M_x is the atomic weight of the nonmetallic element (oxygen, sulfur, etc.).

In some cases, the oxidation rate is constant, which means that the kinetic law is linear:

$$W = k_l t \qquad (20.14)$$

where k_l is the linear rate constant. As will be shown later, the oxidation rate is then governed by an interfacial process such as sorption or reaction at the metal/oxide interface. Using similar nomenclature as for the parabolic rate constants, k_l characterizes the reaction measured by the mass gain per unit surface area during time t and k'_l if the rate is defined by the increase in thickness of the growing scale.

The cubic law ($W^3 = k_c t$) has been observed during the oxidation of several metals, for example, copper, nickel, and zirconium.

Logarithmic laws are observed typically in the case of many metals at low temperatures (generally below 673 K). The initial oxidation rate, corresponding to the growth of oxide layers of thickness generally less than a few tens of nanometers, is quite rapid and then drops off to low or negligible values. This behavior can be described by a direct logarithmic law:

$$W = k_{log} \log(at + 1) \qquad (20.15)$$

or by an inverse logarithmic law,

$$\frac{1}{W} = B - k_{inv} \log t \qquad (20.16)$$

The evaluation of the kinetic parameters in the case of the logarithmic law is, generally, not very precise and this makes it difficult to validate experimentally proposed mechanisms.

The oxidation rate is frequently found to follow a combination of rate laws.

As an example, at low temperatures, a logarithmic law followed by a parabolic rate equation can be observed. At high temperature, oxidation reactions are often described by a parabolic rate equation followed by a linear law ("paralinear" regime) or a linear rate equation followed by a parabolic law [75].

Typical kinetic laws characteristic of the oxidation of a large number of metals as a function of temperature were fully analyzed by Kubaschewski and Hopkins [9], Kofstad [11], and others.

Besides the variation of the kinetic laws with temperature, a change of these rate equations with time can sometimes occur. A typical example characteristic of the changes that may be observed as functions of temperature and time is given in Figure 20.5. At 800°C, for example, the following rate equations are successively observed: parabolic, paralinear, and finally linear after extended oxidation. In many cases, it may be difficult to fit experimental data to simple rate equations, but a first approach can be to plot the $W = f(t)$ curve using double-logarithmic coordinates. In the case of a law of type $W^n = kt$, the slope of the straight line then gives the value of n, that is, 1, 2, and 3 for linear, parabolic, and cubic laws, respectively.

Some authors, using computer software, fit the data to a third-degree polynomial in W:

$$AW^3 + BW^2 + CW + D = t \qquad (20.17)$$

Difficulties in evaluating the proper kinetic law are particularly important in the case of changes in the oxidation behavior during the corrosion process.

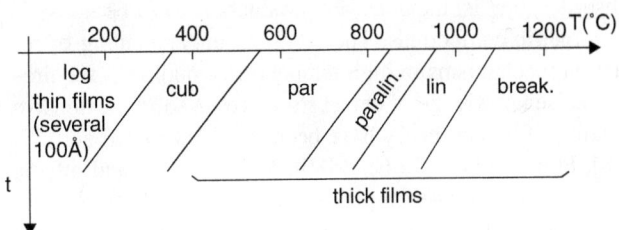

FIGURE 20.5. Successive kinetic laws observed for the oxidation of a given metal as functions of temperature and time.

An elegant method consists of continuously monitoring the kinetic curves with exposure time and calculating, for each experimental point, the rate constant appropriate to the expected model. Deviations from this model can be readily identified [76].

D. DEFECT STRUCTURES

The formation of an oxide scale starts with the adsorption of oxygen gas on the metal surface. During adsorption, oxygen molecules of other gaseous species in the environment dissociate and are adsorbed as atoms. These atoms initially adsorb at sites where the atom is in contact with the maximum number of surface atoms in the metal substrate. Therefore, in polycrystalline materials, grains of preferential orientation exist where the number of adsorbed atoms from the gaseous atmosphere is highest. The result of this process is a two-dimensional adsorption layer. The presence of adsorbed layers may increase the rates of surface diffusion by orders of magnitude compared with those for surfaces with none or small amounts of adsorbate.

When the metal surface which is saturated with adsorbed oxygen atoms or atoms from other gaseous species is further exposed to the gas, the gaseous species may dissolve in the metal and nuclei of the corrosion product are formed on the surface. These nuclei grow laterally and form a continuous film on the surface. Generally nucleation and growth of the nuclei are dependent on the composition of the substrate, the grain orientation, the temperature, and the gas partial pressure. The nuclei grow in thickness and lateral direction and the reaction rate increases with time. As soon as the nuclei impinge on each other, the growth rate decreases. Therefore, the general reaction kinetics can be described by an S-shaped curve [11].

Nuclei of all potential corrosion products can be formed on alloys initially, that is, those that are possible from thermodynamic stability considerations. After the initial stage of oxidation, which is determined by the behavior of the nuclei, growth of the continuous scale occurs in the thickness direction. In dense oxide scales, the growth is determined by solid-state diffusion through the scale. Corrosion products, which include the oxide scales, are ionic structures and diffusion in such structures requires lattice disorder; that is, the corrosion products need to be nonstoichiometric compounds. Therefore, an understanding of reaction mechanisms in high-temperature conditions requires a precise knowledge of defect structures in solids. Extensive studies of defect theory have been provided by Kröger [77, 78], Philibert [79], Kofstad [80], Mrowec [81], and others, but here only an oversimplified discussion will be presented.

Various types of defects may affect scale growth, but we shall only consider the crystalline defects that determine the growth of a compact layer and the three-dimensional defects, such as cracks and pores, that determine the growth of a porous scale. The crystalline defects represent departures from the perfect crystalline array and include point defects (imperfections in the distribution of ions within the lattice), line defects (displacements in the periodic structure of the crystal in certain directions or dislocations), and planar defects or grain boundaries (regions of lattice mismatch).

The point defects comprise either empty crystallographic sites (vacancies) or atoms occupying the interstices between the regular lattice sites (interstitial atoms). Their mole fraction in each sublattice, that is, either the cation or anion sublattice, generally does not exceed 10^{-3}–10^{-2} and is frequently much less. For point defect mole fractions sufficiently high, the defects may associate or cluster to form complex defects such as extended defects or aggregates of point defects (clusters). Point defects strongly influence the growth of compact scales and will be briefly treated hereafter.

D1. Point Defects

D1.1. Real Oxide Structures. A complete development of defect chemistry of inorganic compounds requires a system of notation to describe all the elements of the crystal or "structural elements," that is, not only regular crystallographic sites but also lattice imperfections. The Kröger and Vink notation [78], recommended by the International Union of Pure and Applied Chemistry (IUPAC) because of its great simplicity, will be used here. Thus, in a crystal MO, a structural element of the cation sublattice has a normal charge of $+2$ and, consequently, an effective charge equal to zero. The electronic defects may be considered as structural elements. The electronic defect with positive charge will be written $h^{\bullet}$ (h with a superscript dot). This defect corresponds to the removal of an electron from a regular site of the cation sublattice and can also be written as $M_M^{\bullet}$ (a superscript prime is used for a negative charge). These rules must be followed in writing defects in equilibrium reactions:

- Electroneutrality of the total charges of the structural elements, that is, involving normal charges, actual charges, and effective charges.
- Mass balance, that is, the number of atoms of each chemical species involved in the defect reaction must be the same before and after the defect formation.
- Ratio of regular lattice sites, that is, for a crystal M_pX_q: number of M sites/number of X sites should be equal to p/q.

Following these rules, it is possible to write equilibrium reactions that occur internally without involving the external environment and external equilibria involving mass exchange with the environment. The equilibrium constants will then be evaluated assuming that the activities of atoms on their normal lattice positions can be considered as unity and the activities of point defects will be approximated by their concentration, usually indicated by a double bracket and expressed as the number of moles per mole of compound.

D1.2. Stoichiometry. Alkali halides, silver halides, and several oxides (e.g., Al_2O_3, MgO) are stoichiometric compounds. Some of them are characterized by vacancies and interstitials in one sublattice (e.g., AgBr, with Frenkel disorder); others possess defects in both sublattices (e.g., NaCl, with Schottky disorder). However, it is apparent that neither of these defects can be used to explain material transport during oxidation reactions, because neither defect structure provides a mechanism by which electrons may migrate.

Considering a diagrammatic representation of the oxidation process shown in Figure 20.6, it is seen that either neutral atoms or ions and electrons must migrate in order for the reaction to proceed. In these cases, the transport step of the reaction mechanism links the two phase boundary reactions as indicated. There is an important distinction between scale growth by cation migration and scale growth by anion migration in that cation migration leads to scale formation at the scale–gas interface whereas anion migration leads to scale formation at the metal–scale interface. In order to explain simultaneous migration of ions and electrons, it is necessary to assume that the oxides, for example, that are formed during oxidation are nonstoichiometric compounds. From a macroscopic viewpoint, two alternative classes of nonstoichiometric compounds can be considered:

(i) M_aX_b compounds for which atom number of M/atom number of X $> a/b$. These compounds are ionic semiconductors with metal excess ($M_{a+\delta}X_b$, where δ is the deviation from stoichiometry in comparison with the stoichiometric composition M_aX_b), or with nonmetal deficit ($M_aX_{b-\delta}$). ZnO is a typical n-type

semiconductor with a metal excess; TiO_2, Nb_2O_5, MoO_3, and WO_3 are typical n-type semiconductors with nonmetal deficit.

(ii) M_aX_b compounds for which atom number of M/atom number of X $< a/b$. These include the p-type semiconductors with metal deficit (FeO, NiO, Cr_2O_3, Al_2O_3) or with an excess of nonmetal (UO_2).

In order to allow extra metal in ZnO, it is necessary to postulate the existence of interstitial cations with an equivalent number of electrons in the conduction band. The structure may be represented as shown in Figure 20.7. Here, both Zn^+ and Zn^{2+} are represented as possible occupiers of interstitial sites. Cation conduction occurs over interstitial sites, and electrical conductance occurs by virtue of having the "excess" electrons excited into the conduction band. These, therefore, are called excess or "quasi-free" electrons.

The formation of this defect may be visualized, conveniently, as being formed from a perfect ZnO crystal by losing oxygen, the remaining unpartnered Zn^{2+} leaving the cation lattice and entering interstitial sites and the two negative charges of the oxygen ion entering the conduction band. In this way, one unit of ZnO crystals is destroyed and the formation of the defect may be represented by

$$ZnO = Zn_i^{\bullet\bullet} + 2e' + \tfrac{1}{2}O_2 \qquad (20.18)$$

for the formation of $Zn_i^{\bullet\bullet}$, doubly charged Zn interstitials, or

$$ZnO = Zn_i^{\bullet} + e' + \tfrac{1}{2}O_2 \qquad (20.19)$$

for the formation of $Zn_i^{\bullet}$, singly charged Zn interstitials. The two equilibria shown above will yield to thermodynamic treatment, giving Eq. (20.20) for the equilibrium in Eq. (20.18),

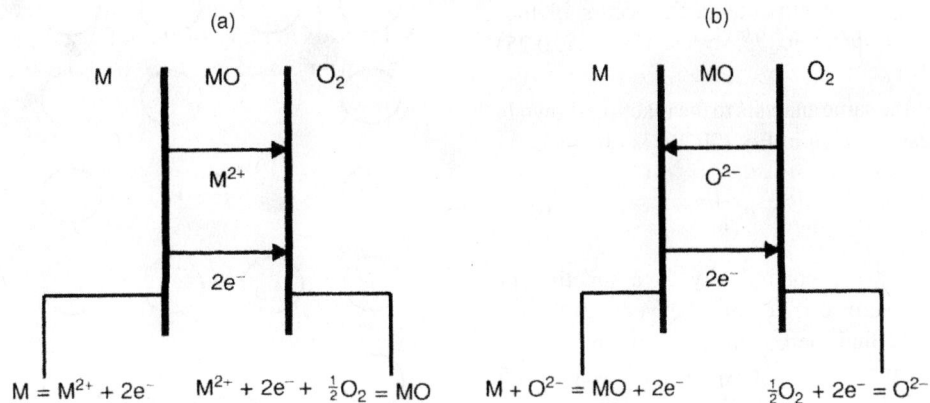

FIGURE 20.6. Interfacial reactions and transport processes for high-temperature oxidation mechanisms: (a) cation mobile; (b) anion mobile [12].

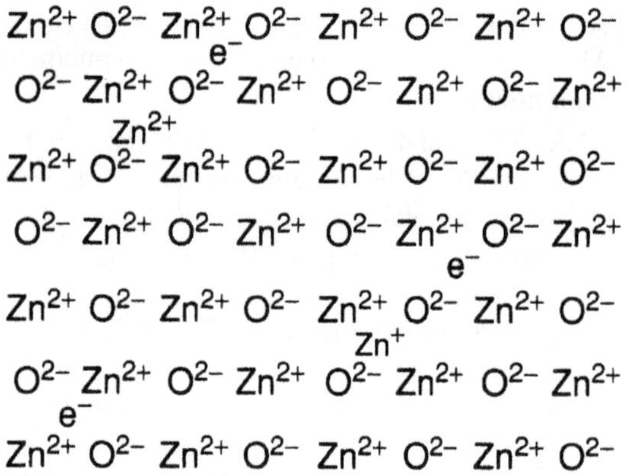

Zn²⁺ O²⁻ Zn²⁺ O²⁻ Zn²⁺ O²⁻ Zn²⁺ O²⁻
                  e⁻
O²⁻ Zn²⁺ O²⁻ Zn²⁺ O²⁻ Zn²⁺ O²⁻ Zn²⁺
     Zn²⁺
Zn²⁺ O²⁻ Zn²⁺ O²⁻ Zn²⁺ O²⁻ Zn²⁺ O²⁻
O²⁻ Zn²⁺ O²⁻ Zn²⁺ O²⁻ Zn²⁺ O²⁻ Zn²⁺
                              e⁻
Zn²⁺ O²⁻ Zn²⁺ O²⁻ Zn²⁺ O²⁻ Zn²⁺ O²⁻
                    Zn⁺
O²⁻ Zn²⁺ O²⁻ Zn²⁺ O²⁻ Zn²⁺ O²⁻ Zn²⁺
e⁻
Zn²⁺ O²⁻ Zn²⁺ O²⁻ Zn²⁺ O²⁻ Zn²⁺ O²⁻

FIGURE 20.7. interstitial cations and excess electrons in ZnO—an *n*-type metal-excess semiconductor [12].

$$K = a_{Zn_i^{\bullet\bullet}} a_{e'}^2 p_{O_2}^{1/2} \qquad (20.20)$$

or, since the defects are in very dilute solution, we may assume that they are in the range obeying Henry's law, when the equilibrium may be written in terms of concentrations $[Zn_i^{\bullet\bullet}]$ and $[e']$ as in Eq. (20.21):

$$K' = [Zn_i^{\bullet\bullet}][e']^2 p_{O_2}^{1/2} \qquad (20.21)$$

If Eq. (20.18) represents the only mechanism by which defects are created in ZnO, then Eq. (20.22) follows:

$$2[Zn_i^{\bullet\bullet}] = [e'] \qquad (20.22)$$

Hence, putting Eq. (20.22) into Eq. (20.21), we obtain

$$K' = 4[Zn_i^{\bullet\bullet}]^3 p_{O_2}^{1/2} \qquad (20.23)$$

or Eq. (20.24) and therefore we obtain Eq. (20.25):

$$[Zn_i^{\bullet\bullet}] = (\tfrac{1}{4}K')^{1/3} p_{O_2}^{-1/6} = \text{const}\, p_{O_2}^{-1/6} \qquad (20.24)$$

$$[e'] \propto p_{O_2}^{-1/6} \qquad (20.25)$$

Similarly, applying the same analysis to the reaction shown in Eq. (20.19), the result shown in Eq. (20.26) is obtained:

$$[Zn_i^{\bullet}] = [e'] \propto p_{O_2}^{-1/4} \qquad (20.26)$$

Measurement of electrical conductivity as a function of oxygen partial pressure carried out between 500 and 700°C [82] indicated that the conductivity of ZnO varied with oxygen partial pressure having exponents between 1/4.5 to 1/5. This result indicates that neither defect mechanism predominates, and the actual structure could involve both singly and doubly charged interstitial cations [80].

Similar approaches can be applied to nonstoichiometric compounds with cation vacancies (Cu$_{2-\delta}$O-type oxide), oxygen interstitials (UO$_{2+\delta}$-type oxide), and so on [80, 82–86].

D1.3. Mass and Electrical Transport. Intragranular or volume diffusion in crystalline compounds takes place through crystal imperfections and mainly through the movement of point defects. Several types of mechanisms may be considered, as shown schematically in Figure 20.8, but mass transport generally occurs by hopping mechanisms from a well-defined site of the crystal into another adjacent site. Consider a one-dimensional flux of particles (atoms, ions, point defects, or electrons) in the *Ox* direction. Let $C(x, t)$ be the defect concentrations (number of particles per unit volume) at the coordinate x and at time t. In a chemical potential gradient and without an electrical potential gradient or other type of driving force, a flux J of particles occurs in the *Ox* direction:

$$J = -\frac{CD}{RT}\frac{\partial \mu}{\partial x} \quad \text{(generalized Fick's law)} \qquad (20.27)$$

where D is the diffusion coefficient of the particle.

Under an additional electrical potential gradient, the particle flux would obey the following general equation:

$$J = -\frac{CD}{RT}\frac{\partial \tilde{\mu}}{\partial x} \qquad (20.28)$$

where $\tilde{\mu}$, termed the electrochemical potential, is related to the chemical potential μ by the equation

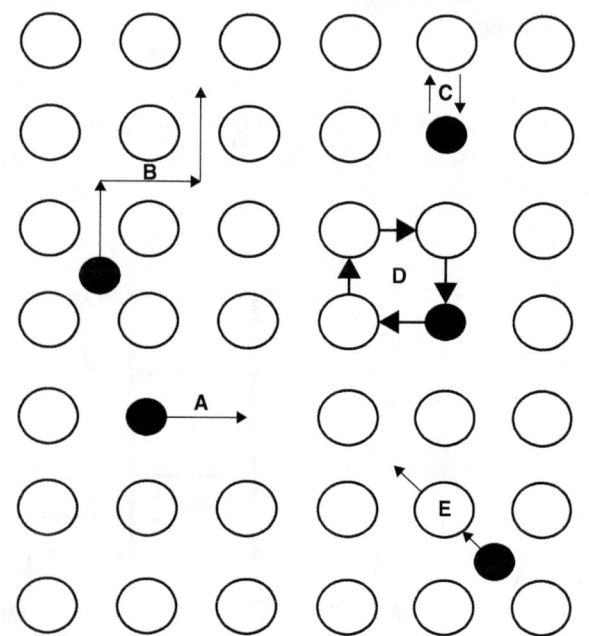

FIGURE 20.8. Schematic illustration of transport mechanisms in crystalline solids [87].

$$\tilde{\mu} = \mu \pm zF\phi \qquad (20.29)$$

where z is the particle charge number, F the Faraday constant, ϕ the electrical potential, and $\pm zF$ the electrical charge.

If the ion movements within one sublattice of the binary compound M_aX_b produce displacements of ions only in that sublattice, M or X diffusion is termed self-diffusion, and the self-diffusion coefficient D_j of component j will obey the following equation:

$$J_j = -\frac{C_j D_j}{RT}\frac{\partial\mu_j}{\partial x} \qquad (20.30)$$

where C_j is the concentration of component j.

Let J_δ be the flux of the defect, δ, in the j sublattice at the coordinate x and in the absence of an electric field. The diffusion coefficient D_δ of the defect δ is defined by the equation

$$J_\delta = -\frac{C_\delta D_\delta}{RT}\frac{\partial\mu_\delta}{\partial x} \qquad (20.31)$$

where C_δ is the concentration of the defect δ.

As a general rule, the relationship between D_j and D_δ may be written as

$$C_\delta D_\delta = C_j D_j \qquad (20.32)$$

If N_δ is the mole fraction of defects δ in the j sublattice, we may write

$$D_j = N_\delta D_\delta \quad \text{since} \quad N_\delta = \frac{C_\delta}{C_\delta + C_j} \approx \frac{C_\delta}{C_j} \qquad (20.33)$$

This relationship shows that the self-diffusion of component j is proportional to the mole fraction of defect δ contained in the j sublattice.

A comparison of the self-diffusion coefficients of anions and of cations may allow identification of the component which provides the majority of mass transport within the crystal. Thus, nonstoichiometric oxides such as NiO, FeO and Cu_2O contain metal vacancies; this observation is in agreement with the order of magnitude of the diffusion coefficients, that is, $D_M \gg D_O$ ($D_M/D_O \approx 10^2$–10^4). In contrast, in some oxides such as TiO_2 that have an oxygen deficit, it has been observed that $D_O \approx D_{Ti}$. This result is not in contradiction with the assumption of several authors who postulate that both oxygen vacancies and titanium interstitials are simultaneously present in this oxide.

Electrical transport in ionic compounds does not necessarily occur by means of point defects. Electrical conductivity due to a charge carrier is given by

$$\sigma = CUzF \qquad (20.34)$$

where C is the molar concentration of charge carriers per unit volume and U is the electrical mobility (expressed in $m^2/s/V$). The mobility U corresponds to the velocity of the charged particles under an electric field equal to unity.

In ionic crystals, the total conductivity σ_t is generally written in terms of ionic and electronic conductivities as

$$\sigma_t = \sigma_{\text{ionic}} + \sigma_{\text{electronic}} \qquad (20.35)$$

Let σ_δ be the partial conductivity relevant to the defect δ, and considering the definition of the electrical mobility U_δ, we obtain

$$\sigma_\delta = zFC_\delta U_\delta = z^2 F^2 \frac{C_\delta D_\delta}{RT} \qquad (20.36)$$

If σ_j is the contribution to the total conductivity of the charged species j, we can write

$$\sigma_j = \sigma_\delta \qquad (20.37)$$

Since

$$C_\delta D_\delta = C_j D_j \qquad (20.38)$$

we obtain the Nernst–Einstein equation

$$\sigma_j = z^2 F^2 \frac{C_j D_j}{RT} \qquad (20.39)$$

In this equation, σ_j is the ionic contribution of species of type j to the total conductivity, D_j is the self-diffusion coefficient of particles j, and C_j is the volume concentration of regular sites in the sublattice that contain species j.

Since the mobility of electronic defects is much higher than that of point defects ($U_e \gg U_\delta$), it can be said that the total electrical conductivity is essentially electronic. Also, it is easy to show that the conductivity varies with oxygen pressure in the same way as does the concentration of the predominant ionized defect (although the current is carried by electronic defects).

The temperature dependence of the conductivity is determined by both the charge carrier mobility and concentration terms.

When ion movements involve jumps between definite sites of the crystal, an energy barrier ΔG_m has to be overcome. The defect mobility then increases strongly with temperature according to an exponential law (*activated process*):

$$U_\delta = U_\delta^0 \exp\left(-\frac{\Delta G_m}{RT}\right) \qquad (20.40)$$

ΔG_m is the free energy of migration of the defect.

The temperature dependence of electron mobility is a function of the electronic structure of the crystal. The electron movement is an activated process and the electronic mobility obeys the equation

$$U_\varepsilon = U_\varepsilon^0 \exp\left(-\frac{E_\varepsilon}{RT}\right) \qquad (20.41)$$

where E_ε is the overall activation energy for polaron migration in the periodic field within the crystal or, in other words, for polaron (electron and distortion field) scattering by lattice vibrations and/or imperfections, also known as polaron hopping and usually treated as a diffusion process.

The determination of the variation of the electronic mobility with temperature may allow us to identify the migration mechanism of electrons in the lattice. Whatever the nature of the charge carrier (ionic or electronic defects), the concentration increases with temperature according to

$$[\text{defect}] = [\text{defect}]_0 \exp\left(-\frac{\Delta G_f}{RT}\right) \qquad (20.42)$$

ΔG_f is the free energy of formation of defects.

Whatever the nature of the conduction mechanism, the electrical conductivity is proportional to the product of the drift mobility and the charge carrier concentration, which varies exponentially with temperature. The crystal conductivity always increases with increasing temperature due to the exponential increase in the number of charge carriers. This characteristic differentiates covalent/ionic compounds from metallic conductors, which exhibit a decrease of the electrical conductivity with increasing temperature [77–81].

D2. Line and Planar Defects

As diffusion along line and surface defects, including dislocations, grain boundaries, and internal and external surfaces, is generally more rapid than lattice diffusion, they are termed high diffusivity or easy diffusion paths. This type of diffusion is often called "short-circuit diffusion." The contribution of grain boundary diffusion to the total diffusion flux decreases as the temperature increases for two main reasons:

- The grain size is larger, the higher the temperature, because of grain growth.
- The activation energy for grain boundary diffusion is less than that for intracrystalline diffusion (from 20 to 30% smaller).

The effective diffusion coefficient may be defined by the Hart equation:

$$D_{\text{eff}} = fD_{\text{gb}} + (1-f)D_v \qquad (20.43)$$

where f is the volume fraction of short-circuit paths, D_v the lattice diffusion coefficient, and D_{gb} the short-circuit diffusion coefficient; D_{eff} may be identified with D_V, the intracrystalline diffusion coefficient, at high temperatures but, at low temperatures, the short-circuit contribution to diffusion can become significant. In general, in accordance with Tamann's empirical law, grain boundary diffusion would be expected to dominate at lower temperatures, say below a transition temperature of between one-half to two-thirds of the absolute melting temperature of the crystal. Conversely, the contribution made by short-circuit diffusion processes will be negligible at higher temperatures. Grain boundary diffusion in growing oxide scales has been reported for NiO [88, 89], Cr_2O_3 [90], Al_2O_3 [91], and other product films.

D3. Three-Dimensional Defects

Stress generation in the oxide layer and the underlying metal may cause through-scale cracking, spalling of the oxide, stratification phenomena, or even detachment of the scale. These phenomena lead to loss of protective properties and faster degradation of metals and alloys. The sources of stress may be either internal (scale growth) or external (mechanical and/or thermal stresses). Often, due to mechanical stresses, a porous layer may develop after the oxide scale has reached a critical thickness. The two main sources of stress are growth stresses, which develop during isothermal formation of the scale, and thermal stresses, which arise from differential thermal expansion between the oxide scale and the metal or alloy during temperature changes.

D3.1. Growth Stresses. Observed stresses depend on the oxidation mechanism and on the physicochemical properties of the alloy and of the oxide. The most important causes of growth stresses include [11, 16] the volume difference between the oxide and the metal (Pilling and Bedworth rule), the oxidation mechanism (e.g., internal or external oxidation of alloys), oxygen dissolution in alloys, epitaxial constraints, physicochemical changes in the alloy or scale during the growth, and specimen geometry.

Two different types of growth stresses can be distinguished, geometrically induced growth stresses caused by the surface curvature of components and the intrinsic growth stresses. As can often be seen in oxidation experiments, the oxide scales crack at the edges of the specimens, initially leading to a locally increased attack at these sites. Such cracking is usually due to geometrically induced growth stresses that arise at edges and corners due to the small surface curvature radius. This situation has been dealt with quantitatively by Manning [92].

With the help of models, the tangential and radial stresses can be calculated for the ideal case of curved surfaces with a constant radius of curvature. Introduction of the oxide displacement vector M is helpful here; M lies perpendicular to the oxide–metal interface and describes the displacement of a reference point in the film resulting from the oxidation. The magnitude and sign of M are incorporated in M, which is calculated as follows:

$$M = PBR(1-a) - (1-V) \qquad (20.44)$$

where a is the fraction of oxide formed on the scale surface, $1-a$ is the fraction of oxide formed at the metal–oxide interface, V is the volume fraction of metal consumed in the oxidation by injecting vacancies into the metal, $1-V$ is the volume fraction of metal consumed in the oxidation which originates directly from the metal surface, and PBR is the Pilling–Bedworth ratio (see later in this section).

Oxidation leads to an increase in the strain in the circumferential direction (tangential strain ε_{Ox}^t) with a rate of

$$\frac{d\varepsilon_{Ox}^t}{dt} = \left(\frac{M}{R_s}\right)\frac{dh}{dt} = \frac{M}{R_s \cdot PBR}\frac{dx}{dt} \qquad (20.45)$$

where R_s is the radius of curvature of the surface (concave $R_s < 0$, convex $R_s > 0$) and h is the metal recession (increase in oxide film thickness $dx = PBR\ dh$). Equation (20.45) allows the tangential stresses σ_{Ox}^t to be calculated assuming linear elastic behavior. The magnitude of the maximum radial stresses, σ_{Ox}^r, is given by

$$\sigma_{Ox}^r = \frac{PBR\ h}{R_s}\sigma_{Ox}^t \qquad (20.46)$$

The relationship between the signs of the tangential and radial strains and stresses in the scale and at the metal–oxide interface, respectively, are

$$\text{sign}\ \varepsilon_{Ox}^t = \text{sign}(-\varepsilon_{Ox}^r R_s) \qquad (20.47)$$

$$\text{sign}\ \sigma_{Ox}^t = \text{sign}(-\sigma_{Ox}^r R_s) \qquad (20.48)$$

A plus sign indicates tensile stress, a minus sign means compressive stress. The sign and level of the stresses in the scale depend on its growth direction and on the radius of service curvature as well as the PBR. The latter was introduced in 1925 in order to explain the formation of growth stresses during oxidation and describes the volume change that is involved in the transition from the metal lattice to the cation lattice of the oxide when only the oxygen anions are diffusing. In other words, the PBR corresponds to the ratio of the volume per metal ion in the oxide to the volume per metal atom in the metal [13]:

$$PBR = \frac{1}{a}\frac{V_{eq}(M_a O_b)}{V_{eq}(M)} \qquad (20.49)$$

It was argued that if the PBR was less than 1, the growth stresses would be tensile and the oxide would crack and not cover the entire metal surface. As indicated in Table 20.1, alkali and alkaline earth metals belong to this class of materials. On the other hand, if the PBR was higher than 1, compressive stresses would develop and the oxide could be protective, at least during the early stages of oxidation. The majority of metals fall into this category.

We now know that the Pilling–Bedworth rule regarding protective behavior exhibits several exceptions. Important examples are tantalum or niobium where, even though the PBR is substantially larger than unity, cracks develop in the oxide scale after extended exposure and these produce nonprotective conditions. Whereas the Pilling–Bedworth paper was a significant advance at the time, it is now recognized that the approach was incomplete and that the influence of the difference between the molar volume of metal and oxide depends on the oxide growth mechanism. However, the Pilling–Bedworth approach may be of great help for the assessment of the geometrically induced growth stresses, as shown earlier.

D3.2. Thermal Stresses. In most applications, high-temperature alloys are subjected to temperature fluctuations even under nominally isothermal conditions. In this case, though, the resultant stresses in the oxide layer resulting from the difference in the coefficient of thermal expansion (CTE) of the metal and oxide (see Table 20.2) are small and may be neglected. This will not be the case, however, for large thermal cycles or during cooling to room temperature when large stresses, perhaps of 1-GPa order, are produced in the oxide layer. Metals have generally a higher coefficient of thermal expansion than oxides (Table 20.2), and consequently, tensile stresses are induced in the oxide scale on heating and compressive stresses during cooling.

The thermally induced stresses can be calculated from the coefficients of thermal expansion according to the equation [94]

TABLE 20.1. Pilling–Bedworth Ratios for Some Metal–Oxygen Systems

Oxide	K_2O	CaO	MgO	CeO_2	Na_2O	CdO	Al_2O_3	ZnO	ZrO_2
PBR	0.45	0.64	0.8	0.90	0.97	1.21	1.28	1.55	1.56

Oxide	Cu_2O	NiO	FeO	TiO_2	CoO	SiO_2	Cr_2O_3	Ta_2O_5	Nb_2O_5
PBR	1.64	1.65	1.7	1.73	1.86	1.9	2.07	2.5	2.7

TABLE 20.2.   Linear Coefficients of Thermal Expansion of Metals and Oxides, °C^{-1} [95]

System	Oxide: $10^6\alpha_{ox}$	Metal: $10^6\alpha_M$	Ratio:α_M/α_{OX}
Fe/FeO	12.2	15.3	1.25
Fe/Fe$_2$O$_3$	14.9	15.3	1.03
Ni/NiO	17.1	17.3	1.03
Co/CoO	15.0	14.0	0.93
Cr/Cr$_2$O$_3$	7.3	9.5	1.30
Cu/Cu$_2$O	4.3	18.6	4.32
Cu/CuO	9.3	18.6	2.0

$$\sigma_{therm} = \frac{-E_{Ox}\Delta T(\alpha_M - \alpha_{Ox})}{\left(1 + 2\dfrac{E_{Ox}}{E_M}\dfrac{d_{Ox}}{d_M}\right)(1-\nu)} \qquad (20.50)$$

where α is the CTE for the metal and the oxide, E is Young's modulus for the metal and the oxide, d is the thickness for the metal and the oxide, respectively, and ν is Poisson's ratio. Here, ΔT stands for the temperature change. The CTEs for technical materials can be found in many of the materials producers' brochures, and those for corrosion products are given in the literature [95]. In most cases, the CTEs can be approximated by linear behavior in the temperature range concerned, but in some cases, where phase changes occur in the scale during the temperature change, nonlinear temperature dependence is found for the CTE. This is, for example, the case for several sulfide layers [96] and is particularly important for magnetite and some iron-based spinels [97], affecting stresses in oxide scales on low-alloy steels. This naturally decisively affects the stress situation in the oxide scales on low-alloy steels [98]. In the temperature range between about 600 and 450°C, the magnetite partial layer is under tensile stress when cooling from 600°C. At lower temperatures, this oxide partial layer may come under compressive stresses, depending on the metallic substrate and its CTE. In the hematite layer, the stresses are always compressive, as the CTE always lies below that of the low-alloy steel (the exception is 9% chromium steel at temperatures below 150°C).

D3.3.   Mechanical Scale Failure.   Growth stresses and thermal stresses may be relieved through various mechanisms that could operate simultaneously:

- Plastic deformation of the oxide scale
- Plastic deformation of the metal substrate
- Spalling of the oxide from the alloy
- Cracking of the scale

When plastic deformation is not sufficient for stress relief, cracking may develop in the scale. It is the more efficient relaxation mechanism but will result in a sudden increase in corrosion rate. The metal oxidation may exhibit repeated regular sequences of cracking and healing of the scale.

Under tensile stresses (heating to temperatures higher than the oxidation temperature or over convex regions of a nonplanar surface) cracks appear as soon as the elastic fracture strain is reached. This critical value will be significantly less than 1% even at high temperatures.

Under compressive stresses, the degradation leads to spallation and the mechanisms are more complex. Two processes are necessary to produce spalling: transverse cracking through the oxide and decohesion along the metal–oxide interface. Two routes of spallation have been identified: the case corresponding to a low cohesive strength of the oxide and a high adhesive strength of the scale on the substrate surface (route 1: cracking of the oxide before decohesion) and the case corresponding to a high cohesive strength of the oxide and poor adhesion of the oxide to the metal (route 2: decohesion before metal cracking). Figure 20.9 illustrates these two distinct mechanisms.

E.   COMPACT SCALE GROWTH

E1.   Elementary Chemical Steps

The overall oxidation reaction of a metal M may be written as

$$a\text{M} + \frac{b}{2}\text{O}_2 = \text{M}_a\text{O}_b \qquad (20.51)$$

The reaction can proceed only if diffusion of matter (oxygen or metal) occurs through the solid scale M_aO_b. If the scale is porous, mass transport occurs by oxygen diffusion; if the scale is compact, mass transport occurs by means of solid-state diffusion. In the latter case, the oxidation mechanism consists of at least four steps [100–104]:

- Surface step—oxygen adsorption on the oxide
- External step—matter exchange at the adsorbed phase/oxide interphase
- Diffusion step—ionic transport through the oxide scale
- Internal step—matter exchange at the oxide/metal interphase

Of course, mass transport by migration of ionized point defects is accompanied by simultaneous electrical transport, which complicates the process. In this section, the kinetics of the oxidation process leading to the growth of a compact scale are analyzed.

Let a chemical elementary step be a local reversible reaction that occurs without the formation of a distinct intermediate product, that is, the reaction proceeds in a single step, and let an interphase elementary step be a chemical

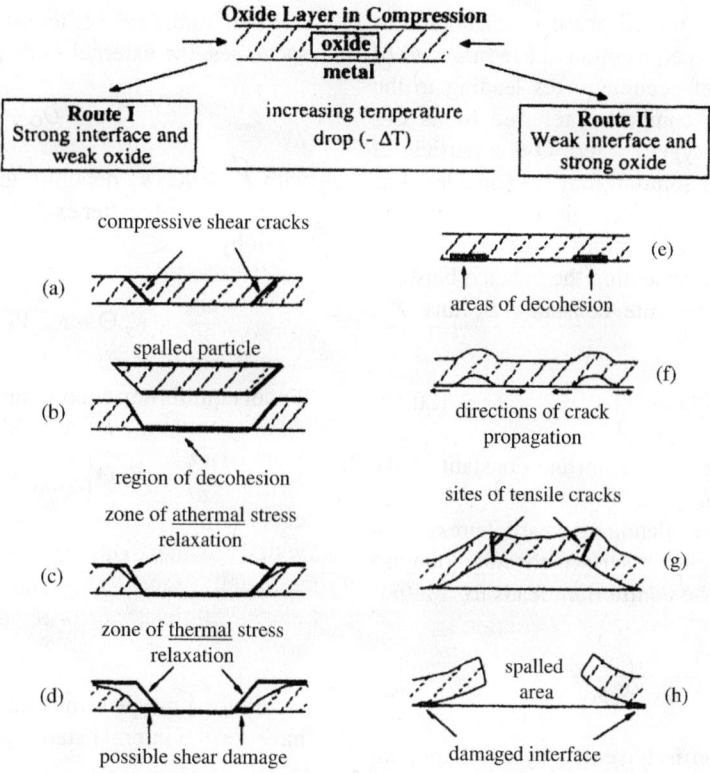

FIGURE 20.9. Cracking and spallation by compressive oxide stresses [99].

process involving matter exchange between two distinct phases. The main difficulty encountered for the formulation of these steps relates to the conditions required for the application of classical theories (Eyring theory) for the calculation of the step rates [105]. We will have to choose, for the adsorbed phase as well as for the oxide, structural models that exhibit ideal behavior for the reacting species in their own phase. Moreover, we will assume that the theory of absolute rates can, under these conditions, be extended to heterogeneous elementary steps involving matter exchange between two different phases.

On a solid surface, the atomic environment is modified in comparison with that in the bulk. The resulting imbalance of the forces in the surface of solids produces attractive forces for gas molecules or atoms. The phenomenon of adsorption can then produce an excess of gas atoms or molecules on the surface compared with the concentration in the adjacent gas phase. For adsorption to occur spontaneously, the process must produce a decrease in free energy, but since that is also a decrease in system entropy, adsorption is always an exothermic process. Consequently, the amount of adsorbed gas at equilibrium at constant pressure (the "adsorption isobar") decreases with increasing temperature.

Depending on the nature of the forces involved, adsorption processes may be classified as physical adsorption (also termed van der Waals adsorption or physisorption) or chemical adsorption, usually abbreviated to chemisorption [106,

107]. Physisorption is generally quasi-instantaneous while chemisorption often proceeds slowly, involving an activation energy E_a. Thus, the chemical adsorption rate becomes appreciable only at sufficiently high temperature.

Many theories and models have been proposed to explain the shape of adsorption isotherms that represent the variation of adsorbed volume as a function of gas pressure or of the p/p_0 ratio (p_0 is the saturation vapor pressure at the experimental temperature). The description of monolayer adsorption can be made using as variable the fraction of the available adsorption sites which are occupied by adsorbed atoms or molecules, $\upsilon = s/s_0$, where s_0 is the number of adsorption sites that are initially available per unit surface area and s is the number of occupied surface sites per surface area unit (thus, υ is the fraction of occupied sites).

Chemisorption involves partial electronic transfer between adsorbed molecules and the substrate. The solid surface appears inhomogeneous and exhibits specific "active" sites on which chemisorption takes place preferentially. When temperature is increased, the amount of adsorbed gas by chemisorption increases because it is an activated process; then the adsorption isobar passes through a maximum because chemisorption is an exothermic process. Since the establishment of high-temperature oxidation requires at least the presence of one monolayer on the surface, it can be assumed that a chemisorbed phase is produced as a surface step.

In this discussion, this adsorbed phase is considered as a two-dimensional solution (i.e., sorption of a monolayer) of free surface sites, s, and of occupied sites leading to the formation of a superficial compound referred to as O-s (atomically chemisorbed oxygen) where O is a particle of the gas phase (O_2). Such a solution may be considered as ideal since it is assumed that no interaction occurs between the free and occupied sites. The sorption process may then be described by an equation representing the balance between two opposite reactions with rate constants K_a' and K_a'', respectively:

$$O_2 + 2s = 2O\text{-}s \tag{20.52}$$

with $K_a = K_a'/K_a''$ denoting the equilibrium constant of the elementary step of sorption.

Under these conditions, if p denotes the partial pressure of the p_{O_2} species, υ^{eq} the fractional surface coverage, the law of mass action applied to the equilibrium leads to

$$\Theta^{eq} = \frac{\sqrt{K_a p_{O_2}}}{1 + \sqrt{K_a p_{O_2}}} \tag{20.53}$$

indicating that the sorption is governed by the Langmuir equation (dissociation occurs on chemisorption, which is likely at high temperature). If $dn_{O\text{-}s}$ is the number of O-s particles formed by adsorption per unit area during the time dt, the rate of the sorption process can be given by the equations

$$\frac{dn_{O\text{-}s}}{dt} = K_a'(1 - \Theta)^2 p_{O_2} - K_a'' \Theta^2 \tag{20.54}$$

or

$$\frac{dn_{O\text{-}s}}{dt} = K_a' \frac{\Theta^2}{K_a} \left[\left(\frac{1-\Theta}{\Theta} \right)^2 K_a p_{O_2} - 1 \right] \tag{20.55}$$

with

$$K_a p_{O_2} = \left(\frac{\Theta^{eq}}{1 - \Theta^{eq}} \right)^2 \tag{20.56}$$

The proposed model to account for the external and internal steps uses the concept of point defects and, for simplicity, it can be assumed that only one defect is predominant in the lattice, that is, either the metal vacancy V_M'' or the oxygen interstitial O_i'' for the case of a p-type semiconductor and either the oxygen vacancy $V_O^{\bullet\bullet}$ or the metal interstitial $M_i^{\bullet\bullet}$ for the case of an n-type semiconductor. The defects are then generated at the external interphase (cationic diffusion, p-type semiconductor) or at the internal interphase (anionic diffusion, n-type semiconductor) (see Fig. 20.6).

Then, in the case of an n-type semiconductor with metal vacancies, the external step can be described by

$$O\text{-}s = O_O + V_M'' + 2h^\bullet + s \tag{20.57}$$

with $K_e = K_e'/K_e''$ denoting the equilibrium constant of the external step of matter exchange. The external reaction rate is given by

$$\frac{dn_e}{dt} = K_e' \Theta - K_e'' [V_M'']_e [h^\bullet]_e^2 (1 - \Theta) \tag{20.58}$$

Under equilibrium conditions, for a given value of υ, we have

$$\frac{dn_e}{dt} = K_e' \Theta \left[1 - \left([V_M'']_e / [V_M'']_e^{eq} \right)^3 \right] \tag{20.59}$$

with

$$[V_M'']_e^{eq} = \frac{(1/4) K_e \Theta}{1 - \Theta} \tag{20.60}$$

Again for a p-type semiconductor with metal vacancies, we have for the internal step

$$V_M'' + 2h^\bullet + M = M_M \tag{20.61}$$

$$\frac{dn_i}{dt} = 4K_i' [V_M'']_i^3 - [V_M'']_i^{eq3} \tag{20.62}$$

with

$$[V_M'']_i^{eq} = \left(\frac{1}{4} \frac{1}{K_i} \right)^{1/3} \tag{20.63}$$

In a thick, growing oxide scale, a positive space charge can appear in contact with the metal balanced by a negative space charge localized near the external interphase. This charge distribution induces, at any point within the scale, an electric field that both accelerates the positively charged ionic defects and slows down the negatively charged electronic defects until no net electrical current flows through the scale. Consequently, stationary concentration profiles are established in the MO scale, the electrically neutral zone extending over practically all of the scale thickness [108, 109].

Taking into account the possible electrical potential gradient through the scale, the particle fluxes are given by

$$J_z = -\left[D_\delta \frac{\partial c_\delta}{\partial x} \pm C_\delta U_\delta \frac{\partial \phi}{\partial x} \right] \tag{20.64}$$

$$J_\varepsilon = -D_\varepsilon \left[\frac{\partial c_\varepsilon}{\partial x} \pm C_\varepsilon U_\varepsilon \frac{\partial \phi}{\partial x} \right] \tag{20.65}$$

In these expressions, C_δ (C_ε) is the volume concentration of point (electronic) defects with z degree of ionization, D_δ (D_ε) is the diffusion coefficient, U_δ (U_ε) is the electric mobility of these charge carriers, and ϕ the electric potential at the coordinate x. The C_δ and $[\delta]$ are linked by the relationship $C_\delta = [\delta]/\Omega$, Ω being the volume of 1 mol of oxide. A positive sign has to be used if the defects are positively charged and a negative sign for negatively charged defects.

A brief kinetic analysis of the four elementary steps for reaction (51) has been presented in this section. The rate expressions of the elementary steps were expressed in a form involving the deviation from equilibrium, that is, within the framework of the thermodynamics of irreversible processes. The general system of equations relating to the growth of an oxide scale MO can then be established by expressing the mass balance at both sides of each interphase in the adsorbed phase and within the scale. These equations, being differential, cannot be solved analytically. However, it is possible to make the simplifying assumption that the concentrations tend to become time independent, that is, a quasi–steady state develops. On this basis, the system of differential equations allowing calculation of the reaction rate can be solved analytically, leading to considerable simplifications. It should be recalled that if one of the rate constants or the diffusion coefficient has a finite value, all the other constants having very large values, we deal with what is called a pure regime [110–113]. In all other cases, it is called a mixed regime [75, 114–116]. Hereafter, we describe pure diffusional regimes.

E2. Diffusion-Controlled Oxidation

Assuming that cationic transport across the growing oxide layer controls the rate of scaling and that thermodynamic equilibrium is established at each interphase, the outward cation flux $J_{M^{2+}}$ is equal and opposite to the inward flux of cation vacancies and we can write

$$J_{M^{2+}} = -J_{V_M''} = D_{V_M''} \frac{[V_M'']'' - [V_M'']'}{y} \quad (20.66)$$

where y is the oxide thickness, $D_{V_M''}$ is the diffusion coefficient for cation vacancies, and $[V_M'']'$ and $[V_M'']''$ are the vacancy concentrations at the scale–metal and scale–gas interphases, respectively.

Since there is thermodynamic equilibrium at each interphase, the volume of $[V_M'']'' - [V_M'']'$ is constant and we have

$$J_{M^{2+}} = \frac{1}{\Omega}\frac{dy}{dt} = \frac{1}{\Omega}\frac{K'}{y} \quad (20.67)$$

where

$$K' = D_{V_M''}\Omega\left([V_M'']'' - [V_M'']'\right). \quad (20.68)$$

Integrating and noting that $y = 0$ at $t = 0$ we obtain

$$y^2 = 2K't \quad (20.69)$$

which is the common parabolic rate law. Furthermore, it has been shown that the cation vacancy concentration is related to the oxygen partial pressure by the equation

$$[V_M''] = \text{const}\, p_{O_2}^{1/n} \quad (20.70)$$

The variation of the parabolic rate constant with oxygen partial pressure can be predicted by

$$K'\alpha\left[\left(p_{O_2}''\right)^{1/n} - \left(p_{O_2}'\right)^{1/n}\right] \quad (20.71)$$

and since p_{O_2}' is usually negligible compared with p_{O_2}'', we have

$$K'\alpha\left(p_{O_2}''\right)^{1/n} \quad (20.72)$$

Clearly, the concentration gradient in the scale never equals zero and, therefore, scale growth never stops.

Originally, Eq. (20.69) was derived by Wagner in a theoretical, detailed analysis of the electrochemical potential situation and the transport conditions in the scale. Figure 20.10 gives a summary of the conditions for which the theory is valid. Assumptions are as follows:

- The oxide layer is a compact, perfectly adherent scale.
- Migration of ions or electrons across the scale is the rate-controlling process.
- Thermodynamic equilibrium is established at both the metal–scale and scale–gas interfaces.
- The oxide scale shows only small deviations from stoichiometry and, hence, the ionic fluxes are independent of position within the scale.
- Thermodynamic equilibrium is established locally throughout the scale.
- The scale is thick compared with the distances over which space charge effects (electrical double layer) occur.
- Oxygen solubility in the metal may be neglected.

This model led to the final equation of the parabolic rate constant, which is [11, 109, 117, 118]

$$K_p' = \frac{1}{2}\int_{p_{O_2}'}^{p_{O_2}''}\left\{\frac{z_c}{|z_a|}D_M + D_O\right\}d\ln p_{O_2} \quad (20.73)$$

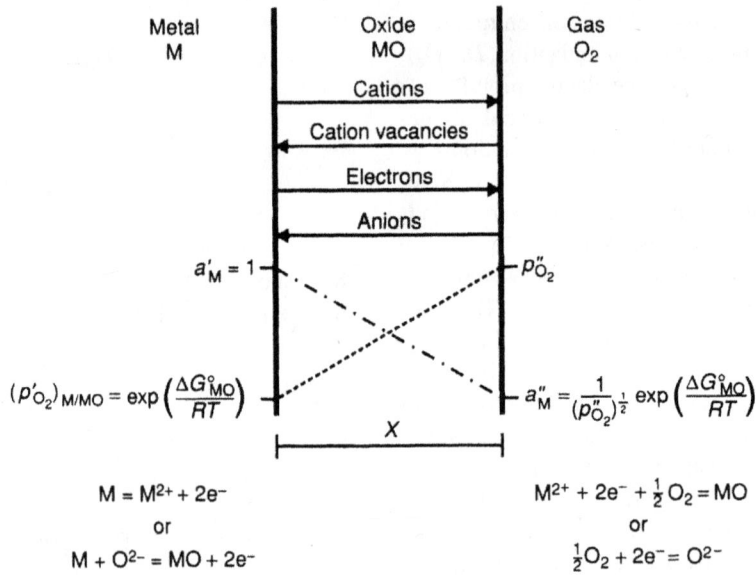

FIGURE 20.10. Diagram of scale forming according to Wagner's model [12].

where K_p', D_M, and D_O are given in units of centimeters squared per second, with D_M and D_O the self-diffusion coefficients for random diffusion of the respective ions (metal and oxygen) and z the valence of the respective ion (anion and cation).

The parabolic rate law that was derived for thickness growth can also be modified for weight gain by oxidation. In this case, K_p' has to be replaced by K_p'' according to the equation

$$K_p'' = 2 \int_z^2 K_p' \qquad (20.74)$$

Values of $\int_z$ are given in Table 20.3 for several oxides. These values are, however, based on the assumption that the scale is free of pores and cavities and consists of only one phase. Under practical conditions, this is not usually the case, and therefore these values should only be used as estimations.

TABLE 20.3.   Calculation Factor $\int_z$ for Conversion of Mass Gain Data into Scale Thickness Data Using Equation (20.74) for Several Oxides and Sulfides [119]

Corrosion Product	$\int_z$ (g/cm³)
FeO	1.28
Fe₃O₄	1.43
Fe₂O₃	1.57
Cr₂O₃	1.64
FeCr₂O₄	1.45
FeS	1.76
FeS₂	2.60

The oxidation rate constant K_p is the most important parameter for describing oxidation resistance. If K_p is low, the overall oxidation rate is low and metal consumption occurs at a very low rate. This is typical for protective oxidation. If K_p is high, metal consumption occurs at a high rate and the case of nonprotective oxidation exists. From K_p the metal consumption rate can also be calculated [120]. This requires knowledge of the stoichiometry of the oxide and the specific weight values as well as the molar weights of the reactants. Then the metal consumption can be calculated.

The real value of Wagner's analysis lies in providing a complete mechanistic understanding of the process of high-temperature oxidation under the conditions set out. The predictions of Wagner's theory for n-type and p-type oxides have been extensively examined by several workers [120–129]. For many systems, the obtained rate constants are generally several orders of magnitude larger than those which one would calculate from lattice diffusion data from Eq. (20.73). This discrepancy indicates that "short-circuit transport" is contributing to growth of the oxide film.

E3. Short-Circuit Diffusion

Lattice diffusion is the dominant process of mass transport at high temperature provided there is a sufficiently high defect concentration, but mass transport can also occur along dislocations or grain boundaries. Thus, the overall transport in polycrystalline oxide scales generally results from two fluxes of matter in parallel: an intragranular flux J_V and a flux along grain boundaries J_{gb}.

The oxidation rate may be expressed as a function of the effective diffusion coefficient defined by Eq. (20.43) (see Section D2). The scale growth rate obeys the differential equation

$$\frac{dy}{dt} = \Omega D_{\text{eff}} \frac{\Delta C}{y} \qquad (20.75)$$

where ΔC is the defect concentration difference through the oxide scale and Ω the molar volume of the oxide. Integration of Eq. (20.75) leads to

$$y^2 = 2D_{\text{eff}} \Delta C\, t = K'_{\text{p(eff)}} t \qquad (20.76)$$

where $K'_{\text{p(eff)}}$ is the effective parabolic constant.

Assuming that the oxide is dense and pure and the grains are spherical, the following rate equation can be obtained [110, 130–132]:

$$y^2 = 2\Omega\, \Delta C \left[D_V t + 4 D_{\text{gb}} \frac{1}{K} (Kt + r_0^2)^{1/2} \right] \qquad (20.77)$$

where r_0 is the initial grain radius and K is a constant which is a function of, among others, the surface energy of grain boundaries, their thickness, and the diffusion coefficient of matter across the grain boundaries. Further numerical development of this simple model has allowed a grain size distribution, different grain boundary widths, or different laws of grain growth to be taken into consideration.

F. MULTILAYERED SCALE GROWTH

The formation of several corrosion scales on metals or alloys is often observed. In order to understand clearly the growth mechanisms of these scales, it is of major importance to know whether they form concurrently or sequentially.

F1. Compact Subscales

The theory of multilayered scale growth on pure metals has been treated by Yurek et al. [133]. The hypothetical system treated is shown in Figure 20.11. It is assumed that the growth of both scales is diffusion controlled with the outward migration of cations large relative to the inward migration of anions. The flux of cations in each oxide is assumed to be independent of distance. Each oxide exhibits predominantly electronic conductivity and local equilibrium exists at the phase boundaries. The total oxidation reaction is

$$2V_{M_2} + M_{M_1} = 3V_{M_1} + M_{M_2} + O_{O_1} \qquad (20.78)$$

The cation vacancies are assumed to be neutral. The cation flux in subscale 1 is

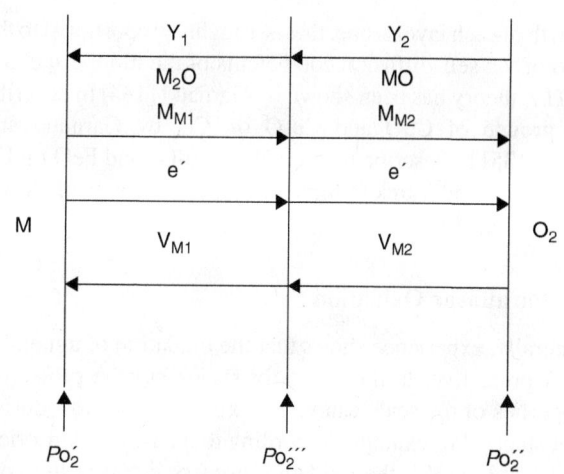

FIGURE 20.11. Schematic diagram of hypothetical two-layered scale.

$$J_M = \frac{1}{\Omega_1} \frac{K'_{p_1}}{y_1} \qquad (20.79)$$

and the amounts Q of metal consumed per unit area to form layers 1 and 2 are, respectively,

$$Q_1 = \frac{2y_1}{\Omega_2} \qquad Q_2 = \frac{y_2}{\Omega_2} \qquad (20.80)$$

The fractions of the cation flux involved in the growth of subscales (1) and (2) are, respectively,

$$J_{M_1} = \frac{Q_1}{Q_1 + Q_2} J_M = \frac{2y_1}{\Omega_1} \cdot \frac{K'_{p_1}/\Omega_1 y_1}{(2y_1/\Omega_1) + (y_2/\Omega_2)} \qquad (20.81)$$

$$J_{M_2} = \frac{Q_2}{Q_1 + Q_2} J_M = \frac{y_2}{\Omega_2} \cdot \frac{K'_{p_2}/\Omega_2 y_2}{(2y_1/\Omega_1) + (y_2/\Omega_2)} \qquad (20.82)$$

with $J_M = J_{M_1} + J_{M_2}$.

With this model, it is not possible to express simply the ratio of the thickness of both subscales. However, the ratio of parabolic rate constants can be obtained as

$$\frac{K'_{p_2}}{K'_{p_1}} = \left(\frac{y_2}{y_1}\right)^2 \frac{1 + (\Omega_1 y_1/2y_2\, \Omega_2)}{1 + (2\Omega_2 y_2/\Omega_1 y_1)} \qquad (20.83)$$

When one of the layers is much larger than the other, this expression simplifies to

$$\frac{y_2}{y_1} = \frac{K'_{p_2}}{K'_{p_1}} \times \frac{\Omega_1}{\Omega_2} \qquad (20.84)$$

for example, if $y_2 \ll y_1$. It appears that this ratio is directly proportional to the ratio of parabolic rate constants for the growth of each layer alone, that is, roughly proportional to the ratio of the self-diffusion coefficients of the mobile species.

This theory has been shown by Garnaud [134] to describe the growth of CuO and Cu_2O on Cu, by Garnaud and Rapp [135] to describe the growth of Fe_3O_4 and FeO on Fe, and by Hsu and Yurek [136] to describe the growth of Co_3O_4 and CoO on Co.

F2. Paralinear Oxidation

Generally, experience shows that the oxidation of a metal is often protective during the early stages but the protective properties of the scale can be partially or totally lost during later stages. For example, according to the Haycock–Loriers model [137, 138], the oxidation process involves the concurrent growth of an inner compact layer of MO, controlled by a diffusion mechanism, and its progressive transformation at its outer interface into an external porous layer MO′. The rate of growth of the inner compact layer controlled by diffusional transport is given by

$$\frac{d(\Delta m_1)}{dt} = \frac{K_p}{2\,\Delta m_1} - K_1 \qquad (20.85)$$

where Δm_1 is the mass of oxygen in the compact scale, K_p is the parabolic rate constant for growth of the layer, and K_1 is the rate constant for its transformation. The growth rate of the outer porous layer controlled by the reaction at its interface with the inner compact layer is given by

$$\frac{d(\Delta m_2)}{dt} = fK_1 \qquad (20.86)$$

where f is the ratio of the oxygen content in the oxide MO′ to that in the oxide MO. If the scale consists of two layers of the same oxide, $f = 1$, and the weight gain Δm is given by the rate equation

$$\Delta m + \frac{K_p}{2K_1} \ln\left[1 - \frac{2K_1}{K_p}(\Delta m - K_1 t)\right] = 0 \qquad (20.87)$$

where the rate constant K_1 is characteristic of a transformation in the solid state that does not depend on the oxygen pressure. The function $\Delta m = \Delta m_1 + \Delta m_2 = F(t)$ can be approximated by the equation

$$\Delta m = \sqrt{K_p t} + fK_1 t \qquad (20.88)$$

which describes paralinear oxidation for which parabolic kinetics predominate during the early stages of oxidation, becoming linear at longer times. This is illustrated in Figure 20.12. Paralinear oxidation is observed during the oxidation of a wide range of metals, especially if mechanical damage occurs to the scale during thermal cycling [139–141].

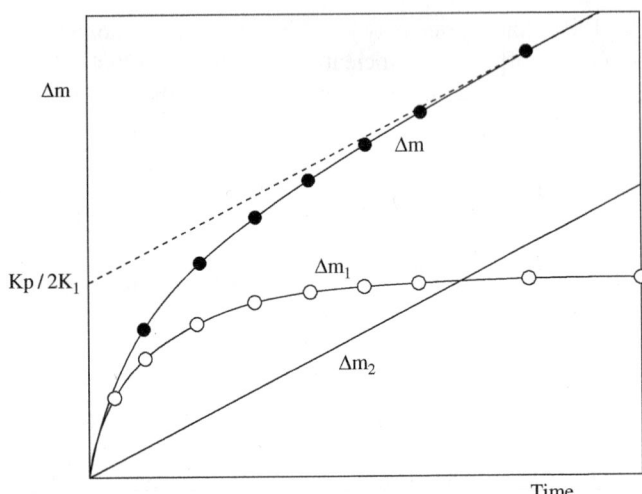

FIGURE 20.12. Schematic representation of total weight gain Δm illustrating paralinear rate equation.

This is more likely to be the case if the thermal expansion coefficient of the oxide is much less or greater than unity. The simultaneous oxidation and evaporation leads to the formation of porous and partially porous scales [59].

F3. Stratified Scales

Several metals, particularly in columns IV and V (Ti, Zr, Nb, Ta) of the periodic table, form stratified scales during their oxidation at high temperatures as a result of periodic cracking of the growing oxide. From a kinetic point of view, such cracking leads to two types of rate laws (Fig. 20.13): The first one is described by successive parabolic or cubic periods

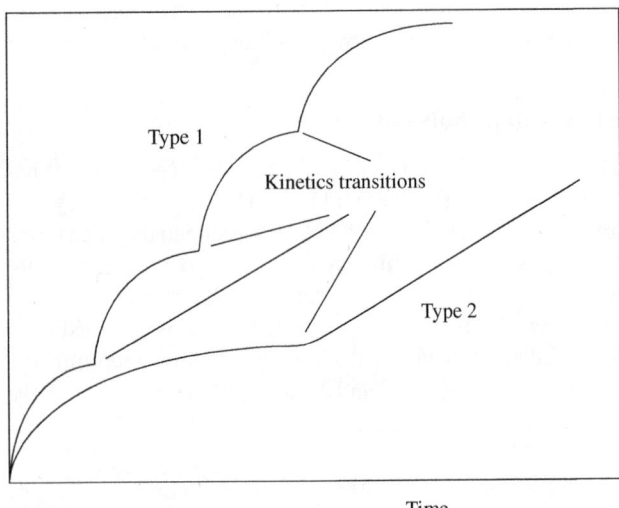

FIGURE 20.13. Schematic rate laws associated with formation of stratified scales.

(type 1), the second one by an initial parabolic or cubic period followed by a near-linear period.

Examination of the scale formed on such specimens, either on fractured or polished cross sections, shows that for both type 1 and type 2 rate laws:

- A compact and adherent scale is formed before the first transition.
- After the first transition, a porous scale exists which consists of well-defined layers formed essentially parallel to the metal surface; these may be separated by isolated cracks, also parallel to the metal surface, or by connected pores.

Mass transport through the oxide occurs by oxygen, mainly by vacancies V_0 (diffusing from the internal to the external interface). Since the scale grows in a confined space at the internal interface, the increase in volume on oxide formation, associated with a Pilling–Bedworth ratio greater than unity, may generate large compressive stresses in the oxide (several gigapascals have been measured). The metal is concurrently submitted to tensile stresses of smaller magnitude (several tens of megapascals) due to its larger thickness. The high compressive stresses in the oxide are probably responsible for the observed tendency for the kinetics to approach a cubic rate law due to an associated reduction in the oxygen diffusion flux. In all cases, oxygen pressure has no influence on the rate constants.

As oxide thickness increases, compressive stresses may also increase and result in localized spallation and/or cracking [142, 143]. Then, free access of the gas to a bare metal surface occurs and reoxidaton results. Two cases can be envisaged. In the first, the bare metal is unchanged compared with that at the beginning of oxidation, in terms of composition, microstructure, or mechanical properties. Reoxidation occurs exactly as in the initial period and the rate law is the exact repetition of the first pretransition curve. Kinetics of type 1 (Fig. 20.13) are then observed with a succession of parabolic-type periods.

In the second example, the bare metal differs from the initial state, for example, dissolution of oxygen may have occurred, leading to increased hardness and lower creep relaxation rates. The second oxide layer that forms does not then reach the thickness of the first since stress increases at a higher rate and early separation occurs from the metal. In the case of TiO_2 growth on titanium at 850°C, for example, the second and all subsequent layers have a thickness of 1–2 μm, whereas the first attains 10–15 μm. The spallation or cracking of these scales does not occur at the same time for all locations on the specimen surface and the resulting law is approximately linear with an increased rate compared to that in the pretransition period. The system can be described as a metal covered by an oxide of statistically constant thickness [8, 144–146].

G. OXIDATION RESISTANCE

The oxidation of alloys and other metallic materials is a complex process consisting of a large number of phenomena which may, themselves, depend on material composition and environment, that is, temperature, mechanical stress, gas composition, and so on. To obtain effective protection of a metallic material, the oxide formed must lead to a continuous scale (external oxidation) and not to precipitates within the alloy (internal oxidation).

Although numerous studies have been devoted to oxidation of alloys, the oxidation mechanisms are often still not fully understood. In fact, the models have usually been developed for the oxidation of the relatively simple example of pure metals and are inadequate to describe alloy oxidation. In order to do this, additional factors have to be taken into account, for example, the different affinity for oxygen of each of the alloy's constituents, dissolution of oxygen into the alloy, solid solubility between the oxides formed, formation of complex oxides, and different mobilities of the various metal ions in the oxide phases. Moreover, alloy oxidation involves complex processes for which the equilibrium state is reached very slowly or may never be reached. It is unrealistic to try to classify all the various types of alloy oxidation using simple criteria. However, several classification systems have been proposed [67, 147] to account for the diverse morphologies of oxidation scales. They use thermodynamic diagrams and kinetic considerations and, for simplicity, we can consider that three methods are available to protect metals and alloys from high-temperature oxidation: (i) control of the atmosphere, (ii) alloying addition of species more easily oxidized than the base metal (chromium, aluminum) and which form a protective scale during oxidation, and (iii) use of protective coatings deposited by various methods (cementation, plasma spraying, ion bombardment, etc.). Some considerations of these three approaches are presented in the following paragraphs.

G1. Atmosphere Control

When a process is considered for use in industry, it must be effective, reliable, and economical. For most applications these requirements rule out systems based on high vacuum and the general practice is to use atmospheres derived from fuels. The gases used are therefore mixtures of N_2, CO, CO_2, H_2, H_2O, and CH_4, which make up the products of combustion of fuels. More recently, atmospheres based on nitrogen have been increasingly used.

Starting from the fuel and air, various types of atmospheres can be produced. The main, or common, differentiation is between "exothermic" and "endothermic" atmospheres. The nomenclature is ambiguous and it is well to be clear about its meaning. An exothermic atmosphere is produced exothermically by burning the fuel with measured amounts of air. This type of atmosphere has the highest oxygen potential. An endothermic atmosphere is produced by heating, by external means, a mixture of fuel gas with air over a catalyst to provide a gas containing reducing species. This atmosphere has a low oxygen potential and heat is absorbed during its preparation; hence the atmosphere is described as endothermic. A glance at the analysis of dried, stripped, exothermic atmospheres will confirm that they are predominantly pure nitrogen. Basically, the fuel has been used to remove oxygen from the air.

The economics of using nitrogen as a controlled-atmosphere source become more attractive when such factors as safety, reliability, and productivity are considered. Furthermore, the present tendency is to move away from oil toward electricity, in which case nitrogen atmospheres will be particularly attractive.

Modern techniques are currently available using carburizing and nitriding systems under vacuum. In these processes of vacuum carburizing and plasma carburizing, the components are heated under vacuum to around 950°C. Methane is leaked into the chamber to a pressure of between 3 and 30 mbar to add carbon to the system. In the absence of a plasma, the methane will only decompose to the extent of about 3%, probably on the surface of the components according to a sequence such as

$$CH_4 \rightarrow CH_3 + H \rightarrow CH_2 + 2H \rightarrow CH + 3H \rightarrow C + 4H \tag{20.89}$$

These reactions may be stimulated to provide 80% decomposition by using a plasma process to excite the methane molecule. In this case, the molecular breakdown may occur in the plasma to produce charged species. Hydrocarbons other than methane may be used as the feedstock. The usual operating sequence involves flushing and evacuation, heating to a temperature under the inert atmosphere, and carburizing for a predetermined time followed by a diffusion anneal in a carbon-free atmosphere. This cycle is designed to provide optimum surface carbon content and carburized depth [9, 10, 12, 13, 148].

The main application of controlled atmospheres is in the area of heat treatment of finished, machined components or of articles of complex shape, which cannot easily be treated subsequently for the removal of surface damage. In this context, the atmosphere is controlled for one of two reasons: to prevent surface reaction or to cause a surface reaction, such as carburizing or nitriding.

Prevention or control of oxide layer formation is primarily a matter of controlling the oxygen partial pressure of the atmosphere at a value low enough to prevent oxidation, as described in Section B1. For a metal that undergoes oxidation according to the reaction shown in Eq. (20.7) (Section C), where MO is the lowest oxide of M, the oxygen partial pressure must be controlled so as not to exceed a value p_{O_2}:

$$p_{O_2} = \exp\left(\frac{2\Delta G^\circ}{RT}\right) \tag{20.90}$$

where ΔG° is the standard free energy of reaction (20.7).

Unfortunately, $(p_{O_2})_{\text{M-MO}}$ is a function of temperature and has lower values at lower temperatures. Thus, if an atmosphere is designed to be effective at high temperatures, it may become oxidizing as the temperature is reduced during cooling. A surface oxide layer may therefore form as the metal is cooled. Although the metallurgical damage to the surface will be negligible, the surface may be discolored, that is, not bright. This condition can be overcome to some extent by rapid cooling or by changing the atmosphere to a lower oxygen partial pressure just before or during cooling.

For alloys, the most critical reaction must be considered when deciding on the composition of the atmosphere to be used. For this purpose, the activities of the alloy components must be known since, if the metal M in Eq. (20.7) exists at an activity a_M, the corresponding equilibrium oxygen partial pressure will be given by p'_{O_2}:

$$p'_{O_2} = \frac{1}{a_M^2} \exp\left(\frac{2\Delta G^\circ}{RT}\right) \tag{20.91}$$

If the metal activities in the alloy are not known, then, by assuming the solution to be ideal, mole fractions may be used instead of activities to give a value of the oxygen partial pressure at which experiments must be performed to establish the correct atmosphere composition.

Low oxygen partial pressures can be provided and, more importantly, controlled by using "redox" gas mixtures. These mixtures consist of an oxidized and a reduced species, which equilibrate with oxygen, for example,

$$CO + \tfrac{1}{2}O_2 = CO_2$$

$$\Delta G^{\circ\prime} = -282200 + 86.7TJ \tag{20.92}$$

from which p''_{O_2} or, more importantly, p_{CO_2}/p_{CO} can be obtained:

$$p''_{O_2} = \left(\frac{p_{CO_2}}{p_{CO}}\right)^2 \exp\left(\frac{2\Delta G^{\circ\prime}}{RT}\right) \tag{20.93}$$

$$\frac{p_{CO_2}}{p_{CO}} = p_{O_2}^{1/2} \exp\left(\frac{-\Delta G^{\circ\prime}}{RT}\right) \tag{20.94}$$

Thus, from Eq. (20.93) the ratio of carbon dioxide to carbon monoxide may be calculated for any oxygen partial pressure and temperature.

Further discussion of protective atmospheres can be found in references [8], [11], and [36].

G2. Alloying

For engineering applications, metals are strengthened and their environmental resistance improved by appropriate alloying. The basic mechanisms operating in pure metal oxidation are also operative in the oxidation of alloys with added complications. These complications include the formation of multiple oxides, mixed oxides, internal oxides, and diffusion interactions within the metals. The effect of alloying on oxidation behavior can be understood first by considering binary alloy model AB consisting of element A, the major component, and B, the minor component [149]. There are two distinct possibilities [150]: (i) One considers element A as more noble and B as more reactive. Thus, at atmospheric pressure of oxygen, B converts to BO, and if the alloy is dilute in B, dispersed precipitates of BO form in A. If the alloy is concentrated in B, a continuous BO scale forms on top of AB. (ii) The other case considers that A and B are both reactive to oxygen with BO more stable than AO. The concentration of B again dictates the oxide morphologies. If the alloy is dilute in B, a stable oxide AO forms as the outer scale. Below the oxide scale, at the AO–alloy interface, the O_2 activity is high enough to oxidize B into BO precipitates. If the concentration of B exceeds the critical level required to form a continuous BO scale, then BO forms on top of AB. The oxide growth rate is parabolic, with activation energy characteristic of the growth of BO. The actual rate depends on how protective the BO scale is and on the presence of additional alloying elements.

Binary alloys of Ni with Cr or Al form the basis of many oxidation-resistant commercial materials. There are three composition (in wt %) regimes, with distinct oxidation characteristics for each family:

Ni–Cr < 10%: Such alloys form an NiO external scale and internal Cr_2O_3.

Ni-30% > Cr > 10%: In such alloys the outer scale consists of NiO on grains and Cr_2O_3 in grain boundaries.

Ni–Cr > 30%: For such alloys the external scale consists of Cr_2O_3, which is maintained because of the large Cr reservoir.

Ni–Al < 6%: Such alloys form an NiO external scale and internal Al_2O_3 and $NiAl_2O_4$.

Ni-17% > Al > 6%: In such alloys the outer scale initially consists of Al_2O_3. However, on continuous exposure, Al depletion occurs in the alloy adjacent to the oxide scale. In the depleted zone, the Al activity is well below the requirement to form continuous Al_2O_3. Therefore, NiO overtakes Al_2O_3. The overall result is the formation of a mixture of NiO, $NiAl_2O_4$ spinel formed by the combination of NiO and Al_2O_3.

Ni–Al > 17%: For such alloys, the external scale of Al_2O_3 is maintained because of the large Al reservoir.

Practical alloys and metallic coatings for high-temperature applications are seldom binary. These alloys are typically Cr_2O_3 and Al_2O_3 formers. Silica (SiO_2) also forms a protective scale, particularly for refractory metals with which it has better thermal expansion matching. However, SiO_2 is not stable at low pressures. It decomposes to gaseous species such as SiO. It also reacts with water vapor at high temperatures, forming $Si(OH)_4$ gas. The use of Cr_2O_3 scale-forming alloys is limited to temperatures below 1000°C.

Volatile CrO_3 forms above this temperature in the presence of oxygen due to the reaction

$$\tfrac{1}{2}Cr_2O_3 + \tfrac{3}{4}O_2 = CrO_3 \qquad (20.95)$$

Thermodynamic analysis of the foregoing reaction shows oxygen partial pressure dependence to be $p_{O_2}^{3/4}$. Thus, the volatilization becomes important at high oxygen partial pressures. Volatilization of Cr_2O_3 has been observed even at lower temperatures, between 850 and 900°C. Also, in the presence of water vapor, $CrO_2(OH)_2$ forms above the Cr_2O_3 scale. The scale rapidly vaporizes, resulting in continuing metal recession. The SiO_2 scales are also subject to water vapor–enhanced volatility. For high-temperature application, above 1000°C, useful alloys are therefore designed to be Al_2O_3 formers. In such alloys, the alumina scale can have one of several allotropic forms, depending on the alloy composition and temperature of oxidation. The allotropes include the transient phases γ, δ, and υ which, on thermal exposure, convert into the stable phase α [151]. The composition of the scale formed and its stability depend on the alloy composition, the temperature, and the cyclic nature of thermal exposure [150, 152–156].

Commercial superalloys contain many alloying elements of significance over and above chromium and aluminum. The oxidation behavior of these alloys is very complex and oxidation resistance varies widely, although the general mechanisms described earlier still apply. The complexities arise from significant influence of the individual elemental constituents. Nickel-based superalloys containing elements such as Co, Cr, Al, Ti, W, and Ta exhibit general behavior similar to simple NiCrAl alloys.

The cyclic oxidation of superalloys consists of several steps. The process involves an initial transient period during which the oxides of the individual constituents form. These include NiO, CoO, Cr_2O_3, TiO_2, and Ta_2O_5. Continued oxidation leads to the formation of the most stable oxide, which is Al_2O_3 for alumina-forming alloys. The parabolic rate constants of the transient oxides are larger than that of the stable oxide. In a cyclic environment the oxide scale cracks and spalls. Aluminum in the alloy diffuses to the oxide–alloy interface to reform the scale. The oxide scale spalling followed by the reformation process continues. An aluminum-depleted zone forms in the alloy below the oxide scale. The thickness of the depleted zone depends on the aluminum content of the alloy. Isothermal oxidation of NiCoCrAlY with 12 wt % Al, for example, exhibits depleted zone thickness three times the thickness of the alumina scale [157]. Because of the loss of aluminum, the depleted zone becomes enriched with the other alloying constituents, some of which, having low solubility in the depleted zone, precipitate out in the form of acicular phases. Additionally, nitrogen, which diffuses into the alloy during oxidation, now exceeds its solubility limit and precipitates out as nitrides of such elements as Ti. The new precipitate phases penetrate into the alloy. The alloy is finally depleted of aluminum to below such a critical level that the regeneration of a continuous scale of Al_2O_3 is no longer feasible. At this stage, for nickel-based alloys, breakaway oxidation starts with the formation of continuous NiO, which is not protective. As a result, oxygen diffuses into the alloy, forming internal oxides of the remaining aluminum, chromium, and other reactive constituents of the alloy. The alloy substrate loses wall thickness and load-bearing capability.

To help maintain oxidation resistance, the following steps can be taken: (i) increase aluminum activity as much as possible; (ii) increase aluminum diffusivity to the alloy surface; (iii) inhibit diffusivity of oxygen into the alloy; and (iv) improve the adherence of the alumina scale to the substrate.

Similar processes and arguments also hold true for cobalt-based alloys, although some differences exist in the detail. For comparable oxidation resistance, cobalt-based alloys need higher combined aluminum and chromium than do nickel-based alloys. Also, once the protective alumina scale fails, CoO forms on cobalt-based alloys, which spalls off catastrophically as opposed to the gradual failure of NiO formed on nickel-based alloys.

The substrate alloy compositions are generally a result of an effort to maximize structural properties such as creep, tensile, and fatigue strength. The alloying constituents, however, affect the oxidation resistance significantly [158–162].

G3. Protective Coatings

The focus on the development of substrate alloys discussed in this chapter is generally to achieve high strength, high ductility, and efficient production. Oxidation resistance may not be consistent with achieving these goals. For example, increased Al and Cr result in improved oxidation resistance; however, beyond a certain level, these elements reduce creep strength of the resulting alloys. To achieve both strength and resistance to environmental degradation, the two functions are separated. The load capability is provided by the application of thin coatings with adequate Al and Cr. The thickness of the coating is controlled so that it does not carry any significant load. Depending on the temperature of use, many high-temperature alloys require coatings compatible with its

composition and structural (modulus) and thermal (coefficient of thermal expansion) properties. Diffusion, overlay, and thermal barrier coatings are discussed in the following paragraphs.

G3.1. Diffusion Coatings. In diffusion coating, the substrate surface is enriched in an element that will provide high-temperature corrosion resistance. Typical elements are chromium (chromizing), aluminum (aluminizing), or silicon (siliconizing). The substrate is involved in the formation of the coating, and substrate elements are incorporated in the coating; in addition, a diffusion zone is developed in the substrate beneath the coating. Such enrichment not only allows a protective scale to form by selective oxidation but also provides a substantial reservoir of the protective element to delay the inevitable breakaway oxidation to substantially longer times. A wide variety of diffusion coatings are used [163–165].

The most common method of aluminizing is pack cementation, which has been a commercially viable process for many years [166, 167].

Pack cementation and vacuum pack coating techniques can be generalized as methods in which a chemical vapor deposition (CVD) process takes place with the substrate surrounded by a mass of the depositing medium. "Cementation" is a misnomer. The substrate is "packed" in a "cement" consisting of a mixture of the master alloy (the source alloy), a salt as activator, and an inert or reducing hydrogen atmosphere. The coating is carried out over a wide range of pressures from a low, near-vacuum 1–20 torr to near-atmosphere, that is, 360 torr, in the enclosed "retorts" (cf. "reactors" for conventional CVD) [168]. The substrate to be coated is surrounded by the pack; various alignments within this principle are possible depending on the substrate requirements.

The first known cementation process was that of Al on steel in 1914 [166, 167]. However, much attention was given to the process, and its variety and development occurred during the mid-1960s to the late 1970s on Ni- and Co-based alloys and iron alloys when the protection of high-temperature gas turbine alloys became paramount together with rocket and space hardware, that is, refractory alloys, mainly Ta, Nb, Mo, and Cr based. The technique itself, in principle, has changed remarkably little since 1914. The composition and quality of the substrate–deposit configuration have been modified as a result of research carried out in the last 10–20 years.

Even now, aluminum stands foremost among pack-coated deposits, closely followed by Cr, Si, and alloys of Al–Cr, and Al–Si can be coated as one- or two-stage packs [169]. Most of the literature referred to here is on Al, Cr–Al, Ni–Al, and Fe–Al systems. Pack coating is particularly suited to treat large substrates either singly or in bulk and can handle intricate shapes as it is not a line-of-sight method. Much of the earlier literature is available in references [169]–[171].

Aluminiding was one of the first high-temperature metallic coating systems. Alumina scale-forming coatings are very protective in high-velocity gas turbines. Uncoated alloys containing Al as a minor constituent are rapidly depleted of Al, which diffuses to the surface to maintain alumina formation due to recurring oxide spallation. Diffusion-coated alloys form a surface alloy layer, for example, NiAl and CoAl, which develop the protective alumina layer under more controlled kinetics and reform on spalling without affecting the substrate alloy properties. A renewal is possible before excessive depletion occurs. Brittle intermetallics can be a danger in this system.

A diffusion barrier layer between the coating and the substrate is necessary to prevent interdiffusion of elements within the coating system. Multicomponent diffusion coatings (Al, B, Cr, Si, Ti, Zr) to protect machines from liquid Al and for brass corrosion have been used [172]. More studies are needed in this area.

On steel, Al coatings are protective up to 500°C, above which brittle intermetallics are formed. Thermal stress-induced cracks then propagate into the substrate metal. Aluminide coatings protect steel from oxidation and corrosion in hydrocarbon and sulfur-containing atmospheres. Aluminized steel is better than stainless steel where oxidation carburization occurs [173].

Chromized steel (diffused in) is resistant to air oxidation up to 700°C. Above 800°C Cr diffuses into the steel, reducing oxidation resistance. At higher temperatures brittle intermetallics form. Good chromized sheet can be bent 180° without damage and is suitable for most firebox and heat exchanger applications up to 600°C. Addition of Al or Si to the chromizing pack process confers oxidation resistance up to 900°C on mild steel, although continuous use at 900°C causes brittle intermetallics and consequent cracking on thermal cycling.

On superalloys, alumina-forming diffusion-bonded coatings provide an Al-rich surface to gas turbine environment. Aluminiding superalloys involve formation of more than one phase in the Ni–Al (and Co–Al) systems. Heat treatment is given to stabilize the NiAl phase. NiAl with some Cr and Ti improves the hot corrosion resistance [174]. Oxide particles (e.g., Y_2O_3) reduce spalling; defects like pinholes, blisters, and cracks may be avoided by a combination of minor additives.

Aluminide coatings lack ductility below 750°C and on thermal cycling undergo surface cracking resulting in spalling of the alumina scale. To overcome these two problems, the coating composition was adjusted to embed the brittle β-NiAl or β-CoAl in a ductile γ solid-solution matrix. Addition of yttrium improved oxide adherence. Improvements in mechanical properties were achieved by HIP (hot isostatic pressing)–densified, argon-atomized prealloyed powder ingots;

tensile ductilities of over 20% were produced using the finer precipitate with its better distribution [175]. Studies in which minor amounts of Si, Fe, and Ti were added to improve the scale resulted in little improvement [176]. Platinum electroplate followed by aluminiding gives improved oxidation resistance which offsets the higher cost [177].

Cobalt-based superalloys for higher temperature but less stressed gas turbine vanes have no Al, and this limits the aluminide coating thickness that can be applied without spalling. Superalloy compositions avoiding sigma and other embrittling phases can be destabilized by coating interdiffusion [178]. Structural strengtheners, such as submicrometer oxides in alloys with oxide dispersion strengthening (ODS) and carbides in eutectic alloys with dispersion strengthening (DS), can also limit coating selection [178].

Further research is needed in these areas as higher performance alloys will restrict the choice of acceptable coatings. In a recent five-year period, over 50 production coatings became necessary to replace an original selection of only one to two compositions [179, 180].

G3.2. Overlay Coatings.
Overlay coatings are distinguished from diffusion coatings in that the coating material is deposited onto the substrate in ways that give only enough interaction with the substrate to provide bonding of the coating. Since the substrate does not enter substantially into the coating formation, in principle, much greater coating composition flexibility is achievable with overlay coatings. Also, elements, such as Cr, that are difficult to deposit into diffusion coatings can be included in overlay coatings. Although Cr is difficult to incorporate into diffusion aluminide coatings, it is readily incorporated into overlay coatings. Overlay coatings based on the Ni–Cr–Al and Co–Cr–Al systems are commonly used to protect superalloys. Also, small amounts of the reactive elements (e.g., Y, Hf) are routinely incorporated into overlay coatings but are difficult (or sometimes impossible) to incorporate into diffusion coatings. The flexibility in composition of the overlay coatings also allows mechanical properties to be tailored for a given application.

Overlay coatings are deposited by physical techniques. The most common are physical vapor deposition (PVD), which includes evaporation, sputtering, and ion plating, and spray techniques (e.g., plasma spraying, flame spraying).

Diffusion-type coatings, used successfully on early gas turbines, were tied to the substrate composition, microstructure, and design. Some changes introduced later were (i) superalloy composition, such as reduction in Cr and increase in other refractory metals; (ii) microstructure, in castings with more segregation; and (iii) design, by air cooling and with thin walls (which introduced higher thermal stresses). These changes required coatings that were much more independent of the substrate. Overlay coatings met this necessity.

Overlay coatings also overcome the process restrictions encountered in diffusion coatings, especially the variants, namely, Cr/Al, Ta + Cr, or the Pt aluminides, all of which give better stability and oxide hot corrosion resistance than Al alone. MCrAlY compositions (M = Ni, Co, Fe alone or in combination) are the main ones in the series of overlay coatings developed by electron beam evaporated physical vapor deposition (EBPVD) for multiple load. This technique is one of the two most important manufacturing techniques widely used for deposition of thermal barrier coatings (TBCs) onto substrates. Due to its feasibility of varying coating structure, it has great potential for multiple load, for example, to produce functionally graded TBCs.

MCrAlY overlays used in gas turbines are usually Ni and/or Co with high Cr, 5–15% Al, and Y addition less than about 1% for stability during cyclic oxidation. They are multiphase alloys with ductile matrix (e.g., γ Co-Cr) containing a high fraction of brittle phase (e.g., β CoAl). The Cr provides oxidation and hot corrosion resistance, but too much Cr affects substrate-phase stability. The success of most overlay coatings is the presence (and perhaps location) of oxygen-active elements, such as Y and Hf, which promote alumina layer adherence during thermal cycling, giving increased coating protection at lower Al levels. Yttrium appears mostly along grain boundaries if MCrAlY is cast but is homogeneous if plasma sprayed. Thus MCrAlY with 12% Al is more protective than the more brittle diffusion aluminides with 30% Al.

Overlay claddings deposited by hot isostatic processing (HIP), electron beam evaporation, or sputtering methods are diffusion bonded at the substrate–coating interface, but the intention here is not to convert the whole coating thickness to NiAl or CoAl. There is, thus, more freedom in coating composition, so that properties can be maximized to the extent required. Compositions based on NiCr, CoCr, NiCrAl, CoCrAl, NiCrAlY, FeCrAlY, and NiCrSi have been used successfully in gas turbine engines. They are generally alumina formers with only 10% Al, unlike the 30% in nickel aluminide coatings. Chromium increases the Al activity allowing this advantage. Higher Al levels cause brittleness and a higher Ductile-Brittle Transition Temperature (DBTT) and the Al levels are generally held below 12% (5–10% preferred). The coatings are also more ductile than NiAl and CoAl and can be rolled and bonded by HIP. In general, NiCrAlY gives best results against high-temperature oxidation whereas CoCrAlY is best for hot corrosion [181].

G3.3. Thermal Barrier Coatings.

Thermal barrier coatings are ceramic coatings that are applied to components for the purpose of insulation rather than oxidation protection. The use of an insulating coating coupled with internal air cooling of the component lowers its surface temperature with a corresponding decrease in its creep and oxidation rates. The use of TBCs has resulted in a significant improvement in the efficiency of gas turbines [12].

The earliest TBCs were frit enamels that were applied to aircraft engine components in the 1950s. The first ceramic TBCs were applied by flame spraying and, subsequently, by plasma spraying. The ceramic materials were alumina and zirconia (MgO or CaO stabilized), generally applied directly to the component surface. The effectiveness of these coatings was limited by relatively high thermal conductivity of alumina and problems with destabilization of the zirconia-based materials. Important developments included the introduction of Ni–Cr–Al–Y bond coats and plasma-sprayed Y_2O_3-stabilized zirconia topcoats in the mid-1970s and the development of EBPVD to deposit the topcoat in the early 1980s. Plasma-sprayed TBCs have been used for many years on combustion liners but, with advanced TBCs, vanes, and even the leading edges of blades, can now be coated. The use of TBCs can achieve temperature differentials across the coating of as much as 175°C [182].

Typical systems consist of a nickel-based superalloy substrate coated with M–Cr–Al–Y (M = Ni,Co) or a diffusion aluminide bond coat, which forms an alumina layer [thermally grown oxide (TGO)]. Onto this is deposited a yttria-stabilized zirconia (YSZ) TBC. The TBC can be deposited by air plasma spraying (APS), or EBPVD. The EBPVD coatings are used for the most demanding applications, such as leading edges of airfoils.

The APS coating consists of layers of *splats* with clearly visible porosity and is microcracked. This microcracking is necessary for strain tolerance. The EBPVD coating consists of columnar grains separated by channels similar to the leaders seen in metallic overlay coatings. These channels are responsible for the high strain tolerance of EBPVD TBCs.

In summary, a wide variety of coatings are used for oxidation protection at high temperature. It is a huge field dealing with different compositions, production methods, properties, degradation, characterization, repair and function of coatings, and so on. These aspects have been surveyed by Bose [183], Khanna [14], Birks et al. [12], Sato et al. [184], Stern [185], Dahotre and Hampikian [186], Gao [179], and Schütze and Quadakkers [180], among others.

H. OXIDATION OF ENGINEERING MATERIALS

There is a large spectrum of engineering materials available for applications subject to oxidation in different temperature ranges. This section focuses on laboratory and field experience of many materials that are being considered for applications in various temperature regimes. They range from carbon and Cr–Mo steels to advanced superalloys.

At temperatures below 550°C, carbon steel in air shows very little weight gain after exposure for nearly one month. As the temperature is increased to 700°C, the oxidation rate is significantly increased, exhibiting a linear rate of oxidation attack. Test results by John [187] showed that carbon steel exhibited about 0.25 mm/year (10 mpy) of oxidation at 604°C; at 650°C, carbon steel exhibited oxidation rates of the order of 1.25 mm/year (50 mpy). The beneficial effects of chromium and silicon additions to carbon steel were reported by Zeuthen [188]. Chromium–molybdenum steels are used at higher temperatures than carbon steel because of higher tensile and creep rupture strengths as well as better microstructural stability. Molybdenum and chromium provide not only solid-solution strengthening but also carbide strengthening.

Silicon is very effective in improving the oxidation resistance of Cr–Mo steels, but the most important alloying element for improving oxidation resistance is chromium.

The superior oxidation resistance of martensitic and ferritic stainless steels to that of carbon and Cr–Mo steels is well illustrated in the open literature [189]. The 25Cr steel (type 446 = S44600) is the most oxidation resistant among the 400 series stainless steels, due to the development of a continuous Cr_2O_3 scale on the metal surface. In Fe–Cr alloys, a minimum of approximately 18 wt % Cr is needed to develop a continuous Cr_2O_3 scale against further oxidation attack [190]. Cyclic oxidation studies conducted by Grodner [191] also revealed that type 446 was the best performer in the 400 series stainless steels, followed by types 430 (S43000; 14–18Cr), 416 (S41600; 12–14Cr), and 410 (S41000; 11.5–13.5Cr). The growth of a thin, adherent $(Fe,Cr)_2O_3$ scale as a function of the accumulated isothermal hold time up to 1000 h was observed by Walter et al. [192] at 650°C in air during cycling from 650 to 300°C.

The 300 series austenitic stainless steels have been widely used for high-temperature components because they exhibit higher elevated-temperature strength than do ferritic stainless steels. Furthermore, they do not suffer 475°C embrittlement or ductility loss problems in thick sections and in heat-affected zones as do ferritic stainless steels. Nevertheless, some austenitic stainless steels can suffer some ductility loss upon long-term exposure to intermediate temperatures (e.g., 540–800°C) due to sigma-phase formation [193]. Moccari and Ali [194] observed the beneficial effects of nickel in improving the oxidation resistance of austenitic stainless steels.

In evaluating materials for automobile emission control devices, such as thermal reactors and catalytic converters, Kado et al. [195] carried out cyclic oxidation tests on various stainless steels. In cyclic oxidation tests performed in still air at 1000°C for 400 cycles (30 min in the furnace and 30 min out of the furnace), types 409 (S40900; 12Cr), 420 (S42000; 13Cr), and 304 (S30400; 18Cr–8Ni) suffered severe attack. Type 420 (S42000; 13Cr) was completely oxidized after only 100 cycles, although the sample did not show any weight changes. Alloys that performed well under these conditions were types 405 (S40500; 14Cr), 430 (S43000; 17Cr), 446 (S44600; 25Cr), 310 (S31000; 25Cr–20Ni), and DIN 4828 (19Cr–12Ni–2Si).

When cycled to 1200°C for 400 cycles (30 min in the furnace and 30 min out of the furnace), all alloys tested except F-1 alloy (Fe–15Cr–4Al) suffered severe oxidation attack. This illustrates the superior oxidation resistance of alumina formers (i.e., alloys that form Al_2O_3 scales when oxidized at elevated temperatures). Their data also illustrate that, for temperatures as high as 1200°C, Cr_2O_3 oxide scales can no longer provide adequate oxidation resistance.

Oxidation data generated in combustion atmospheres are relatively limited. No systematic studies have been reported that varied combustion conditions, such as air-to-fuel ratios. In combustion atmospheres, the oxidation of metals or alloys is not controlled by oxygen only. Other combustion products, such as H_2O, CO, CO_2, N_2, hydrocarbon, and others, are expected to influence oxidation behavior. When air is used for combustion, nitride formation in conjunction with oxidation can occur in combustion atmospheres under certain conditions.

The presence of water vapor can also be an important factor in affecting oxidation behavior of alloys, as discussed later in this section.

Manufacturing processes can greatly influence the surface chemistry of an alloy product. Stainless steels can be finished into the final product by bright annealing (i.e., annealing is performed in a protective atmosphere, such as hydrogen environment or dissociated ammonia environment). This process generally produces a product with minimal depletion of chromium at or near the surface. On the other hand, when the alloy product is finished by black annealing (i.e., annealing is performed in air or combustion atmosphere in the furnace) and followed by acid pickling, there is a good chance that the alloy product may exhibit surface depletion of chromium. This is particularly important for thin-gage sheet products or thin tubular products [196].

Some stainless steel producers may manufacture stainless steels at the bottom of the specification range for key alloying elements, such as chromium, to reduce materials cost. Accordingly, the chromium content can be insufficient to maintain a continuous chromium oxide scale during prolonged service or when subjected to thermal cycling or overheating conditions, thus promoting breakaway oxidation. The oxidation resistance of these "lean" stainless steels can be further degraded by the surface depletion of chromium resulting from manufacturing processes that may involve excessive pickling after "black" annealing (annealing in air or combustion atmosphere), during successive reductions in cold rolling in flat product manufacturing, or pilgering in tubular manufacturing. The chromium concentration at the surface of such a product, particularly a thin-gage sheet or tube, may be too low to form or maintain a continuous chromium oxide scale during service. As a result, iron oxides and isolated nonprotective Fe–Cr oxide nodules can develop on the metal surface, thus resulting in breakaway oxidation, as discussed later.

Some commercial electrical resistance heating elements are made of Fe–Cr–Al alloys, such as Kanthal® alloys, which rely on the formation of the Al_2O_3 scale for applications up to 1400°C [197]. For example, some of the Kanthal alloys that are available in wire, strip, and ribbon product forms are Kanthal A-I (K92500; Fe–22Cr–5.8Al), AF (Fe–22Cr–5.3Al), and D (Fe–22Cr–4.8Al). Since these wrought alloy products are essentially ferritic alloys, they exhibit low creep rupture strengths when the temperature exceeds 650°C and cannot be used for high-temperature structural components. Thus, the electrical resistance heating elements made of these alloys must be properly supported to avoid creep deformation, such as sagging. These Kanthal wires can be used in arc or flame spraying to produce an oxidation-resistant coating or in weld overlay cladding by using a gas metal arc welding (GMAW) process. A powder metallurgy (P/M) process was used to produce a new alloy product, Kanthal APM, reported to exhibit improved creep rupture strength [198]. Other commercial Fe–Cr–Al alloys include ALFA-I™ (Fe–13-Cr–3Al), ALFA-II™ (Fe–13Cr–4Al), and ALFA-IV™ (Fe–20Cr–5Al–Ce) developed by Allegheny Ludlum [199] and Fecralloy® (Fe–16Cr–4Al–0.3Y) developed by Atomic Energy Authority [200].

As the nickel content in the Fe–Ni–Cr system increases from austenitic stainless steels to a group of iron-based alloys with 20–25Cr and 30–40Ni, the alloys become more stable in terms of metallurgical structure and more resistant to creep deformation (i.e., higher creep rupture strengths). In general, this group of alloys also exhibits better oxidation resistance. Some of the wrought alloys in this group are 800H/800HT (Fe–21Cr–32Ni–Al–Ti), RA330 (N08330; Fe–19Cr–35-Ni–1.2Si), HR120 (Fe–25Cr–37Ni–0.7Nb-N), AC66 (N33228; Fe–27Cr–32Ni–0.8Nb–Ce), 353MA (Fe–25Cr–35Ni–1.5Si–Ce), and 803 (Fe–26Cr–35Ni–Al–Ti) [201].

In many Ni–Cr alloys, many alloying elements, such as those for solid-solution strengthening (e.g., Mo, W) and precipitation strengthening (e.g., Al, Ti, Nb), are added to the alloys to provide strengthening of the alloy at elevated temperatures. Many of these alloys are commonly referred to as "superalloys." The superalloys also include oxide dispersion–strengthened (ODS) alloys, which are briefly discussed later.

Similar to Fe–Cr–Al alloys, aluminum is also used as an alloying element in Ni–Cr alloys to improve the oxidation resistance. Although a Ni–Cr alloy generally requires a minimum of 4% Al to form a protective Al_2O_3 scale, the addition of less than 4% Al can significantly improve the oxidation resistance of the alloy. Alloy 601 (N06601), with only about 1.3% Al, shows excellent oxidation resistance [202].

Nickel–chromium alloys containing about 4% Al or higher form a very protective Al_2O_3 scale when heated to very high temperatures; for example, Lai [203] compared alloy 214 (Ni–16Cr–4.5Al–Y) with alloy 601 and alloy 800H in cyclic oxidation tests performed in still air at 1150°C with specimens cycling to room temperature once a day except weekends. Alloy 214 showed essentially no weight loss after 42 days of testing, whereas alloy 601 showed a linear weight loss.

For applications at high temperatures, many superalloys contain numerous alloying elements for increasing the elevated-temperature strength of the alloy. Molybdenum and tungsten are common alloying elements for providing solid-solution strengthening for increasing the creep rupture strength of the alloy. Two iron-based superalloys, Multimet alloy (R30155; Fe–20Ni–20Co–21Cr–3Mo–2.5W–1.0Nb + Ta) and alloy 556 (R30556; Fe–20Ni–18Co–22Cr–3Mo–2.5W–0.6Ta–0.02La–0.02Zr), are good examples. However, the oxides of both molybdenum and tungsten (MoO_3 and WO_3) exhibit high vapor pressures at very high temperatures. Multimet alloy showed rapid oxidation attack at 1150 and 1200°C, with specimens completely consumed at both temperatures. However, formation of the volatile oxides of MoO_3 and WO_3 can be minimized by modification of some key alloying elements in Multimet alloy. The development of alloy 556 was aimed at improving the oxidation resistance of Multimet alloy without losing the elevated-temperature strength by modifying the Multimet alloy composition. The modification involved a slight increase in chromium, a decrease in cobalt, replacement of niobium with tantalum, and addition of a rare earth element, lanthanum, and a reactive element, zirconium, but the concentrations of molybdenum and tungsten were not changed. The result was a much more oxidation-resistant alloy, alloy 556, at 1095 and 1150°C, although rapid oxidation nevertheless occurred at 1200°C.

Cobalt-based alloys with tungsten, such as alloy 188 (R30188; Co–22Cr–22Ni–14W–0.04La), alloy 25 (R30605; Co–20Cr–10Ni–15W), and alloy 6B (Co–30Cr–4.5W–1.2C), showed rapid oxidation at 1205°C. A cobalt-based alloy, alloy 150 (Co–27Cr–18Fe), containing no tungsten also showed rapid oxidation attack at 1205°C.

Again, the oxidation of a cobalt-based alloy can be significantly improved with some modification of alloying elements. Alloy 25 with 15% W exhibits excellent creep rupture strength at high temperatures. However, because of the high level of tungsten, the alloy shows high oxidation rates at very high temperatures, such as 1095 and 1150°C. With a slight increase in chromium and nickel along with the addition of lanthanum, the result of the modification was alloy 188, which has significantly better oxidation resistance than alloy 25 at 1095 and 1150°C.

Oxide-dispersion-strengthened alloys use very fine oxide particles that are uniformly distributed throughout the matrix to provide excessive strengthening at very high temperatures. These oxide particles, typically yttrium oxide, do not react with the alloy matrix, and so no coarsening or dissolution occurs during exposure to very high temperatures, thus maintaining the strength of the alloy. This group of super-alloys is produced using specialty powders that are manufactured by the mechanical alloying process. These powders are essentially composite powders with each particle containing a uniform distribution of submicrometer oxide particles in an alloy matrix. The process of producing these ODS powders involves repeated fracturing and rewelding of a mixture of powder particles in vertical attritors or horizontal ball mills [204]. A vertical attritor is a high-energy ball mill (like the conventional horizontal ball mill) in which the balls and the metal powders are charged into a stationary vertical tank and are agitated by impellers rotating from a central rotating shaft. The shaft turns, but the jar stays put. As the balls are stirred, they fall on the contents and grind whatever is between the balls. Alloy powders are then canned, degassed, and hot extruded followed by hot working and annealing to produce a textured microstructure. Alloys are available in mill products such as bar, plate, and sheet or custom forgings.

Oxidation of alloys can significantly increase under high-velocity gas streams, as in, for example, combustors and transition ducts in gas turbines. These components are also subject to severe thermal cycling, particularly gas turbines in airplane engines. Laboratory burner rigs have been developed to evaluate the type of oxidation, often referred to as "dynamic oxidation," under conditions of very high gas velocities. Some of these dynamic oxidation burner rigs are described elsewhere [204–206].

Hicks [207] performed dynamic oxidation tests with 170 m/s gas velocity at 1100°C with 30-min cycles for several wrought chromia former superalloys and an ODS alumina former (MA956). Alumina former MA956 was found to be considerably better than chromium formers, such as alloys 191, 86, 617, 188, and 263.

MA956, along with some ODS alloys, was studied by Lowell et al. [204] with 0.3 Mach gas velocity at 1100°C with 60-min cycles. ODS alloys included in the study were MA956 (Fe–19Cr–4.4Al–0.6Y_2O_3), HDA8077 (Ni–16Cr–4.2Al–1.6Y_2O_3), TD-NiCr (Ni–20Cr–2.2ThO_2), and STCA264 (Ni–16Cr–4.5Al–1Co–1.5Y_2O_3). Also included in the study was a PVD coating of Ni–15Cr–17Al–0.2Y on MAR-M-200 alloy (Ni–9Cr–10Co–12W–1Nb–5Al–2Ti). MA956 and HDA8077 as well as PVD Ni–Cr–Al–Y coating were found to perform well. No explanation was offered for STCA264, which did not perform as well as HDA8077, although both alloys had similar chemical compositions.

In Fe–Cr, Fe–Ni–Cr, Ni–Cr, and Co–Cr alloy systems, the formation of an external Cr_2O_3 oxide scale provides the oxidation resistance for the alloy. The growth of the Cr_2O_3 oxide scale follows a parabolic rate law as the exposure time increases. As the temperature increases, the oxide scale growth rate also increases. The growth of the Cr_2O_3 scale requires that a continuous supply of chromium from the alloy interior diffuses to the oxide–metal interface. Continued oxidation can eventually deplete chromium in the alloy matrix immediately under the oxide scale. When the chromium concentration in the alloy matrix immediately beneath the oxide scale is reduced to below a critical concentration, the alloy matrix no longer has adequate chromium to reform a protective Cr_2O_3 oxide scale when the scale cracks or spalls due to oxide growth stresses or thermal cycling. Once this occurs, fast-growing, nonprotective iron oxides, or nickel oxides, or cobalt oxides (i.e., oxides of base metal) form and grow on the alloy surface. Breakaway oxidation initiates, and the alloy begins to undergo oxidation at a rapid rate. The alloy thus requires the level of chromium immediately under the chromium oxide scale to reheal [206–211].

To prolong the time for initiation of breakaway oxidation, it is necessary to have an adequate reservoir of chromium immediately below the oxide scale to provide adequate chromium to maintain a protective chromium oxide scale or to reheal the oxide scale after local cracking or failure.

For alumina formers such as Fe–Cr–Al alloys and Fe–Cr–Al- and Ni–Cr–Al-based ODS alloys, breakaway oxidation occurs when aluminum concentration under the Al_2O_3 scale has been reduced to a critical level such that healing of the Al_2O_3 is no longer possible, thus resulting in the formation of nonprotective, fast-growing oxides of base metals (e.g., iron oxides or nickel oxides). The breakaway oxidation due to rapid growth of iron oxides or nickel oxides becomes essentially a life-limiting factor. This critical aluminum concentration was found to be about 1.0–1.3% for Fe–Cr–Al-based ODS alloys (e.g., MA956, ODM751) at 1100–1200°C [210, 211]. These values were obtained from foil specimens (0.2–2 mm thick) tested in still air at 1100–1200°C. For the non-ODS Fe–20Cr–5Al alloy, this critical aluminum concentration was found to be higher (about 2.5%) at 1200°C [210]. Since breakaway oxidation is related to the aluminum reservoir in the alloy, this reservoir is a critical issue when the component is made of thin sheet or foil. Because of excellent oxidation resistance at very high temperatures, there is increasing interest in considering alumina formers for products that require thin foils, such as honeycomb seals in gas turbines, metallic substrates for automobile catalyst converters, and recuperators in microturbines.

For alumina formers to improve their resistance to break-away oxidation, yttrium is frequently used to increase the adhesion of the aluminum oxide scale. Other alloying elements that are known to increase the adhesion of the aluminum oxide scale include zirconium and hafnium. Quadakkers [212] shows that both MA956 (Fe–20Cr–4.5Al–0.5Y_2O_3) and Aluchrom (Fe–20Cr–5Al–0.01Y) exhibited much more cyclic oxidation resistance than Fe–20Cr–5Al when tested at 1100°C in synthetic air with an hourly cycle to room temperature.

Addition of Y_2O_3 to an alumina former has a similar beneficial effect as yttrium added as an alloying element.

Klower and Li [213] studied the oxidation resistance of Fe–20Cr–5Al alloys in 10 different compositions containing various amounts of yttrium ranging from 0.045 to 0.28%. All 10 compositions contained 0.002% S, and 8 compositions contained 0.04–0.06% Zr with two compositions containing no zirconium. The cyclic oxidation tests were performed at 1100 and 1200°C, respectively, with each cycle consisting of 96 h at temperature and rapid air cooling to room temperature. These authors concluded that the yttrium addition of about 0.045% was sufficient to prevent the oxide scales from spalling, and when the yttrium concentration was increased to more than 0.08%, substantial internal oxidation could occur, resulting in rapid metal wastage [213].

Sulfur in the alloy is known to play a very significant role in adhesion of the aluminum oxide scale to the alloy substrate for alumina formers. The role of yttrium is believed to prevent the preferential segregation of sulfur in the alloy to the scale–metal interface to weaken the adhesion of the oxide scale [214–216]. Reducing the concentration of sulfur in a Ni–Cr–Al alloy can significantly improve the oxidation resistance of the alloy. Smeggil [217] compared cyclic oxidation resistance between the normal purity Ni–Cr–Al alloys (approximately 30–40 ppm S) with the high-purity Ni–Cr–Al alloys (approximately 1–2 ppm S), showing a significant improvement in cyclic oxidation resistance when sulfur in the alloy was significantly reduced. Also demonstrated was the beneficial effect of yttrium addition to the normal purity Ni–20Cr–12Al alloy, showing significant improvement in the cyclic oxidation resistance of the alloy without reducing the sulfur content in the alloy. Sulfur was found to segregate to the oxide–alloy interface during oxidation in Fe–Cr–Al alloys [218, 219]. Yttrium is believed to tie up sulfur at the oxide–metal interface, thus improving the oxide scale adhesion [217].

There are some industrial applications that require thin-gage sheet materials or thin foils for construction of critical components. As the component thickness decreases, oxidation becomes a major factor that limits service life. When the component is made of thin foil, prolonging the incubation time before initiation of breakaway oxidation is the controlling factor for extending the service life of the component. Thus, as applications are being pushed toward higher and higher temperatures, alloys that form aluminum oxide scales can offer significant advantages in performance over those alloys that form chromium oxide scales [220–222].

In high-temperature combustion atmospheres, water vapor is invariably present in the environment. The effect of water vapor on the oxidation of alloys is an important factor in the alloy selection process. Most oxidation data are generated in laboratory air, which generally contains low levels of water vapor [223–225]. The effect of water on oxidation, and its detrimental effect, has been established by Onal et al. [226] and many others [2, 4, 12, 16, 170, 183, 227].

As temperature increases, metals and alloys generally suffer increasingly higher rates of oxidation. When the temperature is excessively high, metals and alloys can suffer rapid oxidation. There is, however, another mode of rapid oxidation that takes place at relatively low temperatures, often referred to as "catastrophic oxidation," associated with the formation of a liquid oxide that disrupts and dissolves the protective oxide scale, causing the alloy to suffer rapid oxidation at relatively low temperatures. This phenomenon has been observed by, for example, Meijering [228], Brennor [229], Sawyer [230], Brasunas [231], and Sequeira [232]. The most effective way to alleviate the potential catastrophic oxidation problem is to avoid a stagnant condition of the gaseous atmosphere.

I. CONCLUSIONS

This chapter is a summary of the main factors for determining the nature and extent of gas–metal reactions, the comprehension of which is of paramount importance to understand the subject. Some of these factors are metal structure, oxide structure, metal diffusion, oxygen or metal diffusion at grain boundaries or on the surface, metal and oxide volatility, oxygen solution in metal, cracking, spalling, blistering, sintering, oxide nucleation, electronic conductance, oxide adhesion, and oxide plasticity.

REFERENCES

1. R. Streiff, J. Stringer, R. C. Krutenat, and M. Caillet (Eds.), High Temperature Corrosion of Materials and Coatings for Energy Systems and Turboengines, Elsevier Sequoia, Lausanne, 1987.

2. R. Streiff, J. Stringer, R. C. Krutenat, and M. Caillet (Eds.), High Temperature Corrosion, Vol. 2: Advanced Materials and Coatings, Elsevier, London, 1989.

3. R. Streiff, J. Stringer, R. C. Krutenat, and M. Caillet (Eds.), High Temperature Corrosion and Protection of Materials, Vol. 3, Les Éditions de Physique, Les Ulis, 1993.

4. R. Streiff, J. Stringer, R. C. Krutenat, M. Caillet, and R.A. Rapp,(Eds.), High Temperature Corrosion and Protection of Materials Vol. 4, Trans. Tech. Publications, Zürich, 1997.

5. R. Streiff, I. G. Wright, R. C. Krutenat, M. Caillet, and A. Galerie (Eds.), High Temperature Corrosion and Protection of Materials, Vol. 5, Trans. Tech. Publications, Zürich, 2001.

6. P. Steinmetz, I. G. Wright, G. Meier, A. Galerie, B. Pieraggi, and R. Podor, (Eds.), High Temperature Corrosion and Protection of Materials, Vol. 6, Trans. Tech. Publications, Zürich, 2004.

7. S. Taniguchi, T. Maruyama, M. Yoshiba, N. Otsuka, and Y. Kawahara (Eds.), High Temperature Oxidation and Corrosion 2005, Trans. Tech. Publications, Zürich, 2006.

8. P. Kofstad, High Temperature Oxidation of Metals, Wiley, New York, 1966.

9. O. Kubaschewski and B. E. Hopkins, Oxidation of Metals and Alloys, Butterworths, London, 1967.

10. S. Mrowec and T. Werber, Gas Corrosion of Metals, Foreign Scientific Publications Department of the National Center for Scientific, Technical and Economic Information, Varsovie, 1978.

11. P. Kofstad, High Temperature Corrosion, Elsevier Applied Science, London, 1988.

12. N. Birks, G. H. Meier, and F. S. Pettit, Introduction to the High Temperature Oxidation of Metals, Cambridge University Press, Cambridge, 2006.

13. E. Fromm, Kinetics of Metal-Gas Interactions at Low Temperature—Hydriding, Oxidation, Poisoning, Springer-Verlag, Berlin, 1998.

14. A. S. Khanna, High Temperature Oxidation and Corrosion, ASM International, Metals Park, OH, 2002.

15. P. Sarrazin, A. Galerie, and J. Fouletier, Mechanisms of High Temperature Corrosion: A Kinetic Approach, Trans. Tech. Publications, Zürich, 2008.

16. G. Béranger, J. C. Colson, and F. Dabasi, Corrosion des Matériaux à Haute Température, Les Éditions de Physique, Les Ulis, 1987.

17. G. Y. Lai, High Temperature Corrosion and Materials Applications, ASM International, Materials Park, OH, 2007.

18. L. S. Darken and R. W. Gurry, Physical Chemistry of Metals, McGraw-Hill, New York, 1953.

19. D. R. Gaskell, Introduction to Thermodynamics of Materials, 3rd ed. Taylor and Francis, Washington, DC, 1995.

20. H. I. Aaronson (Ed.), Lectures on the Theory of Phase Transformations, Minerals, Metals and Materials Society, Warrendale, PA, 1999.

21. A. M. Arper (Ed.), Phase Diagrams, Materials Science and Technology, Vol. 1, Academic, New York, 1970.

22. J. Sticher and H. Schmalzried, Zur geometrischen Darstellung thermodynamischer Zustandsgrossen in Mehrstoffsystemen auf Eisenbasis, Report, Clausthal Institute für Theoretischen Huttenkunde und Angewandte Physikalische Chemie der Technischen Universität Clausthal, 1975.

23. S. R. Shatynski, Oxid. Met., **11**, 307 (1977).

24. S. R. Shatynski, Oxid. Met., **13**, 105 (1979).

25. H. J. T. Ellingham, J. Soc. Chem. Ind., **63**, 125 (1944).

26. M. Olette and M.F. Ancey-Moret, Rev. Metall., **60**, 569 (1963).

27. F. D. Richardson and J. H. E. Jeffes, J. Iron Steel Inst., **160**, 261 (1948).

28. F. D. Richardson, and J. H. E. Jeffes, J. Iron Steel Inst., **171**, 165 (1957).

29. M. F. Ancey-Moret, Mém. Sci. Rev. Mét., **70**, 429 (1973).

30. G. R. Belton and W.R. Worrell (Eds.), Heterogeneous Kinetics at Elevated Temperatures, Plenum Press, New York, 1970.

31. C. E. Wicks and F. E. Block, Bulletin 605, Bureau of Mines, U.S. Government Printing Office, Washington, DC, 1963, p. 408.

32. JANAF Thermochemical Tables, J. Phys. Chem. Ref. Data, **4**, 1(1975).

33. E. A. Gulbransen and G. H. Meier, DOE Report on Contract No. DE-AC01-79-ET-13547, University of Pittsburgh, 1979.

34. C. S. Giggins and F. S. Pettit, Oxid. Met., **14**, 363 (1980).

35. R. A. Rapp (Ed.), High Temperature Corrosion, NACE, Houston, TX, 1983.

36. S. A. Jansson and Z. A. Foroulis (Eds.), High Temperature Gas-Metal Reactions in Mixed Environments, American Institute of Mining Metallurgical and Petroleum Engineering, New York, 1973.

37. P. L. Hemmings and R. A. Perkins, Research Project 716-1, Interim Report, Lockheed Palo Alto Research Laboratories for Electric Power Research Institute, Palo Alto, CA, 1977.

38. A. D. Pelton and H. Schmalzried, Met. Trans., **4**, 1395 (1973).

39. K. T. Jacob, D. B. Rao, and H. G. Nelson, Oxid. Met., **13**, 25 (1979).

40. J. Barralis and G. Maeder, Précis de Métallurgie: élaboration, structures-proprietés, normalisation, AFNOR-Nathan, Paris, 1997.

41. R. A. Swalin, Thermodynamics of Solids, Wiley, London, 1972.

42. H. Schick, Thermodynamics of Certain Refractory Compounds, Academic, New York, 1966.

43. O. Kubaschewski and C. B. Alcock, Metallurgical Thermochemistry, 5th ed., Pergamon, Oxford, 1979.

44. A. Yazawa, Met. Trans. B, **10B**, 307 (1979).

45. P. Rocabois, C. Chatillon, and C. Bernard, J. Am. Ceram. Soc., **79**, 1361 (1966).

46. F. S. Pettit, Trans. Met. Soc. AIME, **239**, 1296 (1967).

47. GTT-Technologies, Herzogenroth, Germany.

48. Thermodata–I.N.P.G.–C.N.R.S., Saint Martin d'Héres Cédex, France.

49. G. Eriksson, Chem Soc., **8**, 100 (1975).

50. G. Eriksson and K. Hack, Met. Trans., **B21**, **1013**(1990).

51. E. Königsberger and G. Eriksson, CALPHAD, **19**, 207 (1995).

52. W. C. Reynolds, STANJAN, Version 3, Department of Mechanical Engineering, Stanford University, 1986.

53. B. Sundman, B. Jansson, and J.-O. Andersson, CALPHAD, **9**, 153 (1985).

54. C. B. Alcock and G. W. Hooper, Proc. Roy. Soc., **254A**, 551 (1960).

55. D. Caplan, A. Harvey, and M. Cohen, Corr. Sci., **3**, 161 (1963).

56. D. Caplan and G. I. Sproule, Oxid. Met., **9**, 459 (1975).

57. D. Caplan and M. Cohen, J. Electrochem. Soc., **108**, 438 (1961).

58. H. Asteman, J. E. Svensson, L. G. Johansson, and M. Norell, Oxid. Met., **52**, 95 (1999).

59. C. S. Tedmon, J. Electrochem. Soc., **113**, 766 (1966).

60. E. A. Gulbransen and G. H. Meier, *Proc. 10th Materials Research Symposium, National Bureau of Standards Special Publications, 561, 1639* (1979).

61. E. A. Gulbransen and W. S. Wysong, TAIME, **175**, 628 (1948).

62. E. A. Gulbransen K. F. Andrew, and F. A. Brassart, J. Electrochem. Soc., **110**, 952 (1963).

63. C. J. Wagner, Appl. Phys., **29**, 1295 (1958).

64. C. Wagner, Corros. Sci., **5**, 751 (1965).

65. ASM, Handbook, Corrosion: Fundamentals, Testing and Protection, Vol. 13A, ASM International, Metals Park, OH, 2003.

66. K. Hauffe, Oxidation of Metals, Plenum, New York, 1965.

67. J. Benard, Oxydation des Métaux, Gauthier-Villars, Paris, 1962.

68. N. Cabrera and N.F. Mott, Rept. Prog. Phys., **12**, 163 (1948).

69. F. W. Young, J. V. Cathcart, and A. T. Gwathmey, Acta Met., **4**, 145 (1956).

70. R. K. Hart, Proc. Roy. Soc., **236A**, 68 (1956).

71. M. W. Roberts, Trans. Faraday Soc., **57**, 99 (1961).

72. N. F. Mott, Trans. Faraday Soc., **3**, 472 (1940).

73. K. Hauffe and B. Z. Ilschner, Elektrochem., **58**, 382 (1954).

74. U. R. Evans, The Corrosion and Oxidation of Metals, Edward Arnold, London, 1960.

75. F. S. Pettit and J. B. Wagner, Acta Met., **12**, 35 (1964).

76. D. Monceau and B. Pieraggi, Oxid. Met., **50**, 477 (1998).

77. F. A. Kröger (Ed.), The Chemistry of Imperfect Crystals, North Holland Publishing, Amsterdam, 1975.

78. F. A. Kröger, H. J. Vink, F. Seitz, and D. Turnbull, *Solid State Physics*, Vol. 3, Academic, London, 1956.

79. J. Philibert, Diffusion et Transport de Matière dans les Solides, Monographie de Physique, Les Éditions de Physique, Les Ulis, 1985.

80. P. Kofstad, *Non Stoichiometry*, Diffusion and Electrical Conductivity in Binary Metal Oxides, Wiley, New York, 1972.

81. S. Mrowec, Defect and Diffusion in Solids, Elsevier Science Publications, London, 1980.

82. H. H. von Baumbach and C. Z. Wagner, Phys. Chem., **22**, 199 (1933).

83. I. Branky and N. M. Tallan, J. Chem. Phys., **49**, 1243 (1968).

84. N. G. Eror and J. B. Wagner, J. Phys. Stat. Sol., **35**, 641 (1969).

85. R. Farhi and G. Petot-Ervas, J. Phys. Chem. Solids., **39**, 1169 (1978).

86. M. C. Pope and N. Birks, Corr. Sci., **17**, 747 (1977).

87. C. A. C. Sequeira and D. M. F. Santos, Czech J. Phys., **56**, 549 (2006).

88. A. Atkinson, R. I. Taylor, and A. E. Hughes, Phil. Mag., **A45**, 823 (1982).

89. A. Atkinson, D. P. Moon, D. W. Smart, and R.I. Taylor, J. Mater. Sci., **21**, 1747 (1986).

90. R. J. Hussey and M. J. Graham, Oxid. Met., **45**, 349 (1996).

91. R. Prescott and M. J. Graham, Oxid. Met., **38**, 233 (1992).

92. M. I. Manning, Corros. Sci., **21**, 301 (1981).

93. N. B. Pilling and R. E. Bedworth, J. Inst. Met., **1**, 529 (1923).

94. J. V. Cathcart (Ed.), Stress Effects and the Oxidation of Metals, TMS-AIME, New York, 1975.

95. M. Schütze, *Protective Oxide Scales and Their Breakdown*, The Institute of Corrosion, Wiley, Chichester, UK, 1997.

96. M. Schulte and M. Schütze, Oxid. Met., **51**, 55 (1999).

97. J. Armilt, D.R. Holmes, M.I. Manning, D.B. Meadowcroft, and E. Metcalfe, EPRI Report No. FP 686, Electric Power Research Institute, Palo Alto, CA, 1978.

98. W. Christl, A. Rahmel, and M. Schütze, Oxid. Met., **31**, 1 (1989).

99. H. E. Evans, Int. Mater. Rev., **40**, 1 (1995).

100. F. S. Pettit, R. Yinger and J. B. Wagner, Acta Met., **8**, 617 (1960).

101. J. C. Yang, M. Yeadon, B. Kolasa, and J. M. Gibson, Scripta Mater., **38**, 1237 (1998).

102. J. Romanski, Corros. Sci., **8**, 67, 89 (1968).

103. P. Barret (Ed.), Reaction Kinetics in Heterogeneous Chemical Systems, Elsevier, Amsterdam, 1975.

104. S. Mrowec and A. Stoklosa, Oxid. Met., **3**, 291 (1971).

105. S. Glasstone, K. J. Laidler, and H. Eyring, Theory of Rate Process, McGraw-Hill, New York, 1941.

106. I. Langmuir, J. Am. Chem. Soc., **40**, 1361 (1918).

107. S. Brunauer, P. H. Emmett, and E. Teller, J. Am. Chem. Soc., **60**, 309 (1938).

108. C. Wagner, Z. Phys. Chem. Abt., **B21**, 25 (1933).

109. C. Wagner, Z. Elektrochem., **39**, 543 (1933).

110. M. J. Graham, D. Caplan, and M. Cohen, J. Electrochem. Soc., **119**, 1265 (1972).

111. K. Hauffe, Oxydation von Metallen und Metallegiesungen, Springer, Berlin, 1957.

112. P. Kofstad, Corros. NACE, **24**, 379 (1968).

113. P. Gesmundo and F. Vjani, J. Electrochem. Soc., **128**, 470 (1981).

114. B. E. Deal and A.S. Grove, J. Appl. Phys., **36**, 3770 (1965).

115. U. R. Evans, Trans. Electrochem. Soc., **46**, 247 (1924).

116. K. Fischbeck, Z. Elektrochem., **39**, 316 (1933).

117. C. Wagner, in Atom Movements, ASM, Cleveland, OH, 1951, P. 153.

118. C. Wagner, Prog. Solid State Chem., **10**, 3 (1975).

119. A. Rahmel and W. Schwenk, Korrosion und Korrosionsschutz von Stahlen, Verlag Chemie, Weinheim, 1977.

120. E. Heitz, R. Henkhaus, and A. Rahmel, Corrosion Science—an Experimental Approach, Ellis Horwood, New York, 1992.

121. P. Sarrazin and J. Besson, J. Chim. Phys., **1**, 27 (1973).

122. B. Fischer and D. S. Tannhäuser, J. Chem. Phys., **44**, 1663 (1966).

123. N. G. Eror and J. B. Wagner, J. Phys. Chem. Solids., **29**, 1597 (1968).

124. R. E. Carter and F. D. Richardson, TAIME, **203**, 336 (1955).

125. D. W. Bridges, J. P. Baur, and W. M. Fassell, J. Electrochem. Soc., **103**, 619 (1956).

126. S. Mrowec, A. Stoklosa, and K. Godlewski, Cryst. Latt. Def., **5**, 239 (1974).

127. S. Mrowec and K. Przybylski, Oxid. Met., **11**, 365 (1977).

128. S. Mrowec and K. Przybylski, Oxid. Met., **11**, 383 (1977).

129. R.E. Carter and F.D. Richardson, J. Met., **6**, 1244 (1954).

130. D. Caplan, M.J. Graham, and M. Cohen. J. Electrochem. Soc., **119**, 1205 (1972).

131. J.E. Burke, Trans. AIME, **180**, 73 (1949).

132. N.N. Khoi, W.W. Smeltzer, and J.D. Embury, J. Electrochem Soc., **116**, 1495 (1975).

133. G.J. Yurek, J.P. Hirth, and R.A. Rapp, Oxid. Met., **8**, 265 (1974).

134. G. Garnaud, Oxid. Met., **11**, 127 (1977).

135. G. Garnaud and R.A. Rapp, Oxid. Met., **11**, 193 (1977).

136. H. S. Hsu and G. J. Yurek, Oxid. Met., **17**, 55 (1982).

137. J. Loriers, Comptes Rendus Acad. Sci., **231**, 522 (2006).

138. G. W. Haycock, J. Electrochem. Soc., **106**, 771 (1959).

139. R. A. Rapp and H. Colson, Trans. Met. Soc. AIME, **236**, 1616 (1966).

140. R. A. Rapp and G. Goldberg, Trans. Met. Soc. AIME, **236**, 1619 (1966).

141. S. L. Chang, F. S. Pettit, and N. Birks, Oxid. Met., **34**, 23 (1990).

142. J. Stringer, Acta Metall., **8**, 758 (1960).

143. J. Stringer, Acta Metall., **8**, 810 (1960).

144. S. B. Newcomb and M. J. Bennett, (Eds.), Microscopy of Oxidation, Vol. 2, Institute of Materials, London, 1993.

145. J. A. Roberson and R. A. Rapp, TAIME, **239**, 1327 (1967).

146. Z. A. Munir and D. Cubicciotti (Eds.), High Temperature Materials, Chemistry, Electrochemical Society, New York, 1983.

147. F. Gesmundo and Y. Niu, Oxid. Met., **50**, 1 (1998).

148. H. J. Grabke and D. B. Meadowcroft (Eds.), Guidelines for Methods of Testing and Research in High Temperature Corrosion, Institute of Materials, London, 1995.

149. C. Wagner, Bur. Bunsenges. Phys. Chem., **63**, 772 (1959).

150. J. L. Smialek and G. M. Meier, in High Temperature Oxidation in Superalloy, Vol. 2, C. T. Sims, N. S. Stoloff, and W. C. Hagel, (Eds.), Wiley, New York, 1987, p. 293.

151. K. M. N. Prasanna, A. S. Khanna, R. Chandra, and W. J. Quadakkers, Oxid. Met., **46**, 465 (1996).

152. C. A. Barrett and C. E. Lowell, Oxid. Met., **11**, 199 (1977).

153. C. S. Giggins and F. S. Pettit, J. Electrochem. Soc., **118**, 1782 (1971).

154. H. Hindam and D. P. Whittle, Oxid. Met., **18**, 245 (1982).

155. A. S. Tumarev and L. A. Panyushin, NASA TT F-13221, 1970.

156. G. R. Wallwork and A. Z. Hed, Oxid. Met., **3**, 171 (1971).

157. W. Brandl, H. J. Grabke, D. Toma, and J. Kruger, Surf. Coat. Technol., **68/69**, 17 (1994).

158. C. Sarioglu, M.J. Stiger, J.R. Blachere, R. Janakiram, E. Schumann, A. Ashary, F.S. Pettit, and G.H. Meier, Mater. Corros., **51**, 358 (2000).

159. J. Nicholls (Ed.), High Temperature Surface Engineering, Institute of Materials, London, 1999.

160. H. G. Jung and K. Y. Kim, Oxid. Met., **49**, 403 (1998).

161. A. Mignone, S. Frangini, A. La Barbera, and O. Tassa, Corros. Sci., **40**, 1331 (1998).

162. B. A. Pint, J. Leibowitz, and J. H. Devan, Oxid. Met., **51**, 183 (1999).

163. C. A. C. Sequeira and C. M. E. S. Nunes, Surf. Eng., **3**, 161 (1987).

164. C. A. C. Sequeira, A. M. G. Pacheco, and C. M. E. S. Nunes, Surf. Eng., **3**, 247 (1987).

165. C. A. C. Sequeira, A. M. G. Pacheco, and C. M. E. S. Nunes, Surf. Eng., **4**, 65 (1988).

166. G. Allison and M. K. Hawkins, GEC Rev., **17**, 947 (1914).

167. R. Drewett, Corros. Sci., **9**, 823 (1969).

168. Y. Chen, C. A. C. Sequeira, and X. Song, Corros. Prot. Mater., **24**, 52 (2005).

169. M. Brill-Edwards and M. Epner, Electrochem. Technol., **6**, 299 (1968).

170. E. Lang (Ed.), The Role of Active Elements in the Oxidation Behaviour of High Temperature Metals and Alloys, Elsevier, Amsterdam, 1989.

171. E. Mevrel and R. Pichoir, Mater. Sci. Eng., **88**, 1 (1987).

172. G. V. Samsonov (Ed.), Protective Coatings on Metals, Vol. 5. Consultants Bureau, New York, 1973.

173. R. Sivakumar and E. J. Rao, Oxid. Met., **17**, 391 (1982).

174. C. A. C. Sequeira and F.D.S. Marquis, Mater. Sci. Forum., **514–516**, 505 (2006).

175. P. Lane and N. M. Geyer, J. Met., **18**, 186 (1966).

176. R. G. Ubank, Rev. Int. Htes. Temp. Refr., **14**, 21 (1977).

177. R. G. Wing and I. R. McGill, Pt. Met. Rev., **25**, 94 (1981).

178. B. E. Jacobson and R. E. Bunshah (Eds.), Films and Coatings for Technology, CEI Course, Stockholm, Sweden, 1981.

179. W. Gao (Ed.), Developments in High Temperature Corrosion and Protection of Materials, Woodhead Publ., Cambridge, UK, 2008.

180. M. Schütze and W. J. Quadakkers (Ed.), Novel Approaches to Improving High Temperature Corrosion Resistance, Woodhead Publ., Cambridge, UK, 2008.

181. M. G. Hocking, V. Vasantasree, and P. S. Sidky, Metallic and Ceramic Coatings: Production, High Temperature Properties and Applications, Longman Sci. & Technical, Essex, UK, 1989.

182. J. T. De Masi-Marcin and D. K. Gupta, Surf. Coat. Tech., **68–69**, 1 (1994).

183. S. Bose, High Temperature Coatings, Elsevier, Amsterdam, 2007.

184. Y. Sato, B. Onay, and T. Maruyama (Eds.), High Temperature Corrosion of Advanced Materials and Protective Coatings, Elsevier, Amsterdam, 1992.

185. K. H. Stern, Metallurgical and Ceramic Protective Coatings, Chapman and Hall, London, 1996.

186. N. B. Dahotre and J. Hampikian (Eds.), Elevated Temperature Coatings: Science and Technology, Vol. 3, TMS, San Diego, 1999.

187. R. C. John, Corrosion/99, Paper No. 73, NACE International, Houston, TX, 1999.

188. A. W. Zeuthen, Heating, Piping and Air Conditioning, **42**, 152 (1970).

189. H. E. McGarrow (Ed.), The Making, Shaping and Treating of Steel, United States Steel Corp., Pittsburgh, PA, 1971.

190. I. G. Wright, in Corrosion Metals Handbook, Vol. 13, ASM International, Metals Park, OH, 1987, p. 97.

191. A. Grodner, Weld. Res. Counc. Bull. No. 31, 1956.

192. M. Walter, M. Schütze, and A. Rahmel, Oxid. Met., **39**, 389 (1993).

193. O. D. Sherby, Acta Metall., **10**, 135 (1962).

194. A. Moccari and S. I. Ali, Brit. Corros. J., **14**, 91 (1979).

195. S. Kado, T. Yamazaki, M. Yamazaki, K. Yoshida, K. Yabe, and H. Kobayashi, Trans. Iron Steel Inst. Jpn., **18**, 387 (1978).

196. W. E. Ruther and S. Grunberg, J. Electrochem. Soc., **111**, 1116 (1964).

197. Kanthal Alloys Data Sheet, Sandvik, Sweden, 1995.

198. R. Berglund and B. Jonsson, Ind. Heat, **Oct., 21** (1989).

199. Allegheny Ludlum Data Sheet, Allegheny Ludlum, Pittsburgh, PA, 1990.

200. P. T. Moseley, K. R. Hyde, B. A. Bellamy, and G. Tappin, Corros. Sci., **24**, 547 (1984).

201. F. N. Smith, J. F. McGurn, G. Y. Lai, and V. S. Sastri (Eds.), Applications and Materials Performance, Proc. Nickel-Cobalt 97 International Symposium, The Metallurgical Society of CIM, Montreal, Canada, 1997.

202. P. Ganesan, G. D. Smith, and C. S. Tassen, Corrosion/93, Paper No. 234, NACE International, Houston, Texas, 1993.

203. G. Y. Lai, J. Met., **37**, 14 (1985).

204. C. E. Lowell, D. L. Deadmore, and J. D. Whittenberger, Oxid. Met., **17**, 205 (1982).

205. G. Y. Lai, unpublished results, Haynes International, Kokomo, Indiana, 1988.

206. M. Schütze (Ed.), Corrosion and Environmental Degradation, Wiley-VCM, Weinheim, Germany, 2000.

207. B. Hicks, Mater. Sci. Technol., **3**, 772 (1987).

208. B. Gleeson and M. A. Harper, Oxid. Met., **49**, 373 (1998).

209. H. E. Evans, D. A. Hilton, R. A. Holm, and S. J. Webster, Oxid. Met., **14**, 235 (1980).

210. W. J. Quadakkers and K. Bongartz, Werkst. Korros., **45**, 232 (1994).

211. I. Gurrappa, S. Weinruch, D. Naumenko, and W. J. Quadakkers, Mater. Corros., **51**, 224 (2000).

212. W. J. Quadakkers, Werkst. Korros., **41**, 659 (1990).

213. J. Klower and J. G. Li, Mater. Corros., **47**, 545 (1996).

214. J. G. Smeggil, A. W. Funkenbusch, and N. S. Bornstein, High Temp. Sci., **16**, 163 (1985).

215. A. W. Funkenbusch, J. G. Smeggil, and N. S. Bornstein, Met. Trans. A, **16**, 1164 (1985).

216. J. G. Smeggil, A.W. Funkenbusch, and N.S. Bornstein, Met. Trans. A, **17**, 923 (1986).

217. J. G. Smeggil, Mater. Sci. Eng., **87**, 261 (1987).

218. P. Y. Hou and J. Stringer, Oxid. Met., **38**, 323 (1992).

219. P. Y. Hou, Mater. Corros., **51**, 329 (2000).

220. N. J. Simms, R. Newton, J. F. Norton, A. Encinas-Oropesa, J. E. Oakey, J. R. Nicholls, and J. Wilber, Mater. High. Temp., **20**, 439 (2003).

221. J. Klower, Mater. Corros., **49**, 758 (1998).

222. B. A. Pint, J. Eng. Gas Turbines Power, **128**, 1 (2006).

223. C. W. Tuck, M. Odgers, and K. Sachs, Anti-Corrosion, June 1966, p. 14.

224. K. Segerdahl, J. E. Svensson, and L. G. Johansson, Mater. Corros., **53**, 247 (2002).

225. K. Segerdahl, J. E. Svensson, and L. G. Johansson, Mater. Corros., **53**, 479 (2002).

226. K. Onal, M. C. Maris-Sida, G. H. Meier, and F. S. Pettit, Mater. High Temp., **20**, 327 (2003).

227. A. Rahmel (Ed.), Aufbau von Oxidschichten auf Hochtemperaturwerkstoffen und ihre technische Bedeutung, Deutsche Gesellschaft für Metallkunde, Oberusel, 1982.

228. J. K. Meijering and G. W. Rathenau, Nature, **165**, 240 (1950).

229. S. S. Brennor, J. Electrochem. Soc., **102**, 16 (1955).

230. J. W. Sawyer, Trans. TMS-AIME., **221**, 63 (1961).

231. A. de S Brasunas and N. J. Grant, Trans. ASM, **44**, 1133 (1950).

232. C. A. C. Sequeira (Ed.), High Temperature Corrosion in Molten Salts, Trans. Tech. Publications, Uetikon-Zürich, 2003.

21

THERMOCHEMICAL EVALUATION OF CORROSION PRODUCT STABILITIES FOR ALLOYS IN GASES AT HIGH TEMPERATURE

W. T. THOMPSON

Center for Research in Computational Thermochemistry, Royal Military College of Canada, Kingston, Ontario, Canada

R. C. JOHN

Shell International E&P, Inc. Houston, Texas

A. L. YOUNG

Humberside Solutions Ltd., Toronto, Ontario, Canada

kinetics associated with growth can be treated. Thermochemical analysis of the potential corrosion products can help identify the corrosion mechanisms important for the process conditions and also suggest which alloy types might have sufficient corrosion resistance based upon knowledge of how corrosion products influence corrosion rates. Examples of processes that corrode metals in high-temperature gases include coal gasification, crude oil distilling, steam-methane re-forming, catalytic cracking, fluid bed coal combustion, petroleum coking, combustion, incineration, hydrocracking, naphtha cracking to ethylene, and other energy conversion processes. All involve the presence of gases containing H–O–S–C at temperatures of 600–1300 K ($\sim$300–1000°C)[1].

A. INTRODUCTION

The understanding of corrosion of metals and alloys in hot gases can be assisted by analysis of the thermodynamics of the formation of corrosion product phases involving elements in the alloy and species in the gas. Since the service temperatures and pressures in high-temperature applications may vary over large ranges and the gas phase may involve many species, the stable corrosion products are only known with certainty for applications of specific alloys. It is often helpful to employ a multielement stability (or phase) diagram, which considers the thermodynamic properties of the possible compounds, to establish the most stable corrosion products before the

B. THERMODYNAMICS

B1. Gas Phase

In many processes, the nominal gas composition may not represent the composition present at the corroding surface. In the process of sampling gases containing H, O, S, and C at high temperature and pressure, part of the water vapor may condense during subsequent cooling and pressure reduction. Furthermore, in mixtures rich in CO, there is the possibility of carbon precipitation:

$$2CO = CO_2 + C \qquad (21.1)$$

Uhlig's Corrosion Handbook, Third Edition, Edited by R. Winston Revie

resultant from a shift in the equilibrium with falling temperature given approximately by

$$\log\left[\frac{(P_{CO_2})(a_C)}{(P_{CO})^2}\right] = \frac{8916}{T} - 9.15 \qquad (21.2)$$

Even if the concentrations of the most abundant gases change little with temperature, the partial pressures of the constituent elements such as O_2 or S_2 may vary considerably. It may be necessary to infer, from a gas composition at ambient temperature and pressure, what the gas composition is likely to be at the service condition for the alloy.

Table 21.1 shows the computed equilibrium composition at 900 K (627°C) and 1100 K (827°C) for a gas with an initial composition of 35% CO, 25% H_2, 20% CO_2, 19% H_2O, and 1% H_2S. In both cases, the total pressure is 30 atm. This type of gas composition might be found in coke/coal/oil gasification, Flexicoking (trade name of Exxon Corporation), catalytic cracking in a petroleum refinery, steam-methane re-forming, or instances of fuel-rich conditions of fossil fuel combustion. The nominal (molar) composition may be regarded as the reactants in a chemical process leading to the equilibrium products. In the product equilibrium, the mole fraction is given adjacent to each species considered. The total number of moles of gas appears near the bottom of each listing and is, of course, larger for the higher temperature reflecting the greater degree of dissociation. Although graphite is a potential phase in both figures, it only appears at the lower temperature; at the higher temperature the activity of carbon with respect to graphite is less than 1.

The equilibrium was established by the Gibbs energy minimization [2, 3]. In this versatile computational procedure, the molar concentration of candidate equilibrium species (containing some or all of the elements in the nominal gas) are systematically varied subject to the condition of mass balance for each element. For each such mixture, the total (relative) Gibbs energy, G_t, is calculated by evaluating the following function:

$$G_t = \sum_{i=1}^{\text{gases}} \left(n_i\{\Delta G_i^\circ + RT \ln[(X_i)(P)]\}\right) + \sum_{i=1}^{\text{solids}} n_i\,\Delta G_i^\circ \qquad (21.3)$$

where ΔG_i° are the standard Gibbs energies of formation per mole of each species in the equilibrium gas mixture, R is the gas constant, n_i is the number of moles, X_i is the mole fraction of species in the gas phase at absolute temperature, T, and total pressure, P. The equilibrium is found when successive (and progressively smaller) adjustments in molecular concentrations provide negligible reduction in G_t. The equilibrium partial pressure of the elemental species ($P_{H_2}, P_{O_2}, P_{S_2}$) and the activity of carbon (graphite, a_C) are a by-product of the computation.

B2. Corrosion Product Formation

Having established the equilibrium gas composition at the temperature of interest, it is now possible to consider the reaction of that gas with the elements in the alloy. Consider the case of Inconel 671 (trade name of Huntington Alloys International), which is nearly a binary alloy containing 54% Ni and 46% Cr. This alloy is not widely used but is chosen for this example to illustrate the concepts used to calculate corrosion product stabilities. The most stable chemical phase of nickel can be determined by computing the Gibbs energy change for all possible reactions of nickel with hydrogen, oxygen, sulfur, and carbon, assuming that the composition of the voluminous gas phase is not influenced by reactions with a relatively minor mass of the alloy. The reactions, all based on 1 mol of nickel reacting to form a series of Ni compounds with the general formula $NiH_aO_bS_cC_d$, have the form:

$$Ni + \left(\frac{a}{2}\right)H_2 + \left(\frac{b}{2}\right)O_2 + \left(\frac{c}{2}\right)S_2 + (d)C \rightarrow (NiH_aO_bS_cC_d)$$

$$(21.4)$$

where the constants a, b, c, and d are ≥ 0. When the Gibbs energy change is computed, allowance is made for P_{H_2}, P_{O_2}, and P_{S_2}, and the activity of C appropriate to the gas mixture at the temperature and total pressure involved using the equation

$$\Delta G = \Delta G^\circ - \left(\frac{a}{2}\right)RT \ln P_{H_2} - \left(\frac{b}{2}\right)RT \ln P_{O_2} \qquad (21.5)$$

$$- \left(\frac{c}{2}\right)RT \ln P_{S_2} - (d)RT \ln a_C$$

where ΔG° is the standard Gibbs energy of formation for the nickel-containing compounds (per mole of Ni) obtained from tables or databases of thermochemical properties. The reaction with the most negative Gibbs energy change identifies the most stable compound containing nickel coexisting in direct contact with the gas phase.

In order to evaluate the potential of changes in the gas composition to alter the most stable nickel compound, a diagram may be constructed by systematically varying the partial pressures of two particular elements while holding the partial pressure or activity of the third and fourth elements constant. Such a predominance or phase stability diagram is shown in Figure 21.1 for 1100 K (827°C) [2, 4]. Note that the suggested methodology for the diagram construction does not begin with potentially false suppositions of coexistence of particular combinations of nickel containing phases as a basis for developing equations for each phase boundary. Therein lies the power of the suggested computational procedure. Precision in locating the boundary for graphical purposes simply involves selecting suitably small steps in the variation of the partial pressures for the species used for the

P_{H_2} = 6.17 atm; a_C = 0.838

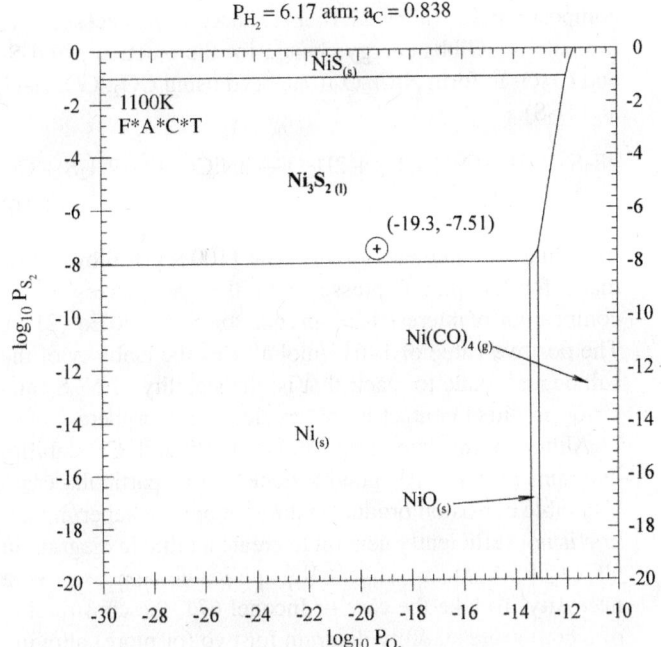

FIGURE 21.1. The **Ni**-H, O, S, C predominance diagram at 1100 K (827°C) prepared using the FACT system [2]. The coordinate ⊕ is consistent with Table 21.1.

$P_{H_2O} = 5.75$ atm; $P_{H_2S} = 0.310$ atm

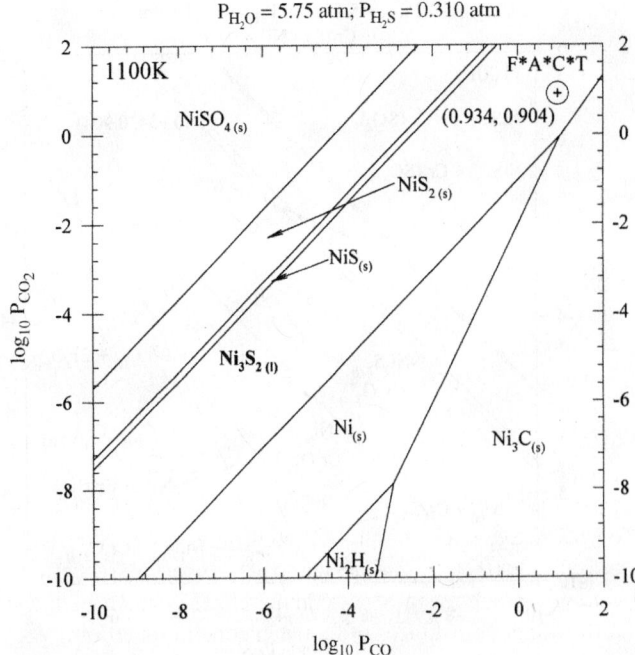

FIGURE 21.2. The **Ni**-H, O, S, C predominance diagram at 1100 K (827°C). The diagram is similar to Figure 21.1 but a different selection of axes and constant conditions has been made. The coordinate ⊕ is consistent with Figure 21.1 and Table 21.1.

axes. Many computations of Gibbs energy can be eliminated following the realization that phase fields must be contiguous. The computational power of a personal computer is more than ample to develop diagrams as in Figure 21.1 in a few seconds [2].

When it is more informative to employ the dominant gas species for the axes and fixed partial pressures in the diagram, a variation on the foregoing methodology is invoked Suppose, for example, log P_{CO_2} is used as oidinate and log P_{CO} as abscissa. The equilibrium constant for the process

$$CO + \tfrac{1}{2}O_2 = CO_2 \qquad (21.6)$$

$$\log\left[\frac{P_{CO_2}}{(P_{CO})(P_{O_2})}\right] = \frac{14,764}{T} - 4.537 \qquad (21.7)$$

enables P_{O_2} to be found for any pair of values of P_{CO} and P_{CO_2}.

Similarly, the carbon (graphite) activity can be found using Eqs. 21.1 and 21.2.

If, for the diagram development, the P_{H_2O} and P_{H_2S} were fixed, then P_{H_2} could be found from the equilibrium constant for

$$H_2 + \tfrac{1}{2}O_2 = H_2O \qquad (21.8)$$

and P_{S_2} from the equilibrium constant for

$$H_2 + \tfrac{1}{2}S_2 = H_2S \qquad (21.9)$$

These preliminary steps reduce the computations used to construct Figure 21.2 to those of Figure 21.1. The two diagrams are consistent in indicating that Ni_3S_2 (liquid) is the most stable phase of nickel in equilibrium with the gas. Comparable points, representative of the equilibrium gas in Table 21.1, are shown on both figures. The two diagrams are *not* topologically the same with respect to triple points and adjacent phase fields since the chemical potentials of all elements in the gas phase are only the same at the coordinates marked.

In the case of corrosion of Inconel 671, it is of course necessary to also consider the chemical compounds of Cr. By procedures similar to that described for Ni, a diagram similar to Figure 21.2 may be developed for Cr. This diagram superimposed on that for Ni is shown in Figure 21.3. The overlapping stability fields give rise to the double labeling of each area on the resultant diagram. For the conditions of temperature, total pressure, and gas compositions covered by Figure 21.3, there are no stable compounds containing both Ni and Cr and little or no mutual solubility of the various nickel- or chromium-containing compounds. Thus the superimposition of Ni and Cr stability diagrams leads to the conclusion, for the gas with the equilibrium composition in Table 21.1 at 1100 K, that liquid Ni_3S_2 and Cr_2O_3 are the most stable corrosion products.

As evidence that Ni_3S_2 and Cr_2O_3 cannot react to form, for example, $NiCr_2O_4$ in contact with the equilibrium gas

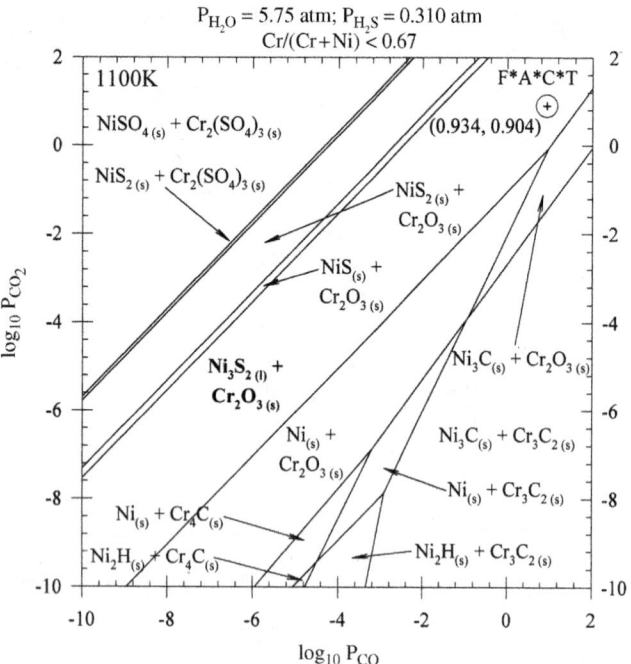

$P_{H_2O} = 5.75$ atm; $P_{H_2S} = 0.310$ atm
$Cr/(Cr+Ni) < 0.67$

FIGURE 21.3. The **Ni, Cr**-H, O, S, C predominance diagram at 1100 K (827°C). The diagram in this case is the superimposition of Figure 21.2 and a similar one for Cr since, within the field of view, there are **no phases that contain both Ni and Cr (such as NiCr₂O₄)**. The placement of the coordinate ⊕ consistent with Table 21.1 shows that **liquid Ni₃S₂** is a stable phase.

TABLE 21.1. Equilibrium Computations for the Initial Gas Composition of 35 Mol CO, 25 Mol H₂, 20 Mol CO₂, 19 Mol H₂O, and 1 Mol H₂S, Shown as Mole Fractions Adjacent to the Given Species

Species	(900 K, 30 atm) X_i	(1100 K, 30 atm) X_i
CO	0.050	0.286
CO₂	0.431	0.267
H₂	0.091	0.206
H₂O	0.335	0.192
CH₄	0.079	0.039
H₂S	0.013	0.010
COS	0.230×10^{-3}	0.455×10^{-3}
H₂CO	0.225×10^{-6}	0.214×10^{-5}
C₂H₄	0.897×10^{-7}	0.821×10^{-6}
C₂H₆	0.189×10^{-5}	0.612×10^{-6}
H₂S₂	0.323×10^{-6}	0.420×10^{-6}
CH₃SH	0.291×10^{-6}	0.228×10^{-6}
S₂	0.323×10^{-8}	0.311×10^{-7}
SO₂	0.217×10^{-9}	0.110×10^{-8}
SO₃	0.171×10^{-19}	0.841×10^{-18}
O	0.209×10^{-23}	0.101×10^{-18}
O₂	0.457×10^{-23}	0.493×10^{-19}
Total moles of gas	75.3	92.8
Activity of C	1	0.838
Moles of C	12.8	0

composition in Table 21.1 at 1100 K, it is necessary to evaluate the Gibbs energy change for the reaction of Ni₃S₂ and Cr₂O₃ to form NiCr₂O₄ (balanced using CO₂, CO, H₂O, and H₂S).

$$Ni_3S_2 + 3Cr_2O_3 + CO_2 + 2H_2O \rightarrow 3NiCr_2O_4 + 2H_2S + CO \quad (21.10)$$

In finding the Gibbs energy change at 1100 K [2], allowance is made for the partial pressures of the gas species at the composition of interest using an equation similar to Eq. (21.5). The positive value of 146 kJ/mol affirms the inability of the sulfide and oxide to react; that is, the stability of Ni₃S₂ and Cr₂O₃ in direct contact in this particular atmosphere.

Although the superimposition of Ni and Cr stability diagrams provides (by good fortune, in this particular case) a suitable corrosion product stability diagram, superimposition is not sufficiently general to create a reliable diagram in all cases. A more universally applicable approach is a necessity. To take the case of Inconel 671, the construction of a composite stability diagram for two (or more) alloying elements can be undertaken by finding the Gibbs energy changes for all reactions of the type:

$$(1-r)\mathbf{Ni} + r\,\mathbf{Cr} + \left(\frac{a}{2}\right)H_2 + \left(\frac{b}{2}\right)O_2 + \left(\frac{c}{2}\right)S_2 + (d)C$$
$$\rightarrow m[\mathbf{Ni}_w\mathbf{Cr}_x(H, O, S, C,)] + n[\mathbf{Ni}_y\mathbf{Cr}_z(H, O, S, C,)] \quad (21.11)$$

In this generalized methodology [5], following the constraints of the phase rule, groups of compounds equal in number to the components in the alloy phase are considered. All the candidate corrosion product compounds in the group must contain some or all of the alloying elements and the molar amounts given by m and n must both of course be nonnegative for the proportion of elements in the alloy given by r. With all equations balanced in terms of 1 mol of alloy with appropriate zero or positive coefficients a, b, c, and d, the Gibbs energy change for each is found in a manner similar to that for reactions of the type given by Eq. (21.4). The corrosion product pair that is most stable is associated with the reaction with the most negative Gibbs energy change. The topology of such diagrams is sensitive to the value of r. Under more oxidizing conditions than those usually associated with coal gasification applications, NiCr₂O₄ may be a stable corrosion product. Clearly, whether NiCr₂O₄ coexists with a Cr- or a Ni-containing compound depends on whether r is greater or less than $\frac{2}{3}$.

C. KINETICS

The qualitative characterization of corrosion rate in high-temperature systems may sometimes be gleaned from a combination of stability diagrams and corrosion rate data. Clearly, the stability diagram for Inconel 671 at 1100 K shown in Figure 21.3 is revealing in terms of indicating liquid sulfide corrosion products, which invariably have high corrosion rates, in contact with the equilibrium gas compositions represented in Table 21.1. Liquid Ni_3S_2, as a stable corrosion product (coexisting with solid Cr_2O_3), would be expected to corrode rapidly in the gas at 1100 K. At the lower temperature (900 K) associated with Figure 21.4, where Ni_3S_2 is solid, somewhat slower corrosion rates might be expected. Thermochemical information is valuable in organizing and interpreting data on corrosion at high temperatures, for design and selection purposes, in combination with corrosion rates measured in actual exposure testing.

To quantitatively represent *isothermal* corrosion testing, it is possible to provide empirical constants in an equation of the form

$$p = kt^n \qquad (21.12)$$

where p is the penetration (or depth of loss of internally oxidation free alloy) and k and n are characteristic parameters established by a statistical treatment of sufficient test data [6]. For a pure metal with only one corrosion product, a value of n near 1 may indicate that the superficial scale is not protective but rather spalls at a constant rate perhaps due to

excessive volume in comparison to the volume of metal replaced (Pilling–Bedworth ratio [7]). In other cases, corrosion rate is limited by arrival of a corrosion species or removal of a volatile corrosion product compound. Protective corrosion products must have a value of $n < 1$. If the growth of the scale is characterized by a constant diffusion coefficient in the alloy or corrosion product compound, n should be $\frac{1}{2}$. This is sometimes termed parabolic growth. Defects in the crystallinity of the scale, particularly as affected by impurities or alloying elements incorporated into the growing scale, may lead to values other than $\frac{1}{2}$. Many corrosion–resistant alloys often have both a value of n of $\frac{1}{2}$ and a small value for k. The parameter k is temperature dependent and is sensitive to the equilibrium partial pressure of the elements in the gas phase as well.

When Inconel 671 is exposed to oxidizing and sulfidizing atmospheres such that Ni_3S_2 and Cr_2O_3 are the *stable corrosion products*, Eq. (21.12) may be generalized [8] to yield

$$\log\left[\frac{p}{(t)}\right] = a + b\left[\log(P_{S_2}) - \frac{3}{2}\log(P_{O_2})\right] + \frac{c}{T} \quad (21.13)$$

Extensive exposure testing for many alloys and corrosive gases has shown that this form provides a good correlation of the corrosion data with the exposure conditions. For the specific case of Inconel 671 exposed to gas with the composition given in Table 21.1 for 900 K and 30-atm total pressure, the penetration is predicted by Eq. (21.13) to be 2.1 mils after 1 year, 4.6 mils after 5 years, and 6.5 mils after 10 years of exposure time [9]. For the same alloy under different service conditions (temperature, pressure, or gas composition), leading to a *different combination* of stable corrosion products, another empirical rate equation of a different form would be required.

D. SOFTWARE

The foregoing approach to alloy selection is computationally intensive and the following logical steps are used:

(a) The equilibrium partial pressures of elements in the corrosive gas must be found.

(b) The stable corrosion product phases on the candidate metal or alloy must be predicted based upon (a).

(c) A suitable empirical corrosion rate expression is selected, as appropriate for the likely corrosion mechanism [i.e., corrosion phases in (b)].

To assist in this manually cumbersome methodology, a user-friendly Windows 95 (trade name of Microsoft Corporation) program called ASSET (Alloy Selection System

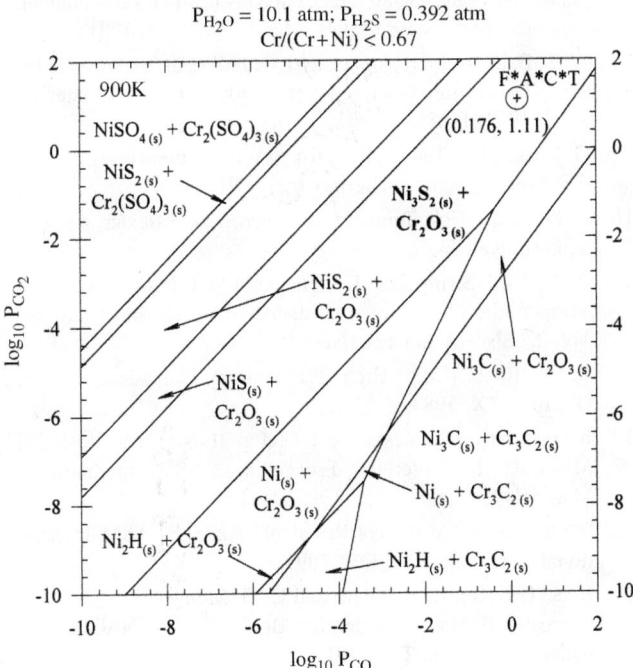

$P_{H_2O} = 10.1$ atm; $P_{H_2S} = 0.392$ atm
Cr/(Cr+Ni) < 0.67

FIGURE 21.4. The **Ni, Cr**-H, O, S, C diagram at 900 K (627°C similar to Figure 21.3. The placement of the coordinate $\oplus$ consistent with Table 21.1 shows that solid Ni_3S_2 is a stable phase coexisting with **solid Cr_2O_3**.

for Elevated Temperatures) has been developed [9]. The thermochemical predictions involved in steps (a) and (b) are performed in a user transparent way with an appropriate subset of programming and databases of Facility for the Analysis of Chemical Thermodynamics (FACT) [2]. Regression analysis is used to produce correlations of alloy corrosion for four different corrosion mechanisms as named for the types of corrosion products that form:

1. Sulfidation
2. Sulfidation/oxidation
3. Oxidation
4. Carburization

A large selection of data has been compiled for these conditions representing nearly 4.7 million h of exposure for 71 commercially available alloys. Details of the program are provided elsewhere [9].

E. CONCLUSION

An engineering approach to the selection of alloys for high-temperature service in many different types of fossil fuel combustion and energy conversion processes, can, with suitable software [2, 9, 10], involve a blend of fundamental science and exposure testing. Thermodynamics is used to compute the equilibrium gas-phase chemical speciation and predict stable corrosion products on commercial alloys. Extensive exposure testing conducted on commercially available alloys is then correlated to those computations to provide a reliable guide to metal loss as a function of time, temperature, gas composition, and pressure. The reader is referred to readily available literature [8, 11–22] for a more comprehensive treatment of the corrosion of metals and alloys in high-temperature gases.

ACKNOWLEDGMENT

The authors thank M. H. Kaye for his assistance in preparing the graphics.

REFERENCES

1. R. C. John, W. C. Fort, and R. A. Tait, Mater. High Temperatures, **11**(1–4) (1993).
2. C. W. Bale, A. D. Pelton, and W. T. Thompson, "Facility for the Analysis of Chemical Thermodynamics," in User Guide, Ecole Polytechnique de Montreal, Montreal, Canada, 1997.
3. G. Eriksson and W. T. Thompson, CALPHAD, **13**(4), 377 (1989).
4. C. W. Bale, A. D. Pelton, and W. T. Thompson, Can. Metall. Q., **25**(3) (July 1986).
5. M. Ahfat and W. T. Thompson, "Computation of Stoichiometric Phase Coexistence in Multicomponent Systems," in Proceedings of the Metallurgical Society of CIM (Computer Software in Chemical and Extractive Metallurgy), Quebec City, Aug. 1993, pp. 99–111.
6. H. H. Uhlig and R. W. Revie, "Corrosion and Corrosion Control," 4th ed., Wiley, New York, 2008.
7. N. Pilling and R. Bedworth, J. Inst. Met., **29**, 534 (1923).
8. R. C. John, "Alloy Corrosion in Reducing Plus Sulfidizing Gases at 600–950°C," in High Temperature Corrosion in Energy Systems, M. F. Rothman (Ed.), Metallurgical Society of AME, Warrendale, PA, 1985.
9. R. C. John, A. L. Young, and W. T. Thompson, "A Computer Program for Engineering Assessment of Alloy Corrosion in Complex High Temperature Gases," in CORROSION 97, National Association of Corrosion Engineers, Houston, TX, 1997.
10. A. D. Pelton, W. T. Thompson, C. W. Bale, and G. Eriksson, High Temp. Sci., **26**, 231 (1990).
11. J. D. Embury (Ed.), High Temperature Oxidation and Sulphidation Processes, Symposium by Metallurgical Society of CIM, 29th Annual Conference of Metallurgists, Pergamon, Elmsford, NY, 1990.
12. A. V. Ley (Ed.), Proceedings Corrosion-Erosion-Wear of Materials in Emerging Fossil Energy Systems, NACE, Houston, TX, 1982.
13. V. L. Hill and H. L. Blade (Eds.), The Properties and Performance of Materials in the Coal Gasification Environment, American Society for Metals, Metals Park, OH, 1981.
14. W. T. Bakker, S. Dapkunas, and W. Hill (Eds.), Materials for Coal Gasification, Conference proceedings of ASM Materials Week 1987, 1988.
15. G. Y. Lai, High Temperature Corrosion of Engineering Alloys, ASM International, Materials Park, OH, 1990.
16. P. Kofstad, High Temperature Corrosion, Elsevier Applied Science, New York, 1988.
17. R. Streiff, J. Stringer, R. Krutenat, and M. Caillet (Eds.), High Temperature Corrosion—Advanced Materials and Coatings, Vol. 2, Elsevier Science, New York, 1989.
18. R. A. Rapp (Ed.), High Temperature Corrosion, NACE, Houston, TX, 1983.
19. K. Natesan, P. Ganesan, and G. Lai (Eds.), Heat-Resistant Materials II, Conference Proceedings, ASM International, Materials Park, OH, Sept. 1995.
20. W. R. Davis (Ed.), Heat-Resistant Materials, ASM International, Materials Park, OH, 1997.
21. R. S. Treseder, R. Baboian, and C. G. Munger (Eds.), NACE Corrosion Engineer's Reference Book, 2nd ed., NACE International, Houston, TX, 1991.
22. H. J. Gabke, Carburization—A High Temperature Corrosion Phenomenon, Materials Technology Institute, St. Louis, MO, 1998.

22

A PROCEDURE TO COMPUTE EQUILIBRIUM GAS-PHASE SPECIATION FOR USE WITH PREDOMINANCE DIAGRAMS

M. H. Piro and B. J. Lewis

Department of Chemistry and Chemical Engineering, Royal Military College of Canada, Kingston, Ontario, Canada

W. T. Thompson

Centre for Research in Computational Thermochemistry, Royal Military College of Canada, Kingston, Ontario, Canada

A. INTRODUCTION

In comparison to corrosion at near ambient temperatures, the oxidation of metals and alloys at high temperatures puts greater emphasis on chemical thermodynamics. Fortunately, the difficulties posed by experimental measurements at high temperatures are offset by more rapid chemical kinetics making computations of equilibrium especially useful as a predictive tool. In some cases, the decreasing stability of metal oxides with increasing temperature also makes exploiting conditions of immunity a practical possibility. A procedure is described in Chapter 21 of this *Handbook* to locate domains of immunity by computational means on isothermal diagrams in which the partial pressures of gases are used as axes on logarithmic scales. These are sometimes called predominance diagrams and have a long history of usage in high-temperature corrosion as well as other matters involving chemical reactions at high temperatures between condensed phases and process gases, such as in smelting and refining. As the gas mixtures become more complex, particularly with respect to the number of elements, predominance diagrams become increasingly difficult to interpret. It is generally not evident how to create in a reacting gas mixture (such as CH_4 and O_2) the equilibrium partial pressures (CO, CO_2, etc.) used as axes on predominance diagrams. Additionally, gas-phase equilibria controlling the speciation in the gas mixture are sensitive to temperature and pressure. Therefore, combinations of alloys and gas mixtures that perform well at one temperature may not do so at another. In these situations, multiphase equilibria states and speciation may be computed through numerical optimization techniques [1–5]. This method is applied to homogeneous gas-phase equilibrium in Section B and heterogeneous equilibrium of a condensed solid phase coexisting with an ideal gas in Section C.

B. HOMOGENEOUS EQUILIBRIUM IN AN IDEAL GAS SOLUTION

High-temperature oxidation and corrosion requires the computation of gas-phase speciation at the temperature and pressure of exposure in a system comprised of several elements. To give focus to the computational methods, consider a hydrocarbon combustion situation. The usual

Uhlig's Corrosion Handbook, Third Edition, Edited by R. Winston Revie
Copyright © 2011 John Wiley & Sons, Inc.

excess of oxygen in relation to a hydrocarbon fuel at high temperature makes it possible to exploit knowledge that the main product species are $CO_{2(G)}$ and $H_2O_{(G)}$, with most of the sulfur appearing as $SO_{2(G)}$. However, with the same collection of elements involved in, for example, coal gasification, the deficiency of oxygen in relation to carbon and the large number of possible chemical compounds makes the computation of molecular concentrations more difficult. A general method to compute gas-phase equilibrium in a system with several elements is therefore a necessity. Even when the gas composition is known from chemical analysis at room temperature, the composition of the gas at the temperature and pressure of exposure must be computed.

Gas-phase speciation is generally discussed with reference to equilibrium constants relating, for example, concentrations of $CO_{(G)}$, $CO_{2\ (G)}$; and $O_{2\ (G)}$:

$$CO_{(G)} + \tfrac{1}{2}O_{2(G)} = CO_{2(G)} \qquad (22.1)$$

where

$$K_{eq} = \exp\left[\frac{-\Delta G^\circ}{RT}\right] = \frac{P_{CO_2}}{P_{CO}\,P_{O_2}^{\ 0.5}} \qquad (22.2)$$

where K_{eq} is the equilibrium constant, ΔG° (J·mol^{-1}) is the Gibbs energy of reaction, R (J·mol^{-1} K^{-1}) is the universal gas constant, T (K) is the absolute temperature, and P_i (atm) is the partial pressure of species i in the gas. In an ideal gas mixture, the fugacity is numerically equivalent to the partial pressure in dimensionless units when the reference pressure is 1 atm. The difficulty with pursuing this method is incorporating the overall atomic proportion of carbon and oxygen in the system to express constraints on the partial pressures. Nonetheless, in a system where experience provides a basis for knowing the dominant gas species, and where the number of those species does not significantly exceed the number of

elements, this traditional method of computing gas-phase equilibrium is quite serviceable. However, it does not lend itself to generalization, even when applied to common combustion gases with several possible elements (C, H, S, O, N), particularly for the case of fuel consumed in a deficiency of air.

B1. Thermodynamic Data

As a basis of discussion that can be related to familiar expectations, consider the binary carbon–oxygen system at 1000 K and 1 atm. The Gibbs energy, G_i° (J·mol^{-1}), of an individual species i may be initially defined as [6]

$$G_i^\circ = \left(\Delta H_i^\circ + \int_{298}^{T} C_{p,i}\,dT\right) - T\left(S_i^\circ + \int_{298}^{T} \frac{C_{p,i}}{T}dT\right) + \int_{1}^{P_i} V_i\,dP \qquad (22.3)$$

where ΔH_i° (J·mol^{-1}) and S_i° (J·mol^{-1}K^{-1}) represent the standard enthalpy of formation and absolute entropy, respectively, at 298 K and 1 atm, $C_{p,i}$ (J·mol^{-1}K^{-1}) is the molar heat capacity at constant pressure, P_i (atm) is the partial pressure, and V_i (m^3·mol^{-1}) is the molar volume occupied by i. Division by the total number of atoms in the formula mass ($M_{i,T}$) provides units of joules per gram-atom (J·g at^{-1}). The use of this latter unit in expressing an extensive property provides a basis for comparing Gibbs energies of compounds with different numbers of atoms per molar mass on an equivalent basis.

Although it is a misnomer, the Gibbs energy in Eq. (22.3), G_i°, is sometimes called the "absolute" Gibbs energy. This nomenclature draws attention to the use of the absolute entropy in the formulation of G_i°. The more usual Gibbs energy of isothermal formation from the most stable form of the elements appears on the extreme right side of Table 22.1.

TABLE 22.1.   Gibbs Energies for Selected Species in C–O System at 1000 K and 1 atm

Species	Mole Fraction of Element		Atoms/ Molecule	"Absolute" Gibbs Energy[a] [7]		Gibbs Energy of Formation[a]	
	$c_{i,C}$	$c_{i,O}$	$M_{i,T}$	(J·g mol^{-1})	(J·g at^{-1})	(J·g mol^{-1})	(J·g at^{-1})
$C_{(G)}$	1	0	1	547,992	547,992	559,952	559,952
$C_{2(G)}$	1	0	2	618,794	309,397	644,114	322,057
$C_{3(G)}$	1	0	3	563,277	187,759	601,255	200,418
$C_{4(G)}$	1	0	4	711,512	177,878	762,151	190,538
$C_{5(G)}$	1	0	5	698,392	139,678	761,690	152,338
$O_{(G)}$	0	1	1	77,353	77,353	187,735	187,735
$O_{2(G)}$	0	1	2	−220,764	−110,382	0	0
$O_{3(G)}$	0	1	3	−119,445	−39,815	211,702	70,567
$CO_{(G)}$	0.5	0.5	2	−323,264	−161,632	−200,222	−100,111
$C_2O_{(G)}$	0.667	0.333	3	29,161	9,720	164,863	54,954
$CO_{2(G)}$	0.333	0.667	3	−629,423	−209,808	−396,000	−132,000
$C_3O_{2(G)}$	0.6	0.4	5	−409,605	−81,921	−150,862	−30,172
$C_{(S)}$	1	0	1	−12,660	−12,660	0	0

[a]In reference to iterations 0 and 1 in Table 22.2

Taking $CO_{(G)}$ as an example, the species of pure carbon with the lowest "absolute" Gibbs energy ($C_{(S)}$) and the lowest for pure oxygen ($O_{2(G)}$) may be deducted from the "absolute" Gibbs energy of $CO_{(G)}$ to yield a Gibbs energy of formation of $-100,111$ (J·g at^{-1}). Table 22.1 deliberately includes several species known to be of little consequence at 1000 K and 1 atm in order to illustrate the unbiased procedure by which the dominant chemical species are found as an initial step in computing the equilibrium composition of the gas phase. The atomic fractions of the elements in each species i are represented by $c_{i,C}$ and $c_{i,O}$ for carbon and oxygen respectively, as shown in Table 22.1 [7].

B2. Identification of Dominant Species

The initial elemental proportions of carbon and oxygen must be specified to identify the dominant gas species. Here, this proportion is arbitrarily set to be 0.65 atom fraction O (equivalently, 0.35 mol $C_{(S)}$ and 0.325 mol $O_{2(G)}$) for discussion purposes. This atomic proportion might be known from the relative flows of oxygen and pulverized carbon in a combustion situation. Temporarily setting aside the effect of concentration (or partial pressure) on the molar Gibbs energy of each species, the overall aim is to iteratively distribute the elements, subject to the constraint on their proportion, to make the Gibbs energy of the gas mixture the most negative. This approach could not be exploited in a practical way until the advent of digital computing using methods generally based on Lagrange multipliers [1–5] as Gibbs himself first recognized [8]. The present approach further exploits knowledge that Gibbs energy is, by nature, a relative quantity since it is not possible to chemically convert one element to another. Only changes in Gibbs energy, which involve processes of chemical conversion from one collection of C- and O-containing species to another, are measurable. Recognition of this thermodynamic principle permits systematic adjustment of the numerical values of the Gibbs energies for species in such a way as to preserve essential differences (as will be demonstrated later in Table 22.3).

The process of adjusting Gibbs energies to preserve Gibbs energy differences for chemical change was coined by Eriksson and Thompson as "leveling" [9]. Leveling is performed by representing the set of Gibbs energies relative to the collection of species assumed to be most stable. The Gibbs energy of species i that is represented relative to a particular assemblage is referred to as the "relative Gibbs energy," ΔG_i^m (J·g-at^{-1}).

The process is best understood using the sequence of diagrams in Figure 22.1 and the adjusted relative Gibbs energies in Table 22.2.

Leveling involves:

(i) Selecting any two species which represent the overall atomic proportion of oxygen to carbon

(ii) Setting the relative Gibbs energy of those species to zero when they are assumed to be the most stable pair

(iii) Readjusting the relative Gibbs energies of all other species relative to the two set to zero

The two species initially selected are the most stable form of each element; that is to say, the form of carbon and oxygen with the lowest Gibbs energy in units of J·g at^{-1}. The use of gram-atom in place of mole (or equivalently gram-mole) permits the direct comparison of equal amounts of monatomic $O_{(G)}$ with diatomic $O_{2(G)}$, as shown on the extreme right-hand side of Figure 22.1. With this initial selection of possible "dominant" species assures the ability to express the overall elemental proportion as represented by the vertical dotted line in Figure 22.1. All species are "leveled" with respect to $C_{(S)}$ and $O_{2(G)}$ in the first iteration, as represented by the horizontal dashed line at $m=1$ in Figure 22.1. The adjustment made to species i as a result of leveling (δG_i) is represented by the following equation for the first iteration:

$$\delta G_i = c_{i,C}\delta G_{C_{(S)}} + c_{i,O}\delta G_{O_{2(G)}} \tag{22.4}$$

By deducting this Gibbs energy evaluated at the composition of each C–O species, the relative Gibbs energies for all species are thereby set relative to $C_{(S)}$ and $O_{2\ (G)}$ in the first iteration, and both elements in their most stable form necessarily become zero. This is depicted graphically at iteration $m=1$ in Figure 22.1 and tabulated in column 3 in Table 22.2.

The next step involves selecting the species with the most negative relative Gibbs energy (i.e., $CO_{2(G)}$), which will

TABLE 22.2. Gibbs Energies Adjusted as Result of Leveling to Identify Dominant Species

Species	Relative Gibbs Energy [ΔG_i^m(J·g at^{-1})] Iteration m			
	0^a	1^a	2	3
$C_{(G)}$	547,992	560,652	560,652	565,097
$C_{2(G)}$	309,397	322,057	322,057	326,502
$C_{3(G)}$	187,759	200,418	200,418	204,863
$C_{4(G)}$	177,878	190,538	190,538	194,983
$C_{5(G)}$	139,678	152,338	152,338	156,783
$O_{(G)}$	77,353	187,735	385,735	383,513
$O_{2(G)}$	$-110,382$	0	198,000	195,777
$O_{3(G)}$	$-39,815$	70,567	268,567	266,345
$CO_{(G)}$	$-161,632$	$-100,111$	$-1,111$	0
$C_2O_{(G)}$	9,720	54,954	120,954	123,177
$CO_{2(G)}$	$-209,808$	$-132,000$	0	0
$C_3O_{2(G)}$	$-81,921$	$-30,172$	49,028	50,806
$C_{(S)}$	$-12,660$	0	0	4,445

^{a}In reference to Table 22.1.

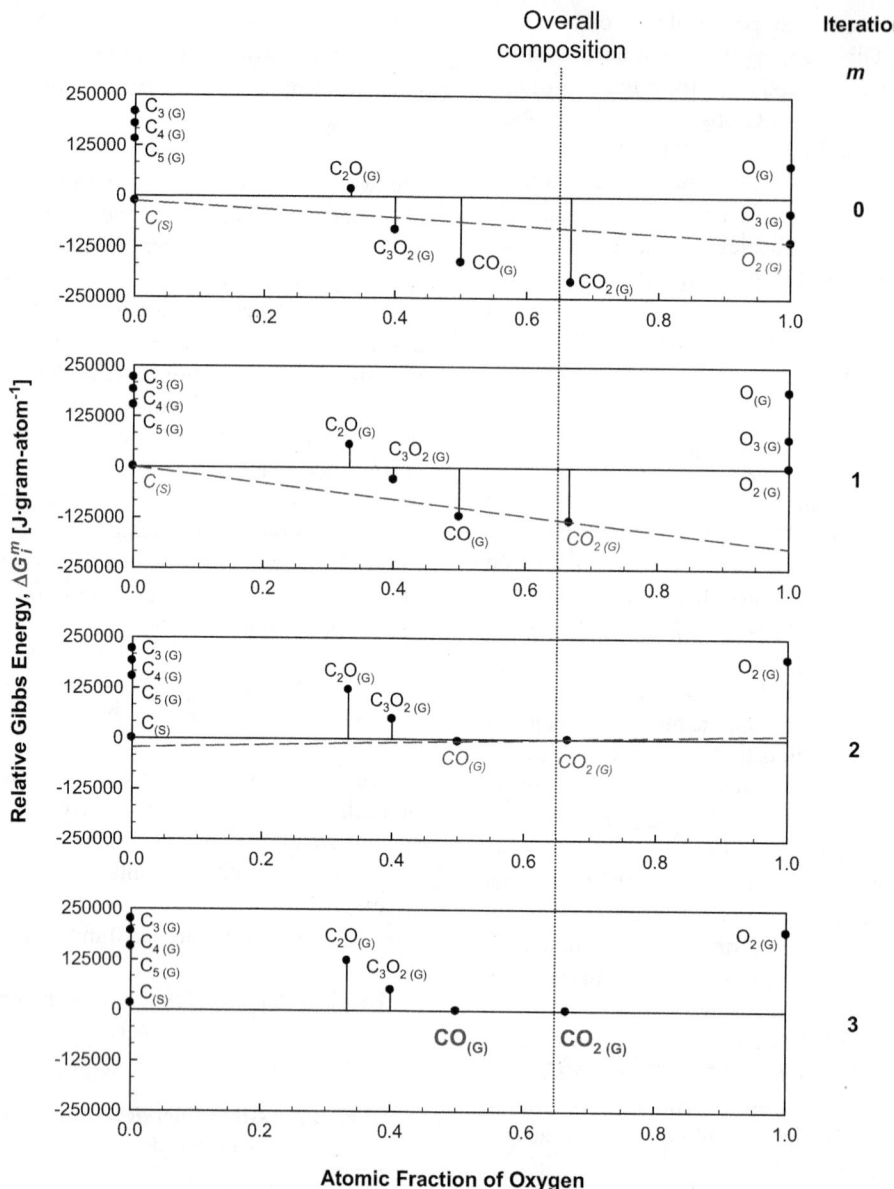

FIGURE 22.1. Illustration of leveling technique applied to C–O system to determine pair of dominant gaseous species at 0.65 atomic fraction of O at 1000 K and 1 atm.

replace one of the species from the first iteration. The choice of the species to be replaced from iteration $m = 1$ is made on the basis of being able to represent the fixed overall atomic fraction of O for the system as a whole. Thus, $CO_{2\,(G)}$ must be paired with $C_{(S)}$ rather than $O_{2\,(G)}$. A combination of $CO_{2\,(G)}$ and $C_{(S)}$ are together more negative than the previous estimated assemblage comprised of $C_{(S)}$ and $O_{2\,(G)}$. The equation representing the horizontal dashed line shown in Figure 22.1 is reestablished and leveling is advanced to iteration $m = 2$ in Figure 22.1 and column 4 in Table 22.2. Species that have positive relative Gibbs energies do not require further consideration as they are correspondingly less chemically stable than the current assemblage. Thus, a large proportion of

possible species are withdrawn from further consideration with every additional iteration. This is a very significant feature when this method is applied to a gas mixture involving many elements and species. Eventually, all relative Gibbs energies are nonnegative, as in iteration $m = 3$ in Figure 22.1 and the extreme right column in Table 22.2. The two species that have zero relative Gibbs energies at the end of leveling ($CO_{2(G)}$ and $CO_{(G)}$) are the dominant gaseous species, and their proportions are as follows:

$$0.35 \text{ mol } C_{(S)} + 0.325 \text{ mol } O_{2(G)}$$
$$= 0.3 \text{ mol } CO_{2(G)} + 0.05 \text{ mol } CO_{(G)} \qquad (22.5)$$

Although the magnitude of Gibbs energies of all species changes throughout the leveling procedure, elemental differences in these energies are preserved at all iterations. The change in relative Gibbs energy for the reaction in Eq. (22.5) at iteration m is thus

$$\Delta G^m_{\text{Reaction}} = \Delta G^m_{\text{Products}} - \Delta G^m_{\text{Reactants}}$$

$$\Delta G^m_{\text{Reaction}}$$

$$= \left(0.3\,\text{mol} \times 3 \times \Delta G^m_{\text{CO}_{2(G)}} + 0.05\,\text{mol} \times 2 \times \Delta G^m_{\text{CO}_{(G)}}\right)$$

$$= -\left(0.35\,\text{mol} \times 1 \times \Delta G^m_{\text{C}_{(S)}} + 0.325\,\text{mol} \times 2 \times \Delta G^m_{\text{O}_{2(G)}}\right)$$

$$(22.6)$$

The preservation of differences in Gibbs energy for the above reaction is demonstrated in Table 22.3 using relative Gibbs energies tabulated in Table 22.2 for all iterations. Differences in relative Gibbs energies of all other conceivable reactions in the system are also preserved.

The thermochemical activity, a^m_i (unitless), of species i is computed with respect to the relative Gibbs energy at iteration m through the following relation:

$$a^m_i = \exp\left[\frac{-\Delta G^m_i M_{i,\text{T}}}{RT}\right] \qquad (22.7)$$

The activity of species i is equivalent to its concentration, $x^m_{i,\text{gas}}$ (and is correspondingly related to its partial pressure), in an ideal gas. $M_{i,\text{T}}$ is the previously defined number of total atoms per formula mass. The estimated concentrations of all species are computed with Eq. (22.7) in Table 22.4 using relative Gibbs energies from iteration $m = 3$. By the nature of this method, the mass constraints imposed on the system are in general not exactly satisfied after leveling. The overall mole fractions of carbon and oxygen in the gas are represented at iteration m by

$$\chi^m_{\text{C,gas}} = \frac{\sum_{i=1}^{N_{\text{gas}}} M_{i,\text{C}} x^m_{i,\text{gas}}}{\sum_{i=1}^{N_{\text{gas}}} M_{i,\text{C}} x^m_{i,\text{gas}} + \sum_{i=1}^{N_{\text{gas}}} M_{i,\text{O}} x^m_{i,\text{gas}}} \qquad (22.8)$$

and,

$$\chi^m_{\text{O,gas}} = \frac{\sum_{i=1}^{N_{\text{gas}}} M_{i,\text{O}} x^m_{i,\text{gas}}}{\sum_{i=1}^{N_{\text{gas}}} M_{i,\text{C}} x^m_{i,\text{gas}} + \sum_{i=1}^{N_{\text{gas}}} M_{i,\text{O}} x^m_{i,\text{gas}}} \qquad (22.9)$$

where $M_{i,\text{C}}$ and $M_{i,\text{O}}$ are the number of carbon and oxygen atoms, respectively, in a molecule of i. The overall mole fraction χ^m_{gas} representing the gas with estimated concentrations at iteration m will differ from the mole fraction that the gas is constrained by, χ'_{gas}. When the system is assumed to be homogeneous and comprised of a gas, the value for χ'_{gas} representing the gas is equal to the overall mole fraction of the system, χ^*. The mass constraint χ'_{gas} imposed on the gas is distinguished from the mass constraint χ^* imposed on the entire system for the condition where the system is heterogeneous and $\chi'_{\text{gas}} \neq \chi^*$. The treatment of heterogeneous systems involving an ideal gas coexisting with a pure condensed phase is addressed in Section C.

Discrepancies between $\chi^m_{\text{C,gas}}$ and $\chi^m_{\text{O,gas}}$ with respect to the fixed mole fractions of carbon and oxygen represented by the gas ($\chi'_{\text{C,gas}}$ and $\chi'_{\text{O,gas}}$) are evaluated on a relative basis. The relative errors representing the overall mass balances are as follows:

$$\xi^m_{\text{C}} = \left|\frac{\chi'_{\text{C,gas}} - \chi^m_{\text{C}}}{\chi'_{\text{C,gas}}}\right| \qquad \xi^m_{\text{O}} = \left|\frac{\chi'_{\text{O,gas}} - \chi^m_{\text{O}}}{\chi'_{\text{O,gas}}}\right| \qquad (22.10)$$

By virtue of the dependency of concentrations of gaseous species on Gibbs energies, the sum of estimated concentrations in the gas, $x^m_{T,\text{gas}}$, will not equal unity. The relative error $\xi^m_{x,\text{gas}}$ representing the deviation of $x^m_{T,\text{gas}}$ from unity is thus defined

$$\xi^m_{x,\text{gas}} = \left|x^m_{T,\text{gas}} - 1\right| \qquad (22.11)$$

The overall mole fraction χ^m_{gas} represented by the gas and the corresponding relative errors are computed in Table 22.4.

Table 22.4 demonstrates that the mass constraints imposed on the gas have not yet been satisfied at iteration $m = 3$. The effect of concentration on the molar Gibbs energy, which was temporarily superceded during leveling, must now be

TABLE 22.3.   Elemental Differences in Gibbs Energies Preserved for All Iterations for Eq. (22.6)

Iteration m	$\Delta G_{\text{Products}}$ (J)		$\Delta G_{\text{Reactants}}$ (J)		$\Delta G_{\text{Reaction}}$ (J)
0	$(0.3)(3)(-209,808) + (0.05)(2)(-161,632)$	$-$	$(0.35)(1)(-12,660) + (0.325)(2)(-110,382)$	$=$	$-128,811$
1	$(0.3)(3)(-132,000) + (0.05)(2)(-100,111)$	$-$	$(0.35)(1)(0) + (0.325)(2)(0)$	$=$	$-128,811$
2	$(0.3)(3)(0) + (0.05)(2)(-1,111)$	$-$	$(0.35)(1)(0) + (0.325)(2)(198,000)$	$=$	$-128,811$
3	$(0.3)(3)(0) + (0.05)(2)(0)$	$-$	$(0.35)(1)(4,445) + (0.325)(2)(195,777)$	$=$	$-128,811$

TABLE 22.4. Estimated Equilibrium Gas-Phase Composition Computed from Adjusted Relative Gibbs Energies from Iteration $m = 3$ in Figure 22.1 and Table 22.2

Species	$M_{i,C}$	$M_{i,O}$	$M_{i,T}$	$\Delta G_i^{m=3}$ (J·g at^{-1})	$a_i^{m=3}$	$x_{i,\text{gas}}^{m=3}$	$M_{i,C} \times x_{i,\text{gas}}^{m=3}$	$M_{i,O} \times x_{i,\text{gas}}^{m=3}$
$C_{(G)}$	1	0	1	564,397	3.30×10^{-30}	3.30×10^{-30}	3.30×10^{-30}	0.00
$C_{2(G)}$	2	0	2	326,502	7.75×10^{-35}	7.75×10^{-35}	1.55×10^{-34}	0.00
$C_{3(G)}$	3	0	3	204,863	7.87×10^{-33}	7.87×10^{-33}	2.36×10^{-32}	0.00
$C_{4(G)}$	4	0	4	194,983	1.82×10^{-41}	1.82×10^{-41}	7.26×10^{-41}	0.00
$C_{5(G)}$	5	0	5	156,783	1.12×10^{-41}	1.12×10^{-41}	5.62×10^{-41}	0.00
$O_{(G)}$	0	1	1	383,513	9.26×10^{-21}	9.26×10^{-21}	0.00	9.26×10^{-21}
$O_{2(G)}$	0	2	2	195,777	3.52×10^{-21}	3.52×10^{-21}	0.00	7.04×10^{-21}
$O_{3(G)}$	0	3	3	266,345	1.82×10^{-42}	1.82×10^{-42}	0.00	5.47×10^{-42}
$CO_{(G)}$	1	1	2	0	1.00	1.00	1.00	1.00
$C_2O_{(G)}$	2	1	3	123,177	4.98×10^{-20}	4.98×10^{-20}	9.96×10^{-20}	4.98×10^{-20}
$CO_{2(G)}$	1	2	3	0	1.00	1.00	1.00	2.00
$C_3O_{2(G)}$	3	2	5	50,806	5.38×10^{-14}	5.38×10^{-14}	1.61×10^{-13}	1.08×10^{-13}
$C_{(S)}$	1	0	1	4,445	5.86×10^{-1}			
Sum						2.00	2.00	3.00
$\chi_{C,\text{gas}}^{m=3}, \chi_{O,\text{gas}}^{m=3}$							4.00×10^{-1}	6.00×10^{-1}
$\xi_{x,\text{gas}}^{m=3}, \xi_C^{m=3}, \xi_O^{m=3}$						1.00	1.43×10^{-1}	7.69×10^{-2}

considered. This matter is the principal cause of the relative errors shown at the bottom right of Table 22.4 (i.e., $\xi_C^{m=3}$, $\xi_O^{m=3}$). The identification of the dominant species by the leveling method provides good initial estimates for further computation.

B3. Computation of Equilibrium Concentrations of Gaseous Species

In many circumstances, depending on the temperature and pressure, the dominant species determined by the leveling algorithm are the only species that contribute significantly to the equilibrium population and all other species are of inconsequential proportions. However, further computation is often required when minor species contribute appreciably to the equilibrium composition in addition to the dominant species.

As a continuation of the concept of leveling, which exploits the fact that Gibbs energy is a relative function, a method is employed that exploits the knowledge that the equilibrium concentration of a solute species dissolved in a solution is related to its Gibbs energy. The objective is to further adjust the Gibbs energies of the pure elements to satisfy mass constraints within an admissible relative error. However, any slight numerical perturbation in the Gibbs energy of the pure elements (i.e., δG_C and δG_O) may result in large transformations in $\chi_{C,\text{gas}}^m$ and $\chi_{O,\text{gas}}^m$. To further complicate matters, changes made to the Gibbs energy of pure carbon will not only result in changes in the estimated number of moles of carbon but also in the number of moles of oxygen, since the dominant portion of molecules in the system are comprised of both carbon and oxygen atoms.

Detailed explanations of the methodologies employed by this technique are discussed by Piro [10].

In view of the nonlinear dependency of $\chi_{C,\text{gas}}^m$ on changes to Gibbs energies of both carbon and oxygen, $\chi_{C,\text{gas}}^m$ is differentiated with respect to changes in the elemental Gibbs energies of both carbon and oxygen. The resulting partial differentials represent the individual contributions to changes in $\chi_{C,\text{gas}}^m$ contingent on changes in elemental Gibbs energies. The numerical problem is reduced to a system of nonlinear equations in Eq. (22.12), which provides a mathematical framework for solving for a set of Gibbs energy adjustments to satisfy mass constraints [10]:

$$\begin{bmatrix} \dfrac{\partial \chi_{C,\text{gas}}^m}{\partial G_C} & \dfrac{\partial \chi_{C,\text{gas}}^m}{\partial G_O} \\ \dfrac{\partial \chi_{O,\text{gas}}^m}{\partial G_C} & \dfrac{\partial \chi_{O,\text{gas}}^m}{\partial G_O} \end{bmatrix} \begin{bmatrix} \delta G_C \\ \delta G_O \end{bmatrix} = \begin{bmatrix} \dfrac{\chi'_{C,\text{gas}}}{x_{T,\text{gas}}^m} - \chi_{C,\text{gas}}^m \\ \dfrac{\chi'_{O,\text{gas}}}{x_{T,\text{gas}}^m} - \chi_{O,\text{gas}}^m \end{bmatrix}$$

(22.12)

The partial differentials in Eq. (22.12) must be computed with respect to Eq. (22.7) in order to solve for the Gibbs energy adjustments for the succeeding iteration. Each term of the square matrix in Eq. (22.12) is computed with the following formulation:

$$\frac{\partial \chi_{j,\text{gas}}^m}{\partial G_k} = \frac{\sum_{i=1}^{N_{\text{gas}}} \left(\dfrac{M_{i,j} M_{i,k}}{RT} \right) x_{i,\text{gas}}^m}{\sum_{i=1}^{N_{\text{gas}}} M_{i,j} x_{i,\text{gas}}^m + \sum_{i=1}^{N_{\text{gas}}} M_{i,k} x_{i,\text{gas}}^m}$$

(22.13)

TABLE 22.5.   Estimated Concentrations of All Species at $m=3$ Used to Compute Coefficients of Eq. (22.12) Using Eq. (22.13)

Species	$M_{i,C}$	$M_{i,O}$	$x_{i,\text{gas}}^m$	$\dfrac{\partial\left(M_{i,C}x_i^m\right)}{\partial G_C}$	$\dfrac{\partial\left(M_{i,C}x_i^m\right)}{\partial G_O}$	$\dfrac{\partial\left(M_{i,O}x_i^m\right)}{\partial G_C}$	$\dfrac{\partial\left(M_{i,O}x_i^m\right)}{\partial G_O}$
$C_{(G)}$	1	0	3.04×10^{-30}	3.66×10^{-34}	0.00	0.00	0.00
$C_{2(G)}$	2	0	7.79×10^{-35}	3.75×10^{-38}	0.00	0.00	0.00
$C_{3(G)}$	3	0	7.90×10^{-33}	8.55×10^{-36}	0.00	0.00	0.00
$C_{4(G)}$	4	0	1.83×10^{-41}	3.51×10^{-44}	0.00	0.00	0.00
$C_{5(G)}$	5	0	1.13×10^{-41}	3.40×10^{-44}	0.00	0.00	0.00
$O_{(G)}$	0	1	9.28×10^{-21}	0.00	0.00	0.00	1.12×10^{-24}
$O_{2(G)}$	0	2	3.53×10^{-21}	0.00	0.00	0.00	1.70×10^{-24}
$O_{3(G)}$	0	3	1.83×10^{-42}	0.00	0.00	0.00	1.99×10^{-45}
$CO_{(G)}$	1	1	1.00	1.20×10^{-4}	1.20×10^{-4}	1.20×10^{-4}	1.20×10^{-4}
$C_2O_{(G)}$	2	1	4.99×10^{-20}	2.40×10^{-23}	1.20×10^{-23}	1.20×10^{-23}	6.00×10^{-24}
$CO_{2(G)}$	1	2	1.00	1.20×10^{-4}	2.41×10^{-4}	2.41×10^{-4}	4.81×10^{-4}
$C_3O_{2(G)}$	3	2	5.39×10^{-14}	5.83×10^{-17}	3.89×10^{-17}	3.89×10^{-17}	2.59×10^{-17}
$C_{(S)}$	1	0					
Sum			2.00	2.41×10^{-4}	3.61×10^{-4}	3.61×10^{-4}	6.01×10^{-4}

Equation (22.13) demonstrates that the adjustments made to the Gibbs energies of the pure elements take into consideration the molecular composition of each species and its estimated concentration from the current iteration. For example, this method recognizes that one molecule of $CO_{2\ (G)}$ contains one carbon atom and two oxygen atoms. The computation of elemental Gibbs energy adjustments incorporates knowledge of the estimated contribution of all gaseous species in the system simultaneously to the overall mass balances of all elements represented by the gas. Table 22.5 exemplifies the computation of individual terms in Eq. (22.12) and (22.13) using estimated concentrations of species from iteration $m=3$ in Table 22.4.

The coefficients of Eq. (22.12) incorporate values from the bottom of Tables 22.4 and 22.5. Components of the vector on the extreme right side of Eq. (22.12) are computed with Eq. (22.8) and (22.9), using mass constraints applied to the gas and the estimated mole fractions of the pure elements at iteration $m=3$. Insertion of computed values from Tables 22.4 and 22.5 into Eq. (22.12) gives

$$\begin{bmatrix} \dfrac{2.41 \times 10^{-4}}{2.00+3.00} & \dfrac{3.61 \times 10^{-4}}{2.00+3.00} \\[2ex] \dfrac{3.61 \times 10^{-4}}{2.00+3.00} & \dfrac{6.01 \times 10^{-4}}{2.00+3.00} \end{bmatrix} \begin{bmatrix} \delta G_C \\[1ex] \delta G_O \end{bmatrix} = \begin{bmatrix} \dfrac{0.35}{2.00}-0.40 \\[2ex] \dfrac{0.65}{2.00}-0.60 \end{bmatrix}$$

$$(22.14)$$

The resulting system of linear equations can be solved with a standard linear equation solver to determine the adjustments made to the Gibbs energies of the pure elements in the next iteration:

$$\begin{bmatrix} 4.81 \times 10^{-5} & 7.22 \times 10^{-5} \\ 7.22 \times 10^{-5} & 1.20 \times 10^{-4} \end{bmatrix} \begin{bmatrix} \delta G_C \\ \delta G_O \end{bmatrix} = \begin{bmatrix} -0.225 \\ -0.275 \end{bmatrix}$$

$$(22.15)$$

The relative Gibbs energies of all species in the system are further altered using Eq. (22.4), analogous to the adjustments made in the leveling method discussed in Section B2. Table 22.6 summarizes the adjustments made to the relative Gibbs energies for all species and the subsequent changes to $x_{i,\text{gas}}^m$, χ_{gas}^m and ξ^m. The process is repeated until the maximum value for ξ^m is less than the specified error tolerance. A value of 10^{-5} has been found acceptable as the error tolerance limit for the majority of thermochemical scenarios. The concentrations of all gaseous species at iteration $m=8$ on the extreme right column in Table 22.6 are representative of the equilibrium composition. The concentration of each gaseous species is equivalent to its partial pressure since the total applied hydrostatic pressure is 1 atm for this example.

The current example has involved a simple two-element system comprised of a few chemical species for discussion purposes. This algorithm can be easily extended to systems of many elements and a correspondingly large number of species. Performance statistics are shown in Figure 22.2 for thermochemical computations of homogeneous systems comprised of 200 species and up to 40 elements using the methodology described. Computation times are (obviously) dependent on many different factors, such as hardware performance, and are thus normalized relative to the computation time of 10 elements.

C. HETEROGENEOUS EQUILIBRIUM INVOLVING A PURE CONDENSED PHASE AND AN IDEAL GAS

The strategy discussed in Section B can be applied to heterogeneous systems involving various combinations of pure condensed phases in equilibrium with an ideal gas. The inclusion of pure condensed phases is illustrated through the

TABLE 22.6. Progression in Estimated Concentrations of Gaseous Species Following Table 22.4 Continued Until the Maximum Relative Error Is Below an Arbitrary Tolerance (10^{-5})

	Iteration									
	$m = 4$		$m = 5$		$m = 6$		$m = 7$		$m = 8$	
Species	ΔG_i^m (J·g at^{-1})	$x_{i,\text{gas}}^m$	ΔG_i^m (J·g at^{-1})	$x_{i,\text{gas}}^m$	ΔG_i^m (J·g at^{-1})	$x_{i,\text{gas}}^m$	ΔG_i^m (J·g at^{-1})	$x_{i,\text{gas}}^m$	ΔG_i^m (J·g at^{-1})	$x_{i,\text{gas}}^m$
$C_{(G)}$	577,568	6.79×10^{-31}	589,086	1.70×10^{-31}	595,141	8.20×10^{-32}	596,177	7.24×10^{-32}	596,173	7.24×10^{-32}
$C_{2(G)}$	338,973	3.88×10^{-36}	350,491	2.43×10^{-37}	356,547	5.66×10^{-38}	357,583	4.41×10^{-38}	357,579	4.41×10^{-38}
$C_{3(G)}$	217,335	8.78×10^{-35}	228,853	1.38×10^{-36}	234,908	1.55×10^{-37}	235,944	1.06×10^{-37}	235,941	1.07×10^{-37}
$C_{4(G)}$	207,454	4.53×10^{-44}	218,972	1.77×10^{-46}	225,028	9.64×10^{-48}	226,064	5.86×10^{-48}	226,060	5.87×10^{-48}
$C_{5(G)}$	169,255	6.25×10^{-45}	180,773	6.14×10^{-48}	186,828	1.61×10^{-49}	187,864	8.63×10^{-50}	187,860	8.65×10^{-50}
$O_{(G)}$	378,316	1.73×10^{-20}	372,469	3.50×10^{-20}	369,224	5.18×10^{-20}	368,626	5.56×10^{-20}	368,615	5.57×10^{-20}
$O_{2(G)}$	190,581	1.23×10^{-20}	184,733	5.03×10^{-20}	181,488	1.10×10^{-19}	180,890	1.27×10^{-19}	180,880	1.27×10^{-19}
$O_{3(G)}$	261,148	1.20×10^{-41}	255,300	9.87×10^{-41}	252,055	3.18×10^{-40}	251,458	3.95×10^{-40}	251,447	3.96×10^{-40}
$CO_{(G)}$	3,638	4.17×10^{-1}	6,473	2.11×10^{-1}	7,878	1.50×10^{-1}	8,097	1.43×10^{-1}	8,090	1.43×10^{-1}
$C_2O_{(G)}$	129,759	4.64×10^{-21}	135,488	5.87×10^{-22}	138,444	2.02×10^{-22}	138,935	1.69×10^{-22}	138,929	1.70×10^{-22}
$CO_{2(G)}$	693	7.79×10^{-1}	634	7.96×10^{-1}	489	8.38×10^{-1}	436	8.55×10^{-1}	427	8.57×10^{-1}
$C_3O_{2(G)}$	56,210	2.09×10^{-15}	60,782	1.34×10^{-16}	63,117	3.28×10^{-17}	63,499	2.61×10^{-17}	63,493	2.62×10^{-17}
$C_{(S)}$	16,917		28,435		34,490		35,526		35,522	
$x_{T,\text{gas}}^m$		1.20		1.01		9.89×10^{-1}		9.97×10^{-1}		1.00
$\chi_{C,\text{gas}}^m$		3.77×10^{-1}		3.58×10^{-1}		3.51×10^{-1}		3.50×10^{-1}		3.50×10^{-1}
$\chi_{O,\text{gas}}^m$		6.23×10^{-1}		6.42×10^{-1}		6.49×10^{-1}		6.50×10^{-1}		6.50×10^{-1}
$\xi_{x,\text{gas}}^m$		1.96×10^{-1}		6.36×10^{-3}		1.14×10^{-2}		2.86×10^{-3}		4.99×10^{-6}
ξ_C^m		7.76×10^{-2}		2.39×10^{-2}		3.23×10^{-3}		5.39×10^{-5}		1.27×10^{-7}
ξ_O^m		4.18×10^{-2}		1.28×10^{-2}		1.74×10^{-3}		2.90×10^{-5}		6.85×10^{-8}

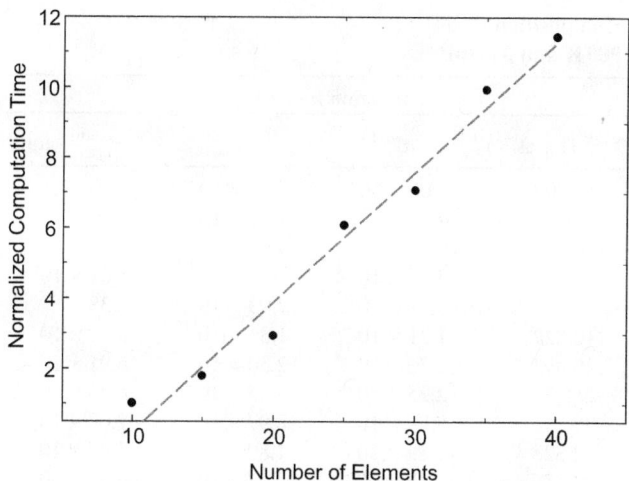

FIGURE 22.2. Performance statistics of thermochemical computations are shown for homogeneous systems (200 ideal species) represented by up to 40 elements. Computation times are averaged by one million unique thermochemical computations by varying composition, temperature, and pressure.

following example applied to the (C, O, H, S) system at 900 K and 30 atm in reference to the example in Chapter 21. The temperature, pressure, and elemental composition reflects coke/coal/oil gasification and several other industrial processes involving corrosion phenomena. The "absolute" Gibbs energies and Gibbs energies of formation computed at

900 K and 30 atm are listed in Table 22.7 for all species of interest in this system.

Leveling is performed using the same methodology described in Section B2 whereby all species are initially treated as pure separate phases. The adjustments made to the elemental Gibbs energies of the system can be interpreted graphically as a hyperplane in four-dimensional Euclidean space, analogous to the horizontal dashed line shown in Figure 22.1 for a binary system. The leveling algorithm determines that $C_{(S)}$, $CO_{2(G)}$, $H_2O_{(G)}$, and $H_2S_{(G)}$ are the dominant species, as shown in Table 22.8 for iteration $m = 4$. Therefore, the system is now assumed to be heterogeneous, consisting of graphite ($C_{(S)}$) coexisting with an ideal gas phase.

The Gibbs energies of the system are further adjusted using Eq. (22.12) to satisfy the mass balances of the system, similar to the previous example for a homogeneous system. Adjustments made to the elemental Gibbs energies are constrained by the relative Gibbs energy of $C_{(S)}$ when it is assumed to precipitate. That is to say that the hyperplane representing adjustments made to the elemental Gibbs energies of the system intersects the point corresponding to the relative Gibbs energy of $C_{(S)}$. The hyperplane pivots about this point and the intercepts correspond to adjustments made to the Gibbs energies of the pure elements.

The mass constraints imposed on the gas (χ'_{gas}) differ from that of the system (χ^*) in regard to a nonzero number of estimated moles of $C_{(S)}$. The number of moles of $C_{(S)}$ is added

TABLE 22.7. Absolute Gibbs Energies and Gibbs Energies of Formation Computed from Thermodynamic Database for (C, O, H, S) System at 900 K and 30 atm[a]

Species	Absolute Gibbs Energy [7]		Gibbs Energy of Formation	
	(J·g mol^{-1})	(J·g at^{-1})	(J·g mol^{-1})	(J·g at^{-1})
$CO_{(G)}$	−274,541	−137,270	−178,659	−89,329
$CO_{2(G)}$	−577,321	−192,440	−395,866	−131,955
$H_{2(G)}$	−103,521	−51,760	0	0
$H_2O_{(G)}$	−399,959	−133,320	−210,866	−70,289
$CH_{4(G)}$	−234,304	−46,861	−16,953	−3,391
$H_2S_{(G)}$	−194,527	−64,842	−58,644	−19,548
$COS_{(G)}$	−339,792	−113,264	−211,548	−70,516
$H_2CO_{(G)}$	−303,908	−75,977	−104,505	−26,126
$C_2H_{4(G)}$	−142,158	−23,693	85,501	14,250
$C_2H_{6(G)}$	−286,433	−35,804	44,747	5,593
$H_2S_{2(G)}$	−220,677	−55,169	−52,433	−13,108
$CH_3SH_{(G)}$	−246,164	−41,027	3,549	592
$S_{2(G)}$	−64,724	−32,362	0	0
$SO_{2(G)}$	−512,325	−170,775	−308,817	−102,939
$SO_{3(G)}$	−624,888	−156,222	−335,808	−83,952
$O_{(G)}$	121,364	121,364	206,937	206,937
$O_{2(G)}$	−171,146	−85,573	0	0
$C_{(S)}$	−10,309	−10,309	0	0

[a]This example is made in reference to Table 21.1 in Chapter 21.

**TABLE 22.8. Equilibrium Compositions Computed for Initial Gas Composition
35 mol CO + 25 mol H_2 + 20 mol CO_2 + 19 mol H_2O + 1 mol H_2S at 900 K and 30 atm**[a]

Species	Iteration $m=4$			Iteration $m=10$			
	$\Delta G_i^{m=4}$ (J·g at^{-1})	$a_i^{m=4}$	$x_{i,gas}^{m=4}$	$\Delta G_i^{m=10}$ (J·g at^{-1})	$a_i^{m=10}$	$x_{i,gas}^{m=10}$	$P_{i,gas}^{m=10}$ (atm)
$CO_{(G)}$	9,637	1.22×10^{-1}	1.22×10^{-1}	11,209	5.00×10^{-2}	5.00×10^{-2}	1.50
$CO_{2(G)}$	0	1.00	1.00	2,097	4.31×10^{-1}	4.31×10^{-1}	1.29×10^{1}
$H_{2\ (G)}$	6,466	2.43×10^{-1}	2.43×10^{-1}	8,980	9.07×10^{-2}	9.07×10^{-2}	2.72
$H_2O_{(G)}$	0	1.00	1.00	2,725	3.35×10^{-1}	3.35×10^{-1}	1.01×10^{1}
$CH_{4(G)}$	1,782	3.77×10^{-1}	3.77×10^{-1}	3,794	7.93×10^{-2}	7.93×10^{-2}	2.38
$H_2S_{(G)}$	0	1.00	1.00	10,822	1.31×10^{-2}	1.31×10^{-2}	3.92×10^{-1}
$COS_{(G)}$	10,699	2.99×10^{-2}	2.99×10^{-2}	20,893	2.30×10^{-4}	2.30×10^{-4}	6.91×10^{-3}
$H_2CO_{(G)}$	26,590	8.90×10^{-6}	8.90×10^{-6}	28,632	2.25×10^{-7}	2.25×10^{-7}	6.76×10^{-6}
$C_2H_{4(G)}$	18,561	5.15×10^{-6}	5.15×10^{-6}	20,236	8.97×10^{-8}	8.97×10^{-8}	2.69×10^{-6}
$C_2H_{6(G)}$	10,443	1.08×10^{-4}	1.08×10^{-4}	12,328	1.89×10^{-6}	1.89×10^{-6}	5.66×10^{-5}
$H_2S_{2(G)}$	12,981	3.42×10^{-3}	3.42×10^{-3}	27,956	3.23×10^{-7}	3.23×10^{-7}	9.70×10^{-6}
$CH_3SH_{(G)}$	12,521	2.71×10^{-4}	2.71×10^{-4}	18,770	2.91×10^{-7}	2.91×10^{-7}	8.72×10^{-6}
$S_{2(G)}$	45,711	4.56×10^{-5}	4.56×10^{-5}	73,147	3.23×10^{-9}	3.23×10^{-9}	9.69×10^{-8}
$SO_{2(G)}$	44,253	4.96×10^{-7}	4.96×10^{-7}	55,494	2.17×10^{-10}	2.17×10^{-10}	6.52×10^{-9}
$SO_{3(G)}$	75,926	3.78×10^{-15}	3.78×10^{-15}	85,140	1.71×10^{-20}	1.71×10^{-20}	5.14×10^{-19}
$O_{(G)}$	404,870	5.94×10^{-20}	5.94×10^{-20}	407,991	2.09×10^{-24}	2.09×10^{-24}	6.27×10^{-23}
$O_{2(G)}$	197,933	1.59×10^{-19}	1.59×10^{-19}	201,066	4.57×10^{-24}	4.57×10^{-24}	1.37×10^{-22}
$C_{(S)}$	0	1.00		0	1.00		
$n_{C_{(S)}}^{m}$			3.00×10^{1}			1.28×10^{1}	
$x_{T,gas}^{m}$			3.78			1.00	
$\chi_{H,gas}^{m}$			5.12×10^{-1}			3.96×10^{-1}	
$\chi_{C,gas}^{m}$			1.30×10^{-1}			1.86×10^{-1}	
$\chi_{O,gas}^{m}$			2.69			4.14×10^{-1}	
$\chi_{S,gas}^{m}$			8.85×10^{-2}			4.40×10^{-3}	
$\xi_{x,gas}^{m}$			2.78			4.97×10^{-6}	
ξ_{H}^{m}						7.26×10^{-6}	
ξ_{C}^{m}						5.79×10^{-6}	
ξ_{O}^{m}						1.49×10^{-6}	
ξ_{S}^{m}						1.78×10^{-6}	

[a]The leveling method ($m = 4$) provides initial estimates for the computation of the final equilibrium state ($m = 10$). The computation of this system is made in reference to Table 21.1 in Chapter 21.

or withdrawn from the system every iteration in a systematic manner. An initial estimate of the number of moles of the pure condensed phase can be computed from the mass balance of the dominant species produced by the leveling algorithm at $m = 4$.

The final equilibrium composition is shown on the extreme right column in Table 22.8, which is referenced to Table 21.1 in Chapter 21. A predominance diagram can be constructed using the methodologies discussed in this chapter by systematically adjusting the partial pressures of two components while maintaining constant temperature, pressure, and activities of the remaining elements. These diagrams are convenient methods to graphically identify the predominant corrosion products over a range of partial pressures of two gaseous species. Figure 22.3 reproduces a predominance diagram from Chapter 21 for the system shown in Tables 22.7 and 22.8. The coordinates corresponding to the partial pressures of $CO_{(G)}$ and $CO_{2(G)}$ in the extreme right column of Table 22.8 determines that $Ni_3S_{2(S)}$ is a stable phase coexisting with $Cr_2O_{3(S)}$.

D. CONCLUSION

The application of computational thermodynamics in determining the equilibrium gas composition and various condensed phases may be utilized for predictive characterization of system speciation and phase behavior. The methodologies described herein are particularly valuable as they do not rely on any prior knowledge of the coexistence of

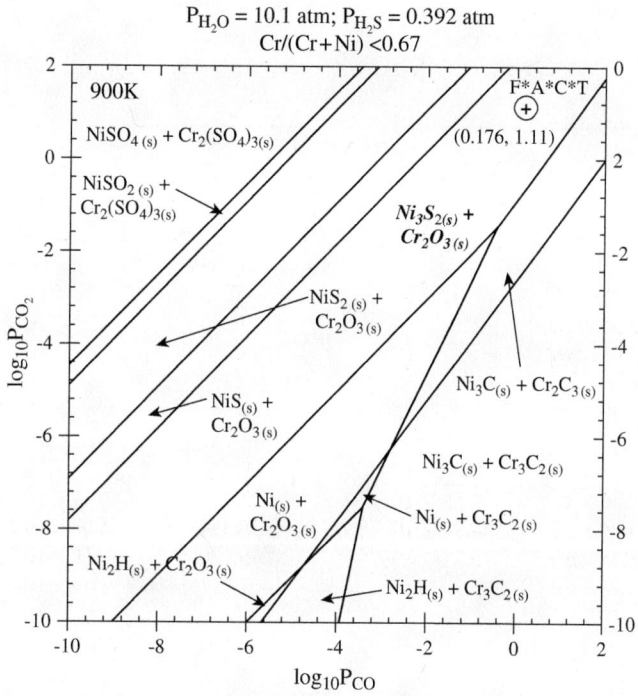

$P_{H_2O} = 10.1$ atm; $P_{H_2S} = 0.392$ atm

$Cr/(Cr+Ni) < 0.67$

FIGURE 22.3. Predominance diagram of the Ni, Cr–H, O, S, C system at 900 K and 30 atm (see Chapter 21). The coordinate $\oplus$ is consistent with computed partial pressures of $CO_{(G)}$ and $CO_{2(G)}$ in Table 22.8.

phases at equilibrium. In this way, it is unnecessary to resort to clumsy empirical correlations of independently generated phase diagram computations with the loss in fidelity from which the empirical approach inherently suffers. This process allows for the capability of dealing in a fundamentally correct manner with a wide range of applications associated with corrosion phenomena.

ACKNOWLEDGMENTS

The authors thank the Natural Science and Engineering Research Council of Canada (NSERC), the University Network of Excellence in Nuclear Engineering (UNENE), and the CANDU Owner's Group (COG) for financial support. They also thank T.M. Besmann of the Surface Processing and Mechanics Group of the Oak Ridge National Laboratory for helpful recommendations.

REFERENCES

1. W. B. White, S. M. Johnson, and G. B. Dantzig, "Chemical Equilibrium in Complex Mixtures," J. Chem. Phys., **28**(5), 751–755 (1958).

2. F. J. Zeleznik and S. Gordon, "An Analytical Investigation of Three General Methods of Calculating Chemical Equilibrium Compositions," Technical Note D-473 NASA, 1960.

3. S. H. Storey and F. van Zeggeren, "Computation of Chemical Equilibrium Compositions," Can. J. Chem. Eng., **42**, 54–55 (1964).

4. F. van Zeggeren and S. H. Storey, The Computation of Chemical Equilibrium, Cambridge University Press, London, 1970.

5. G. Eriksson, "An Algorithm for the Computation of Aqueous Multi-Component, Multiphase Equilibria," Anal. Chim. Acta, **112**, 375–383 (1979).

6. D. R. Olander, General Thermodynamics, CRC Press, Boca Raton, FL, 2007.

7. D. R. Stull and H. Prophet, "JANAF Thermochemical Tables," U.S. Department of Commerce, Washington, DC, 1985.

8. J. W. Gibbs, "On the Equilibrium of Heterogeneous Substances," New York in The Collected Works of J. Willard Gibbs, Vol. 1, Longmans, Green and Co., 1931.

9. G. Eriksson and W. T. Thompson, "A Procedure to Estimate Equilibrium Concentrations in Multicomponent Systems and Related Applications," CALPHAD, **13**(4), 389–400 (1989).

10. M. H. Piro, Ph.D. Thesis, Department of Chemistry and Chemical Engineering, Royal Military College of Canada, to be published.

23

ATMOSPHERIC CORROSION

P. R. ROBERGE

Department of Chemistry and Chemical Engineering, Royal Military College of Canada, Kingston, Ontario, Canada

A. INTRODUCTION

Atmospheric corrosion generally refers to corrosion processes occurring in outdoor or indoor environments. Such a broad definition is associated with very high visibility, for example, rusty bridges, flag poles, buildings (Fig. 23.1) and outdoor monuments (Fig. 23.2) [1]. Many of the parameters used to describe the phenomena associated with atmospheric corrosion are inherited from climatology and describe variables such as rainfalls, wind speeds, pollution level, humidity, and aerosol transport.

Atmospheric corrosion thus refers specifically to the situation of exposure of a component or a structure and, therefore, does not relate to any particular form or type of corrosion that may be developing in such situations. Consider, for example, what has been called brass season cracking. At the turn of the twentieth century, much trouble was caused by the cracking of brass cartridge cases stored in British army depots in India, where the air was hot and contained ammonia from animal waste (horses).

The metallurgical cracking susceptibility of brass in such circumstances was traced down to internal stresses left after fabrication, which were largely tensional in the surface layers of the parts affected, although balanced by compressive stresses in the central zone. The chemical action of ammonia at the grain boundaries allowed the grains, initially in tension, to contract, leaving definite fissures between one and another. This case was studied in many countries, but the extensive work of Moore and Beckinsale in 1920–1923 not only established the conditions favorable to intergranular cracking but also indicated the means of avoiding the trouble by annealing [2].

A more recent example concerned the case of the swimming pool stress corrosion cracking (SCC) of stainless steel supports. In 1985, 12 people were killed in Uster, Switzerland, when the concrete roof of a swimming pool collapsed. The roof was supported by stainless steel rods in tension, which failed due to SCC. A few years later, the suspended ceiling of a municipal swimming pool in the

Uhlig's Corrosion Handbook, Third Edition, Edited by R. Winston Revie
Copyright © 2011 John Wiley & Sons, Inc.

FIGURE 23.1. Noticeable rust on Willow stem plant in downtown Philadelphia. (Courtesy of Kingston Technical Software.)

Netherlands collapsed due to a similar cause. There have been other incidents associated with the use of stainless steel in safety-critical load-bearing applications in the environment created by modern indoor swimming pools and leisure centers [3]. While chloride-induced SCC damage is recognized as a common failure mechanism in stainless steels, a somewhat surprising element of these failures is that they occurred at room temperature. As a general rule of thumb, it had often been assumed that chloride-induced SCC in these alloys was not a practical concern at temperatures below 60°C.

Another incident that has received a high level of attention, particularly by the aircraft industry, was the Aloha incident in 1988, when a Boeing 737 lost a major portion of the upper fuselage in full flight at 7300 m (Fig. 23.3) [4]. This case baffled investigators until it was recognized that the deformation due to the corrosion of aluminum in lap joints of commercial airlines was accompanied by a bulging ("pillowing") between rivets, due to the increased volume of the corrosion products over the original material. The prevalent corrosion product in corroded fuselage joints was found to be hydrated alumina [Al(OH)$_3$], an aluminum corrosion product very stable at low temperatures with a particularly high volume expansion relative to aluminum [5]. The buildup of voluminous corrosion products when hydrated alumina is formed can produce an undesirable increase in stress levels near critical fastener holes (Fig. 23.4) and

FIGURE 23.2. Galvanic corrosion induced on military monument exposed to environments in front of Halifax Citadel. (Courtesy of Kingston Technical Software.)

FIGURE 23.3. Boeing 737 operated by Aloha airlines after it lost major portion of upper fuselage in 1988.

subsequent fracture due to the high tensile stresses resulting from the pillowing.

A similar corrosion situation, described by civil engineers as "pack rust", occurs when wetting and drying cycles produce a manyfold expansion of iron corrosion products within the crevices formed between riveted steel plates (Fig. 23.5). The expanded rust products may eventually build up enough pressure to break rivets used in steel bridge construction. Pack rust can also cause freezing of expansion joints and contribute to an increase in fatigue stress of primary structural components as was noted in the investigative report requested by the Minnesota legislature following the I-35W bridge collapse on August 1, 2007 [6].

The expansion of iron corrosion products is also responsible for spalling of concrete cover on many reinforced concrete structures. An equivalent volume expansion ratio of 3.0–3.2 has been measured in such situations due to the formation of corrosion products on steel bars embedded in concrete [7]. The pressure thus created may be high enough to cause cracks and the degradation of the reinforced concrete material (Fig. 23.6).

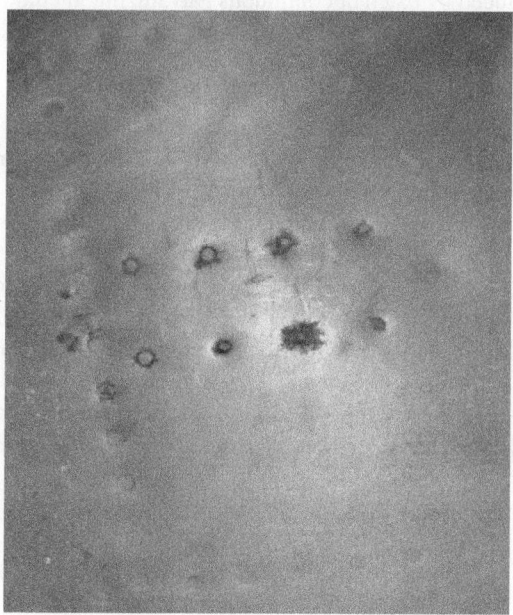

FIGURE 23.4. Advanced stage on belly of Boeing 737, where the corrosion products have expanded to the point where a number of rivets have actually popped their heads. (Courtesy Mike Dahlager Pacific Corrosion Control Corp.)

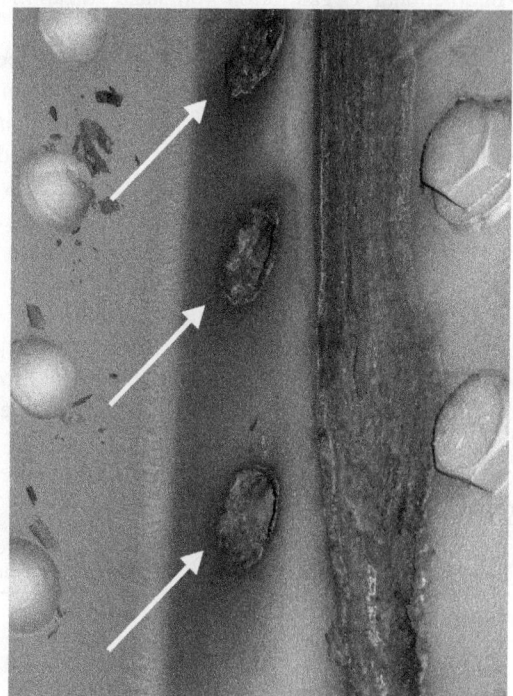

FIGURE 23.5. Effect of "pack rust" that has developed on an important steel bridge under repair. (Courtesy Wayne Senick Termarust Technologies, www.termarust.com.)

FIGURE 23.6. Cracking of concrete cover in marine environment. (Courtesy of Kingston Technical Software.)

B. OUTDOOR ATMOSPHERES

The corrosivity of an atmospheric environment is an important consideration for designing and maintaining systems that are to be exposed to such environments. A widely accepted scheme for the classification of outdoor environments has been developed by a working group (WG 4) of the International Organization for standardization (ISO) corrosion technical committee (TC 156) [8].

In the program nicknamed ISO CORRAG specially prepared coupons of steel, copper, zinc, and aluminum were exposed for 1, 2, 4, and 8 years at 51 sites located in 14 nations in order to generate the necessary data for predicting atmospheric corrosivity from commonly available weather data [9]. Triplicate specimens were used for each exposure. The program was initiated in 1986 and closed in 1998. After a planned exposure, each specimen was sent to the laboratory that had done the initial weighing for cleaning and evaluation. Based on these data, a simple classification scheme of five corrosivity classes was established for each metal (Table 23.1). The environmental and weather data gathered in this program are based on SO_2 and Cl^- deposition rates combined with time of wetness (TOW) measurements at each site.

These five corrosivity categories can be roughly translated into five outdoor situations listed here in decreasing order of corrosivity, that is, industrial, tropical marine, temperate marine, urban, and rural. These outdoor situations have also been used to estimate the zinc coating thickness required to achieve a certain service life with galvanizing (Fig. 23.7) [10]. Service life in Figure 23.7 is defined as the time to 5% rusting of the steel surface.

This graphical method is applicable to zinc-coated steel produced by batch or continuous galvanizing, including hot-dip, electrogalvanized, and thermal sprayed coatings. However, it does not apply to coatings containing more than 1% alloying elements. The method also assumes that the galvanized product is free of significant defects that could accelerate corrosion. Additionally, the service life prediction does not consider issues of water entrapment that can create severe crevice chemistry and degrade a zinc coating very aggressively (Fig. 23.8).

In fact, this last consideration applies generally to most atmospheric corrosion categories since these are usually based on general descriptors that do not consider local variations that can drastically alter the corrosion reactions at an exposed surface. Even the most benign environments may thus become locally extremely corrosive in certain

TABLE 23.1. ISO 9223 Corrosion Rates after One Year of Exposure Predicted for Different Corrosivity Classes

Corrosion Category	Steel (g/m² year)	Copper (g/m² year)	Aluminum (g/m² year)	Zinc (g/m² year)
C1	≤10	=0.9	negligible	=0.7
C2	11–200	0.9–5	=0.6	0.7–5
C3	201–400	5–12	0.6–2	5–15
C4	401–650	12–25	2–5	15–30
C5	651–1500	25–50	5–10	30–60

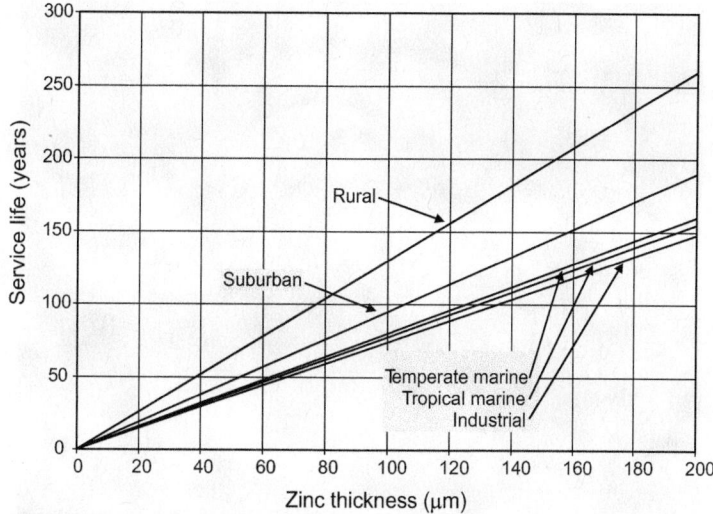

FIGURE 23.7. Service life of hot-dip galvanized coatings as function of zinc thickness and specific environments.

circumstances or conditions, as shown in Figure 23.9 where the statue head may occasionally serve as a perching platform for passing birds, a common situation in many parks where such monuments are erected.

Industrial atmospheres are typically rich in various polluting gases composed mainly of sulfur compounds such as sulfur dioxide (SO_2), a precursor to acid rain, and nitrogen oxides (NO_x), the backbone of smog in modern cities. Sulfur dioxide from burning coal or other fossil fuels is picked up by moisture on dust particles as sulfurous acid. This is oxidized by some catalytic process on the dust particles to sulfuric acid, which settles in microscopic droplets and fall as acid

rain on exposed surfaces. Other corrosive pollutants may also be present such as various forms of chloride, which may be much more corrosive than the acid sulfates. The reactivity of acid chlorides with most metals is more pronounced than the reactivity of other pollutants such as phosphates and nitrates.

Marine atmospheres are laden with fine particles of sea mist carried by the wind to settle on exposed surfaces as salt crystals. The quantity of salt deposited may vary greatly with wind velocity, and it may, in extreme weather conditions, even form a very corrosive salt crust. The quantity of salt contamination decreases with distance from the ocean and is greatly affected by wind currents. In the Tropics, in addition

FIGURE 23.8. Galvanized bolting assembly after 10 years of exposure to deicing salt environment. (Courtesy of Kingston Technical Software.)

FIGURE 23.9. Jesus Breaking Bread statue serving as perching platform. (Courtesy of Kingston Technical Software.)

to the high average temperature, the daily cycle includes a high relative humidity, intense sunlight, and long periods of condensation during the night. In sheltered areas, the wetness from condensation may persist long after sunrise. Such conditions may produce a highly corrosive environment.

The corrosivity of urban and suburban environments is in great part related to the air quality in these environments, which can vary greatly from city to city and from country to country. For more than a century, severe air pollution incidents in cities such as London have shown that breathing dirty air can be dangerous and, at times, deadly. However, it was only in the late 1940s and early 1950s when air pollution disasters on two continents raised an alarm. Both the 1948 "killer fog" in the small town of Denora, Pennsylvania, that killed 50, and the particularly virulent London "fog" of 1952 in which some 4000 died, were associated with widespread use of dirty fuels and were catalysts for government efforts to tackle urban air pollution.

Since then, many nations have adopted ambient air quality standards to safeguard the public against the most common and damaging pollutants. These include sulfur dioxide, suspended particulate matter, ground-level ozone, nitrogen dioxide, carbon monoxide, and lead, all of which are tied directly or indirectly to the combustion of fossil fuels.

Although substantial investments in pollution control in some industrialized countries have lowered the levels of these pollutants in many cities, poor air quality is still a major concern throughout the industrialized world.

Rural atmospheres are, typically, considered to be the most benign from a corrosion standpoint since these atmospheres are free of industrial pollutants, which is usually the case unless one is close to a farm operation where by-products made of various waste materials or the handling of concentrated chemicals such as fertilizers can be extremely corrosive to metals.

C. INDOOR ATMOSPHERES

Normal indoor atmospheres are generally considered to be quite mild when ambient humidity and other corrosive components are under control. However, some combinations of conditions may actually cause relatively severe corrosion problems. Even in the absence of any other corrosive agent, the constant condensation on a cold metallic surface may cause an environment similar to constant immersion for which a component may not have been chosen or prepared. Such systems are commonly encountered in confined areas

FIGURE 23.10. Rusted paint can after a few years in a normally humid basement. (Courtesy of Kingston Technical Software.)

close to ground level or, worse, below ground where high humidity may prevail. Home basements are a good example where such conditions prevail most of the time (Fig. 23.10).

Museums around the world contain all sorts of priceless cultural and historical artifacts that have been collected over centuries. Within such protective environments, surprisingly corrosive microclimates can exist. Corrosive pollutants such as ammonia, formaldehyde, acetaldehyde, formic acid, and acetic acid are often found in higher concentrations indoors than outdoors, inducing an array of chemical, electrochemical, and physical corrosion processes [11].

There can be an abundance of corrosive agents within storage cabinets, display cases, and galleries. The materials making up display cases, paint on the walls, or the carpets on the floors may all be the sources of an artifact's demise. Indeed, even the materials that were used in the manufacture, cleaning, or repair of an object can contribute to corrosion. Perhaps slower in pace and smaller in scale but destructive nonetheless, these corrosion processes mimic the more dramatic and extensive processes outdoors.

Another area of growing concern is related to the ubiquitous presence of electronics and microcomputers. The trend toward miniaturization of technology has led to the development of small personal electronic devices that are now present everywhere, such as pagers, cellular phones, and palm-sized personal organizers and computers [12]. According to the U.S. Census Bureau, the American electronic

manufacturing industries alone shipped \$336 billion worth of electronic components in 1997. As electronic devices become increasingly ubiquitous and robust, the concern over the operating environment seems to lessen, particularly in the personal computer (PC) market.

Materials used in electronic components range from aluminum-based alloys (integrated circuits conductors) to copper contacts electroplated with nickel or gold for improved electrical conductance. Submicron dimensions of electronic circuits, high-voltage gradients, and an extremely high sensitivity to corrosion or corrosion products present a unique set of corrosion-related issues. A major departure from most corrosion situations is the incredibly small volume of the material that can be damaged and lead to a fault. The microchip in an automobile, for example, is not directly subjected to the same environmental hazards as the car body. However, the tolerance for corrosion loss in electronic devices is many orders of magnitude less, that is, on the order of picograms (10^{-12} g). Minimum line width in the state-of-the-art printed circuit boards (PCBs) in 1997 was less than $100\,\mu m$. On hybrid integrated circuits (HICs), line spacings may be less than $5\,\mu m$ [13].

Printed circuit boards can suffer from a variety of problems if the surface is contaminated with electrically conducting materials. When combined with moisture, contamination results in a lowering of the resistance between tracks and pads, which may lead to corrosion of metals. It may also result in the formation of metal filaments or whiskers, which grow between pads, or tracks on rigid or flexible circuits and between oppositely charged metal terminations of components or between the pins of connectors. The essential conditions required for this are a combination of ionic contamination, moisture, and an applied voltage. Not only will the materials degrade, but numerous failure modes may be triggered and become apparent long before any noticeable or detectable changes in the electronic materials themselves. Additionally, these failure modes are quite often intermittent and will only present themselves while exposed to harsh conditions. This makes isolating, identifying, and correcting problems very problematic [14].

Equipment used under dry conditions should not suffer from these problems unless there are large temperature fluctuations that result in condensation occurring on the surface of the circuitry or if the contaminants are hygroscopic and adsorb enough moisture to provide a liquid layer on the surface.

New semiconductor technologies require much more stringent control of the manufacturing environment. From an air quality standpoint, the control of chemical contamination has become as important as the control of particulate contamination for many processes. Levels of airborne molecular contaminants (AMC) in ambient air are high enough to be problematic if introduced into the facility. However, the types and numbers of chemical species used in the

TABLE 23.2. Concentration of Selected Gaseous Air Constituents (ppb) in United States [14]

Gas	Outdoor (ppb)	Indoor (ppb)
O_3	4–42	3–30
H_2O_2	10–30	5
SO_2	1–65	0.3–14
H_2S	0.7–24	0.1–0.7
NO_2	9–78	1–29
HNO_3	1–10	3
NH_3	7–16	13–260
HCl	0.18–3	0.05–0.18
Cl_2	<0.005–0.08	0.001–0.005
HCHO	4–15	10
HCOOH	4–20	20

manufacturing of semiconductor devices also mean that there is a greater chance of chemical contamination from internal sources adversely affecting these sophisticated manufacturing processes [15].

Reactive gases usually have a relatively short life expectancy in the atmosphere and their active presence will last for a few minutes to several days, depending on conditions and type of gas. Many gases combine to form other airborne compounds as well as reacting with or being absorbed by surfaces. Table 23.2 shows a generalized arithmetic mean level of the amount of gaseous pollutants in the air for the United States. Other parts of the world, notably China and Asia, East Central Europe, and large population centers in Latin America have up to four times higher levels of pollutants.

Reactivity monitoring as described in International Society of Automation (ISA) Standard S71.04 has been used for many years to provide accurate environmental assessments to process and control industries. Many of the same types of contaminants that have to be controlled in these industries are those of concern to the semiconductor industry. Because of this, reactivity monitoring has been accepted as a viable environmental monitoring method by a number of semiconductor manufacturers. An air quality classification scheme based on reactivity monitoring has thus been established with wide acceptance throughout the semiconductor industry.

This classification scheme covers airborne contaminants that affect electrical or electronic equipment according to the type of contaminant as shown in Table 23.3. The synergistic effects of all contaminants present are seen in the total thickness of corrosion on the copper reactivity coupons. Then the corrosion damage measured on the reactivity coupons are related to reliability of the electronic equipment in that space.

D. ATMOSPHERIC CORROSIVITY FACTORS AND THEIR MEASUREMENT

As mentioned earlier many of the factors driving atmospheric corrosion are described and monitored as typical meteorological features. The quantity and composition of pollutants in the atmosphere and the variation of these with time are monitored on a regular basis by many governmental organizations for air quality assurance. Temperature, relative humidity (RH), wind direction and velocity, solar radiation, and amount of rainfall are also routinely recorded. Not so easily determined are dwelling time of wetness (TOW) and the surface contamination by corrosive agents such as sulfur dioxide and chlorides. However, methods for these determinations have been developed and are in use at various test stations. By monitoring these factors and relating them to corrosion rates, a better understanding of atmospheric corrosion can be obtained.

Water in the form of rain is not always corrosive. It may even have a beneficial effect by washing away atmospheric pollutants that have settled on exposed surfaces. This effect is particularly noticeable in marine environments. On the other hand levels of humidity causing dew and condensation are always undesirable from a corrosion standpoint if they are not accompanied by frequent rain washing to dilute or eliminate surface contamination.

Temperature plays an important role in atmospheric corrosion in two ways. First, the corrosion activity may double for each 10° increase in temperature. Second, a little

TABLE 23.3. Classification of Reactive Environments [15]

Severity Level	Mild	Moderate	Harsh	Severe
Copper Reactivity Level (in nm)	<30	<100	<200	≥200
Contaminant (gas)	Concentration (ppb)			
H_2S	<3	<10	<50	≥50
SO_2, SO_3	<10	<100	<300	≥300
Cl_2	<1	<2	<10	≥10
NO_x	<50	<125	<1250	≥1250
HF	<1	<2	<10	≥10
NH_3	<500	<10,000	<25,000	≥25,000
O_3	<2	<25	<100	≥100

recognized effect is the temperature lag between metallic objects and ambient air that may follow sudden changes in ambient air temperatures. The period of wetness that may be much longer than the time the ambient air is at or below the dew point is a complex variable that depends on the section thickness and heat capacity of the metal structure and also on air currents, RH, and direct radiation from the sun. This lag may have some dramatic effects in poorly ventilated areas where moisture is trapped during the cold part of a temperature cycle.

As the ambient temperature drops during the evening, for example, metallic surfaces tend to remain warmer than the humid air surrounding them and do not begin to collect condensation until sometime after the dew point has been reached. As the temperature begins to rise in the surrounding air, the lagging temperature of the metal structures will tend to make them act as condensers, maintaining a film of moisture on their surfaces.

Since the dew point of an atmosphere indicates the equilibrium condition of condensation and evaporation from a surface, it is advisable to maintain the temperature some 10–15°C above the dew point to ensure that no corrosion will occur by condensation on a surface that would be colder than the ambient environment.

D1. Relative Humidity, Dew Point, and Time of Wetness

The most important factor in atmospheric corrosion is by far moisture, either in the form of rain, dew, condensation, or high RH. In the absence of moisture, most contaminants would have little or no corrosive effect. A fundamental requirement for atmospheric corrosion processes is the presence of a thin-film electrolyte that may form on metallic surfaces when exposed to a critical level of humidity. While this film is almost invisible, the corrosive contaminants it contains are known to reach relatively high concentrations, especially under conditions of alternate wetting and drying.

In the presence of thin-film electrolytes, atmospheric corrosion proceeds by balanced anodic and cathodic reactions described, respectively, in Eqs. (23.1) and (23.2). The anodic oxidation reaction involves the corrosion attack of the metal, while the cathodic reaction is naturally the oxygen reduction reaction (Fig. 23.11).

$$\text{Anode reaction} \quad 2Fe \rightarrow 2Fe^{2+} + 4e^- \quad (23.1)$$

$$\text{Cathode reaction} \quad O_2 + 2H_2O + 4e^- \rightarrow 4OH^- \quad (23.2)$$

Relative humidity is defined as the ratio of the quantity of water vapor present in the atmosphere to the saturation quantity at a given temperature, and it is expressed as percent. In absence of any particular surface effects, the dew point corresponds to the temperature at which condensation

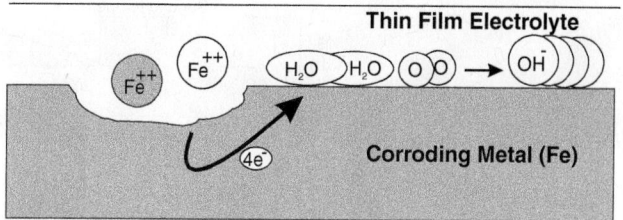

FIGURE 23.11. Schematic description of atmospheric corrosion of iron.

occurs. A high relative humidity would therefore be associated with a dew point close to ambient air temperature. If the relative humidity was 100%, for example, the dew point would be equal to ambient temperature. Given a constant dew point, an increase in temperature will lead to a decrease in relative humidity. Equation (23.3) provides a convenient way to calculate the dew point as a function of temperature to within ±0.4°C [16]:

$$t_d = \frac{B\left(\ln RH + \dfrac{At}{B+t}\right)}{A - \ln RH - \dfrac{At}{B+t}} \quad (23.3)$$

where

$A = 17.625$

$B = 243.04°C$

RH = relative humidity as a fraction (not percent)

t = surface temperature (°C)

t_d = dew point temperature (°C)

Equation (23.3) is valid for $0°C < t < 100°C$, $0.01 < RH < 1.0$, and $0°C < T_d < 50°C$. Figure 23.12 illustrates the relationship between the dew point temperature and relative humidity for selected surface temperatures.

Precise RH measurements can be obtained by comparing the results from wet-bulb and dry-bulb thermometers. The comparison between the two readings provides an indication of water vapor in the air. The readings can be plotted on a chart known as the psychometric chart from where the properties of air vapor mixture such as relative humidity, absolute humidity, and dew point can be directly determined.

The critical humidity level below which there should be no corrosion additionally depends on the nature of the corroding material, the tendency of corrosion products and surface deposits to absorb moisture, and the presence of atmospheric pollutants [17]. It has been shown, for example, that this critical humidity level is 60% for steel when the environment is free of pollutants. In this context, marine environments have typically high RH and are laden with sea salts.

Studies have shown that the thickness of the adsorbed water layer on zinc surface increases with %RH and that corrosion rates increase with the thickness of the adsorbed

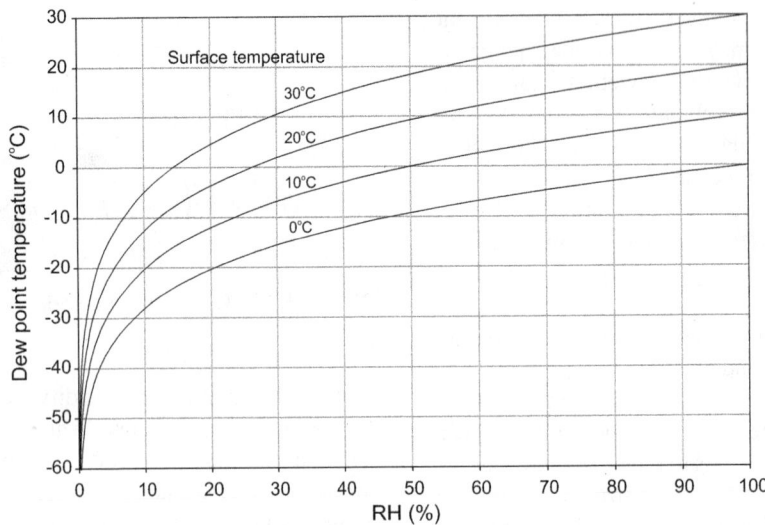

FIGURE 23.12. Relationship between dew point temperature and relative humidity for selected surface temperatures.

layer. There also seems to be a finite thickness to the water layer that, when exceeded, becomes limiting due to oxygen diffusion across the adsorbed water layer [18]. However, when metallic surfaces become contaminated with hygroscopic salts, their surface can be wetted at a lower RH. The presence of magnesium chloride ($MgCl_2$) on a metallic surface can make it wet at 34% RH while sodium chloride (NaCl) on the same surface requires 77% RH to create the same effect [19].

A commonly used compromise between these various observations is the TOW concept, which many corrosion scientists use and report in their studies. TOW is a parameter based on the length of time RH values are greater than 80% at a temperature greater than 0°C. It is commonly reported in hours or days per year or as an annual percentage of time a surface is above this threshold to express the persistence of the wetting water film and its potential consequences.

A method of measuring the TOW has been developed by Sereda and correlated with the corrosion rates encountered in the atmosphere [20]. The moisture-sensing elements in this sensor are manufactured by plating and selective etching of thin films of appropriate anode (copper) and cathode (gold) materials in an interlaced pattern on a thin nonconductive substrate (Fig. 23.13). When moisture condenses on the sensor, it activates the cell, producing a small voltage (0–100 mV) across a 10^7-Ω resistor.

D2. Aerosol Particles

The behavior of aerosol particles in outdoor atmospheres is explained by empirical equations that describe their formation, movement, and capture. These particles are present throughout the planetary boundary layer, and their concentrations depend on a multitude of factors including location,

time of day or year, atmospheric conditions, presence of local sources, altitude, and wind velocity.

Aerosols can either be produced by ejection into the atmosphere or by physical and chemical processes within the atmosphere (called primary and secondary aerosol production, respectively). Examples of primary aerosols are sea spray and windblown dust. Secondary aerosols are produced by atmospheric gases reacting and condensing or by cooling vapor condensation. Once an aerosol is suspended in the atmosphere, it can be altered, removed, or destroyed, and average lifetimes are of the order of a few days to a week, depending on the aerosol size and location.

The highest concentrations are usually found in urban areas, reaching up to 10^8 and 10^9 particles per cubic centimeter, with particle size ranging from a few nanometers to around 100 μm. Size is normally used to classify aerosol because it is the most readily measured property and

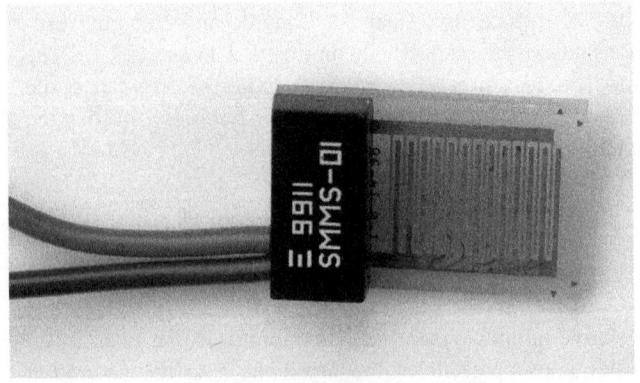

FIGURE 23.13. Interlocking combs of gold and copper electrodes in a Sereda humidity sensor. (Courtesy of Kingston Technical Software.)

other properties can be inferred from size information [21]. The highest mass fraction of particles in an aerosol is characterized by particles having a diameter in the range of 8–80 μm [22]. Some studies have indicated that there is a strong correlation between wind speed and the deposition and capture of aerosols. In such a study of saline winds in Spain, a very good correlation was found between chloride deposition rates and wind speeds above a threshold of 3 m/s or 11 km/h [23].

Aerosol particles have a finite mass and are subject to the influence of gravity, wind resistance, droplet dry-out, and possibilities of impingement on a solid surface. Studies of the migration of aerosols inland of a seacoast have shown that the majority of the aerosol particles are typically deposited close to the shoreline (typically 400–600 m) and consist of large particles (>10 μm diameter), which have a short residence time and are controlled primarily by gravitational forces [22, 23].

Airborne salinity refers to the content of gaseous and suspended salt in the atmosphere. It is measured by the concentration in the air in units of micrograms/cubic meter. Since it is the salt that is deposited on the metal surface that affects the corrosion, it is usually reported in terms of deposition rate in units of milligrams/meter squared/day. Chloride levels can also be measured in terms of the concentration of the dissolved salt in rainwater.

A number of methods have been employed for determining the contamination of the atmosphere by aerosol-transported chlorides, for example, sea salt and road deicing salts. The "wet candle method," for example, is relatively simple [24] but has the disadvantage that it also collects particles of dry salt that might not deposit otherwise. This technique uses a wet wick of a known diameter and surface area to measure aerosol deposition (Fig. 23.14). In this device, the wick is maintained wet, using a reservoir of water or 40% glycol–water solution. Particles of salt or spray are trapped by the wet wick and retained. At intervals, a quantitative determination of the chloride collected by the wick is made and a new wick is exposed.

In reality, the wet candle method gives an indication of the salinity of the atmosphere rather than the contamination of exposed metal surfaces. The technique is considered to measure the total amount of chloride arriving to a vertical surface, and its results may not be truly significant for corrosivity estimates.

Other techniques for measuring surface chlorides in the field are commercially available. The Bresle method, for example, is particularly suited to test for soluble surface contaminants on blasted surfaces. A Bresle patch is attached to the surface where the presence of chlorides is a concern, and distilled water is injected with a syringe in the patch and then extracted and measured with a conductivity meter. An alternate product called CHLOR*TEST, shown in Figure 23.15, uses a proprietary extract solution to retrieve

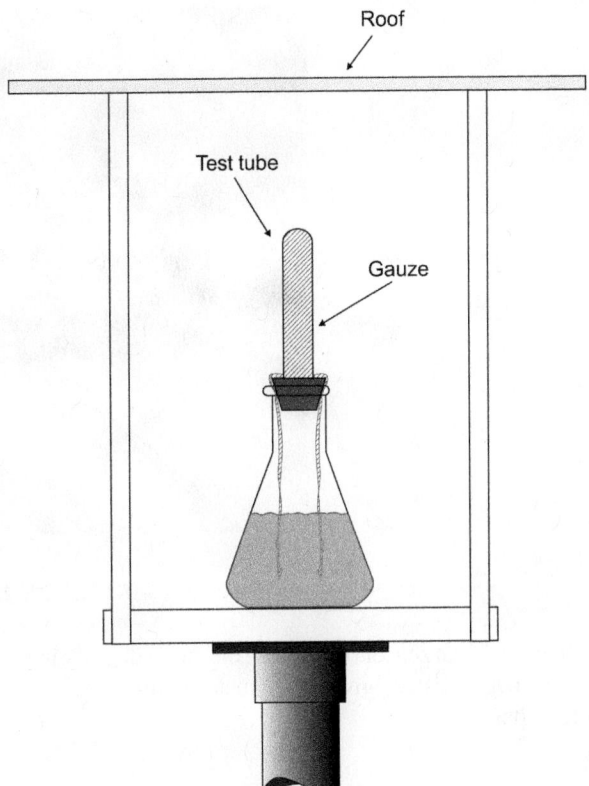

FIGURE 23.14. Schematic of wet candle chloride apparatus.

surface chloride contamination. In the field, this special extract is said to enhance retrieval rates, thereby increasing accuracy.

Another field monitoring technique incorporates a flat sensor to measure the conductivity of a solution from a single drop of sample (Fig. 23.16). Users can either place a sample

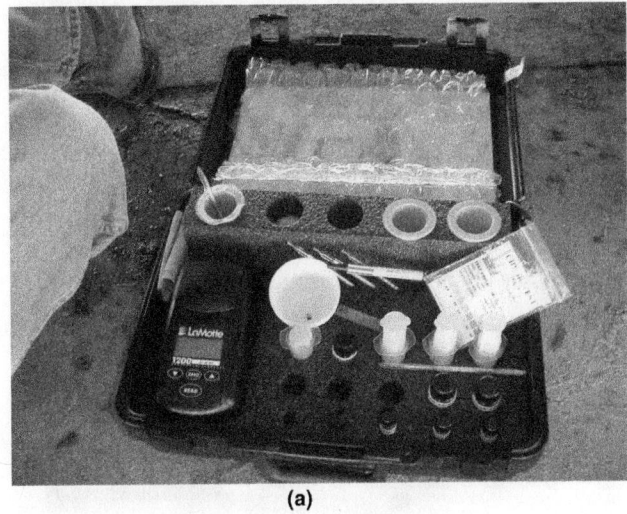

(a)

FIGURE 23.15. (a) Soluble salts detector kit and (b) sample being taken on a surface in preparation. (Courtesy of Termarust Technologies.)

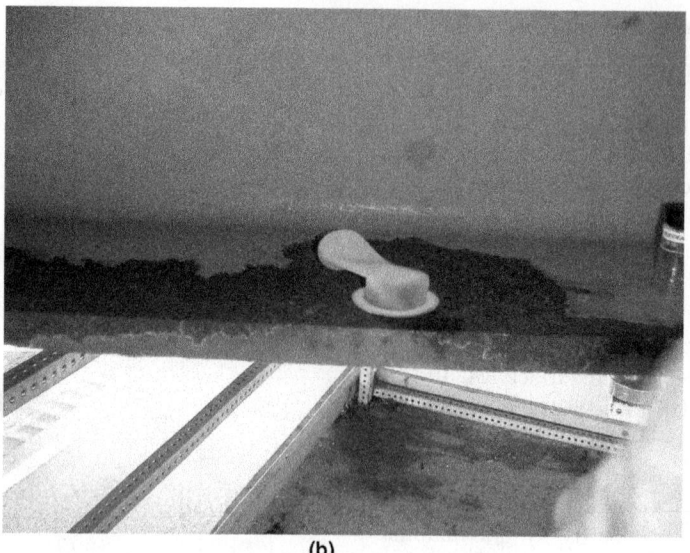

(b)

FIGURE 23.15. (*Continued*)

on the meter's flat sensor or immerse the meter directly in the solution being tested, giving the meter a broad range of applications.

D3. Pollutants

Sulfur dioxide (SO$_2$), a gaseous product of the combustion of fuels containing sulfur such as coal, diesel, gasoline, and natural gas, has been identified as one of the most important air pollutants that contribute to the corrosion of metals.

Less recognized as corrosion promoters are the nitrogen oxides (NO$_x$), which are also products of combustion. A major source of NO$_x$ in urban areas is the exhaust fumes from vehicles. Sulfur dioxide, NO$_x$, and airborne aerosol particles can react with moisture and ultraviolet (UV) light to form new chemicals that can be transported as aerosols. A good example of this is the summertime haze over many large modern cities. Up to 50% of this haze or "smog" is a combination of sulfuric and nitric acids. Sulfur dioxide, NO$_x$, and other urban pollutants are routinely monitored with sophisticated equipment often housed in mobile units (Fig. 23.17) and typically reported in terms of their concentration in air in units of micrograms/cubic meter.

FIGURE 23.16. Portable conductivity meter. (Courtesy of Kingston Technical Software.)

The pollution levels may also be measured by analyzing rainwater for inorganic salts or by measuring its pH. However, this is only indirectly related to the effect of gaseous pollutants on corrosion since only the actual amount of salt deposited on metal surfaces is important.

There are two widely used methods for determining the deposition of SO$_2$ concentration in the atmosphere of interest. Both employ the affinity of lead oxide to react with gaseous SO$_2$ to form lead sulfate. The most common method used in corrosion work is the sulfation plate method. It consists of exposing small disks of lead oxide facing the ground under a small shelter to prevent the reactive paste from being removed by the elements [25]. The disk surface is thus exposed only to gaseous SO$_2$ and protected from particulates. The American Society for Testing and Materials (ASTM) procedure suggests a 30-day exposure, followed by a standard sulfate analysis [26].

The other method sometimes used is the peroxide candle, similar in its function to the chloride candle. In this method, a lead peroxide paste is applied to a paper thimble in the laboratory and allowed to dry thoroughly before exposure. The thimble is then exposed in an instrument shelter to the test yard environment. In both methods, the SO$_2$ deposited results are appropriately reported in terms of deposition rate on the surface in units of milligrams/square meter/day.

D4. Atmospheric Corrosivity

The simplest form of direct atmospheric corrosion measurement is by coupon exposures. Subsequent to their exposure, coupons can be subjected to weight loss measurements, pit density, and depth measurements as well as other types of examination.

FIGURE 23.17. Deployable air quality monitoring unit on Montreal Island. (Courtesy of Kingston Technical Software.)

Battelle Institute has used such coupon exposure as a passive method to monitor the atmospheric corrosivity of Air Force and other sites [27]. The database describing the relative corrosive severity levels of different locations and actual corrosion rates of a variety of metals has now grown to more than 100 sites worldwide. The metals included in that study are three aluminum alloys (A92024, A96061, and A97075), copper, silver, and steel. Figure 23.18(a) shows a closeup view of the coupons before exposure. Once exposed to the environment for a given period of time, the corroded metal strips [Fig. 23.18(b)] are sent back to the laboratory for mass loss measurements following standard methods [28] and further analysis.

Another type of coupon has been developed to provide rapid material/corrosivity evaluations [29]. The helical coil adopted in the ISO 9226 methodology is a high surface area/weight ratio coupon that gives a higher sensitivity than panel coupons of the same material. The use of bimetallic specimens in which a helical A91100 aluminum wire is wrapped around a coarsely threaded bolt enhances further the sensitivity of the device has been described as the Classify Industrial and Marine Atmospheres (CLIMAT) coupon [29, 30].

The mass loss of the aluminum wire of a CLIMAT coupon after 90 days of exposure is considered to be a relative measure of atmospheric corrosivity. However, the results vary greatly between the various combinations of materials suggested in the ASTM standard [30]. The aluminum wire on copper bolts has been found by many to be the most sensitive of the three proposed arrangements in the ASTM standard. The use of triplicate coupons on a single holder additionally provides an indication of the reproducibility of the measure-

ments, and the use of vertical rods can reveal directional information on the corrosive agents as will be illustrated in the following examples.

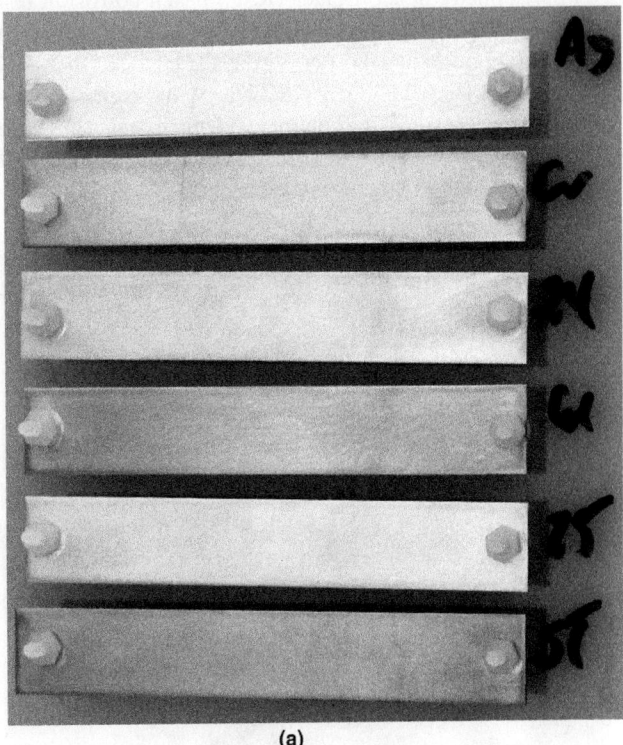

(a)

FIGURE 23.18. Metal coupons (a) before exposure to the environment and (b) after a 3-month exposure in a rural environment. (Courtesy of Battelle.)

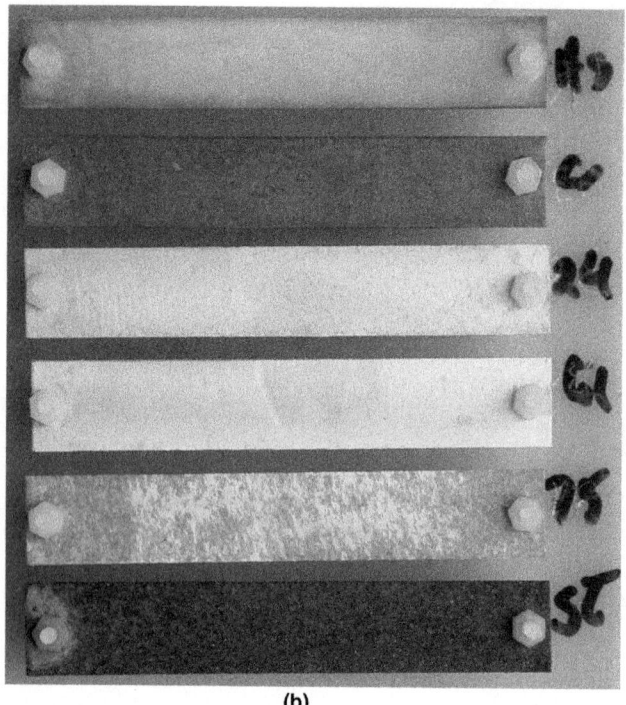

(b)

FIGURE 23.18. *(Continued)*

(a)

(b)

FIGURE 23.20. CLIMAT coupon with three copper rods immediately after it was installed at (a) the Kennedy Space Center beach corrosion test site, (b) after 30 days, and (c) after 60 days.

A CLIMAT coupon with three copper rods installed at the NASA Kennedy Space Center (KSC) beach corrosion test site (Fig. 23.19) is shown immediately after it had been installed [Fig. 23.20(a)], after 30 days [Fig. 23.20(b)], and after 60 days [Fig. 23.20(c)]. KSC having the highest corrosivity of any test site in the continental United States [31], the mass loss recorded even after a shorter exposure than usual can be very high. In the present example it was already

16% of the original aluminum wire after 60 days. The base support of these CLIMATs were purposefully installed parallel to the sea coast, and the directional effect of the marine salts may be illustrated by comparing the front and back of the exposed CLIMATs [Fig. 23.21(a) and (b)].

E. ATMOSPHERIC CORROSION MODELS

One of the principal goals of scientific discovery is the development of a theory or model that can be used to provide explanations and predictions for a specific domain of knowledge. Theory development is a complex process involving three principal activities: theory formation, theory revision, and paradigm shift, as illustrated in Figure 23.22 [32]. A theory is first developed from a collection of known

FIGURE 23.19. Aerial view of NASA Kennedy Space Center beach corrosion test site where atmospheric corrosivity is highest corrosivity of any test site in continental United States.

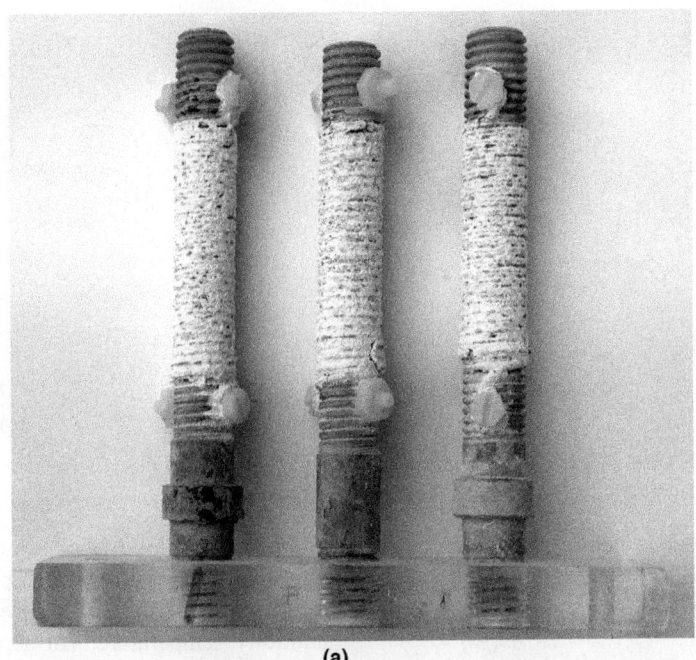

(a)

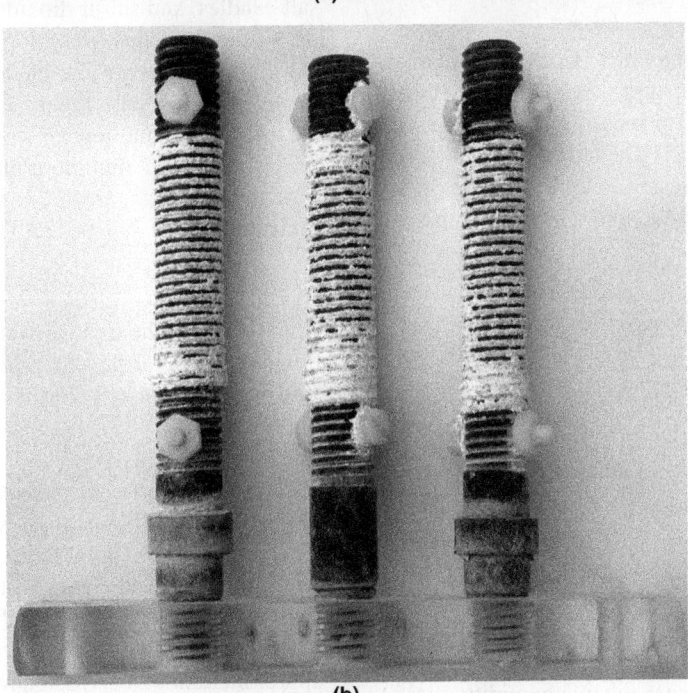

(b)

FIGURE 23.21. Close-up pictures of CLIMAT coupon exposed at KSC for 2 months: (a) seaside and (b) backside.

observations. It then goes through a series of revisions aimed at reducing the shortcomings of the initial model that are often realized when new facts and data become available.

The concept now universally accepted that rusting of iron and other metals in normal atmospheres requires the simultaneous presence of oxygen and water was seriously debated by scientists during the first decades of the twentieth century.

Though the view presented by Dunstan in 1905 was supported by much experimental evidence, many still believed and argued then that the presence of carbon dioxide was required for an atmosphere to be really corrosive [33]. The first models to describe atmospheric corrosion processes and their rates were, however, proposed much later by W.H.J. Vernon following his extensive experimental work with

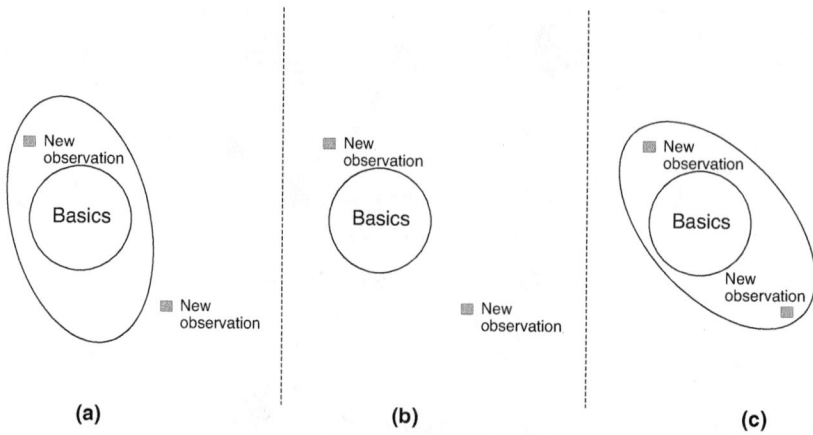

FIGURE 23.22. Theory revision using abduction for hypothesis formation.

copper [34] and iron [35] specimens that had been diligently exposed to various levels of sulfur dioxide, carbon dioxide, and suspended particles [36, 37].

E1. Statistical Models

E1.1. Corrosion Damage Functions.
Corrosion damage functions for aluminum, zinc, and steel were developed based on experiments related to measuring the effects of acid rain during the 1980s [38]. The statistical model described in Eq. (23.4) was presented as a correlation of corrosion damage for aluminum and the environmental variables that were measured:

$$M = 0.206t^{0.987}(0.099 + 0.139f_{90}\text{SO}_2 + 0.0925\text{Cl}$$

$$+ 0.0164\text{H}^+ - 0.0013\text{DUST}) \qquad (23.4)$$

where

M = cumulative mass loss of aluminum per unit area, g/m^2

t = time, years

SO_2 = average SO_2 concentration, μg/m^3

Cl = chloride deposition rate, mg/m^2 day

H^+ = hydrogen ion deposition rate in precipitation, μeq/m^2 day

f_{90} = fraction of time where relative humidity >90% and temperature >0°C

DUST = dustfall deposition rate, mg/m^2 day

The reason for combining the time of wetness term, f_{90}, with the sulfur dioxide term, SO_2, in this equation is that sulfur dioxide requires a water film for absorption and will not deposit onto a dry surface.

E1.2. ISO Corrag.
As mentioned previously, ISO technical committee TC 156 initiated a coupon exposure program that became known as ISO CORRAG. Exposures of flat plates and helixes of mild steel, copper, aluminum, and zinc involved 51 sites in mostly Europe and the United States. The program began in 1986 and ended in 1998. The relative humidity, temperature, salt deposition rate as measured by salt candles, and sulfur dioxide deposition rate as measured by sulfation plates were logged at each site. Statistical correlations between the environmental variables and corrosion rates for these metals have been developed [39, 40]. An example of the regression results for one-year log corrosion rates with aluminum plates are shown in Eq. (23.5):

$$\log(\text{rate}) = -0.739 + 3.26(\text{TOW}) + 5.02(\text{SO}_2) + 6.71(\text{Cl})$$
$$(23.5)$$

where TOW is the time-of-wetness factor, SO_2 is the sulfation rate, and Cl is the chloride deposition rate. This expression was based on data from 32 sites and had an F statistic for regression of 5.9.

E1.3. International Cooperative Program on Effects on Materials.
The International Cooperative Program on Effects on Materials (ICP Materials) was started in September 1987 and involved 39 exposure sites in 14 countries of Europe and North America. The meteorological variables recorded in the data set are listed in Table 23.4 [41]. The first approximation of a corrosion rate, K in Eq. (23.6), consisted of a linear summation of these three contributing factors [42]:

$$K = f(\text{SO}_2) + f(\text{Cl}^-) + f(\text{H}^+) \qquad (23.6)$$

where $f(\text{SO}_2)$ is the sulfur dioxide factor, $f(\text{Cl}^-)$ is the chloride deposition factor, and $f(\text{H}^+)$ is a measure of the acidity of the rain. The sulfur dioxide factor was expanded by multiplying the concentration of sulfur dioxide by the TOW, as expressed in Eq. (23.7):

$$f(\text{SO}_2) = A(\text{SO}_2)^B(\text{TOW})^C \qquad (23.7)$$

TABLE 23.4. Average Annual Ranges Used for ICP Materials Dose–Response Functions

Description	Range	Unit
Time	1–8	Year
Temperature	2–19	°C
Relative humidity	56–86	%
SO$_2$ concentration	1–83	µg/m^{-3}
Ozone concentration	14–82	µg/m^{-3}
Rainfall	33–215	cm
Acid concentration	0.6–130	µg (H$^+$) L^{-1}

The model fitting also revealed the complicated effect of temperature on atmospheric corrosion. In low-temperature regions, the corrosion rate increased with increasing mean temperature. In warm or hot regions a negative temperature effect was observed in the absence of chlorides, which is due to a reduced time of wetness. In marine atmospheres a positive temperature effect was observed also in warm/hot regions due to the presence of hygroscopic salts at the metal surface that prolong the TOW. As the temperature increased the corrosive effect of chloride, the corrosivity in tropical marine atmospheres with high deposition of chlorides can be extremely high.

E1.4. Iberoamerican Atmospheric Corrosion Map Project.

The Iberoamerican Atmospheric Corrosion Map Project (MICAT—Mapa Iberoamericano de Corrosion Atmosferica) was initiated in 1988 with three objectives: (i) construct a corrosion map for Iberoamerica, (ii) provide a better understanding of atmospheric corrosion phenomena, and (iii) identify mathematical models that could predict the corrosion rate of metals as a function of meteorological and pollution variables [43, 44]. Plates of zinc, mild steel, aluminum, and copper were exposed. The environmental variables logged were relative humidity, temperature, number of rainy days per year, sulfur dioxide deposition rate, and chloride deposition rate. One approach was to assume that corrosion rate changes with time according to the parabolic relationship expressed in Eq. (23.8):

$$C = At^n \tag{23.8}$$

where C is the corrosion rate and t is time. The coefficients A and n were obtained by a statistically fit to the measured environmental variables. The regression equations accounted for 83, 62, 59 and 41% of the variance in the annual corrosion data for zinc, steel, aluminum, and copper, respectively. The goodness of fits of data from rural atmospheres was considerably lower. The same data sets were also analyzed with neural networks [45]. This later technique exhibited superior performance in terms of goodness of fit when compared to the classical regression models and reproduced some well-known nonlinear interactions among

the variables of interest. The linear regression model obtained for the corrosion of iron is shown in Eq. (23.9):

$$Fe = b_0 + Cl^-(b_1 + b_2 \times P + b_3 \times RH) + b_4 \\ \times TOW \times SO_2 \tag{23.9}$$

where $b_0 = 6.8124$, $b_1 = -1.6907$, $b_2 = 0.0004$, $b_3 = 0.0242$, and $b_4 = 2.2817$.

E1.5. Topographical Effects on Wind Velocity.

Wind speed, wind direction, and distance from the sea are the most important factors influencing the transport and deposition rate of marine aerosols. A mapping method, based on grids with topographic factors, was developed to assess the influence of various factors on average wind speed at locations near a seacoast [46]. Basically, there are three steps in this method:

- Identify the topographic factors that can affect the flow of wind between a site of interest and the seacoast.
- Determine the topographic factors for each grid and then regress against measured wind speeds.
- Estimate the integrated sea wind for each site.

The concept integrates the geographical texture of the environment in all directions (16 sector resolution) with its effect on airflow. For example, the results of a multiple regression analysis for the annual average wind velocity for sites on the Miura Peninsula (Japan) are expressed in Eq. (23.10):

$$Y = 1.432x_1 + 0.002x_2 - 3.45x_3 - 0.058x_4 - 0.009x_5 \\ - 0.34x_6 + 3.573 \tag{23.10}$$

where
Y = annual average wind velocity, m/s
x_1 = ratio of land to sea per unit area
x_2 = corrugation (effect of height variation within a grid)
x_3 = degree of shielding
x_4 = wind convergence over 2–7 km
x_5 = difference in wind direction
x_6 = wind convergence over 1–2 km

It is interesting to note the least significant factors in the regression analysis (low coefficients) are the corrugation and wind direction effects, possibly due to the presence of important local geographic features. However, the negative values give an indication of the reduction of wind velocity, due to shielding, convergence of wind, and the direction of the wind on the peninsula. The most significant factor affecting sea wind was the degree of shielding.

E2. Mechanistic Models of Corrosivity Factors

E2.1. Pollutant Mass Transfer to a Surface.

In a study on the corrosion of plates of galvanized steel, the pollutant mass transfer rate to the plates was modeled as the product of the pollutant concentration and the deposition velocity. The deposition velocity was modeled from an analogy from momentum transport in Eq. (23.11) [47]:

$$u = \frac{V^{*2}}{V} \tag{23.11}$$

where u is the deposition velocity, V is the average upstream wind velocity, and V^* is the friction velocity. The friction velocity is itself described in Eq. (23.12):

$$V^* = \sqrt{\frac{f}{2}} \tag{23.12}$$

where f is the friction factor, which from boundary layer theory for smooth flat plates is defined in Eq. (23.13):

$$f = \frac{0.03}{(\mathrm{Re}_L)^{1/7}} \tag{23.13}$$

where Re_L is the Reynolds number for the plate as described in Eq. (23.14):

$$\mathrm{Re}_L = \frac{LV}{\nu} \tag{23.14}$$

where L is the length of surface over which the air flows and ν is the kinematic viscosity of air.

E2.2. Marine Aerosol Transport.

A model was developed that accounted for a steady source of marine aerosol particles and their transport fairly near the ground, well within the planetary boundary layer. Aerosol particles were assumed to be transported by convection and turbulent diffusion, while they would be deposited on the ground by turbulent diffusion. The predicted aerosol concentration as a function of distance for 1500 m from a source was consistent with published data on steel corrosion and salinity rates near an ocean [48].

The chloride concentration as aerosol particles was modeled as a function of distance from a source such as a salt-water body [22]. The main assumption was that as wind carries chloride particles from the sea there is a deposition to the ground due to turbulent diffusion. This deposition rate is characterized by a deposition velocity, v_d. The deposition velocity depends upon particle size, ground surface roughness (open terrain, presence of vegetation, vertical projections such as hills, mountains, buildings, etc.), and wind speed. A mass balance of the chloride concentration within a height h above the ground at a distance x from a source of salt aerosols is described in Eq. (23.15):

$$S = S_0^{(-v_d x)/hV} \tag{23.15}$$

where

$S =$ average concentration of chloride in air within a height h above the ground
$S_0 =$ initial chloride conc. at shoreline
$x =$ distance from shoreline
$v_d =$ deposition velocity at ground
$V =$ wind speed
$h =$ height of air layer above ground

This equation was shown to be applicable for the first few hundred meters from a source, but other factors were found to greatly complicate the phenomenon as distance increases:

- The change of particle size distribution with distance as largest particles settle first and therefore changes in v_d.
- A nonuniform vertical profile of salt concentration.
- Decrease of wind speed with downwind distance caused by friction of the ground surface.
- The effect of rainout or even low cloud levels, which reduces the chloride present in the air, is not accounted for.

It was also noted in that study that a minimum wind speed or threshold of approximately 11 km/h is required for the entrainment of marine aerosols over a salt-water body [23].

E2.3. Corrosion under a Droplet.

A detailed model of the corrosion processes within a droplet on a surface was developed [49]. The typical day and night cycle drives changes in condensation and evaporation on metallic surfaces. The model assumed the surface under the droplet functioned as an anode, and the surface at the droplet periphery functioned as a cathode. The effects of changing droplet size and electrolyte concentration at the limiting oxygen reduction current were simulated. The model predicted that there is no significant resistive control in an evaporating droplet under these conditions. However, in a condensate or absorbate of uniform thickness, resistive effects may become significant in layers less than 10 μm thick. In atmospheric corrosion, this has relevance with respect to repeated evaporation and condensation during wetting and drying, to the effects of rainwater washing and surface cleaning.

E2.4. Wind Speed Factor.

In a study focused on the shielding effects of buildings in a marine environment, the directional impact of marine aerosols was revealed by comparing the level of patina on CLIMAT's copper rod exposed for three winter months at facilities on the Pacific coast [50, 51]. What

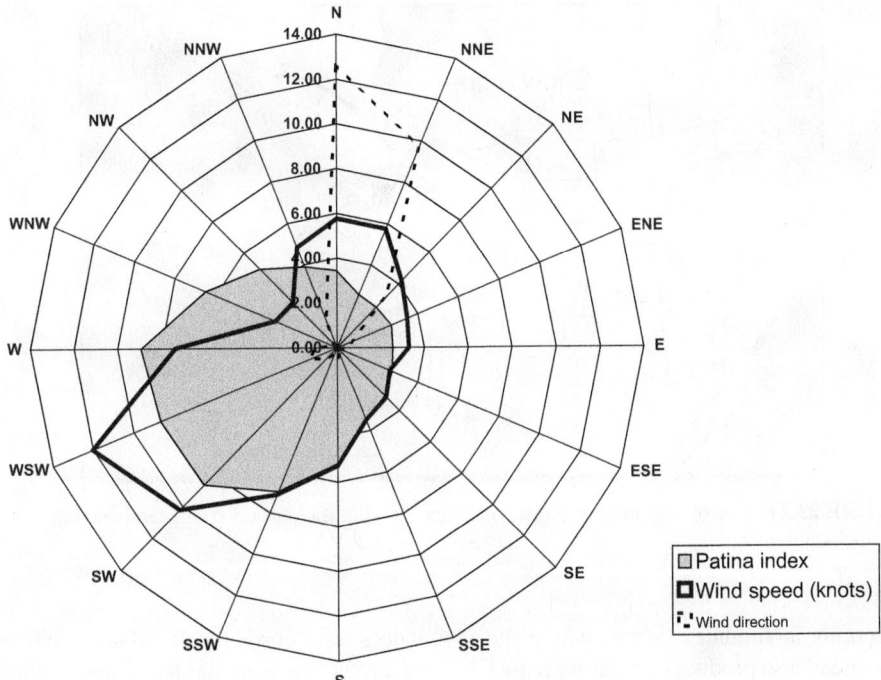

FIGURE 23.23. Average corrosion index for copper rods exposed on rooftop and average wind speed recorded at the local weather station as a function of 16 points of compass.

became evident in that study was that the pattern of the bluish-green patina attributed to the corrosion product $CuCl_2 \cdot 2H_2O$ was not uniformly distributed around the circumference of each copper rod.

A template with the 16 points of the compass was placed onto the outside of each copper rod that was on the most boldly exposed coupons in order to visually quantify the intensity of the bluish-green color. The relative degree of corrosion for each compass point was assessed by assigning a number between 0 and 10 with 0 corresponding to 0 bluish-green patina and 10 corresponding to 100% coverage of the colored corrosion product. The average corrosion index for each of the 16 points of the compass for the copper rods is shown in Figure 23.23.

One attempt to correlate the directional corrosivity observed on the copper rods with weather data was to plot the fraction of time that winds came from the 16 points of the compass during the 3-month exposure period. The dominant direction was the north to northeast. However, the pattern of corrosion product did not correspond to the most frequent wind direction but did correspond to the direction with the highest wind speeds, which were in the west-to-south quadrant at this particular site (Fig. 23.23).

E3. Maps of Atmospheric Corrosivity

Maps are powerful tools for communicating information related to geographical landscapes, and corrosivity maps of various countries have been drawn to illustrate the corrosion severity of regions of these countries [52]. One of the very first atmospheric corrosivity maps was produced to summarize many years of results obtained by exposing bare steel coupons attached to different vehicles in the northeastern United States and Canada (Fig. 23.24) [53]. The results presented in this figure reveal the corrosive effects of using deicing salts since this is the main factor that can explain the higher level of vehicle corrosion in the Snowbelt region when compared to adjacent nonmarine regions.

Similar corrosivity maps have been created from data available in the literature such as the corrosivity map for China (Fig. 23.25), Cuba (Fig. 23.26), and Great Britain (Fig. 23.27) adapted, respectively, from [54], [55], and [56].

F. PREVENTION AND CONTROL

The following sections contain a brief description of methods commonly used to prevent and control the damage due to atmospheric corrosion.

F1. Materials Selection

Only a few metals and alloys are usually boldly exposed to outdoor environments. Among these are all the bronzes and other copper alloys that have been used for centuries for the production of monuments and artifacts. The long life

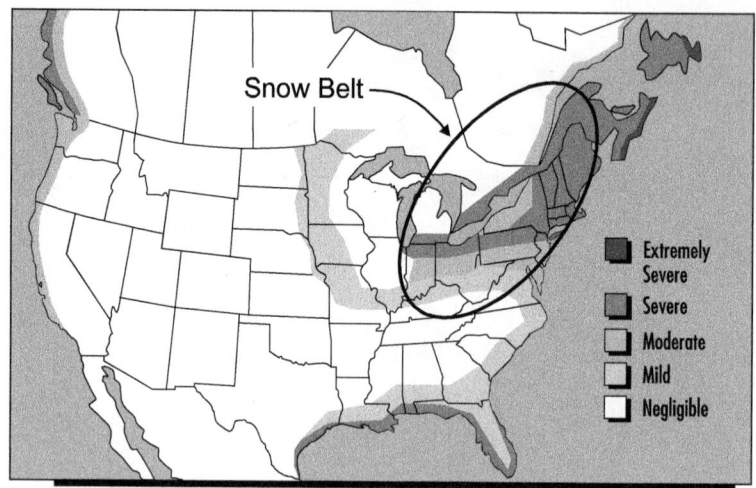

FIGURE 23.24. Corrosivity map of North America showing the particular aggressiveness of Snowbelt region.

provided by copper roofing is another example where the practical usage of such metal also produces a pleasant patina that has become the trademark of many touristic cities. However, most structural metals need some sort of shielding from direct contact with the environment. Aluminum, for example, will typically be anodized and often coated before being put in service. Similarly, steel is typically coated with protective coatings mostly based on organic coatings or galvanized for added protection. There are, of course, some exceptions to this general statement.

Iron, in its various forms, tends to be highly reactive because of its natural tendency to form iron oxide. When it does resist corrosion, it is due to the formation of a thin film of protective iron oxide on its surface by reaction with oxygen of the air. This film can prevent rusting in air at 99% RH, but a contaminant such as acid rain may destroy the effectiveness of the film and permit continued corrosion. Thicker films of iron oxide may act as protective coatings and, after the first year or so, could reduce the corrosion rate, as shown in Figure 23.28.

The Cor-ten high-strength low-alloy (HSLA) weathering steel in Figure 23.28 has shown to be more than 10 times corrosion resistant than carbon steel when freely exposed to mild environments [57]. This is why weathering steel has

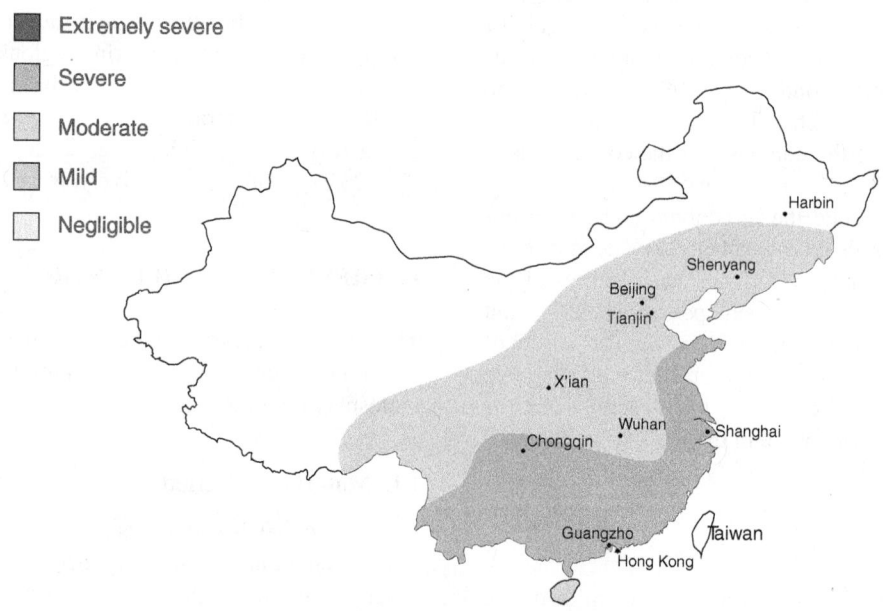

FIGURE 23.25. Corrosivity map of China.

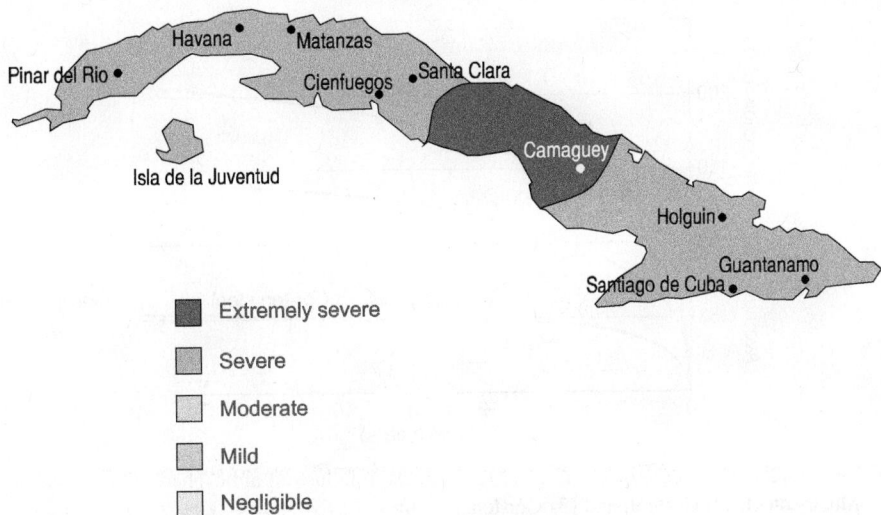

FIGURE 23.26. Corrosivity map of Cuba.

been the choice of many designers for the construction of boldly exposed surfaces, from buildings to utility poles (Fig. 23.29). The formation of a protective rust on weathering steels results from a combination of certain alloying elements, generally totaling <3%, which include Cr, Cu, Ni, Si, and P. Two commonly used weathering steels used for structural purposes are K11430 or Cor-ten B and K12043.

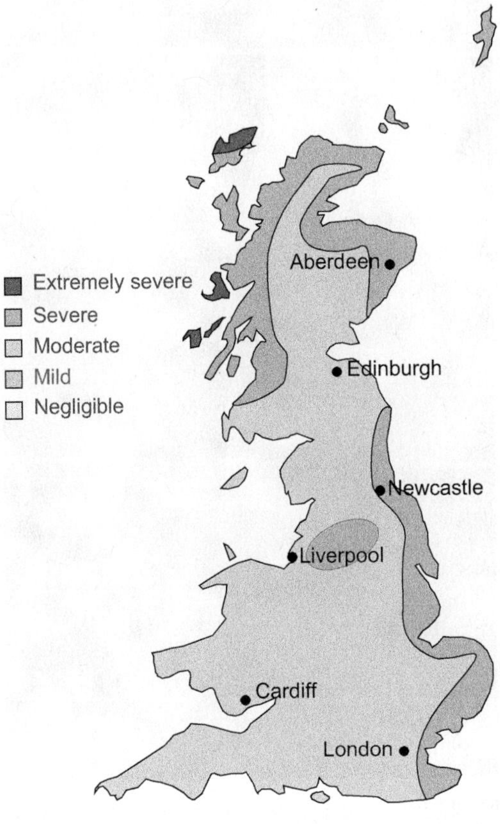

The alloying additions also add strength to weathering steels, thereby enabling further cost and weight savings compared to ordinary structural steels. Regression analyses of long-term atmospheric corrosion tests of hundreds of steels that were conducted at 3 industrial environments lead to conclusions about the effects of 15 elements on corrosion resistance [58]:

- In order of decreasing effect, P, Si, Cr, C, Cu, Ni, Sn, and Mo are beneficial.
- S has a very large and detrimental effect.
- V, Mn, Al, Co, As, and W have no significant effect.

While weathering steels do not require any particular care once installed, it may suffer surprising corrosion attack in crevice areas. As the severity or the physical conditions of exposure change, these weathering steels will show less superiority. In crevices or on the backside of structural forms in a corrosive atmosphere, weathering steel will in fact not perform better than plain carbon steel.

Austenitic stainless steels have become the material of choice of many architects in recent history since they can keep their shiny aspect without tarnishing for many decades as illustrated in The Triad, a tall sculpture erected in the busy part of Toronto in 1984 (Fig. 23.30). Stainless steel has also been used to great advantages in more notorious buildings and monuments as early as 1930.

The Chrysler Building completed in New York City in 1930 was the first high-profile stainless steel application in the world. Type 302 stainless steel was used for the production of six rows of arches topped with a stainless steel spire. Stainless steel gargoyles were installed on the 31st and 61st floors. The present-day condition is estimated to be very good. The exterior of the Empire State Building, completed one year later (1931), was made of stainless steel, gray limestone, and dark gray aluminum. Over 300 metric tons

FIGURE 23.27. Corrosivity map of Great Britain.

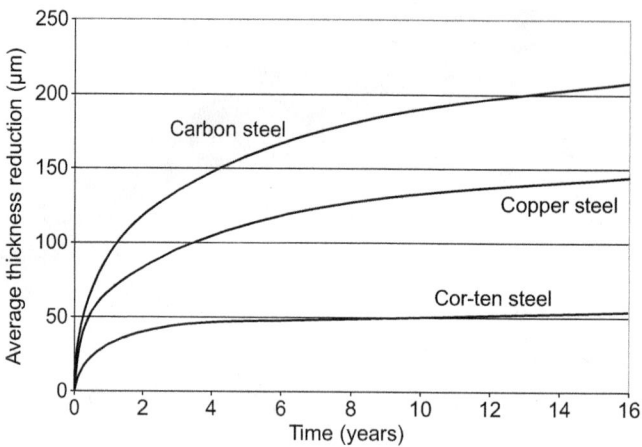

FIGURE 23.28. Time–corrosion curves of three steels in industrial atmosphere, Kearny, NJ: (1) ordinary steel, (2) Custeel, and (3) Cor-ten.

of 1.3-mm-thick Type 302 stainless steel was used for its construction. This historic landmark with a height of 282 m was the world's tallest building for 41 years. The stainless steel was estimated to be in excellent condition 70 years after the construction of this historical landmark. Many other buildings have since then incorporated large amounts of stainless steels in their facades and roofs.

FIGURE 23.29. Thirty-five-meter-high highway lamppost made of weathering steel. (Courtesy of Kingston Technical Software.)

FIGURE 23.30. Triad by Ted Bieler completed in 1984 using stainless steel. (Courtesy of Kingston Technical Software.)

F2. Protective Coatings

The large segment of the paint industry committed to the manufacture and application of products for the protection of metals, as well as the large-scale operations of the galvanizing industry, attest to the importance of controlling atmospheric corrosion. The general subject of protective coatings for the prevention and control of corrosion is the focus of many technical books and much beyond the scope of the present chapter in which only the breakdown of organic coatings exposed to the atmosphere will be discussed.

What happens when moisture and oxygen penetrate the coating? Perhaps nothing, if a truly good bond has been achieved, but a number of reactions may occur. Where the moisture contacts the steel, for example, a corrosion cell comparable to that in a pit on freely exposed iron may then develop. If ionic contaminants such as chlorides and other soluble salts are present at the interface between the coating and the steel, the electrolyte would be more conductive and thus favor a more vigorous cell action.

The rate at which diffusion of water and contaminants is occurring through a coating is largely controlled by its thickness and formulation. The thickness of a coating necessary to resist moisture permeation from the atmosphere and otherwise resist deterioration is approximately 125 μm. If good flowout is not obtained during application, thin spots (particularly at the edges) or actual holidays (holes) may exist in the film and reduce the effective thickness of the coating barrier.

Due to the presence of these imperfections in the coating, the steel substrate is directly exposed to its surroundings and may start corroding as described in Eq. (23.16). In order to maintain electroneutrality within the system, this reaction is balanced by at least one cathodic reaction. In most naturally occurring situations, this reaction will be the reduction of oxygen from ambient air, as illustrated in Eq. (23.17). These two reactions initially take place adjacent to each other but soon separate as the process continues with the cathode moving under the coating.

$$Fe_{(s)} \rightarrow Fe^{2+} + 2e^- \qquad (23.16)$$

$$O_2 + 2H_2O + 4e^- \rightarrow 4OH^- \qquad (23.17)$$

In given conditions of humidity and oxygen permeation, the initial corrosion site may begin to move in a random manner as the corrosion product reduces the oxygen content at the surface and the area becomes highly anodic to the surrounding cathode area of oxygen saturation. The worm track of corrosion that then occurs is termed filiform corrosion (Fig. 23.31).

The ferrous ions produced in Eq. (23.16) carry a positive charge, while the cathode is producing hydroxyl ions (OH^-), resulting in an excess of negative charge. These local charge imbalances are unstable. Additionally, the ferrous ions produced by the corrosion reaction in Eq. (23.16) are not stable in the presence of oxygen and soon react with water to form what is called rust, or $Fe(OH)_3$, following the combination of oxidation and hydrolysis reaction described in Eq. (23.18):

$$Fe^{2+} + 3H_2O \rightarrow Fe(OH)_3 + 3H^+ + 1e^- \qquad (23.18)$$

The electrons produced in Eq. (23.18) are also consumed by the reduction of oxygen. The path from the exposure environment to these initial corrosion sites may be either restricted or blocked completely when the coating adheres well to the substrate. In such cases the corrosion process would be relatively stifled after the initial attack.

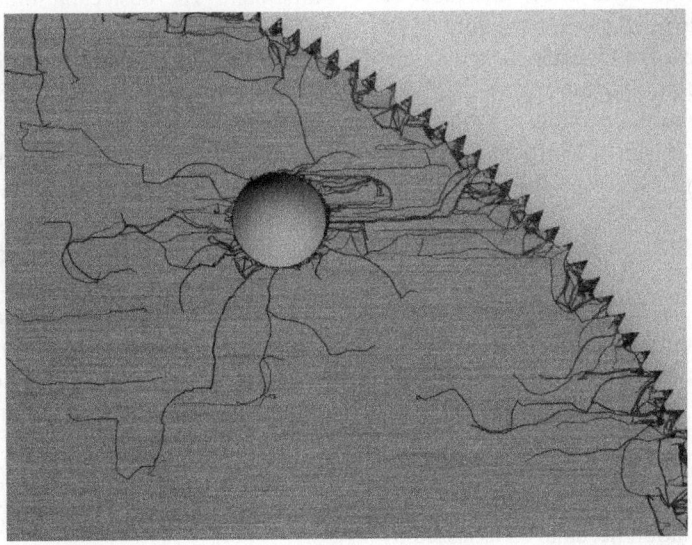

FIGURE 23.31. Filiform corrosion under lacquer protecting surface of new handsaw. (Courtesy Kingston Technical Software.)

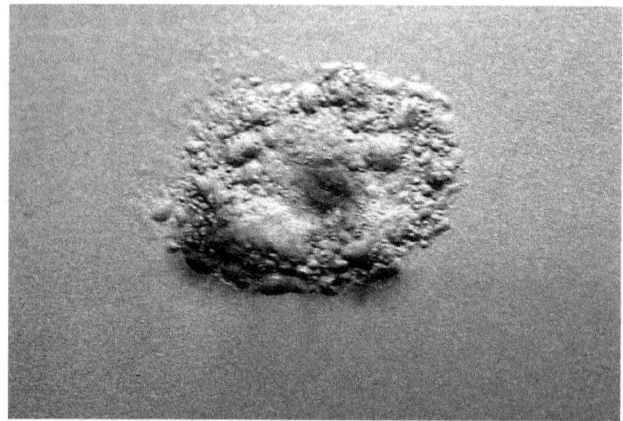

FIGURE 23.32. Corrosion blister formed on a car door months after impact caused break in coating. (Courtesy of Kingston Technical Software.)

For less adherent coatings, much less restrictive pathways such as micronic dust, coating porosity, and holidays are available for counter ions, and the corrosion reaction would be allowed to proceed at a much faster pace. When this happens, a second cathodic reaction may be triggered by the increased acidity at the anodic site, leading to the production of gaseous hydrogen as described in Eq. (23.19) and pry the coating loose (Fig. 23.32):

$$2H^+ + 2e^- \rightarrow H_2 \qquad (23.19)$$

Another aggravating factor is that organic coatings are generally poorly resistant to alkaline conditions and may be attacked by the hydroxyl ion, causing a serious loss of surface adhesion. The reason this alkalinity causes such failure has been variously attributed to either saponification of the coating, dissolution of the oxide layer at the interface, or simply alteration of the ionic resistance of the film [59]. External cathodic currents provided by cathodic protection or internal currents produced by inorganic zinc additives during immersion service, for example, would increase the possibility of failure by the hydrogen or hydroxyl formation because much greater quantities of these cathodic reaction products would be created.

F3. Dehumidification

Corrosion may also be prevented or controlled by modification of the corrosive environment. One method commonly used consists in reducing the water content of the environment by dehumidification to prevent the formation of a continuous film of moisture on metal surfaces. There are a number of different types of dehumidification systems available, and each proposed installation needs to be assessed to determine the most suitable type. Where the item to be protected is in a vapor-sealed enclosure, desiccant beads

such as silica gel may be used successfully. However, if external air is allowed to enter, then the moisture load will soon exceed a desiccant capacity. More efficient dehumidification methods will therefore be required as the complexity and size of the applications increase.

Dehumidification by heating works on the principle that if the air within an enclosed space is heated, then the relative humidity of the air will be reduced. In order for this method to be fully effective, all items in the space, as well as the air itself, need to be heated as humid zones may occur at cold spots on metal surfaces.

Another method commonly employed is the reduction of moisture by passing air through an evaporator coil where it is cooled below its dew point, forcing water to condense on the outside of the coil from where it is drained. The air is then often heated back to ambient temperature. However, cooling to very low temperature makes the refrigeration process impractical, as it requires a great deal of subsequent reheating. The reduction in air temperature is also limited by the freezing point of water condensing on the cooling coil, which in some designs may be offset by the use of a complicated brine spray or a liquid lithium chloride type of system that works on a combination of refrigeration and adsorbent liquid.

The static or dynamic use of desiccants through absorption or adsorption processes also provides a wide spectrum of dehumidification techniques. In the case of absorption processes, desiccants change physically, chemically, or both during the sorption process. Lithium chloride crystals are an example of a solid absorbent. When water is absorbed by this material, it changes to a hydrated state. In liquid sorption dehumidification systems, air is passed through sprays of a liquid sorbent such as lithium chloride or glycol solution. The sorbent in an active state has a vapor pressure below that of the air to be dehumidified and absorbs moisture from the airstream. The desiccant solution, which becomes diluted with moisture during the process of absorption, is regenerated when the solution is subsequently heated.

In the case of adsorption processes the desiccant does not change physically or chemically during the sorption process. Adsorbents such as silica gel, molecular sieve, and activated alumina are normally granular beads or solids with porous structures making these capable of holding large amounts of water on their surface. The principle behind desiccant dehumidification is that the desiccant is exposed to moisture-laden air, from which moisture is extracted. During regeneration, the saturated desiccant is heated and the collected moisture is driven off into the exhaust airstream. Thus a continuous cycle of sorption and regeneration can be set up providing an atmosphere with very low dew point.

F4. Corrosion Prevention Compounds

Corrosion prevention compounds (CPCs) are widely used as a temporary measure to provide cost-effective corrosion

FIGURE 23.33. Fogging of aircraft with commercial CPC. (Courtesy Mike Dahlager Pacific Corrosion Control Corp.)

protection on most metals. CPCs are applied as fluids by wiping, brushing, spraying, or dipping. These compounds are usually immiscible with water but may contain some water displacing components to remove water from surfaces and crevices.

Corrosion prevention compounds are commercially available in a wide range of products. A number of these fluids are based on lanolin and contain various solvents and inhibitors. The evaporation of the solvents leaves either thin soft films, semihard films, or hard resin films that provide varying degrees of short-term protection. The exact compositions of these CPCs are not published for obvious reasons. However, information contained in the Materials Safety Data Sheets (MSDSs) reveal that they may include [60]:

- An oil, grease, or resin-based film former
- A volatile, low surface tension carrier solvent
- A nonvolatile hydrophobic additive
- Various corrosion inhibitors, for example, sulphonates, or surface-active agents

Water-displacing CPCs act by spreading across surfaces, into cracks and crevices where they displace moisture, leaving behind a residue to act as a further barrier after the carrier solvent has evaporated. The hard film CPCs dry to a waxy or hard resinlike finish after application and provide a barrier film to corrosive environments.

Corrosion prevention compounds have been used in many applications where corrosion is a concern. In the aircraft industry, for example, these products are often recommended by the manufacturer in the maintenance manuals as a way to help prevent the onset of corrosion in specific aircraft locations. The application of a CPC is relatively easy by using specially designed high-pressure pump systems to convert it into a dense fog and literally fog the area or cavity that needs protection (Fig. 23.33).

These products are also extensively used on road vehicles to reduce the severity of corrosion, particularly in cold-weather areas where deicing salts are used to break down snow and ice. There are plenty of locations in road vehicles where water can accumulate and accelerate corrosion, leading to premature failure of the equipment or perforation of the sheet metal. The underside of cars is especially vulnerable to rust and corrosion due to the exposure to high levels of moisture.

Since the 1970s, hot-melt wax thermoplastic CPCs have been used extensively during the construction of new vehicles to protect underbody structural components against corrosion and enhance vehicle durability. Hot-melt waxes are usually applied through a dipping process. The wax is preheated to a temperature between 125 and 195°C. Following an alkali cleaning and water rinsing operation, parts are immersed in the molten wax. The thickness of the wax deposited on the parts is controlled through a preheat of the parts prior to dipping and the actual time of immersion in the hot-melt wax.

Thin fluid film CPCs are typically used on road vehicles as a yearly maintenance application for added corrosion protection. The CPC treatment using air-powered spray units is relatively inexpensive and can be provided by various rust-proofing centers (Fig. 23.34). No particular preparation is required. For outdoor spraying a NIOSH-approved oil mist mask, as worn when spraying indoors, is recommended. Gloves are worn for comfort, as the aerosol cans or air-powered spray guns quickly get cold. After treatment, any excess fluid may be wiped up with a cloth or paper towel.

(a)

(b)

FIGURE 23.34. Fogging of (a) car trunk and (b) underbody with a commercial CPC. (Courtesy of Kingston Technical Software.)

F5. Volatile Corrosion Inhibitors

Volatile corrosion inhibitors (VCIs), also called vapor-phase inhibitors (VPIs) are used in close environments where they saturate the ambient atmosphere. In boilers, volatile basic compounds, such as morpholine or hydrazine, are transported with steam to prevent corrosion in condenser tubes by neutralizing acidic carbon dioxide or by shifting surface pH toward less acidic and corrosive values. In closed vapor spaces, such as shipping containers, volatile solids such as salts of dicyclohexylamine, cyclohexylamine, and hexamethylene-amine are used. On contact with the metal surface, the vapor of these salts condenses and reacts with moisture to provide protective ions.

It is desirable, for an efficient VCI, to provide inhibition rapidly and to last for long periods. Both qualities depend on the volatility of these compounds, fast action wanting high volatility, whereas enduring protection requires low volatility. A convenient partial vapor pressure for closed spaces VCIs will lie between 10^{-3} and $10\,Pa$ (Table 23.5) [61].

It is significant that the most effective volatile corrosion inhibitors are the products of the reaction of a weak volatile base with a weak volatile acid. Such substances, although ionized in aqueous solutions, undergo substantial hydrolysis, the extent of which is almost independent of concentration. In the case of the amine nitrites and amine carboxylates, the net result may be expressed by the following reaction:

$$H_2O + R_2NH_2NO_2 \rightarrow (R_2NH_2)^+ : OH^- + H^+ : (NO_2^-)$$

$$(23.20)$$

The nature of the adsorbed film formed at a metal–water interface is an important factor controlling the efficiency of VCIs. Metal surfaces exposed to vapors from VCIs in closed

TABLE 23.5. Saturated Vapor Pressures of Some VCIs

Substance	Temperature (°C)	Vapor Pressure (mm Hg)	(Pa)	Melting Point (°C)
Morpholine	20	8.0	1070	
Benzylamine	29	1.0	130	
Cyclohexylamine carbonate	25.3	0.397	53	
Diisopropylamine nitrite	21	4.84×10^{-3}	0.65	139
Morpholine nitrite	21	3×10^{-3}	0.40	
Dicyclohexylamine nitrite	21	1.3×10^{-4}	0.017	179
Cyclohexylamine benzoate	21	8×10^{-5}	0.010	
Dicyclohexylamine caprylate	21	5.5×10^{-4}	0.073	
Guanadine chromate	21	1×10^{-5}	0.0013	
Hexamethyleneimine benzoate	41	8×10^{-4}	0.110	64
Hexamethyleneamine nitrobenzoate	41	1×10^{-6}	0.00013	136
Dicyclohexylamine benzoate	41	1.2×10^{-6}	0.00016	210

containers typically become covered by a hydrophobic-adsorbed layer as evidenced by an increase of the contact angle of distilled water on treated surfaces with time of exposure.

REFERENCES

1. L. S. Selwyn and P. R. Roberge, Corrosion of Outdoor Monuments, in ASM Handbook, Vol. **13C**, Corrosion: Environments and Industries. D. S. Cramer B. S. Covino (Eds.), ASM International, Metals Park, OH, 2006, 289–305.

2. U. R. Evans, An Introduction to Metallic Corrosion, Edward Arnold, London, 1948.

3. C. L. Page and R. D. Anchor, "Stress Corrosion Cracking in Swimming Pools," Mater. Perform., **29**, 57–58 (1990).

4. D. Miller, "Corrosion Control on Aging Aircraft: What Is Being Done?" Mater. Perform., **29**, 10–1 (1990).

5. J. P. Komorowski, S. Krishna kumar, R. W. Gould, N. C. Bellinger, F. Karpala, and O. L. Hageniers, "Double Pass Retroreflection for Corrosion Detection in Aircraft Structures," Mater. Eval., **54**, 80–86 (1996).

6. Investigative Report to Joint Committee to Investigate the I-35W Bridge Collapse, Gray Plant Mooty, Minneapolis, MN, May 21, 2008.

7. Dissimilar Metals, MIL-STD-889B(3), Army Research Laboratory, Aberdeen, MD, 1993.

8. D. Knotkova, "2005 F. N. Speller Award Lecture: Atmospheric Corrosion—Research, Testing, and Standardization," Corrosion, **61**, 723–738 (2005).

9. C. Leygraf and T. E. Graedel, Atmospheric Corrosion, Wiley, New York, 2000.

10. "Galvanizing for Corrosion Protection: A Special Guide," Report, American Galvanizers Association, Centennial, CO, 1990.

11. J. y Podan, "Corrosion of Metal Artifacts and Works of Art in Museum and Collection Environments," in Corrosion: Environments and Industries, Vol. 13C, D. S. Cramer and B. S. Covino (Eds.), ASM International, Metals Park, OH, 2006, pp. 279–288.

12. G. H. Koch, M. P. H. Brongers, N. G. Thompson, Y. P. Virmani, and J. H. Payer, "Corrosion Costs and Preventive Strategies in the United States," FHWA-RD-01-156, National Technical Information Service, Springfield, VA, 2001.

13. R. P. Frank enthal, "Electronic Materials, Components, and Devices," in Uhlig's Corrosion Handbook, R. W. Revie (Ed.), Wiley-Interscience, New York, 2000, pp. 941–947.

14. D. A. Douthit, "Electronics and Corrosion," Corros. Rev. **21**, 415–432 (2003).

15. "Environmental Conditions for Process Measurement and Control Systems: Airborne Contaminants," ISA Standard ANSI/ISA-S71.04-1985, International Society for Measurement and Control, Research Triangle Park, NC, 1986.

16. M. G. Lawrence, "The Relationship between Relative Humidity and the Dewpoint Temperature in Moist Air: A Simple Conversion and Applications," Bull. Am. Meteorol. Soc., **86**, 225–233 (2005).

17. P. R. Roberge, Handbook of Corrosion Engineering, McGraw-Hill, New York, 2000.

18. S. C. Chung, A. S. Lin, J. R. Chang, and H. C. Shih, "EXAFS Study of Atmospheric Corrosion Products on Zinc at the Initial Stage," Corros. Sci., **42**, 1599–1610 (2000).

19. J. R. Duncan and J. A. Ballance, "Marine Salts Contribution to Atmospheric Corrosion," in Degradation of Metals in the Atmosphere, S. W. Dean and T. S. Lee (Eds.), American Society for Testing and Materials, Philadelphia, PA, 1988, pp. 316–326.

20. P. J. Sereda, S. G. Croll, and H. F. Slade, "Measurement of the Time-of-Wetness by Moisture Sensors and Their Calibration," in Atmospheric Corrosion of Metals, S. W. Dean and E. C. Rhea (Eds.), American Society for Testing and Materials, Philadelphia, PA, 1982, p. 48.

21. G. M. Hidy, Aerosols: An Industrial and Environmental Science, Academic, Orlando, FL, 1984.

22. S. Feliu, M. Morcillo, and B. Chico, "Effect of Distance from Sea on Atmospheric Corrosion Rate," Corrosion, **55**, 883–891 (1999).

23. M. Morcillo, B. Chico, L. Mariaca, and E. Otero, "Salinity in Marine Atmospheric Corrosion: Its Dependence on the Wind Regime Existing in the Site," Corros. Sci., **42**, 91–104 (2000).

24. "Standard Test Method for Determining Atmospheric Chloride Deposition Rate by Wet Candle Method," in Annual Book of ASTM Standards, Vol 03.02, G140-02, American Society for Testing of Materials, West Conshohocken, PA, 2002.

25. H. H. Lawson, Atmospheric Corrosion Test Methods, NACE International, Houston, TX, 1995.

26. "Standard Practice for Monitoring Atmospheric SO_2 Using the Sulfation Plate Technique," in Annual Book of ASTM Standards, Vol 03.02, ASTM G91-97, American Society for Testing of Materials, West Conshohocken, PA, 1997.

27. W. H. Abbott and R. Kinzie, "Corrosion Monitoring in Air Force Operating Environments," 2003 Tri-Service Corrosion Conference, U.S. Department of Defense, Nov. 17, 2003.

28. "Standard Practice for Preparing, Cleaning, and Evaluating Corrosion Test Specimens," Vol 03.02, G1-03, American Society for Testing of Materials, West Conshohocken, PA, 2003.

29. D. P. Doyle and T. E. Wright, "Rapid Method for Determining Atmospheric Corrosivity and Corrosion Resistance," in Atmospheric Corrosion, W. H. Ailor (Ed.), Wiley, New York, 1982, pp. 227–243.

30. Standard Practice for Conducting Wire-on-Bolt Test for Atmospheric Galvanic Corrosion, ASTM G116-99, American Society for Testing of Materials, West Conshohocken, PA, 1999.

31. S. Coburn, "Atmospheric Corrosion" in Metals Handbook, 9th ed, Vol. 1, Properties and Selection, Carbon Steels, L. J. Korb (Ed.), American Society for Metals, Metals Park, OH, 1978, p. 720.

32. P. O'Rorke, S. Morris, and D. Schulenburg, "Theory Formation by Abduction: A Case Study Based on the Chemical Revolution," in Computational Models of Scientific Discovery and Theory Formation, J. Shrager and P. Langley (Eds.), Morgan Kaufmann, San Mateo, CA, 1990.

33. W. R. Dunstan, A. D. Jowett, and E. Goulding, "Rusting of Iron," J. Chem. Soc., **87**, 1548–1574 (1905).

34. W. H. J. Vernon, "A Laboratory Study of the Atmospheric Corrosion of Metals: Part I—Copper," Trans. Farad. Soc., **27**, 255–277 (1931).

35. W. H. J. Vernon, "A Laboratory Study of the Atmospheric Corrosion of Metals: Part II—Iron and Part III—Rust," Trans. Farad. Soc., **31**, 1668–1700 (1935).

36. W. H. J. Vernon, "An Air Thermostat for Quantitative Laboratory Work," Trans. Farad. Soc., **27**, 241–247 (1931).

37. W. H. J. Vernon and L. Whitby, "The Quantitative Humidification of Air in Laboratory Experiments," Trans. Farad. Soc., **27**, 248–255 (1931).

38. F. W. Lipfert, M. Benarie, and M. L. Daum, "Metallic Corrosion Damage Functions," in Proceedings, Vol. **86-6**, The Electrochemical Society, Pennington, NJ, 1986, pp. 108–154.

39. S. W. Dean and D. B. Reiser, "Analysis of Data from ISO CORRAG Program," Paper No. 340, CORROSION 1998, NACE International, Houston, TX, 1998.

40. S. W. Dean and D. B. Reiser, "Comparison of the Atmospheric Corrosion Rates of Wires and Flat Panels," Paper No. 455, CORROSION 2000, NACE International, Houston, TX, 2000.

41. J. Tidblad, A. A. Mikhailov, and V. Kucera, "A Model for Calculation of Time of Wetness Using Relative Humidity and Temperature Data," 14th International Corrosion Congress, (Cape Town, South Africa, Sept.–26 Oct. 1, 1999), Paper No. 337.2, International Corrosion Congress, 1999.

42. J. Tidblad, A. A. Mikhailov, and V. Kucera, "Application of a Model for Prediction of Atmospheric Corrosion in Tropical Environments," in Marine Corrosion in Tropical Environments. S. W. Dean, G. Hernandez-Duque Delgadillo, and J. B. Bushman (Eds.), Philadelphia, PA, American Society for Testing and Materials, 2000, pp. 18–32.

43. S. Feliu, M. Morcillo, and S. Feliu Jr., "The Prediction of Atmospheric Corrosion from Meterological and Pollution Parameters—I. Annual Corrosion," Corros. Sci., **34**, 403–414 (1993).

44. S. Feliu, M. Morcillo, and S. Feliu, Jr., "The Prediction of Atmospheric Corrosion from Meterological and Pollution Parameters—II. Long-Term Forecasts," Corros. Sci., **34**, 415–422 (1993).

45. S. Pintos, N. V. Queipo, O. T. de Rincon, A. Rincon, and M. Morcillo, "Artificial Neural Network Modeling of Atmospheric Corrosion in the MICAT Project," Corros. Sci., **42**, 35–52 (2000).

46. M. Nakajima, "Mapping Method for Salt Attack Hazard Using Topographic Effects Analysis," First Asia/Pacific Conference on Harmonisation of Durability Standards and Performance Tests for Components in the Building Industry, Bangkok, Thailand, Sept. 15, 1999.

47. F. H. Haynie, "Environmental Factors Affecting the Corrosion of Galvanized Steel," in Degradation of Metals in the Atmosphere, S. Dean and T. S. Lee, (Eds.), American Society for Testing and Materials, Philadelphia, PA, 1988, pp. 282–237.

48. R. D. Klassen and P. R. Roberge, "Aerosol Transport Modeling as an Aid to Understanding Atmospheric Corrosivity Patterns," Mater. Design, **20**, 159–168 (1999).

49. S. B. Lyon, C. W. Wong, and P. Ajiboye, "An Approach to the Modeling of Atmospheric Corrosion," in Atmospheric Corrosion, ASTM STP 1239, W. W. Kirk and H. H. Lawson (Eds.), American Society for Testing and Materials, Philadelphia, PA, 1995, 26–37.

50. R. D. Klassen, P. R. Roberge, D. R. Lenard, and G. N. Blenkinsop, "Corrosivity Patterns Near Sources of Salt Aerosols," in Outdoor and Indoor Atmospheric Corrosion, ASTM STP 1421, H. E. Townsend, (Ed.), American Society for Testing and Materials, West Conshohocken, PA, 2002, pp. 19–33.

51. P. R. Roberge, Corrosion Engineering: Principles and Practice, McGraw-Hill, New York, 2008.

52. P. R. Roberge, Corrosion Basics—An Introduction, 2nd ed., NACE International, Houston, TX, 2006.

53. R. F. Steinmayer, "Land Vehicle Management," in AGARD Lecture Series No. 141, NATO, Neuilly-sur-Seine, France, 1985.

54. W. Hou and C. Liang, "Eight-Year Atmospheric Corrosion Exposure of Steels in China," Corrosion, **55**, 65–73 (1999).

55. M. Morcillo, E. Almeida, M. Marrocos, and B. Rosales, "Atmospheric Corrosion of Copper in Ibero-America," Corrosion, **57**, 967–980 (2001).

56. "The Zinc Millennium Map Provides Potential Cost Savings," Anti-Corros. Methods Mater., **48**, 388–394 (2001).

57. F. B. Fletcher, "Corrosion of Weathering Steels," in Corrosion: Materials, Vol. **13B**, D. S. Cramer and B .S. Covino (Eds.), ASM International, Metals Park, OH, 2005, pp. 28–34.

58. H.E. Townsend, "Effects of Alloying Elements on the Corrosion of Steel in Industrial Atmospheres," Corrosion, **57**, 497–501 (2001).

59. D. Greenfield and J. D. Scantlebury, "Blistering and Delamination Processes on Coated Steel," J. Corros. Sci. Engi. (Electronic), **2**, (2000).

60. M. Salagaras, P. G. Bushell, P. N. Trathen, and B. R. W. Hinton, "The Use of Corrosion Prevention Compounds for Arresting the Growth of Corrosion in Aluminium Alloys," Fifth Joint NASA/FAA/DoD Conference on Aging Aircraft, Orlando, FL, Sept. 10–13, 2001.

61. C. Fiaud, "Theory and Practice of Vapour Phase Inhibitors," in Corrosion Inhibitors, Institute of Materials, London, UK, 1994, pp. 1–11.

24

ATMOSPHERIC CORROSION IN COLD REGIONS

G. A. KING*

CSIRO Building, Construction and Engineering, Highett, Victoria, Australia

A. INTRODUCTION

Atmospheric corrosion is an electrochemical process, and moisture must be present on the corroding surface for it to proceed. This can be in the form of a visible liquid layer (rain, dew) or as a thin film formed by condensation at the surface. It has been accepted that such moisture films can form at critical relative humidities, which are determined by the nature of the corroding surface and the chemical species that are present on it. The time that a surface remains "wet" by either of these processes has been defined as the time of wetness (TOW) in the International Standard on Corrosivity of Atmospheres ISO 9223 [1], and it is the critical factor in determining the extent of atmospheric corrosion, together with die presence of pollutants that can be natural (e.g., airborne sea salt) or man-made (SO_x, NO_x). The ISO standard also uses the length of time when the relative humidity exceeds 80% at

a temperature greater than 0°C as an estimate of TOW. However, evidence gathered by several researchers in coastal regions of the Antarctic continent has shown this definition to be inadequate.

There has been a common assumption that there is no corrosion in Antarctica because of the "dry cold," and this myth has persisted even among people with Antarctic experience. Remarkable preservation of metal can be observed in Antarctica, and the excavation of uncorroded metal artifacts from the ice inside the Scott expedition huts in New Zealands's Ross Dependency in the early 1960s probably reinforced the myth. Removal of the ice and snow radically changed the microenvironment and resulted in rapid corrosion of metal items [2]. Notwithstanding the low temperatures, the environment at several historic Antarctic sites was perceived to be severely corrosive and erosive, with very strong winds frequently blowing salt-laden snow off the sea ice and depositing it on huts and equipment. Measurements of corrosivity were made at Cape Evans (the site of Scott's hut) using a standard copper-bearing steel, and the result of 10.8 µm/year (0.43 mil/year) was much higher than expected, considering the site is 77°S, is surrounded by sea ice for 10 months of the year, and has air temperatures rarely above 0°C [3]. At nearby Scott Base (Ross Island), the mean monthly temperature for the coldest month (July) is −29°C, for the warmest month (January) is −5°C, and annually it is −20°C. The precipitation is 150–200 mm of water equiv/year [3].

The myth that corrosion in very cold regions is always low is dispelled. This chapter discusses issues relating to cold climate corrosivity and gives references to studies conducted in these regions.

* Retired.

Uhlig's Corrosion Handbook, Third Edition, Edited by R. Winston Revie
Copyright © 2011 John Wiley & Sons, Inc.

B. INFLUENCE OF TEMPERATURE AND SOLUBLE SALTS

Experimental and theoretical work in the former Soviet Union [4] demonstrated that no significant reduction in corrosion rates occurs until the temperature falls below −25°C; it then falls sharply and approaches zero at −45°C. This was attributed to a retardation of the electrochemical processes at the surface. Laboratory studies in Canada [5] using the potential developed between platinum and zinc electrodes to indicate the presence of an electrolyte showed measured potentials at −20°C, and the conclusion drawn was that there was no freezing point at which corrosion would cease. Furthermore, as discussed above, salt deposition is significant in coastal regions of Antarctica. Rain is very rare and wind-deposited salt is not washed off surfaces. The phenomenon of depression of the freezing point of salt solutions is well established [6], and with appropriate salt compositions involving chlorides of calcium and sodium, the freezing point could be depressed to as low as −50°C. Given the salt deposition regime in coastal regions of the Arctic/Antarctic and the absence of rain washing, a scenario of a saturated salt solution at the surface is definitely feasible in such environments. It is possible also that the long hours of sunlight in the polar summers could raise the surface temperature of dark objects considerably above the surrounding air temperatures, but this temperature rise is inadequate to totally volatilize the electrolyte layer.

C. DEFINITION OF COLD CLIMATE REGIONS

The issue of defining which geographic regions of the world have climates with low temperatures that can significantly influence corrosivity has been addressed by Perrigo et al. [7, 8]. They state that the regions designated Arctic and Antarctic above and below 66°33′ latitude norm and south do not cover this adequately. Neither does the inclusion of the subarctic and subantarctic defined by some as the regions between 60° latitude and the polar circles. In the Northern Hemisphere, Perrigo suggests two regions defined by the U.S. Army Cold Regions Research and Engineering Laboratory (CRREL) [9] in which the average temperature of the coldest winter month is −18°C or less, or between 0 and −18°C. Nonetheless, part of these regions can have "normal" summer temperatures (>20°C) and rainfall, and at these times corrosion proceeds unchecked.

In the Southern Hemisphere, Perrigo has similarly identified two global lines to define the limits of cold climates. The northern most is an isotherm where the mean temperature of the warmest month is 10°C (50°F) [7], and to the south is the Polar Front (previously known as the Antarctic

Convergence [10]), which is a temperature boundary between the near-freezing waters surrounding Antarctica and the warmer waters of the subantarctic. Only the southern tip of the South American continent, the subantarctic islands, and of course Antarctica itself fall within these regions. Heginbottom et al. draw attention to the distribution of permafrost [11] as being another indicator of a cold climate and a factor that will significantly influence cold climate corrosion.

D. ARCTIC CLIMATE AND CORROSIVITY STUDIES

The general pattern of corrosivity in the regions traditionally designated as the Arctic and subarctic can be expected to differ from the cold regions in the Southern Hemisphere, namely, the Antarctic and subantarctic, since the Arctic comprises continental land masses surrounding the North Pole, which is in the middle of the Arctic Ocean and in summer has large areas of open water. Subarctic regions include prairies and forests that can be expected to have low corrosivity, but there are also areas with substantial industrial and mining activity in North America, Scandinavia, and Russia, where higher corrosivity will occur. There are many islands and coastal regions with maritime climates (e.g., Aleutian Islands and Alaska) where marine salts will influence corrosivity. Extensive use is made of deicing salts in North American and Scandinavian cities, which have a profound local effect on corrosivity.

A Canadian site, Norman Wells, some 450 km from the sea in the Northwest Territories has been a benchmark for a site with low corrosivity. The 1-year corrosion rate of cold rolled steel was measured as part of ongoing American Society for Testing and Materials (ASTM) programs and the average rate (for 6 years) was 0.7 μm/year (0.03 mil/year) [12]. This is a surprisingly low result considering that the mean monthly temperature ranges from +6 to +16°C for 5 months of the year with significant summer precipitation and relative humidity (RH) consistently >85% [13].

Divine and Perrigo [14] exposed a range of metals in Anchorage, Alaska, and report a 1-year corrosion rate for carbon steel of 8.1 μm/year (0.32 mil/year). Biefer [15] exposed mild steel wire-on-nylon bolt specimens for a period of 1 year at sites in the Canadian Arctic and subarctic. Corrosion rates as low as 2–5 μm/year (0.08–0.2 mil/year) were obtained at 10 inland sites on the mainland of the western Arctic and the northwest Arctic Islands. At 7 other northern sites, usually within 1 km of the sea, corrosion rates ranged from 21 to 34 μm/year (0.83–1.34 mils/year), comparable to rates measured in southern Canada.

Mikhailov et al. [16] studied corrosivity in eastern Siberia including severe cold climates both on the coast (annual temperatures −6 to −13°C) and inland (−11 to

$-17°C$). They provided corrosivity data for carbon steel and zinc for 29 sites, and this was correlated against TOW, annual average temperatures and relative humidities, and atmospheric pollutants (SO_2, Cl^-). Measured steel corrosion rates ranged from 0.7 to 32 μm/year (0.03–1.26 mils/year). Thirteen references are provided.

There have been a number of studies in Scandinavia concerned with the classification of corrosivity. Kucera et al. [17] exposed specimens of steel and zinc at 32 sites, of which all but one were described as being in a temperate climatic zone. The subarctic station with a yearly average temperature of 1.3°C showed a steel corrosion rate of 6 μm/year (0.24 mil/year). Hakkarainen and Yläsaari [18] measured corrosion with steel and zinc at five sites in southern Finland and correlated the data with TOW and SO_2 levels. Corrosion rates of steel were from 20 to 50 μm/year (0.79–1.97 mils/year). In Norway, a series of exposure programs was reviewed by Atteraas and Haagenrud [19]. Materials included steels, zinc, and aluminum exposed at up to 38 sites mostly close to the coast Measured corrosion rates were generally quite high (e.g., for carbon steel 1-year exposure 17–70 μm/year) (0.67–2.76 mils/year). Seventeen references were provided.

E. SUBANTARCTTC CLIMATIC VARIATION AND CORROSIVITY

Isolated subantarctic islands surround the Antarctic continent. Subantarctic climates are very different from those of Antarctica itself. Characteristically, temperatures are cool, with summer temperatures rarely above +10°C and winter temperatures rarely below 0°C. Formation of sea ice is rare. Winds are characteristically gusty and strong. Being isolated by long sea distances from any industrial activity, pollution is very low. All of the islands have historic sites, and metal artifacts show extensive corrosion. Macquarie Island (an Australian territory at latitude 54°S) was the subject of a small survey (6 sites) to obtain corrosivity data that could assist in conservation measures [20]. The environment is severe marine with annual rainfall of 901 mm, and mean relative humidity at 0900 h is 85–90% throughout the year [21]. Standard low-alloy copper-bearing steel coupons were used, and 1-year corrosion rates were found to be on average 122 μm/year (4.80 mils/year), with a maximum rate of 219 μm/year (8.62 mils/year). Twenty-two references are provided.

F. ANTARCTIC CLIMATE AND CORROSIVITY STUDIES

Corrosion studies conducted in Antarctica and aspects of the Antarctic climate have been reviewed in a paper that also describes a corrosivity survey conducted at nine sites on the continent, including the South Pole [22], Thirty-nine references were provided.

The Antarctic is generally characterized by severe cold, with average temperatures in summer being significantly lower than those in the Arctic [10]. The Antarctic Peninsula is the warmest region of continental Antarctica, but summer air temperatures are still rarely above +5°C. Summer temperatures in the high Arctic are considerably warmer and can exceed +20°C, and dwarf Arctic willow trees grow even at 75°N. Antarctic climates are so cold that there are no trees and even grass only grows in the warmest parts of the Antarctic Peninsula.

The authors of the review recognized that standardized measurements of corrosivity could assist in conservation strategies at historic sites, and clarification was needed to establish why corrosion rates are higher than expected given presumed low TOWs. The 1-year corrosion rate of low-alloy copper steel showed an exceptional range from 27.1 μm/year (1.07 mils/year) at Rothera (Antarctic Peninsula, 67°34′S, on the coast) to 3.4 μm/year (0.13 mil/year) at Mawson (67°36′S, on the coast), 0.87 μm/year (0.03 mil/year) at Vanda (77°35′S, 80 km from the coast [3]), and 0.05 μm/year (0.002 mil/year) at Vostok (78°28′S, 1200 km from the coast). This variation on the Antarctic continent itself represents a factor >500. The influence of TOW, salt deposition, and wind direction on the results was discussed.

Mikhailov et al. [23] studied the corrosion of steel, copper, cadmium, and an aluminum alloy at a Russian coastal station, Mirnyi (66°33′S). A corrosion rate for steel of 7.7 μm/year (0.30 mil/year) was measured and described as comparable with the rate on Ayon Island in the Russian Far East (above the Arctic Circle) and an order of magnitude higher than the corrosion rates measured at two other Russian sites, namely Bilibino and Oimyakon [24]. Climatic parameters and salt levels were measured and the ISO TOW was only 93 h/year in contrast to the range described in the standard [1] of 250–2500 h/year (category τ_3) for dry cold climates and some temperate climates. The authors concluded that in the presence of airborne sea salt, corrosion could proceed at temperatures significantly lower than $-1°C$.

Fahy [25] studied corrosion of aluminum coupons with different surface treatments including mill and anodized finishes. He experienced problems with conducting the exposures since he was unable to travel to Antarctica himself, and some plates were exposed next to diesel generators at Scott Base and pollutants affected results. Exposures were later repeated at unpolluted Arrival Heights that resulted in considerably less pitting of sample coupons.

Rievero et al. [26] used Mössbauer spectroscopy to study the corrosion products on low-alloy steel exposed at a Uruguayan Antarctic island site (Artigas, 62°10′S). The mean temperatures for the coldest month, hottest month, and annual were, respectively, -6.4, 2.2, and $-2.3°C$, and steel corrosion rates (1 year) were $40–66\,\mu m/year$ ($1.57–2.60\,mils/year$).

Resales et al. [27] measured corrosion rates for steel, zinc, copper, and aluminum at Jubany, an Argentinean Antarctic base. The TOW was directly measured using Pd/Ag electrodes on an alumina substrate and compared with the ISO-estimated TOW using climatic data. High corrosion rates were measured for all metals (steel 1-year rates, $36–41\,\mu m/year$) ($1.42–1.61\,mils/year$) and measured TOW (always higher than the ISO estimate) was 24 h/day continuously for 7 months of the year. The minimum values in the coldest months ($-3°C$) still gave a daily mean TOW of 6 h. The authors concluded that in the presence of marine salt, liquid water monolayers could form under ice layers resulting in high corrosion rates at temperatures well below $0°C$

G. CONCLUSIONS

Extensive evidence has now shown that atmospheric corrosion rates can be high in cold regions of the world especially in proximity to the sea. Whereas very low corrosion rates can occur in cold continental climates (especially inland Antarctica), in coastal regions the influence of chloride ion from marine sea salt can enable liquid layers to form at temperatures well below $0°C$ and corrosion proceeds unimpeded. Actual surface temperature in conjunction with the presence of chloride ion determines the rates of corrosion.

REFERENCES

1. International Organization for Standardization (ISO), Corrosion of Metals and Alloys—Corrosivity of Atmospheres, ISO 9223 Classification, ISO, Geneva, Switzerland, 1992.
2. R. Parrish, Corros. Australasia, **8**(3), 2 (1983).
3. G. A. King, G. J. Dougherty, K. W. Dalzell, and P. A. Dawson, Corros. Australasia, **13**(5), 13 (Oct. 1988).
4. A. Dychko and J. Dychko, J. App. Chem. USSR, **30**, 251 (1957).
5. P. Sereda, "Weather Factors Affecting Corrosion of Metals," in Corrosion in Natural Environments. ASTM STP 558, American Society for Testing and Materials, Philadelphia, PA, 1974, pp. 7–22.
6. G. W. Brass, "Freezing Depression by Common Salts: Implications for Corrosion in Cold Climates," in Proceedings of the National Association of Corrosion Engineers, Canadian Region Western Conference, Anchorage, Alaska, 1996, pp. 447–453.
7. L. D. Perrigo, H. G. Byars, and J. R. Divine, Cold Climate Corrosion: Special Topics, NACE, Houston, TX, 1998.
8. L. D. Perrigo, "Summary of a Discussion of Cold Climate Corrosion Issues," presented to the Australasian Corrosion Association Annual Conference, Brisbane, Nov. 12, 1997, available: www.corrprev.org.au.
9. E. A. Wright, CRREL's First 25 Years: 1961–1986, US Army Cold Regions Research and Engineering Laboratory, Hanover, NH, 1986.
10. B. Stonehouse, North Pole–South Pole, Pion, London, 1990.
11. J. A. Heginbottom, J. Brown, E. S. Melnikov, and O. J. Ferrians Jr., Proceedings of the Sixth International Conference Permafrost, Vol. 2, Beijing, China, July 5–9, 1993.
12. ASTM Committee G1 "Corrosiveness of Various Atmospheric Test Sites as Measured by Specimens of Steel and Zinc," in Metal Corrosion in the Atmosphere, ASTM STP 435, American Society for Testing and Materials, Philadelphia, PA, 1968, pp. 360–391.
13. E. A. Pearce and C. G. Smith, The Hutchinson World Weather Guide, Hutchinson, London, 1984.
14. J. R. Divine and L. D. Perrigo, "Atmospheric Corrosion Testing in the Arctic and Subarctic—A Review," Paper No. 389, in Proceedings of the Corrosion 86 Conference, NACE, Houston, TX, 1986.
15. G. A. Biefer, Mater. Perform., **20**(1), 16 (Jan. 1981).
16. A. Mikhailov, M. Syloeva, and E. Vasilieva, Data Base on Atmospheric Corrosivity in Towns and Industrial Centres in the Territory of the Former USSR, Institute of Physics and Chemistry, Russian Academy of Science, Moscow, 1992.
17. V. Kucera, S. Haagenrud, L. Atteraas, and J. Gullman, "Corrosion of Steel and Zinc in Scandinavia with Respect to the Classification of the Corrosivity of Atmospheres," in S. W. Dean and T. S. Lee (Eds.), Degradation of Metals in the Atmosphere, ASTM STP 965, American Society for Testing and Materials, Philadelphia, PA, 1988, pp. 264–281.
18. T. Hakkarainen and S. Yläsaari, *Atmospheric Corrosion Testing in Finland*, in Atmospheric Corrosion, W. H. Ailor (Ed.), Wiley, New York, 1982, pp. 787–795.
19. L. Atteraas and S. Haagenrud, "Atmospheric Corrosion in Norway," in Atmospheric Corrosion, W. H. Ailor (Ed.), Wiley, New York, 1982, pp. 873–891.
20. G. A. King and J. D. Hughes, Bul. Australasia Inst. Conservation Cultural Mater., **8**(3, 4), 25 (1993).
21. N. A. Streten, Papers Proc. R. Soc. Tasmania, **122**(1), 91 (1988).
22. J. D. Hughes, G. A. King, and D. J. O'Brien, "Corrosivity in Antarctica—Revelations on the Nature of Corrosion in the World's Coldest, Driest, Highest and Purest Continent," Paper No. 24, in Proceedings of the Thirteenth International Corrosion Conference, Melbourne, Australia, Nov. 25–29, 1996.
23. A. Mikhailov, Y. Mikhailovski, N. Sharonova, and M. Suloeva, Protection Metals (Zhasheta Metallov, Moscow), **29**(1), 12 (1993) (in Russian).
24. M. Yu. Panchenko, L. A. Shchuvakhina, and N. Yu. Mikhailovski, Protection Metals (Zhasheta Metallov, Moscow), **20**(6), 851 (1984).

25. F. Fahy, "The Corrosion of Architectural Aluminium," unpublished report. Department of Mechanical Engineering, Canterbury University, Christchurch, NZ, 1990.

26. S. Rivera, E. Quagliata, V. Diaz, and L. Morales, "Mossbauer Spectroscopy Study of Corrosion Products Developed on Unalloyed Steel Samples Exposed to Natural Environments," Paper No. 33, in Proceedings of the Thirteenth International Corrosion Congress, Melbourne, Australia, Nov. 25–29, 1996.

27. B. Rosales, A. Fernández, and G. Marietta, "Marine Corrosion of Steel Zinc Copper and Aluminium in Antarctica," Paper No. 31, in Proceedings of the Thirteenth International Corrosion Congress, Melbourne, Australia, Nov. 25–29, 1996.

25

CORROSION BY SOILS

T. R. Jack

University of Calgary, Calgary, Alberta, Canada

M. J. Wilmott

Wasco Coatings Ltd., Kuala Lumpur, Malaysia

A. INTRODUCTION

In the original Uhlig *Corrosion Handbook*, Logan [1] was given the task of describing the state of knowledge regarding the corrosion of metals in soils. Much of the data presented in this study remains pertinent today. Other useful references include Starkey and Wight [2], Booth et al. [3–5] and Flitton and Escalante [6]. Romanoff [7] describes a National Bureau of Standards study in which 37,000 specimens were exposed to 95 types of soil for exposure periods of up to 17 years! Specimens included ferrous and nonferrous metals as well as a selection of coating materials. In 1979, Escalante [8] edited an Amercican Society for Testing and Materials (ASTM) symposium that focused on underground corrosion. Specific articles covered soil surveys, soil corrosivity, soil testing, and mitigation of corrosion in soils. The ASTM *Corrosion and Standards: Application and Interpretation* [9] handbook also contains several sections related to corrosion in soil.

The reason for the abundance of reference material in this area is the economic significance of material degradation in soils. The method of choice for transportation of water, natural gas, oil, and refined hydrocarbons remains that of buried pipelines. In Canada, there are ~580,000 km of buried pipelines, which contribute billions of dollars to the economy. In the United States there are about 2 million km of buried natural gas pipeline and ~280,000 km of petroleum products pipelines. The replacement value for such facilities represents billions of dollars. This mode of transportation remains by far the safest way to ship oil and gas, but deterioration of line pipe steels whether as the result of external corrosion or environmentally assisted cracking

(EAC) can result in failures [10]. Concern over failures has resulted in the development of new direct assessment protocols for external corrosion [11] and stress corrosion cracking [12] that depend in part on characterization of the soil environment [13].

Traditionally, discussions of soil corrosivity have been based on studies where materials including steel were placed in direct contact with the soil environment. Easily measured characteristics such as soil resistivity or moisture content were then correlated with the occurrence of material degradation problems. Such general correlations continue to provide guidance for materials selection, location of structures, route selection, and overall system design but are often too general to allow the accurate prediction or interpretation of damage in specific sites. Soil is a complex material and the many parameters, which affect corrosion by soils, are still not fully understood. Simple models based on the general interpretation of two or three parameters that ignore the specifics of a given situation are not likely to provide a satisfactory explanation of corrosion and EAC processes at specific sites [11, 12]. More sophisticated models are under development. Figure 25.1 shows two scenarios that provide a framework for discussion in this chapter.

Figure 25.1(a) shows the case where a buried steel pipe is in direct contact with the soil environment. The overall system includes the exchange of soil gas with the overlying atmosphere as well as the potential influx of surface water and precipitation to the water table through the unsaturated soil zone. In the unsaturated zone, gas forms the continuous phase between particles in the soil matrix. In Figure 25.1(a), the buried pipe lies partly above the local water table. The steel surface in the unsaturated zone may see a different environment than that lying in the water-saturated soil. Disturbed soil around the buried structure may also lead to a unique environment at the steel surface [6].

Both corrosion and EAC involve local electrochemical processes sustained by a water phase acting as an electrolyte at the metal surface. Soil sets the stage for these processes by controlling the access of agents such as atmospheric oxygen to the steel surface, fostering biological activity, or otherwise altering the chemical composition of the water phase in contact with the metal.

Figure 25.1(b) shows a second situation where the buried steel is shielded from the soil environment by a protective coating or rock shield. Where a protective coating remains effective, corrosion or EAC problems do not occur; however, in situations where water reaches the steel surface, attack on the metal can proceed. In this situation, changes in composition due to Corrosion, microbial activity, cathodic protection, or the incursion of O_2 or CO_2 [14–16] can significantly alter the chemistry of the trapped water.

Cathodic protection is routinely used to protect buried steel structures [17]. This involves installation of a system to impress a negative electrical charge on the steel surface,

effectively preventing dissolution of the metal for thermodynamic reasons. Application of cathodic protection to the two scenarios shown in Figures 25.1(a and b) can lead to different effects. Where bare steel is exposed directly to the soil environment, Figure 25.1(a), protection potentials can be achieved by a properly designed and maintained cathodic protection system. Where effective potentials are not achieved or maintained, corrosion can proceed according to the corrosivity of the local soil environment at pipe depth. Corrosion cells can develop where bare pipe passes through saturated and unsaturated soil zones along its length or where different parts of the pipe surface lie in different soil environments. For the situation shown in Figure 25.1(a), corrosion could result from a differential aeration cell between the top and the bottom of the pipe due to the difference in water saturation of the soil matrix. For steel surfaces exposed to water under disbonded coating, Figure 25.1(b), a very different scenario occurs. If the disbonded coating remains electrically insulating, the underlying pipe surface can be shielded from cathodic protection (CP). Where effective CP potentials are not achieved, corrosion and environmentally assisted cracking (notably stress corrosion cracking) can proceed [18, 19]. Unfortunately, shielding disbondments have proven difficult to detect by surface inspection methods that depend on anomalies in current or potential to identify coating damage [20, 21].

The development of risk models to improve the long-term maintenance of buried structures is an area that has advanced significantly as integrity management plans become a required element for operating pipeline companies.

B. WHAT IS SOIL?

Various schemes have been proposed to categorize soils, but no single scheme spans the diversity of soil environments seen by buried metal structures. Muskegs, peat bogs, and swamps consist primarily of water-saturated decaying organic matter mixed with minor amounts of inorganic material. Thus peat bogs are typically 95% water with more than 17% of the dry weight being carbon [22]. Soils with 20% or more organic content such as those found in the Florida everglades are classed as histosols [23]. In contrast other soils are almost devoid of humic or organic matter, consisting largely of inorganic materials broken and weathered to various particle sizes. The organic matter in a highly weathered sandy soil is only ~0.5%, while poorly drained prairie soils contain ~6% organic matter [23]. Soluble salts precipitated from ancient lakes or oceans dominate the behavior of evaporites. Such large variations in soil environments can support a wide range of potential corrosion mechanisms and influence rates in various ways.

Soils are an admixture of rock disintegrated by physical action and modified by weathering (clastic sediments),

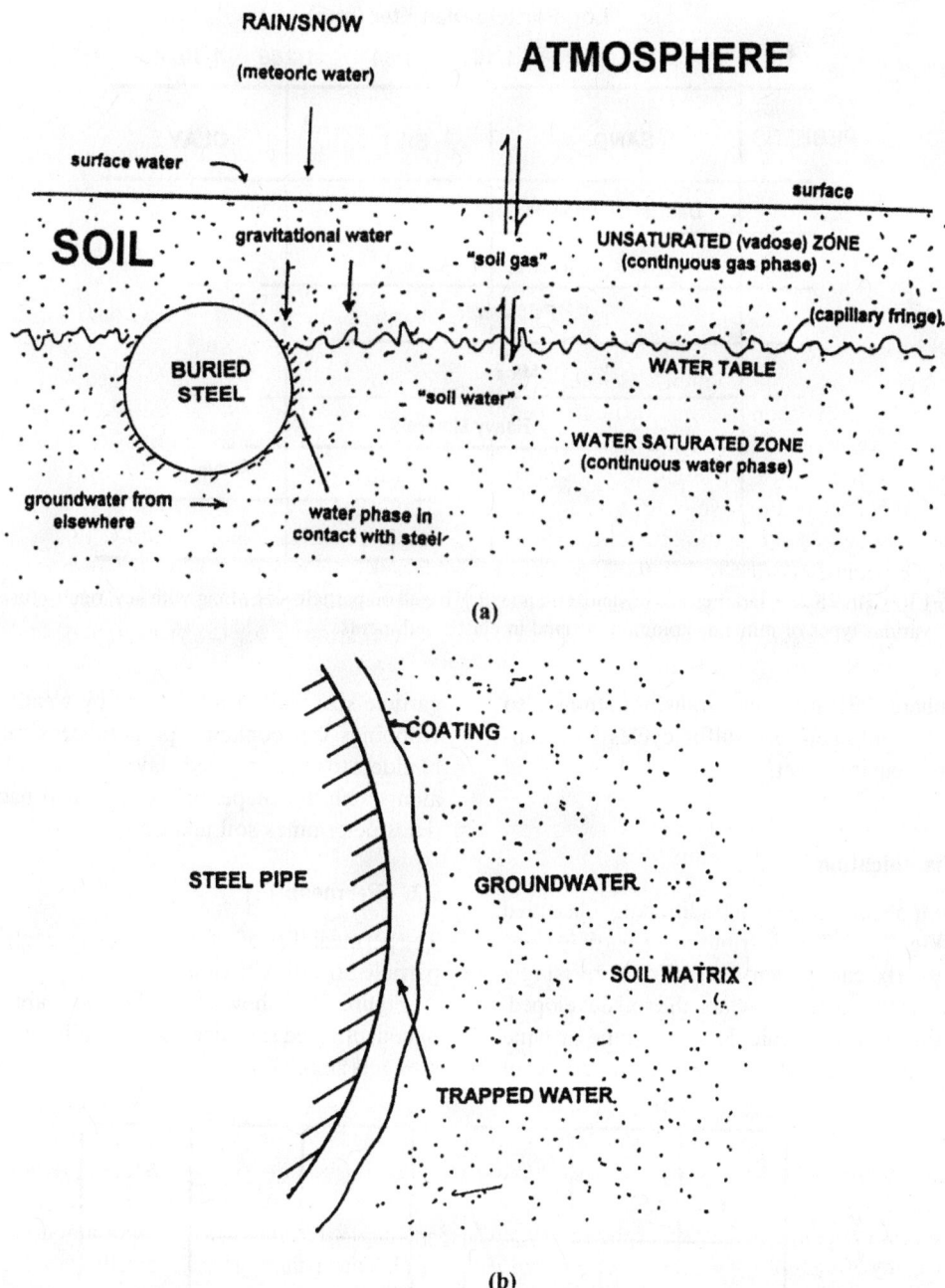

FIGURE 25.1. (a). Exposure of buried line pipe to soil environment. (b). Failure of protective coating can allow water to reach metal surface. Water trapped beneath the coating can develop a composition which is different than groundwater.

materials precipitated chemically from aqueous solution (nonclastic sediments and evaporites), and organic matter (humus). Relative proportions depend on the specific history of soil formation in a location. As much as one half of the volume of a dry soil can be made up of pore spaces between particles. This space is shared by interstitial water and soil gas. Some water is tightly associated with mineral surfaces and cannot be readily removed. Bulk water can flow through the porous matrix as a continuous phase in water-saturated zones and by drainage in unsaturated ones, Figure 25.1(a). In comparison, gas diffuses through unsaturated zones as the continuous phase but must proceed through the saturated zone dissolved in the water phase. Where biological activity by soil organisms is supported by sufficient nutrients, oxygen is removed in the pore space and carbon dioxide builds to concentrations a hundredfold higher than found in the

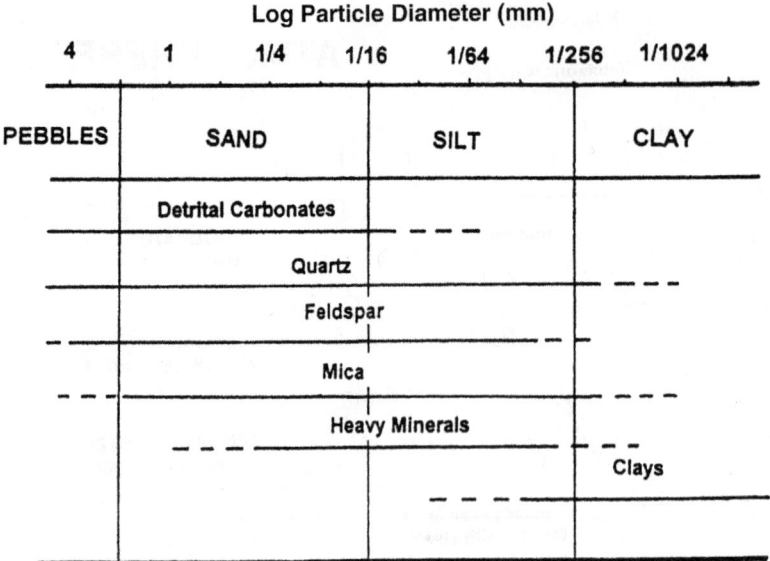

FIGURE 25.2. Classification of sand, silt, and clay based on particle size along with size ranges for various types of minerals commonly found in clastic sediments.

overlying atmosphere [23]. Important transformations also occur as part of the nitrogen and sulfur cycles in a biologically active soil environment.

B1. Textural Classification

Inorganic soil components have been traditionally classified by particle size (Fig. 25.2). The distribution of particle sizes in a given soil matrix can be very broad or surprisingly narrow depending on the manner in which the soil developed.

For example, sand originally laid down as a dune by wind action can be finely sorted with a very narrow range of particle size while soils formed by weathering of rock outcroppings can contain - particle sizes ranging from large boulders to finely divided clays. The particle size distribution along with the shape, orientation, and packing of soil particles determines soil texture.

B2. Permeability

The permeability of a soil matrix is strongly influenced by the particle size distribution.

Figure 25.3 shows how the permeability of well-sorted unconsolidated sand decreases rapidly as the average particle size gets smaller. Permeability is higher in well-sorted sands

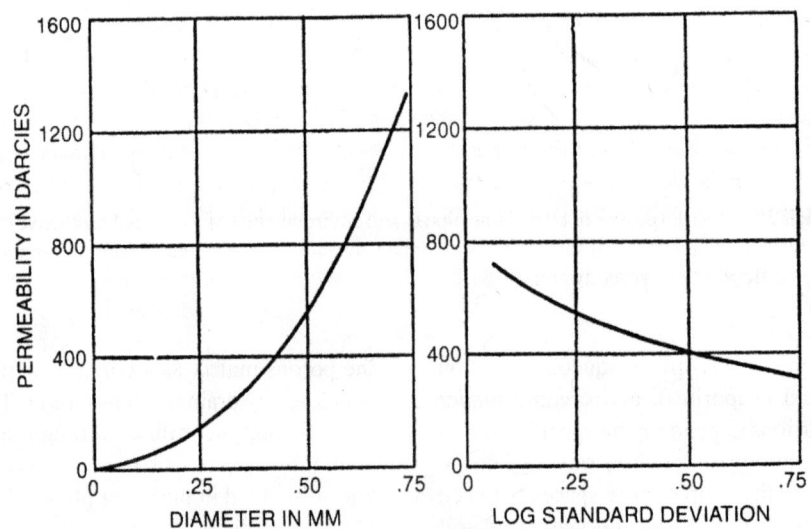

FIGURE 25.3. Variation of sand permeability with average particle size and with breadth of particle size distribution [27].

with a narrow distribution of particular sizes while sands with wider particle size distributions show reduced permeability (Fig. 25.3).

Permeability plays an important role in setting the stage for electrochemical events on a buried steel surface because it controls the relative ease or rate of fluid flow through the soil matrix. Coarse-grained sands allow good drainage and easy access of atmospheric oxygen to structure depth. Fine-grained soils rich in clay are more restrictive. Capillary forces in fine-grained matrices can also draw water up and keep a soil water saturated even during relatively dry conditions by preventing drainage and retarding evaporation [24].

Waterlogged soils are particularly effective at slowing the rate of oxygen ingress from the overlying atmosphere. Consequently, waterlogged soils are often anaerobic at even very modest depths. Control of oxygen access can dictate the basic corrosion scenario seen at a buried steel surface.

Association of inorganic soil particles with organic material can lead to aggregate formation. This influences soil structure and can alter the soil permeability. A well-developed soil structure can enhance water and air movement through the affected soil zone [23].

B3. Soil Color

Soil color can be used to diagnose oxygen permeation. Good drainage and warm temperatures encourage extensive weathering and the formation of iron(III) oxides that give the soil a characteristic red tint. This might be expected in a well-drained sandy soil that is gritty to the touch. Brown and reddish brown hues indicate the presence of organic matter. Yellow or rust colors occur in soils that are slightly wetter due to the iron(III) oxides being left in a hydrated form. This might be expected in silt where a relatively small particle size lowers the permeability of matrix and retains water. Anaerobic soils where iron has not been oxidized to ferric ions are bluish gray, which is characteristic of waterlogged low-permeability clay soils or gley.

Particle size can be roughly correlated with the mineralogy of soils (Fig. 25.2). Techniques such as X-ray diffraction can identify the mineral composition. The diffraction pattern is a fingerprint for the crystal structure of each mineral. Organic material and some inorganic components such as noncrystalline iron sulfide formed by anaerobic microbial action do not show up in this analysis.

B4. Mineralogical Composition

Mineral composition is a key to understanding how a soil can influence the corrosion of buried steel.

B4.1. Clays. Clays are among the most common minerals on earth, constituting perhaps one third of all sedimentary materials [25]. They form through the weathering of parent materials; the chemical transformation of feldspars being one of the most common processes [22, 26]. Clay formation is accelerated where conditions favor weathering and leaching of minerals. In the extreme case, only the most stable clays (e.g., kaolinite) are found. Other clays, being intermediates in the weathering process, dominate the composition where soil has developed under less extreme conditions.

Most clays have notable plasticity when wet and a marked ability to adhere to surfaces. Owing to their small particle size and structure, clays have exceptionally high surface areas and show great surface activity relative to other types of minerals. The surface area of a clay can be 10,000 times that of an equivalent weight of sand [23]. Cations required to charge balance the hydrous aluminum silicate framework in clay minerals provide opportunity for ion exchange and redox chemistry.

In an extensive survey of prairie soils in western Canada, the Alberta Research Council found that the clay content correlated inversely with percent sand present. Concentration of metals including iron, copper, boron, and perhaps cobalt and zinc present in the soil correlated directly with the clay content but calcium, magnesium, and the pH showed no correlation with clay content The plasticity index and liquid limit (% moisture) were also a function of clay content.

Four families of clay minerals are described in Table 25.1.

Characteristics shown in Table 25.1 suggest that the montmorillonite group will have an important influence on the corrosion of buried steel structures. High concentrations of montmorillonite (or smectite) clays are responsible for the sticky yet slippery nature of so-called gumbo soils. Their small particle size leads to very low permeability, extraordinary water uptake and retention, low fluid flow, slow exchange of interstitial water, limited gas migration, and extensive ion exchange with the electrolyte as well as high capacity buffering of pH and poising of the oxidation reduction potential in the soil–water system [26].

Physically, clays of the montmorillonite group such as bentonite can radically change volume through either dehydration/rehydration or ion exchange. This shrinking and swelling can exert force on structures buried in montmorillonite-rich soils leading to potentially detrimental consequences. For example, coating damage and disbondment can occur when a montmorillonite clay adhering to the protective coating on a buried structure changes volume. The ability of clay-rich soils to adhere to surfaces can also aggravate coating damage caused by soil settling around the structure. In extreme cases, shrinkage due to drying can lead to the formation of large-scale cracks through the soil matrix in soils known as vertisols [23]. Swelling conditions can also compromise the permeability of a soil through both the increase in particle volume and the release of particles into fluid flow, resulting in an irreversible plugging of pore throats in the soil matrix [24].

TABLE 25.1. Properties of Major Groups of Clay Minerals[a]

Characteristic	Kaolinite Group	Illite Group	Chlorite Group	Montmorillonite Group
Occurrence	Common	Abundant	Common	Common
Structural type	2 Layer	3 Layer, nonexpandable	3 Layer, nonexpandable	3 Layer expandable
Particle size (μm)	4.0–0.3	0.3–0.1	0.3–0.1	0.2–0.02
Permeability	Large	Moderate	Moderate	Small
Water absorption	Slight	Moderate	Moderate	Very large
Surface area (m²/g)	5–20	100–200		700–800
Cation exchange capacity (meq/100 g)	Slight	Moderate	Moderate	Large
	2.3	16		81
	7–12	18–24	25–30	90–100
Anion exchange capacity (meq/100 g)	7–20	4–17	5–20	20–30
Plasticity	Slight	Moderate	Moderate	Large

[a]This table is generally based on [27]. Cation exchange capacities are from [26] and [25], respectively, anion exchange capacity based on phosphate exchange is from [28], and surface areas are from [29].

B4.2. Sand. Relative to clays, coarse silica sands tend to be relatively permeable, well drained, and inert.

B4.3. Carbonates. Even modest amounts of limestone or dolomite in soils will dissolve over time to saturate associated groundwater. Dissolved carbonate will buffer the solution in the neutral to alkaline pH range [30]. Exposure of this saturated solution to steel surfaces rendered alkaline by electrochemical reactions induced by an effective cathodic protection system will precipitate hard white carbonate scales (e.g., calcite or dolomite) on the metal surface. These scales are strongly adherent and form an impermeable protective layer that indicates both effective cathodic protection and negligible corrosion problems.

B4.4. Evaporites. Evaporation of ancient lakes and oceans has left massive deposits of precipitated salts in some geographic regions. While construction directly in salt beds is a special situation, high salinity salt seeps formed by dissolution of soluble salts into mobile groundwater affect soils more broadly. Consequent increases in the electrical conductivity of the water in the soil system can accelerate electrochemical reactions causing corrosion of steel directly exposed to the soil environment [Fig. 25.1(a)]. In contrast, the increased conductivity for electrolyte under failed coatings can improve the penetration of effective cathodic protection potentials to the underlying steel surface [31] minimizing corrosion or EAC damage in the scenario shown in Figure 25.1(b).

B5. Soil Profiles

Recent geological deposits or those just beginning to develop as soils lack a vertical profile and are called entisols or inceptisols, respectively.

Mature mineral soils are vertically differentiated between the soil surface and underlying strata or bedrock. Three general horizons can often be seen in the vertical profile [25]. At the top is the A or eluvial horizon. The A horizon is in contact with the atmosphere and receives rainfall, nutrients, and so on, from the surface environment. Biological activity and formation of humus arc generally focused here. Leaching by gravitational water removes soluble minerals from this zone, concentrating less soluble, more stable minerals in this layer. The underlying B or illuvial horizon is an intermediate zone that is often low in organics but is enriched by deposition of soluble minerals and secondary transformation products from weathering reactions in the overlying layer. The C horizon at the bottom of the sequence is largely composed of parent material unaltered by weathering and hosts little biological activity. Each of these horizons can be further differentiated into subzones or can be entirely missing from the sequence depending on local conditions and climatic factors.

The principal factors affecting the soil profile are temperature and rainfall. Increased temperature can accelerate biodegradation of organic material, chemical reactions associated with weathering and leaching of soluble minerals. Increased rainfall accelerates the leaching of soluble minerals from the upper horizons in the soil profile. Carbonic acid dissolved in meteoric water drives the conversion of key primary minerals into secondary ones, for example, the conversion of feldspars to clays. The order of decreasing stability of common minerals under weathering conditions is iron oxides, aluminum oxides, quartz, clays (kaolinites being more stable than illites, chlorites, or montmorillonites), muscovite, potassium feldspar, biotite, sodium feldspar, amphibole, pyroxene, calcium feldspar, and olivine [25].

B6. Soils in Various Climates

Soil profiles are developed from parent materials in response to factors associated with climate. The following selected cases illustrate some of the principles and processes outlined above.

B6.1. Arid Regions. In desert regions, low rainfall results in minimal leaching of soluble minerals and low organic loading. Such soils, known as aridisols, tend to be light in color and rich in soluble salts. In general corrosion rates for unprotected steel are very low in these dry soils. Flitton et al. [32] reported corrosion rates less than 12 µm/year and pit depths less than 170 µm for carbon steel 1018 coupons buried in an arid vadose zone for up to 3 years. Where interstitial water is retained in the soil matrix, it can be strongly saline. This can create very aggressive soil conditions locally even in arid regions. Clays tend to be loaded with sodium that is released by ion exchange when water is introduced to yield a solution that ranges from near neutral pH when the prevalent anion is sulfate or chloride to strongly alkaline if carbonate salts prevail [23, 25].

B6.2. Tropical Climates. In contrast tropical or semitropical climates with high rainfalls leach soluble minerals extensively from soil horizons. Warm temperatures promote the degradation of organics thinning the top zone. The result is a red lateritic soil rich in iron and aluminum oxides underlying a thin veneer of humic material. These soils tend to be very acidic and are called oxisols [23, 25].

B6.3. Temperate Zones. In temperate zones, a range of soil profiles can be found that depend in part on the extent of rainfall experienced. Mollisols occur where rainfall is modest, thick top soils accumulate, prairie grasslands being typical. Alfisols are slightly wetter, more acidic soils that occur on the wetter borders of grasslands, while ultisols in even wetter, warmer regions show more clay accumulation in the B horizon.

B6.4. Arctic Conditions. Arctic regions are represented by vast areas of tundra. Freeze–thaw cycles result in the mechanical weathering or disintegration of parent rock. These cycles can also alter soil structure and sort soil components. Low temperatures result in the accumulation of thick organic layers.

Heaving during freeze and thaw cycles can displace buried steel structures. One study has reported effects seen on a length of pipe buried in sand and in silt under controlled conditions [33]. During freezing in silt, ice formed beneath the pipe left a cavity running along the length of the pipe on thawing. This would generate a discrete channel for groundwater flow in the field. In such a situation, the composition of water in contact with exposed steel surfaces might be different to that expected for the local soil environment [Fig. 25.1(a)].

Permafrost found in arctic and antarctic regions underlies up to one quarter of the earth's land area. Permafrost is ground remaining at or below 0°C. It may consist of cold dry earth, cold wet saline zones (called talik), and icy lenses or ice-cemented earth and rock. Unfrozen saline zones may be of special interest in terms of corrosion scenarios seen on buried steel structures. These arise by exclusion of soluble salts from ice as crystals form. This concentrates salts in the remaining unfrozen solution [22, 33]. In coarse soils where migration of soil water is relatively easy, unfrozen, high-salinity zones can appear as thick layers (called cryopegs). In fine-grained matrices where transport is more constrained, near microscopic pockets of unfrozen saline water are often found.

C. INSTALLATION SCENARIOS

The manner in which steel structures are constructed can lead to different scenarios even in similar soil. For example, steel structures can be

Driven into ground (e.g., piles)

Installed in excavations and then buried with backfill (e.g., most pipeline construction)

Inserted into predrilled shafts or horizontal tunnels (e.g., pipeline river crossings)

In the case of driven piles, the soil profile is left largely undisturbed except for compaction around the structural unit being pounded into the ground. This can reduce the permeability of the soil matrix adjacent to the steel, slowing the ingress of oxygen from the atmosphere and forcing an intimate contact between the steel and soil. Romanoff [34] concluded from a study of 19 installations in service for 7–40 years that corrosion on steel pilings driven into undisturbed soil was insufficient to affect the strength or useful life of the load-bearing structures. Soil conditions aggressive to steel buried under disturbed conditions were not corrosive to steel pilings in undisturbed soil. Corrosion was found to be variable but not serious above the water table [34, 35] and negligible below the water table. Correlations between soil properties and corrosion developed for buried steel structures were of no practical value in predicting the corrosion of pilings. Flitton et al. [6] in a review of simulated service testing in soils noted that corrosion in undisturbed soil is always low, regardless of soil conditions, because of the low availability of the oxygen. Recent inspection of facilities that had been in service for decades in very different climates gave a maximum corrosion rate of <19 µm/year for nonaggressive soils and <48 µm/year for an aggressive soil environment with high chloride concentration [36, 37].

In the second installation method, the backfill used to bury the structure can be a jumble of various soil horizons and debris that requires extended periods of time to settle and resume development of a normal vertical profile. Flitton et al. [6] indicated that corrosion of unprotected steel in disturbed soils is strongly affected by soil conditions. This is very relevant to buried pipelines and other structures that are constructed or placed in excavated sites and then buried with backfill. Even after many years, Harris [38] found that only the surface zone of the backfill came to resemble adjacent undisturbed soil. The zone above the pipe remained more permeable, more biologically active, and more heterogeneous than adjacent soil at the same depth. The increased permeability allowed increased oxygen penetration into the soil matrix and provided drainage paths for gravitational water. At and below pipe depth, moisture content in the backfilled trench exceeded that of adjacent soil with free water being found in some cases. This could make measurement of bulk soil properties deceptive in terms of predicting the chemistry occurring at the steel surface. For pipeline construction today, top soils are purposely separated from other horizons during excavation and are restored as the covering layer after burial. This minimizes burial of organic rich soil at pipe depth and allows rapid revegetation of the soil surface. Burial of organic materials can promote bacterial activity in the soil adjacent to buried structures, fostering microbially assisted corrosion scenarios in a zone that would be otherwise deprived of organic nutrients [39]:

The third installation method can leave gaps between the steel surface and soil matrix. The influx of meteoric water or surface water and direct interaction with the atmosphere become possible. In this situation, bulk soil properties again may not reflect conditions at the metal surface.

D. FACTORS THAT INFLUENCE CORROSION IN SOIL

Attempts to correlate soil characteristics with the corrosion damage seen for unprotected buried steel initially relied on generic soil classification schemes developed for agriculture or based on the % clay and silt present in the soil matrix [1]. These schemes were generally based on the A horizon of undisturbed soil in a local area and may or may not have reflected conditions in the backfilled trench around a buried pipe, Fig. 25.1(a). The intent was to extrapolate corrosion information developed for a specific location to areas of similar soil classification within a region to inform design decisions related to setting a corrosion allowance or applying protective coatings in new pipeline construction. An extensive database was developed by the National Bureau of Standards (now the National Institute of Standards and

Technology) in the United States using buried test coupons buried in a wide range of soils for many years for this purpose [7]. A recent attempt to reassess data from these studies using modern statistical techniques [40] concluded that only general trends between soil composition and properties could be supported due to a number of issues with the data. Ricker [40] concluded that more complete, statistically designed data sets based on measurements that reflected properties of the soil and groundwater directly in contact with the exposed steel sample would be needed to enable the useful prediction of corrosion rates and distribution. Numerous factors can affect the corrosion of buried steel.

D1. Soil Type

Many potential consequences of soil properties have already been noted in the foregoing discussion, of soil particle size distribution, organic content, mineralogy, and structure.

In general, the soil structure and particle size distribution determine the physical properties of the matrix such as its permeability. Permeability in turn controls the rate of movement of fluids or gases through the matrix. Hence, soils made up of a broad distribution of small size particles such as clay are restrictive white well-sorted coarse sands allow much greater flux and exchange. Capillary forces in low permeability soils also draw groundwater in to form a water-saturated zone. This effectively raises the local water table and creates an effective barrier to the movement of gases that must dissolve and travel in the continuous water phase. One key consequence is the restriction of oxygen access from the atmosphere to a buried structure. Exchange of water in contact with the buried steel surface with gravitational water or groundwater [Fig. 25.1(a)] can also be limited, allowing the local environment to develop its own composition as events on the steel surface proceed.

Clays of the montmorillonitc or smectite group can exert physical force on underground structures by their ability to adhere to coating or steel surfaces and through volume changes related to the swelling and shrinking caused by hydration effects or ion exchange [26]. As noted previously consequent coating damage can allow water to reach the steel surface [Fig. 25.1(b)], providing the opportunity for corrosion or EAC to develop.

Soil mineralogy plays a key role in the chemistry of associated groundwater. Clays, especially montmoriltonites, support the adsorption and exchange of both cations and anions (notably phosphate). These clays also buffer groundwater pH and poise the oxidation-reduction potential (ORP) of the soil–water system. A detailed discussion of how soil influences groundwater composition can be found in Stumm and Morgan [26].

Soil type and the position of a buried structure in the soil profile influences biological activity. Arid soils, low in water and organic matter, support little microbial activity relative to moist organic rich soils. In the latter case, aerobic microbiological activity is focused in the decaying organic matter on top of the soil and in the rhizosphere around plant roots. Other microbiological processes that use nitrate, snlfate, carbon dioxide, or various metal cations (Fe^{3+}, Mn^{4+}) in place of oxygen are found in more reducing conditions [41]. These tend to develop deeper down in the soil as aerobic processes in overlying zones consume oxygen. Certain microorganisms active in the sulfur cycle, such as the anaerobic sulfate reducing bacteria can be especially important in the corrosion of steel. In mineral soils, organic nutrients become limiting and microbial numbers fall off rapidly with depth [29]. Organic matter mixed in the backfill around a buried structure or even organic components of a protective coating system can become the focal point for microbiological activity [42].

D2. Moisture Content and the Position of the Water Table

The presence of moisture in the soil is an essential requirement for corrosion [43], The position of the water table also influences the nature of the corrosion process as it can determine rates of oxygen transport in the soil [44, 45]. The position of the water table may also vary seasonally, and this in turn may influence the nature of the corrosion process occurring on the buried structure. There are essentially three sources of water in soil [Fig. 25.1(a)]. Gravitational water in soil is derived from precipitation (rainfall or snow), capillary water is held within the capillaries of soil particles, and groundwater, which is the result of accumulation of gravitational water at the water table [43].

The water table is the top of a water-saturated zone where water forms the continuous phase in the soil matrix. It rises and falls seasonally with precipitation. Capillary effects in the soil matrix also influence the height of the water table. High capillary forces associated with low permeability silts and clays draw the water table up nearer to the surface of the soil. Local variations in soil composition mean that the top of the water table is not flat but rises and falls through a zone sometimes referred to as the capillary fringe [Fig. 25.1(a)].

As the soil moisture content increases from 0 to 25%, dramatic changes can be expected in a number of factors that influence corrosion rates. These include an exponential decrease in soil resistivity [7, 46] and the freely corroding potential [46, 47] and polarization resistance [46] for exposed steel. In terms of corrosion Booth et al. [3–5] determined that a soil moisture content of >20% can be potentially aggressive toward corrosion of carbon steel and was likely to result in general corrosion. Moisture contents of <20% would often result in pitting corrosion, while dry soils were not of concern with respect to corrosion.

These observations may be explained as a result of differential aeration process [46, 47]. The moisture content and position of the water table will influence the diffusion of oxygen into the soil. In saturated stagnant soils, anaerobic conditions may become established, oxygen transport will be low, and the corrosion rate of a buried metal (in the absence of microbial influence) will be low.

Terrain surrounding a site of interest can influence conditions in the underground environment. For example, a distant impermeable barrier could block the drainage through a permeable zone creating a perched water table and water-saturated conditions not normally expected for the local soil type. Topography must be considered in assessing a particular location regardless of soil type.

D3. Soil Resistivity

In general, it is believed that as soil resistivity becomes lower (i.e., groundwater becomes saltier and more conductive) corrosion of a buried metal becomes faster. However, this dependency is only true for metals that are not subject to cathodic protection. Cathodic protection also becomes more effective as the resistivity of the soil becomes lower. In the case of buried pipelines coated with polyolefin tape products, it has been shown [31] that penetration of effective cathodic protection potentials under a shielding disbondment is limited in part due to the conductivity of the water trapped under the coating [Fig. 25.1(b)]. As the conductivity of the trapped water increases, cathodic protection will penetrate further under the coating and arrest corrosion on the steel surface.

For unprotected steel exposed to soil, Booth et al. [3–5] suggested that a soil was corrosive if the soil resistivity was less than 2000 Ωcm and noncorrosive if the resistivity was >2000 Ωcm. Palmer [48], as shown in Table 25.2, gave a more refined classification of soil corrosion based on soil resistivity.

Miller et al. [49], Kulman [50], and others [43] proposed similar classifications to those presented in Table 25.2. while all agree that lower resistivity leads to accelerated corrosion, the actual values used to categorize soil corrosivity differ. Others suggest that corrosion by soil can be divided into

TABLE 25.2. Classification of Soil Corrosivity Based on Resistivity[a]

Resistivity Range (Ω cm)	Corrosivity
0–1000	Very severe
1001–2000	Severe
2001–5000	Moderate
5001–10,000	Mild
10,001 +	Very mild

[a]See [48].

microgalvanic cell corrosion leading to general corrosion and macrogalvanic cell corrosion leading to localized corrosion. Low resistivity leads to acceleration of the latter type of corrosion but it does not affect the former type of corrosion.

Resistivity varies with soil texture. For sand and gravel, resistivity varies from 10,000 to 500,000 Ωcm depending largely on how well drained the soil is, while soils rich in silt or clay generally show resistivities in the range from 1000 to 2000 Ωcm or from 500 to 2000 Ωcm respectively, depending on moisture content [46, 51]. Resistivity is also a function of the groundwater conductivity and temperature [46]. Soluble ion content of the soil has a direct impact on the resistivity of the soil. An increased soluble ion content will decrease the soil resistivity, which in turn for unprotected metals will increase the corrosion rate. In some locations, salt seeps can be encountered that can be highly concentrated in sodium, chloride, and sulfate ions. Such locations have very low resistivity and can be potentially very aggressive with respect to metallic corrosion on unprotected steel surfaces.

D4. Water Chemistry

Specific ions and organics dissolved in the groundwater can influence the corrosivity of the soil environment. For example, calcium and magnesium ions can mitigate corrosion by forming protective carbonate deposits on exposed pipe where dissolution of limestone or dolomite minerals has saturated local groundwater. Formation of protective hard white calcareous scale is particularly evident where effective cathodic protection has increased the pH at the metal surface. High levels of soluble chloride can increase the corrosivity of the soil environment by preventing the formation of passivating films on the steel surface so that any disruption will lead to unmitigated pitting. Sulfates are not considered to be directly corrosive to pipeline steel but can support the activity of sulfate-reducing bacteria in anaerobic environments where degradable organics are present [13]. These bacteria reduce the sulfate to sulfide, an anion that can participate in various corrosion and cracking scenarios. Soluble high-molecular-weight organic acids darken the color of groundwater in organic soils such as peat. Dick and Rodrigues [52] have reported that the lower-molecular-weight fulvic acids improved passivation by promoting the formation of surface deposits on the metal surface of API 5LSX65 line pipe steel exposed to test solutions while humic acids promoted pitting, an activity the authors attributed to the formation of soluble iron/humic acid complexes.

D5. Soil pH

The pH of soil will generally fall within the range 3.5–10. Soils containing well-humified organic matter tend to be acidic. Mineral soils can become acidic due to leaching of basic cations (Ca^{2+}, Mg^{2+}, Na^+, and K^+) by rainwater and as the result of dissolving of carbon dioxide into the groundwater.

The corrosion of iron as a function of pH increases considerably at pH values <4, but passivation occurs at high pH values [53]. In contrast to iron, amphoteric metals, such as aluminum, which are protected by oxide films, can be rapidly corroded in alkaline soils with high pH values as well as in acidic environments.

King [54] developed a nomogram that combined the influence of resistivity and pH on the corrosion of steel (Fig. 25.4) but cautioned that the figure should only be used as a guide. The nomogram ignores the influence of both oxidation–reduction potential and microbial activity, key parameters in underground corrosion. It may be best applied to the prediction of corrosion rates in aerobic conditions.

Where acidity is the controlling factor in determining the aggessiveness of a soil environment, the percentage of pipeline failures attributed to corrosion can be correlated to the total acidity of the soil [1].

D6. Oxidation–Reduction Potential

The oxidation–reduction potential (ORP) of soil is the potential of an inert electrode such as platinum, with respect to a reference electrode such as copper/copper sulfate or saturated calomel. It is not a measure of the oxygen concentration but rather an indication of the oxidizing or reducing capacity of the soil. Under aerobic conditions oxygen content of the soil will be high and the ORP will be more positive than that measured for an anaerobic soil. In an anaerobic soil, electron acceptors other than oxygen determine the ORP of the soil. The ORP is somewhat of an abstract value but research and field evidence have indicated that it can be used with some success to help in prediction of soil corrosivity.

Starkey and Wight [2] determined a classification of soil corrosivity based on redox potentials as shown in Table 25.3.

Booth et al. [3–5] developed a similar classification scheme. In this work, soils were considered aggressive if the ORP reading was <0.4 V (NHE at pH 7.0) and nonaggressive if the ORP was >0.4V(NHE at pH 7.0).

The general conclusion made in these systems is that aerobic soil conditions are relatively benign. More severe damage is seen under anaerobic conditions. This observation is at least partly explained by the potential for anaerobic microbial activity in soils with a low ORP.

D7. Role of Microbes in Soil Corrosion

Microbiologically influenced corrosion of iron-based materials in soils has been well documented [55]. While numerous scenarios have been described, two are particularly associated with the corrosion of buried line pipe.

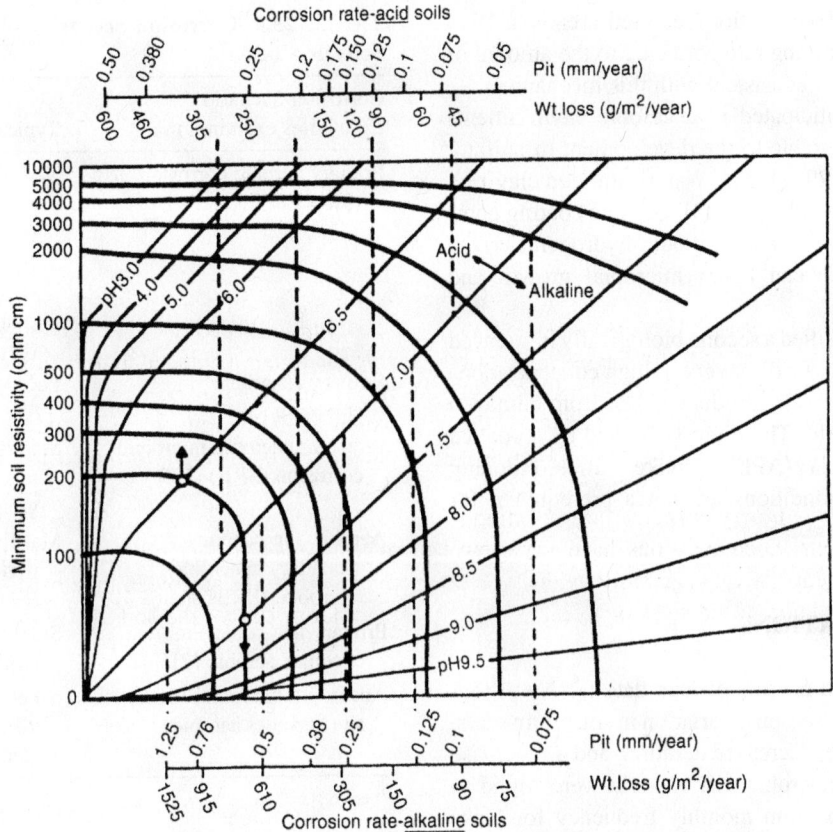

FIGURE 25.4. Nomogram relating soil resistivity, pH, and corrosion rate for steel pipe in soil [54].

As long ago as 1934, sulfate-reducing bacteria were identified as being responsible for severe corrosion damage observed under anaerobic conditions in wet, black soil [56]. These organisms reduce groundwater sulfate to sulfide as part of their anaerobic metabolism and cause the precipitation of black iron sulfides associated with severe corrosion on pipeline systems. These bacteria can derive energy for growth and metabolism by coupling the oxidation of molecular hydrogen to sulfate reduction [57]. Hardy [58] has shown that sulfate reduction can be stimulated by catbodic hydrogen formed on metal surfaces.

A scheme for the corrosion mechanism is shown in Figure 25.5. A galvanic couple formed between steel and

microbially produced iron sulfide is sustained by the removal of electrons possibly in the form of cathodic hydrogen from the iron sulfide matrix by these bacteria. This prevents saturation of the iron sulfide with electrons and leads to further iron sulfide production, which in turn further extends the corrosion cell [20, 59] The amount of iron sulfide present determines the rate of corrosion observed in the lab

TABLE 25.3 Classification of Soil Corrosivity Based on Oxidation–Reduction Potential[a]

ORP Range (mV)[b]	Degree of Corrosion
<100	Severe
100–200	Moderate
200–400	Slight
>400	Noncorrosive

[a]See [2].
[b]The ORP value is corrected to the normal hydrogen electrode (NHE) scale at a pH 7.0.

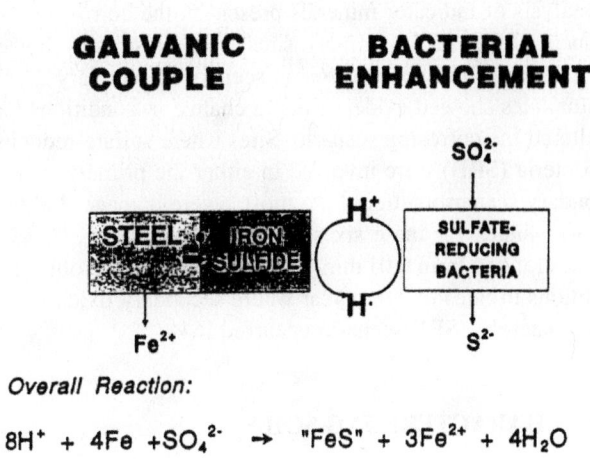

FIGURE 25.5. Corrosion mechanism for microbially influenced corrosion involving anaerobic sulfate-reducing bacteria [20, 59].

experiments [60]. Field observations reported as early as 1940 support an increasing pitting rate corrected to the amount of iron sulfide present [61] consistent with this mechanism.

This scenario is anticipated in anaerobic environments with pH and ORP favorable to the development of sulfate-reducing populations [29, 41, 57]. Water-saturated clay-rich soils provide suitable conditions [43]. Organic coating components and the availability of catbodic hydrogen derived from the steel surface can foster microbial growth and activity [42, 58].

Pope et al. [62] identified a second biologically influenced corrosion scenario with the discovery of high concentrations of organic acids in corrosion products taken from damaged pipe surfaces in the field. These are generated by so-called acid-producing bacteria (APB). Unlike sulfate-reducing microbes, anaerobic conditions are not a prerequisite for activity of these bacteria.

E. FIELD OBSERVATIONS

Kulman [50] compared 5 years of leak data for New York City gas mains with the seasonal variation in soil parameters. Increased soil moisture, decreased aeration, and an increase in the proportion of anaerobic soil bacteria were found to correlate with the maximum monthly frequency for leaks caused by the underground corrosion cycle. In a program of 472 excavations carried out through the 1940s, severe corrosion was found at 185 sites. Of these excavations, 103 of them were in clay soil and ferrous sulfide was found in 81% of the sites. In contrast, soil properties including the concentration of soluble sulfate, the number of sulfate-reducing bacteria, and the soil pH failed to show any correlation with corrosion behavior.

Jack et al. [63] studied the corrosion of line pipe steels over a number of years in a variety of prairie soil in Alberta, Canada. At least six scenarios could be identified from the analysis of indicator minerals present in the corrosion products (Table 25.4), Primary sites were those that showed evidence of only one corrosion scenario. Secondary corrosion sites showed evidence that a change in conditions had altered the corrosion scenario. Sites where sulfate-reducing bacteria (SRB) were involved in either the primary or secondary scenario suffered the most severe damage. Laboratory studies for these six scenarios showed that corrosion rates ranged from 0.01 mm/year in sustained anaerobic conditions to rates of 1 mm/year where secondary oxidation of an anaerobic SRB scenario occurred [64].

F. CHARACTERIZING SOILS

Recently released Recommended and Standard Practices to assess the extent of stress corrosion cracking (SCC) and to

TABLE 25.4. Corrosion Scenarios Identified on Pipelines in Soils in Alberta[a]

Corrosion Scenario (% of sites examined)	Typical Corrosion Deposits
Primary aerobic corrosion (3%)	Various iron(III) oxides including: Fe_3O_4 (magnetite), α-FeOOH (goethite), γ-FeOOH (lepidocrocite), γ-Fe_2O_3, (maghemite), Fe_2O_3 (hematite)
Primary anaerobic corrosion (29%)	$FeCO_3$ (siderite)
Primary anaerobic microbial corrosion (SRB) (27%)	Siderite + amorphous "FeS," mackinawite (FeS), greigite (FeS)
Aerobic site becomes a secondary anaerobic microbial (SRB) site (3%)	Mackinawite (FeS), pyrite (FeS_2;), marcasite (FeS_2 + iron(III) oxides
Primary anaerobic site becomes aerobic (21%)	Siderite + various iron(III) oxides
Anaerobic microbial (SRB) site becomes aerobic (17%)	α-FeOOH (goethite), γ-FeOOH (lepidocrocite) + elemental sulfur

[a]See [63].

assess and reduce the impact of external corrosion on pipeline systems [11, 12] review methods for assessing key characteristics of the soil environment associated with these problems. Specific standard methods have long been available for measuring traditional factors such as soil resistivity (e.g., ASTM G57 and G187) and soil pH (e.g., ASTM G51), but methods in this area are currently in a state of active review and continuous improvement [65, 66]. Task Group 369 of the National Association of Corrosion Engineers International has been assigned to develop a report on aboveground techniques used to identify areas on a pipeline at risk for external corrosion based on the corrosiveness of the environment, but this is likely to be a benchmark report for further development.

F1. Soil Probes

Attention has increasingly shifted over the years from surface observations to measuring key soil characteristics at pipe depth through the use of portable probes that can be inserted directly into the soil to depths of a meter or more.

Starkey and Wight [67] and Deuber and Deuber [68], in association with the American Gas Association, developed the first probes for field measurement of redox potential along pipeline right of ways. Costanzo and McVey [69] developed this probe further. Their probe consisted of two platinum electrodes coupled to a saturated calomel reference electrode. A similar probe for field determination of redox potentials was developed by Booth et al. [3–5]. A more comprehensive probe was patented in Japan [70].

In 1992, Wilmott et al. [71] reported the development of a soil probe that could be used to determine soil pH, resistivity, redox potential, temperature, and pipe to soil potentials. This probe was used in the characterization of buried pipeline environments associated with external corrosion, stress corrosion cracking, and external hydrogen-induced cracking. In most cases, it was found that the soil redox potential and resistivity are important parameters in determining the nature of the scenario that is found at a given location [72].

The seasonal variation in soil parameters can be significant. Soil readings for resistivity, ORP, and moisture content were monitored monthly for 87 sites in the United Kingdom. These readings were then used to classify the sites as aggressive or nonaggressive according to the scheme of Booth et al. [5]. In only about one-half of the sites did the monthly readings consistently result in the same diagnosis. In 5–9% of the sites, an individual monthly reading had a 50% or more chance of giving an incorrect diagnosis relative to long-term site behavior. These observations led to a recognized need to take as many measurements as possible and to use seasonally averaged values to predict the soil corrosivity [5].

One way of characterizing a site more reliably over its seasonal variations is to use permanent soil probes. Gabrys and Van Boven [73] reported such a study. The probe used was based on the principles of that reported by Wilmott et al. [72] but was designed to be permanently buried and can be used in conjunction with corrosion coupons to monitor seasonal variations in cathodic protection, environmental parameters, and associated corrosion rates for pipeline steels, A second approach to soil corrosivity mapping was reported by Wilson [74] in New Zealand who developed a probe, which can be used for measuring corrosion potential, corrosion rate and soil resistivity as a function of time and probe depth. One aim of this research was to monitor seasonal variations in parameters associated with corrosion of buried metals. In 2001, Li et al. [75] reported development of a probe to measure corrosion current density and corrosion potential of metals in soil along with soil resistivity, ORP, and temperature. The last parameter can influence the corrosion rate, ion exchange kinetics with soil components such as clays and coating degradation. Dedicated probes for soil resistivity and corrosion rate measurement (based on linear polarization resistance with compensation for high solution resistance) are available commercially. Electrical resistance probes can be used to measure corrosion rates in very high resistivity soil environments. One application has been the assessment of cathodic protection effectiveness in desert soils where extremely high infrared (IR) drop frustrates other methods

of assessment [76]. A review of corrosion rate probes is available [77] and the U.S. PHMSA has set a research contract [78] to validate their use.

Geotechnical engineers have also developed a number of tools that can be useful in characterizing soil corrosivity. These probes are commonly called cone penetrometers and were initially designed to measure soil mechanical properties. Adaptations of such devices have allowed determination of resistivity, pH, and redox potentials [79–82].

Soil probes provide an essential insight into local conditions adjacent to a buried structure. Disturbed soil, perched water tables, groundwater flow through gaps generated by installation, or freeze–thaw cycles as well as other local effects can preclude the reliable correlation of local soil type with the probability of corrosion behavior at a specific site. Soil probes can characterize the environment at the buried metal surface where the bare metal is in direct contact with the soil [Fig. 25.1(a)] and provide the data needed to assess likely corrosion rates without excavation. Characterization of environments under defective coatings in the scenario illustrated in Figure 25.1(b) is more problematic and will require further research effort to understand how this environment evolves based on background soil and groundwater characteristics.

G. MODELS

Having assayed parameters that can influence corrosion and cracking scenarios on buried structures the question arises as to what to do with the information that is gathered? The goal of soil corrosivity mapping should be to enable development of models that can ultimately be used to identify specific sites that would be susceptible to a given corrosion or cracking scenario.

Traditionally, simple soils models have been used to rank various soil environments in terms of corrosion severity based on the factors described in this chapter [1, 40] with the aim of informing decisions on pipeline design (route selection, corrosion allowance, coating application, remedial backfill, etc.) or prioritizing inspection and maintenance activities. The simple models employed assign a degree of corrosion severity to ranges of values for each factor and then consider the cumulative effect. The approach recommended in ANSI/AWWA C105/A21.5, for example, is to assign points to ranges of values for soil resistivity, pH, ORP, and moisture content and the presence of sulfide. The points assigned to the various factors are then summed to decide whether special action such as the application of a protective coating to the pipe is warranted. The direct measurement of corrosion rates at pipe depth enabled by the new generation of soil probes can be used to prioritize the location and frequency of inspection and maintenance activities, but the cost of collecting such data is sufficient that simple models that rank soil environments along the pipeline right of way may find use in targeting this more sophisticated approach to data collection.

Risk assessment and risk management processes are used in the pipeline industry as a means of prioritizing pipeline hazards and identifying cost-effective maintenance actions. There are a number of components to this process.

Identification of all likely hazards

Determination of the probability of failure from each hazard

Prioritization of pipeline segments based on risk (including consideration of consequences)

Optimization of maintenance activities to reduce overall risk

Recognized protocols are now available to guide this process for external corrosion and stress corrosion cracking [11, 12].

In the case of SCC, Marr and Associates, in conjunction with TransCanada Pipelines, developed a predictive model to identify and rank areas along a pipeline that are most likely to develop SCC [83, 84]. Wilmott and Sutherby [85] reported on a fault tree analysis that can also be used to identify locations of high SCC susceptibility. The factors employed in SCC models [12] include the following;

Type of coating

Year of pipeline construction

Pipeline operating conditions

Soil type, drainage, and topography

In terms of external corrosion of pipelines, risk management has been addressed in a number of articles [86, 87] and a textbook has been written by Muhlbauer [88]. The processes employed in such analyses range from indexing methods [88] through to more complex decision analysis frameworks based on influence diagrams [86].

H. SOIL EFFECTS ON CATHODIC PROTECTION AND COATINGS

Cathodic protection and protective coatings are used to prevent the external corrosion and EAC of buried steel structures [17]. Either method on its own can effectively protect steel in the underground environment. Used together the coating can reduce the cost of cathodic protection while the cathodic protection system can protect steel surfaces exposed to the soil environment through coating damage or defects.

Occasionally, soil conditions can compromise the performance of cathodic protection. For example, the electrical conductivity of the soil is a key factor in the design of an appropriate cathodic protection system. Poor design or an unexpected increase in soil resistivity brought on by drought or other factors can result in a loss of effective protection at buried steel surfaces.

Coating damage or defects can create the opportunity for corrosion or EAC on buried steel surfaces by allowing water to reach the steel surface [Fig. 25.1(b)] in the absence of effective cathodic protection. One example of this occurs where soil stress has pulled a polyolefin tape wrap coating off the steel surface. The electrical insulating ability of the disbonded tape effectively shields the underlying steel surface from effective cathodic protection, allowing metal damage to proceed at essentially the freely corroding potential [31]. This scenario is most common in high clay soils and in locations where soil movement or extensive settling occurs. High temperatures or excessive cathodic protection potentials can accelerate coating disbondment. In another scenario, coating materials including polyvinyl chloride, asphalt, or coal tar can suffer a loss of integrity in service through removal of coating components by water leaching, oxidation, or biodegradation. High temperatures can again accelerate coating deterioration. The resulting increase in coating permeability allows water access to the underlying steel surface. If the permeable coating holds water on the steel surface during dry spells when the surrounding backfill dries out and cuts off effective cathodic protection, corrosion and EAC can occur. Examples of both these scenarios have been seen in the field.

I. SUMMARY

Early work correlating readily measured soil parameters with the degradation of buried materials has provided simple models for soil aggressivity. These models provide general guidance for materials selection and for the location and design of underground systems.

As scientists and engineers learn more about the role of soil in corrosion and environmentally assisted cracking processes it becomes apparent that, while useful generalizations regarding the corrosivity of various types of soil can be made, the underground environment can be complex. A simple, unified theory describing all soil conditions and the resulting corrosion processes is unlikely. Corrosion in soils is a multivariate problem that requires an understanding of chemistry, geology, and materials science to name but a few topics. No single unified process for assessing soil corrosivity has been or is likely to be developed. Direct measurement of soil parameters in disturbed soil adjacent to buried steel structures can improve understanding of site-specific conditions; however, a better understanding of corrosion and EAC mechanisms is needed to refine the linkage between conditions in the surrounding soil and events on a buried steel surface. Where protective coatings have failed in a way that allows water to access the underlying steel surface in the absence of effective cathodic protection, it becomes even more difficult to make this link. An understanding of how coatings fail and the consequences in terms of conditions developed at the steel surface in various soil environments remains a key issue for futurework.

The ability to predict the probability of corrosion or cracking damage on a site-specific basis is important for the ongoing maintenance of a reliable pipeline infrastructure. Risk management tools play a key role in prioritizing effort in both the maintenance and inspection of aging systems to ensure their safety and reliability cost effectively.

REFERENCES

1. K. H. Logan, "Corrosion By Soils," in H. H. Uhlig Corrosion Handbook In "The Corrosion Handbook" H. H. Uhlig (Eds.) Wiley, London, 1948, pp. 446–466.

2. R. L. Starkey and K. M. Wight, "Anaerobic Corrosion of Iron in Soil," American Gas Association Monograph, New York, 1945.

3. G. H. Booth, A. W. Cooper, P. M. Cooper, and D. S. Wakerley, Br. Conos. J., **2**, 104 (1967).

4. G. H. Booth, A. W. Cooper, and P. M. Cooper, Br. Corros. J., **2**, 109 (1967).

5. G. H. Booth, A. W. Cooper, and A. K. Tiller, Br. Corros. J., **2**, 114 (1967).

6. M. K. Adler Flitton and E. Escalante, "Simulated Service Testing in Soils," in Corrosion: Environments and Industries, ASM Handbook, Vol. 13A, ASM International, Materials Park, OH, 2003, pp. 497–500.

7. M. Romanoff, "Underground Corrosion," National Bureau of Standards Circular 579, NBS, Washington, DC, 1957.

8. E. Escalante (Ed.), "Underground Corrosion," ASTM STP 741, American Society for Testing and Materials, Philadelphia, PA, 1979.

9. "Corrosion Tests and Standards: Application and Interpretation," ASTM manual series, MNL 20, American Society for Testing and Materials, Philadelphia, PA, 1995.

10. J. A. Beavers and N. G. Thompson, "External Corrosion of Oil and Natural Gas Pipelines," in Corrosion: Environments and Industries, ASM Handbook, Vol. 13C, ASM International, Materials Park, OH, 2006, pp. 1015–1025.

11. "Standard Practice, Pipeline External Corrosion Direct Assessment Methodology," NACE SP0502-2008, National Association of Corrosion Engineers, Houston, TX, 2008.

12. "Standard Recommended Practice, Stress Corrosion Cracking (SCC) Direct Assessment Methodology," NACE SP0204-2008, National Association of Corrosion Engineers, Houston, TX, 2008.

13. B. N. Leis, E. B. Clark, M. Lamontagne, and J. A. Colwell, "Improvement of External Corrosion Direct Assessment Methodology by Incorporating Soils Data," Contract No. DTRS56-03-T-0003, U.S. Department of Transportation, Research and Special Projects Agency, Nov. 2005.

14. T. R. Jack, "MIC in Underground Environments: External Corrosion in the Gas Pipeline Industry," in A Practical Manual on Microbiologically Influenced Corrosion, Vol. 2 J. G. Stoecker (Ed.) National Association of Corrosion Engineers, Houston, TX, 2001. pp. 6.1–6.12.

15. F. M. Song and N. Sridhar, "Modeling Pipeline Corrosion Under a Disbonded Coating Under the Influence of Underneath Flow," Corrosion, **64**(1), 40–50 (2008).

16. F. M. Song and N. Sridhar, "Modeling Pipeline Crevice Corrosion Under a Disbonded Coating with or without Cathodic Protection Under Transient and Steady State Conditions," Corros. Sci., **50**, 70–83 (2008).

17. A. W. Peabody, "Control of Pipeline Corrosion," NACE, Houston, TX, 1967.

18. T. R. Jack, G. Van Boven, M. Wilmott, R. L. Sutherby, and R. G. Worthingham, "Cathodic Protection Potential under Disbonded Pipeline Coating," Mater. Perform., **33**(8), 17–21, (1994).

19. M. Yan, J. Wang, E. Han, and W. Ke, "Local Environment Under Simulated Disbonded Coating on Steel Pipelines in Solution," Corros. Sci., **50**, 1331–1339 (2008).

20. R. G. Worthingham, T. R. Jack, and V. Ward, "Biologically Induced Corrosion," National Association of Corrosion Engineers, Houston, TX, 1986, p. 33.

21. Y. G. Kim, D. S. Won, and H. S. Song, "Validation of External Corrosion Direct Assessment with Inline Inspection in Gas Transmission Pipeline," Paper No. 08136, CORROSION 2008, New Orleans, LA, National Association of Corrosion Engineers, Houston, TX, 2008.

22. The 1998 Canadian and World Encyclopedia, Multimedia MS, McClelland and Stewart Inc., Toronto, Canada 1998.

23. Compton's Interactive Encyclopedia, SoftKey International Inc., Cambridge, MA, 1995.

24. M. Honarpour, L. Koederitz, and A. H. Harvey, Relative Permeability of Petroleum Reservoirs, CRC Press, Boca Raton, FL, 1986.

25. F. Press and R. Siever, Earth, Freeman, San Francisco, CA, 1974.

26. W. Stumm and J. J. Morgan, Aquatic Chemistry, 2nd ed., Wiley-Iiuerscience, Toronto, Canada, 1981.

27. W. C. Krumbein and L. L. Sloss, Stratigraphy and Sedimentation, 2nd ed., Freeman, San Francisco, CA, 1963.

28. M. J. Wilson (Ed.), A Handbook of Determinative Methods in Clay Mineralogy, Blackie, Glasgow, Scotland, 1987. Anion exchange data for clay cited is originally from R. E. Grim, Clay Mineralogy, McGraw-Hill, New York, 1968.

29. R. M. Atlas and R. Bartha, Microbial Ecology: Fundamentals and Applications, 2nd ed., Benjamin/Cummings, Don Mills, Ontario, Canada, 1987.

30. S. Pawluk and L. A. Bayrock, Some Characteristics and Physical Properties of Alberta Tills, Bulletin 26, Alberta Research Council, Edmonton, Alberta, Canada, 1969.

31. T. R. Jack, G. Van Boven, M. Wilmott, R. L. Sutherby, and R. G. Worthingham, Mater. Perform., **33**(8), **17**, (1994).

32. M. K. Adler Flitton, R. E. Mizia, and C. W. Bishop, "Underground Corrosion of Activated Metals in an Arid Vadose Zone Environment," NACE Paper No. 02531, CORROSION 2002, National Association of Corrosion Engineers International, Houston, TX, 2002.

33. P. J. Williams, T. L. White, and J. K. Torrance, "The Significance of Soil Freezing for Stress Corrosion, Cracking," in the Proceedings of the International Pipeline Conference (IPC'98), June 7–11, 1998, Calgary, Alberta, Canada. American Society of Mechanical Engineers, New York, 1998, pp. 473–484.

34. M. Romanoff, "Corrosion of Steel Pilings in Soil," National Bureau of Standards Monograph 58, U.S. Department of Commerce, Washington, DC, 1962.

35. E. Escalante, "Concepts of Underground Corrosion," in Effects of Soil Characteristics on Corrosion," ASTM STP 1013, V. Chaker and J. D. Palmer (Eds.), American Society for Testing and Materials, Philadelphia, PA, 1989, pp. 81–94.

36. J. B. Decker, K. M. Rollins, and J. C. Ellsworth, "Corrosion Rate Evaluation and Prediction for Piles Based on Long-Term Field Performance," J. Geotech. Geoenviron. Engi., **134**(3), 341–351 (2008).

37. I. Hieng and K. H. Law, "Corrosion of Steel H Piles in Decomposed Granite," J. Geotech. Geoenviron. Engi., **125**(6), 529–532 (1999).

38. J. O. Harris,"Corrosion, 16: 149 (1960)," as discussed in L. L. Shreir, Corrosion, Vol. 1, Corrosion of Metals and Alloys, Wiley, New York, 1963.

39. R. R. Hadley, Petrol. Eng., Mar., 171 (1940).

40. R. E. Ricker, Analysis of Pipeline Steel Corrosion Data from NBS (NIST) Studies Conducted between 1922–1940 and Relevance to Pipeline Management, NISTIR 7415, National Institute of Standards and Technology, U.S. Department of Commerce, Gaithersburg, MD, 2007.

41. J. E. Zajic, Microbial Biogeochemistry, Elsevier, New York, 1969.

42. T. R. Jack, M. M. Francis, and R. G. Worthingham, "External Corrosion of Line Pipe. Part II: Laboratory Study of Cathodic Protection in the Presence of Sulfate-Reducing Bacteria," in S. C. Dexter (Ed.), "Biologically Induced Corrosion" NACE, Houston, TX, 1986.

43. V. Ashworth, C. G. Googan, and W. R. Jacob, Corrosion Australasia, Oct. 10 (1986).

44. C. L. Durr and J. A. Beavers, "Techniques for Assessment of Soil Corrosivity," Paper No. 667, CORROSION 98, NACE International, Houston, TX, 1998.

45. O. E. Picozzi, S. E. Lamb, and A. C. Frank,"Evaluation of Prediction Methods for Pile Corrosion at the Buffalo Skyway," New York State Department of Transportation, Technical Services Division, Feb. 1993.

46. A. Benmoussa, M. Hadjel, and M. Traisnel, "Corrosion Behavior of API 5L X-60 Pipeline Steel Exposed to Near-Neutral pH Soil Simulating Solution," Mater. Corros., **57**(6), 771–777 2006.

47. N. N. Glazov, S. M. Ukhlovtsev, I. I. Reformatskaya, A. N. Pdodbaev and I. I. Ashcheulova, "Corrosion of Carbon Steel in Soils of Varying Moisture Content," Protec. Met., **42**(6), 601–608 (2006).

48. J. D. Palmer, Mater. Perform., **13**(1), 41–46 (1974).

49. F. P. Miller, J. E. Foss, and D. C. Wolf, "Soil Surveys: Their Synthesis, Confidence Limits, and Utilization for Corrosion Assessment of Soil," in Underground Corrosion, ASTM STP 741, E. Escalante (Ed.), American Society for Testing and Materials, Philadelphia, PA, 1979.

50. F. E. Kulman, Corrosion, **9**, 11–18 (1953).

51. P.-J. Cunat, "Corrosion Resistance of Stainless Steels in Soils and in Concrete," paper presented at the Plenary Days of the Committee on the Study of Pipe Corrosion and Protection, Ceocor, Biarritz, Oct. 2001.

52. L. F. P. Dick and L. M. Rodrigues, "Influence of Humic Substances on the Corrosion of the API 5L X65 Steel," Corrosion, **62**(10), 35–42 (2006).

53. H. H. Uhlig and R. W, Revie, Corrosion and Corrosion Control: An Introduction to Corrosion Science and Engineering, Wiley, New York, 1985.

54. R. A. King, "A Review of Soil Corrosiveness with Particular Reference to Reinforced Earths," TRRL Supplementary Report 316, Transport and Road Research Laboratory, Crowthorne, UK, 1977.

55. T. R. Jack, "Biological Corrosion Failures," in Failure Analysis and Prevention, ASM Handbook, Vol. 11, ASM International, Materials Park, OH, 2002, pp. 881–898.

56. C. A. H. von Wolzogen Kuhr and L. W. van der Vlugt, Water, **18**, 147 (1934).

57. L. L. Barton (Ed.), Sulfate-Reducing Bacteria, Plenum, New York, 1995.

58. J. A. Hardy, Br, Corros. J., **18**, 190–193 (1983).

59. R. A. King and J. D. A. Miller, Nature (London), **233**, 491–492 (1971).

60. T. R. Jack, M. J. Wilmott, R. L. Sutherby, and R. G. Worthingham, Mater. Perform., **35**(3), 18–24 (1996).

61. R. F. Hadley, Petrol. Eng., Apr., 112 (1940).

62. D. H. Pope, T. P. Zintel, A. K. Kuruvilla, and O. W. Siebert, "Organic Acid Corrosion of Carbon Steel: A Mechanism of Microbiailly Influenced Corrosion," Paper No. 79, CORROSION 88, NACE, Houston, TX, 1988.

63. T. R. Jack, M. J. Wilmott, and R. L. Sutherby, Mater. Perform., **34**(11), 19 (1995).

64. T. R. Jack, M. Wilmott, J. Stockdale, G. Van Boven, R. G. Worthingham, and R. L. Sutherby, Corrosion, **54**(3), 246 (1998).

65. "Corrosion in the Soil Environment: Soil Resistivity and pH Measurements," TRB Project No. NCHRP 21-06, Transportation Research Board of the U.S. National Academies, Washington, DC.

66. R. G. Wakelin, "Evaluation and Comparison of Soil Resistivity Measurement Techniques," PRCI Report No. PR-262-01114, Pipeline Research Council International, Arlington, VA, 2004.

67. R. L. Starkey and K. M. Wight, "Final Report of the American Gas Association Iron Corrosion Research Fellowship," American Gas Association, New York, 1945. Also U.S. Patent 2,454,952.

68. C. G. Deuber and G. B. Deuber, "Development of the Redox Probe," Research Project PM-20, Final Report to the American Gas Association, New York, 1956.

69. F. E. Costanzo and R. E. McVey, "Development of the Redox Probe Field Technique," Corrosion, **14**, 268T–272T (1958).

70. F. Kajiyama, "Apparatus for Measuring the Corrosivity of Soil," Japanese Patent Application 58-208654, 1982 (in Japanese).

71. M. J. Wilmott, T. R. Jack, J. Geerligs. R. L. Sutherby, D. Diakow, and B. Dupuis, Oil Gas J., **93**(14), 54–58 (1995).

72. M. Wilmott and D. Diakow, "Factors Influencing Stress Corrosion Cracking of Gas Transmission Pipelines: Detailed Studies Following a Pipeline Failure. Part 1 Environmental Considerations," Vol. 1, International Pipeline Conference Calgary, June 1996, ASME, 1996, pp. 507–524.

73. S. Gabrys and C. Van Boven, "The Use of Coupons and Probes to Monitor Cathodic Protection and Soil Corrosivity," NACE Northern Area Western Conference, Feb. 1998.

74. P. T. Wilson"Electrochemical Monitoring of Soil Corrosivity," Paper 30, Progress Corrosion, 1995.

75. M. C. Li, Z. Han, H. C. Lin, and C. N. Lao, "A New Probe for the Investigation of Soil Corrosivity," Corrosion, **57**(10), 913–917 (2001).

76. N. A. Khan, "Using Electrical Resistance Soil Corrosion Probes to Determine Cathodic Protection Effectiveness in High-Resistivity Soils," Mater. Perform., **43**(6), 20–25 (2004).

77. B. S. Covino, Jr., and S. J. Bullard, "Corrosion Rate Probes for Soil Environments," in Corrosion: Environments and Industries, ASM Handbook, Vol. 13C, ASM International, Materials Park, OH, 2006, pp. 115–121.

78. "Validation of External Corrosion Growth-Rate Using Polarization Resistance and Soil Properties," PHMSA Contract No. DTPH56-08-T-000022, U.S. Department of Transportation, Pipeline and Hazardous Materials Safety Administration, Washington, DC, 2010.

79. J. J. Olie, C. C. D. F. Van Ree, and C. Bremmer, Geotechnique, **1**, 13 (1992).

80. R. G. Campanells and I. Weemees, Can. Gcotech. J., **27**, 557–567 (1990).

81. M. R. Horsnell,"The Use of Cone Penetration Testing to Obtain Environmental Data," in Penetration Testing in the UK, Thomas Telford, London, 1988.

82. R. G. Campanella, M. P. Davies, T. J. Boyd, and J. L. Everard, "Geoenvironmental Subsurface Site Characterization using In Situ Soil Testing Methods," Proceedings of the 1st International Congress on Environmental Geotechnics, Edmonton, Alberta, July 10–15, 1994.

83. J. E. Marr, T. Cunningham, and A. Brunton, Pipeline Gas Ind., **81**(3), 27–30, 32–34 (1998).

84. B. Delanty and J. O'Beime. Oil Gas J., **90**(24), 39–44 (1992).

85. M. J. Wilmott and R. L. Sutherby, Corros. Mater., **22**(3), 27–30 (1997).

86. M. J. Stephens and M. A. Nessim, "Pipeline Maintenance Planning Based on Quantitative Risk Analysis," Proceedings of the International Pipeline Conference, Calgary, Canada, 1996.

87. M. Urednicek, R. I. Coote, and R. Courts, "Optimizing Rehabilitation Process with Risk Assessment and Inspection," Proceedings of the International Conference on Gas Pipeline Reliability, Calgary, Canada, June 2–5, 1992.

88. W. K. Muhlbauer, Pipeline Risk Management Manual, 2nd ed., Gulf Publishing Company, Houston, TX, 1996.

26

MICROBIAL DEGRADATION OF MATERIALS: GENERAL PROCESSES

J.-D. Gu

School of Biological Sciences, The University of Hong Kong, Hong Kong, China

T. E. Ford

University of New England, Biddeford, Maine

R. Mitchell

Harvard School of Engineering and Applied Sciences, Harvard University, Cambridge, Massachusetts

A. INTRODUCTION

Microorganisms play a very active and an important role in the biosphere, mediating the decomposition of natural organic materials and recycling essential nutrients. They also have a less recognized negative role to human society, damaging a wide range of natural and engineering materials, including metals [1–4], polymeric materials [4–7], and inorganic minerals and stone [8–12]. Under natural conditions, corrosion of metals is a result of both electrochemical and biological processes operating at the interface between metal surfaces and microorganisms, which can be initiated and accelerated by the presence and the active metabolism of microorganisms [1, 3, 13–16]. Degradation and deterioration of a wide range of polymeric materials are due to the action of microorganisms and other metabolites followed by physical disintegration and damage of the material structures [4, 6, 7]. Concrete and stone are degraded mostly by acidic metabolite attack by the attached microflora. The processes of corrosion, degradation, and deterioration are ambiguously called microbially influenced (induced) corrosion (MIC) in the corrosion and engineering literature without detailed differentiation.

Microbial adhesion and establishment of a complex community (microfouling) are prerequisites for substantial degradation and physical changes of the underlying materials [15, 17–20]. All surfaces, including both nonliving and living, may act as substrata for attachment and then formation of the biofilms [21–27]. Therefore, subsequent attack of materials by microorganisms can take place either directly or indirectly, depending on a combination of factors [6], including the indigenous microflora, material composition, nature of the surface, and environmental conditions. In addition, other specific factors affecting the physical environment also influence the extent of bacterial adhesion on surfaces, including ionic strength of the solution, type of cation, hydrodynamic force, and surface properties (e.g., hydrophobicity or hydrophilicity) [26].

Corrosion of metals is closely associated with the formation of complex microbial biofilms on surfaces [1, 3, 4, 14]. The attached microbial cells on metal surfaces induce the formation of differential aeration cells under aerobic conditions because dissolved oxygen is consumed near and within microbial colonies. At the same time, the decrease in oxygen levels provides an opportunity for anaerobic microorganisms to become established within biofilms. In particular, sulfate-reducing bacteria can become established and cause rapid corrosion of underlying metals. Under conditions that promote microbial corrosion, a variety of minerals have been reported to be associated with different processes [14, 28]. Microorganisms are also capable of extracting electrons from metals under both aerobic and anaerobic conditions [1, 29–31]. Aerobic and anaerobic corrosion of metals are discussed elsewhere [3] as well as in Chapter 39 in this book. In addition, hydrogen produced during microbial metabolism may cause corrosion by affecting the mechanical strength of metals through a process called microbial hydrogen embrittlement [2, 15, 32, 33].

Direct utilization of natural polymeric materials by microorganisms can also proceed at a high rate because of their structural similarities to other biological materials [34–41]. Microbial degradation of complex (high-molecular-weight) polymers is most likely mediated by extracellular depolymerases because the molecular sizes of these polymers are too large to penetrate through cellular membranes for effective assimilation. However, extracellular degradation products can be assimilated by microorganisms as a source of carbon and energy when the polymers are depolymerized. The consequences of degradation are the weakening of polymer backbone linkages and a decrease in mechanical properties due to utilization of the backbone carbon and further damage of the internal bonding between different components of the polymer chain [34, 40, 42]. The process in which a material is decomposed and altered significantly in chemical and physical terms by microorganisms is called biological degradation or deterioration. Mineralization is designated strictly for complete breakdown of polymer carbon to simple inorganic compounds, and the end products during mineralization are CO_2 and H_2O under aerobic conditions and CH_4, H_2S, CO_2, and H_2O under a range of anaerobic conditions [6, 41, 43]. Complete degradation of synthetic and engineering polymer is seldom achieved by microorganisms because part of the substrate carbon will be immobilized in new microbial biomass and recalcitrant organic matter. During degradation, the molecular weight of the polymer decreases under both aerobic and anaerobic conditions.

Corrosive metabolic products from microorganisms have also been found to contribute to the deterioration of a wide range of materials, particularly stone [10, 44], concrete [9, 12, 45], and metals [1, 46–50]. Succession of physiologically different groups of microorganisms on surface of materials and their metabolites contribute significantly to the degradation and deterioration of this class of inorganic materials. Indirect damage to materials by exoenzymes, exopolymeric materials of microbial origin, and other microbial products has drawn increasing attention [46, 48].

A fundamental understanding of bacterial interactions with any kind of material surface requires integration of information from several disciplines, including materials science and engineering, microbial ecology, biochemistry and physiology, and electrochemistry. Basic microbiological knowledge is required for a better understanding of the role of microorganisms in corrosion and deterioration. In this chapter we will discuss microbial diversity and the physiological and biochemical properties of microorganisms with an emphasis on processes relevant to corrosion and degradation of materials.

B. DIVERSITY OF MICROORGANISMS

B1. Evolution and Diversity

Microorganisms are the early life forms preceding all other life on Earth. Mutation and genetic recombination and exchange coupled with natural selection contribute to the divergence of life over the 4.6 billion years of history. Natural selection is the key to the survival of individual species and also provides new opportunities for new ones to evolve. For the diverse life forms on Earth, schemes of classification of the invisible microorganisms have always interested biologists. Prior to the 1960s, biological species were classified solely based on their morphology, biochemistry, and physiology [51–53]. Using these approaches, biologists classified all living organisms into two distinctive kingdoms, the prokaryotes and the eukaryotes [54–59].

Modern molecular biology has contributed significantly to the insights of the fundamental relationships between organisms and the possible events during evolution of life by deciphering the genetic codes in the organism genome. In the case of bacteria, 16S ribosomal RNA (rRNA) has been widely used [60]. The 16S rRNA genes, which are ubiquitous in all biological species on our planet, were proposed to be chronometers for measuring evolutionary changes, and as a result evolutionary relationships between different biological species can be reconstructed by analyzing sequences of nucleotides in the rRNA genes [60].

Currently, the biological world is classified as consisting of Eukarya, Archaea, and Bacteria based on their 16S rRNA [60, 61]. The former prokaryotes were further divided into Bacteria and Archaea as two distinct domains. Since Archaea consists of mostly single cells closely related to early evolutionary life, this domain includes microorganisms thriving in extreme environments, for example, alkaline, acid, high and low temperatures, high pressure, and a range of strictly anaerobic conditions. Further refined classification of Archaea has been made possible by molecular sequencing of 16S rRNA from isolated bacteria and cloned genes from a hot spring pool in Yellowstone National Park. Crenarchaeota, Euryarchaeota and Korarchaeota were proposed for these subdivisions in the phylogenetic tree [62, 63]. As molecular techniques are increasingly used to understand the basic biology and relationships between species, our concept of taxonomic classification and systematics in biology will face constant challenges and revision of the old dogma [64, 65].

B2. Biochemical and Physiological Diversity

Microorganisms also have diverse physiological and biochemical capabilities. For instance, they can be grouped based on their nutritional requirements. While all life forms require certain basic elements, such as nitrogen (N), phosphorus (P), and sulfur (S), they also need a source of carbon and electrons. Microorganisms can be classified based on their sources of carbon and electrons (Table 26.1). An autotroph (lithotroph) is an organism that is capable of utilizing carbon dioxide as the sole source of carbon and obtaining electrons through photosynthesis. At the present time, it is believed that all autotrophs can use organic forms of carbon, but many use carbon dioxide as the primary source. Examples include Fe-oxidizing bacteria. A heterotroph is a microbe that requires reduced forms of carbon for biosynthesis. Carbon dioxide can also be fixed to some extent by heterotrophs, but their primary sources of carbon are more reduced than carbon dioxide, such as glucose, cellulose, and humic materials. This group has probably the largest collection of species.

According to their primary source of energy, electrons and carbon, organisms have a range of nutritional forms (Table 26.1). Early classification schemes used this information exclusively. For example, see the second edition of *The Prokaryotes* [54] and the four-volume set of *Bergey's Manual of Systematic Bacteriology* [55–58]. General microbiology information can be found in a number of textbooks (e.g., Madigan et al. [53] and others [52, 66–68]).

B3. Ecological Diversity

Over the history of life on Earth, microorganisms have left evidence of early life in various unique environments, including deep-ocean thermal vents, hot springs, and sulfate-reducing and methanogenic habitats [59, 69–72]. Based on the environment where a microorganism thrives, microorganisms may be referred to as methanogenic, acetogenic, sulfate reducing (sulfidogenic), alkaliphilic, acidiphilic, barophilic, psychrophilic, and mesophilic. Based on their physiology and biochemistry, these bacteria reflect evolutionary history of life on Earth [73–76].

Contemporary microbiological techniques are incapable of yielding more than approximately 1% of the natural microbial population in laboratory cultures [77]. Difficulties are particularly encountered in isolation of new microorganisms from their natural environments. The majority of the natural microbial community has not been fully explored due to limited understanding of the physiological requirements by the microorganisms. It is commonly accepted that a diverse population of microorganisms is present in the environment, and our inability to culture most microorganisms (~99%) is a direct reflection of our lack of knowledge on the relationship between the microflora of natural habitats and their biochemical, physiological, and nutritional requirements. Corrosion and degradation processes under natural conditions are commonly associated with a mixed population of microorganisms and specific biochemical processes. Molecular cloning and nucleotide probing coupled with the polymerase chain reaction (PCR) provide some advantages in beginning to characaterize nonculturable microorganisms. And, in fact, the emerging field of metagenomics is

TABLE 26.1. Nutritional Types of a Range of Organisms

Nutritional	Organisms	Electron Source	Energy Source	Major Carbon Source
Chemoautotroph (chemoorganotroph)	Most bacteria fungi, protozoa, higher animals	Organic molecules	Electrons from reduced organic molecules	Reduced organic molecules
Chemoautotroph (chemolithotroph)	Hydrogen, iron, Nitrifying, and colorless sulfur bacteria	Reduced inorganic molecules	Electrons from inorganic molecules	Carbon dioxide
Photoautotroph (photolithotroph)	Purple and green sulfur bacteria, blue-green bacteria, algae, higher plants	Reduced inorganic molecules	Light	Carbon dioxide
Photoheterotroph	Purple nonsulfur bacteria, some eukaryotic algae	Organic molecules	Light	Reduced organic molecules

increasingly providing information on the biochemical properties of nonculturable microorganisms [78]. However, a detailed understanding of the physiological and biochemical processes that contribute to biocorrosion and biodegradation remains extremely difficult in the absence of pure cultures of the organisms under study.

C. MICROBIAL BIOFILMS

C1. Fundamental Concepts

Microbial biofilms are a collection of individual bacterial cells and aggregates of cells that are maintained on surfaces by electrostatic forces [26, 79] and/or by adhering exopolymers [21, 48]. Bacteria in planktonic solution are capable of colonizing surfaces of living tissues [80] and inert materials [20, 81, 82]. Formation of bacterial films on surfaces is a result of contact between bacteria and substratum surfaces by short-range forces. The initial contact with a surface is made possible by flagella, Brownian motion of the cells, or hydrodynamic forces. After the initial collision, some cells are freed from the surface by their physical motion or hydrodynamics of the medium, while others will remain on the surface and synthesize expolymeric materials. The production of exopolymers is thought to induce bacterial attachment [83–88]. A pure culture of *Bacillus megaterium* deposited on a membrane filter surface after incubation under laboratory conditions shows extensive space between cells with considerable production of extracellular materials (Fig. 26.1). Results indicate that cell-to-cell signaling molecules, which control the density of bacterial cells in the growth medium, triggers the synthesis of exopolymers when a specific density is reached [89], and more recent information suggests that interspecies communication between

bacteria is also possible and may have considerable ecological implications [90].

Signaling between cells under laboratory conditions [89] induces attachment of *Pseudomonas aeruginosa*. Our understanding of how bacteria attach to surfaces and the initiator of the event is limited under natural conditions. Marshall and co-workers sggested that bacterial adhesion could be separated into two categories, reversible attachment and irreversible attachment [91]. Recently, Dalton and co-workers [92] observed that colonization on hydrophobic surfaces was characterized by the formation of tightly packed biofilms, in contrast to hydrophilic surfaces which exhibited sparse colonization and formation of long chains of more than 100 μm in length. Fletcher [93] and Wiencek and Fletcher [94] further proposed a series of models explaining the selectivity of modified monolayer surfaces of different materials on bacterial adhesion. Heterogeneity of biofilms is evident in both space and time [95–97], and multicellular organization may also have some role in biofilm structure and ecological function [98]. However, the fundamental mechanisms of bacterial adhesion and the behavioral response of bacteria to surface selectivity are still not completely known.

C2. Factors Affecting Biofilm Formation

Formation of microbial biofilms in natural and artificial environments follows a sequence of events: the initial adhesion of a conditioning film consisting of organic molecules on surfaces, recruiting of microorganisms to the surfaces, bacterial colony formation, and growth to a mature biofilm. Biofim growth, senescence, and sloughing off is a continuous process [99–103]. Biofilms are interfacial matrices affecting the physical and chemical conditions of the substratum. The initial events of bacterial attachment to a surface were investigated by Marshall et al. [91]. In their pioneering study,

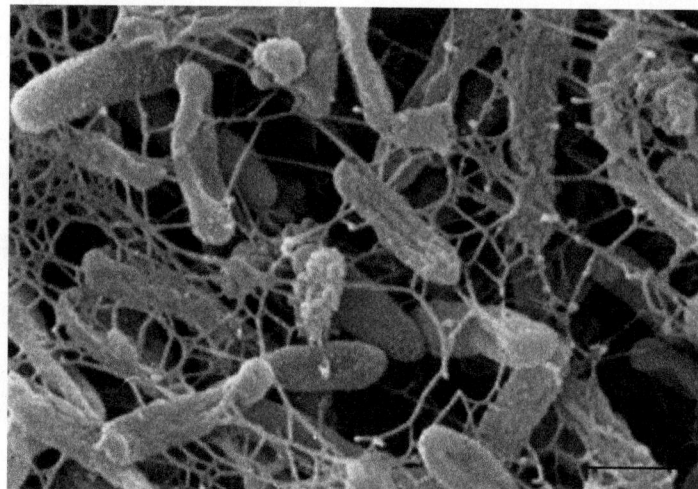

FIGURE 26.1. Scanning electron microphotograph of a pure culture of *Baccilus mageterium* on membrane filter and treated with dehydration and gold-palladium coating before viewing (scale bar, 2 μm).

Pseudomonas sp. adhered to glass slides either reversibly or irreversibly. Flagella promoted irreversible attachment by overcoming the short-range repulsion force between the bacterium and substratum when the distance between the two was very small. Planktonic bacteria may also settle on surfaces due to the presence of molecules that serve as cues for the bacteria [104, 105]. Fletcher [106] concluded that the structure of the surface materials also affects microbial adhesion. Hydrophobicity of a surface favors the formation of biofilms, while hydrophilicity discourages settlement [79, 107–110]. This generalization is not fully understood due to the lack of information about the physical chemistry of interactions between bacteria and surfaces.

A cell-to-cell signal was proposed as a key factor in the development of biofilms [89, 111]. *N*-Acyl homoserine lactone molecules were involved in mediating biofilm formation through a population-dependent process, in which the signaling molecules by bacteria respond to cell density and then induce the synthesis of exopolymers. More recent studies with pure cultures of *Mycobacterium avium* demonstrate that, at least with this organism, addition of the signaling molecule autoinducer-2 triggers an oxidative response mechanism that in turn induces biofilm formation. A similar response was also noted with simple addition of hydrogen peroxide, suggesting that biofilm formation is simply a response to environmental stress [22, 112]. However, bacteria in natural environments differ greatly from those in laboratory studies. Experimental results from the laboratory can be extrapolated only to a limited extent to the natural or artificial environments. After the initial attachment, reproductive growth and colony formation occurs. When the biofilm reaches a maximum thickness, the film may slough off due to hydrodynamic shear forces. The residual cells on surfaces divide and form new colonies; new biofilm will be developed again. During cell division, the daughter cell not attached to the surface may be released into the planktonic phase. The planktonic cells may also settle on newly available surfaces, providing initial colonization for biofilm development.

Physicochemical factors also influence biofilm formation. These include ionic strength, type of cation (valence and charge), concentration of the cation, presence of surfactant, and substratum surface characteristics. Microbiologists largely ignore this information [96, 105, 113]. Marshall discussed the fundamentals associated with this phenomenon [26]. Generally, higher ionic strength, concentration of ions, and higher valence of cations favor the formation of biofilms by decreasing the electric diffuse double layer, which is a barrier against bacterial movement toward a surface. When a bacterium approaches a surface, it must overcome a critical energy level before it can firmly attach. The higher concentrations and valence of cations can compress the diffuse double layer, bringing planktonic bacteria closer to substratum surfaces and facilitating adhesion.

C3. Some Properties of Microbial Biolfilms

C3.1. Biofouling. Biofouling is the undesirable accumulation of microorganisms, their products, and various deposits, including minerals and organic matter, on a substratum surface. The thin film on fouled surfaces usually consists of microorganisms embedded in an organic matrix of biopolymers, which are produced by the microorganisms under natural conditions. Biofilms of various microorganisms can be visible on stone surfaces in tropical climates (Fig. 26.2). In addition, microbial precipitates, minerals, and corrosion products may be present [83, 114]. Industrial fouling is a complex phenomenon involving interactions between chemistry, biology, and materials science. The fouling products may include inorganic particles, crystalline precipitates and scale, and corrosion products. Fouling is common and can be found in purified water systems [115], drinking water filtration and treatment facilities [116], wastewater treatment plants [117, 118], porous media [119–123], and biotechnology processes [124]. In addition, materials immersed in aqueous environments or under high-humidity conditions are also susceptible to biofouling, including both metallic and nonmetallic materials [4, 14, 28, 125–127], urinary catheters [82, 128], water pipes [129], and coatings [125, 130, 131].

C3.2. Changes in Local Environment. Establishment of microbial biofilms on the surface of materials changes the surface chemical properties due to the presence of microbial cells and exopolymeric products. Accumulation of biomass also changes the local environment in terms of water content, which in turn affects expansion/contraction upon heating or cooling [11]. Such physical modifications of the local environment have considerable impacts on the substratum materials and their subsequent susceptibility to further colonization by microorganisms.

C3.3. Heat Transfer Resistance. Biofilms form on a wide range of engineering materials, particularly on equipment used for cooling and transport. They reduce performance and the lifetime of the materials. The thickness of a biofilm may reach between 50 and 100 μm depending on physiochemical conditions of the environment, hydrodynamics of flow, and availability of nutrients. Individual bacterial cells within this layer of gelatinous exopolysaccharides are embedded in the heterogeneous porous matrices, and nutrients may diffuse and be transported in void channels throughout the biofilm matrices [132]. Biofilm establishment on surfaces increases frictional resistance, reduces heat transfer efficiency, and causes pitting corrosion of a variety of alloys. Both aerobic and anaerobic conditions provide opportunities for a metabolically diverse community of microorganisms. When the substratum materials are metals, corrosion may occur, resulting in the buildup of corrosion products on surfaces [133].

FIGURE 26.2. Photograph of natural biofilm community with different colors on sandstone bas-relief of Bayon Temple in Angkor Thom of Cambodia.

Deposited corrosion products may reach a significant thickness that prevents efficient transfer of heat across the cooling pipes and also a significant increase in friction resistance [28, 31, 134]. These changes in surface characteristics affect system performance.

Biofilms may also serve as crucial chemical cues for recruitment of invertebrates [101, 135–137]. Macrofouling is problematic to the power and water industries due to the blockage of pipes and biomass accumulation on surfaces. For example, the freshwater invertebrate zebra mussel (*Dreissena polymorpha*), an accidentally introduced species, has been a serious problems for both the power and drinking water industries in the Great Lakes region of the United States since the 1980s. The combination of heat transfer reduction and corrosion from macrofouling was estimated to be $5 billions for the power-generating industries by 2000 [138].

C3.4. Clogging of Pipes and Porous Media. Another consequence of microbial growth on surfaces, particularly as a result of MIC, is clogging of water distribution systems by corrosion products [30, 139]. Iron-oxidizing bacteria are partially responsible for the clogging under aerobic and microaerophilic conditions. In porous media, bacterial transport and accumulation generally decrease with distance along the flow path, especially if hydrodynamic and shearing forces are low. Bacteria are not easily transported through a tortuous path, and they become attached after a number of collisions on sediment surfaces [140, 141] and synthetic materials. Exopolymers of bacteria also participate in the corrosion of metals and have been shown to solubilize Cu from pipes under soft water conditions [142, 143]. Clogging may also be observed in wastewater treatment facilities

through aggregation of microorganisms in the presence of nucleating agents.

Porous media in subsurface environments are important for bioremediation of contaminated groundwater. Clogging is a common problem in bioremediation. Negatively charged bacteria may readily adhere to surfaces by electrostatic attraction, hydrogen bonding, cation bridging, and/or exopolymers, particularly under low nutrient conditions. Transport of bacteria and collision of bacteria on surfaces are a focus of several recent studies [140, 141]. Bacterial preference for surfaces makes bioremediation *in situ* difficult as exogenous organisms are not readily transported through porous media for effective mobility and bioremediation. In addition, aromatic and polyaromatic pollutants are mostly adsorbed on minerals or organic matter because of their hydrophobicity. Similar problems are also commonly encountered in oil drilling and recovery following seawater injection; clogging may be so severe that the less accessible oil reserves are not recovered. When seawater is introduced, the high ionic strength of the medium promotes microbial production of exopolymeric materials.

C3.5. Corrosion and Deterioration. Activity of microorganisms on surfaces may result in adverse effects on underlying materials, particularly corrosion and deterioration. Corrosion of metals and deterioration of polymers have been reviewed [1, 5, 9, 14, 29]. Corrosion is an electrochemical process induced or accelerated by the presence of microorganisms and their active metabolism [15], while deterioration of polymer is a metabolic process by which microorganisms obtain carbon and energy for their growth. Both corrosion and degradation can take place under aerobic and anaerobic conditions.

Corrosion of metals results in a decrease in tensile strength of the materials and may result in failure of the system. In industrial environments, the consequence of biocorrosion can be catastrophic. Microbial contamination is common even in ultrapure water systems used by pharmaceutical industries and semiconductor manufactures [115]. Several groups of microorganisms have been recognized for their role in corrosion, including the sulfate-reducing bacteria [29, 75], exopolymer-producing bacteria [48], hydrogen-producing and hydrogen-utilizing bacteria [1, 33], acid-producing bacteria, Fe-oxidizing bacteria [144, 145], and fungi [50]. The manganese oxidizing bacterium *Leptothrix discophora* is capable of ennoblement of stainless steel by increasing the open-circuit potential through deposition of MnO_x on surfaces [28, 146]. A correlation between microbial activity and deposition of MnO_x has been established by Dickinson et al. [146], and similar observations have also been reported [147]. Specific mechanisms of corrosion are discussed previously [148] and also in Chapter 39 of this book.

C3.6. Biocide Resistance. Biofilms protect microorganisms from activity of biocides because mass transfer is diffusion limited within the biofilm [149–158]. In order to effectively eradicate biofilms on surfaces, higher concentrations of biocides are often used (Fig. 26.3). Biofilm bacteria are in an environment where gene exchange may occur at high frequency [159–162], and resistance may therefore develop rapidly to a specific chemical, forcing the use of alternative treatments.

Biofilm bacteria respond to biocide concentration levels differently from planktonic bacteria. The efficacy of a biocide should be tested against both planktonic and biofilm bacteria under closely simulated environmental conditions. For example, a biocide for protection of polymeric coatings should be tested in the proposed coating materials and exposed under the relevant conditions [5, 6]. Polymeric materials also contain plasticizers which can be utilized by microorganism as a source of carbon and energy [163, 164]. Such growth-promoting effects of chemicals leaching from these materials can sustain levels of microbial activity and growth not seen in culture solutions. Because of this, extended tests may be needed [165].

D. DIAGNOSIS AND CONFIRMATION

Microbiology as a discipline began when small cells were first observed by Anthonie van Leeuwenhoek under a microscope in 1677 [166, 167]. His observation opened up an unrecognized world. Louis Pasteur investigated microbial contributions to disease in humans and animals and to fermentation of wine and vinegar to improve the quality of these products. Robert Koch advanced the field of medical microbiology and postulated the theory of disease (Koch's postulates) as follows in 1884 [168]:

1. The organism should be present in all animals suffering from the disease and absent from all healthy animals.

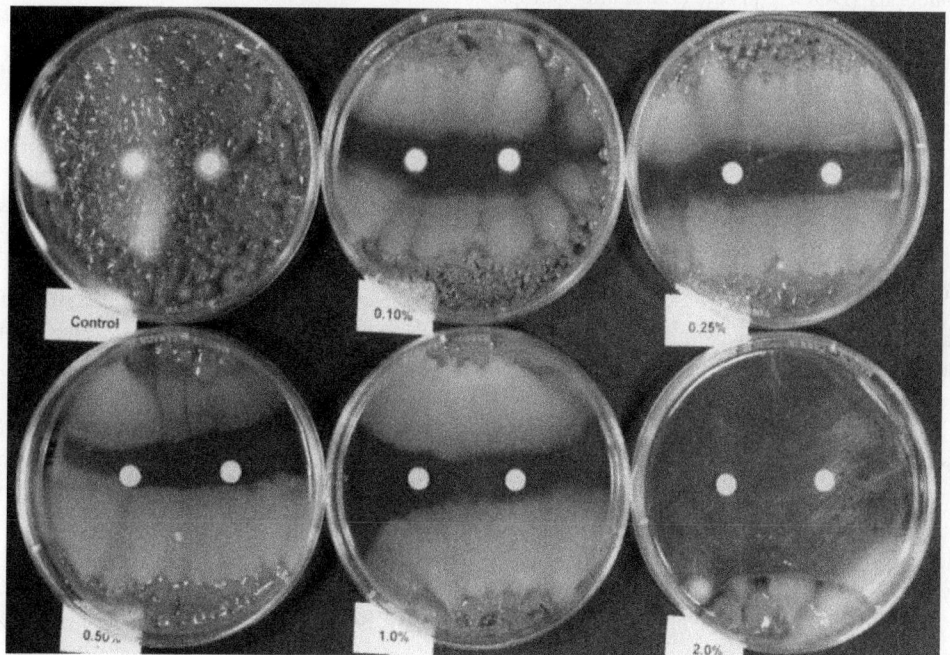

FIGURE 26.3. Photograph showing efficacy of biocide at range of concentrations (0, 0.10, 0.25, 0.50, 1.0, and 2.0%) applied on absorbent filter paper disks on agar plates which were previously inoculated with a testing population of microorganisms on the whole surface.

2. The organisms must be grown in pure culture outside the diseased animal host.

3. When such a culture is inoculated into a healthy, susceptible host, the animal must develop the symptoms of the disease.

4. The organism must be reisolated from the experimentally infected animal and shown to be identical to the original isolate. Today, Koch's postulates remain sound guidelines for medical professionals to prove or disprove the presence of a microbial infection.

Koch's approach can be utilized in research on corrosion and degradation of materials. In this case, organisms should be isolated from the damaged materials to verify their presence. The organisms should be cultured and purified and used to examine corrosion and degradation processes in the laboratory. If corrosion or degradation similar to the initial observation can be duplicated, damage to the material can be linked to the activity of the organisms. Obviously, equipped with molecular fingerprinting techniques and isotope labeling, much higher sensitivity and level of sophistication can be achieved for confirmation of microbial involvement and for understanding the mechanisms.

Uncharacterized mixed cultures have also been used to study degradation and deterioration of polymeric materials [34, 35, 40, 41, 81, 169–171], concrete [45, 172], and metal corrosion [1, 4, 29] and degradation of environmental pollutants [16, 173–175]. Such approaches together with fluorescence *in situ* hybridization (FISH), denaturing gradient gel electrophoresis (DGGE) [176], and stable isotope analysis should allow confirmation of the microorganism's involvement in degradation and deterioration at much higher resolution and sensitivity than currently available with traditional techniques.

E. MATERIAL DEGRADATION PROCESSES

E1. Sources of Carbon and Energy

Organic materials may be utilized by microorganisms as sources of carbon and energy in the presence or absence of molecular oxygen. During degradation, reduced carbon is oxidized, releasing electrons, which are used in bacterial synthesis. Different groups of microorganisms use widely divergent strategies to achieve the goal of decomposition. For example, eukaryotic fungi degrade polymeric materials using extracellular depolymerases. Most bacteria require close proximity to a substratum because the bacteria need to immediately capture the released compounds. This efficient process permits bacteria to maximize their energy conservation. Close proximity allows degradation products to be easily assimilated into cells for further metabolism [5, 6].

Metals can be corroded by bacteria through oxidative processes. Examples are iron- and manganese-oxidizing bacteria [144, 145], which clog drains, water pipes, and wells with deposition of iron oxides and hydroxides [30, 144]. These bacteria are lithoautotrophs utilizing CO_2 from the atmosphere and electrons from metal corrosion for their carbon sources and energy. Most of these bacteria are microaerophilic. Optimal growth is achieved under conditions of reduced oxygen tension/concentrations. *Leptothrix, Gallionella, Metallogenium,* and *Pedomicrobium* are the most commonly described bacteria [30, 177]. New metal-oxidizing species have been isolated recently [145].

Biological corrosion of metals was described more than 60 years ago [178, 179]. Mechanisms of microbial corrosion are discussed in several reviews [1, 28, 29, 31, 134].

E2. Degradation of Polymers

Some polymers support the growth of microorganisms without structural degradation. Additives to polymers are primarily responsible for the observed results [4, 6, 41, 126, 127]. Because bacteria are capable of cometabolism, polymers that are difficult to degrade can be metabolized when other carbon and energy sources are available for growth, altering the mechanical properties and integrity of the polymer. This phenomenon is particularly important in the degradation of polymers used in the electronic, transportation, and infrastructure industries. For example, a slight alteration of the insulation property of a polymer may result in devastating consequences to the proper functioning of integrated circuitry in storing and transmission of data.

E3. Attack of Metals and Concrete by Microbial Metabolites

Damage to materials is frequently incidental to microbial metabolism. For example, organic acids from fungi were found to corrode aluminum alloys in aircraft fuel tanks [50], and H_2 produced by fermenting bacteria may contribute to hydrogen embrittlement of metals under anaerobic conditions [15, 48]. Hydrogen sulfide (H_2S) from sulfate-reducing bacteria attacks both metals [1, 29, 31, 180] and stone materials [4, 10, 45].

Large quantities of exopolymeric materials can be produced by bacteria on surfaces. These materials may bind metal ions and contribute to galvanic corrosion of metals [47] and concrete [45].

F. PREVENTIVE MEASURES

Microorganisms require minimal quantities of water to survive. Active metabolism requires an appropriate relative humidity and temperature. A combination of low humidity and low temperature is the simplest way to control bacterial growth [6, 81]. However, fungi are capable of growing under such conditions. Regular cleaning is good practice to prevent biofilm formation and subsequent biodeterioration.

Microbial biocides are commonly used to prevent unfavorable biofilm formation and biofouling. These chemicals are widely incorporated into products, including toys, filtration systems, coatings, and surface-cleaning agents. Because microorganisms are capable of acquiring resistance after exposure to a particular chemical [160–162], no single chemical can be relied on for long-term use. Frequently, effective eradication of microbial populations is achieved by alternating chemicals or increasing the concentrations of a biocide.

In nature, bacteria normally live on surfaces. However, biocide efficacy tests often involve utilization of planktonic bacterial cultures on agar plate (Fig. 26.3). These test data fail to reflect the actual state of bacteria on material surfaces [6, 81]. More rigorous simulation of environmental conditions is highly recommended. For example, several biocides, including diiodomethyl-*p*-tolylsulfone, were excellent biocides in laboratory testing using planktonic cultures, but active growth of microorganisms was observed when the biocide was incorporated into a polyurethane coating [81]. Some biocides may also contain chemicals that support the active growth of environmental microorganisms, because the active ingredient diffuses out of the biocide formulation on test disks much too rapidly, leaving the formulation chemicals to support microbial metabolism (Fig. 26.4). Microbial resistance to biocides is also of concern to public health because of growth of nuisance organisms and pathogens in, for example, air-conditioning systems, drinking water, and water-heating systems.

G. CONCLUSIONS

Biodegradation of materials is common under both oxic and anoxic conditions. Microorganisms develop strategies to extract electrons and carbon sources from both polymers and metals. During microbial growth on surfaces, metabolites, enzymes, and other exopolymeric materials also affect the underlying materials. Koch's postulates provide an excellent means of associating a degradation process with a causative organism. Degradation may involve the utilization of a polymer for carbon and energy, resulting in mineralization of the materials. Components of a polymer may become substrates for microbial growth, weakening the mechanical properties of the material. Other mechanisms of degradation include corrosive metabolites, for example, H_2S, H_2, organic acids, and exopolymeric materials. Under natural conditions, several of these mechanisms often operate simultaneously. Biofilm formation plays an important role in material degradation, deterioration, and corrosion;

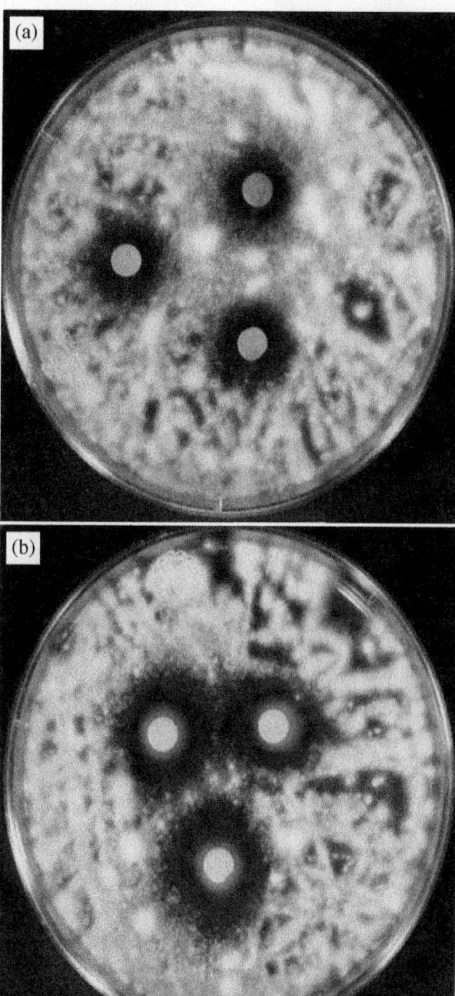

FIGURE 26.4. Photographs showing (a) effective inhibition of test organisms around filter paper disks on surface of agar plate with test chemical embedded in disks and (b) inhibition of test organisms around filter paper disks on surface of agar plate and also active growth of microorganisms on paper disks where test chemical was initially embedded.

microbial films are heterogeneous in both space and time and contain a wide diversity of bacterial species within the gelatinous matrices. Effective eradication of biofilms is difficult because of the phenotypic plasticity and genetic mobility of transferable plasmid genes.

ACKNOWLEDGMENTS

We would like to thank EM Unit at Queen Mary Hospital, The University of Hong Kong, for assistance in taking the scanning electron micrographs used here. We would also like to thank Jessie Lai for helping with the image digitization.

REFERENCES

1. T. Fordand R. Mitchell, in Advances in Microbial Ecology, Vol. **11**, K. C. Marshall (Ed.), Plenum, New York, 1990, pp. 231–262.

2. T. Ford and R. Mitchell, MTS J., **24**, 29 (1990).

3. J.-D. Gu, T. E. Ford, and R. Mitchell, in Uhlig's Corrosion Handbook, 2nd ed., W. Revie (Ed.), Wiley, New York, 2000, pp. 915–927.

4. J.-D. Gu, in Encyclopedia of Microbiology, 3rd ed., M. Schaechter (Ed.), Elsevier, San Diego, California, 2009, pp. 259–269.

5. J.-D. Gu, T. E. Ford, and R. Mitchell, in Uhlig's Corrosion Handbook, 2nd ed., W. Revie (Ed.), Wiley, New York, 2000, pp. 439–460.

6. J.-D. Gu, Int. Biodeter. Biodeg., **52**, 69 (2003).

7. J.-D. Gu, Int. Biodeter. Biodeg., **59**, 170 (2007).

8. R. Gebers and P. Hirsch, in Environmental Biogeochemistry and Geomicrobiology, Vol. **3**, Methods, Metals and Assessment, W. E. Krumbein (Ed.), Ann Arbor Science Publishers, Ann Arbor, MI, 1978, pp. 911–922.

9. J.-D. Gu, T. E. Ford, and R. Mitchell, in Uhlig's Corrosion Handbook, 2nd ed., W. Revie (Ed.), Wiley, New York, 2000, pp. 477–491.

10. R. Mitchell and J.-D. Gu, Int. Biodeter. Biodegr., **46**, 299 (2000).

11. T. Warscheid and J. Braams, Int. Biodeter. Biodegr., **46**, 343 (2000).

12. D. J. Giannantonio, J. C. Kurth, K. E. Kurtis, and P. A. Sobecky, Int. Biodeter. Biodegr., **63**, 30 (2009).

13. H. H. Uhlig and R. W. Revie, in Corrosion and Corrosion Control, 3rd ed., Wiley, New York, 1985.

14. B. J. Little, P. A. Wagner, W. G. Characklis, and W. Lee, in Biofilms, W. G. Characklis and K. C. Marshall (Eds.), Wiley, New York, 1990, pp. 635–670.

15. M. Walch, "Microbial corrosion," Encycloped. Microbiol., **1**, 585 (1992).

16. J.-D. Gu, in Soil Mineral-Microbe-Organic Interactions—Theories and Applications, Q. Y. Huang, P. M. Huang, and A. Violante (Eds.), Springer-Verlag, Berlin, 2008, pp. 175–198.

17. J. A. Breznak, in Microbial Adhesion and Aggregation, K. C. Marshall, Ed., Dahlem Konferenzen, Springer-Verlag, Berlin, 1984, pp. 203–221.

18. A. N. Glagolev, in Motility and Taxis in Prokaryotes, Soviet Scientific Reviews Supplememt Series: Physiocochemical Biology, V. P. Skulachev (Ed.), Harwood Academic, Switzerland, 1984.

19. B. Little, P. Wagner, S. M. Gerchakov, M. Walch, and R. Mitchell, Corrosion, **42**, 533 (1986).

20. B. J. Little, P. Wagner, J. S. Maki, M. Walch, and R. Mitchell, J. Adhesion, **20**, 187 (1986).

21. J. W. Costerton, G. G. Geesey, and K.-J. Cheng, Sci. Am., **238**, 86 (1978).

22. D. E. Caldwell and J. R. Lawrence, Microb. Ecol., **12**, 299 (1986).

23. J. R. Lawrence, P. J. Delaquis, D. R. Korber, and D. E. Caldwell, Microb. Ecol., **14**, 1 (1987).

24. D. R. Korber, J. R. Lawrence, B. Sutton, and D. E. Caldwell, Microb. Ecol., **18**, 1 (1989).

25. G. G. Geesey and D. C. White, Annu. Rev. Microbiol., **44**, 579 (1990).

26. K. C. Marshall, Interfaces in Microbial Ecology., Harvard University Press, Cambridge, MA, 1976.

27. K. C. Marshall, in Adsorption of Microorganisms to Surfaces, G. Bitton and K. C. Marshall (Eds.), Wiley, New York, 1980, pp. 317–329.

28. B. J. Little, P. A. Wagner, and Z. Lewandowski, in Geomicrobiology: Interactions between Microbes and Minerals, J. F. Banfield and K. H. Nealson (Eds.), Mineralogical Society of America, Washington, DC, 1997, pp. 123–159.

29. W. A. Hamilton, Ann. Rev. Microbiol., **39**, 195 (1985).

30. H. H. Hanert, in The Prokaryotes, A Handbook on Habitats, Isolation, and Identification of Bacteria, Vol. **1**, M. P. Starr, H. Stolp, H. G. Trüper, A. Balow, and H. G. Schlegel (Eds.), Springer-Verlag, New York, 1992, pp. 1049–1060.

31. W. Lee, Z. Lewandowski, P. H. Nielsen, and W. A. Hamilton, Biofouling, **8**, 165 (1995).

32. T. E. Ford, P. C. Seaison, T. Harris, and R. Mitchell, J. Electrochem. Soc., **137**, 1175 (1990).

33. M. Walch and R. Mitchell, in Biologically Induced Corrosion, S. C. Dexter (Ed.), National Association of Corrosion Engineers, Houston, TX, 1986, pp. 201–209.

34. R. A. Gross, J.-D. Gu, D. Eberiel, M. Nelson, and S. P. McCarthy, in Fundamentals of Biodegradble Materials and Packaging, D. Kaplan, E. Thomas, and C. Ching (Eds.), Technomic Publishing, Lancaster, PA, 1993, pp. 257–279.

35. R. A. Gross, J.-D. Gu, D. Eberiel, and S. P. McCarthy, in Degradable Polymers, Recycling and Plastics Waste Management, A. Albertson and S. Huang (Eds.), Marcel Dekker, New York, 1995, pp. 21–36.

36. J.-D. Gu, M. Gada, G. Kharas, D. Eberiel, S. P. McCarty, and R. A. Gross, Polym. Mat. Sci. Eng., **67**, 351 (1992).

37. J.-D. Gu, S. P. McCarty, G. P. Smith, D. Eberiel, and R. A. Gross, Polym. Mat. Sci. Eng., **67**, 230 (1992).

38. J.-D. Gu, S. Coulter, D. Eberiel, S. P. McCarthy, and R. A. Gross, J. Environ. Polym. Degr., **1**, 293 (1993).

39. J.-D. Gu, D. T. Eberiel, S. P. McCarthy, and R. A. Gross, J. Environ. Polym. Degr., **1**, 143 (1993).

40. J.-D. Gu, D. Eberiel, S. P. McCarthy, and R. A. Gross, J. Environ. Polym. Degr., **1**, 281 (1993).

41. J.-D. Gu, S. Yang, R. Welton, D. Eberiel, S. P. McCarthy, and R. A. Gross, J. Environ. Polym. Degr., **2**, 129 (1994).

42. G. Odian, Principles of Polymerization, 3rd ed., Wiley, New York, 1991.

43. J.-G. Gu and J.-D. Gu, J. Polym. Environ., **13**, 65 (2005).

44. G. Gomez-Alarcon, M. Munoz, X. Arino, and J. J. Ortega-Calvo, Sci. Total Environ., **169**, 249 (1995).

45. J.-D. Gu, T. E. Ford, N. S. Berke, and R. Mitchell, Int. Biodeter. Biodegr., **41**, 101 (1998).

46. G. Chen, C. R. Clayton, R. A. Sadowski, J. B. Gillow, and A. J. Francis, Paper No. 217, CORROSION/1995, NACE International, Houston, TX, 1995.

47. G. Chen, S. V. Kagwade, G. E. French, T. E. Ford, R. Mitchell, and C. R. Clayton, Corrosion, **52**, 891 (1996).

48. T. Ford, E. Sacco, J. Black, T. Kelley, R. Goodacre, R. C. W. Berkley, and R. Mitchell, Appl. Environ. Microbiol., **57**, 1595 (1991).

49. T. Ford, J. Maki, and R. Mitchell, in Bioextraction and Biodeterioration of Metals, C. C. Gaylarde and H. A. Hector (Eds.), Cambridge University Press, Cambridge, 1995.

50. H. A. Videla, Manual of Biocorrosion, CRC Press, Boca Raton, FL, 1996.

51. A. J. Kluyver and C. B. van Niel, The Microbe's Contribution to Biology, Harvard University Press, Cambridge, MA, 1956.

52. R. M. Atlas, and R. Bartha, Microbial Ecology: Fundamentals and Applications, 4th ed., Benjamin/Cummings, Menlo Park, CA, 1997.

53. M. T. Madigan, J. M. Martinko, P. V. Dunlap, and J. Parker, Brock Biology of Micoorganisms, 12th ed., Pearson Benjamin Cummings, San Francisco, 2009.

54. A. Balow, H. G. Trüper, M. Dworkin, W. Harder, and K.-H. Schleifer, The Prokaryotes: A Handbook on the Biology of Bacteria: Ecophysiology, Isolation, Identification, Applications, Vols. I–IV, Springer-Verlag, New York, 1992.

55. N. R. Krieg and J. G. Holt, Bergey's Manual of Systematic Bacteriology, Vol. 1, Williams and Wilkins, Baltimore, MD, 1985.

56. P. H. A. Sneath, N. S. Mair, M. E. Sharpe, and J. G. Holt, Bergey's Manual of Systematic Bacteriology, Vol. 2, Williams and Wilkins, Baltimore, MD, 1986.

57. J. T. Staley, M. P. Bryant, N. Pfenning, and J. G. Holt, Bergey's Manual of Systematic Bacteriology, Vol. 3, Williams and Wilkins, Baltimore, MD, 1989.

58. S. T. Williams, M. E. Sharpe, and J. G. Holt, Bergey's Manual of Systematic Bacteriology, Vol. 4, Williams and Wilkins, Baltimore, MD, 1989.

59. L. Margulis, Symbiosis in Cell Evolution, W.H. Freeman and Company, San Francisco, CA, 1981.

60. C. R. Woese, Microbiol. Rev., **51**, 221 (1987).

61. C. R. Woese, Proc. Natl. Acad. Sci., **87**, 4576 (1990).

62. N. R. Pace, Science, **276**, 734 (1997).

63. P. Hugenholtz, C. Pitulle, K. L. Hershberger, and N. R. Pace, J. Bacteriol., **180**, 366 (1998).

64. E. Mayr, Proc. Natl. Acad. Sci., **95**, 9720 (1998).

65. C. R. Woese, Proc. Nat. Acad. Sci., **95**, 11043 (1998).

66. G. Gottschalk, Bacterial Metabolism, 2nd ed., Springer-Verlag, New York, 1986.

67. A. G. Moat and J. W. Foster, Microbial Physiology, 2nd ed., Wiley, New York, 1988.

68. O. Ogunseitan, Microbial Diversity, Blackwell, Malden, MA, 2005.

69. T. Fenchel and B. J. Finlay, Ecology and Evolution in Anoxic Worlds, Oxford University Press, New York, 1995.

70. A. N. Nozhevnikova, O. R. Kotsyurbenko, and M. V. Simankova, in Acetogenesis, H. L. Drake (Ed.), Chapman & Hall, New York, 1994, pp. 416–431.

71. G. A. Zavarzin, T. N. Zhilina, and M. A. Pusheva, in Acetogenesis, H. L. Drake (Ed.), Chapman & Hall, New York, 1994, pp. 432–444.

72. S. H. Zinder, in Methanogenesis: Ecology, Physiology, Biochemistry and Genetics, J. G. Ferry (Ed.), Chapman & Hall, New York, 1993, pp. 128–206.

73. J. G. Ferry, Methanogenesis: Ecology, Physiology, Biochemistry and Genetics, Chapman & Hall, New York, 1995.

74. H. Drake, Acetogenesis, Chapman & Hall, New York, 1994.

75. J. R. Postgate, in The Sulphate-Reducing Bacteria, 2nd ed., Cambridge University Press, Cambridge, 1984.

76. J. M. Odom and R. Singleton, Jr., in The Sulfate-Reducing Bacteria: Contemporary Perspectives, Springer-Verlag, New York, 1993.

77. R. I. Amann, W. Ludwig, and K.-H. Schleifer, Microbiol. Rev., **59**, 143 (1995).

78. C. Schmeisser, H. Steele, and W. Streit, Appl. Microbiol. Biotech., **75**, 955 (2007).

79. T. Neu, Microbiol. Rev., **60**, 151 (1996).

80. M. W. Mittelman, in Bacterial Adhesion: Molecular and Ecological Diversity, M. Fletcher (Ed.), Wiley-Liss, New York, 1996, pp. 89–127.

81. J.-D. Gu, M. Roman, T. Esselman, and R. Mitchell, Int. Biodeter. Biodegr., **41**, 25 (1998).

82. J.-D. Gu, B. Belay, and R. Mitchell, World J. Microbiol. Biotech., **17**, 173 (2001).

83. T. J. Beveridge, S. A. Makin, J. L. Kadurugamuwa, and Z. Li, FEMS Microbiol. Rev., **20**, 291 (1997).

84. R. Bonet, M. D. Simon-Pujol, and F. Congregado, Appl. Environ. Microbiol., **59**, 2437 (1993).

85. C. Freeman and M. A. Lock, Limnol. Oceanogr., **40**, 273 (1995).

86. P. Vandevivere and D. L. Kirchman, Appl. Environ. Microbiol., **59**, 3280 (1993).

87. P. Vandevivere, Biofouling, **8**, 281 (1995).

88. C. Whitfield, Can. J. Microbiol., **34**, 415 (1988).

89. D. G. Davies, M. R. Parsek, J. P. Person, B. H. Iglewski, J. W. Costerton, and E. P. Greenberg, Science, **280**, 295 (1998).

90. P. D. Straight and R. Kolter, Annu. Rev. Microbiol., **63**, 99 (2009).

91. K. C. Marshall, R. Stout, and R. Mitchell, J. Gen. Microbiol., **68**, 337 (1971).

92. H. M. Dalton, L. K. Poulsen, P. Halasz, M. L. Angles, A. E. Goodman, and K. C. Marshall, J. Bacteriol., **176**, 6900 (1994).

93. M. Fletcher, in Bacterial Adhesion: Molecular and Ecological Diversity, M. Fletcher (Ed.), Wiley-Liss, New York, 1996, pp. 1–24.

94. K. M. Wiencek and M. Fletcher, J. Bacteriol., **177**, 1959 (1995).

95. M. C. M. van Loosdrecht, J. Lyklema, W. Norde, G. Schraa, and A. J. B. Zehnder, Appl. Environ. Microbiol., **53**, 1893 (1987).

96. M. C. M. van Loosdrecht, J. Lyklema, W. Norde, G. Schraa, and A. J. B. Zehnder, Microbiol. Rev., **54**, 75 (1990).

97. J. W. T. Wimpenny and R. Colasanti, FEMS Microbiol. Ecol., **22**, 1 (1997).

98. G. M. Wolfaardt, J. R. Lawrence, R. D. Robarts, S. J. Caldwell, and D. E. Caldwell, Appl. Environ. Microbiol., **60**, 434 (1994).

99. D. E. Caldwell, G. M. Wolfaaedt, D. R. Korber, and J. R. Lawrence, Adv. Microb. Ecol., **15**, 105 (1997).

100. M. Fletcher and G. I. Loeb, Appl. Environ. Microbiol., **37**, 67 (1979).

101. J. S. Maki, D. Rittschof, A. R. Schmidt, A. S. Snyder, and R. Mitchell, Biol. Bull., **177**, 295 (1989).

102. J. S. Maki, D. Rittschof, M.-O. Samuelson, U. Szewzyk, A. B. Yule, S. Kjelleberg, J. D. Costlow, and R. Mitchell, Bull. Mar. Sci., **46**, 499 (1990).

103. R. Mitchell and J. S. Maki, in Marine Biodeterioration: Advanced Techniques Applicable to the Indian Ocaen, M.-F. Thompson, R. Sarojini, and R. Nagabhushanam (Eds.), Oxford & IBH Publishing, New Delhi, 1989, pp. 489–497.

104. K. Kelly-Wintenberg and T. C. Montie, Appl. Environ. Microbiol., **60**, 363 (1994).

105. J. R. Kirby, Annu. Rev. Microbiol., **63**, 45 (2009).

106. M. Fletcher, in Bacterial Adherence (Receptors and Recognition), Series B, Vol. 6, E. H. Beachey (Ed.), Chapman and Hall, London, 1980, pp. 345–374.

107. H. J. Busscher, J. Sjollema, and H. C. van der Mai, in Microbial Cell Surface Hydrophobicity, R. J. Doyle and M. Rosenberg (Eds.), American Society for Microbiology, Washington, DC, 1990, pp. 335–339.

108. K. C. Marshall, ASM News, **58**, 202 (1992).

109. R. P. Sneider, B. R. Chadwick, R. Pembrey, J. Jankowski, and I. Acworth, FEMS Microbiol. Ecol., **14**, 243 (1994).

110. R. Schmidt, Int. Biodeter. Biodegr., **40**, 29 (1997).

111. R. J. C. McLean, M. Whiteley, D. J. Stickler, and W. C. Fuqua, FEMS Microbiol. Lett., **154**, 259 (1997).

112. H. Geier, S. Mostowy, G. A. Cangelosi, M. A. Behr, and T. E. Ford, Appl. Environ. Microbiol., **74**, 1798 (2008).

113. A. Zachary, M. E. Taylor, F. E. Scott, and R. R. Colwell, in Biodeterioration Proceedings of the 4th International Biodeterioration Symposium, T. A. Oxley, G. Becker, and D. Allsopp (Eds.), Pitman Publishing, London, 1980, pp. 171–177.

114. K. O. Konhauser, S. Schultze-Lam, F. G. Ferris, W. S. Fyfe, F. J. Longstaffe, and T. J. Beveridge, Appl. Environ. Microbiol., **60**, 549 (1994).

115. M. W. Mittelman, in Microbial Biofilms, H. M. Lappin-Scott and J. W. Costerton (Eds.), Cambridge University Press, Cambridge, 1995, pp. 133–147.

116. H.-C. Flemming, G. Schaule, R. McDonogh, and H. F. Ridgway, in Biofouling and Biocorrosion in Industrial Water Systems, G. G. Geesey, Z. Lewandowski, and H.-C. Flemming (Eds.), Lewis, Boca Raton, FL, 1994, pp. 63–89.

117. G. Bitton, in Adsorption of Microorganisms to Surfaces, G. Bitton and K. C. Marshall (Eds.), Wiley, New York, 1980, pp. 331–374.

118. J. D. Bryers and W. G. Characklis, in Biofilms, W. G. Characklis and K. C. Marshall (Eds.), Wiley, New York, 1990, pp. 671–696.

119. A. B. Cunningham, E. J. Bouwer, and W. G. Characklis, in Biofilms, W. G. Characklis and K. C. Marshall (Eds.), Wiley, New York, 1990, pp. 697–732.

120. A. M. Cunningham, W. G. Characklis, F. Abedeen, and D. Crawford, Environ. Sci. Technol., **25**, 1305 (1991).

121. H. H. M. Rijnaarts, W. Norde, E. J. Bouwer, J. Lyklema, and A. J. B. Zehnder, Appl. Environ. Microbiol., **59**, 3255 (1993).

122. B. E. Rittman, Water Res. Res., **29**, 2195 (1993).

123. V. Williams and M. Fletcher, Appl. Environ. Microbiol., **62**, 100 (1996).

124. J. D. Bryers, Colloid Surf. B: Biointerf., **2**, 9 (1994).

125. J. Jones-Meehan, M. Walch, B. J. Little, R. I. Ray, and F. B. Mansfeld, in Biofouling and Biocorrosion in Industrial Water Systems, G. G. Geesey, Z. Lewandowski, and H.-C. Flemming (Eds.), Lewis, Boca Raton, FL, 1994, pp. 107–135.

126. K. E. G. Thorp, A. S. Crasto, J.-D. Gu, and R. Mitchell, in Proceedings of the Tri-Service Conference on Corrosion, T. Naguy (Ed.), U.S. Government Printing Office, Washington, DC, 1994, pp. 303–314.

127. P. Wagner, R. Ray, K. Hart, and B. Little, Mater. Perform., **35**, 79 (1996).

128. R. J. C. McLean, J. C. Nickel, and M. E. Olson, in Microbial Biofilms, H. M. Lappin-Scott and J. W. Costerton (Eds.), Cambridge University Press, Cambridge, 1995, pp. 261–273.

129. J. Rogers, A. B. Dowsett, P. J. Dennis, J. V. Lee, and C. W. Keevil, Appl. Environ. Microbiol., **60**, 1842 (1994).

130. J.-D. Gu, D. B. Mitton, T. E. Ford, and R. Mitchell, Biodegradation, **9**, 35 (1998).

131. K. E. G. Thorp, A. S. Crasto, J.-D. Gu, and R. Mitchell, Paper No. 279, CORROSION/1997, NACE International, Houston, TX, 1997.

132. J. W. Costerton, Z. Lewandowski, D. DeBeer, D. Caldwell, D. Korber, and G. James, J. Bacteriol., **176**, 2137 (1994).

133. W. G. Characklis, in Biofilms, W. G. Characklis and K. C. Marshall (Eds.), Wiley, New York, 1990, pp. 523–584.

134. S. C. Dexter, Biofouling, **7**, 97 (1993).

135. C. Holmström, D. Rittschof, and S. Kjelleberg, Appl. Environ. Microbiol., **58**, 2111 (1992).

136. J.-D. Gu and R. Mitchell, J. Microbiol., **39**, 133 (2001).

137. J.-D. Gu and R. Mitchell, Hydrobiologia, **474**, 81 (2002).

138. J. Ross, Smithsonian, **40**, 24 (1994).

139. R. Lü, Q. Liu, Y. Zhang, and C. Xiao, Acta Microbiol. Sin., **29**, 204 (1989).

140. E. J. Bouwer, in Environmental Microbiology, R. Mitchell (Ed.), Wiley, New York, 1992, pp. 319–333.

141. A. L. Mills and D. K. Powelson, in Bacterial Adhesion: Molecular and Ecological Diversity, M. Fletcher (Ed.), Wiley-Liss, New York, 1996, pp. 25–57.

142. M. W. Mittelaman and G. Geesey, Appl. Environ. Microbiol., **49**, 846 (1985).

143. J. D. A. Miller, Microbial Aspects of Metallurgy, American Elsevier, New York, 1970.

144. H. H. Paradies, in Bioextraction and Biodeterioration, C. C. Gaylarde and H. A. Videla (Eds.), Cambridge University Press, Cambridge, 1995, pp. 197–269.

145. H. L. Ehrlich, Geomicrobiology, 4th ed., Marcel Dekker, New York, 2002.

146. A. Emerson and C. Moyer, Appl. Environ. Microbiol., **63**, 4784 (1997).

147. W. H. Dickinson, F. Caccavo, Jr., B. Olesen, and Z. Lewandowski, Appl. Environ. Microbiol., **63**, 2502 (1997).

148. K. Mattila, L. Carpen, T. Hakkarainen, and M. S. Salkinoja-Salonen, Int. Biodeter. Biodegr., **40**, 1 (1997).

149. K. L. Cargill, B. H. Pyle, and G. A. McFeters, Can. J. Microbiol., **38**, 423 (1992).

150. J. Gillatt, Int. Biodeter., **26**, 205 (1990).

151. X. Liu, F. Roe, A. Jesaitis, and Z. Lewandoski, Biotechnol. Bioeng., **59**, 156 (1998).

152. G. A. McFeters, F. P. Yu, B. H. Pyle, and P. S. Stewart, J. Ind. Microbiol., **15**, 333 (1995).

153. B. H. Pyle, S. C. Broadaway, and G. A. McFeters, J. Appl. Bacteriol., **72**, 71 (1992).

154. H. W. Rossmoore and L. A. Rossmoore, in A Practical Manual on Microbiologically Influenced Corrosion, G. Kobrin (Ed.), NACE International, Houston, TX, 1993, pp. 31–40.

155. C.-T. Huang, G. James, W. G. Pitt, and P. S. Stewart, Colloid. Surf. B: Biointerf., **6**, 235 (1996).

156. P. S. Stewart, Antmicrob. Agents Chemother., **40**, 2517 (1996).

157. P. Stoodley, D. DeBeer, and H. M. Lappin-Scott, Agents Chemother., **41**, 1876 (1997).

158. P. A. Suci, J. D. Vrany, and M. W. Mittelamn, Biomaterials, **19**, 327 (1998).

159. M. L. Angles, K. C. Marshall, and A. E. Goodman, Appl. Environ. Microbiol., **59**, 843 (1993).

160. Y. Wang, P. C. Leung, P. Qian, and J.-D. Gu, Microbes Environ., **19**, 163 (2004).

161. Y. Wang, P. C. Leung, P. Qian, and J.-D. Gu, Ecotoxicology, **15**, 371 (2006).

162. R. Zhang, Y. Wang, and J.-D. Gu, Ant. Leeuvenh. Int. J. Gen. Mol. Microbiol., **89**, 307 (2006).

163. J. Li and J.-D. Gu, Sci. Total Environ., **380**, 181 (2007).

164. Y. Wang, B. Yin, Y.-G. Hong, Y. Yan, and J.-D. Gu, Ecotoxicology, **17**, 845 (2008).

165. J.-D. Gu and L. Pan, J. Polym. Environ., **14**, 273 (2006).

166. A. Leeuwenhoek, Phil. Trans. Roy. Soc. Lond., **11**, 821 (1677).

167. A. Leeuwenhoek, Phil. Trans. Roy. Soc. Lond., **14**, 568 (1684).

168. R. Koch, Mittheilungen aus dem Kaiserlichen Gesundheit-samte, **2**, 1 (1884).

169. J.-D. Gu, T. Ford, K. Thorp, and R. Mitchell, Int. Biodeter. Degr., **39**, 197 (1996).

170. J.-D. Gu, T. E. Ford, and R. Mitchell, J. Appl. Polym. Sci., **92**, 1029 (1996).

171. J.-D. Gu, C. Lu, K. Thorp, A. Crasto, and R. Mitchell, J. Indus. Microbiol. Biotechnol., **18**, 364 (1997).

172. J.-D. Gu, C. Lu, K. Thorp, A. Crasto, and R. Mitchell, Mater. Perform., **36**, 37 (1997).

173. J.-D. Gu and D. F. Berry, Appl. Environ. Microbiol., **57**, 2622 (1991).

174. J.-D. Gu and D. F. Berry, Appl. Environ. Microbiol., **58**, 2667 (1992).

175. J.-D. Gu, D. F. Berry, R. H. Taraban, D. C. Martens, H. L. Walker, Jr., and W. J. Edmonds, Biodegradability of Atrazine, Cyanazine, and Dicamba in Wetland Soils, Virginia Water Resource Research Center, Bull. No. 172, Blacksburg, VA, 1992.

176. G. Muyzer, E. C. de Waal, and A. G. Uitterlinden, Appl. Environ. Microbiol., **59**, 695 (1993).

177. W. C. Ghiorse and P. Hirsch, in Environmental Biogeochemistry and Geomicrobiology, Vol. 3, Methods, Metals and Assessment, W. E. Krumbein (Ed.), Ann Arbor Science, Ann Arbor, MI, 1978, pp. 897–909.

178. R. F. Hadley, in The Corrosion Handbook, H. H. Uhlig (Ed.), Wiley, New York, 1948, pp. 466–481.

179. C. A. H. van Wolzogen Kuhr and I. S. van der Vlugt, Water, **18**, 147 (1934).

180. R. Lü, Q. Liu, C. Xiao, S. Bai, H. Chen, and F. Wang, Acta Microbiol. Sinica, **24**, 243 (1984).

27

CORROSION PROBABILITY AND STATISTICAL EVALUATION OF CORROSION DATA

T. Shibata[†]

Department of Materials Science and Processing, Graduate School of Engineering, Osaka University, Japan

A. INTRODUCTION

Corrosion engineers and scientists are often asked to estimate the corrosion rate of a specific material in a specific environment in order to predict the engineering lifetime. To obtain information to answer such a question, corrosion specialists review all available data in databases accessible to them, as well as in books and in journal literature. Data on general corrosion under commonly encountered conditions, such as steel exposed to the atmosphere, to marine environments, and to soils, are readily available and can be used for lifetime prediction. These data can also be utilized for design purposes, so that the material is made of the thickness required to achieve the desired lifetime, assuming general corrosion is the predominant mode by which degradation will occur during service.

Data on localized corrosion, however, are limited and can be used only for making go–no go decisions on usage of the material in a given environment. For systems that undergo localized corrosion, the corrosion rate cannot be used to predict lifetime. For these systems, the concept of corrosion probability has been introduced for lifetime prediction [1]. The probabilistic concept is essential for quality control, to ensure high-quality products, and in reliability engineering [2], which is a basis for highly efficient production systems in advanced industries.

Corrosion engineering could gain a greater level of confidence if the corrosion probability concept were widely used for estimating lifetime. Probability concepts and statistical procedures, however, can be somewhat difficult for corrosion scientists and engineers to learn, although many textbooks on statistics are available. Although the American Society for Testing and Materials (ASTM) standard G16 [3] is quite helpful in understanding statistical practices for analyzing corrosion data, further insights into corrosion probability, including recent progress, would be very useful. In this chapter, the basic concept of corrosion probability is discussed, along with statistical procedures, based on probability plots, and related topics that have been successfully applied to analysis of corrosion data.

[†]Retired.

Uhlig's Corrosion Handbook, Third Edition, Edited by R. Winston Revie

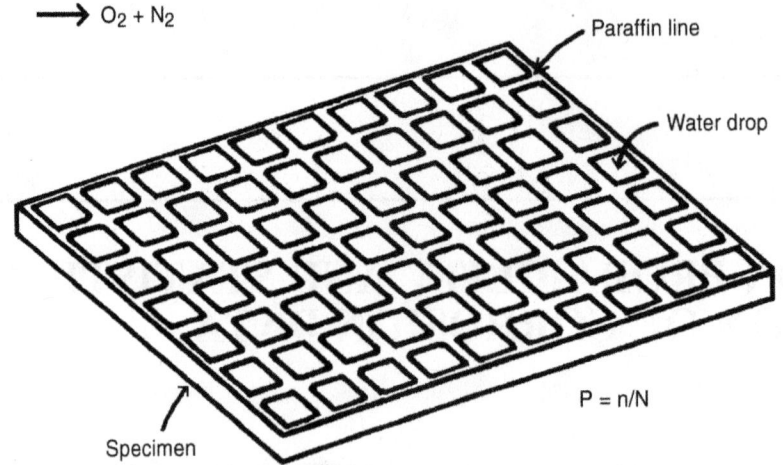

FIGURE 27.1. Demonstration of concept of corrosion probability by Mears and Evans [4].

B. CORROSION PROBABILITY

Mears and Evans [4] designed an interesting experiment for demonstrating the concept of corrosion probability. As illustrated in Figure 27.1, wax lines were drawn on the surface of an iron plate, dividing the surface area into N sections, which were then covered with a thin film of water. This procedure was equivalent to preparing N separated specimens exposed to an identical corrosion condition. After exposure for a fixed time under a mixed-gas atmosphere consisting of oxygen and nitrogen, n squares were corroded. The corrosion probability, P, was calculated as

$$P = \frac{n}{N} \qquad (27.1)$$

The mean weight loss, Q, for a corroded square could be calculated from the total weight loss, W, divided by n. Then the total weight loss of the specimen, W, is given by

$$W = NPQ \qquad (27.2)$$

Mears and Evans [4] demonstrated that both the probability and corrosion rate, equivalent to Q, are functions of the gas composition.

For iron, the corrosion rate increases with increase in the oxygen content in the gas, whereas the corrosion probability changes in the opposite direction, from $P = 1$ to $P = 0$, with increase in the oxygen content, as shown in Figure 27.2. The total weight loss, W, given by Eq. (27.2), shows a maximum at 8% oxygen because of the changes, in opposite directions,

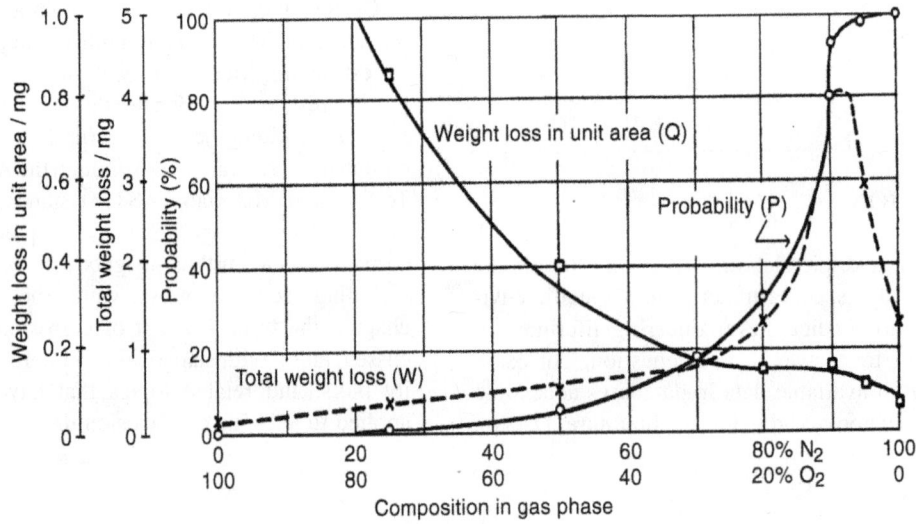

FIGURE 27.2. Corrosion probability, total corrosion loss, and corrosion loss per unit area as function of atmospheric gas composition.

of corrosion probability and corrosion rate with gas composition.

For iron immersed in an aqueous solution of potassium chloride, however, both the corrosion probability and the corrosion rate increase with increase in KCl concentration, resulting in a continuous increase in W, with no maximum. Thus, Mears and Evans [4] succeeded in demonstrating that corrosion probability, rather than the corrosion rate, control the overall weight loss.

B1. Corrosion Probability and Rating Number

For coated steel, galvanized steel, and stainless steel, the surface appearance rather than weight loss is important in many engineering applications. Stain, rust spots, and various types of defects, such as debris and scratches, can degrade the surface appearance. Degradation of surface appearance cannot be judged by weight loss because there is usually no significant loss in weight.

A measure called the rating number (RN) is often used for evaluating the degree of surface degradation. A numerical value of RN is estimated by comparing the surface appearance of the specimen with that of a standard figure, which is provided in various industrial standards. Examples of standard figures for assessing RN are shown in Figure 27.3 [5]. As shown in Figure 27.3, an increase in the number of rust spots and/or in the area of spots causes an increase in the total corroded area, the numerical value of which is shown below each figure. Corresponding to the increase in the total corroded area, the RN, indicated above each figure, decreases. The RN is related to the corroded area, R, of the specimen as follows;

$$RN = a - b \log(R) \qquad (27.3)$$

where a and b are constants. The correlation is illustrated in Figure 27.4. The total corroded area, or the percentage of the

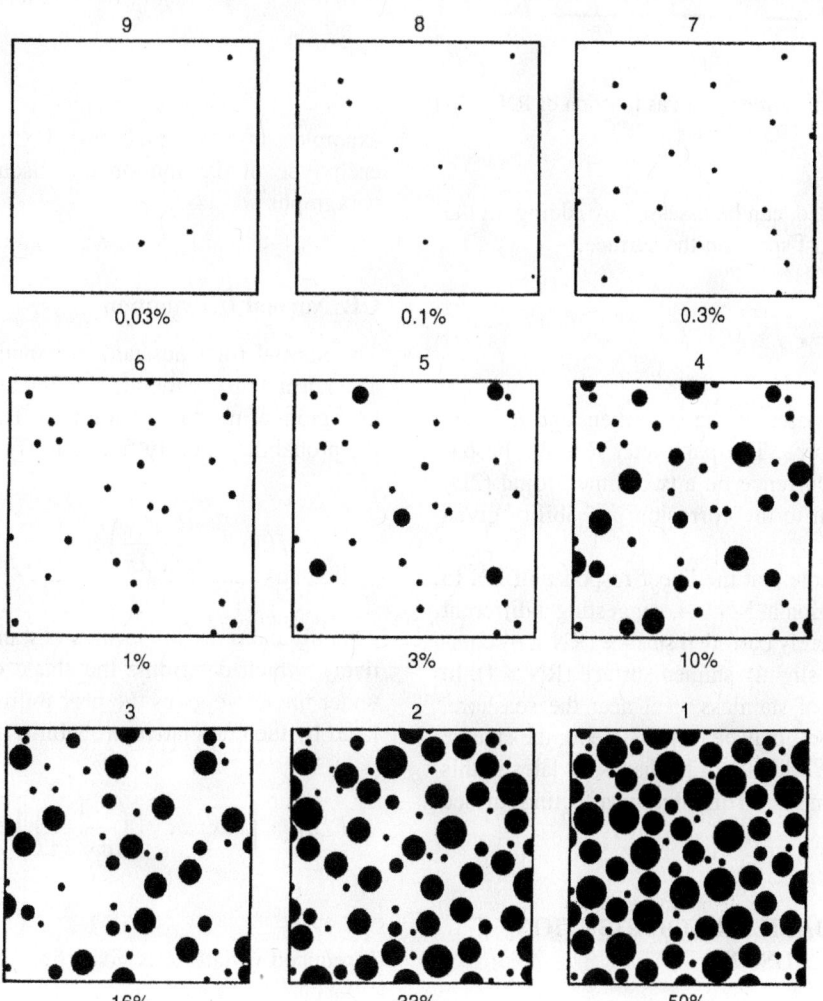

FIGURE 27.3. Examples of model patterns of rust spot distribution with the rating number and total corroded area.

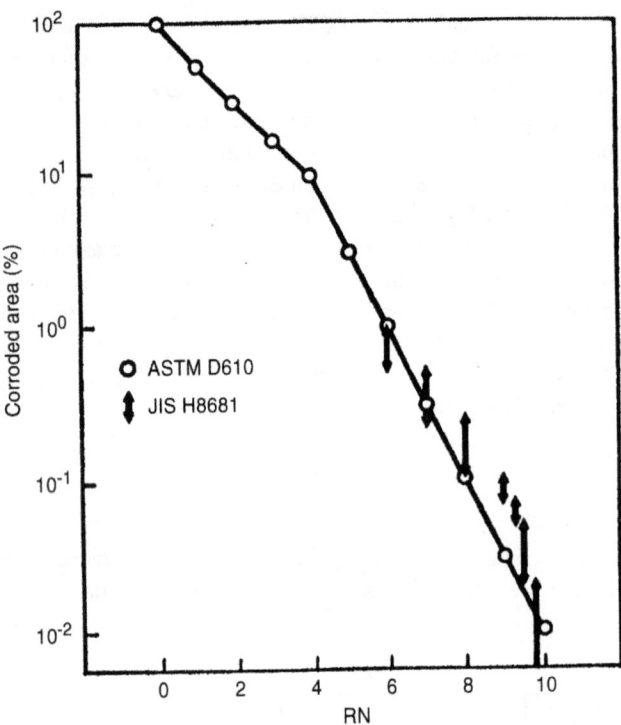

FIGURE 27.4. Total corroded area as function of RN.

TABLE 27.1. Various Probability Distributions Observed in Corrosion

Probability Distribution	Examples in Corrosion	Reference
Normal distribution	Pitting potential	6
Lognormal distribution	SCC[a] failure time	7
Poisson distribution	Two-dimensional distribution of pits	8
Exponential distribution	Induction time for pit generation	9
	SCC and HE[b] failure time	10
Extreme-value distribution		
Gumbel distribution	Maximum pit depth	11
Weibull distribution	SCC failure time	12
Generalized extreme-value distribution	Maximum pit depth	13
	Fatique crack depth	14

[a]Stress corrosion cracking (SCC).
[b]Hydrogen embrittlement (HE).

total area that is corroded, can be assessed by adding all the areas, a_j, covered by rust spots on the surface:

$$R = \frac{\sum a_j}{A} \quad j = 0, 1, 2, 3, \ldots, n \quad (27.4)$$

where A is the surface area of the specimen, and R is the normalized corroded area. The parameter R is the probability of corrosion occurrence on a two-dimensional (2D) scale and is equivalent to the corrosion probability given by Eq. (27.1).

It is interesting to note that the linear response of RN to $\log(R)$ shows an inflection at RN = 4, suggesting a different mechanism on the severely corroded surface (RN < 4) compared with that on the slightly stained surface (RN > 4). In the case of corrosion of stainless steel near the seashore, clustering of the rust spots may be responsible for the change in slope above RN = 4, which will be discussed later in this chapter in the section on spatial distribution affecting surface appearance.

C. TYPES OF PROBABILITY DISTRIBUTION OBSERVED IN CORROSION

The distribution of corrosion data can be reduced to several basic probability distributions, which are listed, along with

examples, in Table 27.1 [6–14]. Characteristic features of each type of distribution are discussed in the following paragraphs.

C1. Normal Distribution

The normal (or Gaussian) distribution is frequently used. The normal distribution is a bell-shaped curve which fits the histogram of most corrosion data. The curve is described by the probability density function, $f(x)$:

$$f(x) = \left(\frac{1}{\sigma\sqrt{2\pi}}\right)\exp\left[-\frac{(x-\mu)^2}{2\sigma^2}\right] \quad (27.5)$$

where μ and σ are the mean and standard deviation, respectively, which determine the shape of the curve. The area under the curve gives the probability of occurrence, calculated by the cumulative probability function, $f(x)$;

$$F(x) = \int f(x)\, dx = \left(\frac{1}{\sigma\sqrt{2\pi}}\right)\int \exp\left[-\frac{(x-\mu)^2}{2\sigma^2}\right] dx \quad (27.6)$$

If reduced variate, s, is given by

$$s = \frac{x-\mu}{\sigma} \quad (27.7)$$

then $F(x)$ can be reduced to a standardized cumulative function, $F(s)$;

$$F(s) = \left(\frac{1}{\sqrt{2\pi}}\right) \int \exp\left(-\frac{s^2}{2}\right) ds \qquad (27.8)$$

Numerical values of $F(s)$ can be found in tables of the cumulative normal distribution.

Graphical analysis using normal probability paper to estimate the distribution parameters of μ and α is simple and useful for engineering applications. Normal probability paper is constructed with values of s on one axis and the cumulative probability, $F(s)$, given by Eq. (27.8) on the same axis. On the other axis, x is plotted on an arithmetic scale. Then the cumulative probability, $F(x)$, obeying the normal distribution can be plotted as a straight line using normal probability paper. To plot the line, the data are arranged in ascending order of value, and the plotting position [3] for cumulative probability is calculated by Eq. (27.9):

$$F(x) = \frac{\left(i - \frac{3}{8}\right)}{\left(N + \frac{1}{4}\right)} \qquad (27.9)$$

where i is the position of the data point in the rank and N is the total number of data points. Instead of Eq. (27.9), another simple equation of $i/(1 + N)$ can be used for calculating the plotting position, but Eq. (27.9) is recommended for use with normal probability plots because it gives almost unbiased estimates of the standard deviation from the slope of the linear plot.

Table 27.2 is a working table for plotting measured values of pitting potential of Type 304 stainless steel on normal probability paper. The first column is the position of the data point. The measured values are arranged in ascending order and are tabulated in the second column. In the third column, the cumulative probabilities given by Eq. (27.9) are listed.

TABLE 27.2. Data Set of Pitting Potential of Type 304 Stainless Steel Measured by the Potential Sweep Method in 3.5% NaCl Solution[a]

i	x_i (V vs. SCE[b])	$F(x_i)(i - \frac{3}{8})/(N + \frac{1}{4})$
1	0.199	0.051
2	0.261	0.133
3	0.263	0.214
4	0.264	0.296
5	0.274	0.378
6	0.275	0.459
7	0.281	0.540
8	0.285	0.622
9	0.286	0.704
10	0.293	0.786
11	0.294	0.867
12	0.295	0.949

[a]Sum: 3.321, mean: 3.321/12 = 0.277, standard deviation: 0.015.
[b]Saturated calomel electrode.

Each set of $[x, F(x)]$, for the pitting potential and the corresponding cumulative probability can be plotted on normal probability paper. The straight line fitting the points, shown in Figure 27.5, indicates that the measured data obey the normal probability distribution. The mean value of μ can be obtained at the 50% point. The standard deviation is the difference between the values at the 50% point and at the 84.13% point, or the slope of line because the slope is proportional to $1/\sigma$. In this case, the values shown in Figure 27.5, $\mu = 0.277$ and $\sigma = 0.015$ are the same values as those obtained by numerical calculation and shown in Table 27.2.

C2. Lognormal Distribution

The probability density function and the cumulative function of the lognormal distribution are given by

$$f(x) = \left(\frac{1}{\zeta x \sqrt{2\pi}}\right) \exp\left[-\frac{(\ln x - \ln \delta)^2}{2\zeta^2}\right] \qquad (27.10)$$

$$F(s) = \left(\frac{1}{\sqrt{2\pi}}\right) \int \exp\left(-\frac{s^2}{2}\right) ds \qquad (27.11)$$

where

$$s = \frac{\ln x - \ln \delta}{\zeta} \qquad (27.12)$$

By introducing the reduced variate s, the cumulative probability function of the lognormal distribution can be converted to the same form as the normal distribution expressed by Eq. (27.8). Then, normal probability paper can be used for plotting the data set, obeying the lognormal distribution by changing the x axis from an arithmetic scale to a logarithmic scale.

Data for failure times by SCC of stainless steel in boiling $MgCl_2$ solution at 154°C are listed in Table 27.3. Plotting the data set of $[\log x_i, F(x_i)]$ on normal probability paper results in a straight line, as shown in Figure 27.6. Again, mean and standard deviation can be determined from data at the 50% point and at the 84.13% point, respectively.

C3. Poisson Distribution

The Poisson distribution is used to describe random phenomena observed in rare events. Pit generation is a good example of a random process that can be described using a Poisson distribution. An example of random occurrence of pitting is shown in Figure 27.7 [15], in which pits in the passive film on Type 304 stainless steel were identified by platinum decoration using a displacement reaction in boiling $MgCl_2$ solution. The spatial distribution of pits on the surface was evaluated by counting the number of pits per unit area of $0.5\,\mu m \times 0.5\,\mu m = 0.25\,\mu m^2$. The results are summarized

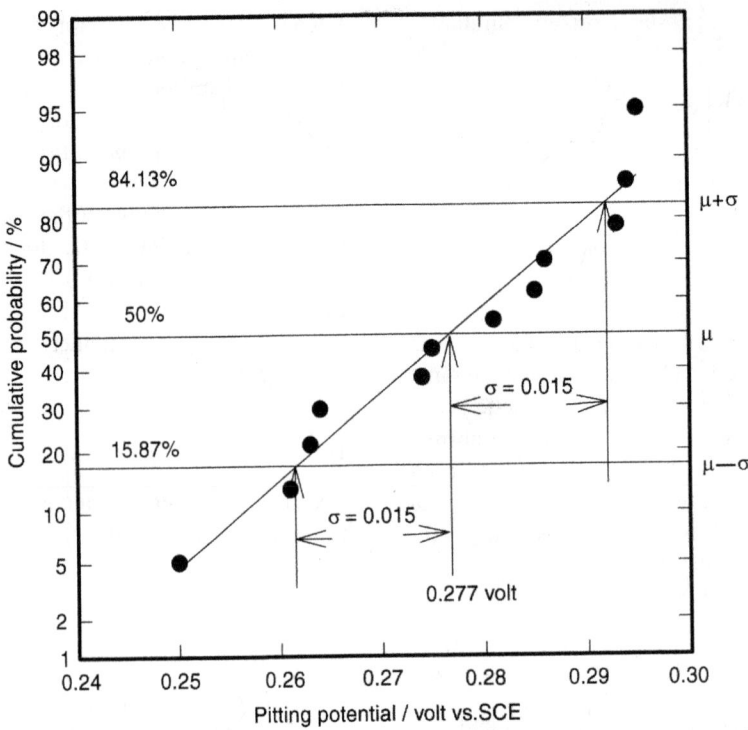

FIGURE 27.5. Probability plot for distribution of pitting potential on normal probability paper, from which mean μ and standard deviation σ can be determined.

in Table 27.4, and the distribution curve is shown in Figure 27.8 [24]. The open circles are observed values, and the closed circles are theoretical values calculated by assuming the Poisson distribution, expressed as,

$$P(x) = \left[\frac{(m)^x}{x!} \right] \exp(-m) \qquad (27.13)$$

TABLE 27.3. Data Set of SCC Failure Times of Stainless Steel in Boiling MgCl$_2$ Solution 154°C

i	x_i (min)	$F(x_i)(i - \frac{3}{8})/(N + \frac{1}{4})$
1	78	0.0439
2	80	0.114
3	96	0.184
4	97	0.254
5	100	0.325
6	101	0.395
7	103	0.469
8	118	0.535
9	123	0.605
10	128	0.675
11	130	0.746
12	141	0.716
13	146	0.886
14	160	0.956

^{a}Sum: 1601, mean: 1601/14 = 114.4, standard deviation: 24.8.

where $P(x)$ is the probability of occurrence of x pits in unit area and m is the mean. Good agreement between the data points and the curve suggests that pit generation in the passive film occurs according to a Poisson process with mutual independent events. Deviation of the observed points from the theoretical curve was found in an A1 alloy and was attributed to a mutual interaction of the pit generation process, as pointed out by Mears and Evans [8].

C4. Spatial Distribution Affecting Surface Appearance

Spatial distribution of rust spots affects the surface appearance, which is sometimes more important than weight loss. Degradation in the surface appearance is usually evaluated by visual inspection. Quantitative evaluation can be achieved by rating the surface appearance as RN, which is simply related to the total corroded area. It should be noticed, however, that a simple RN may not satisfy the observer's intuition because the surface appearance could be affected not only by the total corroded surface area, but also by the distribution pattern of the corroded areas over the total surface.

Figure 27.9 shows patterns of the distribution prepared by Masuko [16, 17] in which nine different patterns with the same RN are demonstrated. Figure 27.9(a), which looks like a lattice of crystal, is called a regular pattern, and

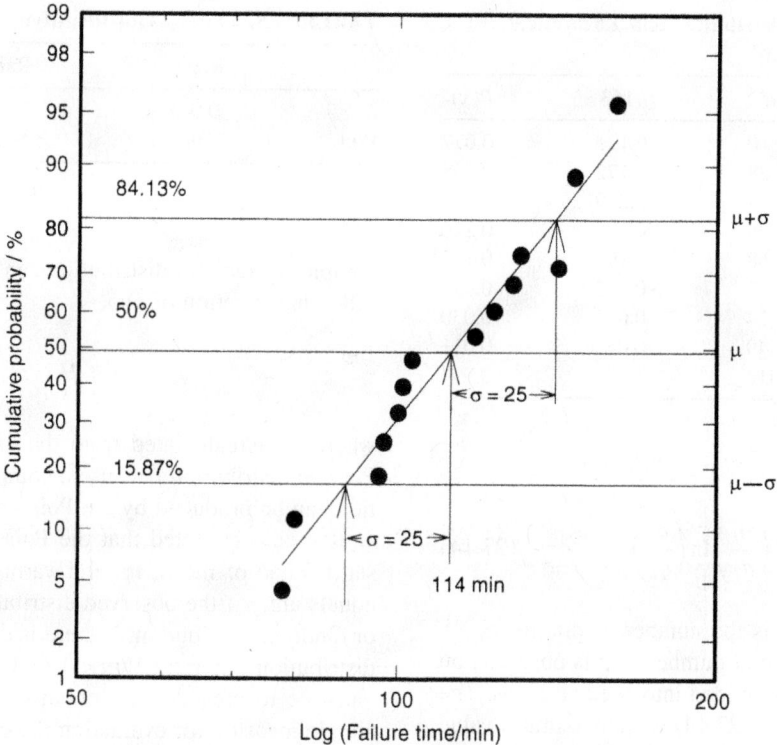

FIGURE 27.6. Probability plot for distribution of SCC failure time on lognormal probability paper, from which mean μ and standard deviation σ can be determined.

Figure 27.9(i) is called a singular or clustered pattern because spots aggregate or localize at separate local sites. Patterns in Figures 27.9(c) and (d), however, show a random appearance. It has to be emphasized, again, that all nine figures have the same RN but provide quite a different visual impression. Another index is required for quantitatively characterizing the patterns. For corroded surfaces, the mode of localization is important for characterizing the distribution pattern. The mode of localization is an important topic in ecology [18] and is analyzed based on spatial analysis theory. In order to describe the mode of localization in corrosion, Masuko [16, 17] introduced a homogeneity function expressed as

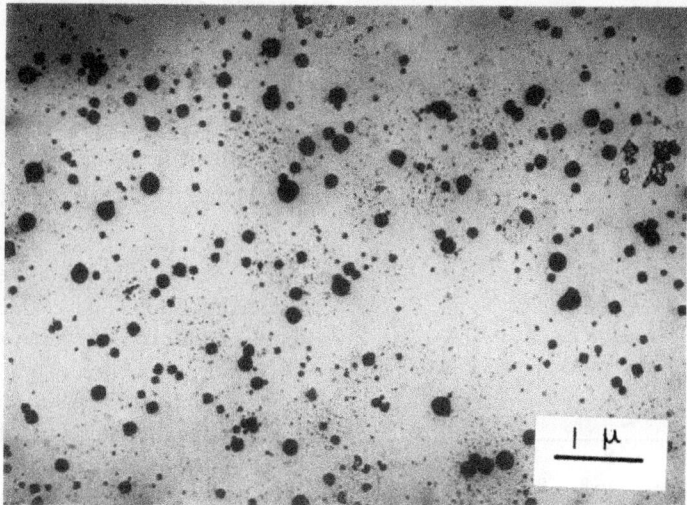

FIGURE 27.7. Transmission electron microscopy (TEM) photograph showing random distribution of pits on surface of Type 304 stainless steel.

TABLE 27.4. Frequency Distribution and Theoretical Probability of Pit Distribution[a]

x	f	xf	$f/163$	$P(x)$
0	21	0	0.129	0.077
1	28	28	0.172	0.198
2	39	78	0.239	0.253
3	32	96	0.196	0.216
4	17	68	0.104	0.138
5	15	75	0.092	0.071
6	4	24	0.025	0.030
7	7	49	0.043	0.011
Sum	163	418	1.000	0.994

[a]Mean = 418/163 = 2.56.

$$H = \sum \left(\frac{q_j}{q_T}\right) \ln \left(\frac{q_j}{q_T}\right) \tag{27.14}$$

where $q_j (j = 1,2,3,\ldots, N)$ is the number of pits in the jth divided area and q_T is the total number of pits observed on the entire surface, which is divided into N equal areas. The proposed function, H, of Eq. (27.14) has a maximum value when q_j is the same in every area, corresponding to a uniform distribution, and $H = 0$ when all pits concentrate in one area, that is, $q_j = q_T$ for one area whereas q_j is zero for all the other areas. A quantitative expression for deviation from a

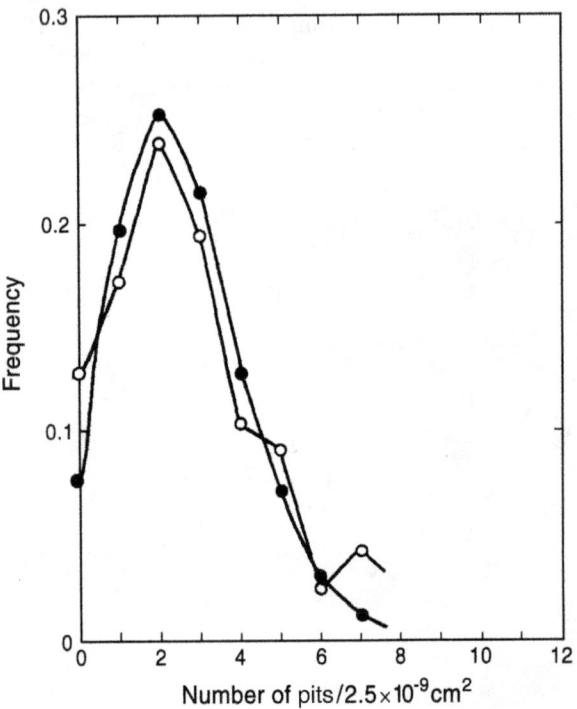

FIGURE 27.8. Distribution curves of pits obeying Poisson distribution.

TABLE 27.5. Indexes for Identifying the Distribution Pattern

	Regular	Random	Cluster
D	$D > 0$	$D = 0$	$D < 0$
V/m	$V/m < 1$	$D/m = 1$	$D/m > 1$

completely random distribution can be calculated using the following function of D:

$$D = H - H_0 \tag{27.15}$$

where H_0 is calculated from the distribution of q_j for the Poisson distribution because a completely random distribution can be produced by the Poisson distribution.

It should be noted that the Poisson distribution has the same value of mean, m, and variance, V. The ratio, V/m, equals unity if the observed distribution obeys the Poisson, or random, distribution. If there is deviation from a random distribution, the ratio $V/m < 1$ or $V/m > 1$. The ratio of the variance to mean, V/m, for an observed distribution is a simple measure for evaluating the deviation from complete randomness. The same argument could be made for H and D functions of Eqs. (27.14) and (27.15). Thus D and V/m can be used to differentiate three basic patterns of localization of pits; the regular, random, and clustered patterns by using the rule shown in Table 27.5. Martin and Mcknight [19] used the variance to mean ratio, V/m, for evaluating clustering of rust spots for painted steel. By using computer-aided image analysis and a program with the V/m ratio [20], it was found that the distribution of rust spots on stainless steel exposed at the seacoast was completely random in the initial stage of the exposure because $V/m = 1$, changing to the regular ($V/m < 1$) and clustered pattern ($V/m > 1$) after corrosion progressed. The change of the distribution pattern occurred at around RN = 4.

D. EXTREME-VALUE STATISTICS

D1. Size and Time Effect in the Corrosion Test

When the required corrosion data are not available, corrosion engineers design a laboratory test that simulates field operating conditions. In some cases, corrosion engineers may be asked to provide reliable data for scaling up the system from the laboratory to the field. They may also be asked to predict the remaining life of apparatus in operation. The size of the test coupon and the duration of testing may change over a wide range depending on the purpose of the test and the requirements of the client. As shown in Table 27.6 [21], tests can be divided into three categories: The first test is carried out in the laboratory by those concerned with materials

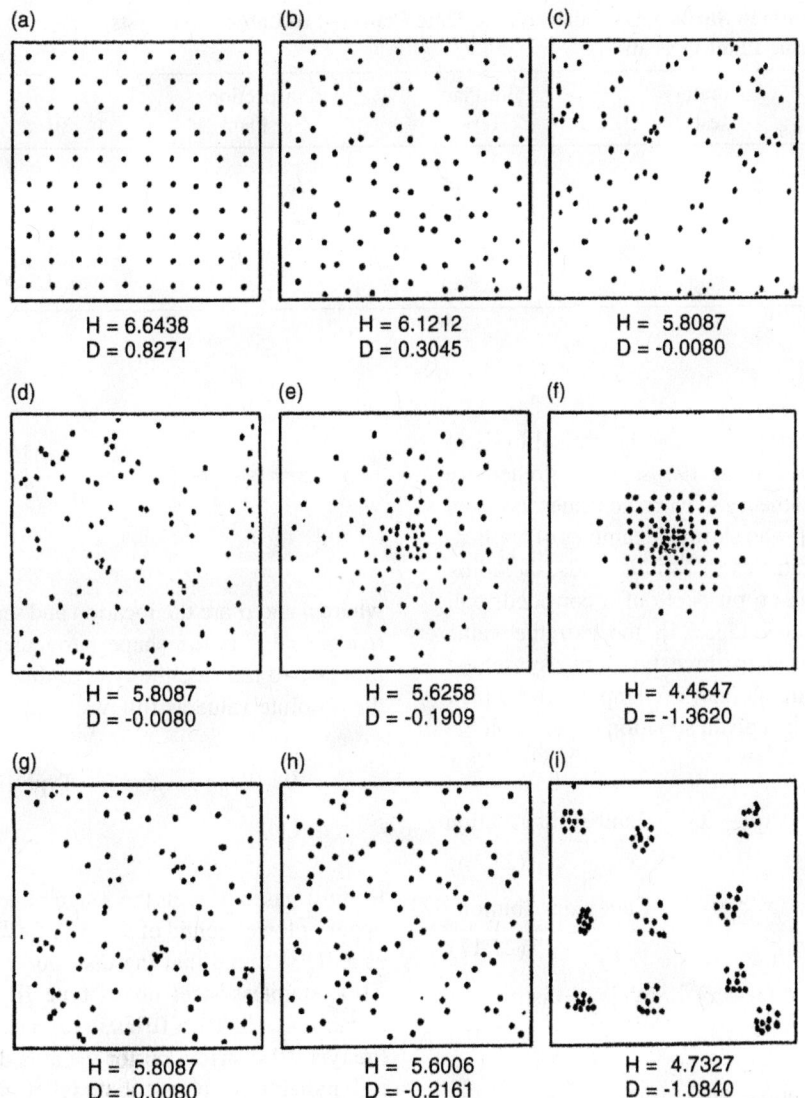

(a)

H = 6.6438
D = 0.8271

(b)

H = 6.1212
D = 0.3045

(c)

H = 5.8087
D = -0.0080

(d)

H = 5.8087
D = -0.0080

(e)

H = 5.6258
D = -0.1909

(f)

H = 4.4547
D = -1.3620

(g)

H = 5.8087
D = -0.0080

(h)

H = 5.6006
D = -0.2161

(i)

H = 4.7327
D = -1.0840

FIGURE 27.9. Model patterns of pit distribution (regular, random, and clustered), which have same rating number.

production; the second is a pilot plant test carried out by those concerned with design and fabrication of the apparatus; and the last is inspection and in-plant examination by those concerned with plant maintenance.

In the laboratory test, coupons of a relatively small area (e.g., $1\,cm^2$) are used and tested in a relatively short time within ~2 weeks under accelerated conditions, whereas larger coupons, several times $10\,cm^2$, are required in pilot plant tests that typically last for at least a few months. During plant operation, a much larger surface of the plant is exposed to the corrosive environment during longer operation times. For instance, the surface area of heat exchangers installed in a plant is in the range of 10^5–$10^6\,cm^2$ depending on the type and design. For such a huge surface area, regular inspection is needed to assess operational reliability and to estimate residual life.

When the coupon size and testing duration in the laboratory test are each equated to unity, the relative size of the coupon and the relative duration of the pilot plant test might be in the range of 10–10^2 and 10–10^3, respectively, as indicated in Table 27.6. Differences in size as well as in duration between the laboratory test and the field examination are extremely large and in the range of 10^4–10^6 and 10^4–10^5, respectively, because plant operation is normally expected to last for >20 years. For this reason, one should be cautious in extrapolating laboratory data directly to design or to prediction of life or residual life of plants [22].

D2. Three Types of Extreme-Value Distribution

Fortunately, a method for bridging the large difference in space and time mentioned in Section D1 is provided by the

TABLE 27.6. Relative Ratio in Surface Area and Failure Time Expected in Laboratory Tests, Pilot Plant Tests, and During Plant Operation

	Laboratory Test	Pilot Plant Test	Inspection at Plant	Failure Life Estimation
Steel maker	a	b		
Plant fabricator	b	a	b	
Plant user		b	a	a
Surface area ratio	1	10–10^2	10^4–10^6	
Failure time ratio	1	10–10^3	10^4–10^6	

[a]Mainly concerned.
[b]Secondary concerned.

statistical theory of extreme values, the theoretical basis of which was well established in the 1950s. A comprehensive treatment of the statistical theory of extreme values has been given by Gumbel [23, 24], who showed examples of applications of extreme value statistics to various fields, including the analysis of the maximum pit depth of a corroded steel pipe. According to Gumbel [23, 24], the extreme value distribution can be reduced to three types of asymptotic distributions for an infinite number of samples, and which of the three types applies in a given situation depends on the initial distribution:

Type I $F(x) \sim \exp[-\exp(-x)]$ Gumbel distribution

$$(27.16)$$

Type II $F(x) \sim \exp[-x^{-k})$ Caucy distribution

$$(27.17)$$

Type III $F(x) \sim \exp[-(\omega - x)^k]$ Weibull distribution

$$(27.18)$$

Each type has two distributions, one for the largest and one for the smallest, so that six asymptotic extreme-value distributions exist. Of these six distributions, however, type I for the largest value and type III for the smallest value are most often observed in corrosion and are called the Gumbel and Weibull distributions, respectively.

D3. Generalized Extreme-Value Distribution

When corrosion data for the maximum or minimum value are collected, one must decide which type of extreme-value distribution should be fitted to the observed data. For this purpose, several methods for testing the closeness of fit are proposed, including the chi-square test, the Komologorov–Simirnov test, the correlation coefficient test, and others.

Recently, a generalized extreme-value (GEV) distribution was introduced for the closeness-of-fit test. The GEV distribution was introduced first [25] in meteorology for analyzing rainfall and other data. The probability function of GEV is given by

$$F(x) = \exp\left\{ -\left[1 - \frac{k(x-u)}{\alpha} \right]^{1/k} \right\}, \quad kx \le \alpha + uk$$

$$(27.19)$$

where u and α are the location and scale parameters, respectively, and k is the shape parameter, which has a unique property to indicate the type of distribution by the sign and the absolute value as follows:

$$k = 0 \quad \text{Type I} \quad k < 0 \quad \text{Type II} \quad k > 0 \quad \text{Type II}$$

$$(27.20)$$

In most cases, k is in the range $-\frac{1}{2} < k < \frac{1}{2}$, and x has an upper or lower bound of $u + \alpha/k$ for $k > 0$ or $k < 0$. Laycock et al. [13] found that the distribution of pit depth on Type 316L stainless steel obeys type III for the largest value, because $k = 0.401 > 0$. The presence of the upper bound in the type III distribution for the area dependence is useful in rationalizing the physical model of pit initiation and growth because the existence of a limiting depth with increase of surface area is more realistic.

D4. Extreme-Value Statistics for Estimating Maximum Pit Depth

Extreme-value statistics using the Gumbel distribution is quite useful for estimating the maximum pit depth and its dependence on surface area [26–30]. A standardized procedure [30] has been proposed for analyzing the maximum pit depth distribution using the Gumbel distribution and the concept of the return period, in order to estimate the maximum depth of pits over the larger surface area from which specimens of small area are extracted.

The Gumbel distribution is expressed as

$$F(x) = \exp\left[-\exp\left(-\frac{x-\lambda}{\alpha} \right) \right]$$

$$(27.21)$$

where $F(x)$ is the cumulative probability of pit depth, x, and λ and α are the location and scale parameters, respectively. The probability density function, $f(x)$, is given by

$$f(x) = (1/\alpha)\exp\{-(x-\lambda)/\alpha - \exp[-(x-\lambda)/\alpha]\} \tag{27.22}$$

The reduced variate, y,

$$y = \frac{x-\lambda}{\alpha} \tag{27.23}$$

is introduced, and then

$$y = -\ln\{-\ln[F(y)]\} \tag{27.24}$$

is used for constructing Gumbel probability paper. The Gumbel probability paper shown in Figure 27.10 is constructed with values of y scaled on the vertical axis and the associated cumulative probabilities, $F(y)$, calculated from Eq. (27.24) on the same axis. Values of the extreme variate, x, are plotted on the horizontal axis using an arithmetic scale. Then the cumulative probability, $F(x)$, of variate, x, obeying the extreme value distribution can be plotted as a straight line on Gumbel probability paper. Plotting position for the cumulative probability can be calculated simply by

$$F(y) = 1 - \frac{i}{1+N} \tag{27.25}$$

where i is the ith position of the ordered values of x, in descending order, and N is the total number of samples. Plotting y as a function of x yields a straight line, and its slope is $1/\alpha$, and the intercept at $y = 0$ gives λ.

For the pit depth distribution, the return period, T, is defined as

$$T = \frac{S}{s} \tag{27.26}$$

where S is the surface area of the object, such as the tank plate to be examined, and s is the area of the small specimens that are sampled randomly from this object. Then the return period, T, is a size factor. The return period is defined as

$$T = \frac{1}{1 - F(y)} \tag{27.27}$$

and y can be correlated with T as follows:

$$y = -\ln\{-\ln[F(y)]\} = -\ln\left[-\ln\left(1 - \frac{1}{T}\right)\right] \tag{27.28}$$

$$= \ln(T) \quad \text{when } T > 18$$

Then the opposite side to the y axis can be scaled as the T axis, as shown in Figure 27.10. The value of x at a given T is the maximum pit depth, x_{max}, for the T times larger surface area, S, compared with the small sample area, s.

An example of plotting the pit depth data is shown in Figure 27.10. In an atmospheric exposure test, 10 sheets of painted steel having 50×50-mm^2 area were exposed for 13 years. After the test, pits with various sizes and depths were found on the sheets. The maximum pit depth was obtained from each sheet and was tabulated in the second column of Table 27.7 in descending order. The third column lists the plotting position calculated by Eq. (27.25). Plotting the data points, $[x_i, F(x)]$, results in the line shown in Figure 27.10. The parameters of the distribution can be assessed from the slope and intercept, $\alpha = 0.139$ and $\lambda = 0.133$, respectively. The maximum depth on the surface area that is 100 times larger than the area of the sheet specimens is estimated to be $x_{max} = 0.773$ mm, which is obtained from the intersection of the linear plot of the distribution and the $T = 100$ line.

Instead of this graphical estimation of the parameters, more reliable estimates of α and λ can be obtained by using the MVLUE (minimum variance linear unbiased estimator) method, the maximum likelihood, and the method of moments. Of these three methods, the MVLUE method, discussed by Lieblein [31], is found to be more efficient and

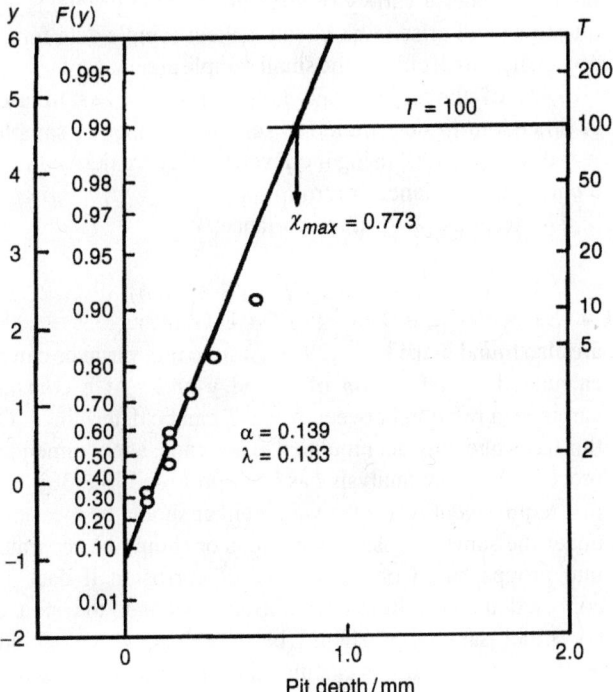

FIGURE 27.10. Probability plot of distribution of pit depth in samples of small area on Gumbel probability paper, from which maximum pit depth for larger surface area can be estimated.

TABLE 27.7. Work Sheet for Plotting Pit Depth Data on Gumbel Probability Paper[a,b]

i	x_i (mm)	$F(x_i) = 1 - i/(1 + N)$	$A_i(N, n)$	$b_i(N, n)$	$a_i x_I$	$b_i x_i$
1	0.6	0.9091	0.00063	0.115279	0.000378	0.069161
2	0.4	0.8182	0.01432	0.11979	0.005728	0.047915
3	0.3	0.7273	0.03046	0.11420	0.009137	0.034261
4	0.2	0.6364	0.04926	0.10060	0.009851	0.020120
5	0.2	0.5455	0.07137	0.07852	0.014274	0.015705
6	0.2	0.4546	0.09790	0.0046027	0.019579	0.009205
7	0.1	0.3636	0.13065	−0.000884	0.013065	−0.00009
8	0.1	0.2727	0.60542	−0.573523	0.060542	−0.05735
9	0	0.1818				
10	0	0.0909				

[a]Pit depth data were obtained from the painted steel sheet after a 13-year exposure test.
[b]Location parameter $\lambda = \sum a_i$; $x_i = 0.1326$. Scale parameter: $\alpha = \sum b_i$; $x_i = 0.1390$.

unbiased for small size samples. The MVLUE estimator can be calculated by

$$\lambda = \sum a_i(N, n)x_i, \qquad \alpha = \sum b_i(N, n)x_i \qquad (27.29)$$

where $a_i(N,n)$ and $b_i(N,n)$ are weights for each sample depending on the sample size, N, and truncated number, n, which are tabulated in the table up to $N = 23$ [29] or the table given by Tsuge [32] up to $N = 45$. Again the example for estimation of the parameters using the MVLUE method is shown in Table 27.7. The fourth and fifth columns list the values of the weight, $a_i(N,n)$ and $b_i(N,n)$, respectively. The sixth and seventh columns list the calculated values of $a_i(N, n)x_i$ and $b_i(N,n)x_i$, respectively, summation of which gives the parameters estimated by the MVLUE method, shown below Table 27.7. There is good agreement between the values obtained by graphical estimation and by the MVLUE estimation.

The mode, λ, of the pit depth distribution for the small specimen is obtained by the MVLUE estimation mentioned above, and then the mode, x_{max}, for the T times larger surface, S, is estimated by

$$x_{max} = \lambda + \alpha \ln(T) \quad \text{if} \ T > 18 \qquad (27.30)$$

The perforation probability, P_p, of the maximum pit through the wall thickness, d, is given by

$$P_p = 1 - \exp\left[-\exp\left(-\left\{ d - \left[\lambda + \frac{\alpha \ln(T)}{\alpha} \right] \right\} \right) \right] \qquad (27.31)$$

The above procedure does not include assessing the closeness of fit of the distribution obtained to the Gumbel distribution, but the closeness of fit can be assessed using

the Kolmogorov–Smirnov or chi-square test if needed. Another convenient method [35], using the GEV of Eq. (27.19), can be utilized for comparing the fit of the distribution to the Gumbel distribution because the shape parameter, k [33], defines the type of distribution, as indicated in Eq. (27.20).

The maximum pit depth, x_{max}, and the perforation probability, P_p, can be estimated using the procedure in the manual method or the MVLUE program of EVAN [34], which operates with MS-DOS, and EVAN-II [35], which operates with Microsoft Excel. If further applications are intended under a variety of conditions, some problems and questions will arise (e.g., how to collect sample data for the object, how to decide on the small sample area, s, for the large area, S, of the object, and how many samples to use). A procedure to determine the suitable number of samples was developed [30] using the T versus N curve derived from a concept of variance control.

The weights, A, B, C, of variance, V,

$$V = \alpha^2[A(N, n)y^2 + B(N, n)y + C(N, n)] \qquad (27.32)$$

are also found in the table [29, 31, 32], so that variance can be calculated as a function of N and y or T. At a constant variance, a relation between N and T can be determined. On this basis and with accumulated experience, a recommended procedure for the analysis has been proposed [30, 36]. The first requirement is that the values either should be measured under the same corrosion conditions or should be separated into groups based on knowledge of corrosion if data are collected under different conditions. Second, the area of the small sample, s, should be chosen so as to include multiple pits. Then T is calculated by $T = S/s$ from S, which is given, and N is decided from the T versus N curves at the given ratio of λ/α. The ratio of λ/α must be known before the analysis but fortunately can be estimated from previously accumulated data.

D5. Weibull Distribution for Analyzing SCC Failure Time

The third type of extreme-value distribution for the smallest value, the Weibull distribution,

$$F(t) = 1 - \exp\left[-\left(\frac{t-\gamma}{\eta}\right)^m\right] \qquad (27.33)$$

can be fitted to the failure life distribution by SCC, where γ, η, and m are the location, scale, and shape parameters, respectively. This third type of asymptotic distribution for the smallest value can be transformed to the first type for the largest value; that is, Eq. (27.21), by changing $1 - F(t)$ to $F(z)$ and by introducing the following reduced variate:

$$X = \ln(t - \gamma) \quad z = \frac{X - \lambda}{\alpha} \qquad (27.34)$$

The same MVLUE method used for Eq. (27.21) can be utilized to estimate the parameters of Eq. (27.33) because the following relations exist between the parameters of both distributions:

$$\lambda = \ln(\eta) \quad \alpha = \frac{1}{m} \qquad (27.35)$$

The above-unified procedure for estimating parameters of the Gumbel and Weibull distribution was coded as a computer program, EVAN [34].

The lifetime during which SCC is the failure mode varies a great deal and, in the past, was analyzed using a lognormal distribution [7, 21] in case of aluminum alloys, stainless steels, and steel wire. More recent analysis, however, has been based on the Weibull distribution [37, 38], including the exponential distribution, because the Weibull distribution can be fitted to various types of distribution by adjusting the shape parameter. The shape parameter, m, is an important parameter because it controls the shape of the probability density function, $f(t)$, and also the failure rate, $\lambda(t)$, as shown below:

$$f(t) = \frac{dF(t)}{dt} = m(t-\gamma)^{m-1}(\eta)^{-m}\exp\left[-\left(\frac{t-\gamma}{\eta}\right)^m\right] \qquad (27.36)$$

$$\lambda(t) = \frac{f(t)}{1-F(t)} = m(\eta)^{-m}(t-\gamma)^{m-1} \qquad (27.37)$$

In reliability engineering [2], a bathtub-shaped mortality curve for describing failure time is widely used to illustrate the failure mode, which is controlled by the shape parameter, as shown in Figure 27.11. The first period, or mode, shows a decreasing failure rate with time and is called an early failure

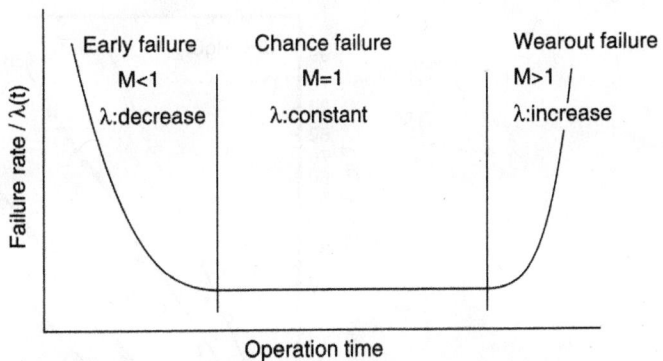

FIGURE 27.11. Bathtub curve for the change of failure rate with time, showing early failure, chance failure, and wear-out failure mode.

mode. The middle part, with a constant and lower failure rate, is a chance failure mode. The final stage, in which the failure rate increases with time, indicates a wear-out mode.

For austenitic stainless steel [39] in boiling $MgCl_2$ solution, applied stress causes transgranular stress corrosion cracking and changes the distribution of failure time, as shown in Figure 27.12. The effects of applied stress on the parameters of the distribution are summarized in Figures 27.13(a) and (b) [39]. With decrease in applied stress, the single distribution changes to a complex one consisting of two distributions, the first distribution of which changes its slope, approaching unity, whereas the second has a constant slope of unity. Both distributions have a trend to approach the chance failure mode with decrease in applied stress.

An opposite shift of the slope with applied stress, however, was reported by Clarke and Gordon [12], who measured the failure time distribution of intergranular stress corrosion cracking (IGSOC) of sensitized Type 304 stainless steel in high-temperature, high-pressure water, simulating the boiling water reactor (BWR) environment, as shown in Figure 27.14 [12]. The Weibull plot shifts to longer time with decreasing applied stress, but the steeper slope occurs at the lower stress. This change in the shape parameter suggests a change in the mechanism of SCC. It is interesting to note that the actual field data of SCC of steam generator rubes in nuclear power plants [40] obeys the Weibull distribution, and the shape parameter was found to be 4.2, as shown in Figure 27.15. This analysis indicates that the SCC failure of the plant was in the final stage, in wear-out mode, so that a remedy such as plugging or replacement was required for safe operation.

E. RELIABILITY ASSESSMENT AND PROBABILITY DISTRIBUTION

As discussed in Section D5, the distribution of failure life of SCC is expressed as a function of experimental conditions,

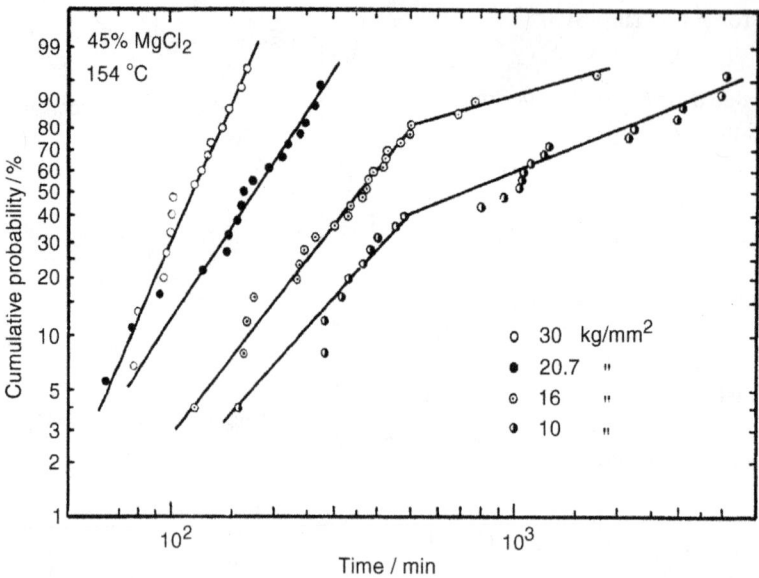

FIGURE 27.12. Probability plots of SCC failure time of stainless steel on Weibull probability paper.

such as applied stress, chloride concentration, oxygen concentration, and so on. If factors that affect failure time are expressed simply as stress, then the shift of the distributions with stress can be illustrated schematically, as shown in Figure 27.16. The laboratory test at the higher stress level produces a series of short failure time data, the distribution of which is located in the left and shorter time region. The distribution of field failure time is located in the extended time region shown on the right side. Usually, the distribution of the field failure life cannot be established because failure time data are very limited. For quantitative estimation of failure life or for assessing reliability in the field, the distri-

bution at each level of stress is estimated. Alternatively, if the distribution is known as a function of stress, extrapolation to lower stress levels can provide the estimated distribution of the field failure time.

Three postulated changes of the distribution plotted on Weibull probability paper are illustrated in the small insert figures in Figure 27.16. Stress corrosion cracking of stainless steels in MgCl$_2$ solution, shown in Figure 27.12, is an example of case (b), and the slope of the distribution likely approaches unity with decrease in stress.

Murata [41] reported the distributions of delayed fracture of high-tension bolts tested in the laboratory as well as in

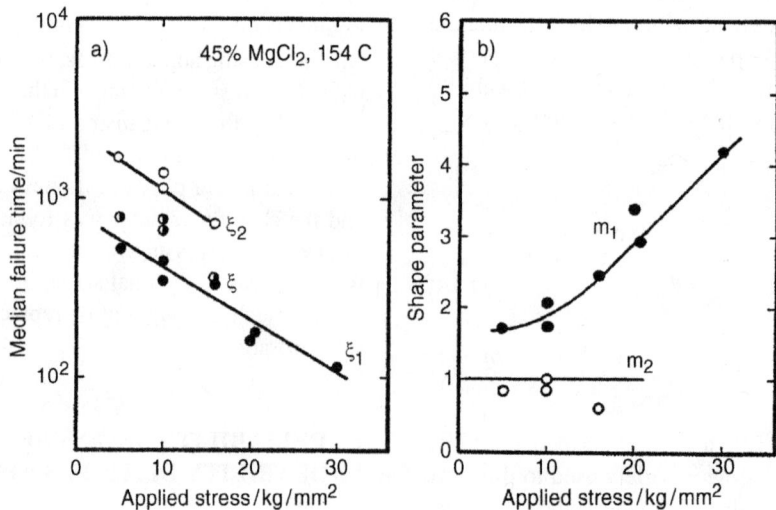

FIGURE 27.13. Median failure time and shape parameter of Weibull distribution as function of applied stress.

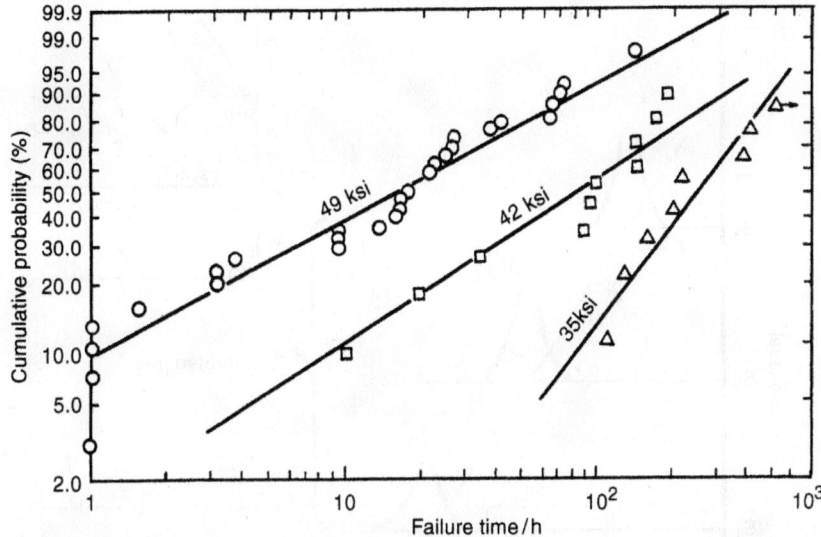

FIGURE 27.14. Probability plots of IGSCC failure time on Weibull probability paper, showing different dependence of shape parameter on applied stress compared with data in Figure 27.13.

atmospheric exposure over a period of 9 years. The results obtained for SNCM 23A (AISI-SAE4320) steel bolts in the exposure test were plotted in Figure 27.17 [41]. Distributions shifted to longer failure times in sea, coastal, industrial, and rural exposure, in that order. The small shape parameter was a specific feature of the sea exposure. In order to compare these atmospheric exposure test results, the laboratory-accelerated tests were carried out using high-temperature and high-humidity environments. The mean failure time of the accelerated test was found to be 10 times shorter than the field exposure test, and the shape parameter was almost the same as in the field exposure test [41].

Atmospheric exposure can be simulated by the high-temperature and high-humidity accelerated test because of the similar shape parameter, but failure in sea immersion cannot be simulated because of the large difference in the shape parameter.

Stress corrosion cracking of stainless steel heat exchanger tubes using industrial water as coolant is a problem in the chemical industry. Failure causes and failure life data have been analyzed [42]. It has been established that both the chloride concentration and the wall temperature are controlling factors causing SCC failures. Conditions to avoid SCC have been recommended based on the data collected. In

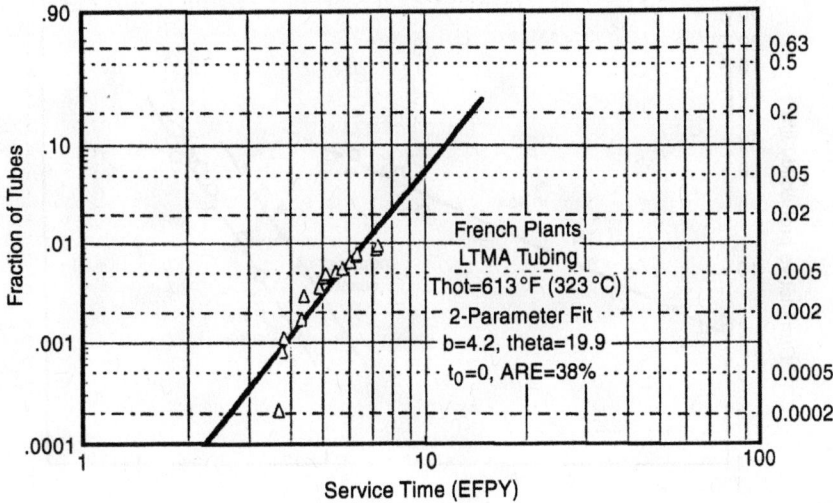

FIGURE 27.15. Weibull probability plot of service life of steam generator tubes operated in French nuclear power plant.

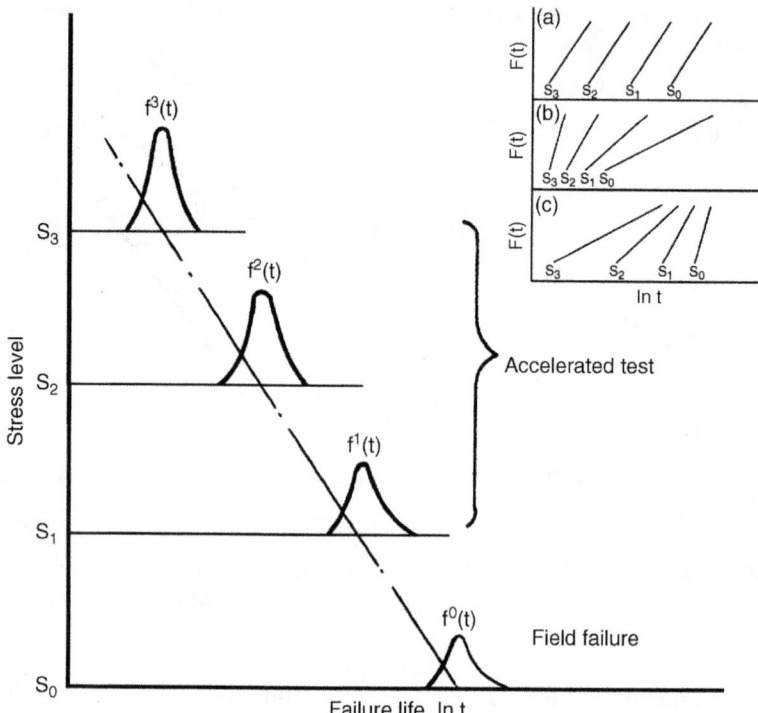

FIGURE 27.16. Distribution curves of failure time observed in various environments, at which stress level increases with severity of environment. Three different modes in the change of the distribution are shown in the insert.

addition, the distribution of failure life of 442 cases reported from 19 plants has been analyzed. The failure life data that were grouped based on wall temperature obey an exponential distribution, as illustrated in Figure 27.18 [42]. The vertical axis is the hazard function, $H(t)$, which is defined as

$$H(t) = \ln[1 - F(t)] \qquad (27.38)$$

or

$$F(t) = 1 - \exp[-H(t)] \qquad (27.39)$$

The hazard function was calculated by summation of the failure rate, $h(t)$, at each time

$$H(t) = \sum h(t) \qquad (27.40)$$

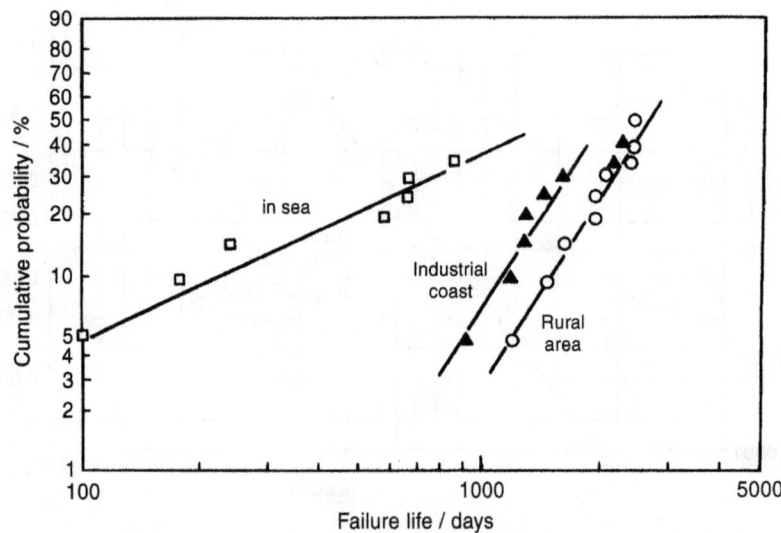

FIGURE 27.17. Weibull probability plots of SCC failure time of high-strength steel bolts exposed under various environmental conditions.

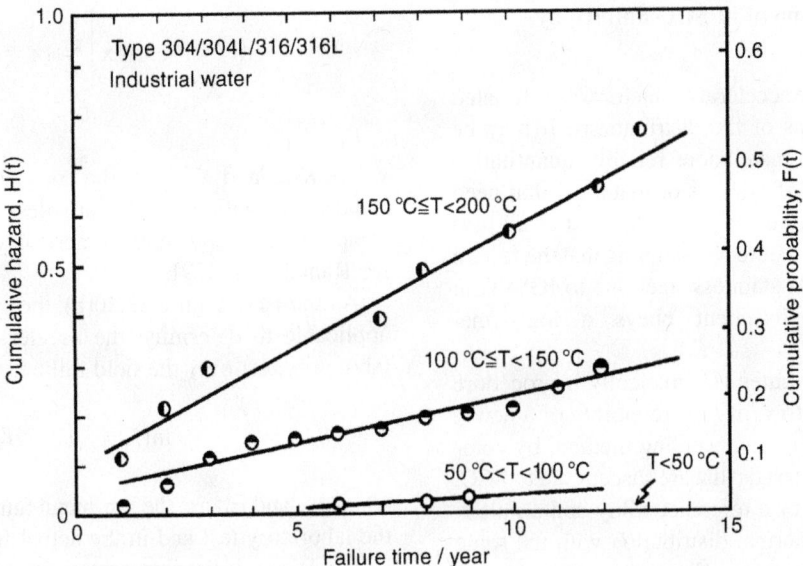

FIGURE 27.18. Hazard plots of SCC failure time of stainless steel tubes for beat exchangers using industrial water in chemical industry.

The exponential distribution has the form of

$$F(t) = 1 - \exp\left(\frac{t - \lambda}{\alpha}\right) \qquad (27.41)$$

which is reduced from the Weibull distribution at $m = 1$. The slope of the line provides the failure rate, which is constant during the lifetime, but depends on the wall temperature, as shown in Figure 27.18.

The field data for the SCC failure time distribution are valuable because they provide a basis for comparison with the laboratory distribution. An example for estimating the

accelerating factor of the laboratory test is shown in Figure 27.19 [43], in which the failure time distributions of Type 304 stainless steel under various conditions, including the field failure time distribution [42], are plotted on Weibull probability paper. Every distribution obeys the Weibull distribution, and the slope decreases from $m = 3.41$ to $m = 1.0$ while the mean failure time increases, indicating the change of failure mode from the wear-out to the chance failure mode. By comparing the medians of the distributions, it can be concluded that the most severe testing condition, using a boiling concentrated $MgCl_2$ solution, accelerates the SCC of stainless steel by a factor of 10^4 compared to the field failure life [42].

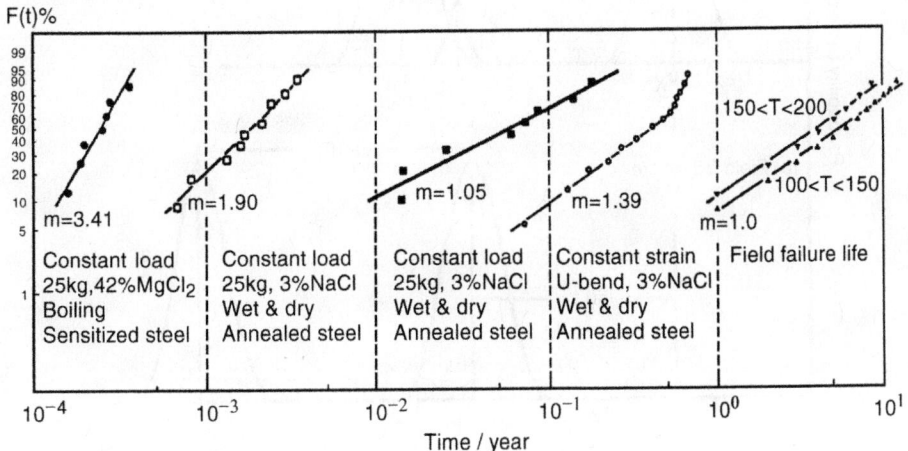

FIGURE 27.19. Comparison of cumulative probability distributions of SCC failure time of Type 304 stainless steel obtained under various conditions.

E1. Quantitative Assessment of SCC Failure by Reliability Test

In Section E, a value of the acceleration factor was estimated by comparing the medians of the distributions. It may be required, however, to estimate more reliably quantitative values for the failure life of systems or materials that need a high level of reliability. For this purpose, Post et al. [44] developed a statistical procedure by assuming that the failure life of sensitized Type 304 stainless steel due to IGSCC in the BWR-simulated environment obeys a lognormal distribution.

Figure 27.20 [38] illustrates schematically a procedure using the accelerated test to verify the reliability of a newly developed alloy or an alternative welding method, by comparing with the current materials that are susceptible to SCC. When both distributions for a reference alloy and an alternative alloy obey the lognormal distribution with the same variance ($\sigma_X^2 = \sigma_A^2$), as shown in Figure 27.20(a), the improvement factor, F, is defined as

$$\ln(F) = \mu_A - \mu_X \qquad (27.42)$$

where μ_A and μ_X are the log-mean failure time of the alternative alloy A and the reference alloy X, respectively. The factor, F, depends on testing time, t, the number, n, of specimens tested, and the level of reliability, β, being expressed as

$$\ln(F) = \ln(t) - \mu_X - \sigma_X \left[K_{\ln A} + \beta \left(\frac{1}{n_X + K_{2nA}^2} \right)^{1/2} \right]$$

$$(27.43)$$

where K_{1nA} and K_{2nA} are the coefficients for the mean and standard deviation of the smallest value distribution of sample size n, and their numerical value can be found in the Rankit table [29].

As shown in Figure 27.20(b), the same equation could be applicable to determine the accelerating factor, L, of the laboratory test over the field failure:

$$\ln(L) = \mu_X - \mu_X' \qquad (27.44)$$

where μ_X and μ_X' are the log-mean failure times of alloy X in the laboratory test and in the actual field, respectively.

For Weibull distributions, a similar procedure could be used for comparing two distributions with the same shape parameter ($m_A = m_X = m$):

$$\ln(F) = \ln(t) - \mu_X + \left(\frac{1}{m} \right) \ln[n_A - \ln\ln(1 - \beta)] \quad (27.45)$$

Akashi [10] found that the distribution of SCC failure times in the laboratory obeys the exponential distribution expressed by Eq. (27.41) and the α/λ ratio is almost constant.

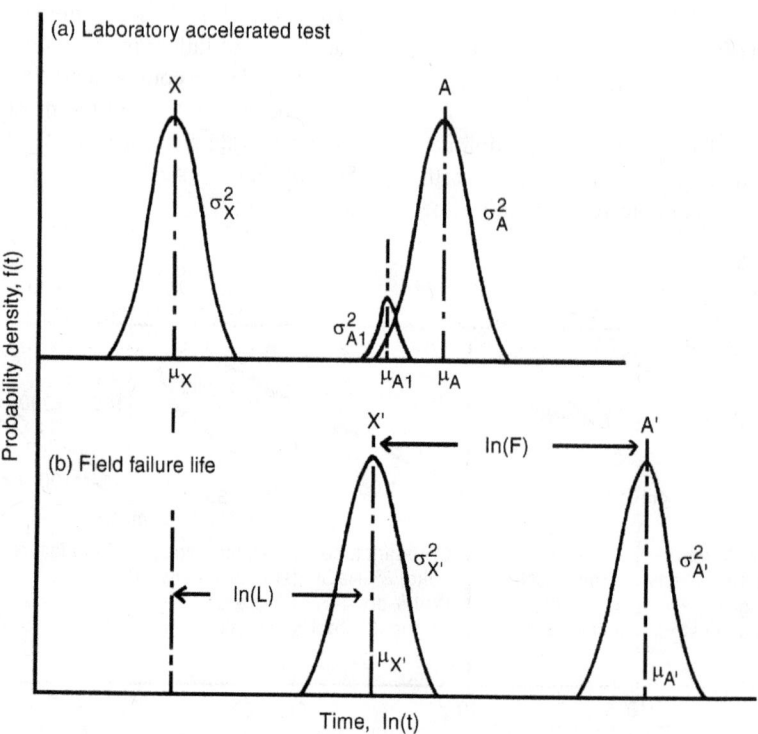

FIGURE 27.20. Schematic illustration of reliability test for assessing acceleration factor based on statistical procedures.

In this case, the condition of $(\alpha_A/\lambda_A = \alpha_X/\lambda_X = \alpha/\lambda)$ is fulfilled and

$$\ln(F) = \ln(t) - \ln(\lambda_X) - \ln\left[1 - \left(\frac{\alpha/\lambda}{n_A}\right)\ln(1-\beta)\right] \quad (27.46)$$

can be used for evaluating the improvement factor, F.

F. CONCLUDING REMARKS

The corrosion probability concept is important and useful for analyzing the highly variable data that are often observed in laboratory corrosion tests and in field failures. Statistical procedures are not very familiar to many corrosion engineers and scientists who have been educated in electrochemical disciplines. Use of statistics is essential when data on localized corrosion must be evaluated. The easiest and fastest way to learn statistical procedures is to plot, on probability paper, your own data that are already stored in your filing cabinet or in your computer.

REFERENCES

1. U. R. Evans, in Localized Corrosion, NACE-3, R. W. Staehle, B. F. Brown, J. Kruger, and A. Agrawal (Eds.), NACE, Houston, TX, 1974, pp. i44–i46.
2. I. Bazovsky, Reliability Theory and Practice, Prentice-Hall, Englewood Cliffs, NJ, 1961.
3. "Applying Statistics to Analysis of Corrosion Data," in Annual Book, ASTM G16-71, American Testing Society for Testing and Materials, Part 27, Philadelphia, PA, 1985.
4. P. B. Mears and U. R. Evans, Trans. Faraday Soc., **30**, 527 (1935).
5. "Standard Method of Evaluating Degree of Rusting on Painted Surfaces," in Annual Book, ASTM D610-85, American Testing Society for Testing and Materials, Part 27, Philadelphia, PA, 1985.
6. T. Shibata and T. Takeyama, Boshoku-Gijutsu, **26**, 25 (1977).
7. F. F. Booth and G. E. G. Tucker, Corrosion, **21**, 173 (1965).
8. P. B. Mears and U. R. Evans, Ind. Eng. Chem., **29**, 1087 (1937).
9. T. Shibata and T. Takeyama, Nature London, **260**, 315 (1976).
10. M. Akashi, in Localized Corrosion, Current Materials Research, Vol. **4**, F. Hine, K. Komai, and K. Yamakawa (Eds.), Elsevier Applied Science, London and New York, 1988, pp. 176–196.
11. M. Aziz, Corrosion, **12**, p. 495t (1956).
12. W. L. Clarke and G. M. Gordon, Corrosion, **29**, 1 (1973).
13. P. J. Laycock, R. A. Cottis, and P. A. Scarf, J. Electrochem. Soc., **137**, 64 (1990).
14. D. C. Buxton and P. A. Scarf, in Life Prediction of Corrodible Structures, Vol. 2, R. N. Parkins (Ed.), NACE, Houston, TX, 1994, pp. 1273–1982.
15. T. Shibata, Boshoku Gijutsu, **23**, 615 (1974).
16. N. Masuko, Proceeding of the 5th ICC, NACE, Houston, TX, 1974, pp. 1051–1055.
17. H. Kawarada, N. Masuko, and H. Yanagida, J. Fac. Eng. Univ. Tokyo, **29**(4), 715 (1972).
18. E. C. Pielou, An Introduction to Mathematical Ecology, Wiley, New York, 1969.
19. J. W. Martin and M. E. Mcknight, J. Coating Technol., **57**(724), 49 (1985).
20. T. Shibata and J. Nakata, in Corrosion Pretension for Industrial Safety and Environmental Control, Proceedings of the 9th Asian Pacific Corrosion Control Conference. Vol. 2, Corrosion Engineering Association of R.O.C., Taipei, Taiwan, 1995, pp. 893–897.
21. T. Shibata, in Life Prediction of Corrodible Structures, R. N. Parkins (Ed.), NACE, Houston, TX., 1994, pp. 64–96.
22. C. C. Nathan and C. L. Dulany, in Proceedings of Localized Corrosion, R. W. Staehle (Ed.), NACE, Houston, TX, 1974, pp. 184–189.
23. E. J. Gumbel, Statistics of Extremes, Columbia University Press, New York, 1958.
24. E. J. Gumbel, Applied Mathematics Series 33, National Bureau of Standards, Washington, DC, 1954.
25. F. Jenkinson, Q. J. Roy. Met. Soc., **81**, 158 (1955).
26. G. G. Eldredge, Corrosion, **13**, 51t (1957).
27. D. E. Hawn, Mater. Perform., **16**(3), 29 (Mar. 1977).
28. T. Shibata, ISIJ Int., **31**, 115 (1991).
29. M. Kowaka, H. Tsuge, M. Akashi, K. Matsumura, and H. Ishimoto, An Introduction to Life Prediction of Industrial Plant Materials—Application of Extreme Value Statistical Method for Corrosion Analysis, Maruzen Publication, Tokyo, Japan, 1984. Translated edition, Allerton Press, New York, 1994.
30. JSCE 60-1 Technical Committee, Boshoku Gijutsu, **37**, 768 (1988).
31. J. Lieblein, "A New Method of Analyzing Extreme-Value Data," NACA TN 3053, NACA, 1954.
32. H. Tsuge, J. Soc. Mater. Sci. Jpn., **36**(400), 35 (1987).
33. J. R. M. Hosking, J. R. Wallis, and E. R. Wood, Technometrics, **27**, 251 (1985).
34. JSCE 60-1 Technical Committee, Working Group, Computer program "EVAN," Maruzen Publication, Tokyo, Japan, 1989.
35. JSCE 95-2 Technical Committee, Working Group, Computer program "EVAN-II," JSCE, 1996.
36. T. Shibata, Corrosion, **52**, 813 (1996).
37. J. H. Harshbarger, A. I. Kemppinen, and B. W. Strum, in Handbook on Corrosion Testing and Evaluation, W. H. Ailor (Ed.), Wiley, New York. 1971, pp. 87–97.
38. T. Shibata, in Localized Corrosion, Current Materials Research, Vol. 4, F. Hine, K. Komai, and K. Yamakawa (Eds.), Elsevier Applied Science, London and New York, 1988, pp. 197–213.

39. T. Shibata and T. Takeyama, Tetsu-to-Hagane, **66**, 693 (1980).

40. R. W. Staehle, J. A. Gorman, K. D. Stavropoulos, and C. S. Welty, Jr., in Life Prediction of Corrodible Structures, R. N. Parkins (Ed.), NACE, Houston, TX, 1997, pp. 1374–1439.

41. T. Murata, in Technology Assesment on Corrosion Protection Technology, Japan Society for Industrial Technology Promotion, Tokyo, 1979, pp. 101–105.

42. Committee of Materials for Chemical Industry Plant, Society of Chemical Engineers, Japan, SCC of Heat Exchanger Tubes-2nd Report on Survey and Failure Life Analysis-, Kagakukogyo-sha, Tokyo, Japan, 1984.

43. E. Sato, Progress Report on Reliability Assessment of Structural Materials, Science and Technology Agency of the Japanese Government, Tokyo, 1988, p. 157.

44. R. Post, J. Lemair, and W. Walker, Proceedings of the Seminar on Countermeasures for Pipe Cracking in BWRs, EPRI Workshop Report, No. WS-79-174, Vol. 1, Paper No. 15, 1980.

PART II

NONMETALS

28

CORROSION OF REFRACTORIES AND CERAMICS

M. RIGAUD

Département de Génie Physique et de Génie des Matériaux, Ecole Polytechnique, Montréal, Québec, Canada

A. INTRODUCTION

The successful use of ceramics to solve material problems involving severely corrosive conditions at high temperatures covers a wide range of applications and industries.

This chapter is focused on the corrosion of (i) industrial refractories for heat and mass confinements in combustion, chemical, metallurgical, materials manufacturing, and related processes and (ii) structural ceramics that substitute for high-temperature metallic alloys in, for example, gas turbine components in the automotive and aerospace industries and in heat exchangers in various segments of the chemical and power generation industries.

Refractories and structural ceramics are generally thought to be inert and corrosion resistant, as compared to metallic alloys, and they are, relatively so, at room temperature, under dry atmosphere, over long time intervals. With increasing temperatures and specific chemical, mechanical, and physical gradients, the propensity to degradation increases rapidly.

Degradation, deterioration, decomposition, and wear are all words used to describe corrosion of these materials. Corrosion of refractories and ceramics is indeed a complex phenomenon to describe and no single word properly fits. Corrosion involves a combination of different mechanisms, such as dissolution and invasive penetration, where diffusion, grain boundary, and stress corrosion may all be present, and oxidation–reduction reactions, where absorption, desorption, and mass transport phenomena all come into play.

Uhlig's Corrosion Handbook, Third Edition, Edited by R. Winston Revie
Copyright © 2011 John Wiley & Sons, Inc.

Exhaustive definition of the material, including its microstructure and surface characteristics, are always needed as in any corrosion study. Here one has to realize that refractory and structural ceramic textures are widely different. Texture means the spatial arrangement of the different phases (minerals and chemicals) from the micrometer level (1 μm) up to 10 mm. In fact, refractories are multiconstituent solids, with varying degrees of crystallinity (in many cases with some glassy phase), with varying degrees of purity (natural and/or synthetic raw materials) with a wide grain size distribution (from 1 to 10,000 μm) with different grain morphology (spherical, flat, or elongated), some with highly anisotropic behavior (many noncubic crystal structures), with inherently high open porosity, as opposed to structural ceramics that are finer grained, purer, and denser.

The exponential increase with temperature in the reactivity of solids and the use of refractories and structural ceramics under pronounced thermal gradients make the study of corrosion of these materials very complex.

A simple, all-encompassing, general theory of corrosion of refractories and structural ceramics does not exist. The interplay between dissolution, penetration, and texture is not sufficient to take into account all the interactions possible with so many parameters involved. This complexity may provide the reason for the shortage of reliable comparative information on the resistance of ceramics to chemical corrosion. Carniglia and Barna [1], on refractories, and McCauley [2] and Lay [3], on technical ceramics, have provided useful compilations. In addition, two more recent reviews on corrosion of refractories have become available [3a, 3b].

The last aspect worth mentioning in this introduction is that corrosion of refractories and ceramics at high temperatures is essentially a chemical rather than an electrochemical phenomenon. Electrochemical dissolution has been considered for the corrosion of refractories in glasses and for the corrosion of oxides (as pure compounds or as minerals) and ceramics in aqueous media at lower temperatures.

B. CORROSION OF INDUSTRIAL REFRACTORIES

Having postulated that corrosion of refractories is essentially the result of a chemical reaction, where the rate-determining step does not involve electronic charge transfer at the reaction interface, this section is divided into two parts: the basic principles and a review of corrosion resistance by broad classes of materials. Corrosion testing of refractories and ceramics is considered in Chapter 83.

B1. Basic Principles

To predict the corrosion of refractories in a given environment, it is worthwhile, first, to consider the concept of acidity–basicity and, second, to estimate the driving force for corrosion using themodynamic laws. This should also be done in two steps: (1) verify the available thermodynamic data for the thermal stability of each constituent and (2) then make the appropriate thermodynamic calculations to estimate the free-enthalpy variations, $(\Delta G)_r$, for all possible reduction or oxidation (redox) reactions that may occur between constituents themselves and between constituents and the environment (gases and/or liquids). To understand the corrosion processes and to select the best refractory for a specific application, kinetic data are required. The principles of penetration, dissolution, and spalling will be presented in the following sections in order to appreciate the particularities of liquids, hot gases, and dusts on corrosion of industrial refractories.

B1.1. Acid–Base Effects. One must consider the chemical nature of the reactants (S and L or S and G), since materials of dissimilar chemical nature, when in contact, will react, especially at high temperatures. The chemical nature of the reactants is best described using the concept of acidity and basicity, a familiar idea but of limited value since a single precise understanding of high-temperature acidity and basicity of all compounds (pure and in solution) has not been developed.

Silica (SiO_2) is the best example of a solid acidic oxide that should be used in applications where the destructive materials (liquid or gaseous) are chemically acidic, for example, coal gasifier ash or iron-making slags or in N_2O_5 or SO_3–SO_2 atmospheres, the most acidic gases. Magnesia and doloma are basic in nature and should be used in applications where slags or gases are generally basic, for example, steelmaking slags and liquid clinker melt in the rotary cement kiln. These generalizations are first approximations that are insufficient in many cases because, in many industrial processes, the corrosive environment changes from acidic to basic during the operation. Nevertheless, the first rule is to make the acid–basic character of the refractory constituents similar to that of the corrosive fluid (liquids and/or gases) to increase the corrosion resistance.

The following table lists the acid–base nature of various compounds:

From most acidic				
Gases	$N_2O_5(g)$	Solids	SiO_2	FeO
	$SO_3(g)$		TiO_2	NiO
	$SO_2(g)$		ZrO_2	MnO
	$CO_2(g)$		Fe_2O_3	MgO
	$B_2O_3(g)$		Cr_2O_3	CaO
	$V_2O_5(g)$		Al_2O_3	Na_2O (s or g)
				K_2O (s or g) *to most basic*

Silica, zirconia, alumina, magnesia, and lime are the most important binary oxides in refractories.

B1.2. Thermodynamic Calculations. The second approach to predict the corrosion resistance of industrial refractories is to make the appropriate thermodynamic calculations, to evaluate the thermal stability of each constituent, to consider the melting and dissolution behavior, and finally to assess redox potentials.

B1.2.1. Thermal Stability Prior to Melting. Several polymorphic transitions may occur that change the microstructural integrity of the solids, as they are being heated, some reversible and some irreversible, with some well-known disruptive transformations, as for silica or zirconia. Other transitions are the decomposition of mixed oxides (e.g., $ZrSiO_4 \rightarrow ZrO_2 + SiO_2$), the devitrification of glassy phases, and the crystallization of high-temperature phases for amorphous or poorly crystallized ones (e.g., the graphitization of carbon). For such modifications, large volume variations may occur, creating sufficiently large compressive, tensile, or shear forces to cause microcracks, and hence porosity, an important aspect of the corrosion of refractories.

Other volume changes or debonding may occur due to thermal mismatch when solids are heated. All noncubic-lattice refractory compounds, for example, Cr_2O_3 (hexagonal), Fe_2O_3 (trigonal), and ZrO_2 (monoclinic), are susceptible to disruptive intercrystalline debonding since they exhibit thermal expansion anisotropy. Another cause of debonding is the juxtaposition of two constituents of different properties, such as different thermal expansion coefficients (e.g., alumina–mullite, mullite–silica, and magnesia–chromite). In fact, in most refractories, phase boundary microcracking can be expected.

B1.2.2. Melting Behavior and the Use of Phase Diagrams. It is still often the case that constituents of refractories are subjected to higher temperatures in service than those attained during their previous history. Frequently, refractories cannot be regarded as being at equilibrium, and it is useful to calculate the amount of liquid that they may contain at a given temperature. For this purpose, one uses thermodynamic principles (e.g., minimization of energy techniques for multicomponent systems) like the SOL/GAS mix protocol and the phase rule, and all data available on the thermodynamics of solutions as well as on the thermodynamic properties of the pure compounds under stable and metastable conditions.

For many years, compositions of refractories were limited to individual minerals as mined. With time, it was recognized that nature did not proportion the oxide contents of the minerals in the most suitable ratio for optimum refractory performance. As pure oxide became available and affordable in terms of cost (first alumina, then magnesia), improvements in performance and, in particular, corrosion resistance became possible. The use of phase diagrams was then recognized as a very good research tool to accelerate such improvements.

Phase diagrams are used to design refractories and to understand the role of slags (of the same nature as the liquid phase formed in the refractories). The early book by Muan and Osborn [4] is a requirement for new researchers in this field and the reviews by Kraner [5], on phase diagrams for fired refractories, and by Alper et al. [6], on fusion-cast refractories, are also important. Phase equilibria in systems containing a gaseous component, in particular the effects of oxygen partial pressure on phase relations in oxide systems, have also been examined. The relative importance of these issues has greatly increased with the advent of the so-called carbon-bonded refractories, magnesia–dolomite and alumina-based systems with carbon and graphite. Also, the roles of oxycarbides, oxynitrides, and sialons as well as metals and alloys in refractories have been documented. These aspects will be treated in the following paragraphs and sections.

In general, the corrosion resistance of a refractory is high if formation of low-melting eutectics and of a large amount of liquid can be avoided. Phase diagrams can be used to predict the formation of these phases. Of course, the better the system is defined, the easier it is to make an accurate prediction. There are, nevertheless, many limitations; for example, phase diagrams are readily available for no more than three constituents, while in practical cases for magnesia or dolomite basic refractories one needs to characterize the CaO–MgO–FeO–Fe_2O_3–Al_2O_3–Cr_2O_3–SiO_2 system. Rait [7] and White [8] offered a comprehensive treatment of the phase assemblages in such a system. A particularly important feature is the CaO/SiO_2 ratio and its effect on the quantities of low-temperature phases to be expected. This ratio is related to the acid–base effect described earlier.

According to Carniglia and Barna [1], the importance in understanding the progressive thermal softening, weakening, and ultimate destruction of refractories cannot be overstressed. It should be clear that the maximum feasible service temperature of a refractory has to be confirmed by (i) its lowest germane eutectic and the melting temperature thereof; (ii) how much liquid is produced at that temperature; (iii) and how rapidly the amount of liquid increases with increasing temperature above this. All these characteristics are linked to the appropriate phase diagrams and the application of the lever rule.

B1.2.3. Oxidation and Reduction Behavior. As for melting, or solid–liquid reactions, the thermodynamic calculations of gas–solid reactions are a powerful tool for describing the stability of refractory materials, in particular, the carbon-containing refractories that are very widely used. Gas–Solid reactions are important not only for dealing with the direct oxidation of carbon in air but also for evaluating the reduction of aggregates (MgO, Al_2O_3, SiO_2, etc.) by carbon (indirect oxidation of carbon) and the role of the antioxidants, being more reducing than carbon in most systems. A powerful use of thermodynamics is to assess the oxidation–reduction behavior of refractories and ceramics by carrying out equilibrium calculations. It is possible to study multicomponent systems using elaborate computer codes, such as Chem Sage or Solgasmix [9, 10]. Graphical displays, such as Ellingham diagrams and volatility and predominance diagrams, are often used to describe the simplest systems of the type $S_1 + G_1 \rightarrow S_2 + G_2$ [11, 12]. For the Ellingham diagrams, all the condensed phases of the reactions are assumed to be pure phases and therefore at unit activity, but deviations from unit activity are very common, and corrections must be applied. Predominance diagrams and volatility diagrams are graphical representations in which the gaseous products are considered at various nonstandard conditions, and these diagrams are used for complex systems involving several gases (e.g., metal–oxygen–carbon–nitrogen under a wide range of partial pressures). These diagrams, the equivalent to Pourbaix diagrams for studying corrosion of a solid in aqueous media, are used to predict the direction of reactions and the phases present.

From thermodynamic calculations, it is possible to explain why and how redox cycling is harmful to any refractory containing iron oxides as impurities, how the reduction of magnesia to magnesium gas and the reoxidation to MgO can be used with great advantage for MgO–carbon bricks (a destruction–reconstruction mechanism that leads to the dense zone magnesia formation theory), how nonoxide refractories (and nonoxide structural ceramics) without exception are subject to high temperature oxidation, and finally how, in the presence of alkalies, refractory chromites (Cr^{3+}) can be oxidized in part to toxic chromates (Cr^{6+}). The alkali chromites ($Na_2Cr_2O_7$ or Na_2CrO_4) are water leachable and present, therefore, an environmental contamination risk in disposal sites.

B1.3. Kinetic Considerations.

Penetration, dissolution, and spalling are the most important phenomena that control the kinetics of the corrosion of industrial refractories. They apply as well to the solid(S) + liquid(L) reactions as to the solid(S) + gas(G) reactions.

Liquids can be either slags or fluxes, molten salts or molten metals, each presenting their own peculiarities. Slags are characterized by their basicity–acidity ratio ($CaO–SiO_2$ ratio either greater or less than 2 : 1) and fluidity or viscosity (very much a function of temperature and overheat, measured by $\Delta T = T_{service} - T_{melting}$ of the lowest melting eutectic compound in the system). Molten salts are known for their low-melting temperatures, high fluidity, and high fugacity (high volatility of alkalies). Molten metals are less reactive toward refractories but are not inert.

The solid refractory material, S, has already been defined as a multiphase material having a texture with intricate surface properties, most often used under a thermal gradient and usually containing some inherent amount of liquid (ℓ)

at the hot face, where ℓ may be equivalent to L in terms of affinity.

Various hot gases, reactive or nonreactive, with or without dusts, are represented by G. The reactivity of G can be determined starting with the acid–base series, noting that the strongest acids and bases are volatile, and by using thermodynamic data in predominance and volatility diagrams.

B1.3.1. Penetration.

It is useful to distinguish between physical penetration and chemical invasion. Physical penetration, without dissolution at all, occurs when a strictly nonwetting liquid is forced into the pores of a solid by gravity or external forces. Chemical invasion occurs when dissolution and penetration are tied together. Both physical and chemical penetration are favored by effective liquid–solid wetting and by low-viscosity liquid. Silicates, particularly silicate glasses, are usually viscous, simple oxidic compounds, and basic slags are less viscous, and halides and elemental molten metals are, in general, the most fluid liquids.

Penetration is the result of an interplay between capillary forces (surface tension), hydrostatic pressure, viscosity, and gravity. Mercury penetration in a capillary glass tube is the best example of physical penetration. The rate of penetration in a horizontal pore, $d\ell/dt$, with a pore radius r is given by the following expression:

$$\frac{d\ell}{dt} = \frac{r\gamma_{\ell g}\cos\theta}{4\eta\ell}$$

where $\gamma_{\ell g}$ is the surface tension of mercury in air, θ the wetting angle of mercury on the glass wall, and η the viscosity of the penetrating liquid (Hg) over a length ℓ at time t. This expression is valid only at constant temperature, whereas $\gamma_{\ell g}$ and η are liquid properties greatly influenced by temperature. Such an equation has often been used to describe the penetration of liquids in refractories without distinction between physical and chemical invasion. However, in the case of chemical invasion, the penetration–dissolution causes changes in composition (which, in turn, affect the values of $\gamma_{\ell g}$ and η) and changes in porosity geometry.

When the pore size distribution is narrow (i.e., pores of the same size), penetration and filling of the porosity by capillarity produce a relatively uniform front moving gradually from the hot face and remaining parallel to it. When pore size distribution is wide (i.e., very large and very small pores) or when open joints, cracks, or gaps between bricks in a refractory wall are accessible at the hot face, rapid and irregular liquid intrusion does occur. There are many penetration paths in a refractory, and the texture of the material is of primary importance; it is important to distinguish between interconnected versus isolated porosity, between open and total porosity, pore sizes, and unbounded boundaries between grains (aggregates and/or matrix) due to thermal mismatches during heating.

For a given temperature gradient, the pertinent eutectic temperature of the penetrating liquid determines its maximum liquid penetration depth.

B1.3.2. Dissolution. The simplest case of pure dissolution is to consider the following reaction: $S_1 + L_1 \rightarrow L_2$, where L_2 is a solution $L_1 + S_1$, S_1 having a continuous surface with no infractuosity.

A general equation useful to describe such a dissolution process is

$$j = \left\{ K \left(1 + K \frac{\delta}{D} \right)^{-1} \right\} (C_{sat} - C_\infty) \qquad (28.1)$$

where j is the rate of dissolution per unit area at a given temperature T; K is the surface reaction rate constant; δ is the thickness of the boundary layer in the liquid phase; C_{sat} is the concentration of the dissolving solid, in the liquid, at the interface; C_∞ is the concentration of the dissolving phase in the bulk of the liquid; and D is the effective diffusion coefficient in the solution for the exchange of solute and solvent.

The parameter δ is further defined by the expression

$$\delta = \frac{C_{sat} - C_\infty}{dc/dy} \qquad (28.2)$$

where dc/dy is the concentration at the interface.

The dissolution rate j may be visualized as the ratio of a potential difference $(C_{sat} - C_\infty)$ divided by a resistance term:

$$K^{-1} \left(1 + \frac{K\delta}{D} \right)$$

Three different cases will be briefly treated.

1. When $K \ggg D/\delta$—that is, when the chemical reaction takes place so rapidly at the solid–solvent interface that the solution is quickly saturated and remains so during the dissolution process. In this case, the dissolution rate, j is controlled by mass transport. Equation (28.1) reduces to Eq. (28.3):

$$j = \frac{D}{\delta} (C_{sat} - C_\infty) \qquad (28.3)$$

The transport process is enhanced by convection due to density differences between bulk and saturated solution (natural convection) and/or by the hydrodynamics of the system under forced convection. Expressions for the boundary layer δ have been derived from first principles for both natural (or free) and forced convection for a variety of simple geometries, and Eq. (28.3) has been validated many times. This process is often called direct dissolution.

2. When $K \lll D/\delta$— that is, when the dissolution rate is phase boundary controlled as opposed to mass transport controlled. Equation (28.1) reduces to Eq. (28.4):

$$j = k(C_{sat} - C_\infty) \qquad (28.4)$$

The phase boundary reaction rate is then fixed by the movement of ions across the interface and hence is governed by molecular diffusion. The effective diffusion length over which mass is transported is proportional to $(Dt)^{1/2}$, and therefore, the change in thickness of the specimen, proportional to the mass dissolved, varies with $t^{1/2}$. This process is often called indirect dissolution.

3. When $K \simeq D/\delta$, both phase boundary and mass transport are controlling. In this case of mixed control, the potential difference $(C_{sat} - C_\infty)$ can be seen as divided into two parts: $C_{sat} - C^*$ is the part that drives the phase boundary condition, and $C^* - C_0$, the part that drives the transport process, so as to keep the dissolution rates for each process equal.

The dissolution rate is extremely temperature sensitive, largely determined by the exponential temperature dependence of diffusion, and can be expressed by the equation

$$j_T = A \exp \left[-B \left(\frac{1}{T} - \frac{1}{T_1} \right) \right]$$

where A is the dissolution rate at temperature T_1 and B is a model constant.

In most practical dissolution problems, the reaction of a solid in a solvent leads to a multicomponent system; no longer is the chemical composition defined by a single concentration nor is there a single saturation composition at a given temperature; instead of pure dissolution the reaction is of the type $S_1 + L_1 \rightarrow L_2 + S_2$, the formation of S_2 causing the interface composition to change.

For porous solids such as refractories, with open porosity and with matrix materials being fine and highly reactive, both dissolution and penetration occur; hence, most slagging situations involve chemical attack of the matrix or low-melting constituent phases, which disrupts the structure and allows the coarse-grained aggregates to be carried away by the slag movement. When penetration is more important than dissolution, another mechanism of degradation needs to be considered: structural spalling.

B1.3.3. Structural or Chemical Spalling. While spalling is a general term for the cracking or fracture caused by stresses produced inside a refractory, chemical or structural spalling is a direct consequence of corrosion penetration. It should not be confused with pure thermal or pure mechanical spalling. Structural spalling is a net result of a change in the texture of the refractory, leading to cracking at a plane of mismatch, at the interface of an altered structure and the unaffected material.

When slag penetration does not cause direct loss of material, the slag does partially or completely encase a volume of refractory, reducing its apparent porosity (by sometimes more than one-half), causing differential expansion with the associated development of stresses. As a result, there is a degradation in material strength and stiffness, the appearance of microcracks, and eventually totally disruptive cracks parallel to the hot face of the lining; this is structural spalling.

Quantification of the mass of spalled materials from the hot face has been attempted by Chen and Buyakozturk [13], who combined slag dissolution and spalling effects into one expression in terms of the residual thickness X and the rate of thickness decrease. They proposed expressions as a function of the hot-face temperature T_H and the maximum depth of slag penetration D_P, referred to the location of the hot face at time t_i during the $(i - 1)$th and ith spalling.

The same authors have developed a very interesting notion to approximate the value of D_p using a critical temperature criterion; they postulated that, for a given system, slag penetration occurs only when temperature reaches a critical value T_c, which could be related to the temperature above which a certain percentage of liquid phase may still exist in the refractory (always under a thermal gradient).

B1.3.4. Corrosion by Gases and Dusts versus Liquids. For corrosion by gases and dusts, the same qualitative relationships are applicable as for gas and liquid penetration into a refractory, except that wetting is not a factor. In gas corrosion, the driving force is the pressure gradient. Gas penetration is less rapid, in terms of mass, than that of wetting liquids; reactions do not necessarily commence at the hot face; and the depth of penetration of gases can be much greater than that of liquids.

Moving down a thermal gradient, gases may condense locally to form a liquid, which then may initiate a dissolution–corrosion process and migrate still further into the solid; alternatively, gases may react chemically to form new compounds (liquid or solid). Once condensation has occurred, the basic criteria for liquid corrosion apply. The condensed liquid fills the pores and debonds the refractory material, creating thermal expansion mismatches, weakening, softening, swelling, and slabbing.

Sulfur dioxide, usually from combustion or smelting operations, provides good examples of the condensation–corrosion patterns experienced in practice. Condensation leads to the formation of sulfate in magnesia at 1100°C, in lime at 1400°C, and in alumina at 600°C. When alkali vapors (Na_2O) are present with SO_2, condensation of Na_2SO_4 overlaps with SO_2 attack. Volatile chlorides also play a role.

When dusts are carried out in hot gases, they may lead to deposition or abrasion. Deposition leads to scaling or caking with either beneficial or life-shortening consequences. Dusts entrained in gases that condense within refractories may facilitate the gradual adhesion of solid particles to the working surface. The resulting buildup can be either a hard scale or lightly sintered cake acting as a protective barrier. In less obvious cases, dust and condensing gases fuse together to create an invasive liquid. In other instances, scaling and caking of refractory walls may not be acceptable to a particular process, and periodic descaling operations can lead to degradation of refractories.

Although no deposition may occur, abrasion can take place when refractory linings are bombarded by dusts. The extent of abrasive wear depends on the impinging particle size, shape, hardness, mass and velocity, angle of impact, fluid dynamics of the system, and viscosity of the eluant gas, among other parameters. Abrasion exacerbates corrosion.

Exposure of a refractory either to a liquid or to its saturated vapor is thermodynamically the same. The differences lie in the transport mechanism. In approximate order of decreasing thermodynamic power, the oxidizing gases are NO_X, Cl_2, O_2, HCl, CO_2, H_2O, and SO_2, and the reducing gases are active metal vapors, NH_3, H_2, C_xH_y, CO, common metal vapors, and H_2S.

B2. Corrosion of Specific Classes of Refractories

Detailed compilations of data on corrosion resistance of industrial refractories are not available because there are too many different products and too many variables to consider. Corrosion is also very seldom the only criterion to consider when selecting a material for a given application; other properties, such as thermal shock resistance and mechanical strength, must also be optimized, and there may be a need to compromise on corrosion resistance.

To assess the corrosion resistance of different types of refractories, a rule of thumb is to consider the temperature limits, above which corrosion is considered to be excessive, established considering the expected durability. From one application to another, this durability does vary greatly as does the mode of degradation: corrosion–dissolution or penetration and structural spalling or both. In Table 28.1, two cases are distinguished, slagging resistance and hot gas corrosion resistance. The first case is represented by a reaction of the type S + L, where L can be either a slag (from ironmaking, steelmaking, or copper smelting) or a flux (from other nonferrous metallurgy), either basic ($CaO/SiO_2 > 2$) or acidic ($Ca/SiO_2 < 2$, on a molar ratio), where S is a

TABLE 28.1. Guideline for Corrosion Resistance Based on the Criterion: Maximum Temperature

	Limit Not Reached for "Normal" Durability			
	Temperature Limits (°C) of Slags and Fluxes[a]		Temperature Limits (°C) of Hot Gases[a]	
Type of Refractories	Basic	Acid	Oxidized	Reduced
1 Magnesia (M)	1700	NR	>2000	1700
Doloma (D)	1700	NR	>1800	1700
(M), (D) or (M + D) + Graphite	1800	1700	800	1700
Magnesia–chrome (MK)	1700	1600	1800	NR
2 High alumina (A)	1600	1600	1900	1900
(A) + Graphite	1700	1700	600	1700
3 Clays + (A) (65% Al_2O_3)	1300	1300	1450	NR
Clays + (A) + C	1550	1550	600	1700
4 Superduty fireclays (F)	1200	1250	1400	NR
Medium–low duty (F)	NR	1100	1200	NR
5 High silica	NR	1400	1750	1600
6 Zircon–zirconia	1450–1600	1500–1700	>2000	1800
Silicon carbide	<1400	1500	1200	1600

[a]NR = not recommended.

given refractory with a given texture. The second case is represented by a reaction of the type S + G, where G can be various hot gases, with or without dusts, either oxidizing (oxygen, air, or CO_2) or reducing (CO, H_2) (see B1.3.4). In Table 28.1, six types of refractories, from basic magnesia to acid silica and neutral zircon–zirconia, are considered.

B2.1. Basic Refractories. Basic bricks are commonly used in the metallurgical industry: steelmaking, copper–nickel, and the cement industry. Magnesia is the most commonly used basic refractory, followed by dolomite. Magnesia bricks are made of various magnesia raw materials (sinters from natural magnesite, sinters from seawater magnesia, or grains of fused magnesia). Many other types of magnesia bricks are used, in combination with chrome (MK), alumina (NA), and doloma (MD). The most corrosion-resistant magnesia bricks are made of fused magnesia grains. Magnesia–chrome bricks were best bricks available before magnesia–carbon-bonded bricks, containing flake graphite additions, were developed for their slagging resistance. Dolomitic refractories are made of burnt natural dolomite, usually low in iron oxide (0.3%) and alumina (0.2%). As for magnesia, very efficient dolomite–carbon bricks have been developed. As shown in Table 28.1, dolomite performance may match that of magnesia for specific applications.

B2.2. High-Alumina Refractories. High-alumina refractories are those that contain more than 65% Al_2O_3, composed of α-alumina (pure Al_2O_3 99.3 + %) and mullite $(Al_2O_3)_3$ $(SiO_2)_2$. For high corrosion resistance, electrofused alumina and mullite products should be considered. As for any type of refractory, there are many different classes of products—fired and unfired—with many different bond systems, a

commonly used one being a phosphate chemical bond system. Corrosion dissolution resistance is improved with increasing alumina content, but slag penetration resistance of such bricks then decreases, and hence, the structural spalling resistance also decreases. Improvements have been achieved with addition of chrome or silicon carbide, and, more importantly, of carbon.

B2.3. High-Alumina Refractories with Clay. High-alumina bricks with clay refractories have alumina content between 45 and 65%; they may contain andalusite or mullite mixed with fire clays and binders. Corrosion resistance varies greatly with mineral composition and relatively low refractoriness (measured by the lowest temperature at which a liquid phase is present in the material). The maximum temperature limit, as indicated in Table 28.1, may not apply for specific applications.

B2.4. Fireclay Refractories. Super duty, medium-duty, and low-duty fireclays are products containing alumina and silica, with alumina content varying between 40 and 45%, 30 and 40%, and 25 and 30%, respectively. They are used in a wide variety of kilns and furnaces because of their low price but are limited to low-temperature applications (1200–1250°C).

B2.5. Silica Refractories. Silica bricks contain typically >93% SiO_2 with minor amounts of lime (CaO < 3.5%), alumina (Al_2O_3 < 2%), and iron oxide (Fe_2O_3 < 1.5%). They have high corrosion resistance to acid slags but are sensitive to high temperatures and to thermal shock even at medium temperature levels. Bricks with reduced porosity and lower alumina content may be used to prevent rapid degradation in the presence of alkalies, when hot gas corrosion prevails.

B2.6. Zircon–Zirconia and Silicon Carbide Refractories.

Among the many available zircon–zirconia refractories, zircon bricks, containing mainly zircon sand ($ZrSiO_4$), have good wear resistance and poor wettability by molten metals at processing temperatures of up to 1450°C.

Zircon is decomposed to zirconia and silica at 1540°C. The most common zircon–zirconia bricks are the very special electrofused A–Z–S bricks containing alumina–zirconia–silica made especially for glass-melting furnaces. Refractory materials with high zirconia content require the use of stabilized zirconia (CaO or MgO in solid solution) to avoid cracking on heating due to monoclinic to tetragonal transformation in the temperature range of 1000–1200°C, with a large volumetric change. Properly prepared zirconia shapes have very high corrosion resistance and are not wetted by metals, even by steels, with different oxygen content.

Silicon carbide refractories are the nonoxide refractories most often used with carbon blocks, usually under reducing conditions. Many different types of silicon carbide refractories are available. They have the following advantages over oxide refractories: high thermal conductivity, superior spalling resistance, superior abrasion resistance, and hence high corrosion resistance against nonoxidizing slag (as in nonferrous metallurgical applications). Properties and costs of SiC refractories vary greatly, depending on the sintering aid used, but they do tolerate much more "impurities" than do structural SiC ceramics, which are discussed in the following sections.

C. CORROSION OF STRUCTURAL CERAMICS

The main differences between industrial refractories and structural ceramics are their purity and texture. Structural ceramics are manufactured from purified synthetic aggregates rather than from as-mined natural minerals. Since the 1960s, the refractory industry has used the available synthetic raw materials, so that the differences between the two groups, although still very significant, are much less than they used to be.

Porosity is a very important factor that influences texture; porosity is < 2% for most structural ceramics, whereas it is 12% or more for most refractories (up to 20% for conventional refractories and 50–70% for insulating refractories). Porosity is a consequence of the manufacturing process route followed, in particular the grain size distribution selected and the sintering conditions. For structural ceramics, mainly very fine grain sizes (in the 0.1–10-μm range) are used to obtain very dense materials. For structural ceramics, the objective is to maximize mechanical properties, whereas for refractories it is to maximize thermal shock resistance and volumetric change, with, of course, good corrosion resistance for both classes of materials.

The fundamental principles of corrosion of industrial refractories also apply to corrosion of structural ceramics, in both the S + L and S + G cases. Nevertheless, as porosity of structural ceramics is much lower, dissolution is more important than penetration. For the S + L case, the corrosion resistance of structural ceramics to molten salts determines their usefulness as components in the chemical industry, such as pumps and heat exchangers. The S + G type of reaction is important for structural ceramics used as kiln furniture in industrial furnaces and as components in gas turbine and diesel engines.

C1. Corrosion by Molten Salts

Comparative information is presented in Table 28.2 on the corrosion resistance to fused salts, alkalis, and low-temperature oxides for two different classes of structural ceramics: oxides and nonoxides.

Values have been compiled from the combination of two main sources [14, 15] and refer to the chemical inertness of pure crystalline materials, with purity >99.5%. For this reason, the data in Table 28.2 are of limited generality. For example, it is to be expected that ceramics with less purity will be attacked more rapidly. The code A, B, C should be used only for comparing the materials considered. It is to be

TABLE 28.2. Corrosion Resistance of Structural Ceramics to Fused Salts, Alkalis, and Low-Melting Oxides

Ceramics at Purity (>99.5%)	NaCl	NaCl + KCl	KNO_3	Na_2CO_3	Na_2SO_4	KOH	Na_2O	V_2O_5
Alumina	A1000	A800	A400	A900	A1000	B500	B500	C800
Zirconia (stabilized)		C800		C900	A1000	B500	C500	C800
SiC (react. bond.)	B900	C800	A400	C900	C1000 (air)	C500 (air)	C600	C800
Si_3N_4 (RBSN)			A400	C750 (air)	C1000 (air)	C500	C500	C800
Si_3N_4 (HPSN)	A800	B900		C900 (air)	B1000 (air)	C500	B500	C800
BN (HP)			A400	C900 (air)	C1000 (air)	C500	C500	

Column header spanning: Fused Salts–Alkalis–Low-Temperature Oxides[a]

[a]A—Resistant to attack up to temperature indicated (°C). B—Some reaction at temperature indicated. C—Appreciable attack at temperature indicated.

assumed that the information is based on short-term observations at ambient pressure. McCauley [2] and Lay [3] provide a more complete coverage of the subject, including attack by glasses, aqueous solutions, and molten metals, but no further generalizations can be made from the specific studies that they reviewed because of limitations in the scope of the studies.

The data in Table 28.2 show that both dense oxides and nonoxides are susceptible to attack at low and intermediate temperatures. The normally protective SiO_2 layer that forms on SiC and Si_3N_4 may explain why such ceramics have a poor corrosion resistance in basic salts.

C2. Corrosion by Hot Gases

The S + G reactions discussed below refer to oxidation at high temperatures in air and oxygen, gaseous corrosion in the presence of condensed deposits, and reduction or oxidation by hot gases, $CO(g)$, $H_2(g)$, $H_2S(g)$, and $H_2O(g)$. High-temperature oxidation in air or oxygen is the most important type of corrosion for carbide- and nitride-based ceramics. Although the corrosion of structural ceramics in hot gases is similar in some aspects to the oxidation of metals and the same theories are used to describe the processes, there are some important differrences:

1. Ceramics have a much higher oxidation resistance than do metals.

2. Because oxidation occurs at much higher temperatures, the corrosion products formed are liquids and gases, as well as solids, which can lead to either weight gains or losses, causing practical experimental difficulties of interpretation.

3. The textures of ceramics are not as homogeneous as those of metals, and so it is very difficult to determine the controlling mechanism of the oxidation process without considering both the processing characteristics and the intrinsic nature of the raw materials used.

Gaseous corrosion in the presence of condensed phase deposits in combustion applications is the second most

important type of corrosion, with sodium sulfate and vanadate being the most corrosive deposits. Nonoxide ceramics are particularly sensitive to this type of environment.

In Table 28.3, the corrosion resistance of dense structural ceramics to hot gases are presented using the same code as in Table 28.2. The oxidation resistance of nonoxide ceramics is clearly less than that of alumina and zirconia.

C3. Corrosion of Specific Classes of Structural Ceramics

C3.1. Alumina. Alumina is considered to be one of the most versatile materials that resist corrosive attack by acids, alkalis, molten metals, molten salts, as well as oxidizing and reducing gases, even at high temperatures. It is widely used in many high-temperature laboratory and pilot plant scale applications, such as crucibles, tubes, and special shapes.

C3.2. Zirconia and Thoria. Thoria is the most thermodynamically stable oxide and the one with the highest refractoriness, with a melting point of about 3300°C. Unfortunately, it is not a readily available material, being used essentially in the atomic energy industrial sector. Zirconia is also a very stable pure oxide, but it has to be used only when stabilized, fully or partially, with CaO, MgO, or Y_2O_3 additions. Such additions make the solid solutions $\langle ZrO_2-X \rangle$ more vulnerable to acid attacks and to alkalis.

C3.3. Silicon Carbide. Silicon carbide has very good resistance to acidic solutions (even HF) and molten metals but is prone to attack under severe alkaline (basic) conditions, in particular in air. The SiC ceramics produced by reaction sintering and containing free silicon are more susceptible to oxidation and chemical corrosion than is single-phase SiC obtained by hot pressing. The SiC is very susceptible to attack by liquid sodium and potassium sulfate. On oxidation, SiC forms a layer of SiO_2, but the possibility of forming $SiO(g)$ increases with temperature and when the partial pressure of oxygen is decreased. The transition between "passive" oxidation (SiO_2 protective layer) and "active" oxidation [$SiO(g)$ without protection] occurs at low pressures

TABLE 28.3. Corrosion Resistance of Structural Ceramics to Hot Gases[a]

Ceramics at Purity (> 99.5%)	Air	Steam	CO	H_2	H_2S	F(g)
Alumina	A1700	A1700	A1700	A1700		
Zirconia (stabilized)	A2400	C1800				
SiC (RB)	A1200	B300	A > 1000	A > 1000	A1000	A > 800
Si_3N_4(RBSN)	B1200	A220	A > 800	—	A1000	
Si_3N_4 (HPSN)	B1250	A250	A > 900	A > 800	A1000	A > 1000
BN (HP)	C1200	C250	A2000	A > 800		

[a]A, B, C, as in Table 28.2.

and high temperatures. SiC ceramics are more sensitive to hot corrosion than is Si_3N_4.

C3.4. Silicon Nitride. Thermodynamic calculations of equilibria in the system Si_3N_4–O_2 revealed that several reactions resulting in the formation of solids SiO_2, Si_2N_2O, SiO, and gaseous N_2, N_2O, NO, SiO, and SiN are possible. Reactions leading to the formation of $SiO_2(s)$ are characterized by the most negative change in free enthalpy, but at high temperatures and low oxygen partial pressure, the possibility of reactions leading to the formation of SiC increases (significant at 1300°C and higher). The notion of passive and active Si_3N_4 oxidation has been well documented, but the controlling factor can be either diffusion of oxygen to the Si_3N_4–SiO_2 interface, diffusion of nitrogen to the SiO_2–air interface, or diffusion of impurity ions and sintering aids from the inner layers to the surface of the ceramic. A detailed survey of the subject is presented in [15].

For hot corrosion to complement the data listed in Table 28.2, only one significant result obtained on reaction-bonding silicon nitride (RBSN) will be recalled. Although Si_3N_4 exhibits high corrosion resistance in the stream of combustion products of pure fuel (10^{-5}% Na and V, 0.5% S) below 1400°C, it corrodes rapidly at 900°C when fuel contains 0.005% Na, 0.005% V, and 3% S. In liquid media, Si_3N_4 resists attack by acids up to $\sim$ 100–200°C. The Si_3N_4 exhibits higher resistance in alkaline solutions than in alkali melts and is rapidly attacked by (Na_2SO_4, Na_2SO_4–NaCl, $Na_2SO_4 + V_2O_5$) melts but have good resistance to chloride melts NaCl + KCl up to 600°C and up to 1100°C in NaF.

C3.5. Boron Nitride. Boron nitride ceramics are usually prepared by hot pressing and normally contain a small proportion of boric oxide, a useful impurity that helps pressing but not its oxidation behavior at high temperatures. The BN ceramics are reported to be attacked by strong acids but are relatively inert toward alkalies. The nonwetting behavior of BN, similar in some ways to pure graphite, may account for this behavior.

Boron nitride starts to oxidize at about 700°C. Oxidation mechanisms differ from Si_3N_4 and SiC because in this case the oxide layer first formed, B_2O_3, is liquid and tends to evaporate to BO(g), BO_2(g), and B_2O_3(g) as temperature increases. Under reducing conditions, pure BN may be usable up to 1800°C. The most valuable characteristic of BN is its resistance to wetting by metals and alloys.

D. PREVENTING CORROSION

Minimizing corrosion of refractories and ceramics under severe conditions at high temperature is a very important task that requires the following three sets of problems to be addressed simultaneously:

1. Material selection, considering both intrinsic and extrinsic characteristics, such as the design and configuration of the component
2. Installation methods and maintenance procedures, considering that each material is a part of a bigger identity
3. Process control to minimize variability and extreme values

In this section, only the first set of problems will be considered, although, for practical solutions, the other two sets of problems should not be overlooked. To select the most appropriate material, the first rule is to make the acid–basic character of the refractory or ceramic constituents similar to that of the corrosive fluids (liquids and/or hot gases) and then to control the penetration–dissolution mechanisms to improve the corrosion resistance. The single, most important factor, in terms of material properties, is the porosity and in a broader sense the texture of the material.

D1. Porosity and Texture

Improvements in the corrosion resistance of refractories have been obtained through texture control resulting from the evolution in manufacturing processes toward much larger size distribution, better mixing, better pressing, and purer raw materials. Total open porosity has decreased to $\sim$12%, a level of porosity that is required to maintain good thermal shock characteristics and acceptable insulating properties. While maintaining the porosity at this value, further improvements are possible through reducing the pore size distribution, so that the larger pores are of a smaller size, and through reducing the permeability of the refractory by using fused grains instead of sintered grains to reduce the openness of the porosity. Other possible ways to minimize penetration in refractories will be discussed in subsequent sections.

For structural ceramics where high strength is more important than thermal shock resistance, porosity can be controlled to very low values (<0.5% in many cases). For these materials, various sintering and densification techniques have been developed, such as reaction-bonded sintering and hot pressing, using very pure, fully densified powder of very small grain size. Porosity control in structural ceramics is nevertheless important. For example, to achieve full density, Si_3N_4 and SiC can be impregnated either with an organosilicon compound that is subsequently decomposed to produce a pore-filling oxide or with a material that is subsequently exposed to a nitriding or a carbiding treatment.

D2. Factors Governing Penetration

Once porosity has been adjusted to its optimal value (fixed for a refractory, nil for a structural ceramic), penetration can be minimized by (1) adjusting the composition of the penetrant fluid; (2) adjusting the wetting–nonwetting characteristics of the same fluid; (3) adjusting the thermal gradient in the material, if at all feasible; and (4) glazing or coating the working face or in some specific cases the back or cold face when oxidation of nonoxide constituents is important.

D2.1. Composition Adjustment of the Penetrant Fluid.

The composition of the penetrant fluid changes during penetration because dissolution and penetration occur simultaneously; if either the melting temperature or the viscosity of the liquid penetrant increases, the depth of penetration can be minimized. This desirable result can be achieved by carefully selecting the matrix additives; for example, a small amount of microsilica can be added to a cement-free alumina-spinel used in some steelmaking applications.

D2.2. Changing the Wetting Characteristics.

In steelmaking applications, wetting characteristics can be changed by adding nonoxides to traditional oxide compositions and using nonfired instead of fired products, for example, by using carbon-bonded refractories and graphite. Insertion of graphite in magnesia-based materials limits penetration and improves corrosion resistance by a factor of 10. Graphite is not wetted by CaO–SiO_2 slags and is therefore a barrier to penetration. Of course, graphite can be oxidized and should not be used in every situation. Impregnation of pitches and tars modifies the bonding phases in refractories and influences penetration phenomena very significantly. Numerous other examples are available in the technical literature, for example, the use of specific additives, such as $BaSO_4$ and CaF_2 in alumina–silica-based castings containing liquid aluminum metal and alloys.

D2.3. Adjusting the Thermal Gradient.

To minimize corrosion, it is important that the penetrating liquid solidify as closely as possible to the hot face, so that the rate of penetration is minimized. For a given hot face temperature, the thermal gradient should be as steep as possible, which can be achieved by vigorously cooling the cold face, by using highly conductive material, and by increasing the thermal conductivity of the lining by conductive backup material.

D2.4. Use of Glazing or Coating.

Sacrificial coatings can be used in some specific situations in which penetration is an important factor. For black (carbon-containing) refractories and for structural ceramics sensitive to oxidation (nonoxide ceramics), the protection of the back face to avoid oxidation has been identified as an important factor. Glazing the hot face using compounds of low-melting temperature has to be analyzed very carefully because glazing can lead to less penetration but more dissolution. Preoxidation treatment to form a protective oxide layer on Si_3N_4 or SiC materials can be beneficial. Uses of multiple layers of new materials, such as "Sialons" (silicon–aluminum oxynitride) and "Simons" (silicon-magnesium oxynitride), have also been studied.

D3. Factors Governing Dissolution

Penetration and dissolution act together and interact, so that the factors controlling them are, to a certain extent, inseparable. Dissolution is controlled by both the chemical reactivity and the specific area of the constituents—the smaller the grains, the faster they dissolve—but in refractories, the smallest grains, by design, bond with the coarser grains (the aggregate), and since this bonding occurs by liquid–solid sintering, the matrix materials (the bonding phases) have the lowest melting temperatures and the highest reactivity. In general, the dissolution of the bonding phases (or the matrix) leads to debonding between the coarser grains, the dissolution of which is inherently slower and usually apparent only at the refractory hot face.

To minimize dissolution of the matrix, the composition can be altered to raise its melting temperature by increasing the chemical purity of the matrix, so as to decrease the importance of segregated impurities. Practical examples include the use of tabular alumina, instead of bauxite grains, and very high purity magnesia grains. Alternatively, to raise the melting temperature of the matrix, finely powdered matrix additives can be included in the refractory mix. Because the matrix is a flux in sintering, these additions require higher sintering temperatures to optimize the bonding and the mechanical properties of the resulting products. The logic of reducing or eliminating additives of low-melting temperature for improved corrosion resistance is compelling, but the consequence is that the sintering temperature must be raised.

A thorough knowledge of the different binder systems (for nonfired refractory products) and the different bonding systems (after firing) goes a long way to explain the different dissolution characteristics of the many refractory materials and ceramics of a given type. The texture and, in particular, the porosity are key parameters to be considered. Dissolution of the coarse grains becomes a life-limiting factor only when penetration and matrix invasion are sufficiently minimized.

D4. Factors Governing Oxidation

Oxidation of carbon-bonded refractories will be considered first. Carbon and graphite oxidation is very detrimental because these material losses lead to porosity; the low specific gravity of carbon causes a 5% material loss to lead to 7–10% porosity. To minimize the rate of carbon and graphite oxidation, new additives, called antioxidants or oxygen inhibitors, have been successfully used. Some general reviews on the role of antioxidants, in such cases, are available, but this topic is outside the scope of this chapter. Among the many additives [16, 17] now available, the metallic aluminum-based alloys are the most commonly used, sometimes in conjunction with various nonmetals, such as silicon carbide and boron compounds. These additives are used to delay carbon debonding and graphite oxidation. In addition, they often act as pore blockers, and many of them have a net positive effect on the mechanical strength of the refractories.

Oxidation of nonoxide structural ceramics can be improved by using protective coatings. The use of a chemically vapor deposited (CVD) layer of the same composition as the substrate is a very promising application of coatings to retard the hot corrosion of structural ceramics [18]. Other methods investigated are cathode sputtering and thin multilayer coating.

As an example, a potential high-temperature oxidation protection scheme could involve a coating system consisting of a refractory oxide outer layer for erosion protection, a silica glass inner layer for an oxygen diffusion barrier and crack sealant, another refractory oxide layer for isolation from the carbon surface, and a refractory carbide inner layer for a carbon diffusion barrier [19]. This system, as with other current or potential protection technologies, involves interfaces between metal oxides and carbides or oxynitrides. The oxidation resistance of nonoxide ceramics is a dominant problem that restricts the wider usage of these materials.

REFERENCES

1. S. C. Carniglia and G. L. Barna, Handbook of Industrial Refractories Technology, Noyes Publication, Park Ridge, NJ, 1992, pp. 220, 243.
2. R. A. McCauley, Corrosion of Ceramics, Marcel Dekker, New York, 1995.
3. L. A. Lay, Corrosion Resistance of Technical Ceramics, HMSO, London, UK, 1983; 3a. J. P. Bennett, K. S. Kwong, G. Oprea, M. Rigaud, and S. M. Winder, "Performance of Refractories in Severe Environments," in ASM Handbook, Vol. 13B, Corrosion: Materials, S. D. Cramer and B. S. Covino (Eds.), ASM International, Materials Park, OH., 2005, pp. 547–564; 3b. D. A. Brosnan, "Corrosion of Refractories," in Refractories Handbook, C. A. Schacht (Ed.), Marcel Dekker, Monticello, NY, 2004, pp. 39–77.
4. A. Muan and E. F. Osborn, Phase Equilibria among Oxides in Steelmaking, Addison-Wesley, Reading, MA, 1965.
5. H. M. Kraner, in Phase Diagrams, Materials Science and Technology, Vol. 6-II, A. M. Alper (Ed.), Academic, New York, 1970, pp. 67–115.
6. A. M. Alper, R. C. Doman, R. N. McNally, and H. C. Yeh, in Phase Diagrams Materials Science and Technology, Vol. 6-II, A. M.Alper (Ed.), Academic, New York, 1970, pp. 117–146.
7. J. R. Rait, Basic Refractories, Their Chemistry and Performance, Iliffe, London, UK, 1950.
8. J. White, in High Temperature Oxides, Part I, A. M. Alper (Ed.), Academic, New York, 1970, p. 77.
9. G. Eriksson and K. Hack, Met. Trans., 21B, 1013 (1990).
10. G. Eriksson, Chem. Scr., 8, 100 (1975).
11. V. L. K. Lou, T. E. Mitchell, and A. H. Heuer, J. Am. Ceam. Soc., 68 (2), 49 (1985).
12. A. H. Heuer and V. L. K. Lou, J. Am. Ceam. Soc., 73 (10), 2789 (1990).
13. E. S. Chen and O. Buyukozturk, Am. Ceam Soc. Bull., 64 (7), 995 (1985).
14. R. Morrell, Handbook of Properties of Technical and Engineering Ceramics, Part I, HMSO, London, UK.
15. Y. G. Gogotsi and V. A. Lavrenko, Corrosion of High-Performance Ceramics, Springer-Verlag, Berlin, 1992.
16. M. Rigaud, in 8th CIMTEC Proceedings, Ceramics: Charting the Future, Vol. 3A, P. Vincenzini (Ed.), Techna Srl., Faenza. 1995, pp. 399–414.
17. A. Watanabe, H. Takahashi, S. Takanaga, N. Goto, K. Anan, and M. Uchida, Taikabutsu Overseas, 7 (2), 17 (1987).
18. J. Desmaison, in High Temperature Corrosion of Technical Ceramics, R. J. Fordham (Ed.), Elsevier Applied Science, New York, 1990, pp. 93–108.
19. J. R. Strife and J. E. Sheehan, Am. Ceram. Soc. Bull., 67 (2), 369 (1988).

29

CORROSION OF GLASS

B. GRAMBOW

La Chantrerie, Laborotorie SUBATECH (UMR 6457), Ecole des Mines de Nantes, Nantes Cedex 3, France

A. INTRODUCTION

Corrosion properties of glass have been investigated since the advent of modern chemistry. Surface reactions of glasses with fluid media, such as adsorption and desorption, chemical reactions, ion exchange, and so on, play an important role both in manufacturing (polishing, cleaning, coating, and production of silica-rich glass) as well as in product properties, such as mechanical strength, reflectance, and catalytic activity. Studies have been performed to qualify glass as a material for reaction vessels and optical instruments, for use in building construction, for cement and plastic reinforcement, for bone replacement, for radioactive waste fixation, and for many other applications.

The corrosion behavior is typically studied in the laboratory using a large variety of test methods. Included are various standard test methods [1–6] allowing comparison of specific dissolution properties of different glass compositions, test methods for quantification of rate law parameters [7, 8], tests for simulating service conditions, and so on. In these tests, powdered or bulk glass samples with or without special surface treatments are exposed to stagnant or flowing aqueous media, distilled water [9], or other corrosive fluids at low or high temperatures in autoclaves [10]. Vapor hydration tests are used by archeologists to understand natural alteration of obsidians [11] or stability of nuclear waste glass in vadose zones. Corrosion properties are characterized by the analyses of detached glass components dissolved or dispersed in these fluids, of solution pH, solid corrosion products or by analyses of corrosion depth (ion implantation or profilometry), concentration gradients of reactants, or glass components at the glass surface. In addition to laboratory testing, features of natural or archeological glasses are analyzed, mainly by surface-sensitive techniques.

Uhlig's Corrosion Handbook, Third Edition, Edited by R. Winston Revie
Copyright © 2011 John Wiley & Sons, Inc.

Probably the most widely studied glasses today are soda lime glasses, vitreous silica, and nuclear waste glasses. By using a multitude of test methods, extensive experimental databases have been created and models have been developed to predict the behavior of these glasses in geological environments for thousands of years. Reviews of the current state of knowledge are available [12–16].

In most cases, it is not possible to use corrosion test results directly to assess behavior under service conditions. A ranking of the relative corrosion resistance of glass compositions often depends on the test method used. To a large extent, this is due to multiply coupled processes, resulting in highly nonlinear corrosion behavior with respect to parameters such as glass composition, pH, solution volume, and so on. Consequently, the prediction of glass performance under service conditions can hardly be based on a single test but requires an extensive test plan. This is particularly obvious when using results from laboratory tests that may be carried out over a period of a few years to assess the performance of nuclear waste glasses intended to immobilize radiotoxic elements for thousands of years. In the laboratory, short-term phenomena and test artifacts become important, while dominant long-term processes are difficult to quantify.

B. STRUCTURE OF GLASS

For understanding basic glass corrosion mechanisms, some fundamental knowledge of glass structure is necessary. Glass structure can be described as a randomly oriented continuous glass network [24, 27]. The amorphous structure is very complex. Neutron-scattering analyses and extended X-ray absorption fine structure (EXAFS) allow three ranges of order for silicate glasses to be distinguished (Fig. 29.1; [17]):

Range I: ordered cation, coordination spheres, as in crystals [18] (with a tendency to smaller coordination number [19, 20]

Range II: cation polyhedra connected by oxygen (in oxide glasses)

Range III: larger units such as rings and chains (e.g., in phosphate glasses)

Depending on formal charge, size, and coordination number, a given cation, such as Si, P, B, or Al, may increase the connectivity of the glass network (network former); others, like the alkali elements, may decrease it (network modifier) or may take an intermediate role [21–24], Glass network connectivity is maintained by mixed ionic–covalent bonds [25]. In silicate glasses the fundamental structural units are silica tetrahedra, which easily interconnect by forming siloxan bridges ($\equiv$Si$-$O$-$Si$\equiv$) [24, 26, 27]. Pure vitreous silica has a high chemical resistance but has various

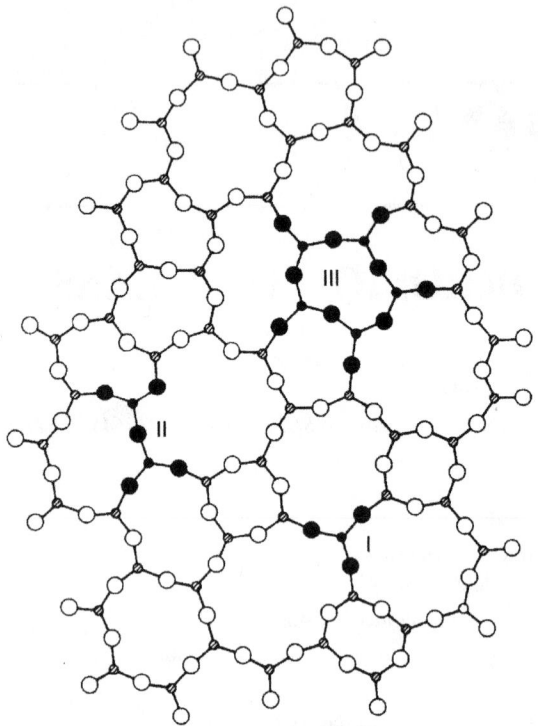

FIGURE 29.1. Glass network according to Zachariasen without far range order but with order in the ranges I, II and III [15].

technical disadvantages (high melting point, small working range), so that it is only used for specific applications. Addition of alkali ions reduce both melting point and corrosion resistance due to disruption of oxo bridges and formation of strongly ionic nonbridging oxygen bonds (NBOs) bonds ($\equiv$Si$-$O$-$Si$\equiv$ + Na$_2$O $\leftrightarrow$ 2$\equiv$Si$-$O$^-$Na$^+$), In case alkali concentrations become very high, one can observe the formation of channels with high mobility for alkali ions. Using ^{29}Si nuclear magnetic resonance (NMR) one can distinguished between Si-tetrahedra with four, three, two, and one oxo bridges (Q^4, Q^3, Q^2, Q^1 sites). The distribution is governed by the alkali–SiO$_2$ ratio. By Coulombic forces, the glass network takes a structure of maximum distance between negative charge centers [28]. Alkali earth cations can increase the crosslink density of the glass network (e.g., Ca^{2+}: bonds $\equiv$Si$-$O$-$Ca$-$O$-$Si$\equiv$). The number of oxo bridges in the glass can also be increased by adding Al to the glass, leading to the formation of anionic aluminate tetrahedra bound to Si-Q^3 sites [29]. The resulting negative charge is compensated by locating alkali ions in the vicinity. At a ratio $R =$ Al–alkali of 1, the charge is compensated entirely and Na ions are very immobile [30].

According to Hinz [32] the coordination number of cations in the glass structure depends on the ratio cation radius/anion radius. Up to a ratio of 0.225, threefold coordination is dominant, to 0.414 tetrahedral, to 0.732 octahedral, and to 1.00 there is cubic coordination. A comparison of

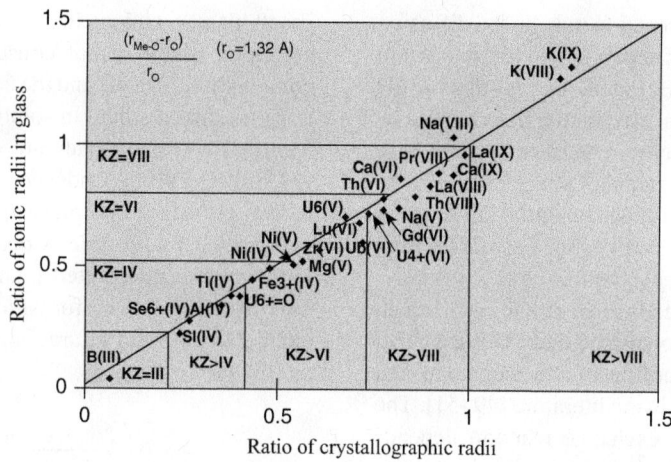

FIGURE 29.2. Comparison of crystallographic ratios of cation to anions with respective ratios for glasses.

cation–anion radii ratios and coordination numbers of various cations in crystalline structures and oxide glasses is given in Figure 29.2. Average crystallographic radii are obtained from bond length data of Shannon [31] (with $r_{oxygen} = 1.32\,\text{Å}$ [32] instead of 1.26 used by Shannon). Figure 29.2 shows a linear relation clearly indicating that bond length in glasses and crystals are similar, but coordination numbers appear to be slightly higher in crystalline structures. According to this diagram, network-forming cations occur up to a ratio of radii of 0.55.

B1. Heterogeneity

When glasses contain multiple phases, typically one phase is more soluble than the other and may control the corrosion resistance or leaching rate. Examples are B-rich borosilicate glasses with an easily leachable B-rich glass phase or soluble alkali molybdate phases formed in ill-defined nuclear waste glass compositions. In contrast, in silica-rich borosilicate glasses, phase separation can lead to an increase in glass durability if a durable Si-rich continuous phase is formed, which includes the B-rich phase in dispersed form [33]. Even in the absence of phase separations, microheterogeneities may occur. In 1921, a glass structure theory was developed [34] based on randomly oriented microcrystals similar to the recent "strained mixed cluster" model [35, 36]. Typical clusters are boroxol rings in B_2O_3–glass, P_8, or P_9 cages in some phosphate glasses or layered Ca octahedra in Ca silicate glass. Cluster boundaries are reactive centers.

C. GLASS CORROSION MECHANISMS

The corrosion rate of glass is not an inherent material property but depends on glass composition, structure, and surface states as well as on environmental conditions, such as

the composition of the corrosive fluid (including pH), hydrodynamic conditions, temperature, stress, strain, vapor pressure, and so on. An overall mechanism of sequential and parallel partial reactions can be formulated for the glass–water reaction, probably applicable to most glass compositions. Different reaction rates and empirical rate laws as well as specific selective leaching properties and pH changes are interpreted as resulting from the different relative importance and consequences of the various partial reactions in the overall reaction scheme. The nature of the rate-limiting reaction in this overall scheme depends on both environmental constraints and glass composition and structure.

The principal partial reactions in the glass–water interaction are glass network hydration, ion exchange, and dissolution of the glass network, resulting in changes of composition of corroding fluids and formation of altered surface layers. Additionally, transport processes, such as access of reactants (e.g., water molecules) and removal of products (e.g., dissolved glass network formers), may become rate limiting under certain conditions.

Glass hydration and alkali–H^+ ion exchange are reactions that occur in parallel with glass network dissolution [37–40], and these reactions are interrelated; for example, in silicate glasses, alkali exchange leads in most cases to pH increase in solution (but not in Mg-rich solution), which increases glass matrix dissolution rates.

The alkali–H^+ ion exchange process is often (not always [41]) diffusion controlled, implying that the rate of selective alkali release from glasses is initially highest and decreases with the square root of time until it becomes equal to the rate of matrix dissolution. The process may be governed either by the electroneutrality coupled, diffusion coefficients of alkali ions, and H^+-bearing species in opposite directions, with the lower mobility of H^+-bearing species such as H_3O^+ ions [42, 43] normally being rate controlling.

Alternatively, the rate-limiting reaction may be the diffusion of water molecules into the glass network [44–46], that is, the hydration of the glass network. Based on findings that the pH response of glass electrodes is essentially determined by exchange equilibria of the outermost surface species with species in solution, it was suggested that ion exchange at the surface is driven by electrochemical potential differences between charged and uncharged surfaces species and the bulk glass [47]. However, the electrical coupling of H^+ and Na^+ migration in the surface layer leads to electrical fields much stronger than those between the surface and the glass [48].

Various models for the coupling of ion exchange and matrix dissolution can be found in the literature [49–51]. The relative importance of the ion exchange reaction depends on the diffusion coefficient D (m^2/s) and the corrosion rate r (m/s). The D depends on glass and solution composition. For borosilicate glasses it can be shown that D is not related to the diffusion coefficient of Na in the dry glass [52]. The ion exchange process is strongly reduced with increasing alkali content of the solutions. The effect increases in the order $Li^+ < Na^+ < K^+ < Cs^+$ [53]. In the time between $t = 0$ and $t \ll r^2/D$, the overall reaction is diffusion controlled [54]. Matrix dissolution becomes dominant for times $t \gg r^2/D$. The ratio D/r gives the average steady-state depth of either alkali diffusion or water diffusion in the glass and the product $r{\cdot}t$ denotes the thickness of corroded glass.

For silicate glasses, it has been shown [55] that the corrosion rate r is proportional to the thermodynamic stability of a hypothetical mechanical mixture of component oxides. This linear free-energy relationship has been confirmed for window glasses, medieval church glasses, natural basalt glasses, and nuclear waste glasses [56], In general, the corrosion rate r is not constant but depends on solution composition. The pH and the concentration of matrix formers (e.g., dissolved silica) in solution are particularly important. Minimal corrosion rates are often observed at neutral pH [57, 58]. The pH dependence is related to the concentration of activated surface complexes [59]. Often, a decrease of corrosion rate r with time is observed. Initially, this decrease with time was interpreted as an effect of protective layers, but later it was shown, for silicate glasses, to be an effect caused by solution saturation, described in a simplified way as [60–62]

$$r = k_+ \left(1 - \frac{C_{\text{Si, actual solution concd.}}}{C_{\text{Si, saturaton concd.}}} \right) \qquad (29.1)$$

with the saturation concentration of silicic acid and the forward rate constant k_+ depending on pH, solution composition, temperature, and ionic strength. Saturation effects were also reported for some phosphate glasses [212]. With the approach of saturation, the affinity [63] of glass dissolution and thus the reaction rate decrease [64–66]. At saturation, the corrosion rate decreases by up to a factor of 1000, but corrosion continues, due to either formation of secondary phase or continued alkali release by the ion exchange reaction. The saturation effect may result in an increase of corrosion rate with increasing solution volume. The effect of surface area to volume ratio (S/V) on glass corrosion rates has been known for more than 20 years [67, 68] and is illustrated with experimental data for a nuclear waste glass in Figure 29.3 [71].

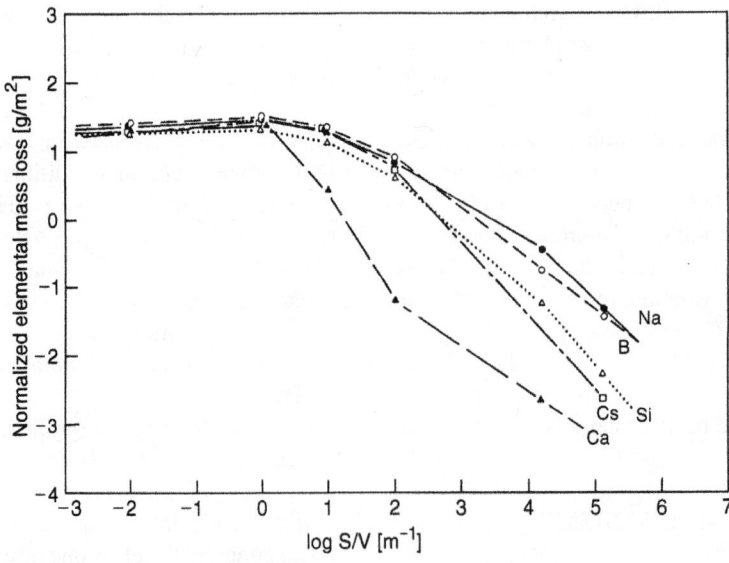

FIGURE 29.3. Dependence of the extent of glass corrosion [normalized elemental mass loss (NL)] on the ratio S/V: PNL–76/86 glass, 90°C, pure water [67]. Divided by the density of the glass (2.7 g/cm^3) NL values can be translated into equivalent reaction layer thickness (in μm).

At the lowest S/V, mass loss per unit surface area, corrosion rates, and corrosion layer thickness are maximum values and independent of S/V. Two further reaction steps can be observed with higher S/V [69]: an intermediate step where the reaction rate decreases to a minimum (as shown in Fig. 29.3) and at very high S/V, where the reaction rate is augmented by secondary phase formation. Whether or not this last step occurs depends on glass composition, particularity on the Al content [70].

Two S/V effects should be distinguished: pH-related effects and saturation effects. Larger S/V values lead to faster saturation and often to pH increase (not always, e.g., not with phosphate glasses, or with silicate glasses in Mg-rich waters). Thus, in a nonbuffered system, the pH-dependent corrosion mechanism [68] varies with the S/V ratio. The saturation-related S/V effect leads to a dependence of the extent of glass corrosion on the product $t. S/V$, as confirmed experimentally [71]; however, simultaneous occurrence of saturation effects and pH variations may complicate this simplistic view [72].

D. EFFECTS OF GLASS COMPOSITION ON CORROSION KINETICS

The stability of the glass matrix against corrosion, ion exchange, and water diffusion as well as saturation concentrations and long-term corrosion mechanism, depend to a large extent on glass composition. The chemical nature and concentration of the glass components are both important. Depending on the interactions with other glass components, a given component may act as either a stabilizing or a destabilizing agent; it may enhance or decrease the relative importance of ion exchange when compared to matrix dissolution.

A comparison of the corrosion rates of important commercial silicate glasses at 90°C and buffered pH 7 is given in Table 29.1 [73].

The table shows that glass components do influence glass stability. However, since effects are not additive, it is difficult to predict the corrosion rates of multicomponent glasses from the corrosion rates of simple glasses. The effect of composition on stability is not an inherent glass property but depends on solution composition, the S/V ratio, and so on. The corrosion stability of a given glass must be experimentally determined if performance is to be predicted for specific service conditions. Nevertheless, some quantitative estimations of glass corrosion effects are possible and are described in the following paragraphs.

There are both theoretical and empirical models describing the effect of glass composition on corrosion rates. Theoretical models are based on semiempirical linear-free-energy relationships (LFER). The most important model is based on studies of Paul [75] showing a relation between corrosion resistance and the hydration enthalpy of a mechanical mixture of component oxides and silicates. This model applies particularly to k_+ in Eq. 29.1 [76]. A list of standard hydration enthalpies $\Delta G^o_{hyd,i}$ is available [77, 78] and can be used to formulate the standard hydration enthalpy of the glass phase in order to calculate corrosion rates caused by solution hydrolysis using the equation

$$\ln r = k_1 \sum X_i \Delta G^o_{hyd,i} + k_2 \qquad (29.2)$$

where k_1 and k_2 are empirical constants that are specific for a given corrosion test. For borosilicate and lead-rich glasses, deviations from linearity were interpreted as resulting from different structural roles of boron in the glass and were corrected [79]. If solution hydrolysis and complexation of

TABLE 29.1. Corrosion Rates of Silicate Glasses at 90°C, pH7[a]

	Composition (wt %)	Linear Rate (μm/d)	E_a (kJ/mol)	Hydration Free Energy (kJ/mol)	References
Sodium-trisilicate	SiO_2 (75), Na_2O (25)	344		−18.5	73
Corning 015 soda lime	SiO_2(72), Na_2O(22), CaO(6)	5.7	84	−19.6	73
PPG soda lime	SiO_2(74), Na_2O(13.3), CaO(8.3), Al_2O_3 (0.06), MgO(3.7)	0.8	68	−13.8	73
R7T7 nuclear waste borosilicate		0.4	70		58
Pyrex borosilicate	SiO_2(82), Na_2O(4), B_2O_3(14), Al_2O_3 (2)	0.2	54	−8.3	73
Kimble R6 soda lime	SiO_2(74), Na_2O(12.9), CaO(8.3), Al_2O_3(1.8), MgO(4.5)	0.02	79	−13.1	73
Obsidian	SiO_2(76), Na_2O(3.8), CaO(0.5), Al_2O_3(13), K_2O(4.8), FeO(0.7)	0.0002	62	3.6	73
Tektite	SiO_2(73), Na_2O(1.5), CaO(1.9), Al_2O_3(13), K_2O(2.4), FeO(4.4)	0.08	80		74

[a]See [73].

dissolved glass constituents as well as surface complexation constants are taken into account, the corrosion rates of silicate glass fibers can be accurately described in the pH range 1–12 [80]. An alternative model is the structural bond strength model [193]. This model assumes that the corrosion resistance is controlled by the average bond strength V in the glass, which can be calculated from the enthalpy of formation of the component oxides, considering the role of a glass component as network former, modifier, or intermediate oxide. For more accurate prediction, empirical models are used that allow interpolation (not extrapolation) in an empirical compositional multidimensional space. These models are based on a multivariate fit of thousands of experimental data to a general equation [81].

E. SURFACE MODIFICATION

After the first contact of a corrosive fluid with the glass, the surface becomes electrically charged, bonds resulting from the population of charge carriers (surface species: $\equiv SiO^-$, $\equiv SiOM$, $\equiv SiOH$, $\equiv POH \cdots$) that belong both to the glass and to the solution. Surface species distribution is responsible for the response function of glass electrodes, and not ion exchange, as is assumed normally [82].

Often, a sequence of reaction layers can be identified, consisting of an inner diffusion layer, followed by a porous gel layer, and one or more laminar layers of precipitated reaction products. The dissolution front of the glass phase is located between the gel layer and the inner diffusion layer. In many cases, there is a smooth transition between the diffusion layer and the gel layer and both together are termed the "gel layer" without distinction [91]. The chemical and mineralogical composition of the assemblage of layers determines the retention capacity for radiotoxic or chemical toxical substances, the reflection properties, and so on. The principal scheme of these layers is given in Figure 29.4

In water-containing fluids, the initial surface modification is glass hydration without loss of the glassy state. If sufficient network bonds are hydrolyzed, a transformed layer [83] (gel) is formed with a clear phase boundary to the glassy phase. The transformed surface layers are porous [84], containing molecular water [85, 86] and allowing for high ionic mobility [87] as well as high water mobility. Molecular water may result from inward diffusion of water carrying H^+ ions or may alternatively, result from the condensation reaction [88, 89], $2 \equiv Si{-}OH \rightarrow H_2O + {\equiv}Si{-}O{-}Si{\equiv}$ bonds. The quantity of water entering the glass depends, among other parameters, on space and strain in the glass network [90].

Ion exchange results in a depletion of alkali ions in the hydrated glass. Potassium-rich silicate glasses show concentration profiles with smoothly decreasing K concentration from the bulk glass to the surface, whereas sodium-rich silicate glass show S-shaped profiles with almost constant Na concentrations between an inflection point and the surface [91]. This difference in shapes was explained by the different structures of Na and K silicate glasses [92], but this is unlikely, as in mixed Na–K lime glasses the same difference in respective concentration profiles was observed [93]. Conradt and Scholze [91] provided a convincing explanation for the different shapes: When ion exchange in the "gel layer" is more rapid than the formation of this layer, the S shape is found, whereas for comparable velocities of the two processes the potassium profile results. In natural tektite glasses [94] and obsidians [95] one has observed water diffusion without alkali release, clearly indicating that ion exchange is a secondary process to water diffusion. Alkali release is also suppressed if the alkali content of the solution is high [96].

The corroded glass may also dissolve entirely (as occurs with some potassium silicate glasses), but in many cases the glass surface becomes enriched at the surface with sparingly soluble glass constituents. This is particularly true if the glass contains large quantities of heavy metals. Depending on glass

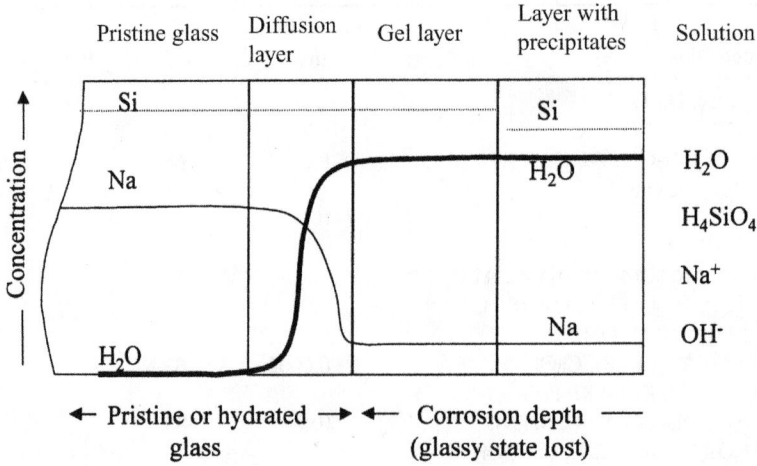

FIGURE 29.4. Schematic representation of different surface regions of corroded silicate glasses.

composition and environmental constraints (temperature, pressure, solution composition, pH, flow rate, vapor pressure, contact time, etc.), amorphous (gel) for crystalline reaction products are formed.

When compared to the original glass, the gel layer is depleted in network-modifying elements (Na, Ca, Li, Cs, etc.) but enriched in Si, Al, transition, and heavy metal contents (Fe, Ti, Nd, U, Zr, etc.) [97, 98]. Under acid conditions, even rare earth elements become depleted [99]. Metal ions from solution may become incorporated into the gel layer (e.g., Mg). The content of molecular water is significantly higher than the content of silanol groups [100] and water may occur phase separated in the structure of the layer [101]. The interface between the diffusion layer and the gel layer often has the character of a phase boundary. The gel layer is attached to the diffusion layer by siloxane bridges ($\equiv$Si$-$O$-$Si$\equiv$) bonds [136] but is often possible to mechanically remove the adhering gel from the glass surface.

The gel layer is a rather stable alteration product whose dissolution rate may become smaller [102] or larger [103] than that of the glass. It is likely [102] that the solubility of the gel determines the saturation effect with respect to the glass [Eq. 29.1].

Released glass components accumulate in solution until solubility limits with respect to pure and mixed secondary phases are achieved. If the solubility is exceeded, precipitates form at the outer surface or even within the gel layer. The composition of the precipitates is a strong function of chemical alteration conditions (e.g., pH, pCO_2, redox potential). In principle, the composition can be predicted both for closed and open systems using models based on equilibrium thermodynamics [104] (geochemical modeling [105]). Successful modeling efforts are reported in the literature, particularly on basalt glass alteration [106–108] and nuclear waste glass alteration [109, 110]. The sequence of secondary phases during silicate glass corrosion is metal (hydrate) oxides $\rightarrow$ clays $\rightarrow$ zeolites $\rightarrow$ SiO_2.

Typical oxides and oxyhydroxides are amorphous or crystalline iron oxyhydrates such as goethite or akaganeite, Tihydroxides, or aluminum hydroxides such as gibbsite. These phases are formed in the first hours of glass/water contact within a layer of few hundred nanometers [111–114]. Clays are among the most abundant crystalline alteration products of both natural glasses and man-made glasses [97, 115–117, 148]. Zeolites are formed mainly at alkaline pH (8–14). Typical zeolites formed are phillipsite [118] and chabazite. Analcime is formed particularly at temperatures between 150 and 250°C [12, 97, 98, 115, 119, 148]. Other crystalline alteration products include calcium–silicate hydrate (CSH) phases $BaSO_4$, anhydrite, molybdates, phosphates (e.g., Apatite [120], $LiPO_4$ [148]), silicates (e.g., uranyl silicates, Weeksite [121], Uranophane, and Haiweeite [122]), and layered double oxides like hydrotalcite [123]. The kinetics of zeolite formation can be described by growth and dissolution reaction processes, closely resembling the Oswald step sequence [124].

With certain glass compositions, secondary-phase formation may also be detrimental to glass corrosion. This is particularly the case if formation of silica-rich zeolites prevents silica concentrations in solution to reach saturation values. Frequently, an increase of corrosion rates is observed once zeolites are formed [125].

F. PROPERTIES OF SURFACE LAYERS

F1. Surface Layers as Protective Surface Films

In order to explain the time-dependent decrease of corrosion rates, it has been assumed [126, 127] that the corrosion layers are protective, but, in many instances, corrosion rates decrease because of saturation effects in solution [131]. Mass balance calculations have shown that borosilicate glass dissolution continues even if tightly adhering surface layers are formed [128]. However, particularly in Mg-rich solutions, a protective effect was observed for borosilicate glass compositions [129], soda lime glasses [130], and natural tektites [131]. For soda lime glass, the addition of Al had a similar effect as Mg in decreasing corrosion rates. Protective effects of surface layers were also observed on bioactive glasses exposed to artificial plasma [132] and phosphate-containing solutions [133]. This effect was attributed to formation of Ca–phosphate crystals (apatite) at the glass surface.

The protective effect of the surface layer may be interpreted by local saturation phenomena, that is, by the formation of a diffusion barrier for dissolved H_4SiO_4(aq) [134, 135]. This implies a protective effect being particularly important if the forward reaction rate of the glass is high. This protective effect becomes unimportant if the bulk solution becomes saturated with respect to dissolved H_4SiO_4(aq) or if, because of tensile stress, fractures are formed in the surface layer [136].

G. EFFECT OF FLUID MEDIUM

G1. Effect of Composition of Aqueous Solution

G1.1. pH The most important parameter governing glass corrosion is probably the solution pH. Solution pH influences the reaction mechanism, the solubility of silica and metal cations, the degree of selectivity of the corrosion process. For example, with borosilicate glasses in acid media, selective extraction of glass modifiers and of B is observed, whereas under alkaline conditions congruent dissolution occurs. Dissolution rates under alkaline conditions increase significantly with pH [137]. If the boron concentration in the glass is sufficiently high to hinder the formation of a continuous three-dimensional Si–O–Si network, smaller (Si–O–Si)$_n$ units may be formed and may be released by a percolation process, leading to congruent release even under acid conditions (pH2) [138].

For evaluating the effect of pH on glass dissolution, it is important to recognize the inverse effect of glass dissolution on pH. A self-accelerating failure mechanism of glass occurs if glass dissolution leads to increase or decrease of pH, and corrosion rates increase because of these pH changes. The interrelationship may be illustrated with some examples:

1. In a nonbuffered static test with limited solution volume, the alkali released from commercial Al-containing soda–lime silicate glass leads to a rapid increase in pH, with increasing dissolution rates. Rates become significantly higher than those of Pyrex glass, the latter showing less increase in pH during corrosion. The situation is entirely different if the pH is buffered at 7. Then the soda–lime glass becomes more stable than Pyrex glass [73]. The slower increase of pH when dissolving Pyrex glass was explained with the smaller Na content of this glass. However, a factor probably even more important is the boron content of the glass. Dissolved boron can act as a very efficient buffer, fixing the pH at a value of ~9. The high stability of borosilicate glasses may, at least in part, be attributed to this buffer effect. A key factor controlling pH changes during glass corrosion is the alkali–B ratio in the glass. If this ratio is smaller than or equal to unity, there will always be sufficient B in solution to buffer the pH.

2. Silicate glass dissolution in Mg-rich solutions leads to decreased pH and increased corrosion rate of Al- or Fe-rich glasses. This effect also occurs during the alteration of natural rhyolitic glasses in arid environments [139]. A particularly detrimental effect of the pH decrease is the mobilization of heavy metals. This effect was observed when actinides were released from borosilicate nuclear waste glasses [140].

3. The phosphoric acid release from phosphate glasses, as well as the fluoride release from heavy metal fluoride glasses, in static tests leads to a rapid decrease in solution pH and to an increase in glass corrosion rates, as discussed in the following paragraphs.

Often, these pH changes occur in standard tests, making data interpretation difficult. Studies on the effect of pH on glass corrosion must therefore be performed in buffer solutions at constant pH, in very high solution volumes, or at high water flow rates. The design of more stable glass compositions may focus more clearly on this self-accelerating mechanism; for example, constituents that provide a pH buffer effect can be added to the glass formulation.

G1.2. Effect of Special Ions in Solution   Highly concentrated alkali salt solutions are known to decrease glass corrosion rates [141]. This effect has been explained by the suppression of the initial ion exchange processes. The corrosion resistance of alkali borosilicate glasses can be enhanced significantly by adsorption of certain ions from solution, such as Al^{3+}, Ca^{2+}, Mg^{2+}, or Zn^{2+} [142]. The effect may in some cases be explained by the formation of protective layers. Also, the solubility of the surface gel layer may be decreased by adsorption of these ions. If the saturation effect of glass dissolution is associated with the solubility of the gel, then the decrease of solubility should have a strong effect on glass corrosion rates. Adsorption of ions from solution was found to decrease in the order Ca > Ba > Ag > Mg [143].

G2. Corrosion by Vapor-Phase Hydration and the Atmosphere

There are many similarities but also significant differences when comparing glass corrosion in liquid water with vapor-phase hydration. In vapor test hydration, only a surface film of adsorbed water molecules exists on the glass surface and dissolved glass constituents cannot be transported away. The thickness of the water film depends on relative humidity (RH) as well as on the hygroscopy of the glass phase and hence, on glass composition. For some glasses, even 30% RH leads to attack of the glass surface. The quantity of water adsorbed increases with time and with the quantity of leached alkali. The humidity resistance increases in the order soda–lime bulb tumblers < soda–lime tubing < commercial plate glass, television panels < commercial window glass, soda–lime bottle glass, soda–lime tubing < commercial light bulb (outside) < alumino silicate, alkali borosilicate tubing < opal tableware, borosilicate tubing [144].

Due to the small thickness of the surface film, one may think of the vapor-phase hydration process as being similar to glass dissolution at high surface area–solution volume ratio (S/V). In such a thin water film on silicate glasses, the pH can rise rapidly. Once the pH is > 9, the corrosion of the glass network may occur, reaction products become deposited, and the glass surface loses its transparency and reflectance. Vertically stacked flat glass sheets may become unusable if stored under humid conditions. On the other hand, in the case of glass windows, the rain always washes these reaction products off, surface pH decreases, and glasses remain stable for longer periods of time.

For corrosion of window glass in buildings under attack of atmospheric agents, four stages can be distinguished [145]:

1. Adsorption of water and CO_2 associated with the formation of alkaline surface pH, leaching and formation of secondary minerals. This surface precipitate can easily be removed by cleaning with water.

2. If humidity contact continues, surface precipitates become thicker ($\sim$1 μm) and calcite forms and the surface becomes colored but can still be cleaned with water.

3. Further contact with humidity leads to strong leaching and enrichment of alkali and alkali earth elements as salt minerals at the surface and the silicate network becomes corroded by forming residual silica-rich structures. After cleaning with water, the surface remains modified, and effective cleaning requires acid treatment or even the use of hydrofluoric acid.

4. In the last stage the surface becomes rough and even a hydrofluoric acid cannot clean the surface.

High S/V ratios are also attained by vapor-phase hydration of stacked window glasses in humid environments, for example, during transport in mountain or maritime regions [145]. On the other hand, at a lower relative humidity of 75% and 22°C it takes $\sim$70 days before significant dealkalization occurs [146]. Alteration phases (e.g., $CaSiO_3$) [147] are rapidly formed due to high surface pH conditions. Also, hydrated carbonate minerals may form upon contact with carbon dioxide from the air. It is recommended to prevent stacked glass sheets to remain for a long time in humid environments, that interleaving materials be used, which may buffer potential pH rises, and that storage areas be well ventilated.

In a corrosion test of borosilicate glass with steam at 200°C (steam generated 0.25 mL liquid water at the bottom of the reaction vessel), water condenses on the glass without dripping and an effective S/V of 4000 m^{-1} was estimated [148]. Higher water volumes led to condensation of water at the glass surface and dripping and thus recycled rinsing of the sample. After an induction period of a few days, corrosion rates in steam were found to be higher (linear rates of 2.5 and 21 μm/day for two similar glasses) than rates (after 30 days 0.01 and 1 μm/days) under hydrothermal conditions at $S/V = 40$ m^{-1} at the same temperature. Corrosion layers in steam were found to be richer in alkali contents and contained more silica–rich crystalline phases (zeolites and feldspars) than the corrosion layers formed under hydrothermal conditions. It was suggested that these phases drive the long-term glass corrosion rate at high S/V ratios caused by uptake of silica and prevention of saturation effects.

Vapor-phase hydration was also studied with phosphate glasses. Here no diffusion control was observed but a linear rate law was observed associated with hydrolysis of the glass network [149].

G3. Hydrofluoric Acid

Etching of glasses by HF is used in many technological and scientific fields, such as glass surface cleaning, glass

strengthening, and fission track edging. Wet chemical etching of silicate glasses in HF solutions has been reviewed [150]. Up to HF concentrations of $\sim$5 wt. %, the etching rate increases linearly with HF concentration, whereas higher HF concentrations lead to an over proportional increase in etching rates. The mechanism of dissolution is governed by adsorption of HF and HF_2^- species and catalytic action of H$^+$ ions [150]. The initial reaction step is HF adsorption, followed by leaching of alkali and alkali earth. The rate-limiting step appears to be the breakage of siloxane bonds both at the outer surface and in the interior of the leached layer [153]. The reaction between HF solutions and silicate glasses was found in certain cases to be transport controlled with respect to the mass transfer of dissolved glass constituents in solution with a typical activation energy of 20–45 kJ/mol [150–152]. Initial corrosion rates in 10 wt % HF solutions were $\sim$100 μm/day at 25°C for fused silica [150]. Corrosion resistance in HF decreased with the addition of Na, Mg [152], Al, and Ca [150] as well as of P or As to the glass, whereas the etching rate decreased after B addition [150]. The relations between glass composition and etching rate are highly nonlinear [153]. The etching of multicomponent glasses by HF does not always lead to complete dissolution. In particular, if the glass contains alkali earth or rare earth elements, the precipitation of sparingly soluble fluoride phases can be observed [150].

G4. Strong Acids

Corrosion of borosilicate glass has been tested in strong nitric acid up to a normality of 16. A maximum corrosion rate of 2500 μm/day (2-h test at 25°C) was observed in 6 N solution, and higher acid strength led to a decrease in reaction rates, attributed to a decrease in water activity [154].

H. BIODEGRADATION OF GLASSES

In contact with bacteria, many glasses show increased corrosion. There is evidence for the existence of "silicate bacteria," defined as chemolithoautrophic bacteria, which gain their energy by deterioration of Si–O bonds in silicates [155]. The effect also depends on the availability of nutrition [156]. Microorganisms, such as molds, are known to affect the stability of optical glasses [157] as well as medieval window glasses [158]. Experimental corrosion measurements of model glasses of medieval glass composition in the *presence* of fungi have shown 5–30 more glass corrosion (gel layer formation) than in the *absence* of fungi [159]. The effect was explained by a combination of an acid attack on the glass surface and a corriplexation reaction.

I. SPECIAL GLASS COMPOSITIONS AND APPLICATIONS

I1. Alkali/Alkali Earth Silicate Glasses— Window Glasses

Among the most important industrial glasses are soda-lime glasses, that is, a special type of alkali/alkali earth silicate glasses. Alkali ions in general decrease the stability of silicate glasses by decreasing the cross-link density of the three-dimensional glass network and by provision of ion exchange sites. In the pH range 1.4–10.9 the reaction of pure alkali silicate glasses with water is characterized by selective alkali release, following initially a square root of time rate law [160]. Divalent or trivalent cations stabilize alkali silicate glasses by formation of O–R–O bridges in the glass structure [161, 162] which hinder the water diffusion/ion exchange process. For a soda–lime glass with the composition SiO_2 75 wt %, Na_2O 15 wt %, CaO 10 wt % exposed to a buffered pH 5.3 solution at 88°C, the diffusion coefficients for H and Na where 6×10^{-16} and 6×10^{-13} cm^2/s, respectively [163], probably governed by diffusion through a transformed surface layer. Diffusion coefficients decreased by about a factor of 5 if an aliquot of 3 wt % of the Na_2O content of the glass is replaced by K_2O, emphasizing a mixed alkali effect [164]. An isotopic effect D/H, as well as ^{18}O update data, is consistent with rate control by breaking an H–O bond (or D–O bond) of indiffusing water molecules and subsequent hydrolysis of Si–O–Si bonds to 2 Si–OH as necessary precursor reactions to Na/H ion exchange [165] Isotopic exchange reactions indicated condensation reactions according to the Scheme $2 \equiv Si–OH \rightarrow \equiv Si–O–Si \equiv + H_2O$ as well as high mobility of water in the leached layer, ~1000 times faster than the rate of leached layer growth. Hence, water diffusion was not rate limiting. Condensation of silanol groups in the leached surface layer was also evidenced by an increase in Q^4 silica groups for a potash–lime glass, analyzed by (magic-angle-spinning nuclear magnetic resonance) ^{29}Si MAS–NMR [166].

While ion exchange and diffusion processes dominate the initial stages of the glass water reaction, glass network corrosion subsequently dominates. For a soda–lime composition for flat glass SiO_2 72.0 wt %, Na_2O 13.9 wt % MgO 4.0 wt %, CaO 8.3 wt % activation energies for forward glass network corrosion rates were found to be 66 kJ/mol [167] at a neutral pH, while increase in activation energy is observed at lower and decrease at higher pH [191]. A primary factor controlling soda-lime glass network corrosion in nonbuffered water is a pH effect associated with the S/V ratio. Under static dissolution conditions the pH rises with the progress of glass corrosion. The higher the temperature or the S/V ratio, the faster this pH rise occurs. If pH > 9, a rapid increase of corrosion rates occurs. This does not happen if the pH is buffered at neutral values [168]. Then slow corrosion rates of about 1 μm/day are observed at 90°C [191].

The rather poor resistance in alkaline solutions results in staining of window glasses at the rainexposed side of buildings if constructions allow for contact with alkaline runoff water from concrete or cement walls [169]. Introduction of nitrogen by adding Si_3N_4 to the glass melt increased durability of window glass considerably (factor of 3 in 2 N NaOH) [170].

For a soda–lime float glass, the composition of top and bottom surfaces are not identical, which is attributed to tin from the bath and Fe enrichment at the bottom. The tin layer can easily be dissolved in 1 M NaOH [171]. The tin-rich layer was found to protect the glass, resulting in about one order of magnitude lower glass surface leach rates [172] and less hydrogen penetration [173] when compared to the tin-free top surface. An addition of 2.5 wt % of tin to soda–lime glass increases glass stability [174].

I2. Alumosilicate Glasses

Addition of Al to soda–lime glasses leads to another class of rather stable glasses, the alumosilicate glasses. The replacement of half of the CaO content of a soda–lime glass by Al_2O_3 (composition SiO_2 75 wt %, Na_2O 15 wt %, CaO 5 wt %, Al_2O_3 5 wt %, leads to a decrease of diffusion coefficients for hydration by H-bearing species by a factor of 5 to a value of 10^{-16} cm^2/s and of Na by a factor of 50 to a value of 10^{-14} cm^2/s [163]. An increase of the Al–Si-ratio of silicate glasses increases the network corrosion resistance [175] by formation of Si–O–Al bridges and immobilization of a fraction of Na ions for reason of charge compensation in the vicinity of these bridges. The ratio of mobile to immobile Na ions in the glass (exchangeable and nonexchangeable Na) depends on the molar Na–Al ratio in the glass. With a molar ratio or Na/Al ≤ 1, an increase in the alkali content of the glass does not lead to an increased glass corrosion rate, because there are no nonbridging oxygen in the glass structure [176]. There are also some cases with no effects of Al on corrosion stability of the glasses [177].

The formation of Si–O–Al bridges stabilizes the glass only if the hydrolytic stability of this bond structure is high. This is the case at neutral to slightly alkaline pH but not at acid pH. Also in strongly alkaline solutions corrosion was observed to increase with increasing Al contents, attributed to the ease of hydrolysis of Al–O bonds under alkaline conditions and the gradual change from tetrahedral to octahedral coordination of Al ions by contact to alkaline solutions [178]. Hence, the stabilizing effect of Al depends also on solution chemistry. In addition, the effect of alkali earth elements on the corrosion resistance of alumosilicate glass depends on the chemistry of the aquatic medium. For example, the partial replacement of Si by alkali earth ions decreases the corrosion resistance of alkali aluminum silicate glasses (18.5 Na_2O–7.4–Al_2O_3–74.1 SiO_2) at neutral pH and increases it at pH 10 [179]. This effect results from the reduced mobility of alkali earth ions at high pH values.

In addition, aluminum plays an important role in controlling the solution concentration of silica. This is important under closed system alteration conditions, where saturation effects may decrease glass corrosion rates. Two cases may be distinguished: (1) control of Si solution concentration by the transformed surface of the dissolving glass, leading to faster slowdown of glass corrosion rates due, to saturation [180, 181], or (2) control of Si concentration by the formation of secondary phases resulting in a hindrance of the slowdown of reaction due to a hindrance of saturation at the glass–solution interface. For example, an Al-rich glass showed a much higher long-term corrosion rate than a glass with much less Al [182], This was due to the reduction of silica concentrations in solution due to the formation of analcime, a zeolite mineral, as secondary solid reaction product.

I3. Borosilicate Glass

Borosilicate glasses with ~4–8% alkali oxide, 2–7% Al_2O_3, 0–5% alkali earth oxides, 70–80% SiO_2, and 7–13% B_2O_3 are particularly water and temperature resistant. Typical samples are Duran or Pyrex glass. Zinc added to borosilicate glasses at concentrations up to 16 mol % leads to increased glass stability [183]. In aluminum silicate glasses without NBO (Na/Al ≤ 1), a replacement of up to 80% of Al_2O_3 by B_2O_3 leads to an increase in corrosion resistance.

The corrosion resistance of borosilicate glass can be understood only on the basis of glass structure. The relation between glass structure, glass composition, and glass stability is described by a model of Dell et al. [184]. Including new results for the ternary system [186] as well as for the binary alkali borate [185] and alkali silicate system, the results of the model are illustrated in Figure 29.5

Besides SiO_4 tetrahedra (Q^n), additional structural units of borosilicate glasses are anionic BO_4 tetrahedra (N_4 sites) and symmetric or asymmetric trigonal groups (N_{3s} or N_{3a} sites), with a distribution among these sites determined by the mol ratio $R = Na_2O/B_2O_3$ and $K = SiO_2/B_2O_3$. The occurrence of tetrahedral groups depends, on the presence of alkali. The maximum tendency for phase separation into Si-rich and B-rich glass phases [186] occurs at $R = 0.19$. The increase of alkali content of the glass increases the solubility of borate groups in the silicate glass network [184] by forming borosilicate groups related to Danburite groups and Reedmergnerite ($NaBSi_4O_{10}$) units [184, 186]. At higher R values, not all Na ions are associated with tetrahedral borate groups but to Si–Q^3 tetrahedras or with further alkali content to Si–Q^2 sites. This leads to an increase of alkali mobility in the glass, resulting in a stronger tendency for ion exchange with contacting aqueous solutions.

High-silica glasses can be fabricated based on phase-separated glasses in the Na_2O–B_2O_3–SiO_2 system (Vycor process [187]). Here, glass corrosion is not detrimental to glass use but is a key design parameter because the borate-rich phase must be dissolved rapidly. The microstructure of the glass determines the rate of dissolution. Rapid dissolution can, for example, be achieved by adding a small amount of high-valency cations such as V, P, or Mo, increasing the size of phase separations in the glass and the rate of borate phase dissolution [188, 189].

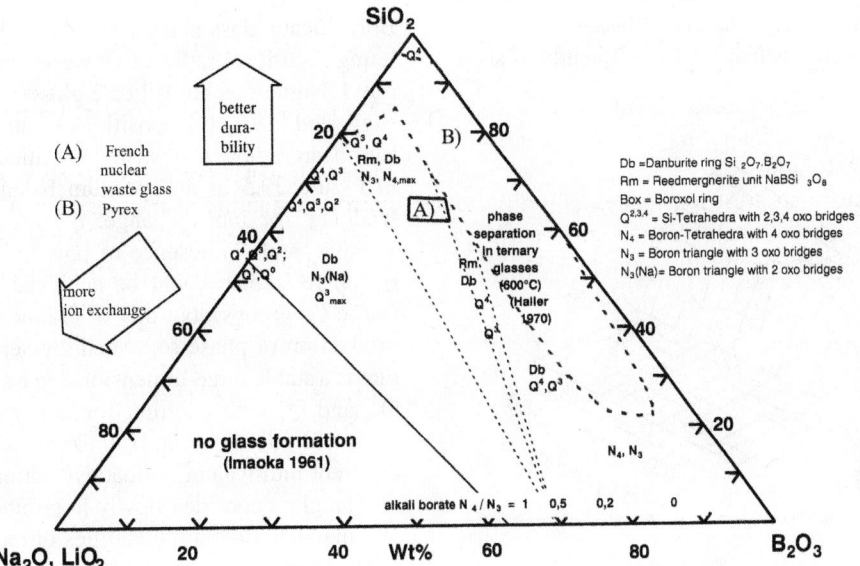

FIGURE 29.5. Ternary Na_2O–B_2O_3–SiO_2 Structure unit stability diagram and relation to glass durability and phase separation.

An industrially important, durable borosilicate glass is Pyrex glass, (SiO_2 80 wt %, Na_2O 4 wt %, Al_2O_3 2 wt % B_2O_3 14 wt %). As expected from Figure 29.5, Pyrex glass is phase separated into a nearly pure silica matrix with a dispersed borosilicate phase with spherical regions of 5–10 nm diameter [190]. Corrosion rates are probably dominated by the dissolution of the boron-rich phase. Corrosion rates were found to be invariant with pH between 0 and 4 and increased by about a factor of 10 until pH 9. At 95°C and pH 9 a linear forward corrosion rate of 0.6 µm/day was measured with pH-independent activation energies of about 60 kJ/mol [191]. At 250–260°C in a closed system, boron release continued with a linear rate of ~4 µm/day, but Na and Si loss reach a temperature-dependent saturation state, with Na/Si ratios in solution equal to those in the glass [192]. Activation energies for hydration were only 17 kJ/mol, much lower than the activation energies for B release. This indicates that hydration and B release are independent processes, and hydration does not occur by filling the holes of phase-separated boron-rich phases.

With simple borosilicate glasses, as well as nuclear waste borosilicate glass compositions, the corrosion resistance is not a linear function of glass composition but shows strong nonlinear behavior. For example, the addition of 0–2 wt % SiO_2 to a waste glass for West Valley (USA) led to only a small increase in corrosion resistance, but 3% addition decreased the corrosion rate by a factor of 10 (Fig. 29.6). A further increase in silica content had a negligible effect [193].

For glass compositions containing redox-sensitive elements, the corrosion resistance becomes a function of the redox conditions during glass production as well as of the redox-state of the aqueous solution. For example, Fe^{III} acts much more as a stabilizing agent than does Fe^{II} (Fig. 29.7) because Fe^{III} may take a similar role in the glass structure as a network-forming element, such as Al. Consequently, if such

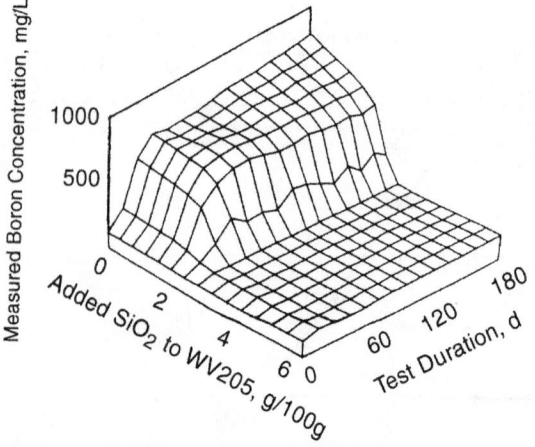

FIGURE 29.6. Dependence of the extent of glass corrosion (release of boron) on the content of silica in the glass [212].

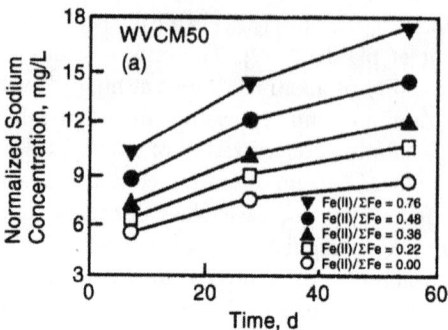

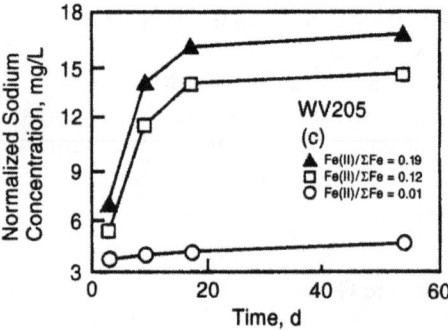

FIGURE 29.7. Corrosion results at 90°C in MCC-3-Test (Na release) of two American radioactive waste glasses as function of redox state ($Fe^{II}/Fe^{(tot)}$) of the glass matrix [212].

glasses are melted under oxidizing conditions, they become more stable when compared to glass melting under reducing conditions [193].

I4. Nuclear Waste Borosilicate Glasses

Borosilicate glass is the principal solid matrix for immobilizing high-level radioactive waste for deep geological disposal. Numerous borosilicate glass compositions have been suggested [194]. Compositions of all industrially produced European radioactive waste borosilicate glasses are located in Figure 29.5 in an optimum balance between (1) low tendency for ion exchange, (2) relatively high corrosion stability, and (3) absence of phase separation. Higher corrosion resistance could be achieved by higher Si content (more Q^4 groups), but this would increase the tendency for production of phase-separated glasses. The resulting structure is a stable three-dimensional glass network with N3, N_4, Q^3, and Q^4 with coordination cages of oxygen of various sizes and coordination, capable of hosting a large suite of different multivalent radioactive cations simultaneously.

The glass corrodes slowly in groundwater and humid air and inevitably certain quantities of radionuclides are mobilized. Various reviews on nuclear waste glass performance have been published [13, 14]. Predicting long-term glass performance is difficult. Disposal sites are located in deep rock with very slow groundwater flow rates. Hence, saturation effects become much more important than for other types of glass applications. Accumulation of dissolved silica and other glass constituents in solution leads in many (not all) cases to a significant decrease in the overall reaction rate [195–197]. Provided the accumulation of Si is not hampered (i.e., by formation of Si-rich crystals such as zeolites, by high water flow), a saturation state will be reached with reaction rates < 1/1000 of the initial rates [198]. This decrease in reactivity has been attributed to decreasing affinity for the rate-limiting dissolution reaction [198–200]. Protective layers were normally not formed. The fundamental disequilibrium between the bulk glass phase and the solution remains, that is, overall affinity remains appreciable and the glass/water reaction will continue as long as an aqueous phase is present. The mechanism of this "long-term reaction" is not yet fully understood. It has been suggested that the dissolution rate may either be driven by the precipitation of secondary phases [198, 201], formation of colloids [196], or resumption of rate control by water diffusion/ion exchange [202].

Glass alteration may or may not be accompanied by transfer of glass constituents (including radionuclides) to a potentially mobile aqueous phase. Sparingly soluble glass constituents become incorporated in or sorbed on solid glass alteration products on the glass surfacer—the gel layer and a precipitated layer—but may also be sorbed on groundwater colloids. The formation of new secondary phases, such as silicates, molybdates, uranates, carbonates, and so on, establishes a new geochemical barrier for reimmobilization of radionuclides dissolved from the waste matrices. As an example, secondary clay minerals (saponite), powellite, and baryte phases formed on the surface of Cogema type HLW-glass during corrosion in brine. The phases formed are normally not pure but are solid solutions of quite complicated composition. Solid-solution formation is a beneficial effect for retention of radionuclides. The clay mineral saponite is important for sorption of trivalent and hexavalent actinides, while powellite and barite are host phases, incorporating either trivalent actinides or divalent radionuclides within their crystal structure. Tetravalent elements appear to become concentrated into thermodynamically very stable zircon or cereanite-type phases. Recently, it has been shown that the retention behavior of trivalent, tetravalent, and hexavalent elements can be described by sorption isotherms in certain pH ranges, whereas solubility/coprecipitation is dominant in other ranges [203].

Radiation damage of the glass phase and transmutation have only a minor effect on glass durability, whereas the effects of radiolysis of water and aqueous species may increase in glass dissolution rate by a factor of 5 [14]. Initial reaction rates may become higher if neutral starting solutions are acidified by radiolysis. Alternatively, for experiments at very high S/V in the presence of nitrogen, it has been shown [204] that dissolution rates of radioactive glasses may be as much as a factor of 40 lower than the corresponding rates of a simulated glass. In this case, the alkaline evolution of solution pH is partly balanced by acidification due to radiolysis. In the absence of nitrogen the initial rate is slightly affected by radiolysis, but there is only little effect on the rate of long-term dissolution at high S/V [205, 206].

I5. Alkali-Resistant Glasses

Glass-fiber-reinforced cement (GFRC or GRC) composites are based on glass fibers with high alkaline durability. Mechanical properties of these cements depend on the chemical durability of the fibers. Suitable alkali-resistant fibers contain high contents of Zr, for example, in CemFIL (16 wt %) or ARG-fiber. The Zr-containing glass fibers (CemFIL-1) were leached at 55°C for a year in Portland cement pore fluids, resulting in only 2 µm of corrosion. Glass stability was caused by the formation of a shell of Zr- and Ca-rich reaction products [207]. In the absence of Ca, no protective effects of Zr-rich surface reaction products were observed. Maximum glass stability was observed with 100 mg/L Ca in solution [208].

Disadvantages of the Zr-rich fibers are the expensive high-melting and processing temperatures. Therefore, a number of alternative alkali-resistant glass compositions have been developed. Glass compositions in the mole percent range CaO < 9, Na_2O 13–15, BaO < 10, ZrO_2 7, MgO 3, SiO_2 65–75 show similar resistance to ChemFIL and 100°C lower manufacturing temperatures [209]. Low corrosion rates, ~0.01 µm/d, of Zn–Al–silicate glass fibers, modified with Fe, Mn, or Ti oxides, were observed at 60°C in $Ca(OH)_2$ solutions at pH 12.6 [210].

I6. Phosphate Glass

Phosphate glasses show a number of advantages when compared to silicate glasses: optical properties, high thermal expansion coefficient, low melting points, and formation of sparingly soluble host phases for radiotoxic elements. However, instability to crystallization and low chemical durability of many phosphate glass composition lead to less widespread use than silicate glasses. Depending on glass composition, the corrosion rates of phosphate glasses cover a range between 1000 and 0.01 µm/day [211].

A particularly stable class of glasses is to be found in the lead–iron phosphate system. A review is given by Sales and Boatner [212].

I7. Glass-Reinforced Plastics

Low specific gravity (<2g/cm^3), low cost, easy repair and shape adaptation, high strength, and so on, make glass-reinforced plastics very popular as structural and corrosion-resistant material for pipes, tanks, and chemical process equipment. There are three main factors controlling corrosion: the fiber, the matrix, and the interface. With respect to the fiber, corrosion stability depends on the same principles as for ordinary glasses. Highly corrosion resistant vitreous silica fibers can, for example, be made from acid-leached E-glass fiber. Stress-induced corrosion processes are important [213]. The fibers may be attached to the resin by silanation [214]. The effect of chemical attack can greatly be reduced by proper choice of the resin and proper fabrication. The following basic principles [215] are useful for resin selection to provide resistance to various chemicals. Aqueous solutions and salts are generally nonaggressive, but strong oxidizing acids attack the resin; organic solvents, such as benzene, attack most polyester or vinyl esters, whereas epoxy esters are better for organic solvent resistance.

I8. Optical Glasses

Corrosion may change the optical properties of glasses during manufacturing, polishing, and so on. The variety of compositions used for optical applications does not allow a generalized treatment of chemical durability. With respect to glass durability, five families of glasses have been distinguished [216], as presented in Table 29.2.

As for alkali silicate glasses, a square root of time rate law was followed by a linear rate law for optical lead silicate glasses (SiO_2 65 mol %, PbO 25–35 mol %, K_2O 0–10 mol %, Al_2O_3 0–2 mol %) under acetic acid static corrosion conditions [217]. Activation energies were ~50 kJ/mol. As for alkali ions in alkali silicate glasses, diffusion-controlled surface depletion profiles were observed for Pb release from the glass [218].

Contact of xBaO/ySiO$_2$ glasses to distilled water led to Ba leaching from surface regions 50–300 nm thick, resulting in decreased density and refractive index of this surface region [219]. Even the initial surface conditions of optical glasses are strongly affected by dissolution effects; for example, mechanical polishing effects were found to be governed by stress-induced chemical surface dissolution [220] or by direct chemical leaching in the polishing solution.

I9. Glass Corrosion and Human Health

Corrosion of glassware may be detrimental to health if toxic elements are released into food to a significant extent. Examples are lead crystal wine decanters, an important potential source of lead exposure (release of 66 μg Pb/L into white wine in 4h) [221]. On the other hand, glasses with 0.2 wt % Cr_2O_3 did not show any release of Cr into solution [222]. For Ni, a beneficial effect may be observed; for example, bottled wine may have significantly less Ni content than the original grapes, attributed to sorption of Ni on the glass surfaces [223].

Bioglass as bone replacement was discovered by Hench et al. [224], characterized to form tight bonds to living bones (bioactivity). Some glasses and glass ceramics are already in clinical use. Bioactivity is based on the ability of the glass or glass ceramic to form, as a reaction product, bonelike apatite during corrosion in body fluids at 37°C. Hence, most glasses contain CaO and P_2O_5 as a main component; however, bioactivity has also been reported for soda–lime glasses, showing apatite formation during corrosion in body fluids [225] and other phosphate-containing solutions [133]. Apatite formation in glass surfaces leads to formation of protective surface layers [132, 133].

Aspirable silicate glass fibers can be detrimental to health if they do not dissolve sufficiently rapidly in physiological solutions. Long-term (120-day) experiments in Gamble solution at 37°C at pH 7.6 have indicated a substantial difference in the residence time of natural mineral fibers and artificial fibers (Table 29.3) [226] expected under in vivo conditions. In another study, 30 glass compositions in the system SiO_2–Al_2O_3–B_2O_3–CaO–MgO–BaO–Na_2O were studied. Predicted corrosion rates in physiological solutions varied between 0.2 and 200 ng/day [227].

I10. Silicate Glass Fibers

Glass fibers and bulk glasses of the sample composition differ in various physical and chemical properties. Faster cooling rates during production make the glass network of fibers more open. Also density, refractive index, and Young's modulus are lower. Consequently, corrosion behavior might

TABLE 29.2.   Optical Glass-type Families in Chemical Durability Tests[a]

Family	Resistance to				
	Climatic Variation	Acids	Alkaline Solutions	Weathering	Staining
Low dispersion		Medium	Medium		
Crown	Good	Good	Good	Good	Very good
Flint	Very good	Very good	Medium or good	Good or very good	very good
Alkaline earth	Medium or good	Medium	Medium or good	Good	Medium or inferior
Rare earth and high diffractive	Very good	Medium or good	Good or very good	Inferior	Very good

[a]See [216].

TABLE 29.3. Overview of Corrosion Results of Various Silicate Fibers in Physiological Solutions at 37°C (Tests for 120 days)[a]

Type of Fiber	Average Diameter (µm) Used	Average Dissolution Rate (nm/day)	Residence Time (years of 1 µm-diameter fiber)
Glass wool			
JM 104 (475)	0.4	0.8	1.7
JM 104 (E)	0.5	0.21	6.5
TEL	3.5	3.4	0.4
Superfine	0.4	1.4	1.0
Mineral wool			
Diabas	4.0	1.15	1.2
Basalt	4.9	1.10	1.2
Schlacke	4.8	0.69	2.0
Fire resistant fibers			
Vitreous silica	0.8	1.1	1.2
Fiberfrax R	1.9	0.27	5.0
Fiberfrax H	1.9	0.28	4.9
Natural fibers			
Chrysotile UICC	0.1	0.005	100
Krokydolite	0.2	0.011	100
Erionit	0.01	0.0002	100

[a]See [226].

also be different. Corrosion of silicate glass fibers have been extensively studied for application fields such as optical waveguides, cement, or plastic reinforcement. A review of the resistance of glass fibers in corrosive environments is given in [228]. A comparative study of the corrosion behavior of glass fibers, in the composition range $(Na.K)_2O$, 0.5–15 mol. %, SiO_2 46–70 mol %, Al_2O_3 0.7–11 mol %, $(Ca,Mg)O$, B_2O_3 0–7 mol %, ZrO_2 0–2 mol % show saturation effects under static corrosion conditions (interpreted as "formation of protective layers") and continued dissolution in dynamic corrosion tests [229]. Specific surface areas of silicon aluminate $(SiO_2 \cdot Al_2O_3)$ glass fibers were found to increase dramatically during corrosion at pH 1 and only little at pH 3. This was interpreted as resulting from Al release, leaving micropores in the glass network [230].

Microphase-separated E-glass fibers SiO_2 52.8 wt %, B_2O_3 10.8 wt %, Al_2O_3 14.4 wt %, CaO 16.7 wt %, MgO 4.9 wt %, Na_2O 0.2 wt %, and K_2O 0.02 wt %) are known to become converted to high-silica glass by leaching in mineral acids [231]. After leaching, the surface area is very high (100–300m^2/g) caused by micropores of <2 nm diameter [232], probably only 0.4 and 0.8 µm, coexisting with larger mesopores of 6 nm [233]. During corrosion, a core-sheath structure is built up with loss of mechanical strength of the outer sheath caused by replacement of Ca and Al (and probably B) by H$^+$ [234]. The extraction of metal ions

proceeds in two steps: (1) dissolution of an interconnected phase exterior to the silica network and (2) dissolution of metal ions form the silica network [235].

Due to formation of voluminous alkali-depleted surface layers, acid corrosion of E glass leads to an increase in the thickness of the fibers and formation of fractures. The thickness of the layer increases with the square root of time [236]. Under alkaline conditions, E glass is not stable. For example, as long as <40% of 10–15-µm-thick glass fibers were dissolved, corrosion rates in saturated $Ca(OH)_2$ solutions (pH 12.6) at 25°C were found to be in the range of 0.08 µm/day [237].

I11. Fluoride Glasses

Heavy metal fluoride (HMF) glasses have been proposed as the next-generation optical waveguides for their transparency to infrared (IR) light. It is particularly important that these glasses show some stability against liquid water and water vapor of common environments. Early studies of BeF_2 glasses have shown highly hygroscopic behavior. A systematic study of chemical durability in multivariate compositional space in the Th–Ba–Al–Zr–F system has been reported [238]. In contrast, ZrF_4-based and BaF_4-ThF_4- based glasses show remarkably high corrosion resistance for atmospheric moisture and temperatures <200°C, but they were not very durable in liquid water.

With congruent matrix dissolution rates at 25°C of 3–30 µm/day BaF_4–ThF_4-based glasses were ~50–100 times more stable in deionized water than Zrf_4 glasses [239]; however, they are still much less durable than typical silicate glasses. Leach rates follow the aqueous solubility of the modifier metal fluorides $(AlF_3, NaF, LiF) > (BaF_2, LaF_3)$ [240]. The high corrosion rates were attributed to evolution of acid pH values in static dissolution tests and corresponding high solubility of glass components, particularly severe with ZrF_4 glasses (pH$_{fin}$ = 2.5). A systematic study of pH effects on HMF glass stability confirmed this interpretation. Indeed, when the pH was fixed at neutral to slightly alkaline values (pH < 10), corrosion rates as low as 0.01–0.1 µm/day were observed after a few hours of testing, even with ZrF_4-based compositions [241]. Corrosion results indicate that the reaction of fluoride glasses with water is controlled by moving boundary diffusion according to a square root of time rate law as a thermally activated process [242]. Corrosion layer formation is accompanied by ion exchange of fluoride ions with OH$^-$ [243, 244]. The existence of this ion exchange reaction leads to a technical application of HMF glasses as fluoride ion-selective electrodes [245]. The release of fluoride ions and consumption of hydroxyl ions thus explains the decrease of pH with the proceeding corrosion reaction. The adherence of surface layers to the glass is rather weak, and thus the layer appears to be nonprotective. Nevertheless, after initial ion exchange, water diffusion into the glass appears to play a significant role in controlling glass corrosion rates [246].

In contrast to HMF glasses, much less work is done on the corrosion stability of other fluoride glasses such as fluoroaluminate glasses. During corrosion, a slight increase in pH is observed. The corrosion mechanism appears to be different from that of the HMF glasses. Congruent leach rates are controlled by metal fluoride solubilities [247].

J. METHODS FOR IMPROVING GLASS SURFACE STABILITY

J1. Polymer Coatings

Successful attempts have been made to protect poorly resistant medieval church window glasses by polymer coatings, such as viacryl resin [248]. Though organic polymers allow for water diffusion to the glass surface beneath, the surface was found unattacked after seven years. This surprising effect was explained by saturation effects of silica in the pore space of the coating, slowing down reaction rates.

J2. Resistant Oxide Glass Coatings

The weathering characteristics of soda–lime glass can be improved by coating with sol–gel derived SiO_2 or nine TiO_2–91 SiO_2 films [249]. The film with Ti is even more protective than pure SiO_2. The durability of soda–lime glasses against alkaline solutions can be enhanced by coating with Ti or Zr [250].

In order to improve the chemical durability of heavy metal fluoride glasses, many efforts have been undertaken to develop different types of coating materials, for example, lead phosphate glass coatings with a transformation temperature t_g of only 227°C and a corrosion rate of $\sim 0.5 \, \mu m$/day [251].

REFERENCES

1. International Organization for Standardization (ISO), Long-Term Leach Testing of Radioactive Waste Solidification Products, International Standard ISO/DIS 6961, UDC 621.039.7, ISO, Geneva, 1979.

2. D. M. Strachan, R. P. Turcotte, and B. O. Barnes, Nucl. Technol., **56**(2), 306–12 (1982).

3. V. M. Oversby, Leach Testing of Waste Forms: Interrelationship of ISO and MCC Type Tests, UCRL-87621, Lawrence Livermore National Laboratory, Livermore, CA, 1982.

4. American Society for Testing and Materials (ASTM), "Standard, Test Method for Determining Chemical Durability of Nuclear Waste Glasses: The Product Consistency Test (PCT)," STM C1285, ASTM, Philadelphia, PA.

5. International Organization for Standardization (ISO), International Standard ISO 719: Glass: Hydrolytic Resistance of Glass Grains at 98°C —Method of Test and Classification, ISO, Geneva, 1985.

6. S. Kimmel, A. Peters, and W. W. Fletcher, Glastech. Ber., **59** (9), 252–258 (1986).

7. A. Barkatt, S. A. Olszowka, W. Sousanpour, T. Choudhury, Y. Guo, Al. Barkatt, and R. Adiga, "The Use of Partial-Replenishment Tests in modeling the Leach Behavior of Glasses", Mater. Res. Soc. Symp. Proc., **212** (Sci. Basis Nucl. Waste Manag. **14**), 133–139 (1991).

8. B. P. McGrail, W. L. Ebert, A. J. Bakel, and D. K. Peeler, J. Nucl. Mater., **249**, 175–189 (1997).

9. International Organization for Standardization (ISO), ISO Soxhlet Leach Test Procedure for Testing of Solidified Radioactive Waste Product, ISO/TC 85/SC 5/WG 5/N 40, ISO, Geneva, pp. 1–6.

10. R. Könneccke and J. Kiesch, EC Static High-Temperature Leach Test Final Report, EUR 9772 EN, 1985.

11. L. Friedman and W. Long, Science, **19**, 347–352 (1976).

12. R. G. Newton, "The Durability of Glass—A Review," Glass Techno. Rev. Pap. **26**(1), 21–38 (1985).

13. W. Lutze, "Silicate Glasses," in Radioactive Waste Forms for the Future, W. Lutze and R. C. Ewing (Eds.), Elsevier Science, Amsterdam, 1988.

14. "High-Level Waste Borosilicate Glass—A Compendium of Corrosion Characteristics," Vols. 1–3, DOE-EM-177, U.S. Department of Energy, Washington, DC, 1994.

15. "Glass as a Waste Form and Vitrification Technology: Summary of an International Workshop," National Research Council, National Academy Press, Washington, DC. (1996)

16. B. Grambow, MRS-Bull., **12** (1995).

17. A. C. Wright, B. Vessal, B. Bachura, R. A. Hulme, R. N. Sinclair, A. G. Clare, and D. I. Grimley, Mat. Res. Soc, Symp. Proc., **376**, (1995).

18. G. N. Greaves, A. Fontaine, P. Lagarde, D. Raoux, and S. J. Gurman, Nature, **293**, 611 (1981).

19. D. A. McKeown, G. A. Waychunas, and G. E. Brown, J. Non-Cryst. Solids, **74**, 345 (1985).

20. N. Binsted, G. N. Greaves, and C. M. B. Henderson, J. Phys. (Paris), **C8**, 18 (1986).

21. K. H. Sun, J. Am. Ceram. Soc., **30**, 277 (1947).

22. A. Dietzel, Z. Elektrochem., **48**, 9 (1942).

23. A. Dietzel, Glastechn. Ber., **22**(3,4), 41 (1948).

24. W. H. Zachariasen, J. Am. Chem. Soc, **54**, 3841 (1932).

25. A. Smekal, Glastechn. Ber., **22**, 278 (1949).

26. J. M. Delaye and D. Ghaleb, J. Non-Cryst. Solids, **195**(3), 239 (1996).

27. B. E. Warren, Z. Kristallogr. Mineralog. Petrogr., **86**, 349 (1933).

28. S. Guman, J. Non-Cryst. Solids, **125**, 151 (1990).

29. G. N. Greaves and K. L. Ngai, J. Non-cryst. Solids, **192–193**, 405 (1995).

30. J. A. Topping and J. O. Isard, Phys. Chem. Glasses, **12**, 145 (1971).

31. R. D. Shannon, "Revised Effective Ionic Radii and Systematic Studies of Interatomic Distances in and Chalcogenide," Acta Crystalog, **A32**, 751–766 (1976).

32. W. Hinz, Silikate,VEB Verlag für Bauwesen, Berlin (1970).

33. P. Taylor, S. D. Ashmore, and D. G. Owen, J. Am. Ceram. Soc, **70**(5), 333–338 (1987).

34. A. A. Lebedev, Trudy Cossud. Opt. Inst., **2**, 57 (1921).

35. C. H. L. Goodman, Phys. Chem. Glasses, **26**(1), 1 (1985).

36. C. H. L. Goodman, Phys. Chem. Glasses, **27**(1), 27 (1986).

37. H. Schröder, Glastechnische Berichte, **26**, 91–97 (1953).

38. K. Zagar and A. Schillmöller, Glastechn. Ber., **33**, 109–116 (1960).

39. R. W. Douglas and T. M. ElShamy, J. Amer. Ceram. Soc., **50**, 1 (1967).

40. R. H. Dormemus, J. Non-Cryst. Solids, **19**, 137–144 (1975).

41. J. Schäfer, and H. A. Schaeffer, U.S.C.V Conference, Brüssels 5, June 2, 1984.

42. W. A. Landford, K. Davis, P. Lamarche, T. Laursen, and R. Groleau, J. Non-Cryst. Solids, **33**, 249–266 (1979).

43. H. Scholze, J. Am. Ceram. Soc, **60**, 186 (1977).

44. R. M. J. Smets and T. P. A. Lommen, Phys. Chem. Glasses, **23**, 83–87 (1982).

45. B. C. Bunker, G. W. Arnold, and E. K. Beauchamp, J. Non-Cryst. Solids, **58**, 295–322 (1983).

46. B. M. J. Smets, M. G. W. Tholen, and T. P. A. Lommen, J. Non-Cryst. Solids, **65**, 319–332 (1984).

47. F. G. K. Baucke, J. Electroanal. Chem., **367**(1–2), 131–139 (1994).

48. T. M. Sullivan and A. J. Machiels, J. Non-Cryst. Solids, **55**(2), 269–282 (1983).

49. Z. Boksay, G. Bouquet, and S. Dobos, Phys. Chem. Glasses, **8**, 140–144 (1967).

50. B. P. McGrail, A. Kumar, and D. E. Day, J. Am. Ceram. Soc., **67**, 463–467 (1984).

51. R. M. J. Smets and M. G. W. Tholen, Phys. Chem. Glasses, **26**, 60–63 (1985).

52. B. P. McGrail, A. Kumar, and D. E. Day, J. Am. Ceram. Soc, **67**(7), 463–467 (1984).

53. X. Feng and I. L. Pegg, Proc. 1992 Int. Symp. Energy, Environ. Inform. Manag., Argonne, II, Sept. 15–18, 1992, pp. 7.9–7.16.

54. R. H. Doremus, Treatise on Materials Science and Technology, Vol. **17**, Academic, New York, 1979, pp. 41–69.

55. A. J. Paul, Mater. Sci., **13**, 2246–2268 (1979).

56. J. Plodinec, C. M. Jantzen, and G. G. Wicks, "A Thermodynamic Approach to Prediction of the Stability of Proposed Radwaste Glasses," DP-MS-82-66, Savannah River Laboratory, Savannah, GA, 1982.

57. K. G. Knauss, W. L. Bourcier, K. D. McKeegan, C. I. Merzbacher, S. N. Ngryen, F. Y. Ryerson, D. K. Smith, H. C. Weed, and L. Newton, Mater. Res. Soc. Symp. Proc, **176**, 371–387 (1990).

58. T. Advocat,These du doctorat, Universite Louis Pasteur, Strasbourg, 1991.

59. E. Wieland, B. Wehrli, and W. Stumm, Geochim. Cosmochim. Acta, **52**, 43–53 (1988).

60. B. Grambow, Glastechn Ber., **56K**, 566–571 (1983).

61. A. Barkatt, P. B. Macedo, B. C. Gibson, and C. J. Montrose, Mater. Res. Soc. Symp. Proc, **44**, 3–14 (1985).

62. K. B. Harvey, C. D. Litke, and A. B. Larocque, Phys. Chem. Glasses, **33**(2), 43–50 (1992).

63. P. Massard, Bull. Soc. Fr. Mineral. Cristallogr., **100**(3–4), 177–184 (1977).

64. T. Advocat, I. L. Crovisier, B. Fritz, and E. Vernaz, Mater. Res. Soc. Symp. Proc, **176** (Sci. Basis Nucl. Waste Manage., **13**), 241–248 (1990).

65. B. Grambow, Mater. Res. Soc. Symp. Proc, (Sci. Basis Nucl. Waste Manage.), **44** 15–27 (1985).

66. W. L. Bourcier, D. W. Peiffer, K. G. Knauss, K. D. McKeegan, and D. K. Smith, "A Kinetic Model for Borosilicate Glass Dissolution Based on the Dissolution Affinity of a Surface Alteration Layer," Report No. UCRL–101107, Lawrence Livermore National Laboratory, Livermore, CA, 1989.

67. L. L. Hench, J. Non-Cryst. Solids, **25**, 343–369 (1977).

68. E. C. Ethridge, D. E. Clark, and L. Hench, Phys. Chem. Glasses, **20**, 35–40 (1979).

69. W. L. Ebert and J. K. Bates, Nucl. Technol., **104**(3), 372–384 (1993).

70. P. Vanlseghem and B. Grambow, Mater. Soc. Symp. Proa, **112**, 63–639 (1988).

71. L. R. Pederson, C. Q. Buckwalter, G. L. McVay, and B. L. Riddle, Mater. Res. Soc. Symp. Proc, **15**, 47–54 (1983).

72. W. L. Ebert and J. K. Bates, Nucl. Techn., **104**(3), 372–384 (1993).

73. G. Perera, R. H. Doremus, and W. Lanford, J. Am. Ceram. Soc, **74**(6), 1269–1274 (1991).

74. P. H. LaMarche, F. Rauch, and W. A. Lanford, J. Non-Cryst. Solids, **67**(1–3), 361–369 (1984).

75. A. Paul, J. Mater. Sci., **12**, 2246–2268 (1977).

76. B. Grambow, Mater. Res. Soc. Symp. Proc, **44**, 15–27 (1985).

77. M. J. Plodinec, C. M. Jantzen, and G. G. Wicks, Mater. Res. Soc. Symp. Proc, **26**, 755–762 (1984).

78. C. M. Jantzen, in D. E. Clark and B. K. Zoitos (Eds.), Corrosion of Glass, Ceramics, and Ceramic Superconductors, Noyes Publications, Park Ridge, NJ. (1992), pp. 153–251.

79. H. Dunken, and R. H. Doremus, Mater. Res. Bull., **22**(7), 863–878 (1987).

80. R. Conradt and P. Geasee, Ber. Bunsen-Ges., **100**(9), 1408–1410 (1996).

81. G. F. Piepel, Spektrum 90, Nucl. and Hazardous Waste Management Topical Meeting, Knoxville, TN, Sept. 30–Oct. 4, 1990, pp. 309–312.

82. F. G. K. Baucke, Ber Bunsen-Ges., **100**(9), 1466–1474 (1996).

83. H. Schnatter, H. Doremus, and W. A. Lanford, J. Non-Cryst. Solids, **102**, 11–18 (1988).

84. B. C. Bunker, T. J. Headly, and S. C. Douglas, Mater. Res. Soc. Symp. Proc, **32**, 41–46 (1984).

85. R. D. Aines, H. C. Weed, and J. K. Bates, Mater. Res. Soc. Symp. Proc., **84**, 547–558 (1987).

86. F. M. Ernsberger, Glastechn. Ber., **56K**, 963–968 (1983).

87. R. H. Doremus, Y. Mehrotra, W.A. Lanford, and C. Burman, J. Mater. Sci., **18**, 612–622 (1983).

88. H. Scholze, Glas–Natur, Struktur und Eigenschaften, Springer-Verlag, Berlin, 1977.

89. D. R. Baer, L. R. Pederson, and G. L. McVay, J. Vac. Sci. Technol., **A2**(2), 738–743 (1984).

90. H. Scholze, J. Non-Cryst. Solids, **102**, 1–10 (1988).

91. R. Conradt, and H. Scholze, "Glass corrosion in Aqueous Media—A Still Unsolved Problem?" Riv. Stn. Sper. Vetro (Murano, Italy), **14**(5), 73–77 (1984).

92. C. A. Houser, J. S. Herman, I. S. T. Tsong, W. B. White, and W. A. Lanford, J. Non-Cryst. Solids, **41**(1), 89–98 (1980).

93. B. M. J. Smets and T. P. A. Lommen, Verres Refract., **35**(1), 84–90 (1981).

94. J. I. Mazer, J. K. Bates J. P. Bradley, C. R. Bradley, and C. M. Stevenson, Nature, **357**, 573–576 (1992).

95. I. Friedmann and W. D. Long, Science, **191**, 247–253 (1976).

96. J. C. Dran, J. C. Petit, L. Trotignon, A. Paccagnella, and G. Della Mea, Mater. Res. Soc. Symp. Proc, **127**, 25–32 (1989).

97. A. Abdelouas, J. L. Crovisier, L. Caurel, and E. Vernaz, C. R. Acad. Sci. Paris, **317**, série II, 1333–1340 (1993).

98. A. Abdelouas, J. L. Crovisier, W. Lutze, R. Miiller, and W. Bemotat, Eur. J. Mineral., **7**, 1101–1113 (1995).

99. H. Scholze, R. Conradt, H. Engelke, and H. Roggendorf, Mater. Res. Soc. Symp. Proc, **11**, 173–180 (1982).

100. R. D. Aines, H. C. Weed, and J. K. Bates, Mater. Res. Soc. Symp. Proc., **84**, 547–558 (1987).

101. T. J. Headly and B. C. Bunker, Am. Ceram. Soc Bull., **63**, 494 (1984).

102. S. Ricol, Ph.D. Thesis, Université P. and M. Curie, Paris, 1985.

103. S. Dobos, Acta Chim. (Budapest), **68**(4), 371–385 (1971).

104. B. Grambow, Mater. Res. Soc. Symp. Proc, **11**, 93–102 (1982).

105. L. N. Plummer, "Geochemical Modeling: A Comparison of Forward and Inverse Methods," in Proc. First Canadian/American Conference on Hydrogeology—Practical Applications of Ground Water Geochemistry, B. Hitchon and E. I. Wallick (Eds), Banff, Alberta, Canada, National Water Well Assoc., Worthington, Ohio, 1984, pp. 149–177.

106. G. Berger, J. Schott, and M. Loubet, Earth Planet. Sci. Lett., **84**(4), 431–445 (1987).

107. J. L. Crovisier, T. Advocat, J. C. Petit, and B. Fritz, Mater. Res. Soc. Symp. Proc., **127** (Sci. Basis Nucl. Waste Manag., **12**), 57–64 (1989).

108. J. L. Crovisier, H. Atassi, V. Daux, J. Honnorez, J. C: Petit, and J. P. Eberhart, Mater. Res. Soc. Symp. Proc., **127**, 41–48 (1989).

109. L. MiChaux, E. Mouche, J. C. Petit, and B. Fritz, Appl. Geochem., Suppl. **1** (Geochem. Radioact. Waste Disposal: Fr. Contrib.), 41–54 (1992).

110. B. Grambow, Adv. Ceram., **8** (Nucl. Waste Manag.), 474–481 (1984).

111. J. L. Crovisier, J. Honnorez, and J. P. Eberhard, Geochimm. Cosmochim. Acta, **51**, 2977–2990 (1987).

112. J. L. Crovisier, H. Attassi, V. Daux, and J. P. Eberhard, C. R. Acad. Sci. Paris, **310**, série II, 941–946 (1990).

113. H. Attassi, Evaluation de la résistance à la corrosion de quelques verres silicatés, Thèse Univ. Louis Pasteur, Strassburg, 1989.

114. J. H. Thomassín, F. Boutonnat, J. C. Touray, and P. Baillif, Eur. J. Mina. **1**, 261–274 (1989).

115. J. Claurel, Altération hydrothermale du verre R7T7. Cinétique de dissolution à 150 et à 250°C, role des phases néoformées, Thése University de Poitier, 1990.

116. J. L. Crovisier, E. Vemaz, J. L. Dussossoy, and J. Caurel, Appl. Clay. Sci., **7**, 47–57 (1992).

117. J. K. Bates, W. L. Ebert, X. Feng, and W. L. Bourcier, J. Nucl. Mater., **190**, 198–227 (1992).

118. J. Honnorez, "Generation of Phillipsites by Palagonitization of Basaltic Glass in Sea Water and the Origin of K-rich Deep-Sea Sediments," in Natural Zeolites: Occurrence, Properties, Use, L. B. Sand and F. A. Mumpton (Eds.), Pergamon, Oxford and New York, 1978, pp. 245–258.

119. J. K. Bates, L. J. Jardine, and M. J. Steindler, Science, **218**, 51 (1982).

120. M. Perez y Jorba, J. P. Dallas, C. Collongues, C. Bahezre, and J. C. Martin, Riv. Stn. Sper. Vetro (Murano, Italy), **14**(5), 121–130 (1984).

121. W. L. Ebert, J. K. Bates, and W. L. Bourcier, Waste Manag., **11**, 205–221 (1991).

122. D. J. Wronkiewicz, L. M. Wang, J. M. Bates, and B. S. Tani, Mater. Res. Soc. Symp. Proc, **294**, 183–206 (1993).

123. A. Abdelouas, J. L. Crovisier, W. Lutze, R. Müller, and W. Bernotat, Mater. Res. Soc. Symp. Proc., **333**, 513–518 (1994).

124. W. E. Dibble, Jr., and W. A. Tiller, Clays Clay Miner. **29**(5), 323–330 (1981).

125. F. Yanagisawa and H. Sakai, Appl. Geochem. **3**(2), 153–163 (1988).

126. G. Malow, Mater. Res. Soc. Symp. Proc, **11**, 25–36 (1982).

127. L. L. Hench and D. E. Clark, J. Non-Cryst. Solids, **28**, 83–105 (1978).

128. V. M. Oversby and D. L. Phinney, J. Nucl. Mater., **190**, 247–268 (1992).

129. B. Grambow and D. M. Strachan, Mater. Res. Soc. Proc, **26**, 623–634 (1984).

130. J. C. Sang, R. F. Jakubik, A. Barkatt, and E.E. Saad, J. Non-Cryst. Solids, **167**(1–2), 158–171 (1994).

131. A. Barkatt, E. E. Saad, R. B. Adiga, W. Sousanpour, Al. Barkatt, M. A. Adel-Hadai, J. A. O'Keefe, and S. Alterescu, Appl. Geochem., **4**, 593–603 (1989).

132. H. Atassi, J. L. Crovisier, A. Mosser, and D. Muster, C. R. Acad. Sci. Paris, **318**, série II, 935–939 (1994).

133. O. H. Andersson and Kaj H. Karlsson, Mater. Sci. Monogr., **69** (Ceram. Substitutive Reconstr. Surg.), 303–311 (1991).

134. D. M. Strachan, Nucl. Chem Waste Manag., **4**, 177–188 (1983).

135. B. Grambow, in Corrosion of Glasses, Ceramics and Ceramic Superconductors, D. E. Clark und B. K. Zoitos (Eds.), Noyes Publ., Park Ridge, NJ, 1992, pp. 124–152.

136. T. A. Michalske, B. C. Bunker, and K. D. Keefer, J. Non-Cryst. Solids, **120**, 126–137 (1990).

137. T. Advocat, J. L. Crovisier, and E. Vemaz, "Corrosion du verre nucléaire R7T7 à 90°C: Passage d'une dissolution séjectiv à congruente par elevation du pH, C. R. Académie des Sciences, Paris, t. 313, Série II, pp. 407–412, 1991, Géochimie/ Geochemistry.

138. M. Kinoshita, M. Harada, Y. Saw, and Y. Hariguchi, J. Am. Ceram. Soc, **74**(4), 783–787 (1991).

139. F Risacher and B. Fritz, Geochim. Cosmochim. Acta, **55**, 687–705 (1991).

140. B. Grambow, A. Loida, L. Kahl, and W. Lutze, Mater. Res. Soc. Symp. Proc., **353** (Sci. Basis Nucl. Waste Manag. **XVIII**, Pt. 1), 39–46 (1995).

141. X. Feng and I. L. Pegg, Phys. Chem. Glasses, **35**(2), 98–103 (1994).

142. J. C. Tait and C. D. Jensen, J. Non-Cryst. Solids, **49**(1–3), 363–377 (1982).

143. N. K. Mitra, M. Banerjee, and S. Pan, J. Inst. Chem. (India), **40** (Pt. 5), 179 (1968).

144. H. V. Walters and P. B. Adams, J. Non-Cryst. Solids, **19**, 183–199 (1975).

145. Y. Godron, "A Systematic Bibliography on the Weathering of Glass Used in Building. Part 4 and Conclusioin," Verres Refract., **30**(5), 635–650 (1976).

146. R. Wuhrer amd A. Ray, Key Eng. Mater., **53–55** (Austceram '90), 372–376, (1991).

147. B. Arman and B. Kuban, Mater. Charact., **29**(1), 49–53 (1992).

148. W. L. Ebert, J. K. Bates, and W. L. Bourcier, Waste Manag., **11**, 205–221 (1991).

149. Y. Tao, X. Bo, and C. Wang, Glass Technol., **30**(6), 224–227 (1989).

150. G. A. C. M. Spierings, J. Mater. Sci., **28**(23), 6261–6273 (1993).

151. S. T. Tso and J. A. Pask, J. Am. Ceram. Soc., **65**(7), 360–362 (1982).

152. G. A. C. M. Spierings and J. Van Dijk, J. Mater. Sci., **22**(5), 1869–1874 (1987)

153. G. A. C. M. Spierings, J. Mater. Sci., **26**(12), 3329–3336 (1991).

154. T. H. Elmer, J. Am. Ceram. Soc, **68**(10), C-273–C-274 (1985).

155. P. Savostin, Z. Pflanzenernaehr. Bodenk., **132**(1), 37–45 (1972).

156. M. Prod'homme "Actipn des micro-organismes sur les surfaces vitreuses," VVII Congress International du Verre, Bruxelles, 1965.

157. G. Theden and V. Kerner-Gang, Glastechn. Ber., **4**, 200–295 (1964).

158. W. E. Krummbein and K. Jens, "Mikrobielle Zerstörungen an mittelalterlichen Kirchenfenstern und antiken Gläsern," DECHEMA-Tagung, 1983, pp. 243–244.

159. R. Drewello, M. Nuessler, and R. Weissmann, Werkst. Korros., **45**(2), 122–124 (1994).

160. C. R. Das, J. Am. Ceram. Soc, **63**(3–4), 160–165 (1980).

161. B. M. J. Smets, M. G. W. Tholen, and T. P. A. Lommen, J. Non-Cryst. Solids, **65**, 319–332 (1984).

162. M. A. Rana and R. W. Douglas, Phys. Chem. Glasses, **2**, 179–195 (1961).

163. T. A. Wassick, R. H. Doremus, W. A. Lanford, and C. Burman, J. Non-Cryst. Solids, **54**(1–2), 139–151 (1983).

164. H. J. Franek, G. H. Frischat, and H. Knoedler, Glastech. Ber., **56**(6–7), 165–175 (1983).

165. P. March and F. Rauch, Glastech. Ber., **63**(6), 154–162 (1990).

166. T. Boehm, J. Leissner, and J. A. Chudek, Glass Sci. Technol. (Frankfurt/Main), **68**(12), 400–403 (1995).

167. Z. Wu and H. Bu, J. Wuhan Univ. Technol., Mater. Sci. Ed., **10** (2), 1–10 (1995).

168. P. Duffer, Am. Ceram. Soc. Bull., **73**(10), 80–83 (1994).

169. H. De Waal, Verres Refract., **35**(2), 303–305 (1981).

170. K. Morita, A. Suganuma, and A. Makishima, J. Mater. Sci., **29** (24), 6587–6591 (1994).

171. M. Hueppauff and B. Lengeler, J. Appl. Phys., **75**(2), 785–791 (1994).

172. V. Gottardi, F. Nicoletti, G. Battaglin, G. Delia Mea, and P. Mazzoldi, Verres Refract., **35**(2), 298–301 (1981).

173. J. E. Shelby and J. Vitko, Jr., J. Non-Cryst. Solids, **38–39**(2), 631–636 (1980).

174. M. Gao, Z. Zhang, L. Li, X. Li, and K. Takahashi, J. Non-Cryst. Solids, **80**(1–3), 319–323 (1986).

175. B. M. J. Smets and T. P. A. Lommen, Phys. Chem. Glasses, **23** 83–87 (1982).

176. J. Tait and D. L. Mandelosi, Limited Report AECL-7803, Atomic Energy of Canada, Pinawa, Canada, 1983.

177. L. A. Chick, G. F. Piepel, G. B. Mellinger, R. P. May, W. J. Gray, and C. Q. Buckwalter, PNL-3188, Pathific Northwest National Laboratory, Richland, WA, 1981.

178. D. Ge, Z. Han, Y. Yan, H. Chen, Z. Lou, X. Xu, R. Han, and L. Yang, J. Non-Cryst. Solids, **80**(1–3), 341–350 (1986).

179. I. O. Isard and W. Müller, Phys. Chem. Glasses, **27**, 55–58 (1986).

180. P. Van Iseghem, W. Timmermans, and R. de Batist, Mater Res. Soc Symp. Proc, **26**, 527–534 (1984).

181. B. Grambow and D. M. Strachan, Mater. Res. Soc. Symp. Proc., **112**, 713–724 (1988).

182. P. Vanlseghem and B. Grambow, Mater. Res. Soc. Symp. Proc., **112**, 631–639 (1988).

183. G. Della Mea, A. Gasparotto, M. Bettinelli, A. Montenero, and R. Scaglioni, J. Non-Cryst. Solids, **84**(1–3), 443–451 (1986).

184. W. J. Dell, P. J. Bray, and S. Z. Xiao, J. Non-Cryst. Solids, **58**, 1 (1983).

185. R. Ota, T. Yasuda, and J. Fukunaga, J. Non-Cryst. Solids, **116**, 46–56 (1990).

186. B. C. Bunker, D. R. Tallant, R. J. Kirkpatrick, and G. L. Turner, Phys. Chem. Glasses, **31**(1), 30 (1990).

187. H. P. Hood and M. E. Norberg, U.S. Patent 2,106,744, 1938.

188. A. Makishima, J. D. Mackenzie, and J. J. Hammel, J. Non-Cryst. Solids, **31**(3), 377–383 (1979).

189. C. Zhang and R. Ye, J. Non-Cryst. Solids, **112**(1–3), 244–250 (1988).

190. R. H. Doremus and A. M. Turkalo, Science, **164**, 418–419 (1969).

191. G. Perera and R. H. Doremus, J. Am. Ceram. Soc., **74**(7), 1554–1558 (1991).

192. S. Yamanaka, J. Akagi, and M. Hattori, J. Non-Cryst. Solids, **70**(2), 279–290 (1985).

193. X. Feng, I. L. Pegg, A. A. Barkatt, P. B. Barkatt, P. B. Macedo, S. J. Cucinell, and S. Lai, Nucl. Technol., **85**, 334–345 (1989).

194. W. Lutze, "Silicate Glasses," in Radioactive Waste Forms for the Future, W. Lutze and R. C. Ewing (Eds.), Elsevier Science, Amsterdam, 1988.

195. L. A. Chick and L. R. Pederson, Mater. Res. Soc. Symp. Proc, **26**, 635–642 (1984).

196. E. Y. Vemaz and J-L. Dussossoy, Appl. Geochem., Suppl., Issue No. 1, 13–22 (1992).

197. B. Grambow, "Influence of Saturation on the Leaching of Borosilicate Nuclear Waste Glasses," Proc. 13th Int. Congr. Glass, Hamburg, July 4–9, 1983, Glastechnische Berichte Vol. 56K, Deutschen Glastechnischen Gesellschaft, Frankfurt 1983.

198. B. Grambow, "Geochemical Approach to Glass Dissolutioni," in Corrosion of Glass, Ceramics, and Ceramic Semiconductors, D. E. Clark and B. K. Zoitos (Eds.), Noyes Publications, Park Ridge, NJ, 1992.

199. T. Advocat, J. L. Crovisier, B. Fritz, and E. Vernaz, "Thermokinetic Model of Borosilicate Glass Dissolution: Contextual Affinity," Mater. Res. Soc. Symp. Proc., **176**, 241–248 1990.

200. W. L. Bourcier, D. W. Peiffer, K. G. Knauss, K. D. McKeegan, and D. K. Smith, "A Kinetic Model for Borosilicate Glass Dissolution Based on the Dissolution Affinity of a Surface Alteration Layer," Mater. Res. Soc. Symp. Proc., **176**, 209–216 (1990).

201. T. Advocat, J. L. Crovisier, B. Fritz, and E. Vernaz, Mater. Soc. Symp. Proc., **176** (Sci. Basis Nucl. Waste Manag., **13**), 241–248 (1990).

202. B. Grambow, W. Lutze, and R. Müller, Mat. Res. Soc. Symp. Proc, **257**, 143–150.

203. B. Luckscheiter and B. Grambow, in Symposium Scientific Basis for Nuclear Waste Management XXI, September 28-October 3, 1997, Davos, Switzerland, Materials Research Society Symposium Proceedings, Vol. 506, I. G. McKinley (Ed.), Materials Research Society, Warrendale, PA, 1998, pp. 925–926.

204. X. Feng, J. K. Bates, C. R. Bradley and E.C. Buck, Mater. Res. Soc Symp. Proc., **294** (Sci. Basis Nucl. Waste Manag. **XVI**), 207–214 (1993).

205. B. Grambow, A. Loida, L. Kahl, and W. Lutze, "Behavior of Np, Pu, Am, and Tc Upon. Glass Corrosion in a Concentrated $Mg(Ca) Cl_2$ Solution," Mater. Res. Soc. Proc, **353** (Sci. Basis Nucl. Waste Manag. **XVIII,** Pt. 1), 39–46, 1995.

206. JSS-1988, JSS-Project Phase V: Final Report, "Testing and Modelling of the Corrosion of Simulated Nuclear Waste Glass Powders in a Waste Package Environment," JSS Project Technical Report 88-02, Stockholm, Sweder, 1988, pp. 86–87.

207. V. T. Yilmaz, J. Non-Cryst. Solids, **151**(3), 236–44 (1992).

208. N. Koshizaki, H. Takayanagi, and K. Kemmochi, J. Non-Cryst. Solids, **95–96** (Pt. 2), 1111–1118 (1987).

209. P. Simurka, M. Liska, A. Plsko, and K. Forkel, Glass Technol., **33**(4), 130–135 (1992).

210. G. Scarinci, D. Festa, G. D. Soraru, B. Locardi, E. Guadagnino, and S. Meriani, J. Non-Cryst. Solids, **80**(1–3), 351–359 (1986).

211. B. C. Bunker, G. W. Arnold, and J. A. Wilder, J. Non-Cryst. Solids, **64**(3), 291–316 (1984).

212. B. C. Sales and L. A. Boatner, "Lead-Iron Phosphate Glassv," in Radioactive Waste Forms for the Future, W. Lutze and R. C. Ewing (Eds.), Elsevier Science, Amsterdam, 1988.

213. E. L. Rodriguez, J. Mater. Sci. Lett., **6**(6), 718–720 (1987).

214. A. Larena, J. Martinez Urreaga, and M. U. De la Orden, Mater. Lett., **12**(6), 415–418 (1992).

215. D. Gurunathan, Chem. Age India, **30**(6), 513–518 (1979).

216. G. Gliemeroth and A. Peters, J. Non-Cryst. Solids, **38–39**(2), 625–630 (1980).

217. S. Wood and J. R. Blachere, J. Am. Ceram. Soc, **61**(7–8), 287–292 (1978).

218. S. Wood and J. R. Blachere, J. Am. Ceram. Soc, **61**(7–8), 292–294 (1978).

219. D. Sprenger, H. Bach, W. Meisel, and P. Guetlich, Surf. Interf. Anal., **20**(9), 796–802 (1993).

220. H. Dunken, J. Non-Cryst. Solids, **129**(1–3), 64–75 (1991).

221. J. H. Graziano, V. Slavkovic, and C. Blum, "Lead Crystal: An Important Potential Source of Lead Exposure," Chem. Speciation, **3**(3–4), 81–85 (1991).

222. K. G. Bergner and G. Braun, "Studies of the Chrormum Contents of Wines. Part 4: Effect of Stainless steel Tanks and Bottles," Mitt. Klosterneuburg, **34**(2), 81–88 (1984).

223. G. Kehry and K. G. Bergner, "Studies on the Nickel Content in Wine," Mitt. Klosterneuburg, **35**(1), 7–15 (1985).

224. L. L. Hench, R. J. Splinter, W. C. Allen, and T. C. Greenlee, J. Biomed. Res. Symp., **2**, 117–141 (1971).

225. H-M. Kim, F. Miyaji, T. Kokubo, C. Ohtsuki, and T. Nakamura, J. Am. Ceram. Soc., **78**(9), 2405–2411 (1995).

226. H. Scholze, "Statement on Chemical Stability of Silicate Glass Fiber" WaBoLu-Hefte (Institut für Wasser-, Boden-, und Lufthygiene, Berlin), (7), 130–133 (1991).

227. R. M. Potter and S. M. Mattson, Glastech. Ber., **64**(1), 16–28 (1991).

228. R. Spaude, Mater. Tech. (Duebendorf, Switz.), **12**(4), 119–126 (1984).

229. H. Tiesler, Glastech. Ber., **54**(12), 369–381 (1981).

230. M. C. Bautista, J. Rubio, and J. L. Oteo, Stud. Surf. Sci. Catal., **87** (C), 449–455 (1994).

231. M. E. Nordberg, "Methods for Making Fibrous Glass Atricles," U.S. Patent No. 2,46,841, Feb. 15, 1949.

232. R. Maddison and P. W. McMillan, Glass Technol., **21**(6) **297–301** (1980).

233. K. Ooi and M. Miyatake, J. Coll. Interf. Sci., **148**(2), 303–309 (1992).

234. B. D. Caddock, K. E. Evans, and I. G. Masters, J. Mater. Sci., **24**(11), 4100–4105 (1989).

235. T. H. Elmer, J. Am. Ceram. Soc., **67**(12), 778–782 (1984).

236. A. K. Bledzki, G. W. Ehrenstein, and A. Schiemann, Kunststoffe, **79**(5), 416–425 (1989).

237. A. Al Cheikh and M. Murat, Cem. Concr. Res., **18**(6), 943–950 (1988).

238. A. Soufiane and M. Poulain, J. Non-Cryst. Solids, **140**(1–3), 63–68 (1992).

239. C. J. Simmons, J. Am. Ceram. Soc, **70**(4), 295–300 (1987).

240. C. J. Simmons and H. Joseph, J. Am. Ceram. Soc, **69**(9), 661–669 (1986).

241. C. J. Simmons, J. Am. Ceram. Soc, **70**(9), **654–661** (1987).

242. P. Tick and S. Mitachi, J. Am. Ceram. Soc., **74**(3), 481–490 (1991).

243. M. Le Toullec, C. J. Simmons, and J. H. Simmons, J. Am. Ceram. Soc., **71**(4), 219–224 (1988).

244. G. H. Frischat and I. Overbeck, J. Am. Ceram. Soc, **67**(11), C238–C239 (1984).

245. D. Ravaine and G. Perera, J. Am. Ceram. Soc, **69**(12), 852–857 (1986).

246. X. Zhao and S. Sakka, J. Mater. Sci., **28**(6), 1622–1630 (1993).

247. Y. Daï, K. Takahashi, and I. Yamaguchi, J. Am. Ceram. Soc, **78**(1), 183–187 (1995).

248. E- Krodl-Kraft, Ostrreichische Zeitschrift für Kunst und Denkmalpflege, **27**, 55–65 (1973).

249. A. Matsuda, Y. Matsuno, S. Katayama, and T. Tsuno, J. Mater. Sci. Lett., **8**(8), 902–904 (1989).

250. G. Carturan, G. Delia Mea, A. Paccagnella, G. D. Soraru, and C. Rizzo, J. Non-Cryst. Solids, **111**(1), 91–97 (1989).

251. M. Hartmann, G. H. Frischat, K. Hoegerl, and G. F. West, J. Non-Cryst. Solids, **184**, 209–212 (1995).

30

MICROBIOLOGICAL DEGRADATION OF POLYMERIC MATERIALS

J.-D. Gu

School of Biological Sciences, The University of Hong Kong, Hong Kong, China

T. E. Ford

University of New England, Biddeford, Maine

D. B. Mitton

Gold Standard Corrosion Science Group, LLC, Boston, Massachusetts

R. Mitchell

Harvard School of Engineering and Applied Sciences, Harvard University, Cambridge, Massachusetts

A. INTRODUCTION

Polymeric materials are used in a wide array of applications in engineering and construction, ranging from basic infrastructure to critical components of aviation and space travel, because of their flexibility, lightness, and high strength. The available literature on deterioration and degradation of organic polymeric materials under natural and/or artificial environments is very limited [1–8], and much less is known about the mechanisms involved in degradation of these materials by microorganisms [4, 6, 7]. Such information is important when assessing material life span and maintenance cost, but available results are widely distributed in different scientific journals due to the highly diverse applications of polymeric materials compared to metals. Because of their versatility in chemical compositions, molecular weights, and applications, issues of polymer deterioration and prevention measures begin to receive more attention as their applications continue to expand.

Polymeric materials differ in their chemical composition, physical forms, processing and molding, mechanical properties, and applications. Variations in their chemical composition and molecular weight of the polymeric monomers are largely due to the versatility of the carbon–hydrogen (–C–C–) bond in the chemical structure and their possible configurations, stereochemistry, and orientation. Their applications include product packaging, insulation, structural components, protective coatings, medical implants, drug delivery carriers, slow-release capsules, electronic insulation, telecommunication, aviation and space industries, sporting and recreational equipment, and building consolidant. In many of these applications, over time of service, they are constantly exposed to a range of natural and artificial conditions that often involve microbial colonization and contamination, resulting in aging, disintegration, discolorization or colorization, and deterioration as well as degradation [9–12].

Evidence of polymeric material degradation by microorganisms was reported in a number of diverse environments, including the Russian Mir space station [6, 8, 13–16], shipping [17], aircraft [18], and sewer pipes [19–21]. In addition, medical devices are also susceptible to microbial contamination [22] and deterioration in vivo [23, 24]. Other materials, including electronic insulation polyimides [8, 25–31], fiber-reinforced polymeric composites [4, 6, 32–36], and corrosion protective coatings [27–29, 36, 37], were also reported for their biodeterioration by microorganisms commonly found in the environment. Degradation and mineralization of polymers used in product packaging were investigated under both aerobic and anaerobic conditions with reproducible methodology [6, 8, 26, 29, 38–46].

Detection of microbial deterioration at the early stage of development is a challenge because of the complex interactions between the materials and their environment, the microbial populations, and the available techniques. Since engineering materials are designed to be highly resistant to change, any physical and chemical change in the material and matrices cannot be observed using traditional techniques because of low sensitivity [6, 8, 29, 46]. Electrochemical impedance spectroscopy (EIS) coupled with microbiological methods enables quantitative and mechanistic analyses of microbial degradation of high-strength polymeric materials within a reasonable period of time [7, 8, 25–31, 33–35, 47]. This technique is particularly useful for detecting the physical failure of polymeric electronic insulating thin films and surface coatings that are water impermeable and may degrade slowly by environmental conditions and/or microorganisms [27, 28, 33, 37]. Furthermore, fiber-reinforced polymeric composite materials (FRPCs) were also successfully assessed using this approach successfully, but the time required to observe any detectable changes in the materials due to microbial growth was at least several months [6, 27–29, 33–35]. With the increasing usage of this class of materials in infrastructure and aviation and space, deterioration, detection, and preventive measures are critical issues that have drawn the attention of end users worldwide.

This chapter is intended to cover selective groups of materials with different properties and applications to illustrate current knowledge and where information and research efforts should be directed. It is also our intention to describe the general mechanisms of degradation and deterioration using selected polymers, the current status of material degradation research, and the strengths and weaknesses of research techniques currently in use. In addition, preventive measures such as incorporation of biocides and implementation of environmental controls are also discussed. Because this chapter deals with biological deterioration of polymers, the wide range of materials is arbitrarily grouped based on their biodegradability for convenience of discussion as (1) relatively degradable, (2) recalcitrant, and (3) completely resistant to breakdown. Our definition of polymeric materials is limited to synthetic industrial polymers, excluding natural polymers, for example, cellulose, chitin, chitosan, lignin, polysaccharides, proteins, and deoxyribonucleic acids.

B. APPLICATIONS OF POLYMERIC MATERIALS

The production and consumption of plastics has increased dramatically worldwide since the 1970s and does not appear to be slowing down. On a volume basis, polymeric materials had already exceeded that of copper, aluminum, and steel by the mid-1980s [48] and they play a major role in our daily life. In 1996, production of polymers reached 3.4 million tons of thermosetting resins, 27.6 million tons of thermoplastic resins, 4.5 million tons of fibers, 2.3 million tons of synthetic rubber, and 4.7 billion liters of paints and coatings [49].

Uses of polymers have constantly expanded to new frontiers, including transportation, aerospace, aviation, medical, infrastructure, electronic, computers, military, and recreation. In addition, demands are also visible in the communication, structural and construction, and protective coatings industries. Polymeric materials may be exposed to natural environments, including soils, seawater, and freshwater under aerobic and strictly anaerobic conditions. In certain cases, artificial or enclosed environments with high acidity or alkalinity (e.g., specialized containment) are encountered in industrial processing, manufacturing, and storage and even within human tissues as implant materials. In most of these incidences, contamination by microorganisms on surfaces of materials is common and, indeed, inevitable, and as a result, susceptibility assessment to microorganisms is not only justified but necessary to provide a better understanding of material performance.

Contact between microorganisms and surfaces of materials is generally unavoidable except under extremely clean conditions where sterility is strictly maintained and reinforced through vigorous filtering and ultraviolet treatment. Information on the extent of microbial biofilm formation on surfaces of materials and subsequent damage to polymers becomes a very important issue in telecommunications, information storage, medicine, aviation, and space travel. This information is invaluable to polymer chemists, engineers, materials scientists, and applied biologists for designing materials with predetermined life spans, stability, and performance under deployed conditions.

C. MICROBIAL BIOFILMS ON POLYMERS

In submerged or humid environments, all surfaces, including metals and polymers, are susceptible to colonization by microorganisms even under nutrient-starved conditions (oligotrophic). A physical layer of microbial cells and their exopolymeric materials may develop on any exposed surfaces when nutrients become readily available for microbial growth. Both sessile microorganisms on surfaces and their microbial exopolymeric materials are collectively called microbial biofilms [7, 50–59]. Biofilms are interfacial communities of microorganisms occurring between the substratum surfaces of materials and the bulk solution, and they may have a very strong influence on the property of the substratum materials, including integrity.

Microbial biofilms are complex communities of microorganisms embedded in a matrix of their exopolymers under natural conditions. When an inert surface of material is introduced into an aqueous environment, the substratum is fouled sequentially with soluble organics and then microorganisms [55, 60–63]. The growth and multiplication of these attached cells increase the thickness of the biofilm and the heterogeneity of the biofilm in terms of O_2 distribution and nutrient profiles. Within the heterogeneous layer of bacteria and their products, both aerobic and anaerobic processes can take place simultaneously over space and time in the highly porous biofilm [8]. Aerobic processes facilitate consumption of more oxygen, which is extremely toxic to the activity of anaerobic microorganisms, and anaerobes benefit from the lowering of oxygen tension by aerobes. Even in nonaqueous conditions, biofilms can form on surfaces [28, 37, 59, 64].

Formation of bacterial biofilms on surfaces provides a strategy for microorganisms to survive environmental stress and predation and to grow, providing an opportunity for invasion and destruction of the colonized materials. It has been widely reported that growth of microorganisms on surfaces may result in corrosion of the underlying metals [65–75] or, in the case of humans, tissue may be infected by pathogens [76]. This detrimental effect occurs particularly in the presence of an active population of sulfate-reducing bacteria in anaerobic corrosion [74, 77–82], during degradation of polymeric materials in contact with soils [38, 39, 42, 44, 45], and in human tissue in contact with polymeric materials, for example, urinary tract catheters [22, 29].

D. DEGRADATION PROCESSES

Polymeric materials may be susceptible to deterioration through physical stress and chemical effects. Microorganisms are ubiquitous and are constantly associated with material surfaces under aerobic and anaerobic conditions depending on the specific application. Some environments can be extreme, such as microgravity in space missions and high levels of irradiation in nuclear power plants or waste depositories. Due to the wide array of environments, simulation studies are often carried out to assess the potential degradability of polymers under either aerobic [26, 33–35, 38, 39, 83, 84] or anaerobic [38, 39, 41, 43, 85] conditions. Other studies emphasize marine environments [86–89], natural soils [90, 91], and river water [86, 92, 93]. In addition, enzymatic assays have been used to assess degradability of a specific polymer under laboratory conditions [86, 94–96] or to detect specific enzyme activity indicative of a degradative process, for example, esterases [12, 97–112]. Experiments in situ always suffer from poor reproducibility because of the broad range of variables involved [83, 84] and the difficulty in interpretation of experimental results and extrapolation of possible mechanisms involved. In contrast, laboratory-simulated conditions with high reproducibility are often used for evaluation of the fate of polymeric materials and investigation of the degradation mechanisms involved [4, 6, 38, 39–45, 113].

Assessment of polymer biodegradability in anaerobic simulation environments can be conducted in laboratory bioreactors. In such systems, the composition of the components and the environmental conditions need to be defined specifically beforehand. The anaerobic Hungate technique [114] can be applied using serum bottles or Bulch anaerobic tubes [115–118]. Degradation of polymer can be determined with either processed thin films [41–43] or powder [119, 120]. However, it should be borne in mind that unprocessed powder does not have the same crystallinity as processed polymers; therefore information about the degradability of a polymer should be interpreted with caution. In addition, aging of the film and the conditions leading to aging also affect the crystallinity of the polymer and then the rate of degradation. A lack of awareness about the physical nature of the polymer results in misinterpretation of polymer degradability in natural environments.

D1. Aerobic Processes

Oxygen is the most favorable electron acceptor in metabolism for a large population of microorganisms. Aerobic conditions are often used to assess polymer biodegradation in many tests; examples are aerobic soil exposure [121, 122], thermophilic composting [42, 44, 123], immersion in surface seawater [88, 89, 122] and freshwater [86, 93], and laboratory batch conditions [92, 95, 96, 124–126]. In addition, the enclosed living environment not only harbors a range of microorganisms; the microorganisms are detrimental to both human health and material integrity [6, 127–138]. Each of these experimental conditions has its weaknesses and strengths and will be discussed later.

Under aerobic conditions, microorganisms capable of utilization of molecular oxygen as an electron acceptor are the dominant species. During degradation, physical disintegration of polymers may subsequently accelerate biological degradation. Mechanisms of polymer degradation by microorganisms usually involve enzymatic attack as the main process. For example, one set of extruded film strips of poly(β-hydroxybutyrate) (PHB) with a molecular weight of 1.46×10^5 (M_w) and a racemic diad of 0.62 were aged at room temperature for two months and buried in flasks containing moist soils incubated at 35°C in the dark. In parallel, a control was prepared by adding antibiotics (streptomycin, nystin) to the moist soil to inhibit microbial growth. During incubation, weight losses of 34 and 66% were observed in the biologically active flasks after 60 and 95 days of incubation, respectively. In contrast, weight loss of the abiotic control was less than 8% during the same period. Dense microbial biofilms were observed on surfaces of the exposed polymer strips, and surface erosion was evident. It is probable that the polymers' physical structure was degraded by exoenzymes excreted by the colonizing microorganisms. *Pseudomonas lemoignei* and *Alcaligenes faecalis* are capable of degrading PHBs through activity of PHB depolymerases [139, 140]. Other polymers, for example, PHB, a copolymer of poly(hydroxybutyrate-*co*-hydroxyvalerate) and poly(d, l-lactide), were also assayed [40, 41].

Thermophilic composting [38–45, 141] and respirometric biometry [39, 43, 46] have been used extensively in quantitatively assessing the fate of polymeric materials. Historically, weight loss was the primary parameter monitored after exposure [32, 40, 41]. For example, in a study of cellulose acetate (CA) degradability in thermophilic composting tests, CA films with an acetyl substitution value of 1.7 were degraded faster than CA with a value of 2.5 [40–45]. It took approximately 14 days for CA with a degree of substitution (DS) of 1.7 to degrade completely, while it took 25 days for DS 2.5 to decompose [42–44]. Biodegradation of CA was confirmed by determining changes in polymer molecular weights before and after the test period as an important indicator of degradation [43, 46]. When necessary, further microbiology work can be conducted to investigate the properties of the microorganism in details. *Pseudomonas paucimobilis* was isolated from CA film surfaces after exposure to aerobic thermophilic microorganisms from compost for two weeks and was found to degrade CA with a DS 1.7 and 2.5 at similar rates [38, 39]. The limitation of this technique is that fragmentation of polymer films may result in nonrecoverable small debris, which may be misinterpreted as degradation, thereby exaggerating the degradation rate.

The respirometry technique is based on quantitative monitoring of O_2 consumed and/or CO_2 produced and can also be used for understanding polymer degradation and more specifically mineralization. In such a test, one set of the test system without polymer amendment serves as a control [40] while another identical set will contain various amounts of polymer powder. In this way, mineralization of polymer can be determined by quantitatively measuring the carbon balance over time without the employment of radiolabeled chemicals. As much as 64% of the substrate polymer carbon was liberated and recovered as CO_2 in NaOH traps during the course of CA degradation under aerobic conditions [38, 39, 42, 46]. It should be borne in mind that complete mineralization of the polymer is not achieved because a fraction of the substrate carbon is always immobilized in microbial biomass and humification materials.

D2. Anaerobic Processes

Simulated anaerobic conditions used to measure polymer degradability include anaerobic soils, landfills [142–146], sewage sludge [147], and laboratory methanogenic or sulfate-reducing environments. These simulations have been widely used in metal corrosion studies involving the sulfate-reducing bacteria (SRB) [74, 75, 78, 79, 81, 82]. Most of the research efforts have been focused on understanding the susceptibility of materials to microorganisms and the mechanisms involved. Polymer biodegradation can be carried out under strictly anaerobic conditions in the laboratory with high reproducibility, but the incubation time is anticipated to be long [38, 39, 40–46]. Other studies were carried out under either natural conditions [148] or semi-natural environments, such as a municipal landfill [144, 148]. The choice of the systems for testing and study is mostly dictated by the objectives and questions to be answered.

Only a very limited number of studies have been conducted on microbial degradation of polymeric materials under strictly anaerobic conditions [39, 41, 119]. Some have also been carried out using natural microbial communities. The powder form of PHB was used in a sludge-amended incubation experiment [119]. Reconstituted microbial consortia containing SRB were used in a study of polymeric composite degradation [149], but this approach may fail to reflect the natural microflora and the relationship between microbial colonization and the possibility of damage to the underlying materials. The mechanisms of SRB corrosion of metals are due largely to the presence of hydrogenase systems causing depolarization of adsorbed hydrogen on metal surfaces and changes to the open-circuit potential [59, 64, 113]. However, the inclusion of a SRB as the major microbial group may not be justified for polymeric materials. As stated earlier, polymer degradation occurs through reactions carried out by depolymerases from microorganisms [139, 150].

E. MECHANISMS OF DEGRADATION

Microbial biofilms are ubiquitous on material surfaces, including polymers. Earlier reports tended to use biofouling as an indication for degradation [151], which should be more carefully stated. Quantitative methods have been introduced coupled with scanning electron microscopic observations to substantiate the extent of deterioration by microorganisms [26–28, 33–35, 37], but elucidation of the mechanisms involved remains a major challenge due to the complexity of the polymeric materials and their chemical constituents. Formation of microbial biofilms on surfaces is the first step and also a prerequisite for subsequent degradation to occur, but it cannot be used as the sole indicator alone. Occasionally, exoenzymes from nonsessile microorganisms may contribute to degradation of polymers, but the extent of such degradation is considerably lower than the decomposition by surface colonizers. When bacteria are attached to a substratum surface, they become protected from predation and may become metabolically more active [61, 152]. They form patches of biofilms on material surfaces. Within biofilms, several groups of microorganisms, including aerobic, facultative, and anaerobic, can coexist in synergistic associations, also called consortia, effectively attacking a wide range of substratum materials [36, 153–158].

Different polymeric materials are degraded to different extents. Some materials are susceptible to mineralization, while in others, specific components are utilized. It is interesting to know that slow-growing bacteria have much high degradation potential than the fast-growing ones [11]. This information is both interesting and important because current methods are biased toward the use of short-term tests and fast-growing microorganisms.

E1. Utilization of Polymers as a Source of Carbon and Energy

Some polymers can be completely utilized as a source of carbon and energy, while others are only partially degraded by microorganisms. Examples of the former include the natural and microbially synthesized materials—poly(hydroxyalkanoates) (PHAs) [121, 150, 159, 160], γ-poly(glutamic acid) [161], cellulose acetates [7, 29, 38, 39, 40–45], polyethers [162], polylactide [40, 41], and polyurethanes [28, 37, 163–168]. A general rule is that biologically synthesized polymers are readily biodegradable in natural environments, while synthetic polymers are either less biodegradable or degrade very slowly, depending on their chemical composition, structural complexity, and molecular weights. However, the rate of degradation is in large part determined by the chemical structure, for example, molecular weights, structure, and configuration of the monomer units. Apart from the intrinsic properties, the environmental conditions, aerobic versus anaerobic, also play a significant role.

Understanding the relationship between the chemical structure of a polymer and its biodegradability is very important for designing a new generation of environmentally degradable polymers. As a general rule, high-molecular-weight polymers are less biodegradable or degrade at a slower rate than those with low molecular weights. Heterogeneity of the monomers also affects degradation rate; the hydrolytic chain scission is dependent on the copolymer composition: poly(3-hydroxybutyrate-co-27% 4-hydroxybutyrate) [P(3HB-co-27% 4HB)] > [P(3HB-co-17% 4HB)] > [P(3HB-co-10% 4HB)] > poly(3-hydroxybutyrate-co-45% 3-hydroxybutyrate [P(3HB-co-45% 3HV)] > [P(3HB-co-71% 3HV)] [121]. Similarly, the sequence of enzymatic hydrolysis follows [P(HB-co-16% HV)] > [P(HB-co-32% HV)] > PHB [169]. The crystallinity of the polymer also affects the rate of degradation but is rarely taken into account [119].

Structural substitution groups, and their number per repeating unit, also affect the degradation kinetics. For example, (CAs) with a lower degree of substitution (DS) values are more quickly degraded than higher substituted ones under both aerobic and anaerobic conditions [38–45, 120]. During degradation of CAs, both molecular weight and degree of substitution showed a decreasing trend, suggesting that deacetylation and decomposition of the polymer backbone proceed simultaneously [44]. Earlier data suggested that CAs with DS values greater than 0.82 are recalcitrant to biodegradation and the limiting step is deacetylation followed by breaking of the polymer carbon–carbon bonds [100].

The general theory of polymer degradation assumes that exoenzymes from microorganisms break complex polymers to yield short units, including oligomers, dimers, and monomers, small enough to be allowed to pass through permeable outer bacterial membranes and subsequently to be assimilated as carbon and energy sources. The process is called depolymerization. The closer a polymer structure is to a natural analog, the easier it is degraded. Cellulose, chitin, pullusan, and PHB are all biologically synthesized and are completely and rapidly biodegradable [170]. The complete decomposition of a polymer to CO_2 and H_2O under aerobic conditions or organic acids, CO_2, and CH_4 under anaerobic conditions is rare. Degradation and mineralization of a polymer substrate can hardly achieve 100% due to the synthesis of microbial biomass and humus matter [171].

Under different conditions, predominant groups of microorganisms differ and degradation pathways vary. In the presence of oxygen, aerobic microorganisms are responsible for destruction of complex materials, with microbial biomass, CO_2, and H_2O as the final products (Fig. 30.1). In contrast, under anoxic conditions, anaerobic consortia of microorganisms are involved in polymer deterioration, and the primary products will be microbial biomass, CO_2, CH_4, and H_2O [142–144] (Fig. 30.1). These conditions are widely

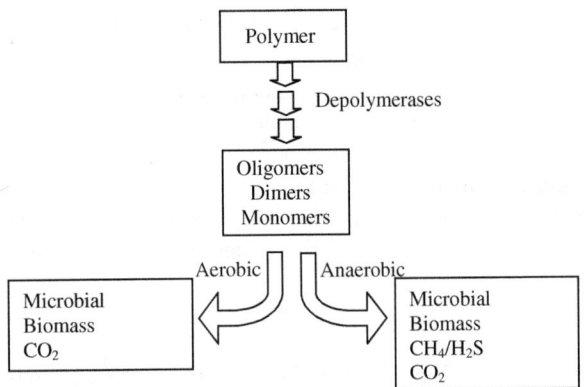

FIGURE 30.1. Schematic diagram showing degradation of polymeric materials under aerobic and anaerobic conditions.

found in natural environments and can be simulated in the laboratory with appropriate inocula.

E1.1. Poly(β-hydroxyalkanoates). Bacterial poly(β-hydroxyalkanoates) are a class of aliphatic polyesters consisting of homo or copolymers of [R]- β-hydroxyalkanoic acids. The polymer is a microbial intracellular inclusion in the cytoplasmic fluid in the form of granules with diameters between 0.3 and 1.0 μm. They can be as much as 30–80% of cellular mass. Unlike other biopolymers, such as polysaccharides, proteins, and DNAs, PHB is a thermoplastic with a melting temperature around 180°C, making it a good candidate for thermoprocessing. Furthermore, a copolyester consisting of 3-hydroxybutyrate and 3-hydroxyvalerate, poly(3HB-*co*-3HV) can be synthesized in cells of *Alcaligenes eutrophus* [172]. Both long side-chain and novel side-chain polymers have been synthesized by the addition of appropriate substrates in the culture medium and with different bacterial strains [173–175]. The copolymers range from thermoplastics to elastomers. The *A. eutrophus* genes encoding PHB synthesis have been transferred into cotton for PHB synthesis in fiber cells for production by agriculture [176].

Homopolymers and copolymers can be degraded in biologically active environments, for example, soils [91, 177], sludge, compost [38, 39, 43], river water [86, 92, 93], and seawater [86, 88, 89]. Extracellular PHB depolymerases have been isolated from *P. lemoignei* [178] and *A. faecalis* [140, 179]. The enzymatic degradation occurred at the surfaces of the polyester film, and the rate of surface erosion was strongly dependent on both the molecular weight (degree of polymerization) and composition of the polyester crystallinity, and the dominant species of bacteria. The stereochemistry of these chemically synthesized polymers also affects their fate after disposal in the natural environment [180, 181].

E1.2. Cellulose Acetates. Cellular acetates are chemically modified natural polymers designed to improve their

mechanical properties for different uses. Generally, CAs with a degree of substitution from 1.7 to 2.5 can be degraded in thermophilic compost in 45 days [38–45] and transformed into constituent chemicals through biologically catalyzed reactions [182]. Increasing the DS value on a repeating unit makes the CA less degradable. CA degradation occurs more rapidly under aerobic conditions. The mechanisms of degradation are deacetylation, which releases the substitution groups, followed by cleavage of the carbon–carbon backbone. The decrease in molecular weight and deacetylation proceed simultaneously during degradation.

E1.3. Polyethers. Polyethers include polyethylene glycols (PEGs), polypropylene glycols (PPGs), and polytetramethylene glycols (PTMGs). They are widely used in pharmaceuticals, cosmetics, lubricants, inks, and surfactants. They frequently contaminate natural water, including coastal waters and streams where wastewater is discharged. Degradability of this class of polymers has been studied under both aerobic [162, 183–186] and anaerobic [187–189] conditions. Polyether degradability is dependent on molecular weight, with molecular weights higher than 1000 considered resistant to biodegradation [162, 190]. However, degradation of PEGs with molecular weights up to 20,000 has been reported. The ability of a microflora to degrade larger PEG molecules is dependent primarily on the ability of a syntrophic association in mixed cultures of bacteria to metabolize the chemicals. For example, *Flavobacterium* and *Pseudomonas* can degrade PEG. After each oxidation cycle, PEG molecules are reduced by a glycol unit.

The central theme of PEG degradation is cleavage of an aliphatic ether linkage. In a coculture of aerobic *Flavobacterium* and *Pseudomonas* species, PEG degradation proceeds through dehydroxylation to form an aldehyde and a further dehydrogenation to a carboxylic acid derivative [184, 185]. Neither of these bacteria can degrade PEG in pure culture. Cellular contact between them seems to be essential for effective cooperation. In the *Flavobaterium* and *Psuedomonas* system, three enzymes are involved in the complete degradation of PEG [162]. PEG dehydrogenase, PEG–aldehyde dehydrogenase, and PEG–carboxylate dehydrogenase (ether cleaving) are required. All three were found in *Flavobaterium*, while only PEG–carboxylate dehydrogenase was present in *Pseudomonas*. However, *Pseudomonas*, though not directly involved in the degradation, utilizes a toxic metabolite that inhibits the activity of the *Flavobacterium*. This appears to be the essential link for their syntrophic association in the degradation of PEG.

E2. Effects on Physical Properties

E2.1. Polyimides. The wide acceptance of polyimides in the electronics industry [191–195] has drawn great attention to the issue of stability of these materials. The National

Research Council in 1987 [3] emphasized the need to develop deterioration preventive measures for polymers used in the electronic industries because data acquisition, information processing, and communication are critically dependent on materials performance and integrity. The interlayering of polyimides and electronics in integrated circuits prompted several studies on the interactions between these two materials [196, 197].

Polyimides are also widely used in load-bearing applications, for example, struts, chasses, and brackets in automotive and aircraft structures, due to their flexibility and compressive strength. They are used in appliance construction, cookware, and food packaging because of their chemical resistance to oils, greases, and fats; microwave transparency; and thermal resistance. Their electrical properties are ideally suited for applications in the electrical and electronics markets, especially as high-temperature insulation materials and passivation layers in the fabrication of integrated circuits and flexible circuitry. In addition, the flammability resistance of this class of polymers may provide a halogen-free flame-retardant material for aircraft interiors, furnishings, and wire insulation. Other possible uses may include fibers for protective clothing, advanced composite structures, adhesives, insulation tapes, foam, and optics operating at high temperatures [195].

Electronic packaging polyimides are particularly useful because of their outstanding performance and engineering properties. However, they are susceptible to degradation by the colonization of fungi (Fig. 30.2, Table 30.1)[25–31, 33–35, 47]. Polyimide degradation occurs through biofilm formation and subsequent physical changes in the polymer. Using electrochemical impedance spectroscopy (EIS) [198, 199], fungal growth on polyimides yields distinctive EIS spectra, indicative of failing resistivity [20, 21, 26]. Two steps are involved during degradation: An initial decline in coating resistance is related to the partial ingress of water and ionic

TABLE 30.1. Polymeric Materials Tested for Their Susceptibility to Degradation and Deterioration by Environmental Microorganisms

Name	Description	References
Adhesive	RTV142 silicone rubber with methyl alcohol	[28]
Insulation foam	Benzophenonetetracarboxylic imide polymert foam	[28]
Cable insulation	Polytetrafluoroethylene, fluorinated ethylene propylene coated polyimides and perfluocarboxyl	[28]
Composites	Fluorinated polyimide/glass fibers, bismaleimide/carbon fibers, epoxy/carbon fibers unidirectional, epoxy/carbon fibers [0,45, 90, -45]2S, poly(ether-ether-ketone)	[28, 34, 35, 36, 200]
	Epoxy/graphite fiber unidirectional	[28]
	Epoxy/carbon fibers, epoxy/glass fibers, bismaleimide/aluminum	[28, 36, 200]
Protective coating	Aliphatic polyurethane coating	[28]
Kapton polymimdes	Pyromellitic dianhydride and 4,4'-diaminodiphenyl ether	[26, 30–32]

species into the polymer matrices. This is followed by further deterioration of the polymer by activity of the fungi, resulting in a large decrease in resistivity. The data support the hypothesis that polyimides are susceptible to microbial degradation resulting in the corrosion of underlying metal. They also confirm the versatility of EIS as a method in evaluation of the biosusceptibility of polymers.

The dielectric properties of polyimides could be altered drastically following growth of microbial biofilms [25, 26, 30, 31, 47], which has wide implications for protection of this class of materials under tropical and subtropical conditions because both humidity and airborne microbial loading are high. This form of deterioration may be slow under ambient conditions in dry and cold regions; however, the deterioration processes can be accelerated in humid conditions or in enclosed environments, for example, submarines, space vehicles, aircraft, and other closed industrial facilities. Very small changes in material properties by the formation of biofilms and trapping of moisture will result in serious functional consequences to the systems.

E2.2. Fiber-Reinforced Polymeric Composite Materials.
The increasing usage of fiber-reinforced polymeric composite materials (FRPCMs) as structural components of public

FIGURE 30.2. Scanning electron micrograph of fungi on deteriorated polyimides (scale bar, 5 μm).

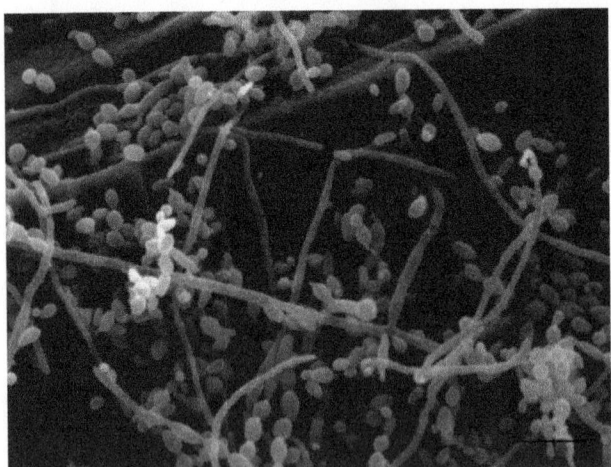

FIGURE 30.3. Scanning electron micrograph of fungi on fiber-reinforced polymeric composite coupon (scale bar, 10 μm).

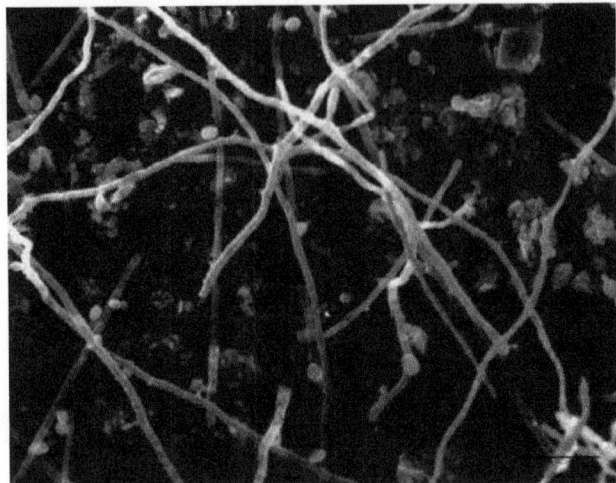

FIGURE 30.4. Scanning electron micrograph of fungi penetrating into matrices of fiber-reinforced polymeric composite coupon (scale bar, 10 μm).

structures and aerospace applications has generated an urgent need to evaluate the biodegradability of this new class of material. It has become clear that FRPCMs are not immune to the colonization by natural microorganisms, including both fungi and bacteria [4–7, 27–29, 37]. Through a series of investigations, impurities and chemical additives in FRPCMs promote microbial growth as they are sources of carbon and energy (Fig. 30.3, Table 30.1). Research has shown that slow-growing microorganisms cause much greater damage to materials than the fast-growing microorganisms [11], but no critical assessment has been made in relation to FRPCMs. At least two groups reported microbial degradation of FRPCs [6, 26, 32–35, 149]. Wagner and her collaborators used a mixed culture of microorganisms, including a sulfate-reducing bacterium, commonly used in corrosion tests. In contrast, Gu and colleagues [8, 26–29, 33–35, 71] used a fungal consortium originally enriched from degraded polymers. This consortium consisted of *Aspergillus versicolor*, *Cladosporium cladosorioides*, and a *Chaetomium* sp. Initial physical and mechanical tests were not sufficiently sensitive to detect any significant physical changes in the bulk materials after 120 days exposure [34, 36, 200]. However, physical penetration of fungal hyphae into the resin matrices has been observed with Scanning electron microscopy (SEM) (Fig. 30.4), indicating that resins were being actively degraded, in turn suggesting that the materials were at risk of failure.

Natural populations of microorganisms are capable of growing on surfaces of FRPC coupons at both relatively high humidity (65–70%) and lower humidity conditions (55–65%) [6, 28]. The accumulation of bacteria on surfaces of composites develops into a biofilm layer, providing some initial resistance to further environmental changes, for example, drying. Since use of a range of mechanical tests did not detect changes in composite coupons after exposure to a fungal culture [200], EIS was applied to examine changes in resistivity of composite materials after fungal exposure. EIS indicated a significant decline in material resistivity and an increase in conductance, providing the first demonstration that this technique can be used in monitoring FRPCMs and electrochemical properties can be used to identify microbial attack [33, 35]. The polymeric composites tested included fluorinated polyimide/glass fibers, bismaleimide/graphite fibers, poly(ether-ether-ketone) (PEEK)/graphite fibers, and epoxy/graphite fibers (Table 30.1) [35]. Graphite fibers are very susceptible to biofilm formation and the surface treatment sizing chemicals are utilized by bacteria as a source of carbon and energy (Fig. 30.5).

A critical question remains about the effect of FRPCM degradation on the mechanical properties and integrity of these composite materials. It is apparent that EIS is sufficiently sensitive to detect resistivity/conductance of com-

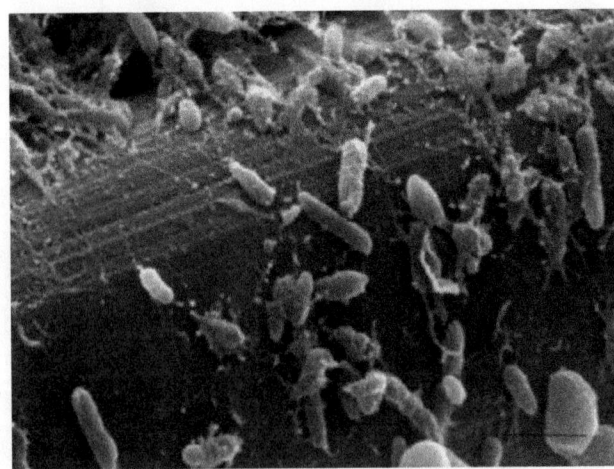

FIGURE 30.5. Scanning electron micrograph of bacteria forming biofilm on surface of graphite fiber (scale bar, 5 μm).

posite coupons, but surface-sensitive analytical techniques are needed to register the minor changes correlating to changes in structure and integrity of polymeric matrices. Acoustic techniques have been proposed as a means of detecting changes in the physical properties of the composite [149], but no comparison has been made between EIS and acoustic techniques.

E2.3. Protective Coatings.

Polymeric coatings are a class of chemicals with increasing production volume and are designed to protect underlying metals from corrosive chemical species and microorganisms. Microbial colonization and formation of biofilms on coatings may accelerate and severely damage the protective coatings and then the underlying metals. Natural bacterial populations were found to readily form microbial biofilms on surfaces of coating materials, including epoxy and polyamide primers and aliphatic polyurethanes (Table 30.1) [6, 28, 36]. Using EIS, both primers and top coatings demonstrated susceptibility to exposure to a mixed culture of fungi compared to sterile controls [28]. Primers were more rapidly degraded in terms of the EIS signal than aliphatic polyurethane coating. A common approach in dealing with coating

life is to incorporate biocides in coating formulations, but addition of biocides to polyurethane coatings did not inhibit bacterial attachment or significantly reduce bacterial growth [28, 37].

E2.4. Packaging Polyethylenes.

High-density and low-density polyethylenes (PEs) are primarily used in product packaging as sheets and thin films. Their degradability in natural environments poses serious environmental problems due to their very slow degradation rate under natural conditions and the hazard they present to freshwater and marine animals. Biodegradation of PEs has been studied extensively [90, 92]. It is believed that polymer additives, such as starch, antioxidants, coloring agents, sensitizers, and plasticizers, may significantly alter the biodegradability of the parent polymers. Degradation rates may be increased by 2–4% following photosensitizer addition.

In one study, extracellular culture concentrates of three *Streptomyces* species were inoculated to starch containing PE films [95, 96], and PE was claimed to be degraded. Degradation of PE by microorganisms may be minimal without taking into account physical and chemical processes, and the data on degradation of PE-containing starch are

TABLE 30.2. Summary of ASTM Methods and Practices and Others for Testing Biodegradation of Polymers and Key Characteristics of Methods

ASTM Code	Purposes	Microorganisms Involved and Key Features	Parameters Monitored	References
D5209-92	Aerobic degradation in municipal sewage sludge	Indigenous microorganisms in sewage sludgre	CO_2 evolved	Cited in [46]
D5210-92	Anaerobic degradation in municipal sewage sludge	Indigenous microorganisms in sewage sludgre	CO_2 and CH_4 evolved	Cited in [46]
D5247-92	Aerobic biodegradability by specified microorganisms	*Streptomyces badius* ATCC39117	Weight loss, tensile strength, elogation and molecular weight distribution	[204]
		Streptomyces setonii ATCC39115		
		Streptomyces viridosporus ATCC 39115		
		or other organisms agreed upon		
D5271-92	Aerobic biodegradation in activated sludge and wastewater	Municipal sewage treatment plant	Oxygen consumption	[205]
D5338-92	Aerobic biodegradation in composting conditions	2–4-month-old compost	Cumulative CO_2 production	[206]
G21-90	Resistance to fungi	*Aspergillus niger* ATCC 9642	Visual evaluation	[207]
		Aureobasidium pullulans ATCC15233		
		Chaetomium globosum ATCC6205		
		Gliocladium virens ATCC9645		
		Penicillum pinophilum ATCC11797		
G22-76	Resistance to bacteria	*Pseudomonas aeruginosa* ATCC 13388	Visual evaluation	[208]
MIL-STD-810E Method 508.4	Resistance to fungi	—	Visual evaluation	[209]

questionable. In addition, microbial metabolites may contaminate the PE surfaces and could be misinterpreted as degradation products of the parent PE when chemical characterization is carried out.

Abiotic degradation of PE is evidenced by the appearance of carbonyl functional groups in abiotic environments. In contrast, an increase of double bonds was observed when polymers showed weight loss resulting from biodegradation [201]. It was then proposed that microbial PE degradation is a two-step process: an initial abiotic photooxidation followed by a cleavage of the polymer carbon backbone. However, the mechanism of the second step needs extensive analysis before plausible conclusions can be drawn.

Lower molecular weight PEs including paraffin can be biodegraded and they undergo hydroxylation oxidatively to form an alcohol group followed by formation of carboxylic acid. At higher temperatures, ketones, alcohols, aldehydes, lactones, and carboxylic acids are formed abiotically within six weeks [201]. PE pipes used in gas distribution systems may fail due to cracking, but it is unlikely that biological processes are involved [202].

E2.5. Polypropylenes. Polypropylenes (PPs) are widely utilized in engineering pipes and containers. Degradation of PPs results in a decrease of their tensile strength and molecular weight. The degradation mechanism may involve the formation of hydroperoxides, which destabilize the polymeric carbon chain to form a carbonyl group [126, 203]. This step has been confirmed by infrared spectroscopy, but no strong evidence is available for the biochemical basis of degradation of PPs.

F. DEGRADATION METHODOLOGY

Traditionally, microbiological degradation of polymeric materials is carried out in microbiological nutrient media for short-term investigations [73, 113, 149]. The American

TABLE 30.3.   Comparison of Several Methods Available for Testing Degradability of Different Polymers and Under Range of Environmental and Simulation Conditions

Methods	Polymer Forms	Inoculum and Degradation Criteria Monitored	Comments	References
Gravimetry	Film or physical intact forms	A wide range of inocula can be used, from soil, waters, sewage, or pure species of microorganisms from culture collections.	This method is robust and also good for isolation of degradative microorganisms from environments of interest. Reproducibility is high. Disintegration of polymer cannot be differentiated from biodegradation.	[40–46, 83, 84, 92, 93]
Respirometry	Film, powder, liquid, and virtually all forms and shapes	Either oxygen consumed or CO_2 produced under aerobic conditions. Under methanogenic conditions, produced methane can be monitored.	This method is most adaptable to a wide range of materials. It may require specialized instruments. When fermentation is the major mechanism of degradation, this method gives an underestimation of the results.	[40–46, 204–209]
Surface hydrolysis	Films or other	Generally aerobic conditions, pure enzymes are used. Hydrogen ions (pH) released are monitored as incubation progresses.	Prior information about the degradation of the polymer by microorganims or particular enzymes is needed to target the specific test.	[180, 181]
Electrochemical impedance spectroscopy	Films or coatings resistant to water	The test polymers should be adhered on the surface of conductive materials and electrochemical conductance is recorded.	Polymer must initially be water impermeable for signal transduction. Degradation can proceed quickly, and as soon as degradation is registered, no further degradation processes can be distinguished.	[26, 28, 30–35]

Society for Testing and Materials (ASTM) offers a range of standard methodologies for assessment of material degradation under specific conditions [6, 46, 204–208] (Table 30.2). These methods are widely accepted by chemists and engineers but few biologists. However, the guidelines (1) are designed for a specific set of conditions which can be reproduced anywhere but may not be relevant to an individual environment; (2) fail to take into account microbiological processes and subsequent biochemical characterization; and (3) are usually qualitative with no or minimal emphasis on the mechanisms involved during degradation (Table 30.3).

Materials used for aerospace applications have been assessed in the United States by application of MIL-STD-810E method 508.4 [209]. In addition to the problems described above, a weakness of this method includes the failure to determine the effects of long-term exposure of the materials, as slow-growing microorganisms appear to be more effective in degrading polymeric materials than fast growers [210]. This method also emphasizes fungi and ignores any participation of bacteria in degradation processes even though both groups of microorganisms may play synergistic roles in the complete degradation of complex polymeric materials.

G. PLASTICIZERS

Polymeric materials are known to contain a wide array of chemicals, including plasticizers to improve their process-ibility and product quality; these plasticizers mostly include phthalate esters, specifically dimethyl phthalate esters (namely *ortho*-dimethyl phthalate ester, dimethyl isophthalate ester, and dimethyl terephthalate ester), di-*n*-butyl phthalate ester, and dibutylbenzyl phthalate ester [12, 101–112]. Since these chemicals are not covalently bound to the polymer resins, they can be utilized by many bacteria isolated from activated sludge [211], mangrove sediments [101–105, 108] and deep-ocean sediment [12, 106, 107] (Table 30.4). Biochemical cooperation is required between two different bacteria, namely *Arthrobacter* species and *Sphingomonas paucimobilis* in utilization of *ortho*-dimethyl phthalate ester (Fig. 30.6) [105, 107], but *Variovorax paradoxus* T4 isolated from deep-ocean sediment of the South China Sea is capable of utilizing dimethyl terephthalate ester as sole carbon and energy source (Fig. 30.7). Di-*n*-butyl phthalate dibutylbenzylphthalate esters can be degraded by *Pseudomonas fluorescence* B-1 isolated from mangrove sediment [109–112]. Results collectively indicated that enzymes involved in the initial cleavage of the two ester bonds can be highly selective.

H. USE OF BIOCIDES

Microorganisms are commonly controlled by application of biocides. While most antimicrobial products are effective against the growth of microorganisms in liquids, biofilm bacteria are more resistant and can rapidly develop resistance after exposure to chemicals [28, 37, 128,

TABLE 30.4. Microorganisms and Degradation of Selective Plasticizers in Class of Phthalate Diester with Source of Bacteria

Substrate	Microorganism(s) Involved	Source of Inoculum	Degradation Intermediates	References
Dimethyl phthalate	*Comamonas acidovorans* fy-1, *Xanthomonas maltophila*, and *Sphingomonas paucimobilis*	Activated sludge	Monomethyl phthalate, phthalic acid	[221, 222]
	Rhodococcus ruber Sa	Mangrove sediment	Monomethyl phthalate, phthalic acid	[101, 102]
	Fusarium sp.	Mangrove sediment	Monomethyl phthalate phthalate	[220]
Dimethyl isophthalate	*Klebsiella oxytoca* Sc and *Methylobacterium mesophilicum* Sr	Mangrove sediment	Monoisophthalate, phthalic acid	[103–105]
	Rhodococcus ruber Sa	Mangrove sediment	Monoisophthalate, phthalic acid	[12,106, 107]
	Variovorax paradoxus	Deep-ocean sediment	Monoisophthalate	[101]
Dimethyl terephthalate	*Rhodococcus ruber* Sa	Mangrove sediment	Monoterephthalate, phthalic acid	[101, 102]
	Pasteurella multocida Sa	Mangrove sediment	Monoterephthalate, phthalic acid	[103]
	Variovorax paradoxus	Deep-ocean sediment	Monoterephthalate,	[12, 106, 107].
	Sphingomonas yanoikuyae	Deep-ocean sediment	?	[107]
Di-*n*-butyl phthalate	*Pseudomonas fluorescens* B-1	Mangrove sediment	Monobutyl phthalate, phthalic acid	[109–111]
butylbenzylphthalate	*Pseudomonas fluorescens* B-1	Mangrove sediment	benzylphthalate, phthalic acid	[109–111]

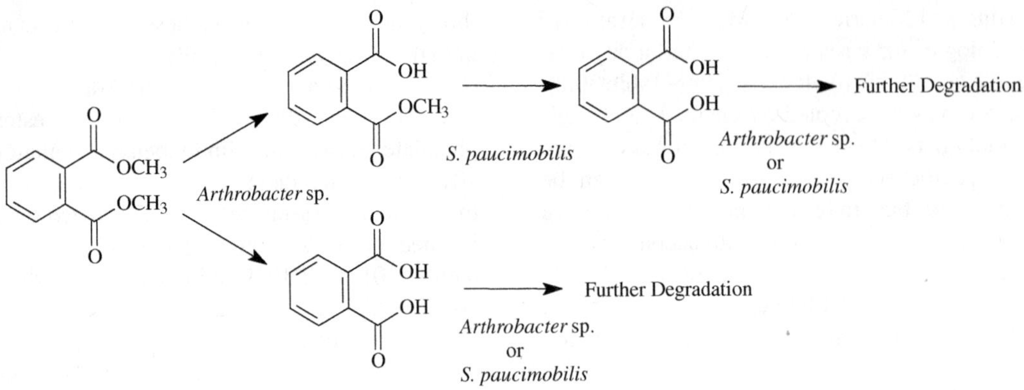

FIGURE 30.6. Biochemical pathway for degradation of *ortho*-phthalate diester by *Arthrobacter* species and *Sphingomonas paucimobilis* through metabolic collaboration.

212–215]. Biofilms reduce the effectiveness of the biocides because of reduced diffusion rates and the binding affinity of exopolymeric materials in the biofilm for biocides. In addition, microorganisms have plasmids which can be effectively exchanged between organisms to rapidly build resistance to the environmental chemicals and biocides [213]. Most industrial tests are conducted in the liquid phase, which does not represent real environmental conditions of exposure. For example, a biocide that effectively inhibits bacterial growth in a test involving liquid culture failed to control growth on surfaces of a polyurethane coating [4, 6, 28, 37]. It is strongly recommended that tests for biocide efficacy should be applied to material surfaces. In addition, compatibility of a biocide with a material should be considered.

It is routine practice that corrosion protective paint formulations are amended with biocides to prolong the shelf-life of products and extend service time [216]. Industrial development of water-based paints and coatings creates the problem of increased microbial contamination and degradation (Fig. 30.8), because these environmentally friendly chemicals may be susceptible to rapid colonization and utilization by microorganisms. Microbial growth deteriorates the quality of products and also provides initial contamination to the product in service. The basic approach to such a problem should emphasize environmental control so that microorganisms are prevented from attaching to surfaces in the first place. Conditions that promote microbial growth should also be avoided so that microbial population growth is kept to a minimum.

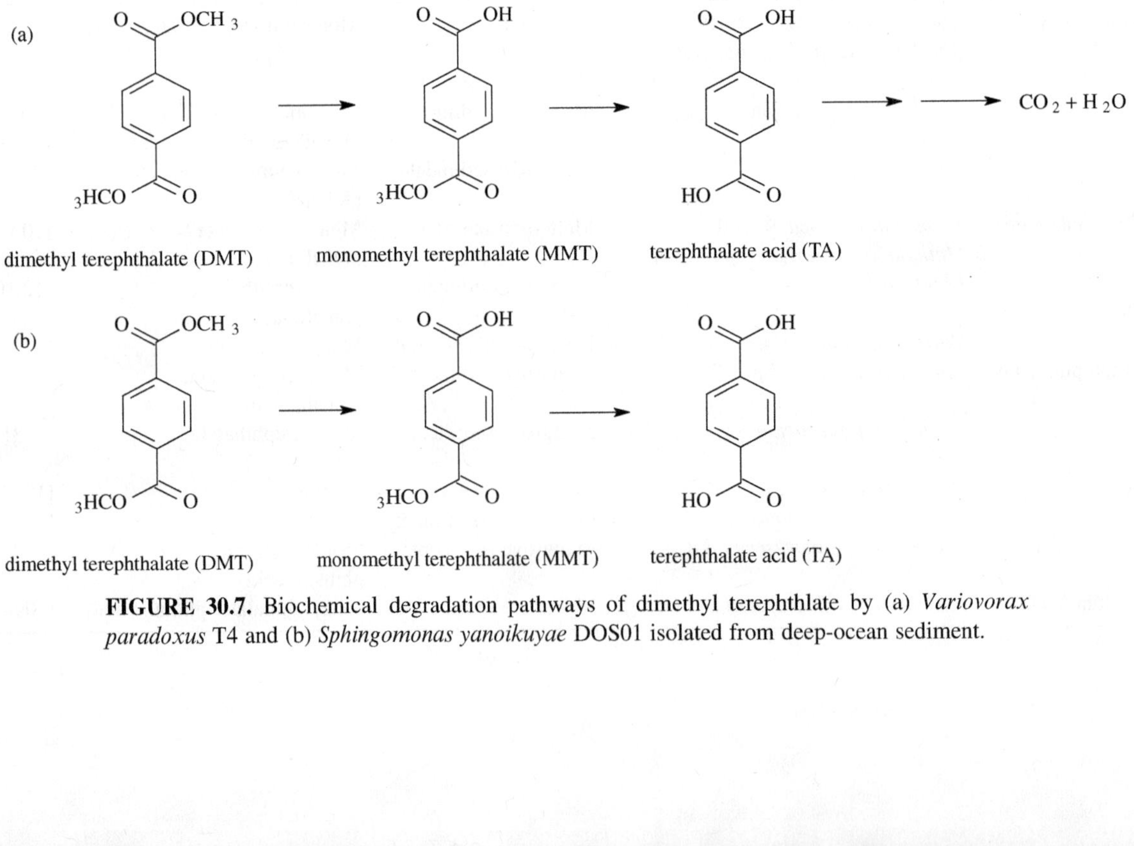

FIGURE 30.7. Biochemical degradation pathways of dimethyl terephthlate by (a) *Variovorax paradoxus* T4 and (b) *Sphingomonas yanoikuyae* DOS01 isolated from deep-ocean sediment.

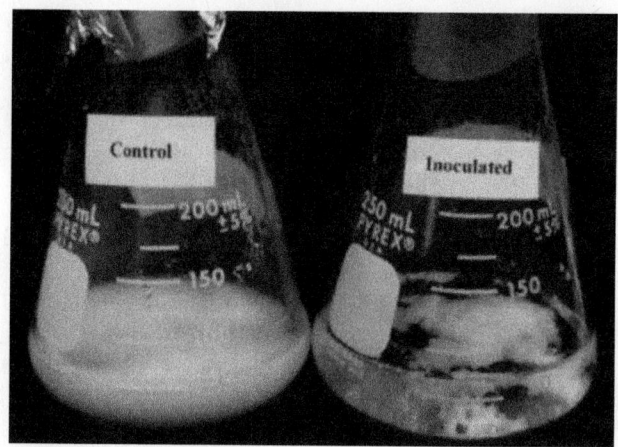

FIGURE 30.8. Photograph of two flasks containing water-based aliphatic polyurethane coating inoculated with soil as inoculums (*right*) and kept sterile throughout incubation period (*left*).

I. BIODEGRADATION AND HEALTH

Polymeric materials are increasingly used not only in engineering and structures, building, and decoration but also in medical implants. Material types cover a wide range of composition, including absorbable sutures and artificial skin. The materials must be thoroughly examined and tested for potential biofilm formation. Some additives, such as PEGs, can promote microbial growth by providing readily available sources of carbon and energy. Other chemicals may serve as attractants for bacteria. In addition, microbial metabolites during active growth, particularly secondary ones, can be allergens or irritants to humans in building materials [217, 218]. Fungi have been studied for their tolerance to lower humidity, where bacteria are barely capable of survival or growth [28]. The growth of fungi may not only damage the materials, integrity but also produce toxic secondary metabolites [8, 219].

J. CONCLUSIONS

Polymeric materials are a diverse class of polymers with multiple applications. Assessment of their biodegradability must address the relationship between their chemical structures and proposed applications. An array of test protocols are available and should be developed to detect early changes in polymer physical and chemical properties. EIS is a sensitive method for detecting early physical changes in polymer films, but the microbial role in deterioration needs to be confirmed with more sensitive methods. In most cases, accelerated testing under specific environmental conditions will be required to determine susceptibility to microbial deterioration and to assess protective measures.

ACKNOWLEDGMENTS

We thank Zhenye Zhao for drawing Figure 30.1; Ed Seling at the Museum of Comparative Zoology, Harvard University, for the scanning electron micrographs taken and used here in Figures 30.2–30.5; and Jessie Lai for assistance in organizing the final figures and archiving.

REFERENCES

1. R. F. Hadley, in The Corrosion Handbook, H. H. Uhlig (Ed.), Wiley, New York, 1948, pp. 466–481.
2. J. A. Lee, in The Corrosion Handbook, H. H. Uhlig (Ed.), Wiley, New York, 1948, pp. 359–365.
3. National Research Council (NRC), Agenda for Advancing Electrochemical Corrosion Science and Technology, Publication NMAB438-2, National Academy Press, Washington, DC, 1987.
4. J.-D. Gu, Int. Biodeter. Biodegr., **52**, 69 (2003).
5. J.-D. Gu, in Handbook of Environmental Degradation of Materials, M. Kultz (Ed.), William Andrew Publishing, New York, 2005, pp. 179–206.
6. J.-D. Gu, Int. Biodeter. Biodegr., **59**, 170 (2007).
7. J.-D. Gu, in Encyclopedia of Microbiology, (Ed.), M. Schaechter 3rd ed., Elsevier, San Diego, California, 2009, pp. 259–269.
8. J.-D. Gu, T. E. Ford, and R. Mitchell, Uhlig Corrosion Handbook, 2nd ed., W. Revie (Ed.), Wiley, New York, 2000, pp. 349–365.
9. J. Lemaire, P. Dabin, and R. Arnaud, in Biodegradable Polymers and Plastics, M. Vert, J. Feijen, A. Albertsson, G. Scott, and E. Chiellini (Eds.), Royal Society of Chemistry, Cambridge, 1992, pp. 30–39.
10. C. G. Pitt, in Biodegradable Polymers and Plastics, M. Vert, J. Feijen, A. Albertsson, G. Scott, and E. Chiellini (Eds.), Royal Society of Chemistry, Cambridge, 1992, pp. 7–17.
11. J.-D. Guand L. Pan, J. Polym., Environ., **14**, 273 (2006).
12. J.-G. Gu, B. Han, S. Duan, Z. Zhao, and Y. Wang, Int. Biodeter. Biodegr., **63**, 450 (2009).
13. O. G. Gazenko, A. I. Grigoryev, S. A. Bugrov, V. V. Yegorov, V. V. Bogomolov, I. B. Kozlovskaya, and I. K. Tarasov, Kosmicheskaya Biologiya i Aviakosmicheskaya Medistina, **23**, 3 (1990). (in Russian).
14. Y. Nefedov, N. D. Novikova, and I. N. Surovezhin, Kosmicheskaya Biologiya i Aviakosmicheskaya Medistina, **22**, 67 (1988). (in Russian).
15. N. D. Novikovaand S. N. Zalogoyev, Kosmicheskaya Biologiya i Aviakosmicheskaya Medistina, **19**, 74 (1985). (in Russian).
16. S. N. Zaloguyev, Kosmicheskaya Biologiya i Aviakosmicheskaya Medistina, **19**, 64 (1985) (in Russian).
17. N. Linsey, R. Conrad, C. Bowles, and M. Kelley, Paper No. 619, Corrosion/96, NACE International, Houston, TX, 1996.

18. B. Zyska, in DECHEMA Monographs, Vol. 133, Biodeterioration and Biodegradation, W. Sand (Ed.), VCH, Frankfurt, Germany, 1996, pp. 427–432.

19. K. Milde, W. Sand, W. Wolff, and E. Bock, J. Gen. Microbiol., **129**, 1327 (1983).

20. J.-D. Gu, T. E. Ford, and R. Mitchell, Int. Biodeter. Biodegr., **41**, 101 (1998).

21. J.-D. Gu, T. E. Ford, and R. Mitchell, Uhlig Corrosion Handbook, 2nd ed., W. Revie (Ed.), Wiley, New York, 2000, pp. 477–491.

22. M. W. Mittelman, in Bacterial Adhesion: Molecular and Ecological Diversity, M. Fletcher (Ed.), Wiley, New York, 1996, pp. 89–127.

23. J. J. Dobbins, B. L. Giammara, J. S. Hanker, P. E. Yates, and W. C. DeVries, in Mater. Res. Soc. Symp. Proc., Vol. 110, Materials Research Society, Pittsburgh, PA, 1989, pp. 337–348.

24. D. F. Williams and S. P. Zhong, Int. Biodeter. Biodegr., **34**, 95 (1994).

25. T. E. Ford, E. LaPointe, R. Mitchell, and D. B. Mitton, in Biodeterioration and Biodegradation, Vol. 9, Institute of Chemical Engineers, Warwickshire, UK, 1995, pp. 554–561.

26. J.-D. Gu, T. E. Ford, and R. Mitchell, J. Appl. Polym. Sci., **92**, 1029 (1996).

27. J.-D. Gu, D. B. Mitton, T. E. Ford, and R. Mitchell, Biodegradation, **9**, 39 (1998).

28. J.-D. Gu, M. Roman, T. Esselman, and R. Mitchell, Int. Biodeter. Biodegr., **41**, 25 (1998).

29. J.-D. Gu, T. E. Ford, and R. Mitchell, in Uhlig Corrosion Handbook, 2nd ed., W. Revie (Ed.), Wiley, New York, 2000, pp. 439–460.

30. B. Mitton, T. E. Ford, E. LaPointe, and R. Mitchell, Paper No. 296, Corrosion/93, National Association of Corrosion Engineers, Houston, TX, 1993.

31. D. B. Mitton, S. Toshima, R. M. Latanison, F. Bellucci, T. E. Ford, J.-D. Gu, and R. Mitchell, in Organic Coatings for Corrosion Control, ACS Symp. Ser. No. 689, G. P. Bierwagen (Ed.), American Chemical Society, Washington, DC, 1998, pp. 211–222.

32. J.-D. Gu and R. Mitchell, Chin. J. Mat. Res., **9** (Suppl.), 473 (1995).

33. J.-D. Gu, T. Ford, K. Thorp, and R. Mitchell, Int. Biodeter. Degr., **39**, 197 (1996).

34. J.-D. Gu, C. Lu, K. Thorp, A. Crasto, and R. Mitchell, Mater. Perform., **36**, 37 (1997).

35. J.-D. Gu, C. Lu, K. Thorp, A. Crasto, and R. Mitchell, J. Ind. Microbiol. Biotechnol., **18**, 364 (1997).

36. K. E. G. Thorp, A. S. Crasto, J.-D. Gu, and R. Mitchell, Paper No. 279, Corrosion/97, National Association of Corrosion Engineers, Houston, TX, 1997.

37. R. Mitchell, J.-D. Gu, M. Roman, and S. Soukup, DECHEMA Monographs, Vol 133, Biodeterioration and Biodegradation, in W. Sand (Ed.), VCH, Frankfurt, Germany, 1996, pp. 3–16.

38. R. A. Gross, J.-D. Gu, D. Eberiel, M. Nelson, and S. P. McCarthy, in Fundamentals of Biodegradble Materials and Packaging, D. Kaplan, E. Thomas, and C. Ching (Eds.), Technomic Publishing, Lancaster, PA, 1993, pp. 257–279.

39. R. A. Gross, J.-D. Gu, D. Eberiel, and S. P. McCarthy, in Degradable Polymers, Recycling and Plastics Waste Management, A. Albertson and S. Huang (Eds.), Marcel Dekker, New York, 1995, pp. 21–36.

40. J.-D. Gu, M. Gada, G. Kharas, D. Eberiel, S. P. McCarty, and R. A. Gross, Polym. Mater. Sci. Eng., **67**, 351 (1992).

41. J.-D. Gu, S. P. McCarty, G. P. Smith, D. Eberiel, and R. A. Gross, Polym. Mater. Sci. Eng., **67**, 230 (1992).

42. J.-D. Gu, S. Coulter, D. Eberiel, S. P. McCarthy, and R. A. Gross, J. Environ. Polym. Degr., **1**, 293 (1993).

43. J.-D. Gu, D. T. Eberiel, S. P. McCarthy, and R. A. Gross, J. Environ. Polym. Degr., **1**, 143 (1993).

44. J.-D. Gu, D. Eberiel, S. P. McCarthy, and R. A. Gross, J. Environ. Polym. Degr., **1**, 281 (1993).

45. J.-D. Gu, S. Yang, R. Welton, D. Eberiel, S. P. McCarthy, and R. A. Gross, J. Environ. Polym. Degr., **2**, 129 (1994).

46. J.-G. Gu and J.-D. Gu, J. Polym. Environ., **13**, 65 (2005).

47. T. Ford, J. Maki, and R. Mitchell, in Bioextraction and Biodeterioration of Metals, C. C. Gaylarde and H. A. Videla (Eds.), Cambridge University Press, New York, 1995, pp. 1–23.

48. F. W. Billmeyer, Jr., Textbook of Polymer Science, 3rd ed., Wiley, New York, 1984.

49. M. S. Reisch, Chem. Eng. News, **74**, 44 (1996).

50. D. E. Caldwell, G. M. Wolfaardt, D. R. Korber, and J. R. Lawrence, in Advances in Microbial Ecology, Vol. **15**, J. G. Jones (Ed.), Plenum, New York, 1997, pp. 105–191.

51. J. W. Costerton, G. G. Geesey, and K.-J. Cheng, Sci. Am., **238**, 86 (1978).

52. M. Fletcher and G. I. Loeb, Appl. Environ. Microbiol., **37**, 67 (1979).

53. H. M. Lappin-Scott and J. W. Costerton, Microbial Biofilms, Cambridge University Press, Cambridge, 1995.

54. B. Little, P. Wagner, S. M. Gerchakov, M. Walch, and R. Mitchell, Corrosion, **42**, 533 (1986).

55. J. S. Maki, B. J. Little, P. Wagner, and R. Mitchell, Biofouling, **2**, 27 (1990).

56. K. C. Marshall, R. Stout, and R. Mitchell, J. Gen. Microbiol., **68**, 337 (1971).

57. H. H. M. Rijnaarts, W. Norde, E. J. Bouwer, J. Lyklema, and A. J. B. Zehnder, Appl. Environ. Microbiol., **59**, 3255 (1993).

58. P. Vandevivere and D. L. Kirchman, Appl. Environ. Microbiol., **59**, 3280 (1993).

59. M. Walch, Encycloped. Microbiol., **1**, 585 (1992).

60. J.-D. Gu, J. S. Maki, and R. Mitchell, in Zebra Mussels and Aquatic Nuisance Species, F. M. D'Itri (Ed.), Ann Arbor Press, Chelsea, MI, 1997, pp. 343–357.

61. D. Kirchman and R. Mitchell, OCEANS, Sept., 537 (1981).

62. J.-D. Gu and R. Mitchell, J. Microbiol., **39**, 133 (2001).

63. J.-D. Gu and R. Mitchell, Hydrobiologia, **474**, 81 (2002).

64. M. Walch, T. E. Ford, and R. Mitchell, Corrosion, **45**, 705 (1989).

65. T. Ford and R. Mitchell, in Environmental Microbiology, R. Mitchell (Ed.), Wiley, New York, 1992, pp. 83–101.

66. T. E. Ford, M. Walch, and R. Mitchell, Mater. Perform., **26**, 35 (1987).

67. T. Ford, J. Maki, and R. Mitchell, in Biodeterioration Vol. 75, D. R. Houghton, R. N. Smith, and H. O. W. Eggins (Eds.), Elsevier Science, New York, 1988, pp. 378–384.

68. T. E. Ford, P. C. Searson, T. Harris, and R. Mitchell, J. Electrochem. Soc., **137**, 1175 (1990).

69. T. E. Ford, J. P. Black, and R. Mitchell, Paper No. 110, Corrosion/90, NACE International, Houston, TX, 1990.

70. H. Ghassem and N. Adibi, Mater. Perform., **34**, 47 (1995).

71. J.-D. Gu, T. E. Ford, and R. Mitchell, in Uhlig Corrosion Handbook, 2nd ed., W. Revie (Ed.), Wiley, New York, 2000, pp. 915–927.

72. B. J. Little, P. Wagner, J. S. Maki, and R. Mitchell, J. Adhes., **20**, 187 (1986).

73. B. Little, P. Wagner, W. G. Characklis, and W. Lee, in Biofilms, W. G. Characklis and K. C. Marshall (Eds.), Wiley, New York, 1990, pp. 635–670.

74. F. Widdel, in Biology of Anaerobic Microorganisms, A. J. B. Zehnder (Ed.), Wiley, New York, 1988, pp. 469–585.

75. F. Widdel, in Biotechnology Focus, Vol. 3, R. F. Finn, P. Prave, M. Schlingmann, W. Crueger, K. Esser, R. Thauer, and F. Wagner (Eds.), Hanser, Munich, Germany, 1992, pp. 261–300.

76. J.-D. Gu, B. Belay, and R. Mitchell, World J. Microbiol. Biotech., **17**, 173 (2001).

77. T. Ford and R. Mitchell, Mar. Technol. Soc. J., **24**, 29 (1990).

78. T. Ford and R. Mitchell, Adv. Microb. Ecol., **11**, 231 (1991).

79. W. A. Hamilton, Ann. Rev. Microbiol., **39**, 195 (1985).

80. K. T. Holland, J. S. Knapp, and J. G. Shoesmith, Anaerobic Bacteria, Chapman & Hall, New York, 1986.

81. W. Lee, Z. Lewandowsky, P. H. Nielsen, and W. A. Hamilton, Biofouling, **8**, 165 (1995).

82. J. R. Postage, The Sulfate-Reducing Bacteria, 2nd ed., Cambridge University Press, New York, 1984.

83. D. F. Gilmore, S. Antoun, R. W. Lenz, and R. C. Fuller, J. Environ. Polym. Degr., **1**, 269 (1993).

84. D. F. Gilmore, S. Antoun, R. W. Lenz, S. Goodwin, R. Austin, and R. C. Fuller, J. Ind. Microbiol. **10**, 199 (1992).

85. P. Wagner, Paper No. 200, Corrosion/95, NACE International, Houston, TX, 1995.

86. A. L. Andrady, J. E. Pegram, and S. Nakatsuka, J. Environ. Polym. Degr., 1, 31 (1993).

87. J. Guezennec, O. Ortega-Morales, G. Raguenes, and G. Geesey, FEMS Microbiol. Ecol., **26**, 89 (1998).

88. B. K. Sullivan, C. A. Oviatt, and G. Klein-MacPhee, in Fundamentals of Biodegradable Materials and Packaging, D. Kaplan, E. Thomas, and C. Ching (Eds.), Technomic Publishing, Lancaster, PA, 1993, pp. 281–296.

89. C. O. Wirsen and H. W. Jannasch, in Fundamentals of Biodegradable Materials and Packaging, D. Kaplan, E. Thomas, and C. Ching (Eds.), Technomic Publishing, Lancaster, PA, 1993, pp. 297–310.

90. A.-C. Albertsson, Eur. Polym. J., **16**, 623 (1980).

91. R. Tsao, T. A. Anderson, and J. R. Coats, J. Environ. Polym. Degr., **1**, 301 (1993).

92. S. H. Imam and J. M. Gould, Appl. Environ. Microbiol., **56**, 872 (1990).

93. S. H. Iman, J. M. Gould, S. H. Gordon, M. P. Kinney, A. M. Ramsey, and T. R. Tosteson, Curr. Microbiol., **25**, 1 (1992).

94. A. J. Anderson and E. A. Dowes, Microbiol. Rev., **54**, 450 (1990).

95. A. L. PomettoIII, B. Lee, and K. E. Johnson, Appl. Environ. Microbiol., **58**, 731 (1992).

96. A. L. PomettoIII, K. E. Johnson, and M. Kim, J. Environ. Polym. Degr., **1**, 213 (1992).

97. R. B. Hespell and P. J. O'Bryan-Shah, Appl. Environ. Microbiol., **54**, 1917 (1988).

98. H. Lee, R. J. B. To, R. K. Latta, P. Biely, and H. Schneider, Appl. Environ. Microbiol., **53**, 2831 (1987).

99. E. Lüthi, N. B. Jasmat, and P. L. Bergquist, Appl. Microbiol. Biotechnol., **34**, 214 (1990).

100. E. T. Reese, Ind. Eng. Chem., **49**, 89 (1957).

101. J. Li, J.-D. Gu, and L. Pan, Int. Biodeter. Biodegr., **55**, 223 (2005).

102. J. Li, J.-D. Gu, and J.-H. Yao, Int. Biodeter. Biodegr., **56**, 158 (2005).

103. J. Li and J.-D. Gu, Ecotoxicology, **15**, 391 (2006).

104. J. Li and J.-D. Gu, Water Air Soil Pollut. Focus, **6**, 569 (2006).

105. J. Li and J.-D. Gu, Sci. Total Environ., **380**, 181 (2007).

106. Y. Wang and J.-D. Gu, J. Human Ecol. Risk Assess., **12**, 236 (2006).

107. Y. Wang and J.-D. Gu, Ecotoxicology, **15**, 549 (2006).

108. Y. Wang, B. Yin, Y.-G. Hong, Y. Yan, and J.-D. Gu, Ecotoxicology, **17**, 845 (2008).

109. X. R. Xu, H. B. Li, and J.-D. Gu, Int. Biodeter. Biodegr., **55**, 9 (2005).

110. X. R. Xu, H. B. Li, and J.-D. Gu, J. Microbiol. Biotechnol., **15**, 946 (2005).

111. X. R. Xu, H. B. Li, and J.-D. Gu, Anal. Bioanal. Chem., **386**, 370 (2006).

112. X. R. Xu, H. B. Li, J.-D. Gu, and X.-Y. Li, J. Hazard. Mater., **140**, 194 (2007).

113. P. A. Wagner and R. I. Ray, in Microbiologically Influenced Corrosion Testing, ASTM STP 1232, J. R. Kearns and B. J. Little (Eds.), American Society for Testing and Materials, Philadelphia, PA, 1994, pp. 153–169.

114. R. E. Hungate, in Methods in Microbiology, Vol 38, J. R. Norris and D. W. Ribbons (Eds.), Academic, New York, 1971, pp. 117–132.

115. C. A. Reddy, T. J. Beveridge, J. A. Breznak, G. A. Marzluf, T. M. Schmidt, and L. R. Snyder, Methods for General and Molecular Microbiology, 3rd ed., American Society for Microbiology, Washington, DC, 2007.

116. J.-D. Gu and D. F. Berry, Appl. Environ. Microbiol., **57**, 2622 (1991).

117. J.-D. Gu and D. F. Berry, Appl. Environ. Microbiol., **58**, 2667 (1992).

118. J.-D. Guin Soil Mineral-Microbe-Organic Interactions—Theories and Applications, Q. Y. Huang, P. M. Huang, and A. Violante (Eds.), Springer-Verlag, Berlin, 2008, pp. 175–198.

119. K. Budwill, P. M. Fedorak, and W. J. Page, Appl. Environ. Microbiol., **58**, 1398 (1992).

120. C. M. Buchanan, R. M. Gardner, and R. J. Komarek, J. Appl. Polym. Sci., **47**, 1709 (1993).

121. Y. Doi, Microbial Polyesters, VCH Publishers, New York, 1990.

122. J. E. McCassie, J. M. Mayer, R. E. Stote, A. E. Shupe, P. J. Stenhouse, P. A. Dell, and D. L. Kaplan, in Fundamentals of Biodegradable Materials and Packaging, D. Kaplan, E. Thomas, and C. Ching (Eds.), Technomic Publishing, Lancaster, PA, 1993, pp. 247–256.

123. M. R. Timmins, D. F. Gilmore, R. C. Fuller, and W. R. Lenz, in Fundamentals of Biodegradable Materials and Packaging, D. Kaplan, E. Thomas, and C. Ching (Eds.), Technomic Publishing, Lancaster, PA, 1992, pp. 119–131.

124. O. Milstein, R. Gersonde, A. Huttermann, M.-J. Chen, and J. J. Meister, Appl. Environ. Microbiol., **58**, 3225 (1992).

125. L. Tilstra and D. Johnsonbaugh, J. Environ. Polym. Degr., **1**, 247 (1983).

126. I. Cacciari, P. Quatrini, G. Zirletta, E. Mincione, V. Vinciguerra, P. Lupattelli, and G. G. Sermanni, Appl. Environ. Microbiol., **59**, 3695 (1993).

127. V. M. Knyazev, V. I. Korolkov, A. N. Viktorov, G. O. Pozharskiy, L. N. Petrova, and V. P. Gorshkov, Kosmicheskaya Biologiya i Aviakosmicheskaya Medistina, **20**, 80 (1986) (in Russian).

128. G. A. McFeters, SAE Technical Paper 911404, Intersociety Conference on Environmental Systems, San Francisco, California, 1991.

129. D. Meshkov, 45th Congress of the International Astronautical Confederation, IAF/IAA-94-G.1.125, Jerusalem, Israel, 1994.

130. D. L. Pierson, M. R. McGinis, and A. N. Viktorov, in Space Biology and Medicine, A. E. Nicogossian, S. R. Mohler, O. G. Gazenko, and A. I. Grigotyev (Eds.), Washington, DC, 1994, pp. 77–93.

131. D. L. Pierson and S. K. Mishra, 43rd Congress of the International Astronautical Confederation, Washington, DC, 1992.

132. G. I. Solomin, Kosmicheskaya Biologiya i Aviakosmicheskaya Medistina, **19**, 4 (1985) (in Russian).

133. M. Stranger-Joannesen, R. Sorheim, D. Zanotti, and A. Bichi, 44th Congress of the International Astronautical Confederation Congress, IAF/IAA-93-G.4.162, 1993.

134. N. D. Novikova, M. I. Orlova, and M. B. Dyachenko, Kosmicheskaya Biologiya i Aviakosmicheskaya Medistina, **20**, 71 (1986) (in Russian).

135. A. N. Viktorov, 45th Congress of the International Astronautical Confederation, IAF/IAA-94-5.165, Jerusalem, Israel, 1994.

136. A. N. Viktorov and V. K. IIyin, 43rd Congress of the International Astronautical Confederation, IAF/IAA-92-0277, Washington, DC, 1992.

137. A. N. Viktorov, V. K. IIyin, and J. Syniak, 44th Congress of the International Astronautical Confederation, IAF/IAA-93-G.4.161, Graza, Austria, 1993.

138. A. N. Viktorov and N. D. Novikova, Kosmicheskaya Biologiya i Aviakosmicheskaya Medistina, **19**, 66 (1985) (in Russian).

139. A. A. Chowdhury, Arch. Mikrobiol., **47**, 167 (1963).

140. T. Saito, K. Suzuki, Yamamoto, T. Fukui, K. Miwa, K. Tomita, S. Nakanishi, S. Odani, J.-I. Suzuki, and K. Ishikawa, J. Bacteriol., **171**, 184 (1989).

141. R. Narayan, in Science and Engineering of Composting: Design, Environmental, Microbiological and Utilization Aspects, H. A. J. Hoitink and H. M. Keener (Eds.), Renaissance, Worthington, OH, 1993, pp. 339–362.

142. M. A. Barlaz, R. K. Ham, and D. M. Schaefer, J. Environ. Eng., **115**, 1088 (1989).

143. M. A. Barlaz, D. M. Schaefer, and R. K. Ham, Appl. Environ. Microbiol., **55**, 55 (1989).

144. M. A. Barlaz, B. F. Staley, and F. L. de los Reyes III, in Environmental Microbiology, 2nd ed., R. Mitchell and J.-D. Gu (Eds.), Wiley-Blackwell, Hoboken, NJ, 2010, pp. 281–299.

145. K. R. Gunjala and J. M. Sulflita, Environ. Sci. Technol., **27**, 1176 (1993).

146. J. D. Hamilton, K. H. Reinert, J. V. Hogan, and W. V. Lord, J. Air Waste Manag. Assoc., **43**, 247 (1995).

147. W. Gujer and A. J. B. Zehnder, Water Sci. Technol., **15**, 127 (1983).

148. V. T. Breslin and R. L. Swanson, J. Air Waste Manag. Assoc., **43**, 325 (1993).

149. P. Wagner, R. Ray, K. Hart, and B. Little, Mater. Perfom., **35**, 79 (1996).

150. K. Nakayama, T. Saito, Y. Fukui, Y. Shirakura, and K. Tomita, Biochim. Biophys. Acta, **827**, 63 (1985).

151. A. Zachary, M. E. Taylor, F. E. Scott, and R. R. Colwell, in Biodeterioration, Proceedings of the fourth International Biodeterioration Symposium, T.A. Oxley, G. Becker, and D. Allsopp (Eds.), Pitman, London, 1980, pp. 171–177.

152. K. C. Marshall, ASM News, **58**, 202 (1992).

153. R. M. Heisey and S. Papadatos, Appl. Environ. Microbiol., **61**, 3092 (1995).

154. J. Jones-Meehan, K. L. Vasanth, R. K. Conrad, M. Fernandez, B. J. Little, and R. I. Ray, in Microbiologically Influenced Corrosion Testing, ASTM STP 1232, J. R. Kearns and B. J. Little (Eds.), American Society for Testing and Materials, Philadelphia, PA, 1994, pp. 217–233.

155. S. Karlsson, O. Ljungquist, and A.-C. Albertsson, Polym. Degr. Stab., **21**, 237 (1988).

156. R. N. Leyden and D. I. Basiulis, in Biomedical Materials and Devices, J. S. Hanker and B. L. Giammara (Eds.), Materials Research Society Symposium Proceedings, Vol. 10, Materials Research Society, Pittsburgh, PA, 1989, pp. 627–633.

157. J. W. McCain and C. J. Mirocha, Int. Biodeter. Biodegr., **33**, 255 (1995).

158. J. Mas-Castellà, J. Urmeneta, R. Lafuente, A. Navarrete, and R. Guerrero, Int. Biodeter. Biodegr., **35**, 155 (1995).

159. E. S. Stuart, R. W. Lenz, and R. C. Fuller, Can. J. Microbiol., **41** (Suppl.), 84 (1995).

160. A. Stenbüchel, in Biomaterials: Novel Materials from Biological Sources, D. Byrom (Ed.), Macmillan, New York, 1991, pp. 127–213.

161. A.-M. Cromwick and R. A. Gross, Can. J. Microbiol., **41**, 902 (1995).

162. F. Kawai, CRC Biotechnol., **6**, 273 (1987).

163. R. C. BlakeII, W. N. Norton, and G. T. Howard, Int. Biodeter. Biodegr., **42**, 63 (1998).

164. J. R. Crabbe, J. R. Campbell, L. Thompson, S. L. Walz, and W. W. Scultz, Int. Biodeter. Biodegr., **33**, 103 (1994).

165. A. H. M. M. El-Sayed, W. M. Mohmoud, E. M. Davis, and R. W. Coughlin, Int. Biodeter. Biodegr., **37**, 69 (1996).

166. Z. Flip, Eur. J. Appl. Microbiol., **5**, 225 (1978).

167. T. Nakajima-Kambe, F. Onuma, N. Kimpara, and T. Nakahara, FEMS Microbiol. Lett., **129**, 39 (1995).

168. M. Szycher, in Materials Research Society Symposium Proceedings, Vol 110, Materials Research Society, Pittsburgh, PA, 1989, pp. 41–50.

169. M. Parikh, R. A. Gross, and S. P. McCarthy, in Fundamentals of Biodegradable Materials and Packaging, D. Kaplan, E. Thomas, and C. Ching (Eds.), Technomic Publishing, Lancaster, PA, 1993, pp. 159–170.

170. D. Byrom, in Biomaterials: Novel Materials From Biological Sources, D. Byrom (Ed.), Macmillan, New York, 1991, pp. 335–359.

171. M. Alexander, Introduction to Soil Microbiology 2nd ed., Wiley, New York, 1977.

172. P. A. Holmes, L. F. Wright, and S. H. Colins, European Patent Application EP 52, 459, 1985.

173. H. Brandl, R. A. Gross, R. W. Lenz, and R. C. Fuller, Appl. Environ. Microbiol., **54**, 1977 (1988).

174. M. H. Choi and S. C. Yoon, Appl. Environ. Microbiol., **60**, 3245 (1994).

175. O. Kim, R. A. Gross, and D. R. Rutherford, Can. J. Microbiol., **41** (Suppl.), 32 (1993).

176. M. E. John and G. Keller, Proc. Natl. Acad. Sci USA, **93**, 12768 (1996).

177. A.-C. Albertsson, S. O. Andersson, and S. Karlsson, Polym. Degr. Stab., **18**, 73 (1987).

178. C. J. Lusty and M. Doudoroff, Proc. Natl. Acad. Sci. USA, **56**, 960 (1966).

179. T. Tanio, T. Fukui, T. Saito, K. Tomita, T. Kaiho, and S. Masamune, J. Biochem., **124**, 71 (1982).

180. J. E. Kemnitzer, S. P. McCarthy, and R. A. Gross, Macromolecules., **25**, 5927 (1992).

181. J. E. Kemnitzer, S. P. McCarthy, and R. A. Gross, Macromol., **26**, 6143 (1993).

182. K. M. Downing, C. S. Ho, and D. W. Zabriskie, Biotechnol. Bioeng., **29**, 1086 (1987).

183. F. Kawai and F. Moriya, J. Ferm. Bioeng., **71**, 1 (1991).

184. F. Kawai and H. Yamanaka, "Biodegradation of Polyethylene Glycol by Symbiotic Mixed Culture (Obligate Mutulism)" Arch. Microbiol., **146**, 125–129 (1986).

185. F. Kawai, H. Yamanaka, M. Ameyama, E. Shinagawa, K. Matsushita, and O. Adachi, Agric. Biol. Chem., **49**, 1071 (1985).

186. H. Yamanaka and F. Kawai, J. Ferment. Bioeng., **67**, 324 (1989).

187. J. Frings, E. Schramm, and B. Schink, Appl. Environ. Microbiol., **58**, 2164 (1992).

188. B. Schink and M. Stieb, Appl. Environ. Microbiol., **45**, 1905 (1983).

189. D. Dwyer and J. M. Tiedje, Appl. Environ. Microbiol., **46**, 185 (1983).

190. L. Pan and J.-D. Gu, J. Polym. Environ., **15**, 57 (2007).

191. G. A. Brown, in Polymer Materials for Electronic Applications, ACS Symposium Ser. No. 184, E. D. Feit and C. W. Wilkins (Eds.), American Chemical Society, Washington, DC, 1982, pp. 151–169.

192. R. J. Jensen, in Polymers for High Technology—Electronics and Phtonics, ACS Symp. Ser. No. 346, M. J. Bouwden, and S. R. Turner (Eds.), American Chemical Society, Washington, DC, 1987, pp. 466–483.

193. J. H. Lai, Polymers for Electronic Applications, CRC Press, Boca Raton, FL, 1989.

194. J. W. Verbicky, in Encyclopedia of Polymer Science and Engineering, Vol. 12, Wiley, New York, 1988, pp. 364–383.

195. T. Verbiest, D. M. Burland, M. C. Jurich, V. Y. Lee, R. D. Miller, and W. Volksen, Science, **268**, 1604 (1995).

196. K. Kelley, Y. Ishino, and H. Ishida, Thin Solid Films, **154**, 271 (1987).

197. P. O. Hahn, G. W. Rubloff, J. W. Bartha, F. Legoues, R. Tromp, and P. S. Ho, in Materials Research Society Symposium Proceedings, Vol. 40, Materials Research Society, Pittsburgh, PA, 1985, pp. 251–263.

198. F. Mansfeld, J. Appl. Electrochem., **25**, 187 (1995).

199. E. P. M. van Westing, G. M. Ferrari, and J. H. W. De Witt, Corrosion Sci., **36**, 957 (1994).

200. K. E. G. Thorp, A. S. Crasto, J.-D. Gu, and R. Mitchell, in Proceedings of the Tri-Service Conference on Corrosion, T. Naguy (Ed.), U.S. Government Printing Office, Washington, DC, 1994, pp. 303–314.

201. A.-C. Albertsson, C. Barenstedt, and S. Karlsson, Acta Polym., **45**, 97 (1994).

202. Z. Zhou and N. Brown, Chin. J. Mater. Res., **9** (Suppl.), 463 (1995).

203. F. Severini, R. Gallo, and S. Ipsale, Polym. Deg. Stabil., **22**, 185 (1988).

204. American Society for Testing and Materials (ASTM), in 1993 Annual Book of ASTM Standards, Vol. 08.03, D5247-92, ASTM Philadephia, PA, 1993, pp. 401–404.

205. American Society for Testing and Materials (ASTM), in 1993 Annual Book of ASTM Standards, Vol. 08.03, D5271-92, ASTM, Philadephia, PA, 1993, pp. 411–416.

206. American Society for Testing and Materials (ASTM), in 1993 Annual Book of ASTM Standards, Vol. 08.03, D5338-92, ASTM, Philadephia, PA, 1993, pp. 444–449.

207. American Society for Testing and Materials (ASTM), in 1993 Annual Book of ASTM Standards, Vol. 08.03, G21-90, ASTM, Philadephia, PA, 1993, pp. 527–529.

208. American Society for Testing and Materials (ASTM), in 1993 Annual Book of ASTM Standards, Vol 08.03, G22-76, ASTM, Philadephia, PA, 1993, pp. 531–533.

209. U.S. Department of Defense (US DoD), Military Standard 810E, Method MIL-STD-810E, Washington, DC, 1989.

210. J.-D. Gu, and L. Pan, J. Polym. Environ., **14**, 273 (2006).

211. Y. Fan, Y. Wang, P. Qian, and J.-D. Gu, Int. Biodeter. Biodegr., **53**, 57 (2004).

212. H. W. Rossmoore and L. A. Rossmoore, in A Practical Manual on Microbiologically Influenced Corrosion, G. Kobrin (Ed.), NACE International, Houston, TX, 1993, pp. 31–40.

213. R. Zhang, Y. Wang, and J.-D. Gu, Antoine van Leeuvenhoek–Int. J. Gen. Mol. Microbiol., **89**, 307 (2006).

214. Y. Wang, P. C. Leung, P. Qian, and J.-D. Gu, Microb. Environogy., **19**, 163 (2004).

215. Y. Wang, P. C. Leung, P. Qian, and J.-D. Gu, Ecotoxicology, **15**, 371 (2006).

216. J. Gillatt, Int. Biodeter., **26**, 205 (1990).

217. I. M. Ezeonu, J. A. Noble, R. B. Simmons, D. L. Price, S. A. Crow, and D. G. Ahearn, Appl. Environ. Microbiol., **60**, 2149 (1994).

218. I. M. Ezeonu, D. L. Price, R. B. Simmons, S. A. Crow, and D. G. Ahearn, Appl. Environ. Microbiol., **60**, 4172 (1994).

219. A. Sunesson, W. H. J. Vaes, C. Nilsson, G. Blomouist, B. Andersson, and R. Carlson, Appl. Environ. Microbiol., **61**, 2911 (1995).

220. Z.-H. Luo, K.-L. Pang, J.-D. Gu, R. K. K. Chow, and L. L. P. Virijmoed, Mar. Pollut. Bull., **58**, 765 (2009).

221. Y. Wang, Y. Fan, and J.-D. Gu, World J. Microbiol. Biotechnol., **19**, 811 (2003).

222. J.-D. Gu, J. Li, and Y. Wang, Water. Sci., Technol., **52** (8), 241 (2005).

31

DURABILITY OF CONCRETE[*]

V. M. MALHOTRA

Consultant, Ottawa, Ontario, Canada

[*]© Her Majesty the Queen in Right of Canada, as represented by the Minister of Natural Resources, 2010.

A. INTRODUCTION

According to the American Concrete Institute (ACI) Committee 201, durability of Portland cement concrete is defined as its ability to resist weathering action, chemical attack, abrasion, or any other process of deterioration; that is, durable concrete will retain its original form, quality, and serviceability when exposed to its environment. If concrete is proportioned correctly and is cast, consolidated, and cured properly, it should be maintenance free for a very long time. But in practice this rarely happens with the result that concrete structures deteriorate prematurely and need constant repairs and maintenance at a considerable cost. This chapter discusses selected aspects of physical and chemical attacks that affect significantly the long-term durability of concrete. The physical attack that causes major distress in concrete structures is the freezing and thawing phenomenon; the chemical attacks include distress caused

Uhlig's Corrosion Handbook, Third Edition, Edited by R. Winston Revie
Copyright © 2011 John Wiley & Sons, Inc.

by the sulfates, alkali-aggregate reactions, seawater, and carbonation of concrete.

Before discussing in some detail the deterioration of concrete by physical and chemical mechanisms, it should be stressed that porosity of hydrated cement paste, and hence concrete, plays a very major role in its durability. As porosity of concrete is directly related to its water–cement (W/C) ratio, any decrease in W/C will reduce significantly the porosity and hence its permeability. The decreased permeability will decrease the transportation of aggressive chemicals into concrete and will also control the moisture content during environmental changes. For example, a decrease in W/C concrete from 0.80 to 0.40 will reduce the coefficient of water permeability from $\sim 130 \times 10^{-12}$ to 10×10^{-12} m/s.

B. DETERIORATION CAUSED BY FREEZING AND THAWING CYCLES

In Canada, northern parts of the United States, northern Europe, Japan, Korea, Russia, and northern parts of China, concrete is subjected to repeated cycles of freezing and thawing. In addition, in several of these countries, deicing chemicals, such as sodium chloride, are routinely used to melt ice and snow on the highways, roads, and sidewalks. The freezing and thawing cycling and the combined action of freezing and thawing cycling and the application of the deicing salts result in considerable deterioration of concrete. In the deicing salt scaling of concrete, the mortar near the surface flakes or peels away. It is primarily a surface phenomenon, in contrast to the internal cracking of concrete in freezing and thawing cycling

when the bulk of the concrete is affected. The concrete made with sound aggregates can be fully protected against freezing and thawing by proper air entrainment; however, air entrainment is found rarely to protect concrete totally from damage due to the combined action of freezing and thawing cycling and the deicing concrete chemicals. It is generally agreed upon that the deterioration caused by the deicing chemicals is mostly physical in nature and that the chemical reactions of the salts with the cement hydration products play only a secondary role in the deterioration mechanisms. Figure 31.1 shows a concrete structure damaged by freezing and thawing cycling.

B1. Mechanisms of Freezing and Thawing Deterioration

B1.1. Hydraulic Pressure. Water in the capillary pores of cement paste in concrete expands $\sim 9\%$ upon freezing. If the increased volume is smaller than the space available, no damage will occur; otherwise the excess water will be expelled by the hydraulic pressure. As cement paste is a permeable material, there is a possibility that excess water can escape from the capillary pores to the nearest air void during the process of freezing if the air void is unfilled. An entrained-air void will be unfilled, that is, filled with air unless a crack has penetrated it. Water forced into it by the mechanism described will exit immediately on thawing, forced out by the compression of the air, and hence no liquid will remain to evaporate and form secondary deposits. The presence of such deposits indicates that the bubble has been penetrated by a crack. Thus, the magnitude of this hydraulic

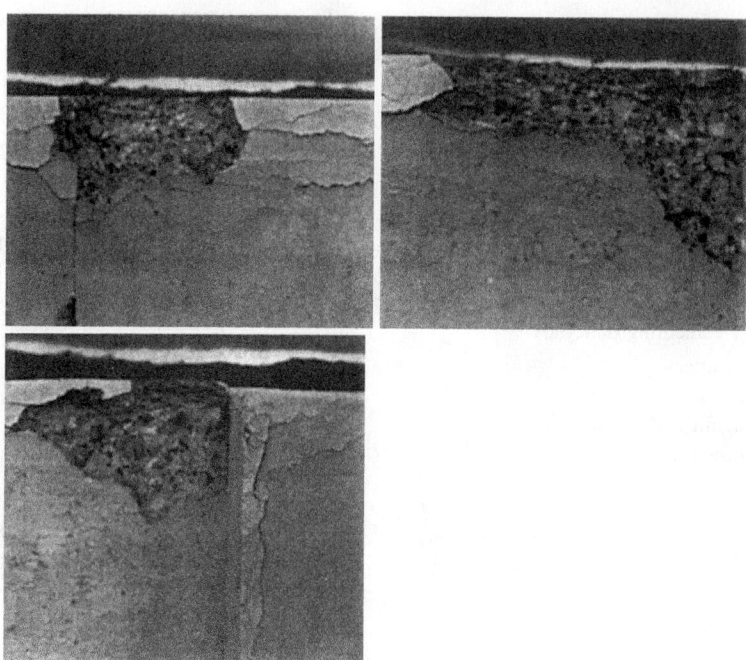

FIGURE 31.1. Damage of a concrete structure caused by freezing and thawing cycling [6].

pressure depends on the permeability of the cement paste, the distance that water must travel to reach the nearest unfilled void, the rate of freezing, the rate of ice formation, and the degree of saturation of the paste. If the pressure is high enough to stress the surrounding paste beyond its tensile strength, it will cause cracking [1, 2].

B1.2. Osmotic Pressure.   In addition to hydraulic pressure caused by water freezing in capillary cavities, the osmotic pressure resulting from partial freezing of the solutions in such cavities can be another source of destructive expansions in cement paste. Water in the capillary cavities is not pure; it contains various soluble substances. Such solutions freeze at slightly lower temperatures than pure water. When solutions of different concentrations of soluble materials are separated by a permeable barrier, the solvent particles move through the barrier toward the solution of greater concentration. The existence of local salt concentration gradients between capillaries is envisaged as the source of osmotic pressure [3].

The hydraulic pressure due to an increase in the specific volume of water on freezing in large cavities and the osmotic pressure due to salt concentration differences in the pore fluid do not appear to be the only causes of expansion of cement paste exposed to frost action [4], but they are believed to be the main ones.

B1.3. Measures to Avoid Damage Due to Freezing and Thawing Cycling.   Air entrainment in concrete has been used as a means of reducing internal damage due to frost action. In air-entrained Portland cement paste, every air void is assumed to be bordered by a zone in which the hydraulic pressure cannot become high enough to cause damage. By reducing the distance between the voids to the point where the protected zones overlap, the generation of disruptive hydraulic pressures during the freezing of water in the capillary cavities can be prevented. In order to avoid damage due to freezing and thawing cycling, concrete in North America is routinely air entrained using chemical admixtures. At a given air content, the protection provided by the air voids against repeated cycles of freezing and thawing is usually greater, the larger the number of voids per unit volume of paste. This implies that voids are more effective when they are close together. It is generally agreed upon that the cement paste is well protected against the effects of frost action if the spacing factor $\bar{L}$ of the air-void system is 200 μm or less, as determined in accordance with American Society for Testing and Materials (ASTM) C 457 test procedure. Field experience has shown that concrete incorporating between 5 and 9% air by volume will generally yield values of $\bar{L}$ which are of the order of 200 μm. When supplementary cementing materials such as fly ash or blast-furnace slag are incorporated into concrete together with superplasticizers, the value of $\bar{L}$ can reach

~230 μm with no detrimental effects on the performance of concrete subjected to freezing and thawing cycling.

B2. Mechanisms of Surface Scaling Due to Combined Action of Freezing and Thawing Cycling and Application of Deicing Salts

In addition to the hydraulic and osmotic pressure theories discussed above, there are a few more phenomena that are associated with the presence of deicing chemicals and are believed to contribute to the surface-scaling type of deterioration of concrete (Fig. 31.2).

B2.1. Layer-by-Layer Freezing.   Differences in the salt concentration in concrete lead to corresponding differences in freezing temperatures of various layers. When ice formation occurs in such a layer-by-layer way, stresses can develop whose extent depends on the dilation difference between the frozen and unfrozen layers [5].

B2.2. Thermal Shock.   Heat is required for the melting of ice and snow. When thawing takes place by the application of deicing chemicals, the heat is extracted mostly from the concrete. The great heat loss causes a rapid temperature drop on the surface of the concrete. The temperature gradient developed can cause stresses of short duration near the surface that may exceed the tensile strength of the concrete and cause damage [5].

B2.3. Supercooling Due to Preventive Salt Application.   The use of deicing chemicals prevents water from freezing

FIGURE 31.2. Surface scaling of concrete caused by deicing chemicals [6].

on the surface of concrete at a temperature near 0°C. When the supercooled water near the surface of the concrete eventually freezes, the destructive effect of the phase transition will be greater than with normal freezing [5].

B2.4. Osmotic Pressure. The use of deicing chemicals will increase the concentration of the chemicals in the capillary pores near the surface of the concrete, which may build up an osmotic pressure high enough to cause a rupture of the cement paste near the surface.

B2.5. Measures to Avoid Damage Due to Deicing Salt Scaling. So far, there is no agreement that any of the above mechanisms are the primary reasons for the deterioration of concrete due to scaling. Consequently, no effective solution has been possible to prevent or to reduce the surface scaling of concrete. Research and practical experience indicate that air entrainment in concrete is effective in preventing internal deterioration; however, the scaling is never completely prevented by the air entrainment, and the critical air-void spacing factor ($\bar{L}$) concept may not be applicable to deterioration due to the scaling. According to Cordon [6], field experience has shown that linseed oil acts as an antiscaling agent. The solutions of this oil and flammable solvents are sprayed on highways, streets, and bridges to reduce damage caused by deicing salts.

C. DETERIORATION CAUSED BY ALKALI–AGGREGATE REACTIONS

For many years, aggregates were believed to be essentially inert and chemically nonreactive in concrete mixtures. Unfortunately, this often is not true, and deleterious reactive aggregates have been found in many parts of the world, including Africa, Australia, Canada, China, England, India, Japan, New Zealand, the Scandinavian countries, and the United States.

C1. Types of Reaction

There are two types of alkali–aggregate reactions:

   (a) Alkali–silica reaction
   (b) Alkali–carbonate reaction

Of these two reactions, the alkali–silica reaction is the most common. For this reason, the alkali–carbonate reaction will not be discussed here.

   The alkali–silica reaction is a reaction in either mortar or concrete between the hydroxyl (OH^-) ions associated with the alkalies (Na_2O and K_2O) from the cement or other sources and certain mineral phases that may be present in the coarse or fine aggregates; under certain conditions,

deleterious expansion and consequent cracking of the concrete or mortar may result [7].

C2. Nature of Reaction and Expansion Processes

The reaction starts when the alkaline hydroxides in the concrete fluid (Na–OH, K–OH) attack the surface of silica minerals in the aggregate. This results in an alkali silica gel and an alteration of the aggregate surface, known as "reaction rim." The resultant gel has a strong affinity for water and, consequently, a tendency to increase in volume. The expanding gel exerts an internal pressure that is sufficient to fracture the surrounding cement paste. Some of the softer gel can leach out through the voids and cracks, but expansion of the solid and semisolid products are more damaging. Figure 31.3 shows a damaged bridge structure due to the alkali–silica reaction, and Figure 31.4 shows expansion in a concrete median barrier due to the alkali–aggregate reactivity.

   Table 31.1 gives a list of minerals and of rock types that have been found reactive in concrete structures. The information given in this table should, however, be used with care. For example, even if some greywackes and granites have been found to be reactive in certain parts of Canada, it does not mean that all greywackes and granites found in Canada are reactive or will react to the same degree. Petrographic analysis will generally help to identify and to determine the

FIGURE 31.3. Cracking in a concrete bridge caused by alkali–silica reaction [8].

FIGURE 31.4. Cracking of a highway medium barrier caused by alkali–aggregate reactivity [8].

TABLE 31.1. Mineral Phases Susceptible to Deleterious Reactions with Alkalies in Cement and Corresponding Rocks[a]

Alkali-Reactive Silica Minerals and Rocks

B1.1 Alkali-Reactive Silica Minerals and Volcanic Glasses (Classical Alkali–Silica Reaction)

Reactants — Opal, tridymite, cristobalite; acid, intermediate, and basic volcanic glasses; artificial glasses; beekite

Rocks — Rock-types containing opal such as shales, sandstones, silicified carbonate rocks, some cherts and flints, diatomites

Vitrophyric volcanic rocks; acid, intermediate and basic, such as rhyolites, dacites, latites, andesites, and their tuffs; perlites, obsidians; all varieties with a glassy goundmass; some basalts

B1.2 Alkali-Reactive Quartz-Bearing Rocks

Reactants — Chalcedony, cryptocrystalline to microcrystalline and macrogranular quartz with deformed crystal lattice, rich in inclusions, intensively fractured or granulated; poorly crystalline quartz at grain boundaries

Rocks — Cherts, flints, vein quartz, quartzite, quartz-arenite, quartzitic sandstones that contain microcrystalline to cryptocrystalline quartz, and /or chalcedony

Volcanic rocks such as in B1.1 (above) but with devitrified, crypto- to microcrystalline groundmass

Micro- to macrogranular silicate rocks of various origin:
Metamorphic rocks: gneisses, quartz-mica schists, quartzites, homfilses, phyllites, argillites, slates
Igneous rocks: granites, granodiorites, charnockites
Sedimentary rocks: sandstones, greywackes, siltstones, shales, siliceous limestones, arenites, arkoses
Sedimentary rocks (sandstones) with epitaxic quartz cement overgrowth

[a]From Canadian Standards Association document CSA 23.1 Appendix B.

proportion of the various potentially reactive rock types in an aggregate sample from a quarried operation or a gravel deposit. However, this information alone will not permit the prediction of the magnitude of the reaction that may occur in the field.

C3. Conditions Conducive to Alkali–Aggregate Reactivity

The rate and extent of expansion due to alkali–aggregate reaction are affected by a large number of factors. These can generally be grouped as follows [8]:

(a) Inherent reactivity of the siliceous material
(b) Total alkali content of concrete
(c) Environmental considerations

C3.1. Inherent Reactivity of Siliceous Material. The nature, amount, and particle size of the siliceous phase within the rock particles play a major role on the actual rate of alkali–aggregate reaction in mortar or concrete. Poorly crystalline forms of silica such as opal and volcanic glass are very reactive because of the nature of their structure through which the alkali hydroxide ions (Na–OH and K–OH) can penetrate quickly. As little as 1% of such reactive components in an aggregate may lead to deleterious reaction and cracking and generally within 10 years of construction.

Quartz is one of the major constituents of the various rocks found in the earth's crust. Quartz has a well-organized three-dimensional crystalline structure that reacts at a much slower rate than opal and volcanic glasses. However, very small particles of quartz, because of their increased surface area, or larger particles, which have been subjected to stress in their geological history and consequently show defects in their crystalline structure, may be fairly reactive.

C3.2. Total Alkali Content of Concrete. The source of alkali for the alkali–aggregate reaction is generally considered to be derived from the Portland cement in concrete. Alkalies may also be contributed by aggregates, admixtures, supplementary cementing materials, and extraneous sources such as deicing salts and seawater. The amount of alkalies contributed by the cement for the alkali–aggregate reaction is calculated by adding the sodium oxide content to 0.658 times the potassium oxide content. The calculation provides the sodium oxide equivalent.

The minimum alkali content at which the alkali–aggregate reaction will occur will vary from one aggregate to another and the conditions to which the concrete is subjected. The alkali–aggregate reaction will generally not occur when the total alkali content is $<3 \, \text{kg/m}^3$ of concrete. There may be special circumstances where one should consider alkali contents $<3 \, \text{kg/m}^3$, such as where concrete is exposed to extraneous alkalies.

Generally, the higher the alkali content, the greater the expansion for a given cement content. In addition, for a given cement composition, finer grinds tend to provide a greater expansion; this may be explained by more complete hydration of the cement causing greater production of alkalies.

C3.3. Environmental Considerations. The progress of alkali–aggregate reaction is also largely influenced by the presence of accessible moisture in the pore space of the concrete, as reactions do not occur in the absence of moisture. Concrete deterioration due to alkali–aggregate reaction has often been observed to be more severe in the portions of structures subjected to cyclic conditions of wetting and drying and is accelerated at higher temperature. For example, in the southwestern Cape Province of South Africa, alkali–aggregate damage typically appears on field concrete structures after ~ 4–7 years, whereas in Canada, where the average annual temperature is much lower than in South Africa, the damage generally appears after ~ 15–20 years for a similar type of reactive greywacke aggregate. Expansion and rates of expansion are known to increase with increasing temperature and humidity conditions, but also in the presence of external alkalies such as deicing salts, seawater and seawater sprays.

C3.4. Preventive Measures Against Alkali–Aggregate Reactivity. The following are the most commonly used preventive measures against alkali–aggregate reactivity:

Use of nonreactive aggregate
Use of low-alkali cement
Limiting the alkali content of the concrete mixtures
Use of supplementary cementing materials

C4. Use of Nonreactive Aggregates

A simple solution to controlling the alkali–aggregate reactions would be to specify and use only nonreactive aggregates. However, the term nonreactive aggregate is a misnomer as most aggregates are reactive to a degree, depending on the exposure conditions and the alkali content of the concrete mixtures. Nevertheless, for very critical structures, such as offshore oil drilling platforms and nuclear reactors, the least reactive aggregate should be used. The past performance record of an aggregate in similar structures and environments as in the proposed structure can provide valuable data on the aggregate performance.

C5. Use of Low-Alkali Cement

Research and field experience have shown that the usage of cements with an alkali content (Na_2O equivalent) $<0.6\%$ will generally arrest alkali–aggregate reaction in many countries; however, the available raw materials may not be conducive to the production of cements with low-alkali contents. The alkali content of cements available in North America for normal construction purposes averages ~ 0.8% equivalent Na_2O.

C6. Limiting Alkali Content of Concrete Mixtures

Alkalies that are available for the alkali–aggregate reaction arise principally from the cement. Other sources of alkali may include aggregate, water, admixtures, and supplementary cementing materials. It is generally considered that for concrete made with potentially reactive aggregates the alkalies in the concrete should not exceed $3 \, kg/m^3$. Critical structures such as dams and nuclear power plants may require even lower alkali contents, closer to 2.0–$2.5 \, kg/m^3$ of concrete.

C7. Use of Supplementary Cementing Materials

Laboratory and field test data available so far have shown that supplementary cementing materials such as fly ash and blast furnace slag, when used in the proper proportions, are efficient in reducing expansion of concrete incorporating reactive aggregates. As a precautionary measure, when there is a possibility that a potentially reactive aggregate may be used, concretes should be made incorporating ~ 20–30% of a fly ash. Any drop in early strength should be compensated for by reproportioning the concrete mixtures. Another approach is to use high-volume fly ash concrete [9].

Silica fume, a byproduct during the manufacture of silicon–boron steels in an arc furnace, is also being used in specific projects where high-strength impermeable concrete is desired. No well-documented data are yet available as to the long-term effectiveness of silica fume in reducing expansion due to alkali–aggregate reaction in concrete. Limited data have shown that, to be effective, one has to use between 10 and 15% silica fume by mass as a replacement for high-alkali Portland cement; however, this will depend on the degree of reactivity of the aggregate and the total alkali content of a concrete mixture.

D. DETERIORATION DUE TO SULFATE ATTACK

One of the major reasons of concrete deterioration is attack by sulfates. This type of chemical attack is widespread and has been reported from many countries in the world, including Canada, the United States, and the Middle East. The U.S. Bureau of Reclamation had cautioned civil engineers of the seriousness of this phenomenon in the early 1930s. The Bureau had emphasized that soluble sulfates in soils ($>0.5\%$) could cause serious damage to concrete [10]. Figure 31.5 shows disintegration of a concrete element caused by sulfate attack.

FIGURE 31.5. Disintegration of concrete caused by sulfate attack [9].

D1. Mechanisms of Sulfate Attack

The sulfates of sodium, calcium, and magnesium present in alkali soils and water react chemically with hydrated lime and hydrated calcium aluminate in the cement paste (hydraulic binder) in concrete to form calcium sulfates and calcium sulfoaluminates respectively. These are expansive reactions and result in degradation of concrete. The chemical reactions involved are as follows [11]:

(A) *Sodium Sulfate Attacks (CaOH)$_2$*

$$Ca(OH)_2 + Na_2SO_4 \cdot 10H_2O \rightarrow CaSO_4 \cdot 2H_2O$$
$$+ 2NaOH + 8H_2O \cdot (gypsum)$$

(B) *Sodium Sulfate Attacks Calcium Aluminate Hydrate*

$$2(3CaO \cdot Al_2O_3 \cdot 12H_2O + 3Na_2OSO_4 \cdot 10H_2O)$$
$$\rightarrow 3C_2O \cdot Al_2O_3 \cdot 3CaSO_4 \cdot 31H_2O + Al(OH)_3$$
$$+ 6NaOH + 17H_2O$$

(C) *Calcium Sulfate Attacks Calcium Aluminate Hydrate.* Calcium sulfate attacks only aluminate hydrate forming calcium sulfoaluminate (3 CaO · Al$_2$O$_3$ · 3 CaSO$_4$ · 32 H$_2$O). This is known as ettringite and causes internal cracking of concrete.

(D) *Magnesium Sulfate Attacks Ca(OH)$_2$, Calcium Aluminate, and Calcium Silicate Hydrates.* The reaction with calcium silicate hydrate takes the form

$$3CaO \cdot 2SiO_2 \cdot (aq) + MgSO_4 \cdot 7H_2O$$
$$\rightarrow CaSO_4 \cdot 2H_2O + Mg(OH)_2 + SiO_2 \cdot (aq)$$

The damage and expansion caused by the magnesium sulfate is more severe than by the calcium or sodium sulfates. Figure 31.5 shows concrete damaged by sulfate attack.

D2. Measures to Protect Against Sulfate Attack

D2.1. General Measures. The best protection against sulfate attack is to make concrete with very low permeability. This can be achieved by the use of low W/C ratio, use of supplementary cementing materials, and superplasticizers, adequate consolidation, and curing of concrete. When the concentration of sulfates is rather high, the use of sulfate-resisting cements can be very beneficial. The recommendations of Mehta and Monteiro [4] on the subject of control of sulfate attack are as follows:

Portland cement containing <5% C$_3$A (ASTM type V) is sufficiently sulfate resisting under moderate conditions of sulfate attack (i.e., when ettringite-forming reaction are the only consideration). However, when high-sulfate concentrations of the order of 1500 mg/L or more are involved, which are normally associated with the presence of magnesium and alkali cations, the ASTM type V Portland cement may not be effective against the cation exchange reactions involving gypsum formation, especially if the C$_3$S content of the cement is high. Under these conditions, experience shows that cements potentially containing little or no calcium hydroxide on hydration perform much better: For instance, high-alumina cements, Portland blast furnace slag cements with >70% slag, and Portland pozzolan cements with at least 25% pozzolan (natural pozzolan, calcined clay, or low-calcium fly ash).

The U.S. Bureau of Reclamation has classified severity of sulfate attacks in four degrees. Based upon this criterion, the American Concrete Institute Building Code 318 stipulates the following requirement for controlling the sulfate attack [4]:

1. *Negligible Attack:* When the sulfate content is <0.1% in soil, or < 150 ppm (mg/L) in water, there shall be no restriction on the cement type and W/C ratio

2. *Moderate Attack:* When the sulfate content is 0.1–0.2% in soil, or 150–1500 ppm in water, ASTM type II Portland cement or Portland pozzolan or Portland slag cement shall be used, with <0.5 W/C ratio for normal-weight concrete.

3. *Severe Attack:* When the sulfate content is 0.2–2.0% in soil, or 1500–10,000 ppm in water, ASTM type V Portland cement, with < 0.45 W/C ratio, shall be used.

4. *Very Severe Attack:* When the sulfate content is > 2% in soil, or >10,000 ppm in water, ASTM type V cement plus a pozzolanic admixture shall be used, with <0.5 W/C ratio. For lightweight-aggregate concrete, the ACI Building Code specifies a minimum 28-day compressive strength of 28 MPa for severe or very severe sulfate attack conditions.

E. DETERIORATION OF CONCRETE IN SEAWATER

In recent years, the performance of concrete in seawater has become of great importance because of the construction of multi-billion-dollar offshore oil drilling platforms and other associated infrastructure. Most seawaters are characterized by the presence of ~ 3.5% soluble salts by weight. In general, the pH of seawater ranges from 7.5 to 8.4. The Mg^{2+} and SO_4^{2-} are present in concentrations of 1400 and 2700 mg/L, respectively, and are the most aggressive in their attack on the products of cement hydration. The concentration of Na^+ and Cl^- ions are ~ 11,000 and 20,000 mg/L.

TABLE 31.2.   Decomposing Action of Seawater on Constituents of Hydrated Portland Cement[a]

Seawater Component That Can Enter into Deleterious Chemical Reactions with Hydrated Portland Cement	Possible Chemical Reactions	Physical Effects Associated with Chemical Reactions
Carbon Dioxide Small quantities of dissolved CO_2, derived mainly from absorption of atmospheric CO_2, always present in seawater. However, decaying vegetable matter can lead to substantially larger and harmful concentrations of dissolved CO_2, which are generally reflected by reduction of the seawater pH to values <8.	$CO_2 + Ca(OH)_2 \rightarrow \underset{\text{Aragonite}}{CaCO_3} + H_2O \xrightarrow{CO_2} \underset{\text{Bicarbonate of calcium}}{Ca(HCO_3)_2}$ $CO_2 + [Ca(OH)_2 + 3CaO \cdot Al_2O_3 \cdot CaSO_4 \cdot 18H_2O]$ $\rightarrow 3CaO \cdot Al_2O_3 \cdot CaCO_3 \cdot xH_2O + \underset{\text{Aragonite}}{CaSO_4 \cdot 2H_2O}$ $3CO_2 + 3CaO \cdot 2SiO_2 \cdot 3H_2O \rightarrow$ $\underset{\cdot \;\text{Aragonite}}{3CaCO_3} + 2SiO_2 \cdot H_2O$	Both calcium bicarbonate and gypsum are soluble in seawater. Loss of material and weakening or mushiness of hardened cement paste can therefore be associated with the formation of these compounds. Since all the hydration products of Portland cement, including the calcium silicate hydrate, can be decomposed by carbonation reactions, permeable concretes in seawater containing larger than normal CO_2 concentration are likely to deteriorate.
Magnesium Salts Typically, seawater contains 3200 ppm $MgCl_2$ and 2200 ppm $MgSO_4$. Regarding cement hydration products, these magnesium salts, even in the small concentrations present, are considered harmful.	$MgCl_2 + Ca(OH)_2 \rightarrow \underset{\text{Brucite}}{Mg(OH)_2} + CaCl_2$ $MgSO_4 + Ca(OH)_2 \rightarrow Mg(OH)_2 + \underset{\text{Gypsum}}{CaSO_4 \cdot 2H_2O}$ $MgSO_4 + [Ca(OH)_2 + 3CaO \cdot Al_2O_3 \cdot CaSO_4 \cdot 18H_2O]$ $\rightarrow Mg(OH)_2 + \underset{\text{Ettringite}}{3CaO \cdot Al_2O_3 \cdot 3CaSO_4 \cdot 32H_2O}$ $MgSO_4 + [Ca(OH)_2 + 3CaO \cdot 2SiO_2 \cdot 3H_2O]$ $\rightarrow 4MgO \cdot SiO_2 \cdot 8H_2O + CaSO_4 \cdot 2H_2O$	$CaCl_2$ and gypsum, being soluble in seawater, lead to material loss and weakening. Formation of ettringite is associated with expansion and cracking. It is reported that the conversion of $3\,CaO \cdot 2\,SiO_2 \cdot 3\,H_2O$ to $4\,MgO \cdot SiO_2 \cdot 8\,H_2O$ is associated with brittleness and strength loss.

[a]From [12].

The decomposing action of seawater on the constituents of hydrated Portland cement as given by Mehta [12] is reproduced in Table 31.2. Figure 31.6 shows the physical and chemical processes responsible for deterioration of reinforced concrete element exposed to seawater.

The presence of large amounts of sulfates in seawater may lead one to consider the possibility of sulfate attack. This is not so because the ettringite formed because of the reaction between sulfate ions and both C_3A and C–S–H is soluble in the presence of chlorides and is leached out and thus does not cause deleterious expansions. This explains why the use of sulfate-resisting cement in concrete exposed to marine environment is not warranted [11].

E1. Requirements of Concrete for Marine Exposure

The selection of concrete for marine exposure must have the lowest possible W/C or water/cementitious ratio to ensure very low permeability. Where possible, the supplementary cementing materials such as fly ash or slag or silica fume and chemical admixture such as superplasticizers should be incorporated in mixture proportioning. The use of sulfate-resisting cement is not essential for concrete exposed to seawater, but the Portland cement should have C_3A content of $<8\%$, when SO_3 content of the cement is $< 3\%$; cements with C_3A contents of up to 10% may be used where the SO_3 content does not exceed 2.5 [10]. Figure 31.7 shows the concrete test specimens at Treat Island (Maine), an outdoor exposure site for determining the long-term performance of concrete exposed to seawater. The concrete blocks cast from well-proportioned concrete are in excellent condition even after 20 years of severe exposure to seawater. Figure 31.8 shows the recently constructed $1 billion Confederation Bridge in eastern Canada. The bridge is about 14 km in length and connects the province of New Brunswick with the province of Prince Edward Island. The prestressed concrete structure is exposed to severe seawater attack and is subject to >100 freezing and thawing cycles a year. The bridge has a design life of 100 years. The high-performance, air-entrained concrete for the bridge was made using silica fume blended cement and incorporates up to 30% fly ash in parts of the foundation elements and piers. Hopefully, the bridge will not suffer major deterioration during its design life.

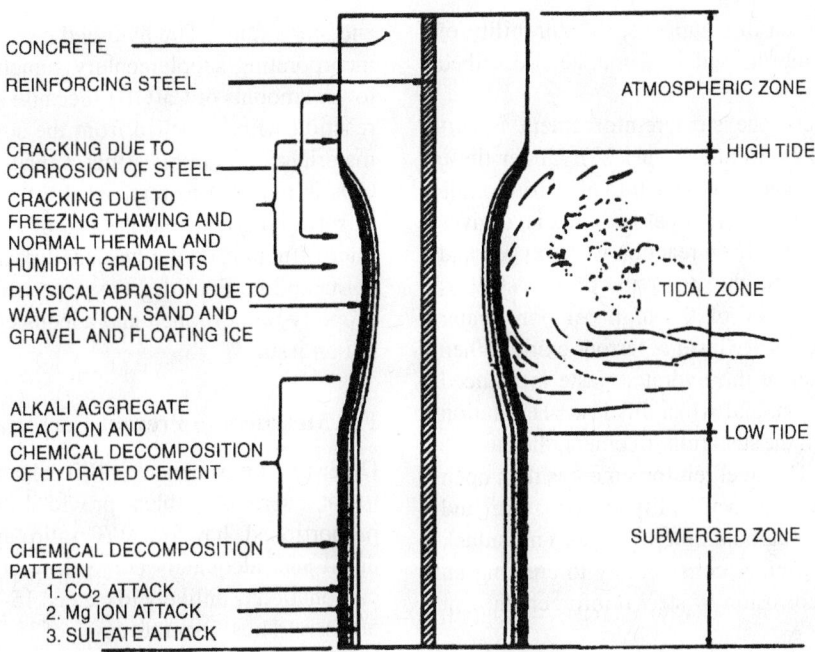

FIGURE 31.6. Physical and chemical processes that cause deterioration of reinforced concrete exposed to seawater [4].

FIGURE 31.7. Treat Island, Maine, exposure site for determining the long term performance of concrete exposed to seawater.

FIGURE 31.8. Confederation Bridge, 14 km long, that connects the Canadian provinces of Prince Edward Island and New Brunswick. Design life of the bridge is 100 years.

F. CARBONATION OF CONCRETE

The term carbonation refers to the reaction of atmospheric carbon dioxide with cement hydrates in concrete. This reaction can have serious implications as regards to durability of reinforced concrete if the concrete is not properly made. The various aspects of carbonation including its mechanism, its rate, and the various factors that influence it are discussed below.

F1. Mechanism of Carbonation

The concentration of CO_2 can vary from 300 ppm (parts per million by volume), that is, 0.03% in nonindustrialized areas

to ~3000 ppm, that is, 0.3% by volume in large cities. The CO_2, even in small concentration, may penetrate into poorly made concrete with high W/C ratio. This is accompanied by the reaction with the cement hydrates, particularly with calcium hydroxide, $Ca(OH)_2$, resulting in the formation of calcium carbonate. The reaction with other hydrates can also take place with the resulting production of silica, alumina, and ferric oxide.

The carbonation of concrete reduces the pH of the pore water in the hydrated cement paste from ~13 to a value approaching ~9. If all the calcium hydroxide is carbonated (this can take many years), the pH of the hydrated paste can reach as low as 8.3. When this happens, the durability of Portland cement is compromised in a manner described below.

In reinforced concrete, the steel reinforcement is surrounded by an alkaline environment (pH 13). Under these conditions, a submicroscopic thin oxide film of ferrous oxide is formed on the steel that renders it passive; that is, it gives the steel complete protection from reaction with oxygen and water. The maintenance of the passivity of the steel is dependent on the availability of the high pH pore water solution surrounding the passive layer of ferrous oxide. When the pH of the pore water of the hydrated paste is reduced because of the carbonation, and when this low-pH solution approaches the surface of the steel reinforcement, the passive oxide film is destroyed. The steel reinforcement is then open to corrosion subject to the availability of moisture and oxygen. Thus, though the carbonation per se does not attack reinforcing steel, it contributes considerably to creating an environment in which corrosion of steel reinforcement can take place.

F2. Factors Affecting Carbonation

The most significant factor affecting carbonation of concrete is the diffusivity of the hardened cement paste in concrete. This, in turn, depends on the composition of concrete, its W/C ratio, and the type of cement used and its curing. All these factors have to be taken into account when investigating the rate of carbonation. Several researchers used the compressive strength of concrete as a parameter for determining the carbonation, primarily because this is a parameter that can be determined easily.

F3. Rate of Carbonation

A most commonly used relationship between depth of carbonation and time depth is as follows [13]:

$$d = k\sqrt{t}$$

where d = depth of carbonation in millimeters (mm)
t = time of exposure in years
k = coefficient of carbonation in millimeters per $\sqrt{\text{year}}$ (mm/year$^{0.5}$)

The values of k can range from <1 for low W/C ratio concrete to >5 for high W/C ratio concrete.

F4. Carbonation of Concrete-Containing Supplementary Cementing Materials

The environmental issues and energy considerations have led to increased use of supplementary cementing materials (SCMs) concrete. These include fly ash, blast furnace slag,

and silica fume. The hydrated cement paste in the concrete incorporating supplementary cementing materials contains lower amounts of $Ca(OH)_2$ because some of it is taken up by reaction with the silica from the supplementary cementing materials. This implies that a relatively smaller amount of CO_2 is required to remove all of the $Ca(OH)_2$. But this is offset by the fact that the use of SCMs results in a denser microstructure of the hardened paste, resulting in reduced permeability. Provided the concrete incorporating SCMs is properly proportioned and adequately cured, carbonation is not an issue.

F5. Measures to Prevent Carbonation

Deterioration of concrete as it relates to carbonation effects is not a serious problem provided that concrete is properly proportioned, has low W/C ratio, and is properly consolidated and adequately cured. In such concrete the depth of carbonation is unlikely to exceed 15–20 mm over a period of 50 years. Thus, concrete cover of 35–50 mm over reinforcing steel should provide sufficient protection against corrosion due to carbonation.

G. CONCLUDING REMARKS

Good-quality concrete that is proportioned properly, is transported and consolidated correctly, is cured adequately, incorporates supplementary cementing materials, and has a low W/C ratio need not deteriorate provided the environment for which it has been designed does not change drastically. Unfortunately, unlike steel, a factory-made product, concrete is normally made at a construction site, and it does deteriorate because all the above requirements are not generally followed. When this happens, concrete cracks giving oxygen and water access to the reinforcing steel results in corrosion of the reinforcing bars and damage to the reinforced concrete structure. As cracking of concrete is very difficult to control, the only alternative then to stop deterioration of reinforced concrete is to use reinforcing bars that do not corrode. No such products are yet available on the market at a reasonable cost. It is therefore imperative that major research efforts be initiated to develop low-cost stainless steel reinforcement or other composite reinforcing materials that have very low tendency for corrosion.

REFERENCES

1. T. C. Powers, J. Am. Concrete Inst., **16**(4), 245 (1945).
2. T. C. Powers, Proc. Highway Res. Board, **29**, 184 (1984).
3. M. H. Zhang, N. Bouzoubaâ, and V. M. Malhotra, in Resistance of Silica Fume Concrete to De-Icing Salt Scaling — A Review, ACI SP-172, V. M. Malhotra (Ed.), American Concrete Institute, Farmington Hills, MI, 1997, pp. 67–102.

4. P. K. Mehta and P. J. M. Monteiro, Concrete: Structures, Properties and Materials, 2nd ed., Prentice-Hall, Englewood Cliffs, NJ, 1993.

5. A. B. Harnic, U. Meier, and A. Rosli, *"Combined Influence of Freezing and Thawing and De-icing Salt on Concretes—Physical Aspects*," Proceedings, First International Conference on Durability of Building Materials and Components, Ottawa, Canada, August 1978.

6. W. A. Cordon, "Freezing and Thawing of Concrete—Mechanism and Control," Monograph No. 3, American Concrete Institute, Farmington Hills, MI, 1966.

7. D Stark, Handbook for the Identification of Alkali-Silica Reactivity in Highway Structure, SHRP-C/FR-91-101, Strategic Highway Research Program, Washington, DC, 1991.

8. W. S. Langley, B. Fournier, and V. M. Malhotra, Alkali-Aggregate Reactivity in Nova Scotia, CANMET, NRCan, Ottawa, Canada, 1993.

9. U.S. Bureau of Reclamation: Concrete Manual, 8th ed., Denver, CO, 1975.

10. A. Bilodeau and V. M. Malhotra, "Concrete Incorporating High Volumes of ASTM Class F Fly Ashes: Mechanical Properties and Resistance to De-Icing Salt Scaling and to Chloride-Ion Penetration," in ACI Special Publication SP-132, Vol. 1, V. M. Malhotra (Ed.), American Concrete Institute, Farmington Hills, MI, 1992, pp. 319–349.

11. A. M. Neville, Properties of Concrete, 4th ed., Prentice-Hall/Pearson Education, Harlow, UK, 1995.

12. P. K. Mehta, Concrete in Marine Environment, Elsevier, New York, 1991.

13. M. G. Richardson, Carbonation of Reinforced Concrete—Its Causes and Management, Citis, New York, 1988.

BIBLIOGRAPHY

ACI Committee 201 "Guide to Durable Concrete," 'Report No. ACI 201-2R-92, American Concrete Institude. Farmington Hills, MI, 1992.

P. Helene, E. P. Figueiredo, T. Holland, and R. Bittencourt, Quality of Concrete Structures and Recent Advances in Concrete Materials and Testing: An International Conference *Honoring V. Mohan Malhotra*, Proceedings, Fourth International Conference, Olinda, Brazil, 2005, ACI Symposium Publication 229, American Concrete Institute, Farmington Hills, MI.

H. K. Hilsdorf and U. Guse, "Frost and Deicing Salt Resistance of High-Strength Concrete," in Proceedings of the RILEM International Workshop, H. Sommer (Ed.), Vienna, Austria, 1994.

D. W. Hobbs, Alkali-Silica Reaction in Concrete, Thomas Telford, London, 1988.

V. M. Malhotra and P. K. Mehta, High Performance High Volume Fly Ash Concrete for Building Sustainable Structures, 3rd ed., Supplementary Cementing Material for Sustainable Development, Ottawa, 2008.

J. Marchand, M. Pigeon, and H. L. Isabelle. "The Frost Durability of Dry Concretes Containing Silica Fume," in Proceedings CANMET/ACI International Workshop on Silica Fume in Concrete, Washington, DC, 1991. (Available from M. Pigeon, Laval University, Quebec City, Canada.)

J. Marchand, M. Pigeon, and H. L. Isabelle, "De-icer Salt Scaling Resistance of Roller-Compacted Concrete Pavements Containing Fly Ash and Silica Fume," in Proceedings CANMET/ACI Fourth International Conference on Fly Ash, Silica Fume, Slag, and Natural Pozzolans in Concrete, Istanbul, Turkey, ACI SP-132, V. M. Malhotra (Ed.), 1992, pp. 151–178.

K. Natesajyer, D. Stark, and K. C. Hover, "Gel Fluorescence Reveals Reaction Product Traces," Concrete Int. **13**(1), 25 (1991).

E. Pazini Figueiredo, T. C. Holland, V. M. Malhotra, and P. Helene, (ed.), Proceedings, (Fifth ACI/CANMET International Conferences on High-Performance Concrete Structures and Materials, ACI Symposium Publication 253, American Concrete Institute, Farmington Hills, MI, 2008.

J. Stark and N. Chelouah, "Freeze-Deicing Salt Resistance of High-Strength Concrete," in Proceedings of the Fourth Weimar Workshop on High Performance Concrete Material Properties and Design, F. H. Wittmann and P. Schwesinger (Eds.), Aedificatio, Freiburg, Germany, 1995, pp. 205–217.

32

MICROBIOLOGICAL CORROSION OF CONCRETE

J.-D. Gu

School of Biological Sciences, The University of Hong Kong, Hong Kong, China

T. E. Ford

University of New England, Biddeford, Maine

R. Mitchell

Harvard School of Engineering and Applied Sciences, Harvard University, Cambridge, Massachusetts

A. INTRODUCTION

Concrete is one of the most widely used materials in construction and has applications in most modern infrastructure, including highways, bridges, and tunnels, and transport of drinking and wastewater, with an approximate consumption rate of 6 billion tons per year. An understanding of the physical performance of concrete materials is vital in order to assess their durability in service under natural conditions. Corrosion or deterioration of concrete has important economic consequences, especially when replacement or repair of infrastructures, such as bridges or municipal sewer systems, is involved [1–8]. However, concrete corrosion is a complex phenomenon involving chemical, physicochemical, electrochemical, and biological processes. Microorganisms have long been implicated in the corrosion of concrete [9–14], but their involvement is probably the least understood among all the processes indicated above. Bacteria in the genus *Thiobacillus* were initially identified as the major culprits through biologically produced sulfuric acid. Other groups of microorganisms involved in the corrosion process have also been implicated, including nitric acid–producing bacteria [15–17], fungi [6, 11], and exopolymer-producing bacteria [2, 11, 18]. Macroscopic fouling organisms also participate in degradation of concrete in submerged structures, but no detailed studies have been reported [19, 20].

Protection against biologically induced corrosion involves modification of the local environment. Corrosion of sewer pipes has been controlled by aeration of stagnant wastewater to decrease the concentrations of H_2S and/or to inhibit populations of sulfate-reducing bacteria (SRB) through addition of specific metal ions, including iron (Fe) compounds [21]. A new generation of technologies using fiber-reinforced resin sheets [22] to coat concrete and addition of polymers [23–25] offer some advantages over conventional methods by providing a physical barrier. However, in addition to concern over cost of these new technologies, polymeric resin–coated concrete was reported to be susceptible to microbial deterioration during exposure in sewer

Uhlig's Corrosion Handbook, Third Edition, Edited by R. Winston Revie
Copyright © 2011 John Wiley & Sons, Inc.

451

systems [26]. Furthermore, microbial degradation of polymeric materials has been reported [12, 27–32]. Fiber-sizing chemicals and plasticizers and the small quantities of organic residues present in composite resins [12, 27, 31, 33–35] can support the active growth of natural populations of microorganisms, including bacteria and fungi [28, 29, 36–38] (also see related chapter 30 in this book). In addition, Slow-growing bacteria cause much greater damage to polymers than the fast-growing ones [35], suggesting that current test methods that focus on fast-growing microbial populations may underestimate the extent of degradation.

B. A BRIEF HISTORY

Corrosion of concrete sewer pipes by hydrogen sulfide, an anaerobic product of sulfur-containing organic compounds and reduction of SO_4^{2-} during microbial metabolism, was identified as the cause of corrosion in 1900 by Olmstead and Hamlin [39]. The corrosion reaction was initially regarded as a purely chemical process [40] in which hydrogen sulfide produced under anaerobic conditions in wastewater is oxidized chemically to sulfuric acid in the presence of molecular oxygen [41]. The sulfuric acid then reacts with calcium in concrete to form $CaSO_4$ or gypsum. Parker and his coworkers established the relationship between acidophilic thiobacilli and concrete degradation by isolating *Thiobacillus concretivorus* (renamed *Thiobacillus thiooxidans*) and another *Thiobacillus* species from corroded concrete (*T. neaopolitanus*) [9, 10, 42–47]. The extent of corrosion correlates positively with the population of these microorganisms, providing sound evidence for their role in the corrosion process.

In the early 1980s, rapid deterioration of newly replaced sewer systems in Hamburg, Germany, renewed the scientific interest in these corrosion processes [8, 26]. In these investigations, positive correlations were observed again between the extent of concrete corrosion and the numbers of *T. thiooxidans*. Sand also observed that *Thiobacillus ferrooxidans* was associated with oxidative activity of H_2S [48]. In the presence of sodium thiosulfate, dominant microorganisms were *Thiobacillus neaopolitanus*, *Thiobacillus intermedius*, and *Thiobacillus novellus*, whereas mercaptan, a S-containing compound, did not support any population of thiobacilli, but heterotrophic bacteria and fungi were commonly found.

Biological corrosion of sewer pipes can be a serious problem in coastal cities due to the high abundance of SO_4^{2-} in the wastewater. At the time of the Hamburg failure, coastal U.S. cities faced similar problems with newly installed concrete sewer systems, especially the city of Los Angeles [7]. A reason for the reemergence of the problem was the advent of the National Pollution Discharge Elimination Systems (NPDES) in 1972, which banned discharge of toxic metals and chemicals into sewers. As a result of this legislation, inorganic and organic toxic wastes are no longer permitted to be discharged directly to sewers, with a resultant increased activity of microorganisms, particularly the SRBs, producing large quantities of H_2S [49–52]. In addition to the sewers, concrete corrosion problems today involve highway bridges, aquaducts, water distribution systems for drinking and wastewater, historic buildings and monuments [53–56], river hydropower dams [57], and nuclear depositories [58, 59] where corrosion rates are unacceptably high. The recognized failure of highways and bridges recently in developed countries has renewed the discussion on concrete corrosion and protection [60, 61].

C. CONTRIBUTIONS BY AIR POLLUTION

Corrosion of concrete is not always caused by microorganisms. Air pollution from industralization and other exhaust systems contributes directly and indirectly to the deterioration of highways and bridges and inorganic building materials such as stone [53, 62, 63], quartz [64, 65], marble [66], mortar, and even glass [67, 68]. Emission of oxides of sulfur and nitrogen contributes to acid rain [69, 70], which dissolves surface materials of concrete, stone, cement, and glass with noticeable effects. Analyses showed that 60–70% of acidic precipitation is due to sulfuric acid, 30–40% to nitric acid, and approximately 5% to hydrochloric acid [71].

Microorganisms participate actively in the degradation process by utilization of atmospheric deposits as nutrient sources for their growth [72, 73]. The predominant groups of microorganisms are those capable of sulfur oxidation [48, 74], nitrification of nitrogen oxides [15, 16, 66, 75–77], and ammonia oxidation [78]. Organic acids, particularly from fungi, are suspected to contribute to concrete corrosion [3, 30]. At the same time, bacterial exopolysaccharides are of equal importance in corrosion [30, 78–80]. Chemical and biological processes interact, resulting in the corrosion of materials [4, 81]. However, a systematic approach integrating the role of both acid deposition and microbial activity to the degradation of inorganic materials has not been attempted in the literature. Data describing stone degradation in the presence of various natural microbial populations are available [13, 24, 25, 82–87]. Pollutants in the air may become a source of microbial substrates in the form of not only sulfurous and nitrogenous oxides but also hydrocarbons and other sulfur- and nitrogen-containing compounds [72].

D. MICROBIAL PROCESSES

The role that microorganisms play in concrete corrosion is linked primarily to the sulfur and/or nitrogen cycle in which biogenically produced inorganic acids, SO_4^{2-} and NO_3^-,

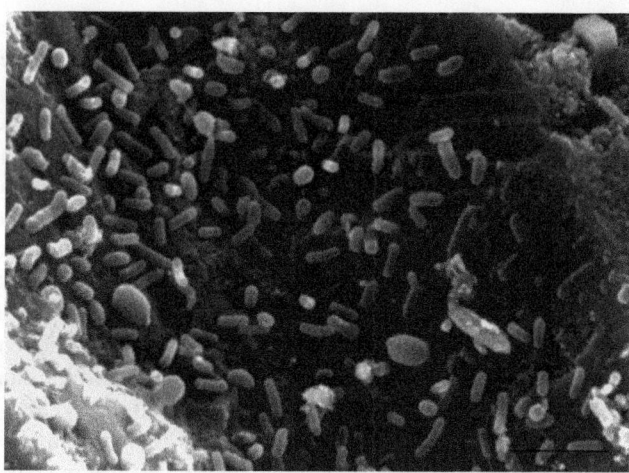

FIGURE 32.1. Scanning electron micrograph of biofilm consisting of *Thiobacillus intermedius* on surfaces of concrete after exposure under laboratory condition for 120 days (scale bar, 5 μm).

FIGURE 32.2. Scanning electron micrograph of a mixed biofilm containing the fungus *Fusarium* species and bacteria before isolation of the *Fusarium* species (scale bar, 10 μm).

are the result of microbial metabolism. In these processes, sulfate reduction and then oxidation and ammonia oxidation and nitrification are the key reactions involved. Nitrifying bacteria have only recently been reported to facilitate the degradation of concrete and building materials under natural conditions [88–90]. Furthermore, fungi also attack concrete at rates comparable to those by sulfur-oxidizing bacteria [11, 84], but no information is available on the contribution of ammonia-oxidizing archaea to corrosion.

Microorganisms are known to form biofilms and produce exopolysaccharides which are also of importance in the corrosion of concrete (Fig. 32.1), specifically through complexation by organic acids and acidic functional groups of the exopolymeric materials. Microbial activity on surfaces and their metabolites affects the susbtratum materials. Degradation and deterioration of materials often occur in the presence of SRBs, sulfur-oxidizing bacteria [91], acid-producing bacteria [92], fungi [11], and microbial exopolymers [4, 78, 79, 81, 93]. Thiobacilli have long been implicated as the culprit for such destruction of concrete materials [1, 26, 57]. For example, *T. intermedius* forms thick biofilms and is capable of etching concrete. The organic acids or multifunctionality molecules may serve as chelating agents, dissolving surface Ca^{2+} into solution. A *Fusarium* species was found to corrode concrete, and the fungal hyphae penetrated into concrete matrices [11] (Fig. 32.2). The mechanism is most likely due to organic metabolites produced by the fungus that form complexes with Ca in the concrete matrix. Milde et al. suggested that a succession of different thiobacilli that are metabolically active under neutral and acidic conditions are responsible for the complete degradation of concrete in natural systems [26]. Similar results were also reported [94]. Other groups of microorganisms have also been implicated for

their role in the biodeterioration of granite and limestone, including lichens [95] and cyanobacteria [96, 97].

D1. Sulfate in Natural Environments

Sulfur exists in various oxidative states from $+6$ in SO_4^{2-} to -2 in H_2S (Table 32.1) and is an essential element for all living cells. Living orgamisms contain approximately 1% S in various proteins. Examples of amino acids containing S are methionine, cystine, and cysteine (Fig. 32.3). Living cells contain these amino acids, which decompose upon death of the organisms. Among the different states of S-containing compounds, including the mercaptan family, hydrogen sulfide is the most noticeable because of the nuisance odor at a concentration as low as 0.2 ppm. Under natural conditions sulfate is widely present in aquatic and marine environments, wastewater treatment facilities, and sewers. The largest reservoir is seawater. Hydrogen sulfide is chemically reactive and the characteristic black color

TABLE 32.1. Oxidation States of Sulfur-Containing Compounds

Form	Formula	Oxidative State (S)
Sulfate	SO_4^{2-}	$+6$
Sulfite	SO_3^{2-}	$+4$
Thiosulfate	$S_2O_3^{2-}$ ($-S-SO_3^-$)	$-2, +6$
Tetrathionate	$S_4O_6^{2-}$	$+2.5$
Thiocyanide	$S-C-N^-$	-2
Trithionate	$-O_3SSSO_3-$	$+6, +2, +6$
Elemental S	S^0	0
Disulfite	HS^-	-2
Sulfide	S^{2-}	-2

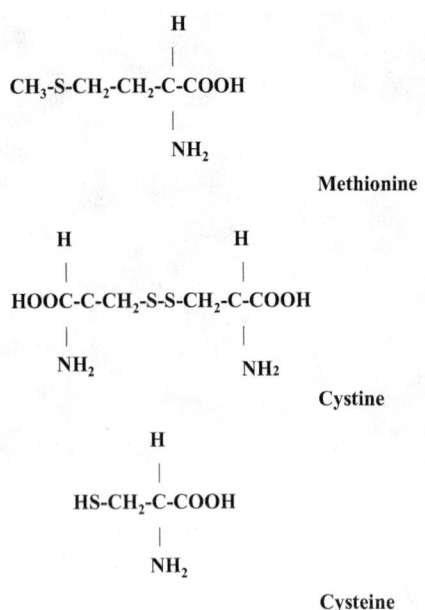

FIGURE 32.3. Chemical structures of three sulfur-containing amino acids.

in sediments is due to the formation of FeS in anoxic zones [98, 99].

Reduction of sulfate in marine environments by SRBs causes serious corrosion problems for shipping, oil-drilling platforms, and other coastal and offshore infrastructures. In addition, production of hydrogen sulfide has a detrimental effect on marine life [100].

D2. Sulfate-Reducing Bacteria

Production of H_2S from SO_4^{2-} in wastewater systems was recognized in 1864 by Meyer [68], who proposed that the reaction was mediated by algae. A few years later, Cohn claimed that the filamentous gliding microorganism *Beggiatoa* was responsible for the reduction of SO_4^{2-} in wastewater to H_2S. In 1895 Beijerinck demonstrated definitively the process of microbial S reduction by isolating *Vibrio desulfuricans* in pure culture (later renamed *Desulfovibrio desulfuricans*). The genus name *Desulfovibrio* is reserved for non-spore-forming sulfate reducers, usually with a curved, motile rod and a limited capability to utilize carbon. Preferred substrates include lactate and pyruvate. Microorganisms in this group also include some species which utilize recalcitrant hydrocarbons.

Other microorganisms capable of sulfate reduction include *Desulfovibrio, Desulfobulbus, Desulfomonas, Desulfobacter, Desulfococcus, Desulfonema, Desulfosarcina, Desulfobacterium, Desulfotomaculum,* and *Thermodesulfobacterium* [51]. *Desulfotomaculum* is the only spore-forming genus of SRB. These SRBs carry out dissimilatory processes during sulfate reduction in which S from SO_4^{2-} is

a source of electrons and energy for microbial growth rather than being assimilated into the cells. At the same time, organic compounds are oxidized.

Sulfate-reducing bacteria are widespread in natural and polluted environments. They are strictly anaerobic microorganisms, and molecular O_2 is toxic to them. Special culturing techniques and laboratory equipment are required for isolation and handling of these bacteria [101]. Their growth covers a wide spectrum of environmental conditions, ranging from psychrophilic (4°C) to extremely thermophilic (90°C and higher). SRBs are found not only in wastewater but also in oil fields, nuclear power plants, and biofilms on surfaces of a wide range of materials. Ecologically, SRBs are important microorganisms scavenging molecular hydrogen produced by anaerobic fermentation because interspecies H_2 transfer is thermodynamically vital to the anaerobic microbial community, as the process essentially removes an inhibitory metabolite (H_2) [102, 103].

D3. Sulfur-Oxidizing Bacteria

As early as 1887, Winogradsky demonstrated that certain groups of bacteria are capable of obtaining their energy for growth from oxidizing S^0 to SO_4^{2-} [104]. The biologically catalyzed reactions are as follows:

$$2H_2S + O_2 \rightarrow S_2 + 2H_2O + 80\,\text{kcal}\ (=334.4\,\text{kJ}) \quad (32.1)$$

$$S_2 + 3O_2 + 2H_2O \rightarrow 2H_2SO_4 + 240\,\text{kcal}\ (=1003.2\,\text{kJ}) \quad (32.2)$$

The microorganisms belong to a group called chemoautotrophs, utilizing CO_2 for carbon and S as a source of electrons. There are eight species in the genus *Thiobacillus*. They differ in their ability to tolerate acidity. *Thiobacillus thiooxidans* and *T. ferrooxidans* are capable of functioning at pH levels as low as 2–3.5, while *Thiobacillus thiopurus* and *Thiobacillus denitrificans* prefer slight acidity for optimal growth. *Thiobacillus ferrooxidans* and *T. thiooxidans* are particularly important for their role in leaching of mining ores.

All thiobacilli are obligate autotrophs except *T. novellus*. Although the heterotrophic bacteria *Arthrobacter, Micrococcus, Pseudomonas,* and *Bacillus,* in addition to some actinomycetes and fungi, can oxidize sulfur compounds, no energy is derived from these transformations [105].

D4. Ammonia-Oxidizing Microorganisms

In the 1870s, Pasteur postulated that oxidation of NH_4^+ to NO_3^- was a microbiological process. The first experimental evidence was presented by Schloesing and Muntz in 1877. Subsequently, Warrington at the Rothamsted Experimental Station in England claimed that the nitrification process was a two-step transformation. Winogradsky then isolated

TABLE 32.2.   Chemoautotrophic Nitrogen Oxidizers

Genus	Species	Habitats
Nitrosomonas	*europaea*	Soil, water, sewage
Nitrosospira	*briensis*	Soil
Nitrosococcus	*nitrosus*	Marine
	oceanus	Marine
	mobilis	Soil
Nitrosovibrio	*tenuis*	Soil
Nitrosolobus	*multiformis*	
Nitrobacter	*winogradsky*	Soil
	(agilis)	Soil, water
Nitrospira	*gracilis*	Marine
Nitrococcus	*mobilis*	Marine
	marina	Marine

nitrifying bacteria in pure culture; however, the purity of his cultures is still debatable [75, 106].

Nitrifying bacteria are chemoautotrophs (Table 32.2). They obtain carbon from CO_2 or carbonate and energy from oxidation of NH_4^+ or NO_2^-. During the processes of nitrification, NH_4^+ is converted to hydroxylamine (NH_2OH) by an endothermic reaction. The next reactant has not been identified but is hypothesized to be a nitroxyl radical. The NO_3^- is proposed to react with this radical to form nitrohydroxylamine before breaking down to HNO_2. Transformation of HNO_2 to NO_3^- requires the presence of oxygen, but the role of O_2 is confined to electron transport [107]. The oxygen in NO_3^- is actually from H_2O. The overall reaction produces energy for growth resulting in a free energy change of -65 and -18.2 kcal/mol for oxidation of 1 mol of NH_4^+ and NO_2^-, respectively according to the following equations:

$$NH_4^+ + 1.5O_2 \rightarrow NO_2^- + H_2O + 2H^+ + 76 \text{ kcal} (= 317.7 \text{ kJ})$$
$$(32.3)$$

$$NO_2^- + 0.5O_2 \rightarrow NO_3^- + 24 \text{ kcal} (= 100.3 \text{ kJ}) \quad (32.4)$$

In addition to chemoautotrophic metabolism, heterotrophs are capable, to a lesser extent, of carrying out nitrification. However, the process does not appear to be ecologically significant.

Bock and Sand [108] found that the nitrifying bacteria play an important role in the degradation of concrete as a result of nitric acid production during nitrification [15]. Nitrifying bacteria were also found to be the predominant contributors to the deterioration of other stone materials [91, 109, 110]. These bacteria differ from the thiobacilli in that the former are capable of growth on nonimmersed surfaces such as buildings, while the latter require an aqueous environment in the presence of sulfate.

More recently, ammonia-oxidizing archaea have been widely detected in environmental samples [78]. This discovery changes our traditional view that bacteria are responsible for the transformation of NH_4^+ available in the environment. However, the limited information currently available also demonstrates some weaknesses in the molecular-based techniques used for detecting this group of archaea. In situ activity of the archaea has not been fully established even though they are widely detected in environmental samples at high abundance.

D5. Role of Fungi

This is a new twist in the deterioration of concrete because fungi differ from bacteria taxonomically and are eukaryotes while bacteria are unicellular prokaryotes. Fungi are well known for their metabolic versatility and are capable of attacking a wide range of organics, including engineered polymeric materials [12, 31, 32]. They are also notorious for their secondary metabolites, the mycotoxins [111]. One of the reactions carried out by fungi is the production of peroxide through peroxidase. This enzyme enables fungi to attack complex organics, including polyaromatics and chlorinated aromatics [112]. In comparison, very little information is available on the role of fungi in the degradation of inorganic materials, particularly stone and concrete. A *Fusarium* species has been isolated that is capable of attacking concrete [3, 11]. The fungus and *T. intermedius* caused almost identical weight loss after 120 days of incubation [11].

Several possible mechanisms of fungal attack of concrete exist. Fungi are known for their production of organic acids, including citric, oxalic, and gluconic acids [85, 86, 113, 114], which react with calcium in the concrete, leading to subsequent dissolution and degradation. They are also physiologically capable of etching the material and extending their hyphae into the interior of the concrete, resulting in enlargement of the damaged area and an increase in porosity. Both bacteria and fungi may form synergistic associations in concrete degradation (Fig. 32.4), and this may be

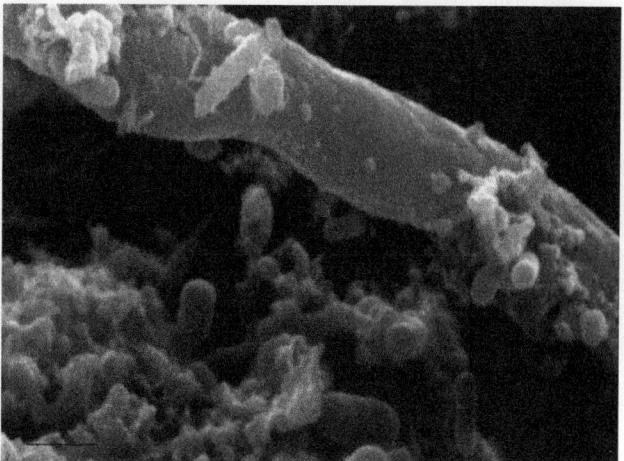

FIGURE 32.4. Scanning electron micrograph of *Fusarium* species hypha with bacteria attached (scale bar, 2 μm).

most pronounced under natural conditions where pure cultures of microorganisms are impossible. Studies of concrete weight loss indicated that precipitates formed with organic acids were responsible for the measured changes in weight. These precipitates, which easily detach, are usually calcium–organic complexes. Fungal activity occurs over a wide range of environmental conditions. Therefore the damage caused by these organisms may be much larger than expected.

D6. Microbial Exopolymers

Microorganisms produce large quantities of exopolymeric materials in late growth phases and/or environments with high carbon/nitrogen (C/N) ratios or during adhesion to and growth on surfaces (Fig. 32.5). These polymeric materials are important carbon and energy reserves and are utilized during periods of nutrient deficiency. Bacterial exopolymeric materials also play an important role in the formation of microbial biofilms [94], subsequent corrosion of metals, and transport of metal ions in porous media [4, 78–81, 115]. They are multifunctional group molecules [93, 116]. The activity of these molecules and their functional groups in chelation and dissolution of calcium from concrete is still poorly understood. Ford and Mitchell [115] and Geesey et al. [117] proposed that bacterial exopolymers bind metals and promote formation of ionic concentration cells, accelerating dissolution and corrosion of metallic materials. Similarly, negatively charged carboxylic and hydroxyl groups of exopolymeric materials from thiobacilli may form complexes with calcium and hence leach the calcium from concrete matrices. This process may contribute to concrete degradation, particularly when biofilms grow in close proximity to the surface. Further research is needed to identifiy a possible role for these exopolymers in concrete degradation.

E. CORROSION OF REBAR

Concrete is formulated for different purposes using steel reinforcement. Reinforcement of concrete with steel bars may increase susceptibility to corrosion because of the growth of SRBs and the resultant production of H_2S, which can penenatre into concrete. The corrosion products from both metals and also concrete constituents may expand in volume, generating stress for the surrounding concrete materials and resulting in formation of cracks [118, 119]. In such situations, both biological and physicochemical processes act in concert. Preventive measures involve the use of polymers to prevent moisture reaching the steel-reinforcing bars and incorporation of biocides. Microbial activity is significantly reduced at low moisture levels [37].

Hydrogen-producing bacteria may also contribute to the stress cracking of high-strength steel. Ford and Mitchell [92] reported that bacteria were found in the cracking areas of a high-strength steel bar under loaded conditions. Cracks did not develop in the absence of hydrogen- and acid-producing bacteria (sterile controls), suggesting that bacterial metabolites could contribute to a significant weakening in the strength of steel rebars due in part to permeation of microbially produced hydrogen into steel [120]. Protection of concrete structures is a significant challenge for both engineers and applied microbiologists because of the multiple applications of reinforced concrete under a variety of different environments (Fig. 32.6).

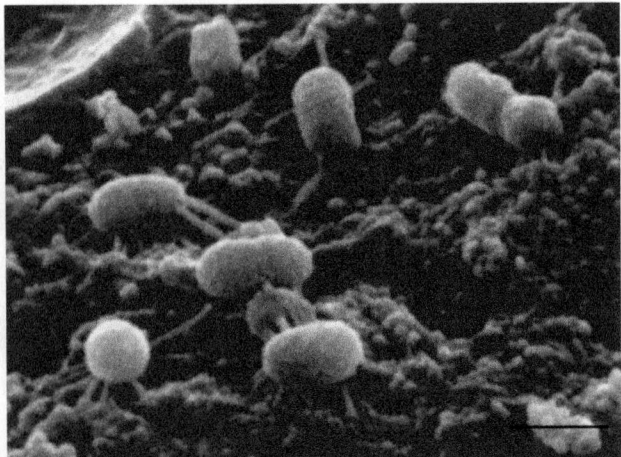

FIGURE 32.5. Scanning electron micrograph of bacteria attached onto substratum through exopolymeric materials (scale bar, 2 μm).

FIGURE 32.6. Photograph of a hydropower station in Guangdong Province in southern China.

FIGURE 32.7. Photograph of an open aquaduct structure under construction for drinking water transport from mainland China to Hong Kong.

F. PREVENTIVE MEASURES

Protection of concrete from biological corrosion has been impeded because of the difficulty in eradicating microorganisms from the materials when the structures are part of the environment and compeletely exposed (Fig. 32.7). In a study conducted by Sand et al. [17], oxygenation of sewer systems was proposed as a means of alleviating the propagation of SRBs in wastewater systems and slowing the subsequent degradation of concrete. Numerous other approaches have been investigated, including precipitation of microbially produced H_2S by ferrous chloride ($FeCl_2$) [21], competition by heterotrophs for displacement of *Thiobacillus* [121], and the use of microbial biocides [122, 123]. However, none of these preventive measures has proven to be successful due to the incomplete understanding of the microbial community under natural conditions [6] and also the plasticity of microorganisms in adapting themselves to the changing environment both physiologically and genetically [124–126].

G. CONCLUSIONS

It is apparent that the microbiology and ecology of concrete corrosion processes are not well understood. Current information on microbial corrosion of concrete is limited to the microorganisms that have been successfully isolated in pure culture from natural environments, particularly *Thiobacillus* species and the nitrifying bacteria. The role of other microorganisms, for example, fungi, acetogenic bacteria, and ammonia-oxidizing archaea, in concrete degradation needs further investigation to improve our knowledge. Fungi are capable of deterioration of concrete, but the synergistic association between bacteria and fungi has not been fully

established. Culture-independent molecular techniques start to show the metabolic diversity of the natural community on concerete and should be utilized in future studies to confirm the specific groups of microorganisms involved in the degradation process. Furthermore, dissolution products from such exposure experiments need to be thoroughly characterized to establish the chemical reactions that are taking place. The roles of many uncultured bacteria shown through molecular biology to be present on deteriorating concrete have yet to be understood.

ACKNOWLEDGMENTS

We thank Ed Seling of the Museum of Comparative Zoology, Harvard University, for taking the scanning electron micrographs used here in Figures 32.1, 32.2, 32.4, and 32.5 and Jessie Lai for assistance in organizing and archiving documents.

REFERENCES

1. T. E. Ford, in Aquatic Microbiology: An Ecological Approach, T. E. Ford (Ed.), Blackwell Scientific, Boston, MA, 1993, pp. 455–482.
2. J.-D. Gu, T. E. Ford, and R. Mitchell, in The H. H. Uhlig Corrosion Handbook, W. Revie (Ed.), Wiley, New York, 2000, pp. 349–365.
3. J.-D. Gu, D. B. Mitton, T. E. Ford, and R. Mitchell, in The H. H. Uhlig Corrosion Handbook, W. Revie (Ed.), Wiley, New York, 2000, pp. 477–491.
4. G. Chen, C. R. Clayton, R. A. Sadowski, J. B. Gillow, and A. J. Francis, Paper No. 217, Corrosion/95, NACE International, Houston, TX, 1995.
5. M. Diercks, W. Sand, and E. Bock, Experientia, **47**, 514 (1991).
6. D. J. Giannantonio, J. C. Kurth, K. E. Kurtis, and P. A. Sobecky, Int. Biodeter. Biodegr., **63**, 30 (2009).
7. F. Mansfeld, H. Shih, A. Postyn, J. Devinny, R. Islander, and C. L. Chen, Paper No. 113, Corrosion/90, NACE International, Houston, TX, 1990.
8. W. Sand and E. Bock, Environ. Technol. Lett., **5**, 517 (1984).
9. C. D. Parker, Aust. J. Exp. Biol. Med. Sci., **23**, 81 (1945).
10. C. D. Parker, Aust. J. Exp. Biol. Med. Sci., **23**, 91 (1945).
11. J.-D. Gu, T. E. Ford, N. S. Berke, and R. Mitchell, Int. Biodeter. Biodegr., **41**, 101 (1998).
12. J.-D. Gu, in Encyclopedia of Microbiology, 3rd ed., M. Schaechter (Ed.), Elsevier, London, 2009, pp. 259–269.
13. D. Nica, J. L. Davis, L. Kirby, G. Zuo, and D. J. Roberts, Int. Biodeter. Biodegr., **46**, 61 (2000).
14. M. Hernandez, E. A. Marchand, D. Roberts, and J. Peccia, Int. Biodeter. Biodegr., **49**, 271 (2002).
15. P.-G. Jazsa, M. Kussmaul, and E. Bock, in DECHEMA Monographs, Vol. 133, Biodeterioration and Biodegradation,

W. Sand (Ed.), VCH Verlagesellschäft, Frankfurt, Germany, 1996, pp. 127–134.

16. P.-G. Jazsa, R. Stüven, E. Bock, and M. Kussmaul, in DECHEMA Monographs, Vol. 133, Biodeterioration and Biodegradation, in: W. Sand (Ed.), VCH Verlagesellschäft, Frankfurt, Germany, 1996, pp. 199–208.

17. W. Sand, B. Ahlers, and E. Bock, in Science, Technology and European Cultural Heritage, N. S. Baer, C. Sabbioni, and A. I. Sors (Eds.), Butterworth-Heinemann, Oxford, 1991, pp. 481–484.

18. T. Gehrke, J. Telegdi, D. Thierry, and W. Sand, Appl. Environ. Microbiol., **64**, 2743 (1998).

19. J.-D. Gu and R. Mitchell, J. Microbiol., **39**, 133 (2001).

20. J.-D. Gu and R. Mitchell, Hydrobiologia, **474**, 81 (2002).

21. R. L. Morton, W. A. Yanko, D. W. Graham, and R. G. Arnold, Res. J. WPCF, **63**, 779 (1991).

22. H. Nakagawa, S. Akihawa, T. Suenaga, Y. Taniguchi, and K. Yoda, Adv. Composite Mater., **3**, 123 (1993).

23. M. F. Azizian, P. O. Nelson, P. Thayumanavan, and K. J. Williamson, Waste Manag., **23**, 719 (2003).

24. N. De Belie, J. Monteny, A. Beeldens, E. Vincke, D. van Gemert, and W. Verstraete, Cement Concrete Res., **34**, 2223 (2004).

25. E. Vincke, E. van Wanseele, J. Moneny, A. Beeldens, N. de Belie, L. Taerwe, D. van Gemert, and W. Verstraete, Int. Bioder. Biodegr., **49**, 283 (2002).

26. K. Milde, W. Sand, W. Wolff, and E. Bock, J. Gen. Microbiol., **129**, 1327 (1983).

27. J.-D. Gu, T. Ford, K. Thorp, and R. Mitchell, Int. Biodeter. Degr., **39**, 197 (1996).

28. J.-D. Gu, C. Lu, K. Thorp, A. Crasto, and R. Mitchell, Mater. Perform., **36**, 37 (1997).

29. J.-D. Gu, C. Lu, K. Thorp, A. Crasto, and R. Mitchell, J. Indus. Microbiol. Biotechnol., **18**, 364 (1997).

30. J.-D. Gu, D. B. Mitton, T. E. Ford, and R. Mitchell, Biodegradation, **9**, 39 (1998).

31. J.-D. Gu, Int. Biodeter. Biodegr., **52**, 69 (2003).

32. J.-D. Gu, Int. Biodeter. Biodegr., **59**, 170 (2007).

33. J.-D. Gu and R. Mitchell, Chin. J. Mater. Res., **9**(Suppl.), 473 (1995).

34. K. E. G. Thorp, A. S. Crasto, J.-D. Gu, and R. Mitchell, in Proceedings of the Tri-Service Conference on Corrosion, T. Naguy (Ed.), U.S. Government Printing Office, Washington, DC, 1994, pp. 303–314.

35. J.-D. Gu and L. Pan, J. Polym. Environ., **14**, 273 (2006).

36. J.-D. Gu, T. E. Ford, and R. Mitchell, J. Appl. Polym. Sci., **92**, 1029 (1996).

37. J.-D. Gu, M. Roman, T. Esselman, and R. Mitchell, Int. Biodeter. Biodegr., **41**, 25 (1998).

38. J.-D. Gu, T. E. Ford, and R. Mitchell, in The H. H. Uhlig Corrosion Handbook, W. Revie (Ed.), Wiley, New York, 2000, pp. 915–927.

39. W. M. Olmstead and H. Hamlin, Eng. News, **44**, 317 (1900).

40. F. M. Lea and C. H. Desch, The Chemistry of Cement and Concrete, Edward Arnold, London, 1936.

41. I. Biczók, Betonkorrosion, betoschutz, Bauverlag, Wiesbaden, Berlin, Germany, 1968.

42. C. D. Parker, Nature, **159**, 439 (1947).

43. C. D. Parker and D. Jackson, in Hydrogen Sulphide Corrosion of Concrete Sewers, Technical Paper No. A.8, Part 6, Melbourne and Metropolitan Board of Works, Melbourne, Australia, 1965, pp. 1–29.

44. C. D. Parker and A. Prisk, J. Gen. Microbiol., **8**, 344 (1953).

45. D. P. Kelley, in The Prokaryotes, M. P. Starr, H. Stolp, H. G. Trüper, A. Balow, and H. G. Schlegel (Eds.), Springer-Verlag, New York, 1982, pp. 1023–1036.

46. D. P. Kelly, in The Nitrogen Cycle, J. A. Cole and S. J. Ferguson (Eds), Cambridge University Press, New York, 1988, pp. 64–98.

47. J. G. Kuenen and O. H. Tuovinen, in The Prokaryotes, M. P. Starr, H. Stolp, H. G. Trüper, A. Balow, and H. G. Schlegel (Eds.), Springer-Verlag, New York, 1982, pp. 1023–1036.

48. W. Sand, Appl. Environ. Microbiol., **53**, 1645 (1987).

49. G. Bitton, Wastewater Microbiology, Wiley, New York, 1994.

50. V. Somlev and S. Tishkov, Geomicrobiology, **12**, 53 (1994).

51. F. Widdel, in Biology of Anaerobic Microorganisms, A. J. B. Zehnder (Ed.), Wiley, New York, 1988, pp. 469–585.

52. J. M. Odom, in The Sulfate-Reducing Bacteria: Comtemporary Perspectives, J. M. Odom and R. Singleton, Jr. (Eds.), Springer-Verlag, New York, 1993, pp. 189–210.

53. F. E. W. Eckhardt, in Environmental Biogeochemistry and Geomicrobiology, Vol. 2, The Terrestrial Environment, W. E. Krumbein (Ed.), Ann Arbor Science, Ann Arbor, MI, 1978, pp. 675–586.

54. M. Islam, N. G. Thompson, D. R. Lankard, and Y. P. Virmani, Paper No. 4, Corrosion/95, NACE International, Houston, TX, 1995.

55. K. K. Jain, A. K. Mishra, and T. Singh, in Recent Advances in Biodeterioration and Biodegradation, K. L. Garg, N. Garg, and K. G. Mukerji (Eds.), Naya Prokash, Calcutta, India, 1993, pp. 323–354.

56. E. May, F. J. Lewis, S. Pereira, S. Taylor, and M. R. D. Seaward, Biodeter. Abstr., **7**, 109 (1993).

57. M. W. Mittelman and J. C. Danko, in 1995 International Conference on Microbial Influenced Corrosion, P. Angell, S. W. Borenstein, R. A. Buchanan, S. C. Dexter, N. J. E. Dowling, B. J. Little, C. D. Lundin, M. B. McNeil, D. H. Pope, R. E. Tatnall, D. C. White, and H. G. Ziegenfuss (Eds.), NACE International, Houston, TX, 1995, pp. 15/1–7.

58. K. Pedersen, Can. J. Microbiol., **42**, 382 (1996).

59. S. Stroes-Gascoyne, K. Pedersen, S. Daumas, C. J. Hamon, S. A. Haveman, T. L. Delaney, S. Ekendahl, N. Jahromi, J. Arlinger, L. Hallbeck, and K. Dekeyser, Microbial Analysis of the Buffer/Container Experiment at AECL's Underground Research Laboratory, AECL-11436, Whiteshell Laboratories, Atomic Energy of Canada Limited, Pinawa, Manitoba, Canada, 1996.

60. G. D. Soraru and P. Tassone, Construct. Build. Mater. **18**, 561 (2004).

61. E. Bastidas-Arteaga, M. Sanchez-Silva, A. Chateauneuf, and M. R. Silva, Struct. Saf., **30**, 110 (2008).

62. J. J. Feddema and T. C. Mererding, in Science, Technology and European Cultural Heritage, N. S. Baer, C. Sabbioni, and A. I. Sors (Eds.), Butterworth-Heinemann, Oxford, England, 1991, pp. 443–446.

63. P. Tiano, in Recent Advances in Biodeterioration and Bio-degradation, K. L. Garg, N. Garg, and K. G. Mukerji (Eds.), Naya Prokash, Calcutta, India, 1993, pp. 301–321.

64. A. M. Lauwers and W. Heinen, Arch. Microbiol., **95**, 67 (1974).

65. E. Lefebvre-Drouet and M. F. Rousseau, Soil Biol. Biochem., **27**, 1041 (1995).

66. P. Hirsch, F. E. W. Eckhardt, and R. J. Palmer, Jr., J. Microbiol. Methods, **23**, 143 (1995).

67. D. R. Fuchs, M. Popall, H. Römich, and H. Schmidt, in Science, Technology and European Cultural Heritage, N. S. Baer, C. Sabbioni, and A. I. Sors (Eds.), Butterworth-Heinemann, Oxford, England, 1991, pp. 679–683.

68. L. Meyer, J. Prakt. Chem., **91**, 1 (1864).

69. K. L. Gauri, Environ. Geol. Water Sci., **6**, 187 (1984).

70. S. S. Yerrapragada, J. H. Jaynes, S. R. Chirra, and K. L. Gauri, Anal. Chem., **66**, 655 (1994).

71. R. P. Webster and L. E. Kukacka, in R. Baboian (Ed.), Materials Degradation Caused by Acid Rain, ACS Symposium Series 318, American Chemical Society, Washington, DC, 1986, pp. 239–249.

72. C. Saiz-Jimenez, Aerobiologia, **11**, 161 (1995).

73. T. Warcheid, M. Oelting, and W. E. Krumbein, Int. Biodeter., **28**, 37 (1991).

74. C. F. Kulpa and C. J. Baker, in Microbially Influenced Corrosion and Biodeterioration, N. J. Dowling, M. W. Mittelman, and J. C. Danko (Eds.), National Association of Corrosion Engineers, Houston, TX, 1990, pp. 4/7–9.

75. E. Bock, H.-P. Koops, and H. Harms, in Nitrification, J. I. Prosser (Ed.), Special Publication of the Society for General Microbiology, Vol. 20, IRL Press, Oxford, 1986, pp. 17–38.

76. W. Sand, E. Bock, and D. C. White, Mater. Perform., **26**, 14 (1987).

77. W. Sand, K. Milde, and E. Bock, in Recent Progress in Biohydrometallurgy, G. Rossi and A. E. Torma (Eds.), Associazione Mineraria Sarda, Iglesias, Italy, 1983, pp. 667–677.

78. M. Könneke, A. E. Bernnhard, J. R. de la Torre, C. B. Walker, J. B. Waterbury, and D. A. Stahl, Nature, **437**, 543 (2005).

79. T. Ford, J. S. Maki, and R. Mitchell, Paper No. 380, Corrosion/87, National Association of Corrosion Engineers, Houston, TX, 1987.

80. T. Ford, J. Maki, and R. Mitchell, in Bioextraction and Biodeterioration of Metals, C. C. Gaylarde and H. A. Videla (Eds.), Cambridge University Press, Cambridge, 1995, pp. 1–23.

81. G. Chen, S. V. Kagwade, G. E. French, T. E. Ford, R. Mitchell, and C. R. Clayton, Corrosion, **52**, 891 (1996).

82. W. E. Krumbein, Zeitschrift fur Allegmeine Microbiologie, **8**, 107 (1968).

83. S. Tayler and E. May, Int. Biodeter., **28**, 49 (1991).

84. M. A. D. Torre, G. Gómez-Alarcón, P. Melgarejo, and J. Lorenzo, in Science, Technology and European Cultural Heritage, N. S. Baer, C. Sabbioni, and A. I. Sors (Eds.), Butterworth-Heinemann, Oxford, 1991, pp. 511–514.

85. M. A. D. Torre, G. Gómez-Alarcón, and J. M. Palacios, Appl. Microbiol. Biotechnol., **40**, 408 (1993).

86. M. A. D. Torre, G. Gomez-Alarcon, G. Vizcaino, and M. T. Garcia, Biogeochemistry, **19**, 129 (1993).

87. J. L. Davis, D. Nica, K. Shields, and D. J. Roberts, Int. Biodeter. Biodegr., **42**, 75 (1998).

88. T. W. Becker, W. E. Krumbein, T. Warscheid, and M. A. Resende, in Investigations into Devices Against Environmental Attack on Stone, G. M. Herkenrath (Ed.), GKSS-Forschungszentrum Geesthacht, GmbH, Geesthacht, Germany, 1994, pp. 147–190.

89. E. Bock, B. Ahlers, and C. Myer, Bauphysik, **11**, 141 (1989).

90. J. J. Ortega-Calvo, M. Hernandez-Marine, and C. Saiz-Jimenez, Int. Biodeter., **28**, 165 (1991).

91. S. Ehrich and E. Bock, in DECHEMA Monographs, Vol. 133, Biodeterioration and Biodegradation, W. Sand (Ed.), VCH Verlagesellschäft, Frankfurt, 1996, pp. 193–198.

92. T. Ford and R. Mitchell, in Microbially Influenced Corrosion and Biodeterioration, N. J. Dowling, M. W. Mittelman, and J. C. Danko (Eds.), University of Tennessee, 1990, pp. 3/94–98.

93. T. Ford, E. Sacco, J. Black, T. Kelley, R. Goodacre, R. C. W. Berkeley, and R. Mitchell, Appl. Environ. Microbiol., **57**, 1595 (1991).

94. D. G. Davies, M. R. Parsek, J. P. Pearson, B. H. Iglewski, J. W. Costerton, and E. P. Greenberg, Science, **280**, 295 (1998).

95. B. Prieto, T. Rivas, B. Silva, R. Carballal, and M. E. Lopez de Silanes, Int. Biodeter. Biodegr., **34**, 47 (1994).

96. F. G. Ferris and E. A. Lowson, Can. J. Microbiol., **43**, 211 (1997).

97. S. Schultze-Lam, G. Harauz, and T. J. Beveridge, J. Bacteriol., **174**, 7971 (1992).

98. B. B. Jorgensen, in The Nitrogen and Sulphur Cycles, J. A. Cole and S. J. Ferguson (Eds.), Cambridge University Press, New York, 1988, pp. 31–63.

99. A. J. B. Zehnder and W. Stumm, in Biology of Anaerobic Microorganisms, A. J. B. Zehnder (Ed.), Wiley, New York, 1988, pp. 1–38.

100. J. R. Postgate, The Sulfate-Reducing Bacteria, 2nd ed., Cambridge University Press, Cambridge, 1984.

101. R. E. Hungate, in Methods in Microbiology, Vol. 3a, J. R. Norris, and D. W. Ribbons (Eds.), Academic, New York, 1971, pp. 117–132.

102. C. Achtnich, A. Schuhmann, T. Wind, and R. Conrad, FEMS Microbiol. Ecol., **16**, 61 (1995).

103. J. Doré, P. Pochart, A. Bernalier, I. Goderel, B. Morvan, and J. C. Rambaud, FEMS Microbiol. Ecol., **17**, 279 (1995).

104. E. A. Paul and F. E. Clark, Soil Microbiology and Biochemistry, Academic, San Diego, CA, 1989.

105. M. Alexander, Introduction to Soil Microbiology, 2nd ed., Wiley, New York, 1977.

106. R. M. MacDonald, in Nitrification, J. I. Prosser (Ed.), Special Publication of the Society for General Microbiology, Vol. 20, IRL Press, Oxford, 1986, pp. 1–16.

107. E. L. Schmidt, in Nitrogen in Agricultural Soils, F. J. Stevenson (Ed.), Agronomy Monograph No. 22, Soil Science Society of America, Madison, WI, 1982, pp. 253–288.

108. E. Bock and W. Sand, in Microbially Influenced Corrosion and Biodeterioration, N. J. Dowling, M. W. Mittelman, and J. C. Danko (Eds.), National Association of Corrosion Engineers, Houston, TX, 1990, pp. 3/29–33.

109. G. Gomez-Alarcon, B. Cilleros, M. Flores, and J. Lorenzo, Sci. Total Environ., **167**, 231 (1995).

110. G. Gomez-Alarcon, M. Munoz, X. Arino, and J. J. Ortega-Calvo, Sci. Total Environ., **167**, 249 (1995).

111. K. F. Nielsen, U. Thrane, T. O. Larsen, P. A. Nielsen, and S. Gravesen, Int. Biodeter. Biodegr., **42**, 9 (1998).

112. J. B. Sutherland, F. Rafii, A. A. Khan, and C. E. Cerniglia, in L. L. Young and C. E. Cerniglia (Eds.), Microbial Transformation and Degradation of Toxic Organic Chemicals, Wiley, New York, 1995, pp. 269–306.

113. M. V. Dutton and C. S. Evans, Can. J. Microbiol., **42**, 881 (1996).

114. A.-L. Sunesson, W. H. J. Vaes, C.-A. Nilsson, G. Blomquist, B. Andersson, and R. Carlson, Appl. Environ. Microbiol., **61**, 2911 (1995).

115. T. Ford and R. Mitchell, in Environmental Microbiology, R. Mitchell (Ed.), Wiley, New York, 1992, pp. 83–101.

116. H. H. Paradies, in Bioextraction and Biodeterioration, C. C. Gaylarde and H. A. Videla (Eds.), Cambridge University Press, Cambridge, 1995, pp. 197–269.

117. G. G. Geesey, M. W. Mittelman, T. Iwaoka, and P. R. Griffiths, Mater. Perform., **25**, 37 (1986).

118. S. G. Millard, K. R. Gowers, and J. H. Bungey, Paper No. 525, Corrosion/95, NACE International, Houston, TX, 1995.

119. H. Saito, Y. Miyata, H. Takazawa, K. Takai, and G. Yamauchi, Paper No. 544, Corrosion/95, NACE International, Houston, TX, 1995.

120. T. E. Ford, P. C. Searson, T. Harris, and R. Mitchell, J. Electrochem. Soc., **137**, 1175 (1990).

121. N. A. Padival, J. S. Weiss, and R. G. Arnold, Water Environ. Res., **67**, 201 (1995).

122. E. Bell, P. Dowding, and T. P. Cooper, Environ. Technol., **13**, 687 (1992).

123. R. D. Wakefield, Scottish Soc. Conserv. Restor. **8**, 5 (1997).

124. Y. Wang, P. C. Leung, P. Qian, and J.-D. Gu, Microb. Environ., **19**, 163 (2004).

125. Y. Wang, P. C. Leung, P. Qian, and J.-D. Gu, Ecotoxicology, **15**, 371 (2006).

126. R. Zhang, Y. Wang, and J.-D. Gu, Antoine van Leeuvenhoek—Int. J. Gen. Mol. Microbiol., **89**, 307 (2006).

33

MICROBIAL DEGRADATION OF WOOD

P. I. Morris

FPInnovations, Vancouver, BC, Canada

A. INTRODUCTION

Before discussing microbial degradation, it should be noted that, in the absence of any one of the critical requirements for decay (described later in this chapter), wood is a remarkably stable material. Some of the longest lived of all organisms are the bristlecone pines of California and these can be >4700 years old. There are numerous examples of temples in Japan with wood components that have remained intact for hundreds of years because they have been kept dry. Treatment of wood with preservatives can extend the service life of wood products for many decades, and this subject is covered in detail in Chapter 34.

It should also be recognized that microbial degradation is only one facet of biodegradation of wood. Biodegradation of wood in terrestrial environments can be caused by a range of microorganisms and insects. In marine environments, there are crustacea and molluscs that will also attack wood. This chapter will focus on microbial degradation of wood in terrestrial environments for a variety of reasons. Damage caused by insects, crustaceans, and molluscs is more readily identifiable with the naked eye than microbial degradation and quite unrelated to corrosion. It typically takes the form of tunneling by biting or abrasion. In contrast, microbial degradation occurs at the microscopic level and is mediated by enzymes and nonenzymic free-radical generation systems, through processes that are more analogous to the corrosion of other materials. Furthermore, while insects, specifically termites and beetles, cause the most damage to wood products in tropical and subtropical regions, microbial degradation is more economically important in temperate climates.

To begin to understand microbial degradation of wood, it is first necessary to understand something about the structure and chemical makeup of wood. The critical requirements for decay will then be detailed and the microorganisms involved will be categorized. The diagnosis of decay types will be covered in some detail and procedures for the elimination of fungal infection will be outlined. Finally, the effects of wood decay on corrosion of metals will be briefly discussed.

B. STRUCTURE AND CHEMISTRY OF WOOD

The structure of wood is related to its functions in supporting the crown of the tree, transporting water from the roots to the crown and storage of food reserves. Wood is a cellular material and most of the water-conducting cells (tracheids and, in hardwoods, vessels) are aligned vertically in the tree (Fig. 33.1), longitudinally in lumber, and in various orientations in an engineered wood products made from veneers, strands, or chips. The cells are connected by pathways

Uhlig's Corrosion Handbook, Third Edition, Edited by R. Winston Revie
Copyright © 2011 John Wiley & Sons, Inc.

FIGURE 33.1. Scanning electron micrograph of a softwood showing cellular structure (×100 approx.)

In the live tree, the cell walls are fully swollen with water, equivalent to ∼ 27% of the dry weight of wood. The sapwood cell voids may also be full of water, resulting in a sapwood moisture content of 150–200%. The heartwood moisture content is typically 30–80%. Shrinkage of wood occurs as a result of drying from 27% to the in-service moisture content. The moisture content of wood inside heated buildings ranges from 4 to 10%. Wood at equilibrium moisture content with exterior air ranges from 12 to 18%. Consequently, wood products must be dried prior to use in construction or allowed to air dry during or after construction if they are to be dimensionally stable and remain free from decay. The critical moisture contents for colonization and breakdown of lignocellulose by microorganisms are considered in the following section.

C. CRITICAL REQUIREMENTS FOR MICROBIAL DEGRADATION OF WOOD

In the forest, microorganisms play a key role in the natural cycles of death, decay, recycling of nutrients, and regrowth. When mankind wishes to use wood as a structural or decorative material, it is necessary to break or delay these natural cycles for the desired life of the structure. This can be done by eliminating one of the critical factors necessary for microbial degradation. Temperature is an important factor, but this mainly affects the speed at which colonization and decay occur. The critical factors are a susceptible wood substrate, the presence of a microorganism capable of decaying wood, oxygen, and water.

Wood-degrading microorganisms have similar temperature optima to humans; consequently, wood in buildings is at an ideal temperature for decay. The fastest growth rates for many wood-inhabiting fungi occur between 20 and 30°C (68–86°F). Most are halted, but not killed, at ∼35°C (95°F). They will, however, be killed after several hours at 60°C (140°F) or minutes at 70°C (158°F) or higher. Moderate growth still occurs down to 15°C (59°F) and growth does not stop until the temperature gets very close to 0°C (32°F).

The susceptibility of the natural wood substrate to microbial degradation depends on the part of the tree it comes from. In the living tree, the inner heartwood is dead, but it is protected to a certain extent by its natural durability. This is provided by natural toxins and blockages laid down by the tree as each annual ring is converted from sapwood to heartwood. This natural durability varies widely among tree species. The sapwood of the living tree transports water from the roots to the crown and stores starches, proteins, and lipids required for future growth. It is protected by the bark and by active response to wounding or invasion. When the tree is converted to wood products, the heartwood retains its natural durability but the sapwood dies and can no longer react to protect itself. Because it contains no natural toxins and has a

through the cell walls that allow for the longitudinal and lateral movement of water. These pathways can also allow the passage of microorganisms from cell to cell. In addition to the longitudinally aligned cells, there are bundles of cells arranged in radial orientation (rays) that conduct soluble materials from the outer to the inner parts of the tree and vice versa. Some of these ray cells are also specialized for storage of food reserves to be used for future growth. The walls of all the cells are constructed from the same basic materials.

Wood cell walls are made up of two major chemical components and many minor ones. The major components are cellulose and lignin, which in combination are referred to as lignocellulose. Cellulose is a long-chain polymer of glucose units and provides the strength properties of wood. Lignin is a three-dimensional resin made up of a range of aromatic monomers. Cellulose chains are aligned into bundles, with crystalline and noncrystalline regions, laid down in a spiral around the walls of each cell. Each cell wall has several layers with a different orientation of the cellulose, resulting in cross-banding. Lignin permeates the entire cell wall but the cellulose–lignin ratio varies between layers. Hemicelluloses, composed of branched chains of sugars, are also part of the matrix.

considerable amount of readily available storage products, the sapwood of all wood species is highly susceptible to fungal colonization and decay. Fortunately, this has been recognized for many years and there are a variety of effective wood preservation technologies to reduce the decay susceptibility of sapwood and enhance the durability of heartwood (see Chapter 34). Some wood species consist largely of heartwood, for example, old growth Douglas fir. Others consist largely of sapwood, for example, plantation-grown southern pine.

Microorganisms reproduce and are disseminated through the production of microscopic spores, smaller than pollen grains, which can be moved short distances by rain splash, carried by insects or blown over long distances by the wind. They can be present in the atmosphere at almost any time of the year but are particularly prevalent in the fall. Wood-degrading organisms can be kept out of the living tree by the barrier effect of the bark, the defense responses of the sapwood, and the natural durability of the heartwood. If they have managed to invade the living tree through wounds, some of these organisms may be able to survive processing into a wood product if the wood remains in a green (wet) condition. However, conventional and high-temperature kiln drying schedules will kill these organisms, as will the temperatures typically encountered in the manufacture of wood-based panels and engineered wood products. Most wood products used for construction are therefore free from infection at the point of manufacture. They will, however, be repeatedly exposed to microorganisms during storage, transportation, and construction and over the entire service life. As a result, the potential for the presence of wood-degrading microorganisms must be assumed.

Oxygen is almost ubiquitously available and many wood-degrading microorganisms are capable of tolerating relatively low oxygen levels. Even in some apparently anaerobic lake sediments, there are bacteria capable of breaking down wood at an extremely slow rate. Attempting to seal the oxygen out of wood is not a successful strategy for preventing microbial degradation.

The requirement for water is by far the most fundamental, and moisture management is by far the best means to prevent microbial degradation. Water is required for fungal cells to remain hydrated and for fungal enzymes, nonenzymic breakdown systems, and breakdown products to move within the wood. Wood products in service can be protected against microorganisms by drying the wood to <20% moisture content and keeping it dry. At moisture contents between 20 and 26%, established decay can continue but it is almost impossible for new infections to be initiated by spores. Some fungi, such as the true dry-rot fungus (see Section E2), can transfer moisture from a wet piece of wood to relatively dry wood, thus creating suitable conditions for decay. Between 27 and 30% moisture content, colonization and decay can occur but the growth of fungi is slow. Above 40% and <80%

are ideal for growth and decay by wood-rotting fungi. Thus, rapid decay of wood is generally reliant on a supply of liquid water (for exceptions see Section E3). The maximum moisture content for rapid decay is between 80 and 120% moisture content depending on density of the wood and the fungus. At these levels, fungal activity is limited by oxygen diffusion rather than by excessive moisture, as such.

In summary,

$$\text{Susceptible wood} + \text{wood-degrading organisms}$$
$$+ \text{ oxygen} + \text{water} = \text{decay}$$

Take any one of these parameters out of the equation and decay does not occur. Preservative treatment can reduce the susceptibility of wood to decay, allowing it to remain in service for many decades. Wet wood can remain sound in a bristlecone pine for thousands of years if microorganisms are kept out. Archeological wood can be recovered from lake sediments after hundreds of years when oxygen is excluded and wood in temples can remain sound for hundreds of years if it stays dry.

D. WOOD-INHABITING MICROORGANISMS AND COLONIZATION SEQUENCE

Discolored and shrunken wood, recognizable as rotten, is almost the last stage of a colonization sequence that begins when sterile wet wood is exposed or dry wood becomes wet. This sequence continues to progress as long as all the critical requirements are in place. Each microorganism changes the environment to make it less favorable for itself, for example, by using up a certain nutrient, and more favorable for subsequent colonizers, for example, by increasing permeability.

The microorganisms that live in wood can be divided into the bacteria and the fungi. Wood-inhabiting bacteria are single-celled organisms (though some grow in chains) which use organic material as a food source. They multiply by cell division and have limited motility so their ability to penetrate wood is very restricted. Wood-inhabiting fungi grow by extension of microscopic multicelled tubes known as hyphae. These hyphae are capable of growing through the existing pathways within the wood and, in some cases, creating new pathways. The entire branching network of hyphae looks something like the root system of a tree and is known as fungus mycelium. The fungi can be further subdivided into molds, staining fungi, soft-rot fungi, and wood-rotting basidiomycetes. The order in which they are listed above is the order in which they invade wood.

The wood-inhabiting bacteria can be separated into early colonizers and wood-degrading bacteria. The early-colonizing bacteria probably develop within hours of a sterile piece of wet wood being newly exposed or a dry piece of wood becoming wet. They live on nonstructural storage products and their only effect is to increase the permeability of wood. However, if other organisms are excluded by unfavorable conditions, wood-degrading bacteria can dominate. These cause extremely slow strength loss over hundreds of years and are a primary cause of failure only in archeological wood and certain types of preservative treated wood.

Mold fungi develop on damp sapwood within days and they also live on nonstructural storage products. They grow rapidly and some produce antifungal antibiotics that help them to compete for the available resources. They have transparent or white mycelium and colored spores. They discolor the surface of the wood with their spores, but these can be wiped off and the effects are hardly noticeable if the wood is dried out. Large quantities of mold spores, particularly of certain species, can cause health problems if they are inhaled by susceptible individuals. Some molds can grow to a limited extent at wood surface moisture contents between 19 and 26%, too low for decay. However, exuberant growth of molds is indicative of conditions that could lead to the next stage of colonization and eventually to decay.

Staining fungi are closely related to mold fungi, and their main distinguishing feature is dark mycelium. They can be further subdivided into those causing sapstain and those causing black-stain in service, although there is considerable overlap between the two groups.

Sapstain fungi can be present in the standing tree and in summer they develop within weeks on felled logs or freshly sawn lumber. Some are capable of growing up to 20 mm/day under ideal conditions. They compete with the molds and bacteria for nonstructural storage products. Their dark brown or black mycelium gives a blue or black color to the wood, and this can penetrate the full depth of the sapwood. Sapstain fungi do disfigure the wood, but they cause little or no structural damage unless the wood stays wet and other organisms are excluded, for whatever reason. Under these circumstances, sapstain fungi can cause detectable reductions in impact bending strength.

The fungi causing blackstain in service also disfigure wood surfaces. They occur on painted and unpainted wood and are particularly problematic when they grow underneath transparent film-forming finishes. Under the latter conditions and on unpainted wood, they may be able to utilize the products released from the breakdown of wood by ultraviolet (UV) light. The aromatic rings in lignin make it a very effective absorber of UV light, but UV light and water produce free radicals which can initiate rapid breakdown of lignin. On unpainted wood, these breakdown products are washed off, leaving delignified cell remnants consisting almost entirely of cellulose. The refraction of light from these delignified cells gives the wood a silvery sheen. Good examples of this appearance can be seen on driftwood. When blackstain fungi and single-celled algae (simple plants) grow

FIGURE 33.2. Soft rot on the surface of an 80-year-old creosoted utility pole below the soil surface.

on such surfaces, they darken the wood to a gray "weathered appearance." The combined effects of UV light, rainfall, and microbial degradation lead to a slow erosion of wood surfaces. This process can be slowed by surface finishes or pressure treatments that exclude one or more of these three factors (see Chapter 34). Conditions conducive to the continued growth of staining fungi are likely to lead to the next stage of colonization.

Soft-rot fungi are closely related to mold and staining fungi. Some do not discolor the wood and others can give blue or black discoloration. They are the first of the colonizers that are capable of using lignocellulose as a food source, but they are commonly prevented from colonizing earlier by competition from the bacteria, molds, and staining fungi. Soft-rot fungi erode or tunnel within the cell wall, causing gradual strength loss. They only cause serious damage if the wood-rotting basidiomycetes are kept out by unfavorable conditions such as preservative treatment or high moisture content (Fig. 33.2). In treated wood in the ground, their rate of attack depends on the type of preservative (see Chapter 34) and the amount in the wood. It is typically measured in millimeters per year starting at the wood–soil interface. In untreated wood exposed above ground the soft-rot fungi are normally replaced very quickly by the wood-rotting basidiomycetes, which can cause much more rapid strength loss.

Wood-rotting basidiomycetes typically take months to become established, probably because they too are outcompeted by the bacteria, molds, and staining fungi. Once they do become established, their effects on the wood are much more rapid and severe than the soft-rot fungi. Wood-rotting basidiomycetes can be divided into two groups: the white-rot fungi and the brown-rot fungi based on their mode of attack on wood. White-rot fungi erode away the surface of the cell walls, causing strength loss faster than soft-rot fungi but much slower than brown-rot fungi. Brown-rot fungi cause rapid strength loss by depolymerizing the cellulose before there are any visible signs of discoloration or shrinkage.

Because of this rapid strength loss and their predominance on structural softwoods, brown-rot fungi have the greatest economic impact of all decay types. Under ideal conditions, which rarely occur in nature, brown-rot fungi can grow up to 10 mm/day. Once established in a piece of sapwood, they can cause up to 40% strength loss in compression parallel to the grain (important for studs and piling) and up to 60% strength loss in compression perpendicular to the grain (important for wall plates and bearing points for horizontal timbers) in 1 week. This means that it is extremely important to stop the colonization sequence by drying the wood or treating it before it gets to the point where wood-rotting basidiomycetes become established.

E. DIAGNOSING DECAY

Determining the type of decay in a wood structure is not a straightforward procedure and is best left to a specialist. Having said this, the most common type of decay in structural softwood products is brown rot and this is the type of decay that most people will recognize.

E1. Major Types of Decay

Bacterial decay typically begins at the surface of wood in contact with soil or fresh water and gradually progresses inward. Visibly, the wood may remain unchanged or it may become paler in color and have the texture of a crumbly English cheese. When the affected wood is dried, it may take on a honeycomb appearance. Some types of bacterial decay can be recognized under the light microscope, but an electron microscope is needed to be certain of this type of decay. The probability of encountering this type of damage is low unless one is dealing with archeological material, old foundation piling, or wood harvested from man-made lakes, decades after flooding.

Soft rot also tends to begin at the surface of wood in contact with soil or fresh water and gradually progresses inward (Fig. 33.2). The appearance of the wood may remain unchanged despite considerable strength loss. The early effects of soft-rot fungi often cannot be diagnosed without looking at thin sections of wood under the microscope. However, there is a tool available that estimates the depth of soft rot by measuring the depth of penetration of a pin fired into the wood with a known force. Some indications of early strength loss can be ascertained by inserting a knife blade at a shallow angle to the surface, at right angles to the grain, and levering up a splinter. A fibrous splinter indicates sound wood and a brash fracture indicates decay (this also works for decay by wood-rotting basidiomycetes). At more advanced stages, soft rot can take on the

appearance of white rot or brown rot. The biochemistry of the breakdown of wood by soft-rot fungi is barely understood compared to our understanding of the other two decay types.

White rot and brown rot are rarely obvious on the surface of a structure until they produce a fruiting structure such as a conk or bracket. In wood in ground contact, they do not compete well with soil fungi so they tend to be confined to deeper within the wood. In wood exposed above ground, the surface is often painted, hiding any underlying damage. Exposed unpainted surfaces may never stay wet long enough to support growth and decay. The first signs of decay in painted wood are usually collapse and cracking of the paint film as the underlying wood shrinks and cracks. When a structure with suspected problems is opened up, such as by removing the cladding or drywall in a building, wood-rotting basidiomycetes can be seen growing on the surface of wood provided the atmosphere in voids within the structure has been at a high relative humidity (RH), close to 100% RH for some time. The fruiting structures and mycelium of white-rot and brown-rot fungi cannot be distinguished without the involvement of a trained mycologist.

White rot and brown rot can be readily distinguished by their effect on the wood. White-rot fungi degrade both lignin and cellulose, leaving the wood with a bleached and often stringy appearance (Fig. 33.3). This can be confused with the effects of chemical bleaching agents (see Chapter 34). White-rot fungi possess a battery of cellulose- and lignin-degrading enzymes, several of which contain metals or are metal dependent. White-rot fungi tend to attack hardwoods, such as aspen and maple, rather than softwoods.

Brown-rot fungi degrade the cellulose and modify the lignin, giving a darkened, almost charred appearance, referred to as cubical cracking (Fig. 33.4). This can also be confused with the electrochemical deterioration that occurs when a galvanic cell forms between two types of metal used

FIGURE 33.3. White rotted wood blocks showing typical bleaching and stringy texture.

FIGURE 33.4. Brown rotted lumber showing typical darkening and cubical cracking.

as fasteners, commonly known as nail sickness (see Chapter 34). There may indeed be a link between the two. It has been suggested that Fenton-type (iron redox) reactions with or without mediation by fungal chelators are involved in the production of free radicals that initiate depolymerization of cellulose and modification of lignin by brown-rot fungi. Brown-rot fungi tend to attack softwoods more than hardwoods. They are the most important fungi causing decay in buildings because softwoods are more commonly used for structural purposes. The brown-rot fungi include those commonly referred to as dry-rot and wet-rot fungi.

E2. Dry Rot, Wet Rot, and Tolerance to Drying

Dry rot is a commonly misused term. It does not refer to wood that is rotted and dried out, nor does it refer to rot that can take place in dry wood. There is, however, a true dry-rot fungus called *Serpula lacrymans*, which is capable of transporting water from wet, rotting wood through mycelial cords to relatively dry wood. When the wood is wetted up, this fungus can colonize and decay it. The true dry-rot fungus is common in Europe, eastern North America, and northern Japan. It is relatively rare in western North America where it is replaced by another fungus, *Meruliporia incrassata*, which also has some water-transporting capabilities. The true dry-rot fungus is particularly prevalent in wood in brick or stone buildings. This may be partly due to a requirement for calcium and partly to a need for relatively constant temperature and humidity. Brick, stone, and concrete tend to retain moisture and slow the drying rate of adjacent wood products. Part of the reason this fungus is so much of a problem is that is can survive in wood hidden within brick, stone, or concrete walls and reemerge to infect new wood if the moisture sources that caused the decay problem in the first place are not completely eliminated.

Wet rot includes decay caused by other wood-rotting basidiomycetes. Two of these, *Gloeophyllum sepiarium* and *Gloeophyllum trabeum*, are particularly tolerant of drying out and of high temperatures. They are particularly prevalent in wood products in above-ground exterior exposure, such as window joinery, siding, and decking. Most of the others require more consistent moisture conditions to thrive. Having said this, it should be noted that many wood-rotting basidiomycetes can survive for years in wood with ~12% moisture content only to become active again if the wood rewets. Simply drying out a structure will not kill the decay fungus.

E3. Identification and Elimination of Wood-Rotting Basidiomycetes

If it becomes important to know which fungus caused the damage, this is best left to a mycologist with specific training in this field. However, with the possible exception of the true dry-rot fungus, it is rarely critical to know which fungus caused the problem; the recommended course of action is the same. Cut out all visibly decayed wood plus 2 ft in the longitudinal direction and the full cross section where the decay occurred. This action is necessary to eliminate wood that may have been colonized by the wood-rotting basidiomycete but not yet visibly changed. Not only will this region have lost considerable strength, but it could act as a source of infection if the moisture sources were not all eliminated or if a new leak arises. The search for moisture sources should consider not only obvious or primary moisture sources, such as a leaky building envelope or a flood, but also possible secondary sources, such as condensation. While wood-rotting basidiomycetes generally need liquid water to become established, they can continue to cause decay supported only by water vapor. Wood-rotting basidiomycetes create their own moisture as they decay the wood. Cellulose is a carbohydrate and the fungus converts it to carbon dioxide and water. If there is the possibility that all infected wood and all sources of moisture have not been eliminated, the remaining wood should be treated with a diffusible preservative and any new wood installed should be pressure treated (see Chapter 34).

E4. Determining Start Date of Decay Problem

When decayed wood is found in buildings or other structures, the question most often asked is, "How long has this been going on?" Unfortunately, this question is almost impossible to answer. The rate of growth and decay by wood-rotting basidiomycetes is affected by the species of fungus, the natural durability of the wood, the wood moisture content, the temperature, the presence and competitive abilities of other fungi, and interactions among these parameters. Too many of these parameters cannot be determined retroactively.

E5. Corrosion of Metals by Wood-Rotting Fungi

The metals, such as manganese and iron, needed by wood-rotting basidiomycetes for their metabolic processes are limiting micronutrients in wood. Both white-rot and brown-rot fungi appear to have specific mechanisms for securing and accumulating these resources. To do this, they produce metal chelators, known as siderophores. Brown-rot fungi also reduce the pH of wood early in the decay process through the production of organic acids. As a result, corrosion of metals can be induced by contact with decaying wood over and above that expected based on the moisture availability.

F. FURTHER READING

More detailed information on the subject of wood biodegradation in general can be obtained from two textbooks by Eaton and Hale [1] and Zabel and Morrell [2] The first is written from a European perspective, whereas the second is written from a North American perspective. A classic textbook in this field was authored by Cartwright and Findlay [3]. More specific information on microbial breakdown mechanisms is covered by Eriksson, Blanchette, and Ander [4]. Inspection of buildings and diagnosis of deterioration problems are covered by Bravery et al. [5] and Levy [6].

REFERENCES

1. R. A. Eaton and M. D. C. Hale, Wood: Decay Pests and Protection, Chapman & Hall, London, 1993, 546 pp.

2. R. A. Zabel and J. J. Morrell, Wood Microbiology: Decay and Its Prevention, Academic, San Diego, CA, 1992, 476 pp.

3. K. St. G. Cartwright and W. P. K. Findlay, Decay of Timber and Its Prevention, Chemical Publishing, New York, 1950, 294 pp.

4. K. E. L. Eriksson, R. A. Blanchette, and P. Ander, Microbial and Enzymatic Degradation of Wood and Wood Components, Springer-Verlag, New York, 1990, 407 pp.

5. A. F. Bravery, R. W. Berry, J. K. Carey, and D. E. Cooper, Recognising Wood Rot and Insect Damage in Buildings, 3rd Edition, Building Research Establishment Report, Department of the Environment, Princes Risborough Laboratory, Princes Risborough, Aylesbury, Bucks, UK, 2003, 120 pp.

6. M. P. Levy, A Guide to the Inspection of Existing Homes for Wood Inhabiting Fungi and Insects, U.S. Department of Housing and Urban Development, Washington, DC, 1979, 104 pp.

34

USE OF CHEMICALS TO PREVENT DEGRADATION OF WOOD

J. N. R. RUDDICK

Department of Wood Science, Forest Sciences Center, University of British Columbia, Vancouver, B.C.,Canada

A. INTRODUCTION

As discussed in Chapter 28, wood is a biodegradable, natural, renewable resource. Over the centuries wood has supported community life in a variety of ways, including construction of buildings for shelter; providing heat, cooking food, and furnishings to improve lifestyle; and as raw material for artisans. When Europeans first came to North America wood was plentiful and could easily meet the demands. In Europe, with the rapid development of shipbuilding in the sixteenth to eighteenth centuries, wood began to be used on such a large scale that Samuel Pepys noted in his diary that there was a shortage of seasoned oak for the navy's ships. The development of large-scale ironworks saw even more destruction of forests for production of smelting charcoal.

Compounding this shortage of wood were the problems of material failure due to rot. In the late-eighteenth century the loss of the Royal George at Portsmouth, with over 800 lives lost, stimulated the Royal Society of the Arts to offer a gold medal for the discovery of the cause of the dry rot in the hulls of wooden ships and a reliable method of eliminating it [1]. Soon several chemicals were considered to enhance wood durability, and in 1770, Sir John Pringle drew up the first list of wood preservatives. By 1810 the *Encyclopedia Britannica* published these lists as a reference. The early preservatives were generally metal salts such as copper sulfate, mercuric chloride, and zinc chloride.

Uhlig's Corrosion Handbook, Third Edition, Edited by R. Winston Revie
Copyright © 2011 John Wiley & Sons, Inc.

One of the problems of these simple salt treatments was their ready loss from treated wood, a problem that was solved when in 1841 Payne was granted a patent for a two-stage treatment involving ferrous sulfate and calcium sulfide which caused the ferrous sulfide to precipitate in the wood. However, lack of efficacy caused the treatment to be abandoned. About the same time, in 1838, John Bethell was granted a patent for his "creosoting" process, which employed solutions of metallic salts and bituminous chemicals. The term creosote was actually used in a patent granted to Moll in 1836 for his process in which wood was injected with extracts of coal tar in closed iron vessels. Moll called the heavier fraction "Kreosot." So in the late-nineteenth century creosote to preserve railway ties and marine timbers began to be used. It is remarkable that the basic creosote preservative and the treatment process proposed by Bethel have remained largely unchanged to the present day. However, as multisalt inorganic-based wood preservatives for utility poles in North America began to be used in the late 1940s alternative processes were required to ensure an effective treatment. The past 20 years have been a number of pressure and nonpressure processes for impregnating timber with preservative, and these will be briefly reviewed in this chapter. Perhaps the biggest single influence on the development of preservatives and processes during the last two decades has been concern over the environmental impact of industrial processes and their products. This has stimulated the industry to create new wood preservatives and process modifications to minimize the environmental impact of treated wood.

B. PRESERVATIVES CURRENTLY STANDARDIZED IN NORTH AMERICA

B1. Creosote

Despite considerable pressure to use alternative preservatives, creosote remains the most important preservative rather than dominant preservative for protection of wood placed in the marine environment and for treatment of wooden railway ties. Information on the properties of creosote can be found in the wood-preserving standards developed by the American Wood Protection Association (AWPA) in the United States [2] and the Canadian Standards Association Wood Preservation Standard 080 in Canada [3]. Over the last two decades, the focus on creosote has been to reduce or eliminate the soluble tar acid fraction and to reduce the benzopyrene content. The tar acid fraction is reduced (eliminated) by washing the creosote and the acids recovered for other uses. The benzopyrene content has been reduced due to its toxic effects on humans. A number of excellent reference papers on creosote and its uses may be found in the proceedings of the AWPA [4–6]. In 1994, $0.26 \times 10^6 \, m^3$ of creosote-treated wood was produced. Since then the trend has been to a

gradually reduce volume of production due to lack of key end uses, such as retail lumber, as well as increased competition from alternative materials for one of the major products—piling. Creosote remains supported for both marine timbers and piles provided that the best management practices guide developed in 1996 by the American Wood Preservers Institute supports its use [7–9].

B2. Pentachlorophenol

Pentachlorophenol was developed in mid-1930s as part of a general interest in the potential of chlorinated phenols [10]. The higher chlorinated phenols were found to have greater efficacy than the trichlorophenol and tetrachlorophenol products. The pentachlorophenol also possessed a higher melting point. Although originally produced either by chlorination of phenol using an aluminium chloride catalyst or by hydrolysis of hexachlorobenzene, the potential for dioxin formation in the latter process has resulted in it being discontinued. Early uses of pentachlorophenol included protection of millwork in which a 5% pentachlorophenol solution in a light organic solvent was applied to wood by either a double-vacuum process in the larger industrial plants, and by spray or dipping in small scale operations. During the 1940s the availability of creosote was limited due to the war, and pentachlorophenol in heavy oil was employed to protect utility poles in North America. Field tests quickly established it to be a very effective broad-spectrum biocide capable of preventing not only decay but also insect attack. It was however unsuitable for protecting wood in the marine environment. It rapidly became the preservative of choice for utility poles and by the mid-1960s it dominated pole protection.

In the mid-1970s the U.S. Environmental Protection Agency (EPA) introduced a review of all wood preservatives, and pentachlorophenol became the focus of attention because of the small amounts of dioxin and furan impurities found in technical-grade material. Concern has also been expressed over the potential for pentachlorophenol-treated wood to produce dioxins in service, although this has not been demonstrated. Perhaps the most important impact of the dioxin impurity is on the disposal of pentachlorophenol-treated waste. Until the last few years the disposal of treated wood recovered from service was not an issue [11]. Since all treated wood as it is still considered to be not hazardous according to EPA. It can be disposed of in a regular landfill. This is now under review and the United States and Canada have limited the procedures that govern the disposal of pentachlorophenol-treated wood can be disposed. Incineration has been proved to be an effective option, provided that the temperature of the process is above 1000°C [12]. Although a "clean" version of pentachlorophenol was developed during the 1980s, cost prevented its use by industry. The importance of the solvent system on the performance of pentachlorophenol has been widely studied and several alternative treatments have been commercialized. These include ammoniacal copper pentachlorophenate and water-soluble forms of pentachlorophenol using emulsifying agents. Perhaps one of the most widely used alternatives was liquefied gas treatment, in which pentachlorophenol was dissolved in an inert liquefied gas such as methylene chloride or butane, often with the aid of a small amount of a cosolvent. The advantage of this treatment was that it could be used to produce treated wood without the color change normally associated with the use of the heavy oil solvent. However, with the passage of time the loss of preservative from wood treated by the liquefied gas processes was found to be too great and the service life of the product was not acceptable. None of these alternative pentachlorophenol-based preservatives or processes are in use today.

In 1994, $0.07\,m^3$ of pentachlorophenol-treated wood was produced. The principal products were utility poles and cross-arms. Its market share of treated wood has been declining since it is not used for retail lumber and many utility companies have changed to using fixed waterborne preservatives. Although initially there was interest in using an oil-borne copper naphthenate in place of pentachlorophenol, the well-publicized failures of copper naphthenate–treated southern pine poles negatively impacted its use.

B3. Chromated Copper Arsenate

Chromated copper arsenate (CCA) was proposed to the AWPA for standardization in 1949 as Greensalt [10]. It was accepted in 1953. The formulation known also as CCA type A was revised in 1968, and in 1969 the formulation was specified in the standards on an oxide basis when it contained 65.5% chromium (expressed as CrO_3), 18.1% copper (expressed as CuO), and 16.4% arsenic (expressed as As_2O_5). A second formulation, known as Boliden K-33, or simply as K-33, had been in use commercially in Sweden since 1950, having been patented in 1947 [10]. It was proposed for inclusion in the AWPA Book of Standards in 1963, as CCA type B. The formulation standardized on an oxide basis in 1969 contained 35.3% chromium (expressed as CrO_3), 19.6% copper (expressed as CuO), and 45.1% arsenic (expressed as As_2O_5). In the mid-1960s the AWPA wished to reduce the complexity of the preservative standards by eliminating obsolete preservatives. A single standard for CCA was proposed which had a formulation which was approximately the average of the two existing CCA formulations [10]. Data were presented to show that such a formulation would have the best leaching resistance. In 1969 a standard for CCA type C was published and the preservative contained 47% chromium (expressed as CrO_3), 19% copper (expressed as CuO), and 34% arsenic (expressed as As_2O_5). From 1970 all three formulations were employed with the industry often switching from one to the other depending on the cost or availability of chromic acid or arsenic acid. However, in the last decade types A and B have

given way to the almost exclusive use of the better balanced type C formulation.

From 1980 to 2003 CCA became the dominant preservative worldwide. In 1994 over $2\times10^6\,m^3$ of CCA-treated wood was produced in the United States. Indeed, over 85% of sawn wood was treated with CCA. This dramatic change in market dynamics from that of the early 1970s reflected the adoption of residential preservative-treated wood which was marketed through retail outlets for use on projects around the home. This market had been taken up exclusively by CCA. With the development of these specialty products for the do-it-yourself market, the use of additives has become much more prevalent. These additives include polyethylene glycol or emulsified oil for improving the climbability of utility poles, dyes and pigments to alter the color of CCA-treated wood, and water-repellant additives to reduce the checking behavior of decking. Despite concerns over surface hardness in CCA-treated southern pine, CCA-treated poles are widely accepted by the industry with a current market share of almost 70%.

In the 1990s the wood-treating industry focused on improving the fixation of CCA before the treated wood leaves the treating plant. Relatively simple tests such as that based upon chromotropic acid were developed which allow both producer and user to confirm that CCA-treated wood does not contain unreacted chromium. Research showed that the rate of reaction of chromium is slower than that of either copper or arsenic so that tests based on the chromium reaction can be used to follow the overall fixation rate. Consistent with the current focus on the environmental impact of wood treatment, greater attention is now given to ensuring that the treated wood is free of surface deposits, and this too has become part of the CSA and AWPA book of standards. The methods used to enhance the fixation rate will be discussed briefly in the following section on treatment processes.

In 2003 the use of CCA dramatically changed. The suppliers of CCA agreed to withdraw its use from the residential treated wood market. It is still allowed for use for industrial products such as poles and piling. It is also allowed for residential uses such as shingles and shakes, plywood, and the preserved wood foundation components. However, for most traditional residential uses, such as fenceposts, decking, fencing, and cladding other preservatives based on copper and organic co-biocides have been introduced and now dominate those markets.

B4. Ammoniacal Copper Arsenate and Ammoniacal Copper Zinc Arsenate

Ammoniacal copper arsenate (ACA, patented by Gordon in 1939) as a preservative was adopted by the AWPA in 1953 [10]. It was formulated as 49.8% copper (expressed as CuO) and 50.2% arsenic (expressed as As_2O_5). This formulation tended to lose unfixed arsenic so that an improved formulation, ammoniacal copper zinc arsenate (ACZA), was developed and replaced ACA in the book of standards [13]. ACZA is formulated as 50% copper (expressed as CuO), 25% zinc (expressed as ZnO), and 25% arsenic (expressed as As_2O_5). Ammoniacal copper-based preservatives have traditionally been preferred for wood species that are refractory to treatment, for example, spruce and Douglas fir. Consequently ACA and ACZA were more common in Canada and the west coast of the United States than in southern United States. Currently, neither ACA nor ACZA are approved in Canada. ACZA remains allowed for use in the United States where its principal uses are the treatment of piles, timber, and utility poles.

B5. Copper and Zinc Naphthenate

Naphthenic acids, by-products of the petroleum industry, can be reacted with copper and zinc salts to produce copper and zinc naphthenates. First used as wood preservatives in Europe, where they were sold under the trade name Cuprinol, they were not introduced into the United States until 1948 [10]. Their principal use was as brush treatments sold at retail outlets to the general public or as dip treatments for commercial products such as fence posts. Their relatively high cost prevented their commercial acceptance for pressure treatment until the last decade, when the utility pole industry became increasingly under pressure to introduce an alternative to pentachlorophenol. It was introduced into the AWPA commodity standards in the early 1990s solubilized in a P9-type oil. However, with the increasing acceptance of fixed waterborne preservatives, its use in this market remains very small. In Europe its main use has been for the double-vacuum treatment of millwork (window joinery and doors), cladding, fencing, and so on, for above-ground exposure. Zinc naphthenate is not used for pressure treatment in North America. The volume of copper naphthenate–treated wood in North America remains relatively small. In Europe the lack of availability of naphthenic acid has stimulated research on the production of synthetic copper and zinc soaps from versatic and octanoic acid. These products have not been introduced into North America.

B6. Bis-tributyltin Oxide

Tributyltin oxide (TBTO) was principally employed for the dip treatment of window joinery and doors. However, following research in the 1970s, which revealed that TBTO rapidly degraded upon impregnation into wood, its use has declined and it is now no longer used.

B7. Copper-8-quinolinolate

Pure copper-8-quinolinolate (oxine copper) is insoluble in light organic solvents. A solubilized form can be produced with nickel hexanoate. Its use remains limited in North America, with its principal advantage being that wood treated with oxine copper at one time could be used in contact with foodstuffs. It was introduced into the AWPA commodity standard for sawn wood exposed above ground but was restricted to the protection of several pine species, including southern pine, western red cedar, and hem-fir [2].

B8. Alkylammonium Compounds

The potential of alkylammonium compounds (AACs) as wood preservatives has been recognized for more than 30 years [14]. However, following early failures of problems of decay in wood treated benzalkonium compounds in New Zealand in the early 1980s their main use has been as a co-biocide in formulations of ammoniacal or amine copper preservatives, such as ammoniacal copper quat (ACQ). They remain standardized as a stand-alone preservative, and if this use is to be realized, it is likely to be formulated with other organic co-biocides for wood exposed above ground. At present only didecyldimethylammonium chloride (DDAC) and DDA didecyldimethylammonium carbonate have been standardized in North America [2]. An important commercial application for DDAC is as a co-biocide in NP-2, a formulation widely used for protecting unseasoned wood in transit and storage. In Europe and Japan other AACs have been commercialized. They include trimethylcocoammonium chloride and alkyldimethylbenzylammonium chloride.

B9. Alkaline Copper Preservatives: Alkaline Copper Quat

Almost universally there has been environmental pressure to reduce the use of CCA in high-volume residential applications. One of the first preservatives to be developed without arsenic or chromium was ACQ (types A and B) where the quat (or quaternary ammonium compound) was DDAC [15]. The preservative consisted of copper (basic copper carbonate) solubilized in ammonium hydroxide and formulated with a quaternary ammonium compound. This preservative combination was commercialized in Scandinavia and was introduced into the AWPA standard in 1992. In 1995, a version of ACQ type B formulated with monethanolamine as the solvent in place of ammonium hydroxide was standardized. The use of amine in place of the ammonium hydroxide provided product improvements including enhanced surface cleanliness and better color to allow penetration of the retail lumber market.

Currently four ACQ formulations are accepted treatments for most commodities in the AWPA [2] and CSA [3] book of standards. The variations include the addition of some ammonia to assist in the penetrability of the preservative in refractory wood species. This preservative has replaced CCA for the treatment of wood for residential applications and is one of two main preservatives commercialized in Canada and the United States for this application.

B10. Alkaline Copper Preservatives: Alkaline Copper Azole

In addition to ACQ a second formulation based on alkaline copper has been widely accepted for use in protecting residential products. Copper azole was standardized in 1995 [16]. This is alkaline copper azole, in which the copper is solubilized in monoethanolamine and formulated with a triazole [2, 3]. The most common triazoles used are tebuconazole and propiconazole. The former is used in North America while the latter has been used in European copper azole formulations. Initially, boric acid was also included in the formulation (CB-A), but this early preservative has since been replaced by one containing only copper and tebuconazole (CB-B).

B11. Micronized Copper Wood Preservatives

Because of concern over the leaching of copper from alkaline copper–treated wood at high retentions as well as enhanced corrosion compared to that from earlier preservatives such as CCA, new innovations continue to be introduced by the industry. The most recent innovation involves the use of small solid particles of copper carbonate which are suspended in aqueous solution together with a co-biocide as a wood preservative [17]. Two formulations have been standardized in North America [2, 3]. They are micronized copper quat (MCQ) and micronized copper azole (MCA). Because the copper is placed into the wood as basic copper carbonate and then is only solubilized by its reaction with the wood components, large concentrations of mobile copper are not present in the treated wood. This is perceived to be an advantage in minimizing the environmental impact and reducing the corrosive effects of the treated wood. In addition, the need to stabilize, or "fix," the preservative after treatment has been eliminated. Currently the micronized copper preservatives are being widely accepted by retail outlets supporting residential use of treated wood. The formulation of the product remains the same as that for the alkaline solvent–based preservatives.

B12. Copper Bis(dimethyldithiocarbamate)

The beneficial value of carbamates has been known for many years and several biocides based on this class of chemical have been commercialized in other industries. In the mid-1990s a new preservative was based on a two-stage treating process involving sequential impregnation of timber with a copper ethanolamine solution and a sodium dimethyldithiocarbamate solution. The two chemicals react in the wood to produce the highly insoluble copper dimethyldithiocarbamate [18]. This preservative has been discontinued in the United States and was never commercialized in Canada.

B13. Ammoniacal Copper Citrate

Ammoniacal copper citrate was standardized [2] for the protection of sawn wood for both in-ground as well as out-of-ground exposures. Its use has been discontinued.

B14. 3-Iodo-2-propynyl Butyl Carbamate

3-Iodo-2-propynyl butyl carbamate (IPBC) is a new generation of organic wood preservatives that has been standardized by the AWPA [2]. Currently its main application is in a formulation NP-2, which is designed to prevent degradation of unseasoned wood during storage and transit. As interest in totally organic-based wood preservative formulations increases, the use of IPBC may well increase to protect wood used in above ground.

B15. Triazoles

Triazoles represent a new group of chemicals that have been shown to provide good protection against decay for timber exposed in above ground. Research has suggested that due to cost and the lack of broad-spectrum activity it is likely that they will be formulated with other biocides to produce an effective preservative, particularly for the protection of residential wood products. Two triazoles have been standardized since 1994: tebuconazole and propiconazole [19]. A third cyproconazole has also been studied both in Europe and North America and has potential for future use in preservative formulations. As with other organic wood preservatives discussed in this chapter, their main use is likely to be for protecting wood used in above ground.

B16. 4,5-Dichloro-2-*N*-octyl-4-isothiazolin-3-one

Isothiazolinones represent a new class of organic wood preservative that has been standardized in both Europe and North America [2]. Like many of the other organic preservatives, the isothiazolinones are likely to find most use for protecting wood exposed in above ground or as co-biocides in formulations to enhance biocidal performance.

B17. Chlorothalonil

Chlorothalonil was introduced into the AWPA standard in 1993 [2]. It is not currently commercially used in wood preservation.

B18. Inorganic Boron

Borate treatment of wood has been well established in New Zealand for the diffusion treatment of unseasoned timber. Indeed, a small amount of borate-treated wood was produced in Canada for supply to the U.K. market for protection against the house longhorn beetle (*Hylotrupes bajulus*). During the past three or four years, there has been increasing interest in the use of borate treatments to prevent insect attack in wood not exposed to liquid water. Examples of such exposures include sill plates and studs in termite-infested regions of the world. Historically, the process involved dipping unseasoned timber in high concentrations of borate followed by long-term storage to allow diffusion to proceed. More recently it has been demonstrated that acceptable treatments can also be achieved by pressure treatment of wood with borate solutions of moderate strength. Consequently, in 1995 inorganic boron was added to the AWPA book of standards [2, 3], and in 1998 borate treatments were added to a number of commodity standards.

B19. Acid Copper Chromate

Acid copper chromate (ACC) was proposed to the AWPA for standardization in 1950 and accepted in 1953 [10]. While it still remains in the book of standards, its use is relatively minor and limited to those industrial products where CCA is not accepted due to the presence of arsenic.

B20. Copper HDO

A recent addition to the array of preservative formulations based on alkaline copper is copper–HDO (where HDO is *N*-cyclohexyl-diazeniumdioxide). The preservative has been widely used in Germany and Scandinavia for over a decade. It was standardized in the United States in 2005 [2] but has not yet been approved in Canada. It is anticipated that its standardization will take place during 2011. The solvent system is monoethanolamine.

B21. Impralit-KDS

Another alkaline copper preservative recently approved and standardized in the United States is Impralit-KDS [2, 20]. This preservative combines the basic copper carbonate and boric acid solubilized in monoethanolamine with a polymeric betaine. The polymeric betaine is a complex borate ester which, once impregnated into wood, breaks down to produce boric acid and a quaternary ammonium like compound. KDS has been used extensively in Europe for more than a decade.

C. PROCESSES

Although several treatment processes have been established in many countries, in North America the main standard-writing organizations have focused on pressure treatment processes. The dip treatment of sawn wood found in many other countries for protection against insect attack is not practiced in North America. (Dip treatment has been widely used for protection of commodities such as window joinery, but neither the AWPA nor CSA standards currently cover it.)

The treatment process for timber can be divided into four main stages [10]: material conditioning, pretreatment steps designed to improve the quality of the protection, the actual pressure process, and the final conditioning/fixation process. For most commodities it is essential to remove the water from the cell lumen to allow the treating solution to enter the wood cell structure. From here it can diffuse into the cell wall. In general, the conditioning of the wood is by air seasoning, although accelerated drying by steam conditioning or kiln drying is increasingly being practiced. Target moisture contents for most treatments are around 25%. An exception to this is where the wood is to be Boultonized prior to treatment, in which case the moisture content may be as high as 35%. Following seasoning, the commodity is cut to the correct size or shape. This is important since during treatment wood is not treated throughout the cross section. Rather the preservative process provides an outer shell of treated wood. The protection of the untreated core is then dependent on maintaining the integrity of this outer shell. To assist in the treatment of refractory heartwood, techniques such as incising are employed. Incising is the creation of fine openings in the wood surface that allows the preservative solution to penetrate immediately to the required depth from the surface. By arranging these incisions in a suitable pattern, the enhanced longitudinal flow of solution allows an integral shell of preserved wood to develop. In round wood techniques such as through boring at the ground line have been commercialized for difficult-to-treat pole species such as Douglas fir.

For oil-borne preservatives such as pentachlorophenol or creosote, the treatment solution is usually applied at high temperature [10]. This has two beneficial effects. First, it sterilizes the wood, eliminating any incipient decay already present. Second, if a vacuum is applied to the treatment system when the timber is immersed in the hot oil, the water present in the wood can be removed. This forms the basis of the Boultonizing of wood, a pretreatment process often practiced with utility poles.

Following the pretreatment processes the products are then subjected to the pressure impregnation stage. While traditionally two main types of processes have been used, these have become modified during the past 10–20 years so that modern processes often depend upon the specific product and chemical combination being considered. The two main processes are the full cell process (sometimes called the Bethel process [10]) and the empty-cell process [10]. In the full cell process the wood is placed in the treatment cylinder and subjected to an initial vacuum of 75 kPa. The preservative solution is then introduced into the cylinder and the pressure then gradually raised to a maximum of 1050 kPa. This pressure is then maintained for 1–16 h depending on the commodity and wood species being treated. Southern pine schedules are generally very short whereas those for refractory species such as Douglas fir tend to be long. Following the pressure phase, the solution is pumped out of the retort and a final clean-up vacuum is applied. This is designed to remove excess preservative solution from the product and prevent dripping of the solution when the product is removed from the cylinder and placed into temporary storage. The full cell process has traditionally been used for waterborne preservatives and creosote treatment of marine piling.

Two versions of the empty-cell process have traditionally been used [10]: the Lowry process and the Rueping process. In the Lowry process, the timber is placed in the treating cylinder and the preservative solution added without any initial vacuum. At the end of the pressure phase the trapped air in the wood assists in recovering much of the creosote or oil-borne preservative, thus preventing overtreatment of the product. For wood species, which has a large amount of sapwood, even the Lowry process can result in excess preservative solution being taken up. In the Rueping process, an initial air pressure is applied to the cylinder containing the timber. The preservative solution is then allowed to enter the cylinder while maintaining the cylinder pressure. As before, the pressure is then increased during the pressure phase. At the end of the pressure phase, the entrapped air then assists in the recovery of the preservative solution, preventing overtreatment.

During the past two decades, the wood-preserving industry has developed several posttreatment processes designed to minimize potential for environmental contamination. These include special sealed concrete surfaces covered from the rain where treated wood can be held before moving into general yard storage. However, some of the most important innovations relate to the treated wood itself. For example, for CCA which undergoes a reaction with the wood components resulting in insolubilization of the components, posttreatment heating is an integral part of the process at many treating plants. This heating stage can be achieved in many ways [21]. Processes based on the application of hot water under pressure, high-temperature steam [22], humid air at moderate temperatures, and the use of radio frequency [23] kilns are just some techniques reported in the literature. Even treating plants using ammoniacal copper preservatives have adopted different posttreatment processes designed to minimize surface deposits on commodities. These include posttreatment washes with ammonia solution.

For oil-borne products including creosote-treated piling, it is recognized that an increase in the moisture content which takes place when the commodity is placed in service causes the fibers to swell and the wood to lose preservative. To counter these effects, posttreatment steaming of oil-borne-treated wood is now recommended to reduce surface loading at the treating plant, thereby reducing the potential for loss to the environment.

While the major focus of the AWPA and CSA standards is on the pressure impregnation of wood, the use of thermal treatments with oil-borne preservatives is also described. In this process the wood is subjected to immersion in a hot preservative solution followed by immersion in a cold preservative treating solution. During the first stage, the air in the wood cells expands and some is lost from the wood surface. In the cold treating solution, the air contracts, creating a partial vacuum which assists the preservative penetration into the wood. The thermal treatment process is restricted to roundwood.

D. STANDARDS

In North America the principal standards governing the production of treated wood are the AWPA [2] and CSA080 [3] standards. These standards have traditionally been based on a "results-type" approach in which a number of pieces in the charge or load of wood are sampled and the acceptability of the treatment depends on the analysis of penetration and retention exceeding set threshold values. The threshold values are established for preservatives through field exposure tests, in which the performance of the new preservative is compared to a reference preservative. During 2009 the AWPA and CSA have begun to explore whether a process specification could be useful for certain types of products. The concept would be to control the process sufficiently to ensure a reliable quality control on the product. Such a process specification would apply to noncritical building components such as fencing boards and deck boards. Currently no process specifications have been standardized.

The AWPA book of standards is updated annually. In earlier versions it contained five sections. The first section described the preservative compositions. The second section was the largest and outlined how different commodities must be treated using pressure processes. This section identified the limits of the various treatment parameters. The objective of this part of the commodity standard is to prevent damage occurring to the product during treatment through the use of excessive pressure or heat. The third section described in detail how treated wood and preservative solution can be analyzed to ensure quality of treatment. A fourth section contained a number of miscellaneous standards, some of which discuss how to establish the quality control of treated wood and of treating plants. The fifth section outlined the procedures that must be followed to evaluate wood preservatives. These include laboratory as well as field tests. Finally, there was a listing of conversion factors.

Until 1998 the AWPA book of standards was based upon Imperial units, but in 1998 the standard was converted to metric units. The Canadian wood-preserving standard converted to metric units in 1989.

A major change in the way the wood preservation commodities are standardized was made in 2003 in North America [2, 3, 24, 25]. Instead of focusing on standards for different commodities with specific details on how to treat wood, the new approach focuses on the different biological hazards that exist. This new system is called the Use Category System (UCS). Under this system different commodities are grouped under a number of categories depending on the degree of biological hazard. Five use categories have been adopted. The first, UC1, covers material not in contact with the ground and protected from exterior weather. UC2 describes interior construction material that is used above ground but which may be expected to become damp in use. The third use category (UC3) is for wood used above ground outdoors and has two subcategories. UC3A covers wood that is coated and not in contact with the ground, for example, windows and cladding. UC3B is for wood that is not coated but is not in ground contact. Typical products might be decks or railings. UC4 covers wood in ground contact. It is divided into several subcategories depending on the degree of criticality. UC4A is for normal ground or freshwater uses, such as fence posts, but excluding wood in critical uses or unusually high hazards. UC4B is for severe ground contact exposure and important construction material that would be difficult to replace, such as preserved wood foundation components. The third subcategory, UCC, is for wood in ground contact but where the biological hazard is unusually high or the end use is structural and failure would be hazardous. Examples of such uses would be freshwater piling and utility poles. The fifth category, UC5, focuses on the marine environment where higher preservative retention is usually required, for example, marine piling. There is in addition a category, UCF, which covers protection of wood against fire rather than biological hazards. This category is further divided into interior and exterior exposure conditions. The intent of this new approach to defining preservative requirements is to make the standards more easily understood by the end user and to ensure consistency between commodities. The adoption of a use category system to describe preservative requirements for commodities is not new, with different versions already in place in Europe, Australia and New Zealand, and Japan.

In the latest versions of the AWPA [2] and CSA [3] book of standards, the structure has been revised to incorporate the information shown previously but centered around the UCS concept. The CSA book of standards mirrors the structurtre of the AWPA book of standards. In the first section of the current AWPA book of standards (U1-08), the user specifications for treated wood are outlined. In addition to describing the various use categories, it also includes a table providing a guide to commodity specifications. The intent is to assist the user in selecting the appropriate treatment type for the product of interest. The table directs the reader to the appropriate "commodity specification" section for additional information. Other tables identify the wood species and treatment combinations standardized by the AWPA (and CSA). It also includes a listing of all preservatives standardized by the AWPA (and CSA). The processing and treatment standard (T1-08) describes all of the technical requirements for successful treatment of wood with preservatives. In a recent development all preservatives have been assigned a separate standard, rather than being grouped together by preservative type, as was previously done. The analytical methods section follows this approach as well, so that each analytical method is now described in its own standard. The evaluation and miscellaneous standards remain as before as separate sections. In general, the analytical and evaluation standards of the AWPA [2] are adopted by the CSA as appendices to the Canadian standard [3].

E. RECOMMENDED SOURCES OF INFORMATION

Several excellent sources of information on wood preservation are available. The Proceedings of the American Wood Protection Association and the Canadian Wood Preservation Association can provide updated information on preservatives and processes. The books of standards produced by the American Wood Protection Association and the Canadian Standards Association are important sources for identifying changes in requirements. A number of texts have been written which can provide more in-depth information on topics covered only superficially in this review. A recommended overall source on wood preservation and biodeterioration from a European perspective is written by Eaton and Hale (1993) [26], while the industrial aspects of preservation have been comprehensively described by Wilkinson (1979) [27]. More detailed information on some of the established preservatives and processes can be found in *Wood Deterioration and Its Prevention by Preservative Treatments* [10]. For up-to-date information on wood preservation research the International Research Group on Wood Protection should be consulted. Internet access is available for many of these organizations (e.g., www.awpa.com, www. irg-wp.com, cwpa.ca).

REFERENCES

1. B. A. Richardson, Wood Preservation, E & FN SPON, London, 1993.

2. American Wood Protection Association (AWPA), Standards, AWPA, Birmingham, AL, 2010.

3. Canadian Standards Association (CSA), Wood Preservation, CSA080-09, CSA, Etobicoke, Ontario, Canada, 2010.

4. D. A. Webb, "Some Environmental Aspects of Creosote," Proc. Am. Wood Preserv. Assoc., **71**, 176–181 (1975).

5. D. A. Webb, "Creosote Its Biodegradation and Environmental Effects," Proc. Am. Wood Preserv. Assoc., **76**, 65–69 (1980).

6. M. J. Wade, M. S. Connor, K. M. Jop, R. E. Hillman, and H. J. Costa, "Summary Evaluation of the Environmental Impact Resulting from the Use of Creosoted Pilings in the Historic Restoration of Pier #2 at the Charlestown Navy Yard," Report, U.S. Department of the Interior, National Park Service, Batelle Ocean Sciences, Duxbury, MA, 1987.

7. Western Wood Preservers Institute (WWPI), Best Management Practices for the Use of Treated Wood in Aquatic Environments, WWPI, Vancouver, WA, 1996.

8. D. Hayward, "Why BMPs for Treated Wood in Aquatic Environments?" Proc. Am. Wood Preserv. Assoc., **93**, 71–78 (1997).

9. K. M. Brooks, "Assessing the Environmental Risks Associated with Creosote Treated Piling Use in Aquatic Environments," Proc. Am. Wood Preserv. Assoc., **93**, 79–103 (1997).

10. D. D. Nicholas, Wood Deterioration and Its Prevention by Preservative Treatments, Vol. 1, Degradation and Protection of Wood, Volume 2, Preservatives and Preservative Systems, Syracuse University Press, Syracuse, NY, 1973.

11. J. N. R. Ruddick, "Wood Preservation and the Environment—A Canadian Perspective," Proc. Can. Wood Preserv. Assoc., **11**, 115–129 (1990).

12. J. N. R. Ruddick, "Disposal of Treated Wood Waste," Proc. CITW Treated Wood Life Cycle Workshop, Can. Inst. of Treated Wood, Ottawa, Ontario, 1994, pp. 135–156.

13. J. Morgan, "The Evaluation and Commercialization of a New Wood Preservative," Proc. Am. Wood Preserv. Assoc., **85**, 16–26 (1989).

14. J. Oertel, "Novel Wood Preservatives of Good Leaching Resistance Based Upon Water Soluble Organic Compounds and Their Potential Uses," Holztechnol, **6**(4), 243–247 (1965).

15. L. Jin and K. J. Archer, "Commercial Development of ACQ in the United States," Proc. Can. Wood Preserv. Assoc., **13**, 43–54 (1992).

16. R. F. Fox and G. R. Williams, "Copper Azole Wood Preservative," Proc. Can. Wood Preserv. Assoc., **14**, 98–119 (1993).

17. J. Zhang and R. Ziobro, "Micronized Copper Preservative Systems," Proc. Can. Wood Preserv. Assoc., **28**, PowerPoint presentation (2008).

18. M. H. Freeman, D. S. Stokes, T. L. Woods, and R. D. Arsenault, "An Update on the Wood Preservative Copper Dimethyldithiocarbamate," Proc. Am. Wood Preserv. Assoc., **90**, 67–87 (1994).

19. W. R. Goodwine, "Suitability of Propoiconazole as a New Generation Wood Preserving Fungicide," Proc. Am. Wood Preserv. Assoc., **86**, 206–214 (1990).

20. H. Hartner and F. Cui, "Performance of Impralit-KDS—New Preservative, New Chemistry," Proc. Can. Wood Preserve. Assoc., **28**, 147–162 (2007).

21. J. F. Lathan, "Practical Implications of Accelerated Fixation of CCA-Treated Wood," Proc. Can. Wood Preserv. Assoc., **14**, 66–76 (1993).

22. S. A. Avramidis and J. N. R. Ruddick, "Effect of Temperature and Moisture on CCA Fixation," Holz als Roh und Werkstoff, **47**(8), 328 (1989).

23. S. Avramidis and J. N. R. Ruddick, "CCA Accelerated Fixation by Dielectric Heating," For. Prod. J., **46**, 52–55 (1996).

24. J. J. Morrell and A. P. Preston, "Use Category: A Proposal for a New Approach to Treating Standards," Proc. Am. Wood Preserv. Assoc., **91**, 63–87 (1995).

25. J. Saur and A. Preston, "AWPA Use Category System—An Update," Proc. Can. Wood Preserv. Assoc., **18**, 13–28 (1997).

26. R. A. Eaton and M. D. C. Hale, Wood. Decay Pests and Protection, Chapman & Hall, London, 1993.

27. J. G. Wilkinson, Industrial Timber Preservation, Associated Business Press, London, 1979.

PART III

METALS

35

METAL–MATRIX COMPOSITES

L. H. HIHARA

Department of Mechanical Engineering, University of Hawaii at Manoa, Honolulu, Hawaii

A. INTRODUCTION

The first metal–matrix composites (MMCs) were created in the 1960s, [1] and new types are still being developed today. MMCs are metals that are incorporated with either discontinuous or continuous reinforcements usually in the form of whiskers (W), particles (P), short fibers (SF), fibers (F), or monofilaments (MF). The reinforcements can be either metal

Uhlig's Corrosion Handbook, Third Edition, Edited by R. Winston Revie
Copyright © 2011 John Wiley & Sons, Inc.

(e.g., tungsten), nonmetal (e.g., carbon, silicon), or ceramic (e.g., silicon carbide or alumina). A variety of MMCs have been developed with matrices such as aluminum, magnesium, lead, depleted uranium, stainless steel, titanium, copper, and zinc for experimental and commercial applications. Aluminum MMCs are the most extensively available [2]. The reinforcement constituents, usually ranging from 10 to 60 vol %, have included boron (B), graphite (Gr), silicon (Si), silicon carbide (SiC), boron carbide (B_4C), titanium diboride (TiB_2), alumina (Al_2O_3), mica, quartz, tungsten, yttria, and zircon. The selection of the matrix metal and reinforcement constituent is usually chosen to increase the specific strength [3] and stiffness [3] of the MMC; however, MMCs have also been developed for other unique properties such as improved thermal conductivity [4], neutron shielding [5], and vibration damping capacity [6] and to reduce thermal expansion [2], friction [7], and wear [7]. Some MMC properties are governed by the rule of mixtures [8], but this does not generally apply to corrosion resistance. In fact, corrosion resistance of MMCs is usually less than that of their monolithic matrix alloys due to the presence of the reinforcements that alter the microstructure, electrochemical properties, and corrosion morphology.

B. MMC TYPES AND APPLICATIONS

MMCs can also be categorized as being discontinuous reinforced (DR) with particulate reinforcements [Fig. 35.1(a)] or continuous reinforced (CR) with fibers or monofilaments [Fig. 35.1(b)]. The MMC notation that is used in this chapter is as follows: matrix type/reinforcement composition/vol % reinforcement type. Hence, $Al/Al_2O_3/50$ P MMC denotes an aluminum matrix MMC reinforced with 50 vol % of Al_2O_3 particles.

DR MMCs generally have isotropic properties and lower reinforcement and fabrication costs in comparison to CR MMCs. For structural-grade DR MMCs, reinforcement content ranges from approximately 15 to 25 vol %, and uniform reinforcement and reinforcement-size distributions are preferred for optimal mechanical properties. These MMCs come in a variety of structural shapes (Fig. 35.2). In one example regarding the catamaran "Stars and Stripes'88" [2], Al/SiC/20 P MMC tubes were used to replace monolithic aluminum tubes to reduce tube weight by 20%. Structural-grade DR MMCs are also being used with greater frequency in the automotive sector for components such as engine pistons, piston-connecting rods, rear-wheel driveshafts, break calipers, cylinder liners, push rods, rocker arms, and valve guides [5]; in the sports industry for high-performance bicycle frames and components, golf clubs, and baseball bats [5]; and in the aircraft industry for fan exit guide vanes in

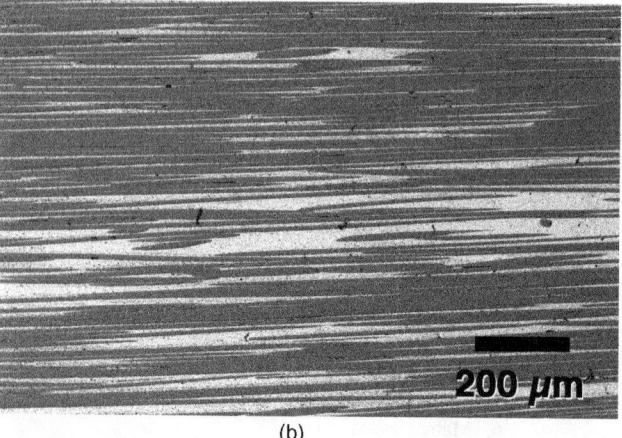

FIGURE 35.1. Micrographs of (a) Al 6092/SiC/40 P-T6 MMC (courtesy of George Hawthorn) and (b) pure $Al/Al_2O_3/50$ F MMC with fibers parallel to surface (courtesy of Shruti Tiwari).

turbine engines, ventral fins, helicopter blade sleeves, and fuel access covers [9]. For electronic-grade DR MMCs used in electronic packaging, reinforcement content generally exceeds 30%. The reinforcement and reinforcement-size

FIGURE 35.2. MMC structural T-section, channel, and tube.

distributions are often not uniform [Fig. 35.1(a)] to maximize the reinforcement volume fraction. The electronic-grade MMCs are usually loaded with ceramics of low coefficient of thermal expansion (CTE) to lower the MMC CTE to levels closer to that of electronic materials such as silicon and gallium arsenide. One MMC example for electronic packaging is the low-CTE Al/Si/43 P MMC, which is also machinable and lightweight. For thermal management applications, copper MMCs are reinforced with milled graphite fibers that have negative CTEs and thermal conductivities exceeding that of copper [10]. DR copper MMCs have also been developed for sliding electrical contact materials [11] and improved machinability for lead-free copper alloys [12]. DR Al/Gr MMCs have been developed for potential use in tribological applications due to good resistance to wear and seizure [13–15]. DR titanium MMCs are candidate materials for gears, bearings, and shafts [16]. Porous DR titanium MMCs have potential use as surgical implant materials, owing to the compatibility of titanium and bone growth [17]. Porous titanium is a good surgical implant material because bone growth is enhanced by the relatively low elastic modulus of titanium, and bone ingrowth is allowed by the porous structure. Porous titanium, however, has poor tribological properties and, therefore, MMCs that incorporate graphite to reduce friction and titanium carbide to improve wear resistance are of interest [17].

CR MMCs are either reinforced with continuous fibers or monofilaments, generally resulting in anisotropic properties with enhancements in the longitudinal direction of the reinforcement [Fig. 35.1(b)]. The reinforcement diameter generally varies from 10 μm for fibers to 150 μm for monofilaments. The reinforcement and fabrication costs of CR MMCs are usually much higher than those of DR MMCs. Examples of CF MMCs are Al/B/MF MMC structural tubes used in the Space Shuttle that resulted in 44% reduction in weight over aluminum alloys in the original design [18] and Al/Gr/F MMCs, which have unique properties such as negative to near-zero coefficient of thermal expansion, that were used as antenna booms in the Hubble space telescope [2]. Al/Al$_2$O$_3$/F MMCs have been used in the automotive industry to replace cast iron components due to MMC properties such as low weight, low thermal conductivity and expansion, good wear resistance, and improved high-temperature tensile and fatigue strengths [19]. Other types of Al/Al$_2$O$_3$/F MMCs with low weight, high strength, and high damping capacity are used for automotive push rods [6]. Structural Ti/SiC/MF MMCs have been used in prototype drive shafts, turbine engine discs, compressor discs, and hollow fan blades and were also candidate materials for the skin of the National Aerospace Plane [20]. CR copper MMCs have been developed for applications requiring reduced weight and high thermal conductivity [21].

C. FACTORS INFLUENCING MMC CORROSION

The nonhomogeneous microstructure of MMCs due to the presence of the reinforcement constituents can lead to higher corrosion rates and different corrosion morphologies in comparison to their monolithic matrix alloys. Corrosion in MMCs may originate from electrochemical, chemical, and physical interaction between MMC constituents due to intrinsic properties or those induced by processing. Corrosion can be accelerated by galvanic interaction between the reinforcement, matrix, and interphases. The interphases and reinforcements may also undergo chemical degradation that are not electrochemical in nature. The presence of the reinforcements may also influence the MMC microstructure by inducing segregation, intermetallic formation, and dislocation generation. Processing deficiencies may result in unexpected forms of corrosion.

The parameters affecting MMC corrosion that will be discussed are (1) electrochemical effects related to the primary MMC constituents, (2) electrochemical effects of the interphases, (3) chemical degradation in MMCs, and (4) secondary effects caused by the microstructure and processing.

C1. Electrochemical Effects of Primary MMC Constituents

One of the major factors in MMC corrosion is galvanic action between the matrix and reinforcement if the reinforcement material is conductive or semiconductive. When using polarization diagrams to depict galvanic corrosion, the intersection of the anodic polarization curve of the matrix metal and the cathodic polarization curve of the reinforcement indicates the magnitude of the galvanic corrosion current density (i_{GALV}) (Fig. 35.3) for a galvanic couple with equal matrix and reinforcement area fractions. When matrix and reinforcement area fractions are not equal, the polarization diagrams must be plotted using the electrode current rather than the current density to account for the unequal area fractions. Galvanic corrosion in MMCs is governed by the reinforcement resistivity, electrochemistry, photoelectrochemistry, and area fraction and the matrix metal, environment, and microstructure.

C1.1. Common Reinforcement Types. Common reinforcement types used in MMCs are summarized in the following paragraphs.

C1.1.1 Boron. Although pure boron is an insulator and cannot support cathodic currents, the conductivity of boron monofilaments (B MFs) is many orders of magnitude greater than that of pure boron due to tungsten and tungsten borides in the MF core [22] (Table 35.1). Boron MFs consist of

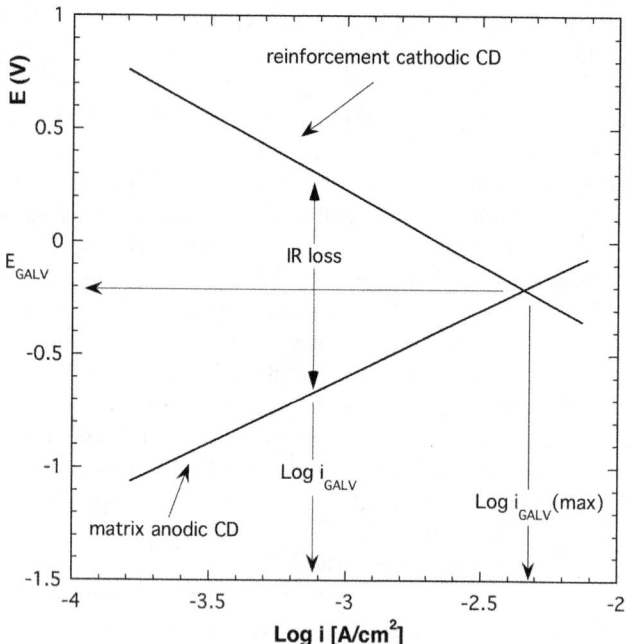

FIGURE 35.3. Anodic and cathodic polarization curves showing galvanic corrosion parameters.

polycrystalline boron with a core of tungsten and tungsten borides [22]. Hence, while pure boron cannot support cathodic currents, B MFs can serve as efficient cathodes.

C1.1.2 Graphite. Graphite reinforcement usually comes in the form of short or continuous fibers. Graphite has relatively

low resistivity (Table 35.1) and therefore serves as very efficient cathodes for oxygen and proton reduction.

C1.1.3 Silicon Carbide. SiC reinforcement comes in the form of particles, whiskers, continuous fibers, or monofilament. The electrical resistivity of SiC increases with purity and can range from approximately 10^{-5} to $10^{13}\ \Omega\cdot$cm [23]. SiC particles can be either of the high-purity green type or low-purity black type. Green SiC, which has higher thermal conductivity (in addition to higher electrical resistivity) than black SiC, is preferred for MMCs used in electronic packaging where high thermal conductivity is needed. SiC fibers have resistivities ranging from 10^3 to $10^6\ \Omega\cdot$cm depending on the fiber grade [24]. SiC MFs have carbon cores and carbon-rich surfaces with resistivities on the order of $10^{-2}\ \Omega\cdot$cm [25] (Table 35.1). Consequently, due to the vast difference in electrical resistivities of SiC, variations in corrosion behavior may result. Hence, galvanic corrosion cannot be ruled out, and SiC can serve as an inert electrode for proton and oxygen reduction.

C1.1.4 Alumina. Alumina reinforcement comes in the form of particles, short fibers, or continuous fibers. The resistivity of Al_2O_3 (99.7 wt % pure) is greater than about $10^{14}\ \Omega\cdot$cm [26], and therefore galvanic corrosion between Al_2O_3 and aluminum is not possible.

C1.2. Reinforcement Resistivity. Reinforcement materials generally fall into the categories of insulators, semiconductors, and conductors. For reinforcements that are insulators,

TABLE 35.1. Resistivities of Reinforcement Materials

Material	Resistivity ($\Omega\cdot$cm)	Temperature. (°C)	Notes	References
Al_2O_3	$>10^{14}$	30	99.7% Al_2O_3	[26]
Mica	10^{13}–10^{17}	—	Muscovite $KAl_3Si_3O_{10}(OH)_2$	[27]
SiC	10^{-5}–10^{13}	—	Function of purity	[23]
SiC fiber	10^3–$>10^6$	25	Nicalon™, dependent on grade	[24]
B	6.7×10^5	25	Pure	[28]
B_4C	10^0	—	—	[29]
Si	10^{-2}–10^5	—	Function of purity	[26]
P100 Gr fiber	2.5×10^{-4}	—	Thornel	[30]
P55S Gr fiber	7.5×10^{-4}	—	Thornel	[30]
SiC MF (ends)	4×10^{-2}	25	The "end" indicates that electrical contact was made with the end of the MF exposing the core; the "circumferential surface" indicates that electrical contact was made with only the MF circumferential surface (excluding the core). SiC MFs have carbon cores and BMFs have tungsten cores.	[31]
SiC MF (circumferential surface)	2×10^{-2}	25		
B MF (ends)	2×10^{-1}	25		
B MF (circumferential surface)	5×10^{-1}	25		

galvanic corrosion is not possible. For semiconductors, the degree of galvanic corrosion will be restricted by the magnitude of ohmic losses (IR) through the reinforcements (Fig. 35.3). The larger the IR loss, the lower will be the galvanic current. Assuming one-dimensional current flow through a reinforcement particle, the IR loss can be approximated as $i\rho l$, where the parameter i is the cathodic current density, ρ is the reinforcement resistivity, and l is the thickness of the reinforcement through which the current is assumed to be flowing. Since reinforcements are usually on the order of several micrometers, their resistivities have to be relatively high to significantly impede galvanic corrosion. For example, for an i_{GALV} of approximately 10^{-4} A/cm^2 and a 5-μm-thick reinforcement having a resistivity of 10^5 Ω·cm, the IR loss will be only 5 mV, which would have a negligible effect on galvanic corrosion. Hence, if IR losses are to limit galvanic corrosion, reinforcements having relatively high resistivity are needed. In addition, resistivity needs to be higher as the particle sizes decrease since ohmic losses decrease with particle size (i.e., IR loss = $i\rho l$). Many semiconductors do not have resistivities high enough to stifle galvanic corrosion. For example, the resistivity of SiC may vary by approximately 18 orders of magnitude depending on its purity [23]. Most particulate SiC-reinforced MMCs utilize black SiC, which has lower cost, purity, and resistivity than green SiC. The resistivities of various reinforcement materials are shown in Table 35.1. The treatment above for the ohmic losses through reinforcement particles should only be considered as an approximation since one-dimensional current flow was assumed. In the actual case, the ohmic drop through the edges of the particle could be much less than through the thickness.

C1.3. Reinforcement Electrochemistry.

Although, as discussed above, the reinforcement resistivity is important, the rate of galvanic corrosion is also highly dependent on the catalytic properties of the reinforcements for proton and oxygen reduction. A low-resistivity reinforcement with poor catalytic properties for proton and oxygen reduction could induce less galvanic corrosion than a higher resistivity reinforcement with facile kinetics for the reduction reactions. For example, in the case of aluminum in aerated 3.15 wt % NaCl (Fig. 35.4) where galvanic corrosion is under cathodic control, the galvanic corrosion rate shows a strong dependency on the reinforcement type. The galvanic corrosion rates between the aluminums and various reinforcements ranked from highest to lowest as P100 Gr > SiC MF (ends exposed) > SiC MF (circumferential surface exposed) > B MF (ends exposed) > hot-pressed (HP) SiC > B MF (circumferential surface exposed) > Si. It should also be noted that ceramic reinforcements may vary in purity and structure, and some reinforcements are in themselves composites. For example, SiC MFs have carbon-rich outer layers and carbon cores, and their polarization diagrams have a stronger

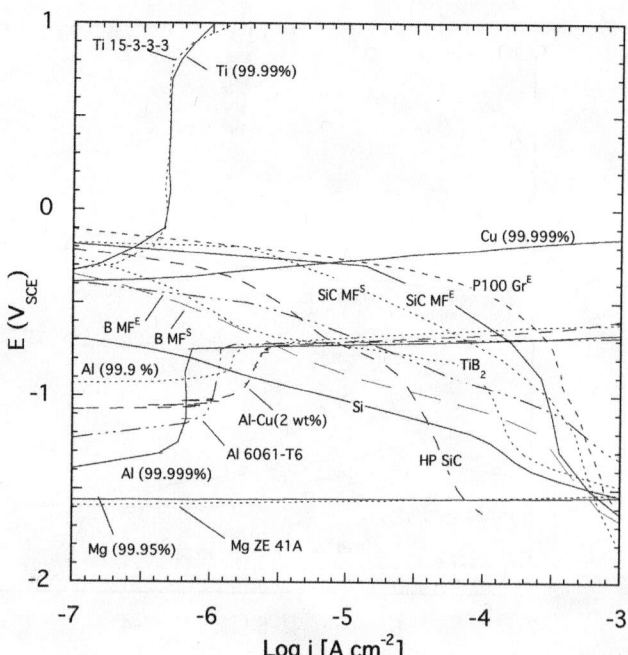

FIGURE 35.4. Collection of anodic polarization diagram of ultrapure Al (99.999%) [32], pure Al (99.9%) [33], Al–2 wt % Cu [34], Al 6061-T6 [32], ultrapure Ti (99.99%) [35, 36], Ti 15-3 [35, 36], ultrapure Cu (99.999%), pure Mg (99.95%) [37], and Mg ZE 41A [37]. Cathodic polarization diagrams of P100 Gr^E [32], HP SiC [32], SiC$_{MF}^E$ [38], SiC$_{MF}^S$ [38], Si [33], TiB$_2$ [32], B$_{MF}^E$ [31], and B$_{MF}^S$ [31] exposed to aerated 3.15 wt % NaCl at 30°C.

resemblance to P100 Gr than HP SiC. In addition, since the MFs often have cores that are of a different material than their surfaces, the orientation of the reinforcements may also affect corrosion behavior. The polarization behavior of MF electrodes with the circumferential surface exposed (MF^S) is different compared to the behavior of MF electrodes with the ends exposed baring the cores (MF^E). Compare cathodic curves for SiC MF^S versus SiC MF^E and B MF^S versus B MF^E in aerated (Fig. 35.4) and deaerated (Fig. 35.5) 3.15 wt % NaCl.

C1.4. Reinforcement Photoelectrochemistry.

If the MMC reinforcements or constituents are semiconductors, galvanic currents between the matrix metal and the semiconductor could be suppressed under illumination if the semiconductor is n type or accelerated if the semiconductor is p type.

An n-type semiconductor is photoanodic and can promote photooxidation reactions under illumination. One such reaction is the oxidation of water. Hence, when an MMC containing n-type semiconductors is wet and under illumination, photogenerated electrons from the n-type semiconductor could polarize the MMC to more negative potentials inducing cathodic protection [39]. Accordingly, anodic

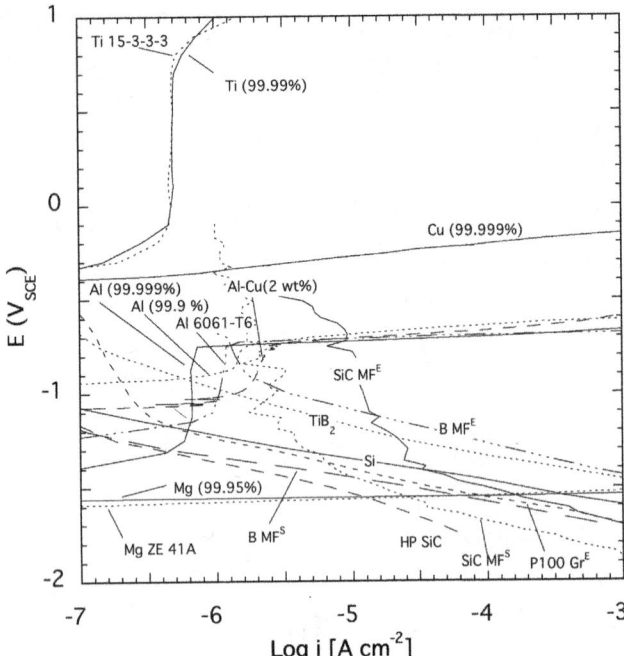

FIGURE 35.5. Collection of anodic polarization diagram of ultrapure Al (99.999%) [32], pure Al (99.9%) [33], Al–2 wt % Cu [34], Al 6061-T6 [32], ultrapure Ti (99.99%) [35, 36], Ti 15-3 [35, 36], ultrapure Cu (99.999%), pure Mg (99.95%) [37], and Mg ZE 41A [37]. Cathodic polarization diagrams of P100 Gr^E [32], HP SiC [32], SiC$_{MF}^E$ [38], SiC$_{MF}^S$ [38], Si [33], TiB$_2$ [32], B$_{MF}^E$ [31], and B$_{MF}^S$ [31] exposed to deaerated 3.15 wt % NaCl at 30°C.

current densities on Al 6092/Al$_2$O$_3$/20 P-T6 MMCs immersed in air-exposed $0.5\,M$ Na$_2$SO$_4$ solutions increased sharply under illumination, which was attributed to photoanodic currents generated by water oxidation on n-type TiO$_2$ particles in the microstructure that were likely introduced with the Al$_2$O$_3$ reinforcements [40]. The open-circuit potentials also decreased upon illumination, indicating that the n-type TiO$_2$ induced cathodic protection. Outdoor exposures of these Al 6092/Al$_2$O$_3$/20 P-T6 MMCs showed that corrosion films were thinner on the topside of specimens exposed to sunlight as compared to the backside of the specimens not exposed to sunlight [41]. Interestingly, MMCs containing p-type semiconductors had thicker corrosion films on the sunlit surfaces as apposed to the shaded surfaces [41].

A p-type semiconductor is photocathodic and under illumination promotes photoreduction reactions. Depending on the electrolyte conditions, proton or oxygen reduction may be enhanced at the p-type semiconductor. In the presence of moisture and illumination on MMCs that contain p-type semiconductors, photoreduction causes the cathodic current to increase, raising the corrosion potential and inducing greater dissolution of the matrix [39]. Accordingly, during cathodic polarization of Al 6092/SiC/P-T6 MMCs [39] that

were immersed in air-exposed $0.5\,M$ Na$_2$SO$_4$ solutions, cathodic current densities and open-circuit potentials increased sharply during illumination. In these MMCs exposed to the outdoors, corrosion films were also thicker on the sunlit surfaces as apposed to the shaded surfaces [41].

C1.5. Reinforcement Area Fraction. If the galvaniccorrosion rate is under cathodic control, galvanic corrosion should increase as the reinforcement area fraction increases. The catchment area principle [42] can be used to determine i_{GALV} as a function of the area fraction of the cathodic reinforcement [32]:

$$i_{GALV} = i_C \frac{X_C}{1 - X_C} \qquad (35.1)$$

where the parameter i_{GALV} is the dissolution current density of the matrix (i.e., I_{GALV}/anode area), i_C is the current density of the cathode, X_C is the area fraction of the cathode, and $1 - X_C$ is the area fraction of the anode. The value of i_C can be set equal to the current density of the cathodic constituents at the galvanic couple potential. For Cu (99.999%), Al (99.999%, 99.9%, Al–2 wt % Cu, and 6061-T6 Al), and Mg (99.95%, ZE41A) coupled to P100 Gr, the i_C values are approximately 10^{-5}, 2.5×10^{-4}, and 7.9×10^{-4}, respectively, in aerated 3.15 wt % NaCl at 30°C (Fig. 35.4). By plotting Eq. 35.1, a graph (Fig. 35.6) was generated from which i_{GALV} of Cu, Al, and Mg can be obtained as a function of the area fraction of P100 Gr^E for exposure to aerated 3.15 wt % NaCl at 30°C. Note that i_C for Cu was in the Tafel regime and that for Al and Mg was in the diffusion-limited oxygen reduction regime (Fig. 35.4). Hence, galvanic corrosion rates for Al and Mg could increase significantly with convection, whereas that for Cu should not. The diagram

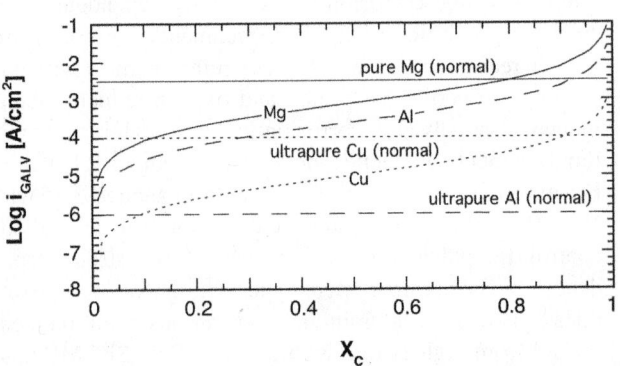

FIGURE 35.6. Graphs showing galvanic corrosion current density i_{galv} of Mg [i.e., pure Mg (99.95 %), Mg AZ41A], Al [i.e., ultrapure Al (99.999%), pure Al (99.9%), Al–2 wt %Cu-T6, or Al 6061-T6], and ultrapure Cu (99.999%) as a function of the area fraction of X_C of P100 Gr^E in aerated 3.15 wt % NaCl at 30°C. Normal corrosion rates of pure Mg, ultrapure Al, and ultrapure Cu are also plotted as horizontal lines.

(Fig. 35.6) also shows the normal corrosion rates of ultrapure Cu, ultrapure Al, and pure Mg. Hence, the area fraction of P100 Gr at which the galvanic corrosion rate would exceed the normal corrosion rate of the matrix metal varies with the type of metal.

C1.6. Matrix Metal. The matrix alloy plays a significant role in galvanic corrosion. In matrix alloys with noble open-circuit potentials, the effect of galvanic corrosion is generally attenuated. Notice that in deaerated 3.15 wt % NaCl (Fig. 35.5), titanium and copper are virtually immune to galvanic corrosion, whereas magnesium is highly susceptible.

C1.7. Environment. The environment can also have a significant effect on galvanic corrosion rates. For matrix alloys that corrode galvanically under cathodic control, galvanic corrosion rates can increase markedly when dissolved oxygen levels increase. The anodic polarization diagrams of several alloys are plotted with the cathodic polarization diagrams of various reinforcements in aerated (Fig. 35.4) and deaerated (Fig. 35.5) 3.15 wt % NaCl at 30°C. Marked increases in galvanic corrosion rates result from aeration (compare Figs. 35.4 and 35.5) for the various types of aluminum when they are galvanically coupled to P100 graphite, SiC MF, B MF, HP SiC, and TiB_2. Notice that galvanic corrosion rates do not increase for the aluminums coupled to Si since the polarization curves intersect in the passive aluminum regime, making galvanic corrosion under anodic control. In halide-free environments such as in 0.5 M Na_2SO_4, aluminum passivates and galvanic corrosion is also under anodic control; hence, aeration (Fig. 35.7) would not increase galvanic corrosion rates higher than that of the passive aluminum current density based on the polarization diagrams. The breakdown of passivity due to the presence of aggressive ions such as chlorides can also significantly increase galvanic corrosion rates (compare Figs. 35.4 and 35.7).

C1.8. Microstructure. The corrosion of the MMC is also affected by the physical presence of the reinforcements, even if they are inert. As the matrix alloy corrodes, the reinforcements are often left in relief, leaving behind a network of fissures that trap corrosion products and exacerbate corrosion. The initiation and propagation of corrosion sites are generally influenced by the electrical resistivity and volume fraction of the MMC constituents, including the reinforcements, interphases, and intermetallics. The corrosion behavior of MMCs in the open-circuit condition can be quite different from what might be expected based on their anodic polarization diagrams. For example, in near-neutral 0.5 M Na_2SO_4 solutions, various aluminum MMCs passivate during anodic polarization [43, 44], but in the open-circuit condition, the same MMCs are susceptible to localized

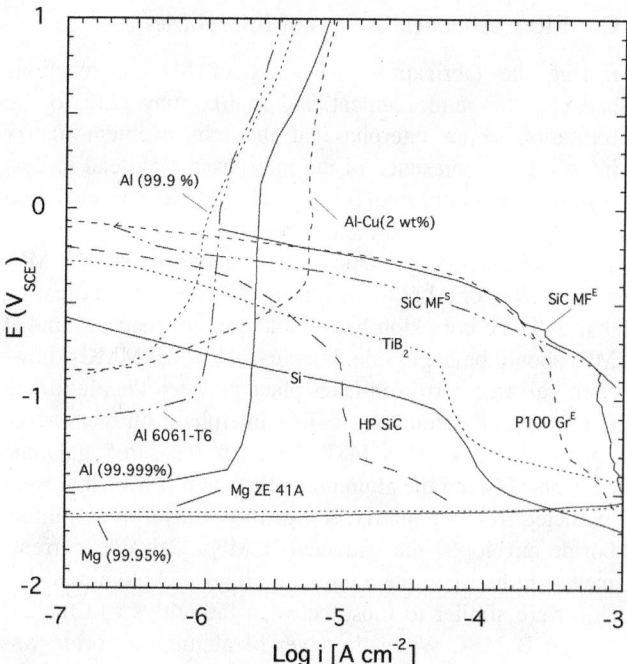

FIGURE 35.7. Collection of anodic polarization diagram of ultrapure Al (99.999%) [32], pure Al (99.9%) [33], Al–2 wt % Cu [34], Al 6061-T6 [32], pure Mg (99.95%) [37], and Mg ZE 41A [37]. Cathodic polarization diagrams of P100 Gr[E] [46], HP SiC [46], SiC$_{MF}$[E] (oxygenated solution) [47], SiC$_{MF}$[S] (oxygenated solution) [47], Si [33], and TiB_2 [46] exposed to aerated 0.5 M Na_2SO_4 at 30°C.

corrosion. At the open-circuit potential, localized cathodic sites can become alkaline, and localized anodic sites can become acidified [45]. If the matrix metal is amphoteric like that of aluminum, dissolution may occur in the alkaline cathodic and acidic anodic regions [45]. Corrosion at alkaline cathodic regions usually occurs in aluminum MMCs when the cathodic constituents are of low area fraction, which concentrates the cathodic current and hydroxide ion buildup over a few cathodic sites. Hence, when predicting MMC corrosion behavior, polarization studies alone may be misleading.

An example of where corrosion was induced by localized cathodic currents was in Al 6092/Al_2O_3/20 P-T6 MMCs. In these MMCs, the Al_2O_3 reinforcement particles are insulators and cannot serve as cathodes; however, the microstructure also contained several types of titanium suboxides with compositions close to that of Ti_6O, Ti_3O, Ti_2O, and TiO; TiO_2 (likely doped); Ti–Zr–Al-containing oxides; and Fe–Si–Al intermetallics [40]. Of these particles, the titanium suboxides, TiO_2, and the Fe–Si–Al intermetallics supported significant cathodic activity [40]. The area fraction of the non-Al_2O_3 particles in the MMCs was estimated to be on the order of 0.01 using image analysis [43]. In the open-circuit condition, these sites were observed to be corrosion initiation sites in aerated 0.5 M Na_2SO_4, where the MMC passivates during anodic polarization.

C2. Electrochemical Effects of Interphases

During the fabrication processing of MMCs, reactions between the reinforcement and matrix may lead to the formation of an interphase at the reinforcement–matrix interface. The presence of the interphase may lead to corrosion behavior different from what might be expected based on virgin MMC constituents. For example, Pohlman [48] could not measure galvanic currents between virgin B MFs and Al 2024 or Al 6061 in 3.5% NaCl solutions, indicating that galvanic corrosion between aluminum matrices and B MFs should be negligible. In actual Al/B/MF MMCs, however, galvanic corrosion takes place between the aluminum matrix and the aluminum boride interphase on the surface layers of the B MFs [48]. Pohlman measured galvanic currents between the aluminum alloys and B MFs that were extracted from the matrix. A 4-μm-thick layer of aluminum boride enveloped the extracted B MFs. Galvanic currents measured between the aluminum alloys and aluminum boride were similar to those between the alloys and the extracted B MFs. When the layer of aluminum boride was removed from the extracted B MFs, the galvanic current ceased, which indicated that the aluminum boride interphase was necessary for galvanic corrosion.

C3. Chemical Degradation of Reinforcements and Interphases

MMCs may also degrade by chemical reactions that cannot be directly assessed by electrochemical methods. Reinforcement phases and interphases may undergo chemical degradation which cannot be detected by polarization techniques.

Muscovite mica $KAl_3Si_3O_{10}(OH)_2$ particles of approximately 70 μm in size [49] used in Al MMCs appeared to have absorbed moisture, swelled, and then exfoliated [50] during exposure to non-deaerated 3.5 wt % NaCl solutions.

The aluminum carbide Al_4C_3 interphase which is present in some aluminum MMCs may hydrolyze in the presence of moisture, forming methane and aluminum hydroxide. The rate of Al_4C_3 hydrolysis was measured to be approximately 1% per hour for hot-pressed Al_4C_3 (78% of theoretical density and porous) exposed to pure water at 30°C [46]. Methane evolution has been detected from Al/Gr MMCs containing Al_4C_3 [51, 52]. Buonanno [52] reported that Al_4C_3 hydrolysis in Al/Gr MMCs leaves fissures at fiber–matrix interfaces. The hydrolysis of Al_4C_3 therefore could result in rapid penetration into the MMC microstructure through reinforcement–matrix interfaces, leading to the formation of microcrevices. Al_4C_3 can form by the reaction of aluminum and carbon [53], SiC [54], or B_4C [55] at elevated temperatures that could be encountered during MMC processing. Hence, if processing conditions are not properly controlled, MMCs containing Al and carbon compounds could contain this deleterious interphase.

C4. Secondary Effects

Corrosion in MMCs is also influenced by factors that are not directly caused by the reinforcement, matrix, or interphase but result due to their presence. For example, the reinforcement phase in the MMC may alter the microstructural features in the matrix metal in ways that do not occur in the monolithic matrix alloy. Examples discussed below are the effects of intermetallic phases which may form around the reinforcements by solute rejection during solidification [56], dislocation generation by the mismatch in CTE between the reinforcement and matrix [57], and the susceptibility of the MMC microstructure to physical damage. In addition, problems related to processing deficiencies are disused.

C4.1. Intermetallics. In MMCs, the presence of the reinforcement phases may affect the normal precipitation characteristics of intermetallic phases in comparison to those of the monolithic matrix alloy. The intermetallics also have their own characteristic electrochemical properties (e.g., corrosion potentials, pitting potentials, and corrosion current densities [58]). Noble intermetallics may induce galvanic corrosion of the matrix, whereas active intermetallics may go into dissolution and leave fissures or crevices. Hence, the difference between the corrosion characteristics of the MMC and its monolithic matrix alloy is also dependent on the extent to which the reinforcement phases affect the amount, distribution, and morphology of the intermetallics in the MMC.

In Al/Al_2O_3 MMCs, Al_8Mg_5 and Mg_2Si intermetallics provided corrosion paths along fiber–matrix interfaces [59]. Pits in Al/Al_2O_3 MMCs exposed to NaCl solutions containing H_2O_2 were attributed to the dissolution of $MgAl_3$, which is rapidly attacked at low potentials [60]. In Al/mica MMCs, a dendritic phase which was probably Mg_2Al_3 or Al_8Mg_5 and spheroidized $CuMgAl_2$ were preferentially attacked in non-deaerated 3.5 wt % NaCl [61].

C4.2. Dislocation Density. The high strength of particulate MMCs in comparison to their monolithic alloys is generated by high dislocation densities caused by a mismatch in the CTE between the reinforcement and matrix and heating and cooling histories [57]. Since cold working, which is the result of generating high dislocation densities, is known to change the corrosion behavior of metals such as steel [62] and aluminum [63], the corrosion behavior of MMCs may also be affected by high dislocation densities [31]. It has been suggested that corrosion near the SiC–Al interface in Al/SiC MMCs could be caused by high dislocation density due to a mismatch of the CTE between SiC and Al [64, 65].

C4.3. Physical Damage. MMCs can also be more susceptible to corrosion initiation at sites of physical damage in comparison to their monolithic alloys. Impacting the MMC

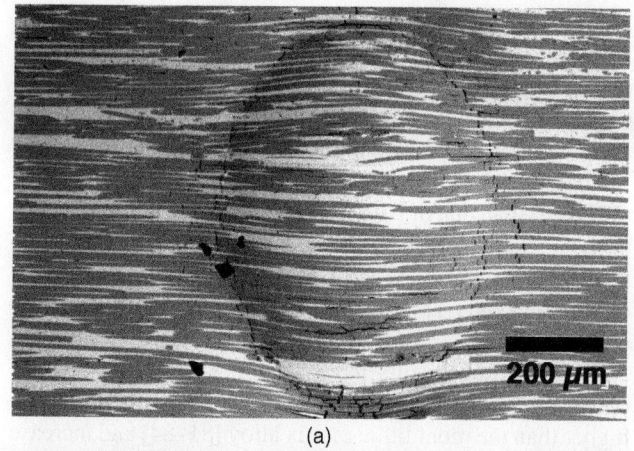

(a)

(b)

FIGURE 35.8. Al–2 wt % Cu/Al$_2$O$_3$/F 50 MMC with localized deformation induced using a 1.58-mm-diameter Si$_3$N$_4$ ball (a) before and (b) after immersion in aerated 3.15 wt % NaCl at 30°C for five days. Notice (a) broken fibers and (b) enhanced corrosion initiation over deformed region. (Photos courtesy of Shruti Tiwari.)

surface may debond the matrix from the reinforcement or fracture the reinforcement (Fig. 35.8) with both cases creating microcrevices that can lead to corrosion initiation sites. In Al-2 wt % Cu/Al$_2$O$_3$/50F MMCs, regions of localized deformation corroded preferentially when the MMCs were exposed to aerated 3.15 wt % NaCl. Significant amounts of Cu precipitated on the surface of the deformed region, where corrosion was exacerabated. For similar levels of localized deformation, pure Al/Al$_2$O$_3$/50F MMCs were much less susceptible to preferred corrosion initiation at locally deformed regions.

C4.4. Processing Deficiencies. Processing-induced corrosion is not inherently caused by the primary components of the MMC system but result from processing deficiencies such as too low or too high processing temperatures, contamination, and so on. Some cases that were observed in various MMCs were (1) low-integrity diffusion bonds that induced crevice corrosion between B MF and the aluminum

matrix [66, 67], (2) microstructural chloride contamination originating from TiCl$_4$ and BCl$_3$ processing gases [68] that induced pitting [69] in some types of Al 6061/Gr/50 F-T6 MMCs in chloride-free environments, and extraneous carbon particles in Al 6092/B$_4$C MMCs [70], exacerbating galvanic effects.

Unlike the case with many types of metal alloys, MMCs cannot be always reprocessed to regain optimal properties. If an MMC had been subjected to high temperatures that lead to interphase formation, reprocessing the MMC to reverse the interphase formation would be unlikely. Also, fractured or damaged reinforcement constituents cannot be repaired. Therefore, it is very important to know the source and history of MMC products.

D. CORROSION OF MMC SYSTEMS

The corrosion behavior of aluminum, magnesium, titanium, copper, stainless steel, lead, depleted uranium, and zinc MMCs is discussed.

D1. Aluminum MMCs

Aluminum is a reactive metal with a high driving force to revert back to its oxide, but it generally has good resistance to aqueous corrosion in near-neutral solutions due to the formation of a passive film [71]. In acidic and basic solutions, the passive film is not thermodynamically stable and thus corrosion rates are high [71]. Aluminum pits in halide-containing solutions and the pitting potential (E_{pit}) is linearly dependent on the logarithm of the halogen anion concentration [72]. However, in order for aluminum to pit in the open-circuit condition, it must be polarized to potentials noble to E_{pit} by a cathodic reaction. Proton and oxygen reduction are two possible cathodic reactions, but in neutral, chloride-containing solutions, oxygen reduction is necessary to initiate pitting. Pits do not nucleate on aluminum in the open-circuit condition if solutions are deaerated [73]. In aerated solutions, ultrapure aluminum (99.999 wt %), which is a poor catalyst for oxygen reduction, does not pit in the open-circuit condition. The slow oxygen reduction kinetics on ultrapure aluminum is believed to be caused by the high resistivity of aluminum oxide which restricts electron migration through the passive film [74, 75]. In aluminum alloys, however, noble precipitates and conducting reinforcements can polarize the alloy to E_{pit} in aerated solutions. Corrosion of aluminum MMCs reinforced with boron, graphite, SiC, Al$_2$O$_3$, and mica will be discussed. Studies on stress corrosion and corrosion fatigue are also discussed.

D1.1. Aluminum/Boron MMCs. Al/B/MF MMCs are usually fabricated by diffusion bonding B MFs between aluminum foils [76]. Aluminum borides (i.e., AlB$_{12}$ and AlB$_2$)

have been found at B MF matrix interfaces [77]. Both the B MF ends and circumferential surfaces can support cathodic currents (Figs. 35.4 and 35.5) and therefore can induce galvanic corrosion. Hence, shielding the MFs from environmental exposure is a likely provision for corrosion control of these MMCs.

D1.2. Aluminum/Graphite MMCs.

Al/Gr MMCs are usually reinforced with either continuous or chopped fibers. The MMCs reinforced with continuous fibers have anisotropic properties with high specific tensile strength and stiffness along the fiber axis, but limitations of shear, compression, and transverse strengths generally excludes their use in structural applications [2]. The high thermal conductivity, negative coefficient of thermal expansion, and high stiffness of graphite, however, has made ultralow expansion Al/Gr MMCs ideal for thermally stable space structures.

The advantages of graphite for enhancements in mechanical and physical properties, however, are tempered with negative characteristics regarding corrosion resistance that graphite imparts due to its low electrical resistivity, catalytic properties for oxygen reduction, and reactivity with aluminum at high temperatures. Severe galvanic corrosion is induced in Al/Gr MMCs in aerated solutions (Fig. 35.4), with the main cathodic reaction being oxygen rather than proton reduction [78]. At elevated temperatures, graphite reacts with aluminum, forming the Al_4C_3 interphase, which readily decomposes in water [46] to produce CH_4 and aluminum hydroxide $Al(OH)_3$ [51]. Also see Section C3.

Al/Gr MMCs should therefore be carefully processed to avoid or minimize Al_4C_3 formation, and microstructures free of chlorides [69] should be ensured (see Section C4.4), either by careful control of processing parameters if chlorinated gases are used or using fabrication methods that do not use chlorinated gases [68]. Due to the reactivity of Al/Gr MMCs, they are more suited for use in dry environments. Certain corrosion-resistant procedures used for monolithic Al such as anodization may not be suitable for Al/Gr MMCs since the graphite fibers can be easily oxidized at positive potentials [79].

D1.3. Aluminum/Silicon Carbide MMCs.

The Al/SiC MMCs have been reinforced with particles, whiskers, fibers, or monofilament. Due to the vast difference in electrical resistivities of SiC, variations in corrosion behavior may result. Hence, galvanic corrosion cannot be ruled out, and SiC can serve as an electrode for proton and oxygen reduction.

The degree of galvanic corrosion is strongly dependent on the type of SiC reinforcement. The anodic polarization diagrams (Fig. 35.4) of ultrapure Al (99.999%), pure Al (99.9%), Al–2 wt % Cu, and Al 6061-T6 with the cathodic polarization diagrams of HP SiC, SiC MF (with either carbon cores or carbon-rich surface exposed) in aerated 3.15 wt %

NaCl shows that the galvanic current density of the aluminums coupled to SiC MF is approximately 15 times that when coupled to HP SiC of equal surface area. The influence of the carbon core and carbon-rich surface of the SiC MF is clearly seen (Fig. 35.4) where the polarization diagram of the SiC MF has a stronger resemblance to that of pitch-based graphite (P100 Gr) than that of HP SiC [25]. Galvanic current between a type of Nicalon™ SiC fiber and an aluminum alloy was also measured in an aerated NaCl solution [80], but the galvanic current was only 15% of that between carbon fiber and the aluminum alloy.

Experimental results have generally indicated that the corrosion rate of particulate and whisker Al/SiC MMCs are higher than the monolithic matrix alloy [81–84] and increase with SiC content [84, 85] in aerated, chloride-containing environments. Weight loss data of 6092/SiC/P-T6 Al MMCs showed an increase in the corrosion rate as the SiC content increased from 5, 10, 20, 40, to 50 vol % for various 90-day humidity chamber tests [86]. At the 50 vol % SiC content level, the corrosion rate for an MMC with high-purity, high-resistivity green SiC was noticeably lower than that of the MMC with low-resistivity black SiC [86]. The black SiC is likely to support more cathodic currents leading to higher corrosion rates in comparison to green SiC. The large variation in resistivity of the SiC may be the cause for conflicting results in the literature on the corrosion of Al/SiC MMCs. For example, no obvious evidence of galvanic corrosion was found in Al 6061/SiC MMCs with 17–27 vol % SiC particles [87].

The presence of SiC particles does not have a significant effect on the aluminum matrix passive current densities [84, 86, 88, 89] and pitting potentials [80, 86–88, 90–95]. Pit morphology, however, is indirectly affected by the presence of SiC particles. Pits on Al/SiC/W MMCs were notably more numerous and much smaller in size [96] compared to pits on wrought and powder-compacted monolithic alloys during anodic polarization in 0.1 N NaCl. The pits nucleated at intermetallic particles (not SiC), which are smaller and more numerous in the MMCs than in the monolithic matrix alloys [96]. SiC whiskers [96] and particles [87] can enhance the precipitation of the intermetallic phases. Pitting has also been observed at dendrite cores [97], near-eutectic silicon [94, 98], and intermetallic particles [98] in various Al/SiC MMCs. In the open-circuit condition, pits have also been observed to initiate at SiC–Al interfaces and could be caused by galvanic action with the SiC particle if the resistivity of SiC is relatively low or the hydrolysis of Al_4C_3 at the SiC–Al interface [43].

Since the corrosion behavior of Al/SiC MMCs is dependent on the microstructure, the corrosion characteristics can be altered by processing. In the MMC, void content [99], dislocation density [100], agglomeration of SiC particles [101], and the precipitation of active phases [102] are affected by processing conditions. Certain solution heat treatments and high extrusion ratios improved the corrosion resistance of an Al 7091/SiC/20 P MMC [99]. Extrusion improved the corrosion resistance of cast MMCs by reducing the amount of pores and agglomerates of SiC particles [101]. Corrosion resistance was also improved by a finer, more homogenous distribution of secondary phases at the T4 temper in comparison to the O and F tempers [103]. It has been suggested that corrosion near the SiC–Al interface could be caused by high dislocation density due to a mismatch of the coefficient of thermal expansion between SiC and aluminum [25, 64, 65], segregation of alloying elements to the SiC–Al interface [104], or the formation of Al_4C_3, which hydrolyzes in water. Aluminum carbide has been identified as a source of corrosion for MMCs reinforced with particles [105] and SiC Nicalon™ fibers [80]. Also see Section C3.

The formation of micro crevices caused by reinforcement particles left in relief as the matrix corrodes also exacerbates corrosion by localized acidification in anodic regions and alkalinization in cathodic regions in Al6092/SiC/P-T6 MMCs [45, 106]. The aluminum matrix loses its ability to passivate when the solution becomes either acidic or alkaline. Also See section C1.8.

D1.4. Aluminum/Alumina MMCs.

D1.4. Aluminum/Alumina MMCs. Particles and both short and continuous Al_2O_3 fibers have been used to reinforce aluminum alloys. Characteristic properties of Al/Al_2O_3 MMCs are low-weight, high-temperature tensile and fatigue strengths, low thermal conductivity and expansion, and superior wear resistance. Galvanic corrosion between Al_2O_3 and aluminum is not possible since Al_2O_3 is an insulator.

The Al_2O_3 reinforcements usually do not have significant effects on pitting potentials [43, 80, 90, 92–94, 107] in chloride solutions. Passive current densities below the pitting potential have been reported to increase with Al_2O_3 content [107], although this may be related to processing since passive current densities for a particulate Al6092/ Al_2O_3/20P-T6 MMC were consistent with other MMCs in sodium sulfate solutions and in chloride solutions under the pitting potential [86]. Microbial corrosion was also reported to be more significant on particulate MMCs in comparison to the monolithic alloy, indicating that the Al_2O_3–Al matrix interface or Al_2O_3 particles may have aided biofilm formation [108].

In Al/Al_2O_3 MMCs, corrosion initiation usually occurs at intermetallic particles or contaminants introduced with the Al_2O_3 reinforcements. Hence, Al/Al_2O_3 MMCs with pure Al matrices usually have excellent corrosion resistance due to minimal amounts of intermetallic precipitates that can serve as cathodic sites [34]. Only slight corrosion damage was observed on pure Al/Al_2O_3/50 CF MMCs exposed to marine atmosphere 0.5 mile from the coastline for an 11-month period [79]. The presence of intermetallics and segregation of alloying elements may contribute to localized corrosion near

reinforcements. Preferential corrosion near fibers [34, 60, 109, 110] and particles [94, 110, 111] is sometimes noticed in chloride-containing solutions. In a 2 wt % Mg aluminum alloy MMC, Fe and high levels of Mg (10 wt %) were detected near fibers [60]. It was suspected that the presence of Mg originated from Mg_2Al_3. Pitting near fibers was attributed to corrosion of Mg_2Al_3, which is rapidly attacked at low potentials [60]. The Al_8Mg_5 and Mg_2Si intermetallics have also been reported to induce corrosion in Al/Al_2O_3 MMCs [59]. In Al–2 wt %Cu/Al_2O_3/50 CF MMCs, corrosion initiation occurred at copper-rich precipitates on the fiber–matrix interface [34]. In Al 6092/Al_2O_3/20 P-T6 MMCs, cathodic sites were identified as Fe–Si–Al intermetallics and low-resistivity Ti oxide or suboxide particles (likely introduced with the alumina reinforcement) [40].

D1.5. Aluminum/Mica MMCs.

D1.5. Aluminum/Mica MMCs. Muscovite ($KAl_3Si_3O_{10}$ $(OH)_2$) mica particles less than about 70 μm in size have been used in Al/mica MMCs [112] for potential use in applications where good antifriction, seizure resistance, and high-damping capacity are required [113]. Since muscovite is an insulator with resistivities that range from approximately 10^{13} to 10^{17} Ω·cm [27], galvanic corrosion should not be a problem. Muscovite is insoluble in cold water [114], but it has also been reported to absorb moisture and then swell [115]. Mica particles were cast in various aluminum alloys [115, 116]. In 3.5 wt % NaCl solution, the Al/mica MMCs had pitting potentials approximately 20–30 mV lower than the monolithic matrix alloys. In addition, intermetallics were preferentially attacked, regions around and away from mica particles pitted, mica–aluminum interfaces corroded, and mica particles exfoliated [50]. Also see Section C3.

D1.6. Stress Corrosion Cracking of Al MMCs.

D1.6. Stress Corrosion Cracking of Al MMCs. There have been only a few studies on stress corrosion cracking and corrosion fatigue of DR and CR Al MMCs.

D1.6.1 Discontinuous Reinforced MMCs.

D1.6.1 Discontinuous Reinforced MMCs. Stress corrosion cracking studies for alternate exposure and immersion in NaCl solutions have been conducted on aluminum MMCs reinforced with Al_2O_3 particles [117] and SiC particles [117–119] and whiskers [117]. The Al2024/Al_2O_3/P MMC was susceptible to stress corrosion cracking while subjected to three-point beam bending and alternate exposure or continuous immersions in a NaCl solution [117]. Under the same conditions, however, the 6061 Al MMCs reinforced with SiC particles and SiC whiskers were not susceptible to stress corrosion cracking [117]. Similarly, Al2024/SiC/P MMCs were not prone to stress corrosion cracking under constant strain at 75% of ultimate tensile strength while exposed to an aerated NaCl solution [118]. Slow strain rate tension testing of Al 2024/SiC/P MMCs indicated that the MMC lost up to 10% of failure strength compared to exposure in air [119].

D1.6.2 Continuous Reinforced MMCs. Stress corrosion cracking studies for the immersed state have been conducted on aluminum MMCs reinforced with unidirectional, graphite fibers [120], boron monofilaments [66], and Nextel 440 (Al_2O_3, SiO_2, B_2O_3) fibers [121]. Al6061/Gr/F MMCs were stressed parallel to the fiber axis in natural seawater. Failure was stress dependent at high stress levels and occurred in less than 100 h. At lower stresses, failure was primarily caused by extensive corrosion and therefore was relatively independent of stress levels. Al2024/B/MF MMCs stressed parallel to the fiber axis at 80% fracture strength in an NaCl solution did not fail in 1000 h but failed after 500 h when H_2O_2 was added to the NaCl solution. Extensive intergranular matrix corrosion and broken filaments at random sites were observed. The monolithic matrix alloy failed within 10 h under similar conditions. For Al2024/B/MF MMCs stressed perpendicular to the fiber axis at 90% yield strength in NaCl and NaCl with H_2O_2 solutions, failure occurred by intergranular matrix corrosion and separation at diffusion-bonded fiber–matrix interfaces. Failure times decreased with increasing B MF content; therefore, the presence of the MF was deleterious when stresses were perpendicular to the fiber axis. For the Al6061/Nextel/F MMCs, specimens were exposed to a pH 2 NaCl solution in the stressed and unstressed states [121]. The composite strength was measured before and after exposure to assess damage. The prevailing mode of failure was attributed to extensive corrosion along the fiber–matrix interface and not stress corrosion cracking.

D1.7. Corrosion Fatigue in Al MMCs.

Corrosion fatigue studies have been conducted on Al MMCs reinforced with graphite fibers [120], SiC whiskers [122–125], and SiC particles [123, 126]. Processing conditions and type of reinforcement affect corrosion fatigue behavior. Unnotched Al6061/Gr/F MMCs were exposed to natural seawater and stressed parallel to the fiber axis. The MMCs were processed with either silica (SiO_2)–coated or TiB_2-coated graphite fibers. For a given stress amplitude, the MMC with TiB_2-coated fibers had the longest corrosion fatigue life, followed by the MMC with the SiO_2-coated fibers and the monolithic matrix alloy. At low stress amplitudes corresponding to longer exposure times, the MMC with the SiO_2-coated fibers suffered premature failure due to extensive corrosion. In Al/SiC MMCs, fatigue crack rates of compact tension specimens are usually higher in NaCl solutions as compared to air [122] or argon [126]. Loading frequency affects corrosion fatigue crack rates [126], but no consistent trends were observed. Fatigue [122] and corrosion fatigue [126] crack rates are influenced by loading and extrusion or rolling direction. The nucleation of a crack was also observed at the bottom of a corrosion pit [125]. The shape of the reinforcement constituent may also have significant effects on stress corrosion and corrosion fatigue, based on modeling that considers crack-tip strain rate [124]. The model predicts that crack rates are reduced by increasing the reinforcement length-to-diameter ratio, which implies that MMCs reinforced with whiskers are more resistant to stress corrosion and corrosion fatigue than those reinforced with particles. This is in agreement with results on Al6061/SiC/W MMCs that were found to have longer corrosion fatigue lives than Al6061/SiC/P MMCs in salt-ladened moist air [123].

D2. Magnesium MMCs

Magnesium is the lightest (density of only 1.7 g/cm³) and most active structural metal in the electromotive series [127]. Therefore, it has a very high driving force for corrosion, making it particularly susceptible to galvanic corrosion if it is coupled to noble reinforcement constituents. The normal corrosion of Mg is generally not affected significantly by dissolved oxygen [127] since the primary cathodic reaction in Mg corrosion is proton reduction. Hence, noble impurity elements that have low hydrogen overvoltage (e.g., iron, nickel, cobalt, and copper) [128] can significantly accelerate the corrosion rate of Mg, which is highly dependent on metallic purity [127]; for example, ultrapure Mg corrodes at the rate of 0.25 mm/y in seawater, but commercial Mg corrodes at about 100–500 times faster due to impurities [127]. If Mg is reinforced with constituents that are catalytic to oxygen reduction, MMC corrosion rates may significantly increase with aeration whereas that of the monolithic matrix alloy may not. Corrosion studies have been conducted on magnesium MMCs reinforced with B MF, graphite fibers, SiC MF, SiC particles, and Al_2O_3 fibers. The stress corrosion cracking behavior of an Mg/Al_2O_3 MMC has also been investigated.

D2.1. Magnesium/Boron MMCs.

Galvanic corrosion between Mg and pure boron is not a concern since pure boron is an insulator [129, 130]. However, tungsten core B MF is not an insulator due to the formation of tungsten boride [130]. Galvanic currents between virgin B MF (tungsten cores either shielded or exposed) and Mg [130] or an Mg alloy [129] were measurable in NaCl solutions. Galvanic currents were higher when tungsten cores were exposed [129, 130] since pure tungsten is an effective cathode [129]. Galvanic current densities increased approximately five times when Mg was coupled to B MF extracted from the matrix [129]. Corrosion rates of actual Mg alloy (MA2-1)/B/MF MMCs in 0.005 N and 0.5 N NaCl solutions were 12.5 and 81.7 g/m² day, respectively, which were about six times the values of the monolithic matrix alloy in respective environments.

D2.2. Magnesium/Graphite MMCs.

The cathodic polarization diagrams of pitch-based graphite (cross section exposed) in aerated (Fig. 35.4) and deaerated (Fig. 35.5) 3.15 wt % NaCl with the anodic polarization diagrams of pure magnesium and Mg ZE41A in deaerated 3.15 wt %

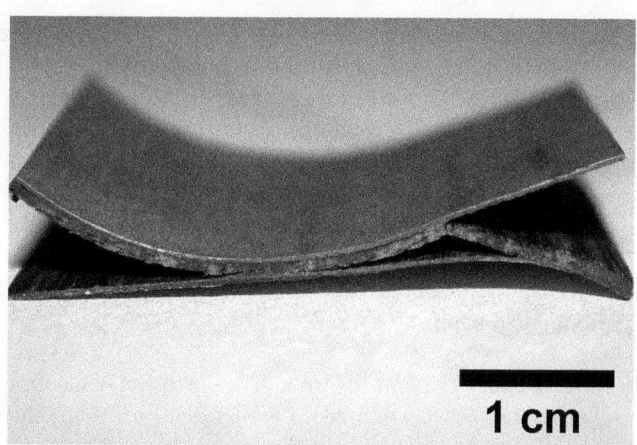

FIGURE 35.9. Exfoliated Mg AZ91C/Gr/12.7 F MMC with AZ31B Mg skins after a 20-year exposure period in an air-conditioned environment.

NaCl show that galvanic corrosion rates with graphite fiber will increase in aerated solutions. In addition, galvanic corrosion of magnesium is cathodically controlled, and therefore galvanic corrosion rates should increase with increasing area fraction of cathodic reinforcement material (Fig. 35.5). Actual MMCs immersed in air-exposed 0.001 N NaCl suffered severe degradation within five days [131]. Comparisons between a Mg AZ91C/Gr/40 P MMC and monolithic Mg AZ91C showed that the MMC open-circuit potential was approximately 0.3 V more noble and corrosion rate 40 times greater than that of the monolithic alloy in a deaerated 50-ppm chloride solution [132]. Even in relatively dry, air-conditioned environments an Mg AZ91C/Gr/12.7 F MMC with AZ31B Mg skins exfoliated over a 20-year period (Fig. 35.9).

D2.3. Magnesium/Silicon Carbide MMCs. Galvanic corrosion between magnesium and SiC depends on the type of SiC reinforcement and on the presence of dissolved oxygen in solution. Anodic polarization diagrams of pure magnesium and Mg ZE41A with cathodic polarization diagrams of HP SiC and SiC MF exposed in aerated (Fig. 35.4) and deaerated 3.15 wt % (Fig. 35.5) show that galvanic corrosion rates (as determined by the mixed-potential theory) are greater in aerated solutions due to oxygen reduction [47, 133]. Galvanic corrosion rates are lower for couples with HP SiC as opposed to the SiC MF which have carbon cores and surfaces. Studies conducted on particulate Mg ZE41A alloy reinforced with 12 vol % SiC particles ranging in size up to approximately 20 μm did not show preferential attack between SiC particles and the matrix in salt spray tests [134]. Instead, macroscopic anodic and cathodic regions developed. Corrosion spread over the MMC surface much more rapidly than on the monolithic alloy, but the local corrosion rates were approximately only three times greater on the MMC. The authors [134] speculated that the higher corrosion rates

on the MMCs could have been caused by iron contamination of the magnesium matrix during processing in a steel crucible. Studies on a model MMC consisting of high-purity magnesium and well-separated SiC particles exposed to 3.5 wt % NaCl also did not show evidence of galvanic corrosion between the particles and matrix [135].

D2.4. Magnesium/Alumina MMCs. Galvanic corrosion should not be expected between magnesium and Al_2O_3 since Al_2O_3 is an insulator. The corrosion rates of an Mg AZ91C/Al_2O_3/CF MMC [136] was approximately 100 times greater than that of the matrix alloy in 3.5 wt % NaCl at 25°C but similar to that of the matrix alloy in distilled water at 20°C. The significantly higher corrosion rates of the composites in the chloride solution appear to be caused by the presence of the Al_2O_3 fibers. Although galvanic corrosion is not expected between magnesium and Al_2O_3, conducting interphases or precipitates could potentially form due to the presence of the Al_2O_3 fibers. The open-circuit potential of a Mg AZ91C/Al_2O_3/CF MMC was more noble than that of the matrix alloy in a 50-ppm Cl$^-$ solution [137], indicating that noble precipitates or interphases could have been present.

Stress corrosion tests [138] of Mg ZE41A/Al_2O_3/CF MMC in an NaCl–potassium chromate (K_2CrO_4) solution showed that notched and unnotched specimens stressed parallel to the fiber axis and exposed for approximately 100–1000 h in the NaCl–K_2CrO_4 solution retained approximately 90% of the strength in air. The matrix alloy and the MMC with the stress direction aligned perpendicular to the fiber axis retained only approximately 40–60% of the strength in air.

D3. Titanium MMCs

Titanium MMCs are being developed for aerospace, commercial, and biomedical applications. Titanium has a density of 4.5 g/cm^3.

D3.1. Titanium/Graphite MMCs. Porous titanium/Gr MMCs were processed and heat treated to fabricate a porous titanium/titanium carbide (TiC)/Gr MMC [17]. Polarization tests were conducted in 0.9 wt % NaCl and lactated Ringer's solution for in vitro use. The anodic polarization current densities of the MMCs were significantly higher than that of pure monolithic titanium, which passivated. The authors attributed the higher corrosion rates of the Ti/TiC/Gr MMC to its porosity, which may have prevented complete passivation of the titanium matrix. Another possibility for the higher current densities could be the oxidation of graphite particles (see Section D3.2).

D3.2. Titanium/Silicon Carbide MMCs. Corrosion studies on titanium alloy Ti–15V–3Cr–3Sn–3Al (Ti 15–3) [36] and titanium aluminide α_2-Ti$_3$Al (14 wt % aluminum, 21 wt %

niobium, balance titanium) [139] reinforced with SiC MF have been conducted. The corrosion behavior of Ti 15-3/SiC/MF MMC was investigated in 3.15 wt % NaCl. There was excellent agreement in the polarization diagrams of the actual MMC and that of a model utilizing the polarization diagrams of the individual constituents and the mixed-potential theory [36]. The matrix passivated and the carbon cores and carbon-rich outer surface of the SiC MF oxidized, likely forming CO_2 similar to that observed in graphite fibers under anodic polarization [69]. In the open-circuit condition, the galvanic current density between Ti 15-3 and SiC MF cannot exceed that of the passive current density of Ti 15-3 based on their polarization diagrams (Fig. 35.4), and zero-resistance ammeter results confirmed that galvanic currents between Ti 15-3 and SiC MF were negligible. The corrosion behavior of the α_2-Ti$_3$Al/SiC/MF MMC [139] was somewhat similar to that of the Ti 15-3/SiC/MF MMC [36], with the exception that the α_2-Ti$_3$Al matrix is less resistant to pitting. During anodic polarization, the α_2-Ti$_3$Al/SiC/MF MMC pitted at approximately 1 V$_{SCE}$ in 0.5 N NaCl, which was approximately 0.5 V less than that of the monolithic matrix alloy. Some matrix pitting and crevice corrosion around the SiC MFs were also observed after anodic polarization. The galvanic current density of the α_2-Ti$_3$Al/SiC/MF MMC was negligible and limited to the passive current density of the α_2-Ti$_3$Al matrix [139].

D3.3. Titanium/Titanium Carbide and Titanium/Titanium Diboride MMCs.

Particulate pure Ti/titanium carbide (pure Ti/TiC/P) and pure Ti/titanium diboride (pure Ti/TiB$_2$/P) MMCs were fabricated by cold isostatic pressing following by sintering [140]. No interphase products were identified in the pure Ti/TiC/P MMC after processing, whereas TiB intephase products were identified in the Ti/TiB$_2$/P MMC after processing. Composites reinforced with 2.5, 5, 10, and 20 vol % TiC or TiB$_2$ were anodically polarized in deaerated 2 wt % HCl in the temperature range of 50–90°C. The passive current density for pure titanium was approximately 10^{-5} A/cm^2 throughout the temperature range. Generally, dissolution currents of the pure Ti/TiC/P and pure Ti/TiB$_2$/P MMCs increased with increasing temperature and reinforcement content, and maximum values were about 20 times and 100 times, respectively, higher than that of pure titanium. For both types of MMCs, microscopy revealed that the titanium matrix was virtually uncorroded, whereas, the TiC particles underwent some degradation, and the TiB$_2$ particles and TiB interphase were corroded significantly.

D4. Copper MMCs

Copper MMCs have been investigated for use in marine, electronic, and thermal applications. Copper is relatively heavy, with a density of 8.96 g/cm^3. Reinforcements are typically chosen to impart strength and stiffness, reduce weight, enhance thermal and electrical properties, improve machinability, and enhance wear resistance. Initial studies [1] were conducted on a wide variety of experimental copper and copper alloy MMCs reinforced with graphite, SiC, TiC, silicon nitride, boron carbide, and Al$_2$O$_3$ for marine applications. The MMCs generally showed corrosion behavior that was similar to that of the monolithic alloys, although corrosion rates were higher for some of the MMCs. Other studies have focused on copper MMCs for electronic, thermal, and tribological applications.

D4.1. Copper/Graphite MMCs.

The corrosion behavior of pure copper MMCs reinforced with 1.2–40 vol % graphite particles and 50 vol % graphite fibers in deaerated and aerated 3.5 wt % NaCl solutions [141] were investigated. The corrosion potential of the particulate-reinforced MMCs became more noble with increasing graphite content in both deaerated and aerated solutions, as would be expected by increasing the content of the noble graphite particles. The corrosion potential of the Cu/Gr/50 F MMC was approximately as noble as the Cu/Gr/40 P MMC in the aerated solution but was significantly more active than the particulate composite in the deaerated solution. This finding could be expected since oxygen reduction kinetics is normally diffusion limited in aerated solutions, but hydrogen evolution kinetics depend on the substrate and could be different on the graphite fibers as compared to on the graphite particles in deaerated solutions. The corrosion potentials of C90300 copper alloy (with 1 wt % titanium additive to increase graphite wettability) MMCs reinforced with 1–10 vol % graphite particles also increased with increasing graphite content in an aqueous solution containing ferric chloride, copper sulfate, and hydrochloric acid [12].

D4.2. Copper/Silicon Carbide MMCs.

The corrosion behavior of pure copper MMCs reinforced with 0, 5, 10, and 20 vol % SiC particles were examined in a 5 wt % NaCl solution [142]. Porosity in the materials ranged from 2.2 to 3.5% and generally increased with increasing SiC content. Corrosion potentials became more active, and corrosion current densities increased with increasing SiC content. Decreasing corrosion potentials would not be expected with increasing SiC content if SiC served as an efficient cathode. The corrosion morphology indicated that there was significant corrosion at SiC–copper interfaces. Voids caused by porosity and SiC–copper interfaces both increased with increasing SiC content. Hence, the decrease in corrosion potential with increasing SiC content is likely to have been caused by an increase in anodic sites at voids and SiC–copper interfaces.

D4.3. Copper/Alumina MMCs.

The corrosion behavior of copper MMCs reinforced with 2.7 vol % Al$_2$O$_3$ was examined in deaerated and aerated 3.5 wt % NaCl [143]. Galvanic corrosion with Al_2O_3 is not expected since Al_2O_3 is an insulator. The corrosion rates of the MMCs were comparable to that of monolithic copper, and the corrosion potentials of the MMC were only 0.01–0.02 V, more active than that of monolithic pure copper.

D5. Stainless Steel MMCs

Sintered, particulate composites consisting of ferritic 434L stainless steel (SS) and Al_2O_3 particles have been developed for potential application in chemical processing plants, turbine blades, and heat exchanger tubes [144–146]. Austenitic 316L SS reinforced with Al_2O_3 and Y_2O_3 have also been investigated for enhanced strength and wear resistance [147].

D5.1. Stainless Steel/Alumina MMCs.

The corrosion behavior of sintered 434L SS/Al_2O_3 MMCs and sintered 434L SS alloy without Al_2O_3 particles was examined [144–146]. The volume percent of Al_2O_3 particles in these materials ranged from 0 to 8%. The effect of small amounts of titanium and niobium alloying elements on corrosion resistance was also investigated. Galvanic corrosion between 434L SS and Al_2O_3 should not occur since the latter is an insulator. In 1 N H_2SO_4 [144–146], there was no strong correlation between Al_2O_3 content and corrosion behavior. One of the few generalities that could be made was that passive-current densities were high and within an order of magnitude of 1 mA/cm^2 for almost all materials. In the 5 wt % NaCl solutions, i_{CORR} of the MMCs was less than 10 μA/cm^2 [145]. Upon polarization, all materials displayed active corrosion behavior in the NaCl solutions.

Particulate 316L SS MMCs [147], fabricated using powder metallurgy, were reinforced with 3, 4, and 5 wt % Al_2O_3 and additions of 2 wt % chromium diboride (CrB$_2$) or 1 wt % boron nitride (BN) for sintering aids. The density of the MMCs ranged from 86 to 96% of the theoretical value. Unreinforced 316L SS specimens were also fabricated using powder metallurgy without sintering aids, resulting in 85% of theoretical density. Less porosity was present in the reinforced MMCs as compared to the unreinforced pure 316L SS specimen. The test samples were immersed in 10 wt % sulfuric acid (H_2SO_4) at room temperature for 24 h, 1 wt % hydrochloric acid (HCl) at room temperature for 24 h, and boiling 10 wt % nitric acid (HNO$_3$) for 8 h. The pure, unreinforced 316L SS specimens passivated in the 10 wt % H_2SO_4 solution, whereas the corrosion rate of the MMC generally increased with increasing Al_2O_3 content to a maximum value of approximately 4 mm/yr. The MMCs performed better than the unreinforced 316L SS specimen in the 1 wt % HCl solution but worse than the unreinforced specimen in the boiling nitric acid solution. There was no strong correlation between Al_2O_3 content in the MMCs and the corrosion rates in 1 wt % HCl and boiling 10 wt % HNO$_3$ solutions.

D5.2. Stainless Steel/Yttria MMCs.

Yttria (Y_2O_3) is an insulator and galvanic effects are not expected. The 316L SS specimens discussed above [147] were also reinforced with 3, 4, and 5 wt % Y_2O_3 and additions of 2 wt % chromium diboride (CrB$_2$) or 1 wt % BN as sintering aids. In all solutions (i.e., sulfuric, hydrochloric, and nitric acid solutions), the Y_2O_3-reinforced MMCs exhibited reduced corrosion resistance as compare to the Al_2O_3-reinforced MMCs. The Y_2O_3 MMCs were sintered to 88–96% of theoretical density, and the Al_2O_3 MMCs were sintered to 86–92% of theoretical density. The Y_2O_3 particles also showed better bonding to the matrix, probably forming a complex YCrO$_3$ oxide, as compared to the Al_2O_3 particles. It is possible that the formation of the reaction layer around the Y_2O_3 particles may have depleted chromium from the matrix, resulting in reduced corrosion resistance, as compared to the Al_2O_3-reinforced MMCs.

D6. Lead MMCs

Lead is a relatively heavy metal with a density of 11.4 g/cm^3. Lead MMCs, therefore, are normally developed for applications where a combination of its structural, physical, and chemical properties is important. The corrosion behavior of pure lead MMCs in simulated lead–acid battery environments has been studied to assess the feasibility of using these composites as positive electrode grids in place of conventional lead-based alloy grid materials. Lead can be alloyed with elements such as arsenic, antimony, or calcium to increase strength and stiffness. These elements, however, reduce corrosion resistance. Monolithic pure lead has very good corrosion resistance in lead–acid battery environments (which consists of sulfuric acid solutions) but is heavy and lacks sufficient mechanical strength. Pure lead, therefore, has been reinforced with strong, lightweight fibers in hopes of achieving the goals of increasing strength, reducing weight, and retaining the corrosion resistance of pure lead [148–150]. For other applications, discontinuous reinforced lead–antimony alloy MMCs were also studied in sodium chloride solutions [151].

To simulate corrosion in lead–acid battery environments, lead MMC reinforced with Al_2O_3, carbon, SiC, and glass–quartz fibers of various volume percents have been [148–150] anodically polarized at 1.226 V (vs. mercury/mercurous sulfate reference electrode) in sulfuric acid solutions (of 1.285 specific gravity) at 50, 60, and/or 70°C. At 1.226 V, lead and water are oxidized to lead dioxide (PbO$_2$) and molecular oxygen (O$_2$), respectively [152, 153]. About one-third of the total anodic current is consumed in the oxidation of lead under these conditions [150]. Poor bonding between Al_2O_3 fibers and the matrix allowed the electrolyte to diffuse into fiber–matrix interfaces, leading to accelerated corrosion [148] and swelling of the composite due to corrosion product buildup [150]. The graphite fibers were also subjected to oxidation [149].

Lead (80 wt %)–antimony (20 wt %) alloy MMCs reinforced with 1–5 wt % zircon ($ZrSiO_4$) particles [151] were exposed to a 1 N NaCl solution. Zircon should not induce galvanic corrosion. Weight loss measurements, made over a 72-h period, showed that the corrosion rate of the MMCs increased with increasing zircon content.

D7. Depleted Uranium MMCs

Depleted uranium/tungsten fiber (DU/W/F) MMCs are the antithesis of the lightweight MMCs and were developed to create high-density materials. Uranium has a density of 18.9 g/cm^3.

DU corrodes galvanically when coupled to tungsten fibers in air-exposed 3.5 wt % NaCl solutions at room temperature [154]. The open-circuit potential of tungsten fiber (-0.25 V_{SCE}) is noble to that of the DU alloy (-0.80 V_{SCE}). The open-circuit potentials of the DU/W/F MMC and galvanic couples consisting of tungsten fiber and DU alloy of equal areas are -0.78 and -0.77 V_{SCE}, respectively, and fall between those of tungsten fibers and the DU alloy. The galvanic corrosion current density measured between equal areas of tungsten fibers and the DU alloy was equal to about 4×10^{-5} A/cm^2. In a 30-day exposure test in the NaCl solution, the DU/W/F MMC lost 43.56 mg/cm^2, which was about 1.3 times that of the DU alloy.

D8. Zinc MMCs

Zinc MMCs have been developed [155] for potential use as bearing materials. Zinc has a density of 7.14 g/cm^3. Zinc alloy ZA-27 MMCs were cast with 1, 3, and 5 wt % graphite particles ranging in sizes from 100 to 150 μm. Zinc alloys are known to have excellent wear and bearing characteristics [156]. The zinc MMCs were resistant to corrosion in SAE 40 grade lubricant that had been in service for six months in an internal combustion engine. In 1 N HCl, the corrosion rates of the MMCs decreased with time.

E. CORROSION PROTECTION OF MMCs

Corrosion of metals can be prevented with the use of protective coatings and inhibitors. The use of impervious, inhibitive, or cathodically protective coatings will depend on the application and substrate. Selecting suitable coatings for MMCs will likely require testing and verification because a proven coating system for an alloy may not be effective for an MMC of that alloy. Poor adhesion and wettability between the coating and reinforcement or differences in the

electrochemical properties of the alloy and MMC may render a good coating system for the alloy ineffective for the MMC. Other coating techniques such as anodization could also be ineffective or even deleterious to the MMC. When an aluminum MMC is anodized, for example, the reinforcements can impede the growth of a continuous aluminum oxide film [157] or the reinforcements can be compromised by oxidation such as in the case of graphite fibers that oxidize to CO_2 [46]. The use of inhibitors is usually reserved for closed systems and, therefore, may not be an option in many practical cases. An inhibitor intended for a monolithic alloy should not be used for an MMC of that alloy until ample examination confirms its effectiveness.

Various studies on the corrosion protection of MMCs utilizing organic coatings, inorganic coatings, anodization, and chemical conversion coatings have been summarized elsewhere [158], and results are given in Table 35.2. The studies have generally shown that the best protection for MMCs that are susceptible to corrosion has been achieved by completely shielding the MMC from the environment utilizing coatings. Undoubtedly, in the future, generally acceptable methods to protect MMCs will likely be developed.

F. SUMMARY

There are many additional concerns regarding the corrosion of MMCs in comparison to their monolithic matrix alloys. Certain MMC systems have inherent corrosion problems. Galvanic corrosion of the matrix may be induced by conductive reinforcements (e.g., Gr, some semiconductors with high impurity levels, and other metals). If reinforcements are semiconductive, n types could suppress galvanic corrosion of the matrix under illumination, while p types could induce galvanic corrosion under illumination. When the reinforcements are insulators, galvanic corrosion of the matrix will not be induced; however, the corrosion behavior of the MMC in comparison to the monolithic matrix alloy may be different. Whether the reinforcements are conductive, semiconductive, or insulating, they can affect the precipitation of intermetallics, induce higher dislocation densities, and react with the matrix-forming interphases. All of these phenomena may alter the corrosion behavior of the MMC in comparison to the monolithic matrix alloy. Corrosion initiation in MMCs may also be more sensitive to physical damage. Impacting the MMC surface may lead to debonding at the reinforcement–matrix interface or fracturing the reinforcement leading to the formation of minute crevices, which may enhance corrosion initiation. The problem could possibly be exacerbated with MMCs having higher strength matrices and higher reinforcement content. Caution should be used in assuming that MMCs belonging to a specific MMC group (e.g., Al/SiC MMCs) all have similar corrosion behavior since variations

TABLE 35.2. Summary of Corrosion Protection Studies on MMCs

MMC Type	Substrate Protected	Coating/Treatment	Environment	Outcome	References
Al/Gr	Surface Al foils on MMC	Organic coatings	Marine and NaCl solution	Protection	[159–161]
	Al/Gr MMC	Inorganic diamondlike coating	NaCl solution	Short-term protection	[162]
	Surface Al foils on MMC	CVD and PVD inorganic coatings	Marine	No protection	[159]
	Surface Al foils on MMC	Electroplated Ni coating (without defects)	Marine	Protection	[159]
	Surface Al foils on MMC	Electroless Ni coating (with defects)	Marine	Accelerated corrosion	[163, 164]
	Al/Gr MMC	Ti and Ni cladding	Marine	Delamination from exposed edges	[159]
	Surface Al foils on MMC	Electrodeposited Al/Mn on electroless Ni coating	Marine	Protection if panel edges were sealed to prevent Al/Gr exposure	[159]
	Surface Al foils on MMC	Anodization with dichromate sealing	Marine	Protection if panel edges sealed	[163]
	Surface Al foils on MMC	Chromate/phosphate conversion coating	Marine	Protection if edges sealed with epoxy	[163]
	Surface Al foils on MMC	Chemical passivation with $CeCl_3$	NaCl solutions	Delayed pitting on surface foils	[165, 166]
Al/SiC	Al/SiC MMC	Organic epoxy coating	Marine and NaCl solution	Protection	[160, 163]
	Al/SiC MMC	Plasma-sprayed alumina coating	Marine	Protection	[163]
	Al/SiC MMC	Flame-sprayed Al coating	Marine	Protection	[163]
	Al/SiC MMC	Anodization	Marine and NaCl solution	Various levels or protection	[167–170]
	Al/SiC MMC	Chemical passivation with $CeCl_3$	NaCl solution	Limited protection	[165, 171]
Al/Al_2O_3	Al/Al_2O_3 MMC	Chemical passivation with $CeCl_3$	NaCl solution	Limited protection	[172, 173]

in the quality of the reinforcement and matrix alloy, manufacturing technique (e.g., powder metallurgy versus casting), thermomechanical processing, and other factors can all alter corrosion behavior. Hence, it will be difficult to obtain generalized and consistent corrosion behavior for a specific MMC group until standards are developed for the manufacture and processing of MMCs.

NOMENCLATURE

B MF	Boron monofilament
B_{MF}^E	B_{MF} electrode with MF ends exposed
B_{MF}^S	B_{MF} electrode with MF circumferential surface exposed
CD	Current density
E_{pit}	Pitting potential
E_{GALV}	Galvanic couple potential
Gr	Graphite
Gr^E	Gr electrode with fiber ends exposed
HP	Hot pressed
I	Current density
i_c	Cathodic current density
i_{CORR}	Corrosion current density
i_{GALV}	Galvanic current density
MMC	Metal–matrix composite
SC	Semiconductor
SiC MF	Silicon carbide monofilament
SiC_{MF}^E	SiC_{MF} electrode with MF ends exposed
SiC_{MF}^S	SiC_{MF} electrode with MF circumferential surface exposed
Superscript E	Electrode with fiber or MF ends exposed (e.g., SiC_{MF}^E)
Superscript S	Electrode with fiber or MF circumferential surface exposed (e.g., SiC_{MF}^S)
T	Thickness
vol %	Volume percent
V_{SCE}	Volts versus a calomel electrode
wt %	Weight percent
x_C or X_C	Cathodic area fraction
ρ	Resistivity

Units

A   Amperes
cm   Centimeter
h   Hours
s   Seconds
V   Volts
Ω Ohm

REFERENCES

1. D. M. Aylor, "Corrosion of Metal Matrix Composites," in Metals Handbook, 9th ed., Corrosion, ASM International, Metals Park, OH, 1987, pp. 859–863.

2. W. C. J. Harrigan, "Metal Matrix Composites," in Metal Matrix Composites: Processing and Interfaces, Academic, New York, 1991, pp. 1–16.

3. J. W. Weeton, D. M. Peters, and K. L. Thomas, Guide to Composite Materials. American Society for Metals, Metals Park, OH. 1987, p. 2–2.

4. G. B. Park and D. A. Foster, International Technical Conference Proceedings, in SUR/FIN'90, July 1990, American Electroplaters and Surface Finishers Society, Boston, MA.

5. DWA Technologies, http://dwatechnologies.com/, Oct. 2007.

6. 3M, Metal Matrix Composites, www.3m.com/market/indus-trial/mmc/.

7. S. V. Prasad and R. Asthana, "Aluminum Metal–Matrix Composites for Automotive Applications: Tribological Considerations," Tribol. Lett., **17**(3), 445 (2004).

8. M. F. Ashby and D. R. H. Jones, Engineering Materials, 2nd ed., Butterworth Heinemann, Oxford, 1998.

9. ALMMC, Aluminum Metal-Matrix Composites Consortium, www.almmc.com. Oct. 2007.

10. FiberNide, http://fibernide.com/copper.html, Oct. 2007.

11. H. L. Marcus, W. F. Weldon, and C. Persad, Technical Report Contract Number N62269-85-C0222, University of Texas at Austin, Austin, TX, 1987.

12. P. K. Rohatgi et al., "Corrosion and Dealloying of Cast Lead-Free Copper Alloy-Graphite Composites," Corros. Sci., **42**, 1553–1571 (2000).

13. V. G. Gurbunov, V. D. Parshin, and V. V. Pamin, Russ. Cast. Prod., 1974, **93**, p. 348.

14. J. Van Muylder and M. Pourbaix, in Atlas of Electrochemical Equilibria in Aqueous Solutions, M. Pourbaix (Ed.), National Association of Corrosion Engineers, Houston, TX, 1974, pp. 449–457.

15. N. A. P. Rao et al., Tribol. Int., **13**, 171 (1980).

16. S. Ranganath, "A Review on Particulate-Reinforced Titanium Matrix Composites," J. Mater. Sci., **32**, 1–16 (1997).

17. D. J. Blackwood et al., "Corrosion Behaviour of Porous Titanium-Graphite Composites Designed for Surgical Implants," Corros. Sci., **42**, 481–503 (2000).

18. M. E. Buck and R. J. Suplinskas, in Engineered Materials Handbook on Composites, ASM International, Metals Park, OH, 1987, pp. 851–857.

19. Saffil, www.saffil.com, May 2002.

20. D. Hughes, Aviat. Week Space Technol., **Nov. 28**, 91 (1988).

21. R. Taylor and Y. Qunsheng, "Thermal Transport in Carbon Fibre-Copper and Carbon Fibre/Aluminum Composites," in ICCM/8, Society for the Advancement of Material and Process Engineering (SAMPE), Honolulu, HI, 1991.

22. A. M. Tsirlin, in Strong Fibres (Handbook of Composites, Vol. 1), W. Watt and B. V. Perov (Eds.), North-Holland, Amsterdam, 1985, pp. 155–199.

23. N. Ichinose, Introduction to Fine Ceramics, Wiley, New York, 1987, pp. 50–52.

24. Nicalon (TM) Ceramic Fiber Brochure, COI Ceramics, Magna, Utah, 2006.

25. L. H. Hihara, "Corrosion of Aluminum-Matrix Composites," Corros. Rev., **15**(3–4), 361–386 (1997).

26. R. E. Bolz and G. L. Tuve, in CRC Handbook of Tables for Applied Engineering Science, CRC Press, Boca Raton, FL, 1973, pp. 262–264.

27. H. R. Clauser, The Encyclopedia of Engineering Materials and Processes, New York, Reinhold Publishing Corporation, 1963, p. 429.

28. N. N. Greenwood and A. Earnshaw, Chemistry of the Elements, Pergamon Press, Oxford, 1984.

29. S. Yamada et al., Ceram. Int., **29**, 299 (2003).

30. J. W. Weeton, D. M. Peters, and K. L. Thomas, Guide to Composite Materials, American Society for Metals, Metals Park, OH, 1987, pp. 5–10.

31. L. H. Hihara, "Corrosion of Aluminum Matrix Composites," Corros. Rev., **15**(3–4), 361 (1997).

32. L. H. Hihara and R. M. Latanision, Corrosion, **48**(7), 546–552 (1992).

33. Z. J. Lin, "Corrosion Study of Silicon-Aluminum Metal-Matrix Composites", in Mechanical Engineering, University of Hawaii at Manoa, Honolulu, 1995.

34. J. Zhu, "Corrosion of Continuous Alumina Fiber Reinforced Aluminum-Matrix Composites," in Mechanical Engineering, University of Hawaii at Manoa, Honolulu, 2008.

35. C. Tamirisa, "Corrosion Behavior of Silicon-Carbide Reinforced Titanium 15-3 Metal-Matrix Composite in 3.15 wt % NaCl," in Mechanical Engineering, University of Hawaii at Manoa, Honolulu, HI, 1993.

36. L. H. Hihara and C. Tamirisa, "Corrosion of SiC Monofila-ment/Ti-15-3-3-3 Metal-Matrix Composites in 3.15 wt.% NaCl," Mater. Sci. Eng. A, **198**, 119–125 (1995).

37. P. K. Kondepudi, "Corrosion Behavior of Magnesium Matrix Composites," in Mechanical Engineering, University of Hawaii at Manoa, Honolulu, 1992.

38. L. H. Hihara and R. M. Latanision, "Corrosion of Metal-Matrix Composites," Int. Mater. Rev., **39**, 245 (1994).

39. H. Ding and L. H. Hihara, A "'Photochemical Corrosion Diode' Model Depicting Galvanic Corrosion in Metal-Matrix

Composites Containing Semiconducting Constituents," ECS Trans., **11**(18), 41 (2008).

40. H. Ding and L. H. Hihara, "Effect of Embedded Titanium-Containing Particles on the Corrosion of Particulate Alumina Reinforced Aluminum-Matrix Composite," ECS Trans., **11** (15), 935 (2008).

41. R. P. I. Adler et al., "Characterization of Environmentally Exposed Aluminum Metal Matrix Composite Corrosion Products as a Function of Volume Fraction and Reinforcement Specie," Paper 06T029, 2005 Tri Serice Corrosion Conference, Orlando, FL, 2005.

42. U. R. Evans, *Metallic Corrosion, Passivity and Protection*, E. Arnold & Co. London, 1937, pp. 513–516.

43. L. H. Hihara et al., "Corrosion Initiation and Propagation in Particulate Aluminum-Matrix Composites, " Tri-Service Corrosion Conference, Orlando, FL, 2005.

44. L. H. Hihara and Z. J. Lin, "Corrosion of Silicon/Aluminum Metal-Matrix Composites," Seventh Japan International SAMPE Symposium & Exhibition, Tokyo, Japan, 1999.

45. L. H. H. Hongbo Ding, "Localized Corrosion Currents and pH Profile over B4C, SiC and Al$_2$O$_3$ Reinforced 6092 Aluminum Composites I. In 0.5M Na$_2$SO$_4$ Solution," J. Electrochem. Soc, **152** (4), pp. B161–B167, (2005).

46. L. H. Hihara, "Corrosion of Aluminum-Matrix Composites," Ph.D. Thesis, Massachusetts Institute of Technology, Cambridge, MA, 1989.

47. L. H. Hihara and P. K. Kondepudi, Corros. Sci., **36**, 1585–1595 (1994).

48. S. L. Pohlman, Corrosion, **34**, 156–159 (1978).

49. D. Nath, R.T. Bhat, and P. K. Rohatgi, J. Mater. Sci., **15**, 1241–1251 (1980).

50. D. Nath and T. K. Namboodhiri, Composites, **19**, 237–243 (1988).

51. K. I. Portnoi et al., Poroshkovaya Metallurgiya, vol **218** (2), 45–49 (1981).

52. M. A. Buonanno, "The Effect of Processing Conditions and Chemistry on the Electrochemistry of Graphite and Aluminum Metal Matrix Composites," Ph.D. Thesis, Massachusetts Institute of Technology, Cambridge, MA, 1992.

53. H. J. Becher, in Handbook of Preparative Inorganic Chemistry, Vol. 1, G. Brauer (Ed.), Academic, New York, 1963, p. 832.

54. T. Iseki, T. Kameda, and T. Maruyama, J. Mater. Sci., **19**, 1692–1698 (1984).

55. A. Grytsiv and P. Rogl, "Aluminum–Boron–Carbon," in Light Metal Systems, Part 1: Selected Systems from Ag-Al-Cu to Al-Cu-Er, Springer, Berlin, 2004.

56. A. Mortensen, J. A. Cornie, and J. Flemings, J. Metals, **40**, 12 (1988).

57. R. J. Arsenault, in Metal Matrix Composites: Mechanisms and Properties, R. K. Everett and R. J. Arsenault (Eds.), Academic, New York, 1991, p. 79.

58. N. Birbilis and R. G. Buchheit, "Electrochemical Characteristics of Intermetallic Phases in Aluminum Alloys: An Experimental Survey and Discussion," J. Electrochem. Soc., **152**(4), B140–B151 (2005).

59. N. K. Bruun and K. Nielsen, in Metal Matrix Composites-Processing, Microstructure and Properties, 12th Riso International Symposium on Materials and Science, Denmark, 1991.

60. J. Y. Yang and M. Metzger, in Extended Abstracts, Abstract No. 155, The Electrochemical Society, Denver, CO, Oct. 1981.

61. D. Nath and T. K. Namboodhiri, Corros. Sci., **29**, 1215–1229 (1989).

62. H. H. Uhlig and R. W. Revie, Corrosion and Corrosion Control, 3rd ed., Wiley, New York, 1985, p. 123.

63. G. Butler and H. C. K. Ison, Corrosion and Its Prevention in Waters, Robert E. Krieger, New York, 1978, 149.

64. Z. Ahmad, P. T. Paulette, and B. J. A. Aleem, "Mechanism of Localized Corrosion of Aluminum-Silicon Carbide Composites in a Chloride Containing Environment," J. Mater. Sci., **35**, 2573–2579 (2000).

65. H.-Y. Yao and R.-Z. Zhu, "Interfacial Preferential Dissolution on Silicon Carbide Particulate/Aluminum Composites," Corrosion, **54**(7), 499–503 (1998).

66. A. J. Sedriks, J. A. Green, and D. L. Novak, Metall. Trans., **2**, 871–875 (1971).

67. A. V. Bakulin, V. V. Ivanov, and V. V. Kuchkin, Zaschita Metallov, **14**(1), 102–104 (1978).

68. L. H. Hihara and R. M. Latanision, Mater. Sci. Eng., **A126**, 231–234 (1990).

69. L. H. Hihara and R. M. Latanision, Corrosion, **47**, 335–341 (1991).

70. T. S. Devarajan, "Corrosion Initiation Sites of Particle Reinforced 6092 Aluminum Metal Matrix Composites," in Mechanical Engineering, University of Hawaii at Manoa, Honolulu, 2005.

71. E. Deltombe, C. Vanleugenhaghe, and M. Pourbaix, in Atlas of Electrochemical Equilibria in Aqueous Solutions, M. Pourbaix (Ed.), National Association of Corrosion Engineers, Houston, TX, 1974, pp. 168–176.

72. J. R. Galvele, in Passivity of Metals, R. P. Frankenthal and J. Kruger (Eds.) The Electrochemical Society, Inc.: Princeton, N.J. 1978, pp. 285–327.

73. W. Hubner and G. Wranglen, in Current Corrosion Research in Scandinavia, IVth Scandinavian Corrosion Congress, Sanoma Osakeyhtio Helsinki, 1964.

74. M. J. Pryor and D. S. Keir, J. Electrochem. Soc., **102**, 605–607 (1955).

75. U. R. Evans, Metallic Corrosion, Passivity and Protection, London, 1937, E. Arnold & Co.

76. M. M. Schwartz, Composite Materials Handbook, McGraw-Hill, New York, 1984.

77. W. H. Kim, M. J. Koczak, and A. Lawley, in Proceedings of the 1978 International Conference on Composite Materials, ICCM/2, The Metallurgical Society of AIME, Toronto, Canada, 1978.

78. D. L. Dull, W. C. J. Harrigan, and M. F. Amateau, "Final Report, The Effect of Matrix and Fiber Composition on Mechanical Strength and Corrosion Behavior of Graphite-Aluminum Composites," The Aerospace Corporation, El Segundo, CA, 1977.

79. L. H. Hihara, "Corrosion of Metal-Matrix Composites," in ASM Handbook, Vol. 13B, Corrosion: Metals, S. D. Cramer and J. B. S. Covino (Eds.), ASM International, Metals Park, OH, 2005.

80. S. L. Coleman, V. D. Scott, and B. McEnaney, "Corrosion Behaviour of Aluminium-Based Metal Matrix Composites," J. Mater. Sci., **29**, 2826–2834 (1994).

81. M. Metzger and S. G. Fishman, "Industrial and Engineering Chemistry," Product Res. Devel., **22**, 296–302 (1983).

82. H. Sun, E.Y. Koo, and H. G. Wheat, "Corrosion Behavior of SiC_p/6061 Al Metal Matrix Composites," Corrosion, **47**(10), 741–753 (1991).

83. O. P. Modi et al., "Corrosion Behaviour of Squeeze-Cast Aluminum Alloy-Silicon Carbide Composites," J. Mater. Sci., **27**, 3897–3902 (1992).

84. G. A. Hawthorn, "Outdoor and Laboratory Corrosion Studies of Aluminum-Metal Matrix Composites," in Mechanical Engineering, University of Hawaii at Manoa, Honolulu, 2004.

85. K. D. Lore and J. S. Wolf, in Extended Abstracts, The Electrochemical Society, Denver, CO, 1981.

86. G. A. Hawthorn and L. H. Hihara, "Out-Door & Laboratory Corrosion Studies of Aluminum Metal-Matrix Composites," U.S. Army Corrosion Summit 2004, Cocoa Beach, FL, 2004.

87. A. J. Griffiths and A. Turnbull, "An Investigation of the Electrochemical Polarisation Behaviour of 6061 Aluminum Metal Matrix Composites," Corros. Sci., **36**(1), 23–35 (1994).

88. P. P. Trzaskoma, E. McCafferty, and C. R. Crowe, J. Electrochem. Soc., **130**, 1804–1809 (1983).

89. S. L. Golledge, J. Kruger, and C. M. Dacres, in Extended Abstracts, The Electrochemical Society, Las Vegas, NV, 1985.

90. Y. Shimizu, T. Nishimura, and I. Matsushima, "Corrosion Resistance of Al-Based Metal Matrix Composites," Mater. Sci. Eng. A, **198**, 113–118 (1995).

91. D. M. Aylor and P. J. Moran, J. Electrochem. Soc., **132**, 1277–1281 (1985).

92. G. W. Roper and P. A. Attwood, "Corrosion Behaviour of Aluminum Matrix Composites," J. Mater. Sci., **30**, 898–903 (1995).

93. C. Monticelli et al., "Application of Electrochemical Noise Analysis to Study the Corrosion Behavior of Aluminum Composites," J. Electrochem. Soc., **142**(2), 405–410 (1995).

94. P. C. R. Nunes and L. V. Ramanathan, "Corrosion Behavior of Alumina-Aluminum and Silicon Carbide-Aluminum Metal-Matrix Composites," Corrosion, **51**(8), 610–617 (1995).

95. G. E. Kiourtsidis, S. M. Skolianos, and E. G. Pavlidou, "A Study on Pitting Behaviour of $AA2024/SiC_p$ Composites Using the Double Cycle Polarization Technique," Corros. Sci., **41**, 1185–1203 (1999).

96. P. P. Trzaskoma, Corrosion, **46**, 402–409 (1990).

97. G. E. Kiourtsidis and S. M. Skolianos, "Corrosion Behavior of Squeeze-Cast Silicon Carbide-2024 Composites in Aerated 3.5 wt.% Sodium Chloride," Mater. Sci. Eng. A, **248**, 165–172 (1998).

98. M. M. Buarzaiga and S. J. Thorpe, "Corrosion Behavior of As-Cast, Silicon Carbide Particulate-Aluminum Alloy Metal-Matrix Composites," Corrosion, **50**(3), 176–185 (1994).

99. R. C. Paciej and V. S. Agarwala, Corrosion, **44**, 680–684 (1988).

100. S. R. Nutt and J. M. Duva, Scripta Metallurgica, **20**, 1055–1058 (1986).

101. M. S. Bhat, M. K. Surappa, and H. V. Sudhaker Nayak, "Corrosion Behaviour of Silicon Carbide Particle Reinforced 6061/A1 Alloy Composites," J. Mater. Sci., **26**(18), 4991–4996 (1991).

102. J. England and I. W. Hall, Scripta Metallurgica, **20**, 697–700 (1986).

103. Z. Ahmad and B. J. Abdul Aleem, "Effect of Temper on Seawater Corrosion of an Aluminum-Silicon Carbide Composite Alloy," Corrosion, **52**(11), 857–864 (1996).

104. W. N. C. Garrard, "The Corrosion Behaviour of Aluminum-Silicon Carbide Composites in Aerated 3.5% Sodium Chloride," Corros. Sci., **36**(5), 837–851 (1994).

105. J. K. Park and J. P. Lucas, "Moisture Effect on SiC_p/6061 A1 MMC: Dissolution of Interfacial Al_4C_3," Scripta Mater., **37**(4), 511–516 (1997).

106. L. H. Hihara, H. Ding, and T. Devarajan, "Corrosion-Initiation Sites on Aluminum Metal-Matrix Composites," U.S. Army Corrosion Summit 2004, Cocoa Beach, FL, 2004.

107. C.-K. Fang, C. C. Huang, and T. H. Chuang, "Synergistic Effects of Wear and Corrosion for Al_2O_3 Particulate-Reinforced 6061 Aluminum Matrix Composites," Metall. Mater. Trans. A, **30A**, 643–651 (1999).

108. R. U. Vaidya et al., "Effect of Microbiologically Influenced Corrosion on the Tensile Stress-Strain Response of Aluminum and Alumina-Particle Reinforced Aluminum Composite," Corrosion, **53**(2), 136–141 (1997).

109. V. S. Agarwala, Abstract No. 15 in Extended Abstracts, The Electrochemical Society, Montreal, Canada, May 1982.

110. L. Bertolini, M. F. Brunella, and S. Candiani, "Corrosion Behavior of a Particulate Metal-Matrix Composite," Corrosion, **55**(4), 422–431 (1999).

111. J. M. G. DeSalazar et al., "Corrosion Behaviour of AA6061 and AA7005 Reinforced with Al_2O_3 Particles in Aerated 3.5% Chloride Solutions: Potentiodynamic Measurements and Microstructure Evaluation," Corros. Sci., **41**, 529–545 (1999).

112. D. Nath, R. T. Bhat, and P. K. Rohatgi, J. Mater. Sci., **15**, 1241–1251 (1980).

113. P. K. Rohatgi, R. Asthana, and S. Das, Int. Mater. Rev., **31**, 115 (1986).

114. R. C. Weast, CRC Handbook of Chemistry and Physics, 67th ed., CRC Press, Boca Raton, FL, 1986, p. B-116.

115. D. Nath and T. K. Namboodhiri, Composites, **19**, 237–243 (1988).

116. D. Nath and T. K. Namboodhiri, Corros. Sci., **29**, 1215–1229 (1989).

117. C. Monticelli et al., "Stress Corrosion Cracking Behaviour of Some Aluminum-Based Metal Matrix Composites," Corros. Sci., **39**(10–11), 1949–1963 (1997).

118. G. E. Kiourtsidis and S. M. Skolianos, "Stress Corrosion Behavior of Aluminum Alloy 2024/Silicon Carbide Particles (SiC$_p$) Metal Matrix Composites," Corrosion, **56**(6), 646–653 (2000).

119. H.-Y. Yao, "Effect of Particulate Reinforcing on Stress Corrosion Cracking Performance of a SiC$_p$/2024 Aluminum Matrix Composite," J. Composite Mater., **33**(11), 962–970 (1999).

120. D. A. Davis, M. G. Vassilaros, and J. P. Gudas, Mater. Perform., 38–42 (1982).

121. D. W. Berkeley, H. E. M. Sallam, and H. Nayeb-Hashemi, "The Effect of pH on the Mechanism of Corrosion and Stress Corrosion and Degradation of Mechanical Properties of AA6061 and Nextel 440 Fiber-Reinforced AA6061 Composites," Corros. Sci., **40**(2/3), 141–153 (1998).

122. S. S. Yau and G. Mayer, Mater. Sci. Eng., **42**, 45–47 (1986).

123. D. F. Hasson et al., in Failure Mechanisms in High Performance Materials, J. G. Early, T. R. Shives, and J. H. Smith (Eds.), Cambridge University Press, New York, 1984, pp. 147–156.

124. R. H. Jones, in Environmental Effects on Advanced Materials, R. H. Jones and R. E. Ricker (Eds.), The Minerals, Metals and Materials Society, Warrendale, PA, 1991, pp. 283–295.

125. K. Minoshima, I. Nagashima, and K. Komai, "Corrosion Fatigue Fracture Behaviour of a SiC Whisker-Aluminum Matrix Composite Under Combined Tension-Torsion Loading," Fatigue and Fracture of Engineering Materials and Structures, **21**, 1435–1446 (1998).

126. R. F. Buck and A. W. Thompson, in Environmental Effects on Advanced Materials, R. H. Jones and R. E. Ricker (Eds.), The Minerals, Metals, and Materials Society, Warrendale, PA. 1991, pp. 297–313.

127. H. H. Uhlig and R. W. Revie, Corrosion and Corrosion Control, 3rd ed., Wiley, New York, 1985, p. 354.

128. G. Butler and H. C. K. Ison, Corrosion and Its Prevention in Waters, Robert E. Krieger, New York, 1978, p. 91.

129. M. A. Timonova et al., Mettallovedenie i Termicheskaya Obrabotka Metallov., **11**, 33–35 (1980).

130. V. F. Stroganova and M. A. Timonova, Metallovedenie i Termicheskaya Obrabotka Metallov, 10, 44–46 (1978).

131. P. P. Trzaskoma, Corrosion, **42**, 609–613 (1986).

132. W. F. Czyrklis, Paper No. 196 in Conference Proceedings of Corrosion 85, National Association of Corrosion Engineers, Boston, MA, 1985.

133. L. H. Hihara and P. K. Kondepudi, Corros. Sci., **34**, 1761–1772 (1993).

134. C. A. Nunez-Lopez et al., "The Corrosion Behaviour of Mg Alloy ZC71/SiC$_p$ Metal Matrix Composite," Corros. Sci., **37**(5), 689–708 (1995).

135. C. A. Nunez-Lopez et al., "An Investigation of Microgalvanic Corrosion Using a Model Magnesium-Silicon Cabide Metal Matrix Composite," Corros. Sci., **38**(10), 1721–1729 (1996).

136. M. Levy and W. F. Czyrklis, in Extended Abstracts, The Electrochemical Society, Denver, CO, Oct. 1981.

137. W. F. Czyrklis, "Corrosion Evaluation of Metal Matrix Composite FP/Mg AZ91C," in 1983 Tri-Service Corrosion Conference, U.S. Naval Academy, Annapolis, MD, 1983.

138. J. M. Evans, Acta Metall., **34**, 2075–2083 (1986).

139. H. M. Saffarian and G. W. Warren, "Aqueous Corrosion Study of α_2-Ti$_3$Al/SiC Composites," Corrosion, **54**(11), 877–886 (1998).

140. B. S. Covino, Jr. and D. E. Alman, "Corrosion of Titanium Matrix Composites," in Proceedings of the 15th International Corrosion Congress, Viajes Iberia Congresos, Madrid, Spain, 2002.

141. H. Sun, J. E. Orth, and H. G. Wheat, "Corrosion Behavior of Copper-Based Metal-Matrix Composites," J. Metals, **Sept.**, 36–41 (1993).

142. Y.-F. Lee, S.-L. Lee, and J.-C. Lin, "Wear and Corrosion Behaviors of SiC$_p$ Reinforced Copper Matrix Composite Formed by Hot Pressing," Scand. J. Metall., **28**, 9–16 (1999).

143. H. Sun and H. G. Wheat, "Corrosion Study of Al$_2$O$_3$ Dispersion Strengthened Cu Metal Matrix Composites in NaCl Solutions," J. Mater. Sci., **28**, 5435–5442 (1993).

144. S. K. Mukherjee, A. Kumar, and G. S. Upadhyaya, Br. Corros. J., **20**, 41–44 (1985).

145. S. K. Mukherjee, A. Kumar, and G. S. Upadhyaya, Powd. Metall. Int., **17**, 172–175 (1985).

146. S. K. Mukherjee and G. S. Upadhyaya, Mater. Chem. Phys., **12**, 419–435 (1985).

147. F. Velasco et al., "Mechanical and Corrosion Behaviour of Powder Metallurgy Stainless Steel Based Metal Matrix Composites," Mater. Sci. Technol., **13**(10), 847–851 (1997).

148. C. M. Dacres et al., J. Electrochem. Soc., **128**, 2060–2064 (1981).

149. J. C. Viala, M. El Morabit, and J. Bouix, Mater. Chem. Phys., **13**, 393–408 (1985).

150. C. M. Dacres, R. A. Sutula, and B. F. Larrick, J. Electrochem. Soc., **130**, 981–985 (1983).

151. K. H. W. Seah et al., "Corrosion Behaviour of Lead Alloy/Zircon Particulate Composites," Corrosion Sci., **39**(8), 1443–1449 (1997).

152. J. Burbank, J. Electrochem. Soc., **106**, 369 (1959).

153. J. Burbank, A. C. Simon, and E. Willihnganz, in Advances in Electrochemistry and Electrochemical Engineering, P. Delahay (Ed.), Wiley Interscience, New York, 1971 p. 157.

154. P. P. Trzaskoma, J. Electrochem. Soc., **129**, 1398–1402 (1982).

155. K. H. W. Seah, S. C. Sharma, and B. M. Girish, "Corrosion Characteristics of ZA-27-Graphite Particulate Composites," Corros. Sci., **39**(1), 1–7 (1997).

156. Smith, W., "Structure and Properties of Engineering Alloys," 2nd ed., McGraw-Hill, New York, 1993.

157. H. J. Greene, *"Evaluation of Corrosion Protection Methods for Aluminum Metal Matrix Composites," Ph.D. Thesis,* University of Southern California, Los Angeles, CA, 1992.

158. L. H. Hihara, "Corrosion of Metal-Matrix Composites," in ASM Handbook, Vol. 13B, Corrosion: Materials, S. D. Cramer and J. B. S. Covino (Eds.), ASM International, Materials Park, OH, 2005, 538–539.

159. J. H. Payer and P. G. Sullivan, in Bicentennial of Materials, 8th National SAMPE Technical Conference, Society for the Advancement of Material and Process Engineering, Seattle, WA, 1976.

160. S. Lin, H. Shih, and F. Mansfeld, "Corrosion Protection of Aluminum Alloys and Metal Matrix Composites by Polymer Coatings," Corros. Sci., **33**(9), 1331–1349 (1992).

161. F. Mansfeld and S. L. Jeanjaquet, Corros. Sci. **26**, 727–734 (1986).

162. B. Wielage et al., "Corrosion Protection of Carbon Fibre Reinforcced Aluminum Composite by Diamondlike Carbon Coatings," Mater. Sci. Technol., **16**, 344–348 (2000).

163. D. M. Aylor and R. M. Kain, "Assessing the Corrosion Resistance of Metal Matrix Composite Materials in Marine Environments," in Recent Advances in Composites in the United States and Japan, ASTM STP 864, J. R. Vinson and M. Taya (Eds.), American Society for Testing and Materials, Philadelphia, PA, 1983, pp. 632–647.

164. D. M. Aylor, R. J. Ferrara, and R. M. Kain, Mater. Perform., **23**, 32–38 (1984).

165. F. Mansfeld et al., Corrosion, **45**, 615–630 (1989).

166. B. R. W. Hinton, D. R. Arnott, and N. E. Ryan, Mater. Forum., **9**, 162 (1986).

167. P. P. Trzaskoma and E. McCafferty, in Aluminum Surface Treatment Technology, R. S. Alwitt and G. E. Thompson (Eds.), The Electrochemical Society, Denver, CO, 1986, pp. 171–177.

168. J. Hou and D. D. L. Chung, "Corrosion Protection of Aluminum-Matrix Aluminum Nitride and Silicon Carbide Composites by Anodization," J. Mater. Sci., **32**, 3113–3121 (1997).

169. C. R. Crowe, D. G. Simons, and M. D. Brown, in Extended Abstracts, The Electrochemical Society, Denver, CO, 1981.

170. S. Lin et al., "Corrosion Protection of Al/SiC Metal Matrix Composites by Anodizing," Corrosion, **48**(1), 61–67 (1992).

171. Z. Ahmad and B. J. A. Aleem, "Degradation of Aluminum Metal Matrix Composites in Salt Water and Its Control," Mater. Des., **23**, 173–180 (2002).

172. P. Traverso, R. Spiniello, and L. Monaco, "Corrosion Inhibition of Al 6061 T6/Al$_2$O$_3$p 10% (v/v) Composite in 3.5% NaCl Solution with Addition of Cerium (III) Chloride," Surf. Interf. Anal., **34**, 185–188 (2002).

173. A. S. Hamdy, A. M. Beccaria, and P. Traverso, "Corrosion Protection of Aluminum Metal-Matrix Composites by Cerium Conversion Coatings," Surf. Interf. Anal., **34**, 171–175 (2002).

36

ENVIRONMENTAL DEGRADATION OF ENGINEERED BARRIER MATERIALS IN NUCLEAR WASTE REPOSITORIES

R. B. REBAK

GE Global Research, Niskayuna, New York

A. INTRODUCTION

All the countries addressing the issue of nuclear waste, mainly produced by nuclear power plants, are considering disposing them in stable geologic repositories. In most of the repositories (e.g., Finland, Sweden, Canada, France) the environments will be reducing in nature, except for the repository in the United States, in which the ingress of oxygen will not be restricted. For the reducing repositories the different national programs are considering materials as carbon steel and copper. For the repository in the United States, some of the most corrosion-resistant commercially available alloys, such as nickel and titanium alloys, are being characterized. This chapter presents a summary of the behavior of the different materials under consideration for the repositories and the current understanding of the degradation modes of the proposed alloys in groundwater environments from the point of view of general corrosion, localized corrosion, and environmentally assisted cracking.

Recent concerns about global warming and the release of greenhouse gases by the fossil fuel power industry have reignited the consideration of alternative sources of energy such as wind, solar, fuel cells, and nuclear power. Currently there are more than 40 nuclear power reactors under construction in the world, most of them in Asia [1]. Because of this, some reports claim that there is a nuclear power renaissance. However, even though nuclear power has been used for more than 60 years, the issue of the toxic radioactive waste generated during energy production still needs to be resolved. Radioactive materials are pertinent not just to nuclear power since they are also used worldwide in other fields, including medical applications and production of weapons. Once the radioactive materials lose their commercial value, they are considered radioactive waste, and they need to be isolated from the environment until the radioactive decay has reduced its toxicity to innocuous levels for plants, animals, and humans. Different types of radioactive waste are produced during commercial and defense nuclear fuel cycles. One type of waste, denoted high-level waste (HLW), contains the highest concentration of radiotoxic and heat-generating species. Because of this factor, the most stringent standards for disposing of radioactive wastes are being placed

Uhlig's Corrosion Handbook, Third Edition, Edited by R. Winston Revie
Copyright © 2011 John Wiley & Sons, Inc.

worldwide on HLW, and the majority of the radioactive waste management effort is being directed toward the HLW problem. One of the most common and most voluminous types of HLW is the spent fuel (SF) from commercial nuclear reactors for power generation.

All of the countries currently studying the options for disposing of HLW have selected deep stable geologic formations to be the primary barrier for accomplishing this isolation. It is postulated that, by the very nature of these geologic sites, they will contain the waste for long times, limiting their spread, for example, through water flow. All the repository designs also plan to delay the release of radionuclides to the environment by the construction of engineered barrier systems (EBS) between the waste and the geologic formation. These barriers will be installed to limit water reaching the repository and to restrict radionuclide migration from the waste. The principal engineered component in this multi-barrier approach is the waste package, which includes the waste itself, possibly a stabilizing matrix for the waste such a glass, and a metallic container that encloses the wasteform. Beyond the metallic containers, other secondary barriers could be added to attenuate the impact of the emplacement environment on the containers. The secondary barriers may include a drip shield such as in the U.S. design or backfilling with bentonite such as in the Canadian and other designs (Table 36.1) [2, 3]. A discussion in detail of the characteristics and significance of the engineered barrier systems in all the planned repositories is given by Bennett et al. [4]

More than 30 nations are currently considering the geologic disposal of HLW [3–6]. A short list of these nations is in Table 36.1. Twenty years ago most of the repository designs specified lifetimes from 300 to 1000 years. Currently, the minimum length of time specified for some repositories has increased to 10,000, 100,000, and even 1,000,000 years [7–11]. The viability of extrapolating degradation data from short-term testing to long-time performance has been addressed by some investigators and the American Society for Testing and Materials (ASTM) [7, 12, 13]. Others have

proposed models to predict the lifetime performance of container alloys [14, 15].

B. NATIONAL PROGRAMS

Table 36.1 lists some countries that are currently considering geologic repositories for nuclear waste. One of the most advanced studies for a repository corresponds to the United States, which is planning to locate its nuclear waste at a remote desert site in Nevada [16]. The container for the waste will be a double-walled cylinder having a 2.5-cm-thick layer of alloy N06022 on the outside and a 5-cm-thick layer of nuclear-grade type 316 stainless steel in the inside. After more than two decades of scientific investigations, the U.S. Department of Energy submitted a license application to build the repository on June 3, 2008, and the Nuclear Regulatory Commission accepted this application on September 8, 2008 [11, 17]. After a formal review process of three to four years, construction may start in late 2011, and the first waste emplacement may not occur until 2020 [7, 18]. Currently, the Yucca Mountain repository project does not seem a high priority of the U.S. Department of Energy [18]. The United Kingdom has also recognized that the geologic disposal of the waste in a mined repository is the best available approach [19]. In June 2008 the U.K. government issued the white paper "Managing Radioactive Waste Safely," where a framework for implementing geologic disposal is outlined. The location for the repository will be defined through geologic screening and community engagement. To complement the permanent repository studies, the U.K. Committee for Radioactive Waste Management has also recommended a robust program of interim storage [19]. The Finnish repository will be located in crystalline bedrock at Olkiluoto Island on the western coast of Finland. The waste containers will be made using nodular cast iron with a 50-mm-thick overpack of copper [20]. The repository in Finland should start operations in 2020 and will continue for

TABLE 36.1.   Selected Countries Considering Repositories for High-Level Waste

Country	Possible Environment, Host Rock	Scheduled Start Operations	Container Materials Being Studied
Belgium	Reducing or anoxic, clay	2035	Carbon steel, cement, stainless steel
Canada	Reducing or anoxic, granite	—	Carbon steel insert, copper
Finland	Reducing or anoxic, granite	2020	Cast iron insert, copper
France	Reducing or anoxic, clay	—	Carbon steel
Germany	Salt dome layer	2030	Carbon steel
Japan	Reducing or anoxic, granite plus bentonite buffer	Late 2030s	Carbon steel, titanium
Sweden	Reducing or anoxic, granite plus clay	2023	Cast iron insert, copper
Switzerland	Reducing or anoxic, clay	—	Carbon steel
United States	Oxidizing, non saturated, volcanic tuff	Initially 1998, now 2020	Ni–Cr–Mo Alloy C-22, titanium Gr 7, 28, and 29

approximately 100 years [21]. Sweden has elected a site for its underground repository in the municipality of Östhammar, 500 m below ground in crystalline rock [22, 23]. The waste will be packed in cast iron baskets inside thick copper canisters surrounded by bentonite clay. Each container is a double-walled cylinder of approximately 1 m diameter and 5 m long [22]. The Swedish repository is scheduled to open in 2023 and it is designed to contain the waste for 100,000 years [22]. The Japanese Final Disposal Plan calls for a repository that will start operating in the late 2030s [24]. The final site for the repository has not been selected yet, but two underground research laboratories have been selected, one 1000 m deep in crystalline rock in the presence of fresh water and the second in sedimentary rock 500 m deep in the presence of saline water [25]. In the final Japanese repository the metal containers will be surrounded by bentonite buffer material. The final material for the container has not been selected yet, but it is reported that a thick steel container surrounded by a bentonite buffer overpack would be robust design [24]. The waste disposal for the French nuclear industry has been outlined in the document Dossier 2005 and calls for the commissioning of a disposal facility by the year 2025 [26, 27]. An important concept in the design of the French repository is its reversibility or design evolution at all steps for at least 100 years [26]. The cylindrical containers for the high-level waste (type C in a glass matrix) will be made of standard steel 5 cm thick, 60 cm diameter, and approximately 1.5 m long. It is estimated that this container will remain leak proof for 4000 years [26]. The spent-fuel container may have a wall thickness of over 10 cm and last 10,000 years. Lithuania has recently announced that it will follow the French design for its own repository. Germany is exploring the possibility of a repository in the salt dome at Gorleben probably using steel containers for the nuclear waste. The waste repository should be stable for a million years and the containers should be retrievable for the entire operation time.

C. ENVIRONMENTS AND MATERIALS

Table 36.1 gives a list of countries that are planning to build repositories and the general characteristics of each repository. As mentioned before, each repository consists of a stable geologic formation with the addition of engineered barriers (e.g., container). The repositories can be divided into two large groups according to the nature of the environment, oxidizing (Yucca Mountain, U.S.) and reducing (the other nations). For the reducing (no-oxygen) repositories, the most common host materials are clay, basalt, salt, and granite [2]. The containers are intended to be placed in alcoves located at varying depths below the water table. The depth of emplacement may vary from country to country, but it is generally assumed to be of the order of 500 m. The United States has the

only nonsaturated (above-the-water-table) repository design, with unrestricted access of oxygen [9].

According to the value of the redox potential, the environment of the world repositories can be categorized as reducing or oxidizing. Under reducing conditions, the cathodic reaction is controlled by the hydrogen evolution reaction. On the other hand, oxidizing conditions are characterized by cathodic reactions such as the reduction of dissolved oxygen. Most of the repositories in the world will be reducing based on redox potentials, since they will rely on depth (where the solubility of oxygen in water is minimal) and a projected backfill with bentonite [3]. One of the intended functions of the backfill is to retard the diffusion of any available oxygen toward the containers and the diffusion of the radionuclides away from the containers. The repository in the United States will not have restrictions regarding the availability of oxygen to contact the containers, that is, the redox potential will be oxidizing in nature, provided an aqueous solution materializes. The groundwaters associated with the rock formations should all be relatively benign to most materials because of their low ionic strengths, near-neutral pH, and low concentrations of halide ions [3, 6]. The corrosivity of the groundwaters could increase if significant vaporization occurs when the containers experience higher temperatures during the early emplacement times. The container temperature will be influenced by the design and loading of the waste package, the density of waste package emplacement, and the thermal properties of the surrounding rock. Because heat is a significant byproduct of HLW decay, the temperature of the waste containers will initially increase and then decrease as the activity of the waste decays. The predicted maximum temperature for waste packages emplaced in a consolidated volcanic ash (tuff) formation in the United States is not expected to be higher than 160–200°C [9]. Typical maximum container temperatures for a number of other repository locations are expected to be lower than 100°C [3, 4].

The International Atomic Energy Agency (IAEA) offers guidelines for the minimum requirements such as tests and evaluations that the waste packages (containers) should undergo during the selection process [28]. The degradation mechanisms that need to be studied include corrosion, microbial activity, and radiation damage. Table 36.1 shows that, except for the United States, most of the recommended materials for the containers will be carbon steel, stainless steel, or copper [2, 4]. Since for most of the nations the environment is rather benign or controlled, the alloys selected for the containers are not in the high end of the scale of the corrosion-resistant alloys. For the U.S. containers, some of the most corrosion-resistant materials currently available have been recommended, including the Ni–Cr–Mo alloy 22 and titanium grade 7. The compositions of some of the candidate materials for the engineered barriers being studied worldwide at this moment are given in Table 36.2.

TABLE 36.2. Approximate Chemical Composition (wt %) for Candidate Alloys

Alloy	UNS	ASTM	Cr	Cu	Fe	Mo	Ni	Ti	Other
Gray cast iron	F10001–F10012	A319–A159	—	—	~95 (bal)	—	—	—	3–3.5 C, 2–2.4 Si, 0.8 Mn
1018 Carbon steel	G10180	A29	—	—	~98 (bal)	—	—	—	0.18 C, 0.5 Mn
4130 Alloy steel	G41300	A29	1.0	—	~97 (bal)	0.2	—	—	0.3 C, 0.5 Mn
2.25Cr–1Mo	K30736	A213	2.25	—	bal	1	—	—	0.05 C, 0.4 Mn, 0.2 V
Type 304	S30400	A182	19	—	~70 (bal)	—	9	—	2 max Mn, 1 max Si
Type 316	S31600	A182	17	—	67 (bal)	2.5	12	—	2 max Mn, 1 max Si
Copper	Various	—	—	~99.9					
Monel 400	N04400	B127	—	~32 (bal)	2.5 max	—	66.5	—	2 max Mn
Incoloy 825	N08825	B163	21.5	2.2	~30 (bal)	3.0	42	0.9	1 max Mn, 0.5 max Si
Inconel 625	N06625	B366	21.5	—	5 max	9.0	~60 (bal)	0.2	4 Nb, 0.5 max Mn
Hastelloy C-4	N06455	B575	16	—	3 max	16	~65 (bal)	—	2 max Co
Hastelloy C-22	N06022	B575	22	—	4	13	~57 (bal)	—	3 W, 2.5 max Co
Ti Gr 2	R50400	B265	—	—	0.3 max	—	—	~99 (bal)	0.25 max O
Ti Gr 7	R52400	B265	—	—	0.3 max	—	—	~98 (bal)	0.2 Pd, 0.25 max O
Ti Gr 16	R52402	B265	—	—	0.3 max	—	—	~98 (bal)	0.06 Pd, 0.25 max O
Ti Gr 12	R53400	B265	—	—	0.3 max	0.3	0.8	~98 (bal)	0.25 max O
Ti Gr 29	R56404	B265	—	—	0.25 max	—	—	~90 (bal)	6Al, 4V, 0.08-0.14 Ru

D. DEGRADATION MECHANISMS

The materials that are being characterized for their corrosion resistance behavior as engineered barriers applications are listed in Tables 36.1 and 36.2. It could be anticipated that most of these materials would suffer several types of corrosion processes, which in general can be grouped as (1) general corrosion, (2) localized corrosion, and (3) environmentally assisted cracking or stress corrosion cracking (SCC). One mode of corrosion that has been extensively investigated in several national programs is microbiologically influenced corrosion (MIC). MIC could affect the three corrosion processes mentioned above mainly by changing the environment in the vicinity of the containers. King provided a decision tree approach to determine if MIC would be an important factor in determining the lifetime of the containers [29]. Figure 36.1 shows schematically how metallurgical and environmental factors may control the occurrence of the three main corrosion processes. If water is present in the repository, the container material would adopt a characteristic potential called the rest potential or corrosion potential (E_{corr}). Localized corrosion such as crevice corrosion or pitting corrosion may happen only above a threshold potential or critical potential (E_{crit}). That is, if the material of the container adopts a potential that is below E_{crit}, only general or uniform corrosion may occur. Environmentally assisted cracking (EAC) may occur at any potential; however, for EAC to occur, the simultaneous presence of three conditions is necessary: (1) susceptible material, (2) specific environment, and (3) tensile stresses. If one or more of these conditions is removed, EAC will not take place. Of the three main corrosion processes listed in Figure 36.1, the least

troubling is general corrosion since the propagation rate by uniform thinning of the container would be low and may not be life limiting for the containers. Localized corrosion and EAC may be more detrimental since these processes would tend to perforate the container at a faster rate at discrete locations allowing the ingress of water and the spreading of the radioactive material without substantial corrosion of the overall container wall. The occurrence of localized corrosion and EAC may be minimized by alloy selection, design, and fabrication.

E. REDUCING OR ANOXIC ENVIRONMENTS

The containers in reducing environments will generally be surrounded by a backfill of bentonite, which will greatly limit the availability of oxygen to the metal surface. The lack of oxygen (or other oxidizing species) will create a redox potential that will be closer to the hydrogen evolution reaction. Elements such as iron (Fe), nickel (Ni), and copper (Cu) are mostly in the range of corrosion immunity at these reducing potentials in the near-neutral pH range [3, 30]. The most common materials under study in typically reducing environments are carbon steel, copper, and titanium [2, 3]. For the least corrosive underground waters, carbon steels could be viable materials; however, for the most saline conditions, titanium alloys are also being studied.

E1. Carbon Steel and Low-Alloy Steel

Carbon steels (and to a lesser extent low-alloy steels) have been tested in several countries in groundwater environments

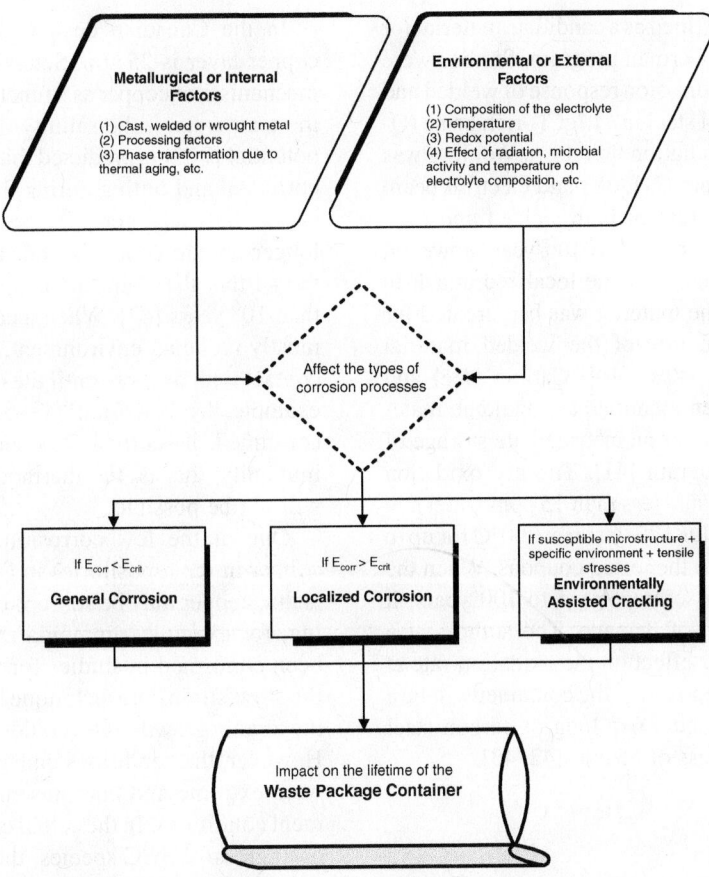

FIGURE 36.1. Schematic representation of the effect of material characteristics plus type of environment on the lifetime performance of the waste package.

for the last 40 years. Carbon steels are rather inexpensive and they tend to suffer general corrosion rather than localized corrosion such as the more expensive stainless steels [31]. Most of the studies find that the corrosion rate of carbon steels and low-alloy steels is low, especially in anoxic conditions. Corrosion rates measured for carbon steels in granitic waters ranged from 1 to 55 μm/year, with one study showing that the rate reached a maximum at around 80°C [32]. The conditions that would lead to localized corrosion of carbon steels are quite specific and unlikely to be present in typical granitic groundwaters [33]. It was proposed that hydrogen embrittlement and hydrogen blistering of carbon steels may be possible if a high rate of hydrogen production exists [33]. However, recent calculations suggested that the failure of buried carbon steel containers by hydrogen damage mechanisms is unlikely [34]. The corrosion behavior of carbon steel was also studied in basaltic water, and it was found that even in oxygenated solutions at 150°C the corrosion rate of all tested carbon steels in basaltic waters was only on the order of 100 μm/year [35].

Under the Swedish program, researchers have studied the anoxic corrosion behavior of carbon steel and cast iron in groundwater at 50 and 85°C and the impact of the presence of copper on the type and mechanical properties of the films formed on the iron alloys [36]. They used a barometric cell filled with a simulated groundwater and monitored the redox potential in the cell at 30°C on a gold electrode. Smart et al. determined that when steel was introduced to the cell, the redox potential decreased rapidly due to the consumption of the residual oxygen by the corrosion of the steel [36].

As part of the Japanese program of nuclear waste disposal, the passive corrosion behavior of steels was found to be dependent on variables such as groundwater pH, temperature, and available dissolved oxygen [37]. Fujiwara et al. have raised the concern that, whenever the corrosion of steel decreases due to a decrease in the oxygen content, the alkalinity in the immediacy of the steel increases. Since higher alkalinity would reduce the free corrosion potential of the steel, the process may increase the rate of hydrogen gas production, which could be detrimental for the stability of the repository [38]. Dong et al. have reported that the corrosion rate of carbon steel is dependent on the amount of bicarbonate (HCO_3^-) present in the water [39]. At bicarbonate levels of $0.1\,M$, similar to the geologic disposal site, the corrosion of carbon steel is inhibited [39].

Carbon steel has been identified as a candidate material for rock salt repositories in the German program. Studies were conducted to determine the corrosion response of welded and nonwelded Fe–1.5 Mn–0.5 Si steel in a $MgCl_2$-rich brine (Q-brine) at 150°C under an irradiation field [40]. Welding was carried out by gas tungsten arc (GTAW) and electron beam (EB). The overall corrosion rate of both welded and nonwelded materials was approximately 70 μm/year; however, the welded materials experienced some localized attack in the weld seam area. When the material was heat treated for 2 h at 600°C, the corrosion rate of the welded material increased by approximately 40% [40]. Carbon steel and low-alloy steel have also been identified as candidate materials to contain nuclear waste for an intermediate storage of 100 years in the French program [41]. The dry oxidation testing of carbon steel in dry air (less than 15 ppm water), in air plus 2% water, and in air plus 12% water at 300°C for up to 700 h showed little damage to the tested coupons. When the depth of the oxide layer was extrapolated to 100 years, it resulted in less than 150 μm of damage. The authors also noted little or no water vapor effect on the oxidation rate at 300°C [41]. For the French repository, the container will be a cylinder 60 cm in diameter and 1.6 m long of carbon steel with an external wall thickness of 55 mm [42, 43].

E2. Copper

The container for the disposition of nuclear waste in Sweden will consist of a 50-mm-thick layer of copper over cast nodular iron, which will provide the mechanical strength. Groundwater in granitic rock (as in the Swedish repository) is oxygen free and reducing below a depth of 200 m. The redox potential is between −200 and −300 mV on the hydrogen scale and the pH ranges from 7 to 9 [8, 44]. The chloride concentration in the groundwater can vary from 0.15 mM to 1.5 M with an equivalent amount of sodium and less calcium. The corrosion of a copper container in this reducing environment is expected to be less than 5 mm in 100,000 years of emplacement [44]. The corrosion of copper is mainly controlled by the availability of oxygen (trapped initially in the pores of the bentonite-based sealing materials), sulfate, and sulfide in the groundwater. The time to failure of the copper layer in the Swedish container has been modeled, and it is predicted that this failure, both by general and pitting corrosion, would be higher than 10^6 years under realistic emplacement conditions [45]. The anodic behavior of copper was also studied as part of the Japanese nuclear waste disposal program using potentiodynamic polarization tests in simulated groundwater at 30°C [46]. The amount of dissolved oxygen as well as different additions of chloride, sulfate, and bicarbonate was controlled. Imai et al. concluded that both sulfate and chloride promote the active dissolution of copper while carbonate is a passivating agent [46].

In the Canadian design, the thickness of the external copper layer is 25 mm. Scientists have modeled the failure mechanism of copper as a function of the oxygen availability, the temperature, the salinity of the solution, and the redox potential [7]. It is predicted that copper will undergo general corrosion and pitting during the initial warm and oxidizing period but only general corrosion during the subsequent longer anoxic cooler period. It has been predicted by this model that the Canadian copper container could last more than 10^6 years [47]. When a copper container is buried in a mostly reducing environment, the metal will initially be in contact with oxygen, until the oxygen is fully consumed, for example, by corrosion [48–50]. When all the oxygen is consumed, its corrosion potential will be in the region of immunity, that is, the thermodynamic oxidation of copper will not be possible.

One of the few corrosion concerns about the use of copper in the repositories in Finland, Sweden, and Canada is that copper may be susceptible to EAC in waters containing, for example, ammonia and nitrite (NO_2^-). This has been confirmed in studies for the Canadian program using the slow-strain-rate technique [51]. It has been reported that the crack growth rate could be as high as 8 nm/s [52]. However, the conditions under which the damage occurred were extreme and unrepresentative of container emplacement conditions. In the actual container, the general absence of aggressive SCC species, the limited applied strain, and the limited supply of oxygen will limit the susceptibility to environmental cracking. In another study, it has been shown that the minimum stress intensity for crack propagation in copper for the Swedish container was 30 MPa$\sqrt{m}$ when tested in a 0.3 M $NaNO_2$ solution [53]. A stress intensity of 30 MPa$\sqrt{m}$ can be considered high for a statically loaded container that may have shallow defects on the surface. That is, the conditions at which copper was cracked in the laboratory are too extreme to be representative of the actual repository conditions.

E3. Stainless Steel and Nickel Alloys

The cyclic potentiodynamic polarization method (ASTM G 61) was used to evaluate the anodic behavior of corrosion-resistant alloys in oxidized Boom clay water (for the repository in Belgium) with varying degrees of added chloride at 90°C [54]. The original Boom clay water is dominated by chloride and sulfate. The alloys studied included 316L SS (also with high Mo and with Ti) (S31603), alloy 926 (N08926), alloy 904L (N08904), alloy C-4 (N06455), and Ti Gr 7 (R52400) (Table 36.2). It was found that both R52400 and N06455 resisted pitting corrosion even at added chloride concentrations of 10,000 ppm and N08926 resisted pitting up to 1000 ppm chloride. The other alloys showed minor pitting at 100 ppm chloride and definite pitting corrosion at the higher tested chloride concentrations [54].

E4. Titanium

Titanium (Ti) alloys have been studied as candidate materials for the containers in Canada, Japan, and Germany. The titanium alloys were selected as a potential alternative because of their excellent performance in more aggressive brine solutions compared, for example, to stainless steels. The corrosion rates for Ti Gr 2 and Ti Gr 12 in both oxygenated and irradiated basalt environments are very low—less than 2 μm/year (0.08 mil/year) [55]. Shoesmith et al. also explained the failure mechanism and a predictive model for the degradation of Ti Gr 2 under the Canadian repository conditions [55]. The model takes into account the crevice propagation rate as a function of temperature and oxygen availability as well as other factors such as the amount of hydrogen absorbed by the alloy during corrosion before a critical concentration for failure is reached. The localized corrosion resistance of titanium alloys has also been investigated extensively as part of the Japanese program [56, 57]. Testing showed that as the temperature and the chloride concentration increased, the repassivation potential ($E_{R, CREV}$) for Ti Gr 1 and Ti Gr 12 decreased to values well below the corrosion potential (E_{corr}) [56]. Ti Gr 12 was more resistant to crevice corrosion than Ti Gr 1. For the other tested alloys, at constant temperature and chloride concentration, $E_{R,CREV}$ increased as the palladium (Pd) content in the alloy increased, rapidly up to 0.008% Pd and then slower between 0.008 and 0.062% Pd [57]. Titanium alloys were also investigated for their resistance to EAC. One way by which titanium alloys may suffer EAC under reducing conditions is by the formation of hydrides due to the slow absorption of hydrogen from the environment. Slow strain rate testing was conducted using Ti Gr 1 in deaerated 20% NaCl at 90°C at an applied potential of − 1.2 V [standard hydrogen electrode (SHE)] [58]. It was confirmed that cracks initiated as deep as the presence of hydrides, that is, the presence of hydride was necessary for cracks to initiate. Based on the critical cracking thickness and the predicted amount of hydrogen generated, the authors dismissed the hypothesis that the titanium-made containers may fail by cathodic EAC [58].

F. OXIDIZING ENVIRONMENTS

The case of the containers for the repository of Yucca Mountain is discussed separately from the other repositories since Yucca Mountain at this time is the only proposed repository that may have an unsaturated environment with unrestricted availability of air (oxygen). The design of the waste package for the Yucca Mountain repository has evolved in the last 15 years [2, 9]. In previous versions of the container, a thick layer of carbon steel was specified for the outer shell of the container and a corrosion-resistant material as the inner shell. However, since 1998, the design of the engineered barriers has not changed significantly, and it currently specifies a double-walled cylindrical container covered by a titanium alloy drip shield. The outer shell of the container will be a Ni–Cr–Mo alloy (N06022) (Table 36.2), with an inner shell of nuclear-grade austenitic type 316 stainless steel (S31600). The function of the outer barrier is to resist corrosion and the function of the inner barrier is to provide mechanical strength and a shield to radiation. The drip shield will be made of Ti Gr 7 and a higher strength Ti alloy (Ti Gr 29) will be used for the internal ribs of the shields. The function of the drip shield is to deflect rock fall and early water seepage on the container [9].

F1. Corrosion Behavior of Alloy 22 (N06022)

The container may suffer corrosion only if water is present in sufficient amount at the repository site. Dry corrosion of alloy 22 is negligible for the emplacement conditions. There are three main modes of corrosion that the container may suffer during its emplacement time (Fig. 36.1): (1) uniform, general, or passive corrosion, (2) localized corrosion (e.g., crevice corrosion), and (3) environmentally assisted cracking (e.g., SCC) [59]. All three types of corrosion may be influenced by the environment, including temperature, solution composition (chloride and nitrate concentration), redox potential, and the presence of microorganisms.

F2. Uniform and Passive Corrosion of Alloy 22

General corrosion (or passive corrosion) is the uniform thinning of the container alloy at its open-circuit potential or corrosion potential (E_{corr}). In the presence of aerated multi-ionic brines, such as those that may be present at the repository site, alloy 22 is expected to remain passive at its E_{corr}. The passive corrosion rates of alloy 22 after 5 years immersion in multi-ionic solutions simulating concentrated groundwaters from pH 2.8 to 10 are extremely low and on the order of 10 nm/year [60–62]. This low corrosion rate was measured at 60 and 90°C for welded and nonwelded alloy 22 at an E_{corr} range from − 100 to + 400 mV SSC (saturated silver chloride) electrode. The low corrosion rates or passive behavior of alloy 22 is because of the formation of a protective inner chromium-rich oxide film between the alloy (metal) and the surrounding electrolyte. This passive film is even stable in the presence of strong mineral acids at temperatures below 60°C [63]. It has been shown that the thickness of this passive film formed in concentrated hot electrolyte solutions could be only in the range of 5–6 nm [64]. The long-term extrapolation of the corrosion rate of alloy 22 has been modeled considering that the dissolution rate is controlled by the injection of oxygen vacancies at the oxide film–solution interface [65]. It has been concluded that it is unlikely that catastrophic failure of

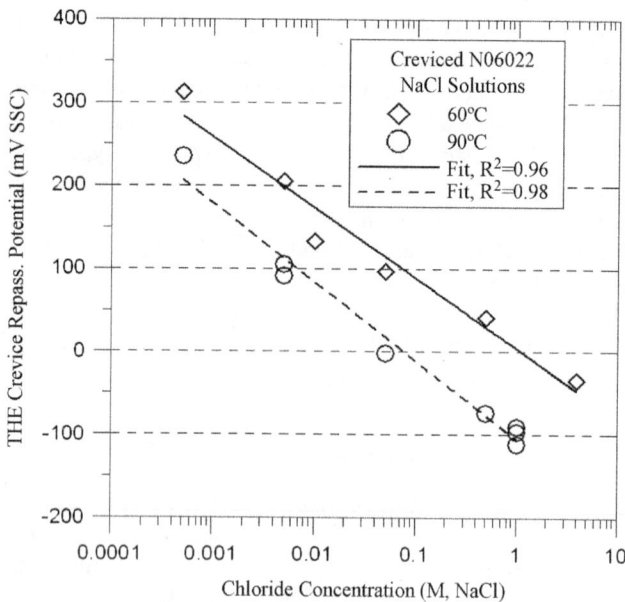

FIGURE 36.2. Effect of temperature and chloride concentration on the crevice repassivation potential of alloy 22 using THE method.

the container may occur due to long-term passive film dissolution [65]. That is, the passive dissolution of alloy 22 is not considered to be the limiting factor for the life performance of the waste container.

F3. Localized Corrosion of Alloy 22

Localized corrosion may be the most detrimental of the degradation modes in Figure 36.1. Localized corrosion (e.g., crevice corrosion) is a type of corrosion in which the attack progresses at discrete sites or in a nonuniform manner. The degradation model assumes that localized corrosion will only occur when E_{corr} is equal or greater than a critical potential (E_{crit}) for localized corrosion [66]. That is, if $E_{corr} < E_{crit}$, only general or passive corrosion will occur. Here, E_{crit} can be defined as a certain potential above which the current density or corrosion rate of alloy 22 increases significantly and irreversibly above the general corrosion rate of the passive metal. The margin of safety against localized corrosion will be given by the value of $\Delta E = E_{crit} - E_{corr}$. The higher the value of ΔE, the larger the margin of safety for localized corrosion. It is important to note here that the values of both E_{corr} and E_{crit} may depend both on the metallurgical condition of the alloy and the environment, such as temperature, chloride concentration, and the presence of inhibitors. The value of E_{corr} is determined by measuring the long-term steady-state value of the open-circuit potential in each environment of relevance and E_{crit} is the crevice repassivation potential measured using electrochemical techniques such as the cyclic potentiodynamic polarization (ASTM G 61) [66].

Localized corrosion was the most extensively studied mechanism of degradation in alloy 22, mainly since 2002 [67, 68]. Alloy 22 is extremely resistant to pitting corrosion but may suffer crevice corrosion, especially in pure chloride solutions and at temperatures higher than 60°C [69, 70]. Figure 36.2 shows the effect of chloride concentration and temperature on the repassivation potential of alloy 22, obtained using the Tsujikawa–Hisamatsu electrochemical (THE) method (ASTM G 192). The higher the temperature and chloride concentration, the lower the resistance of the alloy to localized corrosion [70]. The crevice corrosion susceptibility of alloy 22 is only promoted by the presence of chloride ions, and it can be fully inhibited by the presence of other anions in solution such as nitrate, bicarbonate, sulfate, fluoride, phosphate, and so on [71–73]. The best inhibitor of crevice corrosion in alloy 22 is nitrate [69, 74, 75]. The presence of inhibitor is generally stated using the ratio $R = [\text{inhibitor}]/[\text{Cl}^-]$. The higher is the value of R, the higher is the inhibition effect. Figure 36.3 shows the cyclic potentiodynamic polarization of alloy 22 in three different electrolyte solutions at 110°C. The electrolytes are (1) pure $8\,m$ chloride ($4\,m$ NaCl + $4\,m$ KCl), ($R = 0$) and (2) $8\,m$ chloride with added sodium and potassium nitrate to obtain $R = 0.1$ and $R = 0.5$. Figure 36.3 shows that as R increased, the repassivation potential increased. That is, for $R = 0$, $E_{crit} = -210\,\text{mV SSC}$; for $R = 0.1$, $E_{crit} = -50\,\text{mV SSC}$; and for $R = 0.5$, $E_{crit} = 337\,\text{mV SSC}$. For $R = 0.5$ there was a total inhibition of crevice corrosion (no hysteresis in the reverse scan in Figure 36.3). Since the groundwater at the repository site in Yucca Mountain contains not only chloride but also a variety of anions that may act as inhibitors for crevice

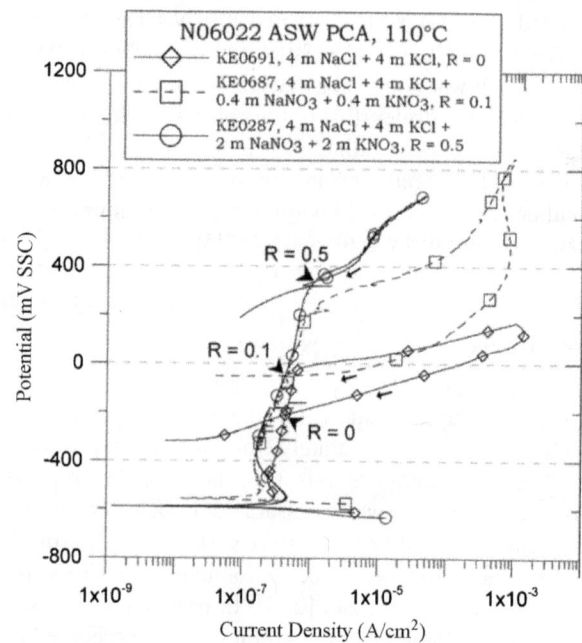

FIGURE 36.3. Effect of nitrate (ratio $R = 0, 0.1, 0.5$) on repassivation potential of alloy 22 in $8\,m$ chloride solution at 110°C.

FIGURE 36.4. Crevice corrosion under a crevice former in alloy 22 under constant applied potential of $+100\,mV$ SSC in $3.5\,m$ NaCl $+ 0.175\,m$ KNO$_3$ solution ($R = 0.05$) at 100°C. Top part of image, interdendritic attack in weld metal; lower part, intergranular attack in base metal.

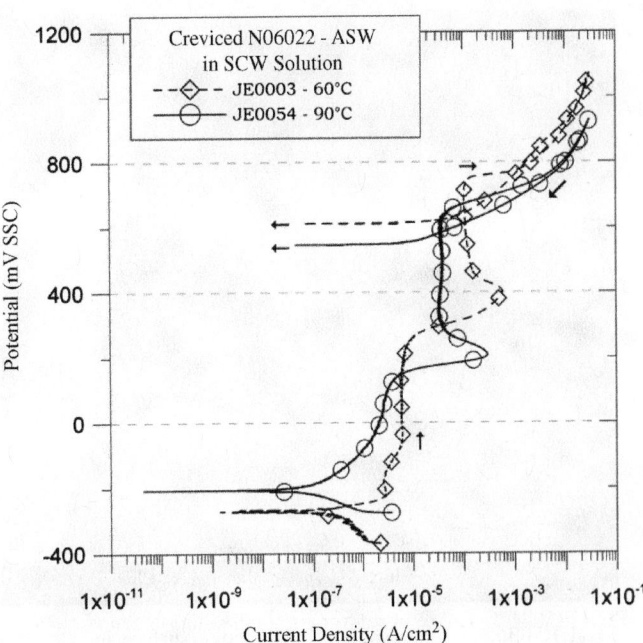

FIGURE 36.5. Anodic polarization curves for alloy 22 in SCW solution at 60 and 90°C, where an anodic peak is observed in the middle on the passive range of potentials.

corrosion, it is unlikely that alloy 22 would suffer crevice corrosion under natural emplacement conditions [76].

Constant potential laboratory tests have shown that crevice corrosion in alloy 22 often repassivates after initiation due to a stifling mechanism [77]. Figure 36.4 shows the crevice corroded area under one crevice former tooth (ASTM G 192) after a week-long test in $3.5\,m$ NaCl $+ 0.175\,m$ KNO$_3$ solution ($R = 0.05$) at 100°C at a constant applied potential of $+100\,mV$ SSC. The test was conducted at approximately 200 mV higher than the crevice repassivation potential of alloy 22 ($-110\,mV$ SSC) in the same conditions [78]. Current measurements showed that crevice corrosion nucleated at 10 min after the potential was applied and it progressed with increasing anodic currents for the next 14 h, after which the anodic current started to decrease, becoming cathodic at hour 79 even though the potential was maintained at $+100\,mV$ SSC during the entire test [78]. Figure 36.4 also shows that crevice corrosion occurred under the entire footprint of the crevice former and that the depth of attack was shallow and even for both the base metal (lower part of the image) and the weld metal (upper part of the image). In the weld metal the attack was interdendritic and in the base metal the attack was intergranular [78].

F4. Environmentally Assisted Cracking of Alloy 22

Wrought mill annealed (MA) alloy 22 is highly resistant to EAC in most environments, including acidic concentrated

and hot chloride solutions. Welded and nonwelded U-bend specimens of alloy 22 and five other nickel-based alloys exposed for more than five years to multi-ionic solutions that represent concentrated groundwater of pH 2.8 to 10 at 60 and 90°C were free from EAC [79, 80]. Even though alloy 22 is resistant to EAC in concentrated hot chloride solutions, it may be susceptible under other severe environmental conditions. Slow strain rate tests were performed using MA alloy 22 specimens in simulated concentrated water (SCW) and other solutions as a function of the temperature and applied potential [81, 82]. SCW has a pH 8–10, and it is approximately 1000 times more concentrated than groundwater. Alloy 22 was found susceptible to EAC in hot SCW solutions and bicarbonate plus chloride solutions at anodic applied potentials approximately 300–400 mV more positive than E_{corr}. The occurrence of EAC was related to the presence of an anodic peak in the polarization curve of the alloy in SCW environments. For example, at ambient temperatures, the peak is not present and EAC does not take place [81]. Currently, the origin of the anodic peak is being investigated [83]. Figure 36.5 shows the polarization curves for alloy 22 in SCW solution, showing the presence of the anodic peak at 60 and 90°C. It was demonstrated that the most aggressive species for EAC in SCW was bicarbonate but that the presence of chloride in the bicarbonate solution enhances the aggressiveness of the environment [82]. Figure 36.6 shows typical transgranular cracking in alloy 22 tested in the laboratory using the slow strain rate technique under anodic polarization. The conditions at which SCC was found

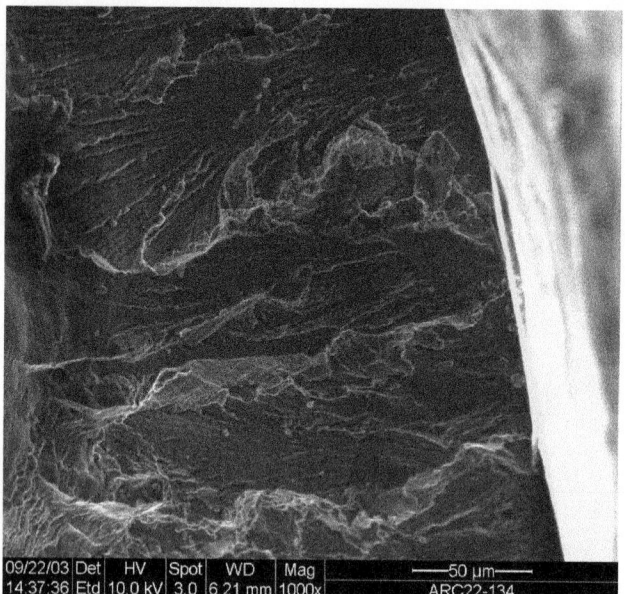

FIGURE 36.6. Transgranular EAC fracture surface in alloy 22 after a slow strain rate test at $1.67 \times 10^{-6} \, s^{-1}$ at an applied potential of $+400 \, mV$ SSC in SCW at 65°C.

in the laboratory in alloy 22 are unrealistic for the natural conditions of the emplacement site in Yucca Mountain.

F5. Corrosion Behavior of Titanium Alloys

Titanium grade 7 (Ti Gr 7 or R52400) was selected to fabricate the detached drip shield for the repository in Yucca Mountain [9]. Other Ti alloys of higher strength such as Ti Gr 29 may also be used for the structural parts of the drip shield. The presence of the drip shield would deflect early water seepage from the containers. This drip shield would also deflect rock fall from the containers. Ti Gr 7 belongs to a family of Ti alloys especially designed to withstand aggressive chemical environments (Table 36.2) [84]. The superior corrosion resistance of Ti and Ti alloys is due to a thin, stable, and tenacious oxide film that forms rapidly in air and water, especially under oxidizing conditions. A detailed review of the general, localized, and environmentally assisted cracking behavior of Ti Gr 7 and other titanium alloys relevant to the application in Yucca Mountain has addressed, among other topics, the effect of alloyed palladium, the properties of the passive films, and the effect of radiation [85]. The presence of fluoride in the groundwater may render Ti Gr 7 more susceptible to general and crevice corrosion under anodic polarization [86].

Weight loss, creviced, and U-bend specimens of Ti Gr 7, 12, and 16 were exposed to three different aerated electrolyte solutions simulating concentrated groundwater for over five years at both 60 and at 90°C in the vapor and liquid phases of these solutions [87]. Ti Gr 7 generally exhibited the lowest corrosion rates irrespective of temperature or solution type

while Ti Gr 12 generally exhibited the highest corrosion rates. Titanium and Ti alloys may be susceptible to EAC, such as hydrogen embrittlement (HE). Embrittlement by hydrogen is a consequence of absorption of atomic hydrogen by the metal to form hydrides. This may happen in service when the Ti alloy is coupled to a more active metal in an acidic solution. A critical concentration of hydrogen in the metal may be needed for HE to occur. Stress corrosion cracking was reported in Ti Gr 7 specimens subjected to constant-load tests in a concentrated groundwater solution pH ~ 10 at 105°C [88]. Results from up to five-year immersion testing at 60 and 90°C of U-bend specimens made of wrought and welded Ti Gr 7 and Ti Gr 16 alloys showed that these alloys were free from EAC in multi-ionic solutions that could be representative of concentrated groundwater [89]. Welded Ti Gr 12 U-bend specimens suffered EAC in SCW liquid at 90°C. Under the same conditions, nonwelded Ti Gr 12 was free from cracking [80, 89].

G. SUMMARY AND CONCLUSIONS

1. The consensus around the world is that high-level nuclear waste should be deposited in stable geologic repositories and several countries are currently developing them.

2. Most of the repositories in the world are planned to be in stable rock formations (e.g., granite) below the water table (saturated). The United States is studying a repository above the water table (unsaturated) and with unrestricted access of oxygen.

3. The repositories will consist of a stable geologic formation within which engineered barriers will be constructed. The most important part of the engineered barrier is the container for the waste

4. The containers are, in general, designed as double-walled metallic cylinders. Some carbon steel containers may be single walled.

5. From the corrosion point of view, most repositories will have reducing or anoxic environments. The U.S. repository will have a natural oxidizing environment.

6. Copper, titanium, and carbon steels were determined to be suitable materials for the reducing or anoxic repositories.

7. High-end materials such as alloy 22 and Ti Gr 7 are being characterized for the mostly dry and oxidizing environment of the U.S. repository.

8. Materials for the engineered barriers are being evaluated for general corrosion, localized corrosion, and environmentally assisted cracking resistance. General corrosion is not generally an important factor that determines the lifetime performance of the containers. Environmentally assisted cracking and localized corrosion are more detrimental and less predictable.

9. The candidate materials for the containers are studied under different metallurgical conditions, such as mill annealed, welded, and thermally aged.

10. Some of the most important environmental variables that may affect the corrosion behavior of the engineering materials include the concentration and type of the aqueous electrolytes at the site, temperature, and redox potential established.

REFERENCES

1. International Atomic Energy Agency, www.iaea.org.

2. R. B. Rebak and R. D. McCright, "Corrosion of Containment Materials for Radioactive Waste Isolation," in ASM Handbook, Vol. 13C, *Corrosion: Environments and Industry*, ASM International, Metals Park, OH, 2006, pp. 421–437.

3. D. W. Shoesmith, *Corrosion*, 62, 703 (2006).

4. D. G. Bennett, A. J. Hooper, S. Voinis, and H. Umeki, "The Role of the Engineered Barrier System in Safety Cases for Geological Radioactive Waste Repositories: A NEA Initiative in Co-operation with the EC," in Proceedings from the XXIX Symposium on Scientific Basis for Nuclear Waste Management, Vol. 932, Materials Research Society, Warrendale, PA, 2006, pp. 43–52.

5. Nuclear Energy Agency, "Engineered Barrier Systems and the Safety of Deep Geological Repositories," Organisation for Economic Co-Operation and Development OECD Publications, Paris, France, 2003.

6. P. A. Witherspoon and G. S. Bodvarsson, "Geological Challenges in Radioactive Waste Isolation—Third Worldwide Review," Report LBNL-49767, Lawrence Berkeley Laboratory, Berkeley, CA, 2001.

7. D. W. Shoesmith, B. M. Ikeda, F. King, and S. Sunder, "Prediction of Long Term Behavior for Radioactive Nuclear Waste Disposal," in Research Topical Symposia—Life Prediction of Structures Subject to Environmental Degradation, NACE International, Houston, TX, 1996, p. 101.

8. L. O. Werme, "Fabrication and Testing of Copper Canister for Long Term Isolation of Spent Nuclear Fuel," Vol. 608, Materials Research Society, Warrendale, PA, 2000, p. 77.

9. G. M. Gordon, *Corrosion*, 58, 811 (2002).

10. R. L. Clark, "Proposed Amendments to the Environmental Radiation Protection Standards for Yucca Mountain, Nevada" in 11th International High Level Radioactive Waste Management Conference Proceedings, Las Vegas, NV, April 30–May 4, 2006, American Nuclear Society, La Grange Park, IL, 2006, pp. 1124–1130.

11. P. N. Swift, K. Knowles, J. McNeish, C. W. Hansen, R. L. Howard, R. MacKinnon, and S. D. Sevougian, "Long-Term Performance of the Proposed Yucca Mountain Repository, USA," in Proceedings from the XXXII Symposium on Scientific Basis for Nuclear Waste Management, Vol. 1124, Materials Research Society, Warrendale, PA, 2009, pp. 3–14.

12. A. A. Sagüés, "Nuclear Waste Package Corrosion Behavior in the Proposed Yucca Mountain Repository," in Scientific Basis for Nuclear Waste Management XXII, Vol. 556, Materials Research Society, Warrendale, PA, 1999, p. 845.

13. "Standard Practice for Prediction of the Long-Term Behavior of Materials, Including Waste Forms, Used in Engineered Barrier Systems (EBS) for Geological Disposal of High-Level Radioactive Waste," ASTM C1174-07, ASTM International, West Conshohocken, PA, 2007.

14. D. D. Macdonald, M. Urquidi-Macdonald and J. Lolcma, J., "Deterministic Predictions of Corrosion Damage to High Level Nuclear Waste Canisters," ASTM Special Technical Publication, ASTM International, West Conshohocken, PA, 1994, p. 143.

15. T. Shibata, "Statistical and Stochastic Aspects of Corrosion Life Predictions," in A Compilation of Special Topic Reports, compiled and edited by F. M. G. Wong and J. H. Payer, Waste Package Materials Performance Peer Review, 2002, May 31, p. 9–1.

16. The Yucca Mountain Project, www.ocrwm.doe.gov.

17. D. J. Duquette, R. M Latanision, C. A. W. Di Bella, and B. E. Kirstein, "Corrosion Issues Related to Disposal of High-Level Nuclear Waste in the Yucca Mountain Repository," in Proceedings from the XXXII Symposium on Scientific Basis for Nuclear Waste Management, Vol. 1124, Materials Research Society, Warrendale, PA, 2009, pp. 15–28.

18. N. J. Zacha, "Yucca Mountain: Dumped and Wasted?" in Radwaste Solutions, The American Nuclear Society, July/Aug. La Grange Park, IL, 2009, pp. 12–18.

19. N. C. Hyatt, S. R. Biggs, F. R. Livens, and J. C. Young, "DIAMOND: Academic Innovation in Support of UK Radioactive Waste Management," in Proceedings from the XXXII Symposium on Scientific Basis for Nuclear Waste Management, Vol. 1124, Materials Research Society, Warrendale, PA, 2009, pp. 77–82.

20. J. Vira, "Further Steps Towards Licensing: Underground Characterisation Started for Spent Fuel Repository in Findland," in Proceedings from the XXIX Symposium on Scientific Basis for Nuclear Waste Management, Vol. 932, Materials Research Society, Warrendale, PA, 2006, pp. 3–12.

21. Nuclear Waste Management in Finland, www.posiva.fi.

22. Radwaste Solutions, The American Nuclear Society, July/Aug. 2009, p. 6.

23. SKB, Technical Report TR-07-12, Sept. 2007, www.skb.se.

24. Nuclear Waste Management Organization of Japan—NUMO, www.numo.or.jp.

25. H. Umeki, K. Shimizu, T. Seo, A. Kitamura, and H. Ishikawa, "The JNC Generic URL Research Program—Providing a Knowledge Base to Support Both Implementer and Regulator in Japan," in Proceedings from the XXIX Symposium on Scientific Basis for Nuclear Waste Management, Vol. 932, Materials Research Society, Warrendale, PA, 2006, pp. 13–22.

26. National Radioactive Waste Management Agency France, www.andra.fr.

27. F. Plas and J. Wendling, "The Geological Research in France—The Dossier 2005 Argile," in Proceedings from the XXX

Symposium on Scientific Basis for Nuclear Waste Management, Vol. 985, Materials Research Society, Warrendale, PA, 2007, pp. 493–504.

28. "Development of Specifications for Radioactive Waste Packages," Report IAEA-TECDOC-1515, International Atomic Energy Agency, Vienna, Austria, Oct. 2006.

29. F. King, "Microbiologically Influenced Corrosion of Nuclear Waste Containers," Corrosion, 65(3), 233–251 (2009).

30. M. Pourbaix, "Atlas of Electrochemical Equilibria in Aqueous Solutions," National Association of Corrosion Engineers, Houston, TX, 1974.

31. N. R. Smart, "Corrosion Behavior of Carbon Steel Radioactive Waste Packages: A Summary Review of Swedish and U.K. Research," Corrosion, 65(3), 195–212 (2009).

32. J. P. Simpson, R. Schenk, and B. Knecht, "Corrosion Rate of Unalloyed Steels and Cast Irons in Reducing Granitic Groundwaters and Chloride Solutions," in Scientific Basis for Nuclear Waste Management IX, Vol. 50, Materials Research Society, Warrendale, PA, 1986, p. 429.

33. G. P. Marsh, K. J. Taylor, I. D. Bland, C. Westcott, P. W. Tasker, and S. M. Sharland, "Evaluation of the Localized Corrosion of Carbon Steel Overpacks for Nuclear Waste Disposal in Granite Environments," in Scientific Basis for Nuclear Waste Management IX, Vol. 50, Materials Research Society, Warrendale, PA, 1986, p. 421.

34. A. Turnbull, "A Review of the Possible Effects of Hydrogen on Lifetime of Carbon Steel Nuclear Waste Canisters," NAGRA Technical Report 09-04, Nagra, Wettingen, Switzerland, 2009, July 2009.

35. R. P. Anantatmula, C. H. Delegard, C. H., and R. L. Fish, Corrosion Behavior of Low-Carbon Steels in Grande Ronde Basalt Groundwater in the Presence of Basalt-Bentonite Packing, Vol. 26, Materials Research Society, Warrendale, PA, 1984, pp. 113–120.

36. N. R. Smart, P. A. H. Fennell, R. Peat, K. Spahiu, and L. Werme, "Electrochemical Measurements during the Anaerobic Corrosion of Steel," in Scientific Basis for Nuclear Waste Management XXIV, Vol. 663, Materials Research Society, Warrendale, PA, 2001, pp. 477 and 487.

37. Y. Fukaya and M. Akashi, "Passivation Behavior of Mild Steel Used for Nuclear Waste Disposal Package," in Scientific Basis for Nuclear Waste Management XXII, Vol. 556, Materials Research Society, Warrendale, PA, 1999, p. 871.

38. A. Fujiwara, I. Yasutomi, K. Fukudome, T. Tateishi, and K. Fujiwara, "Influence of Oxygen Concentration and Alkalinity on the Hydrogen Gas Generation by Corrosion of Carbon Steel," Vol. 663, Materials Research Society, Warrendale, PA, 2001, pp. 497–505.

39. J. Dong, T. Nishimura, and T. Kodama, "Corrosion Behavior of Carbon Steel in Bicarbonate Solutions," Vol. 713, Materials Research Society, Warrendale, PA, 2002, p. 105.

40. E. Smailos, "Influence of Welding and Heat Treatment on Corrosion of the Candidate High-Level Waste Container Material Carbon Steel in Disposal Relevant Salt Brines," Paper 00194, in Corrosion/2000, NACE International, Houston, TX, 2000.

41. A. Terlain, C. Desgranges, D. Gauvain, D. Féron, A. Galtayries, and P. Marcus, "Oxidation of Materials for Nuclear Waste Containers Under Long Term Disposal," Paper 01119, in Corrosion/2001, NACE International, Houston, TX, 2001.

42. M. Hélie, "A Review of 25 Yeast of Corrosion Studies on HLW Container Materials at the CEA," Paper NN8.5 in Scientific Basis for Nuclear Waste Management XXX, Vol. 985, Materials Research Society, Warrendale, PA, 2006.

43. D. Féron, D. Crusset, and J.-M. Gras, "Corrosion Issues in the French High-Level Nuclear Waste Program," Corrosion, 65(3), 213–223 (2009).

44. J. Smith, Z. Qin, D. W. Shoesmith, F. King, and L. Werme, "Corrosion of Copper Nuclear Waste Containers in Aqueous Sulphide Solutions," Paper CC1.12.1 in Scientific Basis for Nuclear Waste Management XXIII, Vol. 824, Materials Research Society, Warrendale, PA, 2004.

45. K. Worgan, M. Apted, and R. Sjöblom, "Performance Analysis of Copper Canister Corrosion under Oxidizing and Reducing Conditions," in Scientific Basis for Nuclear Waste Management XVIII, Vol. 353, Materials Research Society, Warrendale, PA, 1995, p. 695.

46. H. Imai, T. Fukuda, and M. Akashi, "Effects of Anionic Species on the Polarization Behavior of Copper for Waste Package Material in Artificial Ground Water," in Scientific Basis for Nuclear Waste Management XIX, Vol. 412, Materials Research Society, Warrendale, PA, 1996, pp. 589–596.

47. F. King, D. M. LeNeveau, and D. J. Jobe, "Modelling the Effects of Evolving Redox Conditions on the Corrosion of Copper Containers," in Scientific Basis for Nuclear Waste Management XVII, Vol. 333, Materials Research Society, Warrendale, PA, 1994, p. 901.

48. F. King, M. J. Quinn, C. D. Litke, and D. M. LeNeveu, Corros. Sci, 37, 833 (1995).

49. F. King and M. Kolář, "A Numerical Model for the Corrosion of Copper Nuclear Fuel Waste Containers," in Scientific Basis for Nuclear Waste Management XIX, Vol. 412, Materials Research Society, Warrendale, PA, 1996, pp. 555–562.

50. A. Honda, N. Taniguchi, H. Ishikawa, and M. Kawasaki, "A Modeling Study of General Corrosion of Copper Overpack for Geological Isolation of High-Level Radioactive Waste," in Scientific Basis for Nuclear Waste Management XXII, Vol. 556, Materials Research Society, Warrendale, PA, 1999, p. 911.

51. F. King, C. D. Litke, and B. M. Ikeda, "The Stress Corrosion Cracking of Copper Nuclear Waste Containers," in Scientific Basis for Nuclear Waste Management XXII, Vol. 556, Materials Research Society, Warrendale, PA, 1999, p. 887.

52. F. King, C. D. Litke, and B. M. Ikeda, "The Stress Corrosion Cracking of Copper Containers for the Disposal of High-Level Nuclear Waste," Paper 99482, in Corrosion/99, NACE International, Houston, TX, 1999.

53. K. Petterson and M. Oskarsson, "Stress Corrosion Crack Growth in Copper for Waste Canister Applications," in Scientific Basis for Nuclear Waste Management XXIII, Vol. 608, Materials Research Society, Warrendale, PA, 2000, p. 95.

54. F. Druyts and B. Kursten, "Influence of Chloride Ions on the Pitting Corrosion of Candidate HLW Overpack Materials in

Synthetic Oxidized Boom Clay Water," Paper 99472, in Corrosion/99, NACE International, Houston, TX, 1999.

55. D. W. Shoesmith and B. M. Ikeda, "Development of Modeling Criteria for Prediction Lifetimes of Titanium Nuclear Waste Containers," in Scientific Basis for Nuclear Waste Management XVII, Vol. 333, Materials Research Society, Warrendale, PA, 1994, p. 893.

56. M. Akashi, G. Nakayama, and T. Fukuda, "Initiation Criteria for Crevice Corrosion of Titanium Alloys Used for HLW Disposal Overpack," Paper 98158, in Corrosion/98, NACE International, Houston, TX, 1999.

57. G. Nakayama, K. Murakami, and M. Akashi, "Assessment of Crevice Corrosion and Hydrogen Induced Stress Corrosion Cracks of Ti-Pd Alloys for HLW Overpack in Deep Underground Water Environments," in Scientific Basis for Nuclear Waste Management XXVI, Vol. 757, Materials Research Society, Warrendale, PA, 2003, pp. 771–778.

58. N. Nakamura, M. Akashi, Y. Fukaya, G. Nakayama, and H. Ueda, "Stress-Corrosion Crack Initiation Behavior in a-Titanium Used for Nuclear Waste Disposal Overpack," Paper 00195, in Corrosion/2000, NACE International, Houston, TX, 2000.

59. R. B. Rebak and J. C. Estill, "Review of Corrosion Modes for Alloy 22 Regarding Lifetime Expectancy of Nuclear Waste Containers," in Scientific Basis for Nuclear Waste Management XXVI, Vol. 757, Materials Research Society, Warrendale, PA, 2003, pp. 713–721.

60. L. L. Wong, D. V. Fix, J. C. Estill, R. D. McCright, and R. B. Rebak, Characterization of the Corrosion Behavior of Alloy 22 after Five Years Immersion in Multi-Ionic Solutions, Vol. 757, Materials Research Society, Warrendale, PA, 2003, pp. 735–741.

61. J. H. Lee and H. A. Elayat, "A Probabilistic Assessment Model for General Corrosion of Alloy 22 for High Level Nuclear Waste Disposal Container," Paper 04699, in Corrosion/2004, NACE International, Houston, TX, 2004.

62. L. L. Wong, T. Lian, D. V. Fix, M. Sutton, and R. B. Rebak, "Surface Analysis of Alloy 22 Coupons Exposed for Five Years to Concentrated Ground Waters," Paper 04701, in Corrosion/ 2004, NACE International, Houston, TX, 2004.

63. R. B. Rebak and J. H. Payer, "Passive Corrosion Behavior of Alloy 22," in 11th International High Level Radioactive Waste Management Conference Proceedings, Las Vegas, NV, April 30–May 4, 2006, American Nuclear Society, La Grange Park, IL, 2006, p. 493.

64. Y.-J. Kim, P. L. Andresen, P. J. Martiniano, J. Chera, M. Larsen, and G. M. Gordon, "Passivity of Nuclear Waste Canister Candidate Materials in Mixed-Salt Environments," Paper 02544, in Corrosion/2002, NACE International, Houston, TX, 2002.

65. O. Pensado, D. S. Dunn, and G. A. Cragnolino, "Long-Term Extrapolation of Passive Behavior of Alloy 22" in Scientific Basis for Nuclear Waste Management XXVI, Vol. 757, Materials Research Society, Warrendale, PA, 2003, pp 723–728.

66. J. H. Lee, T. Summers, and R. B. Rebak, "A Performance Assessment Model for Localized Corrosion Susceptibility of Alloy 22 in Chloride Containing Brines for High Level Nuclear Waste Disposal Container," Paper 04692, in Corrosion/2004, NACE International, Houston, TX, 2004.

67. R. B. Rebak, "Factors Affecting the Crevice Corrosion Susceptibility of Alloy 22," Paper 05610, in Corrosion/2005, NACE International, Houston, TX, 2005.

68. R. M. Carranza, "The Crevice Corrosion of Alloy 22 in the Yucca Mountain Nuclear Waste Repository," JOM, Jan. 58–65 (2008).

69. R. B. Rebak, "Mechanisms of Inhibition of Crevice Corrosion in Alloy 22," Paper NN8.4, in Scientific Basis for Nuclear Waste Management XXX, Vol. 985, Materials Research Society, Warrendale, PA, 2006.

70. K. J. Evans and R. B. Rebak, "Measuring the Repassivation Potential of Alloy 22 Using the Potentiodynamic-Galvanostatic-Potentiostatic Method," Paper ID JAI101230, J. ASTM Int., 4(9)(2007).

71. G. O. Ilevbare, K. J. King, S. R. Gordon, H. A. Elayat, G. E. Gdowski, and T. S. E. Gdowski, J. Electrochem. Soc., 152(12), B547–B554 (2005).

72. D. S. Dunn, L. Yang, C. Wu, and G. A. Cragnolino, Material Research Society Symposium, Spring 2004, San Francisco, Proc. Vol. 824, Materials Research Society, Warrendale, PA, 2004.

73. D. S. Dunn, Y.-M. Pan, K. Chiang, L. Yang, G. A. Cragnolino, and X. He, "Localized Corrosion Resistance and Mechanical Properties of Alloy 22 Waste Package Outer Containers" JOM, Jan. 49–55 (2005).

74. T. Lian, G. E. Gdowski, P. D. Hailey, and R. B. Rebak, Corrosion, 64, 613–623 (2008).

75. A. K. Mishra and G. S. Frankel, "Crevice Corrosion Repassivation of Alloy 22 in Aggressive Environments," Corrosion, 64 (11), 836–844 (2008).

76. R. M. Carranza and R. B. Rebak, "Anionic and Cationic Effects on the Crevice Corrosion Susceptibility of Alloy 22," in Proceedings of the Materials Research Society (MRS) Scientific Basis for Nuclear Waste Management XXXIII, St. Petersburg, Russia, May 24–29, 2009 Vol. 1193 © 2009 MRS.

77. K. G. Mon, G. M. Gordon, and R. B. Rebak, "Stifling of Crevice Corrosion in Alloy 22," in Proceedings of the 12th International Conference on Environmental Degradation of Materials in Nuclear Power System—Water Reactors, T. R. Allen, P. J. King, and L. Nelson (Eds.), The Minerals, Metals & Materials Society, Warrendale, PA, 2005, pp. 1431–1438.

78. K. G. Mon, P. Pasupathi, A. Yilmaz, and R. B. Rebak, "Stifling of Crevice Corrosion in Alloy 22 During Constant Potential Tests," Paper PVP2005-71174, in Proceedings of the 2005 ASME Pressure Vessels and Piping Division Conference, Denver, CO, July 17–21, 2005, Vol. 7, American Society of Mechanical Engineers, New York, 2005, pp. 493–502.

79. D. V. Fix, J. C. Estill, G. A. Hust, L. L. Wong, and R. B. Rebak, "Environmentally Assisted Cracking Behavior of Nickel Alloys in Simulated Acidic and Alkaline Waters Using U-bend Specimens," Paper 04549, in Corrosion/2004, NACE International, Houston, TX, 2004.

80. R. B. Rebak, "Corrosion Testing of Nickel and Titanium Alloys for Nuclear Waste Disposition," Corrosion, 65(4), 252–271 (2009).

81. K. J. King, L. L. Wong, J. C. Estill, and R. B. Rebak, "Slow Strain Rate Testing of Alloy 22 in Simulated Concentrated Ground Waters," Paper 04548, in Corrosion/2004, NACE International, Houston, TX, 2004.

82. K. T. Chiang, D. S. Dunn, and G. A. Cragnolino, "The Combined Effect of Bicarbonate and Chloride Ions on the Stress Corrosion Cracking Susceptibility of Alloy 22," Paper 06506, in Corrosion/2006, NACE International, Houston, TX, 2006.

83. N. S. Zadorozne, M. A. Rodriguez, R. M. Carranza, and R. B. Rebak (to be published).

84. R. W. Schutz, "Platinum Group Metal Additions to Titanium: A Highly Effective Strategy for Enhancing Corrosion Resistance," Corrosion, 59, 1043 (2003).

85. F. Hua, K. Mon, P. Pasupathi, G. M. Gordon, and D. W. Shoesmith, "Corrosion of Ti Grade 7 and Other Ti Alloys in Nuclear Waste Repository Environments—A Review," Paper 04698, in Corrosion/2004, NACE International, Houston, TX, 2004.

86. C. S. Brossia and G. A. Cragnolino, "Effects of Environmental and Metallurgical Conditions on the Passive and Localized Dissolution of Ti-0.15%Pd," Corrosion, 57, 768 (2001).

87. L. L. Wong, J. C. Estill, D. V. Fix, and R. B. Rebak, "Corrosion Characteristics of Titanium Alloys in Multi-Ionic Environments," in PVP, Vol. 467, American Society of Mechanical Engineers, New York, 2003, p. 63.

88. L. M. Young, G. M. Catlin, G. M. Gordon, and P. L. Andresen, "Constant Load SCC Initiation Response of Alloy 22 (UNS N06022), Titanium Grade 7 and Stainless Steels at 105°C," Paper 03685, in Corrosion/2003, NACE International, Houston, TX, 2003.

89. D. V. Fix, J. C. Estill, L. L. Wong, and R. B. Rebak, "Susceptibility of Welded and Non-Welded Titanium Alloys to Environmentally Assisted Cracking in Simulated Concentrated Ground Waters," Paper 04551, in Corrosion/2004, NACE International, Houston, TX, 2004.

37

CORROSION BEHAVIOR OF ELECTRODEPOSITED NANOCRYSTALS

U. Erb

Department of Materials Science and Engineering, University of Toronto, Toronto, Ontario, Canada

A. INTRODUCTION

This chapter deals with the corrosion behavior of nanocrystalline materials made by electrodeposition. The synthesis method, structure of nanodeposits, and some mechanical and physical properties of these materials will be briefly described. This will be followed by a review of advances made over the past 20 years in the understanding of the corrosion properties of nanocrystalline nickel, cobalt, zinc, and copper and some of their alloys. It will be shown that, contrary to earlier concerns, the high concentrations of intercrystalline defects in nanocrystalline metals do not compromise their corrosion resistance.

Over the past 30 years, numerous synthesis methods have been developed to make nanocrystalline materials in different shapes and forms. From a processing point of view, five basic approaches are used to achieve microstructural refinement down to the 1–100 nm range: vapor-phase processing, liquid-phase processing, solid-state processing, chemical synthesis, and electrochemical synthesis. For each approach there are a variety of individual methods that can make very different or similar materials (Table 37.1). In terms of microstructures, nanomaterials can be classified according to their dimensional structure modulation. In one particular structure scheme [1], zero-dimensional nanomaterials are individual clusters, particles, or fibres with any aspect ratio. One-dimensional nanomaterials are layered (often epitaxially grown) structures with layer thicknesses less than 100 nm. Two-dimensional nanomaterials are thin layers with a grain size less than 100 nm, while three-dimensional nanomaterials are bulk materials with average grain size less than 100 nm.

For bulk three-dimensional nanomaterials there are four basic structure types, depending on the synthesis approach. For materials that are made by the consolidation of precursor nanoparticles (e.g., produced by inert gas condensation or ball milling), the main microstructural defects are grain boundaries (between the consolidated particles) and residual porosity (interparticle voids). In materials made by crystallization from amorphous precursors there are typically several crystalline phases plus, in many cases, a residual amorphous phase separating individual crystals. Nanomaterials made by methods such as severe plastic deformation contain mainly grain boundaries and high densities of dislocations. Finally, in nanomaterials made by techniques such as electrodeposition, the main structural defects are grain boundaries and triple junctions (the lines between three adjoining crystals).

In view of the large number of different synthesis techniques for nanomaterials, their complexity in terms of dimensional structure modulations and the various structure types even within one single group of bulk nanomaterials, it is perhaps not surprising that a comprehensive treatise on the corrosion properties of these exciting new materials is

Uhlig's Corrosion Handbook, Third Edition, Edited by R. Winston Revie
Copyright © 2011 John Wiley & Sons, Inc.

TABLE 37.1. Five Major Processing Routes for Making Nanomaterials with Specific Examples

Processing Route	Specific Examples
Vapor-phase processing	Physical vapour deposition
	Chemical vapour deposition
	Inert gas condensation
Liquid–phase processing	Rapid solidification
	Atomization
	Sonication of immiscible liquids
Solid–state processing	Annealing of amorphous materials
	Mechanical attrition
	Equal channel angular processing
Chemical synthesis	Sol–gel processing
	Precipitation
	Inverse micelle technology
Electrochemical synthesis	Electrodeposition
	Galvanic displacement
	Electroless plating

currently not available. While some earlier studies of sputter-deposited nanocrystalline 304 stainless steel–type Ni–Fe–Cr films showed very promising results in improving the corrosion resistance in 0.3 wt % NaCl solution compared with their conventional polycrystalline counterparts [2], another study using magnetron-sputtered nanocrystalline Fe–8 wt % Al showed both detrimental and beneficial effects of microstructural refinement with respect to corrosion resistance depending on the environment [3].

Two of the most intensively studied subgroups of nanomaterials in terms of their corrosion properties are materials made by (i) crystallization of amorphous precursors and (ii) electrodeposition methods. In the first group of nanocrystallized materials (see e.g., [4–9] for some of the early studies) the interpretation of the electrochemical response of various materials in different environments is often rather difficult because the nanostructured materials may contain several crystallographically and chemically different phases in addition to crystal size refinement and residual amorphous phase. On the other hand, materials made by electrodeposition can be produced both in conventional polycrystalline and nanocrystalline forms without changes in phase composition and without residual amorphous phase. Therefore, the assessment of microstructural refinement down to the nanoscale (i.e., crystal size effects) on the corrosion performance is much easier in this group of materials.

This chapter summarizes our current understanding of the corrosion behavior of nanocrystalline materials made by electrodeposition. Of particular interest will be the following questions:

1. To what extent does grain size reduction influence the electrochemical response of certain groups of pure metals and alloys?

2. What is the effect of grain size on the corrosion morphology?

3. What is the overall gain/loss of corrosion resistance of these materials when we go from conventional polycrystalline structure to nanostructures?

The chapter is organized as follows. First, the synthesis of nanocrystalline materials by electrodeposition will be briefly reviewed. Next, we will look at the unique microstructural features of these materials in comparison with their conventional polycrystalline counterparts. After a brief review of their mechanical and physical properties, the main focus of this chapter will be the corrosion behavior of nanocrystalline nickel, cobalt, copper, and zinc electrodeposits as well as some of their alloys and composites. The last section of this chapter will address some of the applications of these materials and the future outlook for this group of nanomaterials.

B. SYNTHESIS OF NANOMATERIALS BY ELECTRODEPOSITION

Electrodeposition is one of several methods in the general group of electrochemical synthesis of nanomaterials. Other processes using the electrochemical approach include electroless deposition, galvanic displacement, or electrodeposition under oxidizing conditions. In this chapter we will limit our discussion to electrodeposition of metals and alloys from aqueous solutions. Molten salt electrolysis, ionic liquid deposition, or plating of ceramics and semiconductors will not be considered here.

The potential for making metals, alloys, and composites in industrially and economically viable electroplating operations has been established throughout the 1980s and 1990s. The early work focused on nickel and nickel-based alloys. This was followed by expanding the technology to cobalt, copper, and zinc and several of their alloys (see Table 37.2 for examples). The first patents on this technology were issued in 1994 and 1995 [10, 11] and the first large-scale industrial application of an electrodeposited nanocrystalline nickel–phosphorus alloy was reported in 1996 [12]. For a more comprehensive treatment on the synthesis of electrodeposited nanomaterials the reader is referred to recent review articles published on this subject [13–16].

Electrodeposition of a pure metal from aqueous solutions at low pH containing Me^{z+} ions involves the following cathodic reactions:

$$Me^{z+} + ze^- \rightarrow Me \quad \text{(cathodic reaction)} \quad (37.1)$$

$$H^+ + e^- \rightarrow \tfrac{1}{2}H_2 \quad \text{(cathodic reaction)} \quad (37.2)$$

TABLE 37.2. Examples of Electrodeposited Materials with Nanocrystalline Structures

Ni	Co
Ni–P	Co–P
Ni–Fe	Co–W
Ni–Zn	Co–Fe
Ni–Fe–Cr	Co–Fe–P
Ni–Zn–P	Cu
Ni–SiC	Cu–Al$_2$O$_3$
Ni–Al$_2$O$_3$	Zn
Ni–P–BN	Zn–Ni
Ni–MoS$_2$	Fe–Co–Ni
Ni–Al (particles)	Pd
Ni–Carbon nanotubes	Pd–Fe

Alloys can be made by adding two or more species to the plating bath. For example, for Zn–Ni alloys, the bath contains Zn^{2+} and Ni^{2+} ions which come from appropriate metal salts (e.g., sulfates, chlorides) dissolved in the electrolyte in various concentrations, depending on the required composition of the final alloy. In order to maintain the required metal ion solution in the plating bath, two approaches are commonly used. When a dimensionally stable anode (DSA) is used, the bath must be periodically replenished with metal salt additions. On the other hand, when the metal to be deposited is also used as the anode, continuous anode dissolution replenishes the bath with metal ions:

$$Me \rightarrow Me^{z+} + ze^- \quad \text{(anodic reation)} \quad (37.3)$$

Cathodic and anodic reactions can be more complex than shown in Eqs. 37.1–37.3 depending on the metal. Several intermediate steps could be involved, but this is beyond the scope of this chapter.

Electrodeposition can also be used to make composite materials consisting of a metal matrix with a second-phase material embedded in the matrix in various amounts. This is usually achieved by adding the second phase (e.g., particles of SiC, Al$_2$O$_3$ or BN, or carbon nanotubes) to the plating bath, from which it is then codeposited with the metal.

Electrodeposition involves various steps, including diffusion of ions from the bulk of the electrolyte to the cathode and through the Nernst diffusion layer, formation of first adions on the cathode surface, surface diffusion of adions, nucleation of crystals, and growth of crystals. Initially the structure of the cathode is very important (e.g., surface cleanliness, presence of oxides, surface step structures, dislocations, grain boundaries), but with increasing deposit thickness the cathode surface influence diminishes.

Nanocrystalline metals require deposition under conditions that favor nucleation of new crystals and reduce growth of existing crystals. Through several experimental and theoretical studies [17–20] it has been shown that massive nucleation can be maintained during the growth of the deposit under the conditions of high overpotential (i.e., current density during the plating process) and reduced adion mobility. The latter can be controlled within certain limits by the addition of surface-active elements (e.g., saccharin, coumarin, for the case of nickel plating) to the plating bath. However, it must be noted that the breakdown of such additives can result in the incorporation of impurities in the deposit. For example, the use of saccharin as a grain refiner in nickel deposition usually results in sulfur and carbon impurities in the deposit with concentrations on the order of several hundred parts per million [17]. Such impurities can have a significant effect on the corrosion behavior of electrodeposited nanomaterials.

In electrodeposition with conventional direct current, the limiting current density of the system may not allow the application of high enough current densities to induce the massive nucleation required for nanocrystal formation. The limiting current density is reached when the metal ions are plated out at a rate higher than the rate with which they arrive in the Nernst diffusion layer by diffusion from the bulk of the electrolyte. In such cases, cathodic reaction 37.2 dominates and the quality of the electrodeposit deteriorates rapidly with increasing current density. To overcome this problem, two approaches are usually used. In the first approach, extensive agitation of the electrolyte can enhance the replenishment of metal ions in the Nernst diffusion layer. In the second approach, a very high current density is applied for only a short period of time. This is followed by a current-off period during which no plating takes place but which allows metal ions to diffuse back from the bulk of the electrolyte to the Nernst diffusion layer. This approach requires the use of a pulsed current power supply. Pulse plating of nanomaterials is typically carried out with current on and off times on the order of microseconds to milliseconds [10, 17, 18].

Electrodeposition is mainly used to apply a coating on a finished product to enhance certain properties of a part such as wear resistance, corrosion resistance, oxidation resistance, optical appearance, or magnetic properties. For such applications the thicknesses of the coatings are on the order of 0.01–1.0 mm depending on the application. Many industries use electrodeposited coatings for parts in automotive, aerospace, power generation, defense, and consumer products. The same infrastructure that already exists for conventional electrodeposition processes can also be used to make nanocrystalline electrodeposits.

However, electrodeposition is not limited to nanocrystalline coatings. Special processes have been developed to make nanocrystalline thin foils, meshes, wires, thick structural plates, and components for microelectromechanical systems [21–23].

C. STRUCTURE OF NANOCRYSTALLINE ELECTRODEPOSITS

Schematic cross sections showing the microstructures of conventional polycrystalline and nanocrystalline electrodeposits are shown in Figure 37.1. When electrodeposition is carried out at relatively low current densities using direct-current plating, the initial layer close to the substrate contains numerous small crystals. However, with increasing thickness the anisotropy in crystal growth results in a structural transition in which certain crystals grow rapidly while the growth of others is suppressed. Often this results in a columnar cross-sectional structure with grain sizes in the micrometer range. On the other hand, nanocrystalline electrodeposits produced under conditions leading to massive nucleation throughout the entire plating process maintain the nanocrystalline grain structure over the entire thickness. Figure 37.2 shows both a planar view and a cross-sectional view of nanocrystalline nickel produced by pulsed current electrodeposition. The average grain size of this material is on the order of 15 nm. The structure is more or less equiaxed.

The main structural defects in electrodeposited nanomaterials are grain boundaries, separating two crystals with different orientations, and triple junctions, the lines where three crystals meet. These defects are regions in the material with reduced structural order compared with the perfect crystal. Grain boundaries and triple junctions are typically 1 nm thick and vary in their structure depending on the orientations of the adjacent crystals. All polycrystalline materials contain a certain volume fraction of atoms associated with grain boundaries and triple junctions. Detailed calculations [24] have shown that for conventional polycrystalline materials with grain sizes larger than 1 μm, these volume fractions are very small (Table 37.3). On the other hand, for grain sizes less than 100 nm their volume fraction increases very rapidly, reaching close to 50% at a grain size of 5 nm. In other words, in a nanomaterial with 5 nm grain size half of the material is made up of atoms at grain boundaries and triple junctions.

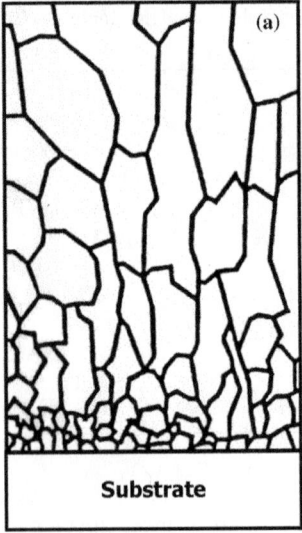

FIGURE 37.1. Schematic cross sections showing microstructures of conventional polycrystalline (a) and nanocrystalline (b) electrodeposits.

TABLE 37.3. Volume Fractions of Atoms Associated with Grain Boundaries and Triple Junctions as Function of Grain Size Assuming Grain Boundary Thickness of 1 nm

Grain Size	Volume Fraction	Grain Size	Volume Fraction
100 μm	2.9×10^{-5}	100 nm	0.003
10 μm	2.9×10^{-4}	10 nm	0.271
1 μm	2.9×10^{-3}	5 nm	0.488

In many polycrystalline materials grain boundaries are preferentially attacked during corrosion because of (i) their enhanced energy compared to the perfect crystal, (ii) the less than perfect atomic structure, and (iii) their susceptibility to segregation of impurities and formation of second-phase particles. For this reason, the initial expectation was that nanocrystalline metals made by electrodeposition would exhibit very poor corrosion properties mainly because of their high-grain-boundary-volume fractions. In Section E, it

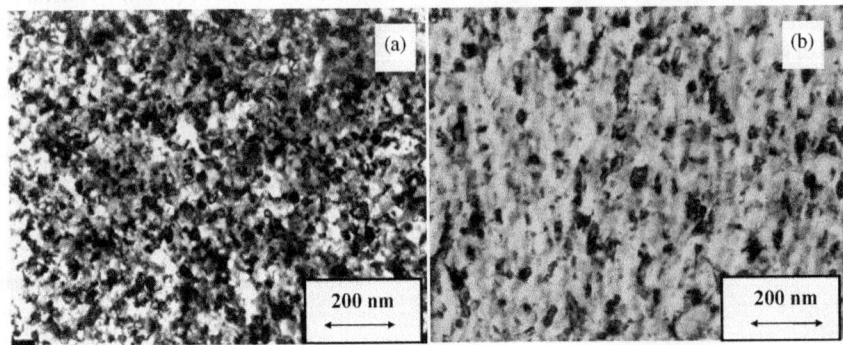

FIGURE 37.2. Transmission electron micrographs of nanocrystalline nickel electrodeposit in planar (a) and cross-sectional view (b).

TABLE 37.4. Effect of Grain Size on Various Properties for Electrodeposited Nickel and Cobalt

Property	Poly Ni (Grain Size 10 µm)	Nano Ni (Grain Size 10 nm)	Poly Co (Grain Size 5 µm)	Nano Co (Grain Size 10 nm)
Hardness (VHN)	140	650	232	525
Yield strength (MPa)	103	800	311	1002
Tensile strength (MPa)	403	1100	805	1865
Young's modulus (GPa)	207	204	207	200
Taber wear index	37	21	40.6	37.0
Thermal expansion ($\times 10^{-6}$/K)	11	10.5		
Electrical resistivity [$\mu\Omega$. cm]	7.5	13.5	7.5	16.2
Saturation magnetization (kA/m)	502	488	1340	1380

will be shown that this is not the case for most electrodeposited nanomaterials studied to date.

D. MECHANICAL AND PHYSICAL PROPERTIES OF NANOMATERIALS

The widespread interest in nanocrystalline materials is largely due to their outstanding mechanical properties and some of their unique property combinations which are not achievable in their conventional or amorphous counterparts. For nanocrystalline metals produced by electrodeposition it is found that grain size reduction to the nanometer range has no major effect on properties such as the Young's modulus, thermal expansion, heat capacity, or saturation magnetization [14]. On the other hand, hardness, yield strength, ultimate tensile strength, wear resistance, and electrical resistivity are strongly affected by grain size [14]. Table 37.4 summarizes some of these properties for nanocrystalline nickel and cobalt in comparison with their polycrystalline counterparts. The substantial increases in their hardness and strength can be understood on the basis of the Hall–Petch effect [25, 26], which describes such increases in terms of grain boundary–dislocation interactions. Higher grain boundary densities hinder easy slip of dislocation, which makes materials harder and stronger.

The Taber wear index is a measure of the material's resistance to abrasive wear; the lower the number, the higher the wear resistance. It has been shown [27] that the Taber wear index of nanocrystalline nickel electrodeposits is directly related to their increased hardness.

Directionally similar results as shown in Table 37.4 for nanocrystalline nickel and cobalt have also been observed for other electrodeposited metals and alloys such as Ni–P, Zn–Ni, Ni–Fe, and Cu.

E. CORROSION PROPERTIES OF NANOCRYSTALLINE ELECTRODEPOSITS

When nanocrystalline metal electrodeposits were initially developed, their corrosion behavior was of great concern

because of their high density of grain boundaries and triple junctions. In conventional polycrystalline materials, these defects are often prone to intergranular and stress corrosion cracking mainly because of the energetic and chemical composition reasons discussed in Section C. This section reviews recent advances in the understanding of the corrosion properties of several pure nanocrystalline metals and some of their alloys. It will be shown that grain size reduction in many of these materials can actually improve their resistance to localized corrosion quite substantially.

E1. Nanocrystalline Nickel and Nickel-Based Alloys

In the early 1990s Rofagha et al. [28, 29] presented the first systematic study on the effect of grain size (100 µm, 500 nm, 50 nm, and 32 nm) on the corrosion behavior of nickel in deaerated $2N$ H_2SO_4 solution at pH 0 using potentiodynamic and potentiostatic polarization tests. Figure 37.3 shows that nanocrystalline nickel exhibited the same active, passive, transpassive behavior as conventional polycrystalline nickel. However, the current density for the nanomaterials in the passive region was about one order of magnitude higher than for polycrystalline nickel. X-ray photoelectron spectroscopy showed that this higher passive current density was the result of a more defective passive film that formed on the nanocrystalline nickel surfaces [30]. Figure 37.3 also shows a positive shift in the corrosion potentials for all nanocrystalline materials which was explained in terms of the catalysis of the hydrogen evolution reaction. Despite the enhanced corrosion rate, Rofagha et al. [28, 29] showed that the corrosion morphology on the nanomaterials was very uniform instead of developing deep localized attack as observed along grain boundaries and triple junctions in polycrystalline nickel.

Wang et al. [31] studied the corrosion behavior of polycrystalline Ni (50 µm) and nanocrystalline Ni (32 nm, 16 nm) in 30 wt % KOH (pH 14.8) at 24°C. Within a wide potential range from the hydrogen evolution reaction to the oxygen evolution reaction, all materials were very inert with low passivation currents. However, as for the tests in $2N$

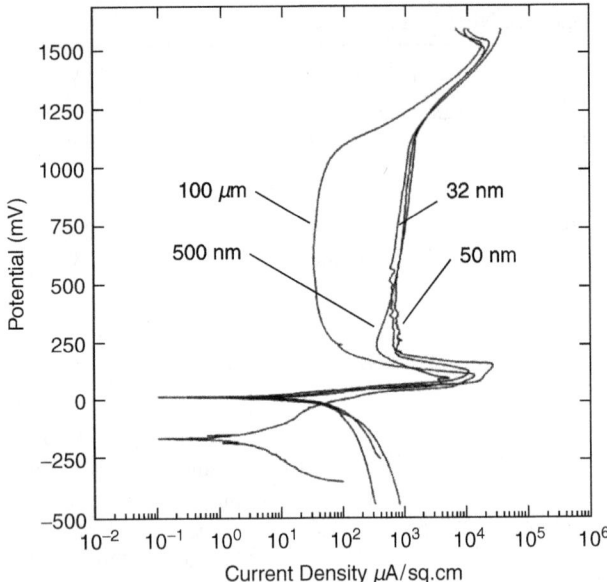

FIGURE 37.3. Potentiodynamic polarization curves for polycrystalline (100 μm) and nanocrystalline (500 nm, 50 nm, and 32 nm) nickel in 2 N H₂SO₄ solution. (Reproduced from [29] with permission.)

H₂SO₄ [28, 29], the current densities for the nanocrystalline materials were somewhat enhanced, about 2–5 times higher than for polycrystalline nickel. The nanocrystalline materials again showed very uniform attack.

Tang et al. [32] compared the corrosion performance of nanocrystalline nickel electrodeposits (grain sizes in the 5–10 nm range) made by different plating methods: pulse plating, direct-current plating, and pulse-reverse plating. Immersion tests were carried out in the following solutions: 7 M nitric acid, 3 M hydrochloric acid, and 20 g/dm³ citric

acid. In addition, samples were exposed to a moist SO₂ environment. Considerable differences in the corrosion performance were observed and the results were interpreted in terms of distinct changes in the crystallographic textures in the different materials.

Nanocrystalline nickel also showed excellent corrosion resistance when exposed to a salt spray environment as per American Society for Testing and Materials (ASTM) B-117 [33]. In this study both nanocrystalline (grain size 10 nm) and polycrystalline (grain size 10 μm) nickel electrodeposits were prepared as 10-μm-thick coatings on mild steel substrates. Percentages of areas covered with red rust were recorded for exposure times up to 250 h. Both the nanocrystalline and the polycrystalline coatings provided the same protection against corrosion of the steel substrate, with area percentages of red rust approaching 30% after 250 h.

In a more recent study, Kim et al. [34] investigated the effects of grain size and sulfur solute segregation on the corrosion behavior of polycrystalline (grains size 100 μm) and nanocrystalline nickel (grain size 20–30 nm), both containing about 1000 ppm by weight of sulfur impurities. Corrosion tests were carried out in 0.25 M Na₂SO₄ solutions at a pH of 6.5. Both materials showed very similar electrochemical behavior in potentiodynamic polarization curves. However, considerable differences were observed in their corrosion morphologies. Figure 37.4 shows that nanocrystalline nickel deposits developed a morphology consisting of numerous shallow corrosion pits (<2 μm deep) evenly distributed over the entire surface. In contrast, polycrystalline nickel shows extensive intergranular corrosion with corrosion attack along some boundaries as deep as 100 μm. This was likely the result of sulfur segregation to the grain boundaries in polycrystalline materials. The tremendous increase in the localized corrosion resistance in the

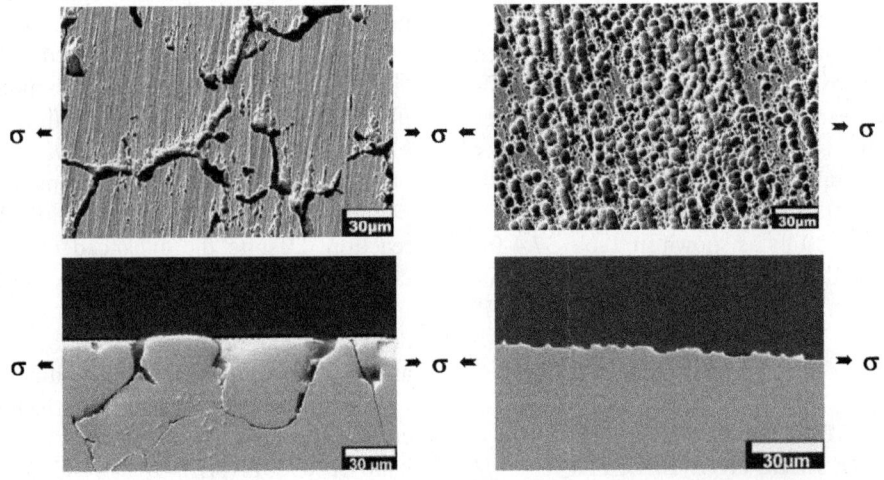

FIGURE 37.4. Scanning electron micrographs showing surface (top) and cross-sectional (bottom) corrosion morphologies of nickel containing 1000 ppm sulfur. Left: polycrystalline Ni. Right: nanocrystalline Ni. Sigma (σ) indicates internal or externally applied stress.

nanocrystalline nickel was interpreted in terms of a solute dilution effect [34]. Because of the large volume fraction of grain boundaries the sulfur content per unit area of grain boundary would be orders of magnitude lower in the nanocrystalline material as compared with the polycrystalline materials.

The problem with the extensive intergranular corrosion observed on polycrystalline nickel is that it can result in (i) excessive material loss due to grain dropping and (ii) unpredictable component failure when stresses are present in a component. For both externally applied and internal component stresses (see Fig. 37.4), the deep corrosion channels will act as stress concentration points, which could result in catastrophic failures. On the other hand, the material loss on the nanocrystalline nickel is much more uniform without deep corrosion channels. Therefore, it is much easier to design a component's long-term corrosion performance based on nanomaterials by choosing an appropriate coating thickness.

A nickel microalloy containing about 3000 ppm by weight of phosphorus (grain size 50–100 nm) was developed for one of the earliest large-scale applications of nanocrystalline Ni: an in situ repair technology of nuclear steam generator tubing (see Section F). Various corrosion tests were performed [35, 36] on this material, including ASTM G28, susceptibility to intergranular attack, and ASTM G35, G36, G44, susceptibility to stress corrosion cracking. The results of these corrosion tests showed that the nanocrystalline Ni–P microalloy is intrinsically resistant to intergranular attack and intergranular stress corrosion cracking. The material was also found to be resistant to pitting attack and only slightly susceptible to crevice corrosion.

Zamanzad-Ghavidel et al. [37] studied the pitting corrosion resistance of nanocrystalline (grain sizes 24, 27, and 31 nm) and microcrystalline (grain size 2 μm) nickel electrodeposits on copper substrates in 3.5% NaCl solution at room temperature. They observed that the breakdown potentials for nanocrystalline coatings were higher than for the microcrystalline nickel. It was further found that the nanocrystalline coating with 31 nm grain size exhibited the highest resistance to pitting corrosion

Gu et al. [38] compared the corrosion performance of nanocrystalline nickel (grain size 40 nm) and two electroless nickel–phosphorus coatings (no grain sizes given) on AZ91D magnesium alloy in 3.0 wt % NaCl solution. Potentiodynamic polarization curves showed that the nanocrystalline nickel coating had the lowest corrosion current density and provided better corrosion protection than the two electroless nickel–phosphorus coatings.

Rofagha et al. [39] studied the corrosion behavior of nanocrystalline Ni–1.4 wt % P (grain size 22 nm), Ni–1.9 wt % P (grain size 8 nm), and Ni–6.2 wt % P (amorphous structure) in 0.1 M H_2SO_4 and compared the results with the corrosion behavior of normal crystalline pure Ni (grain size

100 μm). Potentiodynamic polarization curves showed that, at such high phosphorus contents, the Ni–P alloys were nonpassivating, exhibiting similar polarization curves for the nanocrystalline and amorphous materials. It was concluded that the enhanced corrosion rates of the nanocrystalline materials were due to the high phosphorus content and the nonprotective nature of the surface film on these materials.

Splinter et al. [40] characterized the nature of this film on nanocrystalline and amorphous Ni–P by X-ray photoelectron spectroscopy. They observed an enrichment of elemental P compared to Ni on the surfaces of both materials and suggested that Ni is preferentially dissolved during anodic polarization. The high volume fractions of grain boundaries and triple junctions on the nanocrystalline materials resulted in enhanced dissociative adsorption of oxygen and hydroxyl species from solution. The films were, however, nonprotective because the defective nanocrystalline surfaces also facilitated atom dissolution and oxidation of surface P atoms from hypophosphite to soluble phosphate ions. At higher applied potentials, a thick, porous film formed on the nanocrystalline materials which provided a small kinetic barrier to further dissolution, resulting in slightly lowered anodic current densities as compared with amorphous Ni–P alloys.

Benea et al. [41] carried out a wear/corrosion study comparing pure polycrystalline Ni (grain size > 1 μm) and a nanocrystalline (grain size 100 nm) Ni–SiC nanocomposite coatings using sliding-type wear testing and electrochemical impedance spectroscopy in 0.5 M Na_2SO_4 neutral solution. They showed that the nanocrystalline nickel composite exhibited a higher polarization resistance and a 50% reduced corrosion rate, already in the absence of wear action. Even higher differences in material removal rates were found when the material was subjected to wear/corrosion conditions. For example, at a load of 30 N the nanocomposite materials showed a 90% reduction in material removal rate as compared with polycrystalline nickel.

Peng et al. [42] produced nanocrystalline nickel electrodeposits (grain sizes 42, 38, 31 nm) with varying amounts (0, 4.5, 10.9 wt %) of chromium nanoparticles (average particle size 39 nm) by a codeposition process. Potentiodynamic polarization tests in 3.5% NaCl solution showed that the chromium particles reduced the corrosion potential, increased the breakdown potential, and basically eliminated pitting at the highest chromium concentration. X-ray photoelectron spectroscopy showed that at 10.9 wt % chromium a continuous passive chromium oxide film was formed on the surface.

E2. Nanocrystalline Cobalt and Cobalt Alloys

Kim et al. [43] studied the corrosion behavior of electrodeposited cobalt nanodeposits (grain size 13 nm) in 0.25 M Na_2SO_4 (pH 7) and compared the results with conventional cobalt (grain size 10 μm). Neither the polycrystalline nor the

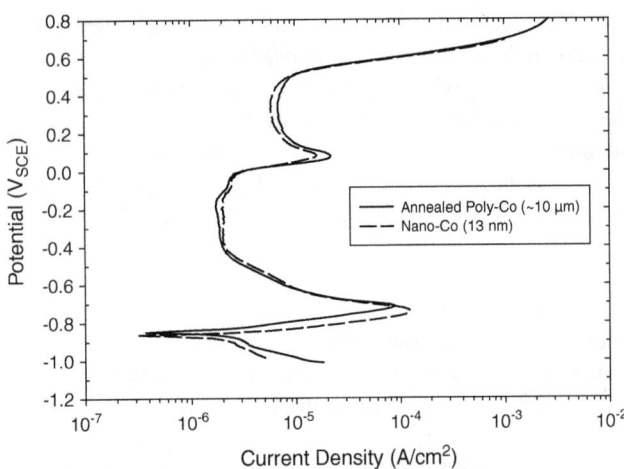

FIGURE 37.5. Potentiodynamic polarization curves in 0.1 M NaOH (pH 13) for polycrystalline and nanocrystalline cobalt.

nanocrystalline cobalt showed passivation in this solution and the potentiodynamic curves for both grain sizes were nearly identical. However, as observed before for the corrosion behavior of nickel, the nanocrystalline cobalt showed high resistance to localized attack.

The same two cobalt materials were also tested in a 0.1 M NaOH solution (pH 13) in which conventional polycrystalline cobalt readily shows passive film formation [44]. Figure 37.5 presents the potentiodynamic polarization curves for both materials in this solution. It can be readily seen that the polarization curve for the nanocrystalline cobalt is very similar to the one for polycrystalline cobalt. In other words, grain size reduction from 10 μm to 13 nm did not substantially change the passivation behavior of cobalt.

Aledresse and Alfantazi [45] also compared the corrosion behavior of polycrystalline (grain size 100 μm) and nanocrystalline (grain size 67 nm) cobalt in 0.25 M Na$_2$SO$_4$ solution. In addition, their study also included a nanocrystalline Co–P alloy with a grain size of 50 nm. None of the materials passivated in this solution.

Jung and Alfantazi [46] performed electrochemical impedance spectroscopy as well as potentiostatic and potentiodynamic polarization tests on microcrystalline cobalt and nanocrystalline cobalt (grain size 20 nm) and cobalt–1.1 wt % phosphorus (grain size 10 nm) in 0.1 M H$_2$SO$_4$ solution. Potentiodynamic polarization results showed that all materials exhibited active dissolution with no transition to passivation. While the polarization curves for nanocrystalline and microcrystalline cobalt were almost identical, a noticeable shift to more positive potential was observed for the nanocrystalline cobalt–phosphorus alloy. In addition, the overpotential for hydrogen evolution was observed to decrease. X-ray photoelectron spectroscopy showed that the enhanced corrosion resistance of the Co–P alloy was due to an enrichment of P of the corroded surface after polarization. However, at higher anodic overpotentials the benefit of phosphorus disappeared due to the formation of a nonprotective surface film which contained elemental phosphorus, hypophosphite, and phosphate species.

Saito et al. [47] studied the corrosion performance of nanocrystalline Co$_{65}$Ni$_{12}$Fe$_{23}$ ternary alloys (grain sizes 10–40 nm) in deaerated 2.5 wt % NaCl solution. Potentiodynamic polarization curves showed that the material passivated in this solution with passivation current densities of ~30 μA/cm^2.

E3. Nanocrystalline Zinc

Youssef et al. [48] compared the corrosion behavior of nanocrystalline zinc electrodeposits (grain size 56 nm) and conventional electrogalvanized zinc (grain size 8–20 μm) in deaerated 0.5 N NaOH solution at 25°C. Both potentiodynamic polarization and alternating current (ac) impedance measurements were used. The polarization curves for both materials were very similar in shape but showed differences in specific electrochemical parameters such as corrosion potential (E_{corr}), passivation potential (E_p), maximum current density (i_m) and passive current density (i_p), as summarized in Table 37.5. Both i_{corr} and i_p were lower for nanocrystalline zinc than for polycrystalline electrogalvanized zinc. It was concluded that the oxide film on nanocrystalline zinc was more protective than the film on polycrystalline film which was supported by ac impedance measurements. The average capacitance value for nanocrystalline zinc was 69 μF/cm^2, as compared to 227 μF/cm^2 for the polycrystalline material. It was further shown that nanocrystalline zinc displayed numerous discrete corrosion pits after potentiodynamic polarization while polycrystalline zinc showed a more uniform corrosion morphology.

The corrosion properties of a series of electrodeposited zinc–nickel alloy coatings on mild steel substrates were

TABLE 37.5. Important Electrochemical Parameters for Nanocrystalline and Polycrystalline Zinc from Polarization Curves in Deaerated 0.5 N NaOH Solution

Material	E_{corr} (mV)	I_{corr} (μA/cm^2)	E_p (mV)	i_m (μA/cm^2)	i_p (μA/cm^2)
Nano Zn	−1470	90	−1362	4503	210
Poly Zn	−1455	229	−1342	3895	828

Source: From [48].

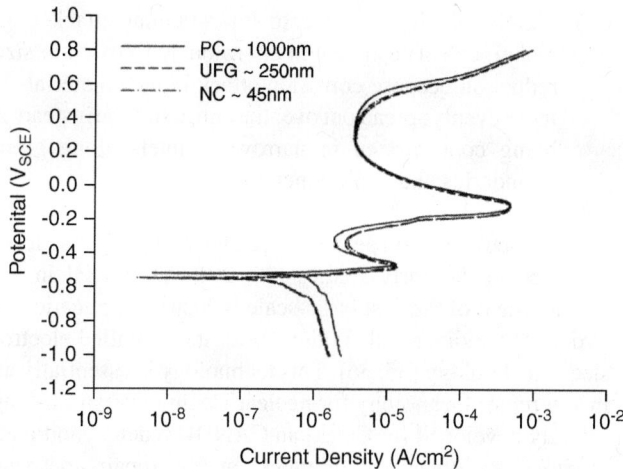

FIGURE 37.6. Effect of grain size in electrodeposited copper on potentiodynamic polarization curves in 0.1 M NaOH solution (pH 13).

studied by Alfantazi and Erb [49]. This study covered both polycrystalline and nanocrystalline materials. It was shown through salt spray testing (ASTM B 117-81) that nanocrystalline Zn-45 w % Ni (grain size 20 nm) and Zn-63 wt % Ni (grain size 2 nm) showed greater resistance to the formation of white and red rust than polycrystalline Zn. However, in this study it was difficult to separate grain size effects from contributions due to chemical composition.

E4. Nanocrystalline Copper

Tao and Li [50] studied mechanical and electrochemical properties of nanocrystalline copper electrodeposits (grain size 56 nm) in comparison with polycrystalline copper deposits (grain size 2 μm). Potentiodynamic polarization curves for both materials in 0.1 M NaOH solution showed two differences. First, the corrosion potential of nanocrystalline copper was shifted slightly to more negative values. Second, the current density in the passive region was lower for the nanocrystalline copper than for the polycrystalline copper. On the other hand, Yu et al. [51] observed that grain size reduction in copper electrodeposits from 1 μm to 45 nm in the same solution (0.1 M NaOH) had no significant effect on the potentiodynamic polarization of copper, as shown in Figure 37.6. After photodynamic polarization tests all materials were covered with a fine needlelike corrosion product (Fig. 37.7) with a morphology similar to what in previous studies on polycrystalline copper was described as the CuO/Cu(OH)$_2$ upper layer [52, 53].

Yu et al. [54] investigated the polarization behavior of nanocrystalline and polycrystalline copper in 3.5 wt % NaCl solution. They observed similar polarization behavior for both materials but noted that the nanocrystalline material had a higher E_{corr} and a lower anodic current density which they attributed to the large density of intercrystalline defects

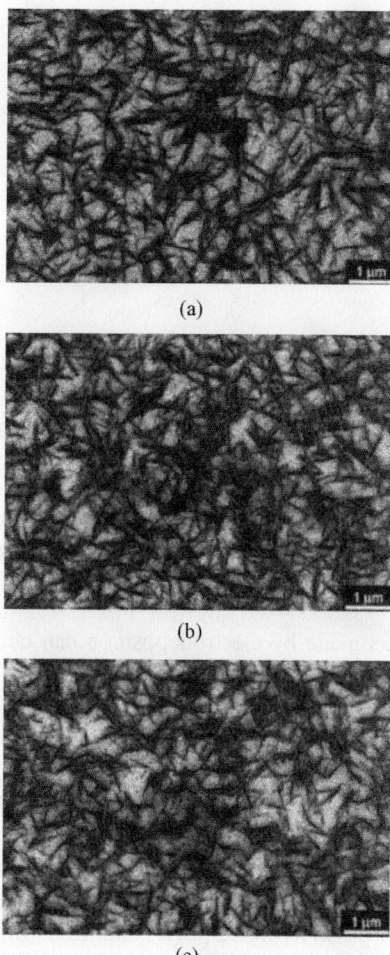

FIGURE 37.7. SEM images after potentiodynamic polarization tests in 0.1 M NaOH solution of polycrystalline and nanocrystalline copper revealed similar "needle like" corrosion product morphologies. Grain sizes: (a) 45 nm, (b) 250 nm and (c) 1 μm.

which act as preferential attack sites and lead to the rapid formation of a passive film. Unfortunately, in their study Yu et al. [54] did not provide any microstructural details on the materials used for the corrosion study. In fact, no grain sizes were given for the two materials.

In a more recent study Yu et al. [55] presented polarization curves for polycrystalline (grain size 1 μm) and nanocrystalline (grain sizes 250 and 45 nm) copper in 0.1 M NaCl solution (Fig. 37.8). Again, the corrosion behavior of copper was not significantly changed by grain size reduction.

F. INDUSTRIAL APPLICATIONS AND OUTLOOK

The results of the corrosion studies over the past two decades on nanocrystalline nickel, cobalt, zinc, copper, and several of

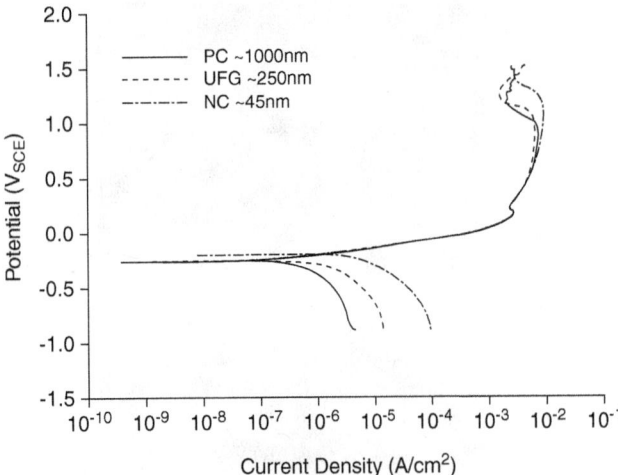

FIGURE 37.8. Potentiodynamic polarization curves of electrodeposited copper foils with varying grain sizes in 0.1 M NaCl (pH 6.5).

their alloys made by electrodeposition can be summarized as follows:

(i) Contrary to earlier concerns, the high-grain-boundary and triple-junction densities in nanocrystalline metals do not compromise their corrosion properties.

(ii) The general shapes of polarization curves obtained in various solutions are not strongly affected by grain size reduction down to the nanometer range. Materials that show passivity in the polycrystalline form usually also exhibit passivity for the nanocrystalline structure. However, with decreasing grain size some changes in specific electrochemical parameters are observed, such as corrosion potential (E_{corr}), corrosion current density (i_{corr}), passivation potential (E_p), maximum current density (i_m), and passive current density (i_p).

(iii) For systems that show clear passivity, different studies found enhanced, reduced, or very similar passive current densities (in comparison with polycrystalline materials), depending on the system. A few discrepancies do exist in different investications for some materials (e.g., for the case of Cu) which would require further studies. Most likely there were some differences in the nanocrystalline and polycrystalline materials used in different studies, such as impurities or crystallographic texture.

(iv) The structure and composition of the passive layer does depend on the grain size of the material. This can be understood on the basis of the high defect concentration (i.e., grain boundaries and triple junctions) intersecting the free surface of nanomaterials. Impurities also have a strong effect on the nature of the passive layer.

(v) Reduced grain size was observed in some studies to enhance the hydrogen evolution reaction.

(vi) Metals which are prone to intergranular attack (e.g, nickel, cobalt) can benefit enormously from grain size reduction because corrosion attack in nanomaterials is more evenly spread out over the entire surface instead of being concentrated in narrow channels along grain boundaries and triple junctions.

The good corrosion performance of nanocrystalline nickel observed in the early studies [28, 29] was crucial in the development of the first large-scale industrial application of structural nanomaterials in the world: the so-called electrosleeve technology [35, 36]. This technology is essentially an in situ repair technology for nuclear steam generator tubing initially developed for Canadian CANDU reactors and later modified for other reactor types. In this repair approach nuclear steam generator tubing which was prematurely compromised by intergranular attack, intergranular stress corrosion cracking, and other forms of corrosion was repaired by the application of an electroformed nanocrystalline Ni–0.3% P microalloy sleeve on the inside of the tubes with a thickness of about 1 mm. The purpose of the sleeve was to (i) seal through-wall cracks to prevent leakage of radioactive water from the core of the reactor and (ii) restore the structural integrity of the tubes. While nanocrystalline nickel microalloy provided the necessary strength, it was the excellent corrosion resistance of nanocrystalline nickel which ultimately made this technology possible. The main purpose of using about 0.3% P in this microalloy as an alloying element was to achieve the long-term thermal stability of the sleeve material in a reactor environment operating at 280°C and a pH of ~11.5. The phosphorus retards grain boundary mobility through a solute drag effect.

The electrosleeve process [56, 57] has been successfully implemented in a CANDU unit in 1994 and a pressurized water reactor (PWR) in 1999. It has been incorporated by the American Society of Mechanical Engineers (ASME) as a standard procedure for pressure tubing repair [58].

Following this early application of electrodeposited nanocrystalline materials numerous applications have been developed over the past 10–15 years. Other applications are currently still at the research and development stage. Table 37.6 summarizes some of these applications, many of which require good to excellent corrosion resistance. For further details on the various applications the reader is referred to references [12–16] and [21–23].

Two of the most successful metallic coating systems used on many manufactured products are chromium, produced by electrodeposition from hexavalent chromium solutions, and cadmium, usually plated from alkaline, cyanide, acid sulfate, or acid fluorobate solutions. Chromium is extensively used for providing surfaces with wear, abrasion, and corrosion resistance or for decorative purposes, while cadmium is mainly used for corrosion resistance on many mass-finished products. However, considerable health risks are associated

TABLE 37.6. Applications of Electrodeposited Nanocrystals

Application	Materials
Armor laminates	Ni, Fe, Co, Ni–Fe
Battery grids	Pb
Catalysts for H_2 evolution	Ni–Mo
Corrosion resistant coatings	Ni, Ni–P, Zn–Ni, Co, Co–P
Chromium-replacement coatings	Co–P
Electromagnetic shielding	Ni–Fe
Electronic connectors	Ni
Electrosleeve	Ni–P
Foil for printed circuit boards	Cu
Free-standing soft magnets	Ni, Co, Ni–Fe, Co–Fe
Hard-facing applications	Ni, Ni–SiC, Ni–Al_2O_3, Co, Co–P
Magnetic recording heads	Ni–Fe
Microelectromechanical systems	Ni, Co, Ni–Fe
Self-lubricating coatings	Ni–MoS_2, Ni–BN
Shaped charge liners	Cu
Structural applications	Ni, Co, Ni–Fe
Transformer core materials	Ni–Fe, Co–Fe
Wear-resistant coatings	Ni, Ni–SiC, Ni–P, Co, Co–P

with both coating systems. Hexavalent chromium baths have been shown to enhance the risk of lung and nose cancer while exposure to cadmium can result in acute respiratory and gastrointestinal effects as well as chronic lung and kidney disease. For this reason several more benign coating systems have been developed to replace chromium and cadmium [15].

Various nanocrystalline-based alloys (e.g., Co–P, Co–Fe–P) are available for chromium replacement, while the best choices for cadmium replacement are Zn–Ni-type nanodeposits.

The outlook for applications of nanocrystalline metals in applications requiring corrosion resistance is excellent. It is the unique combination of the overall good corrosion performance with outstanding mechanical properties which make this relatively new class of materials attractive to many industries.

REFERENCES

1. R. W. Siegel, in Processing and Properties of Nanocrystalline Materials, C. Suryanarayana et al. (Eds.), The Minerals, Metals and Materials Society Warrendale, PA, 1996, p. 3.

2. R. B. Inturi and Z. Szklarska-Smialowska, *Corrosion*, **48**, 398 (1992).

3. W. Zeiger, M. Schneider, D. Scharnweber, and H. Worsch, *Nanostr. Mater.*, **6**, 1013 (1995).

4. R. B. Diegle and J.E. Slater, *Corrosion*, **32**, 155 (1976).

5. K. Hashimoto, K. Osada, T. Masumoto, and S. Shimodaira, *Corros. Sci.*, **16**, 71 (1976).

6. M. Naka, K. Hashimoto, and T. Masumoto, *Corrosion*, **36**, 679 (1980).

7. J. C. Turn and R. M. Latanision, *Corrosion*, **39**, 271 (1983).

8. S. J. Thorpe, B. Ramaswami, and K. T. Aust, *J. Electrochem. Soc.*, **135**, 2162 (1988).

9. P. Bragagnolo, Y. Waseda, G. Palumbo, and K.T. Aust, *MRS. Int. Mtg. Adv. Mat.*, **4**, 469 (1989).

10. U. Erb and A.M. El-Sherik, *U.S. Patent* No. 5,352,266, 1994.

11. U. Erb, A.M. El-Sherik, C. K. S. Cheung, and M. J. Aus, U.S. Patent No. 5,433,797, 1995.

12. C. K. S. Cheung, D. Wood, and U. Erb, in Processing and Properties of Nanocrystalline Materials, C. Suryanarayana et al., (Eds.), The Minerals, Metals and Materials Society, Warrendale, PA, 1996, p. 479.

13. G. Palumbo, F. Gonzalez, K. Tomantschger, U. Erb, and K.T. Aust Plat., *Surf. Fin.*, **90**(2), 36 (2003).

14. U. Erb, K. T. Aust, and G. Palumbo, in Nanostructured Materials, 2nd ed., C. C. Koch, (Ed.), *William Andrew Publ.*, Norwich, NY, 2007, 235.

15. G. Palumbo, J. L. McCrea, and U. Erb, in Encyclopedia of Nanoscience and Nanotechnology, Vol. 1, H. S. Nalwa (Ed.), *American Scientific Publ.*, Stevenson Ranch, CA, 2004, p. 89.

16. U. Erb, in CRC Materials Processing Handbook, J. R. Groza, E. J. Lavernia, J. F. Shackelford, and M. T. Powers, (Eds.), *CRC Press*, Boca Raton, FL, 2007, p. 22–1.

17. A. M. El-Sherik and U. Erb, *J. Mater. Sci.*, **30**, 5743 (1995).

18. R. T. C. Choo, A. M. El-Sherik, J. Toguri, and U. Erb, *J. Appl. Electrochem.*, **25**, 384 (1995).

19. H. Natter and R. Hempelmann, *Z. Phys. Chem.*, **222**, 319 (2008).

20. L. P. Bicelli, B. Bozzini, C. Mele, and L. D'Urzo, *Int. J. Electrochem. Sci.*, **3**, 356 (2008).

21. Erb U, Palumbo G, Aust KT, in Nanostructured Films and Coatings, G.M. Chow et al. (Eds.), *NATO Science Series*, 3 High Technol., **78**, 11 (2000).

22. U. Erb, G. Palumbo, D. H. Jeong, S. H. Kim, and K. T. Aust, in Processing and Properties of Structural Nanomaterials, L. L. Shaw et al., (Eds.), The Minerals, Metals and Materials Society, Warrendale, PA, 2003, p. 109.

23. U. Erb, K. T. Aust, G. Palumbo, J. L. McCrea, and F. Gonzalez, in Processing and Fabrication of Advanced Materials IX, T. S. Srivatsan, et al., (Eds.), ASM International, Materials Park, OH, 2001, p. 253.

24. G. Palumbo, S. J. Thorpe, and K. T. Aust, *Scripta Metall.*, **24**, 1347 (1990).

25. A. M. El-Sherik, U. Erb, G. Palumbo, and K. T. Aust, *Scripta Metall., Mater.*, **27**, 1185 (1992).

26. N. Wang, Z. Wang, K. T. Aust, and U. Erb, *Mater. Sci. Eng.*, **A237**, 150 (1997).

27. D. H. Jeong, K. T. Aust, U. Erb, G. Palumbo, *Scripta Mater.*, **44**, 493 (2001).

28. R. Rofagha, R. Langer, A. M. El-Sherik, U. Erb, G. Palumbo, K. T. Aust, *Scripta Metall. Mater.* **25**, 2867 (1991).

29. R. Rofagha, R. Langer, A. M. El-Sherik, U. Erb, G. Palumbo, and K. T. Aust, *Mater. Res. Soc. Symp. Proc.*, **238**, 751 (1992).

30. R. Rofagha, S. J. Splinter, U. Erb, N. S. McIntyre, *Nanostr. Mater*, **4**, 69 (1994).

31. S. Wang, R. Rofagha, P. R. Roberge, and U. Erb, *Electrochem. Soc. Proc.*, **95-8**, 244 (1995).

32. P. T. Tang, T. Watanabe, J. E. T. Andersen, and G. Bech-Nielsen, *J. Appl. Electrochem.*, **25**, 347 (1995).

33. A. M. El-Sherik and U. Erb, *Plat. Surf. Fin.*, **82**(9), 85 (1995).

34. S. H. Kim, K. T. Aust, U. Erb, and F. Gonzalez, in *Proc. American Electroplaters and Surface Finishers Society (AESF) SUR/FIN 2002*, AESF, Orlando, FL, p. 225

35. F. Gonzalez, A. M. Brennenstuhl, G. Palumbo, U. Erb, and P. C. Lichtenberger, *Mater. Sci. For.*, **225–227**, 831 (1996).

36. G. Palumbo, F. Gonzalez, A. M. Brennenstuhl, U. Erb, W. Shmayda, and P. C. Lichtenberger, *Nanostr. Mater*, **9**, 737 (1997).

37. M. R. Zamanzad-Ghavidel, K. Raeissi, and A. Saatchi, *Mater. Lett.*, **63**, 1807 (2009).

38. C. Gu, J. Lian, J. He, Z. Jiang, and Q. Jiang, *Surf. Coat. Technol.*, **200**, 5413 (2006).

39. R. Rofagha, U. Erb, D. Ostrander, G. Palumbo, and K. T. Aust, *Nanostr. Mater.*, **2**, 1 (1993).

40. S. J. Splinter, R. Rofagha, N.S. McIntyre, and U. Erb, *Surf. Interf. Anal.*, **24**, 181 (1996).

41. L. Benea, P. L. Bonora, A. Borello, and S. Martelli, *Wear*, **249**, 995 (2002).

42. X. Peng, Y. Zhang, J. Zhao, and F. Wang, *Electrochim. Acta*, **51**, 4922 (2006).

43. S. H. Kim, K.T. Aust, U. Erb, F. Gonzalez, and G. Palumbo, *Scripta Mater.*, **48**, 1379 (2003).

44. S. H. Kim, T. Franken, G. D. Hibbard, U. Erb, K. T. Aust, and G. Palumbo, *J. Metast. Nanostr. Mat.*, **15–16**, 643 (2003).

45. A. Aledresse and A. M. Alfantazi, *J. Mater. Sci.*, **39**, 1523 (2004).

46. H. Jung and A. M. Alfantazi, *Electrochim. Acta*, **51**, 1806 (2006).

47. M. Saito, K. Yamada, K. Ohashi, Y. Yasue, Y. Sogawa, and T. Osaka, *J. Electrochem. Soc.*, **146**, 2845 (1999).

48. K. M. S. Youssef, C. C. Koch, and P. S. Fedkiw, *Corros. Sci.*, **46**, 51 (2004).

49. A. M. Alfantazi and U. Erb, *Corrosion*, **52**, 880 (1996).

50. S. Tao and D. Y. Li, *Nanotechnology*, **17**, 65 (2006).

51. B. Yu, P. Woo, and U. Erb, *Scripta Mater.*, **56**, 353 (2007).

52. J. C. Hamilton, J. C. Farmer, and R. J. Anderson, *J. Electrochem. Soc.*, **133**, 739 (1986).

53. S. T. Mayer and R. H. Muller, *J. Electrochem. Soc.*, **139**, 426 (1992).

54. J. K. Yu, E. H. Han, L. Lu, and X. J. Wei, *Mater. Sci.*, **40**, 1019 (2005).

55. B. Yu, P. Woo, and U. Erb, to be published.

56. G. Palumbo, P. C. Lichtenberger, F. Gonzalez, and A. M. Brennenstuhl, *U.S. Patent* Nos. 5,527, 445, 5,516,415, and 5,538,615, 1996.

57. I. Brooks, G. Palumbo, F. Gonzalez, A. Robertson, K. Tomantschger and K. Panagiotopoulos, in *American Electroplaters and Surface Finishers Society (AESF) SUR/FIN 2003 Proc.*, 2003, Orlando, FL, p. 721.

58. ASME Code Case 96-189-BC96-206, Case N-569-Section XI, Division 1: Alternative Rules for Repair by Electrochemical Deposition of Class 1 and 2 Steam Generator Tubing, 1996.

38

CORROSION OF SHAPE MEMORY AND SUPERELASTIC ALLOYS

L. E. Eiselstein

Exponent-Failure Analysis Associates, Inc., Menlo Park, California

A. INTRODUCTION

Shape memory and superelastic alloys are alloys that exhibit thermally recoverable strain (shape memory effect), pseudoelastic (superelastic behavior), or both. This chapter discusses what is known about the corrosion behavior of such alloys. However, most of the information on the corrosion behavior of such materials is focused on the 50 at % nickel–50 at % titanium (atomic percent, at %) (50 at % Ni–50 at % Ti) alloy composition since this material has extensive use in biomedical applications [1–12]. The corrosion data on other shape memory or superelastic materials will be discussed when available. We first describe what makes these alloys unique and what they are typically used for followed by a general description of the various alloy classes. There are also ceramic and polymer materials that exhibit shape memory; however, they are not discussed in this chapter [13–15].

A1. What Are Shape Memory and Superelastic Alloys?

Shape memory and superelastic alloys have unique mechanical properties that make them useful, or even indispensable, for some applications. These properties are known as the shape memory effect and the superelastic or pseudoelastic effect. In 1931 Ölander was the first to point out that the gold–cadmium (Au–Cd) alloy had rubberlike characteristics [16–19]. This "rubberlike behavior" is still not well understood, but as far as the mechanical response, it appears to be superelastic. But unlike normal superelastic alloys in which this effect occurs as a result of stress-induced martensite in an austenitic matrix, the superelastic behavior of these rubberlike materials occur in the martensitic state [18]. The shape memory effect (thermoelastic behavior of the martensite phase) was described more fully in an Au–Cd alloy in 1951 [20] and then in indium–thallium (In-Tl) in 1952 [21]. The discovery of similar behavior in equiatomic nickel– titanium (NiTi) in the early 1960s was the beginning of commercial applications for such materials [22, 23].

Materials that exhibit crystallographically reversible martensitic transformation (or thermoelastic martensitic transformation) are generally called shape memory alloys (SMAs) [24, 25]. Such transformations can be induced by the application of force or straining or from changes in temperature. The martensite transformation is accompanied by a large diffusionless shear-like deformation associated with structural change. This deformation generally amounts to about 20 times the elastic deformation. The martensite is deformable, and it can be induced from the parent austenite phase by loading. A large deformation induced in shape memory alloys can be recovered by heating to temperatures above the reverse transformation finish temperature (A_f) after unloading (the shape memory effect) or simply by unloading at temperatures above A_f (the pseudoelasticity or superelasticity effect) [25]. These transformations can occur in a wide variety of alloys such as Ag–Cd, Cu–Al–Ni, NiTi, Cu–Sn, and Cu–Zn [24]. Rubber-like superelastic behavior has also been observed in Au–Cu–Zn, Cu–Al–Ni, Cu–Zn–Al, and Au–Cd alloys [18].

A2. Shape Memory and superelastic Alloy Classes

The three main commercial types of shape memory alloys are the copper–zinc–aluminum–nickel (Cu–Zn–Al–Ni), copper–aluminum–nickel (Cu–Al–Ni), and nickel–titanium (NiTi) alloys although there are many other SMA families. Examples of various SMAs are shown in Table 38.1. Although shape memory and superelastic alloy classes could be described any number ways, they are broken down into five general classes in this chapter. The five classes we will consider will be NiTi, the founding member class; Ni-free Ti-based alloy composition class; Cu-based, Fe-based, and magnetic shape memory, and shape memory alloys other than Ti-based alloys, a catchall class. Generally only the NiTi, Cu-based, and ferrous families are currently considered materials of commercial interest [25].

B. NiTi-BASED SHAPE MEMORY ALLOYS

The NiTi shape memory alloy was first described by Buehler in the early 1960s [22, 23]. NiTi [50 at % Ni–50 at % Ti, also known as nitinol, since Buehler worked for the Naval Ordinance Laboratory (NiTiNOL)] is the most commonly used superelastic and SMA; thus, it is the alloy composition for which we have the most corrosion data. The Ti–Ni SMAs have become one of the most important metallic biomedical

TABLE 38.1. Various Shape Memory Alloy Compositions

Alloy (Typical Composition, wt %)	References
NiTi Based	
(55–58%)NiTi	[6, 26]
Ni–(45–46%)Ti–(<22%)Cu	[6, 15, 26]
Ni–45% Ti–(<8%)Co	[6]
NiTi–(9%)Nb	[15, 26]
(47–49%)NiTi–(1–3%)Fe	[15]
NiTi–X (X = Pt, Pd, Hf, Zr)	[15, 26, 27]
Ni–Al Based	
Ni–26.5% Al	[6, 27]
Ti Based, Ni Free	
β-Ti	[15, 28–30]
Ti–18 at % Nb–4 at % Sn	[29]
Ti–Nb–Ta–Zr (gum metal)	[30]
Ti–29Nb–13Ta–4.6Zr	[30]
Ti–23 at % Nb–0.7 at % Ta–2 at % Zr–O	[31]
Ti–12 at % Ta–9 at % Nb–3 at % V–6 at % Zr–O	[31]
Cu Based	
Cu–Al	[32, 33]
Cu–44% Al	[6]
Cu–(10–14%)Al–(2–5%)Ni	[15, 25, 33]
Cu–12% Al–5% Ni–2% Mn–1% Ti	[34]
Cu–Sn	[32]
Cu–25% Sn	[6]
Cu–Zn	[32]
Cu–Zn–Al	[25, 33]
Cu–(38–40%)Zn	[6]
Cu–17% Zn–7% Al	[6]
Cu–34.5% Zn–0.9% Si	[6]
(60–85%)Cu–(0–40%)Zn–(0–14%) Al–(0–5%)Si–(0–15%)Mn	[35]
Fe Based	[36]
Fe–(14–30%)Mn–(4–6%)Si	[15, 36]
Fe–Mn–Si–Cr–Ni	[36]
Fe–(~ 25 at %)Pt	[36]
Fe–Ni–Co–Ti	[15, 36]
Others	
Au–(34–36)Cd	[6, 19, 20]
(40–63)Au–(10–27)Cu–(27–33)Zn	[6]
In–Tl	[21]

materials, particularly for percutaneous applications. This occurred primarily as a result of this material's superelasticity properties and good corrosion and biocompatibility properties, which have been utilized to develop various medical devices, such as guide wires, stents, filters, and other unique medical devices [12, 37]. The superelastic properties are used to full advantage in the development of new medical devices that can be delivered through endovascular placement, deployed, and allowed to expand to fully functional size.

Although most corrosion research has been done on "pure" NiTi alloys, NiTi alloys have been alloyed with copper (Cu) and iron (Fe) for other commercial applications. For instance,

NiTi is alloyed with 5–15 wt %* Cu, which reduces the transformation hysteresis to less than one-third that observed for "pure" NiTi, making it similar to those of the Cu-based alloys.

Niobium (Nb) is also alloyed with NiTi at about 9%; however, Nb is not very soluble and is present as a fine dispersion in the NiTi matrix [15]. These precipitates stabilize the martensite phase and the reverse transformation is shifted to higher temperatures, making these alloys good for commercial applications like mechanical couplers [15].

NiTi is also alloyed with Fe to suppress the R-phase and martensite transformation temperature. The R-phase is not suppressed as much and alloys with Fe have good temperature separation between the martensite and R-phase transitions [15].

B1. NiTi Passive Layer

NiTi typically is covered by a naturally formed thin adherent oxide layer known as a passive film [38–40]. This film is very stable and NiTi alloys are resistant to many forms of corrosive attack; however, this passive film is attacked by acidic solutions containing chloride [39]. This has been confirmed by auger and X-ray photoelectron spectroscopy (XPS) as well as through electrochemical impedance spectra. For instance, Pound showed that impedance spectra of NiTi in both phosphate-buffered saline (PBS) and simulated bile show near-capacitive behavior, and the data could be fitted by a parallel resistance–capacitance (as a constant-phase element) circuit [41]. Such behavior is generally associated with a passive oxide film. The thickness of the oxide as determined from the capacitance measured was consistent with surface analytical results reported in the literature. Pound also found compositional differences, if any, between the passive films formed in PBS or simulated bile did not affect the NiTi passive film resistivity [41]. Pound also noted that the passive film on NiTi becomes more defective as the polarization potential is increased. Schroeder found the passive film on NiTi and cathodic-protected (CP) Ti is a TiO_2-based passive film (possibly a hydrated form) containing about the same defect density and can be characterized as an *n*-type semiconductor [38]. The passive film on mechanically polished NiTi is thinner and has a lower film resistance than electropolished or chemically passivated NiTi; however, the oxide continues to grow in aqueous environments. Schroeder also stated that, although the oxide composition and defect density of the passive film are similar for both NiTi and CP Ti and since the diffusion length is shorter for NiTi, one would expect the NiTi film would thicken faster. This however is not the case and Schroeder postulates that the Ti must first diffuse through a Ni-enriched layer before it can

become incorporated in the Ti-rich passive layer. This Ni enrichment has been observed on thermally oxidized NiTi where Ni_3Ti has been found at the metal–oxide interface and islands of metallic Ni have been found in the TiO_2 passive film [38, 42] and on wires with very thick oxide [43]. Undisz et al. have also observed a concentration of nickel (up to 65 at %) beneath a thermally grown oxide [44].

B2. NiTi Uniform Corrosion

Of all the commercially available SMAs, NiTi is by far the most corrosion resistant [32, 40]. Resistance to acids depends on the acid concentration and temperature. Melton and Harrison report boiling concentrated nitric acid (HNO_3) will result in extensive weight loss, whereas exposure to 5% HNO_3 at room temperature (RT) will increase the thickness of the passive film and increase NiTi resistance to corrosion in other media [39]. The SMA Ni–45% Ti–3% Fe shows good resistance to salt fog corrosion after 192 h exposure at 30°C [40].

B3. NiTi Crevice and Pitting Corrosion

All passive alloys, such as the SMA NiTi, will generally corrode as a result of pitting or crevice corrosion when the protective passive film is attacked. This will generally occur either on an unrestricted surface with localized attack (pitting) or within a creviced region where diffusion of reactants and products from the anodic and cathodic process is restricted, resulting in crevice corrosion. Both forms of attack have been observed for NiTi SMA.

B3.1. Pitting of NiTi. Although many have reported good pitting resistance of NiTi in normal-pH chloride-containing environments, repassivation of damage passive films is generally considered difficult and slow [40]. Pitting resistance of NiTi is strongly affected by the surface finish used [41, 43, 45, 46].

The breakdown of a passive film is generally considered a random or stochastic process [46–50]. The comparison of the effect of various surfaces treatments on the pitting resistance of NiTi or the comparison NiTi to other materials has been the source of controversy for many years [51–61]. Using 20 specimens, Zhang et al. evaluated the survival probability of NiTi as a function of polarization voltage and compared it to 317L (Fe–14Ni–19Cr). They also showed pitting potential of the mechanical polished NiTi was generally greater than 317L stainless steel with the same surface treatment and that NiTi has a higher survival probability at higher potentials [47].

Zhang et al. found surface roughness on mechanical polished NiTi did not have an effect on the statistical distribution of breakdown potentials [47]. Melton and Harrison did some of the early work on the pitting resistance of NiTi SMAs [39]. They evaluated the breakdown potential of Ni–50% Ti–3% Fe and Ni–44% Ti–9% Nb NiTi SMA in

---

* All alloy compositions will be given in weight percent otherwise designated (i.e., atomic percent, at %).

natural seawater which was saturated with sodium chloride (NaCl), deaerated with argon, and pH adjusted to various pH levels with hydrochloride (HCl) [39]. Samples were tested in the as-mechanically-polished and HNO_3-passivated condition. Melton and Harrison [62] reported that the breakdown potentials were greater than 1.00 [saturated calomel electrode (SCE)] when the pH was greater than 5. When the pH was lower than 5, there was considerable scatter in the breakdown voltages and there were breakdowns as low as 0.200 V (SCE). Hwang et al. observed the decrease in E_b as pH is lowered [62]. The repassivation potential measured for Ni–50% Ti–3% Fe showed values just slightly lower than the breakdown potential for solutions with a pH greater than 5. For solutions with pH less than 5, the samples did not repassivate. The Nb-containing NiTi SMA had comparatively poor pitting resistance with breakdown potentials lower than 0.400 V (SCE) at a pH of 8 and decreasing with decreasing pH. Melton and Harrison concluded the HNO_3 passivation had leached out small precipitates of nominally pure Nb leaving a prepitted surface. When they anodized this sample, instead of passivating in HNO_3, no breakdown was observed down to a pH level of 4.5 [39].

Electropolishing increased the pitting resistance of NiTi in PBS, NaCl, Hanks' solution, and simulated human bile solutions over what is observed for mechanically polished NiTi surfaces [41, 46, 63]. Surface condition, however, does not have a significant effect on the repassivation behavior of NiTi, as is also the case with CP Ti [63]. Trepanier et al. also showed electropolished NiTi have high breakdown potentials of greater than 800 mV (SCE), whereas nonelectropolished NiTi (described as having a heavily oxidized surface) had breakdown potentials of about 200 mV (SCE). Trepanier also showed thermal oxidation of electropolished surfaces would degrade the breakdown potentials to less than 500 mV (SCE) [2, 45]. Similar results with respect to better pitting resistance of electropolished surface conditions compared to mechanically polished or thermally oxidized surfaces have been observed or reported by others [38, 43, 45, 46, 51, 63–65]. Stephan et al. performed electrochemical tests on thin-film NiTi that demonstrated that as-deposited thin films of

NiTi have breakdown potentials superior to a mechanically polished NiTi and suggested passivation or electropolishing of thin-film NiTi may be unnecessary to promote corrosion resistance in vivo [66].

There is some indication that the passive film on the surface of NiTi, which is responsible for the pitting resistance, is disrupted by the presence of surface-intersecting inclusions [10, 46, 54, 62, 67, 68]. Hwang et al. found Ti–C precipitates within every pit that he looked at and suggested that these pits act as pitting initiation sites as a result of the galvanic couple formed between the NiTi and TiC [62].

It has been observed that the breakdown potential for NiTi generally increases with increased exposure time to physiological solutions, as shown in Figure 38.1 [43, 50]. However both E_b and E_{corr} increase with immersion time and the margin of safety against pitting in vivo (E_b - E_{corr}) remains relatively constant, as shown in Figure 38.2 [43, 50]. Pitting is a stochastic process which results in a fairly wide variation in pitting resistance and SMA NiTi is no different, as shown in Figure 38.3 [46, 47].

B3.2. Crevice Corrosion of NiTi.
NiTi-based alloys are known to be susceptible to crevice corrosion in chloride (Cl^-)–containing environments [39, 69]. Crevice corrosion was observed on creviced plates of Ni–45% Ti–3% Fe after nine months exposure in Langston Harbour, United Kingdom [69]. Melton and Harrison also performed electrochemical crevice testing in artificial seawater and saline of different concentrations at 25°C on the same SMAs. From these tests, Melton and Harrison constructed diagrams showing the critical potential for crevice corrosion and repassivation as a function of total dissolved solids (TDSs). A plot of the critical crevice and repassivation versus the log (TDS) is linear with the crevice and repassivation potential decreasing as TDS increases. They also observed a strong surface finish effect on the crevice corrosion results.

Kaczmarek evaluated the crevice corrosion resistance of NiTi in the as-ground, electropolished, and passivated surface conditions. These tests were done in accordance with the American Society for Testing and Materials (ASTM)

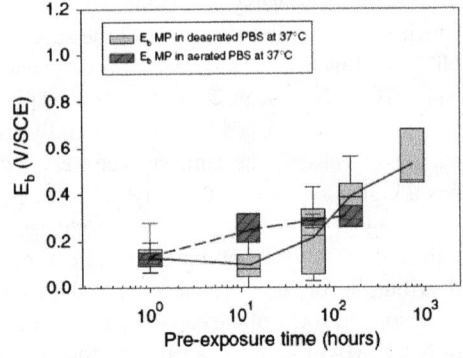
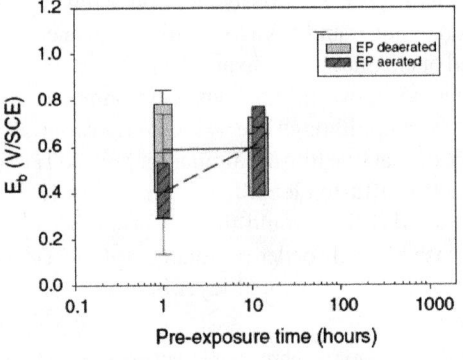

FIGURE 38.1. Increase in E_b upon expose to PBS for mechanical polished NiTi (left) and electropolished (right) NiTi in 37°C PBS during ASTM F2129 polarization testing [46].

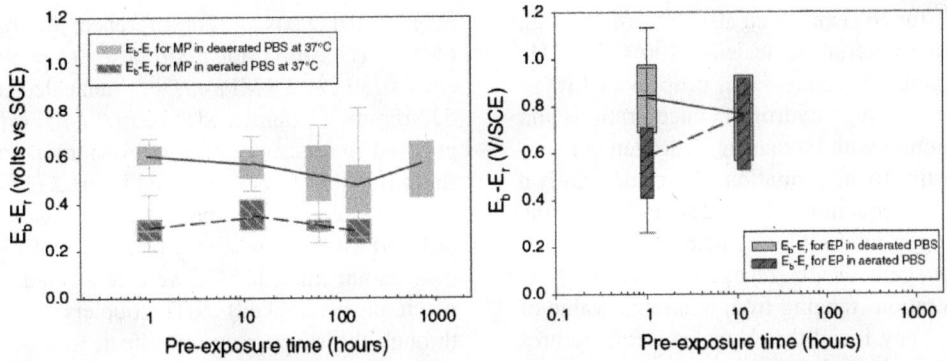

FIGURE 38.2. Margin of safety versus pitting in vivo ($E_b - E_r$) is relative constant with exposure time to 37°C PBS. Mechanical polished NiTi (left) and electropolished (right) [46].

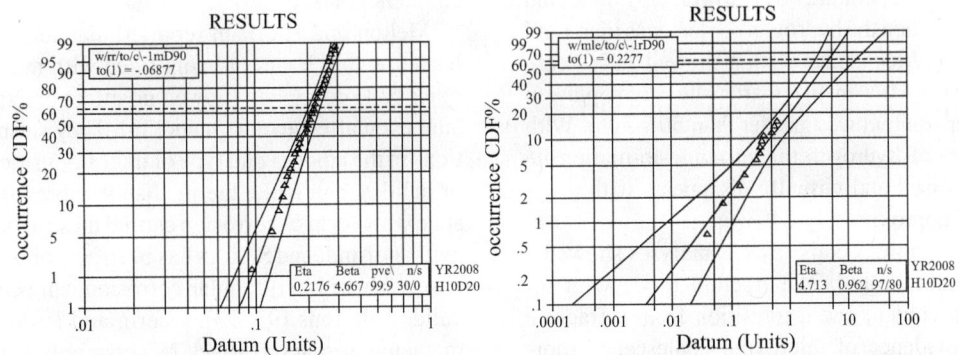

FIGURE 38.3. Margin of safety versus pitting in vivo ($E_b - E_r$) is relative constant with exposure time to 37°C PBS. Mechanical polished NiTi (left) and electropolished (right) [46].

standard F746 [70] in Tyrode's physiological solution at 37°C [71]. The E_{corr} for these three conditions were −248, −193, and −186 mV (SCE) for the as-ground, electropolished, and passivated (boiling water for 1 h) surface conditions, respectively. The critical crevice potential was +450 mV (SCE) for the as-ground surface but was greater than +800 mV for the electropolished and passivated surface conditions. This indicates the surface condition can affect the crevice corrosion resistance of NiTi.

Trepanier et al. showed that overlapped NiTi stents may be susceptible to crevice corrosion [72].

B4. Stress Corrosion Cracking and Hydrogen Effects of NiTi

B4.1. Hydrogen Effects. Hydrogen (H) is known to have detrimental effects on the fracture properties of Ti alloys and NiTi is no exception [73, 74]. Various investigations have shown that under certain conditions severe hydrogen embrittlement (HE) of NiTi SMAs can occur [39, 56, 59, 73, 75–83]. Hydrogen can be absorbed into NiTi during pickling, plating, and caustic cleaning [74]. For instance, NiTi orthodontic arch wires (used in the oral cavity) have been known to show time-dependent failure, which was suspected to result from HE [77]. Relatively short exposures to the hydrogenated pressurized primary reactor water

chemistries can result in hydrogen level in excess of 1000 ppm [74].

Various levels of hydrogen have been considered detrimental to NiTi. For instance, Duerig and Pelton state that hydrogen in excess of 20 wppm* can degrade ductility and more than 200 wppm can severely embrittle NiTi [74]. Others report HE has been observed in NiTi at concentrations of approximately 100 wppm and reported to cause a decrease in the ductility, a loss of the shape memory properties, and a decrease in the fatigue life of NiTi [81, 84, 85]. Studies have shown that the ductility and strength in NiTi are reduced beginning at nominal concentrations (approximately 10–50 wppm) of hydrogen [76, 84].

Asaoka et al. estimated the diffusion constant of hydrogen in NiTi (55.8% Ni) at RT to be $9 \times 10^{-15}\,m^2/s$ [77]. Solubility of hydrogen from 52 to 427°C shows hydrogen is absorbed in NiTi up to a maximum of about 40 at % (13,000 wppm) without hydride formation [86–89]. Schmidt et al. found solubility decreases with increasing temperature and that solubility follows Sievert's law (i.e., solubility is linear with the square root of hydrogen pressure) [89, 90].

He et al. reported hydrogen charging can significantly decrease toughness [91]. He reported the fracture toughness (K_{IC}) for hydrogen-free samples to range between 36

---

* Parts per million by weight (wppm).

and 44 MPa-m$^{-1/2}$ for NiTi annealed at 150°C and as high as 53 MPa-m$^{-1/2}$ for materials annealed at 700°C/$^1/_2$ h. He et al. noted the fracture toughness can drop to as low as 1.3 MPa-m$^{-1/2}$ for very high hydrogen concentrations and the toughness reduction with increasing hydrogen concentration is primarily due to the formation of hydrides. Only a small portion of the reduction (about 2%) is due to the formation of hydrogen-induced martensite [91].

Runciman et al. cathodically charged austenitic NiTi samples to concentrations ranging from a baseline value of 9–650 wppm [84]. They found the A_s and A_f temperatures were lowered by about 8°C when the hydrogen content was increased from 9 to 240 wppm hydrogen. The martensite phase transition temperatures were more strongly affected where the M_s and M_f temperatures were lowered by 80°C and 110°C, respectively when the hydrogen content was increased from 9 to 200 wppm. Mechanical testing showed an increase of approximately 60 MPa in the martensite stress plateau when the hydrogen content was greater than 50 wppm. With increasing amounts of hydrogen the austenite-to-martensite transition is suppressed and virtually disappears with a hydrogen content of approximately 240 wppm.

Scanning electron microscopy (SEM) analyses showed a transition from a ductile-to-brittle fracture mode with increasing hydrogen content and a transition from a fracture surface showing evidence of microvoid coalescence morphology to one which exhibited a featureless, transgranular morphology [75, 84]. The microvoids from the hydrogen-embrittled samples appear to be much smaller than those observed in the nominal, as-received material [75].

Sheriff et al. investigated the effect of hydrogen on the fatigue life in rotating bending [78]. Experimental conditions include load cycling up to 10^7 cycles, a temperature of 23°C, strain amplitudes that ranged from 0.5 to 4.3%, 5–15 samples tested at each load condition, hydrogen contents that ranged from 10 to 80 wppm, and NiTi with $A_f = 13°C$. They did not report the loading frequency. Their data suggested hydrogen concentrations as low as 50 wppm caused a small, yet statistically significant, decrease in fatigue life above 1.4% cyclic strain. However, increasing hydrogen concentration did not appear to affect fatigue life below 1.4% strain up to 80 wppm hydrogen.

B4.2. Stress Corrosion Cracking of NiTi. Given the susceptibility of NiTi to HE, it is not surprising that this alloy can, under some circumstances, exhibit stress corrosion cracking (SCC). We use the term SCC in its broadest sense to include any type of environmentally assisted cracking (EAC), which could include slip dissolution and hydrogen-assisted cracking.

Melton and Harrison reported on the SCC of Ni–50% Ti–3% Fe in both seawater and boiling magnesium chloride [39]. No SCC was observed in precracked cantilever beam specimens of Ni–50% Ti–3% Fe exposed to natural seawater at a stress intensity (K_I) equal to 44 MPa-m$^{-1/2}$

after 2900 h exposure. For reference, the fracture toughness of NiTi (i.e., the critical stress intensity for fracture, K_{IC}) equals 39.2 ± 2.8 MPa-m$^{-1/2}$ (annealed at 150°C) and 53 MPa-m$^{-1/2}$ (annealed at 700°C/$^1/_2$ h) [73]. The specimens exposed to concentrated (double-normal seawater concentration) that had been acidified to pH 2 did crack. The SCC stopped when the threshold stress intensity for SCC (K_{ISCC} = 31 MPa m$^{-1/2}$) was reached. Their SCC tests in boiling magnesium chloride SCC were conducted using smooth (no notch or precracked) NiTi couplers that were clamped, through the shape memory effect, to a solid titanium rod. These stressed couplers were exposed to boiling magnesium chloride at 155°C for 96 h. All the Ni–50% Ti–3% Fe specimens cracked but none of the Ni–44% Ti–9% Nb couplers cracked [39].

Melton and Harrison reported that some NiTi couplings leaked within hours of being filled with methanol, whereas, other NiTi couplings showed no effect of exposure to methanol even after several years [39]. They surmised that variations in the amount and type of trace impurities, such as water or halides, can make the methanol aggressive. They do not state what form of corrosion caused these leaks, but likely it is methanol-induced SCC as has been also observed on other Ti alloys where intergranular corrosion can occur in methanol halide solutions [92, 93]. Duerig and Pelton also note that methanol appears to attack NiTi but only when diluted with low concentrations of water and halides, which leads to pitting and tunneling corrosion similar to what occurs in titanium alloys [74]. Adding as little as 1.5% water is sufficient to suppress the methanol SCC in Ti alloys [92].

Yokoyama et al. showed that NiTi is susceptible to time-dependent initiation and growth of SCC cracks under constant-load conditions [77, 80–83, 94]. Initially, this work was motivated from an orthodontic viewpoint. The constant-load testing was done with smooth wire specimens of NiTi exposed to phosphate solutions containing various levels of fluoride at pH levels as low as 5. The time to failure for these smooth wires exhibited as strong stress dependence with wires subjected to high stresses failing in a few hours and at lower stresses taking 100 h or longer. This is consistent with the time-dependent failures of NiTi orthodontic arch wires noted to break a few months after placement [77]. Yokoyama et al. also noted a discontinuity in the time-to-failure versus stress curves, which he attributed to an increase in hydrogen absorption when the critical stress for martensite transformation is exceeded. They suggested the time-dependent failure may be a result of both HE and active path SCC.

B5. NiTi Galvanic Corrosion

According to Melton, NiTi is slightly more noble in the galvanic series than 316 stainless steel (316 SS) [40]. Venugopalan and Trepanier determined that when NiTi is coupled to stainless steel, Ti and tantalum (Ta) this coupling does not significantly affect its corrosion behavior [95]. However, coupling NiTi to Au, Pt, or Pt–Ir alloys can result in an order-of-magnitude increase in corrosion rate [95]. Hwang suggested there is a galvanic current between the NiTi and Ti–C inclusions at the surface that act to initiate pitting [62].

One example of what may be galvanic in vivo corrosion is described by Kong. He describes an explanted NiTi Amplatzer septal occluder in direct contact with the Pt leads of a pacemaker for 18 months. The occluder showed minor pitting of the passive layer and is believed to be galvanic in nature [96].

B6. NiTi Corrosion Fatigue

Robertson and Ritchie have studied the effect of frequency of loading on the fatigue crack growth rate of NiTi [97]. In their tests they found no difference between the fatigue crack growth rates of NiTi fatigue tested in air or Hanks' balanced saline solution (HBSS) at 37°C. Nor did they find any difference in the fatigue crack growth rates for NiTi tested in HBSS at 37°C at 1 or 50 Hz. Both these observations indicate that corrosion fatigue does not appear to affect NiTi, at least in these environments and for these loading frequencies.

B7. NiTi Fretting Corrosion

The repassivation rate of passive oxides on corrosion-resistant materials is important when such materials are being used where fretting or wear can damage the protective surface oxide when the material is being used in a corrosive environment. For instance, such situations can occur with NiTi used in vivo as a stent and the stent is a woven structure or is rubbing against another stent [2, 38, 61, 72, 96]. The regeneration of a passive surface layer on the Ti surface in in vivo environments is generally slow and Ti–corrosion product has been found in the tissue adjacent to some Ti implants [38, 67, 72].

Asaoka et al. speculated that fretting corrosion may damage the protective passive surface layer of NiTi, allowing water to react with the bare metal surface that could generate adsorbed hydrogen [77].

In implanted medical devices there are concerns regarding fretting corrosion or wear of NiTi at crossover points in braided or knitted wire stents [2]. The U.S. Food and Drug Administration (FDA) required an evaluation of stents that might experience fretting corrosion [98]. Kong examined NiTi Amplatzer septal occluders and no wire fractures were found in vitro after cycle testing with 400 million cycles or in devices taken from the animals and humans [96]. Biochemical studies showed no significant elevation of Ni levels after implantation. These test results imply that fretting corrosion did not occur in these devices in vitro or in vivo during fatigue loading [96].

The wear or fretting between overlapped NiTi stents in PBS has been evaluated in vitro through axial fatigue testing [72]. The overlapped NiTi stents showed about a 300-mV decrease in pitting resistance (E_b) after these fretting tests but were still considered to have sufficient corrosion resistance for use in vivo [72].

B8. NiTi High-Temperature Oxidation

Below about 100°C, NiTi remains shiny in air, but at higher temperatures, the oxide surface slowly thickens, giving interference colors [39, 44]. A gold color is seen for 30 nm oxide (2-min anneal at 540°C), blue color for 45 nm (5-min anneal at 540°C), turning back to golden after annealing to 30 min (125 nm) [44]. Melton and Harrison reported that oxidation above 700°C is more like stainless steel than Ti since no internal oxidation or absorption of oxygen occurs. The oxidation behavior of the NiTi in air at 750–950°C shows parabolic oxidation rate kinetics except for the initial stages. The activation energy for oxidation was determined to be 59 kcal/mol [99]. Titanium is selectively oxidized and the oxide grows from the metal–oxide interface. Satow et al. found the oxidized specimen has a layer structure of TiO_2 (rutile) on the outermost layer, the TiO_2 phase containing $TiNiO_3$ particles, a porous Ni phase containing Ti, the Ni_3Ti phase, and the TiNi phase in sequence from the outer side. They found the rate-determining process for oxidation was the inward diffusion of oxygen ions through the oxide (rutile) layer and the rate of oxidation is slightly faster than that of pure Ti [99].

Undisz et al. investigate the mechanical stability of surface oxide layers on NiTi when subjected to the large reversible pseudoelastic strains (up to 8% strain) [44]. Such cracks in the protective oxide layers could have a significant effect on the corrosion and biocompatibility behavior. They found that the critical strain required to crack the oxide layer was less than 0.6% strain. Furthermore they also note that NiTi undergoes Lüders straining, which can amplify the applied strain. For instance, the application of 1% strain to NiTi can lead to 3% strain at a local level [44]. They found that under tensile straining cracks are formed perpendicular to the loading direction as a result of the longitudinal elongation and parallel to the loading direction as a result of the Poisson contraction of the material. Thin oxide layers (<80 nm) do not form the cracks parallel to the loading direction, and these perpendicular cracks can close during unloading. For oxides layers thicker than 80 nm both types of cracking occur and result in flaking of oxide particles exposing the nickel-enriched layer underneath [44].

B9. Corrosion of NiTi Medical Devices

The use of SMA in medicine has grown rapidly since the early 1980s. The medical device market includes a broad range of equipment and devices used in orthopedic, neurology, cardiology, and interventional radiology [12]. Such applications include stents, guide wires, vena cava filters, and a variety of orthopedic devices. As a result of these applications, considerable research has been done looking into the in vivo and in vitro corrosion resistance.

B9.1. In Vivo Humans.

Hasters et al. used NiTi for various devices in orthopedic surgery for more than 12 years when he wrote his article in 1990 with no known indications of corrosion or biocompatibility [100]. He describes the condition of SMA osteosynthesis staples used for the same applications where Blount's bone staples are used. He describes these staples as having been used in human ankle and knee joint surgeries but does not describe the surface finishing used for these NiTi staples, but it appears they were in an as-ground and polished but not electropolished condition. These staples had been in service for 6–16 months. Staples were surrounded by well-vascularized scar tissue. SEM examination at 1000X magnification showed the indications of the as-ground surface with TiC inclusions. No areas of corrosive attack were detected. For his surgeries, where he used NiTi as a spacer for bone-chip arthrodesis of the spinal column he stated he screened patients first for Ni and Ti allergies [100]. Ninety-five patients were treated with NiTi implant, with exposure times as long as four years, without indication of complications. Lu reported similar results on NiTi staples used for internal fixation [3].

Ries et al. reported on the systemic Ni release after implantation of the NiTi Amplatzer septal occluders [101]. They examined 67 patients with no history of Ni sensitivity with blood samples taken 24 h before and 24 h, 1, 3, and 12 months after occluder implantation. Ni-serum concentrations were measured by atomic absorption spectrometry. They considered the normal Ni concentration to be less than 2 ng/mL. They found the mean serum levels of Ni increased from 0.47 ng/mL before implantation to 1.27 ng/mL 24 h after implantation and ultimately to a maximum of 1.50 ng/mL one month after implantation. During follow-up examinations, the Ni values decreased to those measured before implantation. They speculated Ni was released until a calcium–phosphate layer formed on the passive oxide film of the device or until endothelialization was completed.

Pertile et al. recently measured the open-circuit potential (OCP) of NiTi in vivo in humans [102]. They determined the average in vivo human OCP determined from six independent measurements on human patients in the arterial system was -0.334 ± 0.030 V (SCE). This value was in good agreement with their data from in vitro testing using simulated body fluids [-0.313 ± 0.003 V (SCE) in AFNOR S90-701 artificial saliva, -0.334 ± 0.001 V/SCE in artificial urine, and -0.239 ± 0.007 V/SCE in Ringer's solution]. They noted that, since no E_b values were reported lower than 0.0 V/SCE, NiTi should be considered resistant to pitting in vivo [102].

There have been a few reports of NiTi implants and orthodontic arch wires pitting in vivo [38, 103–106]. Not all agree that this is actually a result of in vivo exposure and may in fact be an artifact of the cleaning and disinfection processes [107]. Additionally, later explant investigations by Major and Guidoin do not report pitting observed on NiTi explants [108].

Carroll and Kelly recently showed the breakdown potential (E_b) of NiTi in Ringer's solution is lower when tested in blood than when tested in Hanks' solution. This implies there are proteins, amino acids, or other components in blood, which improves the corrosion resistance of NiTi over that of Hanks' solution and that corrosion testing in synthetic physiological solutions will give conservative results, that is, does not overpredict pitting resistance [109, 110].

B9.2. In Vivo Animals.

Lu reported Zhang's data on long-term uniform corrosion rate determinations for NiTi implanted in rabbits for up to one year, see Table 38.2 and

TABLE 38.2. Corrosion Rates of NiTi

Media	Corrosion Rate (mm/year)	References
1 wt % NaCl	5.5×10^{-5}	[3]
0.1 wt % NaSO$_4$	6.9×10^{-5}	[3]
1 wt % Lactic acid	5.7×10^{-5}	[3]
Acetic acid (50–99.5% at 30°C to boiling)	2.5×10^{-2}–7.6×10^{-2}	[74]
0.05 wt % HCl	0	[3]
3 wt % HCl at 100°C	9×10^{-3}–8.4×10^{-2}	[74]
10 wt % HNO$_3$ at 30°C	6.4×10^{-4}	[74]
60 wt % HNO$_3$ at 30°C	6.4×10^{-3}	[74]
5 wt % HNO$_3$ at boiling point	5×10^{-2}	[74]
8 wt % FeCl$_3$ at 70°C	2.3×10^{-1}	[74]
Synthetic saliva	2.9×10^{-5}	[3]
Deaerated Hanks' solution	7.85×10^{-5}	[95]
Rabbit (360 days exposure)	2.4×10^{-5}	[3]

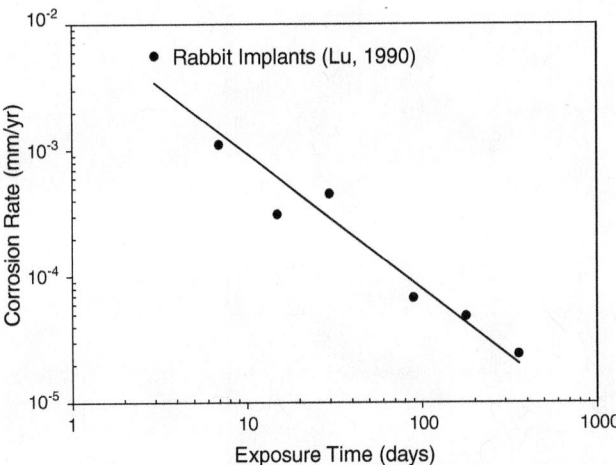

FIGURE 38.4. Plot of Lu's corrosion rate data of NiTi implanted in rabbits [3].

Figure 38.4 [3, 111]. The corrosion rate showed a decreasing corrosion rate with increasing exposure time as would be expected for a passive material. The corrosion rate measured after a one-year exposure is quite small (i.e., 24 nm/year).

Lu reported histological observations on 60 NiTi samples implanted in the femur and subcutaneously in rates for periods of 3–10 months [3]. X-ray radiography after one week showed newly formed bone in contact with the NiTi. No evidence of reaction, rejection, or resorption of bone was observed in the 10-month period. Tissue response formed a thin pseudomembrane around the test specimens and microscopy showed no evidence of inflammatory reaction. They concluded that NiTi had good biocompatibility.

Lu also reported on in vivo animal testing to evaluate Ni release from NiTi [3]. In these tests, NiTi samples were placed in the soft tissue of the dog's right-hind legs and the Ni content of the hair was measured as a function of time. Initial Ni concentration of the hair ranged from 0.5 to 0.6 ppm Ni. After a year, the Ni content ranged from 0.92 to 0.96 ppm, showing a small increase in Ni.

B9.3. In Vitro Cell Culture.
The cytotoxicity of NiTi is comparable with other implantable alloys [3, 8, 52, 112–114]. For instance, Lu describes cell growth inhibition tests in which mouse fibroblast L-cells were grown in Eagle's medium with the addition of 10% calf's serum. The NiTi samples were placed in the culture chamber containing a humidified CO_2 atmosphere at 37°C and a cell suspension was placed on the NiTi for 24 h. The samples were then fixed and compared with a negative control. The results showed the cells grew well on the NiTi [3]. Shabalovskaya et al. showed that cytotoxicity is greatly affected by surface preparation [10].

B9.4. Quality Assurance Testing for Long-Term NiTi Medical Device Implants.
ASTM Standard F2129 provides a quantitative method recognized by the FDA for the accelerated assessment of the corrosion resistance of medical devices [115]. This test determines, among other things, the rest potential* (E_r), breakdown potential (E_b), and repassivation or protection potential (E_p). These values are illustrated in Figure 38.5 for NiTi tested in accordance with ASTM F2129 in PBS [46]. An example of the metal hydroxide plume that forms in the PBS at the pit location is shown in Figure 38.6. An example of a pitted NiTi surface is shown in Figure 38.7.

Currently, there is no universally accepted acceptance criterion for the corrosion resistance of NiTi for medical device implants as measured by ASTM F2129. Results from this test method can be used in a comparative sense to evaluate a new device against a comparable predicate device that has been previously approved for use by the FDA and has been used for several years without any known corrosion failures [46, 64, 65, 116]. This is sometimes quite expensive as predicate devices can be quite expensive. Some companies and researchers suggest a simple criteria in which E_b is specified to be greater than a specified value [2, 117, 118]. For instance, Stoeckel et al. stated Cordis, a Johnson & Johnson company, established $E_b > 500$ mV (SCE) as an acceptance standard for corrosion resistance for implantable NiTi devices [2]. This value was chosen as it corresponds to the corrosion resistance of the stainless steel Palmaz-Schatz stent, the stent with the longest implant history.

Corbett and Rosenbloom suggested a similar criterion where $E_b > 600$ mV (SCE) is acceptable and where $E_b < 300$ mV it is unacceptable. If E_b values range between 300 mV (SCE) and 600 mV (SCE) the medical device is considered to have marginal corrosion resistance and additional testing is required. These methods do not take into account the environment or the material or provide guidance on how many specimens need to be tested. For instance, the margin of safety against pitting is really the difference between the breakdown potential and the rest potential in vivo (i.e., $E_b - E_r$). Eiselstein et al. proposed a quality assurance (QA) acceptance criterion for NiTi for use in long-term medical devices that provides for such a methodology and how many specimens to test [46, 65].

B9.5. Nickel Release.
Implantable medical devices must be able to withstand the corrosive environment of the human body for 10 or more years without adverse consequences. Most research has been focused on developing materials and devices that are biocompatible and resistant to corrosion fatigue, pitting, and crevice corrosion. The biocompatibility studies have generally been done by examining the tissue

---

* The rest potential E_r is also variously known as the OCP, or the corrosion potential (E_{corr} or $\Delta\Phi_{corr}$).

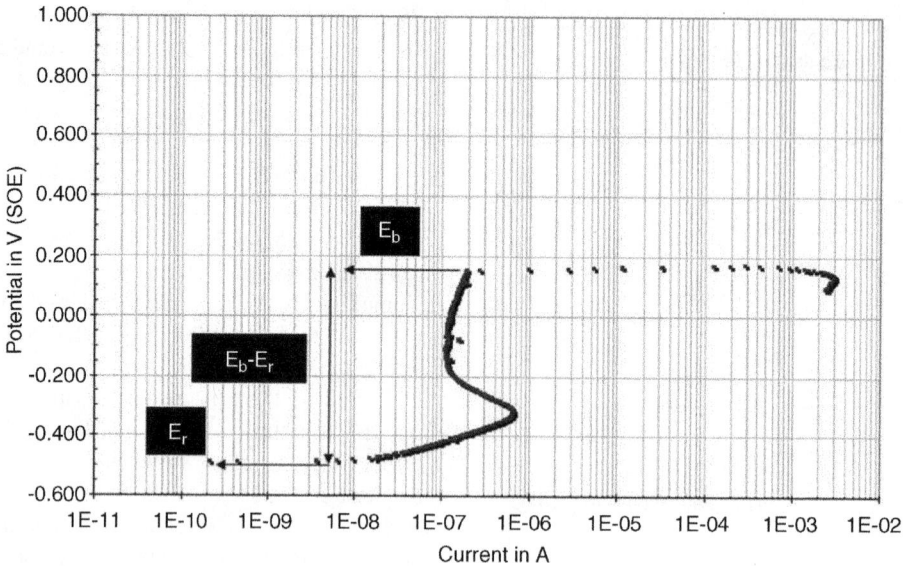

FIGURE 38.5. Typical potentiodynamic polarization curve obtained for NiTi rod tested in PBS solution at 37°C [46].

surrounding devices implanted in animals and how cells respond to the material surface. Although such tests are critical for developing safe medical devices, they provide little help in understanding what it is about a particular material or surface condition that is affecting cytotoxicity or immunological response [119, 120]. The rate at which various alloy elements go into solution, the metal ion leaching rate, is likely to have a direct bearing on the toxicity, biocompatibility, and immunological response of a medical device. Currently, there has been little reported on the metal ion leaching rate for NiTi; however, it appears research into

how various surface treatments affect release rate is increasing rapidly. The release of Ni is of concern as it is a known allergen and carcinogen [121, 122].

The amount of metal ions that are released into the body from an implanted medical device depends on the alloy, its surface treatment, its total surface area exposed to tissue/body fluids (or correspondingly the synthetic physiological solutions used to simulate the in vivo environments), and how long it has been implanted. Much of the currently reported information on Ni ion leaching from NiTi does not report all the required information to evaluate and compare different alloys and surface treatments. For instance, some researchers only report solution concentration (μg/Li of Ni^{2+}) versus exposure time, whereas, the flux (μg/cm^2/week) of Ni^{2+} versus exposure time is actually needed for toxicological comparisons [123]. Nevertheless, some of the reported data clearly illustrate the importance of various variables that affect metal ion flux emanating from an implantable medical device.

Fujita et al. (as reported by Miyazaki) compared the leaching rate of Ni, 304 and 316 stainless steel, and NiTi in the martensite and austenite phase* [1, 124]. He tested these specimens in the as-polished with emery paper condition in 0.9% NaCl at 37°C. Fujita only reported micrograms of material released and does not provide the exposed surface area. Nevertheless, from these data it is clear the metal release rate decreases over time for all the alloys tested and

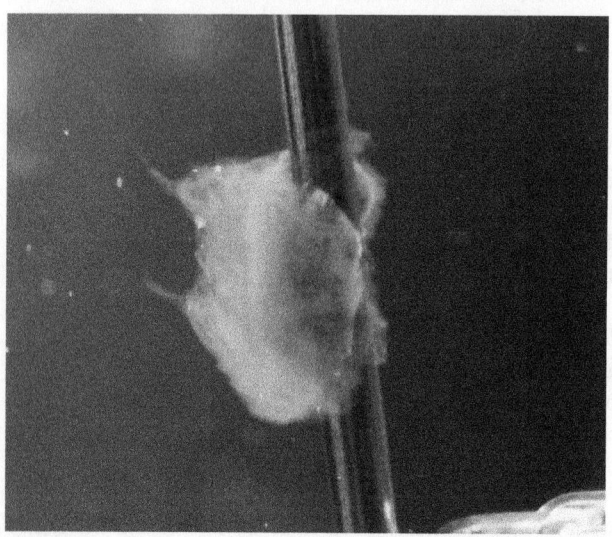

FIGURE 38.6. Pit formed on surface of electropolished superelastic NiTi rod as a result of potentiodynamic cyclic polarization testing in PSB solution at 37°C [46].

* This was done by using alloys with slightly different chemistry or processing which gave M_s temperatures of 0 and 40°C. When tested at 37°C the one with the 0°C M_s was austenitic and the one with the 40°C M_s was partially martensitic.

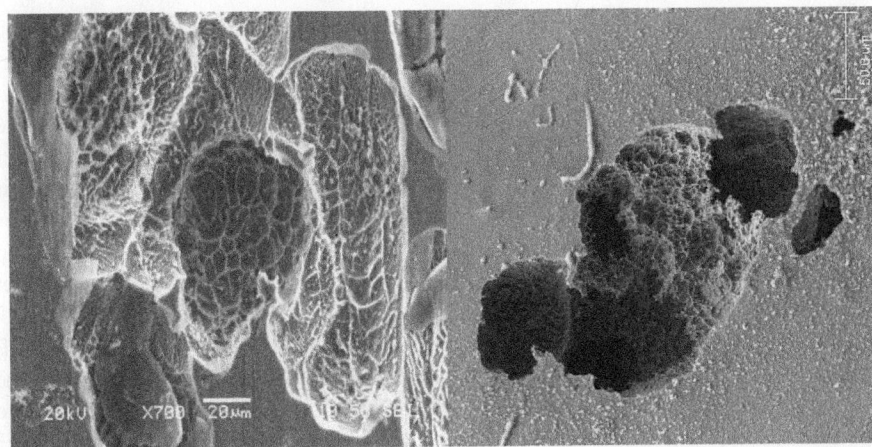

FIGURE 38.7. Example of pits generated in mechanical polished NiTi (left) and electropolished (right) NiTi in 37°C PBS during ASTM standard F2129 polarization testing.

that the Ni released from the NiTi in both the martensite and austenite phases was less than half that measured for the stainless steels tested. These data also indicate that the Ni leaching rate from the austenite was slower than from the martensite phase. Fujita data also showed that passivation treatments will significantly reduce the Ni release rate [1, 124].

Venugopalan et al. showed the decrease in dissolution rate with immersion time. He showed that Ni release from NiTi decreases from well below dietary levels to nearly nondetectable levels in the first few days following immersion in a physiological medium [95]. The data from Trepanier et al. also illustrated the decrease in Ni leaching rate with increasing exposure time and that surface treatment can significantly affect the Ni release rate [123]. They found electropolished NiTi released significantly lower amounts of Ni compared to mechanical polished samples. Trepanier et al. reported the volume of solution used for the leaching but did not report the surface area of the NiTi exposed to Hanks' solution, thereby making these results difficult to compare to other research on Ni leaching from NiTi. Wever et al. reported that the NiTi exposed to deaerated Hanks' solution has a first-day Ni release rate of 0.88 μg/cm^2/week to nondetectable after 10 days [125]. Clarke et al. showed that the surface condition has a significant effect on the Ni leaching at nondetectable levels (less than 5 ppb) for etched wires, etched pickled, or etched–pickled–mechanical polished, whereas Ni leaching rates were significantly higher for wires in the as-drawn condition [43].

In another study on Ni leaching, McLucas, et al. studied leaching as a function of time from NiTi wires given different surface treatments (as drawn, etched, mechanical polished, etched again, and pickled) to determine the effect on human umbilical vein cells (HUVECs) at the transcriptional level by real-time polymerase chain reaction (PCR) measurements of the expression level of three known inflammatory mediators compared to control cells [121]. One of these inflammatory mediators, E-selectin (a marker for endothelial cell injury), was found to be upregulated as a result of Ni release. E-selectin is an adhesion molecule that plays a role in postimplantation reactions and is involved in the initial immunological response [121]. Other experiments showed this upregulation was not a result of surface topography. The Ni release rates reported for 24, 48, and 72 h were determined by exposing wires (total surface area of 3.84 cm^2) to 3 ML of cell culture media at 37°C [121]. The media was then analyzed for Ni^{2+} concentration. Although they only reported the Ni concentration of the media, these values can be converted to an average Ni^{2+} flux over exposure time using the reported surface area and exposure times. The Ni flux ranged from 1.8 μg/cm^2/week to less than 0.009 g/cm^2/week depending on the surface finish. They found that higher Ni concentrations in the passive film layer (as measured by XPS) gave higher Ni leaching rates. The Ni concentrations in the passive films varied from 15 at % to less than 1 at %. Although the average Ni flux over 72 h of exposure was small (less than 0.06 ± 0.007 μg/cm^2/week), it still had a biological effect by upregulating E-selectin.

Fasching et al. have recently reported on how heat treatment, surface condition, and strain levels affect the nickel leaching rate from NiTi surfaces [126]. They found that nickel release rates (μg/cm^2/wk) decreased with increasing immersion times and very low release rates were observed after 30 days. The electropolished sample had the lowest initial nickel release rate of about 7 μg/cm^2/wk, which decreased to less than 0.7 μg/cm^2/wk after one week immersion. Prestrained samples had higher nickel release rates and took longer to reach the low steady-state rates [126].

Okazaki et al. recently showed that the Ni and Ti release rate for mechanically polished NiTi is relatively constant (about 0.1 μg/cm^2/wk) between pH 4 and 8 but increased rapidly with decreasing pH (below pH 4) [127]. The nickel release rate for NiTi is much smaller than that measured for stainless steels and Co–Cr–Mo–Ni–Fe [127]. Although the rapid increases were observed at approximately pH 2, the quantities were even higher than that of Co released from the Co–Cr–Mo and Co–Cr–Mo–Ni–Fe alloys [127].

The Ni leaching rates for NiTi can be compared to other common implant alloys, such as stainless steel and Co-based alloys. For instance, Herting et al. investigated the effect of various simulated physiological solutions [Gamble's solution and artificial lysosomal fluid (ALF)] on the rate of metal ion leaching from various stainless steels [128–130]. They found metal dissolution rates decrease with exposure time, and for stainless steels the metal ion leaching rate is faster at pH 4.5 than at 7.4. For instance, Herting et al. observed the average Ni ion release rate on the first day of exposure of 316L stainless steel in the "as-received" surface condition to ALF solution (pH 4.5) to be 0.33 $\mu g/cm^2$/week, whereas the Ni release rate dropped to 0.08 $\mu g/cm^2$/week after one week of exposure. The average metal ion release rates in pH 7.4 Gamble's solution was at least one order of magnitude lower, 0.006 (Ni) $\mu g/cm^2$/week. Preexposure to Gamble's solution for 8–24 h reduces the total metal ion leaching rate during one week of exposures to the higher pH ALF solutions. Ornberg et al. reported that the metal ion release rate of passivated Ni–Co–Cr alloy 35N LT (36Ni–33Co–20Cr–10Mo) exposed to PBS (pH 7.4) to which 100 mM H_2O_2 was added to simulate immunological response to be 0.55 (Ni) $\mu g/cm^2$/week respectively, after 3 h of exposure [131]. These leaching values are probably higher than observed for stainless steel due to the short exposure time and the oxidizing effect of the peroxide. From this, we can see that the Ni leaching rate of NiTi is comparable to other common implant materials even though the Ni concentration of this SMA is significantly higher than stainless and cobalt-based implant alloys. In addition, the Ni leaching rate can be compared to the daily intake of Ni from food and water, which is estimated to be 200–300 mg/day [95, 122, 132]. The typical Ni leaching rate from medical devices (except for some orthopedic implants with very large surface areas) has Ni release rates several orders of magnitude lower than this daily dietary intake. This, however, should not be taken as evidence that such low release rates are safe, as it has been shown that even low release rates can have biological effects.

B9.6. Effect of Synthetic Physiological Solutions on Pitting. Pound showed that the type of synthetic physiological solution (simulated human bile, salt-only bile, PBS, and Hanks' solution) used had little effect on pitting resistance for NiTi in the electropolished condition with all solutions giving breakdown potentials greater than 1 V (SCE) [41]. However, Pound found that NiTi alloys in the mechanical-polished condition had lower pitting resistance when tested in simulated bile as compared to their pitting resistance when tested in the salt-only bile or PBS as based on $E_b - E_{corr}$ results [41].

Speck and Fraker found pitting resistance of NiTi was not affected when tryptophan or cysteine (amino acids) were added to Hanks' solution at blood-level concentrations [133]. They did find, however, that cysteine did decrease E_b at higher concentrations (4300 times the concentration found in blood). Pound reported results on pitting resistance of NiTi exposed to various amino acids and bovine serum. He found cysteine may increase the pitting susceptibility at blood level concentrations compared to values obtained in PBS. However, he found there was no difference between pitting susceptibility when tests are done in PBS or bovine serum indicating PBS is adequate for simulating in vivo environments for NiTi [134]. Hansen showed that an albumin–fibrinogen combination added to Hanks' solution will decrease E_b and $E_b - E_r$ compared to tests done in Hanks' solution without additives [134, 135].

The pH of the test solution appears to have the greatest effect on corrosion behavior. Speck et al. found that decreasing the pH will decrease breakdown potential and Okazaki et al. observed the amount of Ni leached from NiTi increases dramatically as the pH of the test solution is lowered [127, 133].

C. β-Ti: Ni-FREE Ti-BASED SHAPE MEMORY ALLOYS

Ni-free Ti-based alloys are being developed for shape memory/superelastic biomedical applications [37]. Such materials are important as it is thought they may be more biocompatible than high-Ni-content NiTi compositions most commonly used today for medical devices and implants [7, 29, 30]. SMAs based on β-Ti can exhibit one-way shape memory effects on the order of 3–3.5% strain [15, 30].

Recently, a Ni-free Ti–18 at % Nb–4 at % Sn SMA (elastic strain of 3.5%) was developed to avoid possible allergic or carcinogenic reactions that are usually of concern with Ni-containing alloys [29, 30]. The corrosion behavior of the Ti–Nb–Sn SMA in 0.9% NaCl, 0.05% HCl, and 1% lactic acid in aerated solutions at 37°C was investigated by immersion testing with ICP spectrometry and electrochemical measurements [29]. Takahashi et al. reported this alloy has comparable corrosion resistance to commercial pure Ti and much higher corrosion resistance compared to conventional NiTi SMA [29]. They reported the total metal ion release from Ti–Nb–4Sn in HCl is 1/5th and 1/15th as high as commercial pure Ti and NiTi, respectively.

Saito et al. recently reported another class of superelastic alloys described as "gum metal" [30, 31]. This is a β-type Ti alloy with low modulus and elastic strain of 2.5%; however, its superelasticity is not considered to be the same as for more typical SMAs [31]. Various compositions have been described such as Ti–29Nb–13Ta–4.6Zr [30], Ti–23 at % Nb–0.7 at % Ta–2 at % Zr–O [31], and Ti–12 at % Ta–9 at %

Nb–3 at % V–6 at % Zr–O [31]. Guo reported the corrosion behavior of these newly developed multifunctional β-type Ti–23 at % Nb–0.7 at % Ta–2 at % Zr–O alloy and compared it to Ti–6Al–4V behavior in Ringer's solution. The OCP, potentiodynamic polarization, and XPS techniques were used in this evaluation [136]. The Corrosion property was also measured for comparison. The results showed "gum metal" alloy possessed much better corrosion resistance than Ti–6Al–4V alloy. This alloy had a high corrosion potential and high breakdown potential attributed to the stable and inert passive TiO_2 film modified by the oxides of Nb, Ta, and Zr [136].

Wang et al. studied the corrosion resistance of Ti–22Nb and Ti–22Nb–6Zr alloys [137]. The corrosion behavior was evaluated in 0.9% NaCl at 37°C and neutral pH using OCP, potentiodynamic polarization, and electrochemical impedance spectroscopy (EIS) techniques [137]. The results indicated the addition of Zr played a crucial role in improving the corrosion resistance of the Ti–22Nb–6Zr alloy. The alloys had high-breakdown potential and low corrosion current densities. Both the corrosion current and passive current density decreased with the addition of 6 at % Zr into the Ti–22Nb alloy. Anodic polarization was found to improve corrosion performance [137]. Wang and Zheng also evaluated the corrosion and biocompatibility for a Ti–16Nb low-modulus Ni–free SMA [137]. Their electrochemical testing showed this alloy had a breakdown voltage greater than 2.5 V (SCE) in Hanks' solution at pH 7.4 [137]. XPS showed the passive film contained mainly TiO_2 and Nb_2O_5 and indirect cytotoxicity results showed it to have biocompatibility similar to CP titanium [137].

D. Cu-BASED SHAPE MEMORY ALLOYS

Cu-based SMAs, such as Cu–Zn–Al and Cu–Al–Ni, are available for commercial applications but do not have as good shape memory properties or corrosion resistance as nNiTi; however, they do provide a more economical alternative to nNiTi [32]. Cu-based SMAs typically have body-centered cubic (bcc) crystal structure (e.g., β-brass). Examples include Cu–Zn and Cu–Al alloys in which Zn and Al may be partially replaced by other alloying elements such as Si, tin (Sn), manganese (Mn), or mixtures of these elements [35]. An example of such alloy compositions would be 60–85 wt % Cu with varying amounts of Zn and/or Al in combination with Si and Mn. For instance, alloys having, 0–40 wt % Zn, 0–5 wt % Si, 0–14 wt % Al, and 0–15 wt % Mn form bcc-type structures. Ternary quaternary and more complex alloys of Cu are common [35].

These alloys are generally single phase. Usually the β-phase is obtained by rapidly quenching the alloy from an elevated temperature. If the quenching rate is too slow, a second phase may form which does not undergo the reversible austenite–martensite transformation [35]. An alloy that contains at least 70% β-phase may still possess the same useful properties as the pure β-phase structure. Initially, Cu-based SMA suffered from intergranular failure that was attributed to the large grain size; however, development of fine-grained Cu-based alloys (through grain refining additives such as B, Ce, Co, Fe, Ti, V, and Zr) improved their mechanical properties significantly [32].

In general, Cu–Zn–Al and Cu–Al–Ni SMA are considered to have corrosion performance similar to aluminum bronzes and the performance is rated as poor, low, or fair to excellent [138, 139]. Neither of these Cu-based SMA families are considered to be biocompatible [138].

D1. Cu-Based SMA Pitting

Cu–Ni–Ti shape memory (A_f temperatures of 21°C and 38°C) orthodontic wires have been studied by Pun et al. [140]. They found that the OCP and pitting susceptibility of the Cu–Ni–Ti alloy was significantly higher than that of the NiTi in artificial saliva [140].

D2. Cu-Based SMA Dezincification

Cu–Zn–Al SMAs are susceptible to dezincification which occurs preferentially along surface crevices, grain boundaries, and martensite interfaces [32, 141]. The susceptibility to dezincification strongly depends on the phase and composition. Resistance to dezincification is improved by increasing the Al content [32].

D3. Cu-Based SMA Intergranular Attack and Stress Corrosion Cracking

Cu–Zn–Al alloys are susceptible to intergranular SCC [32, 142, 143]. SCC of Cu–Zn–Al SMAs is insensitive to type of phase present or Al content of the alloy between 4 and 8 wt % [32]. The nitrite anion was found to be the most potent agent for SCC in this Cu–Zn–Al family of SMAs. SCC is also caused by exposure to solutions containing nitrate and sulfate anions and ammonia solutions with pH values between 7.9 and 11 [32].

E. Fe-BASED SHAPE MEMORY ALLOYS

There are three types of martensites with different crystal structures: α' [bcc or body-centered tetragonal (bct) commonly found in Fe–C and Fe–Ni alloys], ε [hexagonal close packed (hcp)], and face-centered tetragonal (fct) martensite [36]. The shape memory effect has been found in each of these three types of ferrous martensites. The iron-based SMA family is composed of Fe–Mn–Si, Fe–25 at % Pt, and FeNiCoTi [15]. The iron-based SMAs are considered to have medium to high corrosion resistance [139].

Although the Fe–(14–30%)Mn–(4–6%)Si alloys do not really undergo a thermo-elastic martensitic transformation, they still exhibit a one-way shape memory effect of several percent and a recovery stress on the order of 300 MPa can be generated [15]. In order to obtain adequate corrosion resistance, this alloy must be alloyed with Cr and Ni, which reduces the shape memory effect slightly; however, a 4% one-way memory effect can be achieved through complex thermomechanical processing. Among the various ferrous SMAs only the Fe–Mn–Si–Cr–Ni alloys appear to have commercial significance [36].

F. MAGNETIC SHAPE MEMORY AND SHAPE MEMORY ALLOYS OTHER THAN TITANIUM BASED

Research into magnetic shape memory alloys has been increasing worldwide [37]. Their shape memory effect is driven by magnetic field instead of temperature variation, allowing faster fast actuation than is possible by temperature change. In addition, there are several other classes of shape memory/superelastic materials, some of which are the copper-based alloys and other alloy systems.

F1. Magnetic Shape Memory

Ferromagnetic shape memory alloys (FSMAs) are actuator materials that deform under magnetic field by the motion of twin boundaries in the martensite phase. The ferromagnetic shape memory effect was first observed in 1996 in Ni–Mn–Ga by Ullakko et al. at MIT [144]. The maximum recoverable strain is about 6%. These FSMAs have the potential to be used as fast-acting actuators and in percutaneously placed implant devices [66]. The corrosion resistance of the Ni–Mn–Ga FSMAs is typically fairly poor. Immersion tests in Hanks' balanced salt solution at 37°C and pH 7.4 for 12 h resulted in the formation of large pits on Ni–Mn–Ga samples while a thin-film NiTi displayed no signs of corrosion [66]. Gebert et al. state that Ni–Mn–Ga FSMAs exhibit a low corrosion rate and stable anodic passivity in weakly acidic to strongly alkaline solutions (pH 5–11) [145]. The passive film formed in pH 5–8.4 solutions was composed of $Ni(OH)_2$, $NiOOH$ and Ga_2O_3, and NiO, MnO_2, and MnO at the surface and film–metal interface, respectively. MnS inclusions which are formed during the materials processing are responsible for the pitting susceptibility of these materials.

REFERENCES

1. S. Miyazaki, "Medical and Dental Applications of Shape Memory Alloys", in Shape Memory Materials, K. Otsuka and C. M. Wayman, (Eds.) Cambridge University Press, Cambridge, 1998, pp. 267–281.

2. D. Stoeckel, A. Pelton, and T. Duerig, "Self-Expanding Nitinol Stents: Material and Design Considerations," Eur. radiol., **14**(2), **292–301** (2004).

3. S. Lu, "Medical Applications of Ni-Ti Alloys in China," in: Eng. Aspects Shape Memory Alloys, T. W. Duerig et al. editors, Jordan Hill, oxford: Butterworth-Heinemann 445–451 (1990).

4. X. Miao and J. Weitao, "Applications of a NiTi Shape Memory Alloy to Medicine and Dentistry," in Proc. of Intl. Symp. on Shape Memory Alloys, Guilin, China, 1986, 411–415.

5. Z. Ming, et al. "Medical Applications of SMA in Beijing General Research Institute for Non-Ferrous Metals, in SMM '94," Beijing, China, 1994, pp. 602–607.

6. K. Mukherjee, "Shape-Memory-Effect Alloys: Biomedical Applications," in Encyclopedia of Materials Science and Engineering, Pergamon, Elmsford, NY, 1986.

7. M. Niinomi, "Recent Metallic Materials for Biomedical Applications," Metall. Mater. Trans. A, **33A**, 477–486 (2002).

8. J. Ryhänen, S. Shabalovskaya, and L. H. Yahia, "Bioperformance of Nitinol: In Vivo Biocompatibility," Mater. Sci. Forum, **394–395**, 131–144 (2002).

9. L. Schetky and M.H. Wu, Issues in the Further Development of Nitinol Properties and Processing for Medical Device Applications, Memory Corporation, Bethel, CT, 2003.

10. S. A. Shabalovskaya, "Surface, Corrosion and Biocompatibility Aspects of Nitinol as an Implant Material," Bio-Med. Mater. Eng., **12**, 69–109 (2002).

11. T. Shusung, et al. "Researches and Applications of Shape Memory Alloys in China," in Proc. of Intl. Symp. on Shape Memory Alloys, Guilin, China, pp. 1–11, 1986.

12. N. Morgan, "Medical Shape Memory Alloy Applications—The Market and Its Products," Mater. Sci. Eng. A, **378**(1–2), 16–23 (2004).

13. M. Irie, "Shape Memory Polymers," in Shape Memory Materials, K. Otsuka and C. M. Wayman (Eds.) Cambridge University Press, Cambridge, 1998, pp. 203–219.

14. K. Uchino, "Shape Memory Ceramics," in Shape Memory Materials, K. Otsuka and C. M. Wayman, (Eds.), Cambridge University Press, Cambridge, 1998, pp. 184–202.

15. J. Van Humbeeck and J. Cederstrom, "The Present State of Shape Memory Materials and Barriers Still to be Overcome," in SMST-94: The First International Conference on Shape Memory and superelastic Technologies, A. R. Pelton, D. Hodgson, and T. Duerig (Eds.), Asilomar, Pacific Grove, CA, 1994, pp. 1–6.

16. Z. Nishiyama, et al. "Martensitic Transformation," in Materials Science and Technology, Academic, New York, 1978.

17. X. Ren and K. Otsuka, "Recent Advances in Understanding the Origin of Martensite Aging Phenomena in Shape Memory Alloys," Phase Transitions, **69**, 329–350 (1999).

18. Otsuka, K. and C. M. Wayman, "Mechanism of Shape Memory Effect and Superelasticity," in Shape Memory Materials, K. Otsuka and C. M. Wayman, (Eds.), Cambridge University Press, Cambridge, 1998, pp. 27–48.

19. A. Olander, "An Electrochemical Investigation of Solid Cadmium-Gold Alloys," J. Am. Chem. Soc., **54**(10), 3819–3833 (1932).

20. L. C. Chang and T. A. Read, "Plastic Deformation and Diffusionless Phase Changes in Metals—the Gold–Cadmium Beta Phase," Trans. AIME, **189**, 47–52 (1951).

21. M. Burkart and T. Read, "Diffusionless Phase Change in the Indium-Thallium System," Trans. AIME J. Metals, **197**, 1516–1524 (1953).

22. W. Buehler, J. Gilfrich, and R. Wiley, "Effect of Low-Temperature Phase Changes on the Mechanical Properties of Alloys Near Composition TiNi," J. Appl. Phys., **34**, 1475 (1963).

23. W. J. Buehler, NITINOL Re-Examination, WOLAA LEAF, WOL Oral History Supplement, White Oak Laboratory Alumni Association, Inc., Olney, MD, www.wolaa.org/files/Nitinol_Oral_History-pdf **VIII**(I), 2006.

24. A. M. Pettinger, "A Regularized Couple Stress Theory and Its Implications on Nucleation and Kinetics of Phase Transformations in Anti-Plane Shear," Department of Mechanical Engineering, Massachusetts Institute of Technology, Cambridge, MA, 1998.

25. S. Miyazaki, "Development and Characterization of Shape Memory Alloys," in Shape Memory Alloys, M. Fremond and S. Miyazaki, (Eds.), Springer-Verlag, New York, 1996, pp. 69–147.

26. T. Saburi, "Ti-Ni Shape Memory Alloys," in Shape Memory Materials, K. Otsuka and C. M. Wayman, (Eds.), Cambridge University Press, Cambridge, 1998, pp. 49–96.

27. S. M. Russell and F. Sczerzenie, "Engineering Considerations in the Application of Ni-Ti-Hf and NiAl as Prictical High-Temperature Shape Memory Alloys," in SMST-94: The First International Conference on Shape Memory and Superelastic Technologies, A. R. Pelton, D. Hodgson, and T. Duerig, (Eds.), Asilomar, Pacific Grove, CA, 1994, pp. 43–48.

28. H. Y. Kim, et al. "Biomedical Ni-Free Ti-Nb Based Shape Memory and Superelastic Alloys," in SMST-06: Proceedings of the International Conference on Shape Memory and Superelastic Technologies B. Berg, M. R. Mitchell, and J. Proft, (Eds.), ASM International Pacific Grove, CA, 2006, pp. 731–740.

29. E. Takahashi, S. Watanabe, and S. Hanada, "Superelastic Behavior in Nickel-Free T-Nb-Sn Alloys," in SMST-2003: Proceedings of the International Conference on Shape Memory and Superelasitc Technologies, ASM International, Metals Park, OH, 2003.

30. M. Niinomi, "Recent Research and Development in Titanium Alloys for Biomedical Applications and Healthcare Goods," Sci. Technol. Adv. Mater., **4**(5), 445–454 (2003).

31. T. Saito, et al. "Multifunctional Alloys Obtained via a Dislocation-Free Plastic Deformation Mechanism," Science, **300**(5618), 464 (2003).

32. M. Wu, "Cu-Based Shape Memory Alloys," in Engineering Aspects of Shape Memory Alloys, T. Duerig et al. (Eds.), Butterworth-Heinemann, London, 1990.

33. T. Tadaki, "Cu-Based Shape Memory Alloys," in Shape Memory Materials, K. Otsuka and C. M. Wayman, (Eds.), Cambridge University Press, Cambridge, 1998, pp. 97–116.

34. K. Sugimoto, K. Kamei, and M. Nakaniwa, "Cu-Al-Ni-Mn: A New Shape Memory Alloy for High Temperature Applications," in: "Engineering Aspects of Shape Memory Alloys", Ed. T. W. Duering et al., Butterworth-Heinemann, London, 1990, pp. 89–95.

35. G. B. Brook, P.L. Brooks, and R.F. Iles, "Mechanical Pre-conditioning Method," U.S. Patent 4,036,669 U.S. Patent and Trademark Office, Raychem, Alexandria, VA 1977.

36. T. Maki, "Ferrous Shape Memory Alloys," in Shape Memory Materials, K. Otsuka and C. M. Wayman (Eds.), Cambridge University Press, Cambridge, 1998, pp. 117–132.

37. S. Miyazaki, "Preface, " in SMST-2007 Proceedings of the International Conference on Shape Memory and superelastic Technologies SMST International Committee, Tsukuba, Japan, 2007.

38. V. Schroeder, "Evolution of the Passive Film on Mechanically Damaged Nitinol," J. Biomed. Mater. Res. A, **90**(1), 1–17 (2009).

39. K. N. Melton and J. D. Harrison, "Corrosion of Ni-Ti Based Shape Memory Alloys," in Proceedings for the First International conference on Shape Memory and Superelastic Technologies A. R. Pelton, D. Hodgson, and T. Duerig (Eds.), SMST International Committee, Asilomar, Pacific Grove, CA, 1994, pp. 187–196.

40. K. Melton, "Ni-Ti Based Shape Memory Alloys," in: "Engineering Aspects of Shape Memory Alloys", Ed. T. W. Duering et al., Butterworth-Heinemann, London, 1990, pp. 21–35 (1990).

41. B. G. Pound, "The Electrochemical Behavior of Nitinol in Simulated Physiological Solutions," J. Biomed. Mater. Res., **85A**(4), 1103–1113 (2008).

42. A. Pelton, et al. "TiNi Oxidation: Kinetics and Phase Transformations," in Solid to Solid Phase Transformations in Inorganic Materials: Phase Transformations in Novel Systems or Special Materials, The Minerals, Metals and Materials Society, San Francisco, CA, 2005, pp. 1029–1034.

43. B. Clarke, et al. "Influence of Nitinol Wire Surface Treatment on Oxide Thickness and Composition and Its Subsequent Effect on Corrosion Resistance and Nickel Ion Release," J. Biomed. Mater. Res., **79A**, 61–70 (2006).

44. A. Undisz, et al. "In Situ Observation of Surface Oxide Layers on Medical Grade Ni-Ti Alloy During Straining," J. Biomed. Mater. Res., Vol 88A: 1000–1009 (2009).

45. C. Trepanier, et al. "Effect of Modification of Oxide Layer on NiTi Stent Corrosion Resistance," J. Biomed. Mater. Res., **43**(4), 433 (1998).

46. L. E. Eiselstein et al., "Acceptance Criteria for Corrosion Resistance of Medical Devices: Statistical Analysis of Nitinol Pitting in In-Vivo Environments," J. Mater. Eng. Perform., **18**(5–6), 768–780 (2009).

47. N. Zhang, X.W.Z. Cao, and Z. Mao, A Statistical Evaluation of Pitting Corrosion on TiNi Shape Memory Alloys, 1995.

48. T. Shibata, "Corrosion Probability and Statistical Evaluation of Corrosion Data," in: Uhlig's Corrosion Handbook, R. W. Revie (Ed.), Wiley, New York, 2000.

49. T. Shibata, "W.R. Whitney Award Lecture: Statistical and Stochastic Approaches to Localized Corrosion," Corrosion, **52**(11) (1996). pp. 813–830.

50. C. Warner, "The Effect of Exposure to Simulated Body Fluids on Breakdown Potentials," J. Mater. Eng. Perform., **18**(5), 754–759 (2009).

51. G. Rondelli and B. Vicentini, "Localized Corrosion Behaviour in Simulated Human Body Fluids of Commercial Ni-Ti Orthodontic Wires," Biomaterials, **20**(8), 785–792 (1999).

52. J. Ryhänen, "Biocompatibility Evaluation of Nickel-Titanum Shape Memory Metal Alloy," Department of Surgery, University of Oulu, Finland, 1999.

53. S. Shabalovskaya, "On the Nature of the Biocompatibility and on Medical Applications of NiTi Shape Memory and Superelastic Alloys," Bio-Med. Mater. Eng., **6**, 267–289 (1996).

54. S. Shabalovskaya, et al. "The effect of Surface Particulate on the Corrosion Resistance of Nitinol Wire," in SMST-2003: The International Conference on Shape Memory and superelastic technologies, 2003, SMST Society, Inc., ASM, Materials Park, OH, pp. 399–408.

55. S. Shabalovskaya, J. Anderegg, and J. Van Humbeeck, "Critical Overview of Nitinol Surfaces and Their Modifications for Medical Applications," Acta Biomater., **4**(3), 447–467 (2008).

56. Shabalovskaya, S. and J. V. Humbeeck, "Analysis of Passivity of Nitinol Surfaces," in SMST-06: Proceedings of the International Conference on Shape Memory and superelastic technologies, B. Berg, M. R. Mitchell, and J. Proft (Eds.), ASM International, Pacific Grove, CA, 2006.

57. S. Shabalovskaya, et al. "Effect of Chemical Etching and Aging in Boiling Water on the Corrosion Resistance of Nitinol Wires with Black Oxide Resulting from Manufacturing Process," J. Biomed. Mater. Res. B Appl. Biomater., **66**(1), 331–340 (2003).

58. S. Shabalovskaya, et al. "Surface and Corrosion Aspects of NiTi Alloys," in SMST-2000: Proceedings of the International Conference on Shape Memory and superelastic Technologies, ASM International, Metals Park, OH, 2000, pp. 299–308.

59. C. Trépanier and A. R. Pelton, "Effect of Temperature and pH on the Corrosion Resistance of Passivated Nitinol and Stainless Steel," in Proceedings of the International Conference on Shape Memory and Superelastic Technologies, ASM International, Baden-Baden, Germany, 2004, p. 361.

60. S. Trigwell, et al. "Effects of Surface Treatment on the Surface Chemistry of NiTi Alloy for Biomedical Applications," Surf. Interf. Anal., **26**(7), 483–489 (1998).

61. L. Tan, R. Dodd, and W. Crone, "Corrosion and Wear-Corrosion Behavior of NiTi Modified by Plasma Source Ion Implantation," Biomaterials, **24**(22), 3931–3939 (2003).

62. W. S. Hwang, K.J. Kim, and W.C. Seo, "Pitting Corrosion of TiNi Shape Memory Alloy in Deaerated Chloride Solution," Paper No. 381, in 3th International Corrosion Congress, NACE International, Houston, TX, 1994.

63. B. G. Pound, "Susceptibility of Nitinol to Localized Corrosion," J. Biomed. Mater. Res., **77A**, 185–191 (2006).

64. A. Nissan, et al. "Effect of Long-Term Immersion on the Pitting Corrosion Resistance of Nitinol," in Proceedings of the International Conference on Shape Memory and Superelastic Technologies, B. Berg, M. R. Mitchell, and J. Proft (Eds.), ASM International, Tsukuba, Japan, 2007, pp. 271–278.

65. L. Eiselstein, et al. "Toward an Acceptance Criterion for the Corrosion Resistance of Medical Devices: A Statistical Study of the Pitting Susceptibility of Nitinol," in Proceedings of the International Conference on Shape Memory and Superelastic Technologies, B. Berg, M. R. Mitchell, and J. Proft (Eds.), ASM International, Tsukuba City, Japan, 2007.

66. L. L. Stepan, et al. "Biocorrosion Investigation of Two Shape Memory Nickel Based Alloys: Ni-Mn-Ga and Thin Film NiTi," J. Biomed. Mater. Res. A, **82**(3), 768–776 (2007).

67. R. Singh and N.B. Dahotre, "Corrosion Degradation and Prevention by Surface Modification of Biometallic Materials," J. Mater. Sci. Mater. Med., **18**(5), 725–751 (2007).

68. W. S. Hwang, K.J. Kim, and W.C. Seo, "Pitting Corrosion of TiNi Shape Memory Alloy in Deaerated Chloride Solutions," in 13th ICC., Paper No. 381 NACE International, Houston, TX pp. 1–7.

69. A. Edwards, "Application of Electrochemical Techniques to the Prediction of Environmental Effects on the Corrosion Behavior of a Ni-Ti Alloy," in Proceedings Marine Corrosion and Fouling, National Technical University, Athens, Greece, 1984, pp. 177–187.

70. American Society for Testing and Materials (ASTM) Standard Test Method for Pitting or Crevice Corrosion of Metallic Surgical Implant Materials, ASTM F746-04, ASTM, West Conshocken, PA, 2004.

71. M. Kaczmarek, "Crevice Corrosion Resistance of NiTi Alloy after Various Surface Treatments," Arch. Mater. Sci. Eng., **29**(2), 69–72 (2008).

72. C. Trépanier, et al. "Effect of Wear and Crevice on the Corrosion Resistance of Overlapped Stents," in SMST-2006: Proceedings of the International Conference on Shape Memory and Superelastic Technologies, B. Berg, M. R. Mitchell, and J. Proft (Eds.), ASM International, Asilomar, Pacific Grove, CA, 2006.

73. L. E. Eiselstein, R.A. Sire, and B.A. James, "Review of Fatigue and Fracture Behavior in NiTi," in ASM Symposium on Materials and Processes for Medical Devices Conference, ASM International, Boston, MA, 2005.

74. T. Duerig and A. Pelton, "Ti-Ni Shape Memory Alloys," in Materials Properties Handbook: Titanium Alloys, R. Boyer et al. (Eds.), ASM International, Metals Park, OH, 1994, pp. 1035–1048.

75. B. James, J. Foulds, and L.E. Eiselstein, "Failure Analysis of NiTi Wires Used in Medical Applications," J. Failure Anal. and Prevention, **5**, 82–87 (2005).

76. T. Asaoka, "Effect of Hydrogen on the Shape Memory Properties of Ni-Ti-Cu," in SMST-94: The First International Conference on Shape Memory and Superelastic Technologies, A. R. Pelton, D. Hodgson, and T. Duerig (Eds.), Asilomar, Pacific Grove, CA, 1994, pp. 79–84.

77. K. Asaoka, K. Yokoyama, and M. Nagumo, "Hydrogen Embrittlement of Nickel-Titanium Alloy in Biological Environment," Metall. Mater. Trans. A, **33A**, 495–501 (2002).

78. J. Sheriff, A. Pelton, and L. Pruitt, "Hydrogen Effects on Nitinol Fatigue," in SMST-04: Proceedings of the International Conference on Shape Memory and Superelastic Technologies," ASM International, Baden-Baden, Germany, 2004, p. 111.

79. M. Wu, "Fabrication of Nitinol Materials and Components," in SMST-01: Proceedings of the International Conference on Shape Memory and Superelastic Technologies, Kunming, China, 2001, pp. 285–292.

80. K. Yokoyama, "Fracture Analysis of Hydrogen Charged Titanium Superelastic Alloy," Mater. Trans., **42**(1), 141 (2001).

81. K. Yokoyama, et al. "Degradation and Fracture of NiTi Superelastic Wire in an Oral Cavity," Biomaterials, **22**, 2257–2262 (2001).

82. K. Yokoyama, et al. "Hydrogen Embrittlement of Ni-Ti Superelastic Alloy in Fluoride Solution," J. Biomed. Mater. Res. **65A**, 182–187 (2003).

83. K. Yokoyama, et al. Delayed fracture of Ni-Ti superelastic alloys in acidic and neutral fluoride solutions. J Biomed Mater Res, 2004. **69A**(1): p. 105–113.

84. A. Runciman, et al. "Effects of Hydrogen on the Phases and Transition Temperatures of NiTi," in SMST-06: Proceedings of the International Conference on Shape Memory and Superelastic Technologies, B. Berg, M. R. Mitchell, and J. Proft (Eds.), ASM International, Pacific Grove, CA, 2006, pp. 185–196.

85. B. L. Pelton, T. Slater, and A. R. Pelton, "Effects of Hydrogen in TiNi," in SMST-97: Proceedings of the Second International Conference on Shape Memory and Superelastic Technologies, A. Pelton, et al. (Eds.), SMST, Pacific Grove, CA, 1997, pp. 395–400.

86. R. Burch and N. Mason, "Thermodynamic Relationships and Structural Transformations in TiCo and TiNi Intermetallic Alloy—Hydrogen Systems," Z. Phys. Chem. Neue Folge (Wiesbaden), **116**, 185–195 (1979).

87. R. Burch and N. Mason, "Absorption of Hydrogen by Titanium–Cobalt and Titanium–Nickel Intermetallic Alloys. II.–Thermodynamic Parameters and Theoretical Models," J. Chem. Soc. Faraday. Trans. I, **75**(3), 578–590 (1979).

88. R. Burch and N. Mason, "Absorption of Hydrogen by Titanium–Cobalt and Titanium–Nickel, Intermetallic Alloys. I. Experimental Results," J. Chem. Soc. Faraday. Trans. I, **75**(3), 561–577 (1979).

89. A. R. Pelton, et al. "Structural and Diffusional Effects of Hydrogen in TiNi," in SMST-2003: Proceedings of the Fourth International Conference on Shape Memory and Superelastic Technologies, A. R. Pelton and T. Duerig (Eds.),

90. SMST Society, Asilomar, Pacific Grove, CA, 2004, pp. 33–42.

R. Schmidt, et al. "Hydrogen Solubility and Diffusion in the Shape-Memory Alloy NiTi," J. Phys. Condens. Matter, **1**, 2473–2482 (1989).

91. J. Y. He, K. W. Gao, Y. J. Su, L .J. Qiao, and W. Y. Chu, "Technical Note—The Effect of Hydride and Martensite on the Fracture Toughness of TiNi Shape Memory Alloy," Smart Mater. Struct., **13**, N24–N28 (2004).

92. R. W. Schutz, "Stress-Corrosion Cracking of Titanium Alloys," in Stress-Corrosion Cracking: Materials Performance and Evaluation, R. H. Jones, (Ed.), ASM International, Materials Park, OH, 1992, pp. 276–284.

93. J. Been and J. Grauman, "Titanium and Titanium Alloys," in Uhlig's Corrosion Handbook, R. Revie (Ed.), Wiley, New York, 2000.

94. K. Kaneko, et al. "Degradation in Performance of Orthodontic Wires Caused by Hydrogen Absorption during Short-Term Immersion in 2.0% Acidulated Phosphate Flouride Solution," Angle Orthodontist, **74**(4), 487–495 (2004).

95. R. Venugopalan and C. Trepanier, "Assessing the Corrosion Behaviour of Nitinol for Minimally-Invasive Device Design," Min. Invas. Ther. Allied Technol., **9**(2), 67–74 (2000).

96. H. Kong, et al. "Corrosive Behaviour of Amplatzer Devices in Experimental and Biological Environments," Cardiol. Young, **12**(3), 260–265 (2002).

97. S. Robertson and R. Ritchie, "In vitro Fatigue–Crack Growth and Fracture Toughness Behavior of Thin-Walled Superelastic Nitinol Tube for Endovascular Stents: A Basis for Defining the Effect of Crack-Like Defects," Biomaterials, **28**(4), 700–709 (2007).

98. U.S., Food and Drug Administration (FDA) Guidance for Industry and FDA Staff: Non-Clinical Tests and Recommended Labeling for Intravascular Stents and Associated Delivery Systems, Material Characterization—Stent Corrosion Resistance," U.S. Department of Health and Human Services, FDA, Center for Devices and Radiological Health, Washington, DC, 2003.

99. T. Satow, T. Isano, and T. Honma, "The High Temperature Oxidation of Intermetallic Compound TiNi," Japan Insti. Met., J., **38**, 242–246 (1974).

100. J. Haasters, G. Salis-Solio, and G. Bensmann, "The Use of Ni-Ti as an Implant Material in Orthopedics," Eng. Aspects Shape Memory Alloys, 426–444 (1990).

101. M. W. Ries, et al. "Nickel Release after Implantation of the Amplatze Occluder," Am. Heart J., **145**(4), 737–741 (2003).

102. L. B. Pertile, et al. "In Vivo Human Electrochemical Properties of a NiTi-Based Alloy (Nitinol) Used for Minimally Invasive Implants," J. Biomed. Mater. Res., **89A**(4), 1072–1078 (2009).

103. C. R. F. Azevedo, "Characterization of Metallic Piercings," Eng. Failure Anal., **2**(4), 255–263 (2002).

104. G. Riepe, et al. "Degradation of Stentor Devices after Implantation in Human Beings," in SMST-2000, S. M. Russell and A. Pelton (Eds.), SMST, Pacific Grove, CA, 2000, pp. 279–283.

105. C. Heintz, et al. "Corroded Nitinol Wires in Explanted Aortic Endografts: An Important Mechanism of Failure?" J. Endovasc. Ther., **8**(3), 248–253 (2001).

106. G. Riepe, et al. "What Can We Learn from Explanted Endovascular Devices?" Eur. J. Vascu. Endovasc. Surg., **24**(2), 117–122 (2002).

107. S. Walak, "Analysis of Nitinol Stent-Grafts After Long Term In-Vivo Exposure," in Materials & Processes from Medical Devices Conference, ASM International, Metals Park, OH, 2005, pp. 290–294.

108. L. Gettleman, "Noble Alloys in Dentistry," Curr. Opin. Dent., **1**(2), 218–221 (1991).

109. W. M. Carroll, M.J. Kelly, and B. O'Brien, "Corrosion Behaviour of Nitinol Wires in Body Enviornments," in SMST-99: Proceedings of the First European Conference on Shape Memory and Superelastic Technologies, SMST International Committee, Antwerp Zoo, Belgium, 1999.

110. W. M. Carroll, and M.J. Kelly, "Corrosion Behavior of Nitinol Wires in Body Fluid Environments," J. Biomed. Mater. Res. A, **67**(4), 1123–1130 (2003).

111. X. Zhang, "A Study of Shape Memory Alloy for Medicine in Shape Memory Alloy 86," in Shape Memory Alloy 86: Proceeding of the International Symposium of Shape Memory Alloys, China Academic Publishers, Beijing, 1986.

112. D. J. Wever et al., "Cytotoxic, Allergic and Genotoxic Activity of a Nickel-Titanium Alloy," Biomaterials, **18**(16), 1115–1120 (1997).

113. J. Ryhänen, "Biocompatibility of Nitinol," in SMST-2000: Proceedings of the International Conference on Shape Memory and Superelastic Technologies, ASM International, Metals Park, OH, 2000, pp. 251–259.

114. J. Ryhänen, et al. "Biocompatibility of Nickel-Titanium Shape Memory Metal and Its Corrosion Behavior in Human Cell Cultures," J. Biomed. Mater. Res., **35**, 451–457 (1997).

115. American Society for Testing and Materials (ASTM), "Standard Test Method for Conducting Cyclic Potentiodynamic Polarization Measurements to Determine the Corrosion Susceptibility of Small Implant Devices," ASTM F 2129-08, in Annual Book of ASTM Standards, ASTM, Philadelphia, PA, 2008, pp. 1684–1690.

116. N. Corlett, et al. "Effect Of Long-Term Immersion on the Localized Corrosion Resistance of Nitinol Wire Under Aerated Conditions," in SMST 2008—International Conference on Shape Memory and Superelastic Technologies, ASM International, Stresa, Italy, 2008.

117. R. A. Corbett, "Laboratory Corrosion Testing of Medical Implants," in ASM Materials & Processes for Medical Devices Conference, ASM International, Anaheim, CA, 2003.

118. S. N. Rosenbloom and R. Corbett, "An Assessment of ASTM F 2129 Electrochemical Testing of Small Medical Implants—Lessons Learned," Paper No. 07674, NACE, Houston, TX, 2007.

119. L. E. Eiselstein, D.M. Proctor, and T.C. Flowers, "Trivalent and Hexavalent Chromium Issues in Medical Implants," Mater. Sci. Forum, **539–543**, 698–703 (2007).

120. L. E. Eiselstein and R.D. Caligiuri,"Ion Leaching from Implantable Medical Devices," in Thermec 2009, Berlin, Germany, 2009. Materials Science Forum, Vols 638–642 (2010), pp. 754–759.

121. E. McLucas, et al. "Analysis of the Effects of Surface Treatments on Nickel Release from Nitinol Wires and Their Impact on Candidate Gene Expression in Endothelial Cells," J. Mater. Sci. Mater. Med., **19**(3), 975–980 (2008).

122. Agency for Toxic Substances and Disease Registry (ATSDR) Toxicological Profile for Nickel, U.S. Department of Health and Human Services Public Health Service, ATSDR, Atlanta, GA, 2005, p. 345.

123. C. Trépanier, et al. "Effect of Passivation Treatments on Nickel Release from Nitinol," in 6th World Biomaterials Congress Transactions, Society for Biomaterial, Mt. Laurel, NJ, 2000 p. 1043.

124. N. Fujita, et al. Report of Osaka Prefectural Industrial Technology Research Institute, No. 86, Osaka, Japan, 1985, p. 32.

125. D. Wever, et al. "Electrochemical and Surface Characterization of a Nickel-Titanium Alloy," Biomaterials, **19**(7–9), 761–769 (1998).

126. A. Fasching, et al. "The Effects of Heat Treatment, Surface Condition and Strain on Nickel-Leaching Rates and Corrosion Performance in Nitinol Wires," in Proceedings of the Materials & Processes for Medical Devices Conference, ASM International, Minneapolis, MN, 2009.

127. Y. Okazaki and E. Gotoh, "Metal Release from Stainless Steel, Co-Cr-Mo-Ni-Fe and Ni-Ti Alloys in Vascular Implants," Corros. Sci., **50**(12), 3429–3438 (2008).

128. G. Herting, "Metal Release from Stainless Steels and the Pure Metals in Different Media," Department of Materials Science and Engineering, Division of Corrosion Science, Royal Institute of Technology, Stockholm, Sweden, 2004.

129. G. Herting, I. Odnevall Wallinder, and C. Leygraf, "Factors That Influence the Release of Metals from Stainless Steels Exposed to Physiological Media," Corros. Sci., **48**(8), 2120–2132 (2006).

130. G. Herting, I. Odnevall Wallinder, and C. Leygraf, "Metal Release from Various Grades of Stainless Steel Exposed to Synthetic Body Fluids," Corros. Sci., **49**(1), 103–111 (2007).

131. A. Ornberg, et al. "Corrosion Resistance, Chemical Passivation, and Metal Release of 35N LT and MP35N for Biomedical Material Application," J. Electrochem. Soc., **154**, C546 (2007).

132. R. D. Barrett, S.E. Bishara, and J.K. Quinn, "Biodegradation of Orthodontic Appliances. Part I. Biodegradation of Nickel and Chromium In Vitro," Amer. J. Orthodont. Dentofac. Orthop. (Official publication of the American Association of Orthodontists, its constituent societies, and the American Board of Orthodontics), **103**(1), 8 (1993).

133. K. M. Speck and A. C. Fraker, "Anodic Polarization Behavior of Ti-Ni and Ti-6Al-4V in Simulated Physiological Solutions," J. Dent. Res., **59**(10), 1590–1595 (1980).

134. B. Pound, Synthetic Human Bile Recipe, 2008.

135. D. Hansen, "The Effect of a Novel Biopolymer on the Corrosion of 316L Stainless Steel and Ti6Al4V Alloys in a Physiologically Relevant Electrolyte," CORROSION 2007, 2007.

136. W. Guo, J. Sun, and J. Wu, "Electrochemical and XPS Studies of Corrosion Behavior of Ti–23Nb–0.7 Ta–2Zr–O Alloy in Ringer's Solution," Mater. Chem. Phys., **113**(2–3), 816–820 (2009).

137. B. Wang, Y. Zheng, and L. Zhao, "Electrochemical Corrosion Behavior of Biomedical Ti-22Nb and Ti-22Nb-6Zr Alloys in Saline Medium," Adv. Funct. Mater., **18**, 23 (2009).

138. J. Van Humbeeck and R. Stalmans, "Characteristics of Shape Memory Alloys," in Shape Memory Materials, K. Otsuka and C. M. Wayman, (Eds.), Cambridge University Press, Cambridge, 1998, pp. 149–183.

139. K. N. Melton, "General Applications of SMA's and Smart Materials," in Shape Memory Materials, K. Otsuka and C. M. Wayman, (Eds.), Cambridge University Press, Cambridge, 1998, pp. 220–239.

140. D. K. Pun and D.W. Berzins, "Corrosion Behavior of Shape Memory, Superelastic, and Nonsuperelastic Nickel-Titanium-Based Orthodontic Wires at Various Temperatures," Dent. Mater., **24**(2), 221–227 (2008).

141. J. Celis, J. Roos, and F. Terwinghe, "Corrosion Behavior of β and Martensitic Al Brasses," J. Electrochem. Soc., **130**, 2314 (1983).

142. F. Terwinghe, J. Celis, and J. Roos, "Stress Corrosion Susceptibility of Beta Aluminium Brasses—I," Br. Corros. J., **19**(3), 107–114 (1984).

143. F. Terwinghe, J. Celis, and J. Roos, "Stress Corrosion Susceptibility of Beta Aluminium Brasses—II," Br. Corros. J., **19**(3), 115–119 (1984).

144. S. Murray, "Magneto-Mechanical Properties and Applications of Ni-Mn-Ga Ferromagnetic Shape Memory Alloy," Dept. of Materials Science and Engineering, Massachusetts Institute of Technology, Cambridge, MA, 2000.

145. A. Gebert, et al. "Passivity of Polycrystalline NiMnGa Alloys for Magnetic Shape Memory Applications," Corros. Sci., **51**(5), 1163–1171 (2009).

39

MICROBIOLOGICAL CORROSION OF METALLIC MATERIALS

J.-D. Gu

School of Biological Sciences, The University of Hong Kong, Hong Kong, China

T. E. Ford

University of New England, Biddeford, Maine

R. Mitchell

Harvard School of Engineering and Applied Sciences, Harvard University, Cambridge, Massachusetts

A. INTRODUCTION

The importance of microorganisms in the corrosion of metallic materials has been recognized for over 70 years, since the first report by von Volzogen Kuhr and van der Vlught [1]. Microbiological corrosion, now also called microbially influenced/induced corrosion (MIC) mostly by corrosion engineers and materials scientists, affects a wide range of industrial materials in multiple applications that include oil fields, offshore platforms, pipelines, pulp and paper industries, armaments, nuclear and fossil fuel power plants, chemical manufacturing facilities, and food industries [2–6]. The terminology of microbiological corrosion was frequently interchangeably used with microbiological fouling, but, strictly speaking, the two are not synonymous. MIC is not clearly defined, and ambiguity and misuse are common.

Corrosion of materials may cause severe economic loss and may have devastating consequences. It was estimated that 70% of the corrosion in gas transmission is due to problems caused by microorganisms, with the American refinery industry losing $1.4 billion a year from microbial corrosion [7]. Different groups of microorganisms are capable of corrosion and degradation: the best-known causative microorganisms include both aerobic and anaerobic bacteria. The sulfate-reducing bacteria (SRB) have become the chosen organisms in a large number of studies on biocorrosion [8–16]. In addition to SRBs, exopolymer (slime)– and acid-producing bacteria were found to participate actively in corrosion processes by a mechanism in which metal ions are complexed with functional groups of the exopolysaccharides, resulting in release of metallic species into solution [17–20]. Similarly, it should be mentioned here that fungi are also involved in the corrosion of aluminum and its alloys by a process in which organic acids of microbial origin attack the

Uhlig's Corrosion Handbook, Third Edition, Edited by R. Winston Revie

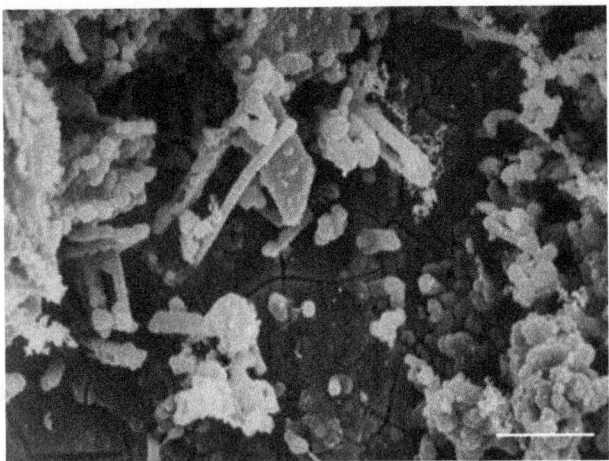

FIGURE 39.1. Scanning electron micrograph showing a biofilm community on the surface of stainless S316 after incubation in marine water from Boston Harbor, Massachussetts. The sample was dehydrated and critical point dried before being coated with palladium and gold (scale bar, 2 μm)

material matrices [21]. Fungi are known in the degradation of concrete [22, 23], stone [15, 24–26], glass [27], and artificial polymers that are widely used for protective purposes against corrosion [28–30].

Microbial involvement in corrosion of materials, including both metallics and polymeric materials as well as stone, is a result of adhesion to and active metabolism on surfaces. Microorganisms are capable of forming complex biofilm communities on surfaces of metals (Fig. 39.1). Microbial associations with surfaces of materials can also contribute to the degradation or corrosion of the underlying materials (e.g., metals). Considering the availability of molecular oxygen, some aerobic microorganisms obtain electrons from metal oxidation and, at the same time, reduce CO_2 for synthesis and growth [31]. The growth of bacteria on these surfaces alters the surfacial microenvironement, yielding acidity, and oxygen depletion, changing the diffusivity of metabolites and nutrients through the biofilm layer. As a result, the anaerobic SRBs may proliferate and produce H_2S, which severely corrodes metals [32]. These corrosion processes are typically mediated by the hydrogenase activity of the SRBs, particularly by the genus *Desulfovibrio*.

A wide range of microbial processes can cause corrosion [10, 11]. In general, the presence of inorganic deposits and differential concentrations of oxygen and chloride are important parameters determining the extent of corrosion. The presence of a microflora and fauna on surfaces of materials alters the local environment, providing appropriate conditions for dissolution of the metal. Some of the processes are summarized below.

B. MICROBIAL BIOFILMS

Microorganisms adhere to nonliving and living (tissue) surfaces under submerged or moist conditions [33–35] and in industrial environments exposed to moisture [28–30]. Adhesion of microorganisms to surfaces of metals changes the electrochemical charateristics of the material. The resultant biofilm can lead to a process called cathodic depolarization due to oxygen depletion near microbial colonies as the result of microbial activity and an increased localized acidity around the microbial colonies. The structure of a biofilm community on any surface is spatially heterogenous in composition [36, 37], reflecting changes in the local environment, nutritional conditions, and selective pressure. Bacterial attachment to surfaces is essential to the initiation of corrosion, recruiting of invertebrate settlement [38], and passivation of metallic surfaces [39]. An understanding of bacterial adhesion processes and the characteristics of biofilms is essential for a better understanding of the initiation and control of corrosion by microorganisms. Information pertinent to microbial biofilm formation is discussed in other relevant chapters in this book.

The effects of biofilms on corrosion are by means of any or a combination of the following factors: (1) direct effects on cathodic or anodic processes, (2) changes in surface film resistivity by microbial metabolites and exopolymeric materials, (3) generation of microenvironments promoting corrosion, including low oxygen concentration and acidic microenvironments, and (4) establishment of ion concentration cells. Because biofilms are present in a wide range of environments, their influence on materials covers a wide range of temperature, humidity, salinity, acidity, alkalinity, and barometric conditions. It should be mentioned here that in some cases biofilms may cause ennoblement rather than corrosion [40].

C. AEROBIC CORROSION

C1. Oxidation Processes

Under aerobic conditions, molecular oxygen (O_2) serves as an electron acceptor to achieve maximal energy for microbial growth. Microorganisms living under natural conditions tend to adhere to surfaces because surfaces concentrate nutrients under oligotrophic conditions and offer protection from predation. Adhesion is by means of long-range and short-range forces operating between the bacterial cells and the surfaces [41, 42]. At a distance, attractive forces dominate. After moving to a critical distance near a surface, repulsion forces become dominant and keep bacteria at a finite distance away from the surface.

The microflora forms random patches on material surfaces, inducing the formation of differential concentration cells on metals. Affter initial adhesion by electrostatic attraction or random collision, the organisms divide and form colonies, which deplete oxygen and release hydrogen ions [41, 43]. The adhesive process may induce the expression of genes responsible for polysaccharide synthesis because exopolymeric materials are important to the formation of a complex biofilm community. Under oligotrophic conditions, microorganisms synthesize large quantities of exopolysaccharides which serve as protectants from dessication and also act as energy reserves. When nutrients are further depleted, the cells can recycle these polymers as a source of carbon and energy through self-initiated depolymerization.

During aerobic corrosion, the area of a metal beneath these colonies acts as an anode, while the area further away from the colonies, where oxygen concentrations are relatively higher, serves as a cathodic site [44]. Electrons flow from the anode to the cathode and the corrosion process is initiated. Concentration of electrolytes affect the distance between the anode and cathode, which are closer at low salt and father apart at high salt concentrations. An electrochemical potential is eventually developed across the two sites and corrosion reactions take place, resulting in the dissolution of metal. Dissociated metal ions form ferrous hydroxides, ferric hydroxide, and a series of Fe-containing minerals in the solution phase, depending on the biological species present and the chemical conditions. It should be noted that oxidation, reduction, and electron flow must all occur for corrosion to proceed. However, the electrochemical reactions never proceed at theoretical rates because the rate of oxygen supply to cathodes and removal of products from the anodes limit the overall reaction [9, 44]. In addition, impurities and contaminants of the metal matrices also stimulate corrosion by initiating the formation of differential cells and accelerated electrochemical reactions.

When aerobic corrosion occurs, corrosion products usually form a structure consisting of three layers called tubercles. The inner green layer is mostly ferrous hydroxide [$Fe(OH)_2$]. The outer one consits of orange ferric hydroxide [$Fe(OH)_3$]. In between these two, magnetite (Fe_3O_4) forms a black layer [13, 14]. The most aggressive form of corrosion is tuberculation caused by the formation of differential oxygen concentration cells on metal surfaces. The overal reactions are as follows:

$$Fe^0 \rightarrow Fe^{2+} + 2e^- \quad \text{(anode)} \quad (39.1)$$

$$O_2 + 2H_2O + 4e^- \rightarrow 4OH^- \quad \text{(cathode)} \quad (39.2)$$

$$2Fe^{2+} + \tfrac{1}{2}O_2 + 5H_2O \rightarrow 2Fe(OH)_3 + 4H^+ \quad \text{(tubercle)}$$
$$(39.3)$$

Initial oxidation of Fe of mild steel at near-neutral pH is driven by dissolved O_2 [45]. Subsequent oxidation of Fe^{2+} to Fe^{3+} is an energy-producing process carried out by a few bacterial species [46]. The amount of free energy from this reaction is small, approximately -31 kJ. Large quantities of Fe^{2+} are oxidized to support the slow microbial growth. Because the Fe^{2+} oxidative reaction is rapid under natural conditions, microorganisms compete with chemical processes for available Fe^{2+}. As a result, biological involvement under aerobic conditions may be underestimated [10, 11].

C2. Microorganisms Involved

Several groups of aerobic microorganisms play an important role in corrosion, including the sulfur bacteria, iron- and manganese-depositing and slime-producing bacteria, fungi, and algae. The so-called iron bacteria include the *Sphaerotilus-Leptothrix*, *Gallionella*, and *Siderocapsa*. Ghiorse and Hirsch [47] also observed that two *Pedomicrobium*-like budding bacteria deposit Fe and Mn on their cell walls. Most of these bacteria are difficult to culture under laboratory conditions [31]. At neutral pH, Fe^{2+} is not stable in the presence of O_2 and is rapidly oxidized to the insoluble Fe^{3+} state. In fully aerated freshwater at pH 7, the half-life of Fe^{2+} oxidation is less than 15 min [48]. Because of this, the only neutral pH environments where Fe^{2+} is present are interfaces between anoxic and oxic conditions. Recently, improved techniques allowed the isolation of new Fe^{2+}-oxidizing bacteria under microaerophilic conditions at neutral pH [49]. Ferric oxides may be enzymatically deposited by *Gallionella ferruginea* and nonenzymatically by *Leptothrix* sp., *Siderocapsa*, *Naumanniella*, *Ochrobium*, *Siderococcus*, *Pedomicrobium*, *Herpetosyphon*, *Seliberia*, *Toxothrix*, *Acinetobacter*, and *Archangium* [46, 47]. Questions remain as to the extent of microbial involvement in specific processes of corrosion involving iron oxidation. A more comprehensive understanding of these mechanisms will likely require an integrated approach that includes microbiology, materials science, and electrochemistry.

Microorganisms in the genus *Thiobacillus* are also responsible for oxidative corrosion. Because metabolically they oxidize sulfur compounds to sulfuric acid, the acid around the cells may attack alloys. Similarly organic acid-producing microorganisms, including bacteria and fungi, are of concern. A number of acid-tolerant microorganisms are capable of Fe oxidation, of which *Thiobacillus* spp. are probably the most common. *Thiobacillus ferrooxidans* oxidizes Fe^{2+} to Fe^{3+}, but the product limits growth of the bacteria [46]. SO_4^{2-} is required by the Fe-oxidizing system in *T. ferrooxidans*. The role of S is probably to stabilize the hexa-aquated complex of Fe^{2+} as a substrate for the Fe-oxidizing enzyme system, with the Fe^{2+} being oxidized at the surface of the bacterirum. The electrons removed from Fe^{2+} are passed to periplasmic cytochrome *c*. The reduced cytochrome *c* binds to the outer plasma membrane of the cell, allowing transport of electrons across the membrane to cytochrome oxidase located in the inner membrane.

Most microorganisms accumulate Fe^{3+} on their outer surface by reacting with acidic polymeric materials. Such mechanisms have important implications not only for corrosion of metals but also to the accumulation of heavy metals in natural habitats. *Aquaspirillum magnetotacticum* is capable of taking up complexed Fe^{3+} and transforming it into magnetite (Fe_3O_4) by reduction and partial oxidation [50]. The magnetite crystals are single-domain magnets and play an important role in bacterial orientation to the two magnetic poles of the Earth in natural environments. However, magnetite can also be formed extracellularly by some nonmagnetactic bacteria [51]. The role of these bacteria in metal corrosion is not fully understood.

Manganese deposition by microorganisms also affects the corrosion behavior of alloys. Gowth of *Leptothrix discophora* resulted in ennoblement of stainless steel by elevating the open-circuit potential to $+375\,mV$ [40]. Further examination of the deposits on surfaces of coupons using X-ray photoelectron spectroscopy (XPS) confirmed that the product was MnO_2. The MnO_2 can also be reduced to Mn^{2+} by accepting two electrons from metal dissolution. The intermediate product is MnOOH [52].

D. ANAEROBIC CORROSION

D1. Anaerobic Processes

In submerged environments or nutrient-enriched conditions, all surfaces are covered with microorganisms and their exopolymeric layers. Adhesion to surfaces provides a strategy for microbial survival and multiplication [42] and an opportunuty for corrosion to occur. The frequency of gene transfer between bacteria in a biofilm community is believed to be high [53]. Within this gelatinous matrix of a biofilm, there are both oxic and anoxic zones, permitting aerobic and anaerobic processes to take place simutaneously. Aerobic processes consume oxygen, which is toxic to the anaerobic microflora, while anaerobes benefit from the lowering of oxygen tension. In the absence of oxygen, anaerobic bacteria, including methogens, sulfate-reducing bacteria, acetogens, and fermenters, are actively involved in corrosion processes. Interactions between these microbial species allow them to coexist under conditions where nutrients are limited.

Sulfate-reducing bacteria are among the most intensely investigated groups of microorgansisms involved in biological corrosion, particularly in oil field and pipeline applications [4, 33, 54, 55]. Under anaerobic conditions, oxygen is not available to accept electrons, but instead SO_4^{2-} or other compounds are used as electron acceptors. Each type of

electron acceptor is unique within a metabolic pathway. When corrosion begins, the following reactions can take place:

$$4Fe^0 \rightarrow 4Fe^{2+} + 8e^- \quad \text{(anodic reaction)} \quad (39.4)$$

$$8H_2O \rightarrow 8H^+ + 8OH^- \quad \text{(water dissociation)} \quad (39.5)$$

$$8H^+ + 8e^- \rightarrow 8H \text{ (adsorbed)} \quad \text{(cathodic reaction)} \quad (39.6)$$

$$SO_4^{2-} + 8H \rightarrow S^{2-} + 4H_2O \quad \text{(bacterial consumption)} \quad (39.7)$$

$$Fe^{2+} + S^{2-} \rightarrow FeS\downarrow \quad \text{(corrosion products)} \quad (39.8)$$

$$4Fe + SO_4^{2-} + 4H_2O \rightarrow 3Fe(OH)_2\downarrow + FeS\downarrow + 2OH^- \quad (39.9)$$

von Wolzogen Kuhr and van der Vlugt (1934) suggested that the above set of reactions is caused by SRBs. This electrochemical generalization has been accepted and is still prevalent. During corrosion, the redox potential of the bacterial growth medium is $-52\,mV$ [32]. After inoculation of a corrosion-testing cell with SRBs, the overall internal resistance decreases from the initial value of $15\,\Omega$ to approximately $1\,\Omega$, while the sterile cell actually shows an increase in resistance.

Hadley described several phases of change in the electrical potential of steel after inoculation with SRBs [32]. Before inoculation, the value is determined by the concentrations of hydrogen ions in the medium. A film of hydrogen forms on surfaces of Fe and steel, inducing polarization. Immediately after inoculation, SRBs begin growth and depolarization occurs, resulting in a drop of 50 mV in the anodic direction. By means of their hydrogenase system, the SRBs remove the adsorbed hydrogen, depolarizing the system. The overall process was described as depolarization based on the theory that these bacteria remove hydrogen that accumulates on the surfaces of iron. Electron removal as a result of hydrogen utilization results in cathodic depolarization and forces more iron to be dissolved at the anode.

The direct removal of hydrogen from the surface is equivalent to lowering the activation energy for hydrogen removal by providing a depolarization reaction. The enzyme hydrogenase, synthesized by many species of *Desulfovibrio* spp., is involved in this specific depolarization process [56]. Under aerobic conditions, the presence of molecular oxygen serves as an electron sink; under anaerobic conditions, particularly in the presence of SRBs, SO_4^{2-} in the aqueous phase can be reduced to S_2^- microbiologically. The biogenically produced S_2^- reacts with Fe^{2+} to form a precipitate of FeS. Controversy surrounding the mechanisms of corrosion includes more complex mechanisms involving both sulfide and phosphide [57–61] and processes related to hydrogenase activity [56]. The addition of chemically prepared Fe_2S and fumarate as electron acceptors also depolarizes. However, higher rates are always observed in the presence of SRBs.

As a result of the electrochemical reactions, the cathode always tends to be alkaline with an excess of OH^-. These hydroxyl groups also react with ferrous irons to form precipitates of hydroxy iron. Precipitated iron sulfites are frequently transformed into minerals, such as mackinawite, greigite, pyrrhotite, marcasite, and pyrite. Lee et al. suggests that biogenic iron sulfides are identical to those produced by purely inorganic processes under the same conditions [14]. Little et al. showed evidence that biogenic minerals are microbiological signature markers [62].

D2. Involvement of SRBs and Their Hydrogenase Systems

The role of anaerobic microorganisms in corrosion was implicated as early as 1910 [63]. Emphasis has traditionally been on SRBs [10–12, 14, 56] and on hydrogenases, in which the hydrogen on metal surfaces is consumed by microbial metabolism. Severe damage by SRBs can be found in oil field drilling steel, materials exposed to deep wells, buried pipelines, and immersed structural materials. However, the involvement of SRBs and their hydrogenases in the corrosion of mild steel is still controversial.

Currently 18 genera of dissimilar sulfate-reducing bacteria have been recognized. They are divided into two physiological groups [64]. One group utilizes lactate, pyruvate, or ethanol as carbon and energy sources, reducing sulfate to sulfide. Examples are *Desulfovibrio, Desulfomonas, Desulfotomaculum,* and *Desulfobulbus.* The other group oxidizes fatty acids, particularly acetate, reducing sulfate to sulfide. This group includes *Desulfobacter, Desulfococcus, Desulfosarcina,* and *Desulfonema.* Some species of *Desulfovibrio* lack hydrogenase. For example, *D. desulfuricans* is hydrogenase negative and *D. salexigens* is positive [65]. Booth et al. observed that the rate of corrosion by these bacteria correlated with their hydrogenase activity [66]. Hydrogenase-negative SRBs were completely inactive. Apparently, hydrogenase-positive organisms utilize cathodic hydrogen, deporalizing the cathodic reaction, which controls the kinetics. In contrast to this theory, it has been suggested that ferrous sulfide (FeS) is the primary catalyst [14, 67].

Other microorganisms should be noted for their role in anaerobic corrosion. They include methanogens [68], acetogens, thermophilic bacteria [67], and obligate proton reducers [69]. More work is needed to elucidate the role of other organisms' contributions to corrosion.

E. ALTERNATING AEROBIC AND ANAEROBIC CONDITIONS

Constant oxic or anoxic conditions are rare in natural or industrial environments. It is more common that the two alternate, depending on oxygen gradient and diffusivity in a specific environment. Microbial corrosion under such conditions is quite complex, involving two different groups of microorganisms and an interface that serves as a transition boundary for the two conditions. Resultant corrosion rates are often higher than those observed under either continuous oxic or anoxic conditions. Microbial activity reduces the oxygen level at interfaces, facilitating anaerobic metabolism. The corrosion products resulting from anaerobic processes, such as FeS, FeS_2, and S^0, can be oxidized when free oxygen is available.

During oxidation of reduced sulfur compounds more corrosive sulfides are produced under anoxic conditions, causing cathodic reactions. The corrosion rate increases as the reduced and oxidized FeS concentrations increase [14]. Cathodic depolarization processes can also yield free O_2, which reacts with polarized hydrogen on metal surfaces.

Corrosion resistance in materials is related to material composition, purity, and surface treatment. For example, stainless steel is more resistant to corrosion than mild steel because a passivation film forms on surfaces. Pitting corrosion of stainless steel can be found, in particular, in areas of welding and crevices [3, 4]. Corrosion of stainless steel American Iron and Steel Institute (AISI) 304 and AISI 316 is often localized in marine habitats.

F. CORROSION BY MICROBIAL EXOPOLYMERS

Bacteria produce copious quantities of exopolymers which appear to be implicated in corrosion [70–74]. These exopolymers are acidic and contain functional groups that bind metal ions. The exopolymers facilitate adhesion of bacteria to surfaces (Fig. 39.2). They are involved in severe corrosion of copper pipes and water supplies in large buildings and hospitals [74, 75]. These materials also play an important role in cueing the settlement of invertebrate larvae and in repelling larvae from surfaces [76]. They primarily consist of polysaccharides and proteins and influence the electrochemical potential of metals [17, 77, 78]. Surface analysis using XPS showed that these functionality-rich materials can complex metal ions from the surface, releasing them into aqueous solution. As a result, corrosion is initiated. Proteins in polymeric materials use their disulfide-rich bonds to induce corrosion.

Bacterial polymers have been found to promote corrosion of copper pipes in water supplies [74, 75, 79] due to the high affinity of the polymeric materials for copper ions [72, 79].

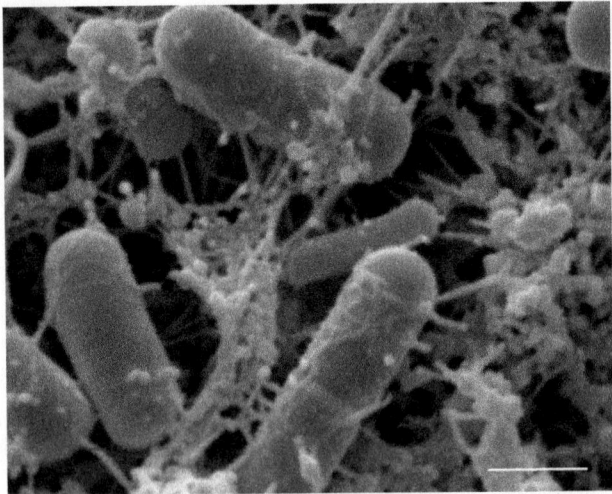

FIGURE 39.2. Scanning electron micrograph of population of aerobic bacteria producing extensive expolymeric materials (scale bar, 0.5 μm)

The corrosion processes are accelerated when the pipes are filled with stagnant soft water.

Cations influence the production of bacterial exopolymers. Polysaccharide production by *Enterobacter aerogenes* is stimulated by the presence of Mg, K, and Ca ions [80]. Toxic metal ions (e.g., Cr) also enhance polysaccharide production. Synthesis is positively correlated with Cr concentration [81].

G. MICROBIAL HYDROGEN EMBRITTLEMENT

The role of bacteria in embrittlement of metallic materials is not fully understood. During the growth of bacteria, fermentation processes produce organic acids and molecular hydrogen. This hydrogen can be adsorbed to material surfaces, causing polarization. Some bacteria, particularly the methanogens, sulfidogens, and acetogens are also capable of hydrogen utilization [82].

Walch and Mitchell [35] and Ford and co-workers [73] investigated a possible role for microbial hydrogen in hydrogen embrittlement. They measured permeation of microbial hydrogen into metal using modified Devanathan cells [83]. In a mixed microbial community commonly found in natural conditions, hydrogen production and consumption occur simultaneously. Competition for hydrogen between microbial species determines the ability of hydrogen to permeate into metal matrices, causing crack initiation [11].

Microbial hydrogen involved in material failure may be explained by two distinctively different hypotheses, pressure and surface energy changes [84]. The kinetic nature of hydrogen embrittlement of cathodically charged mild steel is determined by the competition between diffusion and plasticity. The higher the strength level, the more susceptible the alloy. However, microstructures were also proposed to be the more critical determinant of material susceptibility. Hydrogen permeation may increase the mobility of screw dislocations, but not the mobility of edge dislocations [85].

H. CORROSION BY OTHER MICROBIAL METABOLITES

Fungi have been shown to cause deterioration of a wide array of polymeric materials, including electronic polyimides [22, 28, 86, 87], protective coatings [28], and concrete [22]. Degradation of protective coating needs to be considered because the underlying metals are dependent on the protective properties of the polymer. Fungi produce highly corrosive metabolites, including a wide range of organic acids, and these acids have been shown to corrode fuel tanks [21]. They survive very well at water–fuel interfaces, metabolizing the fuel hydrocarbon as carbon and energy sources. They are also capable of generating corrosive oxidants, including hydrogen peroxide.

I. PREVENTIVE MEASURES

Chlorination is used routinely as a preventive measure to eradicate corrosive bacteria from surfaces and mechanical cleaning is also used to keep the surface clean. The former treatment produces secondary halogenated by-products, which are environmentally unacceptable. In addition, mature biofilms limit diffusion, preventing penetration of the disinfectant, and increasingly higher concentrations of the biocide become necessary to eradicate bacteria. Organic biocides, used to prevent bacterial growth in industrial systems, may selectively enrich a population capable of biocide resistance through selection and gene transfer between cells. No solution to these problems is currently available. However, new environmentally acceptable biocides are being developed with the expectation that at least some of these chemicals will be capable of either preventing biofilm formation or killing microorganisms in previously formed biofilms.

Surface treatment or engineering designs may provide conditions that minimize attachment by bacteria. Smooth surfaces of galvanized iron provide better protection than rough surfaces, because bacteria attach to rough surfaces more readily than smooth ones. Bacteria may align themselves with surface features of materials such as grain boundaries and striations under a scanning laser confocal microscope (Fig. 39.3). Such information is basic and can provide materials-based approaches to controlling establishment and growth of biofilms.

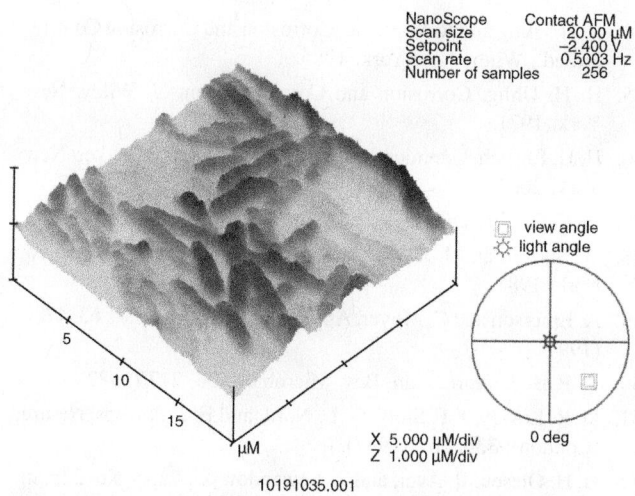

```
NanoScope      Contact AFM
Scan size          20.00 µM
Setpoint            -2.400 V
Scan rate         0.5003 Hz
Number of samples      256
```

view angle
light angle

X 5.000 µM/div
Z 1.000 µM/div   0 deg

10191035.001

FIGURE 39.3. Confocal laser scanning micrograph showing bacteria deposited on rough surface of galvanized steel sheet in three dimensions

Surface treatment with nanotechnology is very attractive, but the toxicological risks to humans from exposure to nanomaterials have not been adequately investigated. Because of this, applications of nanotechnology in food and manufacturing industries need more time and research, particularly in relation to human health and exposure risk.

J. CONCLUSIONS

Microorganisms are involved in the corrosion of metals and alloys under aerobic conditions by oxidation of iron or manganese and by solubilization through acidic metabolites. Under anaerobic conditions, SRBs corrode metals by cathodic depolarization and the formation of FeS or by consuming hydrogen produced by polarization. Other mechanisms have also been implicated in corrosion, including microbial hydrogen embrittlement or complexation of metals by microbial exopolymeric materials. It is clear that we still have a long way to go to fully understand the complex interactions between microflora and metallic surfaces that can lead to corrosion. Modern methods in molecular biology combined with more recently developed techniques in materials science, such as confocal microscopy and stable isotope techniques, should permit us to understand more fully the role and mechanisms of microorganisms in the corrosion of metals and alloys.

ACKNOWLEDGMENTS

We would like to thank Ed Seling at the Museum of Comparative Zoology, Harvard University, for assistance in taking the scanning electron micrograph used in Figure 39.1; the EM Unit at Queen Mary Hospital, The University of Hong Kong, for Figure 39.2; Cheung Kuen Wan for Figure 39.3; and Jessie Lai for assistance.

REFERENCES

1. C. A. H. Von Wolzogen Kuhr and I. S. van der Vlugt, Water, **18**, 147(1934).
2. E. C. Hill, J. L. Shennan, and R. J. Watkinson, Microbial Problems in the Offshore Oil Industry, The Institute of Petroleum, London, 1987.
3. G. Kobrin, A Practical Manual on Microbiologically Influenced Corrosion, NACE International, Houston, TX, 1993.
4. D. H. Pope, D. Duquette, P. C. Wayner, Jr., and A. H. Johannes, Microbiologically Influenced Corrosion: A State-of-the-Art Review, 2nd ed., MTI Publication No. 13, National Association of Corrosion Engineers, Houston, TX, 1989.
5. C. A. C. Sequeira, and A. K. Tiller, Microbial Corrosion, Elsevier Science, Essex, England, 1988.
6. F. Widdel, in Biotechnology Focus, Vol. 3 R. F. Finn, P. Prave, M. Schlingmann, W. Crueger, K. Esser, R. Thauer, and F. Wagner (Eds.), Hanser Publishers, Munich, Germany, 1992, pp. 261–300.
7. J. G. Knudsen, in Power Condenser Heat Transfer Technology, Hemisphere Publishing, New York, 1981, pp. 57–82.
8. S. C. Dexter, Biologically Influenced Corrosion, NACE, Houston, TX, 1986.
9. N. J. E. Dowling and J. Guezennec, in Manual of Environmental Microbiology, C. J. Hurst, G. R. Knudsen, M. J. McInerney, L. D. Stetzenbach, and M. V. Walter (Eds.), American Society for Microbiology, Washington, DC, 1997, pp. 842–855.
10. T. Ford and R. Mitchell, Adv. Microb. Ecol., **11**, 231 (1990).
11. T. Ford and R. Mitchell, Mar. Tech. Soc. J., **24**, 29 (1990).
12. W. A. Hamilton, Ann. Rev. Microbiol., **39**, 195 (1985).
13. J. A. Lee, in H. H. Uhlig (Ed.), The Corrosion Handbook, Wiley, New York, 1948, pp. 359–365.
14. W. Lee, Z. Lewandowski, P. H. Nielsen, and W. A. Hamilton, Biofouling, **8**, 165 (1995).
15. J.-D. Gu, T. E. Ford, and R. Mitchell, in Uhlig's Corrosion Handbook, 2nd ed., W. Revie (Ed.), Wiley, New York, 2000, pp. 439–460.
16. J.-D. Gu, T. E. Ford, and R. Mitchell, in Uhlig's Corrosion Handbook, 2nd ed., W. Revie (Ed.), Wiley, New York, 2000, pp. 477–491.
17. G. Chen, S. V. Kagwade, G. E. French, T. E. Ford, R. Mitchell, and C. R. Clayton, Corrosion, **52**, 891 (1996).
18. C. R. Clayton, G. P. Halada, J. R. Kearns, J. B. Gillow, and A. J. Francis, in Underground Corrosion, Technical Publication 741, J. R. Kearns, and B. J. Little (Eds.), American Society for Testing and Materials, Philadelphia, PA, 1994, pp. 141–152.
19. H. H. Paradies, in Bioextraction and Biodeterioration of Metals, C. C. Gaylarde and H. A. Videla (Eds.), The Biology of World Resources Series 1, Cambridge University Press, Cambridge, 1995, pp. 197–269.

20. H. Siedlarek, D. Wagner, W. R. Fischer, and H. H. Paradies, Corrosion Sci., **36**, 175 (1994).

21. H. A. Videla, Manual of Biocorrosion, CRC Press, Boca Raton, FL, 1996.

22. J.-D. Gu, T. E. Ford, N. S. Berke, and R. Mitchell, Int. Biodeter. Biodegr., **41**, 101 (1998).

23. J.-D. Gu, T. E. Ford, and R. Mitchell, in W. Revie (Ed.), Uhlig's Corrosion Handbook, 2nd ed., Wiley, New York, 2000, pp. 915–927.

24. S. Tayler and E. May, Int. Biodeter., **28**, 49 (1991).

25. T. Warscheid M. Oelting, and W. E. Krumbein, Int. Biodeter., **28**, 37 (1991).

26. R. Mitchell and J.-D. Gu, Int. Biodeter. Biodegr., **46**, 299 (2000).

27. R. Drewello and R. Weissmann, in DECHEMA Monograpgs Vol. 133, Biodeterioration and Biodegradation, W. Sand (Ed.), VCH Verlagsgesellschaft, Frankfurt, Germany, 1996, pp. 17–22.

28. J.-D. Gu, M. Roman, T. Esselman, and R. Mitchell, Int. Biodeter. Biodegr., **41**, 25 (1998).

29. J.-D. Gu, Int. Biodeter. Biodeg., **52**, 69 (2003).

30. J.-D. Gu, Int. Biodeter. Biodeg., **59**, 170 (2007).

31. H. H. Hanert, in The Prokarytes. A Handbook on Habitats, Isolation, and Identification of Bacteria, Vol. I, M. P. Starr, H. Stolp, H. G. Trüper, A. Balow, and H. G. Schlegel (Eds.), Springer-Verlag, New York, 1981, pp. 1049–1060.

32. R. F. Hadley, in H. H. Uhlig (Ed.), The Corrosion Handbook, Wiley, New York, 1948, pp. 466–481.

33. B. J. Little, P. A. Wagner, W. G. Characklis, and W. Lee, in Biofilms, W. G. Characklis, and K. C. Marshall (Eds.), Wiley, New York, 1990, pp. 635–670.

34. T. Ford, E. Sacco, J. Black, T. Kelley, R. Goodacre, R. C. W. Berkeley, and R. Mitchell, Appl. Environ. Microbiol., **57**, 1595 (1991).

35. M. Walch and R. Mitchell, Biologically Induced Corrosion, Proceedings of the International Conference on Biologically Induced Corrosion, in S. C. Dexter (Ed.), National Association of Corrosion Engineers, Houston, TX, 1986, pp. 201–208.

36. J. W. Costerton, G. G. Geesey, and K.-J. Cheng, Sci. Am., **238**, 86 (1978).

37. J. W. Costerton, Z. Lewandowski, D. E. Caldwell, D. R. Korber, and H. M. Lappin-Scott, Annu. Rev. Microbiol., **49**, 711 (1995).

38. J.-D. Gu, J.S. Maki, and R. Mitchell, in Zebra Mussel and Other Aquatic Nuisance Species, F. M. D'Itri (Ed.), Ann Arbor Press, Chelsea, MI, 1997, pp. 343–357.

39. S. C. Dexter, Biofouling, **7**, 97 (1994).

40. W. H. Dikinson, F. Caccavo, B. Olesen, and Z. Lewandowski, Appl. Environ. Microbiol., **63**, 2502 (1997).

41. K. C. Marshall, Interfaces in Microbial Ecology, Harvard University Press, Cambridge, MA, 1976.

42. K. C. Marshall, ASM News, **58**, 202 (1992).

43. J. W. Costerton, Z. Lewandowski, D. DeBeer, D. Caldwell, D. Korber, and G. James, J. Bacteriol., **176**, 2137 (1994).

44. H. H. Uhlig and R. W. Revie, Corrosion and Corrosion Control, 3rd ed., Wiley, New York, 1985.

45. H. H. Uhlig, Corrosion and Corrosion Control, Wiley, New York, 1971.

46. H. H. Ehrlich, Geomicrobiology, 4th ed., Marcel Dekker, New York, 2002.

47. W. C. Ghiorse and P. Hirsch, Arch. Microbiol., **123**, 213 (1979).

48. W. Stumm and J. J. Morgan, Aquatic Chemistry, Wiley, New York, 1981.

49. A. Emerson and C. Moyer, Appl. Environ. Microbiol., **63**, 4784 (1997).

50. R. P. Blakemore, Ann. Rev. Microbiol., **36**, 217 (1982).

51. D. R. Lovley, J. F. Stolz, G. L. Nord, and E. J. Phillips, Nature (London), **330**, 252 (1987).

52. B. H. Olesen, R. Avci, and Z. Lewandowski, Paper No. 275, in Corrosion/98, NACE International, Houston, TX, 1998.

53. M. L. Angles, K. C. Marshall, and A. E. Goodman, Appl. Environ. Microbiol., **59**, 843 (1993).

54. P. Angell, J.-S. Luo, and D. C. White, in 1995 International Conference on Microbial Influenced Corrosion, P. Angell, S. W. Borenstein, R. A. Buchanan, S. C. Dexter, N. J. E. Dowling, B. J. Little, C. D. Lundin, M. B. McNeil, D. H. Pope, R. E. Tatnall, D. C. White, and H. G. Zigenfuss (Eds.), NACE International, Houston, TX, 1995, pp. 1/1–10.

55. J. P. Audouard, N. J. E. Dowling, C. Compere, D. Festy, D. Feron, A. Mollica, V. Scotto, T. Rogne, V. Steinsmo, C. Taxen, and D. Thierry, in 1995 International Conference on Microbial Influenced Corrosion, P. Angell, S. W. Borenstein, R. A. Buchanan, S. C. Dexter, N. J. E. Dowling, B. J. Little, C. D. Lundin, M. B. McNeil, D. H. Pope, R. E. Tatnall, D. C. White, and H. G. Zigenfuss (Eds.), NACE International, Houston, TX, 1995, pp. 3/1–9.

56. R. L. Starkey, in Biologically Induced Corrosion, Proceedings, of the International Conference on Biologically Induced Corrosion, S. C. Dexter (Ed.), National Association of Corrosion Engineers, Houston, TX, 1986, pp. 3–7.

57. W. P. Iverson, in Underground Corrosion, E. Escalante (Ed.), Technical Publication 741 American Society for Testing and Materials, Philadelphia, PA, 1981, pp. 33–52.

58. W. P. Iverson, Mater. Perform., **23**, 28 (1984).

59. W. P. Iverson and G. J. Olsen, Microbial Corrosion, The Metal Society, London, 1983, pp. 46–53.

60. W. P. Iverson and G. J. Olsen, in Petroleum Microbiology, R. M. Atlas (Ed.), Macmillan, New York, 1984, pp. 619–641.

61. W. P. Iverson, G. J. Olsen, and L. F. Heverly, in S. C. Dexter (Ed.), Biological Influenced Corrosion, NACE, Houston, TX, 1986, pp. 154–161.

62. B. J. Little, P. A. Wagner, and Z. Lewandowski, in Geomicrobiology: Interactions between Microbes and Minerals, J. F. Banfield and K. H. Nealson (Eds.), Mineralogical Society of America, Washington, DC, 1997, pp. 123–159.

63. R. H. Gaines, Ind. J. Eng. Ind. Chem., **2**, 128 (1910).

64. M. T. Madigan, J. M. Martinko, P. V. Dunlap, and D. P. Clark, Brock Biology of Micoorganisms, 12th ed., Prentice-Hall, Upper Saddle River, NJ, 2009.

65. G. H. Booth and S. T. Mercer, Nature (London), **199**, 622 (1962).

66. G. H. Booth, L. Elford, and D. J. Wakerly, Br. Corros. J., **3**, 242 (1968).

67. B. J. Little, P. Wagner, and J. Jones-Meehan, in Microbiologically Influenced Corrosion Testing, ASTM STP 1232, J. R. Kearns and B. J. Little (Eds.), American Society for Testing and Materials, Philadelphia, PA, 1994, pp. 180–187.

68. L. Daniel, N. Belay, B. S. Rajagopal, and P. J. Weimer, Science, **237**, 509 (1987).

69. F. A. Tomei, J. S. Maki, and R. Mitchell, Appl. Environ. Microbiol., **50**, 1244 (1985).

70. T. E. Ford, M. Walch, and R. Mitchell, Paper No. 123, in Corrosion/86, NACE, Houston, TX, 1986.

71. T. E. Ford, J. S. Maki, and R. Mitchell, Paper No. 380, in Corrosion/87, NACE, Houston, TX, 1987.

72. T. E. Ford, J. S. Maki, and R. Mitchell, in Biodeterioration Vol. 7, D. R. Houghton, R. N. Smith, and H. O. W. Eggins (Eds.), Elsevier Applied Science, London and New York, 1988, pp. 378–384.

73. T. E. Ford, P. C. Searson, T. Harris, and R. Mitchell, J. Electrochem. Soc., **137**, 1175 (1990).

74. H. H. Paradies, I. Haenssel, W. Fischer, and D. Wagner, "Microbial Induced Corrosion on Copper Pipes," Report No. 44, INCRA, New York, 1990.

75. W. Fischer, I. Haenssel, and H. H. Paradies, Gutachten: Schadensanalyse von Korrosionshäden an Brauchwasserleitungen aus Kupferrohren im Krieskrankenhaus Lüdenscheid/Hellersen, Märkischer Kreis, March, 17, 1987.

76. J. S. Maki and R. Mitchell, Bull. Mar. Sci., **37**, 675 (1985).

77. J.-D. Gu and R. Mitchell, Chin. J. Mater. Res., **9**(Suppl.), 473 (1995) (in English).

78. J.-D. Gu, in Encyclopedia of Microbiology, 3rd ed., M. Schaechter (Ed.), Elsevier, 2009, pp. 259–269.

79. M. W. Mittelman and G. G. Geesey, Appl. Environ. Microbiol., **49**, 846 (1985).

80. J. F. Wilkinson and G. H. Stark, Proc. Roy. Phys. Soc. Edinburgh, **25**, 35 (1956).

81. P. J. Bremer and M. W. Loutit, Mar. Environ. Res., **20**, 249 (1986).

82. G. Gottschalk, Microbial Metabolisms, 2nd ed., Springer-Verlag, New York, 1985.

83. M. A. V. Devanathan and Z. Stachurski, Proc. R. Soc. Lond. A., **270**, 90 (1962).

84. I. M. Bernstein, in The Minerals, A. W. Thompson and N. R. Moody (Eds.), Metals and Materials Society, Warrendale, PA, 1996, pp. 3–11.

85. J.-S. Wang, in Hydrogen Effects in Materials, A. W. Thompson and N. R. Moody (Eds.), The Minerals, Metals and Materials Society, Warrendale, PA, 1996, pp. 61–75.

86. D. B. Mitton, S. Toshima, S. S. Chang, R. M. Latanision, F. Bullucci, T. E. Ford, J.-D. Gu, and R. Mitchell, in Organic Coatings for Corrosion Control, ACS Symposium Series 689, American Chemical Society, Washington, DC, 1998, pp. 211–222.

87. J.-D. Gu, T. E. Ford, and R. Mitchell, J. Appl. Polym. Sci., **92**, 1029 (1996).

40

ELECTRONIC MATERIALS, COMPONENTS, AND DEVICES

R. P. FRANKENTHAL[†]

Bell Laboratories, Lucent Technologies, Murray Hill, New Jersey

L. F. GARFIAS-MESIAS

DNV Columbus, Inc., Dublin, Ohio

A. INTRODUCTION

Although the corrosion mechanisms encountered in electronic materials, components, and devices are similar to those encountered in other materials and structures, several unique differences exist. First, the presence of an applied voltage on many components and devices can cause a small corrosion current to flow between conductors if the resistance of the dielectric separating them is compromised by defects or by mobile ionic impurities (which can be readily found in atmospheric particles). Second, because of the small dimensions encountered in electronic circuits, such as integrated circuits and thin-film devices, only exceedingly small

amounts of contamination or nanosized defects are required to generate corrosion currents sufficient to cause the total failure of a device. Figure 40.1 illustrates the maximum localized corrosion current that can be tolerated to achieve a given lifetime for various cross-sectional areas of a metallic conductor [1]. Similarly, in low-voltage, low-force electrical contacts, tarnish films less than 10 nm in thickness can introduce unacceptable levels of noise and soft errors into circuits.

First, the effects of the environment and contaminants on the corrosion properties of electronic materials, components, and devices are discussed. Then, the corrosion properties and failure mechanisms of integrated circuits (ICs), printed circuit boards, hybrid integrated circuits (HICs) and multichip modules, contacts, and connectors, and their component metals and alloys are considered. The corrosion mechanisms responsible for failures in discrete active devices, such as diodes or transistors, are the same as those for integrated circuits and are not discussed further. The metals commonly found in the above components and devices range from noble metals such as gold, platinum, and palladium to highly reactive metals, such as aluminum and titanium, and include silver, copper, nickel, tin, chromium, and various solders. In Section G, several miscellaneous corrosion failures and failure mechanisms are discussed.

B. ENVIRONMENT AND CONTAMINATION

The relevant factors that affect the corrosion behavior of the various electronic metals, components, and devices include

[†]Retired

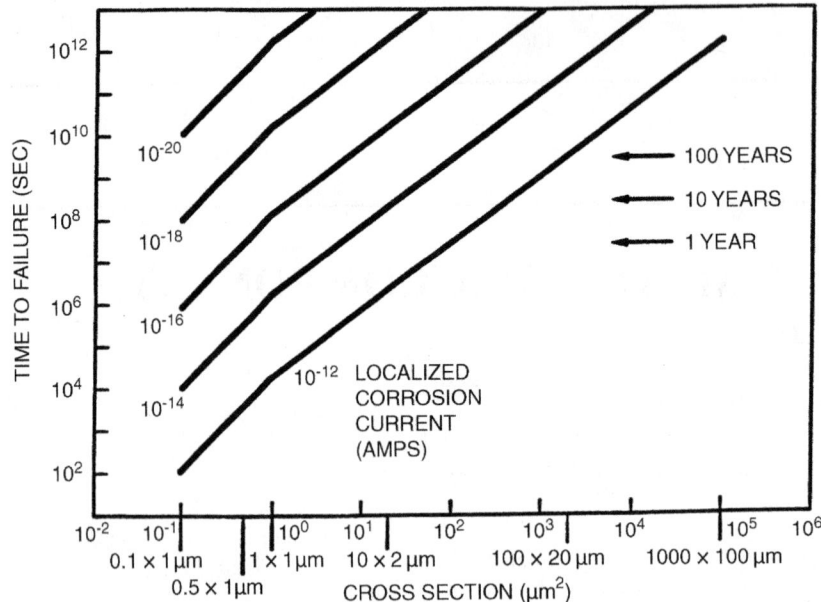

FIGURE 40.1. Illustration of maximum localized corrosion current that can be tolerated and still achieve sufficient lifetime for various cross-sectional areas. (From [1]. Reproduced by permission of The Electrochemical Society, Inc.)

temperature, relative humidity, contamination from processing, handling, and corrosive gases, and applied voltage. The environments to which they are exposed include the many processing steps a device undergoes during manufacture, shipping, and storage conditions as well as the ultimate working environment. The working environment may range from a relatively benign one, for example, room temperature, low relative humidity, and a contamination-controlled office that may be kept under positive pressure, to very aggressive conditions encountered in outdoor environments or industrial sites around the world with high temperatures, high relative humidity, high concentration of dust particles, and aggressive gases that may contain sulfur and chlorine.

An increase in temperature and/or relative humidity of the environment generally increases both the probability of corrosion and the corrosion rate. However, an increase in the device temperature typically lowers the relative humidity of the surface and its surroundings, usually decreasing the corrosion rate. For example, the relative humidity at the surface of a device that generates heat during operation is less than the ambient relative humidity. Consequently, the corrosion rate of a metal on a hot device may be less than the corrosion rate of the same device when it is not powered. Therefore, unpowered (cool) devices may be more susceptible to corrosion (e.g., during shipping) than the same device in service.

Many ionic contaminants (salts) absorb moisture and form electrolyte solutions. The relative humidity at which a given salt begins to absorb significant amounts of moisture is the critical relative humidity (CRH). Each salt has its own distinct CRH. The electrolyte solution formed at or above a salt's

CRH may be corrosive in the absence of an applied voltage, while in the presence of an applied voltage electrolytic corrosion may occur. Corrosion due to the presence of an ionic contaminant generally does not occur below its CRH.

Corrosion that takes place during manufacture of a device most commonly is the result of exposure to corrosive solutions or reactive ion etching [2]. Common corrodants include the dissociation products of plasma etches containing chlorine or fluorine and wet etches containing HF. During shipping, storage, and use, corrosion may be caused by a relatively high-temperature, high-humidity environment, by corrosive gases such as SO_2, H_2S, HCl, and O_3, or by water-soluble ionic particles that are residuals from manufacture or are deposited from the atmosphere.

Common residuals from manufacture are halide salts from wet chemical or reactive ion etches, halide or cyanide salts from plating baths and rinses, and halide salts from solder fluxes. Atmospheric particles exist in two distinct size fractions: Coarse particles (>2 μm in diameter) are usually natural materials generated from rocks, soils, and so on. Because of their size, these can be filtered with high efficiency from the air in indoor environments. Coarse particles usually are not corrosive but can be abrasive or introduce a high resistance into a contact or connector. Fine particles (<2 μm in diameter) are more difficult to remove by filtration and hence may be found in significant concentrations in indoor environments, even those with good air-handling systems. These particles, which typically result from combustion processes, are likely to contain a high proportion of water-soluble ionic compounds that are more corrosive than the natural materials [1, 3–5]. The most common ions found in these particles, NH_4^+ and SO_4^{2-}, are typically present in roughly equal molar proportions [1].

C. INTEGRATED CIRCUITS

The width of or spacing between metal conductors on state-of-the-art commercial silicon ICs was 0.25 μm in 1997. By late 2007, several key electronic manufacturers started mass producing 45-nm chips and the trend continues to scale down the size of the components in an attempt to create smaller devices with higher storage capacity that can perform faster operations. When a voltage is applied to these state-of-the-art circuits, electric fields of 10^5–10^6 V/cm may be generated across their surfaces and in dielectrics, driving ionic motion, and electrochemical reactions. With nanometer dimensions of the conductors and metallic arrays, only nanograms of metal need to corrode to cause total failure of these devices.

Corrosive contaminants such as chlorides may be present during processing or as residual contamination from processing or they can enter plastic packages through microcracks at the package–lead interface. Typical sources of halide contamination from processing and processing materials include wet chemical and reactive ion etches, solder fluxes, and epoxy molding compounds. Moisture can enter through microcracks or by diffusing through the plastic encapsulant.

The most common chip metallization is aluminum alloyed with small quantities of copper and/or silicon. In the presence of copper, Al_2Cu θ-precipitates may form. These precipitates reduce the overvoltage for the cathodic reduction of H_2O or O_2, leading to pitting corrosion. Pits in the range of tens of nanometers in diameter can cause the failure of a device. This is primarily a problem during circuit fabrication, when both wet chemical and reactive ion etch processes can lead to micropitting [2]. The most common corrodants here are the dissociation products of plasma etches containing fluorine or chlorine and wet etches containing HF. When aluminum is deposited onto other metallic conductors to form multilayered metallizations, galvanic corrosion may also occur during fabrication.

Both positively and negatively biased aluminum are susceptible to corrosion. In the presence of moisture and a positive applied voltage, chloride contamination disrupts the passive oxide film and attacks the aluminum metal [6]. In the presence of moisture and a negative voltage, hydrogen generation from the electrolysis of water raises the pH of the adsorbed water sufficiently to cause oxide dissolution,

$$Al_2O_3 + 2OH^- \rightarrow 2AlO_2^- + H_2O \qquad (40.1)$$

Dissolution of aluminum metal follows [7, 8]:

$$Al + 2H_2O \rightarrow AlO_2^- + 4H^+ + 3e^- \qquad (40.2)$$

In both cases failure results from an unacceptable increase in conductor resistance or an open circuit.

A small fraction of ICs uses gold metallization, usually in the form of multilayers of different metals, for example, Ti/Pt/Au or Ti/Pd/Au. Titanium has been traditionally used as a "glue" layer between the silicon or SiO_2 chip surface and the noble metal, and the platinum or palladium layer serves as a diffusion barrier. Positively biased gold metallization may be subject to corrosion [9]. In the presence of moisture and a complexing anion, such as chloride or cyanide, a soluble gold complex forms. These complexing anions usually originate from processing steps. Other sources are the encapsulant and the external atmosphere if microcracks or pin holes exist in the encapsulant or at the encapsulant–lead interface. The soluble complex migrates to the negatively biased conductor, where it is electrodeposited, usually in the form of a dendrite. The dendrite grows in the direction of the positively biased conductor, eventually bridging the conductors and causing a short circuit. Other noble metals, notably silver and copper, are also known to form dendrites when corroded in the presence of a complexing anion and an applied voltage.

In the absence of a complexing anion but with moisture and an applied voltage present, gold may be oxidized to form the hydroxide, which is a voluminous precipitate that spreads across the circuit [9]. Failure results either from a reduced surface resistance of the dielectric separating the conductors or from an open circuit if the positively biased conductor is sufficiently corroded.

Very large scale integrated (VLSI) circuit manufacture involving multilevel metallizations and interconnections has introduced the use of various metal couples, for example, Cu–Cr, Ni–Cr, Ti–Al, TiN–Al, and TiW–Al. These metal couples may be prone to galvanic corrosion, especially during fabrication. Polymers, such as polyimides, may be used as dielectrics between layers. Chemical interactions between metal and polymer may take place, particularly at elevated temperatures during fabrication.

Inorganic dielectrics or passivation layers (e.g., SiO_2) may be deposited over metal conductors for corrosion protection in non–hermetically sealed devices. Small quantities of phosphorus in the deposit are beneficial. However, its concentration should not exceed 4% in undensified films to avoid leaching of corrosive phosphoric acid by moisture [3].

To protect the silicon chip from mechanical damage and contamination, it is encapsulated in a plastic or ceramic package. Plastic packages are most common. However, for high-reliability applications, hermetically sealed ceramic packages may be used. In either case, provision must be made for electrical connection to the chip. The chip is typically mounted on a lead frame, with gold wires acting as the interconnect between an aluminum pad on the chip and the frame. The lead frames may be copper, nickel plate, or a controlled-expansion alloy such as Kovar or Alloy 42 selectively plated with gold for corrosion protection. The two most common failure modes inside the package are (a) galvanic corrosion at the aluminum–gold interface in the presence of moisture and (b) the formation of brittle gold–aluminum intermetallics when the joint is subject to elevated temperatures [10]. Elevated temperatures for longer times may also break down fire-retardant additives in the plastic packaging, releasing bromide ions that can lead to corrosion in the presence of moisture [4]. When the industry replaced bromide-containing fire retardants with phosphorous-containing ones. The phosphorus particles contained in the new materials, in the presence of relatively high levels of humidity, shorted the pins on the ICs and led to failure (either by consuming the anode pin due to corrosion or by "connecting" the two adjacent pins). The frame leads emerging from the package may be subject to electrolytic corrosion in the presence of humidity, ionic contamination, and an applied voltage. This is particularly likely to occur when the Cu frame leads are coated and their ends are cut and exposed to aggressive environments (or "bare toes"). Kovar leads are subject to stress corrosion cracking when bare or at discontinuities in the coating [11, 12]. Tinned leads occasionally are subject to shorting by tin whiskers that grow under the influence of compressive stresses [13–15].

D. PRINTED CIRCUIT BOARDS

Printed circuit boards (PCBs) may be as simple as a single molded plastic board with copper conductors on one or both sides or they may consist of many layers of copper conductors, each separated by a dielectric and interconnected by conductive metal vias. Commercially available state-of-the art minimum line width in 2009 was as small as 75 μm. The plastic boards typically consist of a composite, such as an epoxy resin with layered sheets of woven glass fibers, and currently are available as hard, flexible materials. The dielectric separating the layers of metallization is typically a polymer, such as a polyimide. To maintain its solderability, the exposed copper was typically coated with a tin–lead solder (hot-air solder leveling, or HASL) or protected by an organic inhibitor, such as benzotriazole. Because of the toxicity of the lead employed in the solder, HASL has been slowly replaced by lead-free surface finish products. The surface finishes commercially available are organic solderability preservatives (OSPs), immersion silver (IAg), immersion tin (ISn), direct immersion gold on copper (DIG), electroless nickel/immersion gold (ENIG), electrolytic nickel gold, electroless nickel autocatalytic gold (ENAG) and electroless nickel, electroless palladium, and immersion gold (ENEPIG). More details on the lead-free initiatives are given in Chapter 41.

Components are attached to the PCB with solder or metal-filled conductive adhesives. For additional protection against contamination and moisture and for high reliability, the assembled board may be protected by a covercoat. The covercoat should protect the metallic components, conductive vias, and bare toes on the ICs from the gaseous environment, including high relative humidity.

The most common failure mechanism is corrosion of conductor lines due to ionic contamination, moisture, and an applied voltage. Ionic contamination may originate during processing from solder flux or plating bath residuals, due to incomplete rinsing and to corrosive gases, and particles found in the environment [15a]. High surface conductance resulting from the ionic contamination, humidity, and voltage is the precursor to electrolytic corrosion. Failure may be an open or short circuit, as described in Section C.

Defects, such as pinholes in the covercoat over a conductor or component, can result in leakage currents and, sometimes, arcs between the exposed metal and a nearby conductor. This failure process is prevalent when circuits boards are exposed to relatively high humidities ($>60\%$ RH) and ionic contamination, as is common in many urban, suburban, and industrial environments [4, 5].

Printed circuit boards made from glass fiber–reinforced epoxy may be susceptible to failure caused by the formation of conductive anodic filaments [16, 17]. The first step in this failure mode is the formation of a poor glass/epoxy bond or the delamination of this interface. At sufficiently high relative humidity, moisture is adsorbed by the glass along this interface, reducing the insulation resistance of the circuit board. In the presence of an applied voltage, the positively biased copper conductor becomes the anode and is oxidized in an electrochemical cell. Water is also oxidized at the anode, forming hydronium ions,

$$2H_2O \rightarrow O_2 + 4H^+ + 4e^- \tag{40.3}$$

At the negatively biased copper conductor, water is reduced to form hydroxyl ions,

$$2H_2O + 2e^- \rightarrow H_2 + 2OH^- \tag{40.4}$$

Consequently, a gradient in pH develops between the positively and negatively biased conductors. Copper ions formed at the anode migrate along this concentration gradient until they reach a neutral pH at which they precipitate, forming a filament along the epoxy–glass interface. Eventually, a conductive bridge is formed between the two copper conductors.

Other degradation mechanisms include tarnishing by reactive gases, for example, moist air and sulfur-containing gases, corrosion by sulfur in packaging materials, rubber, and plastics, corrosion by reactive compounds emitted by organic materials, such as adhesives and coatings, during curing or aging [18], galvanic corrosion between dissimilar metals, and electrolytic corrosion caused by the presence of bromide from the decomposition of fire retardants in plastic boards (see Section C).

E. HYBRID INTEGRATED CIRCUITS AND MULTICHIP MODULES

On HICs and multichip modules, layered metallizations may be used instead of aluminum to interconnect individual ICs and other components. Common layered metallizations include chromium/copper, titanium/copper, titanium/palladium/gold, and titanium/palladium/copper/nickel/gold. Silver, tin, and lead are also found in some metallizations. Minimum line widths and line spacings may be 5 µm or less.

The corrosion problems associated with HICs and multichip modules are similar to those described previously for integrated circuit and printed circuit boards. The multilayered metallizations make these circuits more prone to galvanic corrosion, particularly during manufacture.

F. CONTACTS AND CONNECTORS

Contacts and connectors may be characterized by their contact force and the voltage across them. If the contact force is sufficiently high, most tarnish films are penetrated upon closing and good contact is made. Similarly, a sufficiently high voltage across the contact surfaces will destroy most tarnish films and produce good contact. However, for low-force, low-voltage contacts a tarnish film of <10 nm may be sufficient to introduce noise into electronic circuits and somewhat thicker films can have sufficient resistance to cause an effective open circuit. For this reason, most low-force, low-voltage contacts and connectors are coated with a noble metal, such as gold, rhodium, or a palladium–silver alloy. Alloy R156 (60Pd–40Ag) was developed as a low-cost substitute for gold. For high-reliability applications, a thin layer of gold is diffused into the R156 surface. Silver is rarely used for low-force, low-voltage applications because of its propensity to tarnish.

The most frequent corrosion-induced failure mechanisms in low-force, low-voltage contacts and connectors are pore corrosion, corrosion product creep, and fretting. Each of these can raise the contact resistance to an unacceptable level. If the substrate is exposed at the bases of pores or at the edge of a contact, corrosion of the base metal may cause failure if the corrosion product migrates onto the contact surface [18]. Examples here include the sulfidation of silver and the sulfidation or oxidation of copper exposed at the base of pores in gold-plated silver contacts and gold-plated copper contacts [18a], respectively. The corrosion products in both cases creep over the gold contact surface, increasing the contact resistance markedly. The creep rate of the copper corrosion product is a strong function of the relative

humidity [19]. Corrosion product creep has been explained by postulating a local galvanic cell action that is enhanced by the high diffusivity of silver in silver iodide and of the electrons in the substrate metal to the advancing edge of the corrosion product [20]. This hypothesis, however, does not explain why silver sulfide does not creep over rhodium or palladium that has been plated over silver. Fretting corrosion occurs principally on separable connectors having tin or tin–lead solder contacts when paired with each other or with gold [21]. The tin is oxidized in air to form a thin, brittle oxide film. Micromotion of the contact surfaces against each other, due to vibration, cyclic opening and closing of the connector, or even differential thermal expansion and contraction, disrupts the film and leads to the formation of additional oxide. With time the disrupted oxide piles up on the surface producing an open or high-resistance circuit. Stress corrosion cracking (SCC) is also a sporadic problem. Copper or copper–alloy substrates are susceptible to cracking in atmospheres containing ammonia.

Like tarnish films, some organic films produce a high or unstable contact resistance on palladium, platinum, or rhodium contact surfaces after extended service [22, 23]. This results from frictional polymerization of organic vapors during the rubbing that accompanies vibration or the opening and closing of relay contacts.

G. OTHER FAILURES AND FAILURE MECHANISMS

G1. Packaging

Most integrated circuits and many other circuits are protected by plastic packages. All polymers are permeable to moisture. However, moisture cannot condense at the device surface as long as the polymer adheres to the surface. If delamination occurs and moisture condenses, then electrolytic corrosion can take place in the presence of an applied voltage. Failure may be due to any of the mechanisms discussed in the previous sections on ICs and PCBs. Most failure mechanisms are accelerated by the presence of ionic contaminants.

Ceramic packages seal out moisture and contaminants. However, they also seal in those trapped during manufacture. Traces of moisture adsorbed on the wall of the enclosure may desorb at elevated temperatures, condense on the device, and initiate corrosion [24].

G2. Electromagnetic Interference

To avoid electromagnetic interference (EMI), electrical continuity must be maintained in shielding and in ground connections. The integrity of shielding, contacts, and apertures must be maintained by avoiding corrosion and tarnishing. Electrical contact between dissimilar metals can lead to galvanic corrosion. Formation of corrosion films, such as oxides and sulfides, can reduce the thickness and conductivity of the shield below acceptable limits or produce high-resistance connections. Proper design and materials selection is essential to prevent EMI [25, 26].

G3. Aluminum Corrosion in Chlorinated Solvents

Chlorinated solvents corrode aluminum under common processing conditions, for example, cleaning of aluminum-containing components or aluminum–copper interconnections in circuits. Water in the solvent increases the induction time for the onset of corrosion, but it also increases the subsequent corrosion rate. The mechanism of this corrosion reaction is not fully understood [27, 28].

REFERENCES

1. J. D. Sinclair, J. Electrochem. Soc., **135**, 89C (1988).

2. S. Thomas and H. M. Berg, IEEE Trans. Compon. Hybr. Manufact. Technol., **CHMT-IO**, 252 (1987).

3. R. B. Comizzoli, R. P. Frankenthal, P. C. Milner, and J. D. Sinclair, Science, **234**, 340 (1986).

4. R. B. Comizzoli and J. D. Sinclair, in Encyclopedia of Applied Physics, Vol. **6.**, VHC Publishers, New York, 1993, p. 21.

5. R. B. Comizzoli, R. P. Frankenthal, R. E. Lobnig, G. A. Peins, L. A. Psota-Kelty, D. J. Siconolfi, and J. D. Sinclair, Electrochem. Soc. Interf., **2**(3), 26 (Fall 1993).

6. S. C. Kolesar, Annu. Proc. Reliab. Phys., **12**, 155 (1974).

7. W. M. Paulson and R. W. Kirk, Annu. Proc. Reliab. Phys., **12**, 172 (1974).

8. R. B. Comizzoli, RCA Rev., **37**, 483 (1976).

9. R. P. Frankenthal and W. H. Becker, J. Electrochem. Soc., **126**, 1718 (1979).

10. G. G. Harman, Annu. Proc. Reliab. Phys., **12**, 131 (1974).

11. R. G. Baker and A. Mendizza, Electro-Tech., **72**, p. 11 (Oct. 1963).

12. A. J. Raffalovich, IEEE Trans. Parts Hybrids Packag., **PHP-7**, 155 (1971).

13. K. G. Compton, A. Mendizza, and S. M. Arnold, Corrosion, **7**, 327 (1951).

14. S. C. Britton, Trans. Inst. Metal Finish., **52** (Part 3), 95 (1974).

15. C. H. Pitt and R. G. Henning, J. Appl. Phys., **35**, 459 (1964).

15a. M. Reid, J. Punch, L. F. Garfias-Mesias, K. Shannon, S. Belochapkine, and D. A. Tanner, "Study of Mixed Flowing Gas Exposure of Copper," J. Electrochem. Soc., **155**, C147 (2008).

16. J. N. Lahti, R. H. Delaney, and J. N. Hines, Annu. Proc. Reliab. Phys., **17**, 39 (1979).

17. D. J. Lando, J. P. Mitchell, and T. L. Welsher, Annu. Proc. Reliab. Phys., **17**, 51 (1979).

18. R. G. Baker, in Proceedings of the 6th Annual Conference Marine Technical Society, Vol. 2, Marine Technology Society, Washington, DC, 1970, p. 1265.

18a. Y. M. Reid, J. Punch, G. Grace, and L. F. Garfias-Mesias, "Corrosion Resistance of Copper Coated Contacts," J. Electrochem. Soc., **153**, B513 (2006).

19. V. Tierney, J. Electrochem. Soc., **128**, 1321 (1981).

20. C. Ilschner-Gensch and C. Wagner, J. Electrochem. Soc., **105**, 198 (1958).

21. M. Antler, IEEE Trans. Compon. Hybr. Manuf. Technol., **CMHT-8**, 87 (1985).

22. H. W Hermance and T. F. Egan, Bell Syst. Tech. J., **37**, 739 (1958).

23. I. Dietrich and M. Honrath-Barkhausen, Z. Angew. Phys., **11**, 399 (1959).

24. R. P. Frankenthal and G. Y. Chin, in Corrosion of Electronic Materials and Devices, J. D. Sinclair (Ed.), The Electrochemical Society, Pennington, NJ, 1991, p. 296.

25. H. Ott, Noise Reduction Techniques in Electronic Systems, Wiley, New York, 1988.

26. D. C. Smith, C. Herring, Jr., and R. Haynes, in IEEE Symposium on Electromagnetic Compatibility, Chicago, IL, Aug. 1994, p. 224.

27. H. H. Uhlig and R. W. Revie, Corrosion and Corrosion Control, 3rd ed., Wiley, New York, 1985, p. 347.

28. P. A. Totta, J. Vac. Sci. Technol., **13**, 26 (1976).

41

CORROSION OF ELECTRONICS: LEAD-FREE INITIATIVES

M. Reid

Stokes Research Institute, University of Limerick, Limerick, Ireland

L. F. Garfias-Mesias

DNV Columbus, Inc., Dublin, Ohio

A. INTRODUCTION

Lead, mercury, cadmium, hexavalent chromium, and poly-brominated compounds [polybrominated biphenyls (PBB) and polybrominated diphenyl ether (PBDE)] and similar substances contained in modern electronic compounds have been associated with many health risks. The primary health concern has been associated with the migration of lead and other chemicals contained in electronic products from landfill sites into the secondary water sources after disposal [1]. A number of legislative acts have been issued with the aim of reducing the environmental effects due to the use of hazardous substances in electronic equipment. The European Union has issued the most influential of these acts. The European Union adopted the first one, the directive on the restriction of the use of certain hazardous substances in electrical and electronic equipment, 2002/95/EC, commonly referred to as the Restriction of Hazardous Substances Directive (RoHS), in February 2003 [2]. The RoHS directive, which took effect on July 1, 2006, restricts the use of the six hazardous materials indicated above in the manufacture of various types of electronic and electrical equipment. RoHS is closely linked with the Waste Electrical and Electronic Equipment Directive [3] (or WEEE), which aims to prevent the generation of electrical and electronic waste while promoting the reuse, recycling, and recovery of the hazardous substances by reducing the quantity of this type of waste sent to landfills.

The critical substance in the RoHS directive is lead, mainly because the directive requires that the manufacturers of electronic and electrical equipment must replace solders and printed circuit board (PCB) surface finishes on all electronic products sold in the European economic region. Although some exemptions were made for specialized medical, automotive, and telecommunications equipment, those exemptions expire in 2010. Following the European Community directives, other countries, including China and South Korea, implemented their own versions of the RoHS and WEEE initiatives. Although other countries, such as United States and Japan, have not passed similar laws, most of the companies in the United States and Japan have been phasing out electronics that are noncompliant to RoHS and moving toward "environmentally green electronics," mostly because of their exports overseas where RoHS and WEEE

Uhlig's Corrosion Handbook, Third Edition, Edited by R. Winston Revie
Copyright © 2011 John Wiley & Sons, Inc.

laws apply or because of local regulations adopted to protect the community.

In Section B, the main advantages of lead-containing solder are described, particularly its long track record of reliability. In Section C, some of the alternatives for lead-free solder are presented. In Section D, the different lead-free alternatives are described in more detail, along with the manufacturing, properties, and corrosion resistance either when deployed in real environments or in laboratory testing. The difficulties in reproducing creep corrosion in laboratory studies and the effect of flux residues on the new lead-free solders are outlined in Section E. Although flux residues in existing lead-containing solders were not an issue, flux residues in lead-free solders must be investigated. Finally, in Section F, the challenges in the use of lead-free alternatives—corrosion resistance and long-term reliability—are summarized.

B. LEAD-CONTAINING SOLDERS

Traditionally, PCBs are coated with a layer of solder in a process known as HASL (hot-air solder leveling). During HASL, the PCB is typically dipped into a bath of molten solder and all the exposed copper surfaces (including the electronic components) are covered with solder. Excess solder is removed by scraping the PCB with hot air. The HASL process consists of four steps: precleaning, fluxing, hot-air leveling, and postcleaning.

For many years, lead solder has been used and has been proven to be a highly reliable system in most electronic equipment [4]. Even in harsh environments, the long-term corrosion resistance of many devices in the field has been remarkably "unnoticed" until recent years when most companies began transitioning to lead-free solders. The long-term reliability of PCBs and electronic components, either soldered or with a surface finish that include HASL, has been instrumental in the development of many devices that are used daily in communications, transportation, household applications, and so on. In all of these applications, lead solder has demonstrated an excellent corrosion resistance due to its thick coating and the inherent corrosion resistance of the SnPb system [5].

C. LEAD-FREE SOLDERS: CURRENT ALTERNATIVES

Because of the RoHS directives, the move towards environmentally green electronics, and the need for faster and cheaper electronic equipment, the electronic industry is quickly moving toward lead-free PCB surface finishes and high-density circuit boards. The HASL surface finish is increasingly being replaced with other lead-free surface finishes. The new HASL replacements have found applications in several types of electronic equipment, ranging from common household electronic equipment to high-reliability telecommunications equipment (where the expected life spans vary from 2 to 20 years).

At the present time, in early 2010, the most popular choices to replace HASL are immersion silver (IAg), immersion tin (ISn), electroless nickel immersion gold (ENIG), electroless nickel electroless palladium immersion gold (ENEPIG), and organic solderability preservative (OSP). Among these types of lead-free finishes, IAg and OSP are the preferred finishes for many portable applications (e.g., cell phones, personal games, and portable electronics). On the other hand, ISn, ENEPIG, and ENIG are mainly used for applications requiring higher reliability and where higher reliability targets are expected in the electronic equipment.

The alternative PCB surface finishes were developed with the main aim of providing a solderable and coplanar surface for the attachment of both discrete and fine-pitch integrated circuit component packages during assembly. The corrosion resistance of the PCB surface finish, and sometimes of the electronic components attached to it, is one of the main issues that need to be resolved. Since the mid-1990s, suppliers and manufacturers of electronic equipment have invested significant resources in both fundamental and applied research to identify the surface finish that can yield the highest reliability of electronics at a low cost.

As discussed in Chapter 76, airborne particulates plus high relative humidity can result in failures of electronic equipment, in the form of an intermittent signal, high levels of leakage current, or most commonly a short circuit. The other main cause of the failure of electronic equipment is associated with corrosion that disrupts the electrical path; for example, corrosion products may creep and connect a biased gap between conductors.

Extensive testing and solder joint reliability assessments have been performed on the lead-free PCB finishes. However, less attention has been paid to the corrosion resistance of the lead-free PCB finishes. Because of the corrosion of lead-free PCB deployed in natural environments [6] and after less than six months exposure to high-sulfur environments [7], the current lead-free technology needs further development. As discussed in Chapter 76, this further development will be particularly important in very aggressive environments [8] and those that contain high levels of pollutants [9].

D. CORROSION OF ALTERNATIVE LEAD-FREE SOLDERS

In this section, the manufacturing issues, properties, and corrosion resistance of the different technologies that are currently being explored as alternatives to lead-free solder are described based on mixed flowing gas (MFG) testing and from exposure to natural environments.

D1. Immersion Silver Finish

The IAg finish is an attractive lead-free alternative for portable product applications. It combines low cost with a highly conductive planar surface that has been proven suitable for fine-pitch components. In a well-controlled environment, IAg has a shelf life of up to one year [10]. It also does not require special handling during manufacturing. Its low contact resistance reduces in-circuit test (ICT) cycle time after assembly [11]. The IAg process consists of the following steps: cleaning and surface preparation, predipping, silver deposition, postdipping, and finally rinsing and drying. Commercially available IAg surface finishes have silver thicknesses ranging from 0.1 to 0.5 µm. It is important to note that IAg finishes may contain codeposited organic corrosion inhibitors and postplating treatments may be carried out to further enhance the corrosion resistance of the PCB and components. However, atmospheric corrosion of silver is well known to be influenced by the presence of moisture and the presence of oxidizing species (O_3, NO_2, and Cl_2), which enhance the rate of silver sulfidation [12]. In the presence of H_2S, silver can undergo sulfidation [13] through the chemical reaction

$$2Ag + H_2S \rightarrow Ag_2S + H_2$$

Silver is sensitive to Cl_2, but AgCl corrosion products (similar to the reaction of Cu [14] in the presence of Cl_2) are found only in high-chloride environments [15]. Both Ag and Cu corrode [16] in the presence of sulfur-containing species in the atmosphere, such as H_2S, SO_2, and organic acids containing sulfur, resulting in copper sulfide (Cu_2S) and silver sulfide (Ag_2S) [1]. Electronic equipment with IAg exposed to high sulfur environments deployed worldwide have shown to corrode rapidly [4, 6, 7]. In these cases, creeping corrosion between conductive traces or adjacent pins (causing short circuits) was identified as the main failure mode that led to failure of the devices. In general, the susceptibility of IAg to readily tarnish and further corrode in sulfur-containing environments is still regarded as the main reliability concern for this technology [17]. With the exception of creep corrosion, the failure mechanisms of IAg surface finishes have been duplicated using MFG class III tests [11].

D2. Immersion Tin Finish

Immersion tin not only combines good solderability and low contact resistance but also is inexpensive [18]. During soldering, the Sn dissolves in the solder to form part of the solder joint, increasing the lubrication of the surface finish and

improving ICT. The solderability of ISn depends on the amount of free Sn on the surface and the degree of oxidation of the surface finish [19]. The formation of the Cu_6Sn_5 intermetallic, the initial thickness, and dry storage conditions can all influence the shelf life of the ISn surface finish [20]. The (ISn) plating process is similar to the IAg process. However, usually the ISn plating chemistry is more complex and often includes thiourea as a selective complexing agent for the cuprous cations. Since thiourea is a carcinogen, the process is not "environmentally friendly." Unless a new green complexing agent is found, this type of surface finish may become obsolete in the future. Similar to IAg, atmospheric corrosion of Sn requires the presence of moisture on the surface. The most common corrosion products are the Sn oxides, even in the presence of sulfur-containing gases [21]. In the presence of SO_2 and NO_2 gases [22], research has demonstrated that only tin oxides can be formed on the ISn surfaces. The ISn surface finish can undergo low levels of creep corrosion when compared to the IAg surface finish [23]. Although ISn surface finishes may provide slightly better corrosion resistance than IAg at a lower cost, they are susceptible to the formation of Sn whiskers [24], which may lead to failure of the device by bridging adjacent traces or leads in the components in a circuit board.

D3. ENIG and ENEPIG Surface Finish

Electroless nickel immersion gold has been widely used in the electronic industry, mainly in military applications. Its coplanar surface is suitable for fine-pitched device placement, and it can resist multiple reflows, allowing for double-sided assembly of components onto the PCB. ENIG has excellent solderability, which is attractive for most of the modern components with a very low contact resistance [18]. One of the main disadvantages of ENIG is its cost, which is the main reason why ENIG is used primarily on military applications. Commercially available ENIG has a Au plating from 0.1 to 0.125 µm and a Ni layer from 4 to 6 µm [25]. The ENIG plating process is not as simple as the IAg or ISn. In this case, electroless Ni is deposited first and is used mainly as a diffusion barrier between the Cu substrate and the Au layer. It has been demonstrated that a Ni layer deposited with high phosphorus content can increase the corrosion resistance of the ENIG surface finish [26]. The main objective of the Au layer (immersion plated) is to avoid Ni oxidation. Adding a Pd layer to the ENIG surface finish results in an ENEPIG surface finish. The Pd film is also a diffusion barrier, inhibiting Cu migration from the underlying substrate, and typically is under the gold layer, the final layer in this process. The main reason for having the final Au layer on both the ENIG and ENEPIG is to allow the formation of a solderable surface where aluminum wire bonds can be easily attached. Although Ni is an expensive metal, it is very resistant to many types of corrosion. The most typical atmospheric corrosion products of ENIG and ENEPIG are the various forms of nickel sulfates [27]. Initially, the corrosion film formed on nickel in the atmosphere is an inner layer of NiO and an outer layer of $Ni(OH)_2$ [27]. However, as the exposure time increases, a layer of amorphous nickel sulfate is formed, which later develops into a crystalline basic nickel sulfate [28] with slightly higher corrosion resistance. Once the nickel sulfate is formed, it can also transform into hydroxysulfates [27]. The gold layer (typically referred as "flash gold") on the ENIG and ENEPIG surface finish has been subjected to MFG class III testing [29]. Corrosion of the Cu substrate was observed on the surface, and the corrosion products increased in thickness with extended exposure time [29]. Chlorine-containing films were present in the 2-day samples while sulfur became predominant after a 10-day exposure. The chlorine attacked the submicrometer defects in the Au layer. These defects opened and led to the subsequent corrosion of the Ni and Cu alloy substrate layer [30] (as seen in Fig. 41.1). The main corrosion product identified was Cu sulfide (Cu_2S) with chlorine playing an important role at the onset of the corrosion process [31] but without forming a corrosion layer [14]. A reliability concern for ENIG surface finishes is its propensity to display solderability problems (sometimes referred as "Black Pad"), particularly with ball grid array package types [32]. The Black Pad failure mode is typically described as the separation of the solder joint from the nickel surface [32]. Recent reports have found that the gold thickness should not exceed the range of 2 μin. (~50 nm) to 4 μin. (~100 nm) as stated in the IPC-4552 ENIG specification [33] in order to avoid this problem [34].

D4. Organic Solderability Preservative

Using organic and inorganic corrosion inhibitors on Cu exposed to harsh environmental conditions has been a very popular alternative that transformed in the 1990s into the OSP, a low-cost finish with excellent plated-through-hole and surface-mount solderability properties. However, because of its poor contact resistance during in-circuit testing, it usually requires PCBs to be filled with solder in the vias to prevent

continuity problems during electrical testing and damage to the probes. The most common compound that is used as OSP is imidazole, which traditionally has been used as an inhibitor for Cu and its alloys [35]. The corrosion-inhibiting mechanism is based on adsorption of molecules on the Cu surface followed by a formation of a protective complex with Cu [35]. Similar to conventional corrosion-inhibitors of Cu, the combination of imidazole with benzene results in an effective method to increase Cu corrosion inhibition in acidic conditions [36]. Imidazole films decompose 10 times more slowly on Cu under inert atmospheres than under oxidizing atmospheres [37], and since oxygen plays a direct role in the decomposition of the OSP films, it is expected that the OSP layer will decompose in the presence of aggressive environments leading to an increased corrosion rate of the Cu surface [4, 17].

E. LEAD-FREE SOLDERS: CREEP AND FLUX RESIDUES

In this section, two new problems that are frequently encountered in the new lead-free solder technologies are examined. Although creep corrosion is widely encountered in field failures, it has been difficult to design a new experimental method to reproduce this form of corrosion in the laboratory. The flux residues that are left after the deposition constitute a common issue that is prevalent in lead-free solders.

E1. Creep Corrosion

Compared with lead-containing solder, the alternative lead-free surface finishes provide reduced corrosion protection to the underlying copper when exposed to aggressive environments. The high corrosion susceptibility of these alternatives is a reliability concern for products deployed under high levels of sulfur-containing pollutants [4, 6, 7, 9]. Creep corrosion is recognized as the main cause of failure in equipment that failed within a few years (1–5 years) [6].

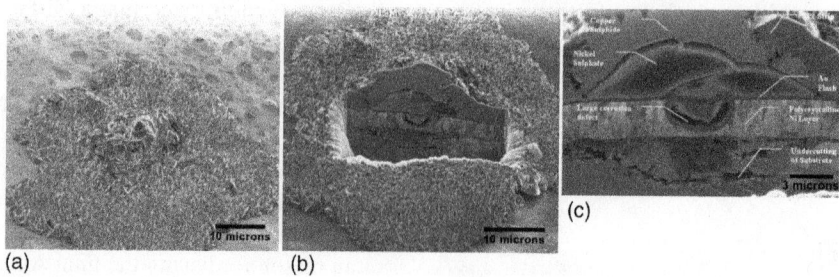

FIGURE 41.1. Scanning electron micrograph of a corrosion site developed on an ENIG plated contact after 30 days exposure to 90% RH, 40°C, and 4 ppm of H_2S: (a) before cross-sectioning with the focused ion beam (FIB); (b) after FIB cross-sectioning; (c) higher magnification image showing the details of the different layers. (Courtesy of The Electrochemical Society [30].)

Such early failures caused by creep corrosion are undesirable, especially if the equipment is expected to survive up to and beyond 20 years in the field [1].

Great interest has been expressed in understanding creep corrosion under the conditions explained above. However, very little progress has been made in reproducing creep corrosion in the laboratories using accelerated corrosion testing. As discussed in Chapter 76, the MFG test is a standard accelerated corrosion test methodology based on the parameters of temperature, relative humidity, and gases that are typical pollutants widely available in the environment (H_2S, SO_2, NO_2, and Cl_2). The MFG class III is widely regarded in the electronic community as the best test to provide realistic simulation of the corrosion behavior of electronic equipment exposed to field environments. However, when the MFG class III was applied to PCBs with lead-free finishes, no creep corrosion was induced. There are clearly other factors to be considered for corrosion products to creep over the surface of components, leads, and PCBs, besides the aggressive environment. Two factors missing in the MFG tests are particulates and, in most cases, a voltage applied to the components, as discussed in Chapter 76. New methodologies should take into consideration all possible factors that can induce exactly the same failure modes that are encountered in the field (including creep corrosion).

E2. Flux Residues

During PCB assembly, the residues that originate from the fluxes used in the process create a new type of surface contamination. These flux residues are typically difficult to clean after the lead-free processing has been completed [38]. Lead-free solder flux formulations require higher molecular weight resins and activators since the more aggressive flux chemistry needs to achieve better wetting. In addition, the higher lead-free reflow temperature induces greater polymerization of the flux constituents, thereby producing a flux residue that is far more difficult to remove (than in traditional fluxes used in lead solders). The flux residue left on the surface of a lead-free PCB assembly is much greater than occurs when lead-containing solders are used.

Flux, necessary in nearly all soldering operations, consists of three major components, solvent, vehicle, and activator [13]. The solvent acts as a carrier for the vehicle and activator and distributes both of them evenly across the surface of the board. The solvent is usually an organic alcohol or water. The solvent used for each flux depends on its ability to dissolve the other flux constituents. The second component, the vehicle, coats the surface to be soldered while the dissolving products (the activators with the surface metal oxides) formed in the reaction act as a barrier between the cleaned metal and the surrounding oxygen-rich atmosphere, preventing further oxidation. Resins are commonly used in traditional fluxes, whereas polyglycols are used in water-soluble fluxes. The final component of the flux, the activator, is the chemical that reacts with the oxides on the surface, exposing the clean metal underneath to be soldered. Typical activators include amine hydrochlorides, dicarboxylic acids (such as adipic or succinic acids), and organic acids, such as malic acid.

The role of flux residues on creep corrosion of the new lead-free solders needs to be addressed in future lab and field testing to establish if alternative flux residues may promote or hinder creep corrosion.

F. SUMMARY

At the present time in 2010, there are several lead-free surface finishes commercially available to the electronics industry. New lead-free surface finishes and improved versions of the existing finishes may be discovered in the years to come. The choice of surface finish may depend on the cost, expected reliability, and lifetime of the equipment. Several studies have shown that these lead-free alternatives need to be improved, particularly when the electronic equipment is going to be deployed in harsh environments.

Furthermore, it is important to develop new testing methodologies (as suggested in this chapter and in Chapter 76) to reproduce the same failure mechanisms that these electronic materials and devices encounter when they are deployed, particularly in very aggressive environments where the pollutant composition exceeds that of typical MFG tests.

Although there is a need for more research on alternative lead-free technologies, it is also important to exchange the information not only on the success stories but also in field failures, where most of the useful knowledge is accumulated.

REFERENCES

1. R. Kellner, "Alternative Surface Finishes—Options and Environmental Considerations," Circuit World, **30**, 30–33 (2004).

2. Directive 2002/95/EC of the European Parliament and of the Council, "On the Restriction of the Use of Certain Hazardous Substances in Electrical and Electronic Equipment," Official Journal of the European Union, L 37, Jan. 27, 2003, pp. 19–23.

3. Directive 2002/96/EC of the European Parliament and of the Council, "On Waste Electrical and Electronic Equipment," Official Journal of the European Union, Article 175(1), Jan. 27, 2003.

4. L. F. Garfias-Mesias, J. P. Franey, R. P. Frankenthal, and W. D. Reents, Gordon Research Conference on Corrosion, New Hampshire, July 26, 2004.

5. S. Yee, and H. Ladhar, "Reliability Comparison of Different Surface Finishes on Cu," Circuit World, **25**(1), 25–29 (1999).

6. R. Schueller, W. Ables, and J. Fitch, "Creep Corrosion on Lead-Free Printed Circuit Boards in High Sulfur Environments,"

SMTA International Conference, Orlando, FL, Oct. 7–11, 2007.

7. P. Mazurkiewicz, "Accelerated Corrosion of Printed Circuit Boards due to High Levels of Reduced Sulfur Gasses in Industrial Environments," in Conference Proceedings from the 32nd International Symposium for Testing and Failure Analysis, Austin, TX, Nov. 12–16, 2006. ASM International, Materials Park, OH, 2006, 469–473.

8. I. S. Cole, W. D. Ganther, J. D. Sinclair, D. Lau, and D. A. Paterson, "A Study of the Wetting of Metal Surfaces in Order to Understand the Processes Controlling Atmospheric Corrosion," J. Electrochem. Soc., **151**, B627–B635 (2004).

9. M. Watanabe, H. Hirota, T. Handa, N. Kuwaki, and J. Sakai, "Atmospheric Corrosion of Cu in an Indoor Environment with a High H_2S Concentration," 17th International Corrosion Congress: Corrosion Control in the Service of Society, Las Vegas, NV, Oct. 6–10, 2008.

10. "Specification for Immersion Silver Plating for Printed Boards," IPC 4553, Institute for Printed Circuits, Bannockburn, IL, May 2009.

11. S. Chada, and E. Bradley, "Investigation of Immersion Silver PCB Finishes for Portable Product Applications," in Proceedings of the SMTA International, Chicago IL, Surface Mount Technology Association, Edina, MN, 2001, pp. 604–611.

12. T. E. Graedel, "Corrosion Mechanisms for Silver Exposed to the Atmosphere," J. Electrochem. Soc., **139**, 1963–1970 (1992).

13. D. W. Rice, P. Peterson, E. B. Rigby, P. Phipps, R. J. Cappell, and R. Tremoureux, "Atmospheric Corrosion of Cu and Silver," J. Electrochem. Soc., **128**, 275–284 (1981).

14. M. Reid, J. Punch, L. F. Garfias-Mesias, K. Shannon, S. Belochapkine, and D. A. Tanner, "Study of Mixed Flowing Gas Exposure of Cu," J. Electrochem. Soc., **155**, C147–C153, (2008).

15. W. H. Abbott, "Effects of Industrial Air Pollutants on Electrical Contact Materials," IEEE Trans. Parts Hybrids Packaging, **10**, 24–27, (1974).

16. S.W. Chaikin, J. Janney, F.M. Church, and C.W. McClelland, "Silver Migration and Printed Wiring", Industrial and Engineering Chemistry, **51**(3), 299–304 (1959).

17. L. F. Garfias-Mesias, J. P. Franey, R. P. Frankenthal, R. Coyle, and W. D. Reents, "Life Prediction and Risk Mitigation for 20 Year Telecom Electronic Equipment in Harsh Environments," in CORROSION 2005, Research in Progress, NACE, Houston TX, Apr. 4–6, 2005.

18. S. Gitachari, "Getting the Lead Out," Circuitree, Jan. 24, 2006, available: http://www.circuitree.com/Articles/Feature_Article/7239c209d5e09010VgnVCM100000f932a8c0.

19. S. Lamprecht, "An Investigation of the Recommended Immersion Tin Thickness for Lead-free Soldering," Circuit World, **31**, 15–21 (2005).

20. "Specification for Immersion Tin Plating for Printed Circuit Boards," IPC 4554, Institute for Printed Circuits, Bannockburn, IL, 2006.

21. H. Strandberg, L. G. Johansson and O. Lindqvist, "Atmospheric Corrosion of Statue Bronzes Exposed to SO_2 and NO_2," Mater. Corros., **48**, 721–730 (1997).

22. V. Brusic, D. D. Dimilia and R. Macinnes, "The Corrosion of Tin, Lead and their Alloys," Corrosion, **47**, 509–518 (1991).

23. R. Veale, "Reliability of PCB Alternate Surface Finishes in a Harsh Industrial Environment," SMTA International Conference, Chicago, IL, 2005, pp. 25–29.

24. B. Z. Lee and D. N. Lee, "Spontaneous Growth Mechanism of Tin Whiskers," Acta Mater., **46**, 3701–3714 (1998).

25. G. Milad, "Surface Finishes: Metallic Coatings Over Nickel Over Cu," in Proceedings of the 1996 Surface Mount International, Technical program, San Jose CA, 794–796, September 10–12, 1996.

26. K. Johal, "Discoloration Related Failure Mechanism and Its Root Cause in Electroless Nickel Immersion Gold (ENIG) Pad Metallurgical Surface Finish," in Proceedings 11th International Symposium on the Physical and Failure Analysis of Integrated Circuits, Taiwan, July 5–8, 2004.

27. C. Leygraf and T. E. Graedel, "Corrosion Mechanisms for Nickel Exposed to the Atmosphere," J. Electrochem. Soc., **147**, 1010–1014 (2000).

28. D. Persson and C. Leygraf, "Initial Interaction of Sulfur Dioxide with Water Covered Metal Surfaces: An *In-Situ* IRAS Study," J. Electrochem. Soc., **142**, 1459–1468 (1995).

29. R. J. Geckle and S. Mroczkowski, "Corrosion of Precious Metal Plated Cu Alloys Due to Mixed Flowing Gas Exposure," IEEE Trans. Comp. Packag. Manufact. Technol., **14**, 162–169 (1991).

30. M. Reid, J. Punch, J. L. F. Garfias-Mesias, G. K. Grace, and S. Belochapkine, "Corrosion Resistance of Cu-Coated Contacts," J. Electrochem. Soc., **153**, B513–B517 (2006).

31. W. H. Abbott, "The Corrosion of Cu and Porous Gold in Flowing Mixed Gas Environments," IEEE Trans. Comp. Hybrids Manufact. Technol., **13**, 40–45 (1990).

32. M. Walsh, "Electroless Nickel Immersion Gold and Black Pad," CircuiTree, available: http://www.circuitree.com/Articles/Feature_Article/9fb0f01e2b7d7010VgnVCM100000f932a8c0, Jan. 1, 2000.

33. "Specification for Electroless Nickel/Immersion Gold (ENIG) Plating for Printed Circuit Boards," IPC-4552, Institute for Printed Circuits, Bannockburn, IL, 2002.

34. G. Milad, "Wiping Out Black Pad: Improved Process Control Can Help the ENIG Process Shine," Printed Circuit Des. Fab., Mar. 1, 2009, p. 873.

35. K. Saeki and M. Carano, "Next Generation Organic Solderability Preservatives (OSP) for Lead-free Soldering and Mixed Metal Finish PWB's and BGA Substrates" Proceedings of the Technical Conference (CDROM) IPC/SMEMA Council APEX 2004 Conference, Anaheim, CA, Feb. 24–26, 2004, pp. S10-2-1-S10-2-9.

36. M. Antonijevic and M. B. Petrovic, "Cu Corrosion Inhibitors. A Review," Int. J. Electrochem. Sci., **3**, 1–28 (2007).

37. R. L. Opila, H. W. Krautter, B. R. Zegarski and L. H. Dubois, "Thermal Stability of Azole-Coated Cu Surfaces," J. Electrochem. Soc., **142**, 4074–4077 (1995).

38. G. C. Munie and L. J. Turbini, "Fluxes and Cleaning," in Printed Circuit Handbook, 6th ed. C. F. Coombs, Jr., (Ed.), McGraw-Hill, New York, 2007.

42

METASTABLE ALLOYS

K. HASHIMOTO

Tohoku Institute of Technology, Sendai, Japan

A. STRUCTURAL CHARACTERISTICS

Amorphous alloys consist of at least two components and have no long-range atomic order. They are produced by a variety of methods based on rapid solidification of the alloy constituents from the gas, liquid, and aqueous phases. Mechanical alloying, that is, solid-state mixing, is also effective for preparation of amorphous alloy powders. Vitrification of metal surfaces is also made by destruction of the long-range atomic order in the surfaces of solid metals.

The formation of the structure with no long-range atomic order is based on the prevention of solid-state diffusion during solidification, and hence the alloys are free of compositional fluctuations formed by solid-state diffusion, such as second phases, precipitates, and segregates. The amorphous alloys are therefore regarded as ideal, chemically homogeneous alloys composed of thermodynamically metastable single-phase solid solutions supersaturated with alloy constituents. This characteristic is particularly suitable in producing new alloys possessing specific properties. Even if amorphous single-phase alloys are not formed, alloys prepared by amorphization methods are often composed of nanocrystalline phases supersaturated with alloying elements. From a corrosion point of view they can be considered as homogeneous alloys.

B. CORROSION-RESISTANT ALLOYS IN AQUEOUS SOLUTIONS

The corrosion behavior of amorphous alloys has received particular attention since the extraordinarily high corrosion resistance of amorphous Fe–Cr–metalloid alloys was reported in 1974 [1]. The preparation of amorphous iron-based alloys by rapid quenching from the liquid state using melt spinning generally requires the alloys to contain large amounts of metalloids, which are mostly close to the eutectic compositions.

The addition of chromium to amorphous Fe–metalloid alloys is particularly effective in enhancing corrosion resistance. For instance, amorphous Fe–8Cr–13P–7C alloy passivates spontaneously even in $2\,M$ HCl at ambient temperature [2]. (The number denoting the concentration of an alloy element in amorphous alloy formulas is expressed as an atomic percent unless otherwise stated.)

Uhlig's Corrosion Handbook, Third Edition, Edited by R. Winston Revie
Copyright © 2011 John Wiley & Sons, Inc.

C. FACTORS DETERMINING THE HIGH CORROSION RESISTANCE OF AMORPHOUS ALLOYS

C1. Passive Films Rich in Cations of Alloying Elements with High Passivating Ability

The spontaneously passive film formed on amorphous Fe–10Cr–13P–7C alloy in $1\,M$ HCl consists of Cr^{3+}, O^{2-}, OH^-, and H_2O, and hence the passive film has been called a passive hydrated chromium oxyhydroxide film $[CrO_x(OH)_{3-2x}\cdot nH_2O]$ [3]. The chromium enrichment occurs in passive films formed not only on amorphous alloys but also on crystalline alloys [4] when the corrosion resistance is based on the presence of chromium. The resistance to passivity breakdown is higher when the chromium content of the passive film is higher. Accordingly, when an alloy has a higher ability to concentrate chromic ion in the passive film, the alloy has a higher corrosion resistance. The concentration of chromic ion in passive films formed on amorphous alloys is far higher than in films formed on crystalline alloys, as shown in Table 42.1. Consequently, amorphous alloys containing strongly passivating elements, such as chromium and tantalum, have a very high ability to concentrate the beneficial ions in their passive films and have the high corrosion resistance based on spontaneous passivation.

C2. Homogeneous Nature of Amorphous Alloys

The high corrosion resistance of amorphous alloys disappears by heat treatment for crystallization [11, 12]. As soon as a nanocrystalline metastable phase is formed in the amorphous Fe–10Cr–13P–7C alloy matrix by heat treatment, the alloy becomes no longer spontaneously passive in $1\,M$ HCl, and the anodic dissolution current continues to increase with increasing time of heat treatment [13]. This change is

TABLE 42.1. Concentrations of Chromic Ion in Passive Films Formed on Amorphous Alloys and Stainless Steels in $1\,M$ HCl at Ambient Temperature

Amorphous Alloy	Cr^{3+}/Total Metallic Ions	Passivation	Reference
Fe–10Cr–13P–7C	0.97	Spontaneous	5
Fe–3Cr–2Mo–13P–7C	0.57	Anodic polarization	6
Co–10Cr–20P	0.95	Spontaneous	7
Ni–10Cr–20P	0.87	Spontaneous	8
Stainless steel Fe–30Cr–(2Mo)	0.75	Anodic polarization	9
Fe–19Cr–(2Mo)	0.58	Anodic polarization	10

due to introduction of chemical heterogeneity into the amorphous alloy consisting of the chemically homogeneous single phase.

A typical example of the detrimental effect of nanocrystalline heterogeneity in the amorphous matrix has been reported for Cr–Ni–P alloys in $6\,M$ HCl at 30°C [14]. Alloys were prepared by melt spinning of liquid alloys composed of mixtures of eutectic Cr–13P and Ni–19P. The Cr–25.31Ni–14.88P, Cr–27Ni–15P, Cr–28.93Ni–15.14P, and Cr–40.5Ni–16P alloys were identified to be amorphous by X-ray diffraction, but the former two alloys showed more than three orders of magnitude higher corrosion rates than those of the latter two alloys. Detailed examination revealed that only a 0.14 at % decrease in the phosphorus content and only a 2.07 at % increase in the chromium content result in precipitation of a nanocrystalline body-centered-cubic (bcc) chromium phase of $\sim10\,nm$ in diameter in the amorphous matrix and that preferential dissolution of bcc precipitates takes place, although phosphorus-containing phases are not corroded. When passivation occurs by the formation of the chromium oxyhydroxide film, the precipitation of the bcc chromium phase is detrimental because chromium itself dissolves actively in $6\,M$ HCl and because a resultant decrease in the chromium content of the spontaneously passive matrix decreases the passivating ability.

The formation of the nanocrystalline phase in the amorphous matrix is not always detrimental. Amorphous Cr–Zr alloys consisting only of corrosion-resistant metals were spontaneously passive, forming a double oxyhydroxide film of Cr^{3+} and Zr^{4+}. Increasing chromium content increases the passivating ability and corrosion resistance. However, as the inherited characteristic from zirconium, the zirconium-rich alloys suffer pitting by anodic polarization. The change in the pitting susceptibility due to structural changes by heat treatment was examined [15]. Specimens were heated with a rate of 4°C/min to the prescribed temperature, kept at the temperature for 30 min, and furnace cooled. These specimens were spontaneously passive in $6\,M$ HCl and their pitting potentials increased with heat treatment temperature. The specimen heated to 500°C exhibited the highest pitting potential, but the specimen heated to 600°C had a lower pitting potential than did specimens heated to 400 and 500°C. The heat treatment at all temperatures resulted in the formation of the less corrosion-resistant hexagonal close-packed (hcp) zirconium precipitates, and the size of precipitates increased with heating temperature. The formation of the hcp zirconium phase led to an increase in the chromium content of the matrix phase and hence to an enhancement of the formation of a more protective chromium-rich passive film covering the entire heterogeneous alloy surface. However, when the average size of the less corrosion-resistant zirconium phase exceeded a critical size ($\sim20\,nm$), the protective chromium-rich passive film could not completely cover the precipitates and the pitting resistance decreased. In this manner, if the precipitation of the nanocrystalline phase enhances the passivating ability of the matrix, the precipitation of the less corrosion-resistant nanocrystalline phase is not always detrimental but sometimes can increase the corrosion resistance.

C3. High Activity of Amorphous Alloys

When the chromium-enriched passive film is formed on amorphous and crystalline Fe–Cr alloys, the composition of the alloy surface just under the chromium-enriched passive film is almost the same as that of the bulk alloy [9]. However, if nickel is contained in alloys, such as austenitic stainless steels, nickel is concentrated in the underlying alloy surface since nickel is not contained in the passive film [16]. Hence, the formation of the chromium-enriched passive film results from selective dissolution of alloy constituents unnecessary for the passive film formation. When an alloy is able to passivate, fast active dissolution of the alloy results in rapid enrichment of beneficial ions. The passivating ability is therefore related to the activity of the alloy.

Because amorphous alloys are thermodynamically metastable, unless passive films are formed, the amorphous alloys dissolve more rapidly than their crystalline counterparts [17]. The high reactivity of amorphous alloys due to their thermodynamically metastable nature is effective in enhancing the accumulation of beneficial passivating elements on the alloy surface. As a result of rapid dissolution of active elements unnecessary for passivation, passivating elements are highly concentrated in the surface, causing passivation to occur rapidly.

D. ALLOYS RESISTANT TO AQUEOUS CORROSION

Special techniques are required for preparation of metastable alloys. Accordingly, unless the resulting metastable alloys have particularly high corrosion resistance, they cannot be used practically. From a corrosion point of view, coating would be the most suitable method. Corrosion-resistant new surface alloys have been tailored, mostly by sputter deposition techniques. Of the various methods for amorphization, sputter deposition is known to form single amorphous phase alloys with the widest composition range. Recent efforts for preparation of extremely corrosion-resistant alloys by sputtering will be summarized hereafter.

D1. Aluminum Corrosion-Resistant Metal Alloys

Alloying of aluminum with refractory metals, such as niobium, tantalum, molybdenum, and tungsten, is a potential method to enhance corrosion resistance. Although the boiling point of aluminum is generally lower than melting points of refractory metals, the sputter deposition technique does not rely on melting to mix the alloy constituents. Therefore, this technique is suitable for forming a single-phase solid solution even though the boiling point of one component may be lower than the melting point of the other components and/or one component may be immiscible with another component in the liquid state.

The sputter deposition method has been applied in preparing corrosion-resistant new aluminum alloys [18–23]. The Al–Ti, Al–Zr, Al–Nb, Al–Ta, Al–Cr, Al–Mo, and Al–W alloys have been successfully prepared in a single amorphous phase over wide composition ranges. Alloying is very useful in enhancing corrosion resistance. Figure 42.1 shows a comparison of corrosion rates of various aluminum alloys with those of conventional corrosion-resistant alloys in $1\,M$ HCl at 30°C. Except for Al–Ti alloys, the corrosion resistance in $1\,M$ HCl increases with increasing alloying additions. The Al–Ti and Al–Cr alloys dissolve actively, but other amorphous aluminum alloys are spontaneously passive even in $1\,M$ HCl. Amorphous Al–Ta and Al–Nb alloys are especially corrosion resistant.

Sputter deposition techniques have been used widely for preparation of corrosion-resistant aluminum alloys in the first one-half of the 1990s. Shaw and co-workers found enhanced passivity of Al–W [25] and Al–Ta [26] alloys over a wide pH range due to the formation of tungsten and tantalum hydroxide films, They attributed high pitting resistance to enrichment of alloying elements with high passivating ability in the underlying alloy surface. High pitting potentials of Al–V, Al–Mn, and Al–W alloys are attributed to suppression of pit growth [27]. Improved pining resistance of Al–Ta alloys is attributed to the formation of thin passive films that are capable of impeding migration of chloride ions through the film [28]. When Al_3Ta is precipitated, passivity breakdown occurs in the dealloyed region of the periphery of the cathodic precipitates [29]. Amorphous Al–(15–45)Mo alloys show stable passivity over a wide potential range in $0.1–1\,M$ NaCl, and their pitting potentials are at least 1.2 V more positive than the pitting potential of aluminum metal [30]. In this manner the formation of single-phase solid solutions of aluminum alloys with corrosion-resistant elements is very effective in enhancing corrosion resistance.

D2. Chromium–Refractory Metal Alloys

Valve metals, such as titanium, zirconium, niobium, and tantalum, are all passivated in strong acids. Chromium also has a very strong passivating ability. Chromium-valve metal alloys have been successfully prepared by sputtering [31–36], Their corrosion resistance to concentrated hydrochloric acids is remarkably high.

Figure 42.2 shows the change in the corrosion rate of Cr–Ti and Cr–Zr alloys in $6\,M$ HCl solution at 30°C and of Cr–Nb and Cr–Ta alloys in $12\,M$ HCl solution at 30°C as a

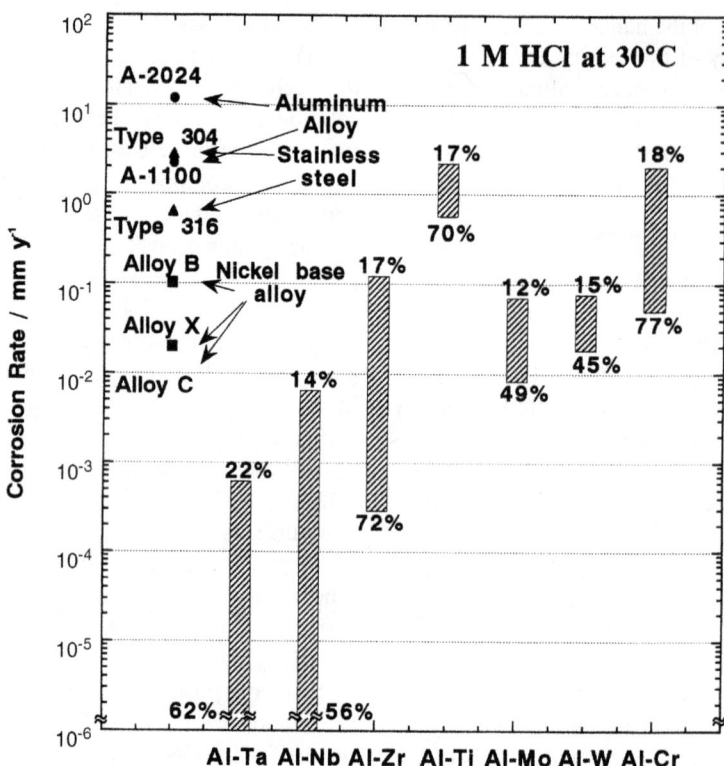

FIGURE 42.1. Corrosion rates of various aluminum alloys and conventional corrosion-resistant alloys in 1 M HCl at 30°C [24].

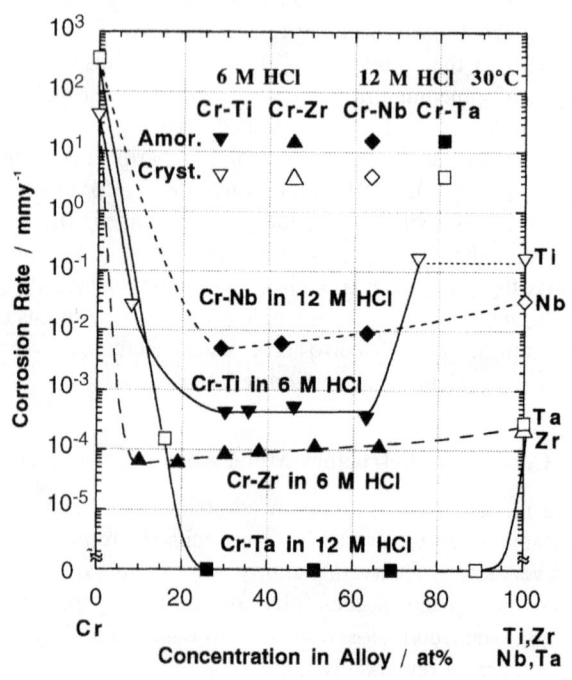

FIGURE 42.2. Corrosion rates of sputter-deposited Cr–Ti [31] and Cr–Zr [33] alloys in 6 M HCl at 30°C and Cr–Nb [34] and Cr–Ta [34] alloys in 12 M HCl at 30°C.

function of valve metal content. In 6 M HCl, chromium and titanium dissolve actively, whereas the Cr–Ti alloys show very low corrosion rates, several orders of magnitude lower than those of the alloy components. Binary Cr–Zr alloys also show low corrosion rates. In spite of the fact that the corrosion rate of chromium metal is five orders of magnitude higher than that of zirconium metal, the corrosion rate of the Cr–Zr alloys decreases with increasing chromium content of the alloy. Amorphous Cr–Nb and Cr–Ta alloys show very high corrosion resistance, higher than that of the alloy components. These results indicate that if both components of binary alloys have a strong passivating ability, the alloys possess even better corrosion resistance than the alloy components. The corrosion rates of Cr–Ta alloys are extremely low and are lower than the level measurable by inductively coupled plasma (ICP) spectrometry, $<10^{-6}$ mm/year. Thus, the Cr–Ta alloys possess the highest corrosion resistance among all known metallic materials in strong acids.

The high corrosion resistance of the chromium-valve metal alloys was due to spontaneous passivation caused by the formation of the passive oxyhydroxide film consisting of both chromium and valve metal cations. X-ray photoelectron spectroscopy (XPS) analysis revealed a charge transfer from chromium ion to valve metal cation in the film [35]. The charge transfer between different cations indicates that these cations are closely spaced and that the passive film does not consist of a simple mixture of chromium oxyhydroxide and valve metal oxide but rather consists of a double oxyhydroxide of chromium and valve metal cations. The resultant double oxyhydroxide films are more protective than chromium oxyhydroxide and valve metal oxide films in these aggressive solutions.

D3. Molybdenum Corrosion-Resistant Metal Alloys

Figure 42.3 shows corrosion rates of sputter-deposited molybdenum–valve metal alloys in $12\,M$ HCl [37–41]. Molybdenum–zirconium alloys become amorphous over a wide composition range, whereas other molybdenum-valve metal alloys were composed of nanocrystalline single bcc phases. Because grain diameters estimated from the full width at half-maxima (FWHM) of X-ray diffraction lines are 5–7 nm, these nanocrystalline alloys are regarded as homogeneous solid solutions from a corrosion point of view. All binary molybdenum-valve metal alloys show significantly higher corrosion resistance than did alloy component elements, regardless of crystalline and amorphous structures of the alloys. The corrosion rates of titanium, zirconium, and niobium are several orders of magnitude higher than the corrosion rate of molybdenum, but the corrosion rate of their alloys decreases with increasing valve metal content of the alloy. However, the corrosion rate of binary Cr–Mo alloys decreases with an increase in the molybdenum content of the alloy and the corrosion resistance never exceeds the corrosion resistance of molybdenum [41].

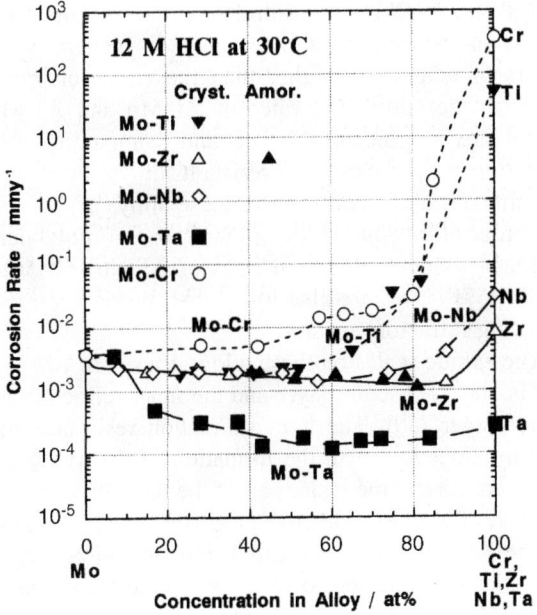

FIGURE 42.3. Corrosion rates of sputter-deposited molybdenum corrosion-resistant metal alloys in $12\,M$ HCl at 30°C [37–41].

The high corrosion resistance of the molybdenum alloys is attributed to the formation of passive double oxyhydroxide films of Mo^{4+} and cations of alloying elements. The molybdenum corrosion-resistant element alloys are spontaneously passive in $12\,M$ HCl, and their corrosion potentials are close to or higher than the corrosion potential of molybdenum. Molybdenum dissolves actively from about -0.8 to $-0.2\,V$ (SCE) and passivates from about -0.2 to $0.2\,V$ (SCE), forming the film consisting of Mo^{4+} [42]. The cathodic activity of passive molybdenum for both proton and oxygen reduction is very high. Accordingly, the corrosion potential of molybdenum in $12\,M$ HCl is very high, and slight anodic polarization results in a sharp current increase due to transpassive dissolution.

The high protective quality of the passive film is attributed to the synergistic effect of two cations forming the double oxyhydroxide film, Even if the alloy is polarized anodically over the transpassive potential of molybdenum, film-forming Mo^{4+} ions are protected by chromium and valve metal cations and are stable. The molybdenum species contributing to the protective quality of the spontaneously passivated film is Mo^{4+} ions that are stable to about 0.5 V (SCE) when the Mo^{4+} ions are protected by chromic ions. When the polarization potential exceeds 0.6 V (SCE), protection by chromic ions is no longer effective and transpassivation of molybdenum occurs, showing a clear increase in the content of Mo^{6+} ions in the film.

D4. Tungsten Corrosion-Resistant Metal Alloys

Since tungsten belongs to the same family as chromium and molybdenum in the periodic table, it is expected that tungsten-valve metal alloys are also extremely corrosion resistant Figure 42.4 shows corrosion rates of tungsten alloys in $6\,M$ and $12\,M$ HCl [43–47]. The corrosion rates of the binary alloys are lower than those of the alloy components. The corrosion rates decrease with increasing alloy concentration, and tantalum-containing alloys show particularly high corrosion resistance. The W–Cr alloys also show higher corrosion resistance than tungsten and chromium, although the corrosion resistance of Mo–Cr alloys does not exceed that of molybdenum. The high corrosion resistance results from spontaneous passivation. The spontaneously passivated films are composed of double oxyhydroxide of W^{4+} and cations of alloying elements.

E. ALLOYS RESISTANT TO SULFIDIZING/OXIDIZING ENVIRONMENTS AT HIGH TEMPERATURES

Amorphous aluminum–refractory metal alloys possess extremely high resistance to high-temperature corrosion in sulfidizing and oxidizing environments. Corrosion of

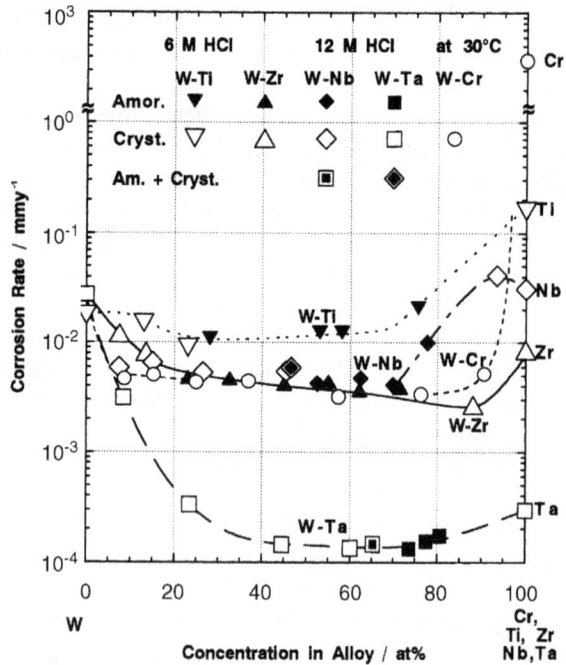

FIGURE 42.4. Corrosion rates of W–Ti alloys [43] in 6 *M* HCl at 30°C and W–Zr [44], W–Nb [45], W–Ta [46], and W–Cr [47] alloys in 12 *M* HCl at 30°C.

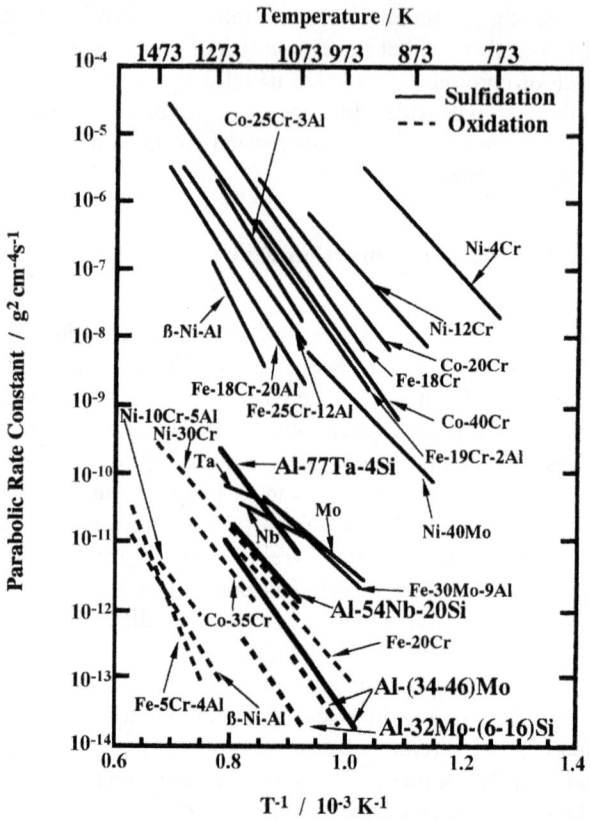

FIGURE 42.5. Sulfidation (solid lines) and oxidation (dotted lines) rate constants for amorphous Al–Mo, Al–Mo–Si, Al–Nb–Si, and Al–Ta–Si alloys as well as several high-temperature alloys [48–51, 56–58].

metallic materials at high temperatures in sulfur-containing atmospheres is much more severe than in purely oxidizing environments. All conventional oxidation-resistant alloys suffer catastrophic corrosion in sulfur-containing atmospheres at high temperatures, because of the poor protective properties of sulfide scales. For instance, the nonstoichiometry of sulfide scales formed on these alloys often reaches 10%. Because of rapid diffusion of cations through the defective sulfide scale, they are sulfidized very rapidly.

Some refractory metals, such as molybdenum, niobium, and tantalum, are resistant to sulfide corrosion and their sulfidation rates are almost comparable to the oxidation rate of chromium. These metals, however, have very low resistance to high-temperature oxidation in spite of the fact that practical sulfidizing atmospheres are often oxidizing. On the other hand, the best alloying element to form a protective scale in oxidizing environments is aluminum, and the second best is chromium. These metals form alumina and chromia scales, respectively. Therefore, aluminum–refractory metal alloys must have the highest resistance to both oxidation and sulfidation.

Sulfidation of sputter-deposited aluminum–refractory metal alloys such as Al–Mo [48–51], Al–Nb [52–56], and Al–Ta [57] alloys follows a parabolic rate law, indicating that the rate-determining step of the overall reaction is the diffusional transport of matter through the sulfide scale formed. Figure 42.5 shows sulfidation (solid lines) and oxidation (dotted lines) rate constants for amorphous Al–Mo

and Al–Mo–Si alloys as well as several high-temperature alloys [48, 49, 51]. The sulfidation rate of conventional oxidation-resistant crystalline alloys is generally many orders of magnitude higher than the oxidation rate. By contrast, the sulfidation rates of Al–Mo and Al–Mo–Si alloys are significantly lower and comparable to the oxidation rate of oxidation-resistant alloys. Furthermore, the sulfidation rate constants of these alloys are more than one order of magnitude lower than those of molybdenum. The steady-state sulfidation rates of Al–Nb [52–56] and Al–Ta [57] alloys are also lower than those of the corresponding refractory metals.

The sulfide scales on these alloys consist of two layers, that is, the Al_2S_3 outer layer and the inner refractory metal sulfide layer [50]. The high sulfidation resistance of the Al–Mo alloys is due to the formation of the MoS_2 phase, which constitutes the major part of the inner barrier layer of the scale. The more protective properties of the sulfide scale on the Al–Mo alloys compared with the MoS_2 scale on molybdenum is attributed to a lower defect concentration in the aluminum-doped MoS_2 phase.

The oxidation rate of Al–Mo alloys is comparable to that of chromia-forming alloys although higher than that of alumina-forming alloys. In particular, the oxidation rate at temperatures >900°C is very high. The scale consists mostly of alumina, but because of the high molybdenum contents of these alloys, molybdenum is also oxidized, forming volatile MoO_3. Since the melting point of MoO_3 is 793°C, the formation of low-melting-point MoO_3 is responsible for the relatively low oxidation resistance of the Al–Mo alloys. By contrast, the ternary Al–Mo–Si alloys have high sulfidation resistance and have higher oxidation resistance than Al–Mo alloys [49, 51]. This high resistance is attributed to the formation of molybdenum silicide, which is stable to oxidation. During sulfidation and oxidation, amorphous alloys are crystallized, forming nanocrystalline intermetallics. The Al–Mo alloys form Al_8Mo_3 and Mo_3Al phases. The molybdenum-rich Mo_3Al phase is readily oxidized, forming volatile MoO_3. Accordingly, when the alumina scale surface on the Al–Mo alloys is analyzed, a low concentration of molybdenum is always found. By contrast, Al–Mo–Si alloys are crystallized to Al_8Mo_3 and Mo_5Si_3 phases without forming the easily oxidizable molybdenum-rich Mo_3Al phase. The Mo_5Si_3 phase is stable to oxidation. Accordingly, any molybdenum and silicon are not detected in the outer surface of the alumina scale. The oxidation resistance of Al–Nb and Al–Ta alloys is also improved by the silicon addition [56, 58].

The chromium–refractory metal alloys have high sulfidation resistance. The sulfidation rates of the alloys containing at least 50 at % refractory metals are almost comparable to those of the corresponding refractory metals [59]. The sulfide scales formed on these alloys consist of two layers: the Cr_2S_3 outer layer and the inner refractory metal sulfide layer. The Cr–Nb and Cr–Ta alloys possess high oxidation resistance nearly comparable to typical chromia-forming alloys.

F. SUMMARY

Almost any kinds of properties can be obtained by preparation of amorphous and nanocrystalline alloys mostly because of the formation of homogeneous alloys due to expansion of solubility limits. A variety of corrosion-resistant materials can be prepared depending upon the application and environmental conditions. Bulk amorphous alloys can be processed if the alloys have a wide gap between glass transition and crystallization temperatures, but their compositions are restricted. Corrosion-resistant amorphous and nanocrystalline alloys can be applied by using surface coatings. Although it is difficult to prepare defect-free perfect coatings at the moment, new surface treatment methods have potential for developing new corrosion-resistant technology. Consequently, investigations of both new surface treatment technology and alloy design that satisfy a variety of demands will exploit a new area in the field of corrosion science and engineering.

REFERENCES

1. M. Naka, K. Hashimoto, and T. Masumoto, J. Jpn, Inst. Metals, **38**, 835 (1974).

2. K. Kobayashi, K. Hashimoto, and T. Masumoto, Sci. Rep. Res. Inst. Tohoku Univ., **A-29**, 284 (1981).

3. K. Hashimoto, T. Masumoto, and S. Shimodaira, in Passivity and Its Breakdown on Iron and Iron Base Alloys, R. W. Staehle and H. Okada (Eds.), National Association of Corrosion Engineers, Houston, TX, 1975, p. 34.

4. K. Asami, K. Hashimoto, and S. Shimodaira, Corros. Sci., **18**, 151 (1978).

5. K. Asami, K. Hashimoto, T. Masumoto, and S. Shimodaira, Corros. Sci., **16**, 909 (1976).

6. K. Hashimoto, M. Naka, J. Noguchi, K. Asami, and T. Masumoto, in Passivity of Metals, Corrosion Monograph Series, R. P. Frankenthal and J. Kruger (Eds.), The Electrochemical Society, Princeton, NT, 1978, p. 156.

7. K. Hashimoto, K. Asami, M. Naka, and T. Masumoto, Boshoku Gijutsu (Corros. Eng.), **28**, 271 (1979).

8. A. Kawashima, K. Asami, and K. Hashimoto, Corros. Sci., **24**, 807 (1984).

9. K. Hashimoto, K. Asami, and K. Teramoto, Corros. Sci., **19**, 3 (1979).

10. K. Hashimoto and K. Asami, Corros. Sci., **19**, 251 (1979).

11. K. Hashimoto, K. Osada, T. Masumoto, and S. Shimodaira, Corros. Sci., **16**, 71 (1976).

12. R. B. Diegle and J. E. Slater, Corrosion, **32**, 155 (1976).

13. M. Naka, K. Hashimoto, and T. Masumoto, Corrosion, **36**, 679 (1980).

14. B.-P. Zhang, H. Habazaki, A. Kawashima, K. Asami, K. Hiraga, and K. Hashimoto, Corros. Sci., **31**, 355 (1991).

15. M. Mehmood, B.-P. Zhang, E. Akiyama, H. Habazaki, A. Kawashima, K. Asami, and K. Hashimoto, Corros. Sci., **40**, 1 (1998).

16. K. Hashimoto and K. Asami, Corros. Sci., **19**, 1007 (1979).

17. K. E. Heusler and D. Huerta, in Corrosion, Electrochemistry and Catalysis of Metallic Glasses, R. B. Diegle and K. Hashimoto (Eds.), The Electrochemical Society, Princeton, NJ, 1988, p. 1.

18. H. Yoshioka, A. Kawashima, K. Asami, and K. Hashimoto, in Corrosion, Electrochemistry and Catalysis of Metallic Glasses, R. B. Diegle and K. Hashimoto (Eds.), The Electrochemical Society, Princeton, NJ, 1988, p. 242.

19. H. Yoshioka, A. Kawashima, K. Asami, and K. Hashimoto, Proceedings of MRS International Meeting on Advanced Materials, Vol. 3, Materials Research Society, Pittsburgh, PA, 1988, p. 429.

20. H. Yoshioka, Q. Yan, H. Habazaki, A. Kawashima, K. Asami, and K. Hashimoto, Corros. Sci., **31**, 349 (1990).

21. H. Yoshioka, H. Habazaki, A. Kawashima, K. Asami, and K. Hashimoto, Corros. Sci., **32**, 313 (1991).

22. H. Yoshioka, H. Habazaki, A. Kawashima, K. Asami, and K. Hashimoto, Electrochim. Acta, **36**, 1227 (1991).

23. H. Yoshioka, H. Habazaki, A. Kawashima, K. Asami, and K. Hashimoto, Corros. Sci., **33**, 425 (1992).

24. K. Hashimoto, N. Kumagai, H. Yoshioka, J. H. Kim, E. Akiyama, H. Habazaki, S. Mrowec, A. Kawashima, and K. Asami, Corros. Sci., **35**, 363 (1993).

25. B. A. Shaw, T. L. Fritz, G. D. Davis, and W. C. Moshier, J. Electrochem. Soc., **137**, 1317 (1990).

26. G. D. Davis, B. A. Shaw, B. J. Rees, E. L. Principle, C. A. Pecile, in Corrosion, Electrochemistry and Catalysis of Metastable Metals and Intermetallics, C. R. Clayton and K. Hashimoto (Eds.), The Electrochemical Society, Princeton, NJ, 1993, p. 1.

27. G. S. Frankel, R. C. Newman, C. V. Jahnes, and M. A. Russak, J. Electrochem. Soc., **140**, 2192 (1993).

28. C. C. Streinz, J. Kruger, and P. J. Moran, J. Electrochem. Soc., **141**, 1126 (1994).

29. C. C. Streinz, P. J. Moran, J. W. Wagner, and J. Kruger, J. Electrochem. Soc., **141**, 1132 (1994).

30. M. Janik-Czachor, A. Wolowik, and Z. Werner, Mater. Sci. Forum, **185–188**, 1049 (1995).

31. J. H. Kim, E. Akiyama, H. Yoshioka, H. Habazaki, A. Kawashima, K. Asami, and K. Hashimoto, Corros. Sci., **34**, 975 (1993).

32. K. Hashimoto, J. H. Kim, E. Akiyama, H. Habazaki, A. Kawashima, and K. Asami, Proceedings of 12th International Corrosion Congress, Vol. 3A, National Association of Corrosion Engineers, Houston, TX, 1993, p. 1102.

33. J. H. Kim, E. Akiyama, H. Habazaki, A. Kawashima, K. Asami, and K. Hashimoto, Corros. Sci., **34**, 1817 (1993).

34. J. H. Kim, E. Akiyama, H. Habazaki, A. Kawashima, K. Asami, and K. Hashimoto, Corros. Sci., **34**, 1947 (1993).

35. J. H. Kim, B. Akiyama, H. Habazaki, A. Kawashima, K. Asami, and K. Hashimoto, Corros. Sci., **36**, 511 (1994).

36. K. Hashimoto, P.-Y. Park, J. H. Kim, H. Yoshioka, E. Akiyama, H. Habazaki, A. Kawashima, K. Asami, Z. Grzesik, and S. Mrowec, Mater. Sci. Eng., **A198**, 1 (1995).

37. P. Y. Park, E. Akiyama, H. Habazaki, A. Kawashima, K. Asami, and K. Hashimoto, Corros, Sci., **37**, 307 (1995).

38. P. Y. Park, E. Akiyama, A. Kawashima, K. Asami, and K. Hashimoto, Corros. Sci., **38**, 397 (1996).

39. P. Y. Park, E. Akiyama, H. Habazaki, A. Kawashima, K. Asami, and K. Hashimoto, Corros. Sci., **38**, 1731 (1996).

40. P. Y. Park, E. Akiyama, H. Habazaki, A. Kawashima, K. Asami, and K. Hashimoto, Corros. Sci., **38**, 1649 (1996).

41. P. Y. Park, E. Akiyama, A. Kawashima, K. Asami, and K. Hashimoto, Corros. Sci., **37**, 1843 (1995).

42. H. Habazaki, A. Kawashima, K. Asami, and K. Hashimoto. The Corrosion Behavior of Amorphous Fe–Cr–Mo–P–C and Fe–Cr–W–P–C Alloys in 6 M HCl Corros. Sci., **33**, 225 (1992).

43. J. Bhattarai, E. Akiyama, A. Kawashima, K. Asami, and K. Hashimoto, Corros. Sci., **37**, 2071 (1995).

44. J. Bhattarai, H. Akiyama, H. Habazaki, A. Kawashima, K. Asami, and K. Hashimoto, Corros. Sci., **39**, 355 (1997).

45. J. Bhattarai, E. Akiyama, H. Habazaki, A. Kawashima, K. Asami, and K. Hashimoto, Corros. Sci., **40**, 19 (1998).

46. J. Bhattarai, E. Akiyama, H. Habazaki, A. Kawashima, K. Asami, and K. Hashimoto, Corros. Sci., **40**, 155 (1998).

47. J. Bhattarai, E. Akiyama, H. Habazaki, A. Kawashima, K. Asami, and K. Hashimoto, Corros. Sci., **40**, 757 (1998).

48. H. Habazaki, J. Dabek, K. Hashimoto, S. Mrowec, and M. Danielewski, Corros. Sci., **34**, 183 (1993).

49. H. Habazaki, J. Dabek, K. Hashimoto, S. Mrowec, and M. Danielewski, in Corrosion, Electrochemistry and Catalysis of Metastable Metals and Intermetallics, C. R. Clayton and K. Hashimoto (Eds.), The Electrochemical Society, Princeton, NJ, 1993, p. 224.

50. H. Habazaki, K. Takahiro, S. Yamaguchi, K. Hashimoto, J. Dabek, S. Mrowec, and M. Danielewski, Corros. Sci., **36**, 199 (1994).

51. H. Habazaki, H. Mitsui, K. Asami, S. Mrowec, and K. Hashimoto, Trans. Mater. Res. Soc. Jpn., **14A**, 309 (1994).

52. H. Mitsui, H. Habazaki, K. Asami, K. Hashimoto, and S. Mrowec, Trans. Mater. Res. Soc. Jpn., **14A**, 243 (1994).

53. Z. Grzesik, H. Mitsui, K. Asami, K. Hashimoto, and S. Mrowec, Corros. Sci., **37**, 1045 (1995).

54. H. Mitsui, E. Akiyama, A. Kawashima, K. Asami, K. Hashimoto, and S. Mrowec, Mater. Trans. JIM, **37**, 379 (1996).

55. H. Mitsui, H. Habazaki, K. Asami, K. Hashimoto, and S. Mrowec, Corros. Sci., **38**, 1431 (1996).

56. H. Mitsui, H. Habazaki, K. Hashimoto, and S. Mrowec, Corros. Sci., **39**, 9 (1997).

57. H. Mitsui, H. Habazaki, K. Hashimoto, and S. Mrowec, Corros. Sci., **39**, 59 (1997).

58. H. Mitsui, H. Habazaki, K. Hashimoto, and S. Mrowec, Corros. Sci., **39**, 1571 (1997).

59. H. Habazaki, K. Ito, H. Mitsui, E. Akiyama, A. Kawashima, K. Asami, K. Hashimoto, and S. Mrowec, Mater. Sci. Eng., **A226–228**, 910 (1997).

43

CARBON STEEL—ATMOSPHERIC CORROSION

I. Matsushima*

Maebashi Institute of Technology, Maebashi, Japan

A. INTRODUCTION

Steel is the most common metallic material used in structures exposed to the atmosphere. Corrosion of steel in the atmosphere therefore constitutes the greatest item in the cost of metallic corrosion in all sectors of engineering. Atmospheric corrosion rates of bare carbon steel depend greatly on the time of wetness and on the atmospheric concentrations of constituents, such as sulfur dioxide and chloride, which vary considerably from location to location; hence, atmospheric corrosion rates vary markedly throughout the world.

On the other hand, steel is seldom used in the bare state. It is usually protected against corrosion by organic, metallic, or other coatings, so that corrosion rate data of unprotected carbon steel are of limited use. However, enormous amounts of atmospheric corrosion data of steel nave been accumulated through exposure tests at many test sites all over the world, because data on corrosion of steel, together with data on

corrosion of zinc, are useful in calibrating the corrosivity of the various atmospheres. Carbon or mild steel specimens are also commonly used as the control specimens in atmospheric exposure tests of other metals and alloys. In fact, long-term exposure test data of carbon steel have been obtained as control data in test programs to determine the effects of alloying additions in low-alloy steels and to demonstrate the superior corrosion performance of low-alloy weathering steels to be used in the atmosphere in the unpainted condition.

B. ENVIRONMENTAL FACTORS

B1. Major Environmental Factors

There are several factors that affect atmospheric corrosion of steel, of which time of wetness and sulfur dioxide and chloride pollution levels are particularly important.

B1.1. Time of Wetness. Water, essential for corrosion to proceed, is provided by meteoric water, dew, and invisible condensed water films from moist air. The corrosion rate of steel depends on the time of wetness, which is defined by the International Organization for Standardization (ISO) standard 9223-1992 [1] as the period during which a metallic surface is covered by adsorptive and/or liquid films of eletrolyte that are capable of causing atmospheric corrosion. The length of time when the relative humidity is $\geq 80\%$ at a temperature $\geq 0°C$ is specified in the ISO standard. As noted in the standard, this specified time does not necessarily correspond to the actual time of exposure to wetness, which is affected by the type of metal, shape, mass and orientation, quantity of corrosion product, nature of pollutants on the surface, and other factors. Nevertheless, this criterion is usually sufficiently accurate for the characterization of atmospheres.

*Deceased

Uhlig's Corrosion Handbook, Third Edition, Edited by R. Winston Revie
Copyright © 2011 John Wiley & Sons, Inc.

B1.2. Sulfur Dioxide. Sulfur dioxide (SO_2) and some SO_3 in the atmosphere, primarily provided by the combustion of sulfur-bearing fossil fuels, react with steel to form ferrous sulfate, probably through the formation of H_2SO_4 [2]. Some ferrous sulfate may be oxidized to ferric sulfate. Schikorr [2] suggested that, on exposure to water, ferrous sulfate is hydrolyzed to regenerate H_2SO_4, which further reacts with steel, accelerating corrosion by means of a cycle. The hygroscopic nature of sulfate, and H_2SO_4 if present, increases the time of wetness by promoting condensation of moisture in the atmosphere. When steel was exposed to the industrial atmosphere of Stuttgart, the corrosion rate in the month of exposure varied directly with the SO_2 content of the atmosphere in that month [2].

With the development of the corrosion product films that act as a barrier to the corrodants in the environment, the action of SO_2 is retarded. However, SO_2 continues to accelerate corrosion by impairing the protective nature of the corrosion product films through the formation of ferrous sulfate nests [3]. For details, see Section C.

The corrosion rate of steel in industrial areas tends to be high because of high concentrations of SO_2 in the atmosphere; the higher its concentration, the higher the corrosion rate.

B1.3. Chloride. Chlorides (Cl^-), carried by the wind from the sea and deposited on the steel surface, are also hygroscopic and promote chemical condensation of moisture in the air. Similar to SO_4^{2-} ions, they decrease the protection of the corrosion product films.

The chloride concentration in the atmosphere is usually high at the seacoast, although the concentration varies greatly from seacoast to seacoast and season to season, depending on the direction and the strength of prevailing winds. It is also affected by the topography of the coast. The chloride content decreases sharply with the distance inland from the coast, but the decay curve is considerably influenced by the geographical features of the area, so that no general data are available. At the exposure test station at Kure Beach, North Carolina, the atmosphere at the 80-ft (25-m) lot is ~3.5 times more corrosive to steel than that at the 800-ft (250-m) lot [4]. Generally speaking, the effect of chloride on corrosion of steel is particularly severe in areas within 0.5–1 km inland from the coast.

B2. Corrosion Rate as a Function of Environmental Factors

Equations have been developed to express the corrosion rate as a function of environmental factors. The research group on the corrosion protection of steel frame structures [5] [Rikujo Tekkotsu Kozobutsu Boshoku Kenkyu Kai (RIKU-BO-KEN), a Japanese group consisting of NKK (steel manufacturer), Ishikawajima Harima Heavy Industries, and Dai Nippon Toryo (paint manufacturer)], based on atmospheric

exposure tests for five years, starting in 1960 at seven sites distributed throughout Japan and on environmental data developed the following equation for carbon steel by multiple regression analysis[1]:

$$\text{Corrosion rate (mdd)}^* = 0.484 \times (\text{temperature,}^\circ C) + 0.701 \\ \times (\text{relative humidity, \%}) + 0.075 \times (Cl^-, \text{ppm}) + 8.202 \\ \times (SO_2, \text{mdd}) - 0.022 \times (\text{rainfall, mm/month}) - 52.67$$

(43.1)

where the values of each factor are based on the annual average in the five-year exposure period. Sulfur dioxide was measured by the amount of SO_2 trapped on a PbO_2–powder coated plate in terms of milligrams per square decimeter per day (mg/dm^2/day or mdd) (close to the procedure of ISO 9225-1992 [6]) and Cl^- by the concentration of Cl^- extracted each month in 1000 mL of water from a 100-cm^2 vinylidene chloride resin film gauze that trapped chlorides. The equation was used to draw a corrosion map of Japan [7].

The group also formulated other equations [8] by using corrosion data for carbon steel in the first year of exposure at 43 locations in Japan compiled from various sources.

For inland and industrial atmospheres,

$$\text{Corrosion rate (mdd)} = 4.15 + 0.88 \\ \times (\text{temperature,}^\circ C) - 0.073 \\ \times (\text{relative humidity, \%}) - 0,032 \\ \times (\text{rainfall, mm/month}) + 2.913 \\ \times (Cl^-, \text{ppm}) + 4.921 \times (SO_2, \text{mdd})$$

(43.2)

and for marine atmospheres,

$$\text{Corrosion rate (mdd)} = 5.61 + 2.754 \times (Cl^-, \text{ppm}) \\ + 6.155 \times (SO_2, \text{mdd})$$

(43.3)

The Tokyo Metropolitan Research Institute for Environmental Protection [9] derived a regression equation for carbon steel based on the exposure tests conducted during 1970–1975 at seven locations in Tokyo:

$$\text{Corrosion rate (mm/month)} = -0.00774 + 3.86 \times 10^{-4} \\ \times (\text{monthly average temperature,}^\circ C) + 2.45 \\ \times 10^{-6} \times (\text{time of wetness, h/month}) + 3.24 \times 10^{-6} \\ \times (\text{conductivity of filtrate of collected dust and rain,} \\ \mu\Omega/cm/L/month/100\,cm^2) \\ + 3.22 \times 10^{-3} \times (SO_3, mg/day/cm^2PbO_2) \\ + 2.4 \times 10^{-3} \times (\text{chloride, mg as NaCl/day/m}^2)$$

(43.4)

---

[1] For steel, to convert the corrosion rate from milligrams per square decimeter per day (mdd) to mm/year, multiply by 0.00464.

It should be noted that the SO_2 concentration in the industrial atmospheres in Japan in the 1960s was much higher than in the 1990s because of the pollution control measures implemented in the 1970s, during which period these equations were obtained. The negative coefficients for rainfall in Eqs. (43.1) and (43.2) show the beneficial effect of rain that washes contaminating SO_2 and its derivatives from the surface. The negative coefficient for relative humidity in Eq. (43.2) was probably caused by die limitation of regression analysis applied due to the small variation, less than 10%, of the annual average relative humidity.

B3. Corrosivity Classification of Atmospheres

Comparative rankings of 45 locations in North America, England, and some other countries in terms of corrosion of carbon steel are listed in Table 43.1 [4]. The rankings are based on the atmospheric corrosion data obtained by two-years exposure during the period from 1960 to 1962. According to Table 43.1, the atmosphere at Dungeness, England, is 326 times more corrosive to steel than that at Norman Wells, NWT, Canada. Although significant yearly variations may exist at each location, the rankings are helpful in estimating comparative corrosivity at different locations.

ISO/TC156 WG 4 [1, 6, 10, 11] has developed a comprehensive corrosivity system, and to substantiate this system, it has carried out an extensive exposure program at 45 sites in 12 countries from 1986 to 1998 (see Chapter 23, Section E1.2). Materials tested were carbon steel, zinc, copper, and aluminum. Time of wetness, SO_2 pollution, and NaCl deposition rates for each year were measured as the environmental data.

A corrosion map based on the corrosion rate of carbon steel plate specimens is shown in Figure 43.1 [12]. The numbers in circles represent the rankings of the corrosion rate, in increasing order (i.e., the lowest corrosion rate results in a ranking of 1).

C. CORROSION PRODUCT FILMS

Corrosion product films on steel formed in the atmosphere tend to be protective, and therefore, the corrosion rate decreases with time, reaching a steady state in a few years, following the relation $p = kt^n$, where p is mass loss, k and n are constants, and t is time [13]. Examples of corrosion–time curves are shown in Figure 43.2 [14].

The higher corrosion rates observed in marine and industrial atmospheres indicate that the corrosion product films formed in these atmospheres are less protective than those in rural atmospheres. In industrial atmospheres that contain appreciable amounts of SO_2, ferrous sulfate, formed by the corrosion reaction, accumulates in small, shallow discrete

TABLE 43.1. Comparative Rankings of 45 Locations Based on Mass Loss of Steel Specimens Exposed for Two Years[a]

Ranking	Location	Mass Loss[b]
1	Norman Wells, N.W.T., Canada	0.73
2	Phoenix, AZ	2.23
3	Saskatoon, Sask., Canada	2.77
4	Esquimalt, Vancouver Island, Canada	6.50
5	Detroit, MI	7.03
6	Fort Amidor Pier, Panama, C.Z.	7.10
7	Morenci, MI	9.53
8	Ottawa, Ont., Canada	9.60
9	Potter County, PA	10.00
10	Waterbury, CT	11.00
11	State College, PA	11.17
12	Montreal, P.Q., Canada	11.44
13	Melbourne, Australia	12.70
14	Halifax (York Redoubt), N.S.	12.97
15	Durham, NH	13.30
16	Middletown, OH	14.00
17	Pittsburgh. PA	14.90
18	Columbus, OH	16.00
19	South Bend, PA	16.20
20	Trail, B.C. Canada	16.90
21	Bethlehem, PA	18.3
22	Cleveland, OH	19.0
23	Miraflores, Panama, C.Z.	20.9
24	London (Battersea), England	23.0
25	Monroeville, PA	23.8
26	Newark, NJ	24.7
27	Manila, Philippine Islands	26.2
28	Limon Bay, Panama, C.Z.	30.3
29	Bayonne, NJ	37.7
30	East Chicago, IN	41.1
31	Cape Kennedy, 0.8 km (0.5 mile) from ocean	42.0
32	Brazos River, TX	45.5
33	Pilsey Island, England	50.0
34	London (Stratford), England	54.3
35	Halifax (Federal Building), N.S.	55.3
36	Cape Kennedy, 55 m (60 yd) from ocean, 60-ft elevation	64.0
37	Kure Beach, NC, 250-m (800-ft) lot	71.0
38	Cape Kennedy, 55 m (60 yd) from ocean, 9 m (30 ft) elevation	80.2
39	Daytona Beach, FL	144.0
40	Widness, England	174.0
41	Cape Kennedy. 55 m (60 yd) from ocean, ground level	215.0
42	Dungeness, England	238.0
43	Point Reyes, CA	244.0
44	Kure Beach, NC, 25-m (80-ft) lot	260.0
45	Galeta Point Beach, Panama, C.Z.	336.0

[a]Reproduced with permission from [4]. Copyright American Society for Testing and Materials (ASTM).
[b]Mass loss: g/4 × 6 in. (101.6 × 152.4 mm) specimen.

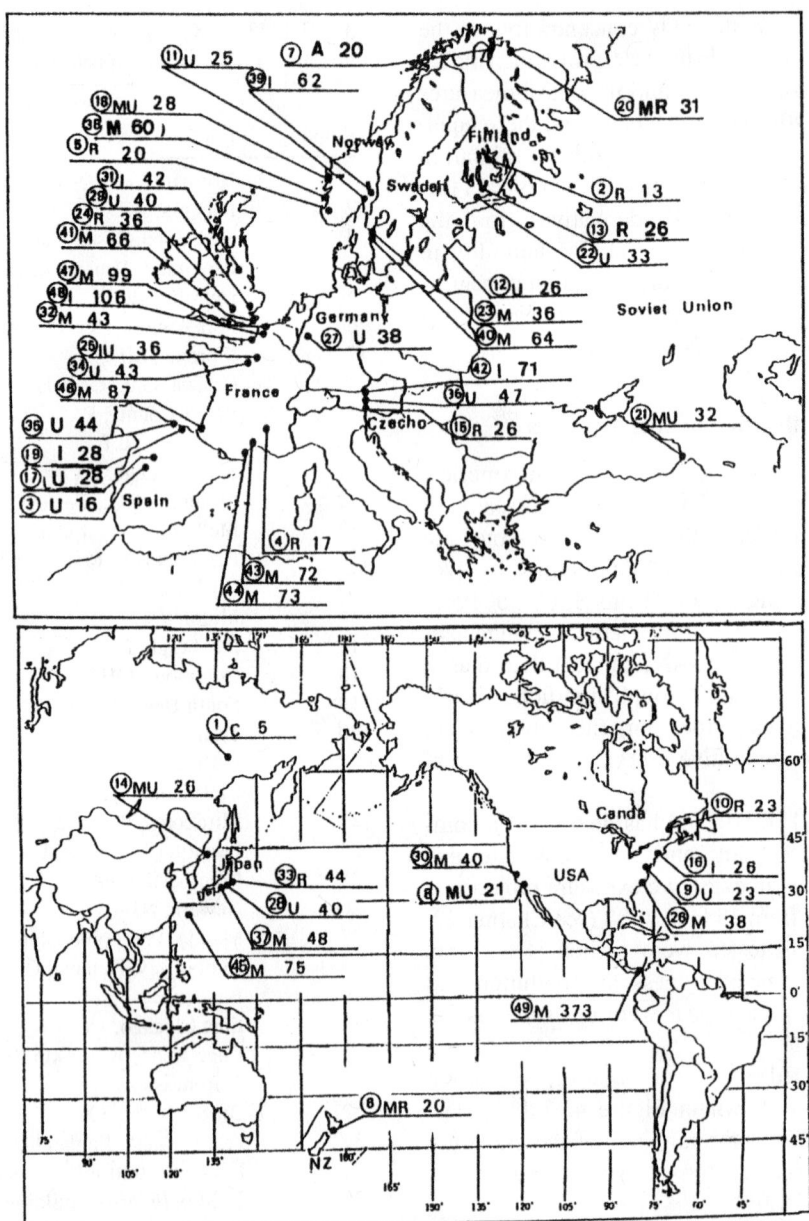

FIGURE 43.1. Corrosion map of the world. Corrosion rates in micrometers per year (μm/y). Based on the average corrosion rates of six flat carbon steel specimens exposed for 1 year starting in spring (in triplicate) and in autumn (in triplicate) [11]. The numbers in circles represent the rankings of the corrosion rate in increasing order. (Reproduced with permission from [12]).

corrosion pits, ~0.5–1 mm in diameter, on the steel surface under mounds of the corrosion product films. The ferrous sulfate agglomerates nesting in pits were named "ferrous sulfate nests" by Schwarz [3], who discovered their existence. Sulfate (SO_4^{2-}) anions migrate to the pits, which are anodic sites, and a greater number of these nests are formed in atmospheres with higher SO_2. Because the protection of the corrosion product films at the nests is poor, the nest areas provide the main paths for further corrosion to proceed. Thus, the corrosion rate is higher with more nests [15].

In marine atmospheres of appreciable chloride content, Cl^- ions probably act in the same way as SO_4^{2-} ions in industrial atmospheres. The presence of chloride nests has been demonstrated recently [16].

The major crystalline constituents in the corrosion product films are γ-FeOOH (lepidocrocite), α-FeOOH (goethite), and Fe_3O_4 (magnetite). The Fe_3O_4 here is not stoichiometric and would be better described as a spinel type oxide. Starting from Fe^{2+} ions, these crystalline compounds are formed, and the relative amounts vary depending on environmental

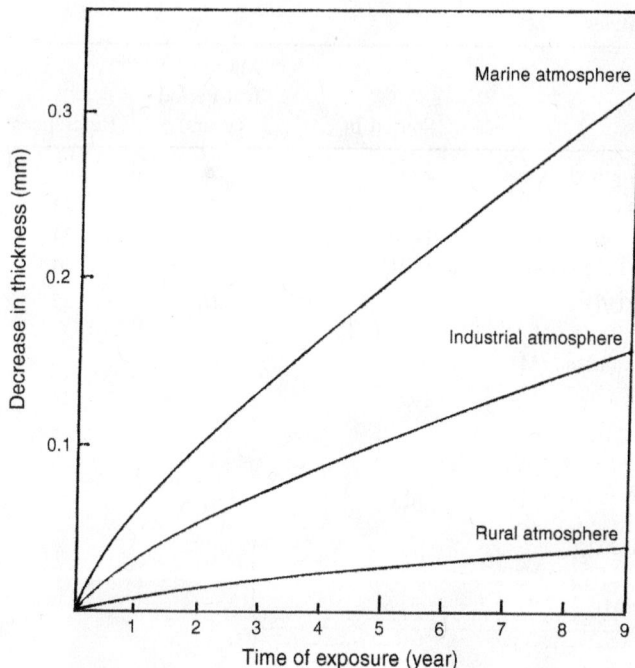

FIGURE 43.2. Atmospheric corrosion of steel as a function of time in different types of atmosperes. (Data taken from [14].)

conditions. Lepidocrocite (γ-FeOOH) is the first stable compound, which changes to α-FeOOH and Fe_3O_4. In marine atmospheres, β-FeOOH (akaganeite) may also be formed. In addition, large amounts of amorphous iron oxides (e.g., 25% in industrial atmospheres [17]) are found in all corrosion product films in the atmosphere. The constituents of corrosion product films formed on carbon steel in various atmospheres are summarized by Misawa [18] in Table 43.2.

Ferric oxides in corrosion product films take part in the cathodic reaction as first proposed by Evans [19]. The final model of this theory [20] is as follows:

$$\text{Anodic reaction:} \quad Fe \rightarrow Fe^{2+} + 2e^- \qquad (43.5)$$

$$\text{Cathodic reaction:} \quad Fe^{2+} + 8FeOOH + 2e^-$$
$$\rightarrow 3Fe_3O_4 + 4H_2O \qquad (43.6)$$

TABLE 43.2. Types of Atmosphere and Constituents of Corrosion Product Films[a]

Type of Atmosphere	Constituents of Iron Compounds in Corrosion Product Films
Industrial	α-FeOOH, γ-FeOOH $\gg Fe_3O_4$; iron sulfates, amorphous rust
Marine	$Fe_3O_4 > \alpha$-FeOOH $> \gamma$-FeOOH; β-FeOOH, amorphous rust
Rural rust	γ-FeOOH, α-FeOOH, Fe_3O_4; amorphous

[a]Reproduced with permission from [18].

$$\text{Oxidation of } Fe_3O_4: \quad 3Fe_3O_4 + \frac{3}{4}O_2 + \frac{9}{2}H_2O \rightarrow 9FeOOH$$

$$(43.7)$$

Despite some objections to this mechanism, particularly about the readiness of reoxidation of Fe_3O_4 expressed in Eq. (43.7), there is no doubt as to the accelerating effect of ferric oxides as cathodic reactants on atmospheric corrosion of steel. Revised models have been proposed by Stratmann [21] and Dünnwald and Otto [22].

D. CORROSION DATA

Long-term atmospheric exposure tests have been conducted at a large number of locations around the world. Because it is not possible to present in this chapter all the data that have been obtained, and because data obtained at one location are not helpful for corrosion engineering at other locations, only a list of major exposure programs is given in Table 43.3.

A survey of the data obtained in these programs shows that the average values of corrosion penetration in the first 10 years are in the following ranges:

Industrial atmospheres	0.1–0.5 mm
Marine atmospheres	0.3–0.8 mm
Rural atmospheres	0.05–0.2 mm

The method of exposure affects the corrosion rate. In most of the tests, specimens in the form of plates or sheets were exposed to the atmosphere at an angle of 30° from the horizontal with the racks facing south. The corrosion rate was measured by mass loss caused by corrosion of both surfaces of the specimen and, using the specific density, converted to the decrease in thickness from one surface so that the value was the average of corrosion of the skyward and groundward surfaces. In polluted atmospheres, corrosion that occurs on the groundward surfaces is greater than on the skyward surfaces, because the former surfaces do not benefit from the washing action of rain and the quick drying action of the sunshine that enhance the protective quality of the corrosion product films. It has been shown [60] that for steel specimens exposed in an industrial atmosphere on the 30° rack the corrosion that occurred on the skyward and groundward surfaces was ~ 40 and 60%, respectively, of the total corrosion.

In some programs, specimens were exposed vertically. LaQue [61] pointed out that the vertical exposure has the disadvantage of poorer reproducibility because a slight variation in the position of the specimen from the vertical greatly affects the washing and drying effects.

Galvanic corrosion data based on atmospheric exposure tests were surveyed by Kucera and Mattsson [62]. In rural and industrial atmospheres, no risk of galvanic corrosion of

TABLE 43.3.   Major Atmospheric Exposure Tests in Various Countries

Country	Organization	Location[a]	Started In	Maximum Test Period (years)	References
International	ISO	12 countries (47 sites)	1987	8	—[b]
Australia	Commonwealth Scientific and Electricity Trust of South Australia (ETSA)/CSIRO	Melbourne (299 sites)	1979	2	23
		South Australia (66 sites)	1990	2	24
		South Australia (475 sites)	1991	2	24
Brazil	Campanhia Siderúrgica Natuinal (CSN)	Volta Redonda (U/I)	1972	16	25
		Volta Redonda (R)			
		Restinga da Marambaia (M)			
	Centro de Pesquisas e Deservolvimento Leopoldo A. Miguez de Mello (CENPES) of Petrobrás	Manaus (E)[c]	1975	2	25
		Belém (E)[c]			
		Fortaleza (M)			
		Aracaju (5 sites, M)			
		Ajacaju (R)			
		Madre de Deus (M)			
		Betim (I)			
		Rio de Janeiro (M)			
		Cubatão (I)			
		S. Mateus (I)			
		Canoas (I)			
		Brasília (U)			
Canada	CANMET	Arctic & Southern Canada (77 sites)	1978	1	26
Czechoslovakia	Akimov State Research Inst. of Material Protection	Prague (U)	Late 1960s	5	27
		Ustí nad Labem (I)			
		Hurbanovo (R/I)			
		Kopisty (I)			
Finland	Technical Research Center (VTT)	Otaniemi (R)	1968	7	28
		Helsinki (U/I)	1975	4	28
		Harmaja (M)			
		Otaniemi (R)			
		Koski (R)			
		Haravalta (I)			
Germany	Staatlichen Materialprüfungsamt, Berlin	Berlin (R)	1914	4	29
		Dortmunt (I)			
		Hörnum (M)			
	Iron and Steel Inst.	Olpe (R)	1956	8	30
		Mülheim (I)			
		Cuxhaven (M)			
	Vereins Deutscher Eisenhüttenleute	Duisburg (I)	1962	16	31, 32
		Gelsenkirchen (I)			
		Mülheim (I)	(about 1970)[d]	8	32
		Cuxhaven (M)			
		Olpe (R)			
Japan	Research Group on Corrosion Protection of Steel Structures	Obihiro (R)	1960	5	5
		Wajima (M)			
		Tokyo (I)			
		Kawasaki (I)			
		Omaezaki (M)			
		Takayama (R)			
		Makurazaki (M)			
	Steel Manufacturers	38 sites in Japan	1960s	12 (max)	33
	Public Works Research Inst.	41 sites in Japan (sheltered)	1981	9	14

TABLE 43.3.   (*Continued*)

Country	Organization	Location[a]	Started In	Maximum Test Period (years)	References
Latin America	Project on "Ibero-American Map of Atmospheric Corrosiveness"	72 sites	(about 1990)[d]	4	34
New Zealand	Building Research Assoc. of New Zealand	168 sites in NZ	1987	1	35
Norway	VERITAS and Norwegian Inst. for Air Research	Sogn (R) Voss (R) Bergen-Bergens Tidende (U) Bergen-VERITAS (U) Bergen-Stend (R) Bergen-Fredriksberg (U) Bergen-Minde (U) Stord 50 (M/R)	1970	8	36
Panama	Naval Research Laboratory	Cristobal (M) Miraflores (Semi-U)	by 1958	16	37, 38
Russia (Formerly USSR)	Inst. of Physical Chemistry	Moscow (I) Zvenigorod (R) Batumi (M/U)	1968	10	39
Scandinavia	Scandinavian Council for Applied Research (NORDFORSC)	32 sites (Norway 19, Sweden 6, Finland 5, Denmark 2)	1975	8	40
Singapore	National University of Singapore	Singapore (M) A raft (M)	1985/86	12 weeks	41
Spain	Ciudad University	South (16 sites) Northwest (21 sites) Center (34 sites)	1976/83	1	42
	University of Barcelona	Catalonia (42 sites, R, U, I, M)	(1980s)[d]	3	43
	Centro Nacional de Investigaciones Metalurgicas	El Escorial (R) Madrid (U) Zaragoza (U) Bilbao (I) Barcelona (M) Cadiz (M) Cabo Negro (M) Alicante (M)	1976	13–16	44
Sweden	Swedish Corrosion Inst.	Ryda (R) Stockholm (U) Borregaard (I) Bohus-Malmön (M)	(1980s)[d]	5	45
Taiwan	China Steel Corp.	Hsinchu (R/U) Kaohsuiung (M) Hsiao Kang (I)	(1980s)[d]	8	46
United Kingdom	Iron Steel Inst.	Calshot (M) Dove Holes Tunnel Llanwrtyd Wells (R) Motherwell (I) Redear (M/I) Sheffield (I) Woolwich (I)	1928	5	47
	BISRA	Sheffield (I)	1937	5	48

(*continued*)

TABLE 43.3.   (*Continued*)

Country	Organization	Location[a]	Started In	Maximum Test Period (years)	References
United States	ASTM (Committee A-5)	Pittsburgh, PA (I)	1916	7.3	49
		Fort Sheridan, IL (R)		12	
		Annapolis, MD (M)		>30	
	U.S. Steel	Kearney, NJ (I)	1938	11.5	50
		Kure Beach, NC (M, 250 m)	1940	7.5	
	INCO	Kure Beach, NC (M, 250 m)	1941	15.5	51
		Block Island, RI (M)		17.1	
		Bayonne, NJ (I)		18.1	
	U.S. Steel	Kearney, NJ (I)	1942	15.5	52
		South Bend, PA (R)			
		Kure Beach, NC (M)			
	ASTM (Committee A-7) (malleable)	Kure Beach, NC (M, 25 m)	1958	12	53, 54
		Newark NJ (I)			
		Point Reyes, CA (M)			
		State College, PA (R)			
		E. Chicago, IN (I)			
	Pacific Gas & Electric Co.	Northern CA (15 sites)	1956	5	55
	Steel Founders' Society of America	Kure Beach, NC (M, 25, 250 m)	1953	7	56, 57
		East Chicago, IN (I)		12	
	Bethlehem Steel	Kure Beach, NC (M, 250 m)	1968	16	58
		Saylorsburg (R)			
		Bethlehem, PA (I)			
	USX	Kearney, NJ (U/I)	Early 1970s	16	59
		Kure Beach, NC (M, 250 m)			
		Potter County, PA (R)			

[a]Rural R, Urban U, industrial I, Marine M.
[b]To be published.
[c]Equatorial.
[d]Not stated in reference and estimated.

carbon steel exists when the steel is in contact with Pb, Zn, Al, Mg, or weathering steel. Increased corrosion results if carbon steel is coupled to stainless steels, Cu, Ni, anodized Al, Sn, or Cr. In marine atmospheres, the effect is essentially the same, except that contact with anodized Al and Cr is not harmful. Unlike the situation in aqueous environments, galvanic corrosion in the atmosphere is usually restricted to a narrow zone in the anodic metal, and the galvanic effects do not extend over several millimeters from the line of contact [63].

REFERENCES

1. International Organization for Standardization (ISO), Corrosion of Metals and Alloys—Corrosivity of Atmospheres—Classification, ISO 9223-1992, ISO, Geneva, Switzerland, 1992.

2. G. Schikorr, Werkst. Korros., **15**, 457 (1964).

3. H. Schwarz, Werkst. Korros., **16**, 93, 208 (1965).

4. American Society for Testing and Materials (ASTM) Task Force on the Calibration of Atmospheric Corrosivity, ASTM STP 435, ASTM, Philadelphia, PA, 1968, p. 360.

5. K. Horikawa, S. Takiguchi, Y. Ishizu, and M. Kanazashi, Boshoku Gijutu (Corros. Eng.), **16**, 153 (1967).

6. International Organization for Standardization (ISO), Corrosion of Metals and Alloys—Corrosivity of Atmospheres—Measurement of Pollution, ISO 9225-1992, ISO, Geneva, Switzerland, 1992.

7. Rikujo Tekkotsu Kozobutsu Boshoku Kenkyu Kai, Memorial Lecture, 1967 Award for Corrosion Protection, Japan Society for Promotion of Science, Tokyo, Japan.

8. Rikujo Tekkotsu Kozobutsu Boshoku Kenkyu Kai, Boshoku Gijutu (Corros. Eng.), **22**, 106 (1973).

9. Study on Metallic Corrosion by Atmospheric Pollutants, 2nd Report, Tokyo Metropolitan Research Institute for Environmental Protection, Tokyo, Japan, 1978.

10. International Organization for Standardization (ISO), Corrosion of Metals and Alloys—Corrosivity of Atmospheres—Guiding Values for the Corrosivity Categories, ISO 9224-1992, ISO, Geneva, Switzerland, 1992.

11. International Organization for Standardization (ISO), Corrosion of Metals and Alloys—Corrosivity of Atmospheres—Determination of Corrosion Rate of Standard Specimens for the Evaluation of Corrosivity, ISO 9226-1992, ISO, Geneva, Switzerland, 1992.

12. Provided by Y. Togawa, based on Y. Togawa, Bousei Kanri (Rust Prevention and Control), 37(2), 7 (1993).

13. R. Passano, Washington Regional Meeting ASTM, March 7, 1934, ASTM, Philadelphia, PA, cited in H. H. Uhlig and R. W. Revie, Corrosion and Corrosion Control, 3rd ed., Wiley, New York, 1985, p. 167.

14. Public Works Research Institute, Kozai Club and Japan Association of Steel Bridge Construction, Joint Research on Application of Weathering Steel to Bridges, Report XVIII, Tokyo, Japan, 1993.

15. I. Matsushima and T. Ueno, Corros. Sci., 11, 129 (1971).

16. Y. Shimizu, K. Tanaka, and T. Nishimura, Zairyo-to-Kankyo (Corros. Eng.) 44, 436 (1995).

17. P. Keller, Werkst. Korros., 18, 865 (1967).

18. T. Misawa, Boshoku Gijutu (Corros. Eng.), 32, 657 (1983).

19. U. R. Evans, Trans. Inst. Metal Finishing, 37, 1 (1960).

20. U. R. Evans and C. A. J. Taylor, Corros. Sri., 12, 227 (1972).

21. M. Stratmann, K. Bohnenkamp, and H-J. Engell, Cor. Sci., 23, 969 (1983).

22. J. Dünnwald and A. Otto, Corros. Sci., 29, 1167 (1989).

23. J. F. Moresby, F. M. Reeves, and D. J. Spedding, in Atmospheric Corrosion, W. H. Ailor (Ed.), Wiley, New York, 1982, p. 745.

24. G. A. King, J. Kapetas, and D. Bates-Brownsword, Paper No. 106, CORROSION/94, NACE, 1994.

25. A. C. Dutra and R. de O. Vianna, in Atmospheric Corrosion, W. H. Ailor (Ed.), Wiley, New York, 1982, p. 755.

26. G. J. Biefer, Mater. Perform., 20(1), 16 (1981).

27. D. Knotková-Čeráková, J. Vičková, and J. Honzák, in Atmospheric Corrosion of Metals, ASTM STP 767, S. W. Dean, Jr., and E. C. Rhea (Eds.), ASTM, Philadelphia, PA, 1982, p. 7.

28. T. Hakkarainen and S. Yläsaari, Atmospheric Corrosion, W. H. Ailor (Ed.), Wiley, New York, 1982, p. 787.

29. O. Bauer, Stahl Eisen, 41, 37, 76 (1921).

30. E. Brauns and U. Kalla, Stahl Eisen, 85, 406 (1965).

31. W. Schwenk and H. Ternes, Stahl Eisen, 88, 318 (1968).

32. G. Burgmann and D. Grimme, Stahl Eisen, 100, 641 (1980).

33. Committee on Steel Corrosion Survey, Kozai Club, Committee Report, A Survey on Corrosion and Corrosion Protection of Steel, Kozai Club, Tokyo, Japan, 1973, p. 18; reproduced in Corrosion and Corrosion Protection Databook, I. Matsushima

(Ed.), Japan Society of Corrosion Engineering, Tokyo, Japan, 1995, p. 6.

34. M. Morcillo, in Atmospheric Corrosion, ASTM STP 1239, W. W. Kirk and H. H. Lawson, (Eds.), American Society for Testing and Materials, Philadelphia, PA, 1995, p. 257.

35. R. J. Cordner, Br. Corros. J., 25, 115 (1990).

36. L. Atteraas and S. Haagenrud, in Atmospheric Corrosion, W. H. Ailor (Ed.), Wiley, New York, 1982, p. 837.

37. C. R. Southwell, J. D. Bultman, and A. L. Alexander, Mater. Perform., 15(7), 9 (1976).

38. C. R. Southwell and J. D. Bultman, in Atmospheric Corrosion, W. H. Ailor (Ed.), Wiley, New York, 1982, p. 943.

39. Y. N. Mikhailovski and P. V. Strekalov, in Atmospheric Corrosion, W. H. Ailor (Ed.), Wiley, New York, 1982, p. 923.

40. V. Kucera, S. Haagenrud, L. Atteraas, and J. Gullman, in Degradation of Metals in the Atmosphere, ASTM STP 965, S. W. Dean and T. S. Lee (Eds.), American Society for Testing and Materials, Philadelphia, PA, 1988, p. 264.

41. S. K. Roy and K. H. Ho, Br. Corros. J., 29, 287 (1994).

42. S. Feliu and M. Morcillo, Br. Corros. J., 22, 99 (1987).

43. E. Brillas, J. M. Costa, and M. Villarrasa, Proceedings of the 11th International Corrosion Congress, Associazione Italiana di Metallurgia (AIM), Milan, Italy, 1990, vol. 2, p. 2. 79.

44. M. Morcillo, J. Simancas, and S. Feliu, in Atmospheric Corrosion, ASTM STP 1239, W. W. Kirk and H. H. Lawson (Eds.), American Society for Testing and Materials, Philadelphia, PA, 1995, p. 195.

45. E. Johansson and J. Gullman, in Atmospheric Corrosion, ASTM STP 1239, W. W. Kirk and H. H. Lawson (Eds.), American Society for Testing and Materials, Philadelphia, PA, 1995, Vol. 2, p. 2.79.

46. Feng-I. Wei, Br. Corros. J., 26, 209 (1991).

47. J. C. Hudson, J. Iron Steel Inst., 148, 161 (1943).

48. J. C. Hudson and J. F. Stanners, J. Iron Steel Inst., 180, 271 (1955).

49. ASTM Committee A-5, Proc. ASTM, 53, 110 (1953).

50. C. P. Larrabee, Corrosion, 9, 259 (1953).

51. H. R. Copson, Proc. ASTM, 60, 650 (1960).

52. C. P. Larrabee and S. K. Coburn, in Metallic Corrosion (Proceedings of the 1st International Congr. Metallic Corrosion), Butterworths. London, 1971, p. 276.

53. G. B. Mannweiler, in Metal Corrosion in the Atmosphere, ASTM STP 435, American Society for Testing and Materials, Philadelphia, PA, 1968, p. 211.

54. C. McCaul and S. Goldspiel, in Atmospheric Corrosion, W. H. Ailor (Ed.), Wiley, New York, 1982, p. 432.

55. H. E. Thomas and H. N. Alderson, in Metal Corrosion in the Atmosphere, ASTM STP 435, American Society for Testing and Materials, Philadelphia, PA, 1968, p. 83.

56. C. W. Wieser, in Metal Corrosion in the Atmosphere, ASTM STP 435, American Society for Testing and Materials, Philadelphia, PA, 1968, p. 271.

57. P. F. Wieser, in Atmospheric Corrosion, W. H. Ailor (Ed.), Wiley, New York, 1982, p. 453.

58. C. R. Shastry, J. J. Friel, and H. E. Townsend, in Degradation of Metals in the Atmosphere, ASTM STP 965, S. W. Dean and T. S. Lee (Eds.), American Society for Testing and Materials, Philadelphia, PA, 1988, p. 5.

59. S. K. Coburn, M. E. Komp, and S. C. Lore, in Atmospheric Corrosion, ASTM STP 1239, W. W. Kirk and H. H. Lawson (Eds.), American Society for Testing and Materials, Philadelphia, PA, 1995, p. 101.

60. C. P. Larrabee, Trans. Electrochem. Soc., **85**, 297 (1944); cited in refs. [51] and [61].

61. F. L. LaQue, Proc. ASTM, **51**, 1 (1951).

62. K. Kucera and E. Mattsson, in Atmospheric Corrosion, W. H. Ailor (Ed.), Wiley, New York, 1982, p. 561.

63. I. L. Rozenfeld, Corrosion and Corrosion Protection of Metals, Izd. Metallurgija, Moscow, Russia, 1970, p. 120.

44

CARBON STEEL—CORROSION IN FRESHWATERS

I. Matsushima*

Maebashi Institute of Technology, Maebashi, Japan

A. INTRODUCTION

Corrosion of steel in freshwater proceeds electrochemically by the action of dissolved oxygen:

$$\text{Anodic reaction:} \quad Fe \rightarrow Fe^{2+} + 2e^- \qquad (44.1)$$

$$\text{Cathodic reaction:} \quad \tfrac{1}{2}O_2 + H_2O + 2e^- \rightarrow 2OH^- \qquad (44.2)$$

$$\text{Overall:} \quad Fe + \tfrac{1}{2}O_2 + H_2O \rightarrow Fe(OH)_2 \qquad (44.3)$$

*Deceased

Uhlig's Corrosion Handbook, Third Edition, Edited by R. Winston Revie
Copyright © 2011 John Wiley & Sons, Inc.

Most natural freshwaters are air saturated, and the concentration of dissolved oxygen at ordinary temperatures is 8–10 ppm. For corrosion of steel to occur, dissolved oxygen must be supplied to the steel surface by diffusion through water and possibly through surface films on the steel surface that act as diffusion barrier layers. Because of its concentration and diffusion coefficient, the supply of dissolved oxygen is slow unless the velocity of water relative to the steel surface is sufficiently high. The corrosion reaction proceeds as rapidly as oxygen reaches the steel surface. Thus, the corrosion is under diffusion control of dissolved oxygen.

In the absence of diffusion barrier layers on the surface, the theoretical corrosion current density, i (A/cm^2), of steel in stagnant air-saturated freshwater can be calculated as follows:

$$i = \left(\frac{DnF}{\delta}\right) C \times 10^{-3} \qquad (44.4)$$

where D is the diffusion coefficient for dissolved oxygen in water (cm^2/s), n is the number of electrons involved in the reaction, F is the Faraday constant (C/mol), δ is the thickness of the diffusion layer (cm), and C is the concentration of dissolved oxygen (mol/L).

Using $D = 2 \times 10^{-5}$ cm^2/s, $n = 4$, $F = 96,500$ C/mol, $\delta = 0.05$ cm, and $C = 8/32 \times 10^{-3}$ mol/L at 25°C, $i = 38.6 \times 10^{-6}$ A/cm^2, which is equivalent to a corrosion rate of 0.45 mm/year [18 mils/year (mpy)].

In hard waters that contain high concentrations of calcium and bicarbonate, the natural deposition of calcium carbonate ($CaCO_3$) on the steel surface provides an effective diffusion barrier to oxygen diffusion, greatly decreasing corrosion. In soft waters, the corrosion rate is higher than in hard waters, but it is lower than the theoretical maximum value because the corrosion product film formed on the surface acts to some extent as a diffusion barrier. The average corrosion rate of steel in stagnant air-saturated soft waters at ordinary temperatures is roughly 0.1 mm/year.

Increase in oxygen concentration, velocity, and temperature, within limits, accelerates corrosion in soft water by increasing the supply of dissolved oxygen to the surface. When oxygen concentration is very high, for example, ~12 mL/L or 17 ppm in distilled water at 25°C [1] or 25–35 ppm [2] in natural waters (e.g., when freshwater is saturated with a gas of high oxygen partial pressure), or water velocity is above a critical value, the corrosion rate drops to a low value. This decrease in corrosion rate is caused by passivation of the steel as a result of oxygen being supplied to the surface in excess of the amount that can be used for the cathodic reaction. More quantitative data regarding the effects of environmental factors are given in Section B.

While the supply of dissolved oxygen controls the overall corrosion rate, the cathodic and anodic reactions do not necessarily occur uniformly over the surface. There are two types of corrosion cells. In macrogalvanic cells, the anodic and cathodic areas are macroscopic, and their locations are fixed, whereas in macrogalvanic cells, the anodic and cathodic sites are microscopic and move randomly with time. When macrogalvanic cells, such as differential aeration cells, are formed on the steel surface and the anodic reaction dissolution of steel takes place predominantly at some limited areas in amount equivalent to the cathodic reaction proceeding around the macroanode, the penetration rate at the anode can be high if the cathode–anode area ratio is large. The extent of the effective cathode area depends on the conductivity and geometry of the system. In waters of high conductance, macrogalvanic cells can operate over long distances.

High corrosion rates of steel are associated with the formation of macrogalvanic cells of some kind, which may not have been anticipated or considered at the design stage for corrosion protection.

B. ENVIRONMENTAL FACTORS

B1. Natural Freshwaters

While corrosion of carbon steel proceeds by the action of dissolved oxygen, the rate of the reaction is affected by other species in water. Among the many species dissolved in natural freshwaters, calcium (Ca^{2+}), bicarbonate (HCO_3^-), and chloride (Cl^-) are very important with respect to corrosion.

The source of natural freshwaters is meteoric water that falls on the earth as rain, snow, sleet, or hail. Rainwater comes in contact with the atmosphere and is saturated with dissolved air. Depending on the kind of atmosphere to which the

rainwater is exposed, it also dissolves impurity gases, such as SO_x, NO_x, NH_3, and HCl, and other matter from suspended atmospheric impurities such as sea salt and dust.

When the precipitation comes into contact with the ground, some of it evaporates, but the rest flows or collects on the surface of the earth as surface water or sinks into the ground as groundwater. The water dissolves materials with which it comes in contact and therefore contains a variety of species, including Ca^{2+}, Mg^{2+}, soluble silica [e.g., $(H_2SiO_3)_n$], HCO_3^-, Na^+, K^+, Cl^-, and SO_4^{2-}.

Carbon dioxide in water reacts with minerals containing $CaCO_3$, such as limestone, marble, chalk, calcite, and dolomite, as follows:

$$CaCO_3 + CO_2 + H_2O \rightarrow Ca^{2+} + 2HCO_3^- \qquad (44.5)$$

Dissolved CO_2 also reacts similarly with minerals containing $MgCO_3$ to form Mg^{2+} and HCO_3^-. The contents of Ca^{2+} and Mg^{2+} as equivalent parts per million $CaCO_3$ are the calcium hardness and the magnesium hardness, respectively, and the sum is the total hardness. In natural waters, the concentration of bicarbonate (HCO_3^-) corresponds to the methyl orange alkalinity (MO alkalinity; the equivalent per liter of titratable base to the methyl orange end point), commonly expressed as parts per million $CaCO_3$. The values of Ca hardness and the MO alkalinity are the primary factors that determine the ease of protective $CaCO_3$ film formation on a metal surface.

The solubility of $CaCO_3$ depends on the dissolved CO_2 concentration. Air-saturated distilled water contains only 0.5 ppm CO_2 at room temperature and can dissolve 53 ppm $CaCO_3$. Rainwater, however, may contain up to 2 ppm CO_2 because of the CO_2 derived from organic matter in the atmosphere. The amount of CO_2 contained in some natural waters, particularly groundwaters, is much higher due to the CO_2 derived from the decay process of organic matter in the soil, and the Ca hardness and MO alkalinity may be over 200 ppm. At a given CO_2 concentration, the amount of $CaCO_3$ dissolution depends on the kind of geological strata to which the water is exposed and the contact time before the water is utilized.

If the dissolution of $CaCO_3$ and $MgCO_3$ is the major source of the total hardness and the MO alkalinity, these two values should be almost identical. In some natural waters, however, the total hardness may be significantly higher or lower than the MO alkalinity. This difference can occur because Ca^{2+} and Mg^{2+} can derive from water-soluble minerals, instead of carbonates, without the action of CO_2, for example, Ca^{2+} from gypsum, alabaster or selenite and Mg^{2+} from epsomite, kainite, picromerite, or loweite. Additional HCO_3^- may be formed by the reaction of dissolved CO_2 with some minerals other than $CaCO_3$ and $MgCO_3$, for example, albite ($NaAlSi_3O_8$). If the MO alkalinity of the water equals or exceeds the total hardness, all of

TABLE 44.1. Typical W

River	Locatio	...ardness[c] ...s CaCO$_3$)					Total Dissolved Solids
		tal	Ca	Cl$^-$	SO$_4{}^{2-}$	SiO$_2$	
		(Concentration in ppm)					
Bandas	Seria	8	1	1	7	6	
Kerteh	Kerteh	15	7	10	1	10	78
Yodo	Sakai	39	29	15	22	6	
Vanda	Helsir	50	31	12	22	8	145
Enim	Tanju	73		12		29	
Changjiang	Wuha	75	53	17	18	5	
Indus	Jamsh	25		15			
Chaopraya	Bang	26	73	109	28	26	
Thachin	Bang	50	100	35	50	20	
Changjiang	Shan	63	98	153	40	10	
Donau	Brat	233	169	23	33	9	
Thames	Lon	262	252	19	27	6	904
Red Deer	Red	290	180	6	37	7	
Saskatchewan	Sask	604	352	476	298	14	1650

[a]Data were supplied from
[b]Water quality is subject
[c]Arranged in increasing t

the hardness is attri
other hand, the MO alkalinity is less than the total hardness, the carbonate hardness equals the MO alkalinity and the balance is the noncarbonate hardness.

The results of analyses of dissolved matter in typical soft and hard river waters are listed in Table 44.1 [3]. Waters with a total hardness of < 100 ppm are soft, those with 100–150 ppm are slightly hard, and those with > 150 ppm are hard. Some waters are very hard, the total hardness being over 300 ppm.

The sources of Cl$^-$ in natural freshwaters are sea salt in the atmosphere, pollution of rivers by sewage and industrial effluents, some geological strata to which the water is exposed, and road deicing salts. The average concentration of Cl$^-$ in rainwater is commonly ~ 1 ppm, but near the seacoast, it can be an order of magnitude higher due to prevailing winds from the sea. The Cl$^-$ content of river waters, about 8 ppm on average, varies in a wide range from a few parts per million to several hundred parts per million except in some unusually high cases.

B2. Effect of Water Quality

B2.1. General. The analytical water quality parameters that may affect corrosivity of natural freshwaters are pH, dissolved oxygen, Ca hardness, MO alkalinity, total dissolved solids (TDS), Cl$^-$, and SO$_4{}^{2-}$. The values of pH, Ca hardness, MO alkalinity, and TDS are factors that determine the saturation index, which is the criterion of whether or not the CaCO$_3$ diffusion-barrier film is formed on a metal

...rface, as will be described in the following section (see ...ction B2.5).

The values of pH and of dissolved oxygen concentration are the fundamental factors that affect the corrosivity of soft waters. They usually do not change the corrosion rate of steel in natural freshwaters, because the pH value remains within a certain range and most waters are air saturated.

The ions Cl$^-$ and SO$_4{}^{2-}$ are always present in natural freshwaters. They affect the penetration rate of localized corrosion by increasing conductivity, and they also affect the critical concentration of oxygen and critical water velocity above which passivation of steel occurs.

B2.2. Saturation Index. For given values of calcium hardness, MO alkalinity and total dissolved salt concentration, a value of pH (pH$_s$) exists at which the water is in equilibrium with solid CaCO$_3$. The deposition of CaCO$_3$ is thermodynamically possible when the pH of the water is higher than pH$_s$.

At equilibrium,

$$K_2 = \frac{[\text{H}^+][\text{CO}_3{}^{2-}]}{[\text{HCO}_3{}^-]} \tag{44.6}$$

$$K_s = [\text{Ca}^{2+}][\text{CO}_3{}^{2-}] \tag{44.7}$$

$$= \frac{[\text{Ca}^{2+}][\text{HCO}_3{}^-]K_2}{[\text{H}^+]} \tag{44.8}$$

Since $-\log[\mathrm{H}^+]$ here is $\mathrm{pH_s}$ and $[\mathrm{HCO_3^-}] \simeq \mathrm{MO}$ alkalinity (alk) in natural freshwaters,

$$\mathrm{pH_s} \simeq \log\left(\frac{K_s}{K_2}\right) - \left\{\log[\mathrm{Ca}^{2+}] + \log(\mathrm{alk})\right\} \quad (44.9)$$

For a more thorough derivation of $\mathrm{pH_s}$, see the textbook by Uhlig and Revie [4], and for a new equation based on a rigorous model of $CaCO_3$ saturation, see the paper by Pisigan and Singley [5].

The value $\mathrm{pH}_{measured} - \mathrm{pH_s}$ is called the saturation index, or the Langelier index, after Langelier [6], who established this index. A positive value of the Langelier index indicates a tendency for the protective $CaCO_3$ film to form and a negative value for it not to form.

In the presence of other dissolved salts, the increase of ionic strength depresses the activities of other ions and increases the values of $\log(K_s/K_2)$ in Eq. 44.9 if K_s and K_2 are based on concentrations rather than activities. On the other hand, the values of $\log(K_s/K_2)$ decrease with temperature, making the saturation index more positive. In practice, the calculation of $\mathrm{pH_s}$ is carried out on the concentration basis with additions of correction factors for ionic strength and temperature to Eq. 44.9.

The values of the Langelier index of representative river waters, calculated based on the analyses provided by Suzuki [3] and others, are shown in Figure 44.1.

Charts have been prepared by Powell et al. [7] for calculating the Langelier index of waters varying widely in relevant dissolved species at various temperatures. The historical charts prepared by Powell et al. [7] and the table arranged by Nordell [8] (Table 44.2) are still in use, but correction factors for ionic strength and temperature have been updated based on recent advances in the solubility chemistry of $CaCO_3$ and solution chemistry (e.g., by Schock [9]).

While the concept of the Langelier index is correct and helpful, it should be emphasized that a positive value of the index can result from waters of totally different quality. As the pH increases, the Ca^{2+} concentration decreases drastically. The corrosion protection characteristics of the resulting $CaCO_3$ film differ accordingly. In other words, waters of different pH, Ca hardness, and MO alkalinity that give the same value of the index have different corrosivity. The buffer capacity and the oxygen concentration of the water and the thickness, composition, and crystalline state of deposits all affect the protectiveness of the deposited $CaCO_3$ film. These facts were pointed out by Stumm [10] as early as 1956 and have not been explained to date either quantitatively or systematically.

Instead of using the thermodynamic prediction of the tendency for $CaCO_3$ precipitation in terms of the Langelier index, an experimental method, the marble test, developed by DeMartini [11] and Hoover [12] can be used. The water is

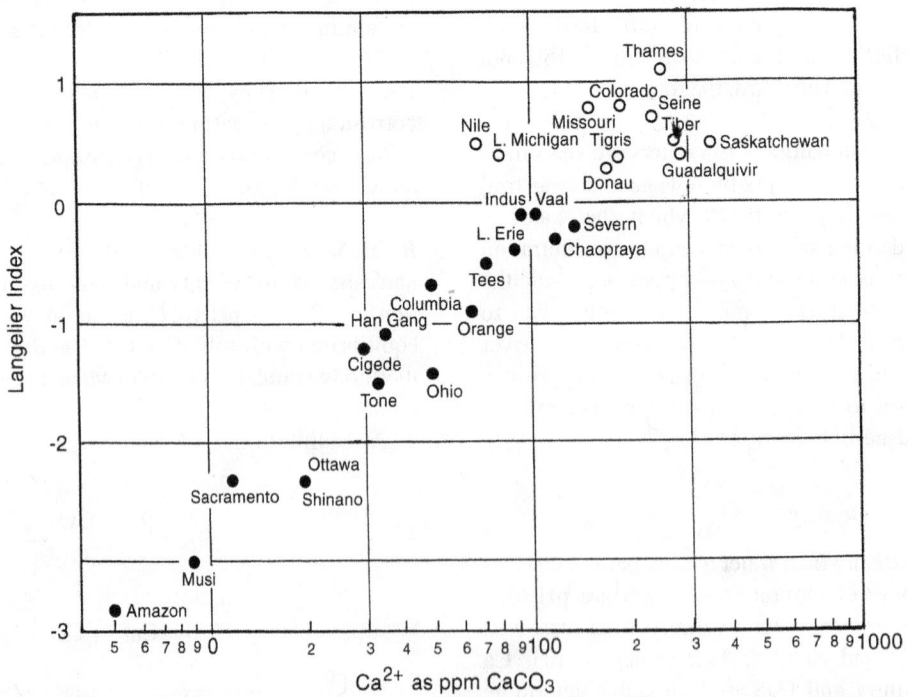

FIGURE 44.1. Values of the Langelier index of representative river waters. Data are primarily from [3].

TABLE 44.2.   Data for Rapid Calculation of the Langelier Index Arranged by Nordell[a]

A		C		D	
Total Dissolved Solid (ppm)	A	Ca Hardness (ppm as $CaCO_3$)	C	MO Alkalinity (ppm as $CaCO_3$)	D
50–300	0.1	10–11	0.6	10–11	1.0
400–1000	0.2	12–13	0.7	12–13	1.1
		14–17	0.8	14–17	1.2
B		18–22	0.9	18–22	1.3
		23–27	1.0	23–27	1.4
Temperature (°C)	B	28–34	1.1	28–35	1.5
		35–43	1.2	36–44	1.6
0–1.1	2.6	44–55	1.3	45–55	1.7
2.2–5.6	2.5	56–69	1.4	56–69	1.8
6.7–8.9	2.4	70–87	1.5	70–88	1.9
10.0–13.3	2.3	88–110	1.6	89–110	2.0
14.4–16.7	2.2	111–138	1.7	111–139	2.1
17.6–21.1	2.1	139–174	1.8	140–176	2.2
21.2–26.7	2.0	175–220	1.9	177–220	2.3
27.8–31.1	1.9	230–270	2.0	230–270	2.4
32.2–36.7	1.8	280–340	2.1	280–350	2.5
37.8–43.3	1.7	350–430	2.2	360–440	2.6
44.4–50.0	1.6	440–550	2.3	450–550	2.7
51.1–55.6	1.5	560–690	2.4	560–690	2.8
56.7–63.3	1.4	700–870	2.5	700–880	2.9
64.4–71.1	1.3	880–1000	2.6	890–1000	3.0
72.2–81.1	1.2				

[a]See [8].
(1) Obtain values of A, B, C and D from the table.
(2) $pH_s = (9.3 + A + B) - (C + D)$.
(3) Langelier saturation index $= pH - pH_s$.

treated with powdered $CaCO_3$, and, after saturation, changes in pH, alkalinity, and concentration of calcium are measured as an estimate of undersaturation. An improved procedure has been proposed by Merrill and Sanks [13].

Another index used in practice for estimating the corrosivity of water is the stability index or the Ryznar index (RI) developed by Ryznar [14]:

$$RI = 2pH_s - pH \qquad (44.10)$$

This index is intended to predict quantitatively the amount of $CaCO_3$ that would be formed and also to predict the corrosivity of waters that are undersaturated with respect to $CaCO_3$. At RI 6.0, $CaCO_3$ is in equilibrium. The deposition of $CaCO_3$ increases proportionately as the index drops < 6, and corrosivity increases as the index rises > 6. Values of 10 or above indicate extreme corrosivity.

Both the Langelier and the Ryznar indexes were derived from thermodymamic considerations. The $CaCO_3$ precipitation does not occur unless a certain degree of oversaturation is indicated by these indexes. A new index, the practical saturation index (PSI), has been proposed by Puckorius and

Brooke [15], which uses what they called an equilibrium pH (pH_e) in the Ryznar equation [Eq. (44.10)]:

$$PSI = 2pH_s - pH_e \qquad (44.11)$$

$$pH_e = 1.485 \times \log (\text{total alkalinity}) + 4.54 \qquad (44.12)$$

The coefficient and the constant in Eq. (44.12) were obtained empirically from the study of hundreds of actual case histories recorded and evaluated by the authors over 12–15 years with concentrated cooling waters, most of which were water treated to a pH level >7.5, often between 8 and 9.5. Because the values of pH_e tend to be smaller than those of the measured pH, making the PSI values larger than the Ryznar index values to compensate the oversaturation required for precipitation to occur [cf. Eqs. (44.10) and (44.11)], the $CaCO_3$ precipitation takes place below exactly PSI 6.0, according to the authors.

For soft waters, a desired level of the Langelier index can usually be attained by the addition of lime, sodium carbonate, or sodium hydroxide to raise the pH to an appropriate value. For very soft waters, however, the required pH may be too

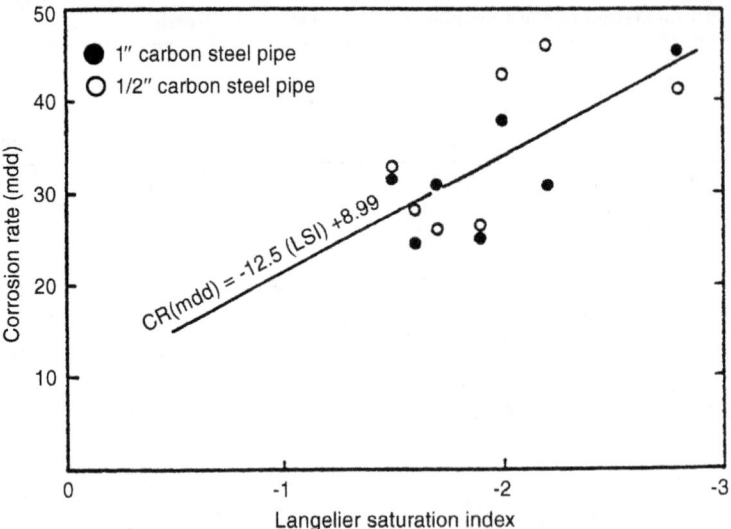

FIGURE 44.2. Relationship between the Langelier index and the corrosion rates of water pipes [18].

high for uses such as tap water unless Ca^{2+} ions are added simultaneously. A Langelier index of $+0.1 \sim +0.5$ is considered to be satisfactory to provide corrosion protection and at the same time avoid excessive deposition or scaling at elevated temperatures [16]. Data by Flentje [17] and Fujii [18] show that water with a less negative Langelier index is less corrosive. Waters with a slightly negative Langelier index may deposit $CaCO_3$ because of pH fluctuations. The relationship between the Langelier index and the corrosion rates of water pipes obtained by Fujii [18] from a field test at a city waterworks are reproduced in Figure 44.2.

Even though the saturation index is positive, the corrosion rate may remain high if the water contains colloidal silica or organic particles, such as algae, because $CaCO_3$ precipitates on them instead of on the steel surface. For waters high in dissolved salt or at high temperature, the $CaCO_3$ film is less protective.

The following sections deal with corrosion in soft waters unless otherwise stated.

B2.3. pH. The pH of freshwaters is determined by the equilibrium of carbonic species:

$$pH = 6.35 + \frac{\log[HCO_3^{-}]}{[CO_{2aq}]} \quad (25°C) \qquad (44.13)$$

where $[CO_{2aq}]$ is the sum of dissolved $[CO_2]$ and $[H_2CO_3]$ in equilibrium with the partial pressure of CO_2 gas in the gas phase. The pH of surface waters ranges from 5 to 9, but that of underground waters exposed to a high CO_2 partial pressure is often < 6.5 and increases on aeration, which eliminates excessive CO_2.

Based on the classic data by Whitman et al. [19], it has generally been considered that, within the pH range of

~ 4–10, the corrosion rate is independent of pH, other factors being equal (Fig. 44.3, line a). Regardless of the bulk pH of water within this range, the steel surface is always in contact with an alkaline solution of saturated hydrous ferrous oxide (pH ~ 9.5).

Some researchers claimed, however, that water quality (i.e., alkalinity, buffer capacity, and concentrations of Cl^{-} and SO_4^{2-}) is the primary factor in corrosion at low-flow velocity and that the influence of dissolved oxygen and pH on corrosion rates is secondary. Skold and Larson [20], by polarization resistance measurements, reported the effect of pH on corrosion, as shown by curves b and c in Figure 44.3, where the corrosion rate increases from pH 7 to a maximum at pH 8. The concentrations of $NaHCO_3$ and $NaCl$ were kept constant at 2.5 and 0.5 mmol/L, respectively, throughout the experiment. Carbon dioxide was used to control pH. A similar curve (Fig. 44.3, curve d) was obtained by Pisigan and Singley [21], who conducted immersion tests in a solution containing 1 mmol/L total carbonate species. These authors pointed out that the increase of the corrosion rate in this pH range corresponds to the decrease of the buffer capacity provided by HCO_3^{-}. The basis of this conclusion is that HCO_3^{-} ions act as an inhibitor (see Section B2.5)

Regarding the effect of pH indicated by curves b–d in Figure 44.3, it should be pointed out that the waters used in these experiments contained no calcium, unlike natural waters.

Fujii et al. [22] observed an appreciable increase in the corrosion rate with increase of pH from 6.5 to 8.5 (Fig. 44.3, curve e) with soft Tokyo tap water (Ca hardness: 51.8 ppm, MO alkalinity: 39.8 ppm, pH 7.1, Langelier index: -1.4) using CO_2 gas or NaOH to control pH. They attributed the increased corrosion rate to the observed enhanced formation

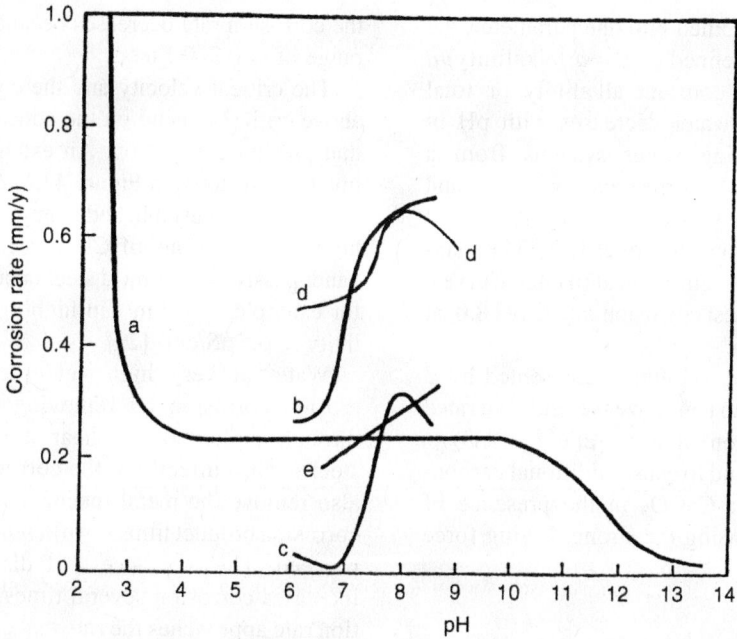

FIGURE 44.3. Effect of pH on corrosion of steel in aerated waters. (a) Soft tap water, Cambridge, MA. pH adjusted by additions of HCl and NaOH, 22°C [19]. (b) Aerated water containing 2.5 mmol/L NaHCO$_3$ and 2.5 mmol/L NaOH. pH controlled by continuous addition of CO$_2$ (22–31°C). Linear polarization method applied after 16 days immersion [20]. (c) Aerated water containing 2.5 mmol/L NaHCO$_3$ and 0.5 mmol/L NaOH. pH controlled by continuous addition of CO$_2$ (19–28°C). Linear polarization method applied after 16 days immersion [20]. (d) Water containing NaHCO$_3$ (alkalinity: 100 mg/L as CaCO$_3$). pH adjusted by introduction of CO$_2$ and air. Room temperature, 3 days [21]. (e) Soft tap water, Tokyo, Japan. pH adjusted by additions of CO$_2$ and NaOH. 0.5 in. (12.7 mm OD) pipe, 2 L/min once-through flow. Room temperature, 32 days [22].

and activity of tubercles at higher pH, where corrosion products are more adherent and the buffer capacity is less, resulting in a lower pH at the anodic areas.

B2.4. Chloride and Sulfate Ions.
As long as diffusion of dissolved oxygen is controlling corrosion, Cl$^-$ and SO$_4^{2-}$ ions in concentrations found in freshwaters have essentially no effect on the overall corrosion rate of steel because they do not affect the solubility of oxygen. Localized attack, however, is accelerated as the conductivity of water increases with increase in the concentration of these ions, expanding the effective cathodic areas.

Above a critical water velocity, passivation occurs, and the corrosion rate of the passivated steel increases with Cl$^-$ concentration. For more details, see Section B3.

Some authors have shown that, at a given HCO$_3^-$ concentration in the absence of Ca^{2+}, increased Cl$^-$ or SO$_4^{2-}$ concentration caused the water to become more corrosive to steel [21, 23, 24]. According to these authors, even in the absence of Ca^{2+}, bicarbonates inhibit the corrosion of steel, and Cl$^-$ and SO$_4^{2-}$ are detrimental to this inhibitive action. For more details, see Section B2.5.

B2.5. Bicarbonate.
Bicarbonate (HCO$_3^-$) is one of the species essential to form a protective CaCO$_3$ film in the presence of Ca^{2+} ions at pH above pH$_s$. Some data indicate the effects of HCO$_3^-$ ions as an inhibitor of corrosion in the absence of Ca^{2+} ions. Although it is not likely that in natural waters an appreciable concentration of HCO$_3^-$ ions is present without a comparable concentration of Ca^{2+} ions, it would be worthwhile to consider this inhibitive action.

With solutions that do not contain Ca^{2+}, Larson and co-workers [23, 24] reported that at a given concentration of Cl$^-$ or SO$_4^{2-}$ the corrosion rate decreased with increase of HCO$_3^-$ and at a given HCO$_3^-$ concentration the corrosion rate increased with increase of Cl$^-$ or SO$_4^{2-}$ and leveled off at a certain concentration of these ions. In other words, the corrosion rate was determined by the Cl$^-$/HCO$_3^-$ or SO$_4^{2-}$/HCO$_3^-$ ratio.

These authors, and also Pisigan and Singley [21], suggested that the corrosion product formed in the presence of HCO$_3^-$ and in the absence of Ca^{2+} was FeCO$_3$, instead of Fe(OH)$_2$, the former being more protective than the latter. The latter authors, having obtained the effect of pH as shown by curve d in Figure 44.3, proposed that the composite effect of pH and alkalinity can be combined into one parameter, the buffer capacity, β which is defined as $\beta = d$ (alkalinity)/d pH. If the pH is varied at constant alkalinity or total carbonate species, β of the water decreases with pH in the range of typical drinking water systems from a maximum to a minimum at pH corresponding to pK_1 and to $\frac{1}{2}(pK_1 + pK_2)$, respectively, where K_1 and K_2 are the first and the second dissociation constants of H_2CO_3. The maximum occurs at pH 6.3 and the minimum at pH 8.3. Curve d in Figure 44.3 shows the highest corrosion rate at pH 8.0, at which β is near its minimum.

The beneficial effect of the alkalinity represented by β was explained by its capacity to neutralize the acid generated at the anode by the oxidation of Fe^{2+} [$2Fe^{2+} + \frac{1}{2}O_2 + 5H_2O = 2Fe(OH)_3 + 4H^+$] and to cause additional carbonate precipitation ($FeCO_3$, or $CaCO_3$ in the presence of Ca^{2+}) at the cathode, decreasing the strong driving force for corrosion.

B3. Effect of Velocity

The effect of water velocity on corrosion of mild steel, from data by Kowaka et al. [25], Fujii et al. [22], Matsushima [26], and Kinoshita et al. [27], is shown in Figures 44.4 and 44.5. According to these data, the critical velocity above which

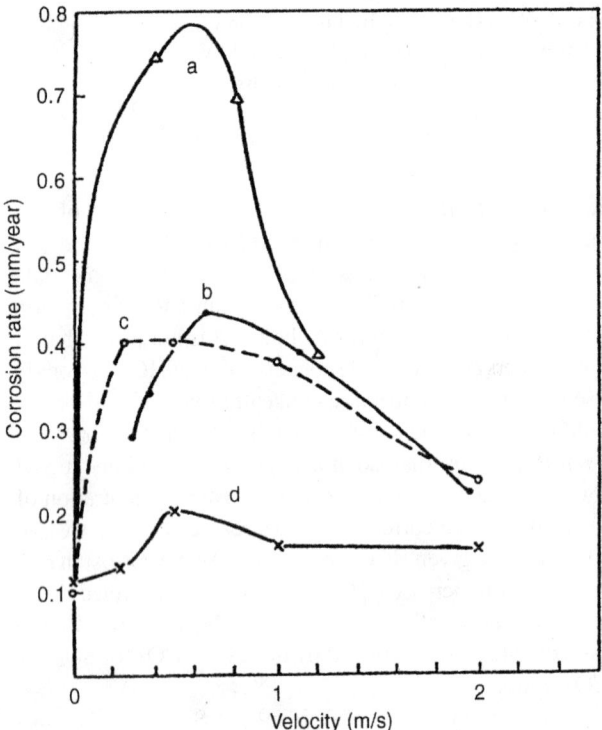

FIGURE 44.4. Effect of water velocity on corrosion of carbon steel (1). (a) Distilled water + 10 ppm Cl^-, 50°C, 14 days [26]. (b) Soft tap water, Tokyo, Japan, room temperature, 67 days [22]. (c) Soft tap water, Amagasaki, Japan, 20°C, 15 days (killed steel) [25]. (d) Soft tap water, Amagasaki, Japan, 20°C, 15 days (rimmed steel) [25].

the corrosion rate decreases because of passivation is in the range of $\sim$ 0.3–0.7 m/s.

The critical velocity and the corrosion rate at velocities above critical depend on the concentration of chloride ions that prevent passivation. An example of data by Matsushima [26] is shown in Figure 44.6. As shown by LaQue [28], passivity is not established at any velocity in the presence of high concentrations of Cl^-, as in seawater. On the other hand, passivation of mild steel occurs at a very low velocity, for example, < 0.1 m/s in high-purity water (e.g., conductivity < 0.5 μS/cm) [29].

Water at very high velocities (e.g., 20 m/s) causes erosion–corrosion by removing corrosion product films through application of shear stress, thereby exposing the steel surface directly to the corrosive environment. It may also remove the metal mechanically, but removal of only corrosion product films is sufficient for erosion–corrosion to proceed. The free access of dissolved oxygen in water increases corrosion several times. The maximum penetration rate approaches the rate that corresponds to the amount of dissolved oxygen supplied from the water without a surface barrier. Erosion–corrosion often occurs at bends or misaligned joints of high-velocity water pipe even if the water does not carry suspended solids.

If conditions of high-velocity water are such that low- and high-pressure areas develop and bubbles form and collapse at the metal–water interface (cavitation), the steel is damaged mechanically by the impact pressure exerted by the collapsing bubbles. This type of degradation is called cavitation erosion or cavitation damage. Cavitation erosion can occur purely mechanically but is accelerated by free access of dissolved oxygen. It occurs typically on rotors of pumps and the trailing side of water turbine blades made of steel and cast iron.

C. LOCALIZED CORROSION

C1. General

Localized corrosion of steel in freshwaters occurs in most cases by the action of macrogalvanic cells. The resulting damage tends to be most critical in water pipes because perforation of the pipe wall by localized corrosion (e.g., pitting) means instant leakage of the water carried by the pipe.

The major types of localized corrosion of steel in freshwater are pitting under tubercles, groove corrosion of electric resistance welded (ERW) pipes, and pitting at discontinuities in corrosion product films. For other types of localized corrosion in water see Chapter 47.

C2. Pitting under Tubercles

Tuberculation or formation of mounds of corrosion products (tubercles) occurs frequently on the steel surface in contact

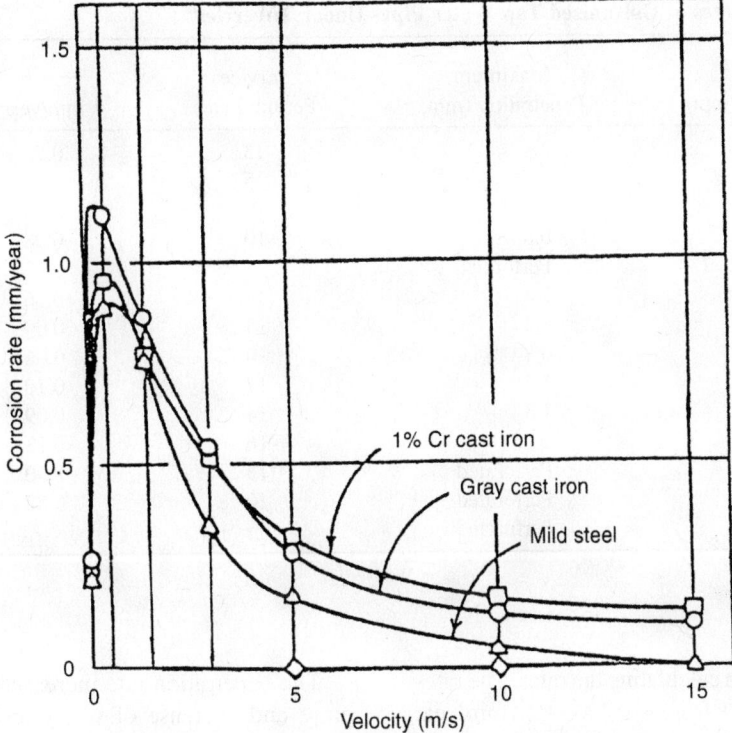

FIGURE 44.5. Effect of water velocity on corrosion of steel (2). 25 ppm Cl⁻, 30°C, 30 days [27].

with water. Differential aeration cells are then formed. The areas under tubercles, where the supply of dissolved oxygen is limited, undergo localized corrosion by the galvanic action of the surrounding cathodic areas that receive free oxygen supply. Galvanized water pipes corrode similarly after the

zinc coating has been lost, as often occurs within several years of service in soft waters.

The penetration rates in millimeters per year (mm/year) and mils per year (mpy) of galvanized city water pipes are listed in Table 44.3 [30]. The periods of protection by the

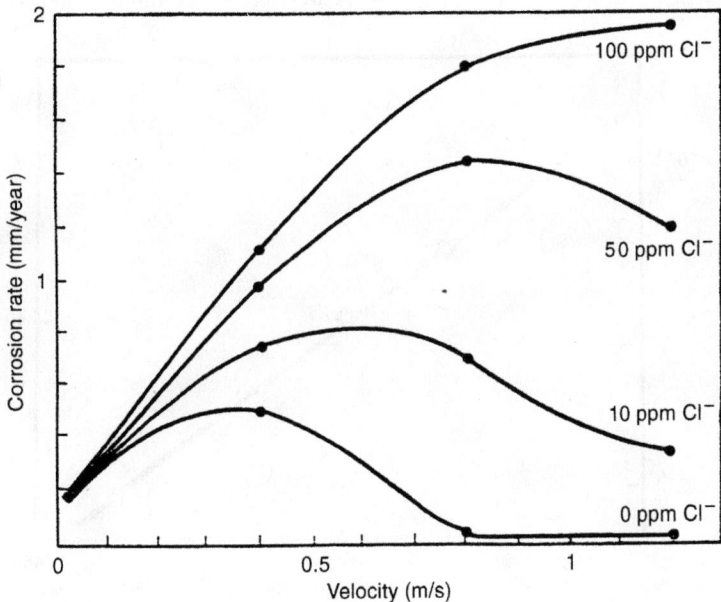

FIGURE 44.6. Effect of water velocity and chloride concentration on corrosion of carbon steel. Distilled water + NaCl, 50°C, 14 days [26].

TABLE 44.3.   Penetration Rates of Galvanized Tap Water Pipes Under Tubercles[a]

Size (in.)	Wall Thickness (mm)	Maximum Penetration (mm)	Service Period (year)	Maximum Penetration Rate	
				mm/year	mils/year (mpy)
1	3.2	3	13	0.23	9
1	3.2	1	5	0.20	8
1	3.2	1.5	10	0.15	6
1¼	3.5	0.8	10	0.08	3
1	3.2	Perforated	14	0.23	9
1¼	3.5	0.8	14	0.06	2
1¼	3.5	1.2	14	0.09	3
2	3.8	0.8	14	0.06	2
1	3.2	1.4	14	0.10	4
1½	3.5	1.3	14	0.09	3
5	4.5	2	16	0.13	5
4	4.5	Perforated	15	0.30	12
1	3.2	Perforated	10	0.32	13
6	5.0	Perforated			

[a]From [30].

zinc coating were neglected in calculating the rates. The rates are not high, ranging generally from < 0.1 to ~ 0.3 mm/year (<4 to ~12 mpy). The highest rate is 0.32 mm/year. Masamura and Matsushima [31] calculated the theoretical penetration rate by assuming the geometry of the corrosion system and the anodic polarization behavior and by solving the Laplace equation for the potential distribution. The rate depended on the resistivity of the water, increasing as resistivity decreased. The penetration rate corresponding to the resistivities of city waters, 5000–8000 $\Omega \cdot$ cm, is about 0.3 mm/year (12 mpy) (Fig. 44.7), in good agreement with the observed highest rate.

The penetration rate increases with increase of temperature and decrease of water resistivity. The highest rate observed by Matsushima [30] for galvanized hot water pipe was about 1 mm/year (40 mpy), except for the cases where the rate was extremely high, for example, 2.8 mm/year (110 mpy) indicating that the localized corrosion was accelerated by reversal of polarity between zinc and steel due to the formation of noble corrosion product films on zinc in waters of particular quality at high temperatures [32]. A penetration rate of 3.8 mm/year (150 mpy) was experienced in black pipe (as rolled steel pipe with mill scale) carrying warm mine water of low resistivity (pH 8.4,

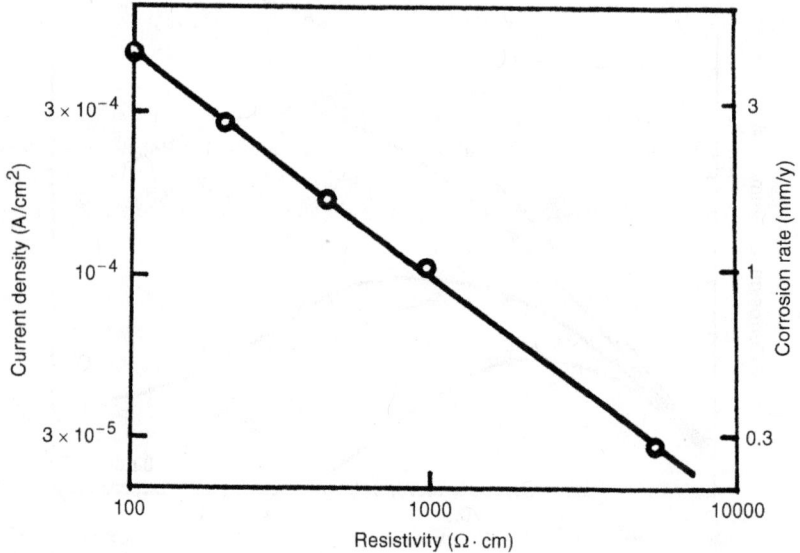

FIGURE 44.7. Effect of water resistivity on theoretical penetration rate of carbon steel under tubercles [31].

TABLE 44.4. Penetration Rates of Steel Pipes Alternately Exposed to Hot Water or Steam and Cold Water[a]

Item	Condition[b]	Penetration (mm)	Service Period (year)	Penetration Rate mm/year	Penetration Rate mils/year (mpy)
Hot water pipe	RT/90 ∼ 95°C	5.8[c]	6	0.97	38
Hot water pipe	RT/max 60°C	3.8[c]	4	0.95	37
Steam/water pipe	6 h cycle	4.2[c]	1.2	3.5	140
Boiler hot water return line near inlet of cold make-up water	48/90°C	4.5[c]	4	1.1	43

[a]See [39].
[b]Room temperature = RT.
[c]Perforated.

Ca hardness 32 ppm, total hardness 72 ppm, MO alkalinity 475 ppm, total dissolved solids 2635 ppm, Cl^- 735 ppm, resistivity 1385 $\Omega \cdot$ cm). The presence of mill scale presumably contributed to the high rate.

It is generally accepted that aerobic iron bacteria greatly accelerate the formation of tubercles by accumulating large amounts of ferric hydrate [33], but tuberculation can also occur in their absence. Sulfate-reducing bacteria flourish in tubercles because of the anaerobic condition and may accelerate corrosion [33]. The corrosion mechanism, however, can be explained solely from electrochemistry [34], and the significance of bacterial action is unknown.

C3. Groove Corrosion of Electric Resistance Welded Pipe

The welded seam of the ERW pipe is heated locally and then cooled rapidly by the ambient air, and its microstructure is much different from that of the parent metal that is not heated during welding. It tends to have more negative, or less noble, potential than the parent metal and, on exposure to corrosive media, such as water and soil, corrodes locally in the form of grooves. This type of localized corrosion does not always occur, but most ERW pipes are potentially susceptible to it

A rate in the range of 1–3 mm/year (40–120 mpy) is common and can be as high as 10 mm/year (400 mpy) [35]. The difference of the open-circuit potential is a maximum of ∼ 70 mV [35]. The more negative, or less noble, potential of the weld seam has been attributed to the formation of unstable MnS inclusions in the weld by the thermal cycle of the welding operation [36, 37]. Butt-welded pipes that have been heated uniformly in the process of shaping and welding are not susceptible to this type of corrosion.

Groove corrosion can be mitigated by minimizing the potential difference between the weld seam and the parent metal. The ERW pipes that are resistant to groove corrosion were developed in the 1970s. The chemical compositions of steels for these new pipes were modified by decreasing sulfur (e.g., $\leq 0.005\%$) and adding up to 0.3% Cu and small amounts of other elements (e.g., Ca, Ni, or Ti). The reduction of sulfur minimizes the formation of unstable MnS, and other elements, such as Ca, stabilize the remaining sulfur. The potential difference between the weld and the parent metal is reduced to almost nil. These ERW pipes are being widely used in diameters ≥ 125 mm, and no occurrence of groove corrosion has been reported. For pipes of smaller diameter, butt-welded pipes are used. For more details, see [38].

C4. Localized Corrosion at Discontinuities in Corrosion Product Films

Pitting-type localized corrosion at penetration rates >1 mm/year (40 mpy) sometimes occurs without the formation of tubercles in pipes alternately exposed to hot water (or steam) and cold water. Examples are shown in Table 44.4 [39].

The inside surface of the pipe is covered by corrosion product films that are tight, dense, relatively smooth, and dark in color. X-ray analysis identifies γ-Fe_2O_3 and Fe_3O_4, unlike the usual iron oxides, α- or γ-FeOOH, Fe_3C_4, and large amounts of amorphous substances that form in aqueous media. The degree of crystallization is much higher, as indicated by high intensities of the diffracted beams in X-ray diffraction.

The mechanism of this type of corrosion is similar to that of the localized corrosion of mill-scaled steel in seawater (see Chapter 47 Section B4). The surface covered by corrosion product films formed during exposure to hot water or steam tends to act as the cathode to the exposed steel at the defects or discontinuities in the films during exposure to cold water, causing localized corrosion. The corrosion potential of the film-covered areas was measured in the range of -590 to -670 mV versus SCE in 3% NaCl in the laboratory. The localized attack was reproduced in an experiment simulating service conditions [40].

REFERENCES

1. H. Uhlig, D. Triadis, and M. Stern, J. Electrochem. Soc., **102**, 59 (1955).

2. S. K. Coburn, in Metals Handbook, Vol. 1, 9th ed., American Society of Metals, Metals Park, OH, 1978, p. 733.

3. T. Suzuki, Private Communication (Database, Kurita Water Industries, Ltd., Tokyo).

4. R. W. Revie and H. H. Uhlig, Corrosion and Corrosion Control, 4th ed., Wiley, Hoboken, N.J., 2008, p. 461.

5. R. A. Pisigan, Jr., and J. E. Singley, J. Am. Water Works Assoc., **77**(10), 83 (1985).

6. W. F. Langelier, J. Am, Water Works Assoc., **28**, 1500 (1936).

7. S. Powell, H. Bacon, and J. Lill, J. Am. Water Works Assoc., **38**, 808 (1945).

8. E. Nordell, Water Treatment, 2nd ed., Reinhold, New York, 1961, p. 287.

9. M. R. Schock, J. Am. Water Works Assoc., **76**(8), 72 (1984).

10. W. Stumm, J. Am. Water Works Assoc., **48**(3), 300 (1956).

11. F. E. DeMartini, J. Am. Water Works Assoc., **30**, 85 (1938).

12. C. P. Hoover, J. Am. Water Works Assoc., **30**, 1802 (1938).

13. D. T, Merrill and R. L. Sanks, J. Am. Water Works Assoc., **70**(1), 12 (1978).

14. J. W. Ryznar, J. Am. Water Works Assoc., **36**, 472 (1944).

15. P. R. Puckorius and J. M. Brooke, Corrosion, **47**, 280 (1991).

16. N. R. Reedy, Mater. Protect. Perform., **12**(4), 43 (1973).

17. M. E. Flentje, J. Am. Water Works Assoc., **53**, 1461 (1961).

18. T. Fujii, Bosei Kanri (Rust Prevention and Control), **27**, 85 (1983).

19. W. Whitman, R. Russell, and V. Altieri, Ind. Eng. Chem., **16**, 665 (1924).

20. R. V. Skold and T. E. Larson, Corrosion, **13**, 139t (1957).

21. R. A. Pisigan and I. E. Singley, J. Am. Water Works Assoc., **79**(2), 62 (1987).

22. T. Fujii, T, Kodama, and H. Baba, Boshoku Gijutsu (Corrosion Eng.), **31**, 637 (1982).

23. T. E. Larson and R. M. King, Corrosion, **10**, 110 (1954).

24. T. E. Larson and R. V. Skold, Corrosion, **14**, 285t (1958). see also R. V. Skold and T. E. Larson, Corrosion, **13**, 139t (1957).

25. M. Kowaka, M. Ayukawa, and H. Nagano, Sumitomo Kinzoku, **21**, 185 (1969).

26. I. Matsushima, in Corrosion Protection Handbook (Boshoku Gijutsu Binran), Japan Society of Corrosion Engineering, ed., Nikkan Kogyo Shinbunsha, Tokyo, Japan, 1985. pp. 176, 179.

27. K. Kinoshita, K. Ichikawa, and N. Kitajima, Boshoku Gijutsu (Corrosion Eng.), **32**, 31 (1983).

28. F. LaQue, in Corrosion Handbook, H. H. Uhlig (Ed.), Wiley, New York, 1948, p. 391.

29. K. Sakai, S. Morishita, T. Honda, A. Minato, M. Izumiya, K. Osumi, and M. Miki, Boshoku Gijutsu (Corrosion Eng.), **30**, 450 (1981).

30. I. Matsushima, paper presented at the Committee on Corrosion and Protection, The Society of Materials Science, Japan, **19**, 94 (1980); reproduced in Corrosion Protection Handbook (Boshoku Gijutsu Binran), Japan Society of Corrosion Engineering, Nikkan Kogyo Shinbunsha, Tokyo, Japan, 1985, p. 181.

31. K. Masamura and I. Matsushima, 23rd Annual Symposium on Corrosion and Its Protection, Japan Society of Corrosion Engineering, Tokyo, Nov. 1978, p. 104.

32. R. W. Revie and H. H. Uhlig, Corrosion and Corrosion Control, 4th ed., Wiley, Hoboken, N.J., 2008, p. 276.

33. F. N. Speller, in Corrosion Handbook, H. H. Uhlig (Ed), Wiley, New York, 1948, p. 496.

34. U. R. Evans, The Corrosion and Oxidation of Metals, Edward Arnold, London, 1960, p. 296.

35. K. Masamura and I. Matsushima, Paper No. 75, CORROSION/81, NACE, Houston TX, 1981.

36. C. Kato, Y. Otoguro, and K. Kado, Boshoku Gijutsu (Corrosion Eng.), **23**, 385 (1974).

37. C. Kato, Y. Otoguro, K. Kado, and Y. Hisamatsu, Corros. Sci., **18**, 61 (1978).

38. I. Matsushima, Low-Alloy Corrosion Resistant Steels, Chijin Shokan Co., Tokyo, Japan, 1995, p. 155.

39. I. Matsushima,"A Study on Localized Corrosion of Steel," Ph.D. Thesis, University of Tokyo, 1982.

40. I. Matsushima, Annual Meeting of Japan Society of Corrosion Engineering, Tokyo, May, 1979, p. 143; reproduced in Localized Corrosion, F. Hine, K. Komai, and K. Yamakawa (Eds.), Elsevier Applied Science, London and New York, 1988, p. 31.

45

CARBON STEEL—CORROSION BY SEAWATER

I. Matsushima*

Maebashi Institute of Technology, Maebashi, Japan

A. Introduction
B. Corrosion by continuous immersion in seawater
    B1. Environmental factors
    B2. Corrosion rate
    B3. Localized corrosion
C. Corrosion of pilings
D. Effect of velocity
References

A. INTRODUCTION

Typical steel structures that are exposed to marine environments are marine piles, offshore structures, vessels, and other structures immersed in seawater. A steel structure in a marine environment may be exposed to five different corrosive zones depending on the position of its parts relative to sea level. The five zones are the atmospheric, splash, tidal, submerged, and seabed-embedded zones. The corrosion characteristics and corrosion behavior of steel are different in each zone. Some kinds of equipment, such as machinery and piping systems that use seawater for industrial purposes, are also subject to seawater corrosion.

While well-established countermeasures are available against marine corrosion, such as application of coatings and cathodic protection, corrosion of bare steel in marine environments is important because some steel structures are used without protection and also because corrosion protection systems in marine environments are particularly susceptible to damage and deterioration, leaving considerable exposure time for corrosion before detection and repair.

*Deceased

B. CORROSION BY CONTINUOUS IMMERSION IN SEAWATER

B1. Environmental Factors

The major variables that affect the corrosion rate of steel are salinity, dissolved oxygen concentration, temperature, pH, carbonate, pollutants, and biological organisms. The characteristics of seawater with respect to these variables are summarized in Table 45.1. Reviews of the variability in seawater at different global locations have been published by Dexter and Culberson [1] and by Dexter [2].

Corrosion of steel in seawater is controlled by the rate of supply of dissolved oxygen to the steel surface, similar to corrosion in freshwaters. The rate of oxygen supply, in turn, is determined by the oxygen concentration in the bulk seawater, the degree of movement of seawater, the diffusion coefficient for oxygen in seawater, and characteristics of corrosion product films on the steel surface as a barrier to oxygen diffusion.

The concentration of oxygen in surface waters is usually near the equilibrium saturation value with the atmosphere, which varies inversely with the temperature and salinity of seawater. Because salinity variations in the surface water are relatively small and do not greatly affect oxygen solubility, the temperature is the major factor affecting oxygen concentration.

The temperature of the surface waters of oceans varies mainly with the latitude, and the range is from about $-2°C$ in the Arctic to $\sim35°C$ in the tropics. Accordingly, the equilibrium saturation concentration of oxygen varies from 11 to 6 ppm [1, 2]. The equilibrium concentration of oxygen in water is shown in Table 45.2 as a function of temperature and salinity.

The diffusion coefficient for oxygen, on the other hand, increases with increase of temperature, and the corrosion

Uhlig's Corrosion Handbook, Third Edition, Edited by R. Winston Revie
Copyright © 2011 John Wiley & Sons, Inc.

TABLE 45.1.   Environmental Factors in Seawater Corrosion[a]

Salinity	1. Open sea: Variation with horizontal location is small, 32–36 parts per thousand (ppt).
	2. Near river outlets: Lower.
	3. Variation with depth: Very small.
Dissolved oxygen concentration	1. Surface water: (1) Near the equilibrium saturation concentration with atmospheric oxygen at a given temperature [6 ppm (in the tropics), 11 ppm (in the Arctic)]. (2) Can be supersaturated due to photosynthesis by microscopic plants (up to 200%) and entrainment of air bubbles (up to ~10%).
	2. Variation with depth: (1) Tends to be undersaturated due to consumption by the biochemical oxidation of organic matter. (2) Goes through a minimum at intermediate depths (400–2400 m deep).
Temperature	1. Surface water: In the open ocean, variations are in the range of −2 to 35°C depending on the latitude, season, currents, and so on.
	2. Variation with depth: Drops with depth. The difference with depth and season may be large or small depending on the location.
pH	1. Surface water: (1) Lies between 7.5 and 8.3 in the open ocean depending on the concentration of dissolved CO_2 determined by air–sea exchange and photosynthesis activity of plants. (2) Microbiological activity affects the pH value; e.g., lower pHs by the formation of CO_2 through the process of biochemical oxidation and higher pH values by the reduction of CO_2 through the process of photosynthesis. (3) Affected by pollutants in the coastal waters.
	2. Variation with depth: Tends to show a profile similar to that of dissolved oxygen (the biochemical oxidation that consumes dissolved oxygen generates CO_2, reducing the pH value).
Carbonate	1. Surface water: Nearly always supersaturated with respect to $CaCO_3$ (200–500%) favored by high pH values and moderate temperatures.
	2. Variation with depth: Saturation state with respect to $CaCO_3$ decreases as the result of lower temperature and pH. Undersaturated in deep waters (e.g., below 200–300 m).
Pollutants	1. H_2S may be 50 ppm or higher in polluted waters in estuaries, harbors, river mouths, and fitting-out basins.
	2. Ammonia may be high in inshore waters and harbors.
Biological organisms	1. Bacteria form bacterial films (slime).
	2. Weeds grow from spores.
	3. Animals (e.g., barnacles, tube worms, and hydroids) adhere.

[a]Prepared based on [2].

rate of steel at a given oxygen concentration approximately doubles for every 30°C rise in temperature [3]. Because of the compensating effects of temperature with respect to the oxygen solubility and the diffusion coefficient, the rate of corrosion of steel in seawater is relatively independent of temperature.

The corrosion product film on the steel surface that more or less serves as a barrier to oxygen diffusion and decreases the corrosion rate of steel contains oxides and hydroxides of iron, and possibly iron sulfide, calcareous deposits, bacterial slime, and macroscopic marine growths. While the film is a physical barrier to oxygen diffusion, oxygen-utilizing bacteria in the film may provide a biochemical barrier by consuming all the oxygen diffusing through the film, resulting in anaerobic conditions at the metal surface. These conditions reduce corrosion by oxygen but may provide a condition where sulfate-reducing bacteria flourish and accelerate corrosion.

TABLE 45.2.   Equilibrium Saturation Concentration of Oxygen in Seawater as a Function of Temperature and Salinity[a]

Temperature (°C)	Oxygen Solubility (ppm) at Indicated Salinity (ppt)					
	0	8	16	24	31	36
0	14.6	13.9	13.1	12.4	11.8	11.4
5	12.5	11.9	11.3	10.7	10.2	9.9
10	10.9	10.4	9.8	9.4	8.9	8.7
15	9.5	9.1	8.7	8.2	7.9	7.7
20	8.5	8.5	7.7	7.3	7.0	6.8
25	7.6	7.2	6.9	6.6	6.3	6.2
30	6.8	6.5	6.2	6.0	5.7	5.6

[a]Calculated based on data from [2].

In spite of wide variations of environmental factors affecting the rate of oxygen supply, differences in the corrosion rate of steel under conditions of continuous immersion in seawater at different locations throughout the world are relatively small. LaQue [4] pointed out that this similarity of corrosion rates in seawater everywhere occurs because the controlling factors change in a compensating way. For example, decrease of oxygen concentration by higher water temperatures is compensated by a large diffusion coefficient. High water temperature, which tends to promote higher reaction rates, also enhances the development of protective calcareous deposits and marine growths, which stifle attack [4]. The action of bacteria that grow beneath the original corrosion product film may simply be to take the place of the excluded oxygen in enabling corrosion reactions to continue at the rates commonly observed [4].

Of course, there are some cases where the corrosion rate is exceptionally high or low. For example, the corrosion rate may be very low because of complete coverage of the surface by macroscopic fouling organisms. Significant salinity variations, caused by river discharge or high evaporation, affect oxygen concentration in seawater and hence the overall corrosion rate. Pollutants in inshore waters (e.g., organic material, hydrogen sulfide, and ammonia) affect corrosion by changing the oxygen concentration and/or the nature of corrosion product films as a barrier to oxygen diffusion.

Intense photosynthesis by macroscopic marine plants growing in the surface layers of the sea can produce high supersaturation of oxygen in the surface waters in the daylight hours. On the other hand, oxygen concentration of seawater below the surface tends to be lower due to consumption by the biochemical oxidation of organic matter. The vertical profile of oxygen concentration depends on the location but commonly goes through a minimum at intermediate depths. The depth of the oxygen minimum ranges from 400 m in the equatorial eastern Pacific to over 2400 m in the central Pacific and the concentration at the depth of the minimum ranges from 0.16 to 6.4 ppm [2].

The pH of open ocean seawater ranges from ~7.5 to 8.3 and has no direct effect on the corrosion rate of steel, but its variation affects the degree of saturation with respect to $CaCO_3$. Spontaneous calcareous deposition is more likely to occur at higher pH values, but the function of the deposits as a barrier to oxygen diffusion is limited because of the presence of organic matter and high salinity of seawater, although enhanced buildup of calcareous deposits under cathodic protection helps to reduce the required protection current, lowering the power consumption.

B2. Corrosion Rate

Overall corrosion rates of steel continuously immersed in quiescent seawater at many locations throughout the world for periods from < 1 year to 40 years collected from various literature [4–14] are plotted in Figure 45.1. The corrosion rates range from 0.02 to 0.37 mm/year (0.8–14.6 mpy), the average rate being ~0.1 mm/year (4 mpy).

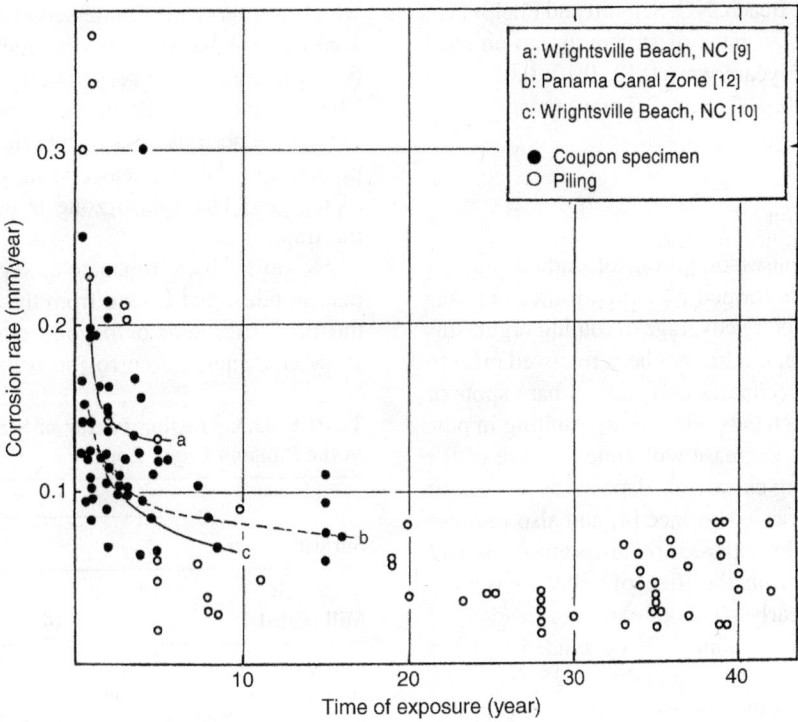

FIGURE 45.1. Average corrosion rates of steel continuously immersed in seawater.

The average of the corrosion rates in Figure 45.1 in the first 5 years is 0.14 mm/year (5.5 mpy) being very close to 0.125 mm/year (5 mpy), which is the value most commonly used as the expected average rate of corrosion of steel continuously immersed in seawater under natural conditions [4, 15]. The average corrosion rates for periods of >5 to 10 years, >10 to 20 years, and >20 years are 0.07, 0.07, and 0.05 mm/year (2.8, 2.8, and 2.0 mpy), respectively, indicating that the corrosion rate decreases with time.

At one time, the corrosion rate of steel in seawater was considered to be linear with time [4]. Larrabee [16] reported in 1962 that the average corrosion rate of submerged sections of 20 H-piles of a steel pier near Santa Barbara, CA, used for offshore oil wells for 23.5 years was ~0.038 mm/year (1.5 mpy) and concluded that the corrosion rate of carbon steel in seawater averages 0.05 mm/year (2 mpy) for the first 20 years, then drops to 0.025 mm/year (1 mpy).

The decrease of corrosion rate with time has been more clearly shown in corrosion tests in which steel was continuously immersed in seawater at a specific location for different periods of time. Southwell and Alexander [12], who conducted an extensive corrosion test in the Panama Canal Zone, showed that the overall corrosion rates [(average penetration)/(time of exposure)] of structural steel in seawater in 1, 2, 4, 8, and 16 years were 0.14, 0.11, 0.094, 0.084, and 0.077 mm/year (5.5, 4.3, 3.7, 3.3, and 3.0 mpy), respectively (curve b in Fig. 45.1). The corrosion rate was 0.15 mm/year (5.9 mpy) after 1 year of exposure and 0.068 mm/year (2.7 mpy) after 16 years of exposure. From the results of corrosion tests of structural carbon steels in seawater at Wrightsville Beach, NC, Schmitt and Phelps [10] reported that the corrosion rates of structural carbon steel in 1.5, 2.5, 4.5, and 8.5 years were 0.12, 0.10, 0.083, and 0.068 mm/year (4.7, 4.0, 3.3, and 2.7 mpy), respectively (curve c in Fig. 45.1).

B3. Localized Corrosion

One of the common causes of pitting of carbon steel is differential aeration cells formed by nonuniformity of corrosion product films or spotty coverage of fouling organisms on the surface. Mill scale, if it has not been removed prior to exposure to seawater, accelerates corrosion at bare spots or breaks in the scale through galvanic action, resulting in pits. Galvanic action tends to decrease with time because of the development of insulating calcareous deposits as a result of cathodic reaction at the scaled surface [4] and also because mill scale is undermined by corrosion products that cause it to be spalled off. For this reason, the effect of mill scale is more pronounced during the early stages of exposure.

From a literature survey, Fink [15] concluded that the pitting attack on bare steel is frequently ~0.25–0.38 mm/year (10–15 mpy) and that the presence of mill scale significantly increases the rate of pitting, the penetration rate being ~0.5 mm/year (20 mpy). Assuming the average corrosion rate of 0.125 mm/year (5 mpy), the pitting factor (the ratio of deepest metal penetration to average metal penetration) is 2–3 for bare steel and 4 for mill-scaled steel. According to LaQue [4], the pitting rate decreases with time, and a normal pitting factor for exposure of ~10 years would be ~2.5 for descaled steel and 3.5 for steel exposed with mill scale. He also stated that, as a general rule, steel exposed with mill scale will be pitted about three times as deeply as descaled steel in a short period of exposure and that this ratio decreases as the exposure is prolonged to become 1.5–1 for a 10-year-exposure period.

The values of the pitting factor found in 15-year tests in seawater at Halifax (N.S., Canada), Auckland (New Zealand), Plymouth (England), and Colombo (Sri Lanka) by the Committee of the Institution of Civil Engineers (London) [17] were 2.0–2.9 for pickled specimens and 2.1–5.6 for specimens with mill scale, in fair agreement with the values indicated by Fink and LaQue.

The pitting factor based on the depth of the deepest pit and the average of the 20 deepest pits found by Southwell et al. [11, 12] in the tropical sea of the Panama Canal Zone for machined and mill-scaled structural steel are summarized in Table 45.3. In these data, the pitting factors are much greater than those summarized by Fink and LaQue.

C. CORROSION OF PILINGS

The corrosion rates in the region between high and low tide, which is alternately immersed and not immersed (the tidal zone), and in the area above it which receives seawater spray (the splash zone) are very high because of constant wetting. The corrosion attack in the tidal zone occurs primarily during the periods of atmospheric exposure because of the presence of the thin seawater film and an abundant supply of oxygen. The splash zone is under such conditions all the time.

However, bare (uncoated) vertical members, such as marine piles, that extend from the submerged zone through the tidal zone into or beyond the splash zone frequently show characteristic corrosion behavior as first reported by

TABLE 45.3. Pitting Factor of Structural Steel in Seawater in the Panama Canal Zone[a]

Surface Finish	Exposure (year)				
	1	2	4	8	16
Machined	12	17	8	6	3[b]
Mill scaled	25	14	10	11	–[c]

[a]Based on data from [11, 12].

[b]Pickled specimen. Pit perforated in machined specimen. The pitting factor for cast steel, which corroded more or less the same as carbon steel, is 3.

[c]Perforated.

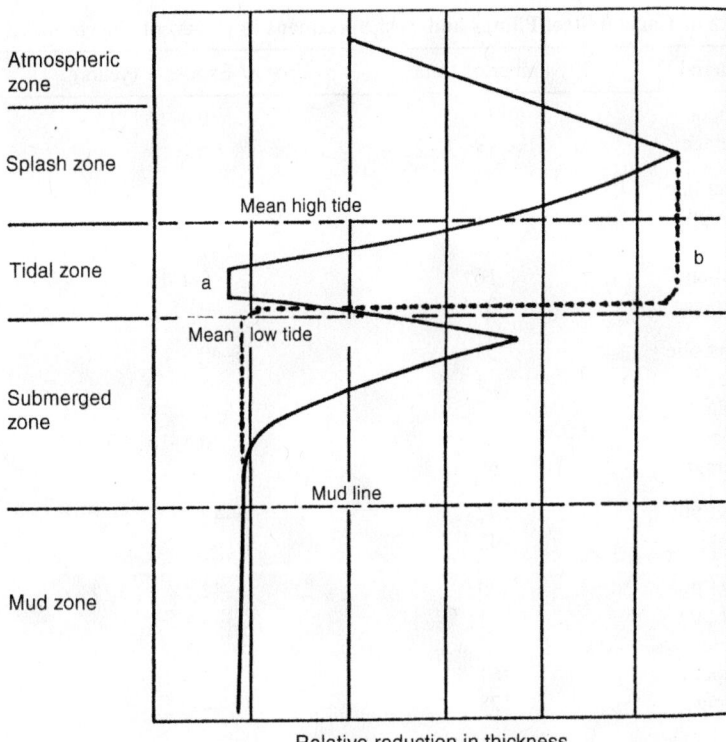

Relative reduction in thickness

FIGURE 45.2. Corrosion profile of steel piling in seawater. (a) Uncoated piling. (Reproduced with permission from [5]. Copyright © NACE International.) (b) At coating defect of piling coated with an electrically insulating substance.

Humble [5] in 1949. Relative corrosion rates of a bare marine pile exposed to the atmospheric, splash, tidal, submerged, and mud (seabed) zones are shown by line a in Figure 45.2. Of these zones, corrosion at the splash zone is the severest, as expected, but corrosion in the tidal zone is relatively mild.

It is generally believed that the low corrosion rate in the tidal zone is the result of galvanic protection by the part near the top of the submerged zone that shows a corrosion peak [5]. The part in the tidal zone acts as the cathode probably because the corrosion products (iron oxides) are oxidized to higher oxidation states during the periods of exposure to the atmosphere, resulting in a more noble corrosion potential. Then, when the part in the tidal zone is submerged during the periods of high tide, it acts as the cathode, with the reduction of the oxides on its surface [18]. If the pile is coated with electrically insulating substances, such as organic coatings, the exposed steel at coating defects in the tidal zone corrodes just as severely as in the splash zone [line b in Fig. 45.2] [19].

High corrosion rates near the top of the submerged zone are not always observed. Zen [14] investigated corrosion of piles used at 43 ports throughout Japan for times of up to 40 years and classified the vertical corrosion profile of bare steel marine pilings (sheet and pipe piles) below the low water level into five types. These types may be rearranged to the following three types:

Type 1 Sharp corrosion peak within several tenths of a meter below the low water level (corrosion rate 0.5–1.0 mm/year, 20–40 mpy) and mild corrosion below this level (corrosion rate $\leq$ 0.1 mm/year, $\leq$ 4 mpy)

Type 2 Corrosion rate highest just below the low water level (0.1–0.2 mm/year, 4–8 mpy), gradually decreasing with depth

Type 3 Low and almost flat corrosion rate below the low water level ($\leq$ 0.1 mm/year, $\leq$ 4 mpy)

Type 1, corresponding to the corrosion profile of line a in Figure 45.2, usually applied to pilings in shallow water, $\leq$ 3 m deep, with the lowest level of concrete coverage staying above the mean water level. When the pilings were covered with concrete below this level or the water depth was $\geq$ 5 m, the sharp corrosion peak at the top of the submerged zone was not found, resulting in the corrosion profile of type 3. At locations where the difference in water level of high and low tides was small, the sharp corrosion peak did not occur. The presence of a freshwater layer at the water surface caused by river discharge, on the other hand, caused the corrosion peak to occur even in pilings in deep water. The corrosion profile of type 2 was intermediate between types 1 and 3.

TABLE 45.4. Corrosion Rates of Carbon Steel Pilings and Test Specimens in Different Corrosion Zones[a,b]

Vertical Position	Material	Number of Data	Time of Exposure (years)	Corrosion Rate (mm/year)
Atmospheric zone	Coupon	19	0.4–16	0.128
	Average	19	—	0.128
Splash zone	Sheet pile	8	6–40	0.112
	Pipe pile	1	8	0.25
	H-pile	2	5–7	0.198
	Coupon	16	0.4–15	0.363
	Average	27	—	0.272
Tidal zone	Sheet pile	35	5–40	0.044
	Pipe pile	4	3–8.5	0.070
	H-pile	2	5–7	0.055
	Coupon	27	0.4–16	0.137
	Average	68	—	0.083
Low water level	Sheet pile	42	5–42	0.047
	Average	42	—	0.047
Immersed zone	Sheet pile	59	5–42	0.039
	Pipe pile	5	3–8.5	0.062
	H-pile	3	5–23.6	0.049
	Coupon	61	0.3–16	0.143
	Average	128	—	0.090
Mud zone	H-pile	2	5–7	0.033
	Coupon	3	3–5	0.103
	Average	5	—	0.075

[a]From [13].
[b]Averages of corrosion rates compiled from literature.

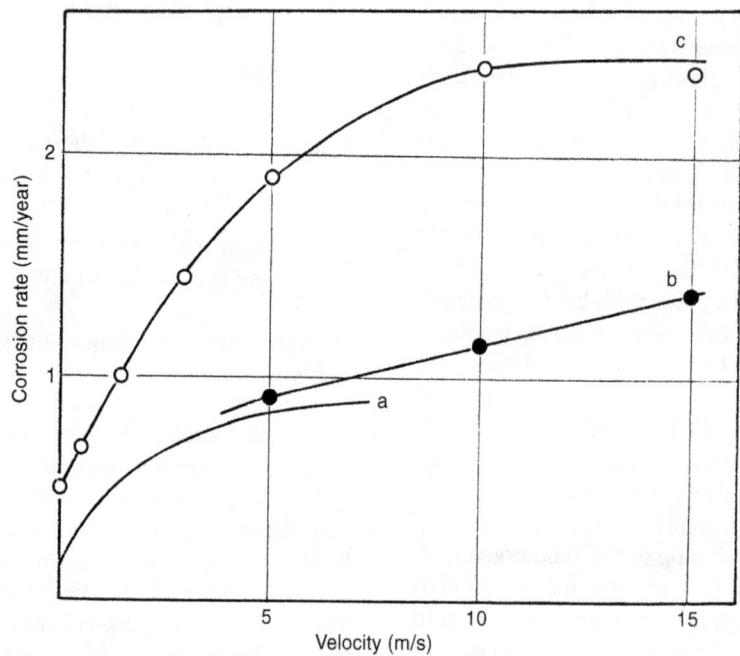

FIGURE 45.3. Effect of velocity on corrosion of steel and cast iron in seawater. (a) Carbon steel tested for 36 days at ~23°C [4]. (b) Carbon steel tested for 30 days at ambient temperature [20]. (c) Cast iron tested for 7 days at 25°C [21].

Corrosion rates of carbon steel pilings and test specimens in different corrosion zones, compiled by a committee of The Kozai Club (organization composed of major steel manufacturers and traders in Japan) [13] from the literature, are listed in Table 45.4. Typical corrosion rates of bare pilings are 0.1–0.2 mm/year (4–8 mpy) in the atmospheric zone, 0.3–0.5 mm/year (12–20 mpy) in the splash zone, 0.1 mm/year (4 mpy) in the tidal zone, 0.5–1 mm/year (20–40 mpy) at the top of the submerged zone (down to 1 m deep) when the corrosion peak appears [otherwise 0.1 mm/year (4 mpy)], 0.1 mm/year (4 mpy) in the submerged zone $\leq$ 1 m, and 0.05 mm/year (2 mpy) in the mud zone. The pitting factor in the tidal and splash zones is in the range of 2–3.

D. EFFECT OF VELOCITY

The corrosion rate of steel by seawater increases with increase of velocity until a critical velocity is reached, beyond which there is little further increase in corrosion rate. The trend of the velocity effect is shown in Figure 45.3 [4, 20, 21]. This behavior is different from that in freshwaters, where corrosion decreases above a critical velocity by passivation (see Sections A and B3, in Chapter 44). In seawater, passivity is not established at any velocity because of the high concentration of Cl^-. The limiting rate of corrosion corresponds to the maximum rate of oxygen supply by diffusion.

When the velocity increases above a critical value, corrosion of steel markedly increases because of erosion–corrosion, even in the absence of solid particles. The critical velocity depends on the state of flow, but it is said to be ~20 m/s [20]. At areas where flow is disturbed (e.g., bends and joints in piping), erosion–corrosion may occur at much lower velocities. The expected maximum corrosion rate of carbon steel piping under velocities up to 4 m/s is ~1 mm/year (40 mpy), but above this velocity, erosion–corrosion would occur at areas of flow disturbance.

REFERENCES

1. S. C. Dexter and C. H. Culberson, Mater. Perform., **19**(9), 16 (1980).
2. S. C. Dexter, in ASM Handbook, Vol. 13, American Society for Metals, Metals Park, OH, 1987, p. 893.
3. G. Skaperdas and H. Uhlig, Ind. Eng. Chem., **34**, 748 (1942).
4. F. L. LaQue, in Corrosion Handbook, H. H. Uhlig (Ed.), Wiley, New York, 1948, p. 383.
5. H. A. Humble, Corrosion, **5**, 292 (1949).
6. J. C. Hudson, J. Iron Steel Inst., **166**, 123 (1950).
7. J. C. Hudson and J. F. Stanners, J. Iron Steel Inst., **180**, 271 (1955).
8. C. P. Larrabee, Proc. ASTM, **44**, 161 (1944).
9. C. P. Larrabee, Corrosion, **14**, 501t (1958).
10. R. J. Schmitt and E. H. Phelps, J. Metals, **22**(3), 47 (1970).
11. C. R. Southwell, B. W. Forgeson, and A. L. Alexander, Corrosion, **16**, 512t (1960).
12. C. R. Southwell and A. L. Alexander, Mater. Prot., **9**(1), 179 (1970).
13. Committee on Corrosion of Marine Structures, Kozai Club, Tokyo, Japan, Report on Corrosion of Marine Structures, 1974.
14. K. Zen, Submerged Marine Structures—Corrosion Survey and Corrosion Protection (in Japanese), Kajima Shuppan Kai, Tokyo, Japan, 1974. Also see Boshoku Gijutsu (Corros. Eng.), **17**, 103 (1968), **18**, 194 (1969), **20**, 414 (1971), **20**, 453 (1971), **21**, 466 (1972), **22**, 55 (1973), **22**, 428 (1973).
15. F. W. Fink, Corrosion of Metals in Sea Water, PB 171344, Battelle Memorial Institute, Columbus, OH, 1960.
16. C. P. Larrabee, Meter. Prot., **1**(12), 95 (1962).
17. Deterioration of Structures in Sea Water, 18th Report of the Committee of the Institution of Civil Engineers, London, 1938; cited in [4].
18. I. Matsushima, Low-Alloy Corrosion Resistant Steels, Chijin Shokan Co., Tokyo, Japan, 1995, p. 122.
19. C. P. Larrabee and F. L. LaQue, the discussion to [5].
20. K. Ichikawa, K. Nagano, S. Kobayashi, and N. Kitajima, Ebara Eng. Rev., No. 85, **2** (1973).
21. M. Miyasaka and N. Takahashi, private communication.

46

CARBON STEEL—CORROSION BY SOILS

I. Matsushima*

Maebashi Institute of Technology, Maebashi, Japan

A. INTRODUCTION

Typical steel structures used underground in contact with the soil are pipelines, utility pipings, and pilings. While most pipelines are protected against corrosion with coatings and cathodic protection, driven pilings are commonly used in the bare condition.

Corrosion rates of steel in soils vary to a marked degree with the kind or type of soil, being affected by many environmental factors such as soil composition, pH, moisture, and so on. Corrosion proceeds basically by the action of water and oxygen, as in water and in the atmosphere, but localized corrosion or pitting is more likely to occur because of nonhomogeneity of the surrounding soils and nonuniform contact of the metal with soils. These effects are more pronounced in disturbed soils than in undisturbed ones, so that pitting in buried pipes is more severe than in pilings driven into undisturbed natural soils.

Soils are not severely corrosive environments, and the overall corrosion rates of buried steel are normally far less than 0.1 mm/year (4 mpy). However, localized corrosion caused by macrogalvanic cells, due to differential aeration and other causes that develop potential differences between different parts of a structure, may increase the rate of penetration by an order of magnitude.

B. DISTURBED SOILS

B1. Overall Corrosion

Major factors that govern corrosivity of a given soil are porosity (aeration), electrical conductivity or resistivity, dissolved salts, moisture, and acidity, or alkalinity [1]. There are mutual relationships among these factors (e.g., a porous soil may retain more moisture and a soil with a high dissolved salt content has a high conductivity). The same factor may accelerate or retard corrosion, for example, a porous and hence well-aerated and moist soil tends to increase the initial corrosion, but the corrosion product films formed in a well-aerated soil may be more protective than those in an unaerated soil, reducing corrosion, particularly pitting, in the long term.

An extensive series of field tests on various metals was conducted by the National Bureau of Standards (NBS; now National Institute of Standards & Technology, NIST) starting in 1910 at many locations in the United States and included almost all types of soils. The results of the tests [2] on 6-in.-(152-mm-) long, 3-in.-(76-mm-) diameter open-hearth steel pipe tested for ~12 years from 1922 at 44 locations are summarized in Table 46.1 and Figure 46.1. The overall corrosion rates range from 0.003 to 0.063 mm/year (0.1 to 2.5 mpy), the average being 0.02 mm/year (0.8 mpy). With respect to the overall corrosion rates, corrosion by soils is relatively mild.

The data on the effects of environmental factors, that is, soil resistivity, internal drainage, and air–pore space, shown

*Deceased.

Uhlig's Corrosion Handbook, Third Edition, Edited by R. Winston Revie
Copyright © 2011 John Wiley & Sons, Inc.

TABLE 46.1. Summary of NBS Field Test Results on Open-Hearth Steel Tested for 12 Years at 44 Locations in the United States[a]

	Overall Corrosion Rate (mm/year)	
Maximum	0.063	Merced silt loam, Buttonwillow, CA.
Minimum	0.003	Everett gravelly sandy loam, Seattle, WA
Average	0.020	44 locations
	Pitting Rate (mm/year)	
Maximum	>0.45[b]	Muck, New Orleans, LA
Minimum	0.033	Everett gravelly sandy loam, Seattle, WA
Average	0.143	44 locations

[a]Original data based on [2].
[b]Perforated.

in Table 46.2 are based on the NBS tests. No single factor controls the overall corrosion rate.

From resistivity measurements of soils of known corrosivity, it was found quantitatively that corrosivity is higher for soils of lower resistivity [3, 4], suggesting that soil resistivity can be used as a rough index of corrosivity of soils. It should be noted, however, that the corrosivity was based on field experience with oil and gas pipelines [3] or gas- and water-distributing systems [4], where localized corrosion was supposedly the main concern. As indicated in Table 46.2, soil resistivity is not a criterion of corrosivity of soils with respect to overall corrosion.

Tables to estimate corrosivity of soils are given in a German Industrial Standard (DIN 50929, Teil 3). One of the tables lists rating indices for each of 12 factors, and by adding the relevant indices and by referring to other tables, the corrosion probability of a given soil can be estimated. The 12 factors include kind of soil (e.g., clay content), resistivity, water content, pH, acidity and alkalinity, sulfide content, and Cl^- and SO_4^{2-} contents.

B2. Localized Corrosion

Localized corrosion, usually in the form of pits, tends to be severe in disturbed soils, and particularly in pipelines, which pass through soils that are different from place to place along the route, thereby enhancing the possibility of the establishment of macrogalvanic cells.

Major causes of the formation of macrogalvanic cells in undergound pipelines are illustrated in Figure 46.2. Pitting, rather than general reduction of thickness, occurs in the anodic areas because corrosion current tends to leave the anodic areas at discrete points of low pipe-to-soil resistance caused by nonuniform contact of the metal with the soil or low resistivity of the soil at localized areas (see Chapter 47, Section B5).

As shown in Table 46.1 and Figure 46.1, the specimens used in the NBS tests were severely pitted, the highest penetration rate being >0.45 mm/year (>18 mpy). The pitting factors ranged from 3.0 to 22.4 (average: 8.5; ~80% of the data points are between ~5 and 10), which are much higher than in waters. These high pitting factors were apparently caused by the formation of differential aeration

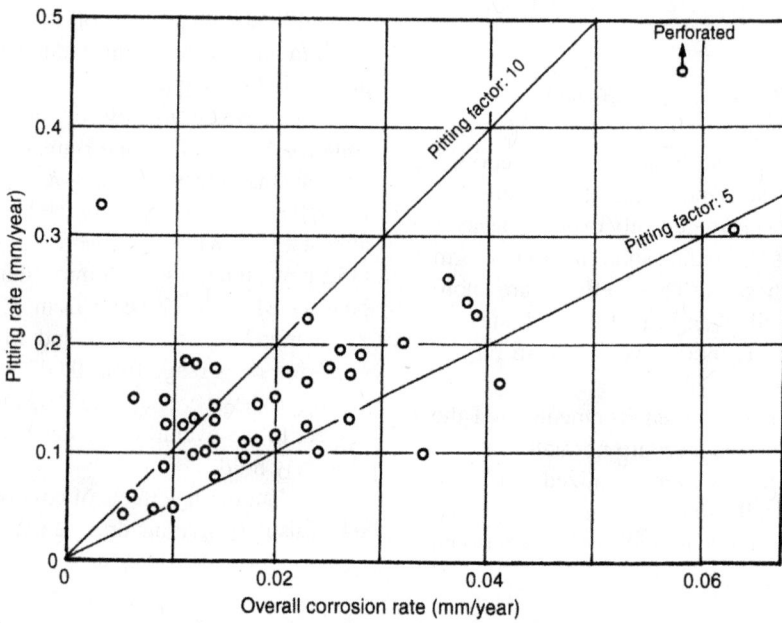

FIGURE 46.1. Overall corrosion rates and pitting rates of open-hearth steel exposed to 44 soils for 12 years. Original data are based on [2].

TABLE 46.2.   Effects of Environmental Factors on Corrosion of Steel in Soils[a]

Environmental Factor	Overall Corrosion Rate (mm/year)			Maximum Pitting Rate (mm/year)		
	Maximum	Minimum	Average	Maximum	Minimum	Average
Resistivity (Ω·cm)						
<1000	0.063	0.018	0.033	0.31	0.11	0.20
1000–5000	0.058	0.006	0.017	>0.45[b]	0.05	0.14
5000–12000	0.033	0.005	0.018	0.23	0.06	0.14
>12000	0.036	0.003	0.014	0.26	0.03	0.11
Drainage						
Very poor	0.058	0.038	0.046	>0.45[b]	0.16	0.28
Poor	0.037	0.010	0.024	0.23	0.05	0.14
Fair	0.063	0.018	0.022	0.31	0.08	0.16
Good	0.022	0.003	0.010	0.18	0.03	0.11
Air–pore space (%)						
<5	0.033	0.010	0.021	0.20	0.05	0.13
5–10	0.063	0.009	0.024	0.31	0.10	0.17
10–20	0.037	0.006	0.017	0.26	0.05	0.15
20–30	0.058	0.012	0.025	>0.45[b]	0.10	0.20
>30	0.038	0.004	0.013	0.23	0.03	0.09

[a]Original data are based on NBS field tests [2] on open-hearth steel for 12 years at 44 locations in the United States.
[b]Perforated.

cells, as suggested by Romanoff [5]. Pitting by differential aeration in operating pipelines would be more severe than in the tests because larger cathodic areas would be available than in the 6-in.-long specimens used in the tests.

The rate at which pits grow in the soil under a given set of conditions tends to decrease with time, as reported by Romanoff [2]. The pit depth–time curves conform to the equation $P = kt^n$, where P is the depth of the deepest pit at time t and k and n are constants that depend on the characteristics of the soil. Examples of the pit depth–time curves obtained in the NBS tests for ferrous metals are reproduced in Figure 46.3 [2].

The most damaging macrogalvanic corrosion cells are formed where there is contact between a pipeline and one or more foreign metallic structures when the pipeline and the structures are not isolated from each other (case A in Fig. 46.2). The damage is catastrophic when buried piping is in contact with reinforcing bars (rebars) in concrete foundations or walls. Being in the alkaline environment of concrete, rebars are passivated and thus have a noble potential constituting a cathode of very large surface area. The penetration rates observed in buried utility piping are shown in Figure 46.4 [6]. A rate of 1 mm/year (40 mpy) is common and the rate reaches over 3 mm/year (120 mpy), being an order of magnitude larger than those found in the NBS tests (see Fig. 46.1).

Pipelines may suffer bacterial corrosion and stray current electrolysis, which are treated elsewhere in this book.

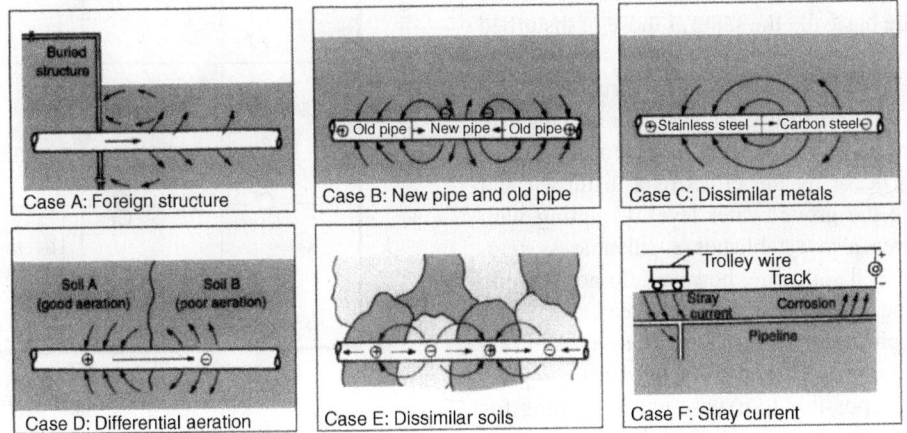

FIGURE 46.2. Major causes of the formation of macrogalvanic cells in underground pipelines.

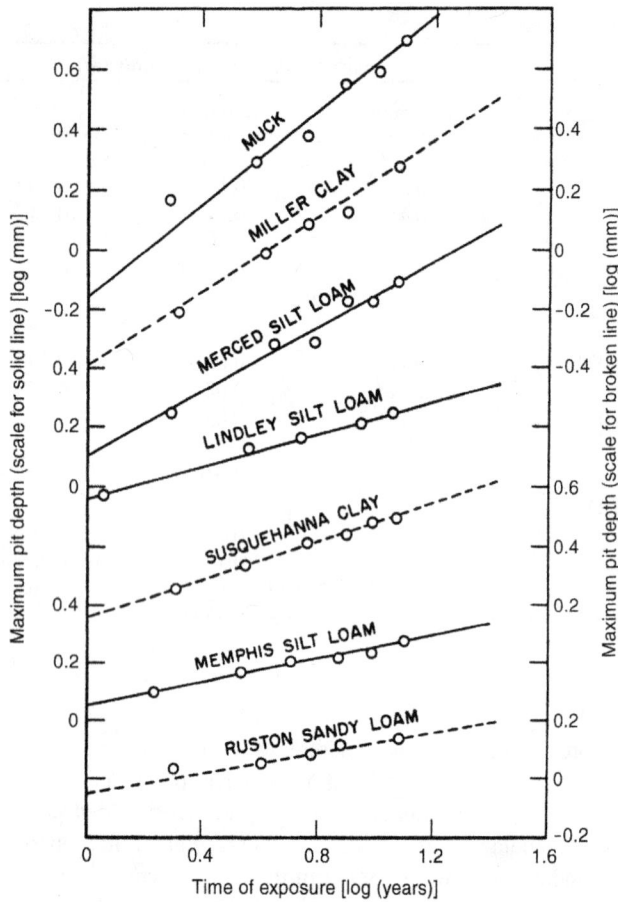

FIGURE 46.3. Pit depth–time curves for ferrous metals exposed to soils of different aeration [2]. (Reproduced courtesy of National Institute of Standards and Technology, Technology Administration, U.S. Department of Commerce.)

C. UNDISTURBED SOILS

Undisturbed soil is the environment of steel piles driven below the water table. Factors affecting the corrosivity of undisturbed soils are basically the same as those of disturbed soils.

When a pile is isolated from subsurface structures, corrosion tends to be low regardless of the nature of the soil because of low availability of oxygen. Corrosion by differential aeration can occur by the effect of a disturbed and aerated zone above the groundwater level. Coupling with subsurface structures also establishes a galvanic system. Piles at the undisturbed soil zone, however, do not corrode severely because the corrosion current is spread over large surface areas. The pitting tendency is less than in disturbed soils because of the uniform metal-to-soil contact.

It is difficult or impossible to expose existing piling for examination, and as a result, corrosion data are scarce. Figure 46.5 has been compiled from limited reported

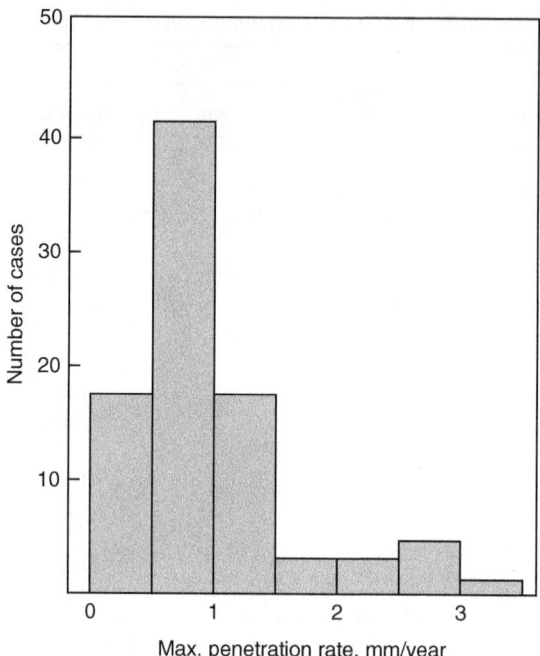

FIGURE 46.4. Maximum pitting rates of buried utility pipings [6].

values [7–10]. The highest rate of penetration for the zone above the water table is 0.37 mm/year (15 mpy), whereas below the water table, the rate is only 0.12 mm/year (4.7 mpy).

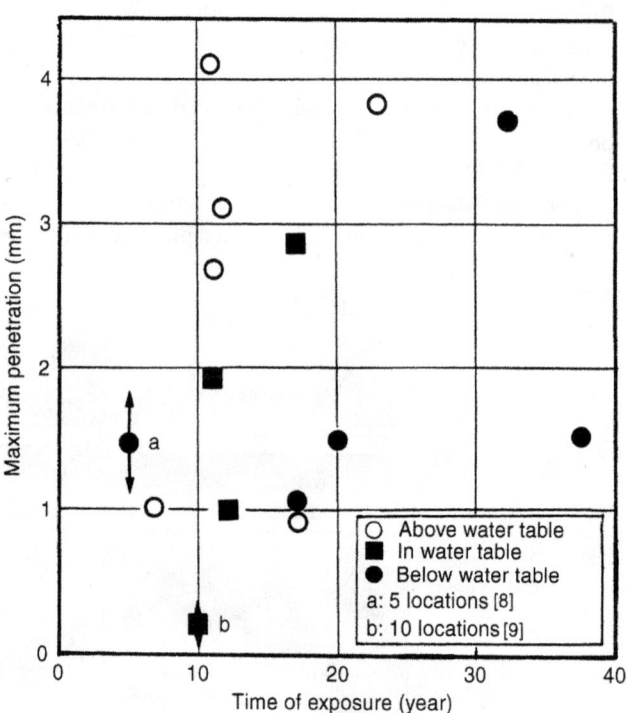

FIGURE 46.5. Maximum penetration of driven steel pilings. Compiled from [7–10].

Note: A source list of 1172 English language journal articles on underground corrosion of steel and other metals as well as its causes and prevention was published in 10 parts by Williams [11].

REFERENCES

1. R. W. Revie and H. H. Uhlig, Corrosion and Corrosion Control, 4th ed., Wiley, Hoboken, N.J., 2008, p. 206.

2. M. Romanoff, Underground Corrosion, National Bureau of Standards Circular 579, National Bureau of Standards, Gaithersburg, MD, 1957. See also: Richard E. Ricker, Analysis of Pipeline Steel Corrosion Data from NBS (NIST) Studies Conducted between 1922–1940 and Relevance to Pipeline Management, NISTIR 7415, National Institute of Standards and Technology, U.S. Department of Commerce, Gaithersburg, MD, May 2, 2007; and J. Res. Natl. Inst. Stand. Technol., **115**(5), 373 (September–October 2010).

3. F. O. Waters, Corrosion, **8**, 407 (1952).

4. G. H. Booth, A. W. Cooper, P. M. Cooper, and D. S. Wakerley, Br. Corros. J., **2**, 104 (1967).

5. M. Romanoff, Underground Corrosion, National Bureau of Standards Circular 579, National Bureau of Standards, Gaithersburg, MD, 1957, p. 11.

6. I. Matsushima, Koatsu Gas, **19**, 557 (1982); cited in Corrosion and Corrosion Protection Databook, I. Matsushima (Ed.), Japan Society of Corrosion Engineering, Tokyo, Japan, 1995, p. 143.

7. M. Romanoff, Corrosion of Steel Pilings in Soils, National Bureau of Standards Monograph 58, National Bureau of Standards, Gaitnersburg, MD, 1962.

8. Tetudo-Denka-Kyokai, A Study on Long-term Corrosion of Steel by Soils, 10th Report, Tokyo, Japan, 1980, p. 11; cited in Corrosion and Corrosion Protection Databook, I. Matsushima (Ed.), Japan Society of Corrosion Engineering, Tokyo, Japan, 1995, p. 31.

9. Y. Ohsaki, Corrosion of Steel Pilings, Kozai Club, Tokyo, Japan, 1980.

10. Y. Mori, Doboku-Gijutsu-Shiryo, **23**(6), 33 (1981); cited in Corrosion and Corrosion Protection Databook, I. Matsushima (Ed.), Japan Society of Corrosion Engineering, Tokyo, Japan, 1995, p. 33.

11. J. Williams, Mater. Perform., **21**(1), 40; (2), 27; (3), 23; (4), 25; (5), 23; (6), 9; (7), 40; (8), 37; (9), 39; (10), 52 (1982).

47

LOCALIZED CORROSION OF IRON AND STEEL

I. MATSUSHIMA*

Maebashi Institute of Technology, Maebashi, Japan

A. GENERAL CHARACTERISTICS

Localized corrosion of iron and steel, typically pitting, usually occurs by the action of macrogalvanic cells. In a macrogalvanic cell, the anodic and cathodic areas are macroscopic, and their locations are fixed, whereas in a microgalvanic cell the anodic and cathodic sites are microscopic and their locations change randomly with time. Relatively small fixed anodic areas surrounded by or connected to relatively large cathodic areas undergo corrosion. Unlike pitting of passive metals, such as stainless steels, where localized breakdown of passivity and the resulting formation of passive–active cells of large potential difference cause deep pits, localized corrosion of iron and steel tends to be shallow in most cases, as the potential difference of various macrogalvanic cells is not as large.

In certain situations, the anodic area of macrogalvanic cells corrodes more or less evenly, resulting in localized corrosion, but in some other situations, specific sites in an anodic area corrode preferentially due to inhomogeneity of the metal surface and/or the environment to which the anodic area is exposed. For example, the anodic areas of differential aeration cells in steel pipe buried in the ground usually form discrete pits at localized sites where the metal/soil contact resistance is lower than at the rest of the anodic areas.

Localized corrosion of iron and steel may occur by the severe corrosive action of the environment concentrated at specific sites without the formation of macrogalvanic cells. The formation of deep pits in cylinder liners (coolant side) of diesel engines by cavitation–erosion under corrosive action is a typical example [1].

B. MACROGALVANIC CELLS

B1. Types of Macrogalvanic Cells

The macrogalvanic cells that frequently cause localized corrosion in steel in service are commonly formed by bimetallic contact, an inhomogeneous steel matrix (typically at welded joints), discontinuous surface films, differential aeration, and differential pH caused by combinations of an alkaline (pH > ~10) and a near-neutral (pH in the range of 5–9) environment that leads to the formation of passive–active cells. An alkaline environment is typically provided by mortar or concrete.

B2. Bimetallic Contact

Bimetallic corrosion of steel that occurs when the steel is coupled to a more noble metal usually takes the form of

*Deceased.

general attack over the steel surface. If the steel in bimetallic contact with a noble metal is coated with organic materials, such as paint, localized corrosion, commonly in the form of severe pitting, occurs at holidays (i.e., defects) in the coating because of the large cathode–anode area ratio.

In one case, 350-mm-diameter sewage piping of 7.9 mm wall thickness, internally coated with coal tar epoxy resin, was perforated in 7 months, and the penetration rate was as high as 13.5 mm/year (530 mdd) [2]. The internal surface of the piping at each butt-welded joint was lined with 400-mm-wide type 304 stainless steel instead of the resin paint coating that would be damaged by welding. The localized attack apparently proceeded at a coating holiday. Thus, the cathode–anode area ratio was extremely large. In addition, the specific resistance of the sewage was low, $60\,\Omega \cdot$ cm, allowing the flow of large galvanic current.

Bimetallic contact tends to accelerate localized attack caused by some other types of macrogalvanic cells. For example, localized corrosion at welds proceeds more rapidly when a more noble metal is coupled, because the resulting shift of the electrode potential of the anodic welds in the noble direction causes a higher corrosion rate of the welds and much deeper localized attack.

B3. Localized Corrosion at Welded Joints

When the weld metal and/or the heat-affected zone (HAZ) are less noble than the steel (parent metal), the former corrodes selectively. Since both are ferrous metals, the difference in the open-circuit potential between the anode and the cathode would not be large, most often a few hundredths of a volt, but the penetration rate could be high because a small shift of the potential of the anode in the noble direction causes a large increase of the dissolution rate due to the small Tafel slope of the anodic polarization curve for steel (see Section D in this chapter).

As mentioned in Section C3 in Chapter 44, the welded seam of electric resistance welded (ERW) pipe is susceptible to groove corrosion. With a potential difference of ~30 mV, the maximum penetration rate could be as high as 10 mm/year (400 mpy), a rate of 1–3 mm/year (40–120 mpy) being quite common. In ERW pipes developed to resist groove corrosion, the potential difference is practically nil [3].

In welded joints, the weld metal and/or HAZ corrodes selectively if it is less noble than the parent metal, which is commonly much larger in surface area than the weld metal and HAZ. The corrosion potentials of the welding materials are almost the same as those of the parent metals but frequently are not exactly the same. Usitalo [4] reported a maximum of 40 mV less noble potential for the welding materials than for the steel plates for shipbuilding. The potential of steel tends to shift in the active direction as a result of the thermal cycle during welding operations, making the HAZ less noble than the parent metal.

The selective attack of weld metal can be prevented by adjusting the weld metal composition so that the potential is slightly positive to that of the parent metal, but not too positive, so as to avoid a detrimental effect on the HAZ. To minimize the potential drop at the HAZ, the chemical composition of the steel plate must be appropriate.

Räsänen and Relander [5] made the welding material for shipbuilding steel [tensile strength: $490\,\text{N/mm}^2$ $(50\,\text{kgf/mm}^2)$] slightly more noble than the parent metal by adding Cu and Ni, which make the potential more noble, and decreasing Si, which makes the potential less noble. The chemical composition of the steel plate was also adjusted by specifying the maximum amount of Cu and the minimum amount of Si to make the steel slightly negative to the weld metal. At the same time, the Si, Mn, and S contents were limited and some Ce was added to minimize the potential drop at the HAZ. Some V and Nb were added to counteract the decrease of tensile strength caused by the low Mn content.

	Chemical Composition (%)					
	Cu	Ni	Si	Mn	S	Others
Welding material (example)	0.6	0.4	0.25	0.8		
Steel plate	0.3/0.5	—	0.2/0.3	$\leqq 1.1$	<0.015	Ce, V, and Nb

In the Canada Centre for Mineral and Energy Technology (CANMET) experimental program [6], it was also found that weld metal corrosion in normalized carbon steel plate was greatly decreased by the presence of Cu (0.3/0.5%) and Ni (0.5/1.6%) in the deposited weld metal. The decrease of Si from 0.42 to 0.15% in the Cu–Ni bearing weld metal was exceptionally effective on steel plate of 0.2/0.3% Si, reducing weld metal corrosion by ~80% in six month tank tests.

Increased amounts of Cu and Ni shift the potential of ferrous metals in the noble direction. Appropriate amounts and ratios of these elements in the parent metal and the weld metal are important in avoiding weld metal corrosion in low-alloy steels. Itoh et al. [7] and Endo et al. [8] reported appropriate distributions of Cu and Ni for $588\,\text{N/mm}^2$ $(60\,\text{kgf/mm}^2)$ high-strength steel for Arctic service and for grade 448 (API X-65) pipeline steel to transport CO_2-bearing wet oil and gas, respectively.

In spite of the susceptibility of welded steel structures to localized corrosion at welded joints, not much attention has been paid to welding practices nor have countermeasures been established, because most such structures are paint coated or otherwise protected from corrosion. Selective attack at welded joints is frequently experienced in bare marine piles. Large petroleum tanks not coated internally tend to suffer corrosion at welds during the hydrostatic pressure test with seawater that may take ~ 1 month. Some studies have been made of ships and offshore structures for Arctic service, where paint coatings are prone to severe mechanical damage by the collision of ice, and of oil and gas pipelines, to which internal coatings are not applied.

B4. Discontinuous Surface Films

A typical example of localized corrosion under discontinuous surface films is pitting of mill-scaled steel plate exposed to water. Mill scale formed on the steel surface by oxidation during hot rolling at the steel mill is not continuous and exposes the underlying steel to the environment at discontinuities or defects. Because the mill-scaled steel surface has a potential much more noble than the exposed steel, macrogalvanic cells are formed between them: the large area of mill-scaled surface being the cathode and the small exposed areas being the anode. When mill-scaled steel is exposed to seawater, the penetration rate at the exposed small areas can be a few millimeters a year. For more details, see Section B3 in Chapter 45.

Steel is seldom used with mill scale, which is removed by either acid pickling or blasting, and the steel is then painted, metal plated, or otherwise protected before use in corrosive environments. When mill-scaled steel is used for large petroleum tanks, pitting may result when it is exposed to the seawater used for the hydrostatic pressure test at the end of construction or to brine precipitated from the stored petroleum.

If mill-scaled steel plates for construction of ships or tanks are stored outdoors for an extended period (e.g., 6 months), pitting corrosion may occur, especially when they are stacked without proper separation between them to allow drying. Rainwater penetrates between the plates and remains there, causing pitting. In one case [9], mill-scaled steel plate used for ship construction corroded along grooves, following the pattern of discontinuities in mill scale resulting from stress concentrations during service.

Steel that is corroded in water and covered with corrosion products may undergo localized corrosion if the corrosion products are removed repeatedly from a fixed area of relatively small size. Corrosion product films protect the underlying steel from corrosion to some extent by acting as a barrier layer against the corrosive environment. If this barrier layer is damaged at a fixed area repeatedly, so that the steel surface is exposed to the environment, the penetration rate at the exposed area increases. Because the surface covered with corrosion products tends to be more noble than the exposed steel, the former acts as the cathode of the macrogalvanic cell. When a piece of mild steel was totally immersed in artificial seawater and the corrosion product film at a fixed small area was removed once a day using a wooden spatula, the penetration at that area in three months was 0.14 mm, while the average penetration was 0.043 mm [10]. More frequent removal of the corrosion product film would have resulted in much larger localized penetration.

The waterside (inside) surface of boiler tubes, reacting with the hot boiler water, develops an Fe_3O_4 film that protects the tubes from further corrosion. Should the film be damaged locally, either chemically or mechanically, pitting corrosion results at the localized areas. Dissolved oxygen, if present, accelerates the localized attack, possibly because the large surrounding areas with intact Fe_3O_4 film act as the cathode.

B5. Differential Aeration

A typical example of localized corrosion caused by differential aeration in water is associated with the formation of tubercles. The steel surface under a tubercle receiving a poor supply of dissolved oxygen is the anode, and the surrounding area of better oxygen supply is the cathode. As stated in Section C2 in Chapter 44, the penetration rate in water pipe carrying city water having a resistivity in the range of 5000–6000 $\Omega \cdot cm$ can be a maximum of 0.3 mm/year [11]. Assuming an average pipe corrosion rate of 0.1 mm/year, the cathode area surrounding a tubercle is about three times that of the anode area. The penetration rate under a tubercle would be larger if the resistivity of the water were lower, although a low resistivity does not necessarily favor the formation of tubercles.

Waterline attack occurs in steel plate or pipe that extends vertically from underwater to air. It is localized corrosion damage at the water–air interface and can be explained in terms of differential aeration caused by limited oxygen supply to the area covered by a thicker corrosion product film at the waterline.

If a buried steel pipe passes through two different soils, one with good aeration, such as sand, and the other with poor aeration, such as clay, a differential aeration cell is set up. The part of the pipe in well-aerated soil is the cathode, and the part in poorly aerated soil is the anode. The corrosion current of the cell leaves the anode at discrete points of low pipe-to-soil resistance, resulting in deep pits. The rest of the anode section of the pipe may corrode, but at lower penetration rates.

In water of a given dissolved oxygen concentration, the supply of oxygen to the metal surface is greater at higher water velocity relative to the metal surface. Corrosion of the cast iron casing of seawater pumps is accelerated by differential aeration near the center, where the water velocity is much lower than at the peripheral area. In one case, the depth of corrosion (graphitic corrosion)* near the center was ~2 mm after 4600 h of operation, corresponding

---

*A type of corrosion that occurs mainly in gray cast iron, composed of iron (ferrite) and graphite flakes. In the corroded layer, corrosion products of iron cement together the residual graphite flakes and form a solid layer of low ductility without apparent reduction in thickness. It occurs in some soils or waters and eventually penetrates the total thickness.

to a corrosion rate of 3.8 mm/year (150 mpy), in contrast to the corrosion rate of 0.8 mm/year (32 mpy) at the peripheral area [12].

B6. Differential pH

Steel is passivated in alkaline environments (above a pH of ~ 10) and assumes noble potentials. If a large passivated area in an alkaline environment is coupled to a small active area in a near-neutral environment, the latter area is attacked by the action of the passive–active cell.

This type of corrosion damage has frequently been observed in underground utility piping leading to steel-reinforced concrete buildings. Because the steel reinforcements are in the alkaline environment of concrete (pH of ~ 12.5), they are in the passive state. If an underground pipe is in contact with the reinforcements, it becomes the anode to the reinforcements and discharges the corrosion current. As the current discharge takes place at limited areas of low pipe-to-soil resistance, the damage is highly localized, usually in the form of pits. Because the surface area of the reinforcements that are connected is much larger than that of the piping, the discharging current densities tend to be large. Thus, the penetration rate is often in the range of 0.5–1.5 mm/year (20–60 mpy) and may be >3 mm/year (120 mpy) [13].

If limited areas of reinforcing bars in concrete structures are neutralized by the action of water penetrating through cracks in the concrete, localized corrosion results at such areas. Along the same lines, piping in constantly wet concrete floors of kitchens in restaurants and in the concrete bottoms of swimming pools may be perforated due to local neutralization by penetrating water. If piping installed in concrete emerges into soil or water, corrosion may occur in the latter environments. The localized attack tends to be concentrated and severe if the piping goes through a thin water layer just outside the concrete, as in the case of a wet concrete floor.

C. LOCALIZED CORROSION BY OTHER CAUSES

C1. Other Types of Macrogalvanic Cells

The presence of chlorides in alkaline environments causes pitting corrosion in passivated steel by locally breaking down passivity. The mechanism is similar to that of pitting corrosion of stainless steels in neutral chloride environments. Localized damage of reinforcing bars in concrete contaminated by chlorides (e.g., use of sea sand in concrete without sufficient washing) and pitting in steel chemical equipment handling crude caustic soda containing chlorides are examples. Pitting corrosion occurs in water (with or without chlorides) inhibited by passivating-type inhibitors (e.g., chromates and nitrites) at insufficient concentrations. Below a certain critical concentration, passivators behave as active depolarizers and increase the corrosion rate at localized areas.

Steel sheets coated with metals more noble than steel (e.g., nickel, silver, copper, lead, or chromium) are attacked at exposed pores by the galvanic effect of the noble coatings. Similar accelerated corrosion occurs in steel coated with thin paint films (e.g., < 100 μm thick) at coating holidays. Steel coated with thin paint films having some degree of ionic conductance exhibits noble electrode potentials whether or not the paint is pigmented with antirusting compounds of the passivator type. The noble potential may be associated with passivity, but its mechanism, particularly when passivating pigments are not used, has not yet been clarified.

C2. Nongalvanic Types

Localized corrosion may occur by the locally concentrated direct action of corrosive environments. The bottoms of pipes or vessels in contact with gasoline containing suspended water are corroded locally by the sedimentary water. The high corrosivity of this water is caused by the good supply of dissolved oxygen from the gasoline, in which the solubility of oxygen is as high as six times that in water.

Steam return lines are pitted by deposited droplets of condensate at locations where condensation starts. Steam condensate is corrosive if it contains CO_2, which is generated in the boiler and contained in the steam. The penetration rate may reach a few millimeters a year [14]. On reaching lower temperatures, steam condensate accumulates at the bottom of the horizontal line to cause elongated general thinning of the bottom areas [14]. Steam lines are pitted similarly if condensate is generated by lowering of temperature. Heat exchanger tubes carrying steam to heat fluid outside the tubing may suffer the same damage.

Stray-current corrosion or electrolysis of underground pipelines commonly takes the form of pitting because the discharging current from the pipeline chooses locally distributed paths of low pipe-to-soil resistance.

Erosion–corrosion causes localized attack at areas where corrosion product films are removed allowing easier access by corrosive species from the environment. Examples are seen at the bottoms of slurry lines, at bends of piping carrying high-velocity water, and at the inlets of heat exchanger tubes handling corrosive fluids.

Cavitation–erosion that occurs typically on diesel engine cylinder liners (cooling water side), on rotors of pumps, and on the trailing side of water turbine blades is accompanied by numerous deep pits. Fretting corrosion caused by oscillatory motion is characterized by the formation of pits.

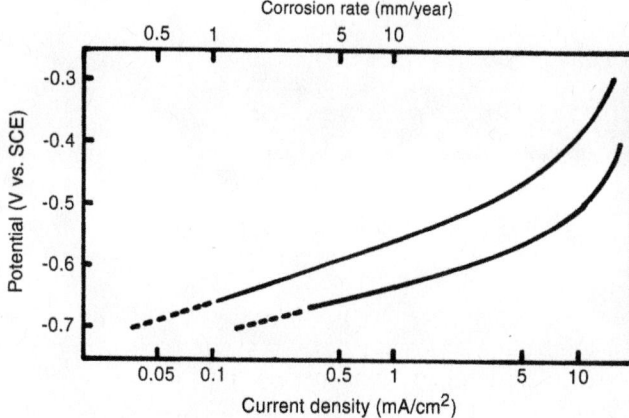

FIGURE 47.1. Steady-state anodic current density as a function of the potential of mild steel in deaerated 3% NaCl at 25 and 60°C [10].

D. MAXIMUM POSSIBLE PENETRATION RATE

The corrosion rate at the anode of a macrogalvanic cell increases as its potential becomes more noble than the open-circuit potential. The shift of the anode potential in the noble direction is greater, the more noble the open-circuit potential of the cathode and the higher the cathode–anode area ratio. If the latter ratio is very large, the potential of the anode is practically the same as that of the cathode, the corrosion rate of the anode, or the penetration rate of localized corrosion being a maximum.

The maximum penetration rate at the anode in a given macrogalvanic cell system can be estimated by the steady-state anodic current density at the potential at which the local anode is maintained The steady-state anodic polarization curves in deaerated 3% NaCl at 25 and 60°C are shown in Figure 47.1 [10]

TABLE 47.1.   Highest Penetration Rates for Various Types of Localized Corrosion in Mild Steel Observed in Practice

Type of Localized Corrosion	Penetration Rate (mm/year)	mpy
Localized corrosion under tubercles in tap water pipe	0.3	12
Localized corrosion under tubercles in hot water pipe	1	40
Localized corrosion in seawater	0.9	35
Localized corrosion by the galvanic action of oxides in steam/industrial water pipe	3.5	140
Localized corrosion of underground service pipe	3.5	140
Groove corrosion at the weld of electric resistance welded water pipe	10	400

The maximum penetration rate at 25°C predicted by Figure 47.1 is ~1 mm/year (40 mpy) at −650 mV and 4 mm/year (160 mpy) at −600 mV (vs. SCE). The actual penetration rates of localized corrosion caused by macrogalvanic cells would be lower than the predicted maximum rates because an incubation period usually exists, during which galvanic cells are formed, or because the anodic area is not always fully active. Unfortunately, data on the corrosion potential during service are not always available. The highest penetration rates observed in some types of localized corrosion mentioned earlier are 0.3–10 mm/year (12–400 mpy), as listed in Table 47.1. Assuming that the anode potentials were in the range of −600 to −650 mV versus SCE, and considering that the actual rates tend to be lower than the steady-state values, these predicted rates are within a reasonable range of the observed values.

REFERENCES

1. F. Speller and F. LaQue, Corrosion, **6**, 209 (1950).

2. I. Matsushima, 32nd Symposium on Corrosion and its Protection, Sapporo, Japan, Aug., 1985, p. 452.

3. I. Matsushima, Low-Alloy Corrosion Resistant Steels, Chijin Shokan Co., Tokyo, Japan, 1995, p. 155.

4. E. Usitalo, Proceedings of the 2nd International Congress on Metallic Corrosion, National Association of Corrosion Engineers, Houston, TX, 1963, p. 812.

5. E. Räsänen and K. Relander, Scand. J. Metall., **7**, 11 (1978).

6. V. Mitrovic-Scepanovic and R. Brigham, Proc. EVALMAT 89, Iron Steel Inst. Japan, 1989, p. 491; also see R. J. Brigham, M. McLean, V. S. Donepudi. S. Santyr, L. Malik, and A. Garner, Can. Met. Q., **27**, 311 (1988).

7. K, Itoh, H. Mimura, T. Inoue, S. Sekiguchi, Y. Horii, and H. Kihara, J. Iron Steel Inst Jpn., **72**, 1265 (1986).

8. S. Endo, M. Nagae, M. Suga, S. Wada, T. Sugino, and T. Nakano, Curt. Adv. Mater. Proc., Iron Steel Inst. Jpn., **4**, 1890 (1991).

9. A. Tamada, Y. Shimizu, and I. Matsushima, Nippon Kokan Giho, No. 17, 271 (1976).

10. I. Matsushima, in Localized Corrosion, F Hine, K, Komai, and K. Yamakawa (Eds.), Elsevier Applied Science, London, 1988, p. 31.

11. I. Matsushima, paper presented at the Committee on Corrosion and Protection, Vol. 19, p. 94 (1980); reproduced in Corrosion Protection Handbook (Boshoku Gijutsu Binran), Japan Society of Corrosion Engineering, Nikkan Kogyo Sinbunsha, Tokyo, Japan, 1985, p. 181

12. M. Miyasaka, Ebara Eng. Rev., No. 137, 1 (1995).

13. I. Matsushima, Boshoku Gijutsu (Corros. Eng.), **25**, 563 (1976).

14. I. Matsushima, Zairyo-to-Kankyo (Corros. Eng.), **44**, 683 (1995).

48

WEATHERING STEEL

T. Murata

Office of Technology Transfer, Innovation Headquarters, Japan Science and Technology Agency, Tokyo, Japan

A. NEED FOR WEATHERING STEEL

Recognition of global warming and environmental degradation has caused the direction of technology to shift toward environmentally conscious and sustainable development, with the result that increased efficiency in the use of limited resources is now a focus (e.g., recycling industrial materials). In addition, the environmental characteristics of industrial products, structures, and materials through their life cycles are being improved by life-cycle assessment (LCA) of the International Organization for Standardization (ISO) 14000s. Although LCA needs further improvement, it is a potentially useful methodology for systematically analyzing a product from the extraction of resources to eventual abandonment, including reuse, recycling, or disposal in terms of energy, in most cases, and materials, emissions of hazardous substances, and wastes.

The LCA approach is being used to clarify quantitatively the important role of steel and steel products in our society in terms of their abundant resources, economic availability in large quantities, workability, reliability, reasonable life-cycle cost, relatively low environmental burden, and recycling possibilities.

Structural steel is expected to remain as one of most environmentally friendly materials available in the twenty-first century. In addition, the recent life-cycle study of automobiles, houses, bridges and so on conducted by the Engineering Academy of Japan has demonstrated that 70~85% of the energy consumption by these products throughout their life cycle occurs during service period, rather than during production or product assembly [1]. This observation indicates that materials performance, durability, and reliability are key considerations in the design of a product or a structure for its life-cycle energy efficiency. Figure 48.1 shows an example of the framework for the systematic assessment of materials performance, including (a) life-cycle cost (LCC) analysis, (b) life-cycle safety (LCS) design, (c) LCA for energy and materials, and (d) diagnosis of life prediction based on the operating mechanisms.

Our living environments consist mainly of freshwater, seawater, the atmosphere, and soil. These four categories include the environments in which the majority of structural steel products and large structures are exposed. Corrosion mechanisms of structural steels and the dominant corrosion parameters in these environments are discussed in other chapters. In this chapter, carbon and low-alloy structural steels resistant to atmospheric corrosion (i.e., weathering steels) are discussed. Weathering steels have been widely accepted in recent years from the viewpoint of saving not only money but energy and resources.

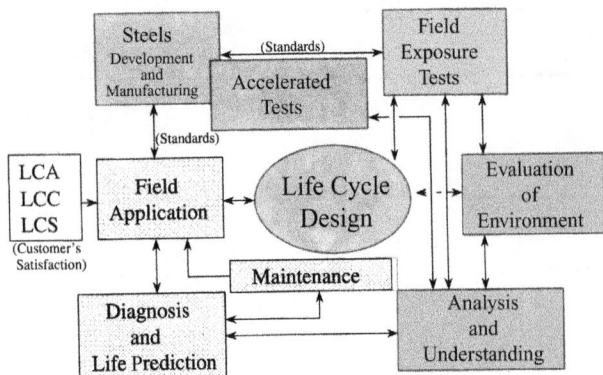

FIGURE 48.1. Framework for the systematic assessment of weathering steel performance.

B. HISTORY

Weathering steel contains Cu, P, Cr, Ni, and Si as alloying elements totaling a maximum of a few percent. During field exposure, an adherent, compact, and protective layer of corrosion products grows under a coarse surface rust layer retarding further corrosion.

Weathering steel was originally developed by U.S. Steel in 1933 as CORTEN, a high-strength, low-alloy structural steel of yield strength >35 kgf/mm^2 and corrosion resistance superior to that of Cu-bearing steel [2]. The initial chemical composition is shown in Table 48.1 [3] with more recent American Society for Testing and Materials (ASTM) specifications [4].

Buck [5] reviewed investigations dating from 1908 which showed the beneficial effects of Cu as an alloying element in steels for atmospheric exposure. In 1916, ASTM took an initiative to conduct long-term exposure tests to evaluate the effect of Cu on atmospheric corrosion resistance of various steels. The final report [6] confirmed the superior corrosion resistance of steels containing Cu up to 0.3%, compared to those without Cu ($<0.03\%$). The additional beneficial effect of P up to 0.09% was also noted. The role of Cu, Cr, Ni, P, Al, and Mo in atmospheric corrosion resistance was assessed in systematic studies in the United States [2], United Kingdom [7], and Germany [8, 9].

Addition of 0.2–0.3% Cu was found to increase the corrosion resistance by a factor of ~2. Weathering steel, with 0.5% Cr in addition to Cu, improved corrosion resistance by an another factor of ~2, as shown in Figure 48.2 [2]. Corrosion resistance is usually assessed by measuring the reduction in thickness from both sides of a steel plate during a specific exposure time.

B1. Practical Applications

The first version of CORTEN was based on the Fe–Cu–Cr–P system whereas later versions were based on the Fe–Cu–Cr–Ni–P system, the basis for the ASTM standard series listed in Table 48.1 [3]. CORTEN B was specially designed with P $< 0.04\%$ (like ASTM 588, registered in 1968) compared with 0.07–0.15% P for CORTEN A, to avoid weld cracking of plates >37 mm (1.5 in.). Nickel was added to minimize plate damage during rolling due to enrichment of Cu in the surface scale [10] and to improve the corrosion resistance in chloride-bearing environments [2].

In the early applications, weathering steel was used for freight trains, trucks, agricultural equipment and so on with various paints. In eight-year atmospheric field exposure tests, painted weathering steel lasted 1.5–4 times longer than on conventional steels [11]. Paints, which retard corrosion initiation and minimize stains from corrosion products, are evaluated for their contribution to life-cycle economy, reduction of environmental stains, and their ease of use.

Bare weathering steel was first used in a full-scale application in the John Deere and Co. office building in Moline, Illinois, in 1964 which stimulated the use of weathering steel in buildings and other structures such as Chicago Civic Center, towers for power transmission lines, and highway bridges. Thus the use of weathering steel expanded in the United States (e.g., by 1993, there were ~2300 bridges in the United States without coatings) [12]. Guidelines were prepared for the removal of mill scale and contaminants, and for the design to install water drainage.

B2. Expansion of the Use of Weathering Steels

Weathering steel was first commercialized in the United States, and the use of weathering steel then has spread to other parts of the world where different environmental conditions exist. In the late 1950s, weathering steel was modified in Japan [3] into Cu–P, Cu–P–Ti, Cu–Cr–Ni–P (CORTEN), Cu–P–Cr–Mo and other systems. The main features included (1) diversification of steels according to P content: high P for high corrosion resistance and low P for better welding properties; (2) three grades of tensile strength covering 41, 50, and 58 kgf/mm^2 (58, 71, and 83 psi, respectively) for welded structures; and (3) a variety of coatings to stabilize the corrosion products early in the exposure period to avoid stains of corrosion products [13–16].

During the 1960s, air pollution in industrial areas in Japan, up to 0.07 ppm SO$_2$, caused by combustion of sulfur-bearing fossil fuel, along with high rainfall and humidity levels, and rather high chloride concentration near the seacoast led to poor performance of weathering steel. The alloy design was modified and coatings were developed to meet the needs for the local applications [3]. Steelmakers collaborated in 12-year exposure tests that

TABLE 48.1. Compositions of Weathering Steels[a]

	C	Si	Mn	P	S	Cu	Cr	Ni		Weight Percent (w/o)
CORTEN	~0.10	0.50–1.00	0.10–0.30	0.10–0.20	—	0.20–0.50	0.50–1.50	0.03		
example [2]	0.09	0.93	0.30	0.16	0.035	0.42	1.1			
1940 CORTEN	≤0.12	0.25–0.75	0.20–0.50	0.07–0.15	≤0.05	0.25–0.55	0.30–1.25	≤0.65		
example [2]	0.06	0.54	0.48	0.11	0.030	0.41	1.0	0.51		
USS CORTEN A	≤0.12	0.25–0.75	0.20–0.50	0.07–0.15	≤0.05	0.25–0.55	0.50–1.25	≤0.65	—	YS ≥ 50 ksi, ASTM-A242 Type 1
USS CORTEN B	≤0.19	0.30–0.65	0.80–1.25	≤0.04	≤0.05	0.25–0.40	0.40–0.65	≤0.40	V0.02–0.10	YS ≥ 50 ksi, ASTM–A588GRA
USS CORTEN B-QT	≤0.19	0.30–0.65	0.80–1.25	≤0.04	≤0.05	0.25–0.40	0.40–0.65	≤0.40	V0.02–0.10	YS ≥ 70 ksi
USS CORTEN C	≤0.19	0.30–0.65	0.80–1.35	≤0.04	≤0.05	0.25–0.40	0.40–0.70	≤0.40	V0.40–0.10	ASTM-A852, YS ≥ 60 ksi, ASTM-A871GR60,65
ASTM A242 (M)	≤0.15	—	≤1.00	≤0.15	≤0.05	≤0.20	—	—	—	
ASTM A588 A	≤0.19	0.30–0.65	0.80–1.25	≤0.04	≤0.05	0.25–0.40	0.40–0.65	<0.40	V0.02–0.10	
ASTM A588 B	≤0.20	0.15–0.50	0.75–1.35	≤0.04	≤0.05	0.20–0.40	0.40–0.70	≤0.50	V0.01–0.10	
ASTM A588 C	≤0.15	0.15–0.40	0.80–1.35	≤0.04	≤0.05	0.20–0.50	0.30–0.50	0.25–0.50	V0.01–0.10	
ASTM A588 D	0.10–0.20	0.50–0.90	0.75–1.25	≤0.04	≤0.05	≤0.30	0.50–0.90	—	Zr 0.05–0.15, Cb ≤ 0.04	
ASTM A588 K	≤0.17	0.25–0.50	0.50–1.20	≤0.04	≤0.05	0.30–0.50	0.40–0.70	≤0.40	Mo ≤ 0.10, Cb 0.005–0.05	

[a]See [3].

623

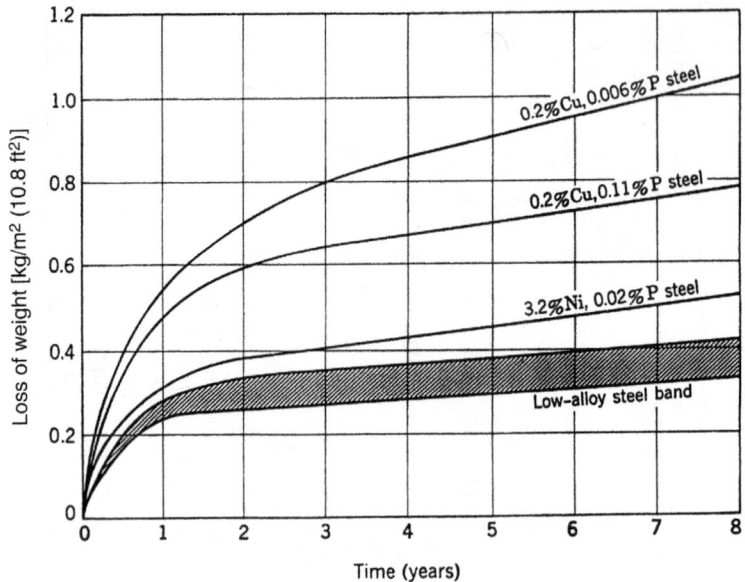

FIGURE 48.2. Atmospheric corrosion of steel as a function of time (industrial atmosphere) [2].

were carried out to classify the severity of the environments in Japan into five zones [17]. The maximum acceptable deposition rate of airborne salt was 0.05 mdd (mg/dm^2/day) in order for the corrosion rate of weathering steel to be <0.3 mm corrosion loss during 50 years [18].

The protective rust layer is the product of corrosion reactions between the steel substrate and the chemical species in the environment, for example, chloride and sulfate ions under wet and dry conditions with temperature variation. Because of these variables, the performance of weathering steel varies with geographic location, such as distance from the seashore, as well as structural location and orientation, such as the angle with respect to incident sunlight. Experience with weathering steel in Japan [3] indicates the need for quantitative assessment of time-dependent environmental parameters, especially concentrations of sulfate and chloride, seasonal humidity and solar radiation, and the effects of these parameters on the formation of protective rust layer.

By 1980, the application of bare weathering steel to highway bridges had increased to 12% of all bridges in the United States [12]. Corrosion of bridges constructed of weathering steel has been observed predominantly on upper decks and joints where deicing salts sprayed during the winter season accumulate [19]. Analysis of the corrosion damage showed that, to ensure the satisfactory performance of weathering steel, it is important to avoid water pools where salts concentrate and degrade the protective rust layer [20]. This experience indicates the importance of design in the use of weathering steel. It is essential to avoid water pool formation and thereby accumulation of salts which could

be deicing salts or airborne sea salts depending on geographic location as described in Section C.

C. ALLOYING ELEMENTS

The role of alloying elements in weathering steel may be divided into (1) effects on the formation of protective rust layer (discussed in this section), (2) enhancement of mechanical strength and toughness, and (3) improvement of weldability. The protective qualities of the layer of corrosion products on steel depend on the continuous growth of the adherent, compact, inner layer, and on the low porosity of the layer. From statistical analysis of data obtained in long-term exposure tests [2, 21–27], the kinetics of atmospheric corrosion were found to obey the equation,

$$C = At^B \qquad (48.1)$$

as shown in Figure 48.3 [2], where C is the corrosion loss, t is time, and A and B are constants. Expressing Eq. (48.1) in a different way the following equation is drawn,

$$\log C = \log A + B \log t \qquad (48.2)$$

and with regression analysis, the two constants, A and B, may be expressed in terms of alloy content or environmental parameters, such as chloride and sulfate ion concentrations [26] depending on the available data. In using statistical analysis, it is essential to establish the reliability of the available data, and then to select a mathematical

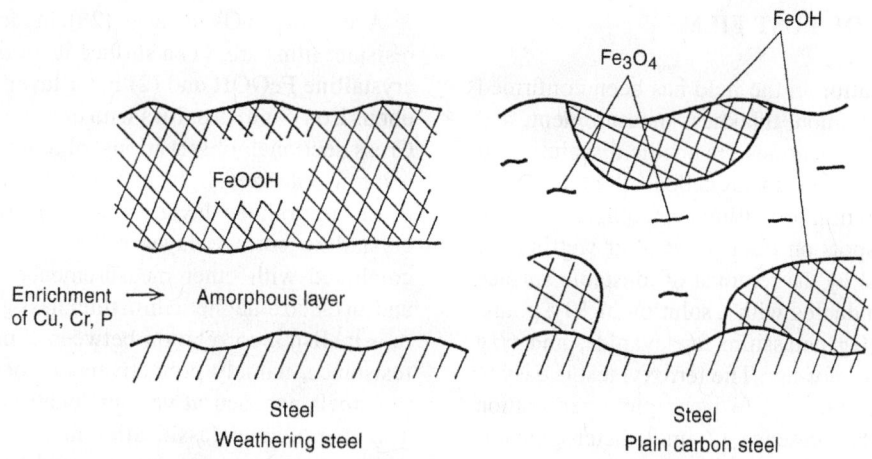

FIGURE 48.3. Schematic illustration of the corrosion product layers identified on steels exposed to rural and marine atmospheres for the periods of up to five years [28].

methodology (e.g., extreme-value analysis or discriminant analysis).

In this way, the beneficial effects of Cu, Cr, and P and the small contribution of Si, Ni, Al, and Mo in the protective layer of corrosion products were identified. The present understanding of the structures of the corrosion product films is described in Chapter 43, pp. 581–583.

The protective nature of weathering steel have been characterized by observing the dark color and limited porosity of the protective rust layer. Study of the cross sections, using polarized light reflection [28] as well as X-ray analysis and electrochemical techniques, identified the inner layer as an amorphous state, spinel-type iron oxide enriched with Cu, Cr, and P, and the outer layer as γ-FeOOH, as schematically illustrated in Figure 48.3. The protective nature of weathering steel was thus attributed to the continuity of the amorphous inner layer. Ferroxyl tests were used to study discontinuities of the amorphous layer caused by porosity or sulfate nests which are spotlike open defects, where chlorides and sulfates accumulate because of localized corrosion [29, 30]. As the sulfate nests become smaller in size and in population, the protective qualities of weathering steel improve. Recently, integrated instrumental analysis, with X-ray diffraction, Raman spectroscopy, and infrared (IR) transmission was utilized to show that Cr-bearing α-FeOOH could be the structure of protective rust layer on the samples exposed for 26 years [31] although identification of this stable phase has not been substantiated for other samples exposed for 35 years [32].

Applications of weathering steel, however, are still very limited in the presence of concentrated sea salt or accumulated deicing salts because of insufficient formation of the protective layer in the presence of chloride nests [33]. One of the proposed solutions to this difficulty is to increase Ni content by as much as 3–5%. Nickel enrichment of the

amorphous inner layer was found to retard corrosion supporting the hypothesis that chloride ions do not penetrate the layer bearing higher Ni contents [34–36].

C1. Mechanistic Aspects

Regarding the role of Cu, Masuko and Hisamatsu [37] and Suzuki et al. [38] found in their studies of artificial rust that addition of Cu ion decreases the size of iron oxyhydroxide (FeOOH) particles, increases its buffer capacity, and stabilizes its amorphous state. The role of Cu was also assessed by Stratmann and Streckel [39] using the Kelvin method, who confirmed the considerable retardation of corrosion by the formation of a compact layer containing Cu. Studies have been carried out on the ion selective properties of corrosion products on local solution chemistry, and on agglomeration of colloidal particles [40, 41]. By observing in situ the formation of colloidal corrosion products with a video-enhanced microscope during the initial stage of corrosion, a phosphate layer was identified beneath which the amorphous layer grows. This mechanism explains the role and location of P at the interface between the inner amorphous layer and the outer FeOOH layer in weathering steel exposed to the field atmosphere for 19 years [42].

Since the corrosion reactions of weathering steel are very slow and have many variables, the precise mechanism of the formation of the protective layer of corrosion products is still somewhat uncertain. Experimental techniques that may be useful in elucidating further the mechanism of corrosion protection of weathering steels include alternating current (ac) impedance with controlled thin water films under wet and dry conditions [35], Kelvin vibrating-capacitor technique [39], colloidal chemical techniques with video-enhanced microscope [42], and high-energy X-ray sources for higher resolution in situ.

D. MONITORING OF RUST FILMS

Protective film formation in the field has been confirmed firstly by visual observation, thickness measurement, and the ferroxyl test, which identifies the anodic dissolution of the Fe^{2+} ion from local active sites called sulfate nests. At these sites, protective film formation is incomplete, as can be noticed by blue spots on filter paper after wetting the steel surface followed by the removal of unstable surface rust. The tests are conducted with a solution of 10 g potassium ferrocyanide, 10 g potassium ferricyanide, and 60 g sodium chloride in 1 L of water. The ferroxyl test is easy to use but can be misleading due to incomplete penetration of test solution in the presence of thick heterogeneous corrosion products.

Kihira et al. [43] developed a portable instrument based on electrochemical ac impedance and harmonic current measurements in order to identify the formation of a protective film on weathering steel. Upon a thin film of $0.1\,M$ Na_2SO_4 solution, two probes were attached to the steel surface which had previously been exposed to a semi-industrial environment for 13 years. By measuring ac impedance and harmonic current characteristics, an appreciable difference in ionic resistance of the films on carbon steel and on weathering steel was found at frequencies of 100 mHz–1 kHz as a noticeable characteristics between two steels.

According to Okada et al. [28], the features of corrosion-resistant films are (1) a surface layer consisting mainly of crystalline FeOOH and (2) inner layer consisting of amorphous iron oxyhydroxides with enrichment of Cu, Cr, and P. Cross-sectional observations of corrosion product films using a polarized light microscope indicate the presence of an amorphous layer, which absorbs light, whereas a crystalline surface layer reflects it. These observations combined with other measurements using ac impedance and an electromagnetic film thickness gage made it possible to establish a correlation between film thickness and film resistance, namely protectiveness, for numerous weathering steels exposed at various locations throughout Japan. The quantitative classification map for the performance of weathering steels was constructed and is shown in Figure 48.4. In the zone designated as A, the film resistance may increase without increasing the thickness where a uniform and protective film is formed. On the other hand, there is a zone C in which thickness increases without increase in the resistance where anomalous rust composed of FeOOH is formed next to the surface, and crystalline Fe_3O_4 is formed in the rest of the film.

The previously described portable instrument called Rust Stability Tester was applied to the 13-year-old bridge located approximately 2 km from the mouth of a river in order to assess the state of the rust films. Figure 48.5

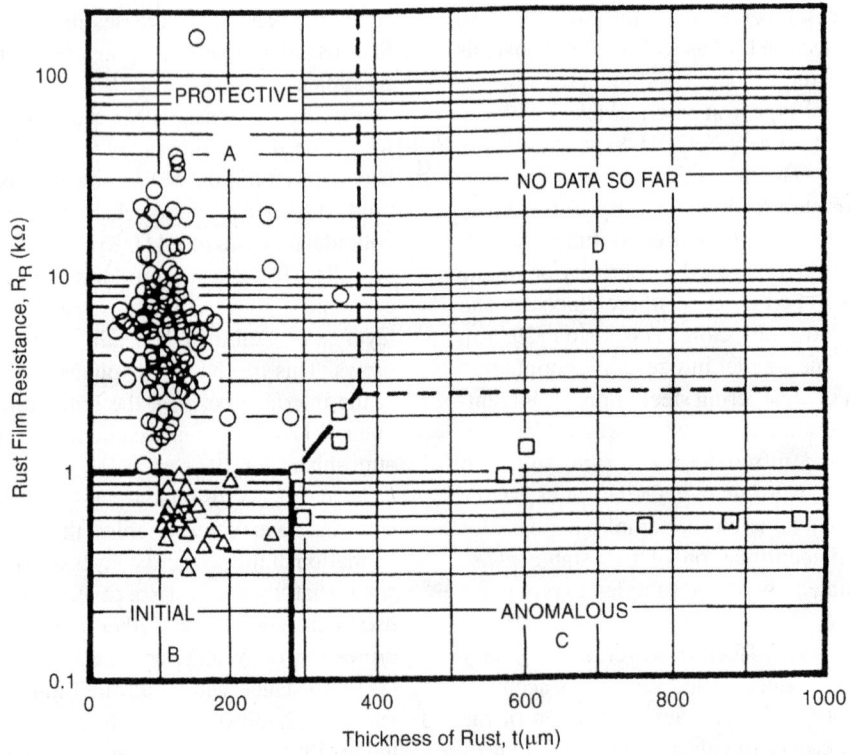

FIGURE 48.4. Quantitative classification of the states of corrosion products on weathering steels based on measurements of film resistance and thickness [43].

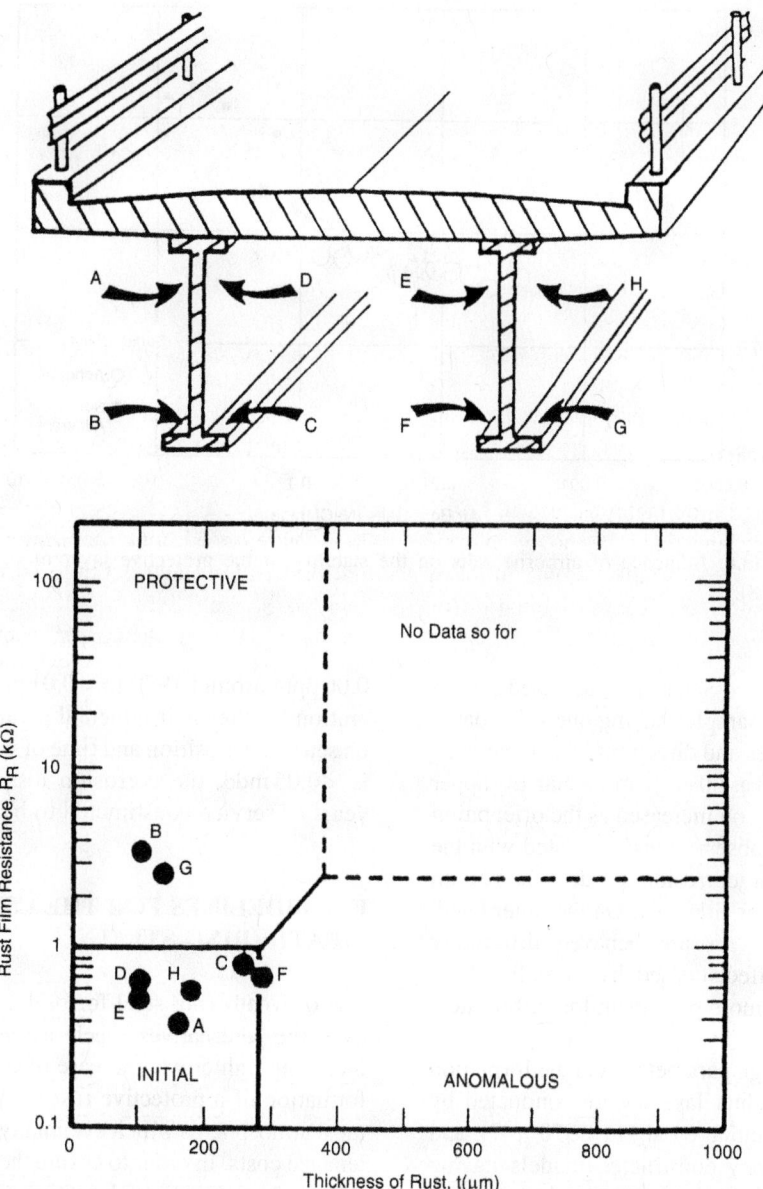

FIGURE 48.5. Distribution of states of corrosion products on a bridge constructed from weathering steel 2 km from the mouth of a river [43].

illustrates the observations. The steel inside the girders (C–D–E–F) and outside the upper girders (A and H) is still covered with the initial rust because of the relatively low humidity. The steel outside the lower girders (B and G) is, however, covered with a protective rust film. The formation of this film may be attributed to washed out effect for deposits by rainfall at this location, in contrast to points C and F, where dust and sea salt accumulate. This nondestructive Rust Stability Tester can be used to assess structures, such as a bridge, so that the formation of a protective film can be monitored and diagnosed. In addition, feedback of the information to the maintenance and design stage becomes very practically useful.

E. DESIGN PARAMETERS

Larrabee [44] and Schikorr [45] noted the importance of orientation and geometric configuration of samples in field exposure tests, that is, the angle with respect to the horizontal, and the upper and lower sides of a sample, due primarily to the time of wetness and amount of deposit. In the study of

splice plates; (6) avoid water leakage at expansion joints; (7) protect the slab from intrusion of water from bridge surfaces; (8) paint the inside surface of the box girders to protect against condensation of water, and paint the girder edge to protect against intrusion of water from expansion joints; (10) blast clean the steel surfaces to facilitate the formation of the protective rust layer; and (11) avoid or minimize foreign substances, such as oil, grease, mortar, and markings.

Minimum maintenance concept for bridges which was proposed by the Ministry of Construction of Japan in 1997 in collaboration with universities and industries will contribute to sustainable development in saving energy, cost and environmental burden.

G. ADVANCEMENT IN WEATHERING STEEL IN THE PAST DECADE

Because of our shared objectives of achieving a sustainable society, life-cycle design for the durability, reliability, and best use of materials and energy is an essential philosophy for social infrastructure such as key bridges and railway stations. The past decade, in this sense, has been a turning period for weathering steel, especially in Japan where a paradigm change in the use of weathering steel has occurred as described in the following paragraphs. This trend is expected to prevail over the environment-sensitive countries.

G1. Life-Cycle Design Specification for Highway Bridges

The Japanese specification for highway bridges was revised in 2002 clearly stating for the first time in its history that degradation of bridge members including those of weathering steels is inevitable during their service lifetime, and therefore, the diagnosis and maintenance must be taken into account at the design stage [53]. This was the first governmental recognition of corrosion degradation of infrastructures during service lifetime followed by necessary maintenance requirements regardless of whether the weathering steels are painted or unpainted.

G2. "Minimum-Maintenance" Concept

For many years, the initial cost has been the main concern for constructing social infrastructures due to budget constraints. Nowadays, it is critical to implement the best performance of infrastructure throughout its service lifetime from the viewpoint of taxpayers as well as future generations, namely social accountability.

Kihira [54] has clearly demonstrated that the design concept for minimizing the maintenance cost is key to

attaining the best life-cycle performance of a bridge, especially that made of unpainted weathering steel. In this concept, three technologies are required:

1. A computational scheme for predicting corrosion of weathering steels, such as those developed by several steelmakers in Japan [55–57] based on rather short exposure test results and quantitative assessment of alloying elements that result in the low corrosion rates, for example less than 0.3 mm per side for 50 years, or 0.5 mm per side for 100 years, under the simulated environmental conditions for a bridge in service.

2. A rust-monitoring system, such as the instrument developed in 1989 as previously described [43] aiming at identifying abnormal rust growth

3. Development of the effective repair methods [54, 58] applicable to identification of the abnormal rust growth during monitoring

G3. Advanced Weathering Steel

After nine years of field exposure tests, an advanced weathering steel, namely 3 mass %Ni–0.4 mass % Cu steel, which is significantly more resistant to airborne salt environments than are conventional steels, was commercialized in 1998 in Japan, as reported elsewhere [59, 60]. Other steelmakers followed [61–63], with the result that the cumulative production of advanced weathering steel exceeded 35,000 tons at the end of the 2004 fiscal year.

The formation mechanism of the protective inner rust layer of 3Ni–0.4Cu-bearing weathering steel was recently assessed by Kimura et al. [64]. Utilizing synchrotron radiation for fine X-ray beams, (1) X-ray diffraction (XRD) Debye rings for crystalline structures, and (2) X-ray absorption fine structures (XAFS) for local linkages around specific elements were studied. Also, (3) electron probe microanalysis (EPMA) for compositional mapping was carried out. They clarified that the protective rust layer on weathering steels formed through wet–dry cycles, which used to be considered from conventional X-ray analysis as an amorphous state, is actually composed of a nanonetwork of $Fe(O, OH)_6$. The formation processes are schematically illustrated in Figure 48.7. Compared with those on conventional weathering and mild steels, the rust formed on advanced weathering steel includes Fe_2NiO_4 where CuO phases are likely to provide nucleation sites for $Fe(O,OH)_6$ nanonetwork to form densely packed fine rust. Fe_2NiO_4 in the nanonetwork of the inner rust layer is considered to increase the negativity of the surface charge of fine iron oxy-hydroxide particles, thereby attracting positively charged sodium ions, which accumulate within the inner protective rust layer increasing the pH adjacent to the steel surface during wet periods. The above proposed mechanism for the protective nature of rust film has to be substantiated by further studies.

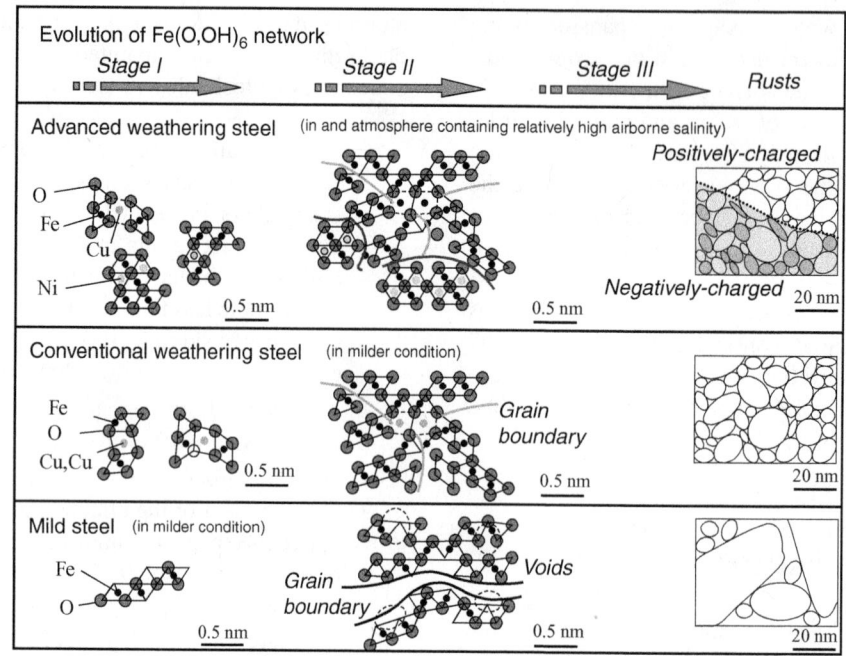

FIGURE 48.7. Schematic diagram for evolution of $Fe(O,OH)_6$ nanonetwork of rust formed on advanced weathering steel in an atmosphere containing relatively high airborne salinity (above), which is compared to conventional weathering steel (middle), and mild steel exposed under milder conditions [64].

REFERENCES

1. K. Yamaji, EAJ Information, No. 74 (1998).

2. C. P. Larrabee and S. K. Coburn, Proceedings, First International Congress Metallic Corrosion, Butterworths, London, 1971, p. 276.

3. I. Matsushima, Low-Alloy Corrosion Resistant Steels, Chijin Shokan, Tokyo, Japan, 1995, p. 21.

4. Standard A588, American Society for Testing and Materials, Philadelphia, PA, l968.

5. D. M. Buck, Ind. Eng. Chem., **5**, 458 (1913).

6. Reports of Committee A5, Proc. ASTM, **53**, 110 (1953).

7. J. C. Hudson, JISI, **148**, 168 (1955).

8. O. Bauer, Stahl Eisen, **41**, 37, 76 (1921).

9. O. Bauer, O. Vogel, and C. Holthaus, Metals Alloys, **1**, 890 (1930).

10. L. Habraken and J. LeComte-Beckers, in Copper, Iron and Steel, I. LeMay and L. M. Schetky (Eds.), Wiley, New York, 1982, pp. 45–81.

11. H. R. Copson and C. P. Larrabee, ASTM Bull., Dec. 1959, p. 68.

12. W. L. Mathay, "Uncoated Weathering Steel Bridges," in Highway Structures Design Handbook, Vol. 1, American Institute of Steel Construction, Pittsburg, PA, 1993, Chapter 9.

13. S. Yoshikawa, "Development of Weather Coat," Bosei Kanri (Corros. Control), **24**, 20 (1980).

14. T. Huga, A. Murao, T. Takeda, and I. Matsushima, Tosou Kogaku (Coating Technology), **18**, 264 (1983).

15. T. Imazu, T. Kurusu, Y. Nakai, T. Hisano, M. Ishiwatari, and T. Sato, Kawasaki Steel Giho, **16**, 123 (1984).

16. T. Watanabe, Bosei Kanri (Corros. Control), **20**, 11 (1980).

17. Technical Report, Corrosion Protection of Steels, Kozai Club, 18 (1986).

18. Technical Report, Guideline for Design and Construction of Bridges by Weathering Steel, Collaboration between National Research Lab. for Civil Engineering, Ministry of Construction, Kozai Club and Japan Association for Construction of Bridges, Publ. Kozai Club, Tokyo, Japan, 1993.

19. Performance of Weathering Steel in Highway Bridges, First Phase Report, American Iron and Steel Institute, Washington, DC, 1982.

20. Guideline, Maintenance Coating of Weathering Steel, Interim Report, FHWA-RD-91-087, U.S. Government Printing Office, Washington, DC, 1992.

21. R. A. Legault and J. G. Dalal, in Atmospheric Corrosion, W. H. Ailor (Ed.), Wiley-Interscience, New York, 1982, p. 285.

22. R. A. Legault and H. P. Leckie, Corrosion in Natural Environments, STP 558, American Society for Testing and Materials, Philadelphia PA, 1974, p. 334.

23. H. E. Townsend and J. C. Zoccola, in Atmospheric Corrosion of Metals, S. W. Dean, Jr., and E. C. Rhea (Eds.), STP 767, American Society for Testing and Materials, Philadelphia, PA, 1980, p. 45.

24. C. R. Shastry, J. J. Friel, and H. E. Townsend, Degradation of Metals in the Atmosphere, S. W. Dean and T. S. Lee (Eds.),

STP 965, American Society for Testing and Materials, Philadelphia, PA, 1986, p. 5.

25. E. A. Baker and T. S. Lee, Degradation of Metals in the Atmosphere, S. W. Dean and T. S. Lee (Eds.), STP 965, American Society for Testing and Materials, Philadelphia, PA, 1986, p. 52.

26. V. Kucera, S. Haagenrud, L. Atteraas, and J. Gullman, "Corrosion of Steel and Zinc in Scandinavia with Respect to the Classification of the Corrosivity of Atmospheres," in Degradation of Metals in the Atmosphere, S. W. Dean and T. S. Lee (Eds.), STP 965, American Society for Testing and Materials, Philadelphia, PA, 1986, pp. 264–281.

27. I. L. Rozenfeld, Effect of Alloy Composition on the Atmospheric Corrosion Rate, in Atmospheric Corrosion of Metals, translated by B. H. Tytell, E. C. Greco (Ed.), National Association of Corrosion Engineers (NACE), Houston, TX, 1972.

28. H. Okada, Y. Hosoi, K. Yukawa, and H. Naito, Tetsu-to-Hagane, **55**, 355 (1969).

29. H. Schwarz, Werkst. Korr., **16**, 93, 208 (1965).

30. I. Matsushima and T. Ueno, Corros. Sci., **11**, 129 (1971).

31. S. Misawa, M. Yamashita, T. Matsuda, H. Yuki, and H. Nagano, Tetsu-to-Hagane, **79**, 69 (1993).

32. A. Usami, Y. Tomita, Y. Tanabe, T. Saito, and K. Masuda, CAMP-ISIJ, **11**, 452 (1998).

33. Y. Shimizu, K. Tanaka, and T. Nishimura, Zairyo-to-Kankyo, **44**, 436 (1995).

34. A. Usami, Y. Tanabe, M. Yamamoto, H. Mabuchi, and H. Inoue, Proc. Fushoku Boshoku '95, D-108 (1995).

35. A. Nishikata, Y. Yamashita, H. Katayama, T. Tsuru, A. Usami, K. Tanabe, and H. Mabuchi, Corros. Sci., **32**, 2059 (1995).

36. K. Tanaka, Y. Shimize, and T. Nishimura, Zairyo-to-Kankyo '98B-115, 1998, p. 67.

37. N. Masuko and Y. Hisamatsu, Boshoku Gijutsu (Corros. Eng.), **17**, 465 (1968).

38. I. Suzuki, Y. Hisamatsu, and N. Masuko, J. Electrochem. Soc., **127**, 2210 (1980).

39. M. Stratmann and H. Streckel, Corros. Sci., **30**, 697 (1990).

40. M. Sakashita and N. Sato, Boshoku Gijutsu (Corros. Eng.), **25**, 450 (1979).

41. H. Kihira, Colloidal Aspects of Rusting of Weathering Steel, in Electrical Phenomena at Interfaces, H. Oshima and K. Furusawa (Eds.), Marcel Dekker, New York, 1998, p. 428.

42. H. Kihira, S. Itoh, and T. Murata, Corros. Sci., **31**, 383 (1990).

43. H. Kihira, S. Itoh, and T. Murata, Corrosion, **45**, 347 (1989).

44. C. P. Larrabee, Trans. Electrochem. Soc., **85**, 297 (1944).

45. G. Schikorr, Z. Electrochem., **43**, 697 (1937).

46. T. Moroishi and J. Satake, Tetsu-to-Hagane, **59**, 125 (1970).

47. J. Satake, T. Moroishi, Y. Nishida, and S. Tanaka, Sumitomo Metals, **22**, 516 (1970).

48. I. Matsushima, Y. Ishizu, T. Ueno, M. Kanazashi, and K. Horikawa, Boshoku Gijutsu, **23**, 177 (1974).

49. Guideline for the Use of Bare Weathering Steel for Bridges, Japan Association for Construction of Bridges, Tokyo, Japan, 1991.

50. M. Kaneko, S. Miyata, S. Fujita, and M. Yasuhara, CAMP ISIJ, **11**, 1111 (1998).

51. Y. Itoh, S. Yamaguchi, K. Masuda, and C. Kato, Zairyo-to-Kankyo, **98B**, 113S (1998).

52. H. Kishikawa, H. Yuki, S. Hara, M. Kamiya, and M. Yamashita, 45th zairyo to kankyo toronkai, (45th Material and Environment Examination Meeting), Zairyo to Kankyo (Corrosion Engineering) **47** (12), 216 (December 15, 1998).

53. Specification for Highway Bridges-Parts I (Common Design Principles) and II (Steel Bridges), Japan Road Association, Pub. Maruzen, Tokyo, Japan, (2002).

54. H. Kihira, Corros. Sci., **49**, 112 (2007).

55. H. Kihira, T. Senuma, M. Tanaka, Y. Fujii, and Y. Sakata, Corros. Sci., **47**, 2377 (2005).

56. T. Nakayama, Feramu, **13**, 804 (2008).

57. I. Kage, Y. Matsuda, K. Shiotani, T. Kamori, and K. Kyono, Materia, **48**, 132 (2009).

58. H. Kihira, A. Imai, M. Hiramatsu, T. Aiga, T. Matsumoto, T. Sato, and M. Nagai, ECS Trans., **16** (43), 115–124 (2009).

59. H. Kihira, A. Usami, K. Tanabe, M. Itoh, G. Shigesato, Y. Tomita, T. Kusunoki, T. Tsuzuki, S. Itoh, and T. Murata, in Proceedings of Symposium on Corrosion and Corrosion Control in Saltwater Emvironments, Honolulu, The Electrochemical Society, Pennington, NJ, 1999, p. 127.

60. H. Kihira, S. Itoh, T. Mizoguchi, T. Murata, A. Usami, and K. Tanabe, Zairyo-to-Kankyo, **49**, 30 (2000).

61. M. Takemura, S. Fujita, S. Shuzuki, and K. Matsui, NKK Technical Report, **171**, 913 (2000).

62. K. Shioya, F. Kawabata, and K. Amano, Kawasaki, Kawaski Steel Technical Report, **33**, 39 (2000).

63. H. Kawano, S. Okano, M. Sakai, T. Nakayama, F. Yuse, and K. Hase, Kobe Steel Technical Report, **52**, 25 (2002).

64. M. Kimura, H. Kihira, N. Ohta, M. Hashimoto, and T. Senuma, Corros. Sci., **47**, 2499 (2005).

49

CORROSION OF STEEL IN CONCRETE

J. P. BROOMFIELD

Corrosion Consultant, London, United Kingdom

A. EXTENT AND COST OF REINFORCEMENT CORROSION

Reinforced concrete is one of the most durable, versatile, and widely used construction materials. It can be used to make small posts, vast bridges, and tall buildings as well as lining tunnels and constructing pipelines. Generally, this composite material is capable of withstanding a wide range of environments, from oil rigs in the North Sea to deserts. However, occasionally it does not give the low maintenance life expected of it. Sometimes this is due to incorrect expectations, sometimes to inadequate specification or construction, and sometimes to more adverse conditions than initially expected. Consequently, there are many structures in the built environment suffering from corrosion-induced damage.

One estimate from the United States [1] is that the cost of damage due to deicing salts alone is between $325 and $1000 million/year to reinforced concrete bridges and car parks. In the United Kingdom, the Department of Transport (DoT) estimates a total repair cost of £616.5 million (~1 billion U.S. dollars) due to corrosion damage to motorway bridges [2]. These bridges represent ~10% of the total bridge inventory in the United Kingdom. The total problem may therefore be 10 times the DoT estimate. There are similar statistics for Australia, Europe, and particularly the Middle East, where the warm marine climate with saline ground conditions increases all corrosion problems. Corrosion control is made more difficult by the problems of curing concrete in hot, drying environments and has led to very short lifetimes for reinforced concrete structures [3]. Deterioration occurs on buildings and other structures as well as bridges.

B. PRINCIPLES OF REINFORCEMENT CORROSION IN CONCRETE

Whereas concrete and concrete deterioration are described in Chapter 31, this chapter is concerned with corrosion of steel in concrete. As concrete is porous and both moisture and oxygen can move through the pores and microcracks in concrete, the basic requirements for corrosion of mild or

Uhlig's Corrosion Handbook, Third Edition, Edited by R. Winston Revie
Copyright © 2011 John Wiley & Sons, Inc.

high-strength ferritic reinforcing steels are present. The reason that corrosion does not occur in most cases is that the pores contain high levels of calcium, sodium, and potassium hydroxide, which maintain a pH of between 12.5 and 13.5. This high level of alkalinity passivates the steel, forming a dense gamma ferric oxide that is self-maintaining and prevents rapid corrosion.

In many cases, any attack on reinforced concrete will be on the concrete. However, there are two chemicals that penetrate the concrete and attack the steel without breaking down the concrete first. The culprits are chlorides and carbon dioxide as these are the main common atmospherically borne species that penetrate concrete without causing significant damage and then promote the corrosion of steel by removing the protective passive oxide layer on the steel, created and sustained by the alkalinity of the concrete pore water.

There are many texts covering the mechanisms of corrosion in concrete and assessment techniques (e.g., [4–7]) as well as specifications and recommended practice documents on how to select and apply repair methods (e.g., NACE SP 0290 and SP 0390, BSEN 12696, BSEN 1504 [7a] and ACI 222R-01; see Bibliography).

Figure 49.1 shows the electrochemical corrosion mechanism common to both carbonation- and chloride-induced corrosion. The separation of anodes and cathodes is an important part of the understanding, measurement, and control of corrosion of steel in concrete. Corrosion of steel in concrete is basically an aqueous corrosion mechanism where there is very poor transport of corrosion product away from the anodic site. This usually leads to the formation of voluminous corrosion product and cracking and spalling of concrete, with delaminations forming along the plane of the reinforcing steel. In the absence of sufficient oxygen at the anodic site, the ferrous ion will stay in solution or diffuse away and deposit elsewhere in pores and microcracks in the concrete, leading to severe section loss without the advanced warning given by concrete cracking and spalling.

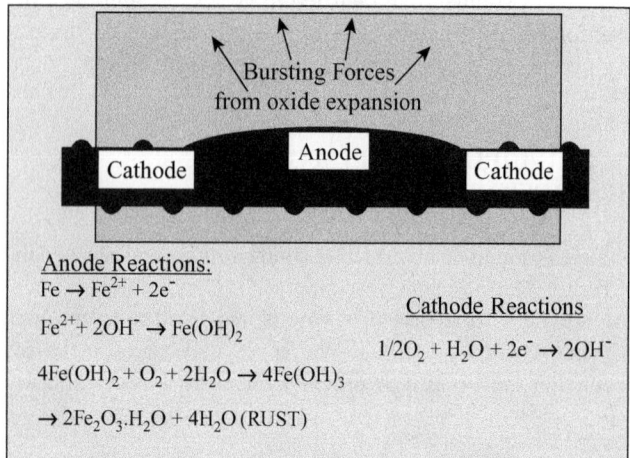

Anode Reactions:
$Fe \rightarrow Fe^{2+} + 2e^-$

$Fe^{2+} + 2OH^- \rightarrow Fe(OH)_2$

$4Fe(OH)_2 + O_2 + 2H_2O \rightarrow 4Fe(OH)_3$

$\rightarrow 2Fe_2O_3 \cdot H_2O + 4H_2O$ (RUST)

Cathode Reactions

$1/2O_2 + H_2O + 2e^- \rightarrow 2OH^-$

FIGURE 49.1. The corrosion process for steel in concrete.

C. CHLORIDE INGRESS AND THE CORROSION THRESHOLD

Chlorides can be present in concrete for a number of reasons. Examples are:

1. Contamination
   a. Deliberate addition of calcium chloride set accelerators
   b. Deliberate use of seawater in the mix
   c. Accidental use of inadequately washed marine sourced aggregates
2. Ingress
   a. Deicing salt ingress
   b. Sea salt ingress
   c. Chlorides from chemical processing

Until the later 1970s, it was widely held that chlorides cast into concrete were largely bound as chloroaluminates and would not cause corrosion. It was then found that large numbers of structures with chloride cast into the mix did suffer from corrosion and that binding was not as effective as initially believed. The ACI Report 222R-01 reviews the national standards and laboratory data. The consensus is that a chloride level of 0.4% chloride by weight of cement is a necessary but not sufficient condition for corrosion and that in variable chloride and aggressive conditions corrosion can occur at lower chloride levels, down to ~0.2% chloride by weight of cement.

A literature review [8] suggested that, whether chlorides are bound or not, the chloroaluminates break down releasing chloride ions for participation in the passivation breakdown process. They suggested that the discussion about the amount of chloride bound in the cement paste is therefore less important than previously thought. They also pointed out that the amount of calcium hydroxide available to maintain the pH has a profound effect on initiation of corrosion. This has implications for cement replacement materials.

Chloride ingress into concrete is generally held to follow Fick's second law of diffusion, forming a chloride profile with depth into the concrete:

$$d[Cl^-]/dt = D_c \cdot d^2[Cl^-]/dx^2$$

where $[Cl^-]$ is the chloride concentration at depth x and time t and D_c is the diffusion coefficient (usually of the order of $10^{-8}\,cm^2/s$). The solution to the differential equation for chlorides diffusing in from a surface is

$$(C_{max} - C_{x,t})/(C_{max} - C_{min}) = erf\frac{x}{(4D_ct)^{1/2}}$$

where

C_{max} = surface or near-surface concentration
$C_{x,t}$ = chloride concentration at depth x at time t

C_{min} = background chloride concentration
erf = error function

The parameter C_{max} must be constant, which is why a near-surface measurement is used, to avoid fluctuations in surface levels on wetted and dried surfaces.

An alternative to the error function calculation is to calculate the progress of a threshold chloride concentration as a parabolic function of time [9, 10] using a similar function to that used for the carbonation front progression.

D. CARBONATION OF CONCRETE AND CONSEQUENCES OF CORROSION

Carbonation is a simpler process than chloride attack. Atmospheric carbon dioxide reacts with the pore water to form carbonic acid. This reacts with the calcium (and other) hydroxides to form solid carbonates. The pH therefore drops from ~pH 13 to ~pH 8. The steel starts to corrode at around pH11.

Carbonation is associated with poor concrete cover, poor concrete quality, poor consolidation, and old age in the absence of chlorides. Carbonated concrete is good quality concrete but it is no longer protective to the reinforcing steel. Carbonation rates generally follow parabolic kinetics:

$$d = At^{0.5}$$

where

  d = carbonation depth
  t = time
  A = constant, generally of the order 0.25–1.0 mm/ year$^{0.5}$

Carbonation rates are a function of the environment where indoor concrete will carbonate faster than outdoor concrete in North American and Northern European environments, but wet/dry cycling in warm conditions can accelerate carbonation in more southern latitudes. As corrosion will not proceed in the absence of water, carbonation rates are usually unimportant inside buildings except in bathrooms, kitchens, and other situations where there is wet/dry cycling and sufficient moisture to cause corrosion after carbonation of the concrete to reinforcement depth.

E. METHODS OF DETECTING AND MEASURING CORROSION AND CORROSION DAMAGE IN CONCRETE

There are several techniques available for assessing corrosion and corrosion damage on reinforced concrete structures. The main ones used in the field are described below.

E1. Visual Inspection

The visual inspection is the first step in any investigation. A properly executed survey consists of a rigorous logging of every defect seen on the concrete surface. The aim of the visual survey is to give a first indication of what is wrong and how extensive the damage is. If concrete is spalling off, then that can be used as a measure of extent of damage. In some cases, weighing the amount of spalled concrete with time can be a direct measure.

The main equipment is obviously the human eye and brain, aided with a notebook, proforma, or hand-held computer, and a camera to record and locate defects. Binoculars may be necessary, but close inspection is preferred if access can be arranged. A systematic visual survey will be planned in advance. Many companies that carry out condition surveys have standardized systems for indicating the nature and extent of defects. These are used in conjunction with customized proformas for each element or face of the structure. It is normal to record date, time, and weather conditions when doing the survey and to note visual observations, such as water or salt rundown and damp areas.

Interpretation is usually based on the knowledge and experience of the engineer or technician conducting the survey. The Strategic Highway Research Program (SHRP) has produced an expert system, HWYCON [11]. This guides the less experienced engineer or technician through the different types of defects seen on concrete highway pavements and structures, including alkali silica reaction, freeze–thaw damage, and corrosion. Further information on bridge surveys can be found in CBDG Report 2 2002 [11a].

The main limitation of visual inspection is the skill of the inspector. Some defects can be mistaken for others. When corrosion is suspected, it must be understood that rust staining can come from iron-bearing aggregates rather than from corroding reinforcing steel. Different types of cracking can be attributed to different causes. The recognition of alkali silica reaction (ASR) is discussed in Chapter 31. Visual surveys must always be followed up by testing to confirm the source and cause of deterioration.

E2. Delamination Surveys

As corrosion proceeds, the corrosion product formed takes up a larger volume than the steel consumed, building up tensile stresses around the rebars. A layer of corroding rebars will often cause a planar fracture at rebar depth prior to spalling of the concrete. The aim of a delamination survey is to measure the amount of cracking between the rebars before it becomes apparent at the surface. It should be noted that this can be a very dynamic situation.

The delamination survey with a hammer is often conducted alongside the visual survey with hollow sounding areas marked directly onto the surface of the structure with a suitable permanent or temporary marker and then recorded on the visual survey proforma. The hammer or chain drag survey is usually quicker, cheaper, and more accurate than alternatives such as radar, ultrasonics, or infrared (IR) thermography. The other techniques do have their uses, for example, in large-scale surveys of bridge decks (ultrasonics and infrared) or waterproofing membranes or other concrete defects (ultrasonics and radar). The reader is recommended to review the literature for further information [12–14].

For interpretation, the experience of the inspector is vital. A skilled, experienced technician will often produce better results than the more qualified but less experienced engineer. Trapped water within cracks, deep cracks (where bars are deep within the structure), and heavy traffic noise can complicate the accurate measurement of delamination.

It is common during concrete repairs for the amount of delamination to be far more extensive than delamination surveys indicate, because of the inaccuracy of the techniques available and also because of the time between survey and repair. Once corrosion has started, delaminations can initiate and grow rapidly. An underestimate of 40% or more is not unusual and should be borne in mind when budgeting for repairs.

E3. Concrete Cover Measurements

The thickness of concrete cover to the reinforcing steel is measured on new structures to see that adequate cover has been provided to the structure in line with design requirements. It is also carried out when corrosion is observed because low cover will increase the corrosion rate by allowing both the agents of corrosion (chlorides and carbonation) more rapid access to the steel and more rapid access of the "fuels" for corrosion, moisture, and oxygen. A cover survey can determine the location of the rebars three dimensionally, that is, their position with regard to each other (X, Y) and depth from the surface (Z). If construction drawings are not available, then it may be necessary to measure rebar diameters as well as locations.

Magnetic cover meters (often known as pachometers in North America) are now available with logging features and digital outputs. With some models, a spacer can be used to estimate rebar diameter. Other alternatives, such as radiography, can be used to survey bridges or other structures.

There is only one standard for cover meters, BS 1881 Part 204, and this standard refers to the measurement on a single rebar. A useful paper discusses cover meter accuracy when several rebars are close together [15]. The paper suggests that different types of head are more accurate in different conditions. The smaller heads are better for resolving congested rebars.

The main problem with cover measurements is the congestion of rebars giving misleading information [15].

Iron-bearing aggregates can lead to misleading readings as they will influence the magnetic field. Different steels also have different magnetic properties (at the extreme end, austenitic stainless steels are nonmagnetic). Most cover meters have calibrations for different reinforcing steel types. See also RILEM TC 189 Report [15a].

E4. Carbonation Depth

There are two principal causes of corrosion: chloride contamination and carbonation of the concrete. Any survey must distinguish between the two. The distinction is important because the type of repair may be determined by the cause of corrosion. The neutralization of the pore solution by the carbonation process leads to a "carbonation front," where there is a transition from ~pH 13 to 8.

Carbonation is easily measured by exposing fresh concrete and spraying with phenolphthalein indicator solution. The measurement can be done either by breaking away a fresh surface (e.g., between the cluster of drill holes used for chloride drilling) or by coring and splitting or cutting the core in the laboratory. The phenolphthalein solution remains clear where concrete is carbonated and turns pink where concrete is still alkaline. The best indicator solution for maximum contrast of the pink coloration is a solution of phenolphthalein in alcohol and water [16]. If the concrete is very dry, then a light misting with water prior to applying the phenolphthalein will also help show the color. Petrographic analysis will also reveal carbonated and partially carbonated zones under an optical microscope.

Sampling can allow the average and standard deviation of the carbonation depth to be calculated. If the carbonation depth is compared with the average reinforcement cover, then the amount of depassivated steel can be estimated. If the carbonation rate can be determined from historical data and laboratory testing, then the progression of depassivation with time can be calculated.

In some aggregates, accurate phenolphthalein readings are difficult to obtain. Some concrete mixes are dark in color and color change may not be visible. Care must be taken that no contamination of the surface occurs from dust and the phenolphthalein-sprayed surface must be freshly exposed or it may be carbonated before testing. High-alumina-cement (HAC) concrete does not show the color change well, and HAC concretes may be highly susceptible to carbonation because of low alkali content (see Chapter 31).

It is also possible for the phenolphthalein to bleach at very high pH (e.g., after electrochemical chloride extraction or realkalization). If the sample is left for a few hours, it will turn pink. There can also be problems on buried structures where carbonation by groundwater does not always produce the clear carbonation front induced by atmospheric carbon dioxide ingress. It should be noted that phenolphthalein changes color at pH 9.5 whereas concrete starts to corrode below pH 11. This could be important in realkalized concrete, where the pH may not reach 11 but may reach 9.5. A new European standard BS EN 14630 on carbonation depth measurements is now available [16a].

E5. Chloride Content

Chlorides can be cast into concrete or can be transported in from the environment. The chloride ion attacks the passive layer even though there is no significant drop in pH. Chlorides act as catalysts to corrosion. They are not consumed in the process but help to break down the passive layer of oxide on the steel and allow the corrosion process to proceed quickly. Chloride contents are measured by sampling the concrete and analyzing a liquid extracted from the sample. This is usually done by mixing acid with concrete dust from drillings or crushed core samples. An alternative is pore water extraction by squeezing samples of concrete or, more usually, mortar. The pore-squeezing technique is frequently used in laboratory experimental work as it is often difficult to extract useful pore water samples from field concrete. A third technique is to crush concrete and boil it in water to extract water-soluble chlorides only.

To measure the chloride profile in the concrete, chloride samples should be collected incrementally from the surface by taking either drillings or sections from cores. The first 5 mm is usually discarded as being directly influenced by the immediate environment and then measurements of chloride content made at suitable increments. For improved statistical accuracy, multiple adjacent drillings are made and the depth increments from each drilling are mixed. There are several ways of measuring the chlorides once samples are taken. Field measurements of acid-soluble chloride can be made using a chloride-specific ion electrode [17]. Conventional titration (e.g., by BS 1881 part 124 and potentiometric titration methods) is also available [18, 18a].

As well as acid-soluble chloride, there are the water-soluble chloride tests [American Society for Testing and Materials (ASTM) C 1218, American Association of State Highway Officers (AASHTO) Test T 260 and Federal Highway Administration Report FHWA RD-77-85]. These techniques use different levels of pulverization of large samples that are refluxed with water to extract the supposedly unbound chlorides. These chlorides are, or can become, free in the pore water to cause corrosion as opposed to the chlorides bound by the aluminates in the concrete or bound up in some aggregates of marine origin. The water-soluble chloride test is rather less accurate than the acid-soluble test because some of the "bound" chlorides can be released, and the finer the grinding, the more will be extracted. However, this test can be useful in showing the corrosion condition where chlorides

have been cast into concrete and particularly where aggregates are known to contain chlorides that do not leach into the pore water.

The corrosion threshold is usually given as $\sim$0.4% chloride by weight of cement, 0.05% chloride by weight of concrete, or 1.2 lb/yd^3 of concrete. This is only an approximation because:

(a) Concrete pH (14-$\log_{10}$ of the OH^- concentration) varies with the cement powder and the concrete mix. A small pH change is a massive change in OH^- concentration, and therefore the threshold moves radically with pH.

(b) Chlorides can be bound chemically (by aluminates) and physically (by absorption on the pore walls). This removes them (temporarily or permanently) from the corrosion reaction.

(c) Some aggregates contain chlorides that cannot be leached into the pore water. They do not play any part in the corrosion threshold but will show up on acid-soluble chloride tests.

(d) In very dry concrete, corrosion may not occur even at very high Cl^- concentration as the water is missing.

(e) In saturated concrete, corrosion may not occur even at a very high Cl^- concentration as the oxygen is missing.

Corrosion can be observed at 0.1% chloride income cases and none seen at >1.0% chloride by weight of cement or more in others. If (d) or (e) is the reason that no corrosion is observed, then a change in conditions may lead to corrosion.

The important questions from chloride measurement are how much of the rebar is depassivated and how will this progress. Points (a)–(c) review how the corrosivity of the chloride can change. If chlorides have been transported in from outside, then the chloride profile can be used along with measurements (or estimates) of the diffusion constant to estimate future penetration rates and the buildup of chloride at rebar depth.

In Europe, it is normal to quote the chloride content as a percentage by weight of cement. In many cases, assumptions must be made about the cement content because the measurement is by weight of sample (concrete) and the true cement content is not known.

E6. Concrete Resistivity Measurements

The four-probe resistivity meter is now used for measurement of concrete resistivity on site. The measurement can be used to indicate the possible corrosion activity if steel is depassivated. The electrical resistivity is an indication of the amount of moisture in the pores and the size and tortuosity of the pore system. Resistivity is strongly affected by concrete quality (i.e., cement content, water/cement ratio, curing, and additives used).

The four-probe resistivity meter used for soil resistivity measurements has been modified for concrete application and is used by pushing pins directly onto the concrete surface with moisture or gels to enhance the electrical contact. Broomfield and Millard (Chapter 5 in [19]) and Andrade et al. (2000) [19a] described two versions of the equipment. Other variations use drilled-in probes or a simpler, less accurate two-probe system.

An alternative approach measures the resistivity of the cover concrete by a two-electrode method using the reinforcing network as one electrode and a surface probe as the other.

Concrete resistivity of the area around the sensor is obtained by the formula

$$\text{Resistivity} = 2RD(\Omega \cdot \text{cm})$$

where

R = resistance by " iR drop" from a pulse between the sensor electrode and the rebar network.

D = electrode diameter of sensor

The interpretation of resistivity measurements with regard to corrosion is empirical. The following interpretation of resistivity measurements has been cited when referring to depassivated steel [20]:

20 k$\Omega \cdot$ cm	Low corrosion rate
10–20 k$\Omega \cdot$ cm	Low to moderate corrosion rate
5–10 k$\Omega \cdot$ cm	High corrosion rate
<5 k$\Omega \cdot$ cm	Very high corrosion rate

Researchers working with a field linear polarization device for corrosion rate measurement have conducted laboratory and field research and found the following correlation between resistivity and corrosion rates using the surface-to-rebar two-electrode approach [21]:

100 k$\Omega \cdot$ cm	Cannot distinguish between active and passive steel
50–100 k$\Omega \cdot$ cm	Low corrosion rate
10–50 k$\Omega \cdot$ cm	Moderate to high corrosion where steel is active
<10 k$\Omega \cdot$ cm	Resistivity is not the controlling parameter

In this method and interpretation resistivity is used along side linear polarization measurements (see below).

The resistivity measurement is a useful additional measurement to aid in identifying problem areas or confirming concerns about poor quality concrete. Measurements can only be considered alongside other measurements. Reinforcing bars will interfere with resistivity measurements.

E7. Electrochemical Potential Measurements

Measurement of the electrochemical potential of the steel with respect to a standard reference electrode gives an indication of the corrosion risk of the steel.

The silver/silver chloride (Ag/AgCl) electrode, mercury/mercury oxide (Hg/HgO), and sometimes the standard calomel electrode (SCE) are commonly used as reference electrodes for steel in concrete. These are now available as extremely robust, double-junction electrodes with gel electrolyte, minimizing the risk of contamination and maintenance requirements. Copper/copper sulfate (CSE) cells are also used but are not recommended because of the maintenance needs, the risk of contamination of the cell, the difficulty of use in all orientations, and the leakage of copper sulfate. A high-impedance digital voltmeter is used to collect the data in the simplest configuration. Other options are to use a logging voltmeter (or logger attached to a voltmeter), an array of cells with automatic logging, or a reference electrode linked to a wheel for rapid data collection.

Interpretation is most reliable for cast in situ, reinforced concrete with chloride-induced corrosion due to diffusion of sea or deicing salts. The ASTM C-876 states that there is a high risk of corrosion if the potential is more negative than −350 mV CSE. If the potential is less negative than −200 mV, there is a low risk of corrosion. In between those values the risk is indeterminate. This interpretation does not necessarily apply to carbonated structures, precast concrete elements, or elements with chlorides cast into the concrete. The readings can also be affected by moisture content, with the base of columns or walls showing more negative potentials regardless of corrosion activity. Stray currents can also influence the readings, which can be used as a diagnostic tool where stray-current corrosion is suspected in the presence of direct current (dc) fields.

Very negative potentials have been measured below the water line in marine environments; however, the lack of oxygen will often slow the corrosion rate to negligible levels.

"Junction potentials" can be created by the change in chemical concentrations within the concrete. This effect was severe in a concrete slab subjected to chloride removal, but that may be due to the treatment, rather than being a real problem under normal conditions. Junction potentials may explain the erratic changes in potential seen in carbonated structures. The potential changes across the carbonation front because of the pH change and because the concentration of dissolved ions changes as carbonated concrete wets and dries quickly because the pores are narrowed by a lining of calcium carbonate. Further information on interpretation can be found in the RILEM recommendation for half-cell potential measurement [21a].

E8. Corrosion Rate Measurement

The corrosion rate is probably the nearest the engineer will get to measuring the rate of deterioration with current technology. Although there are various ways of measuring the rate of corrosion, including ac impedance and electrochemical noise [22], linear polarization, also known as polarization resistance, will be discussed along with various macrocell techniques as they have been shown to be field worthy for measuring the corrosion of steel in concrete.

The rate at which steel dissolves and forms oxide is determined by measuring the electric current generated by electrons liberated by the anodic reaction,

$$Fe \rightarrow Fe^{2+} + 2e^-$$

and recaptured by the cathodic reaction,

$$H_2O + \tfrac{1}{2}O_2 + 2e^- \rightarrow 2OH^-$$

The current is converted to the rate of metal loss by Faraday's law:

$$m = z\mathcal{F}/MIt$$

where

m = mass of steel consumed (g)
I = current (A)
t = time(s)
$\mathcal{F}$ = 96,500 A.s
z = ionic charge (2 for Fe $\rightarrow$ Fe^{2+} + 2e$^-$)
M = atomic weight of metal (56 g for Fe)

This results in a conversion of $1\,\mu A/cm^{-2} = 11.6\,\mu m/$year.

A typical setup is shown in Figure 49.2. The basic system has a rebar connection, a reference electrode, or half cell, an auxiliary electrode through which a small current is applied, and a battery-operated power supply to supply the direct current to change the potential of the steel with respect to the reference electrode. The equation for the corrosion current is given by Stern and Geary [23]:

$$I_{corr} = \frac{B}{R_p}$$

where

I_{corr} = corrosion current.
B = constant related to anodic and cathodic Tafel slopes

and

R_p = polarization resistance, = dE/dI

where

dI = change in current
dE = change in potential

The potential is measured with respect to the reference electrode or half cell. Current, dI, is passed from the surrounding electrode to the steel and the potential shift, dE, is measured. This measurement can be repeated for increasing increments of dI. The potential must be stable throughout the reading so that a true dE is recorded.

The simplest devices do not include a guard ring. Therefore they do not define the area of measurement accurately. At low corrosion rates this can lead to errors by orders of magnitude [24]. A more accurate device has been developed that uses a guard ring controlled by half cells [25, 26]. It has

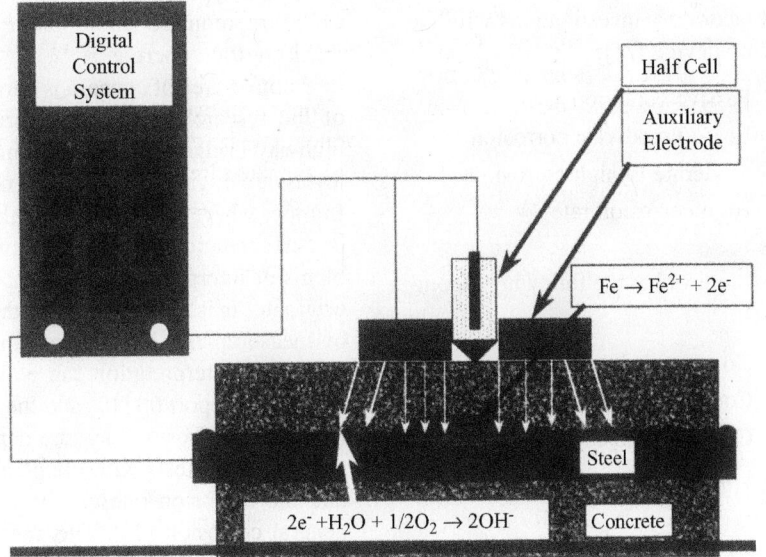

FIGURE 49.2. Schematic of setup for linear polarization measurement.

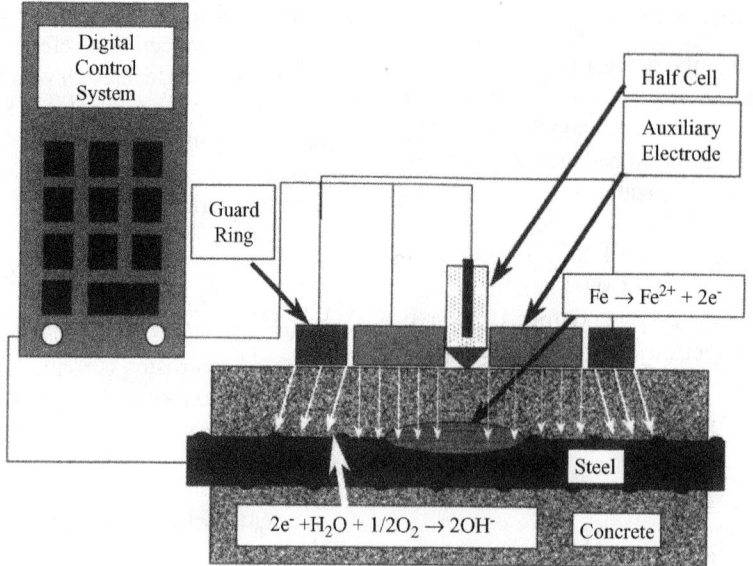

FIGURE 49.3.   Guard ring linear polarization defines area of measurement.

also been used to assess carbonated as well as chloride-induced corrosion. Figure 49.3 shows the system used. Its important features are the two extra reference electrodes that are used to control the guard ring current and define the area of measurement.

In an assessment of three different devices, one without a guard ring, one with a simple guard ring, and the device with the sophisticated half-cell controlled guard ring [24], a good correlation was found between this device and the most sophisticated laboratory measurements, except where the concrete resistivity was very high or cover to the rebar was very deep. Field trials also showed good performance of the device.

The following broad criteria for corrosion have been developed from field and laboratory investigations with the sensor-controlled guard ring device [21]:

$I_{corr} < 0.1\,\mu A/cm^2$	Passive condition
$I_{corr}\ 0.1–0.5\,\mu A/cm^2$	Low to moderate corrosion
$I_{corr}\ 0.5–1\,\mu A/cm^2$	Moderate to high corrosion
$I_{corr} > 1\,\mu A/cm^2$	High corrosion rate

A device without sensor control has the following recommended interpretation [27]:

$I_{corr} < 0.2\,\mu A/cm^2$	No corrosion expected
$I_{corr}\ 0.2–1.0\,\mu A/cm^2$	Corrosion possible in 10–15 years
$I_{corr}\ 1.0–10\,\mu A/cm^2$	Corrosion expected in 2–10 years
$I_{con} > 10\,\mu A/cm^2$	Corrosion expected in 2 years or less

These measurements are affected by temperature and relative humidity, so the conditions at the time of

measurement affect the interpretation of the limits defined above. The measurements should be considered accurate to within a factor of 2.

Although the conversion of I_{corr} to section loss and end of service life has been investigated [28], the loss of concrete is the most usual cause for concern, rather than loss of reinforcement cross section. It is far more difficult to predict cracking and spalling rates, especially from an instantaneous measurement. A simple extrapolation assuming that the instantaneous corrosion rate on a certain day is the average rate throughout the life of the structure often gives inaccurate results [21]. The average rate of section loss and rate of delamination are difficult to predict from this type of measurement because further assumptions are required about oxide volume and stresses for cracking the concrete.

Another area of concern is corrosion due to pitting. Much of the research on linear polarization has been done on highway bridge decks in the United States, where chloride levels are high and pitting is not observed. However, in Europe, where corrosion is localized in run-down areas on bridge substructures and pitting is more common, the problems of interpretation are complicated as the I_{corr} reading originates in isolated pits rather than uniformly from the area of measurement. Further information on methods, equipment, and interpretation can be found in Concrete Society Technical Report 60 [19] and the RILEM recommendations for on-site corrosion measurements [28a].

Laboratory tests with the guard ring device have shown that the corrosion rate can be up to 10 times higher than general corrosion [29]. This means that the device is very sensitive to pits. However, it cannot differentiate between pitting and general corrosion.

An alternative approach, which introduces the theme of long-term corrosion monitoring, is the embedding of macro-cell devices. This includes galvanic couples of different steels [30] or embedding steel in high-chloride concrete to create a corrosion cell, as is popular in cathodic protection monitoring systems. This approach is also popular in laboratory corrosion studies and is an ASTM procedure (ASTM G109 [31]). Concrete prisms are made with a single top rebar in chloride-containing concrete and two bottom rebars in chloride-free concrete. The current flow between the top and bottom bars is monitored.

Another configuration has been used for installation in new structures, for example, the Great Belt Bridge in Denmark [32]. This approach uses a "ladder" of steel specimens of known size installed at an angle through the concrete cover. As the chloride (or carbonation) front advances, each specimen becomes active, and the current flow from the anodic specimen to its adjacent cathodic specimen can be monitored along with the potential.

There are two major problems with interpreting macro-cell systems. The first is that we are not actually measuring corrosion on the reinforcing steel. The extent to which the current represents the reaction on the actual reinforcement partly depends on the care and thought taken about installation of the probes. The second issue is how representative macrocell currents are of the true corrosion currents in the steel. The microcell currents may be more important than macrocell currents. In a comparison with linear polarization, it was found that the macrocell technique underestimates the corrosion rate, sometimes by an order of magnitude [33]. As this was using the ASTM prism technique, it should be considered the most accurate use of the macrocell technique, so if it is an order of magnitude out, field uses of macrocell techniques are probably even less accurate.

The advantage of these techniques is that they are permanently set up and can be used to monitor the total charge passing with time. This approach provides a clearer indication of the total metal loss or total oxide produced than does instantaneous measurement of the corrosion rate.

E9. Permanent Corrosion Monitoring

The technique of corrosion monitoring is well established for cathodic protection of reinforced concrete structures where microprocessor-controlled systems are linked by modem to remote monitoring stations and potentials and current flows are monitored. The sort of system available was described several years ago [20]. Variations on this concept were used in the Middle East and on the U.K. motorway system. Systems have been installed in new and existing reinforced concrete structures to monitor corrosion rate with polarization resistance, resistivity, temperature, and half-cell potentials [34].

The advantage of long-term monitoring is that the progression of changes can be monitored. The growth of anodic areas, potential changes, changes in corrosion rates using linear polarization or macrocell approaches, and changes in concrete resistivity with time are more helpful in predicting long-term durability than the "snapshot" approach of a survey.

F. METHODS OF REPAIRING AND TREATING CORROSION DAMAGE

Repairs are rarely applied in isolation. Most other corrosion control systems are preceded by patch repairing, and patch repairs are frequently followed by coating. No single rehabilitation process is always suitable for all elements of all structures, so there is always a degree of "mix and match" in rehabilitating a structure, by applying the correct repairs and treatments to the different elements in different environments.

F1. Patch Repairing

Almost all corrosion repairs are preceded by patch repairs, since treatments such as cathodic protection are rarely applied until after corrosion damage is observed. The exceptions are discussed below. The degree and extent of patch repairing are determined by what other treatment is being applied.

Patch repairs are usually applied with the intention of making the structure safe, restoring its appearance, and inhibiting further corrosion. Modern patch repair materials produce an extremely durable repair, but the act of patching can actually cause corrosion in adjacent areas by the "incipient anode" effect (Fig. 49.4). If a patch repair is carried out, then an anode is removed, stopping the cathodes from functioning and allowing the development of new anodes around the patch. This phenomenon is commonly observed on structures suffering from chloride-induced corrosion where anodes are large and well defined.

Patch repairs are applicable to most reinforced structures and should be carried out with appropriate materials and skilled operators. The most comprehensive standard for designing and specifying concrete repair systems is BS EN 150, parts 1–10 [7a]. There are other comprehensive guidance documents, for example the Concrete Society Technical Report 38 [35] and the Joint ACI/ICRI Concrete Repair Manual [35a]. Proprietary patch repair systems have the advantage of ensuring that the correct proportions are mixed together from sealed containers of known "use by date." They may include bond coats applied to the parent concrete and primers for the steel and may be applied by hand, flowable grouts into shutters, or spray applied. All systems require cutting the concrete from the corrosion-damaged area of steel back to undamaged concrete and steel,

BEFORE   **AFTER**

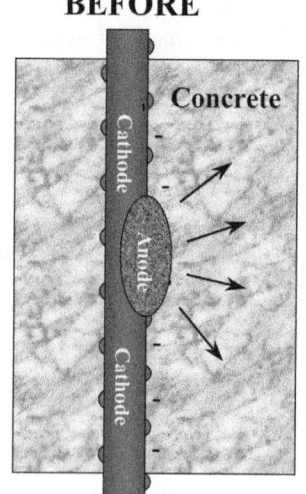

Corrosion at +ve Anode

"Anode" is made
noncorroding -ve cathode

FIGURE 49.4. The formation of incipient anodes after patch repairing.

good surface preparation of the steel, and square edges to the breakout. Materials must be low shrinkage and require good curing to minimize the risk of cracking in the repair.

The length of bar and the amount of the circumference exposed during breakout will depend on the extent and type of corrosion and the other elements of the treatment. If cathodic protection, realkalization, or chloride extraction are being applied, then the repair is mainly cosmetic as the other techniques will stop the corrosion. If only patch repairs are being used, then the repair should extend all round the bar (usually cut out 25–40 mm behind the bar) and at least 100 mm beyond the point where any corrosion damage or rust can be seen along the length of the bar.

Patch repairs are a straightforward, immediate solution to corrosion damage that can improve the condition of a structure and can extend the life of a structure either to give an acceptable repair cycle (e.g., repairs every 10–20 years) or to the desired life of the structure. Patch repairs will not stop corrosion elsewhere in the structure, and may even promote it due to the "incipient anode" effect described above. Nor will patch repairs take the structural load unless specifically designed to do so.

Patches are difficult to match to the existing finish of a structure, and even when that is done, they usually weather differently, so unless an opaque coating is applied, the patches will eventually become visible. Applying patch repairs is noisy, dusty, and dirty. Other treatments often claim the minimization of patch repairing as one of their benefits.

F2. Barriers

The aim of barrier materials is to stop or severely retard the ingress of aggressive agents such as carbon dioxide or chloride ions and to aid the drying out of the concrete. There is considerable literature on anticarbonation coatings and testing for carbonation resistance [e.g., 36]. Penetrating sealers, such as silanes, siloxane, and related compounds and mixtures, are used on highway substructures in many countries to retard chloride ingress without changing the appearance of the structure, or its ability to "breathe," allowing water vapor movement, but being liquid water repellent. The Netherlands, Norway, and the United States are the only OECD (Organisation of Economic Cooperation & Development) countries that do not mandate waterproofing on their bridge decks [37]. Although the membranes themselves require regular replacement throughout the life of the structure, they significantly reduce chloride ingress if good-quality membranes are applied and are properly detailed at edges, corners, drains, joints, and so on.

The principle of barrier materials is to stop ingress of chlorides, carbon dioxide, and/or water to slow down the depassivation and corrosion processes. Any attempt to use coatings once corrosion has initiated requires a careful analysis to see what will happen to the steel after coating the concrete. If chloride levels are well above the corrosion threshold, or if carbonation is well established around the reinforcing steel, then corrosion can continue. However, if the areas of depassivation are small and can be repaired, then coatings can retard or even "freeze" the "corrosion front."

In Europe, it is normal to apply silanes and waterproofing to new structures before chloride ingress occurs. Anticarbonation coatings are applied as part of a rehabilitation process once corrosion is detected. Coatings are frequently applied after electrochemical chloride extraction, realkalization, or inhibitor application to retard further ingress of aggressive agents (and to retain vapor-phase inhibitors). A thorough description of waterproofing membranes for bridge decks is given in [38].

Coatings can be effective in excluding chlorides and carbon dioxide. They will improve the appearance of patch-repaired structures and can be very cost effective if applied appropriately. Coatings, sealers, and membranes are likely to have only marginal effects on the corrosion rate once depassivation has occurred, particularly if there are substantial levels of chloride in the concrete. There is also a risk that, if some oxygen is excluded, then the anodic reaction of oxidation of iron to the soluble ferrous ion proceeds rapidly but there is no oxygen to react with it to form the solid hydrated ferric oxide that expands and cracks the concrete. Severe section loss of rebar without apparent concrete cracking has been observed by the author under failing waterproofing membranes on bridge decks in the United Kingdom and the United States due to this phenomenon. Coatings and membranes require replacement at regular intervals.

F3. Impressed Current Cathodic Protection

Impressed current cathodic protection (ICCP) systems consist of an anode system, a dc power supply, and monitoring probes, with associated wiring as with other cathodic protection systems described in Chapter 69. For steel in concrete, one of the most important decisions is the choice of anode. Anodes consist of conductive coatings on the surface, mixed-metal-oxide-coated mesh, or ribbons under concrete overlay, conductive mortar overlays, coated titanium ribbon in slots, or conductive ceramics or mixed-metal-oxide-coated titanium rods or tubes in holes in the concrete.

The coatings available range from a variety of formulations of carbon-loaded paints and thermal sprayed metals such as zinc (see Broomfield [5]). The overlay systems are either a wet sprayed conductive mortar or titanium grids or mesh fixed to the surface and overlaid with a cementitious overlay that can be sprayed or poured or pumped into oversized shutters. The probe or discrete anodes are rods of coated titanium, coated titanium ribbon, or conductive ceramic tubes in cementitious grout.

The important calculations for designing an ICCP system for steel in concrete are the steel surface area to be protected and hence the current output of the distributed anode and the transformer/rectifier supplying the direct current.

ICCP has been applied to most types of reinforced concrete structures in all types of conditions. The range of anodes means that the technology can be adapted for application to most situations. There is extensive literature on the subject (e.g., [5, 39]) and standards and specifications [40, 41]. ICCP has been applied to most types of reinforced concrete structures, such as:

Buildings (apartment buildings, offices, hospitals, schools, etc.)

Bridges (decks and substructures)

Tunnels

Wharves and jetties

Water towers

Industrial plant

Multistory car parks

It is primarily applied to structures suffering from chloride-induced corrosion due to cast in chlorides and marine or deicing salt ingress. More recently, it has been applied to historic steel framed masonry or brick structures where the stonework was cracking due to expansive corrosion of the steel frame. Cathodic protection has also been applied to new structures, either where chloride ingress is expected or where chlorides have been used in the mix ingredients.

Cathodic protection stops the corrosion process across the whole of the area where anodes are applied. It is extremely versatile and is not limited by the extent and severity of corrosion as long as anodes can be placed in a suitable electrolyte for the current to pass to the corroding steel areas. Using cathodic protection minimizes the amount of breakout for patch repairs, in terms of the need to cut behind the bar and extend along bars. As long as the concrete is sound, the ICCP provides protection. Smaller cutouts may alleviate the need for structural propping during patch repairs. Minimizing patch repairs minimized the disturbance to occupants of offices, apartment buildings, and so on.

The ICCP requires continuous ongoing monitoring for the life of the system/structure by suitably trained and qualified persons, such as described in the European standard BS EN 15257 [41a]. It has a comparatively high initial cost and anodes need replacing on a 10–40-year cycle, depending on the type used.

One of the advantages of the continuous monitoring requirement of ICCP systems is that its effectiveness and the condition of the structure are under continuous scrutiny, so that further intervention if required can be carried out in a timely manner.

Life-cycle cost analysis frequently shows ICCP to be one of the most cost effective repairs for structures with 20 years or more of residual life. Once applied, the system stops the cycle of regular concrete repair associated with structures with high chloride levels.

ICCP cannot easily be applied to elements containing prestressing steel due to the risk of hydrogen embrittlement (see NACE Report 01102 [41b]. There is concern about its application to structures with a high susceptibility to alkali silica reaction (ASR). Also, structures with a lot of electrically discontinuous steel can be expensive to protect as all the steel must be in contact for the current to flow correctly. This situation includes epoxy-coated reinforcing steel, where bars are electrically isolated by their coatings (but see below). Short circuits caused by tie wires and tramp steel touching the anode can also be a problem. Current cannot flow through insulators so coatings and membranes must be removed. Because ICCP anodes evolve gases, gas evolution must be handled if impermeable membranes are being applied over the anodes.

F4. Galvanic Cathodic Protection

Galvanic, or sacrificial anode, systems are attractive in that they cannot be controlled, so they are simpler than impressed current systems. They usually consist of zinc applied to the concrete surface and directly connected to the reinforcing steel. Alloys and other metals such as aluminum–zinc–indium alloys thermally sprayed onto the concrete surface are also used. The principles of galvanic cathodic protection are the same as for impressed current CP, except that the anode is a less noble metal than the steel to be protected and is consumed preferentially, generating the cathodic protection current. The potential difference between anode and cathode is a function of the environment and the relative electrode potentials of the anode and cathode materials. The current is a function of the potential difference and the electrical resistance. As the voltage and current cannot be controlled, the level of protection cannot be guaranteed and a low-resistance environment is required.

The most successful and extensive application has been to the splash and tidal zones of prestressed concrete piles and epoxy-coated reinforcing steel in substructures supporting bridges in the Florida Keys [42]. These consist of thermal sprayed zinc, zinc sheets clamped to the surface, or expanded zinc mesh grouted into permanent formwork on the splash and tidal zone of concrete columns. Recently, there have been applications of a zinc sheet with an adhesive gel on it [43]. A commercial version of a sacrificial anode for installation in patch repairs has been developed to suppress the incipient anode effect as discussed under patch repairs above. Over one million of these units were sold worldwide by 2007.

The simplicity of galvanic CP is very attractive. It requires only the installation of the anode and a direct connection to the steel. In Florida, one method was to clean up the steel exposed by corrosion damage and to arc spray zinc across the steel and the concrete directly connecting them. This procedure provided an added advantage in the cases where the steel had been epoxy coated as it made direct electrical connections. The lower current and voltage meant that the risk of accelerating corrosion in isolated bars was lower than for an ICCP system.

The current flows and voltages in galvanic CP systems are lower than in ICCP systems. Galvanic systems are therefore more readily usable with prestressing because the risk of hydrogen evolution and subsequent embrittlement is lower. The lack of control and the limited voltage and current generated in galvanic systems can be a major limitation; consequently, most galvanic systems are used in splash and tidal zone applications. The anodes are consumed and therefore must be replaced regularly, about every 5–10 years for thermal sprayed zinc in a splash/tidal situation. The anode has been found to work well initially and then the current drops, probably due to rising internal resistance at the anode–concrete interface. A recent development is the use of a chemical agent to increase the moisture around the anode and hence the effectiveness of the galvanic CP system [44]. The trials of this system are underway. Attempts have been made to use other anodes such as aluminum and zinc/aluminum alloys, but there have been application and installation problems [5].

F5. Electrochemical Chloride Extraction

Electrochemical chloride extraction (ECE) is also known as electrochemical chloride removal or desalination. The technique uses a temporary anode and passes a high current ($\sim$1 A/m^2 of steel or concrete surface area) to pull the chlorides away from the steel. A proportion (usually $\sim$50–90%) of the chloride can be completely removed from the concrete with very significant removal immediately around the steel, and a high level of repassivation of the steel is obtained.

ECE can be used in many of the situations where cathodic protection can be applied. It is at its best where the steel is reasonably closely spaced, the chlorides have not penetrated too much beyond the first layer of reinforcing steel, and future chlorides can be excluded. It has been applied to highway structures, car parks, and other structures in Europe and North America. The treatment typically takes 6–8 weeks. On completion the anode is removed and there are no ongoing monitoring requirements. If sufficient chloride is removed and further chlorides excluded, then it is a "one off" treatment. Field and laboratory research showed an improvement in concrete properties such as freeze–thaw, chloride and carbonation resistance, and water uptake resistance for one set of experiments [5]. The treatment time can be too long for some applications, for example, bridge decks that cannot be closed for 6–8 weeks. The problems with isolated steel, prestressing, and ASR mentioned for ICCP are exacerbated for ECE due to the high voltages and current densities. Some concerns with reduction in bond strength have been identified [45], but these only apply to structures containing smooth reinforcement where the mechanical interlock of ribbed steel is not available. A lithium-based electrolyte has been used in trials to mitigate the ASR risk, but results are not yet conclusive on its effectiveness. A U.S. standard for applying ECE is available (NACE SP0107-2007 [46]).

F6. Realkalization

Realkalization can be described as the equivalent of ECE for corroding steel in carbonated concrete structures. It is a shorter term treatment (days instead of weeks), and the proprietary system uses a sodium carbonate electrolyte as an aid to regenerating the alkalinity in the concrete and around the rebar. The same anode systems are used as for ECE. The treatment is most commonly applied to buildings with carbonation damage but with at least moderate cover of the reinforcing steel ($\gtrsim 10$ mm.). The literature is still contradictory about the use of the sodium carbonate electrolyte to maintain alkalinity after treatment and the extent to which it migrates into the concrete. The pH of sodium carbonate is not far above the pH corrosion threshold for steel in concrete. The use of phenolphthalein indicator causes some concern as its color change at pH 9.5 is below the corrosion threshold for steel in concrete at pH 11. A European technical specification on realkalization has been published (CEN/TS 14038-1 (2004) [47]) as well as a U.S. standard for applying realkalization (NACE SP0107-2007 [46]).

F7. Corrosion Inhibitors

The use of calcium nitrite as a corrosion-inhibiting admixture in the concrete mix is well established. However, trials of inhibitor treatments to hardened concrete after corrosion damage has been observed are comparatively recent. Although new materials are being produced and tested, the range of inhibitors presently available can be summarized as follows:

Several proprietary formulations of vapor-phase inhibitors, based on volatile amino alcohols and/or other related organic inhibitors that create a molecular layer on the steel to stop corrosion

Calcium nitrite, an anodic inhibitor in a mixture to aid penetration into concrete

Monofluorophosphate, which seems to create a very alkaline environment as it hydrolyzes in the concrete

In principle, corrosion inhibitors are applicable in any situation. However, for the present understanding of the materials available and the limited field testing the following applications for inhibitors can be considered:

Carbonation or low-to-modest chloride levels (<1% chloride by weight of cement)

Low cover (<20 mm)

Penetrable concrete (carbonated or corrosion, damaged in <20 years)

Barrier coating applied after application

Corrosion monitoring installed in concrete to assess effectiveness with time

Application of an inhibitor to any element with an accessible surface is simple and inexpensive.

Results of field trials on large-exposure slabs by Sprinkel and Oxyildirium 1998 [48] and on long-term field trials on highway structures in the United States by Sohangpurwalla et al. 1997 [49] have shown little benefit. Applications to buildings and car parks have been successful where rigorous and extensive patch repairing was applied [50].

REFERENCES

1. Highway Deicing—Comparing, Salt and Calcium Magnesium Acetate, Special Report 235, Transportation Research Board, National Academy of Sciences, Washington, DC, 1991.
2. E. J. Wallbank, The Performance of Concrete in Bridges, A Survey of 200 Highway Bridges, Her Majesty's Stationery Office, London, Apr. 1983.
3. F. H. Rasheeduzzafar, F. H. Dakhil, M. A. Bader, and M. N. Khan, ACI Mater. J., 89(5), 439 (1992).
4. J. P Broomfield, in "Assessing Corrosion Damage on Reinforced Concrete Structures," in Corrosion and Corrosion Protection of Steel in Concrete, Vol. 1, R. Narayan Swamy (Ed.), Sheffield Academic, Sheffield, UK, 1994, pp. 1–25.
5. J. P. Broomfield, Corrosion of Steel in Concrete—Understanding, Investigation and Repair, 2nd ed., Taylor and Francis, London, 2007.
6. K. Tuutti, Corrosion of Steel in Concrete, Swedish Cement and Concrete Research Institute, Stockholm, Sweden, 1982.
7. C. L. Page and K. W. J. Treadaway, Nature (Lond.) 297, 109 (1982). 7a. "Products and Systems for the Protection and Repair of Concrete Structures". BS EN 1504-10, Parts 1 to 10, British Standards Institute, London.
8. G. K. Glass and N. R. Buenfeld, "Chloride Threshold Levels for Corrosion Induced Deterioration of Steel in Concrete," in Proc., Chloride Penetration into Concrete, RILEM International Workshop, St.-Rémy-lès-Chevreuse, France, October 15-18, 1995, RILEM, Cachan, France, 1997, pp. 428–444.
9. E. Poulsen, "The Chloride Diffusion Characteristics of Concrete: Approximate Determination by Linear Regression Analysis," Nordic Concrete Research, 1 Nordic Concrete Federation, 1990.
10. Germann Instruments, A/S RCT Profile Grinder Mark II Instruction and Maintenance Manual, Germann Instruments, Denmark, 1994.
11. HWYCON—SHRP Report and computer programme, Transportation Research Board, Washington, DC, 1994 11a. Concrete Bridge Development Group, "Guide to Testing and Monitoring the Durability of Concrete Structures," Technical Guide 2, Concrete Society, Camberley, Surrey, UK, 2002.
12. P. D. Cady and E. J. Gannon, Condition Evaluation of Concrete Bridges Relative to Research Reinforcement Corrosion, Vol. 1, State of the Art of Existing Methods, SHRP-S-330, National Council, Washington, DC, 1992.
13. J. H. Bungey (Ed.), Non-Destructive Testing in Civil Engineering, in Proceeding of the International Conference by The British Institute of Non-Destructive Testing, Liverpool, 1993.
14. D. J. Titman, "Fault Detection in Civil Engineering Structures Using Infra-Red Thermography," in Proceedings of the 5th

International Conference on Structural Faults and Repair, Edinburgh University, Vol. 2, 1993, pp. 137–140.

15. J. C. Alldred, "Quantifying the Losses in Cover-Meter Accuracy Due to Congestion of Reinforcement," in Proceedings of the 5th International Conference on Structural Faults Repair, published by Edinburgh University, Vol. 2, 1993 pp. 125–130. 15a. F. Luco, "Comparative Test — Part II — Comparative Test of 'Cover Meters,'" Mater. and Struct. 38, 907–911 (Dec. 2005).

16. "Carbonation of Concrete and Its Effects on Durability," BRE Digest 405, Building Research Establishment, Garston, Watford, UK, 1995. 16a. "Products and Systems for the Repair of Concrete Structures, Test Methods, Determination of Carbonation Depth in Hardened Concrete by the Phenolphthalein Method,". BS EN 14630-06, British Standards Institute, London, 2006.

17. S. E. Herald, M. Henry, I. Al-Qadi, R. E. Weyers, M. A. Feeney, S. F. Howlum, and P. D. Cady, Condition Evaluation of Concrete Bridges Relative to Reinforcement Corrosion, Vol. 6, Method of Field Determination of Total Chloride Content, SHRP-S-328, National Research Council, Washington, DC, 1992.

18. M. G. Grantham, "An Automated Method for the Determination of Chloride in Hardened Concrete," in Proceedings of the 5th International Conference on Structural Faults and Repair, Vol. 2, 1993, pp. 131–136. 18a. "Products and Systems for the Protection and Repair of Concrete structures — Test Methods — Determination of Chloride Content in Hardened Concrete," BS EN 14629:2007, British Standards Institute, London, 2007.

19. "Electrochemical Tests for Reinforcement Corrosion," Technical Report 60, Concrete Society, Camberley, Surry, 2004. 19a. C. Andrade and C. Alonso, with contributions from Gulikers, J. Polder, R. Cigna, O. Vennesland, M. Salta, A. Raharinaivo, and Elsener, B. "Test Methods for On-Site Corrosion Rate Measurement of Steel Reinforcement in Concrete by Means of the Polarization Resistance Method," Construct. Mater., 37, 623–643 (2004).

20. P. Langford and J. Broomfield, Construct. Repair, 1(2), 32 (1987).

21. J. P. Broomfield, J. Rodriguez, L. M. Ortega, and A. M. Garcia, "Corrosion Rate Measurement and Life Prediction for Reinforced Concrete Structures," in Proceedings of the Structural Faults and Repair—93, Vol. 2, University of Edinburgh, Scotland, 1993, pp. 155–164. 21a B. Elsener, C. Andrade, J. Gulikers, R. Polder, and M. Raupach, "Half-Cell Potential Measurements—Potential Mapping on Reinforced Concrete Structures," Mater. Struct. 36, 461–471 (2003).

22. J. L. Dawson, "Corrosion Monitoring in Reinforced Concrete," in Corrosion of Reinforcement in Concrete Construction, A. P. Crane (Ed.), Ellis Horwood, for SCI, London, 1983, pp. 175–192.

23. M. Stern and A. L. Geary, J. Electrochem. Soc., 104, 56 (1957).

24. J. Flis, D. L. Sehgal, Y.-T. Kho, S. Sabotl, H. Pickering, K. Osseo-Assare,and P. D. Cady, Condition Evaluation of Concrete Bridges Relative to Reinforcement Corrosion, Vol. 2, Method for Measuring the Corrosion Rate of Reinforcing Steel,

SHRP-S-324, National Research Council, Washington, DC, 1992.

25. J. P. Broomfield, J. Rodríguez, L. M. Ortega, and A. M. García, "Corrosion Rate Measurements in Reinforced Concrete Structures by a Linear Polarisation Device," Philip D. Cady International Symposium—Concrete Bridges in Aggressive Environments, Special Publication 151, R. E. Weyers (Ed.), Concrete Institute, Published by ACI, American Concrete Institute (ACI), Farmington Hills MI 1994.

26. S. Feliú, J. A. González, C. Andrade, and V. Feliú, Corrosion, 44(10), 761.

27. K. C. Clear, "Measuring the Rate of Corrosion of Steel in Field Concrete Structures," Transportation Research Board Preprint 324, 68th Annual Meeting, Washington, DC, Jan. 1989.

28. C. Andrade, M. C. Alonso, and J. A. Gonzalez, "An Initial Effort to Use Corrosion Rate Measurements for Estimating Rebar Durability," in Corrosion Rates of Steel in Concrete, ASTM STP 1065, N. S. Berke, V. Chaker, and D. Whiting (Eds.), American Society for Testing and Materials, Philadelphia, PA, 1990, pp. 29–37. 28a. C. Andrade and C. Alonso,with contributions from J. Gulikers, R. Polder, R. Cigna, O. Vennesland, M. Salta, A. Raharinaivo, and B. Elsener, "Test Methods for On-Site Corrosion Rate Measurement of Steel Reinforcement in Concrete by Means of the Polarization Resistance Method." Construct. Mater. 37, 623–643 (2004).

29. J. A. Gonzáles, C. Andrade, P. Rodriguez, C. Alonso, and S. Feliú, "Effects of Corrosion on the Degradation of Reinforced Concrete Structures," in Progress in the Understanding and Prevention of Corrosion, Proc. 10th European Corroion Congress, Barcelona, Spain, Public Institute of Materials, London, for European Federation of Corrosion, 1993, pp. 629–633.

30. A. W. Beeby, "Development of a Corrosion Cell for the Study of the Influences of Environment and Concrete Properties on Corrosion," Concrete 85, Conference, Brisbane, Oct. 1985, pp. 118–123.

31. N. S. Berke and L. R. Roberts, "Reinforced Concrete Durability" ASTM Standardization News, Jan. 1992, pp. 46–51.

32. P. Shiessl and M. Ruapach, "Non-Destructive Permanent Monitoring of the Corrosion Risk of Steel in Concrete," in Non-Destructive Testing in Civil Engineering, Vol. 2, J. H. Bungey (Ed.), British Institute of NDT, International Conference, University of Liverpool, Apr. 1993, pp. 661–654.

33. N. S. Berke, D.F. Shen, and K.M. Sundberg, "Comparison of the Linear Polarisation Resistance Technique to the Macrocell Corrosion Technique," in Corrosion Rates of Steel in Concrete, ASTM STP 1065, N. S. Berke, V. Chaker, and D. Whiting (Eds.), American Society for Testing and Materials, Philadelphia, PA, 1990, pp. 38–51.

34. J. P. Broomfield, K. Davies, and K. Hladky, "Permanent Corrosion Monitoring in New and Existing Reinforced Concrete Structures," Mater. Perform., 39(7), 66–71 (July 2000).

35. "Patch Repair of Reinforced Concrete Subject to Reinforcement Corrosion," Technical Report No. 38, The Concrete Society, Slough, UK, 1991. 35a. Joint ACI and ICRI, Concrete

Repair Manual, 3rd ed., American Concrete Institute, Farmington Hills, MI, 2008.

36. H. L. Robinson, "The Evaluation of Coatings as Carbonation Barriers," in Proceedings of the Second International Colloquium on Materials Science and Restoration, Technische Akademie Essingen, Germany, 1986.

37. "Durability of Concrete Road Bridges," Organisation of Economic Co-Operation and Development, Paris, 1989.

38. D. G. Manning,"Waterproofing Membranes for Concrete Bridge Decks," NCHRP Synthesis 220, Transportation Research Board, National Research Council, Washington, DC, 1995.

39. P. Chess (Ed.), Cathodic Protection of Steel in Concrete, E & FN Spon, London, 1998.

40. NACE International Standard Recommended Practice, "Cathodic Protection of Reinforcing Steel in Atmospherically Exposed Concrete Structures," RP0290-90 National Association of Corrosion Engineers, Houston, TX, 1990.

41. "Cathodic Protection of Steel in Concrete," BS EN 12696: 2000, British Standards Institute, London, 2000. 41a. "Cathodic Protection—Competence Levels and Certification of Cathodic Protection Personnel," BS EN15257:2006, British Standards Institute, London, 2006. 41b. "State-of-the-Art Report: Criteria for Cathodic Protection of Prestressed Concrete Structures," NACE Publication 01102, NACE International, Houston, TX, 2002.

42. R. J. Kessler, R. G. Powers, and I. R. Lasa, "Update on Sacrificial Anode Cathodic Protection on Steel Reinforced Concrete Structures in Sea Water," Paper No. 516, CORROSION 95, NACE International, Houston, TX, 1995.

43. R. J. Kessler, R. G. Powers, and I. R. Lasa, "Cathodic Protection Using Zinc Sheet Anodes and an Ion Conductive Gel Adhesive," Paper No. 234, CORROSION 97, NACE International, Houston, TX, 1997.

44. J. Bennett, "Chemical Enhancement of Metallized Zinc Anode Performance," Paper No. 640, CORROSION/98, NACE International, Houston, TX, 1998.

45. J. P. Broomfield, "Electrochemical Chloride Migration—Laboratory and Field Studies in the UK," Paper No. 253, CORROSION/97, NACE International, Houston, TX, 1997.

46. "Standard Practice Electrochemical Realkalization and Chloride Extraction for Reinforced Concrete," NACE SP0107-2007, NACE International, Houston, TX.

47. "Electrochemical Realkalization and Chloride Extraction Treatments for Reinforced Concrete—Part 1: Realkalization." CEN/TS 14038-1, British Standards Institute, London, 2004.

48. M. Sprinkel and C. Oxyildirium, "Evaluation of Exposure Slabs Repaired with Corrosion Inhibitors," in Proc. Intl. Conf.

on Corrosion and Rehabilitation of Reinforced Concrete Structures, Federal Highways Administration, Orlando, FL, 1998.

49. A. A. Sohangpurwalla, M. Islam, and W. Scannell, "Performance and Long Term Monitoring of Various Corrosion Protection Systems Used in Reinforced Concrete Bridge Structures," in Proc. Intl. Conf. on Repair of Reinforced Concrete from Theory to Practice in a Marine Environment, Svolvaer, Norway, Norwegian Road Research Laboratory, May 1997.

50. G. Jones and I. Wood, "Performance of Corrosion Inhibitors in Practice," Aston University Seminar, University of Aston, Birmingham, UK, Jan. 26, 2000.

BIBLIOGRAPHY

ACI 222R-01 "Corrosion of Metals in Concrete," American Concrete Institute, Detroit, MI, 2001.

J. E. Bennett, J. J. Bartholomew, J. B. Bushman, K. C. Clear, R. N. Kamp, and W. J. Swiat, "Cathodic Protection of Concrete Bridges: A Manual of Practice," SHRP-S-372 National Research Council, Washington DC, 1993.

J. E. Bennett and T. Schue, "Chloride Removal Implementation Guide SHRP-S-347," National Research Council, Washington, DC, 1993.

BRE Digest 405 (1995), "Carbonation of Concrete and its Effects on Durability," Building Research Establishment, Watford, UK.

J. P. Broomfield, Corrosion of Steel in Concrete—Understanding, Investigation and Repair," 2nd ed., Taylor and Francis, London, 2007.

BS EN 12696. Cathodic Protection of Steel in Concrete, British Standards Institute, London, 2000.

P. Chess (Ed.), Cathodic Protection of Steel in Concrete, E & FN Spon, London, 1998. Concrete Society Technical Report No. 38, "Patch Repair of Reinforced Concrete Subject to Reinforcement Corrosion," The Concrete Society, Slough, UK, 1991.

NACE SP 0390-2006 Standard Recommended Practice, "Maintenance and Rehabilitation Considerations for Corrosion Control of Atmospherically Exposed Existing Steel-Reinforced Concrete Structures, NACE International, Houston, TX, 2006.

NACE SP290-07, "Impressed Current Cathodic Protection of Reinforcing Steel in atmospherically Exposed Concrete Structures," NACE International, Houston, TX, 2007.

C.L. Page and K.W. J. Treadaway, Nature (London), **297**, 109 (1982).

K. Tuutti, "Corrosion of Steel in Concrete," Swedish Cement and Concrete Research Institute, Stockholm, Sweden, 1982.

50

ETHANOL STRESS CORROSION CRACKING OF CARBON STEELS

J. Beavers, F. Gui, and N. Sridhar

DNV Columbus, Inc., Dublin, Ohio

A. INTRODUCTION

There is significant interest within the pipeline industry in transporting fuel-grade ethanol (FGE) as a result of the increased usage of ethanol as an oxygenating agent for gasoline and interest in ethanol as an alternative fuel. The U.S. Energy Policy Act of 2005 (amended in 2007) established a nationwide renewable fuels standard, starting from 15 billion liters of all biofuels in 2006 to 136 billion liters in 2022. Ethanol will constitute almost 90% of this renewable fuel.

Ethanol is produced in biorefineries and is transported to terminals, where it is blended with gasoline to produce the most commonly used blends; E-10 (10 vol % ethanol–90 vol % gasoline) and E-85 (85 vol % ethanol–15 vol % gasoline). Presently, rail and truck are the predominant means of transporting ethanol from these biorefineries to the population centers in North America, whereas pipeline transportation is the most cost-effective mode of transporting large volumes of ethanol.

There are significant challenges for pipeline transportation of FGE. Most of the current hydrocarbon pipelines move products from the Gulf of Mexico region to the population centers on the east and west coasts and the Midwest. On the other hand, most of the biorefineries in North America are located close to the middle of the continent. Thus, new pipelines will be required to transport the FGE from the biorefineries to the population centers. Ethanol is an excellent solvent so contamination of the FGE is a problem for transportation in existing petroleum pipelines. FGE also is not compatible with some elastomers used in these existing pipelines. Probably the biggest challenge for pipeline transportation of FGE is corrosion, and specifically stress corrosion cracking (SCC), of the carbon steels that are used for pipeline construction.

The early service experience with SCC in FGE was summarized in an American Petroleum Institute (API) publication in 2003 [1]. Documented failures of equipment in FGE distribution terminals and in the end-user gasoline blending and distribution terminals have dated back to the early 1990s. No cases of SCC were reported in ethanol manufacturer facilities, tanker trucks, railroad tanker cars or barges, or following blending of the FGE with gasoline. The survey did not pinpoint what causes ethanol SCC, but the failure history suggests that the SCC may be related to changes in the FGE as it moves through the distribution chain.

This chapter summarizes the present knowledge on the contributing factors in ethanol SCC of carbon steels and identified methods of mitigation.

B. CONTRIBUTING FACTORS IN ETHANOL SCC

A Venn diagram is sometimes used to describe the SCC process. It shows that all three components (a susceptible material, a tensile stress, and a potent cracking environment) are required simultaneously for SCC to occur. How each of these factors influences the SCC process is described further below.

B1. Susceptible Material

API technical reports 939 D [1, 2] summarize the field experience related to SCC in FGE. All of the confirmed SCC failures reported in these documents have been in ethanol storage facilities, blending facilities, or terminals in plate and piping made of carbon steels. Affected grades include hot-rolled American Society for Testing and Materials (ASTM) A 36, ASTM A 53, and A 516 grade 70. SCC of highly cold worked steel springs used in roof hanger assemblies for storage tanks has also occurred, but the steel grade was not identified.

The majority of the cracking has been found at locations near but not in welds. Both axial and longitudinal cracking has been reported. A typical example of a failure of a piping system in a terminal is shown in Figure 50.1. In this case, through-wall cracks occurred near a girth weld in the piping system. There is no evidence that the failures were associated with welding defects or metallurgical defects or deficiencies in the steels. The failures have been associated with steels having compositions and mechanical properties within specifications.

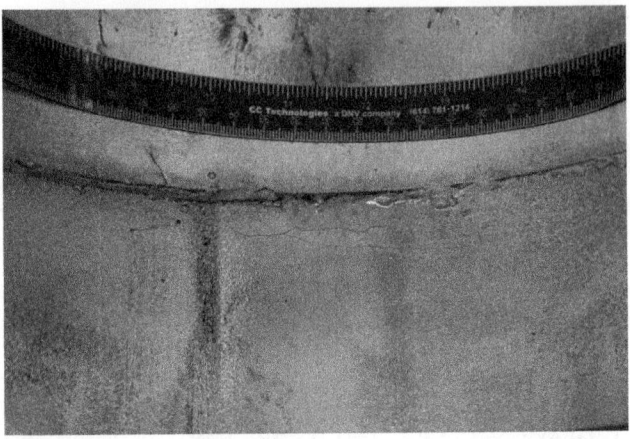

FIGURE 50.2. Photograph of the internal surface of API 5L Grade B piping used in ethanol service following magnetic particle inspection to reveal stress corrosion cracks.

The typical microstructure of these steels consists of ferrite and pearlite and the reported cracking is intergranular, mixed mode, or transgranular. An example of ethanol SCC observed near a girth weld in API 5L Grade B piping is shown in Figures 50.2–50.5. In this example, the failure mode is primarily intergranular.

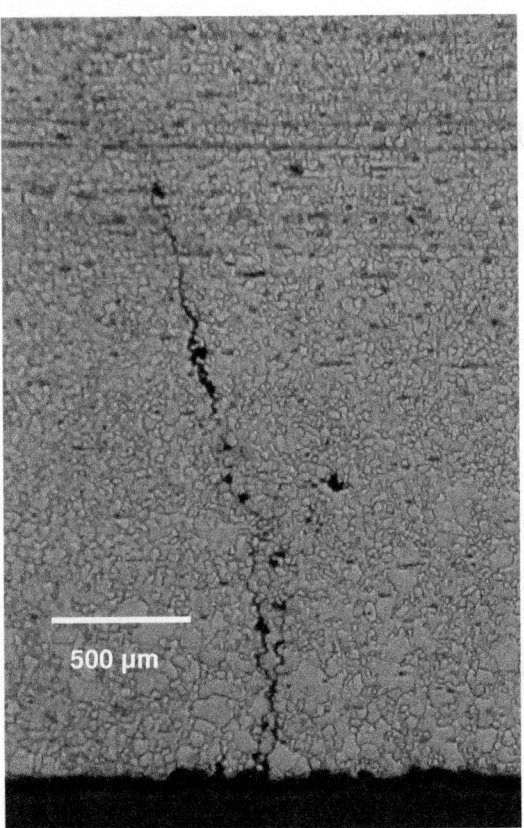

FIGURE 50.3. Light photomicrograph showing circumferential, internal stress corrosion cracks in API 5L Grade B piping used in ethanol service (axial cross section, 4% Nital etchant).

FIGURE 50.1. SCC observed in terminal piping system containing FGE.

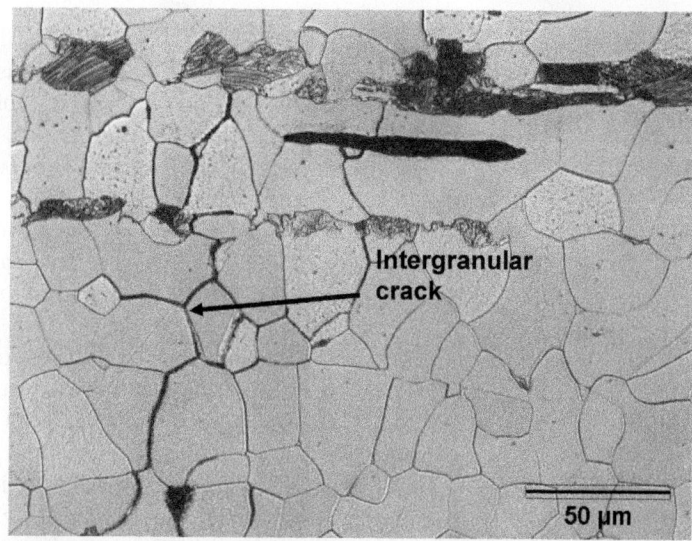

FIGURE 50.4. Light photomicrograph of tip of crack shown in Figure 50.3 (4% Nital etchant).

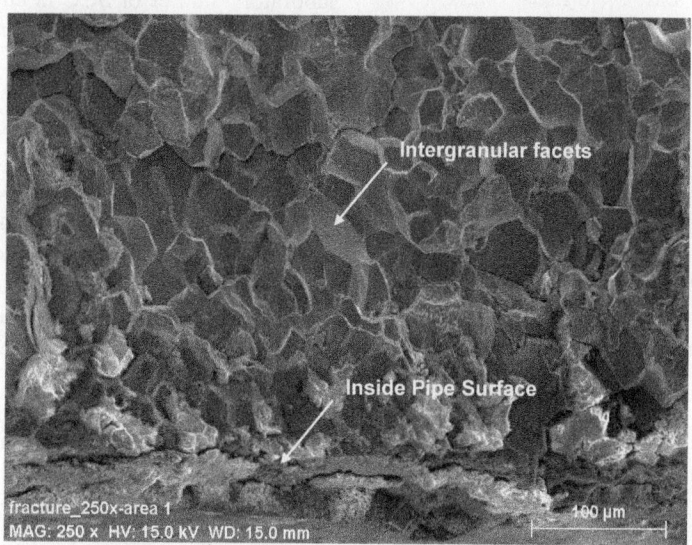

FIGURE 50.5. Scanning electron micrograph of fracture surface of API 5L Grade B piping used in ethanol service showing intergranular SCC.

The Pipeline Research Council International (PRCI) and the Pipeline and Hazardous Materials Safety Administration (PHMSA) sponsored research to investigate the effect of metallurgical factors on ethanol SCC [3]. The majority of testing was performed in simulated fuel-grade ethanol (SFGE) containing reagent-grade ethanol and additions to simulate a FGE meeting ASTM D 4806 specifications [4]; see Table 50.1. In slow-strain-rate (SSR) tests of notched specimens, the severity of SCC of base metal specimens was not highly dependent on steel grade for X-46 double-submerged arc weld (DSAW) pipe material, X-52 high-frequency electric resistance welded (HFERW) pipe material, X-52 low-frequency electric resistance welded (LFERW) pipe material, X-42 seamless pipe material, and cast steel for pumps. However, the cast steel did appear to be somewhat more resistant to cracking; see Figure 50.6. For the X-46 steel DSAW pipe material, the weld metal appeared to be slightly more resistant to SCC than the base metal or heat-affected zone, but all three metallurgies were found to be susceptible to SCC, as shown in Figure 50.7.

Testing of the welds in the electric resistance welded (ERW) pipe material was inconclusive because of the generally poor tensile properties of the welds, although the high ferrite bond line of one ERW line pipe steel appeared to be resistant to SCC [3]. This observation suggests that it may be possible to modify the chemistry/microstructure of these steels to produce ethanol SCC resistance.

TABLE 50.1.   Composition of SFGE

Requirement	ASTM Limits[a]		SFGE
	Minimum	Maximum	
Ethanol, vol %	92.1	—	Balance
Methanol, vol %	—	0.5	0.5
Solvent-washed gum, mg/100 mL	—	5.0	—
Water content, vol %	—	1.0	1.0
Denaturant content, vol %	1.96	5.00	3.75
Inorganic chloride, ppm (mg/L)	—	10 (8)	5.0
Copper, mg/kg	—	0.1	—
Acidity (as acetic acid, CH_3COOH), mass % (mg/L)	—	0.007 (56)	(50)
pH_e	6.5	9.0	—

[a]ASTM D4806-10.

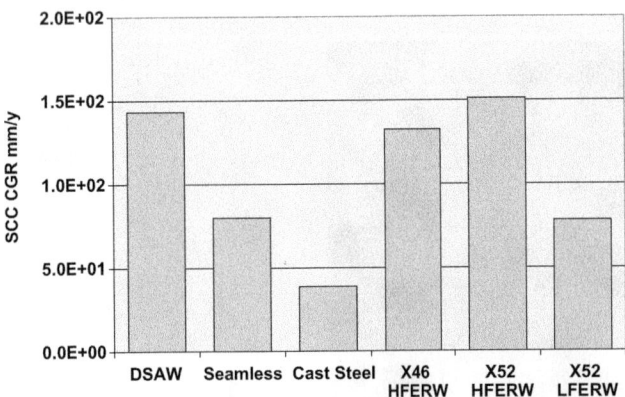

FIGURE 50.6. Crack growth rate for notched base metal (unwelded) specimens of several different steels tested in E-95 (SFGE–gasoline) blend. (After [3].)

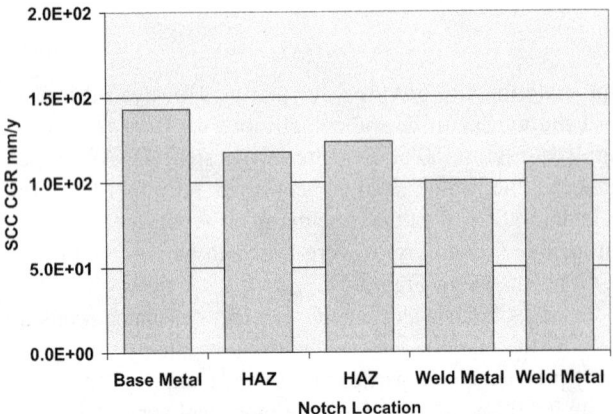

FIGURE 50.7. Crack growth rate as function of notch location for specimens removed from seam weld of X-46 DSAW line pipe steel tested in E-95 (SFGE–gasoline) blend. (After [3].)

B2.   Tensile Stress

The majority of ethanol SCC has occurred in the base metal near welds in low-pressure piping or in tanks, where the primary stresses leading to SCC have been residual welding stresses [1,2]. This cracking was associated with slightly elevated hardness levels, which is most likely indicative of cold working. In analyses of two cases of ethanol SCC, the hardness in the areas of cracking was about 5 Rockwell B hardness (HRB) units higher than in areas away from the cracking [2].

Field experience and laboratory testing also indicate that severe straining is required for ethanol SCC to occur. SCC of ethanol storage tanks has been observed only in severely strained areas associated with non-postweld heat-treated welds and/or in tanks with design/installation issues [1,2]. For example, floor areas that were not adequately supported experienced SCC as a consequence of cyclic loading from filling and withdrawal of ethanol. Some of the earliest laboratory studies of SCC in ethanol were conducted using U-bend specimens. SCC was not observed in these tests unless a poor-quality weld bead perpendicular to the stressing direction and an extremely severe bending mode were included. In SSR tests with unnotched specimens, SCC was observed near the necked region of the specimen [5]. Notched SSR tests exhibited SCC at the notch root [3]. In more recent crack growth tests using compact tension specimens, the presence of a cyclic loading component has been shown to exacerbate SCC [6]. All of these observations suggest that severe plastic deformation and (or) the presence of dynamic plastic strain are necessary for SCC to occur.

B3.   Potent Cracking Environment

B3.1.   Ethanol Composition.
Fuel-grade ethanol can be produced in two forms: anhydrous and hydrous. Because the azeotrope of an ethanol–water binary mixture (the azeotrope is the mixture that boils without any change in composition) corresponds to 95.6% by weight ethanol, it is impossible to produce higher concentration ethanol simply by distillation. The ethanol containing 4–7% water by weight is called hydrous ethanol and has been used extensively as automotive fuel in Brazil since the 1970s. However, because of phase separation and corrosion concerns when blended with gasoline, most countries have adopted ethanol specifications that contain much less water, called anhydrous ethanol. Anhydrous ethanol can be produced by a number of methods, including drying in the presence of calcium oxide, separation using molecular sieves and membranes, distillation using ternary additives, and vacuum distillation. In this regard, it should be noted that even "pure" ethanol (sometimes referred to as 200-proof ethanol) obtained from a supplier of reagent-grade chemicals contains water in the range of 100–700 ppm by weight. The ethanol used as a fuel

TABLE 50.2.   Specifications for Fuel-Grade Ethanol in Various Countries

Constituent	U.S. (ASTM D 4806)	Brazil (Anhydrous)	India (IS 15464-2004)	Europe (EN 15376)
Ethanol, vol %	92.1 (min.)	99.3 min.	99.5 min.	96.7 min
Methanol, vol %	0.5 max.	—	0.038 max	1 max
Solvent-washed gum, mg/100 mL	5 max	—	—	—
Water, vol %	1 max	Not specified (about 0.4 v%)	—	0.3 max
Denaturant content, vol %	1.96–5.00	—	permitted	permitted
Inorganic chloride, mg/L	8 max	—	—	20 max
Copper, mg/kg	0.1 max	0.07 max	0.1 max	0.1 max
Acidity as acetic acid, mg/L	56 max	30 max	30 max	56 max
Sulfur, mg/kg	30 max	—	—	10 max
Sulfate, mg/kg	4 max	—	—	—
Phosphorus, mg/L	—	—	—	0.5 max
pH_e	6.5–9	—	—	Not specified
Appearance	Clear	Clear	—	Clear

additive or a fuel substitute has additional additives and contaminants. A comparison of the different ethanol specifications is shown in Table 50.2.

The denaturant is a compound that is added to render ethanol undrinkable. The United States specifies a certain range of denaturant concentration (1.96–5 vol %) and the only denaturants acceptable for FGE are natural gasoline, gasoline components, or unleaded gasoline [4]. In Europe, there is no specific range, but denaturant is permitted. In addition to these constituents, FGE may contain corrosion inhibitors intended to mitigate corrosion of automotive components. There are other minor differences in the compositions of FGE, as shown in Table 50.2.

The effect of different constituents in ethanol on SCC of carbon steel has been investigated by Beavers et al. [5], Sridhar et al. [7], and Lou et al. [8]. These studies have demonstrated that FGE that meets applicable ASTM standards (Table 50.2 and [4]) can be a potent cracking agent in the presence of oxygen.

B3.2.   Effect of Chemical Factors on SCC.
While reproducible results are generally obtained from SSR tests with SFGE, the potency of actual FGEs can vary significantly [7,9] even with high oxygen concentrations. The FGEs also exhibit an aging phenomenon—with some FGEs, the SCC tendency of steel increases with time while the opposite trend has been observed with other FGEs. In the case of one FGE, the absence of SCC was associated with a very noble corrosion potential [9]. The reason for this behavior is still unclear and is the subject of ongoing research.

B3.3.   Effect of Dissolved Oxygen.
A statistically designed study by Sridhar et al. [7] showed that dissolved oxygen is the most important contaminant affecting SCC in FGE. No SCC occurred under any circumstances without the presence of dissolved oxygen. Other factors, such as acidity, denaturant, and corrosion inhibitor, were not found to be important within the limits allowed by the ASTM D-4806

standard. The effect of different oxygen concentrations in the gas phase on SCC susceptibility is shown in Figure 50.8. In this figure, the corrosion potential of the SSR test specimen measured in the same solution is also plotted. For two ethanol samples, the corrosion potential was rather high even when the sparging gas contained very little oxygen. In these ethanols, very little SCC was observed. This indicates that, perhaps, other redox agents were present in the ethanol to elevate the corrosion potential.

If oxygen is present, then the next most important factor in cracking is contact of steel with the rust layers. Generally, the corrosion potential in SFGE was higher when steel was galvanically coupled to rusted layers [7].

B3.4.   Effect of Chloride.
Chloride concentration appears to be important depending on the test method. In SSR tests of unnotched specimens, only minor SCC was observed in ethanol without any chloride [6]. However, in notched SSR tests, significant SCC was observed even when there was no

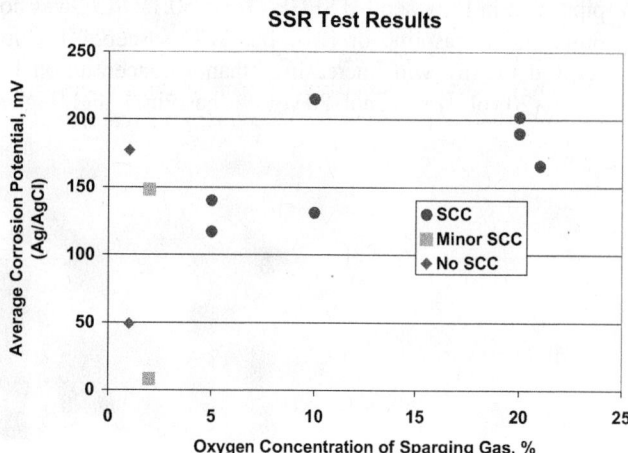

FIGURE 50.8. Effect of oxygen content in the gas phase on the average corrosion potential and SCC severity in one batch of corn-based FGE.

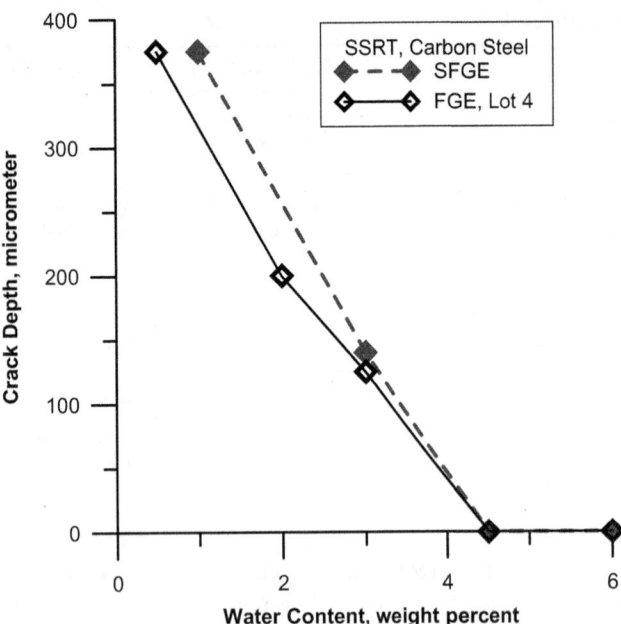

FIGURE 50.9. Effect of water on SCC crack depth of carbon steel in SFGE and FGE from production plant [6].

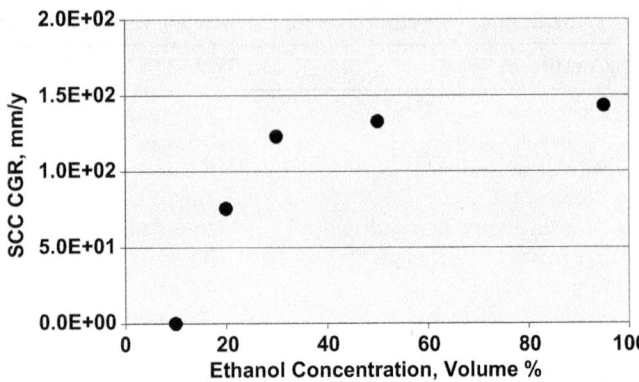

FIGURE 50.10. Crack growth rate as function of ethanol concentration for SSR tests of notched base metal specimens of X46 DSAW pipe material in SFGE–ethanol blends. (After [3].)

addition of chloride. Furthermore, the fracture mode changed from predominantly intergranular to predominantly transgranular as the chloride concentration increased from 0 to 40 ppm [6]. Similar behavior was reported by Lou et al. [8].

B3.5. Effect of Water. Although water within the ASTM D-4806 specification limits was not influential with respect to SCC, above 1% water, cracking susceptibility decreases with increasing water content and water contents above about 4.5% completely inhibit SCC (Figure 50.9).

B3.6. Effect of Ethanol–Gasoline Blends. In a study performed using notched SSR specimens of an X-46 line pipe steel in fully aerated SFGE (Table 50.1), SCC was not observed in gasoline or E-10 but SCC susceptibility increased rapidly with increasing ethanol concentration for E-20 (20 vol % ethanol–80 vol % gasoline) and higher

blends [3]. Surprisingly, E-30 (30 vol % ethanol–70 vol % gasoline) was nearly as potent an SCC agent as FGE, as shown in Figure 50.10. Similar behavior has been observed in crack growth tests performed with precracked compact-type specimens of the X-46 line pipe steel under cyclic loading designed to simulate pressure fluctuations on an operating product pipeline [10].

More recent studies have shown that the 50 vol % ethanol–gasoline mixture (E-50) has a greater propensity to cause SCC than either lower or higher ethanol–gasoline blends [6].

B3.7. Effect of Inhibitors. Sridhar et al. [7] showed that one common corrosion inhibitor added to FGE to protect against corrosion in automotive components (e.g., Octel DCI-11) did not have any effect on SCC of steel. However, Beavers et al. [5] showed that certain amine-type inhibitors had a significant inhibiting effect on SCC. In addition to amine, ammonium hydroxide appears to have a beneficial effect in mitigating SCC. This is illustrated in Figure 50.11 for SSR tests conducted on unnotched specimens in SFGE [6].

The effect of ammonium hydroxide on inhibiting SCC appears to be unrelated to pH effects because an addition of 3.38 mM lithium hydroxide did not inhibit SCC to the same

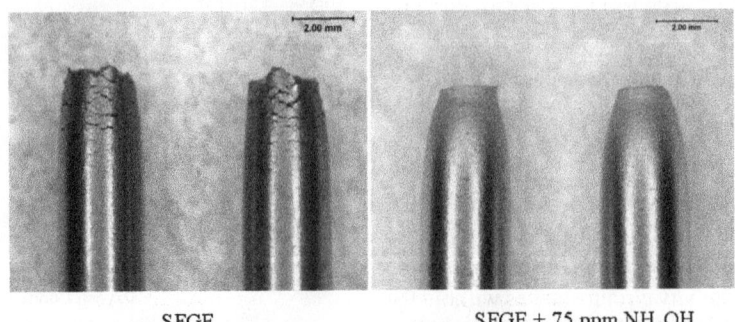

SFGE   SFGE + 75 ppm NH$_4$OH

FIGURE 50.11. Optical photographs of SSR test specimens showing inhibition of SCC on carbon steel in SFGE containing ammonium hydroxide addition [6].

extent as 2 m*M* ammonium hydroxide even though the pH in the lithium hydroxide–ethanol solution was considerably higher [11]. Further research on the mechanism of inhibition of SCC is being pursued to clarify the roles of different species on SCC.

C. MITIGATION OF ETHANOL SCC

The findings from previous field surveys and research specifically addressing ethanol SCC as well as broader industrial experience with other forms of SCC point to potentially effective methods for mitigation of ethanol SCC. SCC of a broad range of materials in potent cracking environments is commonly associated with residual welding stresses. Postweld heat treatment has been shown to be effective in reducing the residual welding stresses, thereby reducing the incidence of SCC near welds. However, the applied operational stresses from, for example, internal pressurization of a pipeline must be taken into consideration. These applied stresses might be sufficient to promote ethanol SCC in the absence of residual welding stresses. Other methods of reducing residual stresses include shot peeing and grit blasting. Grit blasting has been shown to play a role in the mitigation of external SCC of gas transmission pipelines [12]. The compressive residual stress imparted by the grit blasting process effectively overcomes the effects of residual tensile stresses and the tensile hoop stress from internal pressurization. A similar process might be effective for the mitigation of ethanol SCC.

The apparent resistance to ethanol SCC of the high ferrite bond line of one ERW line pipe steel [3] may provide another avenue for mitigation. The newer line pipe steels tend to have lower carbon contents that older steels and these newer steels might be inherently more resistant to SCC. It also might be possible to modify the chemistry/microstructure of the steels to produce ethanol SCC resistance.

Environmental control may be the most effective method for mitigation of ethanol SCC in existing pipeline systems. As previously described, Sridhar et al. [7] demonstrated that dissolved oxygen is the most important contaminant affecting SCC in FGE and that no SCC occurred under any circumstances in the absence of dissolved oxygen. Beavers et al. [5] showed that removing oxygen by chemical (hydrazine at 1000 ppm concentration), mechanical (sparging with nitrogen or vacuum deaeration), or electrochemical methods (reaction with steel wool to reduce oxygen) all resulted in suppression of SCC in SSR tests in a SFGE containing 50 ppm chloride. The oxygen removal was associated with a negative shift in the free corrosion potential, as shown in Figure 50.12.

The application of deaeration for SCC mitigation would create some technical challenges. Once the ethanol is deaerated, any further oxygen ingress must be prevented. For pipeline applications, this could be difficult in that the FGE

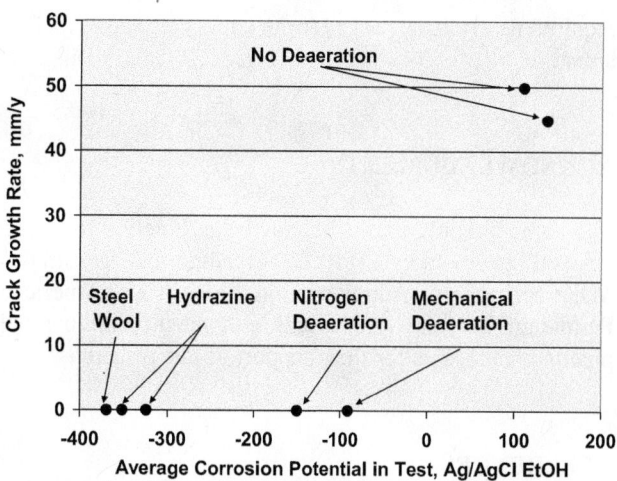

FIGURE 50.12. Crack growth rate versus average potential for SSR tests in SFGE (containing 50 ppm Cl) with various deaeration methods. (After [5].)

likely would be transported to intermediate storage tanks along the pipeline. With the possible exception of chemical deaeration, the deaeration methods would require a sizable capital investment in equipment.

SCC inhibitors also potentially could be effective. As previously described, some amines as well as ammonium hydroxide have been shown to have an inhibiting effect on ethanol SCC in SSR tests. In ongoing research, possible inhibitors or inhibitor packages are being evaluated [6]. In addition to SCC inhibitors, factors being considered include diverse issues such as toxicity, compatibility with combustion engines, and cost. In the SCC tests, candidate inhibitors are being screened using SSR tests of a line pipe steel followed by more realistic crack growth tests with precracked compact-type specimens under cyclic loading conditions. Typical results are shown in Figure 50.13. In this experiment, ammonium hydroxide arrested crack growth in SFGE, but it

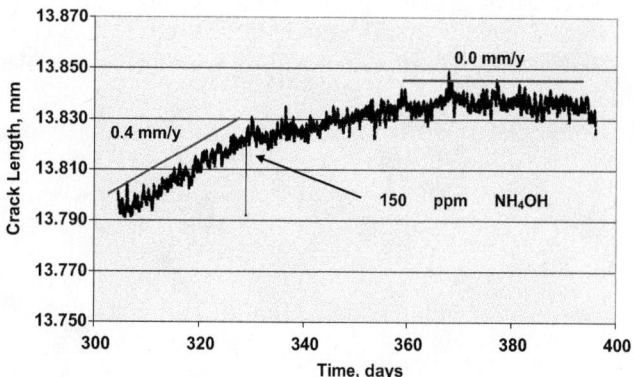

FIGURE 50.13. Crack length and growth rate as a function of time for precracked compact-type specimen of an X-46 line pipe steel cyclically loaded in SFGE (150 ppm of NH₄OH added on day 330) [6].

required nearly four weeks for the stress corrosion crack to arrest.

ACKNOWLEDGMENTS

The authors would like to acknowledge Pipeline Research Council International (PRCI), Pipeline and Hazardous Materials Safety Administration (PHMSA), American Petroleum Institute (API), and a number of individual pipeline companies for their support of this research.

REFERENCES

1. R. D. Kane and J. G. Maldonado, "Stress Corrosion Cracking of Carbon Steel in Fuel Grade Ethanol: Review and Survey," Technical Report 939-D, American Petroleum Institute, Washington, DC, Sept. 2003.

2. R. D. Kane, D. Eden, N. Sridhar, J. G. Maldonado, M. P. H. Brongers, A. K. Agrawal, and J. A. Beavers, "Stress Corrosion Cracking of Carbon Steel in Fuel Grade Ethanol: Review, Experience Survey, Field Monitoring, and Laboratory Testing," Technical Report 939-D, 2nd ed., American Petroleum Institute, Washington, DC, May 2007.

3. J. A. Beavers, N. Sridhar, and C. Zamarin, "Effects of Steel Microstructure and Ethanol-Gasoline Blend Ratio on SCC of Ethanol Pipelines," Paper No. 095465, Corrosion 2009 Conference & EXPO, Atlanta GA, NACE, Houston, TX, Mar. 2009.

4. "Standard Specification for Denatured Fuel Ethanol for Blending with Gasolines for Use as Automotive Spark-Ignition Engine Fuel," ASTM D 4806-10, ASTM International, West Conshohocken, PA 19428, 2010.

5. J. A. Beavers, M. P. H. Brongers, A. K. Agrawal, and F. A. Tallarida, "Prevention of Internal SCC in Ethanol Pipelines," Paper No. 08153, Corrosion 2008 Conference & EXPO, New Orleans, LA, NACE, Houston, TX, Mar. 2008.

6. J. A. Beavers and N. Sridhar, unpublished Results, PRCI/PHMSA Research Program, 2008–2009.

7. N. Sridhar, K. Price, J. Buckingham, and J. Dante, "Stress Corrosion Cracking of Carbon Steel In Ethanol," Corrosion, **62** (8), 687–702 (2006).

8. X. Lou, D. Yang, and P. M. Singh, "Effect of Ethanol Chemistry on Stress Corrosion Cracking of Carbon Steel in Fuel-Grade Ethanol," Corrosion, **65**(12), 785–797 (2009).

9. J. G. Maldonado and N. Sridhar, "SCC of Carbon Steel in Fuel Ethanol Service: Effect of Corrosion Potential and Ethanol Processing Source," Paper No. 07574, Corrosion/2007 NACE International, Houston, TX, 2007.

10. J. A. Beavers, F. Gui, and N. Sridhar, "Recent Progress in Understanding and Mitigating SCC of Ethanol Pipelines," Corrosion/2010 NACE International, Hauston, TX, 2010.

11. F. Gui, N. Sridhar, and J. A. Beavers, "Localized Corrosion of Carbon Steel and Its Role in Stress Corrosion Cracking in Fuel Grade Ethanol," submitted to Corros., J., 2010.

12. J. A. Beavers, N. G. Thompson, and K. E. W. Coulson, "Effects of Surface Preparation and Coatings on SCC Susceptibility of Line Pipe: Phase 1—Laboratory Studies," Paper No. 597, CORROSION/93 New Orleans, LA, NACE, Houston, TX, Mar. 1993.

51

AUSTENITIC AND FERRITIC STAINLESS STEELS*

M. A. Streicher[†]

Revised by J. F. Grubb
Technical & Commercial Center, ATI Allegheny Ludlum Corp., Brackenridge, Pennsylvania

A. DISCOVERY OF STAINLESS STEELS

Harry Brearly was the first not only to recognize the superior corrosion resistance of an iron-based alloy containing chromium but also to put this property to use for making "rustless" cutlery from a 12.8% Cr alloy [1–4] Brearly's independent discovery in 1912, which included the application of a heat treatment to harden the alloy, was the result of the successful exploitation of a chance observation. He had been trying to prevent erosion and fouling in rifle barrels when he alloyed iron with chromium and observed during metallographic work that these steels resisted attack by etchants. Later he gave these ferritic Fe–Cr alloys the name "stainless steel" [1].

This name was then also applied to the austenitic Fe–Cr–Ni compositions that were being developed

---

*Adapted from "Stainless Steel 77," Robert Q. Barr (Ed.), Climax Molybdenum Company.
[†] Deceased.

---

into commercial products at about the same time in Germany [3, 4]. Here, too, it was a chance observation which led to the discovery of the corrosion resistance of the austenitic Fe–Cr alloys with about 8% Ni. In 1912, Maurer [5–8] noticed that some alloys that Strauss had made were impervious to attack after months of exposure to acid fumes in his laboratory. These alloys had been set aside because they could not be worked without cracking. From previous experience on the metallurgy of Fe–Cr–Ni alloys, he devised an annealing heat treatment with a water quench that put the large amount of chromium carbide precipitate into solution and made the alloys ductile [8].

Strauss was the head of the physics section of the research laboratory of the Friedrich A. Krupp Works at Essen, which was founded in 1883. Maurer joined this laboratory in 1909 as its first metallurgist. Patents were immediately applied for in Germany and elsewhere by Strauss and commercial development and production were initiated. By 1914, rapidly increasing quantities of their V2A steel, 20% Cr–7% Ni–0.25% C, were being supplied to the Badische Anilin und Sodafabrik in Ludwigshafen. For Fritz Haber's ammonia synthesis plant, these alloys were the right materials at the right time [8]. Even though the patents for the compositions of austenitic alloys are in the name of Strauss, Maurer's chance observation and heat treatment are recognized [3–10] as the starting points for the industrial application of austenitic stainless steels in chemical plants.*

Advances in production and fabrication techniques led to large-scale applications of both ferritic (17% Cr) and austenitic (18 Cr–8 Ni) stainless steels for ammonia and nitric acid plants in England and the United States as well as in Germany during the years from 1925 to 1935 [11, 12]. Eventually, the volume of production of the austenitic greatly exceeded that of the ferritic alloys by a factor of 2 to 1 in recent years in the United States. By far the greatest effort in research and development has been concentrated on the austenitic grades until recently. As described below, since 1967 there has again been intense activity in the development and commercialization of new ferritic stainless steels.

B. PASSIVE STATE

The first practical applications of a stainless steel originated in England, and so did the most commonly accepted explanation of the mechanism responsible for the superior corrosion resistance of these alloys. On the basis of experiments with iron exposed to nitric acid, Faraday explained what was termed the passive state in 1836 by the Swiss investigator Schönbein [13]. Faraday attributed the resistance of iron in concentrated nitric acid to a protective iron oxide film that forms on the surface of the metal by reaction between the metal and the passivating environment. Once formed, its slow dissolution in this environment then determines the corrosion rate of the metal. From observations [14, 15] made on passive stainless steels exposed to boiling 50% sulfuric acid with ferric sulfate inhibitor, it is apparent that the passive state is not an inert or static state but a dynamic condition in which there is continuous dissolution and repair of the passive film at discrete points in the surface. A similar view of the passive state has also been evolved by Tomashov [16].

Because the techniques for peeling oxide films from passive metals had not been developed in Faraday's time, the invisible protective films could not be detected and identified directly. As a result, their existence was widely disputed even as late as 1908 [13]. In 1930, Evans [17] provided direct evidence of their existence by means of a simple, electrochemical technique that made it possible to peel the film from a passive surface and view it under a microscope.

The factors leading to passivity in various Fe–Cr alloys were first investigated systematically by Monnartz [18] in his doctoral research. He used low-carbon ferrochrome made by the Goldschmidt process for his research and became the originator, not only of a number of methods for enhancing the passivity of these alloys, but also of concepts for understanding this condition. The following is a partial list of his findings of 1911:

Iron–chromium alloys with 12.5% or more chromium are resistant at room temperature to nitric acid at all concentrations. With 14% Cr or more, Fe–Cr alloys resist such solutions at temperatures up to boiling.

In reducing acids, additions of chromium to iron *increase* the rate of corrosion.

Molybdenum additions increase resistance of Fe–Cr alloys in nitric acid containing chloride salts.

Passivity depends on a source of oxygen, either from a compound in solution or from dissolved oxygen gas.

In solutions in which a given Fe–Cr alloy is not passive, for example, nitric acid containing a chloride salt, passivity can be induced by contacting the specimen with a platinum wire, adding platinum as an alloying element, or making the Fe–Cr alloy an anode by means of a cathode and an external electromotive force (emf) (anodic protection). All of these procedures change the potential of the alloy in the noble direction.

With his thesis advisor, Borchers, Monnartz patented a 30–40% Cr alloy with 2–3% Mo. Despite the fact that they used low-carbon chromium derived from the Goldschmidt process, this alloy proved to be too brittle for fabrication.

---

* Chance discoveries also led to the introduction of X-rays (1895), radioactivity (1896), age-hardening aluminum alloys (1906), penicillin (1928), nuclear fission (1939), and the argon–oxygen refining process (1954). See also Appendix A.

The implications of much of this extensive investigation in 1911 were recognized only much later, even though Monnartz pointed out that the chromium-bearing alloys deserved attention by the chemical industry [8].

C. THE ROLE OF ALLOYING ELEMENTS

C1. General Corrosion in Acids

C1.1. Chromiun. When a series of alloys with increasing concentrations of chromium are tested in an acid of constant temperature and concentration, there is frequently a sharp change from high to low corrosion rates within a narrow range of concentration of chromium. The minimum concentration of chromium required for passivity is a function of the type of acid, its concentration, and temperature [19]. Thus, there is no fixed ratio of iron-to-chromium concentration in Fe–Cr alloys that characterizes the passive state for various concentrations, temperatures, and acid compositions [20]. Nor would this be expected if the rate-determining process in the passive state is the rate of dissolution of the protective oxide film in the acid solution. Chemical analyses of oxide films have shown that the ratios of metals are different in the film than in the metal, that is, the rate at which the various metals form hydrated oxides varies [21].

In practice, it has been found that an alloy with 14–18% Cr provides resistance in a number of acid environments, and this has led to a well-established commercial alloy, AISI type 430 (S43000) stainless steel (Table 51.1). Increases in the concentration of chromium above that required to produce passivity can provide significant further reductions in corrosion rates. This had led to the use of

an Fe–24–27% Cr alloy, type 446 (S44600), for some environments. Originally, this alloy was intended to resist high-temperature oxidation.

The effect of the concentration of chromium in iron on corrosion in an oxidizing environment is shown in Figure 51.1. In the boiling 50% sulfuric acid solution with ferric sulfate, the passive rate is reduced by 99.44% as the chromium content is increased from 12 to 25%. Alloys with <12% Cr are active in this solution and dissolve at rates greatly in excess of 600 mm/year. The ferric sulfate in this solution, when present in excess of a certain minimum concentration, which varies with the chromium content of the alloy [14], makes stainless steels passive. The corrosion potential is changed from -0.6 to $+0.6$ V versus a saturated calomel electrode (SCE) (Table 51.2). Even though the rates of corrosion are relatively high and readily measurable, there is no hydrogen evolution and the cathodic reaction consists of the reduction of ferric-ions:

$$Fe^{3+} \xrightarrow{+e} Fe^{2+}$$

This reaction is electrochemically equivalent to the anodic dissolution of iron, chromium, and nickel:

$$\left.\begin{array}{l} Fe \\ Cr \\ Ni \end{array}\right\} \xrightarrow{-e} Fe^{2+} + Cr^{2+} + Ni^{2+}$$

When the available ferric ions are consumed in the cathodic reaction, there is a sudden shift to the active state and, in the case of the boiling 50% sulfuric acid solution, an almost explosive production of hydrogen gas [14].

TABLE 51.1. Composition of Stainless Steels and Related Alloys

UNS No.	Type	Elements, %					
		Chromium	Nickel	Molybdenum	Carbon	Nitrogen	Other
S30400	304	18.0–20.0	8.0–12.0		0.08 max		Mn 2.0 max; Si 1.0 max
S30403	304L	18.0–20.0	8.0–12.0		0.03 max		Mn 2.0 max; Si 1.0 max
S31000	310	24.0–26.0	19.0–22.0		0.25 max		Si 1.50 max
S43000	430	14.0–18.0			0.12 max		Mn 1.0 max; Si 1.0 max
S44600	446	23.0–27.0			0.20 max	0.25 max	Mn 1.50 max; Si 1.0 max
S31600	316	16.0–18.0	10.0–14.0	2.00–3.00	0.08 max		Mn 2.0 max; Si 1.0 max
S31603	316L	16.0–18.0	10.0–14.0	2.00–3.00	0.03 max		Mn 2.0 max; Si 1.0 max
S32100	321	17.0–19.0	9.0–12.0		0.08 max		Ti = 5×%C min
S34700	347	17.0–19.0	9.0–13.0		0.08 max		Cb = 10×%C min
N08020	Carpenter 20 Cb–3	19.0–21.0	32.5–35.0	2.00–3.00	0.06 max		Cb + Ta = 8×%C Cu 3.00–4.00
N10002	Hastelloy alloy C	14.5–16.5	~54	15.0–17.0	0.08 max		Fe 5.5; W4; Co 2.5
S31254	6Mo	19.5–20.5	17.5–18.5	6.0–6.5	0.020 max	0.18–0.22	Mn 1.0 max; Cu 0.50–1.00; Si 1.00 max
N08367	6Mo	20.0–22.0	23.5–25.5	6.0–7.0	0.030 max	0.18–0.25	Mn 2.0 max; Cu 0.75 max; Si 1.0 max

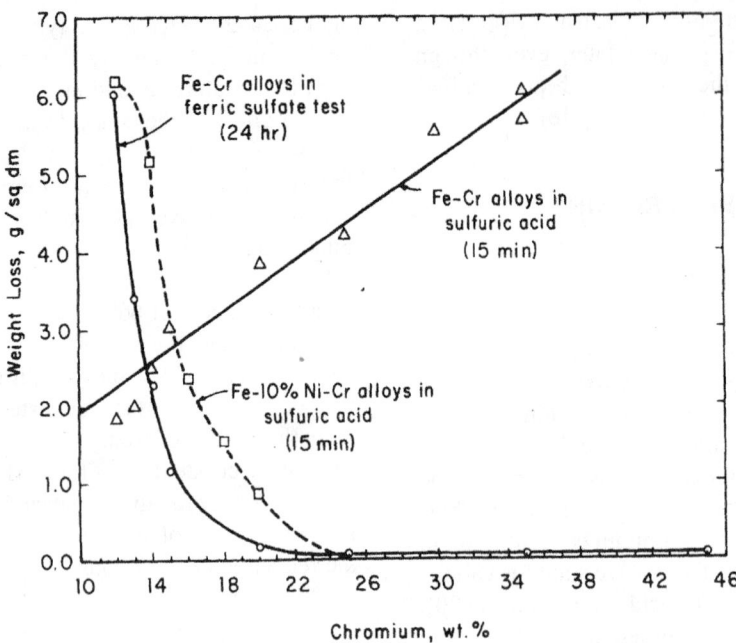

Ferric Sulfate – 50% Sulfuric Acid, + 0.60 V vs. SCE
5% Sulfuric Acid, – 0.60 V vs. SCE

FIGURE 51.1. General corrosion of Fe–Cr and Fe–10% Ni–Cr alloys in sulfuric acid solutions [33]. Ferric sulfate–boiling 50% sulfuric acid with 25 g/600 mL ferric sulfate. 24 h. Sulfuric acid–boiling 5%. 15 min.

Corrosion in the passive state in sulfuric acid solution with ferric sulfate is a function of the acid concentration and of the alloy content of the steel. Large variations in corrosion rates can be obtained by varying the acid concentration (Fig. 51.2) without significantly changing the corrosion potentials, which are near +0.6 V vs. SCE.

The large variation in corrosion rates as a function of chromium content (Fig. 51.1) makes this solution a sensitive

TABLE 51.2. Corrosion Potentials of Stainless Steels in Boiling Acids and in Chloride Solutions

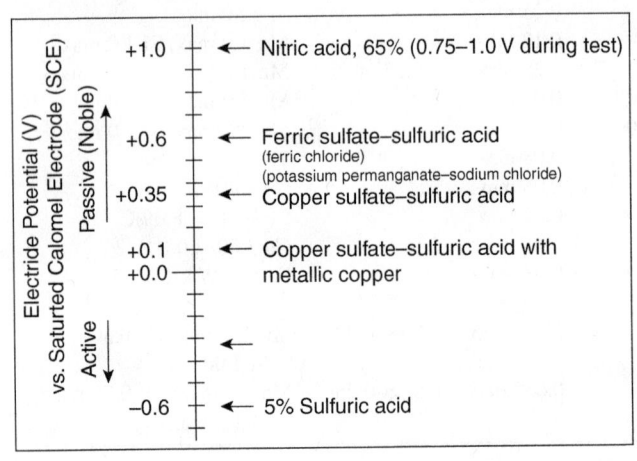

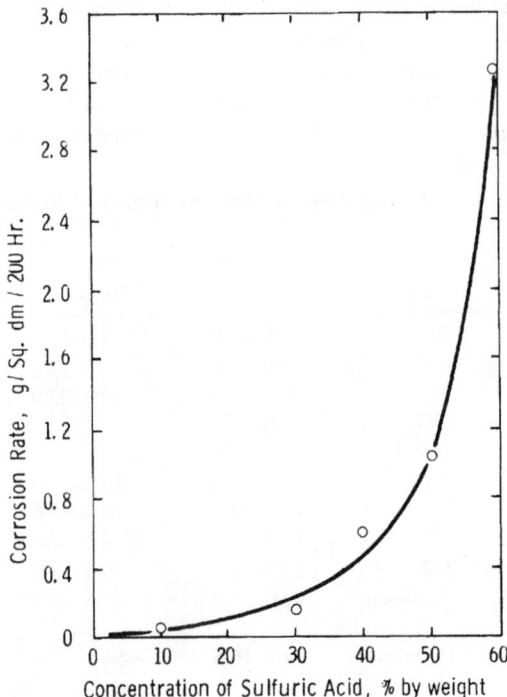

FIGURE 51.2. Effect of concentration of sulfuric acid on corrosion rate of type 304 stainless steel in the passive state. Boiling solutions with 15 g/600 mL ferric sulfate. Boiling points: 10%—102°C (215°F); 50%—123°C (265°F); 60%—140°C (285°F).

tool for detecting the formation of chromium-depleted zones around chromium carbide and nitride precipitates at grain boundaries of stainless steels. These problems are discussed in the section on intergranular corrosion.

Figure 51.1 also shows the results of tests in the active state with hydrogen evolution and a corrosion potential of -0.6 V vs. SCE. The same series of Fe–Cr alloys was exposed to boiling 5% sulfuric acid without ferric sulfate. Only by greatly reducing the acid concentration could tests be made under controlled conditions because of the rapid rate of hydrogen evolution. In agreement with the results of Monnartz [18], it was found that during active corrosion in reducing acids increasing the concentration of chromium in the alloy increases the rate of corrosion. Increasing the chromium content from 10 to 35% increases the dissolution rate by a factor of 3.

The effect of chromium on corrosion by boiling acids is also shown in Table 51.3. For comparison with data on commercial types 430 and 446 alloys, rates are given for carbon steel (0.2% C) and for a low-carbon, low-nitrogen 35% Cr alloy. This high-purity composition was previously investigated by Steigerwald [22]. Even in boiling 65% nitric acid, addition of 16% Cr results in a remarkable decrease in corrosion rate. In acetic acid there is also a decrease in corrosion rate with addition of chromium, leading to complete resistance for the 25% Cr alloy. In all of the other acids, addition of 16% chromium increases the rate of corrosion. Increasing chromium to 25% produces a further increase in rate in sulfuric and hydrochloric acids but decreases the rate in formic acid and in sodium bisulfate. The rates on the 35% Cr alloy add support to the appearance of a maximum in corrosion rate in the range of 16–25% Cr in most of the acids. Passivity in the high-purity alloy may be associated with its low carbon content.

C1.2. Nickel. In the Fe 16–25% Cr alloys, the combination of a body-centered (ferritic) structure and of their normal carbon contents of $\sim$0.1% contributes to a relatively low ductility, which complicates forming operations. These alloys also have a high-notch sensitivity, that is, a high

TABLE 51.3. General Corrosion in Boiling Acids

		Corrosion Rate (mm/year)[a,b]							
UNSNo.	Alloy	Nitric (65%).	50% Sulfuric Acid with Ferric Sulfate	Formic (45%)	Oxalic (10%)	Acetic (20%)	Sodium Bisulfate (10%)	Sulfuric (10%)	Hydrochloric (1%)
G10200	Carbon steel (AISI 1020)	4500		630	62	170	1000	1300	430
S43000	Type 430 (16% Cr)	0.5[c]	7.9[d]	2200	160	80	2300	6400	1500
S44600	Type 446 (25% Cr)	0.2[c]	0.9[d]	250	180	0.0[c]	1600	6900	1900
	Fe–35% Cr (high purity)	0.2[c]	0.2[d]	0.2[d] (1100)	0.0[d] (800)	0.0[c]	0.2[d] (2700)	0.4[d] (5000)	1500
S30400	Type 304 (18 Cr–8 Ni)	0.2[c]	0.6[d]	44	15	0.1[c]	70	400	81
S31600	Type 316 (18 Cr–10 Ni–2.5 Mo)	0.3[c]	0.6[d]	13	2.4	0.1[c]	4.3	22	71
N08020	Carpenter 20 Cb–3 (20 Cr–34 Ni–2.5 Mo–3.5 Cu)	0.2[c]	0.2[d]	0.2[c]	<u>0.2</u>	0.1[c]	<u>0.3</u>	<u>1.1</u>	<u>0.0</u>
N10002	Hastelloy alloy C (16 Cr–54 Ni–16 Mo–4 W)	11.4[c]	6.1[d]	0.1[c]	<u>0.2</u>	0.0[c]	0.2	<u>0.4</u>	<u>0.3</u>
	Titanium	0.3[c]	5.9[d]	22	24	0.0[c]	6.4	160	5.6
S44400	Fe–18 Cr–2 Mo (Ti)	[5.8][d,c]	[4.1][d,e]	10	250	0.0[c]	930	2400	850
S44627	Fe–26 Cr–1 Mo (high purity)	0.1[c]	0.4[d]	<u>0.1</u> (550)	0.2[d] (1800)	0.0[c]	0.0[d] (1800)	3400	0.7[d] (2000)
	Fe–26 Cr–1 Mo (Ti)	0.1[c]	0.3[d]	<u>0.1</u> (350)	0.1[d] (1500)	0.0[c]	0.0[d] (1500)	3200	0.1[d] (1600)
	Fe–28 Cr–2 Mo–4 Ni (Nb)	0.2[c]	0.3[d]	<u>0.1</u>	<u>0.1</u>	0.0[c]	<u>0.0</u>	0.2	<u>0.0</u>
S44700	Fe–29 Cr–4 Mo	0.1[c]	0.2[d]	<u>0.1</u>	<u>0.3</u>	0.0[c]	<u>0.2</u>[d] (500)	1300	<u>0.2</u>[d] (550)
S44800	Fe–29 Cr–4 Mo–2 Ni	0.1[c]	0.2[d]	<u>0.1</u>	<u>0.1</u>	0.0[c]	<u>0.0</u>	0.2	<u>0.2</u>

[a]Acid concentrations in percent by weight. Test specimens, 25 × 25 mm (1 × 1 in.): 600 mL of solution. The length of testing time varied from 5 min for high rates to 10 days for low rates.

[b]0.10 mm = 4.0 mil (thousandths of an inch) per year.

[c]Cannot be activated with an iron rod.

[d]Specimen is passive when immersed, but not self-repassivating when activated by contact with an iron rod while exposed to the solution. Number in parentheses is the rate in the active state. Underlined rates–self-repassivating after activation with an iron rod.

[e]Severe intergranular attack.

transition temperature below which ductile fracture is transformed to brittle fracture. Addition of enough nickel to change the body-centered structure to a face-centered, austenitic, nonmagnetic structure makes the alloy more ductile and provides high impact strengths even at very low temperatures. In the case of the 18% Cr alloy, 8% nickel produces a completely austenitic structure [AISI type 304 (S30400)], and with 25% Cr, an addition of 20% Ni makes it fully austenitic, type 310 (S31000).

Both in oxidizing and in reducing acids, nickel additions may actually increase the corrosion rate of Fe–Cr alloys. An example of this in a reducing acid is shown in Figure 51.1. A comparison of the two plots for corrosion in boiling 5% sulfuric acid shows that the 10% Ni alloys with <16% Cr have higher rates of corrosion than those of the same chromium content without nickel. Only when the chromium content exceeds 16% is there a rapid reduction in corrosion. Thus, in the presence of 10% Ni, increasing the concentration of chromium >16% reverses the deleterious effect of chromium, indicating that at these concentrations there is a synergistic effect on the rate of hydrogen evolution corrosion (active state) between chromium and nickel. This is also shown in Table 51.3 by the data on corrosion in boiling reducing acids, sulfuric, oxalic, and (formic, and in sodium bisulfate. In each of these acids, the corrosion rate of the 18% Cr–8% Ni (type 304) alloy is appreciably lower than that of both the 16% Cr (type 430) and even the 25% Cr (type 446) alloys. Figure 51.3 shows that the first 2% of nickel is more effective in reducing hydrogen evolution corrosion than additions in excess of this concentration.

In oxidizing acids, the effect of nickel additions also depends on the chromium content. Pilling and Ackerman [20]

found that concentrations of nickel up to 30% in alloys with >15% Cr increased the corrosion rate in 5% nitric acid. Only when nickel was added in amounts >30% was there a reduction in corrosion rate.

C1.3. Molybdenum. As mentioned above, Monnartz [18] discovered that molybdenum enhances the passivity of stainless steels in chloride environments. Comparison of the data in Table 51.3 on types 304 (S30400) and 316 (S31600) provides an indication of the effect of ~2.5% Mo on corrosion in acids. Because molybdenum is a ferrite former, the nickel content is increased by ~3% over that of type 304 in order maintain the austenitic structure. This increase in nickel may also contribute to the resistance of type 316 when compared with type 304. Just 2.5% molybdenum significantly reduces hydrogen evolution corrosion in most of the reducing acids but has no significant effect on corrosion in the two oxidizing solutions. It is shown below in connection with the new ferritic alloys that, like nickel, molybdenum converts chromium into a beneficial alloying element in reducing acids. The role of molybdenum in imparting resistance to chloride pitting and crevice corrosion is discussed below in the section on this topic.

The corrosion rates in Table 51.3 for types 430, 446, 304, and 316 in reducing acids merely provide an indication of the effects of various alloying elements. With only three exceptions in acetic acid, the rates are much too high to permit use of these alloys at the temperatures and concentrations shown.

C1.4. Cr–Ni–Mo–Cu [Alloy 20 (N08020)]. To provide resistance in broader ranges of concentrations and temperatures of sulfuric acid solutions, the nickel content of type 316 steel was increased from 11 to 20%, and 2.2% Cu was added in an alloy developed and standardized in 1932 in Germany [23]. Fontana initiated a similar alloy for castings at Du Pont in 1935 [24] with 20% Cr–29% Ni–2.25% Mo–3.25% Cu. Molybdenum enhances resistance at sulfuric acid concentrations between 20 and 70%, while copper contributes to resistance at concentrations <20% and >70%. Increasing the nickel content contributes to resistance over the entire range up to 75% acid [25]. By 1947, improved melting techniques made it possible to produce wrought products from this alloy, and in 1948 niobium was added to combine with carbon and thereby minimize chromium carbide precipitation and susceptibility to intergranular attack [26].

However, in 1960, it was found that this alloy was subject to a unique form of stress corrosion cracking in sulfuric acid solutions in the range of 20–80% [27, 28]. Gräfen [23] attributed this form of cracking to the cathodic action of copper, which after dissolving from the alloy is redeposited on the surface and there accelerates anodic dissolution on adjacent surfaces. This problem was overcome in 1965 by

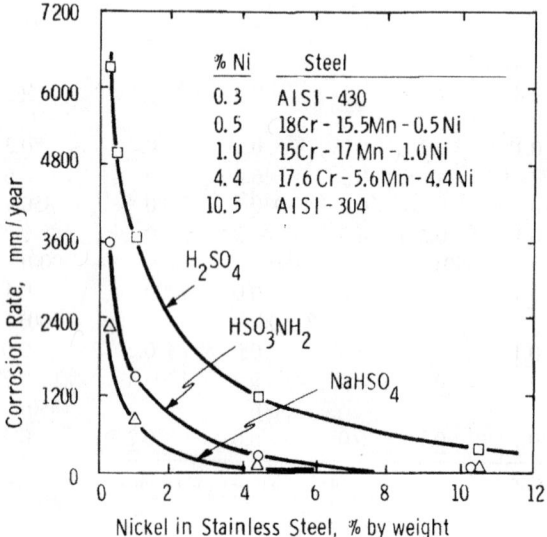

FIGURE 51.3. The influence of nickel on corrosion in the active state [14]. Boiling acids 10% by weight. (HSO₃NH₂ is sulfamic acid.)

increasing the nickel content from 29 to 34% [25, 29] in the Carpenter 20 Cb–3 alloy (Table 51.1).

Data in Table 51.3 show that this alloy does indeed have superior resistance in reducing acids without any decrease in resistance to oxidizing solutions. In fact, the rates for this alloy are low enough to make it usable in all of the environments of Table 51.3. The high nickel content also provides enhanced resistance to chloride stress corrosion cracking, but not to chloride crevice attack. However, these increases in alloy content have substantially increased the price.

The resistance of the nickel-based Hastelloy alloy C(N10002) N10002 has been superceded by N10276, which has similar Cr and Mo contents but reduced contents of C and Si. See the chapter 59 on nickel and nickel alloys. In reducing acids is similar to that of Carpenter 20 Cb–3. Its poor performance in the oxidizing solutions is primarily a result of its high (16%) molybdenum content and lower chromium, 16%, as compared with 18% in type 304. Data on general corrosion of this alloy are given in Table 51.3 for comparison because its resistance to chloride stress corrosion, pitting, and crevice corrosion is superior to all of the stainless steels and to the carpenter 20 Cb–3 alloy.

C1.5. 6Mo Alloys (S31254 and N08367).

While the vacuum-melted superferritic alloys and nickel alloys had demonstrated that stainless-type materials could provide resistance to corrosion by seawater and other high-chloride environments, they were much more expensive than the copper alloys that were traditionally used in such environments. To achieve sufficient pitting resistance to allow an austenitic stainless alloy to be used in seawater, a minimum of about 6% molybdenum is required. The first such alloy to achieve significant commercial success was the AL-6X™* alloy [29a], UNS N08366. It was soon followed by the 254SMO® alloy [29b], UNS S31254, which featured the intentional addition of nitrogen for enhanced corrosion resistance and stability. With the recognition of the beneficial effect of nitrogen, the AL-6X alloy was produced with added nitrogen to create the AL-6XN® alloy [29c], UNS N08367, which replaced the older, N-free alloy. These two alloys, which contain about 20%Cr, 6%Mo, and 0.2%N with 18–24%Ni, are the most widely used of the so-called 6Mo alloys, but numerousothers continue to be developed and used. Many of the newer alloys contain increased Cr contents, typically about 25%, or increased Mo contents of about 7%.

*AL-6X™ is a trademark of ATI Properties Inc. 254SMO® is a registered trademark of Outokumpu Stainless Inc. AL-6XN® is a registered trademark of ATI Properties Inc.

C2. Intergranular Corrosion

C2.1. Effect of Precipitates

C2.1.1. Chromium Carbide and Nitride. A loss in mechanical strength was observed on some of the first stainless steels in use for chemical equipment. It was soon found that corrosion was the cause of this loss. The carbon content of the early austenitic stainless steels was $\sim 0.15\%$. In the very first investigations of this problem by Strauss, Schottky, and Hinnüber [30] and Aborn and Bain [31], it was found that the carbon combines with chromium during exposures of the steel in the range of 425–875°C. Such exposures are involved during welding and, in some cases, during fabrication. A precipitate of chromium carbide, $(Fe, Cr)_{23}C_6$, forms preferentially at the grain boundaries. The high chromium concentration in the precipitate causes depletion of chromium in the metal adjacent to the precipitate. As a result, there is intergranular attack by those environments in which a decrease in chromium content leads to an increase in corrosion rate.

An example of this is shown in Figure 51.4. The specimen of type 446 (25% Cr) steel was heat treated to precipitate chromium carbides at the grain boundaries. Because this commercial alloy also contains a deliberate addition of 0.2% nitrogen, an austenite former, for control of grain size, $\sim 50\%$ of the structure is austenite. In Fe–Cr alloys, nitrogen also reacts with chromium and forms $\beta\text{-}Cr_2N$ [32] at the grain boundaries. As might be expected from Figure 51.1, there is intergranular attack in the oxidizing ferric sulfate–sulfuric acid solution. In contrast, on an identical specimen (Fig. 51.5) exposed in reducing 5% sulfuric acid, there is no

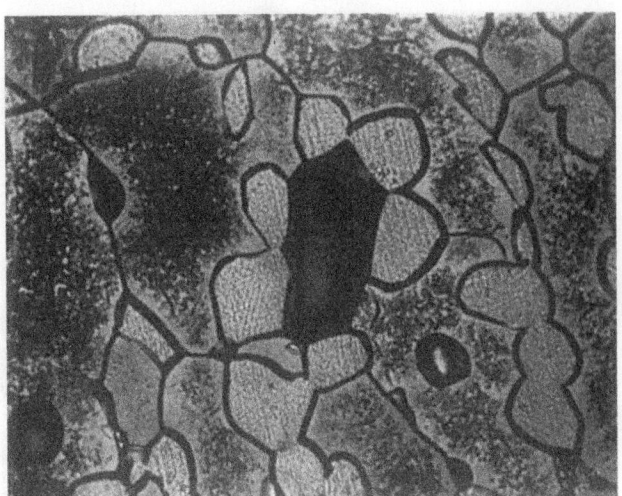

FIGURE 51.4. Initial attack in ferric sulfate/50% sulfuric acid on sensitized type 446 (25 Cr) steel [33] (500×). Steel: Heated 1 h at 1150°C (2100°F). Exposure: 14 h in boiling solution. Structure: Intergranular attack, but not on austenite–austenite boundaries. Austenite islands in ferrite matrix, Grains dislodged.

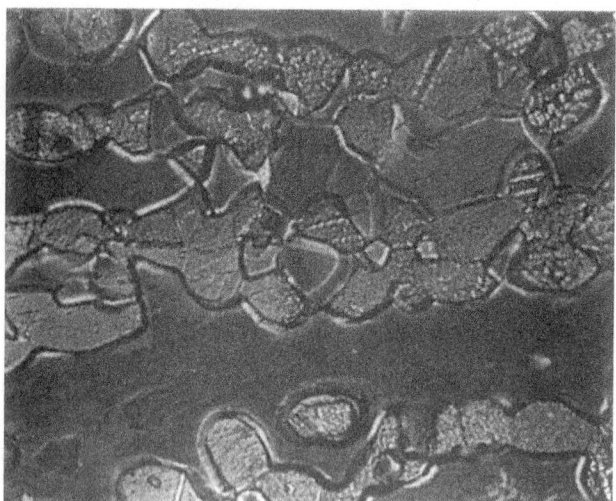

FIGURE 51.5. Initial attack in boiling 5% sulfuric acid on sensitized type 446 steel [33] (500X). Steel: Heated 1 h at 1150°C (2100°F). Exposure: 30 s. Structure: No intergranular attack, preferential corrosion on austenite.

intergranular attack, because in this solution (Fig. 51.1), the corrosion rate of Fe–Cr alloys actually decreases as the chromium content is decreased. The "steps" between the austenite and ferrite grains are a result of higher rates of corrosion on austenite grains in which the concentration of carbon and nitrogen in solid solution is higher than in the ferrite grains [33]. Similarly, Figure 51.1 shows that decreasing the chromium content of iron alloys containing 10% Ni increases the corrosion rate in 5% sulfuric acid. Therefore, on a sensitized 18 Cr–10 Ni alloy (type 304, Fig. 51.6) there is

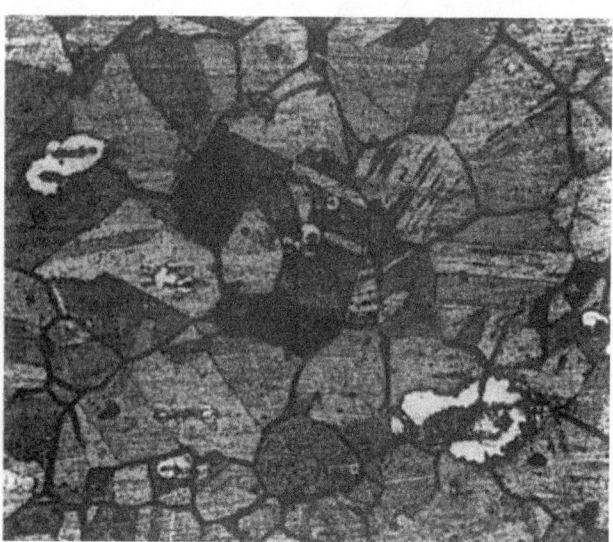

FIGURE 51.6. Initial attack in boiling 5% sulfuric acid on sensitized type 304 stainless steel [33] (250X). Steel: Heated 1 h at 675°C (1250°F). Exposure: 5 min. Structure: Intergranular attack, grains dislodged.

preferential attack at chromium-depleted zones and intergranular corrosion.

The shapes, sizes, and sites of chromium carbide precipitates were determined by Mahla and Nielsen [34] in 1951 by dissolving austenitic stainless steels containing carbide precipitates in a solution of bromine in methanol. This process leaves behind the fine carbide particles that were then examined in the electron microscope. This technique was applied to a specimen which was cooled slowly through the precipitation range to form a large precipitate (Fig. 51.7) [15]. The carbide particles nucleate at grain

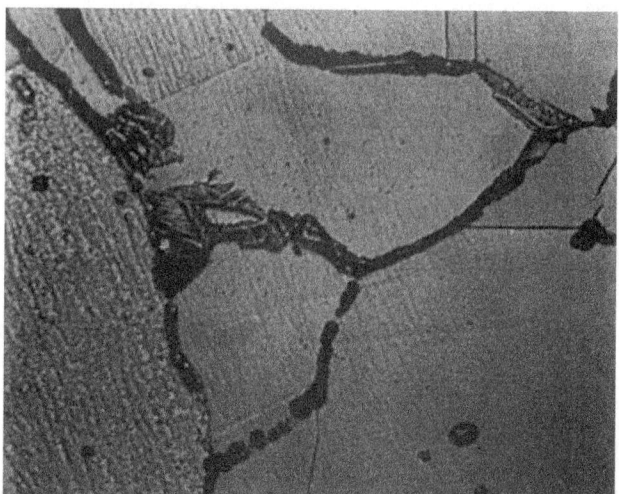

FIGURE 51.7. Precipitation of chromium carbides, (a) Polished 18 Cr–8 Ni–0.05% C stainless steel was heated 24 h at 2000°F. furnace cooled to room temperature and etched electrolytically in oxalic acid. Depressions with dendritic patterns remain where large chromium carbides were dislodged. (b) Electron photomicrograph of dendritic carbide isolated from specimen described in (a) by dissolution in a solution of bromine in methanol (5000X).

boundaries and grow into the metal grains on parallel crystallographic planes.

The original investigations of intergranular corrosion also provided various remedies for this problem.

C2.1.2. Heat Treatments. For austenitic Fe–Cr–Ni alloys annealing treatments at temperatures >1050°C dissolve carbides and put carbon in solid solution. By rapid quenching through the precipitation range of 875–425°C, precipitation of carbides is prevented. Because of the large solubility of nitrogen in austenite, chromium nitrides do not readily form in the austenitic stainless steels.

In the ferritic alloys, the diffusion rates of carbon and nitrogen are much greater than in the austenitic compositions. As a result, even when these alloys are water quenched from temperatures above ∼800°C, precipitation of chromium carbides and nitrides cannot be prevented. The temperature range of most rapid precipitation is 540–600°C. However, because the diffusion rate of chromium is also much higher in ferritic than in austenitic alloys, the chromium depletion zone around the precipitates can be replenished by relatively short heat treatments at 800°C. Thus, in the ferritic alloys, grain boundaries may be immune to intergranular attack even though they contain chromium carbide and nitride precipitates.

C2.1.3. Low Carbon. Reduction of the carbon content to such low concentrations that no precipitates are formed during welding and stress-relieving operations has largely solved the chromium carbide problem in 18 Cr–8 Ni austenitic stainless steels. The maximum carbon content that can be tolerated depends on the heat input during welding and the thickness of the object being welded, that is, on the cooling rate. The solubility of carbon is only 0.007% at 700°C in 18 Cr–8 Ni steel [35] when determined by heating for five weeks at 700°C. For practical purposes, a carbon content of 0.02–0.03%, such as provided in extra-low-carbon grades of type 304 and 316 steels, meets most requirements. Such grades, 304L (S30403) and 316L (S31603), were introduced about 1947 as a result of improvements in the reduction of carbon made possible by blowing oxygen through the melt.

Since 1970 further important improvements in (a) the reduction of the carbon content to 0.01%, (b) the reproducibility of the composition and, (c) the cost of production have been made possible by the introduction of the argon–oxygen decarburization process developed by Krivsky [36]. In this process the starting materials are first melted as usual in an electric arc furnace. The molten metal is then poured into a vessel in which both oxygen and argon are blown into the melt from the bottom. Carbon is removed by reaction with oxygen. By mixing increasing amounts of argon with the oxygen, ever greater amounts of carbon can be removed. The presence of argon reduces the partial pressure of carbon monoxide, which controls the concentrations of carbon in the melt. Even though this process is based on well-known principles of physical chemistry, its discovery was a result of a chance observation. Its industrial success was facilitated by a concurrent drop in the price of argon [36]. This refining process is particularly useful for reducing the carbon content of the new ferritic stainless steels.

Reduction of the carbon content lowers the strength of the 18 Cr–8 Ni alloys and, in some cases, necessitates the design of heavier and more costly sections. To avoid this expense, stabilized steels, types 321 (Ti) (S32100) and 347 (Nb) (S34700), can sometimes be used.

C2.1.4. Use of Stabilizers. Before methods had been developed for reducing the carbon content of commercial melts to low concentrations, titanium [37] and niobium [38] were added to combine with (stabilize) carbon. To prevent the formation of chromium carbides, the mill forms, sheet, tube, and bars must first be heated to ∼1100°C to dissolve all carbides in the austenitic structure and then cooled to 900°C and held for several hours to permit titanium or niobium to react with carbon. As a result, no carbon is available to form chromium carbides during subsequent exposure in the chromium carbide precipitation range of 425–875°C. Stabilized steels were introduced during the early 1930s, and for many years titanium (type 321) was the preferred element in Europe, while in the United States type 347 with niobium has been the most frequently used alloy. Recently, titanium has also been added to type 430 steel [39].

Three of these methods for overcoming susceptibility to intergranular attack are illustrated in Figure 51.8. Four panels were joined by welding and then exposed to a solution that causes intergranular attack on grain boundaries containing chromium-depleted zones surrounding chromium carbide precipitates. In the high-carbon type 304 steel, heat from the liquid weld metal resulted in precipitation of chromium carbides in a zone parallel but not adjacent to the weld metal.

The question of the concentration of nickel in 18% Cr alloys required to convert precipitation of chromium carbides from the ferritic (type 430) to the austenitic (type 304) pattern was investigated by Upp et al. [40]. As expected, type 304 was made susceptible to intergranular attack when heated in the range of 650–760°C. On type 430, this same heat treatment imparts resistance to intergranular corrosion by diffusing chromium into the depleted zones surrounding chromium carbide and nitride precipitates. Quenching from >1040°C provides optimum resistance to type 304 but makes type 430 susceptible to intergranular attack. Additions of only 3.0% nickel to type 430 transformed its response to heat treatments to that of type 304 (18 Cr–8 Ni) steel.

C2.1.5. Sigma Phase. Heating the molybdenum-bearing (2.5% Mo) type 316 stainless steel in the range of 540–1000°C can result in the formation of another precipitate at grain boundaries, an intermetallic compound, sigma phase.

FIGURE 51.8. Weld decay and methods for its prevention. The four different panels were joined by welding and then exposed to a hot solution of nitric/hydrofluoric acids. Weld decay, parallel, but not adjacent to the weld metal, such as shown in the type 304 stainless steel, is presented by reduction of the carbon content (type 304L) or stabilization with titanium (type 321) or niobium (type 347).

It has a tetragonal structure [41] and is rich in molybdenum and chromium [42]. Unlike chromium carbide precipitates, which cause susceptibility to intergranular attack in a large variety of oxidizing and reducing acids, sigma phase only impairs resistance in the highly oxidizing nitric acid environments [43, 44]. This may be a result of the high molybdenum content of sigma phase, which leads to direct attack on these particles rather than on the molybdenum-depleted zones surrounding them. The corrosion potential in boiling 65% HNO_3 is about $+1.0$ V vs. SCE. In the less oxidizing sulfuric acid–Ferric sulfate solution, the corrosion potential is $+0.6$ V vs. SCE and sigma phase in molybdenum-bearing alloys does not lead to intergranular attack [15].

In certain type 316 stainless steels with a very low (e.g., 0.01%) carbon content, it was found that heating for 1 or 2 h at 677°C (1250°F) has no effect on the microstructure (i.e., there were no carbide or visible sigma-phase precipitates at the grain boundaries). Nevertheless, high rates of intergranular attack were observed in nitric acid [15, 43–45]. Because prolonged heating of such alloys at 700°C results in readily visible sigma-phase precipitate and solution annealing at 1060°C with water quenching removes the susceptibility to intergranular attack, it appears that a submicroscopic, preprecipitation form of sigma phase is responsible for the intergranular attack on these alloys. This phenomenon is

TABLE 51.4.   Effect of Chromium Carbide and Sigma Phase on Intergranular Corrosion[a]

Alloy AISI	C (%)	Ferric sulfate–50% Sulfuric Acid (120h)	65% Nitric Acid (240h)	Oxalic Acid Etch Structure (Sensitized Specimen)
304	0.063	11.8	12.8	Ditch
304	0.031	2.1	2.0	Ditch
316L	0.022[c]	1.0	133.0	Step
316L	0.020	1.4	35.6	Step
316	0.046	7.8	19.0	Ditch

[a]See [15].
[b]Rate of sensitized specimen ÷ rate of solution annealed specimen. Sensitizing heat treatment: 1 h at 675°C (1250°F).
[c]Sensitized 1 h at 705°C (1300°F).

discussed further below in connection with data in Table 51.4.

Sigma phase also may form in type 310 and in 16 and 25% Cr (types 430 and 446) stainless steels. However, its rate of formation is so slow that it is of importance only when these alloys are actually used in the temperature range in which sigma phase is formed.

C2.2. Evaluation Tests

C2.2.1. Copper Sulfate. The fact that both the ferritic and austenitic stainless steels could under certain circumstances become subject to rapid intergranular attack made it necessary to evaluate alloys and thereby prevent the use of damaged materials on plants. Tests were proposed in solutions to which these alloys were to be exposed in service. The widely used copper sulfate–sulfuric acid test was derived in 1926 by Hatfield [46] from observation of intergranular attack on austenitic stainless steels in a sulfuric acid pickling tank containing copper sulfate.* Austenitic steels containing chromium carbide that precipitate at the grain boundaries are "sensitized" and, therefore, subject to preferential attack in the boiling solution, while at the grain faces there is only a relatively low rate of attack. The severity of the test is a function of the concentration of sulfuric acid, ~8% in the original composition of Hatfield. This has been increased to 16% [47] and to 50% [48] in more recent modifications. These are now standardized as the American Society for Testing and Materials (ASTM) A262 Practice B and F methods, respectively. The corrosion potential in these solutions is near $+0.35$ V vs. SCE. This value is decreased to $+0.1$ V vs. SCE by immersing metallic copper (sheet or shot) in the test flask along with the stainless steel specimen, as proposed by Rocha [49]. Immersion of metallic copper

_____
*See note in Appendix B.

greatly increases the rate of attack [15]. Sigma phase at grain boundaries of type 316 and 316L steels does not cause intergranular attack in these copper sulfate tests [48].

Hatfield's test solution provided the first quality control method for preventing industrial use of sensitized stainless steels. It was also put to immediate use as a research tool in investigations of intergranular corrosion and for the development of methods for preventing intergranular attack by heat treatments and by alloying (Ti, Nb) additions. Depending on the concentration of the sulfuric acid in the test solution and the alloy content of the steel, evaluation is either by visual examination for fissures on a specimen bent after testing or for grains undermined and dislodged or, in the 50% H_2SO_4 solution, by weight loss determinations.

C2.2.2. Nitric Acid. In connection with the use of iron–chromium alloys for nitric acid plants, Huey [50] described a test procedure in 1930 to determine reproducible corrosion rates and thereby to distinguish differences in the quality of various heats of the new Fe–Cr stainless steels. He proposed five 48-h test periods in boiling 65% nitric acid solution in a glass apparatus. The 65% acid concentration was selected for rapid (240-h) results and to facilitate temperature control because it is near the constant boiling concentration of 68.5%. It was soon found that this test could also be used to detect susceptibility to intergranular attack in Fe–Cr and Fe–Cr–Ni stainless steels.

In numerous comparisons made on identical specimens of various 'sensitized' stainless steels in both the nitric acid and the copper sulfate–sulfuric acid tests, it was established that the nitric acid solution detects susceptibility to sensitization caused not only by chromium carbide but also by sigma-phase precipitates such as found in types 316, 347, and 321 stainless steels [43–45, 48, 51]. The corrosion potential in this highly oxidizing solution is about $+1.0\,V$ vs. SCE. Moreover, this test solution is also unique in that the corrosion products affect the dissolution rate. DeLong [52] reported that as little as 0.004% dissolved chromium appreciably increases the corrosion rate.

Later investigators [53–55] showed that acceleration is due to the action of chromium in the hexavalent state. When chromium dissolves in acids, it enters the solution as divalent, blue ions, which are rapidly converted to the trivalent state, green. These ions are then oxidized more gradually in boiling 65% nitric acid to the hexavalent state, orange. Hexavalent chromium not only increases the rate of general or grain face corrosion and of intergranular corrosion of sensitized stainless steels but also causes intergranular attack on solution-annealed material, which is free of any precipitates at grain boundaries [15]. Therefore, it is essential to limit the accumulation of corrosion products in the nitric acid test, either by changing the test solution frequently (every 48 h) or by distillation designs for test vessels, such as proposed by DeLong [52]. However, not all aspects of this phenomenon

can be controlled. Nitric acid dissolves certain nonmetallic inclusions and, in the case of titanium-stabilized steel, the relatively large concentration of titanium carbide and nitride particles. The pits left behind when these particles have been dissolved limit the movement of dissolved corrosion products and serve to accumulate hexavalent chromium. A self-accelerating dissolution process is initiated which causes rapid growth of the pit by general and intergranular attack. Thus, the interpretation of high corrosion rates in nitric acid tests may be very complex and of limited applicability to other environments.

Only when the high rates in the nitric acid test are a result of a chromium carbide precipitate are the results of more general applicability. The copper sulfate test and several other new methods provide unambiguous ways for detecting susceptibility associated with chromium carbides. Therefore, the nitric acid test is primarily useful as a simulated service test for materials that are to be used for environments involving nitric acid. This is a return to its original purpose. It is somewhat ironic that in the case of Fe–Cr alloys (types 430 and 446), which do not form sigma phase readily, the newer tests described below are much more effective in detecting susceptibility to intergranular attack in the presence of chromium carbides and nitrides than is the nitric acid test (Fig. 51.9). Molybdenum tends to accelerate the rate of corrosion in nitric acid and other highly oxidizing environments. Alloys with more than about 3% Mo do not perform well in this test (ASTM A262 Practice C), and it should not be used to evaluate them except in extraordinary circumstances.

C2.2.3. Oxalic Acid Etch. If intergranular attack in all environments other than nitric acid is associated only with chromium carbide and nitride precipitates (in 304, 304L, 316, and 316L), then the problem of detecting susceptibility to this form of corrosion is reduced to determining the presence of these precipitates. For practical application in quality control by nonspecialists and for unambiguous assessment of results, standardized procedures and simple evaluation criteria are required. To meet these needs, the electrolytic oxalic etch test [56] and the boiling ferric sulfate–50% sulfuric acid test [15, 57] were devised. In the oxalic acid etch test, standardized electrolytic etching conditions are specified along with "acceptable" and "unacceptable" etch structures. Specimens having "acceptable" etch structures [Figs. 51.10(a) and (b)] are immune to intergranular attack and the material they represent can therefore be released for service on the basis of this rapid test method. Only ∼15 min is required to polish, etch, and examine the etch structure. "Unacceptable" etch structures [Fig. 51.10(c)] indicate that the specimen may be subject to intergranular attack. In order to determine the degree of susceptibility on a quantitative basis, the specimen is then tested in one of the boiling acid tests. A weight loss is measured and then converted to a

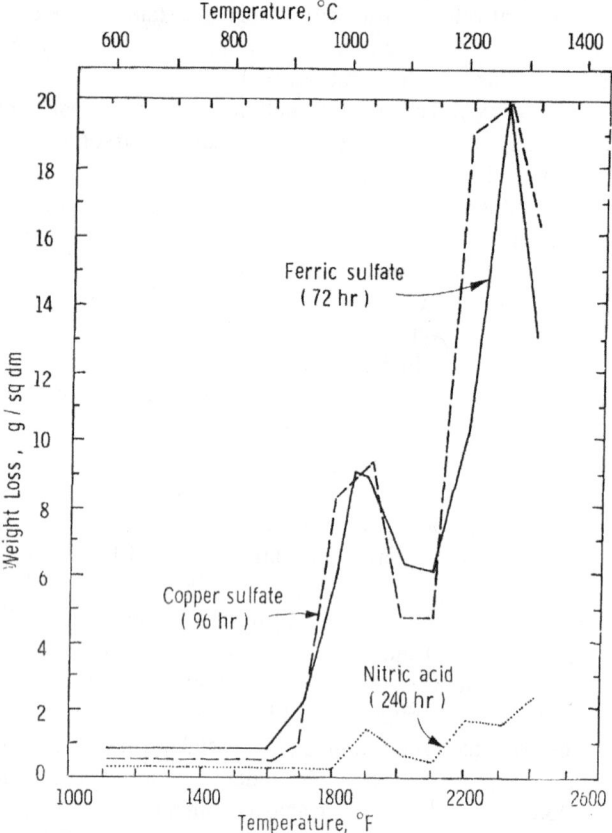

FIGURE 51.9. Effect of heat treatment on corrosion of type 446 stainless steel in boiling acid solutions [33]. Heat treatment: 1 h, water quenched. Acid solutions: 65% nitric; ferric sulfate/50% sulfuric acid; copper sulfate/50% sulfuric with metallic copper.

corrosion rate. Acceptance is determined by its relation to a maximum permissible rate. This rate depends on the test solution, the alloy composition, and any prior "sensitizing" heat treatments. Such heat treatments are applied to alloys to be welded during fabrication or to be stress relieved by heat treatments to assess their response to thermal exposure in the sensitizing range of temperature.

Thus, the oxalic acid etch test is now applied as a rapid screening method for austenitic stainless steels in conjunction with various acid intergranular corrosion evaluation tests [58], copper sulfate–sulfuric acid, nitric acid, and ferric sulfate–sulfuric acid, as specified in ASTM A 262 [47]. The oxalic acid etch test is not applicable to either the old (Fe–Cr) or the new (Fe–Cr–Mo) ferritic stainless steels because the response of sensitized grain boundaries to available etching techniques does not yet provide structures that can be routinely classified for screening purposes.

Oxalic acid etch results are now available for certain ferritic stainless steels. See ASTM A763 [58a] Practice W for a listing of the alloys for which it is applicable.

It should be noted that this etching procedure does not attack the Cr-depleted regions. Instead, it attacks

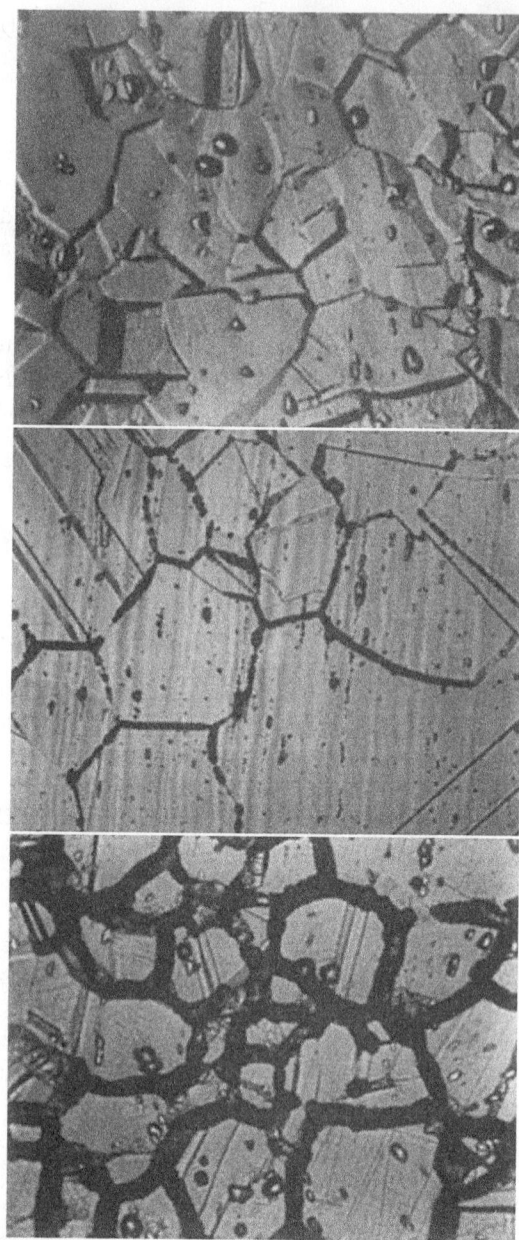

FIGURE 51.10. Oxalic acid etch test structures [56], Etched anodically in 10% oxalic acid at $1.0\,A/cm^2$ for 1.5 min. (a) Step structure, solution anealed (500X). steps between grains, (b) Dual structure, sensitized, ditches at grain boundaries, but not completely surrounding any one grain (250X). (c) Ditch structure, sensitized, ditches at boundaries surrounding grains (500X), one grain or more completely surrounded by ditches.

Cr- and Mo-rich areas such as chromium carbides. This limits the utility of this test for evaluation of high Cr + Mo alloys. It is also the reason why this procedure can be used to accept but not to reject material. Per specification, materials exhibiting unacceptable microstructures must be evaluated by one of the other A262 or A763 procedures.

C2.2.4. Ferric Sulfate. The ferric sulfate–50% sulfuric acid test provides results in 120 h, one-half of the time required for the nitric acid test. It is unaffected by corrosion products, and, therefore, several specimens can be tested simultaneously. Most importantly, it is sensitive only to susceptibility associated with chromium carbide precipitates in types 304, 304L, 316, and 316L stainless steels. It does not detect sigma phase in type 316 and 316L stainless steels. However, in some heats of type 321 (18 Cr–8 Ni–Ti) sigma phase is also formed, and this can increase somewhat the rate of corrosion in the ferric sulfate test [48].

This test has been standardized as ASTM A262 Practice B and A763 Practice X. It is often known as the Streicher test.

The ratios listed in Table 51.4 illustrate the differences between the 120-h ferric sulfate and the 240-h nitric acid tests. For the two heats of type 304 steel, the ratios of the sensitized to the solution-annealed (carbide-free) specimens are essentially the same in both tests. However, for the 316 and 316L steels, there are large differences. The ratios of ~1.0 in the ferric sulfate test for the two 316L heats show that the rates of the solution annealed and of the sensitized specimens are essentially the same. There is no intergranular attack on the specimens heated at 677 and 871°C. However, for both of these steels, the ratios in nitric acid are very high, even though the microstructures of the sensitized specimens showed no carbide- or sigma-phase particles at the grain boundaries. They had a "step structure" (Fig. 51.10) in the oxalic acid etch test. This illustrates the properties of sub-microscopic or invisible "sigma phase." It can be removed or dissolved by annealing the steel at 1070°C and water quenching. Finally, it is apparent from the ratios in the two tests that the type 316 steel of Table 51.4 contains both carbide- and sigma-phase precipitates.

Both the oxalic acid etch test and the ferric sulfate–sulfuric acid test are merely methods for detecting and measuring susceptibility to intergranular attack associated with chromium carbide precipitates. Sensitized materials should not be used in environments that are known from service performance or long-time plant tests to cause intergranular attack in the presence of carbides. Results of extensive long-time tests have been published by Auld [59], Warren [44], and the Welding Research Council [60]. These reports contain extensive lists of environments in which carbides do and do not cause susceptibility to intergranular attack. A major problem in the use of sensitized material even for environments in which data show no intergranular attack is that it is difficult to foresee whether there will be changes in process conditions, solution composition, and temperatures during the lifetime of the stainless steel equipment that will change this from a "safe" to an "unsafe" environment. For environments for which such data are not available, prudence requires that the stainless steel be used in the optimum condition (i.e., free of chromium carbide precipitates).

It is now generally agreed upon that it is the presence of Cr-depleted regions that are usually adjacent to carbide precipitates, rather than the presence of the carbides themselves, that is responsible for sensitization.

Extensive use of evaluation tests for detecting susceptibility to intergranular attack in stainless steels by purchasers of these alloys has contributed not only toward assuring optimum conditions for their use in chemical and other plants but also toward continuing improvements in their compositions (low-carbon grades) and mill processing. When applied for research, they have served as tools for the study of the mechanism of intergranular corrosion and for alloy development. In the United States, the Du Pont Company has been a leader for many years in the development of these test methods and their continuing application as acceptance tests for quality control [15, 21, 24, 32–34, 44, 48, 50–52, 56–59].

C3. Pitting and Crevice Corrosion

As mentioned above, Monnartz [18] observed that chloride salts impair the passive state of iron–chromium alloys and that molybdenum additions increase the resistance in such environments. The molybdenum in type 316 stainless steel, in addition to improving resistance in (organic) reducing acids, also makes this alloy somewhat superior to type 304 steel in chloride pitting and crevice corrosion environments. Further increases in pitting resistance have been obtained by addition of ~2% silicon to type 316 steel [61, 62] and by addition of ~0.2% nitrogen to 23% Cr–4% Ni [63] and 18% Cr–8% Ni [64] steels.

Pitting may be divided into two distinct steps: (a) pit initiation or breakdown of the protective film and (b) pit growth in depth and volume. This division of the pitting process serves as an aid in the analysis of the influence of alloy composition and structure and of the composition and temperature of the environment. Once a pit has been initiated, its growth is a result of an electrochemical process in which the small anode, the pit, is linked electrochemically to a very large cathode, the unpitted area surrounding the pit. The electrolyte is the corroding solution. The corrosion process is self-accelerating because chloride ions migrate into the pit and decrease the pH. Crevices may be viewed as "artificial" pits.

Pit initiation can be studied by anodic polarization of steel specimens in chloride (or other) pitting environments. In a study [65] in which specimens were progressively polarized up to a current density of 3 ma/cm^2 and the number of pits formed per square centimeter was counted, it was found in 18 Cr–8 Ni steels that pit initiation is reduced by decreasing the carbon content, by adding 2% silicon and 0.2% nitrogen (Table 51.5). Surprisingly, addition of 2.5% molybdenum had no effect in pit initiation of pickled specimens. However, it greatly affects the response to passivation. Prior passivation of the surface in a hot nitric acid–bichromate solution greatly

TABLE 51.5. Pit Initiation in Stainless Steels Exposed to 0.1 N NaCl at 25°C (75°F)[a]

| Alloy | Element (%) | | | | | | Pits per cm^{2b} | | Ratio |
	Cr	Ni	Mo	C	N	Si	Pickled[c]	Pickled Plus Passivated[d]	(Pickled to Passivate)
304	18.45	8.90		0.063		0.58	7.4	3.4	2.2
304L	18.30	11.02		0.020	0.033	0.37	4.5	1.6	2.8
302	18.37	8.71		0.10		2.49	6.3	2.8	2.2
302B	17.30	8.62		0.14		2.71	1.5	0.6	2.5
316	17.93	13.50	2.47	0.031		0.31	7.3	0.46	15.8
316L	17.71	11.17	2.44	0.020	0.032		5.0	0.17	29.0
SP-2	18.79	9.24	2.40	0.039	0.23	2.50	0.0	0.0	

[a]See [65].

[b]Pits produced by anodic polarization of 50×50-mm (2×2-in.) specimen to a current density of 3 mA/cm^2 (19 mA/in.^{2}).

[c]Pickled in HNO$_3$–HF–HCl.

[d]Passivated in nitric acid with potassium dichromate.

reduces the incidence of pitting. This phenomenon is also shown in Figure 51.11. Type 304 steel requires much larger concentrations of sodium nitrate inhibitor to prevent pitting by 10% ferric chloride solution than does type 316 steel.

Lowering the carbon content (type 316 vs. 316L) of passivated steels reduces the incidence of pit initiation even further (Table 51.5). In contrast, silicon, which reduces pit initiation, actually promotes pit growth and therefore does not enhance the response to passivation. These findings were incorporated in an alloy composition, SP-2, a type 316L steel with 2.5% Si and 0.23% N. Only when all four factors were incorporated (silicon, molybdenum, low carbon, and 0.23% N) was there a significant improvement in pitting and crevice corrosion resistance over types 304 and 316 (Fig. 51.12).

Even at room temperature, the combination of oxidizing ferric ions, high concentration of chloride ions, and a low pH of 1.6 make the ferric chloride solution a severe pitting medium. For example, the only difference between types

304 and 316 in this test is the length of time before pitting begins, a difference of ~48 h, which is not significant in a 7-month exposure, such as shown in Figures 51.11 and 51.12. Increasing the temperature of the ferric chloride solution results in pitting of the SP-2 alloy in long-time exposures with crevices of the type shown below. These limitations in pitting resistance of even the most resistant, modified 18 Cr–10 Ni stainless steels, along with their susceptibility to chloride stress corrosion cracking, provided some of the incentives for the development of new alloys.

It is now generally acknowledged that pits usually initiate at exposed inclusions, with sulfide inclusions being of greatest concern. It has been suggested that, in addition to creating small pits as MnS and similar sulfides dissolve, their dissolution products modify the local environment to create more aggressive environments [65a]. Additional work has shown that the metal adjacent to these inclusions may have reduced Cr levels, giving it reduced corrosion resistance [65b].

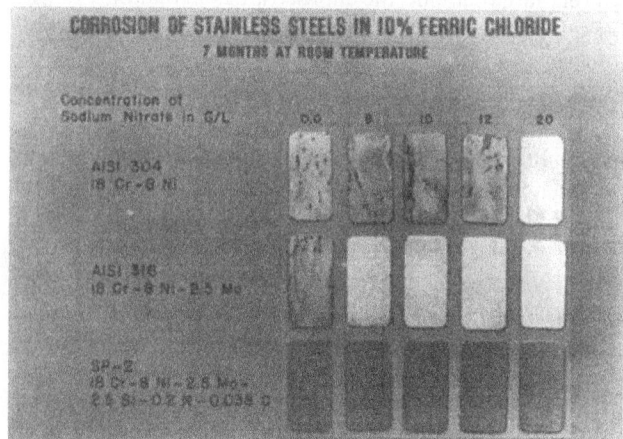

FIGURE 51.11. Inhibition of pitting corrosion.

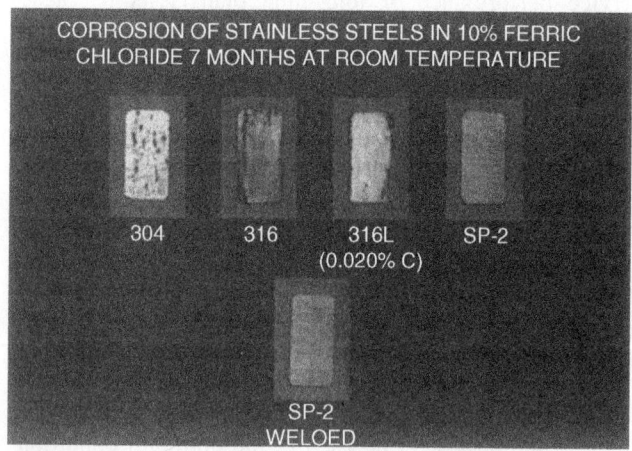

FIGURE 51.12. Effect of composition on resistance to pitting.

Measurements of critical pitting potentials (CPPs) are frequently used to determine resistance to pitting. The specimen is polarized anodically (e.g., in sodium chloride solution) to determine the potential at which there is a precipitous increase in current. The potential at which this occurs, the CPP, can be used to compare alloys. However, such relatively rapid measurements are not reliable guides for predicting immunity to pitting in long-time service in the given environment. Pitting attack has been observed in long-time exposures at potentials below the CPP [65c].

The repassivation potential (E_r), which is significantly less noble than the pitting potential (E_p), has been proposed [65d] as a conservative estimate of the maximum potential at which immunity to pitting will be exhibited in long-term exposure.

C3.1. Pitting and Crevice Corrosion Testing.

The concept of the critical pitting temperature (CPT) was introduced by Brigham and Tozer [65e] in 1973. Their work showed that it was possible to rank a wide range of stainless steels with regard to pitting resistance using temperature as the only parameter. "Temperature" is more easily comprehensible for engineers than the "pitting potential." In most practical situations, chloride crevice corrosion is the limiting factor, rather than pitting. While the details of these corrosion mechanisms differ, they share many basic factors. Ferric chloride solution [65f] provides a low-pH, high-chloride, oxidizing solution that mimics the environment that develops inside crevices in live seawater exposures. It does this quickly, without the extended incubation periods needed to produce attack under "natural" conditions. Use of ferric chloride solutions for pitting and crevice testing was standardized by ASTM as G48 [65g] method A (pitting) and B (crevice) tests. Hydrolysis of the solution limited these procedures to about 50°C maximum. Acidification of the solution limits the hydrolysis and allows testing at higher temperatures. Use of acidified ferric chloride has been standardized as G48 methods C, D, E, and F.

Arnvig [65h] combined this concept with potentiostatic techniques and a crevice-free sample holder to create a procedure for CPT determination. This technique has been standardized as ASTM G150 [65i]. While this procedure provides a rapid, reproducible ranking of alloys, as with CPP, it does not provide an accurate use limit.

C3.2. Ranking and Predicting Chloride Corrosion Resistance.

Chromium and molybdenum act in synergy to provide resistance to chloride pitting. Lorentz [65j] introduced the concept of pitting resistance equivalent (PRE). PRE is established by a linear regression of CPT versus composition and is expressed as $PRE = Cr + 3.3Mo$ (where Cr and Mo are expressed in weight percentages). With the widespread use of nitrogen as an alloying element, the PRE formula was modified to take it into account. The most widely used of these PREN formulas is $PREN = Cr + 16Mo + 16N$, but the

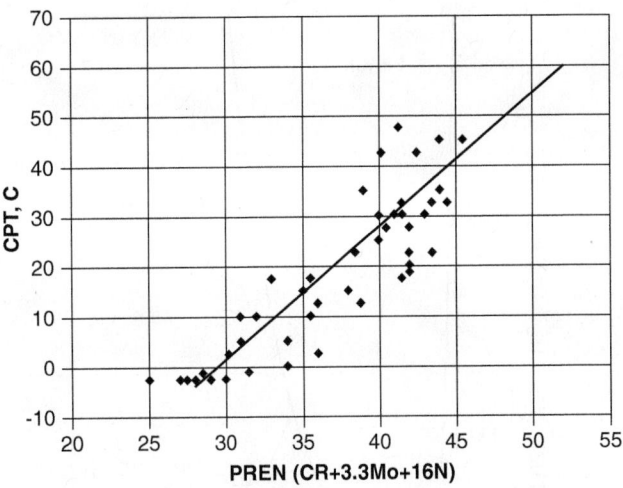

FIGURE 51.13. CPT vs. PREN for austenitic alloys [65l].

formula $PREN = Cr + 16Mo + 30N$ has been cited frequently also. Bauernfeind [65k] has reviewed the various PREN formulations and their extensions to include additional alloying elements. While PREN provides a useful framework for ranking the various corrosion-resistant alloys and has been incorporated into some materials specifications, it remains an empirical relationship. PREN does not account for metallurgical or surface conditions and assumes linearity of effects that are probably not truly linear. In the end, its accuracy is limited by the accuracy of the CPT and composition data used to create it. A comparision of PREN and CPT is shown in Figure 51.13.

C4. Chloride Stress Corrosion Cracking

Beginning at about 1936, structural failures by cracking were observed in the austenitic stainless steels under tensile stresses. By 1940, Hodge and Miller [66] clearly distinguished this type of failure from intergranular corrosion and identified it as stress corrosion cracking (SCC) associated with environments containing chlorides. They also established that cracking can be transgranular and that it can take place in the absence of chromium carbide precipitate. In 1942, Rocha [67] confirmed these findings and by 1944 extensive studies had been carried out showing that aqueous solutions of various chlorides can cause cracking of specimens under tensile stresses of all the common types of austenitic stainless steels. Solutions of concentrated magnesium chloride were introduced for research on this problem [68, 69]. Scheil [68] found that ferritic alloys, types 405 (S40500) (13% Cr) and 430, resist cracking except when they contain some (1.85%) nickel. Nevertheless, even in 1944–1945 it was still thought that the necessary conditions for SCC are not too common and that, while it is an interesting phenomenon, it is not a serious problem from a practical standpoint.

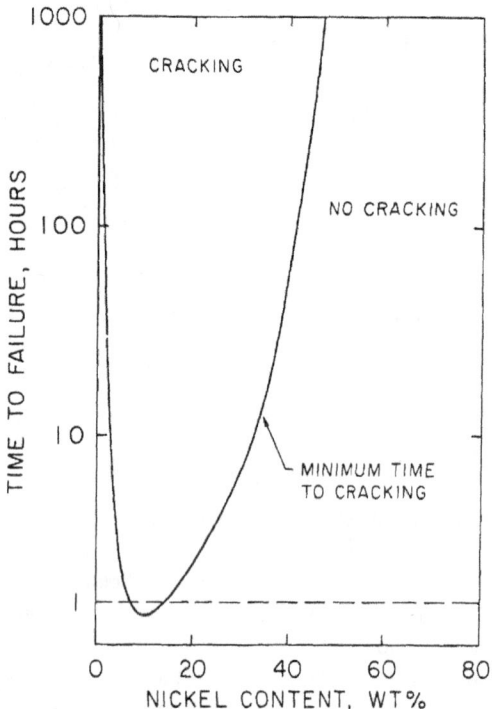

FIGURE 51.14. Copson curve: Effect of Ni content on the susceptibility of SCC of stainless steel wires containing 18–20% Cr in a MgCl₂ solution boiling at 154°C [71].

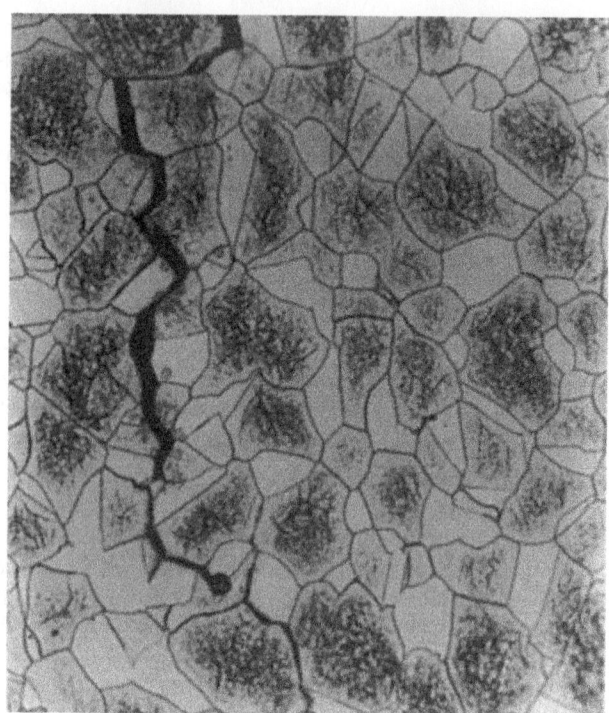

FIGURE 51.15. Stress corrosion cracking of sensitized type 446 stainless steel in boiling sodium chloride solution (250X). Steel: Heated 1 h at 1200°C (2190°F), water quenched. Solution: 50 ppm Cl as NaCl, 100°C (212°F). Exposure: 1610 h as a U-bend. Structure: Intergranular cracking between austenite and ferrite grains (oxalic acid etch).

The intriguing scientific aspects and an increasing number of SCC plant failures stimulated a large number of investigations of cracking in austenitic stainless steels. However, 25 years later, Staehle [70], chairman of an international conference on SCC, was forced to conclude that "there presently is no reliable fundamental theory of SCC in any alloy-environment system that can be used to predict the performance of equipment even in environments where conditions are readily defined. There was an almost uniform conclusion that no unifying mechanisms of stress corrosion cracking exists." This conclusion of 1969 is still applicable today.

From a practical standpoint, the most important finding since 1945 is that increasing the nickel content above or decreasing it below 8% Fe–18% Cr alloys increases the resistance of (austenitic) alloys to stress corrosion cracking and that the 8% Ni alloy actually happens to be the composition that is least resistant [71a] [Fig. 51.14]. It is shown below (Table 51.6) that the 34% Ni in Carpenter 20 Cb-3 alloy increases the time of failure of this alloy in the magnesium chloride test over that of type 304 and 316 stainless steels. Most importantly, this high nickel content makes the alloy *immune* to cracking in certain sodium chloride environments.

By apparently ignoring the early findings of Scheil [68], it was assumed for many years that stainless steels that have an austenitic structure are subject to chloride SCC but that ferritic alloys, because of their structure, are inherently immune to chloride cracking. In 1968 Bond and Dundas [72] clearly established that a 17% Cr ferritic alloy with 1.5% Ni is subject to transgranular cracking in magnesium chloride solution. They concluded that ferritic alloys containing >1 % Ni or 0.5% Cu are subject to this type of chloride cracking and that this is not associated with retained austenite or hydrogen embrittlement.

Furthermore, Renshaw [73] reported that *welded* type 430 is subject to cracking in boiling solutions of sodium chloride containing only 50 ppm Cl. This led to a more detailed investigation [74] on the effect of sensitization in type 430 and 446 steels on their resistance to chloride SCC. These alloys are resistant to cracking when they are free of chromium-depleted zones around chromium carbide precipitates (when chromium is rediffused into the depleted zones during heating) or when carbon is in solid solution. However, when the carbide particles are surrounded by zones depleted in chromium (i.e., severely sensitized), they are subject to SCC—both in the magnesium chloride test and in sodium chloride solution with only 50 ppm Cl. As might be expected, in sodium chloride, cracking is intergranular (Fig. 51.15). But, surprisingly, in the magnesium chloride solution, cracking is transgranular (Fig. 51.16). Because of the 0.2%

TABLE 51.6. Comparison of SCC in 45% MgCl$_2$ and Various NaCl Tests[a]

| | Stress Corrosion Test | | | | |
	MgCl$_2$ Test[b] 155°C (310°F)	NaCl Wick Test[c] 100°C (212°F)	Aerated 26% NaCl[b] 102°C (215°F)	Autoclave Tests[b] in 26% NaCl[d] 155°C (310°F)	200°C (390°F)
Alloy					
AISI 304 18 Cr–8 Ni	Cracked (<3 h)	Cracked (<72 h)	Cracked (72 h)	Cracked (250 h)	Cracked (between 48 and 72 h)
AISI 446 25 Cr Sensitized 1 h at 1150°C (2100°F)	Cracked (<17 h)		Cracked (552 h)	Cracked (<19 h)	Cracked (<23 h)
Carpenter 20 Cb-3 33 Ni–20 Cr 2.3 Mo–3.3 Cu–Cb	Cracked (<40 h)	No cracking[e] (864 h)	No cracking[f] (2544 h)		No cracking (655 h)
W/29/4 29.7 Cr–3.9 Mo C 100 ppm, N 56 ppm	No cracking (2400h)	No cracking (864 h)			No cracking (655 h)
745 Sensitized[g] W/29/4/2 28.5 Cr–4.2 Mo C 210 ppm, N 18 ppm	Cracked (<15 h)	Cracked (72 h)		Cracked (<19 h)	
29.7 Cr–3.9 Mo–2.0 Ni C 90 ppm, N 46 ppm	Cracked (3 h)	No cracking (3360 h)	No cracking[f] (2528 h)	No cracking (487 h)	No cracking[h] (655 h)
775 (high Mo) 29/4 28.5 Cr–7.0 Mo C 20 ppm, N 41 ppm	Cracked (<15 h)	No cracking (864 h)			No cracking (420 h)
28.5 Cr–4.2 Mo[i] With 0.60 Cu	Cracked (432 h)	No cracking (2160 h)			
With 0.15 Ni + 0.15 Cu	No cracking (2544 h)				
With 0.15 Ni + 0.20 Cu	Cracked (432 h)	No cracking (1440 h)			

[a] See [74] and [75].
[b] U-bend specimen, 19 × 76 mm (0.75 in. 3 in.).
[c] U-bend specimen, 51 × 78 mm with 25 mm redius (2 × 7 in.) with 1-in. radius); test solution: 1500 ppm Cl as NaCl.
[d] Air atmosphere in autoclave.
[e] A welded specimen was also tested, with the same results.
[f] pH of MgCl$_2$ test is 4.0 and of NaCl test 7.3. When HCl was added to NaCl test to make pH 4.0, there also was no cracking in 2328 h.
[g] Specimens were sensitized by heating 1 h at 540°C (1000°F). They were subject to severe intergranular attack in the ferric sulfate/sulfuric acid test.
[h] Two specimens were tested.
[i] Welded U-bend specimens.

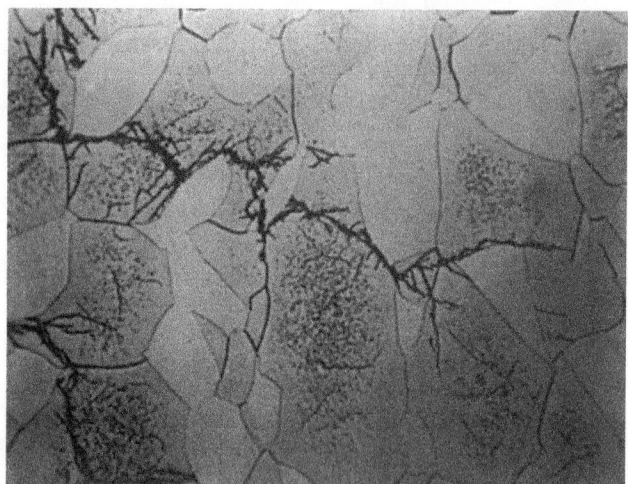

FIGURE 51.16. Stress corrosion cracking of sensitized type 446 stainless steel in boiling 45% magnesium chloride solution (50X). Steel: Heated 1 h at 1200°C (2190°F), water quenched. Solution: 155°C (310°F). Exposure: 17.5 h as a U-bend. Structure: Transgranular cracking in ferrite grains only (oxalic acid etch).

nitrogen in this alloy, there is a major amount of austenite in the structure. Note that, contrary to some expectations, the transgranular crack path is confined to the ferrite phase and avoids the austenite grains.

From the above, it may be concluded that susceptibility to chloride cracking in stainless steels is a function not of the crystal structure of ferrite or of austenite but of the composi-

tions of these phases and the presence of chromium-depleted zones around precipitates in ferritic alloys.

In the work on the new Fe–Cr–Mo ferritic stainless steels discussed below, it is reported [74, 75] that there are important differences in stress corrosion of these alloys depending on the type of chloride, $MgCl_2$ or NaCl, used for testing. These differences are important because the sodium chloride test environments approximate service conditions much more closely than the concentrated magnesium chloride solution.

The wick test (Table 51.6) was originally developed at Du Pont [76, 77] to simulate a hot stainless steel pipe covered with insulation and exposed to rain water. In this process, chlorides in the insulation are leached to the hot stainless steel surface and are concentrated there as water is evaporated. The laboratory apparatus for this simulated service test is shown in Figure 51.17. Its original purpose was to evaluate insulation materials for their tendency to cause SCC of stainless steels. To evaluate alloys, Warren [78] replaced the insulation by glasswool and added chlorides to the water. A relatively large, 17 × 5-cm, U-bend specimen is heated to 100°C by passing an electrical current through it while it is in contact with the glasswool, which is partially immersed in the chloride solution. The wicking action of the glasswool draws the chloride solution to the hot specimen where it is evaporated. Type 304 stainless steel cracks in <3 h in the $MgCl_2$ test and within <3 days in the wick test (Table 51.6).

The test in boiling, saturated (26%) solution of sodium chloride were made on small, 2.0 × 7.5-cm, U-bend

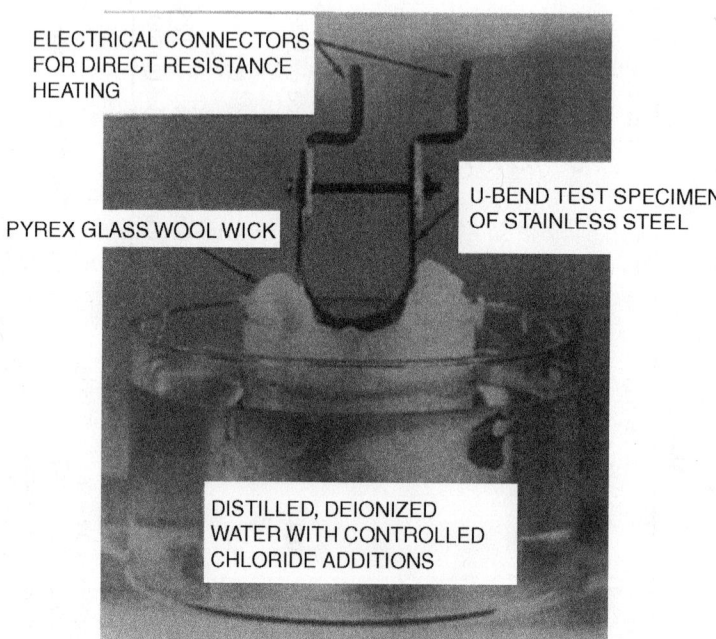

ELECTRICAL CONNECTORS FOR DIRECT RESISTANCE HEATING

PYREX GLASS WOOL WICK

U-BEND TEST SPECIMEN OF STAINLESS STEEL

DISTILLED, DEIONIZED WATER WITH CONTROLLED CHLORIDE ADDITIONS

FIGURE 51.17. Wick test arrangement [76, 77]. Solution: 1500 ppm Cl as NaCl.

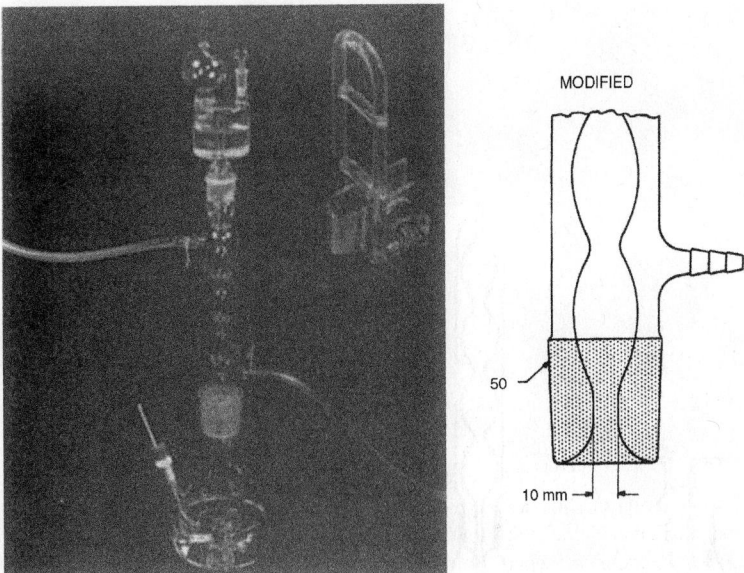

MODIFIED

50

10 mm

FIGURE 51.18. Assembly of glass apparatus for long-time tests in boiling 45% magnesium chloride solution [80]. U-bend stress corrosion specimen. The trap contains a 25% MgCl₂ solution and prevents loss of water vapor from the test solutions. To prevent sudden generation of steam when cold condensate drops into the boiling [155°C (310°F)] solution, the drip tip of the condenser has been replaced by a design that provides an even flow of condensate down the wall of the flask. The result is a continuous flow of preheated condensate into the solution.

specimens to determine whether these simple immersion tests could be used in place of the more complex wick test. To eliminate the effect of the difference in temperatures between the MgCl₂ (155°C) and the NaCl tests (102°C), U-bend specimens were also tested in autoclaves with 26% NaCl at 155 and 200°C. As expected, type 304 steel cracks in all of these NaCl tests, as does the sensitized ferritic type 446 steel. However, the high-nickel Carpenter 20 Cb-3 alloy, which cracks in 40 h in the MgCl₂ test, is resistant in all of the NaCl tests of Table 51.6. The natural pH of the MgCl₂ solution is 4 and that of the 26% NaCl solution is 7.3. To eliminate this difference as a cause of the resistance in NaCl solutions, HCl was added to the NaCl solution to bring the pH to 4.0. Again, there was no cracking on Carpenter 20 Cb-3, (Table 51.6). These results provided a basis for evaluating the stress corrosion resistance of various new ferritic alloys, ASTM G-126-96 [78a].

The concentration of MgCl₂ solutions that boil at 155°C, the test temperature given above, is 45.0% MgCl₂ [79]. These test conditions were proposed in connection with the development of a test apparatus [80] designed to maintain this concentration over long (100-day) test periods (Fig. 51.18). Small losses of water vapor result in a rapid increase of the concentration of MgCl₂ and, consequently, also of the boiling temperature. Both the test condition and the apparatus

have been incorporated in an ASTM recommended practice, G-36 [81].

The even flow of condensate in a large, thin film from the condenser into the boiling (155°C) 45% magnesium chloride solution preheats it and thereby expels dissolved air (oxygen) before it reaches the boiling solution. This is in marked contrast to the dripping of condensate from the tips of Allihn and finger condensers (Fig. 51.19). Preheating is minimized when cold condensate drops fall directly into the solution. Data in Table 51.6A show that in 42–45% magnesium chloride solution there is no condenser effect or effect of sparging the boiling solution with oxygen on stress corrosion of type 304 stainless steel. However, below this concentration and boiling temperature there is a progressively greater effect of condenser design and sparging [81a]. Even the length of the drip tip (1 vs. 5 cm) affects the time to cracking because the longer drip tip results in more preheating of the condensate in the hot vapor space than the shorter drip tip. Similar results have been obtained on Carpenter 20 Cb-3 alloy.

In 26% NaCl solution type 304 fails by stress corrosion in 144 h with a finger condenser and in 248 h with the modified condenser when sparged with oxygen. However, without sparging this alloy does not crack even after 1500 h with the modified condenser. Various types of condensers also affect corrosion rates of stainless steels in 65% nitric acid [81b] and in reducing acids [81c].

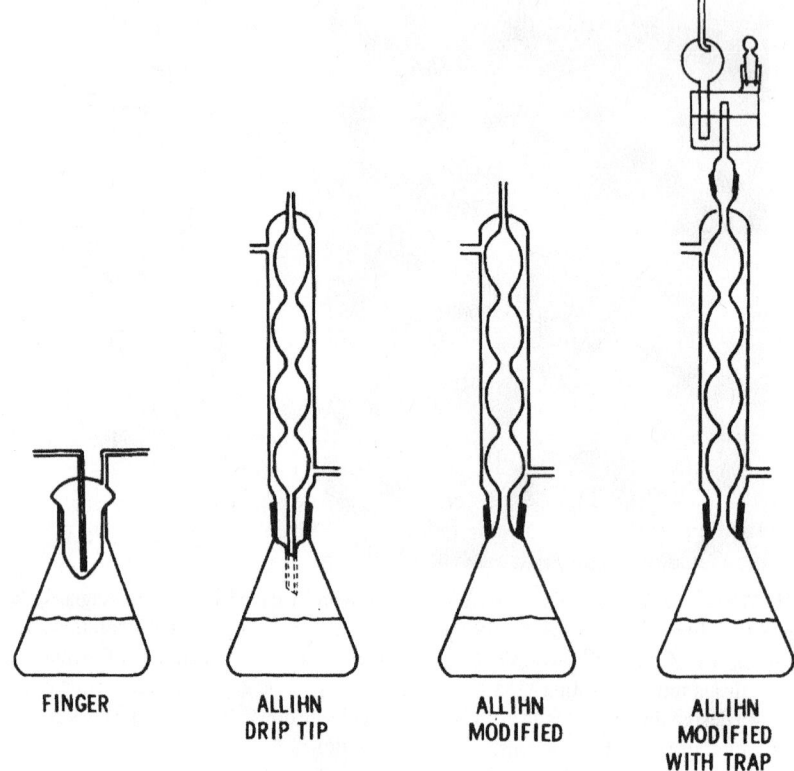

FIGURE 51.19. Types of condensers.

TABLE 51.6A.   Effect of Condenser Design on Time to Cracking of Type 304 SS in Boiling[a] MgCl₂

Solution		Time to Cracking (h)				
			Allihn, Drip-Tip (Length)		Modified Allihn (Sparging)	
Concentration (wt %)	Temperature (°C)	Finger	1 cm	5 cm	(None)	(Oxygen)
13	105	1000				
24	110	63	933	1250[b]	1513	47
28	115	61	548[b]	745[b]	1600[b]	22
32	120	12	28[b]	48[b]	879	16
37	130	5	6[b]	10[b]	21	2
42	145	1[b]	1[b]		1[b]	
45	155	1	1		1	
47	160	1.3[b]			1.3[b]	
50	170	2[b]			2[b]	

[a]U-bend specimens.
[b]Averages of two specimens, all other data are averages of three specimens. Cracking times of <1 h were not determined.

D. NEW FERRITIC STAINLESS STEELS

D1. Origin and Composition

The continuing application of the basic 18% Cr–8% Ni, type 304 stainless steel since 1912 throughout the industrial world indicates that in this alloy, along with its variations, types 304L, 316, 316L, 321, and 347, an optimum balance has been achieved in alloy composition. Contrasting requirements for resistance to corrosion in a wide range of environments, together with requirements for formability and weldability, have been successfully compromised. To obtain a significant improvement in resistance to corrosion by reducing acids and to chloride SCC in NaCl solutions necessitated a major departure from this composition. The nickel content was increased from ~10 to 35% and 3.5% copper was added. However, this large and costly increase in nickel content had no significant beneficial effect on the resistance to chloride pitting and crevice corrosion.

As mentioned above, the ferritic alloys types 430 and 446 are resistant to stress corrosion both in the $MgCl_2$ and various NaCl solutions when in the mill-annealed condition. But welding impairs their ductility as well as their resistance to intergranular corrosion and to SCC. Welded type 430 steel fails by intergranular attack even in tap water [82] and, as noted above, is subject to SCC in boiling water containing only 50 ppm Cl as NaCl [73]. Furthermore, the transition temperature from ductile to brittle fracture is very high for these commercial alloys (e.g., $+120°C$ for type 446 as compared with less than $-240°C$ for type 304).

About 10 years ago, the above limitations in the available ferritic and austenitic stainless steels stimulated several independent and partly simultaneous efforts in the United States and in Germany aimed at developing new ferritic compositions that would overcome these limitations. Also, at this time new improvements in commercial melting and refining methods, along with the previously available vacuum induction and vacuum arc remelt processes, made it possible to produce large heats with low carbon and nitrogen concentrations. These new processes are vacuum-oxygen decarburization (VOD), electron-beam (EB) refining, and the previously mentioned argon–oxygen decarburization (AOD). The latter is now widely employed for the production of stainless steels [36].

In 1951, before the new processes were available, Binder and Spendelow [83] concluded from a study of high-purity Fe–Cr alloys that the decrease in impact strength as the chromium content is raised >20% is due not to chromium but is associated with the carbon and nitrogen contents. When these are low (C + N <0.04%), increasing chromium from 12 to 25% has a toughening influence on these steels. The transition temperature is lowered by about 65°C. Knowledge of the beneficial effects of low carbon and nitrogen concentrations on the mechanical properties and corrosion resistance of Fe–Cr alloys, together with the availability of the necessary commercial processes for achieving these levels, provided the opportunity for the development of new ferritic stainless steels. The following deals with the origin and properties of some of these. The selection is confined to alloys that have already been produced in commercial quantities as mill forms, such as sheet and tubing (Table 51.7).

Ideally, experimental alloys made in the laboratory should be evaluated in plant tests. However, for such tests relatively large amounts of material are required along with testing times of 6–18 months, in some cases, for example, cooling water environments. For these reasons, it is necessary to develop and apply rapid and therefore severe laboratory tests to evaluate experimental alloys for resistance to general, pitting, crevice, intergranular, and stress corrosion. The choice of these tests is critical, because they determine the

validity of the results. By the selection of very severe testing conditions and criteria for evaluation of test specimens, the gap between laboratory data and service performance can be narrowed, but not closed. Ultimate proof of the utility of promising compositions can be provided only by tests in plant environments and service in operating equipment.

All of the new ferritic alloys contain molybdenum for reasons shown below. When they contain nickel, it is in relatively small amounts. This is an advantage during shortages in the supply of this element such as have occurred several times.

D1.1. 18 Cr–2 Mo–Ti (S44400). Three new groups of alloys are shown in Table 51.7, along with the 18 Cr–2 Mo–Ti alloy, which has been in use for some time in Europe and is also produced in the United States. The limits for carbon and nitrogen, (Table 51.7) for this alloy are a result of extensive research on this composition at the laboratories of the Climax Molybdenum Company [84–86]. To maintain the concentrations of carbon and nitrogen as shown in Table 51.7, the alloy must be made by the AOD or VOD process. With the exception described below, it resists stress corrosion cracking in various laboratory tests and its resistance to chloride pitting is approximately on a level with that of type 304 steel. However, in oxidizing acids, there is rapid attack on intermetallic phases and titanium carbides.

D1.2. 26 Cr–1 Mo. In 1970, Schwartz et al. [87] and others [88–91] of Airco Vacuum Metals, Division of Airco Inc., introduced the first high-purity alloy. The E-Brite 26-1[*] (S44627) composition was the result of an effort to improve on the properties of the 26% Cr, type 446 steel. Low carbon and nitrogen concentrations (C + N ≤ 200 ppm) were achieved by a new combination of vacuum induction melting followed by electron-beam continuous hearth refining. Even these low concentrations of nitrogen (150 ppm maximum) resulted in some susceptibility to intergranular attack on weldments. For this reason, niobium, in concentrations from 13 to 29 times the nitrogen content, is being added [92]. However, this small addition is not generally shown in the analyses given for the "high-purity" 26-1 alloy [91]. In addition to its resistance to chloride stress corrosion, this alloy has approximately the corrosion resistance of type 316 stainless steel in acid and in pitting environments. The low carbon and nitrogen concentrations have reduced the transition temperature to $-62°C$ as compared with $+120°C$ for type 446 steel.

When the 26 Cr–1 Mo alloy is made by the argon–oxygen or the vacuum oxygen decarburization processes, carbon and nitrogen cannot be reduced to such low concentrations and, to prevent precipitation of chromium carbides and

[*] Trademark of Airco Inc. Acquired by Allegheny Ludium Industries. June 1977. Now produced by vacuum induction melting.

TABLE 51.7.   New Ferritic Stainless Steels

	Alloy[a]	Limits for Carbon, Nitrogen, and Stabilizers (%)	Melting and Refining Processes	References
I	Fe–18 Cr–2 Mo–Ti UNS S44400	C 0.0250 max N 0.0250 max	Argon–oxygen decarburization or	84, 85, 86,114, 115
		C + N < 0.030 desirable Ti + Nb = 0.20 + 4(C + N) min = 0.80 max	Vacuum–oxygen decarburization	
II	Fe–26 Cr–1 Mo UNS S44627	C 0.0050 max N 0.0150 max	Electron beam hearth refining or	87, 88, 89, 90, 91, 92
		Nb 13–29 (N)	Vacuum induction melting	
II-A	Fe–26 Cr–1 Mo–Ti UNS S44626	C 0.0400 max N 0.0400 max 0.2–1.0 Ti C + N 0.050 typical	Argon–oxygen decarburization	93, 94, 95
II-B	Fe–26 Cr–3 Mo–2 Ni–Ti (UNS S44660)	C 0.030 max N 0.040 max Ti–Nb = 0.20–1.00 Ti + Nb ≥ 6(C + N)	Argon–oxygen decarburization	137
III	Fe–28 Cr–2 Mo DIN 1.4138	C + N ≤ 0.0100 Desirable C + N ≤ 0.050	Vacuum melting follow by arc remelting	97, 98
III-A	Fe–28 Cr–2 Mo–4 Ni–Nb DIN 1.4575	C 0.0150 max N 0.0350 max C + N ≤ 0.0400 with Nb ≥ 12(C + N) + 0.2	Vacuum–oxygen decarburization	99, 101, 102, 116, 120
IV	Fe–29 Cr–4 Mo UNS S44700	C 0.0100 max N 0.0200 max C + N = 0.0250 max	Vacuum induction melting or electron beam refining	74, 75, 103
IV-A	Fe–29 Cr–4 Mo–2 Ni UNS S44800	Same	Vacuum induction melting	104, 110, 119
IV-B	Fe–29 Cr–4 Mo–Ti (UNS S44735)	C 0.030 max N 0.045 max Ti + Nb=0.20–1.00 Ti + Nb≥6(C + N)	Argon–oxygen decarburization	138

[a]Ranges for chromium and molybdenum are shown in Figure 51.18.

nitrides, titanium has been added [93–96]. It has proven to be effective in preventing intergranular attack associated with chromium carbide and nitride precipitates. However, there is an increase in the transition temperature to +40°C and some susceptibility to intergranular attack at grain boundaries of weldments in oxidizing environments associated with intermetallic phases and titanium carbide and nitride precipitates.

D1.3. 28 Cr–2 Mo. A second high-purity alloy was developed at the Deutsche Edelstahlwerke [now Thyssen Edelstahlwerke (TEW)] in Krefeld, Germany, by Oppenheim and Lennartz [97] and Oppenheim and Laddach [98] in 1971.

Their high-purity (C + N ≤ 100 ppm, with 50 ppm desirable) 28% Cr–2% Mo composition requires vacuum melting followed by vacuum arc remelting [99] to give a transition temperature near 0°C. This development was based on previous work by Tofaute and Rocha [100], who reported in 1954 on work carried out about 10 years before. They investigated the effect of nickel, molybdenum, and copper additions to alloys with 20–30% Cr in a search for resistance to sulfuric acid. Because of the high carbon (and nitrogen) content (0.04–07% C) of their alloys, these compositions had high transition temperatures and were difficult to process into mill forms. This prevented their development into commercial products.

The availability of the new processes for melting and refining stainless steels led Oppenheim and Lennartz to reexamine this field with the aim of developing an alloy with resistance to corrosion by sea water up to boiling temperatures. They determined pitting potentials by potentiostatic anodic polarization and concluded that an alloy with 28% Cr–2% Mo met these requirements. In later potentiostatic tests [101] with synthetic crevices, they found that it is subject to crevice corrosion in 3% NaCl solutions at 40°C. Potentiostatic pitting tests are made by using an auxiliary cathode and an external emf for a stepwise increase in the noble direction of the potential of a specimen anode immersed in a chloride solution. The current flow between the specimen and the cathode is observed. The potential at which there is a large increase in this current flow is the pitting potential. The increase in current is a result of pit initiation. High pitting potentials indicate high resistance to pit initiation.

To provide a wider range of resistance to corrosion in sulfuric acid and in sea water, 4% nickel was added to the 28 Cr–2 Mo alloy [102]. Further development work, supported by the German Ministry for Research and Technology, was intended to make it possible to melt this alloy by the vacuum or argon–oxygen decarburization processes in place of the more costly induction melting followed by vacuum arc remelting procedures. About 0.5% niobium was added to combine with the higher carbon and nitrogen contents (C + N ≤ 400 ppm) resulting from this process [99, 101].

D1.4. 29 Cr-4 Mo[*].

The high-purity 29% Cr–4% Mo alloy is the result of a development effort [75, 104] at the Du Pont Experimental Station. It was initiated in 1966. The aim of this investigation was to develop the maximum pitting and crevice corrosion resistance available in the Fe–Cr–Mo system while retaining resistance to chloride SCC. In addition, for industrial applications, formability, toughness, and ductility were required. Note that, unlike the other high-purity alloys in Table 51.7, the total concentration of C + N, which can be tolerated in the 29 Cr–4 Mo alloy is 250 ppm. In particular, the tolerance for nitrogen is 200 ppm. This is important because, while carbon can be reduced to as low as 10–50 ppm in the new refining processes, it is difficult to reduce nitrogen much below 100 ppm. For the high-purity 26 Cr–1 Mo alloy (i.e., in the absence of stabilized elements), the upper limit of 150 ppm N is probably too high by >60ppm when it is essential to avoid intergranular attack on weldments.

Nickel (2%) was added to the 29 Cr–4 Mo composition to increase resistance to reducing acids. This addition makes the alloy self-repassivating in boiling 10% sulfuric acid and in boiling 1% hydrochloric acid. Both the 29 Cr–4 Mo (S44700)

and the 29 Cr–4 Mo–2 Ni (S44800) alloys have been made in commercial quantifies by the Allegheny Ludlum Steel Co. using vacuum induction melting (VIM). The 29 Cr–4 Mo alloy has also been made by electron-beam refining.

D1.5. Stabilized Superferritic Alloys.

The vacuum melting process and the high-purity raw materials required by that process raised the costs of producing the low interstitial alloys like 29-4 and made them too expensive for wide usage. The AOD process allowed the production of high-Cr alloys with interstitial contents low enough to make the production of air-melted, AOD-refined, stabilized superferritic stainless/steels practical. This technology was exploited by Crucible to create the SEACURE® alloy[**], UNS S44660 [104a] Fe–26 Cr–3 Mo–2 Ni–Ti, with was the first seawater-capable, air-melted superferritic stainless steel. Subsequently, Allegheny Ludlum developed an air-melted, stabilized version of the 29-4 alloy that was designated AL 29-4C® alloy,[†] UNS S44735 [104b]. These alloys were not as tough as the vacuum-melted, low-interstitial grades but showed sufficient toughness to allow their widespread use in thin-wall applications. S44735 material has been used up to 1.25 mm thickness, while the nickel addition to S44660 material gives it useful toughness to about 2.5 mm thickness. They have been used as condenser tubing in numerous coastal power stations as well as in other corrosive waters. A notable use for these materials has been in high-efficiency furnace equipment, where they exhibit resistance to the corrosive anions that can be concentrated in flue gas condensate [104e]

D2. Pitting and Crevice Corrosion

The requirements for pitting resistance in the alloy development work at Du Pont [75] were derived from the need for exchanger tubing that would resist crevice, pitting, and stress corrosion when exposed to river waters containing manganese. To prevent clogging by the accumulation of organic slimes in these heat exchangers, the river water was chlorinated for short periods several times a day. Under these circumstances, Carpenter 20 Cb-3 tubes failed in six months of service. The failure by crevice corrosion was attributed to the combined action of permanganate and chloride ions formed when chlorinated water comes into contact with the manganese dioxide water deposit [105]. Simulation of this process in the laboratory led to a solution of 2% $KMnO_4$–2% NaCl with a corrosion potential of + 0.6V vs. SCE and a pH 7.5. Alone, neither of these reagents is corrosive to austenitic stainless steels. However, in combination these reagents cause rapid pitting at *room temperature* on all the common ferritic and austenitic stainless steels, as well as on Carpenter 20 Cbr–3, in <24 h (Table 51.8). When there is pitting

---

[*] In previous publications [75, 103] this alloy was identified as "28-4." For the new ferritic alloys, the center of the concentration range for chromium and molybdenum is being used (Fig. 51.18). Therefore, this practice is also being followed for the 29-4 alloys.

[**] SEA-CURE is a registered trademark of Plymouth Tube Co.
[†] AL 29-4C® is a registered trademark of ATI Properties Inc.

TABLE 51.8. Comparison of Pitting Resistance

Alloy		Element (%)						Pitting Corrosion[a]					
								KMnO$_4$–NaCl[b]				FeCl$_3$[c]	
Stainless Steels	UNS No.	Cr	Ni	Mo	Cu	Si	Cb	RT	50°C (120°F)	75°C (165°F)	90°C (195°F)	RT	50°C (120°F)
AISI 430	S43000	16.2						F				F	
AISI 446	S44600	25.5						F				F	
AISI 304	S30400	18.4	9.1					F				F	
AISI 310	S31000	25	20					F				F	
AISI 316	S31600	17.5	12.8	2.3				F				F	
AISI 316L	S31603	18.4	12.9	3.0				F				F	
Carpenter 20Cb-3	N08020	19.8	34.6	2.3	3.3		0.6	R	R			R	F
SP-2[d]		18	10	2.5		2.5				F			F
Fe–21 Cr–24 Ni–6 Mo–N	N08367	21	24	6			(0.2 N)		R			R	R
Fe–21 Cr–18 Ni–6 Mo–N	S31254	21	18	6			(0.2 N)		R			R	R
Nickel–chromium alloys		*Ni*	*Cr*	*Mo*	*Fe*	*Cu*	*Other*						
Inconel alloy 600	N06600	77	15		7	0.1		R	F			F	
Inconel alloy 625	N06025	59	22	9.0	5	0.4	Cb	R	R	R	R	R	F
Incoloy alloy 800	N08800	32	21		47	0.4		R	F			F	
Incoloy alloy 825	N08825	42	21		30	1.8	Ti	R	R	F		F	
Nickel-base alloys		*Cr*	*Mo*	*Fe*	*Cu*		*Other*						
Hastelloy alloy B	N10001	1	28	5			2.5 Co.	F$_G$				F$_G$	
Hastelloy alloy C	N10002	16	16	6			4W		R	R	R	R	R^e
Hastelloy alloy G	N06007	22	6.5	20	2.5				R	R	R	F	
Titanium								R	R	R	R	R	R^e
New ferritic stainless steels													
Fe–18 Cr–2 MO–Ti	S44400							F	R	F	F	F	F
Fe–26 Cr–1 MO								R	R	F	F	F	F
Fe–28 Cr–2 MO								R	R	F	F	R	F
Fe–28 Cr–2 MO–4 Ni–Nb								R	R	F	F	R	F
Fe–29 Cr–4 MO	S44700							R	R	R	R	R	R
Fe–29 Cr–4 MO–2 Ni	S44800							R	R	R	R	R	R
Fe–27 Cr–3 Mo–1, Ni–Ti, Nb	S44660	27	1	3				R				R	R
Fe–29 Cr–4 Mo–Ti, Nb	S44735	29		4				R	R			R	R

[a] R = No pitting; F = failed by pitting and/or crevice corrosion; F$_G$ = failed by general corrosion.
[b] 2% KMnO$_4$–2% NaCl—no crevices (pH 7.5).
[c] 10% FeCl$_3$·6H$_2$O with crevices (pH 1.6).
[d] With 0.2% N and 0.033% C.
[e] Resistant up to 65°C (150 F).

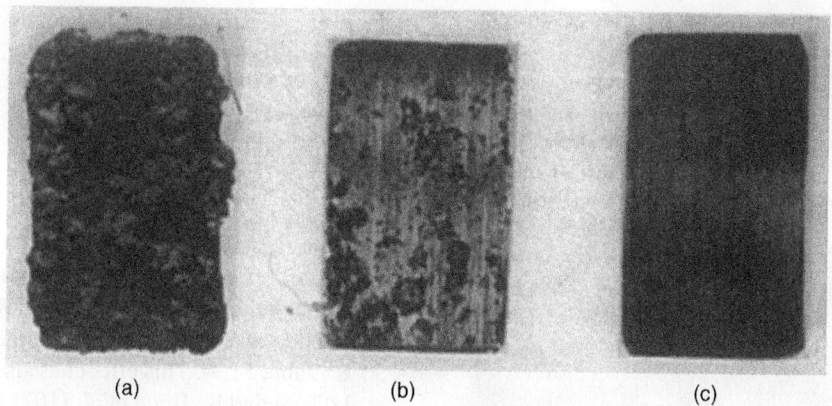

FIGURE 51.20. Permanganate/chloride test specimens [75]. (a) Pitted type 430 specimen with coating of manganese oxides, 15 h at 90°C (195°F). (b.) Same specimen after removal of coating. (c) Typical resistant Fe–Cr–Mo alloy specimen after 16 months at 90°C (195°F).

(anodic sites), the cathodic reaction consists of reduction of soluble permanganate to insoluble manganese oxide at the cathodic areas (Fig. 51.20). A simple, visual demonstration of the electrochemistry involved in corrosion is thus provided.

A lead for the alloy development work was provided by the results on the most pitting-resistan austenitic alloy, SP-2, mentioned above [65]. This experimental alloy, a 316L with 2.5% Si and 0.23% N, proved to be resistant in the KMnO₄–NaCl solution, not only at room temperature but also at 50°C (Table 51.8). In the alloy development program,

a similar alloy without the 10% Ni was made to provide stress corrosion resistance, 20 Cr–2 Mo–2 Si. This was as resistant in the KMnO₄–NaCl solution as the SP-2 alloy. Eventually, because of problems with reproducibility, the silicon was replaced by more molybdenum. To simulate conditions in heat exchangers, the temperature was raised in steps to 90°C in the search for even more resistant compositions. The results are shown in Figure 51.21 [102a].

Because the permanganate–chloride test was new, a comparison was made with results obtained, in the familiar 10%

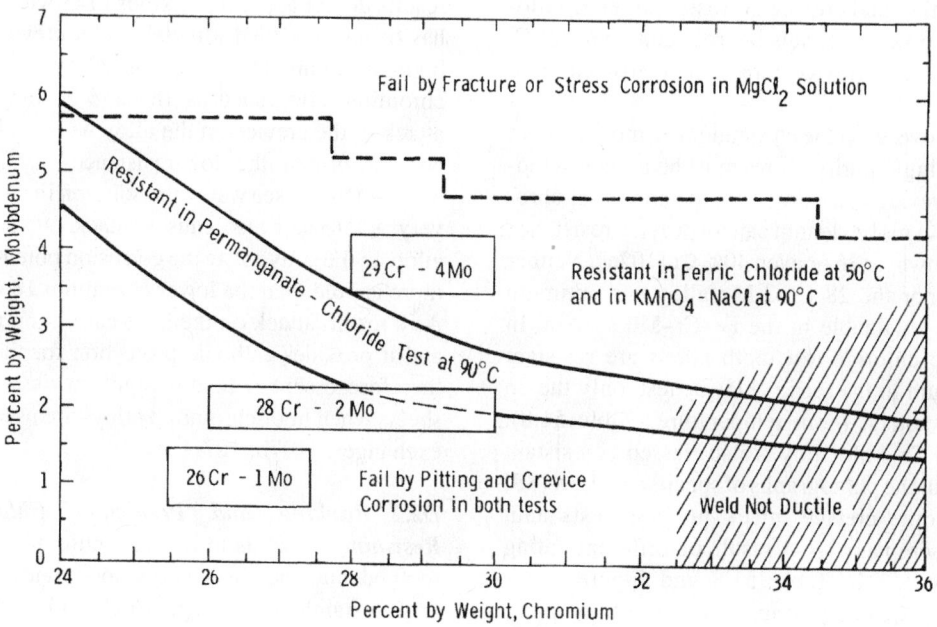

(For 18 Cr - 2 Mo the ranges are 18 to 20 Cr; 1.75 - 2.25 Mo)

FIGURE 51.21. Effect of chromium and molybdenum in Fe–Cr–Mo alloys on pitting, crevice, and stress corrosion.

FIGURE 51.22. Ferric chloride test specimens [75]. Right: Specimen with six crevices ready for testing. Left: After testing with attack at four crevices formed by elastics and under Teflon TM blocks.

ferric chloride solution using a specimen with six crevices (Fig. 51.22) The pH of this solution is 1.6 and the corrosion potential is also about $+0.6$ V vs. SCE. Data in Figure 51.18 and Table 51.8 show that the ferric chloride crevice test at 50°C is consistently more severe than the permanganate–chloride test at 90°C. Data on commercial alloys in Table 51.8 characterize the two pitting tests. Only the nickel–based lnconel alloy 625, Hastelloy alloy C, Hastelloy alloy G, and titanium are resistant in the KMnO$_4$–NaCl solution at 90°C. Of these three materials, only Hastelloy alloy C and titanium resist pitting and crevice corrosion in ferric chloride at 50°C. These two materials are resistant up to 65°C. At 75°C, they too fail by crevice corrosion in ferric chloride solution.

Figure 51.21 shows what the chromium and molybdenum concentrations in high–purity alloys must be to resist chloride pitting and crevice corrosion. It is apparent that chromium alone without molybdenum cannot provide resistance in the two tests shown, at 36 or even 40% Cr [102a]. Neither the 26 Cr–1 Mo nor the 28 Cr–2 Mo alloy has optimum pitting resistance obtainable in the Fe–Cr–Mo system. In the permanganate chloride test, both alloys are resistant only up to 50°C and in the ferric chloride test, only the 28 Cr–2 Mo is resistant at room temperature (Table 51.8). Sometimes the 26 Cr–I Mo alloy is also classed as resistant in the latter test, but the photographs of specimens show [88] that there was some crevice attack in these tests and, therefore, the discrepancy is a result of different rating criteria. Resistant (R) in Table 51.8 and Figure 51.21 indicates that there was no pitting or crevice attack.

The composition of the 29 Cr–4 Mo alloy was selected to provide resistance to crevice and pitting corrosion in ferric chloride solution at 50°C, which means the compositions is also resistant at 90°C in the KMnO$_4$–NaCl

solution. Because high molybdenum contents ($> 4.6\%$) and high chromium contents ($> 32.5\%$) contribute to a loss in ductility, a composition well away from these regions was selected. The 4% molybdenum in this alloy provides additional benefits. It is responsible for the important increase in the tolerance for residual nitrogen discussed below.

Addition of 2% Ni to the 29 Cr–4 Mo composition does not impair resistance to pitting corrosion. However, increasing the nickel content beyond the 2.2% specified for the alloy to 2.5% does reduce resistance in the ferric chloride test at 50°C (Table 51.9). This effect has also been found by Okada et al. [106] and by Bond et al. [107] and is another example of the complex role of nickel in the corrosion resistance of stainless steels.

Both the 29 Cr–4 Mo and the 29 Cr–4 Mo–2 Ni alloys have been tested by immersion in the ocean with severe synthetic (Teflon or Delrin) crevices on the specimens. In addition, during the 9-month exposure a layer of marine organisms 4 cm thick formed on the front and back of each test piece. There was no corrosion, pitting, or crevice attack on either of the two alloys [103].

Similar excellent results were obtained in two other crevice corrosion test programs in natural sea water in 1980. The 29 Cr–4 Mo alloys were part of a group of 36 ferritic and austenitic alloys having a wide range of alloy compositions [107a] and with polymeric (Teflon) crevices.

D2.1. A Precaution. It should be noted, however, that these results do not apply to all kinds of crevices. For example, it has been found that a metal–metal crevice consisting of a high-chromium alloy such as 29 Cr–4 Mo with a lower chromium alloy such as 18 Cr–8 Ni may result in severe attack in the crevice on the alloy with the higher chromium content. When the low-resistance 18 Cr–8 Ni alloy is attacked by the sea water, the solution in the crevice becomes very acidic and also causes attack on the high-chromium alloy. In the active state the corrosion potential of this alloy is more anodic than the lower chromium 18 Cr–8 Ni material. As a result, attack on the high-chromium alloy is increased and it provides cathodic protection for the 18/8 alloy. It is therefore essential to use highly resistant alloys for tube sheets when high-chromium alloys are used for tubes in heat exchangers [107b, 107c].

D2.2. Ranking and Predicting Chloride Corrosion Resistance. Just as in the austenitic alloys, chromium and molybdenum act in synergy to provide resistance to the ferritic stainless steels against chloride pitting. While nitrogen provides no enhancement of the ferritic alloys, the PRE formula provides a useful ranking among the ferritic alloys, but there is a significant offset between the austenitic and ferritic alloys, as shown in Figure 51.23.

TABLE 51.9. Effect of Nickel Addition to 29 Cr–4 Mo Alloy[a]

| Elements(%) | | | Boiling 10% Sulfuric Acid | | Pitting Corrosion[b] | | Stress Corrosion[e] |
Cr	Mo	Ni	State	Corrosion Rate (mm/year)	KMnO$_4$–NaCl[c]	FeCl$_3$[d]	(Not Welded)
28.0	4.0	0.10	Active	1600	R	R	Resistant
28.0	4.0	0.20	Active	1516	R	R	Resistant
28.0	4.0	0.25	Passive	1.4	R	R	Failed
28.0	4.0	0.30	Passive	1.3	R	R	Cracked after 119 h
28.0	4.0	0.40	Passive	0.7	R	R	Cracked after 261 h
28.0	4.0	0.50	Passive	0.6	R	R	Cracked after 16 h
28.5	4.0	1.5	Passive	0.2	R	R	Cracked in < 16 h
28.5	4.2	1.8	Passive	0.3	R	R	Cracked
28.5	4.2	2.0	Passive[f]	0.2	R	R	Cracked in 3 h
28.5	4.2	2.5	Passive[f]	0.3	R	F	—
28.5	4.2	3.0	Passive[f]	0.2	R	F	—

[a]See [103].
[b]R = resistant to pitting and cervice corrosion; F = failed; — = not tested.
[c]2% KMnO$_4$–2% NaCl at 90°C (195°F).
[d]10% FeCl$_3$·6H$_2$O at 50°C (120°F) with cervices.
[e]Magnesium chloride test (45%); resistant = no cracking after 2400 h.
[f]Self-respassivating.

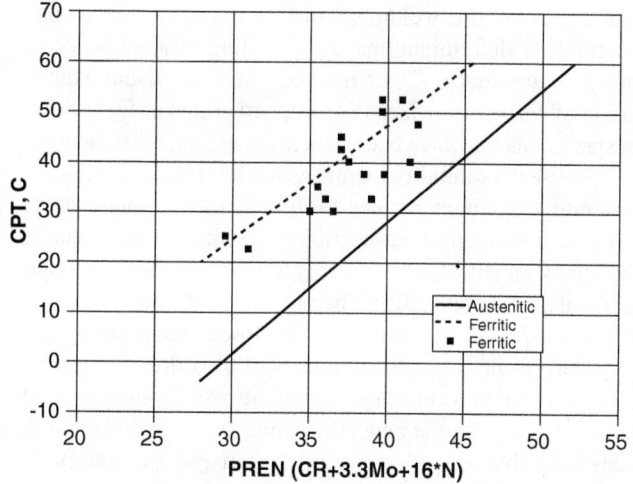

FIGURE 51.23. CPT vs. PREN for ferritic alloys [65].

D3. General Corrosion

Table 51.3 provides a survey of corrosion rates for comparison of the new ferritic stainless steels with each other and with commercially available alloys commonly used for resistance to acids. The acids selected for these tests include a wide range of severely corrosive environments, oxidizing, acids, and organic and inorganic reducing acids. Four types of behavior are represented in Table 51.3:

1. Passive state that is not converted into the active condition by contact with an iron rod while the specimens are immersed in the test solution. Examples are

all of the alloys in concentrated nitric acid and, with the exception of type 430, in acetic acid.

2. Passive state that, after conversion galvanically to active, hydrogen evolution corrosion by contact with an iron rod, is reestablished spontaneously when the galvanic contact is interrupted. This is self-repassivation. Examples are the two nickel-bearing Fe–Cr–Mo alloys in the sulfuric and hydrochloric acid tests.

3. Passive but not self-repassivating. On immersion, the alloy is passive and remains in this state. However, when galvanically activated with an iron rod, there is no repassivation when the rod is removed. Passivity can be restored simply by removing the specimen from the flask and exposing it to the air or rinsing it in water before reimmersing it in the acid solution. In Table 51.3, these are identified by "d" and the rates in both the passive and the active states are listed.

4. Active state with hydrogen evolution corrosion that begins when the specimen is immersed in the boiling solution. Examples are all the tests on type 430 steel in reducing acids.

As already noted above, of all the old alloys in Table 51.3, Carpenter 20 Cb-3 has the best overall resistance to various types of boiling acids. Hastelloy alloy C, because of the deleterious effect of its high molybdenum content (16%) and because of its somewhat low chromium content (15.5%), has the highest corrosion rate of all the alloys in the oxidizing acids [108]. On the new alloys, corrosion in oxidizing acids is a function of the chromium content. The relatively high rate of the 18 Cr–2 Mo–Ti alloy is a result of preferential attack on intermetallic phases and on titanium carbides and nitrides. Similar effects are found on the titanium-stabilized 26 Cr–1 Mo alloy.

The 29 Cr–4 Mo alloy's performance is equal to or superior to that of Carpenter 20 Cb-3 except in inorganic reducing acids. In boiling 10% sodium bisulfate, the alloy is passive, but not self-repassivating, and in boiling 10% sulfuric and 1% hydrochloric acid the alloy is active and has very high corrosion rates. However, addition of only 0.25% Ni to the 29 Cr–4 Mo alloy makes it passive but not self-repassivating in boiling 10% sulfuric acid, (Table 51.9). The amount of nickel required for passivity is a function of both the chromium and molybdenum concentration. For self-repassivation, 2% Ni is required. The 2% nickel also makes the alloy self-repassivating in boiling 1% HCl.

Addition of 4% Ni to the TEW 28 Cr–2 Mo alloy provides resistance comparable to that of the 29 Cr–4 Mo alloy. Both alloy are self-repassivating in boiling 10% sulfuric acid and have corrosion rates that are only one-fifth that of Carpenter 20 Cb-3. But they are not resistant in boiling solutions above 10% [102]. The fact that the higher (4%) nickel content in the 28 Cr–2 Mo alloy does not provide passivity in sulfuric acid at concentrations >10% seems to be a result of the presence of niobium in this alloy. Bond et al. [107] reported that the use of titanium in place of niobium provides passivity up to 25% acid. However, titanium raises the impact transition temperature. The acid corrosion resistance of the two nickel-bearing alloys, 29 Cr–4 Mo–2 Ni and 28 Cr–2 Mo–4 Ni, is essentially equivalent to that of Carpenter 20 Cb-3 in the solutions of Table 51.3.

Because the 26 Cr–1 Mo and 18 Cr–2 Mo alloys do not contain nickel, they are not resistant in the boiling inorganic acids of Table 51.3. Also, because of their lower chromium and molybdenum contents as compared with the 29 Cr–4 Mo alloy, their resistance in the more corrosive organic acids is lower than that of the 29 Cr–4 Mo alloy. In general, the 26 Cr–1 Mo alloy is more resistance than the 18 Cr–2 Mo alloy.

Note that titanium, which has outstanding resistance in the two chloride pitting and crevice corrosion tests and is resistant to chloride stress corrosion, has high resistance only in nitric and acetic acid solutions of Table 51.3. In heat exchangers, for example, titanium provides excellent resistance to corrosion by natural cooling waters, but its range of resistance to process fluids is severely limited as compared with Carpenter 20 Cb-3 and several of the new alloys in Table 51.3. A new titanium alloy (Ti–38A) with 0.3% Mo and 0.8% Ni has recently been introduced to extend the range of resistance to acids. This alloy is now known as titanium Grade 12, UNS R53400. In the solutions of Table 51.3, the corrosion of the new alloy is significantly lower only in boiling 45% formic acid, 22 vs. 0.0 mm/year. In some acids

the rate of the new alloy is higher than that of commercially pure titanium.

The temperatures and acid concentrations of Table 51.3 represent relatively severe conditions. In some cases, alloys that are listed as having high rates of corrosion become passive and self-repassivating at lower temperatures and/or concentrations of acid.

D4. Intergranular Corrosion

D4.1. Chromium Carbides and Nitrides. Susceptibility to intergranular attack in the new ferritic alloys has frequently been determined by testing a specimen containing a weldment without filler metal made by using a shielded tungsten arc to melt the alloy, that is, an autogenous weld. Such a specimen approximates the actual thermal cycles that the material may undergo during fabrication by welding. Because of this, evidence of intergranular attack in one or more of the components of the weldment, weld metal, fusion zone, heat-affected zone, or base metal shows that such material may be subject to intergranular attack when exposed in an environment that produces intergranular attack on sensitized grain boundaries. Preferential attack on a narrow component of the weldment makes it impossible to use weight loss determinations for detecting susceptibility to intergranular attack. Other methods, such as visual examination of corroded surfaces for dropped grains or for fissures on specimens that have been bent after testing, must be used.

Because the solubility for nitrogen is lower in ferrite than in austenite, chromium nitrides form in ferritic stainless steels and also contribute to susceptibility to intergranular attack. The involvement of nitrogen was reported in 1952 by Houdremont and Tofaute [109]. Therefore, the nitrogen content of the Fe–Cr–Mo alloys must be held to low concentrations, not only during melting and refining, but also during welding operations. To prevent absorption of nitrogen by molten weld beads from the air, the new alloys must be protected on both sides by a blanket of argon or helium gas (shielded). In contrast, during heat treatments, there is no problem of nitrogen absorption from air atmospheres by solid metal.

Two copper sulfate tests, the HNO_3–HF test and the ferric sulfate–50% sulfuric acid test, have been used to detect susceptibility to intergranular attack associated with chromium carbide and nitride precipitates in ferritic alloys. As shown in Figure 51.9, the nitric acid test is relatively insensitive for this purpose. In a new ASTM standard, A-763, for evaluating ferritic stainless steels, the number of test methods has been reduced to three, the ferric sulfate test and two versions of the copper sulfate test, with 15% sulfuric acid for alloys containing 17–20% Cr and 50% sulfuric acid for alloys with 25–30% Cr [111].

Intergranular corrosion in Fe–Cr–Mo alloys is a function of the carbon and nitrogen contents, their solubility, and rates of diffusion. Unexpectedly, it has been found that increasing the molybdenum content increases the tolerance for nitrogen, which is difficult to reduce much below 100 ppm by any of the refining procedures. Several investigators [75, 85, 95] have shown that weldments in the 26 Cr–1 Mo alloy containing nitrogen in concentrations greater than ~80 ppm are subject to intergranular attack. Thus, the available refining methods cannot consistently provide a low enough concentration of nitrogen to prevent intergranular attack on weldments in the high-purity 26 Cr–1 Mo alloy.

Because of this problem, small amounts of niobium (0.05–0.08%) have been added to the EB 26-1 alloy [92]. Such additions have been made for ~5 years but have not been shown in recent analysis for which 150 ppm nitrogen is given as the maximum, with 100 ppm as a typical concentration [91]. These additions of niobium have, however, not prevented some susceptibility to intergranular attack. Sweet[*] reported intergranular attack on weldments exposed in the ferric sulfate–sulfuric acid test. The attack was only two layers deep. However, heating of EB 26-1 alloys for 30 min at 593°C (1100°F) resulted in severe susceptibility to intergranular attack throughout the entire cross section. There is a possibility that the reason for this is that the amount of niobium added (0.05–0.08%) is less than that specified (13–29 times the %N) in the patent [92] on this subject. Completely stabilized 26 Cr–1 Mo material, such as 26 Cr–1 Mo–Ti, is resistant to intergranular attack after such a heat treatment at 593°C (1100°F).

Extensive data on the 29 Cr–4 Mo alloy support the conclusion that there is a much higher tolerance for nitrogen, 200 ppm in this alloy than in the 26 Cr–1 Mo composition. Thus, the concentration of ~100 ppm N, which can be achieved by vacuum induction or by electron beam refining, is entirely adequate for the 29 Cr–4 Mo alloy. The limits for carbon and nitrogen in the 29 Cr–4 Mo alloy, $C \leq 100$ ppm, $N \leq 200$ ppm with $C + N \leq 250$ ppm, as defined in the original investigation of this alloy [75], have been completely supported by Nichol and Davis [110]. They tested weldments and specimens that had been heated at 1010, 1121, and 1232°C and either quenched or air cooled. None of the compositions that met the above limits for carbon and nitrogen showed intergranular attack in the four evaluation tests used. These evaluation tests were copper sulfate with 16 and 50% sulfuric acid, HNO₃–HF, and the ferric sulfate test. In the 29 Cr–4 Mo alloy, increasing nitrogen concentrations in excess of 200 ppm impair first mechanical properties and then resistance to intergranular attack and chloride pitting.

Because the carbon contents of all of the new ferritic alloys must be held to low values, pick-up of carbon during processing and forming operations must be avoided. Carbonaceous lubricants must be thoroughly removed before heat treatments.

[*] Report to NACE Task Group 5A-I7, January 21, 1977 (Du Pont Company).

D4.2. Intermetallic Compounds. Thermal exposures in the range of 700–925°C can result in formation of sigma, chi, and perhaps other phases at the grain boundaries of Fe–Cr–Mo alloys. In the high-purity 29 Cr–U Mo and 29 Cr–4 Mo–2 Ni compositions, the rate of formation is so slow that these phases do not form during welding. Even after heating of 1 h at 815°C, the temperature of most rapid formation, only a small amount of sigma phase is formed at grain boundaries [103]. Its presence is without effect on the resistance to intergranular attack, even in the oxidizing ferric sulfate–sulfuric acid test. Nor does it affect the weld-bend ductility. However, it is shown in the next section that there is some reduction in the toughness, the Charpy impact strength. Heat treatments of 100 h at 815°C produce large amounts of chi and some sigma phase. In the 29 Cr–4 Mo alloy, the chi phase contains cracks. Addition of 2% Ni to the 29 Cr–4 Mo composition increases the amount of chi phase and also eliminates the cracks, that is, increases die ductility and toughness [103]. It was found [111] that chi phase is rich in molybdenum and chromium and that there is even greater enrichment in sigma phase.

When titanium and niobium are added to commercial AOD heats of 18 Cr–2 Mo and 26 Cr–1 Mo alloys to stabilize their relatively high carbon and nitrogen contents (C + N = 300–500 ppm, Table 51.7) welded specimens may become subject to intergranular attack in the ferric sulfate–50% sulfuric acid test and in the even more oxidizing nitric acid test. When the same specimens are immune to intergranular attack in the copper sulfate–50% sulfuric acid and the nitric–hydrofluoric tests, this shows that susceptibility to intergranular attack is not associated with chromium carbides or nitrides but is a result either of intermetallic phases rich in molybdenum, titanium, and niobium and/or preferential dissolution of titanium and niobium carbides. The manganese and silicon present in commercial alloys may also promote the formation of these intermetallic compounds. Nichol and Davis [110] tested a series of 26 Cr–1 Mo alloys with a range of Ti/C + N ratios from 3.5 to 18.5. Welded specimens were tested in copper sulfate–50% H₂SO₄ (with metallic copper, +0.1 V vs. SCE) and in ferric sulfate 50% H₂SO₄ (+0.6 V vs. SCE). Compositions having a Ti/C + N ratio of < 9.1 were subject to intergranular attack in both tests because the amount of titanium added was insufficient to prevent formation of chromium carbides and nitrides. On alloys with ratios of 9.1 or greater, there was no intergranular attack in the copper sulfate test, but alloys with high ratios (13.0, 16.6, and 18.5) were subject to intergranular attack in the ferric sulfate test. Recent results of Sweet [112] are in agreement with these findings. By analogy with the effect of sigma phase, in austenitic, molybdenum-bearing stainless steel (316L) as described above (Table 51.4), it may be concluded that the cause of this intergranular attack on 26 Cr–1 Mo with high concentrations of titanium is probably one or more intermetallic, titanium- and molybdenum-rich phases.

Bond and Lizlovs [113] and Dundas and Bond [114] found similar behavior in 18 Cr–2 Mo alloys stabilized with titanium. These results on titanium-stabilized 26 Cr–1 Mo and 18 Cr–2 Mo alloys have led to a recommendation [95] that such alloys not be used in oxidizing environments for which their high chromium contents otherwise make them eminently suitable. Data on 18 Cr–2 Mo–Ti in Table 51.3 support this recommendation.

Some results reported by Troselius [115] on an 18 Cr–2 Mo–Ti alloy that was stressed while exposed at 300°C in water containing 10 and 100 ppm Cl suggest that this problem of second phases or titanium carbide and nitride precipitates may not be confined to nitric acid and ferric sulfate–sulfuric acid solutions. He found intergranular failure that he termed stress-accelerated intergranular attack but which may also have been intergranular stress corrosion cracking. A type 430 specimen did not fail in this environment, which contained 9 ppm oxygen.

Weldments on TEW, niobium-stabilized, 28 Cr–2 Mo–4 Ni sheets 3 mm thick were found to be resistant in all of the acid corrosion tests of ASTM A 262. However, prolonged heating of 15 min and 1 h at 600 and 1200°C was found to increase somewhat the corrosion rate in the nitric acid test [101, 116]. Also, Bond et al. [107] reported that there was preferential pitting attack on weldments of a 28 Cr–2 Mo–4 Ni alloy stabilized with titanium. The alloy was exposed in boiling 25% NaCl solution with an adjusted (HCl) pH of 1.5. There was localized corrosion at the heat-affected zone on an unidentified phase. This appears to be a result of the action of titanium, which may be similar to that found for niobium by Troselius [117]. He observed that niobium has a detrimental effect on the resistance of molybdenum-bearing austenitic steels in chloride-pitting environments.

The stabilization requirements for 29 Cr–4 Mo alloys were investigated by Grubb [117a]. The stabilizing elements Ti, Cb, V, and Zr were investigated. Results of tests on Cb, Ti, and Cb + Ti stabilized alloys showed that the ASTM specification requirements for S44735 are adequate to prevent sensitization. The proposed stabilizing elements V and Zr were investigated and shown to be deficient.

The various intergranular corrosion evaluation tests can be used to determine whether or not the amount of titanium and/ or niobium added was sufficient to prevent formation of chromium carbide and nitride precipitates and whether or not heat treatments applied to react the stabilizers with carbon and nitrogen were effective. If not, chromium carbides and nitrides may be formed, and the unsuccessfully stabilized alloys will be subject to intergranular attack, pitting, and chloride stress corrosion cracking in MgCl2 and NaCl solutions. Enough stabilizer must be present, not only to provide the theoretical amount for combination with the carbon and

nitrogen, but also to ensure a high rate for this reaction and to provide for losses to oxide and sulfide formation. The rate must be high enough so that this stabilizing reaction removes the available carbon and nitrogen during rapid cooling of a weldment through the range of temperature in which the stabilizers react with these elements. The amounts of stabilizers required in relation to the carbon and nitrogen contents are shown in Table 51.7.

D5. Stress Corrosion Cracking

As described above, while ferritic alloys are generally immune to chloride SCC, this property may be impaired in several ways. When ferritic stainless steels contain chromium carbide precipitates surrounded by chromium-depleted zones, they become subject to SCC not only in the MgCl2 test at 155°C but also in various NaCl tests at lower temperatures. This is shown in Table 51.6 for type 446 steel and for an off-grade 29 Cr–4 Mo alloy with 210 ppm C (i.e., more than twice the maximum permissible carbon content of 100 ppm). Also, sensitized 18 Cr–2 Mo and 26 Cr–1 Mo alloys have been found by Dundas to crack when exposed to artificial sea water [85]. It has not been possible to determine the effect of high nitrogen contents on resistance to stress corrosion. Increasing the nitrogen content significantly beyond the 200 ppm limit in the 29 Cr–4 Mo alloy so reduces its ductility that attempts to form U-bends for stress corrosion tests result in fracture of the specimens.

Furthermore, the presence of small amounts of nickel or copper makes ferritic alloys susceptible to transgranular SCC in the *magnesium chloride test*. The effect of nickel has been demonstrated for the 18 Cr–2 Mo, 26 Cr–1 Mo, 28 Cr–2 Mo, and 29 Cr–4 Mo alloys [72, 75, 95, 102]. Residual nickel can be a problem because this element may sometimes be introduced with scrap and from furnace linings. The amount (0.2–1%) which causes cracking in magnesium chloride solution depends on the condition of the metal, wrought, or cast, as in weld metal, the amount of cold work, and the concentration of molybdenum and copper in the alloy (Table 51.10). All of these factors reduce the tolerance of nickel as compared with an annealed, wrought Fe–Cr alloy without molybdenum (Table 51.6). When nickel is added as an alloying element, as in 28 Cr–2 Mo–4 Ni and in 29 Cr–4 Mo–2 Ni alloys, the time to cracking in the MgCl2 test is rapidly reduced as the nickel content is increased to 2% or more (Table 51.9).

Fortunately, nickel as a residual element or an alloying addition does not cause SCC in various *sodium* chloride environments (Table 51.6). There is no cracking even when the temperature is increased to 155 and 200°C in autoclave tests with 26% NaCl solutions and an atmosphere of air [74, 75]. Also, when the natural pH 7.3 of the sodium chloride solution is reduced to 4.0, the pH of the magnesium chloride test, there is no cracking on the 29 Cr–4 Mo–2 Ni or the

TABLE 51.10.   Mechanical Properties of Stainless Steels

Alloy	Yield Strength		Tensile Strength		Elongation (%)	Transition Temperature[a]	Hardness RB
	MPa	ksi	MPa	ksi			
Type 316[b]	206	30	517	75	40	−240°C (−400°F)	95 max
Type 446[b]	310	45	517	75	20	+122°C (+250°F)	95 max
18 Cr–2 Mo–Ti[c]	302	44	468	68	37	+25 to 75°C (+75 to +165°F)	83
26 Cr–1 Mo–Nb[91,92]	345	50	483	70	35	−62°C (−80°F)	83
Electron Beam Refined							
26 Cr–1 Mo–Ti[95]	358	52	517	75	30	+40°C[d](+105°F)	85
28 Cr–2 Mo–4 Ni–Nb[99]	567	82	649	94	26	−5°C (+25°F)	83–98
29 Cr–4 Mo[b]	545	79	614	89	27	+16°C (+60°F)	94
29 Cr–4 Mo–2 Ni[119]	614	89	682	99	24	−7°C[e] (+19°F)	95

[a]Full size, 10 mm (0.39 in.) Charpy impact specimens.
[b]Data from Allegheny Ludlum Steel Corp.
[c]Data from Climax Molybdenum Company.
[d]For 1.91-mm (0.075-in.) sheet, transition temperature −23 to −9°C (−9 to +15°F).
[e]For 1.65-mm (0.065-in.) sheet, transition temperature −90°C (−195°F).

Carpenter 20 Cb-3 alloy in 26% NaCl at 102°C. Alloys with copper and high molybdenum concentrations that fail by cracking in $MgCl_2$ are also resistant in the NaCl tests. Thus, there is a marked difference between stress corrosion caused by chromium carbide precipitates and by nickel, copper, or molybdenum. The sodium chloride test environments are much more like service conditions than the magnesium chloride test solution.

Support for these findings on the difference between $MgCl_2$ and NaCl environments is provided by recent autoclave tests at 200°C with chloride concentrations up to 3000 mg/L in sewage sludge. Welded U-bend specimens were tested [118] and it was found that there was no cracking on the 29 Cr–4 Mo–2 Ni alloy during the 1000-h test period. Type 316 steel cracked in only 48 h in sludge with only 1000 mg/L.

Tests on the TEW 28 Cr–2 Mo–4 Ni alloy gave similar results. This alloy also cracks in the $MgCl_2$ test. However, in stress corrosion tests up to 1000 h in boiling sea water and up to 6000 h in boiling 15% synthetic sea water, there was no cracking, even though types 304 and 316 specimens cracked after 800 and 1800 h, respectively, in the 15% salt solution [101, 116].

D6. Mechanical Properties

A summary of yield and tensile strength, ductility, and transition temperatures is given in Table 51.10 for the new ferritic alloys, along with values for types 316 and 446 stainless steels. New alloys with 26–29% Cr have higher yield strengths than either types 316 or 446, the value for the 29 Cr–4 Mo–2 Ni alloy being the highest of all, 614 N/mm^2 (89 ksi). This alloy also has the highest tensile strength. There is a progressive decrease in ductility from 37 to 24% as

the alloy content of the ferritic compositions is increased from 18 Cr–2 Mo to 29 Cr–4 Mo–2 Ni.

The temperature at which fractures of Charpy V-notch specimens change from a ductile to a brittle appearance is a measure of the notch sensitivity. Reducing the concentration of carbon and nitrogen reduces (improves) the transition temperature [83, 98]. Addition of nickel also lowers this temperature, as does the use of niobium as a stabilizing element [99, 107, 119]. In contrast, addition of titanium raises the transition temperature [99, 107]. Thus the high-purity 26 Cr–1 Mo, 29 Cr–4 Mo–2 Ni, and 28 Cr–2 Mo–4 Ni (Nb) alloys all have relatively low transition temperatures. But both of the titanium-stabilized alloys have a high notch sensitivity (Table 51.10). Low transition temperatures promote the formability of weldments and increase the thickness of sections that can be processed during production. When the transition temperature of a ferritic alloy is too high for processing thick plates, it may be possible to circumvent this problem by cladding thin ferritic alloy sheets onto thick carbon steel tube sheets.

Nichol [119] has made a detailed investigation of the mechanical properties of the 29 Cr–4 Mo–2 Ni alloy. Figure 51.24 contains his data on the effect of specimen thickness on transition temperature for an alloy with 40 ppm C and 146 ppm N. For water-quenched material having a thickness of 1.65 mm (0.065 in.), such as frequently used for heat exchanger tubing, the fracture appearance transition temperature (FATT) is −90°C. The impact strength (toughness) for a standard-size, 10-mm (0.394-in.) Charpy V-notch specimen above the transition temperature (shelf energy) is 250 J (175 ft-lb). Alloys with a higher (3–4%) nickel content, such as the 28 Cr–2 Mo–4 Ni (Nb) alloy, have a lower shelf energy, 160–180 J [99, 107].

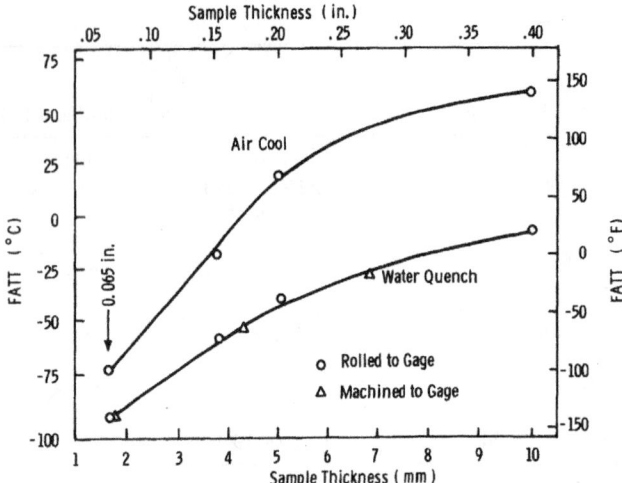

FIGURE 51.24. Effect of sample thickness and colling rate on the fracture appearance transition temperature of 29 Cr–4 Mo–2 Ni alloy [119].

The effect of the temperature of prior heat treatments on the toughness of 29 Cr–4 Mo–2 Ni alloy at room temperature (27°C) is given in Figure 51.25. Prior exposure in two temperature ranges was found to reduce the impact strength, one between 704°C (1300°F) and 871°C (1600°F) and a lower temperature range centered at 482°C (900°F). The higher temperature range corresponds to the range in which chi and sigma phases are formed in this alloy [103]. The lower temperature falls within the range of the 475°C (885°F) embrittlement phenomenon, 400–510°C. All chromium-bearing ferritic stainless steels with > 12% Cr are subject to this type of embrittlement [85, 91, 103, 119, 120]. It limits the prolonged use of the Fe–Cr–Mo alloys to temperatures

below ~350°C. The other new ferritic alloys are probably also subject to a reduction of impact strength after exposure in the range of temperature in which chi and sigma phases are formed, 704°C (1300°F) to 927°C (1700°F).

Detailed investigations of mechanical properties have been reported by Davison [86] for 18 Cr–2 Mo–Ti, by Franson [91] for high-purity 26 Cr–1 Mo, by Pinnow et al. [95] for 26 Cr–1 Mo–Ti, and by Brandis et al. [99] for 28 Cr–2 Mo–4 Ni–Nb).

The oxygen content of ferritic alloys must be kept to low concentrations (~0.001–0.002%) by control of the melting processes to avoid embrittlement and oxide inclusions that impair resistance to pitting attack. To prevent embrittlement by hydrogen, it is essential to avoid diffusion of hydrogen into cold-worked metal during pickling treatments. In cases where absorption of hydrogen has occurred, storage for 24 h at ambient temperature or baking at 100–300°C for several hours permits hydrogen to diffuse out of the alloys and restores ductility [103]. Reaction of ferritic alloys with nitrogen during annealing in cracked ammonia atmospheres must also be avoided.

E. SUMMARY AND CONCLUSIONS

The first use of both the ferritic and austenitic stainless steels can be traced to the skilled application of chance observations by Brearly [1–4] and Maurer [5–8] in 1912. In contrast, the new Fe–Cr–Mo alloys are a result of several simultaneous and deliberate efforts to develop alloys that would meet a number of well-defined objectives. Nevertheless, in the work on these new alloys some unexpected observations were made. Increasing the molybdenum content from 1 to 4% decreases the precipitation of chromium nitrides and,

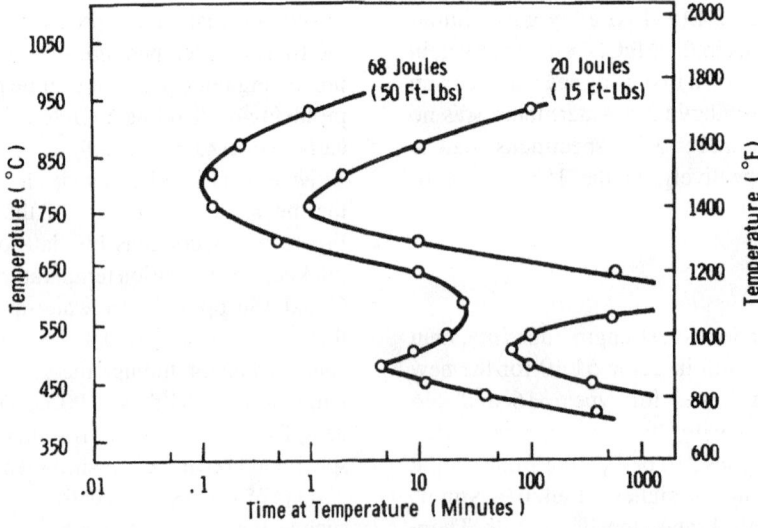

FIGURE 51.25. Effect of temperature and time on impact strength of 29 Cr–4 Mo–2 Ni alloy [119].

TABLE 51.11.   Summary of Properties of Feritic Alloys

Property	18 Cr–2 Mo–Ti	26 Cr–1 Mo (High Purity)	26 Cr–1 Mo–Ti	28 Cr–2 Mo–4 Ni–Nb	29 Cr–4 Mo	29 Cr–4 Mo–2 Ni
			Alloy[a]			
Stress corrosion cracking						
MgCl$_2$ (155°C, 310°F)	R[b,c]	R[b]	R[95,b]	F	R[b]	F
NaCl (103°C, 217°F)	R	R	R	R	R	R
Pitting and crevice corrosion						
KMnO$_4$–NaCl						
Room temperature	F	R	R	R	R	R
50°C (120°F)		R	R	R	R	R
90°C (195°F)		F	F	F	R	R
Ferric chloride						
Room temperature	F	F	F	R	R	R
50°C (120°F)				F	R	R
Corrosion in boiling acids[d]						
Nitric 65%	F[e]	R	F[e]	R	R	R
Formic 45%	F	R	R	R	R	R
Oxalic 10%	F	F	F	R	R	R
Sulfuric 10%	F	F	F	R	F	R
Hydrochloric 1%	F	F	F	R	F	R
Transition temperature[f]	+25 to +75°C (+75 to +165°F)	−62°C (−80 °F)	+40°C (+105 °F)	−5°C (+25 °F)	+16°C (+60°F)	−7°C (+20°F)
Refining processes	AOD or VOD	EB or VIM	AOD	VOD of AOD	VIM of EB	VIM

[a]R = resistant; F = fails by type of corrosion shown; AOD = argon–oxygen decarburization; VOD = vaccum–oxygen decarburization; VIM = vacuum induction melting; EB = electron beam refining.
[b]Copper and nickel residuals must be kept low in these alloys to resist cracking in this solution.
[c]Data from Climax Molybdenum Co. publication. 18 Cr–2 Mo ferritic stainless steel.
[d]R indicates passive or self-repassivating with rates <0.2 mm/y (0.008 in year).
[e]Not recommended for oxidizing solutions.
[f]Full-size Charpy V-notch specimens.

therefore, the susceptibility to intergranular attack in acids. Addition of nickel lowers the impact transition temperature (i.e., increases toughness). Even though nickel additions make the alloys subject to SCC in magnesium chloride solutions, this is not the case in various sodium chloride tests, which simulate actual service environments. When the concentration of nickel exceeds a certain level, resistance to crevice corrosion in oxidizing chloride solutions is reduced.

An overview of the properties of the new ferritic stainless steels is given in Table 51.11. Because of their resistance to stress corrosion cracking in sodium chloride solutions, they can provide a large part of the practical solution to the chloride SCC problems. In addition, depending on their alloy content, they provide resistance to pitting and crevice corrosion in chloride solutions ranging from that roughly equivalent to types 430, 446, and 304 steels for the 18 Cr–2 Mo (Ti) alloy to that equivalent in some cases to the costly Hastelloy alloy C and titanium for the 29 Cr–4 Mo alloys. The latter group of alloys are also the only ones that have been found to be resistant to crevice corrosion during exposure in the ocean.

Note, however, the limitation described under "A Precaution" in Section D2.1: Alloys in metal-to-metal crevices must be of comparable corrosion resistance.

The relatively high chromium contents of these alloys make them resistant in oxidizing acids The only exceptions are the two alloys stabilized with titanium. In oxidizing acids, the numerous titanium carbide and nitride particles are dissolved rapidly. Resistance in organic acids increases with increases in the concentrations of chromium and molybdenum. Addition of nickel is needed to provide resistance to reducing, inorganic acids. Thus, the resistance of the two nickel-bearing alloys to corrosion in acids in the tests of Tables 51.3 and 51.11 is equivalent to that of Carpenter 20 Cb-3, which is the most resistant of all the old alloys of Table 51.3. Weldments of the new alloys are also resistant to intergranular attack in acids, again with the exception of the two titanium-stabilized alloys in oxidizing solutions.

All of the new ferritic alloys have good deep-drawing characteristics and are readily formable with less work hardening than austenitic alloys. For maximum formability of weldments and processing of heavy sections, the alloys with low transition temperatures are preferred.

On the basis of available laboratory data (Table 51.11) the 29 Cr–4 Mo and 29 Cr–4 Mo–2 Ni alloys provide an optimum combination of resistance to chloride SCC, pitting and crevice corrosion, and resistance to general and intergranular corrosion in a broad range of acids, along with excellent strength, ductility, toughness, and weldability. They can also be cast into intricate shapes. Because of the limits for carbon and nitrogen in these two high-purity alloys (100 ppm maximum for C, 200 ppm maximum for N, with C + N ≤ 250 ppm) they must, at present, be melted and refined either by vacuum induction or electron-beam processes that are more costly than the argon–oxygen decarburization process.

These limits for carbon and nitrogen in the 29 Cr–4 Mo alloys are only a relatively small step below the desirable concentrations specified for the *stabilized* 18 Cr–2 Mo, 26 Cr–1 Mo, and 28 Cr–2 Mo–4 Ni alloys. These range from 300 to 500 ppm for C + N in heats made by the lower cost AOD process (Table 51.11). Either titanium or niobium can be used to combine with carbon and nitrogen in the above alloys. However, the properties of the resulting compounds differ depending on the stabilizing element selected. Titanium carbides and nitrides are more rapidly dissolved in oxidizing acid solutions in some cases than the corresponding niobium compounds [113]. In reducing acids the advantage is reversed. The use of titanium for stabilization of the 28 Cr–2 Mo–4 Ni alloy provides for a greater range of concentration of sulfuric acid in which the alloy is passive than does the use of niobium [107], which is specified for this alloy.

To assure rapid formation of titanium carbides and nitrides, as is needed during cooling of welds, more than double the theoretical amount of titanium required for reaction with all the carbon and nitrogen present must be added to 26 Cr–1 Mo and 18 Cr–2 Mo alloys [110, 112, 114]. This is also the case for niobium stabilization of the 28 Cr–2 Mo–4 Ni alloy [99, 101, 102, 116]. Titanium, by raising the impact transition temperature, has a deleterious effect on toughness, whereas niobium, which lowers this temperature, has a beneficial effect. But the joint use of niobium with titanium does not offset the deleterious effect of titanium [121]. Increasing the titanium or niobium contents above the amounts required for stabilization of weldments can result in a decrease in ductility of these weldments. Niobium can cause hot-shortness in weld metal [121]. Both titanium and the more costly niobium promote the formation of intermetallic compounds in molybdenum-bearing alloys. These reduce toughness and have a deleterious effect on corrosion in oxidizing acids. Furthermore, in the 28 Cr–2 Mo–4 Ni alloy, titanium additions result in an unidentified phase in weld metal that reduces resistance to chloride pitting [107]. There is a possibility that niobium has a similar effect.

Thus, for melting of stabilized alloys by the more economical AOD or VOD processes, it is essential not only that the proper stabilizing element be selected to meet specific requirements for corrosion resistance, toughness, and weldability, but also that its concentration be precisely balanced. To prevent susceptibility to intergranular attack on weldments, a minimum amount of stabilizer must be added for rapid reaction during welding with the carbon and nitrogen remaining after the refining of the melt. The upper limit on its concentration is determined by the need to minimize reductions in the ductility of weldments, prevent susceptibility to hot-shortness in weld metal, and retard the formation of intermetallic phases that reduce toughness and resistance to corrosion. The deleterious effects of stabilizing elements on mechanical properties increase with the molybdenum content of the alloy. Addition of titanium and niobium to 29 Cr − 4 Mo alloys containing carbon and nitrogen in excess of the limits for these alloys reduces their ductility and resistance to stress corrosion cracking in the magnesium chloride test. These stabilizers can, therefore, not be used in these alloys to prevent susceptibility to intergranular attack [75]. However, when carbon and nitrogen are kept within the limits specified (C + N ≤ 250 ppm), niobium and aluminum may be added (0.1–03%) for grain refinement. High-purity compositions normally have very large grains which, when it is necessary to prevent an "orange-peel" appearance, can be reduced in size by these additions.

ACKNOWLEDGMENTS

Skilled assistance with experimental work on pitting and stress corrosion was provided by S. J. Kucharsey and A. J. Sweet and on acid corrosion by R. L. Colicchio of the Du Pont Experimental Station, Wilmington, DE.

REFERENCES

1. A. G. Larkin, Metallurgia, Apr. 1966, p. 165.
2. E. Elliott, Metallurgia, May 1971, p. 145.
3. C. A. Zapffe, Stainless Steels, The American Society for Metals, Cleveland, OH, 1949.
4. C. A. Zapffe, Iron Age. Oct. 14, 1948.
5. E. Maurer, Z. Elektrochem., **39**, 820 (1933).
6. W. Köster, J. Inst. Metals, **85**, 113 (1956).
7. E. Maurer, Korros. Metall., **15**, 225 (1939).
8. H. Krainer, Tech. Mitt. Krupp-Werksber., **20**(4), 165 (1962).
9. J. H. G. Monypenny, Stainless Iron and Steel. Chapman & Hall, London, 1951, p. 57.
10. R. Schaefer, Korros. Metall., **13**, 337 (1937).
11. T. H. Chilton, Strong Water, Nitric Acid: Sources, Methods of Manufacture and Uses, The MIT Press, Cambridge, MA, 1968, p. 89.

12. Anon. Metal Ind., Aug. 15, 1963, p. 225.

13. C. Fredenhagen, Z. Phys. Chem., **63**, 1 (1908).

14. M. A. Stretcher, Corrosion, **14**, 59t (1958).

15. M. A. Streicher, J. Electrochem, Soc., **106**, 161 (1959).

16. N. D. Tomashov, Theory of Corrosion and Protection of Metals, Macmillan New York, 1996. p. 346. (translated and edited by B. H. Tytell, 1. Gold, and H. S. Preiser).

17. U. R. Evans, Nature (Lond), **126**, 130 (1930).

18. P. Monnartz, Metallurgie, **8**, 161, 193 (1911).

19. J. H. G. Monypenny, Stainless Iron and Steel. Chapman & Hall, London, 1951, p. 288.

20. N. B. Pilling and D. E. Ackerman, Trans. AIMME, **83**, 248 (1929).

21. T. N. Rhodin, Corrosion, **11**, 465t (1956).

22. R. F. Steigerwald, Met. Trans., **5**, 2265 (1974).

23. H Gräfen, Werkstoffe Korros., **16**, 876 (1965).

24. M. G. Fontana and N. D. Greene, Corrosion Engineering, McGraw-Hill, New York, 1967. p. 166.

25. I. Class and H. Gräfen, Werkstoffe, Korros., **15**, 79 (1964).

26. L. R. Scharfstein, Corrosion, **21**, 254 (1965).

27. BMI-1375. Battelle Memorial Institute. Columbus, OH, Aug. 28, 1959.

28. Y. S. Kuznetsova and G. L. Shvarts, Metallov. term. Obrab. Metall., 1960, No. 8, p. 53. (Brutcher Trans. No. 5483.).

29. L. R. Scharfstein, Chem. Eng. Prog., **60**, 11 (1964).

29a. H. E. Deverell, U.S. Patent 4,007,038, Feb. 8, 1977.

29b. M. Liljas, et al., U. S. Patent 4,078,920, Mar. 14, 1978.

29c. T. H. McCunn et al., U. S. Patent 4,545,826, Oct. 8, 1985.

30. B. Strauss, H. Schottky, and J. Hinnüber, Z. Anong. Allgem. Chem., **188**, 309 (1930).

31. R. H. Aborn and E. C. Bain, Trans. Am. Soc. Steel Treating, **18**, 837 (1930).

32. J. J. Demo, Corrosion, **27**, 531 (1971).

33. M. A. Streicher, Corrosion, **29**, 337 (1973).

34. E. M. Mahla and N. A. Nielsen, Trans. ASM, **43**, 290–322 (1951).

35. S. J. Rosenberg and G. R. Irish, J. Res. NBS, **48**, 40 (1951).

36. W. A. Krivsky, Met. Trans., **4**, 1439 (1973).

37. E. Houdremont and P. Schafmeister, Arch. Eisenhüttenwesen, **7**, 187 (1933).

38. E. C. Bain, R. H. Aborn, and J. J. B. Rutherford, Trans. Am. Soc. Steel Treating, **21**, 481 (1933).

39. R. H. Kaltenhauser, Met. Eng. Q., May 1971, pp. 41–47.

40. J. R. Upp, F. H. Beck, and M. G. Fontana, Trans. ASM, **50**, 759 (1958).

41. L. Menezes, J. K. Roros, and T. A. Read, "Symposium on the Nature, Occurrence and Effects of Sigma Phase," ASTM STP No. 110, American Society for Testing and Materials, Philadelphia, PA, 1950, p. 71.

42. T. P. Hoar and K. W. J. Bowen, Trans. ASM, **45**, 443 (1953).

43. W. O. Binder and C. M. Brown, "Symposium on Evaluation Tests for Stainless Steels," in ASTM STP 93, American Society for Testing and Materials, Philadelphia, PA, 1930, pp. 146–182.

44. D. Warren, Corrosion, **15**, 213t: 22lt (1959).

45. D. C. Buck, J. J. Heger, F. J. Phillips, and B. R. Queneau, "Symposium on Evaluation Testing for Stainless Steels," in ASTM STP 93, American Society for Testing and Materials, Philadelphia, PA, 1930, pp. 56–86.

46. W. H. Hatfield, Discussion of paper by L. B. Pfeil and D. G. Jones, J. Iron Steel Inst., **127**, 380 (1933).

47. "Standard Recommended Practice A 262 for Detecting Susceptability to Intergranular Attack in Stainless Steels," in ASTM Book of Standards, Vol. 01.03, American Society for Testing and Materials, West Conshohoken, PA, 2002.

48. M. A. Streicher, Corrosion, **20**, 57t (1964).

49. H. J. Rocha in discussion of paper by E. Brauns and G. Pier, Stahl und Eisen, **75**, 579 (1955).

50. W. R. Huey, Trans. Am. Soc. Steel Treating, **18**, 1126 (1930).

51. M. H. Brown, W. B. DeLong, and W. R. Myers, "Sympoisum on Evaluation Tests for Stainless Steels," in ASTM STP 93, 1950, pp. 103–120.

52. W. B. DeLong, "Symposium on Evaluation Tests for Stainless Steels," ASTM STP 93, American Society for Testing and Materials, Philadelphia, PA, 1950, pp. 211–216.

53. M. M. Kurtepov and G. V. Akimov, Dok. Akad. Nauk. SSSR, **87**, 93 (1952).

54. A. B. McIntosh, Chem. Ind., **1957**, 687.

55. J. E. Truman, J. Appl. Chem., **4**, 273 (1954).

56. M. A. Streicher, ASTM Bull. No. 188. 1953, p. 35. Werkstoffe Korros, **5**, 363 (1954).

57. M. A. Streicher, ASTM Bull. No. 229, 1958, p. 77.

58. M. A. Streicher, "Advances in the Technology of Stainless Steels and Related Alloys," ASTM STP 369, American Society for Testing Materials, Philadelphia, PA, p. 255, 1965.

58a. "Standard Practices for Detecting Susceptibility to Intergranular Attack in Ferritic Stainless Steels," in Annual Book of Standards, Vol. 3.02, ASTM A 763, ASTM International, West Conshohocken, PA, 1999.

59. J. R. Auld in "Advances in the Technology of Stainless Steels and Related Alloys," ASTM STP 369, American Society for Testing and Materials, Philadelphia, PA, 1965, pp. 183–199.

60. Welding Research Council Bulletin No. 93, 1964, New York.

61. H. A. Smith, Met. Prog., **33**, 596 (1938).

62. G. Riedrich, Metallwirtschaft, **21**, 407 (1942).

63. W. Tofaute and H. Schottky, Arch. Eisenhuttenwesen, **14**, 71 (1940).

64. H. H. Uhlig, Trans. ASM, **30**, 947 (1942).

65. M. A. Streicher, J. Electrochem. Soc., **103**, 375 (1956).

65a. E. G. Webb and R. C. Alkire, J. Electrochem. Soc., **149**, B272 (2002).

65b. D. E. Williams and Y. Y. Zhu, J. Electrochem. Soc., **147**, 1763 (2000).

65c. M. A. Streicher, Mater. Perform., **36**, 65 (1997).

65d. N Sridhar and G. A. Cragnolino, Corrosion, **49**, 885 (1993).

65e. R. J. Brigham and E. W. Tozer, Corrosion, **29**, 33 (1973).

65f. R. J. Brigham, Corrosion, **30**(11), 396–398 (Nov. 1974).

65g. "Standard Test Method for Pitting and Crevice Corrosion Resistance of Stainless Steels and Related Alloys by Use of Ferric Chloride Solution," in Annual Book of Standards, Vol. 3.02, ASTM G 48, ASTM International, West Conshohocken, PA, 2004.

65h. P. E. Arnvig and A. D. Bisgård, Paper No. 96437, Corrosion 96, NACE, Houston, TX, 1996.

65i. "Standard Test Method for Electrochemical Critical Pitting Temperature Testing of Stainless Steels," in Annual Book of Standards, Vol. 3.02, ASTM G 150, ASTM International, West Conshohocken, PA, 1999.

65j. K. Lorentz and G. Medawar, Thyssenforschung, **1**(3), 97–108, (1969).

65k. D. Bauernfeind and G. Mori, Paper No. 03257, Corrosion 2003, NACE, Houston, TX, 2003.

65l. C. W. Kovach and J. D. Redmond, NACE Corrosion 93 paper 93267, 1993.

66. J. C. Hodge and J. L. Miller, Trans. ASM, **28**, 28 (1940).

67. H. J. Rocha, Tech, Mitt. Krupp, Jan. 1942, pp. 1–14.

68. M. A. Scheil, Symposium on Stress Corrosion of Metals, published jointly by ASTM and AIMME, New York, NY. 1945, pp. 395–410.

69. R. Franks, W. O. Binder, and C. M. Brown in Symposium on Stress Corrosion of Metals, published jointly by ASTM and AIMME, New York, NY. 1945, pp. 411–420.

70. R. W. Staehle, Proceedings of Conference on Fundamental Aspects of Stress Corrosion Cracking, NACE, Houston, TX, 1969, p. 1.

71. H. R. Copson in Physical Metallurgy of Stress Corrosion Fracture, Interscience, New York, 1959, p. 247.

71a. M. A. Streicher, Mater. Perform., **36**, 63 (1997).

72. A. P. Bond and H. J. Dundas, Corrosion, **24**, 344 (1968).

73. W. G. Renshaw, Report to the Welding Research Council, Subcommittee on High Alloy Weldments, New York, June 1966.

74. M. A. Streicher, Preprint No. 68, "Stress Corrosion of Ferritic Stainless Steels," NACE, Toronto, Canada, 1975.

75. M. A. Streicher, Corrosion. **30**, 77 (1974).

76. A. W. Dana, ASTM Bull. No. 226, p. 196 (1957).

77. A. W. Dana and W. B. DeLong, Corrosion, **12**, 309t (1956).

78. D. Warren, Proceedings of the Fifteenth Annual Purdue Industrial Waste Conference, Purdue, IN, 1960.

78a. "Standard Test Method for Evaluating Stress Corrosion Cracking of Stainless Alloys with Different Nickel Content in Boiling Acidified Sodium Chloride Solution," ASTM G 123, in Annual Book of Standards, Vol. 03.02, West Conshohocken, PA, 2005.

79. I. B. Casale, Corrosion, **23**, 314 (1967).

80. M. A. Streicher and A. J. Sweet, Corrosion, **25**, 1 (1969).

81. "Standard Recommended Practice for Performing Stress Corrosion Cracking Tests in a Boiling Magnesium Chloride Solution." ASTM G 36, in ASTM Book of Standards, Vol. 03.02, American Society for Testing and Materials, West Conshohocken, PA, 2006.

81a. Y.-L. Chiang and M. A. Streicher, Corrosion, **54**, 740 (1998).

81b. M. A. Streicher, "Effect of Condenser Design on the Corrosion of Stainless Steels in Boiling Nitric Acid," Paper No. 87, CORROSION/80, NACE, Houston, TX, 1980.

81c. M. A. Streicher, "Effect of Condenser Design on Corrosion in Boiling Sulfuric, Acetic and Hydrochloric Acids," Pager No. 86, CORROSION/80, NACE, Houston, TX, 1980.

82. R. H. Espy, "How Composition and Welding Conditions Affect Corrosion Resistance of Type 430 Stainless Steel, Preprint No. 22, NACE, Cleveland, OH, 1968.

83. W. O. Binder and H. R. Spendelow, Trans. ASM, **43**, 759 (1951).

84. M. Semchyshen, A. P. Bond, and H. J. Dundas, "Effects of Composition on Ductility and Toughness of Ferritic Stainless Steels," in Symposium: Toward Toughness and Ductility, Climax Molybdenum Co., Kyoto, Japan, 1971, pp. 239–255.

85. R. F. Steigerwald, A. P. Bond, H. J. Dundas, and E. A. Lizlovs, Corrosion, **33**, 279 (1977).

86. R. M. Davison, Met. Trans., **5**, 2287 (1974).

87. C. D. Schwartz, I. A. Franson, and R. J. Hodges, Chem. Eng., **77**, 164 (1970).

88. R. J. Knoth, G. E. Lasko, and W. A. Matejka, Chem. Eng., **77**, 170 (1970).

89. F. K. Kies, I. A. Franson, and B. Coad, Chem. Eng., **77**, 150 (1970).

90. R. J. Hodges, C. D. Schwartz, and E. Gregory, Br. Corros. J., **7**, 69 (1972).

91. I. A. Franson, Met. Trans., **5**, 2257 (1974).

92. E. Gregory, F. K. Kies, and I. A. Franson, U.S. Patent 3,807,991, Apr. 30, 1974.

93. R. N. Wright, Welding J., **50**, 434S (1971).

94. J. J. Demo, Met. Trans., **5**, 2253 (1974).

95. K. E. Pinnow, J. P. Bressanelli, and A. Moskowitz, Metals Eng. Q, **15**, 32 (Aug. 1975).

96. B. Pollard, Welding J., Welding Res. Supplement, **51**, 222s (1972).

97. R. Oppenheim and G. Lennartz, Chem. Ind., **23**, 705 (1971).

98. R. Oppenheim and H. Laddach, DEW Tech. Ber., **11**, 71 (1971).

99. H. Brandis, H. Kiesheyer, W. Küppers, and R. Oppenheim, TEW Tech. Ber., **2**, 3 (1976).

100. W. Tofaute and H. J. Rocha, Techn. Mitt. Krupp, **12**, 67 (1954).

101. H. Brandis, H. Kiesheyer, and G. Lennartz, TEW Tech. Ber., **2**, 14 (1976).

102. G. Lennartz and H. Kiesheyer, DEW Tech. Ber., **11**, 230 (1971).

102a. M. A. Streicher, Corrosion, **30**, 77 (1974).

103. M. A. Streicher, Corrosion, **30**, 115 (1974).

104. M. A. Streicher, U.S. Patents 3,932,175, 3,929,473, and 3,923, 174 (1975 and 1976).

104a. K. E. Pinnow and J. P. Bressanelli, U.S. Patent 3,957,544, May, 18, 1976.

104b. T. J. Nichol and T. H. McCunn, U.S. Patent 4,456,482, June 26, 1984.

104c. B. Hinden and A. K. Agrawal, In "The Use of Synthetic Environments for Corrosion Testing," ASTM STP 970, American Society for Testing and Materials, Philadelphia, PA, 1988, pp. 274–286.

105. N. A. Long and N. Rice, private communication, Technical Memorandum on Pitting of Stainless Steel, Allegheny Ludlum Steel Co., Wallingford, CT, 1964.

106. H. Okada, Y. Hosoi, and H. Ogawa, J. Iron Steel Inst. (Jpn.), **59**, 155 (1973).

107. A. P. Bond, H. J. Dundas, E. A. Lizlovs, G. Gemmell, and B. Solly, "Corrosion Resistance of Nickel-Bearing Ferritic Stainless Steels," Private communication.

107a. M. A. Streicher, Mater. Perform., **22**, 37 (1983).

107b. J. R. Kearns, M. J. Johnson, and J. F. Grubb, "Accelerated Corrosion in Dissimilar Metal Crevices," Paper No. 228, COROSSION/86, NACE, Houston, TX, 1986.

107c. Y.-H. Yau and M. A. Streicher, Corrosion, **47**, 352 (1991).

108. M. A. Streicher, Corrosion, **32**, 79 (1976).

109. E. Houdreraont and W. Tofaute, Stahl Eisen., **72**, 539 (1952).

110. T. J. Nichol and J. A. Davis, "Intergranular Corrosion Testing and Sensitization of Two High Chromium Ferritic Stainless Steels," in ASTM STP 656, Intergranular Corrosion of Stainless Alloys, R. F. Steigerwald (Ed.), American Society for Testing and Materials, Philadelphia, PA, 1978, pp. 179–196.

111. M. A. Streicher, "Intergranular Corrosion," in Corrosion Tests and Standards—Application and Interpretation, ASTM Manual 20, R. Baboian (Ed.), American Society for Testing and Materials, Philadelphia, PA, 1995, pp. 197–217.

112. A. J. Sweet, "Detection of Susceptibility of Alloy 26–1S to Inergranular Attack," in ASTM STP 656, Intergranular Corrosion of Stainless Alloys, R. F. Steigerwald (Ed.), American Society for Testing and Materials, Philadelphia, PA, 1978, pp. 197–232.

113. A. P. Bond and E. A. Lizlovs, J. Electrochem. Soc., **116**, 1305 (1969).

114. H. J. Dundas and A. P. Bond, "Niobium and Titanium Requirements for Stabilization of Ferritic Stainless Steels," in ASTM STP 656, Intergranular Corrosion of Stainless Alloys, R. F. Steigerwals (Ed.), American Society for Testing and Materials, Philadelphia, PA, 1978, pp. 154–178.

115. L. Troselius, I. Andersson, T. Andersson, S. O. Bernhardsson, J. Degerbeck, S. Henrikson, and A. Karlsson, Br. Corros. J., **10**, 174 (1975).

116. H. Kiesheyer, G. Lennartz, and H. Brandis, Werkstoffe Korros., **27**, 416 (1976).

117. L. Troselius, Corros. Sci., **11**, 473 (1971).

117a. J. F. Grubb, "Stabilization of High-Chromium, Ferritic Stainless Steels," in Procedings of International Conference on Stainless Steels (ISIJ), Chiba, Japan, June 10–13, 1991.

118. T. P. Oettinger and M. G. Fontana, Mater. Perform, **15**(1), 29 (1976).

119. T. J. Nichol, Met. Trans., **8A**, 229 (1977).

120. H. Brandis, H. Kiesheyer, and G. Lennartz, Arch. Eisenhüttenwesen, **46**, 799 (1975).

121. J. M. Sawhill and A. P. Bond, Welding J., Feb. 1976.

52

DUPLEX STAINLESS STEELS

M.-L. Falkland, M. Glaes, and M. Liljas

Outokumpu Stainless AB, Avesta, Sweden

A. DEFINITION OF DUPLEX STAINLESS STEEL

Duplex is a common definition for many dual-phase material systems. Within the stainless steel nomenclature, it can be strictly the definition for combinations of the structures, ferrite, austenite, and martensite. However, in common use duplex stainless steels are defined as steels with an austenitic–ferritic crystal structure, with at least 25 or 30% of the lesser phase. A balance of austenite- and ferrite-stabilizing alloying elements achieves the mixed structure.

B. HISTORY

Bain and Griffiths [1] presented the first duplex stainless steel phase diagrams in 1927. They showed austenitic–ferritic alloys with 22–30% Cr and 1.2–9.7% Ni. No property data were given in this article. The first commercial austenitic–ferritic duplex stainless steels were developed around 1930 [2]. The steels contained essentially 25% Cr, 5% Ni, and 1.5% Mo. Molybdenum was added to improve the corrosion resistance. As with all stainless steels at this time, the carbon content was high. Nevertheless, the duplex grades showed an improved resistance to intergranular corrosion and had equal or better resistance to uniform corrosion than the austenitic grades. The weldability and impact strength

Uhlig's Corrosion Handbook, Third Edition, Edited by R. Winston Revie
Copyright © 2011 John Wiley & Sons, Inc.

were adversely affected by the high carbon content. Even though there were several applications where the duplex grades were successfully used during the first decades of their existence, the real increase in usage came after the development and the introduction of the 22% chromium duplex steel in the late 1970s [3]. This duplex grade, UNS S31803, and more lately modified and designated as UNS S32205, had a better corrosion resistance than 316/3l6L and 317/3l7L. Its useful properties and relatively low cost have made this grade very well known, used, and produced worldwide. In recent years the number of duplex grades has increased rapidly and in the American Society for Testing and Materials (ASTM) A240 standard 20 different grades are listed.

C. METALLOGRAPHY

The chemical composition of the duplex grades is designed to give good corrosion resistance and an adequate phase balance to obtain the desired mechanical and physical properties. The corrosion resistance is, as for all stainless steels, a result of the content of chromium, molybdenum, and nitrogen and also, to some extent, copper and tungsten. Nickel is added in a sufficient amount to give the austenite–ferrite balance, although nickel also has a beneficial effect upon the corrosion resistance in reducing acids, such as dilute sulfuric acid. As a consequence of high nickel price this element is partly replaced by manganese and nitrogen in some newer alloys.

In the solution-annealed condition duplex steels have phase balance of approximately equal amounts of ferrite and austenite. This is achieved by the addition of austenite-stabilizing elements, mainly nickel, manganese, and nitrogen, and by ferrite-stabilizing elements, mainly chromium and molybdenum. The duplex microstructure can be

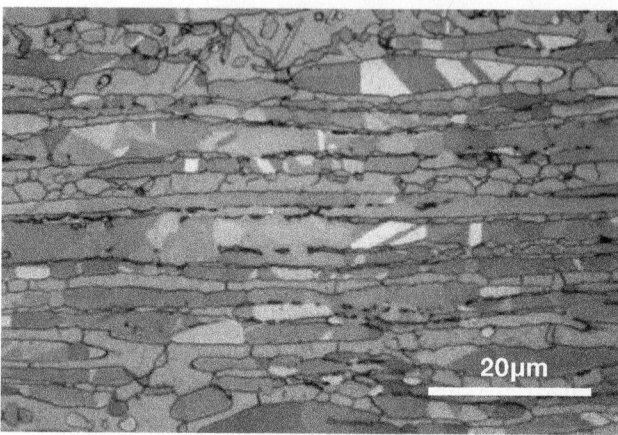

FIGURE 52.1. Typical microstructure of wrought duplex stainless steel. Hot-rolled plate.

described as a continuous ferritic matrix with austenitic islands. Figure 52.1 shows a typical duplex microstructure of a hot-rolled plate material.

If exposed to temperatures in the range of 300–900°C duplex steels may undergo various phase transformations causing changes in properties such as embrittlement and loss in corrosion resistance.

C0.1. 475°C Embrittlement. Duplex steels are known to be susceptible to embrittlement after long-time aging at intermediate temperatures, 325–500°C. This is related to the spinodal decomposition that occurs in the ferrite phase. The transformation is effected by alloy composition and the time of exposure in the critical temperature range (see Fig. 52.2). Because of this risk of embrittlement, duplex grades are not recommended in equipment with design temperatures above approximately 300°C.

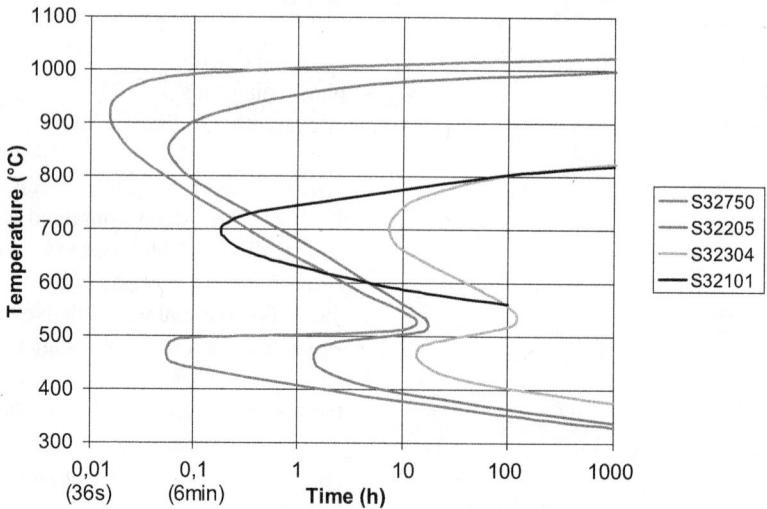

FIGURE 52.2. Curves for reduction of impact toughness to 50% of starting value.

C0.2. Phase Precipitation. At a higher temperature range (600–900°C), duplex stainless steels are susceptible to precipitation of various phases such as carbides, nitrides, and intermetallic phases. Sigma phase is an intermetallic phase that has been of most concern since it is a brittle phase, with high contents of chromium and molybdenum, which adversely affects the mechanical properties as well as the corrosion resistance. The higher alloyed duplex grades are more susceptible to this kind of precipitation (see Fig. 52.2). Compared to spinodal decomposition, the influence on the properties of sigma-phase precipitation is more severe and more rapid. This must be considered on cooling after processing, heat treatment, or welding.

An overview of the metallurgy of the duplex stainless steels has been presented in [4].

D. STEEL GRADES WITHIN THE DUPLEX FAMILY

There are several different stainless steel grades within the duplex family. Many of them are relatively newly developed and therefore also protected by patents and produced by only a few companies. However, the most common grades can be divided into three major groups: lean duplex, duplex, and superduplex. However, there is no agreed-upon clear definition of the individual groups. Recently, a fourth group, hyperduplex, has been introduced. Typical compositions of some duplex stainless steels are listed in Table 52.1.

The lean duplex stainless steels show corrosion resistance on a level with type 304 or 316. UNS32101 is a lean duplex steel that can replace type 304 in many environments, but it can also be an alternative to structural steels. UNS S32304, with no intentional molybdenum addition but higher

TABLE 52.1. Typical Compositions of Some Duplex Stainless Steels

UNS No.[a]	EN No.[b]	Cr	Ni	Mo	N	Other
Lean Duplex						
S32101	1.4162	21.5	1.5	0.3	0.22	Mn 5
S32304	1.4362	23	4.5	0.3	0.1	
Standard Duplex						
S31803	1.4462	22	5.5	3.0	0.15	
S32205	1.4462	22.5	5.5	3.2	0.17	
S32550	1.4507	25.5	6.5	3.1	0.18	Cu 1.6
Superduplex						
S32760	1.4501	25	6.5	3.5	0.25	Cu 0.7, W 0.7
S32750	1.4410	25	7	4.0	0.28	
S32906	—	29	6	2,0	0,35	

[a]Unified Numbering.
[b]European Norm.

chromium content, is an example of a lean duplex grade that favorably can replace type 316.

By far, the most commonly used standard duplex stainless steel is UNS S31803/EN 1.4462 with typically 22% Cr, 5.5% Ni, 3% Mo, and 0.15% N. This grade, developed in the 1970s, is today produced by a number of producers and in several product forms throughout the world. As the benefits of high nitrogen levels have become recognized, UNS S32205, a variation of the 22% Cr duplex, but with slightly higher Cr, Mo, and N levels, has gained popularity. This specification established minimums of 22.0% Cr, 3.0% Mo, and 0.14% N to offset the common tendency of commercial stainless steels to drift toward the low end of their specified composition ranges. Although the 22% Cr duplex is the dominating standard grade, UNS32550, an example of a 25% Cr duplex, also belongs to this group.

Superduplex steels, with high amounts of chromium, molybdenum, and nitrogen, are very corrosion resistant. Several similar but slightly different grades of this group are available. Hyperduplex steels are not listed in the table as they are quite new and to date have limited use.

E. MECHANICAL AND PHYSICAL PROPERTIES

The mechanical and physical properties are the major difference between the duplex stainless steels and some of the more commonly used stainless steels. The mechanical strength is high compared to austenitic or ferritic stainless steels, and the physical properties are different than for the austenitic stainless grades.

E1. Strength at Room Temperature

Relatively high room temperature strength is one of the most typical properties of duplex steels. The duplex microstructure and the addition of nitrogen give the duplex grades very high mechanical strength (see Table 52.2). Higher alloyed grades, particularly with high nitrogen content, are the strongest. The proof strength can be used to advantage for structural applications. Some examples of thinner and lighter construction are given in Section H.

E2. Strength at Elevated Temperatures

The high strength of the duplex grades is maintained up to ~400°C, but at higher temperatures they soften rapidly. In a comparison of the typical proof strength of a duplex grade, an austenitic stainless grade and a carbon–manganese structural grade, the duplex is the strongest up to about 500°C (see Fig. 52.3). Due to the risk of embrittlement, the duplex stainless steels should not be used in applications above about 300°C. The maximum value depends on the grade and the design rules being used. Therefore, design values, listed in Table 52.3, are available up to only 250°C. However,

TABLE 52.2.   Mechanical Properties According to ASTM A 240 and EN 10088 (MPa)

UNS No.	EN No.	Proof $Rp_{0,2}$, (min)		Tensile Rm, (min)		Elongation A_5, (min)		Hardness HB (max)
		ASTM	EN	ASTM	EN	ASTM	EN	
S32101	1.4162	450	450	650	650	30		
S32304	1.4362	400	400	600	630	25	25	290
S31803	1.4462	450	460	620	640	25	25	293
S32550	1.4507	550	490	760	690	15	25	302
S32760	1.4501	550	530	750	730	25	25	270
S32750	1.4410	550	530	795	730	15	20	310
S32906	1.4477	550	550	750	750	25	20	310

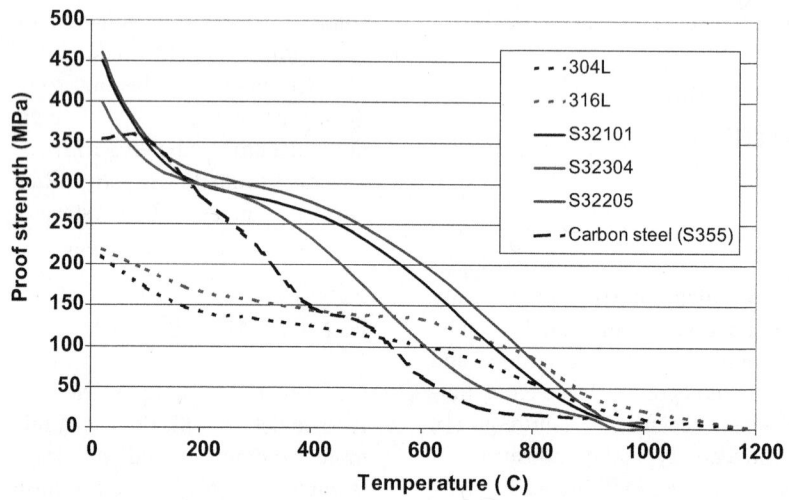

FIGURE 52.3. Proof strengths at elevated temperatures for some·steels.

during short-time exposures, the strength at higher temperature can be of interest. One example is fire resistance in constructions.

E3. Mechanical Properties at Subzero Temperatures

As with most materials the strength of duplex steels increases with lower temperature. Contrary to austenitic stainless steels, duplex steels show a ductile-to-brittle transition behavior. As a rule duplex welds are less tough than the base material.

TABLE 52.3. Design Values at Elevated Temperatures According to EN (MPa)

UNS No.	EN No.	$Rp_{0,2}$ (min)			
		100°C	150°C	200°C	250°C
S32304	1.4362	330	300	280	265
S31803	1.4462	360	335	315	300
S32550	1.4507	450	420	400	380
S32760	1.4501	450	420	400	380
S32750	1.4410	450	420	400	380

However, the toughness is usually sufficient for most applications, except at cryogenic conditions. As an example the European Norm for flat products for pressure purposes, EN 10028-7, allows all listed duplex grades to be used down to − 40°C provided the Charpy impact toughness at this temperature is at least 40 J. The American Society of Mechanical Engineers (ASME) requires Charpy impact tests at the minimum design metal temperature with acceptance criteria expressed as lateral expansion. The material data sheets for duplex steels in the Norsk Sokkels konkuranseposisjon (NORSOK) standard require an impact toughness of 45 J at − 46°C. Fracture mechanics testing of duplex base and weld metals indicate a ductile behavior down to about − 100°C.

E4. Mechanical Properties under Cyclic Load

The high tensile strength of duplex steels also implies high fatigue strength. In many applications, the fatigue strength is of the utmost importance for the life of a structure. The fatigue strength in combination with good corrosion resistance gives good resistance to corrosion fatigue. Table 52.4 shows the result of pulsating tensile fatigue tests

TABLE 52.4. Fatigue Data for Different Stainless Steels in Pulsating Tensile Tests (MPa)

UNS No.	EN No.	$Rp_{0.2}$	Rm	Fatigue Strength
S32101	1.4162	478	696	500
S32304	1.4362	446	689	450
S31803	1.4462	497	767	510
S32750	1.4510	565	802	550
S31603	1.4404	280	578	278

($R = \sigma_{min}/\sigma_{max} = 0.1$) in air at room temperature. The fatigue strength has been evaluated at 2 million cycles and a 50% probability of rupture. The test was made using round polished bars. As shown by the table, the fatigue strength of the duplex steels corresponds approximately to the proof strength of the material. As with all materials the influence of notches has to be considered, and for the design of welded joints the same rules apply as for other types of steels.

E5. Physical Properties

Physical properties depend more on the microstructure than on the content of each alloying element. Thus, the physical properties are similar for all duplex grades. The physical properties of duplex stainless steel are closer to those of ferritic steels than to those of austenitic stainless steels. Compared with austenitic stainless steels, duplex grades have a lower coefficient of thermal expansion and higher thermal conductivity. The physical properties are summarized in Table 52.5.

F. CORROSION RESISTANCE

The corrosion of stainless steel can be described as a two-step process: initiation and propagation. Resistance to the initiation, which is the breakdown of the passive film, depends mainly on the content of chromium and molybdenum. It is the resistance to the initiation of the corrosion that usually determines if a stainless steel is resistant or not. Surface analyses indicate low levels of nickel in the passive layer, but

TABLE 52.5. Physical Properties of Duplex Stainless Steels According to EN 10088

Property	20°C	100°C	200°C	300°C	Units
Density	7.8	—	—	—	kg/dm^3
Modulus of elasticity	200	194	186	180	kN/mm^2
Mean coefficient of thermal expansion between 20°C and T		13.0	13.5	14.0	10^{-6}/°C
Thermal conductivity	15	—	—	—	W/mK
Specific thermal capacity	500	—	—	—	J/kgK
Electrical resistivity	0.8	—	—	—	$\Omega\cdot$mm^2/m

this element appears to be able to improve the corrosion resistance by reducing the active corrosion rate when exposed to media causing uniform corrosion. However, in general, duplex steels show very good corrosion resistance in many acids.

Owing to their high chromium content corrosion resistance of the duplex grades is generally very good. Also, the leanest alloyed duplex grades show a corrosion resistance in the same range as that of the standard austenitic grades. The resistance is especially good in environments where the high levels of chromium and nitrogen are beneficial. Examples of such environments are halide-containing media, oxidizing acids, and hot alkaline solutions.

F1. Resistance in Acids

The type of corrosion normally encountered for stainless steels in acids is uniform corrosion. The resistance to this uniform attack can be described in isocorrosion diagrams. Stainless steel producers often publish isocorrosion diagrams in data sheets or in handbooks [5]. Diagrams comparing duplex and austenitic stainless steels are shown below. Further information can be found in the stainless steel producers' publications.

F1.1. Sulfuric Acid. In pure dilute sulfuric acid, the chromium content is very important. Tests have shown that the best corrosion resistance is demonstrated by grades containing 23–25% chromium—usually duplex grades. At higher acid concentrations increased levels of nickel and copper are beneficial. Accordingly, austenitic grades usually show better performance in the medium- and high-acid concentrations (see Fig. 52.4).

F1.2. Phosphoric Acid. Pure phosphoric acid is very different than the more common wet process phosphoric acid, WPA. The WPAs containing impurities such as chlorides, fluorides, iron, and aluminum are usually the most aggressive. S31803 has shown to be an excellent grade in many of these complicated acids.

Duplex steels are often used in marine chemical tankers, and one of the reasons is the good resistance in WPAs with varying levels of contamination. Figure 52.5 shows the resistance of some duplex and some high-alloy austenitic stainless grades in some typical WPAs.

F1.3. Nitric Acid. Nitric acid is strongly oxidizing at all concentrations. The high chromium content of S32304 makes this steel as resistant as type 304. Duplex grades with molybdenum are less useful in nitric acid

F1.4. Organic Acids. In general, duplex grades have good resistance in organic acids. The isocorrosion diagram in Figure 52.6 shows the resistance in formic acid.

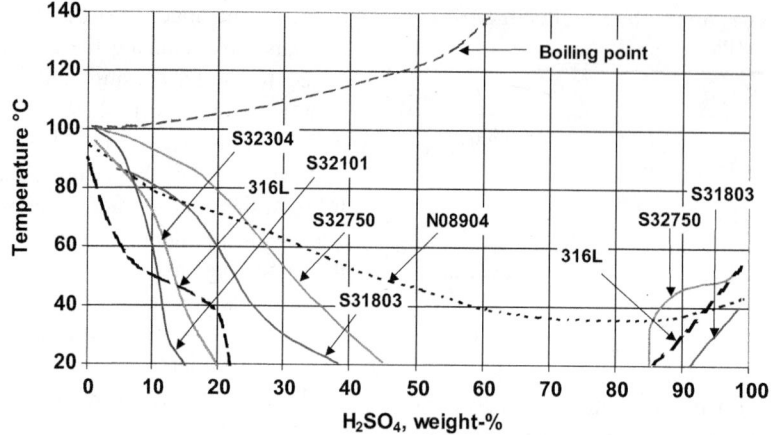

FIGURE 52.4. Isocorrosion curves in sulfuric acid for some steels. Lines represent a corrosion rate of 0.1 mm/year.

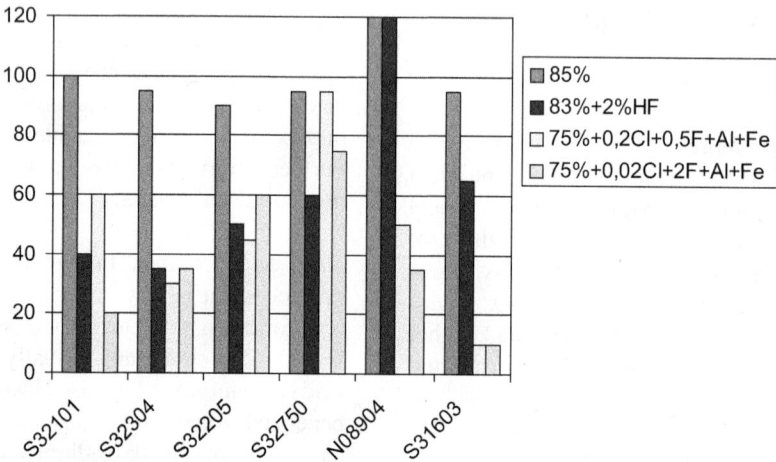

FIGURE 52.5. Critical temperature (°C) causing 0.127 mm/year uniform corrosion in wet process phosphoric acids of different compositions.

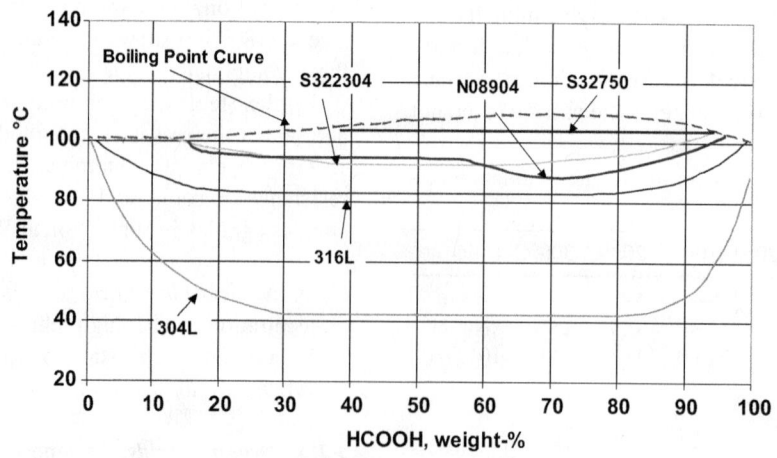

FIGURE 52.6. Isocorrosion curves in formic acid for some steels. Lines represent a corrosion rate of 0.1 mm/year.

TABLE 52.6. Lowest Temperature (°C) at Which Corrosion Rate Exceeds 0.127 mm/year (5 mpy)

Solution	S32101	S32304	S31803/S32205	S32750	S31603	N08904	S31254
1% Hydrochloric acid (HCl)	55	55	85	>100	30	50	70
1% HCl + 0.3% FeCh	20	20	30–40	95	25	50–60	60–95
10% Sulfuric acid (H_2SO_4)	75	65	60	75	50	60	60
60% H_2SO_4 + Nitrogen (N_2) bubbling	≪30	<15	<15	<15	<15	85	40
96.4% H_2SO_4	30	15	25	30	45	35	20
85% Phosphoric acid (H_3PO_4)	100	95	90	95	95	120	110
83% H_3PO_4 + 2% Hydrofluoric acid (HF)	40	35	50	60	65	120	90
80% Acetic acid (CH_3COOH)	>106	>106	>106	>106	>106	>106	>106
50% CH_3COOH + 50% Acetic anhydride [(CH_3CO)20]	105	90	100	110	120	>126	>126
50% Formic acid (HCOOH)	95	80	90	90	40	100	100
50% Sodium hydroxide (NaOH)	85	95	90	110	90	140	115

F2. Resistance in Alkaline Solutions

Alkaline solutions at moderate temperatures are not aggressive to stainless steels in general. Hot alkaline solutions can, however, cause uniform corrosion. The duplex grades, because of their high chromium content, have relatively good resistance in alkaline solutions.

F3. MTI Test Program

The resistance to uniform corrosion can also be evaluated using the method and environments prescribed by the Materials Technology Institute of Chemical Process Industries (MTI). The corrosion testing is carried out as weight loss measurements over a test period of 96 h. The tests are performed until the critical temperature, which is defined as the lowest temperature at which the corrosion rate exceeds 0.127 mm/year (5 mpy), is determined. The interval between test temperatures is 5°C. The critical temperatures for some duplex grades and some austenitic grades are shown in Table 52.6.

F4. Resistance in Halide-Containing Solutions

The most common halide-containing solution is the very common medium water. Water contains chlorides that can destroy passivity of the stainless steel locally and cause pitting corrosion or crevice corrosion. The usually very high chromium content of the duplex steels and the molybdenum- and nitrogen-alloying additions provide good resistance in these media. A common tool for ranking resistance to initiation of pitting is the pitting resistance equivalent (PRE) index. There are several different empirical formulas*; one of the most commonly used for duplex stainless steels was defined by Truman [6]: PRE = %Cr + 3.3×%Mo + 16×%N. The PRE levels for some duplex grades are listed in Table 52.7.

However, this formula ranks the resistance based on the bulk composition of the steel. The surface condition, internal cleanliness, heat treatment, and other factors strongly

TABLE 52.7. Pitting Resistance Equivalents for Some Duplex Steels

Grade	PRE
Lean Duplex PRE < 30	
S32101	25
S32304	26
Duplex PRE = 30–40	
S31803	34
S32205	36
S32550	39
Superduplex PRE > 40	
S32750	42
S32760	42
S32906	40

influence corrosion resistance of stainless steels. A more precise approach is to measure experimentally the resistance of a particular sample in halide-containing media to determine the critical pitting temperature (CPT) or the critical crevice corrosion temperature (CCT) in a defined environment.

Table 52.8 shows some typical intervals of CPT and CCT in the common testing solution 6% FeCl$_3$ (ASTM G48) and in 1 M NaCl (ASTM G 150). The corrosion resistance in seawater of duplex and in particular super duplex stainless steels has been described elsewhere [7].

*The PRE is a relationship derived by statistical regression of the critical pitting temperature as a function of the composition of balanced, fully annealed stainless steels. The Cr, Mo, and N are not truly independent variables and are not mutually substitutable. The PRE may be used to rank performance of grades, but differences of 2 or less are unlikely to be significant.

TABLE 52.8.   Typical Critical Pitting and Crevice Corrosion Temperatures

UNS	EN	PRE	CPT, G48E	CCT. G48F	CPT, G150
S31603	1.4436	26	15	<0	25
S32101	1.4162	25	15	<0	17
S32304	1.4362	26	15	<0	25
S31803/32205	1.4462	34–36	30	20	55
S32550		39	55		
S32750	1.4410	42	65	35	>85
S32760	1.4501	42	65	35	>85

Note: Tests made on ground surfaces (120 grit).

F5.   Corrosion under Mechanical Loading

The environment and the mechanical stresses on or within a stainless steel can cooperate to cause failures in situations where there would not have been any problems in air. The typical corrosion types are stress corrosion cracking (SCC) and corrosion fatigue.

F5.1.   Stress Corrosion Cracking.   Stress corrosion cracking can occur under the simultaneous influence of a critical environment and tensile stresses in the material. The tensile stresses can be residual stresses from the fabrication of a component, that is, welding, forming, or stresses from the structure. There are two typical environments where this might happen. The most common is warm chloride-containing media and the other is in hot alkaline solutions. The duplex stainless steels have good resistance to SCC. Compared to the corresponding austenitic grades, they can be used in a wide range of applications with minimal risk.

Different methods are used to measure and rank the resistance of stainless steel grades to chloride-induced SCC. The ranking depends on the method used. The resistance can be evaluated by laboratory corrosion tests, for example, the drop evaporation test (DET). In this test, a uniaxially stressed specimen is electrically heated to 200°C and then exposed to dripping, dilute sodium chloride solution. The dripping rate is adjusted to let one drop evaporate before the next drop hits the sample. The specimen temperature is ~100°C due to the cooling effect from the sodium chloride solution. The time to failure is measured at different stress levels related to the actual proof strength at 200°C. The threshold values, listed as the stress level that leads to failure after 500 h of testing, are determined. Figure 52.7 shows the threshold values for some grades. The duplex grades show a far better performance than the austenitic steels with similar pitting and crevice corrosion resistance although austenitic grades with high nickel content show better resistance. However, as the duplex steels have higher proof stress, they are also tested at a higher absolute stress level.

F5.2.   Corrosion Fatigue.   The combination of high fatigue strength and, in general, good corrosion resistance results in very good resistance to corrosion fatigue. The resistance to corrosion fatigue can be measured by the same tests as for pure fatigue strength (see Table 52.4), but with the sample exposed to a corrosive medium. The fatigue strength measured in a corrosive medium will generally be lower than that measured in air. A lower test frequency will increase the effect of the corrosive medium and reduce the effect of the mechanical properties.

G.   FABRICATION

There are many similarities in the fabrication of austenitic and duplex stainless steels, but there are also important differences. The duplex microstructure and the high strength of the duplex grades require some changes in fabrication

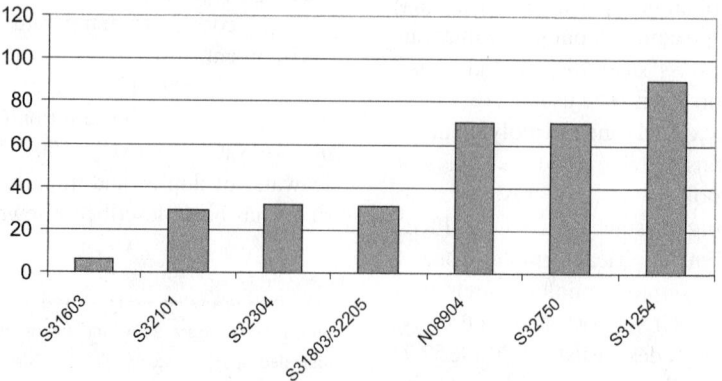

FIGURE 52.7.   Results of stress corrosion cracking tests. Threshold values for the relative stress level at 200°C causing cracking after 500 h.

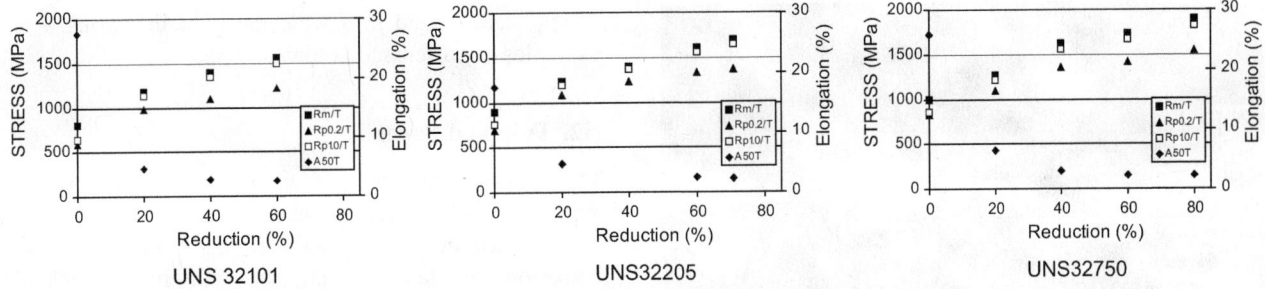

FIGURE 52.8. Effect of cold work on mechanical properties for three duplex steels.

practice. A more detailed discussion of the fabrication of duplex steels is presented in [8].

G1. Forming

The formability of the duplex stainless steels depends on the mechanical property profile. Duplex stainless steels can be cold formed by the same methods as austenitic stainless steels. The high strength implies a high deformation resistance and a greater initial force is required before the steel starts to yield. When this initial force is reached, the material yields without a yield point. Duplex steels strengthen or work harden due to deformation almost as rapidly as the austenitic grades. In Figure 52.8 the work hardening of three duplex grades is illustrated.

The presence of ferrite in the duplex structure signifies less elongation in comparison with standard austenitic stainless steels. As a consequence, forming operations, such as deep drawing, stretch forming, and cold spinning, are more difficult, in the sense of higher strength and the need for more frequent intervening anneals, to perform with duplex steels than with austenitic steels.

Due to the high yield stress of annealed material leading to large elastic deformation of the duplex grades, there is an effect of springback after forming operations such as bending.

The strain and the level of cold deformation due to bending depend on the combination of bending radius and sheet thickness (R/t). Compared to austenitic steels this factor is greater for duplex steels, as a rule 2 for duplex and 1 for standard austenitic steels.

G2. Machining

In general, duplex steels are somewhat more difficult to machine than the conventional austenitic grades, such as type 316. However, the lean duplex grade S32101 shows excellent machinability. The machinability of duplex grades in relation to other stainless steels can be described by a machinability index, as in Figure 52.9. Duplex grades have a different property profile than highly alloyed austenitic stainless steels. The relative reduction in machinability for

the duplex stainless steels is greater for cemented carbide tools than it is for high-speed steel tools, especially for the duplex grades with higher molybdenum and nitrogen contents. The machining index is the result of combining data from several operations. It cannot be strictly applied to a particular application without considering variations in machinery, tooling, lubrication, and operations.

G3. Welding

Duplex steels are readily weldable and can be joined with most welding methods for stainless steels. However, for optimum properties the welding parameters have to be modified compared to those used for austenitic steels. Duplex steels are designed to have approximately equal amounts of ferrite and austenite in the solution-annealed condition. The welding process involves temperature excursions that will change the phase balance to a more ferritic one in both weld metal and heat-affected areas. This could deteriorate weldment properties. Therefore matching filler materials for duplex steels have a more austenitic composition to produce a phase balance similar to that of the base material. Modern duplex alloys also contain sufficient nitrogen to improve the austenite reformation in the heat-affected zone.

If the cooling rate in welding is very high (e.g., low heat input with thick gauges) there is a risk that the ferrite content

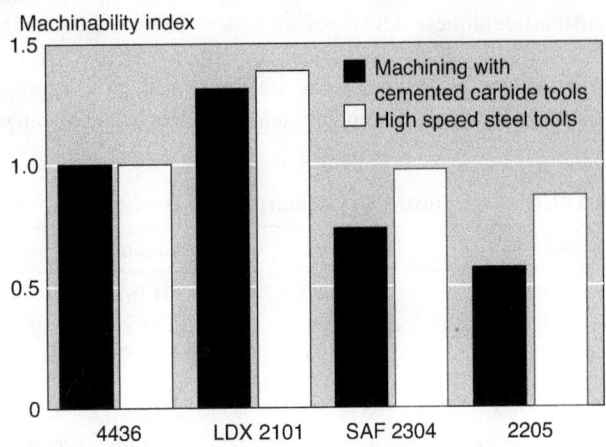

FIGURE 52.9. Machinability of some stainless steels.

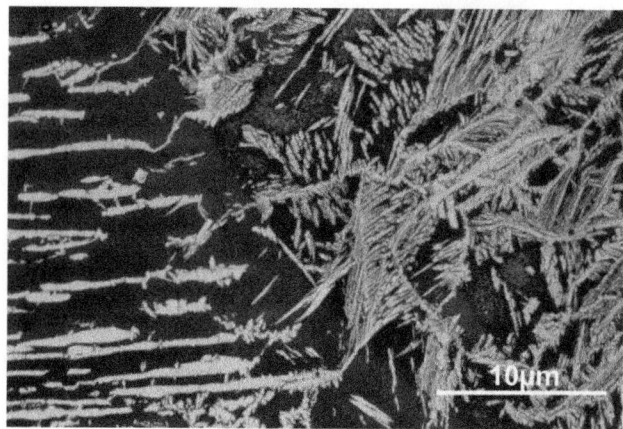

FIGURE 52.10. Microstructure of duplex weld.

will be well above 65%. This high level of ferrite may decrease corrosion resistance and ductility. Similarly, too long exposures at 700–900°C can cause precipitation of intermetallic phases, which impair toughness and corrosion resistance. This has to be considered for high-alloy duplex grades. As with all types of stainless steel weldments, the properties of duplex steel welds differ from those of the base material. However, a properly welded duplex steel shows mechanical and corrosion properties satisfying most requirements. Typical microstructure of a duplex weld is shown in Figure 52.10.

H. APPLICATIONS

H1. General Use

The different groups of duplex grades are in general alternatives to austenitic counterparts, mainly based on their corrosion resistance. An example of how steels can be compared is shown in Table 52.9.

Characteristic benefits of duplex grades compared with austenitic stainless steel types are high resistance to chloride-induced SCC, high mechanical strength, lower thermal expansion, and good erosion and wear resistance. These properties enable the use of duplex stainless steels in a wide range

TABLE 52.9.   Austenitic Counterparts to Duplex Steels

Group	Duplex Grade	Austenitic Grades
Lean duplex	S32101	304 (S30400)
Lean duplex	S32304	316 (S31600)
Duplex	S32205	904L (N08904), 317 (S31700)
Superduplex	S32750	254SMO (S31254), AL6X (N08367), 20-25-6 (N08926)

of applications and in a wide range of industries. A few examples are discussed below.

H2. Pulp and Paper

The pulp and paper industry was one of the first application areas for duplex stainless steel, and these steels are still widely used in several stages of this industry. The good corrosion resistance combined with high mechanical strength make duplex an economical alternative for digesters, blow tanks, oxygen reactors, pulp towers, and chemical recovery liquor tanks. The duplex-grade resistance to SCC is well utilized in chemical recovery evaporation. Suction rolls in paper machines are another application where duplex is used because of the good mechanical properties, particularly superior corrosion fatigue resistance. Figure 52.11 shows a peroxide reactor fabricated in S32101.

H3. Desalination

The need for freshwater has increased the number of desalination plants in the world and duplex stainless steel has become an important substitute for unprotected mild steel and clad plates. Depending on the desalination process a great variety of chloride levels are encountered in a desalination plant. It is possible to select a suitable duplex grade for

FIGURE 52.11. Peroxide reactor in S32101 for pulp and paper mill.

the different hostile environments. The highly alloyed grades like S31803 and S32750 are used in the more aggressive parts of the processes while S32101 and S32304 are used in less severe conditions.

H4. Oil and Gas

A classical application for duplex steel is pipelines for the transport of oil and gas both on- and offshore, but there are many more uses in this industry. In offshore installations superduplex grades such as S32750 are used for pipes, flanges, and fittings in both process systems and in seawater due to their high corrosion resistance. Pump casings and valves are often made of S32205 or S32750. Gravity separators and centrifugal separators in S32205 and S32750 are also common duplex applications where the high fatigue strength is utilized. Another application in the offshore industry where duplex stainless steels are chosen for their high strength is in firewalls and blast walls on the platforms. S32304 absorbs energy rapidly and possesses good resistance to buckling. The wellheads on the bottom of the sea are connected to the drilling platform by flexible pipes. The core part of the flexible pipe is built from profiled stainless steel strip, and it needs to withstand its own weight as well as the pressure of the depth. The lean duplex grade S32101 offers better corrosion resistance and double yield strength compared to S30400, which gives the opportunity of weight savings.

Another important duplex application in offshore industry is umbilicals used for control of subsea systems. Bundles of tubes supply the necessary controls, through hydraulic fluid, electrical cables, and fiber optics, for the subsea field and are often exposed to the same harsh environment as the system it is being used to control and/or monitor. Superduplex steels such as S32750 are mainly used but installations using lean duplex grades with corrosion protection also exist.

H5. Storage Tanks

One of the most important characteristics of the duplex stainless grades is the high mechanical strength compared to that of austenitic stainless steels. This can be used in applications where the strength is controlling the design and an important example is large storage tanks. By selecting a duplex alternative, the wall thickness can be reduced with up to 30% depending on the design standard. Table 52.10

TABLE 52.10. Weight Reductions Replacing Austenitic with Duplex Steel

Standard	API 650	EN 14015
S32101	38200 kg	38200 kg
S30400	48400 kg	52900 kg
Weight savings	21%	28%

FIGURE 52.12. Storage tanks in Barcelona, Spain, constructed of S32101.

shows an example where a standard austenitic grade has been compared to a duplex grade. The tank in this example has a height of 20 m and a diameter of 15 m. In Barcelona, Spain, LDX 2101/S32101 has been used to build 22 tanks for edible oils and a few of them are seen in Figure 52.12.

H6. Bridges

Due to the high strength, duplex is a good choice for different construction purposes. The corrosion resistance together with high strength presents a maintenance-free alternative. The mild environment in the Norwegian mountains makes S32101 the perfect material for the footbridge in Figure 52.13. The construction is so light that it was put in place by helicopter.

FIGURE 52.13. Walking bridge in Norwegian mountains constructed of S32101.

H7. Pipes and Tubes

The use of duplex grades in shell-and-tube heat exchangers is increasing rapidly. Good resistance to SCC and lower thermal expansion relative to the austenitic stainless grades offer a superior performance in tubular heat exchangers. The lower coefficient of thermal expansion can sometimes be used to eliminate expansion bellows and floating heads, resulting in additional cost savings. The good erosion resistance of the duplex grades makes them useful for all systems containing erosive particles, such as water contaminated with sand. The mechanical strength makes the duplex grades very cost effective for high-pressure piping.

H8. Pressure Vessels

An important advantage in constructing pressure vessels is to use a high-strength material as higher design values result in thinner wall sections. How much the mechanical strength can be used is limited by the different pressure vessel codes. Different parts of the world use different rules. As a rule of thumb, the reductions in wall thickness by using a duplex grade instead of an austenitic grade will be 25–50%. This can often make the duplex grades not only an upgrade in corrosion resistance but also more economical when all aspects of design are considered.

One example of a pressure vessel in which duplex grades are being increasingly used is digesters in the pulp and paper industry. Typical design conditions are 15 bars pressure and 200°C. The required wall thickness varies for different pressure vessel codes. In Table 52.11, the dimensions are given according to ASME Section VIII Div. 1 and EN 13445 for such a vessel with a diameter of 4 m. Previously, these vessels were constructed of carbon steel with stainless steel overlay on the inside. The duplex grades permit more aggressive pulp digestion chemistries, and the duplex stainless digesters have a lower total life-cycle cost through reduced maintenance requirements.

TABLE 52.11. Wall Thickness Requirements for Different Steels in Pressure Vessels

Steel Grade	ASME VIII-1 (mm)	EN 13445 (mm)
S30400	24.0	23.0
S32101	17.6	13.7
S32304	20.1	16.2
S32205	18.3	14.3
S32750	14.5	11.3

H9. Chemical Carriers

Stainless steel is a very common material to use in chemical tankers. The corrosion resistance, especially in phosphoric acid, and the ability to clean efficiently between cargos to prevent contamination are two major advantages. Duplex S32205 adds high strength as an advantage and the weight saving that is a result of the higher strength gives the owner the ability to higher payloads. Duplex S32205 has been used in over 200 chemical tankers to date.

REFERENCES

1. E. C. Bain and W. E. Griffiths, Trans. AIME, **75**, 166 (1927).
2. C. Ericsson, Bergsmannen, No. 6, pp. 25–28 (1988), (in Swedish).
3. J. Olsson and M. Liljas, Paper No. 395, Proc. NACE Corrosion 94, Baltimore, MD, 1994.
4. J.-O. Nilsson, Mater. Sci. Technol, **8**, 658 (1992).
5. Outokumpu Stainless Corrosion Handbook, Outokumpu Stainless Steel Oy, Espoo, Finland, 2009.
6. J. E. Truman, in U. K. Corrosion '87, Brighton, U.K., Oct. 26–28, 1987, Institution of Corrosion Science and Technology, Birmingham, UK, 1988, pp. 111–129.
7. B. Wallen, in Proc. 5th World Conf. on Duplex Stainless Steels, Oct. 21–23, 1997, Maastricht, the Netherlands, Stainless Steel World, Zutphen, the Netherlands, 1997, pp. 59–71.
8. Practical Guidelines for the Fabrication of Duplex Stainless Steels, second ed., International Molybdenum Association (IMOA), London, UK, 2009.

53

MARTENSITIC STAINLESS STEELS

J. F. GRUBB

Technical & Commercial Center, ATI Allegheny Ludlum Crop., Brackenridge, Pennsylvania

A. INTRODUCTION

Martensitic stainless steels are iron–chromium alloys with >10.5% chromium and that can be hardened by suitable cooling to room temperature following a high-temperature heat treatment. Although the martensitic stainless steels were invented at about the same time as the ferritic and austenitic stainless steels [1], the mechanical properties of this class of stainless steels have been generally more important than their corrosion resistance. Consequently, the corrosion-resistant properties of these alloys have been less well developed than have the other types of stainless steel. This is a result of the need to restrict their chromium contents to relatively low levels and their often high carbon contents, which together inherently limit corrosion resistance of martensitic stainless steels in comparison with other stainless steels. However, the martensitic stainless steels are hardenable and exhibit high strengths and hardnesses while offering relatively low cost.

The distinction between martensitic stainless steels and other alloys is not sharp. Several nominally ferritic stainless steels, such as American Iron and Steel Institute (AISI) 430 stainless steel (UNS S43000) or the 3CR12 alloy (UNS S41003), can be partially martensitic. Conversely, low-carbon versions of martensitic stainless steel alloys like AISI 410S (UNS S41008) and 416 (UNS S41603) may be substantially ferritic. Lower chromium alloy content alloy steels, like the AISI 500 series heat-resistant steels, also share many of the characteristics of the martensitic stainless steels. Some tool steel alloys are also quite similar to the higher carbon martensitic stainless steels, and some martensitic stainless steels, especially AISI 420, are used as tool (mold) steels.

Uhlig's Corrosion Handbook, Third Edition, Edited by R. Winston Revie
Copyright © 2011 John Wiley & Sons, Inc.

B. STANDARD ALLOYS

Type 410 stainless steel is the prototypical martensitic stainless steel. The standard types of martensitic stainless steels include the following:

AISI Grade	UNS [2]	Description
410	S41000	Basic martensitic stainless steel
403	S40300	Silicon reduced (~0.5%) to improve mechanical properties
414	S41400	Nickel (~2%) added to improve mechanical properties
415	S41500	Carbon decreased and nickel (~4%) added to improve corrosion resistance
416	S41600	Sulfur (0.15–0.30%) added to improve machinability
420	S42000	Higher carbon allows higher hardness for knife blades, etc.
420F	S42020	Sulfur (0.15% minimum) added to improve machinability
420FSe	S42023	Selenium (0.15% minimum) added to improve machinability
431	S43100	Higher chromium plus nickel improve corrosion resistance
440A	S44002	Higher carbon increases hardenability with increased Cr to help maintain corrosion resistance; used for cutting blades and bearings
440B	S44003	Still higher carbon for still higher hardness
440C	S44004	Highest carbon and highest hardness; primary carbides promote wear resistance
440F	S44020	Sulfur (0.10–0.35%) added to improve machinability
440FSe	S44023	Selenium (0.15% minimum) added to improve machinability

In addition to the standard grades, there are many less standard or proprietary versions of the martensitic stainless steels. A few are listed in the below:

Type	UNS	Description
425	S42500	Lower carbon and higher chromium plus nickel improve corrosion resistance
425 Mod	—	Higher carbon allows higher hardness, molybdenum added to improve corrosion resistance
—	S42400	Lower carbon and higher nickel plus some molybdenum improve corrosion resistance

There are also standard grades of cast martensitic stainless steels. These include:

ACI[a] Type	UNS	Similar Wrought Grade
CA-15	J91150	410
CA-15M	J91151	410Mo
CA-40	J91153	420
CA-40F	J91154	420F
CB-6NM	J91650	
CB-6NM	J91540	
CA-28MVW	J91422	422
CB-7Cu-1	J92180	630
CB-7Cu-2	J92110	629

[a]Steel Founders' Society of America, Barrington, IL.

The general compositions (percentages by weight, maximum or range)[*] of these alloys are as follows:

UNS	Type	C	Mn	S	Si	Cr	Mo	Ni	Other
S40300	403	0.15	1.00	0.030	0.5	11.50–13.50			
S41000	410	0.15	1.00	0.030	1.00	11.50–13.50			
S41400	414	0.15	1.00	0.030	1.00	11.50–13.50		1.25–2.50	
S41500	415	0.05	0.50–1.00	0.030	0.60	11.50–14.00	0.50–1.00	3.50–5.50	
S41600	416	0.15	1.25	0.15–0.30	1.00	12.00–14.00			

[*]These are general descriptions only, and the reader should consult the applicable specification for the exact composition limits for the alloy of interest.

UNS	Type	C	Mn	S	Si	Cr	Mo	Ni	Other
S41800[a]	418	0.15–0.20	0.50	0.030	0.50	12.00–14.00		1.80–2.20	2.50–3.50 W
S42000	420	0.15 min.	1.00	0.030	1.00	12.00–14.00			
S42020	420F	0.15 min.	1.25	0.15 min.	1.00	12.00–14.00	0.60		
S42023	420FSe	0.30–0.40	1.25	0.06	1.00	12.00–14.00	0.60		0.15 Se min.
S42200	422	0.20–0.25	1.00	0.030	0.75	11.00–13.50	0.75–1.25	0.50–1.00	0.75–1.25 W, 0.15–0.30 V
S42400	424	0.06	0.50–1.00	0.030	0.30–0.60	12.00–14.00	0.30–0.70	3.50–4.50	
S42500	425	0.08–0.20	1.00	0.010	1.00	14.00–16.00	0.30–0.70	1.00–2.00	
	425 Mod	0.50–0.55	1.00	0.030	1.00	13.00–14.00	0.80–1.20	0.50	
S43100	431	0.20	1.00	0.030	1.00	15.00–17.00		1.25–2.50	
S44002	440A	0.60–0.75	1.00	0.030	1.00	16.00–18.00	0.75		
S44003	440B	0.75–0.95	1.00	0.030	1.00	16.00–18.00	0.75		
S44004	440C	0.95–1.20	1.00	0.030	1.00	16.00–18.00	0.75		
S44020	440F	0.95–1.20	1.25	0.10–0.35	1.00	16.00–18.00	0.40–0.60	0.75	
S44023	440FSe	0.95–1.20	1.25	0.030	1.00	16.00–18.00	0.60	0.75	0.15 Se min.
J91150	CA-15	0.15	1.00	0.040	1.50	11.50–14.00	0.50	1.00	
J91151	CA-15M	0.15	1.00	0.040	0.65	11.50–14.00	0.15–1.00	1.00	
J91153	CA-40	0.20–0.40	1.00	0.040	1.50	11.50–14.00	0.50	1.00	
J91154	CA-10F	0.20–0.40	1.00	0.20–0.40	1.50	11.50–14.00	0.50	1.00	
J91650	CA-6N	0.06	0.50	0.020	1.00	10.50–12.50		6.00–8.00	
J91540	CA-6NM	0.06	1.00	0.030	1.00	11.50–14.00	0.40–1.00	3.50–4.50	
J91422	CA-28MVW	0.20–0.28	0.50–1.00	0.030	1.00	11.00–12.50	0.90–1.25	0.50–1.00	0.90–1.25 W, 0.20–0.30 V
J92180	CB-7Cu-1	0.07	0.70	0.030	1.00	15.50–17.70		3.60–4.60	2.50–3.20 Cu, 0.15–0.35 Nb. 0.05 N
J92110	CB-7Cu-2	0.07	0.70	0.030	1.00	14.00–15.50		4.20–5.50	2.50–3.20 Cu, 0.15–0.35 Nb. 0.05 N

[a]S41800 is only one of the so-called Super 12% Cr alloys [3] used for elevated temperature applications. Others include S41025, S41040, S42300, and AFC-77 alloys.

C. PHYSICAL METALLURGY[†]

In order for an alloy to be a martensitic stainless steel, it must be an iron-based material that contains at least 10.5% (by weight) of chromium (from the definition of a stainless steel), and it must be capable of being substantially transformed through heat treatment to the hard, metastable phase called *martensite*. For this transformation to occur, the alloy must first be transformed by thermal treatment to the stable high-temperature *austenite* phase (often designated *gamma*, γ). The temperature range in which austenite can form depends on the amount of chromium present in iron–chromium alloys. Between 11 and ~12% chromium, binary Fe–Cr alloys go from being capable of forming 100% austenite to being fully ferritic[††] and incapable of forming any austenite [8]. Other elements, notably carbon and nickel, enlarge the so-called *gamma loop*, enabling austenite to be produced in higher chromium steels, but this phenomenon places an upper limit on the chromium content of an austenitic stainless steel of given carbon and nickel content. Once a steel has been austenitized, it is usually transformed to martensite by rapid cooling. However, the high alloy content of martensitic stainless steels gives them great hardenability so that the requirement for rapid cooling can often be satisfied by air cooling; therefore, water quenching is not required.

The influence of chromium, nickel, and especially carbon in depressing the martensite start temperature (M_s) also limits the total alloying element content that can be accommodated in a martensitic stainless steel. The formation of "primary" carbides, which cannot be dissolved during high-temperature heat treatment and that may be undesirable in some applications, may further restrict the maximum carbon content. Other elements are added for specific purposes. Sulfur or selenium are added to improve machinability (usually at the expense of corrosion resistance in some environments). Molybdenum or tungsten is added to

[†] For details on the physical metallurgy of the martensitic stainless steels see [4–7]

[††] The ferritic phase is usually designated as δ (delta) ferrite.

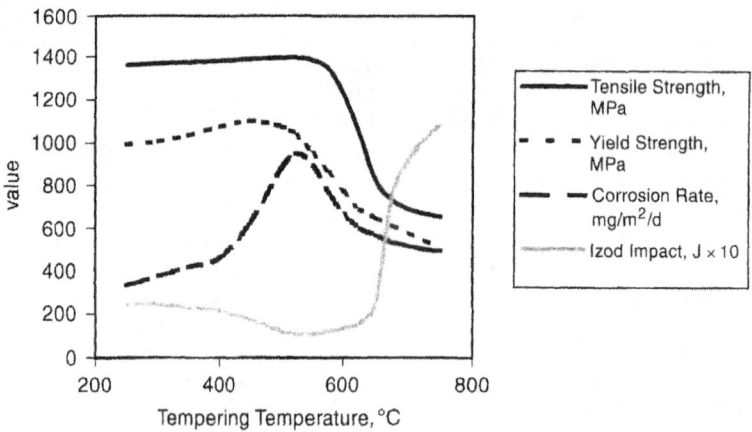

FIGURE 53.1. Effect of tempering (1 h) on the properties of AISI 420 (0.22% C) martensitic stainless steel [9]. Corrosion test run in 3% NaCl at 20°C.

increase resistance to tempering or to improve corrosion resistance.

Although low-carbon martensite can exhibit reasonable ductility and toughness, most freshly formed martensite (with higher carbon) is too hard and brittle for use. Therefore, most of the martensitic stainless steels are usually given the secondary heat treatment called *tempering*. The effect of tempering is controlled by the tempering temperature and time. The effects of tempering temperature on a type 420 stainless steel are shown in Figure 53.1.

The formation of austenite and its subsequent transformation to martensite have a profound grain-refining effect. This can substantially improve the toughness of the martensitic stainless, especially in heavy sections. Multiple heat treatment cycles have been used in some cases to greatly improve toughness. These reactions also involve some volumetric change, which can lead to small dimensional changes and to distortion, especially in thin sheet products. For such reasons, martensitic stainless steels tend to be used more often as bar or plate products and less often as sheet or strip products than the ferritic stainless steels.

D. PRECIPITATION-HARDENING STAINLESS STEELS

To retain the strength of the martensitic stainless steels while increasing their corrosion resistance, many precipitation-hardening stainless steels have been developed. These alloys all[*] derive their strength from the precipitation of a hardening phase within a transformed martensitic matrix. For this reason, some of these alloys are also called *Mar-Aging*

[*] Except for A286 stainless steel. UNS S66286, in which $Ni_3(Ti_2Al)$ phase is precipitated within stable austenite.

stainless steels. Because this hardening precipitate is not carbon based, chromium depletion and loss of corrosion resistance are less of a problem. Precipitation-hardening stainless steels are generally classified as *martensitic*, meaning that they are always martensitic at room temperature, or *semiaustienitic*, meaning that they are usually austenitic and do not become martensitic until suitable heat treatments are applied. Representative precipitation-hardening stainless steels include the following:

Name	AISI Grade	UNS	Type
PH 13-8 Mo	XM^a-13	S13800	Martensitic
15-5 PH	629	S15500	Martensitic
17-4 PH	630	S17400	Martensitic
17-7 PH	631	S17700	Semiaustenitic
PH 15-7 Mo	632	S15700	Semiaustenitic
AM350	633	S35000	Semiaustenitic
AM355	634	S35500	Semiaustenitic
Custom 450	XM-25	S45000	Martensitic
Custom 455	XM-16	S45500	Martensitic

[a]ASTM designation.

The AISI (or ASTM) grade numbers shown above are rarely used and these alloys are typically described by their historical "names," but it should be noted that all of the above names are trademarks. Each of the above trademarks belongs exclusively to one of the various producers. The vast majority of the precipitation-hardening stainless steels were patented, but many of these patents have now expired. For most of these alloys, producers other than the original patent or trademark owner make these alloys and sell them as generic products or under their own (less known) trade names.

The general compositions (percentages by weight, maximum or range) of these alloys are shown in the following table:

UNS	Name	C	Mn	Si	Cr	Mo	Ni	Other
S13800	PH 13-8 Mo	0.05	0.20	0.10	12.25–13.25	2.0–2.5	7.5–8.5	0.90–1.35 Al
S15500	15-5 PH	0.07	1.00	1.00	14.0–15.5		3.5–5.5	2.5–4.5 Cu, 0.15–0.45 Nb
S15700	PH 15-7 Mo	0.09	1.00	1.00	14.00–16.00.	2.0–3.0	6.5–7.75	0.75–1.5 Al
S17400	17-4 PH	0.07	1.00	1.00	15.50–17.50		3.0–5.0	3.0–5.0 Cu, 0.15–0.45 Nb
S17700	17-7 PH	0.09	1.00	1.00	16.00–18.00		6.5–7.75	0.75–1.5 Al
S35000	AM350	0.07–0.11	0.50–1.25	0.50	16.0–17.0	2.5–3.25	4.0–5.0	0.07–0.13 N
S35500	AM355	0.10–0.15	0.50–1.25	0.50	15.0–16.0	2.5–3.25	4.0–5.0	0.07–0.13 N
S45000	Custom 450	0.05	1.00	1.00	14.00–16.00	0.5–1.0	5.0–7.0	1.25–1.75 Cu, 0.10–0.5 Nb
S45500	Custom 455	0.05	0.50	0.50	11.00–12.50	0.5	7.5–9.5	1.25–2.5 Cu, 0.80–1.4 Ti, 0.10–0.5 Nb

Heat treatment for the precipitation-hardening stainless steels varies depending on the alloy and on the properties desired. For the martensitic precipitation-hardening alloys, heat treatment usually comprises three steps:

Austenitizing at high temperature

Rapid cooling to room temperature or below

Tempering plus precipitation hardening

Since 17-4 PH alloy is the prototype of this alloy family, it will be used as an example. Austenitizing is performed at 1040°C (1900°F). The alloy is then cooled to below 32°C (90°F) and held for 2 h. Tempering is then performed for 1–4 h at temperatures ranging from 480 to 620°C (900 to 1150°F). During this tempering process, in addition to stresses being relieved, minute copper-containing particles are precipitated within the martensite, hardening the alloy. The final heat-treated condition is designated "Hxxxx," where xxxx refers to the (Fahrenheit) temperature of tempering. Common heat treatment conditions for 17-4 PH alloy include H900, H1025, H1075, and H1150. The 17-4 PH alloy should not be used in the solution-treated condition because in this condition it has low ductility and low resistance to stress corrosion cracking (SCC).

Heat treatments of the semiaustenitic precipitation-hardening stainless steels are more numerous and more varied, but all conform to a common scheme:

Solution annealing

Austenite destabilization

Martensite formation

Tempering plus precipitation hardening

The 17-7 PH alloy will be used as an example of the various heat treatment schemes. Since material usually is shipped in the solution-annealed condition, that step may sometimes be skipped during final heat treatment.

The first heat treatment to be described is the "RH" (for refrigeration) type heat treatment. Austenite destabilization is accomplished by a "trigger anneal" of 10 min exposure at 955°C (1750°F). This treatment precipitates a small but significant amount of chromium carbide, depleting the austenitic matrix of these two elements and thus raising the martensite start (M_s) temperature. The metal is then cooled to −73°C (−100°F) and held for 8 h to transform it to martensite. Finally, the transformed material is tempered. During this tempering process, a strengthening precipitate of Ni(Al,Ti) is precipitated within the martensite, strengthening it. A typical tempering temperature is 510°C (950°F) for 1 h, which completes the RH950 heat treatment.

An alternative heat treatment is the "TH"-type heat treatment. In this heat treatment, greater austenite destabilization is accomplished by 90 min exposure at 760°C (1450°F). This treatment precipitates a greater amount of chromium carbide than the trigger anneal of the RH treatment. This causes a greater decrease in the carbon and chromium content in the austenitic matrix and thus a greater increase in the M_s temperature. Cooling to only 13°C (55°F) and holding for 30 min becomes sufficient to obtain substantial transformation of the austenite to martensite. Finally, the transformed material is tempered. Tempering at 565°C (1050°F) for 90 min finishes the TH1050 heat treatment.

The marginal stability of the austenite in the semiaustenitic stainless steels also allows their martensitic transformation by way of deformation instead of heat treatment. The CH900 treatment is an example. Cold reducing these alloys ~50% causes substantial transformation to martensite. Tempering at a low temperature with concurrent precipitation completes this process. Tempering for 1 h at the rather low temperature of 482°C (900°F) completes the CH900 treatment, which produces the highest strength.

The differences among the RH, TH, and CH heat treatments have important influences upon the properties of these alloys. The 955°C (1750°F) trigger anneal of the RH heat treatment produces a slight degree of intergranular chromium depletion that slightly sensitizes the material. However, this effect is limited by relatively large values of the solubility product of chromium times carbon and the high rate of diffusion at this temperature. The lower 760°C (1400°F) temperature of the TH heat treatment produces greater Cr depletion and allows for less diffusional healing to occur. The lack of a carbide precipitation step in the CH process prevents this type of sensitization. As a general statement, in many corrosion situations, for a given alloy, the CH condition exhibits the greatest corrosion resistance, followed by the RH conditions, with the TH heat-treated materials exhibiting the lowest corrosion resistance.

E. CORROSION RESISTANCE

The martensitic stainless steels generally exhibit adequate resistance to mild atmospheric corrosion, potable water, and mild chemical environments. Their resistance to chlorides is usually sufficient for short-term contact during use provided that they are cleaned after exposure. They are typically not suitable for continuous immersion in seawater, brackish water, or aerated brines or exposure to marine atmospheres (nor are many of the standard austenitic or ferritic stainless steels). The performance of any given martensitic stainless steel in any given environment may depend on the exact composition of the alloy within the broad range permitted for the alloy, its heat treatment, and its surface finish. For optimum corrosion resistance, martensitic stainless steels should be hardened and tempered. The best corrosion resistance is usually obtained by tempering below 425°C (800°F). Tempering at about 425–540°C (800–1000°F) can lead to increased susceptibility to SCC or hydrogen embrittlement. While tempering >540°C (1000°F) somewhat improves the corrosion resistance (compared with tempering at 425–540°C, 800–1000°F), it generally produces corrosion resistance inferior to that produced by tempering at <425°C (800°F).

A brief discussion of corrosion resistance to specific media follows.

E1. Water

Most of the martensitic stainless steels are resistant to fresh (potable) water. For high-purity water, this resistance extends to high temperatures. Ozaki and Ishikawa [10] examined the SCC behavior of cast 13% Cr martensitic stainless steels in high-purity water (=1 S/cm) at 288°C (550°F) with 8 ppm dissolved oxygen content. Variations in carbon and nickel

contents and in tempering temperatures were examined. They found that the Ni-containing alloys (3.5–5% Ni) had greater susceptibility to intergranular corrosion and intergranular SCC (IGSCC) than alloys with lower Ni content. Hydrogen embrittlement (HE) cracking susceptibility was controlled by the tempering temperature (and resulting strength of the steel) and not by composition within the range studied. Steels with 13% Cr and either Ni = 4% and C = 0.08% or Ni = 2.5% and C = 0.17% when tempered at high temperature (~650°C; ~1200°F) were immune to SCC in high-temperature, high-purity water.

The resistance of the martensitic stainless steels in brackish water, seawater, or aerated brines is more limited. Their resistance to chloride solutions is usually sufficient for short-term contact during use provided that they are cleaned after exposure. They are typically not suitable for continuous immersion in slowly flowing seawater, brackish water, or aerated brines. Corrosion under biofouling deposits should be expected to occur. Type 410 stainless steel was completely perforated through the 6.6-mm specimen thickness after 1 year exposure to tropical seawater in the Pacific Ocean near the Panama Canal [11]. Exposure of type 410 stainless steel in deep seawater sites (715 and 1615 m) also produced significant (up to 15 mils–0.4 mm) crevice attack [12]. The martensitic stainless steels are highly resistant, however, to flow erosion and erosion–corrosion. In situations where rapid flow is the rule, such as boat propellers, propeller shafts, and pump impellers, good performance has been experienced and such products are commercially available. A reasonably large experience base has been developed for 17-4 PH alloy in seawater. For example, when 17-4 PH alloy is aged at temperatures of 550°C (1025°F) or above and used in high-velocity seawater, corrosion is usually avoided [13]. In slowly flowing or stagnant seawater, cathodic protection must be used to prevent pitting and crevice corrosion of the 17-4 PH alloy.

E2. Acid Solutions

The following data compare AISI 409 and 430 annealed ferritic stainless steels with some hardened martensitic stainless steels that were tempered at 204°C (400°F) [14]:

5% Solution at 49°C (120°F)	Corrosion Rate (mm/year)					
	Type 409	Type 410	Type 420	Type 425 Mod	Type 440A	Type 430
Acetic acid	0.022	0.002	0.028	0.122	0.059	0.001
Phosphoric acid	0.002	0.002	0.002	0.015	0.009	0.001

pH 5 Solution at 24°C (75°F)	Pitting Potential (V vs. SCE) in Sodium Chloride					
	Type 409	Type 410	Type 420	Type 425 Mod	Type 440A	Type 430
100 ppm Cl⁻	0.439	0.502	0.581	0.619	0.598	0.590

E3. Petroleum Production and Refining

Some of the martensitic stainless steels have been long used in oil or natural gas production and in oil refining. Use of metallic materials for oil production applications is regulated in many locales by the NACE International Standard MR01-75/ISO 15156 [15]. The concern in this case is the potential for HE cracking (also referred to as sulfide stress cracking) in aqueous service environments containing dissolved H_2S (a common impurity in these applications). The UNS S41000, CA-15, or CA-15M alloy is approved for use by this standard if it is double tempered at 620°C (1140°F) (min) and its hardness is 22 HRC or less. The UNS S42400 or CA-6NM alloy is approved for use by this standard if it is double tempered at ~660 and ~600°C (1220 and 1112°F) and its hardness is 23 HRC or less. S41500 and S42400 are among the first of a growing group often called "supermartensitic stainless steels." These alloys have been used for petroleum pipelines where low-alloy steels have suffered corrosion but where the use of high-alloy materials such as duplex or austenitic stainless steels is not considered to be justified. The UNS S17400 alloy is approved for use by this standard if it is double tempered at 620°C (1148°F) and its hardness is 33 HRC or less. The UNS S45000 alloy is approved for use by this standard if it has been precipitation hardened at 620°C (1148°F) (4 h min) and its hardness is 31 HRC or less. The UNS S66286 alloy is approved for use by this standard if its hardness is 35 HRC or less. As noted in this standard, these materials "have provided satisfactory field service in some sour environments. These materials may, however, exhibit threshold stress levels for sulfide stress cracking in NACE Standard TM0177 that are lower than those of other materials included in this standard." Many martensitic stainless steels are listed as being acceptable for use as well casing or tubing under certain limits of H_2S and chloride concentration, pH, and service temperature that may vary with chemical composition and metallurgical processing or the material. These include API 5CT Grade L-80, API 5CT Grade C-90 Type 1, API 5CT Grade T-95 type 1 (all restricted composition versions of type 420 with 13% Cr), and UNS S42500 alloys. The reader should consult the latest edition of this standard for current information. The CB-7Cu-1 grade in the "H1150 DBL" condition is listed as being acceptable for use in internal valve components when its hardness is 310 HB (30 HRC) maximum.

E4. Atmospheric Corrosion

In general, all stainless steels, even the martensitic stainless steels, exhibit good or excellent resistance to atmospheric corrosion. The increased corrosiveness of marine atmospheres does. however, exhibit significant differences. Type 410 martensitic stainless steel exhibited rusting after a few months exposure to a severe tropical marine atmosphere in the Panama Canal Zone. While weight loss was low, corresponding to an average 0.2-μm/year penetration rate, 0.125-mm-deep pits were observed after eight years exposure [11].

The high strengths attainable with the martensitic and precipitation-hardenable stainless steels greatly increase their susceptibility to SCC, particularly in the presence of chloride ions. For 12% Cr martensitic stainless steels stressed to 75% of their yield strengths in a marine atmosphere, stress corrosion failures were observed in material with strengths above ~1030 MPa (~150 ksi) [16]. Lower strength material similarly tested did not fail in over four years exposure. Tempering between 340 and 540°C (650 and 1000°F) renders these alloys quite sensitive to SCC in marine atmospheres. Tempering below 340°C (650°F) is somewhat better, but for maximum resistance to this problem, tempering >540°C (1000°F) should be employed [16].

The martensitic precipitation-hardening alloys, as represented by 17-4 PH alloy, are more resistant to SCC than the standard martensitic stainless steels and exhibit a threshold yield strength for SCC in marine atmospheres of above 1240 MPa (180 ksi) [10]. This means that 17-4 PH material aged at or above 540°C (1000°F) is essentially immune to this problem while aging at 480°C (900°F) renders it susceptible to cracking in marine atmospheres. Tests of more than 6.6 years produced no failures of specimens aged at 550°C (1025°F) and stressed at 140 ksi or of specimens aged at 620°C (1150°F) and stressed at 105 ksi [17]. Other marine exposures (i.e., in the 25-m lot at Kure Beach) produced failures in 20–322 days, but only for 17-4 PH material aged at 480°C (900°F) [18]. Welding generally increased susceptibility to SCC as evidenced by reduced times to failure.

The semiaustenitic precipitation-hardening stainless steels are also susceptible to cracking in marine atmospheres. Exposures of several of these alloys in the 25- and 250-m lots at Kure Beach produced failures in as little as one day, but other samples survived uncracked for test durations of up to 1100 days [11, 12]. Materials tested included 17-7 PH, PH15-7 Mo, PH 13-8 Mo, and AM355 alloys.

E5. Other Media

The resistance of many alloys to a variety of media is summarized in Corrosion Resistance Tables [19], Corrosion Data Survey [20], and Handbook of Corrosion Data [21]. Corrosion Resistance Tables gives data on 17-4 PH alloy and on "type 400 series" stainless steels, which might include some of the martensitic stainless steels, but more probably refers to type 430 ferritic stainless steel. Corrosion Data Survey reports information for "12 Cr" stainless steel but notes that this might be type 405 ferritic stainless steel or type 410 martensitic stainless steel. Handbook of Corrosion Data [21] has extensive listings for type 410 stainless steel, but information on its heat treatment state is absent from the summaries.

F. OXIDATION/TEMPERATURE RESISTANCE

The chromium content of the martensitic stainless steels gives them good oxidation resistance. They generally offer sufficient oxidation resistance to allow their use to temperatures up to $\sim 700°C$ ($\sim 1300°F$). However, tempering of the martensitic structures and consequent loss of strength usually limits the application of these alloys to lower temperatures.

G. TYPICAL APPLICATIONS

Typical uses for the martensitic stainless steels include the following:

Cutlery

Surgical instruments

Blades, and so on, in turbine engines

Bearings

Aerospace equipment

Petroleum production and refining

Firearms

Valves and stems

Food-processing equipment

REFERENCES

1. R. Castro, "Historical Background to Stainless Steels," in Stainless Steels, P. Lacombe, B. Baroux, and G. Beranger (Eds.), les editions de Physique, Les Ulis, France, 1993.

2. Metals and Alloys in the Unified Numbering System, 11th ed., SAE, Warrendale, PA, 2008.

3. J. Z. Briggs and T. D. Parker, "The Super 12% Cr Steels," Climax Molybdenum Co., New York, 1965.

4. F. B. Pickering, Int. Metals Rev., Dec. 1976.

5. P. T. Lovejoy, "Structure and Constitution of Wrought Martensitic Stainless Steels," in Handbook of Stainless Steels, D. Peckner and I. M. Bernstein (Eds.), McGraw-Hill, New York, 1977, Chapter 6.

6. A. J. Sedriks, Corrosion of Stainless Steels, 2nd ed., Wiley, New York, 1996, pp. 53–58.

7. O. Bletton, "The Martensitic Stainless Steels," in Stainless Steels, P. Lacombe, B. Baroux, and G. Beranger (Eds.), les editions de Physique, Les Ulis, France, 1993, pp. 479–504.

8. T. B. Massalski (Ed.), Binary Alloy Phase Diagrams, American Society for Metals, Metals Park, PA, 1986, p. 822.

9. R. Barker, Metallurgia, Aug. 1967, P. 49.

10. T. Ozaki and Y. Ishikawa, "Intergranular Stress Corrosion Cracking and Hydrogen Embrittlement of Martensitic Stainless Steels in High-Temperature-High Purity Water," in Proceedings of International Conference on Stainless Steels, 1991 Chiba, Japan, The Iron and Steel Institute of Japan, Tokyo, pp. 176–180.

11. A. L. Alexander, C. R. Southwell, and B. W. Forgeson, Corrosion, **17**(7), 345t (1961).

12. A. H. Tuthill and C. R. Schillmoller, "Guidelines for Selection of Marine Materials," paper presented at The Ocean Science and Ocean Engineering Conference, Marine Technology Society, June 14–17, 1965. (Summarized in [13].)

13. F. W. Fink and W. K. Boyd, "The Corrosion of Metals in Marine Environments," DMIC Report 245, Defense Metals Information Center, Battelle Memorial Institute, Columbus, OH, 1970.

14. Allegheny Ludlum Technical Data Blue Sheet Martensitic Stainless Steels types 410, 420, 425 Mod, and 440A, Allegheny Ludlum Corporation, Pittsburgh, PA, 1998.

15. "Petroleum and Natural Gas Industries—Materials for Use in H2S-Containing Environments in Oil and Gas Production," NACE MR01-75/ISO 15156, NACE International, Houston, TX, 2006.

16. E. H. Phelps, "Stress-Corrosion Behavior of High Yield-Strength Steels," Seventh World Petroleum Congress, Mexico City, April 7–9, **9**, 1967, pp. 201–209 (Summarized in [13].)

17. E. E. Denhard, "Stress-Corrosion Cracking of High Strength Stainless Steels," paper presented at Twenty-Fourth Meeting of The AGARD Structures and Materials Panel, Turin, Italy, April 17–20, 1967. (Summarized in [13].)

18. C. J. Slunder, "Stress-Corrosion Cracking of High-Strength Stainless Steels in Atmospheric Environments," DMIC Report 158, Sep. 15, 1961. (Summarized by Fink and Boyd [13].)

19. P. A. Schweitzer, Corrosion Resistance Tables, 4th ed., Marcel Dekker, New York, 1995.

20. Corrosion Data Survey, Metals Section, 6th ed., NACE, Houston, TX, 1985.

21. B. D. Craig (Ed.). Handbook of Corrosion Data. ASM International, Metals Park, OH, 1989.

54

ALUMINUM AND ALUMINUM ALLOYS

E. GHALI

Department of Mining, Metallurgy and Materials Engineering, Laval University, Québec, Canada

Uhlig's Corrosion Handbook, Third Edition, Edited by R. Winston Revie
Copyright © 2011 John Wiley & Sons, Inc.

A. ALUMINUM PROPERTIES AND ALLOYS

Aluminum is second only to iron as the most important metal of commerce. Aluminum is also the third most abundant metal in the crust of the earth, almost twice as plentiful as iron, the fourth most abundant metal. Pure aluminum has a relatively low strength. The density of all alloys (99.65–99.99%) is of the order of 2.7 g/mL, one-third that of steel. In addition to recycling and new smelting processes, aluminum has a relatively low cost, and its alloys provide a high ratio of strength to weight. Salts of aluminum do not damage the environment or ecosystems and are nontoxic. Aluminum and its alloys are nonmagnetic and have high electrical conductivity, high thermal conductivity, high reflectivity, and noncatalytic action [1].

A1. Wrought Alloys

Wrought alloys are of two types: non–heat treatable, of the 1XXX, 3XXX, 4XXX, and 5XXX series, and heat treatable, of the 2XXX, 6XXX, and 7XXX series. Strengthening is produced by strain hardening, which can be increased by solid solution and dispersion hardening for the non-heat-treatable alloys. In the heat-treatable type, strengthening is produced by (1) a solution heat treatment at 460–565°C (860–1050°F) to dissolve soluble alloying elements; (2) quenching to retain them in solid solution; and (3) a precipitation or aging treatment, either naturally at ambient temperature, or more commonly, artificially at 115–195°C (240–380°F), to precipitate these elements in an optimum size and distribution; (4) solution heat treatment and natural aging; (5) air quenched and aged; (6) solution heat treatment and annealed; (7) like entry 6, but overaged; (8) like entry 3, but with accelerated aging; (9) like entry 6 and followed by strain hardening (cold working).

Strengthened tempers of non-heat-treatable alloys are designated by an "H" following the alloy designation, and those of heat-treatable alloys by a "T"; suffix digits designate the specific treatment (e.g., 1100-H14 and 7075-T651). In both cases, the annealed temper, a condition of maximum softness, is designated by an "O" [1]. The temper designation system is used for all forms of wrought and cast aluminum and aluminum alloys except ingot cast materials. Basic temper designations consist of letters; subdivisions of the basic tempers, where required, are indicated by one or more digits following the letter [2]. The nominal chemical compositions of representative wrought aluminum alloys are given in Table 54.1. Typical tensile properties of these alloys in tempers representative of their most common usage are given in Tables 54.2 and 54.3.

All non-heat-treatable alloys have a high resistance to general corrosion. Aluminum alloys of the 1XXX series representing unalloyed aluminum have a relatively low strength. Alloys of the 3XXX series (Al–Mn, Al–Mn–Mg)

TABLE 54.1. Nominal Chemical Compositions of Representative Aluminum Wrought Alloys[a]

Alloy	Percent of Alloying Elements								
	Si	Cu	Mn	Mg	Cr	Zn	Ti	V	Zr
Non-Heat-Treatable Alloys									
1060	99.60%	min Al							
1100	99.00%	min Al							
1350	99.50%	min Al							
3003		0.12	1.20						
3004			1.20	1.0					
5052				2.5	0.25				
5454			0.80	2.7	0.12				
5456			0.80	5.1	0.12				
5083			0.70	4.4	0.15				
5086			0.45	4.0	0.15				
7072[b]						1.0			
Heat-Treatable Alloys									
2014	0.8	4.400	0.80	0.5					
2219		6.30	0.30				0.60	0.10	0.18
2024		4.40	0.60	1.5					
6061	0.6	0.28		1.0	0.20				
6063	0.4			0.7					
7005			0.45	1.4	0.13	4.5	0.04		0.14
7050		2.30		2.2		6.2			
7075		1.60		2.5	0.23	5.6			

[a]Reprinted from [1], pp. 111–145. Courtesy or Marcel Dekker, Inc.
[b]Cladding for Alclad products.

TABLE 54.2.   Typical Tensile Properties of Representative Non-Heat-Treatable Aluminum Wrought Alloys in Various Tempers[a,b]

Alloy and Temper	Strength (MPa)		Percent Elongation	
	Ultimate	Yield	In 50 mn[c]	In 5D[d]
1060 -O	70	30	43	
-H12	85	75	16	
-H14	100	90	12	
-H16	115	105	8	
-H18	130	125	6	
1100 -O	90	35	35	42
-H14	125	125	9	18
-H18	165	150	5	13
3003 -O	110	40	30	37
-H14	150	145	8	14
-H18	200	185	4	9
3004 -O	180	70	20	22
-H34	240	200	9	10
-H38	285	250	5	5
5052 -O	195	90	25	27
-H34	260	215	10	12
-H38	290	255	7	7
5454-O	250	115	22	
-H32	275	205	10	
-H34	305	240	10	
-H111	260	180	14	
-H112	250	125	18	
5456 -O	310	160		22
-H111	325	230		16
-H112	310	165		20
-H116, H321	350	255		14
5083 -O	290	145		20
-H116, H321	315	230		14
5086 -O	260	115	22	
-H116, H32	290	205	12	
-H34	325	255	10	
-H112	270	130	14	

[a]Averages for various sizes, product forms, and methods of manufacture; not to be specified as engineering requirements or used for design purposes.
[b]Reprinted from [1]. pp. 111–145. Courtesy of Marcel Dekker, Inc.
[c]A 1.60-mm-thick specimen.
[d]A 12.5-mm-diameter specimen.

have the same desirable characteristics as those of the 1XXX series, but somewhat higher strength. Almost all the manganese in these alloys is precipitated as finely divided phases (intermetallic compounds), but corrosion resistance is not impaired because the negligible difference in electrode potential between the phases and the aluminum matrix in most environments does not create a galvanic cell. Magnesium, added to some alloys in this series, provides additional strength through solid-solution hardening, but the amount is low enough that the alloys behave more like those with manganese alone than like the stronger Al–Mg alloys of the 5XXX series. Alloys of the 4XXX series (Al–Si) are low-strength alloys used for brazing and welding products and for cladding in architectural products. These alloys develop a gray appearance upon anodizing. The silicon, most of which

is present in elemental form as a second-phase constituent, has little effect on corrosion.

Alloys of the 5XXX series (Al–Mg) are the strongest non-heat-treatable aluminum alloys, and in most products, they are more economical than alloys of the 1XXX and 3XXX series in terms of strength per unit cost. Magnesium is one of the most soluble elements in aluminum, and when dissolved at an elevated temperature, it is largely retained in solution at lower temperatures, even though its equilibrium solubility is greatly exceeded. It produces considerable solid-solution hardening, and additional strength is produced by strain hardening. Alloys of the 5XXX series have not only the same high resistance to general corrosion as other non-heat-treatable alloys in most environments but also, in slightly alkaline ones, a better resistance than any

TABLE 54.3. Typical Tensile Properties of Representative Heat-Treatable Aluminum Wrought Alloys in Various Tempers[a,b]

Alloy and Temper	Strength (MPa)		Percent Elongation	
	Ultimate	Yield	In 50 mm[c]	In 5D[d]
2014 -O	185	95		16
-T4, T451	425	290		19
-T6, T651	485	415		11
2219 -O	170	75	18	
-T37	395	315	11	
-T87	475	395	10	
2024 -O	185	75	20	20
-T4, T351	470	325	20	17
-T851	480	450	6	
-T86	515	490	6	7
6061 -O	125	55	25	27
-T4, T451	240	145	22	22
-T6, T651	310	275	12	15
6063 -O	195	90	25	27
-H34	260	215	10	12
-H38	290	255	7	7
7005 -O	195	85		20
-T63, T6351	370	315		10
7050 -T76, T7651	540	485		10
-T736, T73651	510	455		10
7075 -O	230	105	17	14
-T6, T651	570	505	11	9
-T76, T7651	535	470		10
-T736, T7351	500	435		11

[a]Averages for various sizes, product forms, and methods of manufacture; not to be specified as engineering requirements or used for design purposes.
[b]Reprinted from [1], pp. 111–145. Courtesy of Marcel Dekker, Inc.
[c]A 1.60-mm-thick specimen.
[d]A 12.5-mm-diameter specimen.

other aluminum alloy. They are widely used because of their high as-welded strength when welded with a compatible 5XXX series filler wire, reflecting the retention of magnesium in solid solution.

Among heat-treatable alloys, those of the 6XXX series, which are moderate-strength alloys based on the quasibinary Al–Mg$_2$Si (magnesium silicide) system, provide a high resistance to general corrosion equal to or approaching that of non-heat-treatable alloys. Heat-treatable alloys of the 7XXX series (Al–Zn–Mg) that do not contain copper as an alloying addition also provide a high resistance to general corrosion.

All other heat-treatable wrought alloys have a significantly lower resistance to general corrosion. These include all alloys of the 2XXX series (Al–Cu, Al–Cu–Mg, Al–Cu–Si–Mg) and those of the 7XXX series (Al–Zn–Mg–Cu) that contain copper as a major alloying element. As described later, the lower resistance is caused by the presence of copper in these alloys, which are designed primarily for aeronautical applications, where strength is required and where protective measures are justified [1].

A2. Cast Alloys

Cast alloys are also of two types: non–heat treatable, designated by an "F," for which strengthening is produced primarily by intermetallic compounds, and heat treatable, designated by a "T," corresponding to the same type of wrought alloys where strengthening is produced by dissolution of soluble alloying elements and their subsequent precipitation. Alloys of the heat-treatable type are usually thermally treated subsequent to casting, but in a few cases, where a significant amount of alloying elements are retained in solution during casting, they may not be given a solution heat treatment after casting; thus they may be used in both the F and fully strengthened T tempers (Tables 54.4 and 54.5).

Aluminum casting alloys are produced by all casting processes of which die, permanent mold, and sand casting account for the greatest proportion. Unlike wrought alloys, their selection involves consideration of casting characteristics as well as of properties.

As with wrought alloys, copper is the alloying element most deleterious to general corrosion. Alloys such as 356.0, A356.0, B443.0, 513.0, and 514.0 that do not contain copper

TABLE 54.4. Nominal Chemical Compositions of Representative Aluminum Casting Alloys[a]

Alloy	Percent of Alloying Elements				
	Si	Cu	Mg	Ni	Zn
Alloys Not Normally Heat Treated					
360.0	9.5		0.5		
380.0	8.5	3.5			
443.0	5.3				
514.0			4.0		
710.0		0.5	0.7		6.5
Alloys Normally Heat Treated					
295.0	0.8	4.5			
336.0	12.0	1.0	1.0	2.5	
355.0	5.0	1.3	0.5		
356.0	7.0		0.3		
357.0	7.0		0.5		

[a]Reprinted from [1]. pp. 111–145. Courtesy of Marcel Dekker, Inc.

as an alloying element have a high resistance to general corrosion comparable to that of non-heat-treatable wrought alloys. In other alloys, corrosion resistance becomes progressively less the greater the copper content. More so than with wrought alloys, a lower resistance is compensated for by the use of thicker sections usually necessitated by requirements of the casting process [1].

Other Al-based materials, such as laminates, composites, and ultrafine structures, prepared by conventional or novel techniques are becoming available and their applications will depend in part on their corrosion performance.

Alclad alloys are duplex wrought products, supplied in the form of sheet, tubing, and wire, which have a core of one aluminum alloy and a coating on one or both sides of aluminum or another aluminum alloy. Generally, the core comprises 90% of the total thickness with a coating comprising about 5% of the thickness on each side. The coating is metallurgically bonded to the core over the entire area of

TABLE 54.5. Typical Tensile Properties of Representative Aluminum Casting Alloys in Various Tempers[a]

Alloy and Temper	Type Casting	Strength (MPa)		Percent Elongation
		Ultimate	Yield	In 50 mm[c]
295.O -T6	Sand	250	165	5
336.O -T5	Permanent mold	250	195	1
355.O -T6	Sand	240	170	3
-T6	Permanent mold	375	240	4
-T61	Sand	280	250	3
-T62	Permanent mold	400	360	1.5
356.O -T6	Sand	230	165	3.5
-T6	Permanent mold	255	185	5
-T7	Sand	235	205	2
-T7	Permanent mold	220	165	6
357.O -T6	Sand	345	295	2
-T6	Permanent mold	360	295	5
-T7	Sand	275	235	3
-T7	Permanent mold	260	205	5
360.O -F	Pressure die	325	170	3
380.O -F	Pressure die	330	165	3
443.O -F	Pressure mold	160	60	10
514.O -F	Sand	170	85	9
710.O -F	Sand	240	170	5

[a]Reprinted from [1], pp. 111–145. Courtesy of Marcel Dekker, Inc.

Averages for separate cast test bars; not to be specified as engineering requirements or used for design purposes.

[c]A 1.60-mm-thick specimen.

contact. In the most widely used Alclad materials, the coating alloys are selected so that they will be anodic to the core alloys in most natural environments. Thus, the coating will galvanically protect the core where it is exposed at cut edges, rivet holes, or scratches. Such Alclad alloys are usually more resistant to penetration by neutral solutions than are any of the other aluminum-base alloys [3].

The recent worldwide interest shown in the metal matrix composite "MMC" materials has been fueled by the fact that mechanical properties of light alloys can be enhanced by incorporating reinforcing fibers (usually ceramic). Several manufacturers are marketing a range of particulate reinforced MMC products with different compositions (e.g., 12% alumina, 9% carbon fiber, reinforced A1–12% SiC, particulate SiC/Al ingots). The major reinforcements used in aluminum-based MMCs are boron, graphite, silicon carbide, and alumina.

B. CORROSION BEHAVIOR OF ALUMINUM AND ITS ALLOYS

B1. Description

Alloy 1100 (2S), sometimes known as commercially pure aluminum, contains ~99.0–99.3% aluminum. The rest of the

alloy is made up mainly of iron and silicon with minor amounts of copper. Purer aluminum, containing up to ~99.95% aluminum, is also available; in addition, electrolytic refining has produced a small amount of very pure metal, >99.99% aluminum. The resistance of pure aluminum to attack by most acids and many neutral solutions is higher than that of aluminum of lower purity or of most of the aluminum-base alloys.

Aluminum is an active metal, and its resistance to corrosion depends on the passivity produced by a protective oxide film. In aqueous solutions, the potential–pH diagram according to Pourbaix [4] in Figure 54.1 expresses the thermodynamic conditions under which the film develops. As this diagram shows, aluminum is passive only in the pH range of ~4–9. The limits of passivity depend on the temperature, the form of oxide present, and the low dissolution of aluminum that must be assumed for inertness. The various forms of aluminum oxide all exhibit minimum solubility at about pH 5.

The protective oxide film formed in water and atmospheres at ambient temperature is only a few nanometers thick and amorphous. At higher temperatures, thicker films are formed; these may consist of a thin amorphous barrier layer next to the aluminum and a thicker crystalline layer next to the barrier layer. Relatively thick, highly protective films

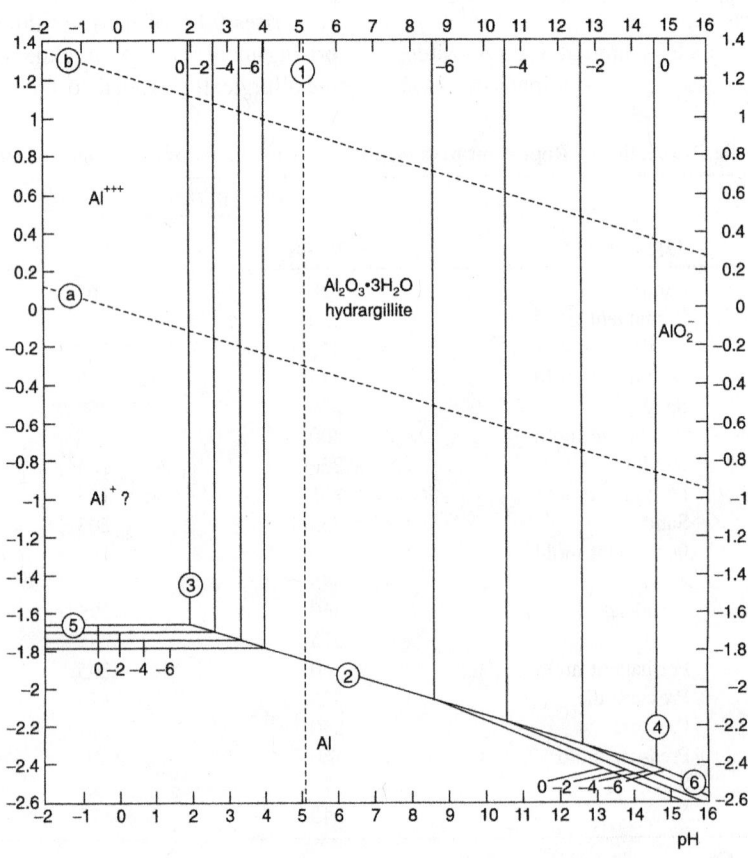

FIGURE 54.1. Potential versus pH diagram for A1/H$_2$O system at 25°C [4]. (Reprinted with permission from [4], NACE International and CEBELCOR.)

of boehmite, aluminum oxide hydroxide AlOOH, are formed in water near its boiling point, especially if it is made slightly alkaline, and thicker, more protective films are formed in water or steam at still higher temperatures.

Since the form of aluminum oxide produced depends on corrosion conditions, its identification is sometimes useful in establishing the cause of corrosion. At lower temperatures, the predominant forms produced by corrosion are bayerite, aluminum trihydroxide $Al(OH)_3$, while, at higher temperatures, it is boehmite $Al_2O_3 \cdot H_2O$. During aging of aluminum hydroxide, which is first formed during corrosion in an amorphous form, still another aluminum trihydroxide, gibbsite, or hydrargilite ($Al_2O_3 \cdot 3H_2O$) may also be formed, especially if ions of alkali metals are present.

Above a temperature of ~230°C (445°F), a protective film no longer develops in water or steam, and the reaction progresses rapidly until eventually all the aluminum exposed in these media is converted into oxide. Special alloys containing iron and nickel have been developed to retard this reaction, and these alloys may be used up to a temperature of ~360°C (680°F) without excessive attack [5].

B2. Effect of O_2 and Some Gases

Oxygen does influence the corrosion of aluminum. The corrosion of aluminum is very slow in deaerated solutions. In the presence of O_2, corrosion is accelerated. In general, high concentrations of dissolved oxygen tend to stimulate attack, especially in acid solutions, although this effect is less pronounced than for most of the other common metals. Hydrogen and nitrogen have no effect, except as they influence the oxygen content [3].

Carbon dioxide and hydrogen sulfide, even in high concentrations, appear to have a slight inhibiting action on the effect of aqueous solutions on aluminum alloys. Aqueous

solutions containing sulfur dioxide etch aluminum, but less than copper or steel. Aqueous solutions of hydrogen chloride are strongly corrosive to aluminum.

B3. Temperature

At low temperatures [4°C (40°F) or below], the action of most aqueous solutions is much slower than at room temperature. However, in many solutions, increasing temperatures above ~80°C (180°F) results in a decrease in the rate of attack. Thus a temperature of 70–80°C (160–180°F) is likely to result in more severe corrosion than temperatures of 20°C (70°F) or 100°C (212°F).

B4. pH

Corrosion of aluminum and its alloys is passive between pH of ~5 and 8.5. Reflecting its amphoteric nature, and as illustrated in Figure 54.1 [4], aluminum corrodes under both acidic and alkaline conditions, in the first case to yield Al^{3+} ions and in the second case to yield AlO_2^- (aluminate) ions. There are a few exceptions, either where the oxide film is not soluble in specific acidic or alkaline solution or where it is maintained by the oxidizing nature of the solution. Two exceptions, acetic acid and sodium disilicate, are included in Figure 54.2 (Alcoa Laboratories). Ammonium hydroxide >30% concentration by weight, nitric acid >80% concentration by weight, and sulfuric acid of 98–100% concentration are also exceptions [1].

There is no general relationship between pH and rate of attack because the specific ions present largely influence the behavior. Thus most aluminum alloys are inert to strong nitric or acetic acid solutions but are readily attacked in dilute nitric, sulfuric, or hydrochloric acid solutions. Similarly, solutions with a pH as high as 11.7 may not attack aluminum

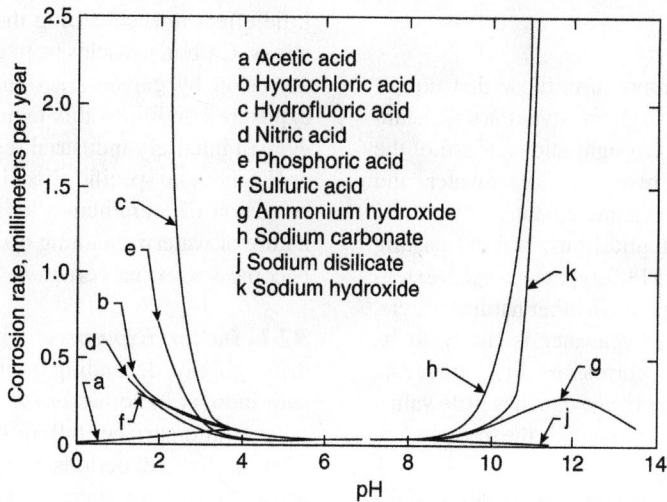

FIGURE 54.2. Relation to pH of the corrosivities toward 1100-H14 alloy sheet of various chemical solutions. (Reprinted from [1], pp. 111–145. Courtesy of Marcel Dekker, Inc.)

alloys, provided silicate inhibitors are present but, in the absence of silicates, attack may be appreciable at a pH as low as 9.0. In chloride-containing solutions, generally less corrosion occurs in the near-neutral pH range, say 5.5–8.5, than in either distinctly acid or distinctly alkaline solutions. However, the results obtained vary somewhat, depending on the specific aluminum alloy under consideration.

B5. Freshwaters

Aluminum-base alloys are not appreciably corroded by distilled water even at elevated temperatures (up to 180°C [350°F] at least). Furthermore, distilled water is not contaminated by contact with most aluminum-based alloys. For this reason, there is a fairly extensive and satisfactory use of aluminum alloy storage tanks, piping valves, and fittings for handling distilled water.

Because natural freshwaters differ so widely in their composition and behavior, it is extremely difficult to make generalizations regarding the resistance of aluminum-base alloys to their action. Most commercial aluminum-base alloys show little or no general attack when exposed to most natural waters at temperatures up to 180°C (350°F) at least. However, certain waters may cause severe localized attack or pitting. Pitting is of most importance where the metal section thickness is small, since the rate of attack at the pits generally falls off with increasing time of exposure. In general, the time necessary to perforate an aluminum alloy sheet 0.10 cm thick or greater is prolonged, as attested to by the wide and successful use of aluminum tea kettles.

The Alclad products are much more resistant to perforation by pitting than are the other aluminum alloys. Therefore, wherever the characteristics of a specific water are not known in advance, it is safer to employ aluminum alloys such as Alclad 3003. Water staining, a type of crevice corrosion of aluminum, can occur.

B6. Seawater

Of the aluminum alloys in common use, those that do not contain copper as a major alloying constituent are resistant to unpolluted seawater. Among wrought alloys, those of the 5XXX series have the highest resistance to seawater, and considering their other desirable characteristics, they are the most widely used for marine applications. Among casting alloys, those of the 356.0 and 514.0 types are used extensively for marine applications [1]. As in other natural waters, any attack that does develop in seawater is likely to be extremely localized (i.e., pitting corrosion). Therefore, rate of attack calculated from weight change data has little value. Measurement of change in tensile strength is the most widely used criterion.

Corrosion of aluminum alloys in seawater is mainly of the pitting type, as would be expected from its salinity and enough dissolved oxygen as a cathodic reactant to polarize the alloys to their pitting potentials. Rates of pitting usually range from 3 to 6 μm/year during the first year and from 0.8 to 1.5 μm/year averaged over a 10-year period; the lower rate for the longer period reflects the tendency for older pits to become inactive. The corrosion behavior of aluminum alloys in deep seawater, judging from tests at 1.6 km, is generally the same as at the surface except that the effect of crevices is greater [6].

B7. Atmospheric Corrosion

B7.1. Outdoor Exposures. The aluminum-base alloys as a class are highly resistant to normal outdoor exposure conditions. The alloys containing copper as a major alloying constituent (over ~1%) are somewhat less resistant than the other aluminum-base alloys, whereas the Alclad alloys are generally the most resistant. Results of typical outdoor exposure tests are based on exposure of machined tensile specimens 103.1 mm (4.06 in.) thick. Loss in tensile strength is generally of the order of 1–2% for the first year depending on the alloy and the atmosphere. An alloy such as 2017T can lose up to 17% in tensile strength during the first year [3].

If the specimens had been thinner, obviously the losses would have been relatively greater, whereas if they had been thicker, the losses would have been smaller. This effect of thickness is especially pronounced in the case of aluminum-base alloys, since the rate of attack greatly decreases with increasing time of exposure.

Specimens were freely exposed to the outdoor locations. If they had been partially sheltered, the rate of attack would have been somewhat greater; if they had been largely sheltered, very little attack would have occurred. Apparently, in the case of aluminum-base alloys, periodic exposure to rain is beneficial probably because the rain washes off corrosive products that settle from the air. Evidently, free exposure to rain is not harmful but, on the contrary, is beneficial.

The gases ordinarily found in industrial atmospheres have little effect in accelerating the corrosion of aluminum-base alloys. Carbon particles from the atmosphere may accelerate corrosion by galvanic action. Under outdoor atmospheric exposure conditions this factor is of secondary importance even in intensely industrial regions. Sulfur compounds such as H_2S have no specific effect in accelerating the tarnishing or corrosion of aluminum alloys. However, the highly acidic nature of water containing dissolved SO_2 or SO_3 causes it to become somewhat corrosive [3].

B7.2. Indoor Exposures. The effects of indoor exposure differ greatly, depending on the exposure conditions. Exposure indoors in homes or offices ordinarily causes, at most, only a mild surface dulling of aluminum-base alloys even after prolonged periods of exposure. In damp locations, especially where there is contact with moist insulating materials, such as wood, cloth, and paper insulation, attack may be more appreciable. In factories or chemical plants, fumes or vapors incident to the operations being conducted may cause a definite surface attack. However, in most indoor atmospheres where pools of contaminated water do not remain in prolonged contact with aluminum alloys, or where extended contact with moist, porous materials is avoided, no appreciable loss of mechanical properties through corrosion will occur. In particular, aluminum alloys are highly resistant to warm, humid conditions where there is appreciable moisture condensation so long as contact with porous materials is avoided. Bare aluminum alloy panels have been used in constructing humidity cabinets that operate just above the dew point at 50°C. After five years of use, there was no corrosion other than minor surface staining [3].

B8. Soil Corrosion

The extent of attack that occurs on aluminum alloys buried underground varies greatly, depending on the soil composition and climatic conditions. In dry, sandy soil corrosion is negligible. In wet, acid, or alkaline soils, attack may be severe. Results of soil corrosion tests in two locations are summarize in Table 54.6 [7]. In both these locations, panels of the various alloys were buried in clay soil of the Aluminum Research Laboratories' properties in New Kensington, PA.

One location was in relatively well-drained soil and the other was in a marshy area <100 ft away. In the well-drained soil, attack on all the aluminum-base alloys, except 2017, was mild after five years. The 2017-T was severely attacked although not as much as the steel.

In the marshy soil, maximum depths of attack on all the uncoated aluminum-base alloys, except Alclad 2024-T, were appreciable and of the same order of magnitude as on steel, although the relative loss in tensile strength was definitely less for most of the aluminum-base alloys than for steel. In the case of the Alclad 2024-T, the attack that occurred was all confined to the coating, as would be expected. Chemical dip and sulfuric acid anodic coatings were definitely protective to 6053-T and presumably to the other aluminum alloys.

B9. Steam Condensate

Condensate from steam boilers, if free from carry-over of water from the boiler, is similarly inert to aluminum-base alloys. Thus, either wrought or cast aluminum alloys are used successfully for steam radiators or unit heaters. Where aluminum alloys are used, it is desirable to install suitable traps in the steam lines, since entrapped boiler water, especially if alkaline water-treating compounds are employed, may be corrosive.

TABLE 54.6. Soil Burial Tests of Five Years Duration with Aluminum Alloy Specimens[a,b]

	Well-Drained Soil			Marshy Soil		
Alloy	Max. Depth of Attack[c] (in.)	% Change in Tensile Strength[d]	Remarks (1 in. = 25.4 mm)	Max. Depth of Attack (in.)	% Change in Tensile Strength	Remarks
1100-$\frac{1}{2}$H	0.0017	−1	Mild general etching	0.0280	−7	Pitted
5052-$\frac{1}{2}$H	0.0007	+1	Mild general etching	0.0140	0	Pitted
6053-T	0.0007	0	Mild general etching	0.0150	0	Pitted
6053-T, Alrok No. 13 coated	0.0006	0	Mild general etching	0.0006	0	Mild general etching
6053-T, aluminite No. 204 coated	0.0003	0	Mild general etching	0.0002	+2	Mild general etching
2017-T	0.0380	−20	Severe pitting	0.0310	−41	Severely pitted
Alclad 2024-T	0.0013	0	Mild general etching	0.0028	−1	General etched
Steel	0.0640	−27	Completely perforated at 3 spots	0.0190	−17	Pitted

[a]Specimens in the form of panels 3 × 9 × 0.064 in. thick were buried to a depth of 61 cm in soil at the property of the Aluminum Research Laboratories in New Kensington, PA.
[b]See [7].
[c]Depth of attack determined by microscopic examination of cross sections.
[d]Change in tensile strength determined by machining tensile specimens from the panels after exposure and comparing their strength with that of unexposed tensile specimens of the same materials.

TABLE 54.7.   Resistance of Aluminum to Aqueous Solutions of Several Gases[a]

Metal	Carbon Dioxide[b] and Water		Sulfur Dioxide,[c] Air, and-Water		Hydrogen Sulfide[d] and Water	
	Av. Wt. Loss (g)	Av. ipy[e]	Av. Wt. Loss (g)	Av. ipy[e]	Av. Wt. Loss (g)	Av. ipy[e]
Aluminum 1100	0.0003	0.00004	0.150	0.0498	0.002	0.00028
Copper	—	—	0.681	0.0701	0.237	0.01030
Steel[f]	0.2153	0.00977	8.583[b]	1.02[b]	1.366	0.06800

[a]See [3].

[b]Metal specimens $1 \times 4 \times 1/16$ in. ($2.5 \times 10.2 \times 0.16$ cm) were partially immersed (to depth of 51 mm).

[c]Metal specimens $2.5 \times 10.2 \times 0.16$ cm thick were partially immersed (to a depth of 2 in.) in distilled water through which air and sulfur dioxide were bubbled. The total period of exposure was 135 h at room temperature.

[d]Metal specimens $1 \times 4 \times 1/16$ in. thick were partially immersed (to a depth of 51 mm) in distilled water through which hydrogen sulfide was bubbled. The total period of exposure was 320 h at room temperature.

[e]Inch/year, this calculation was based on the assumption that all corrosion was confined to the immersed areas of the specimens.

[f]Steel specimen corroded completely through at the water line.

B10. Gases

Most gases, in the absence of water and at or near room temperature, have little or no action on aluminum-base alloys. In the presence of water, the acid gases, such as HCl and HF are corrosive, and wet SO_2 causes corrosion (Table 54.7). Hydrogen sulfide or ammonia, either in the presence or absence of water and at room temperature or slightly above, has negligible action on aluminum-base alloys. Halogenated hydrocarbons, such as dichlorodifluoromethane, dichlorotetrafluoromethane, and monochlorodifluoromethane, are almost completely inert to aluminum. However, methyl chloride and methyl bromide are corrosive and should not be used in contact with aluminum-base alloys.

B11. Chemicals

B11.1. Acids. Acid mine waters are corrosive to aluminum-base alloys. The extent of attack depends on the specific composition of the water. Some use of aluminum pipe has been made in soft coal mines for handling acid mine waters. It has been found that pipe of aluminum alloy 3003 greatly outlasts bare or galvanized steel pipe in this application. Many aluminum-base alloys are highly resistant to nitric acid in concentrations of ~80–99%. Alloys such as 1100, 3003, and 6061 have received the widest use for handling nitric acid of these concentrations. Nitric acid of lower concentrations is more corrosive.

Dilute sulfuric acid solutions, up to ~10% in concentration, causes some attack on aluminum-base alloys, but the action is not sufficiently rapid at room temperature to prevent their use in special applications. In the concentration range of ~40–95%, rather rapid attack occurs. In extremely concentrated or fuming acid, the rate of attack drops again to a very low value.

The action on aluminum (1100) of solutions containing sulfuric acid, nitric acid, and water is illustrated in Figure 54.3. It will be noted that aluminum is most resistant to solutions dilute in both acids, or high in nitric acid concentration, or in 100% sulfuric acid. Hydrofluoric, hydrochloric, and hydrobromic acid solutions, except at concentrations below ~0.1%, are definitely corrosive to aluminum alloys. The rate of attack is greatly influenced by temperature (Fig. 54.4).

Both perchloric acid and phosphoric acid solutions in intermediate concentrations definitely attack aluminum. Dilute (< 1%) phosphoric acid solutions have a relatively mild, uniform etching action that makes them useful for cleaning aluminum surfaces. Boric acid solutions in all concentrations up to saturation have negligible action on aluminum alloys. Chromic acid solutions in concentrations up to 10% have a mild, uniform etching action. Mixtures of chromic acid and phosphoric acid have practically no action on a wide variety of aluminum alloys, even at elevated temperatures. Such mixtures are used for quantitatively removing corrosion products or oxide coatings from aluminum alloys.

Most organic acids are well resisted by aluminum alloys at room temperature. In general, rates of attack are highest for solutions containing ~1 or 2% of the acid. Formic acid, oxalic acid, and some organic acids containing chlorine (such as trichloroacetic acid) are exceptions and are definitely corrosive. Equipment made of aluminum alloys, such as 1100 or 3003, are widely and successfully used for handling acetic, butyric, citric, gluconic, malic, propionic, and tartaric acid solutions.

B11.2. Fruit Acids. Aluminum alloys also have a high resistance to the action of uncontaminated natural fruit acids. Contamination of these substances by heavy metal compounds may cause them to become corrosive. In contrast, the addition of sugar to fruit acids causes them to become even less corrosive [3].

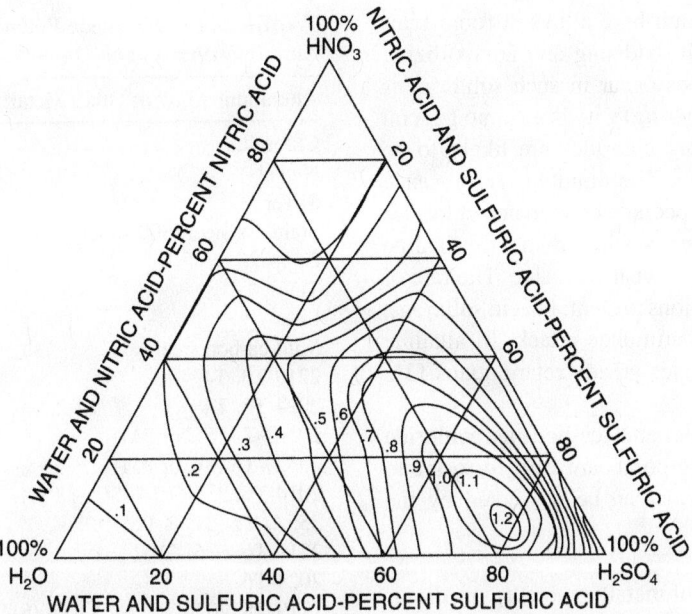

FIGURE 54.3. Action of mixtures of nitric and sulfuric acids on 1100 aluminum 24-h tests, room temperature; contours labeled in in./year (25.4 mm) [3].

B11.3. Organic Compounds. Aqueous solutions of organic chemicals having a substantially neutral reaction are generally not corrosive to aluminum-base alloys, unless these solutions are contaminated with other substances, particularly chlorides and heavy metal salts. At room temperature or slightly above, most organic compounds in the absence of water are completely inert to aluminum-base alloys. This is true for organic–sulfur compounds as well as for other organic compounds. At elevated temperatures, some organic compounds, such as methyl alcohol and phenol, definitely become corrosive, especially when they are completely anhydrous.

B11.4. Alkalis. Solutions of sodium hydroxide or potassium hydroxide in all but the lowest concentrations (<0.01%) rapidly attack aluminum and its alloys. Attack by the very dilute caustic solutions can be inhibited by corrosion inhibitors, such as silicates or chromates, but in more concentrated solutions none of the usual inhibitors are very effective. The alloys of aluminum containing more than ~4% magnesium are somewhat more resistant to attack by alkalis than are the other aluminum-base alloys. Lime or calcium hydroxide solutions are also corrosive, but the maximum rate of attack is limited by the low solubility of these materials.

The aluminum-base alloys are highly resistant to ammonia and ammonium hydroxide. The alloys that contain appreciable magnesium tend to be even less affected by ammonium hydroxide solutions than the other aluminum alloys. The amines generally have little or no action on aluminum alloys. However, a few of the most alkaline do cause definite attack.

B11.5. Salt Solutions. Neutral or nearly neutral (pH from ~5 to 8.5) solutions of most inorganic salts cause negligible

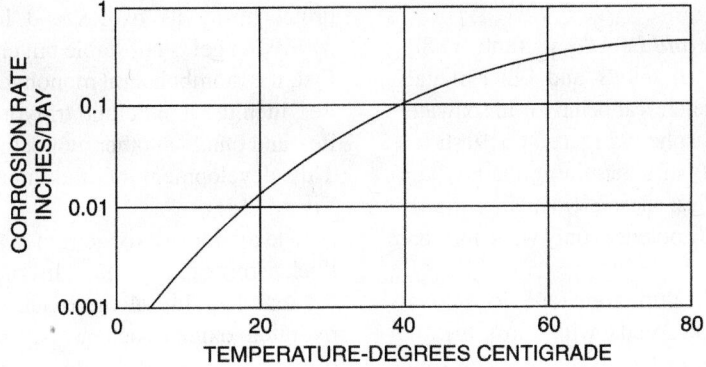

FIGURE 54.4. Effect of temperature on corrosion rate of (6053-T) aluminum in 10% HCl [3].

or minor corrosion of aluminum-base alloys at room temperature. This is true for both oxidizing and nonoxidizing solutions. Any attack that does occur in such solutions is likely to be highly localized (pitting) with little or no general corrosion. Solutions containing chlorides are likely to be more active than other solutions. The simultaneous presence of salts of the heavy metals, especially copper, and chlorides may be very detrimental. Distinctly acid or distinctly alkaline salt solutions are generally somewhat corrosive. The rate of attack depends on the specific ions present. In acid solutions, chlorides, in general, greatly stimulate attack. In alkaline solutions, silicates and chromates greatly retard attack [3].

B11.6. Dry Phenols. Phenols and carbon tetrachloride nearly dry or near their boiling points are very corrosive to aluminum alloys. This behavior can be prevented by the presence of trace water [1].

B11.7. Mercury. The action of metallic mercury on aluminum is unique. It tends to amalgamate readily with aluminum at room temperature to produce an extraordinary corrosion rate in the presence of moisture with the production of voluminous columnar corrosion products. When that reaction is started, the rate of corrosion depends on relative humidity. When dry, metallic mercury reacts only with difficulty because of the oxide film on the aluminum surface. Traces of acidity or halides on the surface cause rapid attack. Solutions containing mercury ions tend to cause rapid pitting of aluminum alloys because mercury plates out in localized areas. In many cases, the amalgamation of stressed aluminum alloy with mercury results in cracking since the mercury penetrates selectively at grain boundaries, thus weakening the material. Mercury can be removed from aluminum surfaces by treatment with 70% nitric acid. Mercury can be distilled away from an aluminum surface by treatment with steam or hot air [8].

C. TYPES AND FORMS OF CORROSION

C1. Uniform Corrosion

C1.1. Potential of Aluminum and Its Alloys. Table 54.8 is a galvanic series of aluminum alloys and other metals representative of their electrochemical behavior in seawater and in most natural waters atmospheres Figure 54.5 [9] shows the effect of alloying elements in determining the position of aluminum alloys in the series; these elements, primarily copper and zinc, affect electrode potential only when they are in solid solution.

As evident in Table 54.8, aluminum (and its alloys) becomes the anode in galvanic cells with most metals, protecting them by corroding sacrificially. Only magnesium and zinc are more anodic and corrode to protect aluminum.

TABLE 54.8. Electrode Potentials of Representative Aluminum Alloys and Other Metals[a,b]

Aluminum Alloy or Other Metal[c]	Potential (V)
Chromium	$+ 0.18$ to $- 0.40$
Nickel	$- 0.07$
Silver	$- 0.08$
Stainless steel (300 series)	$- 0.09$
Copper	$- 0.20$
Tin	$- 0.49$
Lead	$- 0.55$
Mild carbon steel	$- 0.58$
2219-T3, T4	$- 0.64$[d]
2024-T3, T4	$- 0.69$[d]
295.O-T4 (SC or PM)	$- 0.70$
295.O-T6 (SC or PM)	$- 0.71$
2014-T6, 355-O T4 (SC or PM)	$- 0.78$
355.O-T6 (SC or PM)	$- 0.79$
2219-T6, 6061-T4	$- 0.80$
2024-T6	$- 0.81$
2219-T8, 2024-T8, 356.O-T6 (SC or PM), 443,O-F (PM), cadmium	$- 0.82$
1100, 3003, 6061 T-6, 6063-T6, 7075-T6,[c] 443, O-F (SC)	$- 0.83$
1060, 1350, 3004, 7050-T73,[c] 7075-T73[c]	$- 0.84$
5052, 5086	$- 0.85$
5454	$- 0.86$
5456, 5083	$- 0.87$
7072	$- 0.96$
Zinc	$- 1.10$
Magnesium	$- 1.73$

[a]Measured in an aqueous solution of 53 g of NaCl and 3 g of H_2O_2 per liter at 25°C; 0.1 N calomel reference electrode.
[b]Reprinted from [1], pp. 111–145. Courtesy of Marcel Dekker, Inc.
[c]The potential of an aluminum alloy is the same in all tempers wherever the temper is not designated.
[d]The potential varies $\pm$ 0.01–0.02 V with quenching rate.

This type of corrosion can be found in strong acidic or strong basic solutions as illustrated in Figure 54.6 [4, 10]. The rate of corrosion can vary from several micrometers per year to several micrometers per year to several micrometers per hour.

In the range of pH between ~ 4 and 8, aluminum is protected by its oxides and hydroxides. The aluminum hydroxide gel is not stable but crystallizes with time to give, first, the rhombohedral monohydrate ($Al_2O_3 \cdot H_2O$ or boehmite), then the monoclinic trihydrate ($Al_2O_3 \cdot 3H_2O$ or bayerite), and finally another monoclinic trihydrate (hydrargilite). This development of aluminum hydroxide is known as "aging" [4].

The air-formed oxide film is amorphous alumina 2–4 nm thick at room temperature. In contact with wet environments, the external side of the oxide film hydrolyzes to produce hydrated oxides such as bayerite ($Al_2O_3 \cdot 3H_2O$) formed below 70°C and boehmite ($Al_2O_3 \cdot H_2O$ or A100H) formed at > 100°C[11].

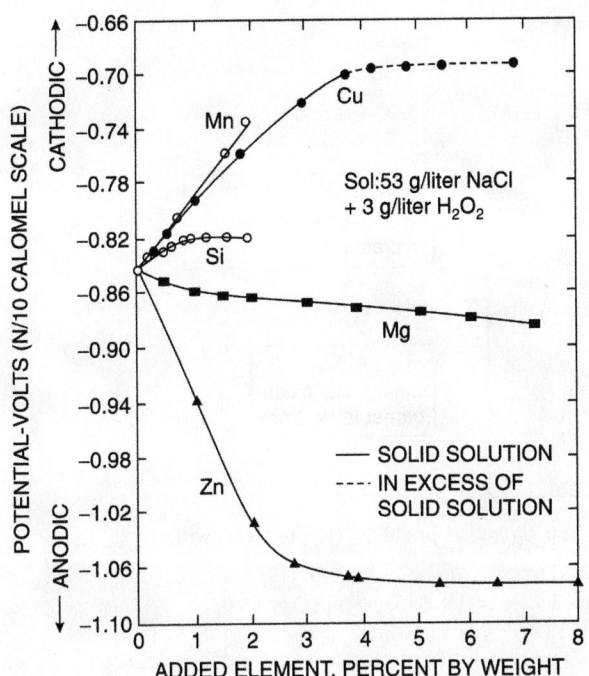

FIGURE 54.5. Effect of alloying elements on the electrode potential of aluminum [9]. (Reprinted with permission from *Metals Handbook, 9th ed.* Vol. 13, *Corrosion*, ASM International, Materials Park, OH, 1984, p. 584.)

Alloying elements such as Li, Mg, and Be, which are more active (less noble) than aluminum, oxidize first, forming poorly protecting oxides at the extreme surface. On the other hand, alloying elements nobler than aluminum, present in solid solution or in the form of small coherent precipitates (size 0.5–50 nm), produce a mixed-oxide film. In contrast, the largest precipitates (size ~1–10 μm) formed from these elements are not coherent with the matrix and often remain unoxidized [12]. Figure 54.7 shows a schematic view of aluminum oxide film on a rolled product.

Some oxides, such as those of Mn and Mg, can improve the resistance to general corrosion when they are partially

introduced into the oxide film, whereas other oxides can deteriorate the protective quality of this film.

Aluminum may corrode because of defects in its protective oxide film. Resistance to corrosion improves considerably as purity is increased, but the oxide film on even the purest aluminum contains a few defects where minute corrosion can develop. In less pure aluminum of the 1XXX series and in aluminum alloys, the presence of second phases is an important factor. These phases are present as insoluble intermetallic compounds produced primarily from iron, silicon, and other impurities and, to a lesser extent, precipitates of compounds produced primarily from soluble alloying elements. Most of the phases are cathodic to aluminum, but a few are anodic. In either case, they produce galvanic cells because of the potential difference between them and the aluminum matrix [1].

C1.2. Galvanic Corrosion. Aluminum-base alloys are anodic to many other common metals and alloys used in structures. Thus, if aluminum-base articles are exposed outdoors or in moist locations in contact with parts made of other metals, galvanic attack of the aluminum surfaces adjacent to the dissimilar metal is likely to occur. Galvanic action is much more pronounced in marine or seacoast atmospheres than in rural or industrial locations.

Contact with copper or copper-base alloys causes more pronounced galvanic attack than does contact with most other metals. In rural or industrial locations, contact with steel does not generally cause a very pronounced acceleration in rate of attack of aluminum-base alloys (especially the Al–Cu alloys such as 2017 and 2024). In seacoast locations, attack may be appreciably accelerated. In certain solutions and in some natural waters this action may be reversed, so that attack of the steel is accelerated and the aluminum is protected.

Contact between stainless steel and aluminum in seawater or other saline solutions usually results in less galvanic action on the aluminum than does contact of aluminum with steel.

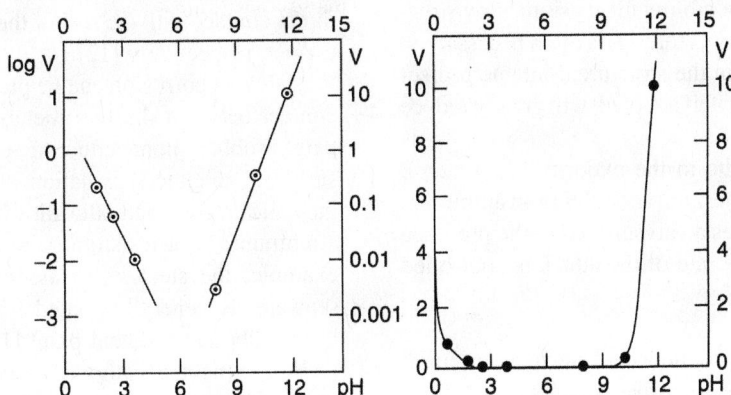

FIGURE 54.6. Influence of pH on the corrosion rate of aluminum. (Reprinted with permission from [4], NACE International and CEBELCOR.)

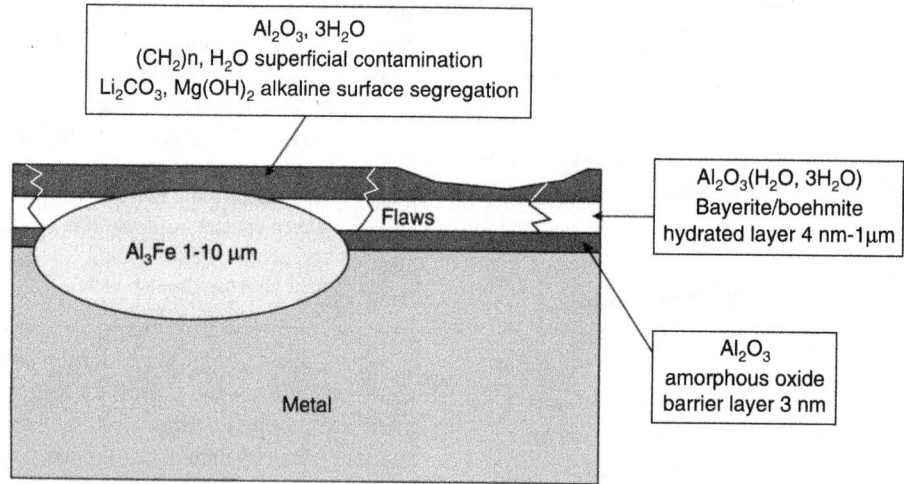

FIGURE 54.7. Schematic view of aluminum oxide film on rolled products. (Reproduced with permission from [12].)

However, no cases of reversal of the stainless steel/aluminum couple, such as occurs with steel, are known. Cadmium has about the same potential as aluminum, and therefore, contact with cadmium usually results in negligible galvanic action. Zinc is anodic to aluminum in most neutral or acid solutions; hence, in such solutions contact with zinc results in protection of the aluminum. In alkaline solutions, the potentials reverse so that, in these media, contact with zinc can cause accelerated attack of aluminum.

Magnesium and its alloys are definitely anodic to the aluminum alloys, and thus, contact with aluminum increases the corrosion rate of magnesium. However, such contact is also likely to be harmful to aluminum, since magnesium may send sufficient current to the aluminum to cause cathodic corrosion in alkaline medium. Cathodic corrosion, as mentioned above, is most likely to be encountered in seacoast locations. Certain aluminum-base alloys (such as 5056) are less affected by contact with magnesium than are other aluminum alloys. For this reason, 5056 rivets have been extensively employed in assembling magnesium alloy structures. In designing outdoor structures, it is often necessary to combine dissimilar metals in the structure. Suitable protective methods are available that, if adopted, will greatly reduce the risk of galvanic corrosion.

Since aluminum is anodic to the majority of structural metals except zinc and magnesium, assembling an aluminum alloy with another alloy gives a galvanic cell in the presence of a corrosive medium. The rate of the attack is controlled by [3]:

The difference of potential between the two structural alloys or metals in the corrosive medium

The electric resistance between the two conductors, which is frequently low

The resistivity of electrolyte; for example, seawater, with a low resistivity (a few Ω/cm^2), is particularly aggressive.

The surface area of the anode as compared to the cathode

Polarization of the two electrodes because of oxide formation and diffusion control

If the surface area of the anode (aluminum or its alloys) is very low with respect to the cathodic surface, the rate of general corrosion will be very high. In some cases, severe localized corrosion occurs, which can lead to perforation, especially if the resistance of the corrosive medium is high and the solution is stagnant.

For natural atmospheric corrosion, aluminum can be compared to other passive alloys in this atmosphere, such as stainless steels. Direct assembly between aluminum and copper should be avoided. This assumption has not been observed for all applications since some electric industries have introduced a bivalent sheet of aluminum and copper in direct contact with the rest of the structure in aluminum and copper, respectively [12].

Galvanic corrosion can be prevented by breaking electric contact between the two metals using, for example, gum, paint, rubber, nonconducting polymers, or washers with sufficient thickness. Isolation of the cathodic metal of the galvanic cell (the cathodic alloy in contact with the structural aluminum) by a resisting paint is a common practice. For example, the steel in an assembly of aluminum–steel in seawater is generally coated with zinc (metalization) and by an adherent resistant paint [13].

Removing the cathodic reactant, galvanic corrosion is reduced because the aluminum is less likely to be polarized to its pitting potential. Thus the corrosion rate of aluminum coupled to copper in seawater is greatly reduced when the seawater is deaerated. In closed multimetallic systems, the corrosion rate, even though it may be high initially, decreases to a low value whenever the cathodic reactant is depleted. Galvanic corrosion is also low where the electrical resistivity is low, as in high-purity water. Some semiconductors, such as graphite and magnetite, are cathodic to aluminum, and in contact with them, aluminum corrodes sacrificially. Galvanic corrosion of aluminum by more cathodic metals in solutions of nonhalide salts is usually less than in solutions of halide ones.

Corrosion of aluminum is controlled by an anodic reaction (oxidation), which leads to metallic dissolution and a cathodic reaction (reduction) of environmental species. The relation where the anodic reaction occurs on aluminum, and thus leads to its corrosion, is shown in Figure 54.8. The anodic polarization curve shown is typical for aluminum and its alloys when they are polarized anodically in an electrolyte free of a readily available cathodic reactant (e.g., in a deaerated electrolyte), whereas the polarization curves for the cathodic reactions are schematic only. The corrosion current developed by the two reactions (which determines the rate of corrosion of the aluminum) is indicated by the intersection of the anodic polarization curve for aluminum with one of the cathodic polarization curves.

The corrosion rate of aluminum when coupled to a more cathodic metal depends on the extent to which it is polarized in the galvanic cell. It is especially important to avoid contact with a mon cathodic metal where aluminum is polarized to its pitting potential because, as shown in Figure 54.8, a small increase in potential produces a large increase in corrosion current.

To minimize corrosion of aluminum in contact with other metals, the ratio of the exposed area of aluminum to that of the more cathodic metal should be kept as high as possible (since such a ratio reduces the current density on the aluminum). Paints and other coatings for this purpose may be applied to both the aluminum and the cathodic metal, or to the cathodic metal alone, but they should never be applied to the aluminum only, because of the difficulty in applying and maintaining them free of defects.

C2. Pitting Corrosion

Fine, white, gelatinous deposits of aluminum hydroxide often cover deep pits. Pitting corrosion is observed when:

Aluminum and its alloys are in the pH range where it is passive. On increasing acidity or alkalinity beyond its passive range of pH, corrosion attack becomes more nearly uniform. Polarization of its potential at least to its pitting potential occurs (Fig. 54.8)

In aerated solutions, the cathodic reaction is oxygen reduction, while the anodic reaction is accelerated by halide

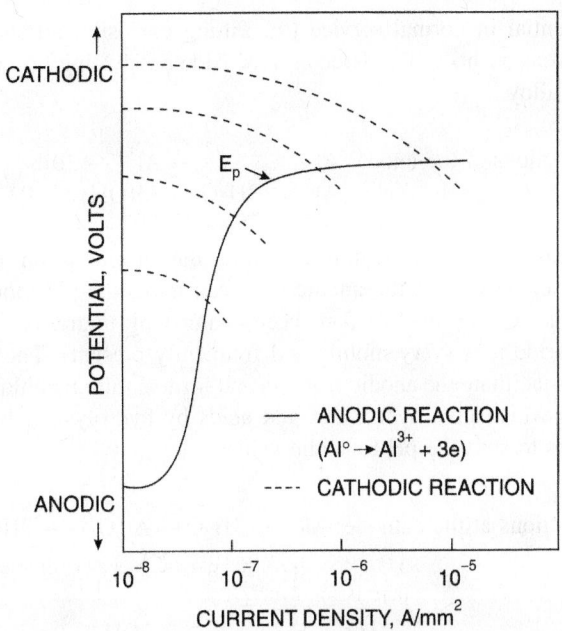

FIGURE 54.8. Typical anodic polarization curve (solid line) for an aluminum alloy in an electrolyte free of a readily available cathodic reactant (commonly oxygen); E_p is the pitting potential of the alloy. E_p for 5XXX and 6XXX alloys at 25°C is of the order of −0.4 to 0.7 V/SCE in a deaerated solution with 0.3–3% Cl at ambient temperature [15]. The intersection of this curve with one of the cathodic polarization curves (schematic) determines the corrosion current of the alloy [14]. (Reprinted with permission from *Metals Handbook, 9th ed.*, Vol. 2, *Aluminum Alloys*, ASM International, Metals Park, OH, 1979, p. 215 (formerly the American Society for Metals, Materials Park, OH.)

ions, of which chloride is the one most frequently encountered in service. Pitting is observed in aerated solutions of halides in the passive region of pH.

The development of pitting can be prevented by removal of the reducible species required for a cathodic reaction. In neutral solutions, this species is usually oxygen. Thus its removal by deaeration prevents the development of pitting in aluminum even in most halide solutions because, in its absence, the cathodic reactions are not sufficient to polarize aluminum to its pitting potential. Metallurgical structure has little effect on the pitting potential of aluminum, nor do second phases in the amounts present in its alloy have significant effect. Severe cold work makes the potential more anodic by a few millivolts, and this change, although small, is sufficient to affect the extent to which pitting develops, for example, more pitting on machined or sheared edges [16].

Four laboratory procedures have been developed to measure E_p– one based on fixed current and the other three on controlled potential [17]. Generally, aluminum does not pit in aerated solutions of nonhalide salts, because the pitting potential is considerably more noble than it is in halide solutions, and aluminum is not polarized to this level of potential in normal service [9]. Pitting corrosion initiates at weak points of the oxide or hydroxide passivating film of the alloy:

Reactions at the anode: $Al \rightarrow Al^{3+} + 3e^-$
$$Al^{3+} + 3H_2O \rightarrow Al(OH)_3 + 3H^+$$

Considering a neutral solution, the consumption of hydroxide ions at the anodic sites can make the pH more acidic, to the level of 3–4, accompanied by migration of chloride ions (very mobile and frequently present). These ions facilitate the anodic reaction and form aluminum chlorides which give hydroxides and acids by hydrolysis. This helps to shift the pH to acidic values.

Reactions at the cathode: $AlCl_3 + 3H_2O \rightarrow Al(OH)_3 + 3HCl$
$$3H^+ + 3e^- \rightarrow \tfrac{3}{2}H_2$$
$$\tfrac{1}{2}O_2 + H_2O + 2e^- \rightarrow 2OH^-$$

The cathodic sites are frequently more alkaline because of the local formation of hydroxides. The presence of oxygen and/ or another oxidant is essential for pitting.

Pitting of aluminum matrix composites 1050 and 2124 each reinforced silicon carbide particles (SiC_p) in the size range 3–40 µm has been studied in 1 N NaCl solution. Pores and crevices at SiC_p/matrix interfaces strongly influence pit initiation, which is further aided by the cracking of large SiC_p during processing. The presence of $CuAl_2$ and $CuMgAl_2$ precipitates in 2124–SiC_p composite also promotes pitting attack at SiC_p–matrix and intermetallic–matrix interfaces [18].

C2.1. Pitting Evaluation.
The pitting density can be obtained for a surface of about $1\,dm^2$ area. The kinetics of perforation is obtained for different periods of immersion, 1, 2, 3, 6, and 12 months. Aziz [19] expressed the pitting kinetics by the following relation:

$$P = K + \tfrac{1}{3}\Gamma$$

Where P is the depth of the pit, K is constant, and Γ is the corrosion time in years. In the case of aluminum, the rate of perforation decreases with time. The average and maximum perforation should be determined. Figure 54.9 gives an expression of the depth of the pit as a function of time. Doubling the sheet thickness can multiply the service life by 8. Although statistical considerations are applied to deduce the resistance of the aluminum alloy in a certain environment, the maximum perforation rate should be used for design and prevention methods [19].

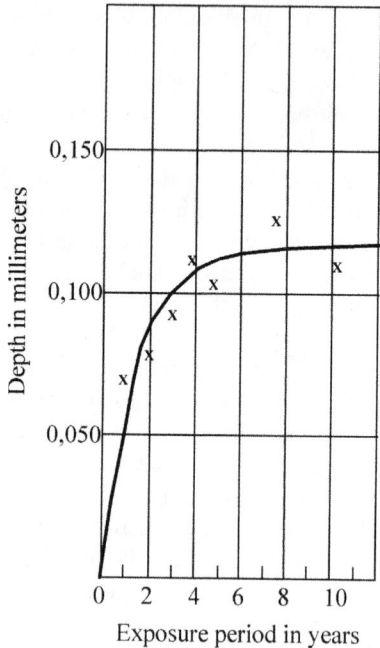

FIGURE 54.9. Pit depth as a function of exposure period. (Reprinted with permission from NACE International, Houston, TX [19]).

C2.2. Temperature.
It has been shown that the pit density of aluminum increases with temperature in water (Figs. 54.10 Figs. 54.11) and that the depth of pits decreases with temperature. It has been suggested that extended contact of the metal with water favors the formation of a protective oxide film [13].

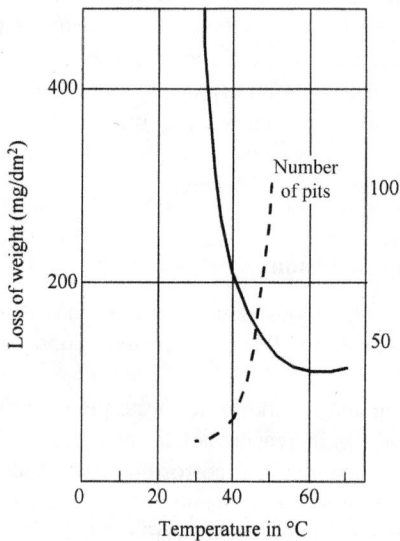

FIGURE 54.10. Influence of water temperature on pitting of aluminum [13]. (Reprinted with permission from Dunod, Paris France, 1976.)

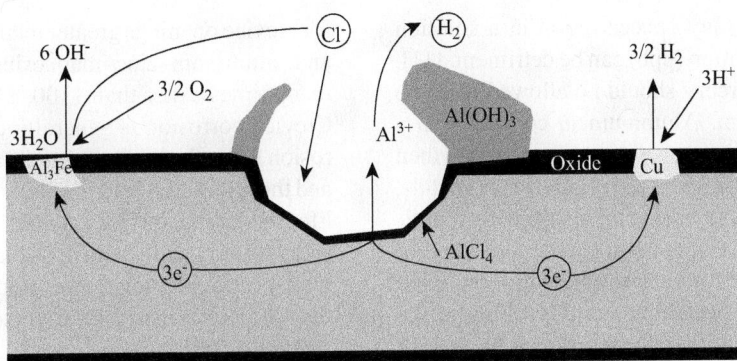

FIGURE 54.11. Pit propagation [22].

C2.3. Pitting Mechanism. Pit initiation in an oxygenated chloride solution is generally controlled by the cathodic reaction Kinetics [20]. Propagating pits require a sufficient Cl^- concentration in the solution contained within the pit [21], and the Cl^- causes the formation of a concentrated $AlCl_3$ solution within active pits (Fig. 54.11).

The corrosion potential exhibits time fluctuations, corresponding to the elementary depassivation–repassivation events. Reboul et al. [22] proposed a 10-step mechanism for the pitting corrosion of aluminum in the presence of chloride ions:

1. The Cl^- adsorption in microflaws of the oxide film, assisted by the high electric field (10^7 V/cm) through the barrier oxide film, resulting form the Al-air corrosion cell (emf 2.9 V).
2. Slow oxygen reduction on the cathodic area, charging the double-layer capacitance ($50\,\mu F/cm^2$).
3. Dielectric breakdown of the oxide film at weak points corresponding to the microflaws.
4. Fast aluminum oxidation of bare aluminum producing soluble chloride and oxychloride complexes at the bottom of flaws.
5. Dissolution of chloride complexes and repassivation of pits.
   These first five steps produce $10^6/cm^2$ micropits of size 0.1–1 μm.
6. Exceptionally, and for some different (often unexplained) reasons, a few micropits propagate. This propagation requires the stabilization of a chloride/oxychloride layer at the active bottom of pits. This layer should be renewed faster than it dissolves, which implies a large enough cathodic area, resulting from the repassivation of the surrounding competitive pits, formed during step 4.
7. Hydrolysis of soluble chlorides/oxychlorides, resulting in the acidification (to pH 3) of the solution within pits.

8. Hydroxide dissolution inside pits and precipitation of aluminum hydroxide outside pits, resulting in the formation of cone-shaped accumulations of corrosion products at the mouths of pits.
9. Aluminum corrosion inside the pits due to the aggressive hydrochloric acid solution.
10. Repassivation and pit death when I_{pit}/r_{pit} (r is the radius of the pit) decreases to 10^{-2} A/cm. The chloride/oxychloride film is dissolved and replaced by a passive oxide film. The solution within the pit reverts to the composition of the bulk solution.

C3. Deposition Corrosion

Deposition corrosion is a particular case of galvanic corrosion that leads to pitting. Aluminum reduces ions of many metals, of which copper, cobalt, lead, mercury, nickel, and tin are the ones encountered most commonly. Reduction of these heavy metal ions leads to corrosion of aluminum and deposition of the more noble metal on the surface of the aluminum. Once this metal is deposited, it leads to serious attack of the alloy due to the galvanic cell that is established [23].

Reducible metallic ions are of most concern in acidic solutions since their solubilities are greatly reduced in alkaline solution. Copper is the heavy metal most commonly encountered in applications of aluminum. A copper ion concentration of 0.02–0.05 ppm in neutral or acidic solutions is generally considered to be the threshold value for initiation of pitting on aluminum.

Ferric ions (Fe^{3+}) can be reduced by aluminum, but do not usually form a metallic deposit. At room temperature, the most anodic aluminum alloys (those with a corrosion potential approaching -1.0 V vs. SCE) can reduce ferrous ions (Fe^{2+}) to metallic iron and produce a metallic deposit on the surface of the aluminum. The presence of Fe^{2+} ion also tends to be rare in service; it exists only in deaerated solutions or in other solutions free of oxidizing agents [24].

In the case of mercury, any concentration in a solution more than a few parts per billion (ppb) can be detrimental [1]. No amount of metallic mercury should be allowed to come into contact with aluminum. Aluminum in contact with a solution of a mercury salt forms metallic mercury, which then readily amalgamates the aluminum. Of all the heavy metals, mercury can cause the most corrosion damage to aluminum [25]. The effect can be severe when stress is present. For example, attack by mercury and zinc amalgam combined with residual stresses from welding caused cracking of the weldment. The corrosive action of mercury can be serious with or without stress because amalgamation, once initiated, continues to propagate unless the mercury can be removed. The corrosive action of mercury can be attributed not only to the galvanic cell but also to the destruction and prevention of formation of aluminum oxide. The corrosion rate can be extremely high, up to 1270 mm/year [26].

C4. Crevice Corrosion

If an electrolyte is present in a crevice formed between two faying aluminum surfaces, or between an aluminum surface and a nonmetallic material, such as a gasket, localized corrosion in the form of pits or etch patches may occur. The oxygen content of the liquid in the crevice is consumed by the film formation reaction on the aluminum surface, and corrosion stops because the replenishment of oxygen by diffusion into the crevice is slow. At the mouth of the crevice, whether it is submerged or exposed to air, oxygen is more plentiful. This difference in oxygen concentration creates a local cell: water with oxygen versus water without oxygen. Localized corrosion occurs in the oxygen-depleted zone (anode) immediately adjacent to the oxygen-rich (cathode) near the mouth of the crevice. Once the crevice attack has initiated, the anodic area becomes acidic and the cathodic area becomes alkaline. These changes further accelerate local cell action.

The amount of aluminum consumed by crevice corrosion is small and is of practical importance only when the metal is of thin cross section or in cases where surface appearance is important. The expansion force of corrosion products produced in a confined space can be more serious. These corrosion products are about five times the volume as the metal from which they were produced, about twice the volume as rust on steel, and can distort even heavy sections of metal [27].

The ratio of the actively corroding surface area in the crevice to the effective external cathode area has been shown by Rosenfeld [28] to be an important factor for submerged crevices. Increasing external cathode areas increases corrosion rate. The corrosion rate of aluminum–magnesium alloy in 0.5 N NaCl increased as the crevice mouth narrowed from 0.14 to 0.04 mm. Similar results were obtained for aluminum and aluminum–manganese alloys. Corrosion rates were low

for crevice openings greater than 254 μm. Aluminum–copper and aluminum–zinc–magnesium–copper alloys corroded many times faster than 1100, 3XXX, or 5XXX alloys [28]. Crevice corrosion is generally critical for atmospheric corrosion when the thickness of the sheet is less than about 1 mm and the required service life is more than ∼ 5 years; for longer life, the faying surfaces should be coated with an inhibitive paint system, and where possible the crevice should be filled with a resilient, moisture-excluding sealant. On thicker sections no provision is usually necessary [27].

C4.1. Crevices in Waters. In most freshwaters, crevice corrosion of aluminum is negligible. In seawater, crevice corrosion takes the form of pitting, and the rate is low. Resistance to crevice corrosion has been found to parallel resistance to pitting corrosion in seawater and is higher for aluminums–magnesium alloys than for aluminum–magnesium–silicon alloys [29].

C4.2. Water Staining. The most common case of aluminum crevice corrosion occurs when water is present in the restricted space between layers of aluminum in close contact, as in packages of sheets or wraps of coil or foil. This may occur during storage or transit because of inadequate protection from rain or be caused by condensation within the crevice when the metal surface temperature falls below the dew point. The color of corrosion products can vary from gray to brown to black. In severe cases, the corrosion product cements the two surfaces together and makes separation difficult.

In some cases, the stain pattern shows a series of irregular rings, like the lines on a contour map. These may indicate the outlines of a receding water pool at various stages of evaporation. The stained areas are not more susceptible to subsequent corrosion; on the contrary, they are more resistant because they are covered with a thickened oxide film. Water staining can be prevented by avoiding exposure to rain and condensation conditions. The metal temperature must be maintained above the dew point, by either providing a low relative humidity or preventing cooling of the metal [27].

C4.3. Filiform Corrosion. Filiform corrosion is another special case of crevice corrosion that may occur on an aluminum surface under an organic coating. It takes the form of randomly distributed threadlike filaments and is sometimes called vermiform or worms track corrosion. The corrosion products cause a bulge in the surface coating much like molehills in a lawn. When dry, their filaments may take on an iridescent or clear appearance because of internal light reflection. The tracks proceed from one or several points where the coating is breached. The surface film itself is not involved in the process, except in the role of providing inadequate zones of poor adhesion that form the crevices in which corrosion occurs upon exposure to moisture with restricted access of oxygen. Filiform corrosion has occurred on lacquered aluminum surfaces in aircraft exposed to marine and other high-humidity environments [27]. Filiform attack is particularly severe in warm coastal and tropical regions that experience salt fall or in heavily polluted industrial areas. Rougher surfaces also experience a greater severity of filiform corrosion.

Aluminum is susceptible to filiform corrosion in the relative humidity range of 75–95%, with temperatures between 20 and 40°C (70 and 105°F). Relative humidities as low as 30% in hydrochloric acid (HCl) vapors have been reported to cause filiform corrosion [30]. Filiforms in aluminum grow most rapidly at 85% relative humidity. Typical filament growth rates average about 0.1 mm/day (4 mils/day). Filament width varies with increasing relative humidity from 0.3 to 3 mm (12 to 120 mils). The depth of penetration in aluminum can be as deep as 15 µm (0.6 mil). Numerous coating systems used on aluminum are susceptible to filiform corrosion, including epoxy, polyurethane, alkyd, phenoxy, and vinyl coatings. Condensates containing chloride, bromide, sulfate, carbonate, and nitrate ions have stimulated filiform growth on coated aluminum alloys.

The filiform cell consists of an active head and a tail that receives oxygen and condensed water vapor through cracks and splits in the applied coating. The cell is driven by a difference of potential between the head and the tail of the order of 0.1–0.2 V. In aluminum, the head is filled with flowing floes of opalescent alumina gel moving toward the tail. Gas bubbles may be present if the head is very acidic. In aluminum, filiform tails are whitish in appearance. The corrosion products are hydroxides and oxides of aluminum. Anodic reactions produce Al^{3+} ions, which react to form insoluble precipitates with the hydroxyl ions produced in the oxygen reduction reaction occurring predominately in the tail.

The mechanisms of initiation and propagation of filiform corrosion in aluminum are the same as for coated iron and steel, as shown in Figure 54.12. The acidified head is a moving pool of electrolyte, but the tail is a region in which aluminum transport and reaction with hydroxyl ions take place. The final corrosion products are partially hydrated and fully expanded in the porous tail. The head and middle sections of the tail are locations for the various initial reactant

ions and the intermediate products of corroding aluminum in aqueous media [31].

In contrast to steel, aluminum has shown a greater tendency to form blisters in acidic media, with hydrogen gas evolved in cathodic reactions in the head region. The corrosion product in the tail is aluminum trihydroxide, $Al(OH)_3$, a whitish gelatinous precipitate. If filiform corrosion is neglected, more serious structural damage caused by other forms of corrosion may develop.

Aluminum is widely used for cans and other types of packaging. Aluminum foil is routinely laminated to paperboard to form a moisture or vapor barrier. If the aluminum foil is consumed by filiform corrosion, the product may be contaminated or dried out because of breaks in the vapor barrier. Typical coatings on aluminum foil are nitrocellulose and polyvinylchloride (PVC), which provide a good intermediate layer for colorful printing inks.

Degradation of the foil-laminated paperboard may occur during its production or its subsequent storage in a moist or humid environment. During the production of foil-laminated paperboard, moisture from the paperboard is released after heating in a continuous-curing oven. Heat curing dries the lacquer on the foil. Filiform corrosion can result as the heated laminate is cut into sheets and stacked on skids while the board is still releasing stored moisture.

As shown in Figure 54.13, the hygroscopic paperboard is a good storage area for moisture. Packages later exposed to humidities >75% in warm areas can also experience filiform attack. Coatings with water-reactive solvents, such as polyvinyl acetate, should not be used. Any solvents entrapped in the coating can weaken the coating, induce pores, or provide an acidic medium for further filament propagation. Harsh curing environments can also result in the formation of flaws in the coating due to uneven shrinkage or rapid volatilization of the solvent. Rough handling can induce mechanical rips and tears.

In aircraft, filiform corrosion was observed on 2024 and 7000 series aluminum alloys coated with polyurethane and other coatings. Two-coat polyurethane paint systems experienced far fewer incidences of filiform corrosion than single-coat systems did. Filiform corrosion rarely occurred when bare aluminum was chromic acid anodized or primed

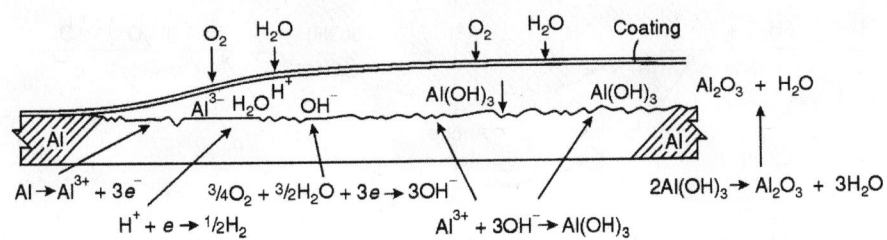

FIGURE 54.12. Filiform corrosion of aluminum [31]. (Reprinted with permission from *Metals Handbook, 9th ed.,* Vol. 13, *Corrosion,* ASM International, Materials Park, OH, 1987, p. 110.)

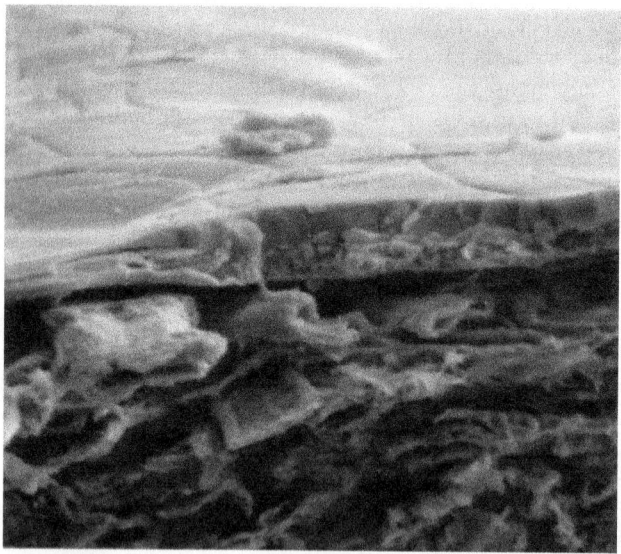

FIGURE 54.13. Cross section of aluminum foil laminated on paperboard showing the expansion of the PVC coating by the corrosion products of filiform corrosion. Note the void spaces between the paperboard fibers that can entrap water. SEM. 650X [31]. (Reprinted with permission from *Metals Handbook, 9th ed.* Vol. 13, *Corrosion*, ASM International, Materials Park, OH, 1987, p. 110).

with chromate or chromate–phosphate conversion coatings [32, 33].

Reducing relative humidity <60%, especially for long-term storage, can prevent filiform corrosion. Also the use of zinc and zinc primers on steel, chromic acid anodizing and chromate or chromate–phosphate coatings have provided some relief from filiform corrosion. Multiple coat systems resist penetration by mechanical abrasion and have a more uniform surface [31].

C5. Biological Corrosion

Biological attack frequently leads to the formation of a tubercle that covers a deep pit. When bacteria are present, the tubercle structure is usually less brittle and less easily removed from the metal surface than when they are absent. The organisms grow either in continuous mats or sludges or in volcano-like tubercles with gas bubbling from the center, as shown schematically in Figure 54.14 [31].

The organisms commonly held responsible are pseudomonas, cladosporium, and desulfovibrio. These are often suspected of working together in causing the attack. *Cladosporium resinae* is usually the principal organism involved; it produces a variety of organic acids (pH 3–4 or lower) and metabolizes certain fuel constituents. These organisms may also act in combination with the slime-forming pseudomonads to produce the corrosion product with oxygen, $Fe(OH)_3$, which, with the biodeposit, forms the wall of the growing tubercle. The outside of the tubercle becomes cathodic, while the metal surface inside becomes highly anodic.

As the tubercle matures, some of the biomass may start to decompose, providing a source of sulfates for sulfur–reducing bacteria (SRB) to use in producing H_2S in the anaerobic interior solution. In some cases, sulfur-oxidizing bacteria may assist in the formation of the sulfates. Depending on the ions available in the water, the tubercle structure may contain some $FeCO_3$ and, when SRBs, are present, some FeS. Finally, if there is a source of chlorides and if the iron-oxidizing bacteria gallionella are present, a highly acidic, ferric chloride solution may form inside the tubercle.

Generally, not all of the above reactions will take place in any single environment. As the individual tubercles on a surface grow under the influence of any combination of reactions, they will eventually combine to form a mass that severely limits flow (or even closes it off altogether), leaving a severely pitted surface underneath.

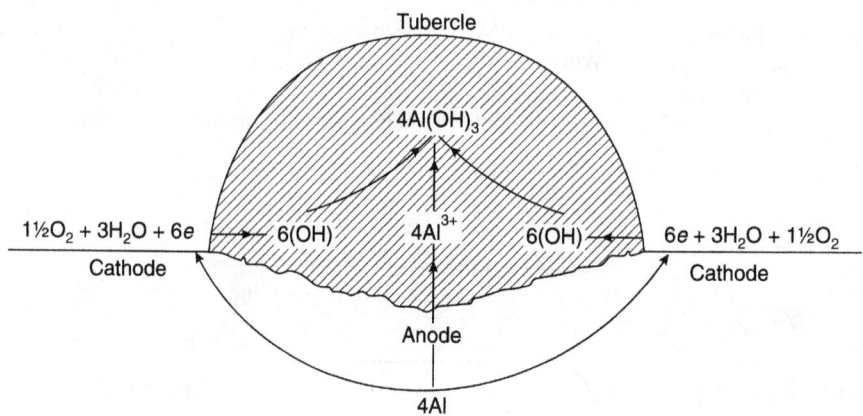

FIGURE 54.14. Schematic of tubercle formed by bacteria on an aluminum alloy surface [31]. (Reprinted with permission from *Metals Handbook, 9th ed.* Vol. 13, *Corrosion,* ASM International, Materials Park, OH, 1987, p. 111.)

Pitting corrosion of integral wing aluminum fuel tanks in aircraft that use kerosene-base fuels is an example of biological attack that has been a problem since the 1950s. The fuel becomes contaminated with water by vapor condensation during variable–temperature flight conditions. Attack occurs under microbial deposits in the water phase and at the fuel–water interface [34].

C6. Erosion–Corrosion

In noncorrosive environments, such as high–purity water, the stronger aluminum alloys have the greatest resistance to erosion–corrosion because resistance is controlled almost entirely by the mechanical components of the system. In a corrosive environment, such as seawater, the corrosion component becomes the controlling factor; thus, resistance may be greater for the more corrosion, resistant alloys even though they are lower in strength. Corrosion inhibitors and cathodic protection have been used to minimize erosion–corrosion, impingement, and cavitation on aluminum alloys [27].

In the case of neutral solutions, the velocity of the solution, up to ~6 m/s, has little effect on the rate of attack. In some cases, increased movement of the liquid may actually reduce attack by assuring greater uniformity of environment. However, increases in velocity decrease the variation in pH that can be tolerated without erosive attack occurring [3].

C7. Intergranular and Exfoliation Corrosion

Intergranular corrosion in aluminum alloys can be caused by direct corrosion of a precipitate that is less corrosion resistant (more active) than the matrix or by corrosion of a denuded zone adjacent to a noble phase.

Although the aluminum alloys are more resistant to intergranular corrosion in the solution-treated condition, avoiding precipitates is not a practical means of avoiding intergranular corrosion in these systems. The precipitates are important to the strengthening of the alloys and are necessary for their performance. Whether or not the alloy will be subject to intergranular corrosion in a particular environment is an important part of the alloy selection process.

Since intergranular corrosion is involved in stress corrosion cracking (SCC) of aluminum alloys, it is often presumed to be more deleterious than pitting or uniform corrosion. Exfoliation is a form of intergranular corrosion that may occur when aluminum alloys have their grains elongated in layers parallel to their surfaces. Intergranular corrosion can occur on the elongated grain boundaries. The corrosion product that forms has a greater volume than the volume of the parent metal, and the increased volume forces the layers apart, causing strips of metal to exfoliate (delaminate).

C7.1. Intergranular Corrosion.
Intergranular corrosion is the selective attack of the grain boundary zone, with no appreciable attack of the grain body or matrix (Fig. 54.15). The mechanism is electrochemical and is the result of local cell action in the grain boundaries. Cells are formed between second-phase microconstituents and the depleted aluminum solid solution from which these microconstituents formed.

The microconstituents have a different corrosion potential than the adjacent depleted solid solution. In some alloys, such as the aluminum–magnesium and aluminum–zinc–magnesium copper families, the precipitates Mg_2Al_3, $MgZn_2$, and Al_x–Zn_x–Mg are more anodic than the adjacent solid solution. In other alloys, such as aluminum–copper, the precipitates ($CuAl_2$ and Al_xCu_xMg) are cathodic to the depleted solid solution. In either case, selective attack of the grain boundary region occurs.

The degree of intergranular susceptibility is controlled by fabrication practices that can affect the quantity, size, and distribution of second-phase intermetallic precipitates. Resistance to intergranular corrosion is obtained by heat treatments that cause precipitation to be more general throughout the grain structure or by restricting the amount of alloying elements that cause the problem. Alloys that do not form second-phase microconstituents at grain boundaries, or those in which the microconstituents have corrosion potentials similar to the matrix (e.g., $MnAl_6$), are not susceptible to

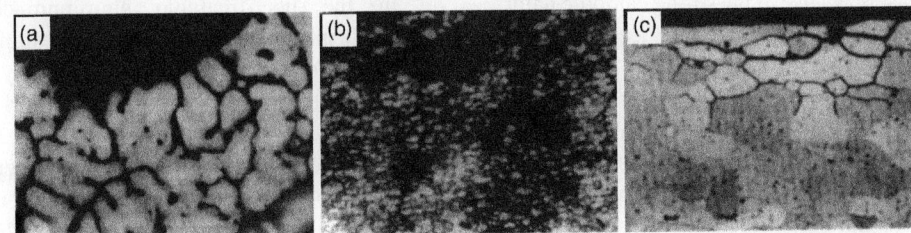

FIGURE 54.15. Various types of intergranular corrosion attack (a) interdendritic (ID) corrosion in a cast structure. (b) Interfragmentary (IF) corrosion in a wrought structure, unrecrystallized. (c) Intergranular (I or IG) corrosion in a recrystallized wrought structure. Keller's etch, (500X) [27]. (Reprinted with permission from *Aluminum: Properties and Physical Metallurgy,* edited by J. E. Hatch, ASM International, Materials Park, OH, 1984, p. 303.)

intergranular corrosion. Examples of alloys of this type are 1100, 3003, and 3004 [27].

In 2XXX series alloys, it is a narrow band on either side of the grain boundary that is depleted in copper. As an example, in the 2024 alloy, $CuAl_2$ precipitates are more noble than the matrix and act as cathodes, accelerating the corrosion of a depleted zone adjacent to the grain boundary. A similar phenomenon is observed for alloy 7075.

In aluminum–copper–magnesium alloys (2XXX), thermal treatments that cause selective grain boundary precipitation lead to intergranular corrosion susceptibility. Many studies have shown that fast cooling or quenching during heat treatment and subsequent aging to peak or slightly overaged strength results in high resistance to intergranular corrosion. Conversely, slow cooling results in intergranular susceptibility.

Aluminum–magnesium alloys (5XXX) containing <3% magnesium are resistant to intersgranular corrosion. In unusual instances, intergranular attack has occurred in the heat-affected zone of a weldment after months or years of exposure at moderately elevated temperatures of 100°C (212°F), in hot, acidified ammonium nitrate solutions of 150°C (300°F), or hot, potable water at 80°C (175°F).

Aluminum–magnesiumalloys that contain >3% Mg (e.g., 5083) may become susceptible to intergranular corrosion because of preferential attack of Mg_2Al_3 (anodic constituent). Intergranular corrosion does not occur when these alloys are correctly fabricated and used at ambient temperatures. These alloys can become susceptible to intergranular corrosion, however, after prolonged exposure to temperatures above 27°C (sensitization). Susceptibility increases with magnesium content, time, temperature, and amount of cold work.

Aluminum–magnesium–silicon wrought alloys (6XXX) usually show some susceptibility to intergranular corrosion. With a balanced magnesium–silicon composition that results in the formation of Mg_2Si constituent, intergranular attack is minor and less than that observed with aluminum–copper (2XXX) and aluminum–zinc–magnesium–copper (7XXX) alloys. When the 6XXX alloy contains an excessive amount of silicon (more than that needed to form Mg_2Si), intergranular corrosion increases because of the strong cathodic nature of the insoluble silicon.

In aluminum–magnesium–zinc alloys such as 7030, the compound $MgZn_2$ is attacked [29]. Intergranular corrosion in aluminum–zinc–magnesium–copper (7XXX) alloys can be affected by thermal treatments. Heat treatment, sometimes in combination with strain hardening, is used to provide good resistance to intergranular corrosion.

C7.2. Exfoliation Corrosion. Exfoliation, also called layer corrosion or lamellar corrosion, is a type of selective subsurface attack that proceeds along multiple narrow paths parallel to the surface of the metal. The attack is usually along grain boundaries (intergranular corrosion), but it has also been observed along striations of insoluble constituents that have strung out in parallel planes in the direction of working. Exfoliation occurs predominantly in relatively thin products with highly cold-worked, elongated grain structures. The intensity of exfoliation increases in slightly acidic environments and when the aluminum is coupled to a cathodic dissimilar metal.

Exfoliation is characterized by leafing, or alternate layers of thin, relatively uncorroded metal and thicker layers of corrosion product of larger volume than the original metal. The layers of corrosion products cause the metal to swell. In an extreme case, an ~1.3-mm- (0.050-in-) thick sheet was observed to swell to one 25 mm (1 in) thick [27].

Exfoliation corrosion may occur on material that has a marked fibrous structure caused by rolling or extrusion. Some authorities regard it as a form of stress corrosion, the stress being either inherent in the metal or produced through the pressure of the larger volume of the corrosion product. It is rare, occurring mainly in copper-bearing aluminum alloys, but can occur in a number of environments, including some regarded as only mildly corrosive. Suitable adjustments of aging treatments and copper content may largely overcome the effect in the higher strength Al–Cu type alloys [35].

Exfoliation usually proceeds inward laterally from a sheared edge, rather than inward from a rolled or extruded surface. In mild cases, it takes the form of blisters that resemble volcanoes, with corrosion products swelling up in the center. In this case, pits occur first and proceed inward until the susceptible layer is encountered. The attack then changes to lateral penetration with generation of less dense corrosion products that cause the blisters to develop. Metallographic examination, visual rating, and weight loss measurements after exposure to corrosive environments (solutions and sprays) at ambient and elevated temperatures can be used to test for exfoliation corrosion susceptibility [27]. The commercial-purity aluminum (1XXX) and aluminum–manganese (3XXX) alloys are quite resistant to exfoliation corrosion in all tempers. Exfoliation has been encountered in some highly cold worked aluminum–magnesium (5XXX) materials–especially in seawater media.

In the heat-treatable aluminum–copper–magnesium (2XXX) and aluminum–zinc–magnesium–copper (7XXX) alloys, exfoliation corrosion has usually been confined to relatively thin sections of, highly worked products with an elongated grain structure. In the 2124-T351 plate, for example the 13-mm plates, was quite susceptible in laboratory and atmospheric tests, while 50– and 100–mm plates with less directional microstructures, did not exfoliate. In extrusions, the surface is often quite resistant to exfoliation because of the recrystallized grain structure. Subsurface grains are unrecrystallized, elongated, and vulnerable to exfoliation.

In aluminum–zinc–magnesium alloys containing copper, such as 7075, resistance to exfoliation can be improved markedly by overaging, designated by the temper designations of T7XXX for wrought products. While a 5–10% loss in strength occurs, improved resistance to exfoliation is provided. In copper-free or low-copper 7XXX alloys, exfoliation corrosion can be controlled by overaging or by recrystallizing heat treatments and can also be controlled to some extent by changes in alloying elements. In aluminum–copper–magnesium (2XXX) alloys, artificial aging to the T6 or T8 condition provides improved resistance [27].

C8. Stress Corrosion Cracking

Only aluminum alloys that contain appreciable amounts of soluble alloying elements, primarily copper, magnesium, silicon, and zinc, are susceptible to SCC. For most commercial alloys, tempers have been developed that provide a high degree of immunity to SCC in most environments.

There is general agreement that for aluminum the electrochemical factor predominates and the electrochemical theory continues to be the basis for developing aluminum alloys and tempers resistant to SCC [36]. The complex interactions among factors that lead to SCC of aluminum alloys are not yet fully understood [37].

According to the electrochemical theory, susceptibility to intergranular corrosion is a prerequisite for susceptibility to SCC, and treatment of aluminum alloys to improve resistance to SCC also improves their resistance to intergranular corrosion. For most alloys, however, optimum levels of resistance to these two types of corrosion require different treatments, and resistance to intergranular corrosion is not a reliable indicator of resistance to SCC.

Stress corrosion cracking in aluminum alloys is characteristically intergranular. This type of corrosion requires a condition along grain boundaries that makes them anodic to the rest of the microstructure so that corrosion propagates

selectively along them. Such a condition is produced by localized decomposition of solid solution, with a high degree of continuity of decomposition products along the grain boundaries. The most anodic regions may be either the boundaries themselves (most commonly, the precipitate formed in them) or regions adjoining the boundaries that have been depleted of solute. In 2XXX alloys, the solute-depleted regions are the most anodic; in 5XXX alloys, it is the Mg_2Al_3 precipitate along the grain boundaries that is anodic. The most anodic grain boundary regions in other alloys have not been identified with certainty [9].

C8.1. Effect of Stress. The SCC in a susceptible aluminum alloy depends on both magnitude and duration of tensile stress acting at the surface. The effects of stress have been established by accelerated laboratory tests, and the results of one set of such tests are shown in the shaded bands in Figure 54.16. Despite the introduction of fracture mechanics techniques for determining crack growth rates, such tests continue to be the basic tools used in evaluating resistance of aluminum alloys to SCC. These tests suggest a minimum (threshold) stress that is required for cracking to develop.

For some alloy/temper combinations, results of accelerated laboratory tests reliably predict stress corrosion performance in service; for example, results of an 84-day alternate immersion test of alloy 7075 and alloy 7178 products correlated well with performance of these products in a seacoast environment [9].

C8.2. Effects of Grain Structure and Stress Orientation. Many wrought aluminum alloy products have highly directional grain structures and are highly anisotropic with respect to resistance to SCC (Fig 54.16). Resistance, measured by magnitude of tensile stress required to cause cracking, is highest when the stress is applied in the longitudinal direction

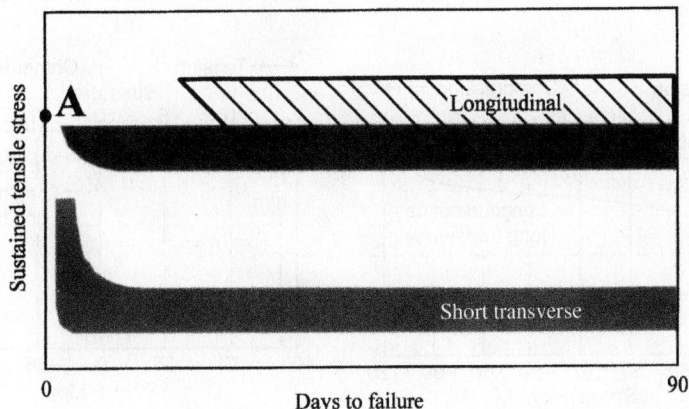

FIGURE 54.16. The SCC of alloy 7075-T651 plate. Shaded bands indicate combinations of stress and time known to produce SCC in specimens intermittently immersed in 3.5% NaCl solution. Point A is minimum yield strength in the long transverse direction for a 75-mm-(3-in.-) thick plate [9]. (Reprinted with permission from *Metals Handbook*, 9th ed. Vol. 13, *Corrosion,* ASM International, Materials Park, OH, 1987, p. 596).

and is lowest in the short-transverse direction and is intermediate in other directions. These differences are most noticeable in the more susceptible tempers but are usually much lower in tempers produced by extended precipitation treatments, such as T6 and T8 tempers for 2XXX alloys and T73, T736, and T76 tempers for 7XXX alloys.

One of the most common practices associated with SCC problems is machining, which leads to high-tensile–stress areas of material. If the exposed tensile stresses are in a transverse direction or have a transverse component and if a susceptible alloy or temper is involved, the probability of SCC is present [38].

C8.3. Effects of Environment.

C8.3. Effects of Environment. Research indicates that water or water vapor is the key environmental factor required producing SCC in aluminum alloys. Halide ions have the greatest effects in accelerating attack. Chloride is the most important halide ion because it is a natural constituent of marine environments and is present in other environment as a contaminant. Because it accelerates SCC, Cl$^-$ is the principal component of environments used in laboratory tests to determine susceptibility of aluminum alloys to this type of attack. In general, susceptibility is greater in neutral solutions than in alkaline solutions and is greater still in acidic solutions [9].

C8.4. Prevention of Stress Corrosion Cracking. The SCC can be greatly retarded, if not eliminated, by polarization to the level of the cathodic protection potential. The heat treatments that provide high resistance to cracking are those that produce microstructures either free of precipitate along grain boundaries or with precipitate distributed as uniformly as possible within grains.

Direction and magnitude of stresses anticipated under conditions of assembly and service may govern alloy and temper selection. For products of thin sections, applied in ways that induce little or no tensile stress in the short transverse (i.e., through–thickness) direction, resistance of 2XXX alloys in T3 or T4 tempers or of 7XXX alloys in T6 tempers may suffice. Resistance in the short transverse direction usually controls application of products that are of thick section or are machined or applied in ways that result in sustained tensile stresses in the short transverse direction. More resistant tempers are preferred in these cases.

Residual stresses are induced in aluminum alloy products when they are solution heat treated and quenched. Figure 54.17 (a) shows the typical distribution and magnitude or residual stresses in thick high-strength material of constant cross section. Quenching places the surfaces in compression and the center in tension. If the compressive surface stresses are not disturbed by subsequent fabrication practices, the surface has an enhanced resistance to SCC because a sustained tensile stress is necessary to initiate and propagate this type of corrosion.

Aluminum products of constant cross section are stress relieved effectively and economically by mechanical stretching. The stretching operation must be done after quenching and, for most alloys, before artificial aging. Note the low magnitude of residual stresses after stretching, Figure 54.17 (b), as compared to the as-quenched material, Figure 54.17 (a). Federal specifications for rolled and extruded products provide for stress relieving by stretching on the order of 1–3%. The stress-relieved temper for heat-treated mill products minimizes SCC problems related to quenching stresses. The stress-relieved temper for most alloys is identified by the designation Tx5x or Tx5xx after the alloy number, for example, 2024-T351 or 7075-T6511 [2].

C9. Hydrogen Embrittlement

Testing in specific hydrogen environments has revealed the susceptibility of aluminum to hydrogen damage. Hydrogen

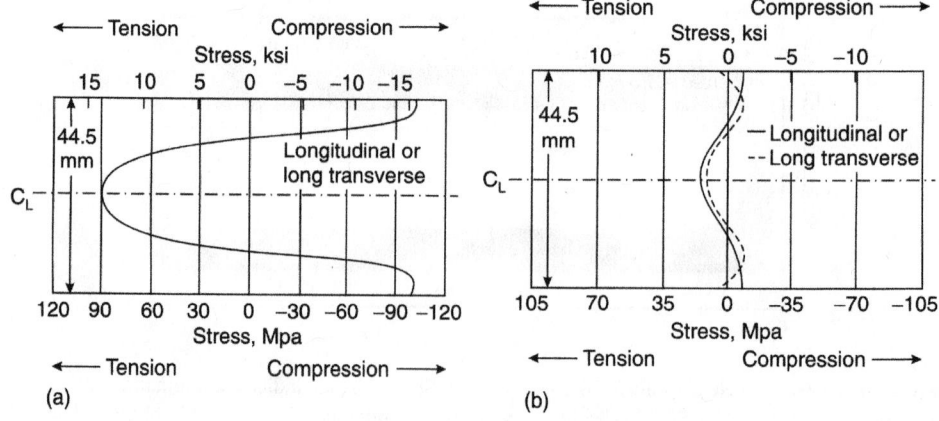

FIGURE 54.17. Comparison of residual stresses in a thick constant-cross-section 7075-T6 aluminum alloy plate before and after stress relief: (a) high residual stresses in the solution treated and quenched alloy; (b) reduction in stresses after stretching 2%. (Reproduced with permission from [9].)

damage in aluminum alloys may take the form of intergranular or transgranular cracking or blistering. Blistering is most often associated with the melting or heat treatment of aluminum where reaction with water vapor produces hydrogen. Blistering due to hydrogen is frequently associated with grainboundary precipitates or the formation of small voids. Blister formation in aluminum is different from that in ferrous alloys in that it is more common to form a multitude of near–surface voids that coalesce to produce a large blister [39].

Hydrogen diffuses into the aluminum lattice and collects at internal defects, most frequently during annealing or solution treating in air furnaces prior to age hardening. Dry hydrogen gas is not detrimental to aluminum alloys; however, with the addition of water vapor, subcritical crack growth increases dramatically (Figure 54.18). The threshold stress intensity for cracking of aluminum also decreases significantly in the presence of humid hydrogen gas at ambient temperature [37].

Hydrogen permeation and the crack growth rate are functions of potential, increasing with more negative potentials, as expected for hydrogen embrittlement behavior. The ductility of aluminum alloys in hydrogen is temperature dependent, displaying a minimum in reduction in area below 0°C; this behavior is similar to that of other face-centered cubic (fcc) alloys [37]. Some evidence for a metastable

aluminum hydride has been found that would explain the brittle intergranular fracture of aluminum(zinc)magnesium alloy (the 7XXX series) in water vapor. However, the instability of the hydride is such that it has been difficult to evaluate. Another explanation for intergranular fracture of these alloys is preferential decohesion of grain boundaries containing segregated magnesium. Overaging of these alloys increases resistance to hydrogen embrittlement in much the same way as for highly tempered martensitic steels [37].

C10. Corrosion Fatigue

Corrosion fatigue failures of aluminum alloys are characteristically transgranular and thus differ from SCC failures that are normally intergranular. Corrosion fatigue is not appreciably affected by stress orientation, and corrosion fatigue failures can be recognized by a characteristic oyster shell pattern on the fractured surfaces. The fatigue strength of aluminum alloys in demineralized water, hard tap water, or brine is almost equal and is relatively half the fatigue strength in air and a quarter of the original ultimate strength of the material. The corrosion fatigue strength of an alloy is not greatly affected by variations in heat treatment. Localized corrosion of an aluminum surface, such as pitting or intergranular corrosion, provides stress concentrations and greatly lowers fatigue life [27].

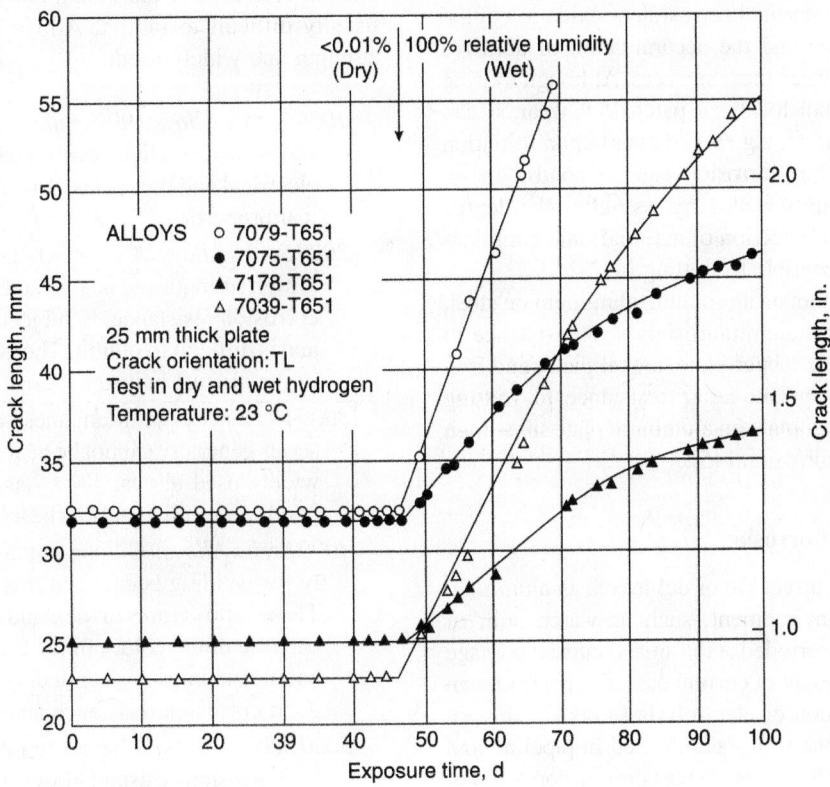

FIGURE 54.18. Effect of humidity on subcritical crack growth of high-strength aluminum alloys in hydrogen gas (TL = transgranular). (Reproduced with permission from [37].)

C10.1. Prevention. Corrosion environments produce smaller reductions in fatigue strength in the more corrosion-resistant alloys, such as the 5XXX and 6XXX series, than in the less resistant alloys, such as the 2XXX and 7XXX series [9]. Some guidelines for selecting aluminum alloys to minimize failure problems have been outlined by Bucci [40].

Stoltz and Pelloux [41] reported the influence of cathodic protection on corrosion fatigue life. Khobaib et al. [42] reported on the development of an inhibitor for corrosion fatigue of high-strength aluminum alloys.

Peening the metal surface increases fatigue life [43] and probably increases corrosion fatigue life as well. Care must be taken not to overpeen the surface to the extent where excessive plastic deformation may cause susceptibility to exfoliation or SCC. Protective surface coatings are also beneficial. Welding lowers both fatigue and corrosion fatigue life, but peening after welding increases the corrosion fatigue life. Paint coatings also increase the corrosion fatigue life, and the highest corrosion fatigue life for welded specimens is achieved by peening followed by coating [27].

C11. Fretting Corrosion

Fretting is the abrasive wear of two touching surfaces subject to cyclic relative motions of extremely small amplitude. Fretting corrosion is an increased degree of deterioration that occurs because of repeated corrosion or oxidation of the freshly abraded surface and the accumulation of abrasive corrosion products between these surfaces. Although fretting is often limited to small localized patches of wear, it can provide a path for leakage (e.g., valve seats) or an initiation site for fatigue. Fretting corrosion can be controlled by lubrication of the faying surfaces, by restricting the degree of movement, or by the selection of materials and combinations that are less susceptible to fretting [44].

Couples like aluminum on aluminum, aluminum on steel, and zinc-plated steel on aluminum show low resistance to fretting corrosion. Zinc, copper plate, nickel plate, and iron plate on aluminum show moderate resistance to fretting corrosion, whereas silver plate on aluminum plate show high resistance to fretting corrosion [45].

C12. Stray-Current Corrosion

Whenever an electric current (ac or dc) leaves an aluminum surface to enter an environment, such as water, soil, or concrete, aluminum is corroded at the area of current passage in proportion to the amount of current passed. This is known as stray-current corrosion or electrolysis (a poorly chosen, ambiguous term, but one firmly entrenched in pipeline and shipping technology). Examples of stray-current corrosion of aluminum have been reported in concrete (electrical conduit), in seawater (boat hulls), and in soils (pipelines and

drainage systems). At low current densities corrosion may take the form of pitting, whereas at higher current densities considerable destruction of the metal can occur. The corrosion rate does not diminish with time [27].

Since the aluminum surface from which the current leaves functions as an anode, oxidation (corrosion) occurs, and the area becomes acidic. The presence of acidity on the surface often provides the clue that reveals unexpected stray-current activity. Local acidity can develop even in an alkaline environment such as concrete.

Stray currents encountered in practice are usually direct current (e.g., from a welding generator) but may also be alternating current. For most metals, ac corrosion is negligible, but with aluminum it can be appreciable. Below a critical small ac current density, no corrosion of aluminum occurs [27].

D. INFLUENCE OF METALLURGICAL CHARACTERISTICS ON CORROSION PERFORMANCE

D1. Metallurgical Aspects

Aluminum alloys containing copper (2000 series) and zinc (7000 series) as major alloying elements are generally less corrosion resistant than those without these elements. For this reason, corrosion of aluminum alloys in these two series is usually difficult to inhibit. Alloys in both series are high strength and widely used.

1000 Series Alloys: 99% pure aluminum or higher. This series has excellent resistance to corrosion and high electrical and thermal conductivities but poor mechanical properties.

2000 Series Alloys: Copper-containing alloys. This series is high strength and heat treatable but has generally low corrosion resistance, is subject to intergranular attack, and is difficult to inhibit. The 2024 alloy is widely used in the aircraft industry.

3000 Series Alloys: Manganese-containing alloys. This series generally cannot be heat treated. One of the most widely used alloys, 3003, has moderate strength and good workability and can be inhibited in certain media.

4000 Series Alloys: Silicon-containing alloys, used mainly for welding because of their lower melting points. These alloys are in demand for architectural uses because of the color effects that can be obtained when anodic coatings are applied. Alloys in this series have good corrosion resistance and can be inhibited.

5000 Series Alloys: Magnesium-containing alloys. These are corrosion-resistant alloys that can be inhibited and are widely used in marine atmospheres, where they exhibit good resistance to corrosion. However, under certain conditions of loading, they are subject to stress corrosion cracking.

6000 Series Alloys: Silicon-and magnesium-containing alloys. The silicon and magnesium are present in the ratio required to form magnesium silicide. These alloys are heat treatable. A major alloy in this series is 6061. These alloys have good corrosion resistance and may be inhibited effectively.

7000 Series Alloys: Zinc-containing alloys. These alloys may also contain smaller percentages of magnesium, copper, and chromium. They are heat treatable and can have very high strengths (e.g., 7075), which is one of the highest strength aluminum alloys. Inhibitors may be used with the 7000 series.

The solution heat-treated tempers are usually more corrosion resistant and more amenable to corrosion inhibition than are the hardened alloys. The strain-or work-hardened alloys are somewhat more readily inhibited than are the alloys hardened by aging treatments [46].

Since corrosion is an electrochemical phenomenon, it might be expected that alloys composed of one homogeneous phase or of two or more phases, all of which have very similar electrochemical (galvanic) potentials, would be more resistant to corrosion than alloys composed of two or more phases with widely different potentials. This expectation is generally correct. Thus pure aluminum or single-phase alloys of aluminum and magnesium or aluminum and silicon are all relatively resistant to corrosion. The Al–Cu alloys heat treated and quenched to retain the copper in solid solution are much more resistant to corrosion than are similar alloys treated so that the copper precipitates out of solution as a constituent, $CuAl_2$, which differs in solution potential from the matrix solid solution and may cause intergranular corrosion. The Al–Mn alloys (such as 3003) are highly resistant to corrosion because the manganese constituent that is present as a separate phase has a potential very similar to that of the matrix.

Metallurgical factors are often important in influencing corrosion rates. They are probably best known in the case of the Al–Cu alloys of the duralumin type. Such alloys contain about 4% copper (alloys 2017-T and 2024-T). This amount of copper is soluble in solid aluminum at elevated temperatures ($>480°C$) but is not entirely soluble at room temperature. After fabrication, such alloys are commonly heat treated at $\sim490°C$ in order to dissolve the copper in the aluminum. They are then immediately quenched in cold water to retain the copper in solution. During aging at room temperature, the hardness and strength of the alloys increase, approaching maximum values after about four days. It is generally assumed that this age hardening is caused by the precipitation of a $CuAl_2$ constituent from the Al–Cu solid solution.

The precipitate particles, if present, are in a very finely divided state and, when in this condition, the alloys, from a corrosion standpoint, behave as if they were substantially single-phase alloys; that is, they are relatively resistant to corrosion. It is in this quenched and room temperature aged condition that they are generally used and, as such, are susceptible only to pitting corrosion with no selective attack at grain boundaries.

If the alloys are quenched more slowly from the heat-treating temperature (i.e., quenched in boiling water instead of cold water), they become susceptible to selective grain boundary attack (intergranular corrosion). This type of attack is attributed to the selective precipitation of relatively large particles of the $CuAl_2$ constituent at the grain boundaries. The Al–Cu solid solution adjacent to the grain boundaries becomes depleted in copper, since the copper is precipitated out of solution. This depleted zone is anodic to the Al–Cu solid solution of the main body of the grain and also to the precipitated particles of $CuAl_2$.

Consequently, the depleted zone corrodes, giving an intergranular form of corrosion. Somewhat similar results occur if the rapidly quenched Al–Cu alloy is heated (artificially aged) to a somewhat elevated temperature ($>120°C$) for a critical period of time. This heating also causes the alloy to become susceptible to intergranular corrosion. However, if the heating is carried out for a sufficiently extended period of time, the susceptibility to intergranular corrosion again disappears, probably because substantially all the copper has precipitated out of solid solution and therefore the zones adjacent to the grain boundaries are no more depleted in copper than are the other areas in the grain boundaries [3].

For many of the other aluminum-base alloys, metallurgical factors have relatively little effect on resistance to corrosion. Alloys such as 1100, 3003, 5052, 6053, and 6061 are relatively insensitive in this respect [3].

In many types of exposure, cold work does not appreciably affect the resistance to corrosion of a wide variety of aluminum-base alloys. In solutions of nonoxidizing acids, however, cold work stimulates corrosion to some extent and also indirectly stimulates corrosion of aluminum alloys containing over about 5% magnesium. With the latter alloys, severe cold work increases the tendency for a magnesium–aluminum constituent to precipitate from solid solution. On exposure to certain media, selective attack of this constituent then occurs [3].

The resistance to corrosion of weldments of aluminum alloys is determined, in part, by the alloy welded, by the filler alloy, and by the welding process. Galvanic cells that cause corrosion may be created because of potential differences among the parent alloy, the filler alloy, and the heat-affected zones where microstructural changes occur. Incomplete removal of fluxes after welding them may also cause corrosion [1].

Weldments in non–heat–treatable alloys generally have good resistance to corrosion. Microstructural changes in the heat-affected region in these alloys have little effect on potential, and the filler alloys recommended have potentials close to those of the parent alloys. In some heat-treatable alloys, however, the effect on potential of microstructural changes may be large enough to cause appreciable corrosion in more aggressive environments; the corrosion is selective, either in the weld bead or in a restricted portion of the heat-affected zone. To a considerable degree, the effect of microstructural changes on corrosion in the heat-affected zone can be eliminated by postweld heat treatment. Stress corrosion cracking in weldments usually is caused by residual stresses introduced during welding, but its occurrence is rare.

Brazed joints in aluminum alloys also have good resistance to corrosion. Excessive corrosion is usually caused by fluxes that are not removed completely or that are removed by a treatment that, together with the fluxes, may cause corrosion. Soldered joints have a resistance to corrosion satisfactory for applications in milder environments, but not for those in more aggressive ones [1].

D2. Composites

Aluminum alloys reinforced with silicon carbide, graphite, alumina, boron, or mica show promise as metal matrix composites with increased modulus and strength and are potentially well suited to lightweight structural applications, including aerospace and military needs. The structures of continuous fiber metal matrix composites (MMCs) are equivalent to those in polymer matrix composites. Industrial applications have emerged recently, for example, reinforced pistons for assembly in light diesel engines, 12% alumina, 9% carbon fiber reinforced Al–12.7% Si MMC cylinder liner [47], and so on.

Generally, long-term tests have shown that the introduction of a reinforcement phase reduces the resistance to corrosion. The extent of this reduction largely depends on the reinforcement species and, form. As with conventional aluminum alloys, the fabrication method and heat treatment influence the corrosion resistance of MMCs and must be carefully controlled. As surface protection will be advisable in certain applications, it is encouraging to see a variety of standard techniques showing promise for MMCs. From the studies performed on the corrosion fatigue of MMCs in saline environments it appears that they are marginally inferior to their matrix alloys [46].

E. CORROSION PREVENTION AND PROTECTION

E1. Design, Alloy Selection, and Joint-Sealing Compounds

During conception, the corrosion specialist should identify the different types of corrosion and prevention methods. Among the most common harmful effects are galvanic action, resulting from direct contact between aluminum and a dissimilar metal, such as copper, and indirect galvanic effects resulting from contact between aluminum and solutions containing reducible compounds of heavy metals. In some cases, design or construction will prevent serious corrosion even though no other factors are altered. Similarly, since the various alloys of aluminum differ widely in behavior, the selection of the most suitable alloy is important [3].

Corrosion can be prevented or reduced by cladding with a more corrosion-resistant alloy, such as high-purity aluminum, a low magnesium–silicon alloy, or an alloy of 1% zinc. All of these cladding materials are frequently employed to give added corrosion protection to the 2000 and 7000 series alloys. The cladding on each side is 2–5% of the total thickness.

Aluminum-base alloys, such as 1100, 3300, 5052, 6053, Alclad 3300, Alclad 1017-T, and Alclad 2024-T, are highly resistant when freely exposed to most natural environments. They will all discolor or darken appreciably in most outdoor exposures but will suffer no structurally appreciable changes in properties unless exposed in relatively thin sections, that is, <0.076 mm (0.03 in.) thick.

Commercial aluminum alloys may contain other elements that provide special characteristics. Lead and bismuth are added to alloys 2011 and 6262 to improve chip breakage and other machining characteristics. Nickel is added to wrought alloys 2018, 2218, and 2618, which were developed for elevated-temperature service, and to certain 3XXX cast alloys used for pistons, cylinder blocks, and other engine parts subjected to high temperatures. Cast aluminum-bearing alloys may contain tin. In all cases, these alloying additions introduce microconstituent phases that are cathodic to the matrix and decrease resistance to corrosion in aqueous saline media. However, these alloys are often and should be used in environments in which they are not subject to corrosion [9].

Joints, depressions, and other areas where moisture and dirt accumulate are more susceptible to corrosion than regions exposed to the atmosphere. Most plastic or semisolid joint-sealing compounds that conform and firmly adhere to adjacent metal surfaces are highly effective in preventing special attack in these regions. Some of these joint-sealing compounds that contain soluble inhibitors are particularly suitable [3].

In some cases, other mechanical factors, such as formability or hardness, may be of great importance in selecting an appropriate alloy for a specific application. Aluminum alloys such as 1100, 3300, or 5052, in the softer tempers, are readily formable and are also highly resistant to corrosion. If greater strength is required, alloy 6061 should be considered. This alloy combines good formability (in the W temper) with relatively high strength and good resistance to corrosion.

E2. Aluminum Thermal Spraying

Aluminum spraying is a current practice to coat less resistant alloys. For some composites, the corrosion behavior is governed by galvanic action between the aluminum matrix and the reinforcing material. Aluminum thermal spraying has been reported as a successful protection method for discontinuous silicon carbide/aluminum composites; for continuous graphite/aluminum or silicon carbide/aluminum, sulfuric acid (H_2SO_4) anodizing has provided protection, as have organic coatings or iron vapor deposited aluminum [48].

E3. Anodic Coatings

Anodizing is an electrolytic process in which the surface of the alloy is made the anode and converted to aluminum oxide, bound as tenaciously to the alloy as the natural oxide film, but much thicker (5–30 μm).

Anodic coatings, particularly those applied in a sulfuric acid electrolyte and suitably sealed, are highly effective in preventing discoloration or surface staining of the aluminum-base alloys mentioned above. In addition, aluminum alloys that are used architecturally are more readily cleaned of atmospheric contaminants if they have been anodically coated. However, anodizing does not provide sufficient protection alone if the alloys themselves are unsuitable for the environment to which they are exposed. Anodic coatings are excellent paint bases.

E4. Inhibitors and Control of the Environment

Current knowledge makes possible the inhibition of aluminum in a wide range of both acidic and alkaline environments. Single materials and combinations have been identified that can be used with considerable confidence, frequently, however, within a narrow range of conditions. Inhibitors may be classified by surface reactivity as adsorptive or surface reactive (where a precipitated film is formed to provide a barrier between the corrosive agent and the aluminum surface). Chromates, silicates, polyphosphates, soluble oils, and other inhibitors are commonly used to protect aluminum. Aluminum is concentration sensitive to chromate solutions as well as to other anodic inhibitors. Combinations of polyphosphates, nitrites, nitrates, borates, silicates, and mercaptobenzothiazole are used in systems that include aluminum and other metals [49].

Composition differences among aluminum alloys often determine whether or not an alloy can be inhibited in a given environment, and so an understanding of the metallurgical variables is important. Investigations into the fundamental reactions at the aluminum–environment interface have added significant new understanding that now permits the selection of inhibitors to be made with greater precision and their application to proceed with fewer trial-and-error adjustments [50].

In a limited number of cases, removing some minor constituent from the contacting liquid or gas can prevent corrosion. For instance, copper compounds, which may make water corrosive to aluminum, can be removed by passing the water through a tower packed with aluminum chips. Finally, the use of periodic cleaning procedures may be highly beneficial in specific cases [3].

E5. Conversion and Organic Coatings

Conversion coatings (chromates or phosphates) are recommended for the preparation of aluminum alloys. For milder environments, paint may be applied on the conversion coating, but a chromated primer should be applied for more aggressive media. Almost any type of paint (acrylic, alkyl, polyester, vinyl, etc.) is suitable [9]. One or two coats of the finish paint should follow the primer. Attention should be made to avoid the discharge of Cr^{VI} in natural environments; otherwise, molybdates can replace chromates in certain applications [51].

E6. Cathodic Protection

Godard [29] cited an early example of aluminum protection using sacrificial zinc anodes and Hatch [27] mentioned the use of impressed current protection systems to protect painted aluminum ship hulls. Cathodic protection requires careful control to ensure that adequate protection is maintained without overprotection, which can lead to alkali attack (cathodic corrosion). Alclad alloys (layered aluminum products with one aluminum alloy integrally bonded to a more noble aluminum alloy core) may be viewed as having a self-contained cathodic protection system [46].

E7. Conventional and Electrochemical Corrosion Testing of Coated Alloys

Evaluation of the protective ability of coatings generally is made in salt spray cabinets, such as those covered in ASTM B 117 and G 85 or in the 3.5% NaCl alternate immersion test (G 44). It is desirable to include ultraviolet (UV) light as part of the cyclic exposure since UV light has a degrading effect on paints and other organic coatings (ASTM G 85). There is considerable interest in potentiodynamic polarization techniques to rapidly assess the durability of coated aluminum surfaces that have been painted or given various polymeric or anodic surface treatments. Corrosion monitoring of epoxy-coated aluminum 2024-T3 was carried out by electrochemical impedance spectroscopy and electrochemical noise measurement (ENM). The corrosive solution was 0.35 wt % ammonium sulfate and 0.05 wt % sodium chloride. It has been concluded that EIS data can be used to monitor the protective quality of the coating as a function of surface treatment or applied voltage, while ENM data were too noisy [52, 53].

E7.1. Corrosion Fatigue of Thermal Spraying of Al as a Coating.
The effect of thermal spray coatings on the fatigue behavior of various substrate materials has attracted increased attention in recent years. The fatigue behaviour of 7075-T651 Al alloy with ductile aluminum thermal sprayed coatings deposited by four different commercial arc spray devices (guns) has been characterized. Coated specimens as well as polished and shot-peened specimens were evaluated under fully reversed uniaxial loading ($R = -1$) at constant amplitude of ± 225 MPa in accordance with ASTM 466-82. A frequency of 20 Hz was selected to avoid potential frequency induced heating with a sinusoidal loading waveform applied via a computer-controlled Materials Testing System (MTS) servohydraulic load frame. While the shot-peening pretreatment was observed to increase the fatigue resistance of polished specimens, application of the coatings subsequently reduced fatigue life to below that of the original polished coupons. Changes in the residual stress state of the shot-peened surface were identified as the most likely source of these reductions. With the absence of microgaps at the coating–substrate interface, even after the fatigue tests and with the lowest coating roughness the equipment provided the best fatigue behavior for the Al-coated 7075-T651 Al alloy in this study [54].

E7.2. Environmentally Assisted Cracking of Metallic Sprayed Coatings.
The thermal spray coating using arc spraying to evaluate the protection against environmentally assisted cracking (EAC) and localized corrosion on aircraft structural 7075 T651 aluminum alloys was examined. EAC and pitting corrosion at the coating–substrate interface are a challenge for thermal spray protective coatings on aluminum alloys under cyclic load and immersion. In this study, EAC has been initiated on polished and shot-peened Al 7075 T651 through a four-point bending test under cycling fluctuation load in 3.5 wt % NaCl solution kept at 25°C in open air [55, 56] (ASTM G39-99). The applied load was kept under the Yield strength (Ys) (503-MPa) of 7075-T6 alloy and oscillated between 24 and 40% Ys in tension ($R = 0.6$) at a frequency of 0.1 Hz. The selected stress level was sufficiently low to avoid premature coating damage and to allow a single failure mode. The failure mode validates the environmentally assisted cracking mechanism such as SCC or fatigue corrosion.

This approach has the benefit to initiate intergranular cracking in the aluminum alloy with a fast response while maintaining the substrate material under elastic deformation. The results had underlined the coating material impact on the interface properties in terms of interface quality (microgap) and adhesive strength. For five different surface properties the Al coating has shown either lower microgap or higher

bond strength than Al–5Mg. Moreover, it is stated that the surface preparation on Al alloy substrate necessitates material removal mechanically (e.g., grit blasting) or by deoxidation [52, 56].

ACKNOWLEDGMENTS

This chapter is based on the pioneering work of R. B. Mears in the first edition of The Corrosion Handbook by H. H. Uhlig and on the valuable contributions of E. H. Hollingsworth and H. Y. Hunsicker in P. A. Schweitzer's Handbook, Marcel Dekker, and in different volumes of Metals Handbook of the ASM International.

REFERENCES

1. E. H. Hollingsworth and H. Y. Hunsicker, in Corrosion and Corrosion Protection Handbook, P. A. Schweitzer (Ed.), Marcel Dekker, New York, 1983, pp. 111–145.
2. Aluminum Standards and Data, Aluminum Association, 8th ed., Washington, DC, 1984, p. 12.
3. R. B. Mears, "Aluminum and Aluminum Alloys," in Corrosion Handbook, H. H. Uhlig (Ed.), sponsored by The Electrochemical Society, Wiley, New York, 1976, pp. 39–56.
4. M. Pourbaix, Atlas of Electrochemical Equilibrium in Aqueous Solutions, 2nd English ed., NACE International (Houston, Texas) and Centre Belge d'Étude de la Corrosion "CEBELCOR," (Brussels, Belgium), 1974, pp. 168–175.
5. M. H. Brown, R. H. Brown, and W. W. Binger, "Aluminum Alloys for Handling High Temperature Water," in High Purity Water Corrosion of Metals, Publication No. 60–13, N. E. Hammer (Ed.), National Association of Corrosion Engineers, Houston, TX, 1960, p. 82.
6. F. M. Reinhart,"Corrosion of Metals and Alloys in the Deep Ocean," Report No. R. 834, U.S. Naval Engineering Laboratory, Port Heceneme, CA, 1976.
7. R. B. Mears and J. R. Akers, Proc. Am. Soc. Brewing Chem., St. Paul, MN, Annual Meeting, 1942.
8. D. J. De Renzo, Corrosion Resistant Materials Handbook, Noyes Data Corporation, Park Ridge, N.J., 1985, p. 621.
9. E. H. Hollingsworth and H.Y. Hunsicker, "Corrosion of Aluminum and Aluminum Alloys", in Metals Handbook, 9th ed., Vol. 13, Corrosion, ASM International, Metals Park, OH, 1987, pp. 583–609.
10. P. Delahay, M. Pourbaix, and Van Russelberghe, Diagramme d'équilibres Potentiel-pH de quelques éléments, C. R. 3e réunion du CITCE, Berne, 1951.
11. R. K. Hart, Trans. Faraday Soc., **53**, 1020 (1957).
12. H. Reboul and R. Canon, "Corrosion galvanique de l'aluminum, mesures de protection," Revue de l'aluminium, **1922**, 403–426, (Aug. 1984).
13. C. Vargel, Le Comportement de l'aluminium et de ses alliages, Dunod, 1976.

14. E. H. Hollingsworth and H. Y. Hunsicker, Corrosion Resistance of Aluminum Alloys, in Metals Handbook, D. Benjamin (Ed.), 9th ed., Vol. 2, American Society of Metals, ASM International, Metals Park, OH, 1979, p. 204.

15. M. Elboujdaini, E. Ghali, R. G. Barradas, and M. Girgis, Corros. Sci., **30**(8/9), 855 (1990).

16. R. B. Mears and R. H. Brown, Ind. Eng. Chem., **33**, 1001 (1941).

17. I. L. Muller and J. R. Galvele, Corros. Sci., **17**, 179 (1977).

18. S. J. Harris, B. Noble, and A. J. Trowsdale, "Corrosion Behaviour of Aluminum Matrix Composites Containing Silicon Carbide Particle," Materials Science Forum, Transtech Publications, Vols. 217–222, Switzerland, 1996, pp. 1571–1576.

19. P. M. Aziz, Corrosion, **12**, 35 (1956).

20. H. Kaesche, "Pitting Corrosion of Aluminum and Intergranular Corrosion of Aluminum Alloys," in Localized Corrosion, B. F. Brown, J. Kruger, and R. W. Staehle (Ed.), NACE International, Houston, TX, 1974, p. 516.

21. N. Sato, J. Electrochem. Soc., **129**(2), 260 (1982).

22. M. C. Reboul, T. J. Warner, H. Maye, and B. Baroux, "A Ten–Step Mechanism for the Pitting Corrosion of Aluminum," Materials Science Forum, Transtech Publications, Vols. 217–222, Switzerland 1996, pp. 1553–1558.

23. S. C. Dexter, J. Ocean Sci. Eng., **8**(1), 109 (1981).

24. E. H. Cook and F. L. McGeary, Corrosion, **20**(4), 111 (1964).

25. M. H. Brown, W. W. Binger, and R. H. Brown, Corrosion, **8**(5), 155 (1952).

26. R. C. Plumb, M. H. Brown, and J. E. Lewis, Corrosion, **11**(6), 277 (1956).

27. J. E. Hatch, Aluminum: Properties and Physical Metallurgy, ASM International, Metals Park, OH, 1984, pp. 248–319, 263, 307.

28. I. L. Rosenfeld, Localized Corrosion, National Association of Corrosion Engineers, NACE International, Houston, TX, 1974, pp. 386–389.

29. H. P. Godard, The Corrosion of Light Metals, Wiley, New York, 1967, pp. 46, 70–73, Chapter 1.

30. W. Slabaugh, W. Deiager, S. Hoover, and L. Hutchinson, Paint, J. Technol., **44**(56), 76 (1972).

31. S. C. Dexter, Localized Corrosion Metals Handbook, ASM International, Corrosion, **13**, 103–122 (1987).

32. W. Ryan, Soc. Adv. Mater. Proc. Eng., **1**, 638 (1979).

33. P. Bijlmer, Adhesive Bonding of Aluminum Alloys, Marcel Dekker, New York, 1985, pp. 21–39.

34. J. J. Elpjick, in Microbial Corrosion in Aircraft Fuel Systems in Microbial Aspects of Metallurgy, J. D. A. Miller (Ed.), American Elsevier, New York, 1970, pp. 157–172.

35. L. L. Shreir, R. A. Jarman, and G. T. Burstein (Eds.), Corrosion, Butterworth-Heinemann, Oxford, UK, 1994, pp. 4–16.

36. V. A. Marichev, Werkst. Korros., **34**, 300 (1983).

37. M. O. Spiedel, "Hydrogen Embrittlement of Aluminum Alloys," in Hydrogen in Metals, I. M. Bernsteinand A. W. Thomson (Eds.), American Society for Metals, ASM International, Metals Park, OH, 1974, p. 249.

38. D. O. Sprowls and E. H. Spuhler, Green Letter, Alcoa, Avoiding Stress-Corrosion Cracking in High Strength Aluminum Alloy Structures, ASM International, Metals Park, OH, Jan. 1982.

39. B. Craig, "Hydrogen Damage," in Metals Handbook, 9th ed., Vol. 13, Corrosion, J. R. Davis (Ed.), ASM International, Metals Park, OH, 1987, pp. 169–170.

40. R. J. Bucci, Eng. Fracture Mech., **12**(3), 407 (1979).

41. R. E. Stoltz and R. M. Pelloux, Metall. Trans. A, **3**(9), 2433 (1972).

42. M. Khobaib, R. M. Pelloux, and F. W. Vahldiek, Corrosion, **37**(5), 285 (1981).

43. N. L. Person, Metal Prog., **120**(21), 33 (1981).

44. L. J. Korb, "Corrosion in the Aerospace Industry," in Metals Handbook, 9th ed., Corrosion, Vol. 13, ASM International, Metals Park, OH, 1987, p. 1082.

45. Society of Automotive Engineers, Aeronautical information, Warrendale, PA, Report 47, 1956.

46. K. A. Lucas and H. Clarke, Corrosion of Aluminium-Based Metal Matrix Composites, Research Studies Press Ltd., Taunton, Somerset, England, Wiley, New York, 1993.

47. A. Walker, Mater. Edge, **34**, 13 (1992).

48. D. M. Aylor and P. J. Moran, "An Investigation of Corrosion Properties and Protection for Graphite/Aluminum and Silicon Carbide/Aluminum Metal Matrix Composites," Paper 202, presented at Corrosion/86, National Association of Corrosion Engineering, Houston, TX, 1986.

49. E. D. Verink, Jr. and D. B. Bird, Mater. Prot., **6**(2), 28–32 (1967).

50. A. H. Roebuck, in Inhibition of Aluminum in Corrosion Inhibitors, C. C. Nathan (Ed.), National Association of Corrosion Engineers, Houston, TX, 1973, pp. 240–244.

51. C. Vargel, Corrosion de l'aluminium, Dunod, Paris, 1999, pp. 154–163.

52. E. Ghali, Corrosion Resistance of Aluminum and Magnesium Alloys: Understanding, Performance and Testing, Wiley, N.Y., 2010, 719 pages.

53. R. L. De Rosa, D. A. Earl, and G. P. Bierwagen, Corros. Sci., **44**, 1607–1620. (2002).

54. B. Arsenault, A. K. Lynn, and D. L. Duquesnay, Can. Metall. Q. J. **44**(4), 495–504 (2005).

55. Y. Z. Wang, R. W. Revie, M. T. Shehata, R. N. Parkins, and K. Krist, in Initiation of Environment Induced Cracking in Pipeline Steel: Microstructural Correlation, International Pipeline Conference, Vol. 1, American Society of Mechanical Engineers, New York, 1998, pp. 529–542.

56. B. Arsenault and E. Ghali, in Prevention of Environmentally Assisted Cracking of Structural Aluminum Alloys by Al and Al-5Mg Thermal Sprayed Coatings Using Different Surface Preparation Techniques, Proceedings of the Aerospace Symposium, 45th Conference of Metallurgists, Montréal, M. Jahazi, M. Elboujdaini, and P. Patnaik, (Eds.), The Canadian Institute of Mining, Metallurgy and Petroleum, Montreal, Canada, 2006, pp. 61–74.

55

COBALT ALLOYS

P. Crook[†]

Haynes International, Kokomo, Indiana

W. L. Silence

Consultant, Fairfield Glade, Tennessee

A. INTRODUCTION

This chapter is concerned with the resistance to aqueous and gaseous corrosion of the chromium-bearing cobalt alloys. These alloys are important because they provide a unique blend of properties, including high strength at elevated temperatures and resistance to many forms of wear. With regard to aqueous corrosion resistance, some of the cobalt alloys are superior to the stainless steels.

The chromium-bearing cobalt alloys have been divided into three distinct groups in this chapter, according to their intended uses, as follows:

1. Alloys designed purely for wear resistance
2. Alloys designed for high-temperature use
3. Alloys designed for aqueous corrosion and wear resistance

All these materials stem from the pioneering work of Elwood Haynes early in the twentieth century. It was he who discovered the corrosion and wear benefits of adding chromium to cobalt (U.S. Patent No. 873,745) and who found that molybdenum and tungsten were outstanding strengthening agents in the cobalt–chromium alloy system (U.S. Patent 1,057,423).

The early alloys were used in cast form or were applied by welding to steel substrates (a technique known as weld overlay or, in the case of the hard alloys, hardfacing). Most of the modern chromium-bearing cobalt alloys designed for wear resistance are also used in the form of castings and weld overlays (wrought processing of most alloys is impractical, given their high-temperature strengths and their lack of ductility).

Those alloys designed for high-temperature use and those designed to resist both aqueous corrosion and wear, on the other hand, are much more ductile and can be forged and rolled into wrought products (sheets, plates, etc.).

Industrial applications of the alloys designed for wear resistance include valve and pump components for the

[†]Retired.

Uhlig's Corrosion Handbook, Third Edition, Edited by R. Winston Revie
Copyright © 2011 John Wiley & Sons, Inc.

chemical process industries, automotive exhaust valves, and hot working tools for the steel industry. The high-temperature alloys are used, in sheet form, for flying gas turbine combustors and afterburners. Those alloys designed to resist both aqueous corrosion and wear are used for applications as diverse as biomedical implants, chemical spray nozzles, and electrogalvanizing rolls (in the steel finishing industry).

B. ALLOYS DESIGNED FOR WEAR RESISTANCE

B1. History

Haynes discovered the inherent corrosion resistance and wear resistance of the cobalt–chromium binary alloys and later the enhanced wear resistance of the cobalt–chromium–molybdenum and cobalt–chromium–tungsten alloys while he was searching for a new spark-plug material. These alloys were found to be hard (even at elevated temperatures), abrasion resistant, and excellent lathe tool materials (their first major application).

One reason for the outstanding abrasion resistance of his alloys was that they contained significant quantities of carbon as an impurity. As a result, carbides of chromium, molybdenum, and tungsten formed in the alloy microstructures. As melting techniques improved, controlled levels of carbon were used in the alloys to promote the formation of carbides.

The extent of the wear resistance of the cobalt alloys did not become known until many years later and was not explained until the 1960s and 1970s when wear became a science (tribology). It was only when the role of microfatigue in wear was established, and the unique atomic structure changes in cobalt and its alloys were understood, that a full appreciation of the attributes of the cobalt alloys was gained.

B2. Metallurgy

Pure cobalt exists in two atomic forms. At temperatures up to 417°C cobalt exhibits a hexagonal close-packed (hcp) structure, and at temperatures above 417°C the structure is face-centered cubic (fcc). Alloying elements such as nickel, iron, and carbon (within its soluble range) suppress the transformation temperature and are known as fcc stabilizers.

Chromium, molybdenum, and tungsten, on the other hand, increase the transformation temperature and are known as hcp stabilizers. Alloys of cobalt, containing both fcc and hcp stabilizers, thus have transformation temperatures that are a complex function of the contents of these elements.

The transformation from fcc to hcp in cobalt and its alloys is very sluggish and does not occur spontaneously upon cooling these materials from high temperatures. Thus castings that have been cooled from the molten state and wrought products that have been cooled after solution annealing typically exist in a metastable fcc form, even if their transformation temperatures are much higher than room temperature. However, partial transformation is easily induced at room temperature by applying mechanical stress. The transformation mechanism during cold work is believed to involve the creation and coalescence of wide stacking faults.

In addition to possessing low-stacking fault energies, which result in stacking fault coalescence and the formation of hcp platelets, metastable fcc cobalt alloys also exhibit mechanical twinning during cold work. These two phenomena result in very high work-hardening rates, and, more importantly from a wear standpoint, accommodate stresses.

With regard to cobalt alloys designed for wear resistance, several compositions are listed in Table 55.1. The STELLITE® materials referred to in this table are the direct descendants of the early Elwood Haynes alloys and contain significant levels of carbon to encourage the formation of carbides in the microstructure during alloy solidification. The two other main alloying elements in the standard STELLITE alloys are chromium and tungsten. Depending on the levels of chromium, tungsten, and carbon, different amounts and different types of carbide form in these alloys. For example, STELLITE 6 alloy (1.1 wt % carbon) contains ~13 wt % carbides, and these are of the chromium-rich M_7C_3 type, whereas STELLITE 3, which has the same carbon content as STELLITE 1 (2.4 wt %), contains ~29 wt % carbides, and these are a mixture of chromium-rich M_7C_3 and tungsten-rich M_6C types [1].

These carbides provide resistance to low-stress abrasion and provide high hardness levels (important if initial deformation resistance is an issue). On the other hand, higher carbide levels are associated with reduced ductilities and lower corrosion resistance, since, in forming the carbides, carbon ties up a portion of the chromium in the alloy.

TABLE 55.1. Nominal Compositions of Wear-Resistant Cobalt Alloys

	Co	Cr	W	Mo	C	Ni	Fe	Si
STELLITE 1	Balance	31	12		2.4	3 max	3 max	1 max
STELLITE 6	Balance	29	4.5		1.1	3 max	3 max	2 max
HAYNES 6B								
STELLITE 12	Balance	30	8		1.4	3 max	3 max	1 max
TRIBALOY T-800	Balance	17.5		28.5		3 max (Ni + Fe)	3 max (Ni + Fe)	3.4

Further references to the STELLITE alloys in this section relate to the standard cobalt–chromium–tungsten–carbon materials, such as alloys 1, 6, 6B, and 12. Nonstandard, cobalt-bearing materials, designed either for wear resistance or for other purposes, are also sold under the STELLITE trademark.

The other alloy in Table 55.1, namely, TRIBALOY® T-800, is one of a group of materials developed by Du Pont in the 1970s. Instead of relying on the formation of carbides in the microstructure, relatively high molybdenum and silicon contents were used to create precipitates of Laves phase (a hard intermetallic compound). A benefit of this approach is enhanced resistance to acids, under active corrosion conditions, since molybdenum ennobles cobalt, and the intermetallic itself has good corrosion resistance. However, the TRIBALOY materials contain large amounts of Laves phase and are thus fairly brittle. Attempts to use them in cast and weld overlay form have met with little success because of their brittleness; however, they have been quite successful as thermal spray deposits, as applied by plasma guns.

B3. Wear Behavior

There are three main categories of wear and several subcategories, as follows:

Abrasion
    Low-stress abrasion
    High-stress abrasion
Erosion
    Solid-particle impingement erosion
    Liquid-droplet impingement erosion
    Cavitation erosion
    Slurry erosion
Metal to metal
    Galling
    Cyclic sliding
    Fretting

Explanations of these wear processes are given in [2]. In essence, they involve either gross deformation and fracture, microfatigue, or wastage due to the continual stripping of surface films.

For those processes involving gross deformation and fracture, such as low-stress abrasion, a network of hard precipitates (e.g., carbides) in the microstructure is advantageous. If the size of the abrading particles is much larger than the size of the microstructural precipitates, then the precipitates can act as outcrops, over which the abrading particles ride. In this case, a high fraction of precipitates in the microstructure is desirable. Often, the need for a high fraction (wt %) of precipitates must be tempered by the need for ductility in castings or weld overlays.

For those processes involving microfatigue, such as liquid-droplet impingement erosion, cavitation erosion, and cyclic sliding, the ability of the cobalt alloys to absorb stress through the transformation and twinning is very important. Thus, it is the cobalt-rich matrix, and not the precipitates, that is critical for many wear applications. Also, under conditions conducive to galling, it is the cobalt-rich matrix that provides resistance to this form of damage, probably due to its atomic bonding characteristics, and because deformation is restricted to the outer layer, which can easily be sheared away, preventing fracture of the bulk material.

B4. Aqueous Corrosion Behavior

The aqueous corrosion behavior of the standard STELLITE alloys is strongly influenced by the main alloying elements, namely, chromium, tungsten, and carbon. As in the stainless steels, chromium is critical to the aqueous corrosion resistance because of its influence on passivation. As already stated, a significant proportion of the chromium in each of the STELLITE alloys is tied up in the form of carbides, so the effective chromium content, from an aqueous corrosion standpoint, is much lower than is evident from the nominal composition.

Tungsten enhances the nobility of cobalt under active corrosion conditions. In this respect, it is like molybdenum, which is used for the same purpose in the austenitic stainless steels and in the corrosion-resistant nickel alloys. Of course, in the STELLITE alloys, tungsten has multiple functions, since it too can take part in the formation of carbides (if the tungsten and carbon contents are high enough) and is used as a powerful matrix-strengthening agent.

In general, the aqueous corrosion resistance of the STELLITE alloys is an inverse function of their carbon contents, and the characteristics of alloys such as STELLITE 6 are broadly similar to those of type 316 stainless steel. For example, like type 316 stainless steel, STELLITE 6 is much more suited to oxidizing acids (such as nitric) than it is to reducing acids (such as hydrochloric). Also, as will be discussed in detail later, the cobalt alloys are prone to stress corrosion cracking (SCC) in chloride-bearing environments, although with alloys such as STELLITE 6, which is not ductile enough to be cold formed, the opportunities for stressing components are limited.

As indicated in Table 55.1, a wrought version of the 1.1 wt % carbon material is also available. This wrought version (HAYNES® 6B) has higher aqueous corrosion resistance than the cast and weld overlay version (STELLITE 6), because hot forging and rolling break down the interconnected carbides into discrete particles. The wrought version also has higher ductility, for the same reason.

Aqueous corrosion data for HAYNES 6B alloy in pure sulfuric, hydrochloric, nitric, and phosphoric acids are presented in Tables 55.2–55.5. Although a corrosion rate of

TABLE 55.2. Corrosion Rates (mm/year) for Haynes 6B in Sulfuric Acid

Concentration, wt %	24°C, 75°F	66°C, 150°F	Boiling
2			0.8
5			2.3
10	<0.1	<0.1	4
20			9.1
30	<0.1	<0.1	>10
50	<0.1	>10	>10
77	<0.1	4.5	>10

0.5 mm/year (20 mpy) is usually regarded as the maximum for industrial use, applications often require corrosion rates of < 0.1 mm/year (4 mpy). From Table 55.2, therefore, it may be concluded that HAYNES 6B is safe to use at room temperature in all concentrations of sulfuric acid but the alloy is suitable for only dilute solutions at high temperatures.

The aggressiveness of hydrochloric acid is apparent from Table 55.3, which infers that HAYNES 6B alloy is only useful in this acid at very low concentrations, even at room temperature. In nitric acid (Table 55.4), HAYNES 6B alloy appears safe for use up to the boiling point, at concentrations of 30 wt % or less, and presumably in higher concentrations at lower temperatures. Finally, Table 55.5 infers that

TABLE 55.3. Corrosion Rates (mm/year) for Haynes 6B in Hydrochloric Acid

Concentration, wt %	24°C, 75°F	66°C, 150°F
2	<0.1	<0.1
5	1.6	>10
10	2.7	>10
20	2.4	>10

TABLE 55.4. Corrosion Rates (mm/year) for Haynes 6B in Nitric Acid

Concentration, wt %	Boiling
10	<0.1
30	0.2
50	>10
70	>10

TABLE 55.5. Corrosion Rates (nun/year) for Haynes 6B in Phosphoric Acid

Concentration, wt %	Boiling
10	<0.1
30	0.1
50	0.5
70	0.6
85	>10

HAYNES 6B is very resistant to pure phosphoric acid and can be used in boiling solutions up to a concentration of 70 wt %.

With regard to the resistance of HAYNES 6B alloy to other media, it is very useful in organic acids. In the presence of water, however, halogen-bearing organic compounds can break down to form inorganic halogen acids, such as hydrochloric, and it is as well to be aware of this possibility.

The presence of chlorides in industrial environments can cause localized attack (pitting and crevice corrosion) and, as already mentioned, stress corrosion cracking (SCC). The resistance to localized attack of HAYNES 6B alloy is much higher than that of type 316L stainless steel.

In boiling, 30 wt % caustic soda (sodium hydroxide), HAYNES 6B alloy exhibits a corrosion rate of 0.3 mm/year (12 mpy). It is therefore assumed that the alloy is suitable for use in this common alkali at lower temperatures and concentrations. It is reported that, at high concentrations and temperatures, caustic stress cracking is a possibility with all the cobalt alloys.

C. ALLOYS DESIGNED FOR HIGH-TEMPERATURE USE

C1. History

The first reported high-temperature applications of the cobalt–chromium alloys were the hardfacing (weld overlay) of engine valves and hot trimming dies in the 1920s [3]. However, the event that led to the evolution of cobalt–chromium high-temperature alloys was the use of VITALLIUM® alloy for the investment casting of aircraft turbocharger blades in the late 1930s. This cobalt–chromium–molybdenum alloy, which was actually designed for aqueous corrosion and wear resistance and was in use as a dental and orthopedic implant material, was found to have the required combination of high-temperature strength, high-temperature microstructural stability (when the carbon content was reduced slightly), oxidation resistance, and castability [4].

To increase the thermal stability of VITALLIUM alloy, 2.5 wt % nickel was added. The resulting composition was renamed STELLITE 21 or modified VITALLIUM alloy. The use of nickel to stabilize the high-temperature cobalt alloys was carried further with the development of cast X-40 (STELLITE 31) and wrought HAYNES 25 alloy (L-605) in the 1940s and of cast MAR-M® 509 and wrought HAYNES 188 alloy in the 1960s. The X-40, HAYNES 25 alloy, and MAR-M 509 contain ~10 wt % nickel, whereas HAYNES 188 alloy contains 22 wt % nickel (Table 55.6).

It is also important that, from the development of X-40 onward, tungsten was preferred over molybdenum as the primary solid-solution strengthening element in the high-temperature cobalt alloys, inferring that it is a more effective

TABLE 55.6.   Nominal Composition of High-Temperature Cobalt Alloys

	Co	Cr	W	Mo	C	Ni	Fe	Si	Others
X-40 (STELLITE 31)	Balance	25.5	7.5	—	0.5	10.5	2	—	B 0.01
MAR-M 509	Balance	24	7	—	0.6	10	—	—	Ta 3.5
									Ti 0.2
									Zr 0.5
HAYNES 25 (L-605)	Balance	20	15	1 max	0.1	10	3 max	0.4 max	Mn 1.5
HAYNES 188	Balance	22	14	—	0.1	22	3 max	0.35	Mn 1.25 max
									B 0.015 max
									La 0.03

high-temperature strengthener than molybdenum, on an atomic percentage basis, which is consistent with its atomic size. Molybdenum, on the other hand, remained the preferred element for aqueous corrosion resistance, as will be discussed.

C2. Metallurgy

The microstructures of the high-temperature cobalt alloys are discussed in detail in [5] and [6]. In the solution-annealed and quenched condition, the wrought materials, HAYNES 25 alloy and HAYNES 188 alloy, exhibit simple microstructures consisting of a sparse dispersion of M_6C carbides in a fcc solid solution. In service, within certain high-temperature ranges, they undergo microstructural changes, the most significant being the precipitation of $M_{23}C_6$ at grain boundaries and along stacking faults and, in the case of HAYNES 25 alloy, the precipitation of Laves intermetallic phase. The precipitation of carbides can provide useful increases in strength. The precipitation of Laves phase is undesirable, since it results in significant loss of ductility.

The cast materials exhibit complicated microstructures. The MAR-M 509, for example, contains both $M_{23}C_6$ and MC carbides in the as-cast condition due to the deliberate addition of active carbide formers, such as tantalum. These carbides provide considerable strength, even prior to extended high-temperature service, during which additional, fine, carbide dispersions are created [5].

C3. Gaseous Corrosion and Molten Metal Corrosion Behavior

One of the main benefits of the high-temperature cobalt alloys is their resistance to sulfidation (hot corrosion) in gas turbine engines. This phenomenon, which is believed to involve the dissolution of the normally protective oxide films by sodium sulfate (created by airborne salts and residual sulfur in gas turbine fuels), is a problem up to ~980°C (1800°F), above which sodium sulfate volatilizes and is relatively innocuous.

HAYNES 25 alloy exhibits good resistance to oxidation up to ~1000°C. HAYNES 188 alloy possesses even higher oxidation resistance, by virtue of the lanthanum addition (0.03 wt %), which reduces the tendency of protective oxide films to spall away during thermal cycling. HAYNES 188 alloy can be used in high-temperature oxidizing environments to ~1150°C.

The cobalt alloys are attacked by some molten metals, notably aluminum and bismuth. However, they are very resistant to zinc and tin alloys and have been widely used in diecasting and zinc galvanizing.

D. ALLOYS DESIGNED FOR AQUEOUS CORROSION AND WEAR RESISTANCE

D1. History

The existence of VITALLIUM in prewar years as a casting alloy for dental and biomedical applications has already been mentioned. This material, the composition of which is given in Table 55.7, is still in use today, having enjoyed 40 years of use in hip joint replacements, initially on its own, then more recently in composite metal/polymer and metal/metal joints. It has been used in cast form, in wrought form, and as made by powder metallurgy processing, in attempts to enhance fatigue resistance. Also to enhance fatigue resistance, nitrogen-bearing derivatives of VITALLIUM alloy have been developed for the biomedical industry.

STELLITE 21, the derivative of VITALLIUM alloy developed for high-temperature use, evolved (in cast and weld

TABLE 55.7.   Nominal Composition of Aqueous Corrosion and Wear-Resistant Cobalt Alloys

	Co	Cr	W	Mo	C	Ni	Fe	Si	Mn	N
VITALLIUM	Balance	28.5		6	0.35 max	1 max	0.75 max	1 max	1 max	
ULTIMET	Balance	26	2	5	0.06	9	3	0.3	0.8	0.08

overlay form) into a candidate material for applications involving aqueous corrosion and wear in particular situations conducive to liquid-droplet impingement erosion and cavitation erosion, such as hydroelectric power turbines and valve seating surfaces, respectively. It was partly the outstanding wear properties of this material, without the benefit of a large amount of carbide in the microstructure, that gave rise to the understanding that, for many wear processes, carbides are not essential and the unique wear characteristics of the cobalt alloys are largely due to the transformation and twinning tendencies under the action of mechanical stress.

With the knowledge that carbides are not essential for wear resistance, ULTIMET® alloy was developed in the late 1980s as a material optimized for aqueous corrosion resistance. Its composition is given in Table 55.7. Primarily a wrought product, ULTIMET alloy has been used also in the form of castings and weld overlays. It is used in a wide variety of chemical process industry situations subject to hot acids and wear, notably in valves, pumps, and nozzles.

D2. Metallurgy

Unlike the cobalt–chromium–tungsten alloys designed for wear resistance, the cobalt–chromium–molybdenum materials designed to resist both aqueous corrosion and wear (i.e., VITALLIUM and ULTIMET alloys) contain few precipitates. The cast VITALLIUM alloy does contain sufficient carbon to cause some grain boundary precipitation of chromium-rich $M_{23}C_6$ carbides; however, the only precipitates in wrought ULTIMET alloy (which requires solution annealing and water quenching) are a few nitrides which occur if reactive residuals are present.

The metallurgical characteristics of cobalt and its alloys, which impart resistance to those wear processes involving microfatigue, have already been discussed. Naturally, VITALLIUM and ULTIMET alloys possess these characteristics. In the case of ULTIMET alloy, the hcp-to-fcc transformation temperature was deliberately lowered by the addition of nickel (at 9 wt %) to reduce the tendency of the material to work harden and thus to make the wrought processing and forming of the alloy into industrial components an easier proposition.

As discussed previously, chromium enhances passivation of the cobalt alloys in the presence of oxygen and is therefore the key ingredient with regard to corrosion resistance in oxidizing media, such as nitric acid. Molybdenum (together with tungsten in the case of ULTIMET alloy) is very beneficial under active corrosion conditions, such as are encountered in hydrochloric acid.

The carbon and nitrogen contents of ULTIMET alloy are critical [7]. During development of the alloy, it was initially thought that carbon should be kept as low as possible, as it is in the austenitic stainless steels and nickel–chromium–molybdenum (HASTELLOY®) alloys, to avoid sensitization (the grain boundary precipitation of carbides during elevated temperature excursions, e.g., during welding). However, carbon was found to be more soluble in the cobalt–chromium–molybdenum system and actually improves SCC resistance when added within the soluble range. Nitrogen is soluble up to ~0.12 wt % (without active residuals), and it enhances strength and localized corrosion resistance (pitting and crevice corrosion) within this soluble range.

D3. Wear Behavior

The results of galling, cavitation erosion, and low-stress abrasion tests at room temperature on ULTIMET alloy (as a representative of the cobalt alloys designed to resist both aqueous corrosion and wear) are given in Figures 55.1–3. For comparison, results are also presented for:

HAYNES 6B
STELLITE 1 and 6 (two-layer gas tungsten arc weld overlays)
C-276 and 625 (corrosion-resistant Ni–Cr–Mo alloys)
316L stainless steel (austenitic)
410 stainless steel (martensitic) in the 23 HRC condition
2205 stainless steel (ferritic/austenitic or duplex)
17-4 PH stainless steel (precipitation hardening)
NITRONIC® 60 (high-silicon, nitrogen-bearing, austenitic stainless steel)

From Figures 55.1 and 55.2, the advantages of cobalt as an alloy base are apparent for situations involving metal-to-metal wear and microfatigue-controlled erosion processes. The only other alloy that comes close to the performance of the cobalt alloys is NITRONIC 60. This alloy also exhibits much higher resistance to galling than the standard stainless steels.

Under low-stress abrasion conditions (Fig. 55.3), the carbide-containing cobalt alloys are outstanding. ULTIMET alloy does possess reasonable resistance to this form of wear, however, indicating that the unique response of the cobalt-rich solid solution to mechanical stress is of some benefit.

The galling test used was a modification of the ASTM G 98 procedure and is described in [8]. The cavitation erosion and low-stress abrasion data were generated using the ASTM G 32 and G 65 test procedures, respectively.

The elevated temperature abrasion properties of ULTIMET alloy, relative to STELLITE 6, 316L stainless steel, and 410 stainless steel, are discussed in [9]. This work indicates that the advantages of cobalt as an alloy base for wear resistance extend to 800°C.

D4. Aqueous Corrosion Behavior

To illustrate the aqueous corrosion characteristics of alloys designed to resist both aqueous corrosion and wear, uniform

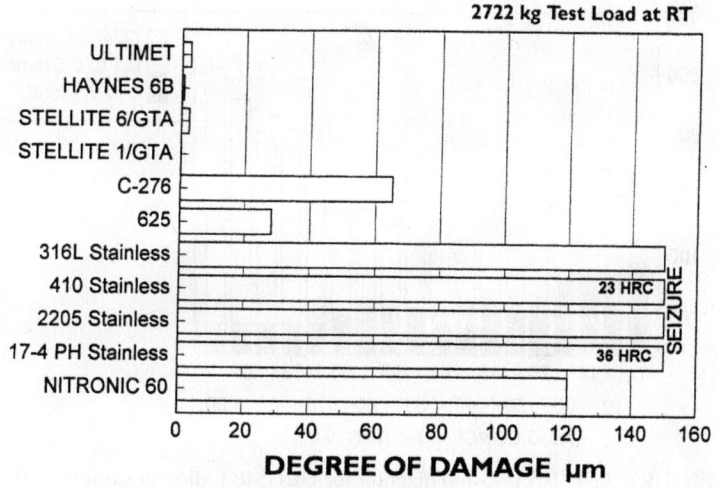

FIGURE 55.1. Comparative galling data (self-coupled).

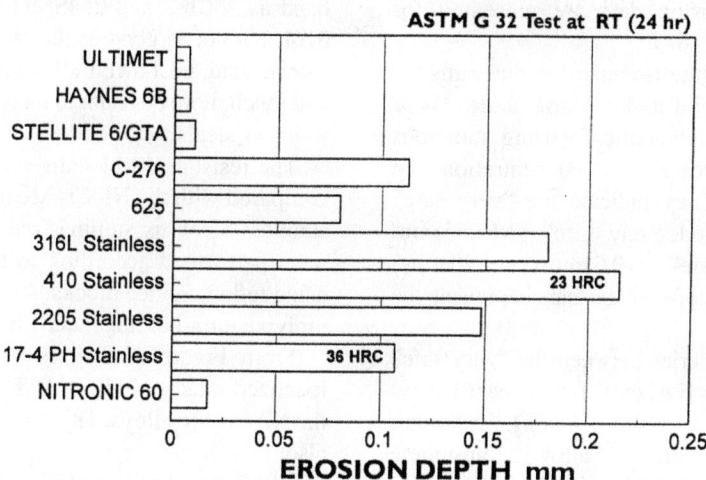

FIGURE 55.2. Comparative cavitation erosion data.

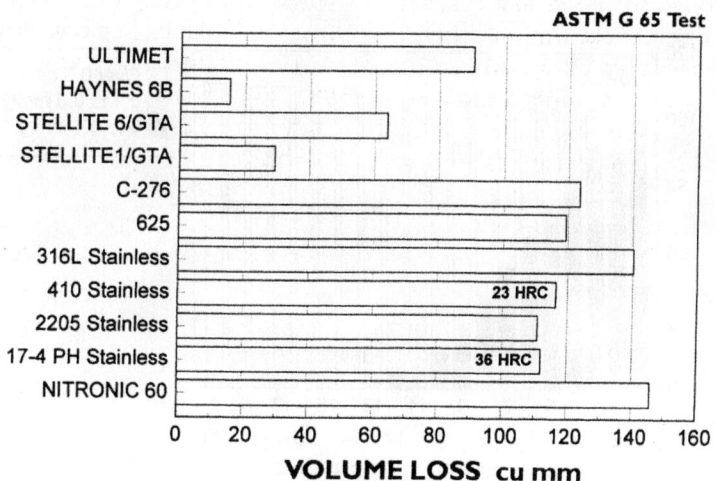

FIGURE 55.3. Comparative low-stress abrasion data.

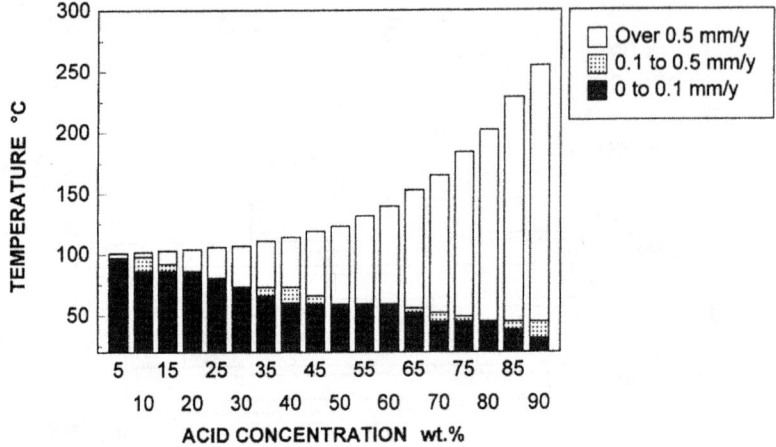

FIGURE 55.4. Isocorrosion diagram for ULTIMET alloy in sulfuric acid.

corrosion and crevice corrosion data are presented for ULTIMET in Figures 55.4–55.8.

Figures 55.4 and 55.5 are the isocorrosion diagrams for ULTIMET alloy in sulfuric and hydrochloric acids. These diagrams were constructed mathematically using numerous corrosion rate values at different acid concentration and temperature combinations. They indicate the "very safe" (0–0.1 mm/year, 4 mpy), "moderately safe" (0.1–0.5 mm/year, 4–20 mpy), and "unsafe" (>0.5 mm/year, 20 mpy) regimes in these acids. The tops of the bars represent the boiling points.

For perspective, the boundaries between the "very safe" and "moderately safe" regimes [i.e., the 0.1-mm/year (4-mpy) lines] are plotted in Figures 55.6 and 55.7 alongside similar lines for 316L stainless steel, 20Cb-3® alloy (an austenitic stainless steel designed to resist sulfuric acid), 254SMO® (one of the 6 wt % molybdenum-bearing stainless steels), and C-2000® alloy, a nickel–chromium–molybdenum material. In sulfuric acid, ULTIMET alloy is in the same performance

band as 20Cb-3 and 254SMO alloys, being far better than 316L but not as good as the Ni–Cr–Mo materials. In hydrochloric acid, ULTIMET alloy is second only to C-2000 alloy and much more resistant to this aggressive acid than the three stainless steels.

The resistance to localized attack of ULTIMET alloy as compared with the Ni–Cr–Mo (C-type) alloys and austenitic stainless steels is summarized in Figure 55.8. These tests were performed according to the ASTM G 48 procedures using teflon crevice blocks. AL-6XN® alloy is also a 6 wt % molybdenum-bearing stainless steel.

From Figure 55.8, it is evident that the resistance to localized attack of ULTIMET alloy is equivalent to that of the Ni–Cr–Mo alloys. This has been confirmed in pitting tests also.

With regard to the SCC resistance of the corrosion and wear-resistant cobalt materials, such as ULTIMET alloy, they appear to be similar to the super austenitic stainless steels in their susceptibility. Reference [8], in which a four-point bend

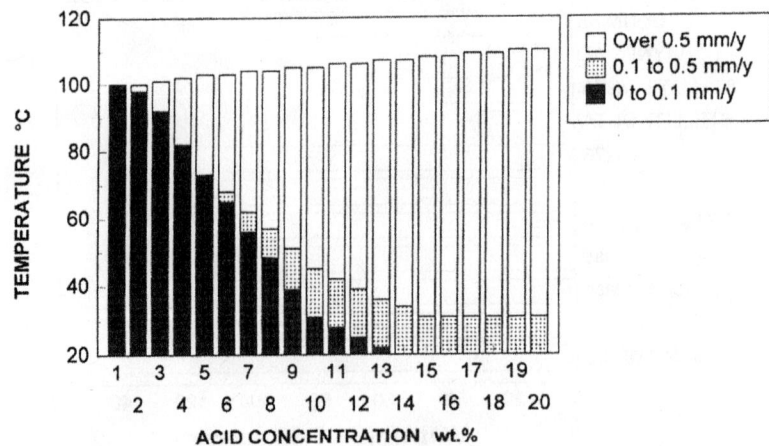

FIGURE 55.5. Isocorrosion diagram for ULTIMET alloy in hydrochloric acid.

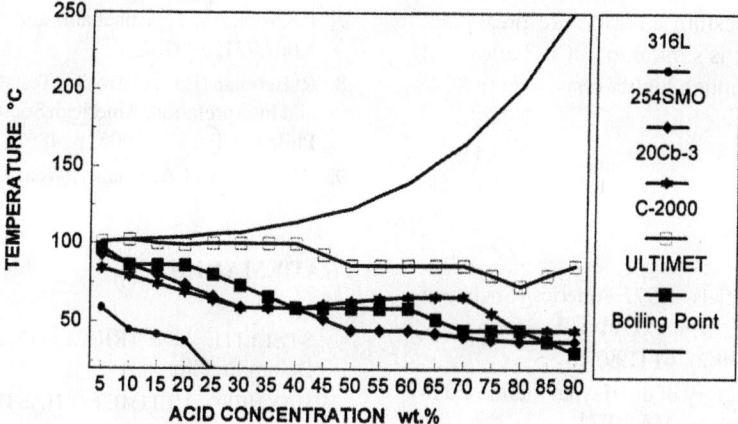

FIGURE 55.6. Comparison of 0.1-mm/year (4-mpy) lines for stainless steels, a Ni–Cr–Mo alloy, and ULTIMET alloy in sulfuric acid.

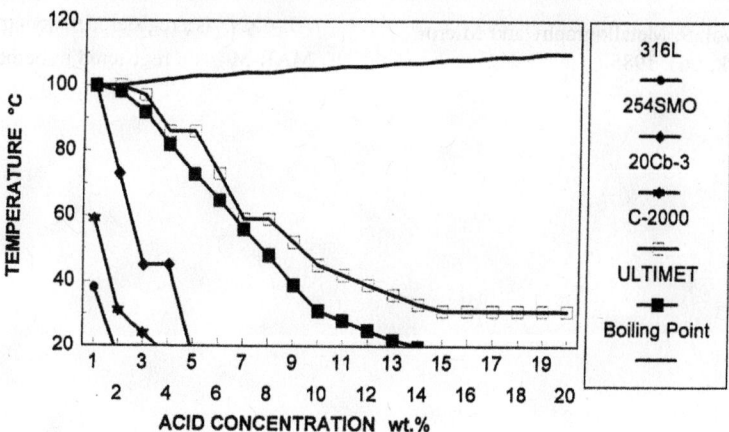

FIGURE 55.7. Comparison of 0.1-mm/year (4-mpy) lines for stainless steels, a Ni–Cr–Mo alloy, and ULTIMET alloy in hydrochloric acid.

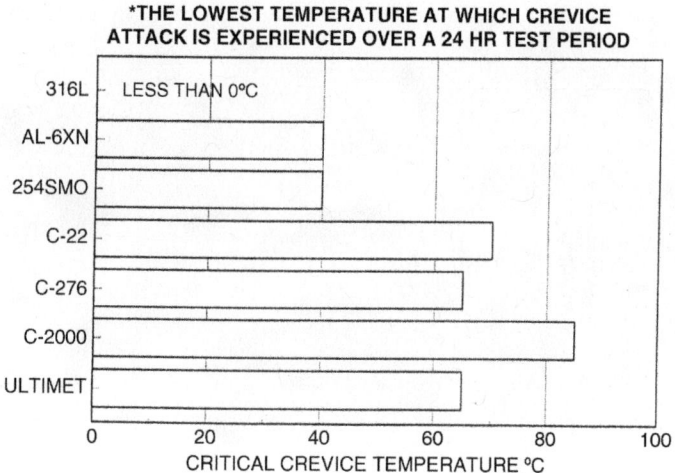

FIGURE 55.8. Critical crevice corrosion temperature (the lowest temperature at which crevice attack is experienced over a 24-h test period) in 6% ferric chloride.

test results in 30 wt % magnesium chloride are presented, indicates that ULTIMET alloy is similar to 20Cb-3 alloy and that both these alloys possess much higher resistance to SCC than does 316L stainless steel.

REFERENCES

1. W. L. Silence, Wear of Materials—1977, American Society of Mechanical Engineers, New York, 1977, p. 77.
2. P. Crook, Mater. Perform., **30**(2), 64 (1991).
3. R. D. Gray, STELLITE—History of the Haynes Stellite Company, 1912–1972, Cabot, Boston, MA, 1974.
4. C. P. Sullivan, M. J. Donachie, Jr., and F. R. Morral, Cobalt-Base Superalloys—1970, Centre d'Information Du Cobalt, Brussels, Belgium, 1970.
5. C. T. Sims and W. C. Hagel, The Superalloys, Wiley, New York, 1972.
6. Metals Handbook, 9th ed., Vol. 9, Metallography and Microstructures, ASM, Metals Park, OH, 1985.
7. P. Crook, A. I. Asphahani, and S. J. Matthews, U.S. Patent 5,002,731, 1991.
8. R. Baboian (Ed.), Corrosion Tests and Standards—Application and Interpretation, American Society for Testing and Materials, Philadelphia, PA, 1995, p. 486.
9. H. Berns and A. Fischer, Wear, **162–164**, 441 (1993).

TRADEMARKS

STELLITE® and TRIBALOY® are registered trademarks of Deloro Stellite.
HAYNES®, ULTIMET®, HASTELLOY®, and C-2000® are registered trademarks of Haynes International, Inc.
VITALLIUM® is a registered trademark of Howmedica.
NITRONIC® is a registered trademark of Armco Steel.
20Cb-3® is a registered trademark of Carpenter.
254SMO® is a registered trademark of Avesta Sheffield.
AL-6XN® is a registered trademark of Allegheny Teledyne, Inc.
MAR-M® is a registered trademark of Lockheed Martin.

56

COPPER AND COPPER ALLOYS

C. A. C. Sᴇǫᴜᴇɪʀᴀ

Instituto Superior Técnico, Lisboa, Portugal

A. INTRODUCTION

Copper is the dominant material for domestic water systems in Europe, North America, Australasia, and many countries of the Commonwealth. The developing world is now rapidly expanding its use of copper tubing as their societies demand increasing quantities of disease-free water. The annual world production of copper water tubing is ~500,000 tonnes, equivalent to ~1.25 billion m or 0.75 million miles. The reasons for this enormous consumption of copper plumbing tube are its excellent corrosion resistance, its ease of fabrication during installation, and hence low-installation costs, and additionally, its contribution to health and the maintenance of healthy water. Further, copper is seen as environmentally friendly due to its potential to be 100% recycled.

Failures of copper water tube from corrosion are rare but in general well-understood events. The high level of quality control exercised in modern tube producing plants plus copper's excellent corrosion resistance gives failure rates of <1 in a million. The fabrication of copper during installation is easy due to copper's malleability and the use of soldered and compression fittings that enable sound jointing to be made quickly. To prevent lead pick-up in soft waters, the industry introduced a range of "potable" integral solder fittings in which no lead is used, the solders being tin–silver or copper–tin.

Uhlig's Corrosion Handbook, Third Edition, Edited by R. Winston Revie
Copyright © 2011 John Wiley & Sons, Inc.

TABLE 56.1.   Generic Classification of Copper Alloys

Generic Name	UNS Number	Composition
Wrought Alloys		
Coppers	C10100–C15760	>99% Cu
High-copper alloys	C16200–C19600	<96% Cu
Brasses	C205–C28580	Cu–Zn
Leaded brasses	C31200–C38590	Cu–Zn–Pb
Tin brasses	C40400–C49080	Cu–Zn–Sn–Pb
Phosphor bronzes	C50100–C52400	Cu–Sn–P
Leaded phosphor bronzes	C53200–C54800	Cu–Sn–Pb–P
Copper–phosphorus and copper–silver–phosphorus alloys	C55180–C55284	Cu–P–Ag
Aluminum bronzes	C60600–C64400	Cu–Al–Ni–Fe–Si–Sn
Silicon bronzes	C64700–C66100	Cu–Si–Sn
Other coppers–zinc alloys	C66400–C69900	
Copper–nickels	C70000–C79900	Cu–Ni–Fe
Nickel silvers	C73200–C79900	Cu–Ni–Zn
Cast Alloys		
Coppers	C80100–C81100	>99% Cu
High-copper alloys	C81300–C82800	> 94% Cu
Red and leaded red brasses	C83300–C85800	Cu–Zn–Sn–Pb (75–89% Cu)
Yellow and leaded yellow brasses	C85200–C85800	Cu–Zn–Sn–Pb (57–74% Cu)
Manganese and leaded manganese bronzes	C86100–C86800	Cu–Zn–Mn–Fe–Pb
Silicon bronzes, silicon brasses	C87300–C87900	Cu–Zn–Si
Tin bronzes and leaded tin bronzes	C90200–C94500	Cu–Sn–Zn–Pb
Nickel–tin bronzes	C94700–C94900	Cu–Ni–Sn–Zn–Pb
Aluminum bronzes	C95200–C95810	Cu–Al–Fe–Ni
Copper–nickels	C96200–C96800	Cu–Ni–Fe
Nickel silvers	C97300–C97800	Cu–Ni–Zn–Pb–Sn
Leaded coppers	C98200–C98800	Cu–Pb
Miscellaneous alloys	C99300–C99750	

Note: UNS = Unified Numbering System.

Pure copper is a very soft, malleable metal. It is alloyed with small quantities of metals such as Be, Te, Ag, Cd, As, and Cr to modify the properties for particular applications, while retaining many of the characteristics of the pure metal. Much larger alloying additions of Zn, Sn, and Ni are made to improve the mechanical properties of the metal, and to retain its excellent corrosion resistance under more arduous service conditions. Nickel permits increased flow rates in water systems; zinc gives increased resistance to sulfide attack. Typical groups of copper alloys that find application in many environments are listed in Table 56.1.

Apart from the wide use of copper in freshwater supply lines and plumbing fittings, copper and its alloys provide superior service in many other applications. They are used for structures open to the atmosphere, for example, in architecture and sculpture. They are used immersed in freshwater and seawater heat exchangers and condensers, as well as in industrial, chemical, and power-generating plants, and buried in the earth for water distribution systems.

In the open air, copper forms a green patina that, in its most stable form, consists of basic copper sulfate, $CuSO_4 \cdot 3Cu(OH)_2$, although in marine environments it may contain chloride, or carbonate in industrial areas. This decorative long-lasting coating makes copper an ideal material for low- maintenance roof coverings and for gutters and channels. A small amount of copper dissolves in water that runs over the metal surface, and this can precipitate on to other less noble metals downstream in the water cycle, leading to galvanic corrosion. Cast iron gutters and pipes used in conjunction with copper roofs benefit from a bituminous or other impervious coating to reduce the possibility of galvanic corrosion. General corrosion or stress corrosion cracking (SCC) may become a problem in industrial areas if ammonium compounds are present in the atmosphere.

Copper and its alloys can safely be buried in most soils, although high corrosion rates have been experienced in those containing cinders or acid peat. If it is expected that corrosion will be a life-limiting factor in the use of the material, it can be protected with bituminous, plastic, or paint coatings. Dezincification can be a problem in brasses with high zinc levels, and it is best to avoid the use of these alloys unless they are specifically required to counter the difficulties that may result from high sulfide levels in the soil.

Copper and its alloys are used extensively in seawater distribution systems and in treatment units, condensers, and heat exchangers where fresh or salt water is used for cooling. Many components in valves, pumps, and taps, as well as pipes and pipe fittings, are made from copper alloys. A distinction is often made between corrosion in freshwater and in seawater. However, the same types of corrosion problem are found in both environments, and such a clear-cut distinction cannot be made. Rather, the change from pure to salt water with varying degrees of pollution should be regarded as a gradually increasing aggressiveness in the environment, aggravated by increases in flow rates and changes in the temperature and oxygen content of the water.

The main problems of copper alloys in water systems are differential aeration corrosion, erosion corrosion, SCC, and demetallification. Differential aeration corrosion is mainly a design problem, although pitting may occur under very slow flow rates, which starve the metal surface of oxygen. Erosion corrosion is a function of flow rate. Pressure changes in a liquid on passing through valves and pumps give rise to cavitation damage, while entrained air or abrasive particles disrupt protective surface films to produce shallow, horseshoe-shaped pits. The deterioration can be very rapid. Ammonia and its salts, together with mercury-based compounds, are the prime cause of SCC.

Dealloying affects many of the alloys, the commonest being dezincification of brasses containing $> 15\%$ zinc, although dealuminification and denickelification have been reported for aluminum bronzes and cupronickels. Dealuminification is most prevalent in aluminum bronzes containing the γ-2 phase in the microstructure, and is most serious when the γ-2 forms a continuous grain boundary network. Rapid cooling from $> 600°C$ ($1115°F$), additions of 1–2% iron, or more than 4.5% nickel, control dealuminification, but microstructural changes that occur during welding can still lead to corrosion problems in the heat-affected zone (HAZ) around the weld. Reheat treatment to remove unsatisfactory postweld microstructures can pose serious problems to the engineer; heating, handling, and quenching large units without introducing distortion while achieving uniform properties throughout the material may be impossible.

Brass fittings may dezincify, especially when the β phase is present. The loss of zinc is accelerated by high-temperature, increased chloride content, low flow rates, and differential aeration. Additions of 1% tin and ~0.04% arsenic, phosphorus, or antimony inhibit dezincification. However, phosphorus can lead to intergranular corrosion, and most manufacturers use arsenic as an inhibitor in brasses. Inhibited α brasses are immune to dezincification in most waters, but the effect of tin and arsenic additions to α/β brasses is not predictable in controlling dezincification. There have been many cases of dezincification in the duplex brasses in both fresh and seawater. In some instances in potable water distribution systems, duplex fittings that have given many years' service suddenly begin to lose zinc when only a slight change in water chemistry occurs.

A protective film of carbonate may be deposited on the metal surface from water containing carbon dioxide and oxygen.

As the flow rate increases, copper and brass tubes become more prone to impingement attack. Aluminum brass and cupronickel offer a greater resistance to higher flow rates, but both have maximum limits that must not be exceeded or the surface film on the metal will be destroyed. While the maximum velocity for inhibited Admiralty brass and aluminum brass is lower than that for cupronickels, they both give better service should sulfide be present in the water, either as a pollutant in rivers or estuaries or in chemical and oil production plants. However, work is in hand to develop coatings for copper-based alloy condenser tubes in land-based and marine systems. These coatings will offer protection against sulfide-polluted water while the natural film is established. Experimental programs have shown that these coatings will also be effective on both cupronickel and aluminum brass tubes [1].

Tin bronzes and phosphor bronzes have good resistance to flowing seawater. The alloys containing 8–12% tin are less susceptible than brasses to SCC and have excellent resistance to impingement attack and to attack in acid waters.

The aluminum oxide film on both aluminum brass and aluminum bronze, which confers the additional corrosion protection, reduces the dissolution rate of copper ions from the alloy and makes them less effective in resisting biofouling.

The resistance of some copper alloys to erosion–corrosion is improved when small quantities of iron are present in the alloy or the water. The iron apparently produces a tougher surface film. This has led to the use of iron sacrificial pieces, in preference to the normal zinc sacrificial anodes, in the water boxes of condensers and heat exchangers that use copper-based tubes and tube plates. The iron ions are absorbed into the film from the water and confer the additional resistance to impingement attack. Zinc ions do not have this beneficial effect.

A very clear distinction can be made between acids that can be safely handled in copper-based equipment and those that cause catastrophic attack. Nonoxidizing acids, such as acetic, phosphoric, dilute sulfuric, and hydrochloric, can be safely handled providing the concentrations of oxidizing agents such as entrained or dissolved air, chromates, and iron(III) ions are kept very low. Oxidizing acids, nitric or concentrated sulfuric, and those containing oxidizing agents must not be handled in copper-based systems. Small additions of oxidizing agents are particularly dangerous in hydrochloric acid, causing a dramatic increase in the rate of metal loss. Before using copper alloys in acid systems, tests should always be made with the particular liquid to be processed, reproducing the actual plant conditions as closely as possible.

In general, copper and its alloys are resistant to attack by alkalis except ammonium hydroxide and those containing ammonium or cyanide ions. Ammonium ions promote SCC and both ions form complex species such as $[Cu(NH_3)_4]^{2+}$ and $[Cu(CN)_4]^{2-}$, which do not allow the double layer to develop to polarize the corrosion cell, hence the corrosion rate remains high.

Iron(III) and tin(IV) salts are aggressive to copper alloys. Copper itself suffers general corrosion, to thin the cross section, in sulfur compounds.

B. PITTING CORROSION

B1. Pitting Mechanisms

Since the first description of pitting of copper tubes in contact with water in 1950 [2], extensive investigations/have since been conducted with various natural waters, reflecting the practical importance of this problem [3–12]. Not only copper, but also copper alloys, such as brasses [13, 14], bronzes [15], some cupronickels (e.g., 70 Cu–30 Ni [16–18]), and other alloys [19–22] can be damaged by pitting. Pitting of copper does not occur exclusively in chloride-containing solutions but in hydrogen carbonate solutions as well [23–25]. Kinetics of passivation and pitting corrosion of copper have been extensively studied using electrochemical techniques [26–30].

These techniques include measurement of redox potentials and of the potentials of corroding electrodes [31], measurement of current density between coupled electrodes, polarization resistance, potentiokinetic curves [32], impedance measurements, and electrochemical noise [30, 33]. Other methods of investigation include eddy current examinations [34], morphological studies [35], and use of chemical and biochemical microsensors [36].

Anodic polarization curves of copper in NaCl solutions show that the breakdown potential is lower the higher the NaCl concentration. When copper specimens are immersed for long periods of time in dilute NaCl solutions, pitting is not observed, and general corrosion occurs. The presence of other halogen anions in NaCl inhibits pitting of copper, but the action of I^- differs from that of Br^- [37]. According to Mor and Beccaria [37], with iodides inhibition is caused by the formation of a CuI layer on the metal surface, whereas KBr acts by reducing the oxygen content of the solution. In very dilute KBr solutions, however, copper corrosion increases.

In hydrogen carbonate solutions, with increasing $NaHCO_2$ concentration, the resistance of the oxide layer toward general corrosion decreases [25], which is probaby caused by increasing soluble copper carbonate complex

formation. However, increasing HCO_3^- oncentration promotes break down of the oxide layer. As the temperature is increased from 25 to 90°C (80 to 197°F), the oxide film is less protective and the pitting potential shifts to less positive values.

Kristiansen [38] studied pitting of copper in distilled water containing 10 mg/L SO_4^{-2} plus 5 mg/L CO_2, with and without Fe^{3+} additions, at temperatures of 45, 50, and 60°C (116, 125, and 143°F). In aerated neutral water ferric ions are reduced on copper surfaces according to

$$Fe^{3+} + Cu = Cu^+ + Fe^{2+}$$

Ferrous ions are reoxidized to Fe^{3+}, causing further copper corrosion. Using radioactive ^{59}Fe, it was established that pits nucleated where iron was present on the metal surface. It was also found that the highest rate of corrosion occurred at 50°C (125°F), which was explained by the decomposition of the basic copper carbonate deposit and formation of a more protective copper oxide at higher temperatures. Pits were also found on copper when no iron had been added, although the presence of iron in the water could not be excluded. The deleterious effect of ferric ions has also been reported by Molvina et al. [39].

The problem of copper tube pitting in supply waters still persists in many countries. This type of localized corrosion was called "nodular pitting" by Campbell [15] because corrosion produces small areas of deep attack covered by small mounds or nodules of corrosion products.

According to Mattsson [40], three types of pitting can be distinguished for copper: Type 1 occurs on annealed or half-hard tubes in cold tap water, caused by a continuous carbon film formed on the inner tube surface during bright annealing; type 2 occurs on hard drawn tubes in hot tap water of low (<7:4) pH and low (< 1) $[HCO_3^-]/[SO_4^{2-}]$ ratio; type 3 occurs on both hard and annealed tubes in cold tap water of high pH, with low salt concentration. This type is not caused by a continuous carbon film, and the reason for its formation is not yet known.

Type 1 pitting usually occurs in supply water from deep wells that is free of organic species. [Natural waters containing organic substances (e.g., rivers and lakes) do not corrode copper. Some undefined organic compounds clearly act as corrosion inhibitors.] There is general agreement that type 1 pitting occurs in hard or moderately hard waters and is favored by water with high sulfate content; failures occur most rapidly in water of fairly low chloride content. Pitting occurs when carbon residues, which arise from the breakdown of lubricants during bright annealing, are present on tube bores [41, 42]. The importance of deposited carbon films in pitting of copper is widely recognized.

Frommeyer [43] studied the composition of corrosion products formed on copper during type 1 pitting by electron spectroscopic methods. Concentration profiles in the coating

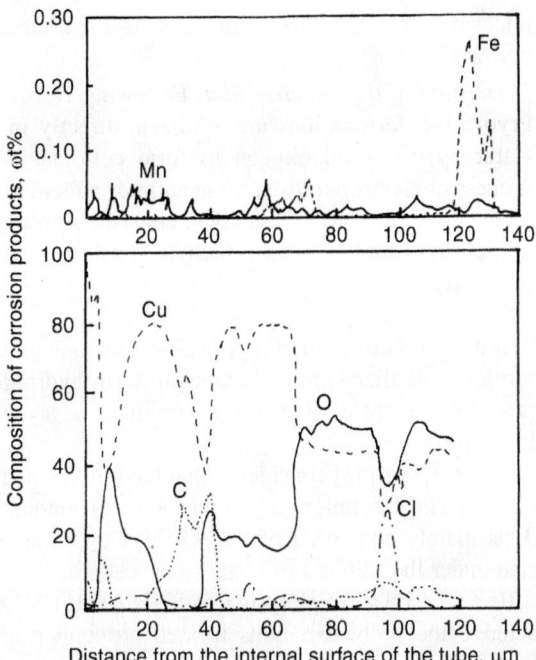

FIGURE 56.1. Concentration profiles in corrosion products on copper. (Reproduction with permission from [43].)

are shown in Figure 56.1 and a schematic representation of copper corrosion products is presented in Figure 56.2. The corrosion regions I and II are below and above the Cu_2O membrane, respectively. Region I forms on a pre-existing inner surface of the tube and is the result of pitting, while region II is a swelling composed of voluminous corrosion products. Electron spectroscopy for chemical analysis (ESCA) studies indicated that carbon does not occur in the form of copper or calcium carbonates, but is derived from the lubricants. The absence of manganese and iron in region I is thought to indicate that only carbon deposited on the copper is responsible for promoting pitting. Because of the detrimental effect of deposited carbon, abrasive cleaning is often

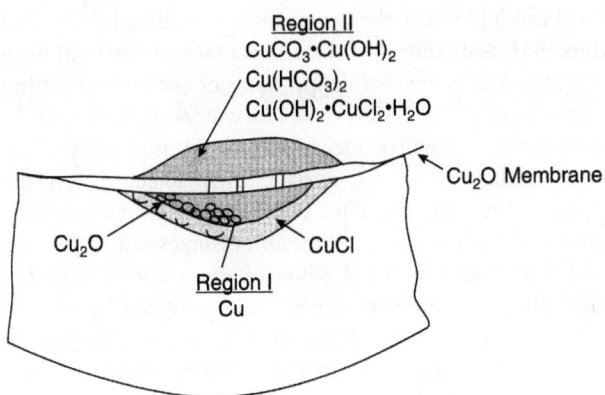

FIGURE 56.2. Schematic representation of layers on corroded copper. (Reproduced with permission from [43].)

used to remove carbon films that form during tube manufacture.

It was found by Cornwell et al. [44] that copper containing 1% Sn or 1% Al (particularly the former) is more resistant to pitting in the presence of carbon residues than ordinary commercial copper. Carbon residue on copper alone, however, is insufficient to produce pitting. Studying the influence of carbon contamination in the bore, Cornwell et al. [41] found that the electrode potential never exceeded the pitting potential in water that did not support pitting, even when high carbon contamination was present. On the other hand, when tube specimens contaminated by carbon were immersed in aggressive water, pitting was observed, but abrasively cleaned specimens did not pit. In the latter case, the electrode did not reach the pitting potential.

Pourbaix [45] first determined the pitting potential of copper. The values measured in Brussels tap water were 170 mV saturated calomel electrode (SCE) for the concave internal surface of copper tubes and 100 mV (SCE) for the surfaces of copper wires. These potentials were only slightly higher than the equilibrium potential of $Cu/CuCl/Cu_2O$ measured inside pits [24 mV (SCE)]. Inside pits, the solution contained 207 ppm Cl^- plus 17 ppm Cu at pH 3.4. Other investigators [15] confirmed the existence of a critical potential for pitting of copper.

Pourbaix [46] observed that the potential of copper specimens immersed in flowing Brussels water shifted in the positive direction during exposure. Three stages were, differentiated. During stage 1, following immersion, the potential is relatively low [−30 to −10 mV(SCE)]; the metal is, covered with a red Cu_2O deposit, and HCl forms as a result of CuCl hydrolysis. During stage 2, potential increases [−10 to +50 mV (SCE)]; greenish malachite appears, and acidification of the solution inside the starting cavity occurs. Stage 3 is characterized by the same processes as stage 2; potential continues to increase irregularly, and pits develop. When equilibrium conditions within the pit have been established, stable CuCl forms on the pit bottom.

Type 2 pitting occurs in water with pH < 7.4 and at temperatures >60°C (143°F) [15]. Mattsson [40] found that pitting occurs in low pH water (generally pH 5–7) and that water that causes pitting has a relatively low HCO_3^- content (≤100 mg/L), whereas no pitting occurs in water with a higher HCO_3^- content (100–300 mg/L). The SO_4^{2-} concentration is 15–40 mg/L. At higher HCO_3^- concentrations, the protective basic copper carbonate forms.

Mattsson [40] observed that in carbonate containing water (15–70 mg/L HCO_3^- plus >0.2 mol/L SO_4^{2-}, a basic copper sulfate crust forms that covers the capillary mouth and thus creates an occluded cell at the anode where the reaction can occur.

$$2Cu + H_2O = Cu_2O + 2H^+ + 2e^-$$

As in type 1 pitting, pits initiate below the metal surface deposit. Mattsson also found that in tubes with pitting, the inner tube wall generally had a greenish gray surface coating, often with a high aluminum content and containing basic copper carbonate; a green, basic copper sulfate crust covered the pits.

Cornwell et al. [44] suggested the following mechanism of copper pitting in aerated supply waters:

$$Cu^+ + Cl^- = CuCl$$

CuCl hydrolyses to form Cu_2O, which is precipitated on the metal surface:

$$2CuCl + H_2O = Cu_2O + 2HCl$$

The cathodic reaction supporting the anodic dissolution process is oxygen reduction:

$$O_2 + 2H_2O + 4e^- = 4OH^-$$

For corrosion to proceed, the hydroxyl ions produced at the cathodic sites must be removed. This occurs more rapidly in acid supply water or water that contains bicarbonate ions:

$$OH^- + HCO_3^- = CO_3^{2-} + H_2O$$

The final reaction causes precipitation of mixed calcium carbonate and basic copper carbonate scale.

B2. Pitting Prevention

There are several ways of avoiding pitting corrosion of copper in waters. Ferrous ion injection has been used for many years in Europe and Japan to protect aluminum, brass, and cupronickel tubes, mainly in polluted waters [47–50]. The beneficial effect of ferrous ion, as $FeSO_4$, can be explained on the basis of either of the following mechanisms:

1. *Electrochemical protection*: Following Cornwell et al. [41], Lecointre et al. [51] and Pourbaix [52] observed that if the tube surface can be maintained at an electrode potential lower than the protection potential of copper, existing pits cannot grow and new pits cannot form. The redox couple Fe^{2+}–FeOOH can maintain the tube surface below this protection potential, and this protection process should remain valid for copper alloys (e.g., brass) and waters containing chloride (e.g., seawater and brackish water). Based on the protection potential for copper alloys in chloride-containing waters and on the electrochemical action of the addition of ferrous ions, a method of control is proposed: The potential of the tubes is measured and ferrous ions are injected to maintain this potential below the protection potential.

2. *Formation of a protective film*: Following North and Pryor [53], ferrous ions are oxidized. directly in the water by dissolved oxygen to form colloids. By a process of electrophoresis, these colloids adhere to the walls of the tubes to form a layer, consisting primarily of FeOOH lepidocrocite, which is considered to be protective.

According to Gasparini et al. [54], ferrous ions precipitate on local alkalized cathodic areas to form hydroxides that are subsequently oxidized to produce a layer of lepidocrocite.

North and Pryor [53] considered that the film formed on copper in a NaCl solution containing a small amount of $FeSO_4$ is mainly composed of η-FeOOH; Cu_2O was also detected under the film.

Castle et al. [55] and Epler and Castle [56] used ESCA to characterize the protective films formed on copper-based condenser tubes protected by ferrous sulfate injection. The film, was found to be composed mainly of two superimposed layers: an inner layer, white, containing hydrotalcite, $Mg_6Al_2(OH)_6CO_3 \cdot 4H_2O$, or other compounds of the hydrotalcite family and paratacamite, $Cu_2(OH)_3Cl$; and an outer layer, brown, porous, which is, a mixture of lepidocrocite (essentially η-FeOOH), paratacamite, iron oxides, and copper oxides. Hydrotalcite can exercise a buffer effect through the effect of Mg^{2+} ions.

In waters containing phosphates, tricalcium phosphate is found in the film. In polluted waters containing sulfides, Cu_2S is found in the external layer of the film. Assuming that the protection is given by a film, one method of monitoring the thickness, porosity, arid other characteristics of the film is to perform polarization tests [52, 53].

Ultraviolet (UV) photochemical decomposition of residual chlorine [57], adequte metal and water treatment (e.g., inhibitors for the solution or alloying additions to the metal tube) [58–63], the application of artificial protective films [64], and control of the redox potential have all been recognized as means of suppressing or preventing pitting corrosion of copper alloy tubes in waters. But, of course, an important effective means of combatting or avoiding pitting attack is by appropriate alloy selection. Of the copper alloys, the most pit resistant are the copper–aluminum alloys known as aluminum bronzes with <8% Al. Tables 56.2 and 56.3 list some of these alloys. Usually, these aluminum bronzes are also not affected by crevice corrosion, a type of attack that is more common in aluminum and chromium-bearing copper alloys, and that can several times be controlled by proper cleaning of the metal surfaces.

TABLE 56.2.   Some Common Wrought Copper–Aluminum Alloys: Chemical Composition

UNS Numbers	Cu	Al	Fe	Ni	Mn	Si	Sn	Others
C–60800	92.5–94.8	5.0–6.5	0.10					0.02–0.35 As 0.10 Pb
C–61000	90.0–93.0	6.0–8.5	0.50			0.10		0.02 Pb
C–61300	86.6–92.0	6.0–7.5	2.0–3.0	0.15	0.10	0.10	0.20–0.50	0.01 Pb
C–61400	88.0–92.5	6.0–8.0	1.5–3.5					0.01 Pb
C–61500	89.0–90.5	7.7–8.3		1.8–2.2				0.015 Pb
C–61800	86.9–91.0	8.5–11.0	0.5–1.5			0.10		0.02 Pb
C–62300	82.2–89.5	8.5–11.0	2.0–4.0	1.0	0.5	0.25	0.60	
C–63000	78.0–85.0	9.0–11.0	2.0–4.0	4.0–5.5	1.5	0.25	0.20	
C–63200	75.9–84.5	8.5–9.5	3.0–5.0	4.0–5.5	3.5	0.10		0.02 Pb

C. STRESS CORROSION CRACKING

C1. Mechanisms

The mechanisms of SCC of copper alloys have been extensively studied and reviewed [65–91]. Several important observations on the mechanisms of SCC are indicated here.

1. Stress corrosion cracking is possible in all alloys but not in pure copper. Alloys with a minimum concentration of solute elements, such as As, Sb, Al, Si, P, Ni, Zn, and Mn, are susceptible to SCC in specific environments under tensile stresses sufficient to produce active slip in microscopic regions of the metal.

2. The environment provides a species that leads to complex ion formation favoring selective removal of solute atoms and film formation to suppress general attack. Complex ion formation is an important feature in the mechanism of SCC of α brass.

3. The regions of complex ion formation and stability of oxide film may be predictable from the potential–pH diagrams if sufficient data are available for the construction of such a diagram. One example of a potential–pH diagram of copper in ammoniacal solution that take into consideration the oxidation of NH_3 to NO_2^- is given in Figure 56.3 [92]. These diagrams extend our understanding of the SCC problem, however, experimental confirmation would still be desirable.

4. There is thus a need to investigate the behavior of stressed specimens over a variety of environments and compositions, as a function of pH, redox potential, Cu concentration, and so on.

5. The ammonium ion is generally considered to be the specific corrodent to produce SCC in Cu-base alloys. Evidence has also been presented that citrates, tartrates, nitrites, sulfur dioxide, carbonates, oxides of N_2 and phosphates, among others, may also produce SCC in some Cu-base alloys. Pyridine and emylenediamine do not produce SCC of brass but do cause intergranular attack. Specificity of the corrodent may depend both on its ability to form a protective film on the metal surface and to lower the energy required to form new surface.

6. Mercurous nitrate (which deposits a coating of Hg on brass) is generally used to determine whether brasses contain high residual stresses that make them susceptible to SCC. Brasses containing appreciably lower stresses than those producing failures in mercurous nitrate solution may fail in an ammoniacal atmosphere that is widely used in the laboratory as a standard environment for SCC tests.

7. Liquid metal embrittlement in contact with molten metal operates by a similar mechanism to that of SCC in respect of lowering the surface energy by alloying and formation of an equilibrium groove angle for grain boundary attack.

TABLE 56.3.   Some Common Cast Copper–Aluminum Alloys: Chemical Composition

UNS Number	Cu	Al	Fe	Ni	Mn	Others
C-95200	86.0	8.5–9.5	2.5–4.0			1.0 total
C-95300	86.0	9.0–11.0	0.8–1.5			1.0 total
C-95400	83.0	10.0–11.5	3.0–5.0	2.5	0.5	0.5 total
C-95500	79.0	10.0–11.5	3.0–5.0	3.0–5.5	3.50	0.5 total
C-95700	71.0	7.0–8.5	2.0–4.0	1.5–3.0	11.0–14.0	0.5 total
C-95800	79.0	8.5–9.5	3.5–4.5	4.0–5.0	0.8–1.5	0.5 total

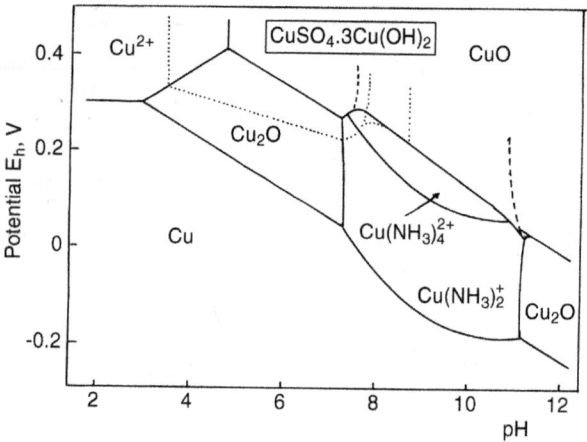

FIGURE 56.3. Potential–pH diagram for the copper–ammonia–water system at 25°C. Total concentration of dissolved copper ions 0.05 *M*, total concentration of dissolved nitrogen-containing species 1 *M*. Dotted lines refer to the formation of $CuSO_4 \cdot 3Cu(OH)_2$. Dashed and dash–dotted lines indicate the situation that would obtain in the absence of the oxidation of NH_3 to nitrite [92]. (Reprinted from T. P. Hoar and G. P. Rothwell, "Potential/pH Diagram for a Copper–Water–Ammonia System: Its significance in the SCC of Brass in Ammoniacal Solutions," *Electrochim. Acta*, **15**, 1032. Copyright 1970 with permission from Elsevier Science.)

8. Stress corrosion cracks may follow intergranular or transgranular paths. In materials having complex slip systems or high stacking fault energies (SEE) and cellular dislocation arrangements, cracks will most probably follow an intergranular path. In materials having low SFE, short-range order (SRO) and planar arrays of dislocations after plastic deformation, cracks may follow transgranular or intergranular paths depending on the corrodent, the composition of the alloy, and the extent of plastic deformation.

9. When polycrystalline α brass is in contact with moist aerated NH_3, the grain boundaries often form preferred sites of electrochemical attack because of the general tendency to form equilibrium grooves, although this condition may be delayed with brass in NH_3 if O_2 supply is limited until protective films have been formed preferentially on the grain faces. Zinc dissolution from the surface can lead to reaction of Cu in the surface with O_2 in solution to form tarnish.

10. Earlier models of SCC of copper alloys [67, 93–108] have now given way to the current models that are mechanistically much more insightful, for example, the film-induced cleavage model of Newman and co-workers [72–75], the surface-diffusion model of Galvele and co-workers [76–79], the corrosion-assisted-cleavage model of Flanagan and co-workers [80–86], and the localized-dissolution/enhanced plasticity model of Magnin and co-workers [87–91].

11. Susceptibility to SCC can be reduced or eliminated by annealing, careful design of the structure, removal of mercury compounds or NH_3 from the environment, control of the pH of the environment, coating with a suitable inorganic or organic material and by using a more resistant metal. A selection of alloys based on SCC tests is, like in the case of pitting corrosion, one of the best preventive measures. The aluminum bronzes listed in Tables 56.2 and 56.3 do not pit and are also not susceptible to SCC. But, of all the copper alloys, the copper nickels, or cupronickels, are the most resistant to SCC in the presence of ammonia and ammoniacal solutions and are highly resistant to SCC in general. Table 56.4 lists some cupronickels used in corrosion engineering.

C2. Effects of Alloy Composition

Thompson and Tracy [108] determined the amounts of a number of elements that, when added singly to Cu, will cause susceptibility to SCC in moist ammoniacal atmospheres. They found that phosphorus when present in residual amounts of only 0.004% make the Cu–P alloy susceptible to SCC in moist ammoniacal atmosphere. Small amounts of As and Sb also increased susceptibility in that medium. The addition of solute elements to Cu produces a concentration of solute in the form of submicroscopic precipitate at the grain

TABLE 56.4.   Some Common Cupronickeles: Chemical Composition

UNS Numbers	Cu	Ni	Fe	Mn	Others
C-70600	Bal	9.0–11.0	1.0–1.8	0.0 max	Pb 0.05 max Zn 1.0 max
C-71500	Bal	29.0–33.0	0.4–0.7	1.0 max	Pb 0.05 max Zn 1.0 max
C-71900	Bal	29.0–33.0 29.0–32.0	0.25 max	0.5–1.0	Cr 2.6–3.2 Zr 0.08–0.2 Ti 0.02–0.08
C-96200	Bal	9.0–11.0	1.0–1.8	1.5 max	Pb 0.03 max
C-69400	Bal	28.0–32.0	0.25–1.5	1.5 max	Pb 0.03 max

boundaries, a richer solid solution or lattice disturbance thus making the grain boundary region more anodic to the grain bodies. Alternatively, formation of a protective film on the exposed surface may leave the grain boundaries highly anodic when insufficient concentration of alloying element is present. Syrett and Parkins [109] found that Sn and As additions both increase the resistance of a 80 Cu–20 Zn brass to SCC in ammoniacal solutions at pH of 7.3–11.3. The effects of alloying elements on the susceptibility of Cu alloys to SCC are well documented in the literature [66–91, 105–113]. In the following sections, a few of the more important alloy systems will be discussed.

C2.1. Aluminum Bronzes. Aluminum bronzes are susceptible to SCC, and many SCC studies of Al bronzes have been reported [114–118]. Element et al. [114] discussed the various possible alloying additions to Al–bronze in relation to their atom size, solubility, and tendency to react with the environment. The addition of 0.2–0.3% Sn or As proved very effective in producing immunity to intergranular cracking of α–Al bronze, but does not affect the relatively mild susceptibility to transgranular cracking that occurs in atmosphere containing appreciable NH_3. Addition of Be and Si was reported to decrease the susceptibility to a lesser degree than Sn or As. Elimination of intergranular SCC depends on decrease or elimination of grain boundary segregation of Al atoms.

C2.2. Copper Nickels. Copper–nickel alloys are not very susceptible, 70 Cu–30 Ni cold-worked tubes do crack on exposure to mercurous nitrate after extremely heavy hollow sinking. Because of its high resistance to SCC, 70 Cu–30 Ni alloys have replaced alloys susceptible to SCC, such as Admiralty brass in environments containing small amounts of NH_3. It has been found that susceptibility to SCC of Cu–Ni–Si alloy in NH_3 atmosphere depends on aging time and on the extent of plastic deformation preceding or following aging, marked improvement in resistance being caused by prior cold working and to a lesser extent by deformation subsequent to aging. The beneficial effect of plastic deformation is attributed to the increase in the number of sites for preferential corrosion and consequently decreasing the rate of penetration [119, 120].

C2.3. Silicon Bronzes. Silicon bronzes are susceptible to cracking in mercurous nitrate solution and in steam atmosphere. Drawn tubes of ternary silicon bronzes (i.e., Cu–Si–Mn, Cu–Si–Sn, and Cu–Si–Zn) are susceptible to intergranular cracking on exposure to NH_3. Allen [121] considers that the factors promoting stress corrosion failure are the composition, high tensile stress, and the presence of moist NH_3 and O_2, not withstanding the outstanding corrosion resistance of these alloys in most outdoor and indoor applications.

C2.4. Tin Brasses. Tin brasses are considerably more resistant to SCC than Cu–Zn alloys [107].

C2.5. Tin Bronzes. If heavily cold worked, tin bronzes crack in NH_3 atmospheres and in mercurous nitrate solution. Tin bronzes are less susceptible than Cu–Zn alloys, and susceptibility increases with Sn content. Alloys that contain >5% Sn are especially resistant to impingement attack. In general, the tin bronzes are noted for their high strength. Their main application is in water service for such items as valves, valve components, pump casings, and so on. Because of their corrosion resistance in stagnant waters, they also find wide application in fire protection systems.

Tables 56.5 and 56.6 list the principal tin bronzes used for corrosion engineering.

C2.6. Nickel Silvers. Nickel silvers are Cu–Zn–Ni alloys. Those containing 5, 10, and 15% Ni did not crack in mercurous nitrate solution after a week, but they can crack if stresses are sufficiently high, usually well above the yield point [122]. The most common nickel silvers are C75200 (65 Cu–17 Zn–18 Ni) and C77000 (55 Cu–27 Zn–18 Ni). They have good resistance to corrosion in both fresh and saltwaters. Primarily because their relatively high nickel contents inhibit dezincification, C75200 and C77000 are usually much more resistant to corrosion in saline solutions than brasses of similar copper content.

C2.7. Aluminum Brasses. Precipitation of $CuAl_2$ at grain boundaries and formation of Cu-depleted zones adjacent to the boundaries causes a potential difference of ≈200 mV in NaCl solution between the boundary and the body of the

TABLE 56.5. Some Common Wrought Copper–Tin Alloys: Chemical Composition

UNS Number	Cu	Pb	Fe	Sn	Zn	P
C-51000	Rem	0.05	0.10	4.2–5.8	0.30	0.03–0.35
C-51100	Rem	0.05	0.10	3.5–4.9	0.30	0.03–0.35
C-52100	Rem	0.05	0.10	7.0–4.9	0.20	0.03–0.35
C-52400	Rem	0.05	0.10	9.0–11.0	0.20	0.03–0.35
C-54400	Rem	3.5–4.5	0.10	3.5–4.5	1.5–4.5	0.01–0.50

TABLE 56.6.   Some Common Cast Copper–Tin Alloys: Chemical Composition

UNS Numbers	Cu	Sn	Pb	Zn	Fe	Sb	Ni	S	P
C-90300	86.0–89.0	7.5–9.0	0.30	2.0–5.0	0.20	0.20	1.0	0.05	0.05
C-90500	86.0–89.0	9.0–11.0	0.30	1.0–3.0	0.25	0.20	1.0	0.05	0.05
C-92200	86.0–89.0	5.5–6.5	1.0–2.0	2.0–5.0	0.25	0.25	1.0	0.05	0.05
C-93700	78.0–92.0	9.0–11.0	0.8–11.0	0.8	0.15	0.55	1.0	0.08	0.15
C-93800	75.0–79.0	6.3–7.5	13.0–16.0	0.8	0.15	0.8	1.0	0.08	0.05
C-93900	76.5–79.5	5.0–7.0	14.0–18.0	1.5	0.4	0.50	0.8	0.08	1.5
C-94700	85.0–90.0	4.5–6.0	0.10	1.0–2.5	0.25	0.15	4.5–6.5	0.05	0.05

grain, and this potential difference is one cause of intergranular cracking of aluminum brasses [123–126]. These alloys are immune to SCC when quenched from the solution heat treatment temperature. Heating the alloy to 190°C resulted in precipitation, at the grain boundaries of $CuAl_2$, which is cathodic to the grain, and formation of a precipitate-depleted zone adjacent to the grain boundaries, thus developing an electrochemical cell [127–130].

C2.8. Manganese Bronzes. Manganese bronzes are susceptible to SCC in NH_3 atmospheres, but less susceptible than Cu–Zn alloys [131]. The time to cracking depends on composition, the minimum in the cracking time versus composition curve lying near 5–6% Mn. The mode of cracking was also altered with changing composition. Alloys containing 4.94% Mn failed intergranularly, whereas those containing 21.38 and 24.5% Mn failed transgranularly. Alloys containing 10.64% Mn showed mixed mode of cracking. With increasing Mn content, alloys became considerably more resistant in seawater and NaOH, but the resistance fell sharply in H_2SO_4 when Mn content exceeded 30% [107]. Alloys containing 11.8 and 24.2% Mn corroded in H_2SO_4 at the same rate as Cu.

C2.9. Brasses. Brasses are extremely susceptible to SCC and have been widely investigated in various environments [132–135]. Brasses containing <15% Zn are considered to be very resistant to SCC. Susceptibility to SCC increases with increase in Zn content up to 40%. Thompson

and Tracy [108] were able to crack brasses containing only 5.18% Zn and Logan [136] easily cracked very large grained brass containing 10% Zn. Elements such as lead, tellurium, beryllium, chromium, phosphorus, and manganese have little or no effect on the corrosion resistance of copper–zinc alloys. These elements are added to enhance such mechanical properties as machinability, strength, and hardness. Table 56.7 lists the compositions of some of the brasses that are used in corrosion engineering. Brasses C44300–C44500, known as Admiralty brasses, are resistant to dealloying because of the tin in the alloy.

Admiralty brass is used mainly in the handling of seawater and freshwater, particularly in condensers. Because these brasses are resistant to hydrogen sulfide, they are used in petroleum refineries.

Red brass, an alloy containing 15% zinc, has basically the same corrosion resistance as copper, but with greater mechanical strength. Most brass piping and fittings are produced from this alloy.

C2.10. Copper–Gold Alloys. Copper–gold alloys belong to a novel but important alloy system with specific sets of properties. The Cu–Au alloys show an anodic polarization behavior that is typical of binary alloy systems with a relatively large difference in the standard potentials of the two alloy components. The current response is the result of copper electrodissolution. Another major advantage of the Cu–Au alloy system for electrddissolution and corrosion studies is the fact that hydrogen cannot be formed on the

TABLE 56.7.   Some Common Brasses: Chemical Composition

UNS Numbers	Cu	Pb	Fe	Zn	Sn	Others
C-27000	63.0–68.5	0.10	0.07	Rem		
C-28000	59.0–63.0	0.30	0.07	Rem		
C-44300	70.0–73.0	0.07	0.06	Rem	0.8–1.2	0.02–0.10 As
C-44400	70.0–73.0	0.07	0.06	Rem	0.8–1.2	0.02–0.10 Sb
C-44500	70.0–73.0	0.07	0.06	Rem	0.8–1.2	
C-46400	59.0–62.0	0.20	0.10	Rem	0.5–1.0	
C-46500	59.0–62.0	0.20	0.10	Rem	0.5–1.0	0.02–0.10 As
C-46600	59.0–62.0	0.20	0.10	Rem	0.5–1.0	0.02–0.10 Sb
C-46700	59.0–62.0	0.20	0.10	Rem	0.5–1.0	
C-68700	76.0–79.0	0.07	0.06	Rem		0.02–0.10 As

surface, or even in pits or cracks. (References [81–86, 137–143] and [137–143] present the pitting corrosion and SCC characteristics of these alloys.)

C3. Specific Corrodents

In a very earlier investigation, brass was tested in 20 different media and was found to crack only in NH_3 and mercury salts; brass was not susceptible to corrosion cracking in air containing nitrogen oxides or solutions of salts of these acids [144]. In contrast to these results, Bobylev [145] reported SCC of brass not only to NH_3 but also nitrites, carbonates, phosphates, and alkalis as well as air containing SO_2 and oxides of N_2. Shreider [146] reported that a Cd-plating solution containing Cd oxide, NaCN, and Ni sulfate induces cracking in stressed α brass. Typical alloy—corrodent combinations in which SCC of copper alloys has been reported is given in [147–150] and others. Data are summarized below for environments in which cracking has been reported.

C3.1. Ammonia System.
The ammonium ion is generally considered to be the specific corrodent causing SCC in Cu-base alloys in service in the presence of water or water vapor. Dioxygen and CO_2 are also considered to accelerate cracking in ammomacal atmosphere [144]. Failure occurs over a wide range of NH_3/air ratios, of the order of 20–80% ammonia. Low relative humidity markedly increases the time required for failure. For stressed specimens containing 60–65% Cu, and exposed to NH_3 vapor, a mixture of intergranular and transgranular cracks developed, whereas for >70% Cu cracks were predominantly intergranular.

Three main environmental factors contribute to the specificity of NH_3 in causing SCC of copper alloys:

1. Corrosion product films reduce the overall anodic activity of the surface while maintaining good electron transport properties to facilitate cathodic process.
2. Enhanced preferential anodic dissolution of Zn at grain boundaries, associated with tarnishing and crystallographic features, leads to "chemical differentiation" of the surface.
3. Stabilization of the Cu(I) valency state by complex formation introduces a very effective cathodic reaction system.

Maximum susceptibility to cracking occurs when all these factors operate together [92, 151].

C3.2. Sulfur Dioxide.
In early research, SO_2/water/air cracked brass [145, 152]. In damp air containing SO_2, SCC of brass occurs if the concentration of SO_2 is between 0.05 and 0.1%, whereas at higher concentration, general overall

corrosion takes place [151, 153]. For stressed brass in SO_2, SCC can be inhibited using benzotriazole (BTAH) [153]. Aqueous potential–H diagrams that were constructed by Tromans [154] for the Cu–BTAH–H_2O and Cu–BTAH–Cl^-–H_2O systems (Figs. 56.4 and 56.5) show the stability regions of Cu BTA salt films. These regions are reasonably consistent with the reported effects of Cu BTA polymer films on corrosion inhibition of copper. These diagrams also provide a basis for understanding the combined

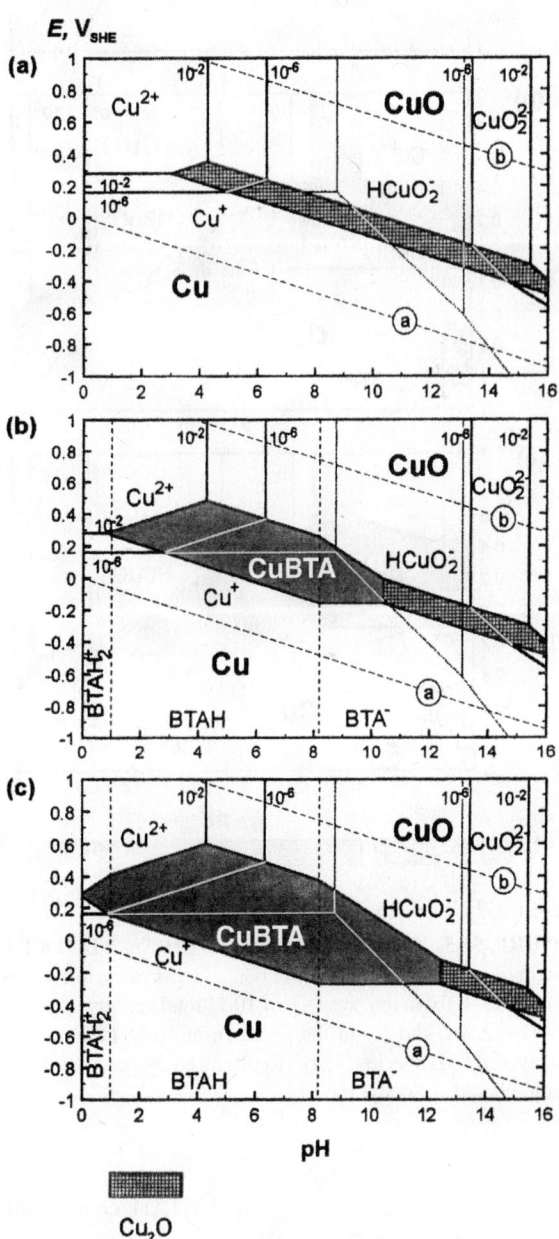

FIGURE 56.4. The E–pH diagrams for the Cu–BTAH–H_2O systems: (a) no BTAH, (b) in the presence of 10^{-4} total activity of dissolved BTAH species, and (c) in the presence of 10^{-2} total activity of dissolved BTAH species [154]. (Reproduced by permission of The Electrochemical Society, Inc.)

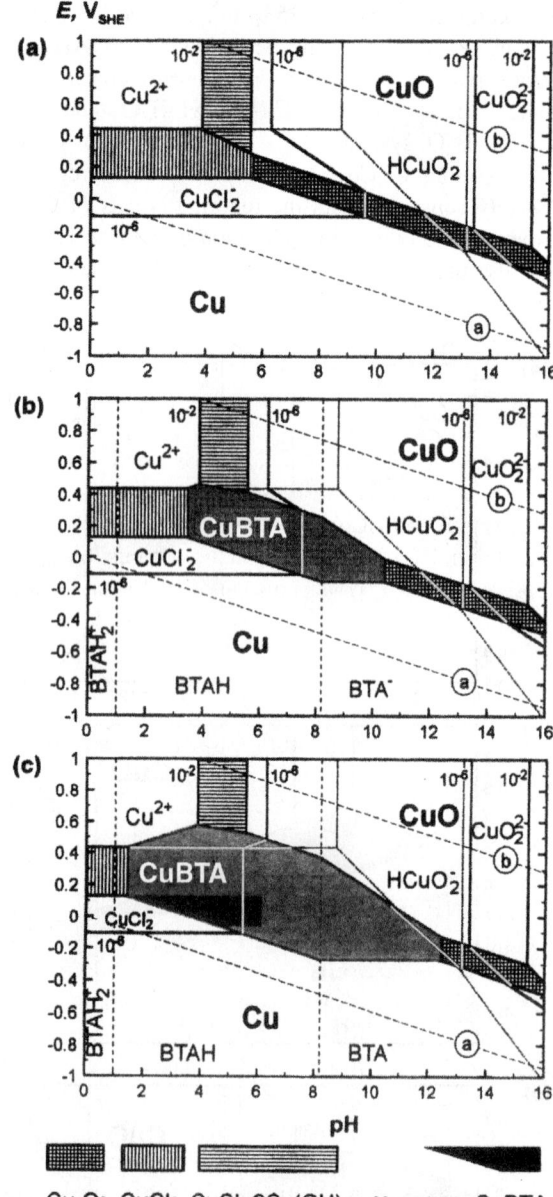

Cu₂O: CuCl: CuCl₂.3Cu(OH)₂: **Most stable CuBTA**

FIGURE 56.5. The E–pH diagrams for the Cu–BTAH–Cl–H₂O systems containing 0.67 activity of [Cl⁻¹] (equivalent to 1 M NaCl): (a) no BTAH, (b) in the presence of 10⁻⁴ total activity of dissolved BTAH species, and (c) in the presence of 10⁻² total activity of dissolved BTAH species [154]. (Reproduced by permission of The Electrochemical Society, Inc.)

effects of potential, pH, Cl⁻, and BTAH concentrations (activities) on corrosion inhibition.

It has been demonstrated by polarization studies of Cu, Fe, and Zn carrying thin condensed moisture films in the presence of O₂ and SO₂ that (a) Cu passivates readily, while Zn remains active, (b) the efficiency of cathodic depolarization increases with decreasing film thickness,

and (c) SO₂ is >10 times more efficient than O₂ as a cathodic depolarizer.

Tromans and Nutting [155] found that cracks in 70 Cu–30 Zn exposed to moist SO₂ were intergranular and grew by linkage of microcracks at grain boundary pits.

C3.3. Citrate and Tartrate Solution. The SCC of brass was found to occur in citrate solution at pH 10.3 after 84 h of immersion in 0.7 M copper citrate and 0.6 M potassium citrate [156]. In tartrate solution intergranular cracking occurred at pH 13 after 31 days immersion in solution containing 0.04 M Cu and 0.5 M potassium tartrate. Cracking is associated with Cu–citrate and Cu–tartrate complexes, the presence of which is controlled by the pH of the solution, concentration of citrate and tartrate ions, and the redox potential of the solution. Alloy C72000 is sensitive to intergranular SCC in citrate solutions containing dissolved copper in the pH range of 7–11 [157].

C3.4. Pyridine and Ethylenediamine Solution. In the presence of moist air, amines can cause intergranular SCC of stress brass, and primary amines are more damaging than secondary or tertiary amines [158]. The corrosion rate of Al–brass exposed to steam condensate containing high concentration of amines is nearly five times greater than that observed in condensate containing no amines [159]. Trace amounts of NH₃ produced by the degradation of the amines might be responsible for the observed high rate of anodic dissolution in the presence of O₂. Alloys C26000 and C68700 are susceptible to SCC in solution of amines [160, 161].

C3.5. Other Environments. Monel (Ni–Cu alloy) is susceptible to SCC in most aerated vapor containing HF. Alloys with 15 and 30% Cu suffered severe SCC with both intergranular and transgranular cracks [107]. The susceptibility to cracking varies with the Ni/Cu ratio. Alloys with 66% Cu resisted cracking in 14-day tests, the only local attack being shallow intergranular corrosion, but not cracking. The insolubility of nickel fluoride keeps the nickel ion concentration low and makes Ni anodic to Cu. Metallic Cu was observed on the specimen surface and also deep within the cracks. Holberg and Prange [162] reported SCC in yellow brass at 64°C (150°F) in HF alkylation in a petroleum refinery.

Failure of a 70 : 30 brass chain occurred in HNO₃ vapor, possibly because of the breakdown of HNO₃ to NH₃ [163]. The SCC of brass in an air conditioning unit was attributed to ammonium ion formed from the oxides of nitrogen produced by the corona discharge in the high-voltage precipitator unit [164].

Neither a pure solution of NaCl nor seawater cause SCC of α brass, while β brass cracks intergranularly in these media. Alloys C26000 [165] and C44300 [166] have lower fracture stresses in NaCl solutions when the metal is anodically polarized.

Failure of Cu–Au alloys in the presence of either aqueous ferric chloride or aqua regia is attributed to the formation and rapture of a mechanically weak surface film consisting of a gold-rich phase that is formed by the preferential dissolution of the less noble elements [107].

Effect of other environments (e.g., cupric acetate, polluted atmospheres, sodium chlorate solutions, sodium formate solutions, sodium hydroxide solutions, sodium nitrate solutions, sodium nitrite solutions, solder, sodium sulfate solutions, sulfide solutions, and sodium tungstate solutions) on SCC of alloys C26000 [167–169], C44300 [166], C70600 [169], and brass [168], Admiralty brass and other copper alloys are well documented in the literature [170–179].

D. ATMOSPHERIC CORROSION

In addition to oxygen, moisture, and gaseous pollutants, such as sulfur dioxide (SO_2), nitrogen oxides (NO_x), chlorine gases (HCl, Cl_2), ammonia (NH_3), and ozone, airborne ionic dust particles strongly affect conditions on corroding surfaces. During atmospheric corrosion of copper, a patina is formed over time. Initially, a layer of Cu_2O, CuO, and CuO_xH_2O forms in which cuprite (Cu_2O) is the main component. Later, a patina with several corrosion products is formed. The basic copper sulfates, posnjakite, $Cu_4SO_4(OH)_6H_2O$, brochantite, $Cu_4SO_4(OH)_6$, and antlerite, $Cu_3SO_4(OH)_4$, are primary constituents. Basic copper chloride and carbonate are also found [180–184]. The mechanism of basic copper sulfate formation has been the subject of several studies that have been reviewed [183, 184].

In metropolitan areas, most of the sulfur acquired by surfaces is not in gaseous form by the reaction with SO_2 but as dry deposition [185, 186]. Outdoor exposures of copper show that the main factors influencing the weight gain rates are relative humidity and concentration of aerosol particles [187]. The most abundant ions found in fine particles are SO_4^{-2} and NH_4^+, with the ratio typically being between that of NH_4HSO_4 and $(NH_4)_2SO_4$ [188].

The effect of submicron $(NH_4)_2SO_4$ particles on the corrosion of copper at varying relative humidities (RHs) and-temperatures was studied by Lobnig et al. [180, 181], who showed that $(NH_4)_2SO_4$ particles lead to the corrosion products found in natural patinas. In a recent study [189], it was investigated whether NH_4HSO_4 and $(NH_4)_3H(SO_4)_2$ may be responsible for formation of basic copper sulfate in the naturally formed patina on copper during atmospheric corrosion. At room temperature, only mixed ammonium copper sulfates, $(NH_4)_2Cu(SO_4)(OH)_2 \cdot xH_2O$, and cuprite were found up to 28 days of reaction time. It was concluded that acid ammonium sulfate is less likely than ammonium sulfate, the sulfur containing species leading to formation of basic copper sulfates in copper patinas.

Basic copper sulfates are the dominant phases in most copper patinas, but particularly at the seacoast, the copper patina also consists of Cl-containing species.

The principal source for chlorine in the atmosphere is sea salt. In marine areas, aerosols are formed with an initial chloride concentration of seawater (0.4 M), but accumulation on surfaces may result in higher concentrations. Sea salt aerosols may be scavenged by water, occurring in rain, snow, fog, and dew, and may be far removed from the coast [182]. Chloride concentrations in precipitation are reported in the range 0.04–4 mM [185]. Another source of chloride is the use of NaCl for deicing, implying that large amounts of chloride may be found on objects close to roads.

Hydrogen chloride is the dominant inorganic gaseous chlorine compound in the atmosphere [185]. The main source is reported to be the reaction of acidic trace gases with sea salt aerosols, for example $NO_2(g)$ and $NaCl(s)$ forming HNO_3 and $HCl(g)$ [190, 191].

In addition, HC1 is emitted from the combustion of fossil fuel and refuse [185]. Concentration of $HCl(g)$ in the atmosphere range from very low values up to 4 ppb [185].

The metal loss of copper coupons in a marine atmosphere has been reported to be 600–700 $\mu g/cm^2$-year, averaged over an 8-year exposure period, the initial corrosion rate being considerably higher [192]. Copper turns brownish during the first year of exposure, while signs of green patina may be observed after 6–7 years in this environment [193]. After decades of exposure, a patchy and streaked appearance with alternating black and light green areas tends to appear. Rain-sheltered areas tend to be black while areas washed by rain tend to be light green. The polymorphs [$Cu_2(OH)_3Cl$], atacamite and paratacamite in combination with different hydroxy sulfates and cuprite (Cu_2O) are dominant in the black sheltered areas, while rain-washed green areas mainly contain brochantite [$Cu_2(OH)_6SO_4$] and cuprite [194]. The patchy black and green appearance is reported to have become especially evident on outdoor objects since the 1960s [195].

Chlorides have been identified in copper corrosion products in marine [196] and inland [197] locations, as well as indoors [198]. Atacamite has been reported to occur more frequently than paratacamite in outdoor environments [194, 196]. Thermodynamic calculations predict atacamite to be stable at the chloride concentrations present in seawater, while it is only marginally stable at concentrations typical of rain and fog (0.4–4 mM) [195]. Jambor et al. [199] recently reported on clinoatacamite, a new mineral polymorph of $Cu_2(OH)_3Cl$. The published Joint Committee on Powder Diffraction Standards (JCPDS) Powder Diffraction File for paratacamite (PDF 25-1427) [194] is only slightly different from the diffraction pattern for clinoatacamite and is claimed to pertain to the latter compound. Thus, previous reports on paratacamite should probably be assigned to clinoatacamite instead.

A few laboratory studies deal with the effect of chloride on the atmospheric corrosion of copper. Working with very high pollutant concentrations (percent levels), Vernon [171] concluded that HCl was more deleterious than SO_2 toward copper at 50% relative humidity (RH). Feitknecht [200] proposed mechanisms for the interaction of chlorides with copper and some other metals. Eriksson et al. [201] reported that the addition of small amounts of sodium chloride caused a marked increase in the corrosion rate of bronze in humid air. More recently, Strandberg and Johansson [202] investigated the effect of NaCl in combination with O_3 and SO_2 on the atmospheric corrosion of copper. Large amounts of cuprite (Cu_2O) formed in all environments at 70 and 90% relative humidity. The corrosive effect of salt was strong in pure humid air and in air containing O3 or SO_2. Corrosion rate was correlated to the amount of chloride applied to the surface and to humidity. In an atmosphere containing a combination of SO_2 and O_3 at 90% RH, corrosion was rapid in the absence of NaCl. In this case, small additions of NaCl resulted in a marked decrease in corrosion rate. In the absence of SO_2, tenorite (CuO), nantokite (CuCl), clinoatacamite [$Cu_2(OH)_3Cl$], and malachite [$Cu_2(OH)_2CO_3$] were identified. In the presence of SO_2, brochantite [$Cu_4(OH)_6SO_4$], soluble sulfate, and an unknown phase occurred, while no tenorite or malachite was formed. The combination of SO_2 and O_3 resulted in the formation of antlerite [$Cu_3(OH)_4SO_4$] and $Cu_{2.5}(OH)_3SO_4 \cdot 2H_2O$ as well.

Although the severity of the atmospheric corrosion of copper alloys depends on the atmospheric contaminants, the corrosion rate usually decreases with time. In general, copper alloys are very suitable for atmospheric exposure. High-copper alloys, silicon bronze, and tin bronze corrode at a moderate rate, and brass, aluminum bronze, nickel silver, and copper nickel corrode at a slower rate. The copper alloys, most widely used in atmospheric exposure are C11000, C22000, C38500, and C75200.

Additional information on the atmospheric corrosion of copper and its alloys is available in [203–212].

E. CORROSION IN WATERS

E1. Copper

The corrosion resistance of copper is due to its being a relatively noble metal. Its satisfactory service in waters depends on the formation of relatively thin adherent films of corrosion products (e.g., cuprous oxide and basic copper carbonate). It has only a weak tendency to passivation, and hence the effect of differential aeration is very slight. However, the influence of copper ion concentration on the potential of copper in solution is very marked. For this reason, when there are varying solution velocities over a copper surface (e.g., when the solution is stirred), the parts exposed to solution with the higher rate of movement become anodes and not cathodes as would be the case with iron, for example.

Cuprous oxide (Cu_2O) is the corrosion product predominantly responsible for protection. This oxide possess semiconducting properties and, in high-temperature corrosion studies (oxidation in the thick-film range) is usually described as a p-type semiconductor in which defects are uniformly distributed in the oxide. In waters, cuprous oxide is produced by anodic oxidation, usually under dynamical flow. That is, it is formed on the metal side and dissolved on the solution side. Under these conditions, the oxide structure is not homogeneous [213–215].

The anodic oxide thickness is very small, typically ~500 nm, in the copper–water system. The coppers–water interface is really a complex copper–oxide–water system, involving solid ionic– electronic thin films able to generate photocurrents [216, 217].

Apart from Cu_2O, there are further types of solid corrosion products that can be formed at the copper–water interface, depending on the composition of the waters. Species more likely to be present in waters are C, CO_2, Cl_2, Cl^-, NO_2, Fe, Na_2SO_4, HSO_4^-, SO_4^{2-}, S, H_2S, HS^-, H_2SO_5, S_8, and SO_2, and these can form copper compounds, such as Cu_2S, Cu_2O, CuO, CuCl, CuOHCl, $Cu(OH)_2$, CuS, and $CuSO_4 \cdot 5H_2O$. Thermodynamic data for most of these compounds are given by Bard et al. [218], Garrels and Christ [219], Wagman et al. [220], and others and are being used in the E–pH graphical form [221] to predict the kinetic behavior of the Cu–O–H, Cu–S–O–H, Cu–C–S–O–H, Cu–F–C–S–O–H and other Cu systems that are relevant to aqueous-phase corrosion processes.

Distilled water is not very corrosive to copper, but over long periods the water will pick up traces of copper. Copper tanks should be tinned for holding distilled water or for water storage on ships.

In soft waters, particularly those containing appreciable amounts of free carbon dioxide, and in carbonated waters in general, the corrosion of copper is much greater. The initial corrosion rate in distilled water may be 0.051–0.16 mm/year (12–7 mdd), but the rate in an aggressive supply water may be as high as 0.26 mm/year (62 mdd). Such waters may be corrosive enough to pick up sufficient copper to form green stains on plumbing fixtures.

Hard waters are seldom corrosive to copper because of the protective film of calcium compounds that soon forms. When waters with an appreciable temporary hardness are softened, however, they can become corrosive, especially if heated above 140°C. The calcium compounds necessary to form a protective film are then absent and the sodium bicarbonate formed breaks down with the release of carbon dioxide.

Copper is very stable in salt solutions and seawater. When constantly immersed, the corrosion rate of pure copper in seawater is 0.02–0.07 rnm/year (5–17 mdd) and under varying immersion, as at half-tide, the rate is 0.02–0.1 mm/year (5–25 mdd). In the latter case, arsenical copper is rather more resistant than tough-pitch copper. Under constant immersion the corrosion resistance of copper is two to five times greater than that of mild steel, and under half-tide immersion the advantage of copper is even more notable. Copper is satisfactory in stagnant seawater or where the velocity is $\lesssim 1$ m/s (3 ft/s). The rate of dissolution of copper is, however, large enough to be toxic to marine boring animals, hence its value as a sheathing material for wooden craft and piling. For the same reason, copper compounds are used in marine antifouling paints.

Copper is not sufficiently corrosion resistant in rapidly flowing seawater to be suitable for condenser tubes in ocean-going ships or in power stations using tidal water. A number of copper alloys have superior resistance under such conditions (see below).

Pitting of copper tubes caused by carbonization of the lubricant used in tube making has been largely eliminated by improving and controlling the tube manufacturing process. Films of manganese oxides, derived by slow deposition from soft moorland waters, can also give rise to localized attack.

A beneficial effect, presumably due to the presence of an organic chemical at very low concentration, is observed in the corrosion of copper in certain waters. The organic agent responsible for this action has not been isolated and the way in which it operates is imperfectly understood.

E2. Copper Alloys

Copper alloys have a long history of successful application in freshwater and saltwater environments. In seawater, in addition to cuprite (Cu_2O), $Cu_2(OH)_3Cl$ forms at the alloy–water interface, independently of alloy composition. Copper alloys (e.g., C44300, C44500, C61300, C68700, C70600, or C71500) are more resistant to corrosion in seawater than pure copper because of the incorporation of relatively corrosion-resistant metals such as nickel, or because of the addition of metals such as iron and aluminum that assist in the formation of protective oxide films. Information is summarized below for the copper alloys more commonly used in water.

E2.1. Brasses. The common brasses are alloys of copper with 10–50% zinc and often a number of other components including tin, iron, manganese, aluminum, and lead. Zinc dissolves in copper up to 39% to give a single phase alloy, α brass, and with zinc contents in the range of 47–50% the alloy is again single phase, β brass, Between these two, that is with 39–47% zinc, the alloy contains both phases, α + β brass. In the corrosion of the α brasses the constituents are dissolved according to the copper/zinc ratio, but as the two-phase alloy is approached, appreciably more zinc dissolves. In the duplex brasses, dezincification takes place by preferential dissolution of the less noble β phase, and in the pure β alloys only zinc appears in the corrosion products, the copper being redeposited as metal.

The stability of brasses in natural waters depends in a complex manner on the dissolved salts, the water hardness, the dissolved gases, and on the formation of protective films. In general, the rates of attack vary from 0.003 to 0.03 mm/year (0.6 to 6 mdd). The corrosion rate is increased by higher concentrations of carbon dioxide and, with the higher zinc brasses, is accompanied by dezincification. Because tin additions are effective in reducing dezincification, naval brass, 63 Cu–36 Zn–1 Sn, is commonly used. For freshwater plumbing, piping of red brass, 85 Cu–15 Zn, is preferred to copper alloys such as Muntz metal, 60 Cu–40 Zn, and leaded yellow brass, 67 Cu–33 Zn, which dezincify in some waters.

The corrosion rate of brass by pure water, such as condensate, is very low, <0.015 mm/year (3 mdd), but this is appreciably increased in the presence of air, carbon dioxide, or ammonia. The scale deposited from hard waters becomes less protective at elevated temperatures owing to the decrease in adhesion.

In seawater, the rate of corrosion is generally small, between 0.008 and 0.12 mm/year (2 and 24 mdd), but may, in some instances, be much greater since it varies with alloy composition and local conditions. Brass with a copper content of ~70% is the most stable of the straight brasses in seawater, and Admiralty brass, 70 Cu–29 Zn–1 Sn, is often used. If the copper content is higher than this, there is a tendency to localized corrosion, particularly at the water line, and if the copper content is lower, there is an increased likelihood of dezincification. The high tensile brasses with a higher zinc content are more resistant to corrosion–erosion and cavitation and are used in the manufacture of propeller screws. Their inclination toward dezincification is reduced by small additions of arsenic, antimony, or phosphorus. The corrosion stability of brass is considerably increased by addition of aluminum and the brass 76 Cu–22 Zn–2 Al is widely used in marine applications [e.g., in condenser tubes (see below)].

Acid mine waters discharged into rivers make the natural water more corrosive owing to the presence of small amounts of sulfuric acid and ferric sulfate. Copper alloys containing tin are the most resistant to this type of water.

E2.2. Copper–Tin Alloy. Copper can form a solid solution with tin containing up to 15.8% tin, the α phase. The actual solubility at room temperature is lower but the decrease in concentration with falling temperature is so very slow that the solubility below 520°C (971°F) for normal alloys can be taken as 15.8%. The bronzes containing up to ~8% tin are used in the rolled form and, at higher tin concentrations, in the wrought form. Alloying additions of small amounts of phosphorus, zinc, and lead are made. Additions of iron, antimony and bismuth are dangerous and are acceptable only up to 0.2–0.5%.

Corrosion by water depends on the oxygen, carbon dioxide, and salt content and on formation of protective layers. Both rolled and wrought bronzes, with and without additions of lead or zinc, may be used. Bronzes are very corrosion resistant to steam, but under more severely corrosive conditions and high steaming rates, corrosion rates as high as 0.8 mm/year (200mdd) can occur. Bronzes are seldom used above 260°C (503°F) and 100-atm pressure.

In seawater, the corrosion rate of bronzes varies from 0.013 to 0.034 mm/year (3 to 8mdd) according to local conditions. In shipbuilding, bronzes with > 5% tin and with additions of lead and zinc are used. These include the well-known Admiralty bronze or Admiralty gun-metal, 88 Cu–10 Sn–2 Zn, and naval bronze, 88 Cu–8 Sn–4 Zn. In certain conditions, the corrosion rate of lead–tin bronzes with tin contents > 10% increases to 0.09 mm/year (22 mdd).

E2.3. Aluminum Bronze.
Aluminum bronzes usually contain not more than 9–10% aluminum and sometimes small additions of manganese and nickel as well. In seawater, they are more corrosion resistant than other copper alloys, with corrosion rate only 10% of that of bronze and 3% of that of brass. The rate of corrosion on an 8.0% aluminum alloy is 0.001–0.003 mm/year (0.3–0.8 mdd) at 30°C (89°F), and 0.01 mm/year (2.3 mdd) at 60°C (143°F). More complex alloys containing up to 11.5% aluminum, 5.5% nickel, 5.0% iron, and 3.5% manganese are casting alloys used in the manufacture of ships' screws. These alloys are able to withstand the severe demands of service in Arctic waters better than manganese brasses and are much more stable toward erosion and cavitation. Alloys C61300 and C63200 are used in cooling tower hardware in which the makeup water is sewage effluent. Aluminum bronzes resist oxidation and impingement corrosion because of the aluminum in the surface film.

E2.4. Silicon Bronzes.
Silicon bronzes contain up to 4.5% silicon and sometimes a number of other additions. The 3.0% Si bronzes are stable in natural waters, including seawater, and are very suitable for hot water equipment and for screws used in marine fittings. Corrosion rates for silicon bronzes are similar to those for copper.

E2.5. Copper-Nickel Alloys.
Alloys based on copper and containing from 5 to 40% nickel have good mechanical properties at moderately elevated temperatures and excellent resistance to corrosion in many environments and in particular in contact with brackish water and seawater [222–225]. They are more stable than the brasses in flowing water, less susceptible to SCC, and are widely used in shipbuilding for installations involving heat exchange. In this group of alloys,

most attention has been directed to 70 Cu–30 Ni and 90 Cu–10 Ni and the improvement in corrosion resistance under flow conditions obtained by small additions of iron and manganese [226]. In both alloys, the optimum performance, even at flow rates up to 5 m/s (16ft/s), is obtained with an iron concentration of 1% (viz., 69 Cu–30 Ni–1 Fe, and 89 Cu–10 Ni–1 Fe). These alloys are recommended for brackish waters, seawaters, and waters with total dissolved solids >2000 ppm (i.e., 0.2%). They have an advantage in that they do not require cathodic protection as is provided for aluminum brass tubes, 76 Cu–22 Zn–2 Al, by iron anodes bolted to the water boxes or by the boxes themselves.

A copper–nickel–iron, 69 Cu–30 Ni–Fe, and also alloys of the Monel type (e.g., 30 Cu–66 Ni with 4% iron and manganese) are used for high-pressure feedwater preheaters. The introduction of copper ions into the boiler is thereby avoided. In recent years, tubes of 69 Cu–30 Ni–1 Fe and 79 Cu–20 Ni–1 Fe alloys have been found to exfoliate—thick layers of corrosion product form and subsequently flake off. Exfoliation is limited to heaters that are operated discontinuously and are used with feedwaters treated with sodium sulfite. Exfoliation may be avoided by using the 89 Cu–10 Ni–1 Fe alloy, which can also be used for equipment evaporating seawater to provide drinking water on board ship.

The 90 Cu–10 Ni alloy and the silicon bronzes, with very good mechanical properties and good weldability, are used in the fabrication of storage tanks for hot freshwater.

The 69 Cu–30 Ni–1 Fe alloy is used for dealing with condensate and feedwater and has a corrosion rate of < 0.08 mm/year (19 mdd). At 70°C (161°F) under air containing carbon dioxide, the corrosion rate is higher, 0.25–0.47 mm/year (64–106 mdd) at 16-atm pressure. This alloy and the corresponding one containing small amounts of manganese are resistant to cavitation.

E3. Microbial Corrosion

Copper and copper alloys in general use, with few exceptions, are susceptible to some form of microbial corrosion that usually arises from the activity of a wide range of microorganisms and their metabolic products. In addition, the microorganisms are ubiquitous and are able to colonize surfaces and, by genetic mutation, acquire the ability to adapt to environmental changes. They constitute a dynamic system that is able to change with time.

The other important feature associated with colonization of copper surfaces is the subsequent formation of biofilms. Adherence of these is brought about by the release of extracellular polymers. Biofilms up to 100 μm thick are not unusual and in nearly all cases contain trapped bacteria. Such films encourage the growth of these bacteria, resulting in the formation of complex biological systems comprising active bacteria, their metabolites, and the chemical changes generated by the system. Because of this complexity, there are numerous microbial corrosion processes and mechanisms, although it must be emphasized that these do not involve any new corrosion process.

Bacterial colonization under aerated conditions usually results in the formation of differential aeration and concentration cells due to the uptake of oxygen by the microbial colony. The oxygen concentration under these colonies becomes depleted and localized pitting corrosion can take place. In some cases, particularly when the iron-oxidizing bacteria are involved, corrosion tubercles develop. Such organisms also facilitate the accumulation of chlorides, resulting deposition of ferric and manganese chlorides.

Tuberculation, like slime films, provides a suitable habitat for the sulfate-reducing bacteria (SRB), which can stimulate corrosion by a number of mechanisms, but, under these conditions, it is likely by the releasing of sulfide ions.

Because copper and copper alloys rely on the presence of a stable oxide film for their corrosion resistance, they are susceptible to corrosion should the film be damaged or the oxygen shielded from the metal by the biofilm.

Under certain circumstances, corrosion can be stimulated by the chemistry within the biofilm and the extracellular polymers that are generated. Chelation of copper ions contributes to the formation of galvanic cells. The mechanism of corrosion may be controlled by the potential difference created between cuprous and cupric ions in the biofilm.

In addition to anodic metabolites, some bacteria are able to generate other substances, such as carbon dioxide and ammonia, the latter being a cause of SCC of copper and its alloys.

All these environments are of concern, but none perhaps is as important as those in which SRB are active. These prolific organisms cause considerable corrosion damage and pollution of the environment. The mechanism of corrosion involving these bacteria is complex and not completely understood. Nevertheless, it is important to emphasize that at present five hypotheses exist to explain the involvement of SRB in the corrosion process. These include (1) the depolarization of the cathode and utilization of molecular hydrogen, (2) corrosion by sulfide ions, (3) galvanic corrosion due to the formation of iron sulfide films, (4) corrosion due to the formation of elemental sulfur, and (5) the production of a corrosive volatile phosphorus compound. Irrespective of these mechanisms, most copper alloys are susceptible to SRB activity.

The rate of corrosion that can arise from SRB activity can be very high particularly when it is controlled by the action of sulfur and its compounds.

Digenite (Cu_5S_9), chalcocite (Cu_2S), and covellite (CuS) have been produced by SRB on copper surfaces [227, 228]. Macdonald et al. [228] reported the formation of djurleite ($Cu_{1.96}S$) from SRB activity on copper alloys. Chalcocite is the most characteristic corrosion product in SRB-induced

corrosion of copper. Baas-Becking and Moore [227] reported that chalcocite cannot be formed abiotically at room temperature and that microbiological formation of chalcocite is a product of digenite, the first species formed during SRB-induced corrosion of copper alloys. Djurleite is important in SRB-induced corrosion of copper alloys because it forms a protective sulfide film [229] and is difficult to synthesize abiotically at room temperature [230].

McNeil and Little [231] analyzed sulfide mineral deposits on copper alloys colonized by SRB in order to identify specific mineralogies that could be used to fingerprint SRB activity. The copper sulfide mineral found in all combinations of copper-containing substrata and cultures was chalcocite. The authors concluded that the presence of chalcocite was an indicator of SRB-induced corrosion of copper. The compound was not observed in sterile controls and its presence in near-surface environments could not be explained thermodynamically.

A range of well-defined types of pitting corrosion of copper were reviewed by Mattsson [40] but none of these was considered to be associated with microbial activity. Since then, two new forms of pitting have been reported that have microbial origins. One of these, termed "pepper-pot pitting" occurred in large institutional buildings in southwestern Scotland, UK [232], whereas the other, frequently termed type $1\frac{1}{2}$ pitting, has been observed in tubes from Saudi Arabia, Germany, and England [233].

Examination of affected copper tubes from buildings in Glasgow showed that there was some superficial corrosion due to deposit attack (differential aeration), but that perforation was at sites containing pepper-pot pits. In cross section the pits had a conical gray cap, often hollow, of corrosion products consisting mainly of copper sulfate and cupric oxide. Removing this cap revealed the pepper-pot cluster of pits.

Pit morphology (i.e., number, size, and shape of pits) varied slightly within a group. In some, but not all, cases, sulfides could be positively detected. The pits were located in a membrane of cuprous oxide or copper and beneath that were crystals of cuprous oxide extending from the perforations outward in a hemispherical fashion. Further examination of these tube bores has shown that a biofilm can always be lifted from the surface and stained using the periodic acid-Schiff stain [234], showing the presence of polysaccharides (by-products of bacteria), which form in copious quantities when biofilms develop. This type of pitting resembles the classical type 3 of pitting [40].

The other new type of pitting occurs in cold, warm, and hot water systems and exhibits features of both type 1 and type 2 pitting (Fig. 56.6) [235]. It resembles type 1 pitting in that the pits are hemispherical and contain soft crystalline cuprous oxide with varying amounts of cuprous chloride under a cuprous oxide membrane. It resembles type 2 pitting in that the oxide on the surface between the

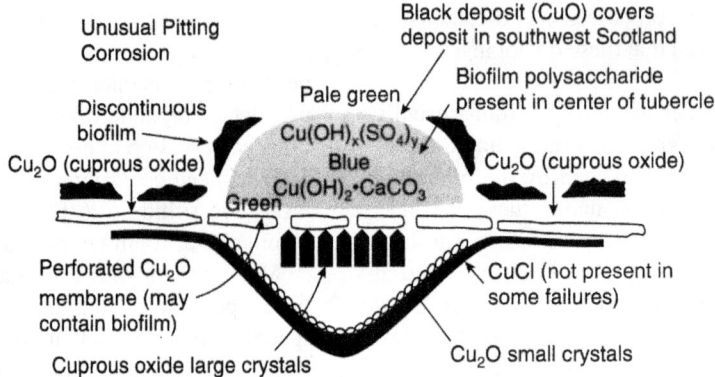

Unusual Pitting Corrosion

Black deposit (CuO) covers deposit in southwest Scotland

Discontinuous biofilm

Biofilm polysaccharide present in center of tubercle

Pale green

$Cu(OH)_x(SO_4)_y$

Cu_2O (cuprous oxide)

Blue $Cu(OH)_2 \cdot CaCO_3$

Green

Cu_2O (cuprous oxide)

Perforated Cu_2O membrane (may contain biofilm)

CuCl (not present in some failures)

Cuprous oxide large crystals

Cu_2O small crystals

FIGURE 56.6. Unusual type of pitting corrosion. (Reproduced with permission from Wiley-VCH [235].)

pits is largely cupric. The mounds above the pits are principally basic copper sulfate, often with a deposit of powdery cupric oxide around the periphery and on parts of the deposit itself.

An interdisciplinary approach for studying type $1\frac{1}{2}$ pitting of copper in potable water systems is reported by Chamberlain et al. [236]. Their electrochemical results are explained on the basis of diffusion processes and multiple Donnan equilibria under the assumption of ion selectivities of mainly anionic exopolymers [237]. Optimizing the alkalinity of potable water and using different tempers and surface conditions are the most promising and economically feasible methods to control type $1\frac{1}{2}$ pitting. Bicarbonate dosing, already used to control several corrosion problems, is beneficial only within a certain optimum concentration range of bicarbonate, as indicated by practical experience and results from limited laboratory experiments. It is also well known (DIN 50930, part 5) [238] that in cold water pitting corrosion damage is enhanced in hard water with high alkalinity, and also at elevated temperature [$>40°C (> 107°F)$] in soft water with low alkalinity. Any measure to improve the performance of different tempers and surface conditions of copper in contact with potable water must not have an adverse effect on tubes already installed and covered by protective layers [238].

In addition to those cases described above, there have been other reports of pitting corrosion in copper and copper alloys exposed to potable water and other industrial water systems; references [35, 107, 239–241] give further information.

E4. Biofouling

Slime algae, sea mosses, sea anemones, barnacles, oysters, and mussels attach themselves to marine structures, such as pilings and offshore platforms, boat hulls and even the insides of pipes and condensers. This phenomenon, called marine biofouling, is common in open waters an estuaries. Either mechanical removal or prevention (e.g., by renewable antifouling coatings) is often required.

Marine organisms adhere to some metals and alloys more readily than they do to others. Copper-based alloys have very good resistance to biofouling, and this property can be used to good advantage. Copper–nickel is used to minimize biofouling on intake screens, seawater pipework, water boxes, mesh cages in fish farming, marine craft, and offshore structures. A prime example of this was demonstrated in 1987 when two early solid copper–nickel hulled vessels, the *Asperida II* and *Copper Mariner*, were inspected after 21 and 16 years, respectively, of service. Neither vessel had required hull cleaning or suffered significant hull corrosion during that time. Alternative antifouling methods for hulls would have required recoating after ~30 months for copper-based paints or 5 years for organotin copolymer coatings.

In the short term, coatings are the less expensive alternative, but based on life-cycle costing, the copper–nickel hull is more cost effective. Cost studies to date have estimated the payback time to be 4–7 years for commercial vessels. With the current environmental concerns and restricted use of tributyltin copolymer paints, the use of copper–nickel to control corrosion and biofouling of ship hulls should be considered. It is also timely to review products incorporating copper–nickel that can be used for protecting marine structures from biofouling.

The 90 Cu–10 Ni alloys were developed from the very successful 70 Cu–30 Ni alloys as a result of conservation of nickel as a strategic metal during World War II [242]. About 1.5% iron increased resistance to erosion–corrosion, but additions >3.5% increase susceptibility to deposit attack and biofouling [243].

Copper–nickel alloys have traditionally been used for pipe work carrying seawater as they exhibit both corrosion and biofouling resistance, but more recently the 90 Cu–10 Ni alloys have been used in marine construction, such as intake screens for seawater-cooled power plants, fish-rearing cages, sheathing for oil and gas platforms, and boats [244]. A number of small boats have been constructed from copper–nickel alloys since the 1940s and have performed well, although at rather low speeds. Trials with a ship travelling at higher velocities are encouraging, especially as the hull material is self-smoothing [244]. Average corrosion rates reported are about 0.02 mm/year. (0.8 mpy, 5 mdd). Although 90 Cu–10 Ni alloys have very good corrosion resistance, at least at flow velocities below 4 m/s, they do corrode slightly. This slow release of copper is responsible for antifouling properties, and if it is halted by a cathodic protection system or galvanic coupling to a less noble metal, then biofouling will occur.

Antifouling properties of 90 Cu–10 Ni depend on the environmental conditions at the exposure site. Deployment in open water with moderate to considerable wave action is likely to result in an essentially clean surface, but exposure in low-water-velocity harbors and basins may result in the intermittent development of thick, complex fouling communities. Although macrofouling is often absent, microfouling usually occurs and is not restricted to prokaryotes but can include unicellular algae, protozoa, and other eukaryotes.

The occurrence of microbial communities on copper alloys has been discussed in detail [222, 242, 245–249]. Efird [250] recorded that "heavy sliming" was present after 12 months exposure in the sea for a range of copper alloys. The development of microbially dominated biofilms on copper–nickel alloys has not really been appreciated by many engineers and corrosion scientists, largely because thin biological films are inconsequential when calculating, for instance, the increased drag and wave loading due to macrofouling or perhaps the increase in weight due to shell-bearing animals. Moreover, it is generally believed that copper ions are lethal to the majority of organisms. Early investigations described microbially influenced corrosion (MIC) of copper alloys [251, 252]. Bacteria isolated on 70 Cu–30 Ni showed considerable copper tolerance, up to 2000 ppm copper [252]. These tests were carried out in suspension culture, and it is now well established that microorganisms residing within a mucilaginous biofilm may show much higher levels of tolerance to toxic compounds, antibiotics and the like. This research showed that MIC occurs on copper–nickel condenser tubes. Mechanisms proposed for MIC of copper–nickel included formation of corrosive metabolic products (e.g., ammonia, organic and inorganic acids, and hydrogen sulfide) that cause cathodic depolarization or catalysis of other corrosion reactions, and anaerobic sulfide production [252].

The effects of sulfide to cause corrosion can be seen long after the source of sulfide has disappeared. Doping a passivating cuprous oxide corrosion product layer by sulfide may lead to very high rates of corrosion once the material is exposed to aerated conditions [253, 254].

E5. Condensers and Heat Exchangers

Apart from the wide range of waters that must be used for cooling and the consequent variability in corrosion characteristics, some of the most corrosive conditions are encountered in power stations and ships that draw their cooling water from estuaries and harbors. These waters are polluted to varying extents depending on location and the time of year. Thus water drawn from an area near the outfall of a sewerage works will be heavily polluted and will cause localized corrosion of condenser tubes.

Corrosion initiated by polluted water, which is sometimes intergranular, may often continue even after changing to a clean water. The concentration of pollutants (e.g., hydrogen sulfide, which arises from the decomposition of organic matter), is likely to be greatest in late summer. Oxygen is consumed by organic matter, and SRB become active.

Hydrogen sulfide is probably the most powerful pollutant of waters and can cause rapid perforation of materials (e.g., brass). Another important pollutant, cystine, is produced by the breakdown of organic material, such as seaweed, and is present in both offshore and in-shore waters, the concentration being particularly large in harbors and estuaries. Cystine is a very efficient cathodic depolarizer and can produce intense attack even in the absence of oxygen. If cystine is present in sufficient concentration, however, a protective film forms on copper alloys consisting of a copper–cystine complex. This film may break down or blister if it becomes too thick (e.g. under impingement) and results in the highly corrosive situation of a small anode and large cathode.

In the construction of condensers and heat exchangers, the metals and alloys used are copper, brasses, bronzes, and cupronickel alloys [255]. Among these alloys C44300, C44400, C44500, C68700, C6130O, C63200, C63000, C95400, C95500, C95800, C70600, C71500, and C12000 provide satisfactory and economical performance for condenser and heat exchanger tubing. The corrosion resistance of these latter materials depends both on the inherent nobility conferred on them by copper and nickel and protection by the layer or film of corrosion product. The properties of this corrosion product, its continuity, adhesion to the metal surface and its ability to form, to be maintained, and to survive the action of erosive forces of high-velocity seawater, determine the alloy chosen for a given use. The effects of velocity are complicated by the presence in water of entrained air bubbles and abrasive materials such as sand, which can cause erosion and impingement attack leading to film breakdown and accelerated corrosion. The incidence of impingement attack on copper–alloy tubes is reduced when entrained air is removed from the cooling water. In other cases, dissolved and entrained air in the water may help to form and maintain protective films on the more resistant materials.

Iron corrosion products also help certain alloys to form protective films. In one case, introduction of ferrous sulfate into the cooling water at a power station reduced the attack on aluminum brass [256].

Pure deoxidized copper is limited to applications where purity of metal is essential. It has good resistance to practically all types of freshwater, but it is not serviceable in salt water and waters polluted with sulfur compounds. [Above ~200°C (395°F) there is a significant loss of strength.] Impingement attack occurs when the velocity of the water is high and no protective scale forms (e.g., in soft water high in free carbon dioxide). Where this type of attack is expected, reinforcement or protection are necessary or a different material should be selected. Arsenical copper containing 0.15–0.5% arsenic has a higher strength and fatigue limit but retains ductility. Its main use is in condensers and heat exchangers for use in freshwater. It is not recommended for waters containing hydrogen sulfide and other sulfur compounds, acid mine water, and salt or brackish waters.

Copper is relatively sensitive to metal ion concentration cells, and so the velocity of liquid across the surface is an important factor in controlling the corrosion rate. The resistance of copper to corrosion by rapidly flowing seawater is unsatisfactory, and it is not suitable for condenser tubes in ocean-going ships or in power stations using tidal waters.

The use of Muntz metal, 60 Cu–40 Zn, is mainly confined to steam condensers operating at low temperatures and using freshwaters from rivers, lakes, and wells. However, it has good resistance to hydrogen sulfide and other sulfur compounds and, with additions of lead, is commonly used in heat exchangers cooled with seawater.

Red brass is also serviceable in heat exchangers using freshwaters and is not susceptible to either SCC or dezincification, but is likely to be severely corroded in the presence of sulfur compounds.

The addition to brass of small amounts of alloying elements is beneficial. Naval brass, 60 Cu–39 Zn–1 Sn, and Admiralty brass, 70 Cu–29 Zn–1 Sn, are used to resist seawater in exchangers if the flow rate is <1 m/s (3 ft/s). The addition of ~0.04% arsenic to Admiralty brass inhibits dezincification, but the alloy is still susceptible to impingement. Nevertheless, with additions of arsenic, antimony, or phosphorus, Admiralty brass is suitable for brackish or saltwater at flow velocities up to 1.5–1.8 m/s (5–6 ft/s). Under more severe conditions aluminum brass, aluminum bronze, or cupronickels are used [257].

The addition of aluminum in aluminum brass, 76 Cu–22 Zn–2 Al, helps to form protective layers resistant to mechanical destruction (e.g., abrasion) and such brasses have a high resistance to corrosion–erosion. Aluminum brass with arsenic may be used in clean or polluted brackish and salt water up to 2.1 or 2.5 m/s (7 or 8 ft/s).

Aluminum bronze, 95 Cu–5 Al, is used at similar velocities in polluted seawater.

Tin bronze is often recommended although its behavior is not always acceptable. It withstands fairly well scouring by abrasives (e.g., sand).

Copper–nickel tubes behave better than aluminum brass in polluted water. The addition of iron to cupronickel helps to form more protective films responsible for excellent performance in seawater. Resistance is highest with the higher nickel and iron contents. In clean and polluted seawater, 90 Cu–10 Ni may be used at flow velocities up to 2.5–3 m/s (8–10 ft/s), but this alloy is not resistant to waters with a high sulfide content. Where long service life under severe conditions is required, the 30% nickel alloy should be used, and has a velocity limit of 3–3.5 m/s (10–12 ft/s).

In clean seawater moving at 3.5 m/s (11.7 ft/s), the rating of the copper alloys in order of decreasing tendency to impingement is Admiralty brass, aluminum bronze, phosphor bronze with 8% tin, followed by the cupronickels. The maximum flow rate of seawater in heat exchange apparatus to avoid corrosion–erosion of various copper alloys is copper, 1 m/s (3.0 ft/s); Admiralty brass, 2 m/s (6.0 ft/s); aluminum brass, 2.5 m/s (8.0 ft/s); and cupronickel, 8 m/s (26.0 ft/s).

Nickel-bearing aluminum bronzes with additions of iron and manganese are used for ship screws and are many times more stable erosion and cavitation than manganese brasses.

Corrosion in condensers and return lines often occurs because of oxygen and carbon dioxide in the condensed water. Although the feedwater to the boiler may be deoxygenated, there may be leakage of oxygen into the circuit external to the boiler. Either the leakage should be eliminated or more resistant alloys should be used [258, 259]. Alternatively, sodium sulfite may be added to the condensate, but a preferable treatment is to raise the pH by adding volatile or film-forming amines.

The addition of sodium silicate, polyphosphates, or oils to condensate have had only partial success. Sodium silicate decreases the corrosive action of acid condensate by forming a protective film of silica on the metal surface or by neutralization of the carbon dioxide by the alkali or both. Addition of oil in inadequate amounts may actually increase the corrosion rate instead of decreasing it. Polyphosphate treatment of return lines requires a long time (weeks) to build up an impermeable protective film on the metal surface. To prevent dissolution of the film, the dosage of polyphosphate must be maintained.

In the past, it was common practice to use Admiralty brass condenser tubes and ferrules in naval brass end plates with ferrous water boxes. The copper alloys were cathodically protected by the iron or steel that underwent enhanced attack, necessitating a corresponding increase in thickness. Alternatively, zinc protectors could be used to protect both the ferrous and copper alloys. This protection also extended into the condenser tubes for a distance of ~2.5 diameters and reduced impingement attack that was most severe there.

If the water boxes are coated or replaced by reinforced plastics, and if the doors are made of gunmetal, steps must be taken to restrict the attack on the end plates and the tube ends.

It was at one time standard practice to use iron protector slabs, which acted as sacrificial anodes, and, in addition, iron corrosion products helped to form a protective film on the condenser tubes.

The replacement of Admiralty brass by aluminum brass and cupronickels has eliminated most of the corrosion problems. Those that do occasionally occur are caused by the accumulation of debris that causes local turbulence, impingement, and deposit attack [260]. Where cast iron or steel doors are used, corrosion has been prevented by using magnesium anodes or impressed current. In coastal power stations, cathodic protection has also been employed for water boxes of condensers that use seawater for cooling. Titanium anodes with automatic current control are particularly suitable for this application. Other remedial and preventative measures for maintaining condenser and heat exchanger integrity have been reported elsewhere [261].

F. CORROSION IN GASES

Copper and its alloys are attacked by hot oxygen, sulfur vapor, sulfur dioxide, hydrogen sulfide, phosphorus, halides, and some acid vapors; they are generally inert toward hot nitrogen, hydrogen, carbon monoxide, carbon dioxide, and reducing gases except so far as cyclic exposure to these may accelerate oxidation. Oxygen-bearing copper is embrittled by hydrogen, dissociated water vapor, and dissociated ammonia (hydrogen disease). Copper alloys annealed without precleaning may develop "red stain."

F1. Oxidation of Copper

When heated in air, copper develops a Cu_2O film (ruby red) covered by a very thin outer layer of CuO (black), which is formed as the film thickness increases [262, 263]. These copper oxides possess semiconducting properties or, in other words, are characterized by mixed ionic/electronic conductivity, an electrical property that can vary in magnitude and in nature depending on the oxide composition. The Cu_2O is a p-type semiconductor, whereas CuO is an intrinsic semiconductor with negligible ionic conductivity [264–267]. The electronic properties of the copper oxides and the variable valency of copper can explain the mechanism of oxide film growth [268] and the severe corrosion attack that can occur.

Around 150°C (305°F), copper has been found to oxidize according to the inverse logarithmic [269, 270] and direct logarithmic [269, 271, 272] relationships, but in the wrong temperature sequence to be accounted for by the Mott–Hauffe–Ilschner mechanism [273, 274].

Between 200 and 1000°C (395 and 1835°F), the oxidation of copper is parabolic. For oxidation in air, Valensi [275] derived and others [262, 276–278] confirmed the following

equations for the temperature dependence of the parabolic rate constants:

300–550°C (575–1025°F): $k_p = 1.5 \times 10^{-5}$ $\exp(-20,140/RT)\ g^2/cm^4$

550–900°C (1025–1655°F): $k_p = 0.266$ $\exp(-37,700/RT)\ g^2/cm^4$

The parabolic plots do not always give straight lines. In particular, the cubic law applies at intermediate temperatures [269, 279] for which the Cabrera–Mott mechanism would account [280].

The effect of moisture in the air is small [277] with a tendency to reduce the oxidation rate slightly [277].

F2. Oxidation of Copper Alloys

Tylecote [262] reviewed the publications prior to 1950 on the effects of alloying elements on the oxidation of copper. At temperatures above ~200°C (1395°F), aluminum, beryllium, and magnesium increase the oxidation resistance of copper considerably [279, 281–283]. At 256°C (496°F), for instance, copper containing 2% Be oxidized ~14 times slower than the pure metal because a very thin, highly protective film rich in BeO forms on the alloy surface [281, 284–287].

The formation of a protective film directly attached to the alloy surface is also assumed to control the oxidation of alloys containing aluminum, but the formation of the protective films depends largely on experimental conditions, such as temperature [288, 289]. Under favorable conditions the oxidation rate of copper can be reduced almost to nil by aluminum, the maximum resistance being obtained at 8% Al [283]. The oxidation resistance of 70/30 brass is also greatly increased by aluminum additions [290].

The beneficial effect of magnesium on the oxidation resistance of copper results from preferencial oxidation [281]. Additions of magnesium also increase the resistance of copper to sulfur dioxide and hydrogen sulfide attack at various temperatures [291].

On binary copper alloys, two distinct oxide layers form, the outer layer consisting essentially of copper oxides, the inner layer containing large proportions of the alloying element [281]. This behavior was clearly recognized with alloys containing Ca, Cr, Li, Mn, Si, or Ti; all these alloys oxidize at approximately the same rate as, or a little slower than-copper.

Oxidation of alloys containing Mn, Ni, Si, Sn, Ti, or Zn resulted in the formation of subscales rich in copper and containing inclusions of the solute oxides that affected the mechanical properties [281]. Similar observations have been reported for Cu–Co–Si [292], Cu–Co [293], Cu–Bi, and Cu–As [294].

Tin bronzes with more than ∼6% Sn oxidize less rapidly than copper. The beneficial effect of tin is independent of concentration above ∼15% Sn [295]. The beneficial effect of tin is partly offset if some phosphorus is also present [281, 282].

The oxidation resistance of copper–nickel alloys with up to 30% Ni is about the same as that of copper, but increases at higher nickel concentrations [296].

Additions of 7.75% iron, 8% cobalt, 1–30% nickel, 2.4% antimony, and other elements have virtually no effect on the oxidation rate of copper [281]; the concentrations of these elements in the scale and the alloy are almost equal [281, 292] only antimony accumulating slightly in the scale.

Precious metals have little effect on the oxidation resistance of copper and are not contained in the scale other than in the metallic form. Early systems investigated have been those with silver [282, 297, 298], gold [298–300], silver and gold [298] platinum [299, 301], palladium [298–301], gold and palladium, and silver and palladium [298].

Tellurium and selenium were found [302] to increase the parabolic rates of oxidation of copper in air and oxygen at 600–1000°C (1115–1835°F) with the exception of some temperatures between 750° (1385°F) and 920°C (1656°F).

Investigations of multicomponent systems have shown that Everdur, for instance (copper containing 3% Si and 1% Mn), has a very low oxidation rate [279, 282] in moist air. Further references to work on multicomponent systems based on copper are found" in the review by Tylecote [262].

Reducing or completely neutral atmospheres prevent oxidation, but atmospheres containing hydrogen, water vapor, or ammonia should not be used with copper containing oxygen except in very short heat treatments at relatively low temperatures [425°C (800°F) maximum].

It is possible to form a subscale upon a dilute alloy in an atmosphere that is nontarnishing for pure copper. The subscale may interfere with buffing and may also to the appearance of surface cracks during subsequent working operations. Either strongly oxidizing or fully reducing atmospheres are to be preferred in such cases.

F3. Corrosion by Gases Other Than Oxygen

Copper is not attacked appreciably by water vapor at temperatures approaching the melting point [303]. The oxidation rate of copper in oxygen at 800°C (1475°F) is independent of water vapor content up to 3.9% [304], although slightly reduced oxidation rate has also been reported [279] for moist compared with dry air. Numerous copper alloys have likewise shown little difference in oxidation rate when investigated at 400° C (755°F) in dry

air and in air containing 10% water vapor; there is generally a slightly decreased attack in moist air, but 2% tin bronze has given the reverse effect [282].

Carbon dioxide and carbon monoxide in dry forms are inert to copper and its alloys. Because some alloy steels are attacked by CO, the high-pressure equipment used to handle this gas is often lined with copper or copper alloys.

Gases containing SO_2 and/or H_2S attack copper in a manner similar to oxygen, leading to the formation of sulfide layers such as Cu_2S or CuS [305–309].

In general, dry gases form very thin corrosion layers, but, in the presence of moisture mixtures of oxide and sulfide scales (e.g., $Cu_2O + Cu_2S$) are formed leading to high rates of corrosion, as happens for C11000 and C23000 copper alloys under hot, wet H_2S vapors.

Dry fluorine, chlorine, bromine, and their hydrogen compounds are not corrosive to copper and its alloys, but these halogen gases are aggressive when moisture is present. Copper and its alloys are also not susceptible to attack by hydrogen unless they contain copper oxide. Hydrogen diffusing inward forms H_2O according to the reaction:

$$Cu_2O + H_2 \Leftrightarrow 2Cu + H_2O$$

and ruptures the metal (hydrogen disease) [310]. Cast tough-pitch copper (containing free Cu_2O) is most sensitive to hydrogen embrittlement. Oxygen-free coppers are used where heating is necessary in the presence of hydrogen, but can become slightly sensitive to hydrogen embrittlement should they be heated in an oxidizing atmosphere.

Additional information on the corrosion of copper and its alloys in gasses and methods of prevention and mitigation are presented in references [311–314].

Studies published in the first decade of this century on the corrosion of copper and copper alloys in atmospheric environments [315–345], in waters [346–352], and in high-temperature media [353–362] confirm the scientific/technical/economic importance of this domain of science and engineering. In particular, the mechanisms of pitting corrosion [363–372], stress corrosion cracking [373–384], microbiologically influenced corrosion [385–395], and high-temperature oxidation [396–400] continue to deserve a lot of concern.

In addition to the forms of corrosion and the environments discussed in this chapter, copper and copper alloys are susceptible to other forms of corrosion and are widely employed in other environments, namely, acid solutions, alkaline solutions, nearly neutral salt solutions, organic compounds, and so on. Additional information on copper and its alloys and on technologies to control corrosion is available in the literature, including the extensive list of references for this chapter.

REFERENCES

1. BNF Press release. BNF Technology Centre, Grove Laboratories, Oxford, UK, Apr. 1985.

2. H. S. Campbell, J. Inst. Met., **77**, 345 (1950).

3. F. M. Al-Kharafi, H. M. Shalaby, and V. K. Gouda, Key Eng. Mater., **20–28**(1), 767 (1988).

4. J. Ehreke and W. Stichel, Werkst. Korros., **40**(1), 17 (1989).

5. O. V. Franque and B. Winkler, Werkst. Korros., **35**(12), 575 (1984).

6. T. Hamamoto, M. Kumagai, K. Kawano, and S. Yamauchi, Sumitomo Light Met. Tech. Rep., **28**(2), 16 (1987).

7. K. Kasahara and S. Komukai, Corros. Eng. (Jpn.), **36**(8), 453 (1987).

8. M. F. Obrecht, W. E. Sartor, and J.M. Kayes, in Proceedings of the 4th International Congress on Metallic Corrosion, N. E. Hammer (Ed.), NACE, Houston, TX, 1972, pp. 576–584.

9. R. Riedl and J. Klimbacher, Int. J. Mater. Prod. Technol., **4**(2), 159 (1989).

10. S. Sato, T. Minamoto, and K. Seki, Boshoku Gijutsu (Corros. Eng.) **31**(1), 3 (1982).

11. P. Tate, Prakt. Metallogr., (6), 311 (1985).

12. E. Triki, M. Smida, J. Labre, and J. Ledion, in Corrosion—Industrial Problems, Treatment and Control Techniques, Pergamon, Oxford, UK, 1987, pp. 667–682.

13. C. Breckon and J. R. T. Baines, Trans. Inst. Mar. Eng., **67**, 1 (1955).

14. V. F. Lucey, Br. Corros. J., **1**, 53 (1965).

15. H. S. Campbell, in Localized Corrosion, R. W. Staehle, B. F. Brown, J. Kruger, and A. Agrawal (Eds.), NACE, Houston, TX, 1974, p. 625.

16. G. L. Bailey, J. Inst. Met., **79**, 243 (1951).

17. A.-M. Beccaria and J. Crousier, Br. Corros. J., **26**, 215 (1991).

18. F. Jones, J. Met., **37**(3), 67 (1985).

19. K. P. Fox, C. N. Tate, G. P. Treweek, R. Trussel, and A. E. Bowers, Report FB86-208717/WMS, Southern California Metropolitan Water District, Los Angeles, CA, 1986.

20. M. Iwata, M. Nishikado, E. Sato, and Y. Itoi, J. Jpn. Inst. Light Met., **34**(9), 531 (1984).

21. A. K. Khachaturov, A. V. Vvedenskii, I. K. Marshakov, I. A. Malakhov, and O. F. Stol'Nikov, Prot. Met. (USSR), **22**(5), 632 (1986).

22. D. M. F. Nicholas, Proc. Conf. CASS 90 "Corrosion Air, Sea, Soil," Australasian Corrosion Association, Auckland, New Zealand, Nov. 1990.

23. F. M. Al-Kharafi, H. M. Shalaby, and V. K. Gouda, Br. Corros. J., **24**, 284 (1989).

24. P. D. Goodman, V. F. Lucey, and C. R. Maselkowsky, in "The USE of Synthetic Environments for Corrosion Testing," STP970, 165-173, ASTM, Philadelphia, PA, 1988.

25. J. G. N. Thomas and A. K. Tiller, Br. Corrs. J., **7**, 256 (1972).

26. H. Baba, T. Kodama, T. Fujii, and Y. Hisamatsu, Boshoku Gijutsu (Corros. Eng.), **30**(3), 161 (1981).

27. M. R. G. de Chialvo, R. C. Salvarezza, D. Vasquez Mol, and A. J. Arvia, Electrochim. Acta, **30**(11), 1501 (1985).

28. T. Fujii, T. Kodama, and H. Baba, Mater. Sci. Forum, **8**, 125 (1986).

29. T. Notoya, Hokkaido Daigaku Kenkyo Hokoku (Bull. Fac. Eng., Hokkaido Univ.), May **146**, 1–10 (1989).

30. S. Smith and R. Francis, Br. Corros. J., **25**, 285 (1990).

31. S. Seri and S. Furutani, Proceedings of the 76th Conference of the Japan Institute of Light Metals Osaka, Japan, 1989, pp. 117–118.

32. H. Baba, T. Kodama, and T. Fujii, Trans. Natl. Res. Inst. Met. (Jpn.), **28**(3), 248 (1986).

33. C. A. C. Sequeira, in Microbial Corrosion—I, C. A. C. Sequeira and A. K. Tiller (Eds.), Elsevier Applied Science, Barking, UK, 1988, pp. 99–118.

34. R. A. Baker and R. S. Tombaugh, Mater. Eval., **48**(1), 55 (1990).

35. H. M. Shalaby, F. M. Al-kharafl, and V. K. Gouda, Corrosion, **45**(7), 536 (1989).

36. C. A. C. Sequeira, M.J. Alyes, and A. C. Costa, in Microbial Corrosion—I, C. A. C. Sequeira and A. K. Tiller (Eds.), Elsevier Applied Science, Barking, UK, 1988, pp. 266–286.

37. E. D. Morr and A.-M. Beccaria, Br. Corros. J., **12**, 243 (1977).

38. H. Kristiansen, Werkst. Korros., **28**, 143 (1977).

39. L. I. Molvina, T. S. Ganzhenko, and V. I. Kucherenko, Izv. VUZ Khim. Khim. Tekhnol, **29**(2), 122 (1986).

40. E. Mattsson, Br. Corros. J., **15**, 6 (1980).

41. F. J. Cornwell, G. Wildsmith, and P. T. Gilbert, Br. Corros. J., **8**, 202 (1973).

42. R. Retief, Br. Corros. J., **8**, 264 (1973).

43. G. Frommeyer, Werkst. Korros., **31**, 114 (1980).

44. F. J. Cornwell, G. Wildsmith, and P. T. Gilbert, in STP576, ASTM, Philadelphia, PA, 1980, p. 755.

45. M. Pourbaix, in Localized Corrosion, R. W. Staehle, B. F. Brown, J. Kruger, and A. Agrawal,(Eds.), NACE, Houston, TX, 1974, p. 12.

46. M. Pourbaix, Tech. Rep. Cebelcor, **100**, 126 (1965).

47. T. W. Bostwick, Corrosion, **17**, 12 (1961).

48. P. T. Gilbert, Trans. Inst. Mar. Eng., **66**, 1 (1954).

49. P. T. Gilbert, in Proceedings of the Conference on Corrosion '81 Paper 194, Houston, Tx, Toronto, Ontario., Canada, NACE, Mar. 1981.

50. A. M. Lockhart, Proc. Inst. Mech. Eng., **179**, 459 (1964–1965).

51. G. Lecointre, V. Plichon, P. Berge, and J. Legrand, Mét.-Corros.-Ind., (620) 35 (1977).

52. M. Pourbaix, J. Electrochem. Soc, **123**(2), (1976).

53. R. F. North and M. J. Pryor, Corros. Sci., **8**, 149 (1968).

54. R. Gasparini, C. Delia Rocca, and E. Isannilli, Corros. Sci., **10**, 157 (1970).

55. J. E. Castle, D. C. Epler, and D. B. Peplow, Corros. Sci., **16**, 145 (1976).

56. D. C. Epler and J. E. Castle, Corrosion, **35**, 123 (1979).

57. K. Kasahara, S. Komukai, and T. Fujiwara, Corros. Eng. (Jpn.), **37**(7), 361 (1988).

58. H. Baba, T. Kodama, and T. Fujii, Boshoku Gijutsu (Corros. Eng.), **34**(1), 10 (1985).

59. G. Cerisola, A. Barbucci, M. Bassoli, L. Fedrizzi, E. Miglio, and P. I. Bonora, in Proc. 7th European Symp. on Corrosion Inhibitors, Ferrara, Italy, Sept. 1990; Ann. Univ. Ferrara, NS, Sez. V, (Suppl. 9), 1990, pp. 1409–1414.

60. T. Fujii, Trans. Natl. Res. Inst. Met. (Jpn.), **30**(2), 81 (1988).

61. T. Hamamoto and M. Kumagai, Sumitomo Light Met. Tech. Rep., **29**(3), 175 (1988).

62. Yu. I. Kuznetsov and I. G. Kuznetsova, Prot. Met. (USSR), **22**(3), 387 (1986).

63. E. Mattsson, Werkst. Korros., **39**(11), 499 (1988).

64. K. Nagata, T. Atsumi, S. Sato, Y. Yamaguchi, and K. Onda, Sumitomo Light Met. Tech. Rep., **25**(3), 144 (1984).

65. E. Mattsson, Electrochim. Acta, **3**, 279 (1961).

66. U. Bertocci and E. N. Pugh, in Proceedings of the 9th International Congress on Metallic Corrosion, Vol. 1, Toronto, Canada, National Research Council of Canada, June 1984, p. 144.

67. H. L. Logan, J. Res. Natl. Bur. Stand., **48**, 99 (1952).

68. H. Leidheiser, Jr., and R. Kissinger, Corrosion, **28**, 218 (1972).

69. R. P. M. Procter and M. Islam, Corrosion, **32**, 267 (1976).

70. N. W. Polan, J. M. Popplewell, and M. J. Pryor, J. Electrochem. Soc., **126**, 1299 (1979).

71. A Parthasarathi and N. W. Polan, Metall. Trans. A, **13 A**, 2027 (1982).

72. R. C. Newman and K. Sieradzki, NATO ASI Ser, Ser. E, **130**, 597 (1987).

73. K. Sieradzki and R. C. Newman, J. Phys. Chem. Solids, **48**(11),1101 (1987).

74. K. Sieradzki, J. S. Kim, A. T. Cole, and R. C. Newman, J. Electrochem. Soc, **134**(7), 1635 (1987).

75. K. Sieradzki and R. C. Newman, Philos. Mag. A, **51**(1), 95 (1985).

76. R. B. Rebak, R. M. Carranza, and J. R. Galvele, Corros. Sci., **28**(11), 1089 (1988).

77. R. M. Carranza and J. R. Galvele, Corros. Sci., **28**(9), 851 (1988).

78. G. L. Bianchi and J. R. Galvele, Corros. Sci., **27**(6), 631 (1987).

79. J. R. Galvele, Corros. Sci., **27**(1), 1 (1987).

80. S. Kim, W. F. Flanagan, B. D. Lichter, and R. N. Grugel, Metall. Trans. A, **24A**(4), 975 (1993).

81. W. F. Flanagan, L. Zhong, and B. D. Lichter, Metall. Trans. A, **24A**(3), 553 (1993).

82. B. D. Lichter, R. M. Bhatkal, and W. F. Flanagan, in Parkins Symp. Fundam. Aspects Stress Corrosion Cracking, S. M. Bruemmer (Ed.), Nashville, TN, 1991; Miner. Met. Mater. Sci., Warrendale, PA, 1992, pp. 279–302.

83. S. Kim, W. F. Flanagan, B. D. Lichter, and R. N. Grugel, in 4th International Conferences on Experimental Methods for Microgravity Materials Science Research, R. A. Schiffman

84. (Ed.), Nashville, TN; Miner.Met. Mater. Soc, Warrendale, PA, 1992, pp. 101–107.

84. B. D. Lichter, W. F. Flanagan, J. B. Lee, and M. Zhu, in Environmental Induced Cracking of Metals, R. P. Gangloff and M. B. Ives (Eds.), NACE-10, NACE, Houston, TX, 1990, p. 251.

85. W. F. Flanagan, P. Bastias, and B. D. Lichter, Acta Metall. Mater., **39**(4), 695 (1991).

86. W. F. Flanagan, J. B. Lee, D. Massinon, M. Zhu, and B. D. Lichter, ASTM Spec. Tech. Publ., **1049**, 86 (1990).

87. Y. Brechet, F. Louchet, and T. Magnin, Mater. Sci. Eng. A, **A164**(1–2), 35 (1993).

88. J. Lepinoux and T. Magnin, Mater. Sci. Eng. A, **A164**(1–2), 266 (1993).

89. T. Magnin, Scr. Metall. Mater., **26**(10), 1541 (1992).

90. T. Magnin, Mem. Etud. Sci. Rev. Metall., **88**(1), 33 (1991).

91. T. Magnin, EGF Publ., **7**, 309 (1990).

92. T. P. Hoar and G. P. Rothwell, Electrochim. Acta, **15**, 1037 (1970).

93. T. P. Hoar and J. G. Ines, J. Iron and Steel Inst., **177**, 248 (1954).

94. T. P. Hoar and J. G. Ines, J. Iron and Steel Inst., **182**, 124 (1956).

95. T. P. Hoar and J. M. West, Nature (London), **181**, 385 (1958).

96. T. P. Hoar and J. M. West, Proc. R. Soc, **A268**, 304 (1962).

97. T. P. Hoar and J. C. Scully, J. Electrochem. Soc, **111**, 348 (1964).

98. C. Edeleanu and A. J. Forty, Philos Mag., **5**, 1029 (1960).

99. A. J. Forty and P. Humble, Philos Mag., **8**, 247 (1963).

100. A. J. Mcevily and A. P. Bond, J. Electrochem. Soc., **112**, 131 (1965).

101. N. S. Stoloff and T. L. Johnston, Acta Met., **11**, 251 (1963).

102. A. R. C. Westwood and M. H. Kamder, Philos Mag., **8**, 787 (1963).

103. G. G. Coleman, D. Weinstein, and W. Rostoker, Acta Met., **9**, 491 (1961).

104. A. R. C. Westwood, Corros. Sci., **6**, 381 (1966).

105. J. H. Johnson, R. T. Kiepura, and D. A. Humphries (Eds.), Metals Handbook, 9th ed., Vol. 13, Corrosion, ASM, Metals Park, OH, 1987.

106. H. H. Uhlig and D. J. Duquette, Corros. Sci., **9**, 557 (1969).

107. E. N. Pugh, W. G. Montague, and A. R. C. Westwood, Corros. Sci., **6**, 345 (1966).

108. D. H. Thompson and A. W. Tracy, Trans. AIME, **185**, 100 (1949).

109. B. C. Syrett and R. N. Parkins, Corros. Sci., **10**, 197 (1970).

110. R. N. Parkins, P. W. Slattery, and B. S. Poulson, Corrosion, **37**, 650 (1981).

111. H. Cordier, C. Dumont, W. Gruhl, and B. Grzemba, Metall (Berlin), **36**, 33 (1982).

112. C-Liang, T. Shinohara, and S. Tsujikawa, Boshoku Gijutsu, (Corros. Eng.) **39**, 309 (1990).

113. E. Iskevitch and R. Chaim, Philos. Mag. Lett., **61**, 209 (1990).

114. J. F. Klement, R. B. Mearsch, and P. A. Tully, Corrosion, **16**, 519t (1960).

115. M. Kamikado, S. Okazaki, M. Kiyoshiga, and K. Tsujimoto, Kushoku Boshoku Bumon Tinkai Shiryo, **26**, 36 (1987).

116. S. I. Pyun, J. K. Choi, and D. J. Hwang, J. Matur. Sci. Lett., **8**, 1402 (1989).

117. A. A. Aksut, Bull. Electrochem., **5**, 3 (1989).

118. S. I. Pyun, T. S. Sue, and H. P. Kim, Werkst. Korros., **38**, 129 (1987).

119. M. Asawa, Corrosion, **46**, 829 (1990).

120. M. Islam, W. T. Riad, S. Al-Kharraz, and S. Abo-Namons, Corrosion, **47**, 260 (1991).

121. A. H. Allen, Metal Prog., 106 (Aug. 1954).

122. A. J. Katkar, J. Balachandra, and K. I. Vasu, J. Electrochem. Soc. India, **29**, 108 (1980).

123. T. Mimaki, H. Fukugana, S. Hashimoto, and S. Miura, Zairyo, **34**, 1327 (1985).

124. R. C. Dorward and K. R. Hasse, Corrosion, **43**, 408 (1987).

125. E. I. Meletis, Mater. Sci. Eng., **93**, 235 (1987).

126. R. Braun and H. Buhl, J. Phys., Colloq., **C3**, 843 (1987).

127. R. C. Durward and K. R. Hasse, Corrosion, **44**, 932 (1988).

128. M. Yamashita, T. Mimaki, S. Hashimoto, and S. Miura, Philos. Mag. A, **63**, 707 (1991).

129. T. Ohmishi and T. Ito, Aluminium (Dusseldorf), **66**, 485 (1990).

130. M. Yamashita, M. Yoshiota, T. Mimaki, S. Hashimoto, and S. Miura, Acta Metall. Mater., **38**, 1619 (1990).

131. A. K. Lahiri and T. Banerjee, Corros. Sci., **5**, 731 (1965).

132. E. T. Meletis and R. F. Hochman, Corros. Sci., **24**, 843 (1984).

133. F. Terwinghe, J. P. Celis, and J. R. Roos, Br. Corros. J., **19**, 107 (1984).

134. M. Froemberg, Korrosion (Dresden), **16**, 323 (1985).

135. M. Takano, K. Teramoto, and T. Nakayama, Corros. Sci., **21**, 459 (1981).

136. H. L. Logan, J. Res. Nat Bur. Std., **56**, 195 (1956).

137. J. S. Chen, M. Salmeron, and T. M. Devine, Scr. Metall. Mater., **26**(5), 739 (1992).

138. J. B. Lee "Studies on De-alloying and Its Relationship to Transgranular SCC in Copper-25 at % Gold Single Crystals," Ph.D. Thesis, Vanderbilt University, Nashville, TN, 1991.

139. M. Zhu, "A Study of Transgranular SCC in Copper-25 at % Gold Single Crystals," Ph.D. Thesis, Vanderbilt University, Nashville, TN, 1990.

140. J. D. Fritz, B. W. Parks, and H. W. Pickering, ASTM Spec. Tech. Publ., **1049**, 76–85 (1990)

141. U. Bertocci, J. Electrochem. Soc., **136**(7), 1887 (1989).

142. J. D. Fritz, B. W. Parks, and H. W. Pickering, Scr. Metall., **22**(7), 1063 (1988).

143. T. B. Cassagne, W. F. Flanagan, and B. D. Lichter, Metall. Trans. A, **19A**(2), 281 (1988).

144. H. Moore, S. Beckinsale, and C. E. Mallison, J. Inst. Met. **25**, 35 (1921).

145. A. V. Bobylev, Zashch. Met., **20**, 631 (1984).

146. A. V. Shreider, J. Appl. Chem. U.S.S.R., **30**, 836 (1957).

147. Y. Suzuki and Y. Hisamatsu, Corros. Sci., **21**, 353 (1981).

148. E. Mattsson, R. Holm, and L. Hassel, ASTM Spec. Tech. Publ., **970**, 152 (1988).

149. T. Shahrabi and R. C. Newman, Mater. Sci. Forum, **44–45**, 169 (1989).

150. M. J. Kanfman and J. L. Fink, Metall. Trans. A, **18 A**, 1539 (1987).

151. T. P. Hoar and C. J. L. Booker, Corros. Sci., **5**, 821 (1965).

152. R. G. Johnson, Sheet Metal Ind., **14**, 1197 (1940).

153. J. B. Cotton, Corrosion, **20**, 202t (1964).

154. D. Tromans, J. Electrochem. Soc, **145**(3), L42 (1998).

155. D. Tromans and J. Nutting, Corrosion, **21**, 143 (1965).

156. H. E. Johnson and J. Leja, J. Electrochem. Soc., **113**, 630 (1966).

157. S. P. Nayak and A. K. Lahire, Indian J. Technol., **10**, 322 (1972).

158. H. Rosenthal and A. L. Jamieson, Trans. AME, **156**, 212 (1944).

159. K. Balakrishnan and P. L. Annamalai, Corrosion, **25**, 22 (1969).

160. S. C. Sircar, U. K. Chatterjee, S. K. Roy, and S. Kisku, Br. Corros. J., **9**, 47 (1974).

161. A. J. Fiocco, Annual Report, 1967–1968, Office of Saline Water, Denver, CO, 1969, p. 241.

162. M. E. Holberg and F. A. Prange, Ind. Eng. Chem., **37**, 1030 (1945).

163. J. P. Fraser, Mater. Protection, **2**, 97 (1963).

164. H. H. Uhlig and J. Sansons, Mater. Protection, **3**, 21 (1964).

165. V. K. Gouda, H. A. El-Sayed, and S. M. Sayed, 8th International Congress on Metallic Corrosion, Frankfurt, Germany, Dechema, Frankfurt am Main, 1981, p. 479.

166. A. Kawashima, A. K. Agrawal, and R. W. Staehle, Stress Corrosion Cracking—The Slow Strain Rate Technique, STP 665, ASTM, Philadelphia, PA, 1979, p. 266.

167. E. Escalante and J. Kruger, J. Electrochem. Soc, **118**, 1062 (1971).

168. R. N. Parkins and N. J. H. Holroyd, Corrosion, **28**, 245 (1982).

169. S. P. Pednekar, A. K. Agrawal, H. E. Chaung, and R. W. Staehle, J. Electrochem. Soc., **126**, 701 (1979).

170. W. H. J. Vernon, Trans. Faraday Soc., **23**, 162 (1927).

171. W. H. J. Vernon, Trans. Faraday Soc., **27**, 255 (1931).

172. R. Chadwick, J. Inst. Met., **97**, 93 (1969).

173. H. W. Pickering and P. J. Byrne, Corrosion, **29**, 325 (1973).

174. J. B. Greer and M. R. Chance, Mater Prot. Perform., **12**(3), 41 (1973).

175. N. G. Goncharov, A. G. Mazel, and S. Y. Golovin, Svar Proizvord, **4**, 21 (1986).

176. T. Magnin and M. Rebiere, J. Phys. Colloq., **C3**, 835 (1987).

177. K. Kon, S. Tsujikawa, and Y. Hisamatsu, Boshoku Gijutsu (Corros. Eng.) **35**, 342 (1986).

178. L. A. Benjamin, D. Hardie, and R. N. Parkins, Br. Corros. J., **23**, 89 (1988).

179. A. K. Agrawal, J. E. Aller, A. R. Hamilton, and J. E. Beavers, Corrosion, **46**, 172 (1990).

180. R. E. Lobnig, R. P. Frankenthal, D. J. Siconolfi, and J. D. Sinclair, J. Electrochem. Soc, **140**, 1902 (1993).

181. R. E. Lobnig, R. P. Frankenthal, D. J. Siconolfi, J. D. Sinclair, and M. Stratmann, J. Electrochem. Soc, **141**, 2935 (1994).

182. T. E. Graedel, Corros. Sci., **27**, 721 (1987).

183. T. E. Graedel, J. P. Franey, and G. W. Kammlott, Corros. Sci., **23**, 1141 (1983).

184. T. E. Graedel, K. Nassau, and J. P. Franey, Corros. Sci., **27**, 639 (1987).

185. T. E. Graedel, Corros. Sci., **27**, 741 (1987).

186. J. D. Sinclair, J. Electrochem. Soc., **135**, 89C(1988).

187. M. Forslund and C. Leygraf, J. Electrochem. Soc., **144**, 113 (1997).

188. J. D. Sinclair and L. A. Psota-Kelty, Atmosph. Environment, **24 A**, 627 (1983).

189. R. E. Lobning and C. A. Jankoski, J. Electrochem. Soc., **145**, 946 (1998).

190. A. Eldering, P. A. Solomon, L. G. Salmon, T. Fall, and G. R. Cass, Atm. Environ., **25A**, 2091 (1991).

191. R. Karlsson and E. Ljungström, J. Aerosol Sci., **26**, 39 (1995).

192. V. Kucera, D. Knotkova, J. Gullman, and P. Holler, in International Congress on Metal Corrosion, Madras, India, Vol. 1, Central Electrochemical Research Institute, 1987, p. 167.

193. R. Holm and E. Mattsson, in Atmospheric Corrosion of Metals, ASTM STP 767, S. W. Dean, Jr., and E. C. Rhea (Eds.), ASTM, Philadelphia, PA, 1982, p. 85.

194. L. Selwyn, N. E. Binnie, J. Poitras, M. E. Laver, and D. A. Downham, Stud. Conserv., **41**, 205 (1996).

195. R. A. Livingston, Environ. Sci. Technol., **25**, 1400 (1991).

196. H. Strandberg, G. Johansson, and J. Rosvall, in ICOM Committee for Conservation 11th Triennial Meeting, Vol. 2, p. 894, Edinburgh, Scotland, Sept. 1996, James and James, London, 1996.

197. S. Oesch and P. Heimgartner, Mater. Corros., **47**, 425 (1996).

198. S. Zakipour and C. Leygraf, J. Electrochem. Soc., **133**, 21 (1986).

199. J. Jambor, J. Dutrizac, A. Roberts, J. Grice, and J. Szymanski, Can. Mineral., **34**, 61 (1996).

200. W. Feitknecht, Chem. Ind., Sept. 5, 1959, p. 1102.

201. P. Eriksson, L.-G. Johansson, and LGullman, Corros. Sci., **34**, 1083(1993).

202. H. Strandberg and L.-G. Johansson, J. Electrochem. Soc., **145**, 1093(1998).

203. D. Behrens (Ed.), Dechema Corrosion Handbook, Vol. 7, VCH Publishers, New York, 1990.

204. T. Drayman-Weisser (Ed.), The Conservation of Bronze Sculpture in the Outdoor Environment, A Dialogue Among Conservators, Curators, Environmental Scientists, and Corrosion Engineers, NACE, Houston, TX, 1992.

205. S. Z. Lewin, Application of Science in Examination of Works of Art, Seminar, Boston, MA, 1970, p. 62.

206. S. W. Dean, Jr. and E. C. Rhea (Eds.), Atmospheric Corrosion of Metals, ASTM-STP 767, ASTM, Philadelphia, PA, 1980.

207. P. N. Cheremiginoff (Ed.), Encyclopaedia of Environmental Control Technology, Gulf Publishing Co., Houston, TX, 1989.

208. C. Pearson, Conservation of Marine Archeological Objects, Butterworths, London, 1987.

209. W. H. Ailor (Ed.), Atmospheric Corrosion, Wiley, New York, 1982.

210. A. Nishikata, Y. Ichihara, and T. Tsuru, Corros. Sci., **37**, 897 (1995).

211. W. W. Kirk and H. H. Lawson (Eds.), Atmospheric Corrosion, ASTM-STP 1239, ASTM, Philadelphia, PA, 1991.

212. F. Mansfeld (Ed.), Corrosion Mechanisms, Marcel Dekker I, New York, 1987.

213. C. A. C. Sequeira and M. I. Panayotova, Sci. Technol. Mater., **9**, 40 (1997).

214. H. Gerischer, J. Electrochem. Soc., **113**, 1174 (1966).

215. W. H. Brattain and C. G. B. Garret, Bell Syst. Tech. J., **34**, 129 (1955).

216. U. Collisi and H. H. Strehblow, J. Electroanal. Chem., **210**, 213 (1986).

217. B. Pointu, M. Braizaz, P. Poncet, and J. Rousseau, J. Electroanal. Chem., **151**, 65 (1983).

218. A. J. Bard, R. Parsons, and J. Jordan, Standard Potentials in Aqueous Solutions, Marcell Dekker, New York, 1985.

219. R. M. Garrels and C. L. Christ, Minerals, Solutions and Equilibria, Harpey and Rowley, New York, 1965.

220. D. D. Wagman, W. H. Evans, V. B. Parker, R. H. Schumm, I. Halow, S. M. Bailey, K. L. Churnay, and R. L Butall, J. Phys. Chem. Ref. Data **11**, Suppl. 2, 392 (1982).

221. C. A. C. Sequeira, Br. Corros. J., **30**, 137 (1995).

222. K. D. Efird and D. B. Anderson, Mater. Perform., **14**, 37 (1975).

223. C. R. Southwell, J. D. Bultman, and A. L. Alexander, Mater Perform., **15**(7), (1976).

224. D. B. Anderson and F. A. Badia, Trans. ASME, **95**(4), (1973).

225. R. D. Schelleng, Technical Publication 949-OP, International Nickel Company, Nov. 1976.

226. C. Pearson, Br. Corros. J., **7**(Mar. 1972).

227. G. M. Bass-Becking and D. Moore, Econ. Geol., **56**, 259 (1961).

228. D. D. Macdonald, B. C. Syrett, and S. S. Wing, Corrosion, **35**, 367 (1979).

229. E. D. Mor and M. Beccaria, Br. Corros. J., **10**, 33 (1975).

230. E. H. Roseboom, Econ. Geol., **61**, 641 (1966).

231. M. McNeil and B. J. Little, Corrosion, **46**, 599 (1990).

232. J. C. Nuttall, in Corrosion and Related Aspects of Materials for Potable Water Supplies, P. McIntyre and A. D. Mercer (Eds.), The Institute of Materials, London, 1993, pp. 65–83.

233. W. Fischer, I. Haenssel, and H. H. Paradies, in Microbial Corrosio, Vol. 1, C. A. C. Sequeira and A. K. Tiller (Eds.), Elsevier Applied Science, London, 1989, pp. 300–327.

234. A. H. L. Chamberlain, P. Angell, and H. S. Campbell, Br. Corros. J., **23**(3), 197 (1988).

235. W. Fischer, H. H. Paradies, D. Wagner, and I. Haenssel, Werkst. Korros., **43**, 56 (1992).

236. A. H. L. Chamberlain, W. R. Fischer, U. Hinze, H. H. Paradies, C. A. C. Sequeira, H. Siedlarek, M. Thies, D. Wagner, and J. N. Wardell, in Microbial Corrosion, A. K. Tiller and C. A. C. Sequeira (Eds.), EFC Publication No. 15, The Institute of Materials, London, 1994, pp. 3–16.

237. C. A. C. Sequeira, A. C. P. R. P. Carrasco, D. Wagner, M. Tietz, and W. R. Fischer, in Microbial Corrosion, A. K. Tiller and C. A. C. Sequeira (Eds.), EFC Publication No. 15, The Institute of Materials, London, 1994, pp. 64–84.

238. D. Wagner, A. H. L. Chamberlain, W. R. Fischer, J. N. Wardell, and C. A. C. Sequeira, Mater. Corros., **48**, 311 (1997).

239. H. Leidheiser, Jr., The Corrosion of Copper, Tin and Their Alloys, Wiley, New York, 1971.

240. T. Burstein, L. L. Shreir, and R. A. Jarman (Eds.), Corrosion, Vol. 1, Butterworth-Heinemann, London, 1994.

241. Handbook of Corrosion Data, 2nd ed., ASM 486, American Technical Publishers, Herts, UK, 1995.

242. W C. Stewart and F. L. LaQue, Corrosion, **8**, 259 (1952).

243. A. H. L. Chamberlain and B. J. Garner, Biofouling, **1**, 79 (1988).

244. B. B. Moreton and T. J. Glover, in 5th Int. Congress on Marine Corrosion and Fouling, Editorial Garsi, 1980, pp. 267–278.

245. D. S. Marszalek, S. M. Gerchakov, and L. R. Udey, Appl. Environ. Microbial., **38**, 987 (1979).

246. M. F. L. De Mele, G. Brankevich, and H. A. Videla, Bri. Corros. J., **24**(3), 211 (1989).

247. H. A. Videla, in Biofouling and Biocorrosion in Industrial Water Systems, G. G. Geesey, Z. Lewandowski, and H. C. Flemming (Eds.), Lewis Publishers, Boca Raton, FL, 1994, pp. 231–241.

248. P. T. Gilbert, Br. Corros. J., **14**, 20 (1979).

249. D. H. Pope, D. J. Duquette, A. H. Johannes, and P. C. Wagner, Mater. Perform., **23**, 14 (1984).

250. K. D. Efird, Mater. Perform., **15**, 16 (1976).

251. G. D. Bengough and R. May, J. Inst. Metals, **32**, 19 (1924).

252. T. H. Rogers, J. Inst. Metals, **75**, 81 (1948).

253. B. C. Syrett, Corros. Sci., **21**, 187 (1981).

254. J. P. Gudas and H. P. Hack, Corrosion, **35**, 67 (1979).

255. R. Francis, in Corrosion in Seawater Systems, A. D. Mercer (Ed.), Ellis Horwood, London, 1990, pp. 65–75.

256. R. Francis, Mater. Perform., **21**(8), 44 (1982).

257. R. Francis, Corros. Sci., **26**, 205 (1986).

258. R. Francis, Br. Corros. J., **20**, 157 (1985).

259. R. Francis, Br. Corros. J., **20**, 175 (1985).

260. W. E. Heaton, in Corrosion in Seawater Systems, D. A. Mercer (Ed.), Ellis Horwood, London, 1990, pp. 99–124.

261. W. E. Heaton, Br. Corros. J., **13**(2), 57 (1978).

262. R. F. Tylecote, J. Inst. Met., **78**, 254 (1950).

263. A. Rönnquist and H. Fischmeister, J. Inst. Met., **89**, 65 (1960).

264. H. Dunwald and C. Wagner, Z. Phys. Chem. B, **22**, 212 (1933).

265. C. Wagner and H. Hammen, Z. Phys. Chem. B, **40**, 197 (1938).

266. C. Wagner and K. Grunewald, Z. Phys. Chem. B, **40**, 455 (1938).

267. K. Hauffe and H. Grunewald, Z. Phys. Chem., **198**, 248 (1951).

268. P. Kofstad, High Temperature Oxidation of Metals, Wiley, New York, 1966.

269. T. N. Rhodin, J. Am. Chem. Soc., **73**, 3143 (1951).

270. F. W. Young, J. V. Cathcart, and A. T. Wathmey, Acta Metall., **4**, 145 (1956).

271. K. Hauffe and A. L. Vierk, Z. Phys. Chem., **196**, 160 (1950).

272. A. H. White and L. H. Germer, Trans. Electrochem. Soc., **81**, 305 (1942).

273. K. Hauffe and B. Ilschner, Z. Elektrochem., **58**, 382 (1954).

274. N. F. Mott, Trans. Faraday Soc., **35**, 1175 (1939).

275. G. Valensi, Rev. Metall., **45**, 10 (1948).

276. D. W. Bridges, J. P. Baur, G. S. Baur, and W. M. Fassell, J. Electrochem. Soc., **103**, 266 (1956).

277. R. F. Tylecote, J. Inst. Met., **81**, 681 (1953).

278. F. De Carli and N. Collari, Chim. Et Industr., **33**, 77 (1951).

279. W. E. Campbell and U. B. Thomas, Trans. Electrochem. Soc., **91**, 623 (1947).

280. N. Cabrera and N. F. Mott, Rep. Progr. Phys., **12**, 163 (1948–1949).

281. K. W. Fröhlich, Z. Metallk, **28**, 368 (1936).

282. A. P. C. Hallowes and E. Voce, Metallurgia, Manchr., **34**, 95, 119 (1946).

283. H. Nishimura, J. Min. Metall., Kyoto **9**, 655 (1938).

284. S. Miyake, Sci. Pap. Inst. Phys. Chem. Res., Tokyo, **31**, 161 (1937).

285. L. De Brouckère and L. Hubrecht, Bull. Soc. Chim. Belg., **61**, 101 (1952).

286. K. Ono, S. Yoshida, and I. Takeda, Am. Chem. Abstr., **47**, 7983 (1953).

287. H. N. Terem, C. A. Acad. Sci., Paris, **205**, 47 (1937).

288. J. P. Dennison and A. Preece, J. Inst. Met., **81**, 229 (1953).

289. P. Spinedi, Metall. Ital., **45**, 457 (1953).

290. J. S. Dunn, J. Inst. Met., **46**, 25 (1931).

291. W. Baukloh and W. W. G. Krysko, Metallwirtschaft, **19**, 157 (1940).

292. F. R. Hensel, E. I. Larsen, and E. F. Holt, Metals Alloys, **13**, 151 (1941).

293. C. S. Smith, Min. Metall. N. Y., **11**, 213 (1930).

294. C. Blazey, J. Inst. Met., **46**, 353 (1931).

295. F. De Carli and N. Collari, Metall. Ital., **44**, 1 (1952).

296. N. B. Pilling and R. E. Bedworth, Ind. Eng. Chem., **17**, 372 (1925).

297. J. A. A. Leroux and E. Raub, Z. Anorg. Chem., **188**, 205 (1930).

298. E. Raub and M. Engel, Z. Metallk. Appendix 83(1938).

299. Õ. Kubaschewski, Z. Elektrochem., **35**, 142 (1929).

300. G. Tamman and W. Rienäcker, Z. Anorg. Chem., **156**, 261 (1926).

301. D. E. Thomas, J. Met. N.Y., **3**, 926 (1951).

302. N. P. Diev, M. I. Kochnev, A. F. Plotnikova, and T. N. Zaidman, Zh. Prikl. Khim., Leningr., **26**, 596, 760 (1953).

303. M. Farber, J. Electrochem. Soc., **106**, 751 (1959).

304. N. B. Pilling and R. E. Bedworth, J. Inst. Met., **29**, 529 (1923).

305. R. V. Chiarenzelli, 3rd Int. Res. Symp. Electrical Contact Phenom., MA, 1966, p. 85.

306. W. H. Abbott, Proc. Holm Semin. Electrical Contacts, 1973, p. 94.

307. K. L. Schiff and N. Harmsen, Proc. Holm Seminar, 1975, p. 37.

308. J. Guinement and C. Fiand, Proceedings of the 13th Conference Electrical Contacts, Lausanne, Switzerland, 1986, p. 383.

309. C. Fiaud, M. Safavi, and J. Vedel, Werkstoffe Korros., **35**, 361 (1984).

310. F. N. Rhines and W. A. Anderson, Trans. Am. Inst. Mining Met. Eng., **143**, 312 (1941).

311. N. Birks and G. H. Meier, Introduction to High Temperature Oxidation of Metals, E. Arnold Publishers, London, 1983.

312. R. A. Rapp (Ed.), Proceeding of the Conference High Temperature Corrosion, NACE, Houston, TX, 1983.

313. D. A. Shores and G. J. Yurek (Eds.), Fundamental Aspects of High Temperature Corrosion, Electrochemical Society, Pennington, NJ, 1986.

314. W. L. Worrell, High Temperature Corrosion, Academic, New York, 1987.

315. E. Garcia-Ochoa, J. González-Sánchez, F. Corvo, Z. Usagawa, L. Dzib-Pérez, and A. Castañeda, J. Appl. Electrochem., **38**, 1363 (2008).

316. S. Syed, Corros. Eng. Sci. Technol., **43**, 267 (2008).

317. F. Sarnie and J. Tidblad, Corros. Eng. Sci. Technol., **43**, 117 (2008).

318. R. Vera, D. Delgado, and B. M. Rosales, Corros. Sci., **50**, 1080 (2008).

319. H. Gil and C. Leygraf, J. Electrochem. Soc., **154**, C611 (2007).

320. F. Samie, J. Tidblad, V. Kucera, and C. Leygraf, Atmospher. Environ., **41**, 4888 (2007).

321. R. Vera, D. Delgado, and B. M. Rosales, Corros. Sci., **49**, 2329 (2007).

322. H. Gil and C. Leygraf, J. Electrochem, Soc., **154**, C272 (2007).

323. Y. C. Sica, E. D. Kenny, K. F. Portella, and D. F. Campos Filho, J. Braz. Chem. Soc., **18**, 153 (2007).

324. F. Samie, J. Tidblad, V. Kucera, and C. Leygraf, Atmospher. Environ., **41**, 1374 (2007).

325. G. A. El-Mahdy and K. B. Kim, Corrosion, **63**, 171 (2007).

326. K. P. Fitzgerald, J. Nairn, G. Skennerton, and A. Strens, Corros. Sci., **48**, 2480 (2006).

327. Z. Y. Chen, D. Persson, and C. Leygraf, J. Electrochem. Soc., **152**, B526 (2005).

328. F. Samie, J. Tidblad, V. Kucera, and C. Leygraf, Atmospher. Environ., **39**, 7362 (2005).

329. Z. Y. Chen, S. Zakipour, D. Persson, and C. Leygraf, Corrosion, **61**, 1022 (2005).

330. Z. Y. Chen, D. Persson, F. Samie, S. Zakipour, and C. Leygraf, J. Electrochem. Soc., **152**, B502 (2005).

331. Z. Y. Chen, D. Persson, A. Nazarov, S. Zakipour, D. Thierry, and C. Leygraf, J. Electrochem. Soc., **152**, B342 (2005).

332. S. Feliú, L. Mariaca, J. Simancas, J. A. González, and M. Morcillo, Corrosion, **61**, 627 (2005).

333. J. Tidblad, T. Aastrup, and C. Leygraf, J. Electrochem. Soc., **152**, B178 (2005).

334. G. A. El-Mahdy, Corros. Sci., **47**, 1370 (2005).

335. I. O. Wallinder, S. Bertling, and C. Leygraf, Metall., **58**, 717 (2004).

336. V. Hayez, A. Franquet, A. Hubin, and H. Terryan, Surf. Interf. Anal., **36**, 876 (2004).

337. A. R. Mendoza, F. Corvo, A. Gómez, and J. Gómez, Corros. Sci., **46**, 1189 (2004).

338. Z. Y. Chen, S. Zakipour, D. Persson, and C. Leygraf, Corrosion, **60**, 479 (2004).

339. I. T. E. Fonseca, R. Picciochi, M. H. Mendonça, and A. C. Ramos, Corros. Sci., **46**, 547 (2004).

340. T. T. M. Tran, C. Fiaud, E. M. M. Sutter, and A. Villanova, Corros. Sci., **45**, 2787 (2003).

341. R. Lobnig, J. D. Sinclair, M. Unger, and M. Stratmann, J. Electrochem. Soc., **150**, A835 (2003).

342. S. Feliu Jr., L. Mariaca, J. Simancas, J. A. González, and M. Morcillo, Rev. Metal. (Madrid), **39**, 279 (2003).

343. M. Morcillo, E. Almeida, M. Marrocos, and B. Rosales, Corrosion, **57**, 967 (2001).

344. T. Aastrup, M. Wadsak, C. Leygraf, and M. Schreiner, J. Electrochem. Soc., **147**, 2543 (2000).

345. M. Wadsak, M. Schreiner, T. Aastrup, and C. Leygraf, Surf. Sci., **454**, 246 (2000).

346. D. A. Lytle and M. A. Schock, J. Am. Water Works Assoc., **100**, 115 (2008).

347. S. Suzuki, Y. Yamada, K. Kawano, and T. Atsumi, Corros. Eng., **54**, 20 (2005).

348. O. Seri, Y. Jimbo, and M. Sakai, Corros. Eng., **55**, 505 (2006).

349. D. B. Harrison, D. M. Nicholas, and G. M. Evans, J. Am. Waters Works Assoc., **96**, 67 (2004).

350. T. Watanabe and S. Komukai, Corros. Eng., **47**, 320 (1998).

351. P. J. Bremer, B. J. Webster, and D. B. Wells, J. Am. Water Works Assoc., **93**, 82 (2001).

352. K. Habib and A. Husain, Desalination, **97**, 29 (1994).

353. Y. Kondo, ISIJ Int., **47**, 1309 (2007).

354. J. Liu and R. Zhang, J. Rare Earths, **24**, 253 (2006).

355. G. Plascensia and T. A. Utigard, Corros. Sci., **47**, 1149 (2005).

356. V. V. Prisedsky and V. M. Vinogradov, J. Solid State Chem., **177**, 4258 (2004).

357. M. Paljević and M. Tudja, Corros. Sci., **46**, 2055 (2004).

358. N. J. Tannyan, G. Plascencia, and T. A. Utigard, Can. Metall. Q., **41**, 213 (2002).

359. S. Akamatsu, T. Senuma, Y. Takada, and M. Hasebe, Mater. Sci. Technol., **15**, 1301 (1999).

360. N. Parkansky, B. Alterkop, S. Goldsmith, R. L. Boxman, Y. Rosenberg, and Z. Barkay, Surfa. Coatings Technol., **120–121**, 668 (1999).

361. U. V. Aniekwe and T. A. Utigard, Can. Metall. Q., **38**, 277 (1999).

362. R. Haugsrud and P. Kofstad, Mater. Sci. Forum, **251–254**, 65 (1997).

363. I.-J. Son, H. Nakano, S. Oue, S. Kobayashi, H. Fukushima, and Z. Horita, Mater. Trans., **49**, 2648 (2008).

364. E. A. Skrypnikova, S. A. Kaluzhina, and I. S. Bocharova, ECS Trans., **6**, 73 (2008).

365. A. Yakuki and M. Murakami, Corrosion, **63**, 249 (2007).

366. A. A. El Warraky, H. A. El Shayeb, and E. M. Sherif, Egyptian J. Chem., **47**, 657 (2004).

367. C. W. Keevil, Water Sci. Technol., **49**, 91 (2004).

368. A. A. El Warraky, H. A. El Shayeb, and E. M. Sherif, Anti-Corros. Methods Mater., **51**, 52 (2004).

369. W. Qafsaoui, Ch. Blanc, J. Roques, N. Pébère, A. Srhiri, C. Mijoule, and G. Mankowski, J. Appl. Electrochem., **31**, 223 (2001).

370. W. Qafsaoui, Ch. Blanc, N. Pébère, A. Srhiri, and G. Mankowski, J. Appl. Electrochem., **30**, 959 (2000).

371. M. Sosa, S. Patel, and M. Edwards, Corrosion, **55**, 1069 (1999).

372. S. Sathiyanarayanan, M. Sahre, and W. Kautek, Corros. Sci., **41**, 1899 (1999).

373. R. Nishimura and T. Yoshida, Corros. Sci., **50**, 1205 (2008).

374. A. Barnes, J. Deakin, and R. C. Newman, Corrosion, **63**, 416 (2007).

375. K. S. Ghosh, K. Das, and U. K. Chatterjee, Mater. Corros. **58**, 181 (2007).

376. H. M. Shalaby, A. Husain, A. A. Hasan, and A. Y. Abdullah, Corros. Eng. Sci. Technol., **42**, 64 (2007).

377. R. G. Song, W. Dietzel, B. J. Zhang, W. J. Liu, M. K. Tseng, and A. Atrens, Acta Mater., **52**, 4727 (2004).

378. D. C. Agarwal, Corros. Eng. Sci. Technol., **38**, 275 (2003).

379. S. Lu, J. Chen, and P. Lin, Corros. Sci. Prot. Technol., **14**, 267 (2002).

380. C.-S. Lee, Y. Choi, and I. G. Park, Met. Mater. Int., **8**, 191 (2002).

381. J. Oñoro and C. Ranninger, Mater. Sci., **35**, 509 (1999).

382. M. J. Robinson and N. C. Jackson, Corros. Sci., **41**, 1013 (1999).

383. M. Puiggali, A. Zielinski, J. M. Olive, E. Renauld, D. Desjardines, and M. Cid, Corros. Sci., **40**, 805(1998).

384. J. Rückert, Mater. Corros., **47**, 71 (1996).

385. J. Liu and I. Neretnieks, Radiochim. Acta, **92**, 849 (2004).

386. E. Zumelzu, C. Cabezas, R. Schöebitz, R. Ugarte, E. D. Rodriguez, and J. Rios, Can. Metall. Quart., **42**, 125 (2003).

387. J. Liu, L. Xu, and J. Zheng, J. Chin. Soc. Corros. Prot, **21**, 345 (2001).

388. R. Pope, B. Little, and R. Ray, Biofouling, **16**, 83 (2000).

389. M. de Romero, Corrosion, **56**, 867 (2000).

390. H. M. Shalaby, A. A. Hasan, and F. Al-Sabti, Br. Corros. J., **34**, 292 (1999).

391. I. B. Beech and C. C. Gaylarde, Rev. Microbiol., **30**, 177 (1999).

392. B. J. Little, J. S. Lee, and R. I. Ray, Electrochim. Acta, **54**, 2 (2008).

393. H. Wang and L. Chenghao, J. Wuhan Univ. Technology, Mater. Sci. Ed., **23**, 113 (2008).

394. Y.-L. Du, J. Li, L.-J. Cui, and Z.-J. Zhao, Corros. Sci. Prot. Technol., **19**, 401 (2007).

395. L. Jing, Z. Jiashen, and X. Liming, Mater. Corros., **52**, 833 (2001).

396. A. S. Khanna, Introduction to High Temperature Oxidation and Corrosion, ASM International, Materials Park, OH, 2002.

397. G. Y. Lai, High-Temperature Corrosion and Materials Applications, ASM International, Materials Park, OH, 2007.

398. M. Schütze (Ed.), Corrosion and Environmental Degradation, Vols. I and II, Wiley-VCH, Weinheim, 2000.

399. W. Gao (Ed.), Developments in High Temperature Corrosion and Protection of Materials, Woodhead Publishing, Cambridge, UK, 2008.

400. P. Steinmetz and I. G. Wright (Eds.), High Temperature Corrosion and Protection of Materials, Vol. 6, Trans. Tech. Publ. Ltd., Stafa-Zürich, 2004.

57

LEAD AND LEAD ALLOYS

F. E. GOODWIN

International Lead Zinc Research Organization, Inc., Research Triangle Park, North Carolina

A. INTRODUCTION

Lead is a malleable, heavy (50% more dense than steel) material that, despite its amphoteric nature, forms insoluble corrosion products under many conditions. This has made it useful in the chemical industry, particularly for its resistance to sulfuric acid, in batteries, and as sheet and pipe for building construction. It enjoys widespread use as solder in the electronics industry because of its low melting point. The low strength of lead often necessitates its support by wooden or steel structures for chemical process equipment or by use of lead clad steel.

There are four common types of lead: (1) pure lead (also called corroding lead) and (2) common lead, (3) chemical lead, and (4) acid–copper lead. The first two contain 99.94% minimum lead while the latter two both contain 99.9% minimum lead. These grades are covered in American Society for Testing and Materials (ASTM) B29. Higher purity (99.99%) is also available in commercial qualities.

Corroding lead refers not to a corrosion characteristic of this grade but rather to a process in which it was formerly used, the Dutch manufacturing process for producing white lead, a pigment. Corroding lead is nearly as corrosion resistant as chemical lead and is the most widely used grade of lead today.

Chemical lead takes its name from its widespread use in the chemical industry. Small amounts of copper and silver are contained, increasing corrosion resistance and mechanical strength. *Common lead* contains higher amounts of silver and bismuth than corroding lead and is used for battery oxide and general alloying. *Acid–copper lead*

Uhlig's Corrosion Handbook, Third Edition, Edited by R. Winston Revie
Copyright © 2011 John Wiley & Sons, Inc.

provides corrosion resistance comparable to that of chemical lead in most applications that require high corrosion resistance. This grade has a higher bismuth tolerance than chemical lead and is used in several types of fabricated lead products.

B. GENERAL CORROSION CHARACTERISTICS

The amphoteric nature of lead makes it susceptible to attack by both acids and alkalis under certain conditions, however, it exhibits excellent resistance in many environments because of the insolubility of its corrosion products that form self-healing protective films. In the electrolyte of a corrosive solution, lead enters the solution at anodic sites as metallic cations or is converted anodically to solid compounds. Both corrosion reactions are represented as

$$Pb \rightarrow Pb^{2+} + 2e^-$$

The standard potential for this reaction is 0.126 V. This oxidation reaction is accompanied by a reduction of some component of the electrolyte at cathodic sites. The rate of corrosion is a function of the current flowing between the anodes and cathodes of the corrosion cell. In the corrosion of a single metal such as lead, local anodes and cathodes may be set up as a result of inclusions, inhomogeneities, stress variations, and differences in temperature. In the case of galvanic corrosion, the different metals will still contain local anodes and cathodes but one of the metals will be more of an anode and corrode in preference to the more noble metal that becomes a cathode. In most environments lead is cathodic to steel, aluminum, zinc, cadmium, and magnesium and thus will accelerate corrosion of these metals. With titanium and passivated stainless steels, lead will be the anode in the cell and will suffer accelerated attack. The rate of corrosion in either case is governed by the difference in potential between the two metals, the ratio of their areas, and polarization characteristics.

Because of the protective films that are readily formed on the surface of lead, the corrosion rate of lead is usually under anodic control. The protective abilities of the different lead surface films vary with their solubility and the extent of mechanically disruptive influences such as agitation. Solubilities of common lead compounds are shown in Table 57.1. Corrosion rates can depend on other factors, for example, the corrosion rate of lead is quite high in weak nitric acid because of the solubility of the nitrate film. However, in concentrated nitric acid the solubility of lead nitrate is much lower, therefore, the corrosion rate of

TABLE 57.1.   Solubility of Lead Compounds[a]

Lead Compound	Formula	Temperature (°C)	Solubility (g/100 mL H_2O)
Acetate	$Pb(C_2H_3O_2)_2$	20	44.3
Bromide	$PbBr_2$	20	0.8441
Carbonate	$PbCO_3$	20	0.00011
Basic carbonate	$2PbCO_3, Pb(OH)_2$		Insoluble
Chlorate	$Pb(ClO_3)_2, H_2O$	18	151.3
Chloride	$PbCl_2$	20	0.99
Chromate	$PbCrO_4$	25	0.0000058
Fluoride	PbF_2	18	0.064
Hydroxide	$Pb(OH)_2$	18	0.0155
Iodide	PbI_2	18	0.063
Nitrate	$Pb(NO_3)_2$	18	56.5
Oxalate	PbC_2O_4	18	0.00016
Oxide	PbO	18	0.0017
Orthophosphate	$Pb_3(PO_4)_2$	18	0.000014
Sulfate	$PbSO_4$	25	0.00425
Sulfide	PbS	18	0.01244
Sulfite	$PbSO_3$		Insoluble

[a]See [1].

lead is also much lower. Other factors such as vibration, agitation, temperature, and aeration can also influence the stability of protective films on lead, increasing its corrosion rate.

The form of corrosion that occurs in each situation also depends on several variables. Lead exposed to normal atmospheric attack will corrode uniformly, however, pitting can occur under conditions of partial passivity or cavitation. Accommodation of corrosion or other conditions such as erosion, fatigue, or fretting can cause more severe damage than any of the individual factors. When lead is placed under continuous load it will creep, continually exposing fresh surface to the corroding environment and therefore increasing corrosion rate.

Lead is generally resistant to sulfuric, sulfurous, chromic, and phosphoric acids along with cold hydrofluoric acid. It is attacked by hydrochloric acid and nitric acid and also organic acids if they are dilute or if they contain oxidizing agents. In caustic solutions, the use of lead is generally limited to concentrations of no more than 10% up to 90°C (195°F). Very little attack is seen in cold, strong amines, however, dilute aqueous solutions will result in attack.

The formation of passive films on the lead surface may make it initially anodic to highly corrosion-resistant nickel-based alloys, but accelerated attack of such alloys can occur after the passive film forms on lead, if the nickel alloy is not electrically insulated from the lead.

B1. Water

The corrosion resistance of lead in water depends on factors that create or destroy the protective lead surface film. Pure water, free from dissolved gases, will not attack lead. Corrosion is dependent on the presence of gases such as oxygen and carbon dioxide, solid impurities such as silicates, carbonates, sulfates, lime, silt, and chlorides, and microorganisms. Agitation and flow also affect lead corrosion.

B2. Natural Water

The corrosion rate of lead in natural water will generally depend on the hardness of the water as shown in Table 57.2. Natural waters of moderate hardness (>125 ppm calcium carbonate) form adequate protective films on lead and therefore attack is negligible. The presence of salts such as silicates increase the hardness and film thickness. Nitrates interfere with protective film formation, causing increased corrosion.

In soft aerated waters, corrosion depends on both hardness and oxygen content. With hardness levels of <125 ppm, natural water behaves more like distilled water, as described below. Therefore, soft waters are usually not suitable for handling such water if potability is required. Drinking water containing >0.015 ppm of lead is generally considered contaminated and unsafe for consumption. Such contamination can be caused even in situations where from a service point of view the corrosion rate of lead is negligible, and can include the use of lead solders in copper pipe.

Corrosion can be accelerated if carbonic acid is present in water, converting calcium carbonate deposits into soluble calcium bicarbonate. Also the presence of organic acids whose lead salts are soluble can promote corrosion. Film-forming lime or sodium silicate can be added to water to lower corrosion rate.

B3. Distilled Water

Distilled water free of oxygen and carbon dioxide does not attack lead. Also such water containing carbon dioxide but no oxygen also has little effect on lead. It is only when lead comes in contact with pure water through which air free of carbon dioxide is bubbled that it quickly oxidizes to form a film of white lead hydroxide. This nonadherent film allows the attack on lead to continue. A yellow crystalline lead oxide forms on the lead surface at or near the water line.

If carbon dioxide is present in addition to oxygen, a basic lead carbonate film can form, protecting lead from further attack. However, once a certain ratio of carbon dioxide to oxygen is reached in the gas, further increases in carbon dioxide level cause the insoluble lead carbonate film to convert to soluble lead bicarbonate, the protective film thus dissolves and corrosion rates increase. The effect of oxygen, with no carbon dioxide present in the gas, on the corrosion of lead in distilled water is shown in Figure 57.1.

Lead pipe coils for steam boiler applications in which a pure water condensate is used give very good service with steam up to 3.4 bars (50 psi) gauge pressure. This is because all condensate is returned to the boiler with negligible use of makeup water; therefore, there is an absence of oxygen and often, carbon dioxide. Lead does not significantly corrode in this application, unless condensate is discarded and fresh feedwater added to the boiler. However, the use of oxygen scavengers such as sodium sulfite or hydrazine can limit such attack.

B4. Demineralized Water

The corrosion rates of chemical lead, Pb–6% Sb and Pb–2% Sn in demineralized water are shown in Table 57.3. These rates were measured after 21 days of immersion. Corrosion rates are observed to be very low; however, the chemical lead did show some waterline attack.

TABLE 57.2. Corrosion of Chemical Lead in Industrial and Domestic Waters[a]

Kind of Water	Temperature		Aeration	Agitation	Corrosion Rate	
	F°	C°			mdd[b]	mpy[c]
Condensed steam, traces of acid	70–100	21–38	None	Slow	6.75	0.85
Mine water, pH 8.3, 110 ppm hardness	68	20	Yes	Slow	2.08	0.26
Mine water, 160 ppm hardness	67	19	Yes	Slow	2.20	0.28
Mine water, 110 ppm hardness	72	22	Yes	Slow	1.98	0.25
Cooling tower, oxygenated Lake Erie water	60–85	16–29	Complete	None	41.7	5.3
Los Angeles aqueduct water, treated by chlorination and copper sulfate	Ambient			0.15 m/s (0.5 ft/s)	2.95	0.38
Spray cooling water, chromate treated	60	16	Yes		2.9	0.37

[a]See [1].
[b]In milligrams/square decimeter/day (mg/dm^2/day).
[c]In mils/year.

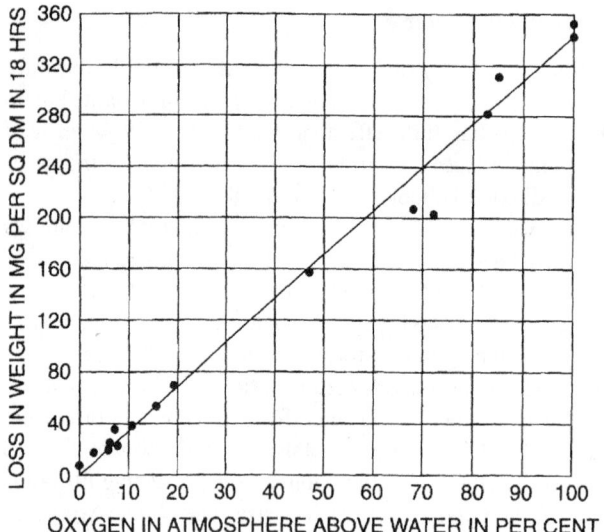

FIGURE 57.1. Effect of oxygen on corrosion of lead submerged in distilled water at 25°C. [Reprinted with permission from H. H. Uhlig (Ed.), *Corrosion Handbook*, Wiley, New York, 1948.]

TABLE 57.3. Corrosion of Lead in Demineralized Water[a]

Lead	mpy	mdd
Chemical lead (ASTM B29)	2.3	18
6% Antimonial lead	0.2	1.6
2% Tin–lead	0.6	4.8

[a]See [1].

B5. Seawater

The corrosion of lead in seawater is relatively low and may be slowed by encrustations of lead salts. Seawater normally has a salt content of ~3–4%. Other factors influencing corrosion

may be pollution, rate of flow, wave action, sand or silt content, temperature, and marine growth. Insufficient data have been produced to reveal the effects of these factors. A summary of the effects of alloying of lead on corrosion rate is shown in Table 57.4.

C. CORROSION IN CHEMICAL MEDIA

C1. Effect of pH on Corrosion

As illustrated in Figure 57.2, the corrosion of lead is accelerated by both acids and alkalis because of its amphoteric nature. Significant corrosion can occur < pH 5 or > pH 10.

C2. Inorganic Acids

The acids to which lead offered good resistance were listed earlier. Lead of 3 mm ($\frac{1}{8}$ in) of thickness or greater usually gives years of service in equipment handling these chemicals. High acid velocities may increase the rate of corrosion. Figure 57.3 shows the effect of velocity of sulfuric acid across a lead base at 25°C.

Lead satisfactorily resists all but the most dilute strengths of sulfuric acid. It performs well with concentrations up to 95% at ambient temperatures, up to 85% at 220°C, and up to 93% at 150°C. Below a concentration of 5%, corrosion rates increase but are still relatively low. At lower concentrations, Pb–Sb alloys are recommended. Similar behavior patterns occur with chromic, sulfurous, and phosphoric acids. The behavior of lead in sulfuric acid is shown in Figure 57.4 and its behavior with air-free hydrofluoric acid in Figure 57.5. Tabular data for the behavior of lead with phosphoric, hydrochloric, and a hydrochloric–ferric–chloride mixture is tabulated in Tables 57.5–57.7. Most concentrations of nitric, acetic, and formic acids corrode lead at a rate high enough to prevent its use; however, nitric acids of 52–70% do not

TABLE 57.4. Corrosion of Lead and Lead Alloys in Seawater[a]

Metal	Days	Depth m	Thickness loss mpy	Weight loss mdd	Pit Depth	Remarks
Chemical lead (ASTM B29)	181	1.5	1.2	9.4	0	Uniform attack
	197	713	0.3	2.4	0	Uniform attack
	123	1719	0.8	6.3	0	Uniform attack
Pb–0.04% Te	181	1.5	1.0	7.9	0	Uniform attack
	197	713	0.3	2.4	0	Uniform attack
	123	1719	1.1	8.7	0	Uniform attack
Pb–6% Sb	181	1.5	1.2	9.4	0	Uniform attack
	197	713	0.3	2.4	0	Uniform attack
	123	1719	0.8	6.3	0	Uniform attack
Pb–33% Sn	181	1.5	3.7	29.1	0	Uniform attack
	197	713	0.5	3.9	0	Uniform attack
	123	1719	0.5	3.9	0	Uniform attack

[a]See [2].

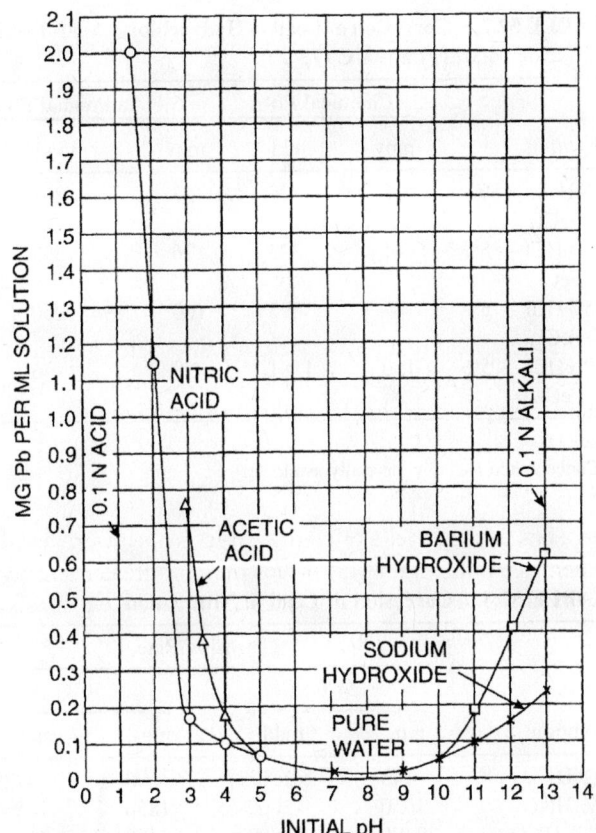

FIGURE 57.2. Corrosion of lead in contact with various acid and alkali solutions. (Reproduced with permission from [2].)

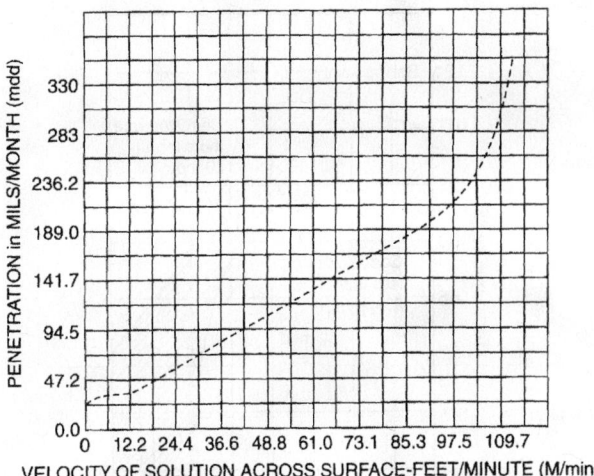

FIGURE 57.3. Effect of velocity across metal face on the corrosion of lead in 20% sulfuric acid at 25°C. (Adapted from [1], used with permission.)

produce significant attack, while dilute concentrations give rapid attack. This pattern is also true for hydrofluoric, acetic, and sodium sulfate acids. Adding sulfuric acid to acids corrosive to lead will often lower the corrosion rate. For example, even though nitric acid of <50% strength is quite corrosive to lead in the presence of 54% sulfuric acid, the corrosion rate for 1 and 5% nitric acid is quite low even at 118°C (245°F). Chemical-grade lead is preferred over Pb–6% Sb for handling such mixed acids. Similar behavior is exhibited in hydrochloric acid when sulfuric acid is present. These data are summarized in Tables 57.8–57.11.

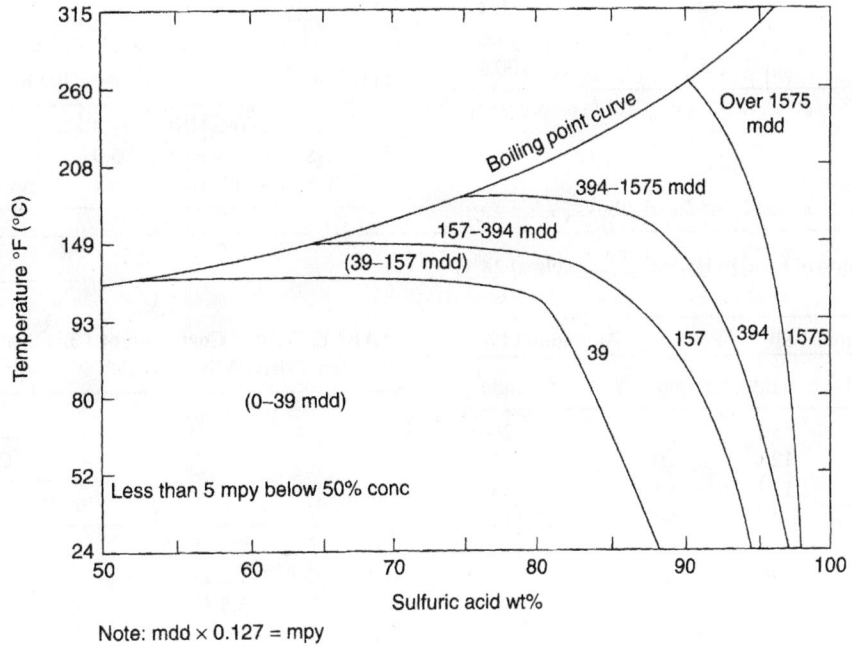

Note: mdd × 0.127 = mpy

FIGURE 57.4. Corrosion rate of lead in sulfuric acid. (Adapted from [1], used with permission.)

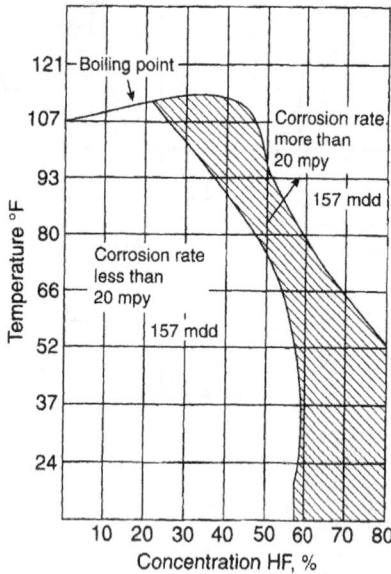

FIGURE 57.5. Corrosion resistance of lead in air-free hydrofluoric acid. (From [1], reproduced with permission.)

TABLE 57.5.   Corrosion of Chemical Lead in Phosphoric Acid at 21°C[a]

	Corrosion Rate	
Solution	mpy	mdd
20% H$_3$PO$_4$ (com'l)	3.4	26.8
30% H$_3$PO$_4$ (com'l)	4.9	38.6
40% H$_3$PO$_4$ (com'l)	5.7	44.9
50% H$_3$PO$_4$ (com'l)	6.4	50.4
85% H$_3$PO$_4$ (com'l)	1.6	12.6
80% H$_3$PO$_4$ (pure)	12.8	100.8

[a]See [1].

TABLE 57.6.   Corrosion of Lead in Hydrochloric Acid at 24°C (75°F)[a]

	Chemical Pb		6% Antimonial Pb	
Solution	mpy	mdd	mpy	mdd
1% HCl	24	189	33	260
5% HCl	16	126	20	158
10% HCl	22	173	43	339
15% HCl	31	244	150	1181
20% HCl	74	583	160	1260
25% HCl[b]	190	1496	200	1575
35% HCl	350	2755	540	1452

[a]See [1].
[b]Concentrated HCl commercially available.

TABLE 57.7.   Corrosion of Lead in Hydrochloric Acid–Ferric Chloride Mixtures at 24°C (75°F)[a]

	Chemical Pb		6% Antimonial Pb	
Solution	mpy	mdd	mpy	mdd
5%HCl + 5% FeCl$_3$	28	220	37	291
10%HCl + 5% FeCl$_3$	41	323	76	598
15%HCl + 5% FeCl$_3$	88	693	160	1260
20%HCl + 5% FeCl$_3$[b]	150	1181	190	1496

[a]See [1].
[b]Concentrated HCl commercially available.

TABLE 57.8.   Corrosion of Lead in Nitric Acid

	Corrosion Rate			
	24°C		50°C	
Solution	mpy	mdd	mpy	mdd
1% HNO$_3$	140	1102	600	4725
5% HNO$_3$	1650	13000	1850	14566
10% HNO$_3$	3400	26800	3490	27500

TABLE 57.9.   Effect of Nitric Acid in Sulfuric Acid on the Corrosion of Lead at 118°C[a]

	Chemical Pb		6% Antimonial Pb	
Solution	mpy	mdd	mpy	mdd
54% H$_2$SO$_4$ + 0% HNO$_3$	7.4	58.2	14	110.2
54% H$_2$SO$_4$ + 1% HNO$_3$	5.9	46.5	22	173.2
54% H$_2$SO$_4$ + 5% HNO$_3$	8.4	66.1	114	897.6

[a]See [1].

TABLE 57.10.   Corrosion of Chemical Lead with Sulfuric–Nitric Mixed Acids[a]

	Corrosion Rate			
	24°C		50°C	
Solution	mpy	mdd	mpy	mdd
78% H$_2$SO$_4$ + 0% HNO$_3$	1	8	2	16
78% H$_2$SO$_4$ + 1% HNO$_3$	3	24	12	95
78% H$_2$SO$_4$ + 3.5% HNO$_3$	3.6	28.3	18	142
78% H$_2$SO$_4$ + 7.5% HNO$_3$	4	32	35	276

[a]See [1].

TABLE 57.11.   Corrosion of Lead in Hydrochloric Acid–Sulfuric Acid Mixtures[a]

Solution	Chemical Lead				6% Antimonial Lead			
	(24°C)		(66°C)		(24°C)		(66°C)	
	mpy	mdd	mpy	mdd	mpy	mdd	mpy	mdd
1% HCl + 9% H_2SO_4	5	39	9	71	5	39	12	95
3% HCl + 7% H_2SO_4	14	110	32	252	21	165	41	323
5% HCl + 5% H_2SO_4	14	110	42	331	21	165	65	512
7% HCl + 3% H_2SO_4	16	126	45	355	22	173	74	583
9% HCl + 1% H_2SO_4	18	142	47	370	30	236	84	661
5% HCl + 25% H_2SO_4	10	79	22	173	22	173	34	268
10% HCl + 20% H_2SO_4	17	134	42	331	80	630	58	465
15% HCl + 15% H_2SO_4	41	323	74	529	90	709	180	1417
20% HCl + 10% H_2SO_4	86	677	120	945	110	866	180	1417
25% HCl + 5% H_2SO4	140	110	160	1260	150	1181	210	1653
5% HCl + 45% H_2SO_4	62	488			53	417		
10% HCl + 40% H_2SO_4	65	511			84	661		
15% HCl + 35% H_2SO_4	66	520			120	945		
20% HCl + 30% H_2SO_4	84	661			130	1024		
25% HCl + 25% H_2SO_4	120	945			210	1654		

[a]See [1].

C3. Organic Acids

The principal organic acids to which lead is corrosion resistant are acetic (concentrated), chloroacetic (limited use), fatty acids (but only in absence of oxygen), hydrocyantic (not in pure acid but with sulfuric acid), or an aqueous solution with cyanide and oxalic and tartaric acids (in the absence of oxygen). Lead is rapidly corroded by dilute acetic or by formic acid in the presence of oxygen. Organic acids harmful to lead can be leached from wood structures, particularly damp timber including western cedar, oak, and Douglas fir. Such corrosion can be prevented by fully drying the wood, treating it to prevent contact with moist air, or inserting a moisture barrier between wood and lead.

C4. Alkalis

The action of strong alkalis on lead is not as rapid as by acids like hydrochloric or acetic but is greater than the action of alkalis on iron. For certain purposes, however, corrosion of lead in contact with NaOH or KOH tip to 30% concentration at 25°C and up to 10% concentration at higher temperatures (90°C) is tolerable. For example, lead has proven useful in the refining of petroleum oil, where a sulfuric acid treatment is followed by an alkaline solution treatment in the same lead-lined tank.

Although $Ca(OH)_2$ (lime) solutions are saturated at ~0.1% $Ca(OH)_2$ at 25°C, they have been found to corrode lead severely when trickling over the surface. This can occur with fresh concrete or with long time seepage of water through cured concrete. In such cases, dissolved oxygen appears to be necessary for corrosion.

C5. Summary of Categorical Data

Categorical data for the corrosion rate of lead in chemical environments are shown in Table 57.12. Four categories of corrosion are shown

A: 51 μm/year (2 mils/year)	Negligible corrosion, lead recommended for use.
B: 510 μm/year (20 mils/year)	Practically resistant, lead recommended for use.
C: 510–1275 μm/year (20–50 mils/year)	Lead may be used for this effect on service life can be tolerated.
D: >1275 μm/year (50 mils/year)	Corrosion rate too high to merit any consideration of lead.

C6. Salts

Lead is generally resistant to corresponding salts of the acids to which it is resistant Nitrates and alkaline salts tend to be actively corrosive.

C7. Gases

Lead resists the actions of chlorine, wet or dry, up to 100°C, but that of bromine only when dry and at lower temperatures. Sulfur dioxide and trioxide, wet or dry, are frequently handled in lead pipes. Hydrogen sulfide, with or without moisture, can be handled with lead providing erosion at high velocities is avoided. Hydrogen fluoride is actively corrosive, therefore, lead is not recommended as a container for this gas.

TABLE 57.12. Corrosion Rate of Lead in Chemical Environments[a]

Chemical	Temp. (°C)	Concentration (%)[b]	Corrosion Class
Abietic acid	24	—	D
Acetaldehyde	24	—	A
Acetaldehyde	24–100	—	B
Acetanilide	24	—	A
Acetic acid	24	Glacial	B
Acetic acid	24	—	C
Acetic anhydride	24	—	A
Acetoacetic acid	24	—	B
Acetone	24–100	10–90	A
Acetone cyanohydrin	24–100	—	B
Acetophenetidine	24	—	B
Acetophenone	24–100	—	B
Acetotoluidide	24	—	B
Acetyl acetone	24–100	—	B
Acetyl chloride	24	—	A
Acetyl thiophene	24–100	—	B
Acetylene, dry	24	—	A
Acetylene tetrachloride	21	Liquid	B
Acridine	24–106	10	B
Acrolein	24–106	10	B
Acrylonitrile	24–100	—	A
Adipic acid	24–100	—	A
Alcohol, ethyl	24–100	10–100	A
Alcohol, methyl	24–100	10–100	A
Alkanesulfonic acid	24	—	D
Alkyl aryl sulfonates	24–100	—	B
Alkyl naphthalene sulfonic acid	93	—	C
Allyl alcohol	24	—	B
Allyl chloride	24	—	C
Allyl sulfide	24	—	D
Aluminum acetate	24–100	10–20	A
Aluminum chlorate	24–100	—	B
Aluminum chloride	24	0–10	B
Aluminum ethylate	24–100	—	B
Aluminum fluoride	24–100	10–20	B
Aluminum fluorosulfate	24	15	A
Aluminum fluosilicate	24	—	B
Aluminum formate	24	—	B
Aluminum formate	100	—	D
Aluminum hydroxide	24–100	10	B
Aluminum nitrate	24	10	B
Aluminum potassium sulfate	24–100	10–20	A
Aluminum potassium sulfate	24–100	20–100	B
Aluminum sodium sulfate	24–100	10	B
Aluminum sulfate	24–100	—	A
Aminoazobenzene	24	—	C
Aminobenzene sulfonic acid	24–100	—	B
Aminobenzoic acid	24–93	—	B
Ammophenol	24–100	—	B
Aminosalicylic acid	100–149	—	C
Ammonia	24–100	10–30	B
Ammonium acetate	25	3.85	B
Ammonium azide	24	—	B
Ammonium bicarbonate	24–100	10	B

TABLE 57.12. (*Continued*)

Chemical	Temp. (°C)	Concentration (%)^b	Corrosion Class
Ammonium bifluoride	24	10	B
Ammonium bisulfate	24–100	—	A
Ammonium carbamate	24–149	—	A
Ammonium carbonate	24–100	10	B
Ammonium chloride	24	0–10	B
Ammonium citrate	100	—	D
Ammonium diphosphate	24–100	10	B
Ammonium fluoride	24	0–20	B
Ammonium fluosilicate	24–52	20	B
Ammonium formate	100	10	C
Ammonium hydroxide	27	3.5–40	A
Ammonium hydroxylamine	20–100	34	B
Ammonium metaphosphate	24	10	B
Ammonium nitrate	24–100	10–30	D
Ammonium oxalate	24	10–30	D
Ammonium persulfate	24–100	10–30	B
Ammonium phosphate	66	—	A
Ammonium picrate	24–100	10	B
Ammonium polysulfide	24–100	10	B
Ammonium sulfamate	24–100	10	B
Ammonium sulfate	24	—	B
Ammonium sulfide	24–100	10	C
Ammonium sulfite	24–100	10–40	B
Ammonium thiocyanate + NH_4OH	24	—	A
Ammonium tungstate	24	10	D
Amyl acetate	24	80–100	B
Amyl chloride	24	—	D
Amyl laurate	24–100	—	B
Amyl phenol	200	—	D
Amyl propionate	24–100	—	B
Aniline	20	—	A
Aniline hydrochloride	24	10	D
Aniline sulfate	24–100	—	B
Aniline sulfite	24–100	—	B
Anthracene	24–100	—	B
Anthraquinone	24–100	—	B
Anthraquinone sulfonic acid	24–100	10–30	B
Antimony chloride	24	—	C
Antimony pentachloride	24–100	90–100	B
Antimony sulfate	100	—	C
Antimony trifluoride	24–100	50–70	A
Arabic acid	24–100	—	B
Arachidic acid	24	—	B
Arsenic acid	24	10	B
Arsenic trichloride	24–100	—	B
Arsenic trioxide	24–100	—	D
Ascorbic acid	24	—	B
Azobenzene	24–100	—	B
Barium carbonate	24	—	D
Barium chlorate	24–100	20	B
Barium chloride	24–100	10	B
Barium cyanide	24	10–70	D
Barium hydroxide	24	10	D
Barium nitrate	24–100	10–30	B

(continued)

TABLE 57.12. *(Continued)*

Chemical	Temp. (°C)	Concentration (%)[b]	Corrosion Class
Barium peroxide	24	10	D
Barium polysulfide	100	—	D
Barium sulfate	24–100	—	B
Barium sulfide	24	10	B
Benzaldehyde	24	10–100	D
Benzaldehyde sulfonic acid	24–100	—	B
Benzamide	24–100	—	B
Benzanthrone	24–100	—	B
Benzene	24	—	B
Benzene hexachloride	24	—	B
Benzene sulfonic acid	24	10–100	B
Benzene sulfonic acid	100	—	D
Benzidine	100	—	B
Benzidine disulfonic acid 2.2	24–100	—	B
Benzidine 3 sulfonic acid	24–100	—	B
Benzilic acid	24–100	10–100	B
Benzobenzoic acid	24–100	—	B
Benzocathecol	24–100	—	B
Benzoic acid	24	—	D
Benzol	24	100	A
Benzonitrile	24–100	—	A
Benzophenone	24–100	—	A
Benzotrichloride	24–100	—	B
Benzotrifluoride	24–100	—	B
Benzoyl chloride	100	—	C
Benzoyl peroxide	24–100	—	B
Benzyl acetate	24–100	—	B
Benzyl alcohol	24–100	—	B
Benzylbutyl phthalate	24–100	—	B
Benzyl cellulose	24–100	—	B
Benzyl chloride	24–100	—	B
Benzyl ethyl aniline	24–100	—	B
Benzylphenol	24–100	—	B
Benzylphenol salicylate	24–100	—	B
Benzylsulfonilic acid	24–100	—	B
Beryllium chloride	100	—	D
Beryllium fluoride	24–100	—	B
Beryllium sulfate	24–100	10–50	B
Boric acid	24–149	10–100	B
Bornyl acetate	24–100	—	B
Bornyl chloride	24–100	—	B
Bornyl formate	24–100	—	B
Boron trichloride	24–100	—	B
Boron trifluoride	24–204	—	A
Bromic acid	24–100	10–30	B
Bromine	24	—	B
Bromobenzene	24–100	—	B
Bromoform	24–100	—	B
Butane	24	—	A
Butanediols	24	—	B
Butyl acetate	24	—	B
Butyl benzoate	24–100	—	B
Butyl butyrate	24–100	—	B
Butyl glycolate	24–100	—	B

TABLE 57.12.   (*Continued*)

Chemical	Temp. (°C)	Concentration (%)[b]	Corrosion Class
Butyl mercaptan	24	—	C
Butyl oxalate	24	—	B
Butyl phenols	24	—	C
Butyl phthalates	24–100	—	B
Butyl stearate	24–100	—	B
Butyl urethane	24–100	—	B
Butyric acid	24	10–100	D
Butyric aldehydes	24–100	—	B
Butyrolactone	24–100	—	B
Cadmium cyanide	24	—	D
Cadmium sulfate	24–100	10–30	A
Calcium acetate	24	20	B
Calcium acid phosphate	24	10–30	B
Calcium benzoate	24–100	—	B
Calcium bicarbonate	24	—	C
Calcium bisulfite	24	—	B
Calcium bromide	24–100	30	B
Calcium carbonate	24	20	D
Calcium chlorate	24	10–30	B
Calcium chloride	24	20	A
Calcium chromate	24–100	10	B
Calcium dihydrogen sulfite + SO_2	24	5	A
Calcium disulfide	24	—	B
Calcium fluoride	24–100	—	B
Calcium gluconate	24–100	—	B
Calcium hydroxide	24	10	D
Calcium lactate	100	10	B
Calcium nitrate	24	10	D
Calcium oxalate	24–100	10	B
Calcium phosphate	100	10	B
Calcium pyridine sulfonate + H_2SO_4	24	20	A
Calcium stearate	24–100	10	B
Calcium sulfaminate	24–100	—	A
Calcium sulfate	24–100	10	B
Calcium sulfide	100	—	C
Calcium sulfite	24–100	10	B
Camphene	24–100	—	B
Camphor	100	—	A
Camphor sulfonic acid	24	20–100	C
Capric acid	24–100	—	B
Caprolactone	24–100	—	B
Capronaldehyde	24	—	A
Capronaldehyde	52–100	—	B
Carbazole	25–100	—	B
Carbitol	25–100	—	B
Carbon disulfide	25–100	—	A
Carbon fluorides	25–100	—	B
Carbon tetrabromide	100	—	C
Carbon tetrachloride (dry)	[b]	100	A
Carbonic acid	24	—	D
Carnallite	24–100	—	A
Carotene	24–100	—	A
Cellosolves	24–100	—	A
Cellulose acetate	24	—	A

(continued)

TABLE 57.12. (*Continued*)

Chemical	Temp. (°C)	Concentration (%)[b]	Corrosion Class
Cellulose acetobutyrate	24–100	—	B
Cellulose nitrate	24–100	—	B
Cellulose tripropionate	24–100	—	B
Cerium fluoride	24–100	—	B
Cerium sulfate	100	—	C
Cesium chloride	24–100	—	B
Cesium hydroxide	24	10	D
Cetyl alcohol	24	—	B
Cetyl alcohol	100	—	C
Chloroacetic acid	24	—	B
Chloral	24–100	—	B
Chloramine	24	10–30	B
Chloranil	24–100	—	B
Chloranthraquinone	24–100	—	B
Chlordane	24–100	—	B
Chlorethane sulfonic acid	100	—	C
Chloric acid	24	10	D
Chlorine	37	—	B
Chlorine dioxide	6	Gas	B
Chloroacetaldehyde	24	—	B
Chloroacetone	24–100	—	B
Chloroacetyl chloride	24	—	B
Chloro–alkyl ethers	24–100	—	B
Chloroaminobenzoic acid	24–100	—	B
Chloroaniline	24–100	—	B
Chlorobenzene + SO_2	18	—	A
Chlorobenzotrifluoride	24–100	—	B
Chlorobenzoyl chloride	24–100	—	B
Chlorobromomethane	24	—	B
Chlorobromopropane	24–100	—	B
Chlorobutane	24	—	B
Chloroethylbenzene	24–100	—	B
Chloroform	24–BP[c]	—	B
Chlorohydrin	24–100	—	B
Chloromethonic ester	24–100	—	B
Chloronaphthalene	24–100	—	B
Chloronitrobenzene	24–100	—	B
Chlorophenohydroxy acetic acid	24–100	—	B
Chlorophenol	24	—	C
Chloroquinine	24	—	C
Chlorosilanes	24–100	—	B
Chlorosulfonic acid	24	—	C
Chlorosulfomc acid + 50% SO_3	14	40	C
Chlorotoluene	24–100	—	B
Chlorotoluene sulfonic acid	24	—	C
Chlorotoluidine	24–100	—	B
Chlorotrifluoro ethylene	24–100	—	B
Chloroxylenols	24	—	C
Chloroxylols	24–100	—	B
Cholesterol	24–100	—	B
Chromic acid	24	—	B
Chromic chloride	24–100	—	B
Chromic fluoride	24–100	—	B
Chromic hydroxide	24–100	—	B

TABLE 57.12.   (*Continued*)

Chemical	Temp. (°C)	Concentration (%)^b	Corrosion Class
Chromic phosphate	24–100	—	B
Chromic sulfate	24–100	10	B
Chromium potassium sulfate	24–100	10	B
Chromium sulfate (basic)	24–100	20–50	B
Chromyl chlorides	24–100	—	B
Critic acid	24–80	10–30	B
Citric acid	24	50–100	D
Cobalt sulfate	24	10–30	B
Copper chloride	24	10–40	D
Copper sulfate	24–100	10–70	B
m-cresol + 10% water	25	Liquid	B
m-cresol + 10% water	BP^c	Vapor	D
o-cresol + 10% water	25	Liquid	B
o-cresol + 10% water	BP^c	Vapor	D
Cresote	24	90	D
Cresylic acid	24	90	D
Cresylic acid	24	100	B
Crotonaldehyde	24–100	—	B
Crotonic acid	24	—	D
Cumaldehyde	24–100	—	B
Cumene	24–100	—	B
Cumene hydroperoxide	24–100	—	D
Cyanamide	24–100	—	B
Cyanoacetic acid	24	—	D
Cyanogen gas	24	—	D
Cyclohexane	24	—	B
Cyclohexanol	24	—	B
Cyclohexanol esters	24–100	—	B
Cyclohexanone	24	—	B
Cyclohexene	24–100	—	B
Cyclohexylamine	24	—	D
Cyclopentane	24–100	—	B
DDT	24	—	B
Dialkyl sulfates	24–100	—	B
Dibenzyl	24–100	—	B
Dibutyl phthalate	24–100	—	B
Dibutyl thioglycolate	24–100	—	B
Dibutyl thiourea	24–100	—	B
Dichlorobenzene	24–100	10–100	B
Dichlorodifluoro-methane–(Freon 12)	24–100	90	A
Dichlorodiphenyldichloroethane (DDD)	24–100	—	B
Dichloroethylene	24–100	—	A
Diethanolamine	24	—	B
Diethyl ether	24	—	B
Diethylamine	24	—	D
Diethylaniline	24–100	—	B
Dielhylene glycol	24–52	—	B
Difluoroethane	24–100	—	B
Diglycolic acid	24	—	D
Dihydroxydiphenylsulfone	24–100	—	B
Diisobutyl	24–100	—	B
Dimethyl ether	24–100	—	B
Dioxane	24–100	—	B
Diphenyl	24–100	—	B

(*continued*)

TABLE 57.12.   (*Continued*)

Chemical	Temp. (°C)	Concentration (%)^b	Corrosion Class
Diphenyl chloride	24–100	—	B
Diphenylamine	24–100	—	A
Diphenylene oxide	24–100	—	B
Diphenylpropane	24–100	—	B
Epichlorohydrin	24	—	A
Ethane	24–100	—	A
Ether	24	—	B
Ethyl acetate	24–80	—	B
Ethyl benzene	24–100	—	B
Ethyl butyrate	24–100	—	B
Ethyl cellulose	24–100	—	B
Ethyl chloride	24–100	—	B
Ethyl ether	24–100	—	B
Ethyl formate	100	—	C
Ethyl lactate	24–100	—	B
Ethyl mercaptan	100	—	D
Ethyl stearate	24–100	—	B
Ethyl sulfonic acid	24	—	B
Ethyl sulfonic acid	100	—	C
Ethylene	24–100	—	A
Ethylene bromide	100	—	B
Ethylene chlorohydrin	24	90	A
Ethylene chlorohydrin	52	100	B
Ethylene cyanohydrin	24	—	A
Ethylene cyanohydrin	52–100	100	B
Ethylene dibromide	24	90	D
Ethylene dichloride	25–100	—	B
Ethylene glycol	–7	50	B
Ethylene oxide	24	—	B
2-Ethylhexoic acid	71	96	C
Ferric ammonium sulfate	24–100	10–20	A
Ferric chloride	24	20–30	D
Ferric ferrocyanide	66–93	—	A
Ferric sulfate	25–80	10–20	A
Ferrous ammonium sulfate	24	10	B
Ferrous chloride	24	10–30	C
Ferrous sulfate	24–100	10	B
Fluoboric acid	24	30	C
Fluocarboxylic acid	24	—	D
Fluorine	24–100	—	A
Fluosilicic acid	45	10	D
Formaldehyde	24–52	20–100	B
Formamide	24–100	—	B
Formic acid	24–100	10–100	D
Furfural	24–100	—	B
Gluconic acid	24	10–100	B
Glutamic acid	24	—	D
Glycerol	24	—	B
Glycerophosphoric acid	24	—	B
Glycol monoether	24–100	—	B
Glycolic acid	24	10–100	B
Glycolic acid	100	10	D
Heptachlorobutene	24	—	B
Heptane	24–100	—	A

TABLE 57.12.   (*Continued*)

Chemical	Temp. (°C)	Concentration (%)[b]	Corrosion Class
Hexachlorobutadiene	24–100	—	A
Hexachlorobutene	24	—	B
Hexachloroethane	24–100	—	B
Hexamethylene tetramine	24–100	10–40	B
Hydrazine	24–100	20–100	D
Hydriodic acid	24–100	10–50	D
Hydrobromic acid	24	10–70	D
Hydrochloric acid	24	0–10	C
Hydrofluoric acid	24	2–10	B
Hydrogen bromide (Anh HBr)	100	—	D
Hydrogen chloride (Anh HCl)	24	100	A
Hydrogen peroxide	24	10–30	D
Hydrogen sulfide	24	90–100	B
Hydroquinine	24–100	10	B
Hydroxyacetic acid	24	—	A
Hypochlorous acid	24	—	D
Iodine	24	—	D
Iodoform	24–100	10	B
Isobutyl chloride	24	—	B
Isobutyl phosphate	24	—	B
Isopropanol	24	—	A
Lactic acid	24	10–100	D
Lead acetate	24	10–30	D
Lead arsenate	24–100	—	B
Lead azide	24–100	—	B
Lead chloride	24–100	—	B
Lead chromate	24–100	—	B
Lead dioxide	24–100	—	B
Lead nitrate	24–100	—	B
Lead oxide	24–100	—	B
Lead peroxide	24–100	—	B
Lead sulfate	24–100	—	B
Lithium chloride	24–100	10	B
Lithium hydroxide	24	—	D
Lithium hypochlorite	24–80	10	A
Lithopone	24	—	A
Magnesium carbonate	24	10	D
Magnesium chloride	24	0–10	C
Magnesium chloride	24	10–100	D
Magnesium hydroxide	24	10–30	D
Magnesium sulfate	24–100	10–60	B
Maleic anhydride	27	10	C
Malic acid	100	—	B
Mercuric chloride	24	10	C
Mercuric sulfate	24–100	10	B
Mercurous nitrate	24	—	D
Mercury	24	100	D
Methanol	30	—	B
Methyl ethyl ketone	24–100	10–100	B
Methyl isobutyl ketone	24–100	10–100	B
Methylene chloride	24–100	—	B
Monochloroacetic acid	24	20–100	D
Monochlorobenzene	24	90	D
Monoethanolamine	171	—	C

(continued)

TABLE 57.12.   (*Continued*)

Chemical	Temp. (°C)	Concentration (%)[b]	Corrosion Class
Naphthalene	24	10–100	B
Naphthalene sulfonic acid + H_2SO_4	88	—	B
Nickel ammonium sulfate	24–100	10	B
Nickel nitrate	24–100	—	B
Nickel sulfate	24–100	10–30	B
Nitric acid	See Table 8		
Nitrobenzene	24–106	—	B
Nitrocellulose	24	—	A
Nitrochlorobenzene	24	—	D
Nitroglycerine	24	—	C
Nitrophenol	24	—	D
Nitrosyl chloride	24	—	B
Nitrosylsulfuric acid	24–80	—	B
Nitrotoluene	24	—	B
Nitrous acid	24	—	D
Oleic acid	24	—	D
Oxalic acid	24	20–100	D
Oxalic acid + 1.5–3% H_2SO_4	52	20–50	A
Pentachlorethane	80	—	B
Perchloroethylene	24	100	B
Persulfuric acid	100	—	C
Phenol	24	90	B
Phenolsulfonic acid	24–100	30	B
Phenyl isocyanate	24	—	B
Phosgene	24–100	—	B
Phosphoric acid	24–93	—	B
Phosphorous acid	27	33	A
Phosphorous chloride	24–148	—	B
Phosphorous oxychloride	24	—	B
Phosphorous pentachloride	24	—	A
Phosphorous pentachloride	52–148	—	B
Phosphorous tribromide	24	—	A
Phosphorous trichloride (dry)	24	10	B
Phthalic anhydride	82	5.25	B
Picric acid	20	25	C
Potassmium aluminum sulfate	26	—	A
Potassium bicarbonate	24	10–30	D
Potassium bifluoride	24–80	10	B
Potassium bisulfate	24–100	10	B
Potassium bisulfite	24–100	10–20	C
Potassium bromide	24–100	10–50	B
Potassium carbonate	24	10–50	C
Potassium chlorate	24	10	B
Potassium chlorate	100	10	D
Potassium chloride	8	0.25–8.0	B
Potassium chromate	24–100	10–40	B
Potassium cyanide	24	10–30	D
Potassium dichromate	24–100	10–60	B
Potassium ferricyanide	24–100	10–60	B
Potassium fluoride	24–80	20	B
Potassium hydroxide	24–60	0–50	B
Potassium hypochlorite	24	10	B
Potassium iodate	24–BP[c]	2–10	B
Potassium iodide	24	30	D

TABLE 57.12. (*Continued*)

Chemical	Temp. (°C)	Concentration (%)[b]	Corrosion Class
Potassium metabisulfite	80	10–30	B
Potassium nitrate	8	0.5–10	B
Potassium permanganate	24	10–40	C
Potassium peroxide	24	10	D
Potassium persulfate	24	10	D
Potassium sulfate	24–100	10–20	B
Potassium sulfite	24	10	B
Propionic acid	24	10–70	D
Pyridine	24–100	10	B
Pyridine sulfate	24	10	B
Pyridine sulfonic acid	24	20	A
Pyrogallic acid	24	—	B
Quinine	24–100	—	B
Quinine bisulfate	24–100	10	B
Quinine tartrate	24–100	—	B
Quinizarin	24–100	—	B
Quinoline	24–100	—	B
Quinone	24–100	10	B
Saccharin solutions	24–100	—	B
Salicylic acid	24–100	—	B
Selenious acid + H_2SO_4 + HNO_3	93	—	A
Silver nitrate	24	10–60	D
Sodium acetate	25	4	B
Sodium acid fluoride	24	10	B
Sodium aluminate	24	10	D
Sodium bicarbonate	24	10	B
Sodium bifluoride	24	—	B
Sodium bisulfate	24–100	10–30	B
Sodium bisulfite	24–100	10	B
Sodium carbonate	24	10	B
Sodium carbonate	52	20	D
Sodium chloride	25	0.5–24	A
Sodium chlorite	24	10	B
Sodium chromate	24–100	10	B
Sodium cyanide	24	10	B
Sodium hydrogen fluoride	71	8	B
Sodium hydrosulfite	24	10–20	A
Sodium hydroxide	26	0–30	B
Sodium hypochlorite	24	1	C
Sodium hyposulfite	24	10	B
Sodium nitrate	24	10	D
Sodium nitrite	24–100	10–60	B
Sodium perborate	24	10	D
Sodium percarbonate	24	—	D
Sodium peroxide	24	10	D
Sodium persulfate	24	10	B
Sodium phosphate	24–100	10–100	B
Sodium phosphate (tri basic)	24	10–20	D
Sodium silicate	24	—	B
Sodium sulfate	24	2–20	A
Sodium sulfide	24–100	10–30	A
Sodium sulfite	24–100	10–30	B
Sodium tartrate	24	—	D
Stannic chloride	24	20	D

(continued)

TABLE 57.12.   (*Continued*)

Chemical	Temp. (°C)	Concentration (%)[b]	Corrosion Class
Stannic tetrachloride (dry)	24	100	B
Stannous bisulfate	24–100	10	B
Stannous chloride	24	10–50	D
Succinic acid	24–100	10–50	B
Sulfamic acid	22	3–20	B
Sulfur dioxide	24–204	90	B
Sulfur trioxide	24	90	B
Sulfuric acid	See Figure 57.4		
Sulfurous acid	60	—	A
Sulfuryl chloride	24	—	B
Tanning mixtures	21	—	B
Tannic acid	24	20–100	D
Tartaric acid	24	30–70	B
Tetraphosphoric acid	24	10–100	D
Thionyl chloride	24–149	—	B
Thiophosphoryl chloride	24	—	B
Tetrachlorethane	63	—	A
Titanium sulfate	24–100	10–30	B
Titanium tetrachloride	24	—	B
Toluene	24–100	—	A
Toluene-sulfochloride	24	—	A
Trichloroethylene	28	—	B
Trichloronitromethane	24	—	C
Triemanolamine	60	0.4	B
Triphenyl phosphite	28	—	A
Turpentine	24	—	B
Vinyl chloride	24	10	D
Zinc carbonate	24	—	B
Zinc fluosilicate	21	30–36	D
Zinc hydrosulfite	24	—	B
Zinc sulfate	35	—	B
Zinc chloride	63	25	B

[a]See [1].
[b]The absence of concentration data is indicated by a dash.
[c]Boiling point = BP.

D. GALVANIC COUPLING

In acid solutions, iron is anodic to lead; therefore, the corrosion of iron is accelerated when coupled to lead. In alkaline solutions, the reverse situation applies and the corrosion of lead is accelerated by coupling with iron. Copper is anodic to lead in strong acid solutions but cathodic in alkaline solutions.

In the handling of sulfuric acid and sulfates in the chemical industries, it is common practice to use Pb–Sb alloys with up to 8% Sb in pumps and valves in electrical contact with sheets and pipes of Te–Pb alloys and chemical lead. Galvanic action is inconsequential because of the formation of an insoluble insulating film of lead sulfate.

Serious galvanic corrosion does not occur under seawater with the use of caulking lead in cast iron pipe joined by bell and spigot types of joints.

E. MECHANICAL AND METALLURGICAL FACTORS

The low melting point of lead allows it to recover and recrystallize at room temperature; therefore, internal stresses or work hardening may dissipate with time at ordinary temperatures. In the corrosion environments for which lead is suited, such factors are usually of minor importance, however, sustained load can result in continuous slow deformation of lead, continually exposing fresh surfaces. In this case, the corrosion rate can increase.

High purity lead (99.99%) may be subject to grain growth at 25°C and all alloys tend to exhibit grain growth at higher temperatures. One or more of the following elements are added to lead alloys as stabilizers or grain growth inhibitors: Cu 0.04–0.08%, Te 0.035–0.05%, Ca 0.02–0.04%. Grain growth becomes an issue if intergranular corrosion is expected. This form of corrosion can cause a significant loss in strength.

F. ATMOSPHERIC CORROSION

Lead has been found to be consistently durable in industrial, rural, and marine environments. The type of corrosion exhibited in each of these three environments is distinct. In rural environments, generally free of pollutants, the humidity, rainfall, and airflow are the only factors influencing corrosion rate. In marine environments, chlorides are entrained in air and can exert a strong effect on corrosion. In industrial environments, sulfur oxide gases and the minerals in solid emissions can influence corrosion behavior. Despite this, lead generally gives satisfactory resistance to smoke-laden air in industrial environments and in ducts for sewer gases. Corrosion data for both lead and lead-coated steel are given in this section.

F1. Lead and Lead Alloy Sheet

In moist air, a dull oxide film or patina forms on the surface of lead. Oxidation generally ceases after 7 days. Several test programs have generated quantitative corrosion data in several natural atmospheres as shown in Table 57.13. The corrosion rates usually decrease with increasing exposure time, and a conservative estimate would use a linear corrosion rate over time. One example of the relationship between corrosion rate and time is shown in Figure 57.6.

TABLE 57.13.   Corrosion of Lead in Various Natural Outdoor Atmospheres[a]

Location	Type of Atmosphere	Material	Duration (years)	Corrosion Rate mdd	Corrosion Rate mpy
Altoona, Pennsylvania	Industrial	Chem Pb	10	0.23	0.029
Altoona, Pennsylvania	Industrial	1% Sb–Pb	10	0.18	0.023
New York	Industrial	Chem Pb	20	0.12	0.015
New York	Industrial	1% Sb–Pb	20	0.10	0.013
Sandy Hook, New Jersey	Seacoast	Chem Pb	20	0.17	0.021
Sandy Hook, New Jersey	Seacoast	1% Sb–Pb	20	0.16	0.020
Key West, Florida	Seacoast	Chem Pb	10	0.18	0.023
Key West, Florida	Seacoast	1% Sb–Pb	10	0.17	0.022
La Jolla, California	Seacoast	Chem Pb	20	0.16	0.021
La Jolla, California	Seacoast	1% Sb–Pb	20	0.18	0.023
State College, Pennsylvania	Rural	Chem Pb	20	0.10	0.013
State College, Pennsylvania	Rural	1% Sb–Pb	20	0.11	0.014
Phoenix, Arizona	Semiarid	Chem Pb	20	0.03	0.004
Phoenix, Arizona	Semiarid	1% Sb–Pb	20	0.09	0.012
Kure Beach, North Carolina 80-ft site	East Coast marine	Chem Pb	2	0.41	0.052
Kure Beach, North Carolina 80-ft site	East Coast marine	6% Sb–Pb	2	0.32	0.041
Newark, New Jersey	Industrial	Chem Pb	2	0.46	0.058
Newark, New Jersey	Industrial	6% Sb–Pb	2	0.33	0.042
Point Reyes, California	West Coast marine	Chem Pb	2	0.28	0.036
Point Reyes, California	West Coast marine	6% Sb–Pb	2	0.20	0.026
State College, Pennsylvania	Rural	Chem Pb	2	0.43	0.055
State College, Pennsylvania	Rural	6% Sb–Pb	2	0.31	0.039
Birmingham, England	Urban	99.96% Pb	7	0.29	0.037
Birmingham, England	Urban	1.6% Sb–Pb	7	0.03	0.004
Wakefield, England	Industrial	99.995% Pb	1	0.58	0.074
Southport, England	Marine	99.995% Pb	1	0.55	0.070
Bourneville, England	Suburban	99.995% Pb	1	0.61	0.077
Cardington, England	Rural	99.995% Pb	1	0.44	0.056
Cristobal CZ	Tropical marine	Chem Pb	8	0.42	0.053
Miraflores CZ	Tropical inland	Chem Pb	8	0.24	0.030

[a]See [1].

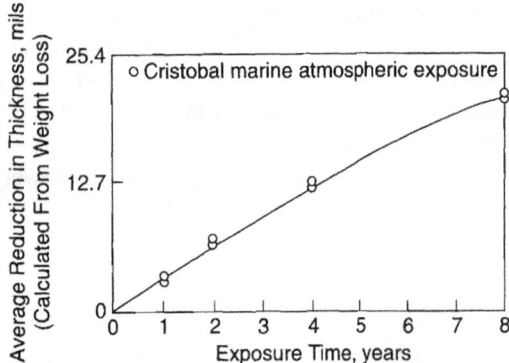

FIGURE 57.6. Relation between corrosion and exposure time for lead exposed to marine atmosphere in Panama Canal zone. (Adapted from [2].)

The Pb–Sb alloys exhibit about the same corrosion rate in atmospheres as chemical lead; however, the greater hardness and strength of this alloy make it more desirable for use in construction applications, especially because of its reduced tendency to buckle over time.

F2. Lead Coatings

Effects of coating porosity on the different reactions that occur at the steel–lead interface can result in variability of corrosion behavior of lead coatings on steel. The corrosion rates of electrodeposited lead coatings in various atmospheres are given in Figure 57.7. Corrosion rates in industrial environments are much lower than the other environments. This is the reverse of experience with other metals. Depositing lead on either nickel or copper-plated steel can enhance corrosion protection. Many terne (Pb–Sn alloy) coated fuel tanks use a nickel precoat for this purpose.

Corrosion resistance can be further enhanced by painting, and this is widely used on terne-coated products.

F3. Installation of Lead Roofing

Design and installation recommendations for lead roofing, flashing, and waterproofing are available in the literature from Lead Industries Association, New York, or LDA International, London. These recommendations provide for accommodation of mechanical factors and free movement of metal when thermal contraction and expansion are anticipated. Patination oil is also available that gives an immediate uniform color to exposed lead, avoiding temporary discoloration that can occur while the lead surface naturally weathers. It is a soybean alkyd resin to which solvents, a special flatting agent and driers have been added. The oil covers $\sim$60 m^2/L and will dry to an even matte finish within 1 h.

G. CORROSION OF BURIED MEDIUM-VOLTAGE POWER CABLE

Medium-voltage (10–50 kV) power cable is customarily placed in underground service, either directly buried in the ground or placed in buried ducts. Although almost all new installations utilize lead-free constructions, there are thousands of kilometers of lead-sheathed cable installed in many cities worldwide that are now at least 20–70 years old. The reliability of these cables, which is in many cases related to the corrosion performance of the lead cable sheathing, is of increasing concern.

Several types of metallurgical or mechanical degradation mechanisms found on failed cables, including grain growth

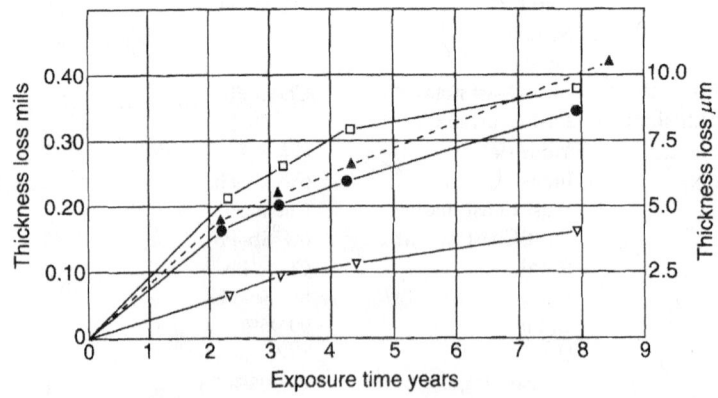

Tests made on 0.002 in (50 μm) thick electro-deposited
lead coating from a fluoborate bath

▲ Tela Honduras-Tropical   ● State College, PA-Rural
□ Kure Beach, NC-Sea coast   ▽ New York, NY-Industrial

FIGURE 57.7. Thickness loss of lead coating versus outdoor exposure. (Adapted from [2].)

and fatigue, can appear to be corrosion damage and in many cases combine with corrosion activity to cause cable failure. In many cases, the only plausible replacement of the failed lead-sheathed cable is with a similar cable, and therefore it is essential to determine the cause of failure. The principal causes of cable failure corrosion are stray-current corrosion and galvanic corrosion. The causes of stray-current corrosion are manmade currents picked up by buried structures and can originate from electrically powered equipment, other power transmission lines, and nearby cathodic protection (CP) circuits. In cities, they include street railway systems, where used, welding equipment, and electroplating shops. The corrosion is localized at the point of discharge of the stray current, wherever the resistance is least for return to the power source. Pits or joined-together pits are observed in the corroded area, but the corroded lead surface is extremely smooth with almost no corrosion products [3]. In the case of stray-current corrosion from a CP circuit, the potential of the sheath is more negative at the point of current pickup and receives the CP current. When the cable passes near the CP source, the current leaves the sheath to return to the negative and the sheath becomes anodic; this is then the site of corrosion. The products of stray-current corrosion, including oxides, chlorides, and sulfates of lead, can be carried away from the immediate reaction site by current flow, resulting in the clean corroded surfaces that are often observed. Stray-current corrosion can be prevented by reducing the electrical resistance of the offending structure or system, including the maintenance of good connections. Stray currents can also be passed from pipelines or other structures operating near high-voltage cables that cause inductive coupling; the currents produced can then be passed to nearly medium-voltage cables [3].

Galvanic corrosion of lead-sheathed cable can occur when it is grounded to a dissimilar metal structure, generally of considerable length. One example is near power stations that use large amounts of copper for grounding. Another is in connection with rusty water pipes; lead is normally cathodic to clean steel but anodic to the oxides comprising rust.

Cathodic corrosion is encountered with lead-sheathed cable buried in soils containing alkali salts. In this case, lead is not removed directly by electric current but dissolved by the secondary action of the alkali produced by the current. Hydrogen ions are attracted to the metal, lose their charge, and are liberated as hydrogen gas. This results in a decrease in the hydrogen ion concentration, and the solution becomes alkaline [4]. This can be observed when a power cable passes between a structure protected by CP and the ground bed of the CP system. Lead becomes very cathodic in close proximity to the DC return, producing a final corrosion product usually consisting of bright orange lead PbO and lead–sodium carbonate.

Because of the nature of their installation, differential aeration corrosion is also common on cable sheaths, but the rate of corrosion is generally rather slow because of the strong tendency of the protective film formation described earlier.

H. PROTECTION MEASURES

The corrosion rate of lead can be reduced by factors that help create or strengthen its protective film. Therefore, the life of lead-protected equipment can be extended, for example, by washing it with film-forming aqueous solutions containing sulfates, carbonates, or silicates. This procedure is suggested for protecting lead when it will be in contact with corrosives that do not form protective films. In water systems, the presence of organic acids whose lead salts are soluble can promote corrosion. In this case, film-forming lime or sodium silicate can be added to the water to lower corrosion rate. A pH 8–9 can be applied. This treatment is also useful to reduce release of lead from solder in copper pipe.

When Portland cement is cast over a lead surface, the free alkali can cause a slow corrosive attack. This can be avoided by use of a suitable underlay such as tar, asphalt, bituminous paint, or a waterproof membrane. After 1 year the free lime in the concrete is usually sufficiently carbonated to eliminate problems; however, continued seepage of water may be a source of long-time corrosion.

In chemical process equipment, the use of acid brick linings will prevent erosion effects and can lower the lead temperature, hereby reducing buckling. Use of automatic steam pressure regulation is recommended for lead coils. This allows for gradual rather than sudden introduction of steam. Similarly, avoiding use of quick shutting valves will prevent failure from water hammering.

When appropriate, galvanic contact between lead and other metals should be prevented by electrical insulation between the metals.

I. LEAD ALLOYS

I1. Lead–Antimony (Pb–Sb)

Antimony is added at levels between 1 and 13% to give greater strength, hardness, and resistance to fatigue. Cast alloys are generally harder man rolled alloys. For sheet and pipe a 6% Sb alloy is frequently employed for its higher strength. Other applications of lead alloys are shown in Table 57.14. The Pb–Sb alloys are usually not used > 95–120°C. Overaging, resulting in loss of properties, can occur at prolonged high temperatures. At compositions > 13% Sb, the alloys become too brittle for engineering uses. The tensile strength of Pb–Sb alloys varies from 17.2 MPa (2500 psi) for pure lead to 50 MPa (7280 psi) for the 13% Sb alloy. At the same time, elongation decreases from 45 to 10%, while hardness increases from 4 to 15.2 BHN (Brinell hardness number).

TABLE 57.14.   Use of Pb–Sb Alloys[a]

Sb (wt %)	Use
1	Electric cable sheathing
2–4	Storage battery connectors and parts
6	Roofing and chemical industry
9	Storage battery grids
6–8	Lead pumps, valves, coatings
15 (+ 5 Sn)	Type metal, bearing metal
15–17 (+ 1 Sn, 1 As)	Bearing metal

[a]See [5].

I2. Lead–Tin (Pb–Sn)

The most popular uses of Pb–Sn alloys are solder and terne plate. Solders can contain between 50 and 95% Sn while terne coatings for steel sheet contain between 3 and 15% Sn. The Pb–Sn alloys have much lower densities and melting points man pure Pb, however, the alloys primarily used for corrosion resistance are low in Sn such as the terne composition. Anodes of 7% Sn are used in chromium plating operations. Ternary Pb–Sn–Sb alloy can have very low coefficients of friction and have high strength. Such alloys are used in special situations such as steam spargers in chemical reactors.

I3. Lead–Calcium (Pb–Ca)

Additions of small amounts of Ca (0.03–0.1% and Sn up to 1.5%) creates alloys with significantly improved mechanical properties and that age harden at room temperature. Moreover these alloys exhibit superior corrosion resistance to Pb–Sb alloys in many applications. Mechanical properties depend on both composition and processing. For example, a 6.4-mm (0.25-in.) section of an air-cooled static cast 0.07% Ca fully aged within 60 days of casting had a breaking strength of 37.9 MPa (5500 psi). The addition of 0.06% Sn increased strength to 51.7 MPa (7500 psi). A 1% Sn level resulted in a strength of 58.6 MPa (8500 psi). Wrought Pb–Ca–Sn alloys have improved properties with breaking strengths between 27.5 and 75.8 MPa (4000 and 11,000 psi). However, a fine grain structure is required, necessitating proper processing. The high corrosion resistance of these alloys is related to their fine grain size. The Pb–Ca alloys with or without Sn are used for battery grids, anodes, and roofing materials.

I4. Lead–Tellurium (Pb–Te)

Additions of 0.04–0.05% Te to chemical lead containing 0.04–0.08% Cu increases corrosion resistance in several environments. These alloys have a refined grain size and exhibit work hardenability, making them useful for steam heating coil applications. The Cu-free alloys have an optimum Te composition of above 0.1%. Such alloys are less resistant to corrosion in some solutions such as low-strength sulfuric acid.

I5. Lead–Silver (Pb–Ag)

The addition of 1% Ag to Pb gives an insoluble anode that exhibits very good corrosion resistance over a wide range of current densities. It is used for production of electrolytic zinc from strong sulfate solutions. Additions of 1% As further increase corrosion resistance, allowing use as an insoluble anode for electrowinning of manganese. The Pb–1% Ag alloy has also given good service in the cathodic protection of ships. The Pb–2.3% Ag alloy has been used as a soft solder; however, unsheltered atmospheric corrosion results in poor performance. Alloys with < 1% Ag (0.002–0.2%) have a higher corrosion resistance than pure lead in certain environments.

I6. Lead–Arsenic (Pb–As)

Arsenic is generally added to Pb–Sb alloys to accelerate age hardening. It also increases resistance to bending and creep of electric power cable sheathing alloys that are exposed to vibration. The most widely used alloy of this type is Pb–0.15% As–0.1% Sn–0.1% Bi.

I7. Lead–Copper (Pb–Cu) and Other Alloys

The Cu additions to Pb are limited to the ranges shown in commercial lead, up to 0.08%. Over this range Cu does not increase the corrosion resistance of lead. Additions of Ag increase corrosion resistance as noted above. Other alloying elements that have been tested for corrosion resistance in lead include Ba, Ca, Co, Au, Li, Mg, Ni, Pa, Pt, Na, Tl, and Zn. Anodes made of Ni and Tl show good performance in sulfuric acid solutions and have been adopted for electrolytic refining in some cases. Ni can be added to lead in small quantities to improve wettability of the steel during lead coating.

REFERENCES

1. Lead for Corrosion Resistant Applications—A Guide, Lead Industries Association, New York, 1974.
2. F. W. Fink and W. K. Boyd, "The Corrosion of Metals in Marine Environments," DMIC Report 245, Defense Metals Information Center, Battelle Institute, Columbus, OH.
3. S. A. Bradford, Practical Handbook of Corrosion Control in Soil, CASTI Handbook Series, CASTI Publishing, Edmonton, Alberta, 1970.
4. R. W. Hymes, "Lead Sheath Cable Corrosion—Cause and Prevention," in Proceedings of the Fortieth Appalachian Underground Short Course, EPRI, Palo Alto, CA, 1995.
5. H. H. Uhlig (Ed.), Corrosion Handbook, Wiley, New York, 1948.

58

MAGNESIUM AND MAGNESIUM ALLOYS

E. Ghali

Department of Mining, Metallurgy and Materials Engineering, Laval University, Québec, Canada

A. PRODUCTION AND FABRICATION OF MAGNESIUM

The perception of magnesium as a rapidly corroding material has been a major obstacle to its growth in structural applications despite its other obviously desirable physical properties. In fact, under normal environmental conditions, the corrosion resistance of magnesium alloys is comparable or better than that of mild steel. It has been the uneducated use of magnesium in wet, salt-laden environments that has given rise to its poor corrosion reputation. Corrosion due to poor design, flux inclusions, surface contamination, galvanic couples, and incorrectly applied or inadequate surface protection schemes are all avoidable and applicable not only to magnesium but to many other metals as well.

Designers and engineers in the magnesium industry have established the correct use of magnesium in corrosive environments and, over the past 30 years, have developed methods to improve the corrosion resistance of magnesium alloys by modifying alloy chemistry and improving surface protection technologies [1].

A1. Physical, Chemical, and Electrochemical Properties

Magnesium is silvery white in approach. It is a divalent metal. The atomic mass is 24.32 and the specific gravity of the pure metal is 1.738 at 20°C (68°F). The structure is closed-packed hexagonal. The lattice structure of magnesium has $c/a = 1.624$ and atomic diameter (0.320 nm) is such that it enjoys favorable size factors with a diverse range of solute elements. The melting point is 649.5°C and the boiling point is 1107°C. The specific heat at 20°C is 1.030 kJ/kg°C and the thermal conductivity at 20°C is f157.5 W/m°C. The electrochemical equivalent is 0.126 mg/°C or 12.16 g/Faraday. The standard electrode potential is

$$E^{\circ}_{Mg^{2+}/Mg} = -2.37\,V$$

Magnesium alloys are used in the aircraft and guided weapons industries and in automotive construction because of their light weight and high strength/weight ratio. New applications are emerging because of required properties, such as high stiffness/weight ratio, ease of machinability, high damping capacity, and casting qualities. Magnesium is used as a canning material for uranium in gas-cooled reactors. Magnesium and its alloys can be used as sacrificial anodes for cathodic protection. Magnesium is itself used for alloying with other metals for different applications.

The corrosion resistance of magnesium and its alloys is dependent on film formation in the medium to which they are exposed. The rate of formation, dissolution, or chemical change of the film varies with the medium and also with the metallic alloying agents, which are impurities present in the magnesium [2].

A2. Magnesium Alloys

Cast magnesium alloys have always predominated over wrought alloys, particularly in Europe, where, traditionally, cast alloys have comprised 85–90% of all magnesium products. The earliest commercially used alloying elements were aluminum, zinc, and manganese, and the Mg–Al–Zn system remains the most widely used for castings. Aluminum, zinc, cerium, yttrium, silver, thorium, and zirconium are examples of widely differing metals that may be present in commercial magnesium alloys. Apart from magnesium and cadmium, which form a continuous series of solid solutions, the magnesium-rich sections of binary phase diagrams show peritectic or, more commonly, eutectic systems. Solubility data for binary magnesium alloys are given in

TABLE 58.1.   Solubility Data for Binary Magnesium Alloys[a]

Element	Solid Solubility at %	Solid Solubility wt %	System[b]
Lithium	17.00	5.50	Eutectic
Aluminum	11.80	12.70	Eutectic
Silver	3.80	15.00	Eutectic
Yttrium	3.75	12.50	Eutectic
Zinc	2.40	6.20	Eutectic
Neodymium	≈ 1.00	≈ 3.00	Eutectic
Zirconium	1.00	3.80	Peritectic
Manganese	1.00	2.20	Peritectic
Thorium	0.52	4.75	Eutectic
Cerium	0.10	0.50	Eutectic
Cadmium	100.00	100.00	Complete SS
Indium	19.40	53.20	Peritectic
Thallium	15.40	60.50	Eutectic
Scandium	≈ 15.00	≈ 24.50	Peritectic
Lead	7.75	41.90	Eutectic
Thulium	6.30	31.80	Eutectic
Terbium	4.60	24.00	Eutectic
Tin	3.35	14.50	Eutectic
Gallium	3.10	8.40	Eutectic
Ytterbium	1.20	8.00	Eutectic
Bismuth	1.10	8.90	Eutectic
Calcium	0.82	1.35	Eutectic
Samarium	≈ 1.00	≈ 6.40	Eutectic
Gold	0.10	0.80	Eutectic
Titanium	0.10	0.20	Peritectic

[a]Reproduced with permission [3, 4], ASM International.
[b]Solid solubility = SS.

Table 58.1; the first 10 elements are those used in commercially available alloys.

Although early Mg–Al–Zn castings suffered severe corrosion in wet or moist conditions, the corrosion performance was significantly improved as a result of the discovery, in 1925, that small additions (0.2%) of manganese gave increased resistance. With this element, iron and certain other heavy metal impurities formed relatively harmless intermetallic compounds, some of which separate out during melting. In this regard, the classic work by Hanawalt et al. [5] showed that the corrosion rate increased abruptly once tolerance limits were exceeded; these tolerance limits are 5, 170, and 1300 ppm for nickel, irons, and copper, respectively. The corrosion rate of pure magnesium as a function of iron content is shown in Figure 58.1, which clearly illustrates the tolerance limit for iron.

Another problem with earlier magnesium alloy castings was that grain size tended to be large and variable, often resulting in poor mechanical properties, microporosity, and, in wrought products, excessive directionality of properties. Values of proof stress also tended to be low relative to tensile strength.

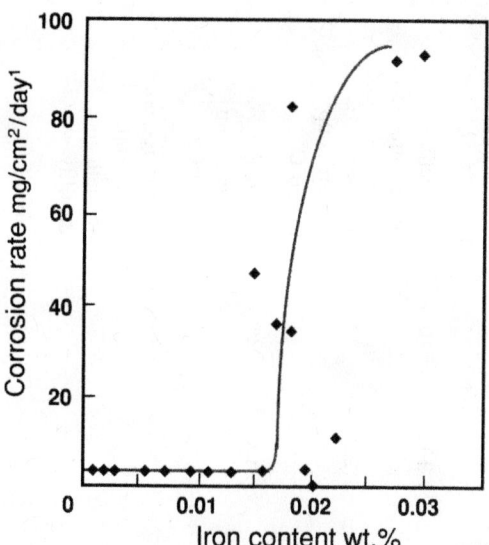

FIGURE 58.1. Effect of iron on corrosion of pure magnesium; alternate immersion test in 3% NaCl [5].

In 1937, it was discovered that zirconium had an intense grain refining effect on magnesium. The lattice parameters of zirconium are very close to those of magnesium. Paradoxically, zirconium could not be used in most existing alloys at that time because it was removed from solid solution owing to the formation of stable compounds with aluminum and manganese. This problem led to the evolution of a completely new series of cast and wrought zirconium-containing alloys with much improved mechanical properties at both room and elevated temperatures. Alloys containing zirconium as a grain-refining agent have the iron content reduced to ~0.004% because impurities separate during the alloying procedure. These alloys are now widely used in the aerospace industries.

A3. Alloy Designations and Tempers

A3.1. Cast Alloys.
Two major cast alloy magnesium systems are available to the designer. The first system includes alloys containing 2–10% Al, combined with minor additions of zinc and manganese. These alloys are widely available at moderate cost, and their mechanical properties are satisfactory from 95 to 120°C (200 to 250°F). At higher temperatures, the properties deteriorate. The second system consists of magnesium alloyed with various elements (rare earths, zinc, thorium, silver, etc.) except aluminum, all containing a small but important zirconium content that imparts a fine grain structure that improves mechanical properties. These alloys generally possess much better elevated temperature properties, but their more costly elemental additions, combined with the specialized manufacturing technology required, result in significantly higher costs. Many of the casting alloys are given simple heat treatments to improve

their properties, while the wrought alloys can be obtained in a number of tempers [6].

An international code for designating magnesium alloys does not exist, although there has been a tendency toward adopting the method used by the American Society for Testing and Materials (ASTM) B275-94 [7–9]. In this system, the first two letters indicate the principal alloying elements according to the following code: A aluminum, B bismuth, C copper, D cadmium, E rare earths, F iron, H thorium, K zirconium, L lithium, M manganese, N nickel, P lead, Q silver, R chromium, S silicon, T tin, W yttrium, Y antimony, and Z zinc. The letter corresponding to the element present in greater quantity in the alloy is used first; if they are equal in quantity the letters are listed alphabetically. Letters are followed by numbers that represent the nominal compositions of these principal alloying elements in weight percent, rounded off to the nearest whole number, (e.g., AZ91 indicates the alloy Mg–9Al–1Zn), the actual composition ranges being 8.3–9.7% Al and 0.4–1.0% Zn. Suffix letters A, B, C are chronologically assigned and usually refer to purity improvement. The X is reserved for experimental alloys. For heat-treated or work-hardened conditions, the designations are specified by the same system as that used for aluminum alloys [8]. Commonly used tempers are T5, alloys artificially aged after casting; T6, alloys solution treated, quenched, and artificially aged; and T7, alloys solution treated and stabilized [10].

Because of the particularly high solid solubility of yttrium in magnesium (12.5% max) and the amenability of Mg–Y alloys to age hardening, a series of Mg–Y–Nd–Zr alloys has been produced, which combine high strength at ambient temperatures with good creep resistance at temperatures up to 300°C [11, 12]. The heat-treated alloys have a resistance to corrosion, which is superior to that of other high-temperature magnesium alloys and comparable to many aluminum-based casting alloys [13, 14]. Since pure yttrium is expensive and difficult to alloy with magnesium because of its high melting point (1500°C) and its strong affinity for oxygen, a cheaper yttrium-containing (~75% Y) mischmetal together with heavy rare earth metals such as gadolinium and erbium could be substituted for pure yttrium (Table 58.2) [15].

A3.2. Wrought Alloys.
Wrought materials are produced mainly by extrusion, rolling, and press forging at temperatures in the range 300–500°C. As with cast alloys, the wrought alloys may be divided into two groups according to whether or not they contain zirconium (Table 58.3). Specific alloys have been developed that are suitable for wrought products, most of which fall into the same categories as the casting alloy already [9]. Examples of sheet and plate alloys are AZ31 (Mg–3 Al–1 Zn–0.3 Mn), which are the most widely used because they offers a good combination of strength, ductility, and corrosion resistance, and thorium-containing alloys such as HM21 (Mg–2 Th–0.6 Mn), which

TABLE 58.2. Nominal Composition, Typical Tensile Properties, and Characteristics of Selected Magnesium Casting Alloys[a]

ASTM Designation	British Designation	Al	Zn	Mn	Si	Cu	Zr	MM	Nd	Th	Y	Ag	Condition	0.2% Proof Stress (Mn/m²)	Tensile Strength[b] (Mn/m²)	El[b] (%)	Characteristics
AZ63		6	3	0.3									As sand cast	75	180	4	Good room temperature strength and ductility
													T6	110	230	3	
AZ81	A8	8	0.5	0.3									As sand cast	80	140	3	Tough, leak tight casting with 0.0015% Be used for pressure die casting
													T4	80	220	5	
AZ91	AZ91	9.5	0.5	0.3									As sand cast	95	135	2	General purpose alloy used for sand and die casting
													T4	80	230	4	
													T6	120	200	3	
													As chill cast	100	170	2	
													T4	80	215	5	
													T6	120	215	2	
AM50		5		0.3									As die cast	125	200	7	High pressure die castings
AM20		2		0.5									As die cast	105	135	10[c]	Good ductility and impact strength
AS41		4		0.3	1								As die cast	135	225	4.5[c]	Good creep properties up to 150°C
AS21		2		0.4	1								As die cast	110	170	4[c]	Good creep properties up to 150°C
ZK51	ZSZ		4.5				0.7						T5	140	235	5	Sand castings, good room temp, strength and ductility
ZK61			6				0.7						T5	175	275	5	As for ZK51

Alloy	BS designation	Zn	Cu	Mn	Ag	RE	Th	Y	Zr	Condition	0.2% PS (MPa)	TS (MPa)	El (%)	Characteristics
ZE41	RZ5	4.2				1.3			0.7	T5	135	180	2	Sand castings, good room temp. strength and castability
ZC63	ZC63	6	3	0.5						T6	145	240	5	Pressure tight castings, good elevated temp., strength, weldable
EZ33	ZREI	2.7				3.2			0.7	Sand cast T5	95	140	3	Good castability, pressure tight, weldable, creep resistant up to 250°C
										Chill cast T5	100	155	3	
HK31	MTZ						3.2		0.7	Sand cast (T6)	90	185	4	Sand casting, good castability, weldable, creep resistant up to 350°C
HZ32	ZTI	2.2					3.2		0.7	Sand or chill cast (T5)	90	185	4	As for HK31
QE22	MSR				2.5	2.5			0.7	Sand or chill cast (T6)	185	240	2	Pressure tight, weldable, high proof stress up to 250°C
QH21	QH21				2.5	1	1		0.7	As sand cast (T6)	185	240	2	Pressure tight, weldable, good creep resistance and proof stress up to 300°C
WE54	WE64					3.25[d]		5.1	0.5	T6	200	285	4[c]	High strength at room and elevated temp, good corrosion resistance, weldable
WE54	WE43					3.25[d]		4	0.5	T6	190	250	7[c]	

[a] Reproduced with permission from [15], Institute of Materials, London, UK.
[b] Mischmetal = MM; El = elongation.
[c] Values quoted for tensile properties are for separately cast test bars and may not be realized in certain parts of castings.
[d] Contains some heavy metal rare earth elements.

TABLE 58.3. Nominal Composition, Typical Tensile Properties, and Characteristics of Selected Magnesium Casting Alloys[a]

ASTM Designation	British Designation	Al	Zn	Mn	Zr	Th	Cu	Li	Condition	0.2% Proof Stress (Mn/m^2)	Tensile Strength (Mn/m^2)	El (%)	Characteristics
MI	AM503			1.5					Sheet, plate/F	70	200	4	Low- to medium-strength alloy, weldable, corrosion resistant
									Extrusions/F	130	230	4	
									Forgings/F	105	200	4	
AZ31	AZ31	3	1	0.3					Sheet, plate/O	120	240	11	Medium-strength alloy, weldable, good formability
				0.2[b]					Sheet, plate/H24	160	250	6	
										130	230	4	
									Extrusions/F Forgings/F	105	200	4	
AZ61	AZM	6.5	1	0.3					Extrusions/F	180	260	7	High-strength alloy, weldable
				0.15[b]					Forgings/F	160	275	7	
AZ80	AZ80	8.5	0.5 0.12[b]	0.2					Forgings/T6	200	290	6	High-strength alloy
ZM21	ZM21		2	1					Sheet, plate/O	120	240	11	Medium-strength alloy, good formability, good damping capacity
									Sheet,	165	250	6	
									Plate/H24	155	235	8	
									Extrusions/F Forgings/F	125	200	9	
ZMC711			6.5	0.75			1.25		Extrusions/T6	300	325	3	High-strength alloy
LA 141		1.2		0.15[b]				14	Sheet, plate/T7	95	115	10	Ultralight weight (specific gravity 1.35)
ZK31	ZW3		3		0.6				Extrusions/T5	210	295	8	High-strength alloy, some weldability
									Forgings/T5	205	290	7	
ZK61			6		0.8				Extrusions/F	210	285	6	High-strength alloy
									Extrusions/T5	240	305	4	
									Forgings/T5	160	275	7	
HK31					0.7	3.2			Sheet,	170	230	4	High creep resistance up to 350°C, short time
									Plate/H24	180	255	4	
									Extrusions/T5				
HM21				0.8		2			Sheet, plate/T8	135	215	6	High creep resistance up to 350°C, short time exposure up to 425°C weldable
									Sheet, plate/T81	180	255	4	
									Forgings/T5	175	255	3	
HZ11	ZTY		0.6		0.6	0.8			Extrusions/F	120	215	7	Creep resistance up to 350°C, weldable
									Forgings/F	130	230	6	

[a]Reproduced with permission from [15], Institute of Materials, London, UK.
[b]Minimum.

show good creep resistance at temperatures of up to 350°C. Magnesium alloys can be extruded at temperatures > 250°C into either solid or hollow sections at speeds that depend on alloy content. Higher strength alloys such as AZ81 (Mg– 8 Al–1 Zn–0.7 Mn), ZK 61 (Mg–6 Zn–0.7 Zr), and the more recent composition ZCM711 (Mg–6.5 Zn–1.25 Cu–0.75 Mn) all have strength/weight ratios comparable to those of the strongest wrought aluminum alloys. The alloy ZM21

(Mg–2 Zn–1Mn) can be extruded at high speeds and is the lowest cost magnesium extrusion alloy available. Again, thorium-containing alloys, such as HM31 (Mg–3 Th–1 Mn), show the optimal elevated temperature properties. Magnesium forgings are less common and are almost always press formed rather than hammer forged.

B. CORROSION FORMS AND TYPES

B1. General Corrosion and Passivation

Magnesium exposed to air is covered by a gray oxide film, which protects the metal from further oxidation. Magnesium can be heated in air to the melting point without burning. However, the fine divided metal reacts vigorously by heating or by contact with water or humid atmospheres. The evolved hydrogen from this reaction can result in an explosive mixture.

In aqueous solutions, magnesium dissociates by electrochemical reaction with water to produce a crystalline film of magnesium hydroxide, $Mg(OH)_2$ [16], and hydrogen gas, a mechanism, which is highly insensitive to the oxygen concentration [17]. Subsequently, all that is needed for rapid corrosion are sites of easy hydrogen discharge [18]. The probable primary overall corrosion reaction for magnesium in aqueous solutions is

$$Mg(s) + 2H_2O(\ell) \Rightarrow Mg(OH)_2(s) + H_2(g)$$

This overall reaction can be described in terms of anodic and cathodic reactions as follows:

Anodic Reaction: $\quad Mg \Rightarrow Mg^{2+} + 2e$

(dissolution of Mg)

and/or

$$Mg(s) + 2(OH)^- \Rightarrow Mg(OH)_2(s) + 2e^-$$

Cathodic Reaction: $\quad 2H^+ + 2e^- \Rightarrow H_2(g)$

(evolution of hydrogen gas)

A subsequent reaction giving OH^- ions can occur and/or

$$2H_2O + 2e^- \Rightarrow H_2(g) + 2(OH)^-$$

In general, the magnesium corrosion products resulting from the anodic reaction depend on the environment and may include carbonate, hydroxide, sulfite, and/or sulfate compounds.

The hydroxide film, brucite, has a hexagonal crystalline structure that is layered, alternating between Mg and hydroxide ions, facilitating easy basal cleavage. Cracking and curling of the film have been noted though it is not clear whether it is from the properties of the film or the evolution of hydrogen gas. The Pilling/Bedworth ratio for $Mg(OH)_2$ is 1.77, which indicates a resistant film in compression. A combination of internal stresses and easy basal cleavage may account for a portion of the cracking and curling of the film. Thus, the structure of the corrosion product directly influences the corrosion behavior of the base metal [19].

Magnesium may form a surface film, which protects it in alkaline environments, and poorly buffered environments where the surface pH can increase. Passivity of magnesium is destroyed by several anions, including chloride, sulfate, and nitrate. Alloying affects the nature of this film, but these effects are poorly understood. The corrosion of magnesium and its alloys is strongly dependent on the absence of impurity elements, some of which have well-defined tolerance levels above which corrosion resistance drops dramatically. For conventional magnesium alloys, these tolerance limits must be observed even if extensive surface treatments are applied [20].

The Pourbaix (potential–pH) diagram [21] shows possible protection of magnesium at high pH values, which may result from $Mg(OH)_2$ formation during the corrosion reaction. Perrault [22] considered the formation of MgH_2 and Mg^+ and assumed that thermodynamic equilibrium cannot exist for a magnesium electrode in contact with aqueous solutions. Such equilibrium is, however, possible if the hydrogen overpotential is about 1 V and the pH is > 5. The following reactions are considered in the E–pH diagram (Fig. 58.2):

$$2H^+ + 2e^- \rightarrow H_2 \tag{58.1}$$

$$MgH_2 \rightarrow Mg^{2+} + H_2 + 2e^- \tag{58.2}$$

$$MgH_2 + 2OH^- \rightarrow Mg(OH)_2 + H_2 + 2e^- \tag{58.3}$$

$$Mg^{2+} + 2OH^- \rightarrow Mg(OH)_2 \tag{58.4}$$

$$Mg^+ \rightarrow Mg^{2+} + e^- \tag{58.5}$$

$$Mg^+ + 2OH^- \rightarrow Mg(OH)_2 + 1e^- \tag{58.6}$$

$$Mg^+ + 2H_2O \rightarrow Mg(OH)_2 + 2H^+ + e^-$$

$$MgH_2 \rightarrow Mg^+ + H_2 + e^- \tag{58.7}$$

Although magnesium has a standard electrode potential at 25°C of −2.37 V, its corrosion potential is more negative than −1.5 V in dilute chloride solution or a neutral solution with respect to the standard hydrogen electrode due to the polarization of the formed film of $Mg(OH)_2$. The oxide film on magnesium offers considerable surface protection in rural and some industrial environments, and the corrosion rate of magnesium lies between that of aluminum and that of low-carbon steels (Table 58.4).

In natural atmospheres, the corrosion of magnesium can be localized. The conductivity, ionic species, temperature of the electrolyte, alloy composition and homogeneity,

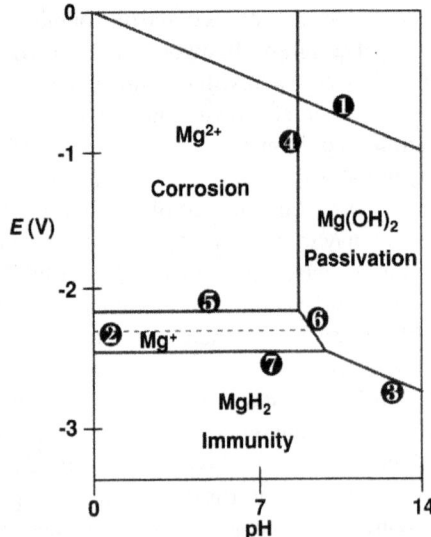

FIGURE 58.2. Equilibria of Mg–H$_2$O system in presence of H$_2$ at 25°C.

TABLE 58.4.   Results of 2.5-Year Exposure Tests on Sheet Alloys[a]

Material	Corrosion Rate (μm/year)	Loss of Tensile Strength after 2–5 years (%)
Marine Atmosphere		
Aluminum alloy 2024	2.0	2.5
Magnesium alloy AZ31	18.0	7.4
Low-carbon steel (0.27%C)	150.0	75.4
Industrial Atmosphere		
Aluminum alloy 2024	2.0	1.5
Magnesium alloy AZ31	27.7	11.2
Low-carbon steel (0.27%C)	25.4	11.9
Rural Atmosphere		
Aluminum alloy 2024	0.1	0.4
Magnesium alloy AZ31	13.0	5.9
Low-carbon steel (0.27%C)	15.0	7.5

[a]Reproduced with permission from [23].

differential aeration, and so on influence the corrosion morphology.

B1.1. Corrosion Prevention. Effective corrosion prevention for magnesium components and assemblies begins at the design stage. General corrosion attack in saltwater exposures can be minimized through the selection of high-purity magnesium alloys cast without introducing heavy metal contaminants and flux inclusions

B2. Galvanic Corrosion

For continuous outdoor use, where magnesium assemblies may be wet or subjected to salt splash or spray, precautions against galvanic attack must be taken. Although corrosive attack from any source can jeopardize the satisfactory performance of magnesium components, attack resulting from galvanic corrosion is probably the most detrimental.

Because magnesium is anodic (or sacrificial) to all other engineering metals, the severe corrosive attack that often occurs with magnesium assemblies in saltwater environments has long been a deterrent to the use of magnesium alloys in structural applications.

Galvanic corrosion of magnesium alloys can generally be attributed to two basic causes: (1) poor alloy quality due to excessive levels of heavy metal or flux contamination and (2) poor design and assembly practices, which can result in severe galvanic corrosion attack.

With the recent development of fluxless melt protection and new high-purity alloys, such as AZ91D, AZ91E, AM60B, and others, a renewed interest in magnesium has developed due to the improved corrosion resistance of these alloys. The new alloys offer no defense against galvanic corrosion attack; however, their improved performance in assemblies can only be realized if proper measures are taken to control the potential for galvanic attack through careful design, selection of compatible materials, and the selective use of coatings, sealants, and insulating materials. The severity of galvanic activity is determined by the galvanic current which flows in the completed circuit. This can be expressed as follows:

$$I = \left(E_k - \frac{E_a}{R_m} + R_e \right)$$

where E_k and E_a are the polarized measured potentials of the cathode and anode, respectively, and R_m and R_e are the resistance of the metal-to-metal contact and the electrolyte portions of the circuit, respectively. The electrochemical reactions are

$$\text{Anode reaction}: \quad \text{Mg(metal)} \rightarrow \text{Mg}^{2+} + 2e^- \quad (58.8)$$

$$\text{Cathode reaction}: 2H_2O + 2e^- \rightarrow H_2 + 2OH^- \quad (58.9)$$

In many practical applications, R_m is negligibly small due to mechanical, electrical, or cost requirements, and R_e (the electrolyte resistance) becomes the controlling factor in the circuit resistance. If the environment is rich in marine mists or deicing salts, the use of drain holes and sealants can help control corrosion by forcing the galvanic current in the electrolyte to flow through a thin and, therefore, highly resistive film (maximizing R_e). In practice, this limits the galvanic activity to an area ~3.2–6.4 mm ($\frac{1}{8}$ – $\frac{1}{4}$ in.) wide on either side of the magnesium–cathode interface. The magnesium alloy can still suffer severely; however, the cathode does not polarize sufficiently to reduce or eliminate the effective potential difference ($E_k - E_a$) [24].

The degree to which the corrosion of magnesium is accelerated by the galvanic couple in a given environment (i.e., a given R_e) depends in part on the relative positions of the two metals in the electrochemical series. Equally important, the polarization that reduces the potential difference of the couple as the galvanic current develops. Because magnesium shows little, if any, anodic polarization in saltwater exposures, the reduction of the potential difference in the galvanic cell typically results from polarization of the cathode, where water is reduced to hydrogen gas and hydroxyl ion [Eq. 58.9]. Some metals, such as iron, nickel, and copper, serve as efficient cathodes in what is thought to be the stepwise process of accepting and reducing hydrogen ion to an atomic form (H) where it then combines to form the evolved hydrogen gas (H_2). These metals have a low hydrogen overvoltage and can consequently cause severe galvanic corrosion of magnesium. Other metals, such as aluminum, zinc, cadmium, and tin, while equally cathodic to magnesium in some environments, serve as much less effective cathodes due to their tendency to inhibit the combination of atomic hydrogen on surfaces to form the hydrogen gas that evolves.

Data on galvanic corrosion of magnesium alloys were compiled in tests at Kure Beach, NC, in which sheets of dissimilar metals were fastened to panels of AZ31B and AZ61A. The dissimilar metals were divided into five groups

(Table 58.5), ranging from the recommended Group 1 to Group 5, metals in Group 5 caused severe galvanic corrosion of magnesium alloys.

Aluminum alloys containing small percentages of copper (7000 and 2000 series and 380 die-casting alloy) may cause serious galvanic corrosion of magnesium in saline environments. Very pure aluminum is quite compatible, acting as a polarizable cathode; but when iron content exceeds 200 ppm, cathodic activity becomes significant (apparently because of the depolarizing effect of the intermetallic compound $FeAl_3$), and galvanic attack of magnesium increases rapidly with increasing iron content. The effect of iron is diminished by the presence of magnesium in the alloy. This agrees with the relatively compatible behavior of aluminum alloys 5052, 5056, and 6061 shown in Table 58.6 [23].

The corrosion of magnesium being largely cathodically controlled, the polarization characteristics of the coupled cathode will largely control the galvanic corrosion. In a highly conducting medium, such as 3% NaCl, most metals will not polarize to the magnesium potential until a relatively high current density is reached. In contact with metals such as steel or nickel, very high corrosion currents are obtained in most

TABLE 58.5.   Relative Effects of Various Metals on Galvanic Corrosion of Magnesium Alloys AZ31B and AZ61A Exposed at the 24.4- and 244-m (80- and 800-ft) Stations, Kure Beach, NC[a]

Group 1 (Least Effect)	Group 2	Group 3	Group 4	Group 5 (Greatest Effect)
Al alloy 5052	Al alloy 6063	AlClad alloy 2024	Zn-plated steel	Low-carbon steel
Al alloy 5056	AlClad alloy 7075	Al alloy 2017	Cd-plated steel	Stainless steel
Al alloy 6061	Al alloy 3003	Al alloy 2024		Monel, titanium
	Al alloy 7075	Zinc		Lead, copper Brass

[a]See [18].

TABLE 58.6.   Corrosion of Mg 6% Al–3% Zn 0.2% Mn Alloy Galvanically Connected to Other Metals in Various Media[a]

	Corrosion Rate (mdd)					
	3% NaCl				Midland Tap Water	Distilled Water
	Separation					
Dissimilar Metal	Close Contact	0.35 cm	2.0 cm	10 cm	0.35 cm	0.35 cm
Steel	23,400	25,500	8300	3900	300	18
Aluminum alloys 2024	12,800	25,700	6800	3200	90	6
Nickel	18,800	22,400	6600		210	19
Aluminum alloys 1100	14,500	15,600	4100		40	4
Copper	8500	8200	3700		90	15
Brass	7100	4000	2500	1700	60	14
Aluminum alloys 5056		1900			10	3
Cd-plated steel	5200	2200	1000		40	14
Zinc	6200	1300	900	700	30	8
Mg–1.5%Mn		50			2	3
Mg–6% Al–3% Zn–0.2% Mn		200			7	3
Size of specimen	$4 \times 1.3 \times 0.2$ cm ($1.5 \times 0.5 \times 0.079$ in.)			Temperature		Room
Relative areas	1:1 (mounted face to face).			Aeration		Nat. convect
Surface preparation	Aloxite 150 ground			Volume of testing solution		100 mL
Velocity	Quiescent					
Duration of test: 3% NaCl, 3 h; midland tap water, 24 h; distilled water, 4 days.						

[a]See [2].

highly conducting media. An exception is the Al–5% Mg rivet alloy that normally polarizes at a very low current density.

The conductivity and composition of the medium in which a couple is immersed are controlling factors in the rate of galvanic corrosion. Equal areas of various cathodic materials and a magnesium alloy were tested by continuous immersion in 3% NaCl, Midland tap water containing approximately 70 ppm chloride, and in distilled water (Table 58.6).

All commonly used metals cause galvanic corrosion of magnesium in a strong chloride electrolyte. Cadmium or zinc plating of the more cathodic metals, such as iron or steel, reduces the galvanic corrosion to one tenth the rate; a reduction in the conductivity, for example, a change from 3% NaCl to tap water, causes an even greater reduction in galvanic corrosion rate.

Under conditions where the corrosion product is not continuously removed or under conditions of high cathodic current density where the surroundings may become strongly alkaline, both the magnesium and an amphoteric contacting metal such as aluminum may suffer severe attack. Aluminum alloys containing appreciable magnesium, such as 5052, 6053, 5056, are least severely attacked in chloride media when galvanically coupled. This fact was observed in galvanic couples of magnesium and aluminum alloys exposed to tide water and in the atmosphere at Hampton Roads, VA.

Zinc, cadmium, or tin plating on steel all reduce galvanic attack of magnesium substantially when compared to that produced by uncoated steel. The relative compatibility of the electroplates in descending order has generally been concluded to be tin, cadmium, and zinc. This is consistent with the data presented at Figure 58.3, where the compatibility of various fasteners (plated on coated steel, plus alternative materials) was determined by the magnesium weight loss in a

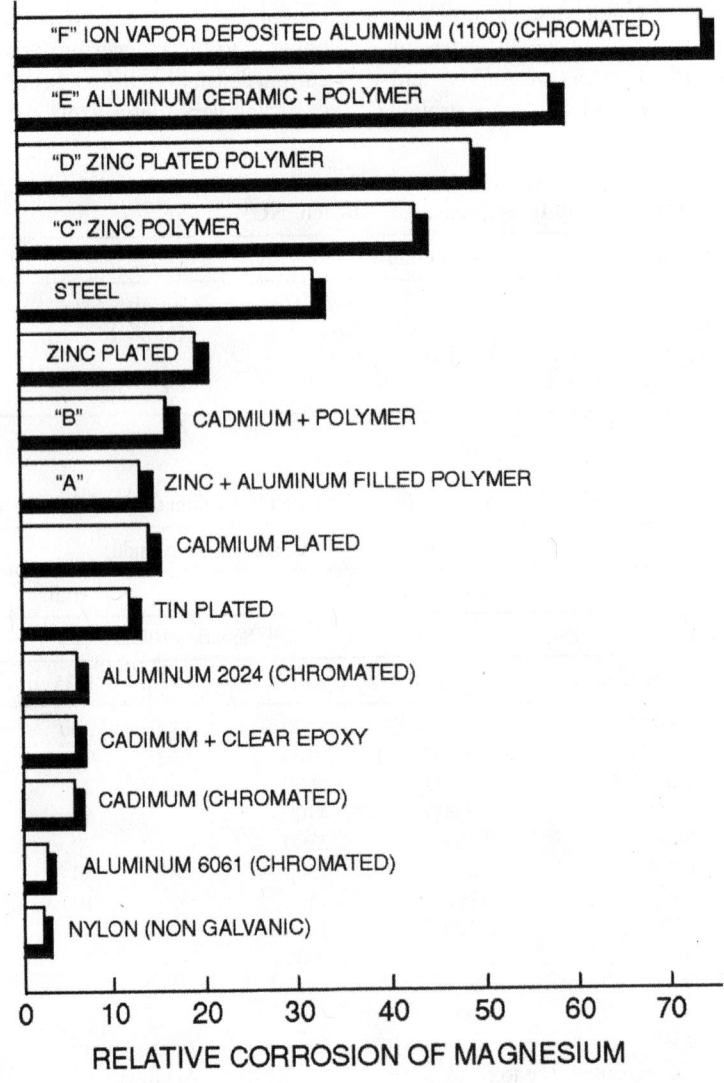

FIGURE 58.3. Galvanic corrosion produced by dissimilar fasteners in AZ91D magnesium alloy. (Reproduced with permission from [24].)

10-day salt spray exposure. Certain zinc and aluminum-filled polymer coatings on steel actually produced more damage to the magnesium than bare steel. This effect may be due to either an increase in the active cathode surface area resulting from the fine metal powders or flake employed in the coatings or it may be due to the presence of a catalytic contaminant on the metal powder surface, such as iron. An inorganic chromate treatment on cadmium electroplate (and perhaps on other electroplates, and on metal surfaces) was as effective in reducing the galvanic attack on magnesium as an epoxy coating. This observation is consistent with the known inhibitive effect of chromates on the cathodic reduction process.

The salt spray test is thought to produce a result biased against zinc due to the rapid cathodic attack on the zinc electroplate produced in this severe test exposure. This attack does not occur in many natural environments. The most compatible fastener coatings are based on zinc plating, with modifications to extend the life of the zinc. These modifications include chromating, silicate treatments, and alloying with tin [25].

In salt spray tests using cast iron disks coupled to AZ91 die cast plates and separated by plastic spacers, it was found that a separation of about 4.45 mm (175 mils) was needed to ensure the absence of galvanic corrosion (Fig. 58.4) [24].

The relative areas of the magnesium anode and the dissimilar metal cathode have an important effect on the galvanic corrosion damage that occurs. A large cathode coupled to a small area of magnesium results in rapid penetration of the magnesium because the galvanic current density at the small magnesium anode is very high, and anodic polarization in chloride solutions is very limited. Painted magnesium should not be coupled to an active cathodic metal if the couple will be exposed to saline or

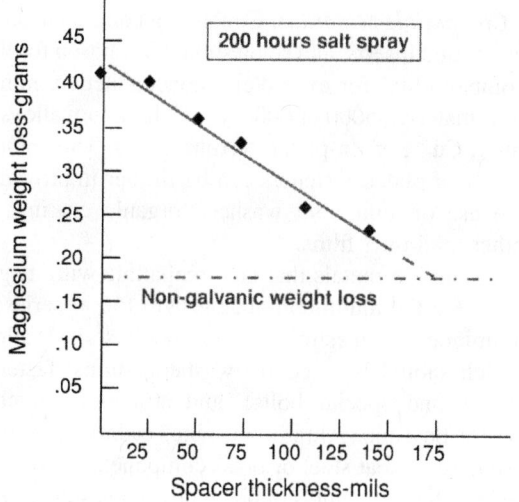

FIGURE 58.4. Effect of spacer thickness on the galvanic corrosion of AZ91 magnesium coupled to cast iron disks through plastic spacers. (Reproduced with permission from [24].)

aggressive environments. A small break in the coating at the junction results in a high concentration of galvanic current at that point. Unfavorable area effects can also be seen in the behavior of some proprietary coatings using aluminum or zinc powder [23].

B2.1. Cathodic Corrosion of Aluminum. Aluminum can be attacked by the strong alkali generated at the cathode when magnesium corrodes sacrificially in static NaCl solutions. Such attack destroys compatibility in alloys containing significant iron contamination, apparently by exposing fresh, cathodic active sites with low overvoltage. The aluminum alloys, having substantial magnesium content (5052 and 5056) are more resistant to this effect but not completely so. A 5052 alloy would meet the essential requirement for a fully compatible aluminum alloy with a maximum of 200 ppm Fe or a 5056 alloy with a maximum of 1000 ppm Fe [23]. Cathodic corrosion of aluminum is much less severe in seawater than in NaCl solution because the buffering effect of magnesium ions reduces the equilibrium pH from 10.5 to ~8.8. The compatibility of aluminum with magnesium is, accordingly, better in seawater and is less sensitive to iron content [26].

Aluminum oxide is amphoteric, that is, soluble in alkaline as well as acid solution. The standard potentials of these two half-reactions are

Acid $$Al^{3+} + 3e^- = Al(-1.66\,V)$$ (58.10)

Alkaline $$H_2AlO_3^- + H_2O + 3e^- = Al + 4OH^-(-2.35\,V)$$ (58.11)

Half-reaction (58.11) has nearly the same standard potential as that for acidic dissolution of magnesium:

$$Mg^{2+} + 2e^- = Mg(-2.37\,V)$$

Commercial aluminum alloys contain several thousand parts per million (ppm) of iron in the form of the intermetallic $FeAl_3$. The mutually destructive galvanic action between magnesium and commercial aluminum alloys in salt water proceeds as follows:

1. Rise in the pH of the liquid in contact with the aluminum member. This is most likely the result of galvanic current flow between the magnesium and the initially passive aluminum.
2. Shift of the aluminum potential in the active direction in accordance with the half-reaction (58.11)
3. Exposure of iron aluminum intermetallic particles (e.g., $FeAl_3$, which then engage in separate galvanic activity with the magnesium). This galvanic current flow accounts for the severe sacrificial corrosion of the

magnesium, and the alkali generated at the cathode ensures continued corrosion of the aluminum in accordance with half-reaction (58.11) [25].

B2.2. Cathodic Damage to Coatings.

Hydrogen evolution and strong alkalinity generated at the cathode can damage or destroy organic coatings applied to fasteners or other accessories coupled to magnesium. Alkali-resistant resins are necessary, but under severe conditions, such as salt spray or salt immersion, which do not simulate adequately a real application, the coatings may be simply blown off by hydrogen, starting at small voids or pores.

B2.3. Prevention of Corrosion and Protection against Galvanic Corrosion [24].

Use Indoors and in Sheltered Outdoor Environments. For indoor use, where condensation is not likely, no protection is necessary. Even in some sheltered outdoor environments, unprotected magnesium components can give good service life providing the absence of water traps, good ventilation, warm component temperature, or the presence of an oil film, and so on.

Seal Faying Surfaces. Sealing compounds, such as non-acidic silicone RTVs, polysulfide, epoxy resins, or plastic tapes, can be employed. If possible the compound or tape should extent beyond the joint interface by 3.2–9.5 mm ($\frac{1}{8}$ – $\frac{3}{8}$ in.) (Fig. 58.5). Iron, and to a lesser extent zinc, based phosphate treatments can replace chromates as inhibitors since the discharge or the formation of CR(VI) in natural water should be avoided.

The joining of two magnesium components invariably involves the use of dissimilar metal fasteners and the formation of a crevice at the joint. Good engineering practice dictates that for corrosive conditions some precautions must be taken (Fig. 58.5). Magnesium faying or mating surfaces should be assembled using "wet assembly" techniques. Inhibited primers or

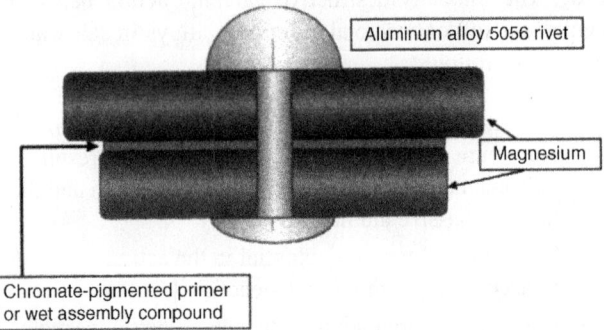

FIGURE 58.5. Schematic of a method to protect faying surfaces in magnesium-to-magnesium assemblies. (Reproduced with permission from [24].)

sealing compounds are placed between the surfaces at the time of assembly. Sealing/jointing compounds of the polymerizing or nonpolymerizing type are preferred as they will remain flexible and resist cracking. Polymerizing-type compounds are also used for caulking operations. In bolted assemblies, the retorquing of bolts a short while after assembly helps to eliminate any joint relaxation problems. For additional protection, mating surfaces can be primed prior to assembly and overpainted after assembly.

Joining Magnesium to Dissimilar Metal Assemblies. Good design can play a vital role in reducing galvanic corrosion (Fig. 58.6). The elimination of a common electrolyte may be possible by the provision of a simple drain or shield to prevent liquid entrapment at the dissimilar metal junction (Fig. 58.7). Alternatively, the location of screws or bolts on raised bosses may also help avoid common electrolyte contact, as would the use of nylon washers, spacers, or similar moisture-impermeable gaskets. The use of studs in place of bolts will, provided the captive ends of the studs are located in blind holes, reduce the area of dissimilar metal exposed by up to 50%.

The use of wet assembly techniques will eliminate galvanic corrosion crevices. Caulking the metal junctions will, by lengthening the electrolytic path, increase the electrical resistance (R_e) of the galvanic couple and so reduce the degree of attack should it occur (Fig. 58.7). Vinyl tapes have also been used to separate magnesium from dissimilar metals or a common electrolyte and so prevent galvanic attack (Fig. 58.8). Finally, overpainting the magnesium and more importantly the dissimilar metal after assembly will effectively insulate the two materials externally from any common electrolyte.

Use Compatible Materials. Contacting components, fasteners and inserts, and so on should be chosen for their compatibility; for example, a nonconductive, nonporous material; 5000 or 6000 series aluminum alloys; or Sn^-, Cd^-, or Zn-plated ferrous alloys. The compatibility of plated fasteners can be further improved by the use of aluminum washers, organic coatings, or other inhibiting films.

Dissimilar metals that are compatible with magnesium are the aluminum–magnesium (5000 series) or aluminum–magnesium–silicon (6000 series) alloys, which should be used for washers, shims, fasteners (rivets and special bolts), and structural members where possible. Aluminum, zinc, cadmium, and tin are used to coat steel or brass components in order to reduce the galvanic couple with magnesium, under mild corrosive environments, but will have minimal effect in corrosive conditions, where additional precautions are required.

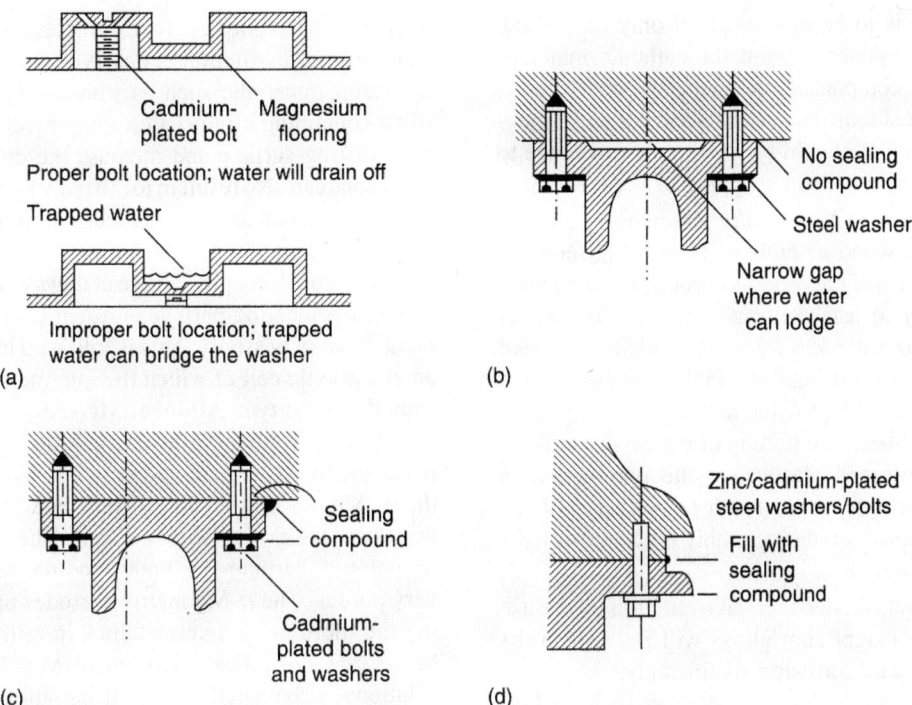

FIGURE 58.6. Design consideration for reducing galvanic corrosion: (a) proper versus improper bolt location, (b) poor sealing practice, (c) good sealing practice, and (d) good sealing practice when direct metal-to-metal contact is required. (Reproduced with permission from [24].)

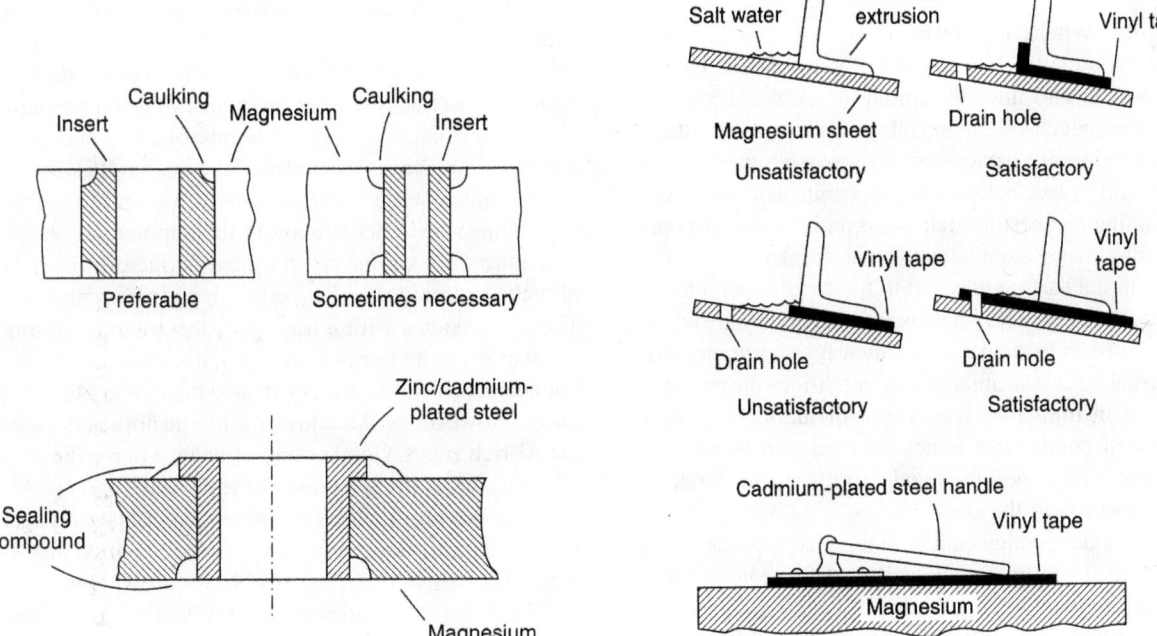

FIGURE 58.7. Examples of good practice for bushing installations. (Reproduced with permission from [24].)

FIGURE 58.8. Proper use of insulating tapes to avoid galvanic corrosion (Reproduced with permission from [24].)

If painting is to be employed on only one of the contacting components, paint the cathodic material. Painting both components is a better practice. Paints employed on cathodic components and complete magnesium assemblies should be chosen for resistance to alkalis in order to prevent stripping of the coating.

Joining Magnesium to Nonmetallic Assemblies. Joining magnesium-to-wood assemblies presents an unusual problem because of the water absorbency of wood and their tendency to leach out natural acids. To protect magnesium from attack, the wood should first be sealed with paint or varnish and the faying surface of the magnesium should be treated as magnesium-to-magnesium assemblies. The joining of magnesium to carbon fiber reinforced plastics, in the presence of a common electrolyte, could result in corrosion of the magnesium unless similar assembly precautions were observed.

High-Purity Alloys. Under corrosive conditions, the use of high-purity magnesium alloys will not reduce the effects of galvanic corrosion significantly.

B3. Localized Attack

General corrosion can lead to localized corrosion, which is favored by a weak electrolyte and small anode/cathode relative area ratios. Localized attack takes the form of pitting, crevice, and filiform corrosion. Intergranular corrosion can also be considered as a localized attack due to metallurgical structure; however, in the case of magnesium alloys, this type of corrosion can be named more properly as granular attack.

B3.1. Pitting.
When corrosion occurs on a smooth machined magnesium alloy surface, this surface is roughened by the chemical action, and after the initial attack the degree of roughness does not change appreciably. In atmospheric attack the roughening is really a microscopic form of pitting. There is a noticeable difference between the appearance of the aluminum-containing magnesium-rich alloys and the zinc/zirconium-containing magnesium alloys. In the former, the microscopic pits in the surface exposed to the weather tend to be narrow and relatively deep, whereas in the latter they are wider and tend to overlap, leading to a slightly wavy appearance [6].

In the usual industrial atmospheric conditions the attack is uniform, but in immersed conditions, including corrosion under pools of condensate, attack may be, and usually is, irregular; some areas become anodic to other areas and, as corrosion proceeds at the anodic areas, a pitting develops. The unequal attack, which occurs in tap water, condensate, and other mild electrolytes, may lead to perforations of thin-gauge sheet and even to deep pitting of castings.

In stronger electrolytes, the effect is variable. In chloride solutions, such as seawater, attack on the metal usually results in pitting of some areas only, for reactive metallic surface, by sand blasting, for example, attack may be so rapid that uniform dissolution is observed [6].

Tramp materials, such as iron-containing shot blast or silica-containing sandblast cleaning media can be entrained on a casting surface and increase the corrosion rate. Flux inclusions can also result in localized attack, but this problem has been eliminated by the current industry practice of fluxless melting.

Stable corrosion pits initiate at flaws adjacent to a fraction of the intermetallic particles present [27] as a result of the breakdown of passivity. This is followed by the formation of an electrolytic cell of which the intermetallic particle is the cathode of the type AlMnFe, $Mg_{17}Al_{12}$, or Mg_2Cu and the surrounding Mg matrix the anode [28]. Hydrogen evolution is the predominant cathodic reaction and where applicable the Fe/Mn ratio within the AlMnFe intermetallics appears to determine the overall corrosion rate. There is no evidence of initiation at particle-free areas, and the resultant surface is very porous. The α-Mg matrix corrodes preferentially leaving the more noble intermetallics in relief along the grain boundaries [29]. The corrosion of Mg–Al alloys in NaCl solutions is characterized by pit initiation and filiform corrosion, which develops into cellular corrosion. Metallographically polished AE alloys with a high Al content exhibited significantly longer induction times for pit initiation than AS, AM, and AZ alloys (ASTM designations, see Section A3.1. at open-circuit condition in the 5% NaCl solution. Pit initiation and growth normally occurred within 1 h of immersion on alloys such as AS41, AM80, and AZ91, whereas AE41, AE42, and AE46 exhibited induction times of a few hours and AE81 more than 24 h.

The AZ-, AS-, and AM-type alloys maintain a bright and shiny appearance in the unattacked part of the corroded surface, whereas the AE alloys tend to become dull due to buildup of a relatively thick hydroxide film and formation of numerous small pits, only a few micrometers in depth. The corresponding backscattered electron image (BEI) and X-ray maps indicate a high chloride concentration in the pits and high aluminum concentration in the unpitted areas.

Figure 58.9 shows the corroded surface of alloy AE81 after the hydroxide film has been stripped off in chromic acid. The grain bodies with a low Al concentration (location B) corrode at a faster rate than the Al-rich regions along the grain boundaries (location A), as can also be seen in other Mg–Al alloys. However, on AE alloys, the pits do not easily penetrate the Al-rich zones. Good pitting resistance of the die-cast AE alloys is, therefore, attributed to the presence of these Al-rich zones, which appear to act as barriers against pit propagation. If these barriers are removed by homogenization heat treatment, the corrosion resistance is reduced. Homogenized AE81 exhibited corrosion rates > 100 times higher than the As cast material during a 3-day immersion test in 5% NaCl solution. It is not yet clear whether this unusual sensitivity of corrosion to heat treatment is related to the absence of Mn in

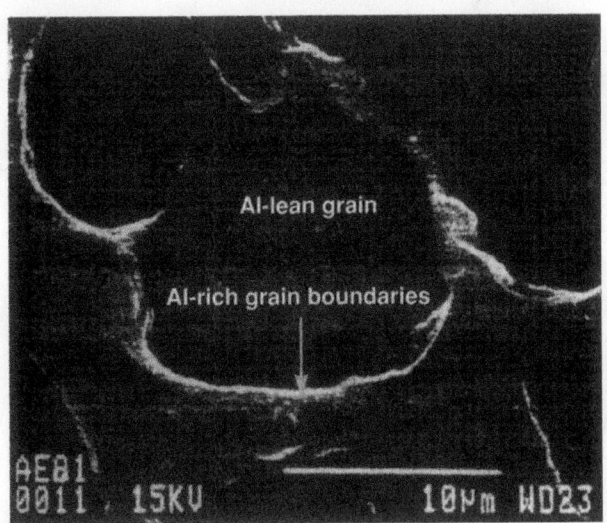

FIGURE 58.9. Morphology of corroded AE81 after removal of the hydroxide film. The grain boundaries with Al-rich areas are more resistant than the Al-lean grain [30]. (Reprinted with the permission from SAE paper No. 930755 © 1993, Society of Automotive Engineers, Inc.)

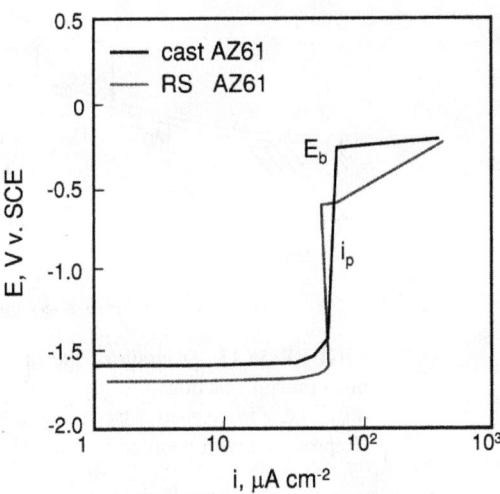

FIGURE 58.10. Anodic polarization scans for cast and rapidly solidified AZ61 (Mg–6Al–1Zn) in pH 10 sodium carbonate–sodium bicarbonate solution with 100 ppm NaCl [31].

this AE81 alloy. The corrosion rate of alloys AM80 and AZ91 were only moderately influenced by a similar heat treatment. In general, homogenized specimens exhibited deeper localized attack than the As cast material [30].

The few studies of pitting of Mg and Mg alloys have been concerned with comparing the pitting behavior of cast to that of rapidly solidified Mg alloys. In these studies, two parameters indicative of pitting resistance were measured: (a) i_p, the passive current density, which is a measure of the protective quality of the passive film, and (b) E_b, the breakdown potential, which indicates the resistance to the breakdown of the passive film that results in pitting attack. The more positive the value of E_b, the more protective the film on the metal surface.

Makar and Kruger [31] showed that rapidly solidified AZ61 (Mg–6 Al–1 Zn) exhibited a breakdown potential that was ~ 200 mV higher than the value found for cast AZ61 in a buffered carbonate solution (pH 10) containing various levels of Cl $^-$ (Fig. 58.10); the higher the value of E_b, the greater the resistance to pitting. In a buffered borate solution (pH 9.2) containing various levels of Cl $-$, there was no improvement in the E_b values observed for the rapidly solidified alloy. However, the pits formed at 1 V below E_b were hemispherical, apparently forming at defects in the black film that is observed when the cast AZ61 surface is at -1.5 V saturated calomel electrode (SCE). No small hemispherical pits were found on the rapidly solidified AZ61 [20].

B3.2. Crevice Corrosion. Although a form of attack that occurs at narrow gaps (crevices) appears similar to crevice corrosion, it is somewhat different because the corrosion

observed is caused by the retention in the crevice of moisture, which, being unable to evaporate, promotes the corrosion of the metal in the narrow recess over extended periods. True crevice corrosion is caused by the development of an anodic region within the crevice because of the exclusion of oxygen and a cathode region outside the crevice where the oxygen concentration is high. Corrosion of magnesium is relatively insensitive to oxygen concentration differences [20]. Corrosion in crevices between Magnox A (Mg–0.18 Al) and mild steel, and between Magnox A and Polytetrafluoroethylene (PTFE) occurred in 200 g/m³ NaOH (pH > 11.5) if the Cl $^-$ concentration was ~ 1 g/m³ or more [32].

B3.3. Filiform Corrosion. Filiform corrosion is typically associated with metal surfaces having an applied protective coating [33]. Its occurrence on bare Mg–Al alloys indicates that highly resistant oxide films can be naturally formed [34]. Filiform corrosion does not occur on bare pure Mg, indicating the strong influence of alloying elements on corrosion products and behavior. The overall variables of significance are temperature, material structure, and polarization of the microgalvanic cell [29]. A diagram showing the mechanism and the products of the filiform corrosion cell of magnesium is presented in Figure 58.11 [35].

After the initiation period of corrosion pits, filiform corrosion dominates the morphology as narrow semicylindrical corrosion filaments project from the pit [36]. Radial propagation is at a much slower rate than that of the filament tips projecting outward. Lunder et al. [37] observed that propagation of the filaments occurs with voluminous gas evolution at the head while the body immediately behind passivates. Electrochemical transport of chloride ions to the head of the filament appears to be an essential component as is precipitation of insoluble $Mg(OH)_2$ by the anodic reaction

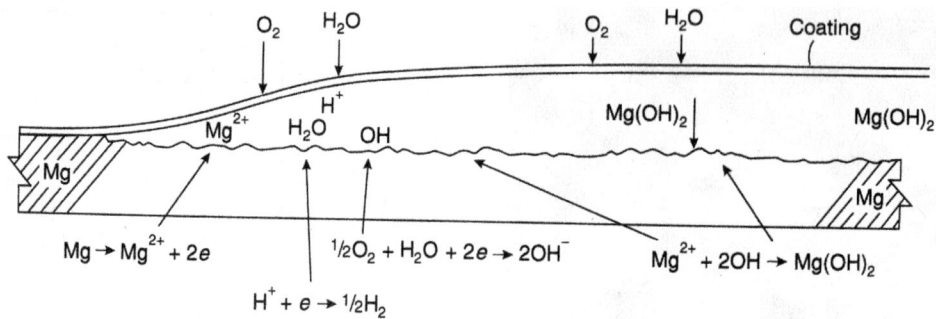

FIGURE 58.11. Diagram of the filiform corrosion cell in magnesium. Corrosion products and predominant reactions are identified. Filiform corrosion is a differential aeration cell driven by differences in oxygen concentration between the head and tail sections about 0.1–0.2 V. [35] (Reproduced with permission from ASM International, Materials Park, OH.)

with Mg^{2+} ions elsewhere along the filament. The corrosion products may vary because they depend on the environment.

Filiform corrosion initiates and then develops into cellular or pitting corrosion. Cellular corrosion occurs when a primary initiation site and secondary pits, formed along the filiform corrosion filaments, coalesce to form a corrosion cell with an epicenter at about the original pit initiation site. The growth rate is at a steady radial rate independent of the material temper. Cellular corrosion continues until the cells impinge on one another, at which point they terminate, thereby forming clearly defined cell boundaries [34]. In the As cast condition, compositional variations orient the growth of filiform corrosion. In homogenized alloys, filiform corrosion propagates transgranularly along crystallographic directions. In Mg–Al alloys, precipitation heat treatment disperses the secondary Mg_{17}–Al_{12} precipitate, which blocks transgranular propagation of filiform corrosion, thereby reducing the corrosion rate [19].

B3.4. Granular Corrosion. Intergranular corrosion of magnesium alloys does not occur because the grain boundary constituent is invariably cathodic to the grain body. Corrosion of magnesium alloys is concentrated on the grains, and the grain boundary constituent is not only more resistant to attack, but is cathodically protected by the neighboring grain.

B3.5. Stress Corrosion Cracking. Pure magnesium is not susceptible to stress corrosion cracking (SCC). The Mg–Al alloys have the greatest SCC susceptibility of all the magnesium alloys, and susceptibility increases with increasing aluminum content. The Mg–Zn alloys have intermediate susceptibility, and the alloys that contain neither aluminum nor zinc are the most SCC resistant. No special heat treatments have been found that will reduce or eliminate SCC [38]. Failures of wrought AZ80 aircraft components resulted from excessive assembly and residual stresses [39–40].

Stress sources likely to promote cracking are weldments and inserts. Welded structures of these alloys require stress-relief annealing. Magnesium castings have been shown to fail in laboratory tests under tensile loads as low as 50% of yield strength in environments causing negligible general corrosion. The apparent low incidence of SCC service failures of castings is attributable to low stresses actually applied or to stress relaxation by yielding or creep when a fixed deflection is imposed.

Although laboratory tests are useful in encouraging conservative design of magnesium alloy structures, results of long-term atmospheric tests of tensile-loaded specimens are considered to be very important. Short-term accelerated tests, such as sodium chloride/potassium chromate (NaCl/K_2CrO_4) tests, do not predict SCC behavior reliably in practice [23].

The SCC in magnesium alloys is usually transgranular with significant secondary cracking (branching). Initiation of these cracks has been found to occur invariably at corrosion pits. Mixed transgranular and intergranular crack propagation, and occasionally totally intergranular cracking, have also been observed during magnesium SCC (Fig. 58.12) [41].

Inhibition, by nitrate or carbonate ions, of SCC in salt—chromate solutions is believed to be associated with the formation of a stronger, more stable, or more readily repaired passive film [38].

The SCC can also occur in many other dilute aqueous solutions, including the following, in order of decreasing severity: NaBr, Na_2SO_4, NaCl, $NaNO_3$, Na_2CO_3, $NaC_2H_3O_2$, NaF, and Na_2HPO_4. *SCC* has also been reported in dilute solutions of KF, KHF_2, HF, KCl, CsCl, NaI, KI, $MgCO_3$, NaOH, and H_2SO_4, HNO3, and HCl acids. When pH is > 12, magnesium alloys become very resistant to SCC.

Increasing temperature accelerates SCC susceptibility of magnesium alloys, but also improves passivation. Creep deformation could improve SCC resistance. Cathodic polarization has been found to reduce SCC in many studies. Anodic potentials increase SCC susceptibility. Fairman and Bray [42] showed that high anodic potentials, which can

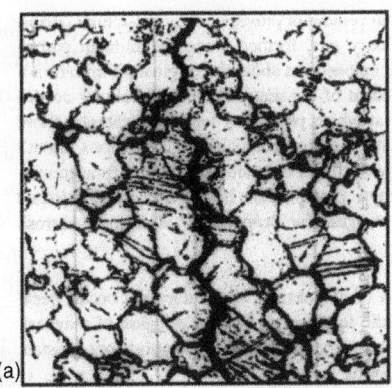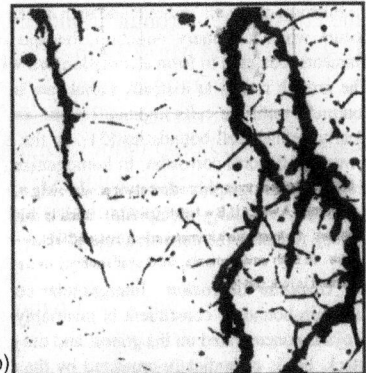

FIGURE 58.12. The SCC in an extruded Mg–6Al–1Zn alloy tested in a salt–chromate solution, showing intergranular crack (a) in the furnace-cooled alloy and transgranular propagation (b) in the water-quenched material [41]. (Reproduced with permission from ASM Transactions, Pittsburgh, PA)

produce a passive film in a single phase, Mg–Al alloys prevent SCC. The $Mg_{17}Al_{12}$ phase promoted pitting and SCC in a multiphase alloy.

It has been recommended that constant stresses applied for long periods of time should be limited to 30–50% of the yield strength to prevent SCC of magnesium alloys in normal atmospheric environments. It has been suggested that the SCC threshold stress is associated with the onset of plastic deformation (i.e., the elastic limit). The 30% yield strength limit recommended for die cast alloy AZ91 correlates with the elastic limit of this material, reported to be approximately one-third of the yield strength [42].

Dissolution models used to interpret transgranular and intergranular SCC include preferential attack, film rupture, or tunneling in specific dissolution processes. Pickering and Swann [43] have proposed corrosion tunnels in certain SCC systems. A mechanically weak, tubular pitted surface is produced along active slip planes. It has been proposed that the role of corrosion is to produce pits or other stress concentrations that cause cracking by cleavage processes, and to remove obstacles that stop the crack. Fairman and Bray [44] proposed that the passage of dislocations on slip planes rupture the surface film, allowing a corrosion pit to develop, which then initiates cleavage.

Liu [45] first suggested that cathodically generated hydrogen be related to magnesium SCC. Experimental evidence supported this model, strengthened by the fact that SCC occurs at crack velocities at which only absorbed hydrogen should be present at the tip [46]. A weak, stress-induced magnesium hydride may form and has been observed on the surface of magnesium SCC fracture [38].

Prevention of SCC is based on avoiding alloys that are susceptible to SCC and environments that cause SCC, and on maintaining the stress below the threshold stress for SCC to occur.

Recommendations to avoid SCC [47] are as follows:

1. The constant stress must be below a threshold level reported to be 30–50% of the tensile yield strength [48].

2. It has been recommended that inserts with a wall thickness greater than 1.25 mm (0.050 in.) be preheated before casting because cast-in inserts may cause SCC due to local residual stresses created in the surrounding magnesium [48].

3. Bolted or riveted joints can also produce high local stresses that can cause SCC, so that attention should be given to proper joint design and construction. Examples include the use of preformed parts, avoiding overtorquing of bolts, and providing adequate spacing and edge margins for rivets [48].

4. Tensile residual stresses from welding were found to be particularly dangerous and, as a result, a low-temperature thermal stress relief treatment has become a recommended practice for welded assemblies.

5. Shot peening and other mechanical processes that create compressive surface residual stresses may also be effective in increasing SCC resistance [49].

6. Cathodic polarization may reduce, or even prevent, SCC of magnesium alloys in aqueous solutions.

7. Coatings have been shown to extend life, but not to totally prevent SCC, with breaks in the coating reducing protection [49]. In one laboratory study, an inorganic coating was found to accelerate SCC of a SCC-resistant alloy under certain conditions [50].

8. Cladding of a susceptible magnesium alloy with a SCC-resistant sheet alloy.

B4. Corrosion Fatigue

There is no endurance limit for magnesium and its alloys in fatigue under corrosive conditions, and the slope of the fatigue curve varies with the corrosive environment and the alloy composition. The Mg–1.5%Mn and Mg–2%Mn–0.5% Ce alloys are more resistant to corrosion fatigue than alloys containing aluminum and zinc; also, media such as 3% NaCl or seawater produce a much more rapid drop in the fatigue curve than does tap water [2]. Substantial reductions in fatigue strength of magnesium alloys are shown in laboratory tests using NaCl spray or drops. Such tests are useful for comparing alloys and heat treatments.

Figures 58.13 and 58.14 show data obtained on 1.6 mm (0.064 in.) sheet alloys Mg–6%Al–l%Zn–0.2%Mn and Mg–3%Al–l%Zn–0.3%Mn tested with plate-type bending fatigue equipment in a chloride-containing spray of 0.01% NaCl. Figure 58.13 also shows data obtained on protected and unprotected sheet to determine the effect of normal laboratory exposure. The two rates of spray shown in Figure 58.13 produced the same decrease in fatigue strength.

Both alloys had approximately the same susceptibility to fatigue under corrosive conditions. Unprotected metal in the laboratory atmosphere had slightly lower fatigue strength than when protected.

Coatings that exclude the corrosive environment are considered to provide the primary defense against corrosion fatigue [23]. The corrosion environment was significantly detrimental relative to the air environment. Quasicleavage fatigue crack growth mechanisms have been identified in corrosion fatigue of AZ91E-T6 cast magnesium alloy in both air and 3.5% NaCl. Final fracture regions of samples in both environments were predominantly quasicleavage with some ductile dimples [52].

Under fatigue loading conditions, microcrack initiation in Mg alloys is related to slip in preferentially oriented grains. Quasicleavage usually occurs in the initial stages of fatigue crack growth, which is common for hexagonal close-packed

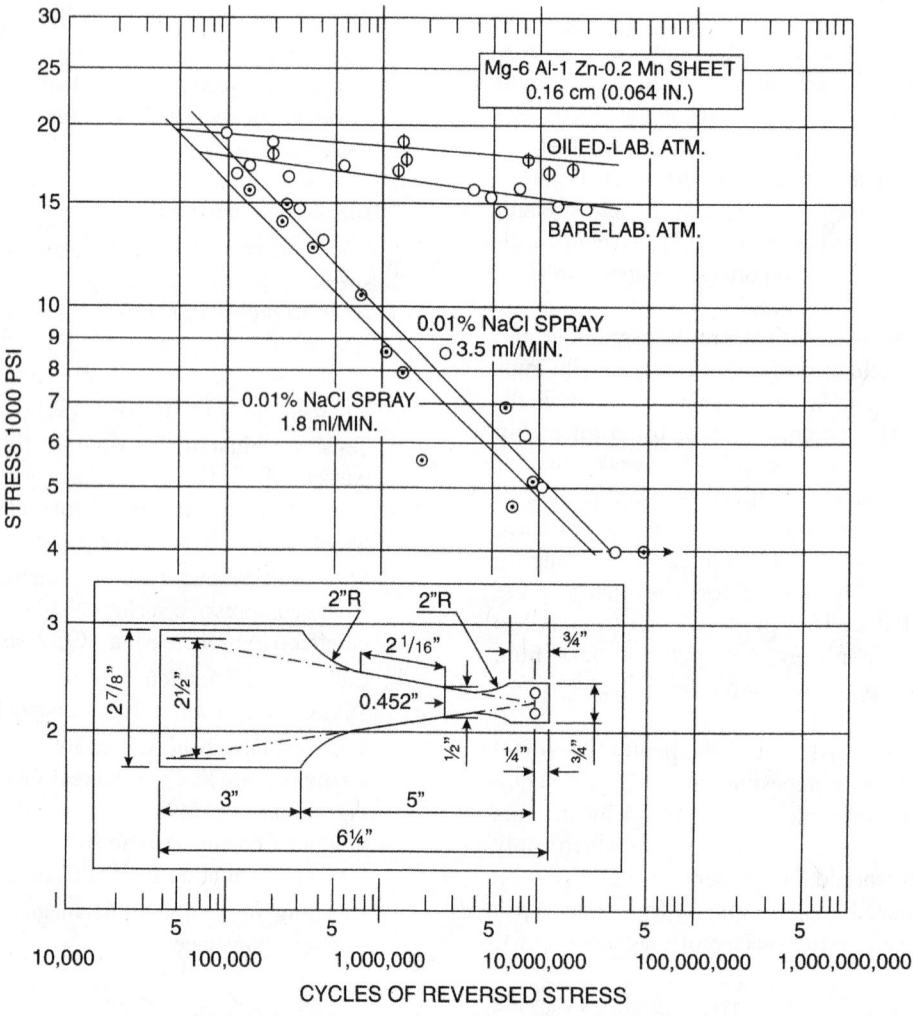

FIGURE 58.13. Effect of spray intensity of 0.01% sodium chloride on the resistance to fatigue of precipitated Mg–6%Al–l%Zn–0.2%Mn sheet [51]. Specimen size—plate-type specimen 1.6 mm (0.064 in.) thick. Surface preparation–aloxite ground. Temperature–about 30°C (90°F).

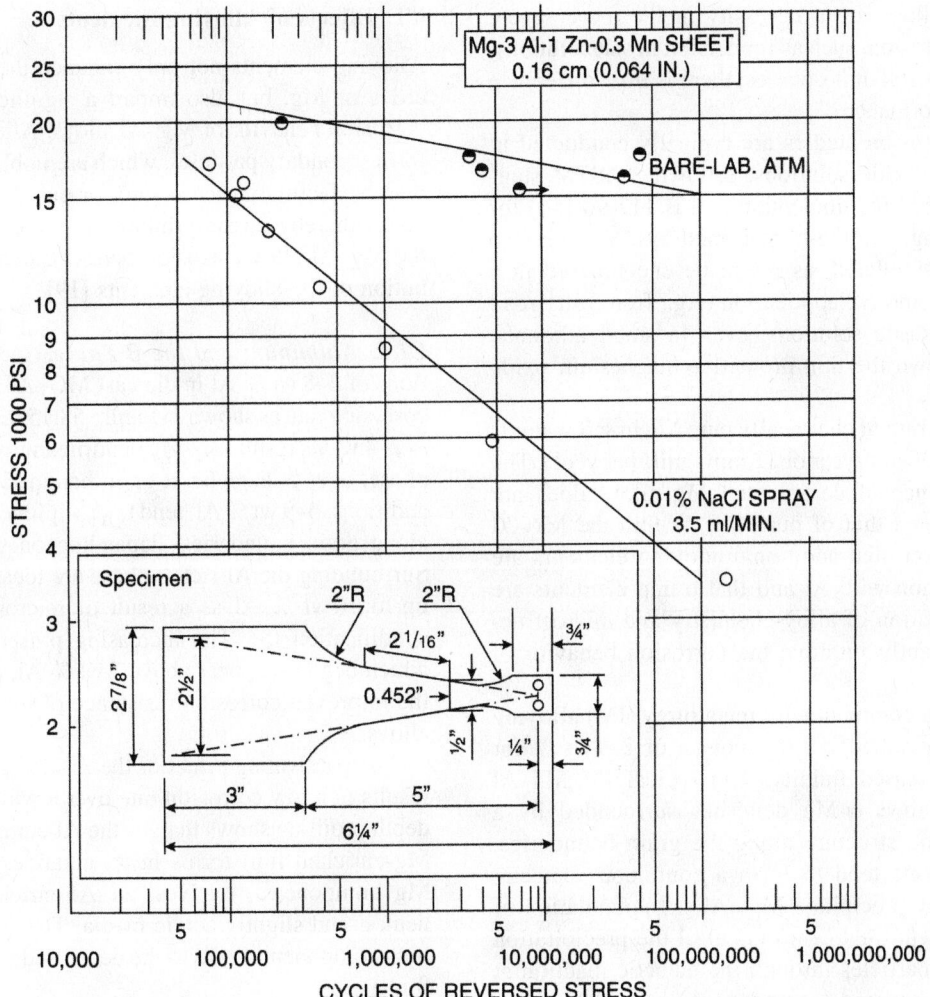

FIGURE 58.14. Effect of 0.01% NaCl spray on the resistance to fatigue of precipitated Mg–3%Al–1 %Zn–0.3%Mn sheet [51]. Specimen size—plate-type specimen 1.6 mm (0.064 in.) thick. Surface preparation—aloxite ground. Temperature—about 30°C (90°F).

cells. Further crack growth micromechanisms can be brittle or ductile and trans- or intergranular, depending on the metallurgical structure and environmental influence [53]. In general, reduction of temperature increases the fatigue life of Mg alloys mainly by lengthening the crack initiation period [54].

Rotating bending fatigue strengths were reduced by 50% in fretting conditions compared to those in air [51]. Oxides and nitrides are often formed on the surfaces of Mg parts subjected to fretting fatigue condition. Surface rolling, sand-blasting, or shot peening can reduce fretting.

C. CORROSION CHARACTERISTICS OF MAGNESIUM ALLOY SYSTEMS

Each group of alloys has its characteristic corrosion behavior that results from metallurgical properties or presence of

certain intermetallics. The general forms and types of corrosion and the specific properties of every alloy should be considered in developing a corrosion prevention strategy. The following groups of alloys can then be identified:

A. Zirconium-free casting alloys

    1. Magnesium–aluminum alloys

    2. Magnesium–zinc

B. Zirconium–containing casting alloys

    3. Mg–Zn–Zr alloys

    4. Mg–Re alloys

    5. Mg–Th alloys

    6. Mg–Ag alloys

    7. Wrought alloys, with the same divisions as cast alloys

    8. Novel alloys

High-purity alloys are a necessity in the more severe conditions of corrosion, such as immersion in salt solutions; however, in industrial atmospheres, there is little difference in corrosion performance.

Saltwater corrosion studies are typically conducted in 3–5% sodium chloride solutions, following ASTM standards G 31-72 [55] for immersion and B 117-90 [56] for salt spray testing. In these test methods, a corrosive environment is simulated, as might be encountered in a marine or an automotive application (e.g., from salty road splash). The chloride solutions, even in small amounts, usually break down the thin protective magnesium oxide film.

The corrosion rate of chemically pure Mg in salt water is in the range of 0.30 mm/year or 12 mpy (mils per year). The corrosion resistance of commercial Mg alloys does not significantly exceed that of pure Mg. Within the Mg–Al alloy system, given that additional alloying elements are used in conjunction with Al and that tramp elements are present, manipulation of alloy chemistry and microstructure can significantly improve the corrosion behavior of these alloys.

Aluminum is a common ingot metallurgy (IM) alloying element typically added in the amounts of 2–9 wt % for strength and increased fluidity. The typical IM Mg–Al microstructure shows α-Mg dendrites surrounded by a two-phase eutectic structure along the grain boundaries. Greater Al contents tend to form a continuous eutectic structure and may precipitate $Mg_{17}Al_{12}$. Manipulation of Al content and heat treatment to control the precipitation of β, $Mg_{17}Al_{12}$, particles through the eutectic reaction at 28 wt % Al can be used to produce a variety of microstructures like precipitated lamellar β phase in heat-treated AZ91 [19].

The slower solidification rates for gravity versus pressure die castings cause increased average grain sizes and increased corrosion rates, although the tolerance levels are not changed. The smaller grain size of the die casting product results in a finer dispersion of the detrimental material, thereby minimizing its effect as a cathode for localized corrosion. This effect is very much evident in rapidly solidified materials.

For die cast Mg–Al alloys in the AM, AS, AZ, and AE series tested by salt spray and by immersion in 5% NaCl solution, the corrosion rate increases when the Al content decreases below ~ 4%. The AE alloys exhibit a high resistance to localized attack because the Al-rich coring along grain boundaries appears to act as an efficient barrier against pit propagation in these alloys. The Fe-rich phases are particularly detrimental, but Al–Mn phases with a low Al/Mn ratio may also have a high cathodic current output. The phases Mg_2Si and Al_4MM (MM = misch metal) appear to be harmless from a corrosion point of view [30].

C1. Effects of Alloying Elements

Alloying elements not only enhance the mechanical properties of Mg, but also impart a significant impact on the corrosion behavior of Mg–Al alloys. Alloying elements can form secondary particles, which are noble to the Mg matrix, thereby facilitating corrosion, or enrich the corrosion products, thereby possibly inhibiting the corrosion rate. Thus, the Mg–Al alloy corrosion behavior depends on the distribution of the alloying elements [19].

C1.1. Aluminum and the β Phase.
Increasing concentrations of 2–8 wt % Al in die cast MG–Al alloys decrease the corrosion rate as shown in Figure 58.15. Low Al additions, of ~ 2–4 wt %, result in α-Mg dendrites surrounded by the two-phase, $\alpha + \beta$, eutectic at grain boundaries, whereas higher additions, 6–9 wt % Al, tend to precipitate distinct β particles along grain boundaries, depending on solidification rates. Surrounding the Al-rich β phase are local concentrations of up to 10 wt % Al as a result of microsegregation during solidification [5]. The increasing presence of β particles, which begin to appear above 2 wt % Al, may cause, in part, the improved corrosion resistance of the higher Al-content alloys.

The passivating effect of the Al-rich β phase, $Mg_{17}Al_{12}$, results in a low corrosion rate over a wide pH range. Auger depth profiling shows that, as the Al component dissolves, a Mg-enriched film forms in an alkaline media, and as the Mg component dissolves, an Al-enriched film forms in neutral and slightly acidic media. The synergistic effect of both components leads to the decreased corrosion rate of the β phase.

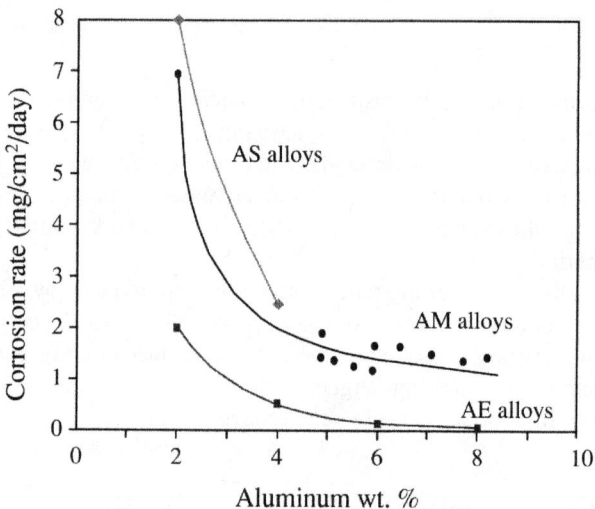

FIGURE 58.15. Corrosion rate of Mg–alloy die cast rods immersed in 5% NaCl solution as a function of Al content [30]. (Reprinted with the permission from SAE paper No. 930755 © 1993, Society of Automotive Engineers, Inc.)

Additions of Al by rapid solidification processing results in decreased corrosion rates without precipitation of the β phase. Faster solidification of IM alloys disperses fine $Mg_{17}Al_{12}$ particles, which increases the corrosion resistance, as does controlled precipitation of the β phase [57]. Increasing Al concentrations have a beneficial effect on the corrosion behavior of Mg–Al alloys, but the specific mechanism depends on the distribution of the Al within the magnesium matrix.

In addition to the alloying ingredients are added certain other metals that usually present in small amounts. In the alloys containing aluminum, for example, iron usually amounts to ∼ 0.02–0.05%. By special techniques and care in melting, this level can be reduced to about one-tenth of this concentration. Such high-purity alloys have much better resistance to salt water than do those of normal purity, but their corrosion behavior in industrial atmospheres is very similar. Furthermore, the practical value of the higher resistance to corrosion is largely offset when components are used in electrical contact with other more cathodic metals. The effect of a steel bolt, for example, even when it has been zinc or cadmium plated, is much greater at the point of contact than that of the local cathodes in the impure alloys. Galvanic corrosion at joints with other metals is not markedly less in the case of the high-purity alloys. Nevertheless, such alloys have their place, and, when they can be used without other metal attachments, provide better intrinsic resistance to corrosion by seawater than the alloys of normal purity [6].

Rare earths (RE) are typically added to Mg–Al alloys as cerium-based misch metal containing lanthanum, neodymium, and praseodymium. A typical composition of MM is 50% Ce, 25% La, 20% Nd, and 3% Pr. These have very low solubilities in Mg (Ce, 0.09; La, 0.14; Nd, 0.10; and Pr, 0.09 at %) [4] and react with Al to form Al_4RE intermetallics [58]. These intermetallics, with their high melting temperature, resist coarsening relative to Mg_2Si and provide enhanced creep resistance at higher temperatures. Compositions of solidified phases are given in Table 58.7 [37].

Corrosion behavior is optimized through alloy chemistry, by minimizing the cathodic sites, which evolve hydrogen gas, or by enriching the corrosion product film, which can inhibit hydrogen gas evolution and decrease the corrosion rate. Microstructural enhancements, which refine the microstructure and homogenize the distribution of alloying elements also, disperse potentially deleterious elements, thereby enhancing corrosion resistance [19]. The potentials of intermetallic phases, prepared synthetically from the pure components by controlled solidification procedures, are given in Table 58.7 [30].

C1.2. Effects of Zn and Si Additions.
Zinc makes the Mg alloy electrochemically more noble, thereby minimizing the corrosion rate [59]. Silicon is intentionally added to only the AS alloys, to combine with Mg, forming Mg_2Si, which

TABLE 58.7. Corrosion Potentials of Synthetically Prepared Intermetallic Phases after 2 h in Deaerated 5% NaCl Solution Saturated with Mg(OH)$_2$ (pH 10.5)^a

Compound	Corrosion Potential (V/SHE)
Al_3Fe	−0.50
$Al_3Fe(Mn)$	−0.71
$Al_6(MnFe)$	−0.76
$Al_6Mn(Fe)$	−0.86
Al_4MM	−0.91
β-Mn	−0.93
$Al_8Mn_5(Fe)$	−0.96
$Mg_{17}Al_{12}(p)$	−0.96
Al_8Mn_5	−1.01
$Al_4Mn(Fe)$	−1.16
Al_4Mn	−1.21
Al_6Mn	−1.28
Mg_2Si	−1.41
Mg 99.99%	−1.42

^{a}Reprinted with permission from SAE, paper No. 930755, © 1993, Society of Automotive Engineers, Inc.)

precipitation strengthens the alloy and is relatively innocuous to the corrosion behavior. The compound Mg_2Si has a corrosion potential of − 1.65 (V SCE), close to the − 1.66/V SCE value for pure Mg in 5% NaCl solution saturated with Mg(OH)$_2$, pH 10.5 [30].

C1.3. Tramp Element Tolerance Levels.
The elements Fe, Ni, and Cu are common tramp elements picked up during melting, handling, and pouring operations. Their influence can be seen in Figure 58.16 for die cast AZ91 corrosion specimens in which the tramp elements were singularly increased.

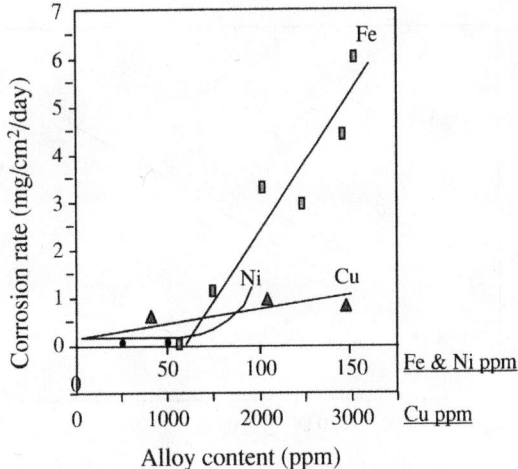

FIGURE 58.16. Die cast AZ91 salt spray performance versus tramp element content. (Reprinted with permission from [60], SAE paper no. 850417 © 1989, Society of Automotive Engineers, Inc.)

TABLE 58.8. Proposed Tramp Element Tolerance Level for Selected Me–Al Die Casting Alloys[a]

Alloy	Fe/Mn	Fe(max)	Cu(max)	Ni(max)
AZ91B	0.032	0.0050	0.030	0.002
AM60B	0.021	0.0050	0.010	0.002
AS41B	0.010	0.0035	0.020	0.002
AE42X1	0.020	0.0050	0.050	0.005

[a]Reproduced with permission from [61]. Copyright ASTM.

The specific ASTM tolerances are given in Table 58.8 (B94-92) [61]. These are typically the same or lower for ingots, as tramp elements are commonly picked up during the melting and pouring operations.

C1.4. Effect of Iron and the Fe/Mn Ratio. The Mg–Fe phase diagram shows a very low solid solubility of Fe in magnesium (9.9 ppm). In the absence of Mn, virtually all the Fe precipitates in magnesium as Al_3Fe, which has a highly cathodic corrosion potential (Table 58.7). Within an appropriate medium, Al_3Fe acts as an effective cathode, catalyzing the reduction reaction, especially hydrogen evolution, which controls the corrosion reaction. Due to the low solubility of Al_3Fe in Mg, increasing additions of Al result in smaller tolerance levels for Fe.

Typically, up to 1 wt % of Mn is added to improve corrosion resistance by reducing the potential difference between iron-containing particles and the matrix. Its beneficial effect is attributed to either Mn combining with the Fe and precipitating to the bottom of the crucible and/or reacting with the Fe left in suspension during solidification [19].

The relationship between the Fe/Mn ratio in the AlMnFe phase and the corrosion rate is shown in Figure 58.17. Mn in

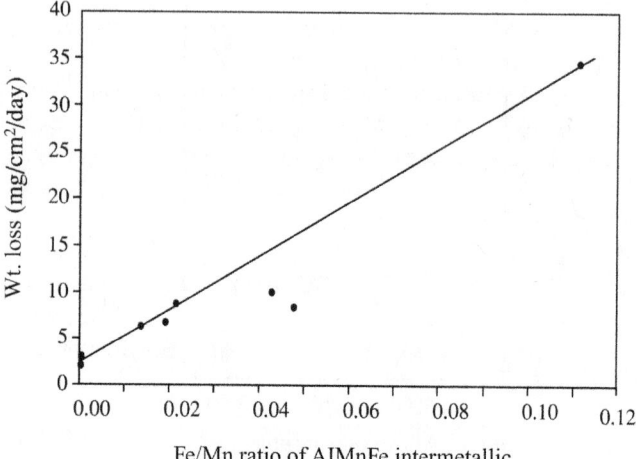

FIGURE 58.17. Relationship between the Fe/Mn ratio in the AlMnFe phase (up to 1 mm in size) and the corrosion rate [27]. (Reprinted with the permission from NACE International.)

excess of that needed to render the Fe content ineffective could be detrimental to corrosion resistance.

C1.5. Summary. In summary [37],

1. The corrosion rate of high-purity die-cast Mg alloys in chloride environment decreases rapidly with increasing aluminum content, up to $\sim$ 4 wt %. Further Al additions, up to $\sim$9 %, gives only a modest improvement in the corrosion resistance. During immersion testing, AE alloys exhibit a lower corrosion rate than AS, AM, and AZ alloys with similar Al content.

2. Intermetallic compounds containing more than a few percent iron are detrimental because they function as efficient cathodes. However, binary Al–Mn phases with a low Al/Mn ratio may also exhibit a relatively high cathodic current output, causing an increase in the overall corrosion rate.

3. The high corrosion resistance of the AE alloys appears to be related to the presence of passive Al-rich zones along the grain boundaries, acting as barriers against pit propagation.

4. Alloying with silicon does not have an important influence on the corrosion properties because the Mg_2Si phase formed is a poor cathode.

5. The Al_4MM phase particles precipitated in AE alloys exhibit a passive behavior and do not affect the corrosion process to a significant extent. A high resistance to localized corrosion is observed for the AE alloys with a high Al content.

C2. Influence of Heat Treatment

Heat treatment can drastically alter the size, amount, and distribution of the precipitated (β phase, $Mg_{17}Al_{12}$, which in turn alters the corrosion behavior of IM Mg–Al alloys. A T4 heat treatment (solution heat treatment only for 16 h at 415°C to homogenize the alloy) increases the corrosion rate slightly compared to the As cast material, as shown in Table 58.9. Aune [57] attributed this increase to the resolution of (β particles and release of elemental Fe. Aune [57] completed two T6 treatments with different aging times and temperatures—the T6 being a T4 followed by a T5 treatment. Both corrosion rates were well below those of samples as cast and

TABLE 58.9. Heat-Treated AZ91 Corrosion Rates[a]

Condition	Corrosion Rate (mg/cm²-day)
As cast	10
T4	11
T6 aged @ 120°C	6
T6 aged @ 205°C	1

[a]Reproduced with permission from [19].

of samples given a T4 heat treatment, as shown in Table 58.9. The lower temperature aging treatment formed a speckled precipitate of (β particles, whereas the higher temperature treatment formed a discontinuous platelike β precipitate with more surface.

In the T6 temper, filiform corrosion follows the same crystalline directionality, but area, and thus, a greater effect to inhibit corrosion usually stops at or close to the grain boundary. Corrosion attacks the Al-depleted region between the β phase lamellae along the grain boundary, but passivates after a few minutes of propagation. The improved corrosion behavior results from the presence of β particles [36].

C2.1. Grain Refinement.
Grain refinement increases the overall grain boundary area, thereby optimizing the distribution and minimizing the size of any possible detrimental intermetallics, such as Fe$_3$Al. The traditional grain refinement method in sand casting is to add an inoculent, which facilitates heterogeneous nucleation during solidification. Additions of strontium to Mg–Al alloys result in reduced grain size and a lower corrosion rate that is attributed not only to the reduced grain size, but also to changes in the oxide layer structure and composition and in the electrochemical properties of the phases present [62].

D. RAPID SOLIDIFICATION

In rapid solidification technologies, including spray or droplet formation, continuous chill casting and in situ melting, typical cooling rates are in the range of 10^5–10^7°C/s [63]. Use of continuous chill casting typically produces a thin ribbon of metal, which is then broken into small particles. Then, as with the material formed by spray or droplet formation, the material is often consolidated and extruded. Improper processing can have a significant impact on corrosion behavior.

The "chunk" effect [64] is caused by surface oxides on powder particles that lead to poor bonding within the final product [65]. Localized corrosion along these prior boundary oxides leads to particle-size pits and high corrosion rates [19].

Corrosion rates for atomized RS alloy are comparable to those of cast AZ91D, though those for melt-spun RS alloys are significantly higher because of the "chunk" effect (Table 58.10).

TABLE 58.10. Corrosion Rates of Selected Materials (mpy)[a]

Material (wt %)	Atomized	Chill Cast	Cast
Mg 7.7 Al 2.9 Zn 6.6 Ce 0.35 Mn	50		
Mg 10.1 Al 2.7 Zn 1.4 Y 0.44 Mn	60		
Mg 10.2 Al 3.2 Zn 5.8 Ce 2.7 Mn		350	
Mg 11.1 Al 2.4 Zn 3.2 Y		430	
AZ 91D			28

[a]Reproduced with permission from [19].

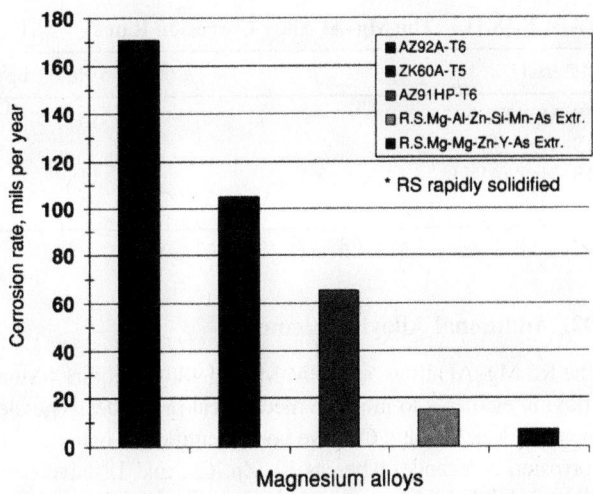

FIGURE 58.18. Corrosion rates (1 mpy ≈ 25 μm/year) of rapidly solidified magnesium alloys tested in 3% NaCl at 21°C compared with some commercial cast alloys (Extr = extruded) [67]. (Reproduced with permission from [15].)

Nonequilibrium phases in RS alloys can influence the corrosion behavior, for example, Makar and Kruger [31] noted an increase in protection against pit initiation in rapidly solidified (RS) AZ61 compared to IM cast AZ61. Corrosion resistance is improved because the more homogeneous microstructures tend to disperse elements and particles that normally act as cathodic centers, and because the extended solubility of various elements may shift the electrode potentials of light alloys to more noble values (Fig. 58.18) [66].

Using rapid solidification processing, a number of magnesium alloys have been produced in the form of melt-spun ribbon, which is then usually mechanically ground to powder, sealed in cans, and extruded to produce bars. Alloy EA55RS (Mg–5 Al–5 Zn–5 Nd) is now available commercially. Microstructures of the bulk products consist of fine grains 0.3–5 μm in size and dispersions of compounds, such as Mg$_{17}$Al$_{12}$, Al$_2$Ca, Mg$_3$Nd, and Mg$_{12}$Ce [62]. Tensile strengths may exceed 500 MNm^{-2}, which compares with maximum values of 250–300 MNm^{-2} for conventionally cast magnesium alloys. Some alloys show improved creep resistance at moderately elevated temperatures, but others undergo accelerated creep deformation [15].

D1. Effect of Aluminum

Hehmann et al. [68], experimentally measured the solid solubility extensions of 22 RS Mg alloys with extension factors ranging from 1.5 to 1000X. The RS Mg–Al alloys with a maximum terminal solid solubility of 23.4 wt % have decreasing corrosion rates with increasing Al contents from 10 to 40 wt %.

TABLE 58.11. The Mg–Al Alloy Corrosion Rates[a]

Material	Corrosion Rate (mpy)
RS–MgAlZnSiMn	15
RS–MgZnAlY	8
RS–MgZnAlNd	11
AZ91HP–T6	82

[a]Reproduced with permission from [19].

D2. Additional Alloying Elements

The RS Mg–Al alloys, as do IM Mg–Al alloys, require further alloying elements to improve mechanical properties. The elements Y, Mn, Nd, and Ce have been identified as beneficial to corrosion resistance, whereas Si, Zn, Ca, and Li have been identified as harmful to corrosion resistance [69]. The Mn acts in the same manner as for the IM alloys, by combining with the Al and Fe to form Al (Mn, Fe) intermetallic [70]. This effect causes the low corrosion rate of RS–MgZnAlSiMn alloy (Table 58.11), despite the presence of detrimental Zn and Si.

D3. Rare Earth Effects

Rare earth alloying elements (Y, Nd, Ce, Pr) result in corrosion rates much lower than those of commercial AZ91HP-T6 alloy (Table 58.11). These elements form stable intermetallic particles in rapidly solidified RS Mg–Al alloys that, similar to the Mg_2Si particles, pin the grain boundaries and result in a refined microstructure. Because of the fast cooling rate, various forms of the intermetallics have been reported such as $Mg_{17}Y_3$, Mg_3RE (RE = Ce, Nd, Pr), and Al_2Nd.

The improved corrosion behavior of these alloys, compared to IM Mg–Al alloys, is attributed to the refined RS microstructure, formation of a protective film on the surface of the RS sample as a result of reaction of the saline solution with the rare earths, and the inertness of the second-phase particles [71].

E. MAGNESIUM FINISHING

To improve corrosion performance and/or for decorative purposes, finishing processes of magnesium and magnesium alloys are carried out, including surface preparation, chemical treatment, and coating. Finishing can also include oil application, wax coating, anodizing, electroplating, and painting. The designer should use the best combination of methods to meet the functional need of the treated part. The degree of superficial corrosion that can be tolerated without affecting performance and the severity of the service environment are determining factors in selecting an optimum finish [72].

A thin oil or wax film is commonly used for storage or shipping. Sand-cast parts are treated before and during machining operations until the final treatments. A dry storage

atmosphere is important. Chemical and electrochemical methods are used for conversion of magnesium surfaces, so that a more corrosion inhibiting and less alkaline to slightly acidic film replaces the natural alkaline hydroxide–carbonate film on magnesium. The converted surface is generally more compatible with organic coatings.

Selected phosphate treatments can be as effective as chromates, even in severe exposures such as marine atmospheric environments. Chemical treatment is strongly recommended for paint formulations, which are based on resins with low resistance to alkaline media. Common chemical treatments alone do very little in aggressive environments and may be unnecessary in mild environments [73, 74].

E1. Cleaning and Surface Preparation

Mechanical cleaning of magnesium alloy products is accomplished by grinding and rough polishing, dry or wet abrasive blast cleaning, wire brushing, and wet barrel or bowl abrading (vibratory finishing). The most frequently used methods of mechanical finishing are barrel tumbling, polishing and buffing, vibratory finishing, fiber brushing, and shot blasting. Chemical cleaning methods for magnesium alloys are vapor degreasing, solvent cleaning, emulsion cleaning, alkaline cleaning, and acid pickling. Acid pickling is required for removal of impurities that are tightly bound to the surface or insoluble in solvents and alkalis. The ferric nitrate pickle deposits an invisible chromium oxide passivating film. The acetic–nitrate and phosphoric acid treatments remove even invisible traces of other metals [75].

During the process of surface protection, corrosion of Mg can occur. Treatment in a boiling dichromate solution (or the equivalent), followed by a slushy oil application, is satisfactory.

E2. Chemical and Electrochemical Finishing Treatments

Chemical and electrochemical finishing treatments can be used alone to provide short-term protection against corrosion and abrasion during shipment and storage, or as pretreatments for subsequent finishing methods.

E2.1. Chrome Pickle and Chrome-Free Phosphate Treatments. Chrome pickle and chrome-free phosphate treatments can be used to provide a base for paint or short-term protection. The steps in a chrome pickle include alkaline cleaning followed by a cold rinse, chrome pickle [180 g/L of $Na_2Cr_2O_7 \cdot 2H_2O$ and 120–180 g/L HNO_3 (sp. gr. 1.42)], holding in air for 5 days, cold rinse and hot rinse. A dichromate seal can be introduced between the cold rinse and hot rinse for better protection. Also, a dichromate treatment can replace the chrome pickle. In a modified chrome pickle treatment, an acid pickle and caustic dip or another acid pickle before the modified chrome pickle solution are used [75]. The modified chrome pickle provides a uniform coating by optimizing the etching and passivating action of the chrome pickle bath and by thorough cleaning and washing.

A number of commercial phosphate treatments provide performance that is comparable to the best chromate-based surface treatments, particularly for new, high-purity die-cast alloys [72, 73].

E2.2. Anodic Treatments.

Anodized coatings have varying degrees of porosity and must be scaled for use in aggressive chloride media. The coatings can be infused with various polymers to produce special properties, including lubricating properties.

Galvanic anodizing is a low-voltage direct-current (dc) treatment that produces a thin black conversion coating, used mainly as a paint base (chemical treatment No. 9) [75]. A source of electric power is not required. Proper galvanic action requires the use of racks, made of stainless steels, Monel or phosphor bronze. The constituents of the anodizing aqueous bath are $(NH_4)_2SO_4$, $Na_2Cr_2O_7 \cdot 2H_2O$, NH_4OH, and the operating temperature is between 49 and 60°C.

More substantial coatings, of 5–30-μm thickness, require anodic polarization by external current. Paint base chemical treatment No. 17 and HAE treatment are currently used. In each of these treatments, a two-phase, two-layer coating is produced. The first layer is about 5 μm thick with a light green or greenish tan color. This layer is covered by a second-phase heavier coating, about 30 μm thick and dark green in color. The second layer is vitreous, relatively brittle, and highly abrasive. As an example, the electrolyte for HAE is composed of KOH, $Al(OH)_3$, K_2F_2, Na_3PO_4, and K_2MnO_4, and the current density is 1.5–2.5 A/dm^2. The terminating potential is 65–70 V after 7–10 min and 80–90 V after 60 min for the HAE treatment. This treatment consists of six or seven steps: alkaline clean, cold rinse, anodize, cold rinse, dichromate bifluoride dip ($Na_2Cr_2O_7 \cdot 2H_2O$ + $NH_4 HF_2$), dry air, and possibly heat humidity aging.

A heavier Cr-22 treatment is a high-voltage process that is commercially available but not currently used. The final potential can be 320 or 350–380 V for heavier coatings. These coatings provide excellent corrosion resistance in mild media and protected unpainted parts of the structure when properly sealed. The sealing posttreatment consists of an immersion for 2 min in a solution of sodium silicate (10% by vol) at 85–100°C [75]. For environmental considerations, two proprietary chrome-free anodizing treatments have been introduced. In a particular military application, these treatments were found to be superior to the two current anodizing treatments (HAE and No. 17) [76–79].

Magoxid-coat is formed in a slightly alkaline bath and results in $MgAlO_4$ and other beneficial compounds on the surface. The innermost, or barrier, layer is extremely thin, followed by a middle ceramic oxide, providing the majority of

corrosion protection since it is almost nonporous. The outermost portion of the coating is a very porous ceramic layer [76].

E2.3. Cathodic Treatments.

Zinc and nickel are the only deposits used commercially as undercoatings upon which other commonly plated metals are deposited. Standard practice for plating magnesium involves surface conditioning, zinc immersion plating (zincate solution), and a cyanide copper strike ($\approx$ 8 μm), followed by a standard plating process. Copper–nickel–chromium plating systems on magnesium satisfy decorative and protective requirements. For interior and mild exterior environments, especially in marine atmosphere, a pore-free deposit is required for satisfactory corrosion resistance. Porosity in the base metal promotes porosity in the deposit [75].

E3. Organic Finishing

For optimum corrosion resistance, a chemical conversion coating or anodizing treatment is required prior to applying the organic finishing system. It has been found that anodized components provide greatly improved corrosion durability if the porosity is sealed with a penetrating resin prior to the application of primer and topcoat [77]. For military applications in the United States, finishing is controlled by specifications [75].

Baking paints are harder and more resistant to attack by solvents and are preferable to air-drying ones for applications in solvents. Primers for magnesium should be based on alkali-resistant vehicles, such as polyvinyl butyral, acrylic, polyurethane, vinyl epoxy, and baked phenolic. Titanium dioxide or zinc chromate pigments are used as inhibitors in these vehicles. Finish coats should be compatible with the primer. Vinyl alkyds are resistant to alkalis; acrylics are resistant to chloride environments; alkyd enamels are used for exterior durability; and epoxies have good abrasion resistance. The following finishes have an increasing temperature resistance in the following order. Vinyls (150°C), epoxies, modified epoxies, epoxy-silicones, and silicones.

Although chemical treatment can retard the natural alkali that forms at any point of the painted film, primers (13 μm thick), which contain an alkali-resistant vehicle, recommended. Oils, alkyds, and nitrocellulose are best avoided, except for mild exposure [73]. Finishing by one ($\approx$13 μm thick) or more coats depends on the corrosive medium and the acceptable corrosion rate. A current example of a paint system can consist of an epoxy-based primer or sealer and an acrylic top coat [73, 74]. With epoxy and epoxy-polyester, electrostatic powder deposition is used successfully on magnesium alloys. Cathodic electrodeposition of a resin from an aqueous emulsion can give a more uniform coverage than spray or dip systems of painting.

In a recently developed process, a hydrogen-rich or magnesium hydride layer is created on the magnesium surface by cathodic electric charging in aqueous solution. This compound is a good base for painting [80].

The following standards on finishing should be considered:

ASTM D1732: Standard Practices for Preparation of Magnesium Alloy Surfaces for Painting; acid and alkaline cleaners, dip treatments, and anodizing described [81].

ASTM D2651: Standard Practices for Preparation of Magnesium Alloy Surfaces for Adhesive Bonding [82].

ASTM D1654: Standard Evaluation of Painted or Coated Specimens Subjected to Corrosive Environments [83].

Accelerated laboratory testing aims to reproduce, in a much shorter time than in the field, natural corrosion and degradation processes of the paint system and the substrate without changing the corrosion/degradation mechanisms occurring in practice. Accelerated corrosion testing that can have different testing modes in different corrosive aqueous media (immersion, alternating immersion, and emersion and spray) as well as in atmospheres with different relative humidity percentages are recommended for corrosion evaluation of coated magnesium alloys. It is always recommended to test the metal or alloy without coating as reference for corrosion rate and corrosion form in the same conditions. Physical and mechanical testing methods such as film hardness are best measured by making microhardness indents on a cross section of the film, but a minimum film thickness of 25 μm is required. This can be increased 10-fold, for example, in the case of an anodized surface layer (keronite process) impregnated with an organic topcoat such as e-coat, PTFE, or powder coat [84].

One of the most effective tests for judging the ability of a paint base to sustain its protectiveness after damage is to scribe the test panels with a sharp instrument that penetrates the paint and coating layer, leaving bare substrate exposed. Once scribed, the panels are placed in salt spray, and evaluations are taken at regular intervals to determine how far corrosion has migrated from the scribe line to undamaged areas (ASTM D1654).

The alternate intermittent immersion in salt water or salt spray is often used to compare the corrosion resistance of magnesium alloys to each other and to other materials. A test period of 28 days has been found to be more severe than a 5-year atmospheric exposure. The humidity, or condensation, test is a simple variation of the salt spray test. Test panels are exposed to a climate with very high relative humidity (RH) and no electrolytes, usually at 40–50°C. Coated panels are exposed so that moisture condenses on the test face [85].

It is admitted frequently that cyclic corrosion testing (CCT) gives the relative corrosion rates, structure, and corrosion morphology quiet close to those seen outdoors, if paint systems of the same type of binder are compared [86]. Simple exposures like prohesion may consist of cyclic events, between salt fog and dry conditions. More sophisticated automotive methods call for multistep cycles that may incorporate immersion, humidity, condensation along with salt fog and dry-off events. The results from an acid rain test (2 h of spraying with acid rain of pH 3.5 plus 1 h drying) did not distinguish the various coated specimens in terms of corrosion behavior. The prohesion test was more corrosive than the acid rain test [84, 86].

ACKNOWLEDGMENTS

This chapter is based on the initial and pioneer contribution of W. S. Loose in the first edition of H. H. Uhlig, *Corrosion Handbook*. Participation and discussions of Dr. I. Nakatsugawa from the Institute of Magnesium Technology, Inc., Sainte-Foy, Québec, Canada, are deeply acknowledged.

REFERENCES

1. D. S. Tawil, "Magnesium Technology," in Proceedings of the Conference of The Institute of Metals, Nov. 3–4, 1986, The Institute of Metals, Ed., London, UK, 1987, pp. 66–74.

2. W. S. Loose, Magnesium and Magnesium Alloys, 1st ed., Corrosion Handbook, H. H. Uhlig (Ed.), sponsored by The Electrochemical Society, Wiley, NY, 1976, pp. 218–251.

3. T. R. Massalski, Binary Phase Diagrams, 2nd ed., Vols. 1–4, ASM International, Metals Park, OH, 1990.

4. A. A. Nayeb-Hashemi and J. B. Clark, "Mg-Si Phase Diagram," in Binary Alloy Phase Diagram, 2nd ed., T. R. Massalski, H. Okamoto, P. R. Subramanian, and L. Kacprzak (Eds.) ASM, Metals Park, OH, 1988, p. 2547.

5. J. D. Hanawalt, C. E. Nelson, and J. A. Peloubet, Trans AIME, **147**, 273 (1942).

6. L. L. Shreir, R. A. Jarman, and G. T. Burstein, "Magnesium and Magnesium Alloys," Corrosion, 1, 4.98–4.115 (1995).

7. ASTM B275-94 "Standard Practice for Codification of Nonferrous Metals and Alloys, Cast and Wrought," in Annual Book of ASTM Standards, Vol. 02.02, Aluminum and Magnesium Alloys, ASTM, Philadelphia, PA, 1994, pp. 282–287.

8. I. J. Polmear, Light Alloys: Metallurgy of the Light Metals, 2nd ed., Edward Arnold, London, 1989.

9. R. S. Busk, Magnesium Products Design, Marcel Dekker, New York, 1987.

10. ASTM B296 (reapproved 1990), "Standard Practice for Temper Designations of Magnesium Alloys, Cast and Wrought," in Annual Book of ASTM standards, Vol. 02.02, Aluminum and Magnesium Alloys, ASTM, Philadelphia, PA, 1994, pp. 288–289.

11. P. F. King, J. Electrochem. Soc, **113**, 536–539 (1966).

12. W. Unsworth and J. P. King, in Proc. Conf: Int. Conf.: Magnesium Technology, The Institute of Metals, London, 1986, p. 25.

13. J. P. King, G. A. Fowler, and P. Lyon, in Proc. Conf: Light Weight Alloys for Aerospace Applications II, E. W. Lee and N. J. Kim (Eds.), TMS, Warrendale, PA, 1991, p. 423.

14. W. Durako and L. Joesten, in Proc. 49th Annual World Magnesium Conference, International Magnesium Association, McLean, VA, 1992, p. 87.

15. I. J. Polmear, Mater. Sci. and Technol., **10**, 1 (1994).

16. S. Krishnamurthy, E. Robertson, and F. H. Froes, "Rapidly Solidified Magnesium Alloys Containing Rare Earth Addition," in Advances in Magnesium Alloys and Composites, Minerals, Metals and Materials Society, H. Paris and W. H. Hunt (Eds.), Warrendale, Pennsylvania, 1988, pp. 77–88.

17. G. L. Makar, I. Kruger, and A. Joshi, "The Effect of Alloying Elements on the Corrosion Resistance of Rapidly Solidified Magnesium Alloy," in Advances in Magnesium Alloys and Composites, H. Paris and W. H. Hunt (Eds.), The Minerals, Metals and Materials Soc., Warrendale, Pennsylvania, 1988, pp. 105–121.

18. M. R. Bothwell, "Magnesium", in The Corrosion of Light Metals, H. P. Godard, W. B. Jepson, M. R. Bothwell, and R. Kane (Eds.), Wiley, New York, 1967, pp. 259–311.

19. B. E. Carlson and J. W. Jones, "The Metallurgical Aspects of the Corrosion Behavior of Cast Mg-Al Alloys," Light Metals Processing and Applications, METSOC Conference, Québec, 1993, pp. 833–847.

20. G. L. Makar and J. Kruger, Int. Mater. Rev., **38**(3), 38 (1993).

21. M. Pourbaix, Atlas of Electrochemical Equilibria in Aqueous Solutions, NACE International and CEBEL-COR (Centre Beige d'Étude de la Corrosion), Houston, TX, 1974, p. 141.

22. G. G. Perrault, Electroanal. Chem. Interfac. Electrochem., **51**, 107 (1974).

23. A. Froats, T. Kr. Aune, D. Hawke, W. Unsworth, and G. Hillis, "Corrosion of Magnesium and Magnesium Alloys," in Metals Handbook, 9th ed., Vol. 13, Corrosion, ASM International, Materials Park, OH, 1987, pp. 740–754.

24. D. L. Hawke, J. E. Hillis, and W. Unsworth, "Preventive Practice for Controlling the Galvanic Corrosion of Magnesium Alloys," Technical Committee, International Magnesium Association, McLean, VA, 1988.

25. D. Hawke, Asbjorn Olsen, Proc., Society of Automotive Engineers SAE, Paper No. 930751, Detroit, MI, 1993, pp. 79–84.

26. M. R. Bothwell, J. Electrochem. Soc, **106**, 1021 (1959).

27. O. Lunder, T. Kr. Aune, and K. Nisancioglu, Corrosion **43**(5), 291 (1987).

28. A. I. Asphahani and W.L. Silence, "Pitting Corrosion," in Corrosion, Metals Handbook, Vol. 13, 9th ed., ASM, Metals Park, OH, 1987, p. 113.

29. W. E. Mercer and J. E. Hillis, "The Critical Contaminant Limit and Salt Water Corrosion Performance of Magnesium AE42 Alloy," Technical Paper No. 920073, Society of Automotive Engineers SAE, Detroit, MI, 1992.

30. O. Lunder, K. Nisancioglu, and R. S. Hansen, "Corrosion of Die Cast Magnesium-Aluminum Alloy," Congress and Exposition, Paper No. 930755 SAE, Detroit, MI, Feb. 26–Mar. 2, 1993, p. 117–126.

31. G. L. Makar and J. Kruger, J. Electrochem. Soc, **13**(2), 414. (1990).

32. C. Kirby, Corros. Sci. **27**(6), 567 (1987).

33. C. Hahin, "Filiform Corrosion," in Corrosion, Metals Handbook, Vol. **13**, 9th ed., ASM, Metals Park, OH, 1987, p. 104.

34. K. Nisancioglu, O. Lunder, and T. Kr. Aune, Corrosion Mechanism of AZ91 Magnesium Alloy, 47th Annual World Mag. Conf., Cannes, France, IMA, 1992, p. 43.

35. S. C, Dexter, "Localized Corrosion," in Metals Handbook, 9th ed. Vol. 13, Corrosion, ASM International, Metals Part, OH, 1987, p. 106.

36. T. Kr. Aune, O. Lunder, and K. Nisancioglu, Microstruct. Sci., **17**, 231 (1989) and [19].

37. O. Lunder, J. E. Lein, S. M. Hesjevik, T. Kr. Aune, and K. Nisancioglu, "Filiform Corrosion of a Magnesium Alloy," 11th Annual Corrosion Congress, Florence, Italy, 1990, pp. 5.255–5.262.

38. W. K. Miller, Stress Corrosion Cracking of Magnesium Alloys: Materials Performance and Evaluation, ASM International, Materials Park, OH, 1991.

39. M. Vialatte, "Study of the SCC Behavior of the Alloy Mg–8% Al," Symposium on the Engineering Practice to avoid Stress Corrosion Cracking, NATO, Neuilly-sur-Seine, France, 1970, pp. 5.1–5.10.

40. M. O. Spiedel, Metall. Trans. A, **6A**, 631 (Apr. 1975).

41. D. K. Priest, F. H. Beck, and M. G. Fontana, ASM Trans, **47**, 473 (1955).

42. L. Fairman and H. J. Bray, Br. Corros. J., **6**, 170, (July 1971).

43. H. W. Pickering and P. R. Swann, Corrosion. **19**(11), 373f (1963).

44. L. Fairman and H. J. Bray, Corros. Sci., **11**, 533, (1971).

45. H. W. Liu, J. Basic Eng., **92**, 633 (Sep. 1970).

46. S. P. Lynch and P. Trevena, Corrosion, **44**(2), 113 (Frb. 1988).

47. W. K. Miller and E. F. Ryntz, SAE Trans, **92**, 524 (1983).

48. E. Groshart, "Magnesium, Part 1–The Metal," Met. Finish. **83**, (10), 17–20 (Oct. 1995).

49. M. A. Timonova, "Corrosion Cracking of Magnesium Alloys and Methods of Protection Against It," in Intercrystalline Corrosion and Corrosion of Metals Under Stress, I. A. Levin (Ed.), translated from Russian, Consultant Bureau, New York, 1962, pp. 263–282.

50. V. B. Yakovlev, L. P. Trutneva, N. I. Isaev, and G. Nemetch, Protection Met., **20**(3), 300 (1984).

51. Beck, A, The Technology of Magnesium and Its Alloys, F. A. Hughes, London, 1940, p. 236.

52. R. I. Stephens, C. D. Schrader, D. L. Goodenberger, K. B. Lease, V. V. Ogarevic, and S. N. Perov, "Corrosion Fatigue and Stress Corrosion Cracking of A291E-T6 Cast Magnesium Alloy in 3.5% NaCl Solution," Technical Paper No. 930752, Society of Automotive Engineers SAE, Detroit, MI, 1993.

53. V. V. Ogarevic and R. I. Stephens, Annu. Rev. Mater. Sci., **20**, 141 (1990).

54. V. A. Serdyuk and N. M. Grinberg, Int. J. Fatigue, **5**(2), 79 (1983).

55. Standard Practice for Laboratory Immersion Corrosion Testing of Metals, 1992 Annual Book of ASTM Standards, ASTM, Philadelphia, PA, 3.02 (G31-72), 1992, p. 102.

56. Standard Test Method of Salt Spray (Fog) Testing, 1992 Annual Book of ASTM Standards, ASTM, Philadelphia, PA, 3.02 (B117-90), 1992, p. 20.

57. T. Kr. Aune, "Minimizing Base Metal Corrosion on Magnesium Products. The Effect of Element Distribution (Structure) on Corrosion Behavior," 40th World Magnesium Conference, IMA, Dayton, OH, June 12–15, 1983, p. 52.

58. L. Y. Wei, "Development of Microstructure in Cast Magnesium Alloys," Ph.D. Thesis, Chalmers University of Technology, Goteborg, Sweden, 1990.

59. J. E. Hillis and S. O. Shook, "Composition and Performance of an Improved Magnesium AS41 Alloy," Technical Paper No. 890205, Society of Automotive Engineers SAE, Detroit, MI, Feb. 1989.

60. K. N. Reichek, K. J. Clark, and J. E. Hillis, "Controlling the Salt Water Corrosion Performance of Magnesium AZ91 Alloy," Technical Paper No. 850417, Society of Automotive Engineers SAE, Detroit, MI, 1985.

61. Standard Specification for Magnesium-Alloy Die Castings, 1992 Annual Book of ASTM Standards, ASTM, Philadelphia, PA, 2.02 (B94-92), Non ferrous Metal Products, 2.02, 1992, p. 55.

62. G. Nussbaum. G. Regazzoni, and H. G. Jestl and, in Proc. SAE International Congress and Exposition, Technical Paper Series 900792, Society of Automotive Engineers SAE, Detroit, MI, Warrendale, PA, 1990.

63. F. Hehmann and H. Jones, Rapid Solidification Processing of Magnesium Alloys, Magnesium Technology, 1986, Institute of Metals, London, UK, 1987, p. 83.

64. A. Joshi, R. D. Adamson, and R. E. Lewis, "Processing and Properties of Rapidly Solidified Mg-Al-Zn-Y (or Ce) Alloys," in Magnesium Alloys and Their Applications, B. L. Moredike and F. Hehmann (Eds.), DGM Informationsgesellschaft, Oberursel, Germany, 1992, pp. 495–502.

65. A. Garboggini and H. B. McShane, "Structural Mg-Al Alloys Produced by Rapid Solidification," Magnesium alloys and Their Applications, in B. L. Mordike and F. Hehmann (Eds.), DGM Informationsgesellsehaft, Ober ursel, Germany, 1992, pp. 503–510 and [19].

66. S. K. Das and C. F. Chang, in 'Rapidly Solidified Crystalline Alloys.' S. K. Das, B. H. Kear, and C. M. Adam (Eds.), Metallurgical Soc. of AIME, Warrendale, PA, 1985, pp. 137–156.

67. S. K. Das and C. F. Chang, Rapidly Solidified High Strength Corrosion Resistant Magnesium Base Metal Alloys," U.S. Patent, 4,853,035 (Aug. 1, 1989) and [15].

68. F. Hehmann, F. Sommer, and B. Predel, Mater. Sci., **A125**, 249 (1990).

69. A. Joshi and R. E. Lewis, "Role of RSP on Microstructure and Properties of Magnesium Alloys," in Advances in Magnesium Alloys and Composites, H. Paris and W. H. Hunt (Eds.), Minerals, Metals and Materials Society, Warrendale, Pennsylvania, USA, 1988, pp. 89–103.

70. C. F. Chang, S. K. Das, D. Raybould, and A. Brown, Met. Powder Pre., **41**(4), 302 (1986).

71. C. F. Chang, S. K. Das, D. Raybould, R. L. Bye, and E. V. Limoncelli, "Recent Developments in High Strength PM/RS Magnesium Alloys-A Review," in Advances in Powder Metallurgy, T. G. Gasbarre and W. F. Jandeska (Eds.), MPIF/APMI, Princeton, NJ, June 3, 1989, pp. 331–346.

72. J. E. Hillis and R. W. Murray, "Finishing Alternatives for High Purity Magnesium Alloys," Paper No. G-T87-003, Society of die Casting engineers, SDCE 14th International, Detroit, MI, May 11–14, 1987.

73. R. W. Murray and J. E. Hillis, "Magnesium Finishing: Chemical Treatment and Coating Practices," Technical Paper No. 900791, Society of Automotive Engineers, Detroit, MI, 1990, SAE technical paper No. 900791.

74. P. L. Hagans, "Surface Modifications of Magnesium for Corrosion Protection," in Proceedings of the International Magnesium Association, London, 1984.

75. J. E. Hillis, "Surface Engineering of Magnesium Alloys," in ASM Handbook, 10th ed., Vol. 5, Surface Engineering, ASM International, Materials Park, OH, 1994, pp. 819–834.

76. C. Jurey, Magoxid-Coat, "A Hard Anodic Coating for Magnesium," in Proceedings of the International Magnesium Association, Washington, DC, 1993, pp. 80–88.

77. J. H. Hawkins, "Assessment of Protective Finishing Systems for Magnesium," in Proceedings of the Inter national Magnesium Association, Washington, DC, 1993.

78. D. E. Bartak, T. D. Schleisman, and E. R. Woolsey, "Electrodeposition and Characteristics of a Silicon Oxide Coating for Magnesium Alloys," North American Die Casting Association Congress and Exposition, Detroit, MI, USA, 1991. Paper No. T91–041.

79. "Corrosion and Protection of Magnesium," Amax Magnesium, Company, Salt Lake City, Utah, 1984.

80. I. Nakatsugawa, J. Renaud, E. Ghali, and E. J. Knystautas, "Electrochemical Formation of Magnesium Hydride and Its Application to Surface Coating," in Magnesium 97, Proceedings of the First Israeli International Conference on Magnesium Science and Technology, 10–12, Nov. 1997.

81. ASTM D1732-67, Standard Practices for Preparation of Magnesium Alloy Surfaces for Painting, Annual Book of ASTM Standards, Vol 02.05, ASTM, Philadelphia, PA, 1990, pp. 713–719.

82. ASTM D2651-79, Standard Practices for Preparation of Metal Surfaces for Adhesive Bonding, Annual Book of ASTM Standards, Vol 15.06, ASTM, Philadelphia, PA, 1990, pp. 165–169.

83. ASTM D1654-79a, Standard Method for Evaluation of Painted or Coated Specimens Subjected of Corrosive Environments, Annual Book of ASTM Standards, Vol 06.01, ASTM, Philadelphia, PA, 243–245.

84. E. Ghali, Corrosion Resistance of Aluminum and Magnesium Alloys, Understanding, Performance and Testing, Wiley, New York, 2010, 719 pages.

85. M. J. Crewdson and P. Brennan, "Outdoor Weathering: Basic Exposure Procedures," JPCL, **12**(9), 17–25 (1995).

86. B. P. Alblas and J. J. Kettenis, "Accelerated Corrosion Tests: Continuous Salt Spray and Cyclic Tests," Protective Coatings Europe, PCE Magazine, Brian Goldie, Editor, Farnham, Surrey, UK, Feb. 2000, pp. 49–59.

59

NICKEL AND NICKEL ALLOYS

D. C. AGARWAL* AND N. SRIDHAR

DNV Columbus, Inc., Dublin, Ohio

*Deceased.

A. INTRODUCTION

Within the chemical process industry, as well as other industries, the 300 series stainless steels have been and will continue to be the most widely used tonnage material after carbon steel. Materials of construction for the modern chemical process and petrochemical industries must not only resist uniform corrosion caused by various corrodents but must also have sufficient localized corrosion and stress corrosion cracking (SCC) resistance. These industries have to cope with the technical and commercial challenges of rigid environmental regulations, the need to increase production efficiency by utilizing higher temperatures and pressures, and more corrosive catalysts and at the same time possess the necessary versatility to handle varied feedstock and upset conditions. Over the past 100 years, improvements in alloy metallurgy, melting technology, and thermomechanical processing, along with a better fundamental understanding of the role of various alloying elements, have led to new nickel alloys, which not only extend the range of usefulness of existing alloys by overcoming their limitations but are also reliable and cost-effective and have opened new areas of applications.

Nickel and nickel alloys have useful resistance to a wide variety of corrosive environments typically encountered in various industrial processes. In many instances, the corrosive conditions are too severe to be handled by other commercially available materials, including stainless and superstainless steels. Nickel by itself is a very versatile corrosion-resistant metal, finding many useful applications in industry. More importantly, its metallurgical compatibility over a considerable composition range with a number of other metals as alloying elements has become the basis for many binary, ternary, and other complex nickel-base alloy systems, having very unique

Uhlig's Corrosion Handbook, Third Edition, Edited by R. Winston Revie

and specific corrosion-resistant and high-temperature-resistant properties for handling the modern-day corrosive environments of chemical process, petrochemical, marine, pulp and paper, agrichemicals, oil and gas, heat treat, energy conversion, and many other industries. These alloys are more expensive than the standard 300 series stainless steels due to their higher alloy content and more involved thermomechanical processing and hence are used only when stainless steels are not suitable or when product purity and/or safety considerations are of critical importance. Corrosion depends on the chemical composition, the microstructural features of the alloy as developed during thermomechanical processing, the various reactions occurring at the alloy/environment interface, and the chemical nature of the environment itself.

In this chapter, the major nickel alloy systems are discussed, including their major characteristics, the effects of alloying elements, and, most importantly, the strengths, weaknesses, and applications of these systems. A few words on fabrication are also included because an improper fabrication may destroy the corrosion resistance of an otherwise fully satisfactory nickel alloy.

B. NICKEL AND NICKEL-BASE ALLOYS FOR CORROSION APPLICATIONS

Nickel-base alloys can be roughly classified into the following family of alloys, each developed for certain corrosion characteristics and mechanical properties:

Alloy System	Type of Environment Designed to Resist Against	Some Major Commercial Alloys
Ni	Caustic	Ni-200/201
Ni–Cu	Nonoxidizing halides, caustic	Alloy 400, K-500
Ni–Mo	Nonoxidizing	Alloys B-2, B-3, B-4, etc.
Ni–Si (some with Cr)	Highly oxidizing, nonhalide	Alloy D, Lewmet, D-205
Ni–Fe	Controlled expansion, nonoxidizing	Invar, Pernifer, etc.
Ni–Cr–Fe	Oxidizing, nonhalide	Alloy 600, 800, 690, etc.
Ni–Cr–Fe–Mo	Oxidizing, moderately localized corrosion environments	825, G-3, G-30, alloy 33, etc.
Ni–Cr–Mo	Oxidizing, highly localized corrosion environments	625, alloy 276, alloy 22, C-2000, alloy 59, etc.
Ni–Cr–Fe–Mo–(Al,Ti,Nb)	Precipitation hardenable, environments similar to those without (Al,Ti,Nb)	X-750, 718, 945, etc.
Ni–Cr–Mo–Fe–Co–C–Al–Ti–Nb–rare earth metals	High-temperature corrosion, wear resistance	X,214, Nimonic, HR160, 230, MP-35N, etc.
Ni–Ti	Shape memory alloys	Nitinol

Commercial nickel production started as early as 1804, and since then Ni-base alloy chemistries have evolved steadily in terms of metallurgical sophistication and their ability to function in a broad range of corrosive environments. The development of Ni–Cr alloys first occurred in the early 1900s at about the same time as stainless steel was being developed. In the 1930s, the Ni–Mo and Ni–Cr–Mo alloys were first commercialized, but they contained high carbon and silicon contents and were not useful in the as-welded condition. The development of the argon oxygen decarburization process in the 1960s and the electroslag remelting process in the early 1980s enabled the manufacturing of low-carbon, low-sulfur Ni–Cr–Mo alloys, such as alloy C-276. Better understanding of the electronic structures of alloys and their effects on the formation of intermetallic phases in these alloys resulted in the development of metallurgically stable variants of these alloys starting in the mid-1970s and further developments have continued through the present.

The Ni–Fe–Cr–Mo corrosion-resistant alloys developed initially were all solid-solution-strengthened alloys, which meant that their strength could be increased only by cold working. This limited the shapes and sizes of parts produced from these alloys. In parallel with the development of solid-solution-strengthened corrosion-resistant alloys, the development of precipitation-hardenable alloys occurred to satisfy the needs of the emerging aerospace and power generation industries. However, most of these early precipitation-hardenable alloys, such as alloys X-750 and 718, did not have sufficient corrosion resistance. Oil and gas production at increasingly deeper zones placed a premium on high-strength, thick-section Ni-base alloys for components such as the tubing hanger. In the last decade this has spurred the development of precipitation-hardenable alloys with quite high corrosion resistance, such as alloy 945.

This chapter provides a concise description of each alloy class with some corrosion data and application areas.

TABLE 59.1.   Alloying Elements and Their Major Effects in Low-Temperature Corrosion Environments

Alloying Elements	Main Effects	Other Effects
Ni	Provides matrix for dissolving other alloying elements; improves thermal stability towards intermetallic phases and fabricability	Improved corrosion resistance to nonoxidizing environments over Fe; resistance to alkalis; increases chloride SCC resistance
Al	Generally added as deoxidizer	In some precipitation-hardening alloys added to form γ' phase
Cr	Improves resistance to oxidizing environments	Increases solubility of nitrogen and carbon; increases susceptibility to precipitation of sigma phase; in combination with Mo, W, and N can improve localized corrosion resistance
C	Increases strength at high temperatures	Increases formation of detrimental grain boundary carbides
Co	Increases resistance to wear	Sometimes an incidental impurity; in high amounts can reduce SCC resistance
Cu	Improves resistance to nonoxidizing environments, such as sulfuric and phosphoric acid; as binary alloying element with Ni can improve seawater corrosion resistance.	Decreases localized corrosion resistance in oxidizing halide environments
Fe	Generally added to Ni-base alloys to reduce cost and enhance scrap utilization; in binary Ni–Fe alloys provides low thermal expansion characteristics; in Ni–Cr–Mo–Fe alloys, higher Fe can reduce tendency to form long-range ordering	Increases tendency to form detrimental intermetallic phases, such as sigma; the higher the Cr, Mo, and W levels, the lower must the Fe addition must be preserve metallurgical stability
Mo	Increases resistance to nonoxidizing environments, such as HCl, H_2SO_4; in combination with Cr enhances resistance to oxidizing halide environments	Provides solid-solution strengthening; increases tendency to form detrimental intermetallic phases, such as μ phase.
Mn	Mainly added as deoxidizer during melting; in levels found in typical Ni-base alloys has no effect on corrosion	High Mn levels could destabilize metallurgical structure by precipitating intermetallic phases
N	In combination with Cr and Mo can increase localized corrosion resistance; increases strength	Higher Cr (at least 25%) is necessary to dissolve N in Ni-base alloys; lower Cr alloys can form deleterious chromium nitrides
Nb (Cb)	Generally added to prevent formation of chromium carbides (stabilization) or to provide precipitation hardening; No effect on corrosion except indirectly through carbide control	May lead to knife-line attack near welds through resolution of carbides and precipitation of chromium carbides; for precipitation hardening through formation of γ'', at least 3% Nb is needed; increases tendency to form detrimental Laves phase
P	Found as tramp element; Detrimental to corrosion resistance	Phosphorus segregation to grain boundaries can lead to enhanced hydrogen embrittlement
S	Found as tramp element; detrimental to corrosion resistance	Sometimes added to improve machinability, but typically not in Ni–base alloys
Sn	Generally undesirable element in manufacturing of Ni-base alloys, especially high-temperature alloys	
Si	Improves resistance to highly oxidizing, nonhalide environments such as concentrated sulfuric acid and red fuming nitric acid; in binary Ni–Si alloys are generally cast and require at least 14% Si; Ni–Cr–Si alloys require much less Si	Sometimes Si is added to cause precipitation hardening through formation of Ni_3Si; Si is detrimental to metallurgical stability; Si increases carbon activity and results in enhanced carbide formation; Si also increases tendency to form sigma phase in presence of Cr and Mo
Ta	Generally added to prevent formation of chromium carbides (stabilization)	May lead to knife-line attack near welds through resolution of carbides and precipitation of chromium carbides
Ti	Generally added to prevent formation of chromium carbides (stabilization) or to provide precipitation hardening; In binary Ni–Ti alloys provides shape memory effect	May lead to knife-line attack near welds through resolution of carbides and precipitation of chromium carbides; enhances precipitation-hardening effect of Al and Nb
W	Similar to Mo in effect, but half as effective because it is twice as heavy as Mo; increases resistance to nonoxidizing environments, such as HCl, H_2SO_4; in combination with Cr and Mo enhances resistance to oxidizing halide environments	Provides solid-solution strengthening; increases tendency to form detrimental intermetallic phases, such as μ phase

TABLE 59.2.   Alloying Elements and Their Major Effects in High-Temperature Alloys

Alloying Elements	Main Feature	Other Features
Cr	Improves oxidation resistance; detrimental to nitriding and fluorination resistance	Improves sulfidation resistance; beneficial to carburization and metal dusting resistance
Si	Improves oxidation, nitriding, sulfidation, and carburizing resistance; detrimental to nonoxidizing chlorination resistance	Synergistically acts with chromium to improve high-temperature degradation
Al	Independently and synergistically with Cr improves oxidation resistance; detrimental to nitriding resistance	Helps improve sulfidation resistance; improves age-hardening effects
Mo	Improves high-temperature strength improves creep strength, detrimental for oxidation resistance at higher temperatures	Helps with reducing chlorination resistance
W	Behaves similarly to molybdenum	
Nb	Increases short-term creep strength; may be beneficial in carburizing; detrimental to nitriding resistance	Improves age hardening
C	Improves strength, helps nitridation resistance, beneficial to carburization resistance, oxidation resistance adversely effected	
Ti	Improves age hardening	Detrimental to nitriding resistance
Mn	Slight positive effect on high-temperature strength and creep; detrimental to oxidation resistance; increases solubility of nitrogen	
Co	Reduces rate of sulfur diffusion, hence helps with sulfidation resistance; improves solid-solution resistance; improves solid-solution strength	
Ni	Improves carburization, nitriding, and chlorination resistance; detrimental to sulfidation resistance	Improves high-temperature mechanical properties
Y & RE (rare earths)	Improves adherence and spalling resistance of oxide layer and hence improves oxidation, sulfidation, carburization resistance	

C. ALLOYING ELEMENTS AND THEIR EFFECTS IN NICKEL ALLOYS

Alloying elements have complex interactive effects on the corrosion resistance and metallurgical stability of Ni-base alloys. The effects of various elements in alloys that are used for wet corrosion resistance and for high-temperature corrosion are presented in Tables 59.2 and 59.3, respectively. Some of the alloying elements are common to both types of alloys but impart different property characteristics in each type. Some elements may be undesirable for wet-corrosion alloys but beneficial for high-temperature corrosion alloys and vice versa.

More detailed information on nickel and nickel alloys is available in the References, including the product literature of the nickel alloy producers. This chapter focuses primarily on the aqueous corrosion alloy systems, with a very brief section on high-temperature alloys and fabrication.

C1. Alloy Systems

C1.1. Nickel

Alloy	UNS No.	Ni[a]	Cu	Fe	Mn	C	Si	S
200	N02200	99.0 min	0.25	0.40	0.35	0.15	0.35	0.01
201	N02201	99.0 min	0.25	0.40	0.35	0.02	0.35	0.01

Note: UNS = Unified Numbering System.
[a]Ni + Co. All values max unless noted otherwise.

The two main alloys, commercially pure alloy 200 and its low-carbon version alloy 201, have good resistance, at low to moderate temperatures, to corrosion by dilute unaerated solution of the common nonoxidizing mineral acids, such as HCl, H_2SO_4, or H_3PO_4. The reason for its good behavior is that the standard reduction potential of nickel is more noble than that of iron and less noble than copper. Because of nickel's high overpotential for hydrogen evolution, there is

TABLE 59.3. Comparison of Some Ni–Cr–Mo Alloys in Various Boiling Corrosive Environments

Media	Uniform Corrosion Rate (mpy)[a]				
	C-276	C-22	686	C-2000	59
$Fe_2(SO_4)_3$ 50% H_2SO_4 + 42 g/L	240	36	103	27	24
ASTM 28B, 23% H_2SO_4 + 1.2% HCL + 1% $CuCl_2$ + 1% $FeCl_3$	55	7	10	4	4
11.5% H_2SO_4 + 1.2% HCL + 1% $CuCl_2$ + 1% $FeCl_3$	26	4	8		5
10% HNO_3	19	2			2
65% HNO_3	750	52	231		40
10% H_2SO_4	23	18			8
50% H_2SO_4	240	308			176
1.5% HCl	11	14	5	1.5	3
10% HCl	239	392			179
10% H_2SO + 1% HCl	87	354			70
10% H_2SO_4 + 1% HCl[b]	41	92	67		3

[a]To convert to millimeters per year (mm/year) multiply by 0.0254.
[b]At 90°C.

no easy discharge of hydrogen from any of the common nonoxidizing acids and a supply of oxygen is necessary for rapid corrosion to occur; hence, nickel can corrode rapidly in nonoxidizing environments in the presence of oxidizing species, such as ferric or cupric ions, nitrates, peroxides, or even oxygen.

Nickel's outstanding corrosion resistance to alkalies has led to its successful use as caustic evaporator tubes. At boiling temperatures and concentration of up to 50% NaOH, the corrosion rate is <0.005 mm/year (0.2 mpy). The iso-corrosion diagram for nickel 200 and 201 in sodium hydroxide clearly shows its superiority and usefulness even at higher concentrations and temperatures [1]. However, when nickel is to be utilized at temperatures above 316°C (600°F), the low-carbon version (alloy 201) is recommended to guard against graphitization at the grain boundaries, which leads to possible loss of ductility causing embrittlement.

Nickel is very resistant to chloride SCC resistance but may be susceptible to caustic cracking in aerated solution in severely stressed conditions. The Ni–Cr–Fe, such as alloy 600, may be more resistant under such conditions. Nickel has a high resistance to corrosion by most natural freshwaters and rapidly flowing seawater. However, under stagnant or crevice conditions, severe pitting attack may occur. While nickel's corrosion resistance to oxidizing acids such as nitric acid is poor, it is sufficiently resistant to most nonaerated organic acids and organic compounds. Nickel is not attacked by anhydrous ammonia or very dilute ammonium hydroxide solution (<2%). Higher concentration causes rapid attack due to formation of soluble (Ni–NH4) complex corrosion products.

Nickel's good resistance to halogenic environments at elevated temperatures, such as in chlorination or fluorination reactions, is utilized in many chemical processes, mainly because the nickel halide films that form on the nickel surface have relatively low vapor pressures and high melting points.

Nickel is used in the production of high-purity caustic in the 50–75% concentration range in the petrochemical industry, in the chemical process industry, in the food industry, and in the production of synthetic fibers. Other applications result from its magnetic and magnetostrictive properties, high thermal and electrical conductivities, and low vapor pressure.

C2. Ni–Cu Alloys

Alloy	UNS No.	Ni(+Co)	C	Mn	Fe	S	Si	Cu[a]	Other
400	N04400	63.0 min	0.30	2.0	2.5	0.024	0.50	31	
K-500	N05500	63.0 min	0.18	1.5	2.0	0.010	0.50	30	Al 2.8[a] Ti 0.60

[a]Typical—all values max unless noted otherwise.

The two main alloys in this system are Monel 400, or alloy 400, and its age-hardenable version, alloy K-500. Alloy 400 was developed near the beginning of the twentieth century and continues to be used in the chemical, petrochemical, marine, refinery, and many other industries. Alloy 400, containing ~31% copper in a nickel matrix, has many characteristics similar to those of commercially pure nickel and other characteristics markedly improved over those of pure nickel. Addition of some iron significantly improves the resistance to cavitation and erosion in condenser tube applications. The main uses of alloy 400 are under conditions of high flow velocity and erosion, for example, in propeller shafts, propellers, pump impeller blades, casings, condenser tubes, and heat exchanger tubes. Corrosion rate in moving seawater is generally less than 0.025 mm/year. The alloy can pit in stagnant seawater, but the rate of attack is considerably less than in commercially pure nickel, alloy 200. Because of its high nickel content, the alloy is generally immune to chloride stress corrosion cracking.

The general corrosion resistance of alloy 400 in nonoxidizing mineral acids is better than pure nickel. However, like pure nickel, it has very poor corrosion resistance to oxidizing media, such as nitric acid, ferric chloride, cupric chloride, wet chlorine, chromic acid, sulfur dioxide, or ammonia.

In unaerated dilute hydrochloric and sulfuric acid solution, the alloy has good resistance up to concentrations of 15% at room temperature and up to 2% at somewhat higher temperature, not exceeding 50°C. Due to this specific characteristic, alloy 400 is also used in processes where chlorinated solvents may form hydrochloric acid due to hydrolysis, which would cause failure in stainless steels.

Alloy 400 has good corrosion resistance at ambient temperature to all HF concentration in the absence of air. Aerated solutions and higher temperature increase the corrosion rate. The alloy is susceptible to SCC in moist aerated hydrofluoric or hydrofluorosilicic acid vapor. This SCC can be minimized by deaeration of the environment or by stress relief anneal of the component.

Neutral and alkaline salt solutions, such as chloride, carbonates, sulfates, and acetates, have only a minor effect on alloy 400, even at high concentrations and temperatures up to boiling point. Hence the alloy is widely used in plants for crystallization of salts from saturated brine.

Alloy K-500, the age-hardenable alloy, contains aluminum and titanium and combines the excellent corrosion resistance features of alloy 400 with the added benefits of increased strength and hardness and maintains its strength up to 600°C. The alloy has low magnetic permeability and is nonmagnetic to −134°C. Some of the typical applications, of alloy K-500 are in pump shafts, impellers, medical blades, and scrapers, oil well drill collars and other completion tools, electronic components, and springs and valves. This alloy is primarily used in marine and oil and gas industrial applications. In contrast, Alloy 400 is more versatile, finding many additional uses, other than those mentioned previously, such as in roofs, gutters and architectural parts on institutional buildings, tubes of boiler feedwater heaters, seawater applications (e.g., sheating), the HF alkylination process, production and handling of HF acid and refining of uranium, distillation, condensation units and overhead condenser pipes in refineries and petrochemical industries, and many others.

The nickel–iron alloys containing 36–80% nickel are commonly applied to take advantage of their special physical properties, such as low coefficient of thermal expansion and/or magnetic properties.

Higher nickel alloys, containing 76–80% nickel with some iron and some molybdenum, have the highest magnetic permeability and are used as inductive components in transformers, circuit breakers, low-frequency transducers, relay parts, and screens. Invar, also known as Pernifer 36, an alloy with 36% nickel, has extremely low coefficient of thermal expansion. Because of its applications in cryogenic environments, this alloy has undergone extensive corrosion testing. The nickel–iron alloys have moderately good resistance to a variety of nonoxiditing industrial environments but are primarily used for their physical characteristics rather than for their corrosion resistance. They can also undergo severe hydrogen embrittlement under same circumstances.

C4. Ni–Si System

Alloy	Ni	Co	Mo	Cu	Cr	Si	Mn	Fe
D	Bal	—	—	3	1 max	9.5	—	2 max
Lewmat 66	Bal[a]	6[a]	0.2	3	31[a]	3[a]	3	16[a]
Alloy D-205	Bal[a]	—	2.5[a]	2[a]	20[a]	5[a]		6[a]

[a]Typical—all other values max unless noted otherwise.

Cast Ni–Si alloys, typically containing 8–10% silicon, were developed for handling hot or boiling sulfuric acid of most concentrations and have also been used to resist strong nitric acid, >50% concentration, and nitric–sulfuric acid mixtures. Unfortunately, cast Ni–Si alloys, such as alloy D, had limited use because they had low fracture toughness and were not corrosion resistant to dilute sulfuric acid environments or environments that contained halides as impurities. The Ni–Cr–Si alloys were developed to provide greater corrosion resistance over a broader range of sulfuric acid environments. Alloy D-205 was especially formulated to be used in wrought form for plate heat exchangers.

C3. Ni–Fe System

Alloy (UNS No.)	Ni	Cr	Co	Mn	Si	C	Fe	Others
36 (K93603) (Invar) Pernifer 36	36[a]	0.2	0.5	0.35	0.2	0.03	Bal	
Magnifer 7904 (Hymu 80)	80[a]			0.50[a]	0.3[a]	0.02[a]	Bal	Mo 5[a]

[a]Typical—all other values max unless noted otherwise.

C5. Ni–Mo System (B Family of Alloys)

Alloy	UNS No.	Ni	Mo	Fe	Cr	C
B-1	N10001	Bal	28	5	0.5	0.03
B-2	N10665	Bal	28	1.5	0.5	0.005
B-3	N10675	Bal	28	1.5	1.5	0.005
B-4	NI0629	Bal	28	3.2	1.3	0.005
B-10	N10624	Bal	24	6	8	0.005

[a]All values are typical unless noted otherwise.

Alloy B, the original alloy in the Ni–Mo family, developed in the 1920s, is susceptible to corrosion of the weld heat-affected zone (HAZ) in nonoxidizing acids (i.e., acetic, formic, and hydrochloric) because of its high carbon content. During the 1960s, argon–oxygen decarburization (AOD) melting technology led to development of alloy B-2. Although this alloy resists HAZ corrosion, it suffers from embrittlement caused by the formation of Ni–Mo intermetallic. This was overcome by the addition of small concentrations of Fe to the alloy. Studies on controlling the composition of alloy B-2 have led to the development of alloy B-3 and Nimofer® 6629 (alloy B-4, UNS N10629), in which the formation of detrimental intermetallic phases were eliminated or reduced, achieving good corrosion resistance behavior and fabricability. Details on the fundamental behavior and understanding of Ni–Mo alloy systems have been presented elsewhere [2]. Alloys B-2, B-3, and B-4 are recommended for service in handling all concentrations of HCl in the temperature range of 70–100°C and in handling wet HCl gas. These alloys also have excellent corrosion resistance to pure H_2SO_4 up to the boiling point in concentrations below 60%. One weakness of the alloy is its lack of chromium and hence its very poor corrosion resistance in the presence of oxidizing species. Alloy B-2 has been successfully used in the production of acetic acid, pharmaceuticals, alkylation of ethyl benzene, styrene, cumene, organic sulfonation reactions, melamine herbicides, and many other products.

Alloy B-4, the improved version of alloy B-2, is used at temperatures between 120 and 150°C in the production of resins, where hydrochloric acid is found as a result of the presence of aluminum chloride. In one chemical company in Spain, alloy B-4 was tested and specified for use in the production of pesticides, where severely corrosive conditions exist because of hydrochloric acid. The "C" family alloys are not acceptable under these conditions. Alloy B-4 has solved both the fabricability problems encountered with alloy B-2 and the susceptibility to stress corrosion cracking in many corrosive environments.

Alloy B-10 [3] (Nimofer 6224) was developed to overcome the limitations of alloys B, B-2, B-3, and B-4 in oxidizing species, where these alloys corrode at unacceptably high rates. The C family alloys, with high chromium contents, such as alloy C-276, or alloy 59 (to a much greater degree), are resistant to corrosion by oxidizing species but lack sufficient molybdenum to resist the reducing conditions in hydrochloric or sulfuric acid. Alloy B-10, an intermediate between the C and B families, has a molybdenum level significantly higher than the C family but somewhat lower than the B family. Also, the chromium and iron levels were increased to 8 and 6, respectively, to resist oxidizing species. This alloy is used, for example, in waste incinerators, where conditions exist that cause crevice corrosion of many alloys.

C6. Ni–Cr–Fe Alloys

Alloy[a]	UNS No.	Ni	Cr	Fe	Al	Si	Others
600	N06600	Bal	16	9	1.4	0.3	
601	N06601	Bal	23	14		0.3	
800H	N08810	32	21	Bal	0.4	0.5	Ti
690	N06690	Bal	30	10		0.3	
602CA	N06025	Bal	25	9.5	2.2		Y, Zr, Ti
45TM	N06045	Bal	27	23		2.7	RE

[a]Typical analysis.

Of the many commercial Ni–Cr–Fe alloys, the major ones are alloys 600 and 601, alloy 800 and its variations, alloy 690 and two new alloys, one being a chromia/alumina-forming alloy (alloy 602CA) and the other chromia/silica-forming alloy (alloy 45TM). These latter alloys are widely used in the chemical, petrochemical, and other process industries, mostly for high-temperature applications, and extensive corrosion test data are available. Improved sulfidation properties, compared to those of alloy 800/800H, have been achieved with a silicon-containing high chromium–nickel alloy, alloy 45TM.

C7. Alloy 600/601/602CA

The high nickel content of alloy 600 imparts excellent resistance to halogens at elevated temperatures, so that this alloy is used in processes involving chlorination. It has good oxidation and chloride SCC resistance. In the production of titanium dioxide, natural titanium oxide (ilmenite or rutile ore) and hot chlorine gases are reacted to produce titanium tetrachloride. Alloy 600 has been successfully used in this process because of its excellent resistance to corrosion by hot chlorine gas. This alloy has found wide usage in furnace and heat treatment applications due to its excellent resistance to oxidation and scaling at 980°C. The alloy is also used in handling water environments, where stainless steels have failed by SCC. It has been used in a number of nuclear reactors, including steam generator boiling and primary water piping systems, although alloy 690 has replaced alloy 600 in some nuclear applications due to its superior stress corrosion cracking resistance. Alloy 600 is used in preheaters and turbine condensers with maximum service temperatures of ∼450°C. However, the low chromium content of alloy 600 prevented its use in applications that require extended exposure to high temperatures and good high-temperature oxidation resistance as well as superior creep and rupture properties. These limitations were addressed by increasing the chromium content to 23% and adding 1.4% aluminum to form alloy 601. The protective oxide that forms on this alloy imparts oxidation and scaling resistance at high temperature, even under severe condition of cyclic exposure to 1100°C. The high nickel and chromium contents of alloy 601 also give it a high degree of resistance to many wet-corrosion environments, along with good resistance to chloride SCC. This alloy has good resistance to carburizing conditions and is used in a wide variety of applications in industries as diverse as thermal and chemical processing, pollution control, and power generation. The alloy has also been used for combustor components and catalyst grid supports in equipment for nitric acid production.

In modern chemical processes, the need to extend the temperature range to 1200°C, while still maintaining good strength characteristics and improved resistance to environmental degradation, led to the development of a new alloy, alloy 602CA. This alloy is a high-carbon, high-chromium, high-aluminum, nickel-base superalloy (Ni Bal, 25 Cr, 2.2 Al, 9.5% Fe, 0.18% C, microalloying additions of titanium, zirconium, and yttrium) with marked improvement in strength and corrosion behavior compared to alloy 601. It has been used in a variety of applications where Alloy 601 was either marginal or inadequate. This alloy is used [4, 5] in the heat treat industry, annealing furnaces, furnace rolls, direct reduction of iron ore technology to produce sponge iron, calciners to produce very high purity alumina, catalytic support systems and glow plugs in the automotive industry, and vitrification of nuclear waste. The alloy is used to combat metal dusting [6] problems in a variety of industries and is finding increased usage against this mode of failure.

C8. Alloy 800/800H/800HT/45TM

Alloy 800 (20% Cr, 32% Ni, and 46% Fe as balance) and its variants are used primarily for their oxidation resistance and strength at elevated temperatures. One of the major benefits is that the alloy does not form embrittling sigma phase even after long time exposure between 650 to 870°C. This behavior coupled with its high creep and stress rupture strength has led to many applications in the petrochemical industry, such as in production of styrene (steam-heated reactors). The alloy exhibits good resistance to carburization and sulfidation and thus has been used in components for coal gasification, for example, heat exchangers, process piping, carburizing fixtures, and retorts. Two major applications are electric range heating element sheathing and extruded tubing for ethylene and steam methane reformer furnaces. It has also found applications as cracking tubes in ethylene dichloride production and in production of acetic anhydride and ketones.

In aqueous corrosion service, alloy 800 is not widely used, because its corrosion resistance is intermediate between types 304SS and 316SS.

Even though alloy 800H is used in coal gasification because of its good carburization and sulfidation resistance properties, it undergoes accelerated attack in some processes. To enhance resistance to these modes of degradation, a higher chromium, silicon-containing nickel alloy was developed,

known as alloy 45TM [4, 7]. This is a chromia/silica-forming alloy with excellent resistance to high-temperature attack in coal gasification and thermal waste incinerators and in refineries and petrochemical industries, where sulfidation has been a major problem.

C9. Ni–Cr–Fe–Mo–Cu Alloys

Alloy[a]	UNS No.	Ni	Cr	Mo	Cu	Fe	Others
825	N08825	Bal	22	3	2	31	Ti
G	N06007	Bal	22	6.5	2	20	Cb + Ta
G-3	N06985	Bal	22	7	2	20	Cb + Ta
G-30	N06030	Bal	30	5	1.5	15	Cb + Ta
20	N08020	35	20	2.5	3.5	Bal	Cb
28	N08028	31	27	3.5	1.0	36	
31	N08031	31	27	6.5	1.2	32	N
33	R20033	31	33	1.6	1.2	32	N
1925hMo	N08926	25	21	6.3	0.9	Bal	N

[a]Typical analysis.

Fortification of Ni–Cr–Fe alloys with molybdenum and copper has resulted in a series of alloys with improved resistance to corrosion by hot reducing acids such as sulfuric, phosphoric, and hydrofluoric acid and acids containing oxidizing contaminants. By maintaining copper content to ~2% or less, chromium between 20 and 33%, and molybdenum between 1.5 and 7.0% and by replacing some of the nickel with iron to reduce cost, a group of alloys were produced with useful corrosion resistance in a wide variety of both oxidizing and reducing acids (except hydrochloric), organic compounds, and acid, neutral, and alkaline salt solutions. These alloys are used extensively in a wide variety of industries to combat corrosion problems.

C10. Alloy 825

Alloy 825 was developed from alloy 800 by increasing the nickel content and by adding molybdenum (3%), copper (2%), and titanium (0.9%) for improved aqueous corrosion resistance in a wide variety of corrosive media. Its high nickel content, about 42%, provides excellent resistance to chloride stress corrosion cracking, although it is not immune to cracking when tested in boiling magnesium chloride solutions. This alloy is an upgrade from the 300 series stainless steels and is used when localized corrosion and SCC are problems. The high nickel, with the molybdenum and copper, provides good resistance to reducing environments, such as those containing sulfuric and phosphoric acids. Laboratory test results and service experience have confirmed the resistance of alloy 825 in boiling solutions of sulfuric acid up to 40% by weight and at all concentrations up to a maximum temperature of 66°C. In the presence of oxidizing species, other than oxidizing chlorides, which may form HCl by hydrolysis, the corrosion resistance in sulfuric acid is usually improved. Hence, the alloy is suitable for use in mixtures containing nitric acid and cupric and ferric sulfates. Alloy 825 is resistant to corrosion in pure phosphoric acid at concentrations and temperatures up to and including boiling 85% acid. The high chromium content imparts resistance to a variety of oxidizing media, such a nitric acid, nitrates, and oxidizing salts. The titanium addition with an appropriate heat treatment stabilizes alloy 825 against sensitization to intergranular attack.

Some typical applications of alloy 825 include various components used in sulfuric acid pickling of steel and copper, components in petroleum refineries and petrochemicals (tanks, agitators, valves, pumps), equipment used in production of ammonium sulfate, pollution control equipment, oil and gas recovery, acid production, nuclear fuel reprocessing, handling of radioactive waste, and phosphoric acid production (evaporators, cylinders, heat exchangers, equipment for handling fluorosilicic acid solution, and many others). Alloy 825 is a versatile alloy for handling a wide variety or corrosive media but is being gradually replaced by newer alloys with superior resistance to localized corrosion, such as the 6% Mo superaustenitic stainless steels, for example, alloy 1925hMo (N08926), alloy 31 (N08031), and the high Cr–Fe–Ni alloy 33, as described later in this chapter.

C11. "G" Family Alloys—G/G-3/G-30

Alloy G was developed during the 1960s from alloy F by adding ~ 2% copper to improve corrosion resistance in both sulfuric and phosphoric acid. This alloy has excellent corrosion resistance in the as-welded condition and can handle the corrosive effects of both oxidizing and reducing agents as well as mixed acids, fluorosilicic acid, sulfate compounds, concentrated nitric acid, flue gases of coal-fired power plants, and hydrofluoric acid. Due to its higher nickel and molybdenum content than alloy 825, the alloy is essentially immune to chloride SCC and has significantly superior localized corrosion resistance. This alloy has been widely used in industries similar to those using alloy 825, but with the added advantage of improved corrosion resistance. Alloy G is now obsolete and has been replaced by alloy G-3.

Alloy G-3 is an improved version of alloy G, with similar excellent corrosion behavior but greater resistance to HAZ attack and with better weldability. Because of its lower carbon content, the alloy has slower kinetics of carbide precipitation, and its slightly higher molybdenum content provides superior localized corrosion resistance. Alloy G-3 has replaced alloy G in almost all industrial applications and alloy 825 in many applications where better localized corrosion resistance is needed.

Alloy G-30 is a modification of alloy G-3 alloy, with increased chromium content and lower molybdenum content. Alloy G-30 has excellent corrosion resistance in

commercial phosphoric acids as well as in many complex and mixed-acid environments of nitric/hydrochloric and nitric/hydrofluoric acids. The alloy also has good corrosion resistance to sulfuric acid. Some typical applications of alloy G-30 are in phosphoric acid service, mixed-acid service, nuclear fuel reprocessing, components in pickling operations, petrochemicals, agrichemicals manufacture (fertilizers, insecticides, pesticides, herbicides), and mining industries.

C12. 6% Mo Alloys

Standard 6Mo alloys, such as Cronifer® 1925hMo, 254SMO®, Inco® 25-6Mo, and A1-6XN®, were derived from alloy 904L metallurgy by increasing the molybdenum content by ~2% and fortifying it with nitrogen as a cost-effective substitute for nickel for metallurgical balance and improved thermal stability. The addition of molybdenum and nitrogen provide the added benefits of improved mechanical properties and resistance to localized corrosion. These alloys are readily weldable with an overalloyed filler metal, such as alloy 625, C-276, or 59, to compensate for the segregation of molybdenum in the interdendritic regions of the weld. They have been extensively used in offshore and marine, pulp and paper, FGD, chemical process industry for both organic and inorganic compounds, and a variety of other applications. The 6Mo family of alloys bridges the performance gap between standard stainless steels and the high-performance nickel- base alloys in a cost-effective manner.

A higher chromium, higher nickel version of 6Mo superaustenites is the alloy Nicrofer®, 3127hMo (Alloy 31, UNS N08031). Its greatly improved corrosion resistance compared with the conventional 6Mo family of alloys and alloy 28 is achieved by increased Cr (27%) and Mo (6.5%) and fortification with nitrogen (0.2%). The corrosion behavior of alloy 31, achieved with only about half the nickel content of alloy 625, makes it a very cost-effective alternative in many applications. The pitting potential of this alloy, as determined in artificial seawater, makes it a suitable alloy for heat exchangers using seawater or brackish water as cooling media. Its corrosion resistance in sulfuric acid in medium-concentration ranges is superior even to that of alloy C-276 and alloy 20.

However, in view of the specific active/passive characteristics of alloy 31, one must be extremely careful when specifying this material for sulfuric acid use at 80% concentration and temperatures above 80°C because these conditions cause active corrosion. Alloy 31 has been extensively used in the most varied applications, including the pulp and paper industry, phosphoric acid environments, copper smelters, sulfuric acid production, pollution control, wastewater treatment in uranium mining, sulfuric acid evaporators, leaching of copper ores, viscose rayon production, and fine chemicals production.

C13. Alloy 20

The first version of alloy 20 was introduced in 1951 for sulfuric acid applications. Later, a columbium-stabilized version was developed as 20Cb, which allowed alloy 20Cb weldments to be used in the as-welded condition without any postweld heat treatment. Further R&D led to the modern version, alloy 20Cb3, with increased nickel content. This alloy is used in many applications because of its superior corrosion resistance in sulfuric acid media and its resistance to stress corrosion cracking. It is used in manufacture of synthetic rubber, high-octane gasoline, solvents, explosives, plastics, synthetic fibers, chemicals, and pharmaceuticals, in the food processing industry, and in many other applications. This alloy contains insufficient molybdenum for resistance to pitting and crevice corrosion in low-pH acidic chloride media.

C14. Alloy 33

Alloy 33 is a chromium-based fully austenitic wrought super stainless steel (33 Cr, 32 Fe, 31 Ni, 1.6 Mo, 0.6 Cu, 0.4 N). This alloy has excellent resistance to acidic and alkaline corrosive media, mixed HNO_3/HF acids, hot concentrated sulfuric acid, localized corrosion, and SCC. Because of its high nitrogen content, this alloy has excellent mechanical properties. Its high pitting resistance equivalent (PRE) makes it a very cost-effective alloy in comparison to the high Ni–Cr–Mo alloys, such as G-3, G-30, and 625. Its localized corrosion resistance is equal to or better than some of the Ni–Cr–Mo alloys in specific environments. Details on the development and properties of this alloy are presented elsewhere [8, 9].

C15. High-Performance Ni–Cr–Mo Alloys

Alloy (UNS No.)	Ni	Cr	Mo	W	Fe	Others
C (N10002)	Bal	16	16	4	6	
625 (N06625)	Bal	22	9		2	Cb
C-276 (N10276)	Bal	16	16	4	5	
C-4 (N06455)	Bal	16	16		2	Ti
C-22/622 (N06022)	Bal	21	13	3	3	
59 (N06059)	Bal	23	16		<1	
686 (N06686)	Bal	21	16	4	2	
C-2000 (N06200)	Bal	23	16		2	Cu
MAT 21 (N06210)	Bal	19	19		<1	Ta

The C family of alloys, the original being Hastelloy® C (1930s), was an innovative optimization of Ni–Cr alloys having good resistance to oxidizing corrosive media and Ni–Mo alloys with superior resistance to reducing corrosive media. The combination resulted in the most versatile corrosion-resistant alloy in the "Ni–Cr–Mo" alloy family with exceptional corrosion resistance in a wide variety of severely corrosive environments typically encountered in the chemical process industry and other industries. The alloy also exhibited excellent resistance to pitting and crevice corrosion in low-pH, high-chloride, oxidizing environments and had total immunity to chloride stress corrosion cracking. These properties allowed this alloy to serve the industrial needs for many year, although it had some limitations. Today, the original alloy C of the 1930s is practically obsolete except for some usage in castings. Alloys with improved corrosion resistance were developed during the 1960s (alloy C-276), 1970s (alloy C-4), 1980s (alloys C-22 and 622), and 1990s (alloys 59, 686, C-2000, and MAT 21). Improvements in corrosion behavior not only overcame the limitations of alloy C but also further expanded the horizons of applications as the needs of the chemical process industries became more critical, severe, and demanding.

C16. Alloy C (1930s–1965)

This section presents the chronology of the development of corrosion-resistant Ni–Cr–Mo alloys during the twentieth century with special emphasis on the evolution in the C family of Ni–Cr–Mo alloys since 1960 and their applications.

The compatibility and optimization between Ni–Cr and Ni–Mo alloys led to the first alloy of the C family, wrought Hastelloy® alloy C in the 1930s, with a typical composition of 55 Ni, 15.5 Cr, 16 Mo, 4 W, and 5 Fe, which exhibited outstanding corrosion resistance in many oxidizing and reducing environments while possessing excellent localized corrosion resistance and resistance to chloride SCC.

This alloy was the most versatile corrosion-resistant alloy available in the 1930s through the mid-1960s to handle the needs of the chemical process industry. However, the alloy had a few severe drawbacks. When used in the as-welded condition, alloy C was often susceptible to serious intergranular corrosion attack in HAZ in many oxidizing, low-pH, halide-containing environments. For many applications, vessels fabricated from alloy C had to be solution heat treated to remove the detrimental weld HAZ precipitates, seriously limiting the usefulness of this alloy. During the late 1940s and 1950s, the chemical process industry was constantly developing new processes that needed an alloy without the requirement of solution heat treating after welding. Also in severely oxidizing media, this alloy did not have enough chromium to maintain passivity and exhibited high uniform corrosion rates.

C17. Alloy 625 (1960s–Present)

One of the "severe oxidizing media" limitations of alloy C was overcome by increasing the chromium content from 15.5 to ~22% in Inconel® alloy 625, an alloy developed in the late 1950s and commercialized in the early 1960s. However, the molybdenum content of the alloy was reduced to 9% and columbium was added for stabilization against intergranular attack, which permitted the use of this alloy in the as-welded condition without the need for solution heat treatment. The increased chromium improved corrosion resistance in a number of strongly oxidizing environments, such as boiling nitric acid and nitric acid mixtures, while at the same time maintaining adequate resistance to a number of reducing acid conditions. The alloy had a good balance of properties but was not as versatile in "reducing acids" as alloy C because of the lower molybdenum level of Inconel® 625.

The alloy is resistant to corrosion and pitting in seawater and has useful resistance to wet chlorine, hypochlorite, and oxidizing chlorides at atmospheric temperatures. It is resistant to all concentrations of hydrofluoric acid, even when aerated, and to such acid mixtures as nitric–hydrofluoric, sulfuric–hydrofluoric, and phosphoric–hydrofluoric acids under most conditions encountered in industrial practice. Alloy 625 has good strength and resistance to scaling in air at temperatures up to 980°C and many of its uses are in high-temperature and aerospace applications.

An alloy with lower nickel content but similar chromium and molybdenum contents, Hastelloy® alloy X (47 Ni, 22 Cr, 18.5 Fe, 9 Mo, 0.10 C, 0.6 W), was developed having good high-temperature strength and resistance to oxidation and scaling up to ~1100°C. It is used mostly for high-temperature applications in the furnace and heat treating fields and for flying and land-based gas turbine components.

Another alloy, Hastelloy® alloy N (69.5 Ni, 7 Cr, 16.5 Mo, 5 Fe), was developed for use with molten fluoride salts at 815°C in nuclear applications. Chromium was reduced to 7% to prevent intergranular attack and mass transfer of chromium in this environment, and the 7% level is adequate to provide strength and oxidation resistance at this temperature.

C18. Alloy C-276 (1965–Present)

To overcome one of the serious limitations of solution annealing in Hastelloy® alloy C after welding, the chemical composition of alloy C was modified by a German company, BASF, which reduced both the carbon and silicon levels by more than 10-fold to very low levels, typically 50 ppm carbon and 400 ppm silicon. This development was possible because of the invention of a new melting technology, the AOD process. The low-carbon and low-silicon alloy came to be known as alloy C-276. It was produced in the United States under a licence from BASF Company, which was awarded a U.S. patent that expired in 1982, at which time other nickel alloy producers started manufacturing this alloy.

The corrosion resistance of both alloy C and alloy C-276 was similar in many corrosive environments, but without the detrimental effects of continuous grain boundary precipitates in the weld HAZ of alloy C-276. alloy C-276 could be used in most applications in the as-welded condition without severe

intergranular attack. The corrosion behavior of both alloy C and alloy C-276 has been adequately covered in the open literature [10–14]. The grain boundary precipitation kinetics and the Time-Temperature-Transformation (T-T-T) diagram for these alloys are also well documented [15–17]. The applications of alloy C-276 in the process industries are extensive, diverse, and versatile due to its excellent resistance in both oxidizing and reducing media, even with halogen ion contamination. However, there are certain process conditions, where even alloy C-276 with its low carbon and low silicon is susceptible to corrosion because it is not adequately stable thermally in regard to precipitation of both carbides and intermetallic phases. Within the broad scope of chemical processing, examples exist where serious intergranular corrosion of a sensitized (precipitated) microstructure has occurred. To overcome this susceptibility, a modification of alloy C-276 was developed in the 1970s, called alloy C-4. Under highly oxidizing conditions, neither alloy C-276 nor C-4 with their 16% chromium content provide useful resistance. This shortcoming was overcome by developments of other alloys, such as alloy C-22/622 and alloy 59.

C19. Alloy C-4 (1970s–Present)

Alloy C-4, in addition to the 10-fold decrease in carbon and silicon of alloy C, had three other major modifications, that is, omission of tungsten from its basic chemical composition, reduction in iron level, and addition of titanium. The above changes resulted in significant improvements in the precipitation kinetics of intermetallic phases when exposed in the sensitizing range of 550–1090°C for extended periods of time, virtually eliminating the intermetallic and grain boundary precipitation of the "mu" phase, which has a $(Ni,Fe,Co)_3(W,Mo,Cr)_2$-type structure, and various other phases. These phases are detrimental to ductility, toughness, and corrosion resistance. The general corrosion resistance of alloys C-276 and C-4 are essentially the same in many corrosive environments, except that in strongly reducing media like hydrochloric acid alloy C-276 is better, but in highly oxidizing media the opposite is true, that is, alloy C-4 is better. Alloy C-4 offers good corrosion resistance to a wide variety of media, including organic acids and acid chloride solution. Details of alloy C-4 development have been documented [10, 18, 19]. This alloy has found greater acceptance in European countries than in the United States, in contrast to alloy C-276, which is more widely used and accepted in the United States. Alloy C-4 is gradually being replaced by another tungsten-free Ni-Cr-Mo alloy, alloy 59, which has superior corrosion resistance.

C20. Alloy C-22 (1982–Present)

Alloy C-22 was introduced to retain the better oxidizing environment capabilities of alloy 625 while maintaining the localized corrosion resistance of alloy C-276. It was found that the "mu" phase control in alloy C-4, accomplished by controlling the "electron vacancy" number by omitting tungsten and reducing iron, was achieved at the expense of reduced corrosion resistance to oxidizing chloride solutions, where tungsten is a beneficial element. In addition, both alloy C-276 and alloy C-4 corrode rapidly in oxidizing, nonhalide solutions because of their relatively low chromium levels of 16%. This alloy C-22 composition with approximately 21% Cr, 13% Mo, 3% W, 3% Fe with balance nickel showed superior corrosion resistance to that of alloy C-276 and alloy C-4 in highly oxidizing and localized corrosion environments. However, its behavior in highly reducing environments and in severe crevice and corrosion conditions is inferior to that of alloy C-276 [20, 21]. Details on the development of alloy C-22 have been described elsewhere [22–24].

C21. Alloy 59 (1990–Present)

Research efforts during the late 1980s at Krupp VDM led to another development within the Ni–Cr–Mo family, alloy 59, which overcame the shortcomings of both alloys C-22 and C-276. It also provided solutions to the most severe and critical corrosion problems of the chemical process, petrochemical, pollution control, and other industries.

As is evident from the composition of the various alloys of the C family, alloy 59 has the highest chromium plus molybdenum content with the lowest iron content, typically <1%. It is one of the highest nickel-containing alloys of this family and is the purest form of a "true" Ni–Cr–Mo alloy without the addition of any other alloying elements, such as tungsten, copper, titanium, or tantalum, which are contained in other alloys in this family. This "purity" and balance of alloy 59 in the ternary Ni–Cr–To system are mainly responsible for the superior thermal stability of this family. The "electron vacancy" number, an important parameter in the "Phacomp calculations" for alloy development and for prediction of occurrence of various phases, now superseded by more precise "New Phacomp" methodology, proposed by Morinaga [25], also lends support to this phenomenon of superior thermal stability of alloy 59.

C22. Alloy 686 (1993–Present)

Alloy 686 is another recent development in the C family of Ni–Cr–Mo–W alloys. This alloy is very similar in composition to alloy C-276, but the chromium level has been increased from 16 to 21%, while maintaining the Mo and W at similar levels. Alloy 686 is overalloyed with the combined Cr, Mo, and W content of around 41%. To maintain its single-phase austenitic structure, this alloy is solution annealed at 1200°C and rapidly cooled to prevent precipitation of intermetallic phases.

C23. Alloy C-2000 (1995-Present)

Alloy C-2000 is another introduction to the C alloy family, in which 1.6% copper has been added to the alloy 59 composition. However, addition of copper results in significantly lower localized corrosion resistance and less thermal stability compared to alloy 59.

C24. Alloy MAT21 (1998–Present)

Alloy MAT21 is a member of the C family of alloys with addition of ~1.8% tantalum to a Ni,19%Cr,19%Mo alloy. Insufficient data are available regarding its localized corrosion behavior and thermal stability.

D. CORROSION BEHAVIOR OF "C" ALLOYS

D1. Resistance to General Corrosion

The uniform corrosion resistance data of the C alloys in various boiling corrosive environments are presented in Table 59.3. The media are both oxidizing and reducing in nature and are used for comparing the relative performance of alloys. As is evident from these data, the overall performance of alloy 59 is better than that of any other C family alloy.

D2. Localized Corrosion Resistance

Localized corrosion has caused more failures in the chemical process and other industries than any other single corrosion phenomenon. Chromium, molybdenum, nitrogen, and to a lesser extent tungsten enhance pitting and crevice corrosion resistance of nickel-base alloys. The critical pitting and crevice corrosion temperatures (CPT, CCT), of these alloys, as evaluated using the ASTM G-48 (10% ferric chloride) test method, are listed in Table 59.4. The lower molybdenum alloy, C-22, has significantly lower critical crevice corrosion temperature. This result is supported by data in Table 59.5, showing that alloy 22 has a slightly lower CCT than alloy 59. In this test, crevice attack of alloy 59 was lower than that of

TABLE 59.4. Critical Pitting and Crevice Corrosion Temperature per ASTM G-48

Alloy	Cr	Mo	PRE[a]	CPT(°C)	CCT(°C)
C-22	21	13	65	>85[b]	58
C-276	16	16	69	>85	>85
686	21	16	74	>85	>85
59	23	16	76	>85	>85

[a]PRE = % Cr + 3.3 % Mo.
[b]Above 85°C the solution chemically breaks down.

TABLE 59.5. Localized Corrosion Resistance in Green Death Solution[a]

Alloy	PRE	CPT(°C)	CCT(°C)	Crevice Depth at 105°C
C-22	65	120	105	0.35 mm
C-276	59	110	105	0.035 mm
59	76	>120[b]	110	0.025 mm
686	74	>120	110	
C-2000	76	110	100	

[a]11.5% H_2SO_4 + 1.2% HCl + 1% $FeCl_3$ + 1% $CuCl_2$.
[b]Above 120°C, the Green Death solution chemically breaks down.

Alloy C-276, showing the beneficial effects of its high PRE number.

D3. Thermal Stability

The superior thermal stability of alloy 59 is shown in Tables 59.6A and 59.6B. The data indicate the detrimental effects of tungsten and copper on the thermal stability of various alloys of the C family. During welding of heavy-walled vessels and/or hot forming of heavy-walled materials, thermal stability is very important in maintaining superior corrosion resistance in the as-welded condition.

Other corrosion resistance data and information on physical metallurgy, fabricability, and weldability of alloy 59 have been adequately covered elsewhere [26–28].

D4. Applications of The "C" Family of Alloys

The C family of alloys has found widespread application in chemical and petrochemical industries producing various chlorinated, fluorinated and other organic chemicals, agrichemicals and pharmaceutical industries producing various biocides, pollution control (FGD of coal-fired power plants, waste water treatment, incinerator scrubbers) [29–32], pulp and paper, oil and gas (sour gas production), marine, and many others.

TABLE 59.6A. Thermal Stability per ASTM G-28A after Sensitization[a]

Sensitization (h)	Corrosion Rate (mpy)[b]				
	C-276	C-22	686	C-2000	59
1	>500[c]	>500[c]	>500[c]	116[c]	40[d]
3	>500[c]	>500[c]	>1000[c]	178[c]	51[d]

[a]At 870°C (1600°F).
[b]To convert to millimeters per year (mm/year) multiply by 0.0254.
[c]Alloys C-276, C-22, 686, and C-2000—heavy pitting attack with grains falling out because of deep intergranular attack.
[d]Alloy 59—no pitting attack.

TABLE 59.6B. Thermal Stability per ASTM G28B After Sensitization[a]

Sensitization (h)	Corrosion Rate (mpy)[b]				
	C-276	C-22	686	C-2000	59
1	>500[c]	339[c]	17[c]	>500[c]	4[d]
3	>500[c]	313[c]	85[c]	>500[c]	4[d]

[a]At 870°C(1600°F).
[b]To convert to millimeters per year (mm/year) multiply by 0.0254.
[c]Alloy C-276, C-22, C-2000, and 686—Heavy pitting attack with grains falling out because of deep intergranular attack.
[d]Alloy 59—no pitting attack.

Between 1966 and 1998, ~60,000 tons of alloy C-276 and C-4 has been used in a variety of industries, some of which are listed in Table 59.7.

Most of the C family alloys are covered under the appropriate ASTM (American Society for Testing and Materials), AWS (American Welding Society), ASME (American Society of Mechanical Engineers), and NACE (NACE International) MR0175 standards and other national and international standards.

E. PRECIPITATION-HARDENABLE ALLOYS

Alloy	UNS No.	C	Cr	Cu	Fe	Mo	Nb	Ni	Other
K-500	N05500	0.2		30	1	—	—	Bal	3 Al, 0.6Ti
X-750	N07750		15.5	—	7.0	—	0.9	Bal	2.5 Ti
718	No7718	0.05	18.0	—	19.0	3.0	5.0	Bal	0.4 max Ti
925	N09925	0.02	21.0	2.0	28	3.0	—	Bal	2.1 Ti
625 Plus	N07716	0.03 max	20.5	—	Bal	8.25	3.5	61	1.3 Ti

The Ni–Cu alloy, K Monel or alloy 500, was used extensively in drill collars and other high-strength components However, many investigators have reported cracking due to hydrogen generated by galvanic coupling with steel [33–35].

The addition of Al, Ti, Nb, and Si causes precipitation of coherent phases upon heat treatment, which increases the strength of the alloys. This characteristic is especially important in manufacturing thick-section or complex-shaped components that cannot be strengthened purely by cold working. These alloys are used in the nuclear industry for components such as hold-down springs in fuel assemblies, high-strength pins, and bolts in reactor cores. They are also used in oil and gas applications for tubing hangers and valve components. A larger number of failures for X-750 have been reported than for alloy 718, and single-aging treatments are

TABLE 59.7. Major Industries Using Alloy C-276/C-4/C-22/59/686

I. Petroleum
    Petroleum refining
    Oils/greases
    Natural gas processing
II. Petrochemical
    Plastic
    Synthetic organic fibers
    Organic intermediates
    Organic chemicals—chlorinated/fluorinated hydrocarbons
      Synthetic rubber
III. CPI—Chemical process industries
    Fine chemicals
    Inorganic chemicals
    Soaps/detergents
    Paints
    Fertilizer—agrichemicals—herbicides/pesticides
    Adhesives
    Industrial gases
IV. Pollution control
    FGD
    Wastewater treatment
    Incineration
    Hazardous waste
    Nuclear fuel reprocessing
V. Pulp and paper
VI. Marine/seawater
VII. Pharmaceuticals
VIII. Sour gas/oil and gas production
IX. Mining/metallurgical

found to yield better SCC resistance than the typical double-aging treatment used in the aerospace industry [36, 37]. Alloy 625Plus has been shown to have greater SCC resistance in the aged condition to sour gas environments than X-750 and alloy 718 [38].

F. HIGH-TEMPERATURE ALLOYS

The need for high-temperature materials is encountered in a wide variety of modern industries, such as aerospace, metallurgical, chemical, and petrochemical, and in many applications, including glass manufacture, heat treatment, waste incinerators, heat recovery, advanced energy conversion systems, and others. Depending on the condition of the chemical makeup and on temperature, a variety of aggressive corrosive environments are produced, which could be either sulfidizing, carburizing, halogenizing, nitriding, reducing, and oxidizing in nature or a combination thereof. All high-temperature alloys have certain limitations and the optimum choice is often a compromise between mechanical property requirements at maximum temperature of operation and requirements for corrosion resistance in the corrosive species present.

Alloys designed to resist high-temperature corrosion have existed since the beginning of the twentieth century. Generally, high-temperature metal degradation occurs at temperatures >540°C, but there are some cases where it can also occur at somewhat lower temperatures. Carbon steel, the "workhorse" material of construction in many industries, is attacked by H_2S >260°C, by oxygen or air > 540°C, and by nitrogen > 980°C. The new technologies of thermal destruction of hazardous and municipal waste, fluidized-bed combustion, coal gasification and chemicals from coal processes, and the use of "dirty feedstock," such as heavy oil and high sulfur coal, coupled with demands for higher efficiency and tougher environmental regulations, have necessitated the use of higher alloy systems of iron-, nickel-, and cobalt-base alloys. Alloy systems must provide reliable and safe performance in a cost-effective manner but must also have sufficient versatility to resist changing corrosive conditions due to feedstock changes. The property requirements in materials of construction for high-temperature applications can be classified under mechanical and high-temperature corrosion resistance as follows:

Mechanical Properties	Corrosion Resistance Properties
High-temperature strength	Oxidation
Stress rupture strength	Carburization
Creep strength	Nitriding
Fatigue strength	Sulfidation
Thermal stability	Halogenation
Thermal shock resistance	Molten salt
Toughness	Liquid metal
Other specific properties	Ash salt deposit
	Others

Requirements will vary for different industries, such as aerospace, heat treating, power generation, and chemical/petrochemical processing.

In nickel-base alloys, the major elements for imparting specific properties or a combination of properties are chromium, silicon, aluminum, titanium, molybdenum, cobalt, tungsten, and carbon. Others, such as yttrium and rare earths, niobium, tantalum, and zirconium, play very specific roles in improving certain high-temperature corrosion characteristics. These alloying elements can also be classified as follows:

Protective scale formers	Cr as Cr_2O_3, Al as Al_2O_3, Si as SiO_2
Solid-solution strengtheners	Mo, W, Nb, Ti, Cr, Co
Age-hardening strengtheners	Al + Ti, Al, Ti, Nb, Ta
Carbide strengtheners	Cr, Mo, W, Ti, Zr, Ta, Nb
Improved scale adhesion (spallation resistance)	Rare earths (La, Ce) Y, Hf, Zr, Ta

References [10, 39–42] provide detailed information on high-temperature alloy systems.

REFERENCES

1. Bulletin CEB-2 "Corrosion Resistance of Nickle and Nickle-Containing Alloys in Caustic Soda and Other Alkalies," International Nickle Co., Inc., Huntington, WV, 1973.

2. D. C. Agarwal, U. Heubner, M. Koehler,and W. Herda, Mater. Perform., 33(10), 64 (1994).

3. M. Köhler, R. Kirchheiner, and F. Stenner, Alloy B-10, "A New Nickel-Based Alloy for Strong Chloride-Containing, Highly Acidic and Oxygen-Deficient Environments," Paper No. 481, CORROSION/98, NACE International, Houston, TX, 1998.

4. D. C. Agarwal and U. Brill, "Material Degradation Problems in High Temperature Environments (Alloys—Alloying Effects—Solutions)," Industrial Heating, Oct. 1994, pp. 55–60.

5. D. C. Agarwal and H. Klein, Applications and Material Performance, Nickel Cobalt 97, Vol. 4, The Metallurgical Society of the Canadian Institute of Mining, Metallurgy and Petroleum, Montreal, 1997, pp. 115–129.

6. H. J. Grabke, E. M. Müller-Lorenz, J. Klöwer, and D. C. Agarwal, "Metal Dusting and Carburization Resistance of Nickel Base Alloys," Paper No. l39, CORROSION/97, NACE International, Houston, TX, 1997.

7. D. C. Agarwal and U. Brill, "NiCr27FeSiRE: A New Alloy for Waste Incineration," Paper No. 209, CORROSION/93, NACE International, Houston, TX, 1993.

8. M. Köhler, U. Heubner, K. W. Eichenhofer, and M. Renner, "Alloy 33, A New Corrosion Resistant Austenitic Material for the Refinery Industry and Related Applications," Paper No. 338, CORROSION/95, NACE International, Houston, TX, 1995.

9. M. Köhler, U. Heubner, K. W. Eichenhofer, and M. Renner, "Progress with Alloy 33, a New Corrosion Resistant Chromium-Based Austenitic Material," Paper No. 428, CORROSION/96, NACE International, TX, 1996.

10. W. Z. Friend, Corrosion of Nickel and Nickel Base Alloys, Wiley, New York, 1980.

11. E. D. Weister, Corrosion, 13, 659 (1957).

12. W. A. Luce, Chem. Eng., 61(3), 254 (1954).

13. R. K. Swandly, in Nickel Base Alloys in Corrosion Resistance of Metal Alloys, 2nd ed., F. L. LaQue and H. R. Copson (Eds.), Van Nostrand Reinhold, Inc., New York, U.S., 1963, pp. 515–552.

14. Metals Handbook, 9th ed., Vol. 13, Corrosion, Metals Park, OH, 1987, pp. 641–657.

15. I. Class, H. Gräfen, and E. Scheil, Z. Metallk, 53, 283 (1962).

16. M. A. Streicher, Corrosion, 19, 272 (1963); 32, 79 (1976).

17. R. B. Leonard, Corrosion, 25, 222 (1969).

18. R. W. Kirchner and F. G. Hodge, Werkst. Korros, 24, 1042 (1973).

19. F. G. Hodge and R. W. Kirchner, Corrosion, 32, 332 (1976).

20. D. C. Agarwal and W. R. Herda, "Alloying Effects and Innovations in Nickel Base Alloys for Combating Aqueous Corrosion," VDM Report No. 23, Krupp VDM, Werdohl, Germany, 1995.

21. V. Yanish, "Corrosion Testing in a Hazardous Waste Incinerator and Waste Heat Boiler," in Proceedings of the Second International Conference on Heat Resistant Materials, Gatlinburg, TN, 11–14 Sept. 1995, ASM, Metals Park, OH, pp. 655–656.

22. P. E. Manning, A. I. Asphahani, and N. Sridhar, "New Developments in Ni–Cr–Mo Alloys," Paper No. 21, CORROSION/83, NACE, Houston, TX, Apr. 1983.

23. P. E. Manning and A. I. Asphaphani, "Advanced Materials Technologies of Interest to the Process Industries," presented at ACHEMA 85, an International Meeting on Chemical Engineering, Frankfurt am Main, Germany, June 9–15, 1985.

24. N. Sridhar, J. B. C. Wu, and P. E. Manning, J. Metals, 37(11), 51 (1985).

25. M. Morinaga, N. Yukawa, H. Adachi, and H. Ezaki, "New Phacomp and Its Application to Alloy Design," Fifth International Symposium Superalloys, Seven Springs, 1984, ASM Metals Park, OH, 1984.

26. R. Kirchheiner, M. Köhler, and U. Heubner, "A New Highly Corrosion Resistant Material for the Chemical Process Industry, Flue Gas Desulfurization and Related Applications," Paper No. 90, CORROSION/90, NACE International, Houston, TX, 1990.

27. D. C. Agarwal, U. Heubner, R. Kirchheiner, and M. Koehler, "Cost Effective Solution to CPI Corrosion Problems with a New Ni-Cr-Mo Alloy," Paper No. 179, CORROSION/91, NACE International, Houston, TX, 1991.

28. U. Heubner, "Nickel Based Alloys," in Materials Science and Technology, A Comprehensive Treatment, R. W. Cahn, P. Haasen, and E. J. Kramer (Eds.), VCH Verlagsgesellschaft, Germany, 1996, Chapter 7.

29. D. C. Agarwal, "Alloy Selection Methodology and Experiences of the FGD Industry in Solving Complex Corrosion Problems: The Last 25 Years," Paper No. 447, CORROSION/96, NACE International, Houston, TX, 1996.

30. W. R. Herda and W. Romer, "Recent Experiences with Alloy 59–UNS N06059 (DIN No. 2.4605) in Waste incineration Plant Construction," Paper No. 557, CORROSION/95, NACE International, Houston, TX, 1995.

31. VDM Case History No. 1, "The Waste Incineration Thermal Power Plant in Essen Kamap," Krupp VDM, Werdohl, Germany.

32. D. C. Agarwal and Miles Ford, "FGD Metals and Design Technology: Past Problems/Solutions–Present Status and Future Outlook," Paper No. 485, CORROSION/98, NACE International, Houston, TX, 1998.

33. J. G. Erlings, H.W. deGroot, and J. F. M. van Roy, "Stress Corrosion Cracking and Hydrogen Embrittlement of High Strength Non-magnetic Alloys in Brine," Mater. Perform., Oct. 1986, pp. 28–34.

34. L. H. Wolfe and M. W. Joosten, "Failures of Nickel/Copper Bolts in Subsea Applications," SPE Production Eng., Aug. 1988, pp. 382–386.

35. K. D. Efird, "Failure of Monel Ni-Cu-Al Alloy K-500 Bolts in Seawater," Mater. Perform., Apr. 1985, pp. 37–40.

36. N. Sridhar and G. A. Cragnolino, "Stress Corrosion Cracking of Nickel-Base alloys," in Stress Corrosion Cracking, R. H. Jones (Ed.), ASM International, Materials Park, OH, 1992.

37. J. Kolts, "Alloy 718 for the Oil and Gas Industry, in Superalloy 718—Metallurgy and applications," E. Loria (Ed.), The Minerals, Metals, and Materials Society, Pittsburgh, PA 1989, pp. 329–344.

38. R. B. Frank and T. A. deBold, "Properties of Age-Hardenable Corrosion-Resistant Nickel-Based Alloy," Mater. Perform., Sept. 1988, pp. 59–66.

39. U. Heubner, Nickel Alloys and High Alloy Special Stainless Steels, 1st ed., Krupp VDM GmbH, Werdohl, Germany, 1987.

40. U. Heubner, Nickel Alloys and High Alloy Special Stainless Steels, 2nd ed. Krupp VDM GmbH, Werdoh, Germany, 1987.

41. U. Brill, "High Temperature Alloys and Their Use In Furnace Construction," VDM Report No. 15, Werdohl, Germany, June 1991.

42. G. Lai, High Temperature Corrosion of Engineering Alloys, ASM International, Materials Park, OH, 1990.

BIBLIOGRAPHY

R. K. Swandly, "Nickle Base Alloys," in Corrosion Resistance of Metal and Alloys, 2nd ed., F. L. LaQue and H. R. Copsen (Eds.), Van Nostrand Reinhold, Inc., New York, U.S., 1963, pp. 515–552.

Publications of The Huntington Alloy Products Division, The International Nickle Co., Inc. Huntington, W VA. Bulletin CEB-1, The corrosion resistance of nickel-containing alloys in sulfuric acid and related compounds, 1983. Bulletin CEB-3, Resistance of nickel and high-nickel alloys to corrosion by hydrochloric acid, hydrogen chloride, and chlorine. Bulletin CEB-4, Corrosion resistance of nickel-containing alloys in phosphoric acid. Bulletin CEB-5, Hydrofluoric acid, hydrogen fluoride, and fluorine. Bulletin CEB-6, Corrosion resistance of nickel-containing alloys in organic acids and related compounds. Individual alloy product brochures of various alloys. High-temperature and corrosion alloys product literature.

V. D. M. Krupp, High Temperature and Corrosion Alloys Product Literature, Krupp VDM GmbH, Werdohl, Germany.

Haynes High Temperature and Corrosion Alloys Product Literature, Haynes International, Kokomo, IN.

Publications of Nickel Development Institute on Nickel Alloys, Nickel Development Institute, Toronto, Canada.

ASTM Book of Standards, Vol 02.04—Non-Ferrous Metals; Vol. 03.02—Wear and Erosion, Metal Corrosion, ASTM, West Conshohocken, PA.

60

TIN AND TINPLATE

T. P. Murphy*

Campion Hall, University of Oxford, Oxford, UK

A. INTRODUCTION

Tin is among the longest-used metals in the service of humankind. Its origins go back millennia, and its combination to form bronze constitutes one of our earliest essays in metallurgy. It may first have been encountered by accident, as a result of the chance reduction of cassiterite in a charcoal fire. Combined with copper, it forms bronze, a metal that helped lift our species out of its primitive condition.

The other component of tinplate, mild steel, is probably the metal most widely used in the modern world. While steel is less venerable than tin and, indeed, than tinplate, its development has shaped our present society. The combination of these two metals has proved valuable for centuries and remains today a material of great economic and environmental importance.

The properties and uses of these metals, the manner in which they may corrode and the factors influencing corrosion, and the consequences of corrosion are reviewed in this chapter. Given the major uses of the metals concerned, their interaction with other metals and with some nonmetallic materials, other than corrodants, will, of necessity, form part of the story.

B. TIN

Tin is a soft, malleable metal with a high luster. It displays allotopic modification, with two forms, α- and β-tin, having densities of 5.8 and 7.3, respectively. The former, known as gray tin, is a brittle, gray, cubic material with little strength. It is the low-temperature modification and, theoretically, may be formed below about 13°C. In practice, temperatures below zero Celsius are required, and the tin must be very pure. Quite small levels of impurity atoms inhibit the transition, and heating to room temperature restores the β form. This, known as white tin, is the familiar, metallic form, ductile and malleable, with a tetragonal structure.

Room temperature represents ∼60% of its absolute melting point at 232°C, so that the mechanical strength of tin is limited. Uses for the metal in its elemental form are thus restricted. It appears, rather, in the form of alloys or coatings. Some use is made of block tin in the brewing industry, but this has largely been superseded by stainless steel. Pewter, an alloy of tin (>90%) with antimony and copper and, nowadays, no lead, is used in decorative work and drinking vessels, and alloys with (decreasingly) lead, bismuth, and other metals find use in solders. Low-melting tin alloys of quite widely differing compositions have served as die-casting materials, and press tools have been made from fusible tin alloys [1]. Solders and tinplate today form the two major uses for tin, with the former having recently

---

*Deceased.

overtaken the latter. Replacement of more toxic materials such as lead in, for example, wine capsules and gunshot has recently provided new applications for this metal.

B1.   Corrosion Behavior

B1.1.   Atmospheric Attack.
Tin is not a noble metal but is not among the most active. It is subject to attack both from the liquid and the vapor phases. The attack depends on the nature of the environment. Tin remains reasonably bright for long periods at ambient temperatures in air free of moisture and pollutants. Oxide formation occurs more quickly as the temperature increases and may produce interference effects. Normal domestic atmospheres have only limited corrosive effects on tin. Pollution changes this, with SO_2 and H_2S capable of producing darkening. The results of such atmospheric attack tend to be formation of oxides and sulfides; the latter are colored so that the attack is obvious. Loss of structural integrity is not a common outcome of atmospheric attack, though the metal is vulnerable to the halogens and vapors of strong mineral acids, and to conditions that lead to formation of films of liquid on the metal. Solderability and electrical resistance can, however, be affected by the formation of surface oxide films, though this is not usually a problem when the tin is sheltered from the weather. Solders are widely used in the electronics industry, with a near-eutectic 60% tin, 40% lead composition chosen for its sharp melting point. Few corrosion problems arise in these sheltered applications, though some care is needed in the choice and use of fluxes, which are, of their nature, corrosive.

Corrosion rates have been measured for exposed conditions. Britton [2] quotes the work of Hiers and Minarcik on American Society for Testing and Materials (ASTM) tests on tin specimens exposed for periods of up to 20 years. Average penetration rates in micrometers per year (μm/year) were as follows:

Rural	0.05
Industrial	0.125–0.175
Marine	0.175–0.275

B1.2.   Aqueous Attack.
The potential–pH (Pourbaix) diagram for tin [3] sets out the regions of stability of the metal and various oxides and hydroxides in the presence of water at 298 K (Fig. 60.1). The diagram does not include data for the metal in the presence of complex-forming materials. We note that tin has no zone of stability corresponding to that of water, so that the formation of oxides and hydroxides is to be expected. The diagram indicates that these are reasonably stable between pH values 3 and 10, so that, in the absence of complexants, attack on the metal should be restricted. Outside this range, general attack is quite likely, with Sn^{2+} or Sn^{4+} ions being formed at lower pH values and stannites

or stannates under more alkaline conditions. Within it, complexants or species capable of causing local attack may cause corrosion.

Thus, in distilled water, attack on tin is limited to the production of an oxide film. This increases in thickness with time and increased temperature and may or may not be protective. The presence of ions like bicarbonate or borate can lead to reinforcement of the film and enhanced protection; other ions can give rise to film weakness and the possibility of localized attack. Some surfactants fall into this category, and chloride is always a potential source of local attack. The formation of oxyhalide films is possible and these can offer some protection. The decisive factor is the solubility of the salts formed. The protective nature of the tin oxide film may be enhanced by the use of oxidizing agents. In practice, few difficulties are encountered in neutral or near-neutral media with tin or high-tin alloys like pewter. The latter has served for many years as a material for drinking vessels.

Tin is then a base metal with a reasonably stable oxide film. It forms complexes with a range of ligands, though the Pourbaix diagram does not show this. These complexes may be critical to the performance of the metal.

Solders in contact with soft water may be attacked, though, as reported by Britton [2], attack on the lead appears to occur preferentially. In recent years, high-tin solders have replaced lead-containing systems in domestic heating and water distribution systems on environmental grounds, with few incidences of failure, although some electrochemical evidence exists for attack on solder in heating systems. A range of inhibitors is available for this application, as also for use in automobiles. The work of Mercer [4] and his colleagues at Teddington remains definitive in this area, where multimetal systems are common. Solders are now less common in packaging systems, and for food packaging those remaining in use are high tin. Can failure by dissolution of solder is rare, though certain inhibitors for automotive antifreezes have been known to attack high-lead solder side seams in tinplate containers. Leidheiser [5] also gives a good general treatment.

Corrosion can sometimes resemble allotropic transformation (tin pest), as shown in Figure 60.2 Here the corrosion is in the specimen on the "outside" of the curve.

C.   TINPLATE

As we pass to a consideration of tinplate, one further aspect of the corrosion of tin should be borne in mind. The foregoing sections have dealt largely with the resistance of tin to attack and oxidative action (i.e., anodic processes). Corrosion needs a cathodic process to consume electrons produced in the anodic step. Tin has a high hydrogen overvoltage; the reduction of protons then takes place on tin only with

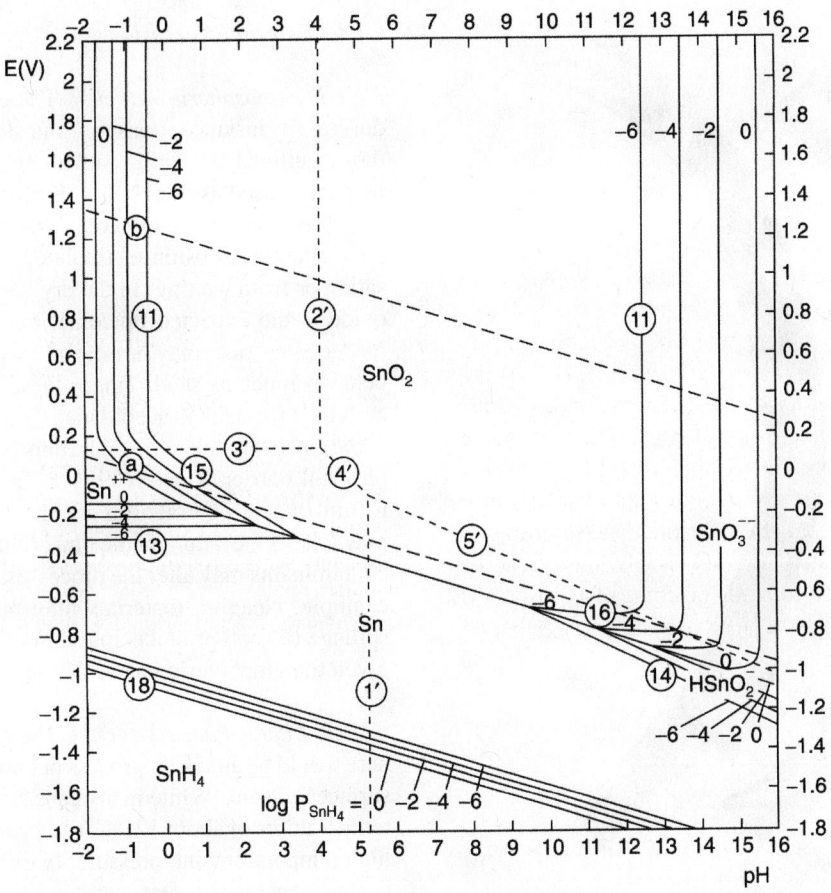

FIGURE 60.1. Potential–pH (Pourbaix) diagram for tin. (Reproduced from [3] by permission of CEBELCOR/NACE.)

difficulty. Corrosion is likely to be slow in the absence of oxygen or other cathodic depolarizers. Contact with a metal having a low hydrogen overvoltage, such as steel, will tend to accelerate the attack. The initial reactions may be summarized as

$$Sn \rightarrow Sn^{2+} + 2e^-$$
$$O_2 + 2H_2O + 4e^- \rightarrow 4OH^- \quad H^+ + e^- \rightarrow 0.5H_2$$

with the first cathodic reaction proceeding readily on tin, while the latter takes place more readily on a material of lower hydrogen overvoltage.

Tinplate itself consists of a sheet of mild, low-carbon steel having controlled levels of copper, phosphorus, and sulfur and coated with a layer of tin. The tin is, nowadays, universally coated by electrodeposition from acid, halogen, or less commonly alkaline baths. Bath chemistry has been the object of much study, and many recipes exist to produce good coatings. Sheet thickness is commonly ~0.25 mm, though a range of ~0.15–0.5 mm may be encountered; the tendency is to thinner stock. The tin coatings are on the order of 0.4 μm, though they are more commonly expressed in terms of

coating mass. Present values range from about 2 g/m² to ~11 g/m² on each surface. After plating, the tin undergoes momentary fusion by inductive or resistive heating followed by quenching in water. The matte, as-deposited coating is converted to a bright, reflective state, and a layer of a tin–iron (FeSn₂) alloy is formed between the tin and the steel substrate. A layer of oxide arises on the tin, and this is usually modified by a "passivation" treatment. This process, using chromic acid or sodium dichromate solutions, sometimes with imposed current, leaves some chromium species in the surface. The object of "passivation" is to control oxidation, suppress sulfide formation in use, and facilitate lacquer coating. A layer of oil, commonly dioctyl sebacate (DOS), is applied to facilitate handling. Thus, from the inside, we have steel, alloy, tin, oxide with chromium, and oil [6].

Tinplate finds its main application in containers for packaging. These cans may be three piece or two piece. The former are produced by forming rectangular blanks round a mandrel and soldering (now increasingly rare), resistance welding, or occasionally cementing the side seam. This produces a cylindrical body to which two ends stamped from tinplate or another sheet material are seamed.

FIGURE 60.2. Allotropic modification and corrosion of tin; corrosion occurred in the specimen on the "outside" of the curve. (From [2]; reproduced by permission of iTRI.)

Aluminum or ECCS, a mild steel coated with a mixed chromium/chrome oxide film, may constitute the end stock. Soldering changes the state of the tin and welding modifies the steel near the weld. Both welding and cementing produce lap joints that may have exposed edges and the possibility of crevices. Two-piece (body and end) containers are made by presswork using circular blanks. The shallow cups formed from these are redrawn with clearance between punch, die, and stock (draw–redraw, or DRD) or without (drawing and ironing, DWI). An end is seamed in place after filling. The DRD generates no new surface, and the area of the can is substantially that of the original blank. Prelacquered or plain stock may be used. In DWI canmaking, the surface undergoes major deformation with disruption of the tin coating, and it is normal to apply lacquers after forming. The canmaking processes can change the nature of the can material and thus influence corrosion [7].

C1. Corrosion Behavior

External and internal attack on containers may be very different. External corrosion will be discussed first.

C1.1. External Corrosion

C1.1.1 Atmospheric Attack.   The effects of condensation during, for instance, transport and storage of containers are those outlined by Uhlig half a century ago [9]. This may be paraphrased as "steel plus air plus moisture equals rust." The formation of a film of moisture is important in the atmospheric corrosion of tinplate; it may arise by condensation or from wetting. In the dry condition, slow growth of oxide is the expected outcome, but if humidity control is inadequate, rust may arise. In this context, the tinplate behaves much as steel. Tin, in these conditions, behaves as a cathode for steel, in accordance with its more noble position in the electromotive series. The only protection it offers is as a physical barrier. Thus, rusting is the expected outcome for neutral or near-neutral films of condensation. Changes in this may arise if the nature of the film is changed. The presence of contaminants may alter the processes. Contamination by, for example, cleaning materials may make the film alkaline; spillage of food products may lower the pH. In either case, given the amphoteric nature of tin, detinning may arise.

C1.1.2 Liquid-Phase Attack.   The most obvious examples here would be attack by process or cooling water on the outer surfaces of cans. While many cans are processed in steam at temperatures of about 121°C, some are processed in water at high temperature and pressure. In either case, rapid cooling of the cans to prevent growth of spoilage organisms is needed, and this will normally be in water. Whether static retorts, hydrostats, or continuous cookers are used, the tinplate will be exposed to water, at temperatures from 40 to 100°C, and in the presence of varying concentrations of air [8]. The situation may be aggravated by spillage of product or residues of cleaning agents, as described above, and other water treatments may play a role. Biological control of such waters is of vital importance for reasons of public health. Again, treatments may be employed to deal with hard waters. The agents used for these purposes may aggravate attack, and the interaction between hardness treatments, microbial control, and corrosion and its prevention is one of the more interesting problems associated with the use of water. As water is often used in a "mains-to-waste" mode, in these applications, the use of inhibitors may be limited by economic factors. Breakthrough of chloride from water-softening systems may, as we might expect, give rise to pitting, and complexation as a result of contamination may accompany pH changes.

In general, few problems arise. Careful control of process temperatures and times and of cooling times is needed to ensure product safety; good practice in this area will normally result in the avoidance of corrosion problems. However, poor drying of cans may not only pose microbiological problems but also initiate corrosion which may only develop later. Poor practice regarding the external packaging of the cans may make things worse. Shrink wrapping wet cans may trap water with the obvious consequences. Uptake of contaminants from cartons may give problems; in particular, chloride may be leached from the board in contact with the cans and may cause local attack. Spillage of product from a damaged or corroded can may cause attack on other cans and may lead to perforation of these, further spillage, and the loss of the whole pallet.

Cathodic disbondment of external lacquers can arise when the cans are in contact with aluminum retort baskets and the latter undergo attack from, say, traces of alkaline cleaning agents. As most food cans are processed externally plain, that is, unlacquered, this is not a widespread problem. Some coated food containers are used, and aerosol containers, which are usually lacquered, are pressure tested in a water bath, giving the possibility of corrosion as for food cans.

To summarize, external attack is likely in many ways to resemble that on steel. Rusting, pitting, and perforation are risks for this metal, while detinning will be possible under some circumstances. Of the problems listed above, can perforation is the most serious. In addition to product loss, there exists the risk of ingress of foreign material into the cans, of which the most serious would be microbes in food cans.

C1.2. Internal Corrosion. Internal corrosion may be very different from external corrosion. Though the material remains the same, the corrodants and conditions are not. For external attack, we are concerned with, essentially, water modified by contaminants in the presence of air, though retorts are purged to reduce air content. Inside the can, the air supply is limited, and the corrodant media may vary enormously. Food and beverages, domestic cleaners, paints, and decorative materials, industrial products, pharmaceuticals, and aerosol products are among the possibilities [10].

These, in turn, may show great variation. Formulated products may vary according to the functional requirements, materials and commercial factors, regulatory constraints, and market needs. Natural products too may vary greatly. Thus, for example, the type and strain of fruit, where it is grown, sunshine, rain, soil condition, use of fertilizers and pesticides, time of harvest, crop treatment, filling, and processing can all affect the corrosive nature of the product. Internally, then, we are faced with a wide range of potentially corrosive media. Let us examine the most important of these product areas, food and beverages.

Of great commercial and humanitarian importance, this area also serves to illustrate the important corrosion characteristics of tinplate. We recall the characteristic corrosion properties of tin as formation of oxides stable over a reasonable pH range, a high hydrogen overvoltage, and the ability to form complexes. These determine tin's behavior inside the can.

The third characteristic, that of forming complexes, changes the electrochemical situation greatly from that obtaining on the exterior of the can. The dissolution reaction presented previously may be modified as follows:

$$Sn \rightarrow Sn^{2+} + 2e^-$$
$$\downarrow\uparrow + L$$
$$Sn^{2+}(L)$$

where L represents a suitable ligand such as citrate. Tin ions readily form complexes with the so-called "fruit acids," such as citric, tartaric, and malic. This produces a diminution in the concentration of free tin ions in solution, which encourages further dissolution and, in accordance with the Nernst equation, depresses the potential of the tin. This becomes sufficiently negative for the "normal" polarity of the tin–iron couple to be reversed, with tin "active" (negative) to the steel. It may thus act as a sacrificial anode. This example of cathodic protection, arising from the complexing action of the natural acids, is vital to the functioning of tinplate cans. The relatively large area of tin exposed in a plain can is an effective protective anode for the small area of steel that is exposed at pores in the tin coating. Should this polarity be reversed, the tin would act as a forcing cathode for attack on the steel. The importance of the complexation has been widely recognized, and many workers have treated this question. In fact, Gouda et al. [11] suggested that this is the most important factor in tinplate corrosion. This view probably underestimates the importance of the initial step, tin oxidation, and the factors controlling this step. In order to form complexes of tin ions, it is helpful to have produced some tin ions. Thus, we must consider what determines the ease with which tin forms its ions.

As with all corrosion processes, the electrons produced in the oxidation step must be consumed if the reaction is to continue. The cathodic process that consumes the electrons is often rate controlling, and this is the case with tin dissolution. For the dissolution and subsequent complexation of tin, an effective cathodic depolarizer is needed. As tin has a high overvoltage for the hydrogen evolution reaction (HER), protons are unlikely to form useful depolarizers, at least on the tin. Reduction of hydrogen can proceed easily on steel, and in the absence of other depolarizers it is this reduction on the steel exposed at pores in the tin coating that controls detinning. As tin acts protectively, the iron dissolution is suppressed and loss of tin is the corrosion reaction encountered. In the situation described of the "well-packed can," in which air has been excluded, the detinning will proceed slowly, and the can will have a useful life of years. As tin dissolves, steel may become exposed, allowing reduction of protons and more rapid detinning, so that the process can become self-accelerating. This is indeed what is observed in practice. As the corrosion progresses, hydrogen gas produced at the steel may build up in the can. This will initially reduce the vacuum in the can arising from processing and later cause positive pressure to build up, which eventually causes the can to swell. A swollen can is unsaleable because a swell as a result of hydrogen buildup (a "hydrogen swell") is not distinguishable in the kitchen from swells produced by gases arising from microbial activity. Thus, the time taken for a can to swell is a measure of its useful life, and an old measure of shelf life was the time taken for 50% of a test batch to swell. Nowadays, the amount of tin dissolved would be a more usual measure. A limit of 200 mg/kg (ppm) seems to enjoy general favor with regulatory bodies, a value consistent with the low oral toxicity of tin. This limit can be reasonably easily met in modern well-packed cans of good quality. Pressure has arisen for much lower limits, perhaps as low as 25 mg/kg, which would be difficult to achieve in a plain can.

Where other depolarizers are possible, the situation changes. If cans are packed with too much residual oxygen in the headspace, as a result of poor packing practice, the oxygen can undergo ready reduction on the tin surface leading to rapid loss of tin, exposure of iron, and premature can failure. Some foods or drinks may contain natural depolarizers. Fruit colors such as anthocyanins may be readily reduced on tin with consequences similar to oxygen. In addition, the reductive bleaching of these materials leads to color changes in the product that may be unacceptable.

In all cases, control is by way of the cathodic process, controlling either by the depolarizers or the reaction surface. Good packing practice with hot filling and steam injection or vacuum closing can reduce the headspace oxygen to an acceptable level; the residual oxygen is quickly consumed and the reaction rate becomes slow. The choice of good tinplate, free of defects and with a high tin coating mass to reduce steel exposure, helps reduce the risk of proton discharge on steel. Can manufacturing and handling practices that reduce the risk of scratches and dents help reduce steel exposure with the same beneficial effects [10]. Reduction of fruit colors is minimized by the use of lacquered tinplate to reduce the tin area available for reduction. This also reduces the relative tin/iron surface area ratio, with some loss in cathodic protection, but a balance has to be struck and, in practice, lacquered cans are very successful.

This happy situation with good cathodic protection of steel by tin is not universal. A number of factors may change it. The relative complexing actions of the fruit acids on tin and iron are not identical. Changes in the acids can influence the corrosion. Thus, in pears, the ratio of citric to malic acids may vary. This may depend on the strain of fruit and upon its degree of ripeness at harvest time. As the citric/malic ratio diminishes, the extent to which the tin potential is negative to that of steel is reduced. This may progress to the stage at which the tin is no longer negative to the steel, and it may in fact become positive. In this case, its cathodic protection

becomes ineffective and its value is as a physical barrier coating on the steel. Should suitable cathodic species be available, the tin may act as a cathode for attack on the steel, causing local dissolution of this metal.

Even in well-packed cans without fruit colors, such depolarizers may arise as a result of farming practices. The use of nitrate fertilizers is common, and residues of this may be present in the fruit or vegetables. Nitrate offers a multi-electron reduction path, starting with nitrate–nitrite and continuing to hydroxylamine and ammonia. This can readily support anodic processes, and nitrate-induced detinning has been a major area of interest in food packaging with studies at Thionville, Parma, Chipping Campden, and in industrial laboratories worldwide. There does seem to be a threshold pH above which the effect does not occur, but when conditions allow, nitrate can produce rapid loss of tin. It may also, of course, support attack on steel when the tin is a cathode in the system.

Pesticide residues may also play a role in steel attack. Dithiocarbamates are known to promote attack on the steel. The mechanism is believed to involve the formation of sulfide species on the surface, causing loss of protection from the tin. This can lead to quite severe attack on the base steel.

Some products are themselves likely to attack steel. Thus, Cola-type drinks, having phosphate as well as citrate ions and a low pH, tend to be iron dissolvers. The electrochemical testing of these materials reveals that protection by tin is doubtful at best and that a situation with steel anodic to tin is common. As these products are normally packed in lacquered cans, the potential effects of a large tin cathode are minimized, so that perforation is uncommon. The prior treatment of the steel can be important. with steels that have been worked more being more vulnerable. Thus, double-reduced steel, which has had extra cold reduction, or DWI steel may be more readily corrodible than normally processed material.

A further form of corrosion that can affect both metals involves the formation of sulfides. Meats, fish, and some vegetables such as peas contain sulfur-bonded protein species. During processing, these may break down to yield hydrogen sulfide. This readily reacts with both tin and steel. The products are colored, tin sulfide being bluish and iron sulfide black. They are among the less soluble compounds of these metals and offer no food contamination problems. However, they are unsightly and the consumer is likely to reject packs showing this defect. The solution is to use internal lacquers pigmented with zinc carbonate or oxide. The sulfide can react with this, but the resulting zinc compound is off-white and causes no concern to the consumer.

C2. Tin–Iron Alloy

The tin–iron alloy in tinplate is an interesting material. It appears to be a true intermetallic and is quite brittle. Though some differences have arisen over its composition, the accepted stoichiometry is $FeSn_2$.

It forms principally during the "flow melting" or refusion process. Examination of non-flow-melted (NFM), as-plated, matte material reveals that some alloy is formed during the plating process. This is present only to the extent of $\sim1\%$ of the level found on flow-melted material.

The alloy tends to adopt a cathodic or positive potential with respect to both tin and steel. It is an effective electrode for proton discharge. Nonetheless, performance of tinplate in packing acid fruits is better with a highly continuous alloy layer at the bottom of the pores in the tin coating.

This arises as a result of its impermeability to atomic hydrogen. The latter is formed during proton reduction and is known to diffuse readily through mild steel. The higher the rate of diffusion, the faster is the potential corrosion. Tin offers a barrier to hydrogen diffusion; so also does $FeSn_2$, hence the value of a continuous alloy coating. Some of the empirical "special property tests" developed by industry recognize this. Thus the alloy tin couple (ATC) test and the aerated medium polarization (AMP) test sought to assess the resistance of the alloy. Serious deformation, as in DWI canmaking, of course, destroys the brittle alloy.

D. CONSEQUENCES OF CORROSION

Two consequences are of major significance: product contamination and package integrity. The first relates, of course, to metal pick-up. Tin and its simple inorganic compounds are, for practical purposes, nontoxic. Cats fed orange juice with some thousands of milligrams per kilogram tin become sick, and human volunteers encounter gastrointestinal upset at some hundreds of milligrams per kilogram. Since 1890, there have been only a few well-documented cases of intoxication by intake of tin in food. All involved thousands of milligrams per kilogram tin, and in each case the problem had arisen by inappropriate use of containers. The effects were, happily, transient in all cases. Nevertheless, a limit of 200 mg/kg tin is applied in most parts of the world. At this level, organoleptic changes become apparent, and visual changes may arise. Tin in beer can give rise to haze, and the color of some foods may be affected. Some foods (e.g., asparagus) benefit from the presence of a little dissolved tin, which preserves the color.

Iron is no more toxic than is tin, and no problems of iron poisoning have been reported. Flavor and color changes are, however, a consequence of iron pick-up and most food producers specify low iron content. Thus, iron at a few milligrams per kilogram can affect the taste of soft drinks, and the manufacturers specify subparts per million levels after 6 months storage. In the case of both tin and iron, it is often the commercial demands rather than the regulations that determine the limits, the former frequently being the more rigorous.

Perforation and loss of package integrity is a more serious matter. The function of the package is to contain the product, protect it from the environment and vice versa, and deliver it to the user, all without costing too much. Perforation of the container, be it through the wall or a seam, allows loss of product. This may, in turn, contaminate or cause attack on other containers. Importantly, perforation may also allow ingress of foreign material, in particular microorganisms. After processing, the contents of a food can are in a state of commercial sterility. This means that they are free of viable forms of microbes having public health significance and of microbes not having public health significance capable of growing and reproducing under normal conditions of ambient storage. Thus, processed foods in intact cans are safe. If microbes gain ingress as a result of perforation, this no longer holds, and a risk of poisoning may arise.

This occurrence is very rare. Only one case comes to mind over the last three decades—of pathogenic contamination as a result of corrosion perforation. The internal steel score of meat cans packed with too much air and in a high-chloride environment failed by perforation, allowing ingress of *Clostridium botulinum*. Normal industrial test procedures detected the situation, and the cans never reached the market. Given the many thousands of millions of food and beverage cans packed each year, the absence of real problems is encouraging.

For nonfood products, the risks to life are less apparent; commercial aspects will be important. Loss of product and market may be significant. Contamination both of other products and the environment may arise. The uses of tin and tinplate mean that the risks of catastrophic failure associated with bridges and oil rigs are diminished. However, tinplate has long been used to manufacture automotive brake reservoirs/master cylinders, and failure here could be disastrous. Again, the failure by corrosion of a tin alloy bearing on a marine drive shaft can have serious consequences, especially if the vessel is an oil tanker on a lee shore.

These two examples of consequential failure of devices or machines as a result of corrosion of a tinplate or tin component reflect the uses and corrosion characteristics of these materials. For obvious reasons tin finds no structural uses, and tin as a coating on structural steel would offer few benefits. Zinc, cadmium, and such protective coatings are more usual, though tin–zinc alloys have some uses here.

The impact of tin corrosion may be out of all proportion to the actual loss of metal. The failure of a small component may give rise to much greater losses. For example, failure of an electronic device in, say, a space probe may vitiate an entire mission. While other causes, for example, tin whiskering, may produce such problems, corrosion and oxidation may play a role.

Tin has been described as a "technologist's metal," because of its critical use in small quantities. Even the usage in tinplate demands only modest tonnage because of the thin coatings used. Nonetheless, the metal has an important impact on our lives. Its corrosion behavior is of great significance, and it is of note that the important function of preserving foods by canning depends to a great extent on the unique corrosion characteristics of tin.

REFERENCES

1. B. T. K. Barry and C. J. Thwaites, Tin, Its Alloys and Compounds, Ellis Horwood, Chichester, 1983.
2. S. C. Britton, Tin Versus Corrosion, iTRI, London, 1975.
3. M. Pourbaix, Atlas d'Equilibres Electrochimiques, Gauthier-Villars, Paris, 1963, Pergamon, Oxford, 1966.
4. A. D. Mercer, Br. Corros. J., **14**(31), 179 (1979).
5. H. Leidheiser, The Corrosion of Tin, Copper and Their Alloys, The Electrochemical Society and Wiley, New York, 1971.
6. W. E. Hoare, E. S. Hedges, and B. T. K. Barry, The Technology of Tinplate, Edward Arnold, London, 1965.
7. E. Morgan, Tinplate and Modern Canmaking Technology, Pergamon, Oxford, 1985.
8. A. Lopez, A Complete Course in Canning, The Canning Trade Inc., Baltimore, MD, 1987.
9. H. H. Uhlig, Corrosion Handbook, The Electrochemical Society and Wiley, New York, 1948.
10. T. P. Murphy, in Progress in the Understanding and Prevention of Corrosion, J. M. Costa and A. D. Mercer (Eds.), Inst. of Materials, London, 1993 pp. 696–711.
11. V. K. Gouda, E. N. Rizkalla, S. Abd-EI-Wahab, and E.M. Ibrahim, Corros. Sci., **21**, 1 (1981).

61

TITANIUM AND TITANIUM ALLOYS

J. Been* and J. S. Grauman
TIMET, Henderson, Nevada

A. INTRODUCTION

Titanium metal became a commercial reality in the early 1950s when its high strength/density ratios were especially attractive for aerospace applications. Titanium's excellent corrosion resistance over a wide range of conditions in many highly corrosive environments has led to a multitude of industrial nonaerospace applications. Some of the first applications in the chemical process industry include wet chlorine gas coolers for chlor-alkali cells, chlorine, and

chlorine dioxide bleach equipment in pulp/paper mills and reactor internals for pressure acid leaching of metal ores [1]. In marine environments, titanium is recognized as the best tube material for seawater power plant condensers with almost 650×10^6 ft installed over the last 40 years and not a single corrosion failure [2].

The market today for titanium is ever expanding. New applications continue to surface as the industry "pushes the process envelope." The relatively high initial cost of titanium is frequently offset by life-cycle costing, reductions in maintenance and operating cost, and costing on a per-unit area basis as opposed to costing on a per-pound basis. A corrosion allowance is generally not required in designs specifying titanium. Usually the only wall thickness criterion is the pressure or structural requirements for that system. Along these lines, the American Society for Testing and Materials (ASTM) and American Society of Mechaning Engineers (ASME) have recently adopted new higher tensile strengths for unalloyed grades 2, 7, and 16. These new grades, designated 2H, 7H, and 16H, offer a 16% improvement in allowable strengths and thus reduced material weight when designing pressure vessels. All other mechanical and chemical requirements are unchanged from the standard grades 2, 7, and 16, allowing material to be dual certified when being applied to ASME pressure vessels fabrications [2a].

Whereas aerospace applications are mainly concerned with mechanical properties, industrial applications place a greater emphasis on the corrosion resistance of titanium [3–7]. Table 61.1 lists the commercially pure and alloy grades most commonly used in industrial service.

Group I contains the commercially pure grades, which differ only in their oxygen and iron content. These grades are highly corrosion resistant, less expensive than titanium alloys, and generally selected when strength is not the main requirement. Increasing oxygen and iron levels improve the

---

*Present Address: Alberta Innovates Technology Futures, Calgary, Alberta, Canada.

Uhlig's Corrosion Handbook, Third Edition, Edited by R. Winston Revie
Copyright © 2011 John Wiley & Sons, Inc.

TABLE 61.1. Titanium and Titanium Alloys Commonly Used in Industrial Applications

Common Alloy Designation	UNS Number	ASTM Grade	Nominal Composition (%)	Minimum Tensile Strength (MPa)	Minimum Yield Strength (MPa)
Group I Commercially Pure Titanium					
Grade 1	R50250	1	0.06 O	240	170
Grade 2	R50400	2	0.12 O	345	275
Grade 2H	R50400	2H	0.12 O	400	275
Grade 3	R50550	3	0.2 O	450	380
Grade 4	R50700	4	0.3 O	550	483
Group II Low Alloy Content Titanium with Pd/Ru Additions					
Grade 2, Pd	R52400	7	0.12 O, 0.15 Pd	345	275
Grade 2H, Pd	R52400	7H	0.12 O, 0.15 Pd	400	275
Grade 1, Pd	R52250	11	0.06 O, 0.15 Pd	240	170
Grade 2, low Pd	R52402	16	0.12 O, 0.05 Pd	345	275
Grade 2H, low Pd	R52402	16H	0.12 O, 0.15 Pd	400	275
Grade 1, low Pd	R52252	17	0.06 O, 0.05 Pd	240	170
Group III Other Alpha and Near-Alpha Alloys					
Ti 3-2.5	R56320	9	3 Al, 2.5 V	620	483
Grade 12	R53400	12	0.3 Mo, 0.8 Ni	483	345
Ti 3-2.5, low Pd	R56322	18	3 Al, 2.5 V, 0.05 Pd	620	483
Ti 3-2.5, Ru	R56323	28	3 Al, 2.5 V, 0.1 Ru	620	483
Ti 5111	R55111	32	5 Al, 1 Sn, 1 Zr, 1 V, 0.8 Mo	689	586
Group IV Alpha-Beta Alloys					
Ti 6-4	R56400	5	6 Al, 4 V	895	828
Ti 6-4 ELI	R56407	23	6 Al, 4 V, 0.13 O max	828	759
Ti 6-4 ELI, Ru	R56404	29	6 Al, 4 V, 0.1 Ru, 0.13 O max	828	759
Group V Beta Alloys					
Beta C	R58640	19	3 Al, 8 V, 6 Cr, 4 Zr, 4 Mo	793	759
Beta 21S	R58210	21	15 Mo, 3 Al, 2.7 Nb, 0.25 Si	793	759

UNS = Unified Numbering System.

material's strength but reduce its ductility. Whereas Grade 2 can be considered the workhorse of the nonaerospace industry, Grade 1 is selected for applications requiring a high formability. The members of group II offer significantly improved corrosion resistance in reducing media through the presence of small concentrations of palladium or ruthenium.

Group III contains other alpha and near-alpha alloys that are characterized by intermediate strength, good ductility, toughness, creep resistance, and weldability. These alloys retain the hexagonal close-packed (hcp) structure characteristic of alpha alloys, which, together with satisfactory strength, make them ideal for cryogenic applications. The presence of molybdenum and nickel in Grade 12, palladium in Grade 18, and ruthenium in Grade 28 improves the corrosion resistance of these grades in reducing acid environments.

The alpha–beta alloys of group IV contain an increase in the percent of beta phase and, hence, the strength level as a result of a higher concentration of vanadium, a beta stabilizer. Heat treatments can be used to control the high room temperature strength. Toughness, ductility, and stress corro-

sion cracking (SCC) resistance can be improved by limiting the level of interstitials such as oxygen, nitrogen, and carbon in extra-low interstitial (ELI) content and very-low interstitial (VLI) content grades.

Beta alloys are readily cold worked in the solution heat treated and quenched condition. They are heat treatable and can be worked and aged to high strengths at some expense of ductility. Because of their high strength/density ratios, beta alloys are predominantly used in the aerospace industry. An excellent account of physical and mechanical properties of titanium alloys can be found in the Materials Properties Handbook: Titanium Alloys [8].

B. TITANIUM OXIDE SURFACES

Titanium is a reactive metal, $E^{\circ}_{Ti/Ti^{+2}} = -1.63 V_{SHE}$ [9], owing its excellent corrosion resistance in many environments to a hard, tightly adherent oxide film which forms instantaneously in the presence of an oxygen source.

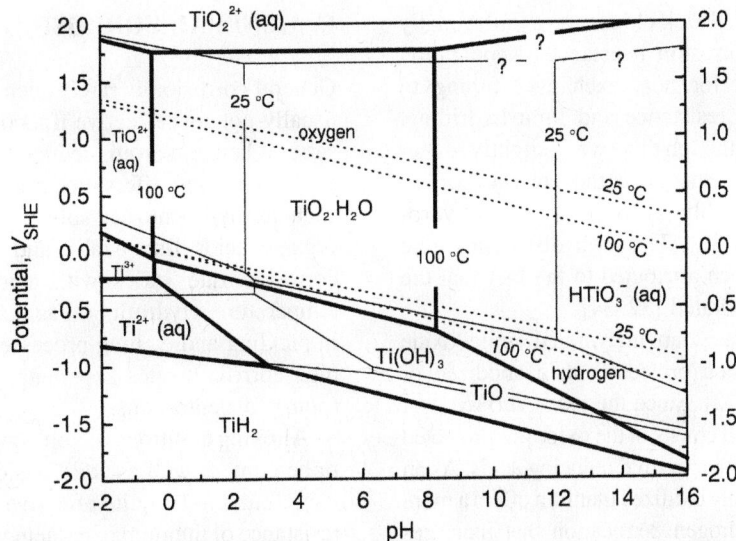

FIGURE 61.1. Phase stability diagram of the Ti–H$_2$O system at 25 and 100°C with a titanium ion activity of 10^{-6} [10].

Figure 61.1 shows the phase stability diagram for the Ti–H$_2$O system [10]. Titanium passivation is present in the stable TiO$_2$·H$_2$O area. At higher pH values, HTiO$_3^-$ dominates in an area that is characterized by corrosion. With increasing temperatures, the area of corrosion extends to lower pH values. The corrosion kinetics are slow within the HTiO$_3^-$ area as indicated by low measured titanium corrosion rates [7]. At potentials between 1.5 and 2 V$_{SHE}$, the unstable peroxide TiO$_3$·2H$_2$O forms [11].

When submerged in a corrosive medium, the overall titanium dissolution rate is very much dependent on the nature and integrity of the oxide. When the oxide is sufficiently thick and stable, electron exchange occurs predominantly with the oxide film. The semiconductive properties of the oxide determine the current/potential behavior of the titanium/oxide system. When the oxide is sufficiently thin (0.4–3 nm), electron exchange occurs between the redox electrolyte and the underlying metal by direct tunneling or resonance tunneling via intermediate states [12, 13]. As a. result of direct tunneling, which consists of electron transfer in one step without loss of energy, electron exchange is under kinetic control with current/potential characteristics that are similar to those of the bulk metal. The oxide functions as a potential energy barrier and the current decreases with increasing oxide thickness. The anodic transfer coefficient becomes smaller with increasing oxide thickness and the cathodic transfer coefficient becomes greater. A cathodic Tafel coefficient of –0.12 V and an anodic Tafel coefficient of ∼ 0.12 V yielded electrochemical corrosion rates that compared satisfactorily with weight loss corrosion rates in acid media [14]. The same cathodic Tafel slope but an anodic Tafel slope of ∼ 0.25 V [15] yielded good results in alkaline peroxide bleaching environments [16].

A freshly abraded titanium surface immediately passivates to form a crystalline rutile and/or anatase oxide layer. Rutile is the more common titanium dioxide (TiO$_2$) and slightly more stable than anatase by ∼12 kJ/mol [17–20]. Anatase is a material with the highest photocatalytic detoxification efficiency in ground and surface water purification [21]. Rutile finds application as a catalyst in organic oxidation reactions [22]. The isoelectric point of TiO$_2$ is ∼6.2, which, together with a high dielectric constant [23], renders titanium oxide waterlike with small electrostatic forces and, consequently, highly compatible as a biomaterial [24]. The titanium oxide gradually decreases in oxygen content from TiO$_2$ at the surface to Ti$_2$O$_3$ and TiO as it approaches the metal oxide interface [25]. Depending on the environment, this oxide may be covered with an amorphous or hydrated surface oxide, giving a two-layer oxide structure. The oxide may be thickened in the presence of oxidizing agents through anodization or thermal oxidation.

In reducing acid environments, severe corrosion can be avoided through the application of anodic protection that aids in the formation of a protective surface. For example, the corrosion rate of a titanium heat exchanger in a 40% sulfuric acid environment can be reduced 11,000 times to a rate of 0.005 mm/year (0.2 mpy) through the application of 2.1 V overpotential [26, 27]. Anodic protection also appears to increase the protective nature of the oxide. Tomashov et al. [28] suggest that this is the result of a decreasing number of defects and decreasing ionic conductivity. Care must be taken not to exceed the repassivation potential of titanium.

Anodization at increasingly higher potentials can thicken the very thin natural oxide from ∼20 Å to several thousand angstroms, depending on the applied potential. As the thickness increases, the oxide progresses through a spectrum of interference colors [29, 30]. Thick oxides were traditionally thought to increase the corrosion resistance. Anodization used to be recommended for heat exchanger tubing to improve crevice corrosion resistance and limit hydriding. Studies later showed that, although there was a slightly higher initial corrosion resistance, the anodized surfaces didn't behave much better than freshly pickled surfaces in hydrochloric acid solutions [31, 32]. The high dissolution rate of the anodized film has been attributed to the fact that the oxide is amorphous and hydrated [32–34].

Thermal oxidation produces an unhydrated rutile oxide which offers greater corrosion protection than anodized or pickled oxide surfaces [31, 32]. Since the oxide surface is at a more anodic potential, small cracks in the oxide are protected by the anodic corrosion potential in mild reducing acids. As an additional benefit, the thermally oxidized titanium offers a more effective barrier against hydrogen permeation that increases with increasing oxide thickness. The more noble oxide potential may, however, increase the driving force for galvanic corrosion.

At elevated temperatures, titanium oxidizes in air to form an oxide scale and an oxygen-rich metal layer. The extent and rate of oxide formation are dependent on the exposure temperature and time. At temperatures below ∼500°C, the oxidation rate of titanium is low and tends to decrease with time [25, 35]. Long-term exposures at temperatures > 650°C will lead to cracking of the brittle oxide scale and rapid continuous oxide growth [36]. Figure 61.2 illustrates how both the oxide scale and the oxygen-rich metal layer on Grade 2 titanium roughly quadruple in thickness as the temperature is increased from 538 to 649°C. The heat resistance of titanium can be increased through alloying [37] or application of oxidation-resistant coatings [38].

C. GENERAL CORROSION

General corrosion is rarely seen in service since titanium is usually not cost effective if a corrosion allowance is necessary. When observed, reducing acids are most often the cause. Titanium offers moderate resistance to mineral acids such as hydrochloric, sulfuric, and phosphoric acid and organic acids like oxalic and sulfamic acid [3, 7]. The corrosion rate varies with acid type, concentration, and temperature. Hydrofluoric acid solutions are routinely used in pickling and etching processes because of the extremely high corrosion rates experienced even at parts per million (ppm) concentrations.

Alloying additions of noble metals such as palladium and ruthenium, as well as additions of molybdenum and nickel, were found to be quite effective in increasing the corrosion resistance of titanium in reducing acid environments [25, 39–41]. Palladium and ruthenium are added in small concentrations, typically 0.05–0.20 wt %, that do not affect the physical and mechanical properties of the titanium alloy but passivate the metal by shifting the corrosion potential into the passive anodic regime [40, 41]. Both nickel and molybdenum reduce the susceptibility of titanium to anodic dissolution. The latter alloying additions increase the alloy strength at the expense of ductility [25, 39].

Small concentrations of oxidizing species effectively increase titanium's corrosion resistance in reducing acids by positively polarizing the metal. Only parts-per-million concentrations of certain multivalent transition metal ions, nitrates, oxychloro anions, noble metal ions, organic compounds, chlorine, and oxygen are required to induce passivity [3]. Inhibitor levels may be present as contaminants in

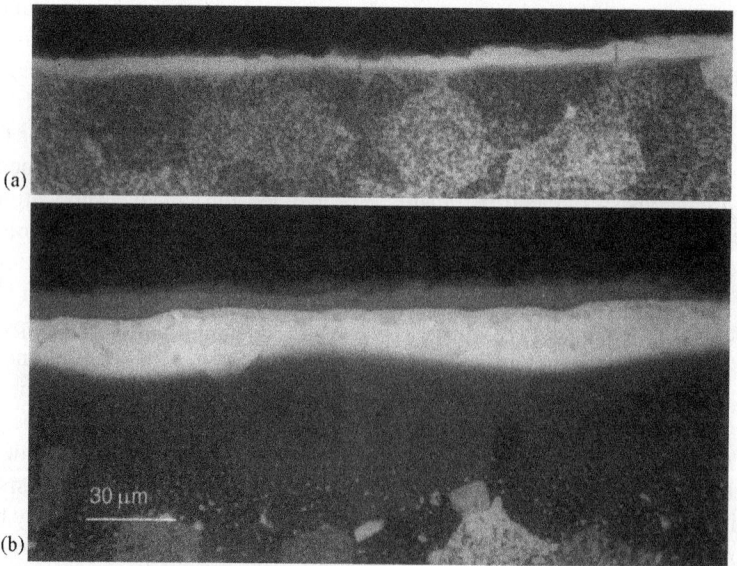

FIGURE 61.2. High-temperature oxidation of Grade 2 exposed for 500 h at (a) 538°C and (b) 649°C. The surface oxide layer covers a layer of oxygen-rich alpha structure, accentuated by a lactic acid etch.

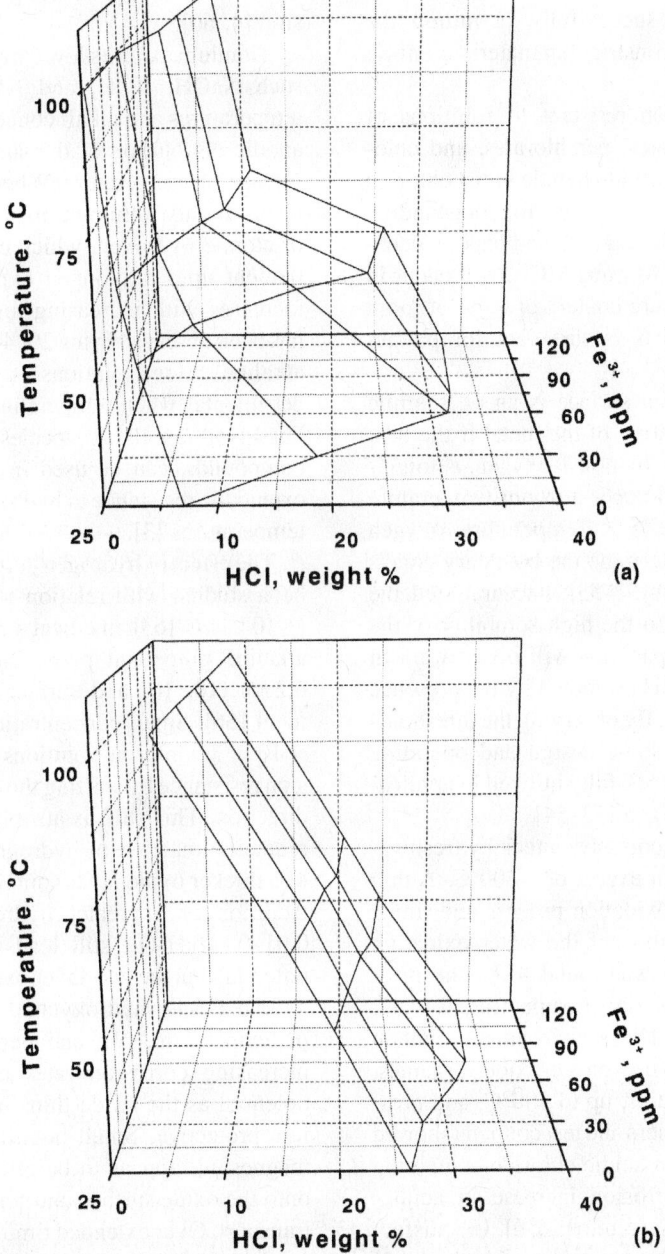

FIGURE 61.3. Effect of ferric ions on the corrosion of (a) Grade 2 and (b) Grade 7 in hydrochloric acid solutions [4]. The isocorrosion plane corresponds to a corrosion rate of 0.13 mm/year.

process streams allowing the safe use of titanium. Figure 61.3 illustrates how hydrochloric acid cleaning solutions can be inhibited by small concentrations of ferric ions.

Titanium is known as one of the most corrosion-resistant metals in oxidizing environments as these conditions generally assure oxide film stability. Highly resistant to oxidizing acids such as nitric and chromic acid at room temperature, general corrosion may occur in nitric acid at boiling temperatures in the 20–70 wt % range [42]. When the solution is

not refreshed, buildup of Ti^{4+} ions leads to a more protective, dehydrated titanium oxide [43]. Other metal ions, such as Fe^{3+}, Ru^{3+}, Rh^{3+}, Ce^{4+}, and Cr^{6+}, and oxidizing ions, such as VO_2^+ and $Cr_2O_7^{2-}$, also inhibit high-temperature corrosion [44]. Only tantalum alloying additions significantly improved the corrosion resistance and yielded titanium alloys that were virtually insensitive to changes in temperature and acid concentration [45]. A pyrophoric reaction may develop in red fuming nitric acid following rapid intergranular attack [46, 47]. The presence of sufficient concentrations of water has successfully permitted the long-term use of titanium as a construction material in nitric acid production plants [48].

Titanium is highly corrosion resistant to solutions of chlorites, hypochlorites, chlorates, perchlorates, and chlorine dioxide [3, 7, 49]. Widely used to handle moist chlorine gas, titanium has earned a reputation for outstanding performance in chlor-alkali cells and pulp and paper bleaching equipment. Rapid ignition, forming $TiCl_4$, will occur in dry chlorine. However, a moisture content of 0.4% at room temperature and 1.2% at 175°C is sufficient for passivation [50, 51].

Oxidation of a fresh titanium surface is an exothermic process which may lead to melting of the metal if the heat cannot be removed fast enough. In <35% oxygen, autoignition of titanium is not likely to occur at room temperature regardless of the total pressure [52]. Temperature, oxygen pressure, and concentration determine the boundary conditions for autoignition of titanium [52, 53]. Once initiated, the reaction is self-sustaining due to the high solubility of the oxide in the molten metal. Propagation will occur in much lower oxygen atmospheres and is promoted by the presence of steam and quenched by water. By observing the thresholds for safe operation and taking some design and operating precautions, titanium can be successfully utilized in applications involving pressurized oxygen [52–54].

Titanium resists all forms of corrosive attack by freshwater and steam to temperatures in excess of ~400°C. A thin rutile oxide layer reduces the oxidation process and limits the uptake of hydrogen, a product of the water oxidation reaction [35]. In fact, titanium was found to be the most corrosion-resistant metallic material in supercritical water oxidation applications [55, 56]. These environments contain supercritical water, oxygen, hydrogen peroxide, organics, and chlorides at high temperatures, up to 600°C, and pressures, up to 40 MPa. Titanium liners and test coupons showed excellent corrosion resistance in acidic chloride containing solutions with a moderate corrosion increase in acidic-sulfate- or phosphate-containing solutions [56]. Unsatisfactory corrosion behavior may be experienced in high-pH (pH ~ 14) environments.

Generally low ozone concentrations, up to 0.6 mg/L, are present in cooling water treatments and ozonated seawater [57]. Titanium is completely corrosion resistant in these environments as well as at higher ozone concentrations of 7–8 mg/L ozone in 10% aqueous NaCl at 50–60°C with no tendency to crevice corrosion [58].

Excellent resistance of titanium to general corrosion in seawater is obtained to temperatures well in excess of 250°C [59]. This includes brackish, polluted, stagnant, aerated, or deaerated water containing contaminants such as metal ions, sulfides, sulfates, and carbonates. Exposure of titanium for many years to depths of over 2 km below the ocean surface has not produced any measurable corrosion [4, 60].

Titanium exhibits low corrosion rates in alkaline solutions such as NaOH, KOH, and NH_4OH over a wide range of temperatures and alkali concentrations [3, 61]. Slow general anodic dissolution of the surface film is accelerated by an increase in temperature. Whereas the corrosion reaction may result in only minimum metal wastage, it is also the source of atomic hydrogen, which could lead to hydrogen embrittlement upon prolonged exposure. Hydrogen pickup also increases with increasing temperatures. Maximum pickup has been observed in the 20–40 wt % NaOH range [62]. In hot alkaline brine solutions, hydrogen uptake can become detrimental when temperatures exceed 80°C and pH > 12. Dissolved oxidizing species, such as chlorate or nitrate compounds, can be used in alkaline cleaning solutions to extend the resistance to hydrogen uptake to somewhat higher temperatures [3].

The effect of hydrogen peroxide on titanium corrosion has been studied with relation to radioactive waste containers ($\sim 10^{-4} M$) [63], medical implants (0.01–0.1 M) [64, 65], alkaline pulp, and paper bleaching environments (~ 0.1–$0.2 M$) [10, 16] and surface etching or bonding pretreatment [66]. Small concentrations of hydrogen peroxide ($\sim 1 \times 10^{-4} M$) in brine solutions strengthen the titanium corrosion resistance by shifting the corrosion potential in the noble direction. This shift is attributed to the additional cathodic reduction reaction of hydrogen peroxide and the formation of a thicker oxide layer containing more stable anatase [63]. Addition of a greater hydrogen peroxide concentration, 0.01–0.1 M H_2O_2, still leads to a more passive corrosion potential, but the two-layer oxide becomes increasingly more hydrated and hydroxylated [64, 65]. Further increases in temperature, pH, and peroxide concentration result in increasing corrosion rates and a more active corrosion potential as the oxide thins becomes more conductive and less protective. Small additions of calcium, silicates, and magnesium appear to be effective inhibitors by adsorbing onto the oxide surface and forming a physical barrier to ion transport. Over extended time periods, however, the effect of calcium additions is questionable as corrosion rates begin to increase again [16]. In pulp bleaching solutions, pulp itself proved to be an effective inhibitor under normal operating conditions [16]. Extremely high titanium dissolution rates may be obtained at extreme conditions. A rate as high as 2300 mm/year was measured at 95°C, 2.5 M NaOH, and $\sim 0.5 M$ H_2O_2 [10]. General corrosion leads to extensive roughening of the surface, which can be attractive for subsequent coating or bonding processes [66].

Titanium is widely used in organic process streams and has been the material of choice for critical areas of terepthalic and adipic acid production. Titanium is highly resistant to solutions of alcohols, aldehydes, esters, ketones, and hydrocarbons [3, 7]. To maintain the integrity of the protective oxide film, some degree of moisture or oxygen should be present. Generally, a moisture content of merely parts per million is sufficient for passivation, a concentration, which experience has shown, can usually be expected in industrial organic processes.

Overall, titanium exhibits excellent corrosion resistance in organic acid solutions [3, 7, 67, 68]. Aeration may be required to maintain passivity. Fully resistant in aerated aqueous formic acid, titanium corrosion rates may become unacceptably high at elevated temperatures in deoxygenated formic acid [7]. A strong oxygen effect has also been observed in urea reactors where corrosion potentials dropped as the supply of oxygen stopped. Titanium is a preferred material of construction since it is not affected much by a temporary lack of oxygen [69]. The corrosivity of urea has been linked to an amino–formic acid intermediate [70].

Negative effects of oxygen have been observed in mixtures of anhydrous acetic acid and acetic anhydride, apparently by facilitating the cathodic reduction reaction that favored the production of the acetate ion. Alloying with palladium did not help in this case [71].

Titanium exhibits a poor resistance to corrosion in propionic acid vapor and has a limited stability in oxalic acid solutions. The corrosion rate in the latter medium increases with increasing temperature and acid concentration [67]. The corrosion resistance in oxalic acid can be improved by molybdenum alloying additions [72] or the addition of oxidizing agents such as antimony(III) [73].

D. PITTING CORROSION

Titanium exhibits remarkable resistance to pitting attack in chloride media with pitting potentials in excess of $+5\,V_{SCE}$ in saturated NaCl at boiling. Thus, titanium generally does not exhibit spontaneous pitting under normal circumstances. Pitting resistance is lower in other halide media; however, potentials still remain at or above $+1\,V_{SCE}$ [74]. Pitting potentials of titanium in sulfate and phosphate media are reported to be in excess of $+80\,V_{SCE}$ [3]. Alloying can lower pitting potentials somewhat, yet even highly alloyed titanium exhibits pitting potentials greater than $+1\,V_{SCE}$ in high-temperature NaCl and HCl environments [75]. Pitting corrosion failures of titanium in service are thus extremely rare.

E. CREVICE CORROSION

Titanium, being a reactive metal and relying on its passive film for corrosion resistance, is susceptible to localized corrosion in much the same manner as other passive film metals like aluminum, stainless steel, and nickel alloys. Crevice attack can occur on titanium in hot halide or sulfate-containing media. Corrosion can be observed in tight gasket-to-metal or metal-to-metal joints or under adherent deposits formed by a process stream. Under normal circumstances, crevice corrosion resistance will probably be the limiting factor for successful use of titanium; thus a thorough understanding of the mechanism, influencing factors, and mitigation techniques will prove invaluable. Several excellent reference sources are available that compliment the information contained herein [1, 3, 8, 76–78].

E1. Mechanism of Crevice Corrosion

As with stainless steels, the mechanism for attack on titanium involves formation of an occluded differential aeration cell, in which slow but finite corrosion depletes the crevice of oxygen through surface oxide formation, as shown in reaction (61.1). Anion migration into the crevice then occurs to preserve mass and charge balance. In the case of chlorides, this results in formation of unstable titanium chloride and oxychloride intermediate compounds that hydrolyze to form free acid, thus lowering the pH in the crevice. At this point, the corrosion reaction becomes selfsustaining as the acid generated from hydrolysis now further attacks the underlying metal. Within the crevice, pH levels of <1 can be obtained, despite having bulk pH levels as high as 7 [79]. This mechanism, first put forth on titanium by Griess [76], has been accepted for many years:

$$2Ti + O_2 + 2H_2O \rightarrow 2TiO_2 + 4H^+ + 4e^- \qquad (61.1)$$

E2. Factors Influencing Crevice Corrosion

Crevice corrosion of titanium has been reported to occur in laboratory tests at solution temperatures of 40–60°C under certain extreme and unusual crevice conditions [78]. However, attack has rarely, if ever, been observed under field and more typical laboratory test conditions at temperatures <70°C. Increasing temperatures have been shown to increase the severity of attack [76]. Once initiated, attack can continue for some time down to temperatures of 25°C, but only when the temperature is gradually decreased. Rapid temperature drop quenches attack [80]. A temperature of 80°C has served well as a conservative upper limit for use of unalloyed titanium in brine environments with a pH equal to or less than 9 [1].

Along with temperature, pH is a critical factor for determining whether crevice attack on titanium can occur. Under alkaline conditions, with a pH > 10, crevice corrosion does not initiate [77]. At a pH < 7, initiation can occur rapidly as the crevice solution pH can drop to <1. At intermediate pH levels between 7 and 10, attack can also occur. However, incubation times may be extended, and frequency and severity of attack are normally lessened.

Studies made on crevice geometry parameters suggest that titanium requires a very narrow, deep crevice for attack to occur. Crevice gaps on the order of 0.001 cm and depths >1 cm are usually required for attack initiation [81, 82]. Fitzgerald and Greene [83] put forth a model relating geometric and electrochemical parameters to crevice corrosion that show titanium requires much deeper crevices for attack than most stainless steel alloys.

In brine media. crevice corrosion has been observed over the range of chloride concentrations from 0.01% up to saturation. Concentration, however, is not as critical a factor as pH and temperature, since attack can occur at any concentration over the minimum value listed above. Factors such as incubation time, severity, and frequency of occurrence can be impacted by concentration [78]. Also, degree of aeration has direct impact on attack and on crevice chloride concentration. Attack will be more severe at higher chloride levels if solution aeration is maintained. However, without aeration, oxygen solubility is so poor at higher chloride concentrations that crevice attack can be stifled. Consistent results are most readily obtained in the laboratory with the use of a 5% NaCl solution at 90°C with constant air sparging of the solution. Resistance to bromides and iodides is similar to chlorides while fluorides tend to be somewhat more aggressive [84]. This is thought to relate to the size of the ion and its ability to diffuse into crevices. Crevice corrosion of titanium in fluorides can only occur with pH levels of ~6 and up. Below a pH of 6, hydrofluoric acid attack will dominate the corrosion process. Titanium is less susceptible to attack in sulfate media, with temperature and concentration minimums increased over those for halide solutions [3].

E3. Detection of Crevice Corrosion

Crevice corrosion is a particularly insidious form of attack since, by its nature, it is a very random process and often can go undetected until complete metal failure occurs.

Monitoring for onset of crevice attack has been studied and appears to be a viable option [85–87]. Titanium is easily polarized and thus one can in effect monitor the activity within a crevice by tracking the potential of the free surface adjacent to the crevice. This technique has been demonstrated in the laboratory for free area–creviced area ratios up to nearly 100:1 [85].

More traditional test techniques have been described elsewhere [3] and basically rely on the formation of a large creviced area and a one-month test period to overcome the randomness of attack. Due to the fact that titanium requires such a deep crevice, the multiple crevice washers routinely used in crevice testing of other materials will not yield accurate test results for titanium and their use should be avoided. A flat 25.4-mm-square gasket former is preferred. Fluorocarbon polymers tend to the most discriminating crevice initiation results and are used extensively in laboratory testing to establish conservative guidelines [77, 81]. However, it is imperative that only virgin material be used. Reprocessed material has been shown to release fluoride ions into the crevice, thereby interfering with test results.

When examining for crevice attack, one must look for tenacious, off-white to gray titanium oxide deposits in the crevice. Normally, the only way to remove these deposits and determine the extent of underlying attack is to lightly (5 s) sandblast the area or metallographically prepare a cross-sectional view of the specimen. Crevice attack will usually appear as multiple, irregularly shaped pits, as shown in Figure 61.4.

Quite often, titanium crevices will display smooth, multicolored oxide interference films after testing. These may range in color from gold to purple to blue, depending on the extent of oxidation the metal has undergone. These oxide films do not represent deleterious attack of the metal and should not be considered as crevice corrosion. On the contrary, the presence of colored oxide films is indicative of

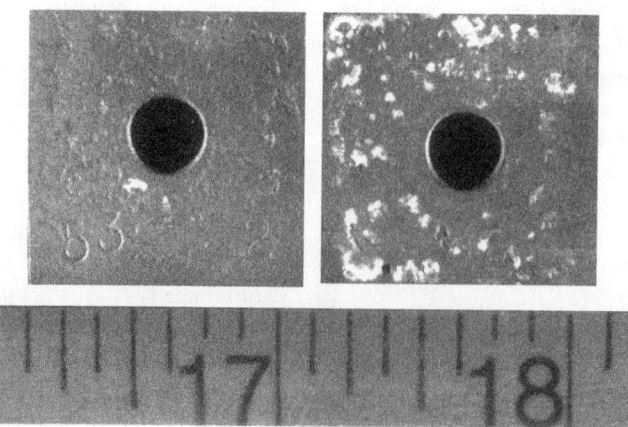

FIGURE 61.4. Crevice corrosion attack on titanium. Attack occurred under polytetrafluoroethylene gasket. Sample on right had some remaining corrosion deposits even after sandblasting.

a passive environment for titanium, one in which the TiO_2 film is stable and increases in thickness [31].

E4. Mitigation of Crevice Corrosion

Crevice corrosion is best mitigated through the use of a more resistant grade of titanium. Several alloying elements have been shown to impart added corrosion resistance to titanium, notably Cr, Ni, Nb, Mo, and precious metals Pt, Pd, and Ru [5]. Precious metal additions are the most effective alloying agents, typically being used in the 0.05–0.20 wt % range. The other elements have been used at levels from ∼ 0.5 to 15 wt %. Enhancement of crevice corrosion resistance can be directly related to the beneficial effect that an alloying addition has on resistance to mineral acid attack [76, 88]. Once sufficient alloying has occurred to passivate the metal in an acid media with a pH of zero, crevice corrosion on titanium can be effectively mitigated [39, 85]. Figure 61.5 illustrates this effect for several different grades of titanium. Grade 12, with small additions of Mo and Ni, offers substantial improvements in crevice corrosion resistance over unalloyed titanium. Clearly, though, the grades alloyed with palladium have a much more dramatic effect on resistance. To achieve an equal effect with molybdenum, a 15 wt % addition is required [39].

In addition to alloying, several other mitigation techniques have been used. Although not as effective as alloying, these methods can often offer remediation from crevice attack once the metal has already gone into service. One method is to coat the surface with a precious metal oxide. This has been shown to offer equivalent resistance as that obtained through alloying with palladium; however, the coating is subject to mechanical damage through scratching or abrasion, rendering it ineffective [89]. Another technique involves placing oxidizing metal ions directly into the crevice. The presence of these ions, such as Cu^{2+}, Fe^{3+}, and Ni^{2+}, can inhibit crevice corrosion in the same manner as they do general corrosion attack (see Section C). This technique has been used successfully in gasket crevices, where the metal oxide paste can be applied directly to the gasket [90]. A more recent innovation utilizes small patches applied directly to the titanium base metal surrounding the crevice in order to ennoblize the occluded titanium within the crevice. This process, referred to as platinum group metal applique, or PGMA, has been shown to offer equivalent crevice performance to Grade 7 and can serve as a retrofit option when unalloyed titanium (Grade 2) has experienced crevice attack [90a]. Enhanced oxide films, obtained through air heating of the titauium at 450–800°C for 2–10 min, have been shown to be beneficial in terms of crevice corrosion resistance [31]. Again, as with the coatings, these oxide films are subject to degradation by mechanical damage, leaving the underlying titanium substrate susceptible to attack. Finally, titanium surfaces may also be protected from crevice attack when coupled to dissimilar metals such as stainless steels, nickel alloys, and copper-containing alloys. It is believed that the release of metal ions from corrosion of the dissimilar alloy acts to inhibit attack of the titanium [77].

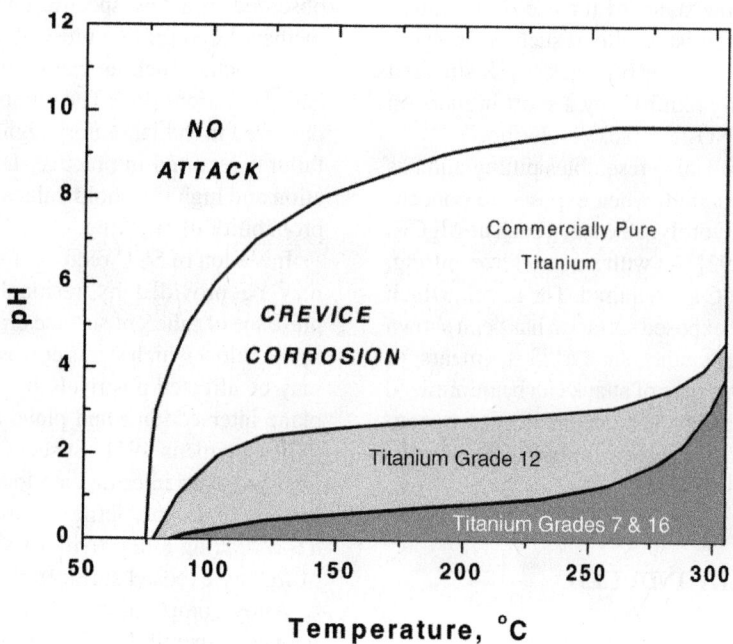

FIGURE 61.5. Temperature/pH crevice corrosion limits for titanium in naturally aerated NaCl brine. The shaded areas represent regimes of susceptibility [1].

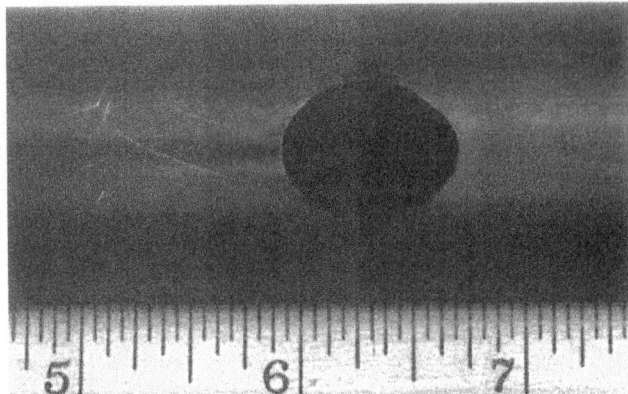

FIGURE 61.6. Titanium Grade 2 tubing exhibiting "smeared iron pitting," a localized corrosion phenomenon.

E5. Specialized Forms of Crevice Corrosion

Two special forms of crevice attack require noting here. The first involves the presence of smeared iron particles on titanium. Unalloyed titanium can undergo attack that resembles pitting when an iron particle embedded in the surface of titanium metal corrodes in a saline environment. The resulting corrosion product (acidic ferric chloride) can induce attack of the titanium at the site where the iron breached the titanium oxide film. This phenomenon, known as "smeared iron pitting," has been well documented [91] and occurs only in concentrated brine solutions when temperatures exceed 80°C. An example of this attack is shown in Figure 61.6. This work resulted in the discontinuing use of steel tooling when handling fabricated titanium parts. Stainless steel tools are now standard for use on titanium. The more crevice corrosion resistant alloys, such as Grades 7, 12, and 16, are immune to this type of attack [4]. If smeared iron is suspected, it is easily removed by a short immersion (2–5 min) in a standard HNO_3/HF pickle solution.

The second form of attack also resembles pitting and has been shown to occur on titanium when exposed to concentrated solutions of certain hydrolyzable salts, such as $MgCl_2$, $CaCl_2$, $AlCl_3$, and $ZnCl_2$ [92]. As with smeared iron pitting, temperatures of at least 80°C are required. The attack, which seemingly occurs on freely exposed surfaces, has been shown to initiate at surface imperfections, such as laps, smears, or grind marks [93]. Again, this type of attack can be minimized by pickling the metal or using a more crevice corrosion-resistant grade of titanium. Guidelines for use of titanium in these salts have been published [1].

F. ENVIRONMENTALLY INDUCED CRACKING

Environmentally induced cracking results from a synergism between tensile stress and a corrosive environment. Corrosion rates are generally low and not apparent. Stress levels that cause cracking are generally below the yield stress. Environmentally assisted cracking includes SCC, corrosion fatigue cracking, and hydrogen embrittlement. Frequently, more than one of the three may be operative, complicating analysis of failures. However, commonalities in the environmental and metallurgical factors responsible for susceptibility of cracking facilitate the determination of appropriate prevention methods.

F1. Stress Corrosion Cracking

The majority of titanium alloys used in the chemical process industry are very resistant to SCC. The latter has been observed in a few specific environments such as absolute methanol, red fuming nitric acid, nitrogen tetroxide, several liquid metals such as cadmium and mercury, and aqueous halide solutions [94]. The number of stress corrosion failures observed in the laboratory significantly outnumbers the few failures reported in practice. Difficult stress corrosion initiation and high threshold values frequently account for a low probability of cracking.

Initiation of SCC requires the presence of a stress riser as may be provided by residual or applied stresses in the presence of inherent surface cracks and flaws. The threshold stress below which stress corrosion cracks will not propagate may be affected positively by cold work, by increasing slip plane intersections, and plane stress conditions as found in thin specimens [95]. Susceptibility to SCC is most pronounced at an intermediate loading rate when a high rate of loading leads to ductile failure before crack initiation and a low loading rate provides sufficient time for repassivation of freshly exposed surfaces [95].

Alloy composition and interstitial content markedly influence susceptibility to SCC. An increase in the interstitial oxygen content of commercially pure titanium results in a slight decrease in K_{ISCC}, the critical stress intensity factor affected by SCC, when exposed to synthetic sea water [96, 97] possibly by increasing the tendency toward planar slip [94]. Interstitial nitrogen and carbon have a similar effect on the susceptibility of titanium. High iron additions ($\sim 0.2\%$ by weight) reduce the α-phase grain size, which leads to some improvement in the resistance to SCC [97]. Aluminum, when present in concentrations > 5 wt % in the α phase, can lead to the formation of Ti_3Al, which lowers K_{ISCC} and increases the velocity of cracking [94]. The presence of tin further decreases the SCC resistance. Whereas SCC in sea water is not a concern for the lower strength commercially pure grades, susceptibility can be lowered substantially in higher strength aluminum-containing alloys by lowering the oxygen content, as, for example, in the case of Ti 6-4 ELI (Grade 23). Furthermore, in Ti 6-4, an acicular structure provides a lower susceptibility to SCC than an equiaxed morphology, which may be related to the mean free path of the susceptible α phase [94, 95] (see Fig. 61.7).

Stress corrosion cracking susceptibility is influenced by the concentration of damaging species, pH, potential, temperature, and viscosity. Addition of halide ions such as Cl^-, Br^-, and I^- may accelerate or induce SCC, an effect that increases with increasing halide concentration [99]. Reducing the pH results in a greater susceptibility to SCC. Crack velocity was found to increase with increasing temperature and decreasing viscosity. Cathodic protection is effective in neutral aqueous halide solutions at potentials more negative than -1 V_{SCE} but ineffective in acid solutions [99].

Stress corrosion cracking of α and α–β alloys takes place by transgranular cleavage of the α phase, where the α phase controls the overall crack propagation rate. Small differences between the main crack propagation plane and the cleavage planes lead to numerous stepwise facets, the result of low-energy ductile rupture. These flutes connect cleavage planes and may exceed the grain size [100].

Intergranular corrosion occurs in methanolic halide solutions through the formation of titanium methoxide. As little as 1.5% water is sufficient to hydrolyze the titanium methoxide and passivate the titanium [101]. Higher strength alloys may require the presence of more water depending on product form and alloying content [102]. Noble metal ion additions such as Pd^{2+} and Au^{3+} facilitate the cathodic process and increase the intergranular corrosion rate. The dissolution process is temperature and viscosity dependent and is accelerated by stress and anodic currents, suggesting a stress-accelerated anodic dissolution or diffusion-controlled mechanism [95]. However, transgranular cracking in medium- to high-strength α and α–β alloys has been attributed to the absorption of hydrogen at the crack tip. Whereas dissolved platinum group metal ions accelerate intergranular corrosion, they have an inhibiting effect on the crack propagation rate, possibly by favoring the recombination of hydrogen atoms as opposed to hydrogen absorption. Since the embrittlement is also dependent on strain rate, it has been suggested that the contribution of stress is, in part, the result of hydrogen embrittlement [103]. A mixed mode of intergranular cracking and transgranular cleavage has been observed in these environments [104].

It has been shown that surface oxide rupture precedes crack initiation with subsequent dissolution and hydrolysis reactions leading to crack tip acidification. The crack tip local pH may be significantly lower than the bulk solution pH.

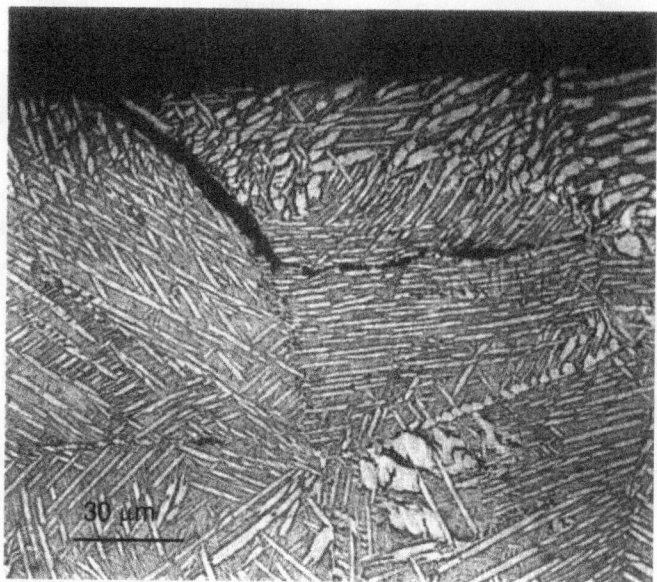

FIGURE 61.7. A hot salt crack changed to a stress unfavorable direction in Ti 6Al-4V with a Widmanstätten microstructure [98].

872 TITANIUM AND TITANIUM ALLOYS

A sharp notch or fatigue precrack appears to be required for hydrogen production to occur at the crack tip. Absorption may then occur at a deforming crack tip surface [105].

Hot salt SCC is of importance in high-temperature applications such as jet aircraft engine components [106] under conditions of high temperature, stress, and exposure to halide salts [59, 107]. Simultaneous cycling of temperature and stress may result in reduced susceptibility to hot salt SCC compared to isothermal, monotonic loading exposure [106]. The mechanism of hot salt cracking resembles that of SCC in aqueous halide solutions, with a fracture process that is associated with hydrogen embrittlement [108].

F2. Corrosion Fatigue

Titanium's superior corrosion resistance renders it an attractive structural material for use in many corrosive environments. In the presence of cyclic loading, the environmental effects on the fatigue properties become important. The excellent corrosion resistance of titanium in seawater and many other aqueous chloride media leads to smooth and notched fatigue run-out stresses, which are virtually unaffected by the environment [8]. Figure 61.8 illustrates the effect of seawater on the fatigue crack propagation rate. Effects of cycle frequency, stress ratio, microstructure, and applied potential have been reported in the literature [8, 109, 111, 112].

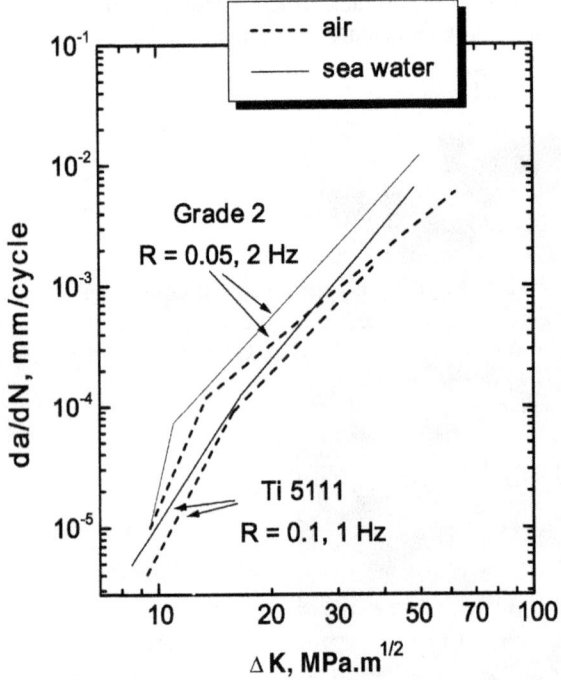

FIGURE 61.8. Trends of fatigue crack propagation rates for commercially pure titanium [109] and the intermediate strength near-α alloy Ti 5111 [110] in air and seawater.

The fatigue crack growth rate was found to be essentially independent of cycle frequency in air and in noncorrosive aqueous sodium sulfate solutions [111]. In aqueous halide solutions, a frequency-related crossover effect can occur at a stress intensity range ΔK, or ΔK_{SCC}, associated with cyclic SCC [111]. Below ΔK_{SCC}, lower frequencies permit more time for repassivation of fresh metal surface at the crack tip, thereby lowering the crack growth rate. Above ΔK_{SCC}, lower frequencies allow more time for hydrogen diffusion and embrittlement, thereby increasing the crack growth rate. In a 3.5% NaCl solution, a crossover effect has been observed with Ti 6 Al–4 V but not with Ti 8 Al–1 Mo–1 V, indicating a relation to alloy chemistry [112]. In methanolic halide solutions, fatigue crack growth rates have been found to increase with decreasing frequency over the whole range of ΔK [111].

In air and saltwater environments, a significant improvement in fatigue crack growth rates in α–β alloys has been associated with a transformed beta microstructure versus an equiaxed microstructure [112]. Under ripple load conditions, in which a small cyclic load is superposed on to a sustained load, an equiaxed microstructure exhibits better cracking resistance [113]. In air, the fracture surface appearance is predominantly ductile. In alcoholic and aqueous environments, an increasing fracture surface roughness appears linked to higher fatigue crack growth rates with cleavage fracture dominating over ductile fatigue striations [111].

Commercially pure titanium and its weld metal have displayed increasing crack growth rates with increasing stress ratio in air and natural seawater [109]. This has been attributed to crack closure effects. At a relatively high ΔK level, an acceleration of the crack growth rate in seawater was more pronounced for weld metal than for base metal indicating a microstructural influence. A small applied cathodic potential reduced the environmental effect by suppressing anodic dissolution when the passive layer failed [109].

F3. Hydrogen-Induced Cracking

The oxide film on titanium is an excellent barrier to hydrogen gas intrusion. Disruption of the oxide film allows easy absorption of hydrogen in high-pressure/temperature anhydrous gas streams. However, small quantities (2%) of moisture or oxygen immediately passivate the surface, forming again an effective barrier at temperatures as high as 315°C and pressures up to 800 psi [114].

At highly acidic or alkaline conditions, corrosion processes affect the integrity of the surface oxide and the corrosion potential drops below the hydrogen evolution potential. Such a drop in the open-circuit potential may also be the result of cathodic protection [115, 116], galvanic coupling, or intense dynamic abrasion. Electrochemically produced atomic hydrogen can now be absorbed. At temperatures <80°C, hydrogen diffusion is very slow and hydrogen will remain on the surface [114, 117]. Surface

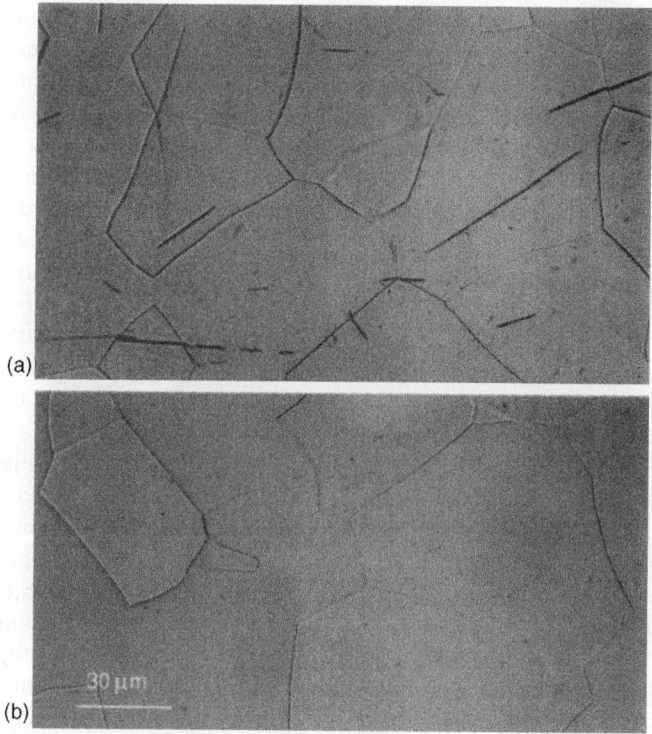

(a)

(b)

30 μm

FIGURE 61.9. (a) Hydride formation at a concentration of 42 ppm of hydrogen affected the formability of Grade 1 titanium. (b) Heating for 4 h in boiling water, followed by an ice water quench, dissolved the hydrides and increased the material's formability temporarily.

hydrogen generally has little effect on the structural integrity of the material. Surface hydriding of Grade 2 power condenser tubes, containing several thousand parts per million of hydrogen in the surface layer, led to only minor decreases in ductility and tensile strength of the in-service tubes [118].

The solubility of hydrogen in α titanium at room temperature is quite low and, influenced by pressure, stress, and alloy composition, is of the order of 20–150 ppm [119]. Once the solubility limit is exceeded, brittle titanium hydride precipitates form in the metal (see Fig. 61.9). These appear as dark, acicular needlelike structures in the metal microstructure [120]. Increasing oxygen content, increasing grain size, and decreasing temperatures facilitate crack initiation of the hydride, restrict slip, and promote cleavage [119]. Hydride embrittlement is also accelerated by increasing hydrogen concentrations, the presence of notches, and increasing strain rates [121, 122].

A loss of impact toughness has been observed in commercially pure titanium at relatively low hydrogen concentrations (∼ 50 ppm) by notched impact testing [121–123]. At high stress intensities, the failure mechanism is thought to be one of hydrogen-assisted localized plasticity at the crack tip, where the presence of hydrogen reduces the stress for plastic flow by enhancing the flow of dislocations [124]. Technically more rigorous toughness tests, following ASTM E 399, showed no effects of hydrogen on fracture toughness

up to 70 ppm [125]. Although impact ductility and toughness are affected by hydride formation at low hydrogen concentrations, only a very small influence on low-strain-rate mechanical properties is observed up to hydrogen concentrations well above 150 ppm, the general ASTM limit for hydrogen.

The hydrogen solubility limit in titanium depends largely on alloying additions. Additions of Al, O, N, and C strengthen and stabilize the α phase. With increasing Al content, absorption of hydrogen gives a supersaturated solid solution corresponding to a greater apparent solubility of hydrogen [121]. Whereas improved resistance to impact embrittlement is observed, plastic strain may result in the strain-induced, diffusion-controlled formation of hydrides with crack propagation through the hydrides at low strain rates [119, 124]. Although iron is a β stabilizer, concentrations of up to 0.15 wt % had no appreciable effect on the impact properties of commercially pure titanium [122, 123].

As a β-stabilizing element, the hydrogen solubility in the β phase is much greater (> 9000 ppm) [119]. As a result of such vast differences in hydrogen solubility in the different α and β phases, strong effects of the microstructure on the hydrogen-induced crack growth susceptibility are observed in near α and α–β alloys. Alloys with a continuous β phase provide a fast diffusion path for hydrogen, and these alloys are more susceptible to hydrogen embrittlement than alloys with a continuous α phase [126]. Similar to slow strain rate embrittlement, a sustained load will promote crack growth in Ti 6-4 with increasing hydrogen levels. However, Ti 6-4 tensile properties are not affected by hydrogen levels up to 300–600 ppm [119]. The high solubility of hydrogen in the β phase renders β alloys rather insensitive to hydrogen embrittlement. Whereas several thousand parts per million may be required for any significant loss of ductility, the hydrogen absorption rate is also much higher, resulting from the much hydrogen diffusion coefficient in the β phase [127].

Hydriding can be avoided if proper consideration is given to equipment design and service conditions. At temperatures ˙80°C, detrimental galvanic couples should be eliminated and impressed cathodic potentials below the hydrogen evolution line should be avoided.

At all temperatures, a pickled surface provides a greater resistance to hydrogen uptake than a sandblasted, abraded, or otherwise damaged surface [114]. Disruption of the surface oxide by smeared iron permits entry of hydrogen at any pH level. In the presence of CO_2, carbonate films reduce hydrogen absorption of titanium, illustrating the importance of the nature of the surface film. Small concentrations of H_2S had no effect [128]. Alloying elements, such as 0.8% Ni in Grade 12, can account for increased hydrogen absorption [128].

At temperatures <80°C, the extent of hydrogen charging under impressed cathodic potentials is a function of potential, pH, temperature, and exposure time [129]. Cathodic protection systems with potentials around −1.0 V vs. Ag/AgCl, as obtained with zinc anodes, will generally not result in titanium hydride problems in natural seawater. However, high stress levels will facilitate hydrogen absorption and hydride formation [130]. The general cathodic potential limit of $-0.75\,V_{SCE}$ has served very well in practice and is possibly somewhat on the conservative side for most applications [116].

G. GALVANIC CORROSION

Titanium is highly corrosion resistant and is often the more noble metal, the cathode, in a galvanic cell [131]. Little or no damage may be observed when the anode area is large in relation to the cathode area.

Rapid accelerated corrosion of the less noble metals magnesium, zinc, and aluminum is likely to occur when these materials are coupled to titanium. In an aluminum structure, localized corrosion around a titanium fastener may result in structural failure. Effective protection against galvanic attack must be provided to copper-based alloys and carbon steel. Titanium may safely be coupled to corrosion-resistant metals and alloys of similar potentials in the galvanic series, such as super duplex stainless steels, 6% molybdenum austenitic stainless steels, alloy 625 (UNS N06625), and alloy C-276 (UNS N10276). Stainless steels

are galvanically compatible in their passive condition. The primary consideration must be to ensure that the selected alloy is appropriate for the service environment. Titanium may safely be coupled to more noble metals and materials such as graphite and carbon fiber composites.

Galvanic corrosion can be avoided by suitable material selection in the design and by protection of adjoining less noble metals in the system. Techniques include (1) reducing the effective cathode/anode area ratio by coating the titanium, (2) electrically isolating the titanium components, and (3) cathodic protection and/or chemical corrosion inhibition of the active metal. It is recommended that impressed cathodic potentials do not fall below $-0.75\,V_{SCE}$ to prevent hydrogen embrittlement.

With the use of titanium seawater piping alongside copper–nickel piping systems on board Navy vessels, several new approaches to the prevention of galvanic corrosion are considered. Current ship designs include thick-walled "waster" pieces that have a high corrosion allowance and serve in physically separating and protecting the dissimilar piping systems. The bielectrode is a device that is positioned between the cathodic and anodic pipes and generates a potential gradient opposing the galvanic potential gradient, thereby eliminating the corrosion driving force [132]. Galvanic corrosion currents are also lowered dramatically by calcareous deposits and parts-per-million concentrations of chlorine.

H. EROSION AND CAVITATION RESISTANCE

Titanium exhibits an outstanding resistance to erosion as a result of its hard adherent surface oxide. In the absence of suspended solids, seawater velocities up to 35 m/s gave a minimal flow-enhanced corrosion rate of ∼0.01 mm/year of Grades 2 and 5 titanium [133]. In sand-laden seawater [134] as well as in coal-washing slurries [135], flow rates as high as 5 m/s gave minimal erosion rates. In the absence of general corrosion, the synergistic effect between erosion and corrosion is small and the metal removal rate can be attributed almost entirely to erosion [136, 137]. As the particle coarseness and velocity are increased to the point that the titanium oxide film is not given sufficient time to re-form, the erosion–corrosion rate increases substantially [134, 135]. Higher slurry velocities can be accommodated to some extent by thermally oxidized [135] or laser-nitrated titanium surfaces [136].

Nitriding is not recommended when the material is subjected to cavitation conditions, as it may lead to brittle fracture [138]. Whereas surface hardness and tensile strength play a dominant role in erosion resistance, material toughness and fatigue are important properties for cavitation resistance [139]. A superb corrosion resistance combined with good toughness and fatigue properties render titanium desirable for cavitation applications, such as the use of Grade 5 for ship propellers [140].

I. MICROBIOLOGICALLY INFLUENCED CORROSION/BIOFOULING

There has never been a reported incidence of microbiologically influenced corrosion attack on titanium [141]. Since titanium is nontoxic, it is susceptible to biofouling when immersed in seawater. The biofilm, however, does not attack the integrity of the underlying oxide and titanium remains resistant to localized corrosion. Water velocities greater than 2 m/s will reduce the extent of biofouling [142]. Chlorination is commonly used for controlling biofouling and can be followed by a dechlorination step. Increasing environmental pressures led to the consideration of nontoxic control methods such as ozone and ultraviolet (UV) irradiation. Both methods have shown positive results when used in titanium piping [142].

REFERENCES

1. J. S. Grauman, "Titanium-Properties and Applications for the Chemical process Industries," in Encyclopedia of Chemical Processing and Design, Vol. **58**, J. J. McKetta, (Ed.), Marcel Dekker, New York, 1998, pp. 123–147.

2. J. A. Mountford and J. S. Grauman, "Titanium for Marine Applications," presented at the 2nd Corrosion Control Workshop, sponsored by the Colorado School of Mines, New Orleans, LA, Feb. 1997.

2a. J.A. McMaster, "Rationalization of Unalloyed Titanium Material Specifications to Current Production Capabilities, Offers new Opportunities for the Titanium Industry," Paper No. 03461, CORROSION 2003, NACE International, Houston, TX 2003.

3. R. Schutz and D.E. Thomas, "Corrosion of Titanium and Titanium Alloys," in Metals handbook, 9th ed., Vol. **13**, J.R. Davis (Ed.), ASM, metals Park, OH, 1987, pp. 669–706.

4. Corrosion Resistance of, Titanium TIMET Brochure, Titanium Metals Corporation, Denver, CO, 1997.

5. R. W. Schutz and J. S. Grauman, "Fundamental Corrosion Characterization of High Strength Titanium Alloys," in Industrial Applications of Titanium and Zirconium: 4th Volume, STP 917, ASTM, Philadelphia, PA, 1986, pp. 130–143.

6. R. W. Schutz, "An Overview of Beta Titanium Alloy Environmental Behavior," in Beta Titanium Alloys in The 1990s, The Minerals, Metals and Materials Society, Warrendale, PA, 1993, pp. 75–91.

7. B. D. Craig and D. S. Anderson (Eds.), Handbook of Corrosion Data, ASM International, Materials Park, OH, 1995.

8. R. Boyer, G. Welsch, and E.W. Collins (Eds.), Materials Properties Handbook, Titanium Alloys, ASM, Materials Park, OH, 1994.

9. R.C. Weast (Ed.), CRC Handbook of Chemistry and Physics, 66th ed., CRC Press, Boca Raton, FL, 1985, pp. D154, F178.

10. J. Been, "Titanium Corrosion in Alkaline Peroxide Environments," Ph.D. Thesis, University of British Columbia, Vancouver, 1998.

11. M. Pourbaix, Atlas of Electrochemical Equilibria in Aqueous Solutions, NACE/Cebelcor, Houston, TX, 1974, pp. 213–221.

12. S. R. Morrison, Electrochemistry at Semiconductor and Oxidized Metal Electrodes, Plenum Press, New York, NY, 1980.

13. W. Schmickler and J. W. Schultze, "Electron Transfer Reaction on Oxide-Covered Metal Electrodes," in Modern Aspects of Electrochemistry, Vol. **17**, J. O'M. Bockris et al (Eds.), 1986, pp. 357–410.

14. Kh. G. Kuchukbaev, V.I. Kichigin, and L.V. kondakova, "Impedance of a Corroding Titanium Electorde in Acid Solutions," Zashchita Metallov, **28**,(2), 202–209 (1992).

15. W. Wilhelmsen and T. Hurlen, Electrochim. Acta, **32**, **1**, 85 (1987).

16. J. Been and D. Tromans, Pulp Paper Can., **100**(1), 50 (1999).

17. J. W. Mellor, A Comprehensive Treatise on Inorganic and Theoretical Chemistry, Vol. **VII**, Longmans, Green, New York, 1952, pp. 27–49.

18. H. E. Boyer and T. L. Gall (Eds.), Metals Handbook–Desk Edition, ASM, Materials Park, OH, 1985, pp. 2–10, 35–59.

19. S. Y. YU, "Mechanisms for Enhanced Active Dissolution Resistance and Passivity of Ti Alloyed with Nb and Zr," Ph.D. Thesis, School of Engineering and Applied Science, University of Virginia, Charlottesville, VA, 1998, p 116.

20. D. R. Stull and H. Prophet, JANAF Thermochemical Tables, NSRDS-NBS 37, National Bureau of Standards, Washington, DC, 1971.

21. M. Lindler, D. W. Bahnemann, B. Hirthe, and W. Griebler, Solar Eng., **1**. 399 (1995).

22. M. A. Barteau, Chem. Rev., **96**, 1413 (1996).

23. Kobe Steel, "On Formation of Thin Oxide Film to Titanium by Anodic Oxidation Method," Technical Note 108, Technical Report R & D, **21**, 2 (1977).

24. P. Tengvall and I. Lunström, Clin. Mater., **9**, 115 (1992).

25. N. D. Tomashov and P. M. Al'tovskii, Corrosion and Protection of Titanium, Government Scientific-Technical Publication of Machine-Building Literature (Russian Translation), Moscow, Russia, 1963.

26. J. B. Cotton, Chem. Eng. Prog., **66**, 10, 57 (1970).

27. J. B, Cotton, Br. Corros. J., **10**, **2**, **66** (1975).

28. N. D, Tomashov, G. P. Chernova, Y. S. Ruscol, and G, A. Ayuyan, Electrochim, Acta, **19**, 159 (1974).

29. B. Seely, Metal Prog., June 1989, pp. 35–37.

30. P. C. S. Hayfield, "Anodic Oxidation of Titanium in Aqueous Solutions," in Titanium Science and Technology, Vol. **4**, R. I Jaffee and H. M. Burte (Eds.), Plenum, New York, 1973, pp. 2405–2418.

31. R.W. Schutz and L.C. Covington, Corrosion, **37**, 10, 585 (1981).

32. T. Fukuzuka, K. Shimogori, H. Satoh, and F. Kamikuba, "On the Beneficial Effect of the Titanium Oxide Film Formed

by Thermal Oxidation," in Titanium' 80 Science and Technology, Vol. **45**, H. Kimura and O. Izumi, (Eds.), AIME, New York, 1980, pp. 2783–2787.

33. T. Ohtsuka, M. Masuda, and N. Satoh, J. Electrochem. Soc., **132**, 4, 787 (1985).

34. L. D, Arsov, C. Kormann, and W. Plieth, J. Electrochem. Soc., **138**, 10, 2964 (1991).

35. T. Moroishi and Y. Shida, "Oxidation Behaviour of Titanium in High Temperature Steam," in Titanium'80, Science and Technology, H. Kimura and O. Izumi (Eds.), AIME, Warrendale, PA, 1980, pp. 2773–2782.

36. J. Stringer, Acta Metall., **8**, 758 (1960).

37. A. M. Chaze, C, Coddet, and G. Beranger, "Dissolution of Oxygen in the Metallic Phase During High Temperature Oxidation of Titanium and Some Alloys," in Sixth World Conference on Titanium, Part IV, P. Lacombe et al. (Eds.), Société Francaise de Métallurgie, France, 1988, pp. 1765–1770.

38. M Li, A Kar, and V. Desai, J. Mater. Sci., **30**, 5093 (1995).

39. J. S. Grauman, "Beta-21S: A New High Strength, Corrosion Resistant Titanium Alloy," in Proceedings of the International Conference on Titanium Products and Applications, Titanium Development Association, Boulder, CO, 1990, pp. 290–299.

40. R. W. Schutz and M. Xiao, "Optimized Lean-Pd Titanium Alloys for Aggressive Reducing Acid and Halide Service Environments," in Proceedings of the 12th International Corrosion Congress, Houston, TX, Sept. 19–24, 1993, NACE, Houston, TX.

41. S. Kitayama and Y. Shida, ISIJ Int., **31**(8), 897 (1991).

42. D. E. Thomas, "Titanium Alloy Corrosion Resistance in Nitric Acid Solutions," in Proceedings of the 1986 International Conference on Titanium Products and Applications, Titanium Development Association, Boulder, CO, 1986, pp. 220–240.

43. E. E. Millaway, R. L. Powell, and S. M. Weiman, "Titanium Behavior in Nitric Acid," NACE 18th National Conference, Chemical Industry Symposium, Kansas City, Mar. 19–23, 1962, NACE, Houston, TX.

44. H. Satoh, F. Kamikubo, and K. Shimogori, "Effect of Oxidizing Agents on Corrosion Resistance of Commercially Pure Titanium in Nitric Acid Solution," in Proceedings of the Fifth International Conference on Titanium, Munich, FRG, Sept., 10–14, 1984, Deutsche Gesellschaft für Metallkunde, Overursel, FRG, 1995, pp. 2649–2655.

45. T. Furuya, H. Satoh, K. Shimogori, Y. Nakamura, K. Matsumoto, Y, Komori, and S. Takeda, "Corrosion Resistance of Zirconium and Titanium Alloy in HNO_3 Solutions," in Proceedings from Fuel Reprocessing and Waste Management, Jackson, WY, Aug. 25–29 1984.

46. L. L. Gilbert and C. W. Funk, Metal Prog., **70**, (11) 93 (1956).

47. D. R. McIntyre and C. P. Dillon, "Pyrophoric Behavior and Combustion of the Reactive Metals," MTI Publication No. 32, NACE, Houston, TX, 1988, pp. 3–9.

48. S. Z. Kostic and B. N. Princip, AT "Corrosion Behaviour of Construction Materials in Nitric Acid Production Plant," Corrosion Prevention & Control, June, pp. 78–80, 1989.

49. A. L. Forrest, "Titanium in the Bleaching Pulp Mill—An Update," in Proceedings of the 1990 International Conference on Titanium Products and Applications, Titanium Development Association, Boulder, CO, 1990, pp. 352–367.

50. E. E. Millaway and M. H. Kleinman, "Investigation of Water Content Required for Passivation of Titanium in Chlorine—II," Timet Progress Report No. 23, Henderson, NY, 1966.

51. E. E. Millaway and M. H. Kleinman, Corrosion, **23**(4), 88 (1972).

52. F. E. Littman, F. M. Church, and E. M. Kinderman, J. Less-Common Met., **3**, 367 (1961).

53. J. D. Jackson, W. K. Boyd, and P. D. Miller, "Reactivity of Metals with Liquid and Gaseous Oxygen," DMIC Memorandum 163, Battelle Memorial Institute, Columbus, OH, Jan. 15, 1963.

54. T.R. Strobridge, J. C. Moulder and A. F. Clark, "Titanium Combustion in Turbine Engines," Technical Report AD A075657, NTIS, Springfield, VA, July 1979.

55. S. Tebbal and R. D. Kane, "Materials Selection in Hydrothermal Oxidation Processes," Paper No. 413, CORROSION 98, NACE, Houston, TX 1998.

56. N. Boukis, C. Friedrich, and E. Dinjus, "Titanium as Reactor Material for SCWO Applications. First Experimental Results," Paper No. 417, CORROSION 98, NACE Houston, TX, 1998.

57. R. Wellauer and M. Oldani, Ozone: Sci. Eng., **12**(3), 243 (1990).

58. T. P. Kashcheeva, L. V. Sologub, and L. Yu. Gadasina, Prot. Met., **17**(2), 161 (1981).

59. J. A. Beavers, G. H. Koch, and W. E. Berry, Corrosion of Metals in Marine Environments, MCIC Report, Columbus, OH, July 1986, Chapter 3.

60. F. M. Reinhart, "Corrosion of Materials in Hydrospace, Part III, Titanium and Titanium Alloys," Technical Note N-921 U.S. Naval Civil Engineering Lab, Port Hueneme, CA, Sept. 1967.

61. M. V. Popa, E. Vasilescu, I Mirza-Rosca, S. Gonzalez, M. L. Llorente, P. Drob, and M. Anghel, "Evaluation of Corrosion Resistance of Ti and Two Titanium Base Alloys in Alkaline Solutions," Vol. **II**. Eurocorr '97, Trondheim, Norway, 1997, pp. 687–692.

62. L. C. Covington and N. G., Feige, "A Study of Factors Affecting the Hydrogen Uptake Efficiency of Titanium in Sodium Hydroxide Solutions," in Localized Corrosion—Cause of Metal Failure, ASTM STP 516, ASTM, Philadelphia, PA, 1972, pp. 222–235.

63. Y. J. Kim and R. A. Oriani, Corrosion, **43**(2), 92 (1987).

64. J. Pan, D, Thierry, and C. Leygraf, J Biomed. Mater. Res., **28**, 113 (1994).

65. J. Pan, D. Thierry, and C. Leygraf, Electrochim. Acta, **41**, 7/8, 1143 (1996).

66. M. Assefpour-Dezfuly, C. Vlachos, and E. H. Andrews, J. Mater. Sci., **19**, 3626 (1984).

67. I. R. Lane, L. B. Golden, and W. L. Acherman, Ind. Eng. Chem., **45**, 1067 (1953).

68. R. A. Clapp, J. J. Kvochak, and B. J. Saldanha, "Corrosion of Titanium and Zirconium in Organic Solutions," Paper 243, CORROSION 95, NACE, Houston, TX 1995.

69. S. K. Bhowmik, Corros. Maint., **8**, **1**, 11 (1985).

70. W. Han, J. Mater. Sci. Technol., **14**, 92 (1998).

71. H. D. Pietka and K. G. Schülze, Werk. Korr., **45**, 325 (1994).

72. O. Radovici and E. Vasilescu, Rev. Roum. Chim., **41**(1–2), 55 (1996).

73. T. A. Krapivkina, T. G. Marchenko, and I. N. Martynova, Prot. Met., **21**(2), 213 (1985).

74. Z. Szklarska-Smialowska, Pitting Corrosion of Metals, NACE International, Houston, TX, 1986, p. 235.

75. R. W. Schutz and J. S. Grauman, "Compositional Effects on Titanium Alloy Repassivation Potential in Chloride media," in Advances in Localized Corrosion, H. S. Isaacs et al. (Eds.), NACE, Houston, TX, 1990, pp. 335–337.

76. J. C. Griess, Jr., Corrosion, **24**, **4**, 96 (1968).

77. R. W. Schutz, "Understanding and Preventing Crevice Corrosion of Titanium Alloys," Paper No. 162, CORROSION 91, NACE, Houston, TX, 1991.

78. M. Kobayashi et al., "Study on Crevice Corrosion of Titanium," in Titanium'80 Science and Technology, H. Kimura and O. Izumi (Eds.), The Metallurgical Society, Warrendale, PA, 1980, pp. 2613–2622.

79. R. W. Evitts et al. "Numerical Simulation of Crevice Corrosion of Titanium: Effect of the Bold Surface," Paper No. 121, CORROSION 96, NACE, Houston, TX, 1996.

80. B. M. Ikeda et al. "The Effect of Temperature on Crevice Corrosion of Titanium," in Proceedings of the 11th International Corrosion Congress, Vol. 5 Florence, Italy, Apr. 1990.

81. H. Satoh et al. "Effect of Gasket Materials on Crevice Corrosion of Titanium," in Titanium—Science and Technology, Vol. **4**, G. Lütjering, U. Zwicker, and W. Bunk (Eds.), Deutsche Gesellschaft für Metallkunde, Oberursel, Germany, 1985, pp. 2633–2639.

82. L-A. Yao et al. Corrosion, **47**(6), 420 (1991).

83. B. J. Fitzgerald and N. D. Greene, "Crevice Corrosion of Active–Passive Metals and Alloys in Acid and Acid Chloride Environments," Paper No. 180, CORROSION 82, NACE, Houston, TX, 1982.

84. G. H. Koch, "Localized Corrosion in Halides Other Than Chlorides," Paper No. 437, CORROSION 93, NACE, Houston, TX, 1993.

85. D. D. Bergman and J. S. Grauman, "The Detection of Crevice Corrosion in Titanium and Its Alloys Through the Use of Potential Monitoring," in Titanium'92 Science and Technology, F. H. Froes and I. Caplan (Eds.), The Minerals, Metals, and Materials Society, Warrendale, PA, 1993, pp. 2193–2200.

86. M. Inman et al. "Detection of Crevice Corrosion in Natural Seawater Using Polarization Resistance Measurements," Paper No. 296, CORROSION 97, NACE, Houston, TX, 1997.

87. R. B. Diegle, "Electrochemical Cell for Monitoring Crevice Corrosion in Chemical Plants," Paper No. 154, CORROSION 81, NACE, Houston, TX, 1981.

88. M. Stern and H. Wissenberg, J. Electrochem. Soc., **106**, 9, 759 (1959).

89. H. Satoh et al. Platinum Metals Rev., **31**(3), 115 (1987).

90. L. C. Covington, "The Role of Multi-Valent Metal Ions in Suppressing Crevice Corrosion of Titanium," in Titanium Science and Technology, R. I. Jaffe and H. M. Burte (Eds.), Plenum, New York, 1973, pp. 2395–2403.

90a. J.S. Grauman, "PGMA—A Novel Corrosion Protection Method for Titanium," in Ti—2003 Science and Technology, Tenth World Conference on Titanium, Vol. IV, G. Lütjering and J. Albrecht (Eds.), Deutsche Gesellschaft für Materialkunde e.V., Frankfurt, Germany, 2003, pp. 2107–2114.

91. L. C. Covington, ASTM STP 576, American Society for Testing and Materials, West Conshohocken, PA, 1976, pp. 147–154.

92. R. W. Schutz and J. S. Grauman, "Localized Corrosion Behavior of Titanium Alloys in High Temperature Seawater Service," Paper No. 162, CORROSION 88, NACE, Houston, TX 1988.

93. J. S. Grauman and R. W. Schutz, "Influence of Surface Condition on the Resistance of Titanium to Chloride Pitting," in Advances in Localized Corrosion, H. S. Isaacs, U. Bertocci, J. Kruger, and S. Smialowska (Eds.), NACE, Houston, TX, 1990, pp. 331–334.

94. M. J. Blackburn, W. H. Smyrl, and J. A. Feeney, in Stress Corrosion Cracking in High Strength Steels and in Titanium and Aluminum Alloys, B. F. Brown (Ed.), Naval Research Laboratory, Washington, DC, 1972, pp. 245–363.

95. D. J. Simbi, Corros. Rev., **14**, 3–4, 343 (1996).

96. R. W. Judy, B. B. Rath, and I. L. Caplan, "Stress Corrosion Cracking of Pure Titanium as Influenced by Oxygen Content," in Sixth World Conference on Titanium, Part IV, P. Lacombe, R. Tricot, and G. Béranger (Eds.), Société Francaise de Métallurgie, France, Les éditiors de physique, Les Ulis, France, 1988, pp. 1747–1752.

97. D.J. Simbi and J. C. Scully, Corros. Sci., **37**(8), 1325 (1995).

98. R. E. Adams, "Effect of Processing Variables on the Hot Salt Stress Corrosion Resistance of Ti 6Al-4V," Technical Report No. 12, TIMET, Henderson, NV, 1968.

99. R. J. H. Wanhill, Br. Corros. J., **10**(2), 69 (1975).

100. D. J. Simbi and J. C. Scully, Corros. Sci., **35**(1–4), 489 (1993).

101. E. G. Haney and W. R. Wearmouth, Corrosion, **25**(2), 87 (1969).

102. R. W. Schutz and J. M. Horrigan, "Stress Corrosion Behavior of Ru-Enhanced Alpha-Beta Titanium Alloys in Methanol Solutions," Paper No. 261, CORROSION 98, NACE, Houston, TX, 1998.

103. A. C. Hollis and J. C. Scully, Corros. Sci., **34**(5), 821 (1993).

104. D. J. Simbi and J. C. Scully, Corros. Sci., **34**(10), 1743 (1993).

105. D. G. Kolman and J. R. Scully, Metall. Mater. Trans. A, **28A**, 2645 (1997).

106. R. L. Fowler and A. J. Luzietti, "Hot Salt Stress Corrosion Behavior of Ti 6-2-4-2 with Alternating Temperature and Stress," Paper No. 18, CORROSION 80, NACE, Houston, TX, 1980.

107. A. J. Hatch, H. W. Rosenberg, and E. F. Erbin, "Effects of Environment on Cracking in Titanium Alloys," in Stress

Corrosion Cracking of Titanium, STP 397, ASTM, Philadelphia, PA, 1966, pp. 122–136.

108. C. F. Clarke, D. Hardie, and P. McKay, Corros. Sci., **26**(6), 425 (1986).

109. R. Murakami and W. G. Ferguson, Fatigue Fract. Eng. Mater. Struct., **16**(2), 255 (1993).

110. J. Been, "Ti 5111 Brings Intermediate Strength, Excellent Toughness, & Corrosion Resistance to Naval Operating Environments," Paper No. 499, CORROSION/99 NACE, Houston, TX, 1999.

111. D. B. Dawson and R. M. Pelloux, Metall. Trans., **5**(3), 723 (1974).

112. G. R. Yoder, L. A. Cooley, and T. W. Crooker, "Effects of Microstructure and Frequency on Corrosion Fatigue Crack Growth in Ti-8Al-1Mo-1V and Ti 6Al-4V," in Corrosion Fatigue: Mechanics, Metallurgy, Electrochemistry, and Engineering, STP 801, T. W. Crooker and B. N. Leis (Eds.), ASTM, Philadelphia, PA, 1983, pp. 159–174.

113. G. R. Yoder, P. S. Pao, and R. A. Bayles, "Ripple-Load Cracking in a Titanium Alloy," Scr. Metall. Mater., **24**, 2285 (1990).

114. L. C. Covington, Corrosion, **35**(8), 378 (1979).

115. L. Lunde and R. Nyborg, "Hydrogen Absorption of Titanium Alloys During Cathodic Polarization," Engineering Solutions to Industrial Corrosion Problems, Sandfjord, Norway, June 1993.

116. R. W. Schutz and J. S. Grauman, "Determination of Cathodic Potential Limits for Prevention of Titanium Tube Hydride Embrittlement in Saltwater," Paper No. 110, CORROSION 89, NACE, Houston, TX, 1989.

117. I. I. Phillips, P. Poole, and L. L. Shreir, Corros, Sci., **14**, 533 (1974).

118. J. P. Fulford, R. W. Schutz, and R. C. Lisenbey, "Characterization of Titanium Condenser Tube Hydriding at Two Florida Power & Light Company Plants," Paper No. 87-JPGC-Pwr-F, presented at the joint ASME/IEEE Power Generation Conference, Miami Beach, FL, Oct. 4-8, 1987.

119. N. E. Paton and J. C. Williams, "Effect of Hydrogen on Titanium and Its Alloys," in Hydrogen in Metals, I. M. Bernstein and A. W. Thompson (Eds.), American Society for Metals, New York, 1974, pp. 409–431.

120. H. Z. Xiao, Scri. Metall. Mater., **27**, 571 (1992).

121. G. A. Lenning, J. W. Spretnak, and R. I. Jaffee, J. Metals, **8**(10), 1235 (1956).

122. D. N. Williams, "Hydrogen in Titanium and Titanium Alloys," TML Report No, 100, Battelle Memorial Institute, Columbus, OH, 1958.

123. K. Rüdinger and H. G. Bitter, "Influence of Temperature on Impact Strength of Notched Samples of Commercially Pure Titanium," in Titanium—Science and Technology, Vol. **3**, G. Lütjering, U. Zwicker, and W. Bunk (Eds.), Deutsche Gesellschaft für Metallkunde, Oberursel, Germany, 1985.

124. D. S, Shih, I. M. Robertson, and H. K. Birnbaum, Acta Metall., **36**(1), 111 (1988).

125. M. L. Wasz, C. C. Ko, F R. Brotzen, and R. B. McLellan, Seri. Metall., **24**, 2043 (1990).

126. N. R. Moody and J. E. Costa, "A Review of Microstructure Effects on Hydrogen-Induced Sustained Load Cracking in Structural Titanium Alloys," in Microstructure/Property Relationships in Titanium Aluminides and Alloys, Y.-W. Kim and R. R. Boyer (Eds.), The Minerals, Metals and Materials Society, Warrendale, PA, 1991, pp. 587–604.

127. B. P. Somerday, N. R. Moody, J. E. Costa, and R. P. Gangloff, "Environment-Induced Cracking in Structural Titanium Alloys," Paper No. 267, CORROSION 98, NACE, Houston, TX, 1998.

128. L. Lunde and R Nyborg. "Hydrogen Absorption of Titanium Alloys During Cathodic Polarization," presented at Engineering Solutions to Industrial Corrosion Problems, Sandefjord, Norway, June 7–9, 1993.

129. J.-I. Lee, P. Chung, and C.-H. Tsai, "A Study of Hydriding of Titanium in Sea Water under Cathodic Polarization," Paper No. 259, CORROSION 86, NACE, Houston, TX, 1986.

130. G. Venkataraman and A. D. Goolsby, "Hydrogen Embrittlement in Titanium Alloys from Cathodic Polarization in Offshore Environments, and Its Mitigation," Paper No. 554, CORROSION 96, NACE, Houston, TX, 1996.

131. D. Peacock, Mater, Perform., **37**(8), 68 (1998).

132. D. A. Shifler, D. Melton, and H. P. Hack, "New Techniques for Galvanic Corrosion in Piping Systems," Paper No. 706, CORROSION 98, NACE, Houston, TX, 1998.

133. G. J. Danek, Naval Eng., J., Oct. 1966, P. 763.

134. Titanium Heat Exchangers for Service in Seawater, Brine and Other Natural Aqueous Environments: The Corrosion, Erosion and Galvanic Corrosion Characteristics of Titanium in Seawater, Polluted Inland Waters and in Brines, Titanium Information Bulletin, Imperial Metal Industries (Kynoch) Limited, Witton, Birmingham, England, May 1970.

135. G. R. Hoey and J. S. Bednar, "Erosion-Corrosion of Selected Metals in Coal Washing Plant Environments," Mater. Perform., Apr. 1983, pp. 9–14.

136. E. Bardal, T. G. Eggen. T. Rogne, and T. Solem, "The Erosion and Corosion Properties of Thermal Spray and Other Coatings," in Proceedings of ITSC'95, Kobe, May 1995, pp. P 645–650.

137. J. Yang and J. H. Swisher, "Erosion-Corrosion Behavior and Cathodic Protection of Alloys in Seawater-Sand Slurries," JMEPEG, **2**(6), 843 (1993).

138. K. S. Zhou and H. Herman, Wear, **80**, 101 (1982).

139. R H. Richman and W. P. McNaughton, JMEPEG, **6**(5), 633 (1997).

140. W. Werchniak and J. P. Gudas, "Seawater Corrosion and Corrosion Fatigue of High Strength Cast Alloys for Propellers and Impellers," Paper No. 199, CORROSION 81, NACE, Houston, TX, 1981.

141. B. Little, P. Wagner, and F. Mansfeld, Int. Mater. Rev., **36**(6), 252 (1991).

142. S. Hoover, J. Jones-Meehan, M. Walch, B. J. Little, and R. W. Erskine, "Evaluation of Nonpolluting Biofouling Control Methods for Titanium Piping at NSWC/Dania, FL." DTIC Report ADA331160.

62

ZINC

X. G. ZHANG

Teck Metals Ltd., Mississauga, Ontario, Canada

A. INTRODUCTION

Zinc, one of the most widely used metals, is silvery blue-gray in color with a relatively low melting point (419.5°C) and boiling point (907°C). The strength and hardness of unalloyed zinc are greater than those of tin or lead but appreciably less than those of aluminum or copper. Pure zinc cannot be used in applications under applied stress because of its low creep resistance. It recrystallizes rapidly after deformation at room temperature and, thus. cannot be work hardened at room temperature. The temperature for recrystallization and the creep resistance can be increased through alloying [1].

The binary zinc alloy systems of most interest for commercial applications are (1) Zn–A1, which at 4% Al forms the basis of the zinc die casting alloys; (2) Zn–Cu, which with zinc up to 45% are brass alloys; (3) Zn–Fe, which includes the phases making up the galvanized coatings; and (4) Zn–Pb, which plays an important role in some pyrometallurgical extraction processes. Ternary and quaternary systems involving these alloys. with additions of such elements as nickel, magnesium, titanium, and cadmium, are also of commercial importance.

The electrochemical properties of zinc are important in the production and application of zinc. For example, electrowinning in zinc refining, electroplating in the production of zinc coating, zinc batteries for energy storage and coatings, and anodes for corrosion protection are all based essentially on its electrochemical properties. These electrochemical properties include the relatively active position in the electromotive force series, fast and reversible dissolution/deposition kinetics, high overpotential for hydrogen evolution, and formation of passive film in slightly alkaline solutions [2].

The uses of zinc can be divided into six major categories: (a) coatings, (b) casting alloys, (c) alloying element in brass and other alloys, (d) wrought zinc alloys, (e) zinc oxide, and (f) zinc chemicals. The use of zinc coatings for corrosion protection of steel structures is the most important application due to the high corrosion resistance of zinc in atmospheric and other environments. Nearly one-half of all zinc produced is consumed for this purpose.

In the past decades, research on various aspects of corrosion of zinc has generated much technical information. This wealth of information has been systematically and critically reviewed in a book entitled *Corrosion and Electrochemistry of Zinc* by Zhang [2]. The information presented here is abstracted from this book. Most of the electrochemical information and much of the specific corrosion data, descriptions, and theories contained in the book are omitted due to limited space.

B. ZINC COATINGS

The many types of zinc and zinc alloy coatings can be classified according to coating composition and production methods [3–15]. According to chemical composition, zinc-based coatings fall into several major categories: (1) pure zinc, (2) Zn–Fe, (3) Zn–Al, (4) Zn–Ni, and (5) zinc composites. In terms of methods, zinc coatings can be produced by hot dipping, electroplating, mechanical bonding, sherardizing and thermal spraying (metallizing). The hot-dip method can be further divided into two processes: continuous hot dip and batch hot dip. Typical applications for zinc and its alloy-coated steel sheet products cover a wide range in the construction, automobile, utility and appliance industries, as shown in Table 62.1.

Hot-dip galvanizing, either continuous or batch, is a process by which an adherent coating of zinc and Zn–Fe, alloys is produced on the surface of iron or steel products by immersion in a bath of molten zinc. In general, an article to be galvanized in a continuous galvanizing process is cleaned, pickled, and fluxed in a batch process or heat treated in a reducing atmosphere to remove surface oxide. It is then immersed in a bath of molten zinc for a time sufficient for it to wet and alloy with zinc, after which it is withdrawn and cooled.

The coating produced in this way is bonded to the steel by a series of Zn–Fe alloy layers, with a layer of almost pure zinc on the surface. The engineering properties of the coating depend on the physical and chemical nature of the Zn–Fe intermetallic layers formed. The thickness and composition of the alloys vary depending on whether it is a batch or continuous process, mainly due to the difference in the immersion time in the molten zinc bath and the bath composition. The coating produced by a batch process is relatively thicker and has clearly distinguishable alloy layers, while that of the continuous process is thinner and has only a very thin and sometimes invisible alloy layer at the coating–steel interface.

In addition to pure zinc coating, there are three major hot-dip zinc alloy coatings galvanneal, Galvalume, and Galfan. Galvanneal is a Zn–Fe alloy coating containing typically 6% Fe that is obtained by annealing the hot-dipped sheet. Galvalume, with 55% Al, 1.5% Si, and 43.5% Zn, has a microstructure consisting of an outer layer and a thin intermetallic layer that bonds the outer layer to the steel. Galfan contains 95% Zn, 5% Al and a small amount of mischmetal and has a multiphase microstructure that is characteristic of its composition, exhibiting a lamellar structure of alternating zinc-rich and aluminum-rich phases.

Electroplating is another common galvanizing method. Owing to its versatility, electroplating has been extensively used to explore new zinc alloy coatings (e.g., Zn, Zn–Fe, Zn–Co, and Zn–Ni),. The plating process generally comprises three stages: (1) degreasing and cleaning, (2) electroplating, and (3) posttreatment. Similar to hot-dip galvanizing, electroplating can be a batch or a continuous process.

C. ELECTROCHEMICAL NATURE OF CORROSION

The corrosion of zinc is an electrochemical process in which zinc is oxidized with simultaneous reduction of hydrogen ions or dissolved oxygen in the electrolyte. The oxidation follows the following reaction:

$$Zn = Zn^{2+} + 2e^- \tag{62.1}$$
$$E° = -0.763 + 0.0295 \log[Zn^{2+}] \quad V_{SHE}$$

TABLE 62.1.   Typical Applications of Zinc-Coated Steel Products[a]

Coatings	Typical Applications
Coatings by Continuous Electroplating and Hot-Data Processes	
Zn and Zn–5Al	Roofing, culverts, housing, appliances, autobody panels and components, nails, guy wire, rope, utility wire, and fencing
Zn–Fe	Autobody panels and structural components
Zn–Ni and Zn–Co	Autobody panels and structural components, housing, appliances, and fasteners
Zn–55Al	Roofing, siding, ductwork, culverts, mufflers, tailpipes, heat shields, ovens, toasters, chimneys, and silo roofs
Hot-Dip Batch Galvanized	

Structural steel for power-generating plants, petrochemical facilities, heat exchangers, cooling coils, water treatment facilities, and electrical transmission towers and poles
Bridge structural members, culverts, corrugated steel pipe and arches
Reinforcing steel for concrete structures
Highway guard rails, lighting stands, and sign structures
Marine pilings and rails
Architectural applications of structural steel, lintels, beams, columns, and related materials

[a]See [3–5].

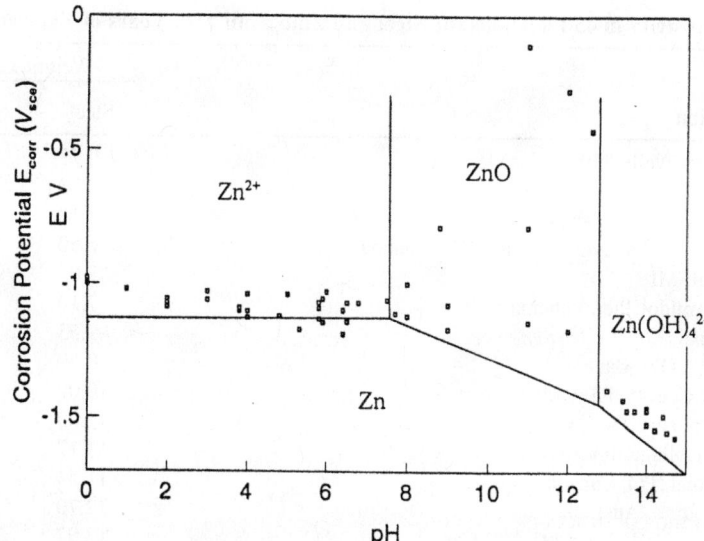

FIGURE 62.1. Corrosion potentials in solutions of various pH values; the solid line indicates the reversible potential which is calculated from Eq. (62.1) assuming $10^{-4}\,M\,Zn^{2+}$ in the solution [2].

The standard potential of this reaction is $-0.763\,V_{SHE}$, which is 0.315 V more negative than iron and 0.9 V more positive than aluminum in the electromotive series.

Figure 62.1 shows the Pourbaix diagram of zinc in aqueous solutions. The solid lines, calculated from Eq. (62.1), assuming $10^{-4}\,M\,Zn^{2+}$ in the solution, define the stability regions for the different solid and dissolved zinc species and, thus, the condition for zinc corrosion and passivation. Also plotted in the figure are the corrosion potentials, reported in different studies on zinc or zinc coatings as functions of pH [2]. It can be seen that the corrosion potentials in the pH range of 4–8 are close to the theoretical values. However, the corrosion potentials in acidic and alkaline solutions are somewhat higher than the reversible potential values, indicating that the zinc electrode at the corrosion potentials is anodically polarized from its reversible value. Part of the deviation of the corrosion potentials from the reversible value may result from a concentration of Zn^{2+} in these solutions being higher than $10^{-4}M$, at least near the surface where the zinc ions from the dissolution may accumulate. The corrosion potentials in slightly alkaline solutions, from pH ~8 to 12, in which ZnO is the stable from, can be much higher than the reversible values due to formation of a surface oxide which, depending on the specific conditions, results in various degrees of passivation.

Since the kinetics of the electrochemical reactions and the surface state determine the rate and form of corrosion in a given circumstance, the corrosion potentials, along with the equilibrium potential values and their variation with pH, shown in Figure 62.1, are the fundamental parameters that can be used to explain the corrosion behavior of zinc under different conditions.

D. CORROSION RESISTANCE

D1. Atmospheric Environments

The high corrosion resistance of zinc in atmospheric environments has resulted in extensive outdoor applications of zinc-coated steels. The corrosion rate is lowest in dry, clean atmospheres and highest in wet, industrial atmospheres. Seacoast atmospheres, not in direct contact with salt spray, are mildly corrosive to zinc. Locations near sea level are subject to salt spray and, hence, the corrosion rate can be much higher. Atmospheres are conventionally defined according to four general types: (1) rural, (2) industrial, (3) Urban and (4) marine. The typical atmospheric corrosion rates of zinc in each of these categories are as follows [6]:

Rural	0.2 to 3 μm/year
Marine (outside the splash zone)	0.5 to 8 μm/year
Urban and industrial	to 16 μm/year

Table 62.2 lists the ranking of the corrosivity of different atmospheres for zinc and steel around the world [7]. The corrosivity of atmospheres from one location to another varies by as much as a factor of 100 for zinc and 500 for steel. The data listed in Table 62.2 show that the corrosion rate of zinc in most atmospheres is at least 10 times lower than that of steel. It is for this reason that zinc is commonly used, through the galvanizing process, to effectively protect steel from corrosion. In addition, Figure 62.2, which is a plot of the data listed in Table 62.2 indicates that the corrosion ratio increases with increasing corrosion rate of steel, suggesting that the more corrosive an atmospheric environment is to steel, the higher

TABLE 62.2.   Ranking of Corrosivity in 45 Locations for Steel and Zinc from Two Years of Exposure[a]

| Ranking | | | Weight Loss (g) | | Loss Ratio |
Steel	Zinc	Location	Steel	Zinc	Steel/Zinc
1	1	Norman Wells, NWT, Canada	0.73	0.07	10.3
2	2	Phoenix, AZ	2.23	0.13	17.0
3	3	Saskatoon, SK, Canada	2.77	0.13	21.0
4	4	Esquimalt, Vancouver Island, BC, Canada	6.50	0.21	31.0
5	15	Detroit, MI	7.03	0.58	12.2
6	5	Fort Amidor Pier, Panama, C.Z.	7.10	0.28	25.2
7	11	Morenci, MI	9.53	0.53	18.0
8	7	Ottawa, ON, Canada	9.60	0.49	19.5
9	13	Potter County, PA	10.00	0.55	18.3
10	31	Waterbury, CT	11.00	1.12	9.8
11	10	State Collage, PA	11.17	0.51	22.0
12	28	Montreal, PQ, Canada	11.44	1.05	10.9
13	6	Melbourne, Australia	12.70	0.34	37.4
14	20	Halifax (York Redoubt), NS, Canada	12.97	0.70	18.5
15	19	Durham, NH	13.30	0.70	19.0
16	12	Middletown, OH	14.00	0.54	26.0
17	30	Pittsburgh, PA	14.90	1.14	13.1
18	27	Columbus, OH	16.00	0.95	16.8
19	21	South Bend, PA	16.20	0.78	20.8
20	18	Trail, BC, Canada	16.90	0.70	24.2
21	14	Bethlehem, PA	18.3	0.57	32.4
22	33	Cleveland, OH	19.0	1.21	15.7
23	8	Miraflores, Panama, C.Z.	20.9	0.50	41.8
24	29	London (Battersea), England	23.0	1.07	21.6
25	24	Monroeville, PA	23.8	0.84	28.4
26	35	Newark, NJ	24.7	1.63	15.1
27	16	Manila, Philippine Islands	26.2	0.66	39.8
28	32	Limon Bay, Panama, C.Z.	30.3	1.17	25.9
29	39	Bayonne, NJ	37.7	2.11	17.9
30	22	East Chicago, IN	41.1	0.79	52.1
31	9	Cape Kennedy, FL (1/2 mile from ocean)	42.0	0.50	84.0
32	23	Brazos River, TX	45.4	0.81	56.0
33	40	Pilsey Island, England	50.0	2.50	20.0
34	42	London (Straford), England	54.3	3.06	17.8
35	43	Halifax (Federal Building), NS, Canada	55.3	3.27	17.0
36	38	Cape Kennedy, FL (60 yards from ocean, 60 ft elevation)	64.0	1.94	33.0
37	26	Kure Beach, NC (800-ft lot)	71.0	0.89	80.0
38	36	Cape Kennedy, FL (60 yards from ocean, 30 ft elevation)	80.2	1.77	45.5
39	25	Daytona Beach, FL	144.0	0.88	164.0
40	44	Widness, England	174.0	4.48	39.0
41	37	Cape Kennedy, FL (60 yards from ocean, ground level)	215.0	1.83	117.0
42	34	Dungeness, England	238.0	1.60	148.0
43	17	Point Reyes, CA	244.0	0.67	364.0
44	41	Kure Beach, NC (80 ft lot)	260.0	2.80	93.0
45	45	Galeta Point Beach, Panama, CZ	336.0	6.80	49.4

[a]Specimen size 150×100 mm (6×4 in.). Copyright American Society for Testing and Materials (ASTM). Reprinted with permission.

the corrosion ratio and the more benefit to use zinc coating for corrosion protection of steel in that environment.

Table 62.3 is a comparison of the corrosion rate of zinc with other common commercial metals in various atmospheres. Zinc has a higher corrosion resistance than iron and cadmium in all atmospheres, higher than copper in industrial atmospheres, and higher than tin and magnesium in marine and rural atmospheres.

The two most important atmospheric factors on the corrosion rate of zinc are time of wetness and level of air

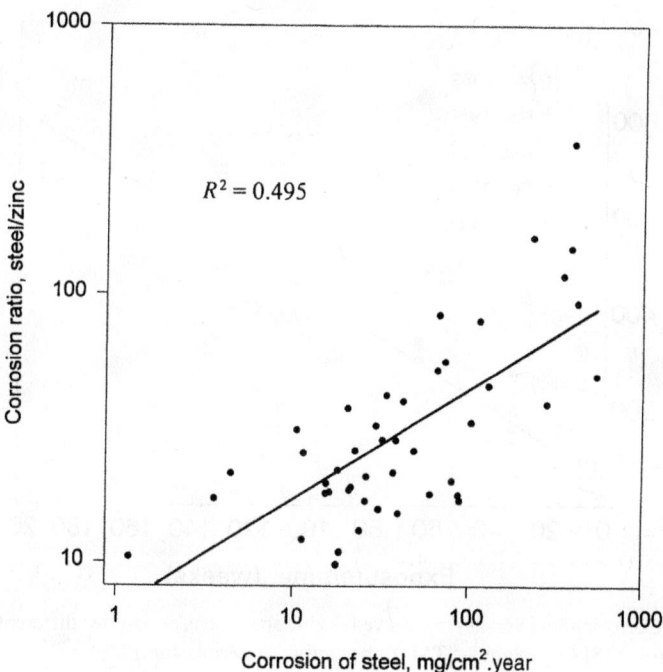

FIGURE 62.2. Corrosion ratio of steel to zinc as function of corrosion rate of steel in atmospheric environments.

pollutants. Time of wetness is determined by relative humidity, temperature, and amount of rain. The corrosion of zinc is negligible when the relative humidity is low but is significant when the surface is wet at a high relative humidity.

Except for the initial few years, the corrosion loss of zinc is generally observed to be almost linear with respect to time, as shown in Figure 62.3 [8]. The variations in the samples exposed at different times of the year are related to seasonal effects that cause the variation in time of wetness and the amount of air pollutants. For a given atmosphere, the yearly averaged corrosion rate may vary because atmospheric conditions, such as the amount of rain or pollution level, change from year to year.

TABLE 62.3. Comparison of Typical Corrosion Rate between Zinc and Other Common Commercial Metals[a]

Metal	Industrial	Marine	Rural
Zn	1	1	1
Cd	2	2	2.4
Sn	0.23	1.6	1.9
Al	0.13	0.3	0.09
Cu	2.4	0.72	0.38
Pb	0.07	0.3	0.28
Ni		0.6	1.1
Sb	0.06		
Mg	0.31	1.8	1.9
Fe	30	50	15

[a]See [2]. Copyright Plenum Press. Reproduced with permission.

The most corrosive pollutant in air is sulfur dioxide, which, in combination with time of wetness, causes abnormally high corrosion rates of zinc [8, 9]. The corrosion rate has been observed to increase roughly linearly with SO_2 concentration [9]. Other air pollutants, such as NO_x, have a relatively insignificant effect on the corrosion of zinc largely due to the much lower content of these species in the air [10–12]. The level of pollution in many developed countries has been considerably reduced over the years because of environmental awareness and regulations; similar trends have generally been found for corrosion rates of many metals. The corrosion rate of zinc was found to be lower in the 1980s than in the 1960s and 1970s [11].

Near the seacoast, the major pollutants are chloride salts. The corrosion rate increases when zinc is exposed closer to seawater. Figure 62.4 shows that the amount of corrosion decreases with distance from the seashore because the salt content in air drops significantly with distance from the shore [13].

The values listed in Table 62.2 and those obtained in many other field-testing programs account for the macroscopic effect of atmospheric environments, namely, the factors determined by the general climate and pollution conditions of a geographic area. Many other factors, namely, the microscopic ones, such as distance from the ground, orientation of the samples, wind or rain shielding, distance to local contaminant sources, and so on, may also significantly affect the corrosion rate. For example, the corrosion rate of zinc

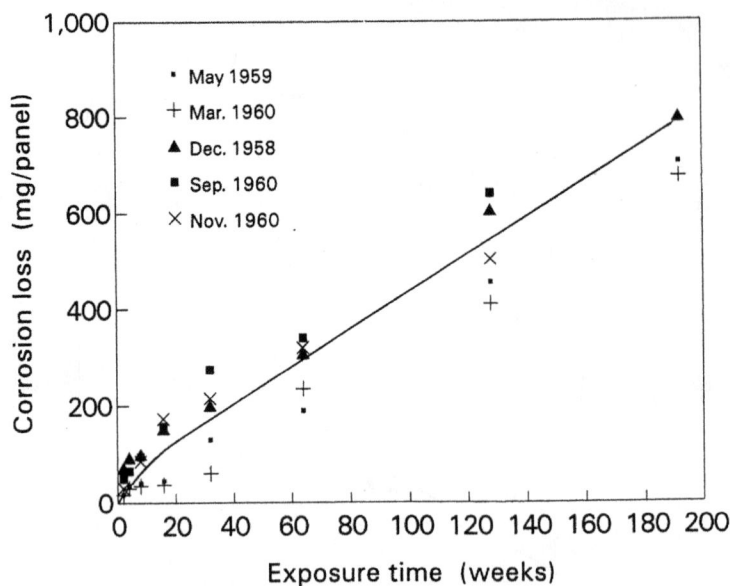

FIGURE 62.3. Corrosion loss versus time curve for zinc specimens exposed at different times of the year. (After Guttman [8] Copyright ASTM, Reprinted with permission.)

measured at 26 sites in one rural area in Spain varied from 0.6 to 3.8 μm/year [14].

Other climatic factors, such as wind and radiation, may also affect condensation and rate of drying as well as the amount of contaminants and corrosion products retained on the surface. The initial climatic conditions at the time of exposure exert marked effects on the corrosion of zinc. Long-lasting rainfall or a relative humidity at or near 100% during the first days tends to cause a higher corrosion rate [2].

The size, shape, and orientation of test samples may affect the corrosion rate of zinc considerably. The corrosion rate is higher on the skyward surface than on the groundward surface, even though the wetting time is longer on the groundward surface. This may be attributed to the effect of rain and larger amount of retention of pollutants on the skyward surface [15].

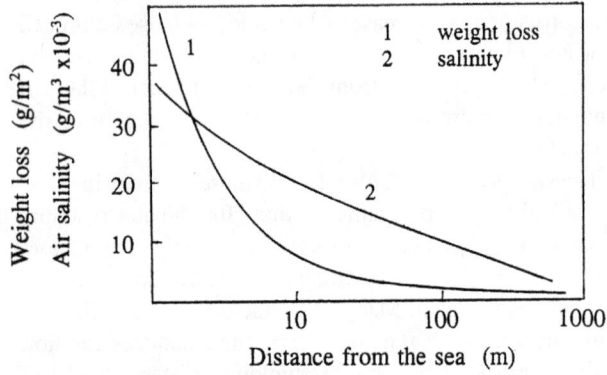

FIGURE 62.4. Corrosion and air salinity as a function of distance from the seashore [13].

The highway environment, experienced by automobiles and highway structures, is particularly aggressive due to the high pollution level from gas exhaustion and, in the winter time, from the deicing salts. The corrosion rate of the zinc coatings in an under-vehicle environment is found to be ~8.5 μm/year, which is comparable to the corrosion rate in a relatively severe marine atmospheric environment [16].

In an indoor atmosphere, the corrosion rate of zinc is very low, typically <0.1 μm/year. Generally, a visible tarnish film forms slowly, starting at spots where dust particles have fallen on the surface. Over a period of time, such films grow gradually until the surface has lost much of its original luster. The appearance and the degree of corrosive attack are related to the relative humidity. Relative humidity of up to ~70% has little influence on corrosion. Above 70%, corrosion activity may occur because moisture precipitates on the surface, especially on that covered with zinc corrosion products and contaminants.

Many zinc compounds can form in each type of atmosphere. However, for a specific atmosphere, only certain compounds dominate. Generally, among the zinc compounds, oxide, hydroxides, and carbonates are most often found in corrosion products [10]. Zinc sulfate ($ZnSO_4 \cdot nH_2O$) and basic zinc sulfate [$Zn_4SO_4(OH_6) \cdot nH_2O$] are also frequently found [10, 17]. In coastal areas, zinc hydroxy chloride [$Zn_5(OH)_8Cl_2 \cdot H_2O$] is also a major compound [17].

The formation sequences of the major zinc compounds found in the corrosion products in four typical types of atmospheres are summarized in Figure 62.5. Initially, the zinc surface is covered quickly with zinc hydroxide. which is gradually converted into zinc carbonate. Within 1 month of

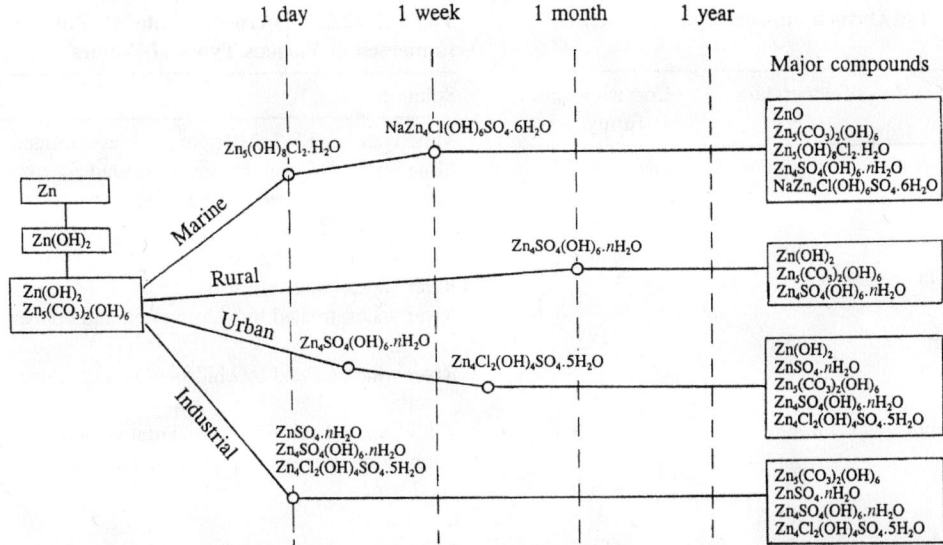

FIGURE 62.5. Formation sequence of the major zinc compounds found in the corrosion products formed in four different types of atmospheres under a sheltered condition. The circles below the compounds indicate the earliest detection of the compounds in the corrosion products. (From [2]. Copyright Plenum Press. Reproduce with permission.)

exposure, almost all major zinc compounds can be detected in the corrosion products. In the more severe atmospheres, such as marine and industrial atmospheres, the formation of chloride and sulfate compounds can be very rapid, occurring within 1 day. As corrosion continues, the various zinc compounds generally increase in quantity but may also disappear due to transformation into different compounds, depending on the specific environmental factors [2].

The corrosion products formed initially are loosely attached to the surface but gradually become more adherent and more dense, resulting from the wetting and drying cycles of the weathering process. After the formation of this corrosion product layer, further corrosion can proceed only within the pores where the zinc surface is not sealed by the corrosion product, while the rest of the surface area, which is sealed by the corrosion products, is protected from corrosion. This process is dynamic. With time, some pores become sealed by newly formed corrosion product while new active pores are opened due to dissolution of the corrosion product. This corrosion mechanism is simplistically illustrated in Figure 62.6 According to this

mechanism, the low corrosion rate observed in atmospheric environments can be described by the following equation:

$$R = ra/A$$

in which R is the observed corrosion rate averaged over the entire surface, r is the actual corrosion rate on active zinc surface unsealed by corrosion products, a is the area of active zinc surface, and A is the area of the entire surface. Because a is very small compared to A, the observed rate R is very low even though the actual corrosion rate r within the pores may be much higher. Thus, the more compact the corrosion product layer, as determined by the corrosion environment, the smaller the active surface area within the pores and the smaller the observed corrosion rate.

D2. Waters and Aqueous Solutions

The typical corrosion rate of zinc in distilled water varies over a wide range between 15 and 150 μm/year [2]. It depends strongly on the amount of dissolved oxygen and carbon dioxide, as shown in Table 62.4 [18]. Figure 62.7 shows that the corrosion rate of zinc in distilled water increases only slightly with temperature up to ~50°C, then increases quickly with temperature, reaching a maximum at ~ 65°C, before decreasing [19]. According to Cox [18], the sharp increase in corrosion rate from 50 to 60°C may be attributed mainly to an abrupt change in the nature of the corrosion products from a protective to a nonprotective state.

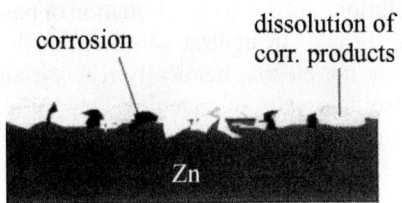

FIGURE 62.6. Schematic illustration of the corrosion process in atmospheric environment.

886   ZINC

TABLE 62.4. Effect of Oxygen on Corrosion of Zinc in Distilled Water[a]

Test Condition[b]	Temperature (°C)	Corrosion Rate[c] (µm/year)
Boiled distilled water; specimens immersed in sealed flasks	Room	25.4
Boiled distilled water; specimens immersed in sealed flasks	40	48.3
Boiled distilled water; specimens immersed in sealed flasks	65	83.8
Oxygen bubbled slowly through the water	Room	218.4
Oxygen bubbled slowly through the water	40	348.0
Oxygen bubbled slowly through the water	65	315.0

[a]See [18].
[b]High-grade zinc specimens, in duplicate, immersed for 7 days.
[c]The corrosion rate was calculated after removal of corrosion products.

TABLE 62.5. Corrosion Rate of Zinc and Zinc Coatings Immersed in Various Types of Waters[a]

Solution	R (µm/year)
Mine water, pH 8.3, 110 ppm hardness, aerated	31
Mine water, 160 ppm hardness, aerated	30
Mine water, 110 ppm hardness, aerated	46
Demineralized water	137
River water, moderate soft	97
River water, moderate soft	61[b]
River water, treated by chlorination and copper sulfate	81
River water, treated by chlorination and copper sulfate	64[b]
Tap water, pH 5.6, 170 ppm hardness, aerated	142
Spray cooling water, chromate treated, aerated	15
Hard water	16
Soft water	15
Seawater in Pacific Ocean	70[c]
Seawater in Bristol Channel	64[d]
Seawater at Eastport, ME	28[c]
Seawater at Kure Beach, NC	28[c]
Seawater at Panama	21[d]

[a]See [2]. Copyright Plenum Press. Reproduced with permission.
[b]Galvanized steel.
[c]One year.
[d]Four years.

In distilled water at room temperature and open to air, zinc corrodes with the formation of pits. The formation of these pits depends on the oxygen content of the water. When the water is depleted of oxygen, there is little corrosion, and when oxygen pressure is high, the corrosion is of a uniform type [2].

As shown in Table 62.5 the corrosion rate in different hard waters can vary significantly. In general, the corrosion rate of zinc is lower in hard water than in soft water or distilled water. Flowing water causes more corrosion than still water.

The presence of trace amounts of copper in water can substantially increase the corrosion of zinc. As little as 0.1 ppm of copper causes a definite increase in corrosion rate [20]. With concentrations of copper of up to ~0.3 ppm, the corrosion rate is proportional to the concentration of copper. The copper appears to deposit small metallic particles on the surface of the zinc. Enhanced corrosion occurs because of the larger cathodic activity generated by the copper particles.

The corrosion rate of zinc in seawater is typically between 20 and 70 µm/year, varying with location, length of exposure, type of zinc, and so on, as shown in Table 62.5. It is generally much higher at the beginning of exposure and decreases with time.

The corrosion processes of zinc in solutions are greatly influenced by the nature of the anions present. Table 62.6 lists the corrosion rates of zinc in solutions of different compositions. The particularly low values in phosphate and chromate solutions are due to the formation of passive films on the zinc surface. In neutral solutions, with chemical agents that are not electrochemically reactive and that do not form insoluble salts or complex ions with zinc, the corrosion rate of zinc is not very different from that in distilled water.

In the absence of reducing or passivating agents, the corrosion of zinc in aqueous solutions is determined primarily by the pH of the solutions. The results in Figure 62.8 show

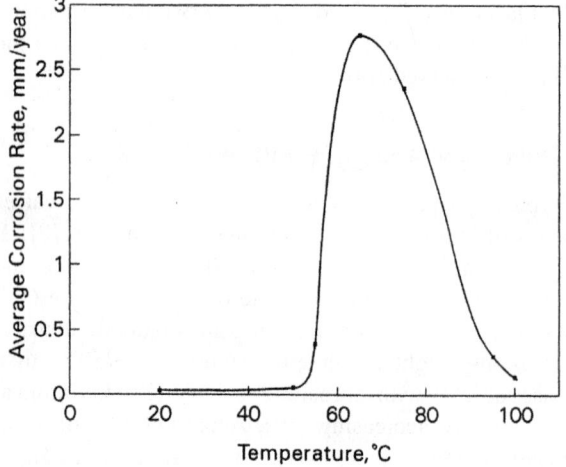

FIGURE 62.7. Corrosion of zinc in air-saturated water as a function of temperature; the test samples were rotated at a speed of 56 rpm. (After Cox [18].)

TABLE 62.6. Corrosion Rate in Different Solutions in Neutral pH Range[a]

Solution	Duration	R (μm/year)
Distilled water	4 weeks	46
Distilled water	1 month	55
0.1 N benzoate	4 weeks	59
0.1 N NaCl	4 weeks	62
0.1 N Na$_3$PO$_4$	4 weeks	1.8
0.1 N Na$_2$ClO$_4$	4 weeks	0.4
5 g/L NaCl	2 months	90
5 g/L KCl	2 months	92
5 g/L NaNO$_3$	6 months	18
5 g/L K$_2$SO$_4$	2 months	52
8% Na$_2$SO$_4$·10H$_2$O	1 month	83
5% NaCl	3 weeks	89
3% Na$_2$SO$_4$	1 day	144

[a]See [2]. Copyright Plenum Press. Reproduced with permission.

that the corrosion rate of zinc in water of pH 6–12 is relatively low [21]. At pH values <6 or >12, the corrosion rate increases substantially. The low corrosion rate at pH values between 6 and 12 is primarily due to the formation of protective corrosion products on the surface of the zinc. According to Roetheli et al. [21], the decrease in corrosion rate at pH near 14, shown in the figure, is due to a decrease in the solubility of oxygen in strongly alkaline solutions.

D3. Soils

The corrosion rate of zinc varies drastically from soil to soil, as shown in Table 62.7, because the chemical and physical properties of soil may vary over extremely wide ranges. For example, the pH of soil may vary from as low as 2.6 to

as high as 10.2, and the resistivity from several tens of ohms to near 100 kΩ [22]. Also, since soil is a highly inhomogeneous environment, both microscopically (e.g., at the dimension of a clay particle) and macroscopically (e.g., at the dimension of a rock), the corrosion in soil is seldom uniform across the metal surface.

The factors that may affect the corrosion of zinc and galvanized steel in soils are numerous, but the correlation between the corrosion behavior and the various factors is, in general, rather poor. The corrosion rates tend to be lower in soils with very high resistivity. There is little correlation between corrosion rate and soil pH, which is an indication of the very complex nature of soil corrosion [2].

Romanoff [22] noted that poorly and very aerated soils are more corrosive to zinc. Soils of fair to good aeration but containing high concentrations of chlorides and sulfates tend to induce deep pitting. Muddy clay and peat (as compared to sand) are, in general, more corrosive to zinc.

D4. Painted Products

The corrosion of painted metals, particularly automotive bodies, is generally characterized as perforation corrosion or cosmetic corrosion. The term "cosmetic corrosion" is applied to an attack that is initiated at the exterior surface, usually at regions where the paint is damaged. Cosmetic corrosion is usually related to (1) red rust, (2) paint creep, and (3) chipping. Corrosion of a painted steel sheet that initiates at an interior surface of a car body panel, penetrates the sheet, and eventually shows through as rust at the exterior exposed surface is known as perforation corrosion [23]. It often occurs at locations which are difficult to clean, phosphate, and coat, such as lapped parts and hem flanges, or at crevices which collect dirt, salt, and moisture.

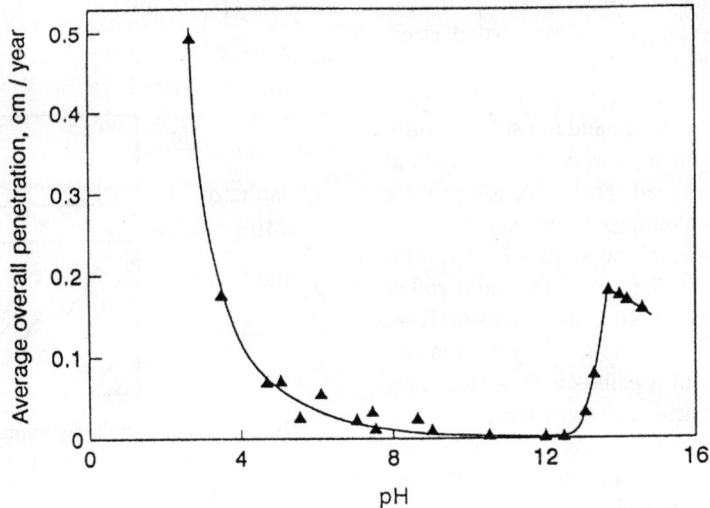

FIGURE 62.8. Corrosion rate in ditilled water as a function of pH (addition of NaOH or HCl for pH adjustment). (After Roetheli [21].)

TABLE 62.7.   Corrosion Rates of Zinc Coating on Steel in Soils of Different Geographic Locations in the United States[a]*

No.[a,b]	Soil Type	ρ (Ω-cm)	pH	R (μm/year)
1	Allis silt loam—Cleveland, OH	1,215	7.0	11.8
2	Bell clay—Dallas, TX	684	7.3	1.5
3	Cecil clay loam—Atlanta, GA	30,000	5.2	1.7
4	Chester loam—Jenkintown, PA	6,670	5.6	7.9
5	Dublin clay adobe—Oakland, CA	1,345	7.0	7.7
6	Everett gravelly sandy loam—Seattle, WA	45,100	5.9	0.5
7	Maddox silt loam—Cincinnati, OH	2,120	4.4	10.8
8	Fargo clay loam—Fargo, ND	350	7.6	3.2
9	Genesee silt loam—Sidney, OH	2,820	6.8	5.0
10	Gloucester sandy loam—Middleboro, MA	7,460	6.6	5.2
11	Hagerstown loam—Loch Raven, MD	11,000	5.3	3.7
12	Hanford fine sandy loam—Los Angeles, CA	3,190	7.1	2.2[c]
13	Hanford very fine sandy loam—Bakersfield, CA	290	9.5	3.7
14	Hempstead silt loam—St. Paul, MN	3,520	6.2	1.1
15	Houston black clay—San Antonio, TX	489	7.5	1.5
16	Kalmia fine sandy loam—Mobile, AL	8,290	4.4	4.2
17	Keyport loam—Alexandria, VA	5,980	4.5	14.8[d]
19	Lindley silt loam—Des Moines, IA	1,970	4.6	2.9
20	Mahoning silt loam—Cleveland, OH	2,870	7.5	4.9

*Average coating thickness, 121 μm.
[a]See [2, 22].
[b]Original soil identification.
[c]Sheet specimens.
[d]Coating corroded completely and the data included the corrosion of steel.

Cosmetic corrosion is most commonly evaluated by measuring the length of underpaint creeping. It is sometimes evaluated also by the extent of rust formation or paint loss. Many factors, such as coating composition, coating thickness, surface treatment, test condition, type of paint damage, and type of paint, can affect the cosmetic corrosion of painted steels. Data from a field survey has shown that, in general, steel panels coated with zinc or zinc alloy coating have much slower uderpaint creeping than cold-rolled steel [24]. The creeping resistance of zinc– and zinc alloy–coated steels increases with coating weight.

The mechanisms of underpaint corrosion for zinc- and zinc alloy–coated steels are complex and are still not fully understood [2]. In general, underpaint corrosion starts at places where the paint is damaged. The corrosion process begins with corrosion of the coating or with paint delamination followed by corrosion of the substrate and, with time, leads to perforation of the steel. For cold-rolled steel, the corrosion products built up at the corrosion front may mechanically delaminate the paint. Delamination can occur at different interfaces in a paint–coating–steel system, depending on the material and environmental conditions, as schematically shown in Figure 62.9. The causes of the delamination at the corrosion front, as reported by different investigators, can be physical, anodic, cathodic, mechanical, or combinations thereof. The predominant cause in a specific corrosion situation can be due to

variations in paints, coatings, phosphates, and test conditions [23–25].

D5. Concrete

The use of galvanized steel rebar as concrete reinforcement has been one of the remedies to alleviate the corrosion problems of steel rebar in concrete caused by water and chloride permeation into the concrete [2]. When galvanized

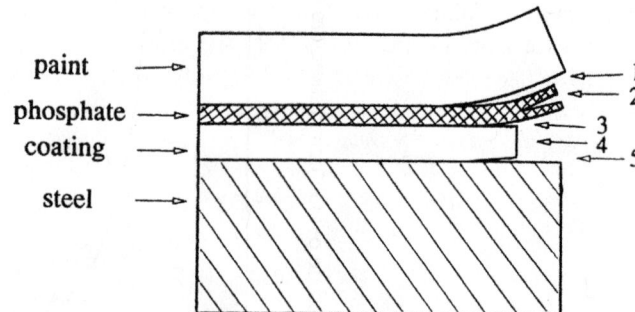

FIGURE 62.9. Possible delamination modes of a painted coated steel: (1) at paint/phosphate interface due to loss of adhesion; (2) within phosphate layer due to mechanical fracture; (3) at phosphate/coating interface due to dissolution of phosphate; (4) dissolution of coating; (5) at coating/steel interface due to mechanical failure. (From [2]. Copyright Plenum Press. Reproduced with permission.)

TABLE 62.8.   Average Corrosion Rate of Zinc Coating Inside Concrete in Marine Environments[a]

Structure	Age (years)	Cover (cm)	Sample Location	Cl⁻ wt % (kg/m³)	R (μm/year)
Longbird bridge	23	10	Above HT[b]	0.19 (4.4)	0.2
St. George dock	12	8	Above HT	0.27 (6.4)	1.1
	10	6	Above HT	0.22 (4.6)	0.5
	7	13	Above HT	0.14 (3.0)	0.29
Hamilton dock	10	5	Near MT	0.08 (1.9)	0.5
	10	6	Above HT	0.14 (3.6)	0.8
Bermuda yacht club	8	7	Below HT	0.16 (3.7)	0.0

[a]See [26].
[b]HT = high tide, MT = mean tide.

steel is covered with a good-quality concrete free of chloride, the corrosion rate is very low [26, 27]. The corrosion rate is higher when concrete contains a high level of chloride salts. According to a field investigation on galvanized-steel-reinforced concrete structures in different marine locations, the corrosion rates of galvanized rebar are, in most cases, <0.5 μm/year, as can be seen in Table 62.8.

There are large amounts of published data on the corrosion performance of galvanized rebar in concrete [2]. Most of these data are obtained from laboratory-simulated testing environments. The data from natural exposure tests, the only reliable information for predicting the life of galvanized reinforced concrete structures, are rather limited. Particularly, there is a lack of data from heavy deicing salt environments. The available field data generally suggest that galvanized steel reinforcement provides longer life compared to black steel; however, this is sometimes in disagreement with the data from the studies using simulated test conditions [28].

D6. Other Environments

Corrosion is also an important issue in some specific applications (e.g., batteries). The corrosion of the zinc electrode in zinc cells and batteries is the main cause for self-discharge, short shelf life, and hydrogen buildup. The corrosion of zinc in a battery environment is extremely complicated because it involves a large number of factors. These factors can be classified into three main groups: (1) electrolyte (composition and physical setting), (2) zinc electrode (form and composition), and (3) operating conditions (temperature, time, current collector, composition of cathode material, rate and depth of discharge, sealing method, and cell geometry) [2].

The corrosion rate of zinc (in some organic solvents) can be much higher than in water while it can be much lower in others. In general, viscosity is an important factor in controlling the corrosion rate of zinc in many organic solvents. Corrosion rate varies only slightly with the molecular weight of the solvent and dielectric constant. For a given solvent, the corrosivity significantly varies with the presence of other chemical agents.

Corrosion in gaseous environments is governed by principles similar to those for atmospheric corrosion, although it also has its own special characteristics. As in normal atmospheres, the amount of moisture plays an important role in gaseous corrosion. Depending on the kind of gases and their concentrations, the critical relative humidity required for corrosion may vary. Also, depending on whether an electrolyte is formed, the corrosion can be electrochemical or chemical in nature.

E. CORROSION FORMS

The corrosion forms which commonly occur on zinc are general corrosion, galvanic corrosion, and wet storage stain [2]. Pitting corrosion and intergranular corrosion are less common. The occurrence of each form of corrosion depends on the specific materials/environmental conditions. The most common form of corrosion encountered by zinc products in various environments, such as atmospheres, soils, and waters, is general corrosion (i.e., uniform corrosion).

E1. Galvanic Corrosion

Galvanic corrosion is a particularly important corrosion form for zinc applications, whether used as a coating, an anode, or a zinc-dust paint [2]. In most situations, unlike many other metals, galvanic corrosion is desirable for zinc because it is required for galvanically protecting the coupled metal, usually steel.

Data in Table 62.9 show the galvanic corrosion rate of zinc coupled to various metals in different atmospheres. Depending on the connected metal and the type of atmosphere, the galvanic corrosion can be as much as five times the normal corrosion of zinc in a rural atmosphere and three times that in a marine atmosphere [29]. Among the metals, mild steel acts as the most efficient cathodic material, largely owing to the voluminous rust that can absorb pollutants and retain moisture and, thus. give rise to an aggressive electrolyte of good conductivity.

TABLE 62.9. Galvanic Corrosion Rate of Zinc Coupled to Other Common Commercial Metals in Different Atmospheres (μm/year)[a]

Coupled Alloy	Rural	Urban	Marine
Zinc freely exposed	0.5	2.4	1.3
Mild steel	3.0	3.3	3.9
Stainless steel	1.1	1.8	2.0
Copper	2.2	2.0	3.2
Lead	1.6	2.4	3.4
Nickel	1.5	1.9	2.8
Aluminum	0.4	1.1	1.1
Anode aluminum	0.9	1.9	1.0
Tin	1.0	2.6	2.4
Chromium	0.7	1.4	1.9
Magnesium	0.02	0.04	1.1

[a]See [30].

Detailed information regarding the principles and factors of galvanic action can be found in Chapter 10 [12]. Also, specific information on galvanic corrosion and protection of zinc/steel couples in various environments is given elsewhere [2].

E2. Pitting Corrosion

Pitting is not a common corrosion form in zinc applications [2]. In atmospheric environments. pitting corrosion has been seldom reported as the main cause of failure of zinc products. In soil, pitting may result from the nonuniform nature of this environment, and the extent varies significantly depending on the soil chemical composition and texture [22].

Pitting may occur in distilled water under an immersed condition. When zinc panels are placed vertically in distilled water, corrosion pits form, often arranged in straight rows of unconnected pits. The local depletion of oxygen was found to be necessary for pitting corrosion. Moving zinc samples in distilled water saturated with oxygen shows no pitting, Also, when distilled water contains very small amounts of dissolved salts, general corrosion, rather than pitting corrosion, occurs.

Pitting is a rather common form of corrosion of zinc in hot water and can be a serious problem for galvanized steel hot water tanks. In hot soft water, pitting corrosion is likely to lead to rapid penetration of galvanized coatings because of the reversal of polarity for zinc/steel galvanic couples in hot water. In hard water, the corrosion is likely to be stifled by the deposition of a protective scale, depending on the heating method. The presence of copper in the water was found to enhance the pitting corrosion of galvanized coatings in hot water [20].

E3. Intergranular Corrosion

Intergranular corrosion, first reported in the first quarter of the twentieth century, has been observed to occur on zinc alloys in different environments: atmosphere, water, solution, and concrete.

Zinc of high purity is not susceptible to intergranular corrosion [3]. The presence of other elements, particularly aluminum, as alloying elements or impurities is necessary to cause intergranular corrosion. Table 62.10 shows the intergranular corrosion rates of Zn–A1 alloys of different compositions in steam or hot water. Intergranular corrosion has also been found to occur in zinc alloys containing only lead or magnesium.

For Zn–Al alloys, intergranular is observed to occur in a concentration range between 0.03 50% A1 [31]. Below 0.03% Al, intergranular corrosion does not occur. The presence of impurities is not required for the occurrence of intergranular corrosion on Zn–Al alloys.

Among the environmental factors, temperature is the most significant. Also, alkaline environments are the most aggressive in intergranular corrosion of Zn–Al alloys [31]. Between pH 5 and 10, the corrosion penetration rate is almost constant. Below pH 5, it decreases with decreasing pH. On the other hand, for pH values >10, increases drastically with increasing pH.

The intergranular corrosion of Zn–Al alloys is attributed to the preferential attack on the aluminum-rich phase at the grain boundaries. The solubility of aluminum in zinc at room temperature is ~0.03% A1. For zinc alloys containing an aluminum concentration >0.03%, the aluminum precipitates

TABLE 62.10. Intergranular Penetration Rates of Some Zinc Alloys in Different Environments[a]

%Al	Zn Purity	Environment	Duration (days)	Rate (mm/day)
0.075	99.999%	95°C water vapor	10	0.02
0.1	99.999%	95°C water vapor	10	0.018
4	99.999%	95°C water vapor	10	0.033
20	99.999%	95°C water vapor	10	0.028
21.1		100% RH at 50°C	42	0.002
0.1	99.99%	95°C tap water		0.1

[a]See [2, 31].

at the grain boundaries and is responsible for the increased corrosion rate at the grain boundaries [31].

E4. Wet Storage Stain

"Wet storage stain" is a term used to describe the zinc corrosion products formed on galvanized steel surface during the period of storage and transportation. It is voluminous, white, powdery, and bulky and is formed when closely packed galvanized articles are stored under damp and poorly ventilated conditions. The crevices formed between the articles can attract and absorb moisture and retain the wetness more readily than the surface area exposed to the open air.

The moisture necessary for the formation of wet storage stain may originate in various ways. It may be present on the galvanized parts at the time of stacking or packing as a result of incomplete drying after quenching. It may result from direct exposure to rain or seawater or from condensation caused by atmospheric temperature changes. Close packing can result in moisture being retained by capillary action between the surfaces in contact because drying is delayed by the lack of circulating air. Under sustained wetting, a fluffy "white rust" is formed. Due to this loose nature, it has little barrier effect on the access of solution to the zinc metal and, also prolongs the time of wetness.

White storage stain discolors the galvanized steel surface and, in some situations, can seriously affect the appearance of the galvanized steel articles. However, it is generally not harmful to the long-term corrosion performance afterward [2]. Wet storage stain can be prevented by properly stacking and storing galvanized products under dry and ventilated conditions [32]. In addition, surface treatments can be applied to freshly galvanized articles to enhance corrosion resistance.

Among surface treatment processes, chromating, in different solution compositions and application methods, has been most widely used in the galvanizing industry as an effective surface treatment to prevent wet storage stain from forming during storage or transportation.

REFERENCES

1. S. W. K. Morgan, Zinc and Its Alloys and Compounds, Wiley, New York, 1985.
2. X. G. Zhang, Corrosion and Electrochemistry of Zinc, Plenum, New York, 1996.
3. H. E. Townsend, "Continuous Hot Dip Coatings," in ASM Metals Handbook, Vol. 5, Materials Park, OH, 1994, pp. 339–348.
4. D. Wetzel, "Batch Hot Dip Galvanized coatings" in ASN Metals Handbook, Vol. 5 Materials Park, OH, 1994, pp. 360–371.
5. S. G. Fountoulakis, "Continuous Electrodeposited Coatings for Steel Strip", in ASM Metals Handbook, Vol. 5, Materials Park, OH, 1994, pp. 349–59.
6. E. Mattsson, "The Atmospheric Corrosion Properties of Some Common Structural Metals—A Comparative Study," Mater. Perform. **21**(7), 9–19. 1982.
7. "Corrosiveness of Various Atmospheric Test Sites as Measured by Specimens of Steel and Zinc," in Metal Corrosion in the Atmosphere, STP 435. ASTM, Philadelphia, PA, 1968, pp. 360–391.
8. H. Guttman, "Effects of Atmospheric Factors on the Corrosion of Rolled Zinc," in Metal Corrosion in the Atmosphere, STP 435, ASTM, Philadelphia, PA, 1968, pp. 223–239.
9. F. H. Haynie and J. B. Upham, "Effects of Atmospheric Sulfur Dioxide on the Corrosion of Zinc," Mater. Protect, Perform. **9**, 35–40 (1970).
10. T. E. Graedel, J. Electrochem. Soc., **136**(4), 193 (1989).
11. P. A. Baedecker, E. O. Edney, P. J. Moran, T. C. Simpson, and R. S. Williams, "Effects of Acid Deposition on Materials," State-of-Science/Technology, Report No. 19, National Acid Precipitation Assessment Program, Government Printing Office, Washington, DC, 1990.
12. X. G. Zhang, "Galvanic Corrosion," Chapter 10 in this book.
13. P. W. Brown and L. W. Masters, "Factors Affecting the Corrosion of Metals in the Atmosphere," in Atmospheric Corrosion, W.H. Ailor (Ed.), Wiley, New York, 1982, pp. 31–49.
14. S. Feliu and M. Morcillo, "Atmospheric Corrosion Testing in Spain," in Atmospheric Corrosion, W. H. Ailor (Ed.), Wiley, New York, 1982, pp. 913–921.
15. R. A. Legault, "Atmospheric Corrosion of Galvanized Steel," in Atmospheric Corrosion, W. H. Ailor (Ed.), Wiley, New York, 1982, pp. 607–613.
16. R. J. Neville, "A Test for Undervehicle Corrosion Resistance," in Automotive Corrosion by Deicing Salts, R. Baboian (Ed.), NACE, Houston, TX, 1981, pp. 182–218.
17. I. Odnevall, "Atmospheric Corrosion of Field Exposed Zinc," Ph.D. Thesis, Royal Institute of Technology, Stockholm, 1994.
18. E. A. Anderson and C. E. Reinhard, "Zinc," in The Corrosion Handbook, H. H. Uhlig (Ed.), Wiley, New York, 1948, pp. 331–346.
19. G. L. Cox, "Effect of Temperature on the Corrosion of Zinc," Ind. Eng. Chem. **23**(8) 902–904 (1931).
20. L. Kenworthy, J. Inst. Metals, **69**, 67 (1943).
21. B. E. Roetheli, G. L. Cox, and W. B. Littreal, Metals Alloys, **3**(3), 73 (1932).
22. M. Romanoff, Underground Corrosion, Circular 579, US. National Bureau of Standards, Washington, DC, 1957.
23. H. E. Townsend, "Coated Steel Sleets for Corrosion-Resistant Automobiles," Paper No. 416, NACE Corrosion '91 Conference, Cincinnati, OH, Mar. 11–15, 1991.
24. Y. Ito, K. Hayashi, C. Kato, and Y. Miyoshi, "A Study on Simulated and Accelerated Corrosion Test Methods for Automotive Precoated Steel Sheet," in Proceedings of the International Conference on Zinc and Zinc Alloy Coated

Steel Sheet, GALVATECH '89, Tokyo, Japan, Sept. 5–7, 1989, pp. 503–510.

25. C. R. Shastry and H. E. Townsend, Corrosion, **45**(2), 103 (1989).

26. K. W. J. Treadaway, B. L. Brown, and R. N. Cox, "Durability of Galvanized Steel in Concrete," in Corrosion of Reinforcing Steel in Concrete, STP 713, ASTM, Philadelphia, PA, 1980, pp. 102–131.

27. K. C. Clear, FHWA/RD-82/028, Federal Highway Administration, Washington, DC, Dec. 1981.

28. P. Hronsky, Corrosion, **37**(3), 161 1981.

29. V. Kucera and E. Mattsson, "Atmospheric Corrosion of Bimetallic Structures," in Atmospheric Corrosion, W. H. Ailor (Ed.), Wiley, New York, 1982, pp. 561–574.

30. L. P. Devillers, "The Mechanism of Aqueous Intergranular Corrosion in Zinc-Aluminium Alloys," Ph.D. Thesis, University of Waterloo, Ontario, Canada, 1974.

31. "Wet Storage Stain," Brochure, American Hot Dip Galvanizers Association, Denver, CO, 1984.

32. C. E. Bird and F. J. Strauss, Mater. Perform., **15**(11), 27 (1976).

63

ZIRCONIUM ALLOY CORROSION

B. Cox[†]

Centre for Nuclear Engineering, University of Toronto, Toronto, Ontario, Canada

A. Introduction
B. Corrosion behavior
    B1. Stress corrosion cracking, delayed hydrogen cracking, and liquid metal embrittlement
    B2. Galvanic coupling
C. Nuclear fuel cladding
References

A. INTRODUCTION

Zirconium alloys are the primary structural metal within the fueled region of the cores of nuclear reactors. They are used for fuel cladding, grid spacers, guide tubes, pressure tubes, calandria tubes, and other minor components. The reasons for this choice are the low-neutron-capture cross section of zirconium, which improves the neutron economy of the reactor; the adequate corrosion resistance in high-temperature water (300–350°C); and the satisfactory mechanical properties. However, the last two properties are not ideal for use under severe in-reactor conditions, so that extensive programs for the development of new alloys with better corrosion resistance and mechanical properties are proceeding worldwide. The alloying additions must also have low-neutron-capture cross sections and be used in their minimum effective concentrations, so that the zirconium alloys have developed along very different paths from the analogous titanium alloys where there are no such restrictions on their use. Unalloyed zirconium is not used in the nuclear industry because of its inadequate strength and corrosion resistance. Because of their good resistance to both acids and bases,

[†]Retired

zirconium alloys, are widely used in chemical plants. The current commercial alloys with their American Society for Testing and Materials (ASTM) specifications are given in Table 63.1 and for nuclear-grade alloys in Table 63.2. Note the tighter impurity levels in the latter. A listing of additional developmental alloys for nuclear applications is given in Table 63.3 [1].

Zirconium occurs naturally as a mixed (Zr, Hf) SiO_4 zircon, and for nuclear uses the 2–3% Hf present must be separated to a level of ≤ 100 ppm because of its high-neutron-capture cross section. Thus, zirconium is available as both a nuclear-grade product (without hafnium) and a commercial-grade product (with hafnium). Hafnium is completely miscible with zirconium and in its pure form has better corrosion resistance than a similarly pure zirconium. Thus, the presence or absence of hafnium has no effect on the corrosion resistance, which is controlled by a very stable oxide that is always present on any zirconium surface (except at high temperature in a good vacuum, when it (dissolves in the metal). At room temperature this passive oxide is 2–5 nm thick and grows to greater thicknesses at elevated temperatures (e.g., in the high-temperature water in-reactor).

B. CORROSION BEHAVIOR

This oxide cannot be reduced electrochemically, as can the passive film on iron-based alloys. It is an amphoteric oxide and thus is soluble in concentrated acids and alkalis, especially at high temperatures. Solubility of the protective oxide in alkalis increases in the order $NH_4OH < KOH < NaOH < LiOH$. It is this that determines the ultimate limits of concentration and temperature at which zirconium can be used in these environments. In general corrosion resistance is good, being relatively better in alkalis than in acids.

Uhlig's Corrosion Handbook, Third Edition, Edited by R. Winston Revie
Copyright © 2011 John Wiley & Sons, Inc.

TABLE 63.1. ASTM Chemical Requirements for Three Commercial Grades of Zirconium (B 551-79)

Element	Composition (%)		
	Grade 702	Grade 704	Grade 705
Zr + Hf, min	99.2	97.5	95.5
Hafnium, max	4.5	4.5	4.5
Fe + Cr	0.20 max	0.2–0.4	0.2 max
Hydrogen, max	0.005	0.005	0.005
Nitrogen, max	0.025	0.025	0.025
Carbon, max	0.05	0.05	0.05
Niobium			2.0–3.0
Oxygen	0.16	0.18	0.18
Tin			1.0–2.0

[a]Reprinted with permission from [5]. Copyright ASTM.

TABLE 63.2. Compositional Ranges of Nuclear–Grade Standard Alloys[a]

ASTM Ref. Name:	R60802 Zircaloy-2	R60904 Zircaloy-4	R60901 Zr–Nb	R60904 Zr–Nb
Alloying Elements (mass %)				
Sn	1.2–1.7	1.2–1.7		
Fe	0.07–0.2	0.18–0.24		
Cr	0.05–0.15	0.070–0.13		
Ni	0.03–0.08			
Nb			2.4–2.8	2.5–2.8
O	Usually 0.1–0.4	Usually 0.1–0.14	0.09–0.13	To be specified
Impurities (ppm)				
Al	75	75	75	75
B	0.5	0.5	0.5	0.5
Cd	0.5	0.5	0.5	0.5
C	270	270	270	150
Cr	Alloy el	Alloy el	200	100
Co	20	20	20	20
Cu	50	50	50	50
Hf	100	100	100	50
H	25	25	25	25
Fe	Alloy el	Alloy el	1500	650
Mg	20	20	20	20
Mn	50	50	50	50
Mo	50	50	50	50
Ni	Alloy el	70	70	35
N	80	80	80	65
Pb				50
Si	120	120	120	120
Sn	Alloy el	Alloy el	50	100
Ta				100
Ti	50	50	50	50
U	3.5	3.5	3.5	3.5
V				50
W	100	100	100	100

[a]See [1].

Unfortunately, many of the common impurities and alloying additions have very low solubilities in α-Zr (the hexagonal phase is stable below $\sim860°C$ (1133 K) when it transforms to the body-centered-cubic (bcc) β-Zr phase). Typical among these impurities are Fe, Cr, and Ni, which have solubilities of ~100 ppm (wt) in unalloyed zirconium at $<600°C$. Iron, in particular, always exceeds this impurity level in unalloyed zirconium produced commercially by the Kroll process and sometimes exceeds this level even in high-purity Zr prepared by the van Arkel process if care is not taken to avoid contamination from the iron vessel in which the production is carried out. Because Fe precipitates as a Zr_3Fe phase in such instances, even a small excess of Fe over the solubility at the final annealing temperature can result in a significant volume fraction of second phase.

The presence of iron intermetallics in the surface of the material provides weak spots in the protective oxide film. Iron has only a low solubility in monoclinic ZrO_2 (the primary phase in the protective oxide) and the iron from the intermetallic therefore tends to form crystallites of an iron oxide [2], which is much more soluble than ZrO_2 in most environments. Thus, chemical attack on the oxide on the intermetallic provides a pit nucleation site. The ZrO_2 film then is undermined rather than dissolved. It may remain bridging the pits when they are small [3, 4], whether the initiation was due to oxide cracking [3] or attack at impurities [4]. However, it eventually breaks up, except in the case of thick thermally formed oxide present on the initial surface. This may remain for some time even when undermined by pitting.

Because transition metal impurities are much more soluble in the β-Zr phase than in the α-Zr, heat treatments in the β phase where the impurities are in solution can be used to control the intermetallic size and distribution. This has been an important factor in the development of nuclear fuel cladding with good corrosion resistance [1]. In this instance, a β quench is usually the first stage in such a treatment. A slow cool from the β phase, however, sweeps all second-phase particles to the Widmanstatten platelet boundaries, where they may form an almost continuous alignment of particles (stringers). This can happen typically at welds, where the mass of the material prevents rapid cooling. Severe pitting of such welds, under conditions where the oxide on the metal matrix is protective, is one common result [5]. A recrystallization heat treatment in the high α-Zr temperature range will usually redistribute the second-phase particles.

B1. Stress Corrosion Cracking, Delayed Hydrogen Cracking, and Liquid Metal Embrittlement

Zirconium alloys have good stress corrosion cracking (SCC) resistance but are susceptible to SCC in many environments (Table 63.4) [6] so that all weldments, even if rapidly cooled, should be heat treated to lower the residual stress at the weld.

TABLE 63.3.   Experimental and Commercial Alloys for Nuclear Applications[a]

Alloy	Typical Weight %								Commercial Application
	Sn	Nb	Fe	Cr	Ni	V	O	Others	
Zr-1	2.5								
Zr-2	1.5		0.13	0.10	0.05		0.11		BWR Clad & structures
Zr-3	0.25		0.25						
Zr-4	1.5		0.21	0.10			0.13		PWR Clad & structures
ZIRLO	1.0	1.0	0.1						
NSF 0.5 (GE)	1.0	1.0	0.5				0.10		
NSF 0.2	1.0	1.0	0.20				0.10		
Valloy			0.15	1.2					
Ozhennite-0.5	0.2	0.1	0.1		0.1				
Scanuk	0.06	0.6	0.04	0.32				0.22 Mo	
Zr-3B	0.5		0.4						
Zr-3C	0.5		0.2		0.2				
Zr-3A	0.25		0.25						
Excel	3.5	0.8						0.8 Mo	
E 110 (Zrl Nb)		1							VVER Clad
Zr2.5 Nb		2.5					0.12		CANDU press. tubes
E635	1.2	1	0.4						
E125		2.5							

[a]See [1].

TABLE 63.4.   Environments Causing Cracking of Titanium and Zirconium Alloys[a,b]

Environment	Temperature (°C)[c]	SCC in Ti	SCC in Zr Alloys		
			Yes/No	Cracking Mode[d]	First Observed References
Hydrogen					
Internal hydrogen	RT–350	Yes	Yes	TG	17
Hydrogen gas	RT–150	Yes	Yes	TG	23
Organic liquids					
Methanol and solutions with I₂, Br₂, NaCl, HCl	RT	Yes	Yes	IG initiation TG propagation	7,9
Higher alcohols with similar additives	RT	Yes	Yes	Mixed IG/TG	9
Iodine solutions in other organic liquids	300–800	Yes	NT		
Chlorinated hydrocarbons (e.g., CCl_4, $CHCl_3$)	RT	Yes	Yes	IG	24
Other halogenated hydrocarbons (e.g., Freons, CH_3I)	RT	Yes	Yes	TG (DCB)	24
Ethylene glycol	RT	Yes	Only in I_2 solution	IG	24
Aqueous solutions					
HCl	25–150	Yes	Yes	IG	12
NaCl, KCl, etc	25–290	Yes	Yes needs polarization	Mixed IG/TG	25
$FeCl_3$, $CuCl_2$	RT	Yes	Yes		10, 11
Hot and fused halides					
NaCl, KCl, CsCl, etc	300–450	Yes	Yes	Mixed IG/TG	7
$KNO_3/NaNO_3$ + KBr/CsI	300–400	Yes	Yes	Mixed IG/TG	26
$KNO_3/NaNO_3$ + NaCl	300–400	Yes	Yes	IG	27

(continued)

TABLE 63.4. (*Continued*)

Environment	Temperature (°C)[c]	SCC in Ti	SCC in Zr Alloys		
			Yes/No	Cracking Mode[d]	First Observed References
Halogen and halogen acid vapors	RT	NT	Yes	TG/(DCB)	24
	250–500	Yes	Yes	Mixed IT/TG	13
Metals (solid, liquid, vapor)					
Hg, Cd, Cs, Zr	RT–400	Yes	Yes	TG	7
Mg	700	NT	Yes	IG	80
Oxides and oxy acids of N_2					
N_2O_4	RT	Yes	NT		
Nitric acid	RT–100	Yes	Yes	T	28, 29
Sulfuric acid	RT	Yes	NT		

[a]A more detailed description of the various environments causing cracking is given in [56].
[b]Reprinted with permission from [6]. Copyright 1990 Elsevier Science. For references see original table in [6].
[c]RT = room temperture.
[d]DCB indicates fractures only in precracked double-cantilever-beam specimens.

Unalloyed Zr, because of its low yield strength, is more SCC resistant. Thus, SCC resistance becomes more important for the higher strength zirconium alloys, such as Zr–2.5% Nb, where in addition to SCC a delayed hydride cracking (DHC) process can occur at highly stressed locations [7–9]. At ambient temperatures even the as-received hydrogen content (which invariably exceeds the room temperature solubility of hydrogen of <1 (parts per million by weight (ppmwt)) is able to cause such cracking. Typical velocities for SCC and DHC cracks are shown in Figure 63.1 [6].

Zirconium alloys are also susceptible to liquid metal embrittlement (LME) in a number of metals (Table 63.5) [10]. Contrary to many other LME phenomena, cracking of zirconium alloys is entirely transgranular in most instances, only in Ga and Cd is the more common intergranular form of cracking observed.

In many chemical plant situations the environment does not fall clearly into a single well-defined category. Mixtures of organic chemicals with inorganic acids and variable valence ions (e.g., Fe, Cu) or halogen (halide) impurities can make it very difficult to determine the primary cause of a corrosion problem. For example, stress cracking in halogens dissolved in organic solutes is very well known. However, where the organic solute is miscible with water in large proportions (e.g., the lower alcohols), a water content higher than a few percent is usually enough to maintain the passivity of the zirconium alloy and prevent SCC [11]. If the system is mixed with sulfuric acid at >50% aqueous content, then occurrence of SCC would not be expected, yet cracking has been observed in such cases. One possible explanation is that acids (such as sulfuric acid) may sequester significant amounts of water so that there is insufficient free water

TABLE 63.5. Metals Known to Cause at Least Some Embrittlement of Zr[a]

Embrittling Element	Zr Alloy	Type of Test[b]	References
Li[c]	Zircaloy-2	ZrB alloy irrad. + UT	24
Cs	Zircaloy-2	SIMFEX, bend UT, DCB	3, 6, 9, 10, 25
Cd	Zircaloy-2	SIMFEX, UT	10, 11, 25
Cs/Cd	Zircaloy-2	SIMFEX, UT, C Tube	10, 25
Cs/Ca	Zircaloy-2	UT	10
Cs/Sr	Zircaloy-2	UT	10
Cs/Y	Zircaloy-2	UT	10
Cs/Zn	Zircaloy-2	UT	10
Ag	Zircaloy-2	Ag spot-welded to Zr, bend	11
Hg	Zirc-2, Zr–2.5 Nb	DCB, CT, bend	6–9, this work
Ga	Zr–2.5 Nb	CT	This work

[a]Reprinted from J. Nucl. Mater., 245, B. Cox and Y-M. Wong, "Liquid Metal Embrittlement of Zr-2.5% Nb Alloy," pp. 34–43. Copyright (1997) with permission from Elsevier Science.
[b]UT, uniaxial tensile test; DCB, double cantilever beam; CT, small compact tension; SIMFEX, simulated fuel expansion test; C tube, compressed tube test.
[c]This test was designed to show the effect of He bubbles produced under irradiation on tensile properties at 300°C. A Zircaloy-2/boron alloy was prepared and irradiated; annealed at 800°C to grow He bubbles; and tested under uniaxial conditions. The fracture surface was brittle, contained few He bubbles, and was almost 100% TG pseudocleavage.

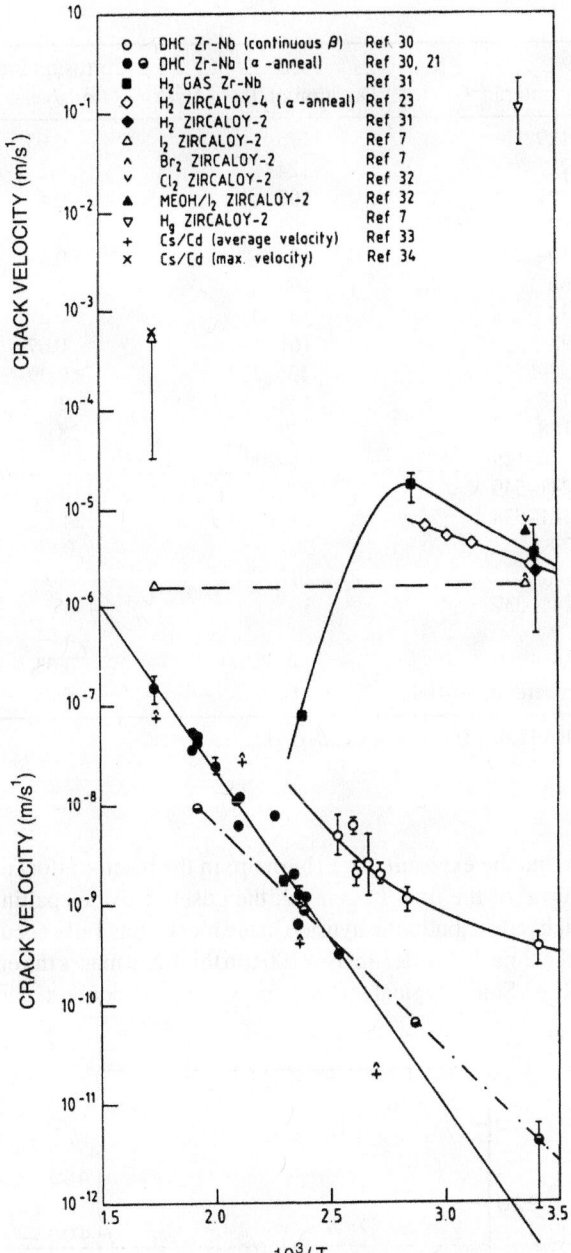

FIGURE 63.1. Collected crack velocities for SCC and hydrogen-induced cracking of zirconium alloys. (From [6]. Reprinted with permission. Copyright 1990 Elsevier Science.)

available in the system to maintain the passive oxide and prevent SCC.

B2. Galvanic Coupling

Such potential interactions in mixed environments together with the effect of variable valence ions such as Fe and Cu in reducing the corrosion resistance in inorganic acids (especially H_2SO_4 and HCl) and traces of halides in other media have a major impact on in-service behavior. Since the oxide on Zr is stable in neutral chlorides in the absence of anodic polarization, the effect of galvanic coupling also needs to be taken into account when comparing tables and plots of corrosion resistance in various inorganic media [12, 13] (such as Tables 63.6 and 63.7 and Figs. 63.2–63.5) with in-service experience. Remember always that the intermetallic particles are the weak points in the protective oxide film and there will be little difference in the volume fraction of these between the Zr_3Fe in many batches of unalloyed zirconium and $Zr(Fe/Cr)_2$ in Zircaloy-4. The commercial alloy specifications allow for large quantities of Fe impurity (Table 63.1), and, in borderline cases where pitting may be a problem, selecting batches of low-Fe material or going to a more expensive van Arkel Zr may offer improvements. There has been no detailed study of these effects for commercial Zr material in the manner that careful control of intermetallic size and distribution is regarded as part of the specification for nuclear fuel cladding [1].

C. NUCLEAR FUEL CLADDING

The behavior of nuclear fuel cladding in a reactor is dependent on a number of variables in addition to alloy composition and fabrication procedure. These factors include the heat flux through the cladding and its evolution with time as a result of the use of burnable poisons; the reactor core loading pattern; the outlet temperature of the reactor core; the thermal hydraulics of the fuel assembly design; and the water chemistry of the reactor type [e.g., differences between pressurised water reactors (PWRs), boiling water reactors (BWRs), Russian water/water reactors (VVERs), and Canadian deuterium uranium reactors (CANDU)]. In any one reactor

TABLE 63.6. Corrosion Data for Zirconium and Its Alloys in HCl[a]

Solution	Temperature (°C)	Duration of Test (days)	Average Corrosion Rate (mils/year)[b]		
			702	704	705
32% HCl	30	91	0.03	0.08	0.03
32% HCl	82	91	0.03	0.2	0.1
10% HCl and 100 ppm $FeCl_3$	30	91	0.9	1.6	0.7
32% HCl and 5 ppm $FeCl_3$	30	91	0.1	1.5	0.1
32% HCl and 50 ppm $FeCl_3$	30	91	c	c	c
32% HCl and 100 ppm $FeCl_3$	30	91	c	c	c

[a]Reprinted with permission from [12, p. 200]. Copyright ASTM International, 100 Barr Harbor Drive, West Conshohocken, PA 19428.
[b]1.0 mil/year = 0.025 mm/year.
[c]Severe pitting and stress cracking observed.

TABLE 63.7. Corrosion Rates of Zirconium in Alkaline Solutions[a]

Solution and Reference No.	Temperature (°C)	Duration of Test (days)	Corrosion Rates (mils/year)[b]
0.6% NaOH, 2% NaClO$_3$, and trace NH$_3$ [13]	129	175	0.05
5–10% NaOH [13]	21	124	0.2
7% NaOH, 53% NaCl, 7% NaClO$_3$, and 80–100 ppm NH$_3$ [12]	191	198	0.4
9–11% NaOH and 15% NaCl [12]	82	207	0.07
10% NaOH, 10% NaCl, and wet COCl$_2$ [13]	10–32	12	0.4
20% NaOH suspended salt, violent boiling [13]	60	196	13
50% NaOH and free Cl$_2$ [12]	38	153	0.2
50% NaOH and 750 ppm free Cl$_2$ [12]	93	161	0.07[c]
50% NaOH [12]	38–57	135–207	<0.09
52% NaOH and 16% NH$_3$ [12]	138	175	3
50–73% NaOH [13]	188	139	28
73% NaOH [12]	110–129	81–200	<2
73% to anhydrous NaOH [12]	241–549	2.5	110
73% to anhydrous NaOH (caustic fusion) [12]	121–538	14	24
13% KOH and 13% KCl [12]	29	207	0.09
50% KOH [12]	27	207	0.06
50% to anhydrous KOH [13]	241–337	3	430
0.2% Ca(OH)$_2$ 14% CaCl$_2$ and 8% NaCl [13]	79	20	0.04
NH$_4$OH, (NH$_4$)$_2$CO$_3$, (NH$_4$)$_2$S, NH$_4$Cl, and NaCl [13]	66	14	0.03
28% NH$_4$OH [14]	Room temperature to 100	14	<0.03

[a]Reprinted with permission from [12, p. 202]. Copyright ASTM International, 100 Barr Harbor Drive, West Conshohocken, PA 19428.
[b]1.0 mil/year = 0.025 mm/year.
[c]Vessel cathodically protected.

type (e.g., PWRs), the actual water chemistry, whether LiOH/ H$_3$BO$_3$ or KOH/NH$_4$OH, and the impurity levels that determine the amount of insoluble steel corrosion products (CRUD) deposited on the fuel cladding also have a major effect on corrosion. Finally, of course, the local fast neutron

flux and the exposure time (burn up) in the reactor affect the survival of the fuel. In general, the onset of oxide spalling, which redistributes the hydride in the metal, generally occurs by the time the oxide film is ~100 μm thick and marks the end of life. Some typical curves for various zirconium alloy

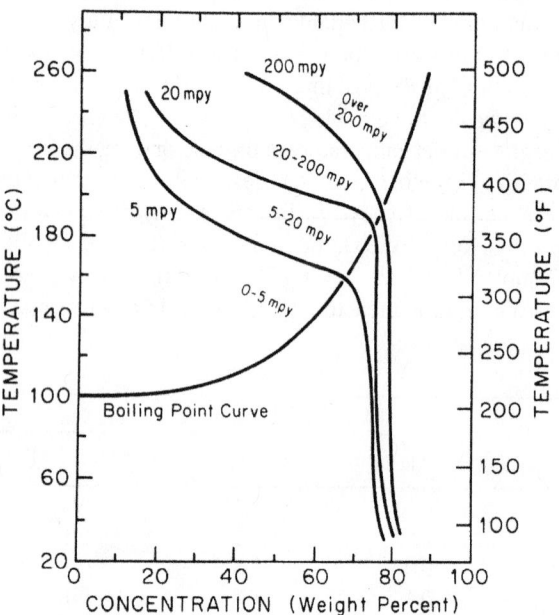

FIGURE 63.2. Corrosion of zirconium in sulfuric acid. (Copyright ASTM International, 100 Barr Harbor Drive, West Conshohocken, PA 19428. Reprinted from [12, p. 194] with permission.)

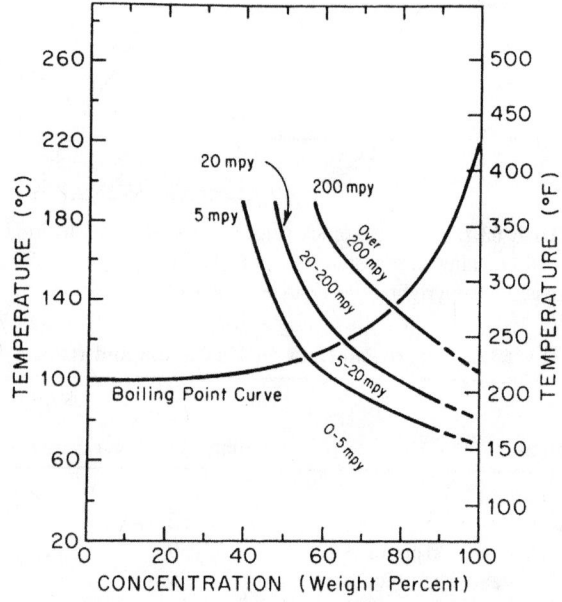

FIGURE 63.3. Corrosion of zirconium in phosphoric acid. (Copyright ASTM International, 100 Barr Harbor Drive, West Conshohocken, PA 19428. Reprinted from [12, p. 198] with permission.)

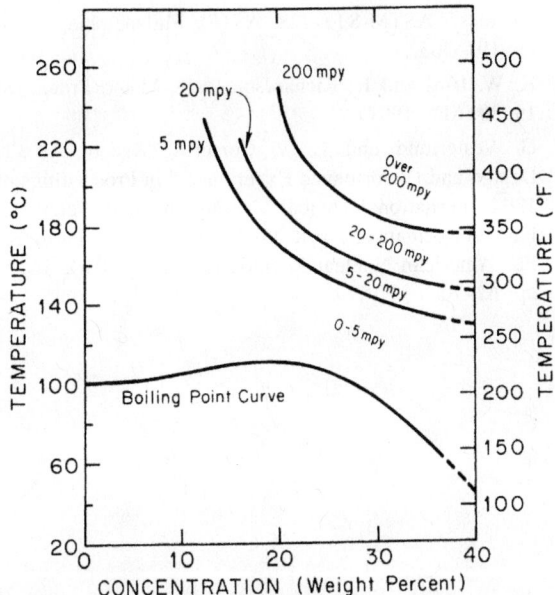

FIGURE 63.4. Corrosion of zirconium in hydrochloric acid. (Copyright ASTM International, 100 Barr Harbor Drive, West Conshohocken, PA 19428. Reprinted from [12, p. 200] with permission.)

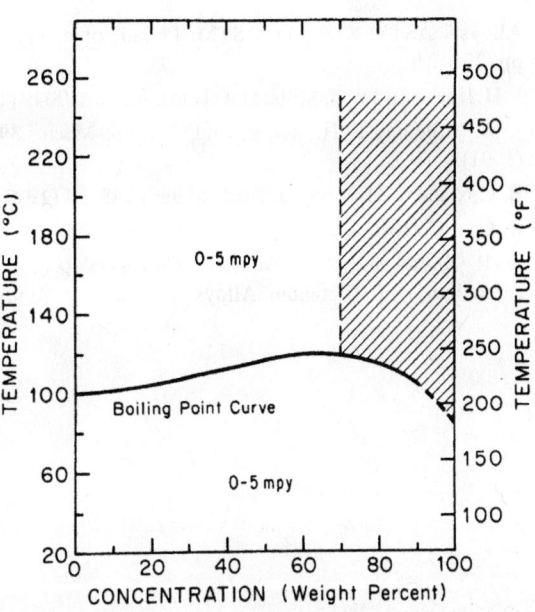

FIGURE 63.5. Corrosion of zirconium in nitric acid. (Copyright ASTM International, 100 Barr Harbor Drive, West Conshohocken, PA 19428. Reprinted from [12, p. 199] with permission.)

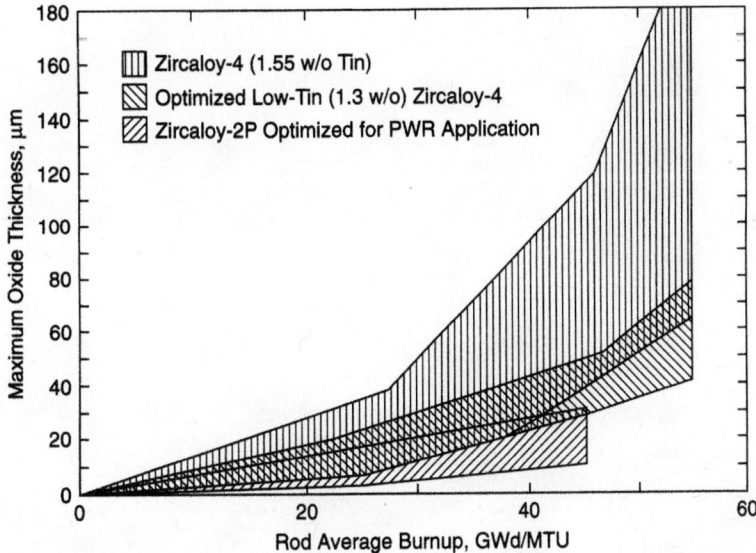

FIGURE 63.6. Maximum oxide thickness as a function of rod average burnup showing the improved performance of optimized alloys. (Copyright 1994 by the American Nuclear Society, La Grange Park, IL. Reprinted from [14] with permission.)

cladding types are shown in Figure 63.6. For a more exhaustive discussion of this complicated problem see [1].

REFERENCES

1. "Waterside Corrosion of Zirconium Alloys in Nuclear Power Plants," IAEA-TECDOC-996, International Atomic Energy Agency, Vienna, 1998.

2. B. Cox and H. I. Sheikh, J. Nucl. Mater., **249**, 17 (1977).

3. B. Cox, Corrosion, **29**, 157 (1973).

4. A. I. Merati and B. Cox, Corrosion, **55**, 388–396 (1999).

5. B. S. Frechem, J. G. Morrison, and R. T. Webster, "Improving the Corrosion Resistance of Zirconium Weldments," STP 728, ASTM, Philadelphia, PA, 1981, pp. 85–108.

6. B. Cox, J. Nucl. Mater., **170**, 1 (1990).

7. B. A. Cheadle, C. E. Coleman, and J. F. R. Ambler, "Prevention of Delayed Hydride Cracking in Zirconium

Alloys," ASTM-STP-939, ASTM, Philadelphia, PA, 1987, pp. 224–240.

8. F. H. Huang and W. J. Mills, Met. Trans. A., **22A**, 2049 (1991).

9. J. W. Mills and F. H. Huang, Eng. Fracture Mech., **39**, 241 (1991).

10. B. Cox and Y-M. Wong, J. Nucl. Mater., **245**, 34 (1997).

11. B. Cox, Corrosion, **28**, 207 (1972); **33**, 79 (1977).

12. D. R. Knittel and R. T. Webster, "Corrosion Resistance of Zirconium and Zirconium Alloys in Inorganic Acids and Alkalies," ASTM-STP-728, ASTM, Philadelphia, PA, 1981, pp. 191–203.

13. K. W. Bird and K. Richardson, Adv. Mater. Proc., **151**(3), 19–20 (Mar. 1997).

14. G. Vesterlund and L. V. Corsetti, "Recent ABB Fuel Design and Performance Experience," in Proceedings of the 1994 International Topical Meeting on Light Water Reactor Fuel Performance, April 17–21, 1994. West Palm Beach, FL, American Nuclear Society, La Grange Park, IL, 1994, pp. 62–70.

PART IV

CORROSION PROTECTION

64

CONTROLLING FLOW EFFECTS ON CORROSION

K. D. Efird

Efird Corrosion International, Inc., The Woodlands, Texas

A. Introduction
B. Occurrence of flow-induced corrosion
C. Methods for controlling flow corrosion
  C1. Minimize fluid turbulence
  C2. Modify flowing fluid
  C3. Minimize flow disruptions
  C4. Modify flow regimes
  C5. Corrosion-resistant alloys
References

A. INTRODUCTION

The effect of flow on corrosion is complex and varied and dependent on both the chemistry and physics of a system. The key variable defining the effect of flow on corrosion is turbulence. High turbulence can result in flow-induced corrosion, erosion–corrosion, or cavitation. Low turbulence can result in corrosion in a separated water phase and allows the occurrence of corrosion under deposits and/or in separated liquid water. *Flow-induced corrosion* is the term used to describe the increase in corrosion resulting from high fluid turbulence due to the flow of a fluid over a surface in a flowing single or multiphase system. *Underdeposit corrosion* is the term used to describe the increased corrosion occurring in a separated water phase beneath deposits of nonmetallic solids on a metal surface resulting from low-flow turbulence.

The key element in controlling flow-induced corrosion is an understanding of the flow characteristics that can produce conditions that favor specific corrosion mechanisms for specific materials and mitigating those conditions by modifying the flow conditions, the material of construction, or the

corrosive environment. The concentration of this chapter is on modification of the flow conditions.

B. OCCURRENCE OF FLOW-INDUCED CORROSION

Knowing when and where flow-induced corrosion could occur in an operating system is a major help in controlling flow-induced corrosion. A logic tree that can be employed to determine the type of flow-induced corrosion expected in an operating system is given in Figure 64.1 [1]. This procedure provides a method to determine whether equilibrium or steady-state corrosion conditions are present.

The first step is to determine if liquid water is present. If it is not, no corrosion can occur. Similarly, if liquid water cannot contact the pipe wall, no corrosion can occur. The determination of the level of turbulence in the water phase from both the flow regimes and the existence of flow disruption then determines whether or not the corrosion reactions are based on equilibrium or steady-state conditions. Steady-state conditions can occur due to either flow disruptions or a nonequilibrium flow regime such as slug flow. Once either steady-state or equilibrium corrosion is confirmed, the expected corrosion rate can be determined. The term *equilibrium* is used to denote a fully developed flow condition for either mass transfer or wall shear stress.

C. METHODS FOR CONTROLLING FLOW CORROSION

A number of methods are available to control flow-induced corrosion. They can involve design modifications, process

Uhlig's Corrosion Handbook, Third Edition, Edited by R. Winston Revie
Copyright © 2011 John Wiley & Sons, Inc.

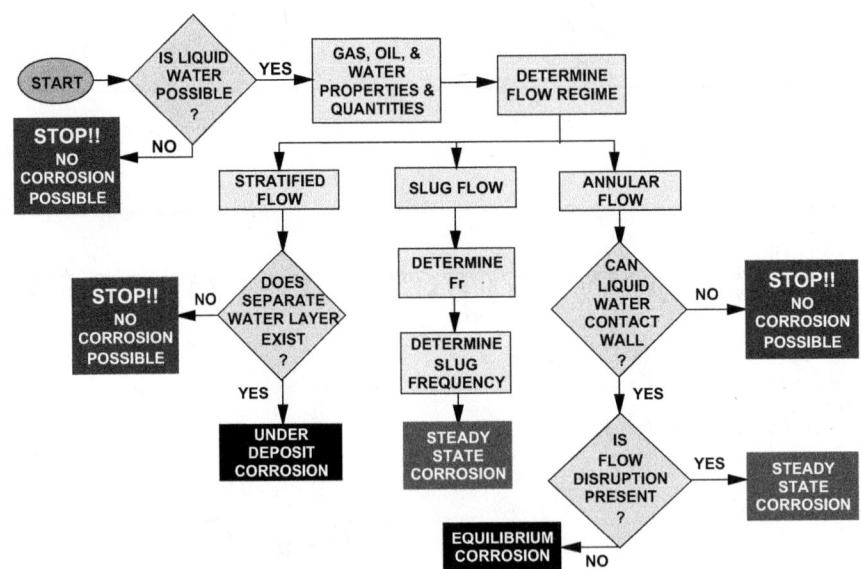

FIGURE 64.1. Logic tree to determine type of flow-induced corrosion in an operating system [1].

changes, and modifications to the corrosive environment. The methods include any one or a combination of the following:

- Minimizing fluid turbulence
- Modification of flowing fluid
- Minimizing flow disruptions
- Modification of flow regimes
- Using a more corrosion resistant alloy

These methods can also be used to control low-turbulence flow corrosion. They also involve design modifications, process changes, and modifications to the corrosive environment.

C1. Minimize Fluid Turbulence

Applied to single-phase systems, lowering the fluid velocity lowers the turbulence intensity (mass transfer and wall shear stress) and hence lowers the corrosion rate. This effect is demonstrated in Figure 64.2 for corrosion of carbon steel in single-phase fluid flow at varying chloride concentrations [2]. Though intuitively obvious, this method of control is not always available due to process and equipment limitations.

Many industries have established "rule-of-thumb" velocity limits for various processes. These are generally based on operational experience regarding flow velocities where corrosion rates increase dramatically. In reality, these "rules" serve to limit the turbulence intensity (mass transfer and wall shear stress), thereby effectively limiting the corrosion rate.

C2. Modify Flowing Fluid

There are two ways to control flow-induced corrosion by modification of the flowing fluid. These are removal of a corrosive species and chemical additives, for example, the addition of corrosion inhibitors. These methods are also used to control corrosion where flow is not a problem, but they can be particularly effective for flow-induced corrosion.

Removal of a corrosive species limits the diffusion of the corrosive species to the metal surface, thereby reducing the corrosion rate. Chemical additives such as corrosion

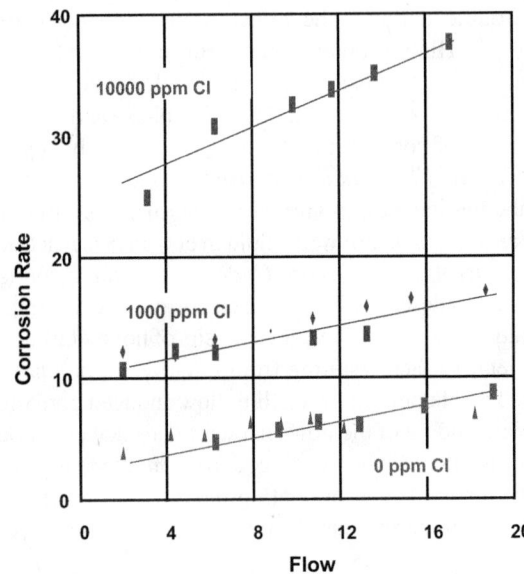

FIGURE 64.2. Effect of single-phase flow on corrosion rate at different chloride concentrations [2].

inhibitors can be employed to modify the surface corrosion products, making them more protective and lowering the corrosion rate. Corrosion inhibitor formulations can also be designed to modify turbulence in the fluid through alterations in viscosity and interfacial surface tension.

Examples of controlling flow-induced corrosion by removal of corrosive species are the removal of oxygen from seawater and raising the pH of a brine solution. The removal of oxygen through either mechanical or chemical means reduces the oxygen concentration in the solution. Consequently, the mass transfer of oxygen to the corroding metal surface is reduced, reducing the corrosion rate. Raising the pH results in a decrease in the concentration of hydrogen ions in a solution. As with oxygen, this reduction results in a decreased mass transfer of hydrogen ion to the corroding metal surface, reducing the corrosion rate.

C3. Minimize Flow Disruptions

Flow disruptions are a major contributor to flow-induced corrosion and resulting failures. As a result, minimizing flow disruptions in equipment design and construction is important. Ways to minimize flow disruptions include the following:

- Minimize the mismatch when welding pipe joints and fittings.
- Grind internal welds smooth, when possible.
- Use long-radius elbows.
- Avoid step changes in pipe diameter.
- Ream and deburr tubing ends before installation.
- Protect stored pipe to prevent internal corrosion before it goes into service.
- Don't use corroded pipe that contains internal pits, even if it still meets the required strength criteria.

C4. Modify Flow Regimes

Flow regime modification is used primarily for pipelines. Flow regimes are modified by changing the gas and/or liquid flow rates in a pipeline. The effects of these changes on flow regime are reflected in the flow regime maps. Resolution of the flow regime existing in various sections of a pipeline can be used to determine where to expect corrosion in the pipeline and consequently where to monitor corrosion.

In the design stage, pipeline sizes can be adjusted to prevent slug flow at the anticipated liquid and gas flow rates. If the anticipated problem is corrosion under deposits in the bottom of the pipeline, the pipe size can be adjusted to eliminate stratified flow and keep the deposits suspended in the fluid.

C5. Corrosion-Resistant Alloys

In some cases, the only alternative available to prevent flow-induced corrosion is the use of a more corrosion-resistant alloy. This is usually the most expensive option and is generally only a last resort. An example is the use of a stainless steel in a situation where high flow rates result in excessively high corrosion rates for carbon or alloy steel.

Corrosion-resistant alloys may also experience turbulence limits in different environments, and the use of a corrosion-resistant alloy should not, a priori, be assumed as the final answer to a flow-induced corrosion problem. In some cases, low flow can be a problem, such as the use of some stainless steels in chloride environments where pitting can occur under deposits.

REFERENCES

1. K. D. Efird, "Disturbed Flow and Flow Accelerated Corrosion in Oil and Gas Production," in Proceedings: ASME Energy Resources Technology Conference, Houston, TX, Feb. 1998, p. 1.
2. J. A. Herce et al., "Effects of Solution Chemistry and Flow on the Corrosion of Carbon Steel in Sweet Production," Paper No. 95111, CORROSION/95, Orlando, FL, Mar. 1995, p. 2.

65

EROSION–CORROSION: RECOGNITION AND CONTROL

J. Postlethwaite[†]

Department of Chemical Engineering, University of Saskatchewan, Saskatoon, Saskatchewan, Canada

S. Nešić

Institute for Corrosion and Multiphase Flow Technology, Ohio University, Athens, Ohio

A. INTRODUCTION

A major prerequisite for the control of erosion–corrosion is the recognition or determination of the relative roles of accelerated corrosion and erosion. Only then can the appropriate action be taken.

If *accelerated corrosion* following damage to protective films is the problem, there are two alternatives:

- Take steps to avoid damage to the film.
- Accept the film damage and use corrosion control methods.

If *erosion* of the underlying metal is a major factor, design and materials selection solutions should be sought.

Recognition of the type of erosion–corrosion is sometimes relatively straightforward. Erosion–corrosion by both single-phase aqueous flow and suspended solids is characterized by the presence of smooth grooves, gullies, shallow teardrop-shaped pits, and horseshoe-shaped depressions most often with an obvious flow orientation. The characteristic pattern of attack often starts at isolated spots on the metal surface and subsequently spreads to a general roughening of the surface [1, 2]. With cavitation and liquid droplet impingement attack, the damage starts in the form of steep-sided pits, which may coalesce into a honeycomb-like structure. The Corrosion Atlas [3] and the NACE International Corrosion Recognition and Control Handbooks [4, 5] contain photographs of all the various forms of erosion–corrosion with suggested control methods built around a large number of case histories.

B. CONTROL OF TURBULENT FLOW ATTACK

In single-phase aqueous flow, the erosion–corrosion of metals such as copper tubing is a process of accelerated corrosion following erosion of the protective film. Control of this type of attack is usually achieved by modifying the design of the system to reduce the hydrodynamic forces and/ or by choosing an alloy with a more erosion-resistant film.

[†] Retired.

Uhlig's Corrosion Handbook, Third Edition, Edited by R. Winston Revie
Copyright © 2011 John Wiley & Sons, Inc.

B1. Design

Design factors, for example, for copper tubing carrying potable water, are the control of the velocity and the minimization of abrupt changes in the flow system geometry. Maximum velocities in the range 0.8–1.5 m/s have been suggested [3, 6]. The tubing should be reamed where it is pushed into fittings such as elbows prior to soldering [7] to mitigate the erosion–corrosion shown in Figure 18.2 (Chapter 18). Recent studies [8] have shown that 1-mm forward- or backward-facing steps are sufficient to initiate film disruption. Reaming in excess of that required to remove the burring on cut tubing may be required to prevent erosion–corrosion. Plastic inserts can be used to solve heat exchanger–tube inlet problems.

B2. Materials

Materials selection could involve, for example, the substitution of copper tubing in hot water distribution systems by stainless steels or plastics [3]. As noted above, only in extremely severe conditions would single-phase aqueous flow damage the passive film on stainless steels leading to erosion–corrosion. The possible pitting and crevice corrosion of stainless steels in the presence of chlorides should be taken into consideration. A range of copper alloys with increasing velocity limits can be used in heat exchangers (Table 18.2). If a very high velocity is required, in a corrosive environment, stainless steel, nickel alloy, or titanium tubes can be considered [9, 10].

B3. Inhibitors

Inhibitors are used in recirculating cooling water systems and steam condensate return lines but find limited application in once-through production systems because of the cost. A notable exception is the extensive use of inhibitors in oil/gas production.

B4. Cathodic Protection

Cathodic protection has very limited throwing power down the inside of pipes and other restricted geometries. Impressed systems can be used in heat exchanger water boxes to protect the important entrance length region for copper alloy tubes [1, 11] carrying seawater. Care must be taken to avoid hydrogen evolution, which can lead to the formation of air/hydrogen pockets at dead zones [1]. In addition to safety problems hydrogen can lead to the embrittlement of titanium and other alloy tubing.

C. CONTROL OF SOLID-PARTICLE IMPINGEMENT ATTACK

Erosion–corrosion problems observed in the presence of solid particles suspended in aqueous solutions are more difficult to solve. The relative role of corrosion and erosion can often only be assessed following testing involving weight loss measurements to determine the total loss with simultaneous electrochemical measurements (polarization resistance) to determine the contribution of corrosion [12, 13]. The relative importance of erosion and corrosion will vary between nondisturbed and disturbed flow and testing should be done in a flow system that simulates both the chemical and the hydrodynamic conditions found in the full-scale process equipment.

C1. Corrosion Control

Corrosion can be controlled by the use of inhibitors and/or solution conditioning. Some *inhibitors* work very well in the presence of very abrasive slurries, including silica, which is one of the most common abrasive components. Chromates and nitrites at high concentrations act as passivating inhibitors [14] whereas chromates at low concentration act as cathodic inhibitors [15, 16] and were used in the first long-distance coal-slurry pipeline [17]. Chromates are, of course, toxic and very low effluent limits < 0.05 ppm have been set in some jurisdictions [18]. Nonchromate inhibitors used in cooling water systems, zinc, sodium tripolyphosphate ($Na_5P_3O_{10}$), and nitrilotris (methylene) triphosphonic acid (NTMP) $N[CH_2PO (OH_2)]_3$, showed little benefit [15] in sand or coal slurries when used alone or along with chromates, in contrast to the synergistic effect normally observed in cooling water.

Solution conditioning involves raising the pH and/or deaeration. Both of these methods of corrosion control have been applied to long-distance slurry pipelines. A problem with raising the pH to control corrosion is the greater likelihood of pitting at elevated pH values where thick scales are more easily formed. Indeed, pitting caused some concern with the Bougainville copper concentrate line [19] which was treated with lime to maintain the slurry at pH > 9.5. Temporary overdosing with lime led to calcium carbonate deposits in the first portion of the Samarco iron ore concentrate line [20] which was maintained at pH ≥ 10. Deaeration can be achieved with oxygen scavengers, sodium bisulfite or hydrazine, or nonchemical steam stripping or vacuum deaeration. The latter two methods of deaeration have not been used with slurry pipelines. They are used extensively, for example, with oil well water injection systems [21].

If the corrosion following the removal of the protective film is liquid-phase mass transport controlled, the flow velocity can be reduced. However, the effect of reducing the velocity will be of secondary importance to controlling the corrosion by inhibitors or solution conditioning. A lower limit is the velocity required to keep the particles in suspension [22]. For example [23], the flow of a 20 vol % silica sand slurry, mean particle diameter 0.43 mm, in a 50-mm horizontal diameter pipe requires a minimum velocity of ∼ 2 m/s

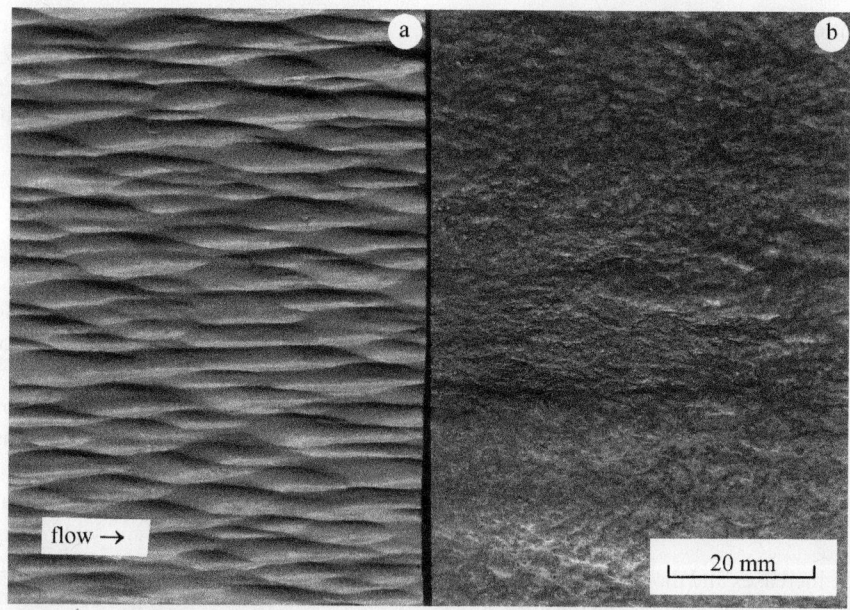

FIGURE 65.1. Variation in erosion–corrosion wear pattern around the circumference of a horizontal 200-mm diameter commercial pipeline carrying an abrasive mineral slurry. (a) Pipe bottom. (b) Pipe top. *Note*: The erosion–corrosion wear pattern in (a) is not characteristic of sliding bed wear. Sliding wear is characterized by long horizontal grooves.

and a 20 vol % slurry of iron ore concentrate mean particle diameter 0.04 mm a minimum velocity of ~1.3 m/s. Lower velocities may result in the *sliding abrasion* of a horizontal pipe bottom. At velocities above the critical velocity to keep the particles in suspension, there is a large concentration gradient from the top to bottom of horizontal pipelines carrying commercial heterogeneous slurries [22]. This can result in a much different erosion–corrosion wear pattern and rate around the circumference of the pipe (Fig. 65.1).

If necessary more corrosion-resistant materials than carbon steel pipe can be chosen, including stainless alloys, ceramics (e.g., cast basalt lined pipe, Mohs hardness ~8), and plastics (e.g., high-density polyethylene pipe or polyurethane linings).

Stainless alloys, with their rapidly healing films, have a greater resistance to corrosion in slurries, but the extra resistance comes at a cost that may not be acceptable. Ceramics and plastics may not have the mechanical and thermal properties for the construction of the particular process equipment. Long-distance pipelines are constructed from carbon steel. More expensive alloys and lined pipe are an option for in-plant operations, including tailings disposal. Titanium alloys perform well in flowing seawater with abrasive solids in suspension [24].

C2. Erosion Control

Erosion can be controlled by design and materials selection. Design involves optimizing the particle size (by grinding,

where there is a choice of size) and the flow velocity [22]. The flow system geometry should be designed to minimize any effects of disturbed flow, for example, by using long-radius elbows, gradual changes in the flow cross section, and specifying maximum weld root protrusion. Other design possibilities are:

- Increasing the thickness of materials in critical areas
- Use of impingement plates to shield critical areas
- Acceptance of a high erosion rate with regular inspection and replacement that may be less costly than using more expensive materials and a practice used extensively in the minerals processing and oil/gas industries
- In some situations pipe rotation, for example, to extend the life of tailing lines

Since there is a major decrease in the erosion rate when the metal surface is harder than the particles, it might be thought to be a simple matter to choose an alloy with a hard surface and eliminate the erosion problem. This is not the case for two reasons:

- The hard alloys that resist erosion are generally cast alloys that are difficult to weld, often brittle, and in general difficult to handle.
- If corrosion is a factor, the alloy must have a suitable corrosion resistance and many alloys that are hardened

by the precipitation of carbides during solidification have a relatively poor corrosion resistance.

Alloyed white cast irons (the most abrasion resistant iron-base alloys), including high-chromium, chrome-moly, nickel–chrome, and pearlitic white irons [25], are noted for their erosion resistance. The relationship between the Cr and C content of the high-chromium cast irons is complex. A high C content is required for the formation of carbides to give erosion resistance, but this leaves less Cr in the matrix for corrosion resistance [26]. The minimum Cr content in the matrix to form a passive film is 12% and the *Cr* required to form carbides is $10 \times \% $ C. The suggested minimum Cr for corrosion resistance is

$$\% \ Cr = (\% \ C \times 10) + 12$$

The erosion resistance increases and the corrosion resistance decreases with an increase in the C content and the alloy must be chosen carefully to accommodate the corrosive and erosive properties of the particular slurry being handled. Alloys containing 20–28% Cr and 2–2.5% C with 2% Mo have good resistance to slurry erosion–corrosion at pH values down to 4. Alloys with less C are required for more corrosive environments [26]. Nickel-hard alloys (4% Ni–2% Cr) find extensive use in abrasive service in sand and gravel pumps handling abrasive but mildly corrosive slurries [27]. Natural rubber is an alternative for pumps to handle abrasive slurries with particles less than 3 mm diameter.

Both Stellite (a cast cobalt alloy) and silicon carbide have excellent erosion resistance in aqueous slurries along with excellent corrosion resistance and can be considered for use under severe service conditions in valves and pumps [3, 28].

D. CONTROL OF LIQUID DROPLET IMPINGEMENT ATTACK

Impingement attack by liquid droplets suspended in high-speed gas flow can be controlled by *design* or *materials selection*.

Design involves optimizing the flow system geometry and the fluid dynamics [29] to reduce the amount of impacting liquid, the angle of impact, and the droplet size. For example, design modifications in steam turbines operating with wet vapor in the low-pressure section have included extracting moisture between blade rows, increasing axial spacing between stator and rotor, and local flame hardening or brazed-on shield of Stellite at the leading edge of the blade. Raising the temperature of the inlet gas above the dewpoint was suggested [4] as a remedy for the impingement attack of a process gas compressor. Impingement plates acting as flow deflectors can also be considered.

The behavior and range of materials utilized to solve high-speed liquid droplet impingement problems are similar to those chosen for resistance to cavitation attack. The "normalized erosion resistance" data shown in Figure 65.2 give an approximate ranking of materials to liquid droplet

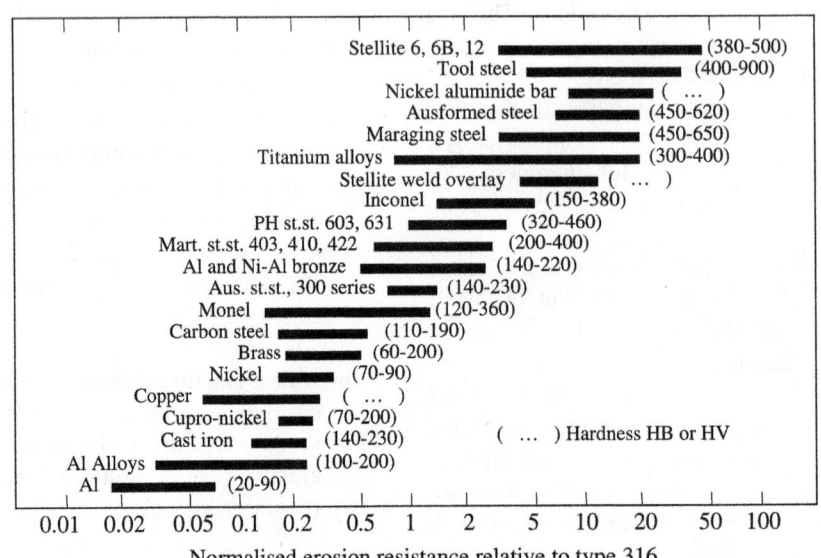

FIGURE 65.2. Normalized erosion resistance of various metals and alloys (the erosion resistance number according to ASTM G73). Selection of the data deduced by Heymann [41] from many sources, including both impingement and cavitation tests. Erosion test data are not very consistent, and the information herein should be used only as a rough guide. (Adapted from Heymann [41].)

impingement and cavitation attack. Such data should be used with great care. Materials selection for turbines includes the use of Stellite mentioned above, 12% chromium martensitic stainless steel, 17 Cr–4 Ni precipitation-hardened stainless steel, and "self shielding" blade alloys that harden under the action of impacts [29].

Liquid droplet impingement attack observed in annular mist flow [Fig. 18.3(b)] in oil/gas production systems is an erosion–corrosion phenomenon, which can be controlled by the use of inhibitors or the use of stainless alloys [30]. Not surprisingly this erosion–corrosion is found to be the most severe under disturbed flow conditions at threaded connections in down-hole tubulars and at elbows, valves, and "Christmas trees" in above-ground facilities.

E. CONTROL OF CAVITATION ATTACK

Cavitation attack is usually controlled by design and materials selection. Tullis [31] has given an extensive discussion of the detailed design of pumps, valves, orifices, and elbows to avoid cavitation problems. Air injection is sometimes used to control cavitation damage. Air injected into the separated flow regime cushions the collapse of the cavities. Cathodic protection has been used, with the protection attributed to cushioning by the hydrogen bubbles evolved in addition to the normal corrosion protection. While this might be satisfactory for a propeller in an open system, the evolution of hydrogen in a closed system could be hazardous as well as leading to hydrogen blistering and embrittlement.

The range of materials available for solving cavitation–erosion problems is similar to those used for liquid droplet impingement, as shown in Figure 65.2. There is a wide range of polymers with good resistance to cavitation–erosion in addition to excellent resistance to corrosion. For example, high-density polyethylene has a cavitation–erosion resistance similar to that of nickel-based and titanium alloys [32].

The major design parameter for centrifugal pumps to avoid cavitation damage is the available *net positive suction head*, $NPSH_A$, the difference between the total pressure (absolute) and vapor pressure at the pump suction, expressed in terms of equivalent height of fluid, or "head," by

$$\text{NPSH}_A = (p/\rho g) + (u^2/2g) - (p_v/\rho g)$$

where

$p =$ static pressure (absolute)
$p_v =$ vapor pressure of the flowing fluid
$\rho =$ density
$g =$ gravitational acceleration
$u =$ flow velocity

The $NPSH_A$ must exceed the value required by the pump, $NPSH_R$. The latter value varies with the flow rate and is a function of the pressure changes as the liquid accelerates over the curved impeller and then decelerates as it approaches the volute. The $NPSH_R$ values, which are supplied by the pump manufacturers, are based on pump efficiency and not on the dangers of cavitation attack. Significant noise and cavitation may occur before the efficiency of the pump begins to decrease. Thus a substantial margin of safety is required if erosion–corrosion is to be avoided. In practice, some cavitation can usually be tolerated and pumps are operated in the NPSH range between cavitation inception and a point where damage is unacceptable [31, 32]. If it is wished to maximize efficiency of the pump operation, more cavitation can be accepted and cavitation attack reduced by the selection of more resistant materials. One very important factor in setting the correct $NPSH_A$ is the relative elevation of the pump and the vessel, from which the liquid is being pumped. Fluid friction in the suction line must also be taken into account, and the suction piping is usually a size bigger than the discharge piping.

F. CONTROL OF FLOW-ENHANCED FILM DISSOLUTION ATTACK

The control of this type of "chemical" erosion–corrosion of carbon steel pipes in power plants may involve one or more of the following [33]:

- Control of the water chemistry: pH and dissolved oxygen concentration
- Materials selection: replacement of carbon steel by low-alloy chromium, >0.1%; low-alloy chromium–molybdenum, 304 ss and in very severe conditions Inconel; or duplex piping with a thin inner layer of stainless steel or other high alloy
- Weld overlay: for protection and repair
- Flame spraying: minimum pipe diameter 600 mm
- Modification of operating conditions: temperature and quality, where wet steam is involved, and flow rate
- Changing the local geometry: for example, installing a larger control valve to reduce downstream turbulence

In general, orifice plates and control valves should be kept well clear, at least 10 pipe diameters upstream of bends, to avoid excessive turbulence and erosion–corrosion at the latter.

The extensive and well-documented Electric Power Research Institute (EPRI) [33] report contains information regarding the detection and control of the problem.

G. PREDICTIVE MODELING

The application of *computational fluid dynamics* (CFD) to the problem of erosion–corrosion in single- and multiphase flow systems under conditions of disturbed flow can help to quantify the effects of the system geometry on rates of erosion–corrosion leading to design improvements. The flow field, including flow separation recirculation and reattachment, and rates of mass transfer can be predicted for a wide range of system geometries [34–39]. In addition the velocities and angles of impact of suspended solid particles with the flow system walls can be determined for application to the calculation of the erosion rate [40–42]. Computer modeling has been applied to the prediction of wear in slurry pumps [43]. Overall, such predictive modeling is still not as accurate as one would hope for [44], even if substantial progress has been made as our understanding of erosion–corrosion processes advanced accompanied by the ever-increasing computing power [45–47].

REFERENCES

1. G. Bianchi, G. Fiori., P. Longhi, and F. Mazza, "Horse Shoe Corrosion of Copper Alloys in Flowing Sea Water: Mechanism, and Possibility of Cathodic Protection of Condenser Tubes in Power Stations," Corrosion, **34**, 396–406 (1978).

2. J. Postlethwaite, B. J. Brady, M. W. Hawrylak, and E. B. Tinker, "Effects of Corrosion on the Wear Patterns in Horizontal Slurry Pipelines," Corrosion, **34**, 245–250 (1978).

3. E. D. D. During, comp., Corrosion Atlas: A Collection of Illustrated Case Histories, Vol. 1: Carbon Steels; Vol. 2: Stainless Steels and Non-Ferrous Materials: "Erosion-Corrosion of Copper Tubing," 06.05.34.01; "Valve Erosion," 04.01.32.01; "Pump Cavitation," 04.11.33.01; Elsevier, Amsterdam, 1988.

4. D. McIntyre (Ed.), Forms of Corrosion, Recognition and Prevention, NACE Handbook 1, NACE, Vol. 2, Houston, TX, 1997, pp. 89, 93.

5. C. P. Dillon (Ed.), NACE Handbook 1, Forms of Corrosion, Recognition and Prevention, NACE, Houston, TX, 1982.

6. A. Cohen, "Corrosion by Potable Waters in Building Systems," Mater. Perform., **32**, 56–61 (1993).

7. ASTM B 828, "Standard Practice for Making Capillary Joints by Soldering of Copper and Copper Alloy Tube and Fittings," in Copper and Copper Alloys, Vol. 02.01, ASTM, West Conshohocken, PA, 1998.

8. J. Postlethwaite, Y. Wang, G. Adamopoulos, and S. Nesic, "Relationship between Modelled Turbulence Parameters and Corrosion Product Film Stability in Disturbed Single-Phase Aqueous Flow," in Modelling Aqueous Corrosion, K. R. Trethaway and P. R. Roberge (Eds.), Klewer Academic, Dordrecht, Netherlands, 1994, pp. 297–316.

9. M. G. Fontana, Corrosion Engineering, 3rd ed., McGraw-Hill, New York, 1986, p. 95.

10. G. J. Danek, Jr., "The Effect of Sea-Water Velocity on the Corrosion Behavior of Metals," Naval Eng. J., **78**, 763–769 (1966).

11. W. Currer and J. S. Gerrard, "Practical Applications of Cathodic Protection," in Corrosion, 3rd ed., L. L. Shreir, R. A. Jarman, and G. T. Burstein (Eds.), Butterworth-Heinemann, Oxford, England 1994, p. 10:112.

12. J. Postlethwaite, M. H. Dobbin, and K. Bergevin, "The Role of Oxygen Mass Transfer in the Erosion-Corrosion of Slurry Pipelines," Corrosion, **42**, 514–521 (1986).

13. B. W. Madsen, "Corrosive Wear," in ASM Metals Handbook, Vol. 18, Friction, Wear and Lubrication Technology, ASM, Metals Park, OH, 1992, pp. 271–279.

14. J. Postlethwaite, "The Control of Erosion-Corrosion in Slurry Pipelines," Mater. Perform., **26**(12), 41–45 (1987).

15. J. Postlethwaite, "Electrochemical Studies of Inhibitors in Aqueous Slurries of Sand Iron Ore and Coal," Corrosion, **35**, 475–480 (1979).

16. J. Postlethwaite, "Effect of Chromate Inhibitors on Mechanical and Electrochemical Components of Erosion-Corrosion in Aqueous Slurries of Sand," Corrosion, **37**, 1–5 (1981).

17. D. R. Bomberger, "Hexavalent Chromium Reduces Corrosion in a Coal-Water Slurry Pipeline," Mater. Protect., **4**, 43–49 (Sept. 1965).

18. V. S. Sastri and M. Malaiyandi, "Spectra Studies on Some Novel Oxygen Scavengers and Their Use in Corrosion Control of Coal Slurry Pipelines," Can. Metall. Q., **22**, 241–245 (1983).

19. R. D. Coale, T. L. Thompson, and R. P. Ehrlich, "Bougainville Copper Concentrate Slurry Pumping System," Trans. SME, AIME, **260**, 289–297 (Dec. 1976).

20. M. E. Jennings, "SAMARCO's 396 km Pipeline: A Major Step in Iron Ore Transportation," Mining Eng., Feb. 1981, pp. 178–182.

21. A. G. Ostroff, Introduction to Oilfield Water Technology, NACE, Houston, TX, 1979, p. 293.

22. E. J. Wasp, J. P. Kenny, and R. L. Gandhi, "Solid-Liquid Flow: Slurry Pipeline Transportation," in Series on Bulk Materials Handling, Vol. 1 (1975/77), No. 4, Trans. Tech. Publications, Clausthal, Germany, 1977, p. 144.

23. J. Postlethwaite, E. B. Tinker, and M. W. Hawrylak, "Erosion-Corrosion in Slurry Pipelines," Corrosion, **30**, 285–290 (1974).

24. R. W. Schutz and D. E. Thomas, "Corrosion of Titanium and Titanium Alloys," in Metals Handbook, 9th ed., Vol. 13, ASM, Metals Park, OH, 1987, pp. 669–706, p. 696.

25. J. H. Tylczak, "Abrasive Wear," in ASM Metals Handbook, Vol. 18, Friction, Wear and Lubrication Technology, ASM, Metals Park, OH, 1992, p. 189.

26. J. Dodd, "High-Chromium Cast Irons," in Corrosion, 3rd ed., L. L. Shreir, R. A. Jarman, and G. T. Burstein (eds.), Butterworth-Heinemann, Oxford, 1994, pp. 3:128–3:137.

27. G. Wilson, "The Design Aspects of Centrifugal Pumps for Abrasive Slurries," in Proc. 2nd Int. Conf. on Hydraulic Transport of Solids in Pipes, BHRA Fluid Engineering, Cranfield, UK, 1972, pp. H2:25–H2:52.

28. E. Heitz, S. Weber, and R. Liebe, "Erosion Corrosion and Erosion of Various Materials in High Velocity Flows Containing Particles," in Proceedings of Symposium on Flow-Induced Corrosion; Fundamental Studies and Industry Experience,

K. H. Kennelley, R. H. Hausler, and D. C. Silverman (Eds.), NACE, Houston, TX, 1991, pp. 5:1–5:15.

29. F. J. Heymann, "Liquid Impingement Erosion," in ASM Metals Handbook, Vol. 18, Friction, Wear and Lubrication Technology, ASM, Metals Park, OH, 1992, pp. 221–232.

30. J. S. SmartIII, "A Review of Erosion Corrosion in Oil and Gas Production," in Proceedings of Symposium on Flow-Induced Corrosion; Fundamental Studies and Industry Experience, K. H. Kennelley, R. H. Hausler, and D. C. Silverman (Eds.), NACE, Houston, TX, 1991, pp. 18:1–18:18.

31. J. Tullis, Hydraulics of Pipelines: Pumps, Valves, Cavitation, Transients, Wiley, New York, 1989, pp. 59, 78.

32. B. Angell, "Cavitation Damage," in Corrosion, 3rd ed., L. L. Shreir, R. A. Jarman, and G. T. Burstein (Eds.), Butterworth-Heinemann, Oxford, 1994, pp. 8:197–8:207.

33. Electric Power Research Institute (EPRI), Flow Accelerated Corrosion in Power Plants, TR-106611, EPRI, Pleasant Hill, CA, 1996, pp. 4:2, 6:25.

34. S. Nesic, G. Adamopoulos, J. Postlethwaite, and D. J. Bergstrom, "Modelling of Turbulent Flow and Mass Transfer with Wall Function and Low-Reynolds Number Closures," Can. J. Chem. Eng., **71**, 28–34 (1993).

35. S. Nesic and J. Postlethwaite, "Relationship between the Structure of Disturbed Flow and Erosion-Corrosion," Corrosion, **46**, 874–880 (1990).

36. S. Nesic and J. Postlethwaite, "Hydrodynamics of Disturbed Flow and Erosion-Corrosion, Part I-A Single Phase Flow Study," Can. J. Chem. Eng., **69**, 698–703 (1991).

37. S. Nesic, J. Postlethwaite, and D. J. Bergstrom, "Calculation of Wall-Mass Transfer Rates in Separated Aqueous Flow Using a Low Reynolds Number $\kappa - \varepsilon$ Model," Int. J. Heat Mass Transfer, **35**, 1977–1985 (1992).

38. J. Postlethwaite, S. Nesic, and G. Adamopoulos, "Modelling Local Mass Transfer Controlled Corrosion at Geometrical Irregularities," Mater. Sci. Forum, **111–112**, 53–62 (1992).

39. J. Postlethwaite, S. Nesic, G. Adamopoulos, and D. J. Bergstrom, "Predictive Modelling for Erosion-Corrosion under Disturbed Flow Conditions," Corros. Sci., **35**, 627–633 (1993).

40. S. Nesic and J. Postlethwaite, "Hydrodynamics of Disturbed Flow and Erosion-Corrosion. Part II—Two Phase Flow Study," Can. J. Chem. Eng., **69**, 704–710 (1991).

41. S. Nesic and J. Postlethwaite, "A Predictive Model for Localized Erosion-Corrosion," Corrosion, **47**, 582–591 (1991).

42. H. Zeisel and F. Durst, "Computations of Erosion-Corrosion Processes in Separated Two-Phase Flows," in Proceedings of Symposium on Flow-Induced Corrosion; Fundamental Studies and Industry Experience, K. H. Kennelley, R. H. Hausler, and D. C. Silverman (Eds.), NACE, Houston, TX, 1991, pp. 9:1–9:21.

43. M. C. Roco, "Wear Mechanisms in Centrifugal Slurry Pumps," Corrosion, **46**, 424–431 (1990).

44. B. Poulson, "Complexities in Predicting Erosion Corrosion," Wear, **233**, 497–504 (1999).

45. A. Keating and S. Nesic, "Numerical Prediction of Erosion-Corrosion in Bends," J. Corrosion, **57**, 621 (2001).

46. S. Nesic, "Using Computational Fluid Dynamics in Combating Erosion-Corrosion," Chem. Eng. Sci. J., **61**(12), 4086 (2006).

47. X. H. Chen, B. S. McLaury, and S. A. Shirazi, "Application and Experimental Validation of a Computational Fluid Dynamics (CFD)-Based Erosion Prediction Model in Elbows and Plugged Tees", Comput. Fluids, **33**, 1251 (2004).

66

USING PLASTICS, ELASTOMERS, AND COMPOSITES FOR CORROSION CONTROL

E. I. DuPont de Nemours & Co. Inc., Wilmington, Delaware

B. Overview of plastics, elastomers, and composite materials

C. Characterization of polymers

D. Comparison of polymeric materials with metals

E. Application of polymers for corrosion control

F. Barrier applications (linings and coatings)
    F1. Introduction
    F2. Selection
    F3. Testing for selection
    F4. Design of vessels to be lined
    F5. Surface preparation
    F6. Thin linings
    F7. Thick linings
        F7.1. Reinforced thermosetting linings
        F7.2. Rubber (elastomer) linings
        F7.3. Thermoplastic linings
            F7.3.1. General-purpose linings
            F7.3.2. Fluoropolymer linings

G. Self-supporting structures: process vessels, columns, and piping
    G1. Thermoplastics
        G1.1. Design of piping
        G1.2. Joining of pipe
        G1.3. Welding of thermoplastics
        G1.4. Welded vessels
    G2. Thermosetting materials: FRP (RTP)
        G2.1. Resins
        G2.2. Reinforcements
        G2.3. Other additives
        G2.4. Curing systems
        G2.5. Selection of resin and laminate composition
        G2.6. Materials for dual-laminate construction
        G2.7. Selection of thermoplastic liners
        G2.8. Design and fabrication of FRP vessels and columns
            G2.8.1 Contact molding (hand layup)
            G2.8.2 Filament wound
        G2.9. Design and fabrication of FRP piping

H. Seals and gaskets
    H1. Gaskets
    H2. Seals
    H3. Selection of seals and gaskets

I. Failures and failure analysis

J. Condition assessment, fitness for service and repairs
    J1. FRP
    J2. Thermoplastic linings
    J3. Elastomeric linings

K. Economic data

L. Conclusion

Bibliography

A. INTRODUCTION

This chapter provides a broad overview of the application of polymers used for corrosion control in the chemical processing industry. The generalizations offered about polymer performance might suggest possible approaches, but they should not be considered as recommendations owing to the complexity of equipment design and the many variables of chemical processing. Consultations with experts, analyses of case histories, and testing are essential to the success of these materials.

Although a great deal of information regarding chemical resistance is available, much of it is not directly useful in selecting an appropriate construction material. It does, however, serve as a first step in determining likely candidates.

Polymer-based materials are attractive candidates due to their chemical resistance and favorable strength/weight ratio. Their use, however, is limited due to their viscoelastic nature and inability to withstand high temperatures. An overview of

Uhlig's Corrosion Handbook, Third Edition, Edited by R. Winston Revie
Copyright © 2011 John Wiley & Sons, Inc.

the use of polymer-based materials for corrosion control is presented in this chapter. All the same, these materials are used very cost effectively in many corrosive and hazardous applications.

B. OVERVIEW OF PLASTICS, ELASTOMERS, AND COMPOSITE MATERIALS

In their final useful form, plastics, elastomers, and composites are polymers made from organic chemicals. Polymers are long-chain molecules consisting of repeating units called monomers. Monomers are predominately manufactured from oil and natural gas, although exceptions exist in naturally occurring rubber and wood. A process of polymerization converts these monomers into polymers. Figure 66.1. illustrates different polymer structures. The polymers are then shaped into final forms using one of several molding processes, such as injection molding, extrusion, transfer molding, and so on. Appropriate fillers can be added during these processes to achieve certain desirable properties such as chemical resistance, strength, and processability. The end results of polymer processing are shapes and forms, which are referred to as plastics, elastomers, and composites. There is a great deal of overlap in these materials and the use of these terms has more to do with the evolution of the terminology than technical accuracy.

To describe polymeric materials more accurately, the following three distinct classifications are used:

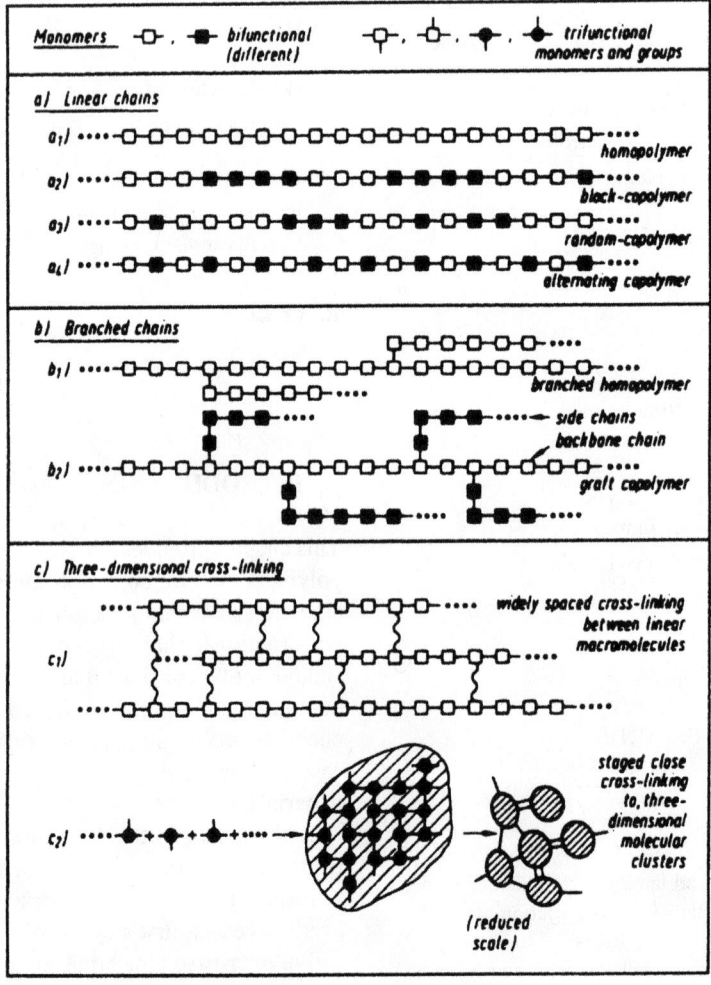

a and **b**: thermoplastics and thermoplastic elastomers
c₁: chemically cross-linked elastomers
c₂: thermosetting materials

FIGURE 66.1. Bi- and trifunctional polymer structures. (Reprinted with permission from the International Plastics Handbook, Hanser Gardner Publishers.)

- Generic nature
- Thermal processing behavior
- Mechanical behavior

Polymers are classified by generic nature according to the organic family to which it belongs. Examples are polystyrene, vinyls, fluoropolymers, and so on. These polymer families typically have several members.

Polymers are also classified by their *thermal processing characteristics* (i.e., thermoplastics or thermosetting). Thermoplastics can be remelted and reprocessed somewhat like ice cubes can be melted and refrozen into new shapes. These materials can be welded and are soluble in certain solvents. Thermoplastics are either amorphous or semicrystalline. Thermosetting materials cannot be remelted or reprocessed and heating them to high temperatures will result in their thermal decomposition and charring. Thermosetting properties are due to their cross-linked structures, which are achieved by the cure process (For elastomers, this process is known as vulcanization). Certain solvents can swell thermosetting materials. Figure 66.2 schematically illustrates thermoplastic and thermosetting materials.

Examples of thermoplastic polymers are:

Acrylics

Fluoropolymers

Vinyls

Polystyrenes

Polyphenelyne oxide

Polysulfones

Polypropylenes

Polybutylenes

Examples of thermosetting materials are:

Epoxies

Melamines

Phenolics

Polyesters

Urethanes (rigid)

The chemical structures of thermoplastic polymers commonly used for corrosion control are shown in Figure 66.3. Polymers can also be classified by their *mechanical properties* as

1. Rigid plastics, elastic modulus > 100,000 psi (7031 kg/cm^2)
2. Semirigid plastics, modulus between 10,000 and 100,000 psi (703 and 7031 kg/cm^2)
3. Nonrigid plastics, modulus < 10,000 psi (703 kg/cm^2)

If semirigid or nonrigid plastics additionally possess properties of high elongation and high recovery, they are known as elastomers or rubbers. Elastomers are defined by the American Society for Testing and Materials (ASTM) as "a macromolecular material that returns rapidly at room temperature to approximately its initial dimensions and shape after substantial deformation by a weak stress and the subsequent release of the stress." Elongation at the breaking point is known to be as high as 900% for many elastomers. Most commonly used elastomers are thermosetting in nature, that is, they are crosslinked. There are a few elastomers which are thermoplastic in nature. They are known as thermoplastic elastomers (TPEs). These, however, do not have applications in chemical handling.

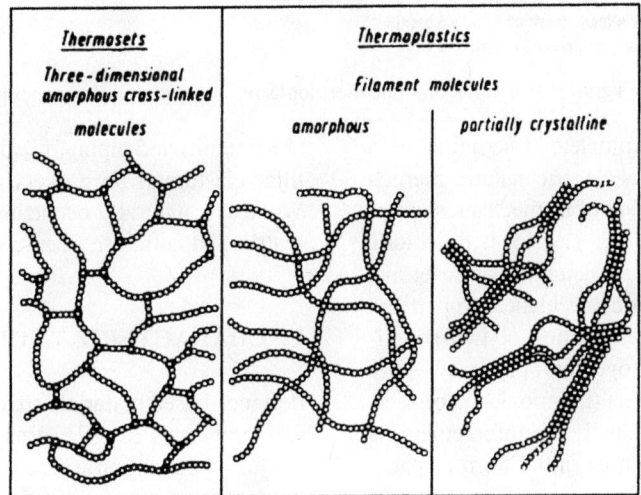

FIGURE 66.2. Structure of plastics-macromolecular arrangements. (Reprinted with permission from the International Plastics Handbook, Hanser Gardner Publishers.)

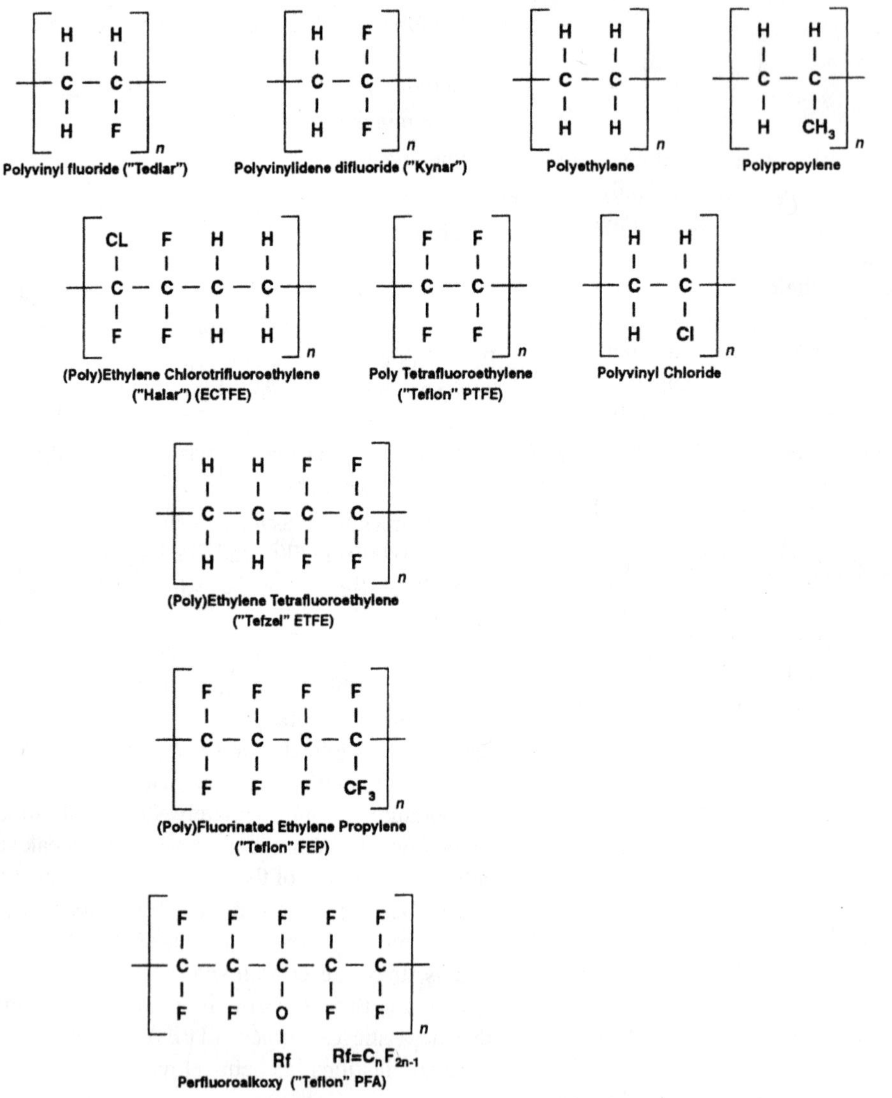

"Tefzel", "Tedlar" and "Teflon" are a trademarks of DuPont
Halar is a trademark of Solvay Solexis
Kynar is a trademark of Arkema

FIGURE 66.3. Chemical structures of some thermoplastic polymers used for chemical handling.

It becomes apparent that a complete description of a polymer material must include its generic nature, thermal processing method, and classification of its mechanical properties. For example, the rubber lining used for hydrochloric acid (HCl) is polyisoprene (chemical nature), thermosetting, (thermal processing), and elastomeric (mechanical properties).

Composite materials are a combination of two generically dissimilar materials brought together for synergy where one phase, the matrix, is continuous (thermoplastic or thermosetting) and the other phase, usually a reinforcement, is discontinuous. Reinforcements can be in the form of particulates, fibers, or cloth. A full description of a composite material must also include information on the reinforcement form. Examples of composites are fiber-reinforced plastic

(FRP) tanks and piping, filled fluoropolymer gaskets, scrim-filled elastomers for gaskets, and impoundment basin liners. Wood is a naturally occurring composite with lignin as the matrix and cellulose fibers as the reinforcing material.

C. CHARACTERIZATION OF POLYMERS

Polymer-based materials are characterized by their mechanical properties (tensile strength, tensile modulus, flexural modulus, elongation at break, impact strength, and hardness), thermal properties (melting point, melt flow index, transition temperature, heat distortion temperature, and coefficient of expansion), electrical properties (dielectric strength), and chemical compatibility (weight gain/loss, swelling, chemical attack such as oxidative or hydrolysis). At a more fundamental level polymers are also characterized by molecular weight and molecular weight distribution, specific gravity, crosslink density, and void content. A list of standard ASTM tests is given in Table 66.1. Manufacturers usually report these properties measured at room temperature. To determine these properties at other temperatures, the user should perform his or her own testing.

Polymer characterization is important from the standpoint of materials selection, processability, long-term performance, and failure analysis.

D. COMPARISON OF POLYMERIC MATERIALS WITH METALS

Compared to metals, most polymeric materials are viscoelastic, that is, they creep. This property has implications in all applications of corrosion control. Polymeric materials also are limited in their ability to withstand higher temperatures. By and large, their use as linings is restricted to 500°F (260°C) and most applications are under 300°F (149°C). Polymer structures such as tanks and piping are rarely used above 200°F (93°C). Polymeric materials are nonhomogeneous as a result of the addition of additives and reinforcements. They absorb liquids more readily than metals and are more permeable than metals. Therefore, unit properties (mechanical, electrical, etc.) of polymers can change over a period of time whereas the unit properties of metals remain the same even after corrosion. Polymer structures and linings are very workmanship sensitive. The supply chain leading up to the final product can be long and complicated. Each step in the chain has an element of workmanship. Polymeric materials do not lend themselves to easy identification once they are in the field. Failure analysis, condition assessment by nondestructive testing (NDT), and accelerated testing for selection are evolving fields. However, there are applications where only-polymeric materials have the required chemical resistance and cost-effective performance. When the end user has stayed involved in all phases of application (e.g., selection, lab and field testing, specification, vendor selection, inspection, maintenance, repair, and replacement), the results have been extremely satisfactory.

E. APPLICATION OF POLYMERS FOR CORROSION CONTROL

Polymeric materials fall into three broad categories from the end-user perspective:

- Barrier applications
- Self-supporting structures (which can be made of composites or solid polymers in tanks, piping, valves, pumps, and other equipment)

- Others such as column internals, seals, gaskets, adhesives, and caulks

Figures 66.4 (a)–(c) show the organization of polymer-based materials used for corrosion control.

F. BARRIER APPLICATIONS (LININGS AND COATINGS)

F1. Introduction

Linings are barriers typically on steel or concrete substrates. In special cases, a lining is also applied to a FRP structure called dual laminates.

Just about all components can be lined for corrosion control. Examples include storage tanks, reaction vessels, columns, piping, valves, pumps, sumps, trenches, and transportation equipment such as railcars and tank trucks. The following polymeric materials are used as linings or coatings:

Thermoplastic: polyvinylchloride (PVC, CPVC), polypropylene (PP), polyethylene (PE), and ethylene copolymers, fluoropolymers (PTFE, FEP, PFA ETFE). Figure 66.3 shows the chemical structures.

Thermosetting: epoxies, phenolics, vinyl esters, elastomeric (natural and synthetic rubbers).

Linings can be classified as follows:

- By method of application: sheet, spray or trowel.
- By thickness: thin [<25 mils (635 μm)] or thick [>25 mils (635 μm)]
- By generic nature

The lining selection is most often determined by the desired thickness. Successful lining applications must take the following steps into account:

Selection of linings
Design of the vessel for lining
Specification writing
Vendor selection
Surface preparation
Application/installations
Inspection
Monitoring

These steps are common to all types and forms of linings. Linings differ in their installation, inspection, and monitoring methods. Linings can be applied both in the shop and field. Where possible, shop lining should be preferred because of better atmosphere control.

TABLE 66.1. Industry Codes and Standards

	Product
ASTM D2310	Standard classification of machine made reinforced thermosetting resin pipe
ASTM D2517	Standard specification for reinforced thermosetting (epoxy) resin pipe
ASTM D2996	Standard specification for filament-wound reinforced thermosetting pipe[a]
ASTM D2997	Standard specification for centrifugally cast-reinforced thermosetting pipe
ASTM D4024	Standard specification for reinforced thermosetting resin flanges
ASTM D3299	Standard specification for filament-wound glass fiber-reinforced thermoset resin chemical tanks
ASTM D4097	Standard specification for contact-molded FRP chemical-resistant tanks
ASTM F1545	Standard specification for PTEE, PFA, FEP and ETFE lined piping
ASTM F492	Standard specification for PE and polypropylene (PP) plastic-lined ferrous metal pipe and fittings
ASTM D1784	Rigid poly(vinyl chloride) (PVC) compounds and chlorinated poly(vinyl chloride) (CPVC) compounds
ASTM D1785	Standard specification for poly(vinyl chloride) (PVC) plastic pipe, schedules 40, 80, and 120
ASTM D3350	Standard specification for polyethylene pipe and fitting materials
ASTM D1248	Polyethylene molding and extrusion material
ASTM D4020	Ultrahigh-molecular-weight polyethylene material
ASTM D1123	Nonmetallic (rubber) expansion joints
ISO 3994	Chemical transfer hose
ASTM D2564	Solvent cements for PVC (CPVC) joints
ASTM D1998	Polyethylene upright storage tanks
ASTM F118	Definition of gasket terms
ASTM F104	Nonmetallic gasket materials
ASTM D1330	Standard specification for rubber sheet gaskets
	Compatibility Testing
ASTM D543	Test methods for determining chemical resistance of plastics
ASTM D471	Effects of liquids on rubbers, test method
ASTM C868	Test method for determining chemical resistance of protective linings
ASTM C581	Test method for chemical resistance of fiberglass-reinforced thermosetting resin
ASTM C3491	Chemical resistance test for rubber linings by Atlas blind flange test
ISO 175	Determination of effects of liquid chemicals on plastic
ISO 4599	Determination of environmental stress cracking (ESC) by the bent-strip method
ISO 4600	Determination of environmental stress cracking (ESC) by the ball or pin impression method
ISO 6252	Determination of environmental stress cracking (ESC) by the constant tensile test method
ISO 8308	Liquid transmission and permeation through hose and tube
ASTM F363	Corrosion testing for gaskets
ASTM F146	Fluid resistance of gasket materials
ASTM D3615	Chemical resistance of thermoset molded compounds used in the manufacture of molded fittings
ASTM D3681	Chemical resistance of reinforced thermosetting resin pipe in a deflected condition
(Not Standardized)	Roberts cell blind flange test for linings exposed to high temperatures and pressures
	Testing for Mechanical Properties
ASTM D638	Standard testing for tensile properties of plastic
ASTM D412	Standard testing for tensile properties of rubbers
ASTM D1415	Test method for indentation hardness of rubbers—international hardness
ASTM D2240	Test for Durometer hardness of rubbers
ASTM D2538	Indentation hardness by Barcol hardness tester for rigid plastics
ASTM F38	Creep relaxation of gasket material
ASTM D395	Compression set of rubbers at ambient and elevated temperatures
ASTM F152	Tension testing of gasket materials
ASTM F36	Short-term compressibility and sealability of gasket materials
ASTM F37	Sealability of gasket material
ASTM F586	Determination of Y and M values of gaskets
ASTM D2290	Apparent tensile strength of tubular (pipe) products by the split-disk method
ASTM D2412	External loading characteristics of plastic pipe by parallel-plate loading
ASTM D695	Compressive properties of plastics
ASTM D429	Adhesion of rubbers to metallic substrates (procedure E1)
ASTM D1781	Climbing drum peel test
ASTM D903	Peel strength of adhesives on metals
ASTM D790	Flexural properties of reinforced and unreinforced plastics

TABLE 66.1.   (*Continued*)

	Product
ASTM D2584	Ignition loss of cured reinforced resin
ASTM D4541	Method of pull-off strength of coatings using portable tester
ASTM D4060	Test method for abrasion resistance of organic coatings
	Recommended Practices and Procedures
SSPC SP-5	Steel structure paint council white metal blast surface preparation
SPI FD118	Proposed test method for pinhole detection by high-voltage spark testing by DC
MTI Project 84	Spark testing practices for linings
NACE PRO 118	Tinker Rasor wet sponge testing for thin linings
ASTM D4787	Practice for continuity testing for linings on concrete
ASTM D4417	Method for measuring surface profile of steel surfaces for coating
ASTM D4258	Cleaning of concrete
ASTM D3486	Installation of vulcanizable rubber tank linings
	Codes
RTP-1	Reinforced Thermoset Plastic Corrosion Resistant Equipment, ASME, NY
Section X	ASME Boiler and Pressure Vessel Code (for fiber-reinforced plastic pressure vessels)
ASME B31.3	Chemical Process, Petroleum Refinery Piping Code

Note: Additional application-specific tests are shown in Table 66.2.
[a]There is no industry standard for contact molded FRP piping.

F2. Selection

Thickness forms the basis of the lining selection. Thickness is in turn dependent on the corrosion rate of the substrate, which is carbon steel (CS) in the majority of cases. Thin linings ≤25 mils 635 µm) are used if the corrosion rate of the CS is <0.25 mm/year (10 mils per year (mpy)). Thin linings are also used for localized corrosion resistance and in the case of thin fluoropolymers for nonstick and product purity. Thick linings >25 mils (635 µm) are used if the corrosion rate of the CS is >0.25 mm/year (10 mpy). Therefore, the corrosion rate of CS is an important determining factor. Linings used for concrete are almost always thick and are of reinforced thermosetting

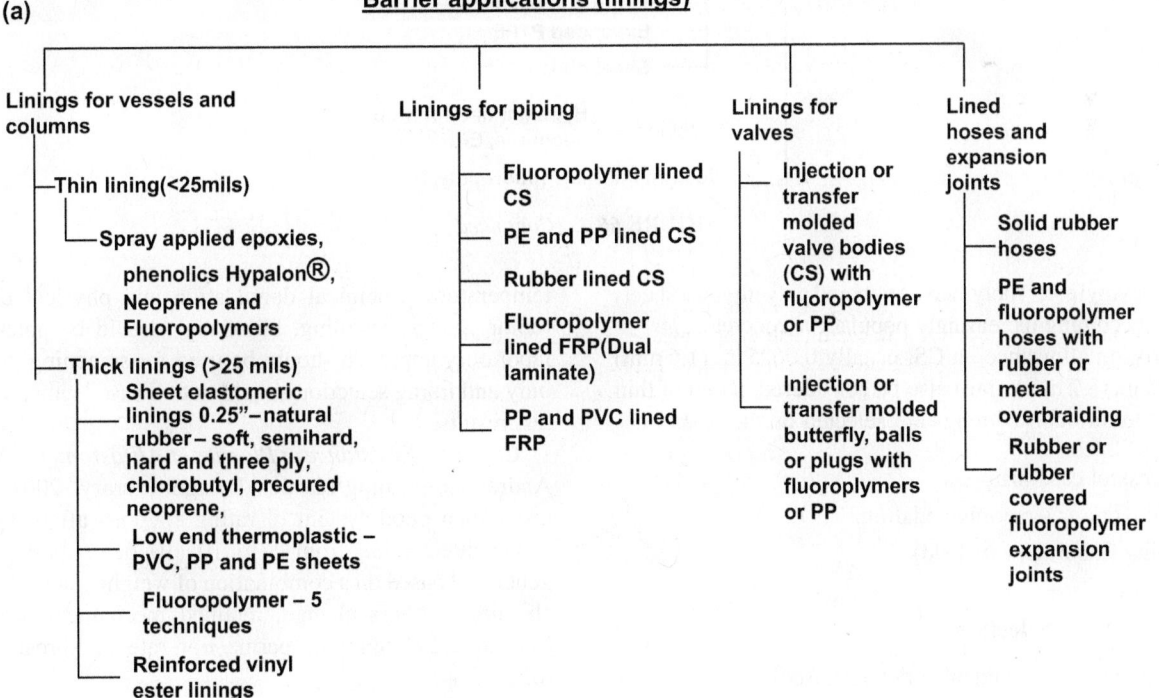

FIGURE 66.4. Overview of polymer-based materials for corrosion control showing classification of polymer-based materials by application: (a) barrier applications (linings); (b) self-supporting structures (vessels, piping, and valves); (c) other applications.

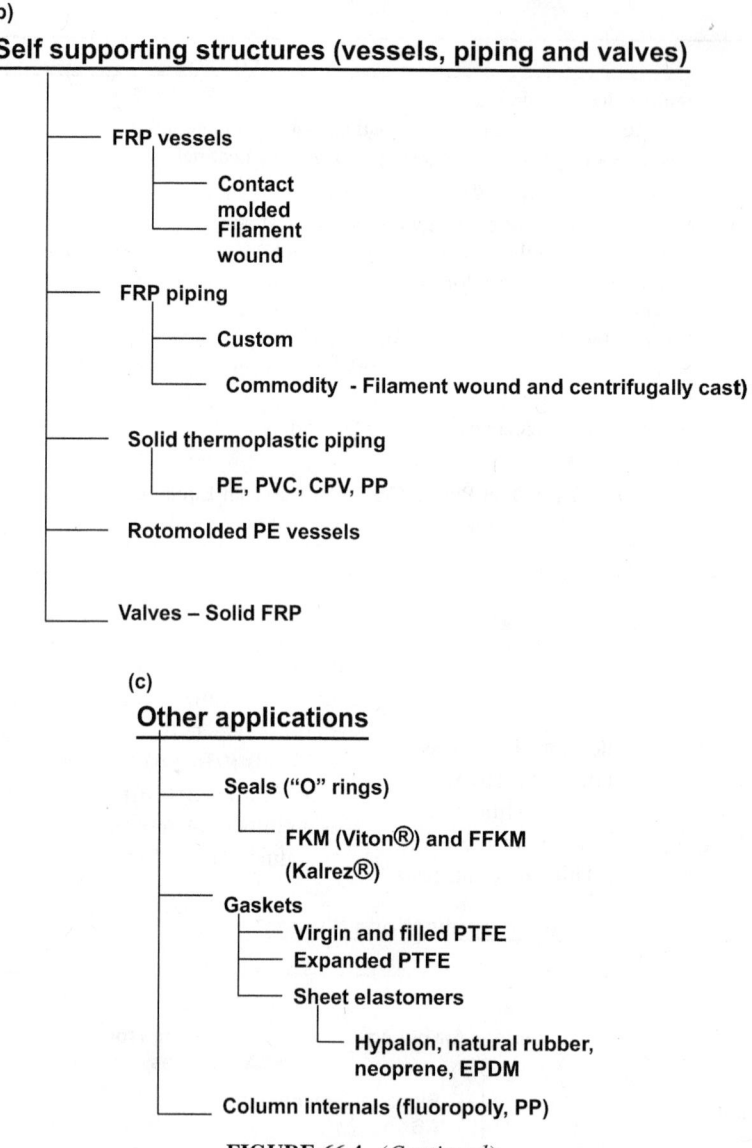

FIGURE 66.4. (*Continued*)

(epoxy or vinyl ester) polymers. Anchored polyethylene sheets are also becoming increasingly popular for concrete.

Corrosion allowance on CS, usually 0.0625 in. (1.6 mm) or 0.125 in. (3.2 mm), must also be considered in lieu of thin linings. Additionally, linings are selected on the basis of:

Successful case histories

Manufacturer's recommendation.

Testing (laboratory or field)

F3. Testing for Selection

Philosophy of testing and other details are covered in another chapter.

Laboratory testing can be simple coupon immersion (both in liquid and vapor) for a period of time at the application temperature. Chemical degradation and physical changes (color change, swelling, blistering) should be noted. This laboratory approach should be used for screening purposes only and lining selection should not be based entirely on the test results.

Chemical Resistance of Plastics and Elastomers (William Andrew Publishing/Plastics Design Library, 2001, 2008) provides a good system of rating coupons after exposure. A weighted value from 1 to 10 (10 being best) can be generated based on a combination of weight change, volume change, hardness change, retained mechanical properties (tensile and elongation), permeation rate, and breakthrough time.

In conjunction with the laboratory-screening test, Atlas blind flange tests (ASTM C868) should be performed [Fig. 66.5(a)]. These tests allow one-sided exposure of

FIGURE 66.5. (a) Atlas cell for one-sided exposure. (b) Manway lined with Teflon® PFA for field testing. (c) Typical atlas cell panel, before and after exposure.

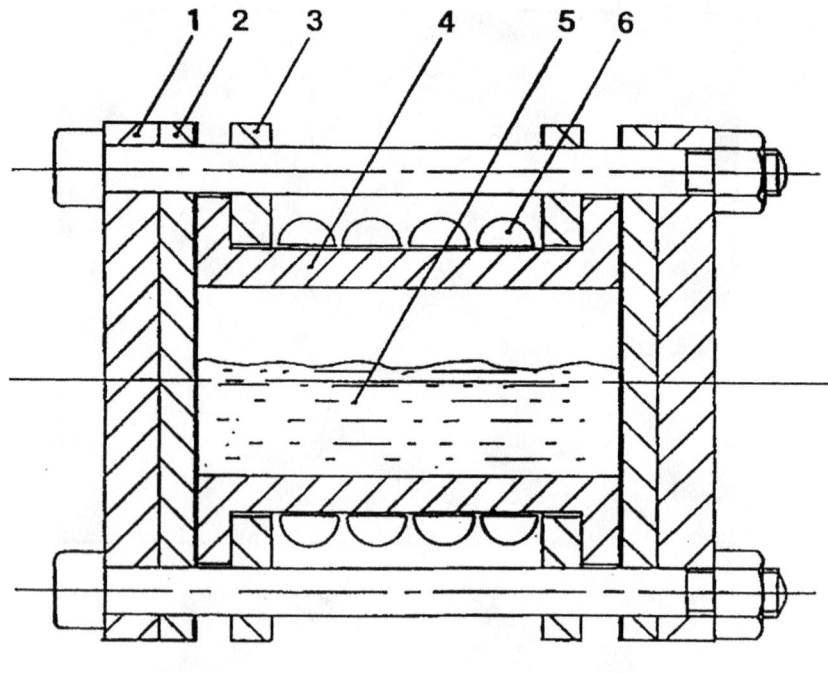

I. Steel Flange   4. Glass Lined Pipe

2. Coated Plate   5. Liquid

3. Loose Flange   6. Electric Ribbon Heater

FIGURE 66.6. Schematic representation of modified Roberts cell.

the lining and simulate the actual field conditions of temperature gradient across the lining. This is an important factor for lining candidates that are susceptible to permeation. Temperature gradient is the main driving force behind permeation. Exposure to liquid above atmospheric boiling point is carried out in a specially designed test cell for higher pressures (Fig. 66.6). Even more realistic than the laboratory test is a blind flange test in the field [Fig. 66.5(b)]. This involves installing a blind flange on a storage tank or a column and monitoring its performance.

Certain thermoplastic linings contain welded seams (e.g., polypropylene and fluoropolymers).The lining on the blind flange must contain a representative seam for accurate evaluation.

Frequent change of test solution is important as is the analysis of the changed liquid to determine pickup of additives from the linings. This testing should be carried out for as long as possible since there is no accelerated testing.

At the end of the test, blind flanges should be observed for discoloration, blistering, swelling, and oxidative attack. Thick linings should be subjected to spark testing before and after testing to detect pinholes in the lining. Thick linings should also be subjected to a peel pull test to determine loss of adhesion, and the substrate should be inspected for obvious signs of corrosion [Fig. 66.5(c)].

Blind flanges with thin linings are a little harder to evaluate. Peel pull tests are not possible, so a qualitative scribe test is employed. This test consists of scribing the lining with an X and lifting the liner. The degree of difficulty is compared with a similar scribing on an unexposed lining area. Instead of a spark test, a low-voltage wet sponge continuity test is employed to detect pinholes.

It is important to note the limitations of the laboratory blind flange test. This test may not be able to simulate the surface area –volume ratio of the actual unit. It is also difficult to simulate thermal shock or thermocycling. It is impossible to simulate other operations such as steaming or other forms of cleaning operations. Field blind flange testing is more meaningful in such cases.

For elastomeric materials it is more important to carry out application-specific testing than coupon screening. Test for elastomer selection listed in Table 66.2.

F4. Design of Vessels To Be Lined

Designing a vessel to be lined is somewhat different from designing a vessel that does not need to be lined. All welds need to be smooth or ground flush, and any weld splatter must be ground off. For field-erected tanks, the shell-to-roof joint must be continuous. Internals should be round pipes

TABLE 66.2. Properties and Test Matrix For Elastomer Selection

	Static Seal "O"-Ring Gasket	Dynamic Seal "O"-Ring Gasket	Diaphragms	Hose	Linings for Vessels, Pipe, Valve, Pumps, etc.	Expansion Joints	Lined Valves
Chemical resistance	ISO 1817 ASTM D471				ASTM C868		
Permeation				ISO 1399 ISO 2782 ISO 2528 ISO 6179 MTI Guide			
Weather and sunlight stability					ISO 4665 ASTM D1171		
Heat aging[a] upper use temperature limit				ISO 188 ASTM D454 ASTM D572 ASTM D573			
Low-use temperature limit				ISO R812 ASTM D746 (brittle temp.) ASTM D1229 (compression set)			
Coefficient of thermal expansion				ASTM D864			
Adhesion to substrate					ISO 814 ISO 813 ASTM 429[b]		ISO 814 ISO 813 ASTM 429[b]
Tensile tests, % modules and elongation at break		ISO 37 ASTM D412				ISO 37 ASTM D412	
Abrasion resistance					ISO 4649 ASTM D22258 (Pico) ASTM D5963		
Tear strength		ISO 34 ISO 816				SO 34 ISO 816	
Stress relaxation	ASTM D6147						
Resistance to compression	ASTM D395 ASTM D1229 (low temp.)						
Hardness				ASTM D1415, ISO 48, (Durometer), ISO 7619			
Fatigue life			ISO 132 ASTM D430 ASTM D843 ISO 133			ISO 132 ISO 133, ASTM D430 ASTM D843	

Note: Properly conducted testing on authenticated samples can afford accurate and reproducible data that can be used to "estimate" seal performance life.
[a]Can be combined with chemical resistance.
[b]Are carried out in conjunction with ASTM C868.

TABLE 66.3.   Fabrication Requirements for Vessels to be Lined

Specification Description	Rubber Linings	Thin Plastic Linings	Glass-Reinforced Plastic Linings	Sheet Linings[a] (Adhesively Bonded) Proprietary
Butt welds permitted	Yes	Yes	Yes	Yes
Lap welds permitted	No	Yes	Yes	No
Welds must be flush	Yes	No	No[b]	Yes
Welds must be smooth	Yes	Yes[c]	Yes[c]	Yes
Weld spatter permitted	No	No	No	No
Maximum nozzle length (in.)				
NPS 1	Not permitted (NP)	NP	NP	Unlimited[d]
NPS 2	6	NP	NP	Unlimited[d]
NPS 3	12	4	4	Unlimited[d]
NPS 4	24	8	8	Unlimited[d]
Minimum number of manholes in field-erected tanks	1	2 1	1 2	2
Grout tank bottom to minimize deflection or "oil canning"	No	Yes	Yes	Yes
Use insulating or structural lightweight aggregate concrete for tank foundation	No	Yes No	No	No

[a]Generally 60 mils thick or greater.

[b]Maximum height of weld bead shall be that permitted by ASME code or $\frac{1}{8}$ in., whichever is less. Welds shall blend into the adjacent surface so that the glass fabric saturated with plastic will be able to follow this contour and not leave a gap or air space.

[c]Weld ripples are acceptable if they are shallow enough so crevices and depressions can be cleaned by sandblasting to remove slag and oxides.

[d]Utilize insert linings. If loose linings are used for nozzles, venting is required.

instead of sharp corner angle irons or I-beams. Table 66.3 lists some lining requirements. Figure 66.7 shows the weld requirements.

F5. Surface Preparation

The extent of surface preparation depends on whether the equipment is new, used (without any lining previously), or to be relined. For new vessels, white metal blast per Steel Structure Painting Council (SSPC SP-5) followed immediately by a primer is needed. For used equipment, a thorough internal inspection is required and repairs or modifications must be performed as needed. Additionally, a chemical or steam cleaning may be needed to remove soluble salts and achieve a neutral pH of the surface. For equipment to be relined, a bake-out is usually required followed by a white metal blast to achieve 2.5–3.5 mils (64–89 μm) anchor profile on steel. In rare cases where stainless steel needs to be lined, a blast profile of 1.5 mils (38 μm) is adequate.

F6. Thin Linings

Classification of linings by thickness rather than lining material is useful because it focuses on both the application technology and lining functionality.

Thin linings <25 mils (635 μm) thick are predominantly based on epoxy and phenolic resins. To a lesser extent, spray-applied elastomeric thin linings (Hypalon® and Neoprene)

are also used. Thin linings are used for situations where the corrosion rate of carbon steel is less than 10 mpy.

Epoxy and modified epoxy resins are cured (crosslinked) using either a chemical curing agent or heat (see Tables 66.4 and 66.5). The curing agent is typically one of several amines (aliphatic, aromatic, or polyamide). Baked phenolic linings based on phenol formaldehyde resin cure by heat alone, typically at 450°F (232°C). Baked phenolic coatings are very brittle and must be applied at the specified thickness only (no more than 8 mils). Epoxy- and phenolic-based linings are used in storage tanks for a variety of chemicals such as wastewater, sodium hydroxide, potable water, fuels, organic compounds, and strong sulfuric acids. Baked phenolic linings have exceptional resistance to a wide spectrum of organic compounds. In general, their, use is limited to 150°F (66°C). Testing is recommended for use at higher temperatures. Epoxy- and phenolic-based linings are very economical when compared to thick linings by installed cost. Although the (National Association of Corrosion Engineers (NACE) specifies three levels of acceptance by the number of allowable pinholes, pinhole-free linings are the most commonly specified.

Epoxy and modified epoxy coatings are relatively easy to repair. Baked phenolic coatings are very difficult to repair. In-service linings are inspected visually and by means of a low-voltage wet sponge test (Tinker Rasor tester).

Spray-applied elastomeric coatings such as Hypalon® and neoprene have good chemical resistance and the ability to withstand thermal shock. However, they are expensive and labor intensive. (More information on elastomers can be found in Section F7.) They are spray applied in multiple coats and are used in hopper car and caustic storage. Spray-applied natural rubber is also used as membranes for acid bricks.

There is a special class of thin linings based on fluoro-polymers (PTFE, polytetrafiuoroethylene; FEP, fluorinated ethylene propylene; PFA, perfluoroalkoxy; ETFF, ethylene tetrafluoroethylene; CTFE, chlorotrifluoroethylene; and PVDF, polyvinyledene fluoride). They are applied to primed surfaces as sprayed waterborne suspensions or electrostatically charged powders. Each coat is baked before

the next is applied. Table 66.6 presents details about these coatings. These linings are expensive and are used mainly for nonstick and product purity applications, particularly at thickness <20 mil (508 μm). They can also be applied at higher thickness as thick linings (see Section F7).

F7. Thick Linings

Thick linings cover a large range of polymers such as polyvinyl chloride (PVC), polyethylene (PE), natural and synthetic elastomers, reinforced vinyl esters, and fluoropolymers. Table 66.7 lists the thick linings commonly used for corrosion control. All the guidelines of selection, vessel

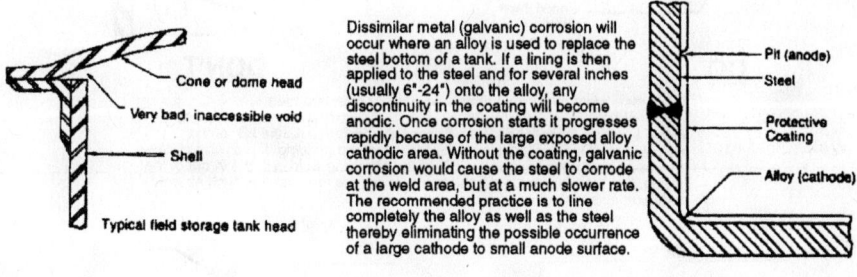

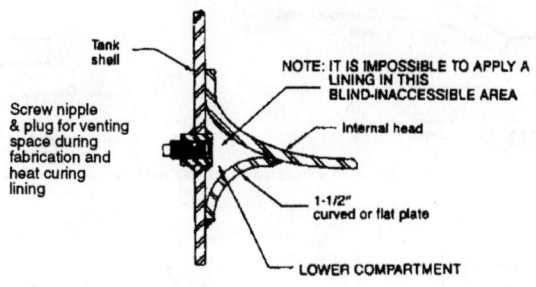

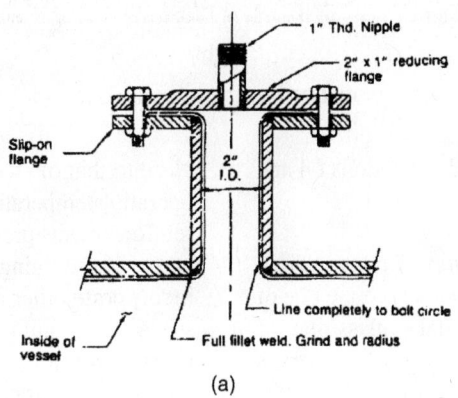

(a)

FIGURE 66.7. Requirements for the quality of welds for lining. (Courtesy of Wisconsin Protective Coatings Co.)

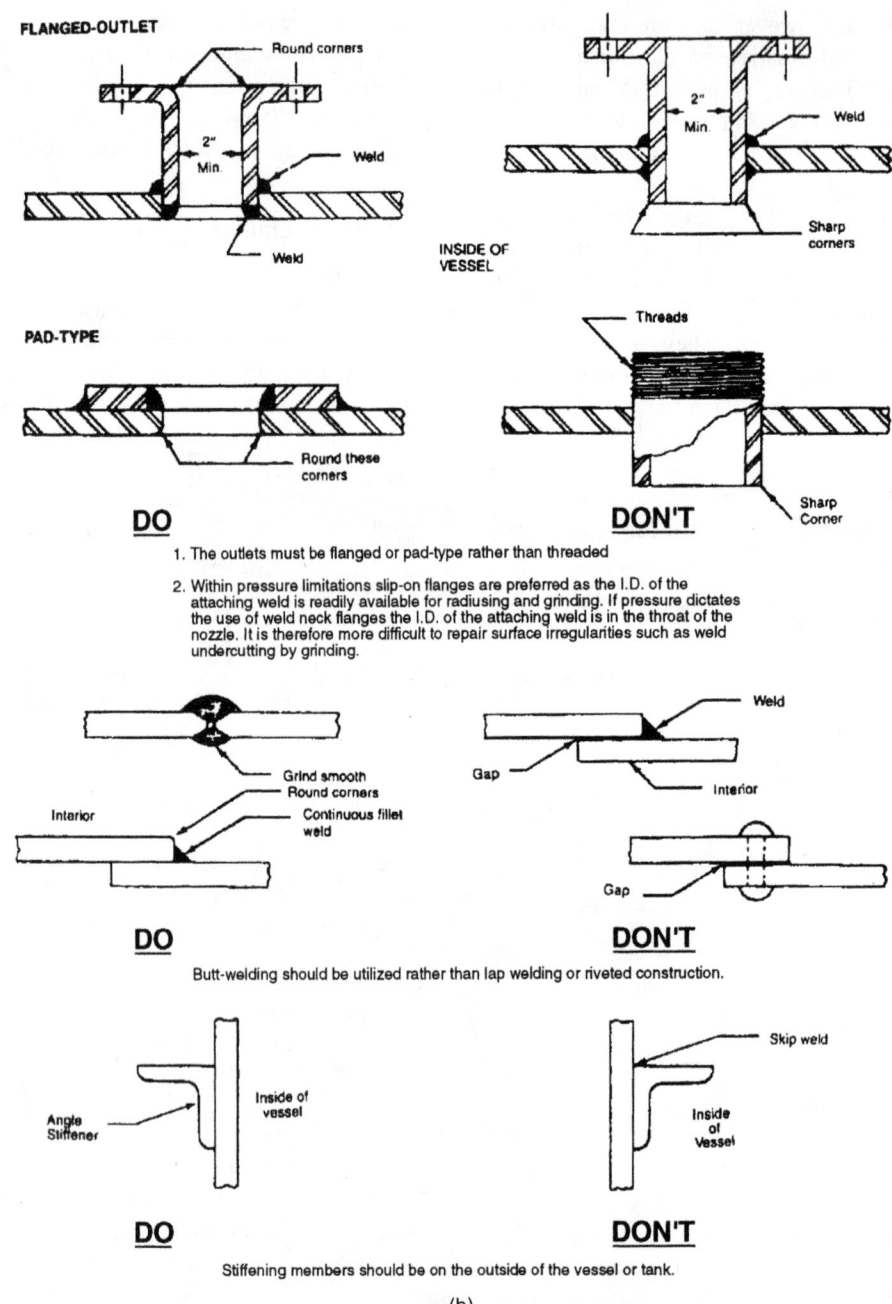

(b)

FIGURE 66.7. (*Continued*)

design, and surface preparation described in Sections F4 and F5 apply equally to thick linings.

F7.1. Reinforced Thermosetting Linings. Epoxy-, vinyl ester–, and polyester-based lining systems incorporate one of three types of reinforcements (i.e., glass flake, glass mat, or woven roving, see Section H3.2.2 for vinyl esters). Each serves a specific purpose. In general, glass or other forms of flake offer the maximum level of permeation resistance. Mat and woven cloth lower the coefficient of thermal expansion of the resin to close to that of steel. This is important for service where the operating temperature is higher than ambient. In addition, these reinforcements prevent cracking due to shrinkage as the resin cures. These linings are applied in two to three coats, which incorporate other additives such as other inert flakes (mica, silica, aluminum oxide) and pigments. These additives impart additional properties such as abrasion resistance, identification of coats, and so on. Figure 66.8 illustrates the components of a reinforced lining system. Figures 66.9 (a)–(d) illustrate the multilayer nature of these linings.

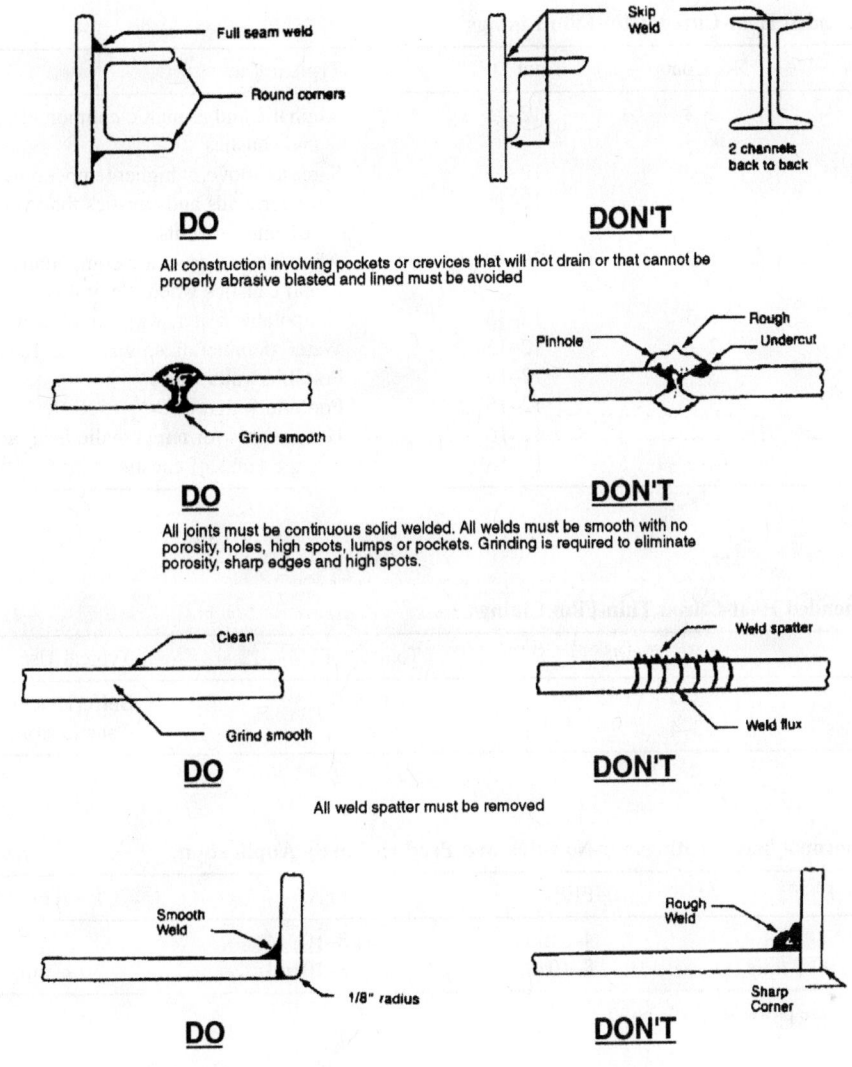

FIGURE 66.7. (*Continued*)

These linings are either trowelled or spray applied. They are applied on concrete as well as steel. Crosslinking is by a chemical curing agent and additional heat is rarely required. Typical thickness is 0.125 in. (3.2 mm) although higher thicknesses of up to 0.187 in. (4.8 mm) are also used. Maximum service temperature is limited to 170°F (77°C). They can be applied in the shop as well as in the field, which makes them attractive for a variety of applications such as waste treatment, ground water remediation, and hydrochloric acid. Table 66.8 is a general guide listing the chemical resistance rating of reinforced linings. For specific cases testing is recommended (see Section F3).

Vinyl ester–based linings are used for vessels, large-diameter columns, and concrete sumps and trenches.

Plasticized PVC is another useful thick lining material for service with many acids and bases. The maximum service temperature is ~150°F (66°C). It is sometimes used in combination with unplasticized PVC. Sheets are bonded with adhesive to primed surfaces.

F7.2. Rubber (Elastomer) Lining. Thermosetting rubber linings, either natural or synthetic, are well established in the chemical processing industry (CPI). Their highly elastic nature permits use as linings for vessels, column, piping, and valves. Typically, the minimum thickness is 0.25 in. (6.4 mm) for vessel linings and 0.125 in. (3.2 mm) or 0.1875 in. (4.8 mm) for piping, depending on size. These linings are installed as adhesive-bonded sheets in vessels and are heat cured (vulcanized) after installation. Loose linings for impoundments and spray-applied rubber coatings for some applications are also known.

TABLE 66.4.   Recommended Heat-Cured Thin-Film Linings

Generic Type	No. Coats	Total DFT[a]	Typical Uses
Epoxy phenolic	2–3	12–15	Aliphatic and aromatic compounds, waste treat, weak acids and caustics
Epoxy phenolic glass flake	2–3	12–15	Same as above, at higher temperature and abrasion resistance
Epoxy Novalac	2–3	12–15	Stronger acids and caustics than above process affluent streams, solvents
Epoxy polysulfide	1–2	20–40	Aliphatic and aromatic compounds, waste treat, weak acids and caustics where flexibility is required
Epoxy amine	2–3	12–15	Nonpotable water, waste treatment
Epoxy amine	2–3	12–15	Water, demineralized water 82–121°C (180–250°F)
Epoxy polyamide	3	12–15	Portable water, fuels
Epoxy polyamide	3	12–15	Portable water, rules
Hypalon®	2–3	12–16	Hopper cars for terephthalic acid, adipic acid
Neoprene	2–3	12–16	Storage tanks of caustic

[a]Dry film thickness.

TABLE 66.5.   Recommended Heat-Cured Thin-Film Linings

Generic Type	No. Coats	Total DFT	Typical Uses
High baked phenolic	3–4	6–8	Sulfuric acid rail cars, organic waste
High baked epoxy	3–4	12	Caustic storage

TABLE 66.6.   Thin Fluoropolymer Coatings for Nonstick and Product Purity Applications

	ETFE	FEP	PFA	E-CTFE	PVDF
Dispersion		4–5 mils	8–10 mils		25 mils
Powder	90 + mils	8–10 mils	8–10 mils	50 + mils	25 mils

Note: For thicker versions of these please see Table 66.17

TABLE 66.7.   Thick (>25-mil) Linings

Reinforced vinyl ester or epoxy linings
  Reinforced with glass cloth, mat or woven
  Spray or trowel applied
  Shop or field applied
  Chemically cured (no baking required)
  Relatively easy to repair

Sheet rubber linings
  Natural rubbers (soft, semihard, butyl)
  Standard thickness $\frac{1}{4}$ in.
  Shop or field applied
  Steam or autoclave curing needed
  Relatively easy to repair

Thermoplastic sheet lining
  General purpose (PVC, PP) for steel and PE (anchored) for concrete
  High performance (fluoropolymers—PVDF, FEP, PFA, ETFE, ECTFE) for steel

Elastomers are classified as either natural or synthetic and as general purpose, or high performance (Tables 66.10A and B). Rubber linings for chemical handling are usually of general -purpose type. High-performance rubbers are used principally for seal applications. In the past, rubber linings were also used for piping.

Natural rubber is manufactured from naturally occurring rubber gum mixed with carbon, antioxidant, accelerator, and crosslinking agent (Table 66.9 shows a typical recipe). The ingredients are mixed in a Banbury mixer and sheets are calendered with multiple layers. These sheets are applied as a lining to the interior of vessels, piping, and pumps. The lining is applied over a sandblasted and primed surface. Adhesive is applied to both the metal side and the rubber sheet. Joints are made by overlapping two adjoining sheets and "skiving," that is, either up skiving or down skiving. For semihard, hard, and chlorobutyl rubbers, a soft rubber backing material is used to attain better adhesion [Figs. 66.10(a) and (b)].

Elastomeric linings for chemical handling are natural rubbers. Occasionally synthetic rubbers such as ethylene

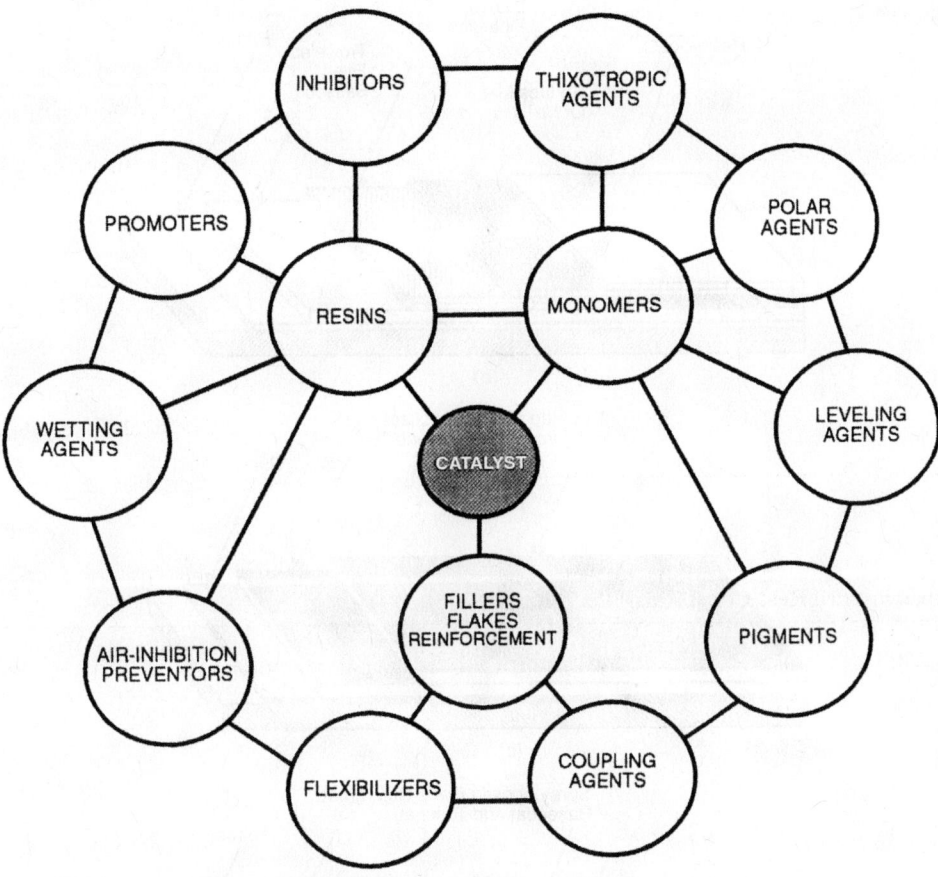

LINING COMPONENTS

FIGURE 66.8. Elements of a reinforced lining system. (Courtesy of Ceilcote Company.)

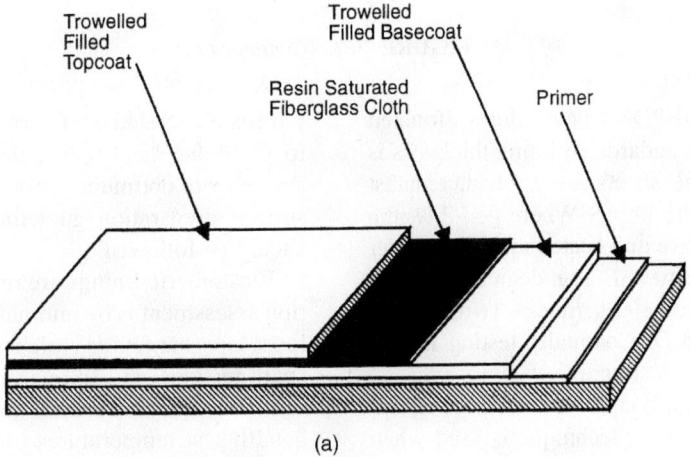

(a)

FIGURE 66.9. (a). Fiberglass cloth-reinforced lining with filled basecoat and topcoat. (Courtesy of Ceilcote Company.) (b). Fiberglass mat-reinforced lining with silica-filled basecoat. (Courtesy of Ceilcote Company.). (c). Trowelled glass flake lining. (Courtesy of Ceilcote Company.). (d). Spray-applied flake-reinforced lining. (Courtesy of Ceilcote Company.)

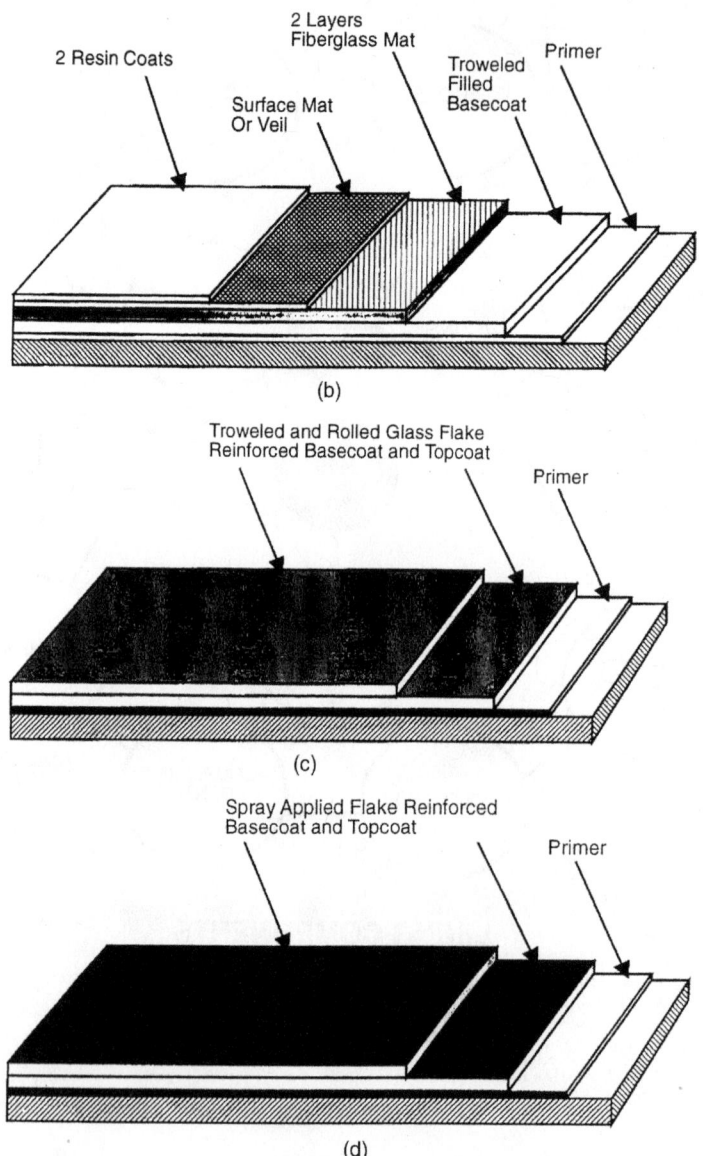

FIGURE 66.9. (*Continued*)

propylene diene monomer (EPDM) and chlorosulfonated polyethylene are used. The standardized lining thickness is 0.25 in. (6.35 mm). Once the sheets are applied, exhaust steam heat is used to cure the lining. Where possible, the vessel is placed in an autoclave and heated until the proper degree of crosslinking is achieved. The degree of crosslinking is determined by Durometer hardness. Two types of Durometers are used: A and D. Continuity testing is performed by high-voltage spark testing.

Cure is also achieved by applying sulfur-bearing organic compounds. This chemical curing technique is used when steam or autoclave curing is not possible, as in the case of repairs.

Because of the flexible nature of rubber, installation and use are straightforward. It is common to see rubber on bolted joints and on relatively sharp surfaces. It is also common not to grind the steel welds flush but to simply grind them smooth. For optimum performance, however, the design and surface preparation guidelines (see Sections F4 and F5) should be followed.

Elastomeric linings are relatively easy to repair. Condition assessment is by internal visual inspection, looking for loose laps, hardening surfaces, cracks, blisters, pinholes (by spark testing), and so on. Various rubber formulations are available for resistance to chemicals and food-grade handling at temperatures up to 180°F (82°C). Successful applications include storage vessels and railcars for HCl, HF, and abrasive slurries. Table 66.11 shows the ASTM classification of rubber and Table 66.12 is a chemical resistance chart.

TABLE 66.8. Chemical Resistance Rating for Polyester and Vinyl Ester Linings

Chemical	Generic Resin Type					Resin Abbreviations	
	ISO	BIS-A	CHLOR	VINYL-E	NOV-VE		
Acetic acid, 10%	D1	C1	C1	C1	A1	ISO	Isophthalic
Acetic acid, 100%	E2	D2	D2	D2	D1	BIS-A	Bisphenol-A fumarate
Acetone	N	N	N	N	C2	CHLOR	Chlorinated
Ammonium hydroxide — 20%	N	E1	N	E1	E1	VINYLE-E	Vinyl easter
Ammonium nitrate	A1	A1	A1	A1	A1	NOV-VE	Novolac vinyl easter
Benzene	D2	D1	D1	E2	D1		
Chromic acid, 10%	N	E1	C1	N	E1	Ratings	
Formaldehyde	A1	A1	A1	A1	A1		
Formic acid	E2	D1	D1	E1	D1		
Gasoline, unleaded	A1	A1	A1	A1	A1	1	Good for immersion or constant flow
Hydrochloric acid, 10%	A2	A1	A1	A1	A1		
Hydrochloric acid, 37%	E2	D2	D2	D1	D1		
Hydrofloric acid, 10%	E1	D1	D1	D1	D1	2	Limited to spillage or secondary containment
Kerosene	A1	A1	A1	A1	A1		
Methene Cloride	N	N	N	N	E2		
Methyl ethyl ketone	E2	E2	E2	E2	D2	A	Good to 200° F
Nitric acid, 10%	D2	B1	A1	CI	A1	B	Good to 180°F
Nitric acid, 60%	N	D1	D1	N	D1	C	Good to 140° F
Oils	A1	A1	A1	A1	A1	D	Good to 120°F
Phosphoric acid, 85%	A1	A1	A1	A1	A1	E	Good to 100°F
Sodium chlorate	B1	A1	A1	A1	A1	N	Not recommended
Sodium hydroxide, 10%[a]	N	D1	N	D1	D1		
Sodium hydroxide, 50%[a]	N	C1	N	C1	C1		
Sulfuric acid, 50%	B2	B1	A1	B1	A1		
Sulfuric acid, 75%	E2	E1	E1	E1	E1		
Toluene	N	E2	E2	E2	E1		
Trichloroethylene	N	E2	E2	E2	E1		
Vinegar	A1	A1	A1	A1	A1		

Note: The above ratings are typical for polyester and vinyl ester linings and reflect the maximum recommended temperatures. Maximum temperature for any given lining type may be lower depending on lining thickness and permeation resistance. Courtesy of Ceilcote Co.
[a]Exposed lining surface requires carbon filler or synthetic veil.

F7.3. Thermoplastic Linings.

Thermoplastic linings fall into two broad categories: general purpose and high performance. General-purpose linings consist of PVC, CPVC (chlorinated polyvinyl chloride), PE, and PP (polypropylene). High-performance thermoplastic lining materials include fluoropolymers such as PVDF, CTFE, ETFE, polytetrafluoroethylene (PTFE), FEP and perfluoroalkoxy (PFA).

Thermoplastic linings are used for vessels, piping, valves, and pumps, Common lining methods include sheet linings for vessels, extruded loose linings for piping, and injection molded linings for valves and pumps. In some cases, spray-applied linings are also used. Examples are flame spray coatings of ethylene methacrylic acid and its ionomer. Spray and baked fluoropolymer linings will be discussed later.

F7.3.1. General-Purpose Linings.

General-purpose linings are limited by their maximum temperature use and chemical resistence. Usually, their use is limited to 150°F (66°C). Many organic solvents easily attack general-purpose linings, but due to their relative low cost, they can be successfully used for many inorganic acids.

Among the commonly used thermoplastics materials in this category are PP (homopolymer and impact-grade co-polymer), PVC (rigid and plasticized), CPVC, and PE. Table 66.13 summarizes these linings for vessels.

TABLE 66.9. Formulation of a Soft Natural Rubber

Ingredients	Parts by Weight
Natural rubber gum	100
Sulfur powder	1–2
Accelerator	1
Carbon black	25–50
Internal plasticizer (oils and resins)	1–3
Antioxidants	1

TABLE 66.10A.   Chemical Structures of Commonly Used Elastomers

Name[a]	Chemical Name	Structure
Natural rubber	*Cis* –1,4-Polyisoprene	$[CH_2-C-CH-CH_2]_n$ $\quad\quad\; \vert$ $\quad\quad CH_3$
Nitrile	Poly(butadiene-*co*-acrylonitrile)	$[(-CH_2-CH=CH-CH_2-)_3(-CH_2-CH-)]_n$ $\quad\quad\quad\quad\quad\quad\quad\quad\quad\quad\quad\quad\;\; \vert$ $\quad\quad\quad\quad\quad\quad\quad\quad\quad\quad\quad\quad CN$
Butyl	Poly(isobutylene-*co*-isoprene)	$\quad\quad CH_3$ $\quad\quad\;\; \vert$ $[-(CH_2-C-)_m(-CH_2-C=CH-CH_2-)]_n$ $\quad\quad\;\; \vert \quad\quad\quad\quad \vert$ $\quad\quad CH_3 \quad\quad\; CH_3$
Neoprene	Polychloroprene	$[-CH_2-C=CH-CH_2-]_n$ $\quad\quad\quad \vert$ $\quad\quad\quad Cl$
EPDM (Nordel®)	Poly(ethylene-*co*-propylene-*co*-diene)	$[-(-CH_2-CH_2-)_{37}(-CH_2-CH)_{13}-diene-]_n$ $\quad\quad\quad\quad\quad\quad\quad\quad\quad\quad\quad \vert$ $\quad\quad\quad\quad\quad\quad\quad\quad\quad\quad CH_3$
Chlorosulfonated Polyethylene (Hypalon®)		$[-CH_2-CH-CH_2-CH_2-CH_2-CH-CH_2-]_n$ $\quad\quad\; \vert \quad\quad\quad\quad\quad\quad\quad\quad \vert$ $\quad\quad\; Cl \quad\quad\quad\quad\quad\quad\quad SO_2Cl$

[a]Nordel and Hypalon are registered trademarks of DuPont Dow Elastomers Company.

These materials are also used as linings for valves, pumps, and piping. Ultrahigh-density polyethylene is used as valve linings for its exceptional abrasion resistance. High-density PE is used as lining for complex shapes using the rotolining process. Polypropylene is used as lining for valves using the injection molding process. Extruded PP liners are used for lined pipe.

F7.3.2. Fluoropolymer Linings. Fluoropolymers, also known as high-performance thermoplastics, are well-proven materials used as thick linings to control corrosion of process equipment. They resist a broader range of chemicals than other polymers and generally have higher service temperatures as well. They typically compete with high-nickel alloys as well as titanium, tantalum, and zirconium.

TABLE 66.10B.   Specialty Elastomers—Fluoroelastomers

Name	Tradenames[a]	Structure
Vinyledene fluoride/hexafluoropropylene	Viton®, Fluorel®, Technoflon®	$\quad\quad\quad\quad\quad CF_3$ $\quad\quad\quad\quad\quad\; \vert$ $[(CH_2-CF_2)_x(CF_2-CF)]_y$
Vinyledene fluoride/hexafluoropropylene/ tetrafluoroethylene/terpolymer	Viton®, Fluorel®, Technoflon®, Dai-El®	$(CH_2-CF_2)_x(CF_2-CF)_y(CF_2-CF_2)_z$ $\quad\quad\quad\quad\quad\quad\quad \vert$ $\quad\quad\quad\quad\quad\quad\; CF_3$
Tetrafluoroethylene/propylene copolymer	Aflas®	$(CF_2-CF_2)_x(CH_2-CH)_y$ $\quad\quad\quad\quad\quad\quad\quad \vert$ $\quad\quad\quad\quad\quad\quad\; CH_3$
Vinylidene fuoride/chlorotrifluoroethylene copolymer	Kel-F®	$(CF_2-CH_2)_x(CF_2-CFCl)_y$
Perfluorocarbon rubber (tetrafluoroethylene copolymer)	Kalrez®	$(CF_2-CF_2)_x(CF_2-CF)_y(CF_2-CF)_z$ $\quad\quad\quad\quad\quad\quad \vert \quad\quad\quad\; \vert$ $\quad\quad\quad\quad\quad OCF_3 \quad\quad X$

[a]Viton® and Kalrez® are registered trademarks of DuPont Dow. Fluorel® is a registered trademark of 3 M Company. Aflas® is a registered trademark of Green Tweed Company.

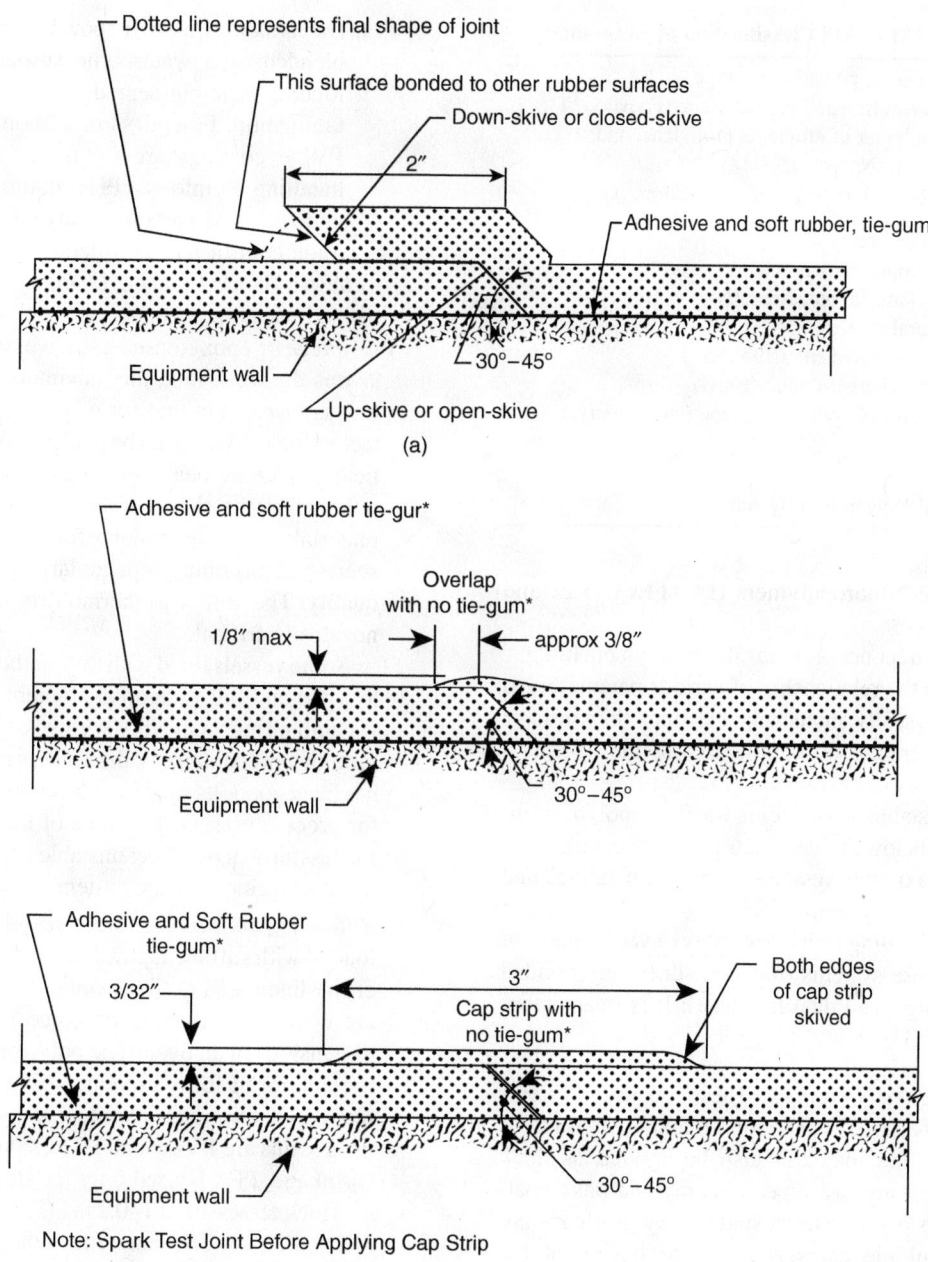

Dotted line represents final shape of joint

This surface bonded to other rubber surfaces

Down-skive or closed-skive

2″

Adhesive and soft rubber, tie-gum

Equipment wall

30°–45°

Up-skive or open-skive

(a)

Adhesive and soft rubber tie-gur*

Overlap
with no tie-gum*

1/8″ max

approx 3/8″

Equipment wall

30°–45°

Adhesive and Soft Rubber
tie-gum*

3/32″

3″

Cap strip with
no tie-gum*

Both edges
of cap strip
skived

Equipment wall

30°–45°

Note: Spark Test Joint Before Applying Cap Strip

(b)

FIGURE 66.10. (a) Skived joint for semihard or hard rubber lining (e.g., code B) having a thin, soft-rubber, tie gum surface for bonding to steel. (b) Low-profile joints as typically called for under brick linings. (Courtesy of Dupont Co.)

Fluoropolymers are widely specified when contamination must be minimized.

There are two general categories of fluoropolymers: fully and partially fluorinated. The fully fluorinated materials PTFE, FEP, and PFA have higher service temperatures and broader chemical resistance than other fluoropolymers. The partially fluorinated fluoropolymers ETFE ethylene chlorotrifluoro, thylene (ECTFE), and PVDE have higher tensile strength and stiffness and less extensive chemical and thermal resistance.

Table 66.14 lists key properties of fluoropolymers. It should be noted that service temperatures vary with a number of factors, including the mechanical requirements of lining systems, and may be significantly lower than the maximum service temperatures shown.

Fully fluorinated fluoropolymers can be used with most aggressive corrosives. Partially fluorinated types can perform well, depending on the chemicals involved. Table 66.15 summarizes the chemical resistance of

TABLE 66.11.   ASTM D1418 Classification of Elastomers

M:	Saturated chain of polyethylene type
	CSM: chlorosulfonated polyethylene (Hypalon®)
	EPDM: terpolymer of ethylene, propylene, and a diene (Nordel®)
	FKM: fluoroelasters (e.g., Viton®, Kalrez®)
R:	Unsaturated carbon type
	CR: chloroprene (neoprene)
	NBR: acrylonitrile butadiene (Nitrile, Buna N)
	NR: natural rubber
	SBR: styrene butadiene (Buna S)
	IIR: isobutylene isoprene (Butyl)
	CHR: chloro-isobutylene-isoprene (chloro-butyl)
U:	Polyurethane rubbers
Q:	Silicon- and oxygen-bearing rubbers

partially fluorinated fluoropolymers (ECTFE, ETFE, and PVDF).

Permeation is a concern with fluoropolymer linings. Table 66.16 shows the relationship of various parameters to permeation rate. Usually increasing the liner thickness or reducing the temperature differential across the liner is a practical approch.

There are six established systems for fluoropolymer linings as described below. Table 66.16 provides details such as thicknesses, maximum vessel size, vacuum rating, and repairability for these systems.

Thermoplastic welding technology plays a very important role in three of these systems (namely, adhesively bonded linings, loose linings, and dual laminate). It is treated separately in Section G.

1. *Spray-and-Bake Coatings.* These coating systems were described earlier (Section F6 and Table 66.6) as thin linings, but they can also be applied as thick linings. There are two types of spray-and-bake coatings: dispersion and electrostatic spray. Both are applied in multiple coats and require baking at the melting temperature of the polymer. These coatings frequently require the use of appropriate primers to promote adhesion between the fluoropolymer and the substrate. Since the coating is applied as a spray in both cases, the design of the vessel must include body flanges for direct access to the surface. Since baking is performed in ovens, the largest size oven available in the industry determines the largest size vessel that can be coated. For this reason, this technology is frequently used for small parts such as agitators, pump housings, small open-top containers, and dip tubes. Spray-and-bake coatings can also be used for large units. However, great care must be taken to ensure that there is proper hot air circulation in the oven.

1a. *Dispersion.* Polymer powder of 30–60 μm is blended as a waterborne suspension. Often the topcoat is unpigmented to prevent product contamination. Fiberglass or carbon cloth-reinforced PVDF coatings are frequently used for H_2SO_4 handling. Reinforced PFA coatings with wire mesh and activated carbon are also used where permeation control is a consideration.

2. *Adhesive-Bonded Sheet Linings on Steel.* Fabric-backed sheets are bonded to steel vessel walls with neoprene or epoxy-based adhesive. Because the adhesive is the weak link, the maximum service temperature is lower than that for other fluoropolymer lining technologies. Work can be performed in the shop or the field. Joints are heat welded using a rod of the same polymer as the sheets. Table 66.16 lists the range of materials currently available for this lining system. The success of this lining depends largely on the fabrication quality. The ability to thermoform the head and flare nozzles is critical.

Some vessels lined with this method have been rated to withstand full vacuum at ambient temperature. The vacuum ratings may vary with vessel size. When used for transport containers, the success rate for this system has been good. Its success rate has been fair when used for process vessels. This type of lining has been used successfully in handling unstable sodium hypochlorite in the chlorine destruct systems.

3. *Rotolining on Steel.* A steel vessel or component is loaded with sufficient powdered fluoropolymer to cover the lined area at the specified thickness. Openings are covered, and then the piece is rotated in three dimensions in an oven. The polymer melts and covers the interior to form a seamless lining. Rotolining is a relatively new process for fluoropolymers. Commonly used resins are PVDF and ETFE. Due to a high rate of shrinkage, PFA is used only for small sizes.

Thicknesses of 0.1–0.2 in. (2.54–5.1 mm) are the most common. The size of the coated piece is limited by the oven size. The largest oven today is 8 ft (2.4 m) in diameter and 22 ft (6.7 m) long.

Rotolining is very effective for sections of columns and for components like complex piping where its use allows eliminating flanged connections. It has been used to line scrubbers and neutralization equipment used to manufacture chlorofluorocarbons.

4. *Loose Linings.* Fluoropolymer sheets are welded into lining shapes, folded, and slipped into the housing. The lining is flared over the body and nozzle flanges to hold it in place. The American Society of Machanical Engineers (ASME) code allows weep holes to release permeants. FEP, PFA, and modified PTFE have been successfully used as loose linings.

TABLE 66.12. Chemical Resistance of Rubber Linings

Chemicals	Remarks	Neoprene	Soft Natural Rubber	Semihard and Hard Rubber	Butyl	Hypalon
Acetic acid, dilute		NR	NR	NR	150	NR
Acetic acid, glacial		NR	NR	NR		NR
Acetic anhydride		NR	NR	NR		
Acetone		NR	NR	NR	NR	NR
Alum: ammonium		200	150	200	150	175
Alum: chrome		200	150	200		
Alum: potassium		200	150	200	200	
Aluminum chloride	pH > 6	150	150	200	200	175
Aluminum fluoride		200	125			
Aluminum hydroxide		200	150	200	185	150
Aluminum nitrate	pH > 6.5	150	150		185	150
Aluminum sulfate		200	150	200	200	175
Ammonia: aqua 18–25%		NR	NR	NR	150	
Ammonia: gas (dry)		NR	NR	NR	NR	NR
Ammonia thiocyanate						
Ammonia water (household ammonia)		NR	NR	NR	150	NR
Ammonium acetate, 10%	pH > 6	NR	100	150	ID	100
Ammonium bifluoride						
Ammonium bromide			100	175		
Ammonium carbonate		150	150	185	185	
Ammonium chloride	pH > 6	180	150	200	200	175
Ammonium fluoride		NR	NR	NR	150	ID
Ammonium hydroxide		NR	NR	NR	150	NR
Ammonium nitrate	pH > 6.5	200	100	150	200	175
Ammonium phosphate		150	125	200	150	
Ammonium sulfate		200	150	200	185	175
Ammonium sulfide			100	150		
Ammonium thiocyanate		ID	ID	ID	ID	ID
Amyl alcohol			100	150	150	
Aniline and aniline oil		NR	NR	NR	NR	NR
Aniline hydrochloride		NR	NR	NR	ID	ID
Aromatic hydrocarbons		NR	NR	NR	NR	NR
Arsenic acid		125	150	150	150	
Arsenous acid				150		
Barium carbonate		150	100			200
Barium chloride	pH > 6	175	150	200	185	200
Barium hydroxide		200	150	150	ID	
Barium nitrate	pH > 6.5		NR			200
Barium sulfate		200	100	200		200
Barium sulfide			150	175		
Barium sulfite		NR	150	200		
Benzene (coal tar)		NR	NR	NR	NR	NR
Benzine (gasoline type)		NR	NR	NR	NR	NR
Benzoic acid		150	150	150		
Black liquor (sulfate)						
Bleach		NR	NR	NR	100	
Borax		200	150	200		200
Boric acid		200	150	200	185	200
Brine solution		200	150	200	200	200
Bromine		NR	NR	NR	NR	NR
Butane		NR	NR	NR	NR	NR
Butyl acetate		NR	NR	NR	NR	NR

(continued)

TABLE 66.12.   (*Continued*)

Chemicals	Remarks	Neoprene	Soft Natural Rubber	Semihard and Hard Rubber	Butyl	Hypalon
Butyl alcohol (butanol)		NR	100	150	ID	NR
Butyric acid		NR	NR	100	NR	NR
Cadmium cyanide		150	100	150		150
Calcium acetate			NR	NR		150
Calcium bisulfate		150				150
Calcium bisulfite		NR	100	150	120	
Calcium bleach	(Calcium hypochlorite)		NR	NR	125	125
Calcium carbonate		200	100	200		
Calcium chloride	pH > 6	175	150	200	185	200
Calcium hydroxide		200	150	175		
Calcium hypochloride		NR	NR	NR	125	
Calcium nitrate	pH > 6.5	200	150	200	185	200
Calcium oxide, dry		200	150	200		
Calcium sulfate		150	150	200		175
Carbolic acid (phenol)		NR	NR	NR	100	NR
Carbon bisulfide		NR	NR	NR	NR	NR
Carbon dioxide (wet)		200	150	175	175	200
Carbon dioxide (dry)		200	150	175	175	200
Carbon tetrachloride		NR	NR	NR	NR	NR
Carbonic acid		200	150	200	185	200
Castor oil		ID	ID		150	
Caustic soda	(Sodium hydroxide)	200	150	200	185	200
Chloracetic acid		NR	NR	100	120	
Chlorinated hydrocarbons		NR	NR	NR		
Chlorine, dry		NR	NR	SC	NR	NR
Chlorine, wet		NR	NR	SC	NR	NR
Chlorine dioxide		NR	NR	NR	NR	NR
Chromic acid		NR	NR	NR	NR	125
Citric acid		150	ID	150	ID	100
Copper carbonate				200		
Copper chloride				100	185	
Copper cyanide			125	150		
Copper nitrate	pH > 6.5	150	NR	150	185	150
Copper sulfate		200	150	150	185	200
Cottonseed oil		NR	NR	NR	100	NR
Cresylic acid		NR	NR	NR	NR	NR
Ethanol	(Ethanol)[a]	100	100	100	100	ID
Ethers		NR	NR	NR		NR
Ethyl acetate		NR	NR	NR		NR
Ethyl alcohol	(Ethanol)[a]	100	100	100	100	ID
Ethyl chloride		NR	NR	NR	NR	NR
Ethylene glycol		100	100	200	185	150
Fatty acids		NR	NR	NR	NR	NR
Ferric chloride	pH > 6	85	150	200	185	100
Ferric hydroxide			100	150		
Ferric nitrate	pH > 6.5		NR	150		125
Ferric sulfate			150	200	185	125
Ferrous ammonium sulfate		200	150			200
Ferrous chloride	pH > 6		100	200	185	175
Ferrous hydroxide			100	200		
Ferrous nitrate				150		
Ferrous sulfate		150	100	200	185	175
Fluoboric acid		100	150	200		125

TABLE 66.12.   (*Continued*)

Chemicals	Remarks	Neoprene	Soft Natural Rubber	Semihard and Hard Rubber	Butyl	Hypalon
Fluorine gas (wet)		NR	NR	NR	NR	NR
Fluorine gas (dry)		NR	NR	NR	NR	NR
Fluosilicic acid		100	150	200	185	125
Formaldehyde, 5%		NR	NR	150		
Formaldehyde, 40%		NR	NR	150	125	NR
Formic acid		NR	NR	NR	150	NR
Gasoline		NR	NR	NR	NR	NR
Glauber's salts	(Sodium sulfate)	200	150	200	150	200
Glycerine			100		150	
Hydrobromic acid		NR	100	200	150	NR
Hydrochloric acid		NR	150	200	NR	NR
Hydrofluoric acid		NR	NR	NR	150	NR
Hydrofluosilisic acid		100	100	200	185	125
Hydrogen peroxide		NR	NR	NR	NR	NR
Hydrogen sulfide		NR	NR	NR	NR	NR
Hydrogen sulfite, dry		NR	NR	NR		
Hydrogen sulfite, wet		NR	NR	NR		
"Hypo" photographic solution	(Sodium thiosulfate)	100	150	200		
Hypochlorous acid		NR	NR	NR	NR	NR
Kerosene		NR	NR	NR	NR	NR
Lacquer solvents		NR	NR	NR	NR	NR
Lactic acid	Pure	80	120	150	150	ID
Lead chloride	pH > 6	200	100	150		150
Lead nitrate	pH > 6.5		100		150	
Lead sulfate		100	100	150	180	175
Lime, dry	(Calcium oxide)	200	150	200		
Lime, slaked	(Calcium hydroxide)	200	150	175	180	
Linseed oil		NR	NR	NR	150	
Lithium chloride	pH > 6	200				
Lye	(Sodium hydroxide)	200	150	200	158	NR
Magnesium carbonate	(Basic)	200	100			150
Magnesium chloride	pH > 6	200	150	200	185	175
Magnesium citrate				150		
Magnesium hydroxide		200	150	200	185	150
Magnesium nitrate	pH > 6.5	200	150	200	185	150
Magnesium sulfate		200	150	200	185	175
Maleic acid						
Malic acid		ID	NR	150	NR	80
Manganese ammonium sulfate						
Manganese chloride						
Manganese sulfate		200	150	150	180	175
Mercuric chloride	pH > 6	NR	100	200	150	ID
Mercuric cyanide		NR	100	150	150	ID
Mercuric nitrate				150	ID	ID
Mercurous nitrate		NR	NR	150	ID	ID
Methyl alcohol	(Methanol)[a]	ID	80	150	180	80
Methyl chloride		NR	NR	NR	NR	NR
Mineral oils		NR	NR	NR	NR	NR
Muriatic acid	(Hydrochloric acid)	NR	150	180	NR	NR
Nickel acetate	pH > 6	ID	100	200	150	125
Nickel ammonium sulfate						
Nickel chloride	pH > 6	200	150	200	150	175
Nickel nitrate	pH > 6.5	100	100	200	ID	150
Nickel sulfate		200	150	200	185	200

(continued)

TABLE 66.12. *(Continued)*

Chemicals	Remarks	Neoprene	Soft Natural Rubber	Semihard and Hard Rubber	Butyl	Hypalon
Niter	(Potassium nitrate) pH > 6.5	200	150	180	ID	200
Nitric acid, 5%		NR	NR	NR	150	100
Nitric acid, 10%		NR	NR	NR	125	NR
Nitric acid, 25%		NR	NR	NR	100	NR
Nitric acid, 40%		NR	NR	NR	NR	NR
Nitrous acid		NR	NR	150		NR
Oleic acid		NR	NR			
Oleum	(Fuming sulfuric acid)	NR		NR		NR
Oxalic acid		150	NR	150	150	175
Palmitric acid		NR	NR	NR		
Paraffin				NR		
Perchloric acid	(Dihydrate)	NR	NR	NR	ID	NR
Peroxide bleach	(Sodium perborate)					
Petroleum oils, crude		NR	NR	NR	NR	
Phenol	(Carbolic acid)	NR	NR	NR	NR	NR
Phosphoric acid, 85%	(Over 85% use butyl)	150	150	200	185	
Plating solution, brass				200		
Plating solution, cadmium				200		
Plating solution, chrome		NR	NR	NR	NR	NR
Plating solution, copper				200		
Plating solution, gold				200		
Plating solution, lead				200		
Plating solution, nickel				200		
Plating splution, silver				200		
Plating solution, tin				200		
Plating solution, zinc				200		
Potassium acetate	pH > 6					
Potassium acid sulfate						
Potassium alum		200	150	200	180	200
Potassium aluminum sulfate	(Alum)		150	200	175	200
Potassium antimonite				150		
Potassium auricyanide				150		175
Potassium bicarbonate		200	100	180		
Potassium bichromate	pH > 6	150	NR	NR		
Potassium bisulfate		150	150	200		175
Potassium bisulfite		NR	NR	200		ID
Potassium borate			100	200		
Potassium bromide			100	200		
Potassium carbonate		200	100	200		200
Potassium chlorate				200		200
Potassium chloride	pH > 6	150	150	200	180	200
Potassium chromate	pH > 6		NR	NR	NR	NR
Potassium cyanide		150	125	200	150	150
Potassium cyprocyanide			125	200		
Potassium dichromate	pH > 6	NR	NR	NR	NR	175
Potassium ferricyanide		NR	100	NR	ID	ID
Potassium hydroxide, 25%		220	150	200	185	NR
Potassium hydroxide saturated	pH > 25%	220	150	150		NR
Potassium iodide	pH > 6.5		150	200		150
Potassium nitrate	pH > 6.5	200	150	180	180	200
Potassium nitrite	pH > 6.5					
Potassium oleate						
Potassium permanganate	pH > 7.0		NR	150		NR

TABLE 66.12. *(Continued)*

Chemicals	Remarks	Neoprene	Soft Natural Rubber	Semihard and Hard Rubber	Butyl	Hypalon
Potassium phosphate	Mono-di- or tribasic	200	150	200		200
Potassium salicylate				200		
Potassium silicate		200	100	200		
Potassium sulfate		200	150	200	105	200
Potassium sulfide		200		200		
Potassium sulfite	pH > 6	150	150	200		
Potassium thiosulfate		AMB	100	200		
Propane		NR	NR	NR	NR	NR
Propionic acid (dilute)			NR	NR		
Propyl alcohol			100	150		
Rochelle salts	(Potassium sodium tartrate)		100	200		200
Sludge acids						
Soap solutions			100	180		
Sodium acid sulfate						
Sodium aluminum sulfate		200				
Sodium antimonate				200		
Sodium bicarbonate		200	100	200		200
Sodium bichromate	pH > 6	150				
Sodium bisulfate		170	150	200		
Sodium bisulfite						
Sodium borate			100	200		
Sodium carbonate		200	150	200		200
Sodium chlorate						
Sodium chloride	pH > 6	200	150	200	185	200
Sodium cuprocyanide						
Sodium cyanide		150	100	200		150
Sodium dichromate	pH > 6		NR	NR	150	
Sodium ferricyanide		NR	100			ID
Sodium fluoride			NR	NR		
Sodium hydroxide, 25%		200	150	200	185	NR
Sodium hydroxide, saturated	pH > 25%	200				
Sodium hypochlorite	pH > 9	NR	NR	NR	100	
Sodium nitrate	pH > 6.5	200	150	180	180	200
Sodium nitrite	pH > 6.5	150				
Sodium oeate						
Sodium perborate		200	150	150		
Sodium permanganate	pH > 7.0		NR	200		NR
Sodium peroxide			NR	NR		
Sodium phosphate	Mono-di- or tribasic	200	150	200	185	200
Sodium salicylate				200		
Sodium silicate		200	150	200		
Sodium sulfate		200	150	200	150	200
Sodium sulfide			100	200		
Sodium sulfite	pH > 6	150	150	200	150	150
Sodium thiosulfate			150	200		
Soft drink syrups and concentrates	NR	NR	NR			
Stannic chloride	pH > 6		150	200		125
Stannous chloride	pH > 6	NR	150	200		150
Stearic acid		NR	NR	NR		
Sulfate liquors						
Sulfite liquors						
Sulfur dioxide, wet				150		

(continued)

TABLE 66.12. (*Continued*)

Chemicals	Remarks	Neoprene	Soft Natural Rubber	Semihard and Hard Rubber	Butyl	Hypalon
Sulfuric acid, 5%		180	100	185	185	175
Sulfuric acid, 25%		170	100	175	175	175
Sulfuric acid, 50%		75	NR	150	150	150
Sulfuric acid 75%		NR	NR	NR	NR	
Sulfurous acid		NR	NR	200	150	
Tannic acid		1D	100	150		
Tartaric acid		100	150	200	175	125
Tin chloride	Ph > 6					
Trichloroethylene		NR	NR	NR	NR	NR
Triethanolamine						
Trisodium phosphate	pH > 6	200	150			
Turpentine		NR	NR	NR	NR	NR
Urea						
Urea ammonia liquor						
Urec acid						
Vegetable oils			NR			
Water, acid mine		150	100	150		
Water, fresh			150	185		100
Water sea or salt		200	150	200	200	200
White liquor						
Wood pulp and pulp liquors						
Zinc acetate						
Zinc chloride	pH > 6	150	150	200	185	150
Zinc nitrate						
Zinc sulfate		150	130	200	185	175

[a]No harmfull impurities.

Loose linings are fabricated by making cylinders by using butt fusion machines. Axial welding of the cylinders is performed by hot air welding. Nozzles are welded either by butt fusion or hot air welding. Flaring of the nozzle is performed after the liner is inserted in the housing.

Performance for loose linings has been fair in larger vessels. For obvious reasons this system is not used for vacuum service. Such linings have been used in service with fluorobenzene.

5. *Molded Linings.* For components like pumps and valves, melt-processible fluoropolymer linings are often formed by transfer molding. Sizes are limited by the capacity of the transfer molding equipment, polymer rheology, and other factors.

Special molding technology is used to form PTFE linings for components and piping. Although PTFE is a thermoplastic, it is so viscous in the melt that conventional molding techniques cannot be used. To line components, the resin is pressed in housings, usually by isostatic pressure applied by inflating rubber tooling. The resulting "green form" is sintered at high temperature so that it coalesces into a solid shape.

Isostatic molding is also used to form PTFE shapes for pumps and other components. In most cases, these shapes require machining to meet required tolerances.

Some PTFE pipe linings are formed by a paste extrusion process. The resin is mixed with a hydrocarbon extrusion aid and extruded as tubing. The extruded shape is sintered to drive off the hydrocarbon and coalesce

TABLE 66.13.   General-Purpose, Thick Thermoplastic Sheet Linings[a]

Material	Forum	Description
Polypropylene	Rubber-backed sheet	0.060-in.-thick Polypro on 0.060 in. rubber
Plasticized PVC	Sheet	$\frac{1}{16}$ in., $\frac{3}{32}$ in and $\frac{3}{16}$ in. thick
Unplasticized PVC on plasticized PVC	Laminated sheet	0.085 in. thick

[a]Root pass welding followed by capstrip covers is used.

TABLE 66.14. Fluoropolymer Physical Properties[a]

Property	Test Method	Units	PTFE	FEP	PFA	ETFE	ECTFE	PVDF
Tensile strength	D638	psi × 1000 (MPa)	3–5 (21–35)	3.4 (23)	3.6 (25)	5.8–6.7(40–47)	4.2–4.3 (28.9–29.6)	6.8–8.0(46.9–55.2)
Elongation	D638	%	300–500	325	300	150–300	200	50–250
Flexural modulus	D790	psi × 1000 (MPa)	72 (500)	85 (600)	85 (600)	170 (1200)	240–245(1655–1689)	165–325(1138–2241)
Hardness	D2240	Shore D	50–65	56	60	63–72	75	77
Melting point	DTA, E-168	°F (°C)	621 (327)	500 (260)	582 (305)	473–512 (245–267)	464 (240)	320–338 (160–170)
Upper service temp.	UL746B	°F (°C)	500 (260)	400 (204)	500 (260)	300 (150)	300 (150)	275 (135)

Arkema sells PVDF as Kynar®.

[a]Sources: DuPont (PTFE, FEP, PFA, ETFE) and Solvay Solexis (ECTFE, PVDF). DuPont sells PTFE, FEP, and PFA as Teflon® fluoropolymer resins and ETFE as Tefzel® fluoropolymer resin. Solvay Solexis sells ECTFE as Halar® and PVDF as Hylar®.

TABLE 66.15. **Chemical Resistance of Partially Fluorinated Fluoropolymer Manufacture Recommended Maximum Service Temperature**

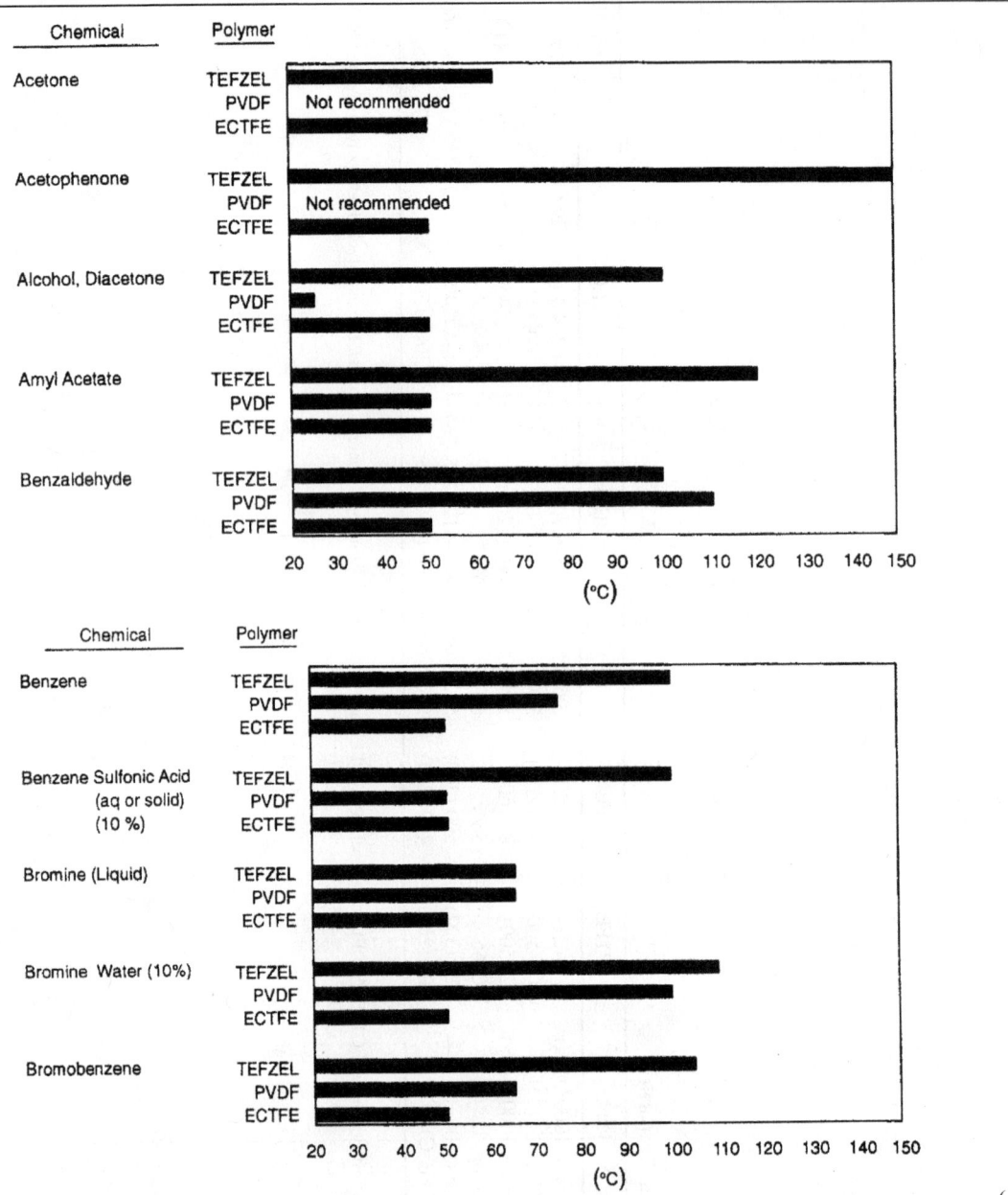

(continued)

the resin. Other processes for PTFE are ram extrusion and tape-wrap, where layers of tape are built up on a mandrel to the required thickness before sintering.

6. *Dual Laminates.* Structures are made of FRP built up on fluoropolymer sheeting that serves as a lining. Dual-laminate structures combine the technology of fluoropolymer fabrication with the technology of FRP (see Section G for self-supporting structures). This lining is fabricated by machine and hand welding of fabric-backed fluoropolymer sheets like those described above for adhesive-bonded and loose

linings. The fabric aids in bonding the sheeting to the FRP.

Dual laminate construction involves fabricating the thermoplastic shell using the welding techniques described earlier for adhesive bonds and loose linings. The FRP shell is then wound on the thermoplastic structure followed by welding of the nozzles and an FRP overlay on the joints.

Some commonly used liners are PVDF, ECTFE, ETFE, FEP, and PFA.

TABLE 66.15. (*Continued*)

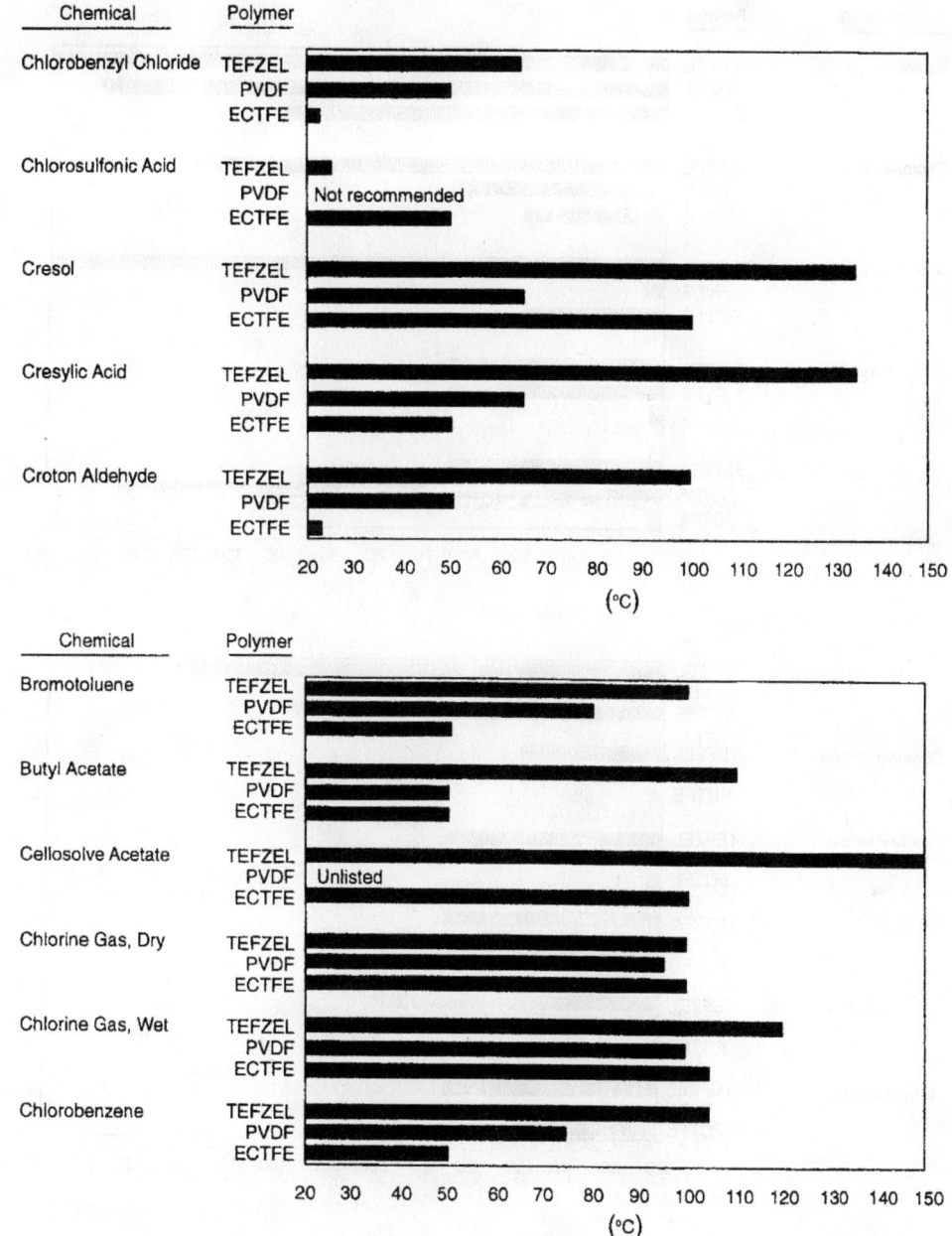

(*continued*)

Table 66.17 compares the five technologies of Fluoropolymer lining.

G. SELF-SUPPORTING STRUCTURES: PROCESS VESSELS, COLUMNS, AND PIPING

G1. Thermoplastics

The use of thermoplastic polymers in self-supporting structures is limited to piping. The exceptions are rotomolded polyethylene and welded polypropylene tubs. Materials used for piping are PVC, CPVC, PE, and PP. On rare occasions PVDF piping is also encountered. These materials are extruded for piping that is available up to 356 mm (14 in.) diameter.

G1.1. Design of Piping. Thermoplastic piping is designed on the basis of the hydrostatic design basis (HDB), which involves carrying out short-term burst tests at various pressures and recording the time to failure. The stress versus time to failure is extrapolated to an arbitrarily chosen long-term

TABLE 66.15. (*Continued*)

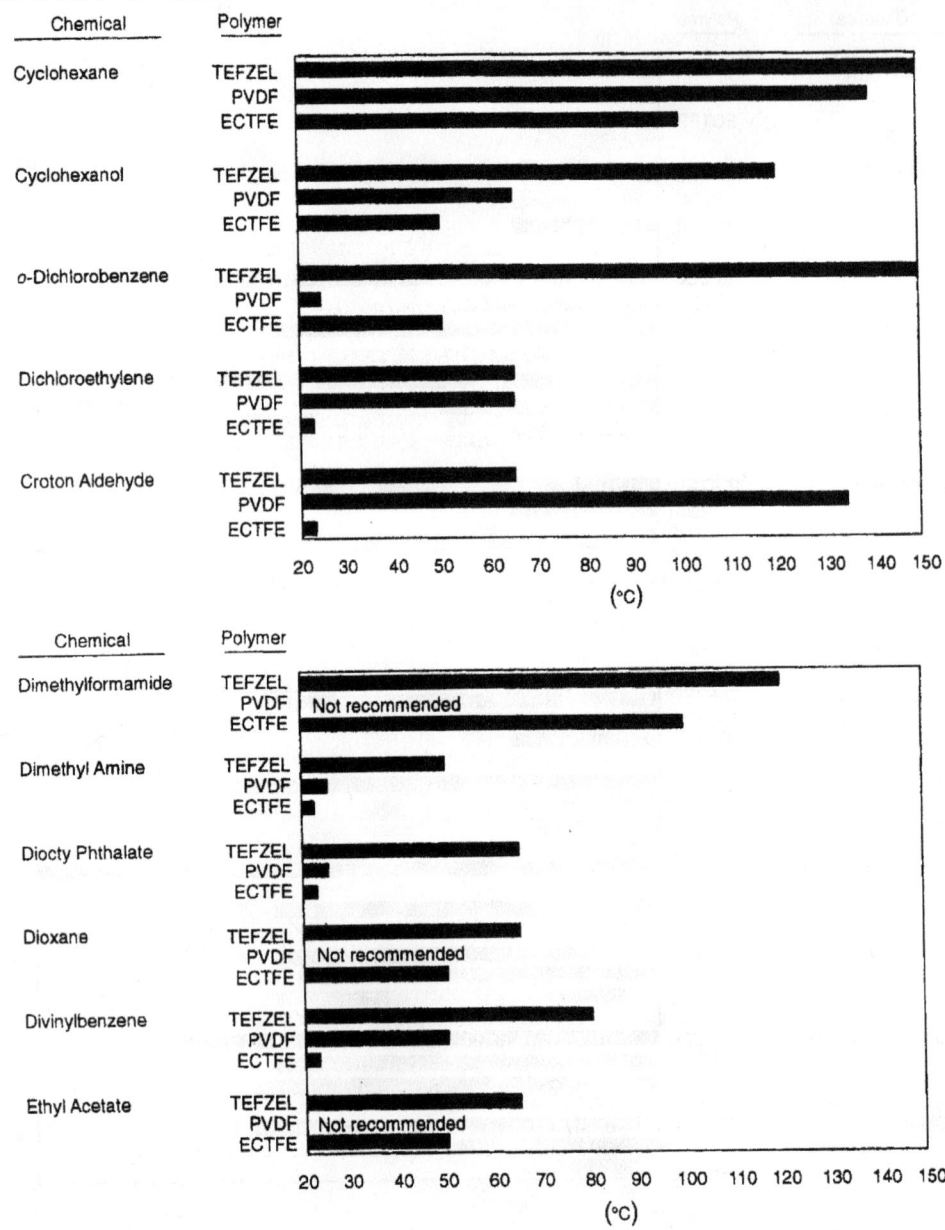

(*continued*)

value called the HDB. A factor of safety is applied to calculate the maximum allowable stress and pressure.

G1.2. Joining of Pipe. Joining of piping products such as lengths and fittings is done by welding (for PE, PP, and PVDF) and solvent cementing for PVC. Solvent converting is a quick and easy way to join pipe lengths and fittings through couplings. Welding requires qualified personnel using standardized industry practices.

G1.3. Welding of Thermoplastics. There are three principal methods of welding thermoplastic materials. Welding is needed to join piping ends on edges of sheet such as in lining:

Hot air (hand) welding

Extrusion welding

Butt fusion (hot plate or heated tool) welding

Flow fusion welding

Cap strip welding

TABLE 66.15. (*Continued*)

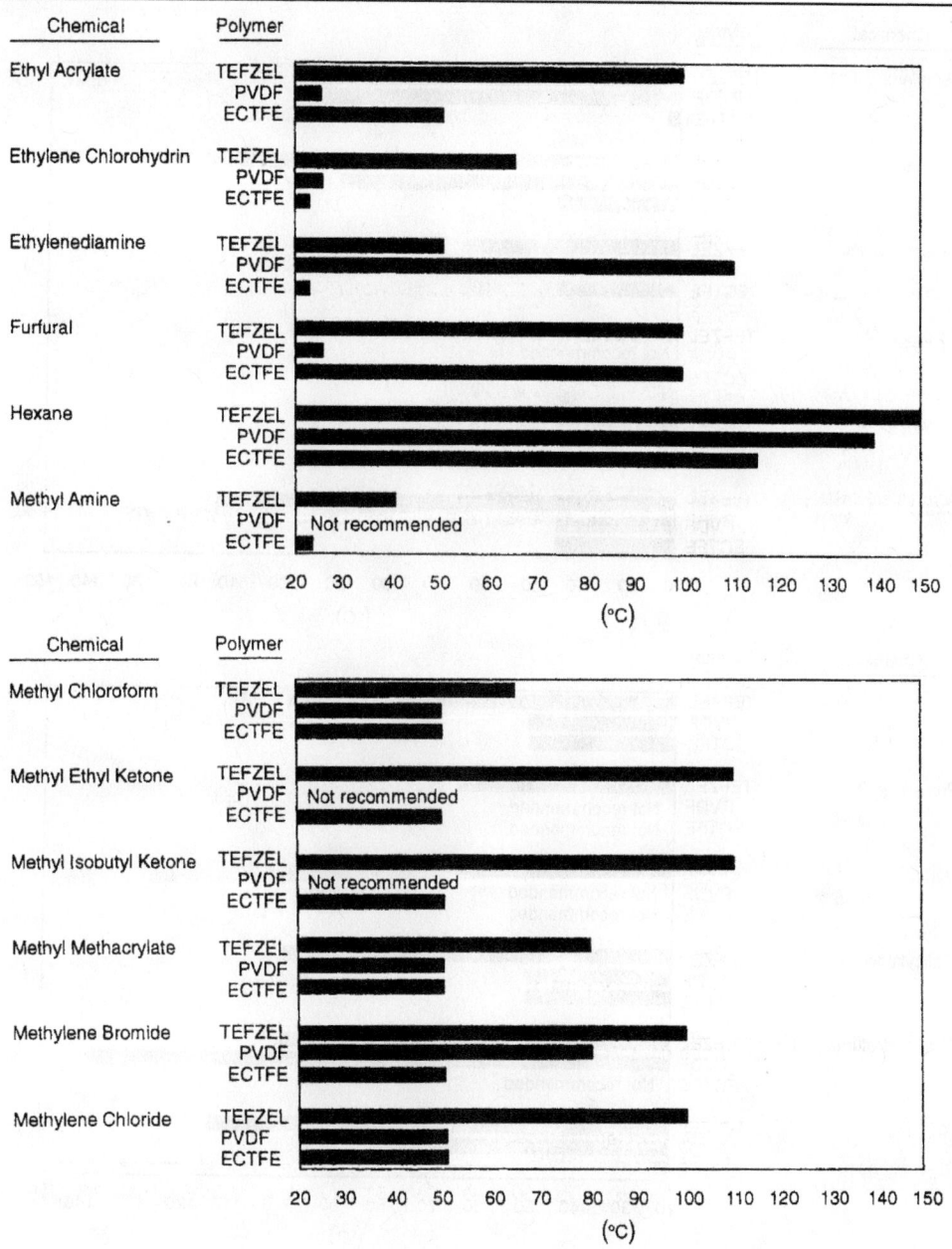

Hot air and extrusion welding are similar except that extrusion weld is used for sheets thicker than 1/8 in. In both cases a weld rod is fused between the two sheets to be joined. In the case of hot air welding the weld rod is manually pressed in a V-groove while hot air or nitrogen melts the skin of the sheet and the rod. Sheet linings for fluoropolymers, adhesively bonded, PP and PVC use this technique. In extrusion welding the filler rod is extruded in the groove by an extruding screw.

Butt fusion welding is most used for welding pipe ends of both single- and double-wall piping. A special machine capable of properly aligning the pipe ends and melting the ends and pressing them together is used. This technique also used for axial welds of sheets for linings. How fusion welding is a variation of the butt fusion welding where a special machine is used which melts and fuses the edges of sheets together. This technique is used for making longitudinal welds in loose fluoropolymer linings as well as dual-laminate liners.

(*continued*)

TABLE 66.15. (*Continued*)

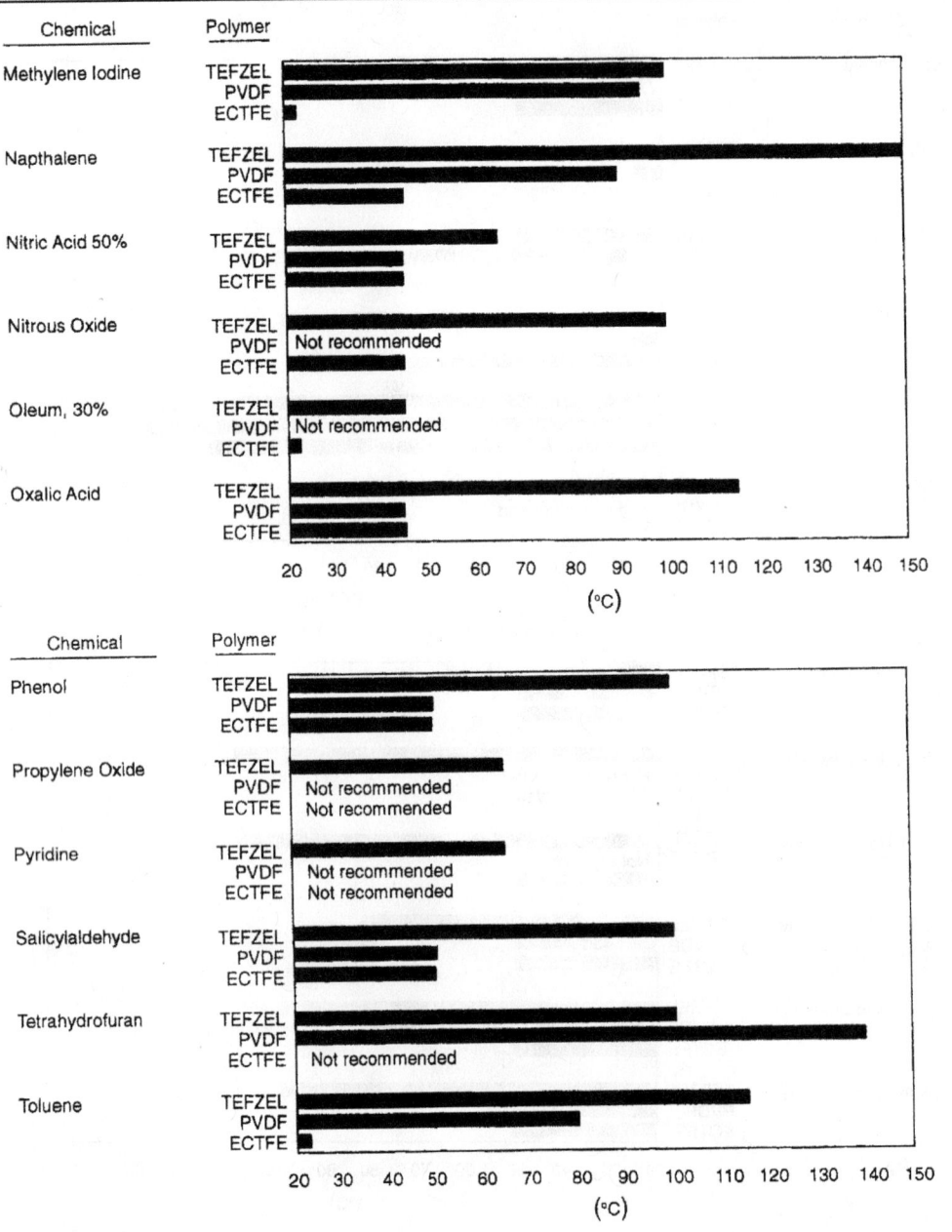

(*continued*)

Cap strip welding is used to cover hand welds in the bonded linings (fluoropolymers, PVC, and PP) as well as dual laminate. It is similar to hand (hot gas) welding except a strip is used instead of a rod.

Regardless of the welding technique used, successful welds have the following in common: controlled heat input, consistent and right level of pressure, and the time for which heat and pressure are maintained. Figure 66.11 shows the schematic of a welding gun used for any free hand welding including extrusion welding. Figure 66.12 shows how free hand welding is done using a gun and the weld rod. Fluoropolymers are welded with an electric weld gun using a speed welding tip. (Fig. 66.13). Figure 66.14 shows root pass welding followed by cap strip welding.

Currently the industry follows AWS (American Welding Society) and DVS (German Welding Institute) standards

TABLE 66.15. (*Continued*)

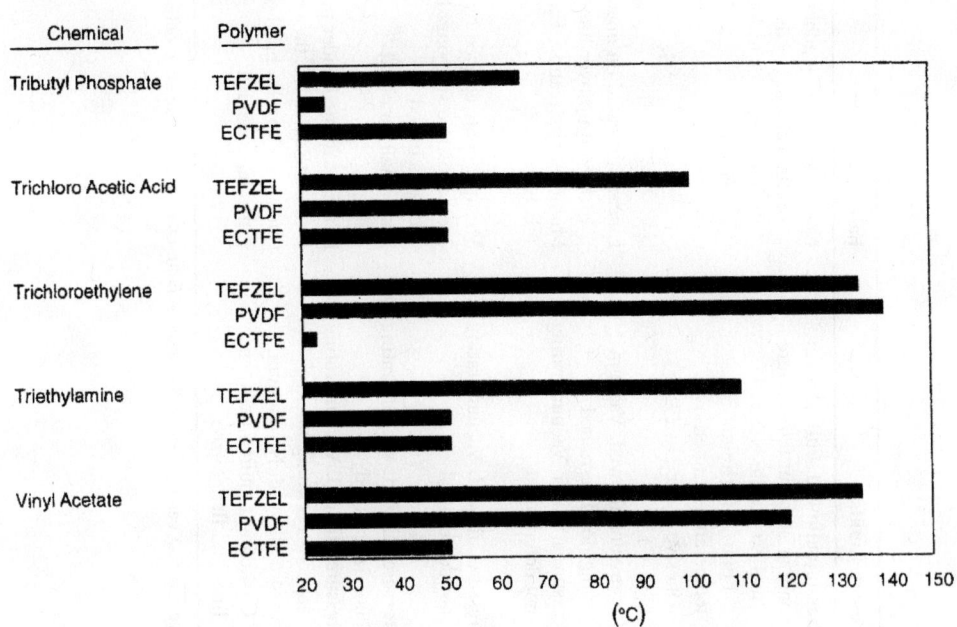

for weld quality and qualification. Following are some of the standards used:

Weld qualification

    DVS certification program per DVS 2212

    AWS B 2.4 (2006)

Weld evaluation

    DVS 2203 Part 5 (bend test)

    DVS 2203 part 4 (tensile creep test)

    AWS G 1.10M (2001)—visual examination

    EN 12814-7, Tensile test with waisted test specimen

G1.4. Welded Vessels. Using the hand (hot gas) welding techniques PP tubs are fabricated and are widely used for

TABLE 66.16.   Permeation Rules of Thumb

	Change	Permeation
Voids in polymer	⇓	⇓
Polymer crystallinity	⇑	⇓
Polymer chain stiffness	⇑	⇓
Polymer interchain forces	⇑	⇓
Polymer temperature	⇓	⇓
Permeant size/shape	⇑	⇓
Permeant concentration	⇓	⇓
Polymer thickness	⇑	⇓
Permeant temperature	⇓	⇓
Voids in polymer	⇑	⇑
Permeant/polymer chemistry	⇓ Similarity	⇓

pickling operations. By far the most commonly used vessels are rotomolded PE (HDPE as well as crosslinked) per ASTM D 1198. These are economical and find use in HCl and H_2O_2 storage.

G2.   Thermosetting Materials: FRP (RTP)

The bulk of self-supporting structures are made of reinforced thermosetting materials commonly known as fiber-reinforced plastics (FRPs or reinforced thermosetting plastics (RTPs.) Resin constitutes the continuous phase (matrix) and fibers (continuous or chopped strand as well as mat or woven roving) form the reinforcement. Resin provides the needed chemical resistance while the reinforcement provides the needed mechanical and physical properties such as strength, elongation, and insulation.

G2.1. Resins. The resins used are generally known as polyesters. In reality, resins belong to the following broad classifications:

Bisphenol A fumarate

Chlorendic anhydride

Epoxy vinyl esters

    Epoxy Novalac–based vinyl esters

    Modified vinyl esters

Furan

Table 66.18 shows the building blocks of these resins.

TABLE 66.17. **Fluoropolymer Lining Systems**[a]

Lining System	Lining Materials	Maximum Size	Design Limits	Fabrication
Adhesive bonding	Fabric-backed PVDF, PTFE, FEP, ECTFE, PEA	No limit	Pressure allowed. Full vacuum only at ambient temperature. Smallest nozzle is 2 in. (51 mm). Maximum temperature limited by adhesive, typically 275°F (135°C)	Neoprene or epoxy adhesive, sheets welded. Heads thermoformed or welded
Dual laminate	Same as adhesive bonding	12 ft (3.7 m) dia.	No pressure allowed. Vacuum rating not determined	Linear fabricated on mandrel. FRP built up over liner
Sprayed dispersion	FEP, PFA, PFA w/mesh and carbon, PVDF, PVDF w/glass or carbon fabric	8 ft (2.4 m) dia., 40 ft (12.2 m) length	Pressure allowed. Vacuum rating not determined	Multicoat application. Each coat is baked
Electrostatic spray—powder	ETFE, FEP, PFA, ECTFE, PVDF	8 ft (2.4 m) dia., 40 ft (12.2 m) length	Pressure allowed. Vacuum rating not determined	Multiple coats applied by electrostatic spraying. Each coat is baked
Rotolining	ETFE, PVDF, ECTFE	8 ft (2.4 m) dia., 22 ft (6.7 m) length	Pressure allowed. Vacuum rating not determined	Rotationally molded
Isostatic molding, paste extrusion	PTFE		Pressure allowed. Vacuum rating depends on lining thickness	PTFE performed under pressure or paste extruded, then sintered
Loose lining	FEP, PFA	Determined by body flange	Pressure allowed. No vacuum. Gasketing required between liner and flange face	Liner with nozzles hand or machine welded

[a]Nondestructive spark testing should be used, along with visual inspection for all systems except loose linings. Adhesive bonding can be done in the shop or field; other systems are shop only.

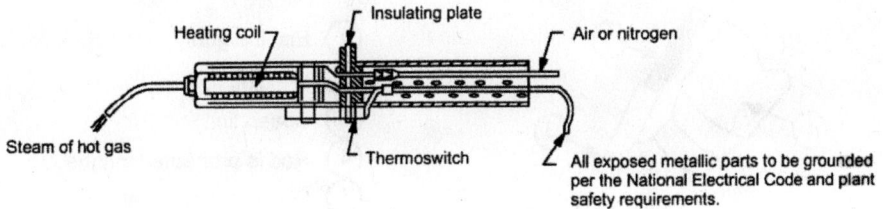

FIGURE 66.11. Electrically heated welding gun.

Styrene is added to these resins as a reactive diluent. Styrene serves two purposes: First, it helps in adjusting the viscosity of the resin mix for ease of fabrication and, second, it lowers the cost of the resin. The final structure is a copolymer of styrene and the polyester resin.

Choice of resin will depend on chemical resistance, temperature resistance, flexibility, and ease of fabrication. Resin manufacturers routinely publish chemical resistance information as well as mechanical properties. Bis A fumarate resins offer the best resistance to caustic solutions but lack the flexibility of modern resins. These resins were the early workhorse in CPI and have been virtually replaced by the more versatile vinyl ester resins which offer a variety of combinations of chemical resistance, flexibility, high-temperature resistance, and ease of fabrication. The chlorendic anhydride family is principally used for chlorine-handling applications while furans are used for solvent resistance. Currently, vinyl ester resins account for the largest number of applications in the CPI.

G2.2. Reinforcements. The FRP laminates are heterogeneous (i.e., the type and the amount of reinforcement varies across the thickness.) The surface against the chemical is reinforced by a glass or synthetic veil (polyester or carbon) and is resin rich (90% resin, 10% glass). Next are two or four layers of chopped strand mats with a somewhat lower resin/glass ratio (70% resin). Together, the veil(s) and mats form the corrosion barrier. The structural thickness that follows is either by alternating layers of mat and woven roving known as contact molding (hand layup) or continuous strand roving

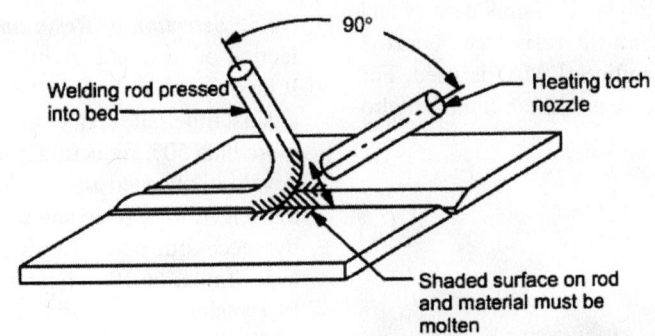

For welding polyethylene and polypropylene

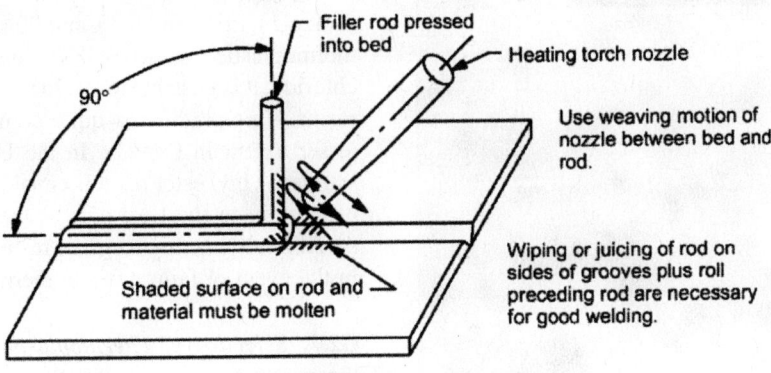

For welding PVC and CPVC

FIGURE 66.12. Welding positions.

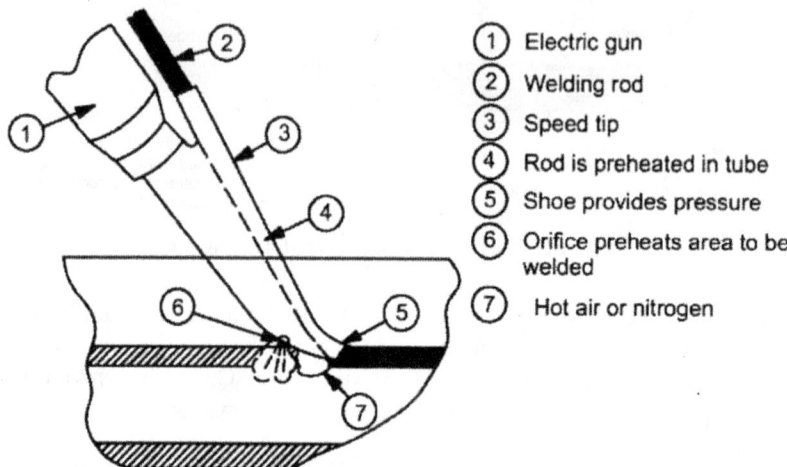

FIGURE 66.13. Electric gun with speed welding tip.

interspersed with uniaxial fabric and spray up (filament wound construction). [see Figs. 66.15(a) and (b)].

G2.3. Other Additives. For fire retardancy, resins are brominated. Antimony trioxide is also added.

G2.4. Curing Systems. There are three principal curing systems used for FRP. Methyl ethyl ketone peroxide (MEKP) is by far the most common. For higher crosslink density and the consequent increase in chemical resistance, benzoyl peroxide (BPO) with dimethylaniline (DMA) is used. For thicker sections, where a slower cure is desired, cumine hydro

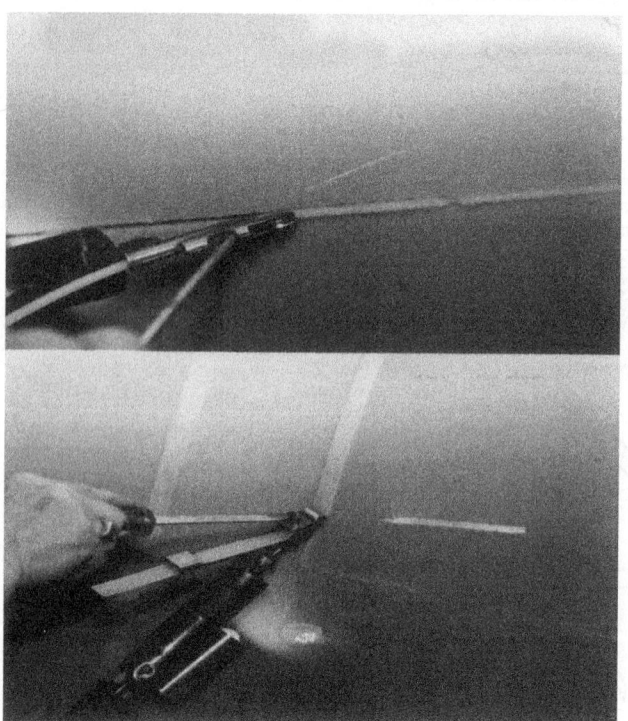

FIGURE 66.14. Root pass and cap strip welding for fluoropolymer lining.

peroxide (CHP) is used. Curing refers to the resin crosslinking process caused by free radicals formed by the curing agents.

Sometimes postcuring is required to increase the degree of crosslinking which enhances the chemical resistance. Hardness measurements by the Barcol tester or the acetone sensitivity test measure the degree of crosslinking and the chemical resistance.

G2.5. Selection of Resin and Laminate Composition. Selection of resin is currently done by the ASTM C581 test. It involves plotting the retained flexural strength of coupons against time. The resin is accepted if the curve levels off at no more than 50% reduction. Currently, efforts are underway to develop enhanced resin selection techniques. Quite often, the selection of resin and the laminate design is determined by successful case histories in the industry or actual field tests. Table. 66.19 indicates general chemical resistance of FRP resins.

G2.6. Materials for Dual-Laminate Construction. Dual laminate refers to FRP structure with a bonded thermoplastic liner on the inside. The liner materials range from low-end PVC through PP, PVDF (Kynar®), ECTFE (Halar®), ETFE (Tefzel®), FEP (Teflon®), and PFA. When using the low-end thermoplastics such as PVC and chlorinated polyvinyl chloride, it is possible to use low-grade isophthalic polyester resins. This practice is quite common in Europe and to a lesser extent in Canada. In the United States, the cost of standard vinyl ester resin is considered to be insignificant in proportion to the total cost of the dual-laminate structure. Additionally, the higher grade resin provides some protection in the event of failure of the thermoplastic liner.

G2.7. Selection of Thermoplastic Liners. Selection of the thermoplastic depends on the chemical and temperature resistance required, weldability of the thermoplastic, and cost. Dual laminates must be considered in conjunction with

TABLE 66.18.   Building Blocks for FRP Resins

Generic	Acid	Glycol	Unsaturated Acid	Trade Names[a]
ISO	Orthophthalic Isophthalic Terephthalic	Ethylene Propylene Diethylene Dipropylene Neopentyl	Fumaric Maleic anhydride	Atlac 400
Bis A furmarate	Bisphenol A	Ethylene oxide Propylene oxide	Fumaric acid	Atlac® 382
Vinyl ester	Bisphenol A	Epichlorohydrin[b]	Methacrylic acid	Atlac 580 Derakane® 411 Hetron® 922
Epoxy Novoloc Vinyl	Phenol formaldehyde	Epichlorohydrin[b]	Methacrylic acid	Derakane® 470 Hetron® 970
Ester Chlorendic anhydride	Chlorendic acid	Ethylene oxide Propylene oxide	Maleic anhydride	Hetron® 92

[a]Atlac® is a registered trade name of Reichhold Co. Hetron® and Derakane® are registered trade names of Ashland Chemical Co.
[b]Not a glycol.

other alternatives such as metals and ceramics. Manufacturers of thermoplastics publish chemical resistance data. In the event that testing is needed, dunk tests of thermoplastic materials are carried out for weight change and retained mechanical properties. More appropriately laboratory (ASTM C868) or field blind flange tests should be carried out. Loss of adhesion from FRP, color change, embrittlement, and stability of welds should be examined.

G2.8. Design and Fabrication of FRP Vessels and Columns. The FRP structures are designed on the basis of operating condition temperatures and pressures (imposed as

well as hydrostatic) as well as other loads such as seismic, wind, snow, attached piping, and agitator. Design is carried out by established industrial practices and standards such as ASTM D3299, D4097, RTP-1, and ASME Section X. Complete design includes shell and head thicknesses, type of heads, joints for shell sections, nozzles to shell, and shell to heads. Design of lifting, support lugs, and anchors is also included.

The FRP vessel fabrication is performed by one of the following two techniques.

G2.8.1. Contact Molding (Hand Layup). This technique involves sequential layup of layers saturated with resin on

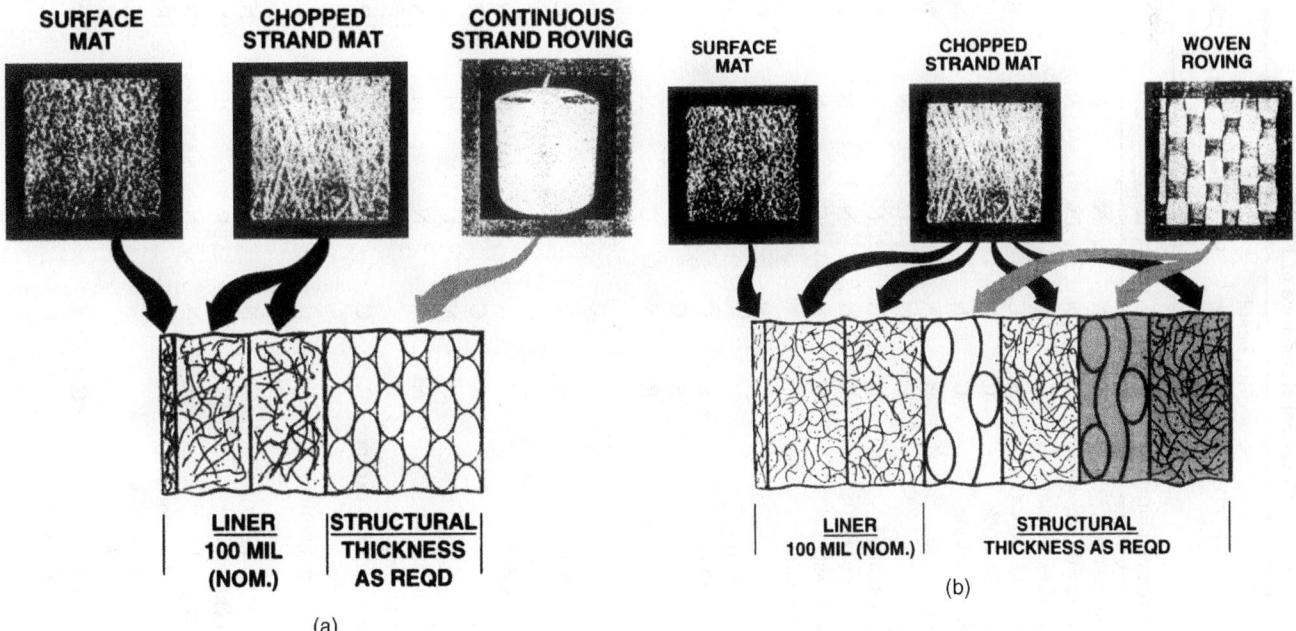

FIGURE 66.15. Chemical resistant structural laminate: (a) Filament wound; (b) hand lay up.

TABLE 66.19. Chemical Resistance of FRP

Up to Temperature (°F)	Bisphenol A Epoxy, Amine Cured 120	210	Bisphenol A Epoxy, Anhydrous Cured 120	210	Bisphenol A Vinyl Ester 120	210	Novolac VE 120	210	Bisphenol A Fumarate Polyester 120	210	Chlorendic Acid Polyester 120	210	Furan 120	210	Isophthalic Acid Polyester 120	180	Orthophthalic Acid Polyester 120	150
Acetaldehyde	N	N	N	N	N	N	R	N	N	N	N	N	R	R	N	N	N	N
Acetaldehyde, aq. 40%	N	N	N	N	N	N	R	N	N	N	N	N	R	R	N	N	N	N
Acetic acid, glacial	C	C	N	N	N	N	C	N	N	N	C	N	N	N	N	N	N	N
Acetic acid, 20% (25)	R	C	R	C	R	R	R	R	R	R	R	R	C	C	R	N	N	N
Acetic acid, 80%	C	N	N	N	C	N	R	N	N	N	N	N	C	C	N	N	N	N
Acetic anhydride	C	N	N	N	N	N	C	N	N	N	C	N	R	R	N	N	N	N
Acetone, 10%	N	N	N	N	N	N	R	N	N	N	N	N	R	N	N	N	N	N
Adipic acid	C		C		R		R											N
Alcohol, allyl	C	N	N	N	N	N	C	N	N	N	N	N			N	N	N	N
Alcohol, benzyl	C	N	R	N	N	N	C	N	N	N	N	N	R		N	N	N	N
Alcohol, butyl (n-butanol)	C	N	C	N	C	N	R	N	N	N	N	N	R		N	N	N	N
Alcohol, butyl (2-butanol)	C		N		C	N	R		N		N		R		N		N	
Alcohol, ethyl	R	C	C	C	C	N	C	N	C	N	N	N	R	R	R	N	N	N
Alcohol, hexyl	R	C	R	C	C	N	R	R	N	N	N	N	R		N	N	N	N
Alcohol, isopropyl (2-propanol)	R	N	C	N	R	N	R	N	N	N	N	N	R	N	C	N	N	N
Alcohol, methyl	C	N	N	N	N	N	C	N	N	N	N	N	R	R	C	N	N	N
Alcohol, propyl (1-propanol)	R	N	R	N	C	N	R	N	N	N	N	N	R		N	N	N	N
Allyl chloride	N	N	N	N	N	N	N	N	N	N	N	N	R	R	N	N	N	N
Alum	R	C	R	C	R	R	R	R	R	R	R	R	R	R	R	R	N	N
Ammonia, gas	C	N	C	N	C	C	C	C	N	N	N	N	R	R	N	N	N	N
Ammonia, liquid	N	N	N	N	N	N	N	N	N	N	N	N			N	N	N	N
Ammonia, aq. 20%			N		R	N	R	N	C	N	N	N			N	N	N	N
Ammonium salts, except fluoride,	R	C	R	C	R	R	R	R	R	R	R	C	R	R	R	R	R	R
Ammonium fluoride, 25%	R	N	R		R		R		R		R		N	N	N		N	N
Amyl acetate	C	C	N	N	N	N	R	N	N	N	N	N			N	N	N	N
Amyl chloride	R	N	R	N	N	N	R	N	N	N	N	N	R	R	N	N	N	N
Aniline	N	N	N	N	N	N	N	N	N	N	N	N	R	R	N	N	N	N
Aniline hydrochloride	R	N	R	N	R	N	R	N	R	N	N	N	R	R	R	N	N	N
Antimony trichloride					R	C			N	N	R	C	R	R	R	N	N	N
Aqua regia					N	N			R	N	R	N						
Arsenic acid, 80%	C	N	C	N	C	N	C	N	N	N	N				N	N		
Aryl-sulfonic acid	R	R	R	R	R	R	R	R	R	R	N	N				N	N	

Barium salts	R	C	R	C	R	R	R	R	R	R	R	C	R	N	R	N	N	N
Beer	R	N	C	N	R	N	N	N	N	N	N	N	N	N	R	N	N	N
Beet sugar liquor	R	N	R	N	R	N	R	N	R	N								
Benzaldehyde, 10%					N	N			N	N	N	N	R	N	N	N	N	N
Benzaldehyde, 10–100%	N	N	N	N	N	N	N	N	N	N	N	N	R	N	N	N	N	N
Benzene (benzoil)	R	N	C	N	N	N	C	N	N	N	N	N	N	N	N	N	N	N
Benzene sulfonic acid, 10%	R	C	R	R	R	R	R	R	R	C	R	C	R	N	R	N	N	N
Benzene sulfonic acid, 50%	R	N	C	N	R	N	R	N	R	R	N	N	R	N	N	N	N	N
Benzoic acid	R	R	R	R	R	R	R	R	R	R	R	R	R	R	R	N	N	N
Black liquor—paper	R	R	R	C	R	R	R	R	R	R	N	N			N	N		
Bleach, 12.5% active chlorine	C	N	C	N	R	N	R	N	R	N	N	N	N	N	N	N	N	N
Bleach, 5.5% active chlorine	C	N	C	N	R	N	R	N	R	N	N	N	N	N	R	N	N	N
Borax	R	R	R	R	R	R	R	R	R	R	R	N	R	N	R	N	N	N
Boric acid	R	R	R	R	R	R	R	N	R		R	N	R	N	R	N	N	N
Brine	R	R	R	N	R	R	R	N	R	N	R	N	R	N	R	R	R	N
Bromic acid, <50%	C	N	N	N	R	N	R	N		N		N				N		N
Bromine, liquid	N	N	N	N	N	N	N	N	N	N	N	N	N	N	N	N	N	N
Bromine, gas, 25%	N	N	N	N	N	N	N	N	N	N	N	N			N	N	N	N
Bromine, aq.	C	N	N	N	R	N	R	N										
Butane	R	R	R	R	R	R	R	R	R	R	R	R	R	R	R	R	R	R
Butanetriol (erythriol)					R	C	R	R	R	C	R	N				N	N	N
Butanediol					N	N	R	R	N	N	N	N	R	R	N	N	N	N
Butyl acetate	C	N	N	N	N	N	N	N										
Butyl phenol	N	N	N	N	N	N	N	N	N	N	N	N			N	N	N	N
Butyric acid, <50%	R	R	R	R	R	R	R	R	N	N	N	N	N	R	N	N	N	N
Calcium salts, aq.	R	R	R	R	R	R	R	R	R	R	R	R	R			N	N	N
Calcium hypochlorite	R	N	C	N	R	N	R	N	R	N	R	N	N	N	R	N	N	N
Calcium hydroxide, 100%	R	R	R	R	R	R	R	R	R	N	R	N	N	N	R	N	N	N
Cane sugar laiquors	R	C	R	N	R	N	R	C										
Carbon disulfide	C	N	N	N	N	N	N	N	N	N	N	N	N	N	N	N	N	N
Carbon dioxide	R	C	C	C	R	R	R	R	R	R	R	R	R	R	R	N	N	N
Carbon dioxide, aq.	R	C	C	C	R	R	R	R	R	N	R	R			R	R	R	N
Carbon monoxide	R	C	R	C	R	R	R	R	R	R	R	R	R	N	R	R	R	R
Carbon tetrachloride	R	N	R	N	R	N	R	N	N	N	R	N	R	R	N	N	N	N
Casein	R	R	R	R	R	R	R	R	R	R	R	R	R	R	R	R	R	R
Castor oil	R	N	R	N	R	N	R	N										
Caustic potash (KOH)	R	N	C	N	R	N	R	N	R	N	N	N	N	N	N	N	N	N
Caustic soda (NaOH)	R	C	C	N	R	N	R	N	R	N	N	N	N	N	N	N	N	N
Chlorine, gas, dry	R	C	R	C	R	R	R	R	R	N	R	N			R	N	N	N
Chlorine, gas, wet	N	N	N	N	R	R	R	R	R	N	R	R			N	N	N	N
Chlorine, liquid	N	N	N	N	N	N	N	N	N	N	N	N	N	N	N	N	N	N
Chlorine, water	C	N	C	N	R	N	R	R	C	N	C	N			N	N	N	N
Chlorocetic acid	R	N	R	N	R	N	R	N	N	N	N	N	N	N	N	N	N	N

(*continued*)

TABLE 66.19. (Continued)

Up to Temperature (°F)	Bisphenol A Epoxy, Amine Cured 120	210	Bisphenol A Epoxy, Anhydrous Cured 120	210	Bisphenol A Vinyl Ester 120	210	Novolac VE 120	210	Bisphenol A Fumarate Polyester 120	210	Chlorendic Acid Polyester 120	210	Furan 120	210	Isophthalic Acid Polyester 120	180	Orthophthalic Acid Polyester 120	150
Chlorobenzene	C	N	N	N	N	N	C	N	N	N	N	N	R	R	N	N	N	N
Chloroform	N	N	N	N	N	N	N	N	N	N	N	N	N	N	N	N	N	N
Chlorosulfonic acid, 10%	N	N	N	N	N	N	N	N	N	N	N	N	N	N	N	N	N	N
Chromic acid, 10%	N	N	N	N	R	N	R	N	N	N	R	N	N	N	N	N	N	N
Chromic acid, 30%	N	N	N	N	N	N	N	N	N	N	N	N	N	N	N	N	N	N
Chromic acid, 40%	N	N	N	N	R	N	R	N	N	N	R	N	N	N	N	N	N	N
Chromic acid, 50%	N	N	N	N	N	N	R	N	N	N	R	N	N	N	N	N	N	N
Citric acid	R	R	R	R	R	R	R	R	R	R	R	R	R	R	R	R	R	N
Coconut oil	R	R	R	N	R	N	R	R	R	R	R	R	R	N	R	R	R	N
Copper salts, aq.	R	R	R	R	R	R	R	R	R	R	R	R	R	R	R	R	R	N
Corn oil	R	R	R	C	R	C	R	R	R	N	R	R	R	R	R	R	C	N
Corn syrup	R	R	R	R	R	R	R	R	R	R	N	N	R	R	R	R	N	N
Cottonseed oil	R	R	R	R	R	R	R	R	R	R	N	N	R	R	R	R	N	N
Cresylic acid, 50%	N	N	N	N	N	N	R	N	R	N	N	N	N	N	N	N	N	N
Crude oil	R	R	R	R	R	R	R	R	R	R	R	R	R	R	R	R	R	N
Cyclohexane	R	N	R	R	R	N	R	R	R	N	R	N	R	R	R	R	R	N
Cyclohexanol	R	N	R	N	R	N	R	N	R	N	R	N	R	N	R	R	N	N
Cyclohexanone					N	N			R	N	R	N			N	N	N	N
Diesel fuels	R	R	R	N	R	R	R	R	R	R	R	R	N	N	R	R	R	C
Diethyl amine	N	N	N	N	N	N	N	N	N	N	N	N	N	N	N	N	N	N
Dioctyl phthalate	R	R	R	C	R	N	R	R	R	N	R	N	R	R	R	R	N	N
Dioxane-1,4			N	N	N	N	N	N	R	N	N	N	R	R	N	N	N	N
Dimethylamine	N	N	N	N	N	N	N	N	N	N	N	N	N		N	N	N	N
Dimethyl formamide	N	N	N	N	N	N	N	N	N	N	N	N	N	N	N	N	N	N
Detergents, aq.	R	R	R	R	R	R	R	R	R	R	R	R	R	R	R	R	R	N
Dibutylphthalate	R	C	R	N	R	N	R	R	R	N	R	N	R	R	R	R	N	N
Dibutyl sebacate	R	N	R	N	R	N	R	N	R	N	R	N	R	R	R	R	N	N
Dichlorobenzene	C	N	N	N	N	N	R	N	N	N	R	R	R	R	N	N	N	N
Dichoroethylene	C	C	N	N	R	R	R	R	R	N	R	R			R	R	N	N
Ether (diethyl)	C	C	N	N	N	N	R	R	R	N	N	N			N	N	N	N
Ethyl halildes	C	C	N	N	N	N	N	N	N	N	N	N			N	N	N	N
Ethylene halibes	R	N	R	R	R	N	R	R	N	N	R	R	R	R	R	R	N	N
Ethylene glycol	R	R	R	R	R	R	R	R	R	R	R	R	R	R	R	R	R	N
Ethylene oxide	N	N	N	N	N	N	N	N	N	N	N	N	R	R	N	N	N	N
Fatty acids	R	C	C	R	R	R	R	R	R	R	R	R	R	R	R	R	N	N
Ferric salts	R	R	R	R	R	R	R	R	R	R	R	R	R	R	R	R	N	N
Fluorine, gas, dry	N	N	N	N	N	N	N	N	N	N	N	N	R	R	N	N	N	N
Fluorine, gas, wet	N	N	N	N	N	N	N	N	N	N	N	N	R	C	N	N	N	N

956

Chemical															
Fluoroboric acid, 25%	N	N	N	N		R	N	N	N	R	R	R	R	R	R	R
Fluorosilicic acid, 10%	N	N	N	N		R	N	N	N	N	R	R	R	R	C	C
Formaldehyde	N	N	N	R	N	N	R	R	N	N	R	N	R	R	C	C
Formic acid	N	N	N	N	N		N	N	N	N	C	N	R	N	C	C
Freon, F11, F12, 113, 114		N	N	N		N	N	N	N	N	N	N	N	N	N	C
Freon, F21, F22	R	N	N	R	N	R	N	N	N	N	N	N	C	C	N	R
Fruit juices and pulps	R	N	N	R	R	R	R	R	C	R	R	C	R	C	R	R
Fuel oil	C	R	N	N	R	C	R	R	R	R	R	R	R	R	R	N
Furfural	R	N	R	R	C	R	R	R	R	R	R	R	R	R	R	R
Gas, natural, methane	R	N	N	R	R	R	N	C	C	C	R	R	C	C	C	R
Gasoline	C	N	N	R	R	C	R	R	R	R	R	R	R	R	C	R
Gelatin	N	N	N	R	C	N	R	R	R	R	R	R	R	R	C	C
Glycerine (glycerol)	R	N	N	R	N	R	N	N	N	R	C	C	C	C	R	R
Glycols	R	R	R	N	R	R	R	R	R	R	R	R	R	R	R	R
Glycolic acid	C	N	R	R	R	R	N	R	R	R	R	R	R	R	R	R
Green liquor-paper	R	N	N	R	N	R	N	N	N	N	C	R	N	N	C	R
Haptane	R	N	N	C		R	R	R	R	R	R	R	R	R	R	C
Hexane	R	N	N	N	N	N	N	N	N	N	R	C	C	C	C	R
Hydrobromic acid, 25%	C	N	N	N	N	N	N	N	N	N	N	C	C	C	C	C
Hydrochloric acid	R	N	N	N	N	N	N	N	N	N	N	C	C	C	C	R
Hydrofluoric acid, 10%	R	N	N	N	R	R	R	R	R	R	R	R	R	R	R	R
Hydrofluoric acid, 60%	R	N	N	R	R	R	N	R	R	R	R	R	R	R	R	N
Hydrofluoric acid, 100%	N	N	N	R	N	N	N	N	N	R	N	N	N	N	N	R
Hydrocyanic acid	N	N	N	N	R	R	N	R	R	R	R	N	R	R	R	N
Hydrogen peroxide, 50%	R	N	N	R	R	R	N	R	R	N	R	R	N	N	N	N
Hydrogen peroxide, 90%	R	N	N	R	R	R	N	R	R	R	R	R	N	N	N	R
Hydrogen sulfide, dry	N	N	N	N			R	N	N	R	N	R	R	R	R	R
Hydrazine	R	N	N	R	R	R	N	R	R	R	R	R	N	R	R	R
Hypochlorous acid, 10%	R	N	N	R		R	N	R	R	R	R	R	R	R	R	R
Jet fuel, JP4 and JP5	R	N	N	R	N	N	N	R	R	R	R	R	R	R	R	R
Kerosene	R	N	N	R	R	R	N	R	R	N	R	R	R	R	R	R
Lactic acid, 25%	R	N	N	R	R	C	R	R	R	R	R	R	N	R	R	R
Lauric acid	R	N	N	R	R	N	N	R	R	R	R	R	R	R	R	R
Lauryl chloride	R	N	N	R	R	R	N	R	R	R	R	R	R	R	R	R
Lauryl sulfate	R	N	N	R	R	R	N	N	R	R	R	R	R	R	R	R
Lead salts	R	N	N	R		R	R	R	R	R	R	R	N	R	R	R
Linoletic acid	R	N	N	R	R	R	N	N	R	R	R	R	R	R	R	R
Linseed oil	R	N	N	R		R	N	N	N	R	R	R	N	R	R	R
Lithium salts	R	N	N	R	N	N	N	N	N	R	R	R	N	R	R	R
Lubricating oils	R	N	N	R	N	R	N	R	R	R	R	R	R	R	R	R
Machine oil	R	N	N	R	R	R	N	R	R	N	R	R	R	R	R	R
Magnesium salts	R	N	N	R	R	R	N	N	N	R	R	R	R	R	R	R
Maleic acid	R	N	N	R	R	R	N	R	R	R	R	R	R	R	R	R
Manganese sulfate	R	N	N	R	R	R	N	R	R	R	R	R	R	R	R	R

(continued)

TABLE 66.19. (Continued)

Up to Temperature (°F)	Bisphenol A Epoxy, Amine Cured		Bisphenol A Epoxy, Anhydrous Cured		Bisphenol A Vinyl Ester		Novolac VE		Bisphenol A Fumarate Polyester		Chlorendic Acid Polyester		Furan		Isophthalic Acid Polyester		Orthophthalic Acid Polyester	
	120	210	120	210	120	210	120	210	120	210	120	210	120	210	120	180	120	150
Mercuric salts	R	R	R	R	R	R	R	R	R	R	R	R	R	R	R	R	N	N
Mercury	R	R	R	R	R	R	R	R	R	R	R	R	R	N	R	R	N	N
Methane	R	R	R	N	R	R	R	R	R	R	R	R	R	R	R	R	R	R
Methyl acetate	N	N	N	N	N	N	N	N	N	N	N	N	R	N	N	N	R	R
Methyl bromide (gas)	N	N	N	N	N	N	N	N	N	N	N	N			N	N		
Methyl cellosolve					N		N		N		N		R	N	N		N	
Methyl chloride	N	N	N	N	N	N	N	N	N	N	N	N			N	N	N	N
Methyl chloroform	N	N	N	N	N	N	N	N	N	N	N	N			N	N	N	N
Methyl cyclohexanone	N	N	N	N	N	N	N	N	N	N	N	N	N	N	N	N	N	N
Methyl methacrylate	N	N	N	N	N	N	N	N	N	N	N	N		N	N	N	N	N
Methylene bromide	N	N	N	N	N	N	N	N	N	N	N	N	N	N	N	N	N	N
Methylene chloride	N	N	N	N	N	N	N	N	N	N	N	N	R	N	N	N	N	N
Methylene iodide	N	N	N	N	N	N	R	N	N	N	N	N			N	N	N	N
Milk	R	R	R	R	R	R	R	R	R	N	R	R	R	N	R	N	N	N
Mineral oil	R	R	R	R	R	R	R	R	R	R	R	R	R	R	R	R	R	R
Molasses	R	R	R	N	R	R	R	N	R	R	R	R	R	R	R	R	N	N
Monochlorobenzene	N	N	N	N	N	N	C	N	N	N	N	N	N	N	N	N	N	N
Monoethanolamine	N	N	N	N	N	N	N	N	N	N	N	N	R	N	N	N	N	N
Motor oil	R	R	R	R	R	R	R	R	R	R	R	R	R	R	R	R	N	N
Nephtha	R	R	R	N	R	R	R	R	R	R	R	R	R	R	R	R	N	N
Nephthalene	R	R	R	R	R	R	R	R	R	R	R	R	R	R	R	R	N	N
Nickel salts	R	R	R	R	R	R	R	R	R	R	R	N	R	R	R	R	N	N
Nitric acid, 0–20%	C	N	N	N	R	N	R	N	R	N	R	R	N	N	R	N	N	N
Nitric acid, 21–100%	N	N	N	N	N	N	N	N	N	N	N	N	N	N	N	N	N	N
Nitric acid, fuming	N	N	N	N	N	N	N	N	N	N	N	N	N	N	N	N	N	N
Nitrobenzene	C	N	N	N	N	N	C	N	N	N	R	N	R	R	N	N	N	N
Nitrous acid	R	R	R	N	R	N	R	N	R	N	R	R	R	R	R	N	N	N
Oleic acid	R	R	R	R	R	R	R	R	R	R	R	R	R	R	R	R	R	R
Oleum	N	N	N	N	N	N	N	N	N	N	N	N	N	N	N	N	N	N
Olive oil	R	R	R	R	R	R	R	R	R	R	R	R	R	R	R	R	R	R
Oxalic acid	R	R	R	R	R	R	R	R	R	R	R	N	R	R	R	R	N	N
Ozone, gas, 5%	C	N	C	N	R	N	R	R	R	N	R	R	R	N	R	N	N	N
Palmitic acid, 10%	R	R	R	R	R	R	R	R	R	R	R	R	R		R	R	R	R
Palamitic acid, 70%	R	R	R	R	R	R	R	R	R	R	R	R	R	N	R	R	R	R
Paraffin	R	R	R	R	R	R	R	R	R	R	R	R	R	R	R	R	R	R
Pentane	R	N	R	N	R	N	R	N	R	N	R	R	R	R	R	R	R	R
Perchloric acid, 10%	R	C	R	C	R	N	R	N	N	N	R	R	N	N	R	R	R	R
Perchloric acid, 70%	R	C	R	C	N	N	R	N	N	N	N	N		N	N	N	N	N
Perchloroethylene	R	C	R	C	N	N	R	N	N	N	N	N	N		N	N	N	N

958

Chemical																
Petroleum, sour	N	N	N	R	R	R	R	N	R	R	R	R	R	R	R	R
Petroleum, refined	R	R	R	R	R	R	R	N	R	R	R	R	R	R	R	R
Phenol, 88%	N	N	N	N	N	N	N	N	N	N	N	N	N	N	N	N
Phenylcarbinol	N	N	N	R	N	N	N	N	N	N	N	N	N	N	N	N
Phenylhydrazine	N	N	C	R	R	N	N	N	N	N	N	N	C	C	C	C
Phosphoric acid	N	N	R	N	R	R	R	R	R	R	R	R	R	R	R	R
Phosphorus, yellow	N	N	N	N	N	N	N	N	N	N	N	N	N	N	N	N
Phosphorus, red	N	N	N	N	N	N	N	N	N	N	N	N	N	N	N	N
Phosphorus trichloride	N	N	R	N	R	N	N	N	N	N	N	N	N	N	N	N
Phthalic acid	N	R	R	R	R	R	R	R	R	R	R	R	R	R	R	R
Potassium salts, aq.	R	C	N	R	R	R	R	R	R	R	R	R	R	R	R	R
Potassium permanganate, 25%	R	R	R	R	R	N	R	R	N	N	N	N	R	R	R	C
Propane	N	N	R	R	R	R	R	R	R	R	R	R	R	R	R	R
Propylene dichloride	N	N	N	N	N	N	N	N	N	N	N	N	N	R	R	R
Propylene glycol	N	N	N	R	R	R	R	R	R	R	R	R	R	R	R	R
Propylene oxide	N	N	N	N	N	N	N	N	N	N	N	N	N	N	N	N
Pyridine	N	N	N	N	N	N	N	N	N	N	N	N	N	R	R	R
Rayon coagulating bath	R	R	R	R	R	R	R	R	R	R	R	R	R	R	R	R
Sea water	R	R	R	R	R	R	R	R	R	R	R	R	R	R	R	R
Salicyclic acid	R	R	R	R	R	N	R	N	N	N	N	R	R	R	R	R
Sewage, residential	R	C	R	N	C	N	C	C	C	C	C	R	R	R	R	R
Silicic acid	R	R	R	R	R	R	R	R	R	R	R	R	R	R	R	R
Silicone oil	R	R	R	R	R	R	R	R	R	R	R	R	R	R	R	R
Silver salts	R	R	R	R	R	R	R	R	R	R	R	R	R	R	R	R
Soaps	R	R	R	R	R	R	R	R	R	R	R	R	R	R	R	R
Sodium hydroxide	N	R	R	N	C	N	R	C	N	R	C	R	R	C	C	R
Sodium salts, aq. except	R	R	R	R	R	R	R	R	R	R	R	R	R	R	R	R
Sodium chlorite, 10%	N	R	R	R	R	N	R	N	R	R	R	R	R	R	R	R
Sodium chlorate	R	R	R	R	R	R	R	R	R	R	R	R	R	R	R	R
Sodium dichromate, acid	R	R	R	R	R	R	R	R	R	R	R	R	R	R	R	R
Stanic chloride	R	R	R	R	R	R	R	R	R	R	R	R	R	R	R	R
Stanous chloride	R	R	R	R	R	R	R	R	R	R	R	R	R	R	R	R
Stearic acid	R	R	R	R	R	R	R	R	R	R	R	R	R	R	R	R
Sulfite liquor	N	R	R	R	R	R	R	R	R	R	R	R	R	R	R	R
Sulfur	N	N	N	N	N	R	N	R	R	R	N	N	R	R	R	R
Sugars, aq.	N	N	R	R	R	R	R	R	R	R	R	R	R	R	R	R
Sulfur dioxide, dry	R	R	R	R	R	R	R	R	R	R	R	R	R	R	R	R
Sulfur dioxide, wet	N	N	N	N	R	R	R	R	R	C	C	C	R	C	C	C
Sulfur trioxide, gas, dry	N	N	N	N	R	N	R	R	R	R	R	R	R	R	R	R
Sulfur trioxide, wet	N	N	N	N	N	N	N	N	N	N	N	N	N	N	N	N
Sulfuric acid, < 26%	N	N	R	N	R	N	R	N	R	R	R	R	R	C	R	C
Sulfuric acid, 26–80%	N	N	R	N	R	N	C	N	N	N	N	R	R	C	C	C
Sulfuric acid, 81–100%	N	N	N	N	N	N	N	N	N	N	N	N	N	N	N	N
Sulfurous acid, 10%	N	N	N	N	R	N	R	R	R	N	R	N	R	R	R	R

(continued)

959

TABLE 66.19. (*Continued*)

Up to Temperature (°F)	Bisphenol A Epoxy, Amine Cured		Bisphenol A Epoxy, Anhydrous Cured		Bisphenol A Vinyl Ester		Novolac VE		Bisphenol A Fumarate Polyester		Chlorendic Acid Polyester		Furan		Isophthalic Acid Polyester		Orthophthalic Acid Polyester	
	120	210	120	210	120	210	120	210	120	210	120	210	120	210	120	180	120	150
Tall oil	R	R	R	R	R	R	R	R	R	N	R	N	R	N	R	N	N	N
Tannic acid	R	R	R	R	R	R	R	R	R	R	R	R	R		R	R	N	N
Tartaric acid	R	R	R	R	R	R	R	R	R	R	N	R	R	R	R	R	R	R
Tetrachloroethane	C	N	C	N	N	N	R	N	N	N	N	N	R	R	N	N	N	N
Tetrahydrofuran	N	N	N	N	N	N	N	N	N	N	N	N	N	N	N	N	N	N
Thionyl chloride	N	N	N	N	N	N	N	N	N	N	N	N	R	N	N	N	N	N
Thread cutting oil	R	R	R	R	R	R	R	R	R	R	R	N	R	N	R	R	N	N
Terpineol	R	R	R	R	R	R	R	N	R	R	R	R			R	R	N	N
Toluene	R	C	C	N	N	N	R	N	N	N	N	N	R	R	N	N	N	N
Tributyl phosphate	R	N	R	N	R	N	R	N	N	N	N	N			N	N	N	N
Tricresyl phosphate	R	N	R	N	R	N	R	N	N	N	N	N			N	N	N	N
Trichloracetic acid	C	C	C	C	R	R	R	N	R	N	N	N			N	N	N	N
Trichloroethylene	N	N	N	N	N	N	N	N	N	N	N	N	R	N	N	N	N	N
Triethanolamine	R	N	R	N	R	N	R	N	N				N	N	N	N	N	N
Triethylamine	R	N	C	N	R	N	R	N							N	N	N	N
Turpentine	R	N	R	N	R	R	R	R	N	N	R	N	R	N	N	N	N	N
Urea, 50%	R	N	R	N	R	N	R	N	R	N	R	N	R	R	N	N	N	N
Urine	R	N	R	N	R	N	R	N	R	N	R	N	R	R	R	N	N	N
Vaseline	R	R	R	R	R	R	R	R	R	R	R	N	R	R	R	R	R	R
Vegetable oils	R	R	R	R	R	R	R	R	R	R	R	N	R	R	R	R	R	R
Vinegar	R	R	R	R	R	R	R	N	R	N	R	N	R	R	N	N	N	N
Vinyl acetate	N	N	N	N	N	N	N	N	N	N	N	N			N	N	N	N
Water, distilled	R	C	C	N	R	R	R	R	R	N	R	N	R	N	R	R	N	N
Water, fresh	R	C	R	N	R	R	R	R	R	N	R	N	R	N	R	R	R	R
Water, mine	R	R	R	N	R	R	R	R	R	N	R	N	R	N	R	R	R	R
Water, salt	R	R	R	N	R	R	R	N	R	N	R	N	R	N	R	R	R	R
Water, tap	R	C	R	N	R	R	R	N	R	N	R	N	R	N	R	R	R	R
Whiskey	R	C	R	C	R	R	R	R	R	N	R	N	N	N	R	R	R	R
Wines	R	C	R	C	C	R	R	R	R	N	R	N	R	N	R	R	R	R
Xylene	R	N	C	N	N	N	R	N	N	N	N	N	R	R	N	N	N	N
Zinc salts	R	R	R	R	R	R	R	R	R	R	R	N	R	R	R	R	N	N

R = Generally resistant, N = generally not resistant, C = less resistant than R but still suitable for some conditions.

960

TABLE 66.20. Comparison of Custom and Commodity Piping

	Custom	Commodity
Resin	Large choice of resin	Usually vinyl ester only
Manufacturing	Filament wound or hand layup	Centrifugal cast, filament wound
Corrosion barrier	100 mils, min w/choice of veil	Varies from 50 mils. May not have a corrosion barrier
Joints	Butt and wrapped	Bell and spigot, socket
Fittings	Filament wound or hand layup	Compression molded

a mandrel. A typical sequence is application of mold release agent, polyester film, buildup of corrosion barrier (veil and mat) followed by the structural thickness (mat and woven roving). A pigmented final coat for ultraviolet (UV) protection completes the fabrication. Contact molding is used for complex shapes as well as simple cylindrical forms such as tank shells. This process makes vessel heads, flanges, nozzles, pads, and secondary bonds.

G2.8.2. Filament Wound. Filament winding consists of using a winding machine on which resin-saturated continuous-strand roving is applied. Wind angle varies from 54° (considered to be ideal) to almost 90°. Structures wound at higher angles usually require axial reinforcement. The corrosion barrier is still made by the contact molding process.

Filament-wound laminate is stronger than hand layup and is easier and quicker to make. However, due to the absence of continuous paths, hand layup laminate is more forgiving of flaws.

G2.9. Design and Fabrication of FRP Piping. The FRP piping is classified into two broad classes: custom made and commodity (also known as off-the-shelf or machine made). Custom-made piping is specified completely by the user. Resin, cure system, composition of the corrosion barrier, and so on, are determined by the user. Commodity piping is designed by the manufacturer and sold as a product line with trade names. Commodity piping usually has smaller corrosion barrier thicknesses and incorporates flexibilizing agents for ease of fabrication. Usually the fittings are compression molded with bulk molding compound. For these reasons, it is important to verify the chemical resistance of these products.

The design of FRP piping (both custom and commodity) follows the same rules of HDB. For custom piping, the maximum allowable stress is determined to be 1500 psi (150 kg/cm^2) per ASTM D2992. For commodity piping the maximum allowable stress is 0.5–1.0 times the HDB determined per ASTM D2992 by the manufacturer by short- and long-term tests. The factor 0.5 is used for static applications and 1.0 for cyclic applications.

Table 66.20 shows a comparison of custom and commodity piping. Assuming that the laminate is properly chosen, failure of FRP piping is mostly at the joints, particularly for commodity piping. The economic advantages of commodity piping may be offset by a variety of joint problems. The result is that commodity piping is generally not used for severe service.

Some reasons for commodity piping joint failure are listed below:

Contamination of joining surfaces
Excessive adhesive thickness
Inadequate cure of adhesives
Exceeding the pot life of adhesive
Exceeding the shelf life of adhesive

It is a common experience that more failures occur at field joints than at shop joints.

H. SEALS AND GASKETS

Seals and gaskets are used for flange joints and mechanical seals. Seals usually refer to nonflat sealing devices such as O rings. Two broad classes of materials are used: elastomers and fluoropolymers.

H1. Gaskets

Gaskets are classified in three broad categories: metallic, semimetallic, and nonmetallic. Metallic gaskets are not included in this chapter.

Semimetallic gaskets include envelope and spiral-wound types. The envelope is always a fluoropolymer, usually PTFE. The reinforcing can be metallic, elastomeric, or composites. Spiral-wound gaskets are typically constructed from metals incorporating a fluoropolymer material.

Nonmetallic gaskets usually include sheet gaskets. Materials include elastomeric and thermoplastics. Elastomeric sheet gaskets are used either as plain sheets or scrim-reinforced sheets. Commonly used thermoplastic gaskets include expanded, filled, and virgin fluoropolymers. The most frequently used fluoropolymer is PTFE. Fillers include calcium sulfate, barium sulfate, and graphite. Each helps to enhance creep performance of the gasket.

The success of a bolted flange connection using gaskets depends on the right choice of gasket, type and quality of mating flanges, and bolts used. Additionally, the right bolt torque to achieve an appropriate level of gasket stress is very important. The gasket stress needs to be adequate to achieve proper seal yet not be so high that the flanges or the gasket are damaged. In this regard there are two distinct approaches, one for fluoropolymer gaskets and the other for others such as the elastomeric gaskets.

Fluoropolymer gaskets require higher compressive stress to achieve a seal. These materials also creep, which results in stress relaxation. To maintain a leak-tight joint, retorqueing after 24 h is necessary. The Pressure Vessel Research Council (PVRC) has established guidelines whereby fluoropolymer gaskets can be compressed to achieve the tightness parameter required. Torque tables are also published by the manufacturers of gaskets which show torque values for pipe sizes for a given flange.

Elastomeric gaskets do not creep and require a much lower level of preload to achieve a seal. Usually 20–25% compression is adequate to achieve the required level of tightness. Elastomeric gaskets are usually used on full-face flanges and cannot be used on raised-face flanges.

H2. Seals

The word "seals" refers to compression molded T seals, O rings, V rings, and U cups. High-performance elastomers (fluoroelastomers) and PTFE are usually the materials of choice for such seals where high temperatures and corrosive or hazardous chemicals are involved. These are most commonly used in mechanical seals, valves, and pumps.

Design of the groove for the O ring is just as important as the selection of the material. Proper groove design enables optimum compression for a tight seal.

H3. Selection of Seals and Gaskets

Compression set (sometimes known as permanent set) is the key data by which seals and gaskets are selected. These numbers are, however, generated in air and cannot be translated to a situation where the seals and gaskets are exposed to chemicals. The material is thus selected by immersing coupons in chemicals under operating conditions and evaluating properties such as tensile strength, 100% modulus, elongation, volume swell, weight gain, and hardness. Combining this information with the compression set data selection is made. Table 66.22 gives guidance on the possible damage mechanisms based on the combination of property changes.

Attempts are being made to expose gaskets and O rings to chemicals while under compression using the apparatus shown in Figure 66.16. Multiple samples can be fitted between the parallel plates such that exact compression is achieved. The whole apparatus is immersed in the chemical.

Table 66.10B shows the chemical formulas for high-performance rubber seals and gaskets. Tables 66.21A and 66.21B list their properties and general chemical resistance.

I. FAILURES AND FAILURE ANALYSIS

Failure of a unit is defined as a condition where it is not able to function due to some form of damage to the structure or the internal or external surfaces. Failure is also defined as a condition where the continued operation of the unit is likely to lead to a release of a chemical or creating some form of safety or environmental hazard. In either case the interaction of the unit and the exposure forces have resulted in a condition which makes it impossible for continued operation. Under such conditions the unit has to be shut down and the cause is investigated.

In general failures occur due to the following factors:

Improper selection—material, thickness, and fabrication technique

Inadequate attention to supplier selection

Lack of quality assurance program during fabrication

Operating beyond the parameters

Permeation, cycling (pressure and temperature)

Absence of a condition assessment (in-service inspection) program

Failure analysis is based on a methodology as well as data on known polymer–chemical–stress interactions. Failure analysis begins with visual observations of the part. These observations are common to all materials and are listed in Table 66.23 along with the possible causes. Specific failure modes and damage mechanisms for FRP are listed in Table 66.24 Following also are some specific manifestations of failures of some other components:

Piping. Sagging or general deformation is due to inadequate pipe supports. This is particularly true for solid thermoplastic piping such as PE, PP, and PVC. Leaky joints in FRP—a frequently occurring problem—is usually due to improperly made butt and wrap joint. Lined piping, especially fluoropolymer piping, is susceptible to liner collapse due to temperature cycling.

Valves. The most commonly occurring problem in lined valves is swelling of the liner leading to seizing of the ball or plug. Swelling is caused by permeation and is usually countered by a different type of valve. Butterfly valves usually fail due to the elastomer backing being attacked by the permeants. Swelling of the plate liner becomes evident.

Hoses. The most common problem with hoses is the failure of the end connection to the liner–carcass assembly. Qualification of the assembler and rigid quality assurance usually counters this problem.

Expansion Joints. Expansion joints fail due to exceeding manufacturers' limits on linear, bending, and torsional movement. Tearing in the body of the joint results.

FIGURE 66.16. Apparatus for immersing seals and gaskets while under compression.

If analysis beyond visual observation is needed, a variety of tools are available. The methodology of correlating such information to the failure mode is not very well established.

The first step in the analysis is identification of the polymer backbone, if it is not already known, through infrared (IR) spectroscopy along with thermogravimetric analysis and melting point determination where applicable.

TABLE 66.21A.   Mechanical Properties of General Chemical Resistance of High-Performance Elastomer Sealing Materials

ASTM Designation	FKM	TFE/P	FVMQ	FFKM
Trade Name[a]	Viton®	Aflas®	Fluorosilicone	Kalrez®
Common Name	Fluorinater	PTFE/Polypropylene		
	Hydrocarbon	Copolymer	Fluorosilicone	Perfluoroelastomer
Min. continuous-use temp (°F)	−20 to −70	−20	−90 to −112	−36
Max. continuous-use temp(°F)	440	400–446	450	554–600
Tensile strength ($\times 10^3$ psi)	0.5–3.0	2.0–3.2	0.5–1.4	0.5–1.5
Tensile modulus at 100% elongation	200–2000	900–2500	900	0.9–1.9×10^3
Hardness (durometer)	50A–95A	60–100A	35A–80A	65A–95A
Compression set	9–16, 70h	25, 70h	17–25 22h	20–40, 70h
(ASTM D395,3%)	at75°F	at 200°F	at 300°F	at 70°F
Elongation %	100–500	50–400	100–480	60–170

[a]Viton® and Kalrez® are trademarks of DuPont Dow. Aflas® is a trademark of the Green Tweed.

TABLE 66.21B.   Mechanical Properties of General Chemical Resistance of Midperformance Elastomer Sealing Materials

ASTM Designation Trade Name[a] Common Name	EU Adiprene Polyurethane	CR Neoprene Polychloroprene	CSM Hypalon® Chlorosulfonated Polyethylene	EPR Nordel® Ethylene Propylene
Min. continuous-use temp. (°F)	− 65	− 80	− 65	− 75
Max. continuous-use temp. (°F)	+ 250	+ 300	+ 275	+ 325
Tensile strength ($\times 10^3$ psi)	0.3–3.5	0.5–4.0	—	0.3–3.5
Tensile modulus at 100% elongation	100–3000	100–3000	450–500	100–3000
Hardness (durometer)	30A–90A	15A–95A	40A–100A	30A–90A
Compression set	20–60,70h at 212°F	20–60h at 212°F	38–80, 22h at 212°F	20–60, 70h at 212°F
Elongation %	100–700	100–800	100–700	100–700

[a]Hypalon® and Nordel® are trademarks of DuPont.

TABLE 66.22.   Interpretation of Test Data Results

	Root Cause of Property Damage				
	(a)	(b)	(c)	(d)	(e)
Property change	Process Medium absorption	Extraction of compound ingredients	Medium attack on filler system	Attack and degradation of crosslinks and/or polymer backbone	
Hardness	Decrease	Increase	Usually decreases	Increase (hard/brittle)	Decrease (soft/gummy)
Volume	Increase	Decrease	Increase	Increase	Increase
Tensile stress at break	Decrease	Increase	Decrease	Decrease	Decrease
Modulus	Decrease	Increase	Decrease	Increase	Decrease
Elongation at break	Increase	Decrease	Increase	Decrease	Increase or decrease

Notes: 1. Dependent upon the elastomer compound, process media, and conditions, more than one of these phenomena may occur simultaneously.
2. Property changes (increase or decrease) represent trends often observed for most families of elastomers. There may be exceptions.

For composite materials, a burnout test will define the proportions of resin and reinforcement or filler.

Failed linings should be examined for loss of adhesion and discoloration or other evidence of permeation and absorption. The contents of a blister can be removed for analysis with a hypodermic needle. Useful nondestructive measure-

TABLE 66.23.   Visual Evidence of Failure and Possible Causes

Defect	Possible Causes
Swelling	Sorption, permeation
Blistering	Permeation, localized polymerization
Discoloration	Chemical reaction with additives, permeation
Cracking	Stress cracking and environmental stress cracking
Deformation	Creep, stress overload
Debonding from a substrate	Permeation and adhesion failure
Loss of thickness	Chemical attack, abrasion
Spalling	Lack of adhesion, thermal shock
Pinholes	Lack of proper curing or coverage
General degradation	Chemical attack—oxidative, hydrolysis, de-crosslinking, depolymerization

ments include hardness and spark testing. Measuring tensile properties of a sample of the lining is an excellent means of determining its condition.

Visual analysis is still the most useful method for studying failures. Surface etching or cracking indicates that the polymer has been chemically attacked. Swelling is evidence of absorption of organics. How well the substrate surface was prepared prior to lining can be determined by microscopic examination of the back of the lining specimen.

After a failure has been analyzed, repair may be considered. The degree of repair difficulty depends on the age and quality of the material and the severity of the damage. For linings, a test patch should be spark and adhesion tested. If thermoplastic welding is required, as for fluoropolymers, testing for weldability is necessary. Damaged FRP can be repaired by exposing an undamaged surface and laying up new material.

J. CONDITION ASSESSMENT, FITNESS FOR SERVICE AND REPAIRS

Condition assessment of polymeric materials in chemical handling is an evolving field. Condition assessment implies

TABLE 66.24.   FRP and Chemical Interaction

Chemical	Appearance (Failure Mode)	Nature of Interaction (Damage Mechanism)
Sodium hydroxide	Swells, resin disappears	Glass attack and hydrolysis (de-esterification) of polymer backbone
HF	Resin disappears	Glass attack only (higher temperature may cause resin attack?)
HCL	May turn greenish color, blisters, may become brittle	Glass attack only (higher temperature may cause resin attack?)
H_2SO_4	May turn reddish, blister depending on concentration, >80% becomes black, swells	Oxidative attack of resin
NaOCl	Resin disappears	Oxidative attack of resin; glass attack and hydrolysis (de-esterification) of polymer backbone
HNO_3	Resin disappears or reaction products	Oxidative attack of resin
O_3	Resin disappears	Oxidative attack of resin
H_2O_2	Resin disappears	Oxidative attack of resin
Chromic acid	Severe cracking	Oxidative attack of resin
DI water	Blistering, severity depends on temperature	Hydrolysis (de-esterification) of resin
Gasoline	See comments for aliphatic solvents, below; dependent on components and amount	Glass resin interface physical attack due to permeation
Aliphatic solvents	Swelling dependent on temperature and size of molecule; Derakane 411 recommended up 160°F for 100% hexane	Glass resin interface physical attack due to permeation
Chlorobenzene	Swelling leading to resin tearing (cracking)—excessive weight gain	Glass resin in interface physical attack due to permeation
Methanol	Swelling leading to resin tearing (cracking)—excessive weight gain	Glass resin interface physical attack due to permeation

Courtersy of Dupont Company.

online monitoring and inspection/evaluation. In both these areas tools and techniques that give measurable quantitative results are rare and the end user is limited to a great deal of visual observation and experiences in on-line monitoring. Two systems have been evaluated one of which consists of embedding probes at different thickness in the laminate. The changing electrical resistance is measured at different times. The second system consists of using a conductive fabric in the laminate usually at the end of the corrosion barrier. This conductive layer is connected to a measuring device which measures the changing electrical resistance of the corrosion barrier.

Nondestructive testing (NDT) techniques applicable to metals, such as ultrasonics, radiography, magnetic particles, and eddy current, are not applicable for polymeric materials. Acoustic emission testing has proved to be the most effective for FRP. In many applications destructive testing is used to assess the condition of the material. Examples are pull testing of linings for loss of adhesion and mechanical property retention of FRP and other structural materials. Destructive testing is usually carried out on easily replaceable parts such as a manway cover, a blind flange, or a pipe spool. Interpretation of such results for remaining life prediction is done on the basis of experience.

Condition assessment for the purpose of determining the fitness for service of a component consists of inspection of the unit and recording the visual observations followed by microscopic where appropriate. Where possible, other NDT techniques are employed.

J1. FRP

Acoustic emission (AE) is used to determine the general integrity of a vessel. This technique involves installing sensors on the outside of the vessel and subjecting it to a hydrostatic load. Resin or fiber breaks are detected as events which are analyzed for type and intensity. Accurate locations of the breaks are also possible. Repairs are carried out as needed or the vessel is rerated for a lesser capacity.

Microwave probes have been used to detect blisters in FRP structures with limited usefulness. Blisters containing water in HCl storage vessels have been detected with good confidence. For other type of damage more work needs to be done.

Laser shearography offers great potential, but more work needs to be done. This technique involves stressing the tank either by internal pressure or by some other means and observing the stress patterns of the structure by a laser camera. Localized damaged areas are detected and repaired. Although this technique is successfully used in FRP pressure containers for the defense industry, it has not been evaluated for chemical handling.

FRP damage is described at various levels starting with visual descriptions (blisters, discoloration, cracking, chemical attack, etc.) followed by determining whether the damage is occurring to resin, glass, or the resin–glass interface. Finally the nature of the chemical or physical interaction between the laminate and the liquid being contained is described (oxidative, hydrolysis etc). Table 66.23 describes the current method of describing such interactions for some commonly used chemicals.

Destructive testing of the FRP laminate is carried out in an extreme case where visual or acoustic emission does not yield satisfactory information. A blind flange or a manway cover is chosen and is destructively tested for retained mechanical properties, extent of permeation, and loss of corrosion barrier.

FRP linings are inspected for blisters, spalling, cracking, and discoloration. Retained adhesion to the steel substrate is determined by a peel pull test. Corrosion of the substrate is evaluated by ultrasonic testing from the outside.

Current generally accepted guidelines for the type and frequency of internal examinations are as follows. However, a better approach would be to use the risk-based inspection strategy for FRP developed by the Materials Technology Institute, St. Louis, Missouri.

All equipment should be inspected after one year of operation:

After the first inspection: every 18 months for:

- Equipment handling severe chemicals (e.g., NaOCl, HCl, HNO$_3$, H$_2$SO$_4$)
- Equipment operating above 180°F
- Equipment handling abrasives
- Critical pieces (stacks, scrubbers, etc.)
- Dual-laminate structures

After the first inspection: every 36 months for:

- Storage vessels (50,000 gal and above) for waste
- Equipment for organic solvents
- Equipment operating over 150°F

After the first inspection: every 48 months for:

- Waste wells
- Agitated vessels
- Equipment subjected to high-temperature differentials

Repairs of FRP are done after careful engineering evaluation for structural integrity followed by determination of the required layup for strength and corrosion resistance.

A bond test for adhesion is essential as is proper cleaning and decontamination.

J2. Thermoplastic Linings

Thermoplastic linings are mainly examined from the inside of the vessel for tears, particularly in the welds if the liners are of welded construction. Bulging and blisters are also noted. Because these materials have high elongation and are not generally subject to embrittlement, blisters and bulges are not considered failures unless they interfere in the functioning of the unit. Ultrasonic testing of the substrate from the outside should be carried out when concerned with corrosion of the substrate. Spark testing should be done only as a last resort and that too at a reduced voltage. Attempts are being developed to test the lining by helium leak test under the lining with good promise.

Repairs of the linings usually require some form of welding. Use of qualified welders (AWS or DVS) and qualified welding procedures is essential. Equally importantly test for weldability on a used surface is important. This usually takes the form of running a weld bead and pulling it to test its bond. Inadequate bond may mean decontamination of the surface or relining.

J3. Elastomeric Linings

Rubber linings are subject to progressive hardening due to exposure to certain acids. Organic solvents or wastewaters containing organic solvents can soften these materials. Internal examinations should look for surface cracking, blistering, and hardening of the rubber. Spark testing is also recommended. However, unless there is reason to believe that there is a problem, it is best to leave the rubber lining alone. In an extreme situation where the bond of the lining to the substrate is in question, a peel pull test of the lining is undertaken. Usually a manway cover is selected for the test and a $\frac{1}{2}$-in. wide strip is pulled at 180° and the force to separate is measured. The type of separation (adhesive, cohesive) is noted. A break in the lining itself (cohesive failure) is preferred. An arbitrary width of 10 lbf per lineal inch is agreed upon as the minimum required bond strength.

Figure 66.17 shows a typical risk-based inspection logic diagram for rubber lining in HCl service.

Repair of the rubber lining is relatively easy. After removing the damaged rubber a fresh patch is applied and chemically cured. Steam vulcanizing is not considered necessary. Durometer hardness for an appropriate level of hardness and spark test should be done.

K. ECONOMIC DATA

Tables 66.25A, B, and C provide guidelines for cost estimation purposes.

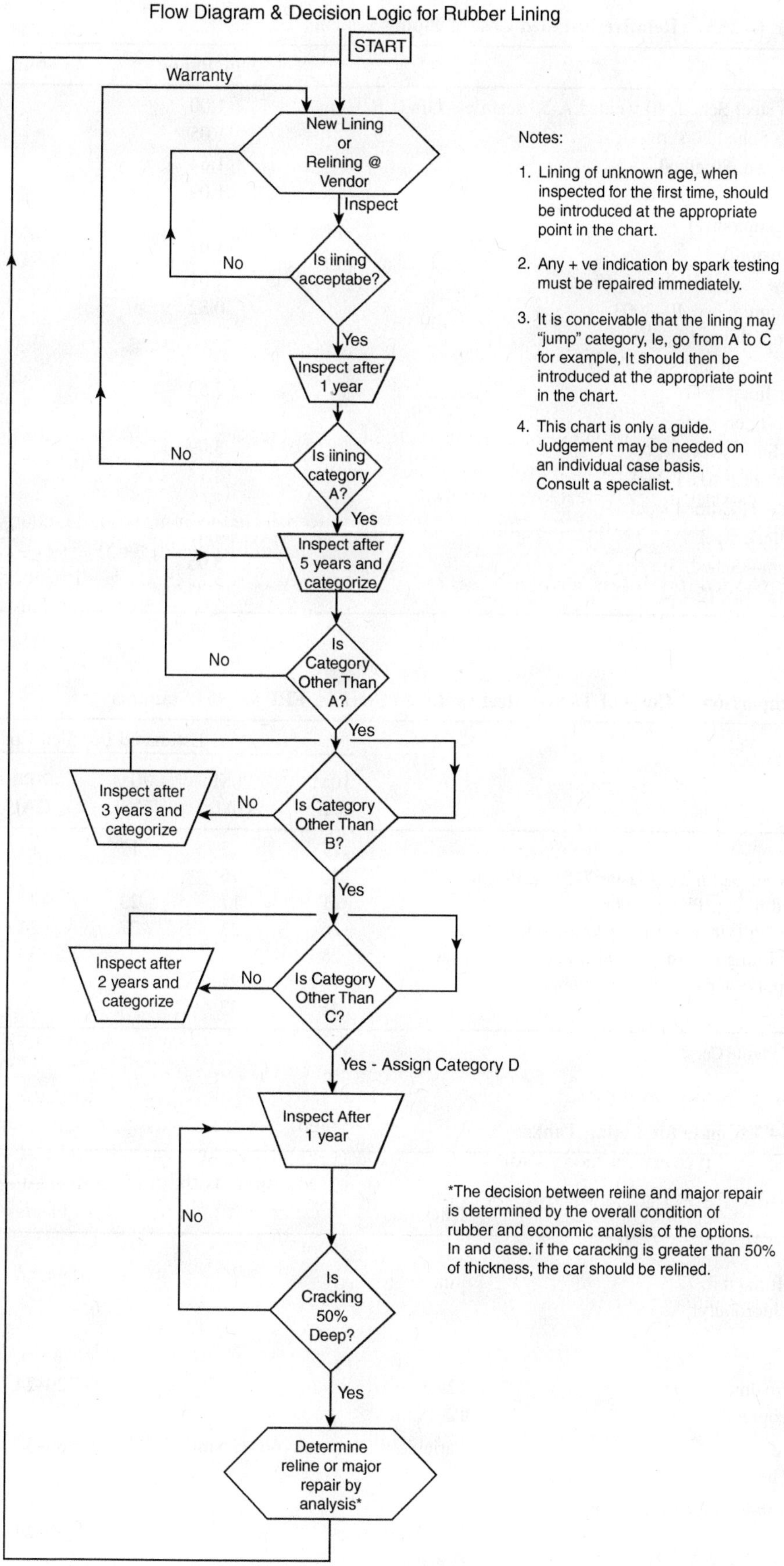
Flow Diagram & Decision Logic for Rubber Lining
START
Warranty
New Lining or Relining @ Vendor
Inspect
Is iining acceptabe?
No
Yes
Inspect after 1 year
Is iining category A?
No
Yes
Inspect after 5 years and categorize
Is Category Other Than A?
No
Yes
Inspect after 3 years and categorize
No
Is Category Other Than B?
Yes
Inspect after 2 years and categorize
No
Is Category Other Than C?
Yes - Assign Category D
Inspect After 1 year
No
Is Cracking 50% Deep?
Yes
Determine reline or major repair by analysis*
Notes:
1. Lining of unknown age, when inspected for the first time, should be introduced at the appropriate point in the chart.
2. Any + ve indication by spark testing must be repaired immediately.
3. It is conceivable that the lining may "jump" category, Ie, go from A to C for example, It should then be introduced at the appropriate point in the chart.
4. This chart is only a guide. Judgement may be needed on an individual case basis. Consult a specialist.
*The decision between reiine and major repair is determined by the overall condition of rubber and economic analysis of the options. In and case. if the caracking is greater than 50% of thickness, the car should be relined.

FIGURE 66.17. Flow diagram and decision logic for in-service inspection of rubber lining.

TABLE 66.25A. Relative Installed Cost of Piping Systems

	2-in. Piping	6-in. Piping
Carbon steel Sched. 40 welded A-53 seamless Grade B	1.00	1.00
Polypro Sched 80 screwed	1.05	
PVC Sched. 80 glued	1.64	0.78
CPVC Sched. 80 glued	1.09	
FRP (commodity)	1.53	1.42
FRP (custom)	1.91	1.81
Stainless steel 304 Sched, 10	1.51	1.11
Aluminum Sched 40–3003	0.82	0.89
Cast iron, flanged		
Cast iron, bell and spigot		
Rubber-lined steel	2.52	1.932
Polypro-lined steel	2.43	2.673
PVDF-lined steel	3.09	4.02
"Teflon" FEP-lined steel	4.15	5.57
"Teflon" TFE-lined steel		
Glass-lined steel	4.43	5.27
Titanium—Sched. 5	5.95	9.31
Hastelloy B or C	7.83	

TABLE 66.25B. Comparison of Costs of Tanks; Steel vs. Lined Steel vs. FRP vs. 304 Stainless

Material for Tank[a]	Estimated Installed Cost, $ ft^2					
	10M GAL	20M GAL	70M GAL	70M GAL	05M GAL	305M GAL
Carbon steel (C/S) 1/4 in.	17	13	17	14	16	16
C/S + thin chemically cured lining (Plasite 7155 or Plasite 7122)	20	16	21	21	18	20
C/S + thick heat cured lining (Plastic 3066)	21	17	23	23	20	22
C/S + thick plastic lining (1/8 in. Ceilcote lining 74)	26	23	27	24	26	26
C/s + thick elastomer lining (1/4 in. natural rubber)	28	24	23	30	32	32
Glass fiber-reinforced polyester (std. filament wound)	13	21				
Type 304 stainless steel	45	33				25

[a]Plasite is a trademark of Plasite Corp.

TABLE 66.25C. Cost Estimate for Lining Tanks

Lining	Thickness	Installed Cost, $/ft^2	
		Field	Shop
1. Elastomeric sheet			
• Natural rubber: soft natural, semihard, triflex, chlorobutyl	$\frac{1}{4}$ in.	25–30	20–24
• Synthetic			
Precured neoprene	$\frac{1}{4}$ in.		
Spray applied hypalon	12–20 mils	20–24	20–24
spray-applied neoprene	12–15 mils		
2. Fiberglass-reinforced vinyl ester (spray or trowel applied)	Variable thicknesses (60–125 mils)	30–34	25–30
3. Epoxy or phenolic spray-applied thin linings			
• Chemically cured	10–12 mils	20–24	18–20
• Heat cured	6–8 mils		

TABLE 66.25C. (*Continued*)

Lining	Thickness	Installed Cost, $/ft^2	
		Field	Shop
4. PVDF dispersion coatings			
• Nonreinforced	25 mils	N/A	40
• Glass fabric reinforced	40 mils	N/A	60
• Carbon cloth reinforced	40 mils	N/A	75
5. Powder coatings: ETFE (Tefzel), ECTFE (Halar), PFA (Teflon PFA)	40–80 mils	N/A	40–80
6. Wire mesh reinforced PFA (Fluoroshield)	80 mils	N/A	200
7. Ebonite with fluoropolymer	210 mils	N/A	190
8. Fabric-backed sheet linings			
• PVDF (Kynar)	—	100–120	75–80
• Polypropylene	—	60–64	55–60
• ECTFE (Halar)	—	62–66	64–68
• ETFE (Tefzel)	60–90 mils	72–76	70–74
• Modified PTFE			
• FEP			
• PFA			
9. Loose sheet: PTFE, modified PTFE	120 mils	175–185	150–165

Notes: 1. All prices are intended as general guidelines. Pricing per square feet of area will vary significantly depending on size of tanks, complexity of tank, and labor situation on job site (in case of field work.)
2. Field installation cost will vary significantly depending on number of square feet over which to amortize mobilization and travel costs.

L. CONCLUSION

Polymeric materials have proved to be very effective in corrosion control. In many applications they are the only acceptable materials due to their chemical inertness. However, developments in the area of accelerated testing and condition assessment need to take place to meet the challenges of increasingly stringent environment and safety restrictions.

BIBLIOGRAPHY

1. J. Alexander, P. Khaladkar, B. Moniz, W. Stahl, and T. Taylor, Environmental Performance of Elastomers, ASM Handbook, Vol. 13B, ASM International, Materials Park, OH, 2005.

2. R. O. Babbit (Ed.), The Vanderbilt Rubber Handbook, R. T. Vanderbilt Company, Norwalk, CT, 1978.

4. L. W. Buxtonand D. R. Goldsberry, "Fluoropolymer Lined Chemical Systems and Permeation," in Proceedings of Managing Corrosion with Plastics Symposium, NACE International, Houston, TX, Nov. 1993.

5. Chemical Resistance, Vols. 1 and 2, 2nd edition, PDL Handbook Series, Plastics Design Library, New York, 1994.

6. S. Ebnesajjad, Fluoroplastics, Vol. 1, Non-Meltprocessibles Fluoroplastics, PDL Handbook Series, PDL Design Library, New York.

7. S. Ebnesajjad, Fluoroplastics, Vol. 2, Melt Processible Fluoroplastics, PDL Handbook Series, PDL Design Library, New York.

8. S. Ebnesajjad and P. Khaladkar, Fluoropolymer Applications in Chemical Processing Industry, PDL Handbook Series, William Andrew, Norwich, New York, 2005.

9. Fiberglass Pipe Handbook, SPI Composites Institute, New York, 1992.

10. G. A. Glein, "Dual Laminate for Difficult Corrosion Problems — Selection Criteria and Techniques," in Proceedings of Managing Corrosion with Plastics Symposium, NACE International, Houston, TX, Oct. 1991.

11. G. A. Glein, "Dual Laminate Selection Criteria and Techniques," Mater. Perform., **31** (3), 44 (Mar. 1992).

12. N. L. Hall, "Fluoropolymer-Lined Reinforced Thermosetting Plastic Chemical Process equipment." in Proceedings of the Society of the Plastics Industry, Western Section, Composites Institute, Las Vegas, NV, Apr. 1988.

13. N. L. Hall, "Manage Corrosion with Fluoropolymer Dual Laminates," Chem. Eng. Prog. Mag., June 1994.

14. Handbook of Plastics, Elastomers and Composites, 4th edition, C.A. Harper (Ed.), McGraw-Hill, New York, 2002.

15. D. K. Heffner, "Fluoropolymer Linings in the Transportation Industry," Mater. Perform. **31** (7), 33 (July 1992).

16. J. F. Imbalzano, D. N. Washburn, and P. H. Mehta, "Permeation and Stress Cracking of Fluoropolymers", Chem. Eng. Mag., Jan. 1991.

17. P. Khaladkar, "A Comparison of Fluoropolymer Linings," Mater. Perform., **33** (2), 35 (Feb. 1994).

18. P. Khaladkar, "Fight Corrosion with Plastics," Chem. Eng. Mag., Oct. 1995.

19. P. Khaladkar, Modern Fluoropolymers, Wiley, New York, 1997.

20. M. Morton (Ed.), Rubber Technology, 3rd ed., Van Nostrand Reinhold, New York, 1987.

21. MTI Publication, "Fiber Reinforced Plastic Flange Design," Materials Technology Institute, St. Louis, MO, 2007.

22. MTI Publication, "Practical Guide to Field Inspection of FRP Equipment and Piping," Materials Technology Institute, St. Louis, MO, 2001.

23. MTI Publication, "User's Guide to ASME Standards for Fiberglass Tanks and Vessels," Materials Technology Institute, St. Louis, MO, 1996.

24. MTI Publication, "Materials Selection for the Chemical Process Industry," Materials Technology Institute, St. Louis, MO, 2004.

25. MTI Report, "Guide to Elastomers Testing for Chemical Resistance Applications," Materials Technology Institute, St. Louis, MO, 2004.

26. MTI Report, "FRP Self Help Guide," Materials Technology Institue, St. Louis, MO, 2001.

27. MTI Report, "User's Guide for Evaluating New Polymer Systems," Materials Technology Institute, St. Louis, MO, 2009.

28. MTI Report, "Permeation Testing Protocol," Materials Technology Institute, St. Louis, MO, 2005.

29. MTI Report, "Nondestructive Evaluation (NDE) Methods for Inspecting Polymer Vessels Used in Process Industries - A Review," Materials Technology Institute, St. Louis, MO, 1998.

30. NACE Coatings and Linings Handbook, National Association of Corrosion Engineers, Houston, TX, 1978.

31. Rapra Technology, MTI Publication No. T-4: Prediction of Service Performance of Equipment Made of or Lined with Polymeric Materials, published for the Materials Technology Institute of the Chemical Process Industries, Inc., by NACE International, Houston, TX, 1993.

31A. W. Slama, "Polyester and Vinyl Ester Coatings," J. Protec. Coat. Linings, May 1996.

32. L. P. Smith, The Language of Rubber: An Introduction to the Specification and Testing of Elastomers, Butterworth–Heinemann, London, 1993.

33. R. E. Tatnall, MTI Project 84: Review of Spark Testing Practices, Materials Technology Institute of the Chemical Process Industries, 1994.

34. R. E. Tatnall, MTI Project 99: Manual on Inspection of Linings, Materials Technology Institute of the Chemical Process Industries, 1996.

35. Webster-Atkinson Associates, Furan Reinforced Thermoset Plastics for Chemical Process Equipment, MTI Publication No. 21, Materials Technology Institute of the Chemical Process Industries, Columbus, OH, 1986.

36. W. Woebcken, Saechtling International Plastics Handbook: for the Technologist, Engineer, and User, 3rd ed., translated and edited by J. Haim and D. Hyatt, Hanser-Gardner, Cincinnati, OH, 1995.

67

CORROSION CONTROL OF STEEL BY ORGANIC COATINGS

C. H. Hare[†]

Coating System Design Inc., Lakeville, Massachusetts

A. INTRODUCTION

The application of paint films to metal is the oldest means of corrosion control and dates back to antiquity. Not until the second quarter of the twentieth century, however, was the role of such paint films understood and their mechanism of protection suitably categorized in terms of corrosion science. Up until that time, it was assumed that protection with paint films was based simply on insulation of the metal from the corrosive environment, specifically water and oxygen, thereby depriving the cathode reaction of fuel. Although certain pigments, such as red lead, had long been known and valued for enhancing the protection of iron and steel, the role of such inhibitors remained ill defined.

The fallacy of the insulation assumption was revealed in the late 1940s. Researchers determined that most practical paint films (including those known to provide protection in the field) allowed far more moisture (and, in most instances, more oxygen) through their continuum to the metal than those levels of such reactants that would be necessary to sustain the cathode reaction (Table 67.1) [1, 2].

During this same period, the introduction of new anticorrosive pigments (such as zinc potassium chromate) and tentative experiments with high loaded zinc (rich) primers fulfilled predictable enhancements in protection and these advances gradually led to a more organized understanding of control by coatings in terms of corrosion science [3, 4].

[†]Retired.

Uhlig's Corrosion Handbook, Third Edition, Edited by R. Winston Revie
Copyright © 2011 John Wiley & Sons, Inc.

TABLE 67.1.   Transmission of Water and Oxygen through Protective Coating Films

	g of Water/m²/day 25 μm film @95% RH and 38°C	mL of Oxygen/m²/day 100 μm film @85% RH, 23°C and 1 atmos O₂	
Threshold quantity of water necessary to support corrosion rate of 70 mg Fe/cm²/year[a]	0.93		
Chlorinated rubber primer	20 ± 3	30 ± 7	Chlorinated rubber primer
Chlorinated polymer	26 ± 5	33 ± 2	Chlorinated polymer
Coal tar epoxy	30 ± 1	213 ± 38	Coal tar epoxy
Aluminized epoxy mastic	42 ± 6	110 ± 37	Aluminized epoxy mastic
		575	Threshold quantity of oxygen necessary to support corrosion rate of 70 mg Fe/cm²/year[a]
Titanium dioxide pigmented alkyd	258 ± 6	595 ± 49	Titanium dioxide pigmented alkyd
Red lead/linseed oil primer	214 ± 3	734 ± 42	Red lead/linseed oil primer

[a]Corrosion rate of unprotected mild steel in typical industrial environment (Motherwell and Sheffield, U.K.).
Data from J. C. Hudson, The Corrosion of Iron and Steel, Chapman and Hall, London, 1940, p. 66.

B. FUNDAMENTAL MECHANISMS OF CORROSION CONTROL BY COATINGS

Corrosion control by coatings may be subclassified into control by any of three (possibly four) techniques [5]:

1. Barrier coatings that protect by one or more of two mechanisms:
   a. Resistance inhibition: the paint film acting as an ionic filter and ensuring that any moisture accessing the metal through the paint film continuum is, at the paint film–metal interface, of a sufficiently high electrical resistance to mitigate current transfer between anodic and cathodic sites [1, 2, 6].
   b. Oxygen deprivation: where suitably formulated paint films are able to exclude sufficient oxygen from the metal to impede the cathodic reaction [7–10].
2. Cathodic protection, wherein the paint film protects a metal substrate by preventing current discharge from the metal to the environment [11, 12]. This effect is accomplished by applying to the metal a film pigmented with a more anodic metal (usually zinc, for protecting iron and steel). Loadings of the anodic zinc dust pigment must be high enough to ensure a continuous current flow through the film itself and across the interface [11, 13] (i.e., between anodic film and cathodic metal substrate). All discharge occurs at the paint film, and as long as the conductivity of the model is sufficient to just sustain current flow from the film to the environment, the steel remains protected.
3. Inhibitive primers that control corrosion by modifying the interfacial (primer/metal) environment so that

passivation (or inhibition) of the substrate metal may be achieved and maintained [14, 15]. In this case, the design device of the formulation encourages moisture access into the film, so that soluble inhibiting moieties supplied from the coating film may be carried to the metal where they may stimulate the establishment of passive (or inhibitive) films.

Each design device dictates its own requirements in terms of the necessary formulatory response (the selection and apportionment of desired pigments, binders, additives, and solvents). These responses are generally quite specific, and formulatory approaches toward optimized corrosion resistance via any one mechanism may be entirely incompatible with (even counterproductive to) approaches necessary for protection via a second mechanism. The use of hydrophilic inhibitive pigments in barrier primers, for example, is quite counterproductive. In addition, the formulation of primer coats, intermediates, and finishes that are used together to make composite paint systems may or may not themselves have conflicting requirements.

We now review each of the three types of coatings in more detail.

C. BARRIER COATINGS

Whether protecting by resistance inhibition or oxygen deprivation (Fig. 67.1), the fundamental requirements of the barrier system are that the coating should (1) be impermeable to ionic moieties [6, 7, 15] and, if possible, to oxygen [7–9, 16] and (2) maintain adhesion (to the metal) under wet service conditions [17, 18]. As noted above, sufficient

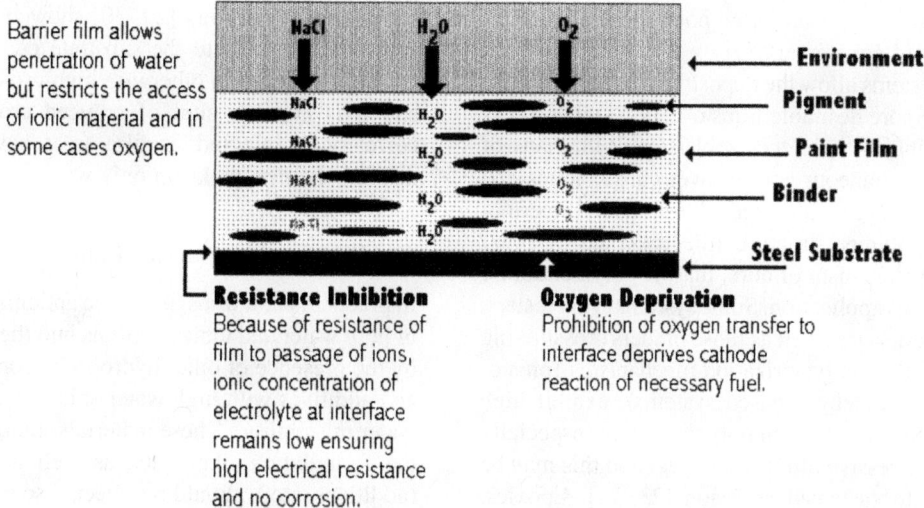

Barrier film allows penetration of water but restricts the access of ionic material and in some cases oxygen.

→ **Environment**
→ **Pigment**
→ **Paint Film**
→ **Binder**
→ **Steel Substrate**

Resistance Inhibition
Because of resistance of film to passage of ions, ionic concentration of electrolyte at interface remains low ensuring high electrical resistance and no corrosion.

Oxygen Deprivation
Prohibition of oxygen transfer to interface deprives cathode reaction of necessary fuel.

FIGURE 67.1. Corrosion control by barrier coatings.

impermeability to water is not possible except in very thick films (>20 dry mils). Impermeability to ionic solutions and oxygen is an entirely more practical objective, however, and this factor is thought to be rate determining for corrosion beneath barrier films. Permeability of the film to water is generally thought to have greater direct consequence on deadhesion, leading subsequently to corrosion.

C1. Binders for Barrier Coatings

C1.1. Thermoplastic Binders. Impermeability is ensured primarily by the molecular structure of the binder [19, 20], although the selection of pigments and other components of the coating (volatiles as well as nonvolatiles) can be critical [21, 22]. Dense, molecularly tight films (thermosetting films of uniform high crosslink density or relatively crystalline thermoplastics) that are hydrophobic and give optimized resistance to the ingress of water, oxygen, and ionic material are the preferred binders.

High-molecular-weight hydrocarbons, particularly halogenated species (fluoropolymers, polyvinyl chloride, etc.), are particularly suited to this employment. Unfortunately, because of the high crystallinity of these systems, the polymers are not soluble in conventional solvent systems, and, where they are, the low concentration of the polymer at practical solution viscosities excludes their use in many applications where restrictions apply to the amount of solvent which may be employed. Their use in the form of dispersions from water and as powder coatings has been widely practiced in coil coatings, container coatings, and so on, although these systems cannot readily be employed in field applications because of thermal curing requirements.

Such is the cohesive strength of many of these polymers (e.g., polyvinyl chloride) that adhesion under conditions of

field-applied stress may be tentative unless the polymer is suitably modified. At some compromise with performance, chemically engineered modifications to the polymer (plasticization, copolymerization, introduction of carboxylic acid groups, etc.) and reductions in molecular weight improve both solubility and adhesion. These modifications enable coatings based on such polymers to be applied as lacquers and dried in the field by evaporation of the solvent carrier. Nevertheless, these systems often fail to satisfy modern environmental requirements that limit the amount of volatile solvents in the coating. Additional difficulties with chlorinated polymers are related to the propensity of such materials to undergo heat- and light-induced dehydrochlorination, producing HCl and chlorides at the substrate that effectively short circuit the barrier properties of the film. Finally, solvent retention in applied thermoplastic lacquers may adversely affect the impermeability of the film to water, where retained solvents are hydrophilic. Such hydrophilic solvent entrapment will reduce the glass transition temperature T_g and may induce osmotic blistering failure.

C1.2. Thermosetting Binders. In consequence of all these things, with the possible exception of high build bituminous cutbacks, most modern barrier systems are based on thermosetting polymers (epoxies, polyesters, vinyl esters, and polyurethanes) [4]. These take the form of two-pack systems that are polymerized in situ from low-molecular-weight oligomeric and monomeric precursors or single-pack thermosets (epoxy/phenolics, amino formaldehyde resin crosslinked acrylics, etc.) applied as premixed systems and polymerized by baking after application. Single-pack, thermally cured, thermosetting powder coatings may also be used, where practical, although the corrosion resistance properties of these systems are often poorer than their two-pack, solvent-based counterparts.

All of these systems allow the deposition of higher solids, environmentally more desirable films of much higher thickness than is possible with the thermoplastics. With many of these systems, simultaneous control over cure kinetics and solvent release allows for enhanced adhesion and impermeability as does polymer structure (the presence of polar groups) and the mechanism of cure, that is, polymerization after (but not before) application. Some systems (vinyl esters, unsaturated polyesters), as well as those binders crosslinking via condensation-type polymerization mechanisms (phenolics and other formaldehyde-based systems), exhibit high internal stress from shrinkage on polymerization (especially when applied at excessive film thicknesses) and this may be disadvantageous to sustained adhesion [23, 24]. Epoxies, however, curing by addition polymerization, shrink less and exhibit much improved adhesion, making very valuable binders for barrier coatings. Some thermosets (epoxies and polyurethanes) are modified with crude, highly hydrophobic, bituminous-based thermoplastics (coal tar resins) and give adhesive and highly impermeable multicoat systems for optimum performance under immersion and below grade service [4, 25].

With all of these thermosets, as with the prepolymerized thermoplastics, design objectives are to create a dense hydrophobic molecular matrix through which the transport of water and oxygen is minimized and that of ionic species is entirely prevented [26, 27]. These requirements are at best satisfied by polymeric films having high uniform crosslink density, high T_g, and molecular structures that are either nonpolar or derived from non-water-attracting moieties [28].

Mayne and coworkers [29, 30] showed that in thin barrier films ionic ingress to the substrate occurs at areas of low crosslink density (in otherwise high-crosslink-density films) and it is at these sites of reduced crosslink density that corrosion is initiated. In film areas having high crosslink density, the film picks up only water.

C2. Pigments for Barrier Films

In practical paint films, it may be anticipated that the transfer of both water and ionic solutions into the film is also favored by the presence of other hydrophilic components (pigments and additives with high water solubility, solvents with high water miscibility). These materials should be excluded from the formulation. Pigments, as well as other components (additives, etc.), should be selected so as to enhance hydrophobicity. Wherever possible, adhesion across pigmentary/binder interfaces should also be maximized and pigment flocculation minimized. Impermeability is enhanced by the use of flat platey pigmentations, especially aluminum flake, glass flake, stainless steel flake, and micaceous iron oxide. In wet films, these pigmentations orient themselves parallel to the substrate and both laterally reinforce the film and present a more tortuous path for the penetration of any corrosive agent through the film [20]. Pigmentation levels for barrier coatings are less critical than are pigmentation levels for zinc and inhibitive based primers, being normally much lower (see Fig. 67.2). Excessive pigmentation levels should be avoided, especially where the metallic pigmentation is cathodic to the metal substrate. This maintains the electrical resistance of the film and prevents pitting at uncoated pinholes, which act as anodes.

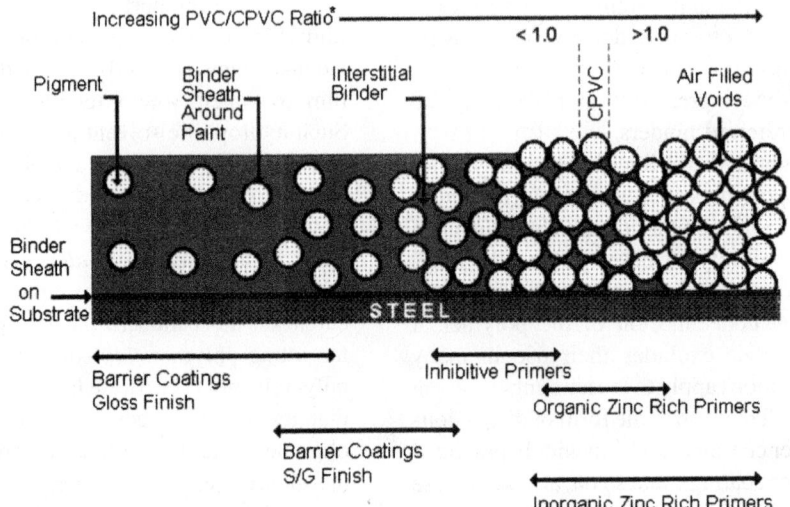

*Ratio of actual pigment volume concentration in total dry film volume (PVC) to the critical pigment volume cocentration at maximum pigment packing (CPVC).

FIGURE 67.2. Pigment volume criteria for classes of anticorrosive coatings.

C3. Solvent Systems for Barrier Films

Solvent systems should be designed so that lateral orientation of the pigment in the film is optimized but so hydrophilic solvents evaporate and diffuse from the film rapidly, if possible, before cure is complete and certainly before the film is put into service. Slow solvents that readily wet the substrate and provide enough motility in the wet film to allow the flat, platelike, barrier pigments to settle flat against the substrate are to be preferred. However, slow-drying hydrophilic solvents, especially those having high affinity for the binder or those which have nonplanar complex structures, tend to be held within the drying film for long periods. They may, in service, attract water into the film osmotically, reducing the electrical resistance of the interfacial area and leading to blistering. The presence of water and/or associations of solvent and water within the film tend to plasticize and swell the film, reducing T_g and thereby facilitating the ingress of additional water. This causes further decrease in T_g. Optimally, therefore, high-boiling solvents that are last to leave the film should be hydrophobic, free of functional groups, and planar in structure. Here, some compromise with possible conflicting requirements, such as binder solubility and the need for defect-free films, may be necessary. Care should also be taken in the selection of nonvolatile additives (pigment dispersants, surfactants, thixotropes, and flow control agents) as many of these materials may be excessively hydrophilic.

C4. Wet Adhesion

It has been noted that corrosion beneath a barrier film can begin only after deadhesion has taken place, and if adhesion can be maintained under wet service conditions, then protection is assured [31]. Adhesion is also optimized by polymer design. Most coatings must rely on secondary valency bonding for adhesion. A limited number of polymer types form primary valency bonds with the metal. In both cases, adhesion may be optimized when the polymeric binder is richly endowed with polar groups (hydroxyls, carboxylic acid groups, etc.) on their molecule. It is because of this that epoxies, alkyds, and polyesters show their characteristic high levels of adhesion to metals. These groups readily associate with metal oxide groups on the exposed metal surface.

In designing systems for optimized impermeability and maximized adhesion, there is some dichotomy of purpose [31, 32]. As noted above, those polymers bearing polar groups that encourage adhesion (hydroxyls, carboxylic acid groups) tend to be hydrophilic and attract water into the film. Polar ester groups (e.g., in alkyds and other oxidizing systems) and, to a lesser extent, amide groups (urethanes) are also vulnerable to alkali-induced hydrolytic cleavage and may in consequence be attacked by alkalinity developed at the cathodes of corrosion cells. The effect is seen in the peripheral areas around corrosion spots and breaks in these sensitive films in service, where both cohesive loss (via saponification) and adhesive loss at the coating–metal interface will result.

The dichotomy of need for increased adhesion and minimized water takeup and alkali resistance is resolved to some extent by the use of multicoat systems. Here adhesion becomes of primary importance in barrier primers (perhaps at some compromise to maximized impermeability) while impermeability (and the use of less polar components) becomes paramount in barrier-type finish coats [28]. Adhesion of topcoats to primers is, if recoating is done soon after priming, more readily secured than is the adhesion of the organic primer to the metal substrate. Ester group sensitivity is best resolved by avoiding polymers based on these types of binders or at least in seeking to maximize alkali resistance via the selection of polymers with suitable polymeric architecture. Not all ester-based coatings have the same degree of vulnerability to hydrolysis as do alkyds and oxidizing systems. Polyesters, vinyl esters, and ester-based polymers prepared from polyols bearing no hydrogen on the carbon atom that is beta to the hydroxyl are very often quite resistant enough for successful service in alkaline environments.

If the hydroxyl group of a barrier coating binder is bound to a relatively rigid polymeric backbone, molecular mobility after cure is much reduced. Optimally, the hydroxyls in the interfacial layer of the primer would orient themselves against the metal (associating with metal oxides on the substrate) so that they are less available for hydration, and the substrate is therefore more protected. Moreover, such rigid systems are molecularly too fixed to facilitate readily the transfer of water to the substrate because of the reduced kinetic energy of the molecule [32]. The opportunity available for access of water to the substrate is therefore limited.

On the other hand, during application and wetting (while the paint film is still heavily solvated) the same hydrophilic groups should display high mobility so that substrate wetting is maximized and molecular orientation with available metal oxide groups on the surface of the metal is favorably enhanced. Thus, barrier film binder design is optimized in systems having high molecular mobility in the wet stage but high immobility when the film is cured. These characteristics may be maximized where low-molecular-weight polar monomers and oligomers are applied in wet films that may soak into the substrate before converting to high T_g films of limited mobility after cure [33].

C5. Effects of Structure on Oxygen Impermeability

It has been shown that polymers rich in hydroxyls have greater impermeability to oxygen than do nonpolar polymers [19, 20]. As the control of oxygen transport would appear more readily achievable in practical systems than the control of water transport, it may be argued that the presence of hydroxyl groups is not entirely incompatible with good corrosion control where oxygen deprivation is the desired mechanism.

C6. Film Thickness

One of the most effective strategies by which good barrier protection may be assured is found in the application of additional thicknesses of paint. This may be done as high-build single-coat applications, as are often employed in protection of buried structures with bituminous cutbacks, or more effectively in multicoat systems. In the latter case, finish coats applied over primers and primer/intermediate coat combinations render additional service, that is, either enhanced impermeability (coal tar epoxies) or for other reasons (e.g., light-stable aliphatic urethanes over high-build epoxy primers for improved weathering resistance). As film thicknesses increase, so the transfer of moisture, oxygen, and ionic species to the substrate is greatly diminished, if not the absorption of such penetrants into the film.

High film thicknesses may be one of the primary defenses for corrosion where engineering structures must be buried or subject to long-term immersion. In these applications, coatings are often used in tandem with cathodic protection systems, and the coatings effectively reduce the area of the exposed metal requiring cathodic protection. Thus, cathodic protection current requirements are lowered. In these systems especially, high levels of alkali resistance are necessary if adhesion of the coating is to be maintained while the coated metal remains under cathodic protection.

Light-stable barrier finishes that are resistant to ultraviolet light (UV)–induced changes (degradation) in the mechanical properties of the film also prevent gloss loss, microcracking, and other age-related film defects, which lead to reduced long-term corrosion resistance of the total system.

Multicoats have practical advantages over single coats in that the superimposition of several films does much to eliminate the possibility that pinholes, holidays, and misapplications of single coats may allow direct access of the environment to the metal. It is statistically unlikely that holes in one coat will coincide with similar faults in the next, although pinholes can telegraph through two coats.

Perhaps, more importantly, certainly with thermosetting (chemically curing systems) internal stress development on drying, arising from solvent release and/or from polymerization, will increase as film thickness increases [24, 34]. The stress produced by the conversion of thick films (even strongly adherent epoxies) is capable of overcoming the adhesion of the coating, even over abrasive blasted surfaces. In less extreme cases, residual internal strain left in a coating after curing will inevitably reduce the amount of tensile stress from other sources (service stresses, thermal stresses on cooling, etc.) that the film is able to withstand before undergoing brittle failure (either cracking or delaminating) at some later epoch in the life of the system. These untoward stress conditions may be ameliorated by reducing the amount of film applied in any single coat [24]. Flat, platelike pigments also tend to reduce the amount of internal stress buildup in coating films [23]. These considerations become singularly more critical when the original adhesion of the coating to the metal is in any way compromised. This is one of the primary motivations for metal scarification and the establishment of a viable anchor pattern before the primer is applied. On smooth surfaces (particularly thin gauged substrates where flexing and thermal distortions are maximized), high film thicknesses should not be applied. In these cases (coil coatings, aircraft skins, container coatings, etc.) thin-film systems based on barrier primers (or inhibitive primers) are preferred to high-build barrier systems.

D. SACRIFICIAL COATINGS

Perhaps the most effective corrosion control coatings are the zinc-rich primers [12, 35, 36]. As this type of composition requires intimate contact with the steel being protected, these coatings are used exclusively as primers and, although many types are self-recoatable, they can never be employed over other types of coatings. Protecting steel cathodically (Fig. 67.3), these primers require high volumes of zinc pigment in order to ensure an electrical pathway across the film (the anode of the artificial galvanic cell) as well as between the film and the substrate (the cathode). The zinc pigment employed in these coatings is in the form of a spherical dust of between 3 and $20\,\mu m$ in diameter (usually $\sim 7\,\mu m$). Contact necessary to maintain current transfer is thus tangential (Fig. 67.4) between individual spheres of pigment and between the particles of pigment and the metal surface.

In order to sustain the electrical continuity requirements of the primer, the spherical zinc dust particles must be packed sufficiently close together to ensure this contact [11]. This configuration allows for little binder, and formulations are quite critical in order to maintain electrical conductivity without depriving the film of the level of binder necessary to maintain adhesion and acceptable physical properties.

Two fundamentally different classes of zinc-rich primer have been developed, the organic zinc-rich primer and the inorganic zinc-rich primer [36].

D1. Organic Zinc-Rich Primers

Organic zinc-rich primers are based on a variety of resin systems, including epoxy/polyamides, high-molecular-weight linear epoxies, moisture-cured urethanes, high-styrene resins, chlorinated rubbers, and epoxy esters, and the chemical and physical properties of the film in part reflect

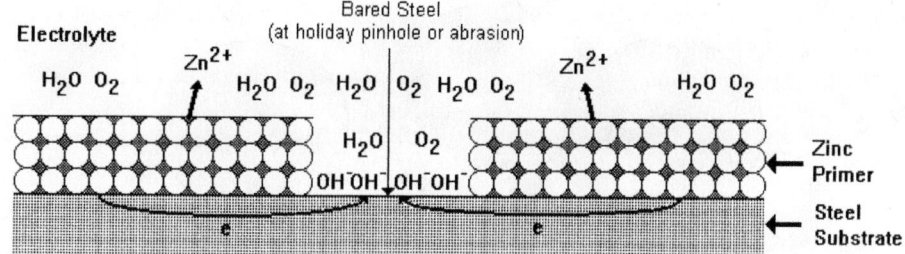

The presence of strongly electronegative zinc-pigmented coating short circuits all local cell activity on steel. The steel becomes totally cathodic to the anodic zinc coating. The zinc corrodes, but the steel will not corrode even at bare spots. It is mandatory that the zinc coating be in electrical contact with the steel surface; therefore, the steel must be stripped of all contamination.

FIGURE 67.3. Fundamentals of zinc-rich protection.

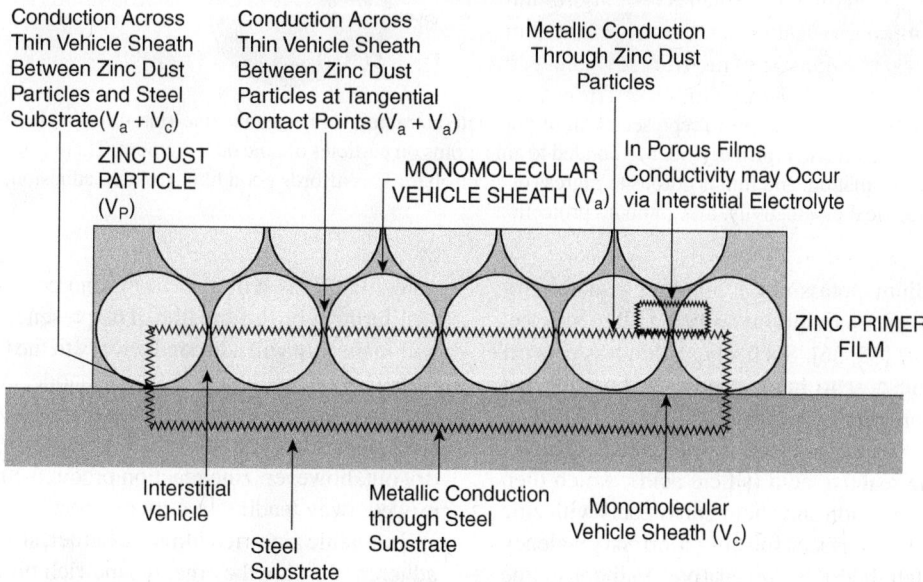

FIGURE 67.4. Electrical conduction in zinc dust pigmented primer.

the resin system used. Such primers, employing an organic binder, are not, in fact, radically different from other organic paint films except for their very high volumetric concentration of pigment ($\sim$65% zinc by volume of the total dry film volume). The formulation of organic zinc-rich primers has been treated at length by several authors [13, 37, 38]. In this type of primer, each individual sphere of zinc is lightly encapsulated with a monomolecular layer of binder that facilitates film cohesion and provides adhesion to the substrate. While it is necessary that all zinc particles bear at least a monomolecular layer of binder on their surface to ensure adhesion and film cohesion, it is important that the presence of the binder sheath around each particle of paint does not introduce levels of electrical resistance that are too high. Conductivity must be high enough to support the necessary unidirectional flow between cathode and anode. Control of

the volumetric ratio of conductive pigment to dry nonconductive binder ensures that the binder sheath around each particle of pigment (zinc) does not become so thick that the current transfer is too greatly reduced or so depleted that cohesion and film strength are lost [37]. Optimum loadings of zinc will actually depend on the specific formulation, the presence of auxiliary pigments (thixotropes, coloring agents, inhibitors, etc.), and the geometries of the pigment packing within the dried film [11, 13].

D2. Inorganic Zinc-Rich Primers

The inorganic zinc-rich primers are fundamentally different from the organic [11]. Instead of employing a binder which encapsulates the zinc, these materials employ a reactive binder, usually an inorganic silicate (either an alkaline

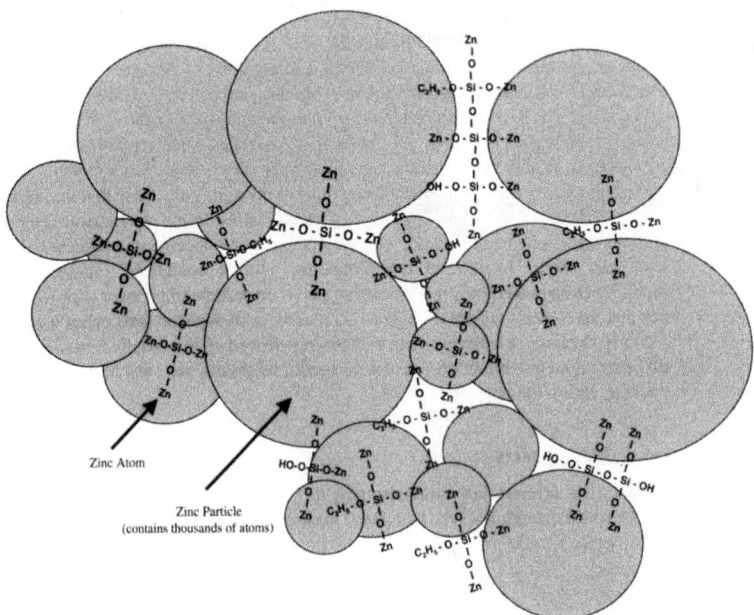

FIGURE 67.5. Stylistic representation of postulated structure of inorganic zinc-rich film in which silicate vehicle is primary valency bonded to zinc atoms on particles of zinc dust pigment. There is no encapsulation, and film is porous to ingress of electrolyte. This affords good film strength, adhesion, electrical conductivity, and cathodic protection.

silicate, such as sodium, potassium, lithium, or a quaternary ammonium silicate, or a partially hydrolyzed alkyl silicate, such as ethyl silicate) [35, 36]. Such silicate binders convert on mixing with zinc dust to hard, cohesive films that are virtually wholly inorganic in nature.

Reactions involve loss of solvent or water and hydrolysis of the silicate to the reactive acid (silicic acid), which then undergoes self-condensation and chemically reacts with zinc atoms on the zinc dust surface, forming a primary valency bonded matrix of tetrahedrally linked (poly) silicon oxide bridges between adjacent particles of zinc (Fig. 67.5) [35]. Although the reaction paths are somewhat different depending upon the silicate type used (alkaline or alkyl silicate), the eventual structures of all films are probably quite similar. For specifics on the reactions involved, the reader is referred to [12, 35, 36]. (A bonding similar to that which occurs between the silicate and zinc is also thought to occur across the interface with the iron in the steel surface [35], although this has never been definitively proven.)

The resultant film matrix is entirely inorganic, open, and much more porous than the organic zinc film. In practice, films may also contain unhydrolyzed material (alkyl silicate or alkaline silicate), which may remain within the film for months after the material has originally dried. Subsequent reaction of residual binder with carbonic acid from the atmosphere probably completes the conversion of residual silicate to silicic acid and adds to film cohesion. In service, the open film may become saturated with corrosive electrolyte, such as brine, which is readily absorbed into the film

porosities. This will also form zinc corrosion products that will further convert the film. The presence of brine solutions within the film will add conductivity to the film, and, as long as sufficient zinc remains to act as an anode, even in the presence of chloride solution, the primer will retain its ability to protect steel at least in the short term. Unless given the opportunity to dry out, however, zinc reaction products may be dissolved or eroded away leading to a more linear loss of zinc.

Inorganic zinc-rich films are harder, stronger, and far more adherent than are the organic zinc-rich films [35]. As the film matrix is entirely inorganic, these primers also have better resistance to solvent and to heat than the organic zinc-rich primers and may be used for tank linings and applications involving temperatures as high as 400°C (750°F).

The lack of encapsulating binder in inorganic zinc-rich primers means that zinc loading concentrations are not as critical to galvanic conductivity as they are in the case of organic zinc-rich primers, although reduced anode loadings will inevitably produce some reduction in performance. For the most part, performance levels realized from the inorganics are superior to those realized from the organics, although surface preparation and application requirements (especially application of the alkaline silicate-based materials) are more exacting. Also vulnerable is the security of intercoat adhesion between organic topcoats and the inorganic primers, especially in the case of the alkaline silicates. Alkaline silicates based on sodium silicate, in the absence of high baking temperatures, are cured by the postapplication of amine phosphate solutions, residues of which must be carefully removed if subsequent deadhesion of topcoats is to be avoided.

D3. Postgalvanic Protection Phenomena

In both organic and inorganic zinc-rich primers, the duration of cathodic activity is finite, and the effects of zinc polarization and the generation of zinc corrosion product gradually convert protection from the galvanic mechanism to one that is, at least in part, a barrier mechanism [11, 12]. The corrosion product blocks and seals the porosities of the film with a dense inorganic coat. In the case of the organic, this polarizing film is most usually a surface phenomena. In the more porous inorganic, zinc corrosion product occurs within the interparticulate spaces, building up and more completely sealing the film. Other postgalvanic phase mechanisms have also been suggested, including inhibition from zinc products [39], locally elevated pH, and control of oxygen reduction [40]. Most authorities, however, ascribe long-term protection to barrier effects [4, 11, 12, 38]. The interval of galvanic protection before conversion to the secondary protecting mechanism depends on the type and composition of system and the nature of the environment. Galvanic activity in alkyl silicates has been recorded after three years under atmospheric conditions. In epoxy zinc-rich primers, the cathodic protection phase is briefer [41].

E. INHIBITIVE PRIMERS

E1. Pigmentation

Inhibitive primers are typified by traditional compositions based on red lead and linseed oil, alkyd resins pigmented with zinc yellow (zinc potassium chromate), and, more recently, the many proprietary epoxy, alkyd, urethane, and latex systems that rely on modified phosphate, borate, and molybdate pigments [15, 42, 43]. While traditional chromate pigments are still being employed where inhibitive primers are used for the most demanding service (aircraft coatings, automotive primers, coil coating primers, etc.), the press of environmental and toxicological concerns threatens to eventually eliminate their employment in all coatings in the same manner that led to the removal of lead-based pigmentations.

All of the above pigment types either have some direct limited solubility in water [44] or form reactive products (of limited solubility) with certain binders or their degradation products [45]. Several reviews of the many traditional and newer offerings of these types of pigments have been published [14, 15, 42, 46]. Nevertheless, the exact mechanism whereby these various inhibiting species inhibit corrosion remains imprecisely understood and is probably not the same for all inhibitors.

It is, however, believed that chromates, and possibly molybdates, function by decreasing the oxidizing threshold at which passive films are naturally formed [44]. While oxygen alone will effectively passivate steel at high enough concentrations, those concentrations required for the passivation of steel (unlike the case with aluminum) exceed the levels of oxygen that dissolve in water at neutral pH. Oxidation levels high enough to induce passivation on steel are possible using oxidizing inhibitors, such as the chromates, however, which are thought to form passive films of complex chromic and ferric oxides on the metal [47].

Auxiliary pigmentations which increase the basicity of the film (zinc oxide, wollastonite, etc.) increase the pH of the interfacial environment, and at these levels of alkalinity, oxygen concentrations necessary for passivation fall nearer to those levels of oxygen that will dissolve in water [48]. It is thought, therefore, that these auxiliaries act in tandem with the inhibitor, reducing the amount of inhibitor necessary to achieve passivation under any given set of conditions. In many of the more popular binder systems for this type of product (e.g., alkyds, oils, and other fatty acid–based systems), these same basic pigments (both inhibitive and auxiliary) may react with acid groups on the binder, acting as pigmentary "crosslinking centers," introducing unwanted embrittlement into the film. These effects must be controlled by judicious balancing of the formulation.

Active ions from other inhibitors are thought to act to reinforce the naturally occurring oxide layer rather than establish a passive film themselves. Lead cations from lead-based inhibitors, as well as the modified zinc phosphate inhibitors, are thought to plug up discontinuities in the natural oxide layer, reinforcing it and thereby preventing ionic egress from the metal to the electrolyte. In spite of their long history, inhibition from lead-based pigments is itself incompletely understood. It is most generally thought to be attributed to the reaction of certain acidic moieties (azeleic acid, pelargonic acid, and other long-chained mono- and dicarboxylic acids) from oxidizing binders with lead monoxide giving soluble azelates and pelargonates which serve as a continuous source of inhibitive lead cations [45, 49]. Binders that produce higher concentrations of lower molecular weight acids (e.g., acetic acid), in lieu of the azelates, give much less effective protection [49], as do pigments such as lead dioxide [15]. These components appear to be corrosive rather than inhibitive.

E2. PVC/CPVC and Controlled Permeability

Inhibitive metal primers are designed with relatively high pigment volume concentrations. This design facilitates sufficient water absorption into the film, so that soluble inhibitive ions may be released by the pigment (or its salts with the binder) and carried to the metal surface beneath

Film of controlled porosity allows water through continuum to leach inhibitive ions from soluble pigment, rendering water inhibitive.

Film is less porous to corrosive ions which largely remain outside film. At the substrate, ratio of inhibitive ions to corrosive ions is conducive to formation and preservation of passive condition.

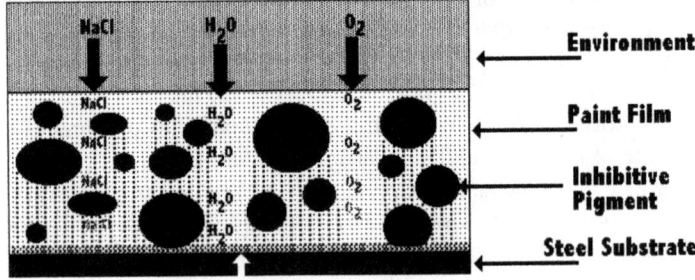

Passive Film
Surface of steel beneath inhibitive primer is comprised of a passive layer which prevents metal ions from entering electrolyte.

FIGURE 67.6. Corrosion protection by inhibitive primer.

the primer where passive films may be established. Formulation devices are designed so that enough water is able to penetrate the film and dissolve sufficient inhibitor to provide ionic concentrations high enough to achieve passivation at the subfilm metal (Fig. 67.6). The permeability to ionic solutions must not, however, be so great as to allow depassivating ions (chloride, sulfates, etc.) into the film from the environment. These depassivators raise the inhibitor threshold necessary to achieve passivation [50]. Where depassivator concentrations are high enough, it may not be possible to achieve passivation with even the strongest inhibitor. Fortunately, in normal primer films, the relative ionic impermeability of the coating films and the presence of the source of passivating ion within the film encourage favorable passivator/depassivator ratios at the substrate and preserve protection.

Selection of the pigment/binder composition should be such that the inhibitor is not so soluble or the film so permeable that excessive release of soluble inhibitive ions leads to the premature exhaustion of the film's available inhibitive reservoir and, in consequence, loss of primer effectiveness [51]. (Additional dangers with the use of very soluble inhibitors in paint films are the propensity of such films to exhibit osmotic blistering in high-humidity and freshwater conditions, especially when inhibitive primers are employed beneath lower permeability finishes.) Alternatively, the binder should not be so impermeable (or the pigment not so insoluble) that insufficient inhibitive ions are released to establish or sustain protection. This condition will lead to little effectiveness as an inhibitive primer,

although such systems may make rather inefficient barrier films.

There remain unanswered questions concerning these mechanisms, for it is difficult to understand how passivating films may form on metals bearing tightly adherent organic films. It has been suggested that inhibition may not become established until some initial adhesive breakdown in a quasi-barrier protection has occurred [32].

E3. Water-Based Inhibitive Primers

Recent emphasis on minimizing volatile organics released from the application of coating films has encouraged the use of water-based systems. With some important exceptions [9], the films deposited using such technologies are more water sensitive than are those deposited from solvent-based systems. Thus the films do not generally make good barrier coatings. They do, however, make more suitable binders for inhibitive primers. Water-based primers of this type are based on acrylic latexes, epoxies, and water-borne alkyds. In these systems, the inhibitor exists in contact with the water carrier in the can prior to the application of the paint, and this constitutes a special case. Here, the inhibitive ions are already in solution at the time of application, and, where passivating films may form on the steel before the wet primer film dries, the initial passive film may be established without the access of water through the coating in the field [51]. Thus, classical inhibition from the access of external water to the pigment in service is presumably necessary only to sustain the passive film.

F. SURFACE PREPARATION

Although the primary foci of this chapter are the mechanisms by which coating films may be employed to control metallic corrosion, such discussions cannot be considered complete without some reference to the preparation of the metal surface prior to coating application. The relevance of good surface preparation to the effectiveness of applied coating systems on steel has been well recognized since the work of Hudson in the 1930s, 1940s, and 1950s [52]. Since then, numerous other researchers have reaffirmed the singular importance of good practice in this regard, whether we are considering the sandblasting of structural steel, the anodizing of aluminum, or the pickling of nickel surfaces [53–55]. Morcillo [56] emphasizes the particular importance of good surface preparation to the performance of modem corrosion-resistant coating systems on steel.

The contribution of surface preparation to coating system performance may be reduced to one or more of the factors described in the following paragraphs.

F1. Removal of Interference Material

The presence of numerous and diverse conditions and contaminations on practical metal surfaces is virtually axiomatic. Oils, greases, and other organics may reduce the surface energy of normally high-energy metal surfaces from values near 400 dyn/cm to values <20 dyn/cm, thereby reducing wetting and impeding or even eliminating adhesion of the film. Insoluble inorganics (dirts, soils, and other powders) prevent the necessary close association between the binder and metal surface necessary for good adhesion, regardless of the nature of the bond (primary or secondary valency

bonding). The required distance between attracting moieties on the metal (oxides and hydroxides) and on the coating polymer (acidic groups, hydroxyls, etc.) is very small [of the order of ~0.5 nm (5 Å)] [57]. On this scale, dirt particles present very large obstructions. Soluble inorganics (e.g., salts), which also interfere with adhesion, may have more serious consequences [58]. Chlorides and sulfates, for example, act as depassivators and shut down passivation from inhibitive primers. Beneath barrier primers [50], the same salts dramatically decrease the electrical resistance of the interfacial region, therefore nullifying resistance inhibition [5]. In both types of systems, the presence of these salts beneath a film in freshwater immersion or in high-humidity environments will result in osmotic blistering failure.

Removal of these contaminants by one of the multiple cleaning techniques noted in Table 67.2, or by scarification techniques, such as abrasive blasting and pickling, eliminates these conditions.

F2. Activation of Metal Sites for Subsequent Adhesion of Coatings

On exposure to the environment, metal surfaces react rapidly with components of that environment, producing surface conditions in which these sites are blocked by reaction or corrosion products, for example, oxide scale (mill scale) which forms on new steel during hot rolling processes. In many instances, the corrosion products that saturate the reactive sites on the metal are poorly adhesive, insufficiently cohesive, or otherwise unstable.

If a coating system is to achieve full adhesion to the metal, these saturating moieties must be removed from the metal surface before the coating is applied. Various techniques are

TABLE 67.2. Prepainting Cleaning and Surface Preparation Processes

Solubilization[a]	For removal of soluble inorganics with water and soluble organics with solvent
Emulsification[a]	For removal of insoluble and/or water-immiscible inorganics by emulsification with detergent solutions
Saponification[a]	For removal of ester and amide based organics and inorganic salts by chemical hydrolysis with alkaline solutions
Chelation and sequestration[a]	For removal of metal ions from solutions and surfaces by chelation or complexation reactions
Deflocculation[a]	For wetting and dispersion of soils with surfactants, suspension of soil residues in order to prevent resedimentation, and recontamination on metal surface
Pickling[b]	For removal of surface contaminations, rust scale, mill scale, and other bound moieties (including surface layers of metal itself) by chemical dissolution with acids or alkaline deoxidation with or without the application of an electric current
Mechanical scarification[c]	For removal of surface contamination, rust scale, mill scale, and other bound moieties (including surface layers of metal itself) by wet or dry abrasion, erosion, and/or cavitation processes

[a]Cleaning processes may involve spray, dip, wipe, and brush application. Solvent cleaning by vapor degreasing is very common and efficient. Processes generally become more efficient with increasing temperature and pressure (e.g. steam cleaning). Surface neutralization may be necessary with some processes (saponification). Final rinsing step with clean water or solvent is mandatory.
[b]Neutralization and/or rinsing with clean water is always necessary.
[c]Abrasive blasting (air nozzle or centrifugal) is preferable to hand tool grinding, brushing, and so on, which are slow and inefficient. Wet blasting without abrasive does not produce an anchor pattern.

employed for this task, many employing mechanical or chemical scarification of the metal itself. Abrasive blasting (in the field) and (in the shop) acidic deoxidation (pickling) are typical of such procedures. The coating, if applied sufficiently rapidly after such removal, may itself satisfy a sufficient number of these newly freed sites to maximize adhesion. Such coatings may be prepaint conversion coatings such as phosphate coatings on steel, anodized coatings or chromate conversion coatings on aluminum, or amorphous oxide coatings on zinc, or they may be the organic primer itself. The conversion coating (often more porous than the metal) improves the metal surface for subsequent adhesion of the organic coating film.

Unremoved, the original species, contaminating the metal surface, may not only reduce the adhesion of the organic paint primer but, depending on its nature, may also interfere with subsequent performance of the applied system. Mill scale, noted above, is cathodic to steel and can induce corrosion of the metal where the mill scale is ruptured [4]. It is also smooth, so that coatings tend to delaminate under stress more easily than with a rougher sandblasted surface. Mill scale may also, under stress, delaminate from the metal itself (thus carrying away the coating system as it delaminates).

Where steel surfaces bear reacted salts (e.g., ferrous sulfate), these materials, if not removed, will not only interfere with adhesion but may, if trapped beneath a barrier film, be sufficiently water soluble to short circuit all resistance inhibition, leading to the establishment of conductive electrolytes beneath the coating system. With inhibitive systems, these same salts, unless removed, will act as powerful depassivators, changing the ratio of passivating ions (from the film) to depassivating ions (from the substrate) in an unfavorable direction.

F3. Increased Surface Area

In scarifying a metal surface prior to coating, whether by mechanical or chemical means, the surface is roughened [59]. It is changed from one in which the real (microscopic) surface area equates fairly closely with the apparent (macroscopic) area to one in which the real surface area is very much greater than the apparent area. In accomplishing this roughening, the number of active sites on the metal surface (which may subsequently serve as reaction sites for coating film adhesion) is greatly increased. Initial adhesion is thus much improved by scarification. As the lower areas of the film are also mechanically "reinforced" by the peaks and irregularities of the induced metal profile, the long-term adhesion and resistance to stress-induced delamination is also greatly improved. Under extreme stress, it is more likely that the cohesive integrity of the system above the peaks of the anchor pattern will fail before actual deadhesion of the system at the interface will occur.

F4. Surface Normalization

In designing a coating system for corrosion control on metal, the design engineer assumes that the metal surface will approximate that surface he or she conceives. In practice, this may or may not be the case, for numerous anomalous conditions are inevitable on practical surfaces, derived from fabrication mispractice to contamination incurred in transportation, storage, processing, and/or erection. This contamination, moreover, may not be entirely uniform across the entire surface. Surface preparation, in removing all of the contamination that may be present and in establishing a pristine surface which is, from an engineering standpoint, greatly enhanced, also produces a substrate that is at once a more uniform and a better replicate of that surface originally conceived by the design engineer. The entire coating system (substrate/paint film) thus becomes more consistent with the theoretical model, and performance is therefore more predictable.

REFERENCES

1. J. E. O. Mayne, Official Digest, **24**(127), 127 (1952).
2. R. C. Bacon, J. J. Smith, and F. M. Rugg, Ind. Eng. Chem., **40**(1), 161 (1948).
3. J. E. O. Mayne, Br. Corros. J., **5**, 106 (1970).
4. R. A. Hartley, "Coatings and Corrosion," J. Mater., **7**(3), 361 (1972).
5. C. H. Hare, "Barrier Coatings," J. Protect. Coat. Linings, **15**(4), 17 (1998).
6. E. M. Kinsella and J. E. O. Mayne, Br. Polym. J., **1**(4), 173 (July 1969).
7. H. Haagen and W. Funke, JOCCA, **58**(10), 359 (1975).
8. S. Guruviah, JOCCA, **53**(8), 669 (1970).
9. N. Thomas, JOCCA, **74**(3), 83 (1991).
10. M. Morcillo et al., JOCCA, **73**(1), 24 (1990).
11. C. H. Hare, J. Protect. Coat. Linings, **15**(7), 17 (1998).
12. C. G. Munger, "Corrosion Resistant Zinc Rich Coatings," in Corrosion Control by Protective Coatings, NACE International, Houston, TX, 1984, p. 129, Chapter 6.
13. M. Leclerq, J. Protect. Coat. Linings, **7**(3), 57 (1990).
14. H. Leidheiser, J. Coat. Technol., **53**(678), 29 (1981).
15. M. Svoboda and J. Mleziva, Prog. Org. Coat., **12**(3), 251 (1984).
16. K. Baumann, Plaste Kautschuk, **19**(9), 694 (1972).
17. H. Leidheiser and W. Funke, JOCCA, **70**(5), 121 (1987).
18. W. Funke, JOCCA, **68**(9), 229 (1985).
19. A. S. Michaels, Official Digest, **37**(485), 638 (1965).
20. N. Thomas, Prog. Org. Coat., **9**, 101 (1991).
21. C. H. Hare, J. Protect. Coat. Linings, **7**(2), 53 (1990).
22. W. Funke, Prog. Org. Coat., **9**, 29 (1981).
23. K. Sato, Prog. Org. Coat., 143 (Aug. 1980).

24. S. G. Croll, J. Coat. Technol., **53**(672), 85 (1981).

25. Anonymous, "Coal Tar Epoxy Coatings—State of the Art Review," TPC-12, NACE International, Houston, TX, 1991.

26. W. Funke, JOCCA, **62**(2), 63 (1979).

27. M. Yasseen, "Permeation Properties of Organic Coatings in the Control of Metallic Corrosion," in Corrosion Control by Protective Coatings, H. Leidheiser, Jr. (Ed.), NACE, Houston, TX, 1981, p. 24.

28. C. H. Hare, Coatings World, **2**(4), 24 (1997).

29. D. J. Mills and J. E. O. Mayne, "The Inhomogenous Nature of Polymer Film and Its Effect on Resistance Inhibition," in Corrosion Control by Protective Coatings, H. Leidheiser, Jr. (Ed.), NACE, Houston, TX, 1981, p. 12.

30. C. C. Maitland and J. E. O. Mayne, Official Digest, **34**(452), 972 (1962).

31. W. Funke, JOCCA, **68**(9), 229 (1985).

32. W. Funke, JCT, **55**(705), 31 (1983).

33. C. H. Hare, J. Protect. Coat. Linings, **13**(7), 79 (1996).

34. D. Perera, "Stress Phenomena in Organic Coatings," in Paint and Coatings Testing Manual, 14th edition of Gardner-Sward Handbook, J. V. Koleske (Ed.), ASTM, Philadelphia, PA, 1995, Chapter 49, p. 585.

35. C. G. Munger, Mater. Perform., **14**(5), 25 (1975).

36. D. M. Berger, J. Protect. Coat. Linings, **1**(2), 20 (1984).

37. C. H. Hare, M. J. O'Leary, and S. J. Wright, Modern Paint Coat., **73**(6), 30 (1983).

38. A. C. Elm, "Zinc Dust Metal Protective Coatings," New Jersey Zinc Company Publication 5-68-2M, New Jersey Zinc Co., New York, 1968.

39. S. A. Lindquist, L. Meszaros, and L. Svenson, JOCCA, **68**(1), 10 (1985).

40. R. Romagnoli and V. F. Vetere, JOCCA, **76**(5), 208 (1993).

41. S. Feliu, Jr., M. Morcillo, J. M. Bastidas, and S. Feliu, J. Coat. Technol., **65**(826), 43 (1993).

42. G. Salensky, "Corrosion Inhibitors," in: Handbook of Coatings Addition, L. J. Calbo (Ed.), Marcel Dekker, New York, 1987, Chapter 12, p. 307.

43. M. J. Austin, "Anti Corrosive Inorganic Pigments," in Surface Coatings, Vol. I. P. Parson, (Ed.), Chapman and Hall, London, 1993, Chapter 25, p. 409.

44. C. H. Hare, Paint Coat. Ind., **XIII**(8), 50 (1997).

45. A. J. Appleby and J. E. O. Mayne, JOCCA, **50**(10), 897 (1967).

46. H. Weinard and W. Ostertag, Mod. Paint Coat., **74**(11), 38 (1984).

47. Z. W. Wicks, "Corrosion Protection by Coatings," Federation Series on Coatings Technology, F.S.C.T., Philadelphia, PA, 1987.

48. C. H. Hare, Paint Coat. Ind., **XIV**(3), 74 (1998).

49. A. J. Appleby and J. E. O. Mayne, JOCCA, **59**(2), 69 (1976).

50. R. W. Revie and H. H. Uhlig, Corrosion and Corrosion Control, 4th ed., Wiley, Hoboken, NJ, 2008, Chapter 17, p. 303.

51. C. H. Hare, J. Protect. Coat. Linings, **15**(5), 48 (1998).

52. J. C. Hudson, J. Iron Steel Inst., *JISIA*, **168**(1), **165**, (1951).

53. Anonymous, "Effects of Surface Preparation on Service Life of Protective Coatings," Interior Statistical Report by NACE Tech. Committee T-6H-15, NACE, Houston, TX, Dec. 1977.

54. S. Spring, Preparation of Metals for Painting, Reinhold, New York, 1965.

55. B. M. Perfetti, "Metal Surface Characteristics Affecting Organic Coatings," Federation Series on Coatings Technology, Federation of Societies for Coatings Technology, Blue Bell, PA, 1994.

56. M. Morcillo, "Minimum Film Thickness for Protection of Hot Rolled Steel: Results after 23 Years of Exposure at Kure Beach, North Carolina," in New Concepts for Coating Protection of Steel Structures, ASTM, STP 841, D. M. Berger and R. F. Wint (Eds.), American Society for Testing and Materials, Philadelphia, PA, 1984, p. 95.

57. C. H. Hare, Constitution Specifier, 27 (Mar. 1973).

58. L. Igetoft, "Surface Cleanliness and Durability of Anticorrosive Paint," in Proceedings of Second World Bridge Congress, Oct. 26/27 1982, New York, University of Missouri Rolla, Rolla, MO, 1983, p. 5.

59. J. D. Keane, J. A. Bruno, and R. E. F. Weaver, "Surface Profile for Anti-Corrosive Paints," SSPC Report, Steel Structures Painting Council, Philadelphia, PA, 1976.

68

SELECTION AND USE OF COATINGS FOR UNDERGROUND OR SUBMERSION SERVICE

R. Norsworthy

Lone Star Corrosion Services, Lancaster, Texas

A. INTRODUCTION

Coatings have been used for many years to provide corrosion protection for metal structures that are buried or submersed. Unlike most above-ground coatings these coating materials are normally used for corrosion control and therefore do not have to meet appearance requirements. These coatings are normally applied much thicker than on above-ground structures and must be able to withstand an entirely different environment. The amount of corrosion protection provided by a particular coating system depends on many variables.

Various types of pipelines, underground storage tanks, and subsea structures made from a variety of metals make

Uhlig's Corrosion Handbook, Third Edition, Edited by R. Winston Revie
Copyright © 2011 John Wiley & Sons, Inc.

up the large percentage of structures that are normally coated for corrosion control. Many tanks, pipelines, and other structures are also internally coated for corrosion protection, but these coatings will not be discussed in this section (even though many of the same principles are used for these coatings). When selecting a corrosion coating for the external surfaces of structures, there are many criteria to be considered. The following is a brief discussion of some of these criteria.

B. CRITERIA FOR SELECTION OF COATINGS FOR UNDERGROUND OR SUBMERSED STRUCTURES

B1. Safety

Safety should always be the first consideration when selecting any material. The application of many coating systems involves potentially hazardous or dangerous processes. Some coatings emit hazardous fumes or particles. Others involve high temperatures, hot liquids, open flames, or dangerous moving equipment. The type of surface preparation needed for a particular coating can be a safety problem. Safe removal and disposal of an existing coating system or disposal of waste materials after application must also be considered. Designing a coated structure that will operate safely for the life of the structure without leaks or failures protects the surrounding community and the environment.

B2. Cost

Cost is always an important part of any selection criteria, but it is important to look past the initial cost of providing a coating for the structure. The coating cost for most projects is minor when compared to the total project cost, yet frequently a "cheaper" coating or surface preparation is selected to save money. A coating must be selected that will perform in that environment for at least the projected life of the system. Many systems operate well beyond their projected life. For example, there are several pipeline systems in use today that have operated many years beyond the projected life. Systems that use the right coating for the environment, ensure proper installation and handling of the coated structure, and do not exceed the operating requirements for the coating can operate safely for many years beyond the design life. Other systems fail before the design life because the coating did not perform properly. The reasons for coating failures include improper coating selection for the environment, poor surface preparation, and poor coating application. Failed coating systems require continuous maintenance on the structure and the coating. For these reasons, the long-term cost must be considered when determining the coating system to be used.

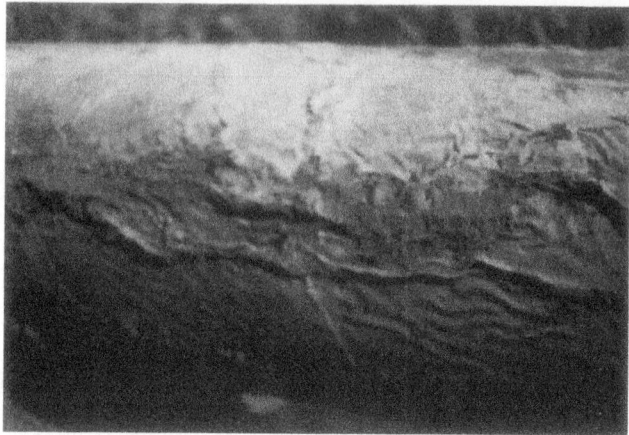

FIGURE 68.1. Soil stress effects on a coal tar coating. (Photo furnished by Lone Star Corrosion Services, Lancaster, TX.)

B3. Environment

The type of environment the coating system will be placed in is very important. When placed in soils, the most critical factor to consider is soil stress. Heavy clays and gumbo-type soils can cause severe coating damage, especially on pipeline, as shown in Figure 68.1. Coatings that stretch easily can wrinkle or move on the pipeline. Elongation, shear strength, and pull-off tests all help to determine if a coating will be affected by soil stress. Coatings with little or no elongation but with good shear strength and adhesion to the pipe surface are less susceptible to soil stress. Twenty-eight U.S. and Canadian operators who responded to a 1989 survey said that disbondment (39% of total responses) and soil stress (27% of total responses) were their greatest concerns relative to coal tar, tape, and asphalt coatings [1].

Sandy soils in hot environments can cause some types of coatings to sag or move. Wet–dry environments, such as "subka sand," can affect some types of coatings. Sea mud and brine soils can cause other coatings to fail. Moving water, such as in rivers, or tidal waters carry debris and other substances that can erode and damage many types of coatings. This has been a serious problem in the "splash zone" of offshore structures, as shown in Figure 68.2. Areas such as bogs, tundra, and swamps have bacteria that may use some coating materials for a food source, thereby causing coating failure. Cold or arctic conditions cause some coatings to perform poorly. Contaminated soils and other environments must be considered when selecting a coating because some types of contamination have an adverse effect on certain types of coatings.

B4. Operating Temperature

The operating temperature of the system is critical in selecting a coating system. Maximum and minimum operating temperatures of the system must be considered. The data and

FIGURE 68.2. Splash zone coating damage and corrosion on offshore platform. (Photo furnished by Lone Star Corrosion Services, Lancaster, TX.)

FIGURE 68.3. Blistering of FBE coating on hot pipeline (180°F). (Photo furnished by Lone Star Corrosion Services, Lancaster, TX.)

test results of each coating to be considered for a particular application must be assessed and compared at these temperatures. Coatings that perform well at higher temperatures usually perform well at lower temperatures but may not perform well in very cold temperatures. The same reasoning applies to coatings that may perform well in very cold temperatures but may not perform well at high temperatures.

Operating temperatures >65°C (150°F) can cause most adhesives, which some coatings use as the main bonding mechanism, to flow and move toward the lower parts of the structure. When this happens on pipe, a loose shell of the outer material is left on the top half of the pipe with no bond to the pipe. Many tapes, shrink sleeves, and extruded polyolefins are susceptible to this problem. This process may take only a few months to several years to occur, according to the temperature and the type of coating. If water penetrates this space, corrosion becomes a problem. When high temperatures are encountered, polypropylenes instead of polyethylene are normally used. Coal tar and asphalt based coatings can slowly flow with gravity, leaving the top part of the structure exposed.

Fusion-bonded epoxy (FBE) coatings can absorb more water than normal at higher temperatures but do not flow or move on the structure. If underfilm contaminants are present, FBE coatings can blister or disbond, as shown in Figure 68.3.

B5. Surface Preparation

It is widely recognized that surface preparation is the most important single factor in coating performance [2]. Many in the coating industry agree that two-thirds the cost of any coatings job should be spent on surface preparation of the structure to be coated. One should always prepare the best surface possible for a given coating situation. Blasting using

the correct material and method not only cleans the metal surface but also provides a surface profile (anchor pattern) for coating adhesion. Other methods of surface preparation include the use of hand tools, power tools, water blasting, and wire brushes. The Steel Structures Painting Council and NACE International are both excellent resources for information on surface preparation in the coating industry.

Field repairs or coating replacements do not always allow for the best surface preparation. When the proper surface preparation is not possible, replacement coatings are selected that are more tolerant to poor surface preparation.

Coating tests performed on a variety of prepared surfaces, as well as past experiences with coatings that performed well in a particular condition, should be used to help select the best coating for the situation. For field repairs, leaving the structure uncoated (if adequate Cathodic Protection (CP) is available) may be a better choice than to apply a coating that may not bond to the structure because of poor surface conditions. This is especially true during cold or damp weather or if the structure is sweating or cold. Hot applied coatings should never be applied on a cold structure. Petrolatum-based coatings are sometimes used on cold or wet structures. If a structure is left uncoated or if a poor coating condition exists, this part of the structure should be monitored and properly coated when the conditions are more suitable.

Most plant-applied coatings require more stringent surface preparations. Surface preparations are much easier to control and perform in plant setting, as shown in Figure 68.4. Some of the common surface preparation mistakes made in plants are:

Using the wrong type of blast material
Using contaminated blast material
Improperly using surface treatments
Using rinse water that contaminates the steel surface

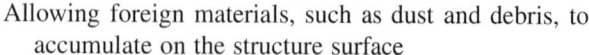

FIGURE 68.4. Inspection for and grinding of surface defects after blasting in FBE. (Photo furnished by EB Pipe Coating, Inc., Panama City, FL.)

FIGURE 68.5. Proper handling of FBE coated pipe with padded forks. (Photo furnished by EB Pipe Coating, Inc., Panama City, FL.)

Allowing foreign materials, such as dust and debris, to accumulate on the structure surface

Using air that may be contaminated with oil or water; debris, tape, and other contamination on conveyors, wheels, and supports

Some coating processes require that a chromate and water mixture be applied to the structure before the corrosion coating is applied. Chromate is a conversion coating that reacts with steel to change the surface chemistry. Too much chromate can cause a buildup that may affect the coating bond.

Weather conditions should always be considered whether in the field or a plant. Flash rusting after surface preparation but before coating application can occur if the humidity is too high. Flash rusting can cause some coatings to fail. Oil or grease contamination must be removed with proper solvents. Blasting will only spread the oil or grease on the structure surface. The original surface condition of the structure must be considered. New pipe or structural steel may have mill lacquers, mill scale, or other surface contaminants. Stored or in-service structures may have considerable corrosion, scale, or other contamination. Leaking product can contaminate surfaces of in-service pipes or tanks in leak locations.

B6. Application Methods

Field and plant application techniques vary considerably for each coating system. Plant applications are usually much faster, and each step can be controlled more easily. The coating itself is usually applied with automatic equipment on a production line. Inspections and more sophisticated testing can easily be performed. Storage, handling, and transportation can present a problem for plant-coated structures, as shown in Figure 68.5.

Field applications are by hand or machine but are usually labor intensive. Applying coatings in the field requires training and practice with the particular coating and application method chosen.

Tape coatings for pipelines appear to be easily applied. Too many times, tape coatings are applied over the wrong or uncured primer with improper tension or not enough overlap. Shrink sleeves can be over- or under heated, be unevenly heated, leave air pocket, or allow debris to be blown between the sleeve and the pipe.

Brushable or sprayed coatings must be allowed to properly cure before applying the second coat or before the structure is back-filled or submersed. Frequently, these coatings are applied too thin or thick. Liquid coatings must be mixed properly, especially two-part coatings, such as epoxies. Urethanes and polyurethanes are very susceptible to moisture and humidity and require specialized equipment and training for application.

Any coating applied in the field must be compatible with any existing coating. The transition from one coating to another is a very critical area. Many coating failures have occurred at the transition area.

B7. Weld Area, Joint, and Additional Component Coatings

Whether the main coating is plant or field applied, there are areas around girth welds (called field joints on pipelines), flanges, bolts, and other joined areas where the structure is connected together that must be considered when choosing the primary coating. In some cases, a very good coating is used on the main structure, but then an inferior or poorly applied coating is used in the areas where connections are made in the field. Field connections are the most critical area on many systems. Heat-affected zones and other stressed areas around welds can corrode quickly when exposed to an electrolyte. Bolted or riveted connections can have crevices and other areas that can easily corrode or cause stress points where cracking can occur.

The best field connection coating for a particular system is usually the same as or better than the coating used on the main structure. Any field connection coating must be compatible with the previously applied coating. The field connecting process should cause very little or no damage to the existing coating. Prequalification testing of the applicator will help to ensure the crew is capable of applying the coating as specified.

B8. Repairs

Any coating selected must be easily repaired with a material that is compatible with, or the same as, the parent coating [3]. Many times repair coatings are improperly applied, not allowed to cure, or incorrectly mixed. Many of the problems mentioned above for field connections and field-applied coatings are the same for repair coatings as well. Once again, testing and experience are important when choosing a repair coating.

B9. Cathodic Protection

Since corrosion always requires the presence of an electrolyte (moisture) in contact with the metal, if a metal could be coated with a material which was absolutely waterproof and absolutely free from holes, all attack would be stopped [4]. Always assume "there is no perfect coating"! Even if it were possible to apply and install a "perfect" coating, deterioration, soil stress, environmental factors, and damage from outside forces would soon cause portions of the coating to fail. For this reason, cathodic protection must be applied and properly maintained. This synergistic relationship is well proven and documented.

Cathodic protection can affect all coatings. The hydrogen, hydroxyl ions, and other electrochemical reactions caused by cathodic protection currents at the cathode may cause blistering or cathodic disbondment of the coating. Hydroxyl ions are one of the most aggressive chemical species and nearly all organic binders are capable of reacting with them [5]. Cathodic protection can cause loss of adhesion between the coating and the metal, leaving a void between the coating and the structure. The surrounding environment and the level of cathodic protection at a particular site determine the extent of cathodic disbondment. The amount of cathodic protection needed to cause cathodic disbondment is very difficult to determine. There are several rules of thumb but little data to actually use for determining the level that affects a particular coating in a certain environment.

The amount of cathodic protection needed to achieve adequate protection varies greatly for each type of coating. The effective electrical strength of a coating is often expressed as the resistance per average square foot of coating [6]. Determining the actual amperes per square foot needed to protect the coated pipe after installation can provide valuable information for future projects. There are charts, tables, and computer programs used for determining the amperes of current per square foot needed for a particular coating system. Current requirements or current density measurements are related to coating conductance [7]. One must also consider the coating breakdown factor. How long will it take the coating to deteriorate before cathodic protection has to be increased? At 10, 20, or 30 years after installation, what percent of the structure will be exposed because of deteriorated or damaged coating?

Coatings that incorporate an electrically insulating outer layer, such as polyolefin tape coatings, can preclude effective CP from reaching the pipe surface in disbonded areas [8]. Some research and considerable debate have not fully answered the question of whether cathodic protection can be provided under disbonded coating. There is some information available that indicates some protection can be achieved when the electrolyte present under the disbonded coating has very low resistance (e.g., saltwater environments). The use of pulsed cathodic protection has also been studied to determine its benefits (if any) for providing cathodic protection under disbonded coatings.

When selecting a coating system to be used with cathodic protection, the "nonshielding" or "fail-safe" characteristics of that coating may be more important than other issues that are normally considered [8a]. Nonshielding means if the coating system adhesion fails and water penetrates, corrosion on the metal is significantly reduced or eliminated when adequate CP is available. Fusion-bonded epoxy is a nonshielding coating.

B10. Handling, Storage, and Transportation of Coated Pipe

Coating materials selected must be able to withstand the rigors of handling, storage, and transportation for a particular project. Coated pipes and other structures should be handled with padded forks, hooks, or other equipment that will not damage the coating. Stacking and storage should be according to structure weights, with the proper separation materials between each piece. The separation material should not damage the coating in any way. When coated structures are stored outside, separation helps to protect the coating by not letting debris settle in openings and crevices.

During transportation, the coated structures must be properly strapped down with soft straps, not chains or bands that may damage the coating. Coated structures should also be loaded and fastened in a manner that does not allow the coated structures to shift or slide during transportation. Once again, proper separation and support should be used between the structures. Rocks and other road debris can damage coatings during transportation. Mud flaps and protective shields help eliminate most of these problems during truck or rail transportation.

Outside storage can cause problems with all coatings. Ultraviolet (UV) rays damage most materials with time. Polyolefin coatings become brittle and crack. The FBE coatings can chalk. This chalking is usually only 1 or 2 mils in depth, but rain can wash off the "chalked" layer and expose new FBE to the sun. Coal tar coatings tend to crack with exposure. Heat and cold can cause some coatings to contract or expand or cause the adhesive to flow.

B11. Specialty Coating and Overcoats

Coatings may be specific for a particular situation. One may involve weight coatings. Concrete is applied over coated pipe to provide enough weight to sink the pipe or provide it with mechanical protection. Other "overcoats" are applied to provide mechanical protection when pipe or other structures are pulled through road or river bores, put inside casings, or installed in rocky terrain or for other unusual circumstances. Other special "overcoat" material may be used in splash zone areas of subsea structures. These overcoats can shield the cathodic protection if they disbond from the metal surface. If the reinforcement in concrete coated pipe or structures contacts the pipe, the cathodic protection will protect the reinforcement but not the pipe.

C. COATING TYPES FOR UNDERGROUND OR SUBMERGED STRUCTURES

This discussion covers the variety of coatings available to industry at this time. The information provided about each system comes from numerous tests, specification writing and reviews, coating inspections, field applications and evaluations, literature reviews, and plant applications. Each coating being considered for use on an underground or submerged system should be evaluated using the above criteria and any other criterion that may affect the performance of the coating.

C1. Coal Tar Based Coatings

C1.1. Application. Coal tar based coatings were one of the first types to be successfully used as a corrosion coating. These coatings are made of coal tar mixed with various blends or formulations of fillers and extenders. As with most types of coatings, there are several grades of coal tar coatings available. One must study the different types available to ensure the best choice.

Coal tar mixtures have been widely used on steel underground storage tanks. Most are applied at the tank manufacturing facility by spraying or brushing the coal tar epoxy or mastic onto clean, primed surface. Surface

preparation normally involves brushing or blasting. A reinforcing wrap may be used to add mechanical strength. These tanks are not always carefully handled during transportation and construction. Coating damage is easily repaired by cleaning the area and applying a coal tar based epoxy or mastic.

Coal tar coatings have also been successfully used on subsea structures, such as offshore platforms and pipelines, submersed bridge structures, docks, and other submersed metal structures.

Plant-applied coal tar enamels have been used for many years on underground pipelines, Coal tar is applied to a blasted and primed pipe surface. The surface is usually grit blasted in accordance with SSPC-SP6 after any dirt, oil, grease, and so on, are removed. The pipe is preheated to remove any moisture and keep the steel temperature above the dewpoint. A heated [50°C (120°F) or less] type A or type B primer is applied by spraying as the pipe rotates down the conveyor line. The type A primer is made from coal tar. Type B is a fast-drying synthetic and is normally specified.

The coal tar normally used for plant application contains fully plasticized enamels. Hot coal tar enamel flows onto a prepared pipe as it rotates down a conveyor. After the coal tar flows onto the pipe, a felt wrap is applied to reinforce the coating. This is normally topped with whitewash or kraft paper to provide ultraviolet protection. The thickness of the coal tar and felt wrap is normally $\sim 2.4\,\mathrm{mm}\left(\frac{3}{32}\ \mathrm{in.}\right)$ $\left(\pm 0.8\,\mathrm{mm}\left[\frac{1}{32}\ \mathrm{in.}\right]\right)$ Early felt wraps contained asbestos, but in the 1980s most of the industry changed to felt that is fiberglass reinforced. There are other methods for plant application of coal tar, but this method is the most common. Coal tar coatings are not as popular as they once were but are still a major part of the coating industry.

Variations of coal tar enamels, epoxies, mastics, and urethanes are used for field applications and repairs. Coal tar enamels are hot-applied materials that are usually applied with wraps containing coal tar coated fiberglass felt. Coal tar epoxies are solvent-cured coatings that are usually brushed or hand applied in multiple layers. Coal tar urethanes are multicomponent applied mixtures that are normally used for large-scale rehabilitation projects. Coal tar mastics are hand applied and are used for a variety of irregular shaped structures.

C1.2. Pluses. Coal tar enamel coatings have a long history of corrosion protection. Many coal tar coated pipelines have been in service for over 50 years and are still in very good condition. Coal tar coatings are easily repaired with field-applied coal tar enamel, coal tar epoxy, or tape coatings. Coal tars are thick coatings that have excellent electrical insulating properties, have low water permeation properties, resist bacterial attack, and have the solvent action of petroleum oils.

C1.3. Minuses. Even though many pipelines are well coated and still perform as intended, many pipelines have major coal tar coating failures. Much of the failed coal tar was applied over the ditch, on pipe surfaces that were not well prepared, or in undesirable weather conditions. In many cases, plant-applied coal tar was not applied using an acceptable method. Handling during transportation, storage, and construction also contributed to these failures.

If the temperature is not properly maintained, both the application and future performance of coal tar enamel coating are compromised. Coal tar epoxy is a multicoat application that requires time for solvents to evaporate before burying or submersing. Coal tar urethanes require sophisticated equipment and well-trained individuals for application. Application of urethanes is not recommended in high-moisture situations.

Coal tar enamel coatings are subject to soil stress that may cause the coating to wrinkle, crack, disbond, and expose steel surfaces. If water penetrates under the disbonded coating, cathodic protection shielding can become a problem. Cathodic protection requirements normally increase as the coal tar coatings age. Operating temperatures of pipelines coated with coal tar are normally limited to 65°C (150°F) or less, Coal tar enamels can "cold flow," leaving the top of the pipe without adequate coating, especially on high-temperature pipelines.

There are safety concerns with hot-applied coal tar enamels. Not only are there the problems associated with the dangers of hot application, but, in addition, the fumes may be toxic. The early use of asbestos felt has been a problem for those removing and disposing of these coatings. Each of these problems can be studied and handled with proper education and precautions.

C1.4. Improvements.
During the 1990s, there have been improvements in coal tar enamel coatings, including an effort to make these coatings more tolerant to higher temperature operations.

C2. Fusion-Bonded Epoxy

C2.1. Application.
Fusion-bonded epoxy pipeline coatings have been used successfully since the late 1960s. A typical FBE formulation consists of epoxy resins, curing agents (hardeners), catalysts and accelerators, prime and reinforcing pigments, control agents (for flow and stability), and specialty ingredients [9]. The first FBEs were applied over a primer, but later developments allowed them to be used without a primer. Originally, FBEs were applied much thinner than the 12–16 mils normally specified in 1999. The FBE has been used to coat rebar used in bridge, road, and building construction to help prevent corrosion of the rebar in concrete.

The application process for FBEs is one of the most stringent and complicated in the industry. Because of the thin film and potential for water absorption, the surface cleanliness must be very good. The pipe is moved through the plant on a series of inspection racks and conveyors. Oil, grease, or other contaminants are removed from the pipe. The pipe is then preheated to keep it above the dewpoint during the surface preparation phase. After blasting to SSPC-SP10 (Near White), most companies now require a phosphoric acid wash and rinse to further clean, etch, and provide surface energy to the pipe that helps to attract and bond the FBE to the steel surface. Many companies (especially in Europe) require the application of a chromate solution for an additional surface treatment.

After the surface preparation, the pipe is heated to a range of 232°C (450°F) to 260°C (500°F) by either heat induction coils or gas-fired heaters. The FBE powder is applied to the hot pipe by a dry air spray system. Electrostatic gun are used to attract the optimum amount of powder to the pipe surface. The FBE powder melts, gels, and cures as the pipe rotates down the conveyor line. Gel times are usually in a range of 10–30 s at these application temperatures. Once the powder has gelled, the FBE on the pipe is hard enough to support the weight of the pipe on the conveyor tires. After the FBE is cured (usually 30–100 s) a water-quenching system cools the coated pipe. The coated pipe is then inspected for acceptance, rejected, or placed on hold for necessary repairs.

C2.2. Pluses.
The FBE coating has the best bond (adhesion) to steel of any pipe coating. The bond is mechanical as well as chemical. The FBE is very flexible and allows for field bending of pipelines or rebar. The FBE mostly cures by the time it is in the quench, so it can be handled immediately after the process. Repairs are made easily with two part epoxies or patch sticks. The water absorption of FBEs is a plus when disbondment occurs. Unlike thick coatings that have high electrical resistance, FBE (because of the water absorption) will allow enough cathodic protection current through the film to protect the pipe under the coating. Therefore, corrosion and pitting are rarely encountered under disbonded FBEs if adequate cathodic protection is provided. Unless severe failure occurs on an FBE coated structure, the cathodic protection requirements normally do not increase significantly as the FBE ages. Stress corrosion cracking has been studied extensively and has never been observed on FBE-coated pipelines [10].

The FBE coatings handle well during transportation and construction activities. Comments have been made by some in the industry that they do not use FBE because it is easily damaged during construction, compared to thick coatings. When FBE coatings are damaged, the damage is easily found and repaired. The thicker coatings may "hide" damage that has occurred and disbondment may not be seen in some cases until the pipe fails.

Multilayer coatings with FBE as the base or primer coat have been very successful when properly specified and applied. These coatings have been used successfully on pipelines with internal operating temperatures up to 150°C (300°F) or pipelines in critical areas. These systems are discussed further in Section C3.

C2.3. Minuses. The FBEs require more stringent surface preparation and application techniques than most plant-applied coatings, thus requiring more attention during the coating process. Slivers and other steel imperfections can cause "holidays" that may not be a problem on thicker coatings. Field joints are expensive if FBE is used for the field joint coating (even though FBE is the preferred coating). Higher temperatures can cause water to be absorbed more quickly and cause disbondment if surface contaminants are present. For pipelines operating at temperatures between 65°C (150°F) and 85°C (185°F), thicker, up to 800 μm (32 mils), FBE should be used. Flexibility during bending can be a problem when using thicker FBE coatings. When stored outside, UV rays can cause the FBE to "chalk" and become grainy in the first few mils of coating. Multilayer coatings with FBE as a base coat are expensive but offer an excellent coating.

C2.4. Improvements. Recent changes in FBE coatings include the development of FBEs that will perform better in high-temperature operations. Each manufacturer offers several formulations of FBE. Each formulation has specific uses and they usually vary in gel, cure time, application technique, and test results. The FBE-based multilayer systems have become an important part of the FBE market.

New formulations of FBEs are being tested that can be applied and cured at much lower temperatures to reduce the amount of energy required and to accommodate the higher strength steels that may be affected by higher application temperatures.

C3. Polyolefin Coatings

Plant-applied polyolefin coatings have been used for pipeline coatings since the early 1960s. Because of application techniques, plant-applied polyolefins are not normally used on other structures. Since polyolefins do not bond well to steel, it is normally applied over an adhesive or other product that provides the bond to the steel. Polyethylene has been the polyolefin of choice for most extruded coatings, but polypropylene is normally used for higher temperature [>65°C (150°F)] operations. For many extruded polyolefins, an adhesive that is usually a rubber-modified asphalt adhesive or a butyl rubber compound provides the bond to the steel. Three-layer systems use an epoxy primer or a coating quality FBE for a base coating to the steel; then an ethylene copolymer or terpolymer adhesive is used for the "tie" layer. Some of these multilayer systems are now using chemically modified polyolefins for the topcoat or the tie layer. These

chemically modified polyolefins chemically and mechanically bond to the FBE base coat.

C3.1. Polyolefin over Adhesive. Polyolefins essentially have no adhesion to steel [11]. When adhesives are used as the bond to steel, the process usually involves heating the pipe to a temperature above the dewpoint. The pipe is then blasted and the heated adhesive is applied by flood coating onto the pipe. One method uses a crosshead extrusion die system that extrudes the melted polyolefin over a rubber-modified asphalt adhesive coated pipe as it travels down a nonrotating conveyor system.

Another method uses a rotating conveyor system and side extruders to apply the adhesive and the polyolefin. After preheating and blasting, the pipe rotates down a conveyor where the molten butyl rubber adhesive is spirally applied by the side extruder. The molten polyolefin is then side extruded over the adhesive. This process allows the layers of adhesive and polyolefin to fuse together and produce a theoretically seamless pipe coating.

C3.2. Polyolefin over Epoxy Primer. Three-layer systems using a liquid epoxy primer or powder epoxy primers also use a rotating conveyor system for the application process. These primers are sprayed onto the hot pipe as it rotates down the conveyor. The co-terpolymer is then side extruded over the primed pipe as the "tie" layer between the epoxy and polyolefin. The polyolefin is then side extruded over the co-terpolymer. This system has been used in Europe and the Middle East for over 20 years with excellent results. The three-layer coating evolved from the extruded two-layer (co-terpolymer adhesive plus polyethylene) polyethylene coating first used in Europe in about 1960 [12]. The addition of the epoxy primer improved the cathodic disbondment characteristics of these coatings. Many three-layer systems now use powdered epoxies or FBE for the primer coating.

C3.3. Polyolefin over FBE. Today, the trend for many multilayer coatings is to use a thicker 300–625 μm (12–25-mil) FBE base coat. The FBE is then topped with a chemically modified polyolefin (CMP) as either the topcoat or the middle layer of a three-layer system. This is an improvement over the traditional three-layer systems because of the thicker FBE quality base coat and use of the CMP. The bond of the CMP to the FBE is normally much stronger than the bond of the co-terpolymer to the FBE. The bond of the CMP to the FBE is both chemical and mechanical and will not allow a separation between the two when properly applied.

When a three-layer system is used, the hot CMP will bond well to the top layer of polyolefin, making a coating system that is excellent. Some of these coating systems now use a powdered CMP for the middle layer of a three-layer system or top layer of a two-layer system. These have been very successful coatings when applied properly.

C3.4. Pluses. Plant-applied polyolefin coatings are overall excellent coatings. Polyolefins are among the best materials to prevent water permeation. The "slick" surface of plant-applied polyolefins is not normally affected by soil stress. These systems handle easily during transportation and construction activities. Polyolefins are resistant to many chemicals, environments, and bacterial attack. The polyolefin over FBE systems is an excellent pipeline coating system for high-temperature service. Test results show that coating systems consisting of polypropylene (PP) over FBE can be used successfully when internal operating temperatures reach 150°C (300°F) [13]. Even though polyolefin over FBE coating systems can be more expensive than other coatings, they provide an excellent overall coating system when specified and applied properly.

C3.5. Minuses. One of the problems with polyolefins over adhesive coatings is that the adhesive layer provides the only bond to the steel. Adhesive can allow the polyolefin to shrink and expand on the pipe surface as the temperature changes. If this movement occurs, steel can be exposed or it can cause problems with field joint coatings. If the polyolefin is damaged while in service, it can disbond or split and allow water penetration between the polyolefin and pipe (Fig. 68.6.) Since the polyolefin has very high electrical resistance, cathodic protection is shielded. Underfilm corrosion can then become a significant problem, since the primer does not offer much corrosion protection. External damage can allow water to permeate the polyolefin more easily even though the film is not broken enough for the holiday detector voltage to penetrate. This external damage can also break the bond between the adhesive and the polyolefin. This is especially true of the crosshead die extrusion process.

The co-terpolymer tie layer for some polyolefin over epoxy primer coatings also allows the top layer to shrink and expand as the temperature changes. Many times, the epoxy primer used is very thin, 80–130 μm (3–5 mils), and may not provide corrosion protection if the top layers are damaged. If water were to penetrate between the polyolefin and the primer, the thick polyolefin would shield the cathodic protection.

One of the major disadvantages of all polyolefin coatings is coating selection for use on repairs, field joints, valves, and other components. Most tapes and shrink sleeves have problems with soil stress, do not bond well to the polyolefin, or have application problems that may cause failures at the repair or field joint. It is difficult to coat bends and other components with most tapes and shrink sleeves. Liquid epoxies, mastics, and other such coatings do not bond well to the polyolefin, possibly allowing water to penetrate at the junction of the two coatings. One system coats the bends and many of the other components with an FBE covered with a powdered chemically modified polyolefin. This system works well with the polyolefin over FBE two-and three-layer systems. Since high temperature is required for this application, the heat can affect the plant-applied coating if this process is used for the field joints. These coatings are expensive when compared to most other plant-applied coatings.

C3.6. Improvements. Recent developments include providing better repair and field joint coating systems for the three-layer systems. The use of thicker, high-quality FBE for the first layer and chemically modified polyolefins as the tie layer has improved the overall quality of these coatings. Flame spraying of thermoplastics and other materials is being developed to provide another method of applying these coatings to field joints, bends, and other components.

C4. Tape Coatings

Several types of hand- and plant-applied tapes are used for corrosion protection on underground or submersed metals. Though some systems were developed earlier, most were developed in the late 1950s and early 1960s. Most have been used on buried piping systems, but some have been used for other applications. Tapes are normally applied over a primed surface, but some companies are promoting a "primerless" tape that does not require a primer. Polyethylen-backed tape systems have been the most widely used type of tapes in the pipeline industry. Other types of tapes are made from coal tar with special fibers, petrolatum, polyvinyl chloride (PVC), or polyolefin fiber mesh with compound. The proper selection and use of tape product depend greatly on the environment (e.g., soil conditions), size of pipe, and operating temperature.

FIGURE 68.6. Damage to extruded polyethylene coat with adhesive only. Also note the movement of the coating from the field joint area. (Photo furnished by Lone Star Corrosion Services, Lancaster, TX.)

Polyethylene systems were developed from electrical and general use industrial tape applications. These tapes normally consist of one layer of polyethylene backing and an adhesive layer bonded to the polyethylene backing. The polyethylene backing layer in most cases is a low-density or blend of high- and low-density polyethylene. Adhesive compounds usually consist of elastomer butyl rubber, natural rubber, rubberized bitumen, or coal tar derivatives. Processing oils, fillers, tackifiers, and stabilizers are added to provide adhesion, shear resistance, and thermal and chemical resistance. Pigments and stabilizers are added to the polyethylene to provide color, UV resistance, and thermal stability and to improve aging. Polyethylene tapes come in several thicknesses of polyethylene and adhesive compounds. These tapes are usually made by one of three different methods: extrusion, coextrusion, or calendering. A polyethylene film is formed in a continuous sheet and the adhesive is laminated to it as the system is produced in large rolls. These rolls are then cut into the required widths and lengths.

Some tapes use a woven polyolefin geotextile fabric for the backing. This product has several advantages over the solid polyethylene backed tapes that will be discussed later. These tapes use the same type of compounds for adhesion to the metal. The woven fabric provides the mechanical protection and the compound provides the corrosion protection in these coating types.

Another type of tape is composed of coal tar base coating material supported on a fabric of organic or inorganic fibers. The fabric is covered on both sides by the coating materials. These tapes must be pliable enough to unwind from the roll during application. They are applied to the structure by heating the tape (usually with an open flame) on the structure side surface until the tape becomes liquid. The tape is then wrapped or applied to the primed surface.

Many tape systems require that an outer wrap be placed over the tape product for buried service. These outer wraps may help to alleviate soil stresses on the coating and provide additional mechanical protection or insulation. In all these cases, outer wraps should be used that do not shield cathodic protection from the metal if the coating is damaged.

C4.1. Pluses. Tape coatings are used in the field for repair, replacement, or base coatings. Most tape products can be quickly back-filled or immersed after application. Ease of application is another advantage of most tape products. Tapes can be applied by hand or with a tape machine. Even though there are a variety of tape types and prices, tape products usually cost less per square foot to apply than other coating products. Plant-applied systems are fast and handle easily. When properly selected for the environment and properly applied, tape coatings can provide an economical, easily applied coating system.

Tape systems can be applied in several layers to provide a thick coating when needed for certain environments. Tape coatings are normally very flexible and can be formed to fit many irregular shapes. Tape coatings normally have high electrical resistance (or low conductance) and therefore low cathodic protection requirements, especially for the polyethylene backed tapes.

C4.2. Minuses. Tape systems have been widely used for coating pipelines and other structures, but there have been major failures with the solid film backed tape products. Soil stress, poor surface preparation, and poor application techniques have been reasons for most tape coating failures. Other major reasons for tape failures have been the use of an incorrect primer, not using primer when required, or not allowing the primer to cure properly before tape application. There have been cases of bacterial attack on some butyl rubber compounds used for the adhesive. At this time, tapes have a very limited use on structures that operate at temperatures >65°C (150°F).

Soil stress effects on geotextile backed tape is much less than on most polyethylene backed tapes. The geotextile fabric has approximately a 20% stretch compared to the polyethylene backings, which can have up to a 600% stretch. When soil compresses around the pipe and forces downward on each side of a pipe, the force can cause the tape to wrinkle. Electrical resistance of polyethylene coatings and their susceptibility for unbonded installation create a serious problem on pipelines [14]. The electrical shielding comes primarily from the polyethylene backing. Even with the compound present, if water were to penetrate between the polyethylene backing and the metal, corrosion can occur because cathodic protection currents are shielded from the metal. On the woven fabric backed tape, soil can compress the compound and possibly move it to expose the metal surface, but the woven fabric will not shield cathodic protection currents.

C4.3. Improvements. Recent improvements with tape coatings involve the use of materials that will function effectively at higher temperatures. The success and technology of the fiber mesh backed tapes provide a superior product because their stretch is minimal, and therefore, soil stress is not as significant a problem. The woven mesh backing will not shield cathodic protection currents even if the compound is compressed and the metal is exposed. One company has developed a mesh backed tape that is nonshielding to cathodic protection should there be a disbondment and water penetrate between the coating and the pipe. After 20 years of use this coating system has had no reports of corrosion under the coating as with other types of coatings, such as solid film backed tape and shrink sleeves, even when improperly applied [14a].

C5. Shrink Sleeves

Shrink sleeve coatings are normally made from cross-linked heat-shrinkable thermoplastic backing (usually polyethylene) that serves as a tough permanent outer layer. Because of radiation cross linking these materials have an elastic memory that allows the product to be supplied in an expanded state. When heated, the material shrinks. Similar technology is used in the electrical connector industry. Special adhesives, from soft sealants to highly crystalline hot melts, are applied (similar to tapes) to the backing material. This technology was developed in the early 1960s and continues today.

These products were introduced to compete with cold-applied tapes. Early shrink sleeve applications had serious adhesion problems. Later versions have major improvements in adhesion and application procedure. Even with recent improvements, application procedures for shrink sleeves are very critical to their performance. During inspection, poor adhesion at the interface dual epoxy/sleeve has been detected mainly because of the insignificant amount of heat applied to the sleeve [15].

There are basically two types of shrink sleeve on the market. One is a sleeve that is installed over a pipe end and slid into place. This type requires the sleeve to be installed before making connections. The other type is a wrap-around sleeve with a closure strip that is normally preattached to the sleeve for easy application. The closure has an adhesive to help hold the sleeve in place until it is shrunk to the structure. These sleeves have the advantage that they can be applied on any structure (normally pipes) even after construction.

C5.1. Pluses. Shrink sleeves are quick and relatively easy to use for field joint and repair coatings. They provide a tough, durable coating when properly selected and applied. Shrink sleeves are compatible with a variety of plant-applied coating systems. They work best when used on new construction or systems where the metal surface can be heated to the desirable application temperature before the shrink sleeve is applied.

C5.2. Minuses. As mentioned above, shrink sleeves must be properly applied. Even though the process appears to be easy, instructions must be followed. The most frequent mistakes involve not heating the metal surface properly before application or not using the proper heat to apply the sleeve. Metal on in-service pipelines that are operating at low temperatures or structures in cold weather regions may not allow for proper metal heating before applying the sleeve. If the structure is too cold, the adhesive next to the metal may not melt enough to allow proper adhesion, even though the sleeve will shrink and appear to be properly installed. Improper heating usually occurs because the wrong type of torch is used or certain areas are heated either too much or not enough. Overheating can cause the sleeve to overshrink and

possibly cause the sleeve to crack or split. If not heated sufficiently, the sleeve may not properly bond to the metal. Shrink sleeves can have the same problems with soil stress, cathodic shielding, and temperatures >65°C (150°F) as polyolefin backed tapes.

C5.3. Improvements. Shrink sleeves are being developed for use at high temperatures, which will make them easier to apply and less affected by soil stress. The use of infrared heaters for applying shrink sleeves has improved the process by allowing for more uniform heating of the sleeve and pipe.

C6. Wax Based Coatings

Waxes have been successfully used for coating materials on a variety of structures. Petrolatum tape systems are made from a combination of saturated petrolatum impregnated with a neutral compound covering a synthetic fabric carrier for strength. Another tape uses pure wax in combination with synthetic fibers. As with most tapes, the proper primer must be used. Wax based tapes are easily formed to fit irregular shapes. Hot applied wax is normally composed of microcrystalline wax, fillers, and sometimes wetting agents and corrosion inhibitors. These coatings are applied by first melting the wax at temperatures in a range of 94–260°C (200–500°F) and flood coating the structure. An outer wrap is normally applied over the wax to provide mechanical protection to the pliable wax material.

Another coating is a hand-applied coal tar wax mastic that typically consists of microcrystalline or petrolatum compounds. After hand application, the mastic is usually covered with a protective wrap to help prevent mechanical damage.

C6.1. Pluses. Wax coatings are easy to apply and work well on irregular surfaces. When selected and applied properly, these coatings can provide excellent corrosion protection.

C6.2. Minuses. Wax based coatings are soft and sensitive to high temperatures. As with any hot-applied coating, there are safety concerns.

C7. Asphalt

Asphalt coatings were derived from petroleum and have been used similarly to coal tar coatings on a variety of structures. Asphalt coatings were cheaper than coal tar coatings, but they did not have the same chemical resistance and bendability and were not effective at higher temperatures. Application was similar to that of coal tar. Asphalt mastics are used on a variety of irregular structures. They are normally hand applied, and, if allowed to cure properly, provide a good coating system for certain environments.

Asphalt coatings are not widely used at this time for underground or submersed metal corrosion protection.

C8. Liquid Coatings

Liquid coatings for underground or submerged structures are very diverse but must be selected with caution. Thus a coating has worked well above ground or in certain types of submersion service does not mean this same coating will perform well in conjunction with cathodic protection when buried or submersed. Extensive testing should be performed to ensure that the coating will provide the required protection in that environment.

These coatings can be applied in plants or at construction sites to a variety of structures, valves, and other components that are not easily coated by other coating methods. Many of these coatings are applied in the field for replacement or repair coatings. The application methods include, but are not limited to, brushing, air spray, airless spray, and plural component spraying. Since the list is extensive, only the most popular and most promising will be discussed.

C8.1. Epoxies. Coal tar epoxies, as mentioned above, have been used for pipeline, tank, and structure coatings for many years. They have been fairly successful coating but are not flexible, easily crack, and are affected by temperatures over 65°C (150°F).

There are several types of 100% solids epoxy coatings formulated for burial or submersed use in conjunction with cathodic protection. These are excellent coatings when properly applied and used in the correct environment. These coatings are fast-cure, high-build materials with good adhesion. Epoxies of 100% solids are environmentally safe and do not shield cathodic protection to the extent that thicker coatings do. Some of these epoxies are formulated to perform well on systems operating at temperatures up to 100°C (212°F). Epoxies must be mixed in the correct ratio for the coating to cure properly.

C9. Urethanes and Urethane Blends

Urethane and urethane blends are becoming more popular as coatings for burial or submersed structures. These coatings have been successfully used over FBE coatings to provide mechanical protection during construction, in rocky areas, and for road and river crossings. The bond to the metal is not as strong as some other liquid coatings, but improvements are being made. These are very fast curing, 100% solids that are normally applied with plural component spray equipment. Urethanes used for underground or submersion service are normally adversely affected by moisture or high humidity.

C10. Epoxy Phenolic

One coating that has been used in other industries for years is modified epoxy phenolic [16]. Epoxy phenolics have been successfully used in some applications, such as the internals of tanks and other systems that operate at temperatures up to 121°C (250°F) with some success. Cathodic protection may damage some of these coating applications at these temperatures. More study and field experience are required at this time.

C11. Polyurea

Polyurea coatings are new to the underground and submersion coating industry. At this time, most of these are modified polyurea systems. They are very rapid cure even at temperatures less than -29°C (-20°F) but have high thermal stability, even at temperatures greater than 150°C (300°F). Polyureas are moisture insensitive and have high abrasion resistance. These systems are normally applied using plural component, high-temperature/high-pressure impingement mix application equipment. At this time, polyureas have not performed well in cathodic disbondment and moisture permeation testing, but these coatings do hold promise if the adhesion to the metal surface can be improved.

C12. Esters

Esters are normally used as polyester or vinyl esters. They are used in areas where high temperature is a concern. Esters must be applied to a very well prepared surface with a very good anchor pattern to have adequate adhesion and good cathodic disbondment and moisture permeation results. This material is normally applied in two coats with relatively short recoat time intervals. Mixing must be monitored closely since large component mixing ratios (up to 64 : 1) are normally used.

D. TESTS TO EVALUATE UNDERGROUND OR SUBMERSION COATINGS

There are many different standard tests for evaluating and comparing the various coatings mentioned above. The tests most often used and some of the various organizations that provide the standards are given in this section. Frequently, these standard tests are modified to provide test information for particular environments or conditions. Short-term laboratory testing is limited in providing longterm performance information but is very valuable in providing information about differences between various types of coatings. The longer the test, the more meaningful the information. Short-term tests are performed at coating plants during production to determine if the application process is providing the specified product.

Sample preparation for each type of test is very important. When possible, samples should be taken from the actual production or field application. When this is not possible, the application process should match the production or field application as much as possible. One should not rely totally on one set of test results to select a coating system, because all

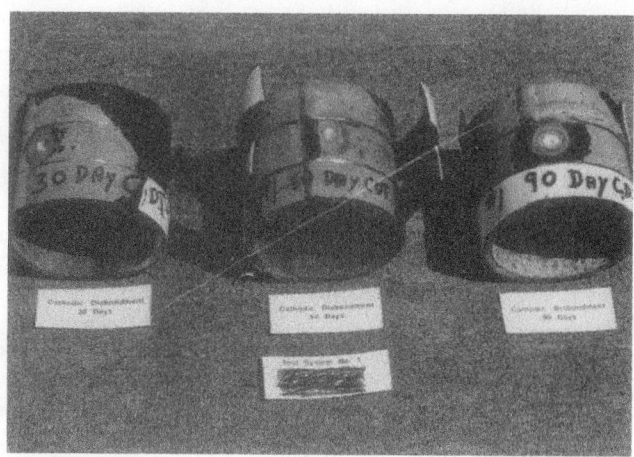

FIGURE 68.7. Cathodic disbondment test of a tape coating system. This particular system failed the test (Photo furnished by ITI Anti-Corrosion, Inc., Houston, TX.)

the performance parameters of a coating system cannot be obtained using only one type of test. One must consider all the various conditions of application and service to select the test criteria for a particular coating system.

Tests normally performed for all underground or submersion coatings:

Cathodic disbondment test (CDT) as shown in Figure 68.7
Moisture permeation (sometimes called hot water soak or adhesion testing)
Resistance to impact
Flexibility (sometime called bend test)
Chemical resistance
Adhesion (strength of bond of the coating to the metal)

Fusion bonded epoxies are additionally tested for:

Porosity (through film and interface)
Surface contamination (amount of contamination on the back side of the coating surface)
Thermal characteristics by differential scanning calorimetry (DSC)

Tape coatings are additionally tested for:

Adhesion or peel strength
Shear, stretch, or soil stress characteristics

One should be familiar with the various organizations that publish standard procedures for testing coatings. Some of these are:

NACE International
ASTM (American Society for Testing and Materials)

CSA (Canadian Standards Association)
SSPC (Steel Structures Paint Council)
API (American Petroleum Institute)

The most valuable information is that taken from inspections performed on in-service coatings. Any time a coating is exposed after being in service for a length of time, the coating system should be inspected and tested using many of the same tests mentioned above. One should inspect for blisters, disbondment, adhesion strength, discoloration, and any other adverse effects to the coating.

Results of cathodic disbandment tests are presented in Table 68.1.

When a coating is used in conjunction with cathodic protection, the pH should be checked under any blisters or disbondments. This will provide information on the effectiveness of the CP system to provide protection in these areas. This is discussed in more depth in the next section.

E. SHIELDING VERSUS NONSHIELDING COATINGS

When using coating with cathodic protection, it is critical for the end user to consider what potential problems could exist if the coating system fails and electrolyte penetrates between the coating and the metal being protected. Shielding of the cathodic protection current by disbonded external coatings on pipelines causes more external corrosion than any other process in today's pipeline industry. The resulting corrosion cells cannot be protected by the cathodic protection system, so corrosion will develop, as illustrated in Figure 68.8. There have been many articles written about the problems of external corrosion caused from disbonded and shielding pipeline coatings [17–23].

When selecting a pipeline coating, the nonshielding characteristics of the coating system may be more important than other issues normally considered. To adequately protect underground pipelines, a coating must conduct CP current when disbondment occurs [24]. Nonshielding means if the coating system adhesion fails and water penetrates corrosion on the metal is significantly reduced or eliminated when adequate CP is available, Fusion-bonded epoxy is a nonshielding coating (Fig. 68.9). Corrosion and stress corrosion cracking (SCC) have not been an issue under disbonded FBE in over 40 years of service or the mesh backed tape after over 20 years of service (Fig. 68.10). Other coatings may also provide this property but should be studied and tested to ensure this property. This has been proven by the use of the various in–line inspection (ILI) tools that are used to find corrosion and other defects in pipelines.

With the use of the external corrosion direct assement (ECDA) methods now being used, companies can evaluate

TABLE 68.1.   Cathodic Disbondment Test Results Comparing Tape Coatings and Shrink Sleeves[a]

System	30 days	60 days	90 days
1. Two-layer tape/primer	41 mm	45 mm	58 mm
2. Shrink sleeve/two-part primer	25 mm	>65 mm	>65 mm
3. Shrink sleeve/no primer	0 mm	5.6 mm	0 mm
4. Shrink sleeve/two-part primer	>50 mm	>50 mm	>50 mm
5. Tape/primer	Complete failure	Complete failure	Complete failure
6. Tape/primer	>50 mm	>50 mm	>50 mm
7. Shrink sleeve/two-part primer	Complete failure	Complete failure	Complete failure
8. Tape/two-part primer	8.3 mm	6.3 mm	8.9 mm
9. Shrink sleeve/two-part primer	Complete failure	Complete failure	Complete failure
10. Shrink sleeve/primer	Complete failure	Complete failure	Complete failure
11. Shrink sleeve/primer	Complete failure	Complete failure	Complete failure
12. Mesh-backed tape	0 mm	0 mm	0 mm
13. Two-layer tape/primer	10.7 mm	12.3 mm	12.5 mm
14. Two-layer tape/primer	38.3 mm	>50 mm	>50 mm
15. Two-layer tape/primer	>75 mm	>75 mm	>75 mm

[a]Provided by Lone Star Corrosion Services, Ennis, TX.

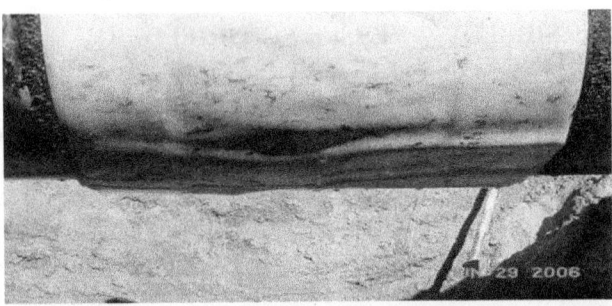

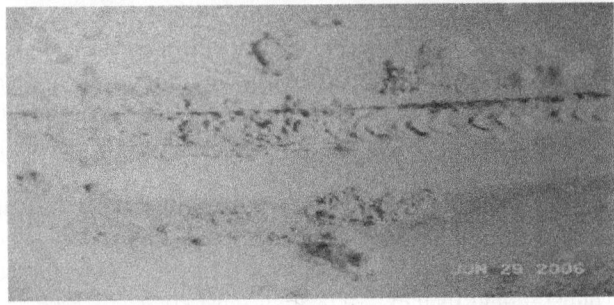

FIGURE 68.8. Shrink sleeve applied in 1997 and resulting corrosion found in 2006. Shrink sleeve shielded the CP. Pipe potential met all NACE criteria.

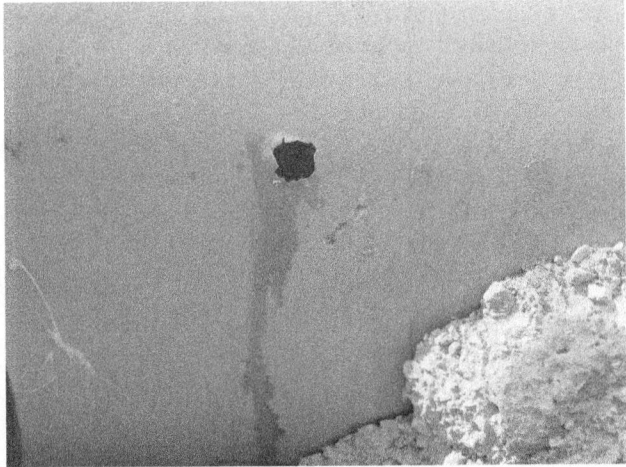

FIGURE 68.9. Water under blisters on FBE coated pipe used for gas transmission in central United States. Water under the blisters had a pH of 12 with no metal loss showing nonshielding properties of FBE. (Photo furnished by Lone Star Corrosion Services, Ennis, TX.)

the condition of the external coatings. When doing the ECDA diagnosis of any failed coating and potential corrosion, one of the most critical evaluation methods is to check the pH of any water under the coating. The electrochemical process of cathodic protection will cause an alkaline area where the current enters the metal. External corrosion may occur for other reasons, such as AC or DC interference or shielding from other materials that may shield the CP. The cause of external corrosion must be properly evaluated to distinguish between shielding and other causes.

If the coating system is nonshielding to the CP current, the pH will typically be 9 or above, indicating that CP is being effective even under the disbonded coating. At this pH level, corrosion on steel surfaces will be limited or nonexistent since steel is protected at these levels of pH. The high pH also indicates adequate CP is available under the coating. Disbonded coatings that are shielding CP current will typically have a pH of 7 or less. Many times significant corrosion will develop under these coatings because CP cannot be effective under these coating systems, yet the CP potentials taken above ground will meet or exceed the various CP criteria.

FIGURE 68.10. Mesh backed tape applied to pipe with condensation on one side. After three years in service, adhesion was very good on the dry side, but there was no adhesion on wet side, but the pH of the water under the coating was 11 with no corrosion on the pipe, proving the nonshielding properties of this coating system. (Photo provided by Polyguard Products, Inc., Ennis, TX.)

F. CONCLUSIONS

There may be other selection criteria that are important for a particular situation, but the ones given in this chapter are always important. Testing of materials and judgment of experienced individuals will help to determine the correct coating for a particular structure and situation. Clear, precise specifications are very important. Specifications should be continually updated to reflect changes in the industry. It is essential to have well-trained and qualified inspectors who have a real passion for ensuring the best coating possible under the circumstances.

If coatings are plant applied, preproduction meetings, coating, and testing help to ensure that the plant understands the specifications and can produce coated material as specified. For field-applied coatings, well-trained applicators should be employed. Once again, preproduction meetings, coating, and testing are advised. Company employees who apply coatings should be trained for each particular coating type to be used.

Many factors must be considered when selecting a coating for a particular system. Short-term testing will help to determine which coatings will perform best in certain conditions. Past experience with a particular coating system used in the same environment should always be considered. Good coating performance in one environment does not mean the same coating will perform well in a different environment. Once selected, the coating must be properly applied to a clean, well-prepared surface. The coated product must then be handled, transported, and stored properly during

the construction phase. After installation, the parameters of the coating, such as temperature and cathodic protection limits, should not be exceeded.

Well-written coating specifications and inspection in the field or plant by inspectors who are well trained and passionate about what they are doing will help to ensure the best possible coating system under the circumstances.

When selecting a pipeline coating, the shielding and nonshielding characteristics of the coating system may be more important than other issues normally considered. Most of the external corrosion on pipelines today is caused from coating that have disbanded (for a variety of reasons), allowed water to penetrate between the coating and the pipe, and then shield the CP current allowing corrosions to develop. More CP does not solve this problem; only recoating the pipe properly will solve the problem.

REFERENCES

1. A. C. Coates, "Pipeline Recoating—A Cover Up Story," Pipeline Digest, Apr. 1991, p. 11.

2. Good Painting Practice, Steel Structure Painting Manual, Vol. 1, 3rd ed., Steel Structure Painting Council, 1993, p. 19,

3. R. Norsworthy, "Select Effective Pipeline Coating," Hart's Pipeline Digest **34**(4), 17 (Feb. 1997).

4. Corrosion Basics, An Introduction, National Association of Corrosion Engineers, NACE International, Houston, TX, 1984, p. 213.

5. G. B. Byrnes, Mater. Perform., Sep. 1989.

6. Control of Pipeline Corrosion, National Association of Corrosion Engineers, (NACE International), Houston, TX, 1967, p. 18.

7. J. L. Banach, "Evaluation Design and Cost of Pipe Line Coatings," Pipe Line Ind., Mar. 1998, p. 62.

8. D. A. Diakow, G. J. Van Bovan, and M. J. Wilmott, Mater. Perform., **37**(5), 17, (1998).

8a. R. Norsworthy, "Is Your Pipeline Coating 'Fail Safe," Pipeline Gas J., **233**(10) 62 (Oct. 2006).

9. T. A. Pfaff, "FBE Serve A Broad Market," Hart's Pipeline Digest, Oct. 1996, p. 20.

10. D. Neal, "Fusion-Bonded Epoxy Coating: Aging and Below-Ground Performance," Hart's Pipeline Digest, **34**(13), 20 (Sep. 1997).

11. D. Neal, Pipeline Gas Ind., **81**(3), 43 (1998).

12. S. J. Lukezich, J. R. Hancock, and B. C. Yen, "State-of-the-Art for the Use of Anti-Corrosion Coatings on Buried Pipelines in the Natural Gas Industry," Topical Report, Document No. 92/0004, Gas Research Institute, Chicago, Apr. 1992.

13. R. Norsworthy, "High Temperature Pipeline Coating Using Polypropylene Over Fusion Bonded Epoxy," in International Pipeline Conference, Vol. 1, ASME, New York, 1996.

14. C. D. Tracy, Pipeline Gas Ind., **80**(2), 27 (1997).

14a. C. Hughes and R. Norsworthy, "Proven Protection," World Pipelines, Oct. 2007, p. 49.

15. V. Rodriquez, E. Perozo, and E. Alvarez, Mater. Perform., **37**(2), 44 (1998).

16. R. Norsworthy, Pipeline Gas J., **225**(3), 44 (1998).

17. J. A. Beavers and N. G. Thomson, "Corrosion Beneath Disbonded Pipeline Coatings," Mater. Perform. Apr. 1997, p. 19.

18. "Coatings Used in Conjunction with Cathodic Protection," NACE International, Technical Committee Report, Item No. 24207, Houston, TX, 2000, p. 2.

19. S. Papavinasam, M. Attard, and R. W. Revie, "External Polymeric Pipeline Coating Failure Modes," Mater. Perform., Oct. 2006, p. 28.

20. F. M. Song, D. W. Kirk, D. E. Cormack and D. Wong, "Barrier Properties of Two Field Applied Pipeline Coatings," Mater. Perform., Apr. 2005, p. 26.

21. J. A. Beaversand N. G. Thompson, "Corrosion Beneath Disbonded Pipeline Coatings," Mater. Perform., Apr. 1997, p. 13.

22. G. R. Ruschau and Y. Chen, "Determining the CP Shielding Behavior of Pipeline Coatings in the Laboratory," Corrosion 2006, Paper No. 06043, NACE International, Houston, TX, 2006.

23. J. Alan Kehr, Fusion Bonded Epoxy (FBE)—A Foundation for Pipeline Corrosion Protection, NACE International, Houston, TX, 2003.

24. D. P. Moore, "Cathodic Shielding Can Be a Major Problem After a Coating Fails," Mater. Perform., **39**(4), 44 (2000).

69

ENGINEERING OF CATHODIC PROTECTION SYSTEMS

J. H. Fitzgerald III

Grosse Pointe Park, Michigan

A. INTRODUCTION

This chapter presents the practical application of cathodic protection. Various uses are shown along with the principles of design for galvanic anode and impressed current systems. This chapter also presents recommended steps and formulas as well as overall information that will guide the engineer in the choice and design of the most appropriate cathodic protection for the structures involved. Standards and other publications containing information on various facilities are referenced.

Engineering of cathodic protection involves not only the calculation of current requirements, resistances, and voltages but also an understanding of the type and configuration of cathodic protection to be used. The designer must consider its

Uhlig's Corrosion Handbook, Third Edition, Edited by R. Winston Revie
Copyright © 2011 John Wiley & Sons, Inc.

practicability, life-cycle costs, and maintenance and operational requirements.

B. OPERATION OF CATHODIC PROTECTION

Essentially, cathodic protection involves the application of a direct current (dc) from an anode through the electrolyte to the surface to be protected. This is often thought of as overcoming the corrosion currents that exist on the structure. That is not really what happens as there is no flow of electrical current (electrons) through the electrolyte. There is, of course, a flow of ionic current in the electrolyte as explained in earlier chapters on electrochemistry.

Cathodic protection eliminates the potential differences between the anodes and cathodes on the corroding surface. A potential difference is then created between the cathodic protection anode and the structure such that the cathodic protection anode is of a more negative potential than any point on the structure surface. Thus, the structure becomes the cathode of a new corrosion cell. The cathodic protection anode is allowed to corrode; the structure, being the cathode, does not corrode.

C. USES OF CATHODIC PROTECTION

Cathodic protection is widely used to protect many structures. Among these are underground structures, on-grade tank bottoms, marine facilities and ship hulls, and water storage tanks. It can also be used for stray-current corrosion control. See Kumar [1] and [2] and [3, Chapter 6–8] for further information.

Effective cathodic protection may or may not require electrical isolation of the structure. Where the protected structure is not electrically isolated, protective current requirements are increased as some current will be lost to interconnected facilities. Many structures such as gas distribution piping, transmission lines, and small piping systems can usually be easily isolated. Other facilities such as large tank farms, wharves, foundation or sheet piling, and complex piping networks often cannot be isolated economically or with any degree of certainty.

The designer must consider the feasibility of isolation in the overall current requirements and in the layout of cathodic protection. The presence or absence of isolation may also affect the type of cathodic protection that the designer chooses.

D. TYPES OF CATHODIC PROTECTION

There are two types of cathodic protection: galvanic anode and impressed current. Typical uses and selection recommendations are discussed later.

D1. Galvanic Anodes

Galvanic anode protection is often called "sacrificial" because the anode is thought of as "sacrificing" itself to protect the structure. This type of protection utilizes a galvanic cell consisting of an anode made from a more active metal than the structure. The anode is attached to the structure, either directly or, to permit measurement of the anode output current, through a test station.

Magnesium and zinc are the most common galvanic anodes for underground use. In salt water, zinc anodes and aluminum alloy anodes are commonly used. In freshwater, magnesium is frequently used.

For underground use, magnesium anodes are packaged in a backfill consisting of 75% gypsum, 20% bentonite, and 5% sodium sulfate. The purpose of the backfill is to absorb products of corrosion and to absorb water from the soil to keep the anodes active. Magnesium and zinc are also available in ribbons and extruded rods.

Galvanic anodes require no external power. The protective current comes from the electrochemical cell created by the connection of the anode material to the more noble or electrically positive metal of the structure.

D2. Impressed Current

Impressed current protection provides dc from a power source. The current is delivered to anodes made of a material having a very low or essentially inert dissolution rate. The anodes serve simply to introduce the protective current into the electrolyte.

D2.1. Power Sources. The most common power source for impressed current protection is the transformer rectifier. This unit, commonly called simply a rectifier, reduces incoming alternating current (ac) voltage and rectifies it to dc. There are also solid-state "switchmode" rectifiers that perform similar functions without the use of transformers. Rectifiers can be provided with constant voltage, constant current, or structure-to-electrolyte potential control.

In areas where electrical power is not readily available, solar power– and wind-driven generators coupled with storage batteries are used, There is also some use of thermoelectric cells, in-line turbine generators (in gas or oil pipelines), and internal combustion engine–driven generators.

D2.2. Anodes. A variety of materials are used for impressed current anodes, Among the oldest are high silicon, chromium-bearing cast iron, graphite, and junk steel. Magnetite and lead–silver anodes are also used, with lead–silver being confined to use in seawater.

Among newer materials are "dimensionally stable anodes," so-called because the anode itself consists of a deposit on an inert substrate. This deposit may be consumed, but the anode shape tends to remain stable. Included in this category are platinized niobium or titanium and mixed-metal oxide (MMO)/titanium anodes.

Underground impressed current anodes are usually back-filled in a carbonaceous material such as metallurgical or calcined petroleum coke. The purpose of the backfill is to increase the effective size of the anode, thus reducing its resistance to earth, and also to provide a uniform environment around the anode, increasing its life. Anode life is extended by the large coke backfill column since the current is discharged from the coke column as opposed to being discharged from only the anode. Another advantage to increasing the size of the anode is that the resultant reduction in anode current density reduces acidity in the vicinity of the anode.

E. GALVANIC ANODE DESIGN

E1. Typical Uses

Galvanic anode systems are typically used where protective current requirements are relatively low, usually in the range of several hundred milliamperes to perhaps 4 or 5 A. Offshore structures, having a current requirement of many hundreds of amperes can also be protected by large galvanic anodes weighing anywhere from 135 to 635 kg.

Common places to use galvanic anode protection include well-coated, electrically isolated structures, offshore structures, ship hulls, hot-spot pipeline protection, heat exchanger water boxes, and in environments of resistivity below $\sim$10,000 Ω cm, although they can be used in much higher resistivities, especially if coating and isolation conditions are favorable.

E2. Advantages and Limitations

There are advantages and limitations to galvanic and impressed current systems. The designer needs to assess the engineering and economic aspects of each making the type of protection system to use:

Advantages of galvanic systems
- No external power required
- Little maintenance
- Relatively easy installation
- Little chance of cathodic interference
- Less inspection and recordkeeping than with impressed current systems

Limitations of galvanic systems
- Lack of adjustment without resistors in anode circuits
- Limited current output

- Possible high replacement costs
- Need for good coating
- Need for electrical isolation of protected structure

E3. Design Process

The following will serve as a guide to the designer for the steps to follow during the design process. The example is based on an underground pipeline. Later in this chapter the reader is referred to publications and standards that will provide information on other structures. Refer also to [5], Chapter 6] for additional information on galvanic anode design.

E3.1. Design Parameters. To begin the design, establish the electrolyte resistivity, the protective current requirement, the desired life of the anodes, whether or not the structure is electrically isolated, whether or not there are any stray-current concerns, and the physical configuration desired for the anode.

E3.2. Current Requirement. There are two ways of establishing current requirement. If the structure is in place, one can test it as shown in Figure 69.1. In this procedure, a test current is applied to the structure and the resultant change in structure-to-electrolyte potential is measured. From the data, the current requirement can be calculated. Details on current requirement testing are given by Peabody [4] and in [5, Chapter 5].

In new construction, the current requirement is often calculated from estimates such as those in Table 69.1. The amount of bare steel depends on the assumed quality of the coating, both to begin with and after several years of operation. A coating efficiency, for example, of 95%, equals 5% bare metal.

The current requirement is then the total area of the pipe times the percent bare (as a decimal) times the current per square meter, or

$$I_{req} = A \times \%bare \times mA/m^2 \qquad (69.1)$$

Where I_{req} is the total, current requirement in milliamperes and A is the total area of the structure in square meters.

E3.3. Anode Selection. Table 69.2 lists the characteristics of several types of galvanic anodes. There are two types of magnesium anode alloys, standard (H-1) alloy and high-potential alloy. Generally speaking, high-potential anodes are desirable if the electrolyte resistivity exceeds 8000 Ω cm. There are two grades of zinc anodes, one for seawater use and one for underground use. There are also two grades of aluminum alloy anodes for saltwater use.

If zinc anodes are chosen, it is important that they meet the purity required by the American Society for Testing

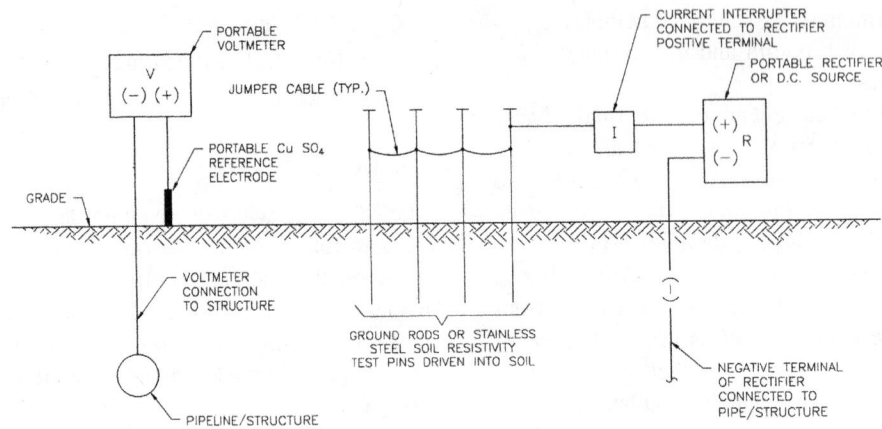

FIGURE 69.1. Typical current requirement test setup. (From Advanced Course Text, Appalachian Underground Corrosion Short Course [5]. Used by permission.)

TABLE 69.1. Typical Current Requirement for Cathodic Protection of Bare Steel[a]

Environment	(mA/m²)
Neutral soil	4.5–16.0
Well-aerated neutral soil	21.5–32.0
Highly acid soil	32.0–160.0
Soil supporting sulfate-reducing bacteria	65.0–450.0
Heated soil	32.0–270.0
Stationary freshwater	11.0–65.0
Moving, oxygenated freshwater	54.0–160.0
Seawater[b]	32.0–110.0

[a]Data from Air Force Manual 88–9, Corrosion Control, Chapter 4, p. 203, and [1, Table A-1].
[b]May be as high as 160.0–430.0 in cold and Arctic waters.

and Materials (ASTM) standard B-418 Type I (MILL-Spec A-18001) for seawater use and ASTM B-418 Type II for underground use, Lower purity anodes will not function properly. Zinc anodes are generally limited to environments of resistivity below 1500 Ω cm because of their low driving voltage. Zinc anodes are used successfully in higher resistivity soils, however, on some very well coated, electrically isolated facilities such as underground storage tanks.

E3.4. Anode Requirement. This step involves calculation of the resistance of the cathodic protection circuit, the potential difference between the anode and the structure and, from these numbers, the anode output.

TABLE 69.2. Galvanic Anode Characteristic[a]

Material	Theoretical Output (A-h/kg)	Actual Output (A-h/kg)	Efficiency	Consumption Rate (kg/A-year)	Potential to CDE[b]
Zinc					
type I	860	781	90%	11	1.06
type II	816	739	90%	12%	1.10
Magnesium					
H-1 alloy	2205	551–1279	25–58%	6.8–16	1.40–1.60
Magnesium					
High					
Potential	2205	992–1191		7.3–8.6	1.70–1.80
Al/Zn/Hg	2977	2822	95%	3.1	1.06
Al/Zn/In	2977	2591	87%	3.3	1.11

[a]Data adapted from Advanced Course Text. Chapter 6. Appalachian Underground Corrosion Short Course, West Virginia University, Morgantown, WV, 2008, and from [3, p. 139].
[b]Copper–copper sulfate electrode.

E3.4.1. Cathodic Protection Circuit Resistance. The resistance of a vertical anode to ground can be calculated from the following equation (after H. B. Dwight [6]), based on the dimensions of the anode package (Eq. (69.2) [4, p. 134] and on accompanying computer disks):

$$R_v = \frac{\rho}{2\pi L}\left(\ln\frac{8L}{d} - 1\right) \quad (69.2)$$

where R_v is the resistance to earth in ohms, ρ is the soil resistivity in ohm centimeters, L is the anode length in centimeters, and d is the anode diameter in centimeters.

If the designer decides to use horizontal anodes, the resistance to earth is different from the vertical anode and is given by (see [6] and [1, pp. 1–8])

$$R_h = \frac{\rho}{2\pi L}\left(\ln\left[\frac{4L^2 + 4L\sqrt{(2h)^2 + L^2}}{2dh}\right] + \frac{2h}{d} - \frac{\sqrt{(2h)^2 + L^2}}{L} - 1\right) \quad (69.3)$$

where R_h is resistance to earth in ohms, ρ is the soil resistivity in ohm centimeters, L is anode length in centimeters, d is anode diameter in centimeters, and h is depth below the surface to the center of the anode in centimeters.

The designer will find anode dimensions given in catalogs prepared by anode manufacturers. The packaged dimensions are used in the calculation.

The anode is conected to an underground structure with a lead wire. The resistance of the wire (R_w) is very small but should be considered. Wire and cable data appear in Table 69.3.

The resistance of the structure (R_s) to ground mayor may not be significant. In many cases, this resistance is small enough to be omitted from the calculation. Peabody [4, p. 135] provides additional information on calculating structure to ground resistance.

The total cathodic protection circuit resistance (R_t in ohms) then is

$$R_t = R_a + R_w + R_s \quad (69.4)$$

where R_a is the resistance of the anode to ground from Eqs. (69.2), (69.3) or (69.9), R_w is the wire resistance, and R_s is the structure-to-ground resistance.

E3.4.2. Anode Output. The anode output depends on the anode circuit resistance and the potential difference between the anode and the structure (the driving potential). The potential difference between the anode and the structure, ΔV_g in volts, is the difference between the open-circuit potential of the anode (V_{ga} vs.) and the desired polarized

TABLE 69.3. Copper Cable Data[a]

Size AWG	dc Resistance 20°C Ω/1000 m	Max. dc Current Capacity (A)
14	8.4650	15
12	5.3152	20
10	3.3466	30
8	2.0998	45
6	1.3222	65
4	0.8334	85
3	0.6595	100
2	0.5217	115
1	0.4134	130
1/0	0.3281	150
2/0	0.2608	175
3/0	0.2070	200
4/0	0.1641	230
250 MCM	0.1388	255

[a]Data adapted from Advanced Course Text, Chapter 5, Appalachian Underground Corrosion Short Course, West Virginia University, Morgan town, WV 2008.

potential of the Structure (V_{gs} vs.), or

$$\Delta V_g = V_{ga} - V_{gs} \quad (69.5)$$

The current output, I in milliamperes, is then calculated from Ohm's law:

$$I = \frac{\Delta V_g}{R_t} \times 1000 \quad (69.6)$$

E3.4.3. Number of Anodes Required. This is calculated from the current requirement divided by the individual anode output:

$$\text{Anodes required} = \frac{\text{current requirement}}{\text{anode output}} \quad (69.7)$$

Anodes are usually spaced evenly along or around the structure. It is also good practice to place anodes near isolating fittings, building walls, or other locations where an inadvertent, although likely high-resistance, contact to another structure might occur.

E3.5. Anode Life. Calculate the life of the anode from [5],

$$L = \frac{\text{Th} \times W \times E \times \text{UF}}{h \times I} \quad (69.8)$$

where L is life in years, Th is the theoretical A·h/kg output (Table 69.2), W is the anode weight in kilograms, E is current efficiency (Table 69.2), UF is the utilization factor, h is hours per year (8766), and I is anode output in amperes. The utilization factor is usually chosen as 0.85 (85%); this means that once the anode is 85% consumed, its resistance to earth

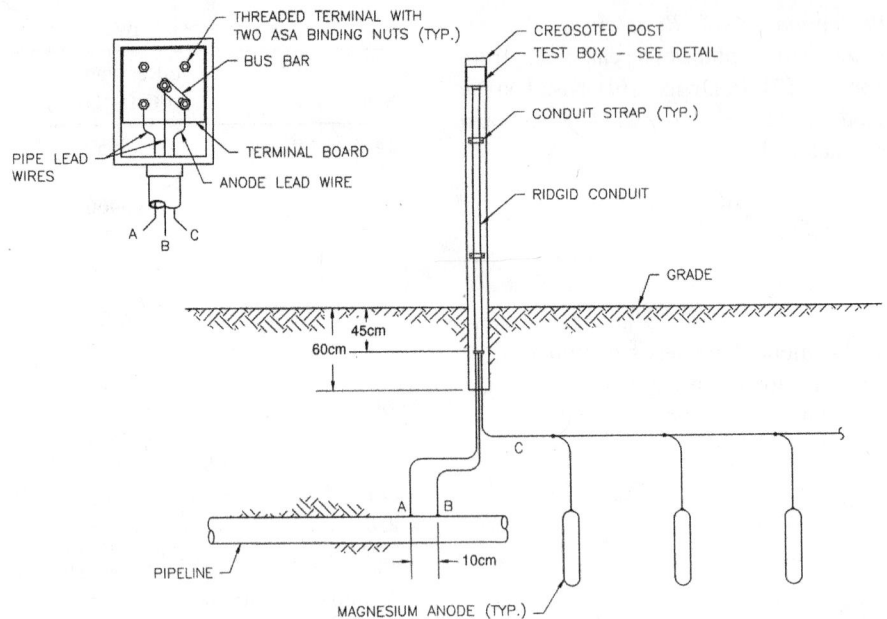

FIGURE 69.2. Typical galvanic anode system installation. (Courtesy of Advanced Course Text, Appalachian Underground Corrosion Short Course [5]. Used by permission.

begins to increase to the point that its output is reduced significantly.

If the calculated life is insufficient, then the designer needs to choose a heavier anode or perhaps use more anodes than the design calculations require. Also, Eq. (69.8) can be solved for total required anode weight for a given life. The total weight can then be divided the individual anode weight to obtain the required number of anodes.

E4. Grouped Anodes

There are times when the designer may desire to group several anodes together in a bank. The anodes may then be connected to the structure through a test station. This is particularly useful when replacing anodes to minimize the number of excavations required.

When placed in groups, the individual anode outputs are reduced due to a mutual interference resistance proportional to the anode spacing. This resistance decreases as anode spacing increase. The resistance to earth of a group of vertical anodes may be calculated from the Sunde equation (see [4, p.134] and [7]):

$$R_n = \frac{\rho}{2pNL}\left(\ln\frac{8L}{d} - 1 + \frac{2L}{S} \times \ln 0.656N\right) \qquad (69.9)$$

where R_n is resistance of the anodes to earth in ohms, ρ is soil resistivity in ohm centimeters, L is anode length in centimeters, d is anode diameter in centimeters, N is number of anodes, and S is center-to-center spacing of anodes in centimeters.

E5. Anode Installation

Figure 69.2 shows a typical installation. It is good practice to use test stations wherever possible. Test stations are necessary to evaluate the effectiveness of cathodic protection on underground structures; placing at least some anodes at test stations permits measurement of the output current and "instant off" polarized potentials at that location. Knowledge of anode current aids in evaluating anode performance and in calculating eventual anode life. Figure 69.3 shows anodes on a distribution pipeline.

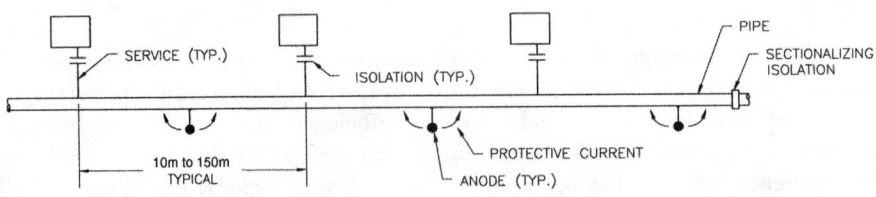

FIGURE 69.3. Typical galvanic anode cathodic protection.

F. IMPRESSED CURRENT DESIGN

F1. Typical Uses

Impressed current cathodic protection has wide application. It is especially applicable when current requirements are large, in some cases as high as 500 A or more. Then, too, a small impressed current system putting out less than an ampere might be used to replace a dissipated galvanic anode system.

Common uses of impressed current include long transmission pipelines, complex underground structures, pilings, marine structures and ship hulls, replacement for dissipated galvanic systems, large condenser water boxes, reinforcing steel in concrete, bare or poorly coated structures, unisolated structures, and water storage tank interiors.

F2. Advantages and Limitations

As discussed in Section E, there are advantages and limitations to galvanic and impressed current systems. The designer needs to assess the engineering and economic aspects of each in making the choice of the type of protection system to use:

Advantages of impressed current systems
- Adjustable output
- Large current available
- Applicable to poorly coated or bare structures
- Applicable to nonisolated structures
- Low-cost method of replacing spent galvanic anodes

Limitations of impressed current systems
- Constant power required
- More maintenance, inspections, and recordkeeping than with galvanic systems, especially for regulated tanks or pipelines
- Experienced electrical personnel may be needed for installation
- Possibility of cathodic interference

F3. Design Process

The following will serve as a guide to the designer for the steps to follow during the design process. This example is based on an underground pipeline. Later in this chapter the reader is referred to publications and standards that will provide specific information on other structures. Refer also to [5, Chapter 5] for additional information on impressed current design.

F3.1. Design Parameters.
To begin the design, establish the electrolyte resistivity, the protective current required, the desired life of the groundbed (anodes), whether or not the structure is electrically isolated, and the physical configuration desired for the groundbed.

F3.2. Current Requirement.
The current requirement can be determined by testing if the structure is in place as explained under galvanic anode protection and as shown in Figure 69.1 and further described by Peabody [4].

In new construction, the current requirement is often calculated from estimates such as those in Table 69.1. The amount of bare steel depends on the assumed quality of the coating both to begin with and after several years of operation. A coating efficiency, for example, of 95% equals 5% bare metal.

As described in Section E3.2 for the current requirement in a galvanic anode system, the current requirement for an impressed current system is also calculated by multiplying the total area of the pipe, the percent bare (as a decimal), and the current per square meter, or

$$I_{reg} = A \times \% \text{ bare} \times mA/m^2 \qquad (69.1)$$

where I_{req} is the total current requirement in milliamperes and A is the total area of the structure in square meters.

F3.3. Anode Selection.
The designer has a variety of anodes from which to choose and with experience will learn which anodes perform best for the situation at hand. High silicon, chromium-bearing cast iron, and graphite have similar characteristics. Dissipation rate varies with the environment, but 0.5 kg/A-year is typical; steel dissipates at 9.1 kg/A-year. Kumar et al. [1] present data on MMO anodes. For characteristics of other anode materials, refer to the manufacturer's literature.

F3.4. Anode Requirements and Life.
In determine the number of anodes required for an impressed current groundbed, it is best to start with the calculation of anode life. While the dissipation rate of high SiCrFe and graphite anodes is actually ~0.5 kg/A-year or less, it is common practice to use 1.0 kg/A-year to allow for a safety factor (similar to the utilization factor used in galvanic anode design). For other anode materials, the designer should follow the dissipation information given by the manufacture. The total anode weight (W_t) for the desired life then is

$$W_t = D_r \times I_{req} \times L \qquad (69.10)$$

where W_t is the total weight required in kilograms, D_r is the dissipation rate in kg/A-year, I_{req} is current requirement in amperes, and L is life in years.

The number of anodes required is simply the total weight required divided by the individual anode weight.

F3.5. Groundbed Resistance

F3.5.1. Anode-to-Ground Resistance.   Since multiple anodes are usually involved, Eq. (69.9) is used to calculate the resistance of the anode bed to earth (for vertical anodes), as discussed in Section E4 for groups of anodes in a galvanic system.

The dimensions of the coke column are used for L and d in the equation.

F3.5.2. Cable Resistance.   Figure 69.4 shows a typical surface point groundbed layout. There are three cable resistances to consider:

Negative cable to the pipeline
Positive cable to the groundbed
Positive cable in the anode groundbed

Since the current flow through the anode portion of the groundbed cable drops as each anode is encountered, the effective resistance of the groundbed cable is usually taken as one-half of its length. Thus the total effective cable resistance in ohms becomes

$$R_c = R_{(-)} + R_{(+)} + R_{gb} \qquad (69.11)$$

where R_c is the total cable resistance, $R_{(-)}$ is the negative cable resistance, $R_{(+)}$ is the positive cable resistance, and R_{gb} is one-half of the resistance of the total length of the anode portion of the groundbed. Cable resistances are given in Table 69.3.

F3.5.3. Structure-to-Ground Resistance.   As discussed in Section E, the pipe-to-ground resistance is frequently negligible. Peabody [4, p. 135] provides additional information on calculating structure-to-ground resistance.

F3.5.4. Total Circuit Resistance.   The total circuit resistance in ohms (R_t) is the sum of the anode to ground (R_a), cable (R_c), and structure-to-ground (R_s) resistances, or

$$R_t = R_a + R_c + R_s \qquad (69.12)$$

F3.6. Rectifier Selection.   Required driving voltage (E) is determined by Ohm's law:

$$E = \frac{I_{req}}{R_t} \qquad (69.13)$$

The designer should allow for some increase in current requirement over the years; 25% is a reasonable figure, but the designer should also rely on experience. It is prudent also to allow for some circuit resistance increase in the future, so the required rectifier voltage is usually multiplied by 1.5. The designer then chooses a commercially available unit meeting the design requirement.

F4. Other Groundbed Configurations

A distributed anode system, shown in Figure 69.5, consists of anodes spaced at intervals along the structure to be protected. Typical applications include pipelines, sheet pilings, large tank farms, and other complex networks.

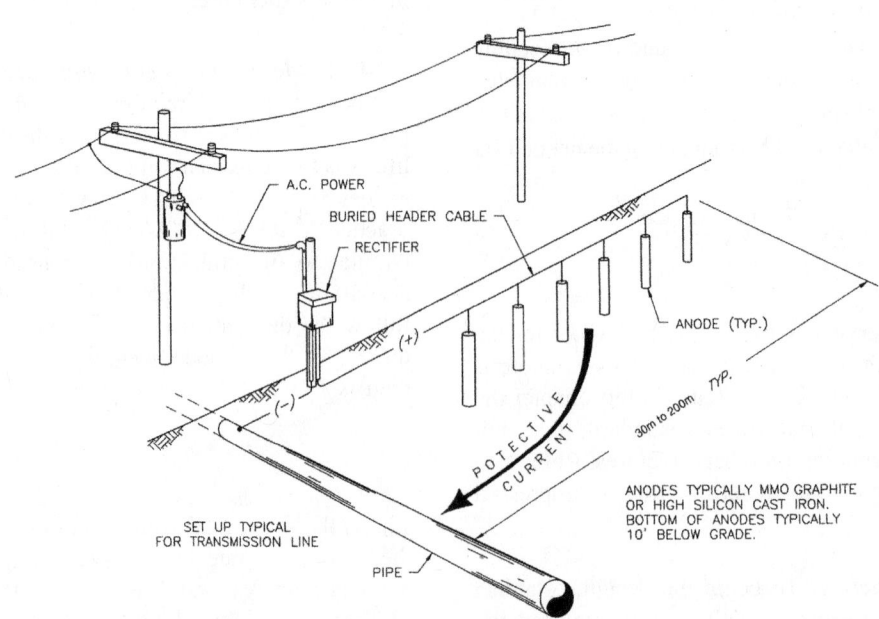

FIGURE 69.4. Impressed current cathodic protection using surface point groundbed.

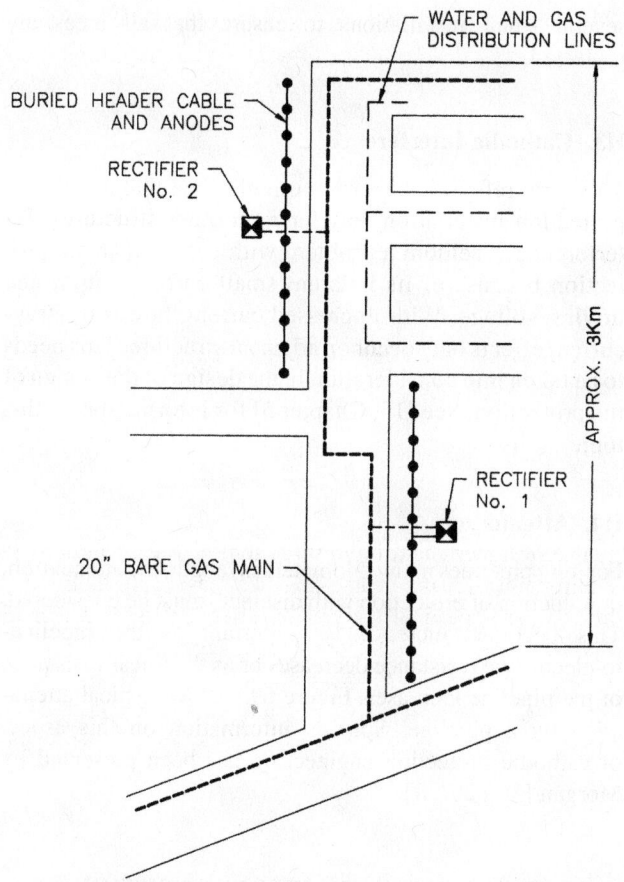

FIGURE 69.5. Impressed current cathodic protection using distributed groundbed.

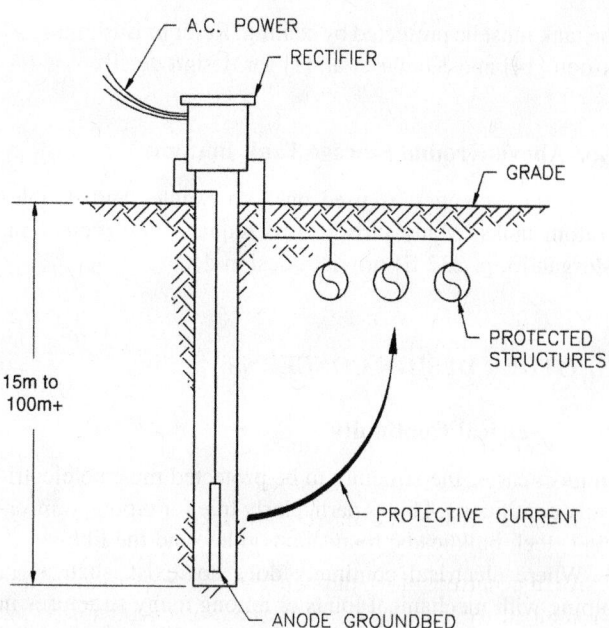

FIGURE 69.6. Impressed current cathodic protection using a deep anode groundbed.

Figure 69.6 shows a deep anode bed. This installation utilizes anodes placed in a coke column vertically in the earth. Typical depths are 60–90 m. The resistance to earth is based on the dimensions of the coke breeze column and may be calculated from Eq. (69.2). Lewis [8] presents detailed information on the design and use of deep anode groundbeds.

G. DESIGN OF OTHER CATHODIC PROTECTION SYSTEMS

Not all uses of cathodic protection can be explained in this chapter. The following sections provide references for various applications.

G1. Marine Structures

Design of cathodic protection for marine structures in both fresh and salt water require special techniques. Galvanic systems usually employ zinc or aluminum alloy anodes. Impressed current systems frequently use high silicon, chromium-bearing iron, platinized niobium, or mixed-metal oxide/titanium anodes.

The structure being protected affects the design. Stationary facilities such as bulkheads and support piles require different techniques from ship hulls. Morgan [3, Chapter 7] discusses this in detail.

G2. Heat Exchangers

Cathodic protection is often used to overcome corrosion in heat exchanger water boxes. It is especially useful in controlling galvanic corrosion between steel tube sheets and copper or other noble metal tubes such as titanium or stainless steel. Morgan [3, p. 398 ff] and Lane [9] present information on this application.

G3. Steel in Concrete

Bridge decks, parking structures, and other reinforced concrete structures lend themselves to cathodic protection, particularly when the concrete is contaminated with chlorides from deicing salts or a marine environment. Special techniques are required as described by Rog and Swiat [10], NACE [11, 12] and Morgan [3, p. 242 ff].

Prestressed concrete pipelines occasionally require cathodic protection. Protection must be done carefully to avoid damage to the prestressing wire from hydrogen embrittlement or stress corrosion cracking (SCC). Helpful advice is found in [13].

G4. Water Storage Tanks

The wetted surfaces of water storage tank interiors can easily be protected by cathodic protection. Above the water surface, the tank must be protected by coating. Refer to Bushman and Kroon [14] and Kumar et al. [1] for design details.

G5. Above-Ground Storage Tank Bottoms

Cathodic protection is used on both Single- and double-bottom tanks. Many jurisdictions require such protection. Morgan [3, p. 232 ff] provides design details.

H. OTHER DESIGN CONCERNS

H1. Electrical Continuity

In most cases, the structure to be protected must be electrically continuous. This is particularly true for piping, reinforcing steel, bulkheads, foundations piles, and the like.

Where electrical continuity does not exist, such as in piping with mechanical joints or among many structures in a complex network, joint bonds, consisting of insulated copper cable, are necessary to ensure that all of the structures to be protected are electrically continuous.

There are times where electrical continuity is not required. One example is the installation of galvanic anodes on existing ductile iron pipe where anodes can be installed on individual lengths of pipe.

H2. Shielding

In complex situations such as tank farms, industrial plants, and other large underground and underwater structures, it is important to lay out the anodes so that all structures receive protection. This may require a distributed anode system or several small installations to ensure that all areas are protected.

H3. Cathodic Interference

This term refers to the stray-current effect that a cathodic protection installation may have on other structures. Interference is seldom a problem with galvanic anode protection because of its inherent small current output and driving voltage. With impressed current, however, stray-current effects may occur on adjacent structure. This needs to be taken into consideration in the design of the design of the protection. See [15, Chapter 5] for information on this topic.

H4. Attenuation

For long pipelines many kilometers in length, the attenuation, or reduction of protection with distance, must be considered. This becomes increasingly important as the pipeline-to-electrolyte resistance decreases or as the linear resistance of the pipeline increases. Figure 69.7 shows typical attenuation on a pipeline. Specific information on this aspect of cathodic protection engineering has been presented by Morgan [3, p.201 ff].

I. CRITERIA FOR CATHODIC PROTECTION

The designer must understand the applicable criterion and the proper test techniques. Various criteria for underground and submerged structures are listed below. The reader must study these criteria and associated testing techniques in detail.

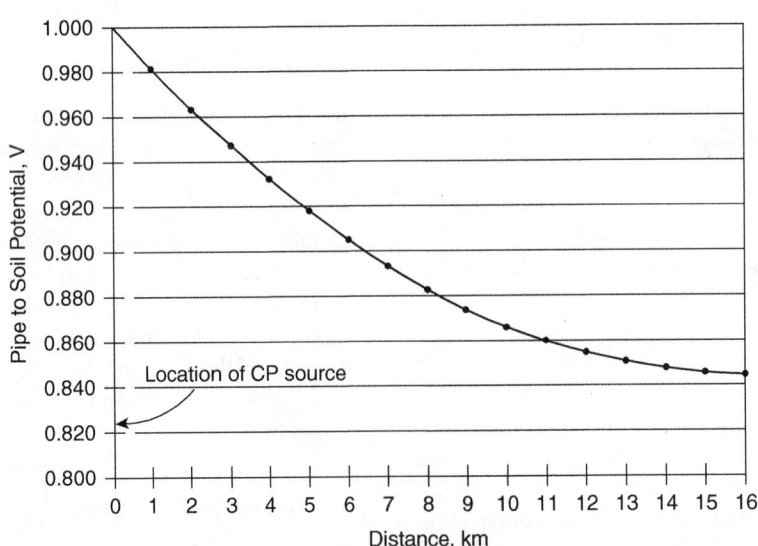

FIGURE 69.7. Typical attenuation on a pipeline.

See [16] regarding underground and submerged structures and Section 2 of [12] for steel in concrete.

I1. Steel and Cast Iron

A negative potential of at least 850 mV to a saturated copper–copper sulfate reference electrode (CSE) with the cathodic protection current applied. Voltage drops other than those across the structure to electrolyte boundary must be considered.

A negative polarized potential of at least 850 mV with respect to CSE.

A minimum of 100 mV of cathodic polarization between the structure and a stable reference electrode. This criterion also applies to steel in concrete.

I2. Aluminum

A minimum of 100 mV of cathodic polarization between the structure and a stable reference electrode. Precautions must be taken to prevent overprotection of aluminum.

I3. Copper

A minimum of 100 mV of cathodic polarization between the structure and a stable reference electrode.

REFERENCES

1. A. Kumar, J. B. Bushman, J. H. Fitzgerald, A. E. Brown, and T. M. Kelly, Impressed Current Cathodic Protection Systems Utilizing Ceramic Anodes, U.S. Army Corps of Engineers Construction Engineering Research Laboratories, Champaign, IL, 1990.

2. Basic Course, Text, Appalachian Underground Corrosion Short Course, West Virginia University, Morgantown. WV, 2008.

3. J. H. Morgan, Cathodic Protection, 2nd ed., NACE International, Houston, TX, 1993.

4. A. W. Peabody, Control of Pipeline Corrosion, 2nd ed., NACE International, Houston, TX, 2001.

5. Advanced Course Text, Appalachian Underground Corrosion Short Course, West Virginia University, Morgantown, WV, 2008.

6. H. B. Dwight, "Calculation of Resistance to Ground," Electric. Eng., **55**(12), 1319 (1936).

7. E. D. Sunde, Earth Conduction Effects in Transmission Systems, D. Van Nostrand, New York, 1949.

8. T. H. Lewis, Deep Anode Systems, NACE International, Houston, TX, 2008.

9. R. W. Lane, Control of Scale and Corrosion in Building Water Systems, McGraw-Hill, New York, 1993.

10. J. W. Rog and W. J. Swiat, "Guidelines for Selection of Cathodic Protection Systems for Reinforced Concrete," paper presented at Corrosion 87, San Francisco, NACE International, Houston, TX, 1987.

11. V. Chaker (Ed.), Corrosion Forms and Control for Infrastructure, NACE International, Houston, TX, 1992.

12. Standard Practice SP02-90-2007, "Impressed Current Cathodic Protection of Reinforcing Steel in Atmospherically Exposed Concrete Structures," NACE International, Houston, TX, 2007.

13. M. Szeliga (Ed.), Corrosion in Prestressed Concrete: Pipes, Piles and Decks, NACE International, Houston, TX, 1995.

14. J. B. Busman and D. H. Kroon, J. AWWA, Jan. 1984.

15. Intermediate Course Text, Appalachian Underground Corrosion Short Course, West Virginia University, Morgantown, WV, 2004.

16. Standard Practice SP01-69-2007, "Control of External Corrosion on Underground or Submerged Metallic Piping Systems," Section 6, NACE International, Houston, TX, 2007.

70

STRAY-CURRENT ANALYSIS*

J. H. Fɪᴛᴢɢᴇʀᴀʟᴅ III
Grosse Pointe Park, Michigan

A. INTRODUCTION

The causes and common means of detecting and mitigating stray-current interference effects from direct-current (dc) sources are reviewed in this chapter. Alternating current (ac), which can create potential safety hazards has also been shown to contribute to corrosion of ferrous structures. Alternating ac stray current is discussed in Section E.

Stray currents are defined as electrical currents flowing through electrical paths other than the intended paths. Stray currents, or interference currents, are classified as either static or dynamic.

Static interference currents are those that maintain constant amplitude and constant paths. Examples of typical sources are railroad signal batteries, high-voltage direct-current (HVDC)

* Adapted with permission from the Appalachian Underground Corrosion Short Course [1, 2].

ground electrodes, and cathodic protection system rectifiers. Dynamic interference currents are those that continually vary in amplitude, magnitude, and electrolytic paths. These currents can be man made (e.g., dc welding equipment, dc railway systems, chloride plants, and aluminum plants) or caused by natural phenomena. Natural sources of dynamic stray currents, called tellurics, are caused by disturbances in the earth's magnetic field from sun spot activity. Telluric effects may contribute to corrosion and, in addition, can create measurement difficulties and interfere with the ability to assess cathodic protection system performance.

The lower the resistivity of the soil, the more severe the effects of stray currents may be, If there is a current flowing in the earth and a potential difference exists between points where a metallic conductor, such as a pipeline or cable, is located, then the conductor will readily acquire a part of the current that is flowing. Thus, metallic pipelines and cables can become conductors of stray currents in the earth environment.

B. DETECTION OF STRAY CURRENTS

Static stray currents on a pipeline can be detected by analyzing pipe-to-soil potentials. The graph in Figure 70.1 shows a pipeline with no interference. Figure 70.2 shows a potential plot for a coated pipeline with stray-current interference. Interference may be suspected if:

The voltage curve profile shows abnormal variation from previous survey graphs.

High negative values are noted remote from any cathodic protection system on the surveyed line.

Unusual current flows are measured along the pipeline.

Low negative or positive voltages are measured.

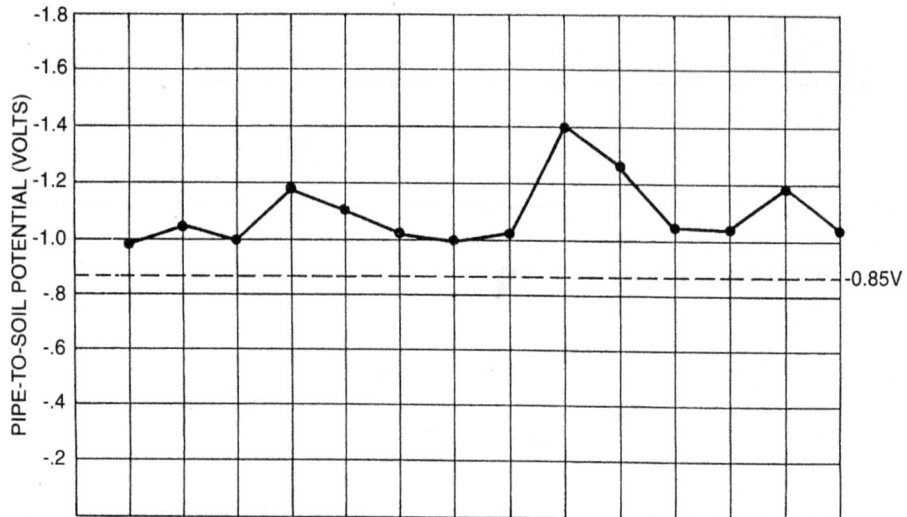

FIGURE 70.1. Potential versus distance plot, no interferance.

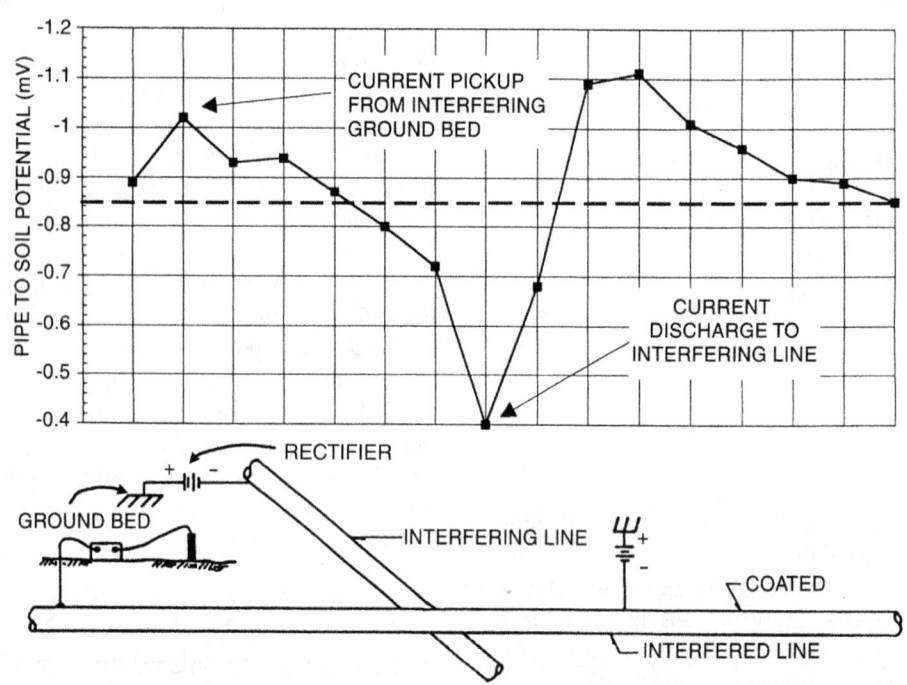

FIGURE 70.2. Potential plot with interference; coated pipeline.

In practice, if the positive shift does not cause a potential less negative than -0.850 vs. $Cu/CuSO_4$ (free of voltage IR drop caused by current flow through ohmic resistance), then negligible corrosion can be expected.

Dynamic stray currents are more easily detected than static ones. Dynamic stray currents are present if the structure-to-soil potential is continually fluctuating while the reference electrode is kept in a stationary position in contact with the soil. These potential changes are the result of current changes at the source of the interference.

C. LOCATING SOURCE OF INTERFERENCE

C1. Static Interference

The path of current flow in the earth can be tracked to its source by measuring the currents in the earth using two identical portable reference electrodes and a digital voltmeter. By measuring the potential difference between the two electrodes spaced about 8 m (25 ft) apart, as shown in Figure 70.3, the direction of current flow can be determined

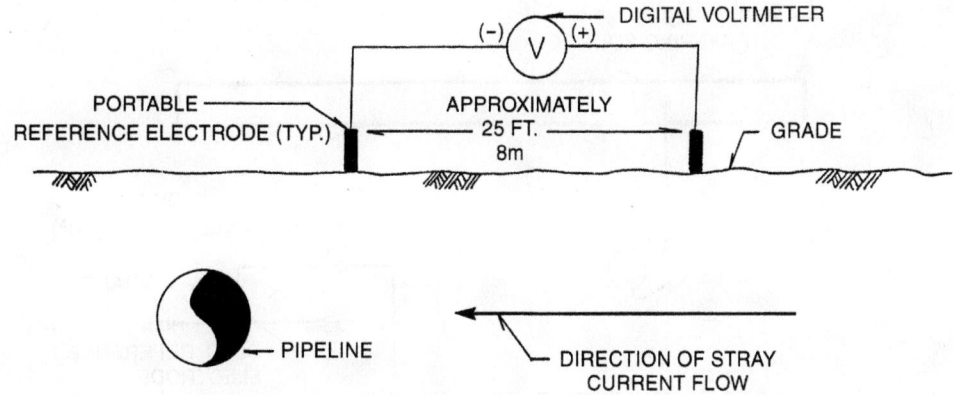

FIGURE 70.3. Tracing stray-current flow.

and its path traced. With the voltmeter connected as shown, positive potential readings would indicate current flow toward the pipeline. Negative potential readings would indicate that current is flowing away from the pipeline.

If there are test facilities for measuring current flow along the pipeline, a line current survey can be conducted to determine the areas of current pick-up and discharge. This information can also be used to track stray currents toward their source.

To confirm the cause of abnormal pipe-to-soil potential readings, the effect of interrupting the suspected current source on a potential survey along the protected pipeline or structure should be assessed. If there is no effect on the potential readings of the interfered structure, the search for the current source must continue until the actual interfering current source is located.

C2. Dynamic Stray Currents

After identifying possible sources, such as dc electrical railway systems, mines, or industrial plants, such as aluminum and chlorine, the current flow should be traced along the interfered structure to its source. One method is to observe the current flow at intervals along the structure using millivolt drop test station lead wires to determine the direction of current flow, as illustrated in Figure 70.4.

Assume that a situation exists where a single source is causing interference problems. A voltmeter connection, to be used for pipe-to-soil potential measurements, is made between the pipeline (interfered structure) and a reference electrode within the earth current pattern of the source and its load. Observing the fluctuating potential readings at this point alone would not enable one to determine if the readings are being taken at a point where the pipeline is picking up or

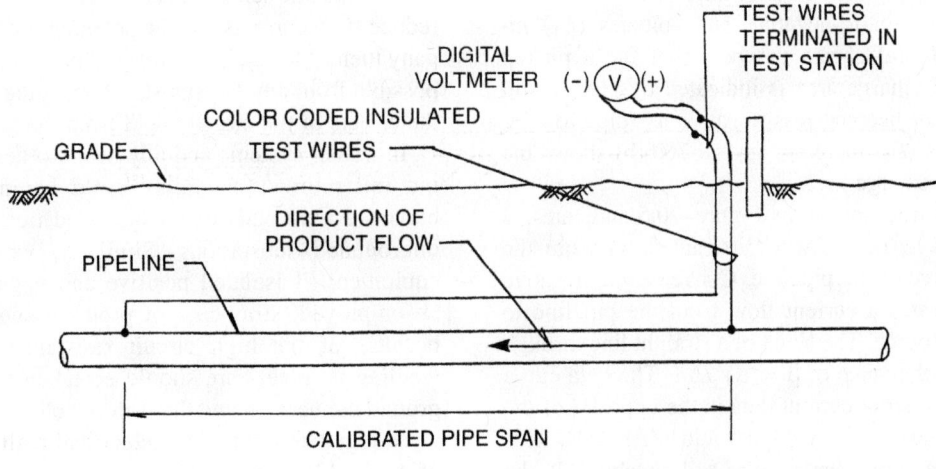

NOTE: POSITIVE POTENTIALS MEASURED WITH METER HOOK-UP AS SHOWN INDICATES CURRENT FLOW IN DIRECTION OF PRODUCT FLOW

FIGURE 70.4. Determining direction of line current flow.

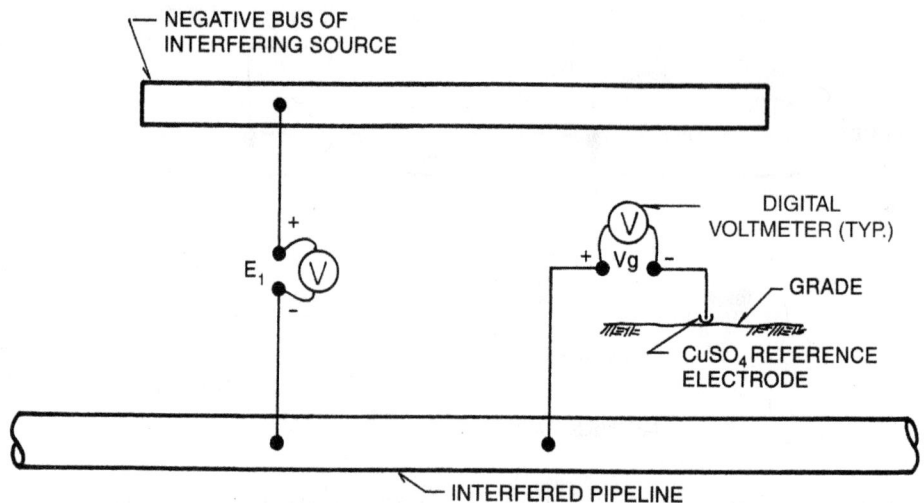

FIGURE 70.5. Typical test setup used to locate point of maximum exposure.

discharging current. If measurements are being taken at a point of current pickup, a negative potential swing would be indicated. A positive swing would indicate a decrease of current flow and a condition of the pipe returning to its steady-state condition. Readings observed at a discharge point swinging in the positive direction would indicate an increase in current leaving the line and a negative swing would indicate a decrease in current discharge.

Test locations can be identified as pick-up or discharge points by correlating the open-circuit voltage between the interfered pipeline and the stray-current source and the pipe-to-soil potential of the pipeline. A plot of these data is called a beta plot. Figure 70.5 illustrates a typical experimental setup to obtain a beta plot. A current pickup area is indicated by pipe-to-soil potentials (V_g) that become more negative as pipe-to-negative bus voltages (E_1) increase; Figure 70.6(a) shows the beta plot for a pickup area. A current discharge area is indicated by pipe-to-soil potentials (V_g) that become less negative as (pipe-to-negative bus voltages (E_1) increase; Figure 70.6(b) shows the beta plot for a discharge area.

As the connection shown in Figure 70.5 indicates, a positive value of E_1 occurs when current flows from the stray-current source to the pipeline. Conversely, a negative value of E_1 indicates a current flow from the pipeline to the stray-current source. The slope of a straight line through the data points is the value of $\beta = \Delta V_g / \Delta E$. The beta curve (slope) of dynamic stray-current data is the opposite of the geometric slope; the geometric slope would be $\Delta E_1 / \Delta V_g$. This type of testing requires that pipe-to-soil potential V_g be measured at two or more locations. Many readings should be taken at these locations simultaneously and the meters used must be identical or comparison of the sets of readings will be difficult. In most cases, a dual-channel recorder, such as an X–Y plotter, or a multichannel data logger is used.

If the points plotted form a vertical line, a neutral curve, there is no influence on V_g by the output fluctuations of the current source. The point of maximum exposure to stray currents can be determined from a plot of slopes of beta curves versus distance along the pipeline. The location of maximum discharge area slope (the line closest to the horizontal) is the location of maximum exposure.

D. MITIGATION OF STRAY-CURRENT CORROSION

D1. Controlling Stray Currents at Source

Groundbed site selection can be used to eliminate or greatly reduce stray currents and the potential gradients that accompany them. Ideally, groundbeds should be installed as far as possible from any foreign structure in the area to minimize the effects of the electric field from the groundbed.

In transit systems and any of the other systems involving rail returns, the rails should be installed on well-ballasted road beds or on insulated ties or padding with ungrounded substations. Similarly, when dealing with equipment, if isolated positive and negative circuits can be employed, stray-current problems will be minimized because of the high circuit resistance to earth. When welding is done, care should be taken to ensure that the ground connection and the welding electrode are relatively close together and that the electrical path between them is of negligible resistance.

D2. Static Stray Currents

Mitigation bonds are used to mitigate the effects of stray-current corrosion on a structure. The purpose of the miti-

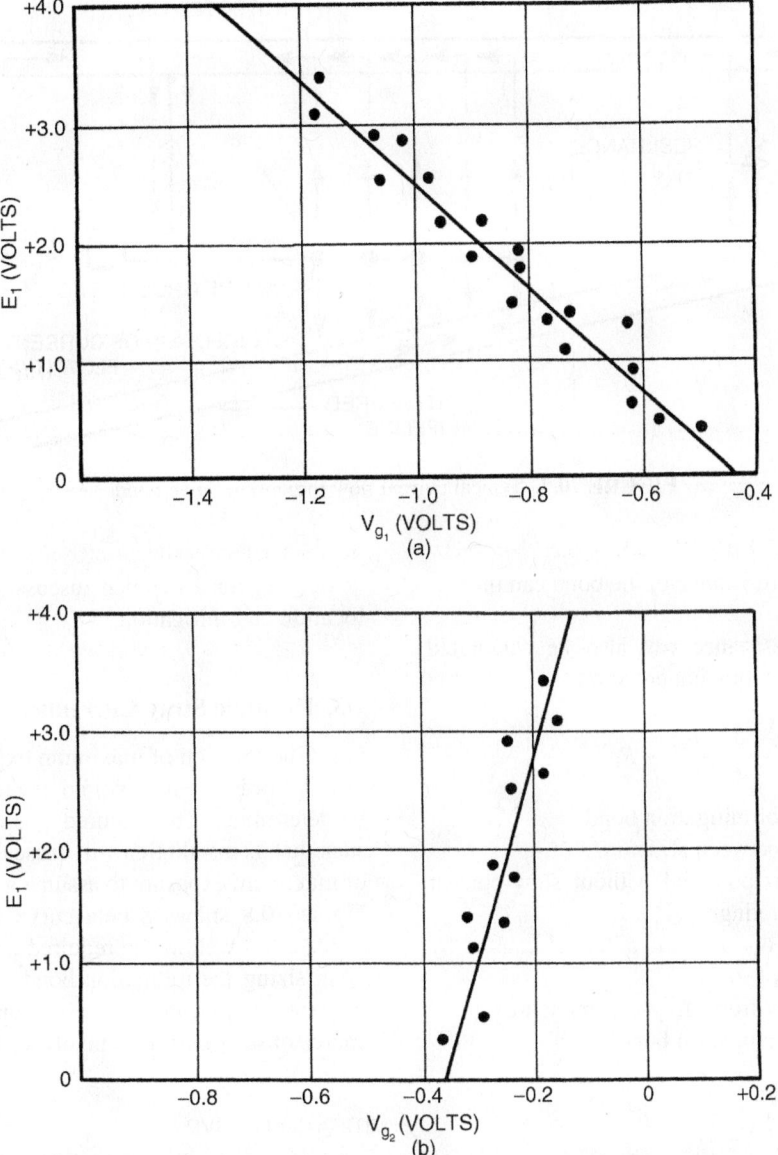

FIGURE 70.6. Typical beta curve: (a) pickup area; (b) discharge area.

gation bond is to eliminate current flow from a metallic structure into the earth by providing a metallic return path for the current. This bond allows the stray current flowing from a groundbed to the interfered structure to flow through the structure and back to the protected structure through the bond. The typical current flow when a drainage bond is installed is shown in Figure 70.7. Corrosion will occur only if the current flows from the metal surface into the earth.

To size the mitigation bond, the point of maximum current flow between the two affected structures must be located. Typically, this point is situated near the point of pipeline crossing, where the circuit resistances are the lowest, but it can be located some distance away, particularly

with well-coated pipelines at areas of coating failure or damage.

A "trial-and-error" method can be used to determine the correct bond resistance and current for a solution to static stray currents. A reference electrode is placed at the point of maximum stray-current exchange to monitor the potential-to-soil of the interfered structure. With the current source operating, a variable resistor is placed between the two pipelines. When the potential-to-soil of the interferred line with the current source operating and the bond installed equals the potential-to-soil of the interferred line with the current source deactivated and with the bond disconnected, the correct bond resistance is determined. The current flow

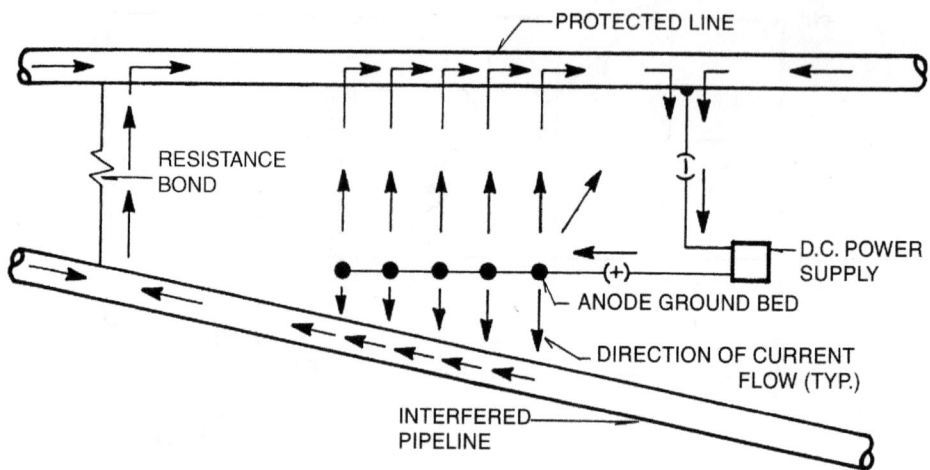

FIGURE 70.7. Typical current flow through resistive bond.

through the bond and the resistance of the bond can then be measured.

The required bond resistance can also be calculated mathematically using the following equation:

$$R_b = \frac{E_0 + \Delta E / I_{SC}}{I_b} - R_I$$

where R_b = resistance of mitigation bond
R_I = resistance between structure
E_0 = open-circuit potential without stray-current source operating
ΔE = change in open-circuit potential caused by stray-current source
I_{SC} = current flow from stray-current source
I_b = current in mitigation bond

See [1] for a detailed discussion of static stray-current location and mitigation.

D3. Dynamic Stray Currents

Once the location of maximum exposure is determined and its slope or beta curve plotted, the resistance of the bond can be determined. The required size of the resistance bond is such that its installation will cause the beta curve at the point of maximum exposure to assume a neutral or pick-up slope. Figure 70.8 shows a beta curve at a point of maximum exposure as well as the required mitigation curve.

In sizing the mitigation bond, a trial-and-error solution may be possible in relatively simple cases where a single source of stray current is involved. The size of the resistance

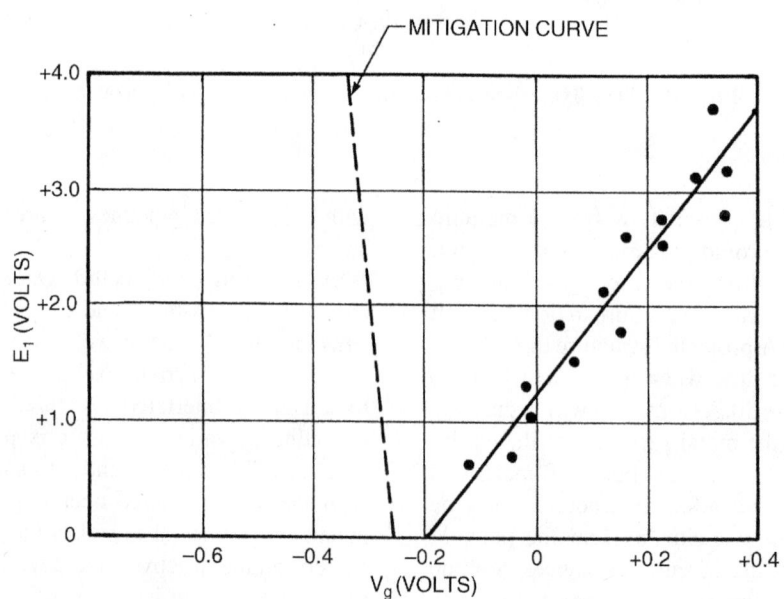

FIGURE 70.8. Beta curve—discharge area (mitigation current).

bond can be determined by installing a temporary variable resistance and determining when stray-current corrosion has been mitigated. The procedure is similar to that described for static stray currents, except that a mitigation curve, such as that shown in Figure 70.8, must be obtained.

Where a more complex interference problem exists, which precludes the use of the trial-and-error method, a mathematical method can be used. In this method, the following equation is used to obtain the required resistance of the mitigation bond:

$$R_b = \frac{\Delta V_g / I_b}{\beta} - R_I$$

where
R_b = resistance of mitigation bond, Ω
R_I = resistance between structures, Ω
I_b = bond current, A
β = beta slope
ΔV_g = change in pipe-to-soil potential caused by I_b

Using values determined experimentally by field measurements, R_b, the resistance of the bond, is calculated. For example, if

$$\Delta V_g / I_b = 0.00169 \text{ V/A} \qquad \beta = 0.017 \qquad R_I = 0.070 \, \Omega$$

then

$$R_b = 0.0294 \, \Omega$$

The bond resistance value, $0.0294 \, \Omega$ for this example, can be a simple cable or a combination of cable and variable or fixed resistors. The bond must also be sized to permit the maximum current to flow and remain in the current range for the bond. In a dynamic stray-current situation, the maximum current can be calculated once the maximum open-circuit voltage between the structures is known. This value is usually obtained by measuring the park value over several operating cycles of the stray-current source. For most stray-current sources, the typical cycle is 24 h.

For example, if the maximum value of open-circuit voltage E_1 (Fig. 70.5) is 12.0 V, the value of the maximum stray current through the bond can be calculated as follows:

$$E_1 = I_b(R_I + R_b) \qquad I_b = 120.9 \text{ A}$$

See [2] for detailed analysis and calculations.

In many instances of stray current where rail transit systems are involved, there may be locations where drainage bonds are required in areas where reversals of potential could occur. Therefore, it is often necessary to install an electrolysis (reverse-current) switch or silicon diode into the circuit to prevent current flow from the substation back onto the pipeline through the bond. The resistance to the forward flow of current created by these devices must be included in the sizing of the bond.

D4. Galvanic Anodes

Galvanic, or sacrificial, anodes may be used to mitigate stray-current effects in situations where small current flows or small voltage gradients exist. In effect, a potential gradient produced by the galvanic anode(s) counteracts the interference current, with a resulting net current flow to the interfered-with structure. Because galvanic anodes produce limited voltages, they can overcome only limited stray-current voltages. In addition, resistance of galvanic anodes to ground increases as the anodes are consumed during their lifetime, leading to reduced current flow from the anode and decreased voltage gradients. Galvanic anodes should be sized to provide a sufficient anticipated life span and should be monitored carefully.

Galvanic anode drains are commonly used in lieu of bonds where small drain currents are involved.

D5. Impressed Current Systems

When the magnitude of stray currents is beyond the ability of galvanic anodes to counteract, impressed current systems can sometimes be utilized. Impressed current systems have much higher voltage capacities than galvanic anodes and a greater life per kilogram of anode material. Built-in control and monitoring circuits in the impressed current rectifier can be used to adjust the protective current output based on the voltage-to-earth fluctuations of the interfered-with pipeline.

The design of galvanic or impressed current mitigation systems is done by the trial-and-error method. Simulated systems are placed in the field and the results measured. Based on the results, a full-scale system can be designed.

E. STRAY CURRENT FROM AC SOURCES

Stray current from ac sources is a well-known safety hazard, particularly in the form of ac voltages that can occur on pipelines paralleling or crossing high-voltage ac electrical tower lines. Recent work, however, is leading to the understanding that ac discharge from a pipeline may indeed cause corrosion [3]. Both steady-state and dynamic ac stray current may be found. See [4] for detailed information.

Steady-state ac is commonly associated with pipelines laid in close proximity or paralleling high-voltage ac electrical transmission lines. Dangerous voltages can be created on such pipelines. Mitigation of the voltage as well as safety grounding of test stations and above-ground appurtenances to buried equipotential mats are recommended if ac pipe-to-soil potentials exceed 15 V.

Dynamic ac stray currents may be generated by ac welding, bad building grounds, or ac electrified railroads.

REFERENCES

1. Intermediate Course Text, Appalachian Underground Corrosion Short Course Omit, West Virginia University, Morgantown, WV, 2008, Chapter 5.

2. Advanced Course Text, Appalachian Underground Corrosion Short Course, West Virginia University, Morgantown, WV, 2008, Chapter 4.

3. R. Gummow, "AC Corrosion—A Challenge to Pipeline Integrity," Mater. Perform. **38**(2) (Feb. 1999).

4. NACE: Standard Practice SP01-77-2007, "Mitigation of Alternating Current and Lightning Effects on Metallic Structures and Corrosion Control Systems," NACE International, Houston, TX, 2007.

71

CORROSION INHIBITORS

S. Papavinasam

CANMET Materials Technology Laboratory, Hamilton, Ontario, Canada

A. INTRODUCTION

A corrosion inhibitor is a chemical substance which, when added in small concentrations to an environment, minimizes or prevents corrosion [1].

Corrosion inhibitors are used to protect metals from corrosion, including temporary protection during storage or transport as well as localized protection, required, for example, to prevent corrosion that may result from accumulation of small amounts of an aggressive phase. One example is brine in a nonaggressive phase such as oil. An efficient inhibitor is compatible with the environment, is economical for application, and produces the desired effect when present in small concentrations.

Inhibitor efficiency P is given as

$$P = \left(\frac{w_0 - w}{w_0}\right) \times 100 \qquad (71.1)$$

where w_0 is the corrosion rate in the absence of inhibitor and w is the corrosion rate in the same environment with the inhibitor added.

B. CLASSIFICATION OF INHIBITORS

Inhibitor selection is based on the metal and the environment. A qualitative classification of inhibitors is presented in

Uhlig's Corrosion Handbook, Third Edition, Edited by R. Winston Revie
Copyright © 2011 John Wiley & Sons, Inc.

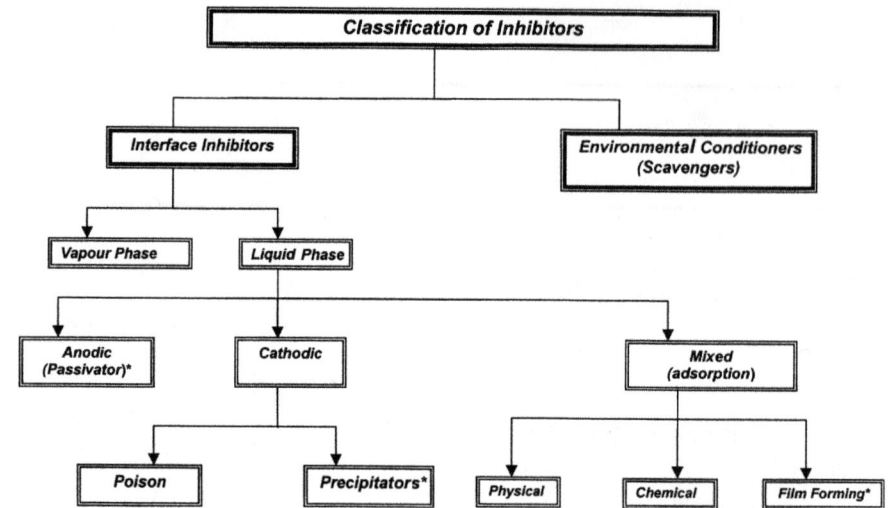

FIGURE 71.1. Classification of inhibitors: (∗) Form three-dimensional layers at the interface, so they are classified collectively as interphase inhibitors.

Figure 71.1. Inhibitors can be classified into environmental conditioners and interface inhibitors.

B1. Environmental Conditioners (Scavengers)

Corrosion can be controlled by removing the corrosive species in the medium. Inhibitors that decrease corrosivity of the medium by scavenging the aggressive substances are called environmental conditioners or scavengers. In near-neutral and alkaline solutions, oxygen reduction is a common cathodic reaction. In such situations, corrosion can be controlled by decreasing the oxygen content using scavengers (e.g., hydrazine [2]).

B2. Interface Inhibitors

Interface inhibitions control corrosion by forming a film at the metal/environment Interface. Interface inhibitors can be classified into liquid- and vapor-phase inhibitors.

B2.1. *Liquid-Phase Inhibitors.*
Liquid-phase inhibitors are classified as anodic, cathodic, or mixed inhibitors, depending on whether they inhibit the anodic, cathodic, or both electrochemical reactions.

B2.1.1 Anodic Inhibitors. Anodic inhibitors are usually used in near-neutral solutions where sparingly soluble corrosion products, such as oxides, hydroxides, or salts, are formed. They form, or facilitate the formation of, passivating films that inhibit the anodic metal dissolution reaction. Anodic inhibitors are often called passivating inhibitors.

When the concentration of an anodic inhibitor is not sufficient, corrosion may be accelerated, rather than inhibited. The critical concentration above which inhibitors are effective depends on the nature and concentration of the aggressive ions.

B2.1.2 Cathodic Inhibitors. Cathodic inhibitors control corrosion by either decreasing the reduction rate (cathodic poisons) or by precipitating selectively on the cathodic areas (cathodic precipitators).

Cathodic poisons, such as sulfides and selenides, are adsorbed on the metal surface, whereas compounds of arsenic, bismuth, and antimony are reduced at the cathode and form a metallic layer. In near-neutral and alkaline solutions, inorganic anions, such as phosphates. silicates, and borates, form protective films that decrease the cathodic reaction rate by limiting the diffusion of oxygen to the metal surface.

Cathodic poisons can cause hydrogen blisters and hydrogen embrittlement due to the absorption of hydrogen into steel. This problem may occur in acid solutions, where the reduction reaction is hydrogen evolution, and when the inhibitor poisons, or minimizes, the recombination of hydrogen atoms to gaseous hydrogen molecules. In this situation, the hydrogen, instead of leaving the surface as hydrogen gas, diffuses into steel causing hydrogen damage, such as hydrogen-induced cracking (HIC), hydrogen embrittlement, or sulfide stress cracking.

Cathodic precipitators increase the alkalinity at cathodic sites and precipitate insoluble compounds on the metal surface. The most widely used cathodic precipitators are the carbonates of calcium and magnesium.

B2.1.3 Mixed Inhibitors. About 80% of inhibitors are organic compounds that cannot be designated specifically as anodic or cathodic and are known as mixed inhibitors. The effectiveness of organic inhibitors is related to the extent to which they adsorb and cover the metal surface. Adsorption depends on the structure of the inhibitor, on the surface charge of the metal, and on the type of electrolyte.

Mixed inhibitors protect the metal in three possible ways: physical adsorption, chemisorption, and film formation.

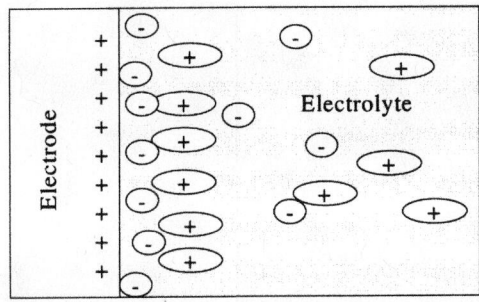

FIGURE 71.2. Adsorption of negatively charged inhibitor on a positively charged metal surface.

Physical (or electrostatic) adsorption is a result of electrostatic attraction between the inhibitor and the metal surface. When the metal surface is positively charged, adsorption of negatively charged (anionic) inhibitors is facilitated (Fig. 71.2).

Positively charged molecules acting in combination with a negatively charged intermediate can inhibit a positively charged metal. Anions, such as halide ions, in solution adsorb on the positively charged metal surface, and organic cations subsequently adsorb on the dipole (Fig. 71.3). Corrosion of iron in sulfuric acid containing chloride ions is inhibited by quaternary ammonium cations through this synergistic effect [3].

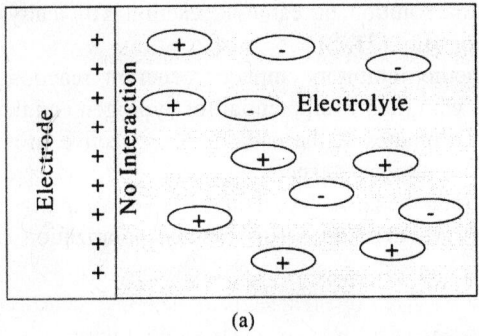

(a)

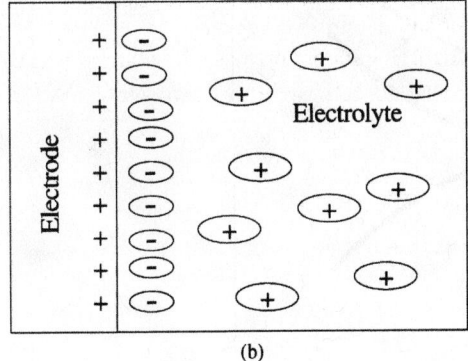

(b)

FIGURE 71.3. (a) Positively charged inhibitor molecule does not interact with positively charged metal surface. (b) Synergistic adsorption of positively charged inhibitor and anion on a positively charged metal surface.

Physically adsorbed inhibitors interact rapidly, but they are also easily removed from the surface. Increase in temperature generally facilitates desorption of physically adsorbed inhibitor molecules. The most effective inhibitors are those that chemically adsorb (chemisorb), a process that involves charge sharing or charge transfer between the inhibitor molecules and the metal surface.

Chemisorption takes place more slowly than physical adsorption. As temperature increases, adsorption and inhibition also increase. Chemisorption is specific and is not completely reversible [4].

Adsorbed inhibitor molecules may undergo surface reactions, producing polymeric films. Corrosion protection increases markedly as the films grow from nearly two-dimensional adsorbed layers to three-dimensional films up to several hundred angstroms thick. Inhibition is effective only when the films are adherent, are not soluble, and prevent access of the solution to the metal. Protective films may be nonconducting (sometimes called ohmic inhibitors because they increase the resistance of the circuit, thereby inhibiting the corrosion process) or conducting (self-healing films).

B2.2. Vapor-Phase Inhibitors. Temporary protection against atmospheric corrosion, particularly in closed environments, can be achieved using vapor-phase inhibitors (VPIs). Substances having low but significant pressure of vapor with inhibiting properties are effective. The VPIs are used by impregnating wrapping paper or by placing them loosely inside a closed container [5]. The slow vaporization of the inhibitor protects against air and moisture. In general, VPIs are more effective for ferrous than nonferrous metals.

C. MECHANISTIC ASPECTS OF CORROSION INHIBITION

C1. Environmental Conditioners (Scavengers)

In near-neutral solutions, the common cathodic reaction is oxygen reduction:

$$O_2 + 2H_2O + 4e \rightleftarrows 4OH^- \qquad (71.2)$$

Scavengers deplete the oxygen by chemical reaction; for example, hydrazine removes oxygen by the following reaction [2]:

$$5O_2 + 2(NH_2 - NH_2) \rightleftarrows 2H_2O + 4H^+ + 4NO_2^- \quad (71.3)$$

C2. Anodic Inhibitors (Passivators)

The mechanism of anodic inhibition can be explained using the polarization diagram of an active–passive metal (Fig. 71.4) [6].

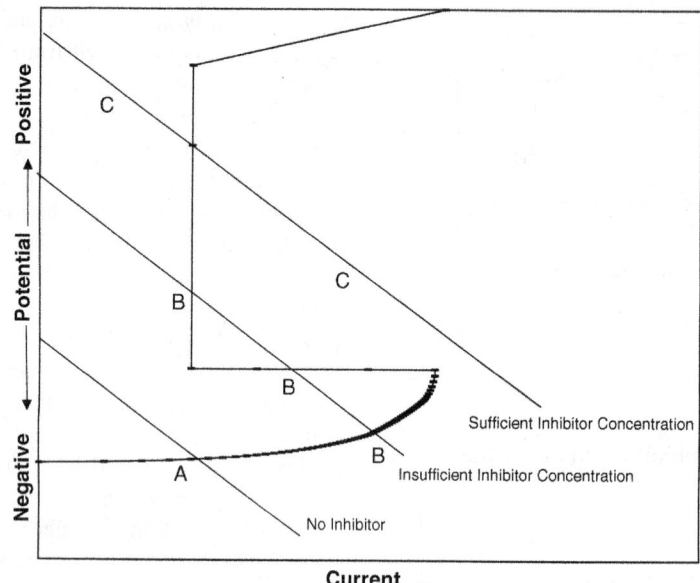

FIGURE 71.4. Polarization diagram of active–passive metal showing dependence of current on concentration of passivation-type inhibitor [7].

In the absence of inhibitors, the metal corrodes in the active state at a rate corresponding to point *A* in Figure 71.4. As the concentration of inhibitor is increased, the corrosion rate also increases until a critical concentration and a critical corrosion rate (point *B*, Fig. 71.4) are reached. At the critical concentration, there is a rapid transition of the metal to the passive state, and the corrosion rate is decreased (point *C*).

C3. Cathodic Inhibitors

In acid solution, the cathodic reaction is, typically, the reduction of hydrogen ions to hydrogen atoms, which combine forming hydrogen molecules:

$$H^+ + e^- \rightleftarrows H \qquad (71.4)$$

$$2H \rightleftarrows H_2 \qquad (71.5)$$

In alkaline solution, the cathodic reaction is typically oxygen reduction [Eq. (71.2)].

Cathodic inhibitors impede reduction reactions. Substances with high overpotential for hydrogen and those that form precipitates at the cathode are effective in acid and alkaline solutions, respectively. The effect of a cathodic inhibitor on the polarization curves is shown in Figure 71.5. In this case, the slope of the anodic polarization curve is

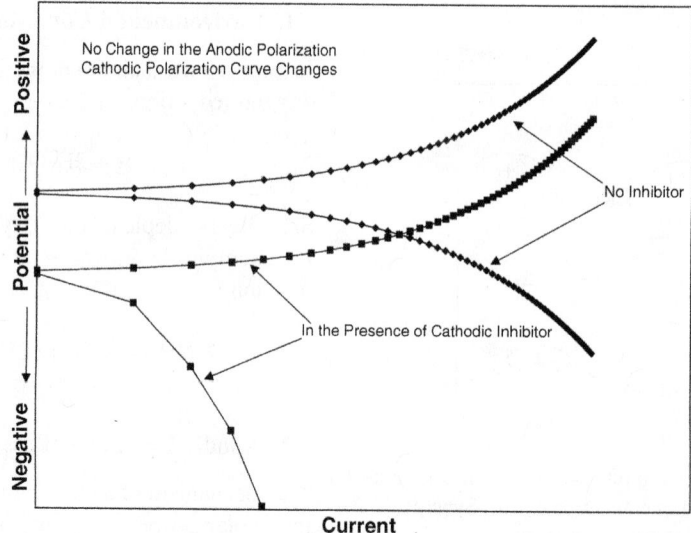

FIGURE 71.5. Polarization curve in the presence of cathodic inhibitor [7].

unaffected, but the slope of the cathodic polarization curve is changed [7].

C4. Mixed Inhibitors (Adsorption)

Adsorption occurs as a result of electrostatic forces between the electric charge on the metal and the ionic charges or dipoles on the inhibitor molecules.

The potential at which there is no charge on the metal is known as the zero-charge potential (ZCP) (Table 71.1) [8]. The charge on a metal surface in a given medium can be determined from the corrosion potential (E_{CORR}) and zero-charge potential. When the difference $E_{CORR} - ZCP$ is negative, the metal is negatively charged and adsorption of cations is favored. When $E_{CORR} - ZCP$ is positive, the metal is positively charged and adsorption of anions is favored.

The charge on inhibitors depends on the presence of loosely bound electrons, lone pairs of electrons, π-electron clouds, aromatic (e.g., benzene) ring systems, and functional groups containing elements of group V or VI of the periodic table. Most organic inhibitors possess at least one functional group, regarded as the reaction center or anchoring group. The strength of adsorption depends on the charge on this anchoring group [rather on the hetero atom (i.e., atoms other than carbon, including nitrogen and sulfur) present in the anchoring group]. The structure of the rest of the molecule influences the charge density on the anchoring group [4].

Water molecules adsorb on the metal surface immersed in an aqueous phase. Organic molecules adsorb by replacing the water molecules:

$$[\text{Inhibitor}]_{\text{soln}} + [n\text{H}_2\text{O}]_{\text{adsorbed}} \rightleftarrows [\text{Inhibitor}]_{\text{adsorbed}} + [n\text{H}_2\text{O}]_{\text{soln}}$$
(71.6)

where n is the number of water molecules displaced by one inhibitor molecule.

The ability of the inhibitor to replace water molecules depends on the electrostatic interaction between the metal and the inhibitor. On the other hand, the number of water molecules displaced depends on the size and orientation of the inhibitor molecule. Thus, the first interaction between inhibitor and metal surface is nonspecific and involves low activation energy. This process, called "physical adsorption," is rapid and, in many cases, reversible [9].

Under favorable conditions, the adsorbed molecules involved in chemical interaction (chemisorption) form bonds with the metal surface. Chemisorption is specific and is not reversible. The bonding occurs with electron transfer or sharing between metal and inhibitor. Electron transfer is typical for transition metals having vacant, low-energy electron orbitals.

Inhibitors having relatively loosely bound electrons transfer charge easily, The inhibition efficiency of the homologous series of organic substances differing only in the heteroatom is usually in the following sequence: $P > S > N > O$. An homologous series is given in Figure 71.6. On the other hand, the electronegativity, that is, the ability to attract electrons, increases in the reverse order.

Adsorption strength can be deduced from the adsorption isotherm, which shows the equilibrium relationship between concentrations of inhibitors on the surface and in the bulk of the solution. Various adsorption isotherms to characterize inhibitor efficiencies are presented in Table 71.2 [10–13]. To evaluate the nature and strength of adsorption, the experimental data (e.g., corrosion rate) are fitted to the isotherm, and from the best fit, the thermodynamic data for adsorption are evaluated.

TABLE 71.1. Values of Zero-Charge Potentials[a]

Metal	Zero-Charge Potential, mV (SHE)
Ag	− 440
Al	− 520
Au	+ 180
Bi	− 390
Cd	− 720
Co	− 450
Cr	− 450
Cu	+ 90
Fe	− 350
Ga	− 690
Hg	− 190
In	− 650
Ir	− 40
Nb	− 790
Ni	− 300
Pb	− 620
Pd	0
Pt	+ 20
Rh	− 20
Sb	− 140
Sn	− 430
Ta	− 850
Ti	− 1050
Tl	− 750
Zn	− 630

[a]See [8].

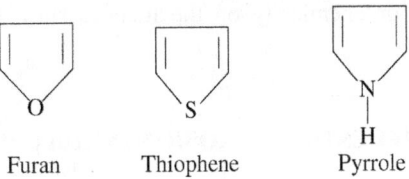

FIGURE 71.6. Homologous series of organic molecules (the molecules differ only in the heteroatom.)

TABLE 71.2. Adsorption Isotherms[a]

Name	Isotherm[b]	Verification Plot
Langmuir	$\theta/(1/\theta) = \beta \cdot c$	$\theta/(1-\theta)$ vs. log c
Frumkin	$[\theta/(1-\theta)]e^{f\theta} = \beta \cdot c$	θ vs. log c
Bockris– Swinkels	$\theta/(1-\theta)^n \cdot$ $[\theta+n(1-\theta)]^{n-1}/$ $n^n = c \cdot e^{-\beta}/55.4$	$\theta/(1-\theta)$ vs. log c
Temkin	$\theta = (1/f)\ln K \cdot c$	θ vs. log c
Virial Parson	$\theta \cdot e^{2f\theta} = \beta \cdot c$	θ vs. log (θ/c)

[a]See [10–13].

[b]θ, %P/100, surface coverage; β, $\Delta G/2.303RT$; ΔG, free energy of adsorption; R, gas constant; T, temperature; c, bulk inhibitor concentration; n, number of water molecules replaced per inhibitor molecule; f, inhibitor interaction parameter (0, no interaction; $+$, attraction; and $-$, repulsion); K, constant; and $\%P = 1 -$ inhibited corrosion rate/uninhibited corrosion rate.

The principle of soft and hard acids and bases (SHAB) has also been applied to explain adsorption and inhibition [14]. The SHAB principle states that hard acids prefer to coordinate with hard bases and that soft acids prefer to coordinate with soft bases. Metal atoms on oxide-free surfaces are considered to be soft acids, which in acid solutions form strong bonds with soft bases, such as sulfur-containing organic inhibitors. By comparison, nitrogen or oxygen-containing organic compounds are considered to be hard bases and may establish weaker bonds with metal surfaces in acid solutions.

Whatever may be the mechanism of adsorption, the electron density of the functional groups, polarizability, and electronegativity are important parameters that determine inhibition efficiency.

C5. Vapor-Phase Inhibitors

The process of vapor-phase inhibition involves two steps: transport of inhibitor to the metal surface and interaction of inhibitor on the surface. Either a VPI may vaporize in the molecular form or it may first dissociate and then vaporize. Amines vaporize in the undissociated molecular form. Subsequent reaction with water, present as moisture at the surface, dissociates the inhibitor. On the other hand, dicyclohexylamine nitrite dissociates liberating amine and nitrous acid, which deposit on the metal surface [15]. Both in molecular and in dissociated forms VPIs adsorb either physically or chemically on the metal surface to inhibit corrosion.

D. EXAMPLES OF CORROSION INHIBITORS

Inhibitors used in practice are seldom pure substances but are usually mixtures that may be byproducts, for example, of

TABLE 71.3. Some Anchoring (Functional) Groups in Organic Inhibitors

Structure	Name	Structure	Name
—OH	Hydroxy	—CONH$_2$	Amide
—C≡C—	-Yne	—SH	Thiol
—C–O–C—	Epoxy	—S—	Sulfide
—COOH	Carboxy	—S=O	Sulfoxide
—C–N–C—	Amine	—C=S—	Thio
—NH$_2$	Amino	—P=O	Phosphonium
—NH	Imino	—P—	Phospho
—NO$_2$	Nitro	—As—	Arsano
—N=N–N—	Triazole	—Se—	Seleno

some industrial chemical processes for which the active constituent is not known. Commercial inhibitor packages may contain, in addition to the active ingredients for inhibition, other chemicals, including surfactants, deemulsifiers, carriers (e.g., solvents), and biocides.

The active ingredients of organic inhibitors invariably contain one or more functional groups containing one or more heteroatoms, N, O, S, P, or Se (selenium), through which the inhibitors anchor onto the metal surface. Some common anchoring groups are listed in Table 71.3. These groups are attached to a parent chain (backbone), which increases the ability of the inhibitor molecule to cover a large surface area. Common repeating units of the parent chain are methyl and phenyl groups. The backbone may contain additional molecules, or substituent groups, to enhance the electronic bonding strength of the anchoring group on the metal and/or to enhance the surface coverage. The outline of the constitution of an organic inhibitor is presented in Table 71.4.

D1. Inhibitors Containing Oxygen Atom

Benzoic acid and substituted benzoic acids are widely used as corrosion inhibitors [16]. Adsorption and inhibitor efficiencies of benzoic acids depend on the nature of the

TABLE 71.4. Constitution of an Organic Corrosion Inhibitor

Anchoring Group[a]	Backbone	Substituent Groups[a]
Binds onto the metal	Bears anchoring and substituent groups Provides surface coverage	Supplements electronic strength and surface coverage

[a]Anchoring and substituent groups are interchangeable, that is, the substituent group through which the inhibitor anchors onto the metal surface depends on the electron density, charge on the metal, and orientation of the molecule in a particular environment.

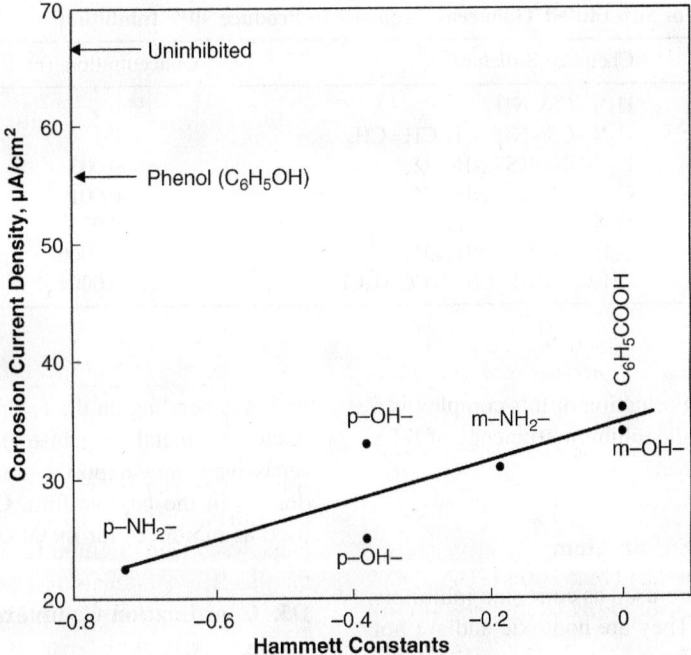

FIGURE 71.7. Variation of inhibitor efficiency as a function of substituents (benzoic acid) (substituents with negative Hammett constants will attract electrons from the anchoring –COOH group, thereby decreasing the efficiency) [16]. (Hammett constant is a measure of ability of the substituents to attract or repel electrons.)

substituents. Electron-donating substituents increase inhibition by increasing the electron density of the anchoring group (–COOH group); on the other hand, electron-withdrawing substituents decrease inhibition by decreasing the electron density. Percent inhibition as a function of substituents is presented in Figure 71.7.

D2. Inhibitors Containing Nitrogen Atom

Benzotriazole (BTA) and its derivatives are effective inhibitors for many metals, especially copper, in a variety of conditions [17]. At low concentrations, BTA is adsorbed slightly on the surface. At sufficiently high solution concentrations, bulk precipitation of the complex on the surface

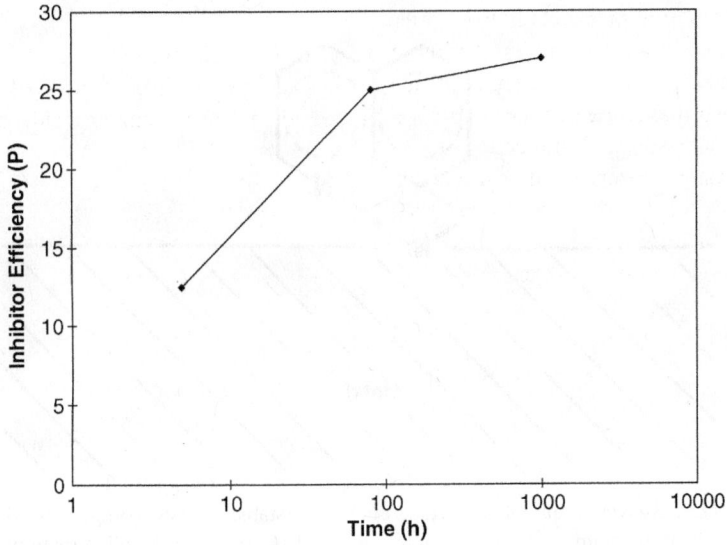

FIGURE 71.8. Dependence of inhibitor efficiency on time (benzotriazole on copper) [17].

TABLE 71.5. Concentration of Substituted Thioureas Required to Produce 90% Inhibition[a]

Inhibitor	Chemical Structure	Concentration, (mol/L)	Molecular Weight
Thiourea	$H_2N-CS-NH_2$	0.1	76.13
Allyl thiourea	$H_2N-CS-NH-CH_2CH=CH_2$	0.1	116.19
N,N-Diethyl thiourea	$C_2H_5HN-CS-NHC_2H_5$	0.003	132.23
N,N-Diisopropylthiourea	$C_3H_7HN-CS-NHC_3H_7$	0.001	160.28
Phenyl	$H_2N-CS-NH-C_6H_5$	0.001	152.21
Thiocarbamide	$C_6H_5HN-CS-NHC_6H_5$	0.0006	228.38
Symdiotolylthiourea	$CH_3C_6H_4NH-CS-NHC_6H_4CH_3$	0.0004	256.35

[a]See [18].

occurs, inhibiting corrosion. Formation of this complex is a slow process and, as a result, the inhibitor efficiency of BTA increases with time (Fig. 71.8).

D3. Inhibitors Containing Sulfur Atom

Thiourea and its derivatives are used as corrosion inhibitors for a variety of metals [18]. They are nontoxic and are not an environmental hazard. The variation in the inhibitor efficiencies of various derivatives of thiourea depends on the molecular weight. By using lower concentrations of large molecules, higher inhibitor efficiencies can be obtained (Table 71.5.)

D4. Electronically Conducting Polymers

In situ polymerization of heterocyclic compounds, such as pyrrole and thiophene (structures in Fig. 71.6) and aniline, produces homogeneous, adhesive films on the metal surface [19]. These films are electronically conducting and have the advantage of tolerance to microdefects and minor scratches. Conductivity can be up to 100 S/cm and can be

varied depending on the extent of oxidation, from semiconductor to metal. Because of these properties, the films repassivate any exposed areas of metal where there are defects in the passive film. Conducting polymers are now used as inhibitors for metal corrosion.

D5. Coordination Complexes

A variety of chelants provide either corrosion inhibition or corrosion acceleration, depending on the structure and functional groups. The chelants displaying high surface activity and low solubility in solution are effective corrosion inhibitors. If they do not have these characteristics, they stimulate corrosion.

The 8-hydroxyquinoline molecule satisfies the structural requirements for surface chelation, but formation of a nonadherent film is a distinct disadvantage (Fig. 71.9). On the other hand, pyrocatechols (Fig. 71.10), forming adherent chelants on the metal surface, are effective inhibitors. Inhibition efficiency can be increased by decreasing the solubility through alkylation (increase in chain length) (Table 71.6) [20].

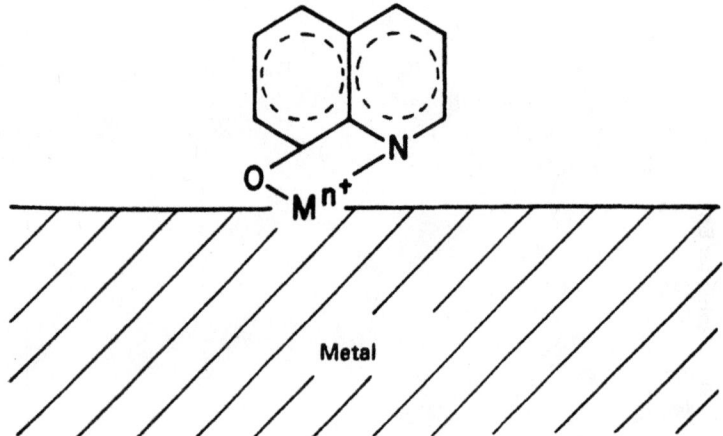

FIGURE 71.9. 8-Hydroxyquinoline surface chelation (stable chelate complex is formed but is soluble in aqueous medium —no corrosion inhibition [20]. (Copyright NACE International. Reprinted with permission.)

FIGURE 71.10. Pyrocatechols (forms insoluble chelate complex with the metal. Efficient corrosion inhibitor; also refer to Table 71.6 [20].

TABLE 71.6. Dependence of Inhibitor Efficiency of Pyrocatechol on Chain Length[a]

Inhibitor	Substituent (R value in Fig. 71.10)	% Inhibition
Pyrocatechol	–H	– 14
4-Methylpyrocatechol	–CH$_3$	84
4-*n*-Butylpyrocatechol	–(CH$_2$)$_3$CH$_3$	93
4-*n*-Hexylpyrocatechol	–(CH$_2$)$_5$CH$_3$	96

[a]See [20]. Copyright NACE International. Reproduced with permission.

E. INDUSTRIAL APPLICATIONS OF CORROSION INHIBITORS

E1. Petroleum Production

Corrosion in the hydrocarbon industries may be divided into two types, "wet corrosion" and "dry corrosion." At low temperature (i.e., below the boiling point or dewpoint of water), material corrodes due to the presence of an aqueous phase (wet corrosion). At higher temperature (above the boiling point of water), corrosion occurs in the absence of an aqueous phase (dry corrosion).

Wet corrosion is influenced by pressure, temperature, and compositions of aqueous, gaseous, and oil phases. In refineries and petrochemical plants, the amount of water is usually small, but the corrosivity is high and is localized at regions where the aqueous phase contacts the metal. The water may contain dissolved hydrogen sulfide (H$_2$S), carbon dioxide (CO$_2$), and chloride ions (CI$^-$). Corrosion may occur even when the produced water content is as low as 0.1%, or corrosion activity may not begin until after several years of production.

Refineries and petrochemical industries employ a variety of film-forming inhibitors to control wet corrosion. Most of the inhibitors are long-chain nitrogenous organic materials, including amines and amides. Water-soluble and water soluble–oil dispersible inhibitors are continuously injected, or oil-soluble and oil soluble–water dispersible inhibitors (batch inhibitors) are intermittently applied to control corrosion.

Film-forming inhibitors anchor to the metal through their polar group. The nonpolar tail protrudes out vertically. The physical adsorption of hydrocarbons (oils) on these nonpolar tails increases film thickness and the effectiveness of the hydrophobic barrier for corrosion inhibition [21].

Because inhibitors are interfacial in nature, they are active at liquid–liquid and/or liquid–gas interfaces and can lead to emulsification. As a result, foaming is sometimes experienced in the presence of inhibitors.

E2. Internal Corrosion of Steel Pipelines

Gathering pipelines, operating between oil and gas wells and processing plants, have corrosion problems similar to those in refineries and petrochemical plants. The flow regimes of multiphase fluids in pipelines influence the corrosion rate. At high flow rates, flow-induced corrosion and erosion–corrosion may occur, whereas at low flow rates, pitting corrosion is more common.

Corrosion is related to the amount and nature of sediments. High-velocity flow tends to sweep sediments out of the pipeline, whereas low velocity allows sediments to settle at the bottom, providing sites for pitting corrosion. Internal corrosion of pipelines is controlled by cleaning the pipeline (pigging) and by adding continuous and or batch inhibitors.

E3. Water

Potable water is frequently saturated with dissolved oxygen and is corrosive unless a protective film, or deposit, is formed. Cathodic inhibitors, such as calcium carbonate, silicates, polyphosphates, and zinc salts, are used to control potable water corrosion.

Water is used in cooling systems In many industries. In a recirculating system, evaporation is the chief source of cooling. As evaporation proceeds, the dissolved mineral salt content increases. Cooling systems may consist of several dissimilar metals and nonmetals. Metals picked up from one part of the system can be deposited elsewhere, producing galvanic corrosion. Corrosion is controlled by anodic (passivating) inhibitors, including nitrate and chromate, as well as by cathodic (e.g., zinc salt) inhibitors. Organic inhibitors (e.g., benzotriazole) are sometimes used as secondary inhibitors, especially when excessive corrosion of copper occurs [22].

E4. Acids

Acids are widely used in pickling, in cleaning of oil refinery equipment and heat exchangers, and in oil well acidizing. Mixed inhibitors are widely used to control acid corrosion.

E5. Automobile

Inhibitors are used in an automobile for two reasons: (1) to reduce the corrosivity of fluid systems (internal corrosion) and (2) to protect the metal surfaces exposed to the atmosphere (external corrosion). Internal corrosion is influenced by the coolants, flow, aeration, temperature, pressure, water impurities and corrosion products, operating conditions, and maintenance of the system. Some common inhibitors dissolved in antifreeze are nitrites, nitrates, phosphates, silicates, arsenates, and chromates (anodic inhibitors); amines, benzoates, mercaptans, and organic phosphates (mixed inhibitors); and polar or emulsifiable oils (film formers) [23].

Atmospheres to which automobiles are exposed contain moist air, wet SO_2 gas (forming sulfuric acid in the presence of moist air), and deicing salt (NaCl and $CaCl_2$). To control external corrosion, the rust-proofing formulations that are used contain grease, wax resin, and resin emulsion, along with metalloorganic and asphaltic compounds. Typical inhibitors used in rust-proofing applications are fatty acids, phosphonates, sulfonates, and carboxylates.

E6. Paints (Organic Coatings)

Finely divided inhibiting pigments are frequently incorporated in primers. These polar compounds displace water and orient themselves in such a way that the hydrophobic ends face the environment, thereby augmenting the bonding of the coatings on the surface. Red lead (Pb_3O_4) is commonly used in paints on iron. It deters formation of local cells and helps preserve the physical properties of the paints. Other inhibitors used in paints are lead azelate, calcium plumbate, and lead suboxide [24].

E7. Miscellaneous

Inhibitors are used to control corrosion in boiler waters, fuel oil tanks, hot chloride dye baths, refrigeration brines, and reinforcing steel in concrete, and they are also used to protect artifacts.

F. OTHER FACTORS IN APPLYING INHIBITORS

Some factors to be considered in applying inhibitors are discussed in the following paragraphs.

F1. Application Techniques

A reliable method should be applied for inhibitor application. A frequent cause of ineffective inhibition is loss of the inhibitor before it either contacts the metal surface or changes the environment to the extent required. Even the best inhibitor will fail if not applied properly.

If the inhibitor is continuously applied in a multiphase system, it should partition into the corrosive phase, usually the aqueous phase. This partitioning is especially important when using water-soluble, oil-dispersible inhibitors.

In batch treatment, the frequency of treatment depends on the film persistency. It is important that the corrosion rates are measured frequently to ensure that a safe level of inhibition is maintained. It is also important that the inhibitor contacts the entire metal surface and forms a continuous persistent film.

When using volatile inhibitors, care must be taken in packaging to prevent the loss of inhibitor to the outside atmosphere.

Inhibitors are added to the primers used in paint coatings. When moisture contacts the paint, some inhibitor is leached from the primer to protect the metal. The inhibitor should be incorporated in such a way that it protects the areas where potential corrosion can take place and not leach completely from the primer during the service life.

F2. Temperature Effects

Organic molecules decompose at elevated temperatures. In general, film-forming inhibitors that depend on physical adsorption become less effective at elevated temperatures, so that larger treatment dosages may be required to maintain protective films. Chemisorption, on the other hand, increases with temperature due to the strengthening of chemical bonds. As a result, inhibitor efficiency increases with temperature up to the temperature at which decomposition of the inhibitor occurs.

F3. Poisoning

Inhibitors for hydrogen damage should reduce not only the corrosion rate but also the rate of absorption and permeation of hydrogen into the steel. For example, the corrosion rate of steel in sulfuric acid is decreased [25], while hydrogen flux through a steel membrane is increased by adding thiourea [26]. Although thiourea inhibits corrosion, it poisons the hydrogen recombination reaction, so that much of the hydrogen produced at the steel surface enters the steel, causing hydrogen damage, rather than recombining with other hydrogen atoms to form bubbles of hydrogen that escape from the system.

F4. Secondary Inhibition

The nature of the inhibitor initially present in acid solutions may change with time as a consequence of chemical or electrochemical reactions. Inhibition due to the reaction products is called secondary inhibition. Depending on the effectiveness of the reaction products, secondary inhibition may be higher or lower than primary inhibition. For example, diphenyl sulfoxide undergoes electrochemical reaction at the metal surface to produce diphenyl sulfide, which is more effective than the primary compound [27]. On the contrary,

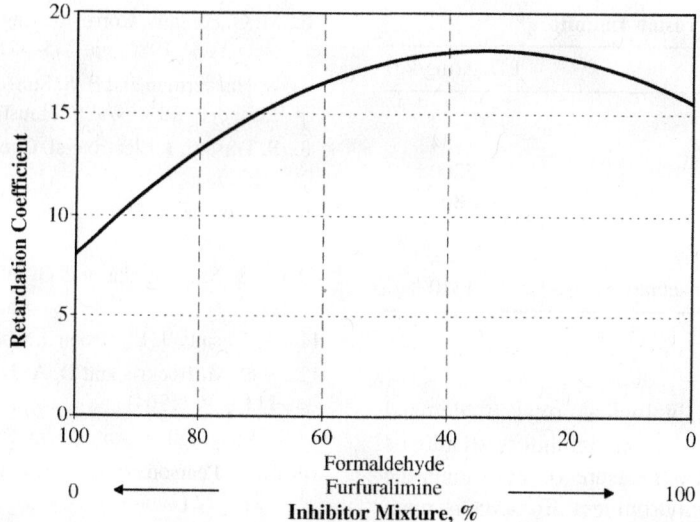

FIGURE 71.11. Synergistic effect of mixing formaldehyde and furfuralimine [28].

the reduction of thiourea and its alkyl (e.g., methyl, ethyl) derivatives gives rise to HS⁻, which accelerates corrosion.

F5. Synergism and Antagonism

In the presence of two or more adsorbed species, lateral interaction between inhibitor molecules can significantly affect inhibitor performance. If the interaction is attractive, a synergistic effect arises, that is, the degree of inhibition in the presence of both inhibitors is higher than the sum of the individual effects. For example, because of this synergistic effect, the inhibition efficiency of a mixture of formaldehyde and furfuralimine is higher compared to the inhibition

efficiency when these inhibitors are used separately (Fig. 71.11). On the other hand, when narcotine and thiourea are used as mixed inhibitors, there is an antagonistic effect and a decrease in inhibitor efficiency compared to that which exists when these inhibitors are used separately (Fig. 71.12) [28].

F6. Green Inhibitors

Environmental concerns worldwide are increasing and are likely to influence the choice of corrosion inhibitors in the future. Environmental requirements are still being developed, but some elements have been established.

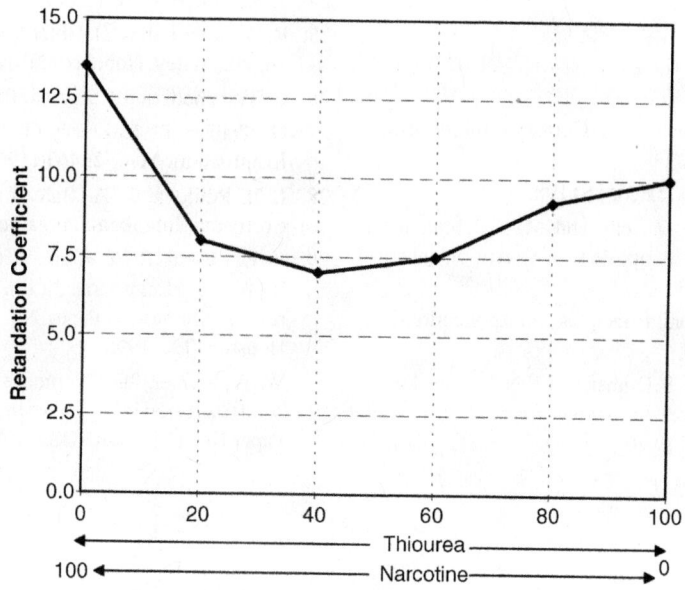

FIGURE 71.12. Antagonistic effect of mixing narcotine and thiourea [28].

TABLE 71.7. Toxicity of Corrosion Inhibitors[a]

Compound	LD_{50}, (mg/kg)
Propargyl alcohol	55
Hexynol	34
Cinnamaldehyde	2200
Formaldehyde	800
Dodecylpyridinium bromide	320
Naphthylmethylquinolinium chloride	644
Nonylphenol–ethylene oxide surfactants	1310

[a]See [30].

The biodegradation, or biological oxygen demand (BOD), should be at least 60%, and inhibitors should be nontoxic [29]. The BOD is a measure of how long the inhibitor will persist in the environment. Toxicity is measured as LC_{50} or EC_{50} [30]. LC_{50} is the concentration of the inhibitor needed to kill 50% of the total population of the test species. The results are quoted as milligrams of chemical per liter of fluid (or LD_{50}, milligrams per kilogram) for exposure times of 24 and 48 h. The EC_{50} is the effective concentration of inhibitor to adversely affect 50% of the population. In general, EC_{50} values are lower than LC_{50} values because the former are the concentrations required to damage the species in some way without killing it. Some chemicals are excellent inhibitors but are quite toxic and readily adsorbed through the skin. Toxicity of some inhibitors is presented in Table 71.7.

There is a growing demand for corrosion inhibitors that are less toxic and biodegradable compared to current formulations. Green inhibitors displaying substantially improved environmental properties will be the inhibitors most widely used in the future.

REFERENCES

1. O. L. Riggs, Jr., in C. C. Nathan (Ed.), Corrosion Inhibitors, NACE, Houston, TX, 1973, p. 11.
2. M. G. Noack, Mater. Perform., 21(3), 26 (1982).
3. A, Frignani, G. Trabanelli, F. Zucchi, and M. Zucchini, in Proceedings of 5th European Symposium of Corrosion Inhibitors, Ferrara, Italy 1980, p. 1185.
4. V. S. Sastri, Corrosion Inhibitors: Principles and Applications, Wiley, New York, 1998, p. 39.
5. S. A. Levin, S. A. Gintzbergy, I. S. Dinner, and V. N. Kuchinsky, in Proceedings of Second European Symposium on Corrosion Inhibitors, Ferrara, Italy 1965, p. 765.
6. M. G. Fontana, Corrosion Engineering, 3rd ed., McGraw-Hill, New York, 1986, pp. 445–481.
7. N. Hackerman and E. S. Snavely, in Corrosion Basics, L. S. V. Delinder (Ed.), NACE, Houston, TX, 1984, pp. 127–146.
8. S. Trasatti, J. Electroanal. Chem. Interf. Electrochem., 33, 351 (1971).
9. W. Lorenz, Z. Phys. Chem., 219, 421 (1962); 224, 145 (1963); and 244, 65 (1970).
10. Z. S. Smialowska and G. Wieczorek, Corros. Sci., 11, 843 (1971).
11. S. Trasatti, J. Electroanal. Chem., 53, 335 (1974).
12. J. O. M. Bockris and D. A. J. Swinkels, J. Electrochem. Soc., 111, 736 (1964).
13. I. Langmuir, J. Am. Chem. Soc., 39, 1848 (1947).
14. R. G. Pearson, J. Am. Chem. Soc., 85, 3533 (1963); Science, 151, 172 (1966).
15. I. L. Rosenfeld, V. P. Persiantseva, and P. B. Terentief, Corrosion, 20, 222t (1964).
16. A. Akiyama and K. Nobe, J. Electrochem. Soc., 117, 999 (1970).
17. P. G. Fox, G. Lewis, and P. J. Boden, Corros. Sci., 19, 457 (1979).
18. I. Singh, Corrosion, 49, 473 (1993).
19. T. A. Skotheim (Ed.), Handbook of Conducting Polymers, Vols. 1 and 2, Marcel Dekker, New York, 1986.
20. A, Weisstuch, D. A. Carter, and C. C. Nathan, Mater. Perform., 10(4), 11, (1971).
21. C. C. Nathan (Ed.), Corrosion Inhibitors, NACE, Houston, TX, 1973, p. 45.
22. S. Matsuda and H. H. Uhlig, J. Electrochem. Soc., 111, 156 (1964).
23. Snow and Ice Control with Chemical and Abrasives, Bulletin 152, Highway Research Board, Washington, DC, 1960.
24. T. Rossel, Werkstoffe Korrosion, 20, 854 (1969).
25. R. W. Revie and H. H. Uhlig, Corrosion and Corrosion Control, 4th ed., Wiley, Hoboken, NJ, 2008, p. 312.
26. G. Trabenelli and F. Zucchi, Rev. Coat. Corros., 1, 97 (1972).
27. G. Trabanelli, F. Zucci, G. L. Zucchini, and V, Carassiti, Electrochim. Met., 2, 463 (1967).
28. I. N. Putilova, S. A. Balezin, and V. P. Barannik, Metallic Corrosion, Inhibitors, Pergamon, New York, 1960, pp. 17–24.
29. R. L. Martin, B. A. Alink, T. G. Braga, A. J. McMahon, and R. Weare, "Environmentally Acceptable Water Soluble Corrosion Inhibitors," Paper No. 36, CORROSION/95, NACE, Houston, TX, 1995.
30. W. W. Frenier, "Development and Testing of a Low-Toxicity Acid Corrosion Inhibitor for Industrial Cleaning Applications," Paper No. 152, CORROSION/96, NACE, Houston, TX, 1996.

72

COMPUTER TECHNOLOGY FOR CORROSION ASSESSMENT AND CONTROL

S. Srinivasan

Advanced Solutions—Americas, Honeywell International, Inc, Houston, Texas

A. INTRODUCTION

Information and knowledge are the foundations on which all technology has been built. Computers have become ubiquitous, key components of automation and optimized problem solving in most domains of science and engineering, and corrosion is no exception. Computer-based information systems and computer models of corrosion processes have revolutionized our approach to problem solving, information access, and knowledge processing in significant ways. It would not be superfluous to say that in the current-day environment every aspect of corrosion data analyses, modeling, and production is managed through computer-based tools and technologies.

This chapter provides an overview of the types of computer tools utilized for solving corrosion-related problems, data storage, and data analyses. An introduction to computer-based corrosion problem solving is followed by a description of types of computer programs employed in the domain of corrosion, with an overview of early systems leading to current-day models for thermodynamic analyses, corrosion prediction, and material selection. Brief descriptions of different types of computer applications for corrosion, including expert systems, neural networks, and object-oriented software systems, is also included for purposes of both completeness and relevance. A description of currently available computer tools for modeling corrosion and cracking problems, selection of materials/equipment, as well as corrosion management, monitoring, and control is also provided. Aspects relevant to the role of emerging technologies and the availability of computer-based tools for corrosion analyses from Web/Internet-based systems to cloud computing (data shared on Internet servers) are discussed.

B. COMPUTER-BASED CORROSION PROBLEM SOLVING: CLASSIFICATION AND BACKGROUND

Using computer tools to model and represent corrosion processes is a challenging task since characterizing corrosion processes requires a fundamental understanding of principles underlying multiple disciplines, from electrochemistry, thermodynamics, and fluid mechanics to material science and engineering. The complexity of characterizing corrosion-related tasks has necessitated use of computer tools in corrosion science and engineering, from thermodynamic and phase behavior modeling to data acquisition and analysis. Computers, in the current-day environment, are an intrinsic part of both data representation and automated problem solving. In this context, computer-based corrosion problem-solving systems may be classified as:

Uhlig's Corrosion Handbook, Third Edition, Edited by R. Winston Revie
Copyright © 2011 John Wiley & Sons, Inc.

Systems for modeling corrosion/cracking processes

Material selection and equipment specification programs

Computer-based corrosion monitoring systems

Computer-based systems for control of corrosion testing equipment

Databases and hypertext systems

Internet-based databases and software programs

A large number of early programs in corrosion were billed as expert systems, primarily because the programs typically attempted to capture human expertise in corrosion [1], and these programs represented research-based development efforts normally lacking rigorous software engineering foundations necessary for commercial distribution. Most of these programs were developed using software platforms called shells [2] that supported easy implementation of heuristic rules (rules of thumb) and representation of common concepts of reasoning. Table 72.1 lists a few of the early, well-known computer programs in corrosion developed in the late 1980s and early 1990s [3–8]. It is interesting to note that *none* of these early systems were implemented in commonly used programming languages (such as C++, Fortran, etc.) or current programming languages/environments (JAVA, C#, .Net, etc.), and many were implemented by corrosion/materials specialists with little or no formal training in software development [9].

Most current-day applications for corrosion may be broken into different categories as follows:

Programs for prediction of corrosion for different industrial applications, such as oil and gas production, transmission, and refining

Applications for characterizing thermodynamics and phase behavior as well as system speciation for a range of process/plant applications (broadly termed *ionic modeling* and *process modeling* tools)

Applications for selection of materials for corrosion resistance as well as resistance to stress corrosion cracking

Applications for risk-based inspection and risk-based integrity characterization

The single most popular application of computers in corrosion stems from programs built for modeling and predicting corrosion for different applications related to oil and gas. Such programs (currently commercially available) include Predict® [9, 10] and NORSOC® [11] as well as programs such as HydroCorr from Shell and Cassandra from BP, all of which focus on predicting corrosion in CO_2/H_2S oil/gas production environments. Numerous other systems for corrosion prediction include Predict-Pipe [12] for internal corrosion direct assessment (ICDA) automation for dry-gas pipelines and Corrosion Analyzer [13] for characterizing thermodynamics/phase behavior/speciation in corroding systems. Some early systems, such as Auscor [14], provided assessment of stress corrosion cracking in austenitic stainless steels, and legacy programs for general corrosion prediction included LipuCor [15] and Cormed [16]. A more detailed description of some of the current-day corrosion prediction models are provided in a subsequent section in this chapter.

Several corrosion-related applications and databases are available to assist with material selection and equipment specifications. The CORIS system [17] is a legacy expert system database that integrates corrosion problem-solving expertise with a comprehensive database on corrosion, materials, and thermodynamic properties of corrosive media. Other programs for material evaluation and selection, some of which have had wide commercial application, that provide an automated basis for materials selection include SOCRATES [18, 19] for oil and gas production, Predict®-SW [20, 21] for refinery sour water corrosion prediction and material selection and Selmatel [22] for selection of materials exposed to elevated temperatures in refinery furnace tubes.

Early applications of computers to corrosion (in the 1990s) focused on development of databases integrated with material evaluation heuristics for corrosion and materials data. The Materials Technology Institute [1] in the late 1980s and early 1990s developed a series of systems called ChemCor relevant to evaluation of materials applicable to different segments of the chemical processing industries. CORSUR [23] was another large database application on the corrosion behavior of metals and nonmetals in over

TABLE 72.1. Early, Legacy Software (Expert) Systems in Corrosion

Name	Application	Country of Development	Shell Used
Achilles	Diagnosis and prediction of localized corrosion	United Kingdom	Spices
Aurora	Prediction of localized corrosion in austenitic steels	Finland	Level 5
Auscor	Prediction of corrosion of austenitic stainless steels	United Kingdom	Savior
ChemCor	Materials selection for hazardous chemical service	United States	KES/Level 5
DIASCC	Evaluation of cracking in stainless steels	Japan	OPS83
KISS	Material selection	Germany	Nexpert Object
Prime	Materials selection for chemical process industry	Belgium	KEE
Suscept	Evaluation of SSC in steels	United States	PCPLUS

700 chemical environments. However, both ChemCor and Corsur have become obsolete, since MTI has not pursued further development and commercialization of these programs beyond the 1990s.

C. TYPES OF COMPUTER PROGRAMS FOR CORROSION CONTROL

Different types of software technologies have found application in corrosion engineering and science. These programs can be broadly classified as follows:

Conventional software systems (structured software systems)

Artificial intelligence and expert systems applications

Object-oriented software systems (includes current-day component-based development frameworks such as .Net and Java)

Neural networks

Hybrid systems that utilize one or more of the above technologies

Most early computer programs and legacy systems (meaning older systems whose technology is currently obsolete) developed in the 1970s and 1980s used high-level programming languages such as Basic, Fortran, and Pascal. Typically, such systems provided a front end for data modeling and analysis that was computation intensive but provided little or no support for representation of heuristic concepts and manipulation of symbolic information.

The advent of expert systems in the late 1980s drastically changed the direction of computer programs for corrosion applications. *Expert systems* or *knowledge-based systems,* defined as intelligent computer programs that use expert knowledge to attain high levels of performance in narrow problem domains [24], became quite common and prominent. The term *expert* in the expert systems implied the narrow specialization and competence of a human expert embodied in the system. Expert systems typically had a specific structure that distinguishes them from conventional computer programs in that the domain knowledge (knowledge base) is usually separated from the reasoning process (inference engine), as shown in Figure 72.1.

The knowledge base housed the expertise that is embodied in an expert system. The inference engine controlled the manner or the logical path used by the expert system to access the information (facts and rules) in the knowledge base to make decisions. *Knowledge representation* in an expert system referred to the scheme used to represent a given piece of information in the system. Decision making in an expert system was represented through interconnected rules (rules of thumb that correlate facts, data, and expertise) and objects (descriptions of different system components).

Expert systems represent by far the most widely used segment of computer applications in corrosion. Numerous expert systems have been developed to address different aspects of corrosion [1] and include systems for applications in cathodic protection [25], assessment of stress cracking in light water nuclear reactors [26], and prediction of localized corrosion of stainless steels [27].

With the advent of object-oriented systems and the concept of reusable entities that contain both the data and the procedures relevant to an object, the need for separation of knowledge and reasoning as found in expert systems vanished. Object-oriented systems provided a framework for modeling through simulation of behavior of real-life systems. An object in this context is any entity characterized by data (properties) and procedures (methods) for manipulating that data. In an object-centric view of the world,

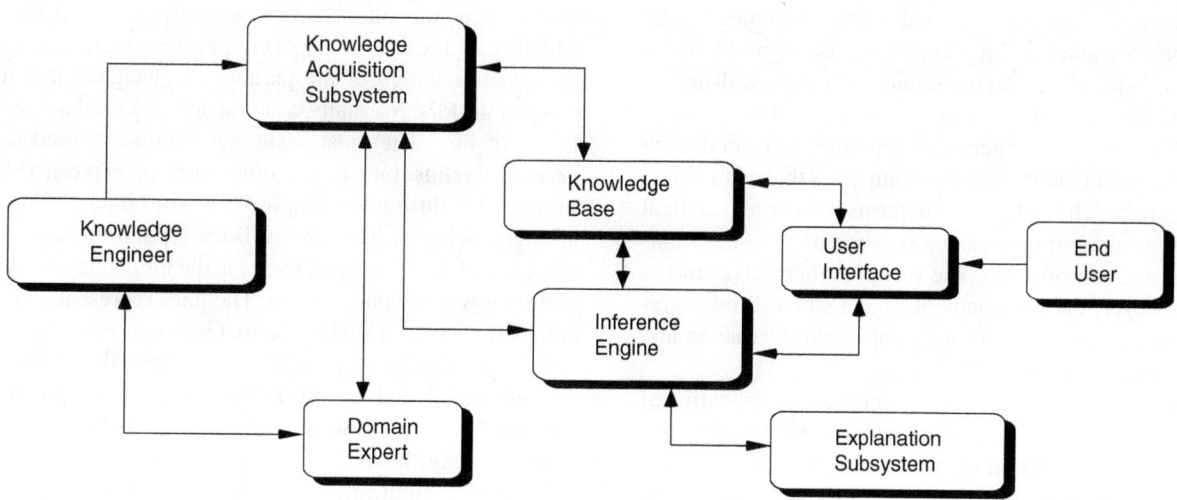

FIGURE 72.1. Expert systems: a simple schematic.

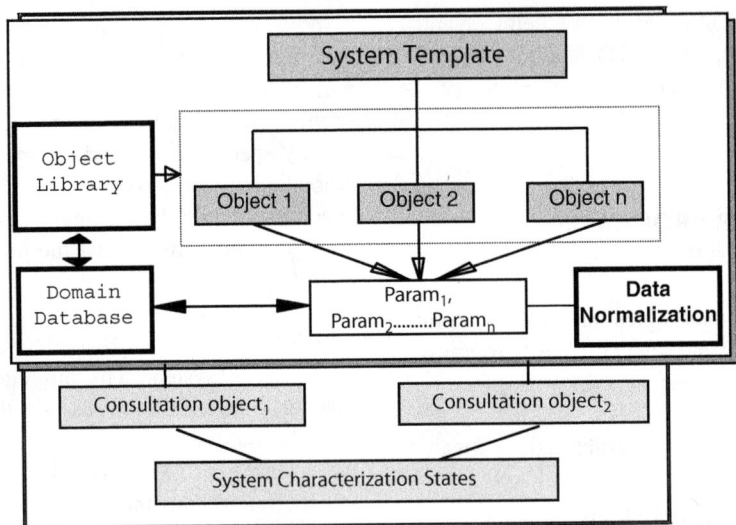

FIGURE 72.2. Structural framework of a generic development system.

computation is behavior simulation [28]. Objects represented as computational abstractions are simulated complete with the characteristics of the simulated objects. Once the abstraction has been adequately characterized, other objects can be derived from the abstraction thus created and such *derived* objects may be allowed to *inherit* the properties already specified for the abstraction. Most current-day programs and applications are built using object-oriented frameworks (such as .Net) or languages (C++, C#, or Java).

Genera [29] is an object-oriented framework for development of problem-solving systems in corrosion and materials and has been used to develop numerous problem-solving systems. The Genera system uses the following framework for implementation, a schematic representation for which is given in Figure 72.2: Material and corrosion systems can be represented as objects whose states can be defined in terms of critical parameters and interrelationships between these parameters. States of different properties of objects are encapsulated within the objects. The object-based framework represents a class of problems in terms of abstractions of commonalities between different domains. For example, an environment could be an environment relevant to refineries or pipelines or a producing well. An abstract view of the environment is that of an object which can be characterized in terms of certain critical variables or operating parameters, such as pressure, temperature, and composition. The variables themselves might vary. However, the environment object can still be represented as a function of the states of certain variables and their interrelationships.

Neural networks represent another exciting application of computing technology to corrosion. Evolving from neurobiological insights, neural network technology gives a computer system the capability to actually learn from input data. Artificial neural networks, as neural networks are commonly

referred to (because neural networks by themselves mean biological neural systems found in humans and other carbon-based organic life forms), have provided solutions to problems normally requiring human observation and thought processes. Some real-world applications in corrosion include corrosion data modeling and prediction [30], corrosion data reduction [31], and electrochemical impedance spectroscopy data analysis [32].

Neural networks perform computation in a manner quite different from that used by conventional computers, where a single central processing unit sequentially dictates every segment of activity. Neural networks are built from a large number of very simple processing elements that individually deal with pieces of a big problem. A processing element (PE) simply multiplies an input by a set of weights, and a nonlinearity function transforms the result into an output value. The principles of computation at the PE level are deceptively simple. The power of neural computation comes from the massive interconnection among the PEs which share the load of the overall processing task and from the adaptive nature of the parameters (weights) that interconnect the PEs. Normally, a neural network can have several layers of PEs. The most basic and commonly used neural network architecture is the multilayer perceptron (MLP). Figure 72.3 illustrates a simple MLP. The circles are the PEs arranged in layers. The top row is the input layer, the middle row (there can be many of these) is the hidden layer, and the bottom row is the output layer. The lines represent weighted connections (i.e., a scaling factor) between PEs.

The performance of an MLP is measured in terms of a desired signal and an error criterion. The output of the network is compared with a desired response to produce an error. An algorithm called back-propagation is used to adjust the weights a small amount at a time in a way that reduces the error. The network is trained by repeating this process many

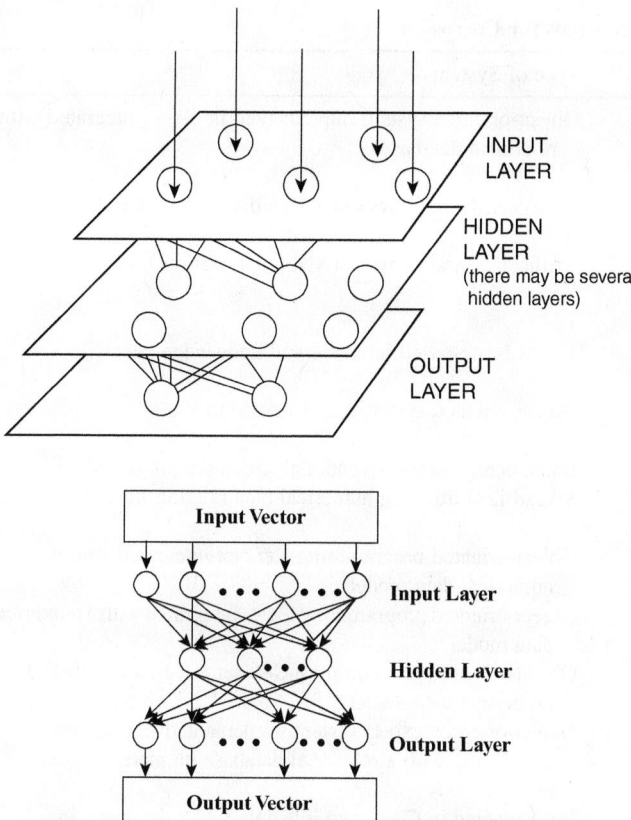

FIGURE 72.3. Schematics for a multilayered perceptron of an artificial neural network.

times. The goal of the training is to reach an optimal solution based on the performance measurement.

Current-day computer programs for corrosion assessment and control utilize one or more of the computing technologies described above. For instance, the Strategy [33] programs for evaluation of hydrogen-induced cracking in steels combine principles of expert systems and object-oriented programming, and the work done by Silverman et al. [30] integrates expert systems with neural networks.

The next section provides a brief description of specific computer programs utilized for corrosion modeling, prediction, monitoring, and control.

D. COMPUTER TOOLS: EVOLUTIONARY DEVELOPMENT TRENDS

This section will focus on examples of current-day commercial applications in corrosion that have find widespread use and application. It is important to note that numerous applications of computers and software technology have been reported over the last two decades. Just in the area of expert system development, over 57 systems were reported by two surveys [34] conducted by the Materials Technology Institute and the European Federation of Corrosion (EFC) in the late 1990s. If databases and other types of applications are included, the number of computer programs/software systems for corrosion prediction and control extend well into the hundreds. However, it would be instructive to know about systems that define the evolving trends and those that have contributed to the continuous growth of computing technology to corrosion applications. Table 72.2 provides a listing of different types of representative computer applications in corrosion currently available in the industry.

In this section, specific examples of current computer systems from different types of computing applications in corrosion are provided. These include:

Cormed, a program for corrosion prediction and assessment

SOCRATES, a software system for material selection for oil and gas applications

Predict-SW, a program for corrosion prediction and material selection for refinery sour water applications (an example case study demonstrating integration with a process modeling system) and a framework for real-time risk-based integrity and corrosion Analysis

Programs for online corrosion monitoring and control

Corsur, a database system for evaluation of metals and nonmetals

An application of neural networks to corrosion data reduction

The Cormed corrosion prediction model [16] developed by Elf (a French oil and gas company) predicts the probability of corrosion in wells. It uses a detailed analysis of field experience on CO_2 corrosion as well as data from other published computer models of corrosion. The model uses only a limited number of factors, such as CO_2, in situ pH, or bicarbonates for assessing the risk of tubing perforation in production wells. Other comprehensive corrosion prediction computer models include Predict [9, 10], LipuCor [15], and the USL [35] model, systems that examine a larger number of parameters in predicting corrosive damage.

The SOCRATES system, developed by InterCorr International, and now by Honeywell International, provides access to the material selection decisions and decision logic of a domain expert and significant experience on utilization and selection of corrosion-resistant alloy (CRA) materials and steels. The primary basis for the system data comes from two large Joint Industry Projects (JIP) conducted by Honeywell for CRA data development, which is embodied in the rules built into SOCRATES. SOCRATES also incorporates data sources such as published literature on as well as field experience related to oil and gas field service.

TABLE 72.2. Representative Sample of Current-Day Software Programs for Corrosion

Application	Type of System
Corrosion analyzer: Faciliates simulation of aqueous and nonaqueous multicomponent systems from a standpoint of thermodynamic behavior and speciation	Object-oriented system implemented in C++ integrated with relational databases
CP diagnostic: Troubleshooting and diagnosis of sacrificial anode and impressed current cathodic protection systems	Shell-based expert system with a database on CP data
EIS data extrapolation: Uses neural networks to train on electrochemical impedance spectroscopy data for extrapolation	Artificial neural network (ANN) application
Filter debris analysis (FDA) expert system: condition monitoring of aircrafts	Visual Basic-based interface and knowledge base
Genera: Generic problem-solving framework for characterizing corrosion and materials problems	Object-oriented system implemented in C++
LipuCor: Prediction of corrosion in oil and gas systems	Implemented as a conventional structured program
NORSOK: Prediction of CO_2 corrosion in oil and gas production and transmission environments	Spreadsheet utilizing numerical data relationships
Predict: Prediction of corrosion in CO_2/H_2S multiphase oil and gas production/transmission environments	Object-oriented programming C#/.Net integrated with a numerical data model
Predict-Pipe: Automation of internal corrosion direct assessment for gas transmission pipelines	Object-oriented programming C++ integrated with a numerical data model
Predict-SW: Corrosion prediction and material selection for refinery sour water applications	Object-oriented programming C#/.Net integrated with a numerical data model
Socrates: Selection of materials for oil and gas production service	Object-oriented expert system implemented in C#/.Net interfacing with a relational database on materials and compositions
Strategy: Programs for evaluation of cracking in steels used in pipelines and refineries	Implemented in C++ and integrated with databases in Microsoft Access
USL corrosion model: Program for prediction of corrosion in gas condensate wells	Implemented as an expert system in Visual Basic

The flow of data in the Socrates system is shown in Figure 72.4. Even though a specific order of decision making is indicated by the hierarchy shown in Figure 72.4, the object-oriented implementation ensures that data specification does not have to proceed in any specific sequence. At level 1, the initial set of applicable materials is obtained by determining the application for which the CRA is to be selected. If the application is not known, all classes of materials known to the system become part of the solution set. The system characterizes the environment for severity of general corrosion, localized corrosion, stress corrosion cracking, and other relevant mechanisms using common environmental parameters such as H_2S and CO_2, bicarbonates, pH, operating temperatures, chlorides, aeration/sulfur, water-to-gas ratio/water cut and gas-to-oil ratio as well as metallurgical parameters.

Current-day applications have evolved toward availability of real-time, online corrosion prediction and material selection, as opposed to the conventional offline approach. A framework for real-time availability of applications is shown in Figure 72.5. An online, real-time framework ensures that critical decisions related to corrosion problems and upset conditions may be made before the issues becomes a problem.

The framework shown in Figure 72.5 is possible only through availability of integrated corrosion prediction (Predict®-SW) and process modeling (Unisim Design) applications [36].

Predict®-SW ensures that process equipment is properly specified to deliver desired product throughput and specifications consistent with demands of the operating parameters and potential for material/corrosion damage. The tool helps evaluate effects of feed changes, upsets and equipment downtime on process safety, reliability, and profitability. It also improves plant control, operability, and safety using dynamic simulation (sensitivity analyses) of planned and existing plants.

Predict®-SW quantifies corrosion rates for the most common construction materials of construction utilized in refineries allowing planning for complex and difficult-to-quantify refinery processes. Predict®-SW utilizes a quantitative engineering database and decision support model to predict corrosion in alkaline sour water systems as a function of critical environmental parameters, such as NH_4HS concentration, H_2S partial pressure, temperature, hydrocarbon content, and chemical treatments integrated with characterization of flow regimes and wall shear stress.

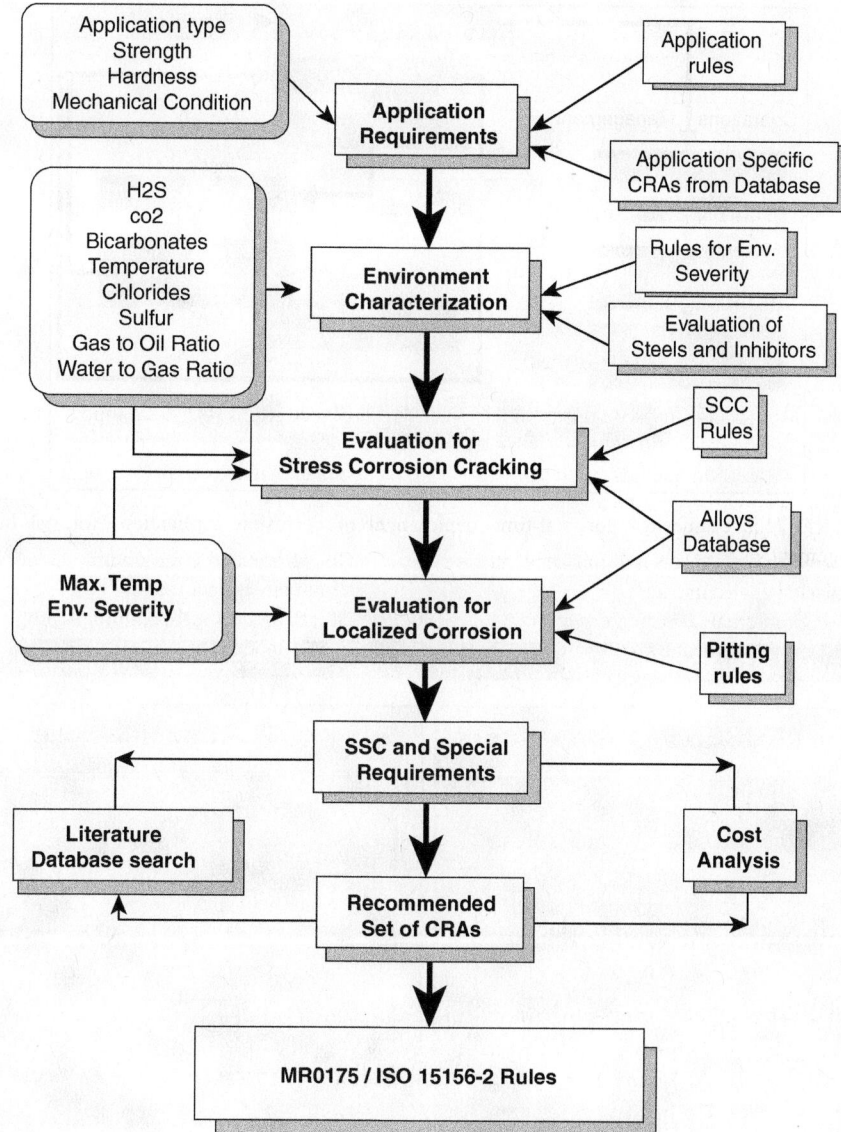

FIGURE 72.4. Flow of data in the SOCRATES system for selection of corrosion-resistant alloys.

Predict®-SW is the resultant product of this extensive research and can:

- Predict corrosion rates for a wide range of applicable conditions for 14 commonly used materials, including carbon steel, stainless, Monel, and C-276.
- Perform flow modeling, compute wall shear stress, and analyze flow regimes in multiphase flow.
- Evaluate parametric effects with a sensitivity analysis tool.
- Account for effects of light and heavy hydrocarbons.
- Correlate flow effects with corrosion rate based on extensive laboratory data and flow modeling.
- Access data used to support system decision making and analyses.

- Quantify, characterize, and analyze sour water systems helping to prevent unscheduled shutdowns.
- Facilitate multipoint analysis and data sharing using Microsoft Excel, Microsoft Word, or email.
- Quantify corrosion in alkaline sour water systems as a function of NH_4HS concentration, velocity (shear stress), H_2S partial pressure, temperature, and various additional parametric variables.
- Seamlessly share data with other analysis tools and modeling systems.
- Share data with OPC-compatible control systems such as Honeywell's Experion® Process Knowledge System (PKS), as shown in Figure 72.6. The figure details functional interactions in real-time *process modeling* using UniSim® Design and *process intelligence*

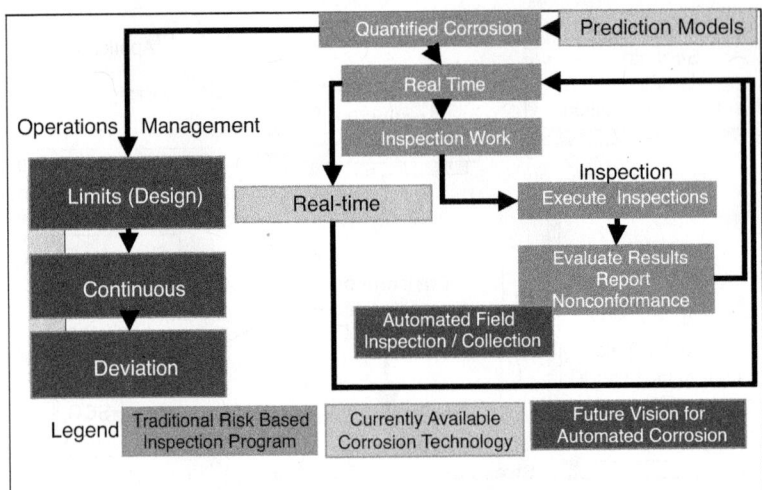

FIGURE 72.5. Framework for real-time deployment of corrosion applications for risk-based management.

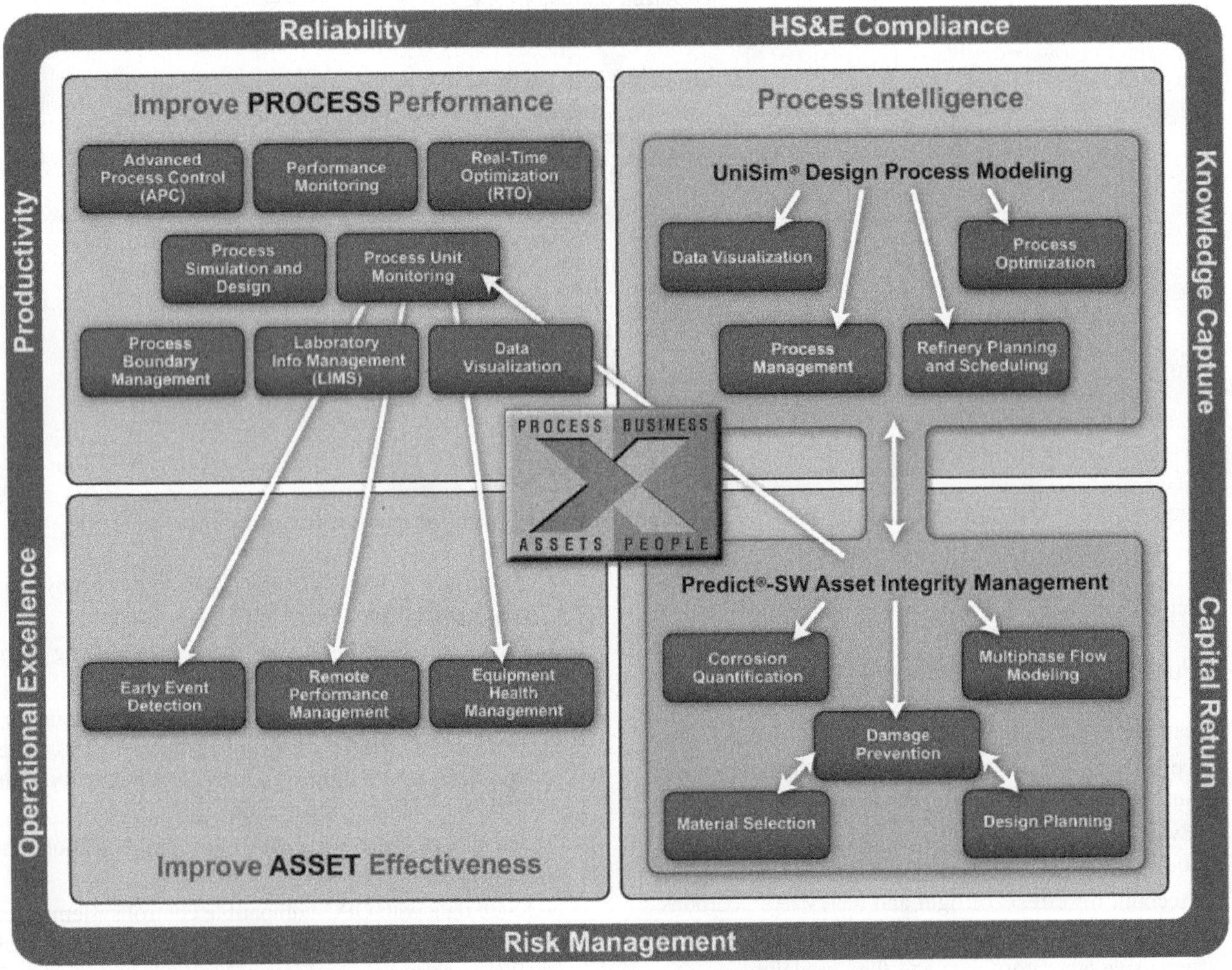

FIGURE 72.6. Automated asset integrity framework reliability and risk management in process plants.

provided by Predict®-SW as a means to *enhanced asset integrity and value retention.*

- Easily customize Predict®-SW (through Honeywell) and deploy the system on a Web-based framework.

One of the most prominent neural network applications in corrosion facilitates electrochemical impedance spectroscopy (EIS) data extrapolation [32]. EIS is reported to be by far the most popular technique for obtaining instantaneous corrosion rate information. However, one limitation of EIS data is the inability of conventional techniques for validation and extension of the data frequency range. Here, artificial neural networks have been used to extrapolate impedance data to a wider range of frequencies. Another important application of neural networks is a system that identifies the problems associated with the inherent variability of corrosion data and attempts to develop solutions to these problems [31]. These solutions are considered to be general advances in neural network technology for use with noisy or poor-quality data. In essence, they depend on the production of two further output parameters in addition to the corrosion rate. The first of these is the variance and is essentially the same as the conventional statistical parameter. The second has been termed the confidence and is an indicator of the reliability of the prediction of the neural network for a given set of input parameters. he Ecorr (acronym for engineering corrosion) [36] computer-aided learning (CAL) package represents a novel application of computing to corrosion education. This system uses advances in multimedia technology to promote interactive, computer-aided learning.

In the Ecorr approach, the student is presented with a series of problems to solve by reference to supporting information. The program itself is built as a string of books wherein information is presented in terms of "objects" such as text, photographs, video, or sound objects on a page. The system consists of a theory base, case study modules, a glossary, and a control center to coordinate navigation and information flow between the user and different modules. The theory base contains basic corrosion information that a student can use while working with a specific case study module. The organization of the system facilitates learning by users with different levels of corrosion proficiency due to the inherent ease of navigation. Currently the program is being designed to provide 12 case study modules at two levels of proficiency (basic/level 1 and advanced/level 2). Real-time computer-based control systems represent an important development in the application of computing to corrosion monitoring and control. Figure 72.7 shows the structure of an automated constant extension rate system (ACert) used for conducting corrosion and cracking evaluation tests. This system facilitates test initiation and conduct and data collection and analysis from a software system integrated with a data acquisition control system and data measurement devices. Corrosion tests can be run with minimal human intervention, and such systems promote corrosion evaluation of difficult-to-simulate environments in an automated manner.

One of the problems in conducting conventional laboratory tests stems from the fact that the time required to evaluate cracking can take months of exposure. Slow-strain-rate (SSR) tests (also referred to as constant-extension-rate tests) may be used to mechanically accelerate testing with the application of a slow constant extension rate. Such constant-extension-rate tests promote speeding up the crack initiation process and promote evaluation of material to stress corrosion cracking in a matter of hours or days. The Acert Constant Extension Rate Tester is a computer-controlled closed-loop feedback system facilitating testing over a wide range of testing speeds without the need of cumbersome gear changes. It also has an anti-backlash feature for application of cyclic loads and stresses. An Acert system can be programmed to conduct a variety of tests, including

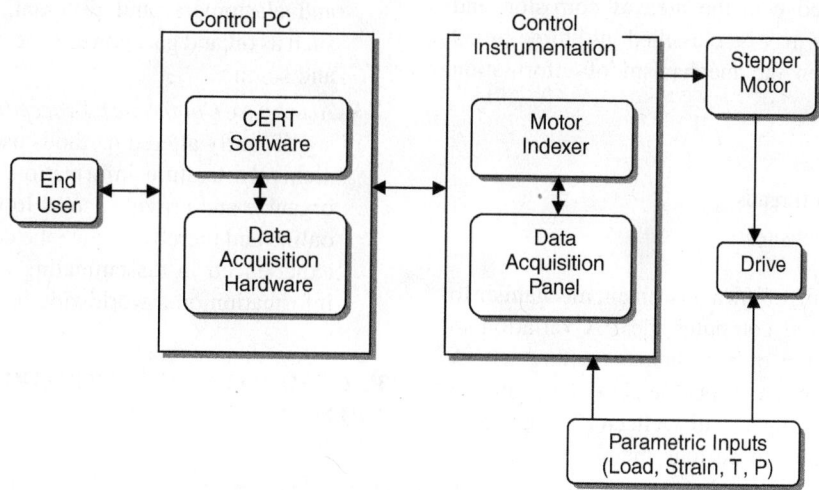

FIGURE 72.7. Schematic representation of automated constant-extension-rate tester.

constant-extension-rate tests, cyclic slow-strain-rate tests, fracture mechanics tests, and crack growth tests.

Hitherto, computing and its applications in corrosion typically dealt with individual or group-based development efforts. The advent of the Internet and the World Wide Web has fundamentally altered computing, communication, and human interaction on a global basis. Now, we have access to a new medium where we are not bound by barriers of differing media in sharing information. Given its ability to combine the best of different media such as printed information, sound, video, and computer files of different types, the Internet has truly revolutionized the manner in which corrosion engineers and scientists across the globe access, process, and share data and knowledge. The next section looks at the impact of the Internet on corrosion computing and the new directions for applications of computers in corrosion.

E. INTERNET AND WORLD WIDE WEB: NEW FRONTIERS IN COMPUTING

Since 1996, there has been significant growth in the use of the Internet in sharing and exchanging technical information [37–39]. The growth in terms of utilization of the Internet has been so compelling that the Internet has become the default medium for people to access information, knowledge, and applications [40]. The resources available to the materials/corrosion engineer/scientist fall into different categories and span a wide range and type of relevant topics, from simple descriptions of fundamental mechanisms of corrosion and cracking to technical databases focused on specialized material problems to problem-solving software tools that can be accessed and used on the Internet.

E1. Internet Resources for Corrosion Engineer

The Net offers a variety of resources to promote exchange of information and knowledge in the area of corrosion and materials. The resources may be classified into three broad categories, depending on the mechanism of information exchange:

> E-mail-based resources
>
> Blogs and discussion threads
>
> Technical resource websites

Electronic mail (e-mail) offers a convenient mechanism to exchange information and computer files. A variation to using e-mail is the notion of a technical mailing list. A mailing list represents a group of people communicating on a specific topic using a simple mail "reflector" program to distribute e-mail messages to various members that have subscribed to the mailing list. A typical example is the CORROS-L [38] list, which is a forum for people interested

in corrosion operated through the corrosion server at the Corrosion and Protection Centre at the University of Manchester at e-mail: corros-l@listserv.rl.ac.uk.

Blogs and discussion threads online evolved from the early news groups, which were the equivalent of a world wide bulletin board (or a class room) where questions and messages could be posted, comments made, discussions monitored, and answers given. There are currently countless blogs, discussions threads, and websites dedicated to corrosion/materials information Two examples of websites providing dedicated corrosion content are corrosionsource.com and corrosiondoctors.org.

The Internet has altered most conventional methods of information access in corrosion science and engineering. Already, significant technical resources are available online, including:

- *Different Types of Technical Databases.* This includes corrosion data on general, localized, galvanic, intergranular and erosion corrosion and cracking data relevant to stress corrosion cracking (SCC), Sulfide stress cracking (SSC), Hydrogen embrittlement cracking (HEC), and so on, as well as materials data on steels, stainless steels, corrosion-resistant alloys, plastics, composites, and so on, all available online
- *Corrosion and Materials Problem-Solving Software Tools.* It is now possible to run different software tools for corrosion problem solving from the Internet on a global basis. These programs provide access to state-of-the-art tools without any of the overheads associated with software procurement and distribution.
- *Archival Laboratory Testing Data and Reports.* There is already a significant trend toward providing corrosion scientists and engineers access to archival data and reports online. Several websites provide well-organized indices to data and reports on specific topics, such as general and localized corrosion, different types of steels and elastomers, and different application segments, such as oil and gas, power generation, nuclear industry, and so on.
- *Access to Conference Proceedings.* The Internet has significantly altered methods used by people to access archival literature information. It is now possible to organize and provide access to valuable literature data online and thereby obviate the conventional difficulties experienced in disseminating current technology and information on a worldwide basis.

F. COMPUTERS IN CORROSION CONTROL: CRITICAL EVALUATION

The last two decades have witnessed explosive growth in the utilization of computers to generate, analyze, and interpret

PART V

TESTING FOR CORROSION RESISTANCE

73

PRINCIPLES OF ACCELERATED CORROSION TESTING

D. L. Jordan

Ford Motor Company, Dearborn, Michigan

A. INTRODUCTION

Few topics in science and engineering incite as much spirited debate as do the selection, execution, and interpretation of accelerated corrosion tests for raw materials and consumer products. The ongoing dialogue spurred by LaQue's seminal 1952 paper [1] and the subsequent discussion [2] on salt fog testing has continued for a half-century [3–13] and has been expanded to include the plethora of accelerated corrosion tests with "improved correlation" that have been developed over the decades [14–20]. Despite the huge amount of corrosion research and improved understanding that have been accomplished, one of LaQue's oft-quoted observations still rings true: "... sooner or later one's judgement may be questioned if continued use of an unreliable test in place of the exercise of good judgement in some other way should lead to a series of poor decisions" [1, p.78]. The fact remains that engineers and technologists need rapid methods to determine the corrosion characteristics and behaviors that take a long time to mature in the natural service environment. The challenge is to develop and exercise good judgment in the selection, conduct, and interpretation of accelerated corrosion tests.

The purpose of this chapter is to encourage the reader to study and understand specific in-service environmental and exposure conditions relevant to particular situations and how those conditions affect various aspects of the overall corrosion behavior of materials in a given application. Only then can we hope to derive benefits from an accelerated corrosion test that includes all relevant stressors and degradation mechanisms.

Rather than provide a glossary of popular accelerated corrosion test methods, many of which are discussed in other chapters of this handbook, this chapter introduces some of the many considerations necessary to understand why an accelerated corrosion test may be used for a particular set of conditions. Clearly, the definition of "particular set of conditions" can vary dramatically from case to case and will involve unique definitions or interpretations of the terms *test material, environment, corrosive species, protective films,* and others, but the principles remain the same.

In keeping with the theme of this handbook, "corrosion" is defined as "the degradation of a material by its environment." The word "degrade" in this context implies that corrosion is a process (i.e., a means of changing from one state to another by way of one or more steps), not a property (i.e., a fixed characteristic that defines a specific state of material). Finally, "accelerated corrosion test" is intended to represent any of a number of cyclic cabinet tests, immersion tests, or electrochemical measurements, all of which are covered in detail by standards organizations [21] and elsewhere in this handbook.

B. PURPOSES AND PITFALLS OF TESTING

The primary goal of accelerated corrosion testing is to obtain desired information about the corrosion behavior of a

Uhlig's Corrosion Handbook, Third Edition, Edited by R. Winston Revie
Copyright © 2011 John Wiley & Sons, Inc.

material or material system in a particular application in a short period of time. Accelerated corrosion tests are conducted with a variety of purposes, including:

Quality control: Determination of batch-to-batch variability.

Quality control: Comparison with a reference standard or other acceptance criteria.

Study corrosion mechanisms: Probe total system with known factors and measure results.

Material selection: Determine the performance ranking of competing materials in a given environment.

Development of new corrosion-resistant products: Determine effects of one or more changes to the corrosion protection design, including new materials of construction and corrosion inhibitors.

Prediction of service life in specific environments: Test design characteristics and determine safety factors vis-à-vis results of comparable in-service performance characterizations.

Simulate atmospheric load on a material or object: Avoid time and expense of long-term exposure when specific individual stressors or combinations of stressors are of interest.

Simulate specific degradation mechanisms: Determine if hypothesized changes to the engineering design had the expected effect on the degradation mechanism, including the effect on the test environment.

Stimulate specific degradation mechanisms: Intentionally introduce or intensify specific stressors and combinations of conditions expected to affect the performance of the engineered system.

In all cases, the intent is to obtain approximate or directional data quickly when more exact data take an unacceptably long period of time to develop. It is clear that we must be prepared to provide an accurate description of the operable in-service corrosion mechanisms before we can aspire to reproduce those mechanisms and control their rates masterfully in a meaningful accelerated corrosion test. If that can be accomplished, we may be confident that the accelerated corrosion test provides information that is consistent with the outcome of exposure to some or all of the actual service conditions and is worthy of being used as a tool for material selection and decision making.

The difficult task is to find ways to increase the rate of the relevant corrosion processes without introducing extraneous mechanisms or altering (including completely neglecting) the specific degradation mechanisms that occur in the service environment of interest. The "environment" in this context includes not only the natural exposure environment but also all parts of the natural and engineered system that

may directly or indirectly affect the corrosion process in any way.

Accelerated corrosion tests that have no proven relationship to the service environment of interest are often irrelevant and can be misleading. With that in mind, there is more to accelerated corrosion testing than just subjecting a material to a standard test and reporting the results. For example, one should not expect information that is immediately applciable to a sulfate exposure to be provided by an accelerated corrosion test that features chloride. A relevant accelerated corrosion test may result from the careful analysis of the corrosion process in the expected service environment and the development and testing of a hypothesis regarding the behavior of the material in a test sequence that contains the appropriate stressors. These stressors include not only elements of the exposure environment itself and all parts of the engineered system but also items that may be inadvertently overlooked [e.g., metallic and nonmetallic fasteners, sheared (work-hardened) edges, identification labels, shipping oils, or any of a number of in-service contaminants].

C. DEVELOPMENT, SELECTION, AND EXECUTION OF MEANINGFUL ACCELERATED CORROSION TESTS

C1. Types of Information

There are numerous opportunities to measure meaningful quantitative and qualitative parameters that provide a glimpse into the corrosion behavior of the engineered material system at any given point during the corrosion process. The corrosion process that occurs during an accelerated corrosion test may provide reasonable estimates of some, all, or none of the measurable parameters provided by any given in-service environments, which may include polluted and nonpolluted versions of the following:

Indoor atmospheres

Outdoor atmospheres

Total immersion in chemicals, processing fluids, and various waters

Intermittent immersion in chemicals, processing fluids, and various waters

Practically infinite combinations of the above, with various contributions of temperature, pressure, soluble gases, agitation, and mechanical inputs

It is important to decide which corrosion behavior characteristics are of interest for a typical application and test with those in mind. *The time to decide is long before conducting the accelerated corrosion test, not during or after.*

Typical measurable parameters in accelerated corrosion tests include:

Visual appearance of the raw material or consumer product at the end of the test

Time to develop the first visual sign of undesirable corrosion products

Weight loss

Weight gain

Corrosion rate (in mm/y, mpy, gmd, or A/m^2)

Pit depth

Pit density

Concentration of metal ions in solution

Dissolved oxygen consumption in local or bulk electrolyte

pH change of local or bulk electrolyte

Mechanical or physical properties (e.g., tensile strength, electrical conductivity, toughness)

Electrochemical parameters (e.g., corrosion potential, current density, Flade potential)

The selection of experimental factors or conditions to include in an accelerated corrosion test depends on what is known about the stressors and corrosion mechanisms in the service environment and what decisions are to be made based on the test results. In addition, the initial, intermediate, and "final" characteristics of corrosion products have a tremendous effect on the corrosion rate in many applications and should be considered to be part of the ever-changing local exposure environment during the course of the test [22, 23].

Specific in-service conditions, including geometric configuration, orientation of the corroding surface with respect to rainfall or process fluid flow, soiling by atmospheric fallout and wandering wildlife, proximity to emissions of other corroding or otherwise reactive material systems, exposure to mechanical damage by way of external loading or internal stresses, and differential sun loading, all may seem to be outside the scope of accelerated corrosion testing but could hold the key to the establishment of a meaningful and informative test upon which knowledgeable corrosion-resistant material selection decisions may be made. *While it may not be immediately practicable to reproduce certain naturally occurring conditions in the laboratory, there is no reason to ignore their existence while testing and especially during data analysis.*

Similarly, the interaction of mechanical and environmental stresses during some in-service exposures may require a specific sequence of events during testing. For example, exposure of a specimen to a corrosive environment while under tensile loading (resulting in a yawning crack) will give different results than such an exposure in the nonloaded condition due to the differing abilities of the corrosive

environment to freely enter, react in, and exit the crack. The exposure conditions within the crack are different in the two loading conditions, so it is reasonable that the corrosion behavior within the crack will differ accordingly. As another example, metals with naturally occurring passive films may be unfairly evaluated when nonrepresentative specimen preparation inadvertently compromises the passive film prior to accelerated corrosion testing.

If the in-service conditions are well defined and understood, acceleration of degradation may be realized by exposing thoughtfully prepared specimens to purposefully manipulated combinations of temperature, type and concentration of corrosive species, or time of wetness. On the other hand, incompletely understood service conditions can lead to the selection of inappropriate exposures and acceleration factors that may reduce total testing time but may introduce irrelevant or even misleading results. For example, increasing temperature often results in decreased oxygen solubility in aqueous electrolytes, which in turn leads to unexpectedly low corrosion rates when oxygen reduction is the primary cathodic reaction [4].

Carefully collected and appropriately analyzed field data from a range of typical in-service conditions provide the best input to the development of a relevant accelerated corrosion test for future generations of raw materials or consumer products for similar service. *There is no substitute for knowing how your product performs when in the hands of your customer.*

C2. Physical and Experimental Considerations

When using an accelerated corrosion test to evaluate modifications to existing corrosion-resistant raw materials or consumer products, it may be tempting to expect an A : B comparison that is always relevant to the service condition to result. The reality is that accelerated corrosion tests with demonstrable and nontrivial correlation with in-service corrosion performance are rare [9, 16, 17], usually due to an incomplete understanding of the in-service degradation mechanisms.

The difficulty here is that, for example, an accelerated corrosion test that simply increases temperature and aggressive ion concentration should not be expected, *a priori*, to highlight differences in coating adhesion, diffusivity of specific anions though a coating, charge transfer at a modified surface, or any of a number of other corrosion-resistant product development strategies that can improve corrosion performance but are much better characterized by property-specific tests or measurements that are not normally characterized as "corrosion" tests. Relying on one of the many standard accelerated corrosion tests to reveal improved behavior resulting from a strategy change that is not specifically probed by the standard test can result in the unfair elimination of promising ideas for improved in-service performance or the unwitting acceptance of improvement ideas that look good in the test but ultimately disappoint in the field. *Know what specific material properties or characteristics are being tested or measured and why.*

The selection of aggressive ions for accelerated corrosion tests may seem to be a straightforward task. Consider the introduction of the chloride ion to a corrosion test environment. Sodium chloride is an inexpensive, widely studied, and well-accepted source of the frequently destructive chloride ion in aqueous corrosion testing solutions. In addition, sodium chloride is pervasive in daily life and is encountered in the form of road deicers, seawater, and various process fluids that contribute to undesired corrosion events in automotive, aerospace, construction, buried pipeline, and other industries. It seems natural that sodium chloride should be part of many accelerated corrosion tests.

Magnesium chloride and calcium chloride are also used a road deicers, and both are more hygroscopic than sodium chloride. Consequently, the time of wetness for a product exposed to the more hygroscopic road deicers will be higher than that of a product exposed to only sodium chloride. In turn, the in-service corrosion rate for materials exposed to the more hygroscopic road deicers may be expected to be higher, all other considerations equal. Accordingly, a continuously wet accelerated corrosion test would not be expected to distinguish the differences in hygroscopicity, time of wetness, or corrosion rate and would fail to detect the true in-service differences in corrosion rates. Not all sources of chloride ions give the same rank order in service as they do in accelerated corrosion tests. The same may be said for other components of accelerated corrosion tests.

In accelerated corrosion tests, only certain aggressors are delivered. Some critical aggressors (e.g., thermal cycling to expand and contract surface cracks or to compromise otherwise adherent and protective oxide films) may be key components of in-service conditions but may not be represented at all in an accelerated corrosion test. A limited subset of aggressive ions or species, some of which participate in the initial corrosion reactions and some of which participate in subsequent reactions, and the possible absence of other necessary but not sufficient reactants will allow the painting of only an incomplete picture. Furthermore, some products of initial reactions themselves act as reactants in subsequent reactions [22, 23]. *Development of nonrepresentative corrosion products can propagate a series of errors and lead to poor decisions.*

The presence of a chemical or biological species in the exposure environment or in the resultant soluble and insoluble products of the initial corrosion reactions does not imply that the given species actually participates in the overall reaction. A given species may be an innocent bystander or, at worst, a possible "inhibitor" to the acceleration of the reactions of interest if present in inappropriate concentrations or introduced at an inopportune moment in the overall degra-

dation process (e.g., while conditions are such that a previously yawning crack is held closed).

Another practical example of this is where calcium was detected in corrosion products in an automotive body application and a calcium compound was introduced into a developmental accelerated corrosion test. The undissolved portion of the calcium compound clogged nozzles in the test chamber and actually decreased the corrosion rate of materials in the test. The clogging issue was overcome and the intended corrosivity was restored [18], but the exact role of the calcium ions in the test has not yet been established.

All is not lost. The point is not to discourage the reader from using accelerated corrosion tests to study and understand corrosion behavior of engineering material and environmental systems but rather to urge careful consideration of how the particular stressors characteristic of the selected accelerated corrosion test relate to every aspect of the corrosion behavior in the intended service application.

Test parameters and methods must be adjusted to suit the numerous materials and corrosion protection strategies associated with exposure to gaseous, liquid, and solid environments. A nonexhaustive list of possible factors to consider in an accelerated corrosion test for any given in-service condition follows:

Reactive ionic and nonionic chemical or biological species in aqueous solutions

Reactive ionic and nonionic chemical or biological species in nonaqueous solutions

Nonreactive but possibly participative ionic and nonionic chemical or biological species in aqueous or nonaqueous solutions

Oxygen gas

Ozone

Solar radiation of various wavelengths

Particles that abrade the material surface

Particles that temporarily adhere to the material surface

Particles that permanently adhere to the material surface

Temperature extremes and cycles

Pressure extremes and cycles

Rate of replenishment of reactants to the material surface

Rate of replenishment of nonreactants to the material surface

Level of agitation (or stagnation) of environment

Initial surface chemistry and condition of the test material

Initial, intermediate, and final reaction products (soluble)

Initial, intermediate, and final reaction products (insoluble, including biofilms and gels)

Initial, intermediate, and final reaction products (insoluble and adherent, including biofilms and gels)

Over the years, the most important insight that has come to light is that the highest rate of corrosion damage occurs during the transition from wet to dry, not when a material is fully immersed in the corrosive environment [23]. During the drying event, the concentrations of all reactants become optimum for rapid metal dissolution. Reproducing such an event in a practical laboratory accelerated corrosion test for consumer products continues to be an elusive endeavor.

C3. Statistical Considerations

The topic of experimental design, particularly the selection of sample size or replication, is a bitter rival to accelerated corrosion testing for the spirited scientific debate crown. As can be expected, mixing the two can result in ill-advised decisions or gridlock if not handled judiciously.

Replication, randomization, and blocking are the cornerstones of all experimental designs and should receive full consideration before embarking on an accelerated corrosion testing program. Plan for the final statistical analysis long before preparing specimens for testing. Decide what will be measured (and with what instruments) so that the measuring system may be evaluated for repeatability, reproducibility, accuracy, and precision long before estimates of experimental error are needed. Decide what will be measured (and with what tools), estimate the standard deviation of the values to be measured, commit to a statistical level of significance to be met for both type I and type II errors, and determine the difference between two values that is deemed to be practically significant. With that information in hand, one can calculate the sample size (replication) necessary to achieve the stated goals. Operating characteristic (OC) curves are useful in this regard [24]. Many promising product ideas have been abandoned due to inadequate replication, resulting in the inflation of type I and type II errors during initial screening experimentation. Numerous statistical texts outline the process of experimental design and statistical data analysis [24–27].

D. CONCLUDING REMARKS

An athlete may be judged in initial tryouts by his or her ability to run a given distance in a short period of time, to lift a prescribed weight numerous repetitions, or to accomplish any of a number of astonishing physical feats, but the true test of the athlete's ability to perform comes when all of the conditions of the particular sport are present in the correct proportions and under relevant circumstances, that is, during game or match conditions. Particular physical skills, as independently measured, are mere predictors or enabling attributes that contribute to the athlete's ability to play a given sport and may have little relevance outside of the testing environment.

Similarly, an accelerated corrosion test or series of tests and measurements cannot, *a priori*, be expected to tell a complete story. Expert consideration, coaching if you will, is needed to derive the desired information from an imperfect set of raw data.

Don't just toss specimens into a test cabinet and ignore them until the test is completed—observe the degradation process by whatever means available to illuminate the characteristics that give you the information needed to make a decision.

Corrosion is a process, not a property. Development, selection, conduct, and interpretation of accelerated corrosion tests benefit greatly if that fact is remembered.

REFERENCES

1. F. L. LaQue, "A Critical Look at Salt Spray Tests," Mater. Methods, **35**(2), 77–81 (1952).
2. F. L. LaQue, "A Critical Look at Salt Spray Tests," Mater. Methods, **35**(3), 77–81, 156, 158, 160, 162, 164, 166, 168 (1952).
3. W. D. McMaster, The Accelerated Corrosion Testing of Metals, First International Congress on Metallic Corrosion, Butterworths, London, 1962, pp. 679–684.
4. L. Schlossberg, "Corrosion Theory and Accelerated Testing Procedures, Part 1," Met. Finish., **62**(4), 57–63 (1964).
5. L. Schlossberg, "Corrosion Theory and Accelerated Testing Procedures, Part 2," Met. Finish., **62**(5), 93–99 (1964).
6. J. Mazia, More Metal Finishing Myths (MFM's). Metal Finishing **75**(5), 77–81 (1977).
7. T. S. Lee and K. L. Money, "Difficulties in Developing Tests to Simulate Corrosion in Marine Environments," Mater. Perform., **23**(8), 28–33 (1984).
8. W. Funke, et al., "Unsolved Problems of Corrosion Protection by Organic Coatings: A Discussion," J. Coat. Technol., **58**(741), 79–86 (1986).
9. Cleveland Society for Coatings Technology Technical Committee, "Correlation of Accelerated Exposure Testing and Exterior Exposure Sites," J. Coat. Technol., **66**(837), 49–67 (1994).
10. J. Maxted, "Short Term Testing and Real Time Exposure," J. Corros. Sci. Eng., **2**(15) (1999).
11. J. Repp, "Accelerated Corrosion Testing—Truths and Misconceptions," Mater. Perform., **41**(9), 60–63 (2002).
12. J. Guthrie, B. Battat, and C. Grethlein, "Accelerated Corrosion Testing," AMPTIAC Q., **6**(3), 11–15 (2002).
13. G. J. Jorgensen, "A Phenomenological Approach to Obtaining Correlations between Accelerated and Outdoor Exposure Test Results for Organic Material," J. Test. Eval., **32**(6), 494–499 (2004).
14. K. Barton, Acceleration of Corrosion Tests on the Basis of Kinetic Studies of the Rate Controlling Combination of Factors., First International Congress on Metallic Corrosion, Butterworths, London, 1962; pp. 685–690.

15. D. Grossman, "More Realistic Tests for Atmospheric Corrosion," J. Protec. Coat. Linings, **13**(9), 40–45 (1996).

16. B. Boelen, B. Schmitz, J. Defourny, and F. Blekkenhorst, "A Literature Survey on the Development of Accelerated Laboratory Test Methods for Atmospheric Corrosion of Pre-coated Steel Products," Corros. Sci., **34**(11), 1923–1931 (1993).

17. A. Forshee, "Accelerated Corrosion Testing, Part 1: An Overview of 20 Possible Accelerated Corrosion Tests," Met. Finish., **91**(9), 51–54 (1993).

18. SAE, J2334, Laboratory Cyclic Corrosion Test, SAE International, Warrendale, PA, 2003.

19. F. Altmayer, "Choosing an Accelerated Corrosion Test," Met. Finish., **100**(1A), 572, 574–578 (2002).

20. H. D. Hilton, "Selecting a Cyclic Corrosion Test Cabinet," Mater. Perform., **42**(2), 76–79 (2003).

21. R. Baboian, (Ed.), Corrosion Tests and Standards: Application and Interpretation, Manual MNL 20, American Society for Testing and Materials, Philadelphia, 1995.

22. D. L. Jordan, "Location and Identity of the Cathodic Reaction During Underfilm Corrosion of Painted Galvanized Steel", Dissertation, Illinois Institute of Technology, Chicago, IL, 1996.

23. M. Stratmann and H. Streckel, "On the Atmospheric Corrosion of Metals Which are Covered with Thin Electrolyte Layers—II. Experimental Results," Corros. Sci., **30**(6/7), 697–714 (1990).

24. M. G. Natrella, Experimental Statistics, Handbook 91, National Bureau of Standards, Washington, DC, 1963.

25. W. Nelson, Accelerated Testing, Wiley, New York, 1990.

26. D. C. Montgomery, Design and Analysis of Experiments, 2nd ed., Wiley, New York, 1984.

27. G. E. P. Box, J. S. Hunter, and W. G. Hunter, Statistics for Experimenters, 2nd ed., Wiley, Hoboken NJ, 2005.

74

HIGH-TEMPERATURE OXIDATION—TESTING AND EVALUATION

C. A. C. Sequeira

Instituto Superior Técnico, Lisboa, Portugal

A. Introduction
B. Spectroscopy
C. Conclusions
References

A. INTRODUCTION

One of the key parameters in high-temperature oxidation is the parabolic rate constant. This is true as long as protective oxidation determines the material behavior. Consequently, measurement of the weight change of the specimen is the key experimental technique in high-temperature oxidation. Generally, there are two possibilities. The first is to take a number of specimens of the same type, expose them to the respective atmosphere in a closed furnace in which defined atmospheres can be established, and take the specimens out after different oxidation states. Before and after the tests, the specimens are weighed on a high-resolution laboratory balance and the weight change Δm provided by the surface area of the specimen, A, is plotted versus time. Therefore, before exposure, the surface area A of the specimens has to be determined accurately. If the results are plotted in a parabolic manner, that is, $\Delta m/A$ is squared while t remains linear, the rate constant can be determined directly from the slope, as discussed in Chapter 20.

A more elegant method is to use continuous thermogravimetry [1]. In this case, a platinum or quartz string is attached to a laboratory balance that extends down into a furnace. At the lower end of the string, a coupon specimen is attached for which the surface area must be determined before the test. A movable furnace can also be installed that allows thermocyclic oxidation testing by periodically moving the furnace over and away from the specimen area. Ideally, the specimen should be in a quartz chamber or a chamber of another material that is highly corrosion resistant so that defined gas atmospheres can be used in the tests. The interior of the microbalance must be shielded against the aggressive gas atmospheres, usually by a counterflux of a nonreactive gas, such as argon. In a more sophisticated type of thermobalance, acoustic emission measurements can be made, for example, by acoustic emission thermography (AET) [2]. This becomes possible if a waveguide wire is attached to the string hanging down from the balance, to which the specimen is attached. In particular, under thermocyclic conditions, acoustic emission measurements allow the determination of the critical conditions under which the oxide scales crack or spall [3]. If this type of scale damage is accompanied by mass loss due to spallation of the scale, then this is directly reflected in the mass change measurements and can be correlated with acoustic emission results.

In some situations, internal oxidation or corrosion may occur which cannot be detected directly by thermogravimetric measurements. Therefore, it is necessary to perform metallographic investigations as well. In particular, for continuous thermogravimetric testing, at the end of each test a metallographic cross section should be prepared in order to check whether the mass change effects measured in the tests are caused by surface scales alone or whether the metal cross section has been significantly affected. Furthermore, if the kinetics of internal corrosion are to be determined, it is necessary to perform discontinuous tests where specimens are taken out of the test environment after different testing times and then investigated by metallographic techniques [1, 4–7].

Uhlig's Corrosion Handbook, Third Edition, Edited by R. Winston Revie
Copyright © 2011 John Wiley & Sons, Inc.

Standard high-temperature corrosion investigations usually also include an analysis of the corrosion products formed in the tests or under practical conditions because this allows conclusions as to which are the detrimental species in the environment and whether protective scales had formed. In most cases, this is done either in the scanning electron microscope by using energy-dispersive X-ray analysis or with metallographic cross sections in the electron probe microanalyzer (wavelength-dispersive X-ray analysis). Another tool may be X-ray diffraction by the grazing angle (GAXRD) technique, which allows analysis of the composition of thin layers.

Common experimental investigation techniques used to assess corrosion morphology, identify corrosion products, and evaluate mechanical properties are the following [1, 4, 8–11]:

(a) Corrosion morphology assessment

Optical microscopy
Scanning electron microscopy (SEM)
Electron probe microanalysis (EPMA)
Transmission electron microscopy (TEM)

(b) Corrosion product identification

X-ray diffraction (XRD)
Energy-dispersive X-ray analysis (EDX)
Secondary ion mass spectroscopy (SIMS)
X-ray photoelectron spectroscopy (XPS)
Auger electron spectroscopy (AES)
Laser Raman spectroscopy (LRS)

(c) Mechanical properties evaluation

Creep rupture
Postexposure ductility
Modulus of rupture (MOR)

In general, creep rupture, hardness, and MOR have been used equally to assess the mechanical properties of corroded test pieces. When the material is difficult to grip (as is a ceramic), its strength can be measured in bending. The MOR is the maximum surface stress in a bent beam at the instant of failure [International System (SI) units, megapascals; centimeter–gram–second (cgs) units, $10^7 \, dyn/cm^2$]. One might expect this to be exactly the same as the strength measured in tension, but it is always larger (by a factor of about 1.3) because the volume subjected to the maximum stress is small and the probability of a large flaw lying in the highly stressed region is also small. (In tension all flaws see the maximum stress.) The MOR strictly applies only to brittle materials. For ductile materials, the MOR entry in the database is the ultimate strength [1].

Spectroscopic techniques used for chemical analysis of oxidation problems and characterization of thin layers of corrosion scales are of considerable importance, as discussed in the following paragraphs.

B. SPECTROSCOPY

Chemical analysis by spectroscopy has made rapid advances in high-temperature studies and almost always includes equipment for high-resolution microscopy. Several books and monographs are available, including most of the old and newly developed techniques [10, 12–14]. Glow discharge spectroscopy (GDMS) is fast, sensitive, accurate, simple, and reliable and can be used for surface analysis if the specimen can be attached to a vacuum cell [15].

TABLE 74.1. Methods of Material Characterization by Excitation and Emission

Primary Excitation	Detected Emission	Methods of Analysis: Name and Nomenclature
Photons, optical	Optical	Spectroscopy
		AA: Atomic absorption)
		IR: Infrared)
		UV: Ultraviolet)
		Visible)
	Electrons	UPS: Vacuum UV photoelectron spectroscopy *Outer shell*
Photons,	Electrons	XPS: X-ray photoelectron spectroscopy
X-rays		*Inner shell*; also called ESCA
		ESCA: Electron spectroscopy for chemical analysis
	X-rays	XFS: X-ray fluorescence spectrometry
		XRD: X-ray diffraction
Electrons	X-rays	EPMA: Electron probe microanalysis
	Electrons	SEM: Scanning electron microscopy
		TEM: Transmission electron microscopy
		STEM: Scanning transmission electron microscopy
		SAM: Scanning Auger microanalysis
		AES: Auger electron spectroscopy
Ions	Optical	SCANIIR: Surface composition by analysis of neutral and ion impact radiation
	X-rays	IIXA: Ion-induced X-ray analysis
	Ions (±)	ToFMS: Time-of-flight mass spectrometry
		SIMS: Secondary ion mass spectrometry
		IPM: Ion probe microanalysis
		ISS: Ion scattering spectrometry
		RBS: Rutherford backscattering spectrometry
Radiation	Optical	ES: Emission spectroscopy
	Ions (±)	SSMS: Spark source mass spectrography

TABLE 74.2. Summary of Various Characteristics of the Analytical Techniques

Characteristic	AES	XPS	ISS	SIMS	RBS	NRA	IXX
Sample alteration	High for alkali halogen organic Insulators	Low	Low	Low	Very low	Very low	Very low
Elemental analysis	Good	Good	Good	Poor	Fair	Fair	Good
Sensitivity, variation, resolution	Good	Good	Fair	Good	Fair	Good	Good
Detection limits	0.1%	0.1%	0.1%	$10^{-4}\%$ or higher	$10^{-3}\%$ or higher	$10^{-2}\%$ or higher	$10^{-2}\%$ or higher
Chemical state	Yes	Yes	No	Yes	No	No	No
Quantification	With difficulty, req. standards	With difficulty, req. standards	With difficulty, req. standards	Very difficult, req. standards	Absolute, no standards	Absolute, no standards	Absolute, no standards
Lateral resolution	200 nm	2 mm	100 m	100-1 m	1 mm	1 mm	1 mm
Depth resolution	Atomic layer to	Atomic layer to	Atomic layer to	Atomic layer to	10 nm	10 nm	None
Depth analysis	Destructive, sputter	Destructive, sputter	Destructive, sputter	Destructive, sputter	Non destructive	Non destructive	Very difficult

TABLE 74.3.   Outline of Some Important Techniques to Study Metallic Surfaces

Technique	Abbreviation	Information	Comments
Optical microscopy	OM	Surface topography and morphology	Inexpensive but modest resolving power and depth of field
Transmission electron microscopy	TEM	Surface topography and morphology	Very high resolution but requires replication; artifacts can be a serious problem
Scanning electron microscopy	SEM	Surface topography and morphology combined with X-ray spectroscopy gives "bulk" elemental analysis	Resolving power >> optical microscopy; preparation easier than TGEM and artifacts much less likely
X-ray photoelectron spectrometry	XPS (ESCA)	Chemical composition, depth profiling	Especially useful for studying adhesion of polymers to metals
Secondary ion mass spectroscopy	SIMS	Elemental analysis in "monolayer range," chemical composition and depth profiling	Extremely high sensitivity for many elements
Auger electron spectroscopy	AES	Chemical composition, depth profiling, and lateral analysis	High spatial resolution which makes the technique especially suitable for composition-depth profiling
Contact angle measurement	—	Contamination by organic compounds	Inexpensive; rapid

The resolving power in depth profiling is similar to AES and SIMS. A 1-kV glow discharge causes ion bombardment and surface erosion which is fed (optically) to a multichannel spectrometer for elemental analyses. Other long and complex methods of surface analysis, such as AES, SIMS, XPS, ion scattering spectroscopy (ISS), Rutherford backscattering (RBS), nuclear reaction analysis (NRA), ion-induced X-ray emission (IIXA), and ESCA, are difficult for field use. Several authors have reviewed these methods [16–28]. Tables 74.1–74.4 compare the techniques, and Figure 74.1 shows the relative sizes of areas analyzed using these techniques.

Commonly used methods are optical and scanning electron microscopy for surface studies. Transmission electron microscopy of interfaces has been explored. Selected area diffraction patterns show the orientation between different grains. In a ceramic coating, the interface between different phases can be coherent, semicoherent, or incoherent. Coherent phases are usually strained and can be studied by TEM contrast analysis. Other aspects of analytical electron microscopy analysis are discussed [29, 30]. TEM resolution is better than 1 nm and selected volumes of 3 nm diameter can be chemically analysed. Methods of preparing thin TEM transparent foils are described [16, 31–33].

TABLE 74.4.   Types of Samples and Techniques Nearly Appropriate for Their Analysis

Required Sample Analysis	Appropriate Technique
Depth profiling lower Z elements and thin films; trace or minor analysis of light elements; quantitative analysis	NRA
Depth profiling of higher Z elements and thin films; trace analysis of heavy elements in light matrix; quantitative analysis	RBS
Trace, minor, and major element analysis in thicker samples; quantitative analysis	IIX
Minor and major elements at surface or interface of small samples	AES
Trace elements at surface or interface of medium to small size samples; analysis of insulators; sputter profiling of light elements	SIMS
Chemical state analysis; analyses of organics, insulators	XPS
Analysis of outer atom layer; analysis of insulators	ISS

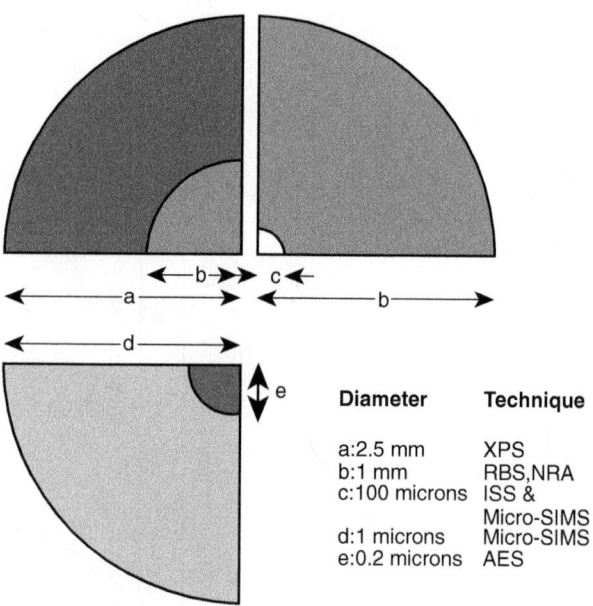

Diameter	Technique
a:2.5 mm	XPS
b:1 mm	RBS,NRA
c:100 microns	ISS & Micro-SIMS
d:1 microns	Micro-SIMS
e:0.2 microns	AES

FIGURE 74.1. Schematic illustrating relative sizes of areas scanned by spectrometric analytical techniques.

Photoemission with synchrotron radiation can probe surfaces on an atomic scale [34, 35], but this method requires expensive equipment. Complex impedance measurements can separate surface and bulk effects, but problems of interpretation need to be resolved [10, 34]. X-ray and gamma radiographs, as used in weld inspection, can be used to inspect coatings for defects. The method has been discussed by Helmshaw [36]. Inclusions, cracks, porosity, and sometimes lack of fusion can be detected. Surface compositions of ion-implanted metals have been studied by RBS [17, 35]. In this nondestructive way a microanalysis of the near-surface region is obtained. Interpretation is relatively easy. Assessment of radiation damage in ion-implanted metals by electron channelling is described using SEM [35]. NRA for the characterization of surface films is described [10, 35].

AES and XPS analyze the top of the surface only and erosion by ion bombardment or mechanical tapering is needed to analyze deeper regions. AES detects 0.1% of an impurity monolayer in a surface. Auger electrons are produced by bombarding the surface with low-energy (1–10 keV) electrons. In XPS the surface is exposed to a soft X-ray source and characteristic photoelectrons are omitted. Both AES and XPS electrons can escape from only 1 nm depth from the surface, and so these are surface analytical methods [10, 17].

It is most important to avoid surface contamination during preparation for surface analysis. Semiquantitative in situ analysis by AES has been reported [37]. Nitrides and other compound refractory coatings are frequently analyzed by AES and RBS methods. Depth and crater edge profiling have been done for direct-current (dc) magnetron sputtered and ARE (activated reactive evaporation) samples of (Ti, Al) N, TiN, and TiC coatings [38, 39]. Round-robin tests of characterization by including a range of analyses such as XPS, EPMA, XRS, AES, Atom Probe Microanalysis (APMA), and XRD are not uncommon. Among these, XRD was felt to be unreliable [40].

Ion spectroscopy is a useful technique for surface analysis [10, 17]. ISS uses low-energy back-scattered ions [41] and has a high sensitivity. SIMS has the possibility of sputter removal of layers allowing depth profiling [17]. It can act as a stand-alone single system to solve surface analysis. Three-dimensional SIMS of surface-modified materials and examination of ion implantation is reported [42]. Lattice vacancy estimation by positron annihilation is another approach [10, 43]. Transmission and scanning electron microscopy (TEM and SEM) are valuable techniques, and replication methods, using, for example, acetate replicas, can nondestructively reveal surface features of specimens too thick for TEM [1, 17]. ARE coatings of V–Ti in C_2H_2 give wear-resistant (V,Ti)C coatings. The hardness is related to grain size, stoichiometry, free graphite, and cavity networks. SEM and XRD analysis could not be used to explain the large hardness variations obtained by varying temperature and gas pressure, but TEM revealed

microstructural changes [1, 10, 16]. Beta backscatter and X-ray fluorescence have low sensitivity (0.5 cm²/min and 1 cm²/h, respectively). Thickness and uniformity of silica coatings on steel have been determined by X-ray fluorescence measurements of Si concentrations along the surface [16, 44]. Round-robin tests for microstructure and microchemical characterization of hard coatings have included XPS, UPS, AES, EELS (electron energy loss spectroscopy), EDX, WDX (wave-dispersive X-ray analysis), RBS, SIMS, TEM, STEM, and XTEM (X-ray transmission electron microscopy) [1, 44–46]. Field emission STEM has been applied for profiling Y across a spinel–spinel grain boundary [47].

C. CONCLUSIONS

In studying oxidation behavior at high temperatures, the foremost requirement is to monitor the extent and kinetics of attack. To obtain a complete mechanistic understanding, such data have to be augmented by precise details of all the processes involved, starting with the chemical reaction sequence leading to the formation of gaseous products and solid products at the reacting surface. The development and failure of protective surface scales crucially govern the resistance of most materials in aggressive environments at elevated temperatures. Knowledge is also essential on the changes throughout the exposures of the scale chemical composition, physical structure (including topography), stress state, and mechanical properties as well as on the scale failure sequence (e.g., by cracking and spallation).

All these processes involved in high-temperature oxidation are dynamic. Therefore, to obtain unambiguous information, the main experimental approach in research should be based on in situ methods. These can be defined as being techniques which either measure or observe directly high-temperature oxidation processes, as they happen, in real time. Although numerous in situ methods have been developed, to date, with several notable exceptions, the most important being controlled atmosphere thermogravimetry, the deployment of these techniques often has been limited. This may be attributed largely to experimental difficulty and also to the lack of suitable equipment. Current understanding of the chemical and physical characteristics, stress state, and mechanical properties of oxidation scales largely derives from postoxidation investigations. In fact, certain detailed aspects, for example, variations in mechanical properties and microstructure through scales, can be revealed only by postoxidation studies. The two main experimental approaches, in situ oxidation and postoxidation, are not mutually exclusive, as they complement and augment each other. Nevertheless, at the current state of mechanistic knowledge of high-temperature oxidation, further understanding of many critical facets (e.g., the breakdown of protective oxide scales) will be revealed only by real-time experimentation. These requirements taken in conjunction with recent advances in both commercial and experimental equipment design/capabilities and in data storage/processing make it imperative that all investigators in this field be fully aware of the available in situ experimental test methods.

The purpose of this chapter is to provide a very brief summary of the main techniques in current use, their main limitations, and scope for development. Information on the detailed methodology of any technique or the complete results of any specific study using any such technique should be obtainable from the references given to published papers.

REFERENCES

1. H. J. Grabke and D. B. Meadowcroft (Eds.), Guidelines for Methods of Testing and Research in High Temperature Corrosion, Institute of Materials, London, 1995.

2. M. Walter, M. Schütze, and A. Rahmel, Oxid. Met., **39**, 389 (1993).

3. M. Schütze, Protective Oxide Scales and Their Breakdown, The Institute of Corrosion, Wiley, Chichester, 1997.

4. N. Birks, G. H. Meier, and F. S. Pettit, Introduction to the High Temperature Oxidation of Metals, Cambridge University Press, Cambridge, 2006.

5. Y. Wouters, A. Galerie, and J. P. Pettit, Solid State Ionics, **104**, 89 (1997).

6. B. Glaser, K. Rahts, M. Schorr, and M. Schütze, Sonderband des Praktischen Metallographie, **25**, 75 (1994).

7. H. Baboian (Ed.), High-Temperature Gases (Corrosion Tests and Standards: Applications and Integration), American Society for Testing and Materials, Washington, DC, 1995.

8. S. Taniguchi, T. Maruyama, M. Yoshiba, N. Otsuka, and Y. Kawahara (Eds.), High Temperature Oxidation and Corrosion 2005, Trans. Tech. Publications, Zürich, 2006.

9. A. Rahmel (Ed.), Aufbau von Oxidschichten auf Hochtemperaturwerkstoffen und ihre technische Bedeutung, Deutsche Gesellschaft für Metallkunde, Oberursel, 1982.

10. P. Marcus and F. Mansfeld (Eds.), Analytical Methods in Corrosion Science and Engineering, CRC Press, Taylor & Francis Group, Boca Raton, FL, 2006.

11. C. A.C. Sequeira, Y. Chen, D.M.F. Santos, and X. Song, Corros. Prot. Mater., **27**, 114 (2008).

12. P. A. Psaras and H. D. Langford (Eds.), Advancing Materials Research, National Academy Press, Washington, DC, 1987.

13. O. Kubaschewski and B. E. Hopkins, Oxidation of Metals and Alloys, Butterworths, London, 1967.

14. P. Kofstad, High Temperature Corrosion, Elsevier Applied Science, London, 1988.

15. R. Berneron and J. C. Charbonnier, in Proceedings 7th ICVM Conference, Tokyo, 1982, p. 592.

16. E. Lang (Ed.), Coatings for High Temperature Applications, Applied Science, New York, 1983.

17. D. M. Brewis (Ed.), Surface Analysis and Pretreatment of Plastics and Metals, Applied Science, London, 1982.

18. H. Nickel, Y. Wouters, M. Thiele, and J. Quadakkers, J. Anal. Chem., **361**, 540 (1998).

19. L. Pfeil, J. Iron Steel Inst., **119**, 501 (1929).

20. L. Pfeil, J. Iron Steel Inst., **183**, 237 (1931).

21. J. Mougin, M. Dupeux, A. Galerie, and L. Antoni, Mater. Sci. Technol., **18**, 1217 (2002).

22. J. L. Liu and J. M. Blakely, Appl. Surf. Sci., **74**, 43 (1994).

23. W. D. Jennings, G. S. Chottiner, and G. M. Michal, Surf. Interf. Anal., **ii**, 377 (1988).

24. W. J. Quadakkers, J. Jedlinski, K. Schmidt, M. Krasovec, G. Borchardt, and H. Nickel, Appl. Surf. Sci., **47**, 261 (1991).

25. H. J. Grabke, M. Steinhorst, M. Brumm, and D. Wiener, Oxid. Met., **35**, 199 (1991).

26. D. Clemens, K. Bongartz, W. Speier, R. Hussey, and W. J. Quadakkers, Fresenius J. Anal. Chem., **346**, 318 (1993).

27. H. Viefhaus, K. Hennesen, M. Lucas, E. M. Muller-Lorenz, and H. J. Grabke, Surf. Interf. Anal., **21**, 665 (1994).

28. H. Bohm, Metalloberflache, **46**, 3 (1993).

29. M. J. Thoma, Vac. Sci. Technol., **A4**, 2633 (1986).

30. H. Hansmann and J. Mosle, Adhesion, **26**, 18 (1982).

31. B. E. Jacobson and R. E. Bunshah (Eds.), Films and Coatings for Technology, CEI Course, Stockolm, Sweden, 1981.

32. J. Doychak, J. L. Smialek, and T. E. Mitchell, Met. Trans. A, **20A**, 499 (1989).

33. A. Strecker, U. Salzberger, and J. Mayer, Practical Metallogr., **30**, 482 (1993).

34. J. Pask and A. Evans (Eds.), Surfaces and Interfaces in Ceramic Metal Systems, Plenum, New York, 1980.

35. V. Ashworth, W. A. Grant, and R. P. M. Procter (Eds.), Ion Implantation in Metals, Pergamon, London, 1980.

36. R. Helmshaw, Industrial Radiology, Applied Science, London, 1982.

37. A. Bosseboeuf and D. Bouchier, Surf. Sci., **162**, 695 (1985).

38. H. A. Jehn, S. Hofman, and W.-D. Munz, Thin Solid Films, **153**, 45 (1987).

39. N. Kaufherr, G. R. Fenske, D. E. Busch, P. Lin, C. Despandey, and R. F. Bunshah, Thin Solid Films, **153**, 149 (1987).

40. A. J. Perry, C. Strandberg, W. D. Sproul, S. Hofmann, C. Erneberger, J. Nickerson, and L. Cholet, Thin Solid Films, **153**, 169 (1987).

41. A. W. Czanderna (Ed.), Methods of Surface Analysis, Elsevier, London, 1975.

42. R. H. Fleming, G. P. Meeker, and R. J. Blattner, Thin Solid Films, **153**, 197 (1987).

43. J. Brunner and A. J. Perry, Thin Solid Films, **153**, 103 (1987).

44. M. J. Bennett, J. Vac. Sci. Technol., **B12**, 800 (1984).

45. J. E. Sundgren, A. Rockett, J. E. Greene, and U. Helmersson, J. Vac. Sci. Technol., **A4**, 2770 (1986).

46. S. Bose, High Temperature Coatings, Elsevier, Amsterdam, 2007.

47. M. J. Bennett and A. T. Tuson, Mat Sci. Eng., **92**, 180 (1989).

75

TESTING FOR FLOW EFFECTS ON CORROSION

K. D. EFIRD

Efird Corrosion International, Inc., The Woodlands, Texas

A. Introduction
B. Correlation of test data to operating facilities
C. Flow effect test techniques
D. Low-turbulence flow corrosion test methods
References

A. INTRODUCTION

Fluid velocity was long used as the primary parameter for scaleup of laboratory test results to field applications, but this concept began changing in the 1970s. Data relating the calculated hydrodynamic parameter of wall shear stress to corrosion were first published by Efird in 1977 for copper alloys in flowing seawater [1]. Corrosion science now understands that fluid flow must be expressed in terms broadly related to fluid flow parameters common to all hydrodynamic systems to allow application of laboratory test data to field operations [2–11]. These hydrodynamic parameters are calculated from empirical equations developed to characterize fluid flow. The hydrodynamic parameters employed are wall shear stress (τ_w) and mass transfer coefficient (k).

B. CORRELATION OF TEST DATA TO OPERATING FACILITIES

The two primary considerations in the correlation of laboratory data to corrosion in operating facilities are the material

tested and the laboratory test procedure. The material tested must correspond as closely as possible to the material used in the facility in both alloy chemistry and metallurgical structure. The laboratory corrosion tests for flow-induced corrosion must be conducted in a manner that allows calculation of the mass transfer coefficient or wall shear stress [2–11]. Experimentally determined corrosion rates are then applied to the operating facilities for identical calculated hydrodynamic parameters. The basic assumptions are as follows:

- Calculation of the hydrodynamic parameters is valid.
- Calculated hydrodynamic parameters are those controlling corrosion or are intimately related to them.
- Scaleup of these hydrodynamic parameters to field operations with respect to corrosion is valid.

Both wall stress (τ_w) and mass transfer coefficient (k) meet these basic criteria. The process of correlating laboratory data to field operations is outlined in Figure 75.1 [12]. The operating flow rate allows calculation of the wall shear stress, which is then correlated through the wall shear stress to the measured laboratory corrosion rate.

The existing applicable equations for both wall shear stress and mass transfer coefficient assume equilibrium conditions. As discussed in Chapters 17 and 64, both equilibrium and disturbed-flow, steady-state conditions exist in operating systems. While the existing equations and test methods are effectively used for equilibrium flow conditions in operating systems, they may not be applicable to disturbed-flow conditions.

The basic methodology for relating an experimentally determined corrosion rate to allow prediction of the corrosion

Uhlig's Corrosion Handbook, Third Edition, Edited by R. Winston Revie
Copyright © 2011 John Wiley & Sons, Inc.

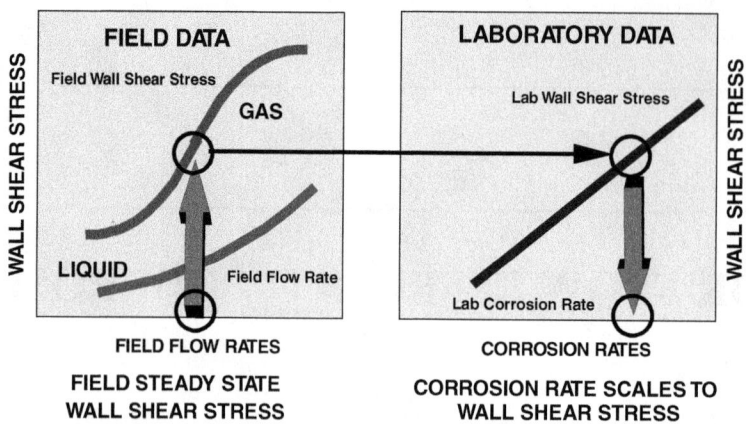

FIGURE 75.1. Process for relating laboratory data to facility operating systems using hydrodynamic conditions [12].

rate for flow-induced corrosion in an operating system is outlined in Figure 75.2 [13]. This provides a procedure to relate the results of a corrosion test directly to corrosion in an operation system where flow-induced corrosion conditions are expected or must be considered.

The steps in the process areas follows:

- The steady-state mass transfer, as defined by the limiting diffusion current density, is measured as a function of wall shear stress using a reversible, well-characterized electrochemical reaction.
- The equation relating the wall shear stress or mass transfer coefficient to the limiting diffusion current density is defined.
- The experimental apparatus for the corrosion measurements is calibrated by measuring the limiting diffusion

current density for the same reaction as a function of a convenient flow parameter.

- Applying the equation relating the wall shear stress or mass transfer coefficient to the limiting diffusion current density allows calibration as a function of the flow parameter.
- The corrosion rates obtained in the calibrated test apparatus are related to the calculated disturbed-flow wall shear stress or mass transfer coefficient in the operating system, allowing estimation of the expected corrosion rates in the operating system for disturbed flow.

C. FLOW EFFECT TEST TECHNIQUES

A number of techniques are available for testing the effects of fluid flow on materials. These techniques fall into two categories, rotating systems and flow systems. They are distinguished by the means used to induce flow across the test specimen. Basically, rotating systems move the test specimen in the fluid and flow systems move the fluid across the test specimen. Techniques in use are given in Table 75.1. A comparison of the various techniques for testing the effect of flow on materials is given in Table 75.2. Volume 8 of the NACE International series Corrosion Testing Made Easy is available for those interested in conducting tests for

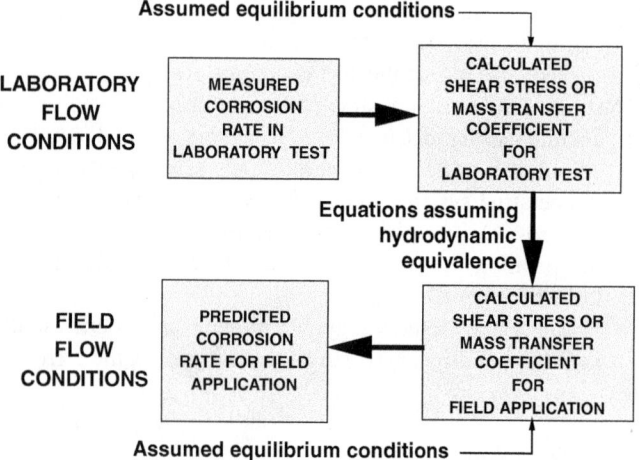

FIGURE 75.2. Relating laboratory data to field applications using equilibrium conditions [12].

TABLE 75.1. Techniques for Testing Effect of Flow on Materials

Rotating Systems	Flow Systems
Rotating cylinder	Large-diameter flow loop (≥ 4 in.)
Rotating cage	Small-diameter flow loop (≤ 1.5 in.)
Rotating disk	Jet impingement

TABLE 75.2.   Operational Comparison of Techniques for Testing Effect of Flow on Materials[a]

Criteria	Small Flow Loop	Rotating Cylinder	Jet Impingement	Rotating Cage	Large Flow Loop
Fluid requirements	High	Low	Medium	Low	Very High
Construction cost	High	Low	Medium	Low	Very High
Operating cost	High	Low	Medium	Low	Very High
Test work effort	Medium	Low	Medium	Low	High
k and/or τ?	Yes	Yes	Yes	No	Yes
Multiphase testing?	Yes	No	Yes	Yes	Yes
Scale up—liquid?	Yes	No	Yes	No	Yes
Scale up—gas?	No	No	No	No	Yes
Use coupons?	Yes	Yes	Yes	Yes	Yes
Use ER?	Yes	Yes	Yes	No	Yes
Use LPR, EIS?	Yes	Yes	Yes	No	Yes

[a]Not including rotaing disk.

the evaluation of the effects of flow on the corrosion of materials [14]. This volume provides working details for the various test techniques as well as a comparison of strengths and weaknesses.

D. LOW-TURBULENCE FLOW CORROSION TEST METHODS

There are no standardized corrosion test methods or test protocols for low-flow corrosion. However, based on knowledge of the low-flow conditions that can result in low-flow

corrosion, certain criteria for test protocol design are required.

The objective is to model the conditions that exist in the system stratified flow or liquid holdup locations. In general, this can involve exposure of the test material to solid, fluid, and gas phases that could be present as well as the respective interfaces. This can be as simple as a horizontal coupon with solids placed on the coupon or as complex as a single coupon that passes through all the existing phases. This test arrangement has the benefit of including any galvanic effects that might be present. An example of this setup is shown in Figure 75.3.

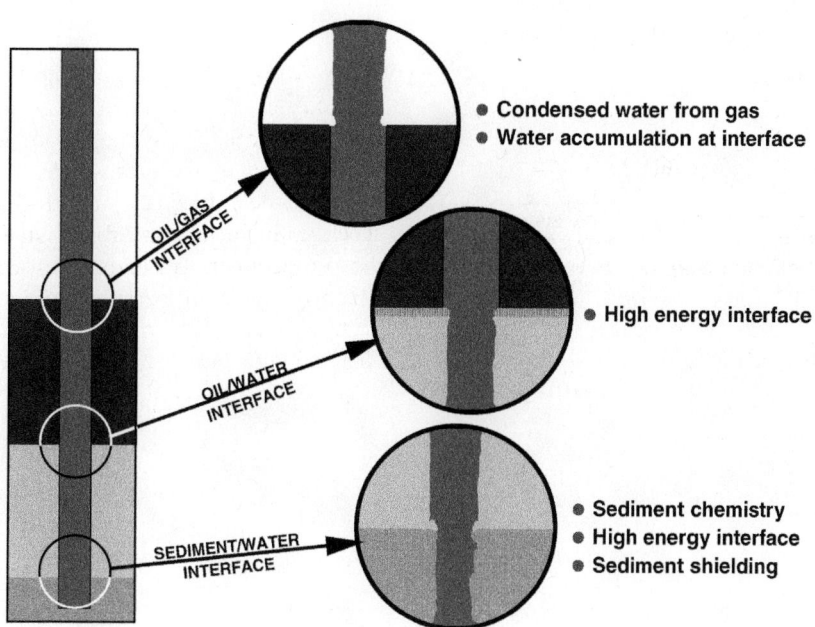

FIGURE 75.3.   Example of a low-flow test using a single coupon that passes through all the existing phases.

REFERENCES

1. K. D. Efird, "The Effect of Fluid Dynamics on the Corrosion of Copper Base Alloys in Seawater," Corrosion, **33**(1), 3–8 (Jan. 1977).

2. T. Y. Chen, A. A. Moccari, and D. D. Macdonald, "The Development of Controlled Hydrodynamic Techniques for Corrosion Testing," Paper No. 292, CORROSION/91, National Association of Corrosion Engineers, Cincinnati, OH, Mar. 11–15, 1991.

3. E. Heitz, "Chemo-Mechanical Effects of Flow on Corrosion," MTI Publication No. 23, MTI Project No. 15, Materials Technology Institute, Columbus, OH, 1986.

4. B. T. Ellison and C. J. Wen, "Hydrodynamic Effects on Corrosion," in Tutorial Lectures in Electrochemical Engineering and Technology, R. Alkire and T. Beck (Eds.), AIChE Symposium Series, Vol. 77, AIChE, New York, 1981, pp. 161–169.

5. K. D. Efird et al., "Experimental Correlation of Steel Corrosion in Pipe Flow with Jet Impingement and Rotating Cylinder Laboratory Tests," Corrosion, **49**(12), 992 (Dec. 1993).

6. K. D. Efird et al., "Wall Shear Stress and Flow Accelerated Corrosion of Carbon Steel in Sweet Production," in Proceedings: 12th International Corrosion Congress, Houston, TX, Sept. 19–24, 1993.

7. J. A. Herce et al., "Effects of Solution Chemistry and Flow on the Corrosion of Carbon Steel in Sweet Production," Paper No. 95111, CORROSION/95, Orlando, FL, Mar. 1995.

8. K. G. Jordan and P. R. Rhodes, "Corrosion of Carbon Steel by CO_2 Solutions: The Role of Fluid Flow," Paper No. 95125, CORROSION/95, Orlando, FL, Mar. 1995.

9. C. DeWaard, U. Lotz, and A. Dugstad, "Influence of Liquid Flow Velocity on CO_2 Corrosion: A Semi-Empirical Model," Paper No. 95128, CORROSION/95, Orlando, FL, Mar. 1995.

10. B. F. M. Pots, "Mechanistic Models for the Prediction of CO_2 Corrosion Rates under Multi-Phase Flow Conditions," Paper No. 95137, CORROSION/95, Orlando, FL, Mar. 1995.

11. W. P. Jepson, S. Bhongale, and M. Gopal, "Predictive Model for Sweet Corrosion in Horizontal Multiphase Slug Flow," Paper No. 96019, CORROSION/96, Denver, CO, Mar. 1996.

12. K. D. Efird, "The Effect of Disturbed Flow on Flow Accelerated Corrosion," EPRI Corrosion and Degradation Conference, St. Petersburg Beach, FL, June 2–4, 1999.

13. K. D. Efird, "Disturbed Flow and Flow Accelerated Corrosion in Oil and Gas Production," J. Energy Resources Technol., **120**(1), 72–77 (Mar. 1998).

14. P. Roberge, "Erosion-Corrosion," in Corrosion Testing Made Easy, Vol. 8, B. C. Syrett (Ed.), NACE International, Houston, TX, 2004.

76

ACCELERATED TESTING OF ELECTRONICS TO SIMULATE LONG-TERM WORLDWIDE ENVIRONMENTS

L. F. GARFIAS-MESIAS

DNV Columbus, Inc., Dublin, Ohio

M. REID

Stokes Research Institute, University of Limerick, Limerick, Ireland

A. Introduction
B. Mixed flowing gas testing
C. Corrosion of electronic equipment worldwide
D. Accelerated corrosion testing to simulate worldwide corrosion of electronics
E. Summary
References

A. INTRODUCTION

Because of the very high rate at which new commercially available electronic materials and devices have been introduced into the market and phased out in the past 50 years, it has been difficult to understand the mechanisms and factors that influence their corrosion performance. In the late 1990s, research to develop new electronic materials and devices typically took one to two years, followed by a very fast ramp-up to production and worldwide deployment that typically took another year. The equipment that contained these devices became obsolete by their fifth year in service, since new, faster and higher capacity equipment was needed to keep up with the demands for modern communications. In this very fast cycle of supply and demand of new equipment with high-performance electronic devices, failure of electronics was traditionally regarded as nonexistent, mainly because suppliers and users were willing to exchange dated technology that showed poor performance for the latest advances in technology. As a consequence, there was no time to investigate the failure of devices and components. This hindered the improvement of the design of the new generations of electronic devices.

Another important issue in the fast-changing world of electronics was globalization. In the 1960s and 1970s, most electronic equipment had to be tested to stringent requirements. Qualification of electronic devices took at least a couple of years, and companies heavily promoted the good performance and reliability of their equipment. However, by the late 1990s, the need to produce and export new electronic equipment worldwide together with increasing levels of competition (that drove prices to unimaginably low levels) pushed most companies to lower the number of tests to qualify new devices and in some cases to compromise long-term reliability.

These events promoted the creation of new standardized tests that could accelerate (and mimic) the environmental challenges that the equipment would see in its short life span. Mixed flowing gas (MFG) testing has been used in North America and Europe for several years to simulate the corrosion of electronics in harsh environments. This type of laboratory testing with a controlled environment is briefly explained in Section B. A few of the challenges that the industry has found, particularly when trying to deploy electronic equipment worldwide in environments that are more aggressive than the MFG Class III test, are described in Section C. Finally, in Section D, the future of accelerated testing in this ever-changing world of electronics is discussed, along with the attempts to mimic the very aggressive environments found in developing countries.

B. MIXED FLOWING GAS TESTING

In the early 1980s, with the discovery of significant printed circuit board (PCB) and component failure modes (mainly due to corrosion), a number of firms and laboratories set out to develop accelerated corrosion test methods with known acceleration factors. The aim of such efforts was to shrink years of service in the worst type of environment into days of testing in the laboratories. The main goal of this testing was to prove that the field failure modes encountered during service could be replicated during the laboratory tests. To simulate the environment under operational conditions, the PCB and its components were exposed to different gas mixtures, temperatures, and relative humidity. IBM, AT&T, and Battelle laboratories participated in this effort [1]. The result of this work was the development of a MFG test, which is primarily a laboratory test in which the temperature, relative humidity, and concentration of gaseous pollutants are carefully defined, monitored, and controlled [2]. Since then, several internationally recognized standards, including American Society for Testing and Materials (ASTM) [3] and Telcordia Technologies [4], based on the MFG testing have been created. The standards are readily available and describe in detail the hardware as well as the protocol for testing and calibration. Figure 76.1 shows a typical design of a MFG testing chamber. In order to calibrate the environmental chamber and the test procedure, a set of copper coupons is exposed inside the chamber, and the weight gain of the coupons and thickness of the corrosion products are compared with the standard. This is probably the most critical part of the MFG testing and should be undertaken prior to testing.

A set of four environmental corrosion classes were defined by the original group that developed the MFG tests, so that the corrosion in the field could be compared with the

TABLE 76.1. Nominal Test Conditions for Standard MFG Exposure

Class	H_2S (ppb)	Cl_2 (ppb)	NO_2 (ppb)	SO_2 (ppb)	% RH	T (°C)
I						
II	10	10	200	200	70	30
III	100	20	200	200	70	30
IV	200	50	200	200	75	50

corrosion under the laboratory test conditions in the chambers [1, 2]. The Class I environment corresponds to a benign environment in which the corrosion film on Cu in one year will not exceed 35 nm. This class is typically encountered in a central office with very good environmental control (mainly constant temperature and constant relative humidity as well as good filtering of the outdoor air entering the building). Class II is a mildly corrosive environment that ranges from a poorly controlled central office to a rural environment in which the environmental conditions are mild enough to produce a corrosion film on Cu, in one year, between 40 and 70 nm. Class III is a harsh corrosive environment typical of an outdoor cabinet containing electronic equipment without any control of its surroundings in a populated urban environment, in which the corrosion film on Cu in one year would reach between 80 and 400 nm. Finally, Class IV is a severe corrosive environment, such as in industrial areas or cities with a polluted environment, where the corrosion film on Cu would exceed 500 nm in one year. Table 76.1 summarizes the accepted test conditions for standard MFG exposure ranked in order of severity. Evidently, if the electronic equipment would be deployed in a Class I environment, there is no need for testing as the corrosion rate is significantly low that no failures due to corrosion are expected throughout the life span of the equipment.

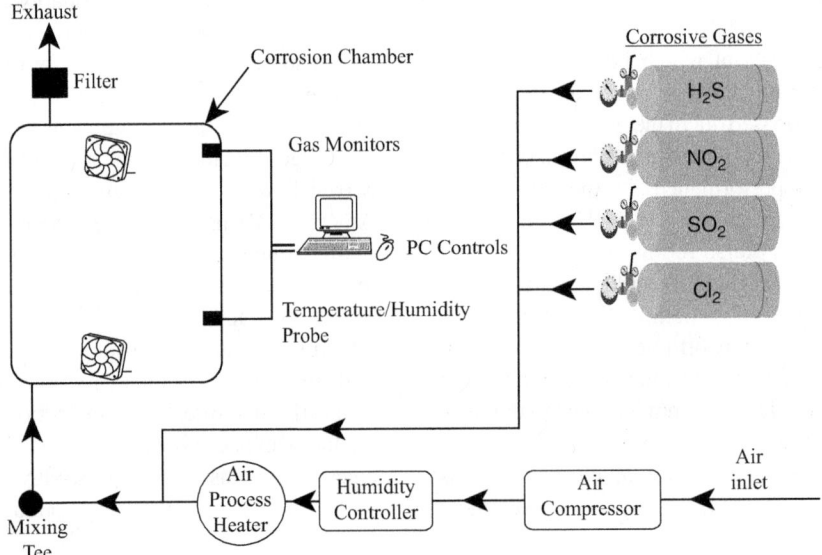

FIGURE 76.1. Schematic diagram of MFG test chamber setup. (Courtesy of The Electrochemical Society [23].)

Numerous authors have studied the effect of MFG (Class III) exposure on different parts of electronic components [5], including plated Cu contacts with precious metals [6] and the effect of exposure to MFG Class III [7] on contact resistance. Other types of studies include the diffusion of gases through the encapsulants [8] and creep corrosion on lead frames [9].

C. CORROSION OF ELECTRONIC EQUIPMENT WORLDWIDE

Environmental data from different parts of the world suggest that more aggressive atmospheres exist in Asia [10], Australia [11], and South America [12], compared to the environment found in developed countries such as the United States [13] and Europe [14]. Typically, in those countries with harsher environmental conditions, the local regulations are not as strict as in the developed world. Contrary to what is expected [15], some cities in the developing world, particularly in rural areas with low population and low industrial activity, may have similar environmental conditions to those in cities in the developed world. However, in developing countries, the extremely aggressive environment in cities is caused mainly by the large concentration of population and/or high density of industrial activity [16].

The most common cause of failure in electronic components and devices (cited in the literature) is the high corrosion rates of metals due to the (relatively high) concentration of corrosive gases, typically sulfur-containing and/or chloride-containing gases. Field studies and laboratory corrosion tests have been carried out to identify not only the corrosion products of pure metals (such as Cu [17] and Ni [18]) but also the corrosion products of relevant engineering alloys (such as stainless steels [19] and Cu–Au–Sn [20]). However, few studies have been undertaken to explore the corrosion of electronic equipment and devices deployed in harsh environments (typically exposed to Class III or harsher environments). In the majority of cases, the corrosion rate has been expressed in terms of film thickness accumulated during one year on a given metal (this value can be obtained by extrapolating from measurements made during a long period of time through the year). Although it has been found that the corrosion films on different metals may be composed of several layers of different compositions, typically it has been accepted that there is a predominant film (or corrosion layer) that thickens as the metal is exposed to the aggressive environment. The corrosion layers of copper (one of the most common metals used in electronic equipment) has been investigated during exposure to both field and laboratory environments [11–14, 21]. Clearly, the most relevant studies are those in which metals are exposed to laboratory environments that mimic field environments, so that the corrosion data obtained in the short-term laboratory test can be extrapolated to a longer term in order to estimate the service life in the field environment. Figure 76.2 shows the corrosion film thickness (measured in nanometers of Cu_2S per year) formed on Cu coupons exposed indoors and outdoors in 10 selected cities in Asia. As expected, the Cu_2S film formed on the coupons outdoors is typically thicker, in some cases 10 times thicker, than the film formed on coupons exposed indoors.

Unfortunately, most of the useful data related to corrosion of electronic equipment and devices in real environments have not been published, mainly because companies that

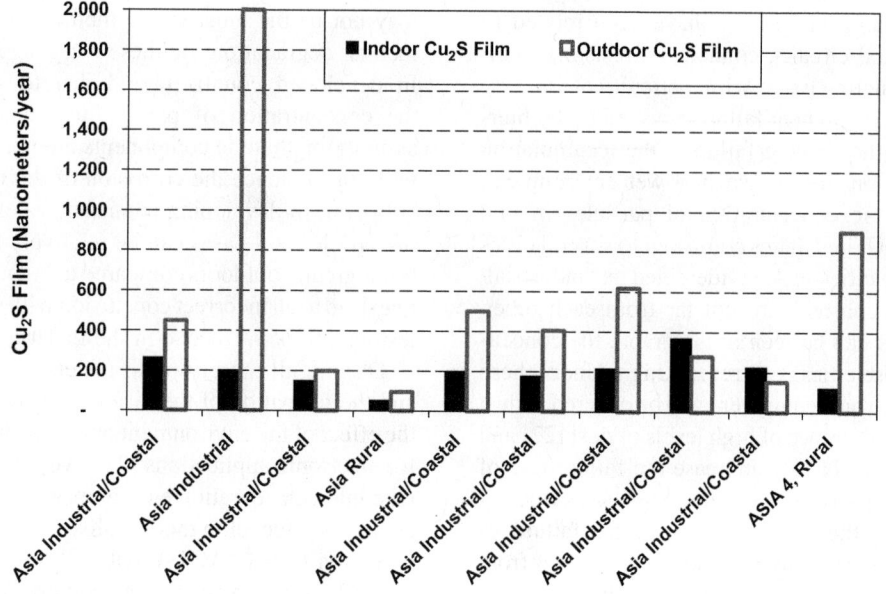

FIGURE 76.2. Corrosion film thickness (nm/year) on copper exposed in indoor and outdoor environments in 10 selected cities in Asia.

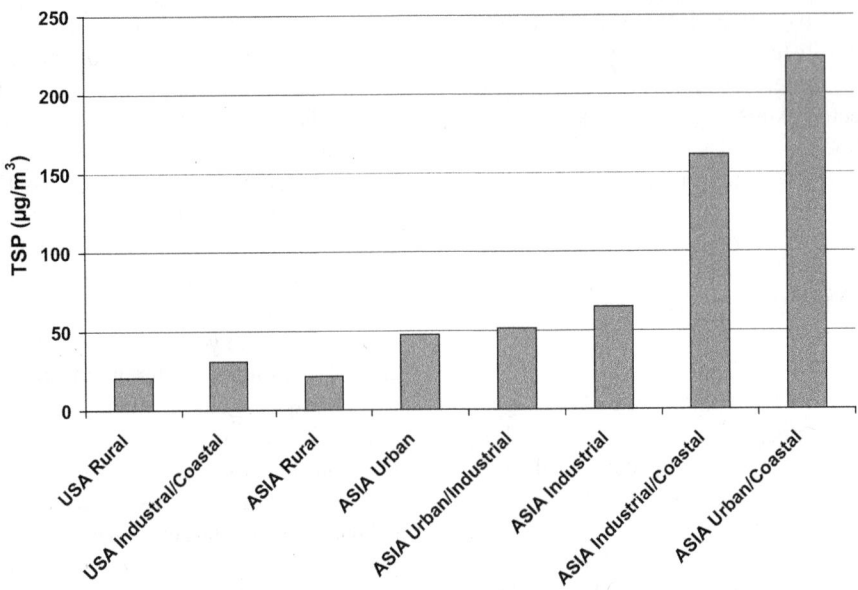

FIGURE 76.3. Total suspended particles ($\mu g/m^3$) found in the outdoor environment of two selected U.S. cities compared with several selected cities in Asia.

experience failures use the knowledge gathered during the failure investigations to improve the design of future products. This gives them a competitive advantage while keeping all the knowledge inside their R&D laboratories. The most common type of data that has been published is mainly in coupons made of the same materials that are typically used in the electronic industry. However, the corrosion rate of coupons (as discussed in the next section) is very different from a device that conducts an electrical signal and that is adjacent to a dissimilar metal (most likely fabricated under different circumstances and with different materials).

Most of the data available in the literature regarding failure of electronic equipment and devices have been related to corrosion of integrated circuits, connectors, resistors, transistors, or even conducting traces within circuit boards. Few studies on electronic equipment failure reported in the literature have dealt with the effect on failure of the accumulation of hygroscopic dust on the surface in a wet environment. Figure 76.3 shows the concentration of particles in two selected cities in the United States compared to several cities in Asia. The last two cities in Asia, identified as "industrial/coastal" and "urban/coastal," are not far from each other. More importantly, as can be seen in the graph, the concentration of dust in those cities is extremely high. Under these conditions, most electronic equipment will be covered in dust in a few months. The presence of high levels of dust [22] and high relative humidity [11] can increase the failure rate of electronics. In the ideal case where there are no gases (such as H_2S, SO_2, and Cl_2), the main mechanism for failure of electronics may be attributed to the adsorption of water from the environment on the surfaces covered with dust, facilitating a short circuit in the powered components or even in

adjacent pins (or leads). Although some electronic devices may be protected with a conformal coating (that can retard the diffusion of water and gases through the coating), this protection may be only temporary (depending on the coating) because the device may be subjected to long periods of high humidity due to the highly hygroscopic nature of the dust layer, absorbing the water from the environment. In reality, every electronic device is surrounded by an environment containing, to some extent, hygroscopic particulates, corrosive gases, and atmospheric water (daily wet and dry cycles).

The simple effect of the gases alone (together with the conservative values of temperature and relative humidity) may not be the most viable method to mimic the environmental degradation in these very aggressive conditions. In developed countries, and due to stricter regulations, the concentration of particulates (dust) is low, and the assumption that the components are not covered in dust and will not influence the corrosion of the metals is justifiable. However, in developing countries, where the concentration of particles and gases can be relatively high (sometimes 10 times greater outdoors compared to indoors), this assumption may lead to an incorrect conclusion with respect to laboratory testing and performance in the in-service environment.

Despite all the efforts to understand the corrosion rates and the formation of the different corrosion layers [23] and the effect of the environment on those materials (particularly for electronic applications [24]), very little has been done to take into consideration environmental data from developing countries to develop more realistic accelerated atmospheric corrosion tests (AACTs) that will mimic the environments typically found in developing countries. This subject will be discussed in more detail in the next section.

D. ACCELERATED CORROSION TESTING TO SIMULATE WORLDWIDE CORROSION OF ELECTRONICS

During the decade of the 1990s, professional organizations, including ASTM, Electronic Industries Association (EIA), International Electrotechnical Commission (IEC), and Telcordia, began to standardize these test methods and published corresponding documents as guidelines [3, 4]. The standards developed for equipment deployed in the North America Region (NAR) are considered to provide accelerated aging conditions for NAR- and European-type environments only. Typically these environments are Class I or Class II and in some extreme circumstances Class III.

The MFG Class III is a reliable test to accelerate these types of environmental conditions. Typical methods to evaluate the corrosion rate of metal coupons (or electronic devices) after testing include weight loss (mass change), cathodic reduction, and microscopy of the different corrosion layers. Figure 76.4 shows a comparison of the results from Cu coupons that were exposed to MFG Class III and were further evaluated using these three different methods.

Figure 76.5 shows focused ion beam (FIB) images of the corrosion films on Cu coupons that were exposed to MFG Class III and further evaluated using these three different methods. Notice that the oxide film thickens with time. Although the MFG Class III test is a reliable test, care should be taken especially when selecting the test methodology to evaluate the coupons, tested devices, or field failures.

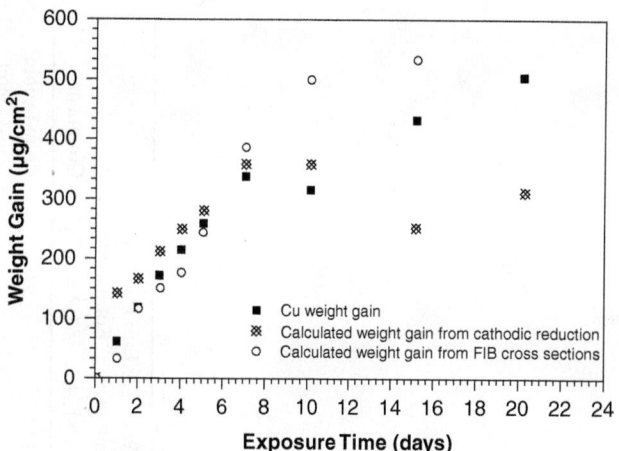

FIGURE 76.4. Weight gain as a function of time for Cu samples exposed to MFG Class III environment. Comparison of the weight gain measured in the Cu coupons after exposure with the weight gain calculated using the data from cathodic reduction and data observed using cross-sectional examinations (using focused ion beam). (Courtesy of The Electrochemical Society [23].)

Figure 76.4 shows that the calculated cathodic reduction corrosion layer thickness, the FIB cross-sectional measurements, and the Cu weight gain are in good agreement within the first week of exposure (when the corrosion layer is thin and well adhered to the substrate). However, after one week of exposure (or 1-μm corrosion layer) [23], the thicknesses determined by the three methods show significant differences. The discrepancies between the measurements and the

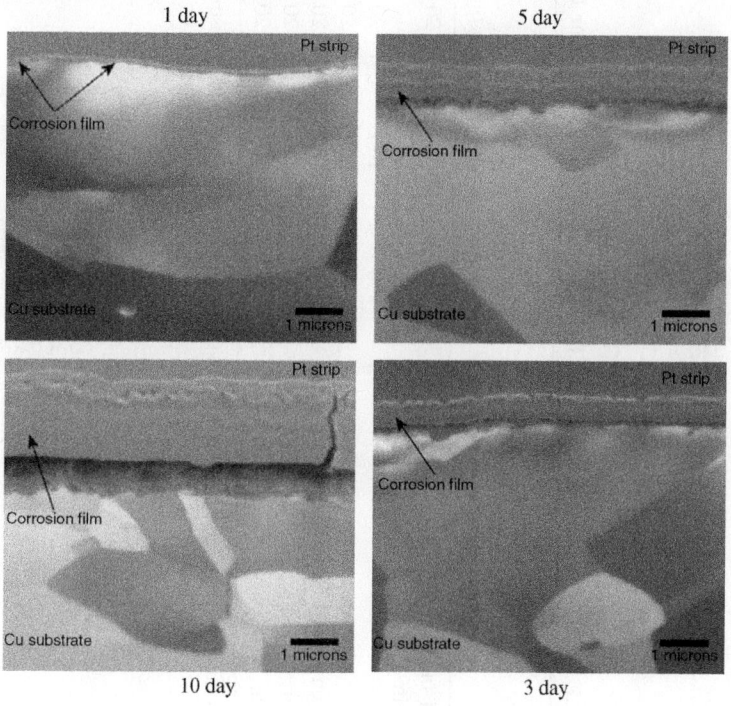

FIGURE 76.5. Cross sections of corrosion films formed on Cu coupons after exposure to MFG Class III for 1, 5, and 10 days. (Courtesy of The Electrochemical Society [23].)

TABLE 76.2. Typical Setup of Mixed Flowing Gas (MFG) for Outdoor Environments: Level 3MFG Test

	Temperature (°C)	RH (%)	H$_2$S (ppb)	Cl$_2$ (ppb)	NO$_2$ (ppb)	SO$_2$ (ppb)	VER (Times/h)	Test Duration (days)	Life Prediction Equivalent (years)	Comments
AACT in aggressive environments	40	90	2000–4000	20	200	200	4 or 5	14	20	Round-robin tests required!
Telcordia	30	70	100	20	200	200	20	10	15	
Telcordia	30	70	100	20	200	200	20	14	20	
EIA 1997	30	75	100	20	200	200	N/A	N/A	N/A	SO$_2$ (for test IIIA)
IEC 1995	30	75	100	20	200	0	3.2	10	10	
Battelle	30	75	100	20	200	0	6	14	7	

Abbreviations: VER, volume exchange rate inside the chamber; EIA, Electronic Industries Alliance; IEC, International Electrotechnical Commission.

calculated values (using cathodic reduction thickness) may be attributed to a number of factors. First, variations in corrosion product thickness caused by localized differences (and stress, particularly around the coupon edges), which are not taken into consideration during FIB cross-sectional analysis, are very important. Second, in an attempt to compare the different corrosion products and film thicknesses after different exposure times, a current density value ($0.35 \, \text{mA/cm}^2$ in the case of Fig. 76.4) has to be chosen. This relatively high current density may have attributed to the film reduction proceeding at an unreasonable rate with disbonding of thick corrosion products during the cathodic reduction test. Additionally, flaking of the thick corrosion product after 10 or more days (corrosion products larger than $1 \, \mu\text{m}$ thick) does not allow full reduction using cathodic reduction, particularly when the disbonded film is not in electrical contact with the substrate. Finally, in some cases, the corrosion products with the lowest reduction potentials are reduced first; in the case of copper, CuO is reduced first, followed by Cu_2O, CuS, and finally Cu_2S [25]. As copper has a naturally thin oxide film on its surface, it is reasonable to assume that for thicker films with a multilayered structure the reduction of the inner Cu oxides will invariably be reduced first. This may reduce the stability of CuS and Cu_2S layers and may cause unnecessary flaking; also this effect is more likely to be more pronounced after more prolonged exposure, where the corrosion products are thicker.

In most developing countries, urban cities, where most of the electronic equipment is deployed, fall into Class III and in some cases exceed Class IV [10, 15, 16]. The environmental and corrosion data as well as the few reported failures of electronic equipment in the field in different parts of the world suggest that current AACTs do not mimic the environments typically found in developing countries. In those countries, harsher environmental conditions have caused high corrosion rates of metals, leading to failure of devices and electronic equipment. In some regions of the world, corrosivity has been described in a corrosivity map [26], which indicates, by color, the typical corrosion rates, from negligible to mild and severe, measured in different regions indicated on the map. The levels of H_2S (one of the most common gases found in very aggressive environments in developing countries) in the MFG for the outdoor tests (Class II or Class IV) are considerably below those levels typically found in some Asian countries [10, 15, 16]. In a recent study, Wattanabe and co-workers [10] measured the concentration of H_2S near the Noboribetsu hot springs area of Hokkaido, Japan, and found that the H_2S concentration ranged from 160 to 1770 ppb. Similarly, it has been reported that in parts of China, Taiwan, and Mexico the levels of H_2S were as high as 850 ppb [15]. These findings strongly suggest the importance of increasing the H_2S concentration in order to use MFG testing to qualify electronic equipment for use in aggressive environments. Table 76.2 shows the typical concentrations used for most mixed flowing gases according to the different standards. Based on the findings of recent studies, it is suggested that the levels of H_2S when using the MFG test to qualify electronic equipment that will be deployed in very aggressive outdoor environments should be increased taking into consideration the real environment (see Table 76.2).

It is clear that the suggested temperature for the MFG test (30°C) may be adequate in most cases, except when the temperature of the city (or cities) where the equipment is going to be deployed will be significantly different. In tropical cities, where the temperature year round is high, the 30°C temperature for the MFG tests seems reasonable. However, in colder areas, where there is the possibility that the temperature fluctuates significantly during the day (hot) and the night (when condensation may occur), it is recommended that these cycles be considered in the design of the test.

Because the relative humidity in aggressive environments around the world ranges from low values in the northern hemisphere to high values in the tropics, the relative humidity used in the test depends on whether the objective of the test is to mimic indoor or outdoor environments. Because the outdoor environment is more humid near tropical areas in the coastal cities, the relative humidity inside the chamber should be increased; values around 90% may be acceptable if the objective is to mimic the outdoor environment.

Other variables for the test should also be considered, for example, different gases, the volume exchange ratio in the chamber, and test duration sufficient for lifetime prediction. Although these variables are important, they may need to be adjusted after the values of the key variables (H_2S content, temperature, and relative humidity) have been fixed.

One additional variable that has not been considered during most of the AACT studies is the voltage across the electronic devices. Typically, coupons and electronic devices have been tested without power. The devices are subjected not only to particulate, gases, temperature cycles, and high relative humidity but also to a voltage. A common practice in the industry is to use coupons to measure the leakage current across the traces (lines) while the coupon is subjected to a voltage. The two voltages usually selected are 24 and 48 V because these voltages are widely used in the electronics industry.

E. SUMMARY

A review of the currently available literature and common industry practices suggests that MFG testing for both indoor (typically Class II) and outdoor (normally associated with Class III) environments is the most common testing used to qualify materials and devices for use in aggressive environments.

In order to mimic the overall environment in more aggressive areas of the world (e.g., in developing countries or near industrial sites in developed countries), addition of higher levels of pollutants (including atmospheric particles and gases) should be considered for tests that are used to qualify electronic materials and devices.

"Dusting" followed by AACT that includes electrical testing of interdigitated coupons may also be used as a qualification method. This may have a considerable impact when trying to reproduce field failures (e.g., creep corrosion).

At the same time, it is extremely important to reproduce (and induce) during testing the same failure modes encountered in electronic devices. A comparison of available data from the United States and developing countries shows that different testing criteria should be developed (if feasible) during qualification testing for each particular environment. The ultimate goal should be to develop a meaningful AACT that can mimic the real, in-service environment while achieving an adequate acceleration factor. The available data and field experience suggest that more aggressive testing may be desirable in some situations.

REFERENCES

1. W. H. Abbott, "The Development and Performance Characteristics of Mixed Flowing Gas Environment," IEEE Trans. Components Hybrids Manufact. Technol., **11**, 22–35 (1988).

2. W. H. Abbott, "The Corrosion of Copper and Porous Gold in Flowing Mixed Gas Environments," IEEE Trans. Components Hybrids Manufact. Technol., **13**(1), 40–45 (1990).

3. ASTM B 845-97. Standard Guide for Mixed Flowing Gas Tests for Electrical Contacts, ASTM International, West Conshohocken, Pennsylvania, 2003.

4. NEBS™ Requirements: Physical Protection, Generic Requirements GR-63-CORE, Method 5.5.2, Telcordia Technologies, Chester, 2002.

5. D. C. Abbott, "Nickel Palladium Finish for Leadframes," IEEE Trans. Components Packaging Technol., **22**, 99–103 (1999).

6. R. J. Geckle and R. S. Mroczkowski, "Corrosion of Precious Metal Plated Copper-Alloys Due to Mixed Flowing Gas Exposure," IEEE Trans. Components Hybrids Manufact. Technol., **14**, 162–169 (1991).

7. R. Martens and M. G. Pecht, "An Investigation of the Electrical Contact Resistance of Corroded Pore Sites on Gold Plated Surfaces," IEEE Trans. Adv. Packaging, **23**, 561–567 (2000).

8. C. Hillman, B. Castillo, and M. Pecht, "Diffusion and Absorption of Corrosive Gases in Electronic Encapsulants," Microelectron. Reliabil., **43**, 635–643 (2003).

9. P. Zhao and M. Pecht, "Mixed Flowing Gas Studies of Creep Corrosion on Plastic Encapsulated Microcircuit Packages with Noble Metal Pre-plated Leadframes," IEEE Trans. Device Mater. Reliabil., **5**, 268–276 (2005).

10. M. Watanabe, H. Hirota, T. Handa, N. Kuwaki, and J. Sakai, "Atmospheric Corrosion of Cu in an Indoor Environment with a High H_2S Concentration," 17th International Corrosion Congress: Corrosion Control in the Service of Society, Las Vegas, NV, Oct. 6–10, 2008.

11. I. S. Cole, W. D. Ganther, J. D. Sinclair, D. Lau, and D. A. Paterson, "A Study of the Wetting of Metal Surfaces in Order to Understand the Processes Controlling Atmospheric Corrosion," J. Electrochem. Soc., **151**, B627–B635 (2004).

12. J. R. Vilche, F. E. Varela, E. N. Codaro, B. M. Rosales, G. Moriena, and A. Fernandez, "A Survey of Argentinean Atmospheric Corrosion: II—Cu Samples", Corros. Sci., **39**(4), 655–679 (1997).

13. D. W. Rice, P. Peterson, E. B. Rigby, P. Phipps, R. J. Cappell, and R. Tremoureux, "Atmospheric Corrosion of Cu and Silver", J. Electrochem. Soc., **128**(2), 275–284 (1981).

14. J. Tidblad and C. Leygraf, "Atmospheric Corrosion Effects of SO_2 and NO_2, a Comparison of Laboratory and Field-Exposed Copper," J. Electrochem. Soc., **142**(3), 749–756 (1995).

15. L. F. Garfias-Mesias, J. P. Franey, R. P. Frankenthal, and W. D. Reents, Gordon Research Conference on Corrosion, New London, NH, July 26, 2004.

16. L. F. Garfias-Mesias, J. P. Franey, R. P. Frankenthal, R. Coyle, and W. D. Reents, in CORROSION 2005, Research In Progress, NACE, Houston, TX, Apr. 4–6 2005.

17. M. Lenglet, J. Lopitaux, C. Leygraf, I. Odnevai, M. Carballeira, J.-C. Noualhaguet, J. Guinement, J. Gautier, and J. Baissel, "Analysis of Corrosion Products Formed on Copper in $Cl_2/H_2S/NO_2$ Exposure," J. Electrochem. Soc., **142**(11), 3690–3696 (1995).

18. S. Zakipour, J. Tidblad, and C. Leygraf, "Atmospheric Corrosion Effects of SO_2, NO_2, and O_3, A Comparison of Laboratory and Field Exposed Nickel" J. Electrochem. Soc., **144**(10), 3513–3517 (1997).

19. G. Herting, I. Odnevall Wallinder, and C. Leygraf, "A Comparison of Release Rates of Cr, Ni, and Fe from Stainless Steel Alloys and the Pure Metals Exposed to Simulated Rain Events," J. Electrochem. Soc., **152**(1), B23–B29 (2005).

20. S. Zakipour and C. Leygraf, "Evaluation of Laboratory Tests to Simulate Indoor Corrosion of Electrical Contact Materials," J. Electrochem. Soc., **133**, 21–30 (1986).

21. M. Reid, J. Punch, C. Ryan, L. F. Garfias, S. Belochapkine, J. P. Franey, G. E. Derkits, and W. D. Reents, J. Electrochem. Soc., **154**, C209 (2007).

22. R. B. Comizzoli, C. A. Jankoski, G. A. Peins, L. A. Psota-Kelty, D. J. Siconolfi, J. D. Sinclair, W. Chengen, and M. Gao, in Corrosion and Corrosion Protection, J. D. Sinclair, R. P. Frankenthal, E. Kalman, and W. Plieth (Eds.), PV 2001-22, Proceedings Series, The Electrochemical Society, Pennington, NJ, 2001, pp. 691–705.

23. M. Reid, J. Punch, L. F. Garfias, K. Shannon, S. Belochapkine, and D. A. Tanner, "Study of Mixed Flowing Gas Exposure of Cu," J. Electrochem. Soc., **155**(4), C147–C153 (2008).

24. M. Reid, J. Punch, L. F. Garfias, G. K. Grace, and S. Belochapkine, "Corrosion Resistance of Cu-Coated Contacts," J. Electrochem. Soc., **153**(12), B513–B517 (2006).

25. S. J. Krumbien, B. Newell, and V. Pascucci, J. Test. Eval., **11**, 357 (1989).

26. W. Hou and C. Liang, "Eight-Year Atmospheric Corrosion Exposure of Steels in China," Corrosion, **55**(1), 66 (1999).

77

TESTING FOR ENVIRONMENTALLY ASSISTED CRACKING

R. D. KANE

i Corrosion LLC, Houston, Texas

A. INTRODUCTION

The study of environmentally assisted cracking (EAC) involves the consideration and evaluation of the inherent compatibility between a material and the environment under conditions of either applied or residual stress. However, this is a very broad, encompassing topic with many possible combinations of materials and environments in which EAC has been investigated and documented. EAC is also a critical problem because equipment, components, and structures are intended to be used while exposed to various environments and conditions of stress and must resist EAC over prolonged periods of service. Furthermore, the materials used in construction typically have a multitude of manufacturing and process variables that may affect their metallurgical condition and structure which, in turn,

influence resistance to EAC. Testing for resistance to EAC is one of the most effective ways to determine the interrelationships among material, environmental, and mechanical variables on the process of EAC.

The proportions of this subject immediately limit attempts to make simplistic use of only a single method of testing for all cases. Factors such as material type, process history, product form, active cracking mechanism(s), loading configuration and geometry, and service environment all can have a major impact on the type of specimen and test condition to be utilized for the evaluation of EAC. The prudent approach to selection of testing methods is usually to start with a survey of previous experiences from prior investigations conducted on similar classes of materials and types of environments found by surveying the published literature. Additionally, an extensive amount of information on standardized stress corrosion cracking (SCC) testing methods, evaluation procedures, and experimental techniques is available in International Organization for Standardization (ISO) 7539 Parts 1–8 [1], American Society for Testing and Materials (ASTM) and NACE standards and existing reviews published in the literature [2].

B. BACKGROUND

To better establish a basis for understanding the role that each of the various cracking processes plays in the selection of EAC test methods on a particular material, the applicable EAC processes must first be identified for the specific material and environment under consideration. Mechanisms of EAC have been debated for decades and there are still controversies over the use of specific terms and definitions. Therefore, in this chapter the discussion of the subject will

Uhlig's Corrosion Handbook, Third Edition, Edited by R. Winston Revie
Copyright © 2011 John Wiley & Sons, Inc.

be limited to basic phenomenological descriptions of the various cracking processes with an attempt to discuss them in simple but description terms and in relation to important testing-related variables.

B1. Stress Corrosion Cracking

Stress corrosion cracking is the formation of embrittlement whereby cracks form in a normally sound, ductile material through the simultaneous action of a tensile stress and a corrosive environment. In most cases, SCC has been associated with the process of active path corrosion (APC) whereby the corrosive attack or anodic dissolution initiates at specific, localized sites and is focused along specific paths within the material. Crack initiation often occurs at sites of local anodic attack (e.g., pits or local crevice corrosion). In some cases, crack propagation is along grain boundaries (i.e., intergranular SCC, or IGSCC); in other cases, the path is along specific crystallographic planes within the grains (i.e., Transgranular SCC). Quite often, SCC is strongly affected by alloy composition, the bulk concentration of specific corrodent species or the local concentration on the metal surface, and usually, to a lesser degree, the stress intensity. In some cases, this latter point may make use of test methods based on fracture mechanics concepts difficult to utilize effectively due to multiple crack initiations, excessive crack branching, and tendencies for nonplanar propagation of cracks.

Furthermore, corrosion film characteristics (i.e., passivation) and local anodic attack (i.e., depassivation) often serve as controlling factors in SCC crack initiation and growth. Therefore, localized corrosion can promote SCC, making exposure geometry and specimen design important factors. In many cases, mechanical straining or electrochemical inducements such as crevices or controlled potential are utilized to overcome the problems and uncertainties of SCC initiation so that the inherent resistance of the material to SCC can be obtained at reasonable test duration (see Table 77.1) [3].

B2. Hydrogen Embrittlement

Hydrogen is often a byproduct of corrosion and electrochemical processes and may also be a major constituent in various service environments. During electrochemical reactions in aqueous environments, it is common for a hydrogen ion (H^+) to combine with an electron (e^-) to form atomic hydrogen on the surface of the material. The effects of hydrogen on cracking are contrasted to those of local metallic dissolution in Figure 77.1 [4]. Depending on the solution and interfacial characteristics, the hydrogen atoms formed by the corrosion process may recombine to form molecular hydrogen that can simply bubble off of the specimen surface without any further complications. However, under certain circumstances, when hydrogen

TABLE 77.1. Applied Potential for SCC in Steel Exposed to Various Service Environments

Environment	Potential Range (mV SCE)
Nitrate	− 250 to 1200
Liquid ammonia	− 400 to > + 1500
Carbonate	− 650 to − 550
Hydroxide	− 1100 to − 850 and + 350 to + 500

recombination poisons (e.g., S, P, As, Sn) are present in the environment, hydrogen recombination is retarded, promoting the absorption of atomic hydrogen into the material. Once inside the material, hydrogen can affect the mechanical performance of materials in several ways:

1. The formation of internal hydrogen blisters or blister-like cracks at internal laminations or at sites of nonmetallic inclusions in low-strength materials. These internal cracks may propagate by a process called hydrogen-induced cracking (HIC) or hydrogen blistering. No external stress on test specimens is usually required to examine this type of cracking behavior. In some cases, however, these blister cracks may take on an alignment caused by the presence of residual or applied tensile stresses, which is referred to as stress-oriented hydrogen-induced cracking (SOHIC).

2. The process of hydrogen-assisted microvoid coalescence can occur during plastic straining. This can reduce the ductility of normally ductile engineering materials while not inducing brittle cracking.

3. An extreme case of ductility loss from hydrogen is the brittle fracture of susceptible materials under applied or residual tensile stresses. This form of cracking typically results in either transgranular or intergranular cracks, depending on the material type and condition, yield strength, and processing variables and is normally referred to as HEC.

With respect to HEC, most susceptible materials show a major effect of stress concentration (i.e., notches) and level of stress intensity and tend to produce failures in a relatively short time (i.e., <1000 h). Therefore, tension, notched, and precracked specimens and fracture methods are widely utilized in the evaluation for HEC. Once hydrogen has entered a material, it can produce delayed failure (i.e., fracture resulting well after application of a constant load on the specimen). Additionally, slow dynamic straining can be used to accelerate crack initiation and propagation.

An ancillary procedure that can greatly extend the utility of testing for hydrogen embrittlement is the use of hydrogen flux monitoring. Methods are provided in ASTM G148 [5] and F1113 [6] for determining the severity of hydrogen

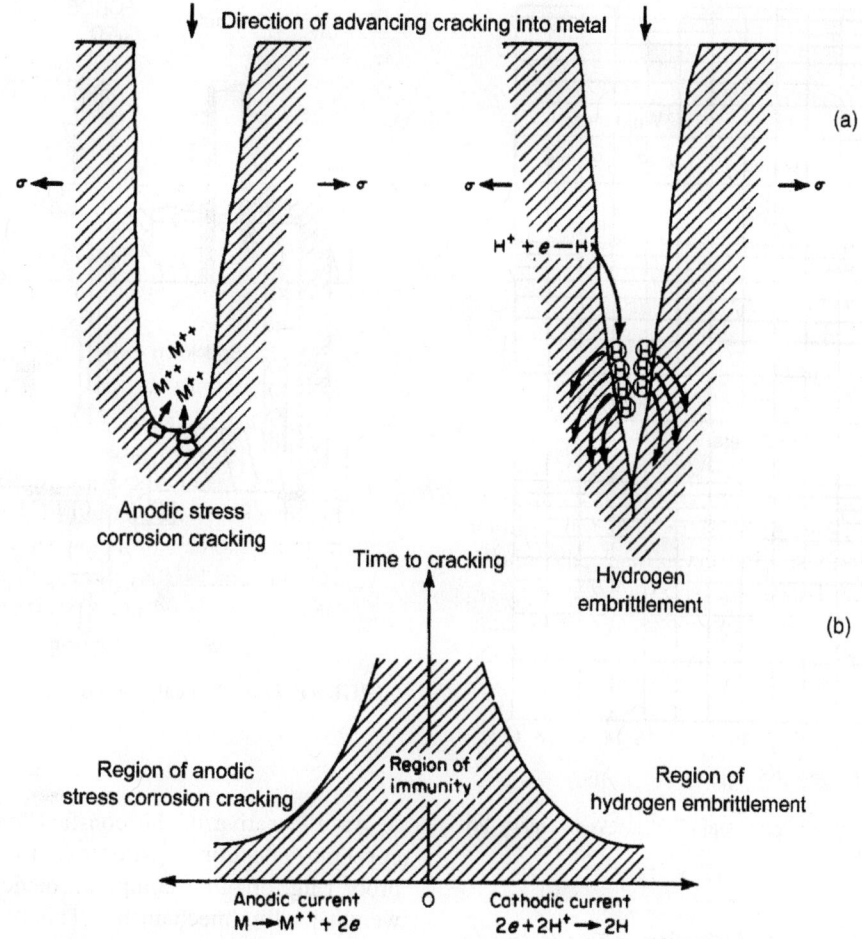

FIGURE 77.1. Schematic comparison of (a) anodic SCC and (b) hydrogen embrittlement cracking (HEC) mechanisms [3].

charging in materials from exposure to various environments [7, 8]. Most importantly, data from hydrogen permeation tests can often be correlated to the results of HIC, SOHIC, and HEC tests and, in some cases, can even be extended using analytical methodologies [9] to predict the extent of cracking.

B3. Liquid-Metal Embrittlement

Certain materials exhibit general and/or localized corrosion and embrittlement when in contact with certain liquid metals. Liquid-metal embrittlement (LME) shows many of the characteristics of both SCC and HEC. For example, LME is often preceded by an incubation period required for the liquid metal to penetrate surface oxide or passive layers on the material, which is analogous to the local, depassivation or pitting prior to SCC. However, in many cases, LME shows a very strong effect of stress intensity and a rapid transition from slow to rapid crack growth (see Fig. 77.2) [10] which makes it similar to HEC. Therefore, it is common in LME

tests to utilize surface-active agents or dynamic strain to promote surface attack and thereby reduce the incubation time required to initiate cracking, thus allowing the test to focus on the cracking resistance of the substrate material. Second, tension, precracked, or notched specimens and fracture mechanics methods are also utilized extensively in LME testing so that the cracking response of the material can be more precisely investigated.

C. CONSTANT-LOAD/DEFLECTION TECHNIQUES

C1. Tension Tests

One of the most common and straightforward methods utilized in EAC tests is the use of an applied load that acts as the driving force for EAC. Typically, a tension specimen is employed and specimens are loaded to various levels of applied stress as defined by ISO 7539-Part 4 [1], ASTM G-49 [11], and NACE TMO177 [12]. A typical tension specimen and exposure cell are

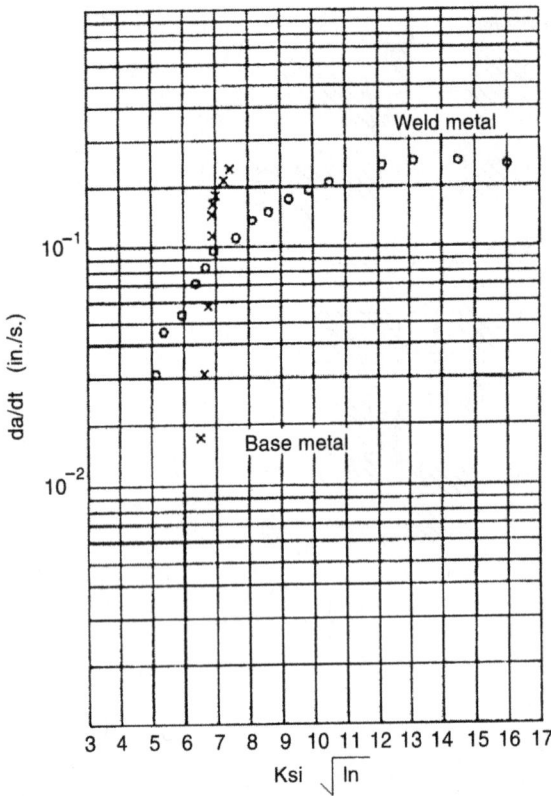

FIGURE 77.2. Crack growth versus applied stress intensity for LME of aluminum in Hg.

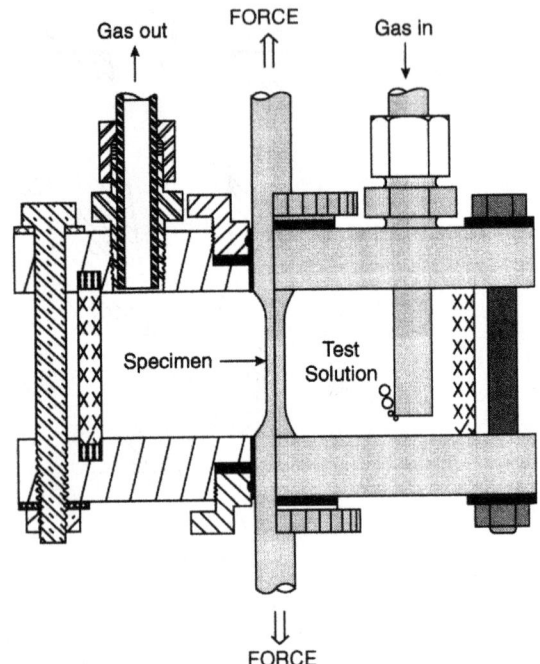

FIGURE 77.3. Typical smooth tension specimen in test cell.

shown in Figure 77.3 [12]. A distinction is usually made between the procedure of running these tests with regard to the methods employed for applying the load. The constant load is usually applied using a dead-weight fixture. In its simplest form, a hanging weight is suspended from the specimen. This usually works well if the loads required are relatively low. Alternately, for higher loads, a simple lever can be utilized to magnify the applied load, similar to the type of apparatus used to perform creep or stress rupture tests.

The stress (S) on a smooth, uniaxial tensile specimen is calculated with the following formula: $S = P/A$, where $P =$ load on the specimen and $A =$ specimen cross-sectional area. For the case of dead-weight loading, a constant load is produced on the specimen. However, once cracking initiates in the specimen, the cross-sectional area is reduced so the applied stress actually increases as the crack increases in length. Therefore, in this type of test performed on a susceptible material, the specimen often fails soon after initiation of cracking and little information on crack propagation is obtained. The effects of corrosion and stress can be focused at a single location by the use of a single notch in the specimen gauge section. In many cases, multiple specimens and stress levels are utilized to determine a threshold stress curve, as shown in Figure 77.4 [12].

An alternative to the constant-load tensile test is the constant-deflection tensile test. In this case, a spring, proof ring, or other compliant device replaces the dead-weight loading mechanisms. This type of loading configuration is usually much simpler and easier to set up. It also allows for more specimens to be tested in a limited area, such as a laboratory exhaust hood, if necessary. In the case of a constant-deflection test, the load will decrease as the crack propagates through the specimen according to the compliance of the test fixture. If insufficient compliance is available in the loading fixture, the crack will stop prior to specimen fracture. Therefore, for deflection-controlled tests, it is important to obtain information regarding the load/deflection relationship for the particular geometry of fixture used and, where possible, select test fixtures having a reproducible and reasonably high level of compliance. Normally, the important aspects of testing are as follows:

1. Check deflections before, during, and after tests.
2. Examine the specimens closely for subcritical crack growth.
3. Calibrate the fixture periodically for load versus deflection.
4. If only limited compliance is available, break specimens in air after testing to locate any subcritical cracks and determine the reduction in load-carrying capacity they cause on the test specimen.

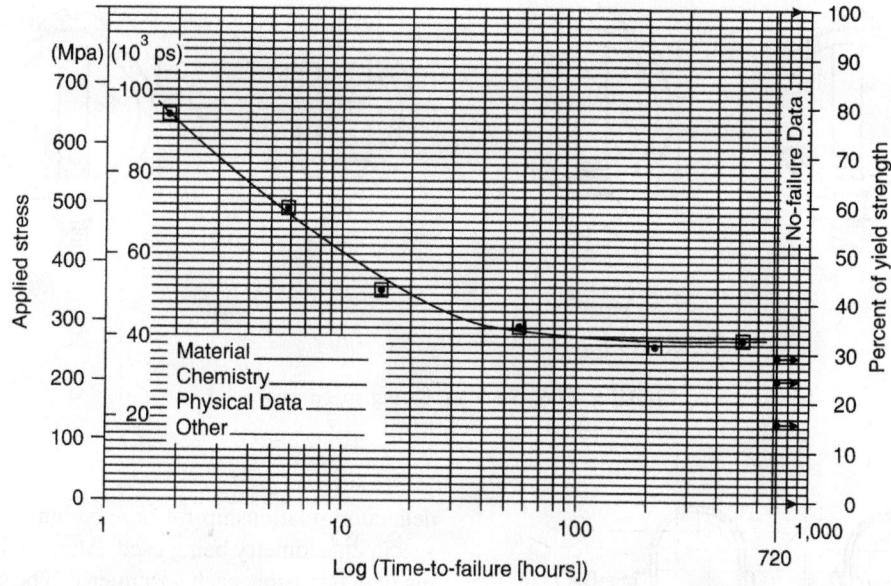

FIGURE 77.4. Typical applied stress versus time to failure for curve.

This latter approach is usually referred to as the breaking load test, whereby after completion of the intended exposure period the nonfailed specimens are pulled to failure and the load at failure recorded and the fracture surface examined for evidence of EAC (see ASTM G139) [13]. Another concern for constant-deflection specimens occurs at high levels of applied stress relative to the yield strength of the material (i.e., above the elastic limit). In these cases, time-dependent deformation can result in creep and a reduction in the applied load with time. Therefore, step 1 above is critical when constant-deflection tests are conducted.

C2. Other Constant-Deflection Specimens

There are a variety of specimens that can be utilized for constant-deflection tests. These include:

1. Bent-beam specimens (two-, three-, and four-point loading) per ISO 7539 Part 2 [1], ASTM G38 (see Fig. 77.5) [14], and NACE TM0177 Method B [12]
2. C-ring specimens per ISO 7539 Part 5 [1], ASTM G38 (see Fig. 77.6) [15], and NACE TM 0177 Method C [12]
3. U-bend specimens per ISO 7539 Part 3 and ASTM G30 (see Fig. 77.7) [16].

Each type of specimen has a compliance dictated by the specimen geometry and dimensions and the modulus of elasticity of the material being tested. For purely elastic deflections, simple stressing equations can be utilized to relate applied tensile stress to specimen deflection (see the above-mentioned test methods for guidance). Another

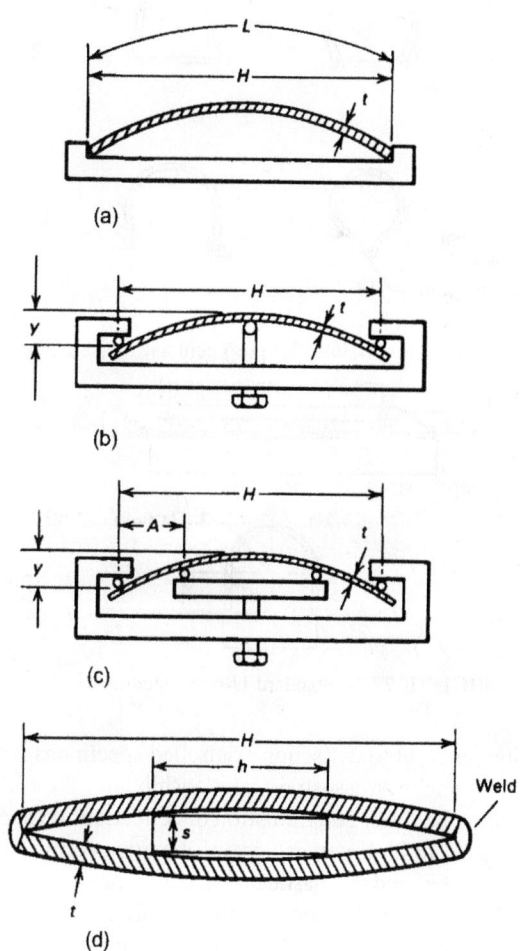

FIGURE 77.5. Standard bent-beam specimens.

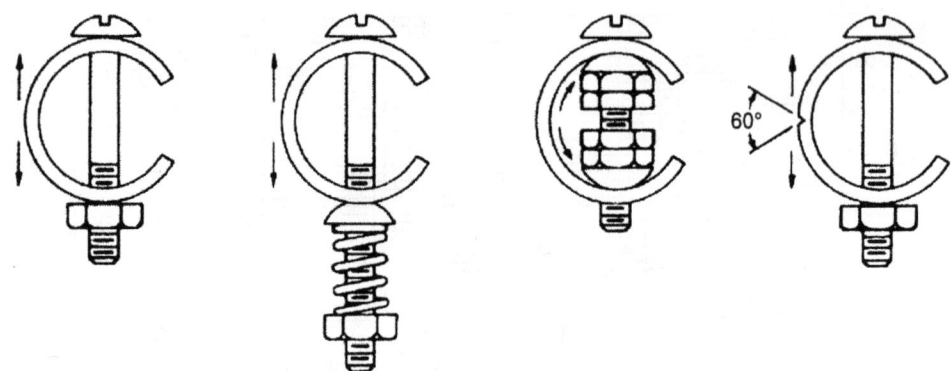

FIGURE 77.6. Standard C-ring specimens.

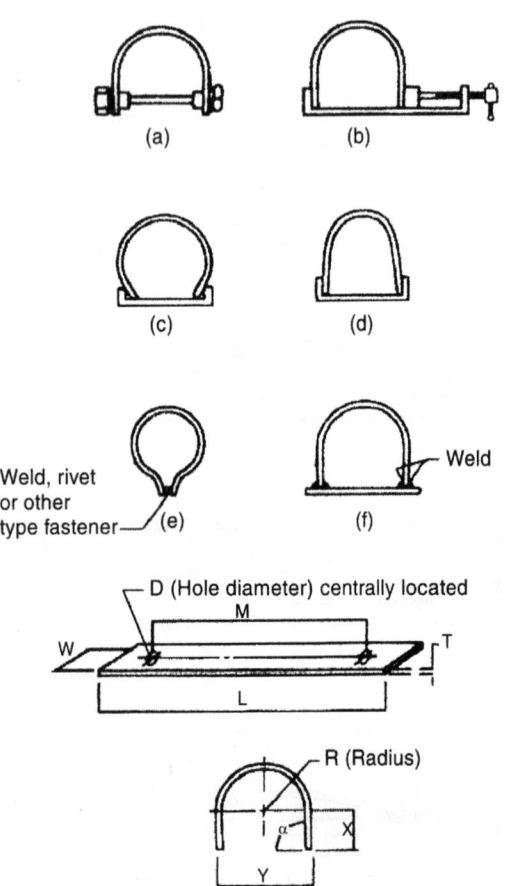

FIGURE 77.7. Standard U-bend specimens.

limitation inherent to deflection-controlled specimens is the nonlinearity in the stress–strain relationship once the elastic limit is surpassed. For determination of deflections on specimens stressed to high percentages of the engineering yield strength (i.e., beyond the elastic limit), it is common to utilize a strain-gauged calibration specimen having the same geometry, dimensions, and modulus as being used in the test. This specimen can be used to measure the exact stress/deflection relationship for the combination of material and specimen geometry being used. Alternately, for tests involving high precision, each specimen can be strain gauged prior to loading and the strain gauge can be removed before exposure of the specimen to the test environment.

Once the stress/deflection relationship for the specimens to be used is determined, it is typical to stress several specimens at various levels of stress. This allows for assessment of the susceptibility to cracking as a function of applied stress (Fig. 77.4). This can be performed either by monitoring time to failure as a function of applied stress or by evaluation of the failure/no-failure performance at a fixed test duration and level of applied stress. The latter technique is particularly useful in quality assurance or lot release testing where the combination of service experience and laboratory testing has provided information regarding the level of laboratory performance required to determine acceptable service performance. In some cases, constant-deflection specimens, such as bent beams and C rings, are not as severe as dead weight–loaded specimens. This is usually due either to variations in susceptibility with orientation or to limited compliance reducing the driving force for crack propagation. Additionally, it is often observed that bent-beam or C-ring specimens are not as severe as uniaxial tension specimens (Fig. 77.8) [17]. This effect is usually explained in terms of the changes in stress state produced in these specimens as a result of the crack propagation.

U-bend specimens are constant-deflection specimens but are normally not stressed to various levels of deflections or load. They are severely plastically deformed by bending a strip of material around a mandrel to form the U shape. Usually, multiple specimens are utilized and failure/no-failure is monitored after various durations of exposure. The plastic deformation in most cases provides a mechanical inducement to the initiation of SCC. Therefore, it can accelerate SCC in some systems that normally require an unacceptably long time of crack initiation on other types of specimens.

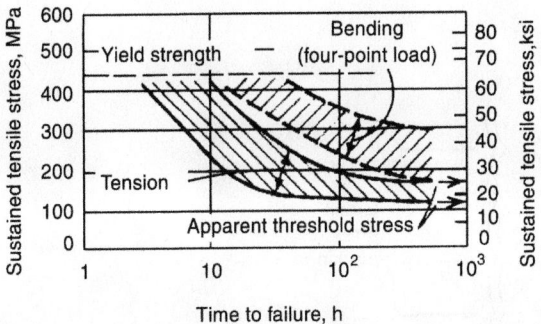

FIGURE 77.8. Comparison of SCC data for Al alloy in tension and bent beam.

D. DYNAMIC TESTS

Slow-strain-rate (SSR) tests, also known as constant-extension-rate tests (CERTs), are a modification of the constant-load tension test as shown schematically in Figure 77.3. In this case, the constant load on the test specimen has been replaced by a slow extension of the specimen that produces a ramping load (increasing stress) on the specimen until failure occurs (see Fig. 77.9). More detailed descriptions of these test procedures are given in ASTM G129 [18], NACE TM0198 [19], and ISO 7539-7 [I]. The benefit of SSR testing is that it produces a result in a reasonably short time, usually within 1–2 days. Even at the slowest rates, the test is usually complete in no more than a week. The dynamic straining reduces incubation time to the onset of cracking in susceptible materials and sustains the cracking process. Therefore, in applications where the available testing time is limited and rapid screening is an utmost consideration, SSR techniques can provide significant benefits.

The plastic strain used in the SSR procedure causes an accelerated disruption of surface films, thereby overcoming the initial period of incubation that can result in unacceptably long test durations prior to the onset of cracking (Fig. 77.10) [20]. It also provides dynamic straining at the

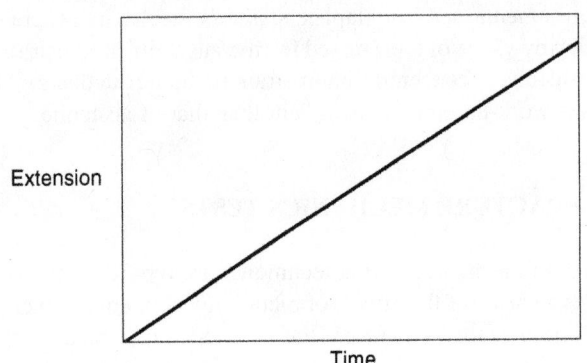

FIGURE 77.9. Schematic representation of constant-extension-rate test (load vs. time).

crack tip after initiation that helps to sustain the further growth of the crack. However, in some cases, a major concern of the SSR technique is that the plastic strain can add complications to the interpretation of the test results because most materials are not prone to this degree of straining in actual service. Furthermore, some material may actually show vastly increased cracking susceptibility as a result of the plastic straining compared to tests conducted at constant load or strain.

Another benefit of the SSR technique is that it allows the evaluation of the effects of metallurgical variables, such as alloy composition, heat treatment, and processing and/or environmental parameters (e.g., aeration, concentration, and inhibition) in a relatively short period of time resulting from the short exposure period usually required versus conventional constant-load or constant-stress specimens. Extension rates utilized for SSR testing are typically ~ 1–4×10^{-6} in./s (2.5×10^{-5}–1×10^{-4} mm/s). At this extension rate, the testing speed is ~ 0.25–1% strain per hour on a 1-in.- (2.5-mm-) long gauge section and failure of most engineering materials will occur within a few days. In some cases, slower strain rates are required to produce the necessary degree of sensitivity with the SSR technique (see Fig. 77.11). However, in other cases, slower stain rates reduce the measured cracking susceptibility [19]. Therefore, tests at multiple strain rates are usually desired.

In some cases, longer exposure periods prior to testing may be necessary if longer term formation of corrosion films is a critical step in the cracking process. The in-service corrosion potential of metallic components may also change with time and eventually move into a range of potential where the material is susceptible to cracking. Therefore, it may also be necessary to evaluate the electrochemical potential to define a specific range where susceptibility to cracking can occur; for example, film formation of water scales on austenitic stainless steels can exacerbate SCC in chloride-containing waters as found in some heat exchange applications [21]. Additionally, Figure 77.12 shows the influence of electrochemical potential on cracking of steels in caustic environments, which is a major factor in SCC of chemical process equipment [22].

The SSR evaluation for susceptibility to EAC is normally obtained through comparison of the results of tests conducted in a corrosive environment versus corresponding data obtained in an inert environment (see Fig. 77.13). An inert environment is one that has been shown not to promote EAC or significant corrosion in the material being tested. In most cases, these baseline tests are conducted in air. Direct examination of the specimen gauge section for EAC and documentation of fracture mode is also important to a full interpretation of the SSR test results. The SSR test results that are used include time to failure, plastic elongation to failure, reduction in area, ultimate tensile strength, and fracture energy as measured by the area under the stress–strain curve.

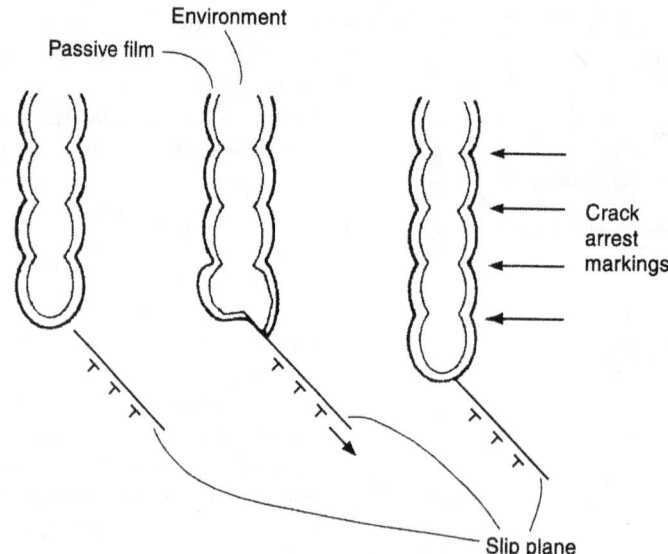

FIGURE 77.10. Disruption of surface films at crack tip by plastic deformation.

For notched tensile SSR tests, the notched tensile strength is usually involved in the evaluation. These data are usually presented in terms of their ratios versus the corresponding value from a test conducted in the inert environment. Ratios in the range 0.8–1.0 normally denote high resistance to EAC, whereas low values (i.e., < 0.5) show high susceptibility. In some cases, hydrogen can cause loss in ductility without indication of brittle cracking in the specimen. The slight loss of ductility is usually a less important situation than where extreme loss of ductility (i.e., embrittlement) has been observed, particularly if the material still exhibits a high-tensile-strength ratio.

In order to overcome the problems associated with cumulative plastic strain in conventional SSR tests, a novel alternative testing methodology is the use of a cyclic slow-strain-rate (CSSR) testing procedure. In this case, the extension rates are still in the range cited previously for the SSR technique. However, for CSSR testing, the specimen is loaded to a relatively high percentage of its yield strength in tension and the stress is then varied above and below this value. Whereas the magnitude of the cyclic component of the stress can vary, in many cases a range of at least ±10% of the mean stress is used (see Fig. 77.14). The frequency of the cyclic loading will vary with both the load limits and the strain rate. The benefit of this technique is that the amount of cumulative plastic strain received by the specimen is low relative to that in the conventional SSR technique (see Fig. 77.15) [23]. It also retains the dynamic straining of the specimen that provides the mechanical acceleration of crack initiation through its influence on disruption of protective surface films. The number of cycles used in the evaluation are typically on the order of 100–200 and a test can take from a few days to two weeks depending on the actual number of cycles used. This type of test is better than the conventional SSR technique for material/environment combinations where excessive straining produces a large increase in susceptibility to cracking. Examples of this type of behavior are the case of ferritic and martensitic steels in hydrogenating environments and for duplex stainless steels, in general, which by virtue of their mixed ferritic/austenitic microstructure receive a concentration of strain in the ferrite due to its lower work-hardening coefficient than that of austenite.

E. FRACTURE MECHANICS TESTS

Fracture mechanics testing techniques are typically utilized for evaluation of the effects of metallurgical or environmental variables on EAC where the specimen contains a sharp crack. Figure 77.16 [24] shows many of the possible fracture mechanics test configurations. One of the most common and relatively simple techniques for incorporation of fracture

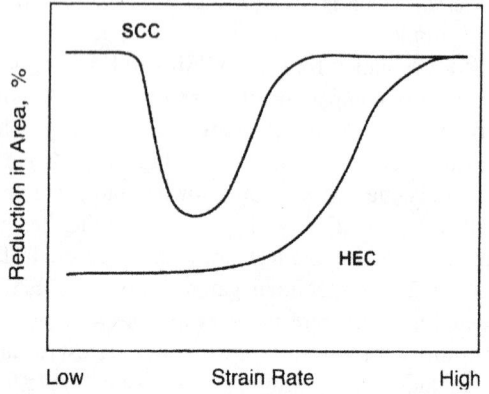

FIGURE 77.11. Schematic representation of the influence of strain rate on SCC and HEC.

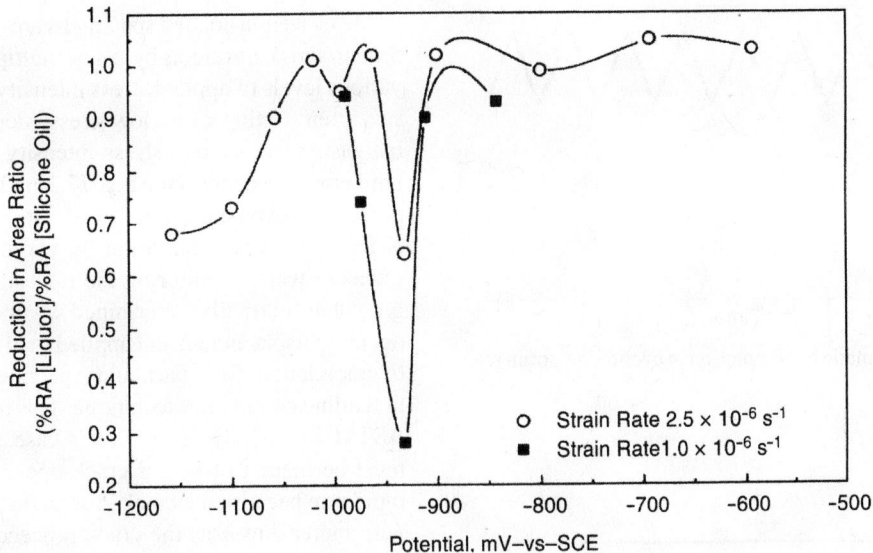

FIGURE 77.12. Influence of electrochemical potential on SCC of steel in caustic (NaOH) solution. Note strain rate effect.

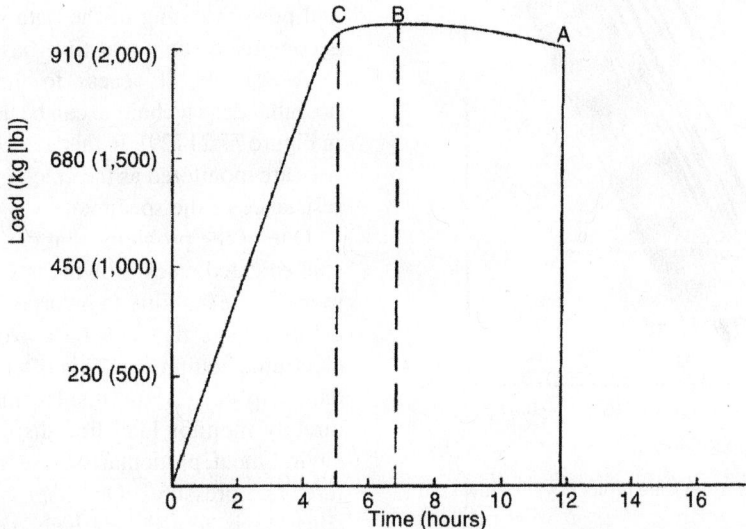

FIGURE 77.13. Comparison of inert environment CERT test (solid line) to those in corrosive environments (dotted line).

mechanics techniques for the evaluation of EAC is through the use of constant-load or constant-deflection specimens in combination with a precracked specimen. In the case of constant-load specimens, a load is applied to a fracture mechanics specimen using a dead-weight load, by a hydraulic cylinder, or through a pulley or lever system to magnify the dead-weight load. These methods are generally analogous to those used for constant-load tensile specimens discussed previously. The most common types of specimens utilized for evaluation of EAC are the compact tension (CT), precracked double beam (PDB—also referred to as double-cantilever beam, or DCB), or single-edge notched bend

(SENB) specimens (see Figs. 77.17–77.19). Normally, they are fatigue precracked or, in some cases, slotted using controlled electrodischarge machining (EDM) procedures prior to exposure to the environment to produce a cracklike defect that can successfully initiate EAC at high levels of stress intensity. The fatigue precracking must be performed at a low enough stress intensity to minimize the plastic zone ahead of the crack. This is usually accomplished through load-shedding techniques whereby an initially high peak load is used to initiate the fatigue precrack and the cyclic load is then decreased as the precrack approaches its desired length. In the testing of precracked specimens, it has been

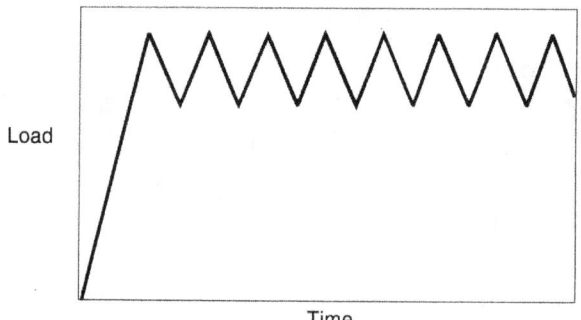

FIGURE 77.14. Schematic representation of cyclic slow strain rate test (load vs. time).

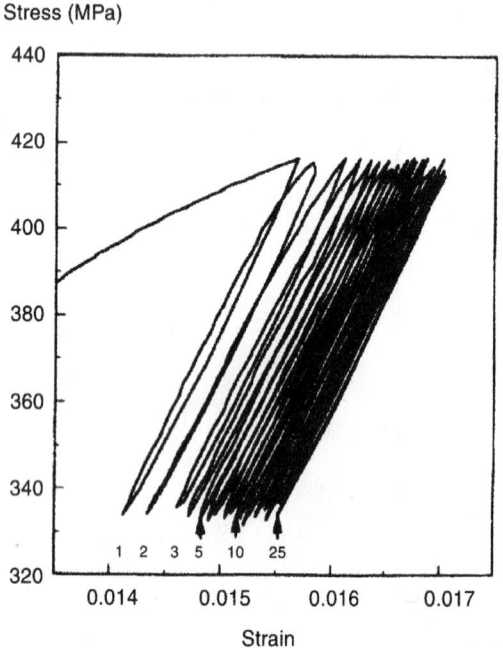

FIGURE 77.15. Stress versus strain for a typical cyclic slow strain rate test.

found that excessive initial stress intensity of precracked specimens can produce a large plastic zone that can act as a barrier to EAC initiation. In these cases, the effect will be to produce nonconservative data.

The stress intensity at the tip of the crack can be calculated using standard equations as given in ASTM E399 [25] for CT and SENB specimens and in ASTM GXXX [26] or NACE TM0177- Method D [12] for the PDB specimen. As shown for the PDB specimen, side grooves can be utilized to assist in keeping the crack growing in a planar fashion under plane strain conditions. In some cases, the crack will tend to grow out of plane resulting in an invalid test. The important consequence of using side grooves is that the equations for the CT or PDB specimens must contain a correction factor that accounts for the geometry and dimension of the side grooves.

Dead-weight-loaded specimens are often used to measure time to crack initiation by using multiple specimens having various levels of applied stress intensity or by taking a single specimen starting with a low stress intensity and periodically increasing the applied stress intensity in incremental steps until cracking occurs (see Fig. 77.20) [27]. Alternatively, it is possible to evaluate crack growth rate versus stress intensity K following crack initiation by varying the applied stress intensity while monitoring the rate of crack growth. Crack growth is normally determined by measurement of crack opening displacement and applied load, which can be related to crack length for a particular specimen geometry using an unloading compliance technique. This technique is defined in ASTM E813 [28]. In the latter case, however, provisions must be made to monitor crack opening displacement at a rapid rate because the crack growth rate will tend to increase with increasing K as the crack proceeds through the specimen. This is usually accomplished by integrating load cell and displacement gauge signals through a high-speed data logger or computer-based data acquisition system. The latter system also has the potential for combining data acquisition and postprocessing of the data so that data display can be accomplished on a real-time basis.

Alternately, if access to the specimen is difficult, a potential drop technique can be used as shown schematically in Figure 77.21 [29]. In this procedure, changes in the current flow are monitored as the crack grows, thereby changing the resistance of the specimen.

One of the problems that can occur in conducting these sophisticated crack growth tests in some corrosive environments is the inability to incorporate electronic equipment in environments that can be corrosive to materials used in electronic equipment. This often precludes the use of standard clip gauges and displacement monitoring devices to directly monitor load line displacements in the corrosive environment, particularly those involving elevated temperatures and pressures. One approach used effectively in this situation is to use relatively simple mechanical devices made from corrosion-resistant material (but electrically isolated from the specimen) that can be used to monitor displacements directly on the test specimens. These mechanical devices then produce a relative displacement outside of the test chamber where the measurement can be made using conventional electronic devices. The main areas for concern when using these techniques are that electrical isolation be maintained between the test specimens and any dissimilar metals. This can be accomplished using nonmetallic components or coatings (e.g., plastics or ceramics). When used in a load-bearing manner, these materials must remain rigid and not result in excessive compliance or time-dependent deformation. Finally, the seals through which these devices must react must be able to handle the pressure and temperature constraints of the system while providing minimal frictional loading.

Precracked specimen configurations for stress-corrosion testing

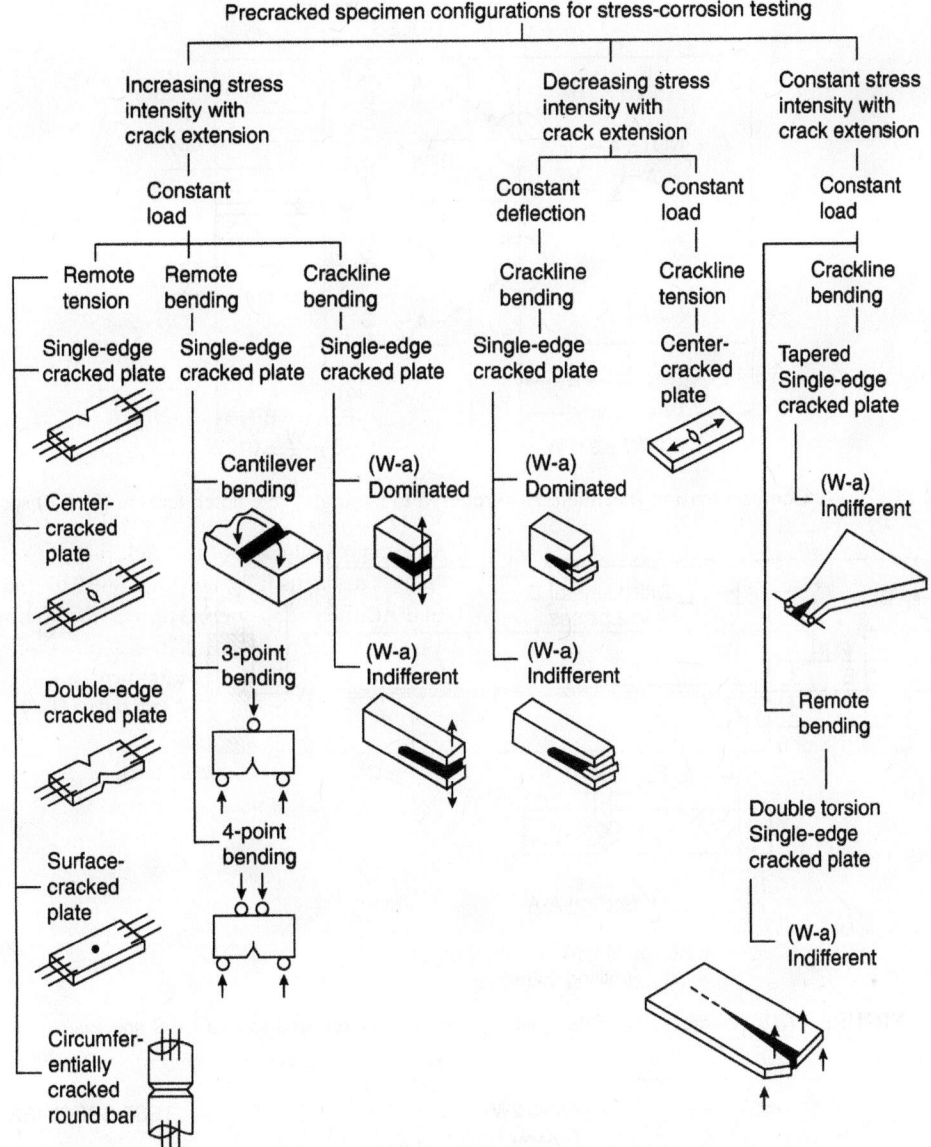

FIGURE 77.16. Fracture mechanics test configurations.

In some materials, another problem that can be encountered associated with conducting tests on precracked specimens is that of corrosion product wedging. Where excessive insoluble corrosion products can buildup on crack surfaces, additional mechanical loading (over that provided intentionally) can result. In the case of precracked specimens, the corrosion product buildup forces the arms of the specimen apart, thereby increasing the applied stress intensity on the crack trip. Figure 77.22 [30] shows some typical cases with and without corrosion product wedging. It is usually manifested in cases where the crack growth rate does not decrease with the applied stress intensity or where apparent threshold stress intensity has been reached but then an increase in crack growth activity occurs. This is a problem seen in aluminum alloys in some environments

where a voluminous alumina (Al_2O_3) corrosion product is generated.

An attractive alternative to dead-weigh-loaded specimens is constant-deflection specimens. In this situation, either CT or PDB specimens are loaded to an initial level of crack tip stress intensity by deflection of the arms of the specimen. This deflection is obtained either by inserting the wedge into the specimen or by tightening a bolt arrangement that deflects the arms of the specimen. The initial stress intensity must be above the threshold stress for EAC, which allows cracking to initiate. Once cracking initiates, it proceeds while the stress intensity decreases as the crack propagates through the specimen. Thus, this type of test is commonly referred to as a decreasing K test and is extensively utilized for evaluation of EAC in its various forms. Once the stress

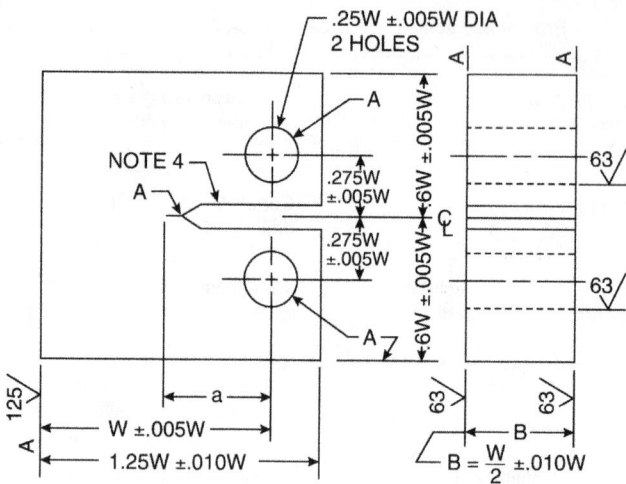

FIGURE 77.17. 1-T Compact tension specimen. Also referred to as single-edge notch tension (SENT) specimen.

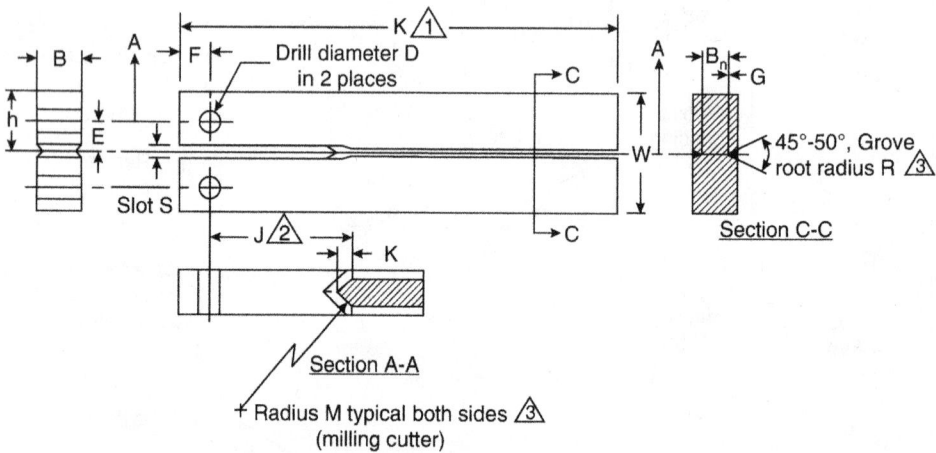

FIGURE 77.18. Precracked double-beam specimen. Also referred to as a DCB specimen.

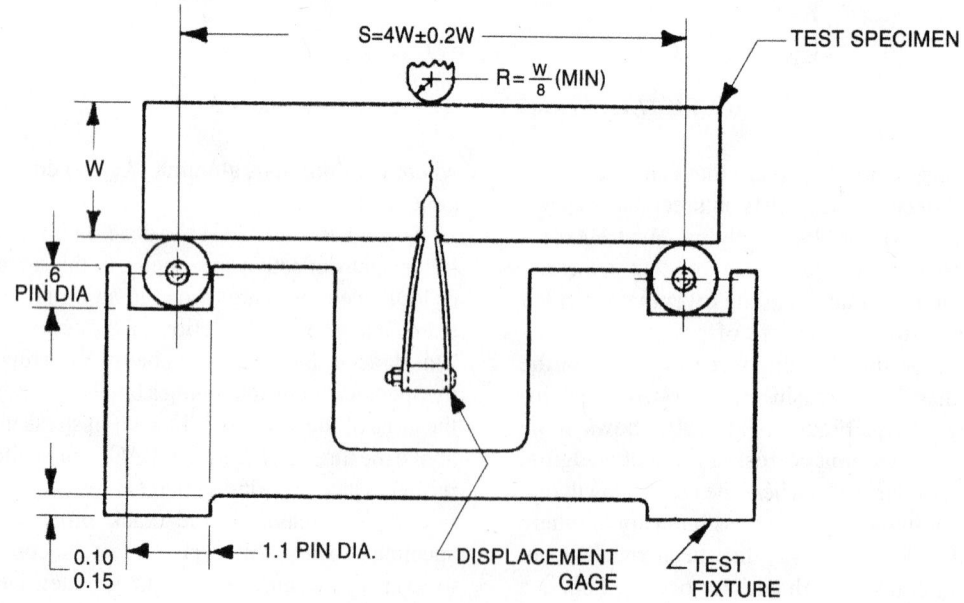

FIGURE 77.19. SENB specimen.

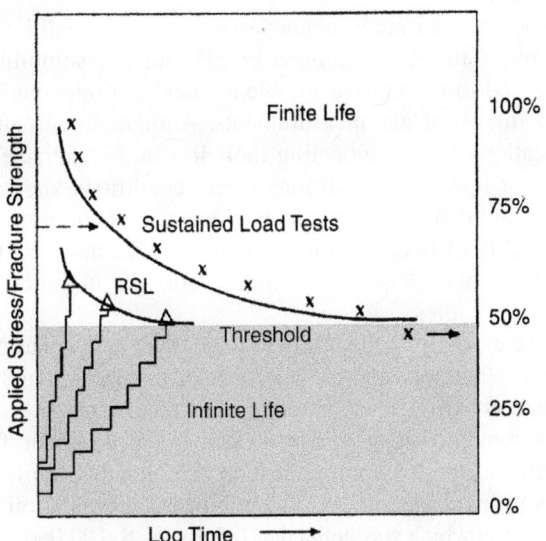

FIGURE 77.20. Comparison of test data for HEC from sustain-loaded precracked specimens and rising-step-loaded (RSL) specimens.

intensity at the crack tip reaches a value insufficient to sustain crack growth, crack growth will stop. Therefore, the final conditions of load and crack length can be used to define the threshold stress intensity using the appropriate equations for either the CT or DCB specimen.

Sometimes the period required to run a decreasing K test is very long. An alternative is to use a rising load test whereby the fracture mechanics specimens are subjected to an increasing load in a similar manner as used in conventional CERT testing. In this case, the crack open displacement and load are monitored simultaneously and the results are analyzed as in conventional fracture mechanics tests. One of the difficulties in the interpretation of rising-load tests is that the threshold stress intensity obtained by this method often differs from that determined by decreasing K tests (see Fig. 77.20). The dynamic strain rate in the rising-load test can complicate the interpretation of the test results, particularly if HEC or other operable cracking mechanisms are strongly dependent on strain rate, diffusion rate of particular corrosive species or film formation, or repassivation rates. In these cases, a loading rate is

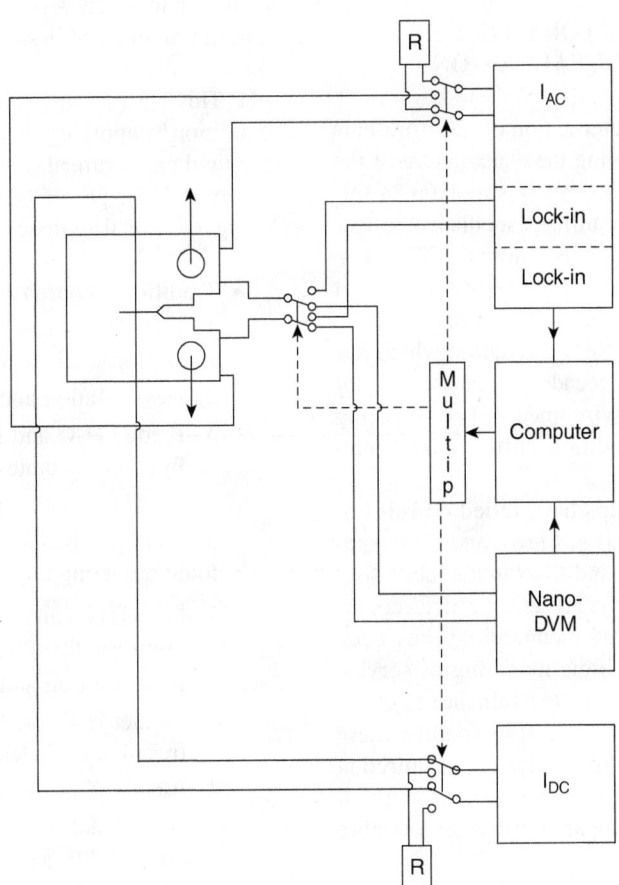

FIGURE 77.21. Schematic representation of a potential drop set up for measurement of crack growth in a fracture mechanics specimen.

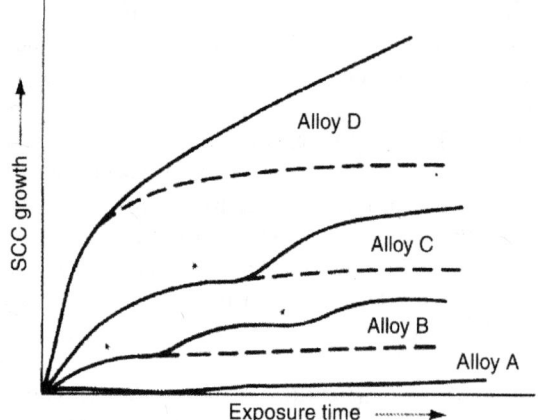

FIGURE 77.22. Influence of corrosion product wedging on results from precracked test specimen. Curves with an asterisk show increase in SEE growth resulting from corrosion product wedging.

usually selected that results in crack growth rates and fracture morphologies that correlate with those obtained from applicable service experience, field testing, or laboratory tests conducted under simulated service conditions.

F. DEFINITION OF LABORATORY TEST ENVIRONMENTS FOR EAC EVALUATION

Laboratory testing for the characterization of EAC presents an ongoing challenge [31]. Defining the exact nature of the laboratory test environment that correctly simulates in-service process conditions presently utilizes sophisticated theoretical (ionic/phase behavior) models. These models translate in-service conditions to laboratory test conditions so that relevant EAC evaluations can be performed. This approach has been utilized for laboratory EAC evaluation which has progressed over the past two decades from the use of relatively simple standardized environments to emphasizing simulated service or process conditions, often at high temperature and pressure.

Traditional experimental setups have relied on rules of thumb, simple relationships (inert gas law), and past experience to achieve certain target conditions in the laboratory that may or may not accurately represent the field conditions. However, a state-of-the-art methodology has been advanced that includes the use ionic modeling of service conditions to provide an accurate path to attain the target test conditions in the laboratory. Even more importantly, these models provide the correct amounts of species required at loading conditions (often at room temperature) which will result in the target test conditions at elevated temperature and pressure.

The basic difference between the traditional laboratory approaches and modern test methodologies is the use of an ionic modeling tool. Such modeling tools also provide a

basis to accurately characterize service conditions in terms of in situ conditions of pH and gas solubility as opposed to normally available ambient or "open-cup" pH and dissolved gas measurements. Another important application of ionic modeling tools lies in the characterization of operating conditions where conditions known for EAC in laboratory tests can be documented and translated into field or plant operating conditions for more accurate prediction of EAC susceptibility for use in fitness-for-service studies.

An example of the utility of ionic modeling versus traditional laboratory approaches for EAC testing is found in a situation where the laboratory evaluation of candidate alloys was required in a simulated service environment for a petroleum production application. The goal of the testing was to investigate EAC susceptibility at a typical oil well condition which was defined as 0.5 psia H_2S, 100,000 mg/kg Cl^-, and an in situ pH of 4.5. The evaluation of EAC susceptibility included testing for an HEC phenomenon known as sulfide stress cracking at 23°C and also the susceptibility to anodic SCC at an elevated temperature of 204°C. Ionic modeling enabled the calculation of the particular H_2S and CO_2 gas additions and the necessary amount of buffer ions (HCO_3^-) to achieve the target pH and H_2S partial pressure and in situ pH at temperature without relying on acid or base additions to the solution to adjust the solution pH. This ensured an accurate representation of the service condition by applying only species that are actually present in the field environment.

For the case presented above, a comparison of the results using traditional approaches and ionic model is shown below:

- Traditional approach
  1. Load H_2S (excess liquid) based on calculations from real or ideal gas law.
  2. Heat solution and H_2S to 204°C.
  3. Predict H_2O and H_2S partial pressure contributions from steam tables.
  4. Bleed off excess pressure.
  5. Final pH, taken as an "open-cup" measurement.
- Ionic modeling approach
  1. Load H_2S (calculated amount of liquid H_2S added by gram weight).
  2. Heat solution and H_2S to 204°C.
  3. Predict H_2O and H_2S partial pressure contributions from ionic model.
  4. Bleed off excess pressure (if necessary).
  5. Use model to calculate the in situ pH and obtain dissolved H_2S content in solution as well as final pH adjusted for corrosion rate of the material and consumption of reactive species during the test duration.

TABLE 77.2. Results of Ionic Modeling of EAC Test Environment

Species	Total (g)	Aqueous (g)	Vapor (mol)	ppm	Mol%	Pressure (psia)
Water	3000	2987.38	0.700586			
Carbon dioxide	69.0483	10.9882	1.31925		39.51	500.0
Hydrogen sulfide	51.2501	6.2837	1.31936		39.51	500.0
Hydrogen chloride	2.03E–04	3.25E–05	4.67E–06			
Sodium bicarbonate	3.17E–04	3.17E–04	0.00E+00			
Bicarbonate ion (−1)	2.18E–03	2.18E–03	0			
Carbonate ion (−2)	8.41E–08	8.41E–08	0			
Chloride ion (−1)	6.08E+02	6.08E+02	0	151,728		
Hydrogen ion (+1)	1.51E–03	1.51E–03	0			
Hydrogen sulfide ion (−1)	4.84E–02	4.84E–02	0			
Hydrogen ion (−1)	1.52E–06	1.52E–06	0			
Sodium carbonate ion (−1)	1.15E–08	1.15E–08	0			
Sodium ion (+1)	3.94E–02	3.94E–02	0			
Sulfide ion (−2)	2.97E–09	2.97E–09	0			
Total (by phase)	4.12E+03	4.01E+03	3.3392			1265.65

Note: The following calculations were made with the ionic model: (a) The gram weight of H_2S and CO_2 to be added in liquid form at the start of the test, (b) the contribution of each species to the total system pressure, (c) partitioning between aqueous solution and vapor phase, and (d) the amount of each of the dissolved species in the heavy brine test solution at the test temperature.

A second example identifies the errors of traditional laboratory approaches for EAC evaluation when simulating extremely sour, heavy brine conditions. A study focused on the EAC testing of alloys in an environment consisting of 500 psia H_2S, 500 psia CO_2, and 151,700 mg/L Cl^- in aqueous solution at a temperature of 204°C. Traditional laboratory approaches rely on steam tables to calculate the contribution of H_2O to the total pressure. This calculated value can be significantly inaccurate due to the contribution of high chloride content on depression of the vapor pressure of H_2O and the interaction of multiple soluble gases in the system. Using traditional methods leads to conditions in the laboratory that are actually very different from the target conditions (i.e., much more H_2S was dissolved in the test solution than needed to accurately simulate the service conditions in question). As shown in Table 77.2, the ionic modeling calculated the contributions of each gas species to the overall pressure at the test temperature while also accounting for the solubility of each gaseous component in the heavy brine solution, thus facilitating a much more accurate environmental simulation.

It should be noted that ionic models and associated software are available from a number of commercial and academic sources. These are computer-based computational models that are based on documented thermodynamic relationships for various chemical species. The user should be aware that the models have varying content in terms of the ionic species that they contain with some containing information on over 2500 chemical species. Just as importantly, different models may vary in terms of the ranges of conditions where they can be successfully applied depending on the source of their thermodynamic data and relationships.

G. SUMMARY

It can be said that there is no single perfect testing technique for the evaluation of EAC. Therefore, evaluation of materials typically involves the use of the specimen and testing technique that takes into account as many relevant aspects as possible for the particular material, environment, and cracking mechanism under consideration. In some cases, this may mean the use of:

1. More than one type of test specimen
2. Various alternative configurations of the same specimen
3. Alternative test techniques with the same specimen (e.g., crevices or applied potential, constant load, and slow strain rate)

Most of all, it is important to provide linkage between the results of laboratory evaluations and real-world service applications. This is often developed through studies involving:

1. Integrated laboratory and field or in-plant tests
2. Correlation of laboratory data with service experience
3. Reviews of published literature on the service performance of similar materials

The evaluation of EAC susceptibility using laboratory testing methods provides data that can help the investigator better define cracking mechanisms as well as the possible service performance of materials of construction. Consequently, this information can provide a better technical basis for using materials in engineering structures and operating equipment resulting in an increased confidence level when materials selection decisions need to be made. This, in turn, also leads to optimization of the materials of construction by reducing the allowance for unpredictable service behavior resulting in a lower material cost, less downtime, and a reduction in the number of costly failures and associated loss production.

REFERENCES

1. ISO 7539 Corrosion of Metals and Alloys—Stress Corrosion Testing, International Organization for Standardization, Geneva, Switzerland, 1989.

2. A. Turnbull, Br. Corros. J., **17**, 271 (1992).

3. M. G. Fontana, Corrosion Engineering, McGraw-Hill, New York, 1986, p. 148.

4. A. Turnbull, Br. Corros. J., **17**, 148 (1992).

5. ASTM Standard Designation, Standard Practice for Evaluation of Hydrogen Uptake, Permeation, and Transport in Metal, by an Electrochemical Technique, ASTM, West Conshohoken, PA, 1997.

6. ASTM, Standard Designation, Standard Test Method for Electrochemical Measurement of Diffusible Hydrogen in Steels (Barnacle Electrode), ASTM, Philadelphia, PA, 1994.

7. M. A. V. Devanathan and Z. Stachurski, Proc. R. Soc., **A270**, 90 (1962).

8. J. McBreen, L. Nanis, and W. Beck, J. Electrochem. Soc., **113**, 1218 (1966).

9. M. S. Cayard, R. D. Kane, C. J. B. Joia, and L. A. Correia, "Methodology for the Application of Hydrogen Flux Monitoring Devices to Assess Safety Margins on Equipment Operating in Wet H_2S Service," Paper No. 394, CORROSION/98, NACE International, Houston, TX, 1998.

10. R. D. Kane, Slow Strain Rate Testing for Evaluation of Environmentally Induced Cracking: Research and Engineering Applications, STP 1210, ASTM, West Conshohoken, PA, 1993, p. 187.

11. G49, Standard Practice for Preparation and Use of Direct Tension Stress-Corrosion Test Specimens, ASTM, West Conshohoken, PA, 1995.

12. TM0177-96, Testing Of Metals For Resistance To Sulfide Stress Cracking at Ambient Temperature, NACE International, Houston, TX, 1996.

13. G139, Standard Test Method for Determining Stress Corrosion Cracking Resistance of Heat-Treatable Aluminum Alloy Products Using Breaking Load Method, ASTM, West Conshohoken, PA, 1996.

14. G39, Standard Practice for Preparation and Use of Bent-Beam Stress-Corrosion Test Specimens, ASTM, West Conshohoken, PA, 1999.

15. G38, Standard Practice for Making and Using C-Ring Stress-Corrosion Test Specimens, ASTM, West Conshohoken, PA, 1995.

16. G30, Standard Practice for Making and Using U-Bend Stress-Corrosion Test Specimens. ASTM, West Conshohoken, PA, 1997.

17. D. Sprowls, ASM Handbook, Vol. 13, Corrosion, ASM International, Materials Park, OH, 1987, p. 247.

18. G, 129, Standard Practice for Slow Strain Rate Testing to Evaluate the Susceptibility of Metallic Materials to Environmentally Assisted Cracking, ASTM, West Conshohoken, PA, 1995.

19. TM0198-98, Slow Strain Rate Test Method for Screening Corrosion Resistant Alloys (CRAs) for Stress Corrosion Cracking in Sour Oilfield Service, NACE International, Houston, TX, 1998.

20. R. W. Staehle, Stress Corrosion Cracking and Hydrogen Embrittlement of Iron Based Alloys, NACE International, Houston, TX, 1977, p. 180.

21. J. A. Beavers and G.H. Koch, Limitations of the Slow Strain Rate Test for Stress Corrosion Cracking Testing, MTI Publication No. 39, Materials Technology Institute of the Chemical Process Industries, St. Louis. MO, 1995, p. 5.

22. D. Singbeil and A. Garner, "Stress Corrosion Cracking of Kraft Continuous Digesters," International Congress on Metallic Corrosion, Vol. 1, Pu 61. National Research Council of Canada. Ottawa, Canada, 1984, p. 187.

23. W. J. R. Nisbet, R.H.C. Hertman, and G. vd Handle, "Rippled Strain Rate Test for CRA Sour Service Materials Selection," Paper No. 58, CORROSION/97, NACE International, Houston, TX, 1997.

24. W. Barry Lisagor, in Corrosion Tests and Standards—Application and Interpretation, R. Baboian (Ed.), ASTM, West Conshohoken, PA, 1995, p. 246.

25. E399, Standard Test Method for Plane-Strain Fracture Toughness of Metallic Materials. ASTM, West Conshohoken, PA, 1997.

26. G168, Standard Practice for Making and Using Precracked Double Beam Stress Corrosion Specimens, ASTM, West Conshohoken, PA. (To be published 2006).

27. L. Raymond. ASTM Standardization News, **24**(10), 42 (1996).

28. E813, Standard Test Method for J_{IC}, A Measure of Fracture Toughness, ASTM, West Conshohoken, PA, 1998.

29. W. Dietzel and K. H. Schwalbe, Zeitscrift Für Materialprüfung, **28**(11), 369 (1986).

30. D. Sprowls, ASM Handbook, Vol. 13, Corrosion, ASM International, Materials Park, OH, 1987, p. 268.

31. B. D. Chambers, V. Lagad, S. Srinivasan, and M. Yunovich, "Ionic Modeling of Field Conditions: Improvements over Traditional Laboratory Approaches," Presentation 8703, Research in Progress Symposium, Corrosion 2009, NACE International, Houston, TX, Mar. 2009.

78

TEST METHODS FOR WET H₂S CRACKING

TEST METHODS FOR WET H_2S CRACKING

M. ELBOUJDAINI

CANMET Materials Technology Laboratory, Ottawa, Ontario, Canada

A. INTRODUCTION

Much of the equipment in petroleum operations is exposed to aqueous environments containing H_2S and is fabricated from carbon steel that is susceptible to cracking in wet H_2S. Standards on material requirements for resistance to cracking in wet H_2S and test methods to assess the resistance of materials to H_2S cracking were developed by NACE International.

The development of testing methods for sulfide stress cracking (SSC), hydrogen-induced cracking (HIC), and stress-oriented hydrogen-induced cracking (SOHIC) was driven by industry requirements to evaluate and qualify materials for sour service. The NACE Standard TM0177 [1], "Laboratory Testing of Metals for Resistance to Specific Forms of Environmental Cracking in H_2S Environment," was developed to assess SSC resistance, and another NACE standard, TM0284 [2], was developed to evaluate pipeline and pressure vessel steels for resistance to HIC. The methodology specified in NACE standard TM0284 has been successfully used to evaluate the effects of chemical composition, structure, materials processing, and orientation on the HIC resistance.

© Her Majesty the Queen in Right of Canada, as represented by the Minister of Natural Resources, 2010.

Uhlig's Corrosion Handbook, Third Edition, Edited by R. Winston Revie
Copyright © 2011 John Wiley & Sons, Inc.

The double-beam specimen configuration described in American Society for Testing and Materials (ASTM) G-39 [3] has been used to study weldments and parent steels under applied tensile stress in order to study SOHIC.

B. TEST PROCEDURES

The test methods that have been developed are used for developing improved alloys for sour service and for selecting materials for application in specific sour environments. Variables that influence cracking behavior include alloy composition and microstructure, hardness, total stress (applied stress plus residual stress) and environmental parameters, such as pH and corrosivity. For example, Figure 78.1 shows the effect of hardness [4] of two materials, American Iron and steel Institute (AISI) 4130 low-alloy steel and 12% Cr stainless steel, on the threshold stress, or critical stress, above which SSC occurs. The conditions of a laboratory test technique, the "BP Test," for studying HIC in a wet H_2S environment, originally developed by Cotton and subsequently included as part of NACE standard TM0284, are listed in Table 78.1.

The specimen location and orientation for seamless and welded pipe are illustrated in Figure 78.2. As shown, the coupons of both parent and weld metal are 20×100 mm $\times$ wall thickness. Three unstressed specimens of each material are immersed in the H_2S-saturated solution. After the test, three sections of each coupon, oriented as shown in Figure 78.3, are polished and examined for HIC using the optical microscope at a magnification of $100\times$. As indicated in Figure 78.4, HIC is assessed quantitatively by determining the following three ratios:

1. Crack length ratio (CLR), representing the amount of cracking in the rolling plane

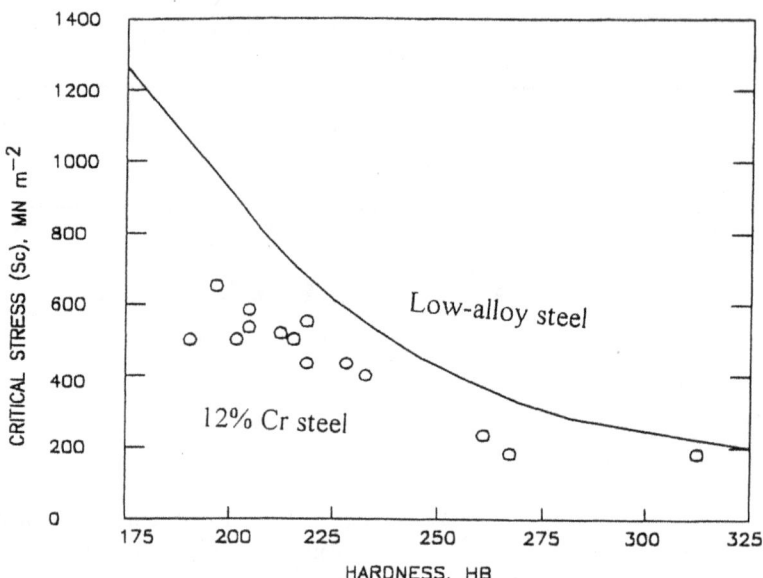

FIGURE 78.1. Comparison of SSC threshold data (σ_{th}) for low-alloy (AISI 4130) and AISI 410 steels [4]. (Copyright by NACE International; reprinted with permission.)

2. Crack/thickness ratio (CTR), representing the amount of cracking in the through-thickness direction of the steel wall
3. Crack/sensitivity ratio (CSR), representing a combination of CLR and CTR

The CLR is defined as the ratio of the sum of the individual crack lengths, a_i, to the width of the section observed, whereas CTR is the ratio of the sum of the individual crack thicknesses, b_i, to the thickness of the specimen.

The CLR, CTR, and CSR can be calculated as percentages as follows:

Crack/thickness ratio:
$$\text{CTR} = \sum_i \left(\frac{b_i}{t} \right) \times 100\%$$

Crack/length ratio:
$$\text{CLR} = \sum_i \left(\frac{a_i}{W} \right) \times 100\%$$

Crack/sensitivity ratio:
$$\text{CSR} = \sum_i \left(\frac{a_i b_i}{tW} \right) \times 100\%$$

TABLE 78.1. Standard Test Conditions for NACE TM0284[a]

Test period	96 h (4 days)
Temperature	$25 \pm 3°C$
H₂S concentration	Saturated (>2300 ppm)
H₂S flow rate	100 mL/min-L of solution
pH	5.1–5.4
Test solution	Synthetic seawater (ASTM D1141-52 stock solution No. 1 or 2)
Loading stress	Not stressed

[a]See [2].

where t is specimen thickness, W is specimen width, a is crack length, and b is crack thickness. As an alternative to metallographic examination, an ultrasonic C scan can be used along with a quantitative image analysis system to quantify HIC [5, 6].

The wet-fluorescent magnetic-particle inspection (WFMPI) technique can be used to detect cracks in welds and adjacent heat-affected-zones (HAZs) on the inside surfaces of pipes or vessels. This technique was found to be more sensitive in locating HIC cracks than other inspection techniques, such as radiography or dry magnetic-particle examination.

B1. Sulfide Stress Cracking

The standard test conditions for the NACE TM0l77 [1] test are listed in Table 78.2. This test is frequently carried out by immersing specimens in an aqueous solution containing 5% sodium chloride (NaCl), 0.5% acetic acid (CH₃COOH), saturated with H₂S gas at ambient temperature and pressure. Round tensile specimens are commonly used, but, as described in Chapter 77, other types of specimens are also used. Specimens are loaded to various stress levels using calibrated proof rings or dead-weight testers. Time to failure is monitored for the 720-h test duration; "no failure" is recorded if failure does not occur within this period.

After characterizing the tensile properties of a material, SSC testing is carried out under constant load at various load levels, and the times to failure are plotted in terms of applied stress versus time to failure (see Fig. 77.4 in Chapter 77). In general, an SSC threshold stress can be determined and is used as a measure of cracking susceptibility. Susceptibility is

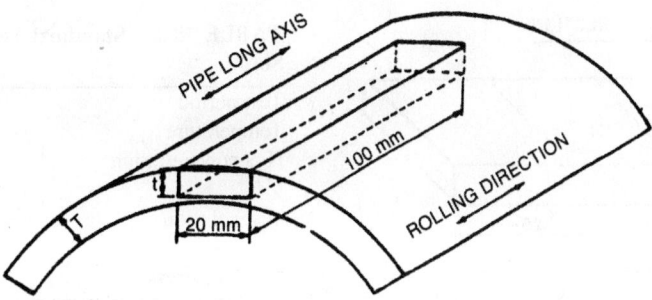

(a)   Orientation of test coupons taken from seamless pipe and
from the parent material of longitudinally welded pipe

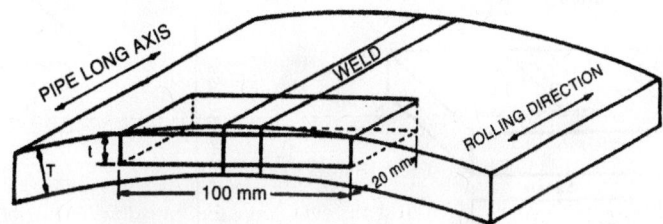

(b)   Orientation of test coupons taken from the
weld area of longitudinally welded pipe

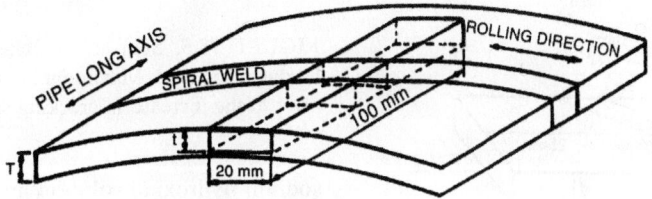

(c)   Orientation of test coupons taken from the
weld area of spiral welded pipe

FIGURE 78.2. HIC samples from pipe material.

strongly influenced by steel strength and hardness, as shown in Figure 78.5. A sharp increase in cracking susceptibility with increased hardness is often observed. Heat treating steels to hardness levels below that at which increased cracking susceptibility occurs has been effective in minimizing field failures in sour service.

B2. Determination of Dissolved Hydrogen Concentration Causing HIC of Linepipe Steels

After the HIC immersion test, some investigators immerse test coupons in glycerin or mercury-filled collectors held at 45°C. The "diffusible hydrogen at 45°C" which is evolved is collected and usually reported as milliliters of H_2 (NTP)/ 100 g of steel. A number of researchers have used measurement of diffusible hydrogen as a means of assessing HIC susceptibility [7–10] and have correlated diffusible hydrogen measurements with CLR and CTR [11]. The amount of

hydrogen absorbed by a test coupon at pH_{th}, the pH just low enough to cause HIC, is the value of C_{th}. Ikeda et al. [10] and Hoey et al. [8] have shown that there is a critical concentration of dissolved hydrogen in a steel, termed C_{th}, that must be attained in order for HIC to occur.

C. HYDROGEN DIFFUSIVITY MEASUREMENTS

A well-established approach to measuring hydrogen diffusion in steels involves the use of a dual electrochemical cell with the steel specimen between the charging and oxidation cells, as shown schematically in Figure 78.6. Hydrogen atoms are generated on the side of the specimen exposed to the environment of interest in the charging cell, diffuse through the steel, and are oxidized electrochemically on the opposite side of the specimen. This side is exposed to a

ROLLING DIRECTION

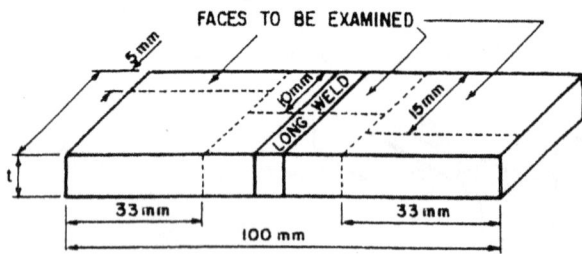

(a)   Sectioning procedure for cutting test coupons from seamless pipe and the parent material welded pipe.

FACES TO BE EXAMINED

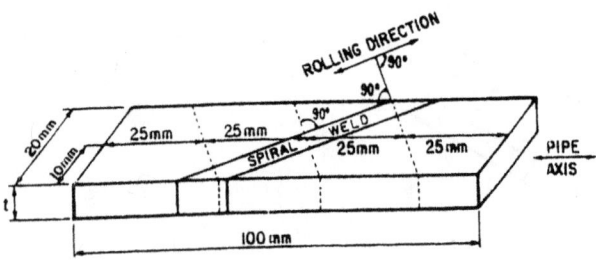

(b)   Sectioning procedure for cutting test coupons from the weld area of longitudinally welded pipe.

(c)   Sectioning procedure for cutting test coupons from the weld area of spiral welded pipe.

FIGURE 78.3. Geometry of specimens cut from plate and from pipe.

TABLE 78.2. Standard Test Conditions for NACE TM0177 Solution[a]

Test period	720 h (30 days)
Temperature	$25 \pm 3°C$
H₂S concentration	Saturated (>2300 ppm)
pH	Start pH 2.7, end pH 4.5
Test solution	5% NaCI + 0.5% acetic acid (CH₃COOH) saturated with H₂S
Loading stress	Up to the yield strength

[a]See [1].

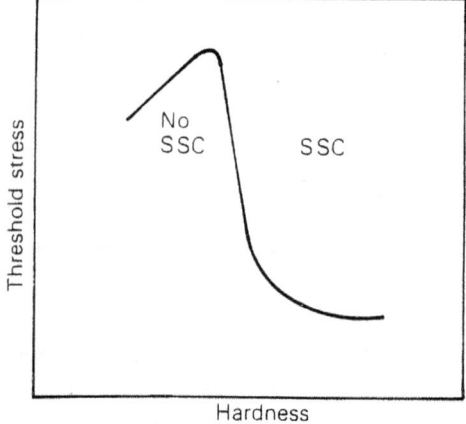

FIGURE 78.5. Sour gas cracking threshold stress variation with hardness for steel. Maximum hardness can be selected as that just prior to the increase in cracking susceptibility.

sodium hydroxide solution and is maintained at a constant potential sufficiently anodic to oxidize the hydrogen atoms that diffuse through the steel [12]. The current in the potentiostatic circuit on the oxidizing side is a direct measure of the instantaneous rate of hydrogen permeation and, by continuous recording of that current, the effective diffusion coefficient of atomic hydrogen can be determined as well as the extent of hydrogen trapping in the steel.

The HIC develops when hydrogen concentration, C_0. in the steel matrix exceeds the threshold hydrogen

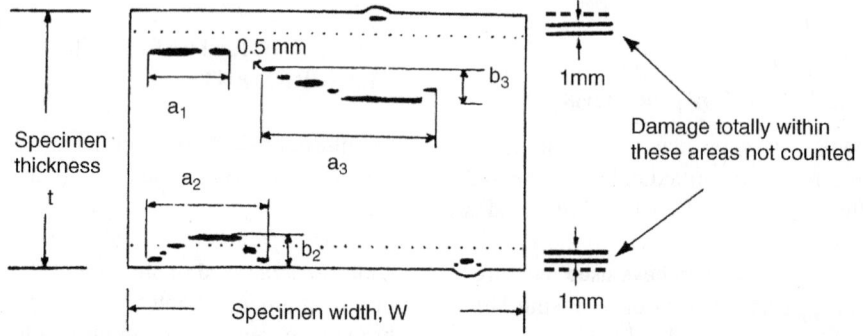

FIGURE 78.4. The HIC susceptibility parameters: cross section perpendicular to rolling direction.

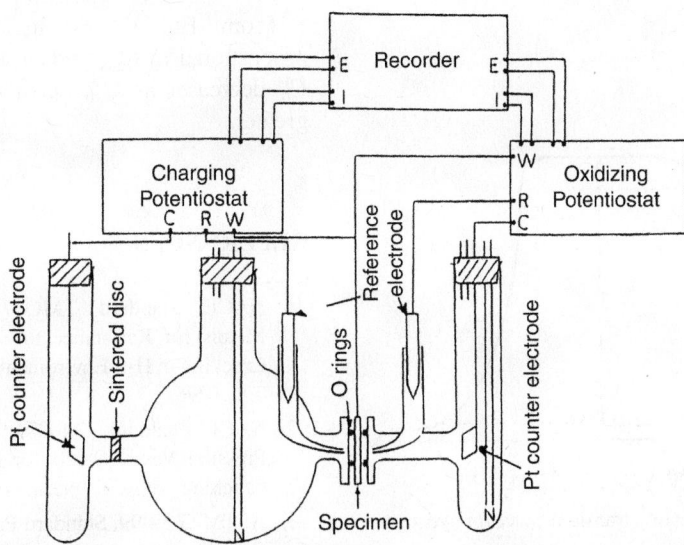

FIGURE 78.6. Schematic of the experimental apparatus for studying hydrogen permeation.

concentration, C_{th} [13]. The C_0 depends on alloy composition, H_2S partial pressure, and pH. The C_{th} depends on inclusions and segregation in the matrix [10].

The distribution of hydrogen atoms is derived from the solution of Fick's second law. In an operating pipeline, hydrogen atoms entering the steel at the internal surface diffuse through the wall and exit at the external surface, where they form hydrogen gas molecules. The driving force for this flux of diffusing hydrogen atoms is the concentration gradient between the internal surface [10, 14, 15], where $C^H = C_0^H$, and the external surface, where

$C^H = 0$, as shown schematically in Figure 78.7. The concentration of hydrogen atoms in an operating pipeline is assumed to decrease linearly with distance through the pipe wall. At midwall, $C^H = 0.5 C_0^H$.

A schematic graph of permeation flux versus time is shown in Figure 78.8. In order to calculate the hydrogen concentration (C_0^H) on the inside wall of the pipe, the hydrogen diffusion coefficient, D, must first be calculated, either using the half-rise time ($t_{1/2}$) formula or using the breakthrough time (t_b) formula given by Devanathan and Stachurski [12].

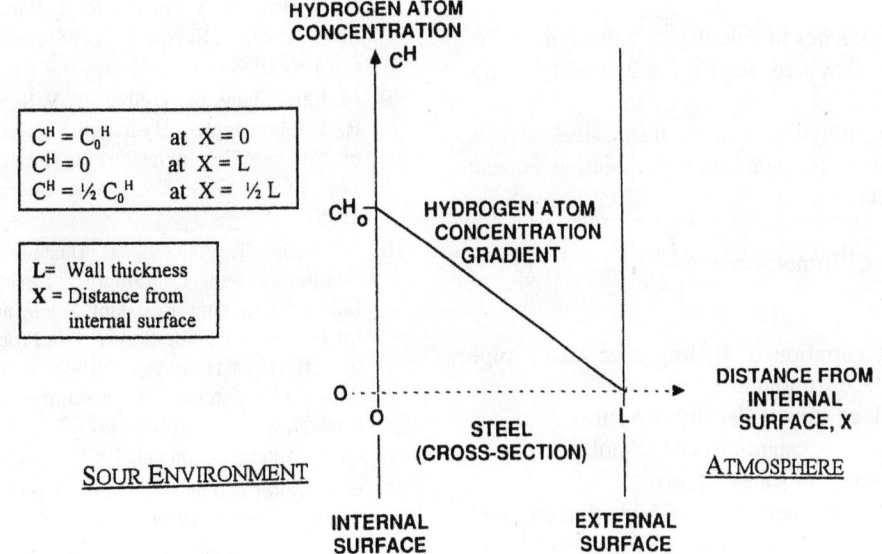

FIGURE 78.7. Hydrogen atom concentration gradient in steel wall.

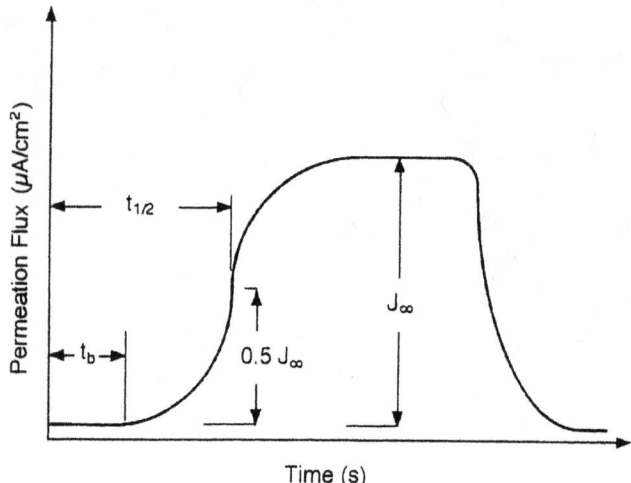

FIGURE 78.8. Schematic permeation transient curve for hydrogen permeation test.

The diffusion coefficient D using the half-rise time ($t_{1/2}$) formula is calculated as follows:

$$D(\text{cm}^2/\text{s}) = \frac{L^2}{7.2 t_{1/2}} \qquad (78.1)$$

where L = thickness(cm) of steel where permeation probe was attached

$t_{1/2}$ = time (s) needed to obtain one-half of the steady-state current, J_∞ (see Fig. 78.8)

The diffusion coefficient can also be calculated using the breakthrough time (t_b) expression (see Fig. 78.8):

$$D = \frac{L^2}{15.3 t_b} \qquad (78.2)$$

There are other approaches to calculating values of D [16, 17], but Eq. (78.1) is often used, and Eq. (78.2) usually results in similar values.

The concentration of hydrogen on the inside steel surface, C_0^H, is calculated from the maximum permeation current density using the relationship

$$\text{Peak } C_0^H (\mu\text{mol/cm}^3) = \frac{I_{\max} L}{DF} \qquad (78.3)$$

where C_0^H = concentration of hydrogen at inside pipe surface (μmol/cm³)

$I_{\max}$ = peak of current density (μA/cm²)

F = faraday constant (96,487 C/mol)

L = thickness of sample (cm)

D = diffusion coefficient of hydrogen in steel (cm²/s) calculated from Eqs. (78.1) and (78.2)

From Eq. (78.3), hydrogen concentration C_0^H is proportional to $I_{\max}$ and inversely proportional to D; thus, C_0^H decreases as $I_{\max}$ becomes smaller and as D becomes greater.

REFERENCES

1. NACE Standard TMO177-96, "Laboratory Testing of Metals for Resistance to Specific Forms of Environmental Cracking in H₂S Enviroments," NACE International, Houston, TX, 1996.

2. NACE Standard TM0284-96, "Evaluation of Pipeline and Pressure Vessel Steels for Resistance to Hydrogen-Induced Cracking," NACE International, Houston, TX, 1996.

3. ASTM G 39-99, Standard Practice for Preparation and Use of Bent-Beam Stress-Corrosion Test Specimens, ASTM Volume 03-02, ASTM, West Conshohoken, PA, 1999.

4. R. S. Treseder and T. M. Swanson, Corrosion, **24**(2), 27 (1968).

5. M. Elboujdaini, M. T. Shebata, and R. W. Revie, "Performance of Pipeline Steels in Sour Service," in Proceedings, Materials for Resource Recovery and Transport, L. Collins (Ed.), The Metallurgical Society of CIM, Calgary, August 1998, CIM, Montreal, Canada, 1998, pp. 109–127.

6. A. Ikeda, A. Nakamura, and J. Kushida, "Evaluation Method of Hydrogen-Induced Crackin Susceptibilities of Steels for Pipe Lines and Pressure Vessels," in Proc. Int. Conf. Interaction of Steels with Hydrogen in Petroleum Industry Pressure Vessel Service, Paris, France, 1989, Martin Prager (Ed.), Publ. Materials Properties Council, New York, 1993, pp. 289–304.

7. R. W. Revie, V. S. Sastri, M. Elboujdaini, R. R. Ramsingh, and Y. Lafrenière, Report MTL 91-60(TR), CANMET, Ottawa, Canada, 1991.

8. G. R. Hoey, R. W. Revie, R. R. Ramsingh, D. K. Mak, and M. T. Shehata, Report MTL 88-41(TR), CANMET, Ottawa, Canada, 1988.

9. M. Elboujdaini, M. T. Shehata, V. S. Sastri, R. W. Revie, and R. R. Ramsingh, "Hydrogen-Induced Cracking and Effect of Non-metallic Inclusions in Linepipe Steels," Paper No. 748, CORROSION/98, NACE International, Houston, TX, 1998.

10. A. Ikeda, T. Kaneko, T. Hashimoto, M. Takeyama, Y. Sumitomo, and T. Yamura, "Development of Hydrogen Induced Cracking Resistant Steels and HIC Test Methods for Hydrogen Sulfide Service," in Procedings, Symposium on the Effects of Hydrogen Sulfide on Steel, Paper No.6, 22nd Annual Conference of Metallurgists, Edmonton, Alberta, Canada, S. A. Bradford (Ed.), The Metallurgical Society of CIM, Montreal, Canada, 1983.

11. G. J. Biefer and M. J. Fichera, Report 84-13(TR), CANMET, Ottawa, Canada, 1984.

12. M. A. V. Devanathan and Z. Stachurski, Proc. R. Soc., **290A**, 220 (1965).

13. A. Ikeda, Y. Morita, F. Terasaki, and M. Takeyama, Proceedings, 2nd International Congress on Hydrogen in Metals—Paris, **4A**, 7 (1977).

14. K. Van Gelder, M. J. J. Simon Thomas, and C. J. Kroese, Corrosion, **42**(1), 36 (1986).

15. M. G. Hay, "An Electrochemical Device for Monitoring Hydrogen Diffusing through Steel." CIM Symposium on "The Effects of Hydrogen Sulfide on Steel," Paper 11, Edmonton, Canada, Aug. 1983, CIM, Montreal, Canada.

16. D. Noel, C. P. Vijayan, and J. J. Hechler, Surface Coatings Technol., **28**, 225 (1968).

17. J. Brogan, I. M. Austen, and E. F. Walker, "Method of Calculating the Hydrogen Diffusion Coefficient in Steel from Hydrogen Permeation Data," British Steel Corporation, London, 1987.

79

ATMOSPHERIC CORROSION TESTING

D. L. Jordan

Ford Motor Company, Dearborn, Michigan

A. INTRODUCTION

Spectacular corrosion failures occur frequently as a result of exposure of engineering materials to aggressive aqueous and nonaqueous process fluids. Many materials of construction are not exposed to such conditions but, rather, are used in the naturally occurring atmosphere and are subject to degradation processes that limit their ability to function in the desired manner. While often plain in appearance, corrosion in the atmosphere is recognized as the single most severe form of corrosion on a tonnage basis. Knowledge of the expected corrosion performance of a material in the atmosphere is needed when designing engineered components and assemblies. The purpose of this chapter is to provide information to allow design and materials selection engineers to decide if atmospheric exposure tests are a necessary part of their function and, if so, to provide them with guidelines to conduct a successful test or to interpret existing test data.

B. PURPOSE OF TESTING

Before committing to the effort and expense of long-term corrosion testing, it is important to understand and record the reasons why test data are needed. The specific application, the material attributes important to that application, and the desired form of the final data must all be considered. Atmospheric corrosion test data typically are required to determine or to predict the following types of information:

Material lifetime in a specific atmosphere:
> How long will it last before failing to satisfy a key performance indicator?

Pick the winner of a group of materials in a specific atmosphere:
> Is material A better than material B?

Corrosion rate in a specific atmosphere:
> What is the rate of material thickness loss?

Atmospheric corrosivity comparisons:
> Is atmosphere A more aggressive than atmosphere B?

Examples of material applications that may require the collection or interpretation of atmospheric corrosion data are as follows:

Bridge work
Highway guardrails

Uhlig's Corrosion Handbook, Third Edition, Edited by R. Winston Revie
Copyright © 2011 John Wiley & Sons, Inc.

Utility poles

Power lines

Electrical switchgear cabinetry

Automatic teller machine cabinetry

Building siding and roofs

Automobile bodies

Airplane skins

Refrigerator wrappers or back plates

Outboard motor housings

Oil derricks

Mail boxes

Chain link fences

Stadium seats and decks

The material attributes or performance indicators that may be measured to provide information for proper materials selection and life prediction include:

Weight loss

Thickness loss

Loss of impact strength

Loss of tensile strength

Loss of ductility

Pitting

Perforation

Coating life

Discoloration or other measure of appearance

Contact resistance

The data should be reported in a format that allows it to serve the desired function of the test; comparison with existing or future data is frequently, but not always, required.

C. TYPES OF ATMOSPHERES

The wide variety of possible climatic conditions, combined with a multiplicity of geometric and spatial arrangements of materials, provides an unlimited assortment of atmospheres to which materials of construction may be exposed. A given material performance in one atmosphere or application should in no way be interpreted to mean that said performance would be repeated for other conditions. Many examples that demonstrate this principle exist and provide the impetus and justification for continued testing of existing and new materials in known and untested atmospheres and applications.

Despite the poorly defined target, standards organizations have used different procedures to define general classifications of various atmosphere types [1, 2]. General atmospheric

classifications are needed in order to limit the number of candidate materials for an application. Classifications provide qualitative or quantitative descriptions of the key characteristics of the atmosphere that contribute to corrosion behavior:

Time of wetness

Sun load [exposure to ultraviolet (UV-B) radiation]

Seasonal temperature and humidity cycles

Natural contaminants (sea salt, particulates)

Man-made contaminants (SO_x, NO_x, road deicing salt, tropospheric ozone, particulates)

Other characteristics that may be unique to specific locations, geometries, or applications are as follows:

Proximity to contaminant point sources

Shelter from direct rainfall or sun load

The simplest atmospheric classification scheme is based on the primary aggressors for most metallic materials in atmospheric corrosion: moisture, fallout of industrial pollution, and the chloride ion. Accordingly, the simplest classifications are as follows:

Rural (very little fallout)

Industrial (primarily sulfur-containing fallout)

Marine (chloride-containing fallout)

A refinement of the method includes some of the location characteristics cited earlier but provides better discrimination [3]:

Dampness	Temperature	Contaminants
Dry	Tropical	Rural
Humid	Temperate	Urban
Marine	Arctic	Industrial

One entry from each of the three columns provides a reasonable qualitative description of most common atmospheres.

The international program ISO CORRAG (International Organization for Standardization Technical Committee 156, Working Group 4) comprehensively characterized and classified dozens of atmospheric locations globally by two semiquantitative methods. One concentrates on the predicted corrosivity based on time of wetness and contaminant measurements [2, 4] while the other relies upon actual corrosion rate measurements on standard metal specimens [2, 5, 6]. The following descriptions are now standard:

Category	Time of Wetness (%)
τ_1	<0.1
τ_2	0.1–3
τ_3	3–30
τ_4	30–60
τ_5	>60

Category	Pollution ($SO_2\ \mu g/m^3$)
P_0	<12
P_1	12–40
P_2	40–90
P_3	90–250

Category	Chloride (mg/m^2-day)
S_0	<3
S_1	3–60
S_2	60–300
S_3	300–1500

Combinations of the τ, P, and S categories are used to estimate the corrosivity category, C, for carbon steel, zinc, copper, and aluminum based only on measurements of the environmental characteristics, as summarized here for a one-year exposure [2]:

Category	Corrosion Rate ($\mu m/year$)			
	Steel	Zinc	Copper	Aluminum
C_1	<1.3	<0.1	<0.1	Negligible
C_2	1.3–25	0.1–0.7	0.1–0.6	<0.6
C_3	25–50	0.7–2.1	0.6–1.3	0.6–2
C_4	50–80	2.1–4.2	1.3–2.8	2–5
C_5	80–200	4.2–8.4	2.8–5.6	5–10

The American Society for Testing and Materials (ASTM) provides atmospheric classification methodologies similar to those of ISO. Standards exist for site characterization [1], time of wetness monitoring [7], and pollution monitoring [8, 9]. In addition, a fast and convenient method to compare atmospheric corrosivities is the CLIMAT (**CL**assification of **I**ndustrial and **M**arine **AT**mospheres) test [10]. It consists of a number of aluminum wires wound tightly around threaded bolts of different materials, usually nylon, steel, or copper. Exposure of the wound wire-on-bolt assembly in an atmospheric environment of interest will result in various amounts of uniform or galvanic corrosion of the aluminum wire, which may be used as a corrosivity index for that atmosphere.

It is critical to remember that the most important material and atmospheric data are not those reported by ISO COR-RAG, ASTM, or any other source, but those data that are relevant to the application of interest. For instance, it has been shown that the ISO atmospheric corrosivity categories do not pertain to many sites throughout the world, including very cold areas and those where dew formation may occur in a relative humidity of <80% [11].

D. STATISTICAL CONSIDERATIONS

The long-term time commitment (several months to several decades) required to obtain meaningful atmospheric corrosion data is sufficient justification to be particularly careful to perform the tests in the desired manner the first time. The expense of repeating a failed testing program due to an ineffective experimental design is unfortunate enough but pales in comparison to the unrecoverable time lost.

The type of experimental design is dependent on the purpose of the testing and the desired use of the data. The time to decide what type of statistical techniques to use for the experimental design and data analysis is long before any material is secured for testing. In any event, careful observance of the basic principles of replication, randomization, and blocking is a minimum requirement.

Attrition of specimens is a particular problem for long-term programs. It is important to adjust replication accordingly or to be prepared to decrease the expectations of the program postmortem. Long-term programs often gain additional expectations as they mature; extra replicates to accommodate future brainstorms are rarely wasted. It is a common misconception that specimens of a given type that are part of a periodic removal schedule should be counted as replicates for that material. In reality, only those specimens that are evaluated, either destructively or nondestructively, at a given inspection interval can be counted as replicates for that material type; subsequent evaluations comprise a different response characteristic and, thus, cannot be pooled to create a larger number that enhances the data artificially. In addition, the practice of measuring a given response at different locations on the same test specimen may be expedient but does not provide independent measurement of the response and thus represents only one replicate in the statistical analysis.

Randomization of specimen mounting locations within the given atmospheric corrosion test site and on the individual racks and frames should be practiced when possible to reduce the possible detrimental effects of nonuniform time of wetness at different frame locations (top-to-bottom of frame, edge-to-center of site) and nonuniform contaminant fallout (animal droppings are a noteworthy problem in this regard). This randomization of specimens within the atmospheric corrosion test site is, of course, in addition to the random selection of specimens from the population of material under study.

In cases where randomization is not possible, it is permissible (or even desirable) to create statistical blocks that are easily handled statistically. Replication, randomization, and blocking are covered in detail in a variety of statistics texts and handbooks [12–14].

E. HARDWARE, MAINTENANCE, AND PRACTICAL CONSIDERATIONS

Several excellent documents and manuals that give detailed plans and drawings for atmospheric corrosion test sites, hardware, and administration are readily available and should be consulted before constructing a site [15–23].

Many existing sites contain instrumentation and maintain historical databases on temperature, relative humidity, time of wetness, chloride, and sulfur dioxide, all of which have direct and usually dramatic influences on the corrosivity of the atmosphere. While it is not critical to have such data for a single test in a single atmosphere, comparison of several tests in several atmospheres certainly would be facilitated.

In general, an atmospheric corrosion test site should be located in a well-secured area, not subject to frequent disturbance by human or other passersby. Racks should be constructed of a corrosion-resistant material that has been shown to last at least as long, preferably much longer, as the expected test in the given atmosphere. The bottom edge of the rack should be 750 mm or more from the ground to reduce the possible interference of plant growth. Herbicides, fungicides, and insecticides should never be used on or around the atmospheric corrosion test site due to the possible corrosion inhibitive nature of some of their contents. Many sites are covered with thick polyethylene and up to 200 mm of crushed rock or gravel to discourage plant growth and to eliminate ongoing maintenance. Be it grass, concrete, sand, or gravel, the ground should be flat and of uniform composition throughout the site so that any influence of moisture evaporation is distributed evenly throughout. Periodic inspection of the mounting of each and every specimen is recommended due to loosening by windstorms, breakage of insulators, and dissolution of the specimens themselves.

Attachment to the rack depends on the type of specimen but should always be done with insulated hardware to eliminate galvanic effects between the specimens and the racks. Metal sheet specimens are usually mounted so that they face skyward at 30° to the horizontal, toward the predominant direction of the sun (facing south in the Northern Hemisphere). In seacoast exposures, the specimens usually face the surf regardless of compass direction. Painted metal sheet specimens are exposed in the same manner as their unpainted counterparts but typically are placed at a 45°

angle. It is common to analyze the skyward and groundward surfaces together in performance determinations, but it is important to ensure that the groundward surface is not shielded from the ground by rack support brackets. If determination of the skyward and groundward behaviors is required, appropriate masking with organic materials may be used with the recognition that diligent periodic maintenance will be required. Specimens with unusual shapes should be oriented in a manner consistent with their in-service application; however, mounting unusually shaped specimens in a variety of orientations may provide unexpectedly useful performance data. The U-bend specimens for stress corrosion cracking (SCC) susceptibility [24] should be mounted in such a way as to approximate the expected service orientation. Racks may include covers of various types to simulate sheltered conditions; periodic inspection of the covers is necessary to detect and repair leaks.

Specimen preparation usually involves trimming to a uniform size and removing any processing fluids with an inert cleaner. Specimens should be marked in a manner that is not obliterated by corrosion. Drilled holes or cut notches have proven to be effective for long-term exposures but add complications where coatings are breached. Various paints, inks, and plastic labels may be appropriate for very short-term exposures or in cases where frequent periodic maintenance of the labels is planned.

Meticulous record keeping is a necessity. It is common for a long-term testing program to have several caretakers over its lifetime. An accurate description of every experimental detail and a vivid account of reasoning behind any decision are always useful to whomever is given the task of maintaining a periodic removal schedule and reporting the long-term data. Standard forms for data to be collected before, during, and after exposure are useful [15, 18, 19].

It is common to calculate the weight loss of exposed specimens by subtracting the weight of the individual specimens after cleaning [16] from their weight before exposure. Dividing the weight loss by the exposed surface area and dividing that quotient by the material density provides the uniform material thickness loss. This practice provides useful information for loss of load-bearing capacity but must be interpreted carefully. The conversion of weight loss to thickness loss assumes uniform material density and uniform corrosion over the entire specimen. Pitting corrosion results in minimal weight loss but dramatic localized thickness loss that frequently is of catastrophic consequences. In addition, different layers of coated materials may be of different densities or may corrode at different rates.

REFERENCES

1. ASTM G92, "Standard Practice for Characterization of Atmospheric Test Sites," ASTM, West Conshohocken, PA, 2003.

2. ISO, 9223, "Corrosion of Metals and Alloys—Corrosivity of Atmospheres—Classification," ISO, Geneva, Switzerland, 1992.

3. J. F. Stanners, "Control of Corrosion in Common Environments," in Corrosion and Protection of Metals, The Institution of Metallurgists, Iliffe Books, London, 1965, p. 88.

4. ISO 9225, "Corrosion of Metals and Alloys—Corrosivity of Atmosphers—Measurement of Pollution," ISO, Geneva, Switzerland, 1992.

5. ISO 9226, "Corrosion of Metals and Alloys—Corrosivity of Atmospheres—Determination of Corrosion Rate of Standard Specimens for the Evaluation of Corrosivity," ISO, Geneva, Switzerland, 1992.

6. ISO 9224, "Corrosion of Metals and Alloys—Corrosivity of Atmospheres—Guiding Values for the Corrosivity Categories," ISO, Geneva, Switzerland, 1992.

7. ASTM G84, "Standard Practice for Measurement of Time-of-Wetness on Surfaces Exposed to Wetting Conditions as in Atmospheric Corrosion Testing," ASTM, West Conshohocken, PA, 1993.

8. ASTM, G91, "Monitoring Atmospheric SO_2 Using the Sulfation Plate Technique," ASTM, West Conshohocken, PA, 2004.

9. M. R. Foran, E. V. Gibbons, and J. R. Wellington, Chem. Can., **10**, 33 (1958).

10. ASTM, G116, "Standard Practice for Conducting Wire-on-Bolt Test for Atmospheric Galvanic Corrosion," ASTM, West Conshohocken, PA, 2004.

11. G. A. King and J. R. Duncan, Corros. Mater., **23**(1), 8, 22 (1998).

12. G. E. P. Box, W. G. Hunter, and J. S. Hunter, Statistics for Experimenters, Wiley, New York, 1978.

13. D. C. Montgomery, Design and Analysis of Engineering Experiments, 2nd ed., Wiley, New York, 1984.

14. M. G. Natrella, Experimental Statistics, Handbook 91, National Bureau of Standards, Washington, DC, 1963.

15. H. H. Lawson, Atmospheric Corrosion Test Methods, NACE International, Houston, TX, 1995.

16. ASTM G1, "Standard Practice for Preparing, Cleaning, and Evaluating Corrosion Test Specimens," ASTM, West Conshohocken, PA, 2003.

17. ASTM G50, "Standard Practice for Conducting Atmospheric Corrosion Tests on Metals," ASTM, West Conshohocken, PA, 2003.

18. ASTM G33, "Standard Practice for Recording Data from Atmospheric Corrosion Tests of Metallic-Coated Steel Specimen," ASTM, West Conshohocken, PA, 2004.

19. ASTM, G107, "Standard Guide for Formats for Collection and Compilation of Corrosion Data for Metals for Computerized Database Input," ASTM, West Conshohocken, PA, 2008.

20. S. W. Dean, Jr., Mater. Perform., **27**(10), 56 (1988).

21. S. W. Dean, Jr., Mater. Perform., **27**(11), 64 (1988).

22. S. W. Dean, Jr., Mater. Perform., **27**(12), 35 (1988).

23. S. W. Dean, Jr., Mater. Perform, **28**(1), 52 (1989).

24. ASTM G30, "Standard Practice for Making and Using U-Bend Stress-Corrosion Test Specimen," ASTM, West Conshohocken, PA, 2003.

80

GALVANIC CORROSION TESTING

X. G. ZHANG

Teck Metals Ltd., Mississauga, Ontario, Canada

A. General
B. Standards
References

A. GENERAL

Galvanic corrosion is very complex because it involves multiple material factors as well as environmental and geometric factors (as shown in Fig. 10.1 in Chapter 10). It is thus important in a galvanic test to identify clearly and to control those factors that are significant to the system. This requires careful consideration of the testing objectives. Galvanic corrosion tests are often carried out for two basic objectives: to assess materials compatibility in terms of the polarity and rate of galvanic corrosion of bimetallic couples and to predict the extent and spatial distribution of corrosion damage.

Galvanic action is governed by the specific potential distribution on the entire surface of the coupled metals, which in turn is determined by the geometry of the metal surfaces. Because of specific requirements of the geometry in galvanic corrosion testing, it is generally not feasible to use a universal geometry for different situations. A galvanic corrosion test is, thus, often distinctive in its cell design [1].

Geometry should also be considered when the intent is to test nongeometric factors; for example, to test metallurgical factors, such as alloying or mechanical working, on reaction kinetics, the geometry of the design should result in a uniform potential distribution on the surface of the anode and cathode; to meet this requirement, the distance between the two metals

should be larger than the dimensions of the samples (Chapter 10). To test galvanic corrosion in the atmosphere, the two metals should be in close contact and have a small dimension in the direction perpendicular to the contact line because the electrolyte formed under an atmospheric environment is very thin, and the galvanic action generally does not extend beyond a few millimeters from the contact line [2, 3]. The various cell designs and procedures in galvanic corrosion testing have been summarized by Hack [1].

Electrochemical and nonelectrochemical methods can be used to measure the parameters of interest. Electrochemical methods include measurement of potentials, under coupled or noncoupled conditions, to provide information on the polarity of a bimetallic couple and extent of anodic and cathodic polarization; or measurement of galvanic current to indicate quantitatively the intensity of galvanic corrosion; or measurement of the current–potential relationship of each metal with a potentiostat to understand the kinetics. The nonelectrochemical methods include weight loss or thickness loss determination and visual or instrumental examination of the corroded surface. Each method has its advantages in providing information about galvanic corrosion and, depending on the specific needs in each circumstance, one or a combination of several methods are needed to provide sufficient information. Detailed descriptions of each of these methods can be found in the literature [1].

B. STANDARDS

The many specific requirements for a galvanic corrosion test make standardization very difficult. Few test methods for galvanic corrosion have been standardized [1]. There are two American Society for Testing and Materials (ASTM)

Uhlig's Corrosion Handbook, Third Edition, Edited by R. Winston Revie
Copyright © 2011 John Wiley & Sons, Inc.

guidelines for galvanic corrosion testing: ASTM standard guide G71 [4] and ASTM standard G82 [5] on "Development and Use of a Galvanic Series for Predicting Galvanic Corrosion Performance." Standard tests have been developed for this situation because galvanic corrosion in atmospheres results mainly from the thin layer of electrolyte that limits galvanic action to an area a few millimeters normal to the contact line, so that the effect of many geometric factors can be neglected.

Under atmospheric conditions, two different types of testing methods have been standardized for determination of the weight loss due to galvanic corrosion: International Organization for Standardization (ISO) 7441-84 and ASTM G104-89 for plate test and ASTM G116-99 for wire-on-bolt test [6–8]. In the plate type of assembly, a strip of one metal is attached by bolts to a panel of another metal. The bolts are insulated from the strip and panel. The galvanic corrosion is evaluated by visual examination or by weight loss measurement of the strip or panel. In the wire-on-bolt type of assembly, a wire of the metal to be tested is tightly wound around the threads of a bolt of the other metal in the couple. The galvanic corrosion can be quantitatively estimated by comparing the weight loss of the coupled wire to that wound on the threads of a plastic bolt.

REFERENCES

1. H. P. Hack, "Galvanic Corrosion Test Methods," NACE International, Houston, TX, 1993.

2. X. G. Zhang, Paper 743, NACE'98, Mar. 1998.

3. X. G. Zhang, Corrosion and Electrochemistry of Zinc, Plenum, New York, 1996.

4. ASTM, G71-81 (Reapproved 2009), "Standard Guide for Conducting and Evaluating Galvanic Corrosion Tests in Electrolytes," American Society for Testing and Materials, West Conshohocken, PA, 2009.

5. ASTM, G82-98 (Reapproved 2009), "Standard Guide for Development and Use of a Galvanic Series for Predicting Galvanic Corrosion Performance," American Society for Testing and Materials, West Conshohocken, PA, 2009.

6. ASTM, G116-99 (Reapproved 2004), "Standard Practice for Conducting Wire-on-Bolt Test for Atmospheric Galvanic Corrosion," American Society for Testing and Materials, West Conshohocken, PA, 2004.

7. ISO, 7441-84, "Corrosion of Metals and Alloys—Determination of Bimetallic Corrosion in Outdoor Exposure Corrosion Tests," ISO, Geneva, Switzerland, 1984.

8. ASTM, G104-89, "Standard Test Method for Assessing Galvanic Corrosion Caused by the Atmosphere," American Society for Testing and Materials, Philadelphia, PA, 1976.

81

TESTING OF ALUMINUM, MAGNESIUM, AND THEIR ALLOYS

E. Ghali

Department of Mining, Metallurgy and Materials Engineering, Laval University, Québec, Canada

A. ALUMINUM AND ALUMINUM ALLOYS

At 25°C and atmospheric pressure, the value of the pitting potential of Al 1199 (99.99% Al) in sodium chloride solutions can vary from ~ -0.35 to $0.54\,V$. versus saturated hydrogen electrode (SHE) depending on the activity of chloride ions. The critical pitting potential E_{SCE}, which lies between the breakdown potential and the protection potential, should be also considered [1]. It is current practice to test the performance of aluminum and aluminum alloys for intergranular corrosion, stress corrosion cracking (SCC), and corrosion fatigue.

A1. Intergranular Corrosion Testing

Testing for intergranular susceptibility varies with the alloy family. Metallographic examination after exposure to a $NaCl–H_2O_2$ corrosive solution [2] is used primarily for aluminum–copper–magnesium (2XXX) alloys. It has also been used with aluminum–magnesium–silicon (6XXX) and aluminum–zinc–magnesium–copper (7XXX) alloys. A practice for measurement of corrosion potentials of aluminum alloys has been established [3]. With aluminum–copper–magnesium (2XXX) and aluminum–zinc–magnesium–copper (7XXX) alloys, limited use has been made of electrochemical methods for predicting intergranular corrosion susceptibility. However, confirmation by metallographic examination is still considered necessary [2].

A2. Stress Corrosion Cracking Testing

Several aluminum alloy product specifications require defined levels of performance with respect to resistance to SCC. Standard tests used to measure such performance are described in standard methods and are referenced in materials specifications. Among these are tests for evaluating resistance to SCC of 2XXX alloys and 7XXX alloys that contain copper, by alternate immersion in 3.5% NaCl solution [4, 5]. Lot acceptance criteria for products of 7XXX copper-containing alloys in T76, T73, and T736 tempers are based on combined requirements for tensile strength and electrical conductivity [6]. A system of rating SCC resistance of high-strength aluminum alloy products has been developed by a joint task group of the American Society for Testing and Materials (ASTM) and the Aluminum Association to assist alloy and temper selection and has been incorporated into [7]. The mentioned stress levels are not to be interpreted as threshold stresses and are not recommended for design.

A3. Fatigue and Corrosion Fatigue Testing

Fatigue tests are carried out in ambient air with laboratory apparatus that subjects the specimen to cyclic stress. The cyclic stress may be axial, torsional, or reverse bending. When testing polished cylindrical specimens rotating under flex in a Moore machine, the stress just below that which causes failure at 500×10^6 cycles is taken as the fatigue strength. For sheet specimens subjected to reverse bending in a Krouse machine, the value is usually taken for survival at 50×10^6 cycles.

Corrosion fatigue tests are carried out with the same apparatus, with provisions made to subject the specimen to a corrosive environment by spraying or dripping a solution onto the specimen, creating a mist in an enclosure about the specimen, or immersing the specimen in a solution. The corrosive environments used most frequently are distilled or demineralized water, tap water, and brines (including natural or synthetic seawater). Because of the complexity of service environments, a good correlation of normal laboratory corrosion fatigue data with actual service performance is difficult [8]. Low-stress long-duration laboratory tests lower the fatigue strength of aluminum alloys.

B. MAGNESIUM AND MAGNESIUM ALLOYS

The testing which has been published has been focused on automotive applications, where the standard testing has been ASTM B117 [9] salt spray and automotive cycle tests. The cycle tests are specific to each of the autumotive companies but involve cyclic exposure to salt water, drying, and high humidity. These are primarily targeted at generating a more realistic evaluation than standard salt spray exposure alone for coating adhesion on ferrous metal. The high-purity magnesium alloy components typically perform very well in such tests, since differential aeration does not appear to play a role in magnesium corrosion [10].

Quantitative control tests of coatings on magnesium alloy surfaces for painting are desirable. These generally consist of exposures, with and without paint, to salt spray, humidity, or natural environments. Precautions are taken to remove surface contamination before painting, generally to a depth of at least 25 µm (1 mil) per side by acid pickling [11].

All organic materials should be degreased thoroughly and if desired a 10% caustic solution is frequently used for cleaning magnesium at temperatures up to the boiling point before testing. Soak cleaners are used as alkaline cleaning in concentrations of 30–75 g/L (4–10 oz/gal) and at 71–100°C [12]. In addition to weight loss, potentiody-

namic polarization methods, an estimate of the polarization resistance R_p, and volume measurement of evolved hydrogen can all be used to assess the instantaneous corrosion rate as well as the kinetics and evolution of the corrosion behavior of magnesium and its alloys. These types of measurements can all be carried out at controlled pH.

The alternate intermittent immersion in salt water or salt spray is often used to compare the corrosion resistance of different magnesium alloys [1].

B1. Electrochemical Methods of Testing

B1.1. Cyclovoltammetric Sweeps. During cyclovoltametric sweeps, changes in the passive films, the solution composition at the interface metal/electrolyte (e.g., hydroxide formation and pH increase), as well as development of localized forms of corrosion such as pitting take place. The advantage of polarization measurements with the Rotating Disc Electrode (RDE) as compared to the more commonly used salt spray test is the ease of obtaining an instantaneous corrosion rate and the removal of water-soluble and not strongly adherent corrosion products from the exposed surface [1, 13].

B1.2. Electrochemical Noise Measurements (ENM). The corrosion rate can be obtained from an estimate of the polarization resistance, R_p, which is inversely related to the linear corrosion rate by the Stern–Geary approximation. Three approaches are possible for obtaining instantaneous values of R_p from the Electrochemical Noise (EN) current and potential measurements, namely noise resistance, spectral noise resistance, and self-linear polarization resistance (SLPR) [1, 14].

B1.3. Electrochemical Impedance Spectroscopy (EIS). Over a period of 72 h immersion of the die cast AZ91D in ASTM corrosive water saturated with magnesium hydroxide (pH 10.6), the R_t (transfer resistance) increased gradually due to the formation of a protective corrosion film. Impedance measurements are usually performed at open-circuit potentials (E_{oc}) and under potentiostatic conditions. Generally, high- and low-capacitive loops are seen. The high-frequency capacitive loop can be attributed to charge transfer reactions and the diameter of the loop to the transfer resistance. The capacitance values for this loop were always below 50 µF/cm^2 and can be attributed to the double-layer capacitance (c_{dl}) of the partially covered surface. The C_{dl} values were on the order of 5–10 µF/cm^2 and increased progressively to about 20 µF/cm^2. The second capacitive loop is generally attributed to the diffusion of ions through the hydroxide or oxide coating [1, 12, 15].

B1.4. Scanning Reference Electrochemical Technique (SRET). The term quasi-Electromotive force (QEMF), corresponding to the difference between the most active cathode potential and the most active anode potential, is used to interpret the SRET results. It is believed that the corrosion rates and tendency of metals and alloys can be related at least indirectly to the evolution of the electromotive force (EMF) of the corrosion cell. This difference can be normally determined every complete scan of 12 min for a surface area of $1\,cm^2$, for example, and should have the same trend as the EMF of the most active corrosion cell [16].

B2. Testing Solutions for Magnesium Alloys

Most solutions for testing magnesium include chloride or sulfate ions in neutral or alkaline media and are used to evaluate general and localized corrosion. A sodium chloride solution (3%) is commonly used. Polarization curves obtained in $1\,N$ NaCl solution adjusted to pH 11 with NaOH showed different pitting potentials as a function of the microstructure. With pH increasing above 10.2 (the point at which magnesium hydroxide is formed), the effect of impurities both in the metal and in the solution medium is apparently overshadowed by the high tendency of film formation. Buffer solutions and saturation of the solution with magnesium hydroxide as well as surface cleaning have been adopted frequently [1, 12].

A $2\,N$ NaCl solution alone has also been used, but it seems to be aggressive especially for magnesium composites. Solutions of $0.1\,M$ NaOH with additions of 0.005, 0.01, 0.02, and $0.03\,M$ NaCl are used for determination of pitting or filiform corrosion during a 24-h immersion period. Passivation and reproducible results are obtained by using an addition of 10 mL hydrogen peroxide (30%) in corrosion pitting studies, preferably with pH adjustment to 11. A $1\,N$ sodium hydroxide solution saturated with Mg_2SO_4 can show the influence of the aggressive sulfate ion with reference to industrial media [18]. Corrosive water according to ASTM D 1384-05 contains 100 ppm, each of sulfate, chloride, and bicarbonate ions introduced as sodium salts. It has been used, either alone (moderately aggressive) or saturated with magnesium hydroxide (less aggressive), to simulate the interface Mg or its alloy/solution interface that may contain important concentrations of magnesium hydroxide in stagnant solutions [1, 17].

B3. Open-Circuit and Pitting Corrosion Potentials

In sodium chloride air-saturated solution without peroxide additions, the free corrosion potential of Mg and that of some of its alloys were found in the range of -1500 to $-1600\,mV$ versus saturated calomel electrode (SCE) for chloride concentrations of 5×10^{-3}–$5 \times 10^{-4}\,M$. In many magnesium alloys, the open-circuit potential is higher than the pitting potential, and thus, the pitting potential cannot be determined using the traditional methods. Borated boric acid buffer pH 8.4 containing $0.001\,N$ NaCl is advantageous for materials that are susceptible to pitting. The low percentage of sodium chloride and sodium hydroxide can give a controlled attack and reproducible results. Higher quantities of NaCl can also give reproducible results for situations less sensitive to agitation [17–19].

B4. Stress Corrosion Cracking Testing

Magnesium castings have been shown to fail in laboratory tests under tensile loads as low as 50% of yield strength in environments causing negligible general corrosion. Under certain conditions of testing, stressed specimens may (1) be as resistant to attack as unstressed specimens, (2) show acceleration of corrosion when stressed, and (3) develop stress corrosion cracks without showing appreciable acceleration of general corrosion. The strain rate chosen for the tests indicating maximum susceptibility to cracking was about $2 \times 10^{-6}\,s^{-1}$ for a solution containing 5 g/L sodium chloride and 5 g/L $KCrO_4$ [1, 20, 21].

The two frequently used laboratory methods for characterizing SCC and mechanism considerations are the constant-extension-rate test (CERT) and the linearly increasing stress test (LIST). In the CERT and LIST, the nominal load and extension are increased until specimen fracture, respectively. They allow rapid characterization of SCC susceptibility and permit direct measurement of critical parameters, such as the threshold stress (σSCC) and crack velocity (VC) when coupled with equipment to measure crack extension. However, the constant-load tests are generally valuable since they reflect better the initiation and propagation cracking processes of SCC, caused by the evolution of electrochemical properties of the interface and the physical and mechanical properties of the corroded metal [22].

REFERENCES

1. E. Ghali, Corrosion Resistance of Aluminum and Magnesium Alloys: Understanding, Performance and Testing, Wiley, Hoboken, NJ, 2010.

2. J. E. Hatch, "Aluminum: Properties and Physical Metallurgy," American Society for Metals, Metals Park, OH, 1984, pp. 248–319.

3. G69-97 (Reapproved 2009), Annual Book of ASTM Standards, Standard Test Method for Measurement of Corrosion Potentials of Aluminum Alloys, American Society for Testing and Materials, West Conshohocken, PA, 2009.

4. G44-99 (Reapproved 2005), Annual Book of ASTM Standards, Standard Practice for Exposure of Metals and Alloys by Alternate Immersion in Neutral 3.5% Sodium Chloride Solution," American Society for Testing and Materials, West Conshohocken, PA, 2005.

5. G47-98 (Reapproved 2004), Annual Book of ASTM Standards, Standard Test Method for Determining Susceptibility to Stress-Corrosion Cracking of 2XXX and 7XXX Aluminum Alloy Products, American Society for Testing and Materials, West Conshohocken, PA, 2004.

6. E. H. Hollingsworth and H. Y. Hunsicker, "Corrosion of Aluminum and Aluminum Alloys," in Metals Handbook, 9th ed., Vol. 13, Corrosion, J.R. Davis, Senior Ed. ASM International, Metals Park, OH, 1987, pp. 583–609.

7. ASTM, G64-99 (Reapproved 2005), Annual Book of ASTM Standards, "Standard Classification of Resistance to Stress-Corrosion Cracking of Heat-Treatable Aluminum Alloys," American Society for Testing and Materials, West Conshohocken, PA, 2005, pp. 3–10.

8. P. J. E. Forsyth, J. of the Society of Environmental Engineers, **19**-1(84) (1980).

9. ASTM B117-07a, Annual Book of ASTM Standards, Standard Practice for Operating Salt Spray (Fog) Apparatus, American Society for Testing and Materials, West Conshohocken, PA, 2007.

10. J. E. Hillis and R. W. Murray, "Finishing Alternatives for High Purity Magnesium Alloys," Paper No. G-T87-003, Society of Die Casting Engineering, Toronto, Ontario, May 1987.

11. ASTM D1732-67, Annual Book of ASTM Standards, Standard Practices for Preparation of Magnesium Alloy Surfaces for Painting, American Society for Testing and Materials, Philadelphia, PA, reapproved 1984.

12. E. Ghali, W. Dietzel, and K.-U. Kainer, J. Mater. Eng. Perform., **13**(5), 517–529 (2004).

13. T. Trossmann, K. Eppel, M. Gugau, and C. Berger, in Investigation of the Passivation Behavior of Magnesium Alloys by Means of Cyclic Current-Potential-Curves, the 7th International Conference on Magnesium Alloys and Their Applications, Weinheim, Germany, 2006, K.-U. Kainer (Ed.), Wiley-VCH, Weinheim, Germany, 2006, pp. 802–808.

14. R. D. Klassen, P. R. Roberge, A.-M. Lafront, M. Ö. Öteyaka, and E. Ghali, in Comparison of the Corrosion Behaviour between Two Magnesium Alloys by Electrochemical Noise and Scanning Reference Electrode Technique, K. U. Kainer (Ed.), Deutsche Gesellschaft fur Materialkunde e. V. (DGM), Weinheim, Germany, 2003, pp. 655–660.

15. S. Mathieu, C. Rapin, J. Hazan, and P. Steinmetz, Corros. Sci., **44**, 2737–2756 (2002).

16. S. Jin, E. Ghali, C. Blawert, and W. Dietzel, "SRET Evaluation of the Corrosion Behavior of Thixocast AZ91 Magnesium Alloy in Dilute NaCl Solution at Room Temperature," Proceedings 210th ECS 2006 meeting, Cancun, Mexico, N. Missert, A. Davenport, M. Ryan, and S. Virtanen (Eds.), Pennington, NJ, 2007, pp. 295–311.

17. V. Mitrovic-Scepanovic and R. J. Brigham, "Localized Corrosion Initiation on Magnesium Alloys," Corrosion, **48**(9), 780–784 (1992).

18. G. L. Song and A. Atrens, Adv. Eng. Mater., **1**(1), 11–33 (1999).

19. P. P. Trzaskoma, Corrosion J. (USA), **42**(10), 609–613 (1986).

20. B. W. Lifka, "Aluminum (and Alloys)," in Corrosion Testing and Standards: Application and Interpretation, R. Baboian, (Ed.), American Society for Testing and Materials, Philadelphia, PA, 1995, pp. 447–457.

21. K. Ebtehaj, D. Hardie, and R. N. Parkins, Corros. Sci., **28**, 811–829 (1988).

22. N. Winzer, A. Atrens, W. Dietzel, V. S. Raja, G. Song, and K. U. Kainer, Mater. Sci. Eng. **488**, 339–351 (2008).

82

TESTING OF POLYMERIC MATERIALS FOR CORROSION CONTROL

B. THOMSON AND R. P. CAMPION

MERL Ltd., Wilbury Way, Hitchin, UK

A. INTRODUCTION

This chapter reviews the material properties that influence the selection of a polymer for corrosion control or limitation duty. Mechanical integrity, thermal resistance, fluid perme-ation, chemical aging, and fluid resistance are some of the factors which should be considered and, if possible, measured under service conditions using realistic samples. Tests developed to assess these and other properties are discussed below, accompanied by background theory where relevant. The extent to which a polymer resists service fluids, hostile or otherwise, largely determines its suitability as a protective barrier in static applications (e.g., a chemical storage tank liner). Considerable importance is attached in this chapter to permeation, the phenomenon which underpins the ultimate success (or failure) of a polymeric material employed to separate a metal surface from contained fluids. The polymer must have sufficient strength to withstand the mechanical stresses of the application, and for barrier materials operating in a dynamic service environment, additional testing to evaluate stress and cyclic loading effects on material performance must also be undertaken.

The selection of a polymer for anticorrosion purposes depends primarily on service conditions, but cost and processability can also be factors. No polymer is entirely immune to interaction (physical and/or chemical) with contacting fluids (liquids, high-pressure gases) but, with correct selection and design, the permeation rate can be extremely slow—hence, the widespread use of polymers to limit the access of hostile fluids to the surfaces of underlying metallic structures. Herein lies the dilemma for the evalua-tion of anticorrosion materials—in many applications they are expected to have multiyear service lives, but testing in the laboratory over such periods is clearly unrealistic.

To resolve the dilemma as best as possible, accelerated procedures are necessary. However, their application re-quires knowledge of the basic mechanisms involved, whether chemical, physicochemical, or physical. As an example, and in the interests of covering as much ground as possible

Uhlig's Corrosion Handbook, Third Edition, Edited by R. Winston Revie
Copyright © 2011 John Wiley & Sons, Inc.

in this chapter, the upcoming discussion will outline one specific, and commercially important, polymeric anticorrosion application to illustrate the scope of testing requirements. The example selected is the internal pressure sheath (or liner) of unbonded flexible pipe used to transport crude oil in offshore oil production systems, introduced below and referred to at intervals throughout the chapter.

A1. Example of Polymer Barrier Application

Unbonded flexible pipes are employed offshore to convey pressurized hot crude oil and gas from the wellhead on the seabed to the surface for transport to processing facilities. In addition, at intervals they carry a variety of injection and service fluids, which are necessary to maintain optimum production rates and are subjected to wide variations in fluid pressure and temperature. Some of the fluids can be very hostile to certain polymers, particularly at elevated temperatures. "Flexible" is a relative term: These pipes are enormously stiff compared with a garden hose but are flexible compared with steel pipes of similar dimensions.

A typical unbonded flexible pipe comprises discrete metallic and polymeric layers that are allowed a degree of slip relative to one another (Fig. 82.1). In addition to the

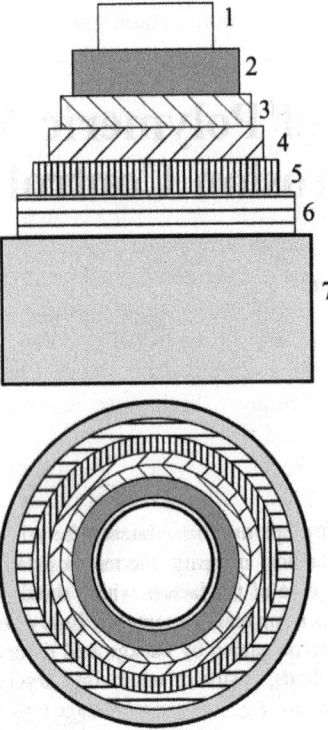

FIGURE 82.1. Schematic representations showing the composite layer structure of a typical unbonded flexible pipe: (1) carcass; (2) internal pressure sheath; (3) pressure armor; (4) back-up pressure armor; (5) inner tensile armor; (6) outer tensile armor; (7) outer sheath.

illustrated layers, between the armor windings there may also be included thin (usually polymeric) antifriction layers. The internal pressure sheath is a critical component of the pipe [1] and acts to contain production fluids, transmit internal fluid pressure to the surrounding layers of metallic armor reinforcement, and prevent corrosion of the armor wires by the transported fluids. The armor layers provide mechanical resistance to pressure and tension. The internal diameter of a typical current pressure sheath might be ∼25 cm, with a wall thickness of 6–10 mm; a requirement of the industry is the development of ever larger diameter pipes. Bonded flexible pipe technology is also in use. In bonded flexile pipes, the inner polymer layer, usually an elastomer (a rubbery polymer), is bonded to the underlying metal; the whole pipe is a composite of bonded metal cord and polymer layers. Bonded pipes have their own requirements for testing; in particular, blistering/splitting of the inner elastomer as a result of rapid gas decompression events assumes great importance. This phenomenon is also possible in unbonded flexible pipes based on thermoplastics.

For risers (those flexible pipes that leave the seabed to rise to the rig, as opposed to flowlines running from wellheads along the sea floor), the transport of multiphase production fluids in the manner described above is clearly a dynamic service environment, and the flexible pipe experiences loads from vessel motions, waves, currents, self-weight, and internal fluid pressure. With required service lives of up to 20 years, almost continuous contact with a range of aggressive fluids at high temperatures and pressures, numerous decompression cycles, sporadic combinations of bending, tensile, and compressive strains, and no opportunities for repair should failure occur, the requirements for a sheath polymer are stringent. Needless to say, comprehensive evaluation of candidate materials for such a role is absolutely essential prior to deployment offshore [2].

For various reasons, thermoplastic polymers (Section B1) are most suited to this application, with the choice of thermoplastic governed to a large extent by service temperature. At moderately elevated temperatures (up to 90°C 194°F) PA 11 (nylon 11) and crosslinked polyethylene (PE) are suitable pressure sheath materials. At higher temperatures, up to 120°C, plasticized and modified unplasticized types of poly(vinylidene fluoride) (PVDF) are current preferences. For the future, there is a need to evaluate polymers that will operate safely at temperatures as high as 200°C and pressures up to 10000 psi to enable the exploitation of deeper wells. The investment necessary to select and test suitable materials is currently being made by sections of the offshore oil and gas industry. The testing requirements for a dynamic flexible pipe pressure sheath are clearly more extensive and involved than those for a "simple" polymeric lining for a chemical storage tank.

The means of accelerating testing to allow residual product life to be estimated is of considerable current interest.

Arrhenius plots can be used to assess aging effects and fluid permeation rates, but only if a single activation energy is involved. Chemical kinetic effects should also be examined. In addition, sudden changes can occur at critical temperatures for physicochemical reasons: for example, a severe deterioration of PVDF in methanol at vapor pressure and 140°C has been observed after relatively short periods [2].

B. POLYMERS FOR CORROSION CONTROL

Before discussing the testing of polymers, a summary of their status as materials is essential. With the exception of natural rubber and a few other materials, all polymers are synthetic (i.e., man made) and are usually derived from oil cracking products. Polymers [3–6] can be divided into three main categories: elastomers (rubbers), thermoplastics, and thermosets. In general terms, thermosets resist deformation whereas elastomers elongate readily under a small applied stress, with most of the energy being recoverable—they are elastic. Thermoplastics exhibit intermediate stress–strain characteristics, are nonelastic, and can be further subdivided into amorphous and semicrystalline types. The majority of the latter will soften when heated (crystalline melting) and can thus be rapidly processed into useful shapes. However, service temperatures are limited by the melting point. Elastomers and thermosets, on the other hand, are characterized after processing and curing at, say, 130–180°C by the presence of permanent crosslinks within the polymer matrix; these tie individual macromolecules into a three-dimensional network that will not soften when heated (especially thermosets) and in the case of elastomers is capable of tolerating large reversible extensions. A more recent innovation is the thermoplastic elastomer (TPE), the melt processability of which arises from the existence of thermally labile crosslinks, but it is more prone to high-temperature service creep/extrusion than normal elastomers. The means of deterioration differs between the above polymer classes. For example, thermosets resist swelling better than elastomers but have poorer impact strength.

The choice of polymer to function in a particular anticorrosion application depends on the service conditions. In static applications, it may not matter which polymer type is selected as long as the fluid is safely contained. However, other factors are relevant, including ease of manufacture and cost.

For elastomeric tank linings, it is common to coat the metal surface requiring protection with bonding agent (adhesive) after suitable preparation to remove grease and debris, dry it, and then apply an uncured rubber layer. When the tank is sealed, superheated water can be used to apply some consolidating pressure and heat; for this arrangement, active recipes which bring about curing at temperatures as low as 100°C must be used. Work has been performed to show that strong, well-bonded rubbers can be formed at temperatures down to 40°C with the correct procedures [7].

B1. Types of Polymeric Anticorrosion Barrier

A polymer coating is similar in one respect to the continuous oxide layer that develops on some metal surfaces when exposed to air; it imparts protection against further corrosion or at least acts to slow the corrosion rate. Rubber (e.g., halobutyl) tank linings are common, fabricated as outlined above. However, for the most aggressive fluids, the high chemical resistance shown by fluorinated thermoplastics makes them preferable.

There are two general categories of polymer protection [8, 9]. Barrier coatings/linings, which are classified as thin (<0.64 mm) and thick (>0.64 mm), can be applied in a number of ways to a metal (or other) substrate; examples of application techniques include thermal/chemical curing, spray-and-bake and electrostatic powder coating, rotolining, and adhesive bonding. The choice of a thin or thick lining depends on the expected rate of corrosion rate of the steel to be protected. The second form in which polymers are deployed in an anticorrosion role is as self-supporting structures. Since a large number of thermoplastics are suitable for extrusion and/or molding, components such as pipes, valves, and fittings can be fabricated to carry hostile fluids or to protect metallic structures. Thermosets on their own are not employed as thick (>0.65 mm) coatings for pipes and tanks, but epoxy and phenolic resins are utilized as thin (<0.3 mm) protective linings. Fiber-reinforced plastics (FRPs) are widely employed to fabricate piping and tanks for protection against corrosion and aggressive chemicals. These are composite materials, a combination of thermosetting resin and fiber reinforcement (usually glass), and are outside the scope of this chapter. Of the thermoplastics, it is the perfluoropolymers, in which all C–H bonds that can be "replaced" are fluorinated, which possess the highest combined levels of chemical and thermal resistance; examples include PTFE (poly(tetrafluoroethylene)), PFA (perfluoroalkoxy), MFA (copolymer of TFE and perfluoromethyl-vinylether), and FEP (fluorinated ethylene-propylene). Partially fluorinated polymers such as PVDF, ETFE (ethylene-tetrafluoroethylene), and ECTFE (ethylene-chlorotrifluoroethylene) are also widely used by the chemical process industry [10].

Polymer coats that are chemically bonded to the metal via an adhesive system [11] or that intimately contact the metal as a result of the application method (e.g., dipping, spraying, electrostatic spraying, vacuum coating [12]) generally provide better corrosion protection (because air and water tend to be excluded from the polymer–metal interface during fabrication) than polymeric liners that operate by being in close physical contact with the metal surface. An example of the latter is the lining of steel pipe with PE pipe [13, 14]. In this case, fluid molecules could slowly gather in this interfacial region, following permeation through the polymer, possibly allowing metal corrosion to commence. For elastomers bonded to metal, the bond joining the two material types is often stronger and more durable than the rubber itself [15, 16].

C. RELEVANT POLYMER PROPERTIES

There are numerous factors to be considered when selecting a polymer to control corrosion. Of the utmost importance is the nature of the service fluid and the conditions likely to be experienced by the polymer during service (temperature, pressure, flow, stress, etc.). The criteria which influence the choice of polymer for a particular fluid environment have been outlined in Chapter 66.

In an amorphous polymer at sufficiently low temperature, all large-scale chain movement is "frozen" and no long-range order exists; the polymer is said to be in a glassy state. With heating, a temperature is reached at which segmental chain motion becomes possible, and the polymer stands on the threshold of a rubbery condition. The apex of this region, which varies with polymer type, is known as the glass transition temperature (T_g); the vast majority of polymers possess a T_g. In general, the more flexible a polymer chain (a function of its chemical structure), the lower its T_g. During dynamic motions, there is a high-energy absorption range in the vicinity of the T_g, envisaged as the "internal friction" between the glassy and rubbery regions.

C1. Polymer/Fluid Compatibility

A polymer selected to protect a metal surface should be incompatible, in the chemical sense, with the contacting fluid; that is, the fluid should not interact with the polymer to a significant degree. The chemical compatibility of a liquid and a polymer can be estimated empirically from the proximity of their solubility parameters (symbol δ; units usually $cal^{1/2}/cm^{3/2}$; multiplying by 2.05 changes them to MPa) [17–19]. This is a measure of the energy of attraction between the different molecular species involved. If the δ values differ significantly, there is no chance of swelling, but if they are similar, within ~2 units, swelling may occur. For example, a nitrile rubber with a low acrylonitrile content ($\delta \sim 9$) would not be considered an appropriate lining for a tank containing toluene (δ also ~9) as considerable swelling (and therefore weakening) of the rubber could result. Structural (entropic) considerations are important here and, particularly with elastomers, there is scope to counteract the swelling expected from solubility parameter similarities at the compounding stage of processing. Increased filler loadings and/or increased crosslink density will reduce the propensity of a given elastomer to swell in a compatible liquid; in the first instance, the volume of rubber present in the compound is

lowered, in the second, tighter network formation reduces free volume (as would starting with a higher T_g base rubber) and restricts the volume of solvent that the material can accommodate; passage of solvent through the material is also impeded. Increasing the crystallinity of a semicrystalline thermoplastic, perhaps by altering the molecular weight distribution and/or processing conditions, has the same effect. The viscosity of the contained fluid should also be considered, with more viscous liquids ultimately absorbed in smaller amounts and at a slower rate than low-viscosity solvents.

C2. Mechanical Strength and Fatigue Resistance

The mechanical strength of semicrystalline thermoplastics originates from their crystalline domains; crystallites are the strength-giving elements, literally holding the material together above T_g and furnishing the polymer with reasonable mechanical properties below the melting point. Crystallites, dense regions of highly ordered polymer chains, are dispersed in a less dense amorphous (randomly oriented or disordered) polymer phase: some polymer molecules, which are very long, run in and out of the two phases. Crystallites will not admit solvent molecules and diffusing species must circumvent such obstacles, increasing their pathlength (tortuosity) and residence time in the polymer. In articles fabricated by extrusion of a polymer melt (e.g., flexible pipe internal pressure sheath), polymer chains tend to orient in the flow direction. Subsequent cooling of the extruded melt means that developed crystallites also tend to lie in the flow direction. In other words, properties of the extruded article may be anisotropic. The degree of anisotropy depends on many factors, including polymer type and processing condition, and can range from being extreme (as in liquid crystalline polymers) to being insignificant.

In the example of Section A1, the flexible pipe concept requires the correct balance between the compact self-organization of crystalline regions and chain mobility within amorphous domains to provide strength together with some flexibility. Mechanical property requirements depend on service conditions, particularly temperature, but there is scope at the pipe design stage to accommodate a range of material attributes. For example, if a rigid (high-modulus) polymer is being assessed as a potential pressure sheath, a reduction in pipe wall thickness may be one way in which to compensate for the increased material stiffness.

C3. Thermal Stability

Although thermoplastics are generally less permeable than elastomers as a direct result of structural differences, destruction of the crystallites by melting weakens the material further and removes these barriers to solvent passage—hence the requirement that service temperature not come close to the melting point or other transition point. For instance, the continuous-use temperature (CUT) for the PVDF example following is brought about by a transition of crystalline form at 150–155°C. In fact, the optimum CUT of a semicrystalline thermoplastic is typically well below its melting point and varies with exposure conditions. For example, the CUT of plasticized PVDF in air is 145–150°C, but this falls to 120–130°C in contact with oilfield production fluids. Another factor that must not be overlooked is increase in size experienced by most materials with temperature. Since polymers generally have higher coefficients of thermal expansion than do metals, glass, or ceramics, thermal stresses resulting from mismatched joined materials should be taken into consideration at the design stage.

C4. Chemical Resistance and Degradation

The next question concerns interaction, that is, does the fluid or any of its components chemically react with the polymer? Polymers do not corrode in the sense that corrosion is seen as the deterioration of metals [12], but nevertheless they can degrade chemically in contact with hostile chemicals. The main point of issue is that, to retain a barrier function, a polymer lining must be affected less rapidly than the metal being protected from a particular hostile fluid. In addition, it is necessary to know the residual life of the polymer to avoid fluid eventually reaching the metal. Degradation (by definition) chemically alters polymer structure at the point of contact. The chemical changes are usually irreversible and surface alterations may or may not enhance the entry of fluids into the polymer bulk. Properties (physical, thermal, and mechanical) are thus changed. For example, the discoloration and embrittlement of PVDF by *n*-butylamine is well known. Although much information about the effect of specific chemicals on polymers has been published, it is by no means complete, especially for complex formulations such as the corrosion inhibitors, scale inhibitors, biocides, and drilling fluids used in offshore oil and gas production. These fluids contact the flexible pipe inner sheath, some for short periods neat, others continuously in trace amounts. Appropriate assessments of their impact on polymeric components (pipes, seals) must be made before deployment.

C5. Permeation

Because of its relevance to all anticorrosion linings and barriers, this section is devoted to a discussion of permeation and its measurement. Other properties and test methods germane to the flexible pipe example are outlined briefly in Section D.

Permeation is a molecular phenomenon involving the passage of a fluid through a material [20]. Due to their nature, a degree of permeation is an unavoidable consequence of polymer contact with a liquid. The driving force is the chemical potential (μ), which is normally quantified as a concentration gradient. When the fluid is a gas or vapor, the concentration becomes pressure dependent (i.e., the driving force is now the applied partial pressure). In all polymeric barrier/lining applications, including the unbonded flexible pipe inner sheath, the contained service fluid initially contacts only one polymer surface.

Permeation and degradation are unrelated properties but may interact; for instance, the components of the permeation process (solvation and diffusion) give hostile chemicals a route into the polymer bulk, where they may (or may not) interact in a degradative way with the material. If an impermeable polymer existed, any degradative chemical reaction would be confined to the surface. While there is only one permeation phenomenon, other physical features of polymers can affect the permeation rate. The rate depends on a balance of factors; polymer and fluid type, fluid concentration (may apply as a pressure differential), temperature, exposed surface area, and thickness. For any one fluid, permeation is an inherent material property, that is, different polymers display different permeation behavior in a particular liquid.

C5.1. Permeation Defined. Permeation is a two-stage process involving, first, adsorption/evaporation of the fluid by the polymer surfaces (quantified by the surface solubility coefficient *s*) and, second, diffusion (quantified by coefficient *D*) of the fluid through the polymer bulk. (A related total-immersion phenomenon—absorption—possesses no evaporation stage.) Both *D* and *s* are inherent material properties and the permeation coefficient *Q* is the product of these other two coefficients, that is,

$$Q = Ds$$

At constant conditions, different fluids will permeate at different rates through a given polymer, with the rate raised by increasing the exposed area and/or decreasing specimen thickness. As already noted, the concentration gradient is the main driving force for permeation of a single fluid through a polymer membrane. For a gas, this is used most conveniently by considering the difference between its partial pressures on either side of the membrane: if the gas is the only permeant and there is no restriction to evaporation when through the membrane, this difference will equal its applied pressure.

With the units in commonplace use (cm^2/s/atm), *Q* is defined as the permeation rate for a fluid passing right through a polymer cube of dimension 1 cm driven by a pressure of 1 atm. Similarly, *D* is the diffusion rate across the cube and *s* is the fluid concentration in the polymer surface, both at atmospheric pressure. Hence, for a given fluid, all these coefficients can be measured from a single permeation experiment. The temperature dependencies of *Q*,

D, and s are each given by Arrhenius relations of the form (for Q)

$$\ln Q = A \exp\left(\frac{E_a}{RT}\right)$$

were A is a constant and E_a is the activation energy associated with permeation. By performing accelerated testing at elevated temperatures, a plot of $\ln Q$ (or D or s) versus reciprocal temperature, followed by extrapolation to the service temperature, allows service permeation, and so on, characteristics to be estimated. It should be noted that at high pressures these coefficients can become pressure dependent. Permeation can occur simultaneously in opposite directions if, for example, a polymer membrane separates two different fluids. It is also worth stressing that permeation characteristics can change over time if other factors that affect the polymer are altered.

The parameter Q provides insight on how much fluid passes right through a polymer at steady-state conditions, whereas D is used to estimate the time to breakthrough and s governs the amount of fluid initially available for migration. Predictions stemming from Q should be applied to the correct service geometry. In general, permeation rates for thermoplastics are an order of magnitude below those typical of elastomers, reflecting the different molecular architecture prevalent in each material class. Most elastomers are amorphous with relatively high free volume content (i.e., space for diffusing small molecules to occupy) and exist above T_g at room temperature. Thermoplastics possess crystalline zones which, as already described, do not admit diffusing small molecules; the path of diffusing molecules is thus distorted and lengthened.

C5.2. Permeation Measurement.

As far as testing is concerned, permeation ("transmission") tests are best for gases (and perhaps water [21]), while absorption ("total immersion") tests are more suitable for liquids. A variety of arrangements exist to measure gas permeation at high pressures and temperatures. One setup, developed by MERL [22], can perform at pressures up to 15,000 psi and temperatures in excess of 200°C and is shown schematically in Figure 82.2. The centerpiece is a permeation cell in which the polymer membrane is supported by a smooth, porous, stainless steel sinter; a gauge is incorporated to measure sample thickness changes while under pressure. Band heaters and good insulation provide the required test temperature. After permeation through the polymer sample, gas on the lower pressure side is collected in a fixed-volume reservoir. All pressures and temperatures are continually monitored using pressure transducers and thermocouples, respectively, and logged by a computer. The increase in low pressure with time is converted to a rate of gas permeation at standard conditions.

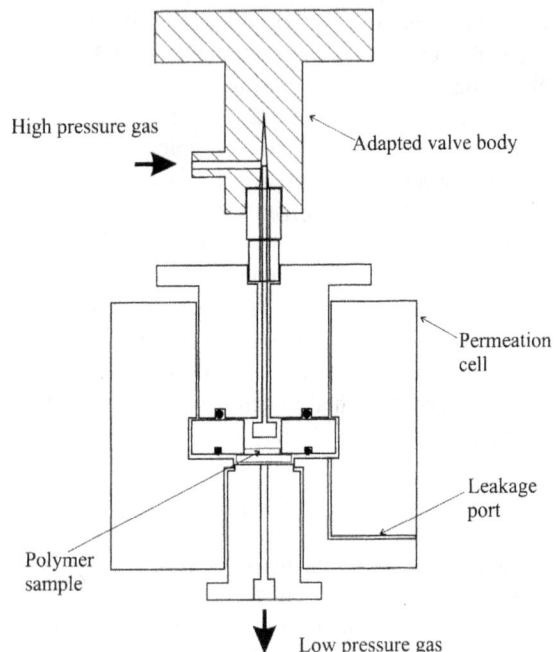

FIGURE 82.2. Schematic of high-pressure permeation test cell assembly showing location of elastomer sample and the developed in situ thickness gauge.

With oriented thermoplastics (e.g., some extruded polymers), anisotropic effects have led to the suggestion that the rate of permeation through the exposed ends of the article, say a pipe, can greatly exceed that of the same fluid through the curved surface. The passage of migrating fluid molecules between crystallites along the length of the pipe is often less impeded than that of molecules passing through the pipe in a direction orthogonal to the extrusion axis. Thus the permeation rate obtained by testing an unsealed pipe section may be misleading (too high) if such orientation effects are present; after all, the exposed pipe cross section is not a surface relevant to pipe service conditions. A sealed edge/unsealed combination of tests is required. For aqueous acids, permeation tests detect acid by titration or electrochemical means; results should be validated by a series of absorption tests.

It is recommended that laboratory permeation tests be performed on samples removed from actual structures. Failing that, samples should be manufactured under conditions that mimic actual production as closely as possible; this is less important for elastomers.

A special case is that of PTFE linings and barriers. As a result of the special processing techniques required for its fabrication, a consequence of the lack of melt processability due to the extremely high molecular weight of commercial grades, this perfluoroplastic invariably contains ultrafine porosity within its bulk. These special techniques (e.g., ram extrusion) involve the fusion of partially molten PTFE particles, after the required shape has been formed, by sintering; pores remain where the process is incomplete. If PTFE contains open pores, then testing with a liquid at several pressures should resolve the issue; with open pores, the "permeation" rate will be high and pressure dependent. If the pores are closed, the rate will be low and generally pressure independent The presence of pores in PTFE can lead to applications in filter systems.

C5.3. Other Factors That Influence Permeation. The permeation rates characteristic of elastomers are generally an order of magnitude greater than those measured for thermoplastics [6]. Hydrophilic residues in polymers [e.g., residual protein in natural rubber (NR)] lead to higher than expected amounts of water being absorbed. Increasing temperature permits greater thermal motion of polymer chains, thereby easing the passage of diffusants. If a polymer surface is degraded by contact with hostile chemicals, access to the interior may be facilitated; the permeation characteristics change accordingly. Application of mechanical stresses can, by altering chemical potential factors, also affect permeation rates.

D. TEST METHODS

In this section, available test methods for measuring the properties outlined in Section C are discussed, with emphasis on the flexible pipe example [3]; the methods described below are not relevant to composites.[*] In evaluating polymeric materials for barrier applications, test pieces should ideally be cut/machined from actual liner or pipe or from articles fabricated under similar processing conditions.

D1. Tensile Properties

Mechanical properties can be determined by tensile testing [e.g., American Society for Testing and Materials (ASTM) D638] using standard test pieces, at a series of service-related temperatures, to obtain values for Young's modulus, tensile stress at yield, percent elongation at yield, tensile (ultimate) strength at break, and percent elongation (strain) at break. Such measurements should be repeated using (standard) test pieces that have been subjected to relevant chemical aging procedures in order to determine changes after specified times of exposure to chemicals of interest. Limitations of tensile strength include its rather strong dependence on the method of testpiece preparation and its failure to test notch sensitivity.

---

[*]For composite materials, relevant test methods are the remit of the following ASTM committees: committee D20 on Plastics, committee D30 on Composite Materials.

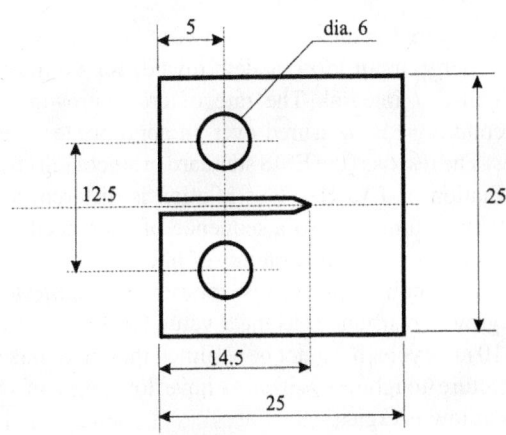

FIGURE 82.3. Side view of compact tension test piece showing the slot where the crack tip is added and the holes for loading the test piece; all dimensions in millimeters.

D2. Fracture Toughness

The basis of this test is given in a draft European Structural Integrity Society (ESIS) protocol [23]. It provides a measure of fracture toughness (J) by relating the total work done in displacing the compact tension test piece (Fig. 82.3) by a preselected amount to the depth of the crack that results. After the test, the fracture surfaces are exposed and the crack depth measured. It is necessary for compact tension test pieces (described in the protocol) to be accurately machined and for thick (6-mm minimum) samples to be used to ensure that only plane strain conditions apply during testing. The test piece has a lateral slot at the center of one side into which a sharp crack tip is added with a razor blade just before testing. A useful form in which to present data is as a plot of fracture toughness (resistance) versus crack growth constructed using the results from several separate tests in which different levels of deflection are employed. Fracture toughness J is calculated from an appropriate equation for the compact tension test piece. In summary, the test measures the intrinsic fracture toughness of the polymer immediately prior to crack growth, that is, the capacity of the polymer to withstand crack propagation. This is a more fundamental property than tensile strength and can also determine notch sensitivity.

D3. Crack Growth Fatigue

This test extends the fracture toughness concept by providing stress–strain cycles on the same test piece type to illustrate the fatigue behavior of thermoplastic materials. Dynamic fatigue occurs under cyclic loading where a small amount of crack growth appears during each load cycle, initially from the crack introduced into the material. Dynamic fatigue resistance is measured using a compact tension test piece that contains a crack of defined dimension. The material property required is the relation between crack growth rate and the energy input level as determined, for example, by the so-called J integral. The rate of crack growth under cyclic conditions is measured over an appropriate range of J values. The test uses the ESIS standard protocol [23] for the determination of J values. This relation is different at different temperatures and so a sequence of tests needs to be carried out at service temperatures of interest. A convenient way of representing the results is empirically to define a term "crack growth resistance" as the J value for a crack growth rate of 10 nm/cycle. It cannot be assumed that materials with high fracture toughness will also have low rates of crack growth at low energies.

D4. Stress Relaxation

Stress relaxation provides a measure of the rate of decrease in stress during a constant state of strain. A linearity of applied stress with log time is the norm for viscoelastic materials, providing that no chemical (air) aging occurs during the test. For semicrystalline and/or glassy polymers, with less viscous flow, linearity of stress with another function of time may apply. Any changes caused by oxidation are only likely to occur (if at all) at high test temperatures. If no such chemical aging occurs during service, the same relationship can be extrapolated to indicate stress retention during service life. If measurements with suitably aged samples should indicate a chemically induced change in relaxation rate, appropriate corrections to the simple extrapolation might be estimated.

The relation of stress is important in end-fitting regions of flexible pipe. At the end of the pipe, the pressure sheath is terminated by a swaging process involving metal elements to isolate the bore from the outer armor layers. This ring supports the entire weight of the internal sheath during periods when the production schedule requires the pipe to be decompressed. Hence, this test is necessary to measure the rate of loss of applied stress when a polymeric material is compressed to a fixed deformation. If the rate of stress relaxation is high, then it is possible that the loss of sealing force could lead to leakage, or, in the extreme, the end fitting could loosen.

D5. High-Pressure Gas Permeation

As described in Section C5, this test involves applying gas at a predetermined pressure to a suitably designed and sealed test piece mounted in an appropriate cell. The gas permeation coefficient (Q), diffusion coefficient (D), solubility coefficient (s), and concentration (c) are all determined from the test: c is the product of s and applied pressure (Henry's law). If the test is repeated for a range of pressures and temperatures, a predictive model of these coefficients, based on Arrhenius plots, can be established.

D6. Rapid Gas Decompression

It is a crucial requirement for any polymeric lining subject to prolonged exposure to high-pressure gas (>5000 psi) that rapid decompression of the gas should not cause internal fracturing/blistering of the polymer [24–27]. This can be caused by dissolved gas no longer being in equilibrium, from Henry's law. A full characterization of resistance to rapid decompression should include the effect of appropriate mechanical constraints and also the effect of temperature, pressure, and relevant gas type. The test consists of exposing samples, preferably cut from pipe/liner, to the test gas for sufficiently long as to allow them to become saturated with gas. Knowledge of permeation behavior (Section C5) is highly relevant here, involving D and c, allowing the calculation of the soak period and the amount of gas absorbed. The pressure is then released at a controlled rate (e.g., 1000 psi/min). A representative number of repeat cycles are performed and, after a suitable degassing period, samples are removed and inspected for damage. Ideally, thermoplastic materials likely to best resist rupturing will be characterized by low gas solubility in the relevant gas and a high diffusion coefficient; the latter is essential if the gas is to exit the polymer rapidly after/during the decompression.

D7. Liquid Compatibility

When evaluating the compatibility of any polymeric lining or barrier with the liquid it is to contact during service, testing must be undertaken at realistic temperatures and pressures using samples cut from actual liner or similarly processed material. In the case of the flexible pipe pressure sheaths, which carry a variety of different crude oils, depending on the field location, model reference oils can be developed that have the same solubility parameter as the crude and a similar ratio of aliphatic/naphthenic/aromatic constituents [19]. Other solvents can be included to represent oil field production additives, for example, corrosion inhibitors. Measurements made during exposures can lead to the equilibrium mass uptake and volume swell for the polymer/liquid system and provide diffusion and permeation data. If tensile test pieces are used, the change in tensile properties as a function of time in the fluid of interest can be monitored.

D8. Chemical Deterioration/Aging

The aging effects of hostile service fluids are assessed in a similar way; test vessels used tend to be pressure cells. Chemicals that may be hostile to the polymeric flexible pipe inner sheath are constituents of injected fluid mixtures such as biocides, corrosion inhibitors, and scale inhibitors; these include quaternary ammonium compounds, lower alcohols, and glycols. Hydrogen sulfide is present at low levels in many oil wells (sour wells) and steel surfaces are vulnerable to attack by this corrosive gas. When including H_2S in laboratory aging fluids, special precautions are necessary to safely contain, monitor, and dispose of liquids containing this dangerous chemical. Some flexible pipes are employed in fields requiring that they be exposed for short periods to strong acids and neat alcohols; resistance should be assessed. In total, there could be as many as a dozen different fluids to which the polymer is exposed, long term, during service, and hence should be so during realistic testing. These tend to be longer term (typically one year) tests, designed to evaluate the effect on relevant polymer properties (see above) of exposure to service fluids. The effects of aging are often visible after exposures: changes in color, changes in volume, presence of surface fractures, or complete disintegration in extreme cases.

D9. Environmental Stress Cracking

When mechanical stress is combined with a hostile chemical environment, faster rates of failure can occur than would occur due to the stress or the environment alone. The cause can be physicochemical (i.e., high local swelling in solvents with certain solubility parameters) and/or truly chemical (e.g., ozone cracking of white-walled tyres in very sunny climates). The stress required to induce cracking is invariably lower than it would be in air, and hence the need to assess relevant polymer/fluid combinations. Temperature, stress level, time, and nature of the fluid all influence the process. In general, the extent of the damage incurled increases with stress level and temperature. The effect of dilution is less clear-cut. For polymers, increasing molecular weight can reduce susceptibility to this problem. Testing is straightforward: samples are held strained at the appropriate level in rigs, and these are immersed in the fluid of interest. MERL has developed special test methods for characterizing the combined effects of material stress and aggressive chemical environment, including high fluid pressure.

E. PREDICTION OF POLYMER SERVICE LIFE

If service temperatures are low (e.g., 23°C) the permeation rate, especially for thermoplastics, is often very slow. In testing, therefore, an element of acceleration (usually higher temperature) is often introduced. As already noted in Section C5.1, because diffusion follows Arrhenius-type kinetics, testing at (typically) three different higher temperatures allows Arrhenius plots to be constructed with extrapolation giving an estimate of the permeation rate at the service temperature. If service temperature is high, property measurements should be made at this temperature.

Predictions are also possible for the consequences of chemical aging using a similar technique to that for permeation. For these, it is necessary to identify a property change or property level that marks the limit of being acceptable. Then, by accelerated testing at elevated temperatures, in each case the aging time $t\alpha$ necessary to bring about the defined changes (or reach the defined level) is recorded. Because reciprocal time is a rate, from the original Arrhenius concept, $\ln(1/t\alpha)$ versus $1/T$ plots will be linear (T in kelvin). Once again, extrapolation will lead to the time to reach the limiting point at service temperature. Practically, for partially accelerated tests at slightly elevated temperatures, an extrapolation of property-versus-time plots may be necessary to obtain a value for $t\alpha$: a knowledge of the chemical kinetics (i.e., first order, second order) can improve the quality of this extrapolation.

F. CONCLUSIONS

It should be apparent from the preceding discussion that the largest single factor in using a polymer to restrict, or even prevent, a potentially corrosive fluid reaching a metal surface is permeation—hence the focus on this topic and its measurement in polymeric materials. The challenge in qualifying polymers for corrosion control is to apply (or if necessary develop) relevant tests to measure appropriate permeation characteristics for the particular fluid in question.

However, other factors also apply. A flexible pipe example has been used to illustrate the complexity and breadth of testing required to validate a polymeric material for duty in hostile dynamic service conditions. Properties besides permeation requiring measurement are selected as appropriate from mechanical, fracture toughness, crack growth fatigue, stress relaxation, rapid gas decompression, environmental stress cracking, and liquid compatibility and chemical aging. Although extreme in the number of different property/fluid combinations requiring assessment, the flexible pipe case does highlight the need to realistically test polymers for critical applications.

Clearly, in relatively simple static applications (e.g., a tank liner operating at low temperature and pressure, where the polymer coating is expected to protect the tank surface from a single fluid type), less testing is required. Compatibility with and aging in the relevant service fluid must be appraised as well as permeation issues and general strength properties, but dynamic testing, rapid gas decompression, stress relaxation, and crack growth fatigue are not relevant for such an end use. In the case of widely used barrier materials, such as fluoroplastics, extensive technical information is available from suppliers and the open literature. However, this is no substitute for properly targeted testing if the slightest doubt exists about the suitability of any polymer/fluid combination under service conditions.

REFERENCES

1. F. Grealish, G. McGuinness, and P. O'Brien, in Proceedings of the Conference on Oilfield Engineering with Polymers, London, Oct. 28–29, 1996, MERL, Hertford, UK 1996, pp. 47–55.

2. R. Campion, M. Samulak, A. Stevenson, and G. Morgan, in Proceedings of the Conference on Oilfield Engineering with Polymers, London, Oct. 28–29, 1996, MERL, Hertford, UK 1996, pp. 60–80.

3. P. Painter and M. Coleman, Fundamentals of Polymer Science, Technomic, Lancaster, UK, 1994, Chapters 1, 7, 8.

4. G. Odian, Principles of Polymerization, Wiley, New York, 1991, pp. 1–37.

5. K. Saunders, Organic Polymer Chemistry, Chapman and Hall, London, 1988, pp. 1–44.

6. J. Brydson, Plastics Materials, Butterworth Heinemann, Oxford, 1995, pp. 19–100.

7. R. P. Campion, in Proceedings of the Conference on Rubbercon 92, Brighton, June 9–12 1992, Plastics and Rubber Institute, London 1992, pp. 411–418.

8. P. R. Khaladkar, Mater. Perform., **33**, 35 (1994).

9. P. Khaladkar, Chem. Eng., **102**, 94 (1995).

10. W. A. Miller, Chem. Eng., **100**, 163 (1993).

11. B. G. Crowther, RAPRA Rev. Rep., 80, **7**, 3 (1995).

12. L. L. Shreir (Ed.), Corrosion, Vol. 2, Corrosion Control, Newnes-Butterworths, London, 1976, Chapter 17.2.

13. J. Lovell, T. Matthews, R. Weaver, and P. J. DeRosa, in Proceedings of the Conference on Oilfield Engineering with Polymers, London, Oct. 28–29, 1996, MERL, Hertford, UK 1996, pp. 114–120.

14. D. Hill, K. A. Wilson, and A. Maclachlan, in Proceedings of the Conference on Oilfield Engineering with Polymers, London, Oct. 28–29, 1996, MERL, Hertford, UK 1996, pp. 121–134.

15. R. P. Campion, Mater. Sci. Technol., **5**, 209 (1989).

16. J. F. E. Ruffell, J. Inst. Rubber Ind., **3**, 166–169 (1969).

17. P. C. Painter, M. M. Coleman, and J. F. Graf, Specific Interactions and the Miscibility of Polymer Blends, Technornic, Lancaster, UK, 1994, Chapters 1, 2.

18. P. I. Abrams and R. P. Campion, Plastics, Rubber and Composites Processing Appl., **22**, 137 (1994).

19. R. Campion, "Model Test 'Oils' Based on Solubility Parameters for Artificial Ageing of Polymers," in Proceedings of the Conference on Polymer Testing '96, Shawbury, Sept. 9, 1996, RAPRA, Shrewsbury, UK 1996, pp. 1–5.

20. J. F. Imbalzano, D. N. Washburn, and P. M. Mehta, Chem. Eng., **98**, 105 (1991).

21. M. L. Lomax, Permeability Review, RAPRA Members Report No. 33, publication RAPRA, Shawbury, UK, 1979.

22. R. P. Campion and G. J. Morgan, Plastics, Rubber and Composites Processing and Appl., **17**, 51 (1992).

23. G. E. Hale, "A Testing Protocol for Conducting J-Crack Growth Resistance Curve Tests on Plastic," ESIS Draft Protocol, May 1994.

24. R. P. Campion, Cellular Polym., **9**, 206 (1990).

25. A. N. Gent and D. A. Tompkins, J. Appl. Phys., **40**, 2520 (1969).

26. A. Stevenson and G. J. Morgan, Rubber Chem. Tech., **68**, 197 (1995).

27. B. J. Briscoe, T. Savvas, and C. T. Kelly, Rubber Chem. Tech., **67**, 384 (1994).

83

CORROSION TESTING OF REFRACTORIES AND CERAMICS

M. Rigaud

Département de Génie Physique et de Génie des Matériaux, Ecole Polytechnique, Montréal, Québec, Canada

A. INTRODUCTION

In an echo to Lord Kelvin, who said, "When you cannot measure what you are speaking about or cannot express it in numbers, your knowledge is of a meager or unsatisfactory kind,"much effort has been expended in developing tests to measure the corrosion resistance of refractories under slag-.ging conditions and of structural ceramics under hot gas corrosion and oxidation at high temperatures, especially for nonoxide materials. Numerous methods have been tried and some reasonable correlations have been obtained for very specific conditions, but very few methods have reached the status of standard operating practices and none have yet been accepted for universal use.

The main reason is that corrosion resistance data obtained in a laboratory environment very rarely simulate the conditions that prevail in service: sample size and geometry, state of stresses in the lining, thermal gradient and thermal cycling, as well as time, which are very difficult to be scaled down to fit with acceptable laboratory test conditions. It must always be remembered that accelerated tests, specially those done using very severe conditions, can lead to erroneous predictions.

Compared with laboratory testing, field trial testing is of course much more costly and, in some instances, unsafe. It

may then be worthwhile to test small panels rather than to carry out full-size testing; the larger the installation, the more confidence one will have in the selection of the proper material to use.

Postmortem examination of in-service trials also provides very useful insights to understand and determine the controlling mechanisms in the degradation of ceramics. Detailed investigations include the use of a wide variety of characterization methods, including chemical analysis, X-ray diffraction, mineralogical analysis, and scanning electron microscopy (SEM)/energy dispersive spectroscopy. Selecting samples and carrying out sample preparation are often very challenging. Observations of the uncorroded part often yield clues as to what may have happened in the corroded part.

Crescent and Rigaud [1] reviewed and classified some 106 different experimental setups, all falling into one of 12 categories, as shown schematically in Figure 83.1. Among the nonstandard tests, the most commonly used are the crucible test, the rotary disk, and the finger test.

The Refractory Committee C8 of the American Society for Testing and Materials (ASTM) [2] has defined two standard practices dealing with slags (C-768 and C-874); two others are designed to measure corrosion resistance of refractories to molten glass, one under isothermal conditions (method C-621), the other standard practice using a basin furnace to maintain a thermal gradient through the refractory. These methods could be applicable to other liquids as well. In addition, there are other ASTM methods dealing with the disintegration of certain refractories in carbon monoxide (C-288), by alkali for carbon refractories (C-454), and by alkali vapor for glass furnace refractories (C-863). Only the three standard practices, C-768, C-874, and C-863, will be described here.

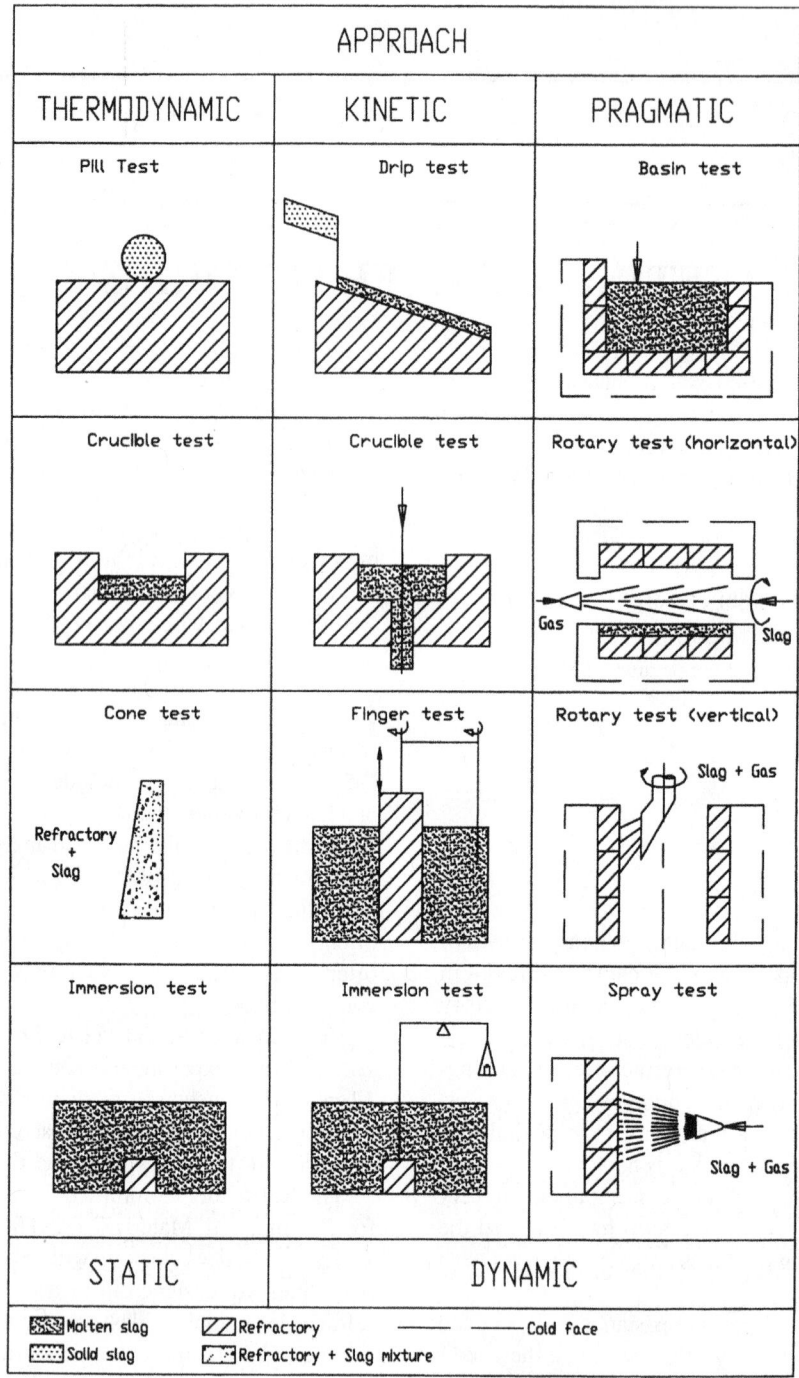

FIGURE 83.1. Different approaches to corrosion testing in slags. (From [1]. Reprinted with permission from the Canadian Institute of Mining, Metallurgy and Petroleum.)

B. DRIP SLAG TESTING (ASTM C-768)

The standard practice on drip slag testing covers the determination of the relative resistance of various refractory brick (and structural ceramics) to the action of molten slag dripped continuously onto test specimens within a heated furnace.

Test samples are mounted into the wall of the furnace with their top surface sloping down at a 3° angle. Rods of slag are placed through a hole in the furnace wall so that the melting slag drips onto the specimens. Slag is fed continuously to maintain consistent melting and dripping. The bottom of the furnace should be constructed to provide a base for the specimen supports and a sand pit to receive the spent slag. The testing temperatures can be adjusted to the conditions, as can the duration of the test (from 5 to 8 h and from 300 to 800 g of slag are the usual ranges considered). The amount of corrosion is determined by measuring the amount of modeling clay or fine sand required to fill the corroded cavity. In addition, the depth of penetration of slag into the refractory may be determined by cutting the sample in half, perpendicular to the slag cut where it is deepest. Since the entire specimen is at the test temperature, penetration of the slag into the specimen is promoted and may exceed that observed under thermal gradients that often exist in refractory applications. Oxidation of carbon-containing materials or of oxidation-prone structural ceramics may occur during such a test since the entire sample is not covered by slag and since the furnace atmosphere is air.

C. ROTARY-KILN SLAG TESTING (ASTM C-874)

The standard practice for rotary-kiln slag testing is used to evaluate the relative resistance of refractory materials against slag erosion. The furnace is a short kiln, a cylindrical steel shell mounted on rollers and motor driven with adjustable rotation speed and tilt angle, heated with a gas–oxygen torch capable of heating up to 1750°C (3182°F), if needed. The kiln is lined internally with the test specimens. Normally, six test specimens 228 mm long and two plugs to fill the ends of the shell constitute a test lining. In principle, the furnace is tilted at a 3° angle and rotated at 2–3 rpm. An initial amount of slag is first fed into the kiln, which coats the lining and provides a starting bath. Extra amounts of slag are then added at regular intervals of time at a rate of ~1kg/h for at least 5 h. The test atmosphere is usually oxidizing, but neutral conditions may be obtained by using a reducing flame or by adding carbon black to the slag mixture. The tilt angle is adjusted axially toward the burner end so that the molten slag washes the lining and drips regularly from the lower end after each addition of new slag at the higher end of the kiln. One method of assessing erosion is to measure the profile of the refractory thickness along a surface cut in the middle of the exposed surface of the specimen and measure the eroded area using a planimeter. Results can be reported as percent area eroded

from the original specimen area. It is to be noted that, for this practice, the thermal gradient through the test specimen is controlled by the thermal conductivity of the specimen and the backup material used at the cold face of the lining. The slag is constantly renewed so that a high rate of corrosion is maintained. The flow of slag can cause mechanical erosion of materials. Care must be taken to prevent oxidation of carbon-containing refractories and oxidation-prone structural ceramics during heat up. A reference refractory specimen should be used for comparison in each consecutive test run. Caution should be exercised in interpreting results when materials of vastly different types are included in a single experiment. A full discussion on the advantages and shortcomings of the method has been published [1].

D. OXIDATION RESISTANCE AT ELEVATED TEMPERATURES (ASTM C-863)

ASTM standard practice C-863 covers the evaluation of the oxidation resistance of silicon carbide refractories in an atmosphere of steam. The steam is used to accelerate the test. Oxidation resistance is the ability of SiC to resist conversion to SiO_2. The volume changes of cubes (64 mm side) are evaluated at a minimun of three temperatures between 800 and 1200°C (1472 and 2192°F). The duration of the test at each temperature is normally 500 h. In addition to the average volume change (based upon the original volume), weight, density, and linear changes are also reported as supplementary information.

More information on corrosion testing of ceramics can be found in McCauley [3] and ASTM Volumes 2.05, 2.06, 4.01, and 15.02 [2].

REFERENCES

1. R. Crescent and M. Rigaud, in Advances in Refractories for the Metallurgical Industries, Proceedings of the 26th Annual Conference of Metallurgists, CIM, Montreal, Canada 1988, pp. 235–250.

2. American Society for Testing and Materials, Vol. 15.01, West Conshohocken, PA, published annually.

3. R. A. McCauley, Corrosion of Ceramics, Marcel Dekker, New York, 1995, pp. 109–128.

84

EVALUATION AND SELECTION OF CORROSION INHIBITORS

CANMET Materials Technology Laboratory, Hamilton, Ontario, Canada

Uhlig's Corrosion Handbook, Third Edition, Edited by R. Winston Revie
Copyright © 2011 John Wiley & Sons, Inc.

A. INTRODUCTION

In principle, any method to determine the corrosion rate can be used to test a corrosion inhibitor. The primary criterion for evaluation is inhibitor efficiency. Inhibitors having sufficiently high efficiency are tested for "side effects," which include environmental compatibility, emulsion formation, viscosity, and pour point density. Finally, the inhibitor formulation is tested in the field. In general, the process of evaluation and selection of corrosion inhibitors involves three steps: (1) laboratory evaluation, (2) evaluation of compatibility (including cost), and (3) field evaluation.

B. LABORATORY METHODOLOGIES

The laboratory test methods should be carried out under conditions that simulate operational conditions in the field, including composition of material and environment, temperature, flow, pressure, and the method by which the inhibitor is added, continuous or batch. Screening of inhibitors in the laboratory requires two parts: methodology and measuring techniques. Methodology is defined as an experimental setup to generate corrosion, and measuring techniques are used to determine the corrosion rate and inhibitor efficiency.

The laboratory methodology should simulate the variables that influence inhibitor performance. These variables are classified into two categories: direct and indirect. The direct variables include composition (environment and metallurgy), temperature, and pressure. If the direct variables of a system are known, then the laboratory experiments can be carried out under these conditions. On the other hand, simulation of indirect variables in a laboratory experiment in not straightforward. One common indirect variable is flow.

To simulate flow effects in laboratory experiments, hydrodynamic parameters, including mass transfer coefficients, wall shear stress, and Reynolds number, are used, as described in Chapters 17 and 18. The merit of a laboratory methodology is generally judged by the ability to control and determine these hydrodynamic parameters. Some of the laboratory methodologies to evaluate inhibitors are discussed in the following sections.

B1. Wheel Test

The wheel test is performed by adding the corrosive fluids and inhibitor to an ~7-oz (~207-cm³) bottle with a metal coupon, purging with a corrosive gas, and capping the bottle [1]. The bottle is then agitated for a period of time by securing it to the circumference of a "wheel" and rotating it. In the high-temperature, high-pressure wheel test, autoclaves substitute for the conventional bottle and allow different partial pressures of corrosive gases, such as CO_2 and H_2S, to be used. The wheel test is best regarded as a screening test in a preliminary stage of inhibitor evaluation because it may discriminate poor inhibitors from good ones but not necessarily the best inhibitor among several good ones [2]. There is no theoretical or hydrodynamic method to determine the flow patterns in a wheel test.

B2. Bubble Test

The bubble test is also known as the stirred corrosion test or the kettle test. It is a flexible laboratory procedure for monitoring corrosion rates and inhibitor performance. The test is performed in a 1-L glass beaker or conical flask. The glass lid of the vessel has access ports for the working, counter, and reference electrodes, thermometer, and gas inlet and outlet tubes (Fig. 84.1 [3]). In the bubble test, the composition and temperature of the field can be simulated. The only fluid movement is generated by the bubbles of the purging gas. No hydrodynamic equation exists to describe the flow conditions in the bubble test.

B3. Static Test

The static test is used to evaluate inhibitor performance in the absence of flow. The apparatus described for the bubble test can also be used for the static test. The corrosion rates can be measured by weight loss or by electrochemical methods.

B4. Rotating-Disk Electrode

The rotating-disk system is simple and provides information quickly and inexpensively. For these reasons, the rotating-disk electrode (RDE) is very popular in electrochemical studies [4–6]. A typical RDE apparatus consists of a rotating unit driven by a motor that is attached to a sample holder.

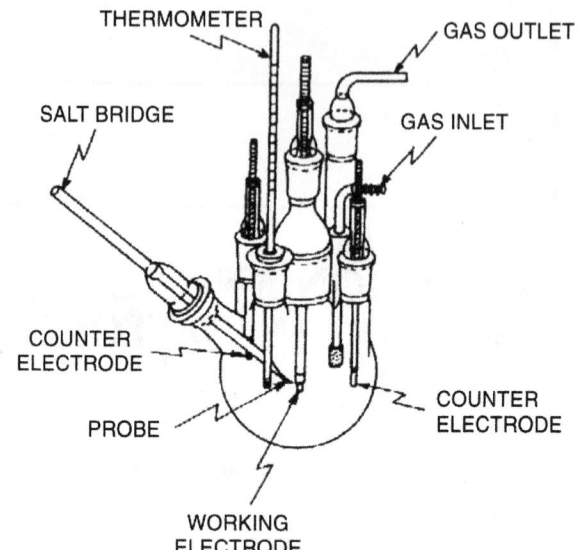

FIGURE 84.1. Apparatus for the bubble test. (From [3]. Copyright ASTM. Reprinted with permission.)

Electrochemical connections to the electrodes are made using brush contacts. The corrosion rates are measured using conventional electrochemical instruments.

The RDE surface is uniformly accessible. It is one of the few convective systems for which the equations of fluid mechanics have been solved rigorously for steady-state conditions. The limiting current density, i_d, at the RDE is given by [7]

$$i_d = 0.62n\mathcal{F}CD^{2/3}\nu^{-1/6}\omega^{1/2} \qquad (84.1)$$

where n is number of electrons, $\mathcal{F}$ is the Faraday constant, C is concentration of reactant (or product), D is the diffusion coefficient of the reactant (or product), ν is the kinematic viscosity, and ω is the angular velocity.

Equation (84.1) can be applied only under conditions of laminar flow. At higher rotation speeds, the flow at the RDE changes from laminar to turbulent. Hydrodynamic relationships have been derived to correlate RDE and other systems, for example, pipe flow [5] [Eq. (84.2)]. The relation between the RDE under laminar conditions and under pipe conditions is

$$\mathrm{Re_p} = 79\left(\frac{d}{r_o}\right)\left(\frac{e}{d}\right)^{-0.15}\mathrm{Sc}^{-1/6}\,\mathrm{Re_D^{1/2}} \qquad (84.2)$$

where $\mathrm{Re_D}$ is the Reynolds number for the rotating disk, $\mathrm{Re_p}$ is the Reynolds number for pipe, d is the diameter of pipe, e is the roughness of the pipe wall, r_o is the radius of the rotating disk, and Sc is the Schmidt number. Only a few experimental verifications of this equation have been performed, and, for this reason, RDE has not been widely used for inhibitor evaluation. Rotating electrode systems are now available for studies under flowing conditions at elevated temperature and pressure [8].

B5. Rotating-Cylinder Electrode

The rotating-cylinder electrode (RCE) test system is compact, relatively inexpensive, and easily controlled [9]. It provides stable and reproducible flow in relatively small volumes of fluid. It operates in the turbulent regime over a wide range of Reynolds numbers. The design of the RCE has several features in common with the RDE, including control of the rotation speed that is continuously variable at high and low speeds.

For RCE, the reaction rates may be mass transport controlled. Provided the ohmic polarization (IR drop) is constant in the cell, the current distribution over the electrode surface may be uniform, and concentration changes may be calculated even though the fluid flow is generally turbulent. Laminar flow is limited because in the conventional arrangement the RCE is enclosed within a concentric cell and $Re_{crit} \sim 200$, corresponding to rotation speeds of <10 rpm. Notwithstanding the instability of turbulent motion, the RCE has found a wide variety of applications, especially when naturally turbulent industrial processes must be simulated on a smaller scale or when mass transport must be maximized.

Dimensionless group correlation for turbulent flow in RCE has been described [10–14]. When the wall shear stresses are equal in the RCE laboratory test and in the pipe in the fluid, similar hydrodynamic conditions (e.g., turbulence) are maintained. Under these conditions, the corrosion mechanism (not the rate) is hypothesized to be the same in two geometries (e.g., RCE and pipe).

The wall shear stress of RCE, τ_{RCE}, is given as [10–12]

$$\tau_{RCE} = 0.0791\,Re^{-0.3}\rho r^2 \omega^2 \qquad (84.3)$$

where Re is the Reynolds number, ρ is the density, ω is the angular velocity, and r is the radius of the cylinder.

In the absence of other correlations, Eq. (84.3) can be used as a first approximation to establish the appropriate RCE velocity for modeling the desired system when trying to examine corrosion accelerated by single-phase flow. The American Society for Testing and Materials (ASTM) G170 and G185 provide guidelines to operate RCE [15, 16].

B6. Rotating Cage

The rotating cage (RC) has been reported in the literature as one of the promising laboratory methodologies to evaluate corrosion inhibitors [17–19]. Several results have been reported using this methodology, but methods to calculate the hydrodynamics of this system, to optimize apparatus dimensions, and to assess their effects on flow have not been clearly defined.

FIGURE 84.2. Photograph of rotating cage (note the vortex formation).

Figure 84.2 shows a rotating-cage system. Several coupons (8 or 10) are supported between two Teflon disks mounted a fixed distance apart on a stirring rod.

When the rod is rotated, a vortex is formed and the dimensions of the vortex (both length and width) increase with rotation speed until the width reaches the side walls of the container. The flow patterns in the rotating cage can be qualitatively divided into four zones that depend on the rotation speed and the volume of the solution and of the container [20]:

1. *Homogeneous zone*: Vortex dimensions (length and width) increase with rotation speed.
2. *Side-wall affected zone*: Vortex length increases, but the width has reached the side and collides with the wall.
3. *Turbulent zone*: Vortex length penetrates into the rotating-cage unit and creates turbulent flow.
4. *Top-cover affected zone*: The liquid level oscillates and rises to the top cover, which restricts the development of the vortex.

The rotating cage illustrated in Figure 84.2 is in the homogeneous zone. As a first approximation, the rotating-cage wall shear stress can be calculated using the equation

$$\tau_{RC} = 0.0791 \, Re^{-0.3} \rho r^2 \omega^{2.3} \qquad (84.4)$$

where r is the radius of the rotating cage. ASTM standards G170, G184, and G202 provide guidelines to operate RC [15, 21, 22].

B7. Jet Impingement

The jet impingement (JI) test can simulate reliably and repeatedly high-turbulence conditions at high temperature and pressure for gas, liquid, and multiphase turbulent systems. It requires relatively small volumes of test fluids and is controlled easily. Jet impingement is a new methodology to evaluate corrosion inhibitors.

For a circular jet impinging on a flat plate with the central axis of the jet normal to the plate, a stagnation point exists at the intersection of this axis with the plate, and the flow is axisymmetric. Only the flow and fluid properties in the radial plane normal to the disk are considered (Fig. 84.3) [23–25].

Region A in Figure 84.3 is the stagnation zone. The flow is essentially laminar near the plate, and the principal velocity component is changing from axial to radial, with a stagnation point at the center. Region A extends from the central axis to the point of maximum velocity and minimum jet thickness at $r/r_o = 2$.

Region B in Figure 84.3 is a region of rapidly increasing turbulence, with the flow developing into a wall jet (i.e., the primary flow vector is parallel to the solid surface). This region extends radially to $r/r_o = 4$. The flow pattern is characterized by high turbulence, a large velocity gradient at the wall, and high wall shear stress. Thus, region B is of primary interest for studying fluid flow effects on corrosion in high-turbulence areas. This region has not been rigorously characterized mathematically, but results of some research indicate that wall shear stress is proportional to the velocity squared [11]:

$$\tau_J = 0.179 \rho U_o Re^{-0.182} \left(\frac{r}{r_o} \right)^{-2.0} \qquad (84.5)$$

$$Re = \frac{2 r_o U_o}{\nu} \qquad (84.6)$$

where τ_j is the wall shear stress of the jet, ρ is the density of the fluid, U_o is the flow rate, Re is the Reynolds number, r is the jet radius, r_o is the radial distance, and ν is the kinematic viscosity.

In region C in Figure 84.3, the bulk flow rate and turbulence decay rapidly as the thickness of the wall jet increases, momemtum is transferred away from the plate, and the surrounding fluid is entrained in the jet. This region is amenable to mathematical characterization, but the flow cannot be related to other flow conditions (e.g., pipe flow), because momentum transfer and fluid entrainment are in the opposite direction from pipe flow.

B8. Humidity Chambers

Inhibitors to prevent corrosion in the vapor phase (vapor-phase inhibitors, VPI) are tested in humidity chambers in which wet–dry cycles are carried out and parameters such as temperature and relative humidity are measured. Various accelerated procedures have been developed to simulate different conditions (e.g., tropical, marine, and industrial

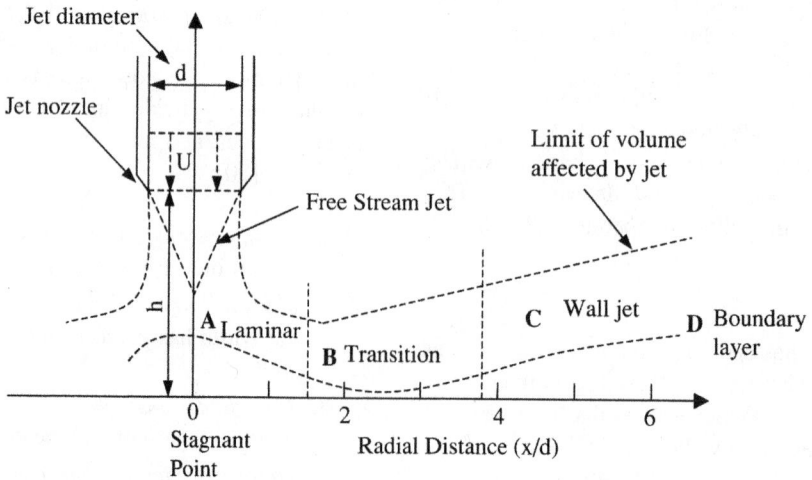

FIGURE 84.3. Schematic representation of fluid flow characteristics of jet impingement. (Adapted from [11].)

[26–29]. Observations are made after a predetermined time or number of cycles (for wet–dry tests). Corrosion of ferrous metals is assessed in terms of the percentage of the surface area covered by rust, whereas corrosion of nonferrous metals is assessed from changes in the state of the surface, as reflected by color changes.

C. MEASURING TECHNIQUES

The measuring techniques used to determine corrosion rates can be broadly classified into two categories: nondestructive and destructive. A measuring technique is destructive if it alters the corrosion process during the measuring process (e.g., potentiodynamic polarization) or if the material is physically removed from the environment (e.g,. weight loss measurements). Nondestructive techniques, including linear polarization resistance (LPR), electrochemical impedance spectroscopy (EIS), and electrochemical noise, can be used to make repeated measurements at different time intervals.

C1. Weight Loss

Determination of weight loss of the coupons in a corrosion experiment is one of the common methods to calculate corrosion rates. The coupons are cleaned and weighed before and after the experiments to remove the surface and/or corrosion products. Corrosion rates in wheel tests and in rotating-cage experiments can be determined only by weight loss methods. This method can, in principle, be used in all laboratory methodologies and for monitoring corrosion rates in the field using coupons.

From the weight loss measurements, the corrosion rates are calculated using the equation

$$\text{Corrosion rate (mm/year)} = \left(\frac{w \times 10^4}{a}\right)\left(\frac{0.365}{\rho}\right) \quad (84.7)$$

where w is weight loss (g), a is the surface area (cm^2), and ρ is the density (g/cm^3).

The inhibitor efficiency is usually given as percentage inhibition (% P). Percentage inhibition can be calculated as

$$\%P = \frac{X_o - X_i}{X_o} \times 100 \quad (84.8)$$

where X_o and X_i are weight loss (or corrosion rate or corrosion current) in the absence and presence of inhibitor, respectively.

C2. Electrochemical Methods

Linear polarization resistance, EIS, electrochemical noise (all nondestructive), and potentiodynamic polarization

(destructive) are all used in measurements to assess inhibitors. These methods are discussed in detail in Chapters 85 and 86.

C3. Solution Analysis

In this method, determinations are made of changes with time in the content of metal ions measured in the process liquid resulting from the corrosion process; inhibition will be reflected in the analytical data. The method is clearly of most use when the corrosion products are soluble and is of less value when the corrosion, or inhibition, process leads to the formation of insoluble products.

A recent variation of this method is the technique of thin-film activation analysis. In this technique, metal components or special test coupons are irraliated to produce a thin layer of radioactive isotopes on the metal surface. The extent of inhibition can then be determined from changes in the radioactivity of the medium or the specimen [30].

D. EVALUATION OF COMPATIBILITY

D1. Cost

A philosophy for using corrosion inhibitors is to define the benefits in terms of reduced costs of operation, shutdown, inspection, reliability, and maintenance by using a cheaper material with inhibitor versus using a costly corrosion-resistant material without inhibitor. Economic considerations and factors such as plant life dictate the material chosen and the corrosion prevention strategies used. Various methods (e.g., discounted cash flow discussed in Chapter 3) can be used to determine the benefits of using corrosion inhibitors.

D2. Environmental Issues

There has been increasing concern about the toxicity, biodegradability, and bioaccumulation of inhibitors discharged into the environment. Standardized environmental testing protocols are being developed [31–33].

D3. Quality Control

Spectroscopic techniques, including carbon-13 nuclear magnetic resonance (^{13}C NMR) and Fourier transform infrared spectroscopy (FTIR), are used as quality control methods to ascertain whether the correct inhibitor ingredients are blended into the product and whether the ingredients are added in the correct proportions [34].

D4. Emulsion Formation

The effect that an inhibitor has on the emulsion-forming tendency of well fluids is critical in the selection of the proper inhibitor. Test methods available to determine emulsion formation include the Setzer bottle test and high-pressure cell test [35].

E. FIELD EVALUATION

Corrosion inhibitor monitoring in a field may be needed at several points to ensure that the inhibitor has reached all locations, has formed films and/or adsorbed, is persisting, and more importantly has reduced the corrosion rates. Techniques for the field should satisfy the following requirements: high resolution/short response time; high reliability; no regular maintenance or long maintenance intervals (years); and simple signal collection, processing, and transmission. Techniques for field monitoring are discussed in the following sections.

E1. Weight Loss Coupons

Corrosion monitoring using weight loss coupons is one of the common practices in many industries. Preweighed coupons are exposed for a fixed duration and then retrieved. Equipment is now available to place and remove coupons without shutting down the plant or reducing pressure, temperature, or flow. However, coupon measurements are used to calculate only time-averaged values of corrosion rate and may not adequately represent corrosion rates throughout the entire system.

E2. LPR Probes

The presence of a conducting solution (e.g., NaCl solution) is necessary for LPR probes. In systems where there is no water or the probe is not exposed to water, LPR probes do not give meaningful results.

E3. Mass Counts

Historical metal count data are useful for determining corrosion trends. The method of collecting metal count data is difficult to apply as a tool for controlling the day-to-day dosage of inhibitors.

E4. Pipeline Integrity Gauge (PIG)

Internal inspection of pipelines with intelligent (containing sensors) pigs defines the corrosion status of the pipeline for scheduling the inhibitor treatment frequency. This method is relatively expensive and, for this reason, cannot be used frequently [36].

E5. Field Signature Method (FSM)

FSM technology is a relatively new, nonintrusive electrical resistance method for measuring wall thickness on a semicontinuous basis [37]. In this method, a large number of electrical resistances measured on a short section of the equipment are used to generate a map of the monitored area in terms of the changes in wall thickness over time as compared to the original readings. Even though this technique does not directly measure inhibitor efficiency, it can be used to monitor corrosion, erosion, and cracking.

E6. Visual Inspection

Visual inspection by permanently installed video cameras is now a possibility. A very small color video camera for borehole inspection that can operate at temperatures up to 150°C and at pressures up to 10,000 psi (69 MPa) is available. For certain critical equipment, for example, subsea oil production equipment, and at critical locations such an internal visual monitoring system can be valuable.

E7. Hydrogen Permeation

One of the common cathodic reactions during corrosion in acid solution is hydrogen reduction:

$$H^+ + e \rightarrow H \tag{84.9}$$

The atomic hydrogen that can form, for example, inside a pipeline carrying wet sour gas can diffuse and recombine on the outside to form molecular hydrogen. Monitoring of this hydrogen gas on the outside surface can provide an indirect measurement of the corrosion inside the equipment [38]. This principle is used in assessing corrosion and inhibitor efficiency using externally mounted hydrogen foils. Two different types of hydrogen monitoring probes are used: electrochemical and vaccum probes [39–42].

E8. Electrochemical Techniques

Electrochemical techniques are reviewed in Chapters 85 and 86.

REFERENCES

1. Publication 1D182, Wheel Test Method Used for Evaluation of Film Persistent Inhibitors for Oilfield Application, T-ID-8, Technical Practices Committee, NACE, Houston, TX, 1982.
2. R. H. Hausler, D. W. Stegmann, and R. F. Stevens, "The Methodology of Corrosion Inhibitor Development for CO_2 Systems," Paper No. 360, CORROSION/88, NACE, Houston, TX, 1988.
3. N. D. Greene, Experimental Electrode Kinetics, Rensselaer Polytechnic Institute, Troy, NY, 1965; "Standard Reference Test Method for Making Potentiostatic and Potentiodynamic Anodic Polarization Measurements," G5, 1994 (Reapproved 2004), 2009 Annual Book of ASTM Standards, Vol. 03.02, American Society for Testing and Materials, West Conshohocken, PA, 2009, p. 45.

4. Yu. V. Pleskov and Y. Yu. Filinovskii, The Rotating Disc Electrode, translated by H. S. Wroblowa, H. S. Wroblowa and B. E. Conway (Eds.), Consultants Bureau, New York, 1976.

5. G. Liu, D. A. Tree, and M. S. High, Corrosion, **50**, 584 (1994).

6. C. G. Law and J. Newman, "Corrosion of a Rotating Iron Disk in Laminar, Transition, and Fully Developed Turbulent Flow," J. Electrochem. Soc., **133**, 37 (1986).

7. V. G. Levich, Physicochemical Hydromechanics, Prentice-Hall, New York, 1962; Acta Physicochim. URSS, **17**:257 (1942); Zh. Fiz. Khim., **18**, 335 (1944).

8. S. Papavinasam and R. W. Revie, "High-Temperature, High-Pressure Rotating Electrode System," International Pipeline Conference, Vol. 1, ASME, New York, 1998, p. 341.

9. D. R. Gabe, J. Appl. Electrtchem., **4**, 91 (1974).

10. J. S. Newman, Electrochemical Systems, Prentice-Hall, Englewood Cliffs, NJ, 1973.

11. K. D. Efird, E. J. Wright, J. A. Boros, and T. G. Hailey, Corrosion, **49**, 992 (1993).

12. D. C. Silverman, Corrosion, **40**, 220 (1984).

13. D. C. Silverman, Corrosion, **44**, 42 (1987).

14. M. Eisenberg, C. W. Tobias, and C. R. Wilke, J. Electrochem Soc., **101**, 306 (1954).

15. ASTM G170, 2006, Standard Guide for Evaluating and Qualifying Oilfield and Refinery Corrosion Inhibitors in the Laboratory, Annual Book of ASTM Standards, Vol. 03.02, American Society for Testing and Materials, West Conshohocken, PA, 2009, p. 693.

16. ASTM G185, 2006, Standard Practice for Evaluating and Qualifying Oil Field and Refinery Corrosion Inhibitors Using the Rotating Cylinder Electrode, Annual Book of ASTM Standards, Vol. 03.02, American Society for Testing and Materials, West Conshohocken, PA, 2009, p. 758.

17. R. H. Hausler, D. W. Stegmann, C. I. T. Cruz, and D. Tjancroso, "Laboratory Studies on Flow-Induced Localized Corrosion in CO_2/H_2S Environments, II. Parametric Study on the Effects of H_2S, Condensate, Metallurgy, and Flow Rate," Paper No. 6., CORROSION/90, NACE, Houston, TX, 1990.

18. G. Schmitt, W. Bruckhoff, T. Faessler, and G. Bluemmel, "Flow Loop Versus Rotating Probes—Correlations between Experimental Results and Service Applications," Paper No. 23, CORROSION/90, NACE, Houston, TX, 1990.

19. G. Schmitt and W. Bruckhoff, "Relevance of Laboratory Experiments for Investigation and Mitigation of Flow-Induced Corrosion in Gas Production. " Paper No. 357, CORROSION/88, NACE, Houston, TX, 1988.

20. S. Papavinasam, R. W. Revie, M. Attard, A. Demoz, H. Sun, J. C. Donini, and K. Michaelian, "Inhibitor Selection for Internal Corrosion Control of Pipelines: 1. Laboratory Methodologies," Paper No. 99001, CORROSION/99, NACE, Houston, TX 1999.

21. ASTM G 184, 2006, Standard Practice for Evaluating and Qualifying Oil Field and Refinery Corrosion Inhibitors Using Rotating Cage, Annual Book of ASTM Standards, Vol. 03.02, American Society for Testing and Materials, West Conshohoken, PA, 2009, p. 752.

22. ASTM G202, 2009, Standard Test Method for Using Atmospheric Pressure Rotating Cage, American Society for Testing and Materials, West Conshohocken, PA, 2009.

23. J. L. Dawson and C. C. Shih, "Electrochemical Testing of Flow Induced Corrosion Using Jet Impingement Rigs," Paper No. 453, CORROSION/87, NACE International, Houston, TX, 1987.

24. F. Giralt and D. Trass, J. Fluid Mech., **53**, 505 (1975).

25. F. Giralt and D. Trass, J. Fluid Mech., **54**, 148 (1976).

26. E. G. Stroud and W. H. J. Vernon, J. Appl. Chem., **2**, 178 (1952).

27. A. Wachter, T. Skei, and N. Stillman, Corrosion, **7**, 284 (1951).

28. S. Z. Levin, S. A. Gintzberg, I. S. M. Dinner, and V. N. Kuchinksy, Ann. Univ. Ferrara (Nuove Serie), Sezione 5 (1966), Suppl. 4, 765, in Proceedings of Second European Symposium on Corrosion Inhibitors (SEIC), 1965.

29. A. D. Mercer, Br. Corros. J., **20**, 61 (1985).

30. S. Sastri, Corrosion Inhibitors: Principles and Application, Wiley, Chichester, UK, 1998, p. 82.

31. W. M. Hedges and S. P. Lockledge, "The Continuing Development of Environmentally Friendly Corrosion Inhibitors for Petroleum Production," Paper No. 151, CORROSION/96, NACE, Houston, TX, 1996.

32. W. W. Frenier, "Development and Testing of a Low-Toxicity Acid Corrosion Inhibitor for Industrial Cleaning Applications," Paper No. 152, CORROSION/96, NACE, Houston, TX, 1996.

33. P. Prince, A. R. Naraghi, and C. E. Saffer, "Low Toxicity Corrosion Inhibitors," Paper No. 153, CORROSION/96, NACE, Houston, TX, 1996.

34. J. A. Martin and F. W. Valone, Corrosion, **41**, 465 (1985).

35. J. D. Garber, R. D. Braun, J. R. Reinhardt, F. H. Walters, J. H. Lin, and R. S. Perkins, "Comparison of Various Test Methods in the Evaluation of CO_2 Corrosion Inhibitors for Downhole and Pipeline Use," Paper No. 42, CORROSION/94, NACE, Houston, TX, 1994.

36. M. W. Joosten, K. P. Fischer, R. Strommen, and K. C. Lunden, Mater. Perform., **34**(4), 44 (April, 1995).

37. A. M. Pritchard, "Use of FSM Technique in the Laboratory to Measure Corrosion Inhibitor Performance in Multiphase Flow," Paper No. 261, CORROSION/97, NACE, Houston, TX, 1997.

38. H. B. Freeman, "Hydrogen Monitoring Apparatus," U.S. Patent 5,279,169, 1994.

39. O. Yépez, J. R. Vera, and R. Callarotti, "A Comparison between an Electrochemical and Vaccum Loss Technique as Hydrogen Probes for Corrosion Monitoring," "Paper No. 272, CORROSION 97, NACE, Houston, TX, 1997.

40. W. J. D. Shaw, D. M. Jayasinghe, and D. Matei, "Correlations between Field and Laboratory Results for Hydrogen Flux Measurements Using a Vacuum Foil Technique," Paper No. 391, CORROSION 98, NACE, Houston, TX, 1998.

41. Betafoil™ Operating Manual, Beta Corporation, Calgary, Canada, Jan. 1995.

42. D. R. Morris, V. S. Sastri, M. Elboujdani, and R. W. Revie, Corrosion, **50**, 641 (1994).

85

PRACTICAL CORROSION PREDICTION USING ELECTROCHEMICAL TECHNIQUES

D. C. SILVERMAN

Argentum Solutions, Inc., Chesterfield, Missouri

A. INTRODUCTION

Corrosion evaluations usually have two main objectives: to predict compatibility before a material is used in an environment and to aid in understanding why a certain material—environment interaction has been observed. Very often there is a need to make this evaluation quickly as, for example, when moving a product or process very quickly from the laboratory to full-scale production or when monitoring corrosion in real time either in a process or in the laboratory. Instances arise requiring corrosion evaluations to be completed in 1–3 days. Coupon immersion tests often cannot provide the necessary information in this short of a time period.

Under these circumstances, electrochemical techniques provide a viable alternative for rapid prediction or evaluation of alloy corrosion. Their applicability arises from the fact that most aqueous corrosion processes involving metals require the transfer of charge across the alloy–solution interface. These techniques can be loosely pictured as placing a measuring device directly in the electrical circuit creating the corrosion process. Widespread use of these techniques arises from their relative ease of implementation requiring instrumentation that, today, is relatively inexpensive while being completely automated. Electrochemical techniques can be used both for laboratory and field evaluations though, in this chapter, the emphasis is on their use in the laboratory. While the techniques are usually fairly easy to use, the interpretation of the results can be difficult. The interpretation is usually made by examining the measured response (current or voltage) to an input perturbation and then by comparing the characteristics of the response to those that might be expected for different types of corrosion phenomena. That expectation ranges from agreement with highly sophisticated mathematical models to little more than comparison of output plots of current, voltage, and resistance to those expected for given types of behavior. The difficulty arises from the fact that one relationship (voltage or current) is being used to dissect a rather complex set of phenomena. If only one technique is being used, the analogy is similar to attempting to solve for several unknowns by using only one equation. The best approach is to apply several independent techniques, some of which may be nonelectrochemical, to the same problem. Thermodynamics, various electrochemical techniques, theoretical calculations, various types of spectroscopies, and traditional coupon immersion tests have been combined to provide the insights needed to make reliable corrosion predictions [1–4].

The purpose of this chapter is to provide practical examples of how a number of electrochemical techniques might be used to predict corrosion behavior in practical situations. The emphasis is on simple laboratory tests for rapid corrosion testing. The examples do not provide step-by-step procedures that cover all possibilities. Nor does the discussion focus on how to set up and run the electrochemical corrosion experiments. Such is provided in the references that the practitioner is urged to read. The examples are presented to give a flavor for the approaches that have been taken when tight time and budget constraints limit the number of tests that can be run in order to make the prediction. The practitioner must make the final judgment about which techniques and procedures are most applicable to the situation. Detailed discussions in the references should help. The techniques discussed here are the following:

1. Thermodynamic potential–pH diagrams
2. Cyclic potentiodynamic polarization scans (especially for predicting localized corrosion)
3. Polarization resistance technique
4. Electrochemical impedance spectroscopy
5. Velocity sensitivity testing with emphasis on the rotating-cylinder electrode in single-phase fluids

Methods for estimating the onset and propagation of stress corrosion cracking and use of electrochemical noise are not considered in this chapter. Many times, the above five techniques are combined to provide greater depth of understanding and more reliable predictions. Wherever applicable, such examples of combining techniques are discussed.

B. THERMODYNAMIC POTENTIAL–pH DIAGRAMS

Predicting alloy corrosion requires information in two areas, the expected state of the alloy (e.g., nonoxidized metal, oxidized metal surface, oxide, and ion) and the rate at which the alloy proceeds to that state. Thermodynamics can address the first area. The potential–pH diagram originally developed by Pourbaix [5] is a pictorial method of displaying metal stability (the lowest free-energy state) as a function of hydrogen ion activity (pH) and equilibrium potential (oxidizing power). Though these diagrams tend to be used to aid in explaining corrosion mechanisms by identifying possible corrosion products, they nave an equally important ability to act as corrosion "road maps." If all components important to corrosion are included and the thermodynamic data used represent that expected for the interface and the reactions occurring there (a nontrivial assumption), the diagrams can, in principle, show under what circumstances of pH and potential does corrosion *not* occur (region of immunity). The diagrams show what pH and potential conditions *might* cause the metal to transform to an ion (metal loss) or an oxidized solid (possible passivity). The important point is that corrosion penetration rates of micrometers per century or kilometers per second would both be depicted as regions of corrosion. Kinetic experiments are required to show what would, indeed, happen. If the measured corrosion rate is high, the potential–pH diagram might provide guidance as to how this rate might be suppressed (change pH, change potential, change concentration, etc.). The diagrams tend to be generated for pure metals because appropriate thermodynamic property values for alloys usually do not exist.

B1. Limiting Amount of Corrosion Testing

The ability of potential–pH diagrams to narrow the necessary amount of corrosion testing is illustrated by the following example [6]. The purpose of this discussion is to provide a practical example of how potential–pH diagrams might be applied to analyzing corrosion in a practical system. The reader is strongly urged to consult the chapters on aqueous thermodynamics and construction of Pourbaix diagrams elsewhere in this book as well as the literature referenced in [5] and [6] for further applications of this approach. The question arose as to the lowest possible pH (greatest acidity) titanium could withstand when exposed to a process stream somewhat depleted in oxygen and containing a high concentration of sodium chloride, hydrochloric acid, and an amino acid containing two carboxylic acid groups in addition to the amine group. The equipment involved was made from Unified Numbering System (UNS) R50400 (Grade 2 titanium). The temperature was 75°C (167°F). The only information available was that below a certain pH titanium was known to corrode catastrophically. The point at which such corrosion begins to occur was unknown. With little a priori knowledge, many costly experiments might have been run to define the threshold pH. The situation is an excellent example of where potential–pH diagrams can be most helpful in providing the guidance to minimize the number and complexity of the experiments.

Practical use of the potential–pH diagram requires placement of the coordinate of hydrogen ion activity and equilibrium potential on the diagram. Though the abscissa is actually the logarithm of the inverse hydrogen ion activity, under most circumstances, that quantity is approximated reasonably well by the measured pH of the solution. This example is somewhat different. It is provided to summarize one approach for estimating the "pH" under extremely acidic conditions with a large amount of additional salt. In this case, since the desire is to find the lowest pH or highest hydrogen ion activity that titanium might withstand, some values of hydrogen ion activity would be expected to be greater than one meaning an effective pH < 0. Most hydrogen ion electrodes have difficulty responding to the high concentrations of hydrogen ion expected here (hydrogen ion molality between ~1 and 10). Note that under these conditions the activity of water may not be 1 or be independent of pH.

The hydrogen ion activity was estimated from the compositions of the solutions provided for this study. The method is discussed in detail elsewhere [6] and only outlined here to give a flavor for the approach required because of space considerations. The expected process streams that might exist at various levels of acidity were used to estimate the hydrogen activities for each case. The only information known about the sodium chloride was that it was at saturation in all solutions, meaning that not all of the material was in solution. The molality at saturation was estimated for each solution using the correlation of Potter and Clynne [7]. That reference should be consulted for details on the procedure and equations. No account was taken of possible "salting out."

The organic acid was assumed to have a lowest acidity constant equivalent to aspartic acid and iminodiacetic acid (~1.9) [8]. The equations that relate the acid–base equilibrium constants of water and the organic components to the individual concentrations, the charge balance equation that equates the sum of the concentration of positive and negative charges, and the mass balance equations that relate the initial and equilibrium concentrations of the components were solved simultaneously for the hydrogen ion concentration. Ideality was assumed for this portion of the calculation for simplicity. The activity coefficient of hydrogen chloride in

these solutions was estimated by the methods outlined by Kusik and Meissner [9] and should be consulted for details. The carboxylate moiety on the organic acid was assumed to affect activity coefficients in a manner similar to chloride. Errors in the hydrogen activity are expected. But the effect of those errors on the predictions by the potential–pH diagram are somewhat deemphasized by the fact that the abscissa is actually the logarithm of the hydrogen ion activity, not the activity itself. The calculated activity is the abscissa portion of the (hydrogen ion activity, potential) coordinate to be placed on the diagram.

The corrosion potential is assumed in place of the equilibrium potential for the potential portion of the coordinate because of ease of measurement. This potential is usually the only one available for placement on the diagram. Judging the magnitude of the error introduced by this substitution is impossible without knowing the electrochemical mechanism. Once again, some solace can be gained by noting that the voltage ordinate is actually a logarithm of the "electron activity" so only the logarithm of the error is included in the coordinate.

Figure 85.1 shows the calculated potential–pH diagram for titanium at 75°C (167°F). Chloride ion activity was assumed to be 10. Sources of the thermodynamic values for the various compounds and the computer algorithm are discussed in detail elsewhere [6]. The important point is that the thermodynamic data for the hydrolysis products must be self-consistent, that is, consistent with measured equilibrium constants between various forms of the species. The coordinates of pH and corrosion potential have been placed on Figure 85.1. The diagram predicts that the coordinates for the first three solutions lie in the region of stability of titanium dioxide (TiO_2), an oxidized solid. The coordinates for the last two solutions lie in a region of stability of titanium as either Ti^{3+} or $TiCl^{2+}$, both of which are ions. The results suggest that exposure of titanium would be resistant to the first three solutions because the solid oxide has the lowest free energy. Dissolution of titanium might be expected in the last two solutions since an ion (a soluble corrosion product) has the lowest free-energy state. That metal dissolution occurs must be verified experimentally.

As mentioned previously, the potential–pH diagram can only answer the question, "Is corrosion possible?" The next question is "What is the corrosion rate?" That can only be answered by kinetic measurements. But the thermodynamic prediction has greatly simplified the required experiments in two ways. First, only the region ~ pH 0 needs to be examined. Second, since an ion is expected, corrosion rates might be expected to be large. Experiments of short duration might suffice.

Static immersion tests were run in solutions similar to those used in the calculation but concentrating in a narrower region of acidity. As predicted, the original solution near pH 2 showed effectively no corrosion. As the acidity increased

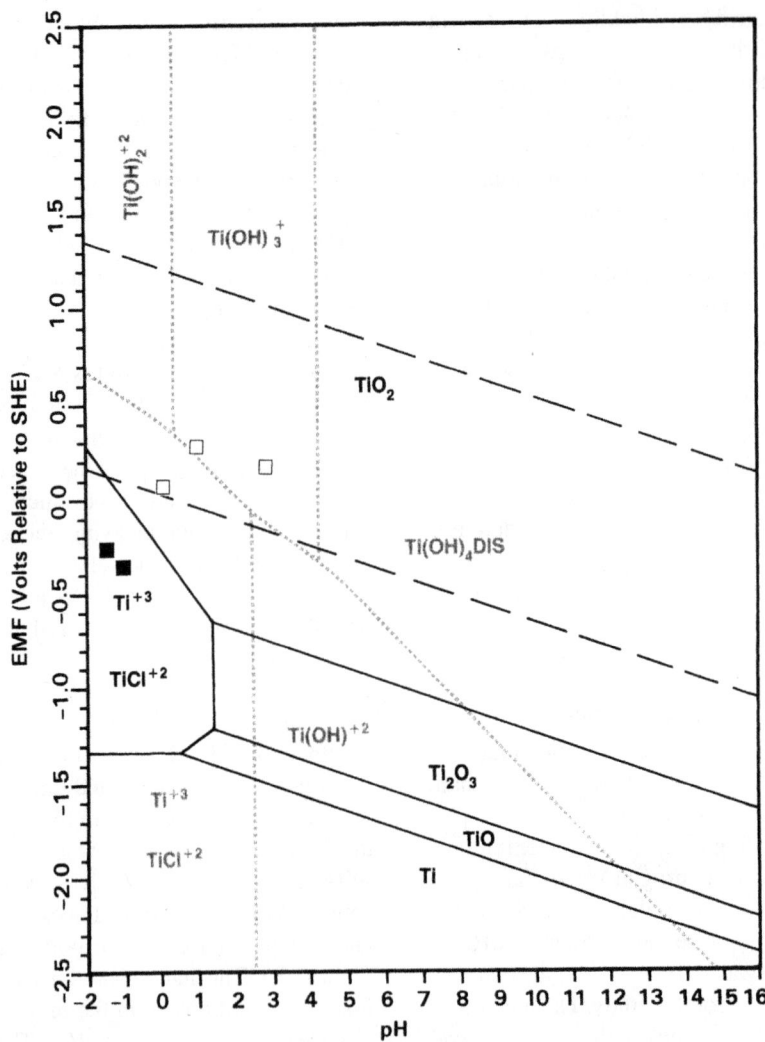

FIGURE 85.1. Potential–pH diagram of titanium at 75°C. All dissolved titanium species are at 10^{-6} activity. Chloride ion is at an activity of 10. Symbols correspond to various pH values. Open symbols correspond to no corrosion. Closed symbols correspond to corrosion. (From [6]. Copyright NACE International. Reprinted with permission.)

above a hydrogen ion activity of 1, the corrosion rate increased dramatically, rising from ∼0.025 mm/year (1 mpy) at pH 0 to 2.5 mm/year (100 mpy) for a hydrogen ion activity of 10. The conclusion is that in this situation a hydrogen ion activity of between 1 and 2 is probably about the maximum that can be tolerated before corrosion rates become exorbitant. The important lesson from this exercise is that, while thermodynamic potential–pH diagrams in and of themselves cannot predict corrosion, they can provide experimental guidance and some mechanistic understanding so that subsequent experimentation is simplified and more effective.

B2. Thermodynamic Calculations

Before leaving this subject, mention should be made of the ease with which these diagrams can be constructed and how

they can be modified for a given situation. With respect to construction, computer algorithms are available either in the literature (see the chapter 7 on constructing Pourbaix diagrams in this volume and the discussion in [10]) or as "ready-to-run" packages ([11] is one example). Reasonably self-consistent thermodynamic databases have been compiled for inorganic solids and ions [12–15]. Methods for ensuring that the database used for the diagrams is self-consistent have been described in detail [6]. For many simpler applications, the database associated with the commercial software package is adequate. More recently, generation of potential–pH diagrams has been shown to be feasible as a Web-based application on the Internet [16]. This approach circumvents existing compatibility issues that exist with present applications run on individual computer operating systems. By using the Web browser, generation of the diagrams becomes platform independent. This approach is probably the future direction for many corrosion computer applications.

Often an environment of interest contains organic ions in addition to inorganic species. Databases for stability constants of metal and organic ion adducts exist [8]. The information in these databases is usually adequate but is often incomplete. In addition to using these sources, the user should perform a literature search because new or updated data are continually being published. These adducts can be incorporated in the diagrams but to do so requires a minimal background with thermodynamic theory and the understanding that thermodynamic properties are a function of state, not path. This concept enables data to be estimated by constructing "calculation pathways" from initial reactants for which data are available to final products for which data are not available. These pathways may include virtual states as intermediate states. The reason for using this approach is that, though acid–base constants and stability constants are available for many organic and organometallic ions, the free energy of such species is usually not available. These data can be obtained by constructing the molecule in the "virtual" vapor state and then moving from this vapor state to an ideal liquid state. One such method described in detail elsewhere [6, 16, 17] utilizes the fact that group contribution methods can be used to estimate the free energy of the organic species in the vapor state and in moving from the vapor state to the liquid state. Then, the stability constants are used to estimate the free energies of the various ions. This type of procedure is only required when considering adducts between organic ions and metals. The interested reader is directed to [6, 16, 17] for a more thorough discussion of this methodology.

C. CYCLIC POTENTIODYNAMIC POLARIZATION SCANS FOR PREDICTING LOCALIZED CORROSION

The cyclic potentiodynamic polarization technique for corrosion studies was introduced in the 1960s and refined especially during the 1970s into a fairly simple technique for routine use. It tends to be most useful for the so-called self-passivating alloys susceptible to localized corrosion such as austenitic stainless steels, nickel-based alloys containing chromium, and reactive alloys such as titanium and zirconium. The technique is built on the idea that corrosion can be predicted by observing the response to a controlled upset from steady-state behavior. The upset can be created by application of voltage or current. In most cases the voltage is ramped in a cyclic manner from the corrosion potential and the characteristics of the current generated during the cycle are used to predict possible behavior at the corrosion potential.

The voltage applied between an electrode made from the alloy under study and an inert counter electrode is ramped at a continuous, often slow, rate relative to a reference electrode using a potentiostat. The voltage is first increased in the anodic or noble direction (forward scan). The voltage scan direction is reversed at some chosen current or voltage and progresses in the cathodic or active direction (backward or reverse scan). The scan is terminated at another chosen voltage, usually either the corrosion potential or some potential active with respect to the corrosion potential. The potential at which the scan is started is usually the corrosion potential measured when the corrosion process reaches steady state. Sometimes the sample has to be immersed in the environment for 24 h or more for the potential to become constant indicating steady state. The corrosion behavior is predicted from the structure of the polarization scan. An American Society for Testing and Materials (ASTM) standard exists for this procedure [18] though the procedure as outlined there is not the only or even necessarily the best way to generate the polarization scan in all situations. Several vendors offer software with their potentiostats that enable the scans to be generated automatically under computer control.

Though the generation of the polarization scan is simple, its interpretation can be difficult. The interpretation is derived from the relationship between the current and voltage and differences in that relationship between the forward and the reverse portions of the scan. Certain characteristics were identified as being important very early in the development of this technology. Characteristic potentials identified as important for determining the propensity for localized corrosion were the "protection" or "repassivation" potential and "pitting" potential [19]. Over the years, investigators have reexamined this technique to determine the relevancy of these and other parameters for the routine prediction of localized corrosion. The practitioner should note that these parameters have not been theoretically derived and so are empirical in nature. For example, as discussed later, the critical pitting potential and the repassivation or protection potentials are not fundamental properties but can change with such experimental variables as scan rate and point-of-scan reversal. Pitting may be observed at potentials below the measured pitting potential. This empiricism does not mean that the technique is not useful for practical corrosion screening. What the empiricism does mean is that care is required to ensure that experimental variables are chosen so that differences among polarization scans reflect actual differences in corrosion behavior. No single set of experimental parameters can be recommended though guidelines can be given. The following discussion is meant to provide background so that the practitioner can make his/her own procedure relevant to the system under study.

The polarization scan should not be used to estimate the general or uniform rate of corrosion. Making such an estimate from polarization scans would usually require the assumption of a mechanism and the curve fitting of the scan to equations describing that mechanism over a significant potential range (e.g., several hundred millivolts). Assuming that the corrosion mechanism does not change over this potential range may not be valid. Better technologies exist for estimating corrosion rates (e.g., electrochemical impedance spectroscopy, polarization resistance technique, and coupon immersion testing).

C1. Features Used for Interpretation

The features used for the interpretation should be consistent across a broad range of alloys and conditions assuming that the experiments are run properly. Table 85.1 shows five such generally applicable features. That these features provide such consistency is supported by the successful development of an expert system in which the scan features could be trained to predict the type of corrosion by an artificial neural network [20]. Such training would be possible only if the "input" features and "output" corrosion are internally consistent. The work originated from the realization that interpretation of polarization scans is akin to recognizing the pattern that the polarization scan presents and from which the interpretation is made. The features listed in Table 85.1 provide much of the necessary consistency between observation and prediction and were those that enabled the artificial neural network to be trained with reasonable success. The features cannot be used separately. Only when used together and generated under a consistent set of conditions do they present the pattern that is interpretable. Also, the values of some of the features can be a function of experimental variables as discussed later. Care is required to maintain consistency across polarization scans when screening corrosion.

Figures 85.2–85.5 show some generalized polarization scans that are often observed in practice with the features listed in Table 85.1 labeled on them. These figures represent different types of corrosion phenomena as discussed in their captions. The figures are drawn assuming an arbitrary minimum recorded current (e.g., $0.1\,\mu A/cm^2$) that could lie above the actually measured minimum current (e.g., $0.01\,\mu A/cm^2$) sometimes observed in an experiment. Hence, the lowest current portion of the scan may sometimes overlay the voltage axis. Some of the features that were found to be relevant are shown on the figures.

The methodology for choosing the values of these parameters is described in detail elsewhere [21] and is summarized here. The propensity for localized corrosion in

the form of pitting and crevice corrosion may be deduced by proper consideration of the parameters shown in Table 85.1. The relative position of the "pitting" and "repassivation" potentials with respect to the corrosion potential is one important observable. For this reason, the scan should be started from the corrosion potential when steady state has been established, which may take at least 24 h. A traditional interpretation is that pitting would occur if the hysteresis between the forward and reverse scan appears as in Figure 85.2 and the corrosion potential is ~100–200 mV more active than or is anodic (noble) with respect to the pitting potential. Preformed pits (e.g., crevices) might be expected to activate and grow if the corrosion potential lies between the pitting and repassivation potentials. The alloy would be expected to resist localized corrosion if the corrosion potential lies cathodic (active) with respect the repassivation potential or if the polarization scan appears as in Figure 85.3. A useful rule-of-thumb is to require that the corrosion potential be some value (e.g., 200 mV) more active than the repassivation potential for the risk of crevice corrosion to be negligible. In that case, the pitting potential either should not be present or should be somewhat more noble (e.g., 100–200 mV) to the repassivation potential. These rules of thumb are recommended because of the influence that experimental variables can have on the value of the repassivation potential.

The pitting potential is a function of the scan rate, becoming less noble (more active) as the scan rate is decreased. The difference between the pitting potential and repassivation potential has been found at least in some environments to be a measure of the extent of crevice corrosion suffered by the electrode [22]. Extrapolation of reported results suggests that at low scan rates the pitting potential and repassivation potential approach each other [23]. Such convergence suggests that the repassivation potential and pitting potential might become equal at zero scan rate. This concept has not been proven. The concept does imply that the corrosion predictions deduced from such measured repassivation potentials would be conservative. The practical conclusion is that the scan rate should be as slow as possible, consistent with the timing required to obtain an answer. Often 0.5 mV/s is a reasonable choice.

The hysteresis refers to a feature of the polarization scan in which the forward and reverse portions of the scan do not overlay each other. The hysteresis is shown in Figures 85.2

TABLE 85.1. Features for Interpreting Polarization Scans[a]

Feature	Value or Quality of Feature
Repassivation or protection potential	(repassivation potential − corrosion potential)
Pitting potential	(pitting potential − corrosion potential)
Potential of anodic-to-cathodic transition on reverse portion of scan	(potential of anodic-cathodic transition − corrosion potential)
Hysteresis	Positive, none, negative
Active–passive transition	Present, absent

[a]See [21].

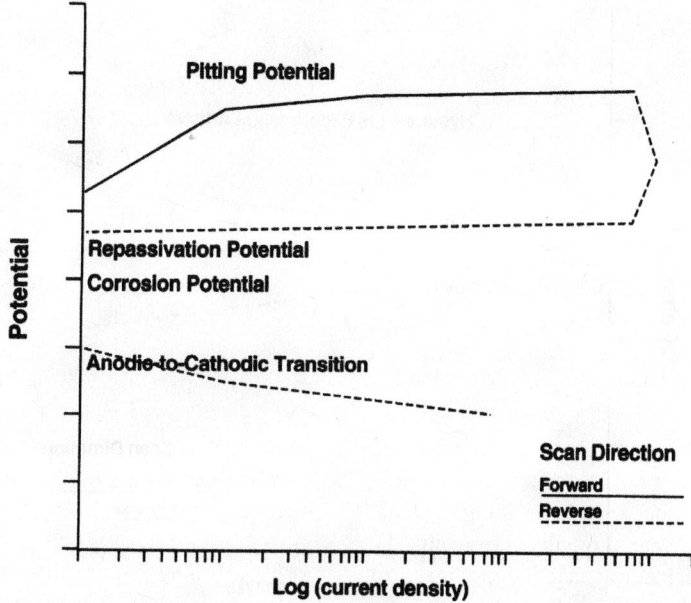

FIGURE 85.2. Typical potentiodynamic polarization scan for an alloy suggesting a significant risk of localized corrosion in the form of crevice corrosion or pitting depending on relative values of characteristic potentials. General corrosion is not likely. (From [21]. Copyright NACE International. Reprinted with permission.)

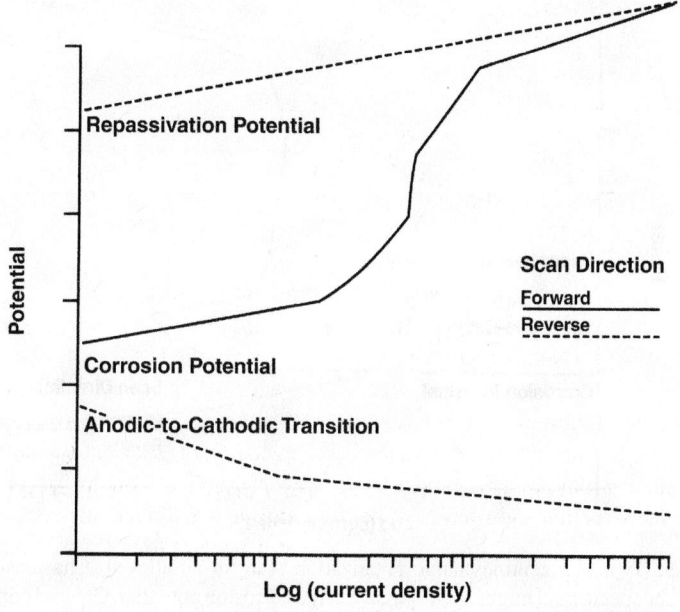

FIGURE 85.3. Typical potentiodynamic polarization scan for a completely passive alloy suggesting little risk of crevice corrosion, pitting, or general corrosion. (From [21]. Copyright NACE International. Reprinted with permission.)

and 85.3 and is created by the current density difference between the forward and reverse portions of the scan at the same potential. The difference is a result of the disruption of the passivation chemistry of the surface by the increase in potential and reflects the ease with which that passivation is restored as the potential is decreased back toward the corrosion potential. For a given experimental procedure, the greater the current on the return portion relative to the forward portion, the greater the disruption of surface passivity, the greater the difficulty in restoring passivity, and, usually, the greater the risk of localized corrosion. This type of hysteresis is labeled in this chapter as "negative hysteresis." Figure 85.2 shows negative hysteresis. In this case, the current decreases much more slowly during the

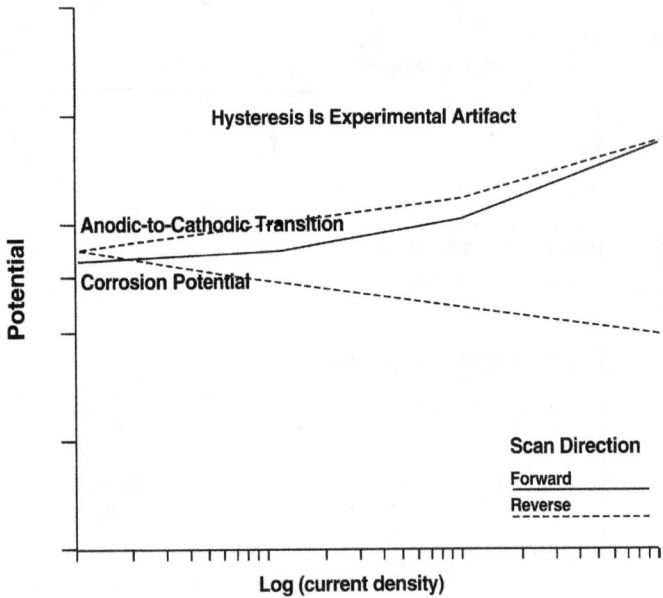

FIGURE 85.4. Typical potentiodynamic polarization scan for an alloy that might suffer general corrosion, possibly at a high rate of penetration. (From [21]. Copyright NACE International. Reprinted with permission.)

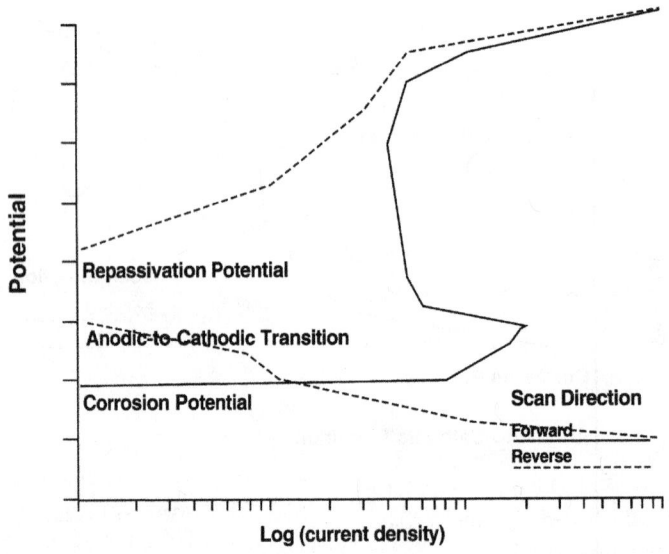

FIGURE 85.5. Typical potentiodynamic polarization scan for an alloy that has easily oridizable/reducible surface species and might not be passive at the corrosion potential. General corrosion may be measurable. (From [21]. Copyright NACE International. Reprinted with permission.)

reverse scan and indicates an alloy that has difficulty in repassivating. Such behavior often suggests a lack of resistance to localized corrosion. In many experiments involving passive alloys, the voltage is ramped until the pitting potential is exceeded. Figure 85.3 shows the opposite type of hysteresis labeled as "positive hysteresis." In this case, the current decreases very rapidly during the reverse scan and indicates an alloy that passivates very easily. Such behavior often suggests a resistance to localized corrosion. Such localized corrosion is not expected in the absence of the

type of hysteresis as shown in Figure 85.4. Such a polarization scan often indicates an active surface with a possibly large corrosion rate. The corrosion rate should be determined by the techniques listed previously.

The active–passive transition or "anodic nose" reflects the characteristic in which the current increases rapidly with increasing potential in the anodic direction near the corrosion potential, goes through a maximum value, and then decreases to a low value. Iron or some austenitic alloys may demonstrate this type of behavior in acidic environments, for example. The decrease in current may suggest an alloy surface undergoing some type of passivation process or valency change (Fe^{II} to Fe^{III}) as the potential is increased. Figure 85.5 shows such an example. This feature often implies that the alloy has a finite, possibly small, general corrosion rate at the corrosion potential.

The "anodic-to-cathodic transition potential" at which current changes from anodic to cathodic current during the reverse portion of the scan is assumed to be the potential of the anodic-to-cathodic transition. The difference between this potential and the corrosion potential is an additional feature useful for screening. If the polarization scan appears as in Figures 85.2 and 85.3, this potential still exists, but the current at the transition is lower than the lowest recorded value of the current density. Under these circumstances, this potential might be assumed to be the potential at which the cathodic current rises above the lowest recorded value. Note that in Figure 85.5 the feature labeled as anodic nose might not appear. Since this feature results from a potential induced change in surface chemistry that makes the alloy surface suddenly more passive, it often is further evidence of a somewhat easily reducible surface.

The difference between this anodic-to-cathodic transition potential and the corrosion potential would provide an additional indication of persistence of passivity. In alloys that can passivate either by a change in oxidation state (ferrous to ferric) or a change in the passive layer (greater enrichment of chromium oxide, e.g.), polarization to more noble potentials relative to the corrosion potential might place the surface in a more passive state than at the corrosion potential, at least until the transpassive region is reached. If in the reverse scan this potential is cathodic or active with respect to the corrosion potential, the suggestion would be that passivity persists as the scan returns through the corrosion potential (Figs. 85.2 and 85.3). Following that reasoning, the passive layers that would normally develop on the alloy in the environment would be considered to be very stable. General corrosion should be low. If this potential is more noble than the corrosion potential, the passive layers formed by oxidizing the surface are not stable at the corrosion potential (Fig. 85.5). Under these circumstances, the general corrosion rate should be measured by the techniques mentioned before.

C2. Artifacts Introduced by Experimental Details

Numerous examples of how these polarization scans might be used to predict localized corrosion abound in the literature [4, 21–27]. The technique can be used in the laboratory or in actual process equipment. The reader is recommended to read these and other references on how the technique has been applied in practice. The focus here is on three experimental artifacts that can arise even if the physical equipment and cell arrangement are verified to be appropriate: excessive resistivity in the fluid, inappropriate scan rate, and inappropriate point of reversal of the polarization scan.

C3. Solution Resistivity

The typical arrangement in an electrochemical cell is for the sensing point of an external reference electrode to be brought close to the working electrode by means of a Luggin–Haber capillary [28]. A rule of thumb is that the sensing point can be no closer than about two outer diameters of the capillary. Sometimes, such close proximity is not possible. The potentiostat can compensate for the voltage drop between the counter electrode and the sensing point of the reference electrode. The potentiostat is blind to the voltage drop (resistance) between the sensing point and the working electrode. This resistance includes the solution resistance, any passive film resistance, and the electrical resistance of the electrode and leads. This uncompensated resistance may change with changing applied potential.

Often in well-controlled corrosion studies the uncompensated voltage drops are not significant enough to warrant attention. But a large solution resistance might be encountered in routine corrosion testing where the practitioner may have less control of the environment and possibly the cell arrangement. The problems are that (1) the potential affecting the alloy surface region can be different from the potential applied by the potentiostat and (2) the effective scan rate at the surface can be different from the scan rate applied by the potentiostat [29]. If the uncompensated voltage drop becomes significant, these offsets can become so large that they drastically affect scan appearance and the subsequent interpretation. The applied potential (which is plotted on the polarization curve) can become much greater than the voltage that is affecting the corrosion processes (the voltage that is assumed to be plotted on the polarization curve). The applied scan rate can be much greater than the effective scan rate. The differences will become larger with increasing current.

An example can best show how this phenomenon might manifest itself. A rotating-cylinder electrode was used to assess the effect of velocity on corrosion of steel for a plant solution of acetone cyanohydrin [30]. The solution conductivity was 100 μmho/cm. Mass loss of the electrode measured as a function of fluid velocity demonstrated that the corrosion rate of steel in this environment was equal to the mass transfer rate of iron from the surface, as shown in Figure 85.6. Under these circumstances, the polarization scans would have been expected to have an anodic current density that increases with flow rate while maintaining a constant primary passivation potential. Two of the polarization scans as generated at different rotation rates are shown in Figures 85.7 and 85.8.

In both figures, the mass transfer limiting current would appear to be independent of rotation rate while the "measured" primary passivation potential seems to increase with velocity (1.0 V at 500 rpm and 1.8 V at 5000 rpm). These observations are artifacts of the uncompensated resistance. The voltage drop is so great that the compliance of

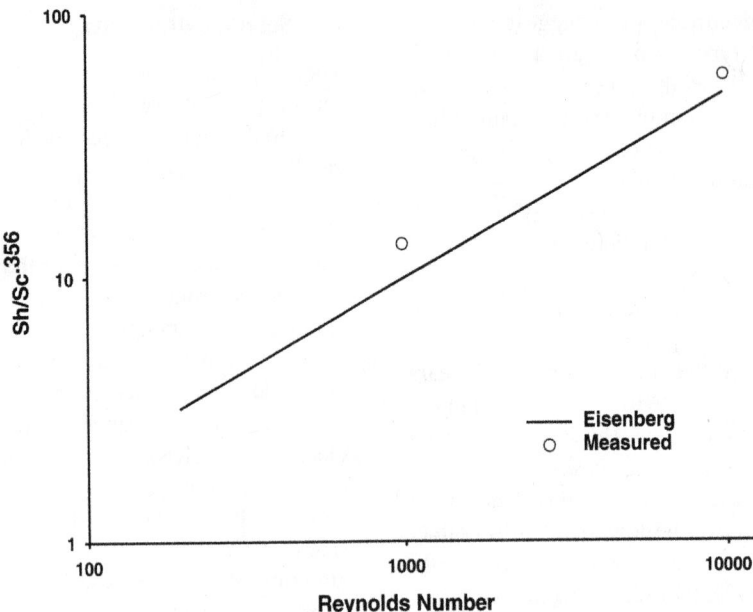

FIGURE 85.6. Sherwood number versus Reynolds number relationship calculated from mass-loss-derived corrosion rate of steel in process acetone cyanohydrin versus that expected in the rotating-cylinder electrode apparatus.

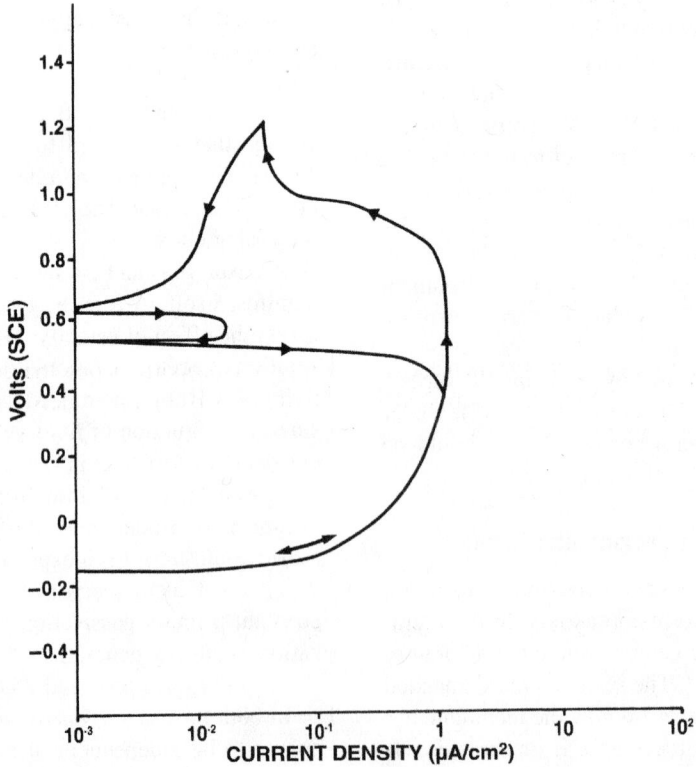

FIGURE 85.7. Potentiodynamic polarization scan for steel in process acetone cyanohydrin at 500 rpm in rotating-cylinder electrode apparatus. (From [30]. Copyright NACE International. Reprinted with permission.)

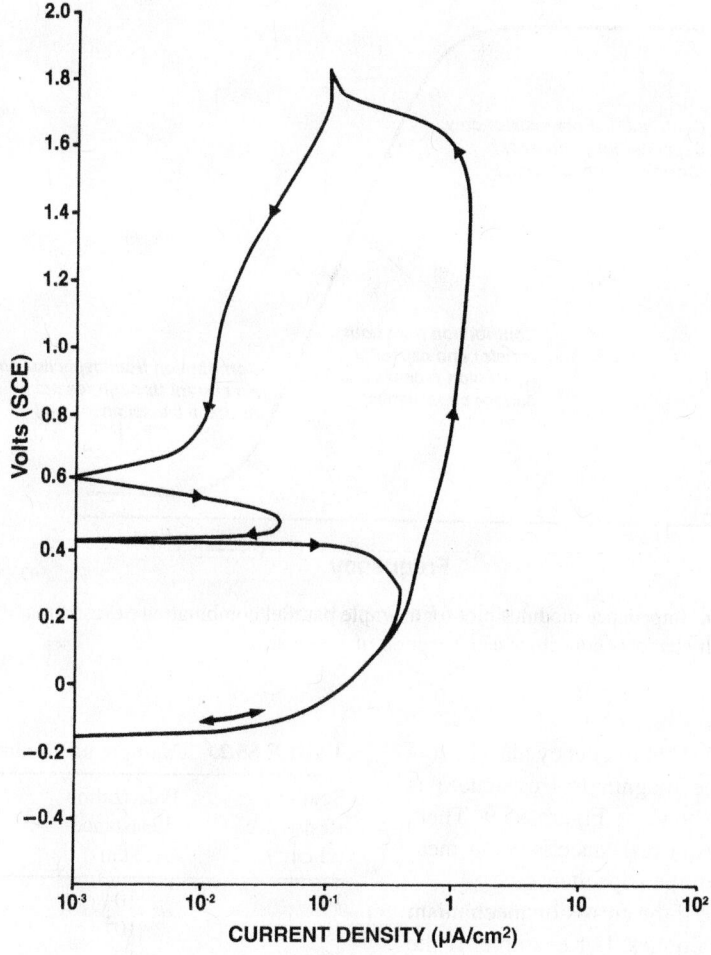

FIGURE 85.8. Potentiodynamic polarization scan for steel in process acetone cyanohydrin at 5000 rpm in rotating-cylinder electrode apparatus. (From [30]. Copyright NACE International. Reprinted with permission.)

the potentiostat has most likely been reached, limiting the current. The observed primary passivation potential is a function of rotation rate because the difference between the applied potential and the potential driving the corrosion process (the effective potential) increases with the increased voltage drop caused by the increased rotation rate (increased current). This point is confirmed by the independence of the characteristic potentials found on the reverse portion of the scan when the current densities are much lower. Had the effect of uncompensated resistance not been realized, the *erroneous* interpretation might have been that the corrosion rate is not sensitive to fluid velocity and the passivation mechanism depends on velocity.

C4. Scan Rate

The rate at which the potential is changed, the scan rate, is an experimental parameter over which the user has control. If not chosen properly, the scan rate can alter the scan and cause a misinterpretation of the features. The problem is best

understood by picturing the surface as being modeled by a simple resistor and capacitor in parallel. As an example, the capacitor could represent the double-layer capacitance and the resistor could represent the polarization resistance (inversely proportional to the corrosion rate). The goal is for the polarization scan rate to be slow enough so that this capacitance remains fully charged and the current/voltage relationship reflects only the interfacial corrosion process at every potential. If not, some of the current being generated would reflect charging of the surface capacitance in addition to the corrosion process. The result is that the measured current would be greater than the current actually generated by the corrosion reactions. The measured current would not reflect the corrosion process.

The question is, "What is that proper scan rate?" No recognized method exists to estimate this scan rate because the capacitance and resistance would be functions of the applied voltage and change during the course of the scan. But an estimate can be derived using the premise that the scan rate (rate of change of voltage) is analogous to a frequency at

1140   PRACTICAL CORROSION PREDICTION USING ELECTROCHEMICAL TECHNIQUES

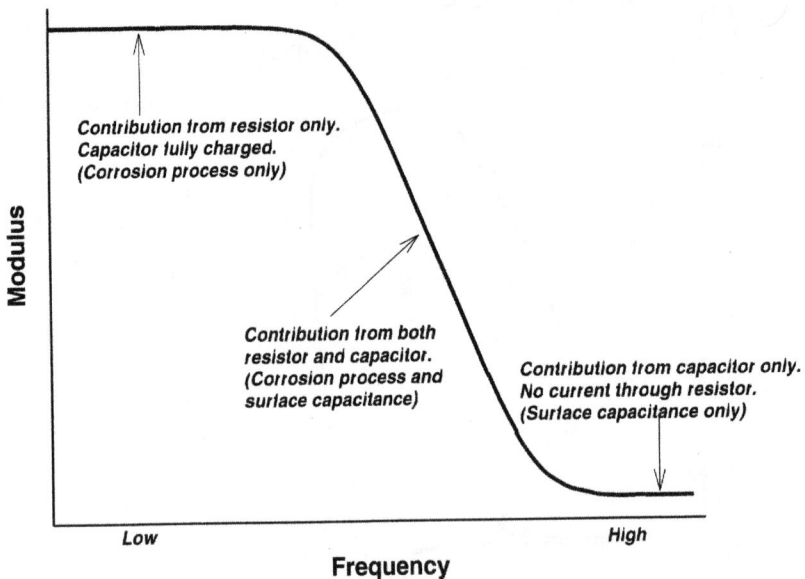

FIGURE 85.9. Impedance modulus plot for a simple parallel combination of resistor and capacitor showing which elements contribute as a function of frequency.

every applied potential [31, 32]. That frequency must be low enough so that the impedance magnitude (resistance) is independent of frequency, as shown in Figure 85.9. Then, the polarization or charge transfer resistance is being measured with no interference from the capacitance.

This rate might be estimated if the corrosion mechanism can be modeled by circuit analogues. For example, if the corrosion mechanism can be modeled as a resistor (solution resistance) in series with a parallel combination of a resistor and capacitor (polarization resistance and double-layer capacitor, for instance), then the frequency corresponding to the inflection or break point of Figure 85.9 can be described by[*]

$$f_b = \frac{1}{4\pi R_\Omega C_{dl}} \left\{ 1 - \frac{1}{R_p} \sqrt{R_p^2 - 4R_\Omega R_p - 4R_\Omega^2} \right\} \quad (85.1)$$

Following Mansfeld and Kendig [32], to obtain a frequency in the horizontal portion of Figure 85.9, the frequency f_b in Eq. (85.1) should be divided by 10. That is, the applied frequency should be about an order of magnitude lower than the breakpoint frequency. The maximum scan rate is then estimated by

$$S_{max} = \left[\pi V_{pp} f_{max} \right] < \left[\frac{\pi V_{pp} f_b}{10} \right] \quad (85.2)$$

Table 85.2 shows estimated maximum scan rates for several polarization resistance values, solution resistance

[*] Parameters are defined in Section H.

TABLE 85.2. Example of Maximum Scan Rates[a]

Solution Resistance ($\Omega \cdot cm^2$)	Polarization Resistance ($\Omega \cdot cm^2$)	Surface Capacitance ($\mu F \cdot cm^2$)	Maximum Scan Rate (mV/s)
10	10^3	100	5.1
10	10^4	100	0.51
10	10^5	100	0.05
10	10^6	100	0.005
100	10^3	100	6.3
100	10^4	100	0.51
100	10^5	100	0.05
100	10^6	100	0.005
10	10^3	20	25.
10	10^4	20	2.5
10	10^5	20	0.25
10	10^6	20	0.025
100	10^3	20	50.
100	10^4	20	2.6
100	10^5	20	0.25
100	10^6	20	0.025

[a]See [21].

values, and capacitance values sometimes encountered. For example, many passive alloys have estimated polarization resistance values of 10^4–$10^6 \, \Omega \cdot cm^2$ and capacitance values of the order of 10–100 $\mu F/cm^2$. The estimates are approximate. They do suggest that the maximum permissible scan rates at which the capacitive contribution is eliminated are fairly low for very passive types of alloy–environment interactions. Even scan rates as low as

0.5 mV/s suggested earlier as being reasonable may not allow for complete charging of the surface capacitance. This observation does not mean that the cyclic polarization scans cannot be used for screening or that the scan rates must be so low so as to make them impractical. What the observation does mean is that the polarization scan is probably not measuring only the charge transfer process. From a practical standpoint, consistent interpretations can still be made if the scan rates are kept low and the same (e.g., 0.5 mV/s) when screening alloys or environments (with comparable uncompensated resistance).

C5. Point of Scan Reversal

The current density (potential) at which the polarization scan is reversed can play a significant role in the appearance of the polarization scan and the value of the repassivation or protection potential. The reason is that the value of the repassivation potential is dictated by the amount of prior damage done to the passive surface. The farther the polarization scan is generated in the anodic direction, the greater tends to be the degree of upset of the surface region. The influence of the point of reversal on repassivation potential is especially pronounced if the pitting potential is exceeded or some other electrochemical transformation is precipitated, especially if it does not reflect the behavior at the corrosion potential. The result can be an erroneous prediction of corrosion behavior. There is no "best" recommendation because the amount of upset of the surface required for a prediction is somewhat dictated by the information desired One rule of thumb that has been suggested is to reverse the potentiodynamic polarization scan when $100\,\mu A/cm^2$ is reached so that the surface is perturbed but not overly perturbed [21]. As shown in the example below, that rule of thumb cannot be followed blindly. Maintaining a constant reversal point can be most important if alloys are being screened in a constant environment or if a single alloy is being evaluated across a number of environments but the choice must be dictated by the environment and application.

The following example shows how this artifact can affect the polarization scan and the prediction from it [21]. The polarization scan shown in Figure 85.10 was generated for UNS N08825 during a test program in a low-pH (pH 1–2) environment. The scan shows a large hysteresis with the repassivation potential lying about 50 mV more active than the corrosion potential. The conclusion would be that localized corrosion in the form of crevice corrosion has a reasonable risk of occurring. No such attack was ever found in practice, contrary to the prediction.

Figure 85.11 shows a polarization scan in the same environment but generated so as to avoid potentials in the region of the Cr^{III}–Cr^{VI} transformation above the 0.8–1.0 V saturated calomel electrode (SCE) range as deduced from a

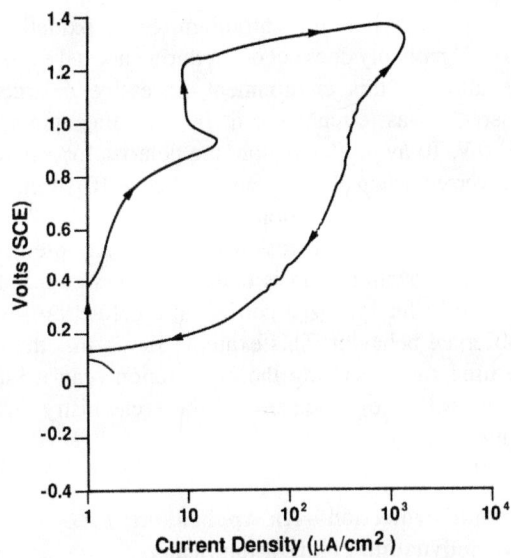

FIGURE 85.10. Potentiodynamic polarization scan of UNS 08825 in process stream sample. Maximum potential and current density at scan reversal are excessive for this system. (From [21]. Copyright NACE International. Reprinted with permission.)

potential–pH diagram for chromium. This potential–pH diagram was used because in the strongly acidic solution UNS N08825, which has a fairly high bulk chromium concentration, was believed to have a surface enriched in chromium, a surface that might be modeled by the

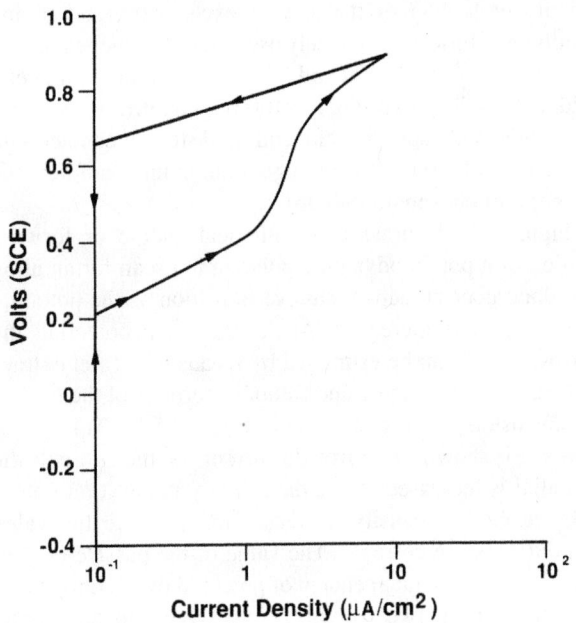

FIGURE 85.11. Potentiodynamic polarization Scan of UNS 08825 in process stream sample. Maximum potential and current density at scan reversal are appropriate. (From [21]. Copyright NACE International. Reprinted with permission.)

potential–pH diagram for chromium. Such oxidation of Cr^{III} to Cr^{VI} probably does not occur during normal exposure of this alloy in this environment Excessive destruction of passivity was circumvented by avoiding potentials >0.8–1.0 V. To avoid this region, the polarization scan had to be reversed when the current reached $\sim 10\,\mu A/cm^2$. In Figure 85.11, the repassivation potential lies $\sim$ 0.5 V above (more noble than) the corrosion potential. The prediction from this polarization scan is that the alloy would not be expected to suffer localized corrosion, a prediction in line with observed behavior. This example shows that the procedure used for generating the polarization scan must be consistent with the anticipated surface chemistry in the application.

C6. Anodic Protection—An Application of Potentiodynamic Polarization Scans

The potentiodynamic polarization scan can be used to estimate if a technology called anodic protection might be used for corrosion control. The principle of anodic protection is straightforward. In certain alloy–environment systems, a potential region exists anodic or noble with respect to the corrosion potential such that if the voltage is controlled in that region the corrosion rate of the alloy can be drastically reduced relative to what it is at the corrosion potential. This technology has been applied to a number of practical systems [33]. The reader is encouraged to consult [33]. One very common application is the protection of carbon steel and stainless steel in various concentrations of sulfuric acid. Anodic protection of tanks, heat exchangers, and piping handling sulfuric acid is widely used. Several other examples are protection of S31600 and S31700 in mixtures of acetic acid and acetic anhydride, S32100 in phenol, various stainless steels in phosphoric acid, and mild steel in contact with fertilizer solutions containing ammonium sulfate and ammonium phosphate [34–36].

Figure 85.12 shows a hypothetical anodic or forward portion of a potentiodynamic polarization scan for an alloy that undergoes an active–passive transition as the potential applied to it is increased. At the corrosion potential, the corrosion rate can be estimated by means of a Tafel extrapolation [1] of the anodic and cathodic portions of the scan to the corrosion potential, point A in Figure 85.12. This corrosion rate is shown as "corrosion current" on the figure. If the potential is increased above the primary passivation potential, the current density decreases markedly to the value marked "passive current." The value of the passive current may then become independent of potential over a fairly wide potential range between points B and C on Figure 85.12. The passive current can be several orders of magnitude lower than the corrosion current (note that current is on a logarithmic scale). The concept of anodic protection is to control the potential of the alloy between points B and C.

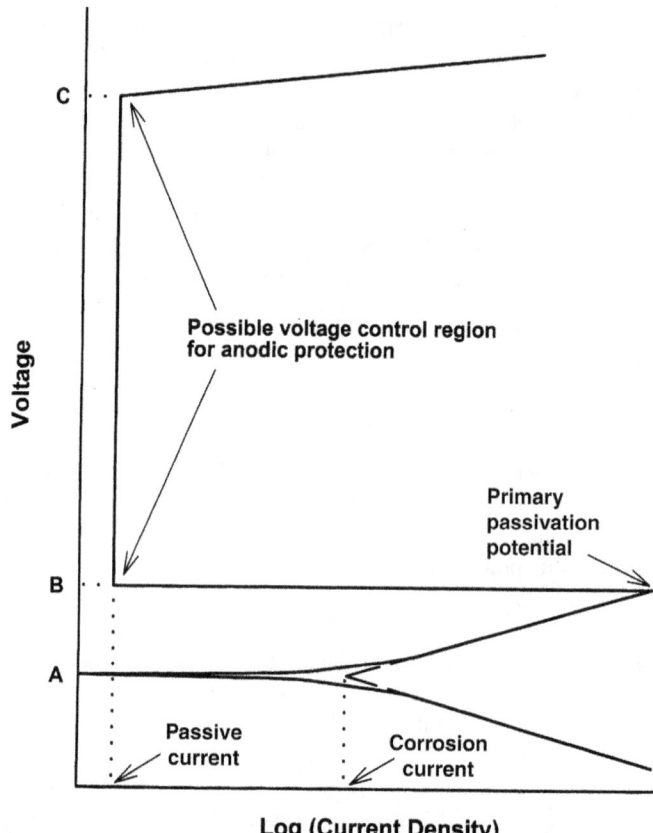

FIGURE 85.12. Anodic portion of a potentiodynamic polarization scan that shows an active–passive transition. The corrosion potential is at point A. The passive region lies between points B and C.

The polarization scan illustrates some of the complications of anodic protection. Under steady-state conditions, the current requirements can be very low as suggested by the passive region BC in Figure 85.12. But if the alloy is initially at the corrosion potential (point A), the current must follow the curve as the potential is increased into the passive region between points B and C. The required maximum current density to establish steady state can be several orders of magnitude higher than the steady-state current The practical significance is that the power supply must be significantly oversized relative to that needed to maintain the steady-state current or the equipment to be protected must be prepassivated or passivated at a lower temperature.

Under some conditions, the corrosion potential is noble with respect to the primary passivation potential (B), lying between points B and C. But the initial current is actually greater than that found at steady state [37, 38] or greater than that predicted from the polarization scan. The actual cause is unknown. One theory is that in the case of sulfuric acid the surface undergoes a reconstruction into, for example, an iron oxide–iron sulfate matrix that ultimately functions to increase electrical resistance across the interface. No matter the cause, this phenomenon translates into a required power supply that is greater than that which would be deduced by using the current between points *B* and *C*, the region of the initial potential.

One last point should be mentioned. Anodic protection might not be a viable corrosion control technology if the polarization scan appears as in Figure 85.2. The reason is that should the potential stray above the repassivation potential crevice corrosion and ultimately pitting could result. The best systems for which anodic protection might be considered are those with polarization scans as shown in Figures 85.3 and 85.5. Longer term controlled potential tests must be used to ensure that all currents used for specifying electronic equipment are at steady state.

D. POLARIZATION RESISTANCE TECHNIQUE FOR CORROSION PREDICTION

Predicting rates of general corrosion is far easier than predicting the risk of crevice corrosion or pitting. Several electrochemical techniques are very useful in predicting such corrosion rates. One such technique is the linear polarization technique or, as it should be mere properly named, the polarization resistance technique. The technique is rapid, simple, and relatively inexpensive. It can be applied without detailed knowledge of the controlling electrochemical parameters. The technique can be used in complex, poorly defined electrolytes. The methodology for generating the potentiodynamic sweep in the vicinity of the corrosion potential is adequately described in ASTM G59 [39]. This standard provides a test circuit and standard solution useful for determining if a setup is functioning properly. The name "linear polarization technique" is actually a misnomer because the inverse relationship between the polarization resistance (slope of the voltage versus current curve at the corrosion potential) and the corrosion current (corrosion rate) exists even though the curve itself may not necessarily be linear at the corrosion potential [40].

The polarization resistance technique is very well established for routine use both for corrosion prediction and corrosion monitoring. During routine use, the realistic limit for corrosion rate estimation is about 0.001 mm/year ($\sim$0.1 mpy) mostly because of limitations in estimating Tafel slopes. Under quiescent conditions and with no outside electrical or other interference, a polarization resistance of greater than $10^6 \, \Omega \cdot cm^2$ may be measured reasonably reliably. Such values would translate to corrosion rates of $\sim$$10^{-4}$ mm/year, or $\sim$0.01 mpy. Reviews by Mansfeld [40, 41] provide an overview of the applications of this technique and assumptions inherent in its use. Sensors are available for online corrosion monitoring [42] and are available from several commercial suppliers.

D1. Assumptions Behind Corrosion Rate Estimate

The data are usually analyzed by assuming that the relationship between the current and voltage in the polarization curve is given by

$$\frac{i}{i_{corr}} = \exp\left[\frac{2.303(V - V_{corr})}{b_a}\right] - \exp\left[\frac{-2.303(V - V_{corr})}{b_c}\right] \tag{85.3}$$

The corrosion current can be related to the Tafel slopes by

$$i_{corr} = \left(\frac{1}{2.303 R_p}\right) \frac{(b_a b_c)}{(b_a + b_c)} \tag{85.4}$$

No assumptions beyond fulfilling Eq. (85.3) are needed to derive Eq. (85.4). The derivation of Eq. (85.4) is completely mathematical. The assumption of linearity between voltage and current is not needed [43]. Substituting Eq. (85.4) into Eq. (85.3) enables Tafel slopes and the polarization resistance to be extracted from a curve fitting of Eq. (85.3) to the actual data.

As shown in Figure 85.13, the polarization resistance R_p is the reciprocal of the slope of the polarization curve at the corrosion potential when plotted with current density on the ordinate and voltage on the abscissa. The polarization resistance is inversely proportional to the corrosion current density, which can be transformed to a corrosion or penetration rate for uniform corrosion. From Eq. (85.4), all that would be needed in principle to estimate the corrosion rate is to have the values of the two Tafel slopes and the polarization resistance, *all measured at the corrosion potential*. The value of the technique lies in the fact that, in most instances, the method of making the measurement does not interfere with the quantities being measured as long as the polarization is in the vicinity of the corrosion potential ($\sim \pm 30$ mV or less offset) and the measurement can be made very quickly, usually in a matter of minutes. Notice also that in Figure 85.13 the curve is not linear in the vicinity of the corrosion potential. In addition, a polarization curve is symmetrical in the vicinity of the corrosion potential only when the two Tafel slopes are equal [40].

Using Eq. (85.3) to quantify the corrosion process and estimate corrosion rates from a polarization curve, such as that in Figure 85.13, requires assumptions such as those summarized below:

1. The reaction rate (corrosion current) can be expressed as being proportional to the exponential of the voltage offset from the corrosion potential for one oxidation (anodic) and one reduction (cathodic) reaction.

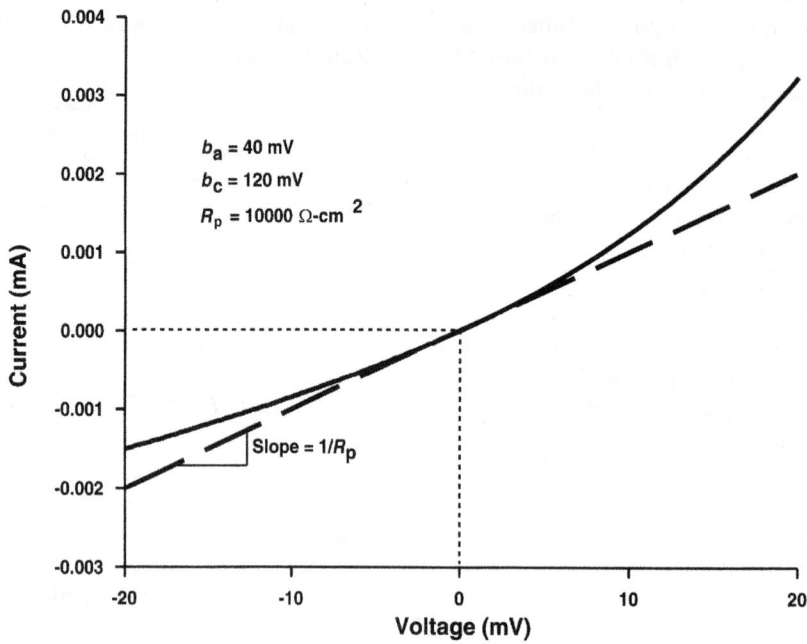

FIGURE 85.13. Theoretical polarization resistance curve showing the relationship between the polarization resistance and die curve structure [44]. (Reprinted from Proceedings of the 1995 Water Quality Technology Conference, by permission. Copyright © 1996, American Water Works Association.)

2. Uncompensated resistance in the electrolyte is either absent or much smaller than the polarization resistance. The resistance estimated by the technique as the polarization resistance contains all contributions to the total resistance.

3. For the full form of Eq. (85.4) to be used, mass transfer is not the controlling or rate-limiting step and both the anodic and cathodic reactions are under activation control. Otherwise, the Tafel slope corresponding to the mass-transfer-controlled process is infinite. The technique itself can still be used to estimate corrosion rates under these circumstances. Nonlinear regression of the combined Eqs. (85.3) and (85.4) would result in an extremely large Tafel slope for the subprocess under mass transfer control.

4. The corrosion potential does not lie close to the reversible potentials for the oxidation and reduction reactions. Being 25 mV from the reversible potential should be sufficient [43].

5. To estimate the rate of uniform corrosion from the polarization resistance, the entire electrode surface must function simultaneously as a cathode and an anode. The anodic and cathodic reactions must not occur on different sites. Otherwise, corrosion will be localized and the corrosion rate calculated using Eqs. (85.3) and (85.4) is not the rate of uniform corrosion. If such localized attack is severe, the method might be used as a sensitive detector of such corrosion, but that use is beyond the scope of this discussion [42].

6. No additional electrochemical reactions are occurring.

Assessing how well these assumptions are fulfilled requires some knowledge of the corrosion process. The polarization resistance technique, like all electrochemical techniques, cannot be used blindly. Fulfilling the above assumptions means that using the polarization resistance technique to estimate the corrosion rate is valid for that corrosion process. Many practical systems are often poorly characterized so assessing how well these criteria are fulfilled can be difficult. Some degree of imprecision must be associated with the estimated corrosion rate under these conditions. Additional sources of error can arise when the technique is applied in practice.

D2. Sources of Error

D2.1. Voltage Scan Rate. The rate at which the voltage is ramped can affect the slope of the polarization curve at the corrosion potential and, hence, the polarization resistance [32, 40, 44] for the same reasons as discussed previously with respect to the potentiodynamic polarization scan. When one ramps the voltage in one direction (e.g., −20 to + 20 mV relative to the corrosion potential) and then in the other direction (+ 20 to −20 mV relative to the corrosion potential), the two curves should overlay each other. Otherwise, the slope of either curve may not represent the true slope. If the scan rate is too large, these curves will not overlap. The methodology outlined with respect to Eqs. (85.1) and (85.2) and as shown in Table 85.2 could be used to estimate the maximum scan rate. Note that, as in the case of the potentiodynamic polarization scan, the applied potential may not be equal to the effective potential if the scan rate is too high.

D2.2. Uncompensated Solution Resistance.
The resistance calculated from Eq. (85.4) is the sum of the actual polarization resistance and the uncompensated resistance between the sensing point of the reference electrode and the working electrode. The lower the conductivity of the solution, the greater the uncompensated resistance and the greater the possible error in the estimated polarization resistance. The slope of the polarization curve at the corrosion potential as plotted in Figure 85.13 would make the estimated R_p too large and the estimated corrosion rate too small. Often, the resistance estimated at high frequency (e.g., several thousand hertz) as measured by electrochemical impedance spectroscopy can be used as the solution resistance. That number would be subtracted from the measured polarization resistance to provide the "true" polarization resistance.

D2.3. Nonlinearity in Vicinity of Corrosion Potential.
Sometimes, the slope of the voltage-versus-current curve in the vicinity of the corrosion potential is assumed to be independent of applied potential. Devices exist that apply two voltages, one at -10 to $-20\,\text{mV}$ and the other at $+10$ to $+20\,\text{mV}$, both relative to the corrosion potential. The voltage difference is divided by the current difference and the result is assumed to be the polarization resistance. Other devices exist that measure the current at discrete voltage increments and determine the curve by linear regression. Figure 85.13 shows how a true polarization curve might appear. For Eq. (85.3), the second derivative is not zero at the corrosion potential so curvature of the polarization curve might be expected at that point [43, 44]. The amount of curvature will depend on i_{corr}, which itself depends on the Tafel slopes and polarization resistance. Invoking the assumption of linearity where the curve is actually nonlinear has been estimated to result in errors as high as 50% from this source alone [40]. Such errors may be acceptable because corrosion rates estimated from mass loss can also be in error by 100%. During screening, rates differing by a factor of 2 or 3 may be considered to be the same. Then, linearity may be assumed for simplicity. If more accuracy is needed, account should be taken of the full nonlinearity in the polarization curve near the corrosion potential. Curve-fitting by nonlinear regression of the data against an equation such as (85.3) is a reasonable way to extract the polarization resistance. One straightforward way to curve fit the data is place the voltage–current pairs into a spreadsheet program such as Excel (trademark of Microsoft Corporation) and to subtract the corrosion potential from the each voltage. If Excel is the spreadsheet program, the option "Solver" can be used to obtain the Tafel slopes and polarization resistance by nonlinear regression of the data against Eq. (85.3) with Eq. (85.4) substituted for the corrosion current.

D2.4. Errors in Tafel Slopes.
A plethora of methods exist for estimating the corrosion current (polarization resistance) and Tafel slopes that do not assume linearity in the relationship between voltage and current. A number of these methods have been developed that tend to use a regression against the actual polarization curve to calculate the Tafel slopes and the polarization resistance followed by calculation of the corrosion current. Differences between actual and calculated Tafel slopes can cause large errors in estimated corrosion rates [45–47]. Regression techniques can sometimes lead to nonunique solutions by locating a local and not a global minimum in the response surface. Care must be used when trying to extract the corrosion rates from the polarization curve shown as in Figure 85.13. One simple technique useful during screening is to assume that the corrosion current is equal to the reciprocal of the polarization resistance multiplied by 0.025 V [48].

D2.5. Varying Corrosion Potential.
The corrosion potential is the potential of a corroding surface in an electrolyte relative to a reference electrode measured under open-circuit conditions. This potential is created by all of the electrochemical reactions occurring on the corroding surface. One of the requirements of the polarization resistance technique is that the electrochemical reactions must be at steady state or at least constant during the measurement. Such a condition is identified by a constant corrosion potential. If the corrosion potential is varying, the current–voltage relationship defining the polarization curve may not reflect the same corrosion phenomena at all points of the curve.

D2.6. Nonuniform Current and Potential Distributions.
The Wagner number W is useful for qualitatively predicting if a current distribution is uniform or nonuniform [44, 49]. The parameter W is dimensionless and is given by

$$W = \left(\frac{\kappa}{L}\right)\left[\frac{\partial V}{\partial i}\right] \qquad (85.5)$$

This number can be considered as the ratio of the resistance to electron transfer across the interface to the resistance of the solution. For practical purposes, Eq. (85.5) can be represented by

$$W = \frac{R_p}{R_\Omega} \qquad (85.6)$$

One rule-of-thumb proposed is that if W is less than 0.1 the current distribution is likely to be nonuniform unless precautions are taken to ensure that the cell geometry is ideally symmetric [44]. The solution resistance will be the dominant factor. In this case, the voltage must be corrected for IR drops. Reference [50] provides a significant amount of information on the measurement of and correction for the uncompensated resistance.

D2.6.1 Example of Use of Technique (Nickel in Strong Acid).

The corrosion rate of nickel had to be estimated in several strongly acidic phosphoric/phosphorous acid solutions, the total acid content being ~50 wt %. The polarization resistance method was used because of the ability of this technique to provide corrosion rates fairly quickly as a function of time over a 24-h period. Several solutions were examined.

The procedure was to scan the potential between − 20 and + 20 mV at 0.1 mV/s after about 1 h, after about 4 h, and finally after 24 h of exposure. Voltage–current data were recorded at every 0.2 mV. The corrosion potential was stable during the generation of replicates at each time period. In several solutions, the potential did change by ~50 mV over the 24-h period.

Figure 85.14 shows a typical polarization resistance plot for these systems. One feature is that the curve is not completely linear over this voltage range. Even between −10 and + 10 mV, the curve is not completely linear. A nonlinear regression routine was used to curve fit Eq. (85.3) to the data in all cases. The two Tafel slopes and the polarization resistance were used as independent variables whose values were determined directly by the regression analysis. A number of variations of this procedure have been

described in the literature [40, 41, 51–53]. A simple method using a spreadsheet was briefly described in the previous section. Corrosion rates calculated from these parameters were compared to corrosion rates estimated from the mass loss of the nickel electrode in each solution. This last comparison is important because it provides a check against the assumption that Eq. (85.3) is valid. An ASTM standard provides a methodology for how to make such calculations for all alloys [54]. Also, the electrode itself was examined under a microscope for localized corrosion. If such appears, the assumption about uniform corrosion might be violated, providing another source of error.

Table 85.3 shows the calculated Tafel slopes and the corrosion rates estimated from the regression analysis using Eq. (85.3) and from mass loss. The Tafel slopes and polarization resistance values were averaged across the runs over the 24-h period because the corrosion rate did not change over that period. The error shown is the standard deviation. As shown in the first two cases, good agreement is possible between corrosion rates calculated by mass loss and those calculated from Eq. (85.4).

The difference between corrosion rates in the last two cases touches on several areas that can contribute to errors. First, the solution resistance was ~1–2 Ω·cm² as measured by electrochemical impedance spectroscopy at 5000 Hz. The error introduced by ignoring solution resistance is very small in the first two cases, but that error can account for at least 10% of the value in the last two. Second, though the standard deviation in the measurement of the polarization resistance is only ∼ 20% of the average value, the difference between the

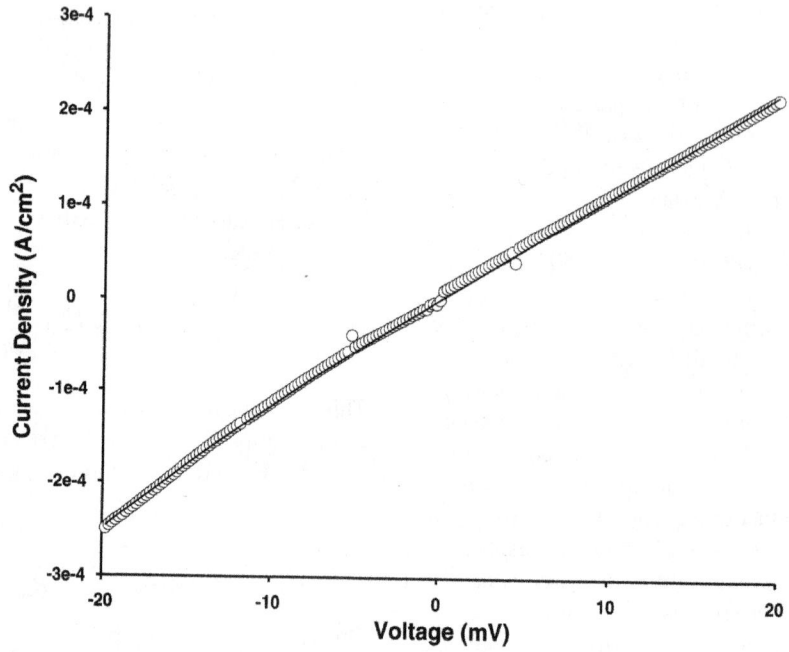

FIGURE 85.14. Polarization resistance curve for nickel in a 50–60 wt % mixture of phosphoric and phosphorous acid.

TABLE 85.3. Nickel in Strong Acid Tafel Slopes, Polarization Resistances, and Corrosion Rates

Solution	Tafel Slope, b_a (mV)	Tafel Slope, b_c (mV)	Polarization Resistance ($\Omega\cdot cm^2$), R_p	Corrosion Rate (mm/year)	
				Corrosion Current	Mass Loss
1	47($\pm$25)	65($\pm$24)	83($\pm$15)	1.55	1.80
2	62($\pm$16)	82($\pm$20)	116($\pm$18)	1.58	1.62
3	50($\pm$10)	54($\pm$9)	24($\pm$5)	5.03	22.3
4	59($\pm$9)	44($\pm$10)	16($\pm$5)	7.30	29.5

extremes in the polarization resistance (measurement + standard deviation) − (measurement − standard deviation) is ~100% for those two cases. Small errors in small polarization resistance values can lead to large errors in estimated corrosion rates. Third, the corrosion potential for those two cases is about − 250 mV (SCE), very close to the reversible potential for hydrogen under the extremely acidic conditions. Hydrogen evolution in the form of bubbles was observed, suggesting that the corrosion potential might have been close to the reversible potential for the hydrogen evolution reaction in this system. Such proximity could have led to errors because Eq. (85.3) would not have completely described the electrochemistry [43].

D3. Steel in Inhibited Water System (Effect of Experimental Artifacts)

In view of the plethora of literature on successful applications, a discussion showing the influence of experimentally induced artifacts on the estimates may be useful. Reasonable values of the polarization resistance may be obtained even if the Tafel slope estimates are poor or if extraneous experimental artifacts may be interfering with the measurement. The results of a corrosion evaluation of steel (UNS G10180) in the presence of aminotrimethylene phosphonic acid are used for this purpose. Details of the techniques, materials used, and results are given elsewhere [55]. The aqueous environment was an aggressive water to which the phosphonate was added.

Figures 85.15–85.17 show polarization resistance curves generated between − 20 and + 20 mV from the corrosion potential as measured at the time the polarization was performed. They were chosen because they demonstrate different experimental problems that can arise because of electrical interference. The scan rate was 0.1 mV/s. Figure 85.15 shows a polarization resistance curve for steel generated under static conditions under mildly heated conditions at 35°C. Figure 85.16 shows a polarization

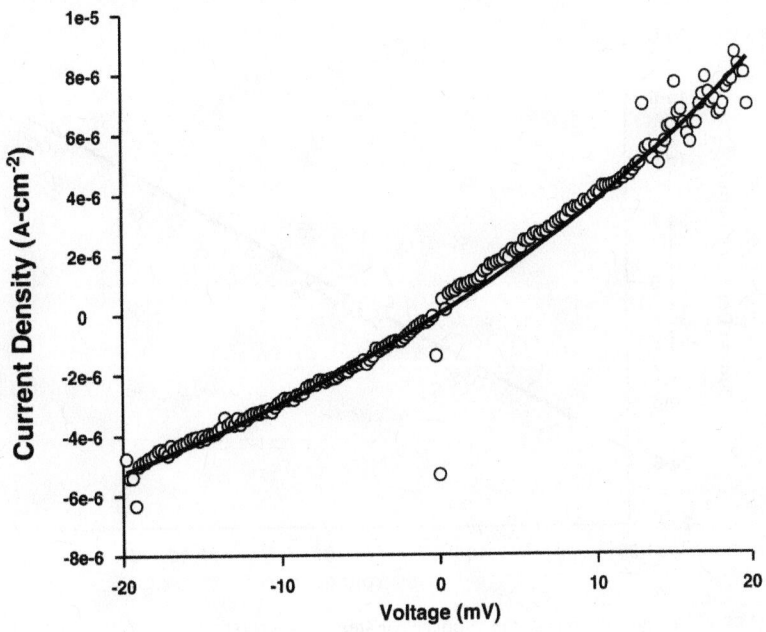

FIGURE 85.15. Polarization resistance curve for steel in aggressive water with 50 ppm DEQUEST 2000 at 32°C under state conditions [44]. (Reprinted from Proceedings of the 1995 Water Quality Technology Conference, by permission. Copyright © 1996, American Water Works Association.)

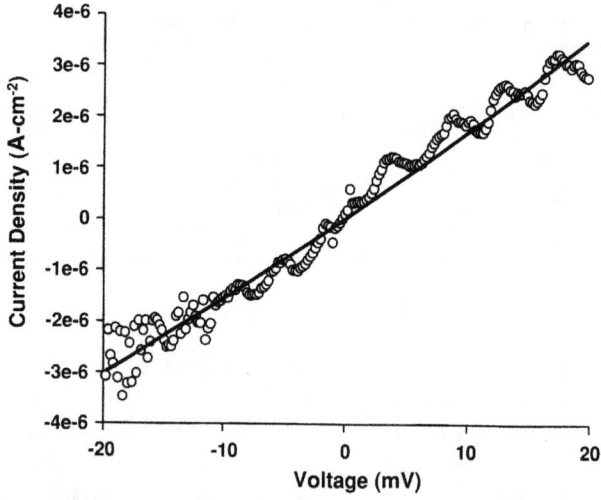

FIGURE 85.16. Polarization resistance curve for steel in aggressive water with 50 ppm DEQUEST 2000 at 35°C at 200 rpm in rotating-cylinder electrode [44]. (Reprinted from Proceedings of the 1995 Water Quality Technology Conference, by permission. Copyright © 1996, American Water Works Association.)

resistance curve for steel under dynamic conditions of 200 rpm using a rotating-cylinder electrode at 35°C (95°F). Figure 85.17 shows a polarization resistance curve for steel under dynamic conditions of 200 rpm using the same rotating-cylinder electrode but at 80°C (144°F). No attempt was made to compensate for solution resistance or to isolate the electrical interference.

The polarization curve was fit by computer to Eq. (85.3) using nonlinear regression. Table 85.4 lists the anodic and cathodic Tafel slopes b_a and b_c and the polarization resistance, R_p calculated from these curves. Tafel slopes in the range of 30–120 mV can be explained theoretically [40, 41] in terms of single electrochemical processes that allow for the application of Eqs. (85.3) and (85.4). Values much beyond this range raise warning flags. In the case of mass transfer control, one of the Tafel slopes should be infinite.

In terms of the values in Table 85.4, the anodic Tafel slope corresponding to the curve calculated in Figure 85.15 is reasonable. The cathodic Tafel slope most likely corresponds

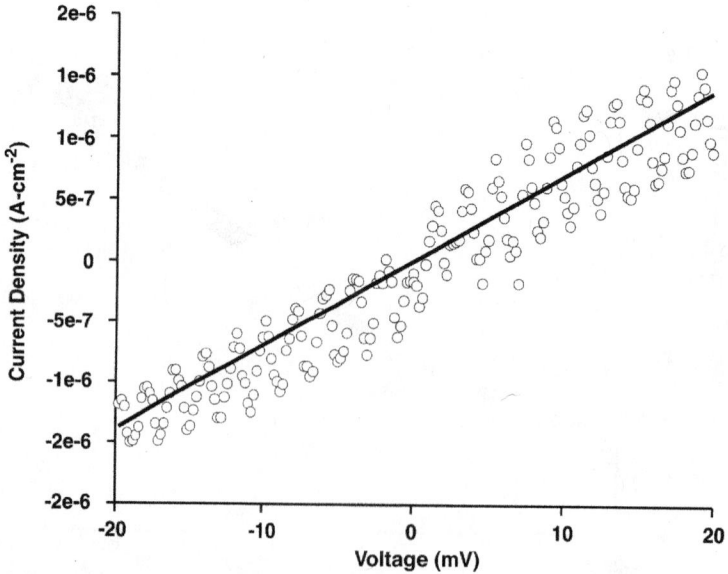

FIGURE 85.17. Polarization resistance curve for steel in aggressive water with 50 ppm DEQUEST 2000 at 80°C at 200 rpm in rotating-cylinder electrode [44]. (Reprinted from Proceedings of me 1995 Water Quality Technology Conference, by permission. Copyright © 1996, American Water Works Association.)

TABLE 85.4. Electrochemical Parameter From Curve Fit of Figures 85.14–85.16

Figure Number	Polarization Resistance ($\Omega \cdot cm^2$)	Tafel Slopes (V)	
		Anodic (b_a)	Cathodic (b_c)
85.15	3.01×10^3	9.3×10^{-2}	1.5×10^7
85.16	6.14×10^3	2.9×10^{-1}	2.3×10^5
85.17	1.44×10^4	8.5×10^6	1.8×10^9

to the mass-transfer-controlled process of oxygen reduction under the static conditions in the aerated, inhibited water. The appearance of the small sinusoidal variation superimposed on the curve in Figure 85.16 can make estimating the Tafel slopes and polarization resistance more difficult. In addition, the polarization resistance values are larger, suggesting a lower corrosion rate than above. Figure 85.17 shows the effects of still greater noise on the estimation of the Tafel slopes. Assuming the 25-mV proportionality, a polarization resistance of $>10^4 \, \Omega \cdot cm^2$ corresponds to a corrosion rate of under 2.5×10^{-2} mm/year (1 mpy). The anodic Tafel slopes estimated from Figure 85.16 and especially Figure 85.17 are large.

The corrosion currents estimated using the Tafel slopes are much greater than the corrosion currents of 10^{-6} A/cm^2 suggested by the polarization resistance values. The difference clearly shows that one should not curve fit Eq. (85.3)

to experimental data or substitute calculated Tafel slopes into Eq. (85.4) blindly. The practitioner should ensure that experimental artifacts are not interfering with the measurement.

One additional question is, "Are large errors in the estimated Tafel slopes necessarily associated with large errors in the polarization resistance?" The polarization resistance estimated using the polarization resistance technique was compared to that estimated from electrochemical impedance spectroscopy. The impedance spectra were generated on the same electrode used for the polarization resistance measurements. The analysis is described in detail elsewhere [55]. The results in Table 85.5 show that the polarization resistance values are in good agreement between the two techniques even though the Tafel slopes might be in error. Though significant error can be associated with Tafel slope estimates, the polarization resistance values may be reasonably accurate.

Changes in the polarization resistance may themselves be used as indicators of changes in the corrosion rate. Assuming a constant Tafel constant B [calculated as $(1/2.303)[b_a b_c/(b_a + b_c)]$ provides a reasonable way of estimating a corrosion rate. The value of 25 mV is a reasonable average value (± 15 mV) to use [40]. This approach has been used in commercial instruments using this technique for corrosion monitoring. Corrosion rates calculated by this method can differ from that calculated from mass loss by greater than a factor of 2.

TABLET 85.5. Comparison of Polarization Resistance Measurements^a

Environment and Temperature	Rotation Rate (rpm)	Estimated Polarization Resistance ($\Omega \cdot cm^2$)	
		Impedance	dc Polarization
Uninhibited, 35°C	200	8.78×10^2 (2 time constants)	9.30×10^7
	200	7.63×10^2 (2 time constants)	7.95×10^2
	200	7.55×10^2 (2 time constants)	1.02×10^2
	200	1.18×10^3 (2 time constants)	1.02×10^2
Uninhibited, 80°C	200	2.05×10^2	1.87×10^2
	200	2.32×10^2	2.08×10^2
	1000	1.38×10^2	1.28×10^2
	200	2.15×10^2	2.15×10^2
30 ppm DEQUEST 2000, 35°C	200	1.16×10^4	1.13×10^4
	200	1.78×10^4	1.44×10^4
	1000	2.02×10^4	1.68×10^4
	200	2.52×10^4	1.72×10^4
	1000	2.43×10^4	1.89×10^4
30 ppm DEQUEST 2000, 80°C	200	3.11×10^4	3.75×10^3
	200	1.12×10^4	6.14×10^3
	1000	1.33×10^4	6.59×10^3
	200	1.43×10^4	7.18×10^3
	1000	1.17×10^4	6.36×10^3

^{a}See [55]. Results correspond to exposure at increasing times. The first 200 rpm result is after 3–5 h of exposure. Afterward, spectra were generated at 200 rpm every 24 h. The spectrum at 1000 rpm was generated 1 h after the preceding spectrum at 200 rpm.

E. CORROSION PREDICTION USING ELECTROCHEMICAL IMPEDANCE SPECTROSCOPY

Electrochemical impedance spectroscopy (EIS) has become a routine tool for practical corrosion prediction. A number of reviews and tutorials have been written discussing the experimental setup, methodology for making the measurement, and methods for analyzing the data [4, 56–62]. Generating the spectra is now very easy with several "turnkey" commercial systems available. Some types of electrical interference as shown with the polarization resistance technique can have much less influence on the EIS measurement, especially when a good frequency response analyzer is used. Several software packages are available for fitting the spectra to analogous circuits [63, 64], a technique often used to analyze the data, and others have been reported in the literature [65, 66]. The technique has been established to the point that an ASTM standard has been written to provide the practitioner with a test method for ensuring that the electronic equipment, algorithm for generating the impedance spectra, and electrochemical cell are functioning properly [67]. This standard is geared to relatively low values of the polarization resistance so it may not be appropriate for ensuring that the equipment can examine coatings or other systems exhibiting an extremely high impedance.

The areas that have been demonstrated as appropriate for using EIS for corrosion measurements are:

Rapid estimation of corrosion rates (within 30 min to 24 h)

Estimation of extremely low corrosion rates and metal contamination rates ($<10^{-4}$ mm/year, <0.01 mpy)

Estimation of corrosion rates in low-conductivity media

Rapid assessment of corrosion inhibitor performance in aqueous and nonaqueous media

Rapid evaluation of coatings

Following are examples of practical applications of EIS encountered in industry. The emphasis is on situations in which either the environment is poorly defined or characterization of the corrosion mechanism could not be done beyond that needed for making a practical prediction. Application of EIS to coatings is not being considered in this section. References [58, 60] and [68–73] are provided as places to start for information on coating evaluation.

E1. Validity of Impedance Spectra

Before attempting to model impedance spectra, one must be assured that the spectra themselves are valid. When a sine wave is used as the perturbation, the relationship between the current and applied voltage can be characterized by the ratio between the amplitudes of the voltage and the current and the

phase shift between the rotating vectors which represent the instantaneous voltage and current. These two quantities are the modulus and phase shift of the vector (a complex number) representing the impedance [56, 62]. In mathematical terms, the impedance is a transfer function relating (the Laplace transform of) a response (e.g., current) to (the Laplace transform of) a perturbation (e.g., voltage). The transfer function can only become an impedance when the following four conditions are fulfilled [73]:

1. *Causality:* The response of the system must be a result only of the applied perturbation.
2. *Linearity:* The relationship between the perturbation and response is independent of the magnitude of the perturbation.
3. *Stability:* The system returns to its starting state after the perturbation is removed.
4. *Finite valued:* The transfer function (impedance) must be finite as the frequency approaches both 0 and ∞ and is continuous and finite valued at all intermediate frequencies.

Causality is difficult to verify. Agreement with the results in ASTM G-106 [67] to check equipment and algorithm would provide confidence that the signals are not arising from extraneous sources in the equipment and cell. Linearity is very easy to verify. The impedance spectra should be independent of the amplitude of the applied voltage (or current). Generating spectra with excitation amplitudes greater and less than that used and verifying that the modulus and phase angle values extracted from the spectra do not change with excitation amplitude at each frequency comprise the simplest method for checking linearity. Stability can be verified by generating spectra from high to low frequency and repeating from low to high frequency. Since the lowest frequency takes the longest to generate (e.g., 0.001 Hz requires $\sim$15 min for the measurement), this frequency would have the greatest chance of upsetting the system. If the spectra are the same, stability has not been violated.

One of the techniques that has been proposed to validate the polarization resistance is to use the Kramers–Kronig transformations. They enable calculation of the real contribution from the imaginary contribution and vice versa [74, 75]. In theory, as long as the four criteria cited above are valid, the contributions calculated by the transforms will equal the real and imaginary contributions as measured. The calculated and measured spectra will be identical. Unfortunately, if the impedance measured at the lowest frequency is far from the value that it would have at zero frequency and additional time constants are present in that region, the transforms cannot be applied easily [76]. Methods have been introduced that help the practitioner to assess the error structure of the measurement and give direction on the need to minimize stochastic "noise" and bias [77]. The reader is directed to the above references and those contained in those articles for more information about these techniques.

E2. Modeling of Impedance Spectra

Typically, the impedance spectra are modeled by assuming a circuit made up of resistors, capacitors, and inductors and then fitting that circuit to the spectra to extract values of the circuit elements. The values may then be related to physical phenomena to try to verify that the circuit model is a reasonable representation of the corrosion process. A constant-phase element is often introduced in the imaginary (capacitance) terms. It has the form $A(j\omega)^{\alpha}$. Some workers have tried to relate A to a capacitance and α to a degree of surface roughness or cell geometry. From a practitioner's standpoint, the constant-phase element serves as a way of ensuring that the capacitive contributions fit the data. From a mathematical standpoint, this parameter is actually two adjustable parameters, A and α.

Nonlinear regression is used to curve fit the analogous circuit model to the spectra. Care must be taken to ensure that the number of circuit elements (independent variables) does not exceed the number of data points or even approach the number of data points. Usually, the simpler the model that fits the data, the more likely it represents the physical process. Two types of circuits that are often used to fit spectra are shown in Figure 85.18 [(1) and (2)]. The first analogous circuit represents a single charge transfer reaction. The second represents an imperfectly covered electrode as might be found with an inhibitor, an adsorbed intermediate, or an imperfect coating. Pseudoinductance will be discussed in a later example. Note that the maximum number of adjustable parameters is kept small so that agreement between calculated and measured impedance spectra would hopefully reflect the adequacy of the model. Four adjustable parameters are present in the first model and seven in the second.

A word of caution is in order. The tendency might be to assume that good agreement between the measured impedance spectrum and that calculated from the modeling circuit means that the model used is the best representation of the corrosion process and provides an explanation for it. One cannot, however, assume the uniqueness of a circuit model merely on the basis of a good fit with the observed spectrum. Tables of mutually degenerate networks exhibiting identical total impedance over the entire frequency spectrum have been published [78]. These provide insight into the large number of circuits that can produce the same spectrum.

Another reason for two analogous circuits seeming to fit the same spectrum is that signal noise can sometimes prevent nonlinear regression from differentiating between two models because of the structure of the equations used for the models [79]. This phenomenon was recently

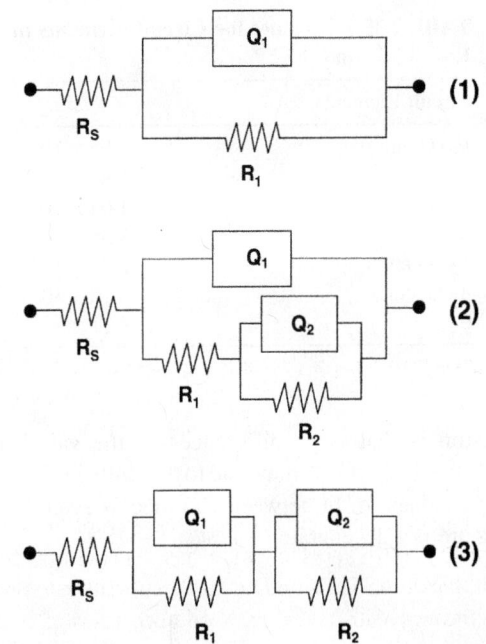

FIGURE 85.18. Simple circuit models for analyzing EIS spectra: (1) single-resistor and constant-phase element in parallel, (2) nested parallel combination of resistors and constant-phase elements, and (3) series of two parallel combinations of single-resistor and constant-phase element. See discussion for meaning of models.

demonstrated for the second and third models in Figure 85.18. Both models were curve fit to the same data using EQUIVCRT [63]. The resulting values for the elements are shown in Table 85.6. Both models yielded the same element values. Figures 85.19(a) and (b) show the measured and calculated impedance spectra from the two circuits. The reason for the similarity can be found by examining the structure of the equations for the two circuits. The equation for circuit 2 is

$$Z = R_\Omega + \left[\frac{R_1 R_2 Q_1 + R_1 Q_1 Q_2 + R_2 Q_1 Q_2}{R_1 R_2 + R_1 Q_2 + R_2 Q_2 (1 + Q_2/Q_1) + Q_1 Q_2} \right]$$

(85.7)

and for circuit 3 is

$$Z = R_\Omega + \left[\frac{R_1 R_2 Q_1 (1 + Q_2/Q_1) + R_1 Q_1 Q_2 + R_2 Q_1 Q_2}{R_1 R_2 + R_1 Q_2 + R_2 Q_1 + Q_1 Q_2} \right]$$

(85.8)

The only difference between these two equations is the location of the term $1 + (Q_2/Q_1)$. If $Q_2/Q_1 \ll 1$, then the term is $\simeq 1$. The magnitude of Q is the inverse of the value of A reported in Table 85.6. In this example, the quotient $A_1/A_2 = 0.017$, a number $\ll 1$. Since the experimental data have scatter, the effect of the small quotient is probably buried in the numerical "noise" of the regression. The

TABLE 85.6. Values for Circuit Elements in Eqs. (85.7) and (85.8)[a]

Circuit Element	Value
R_s ($\Omega \cdot cm^2$)	1.14×10
R_1 ($\Omega \cdot cm^2$)	1.87×10^2
A_1 (F/cm^2)	1.04×10^{-4}
α_1	8.88×10^{-1}
R_2 ($\Omega \cdot cm^2$)	3.72×10^4
A_2 (F/cm^2)	6.12×10^{-3}
α_2	8.91×10^{-1}

[a]See [79].

conclusion is that large differences in the values of the constant-phase elements may lead to difficulty in distinguishing the goodness of fit between two models even if the two models are not degenerate.

E3. Example of Pseudoinductive Behavior

Impedance spectra possessing pseudoinductive characteristics have been observed since about the time that the modern version of the technique was reported for corrosion prediction. Such behavior is characterized by a portion of the impedance spectrum appearing in the fourth quadrant when plotted in Nyquist format or when the impedance modulus decreases at low frequency coupled with negative phase angles when plotted in Bode format.

Deducing corrosion mechanisms in the presence of this behavior is controversial and requires care because phenomena other than corrosion might be causing the effect. For example, impedance spectra for iron in 1 M sulfuric acid with small amounts of propargyl alcohol exhibited such pseudoinductive behavior [80]. Later, this behavior was shown to be a result of a nonlinearity in the response to

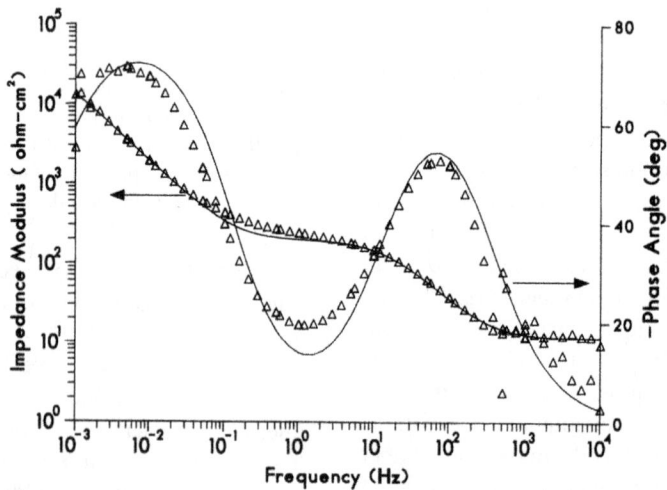

FIGURE 85.19a. Measured EIS spectrum versus that calculated using circuit 2 in Fig. 85.18 and parameters from Table 85.6. (From [79]. Copyright NACE International. Reprinted with permission.)

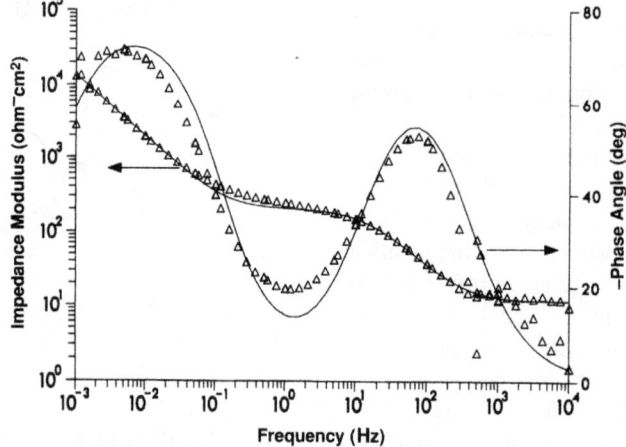

FIGURE 85.19b. Measured EIS spectrum versus that calculated using circuit 3 in Fig. 85.18 and parameters from Table 85.6. (From [79]. Copyright NACE International. Reprinted with permission.)

the voltage perturbation [81]. An irreversible desorption of the inhibitor occurred in the vicinity of the corrosion potential [82]. This particular system violated the criteria of linearity and stability. The pseudoinductive behavior was caused by violation of two of the criteria required for the quotient of the applied voltage divided by the measured current to be the impedance.

Corrosion rates estimated from the low-frequency limit of the impedance modulus might be in significant error because the decrease in the modulus can be large when pseudoinductive behavior is observed. An example of the care required is shown in the following example. In this case, an estimate of the corrosion rate was needed for steel handling a waste stream containing iminodiacetic acid which could form complexes with steel under process conditions [83] Figures 85.20(a) and (b) show a typical impedance spectrum from that study. The spectra were analyzed using Eq. (85.9)

$$Z = R_\Omega + \left[\frac{1}{R_t}\left(1 + (j\omega\tau_1)^\alpha\right) + \frac{1}{\rho(1+j\omega\tau_2)} \right]^{-1} \quad (85.9)$$

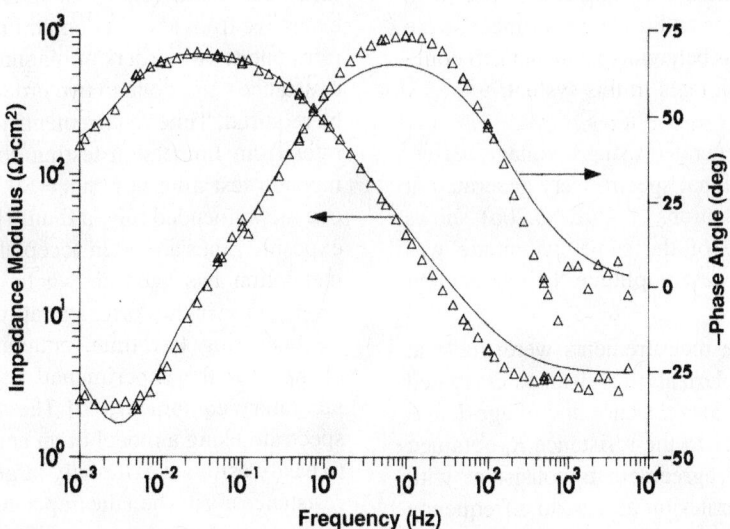

FIGURE 85.20a. The EIS spectrum of waste stream demonstrating pseudoinductive behavior. Spectrum generated at 5-mV perturbation peak height. (From [83]. Copyright NACE International. Reprinted with permission.)

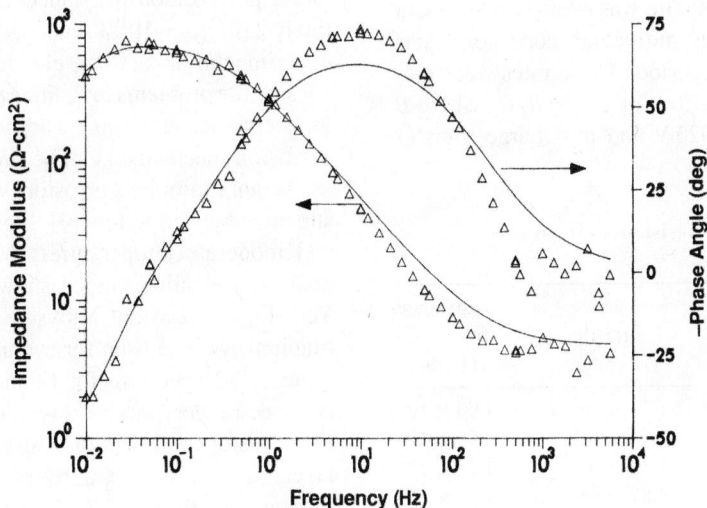

FIGURE 85.20b. The EIS spectrum of system in Fig. 85.19(a). Spectrum generated al 2-mV perturbation peak height. (From [83]. Copyright NACE International. Reprinted with permission.)

This equation was proposed to describe the impedance of iron in a solution containing 0.1 M sulfuric acid and 0.9 M sodium sulfate [84]. The derivation of Eq. (85.9) assumed that an adsorption process with electron transfer at equilibrium preceded the rate-determining step. Note that this analysis uses an equation different from that in the more generally available software. Such an approach does not mean the available software cannot be used for analyzing impedance spectra with pseudoinductive behavior. The example does indicate that, at times, the practitioner should look beyond that software.

Verifying that the criteria of stability and linearity are not violated is needed before attempting to use the spectra to estimate corrosion rates. Three additional experiments were run to aid in verifying that this behavior is real and EIS could be used to estimate corrosion rates in this system:

1. Figure 85.20a) was generated using a voltage perturbation of 5 mV. Additional spectra were generated at 2- and 10-mV perturbations. Figure 85.20b) shows that pseudoinductance of the same magnitude was observed even at the lowest amplitude. This agreement implies linearity.

2. Polarization resistance measurements were made at the same time. The polarization resistance estimated by curve fitting Eq. (85.3) to the current–voltage data is compared in Table 85.7 to the resistance R_t obtained using Eq. (85.9). The agreement is consistent with fulfilling the stability criterion at least to a frequency of ~0.001 Hz, equivalent to a scan rate of ~0.1 mV/s according to Eq. (85.2).

3. The experiments were run over a 24-h period and the corrosion rates were such that mass loss was measurable. Then corrosion rates calculated from the EIS analysis could be compared to those calculated from mass loss. The corrosion rate was estimated bom EIS by time averaging the individual corrosion rates determined during that period. These rates were calculated by using the relationship $i_{corr} = B/R_t$, where B was assumed to be 0.025 V and the charge transfer

process was assumed to involve two electrons. These rates showed good agreement with those estimated from mass loss of the same electrode (see reference for details). The agreement in rates as determined by two independent techniques supports that the corrosion process was, indeed, being measured by EIS.

E4. Low-Corrosion-Rate Estimation

Very often, concern is not with integrity of equipment but with contamination of the product by corrosion of the containment vessel. Corrosion rates acceptable from a structural standpoint (e.g., 0.01 mm/year, or 0.4 mpy) may be excessive from a metal contamination standpoint. Corrosion rates one to two orders of magnitude lower and polarization resistance values one to two orders of magnitude higher may be required. Time requirements to obtain accurate corrosion rates from Immersion testing could be excessive (e.g. immersion test time in hours = 2000/corrosion rate in mpy is one recommended rule of thumb [85] though shorter implied exposure times are often acceptable). This particular recommendation has been shown to guarantee negligible error propagation in the corrosion rate calculation caused by errors in measuring test time, coupon dimensions, and weight change for the experimental accuracy routinely found in laboratory equipment [86] The ability to fit the impedance spectrum using a model of an analogous circuit enables one to easily estimate extremely large values of the polarization resistance even when the impedance value at lowest frequency measured still has a significant capacitive contribution. EIS has been demonstrated to enable such rates to be estimated within 24 h, the delay being determined by the time required for the corrosion process to reach steady state, the corrosion potential to become constant [66]. The advantages over the polarization resistance technique are that much larger polarization resistance values can be measured routinely (10^7 to ~10^{10} $\Omega \cdot cm^2$ depending on equipment and experimental apparatus), electrical interference of the type causing the problems in Figures 85.16 and 85.17 can be easier to overcome, and some additional information about the corrosion mechanism can be provided very quickly.

As an example, corrosion was examined in a process stream subjected to low pH, higher chloride concentrations, and moderate temperatures (60–75°C) (116–137°F). The goal was an alloy choice showing virtually no corrosion. Very highly resistant alloys (e.g., UNS R52400, Grade 7 titanium) were chosen for evaluation. Figure 85.21 shows a typical EIS spectrum for titanium in the process streams. All spectra were modeled by a simple parallel combination of a resistor (polarization resistance) and capacitor (double layer) in series with a resistor (uncompensated solution resistance). Table 85.8 shows some of the polarization resistance values for several of the alloys, all measured after 24 or so hours of exposure. A value of 0.025 V was assumed

TABLE 85.7. Comparison of Resistances from EIS and DC Measurements[a]

Experiment No.	Exposure (h)	Resistance from EIS ($\Omega \cdot cm^2$)	Resistance from DC ($\Omega \cdot cm^2$)
1	4	7.21×10^2	4.90×10^2
	24	4.09×10^2	3.05×10^2
2	4	5.10×10^2	4.36×10^2
	24	4.87×10^2	4.10×10^2
3	24	2.82×10^2	2.27×10^2

[a]See [83].

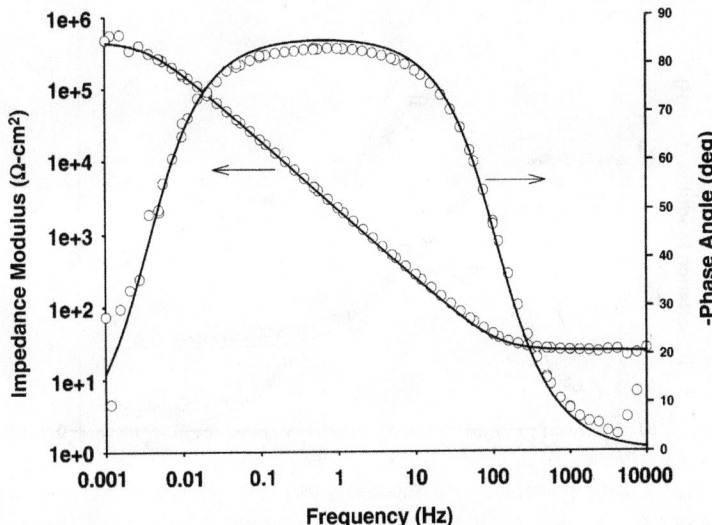

FIGURE 85.21. The EIS spectrum of UNS R52400 (Grade 7 titanium) in process stream demonstrating ability to estimate extremely low corrosion rates. Spectrum modeled by circuit 1 in Fig. 85.18.

TABLE 85.8. Low-Corrosion-Rate Estimation by EIS

Alloy and Process Stream	Polarization Resistance ($\Omega \cdot cm^2$)	Corrosion Rate [mm/year (mpy)]
Tantalum slurry product	1.7×10^6	1.0×10^{-4} (0.004)
Grade 7 titanium slurry product	5.2×10^5	4.2×10^{-4} (0.02)
Zirconium 702 slurry product	4.2×10^6	6.8×10^{-5} (0.003)
Grade 7 titanium condensate	4.5×10^5	4.8×10^{-4} (0.02)

TABLE 85.9. Order-of-Magnitude Estimates of Corrosion Rates for Iron- and Nickel-Based Alloys

Polarization Resistance ($\Omega \cdot cm^2$)	Corrosion Rate [mpy (mm/year)]
10	1×10^3 (25)
10^2	1×10^2 (2.5)
10^3	1×10 (0.25)
10^4	1 (0.025)
10^5	1×10^{-1} (0.0025)
10^6	1×10^{-2} (0.00025)
10^7	1×10^{-3} (0.000025)

for the proportionality between corrosion current and the inverse of the polarization resistance because the Tafel slopes could not be determined. Corrosion rates this low are virtually impossible to measure by mass loss. Even if the error in the estimates is significant (an order of magnitude), the results do provide guidance for the types of metal loss and contamination that might be observed. Table 85.9 shows order-of-magnitude estimates of corrosion rates as a function of polarization resistance values for iron- and nickel-based alloys assuming a value of 0.025 V for "B" and a two-electron process. This approach can be used to provide reasonable *estimates* of corrosion rates when Tafel slopes are unavailable or not easily obtained.

E5. Rapid Estimation of Corrosion Rates in Complex Systems

The use of passive circuit elements to curve fit impedance spectra so that the polarization resistance can be estimated

means that, in principle, corrosion rates can be estimated rapidly. That rather simple models can be fit to spectra from systems exhibiting complex chemistry or in systems for which the chemistry is unknown means that EIS allows for rapid corrosion estimation in complex, poorly characterized systems [87]. The extreme example of a poorly characterized environment is a waste stream which contains all of the unknown components in a plant. The example below is a corrosion study in such environments made even more complicated because the waste streams could be combined [88].

Seven different waste solutions were examined. The effect of fluid velocity was examined by coupling the EIS measurement with a rotating-cylinder electrode and estimating polarization resistance as a function of rotation rate. Figures 85.22 and 85.23 show EIS spectra for two waste solutions. The spectrum in Figure 85.22 was modeled using a parallel combination of a resistor and capacitor in series with a resistor. That in Figure 85.23 was modeled using Eq. (85.9). EIS spectra were generated daily over several days. The

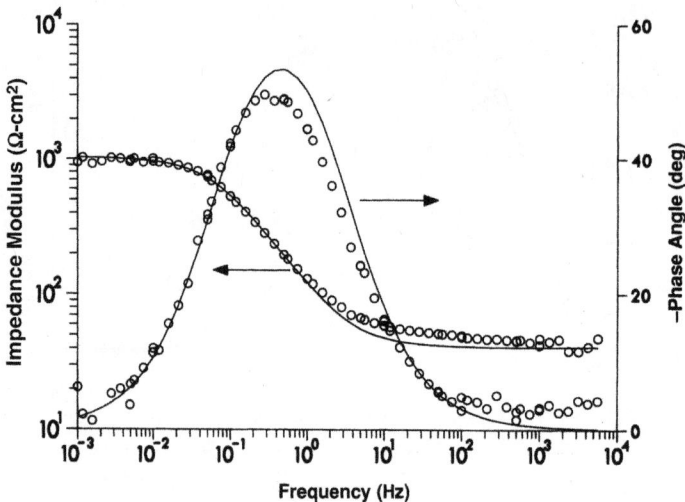

FIGURE 85.22. The EIS spcctram of waste solution 7 in Table 85.10 at 200 rpm. Spectrum modeled by circuit 1 in Fig. 85.18 [88]. (Reprinted from Electrochimica Acta, Vol. 38, D. C. Silverman, "Corrosion Prediction in Complex Environments Using Electrochemical Impedance Technique," p. 2075, Copyright 1993, with permission from Elsevier Science.)

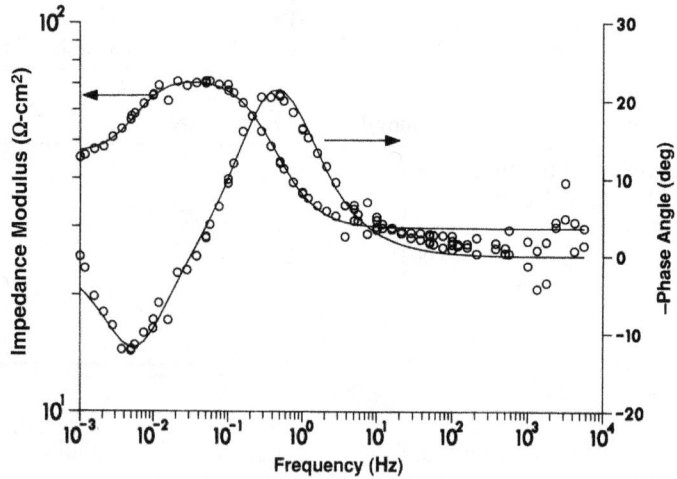

FIGURE 85.23. The EIS spectrum of waste solution 4 in Table 85.10 at 200 rpm. Spectrum modeled using Eq. (85.9) [88]. (Reprinted from Electrochimica Acts, Vol. 38, D. C. Silverman, "Corrosion Prediction in Complex Environments Using Electrochemical Impedance Technique," p. 2075, Copyright 1993, with permission from Elsevier Science.)

corrosion rates were estimated by assuming a proportionality constant of 0.025 V between the polarization resistance and the corrosion current because of difficulties in estimating reasonable Tafel slopes under the dynamic conditions in these complex systems. The time-averaged corrosion rate was compared to that estimated from mass loss of the electrode. The latter measurement is useful because it provides an independent check on the corrosion rates estimated by EIS, especially important when making measurements in complex, poorly characterized environments or where pseudoinductance is observed. The direct current (DC) measurements

were also made, with the polarization curves being modeled by Eq. (85.3) to provide yet another check on the results.

Table 85.10 shows the time-averaged corrosion rates from EIS compared to those from mass loss. The agreement provides strong support that the polarization resistance values represent the corrosion process. The agreement means that the pseudoinductance in Figure 85.23 is not an artifact and is caused by the corrosion process. Further, this agreement means that polarization resistance values measured as a function of rotation rate are valid and might provide information on velocity sensitivity of corrosion. Figure 85.24

TABLE 85.10. Comparison of Time-Averaged Corrosion Rates as Function of Waste Solution[a]

Waste Solution	Time-Averaged Corrosion Rate			
	Impedance		Mass Loss	
	mm/y	mpy	mm/y	mpy
1 (high tar)	0.36	14.0	0.30	12.0
2 (low tar)	0.44	17.0	0.38	15.0
3 (low tar)	0.12	4.7	0.30	12.0
4 (high tar)	8.5	335.0	7.4	291.0
5 (high tar)	1.6	63.0	1.8	71.0
6 (high tar)	0.15	5.9	0.19	7.4
7 (high tar)	0.30	12.0	0.43	17.0

[a]See [88].

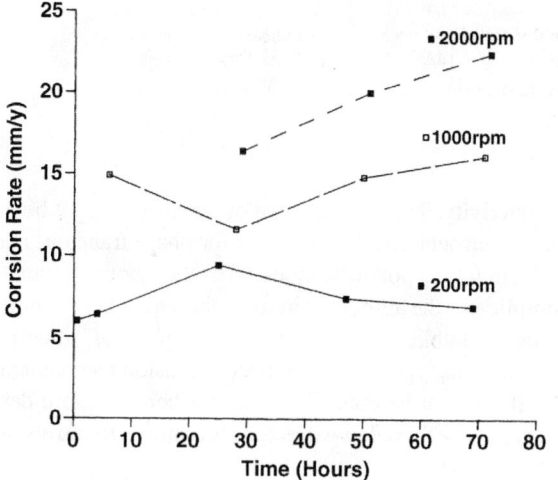

FIGURE 85.24. Corrosion rote as a function of time and rotation rate for solution 4 in Table 85.10. (Reprinted from Electrochimica Acta, Vol. 38, D. C. Silverman, Corrosion Prediction in Complex Environments Using Electrochemical Impedance Technique, p. 2075, Copyright 1993, with permission from Elsevier Science.)

shows the corrosion rate as a function of rotation rate measured at various times during immersion for the environment of Figure 85.23. Fluid motion must be taken into account when considering corrosion of steel in this environment. Section E6 provides some information on using electrochemical measurements to estimate the influence of fluid velocity on corrosion.

One of the needs for the impedance spectra to be valid is steady-state conditions. Monitoring of the corrosion potential can aid in determining if and when steady state has been achieved. "The question is, "How close to steady state must one be in order to use EIS for practical corrosion screening?" Often, rapid screening forces one to try to generate spectra

prior to reaching actual steady state. Under these circumstances, the concern is if the electrochemistry and reaction rates change enough during generation of the spectrum so that the lower frequency portion represents chemistry different from the higher frequency portion. If this inconsistency occurs, the higher frequency portion of the spectrum might be extrapolated to low frequencies to estimate the polarization resistance if no additional low-frequency time constants are present. This type of extrapolation was required for some of the spectra obtained after shorter exposure times. An alternative approach might be to use DC polarization at short exposure times because this technique requires less measurement time. EIS would be used only after longer exposure times. Though phase angle information would be absent, this approach would enable estimation of the polarization resistance after short exposure times.

E6. Evaluation of Corrosion Inhibitors

Corrosion inhibitors are prevalent throughout industry functioning in many diverse applications. EIS offers a powerful technology for evaluating these compounds, often in real time. The literature is saturated with studies of corrosion inhibitors. Corrosion inhibition under near-neutral pH conditions offers certain challenges not found under more acidic conditions. The reason is that this type of inhibition has been characterized, at least for low-alloy steels, as interphase inhibition [89, 90]. This type of inhibition is characterized by a three-dimensional protective layer between the metal and the electrolyte. Corrosion behavior depends on exposure time and conditions (e.g., fluid motion). The three-dimensional matrix takes time to develop, resulting in a slow drift in corrosion potential. Fluid motion can affect the rate at which the inhibitor or oxygen reaches the surface. In addition, one side benefit of using EIS for this system is that use of a frequency response analyzer can sometimes circumvent the electrical noise that causes DC curves of the type shown in Figures 85.16 and 85.17 resulting in more accurate values of the polarization resistance.

As discussed in Section D and shown in Table 85.4, though estimating the Tafel slopes and the value for B can be difficult, reasonable agreement can be achieved between the polarization resistance values determined by EIS and by *DC* polarization. The question becomes how to estimate corrosion rates from only the polarization resistance on a routine basis. One approach to obtaining a corrosion rate number is to assume some reasonable value for the proportionality constant B (e.g., 0.025 V) and use it for all measurements as discussed previously, understanding that the error could be 50% or greater.

Another approach is to plot the value of $1/R_p$ (or $1000/R_p$ to make the ordinate usually lie between 0.1 and 10) versus the variable under consideration [90], This procedure provides a practical way to track the inhibitor behavior with time

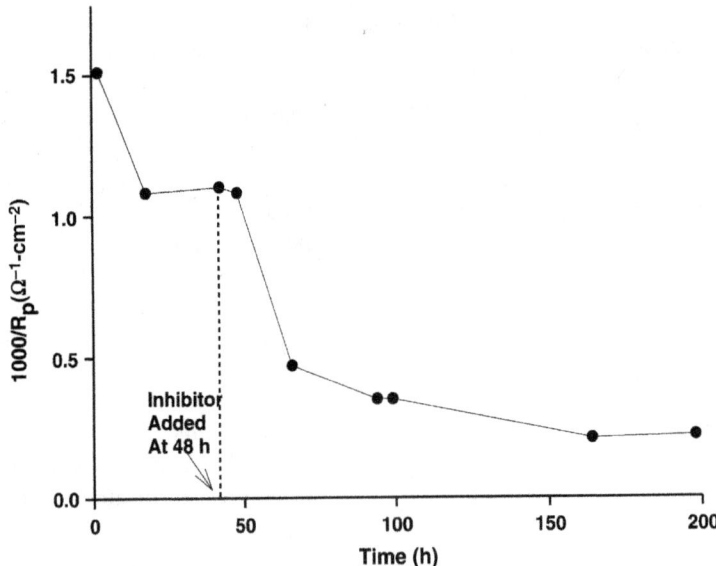

FIGURE 85.25. Reciprocal of the polarization resistance versus time for steel in an aggressive water solution under static conditions. The inhibitor was added after 48 h [44]. (Reprinted from Proceedings of the 1995 Water Quality Technology Conference, by permission. Copyright © 1996. American Water Works Association.)

as a function of variables such as inhibitor and velocity. An example is tracking inhibitor behavior on a surface that has already suffered corrosion. Figure 85.25 shows such a plot for the polarization resistance as measured by EIS for corrosion of steel prior to and after addition of an inhibitor. This figure demonstrates the corrosion inhibitor monitoring capability with EIS when the polarization resistance values are near the upper limit for polarization resistance measurements for the particular experiment. Since the Tafel slopes might have changed after addition of the inhibitor, the absolute differences may be somewhat approximate. But the procedure is no worse than assuming a constant value for B and provides a practical way to assess relative differences in inhibitor efficacy.

F. TESTING FOR VELOCITY-SENSITIVE CORROSION—ROTATING CYLINDER ELECTRODE

Fluids are very rarely stagnant. Under appropriate conditions, fluid flowing past a metallic surface may accelerate its rate of corrosion. Four general mechanisms have been proposed: mass transport–controlled corrosion, phase transport–controlled corrosion; erosion–corrosion usually involving suspended solid particulates, and cavitation corrosion [91]. Erosion–corrosion and cavitation corrosion are beyond the scope of this chapter. Phase transport–controlled corrosion involves corrosion in the presence of more than one phase in which each phase has differing corrosivity and often differing

conductivity. Though this type of corrosion might be examined electrochemically, the need for phase transport and not molecular transport to be modeled by the laboratory apparatus complicates the ability to simulate the actual field conditions. Another publication provides examples of attempts to examine this type of flow-induced corrosion mechanism [92] and it will not be considered further here. A more detailed discussion of fluid flow–affected corrosion is found elsewhere in this volume.

A mechanism very suitable to evaluation and prediction by electrochemical techniques is mass transport affected corrosion in single-phase fluids. It is the only one considered in this section. Numerous examples exist, corrosion of steel in sulfuric acid [93, 94] and steel in oxygenated water without inhibitors [95] being two. The question is how best to simulate this type of corrosion. Under some conditions the exact geometry must be modeled. Under others, one geometry can be used to simulate the mechanism of another. A recent state-of-the-art report compares in detail four techniques for examining flow-induced corrosion in the presence of a single–phase fluid [96], The techniques described are the rotating-disk electrode, rotating cylinder electrode, impinging jet (in stagnation region), and flow loop. Each of the techniques can be equipped to measure corrosion electrochemically. Table 85.11, taken from [96], shows the strengths and weaknesses of each of the techniques for this type of study.

Choice of the appropriate laboratory geometry to use to model the field geometry suffering velocity–sensitive corrosion depends on both geometries and their fluid mechanics. The reader should consult the references provided at the end

TABLE 85.11.   Summary of Characteristics of Systems for Studying Single-Phase Flow-Induced Corrosion[a]

Characteristic	Rotating Disk	Rotating Cylinder	Impinging Jet	Flow Loop
Flow regime usually tested	Laminar	Turbulent	Stagnation	Laminar and turbulent
Well-characterized flow pattern	Yes	Yes	Yes	Yes (beyond entrance length)
Uniform mass transfer	No	Yes	Yes	No
Uniform primary cunent distribution	No	Yes/No	No	No
Uniform shear distribution	No	Yes	No	No
Possibility of noise at rotating contacts	Yes	Yes	No	Yes
Possibility of wobble-induced errors	Yes	Yes	No	No
Requires pump	No	No	Yes	Yes
Potential for leaks	Low	Low	High	High
Suitable for application of in situ surface analysis	No	No	Yes	Yes
Suitable for well-defined studies of mass transfer	Yes	Yes	Yes	Yes
Suitable for well-defined studies of shear	No	Yes	Yes	No
Suitable for qualitative screening studies	Yes	Yes	Yes	Yes

[a]See [96].

of this chapter and the chapter on fluid flow–affected corrosion elsewhere in this volume as starting points to determine which type of geometry might best model the system being investigated [97–102]. Using the rotating cylinder electrode as the example, the following discussion is meant to provide an overview of how proper flow conditions might be constructed in the laboratory to model the corrosion mechanism, if not the corrosion rate, in the field situation.

How conditions are chosen depends on how detailed the information has to be. Sometimes, the answer only has to be qualitative. For example, corrosion of steel in an aqueous solution tends to be sensitive to fluid velocity because of mass transport of oxygen or inhibitors or shear-induced removal of inhibitor films. The need may be only to demonstrate that such a sensitivity exists. Under these circumstances, the experimenter need only measure the corrosion rate at several velocities. Figure 85.24 is an example of this qualitative approach. Plotting the reciprocal of the polarization resistance versus time and rotation rate as in that figure only shows that corrosion is affected by fluid velocity but does not indicate if the mechanism is one of mass transfer control. To define the mechanism requires estimation of the mass transfer coefficient from the corrosion current and comparison to the expected mass transfer coefficient for that fluid velocity. Such a calculation requires the determination of the concentration difference of the species being transferred between the surface and bulk. The unknown nature of the environments in the previous example makes such a determination difficult, forcing the more qualitative approach.

Following is a methodology that might be used if the corrosion mechanism is known or suspected to be under mass transfer control. The fluid boundary layer is assumed to be fully developed in both geometries and the surface is assumed to be hydraulically smooth. A hydraulically smooth surface is one in which all protuberances on the surface caused by nonuniformity are smaller than the thickness of the laminar sublayer portion of the boundary layer immediately adjacent to the wall. The assumption in this discussion is that the rotating cylinder electrode is the laboratory apparatus. Equations in Table 85.12 have been shown to relate velocities in the rotating cylinder electrode to several other geometries so that the mass transfer coefficients are the same in both. If the flow rate is known in the field geometry, then the appropriate velocity could be used in the rotating cylinder electrode to maintain equal mass transfer coefficients. Then, the polarization resistance technique or EIS could be used to obtain the polarization resistance. The corrosion current would be estimated from that parameter ($i_{corr} = B/R_p$). This current would be expected to be the same in the two geometries because the mass transfer coefficients are the same and the corrosion mechanisms are the same. The corrosion current is related to the mass transfer coefficient by

$$i_{corr} = nFk_{mt}(C_{Bulk} - C_{wall}) \qquad (85.10)$$

Often the concentration either at the wall or in the bulk is assumed to be zero, the other being at saturation.

Rearranging Eq. (85.10) so that the mass transfer coefficient is estimated from the corrosion current would enable an assessment to be made of the degree of mass transfer control. The logarithm of the Sherwood number ($Sh = k_{mt}D/d$) when plotted as a function of the logarithm of the Reynolds number ($Re = \rho vd/\mu$) should be a straight line consistent with the Sherwood–Reynolds relationship for that geometry. Equation (85.11) has often been used to determine if data generated in the rotating cylinder electrode follows complete mass transfer control [106]:

$$Sh = 0.079 \, Re^{0.7} \, Sc^{0.356} \qquad (85.11)$$

As an example, Figure 85.26 shows actual data from a study of UNS S44627 (26 wt % chromium and about 1 wt %

TABLE 85.12. Rotating Cylinder Electrode Velocity Versus Geometry to Maintain Equal-Mass-Transfer Coefficients

Geometry	Relationship	Source
Pipe	$u_{cyl} = 0.1185 \left[\left(\dfrac{\mu}{\rho} \right)^{-0.25} \left(\dfrac{d_{cyl}^{3/7}}{d_{pipe}^{5/28}} \right) Sc^{-0.0857} \right] u_{pipe}^{5/4}$	Silverman [98]
Pipe	$u_{cyl} = 0.1066 \left[\left(\dfrac{\mu}{\rho} \right)^{-0.229} \left(\dfrac{d_{cyl}^{3/7}}{d_{pipe}^{1/5}} \right) Sc^{-0.0371} \right] u_{pipe}^{1.229}$	Nesic et al. [103]
Pipe	$u_{cyl} = 0.1786 \left[\left(\dfrac{\mu}{\rho} \right)^{-0.211} \left(\dfrac{d_{cyl}^{0.346}}{d_{pipe}^{0.135}} \right) Sc^{-0.0808} \right] u_{pipe}^{1.211}$	Silverman [27]
Annulus	$u_{cyl} = 0.1185 \left[\left(\dfrac{\mu}{\rho} \right)^{-0.25} \left(\dfrac{d_{cyl}^{3/7}}{d_{annulus}^{5/28}} \right) Sc^{-0.0857} \right] u_{annulus}^{5/4}$	Silverman [98]
Wall jet region of impinging jet	$u_{cyl} = 0.7939 \left[\left(\dfrac{\mu}{\rho} \right)^{-0.2} \left(\dfrac{d_{cyl}^{3/7}}{d_{jet}^{0.229}} \right) Sc^{-0.0329} \left(\dfrac{x}{d_{jet}} \right)^{-1.71} \right] u_{jet}^{1.2}$	Silverman [105]

molybdenum, the balance iron) in several higher concentrations of sulfuric acid [94]. The mass transfer coefficients were calculated from both current densities on polarization scans at the anodic current limit and mass loss. The mechanism was assumed to have a rate-limiting step of iron mass transfer from a surface saturated in iron sulfate similar to that for steel [94]. The agreement with Eq. (85.11) is reasonable. Under conditions of complete mass transfer control the

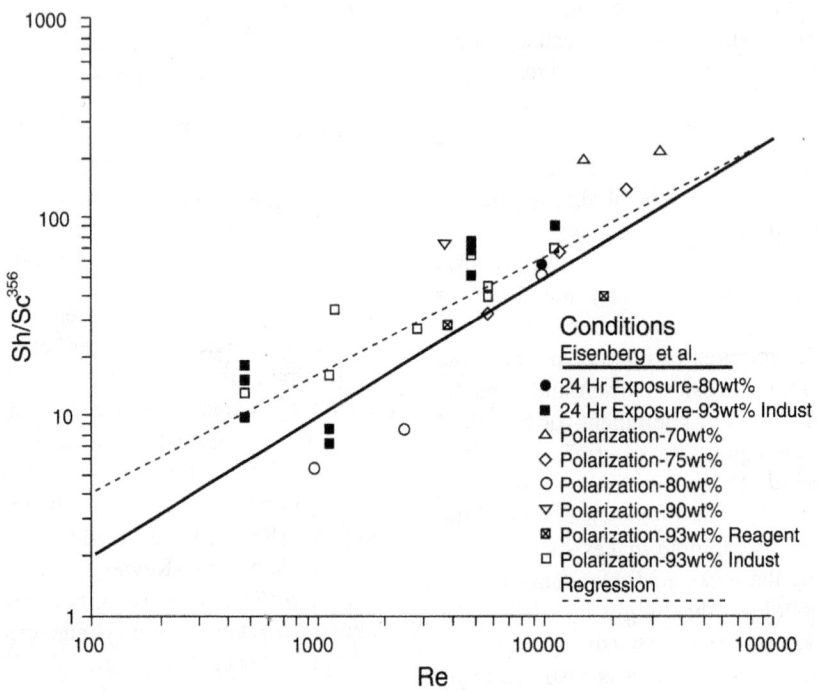

FIGURE 85.26. Sherwood number versus Reynolds number relationship for UNS S44627 (26 wt % chromium and ~1 wt % molybdenum, the balance iron) in various concentrations of sulfuric acid. The Eisenberg et al. [106] correlation is included for comparison ref [94]. (Copyright NACE International. Reprinted with permission.)

TABLE 85.13.   Reported Sherwood Number Versus Reynolds Number Relationships for Hydrodynamically Smooth Cylinder

$Sh = a\,Re^b\,Sc^c$	Experimental Conditions	Reference
$Sh = 0.0791\,Re^{0.7}\,Sc^{0.356}$	Ferricyanide–ferrocyanide on nickel electrode, $10^3 < Re < 10^5$	Eisenberg, Tobias, and Wilke [106]
$Sh = 0.0964\,Re^{0.7}\,Sc^{0.356}$	Ferricyanide–ferrocyanide on platinum electrode, $10^3 < Re < 2 \times 10^4$	Morrison, Striebel, and Ross [107]
$Sh = 0.219\,Re^{0.645}\,Sc^{0.27}$	Theoretical calculation compared to data from ferricyanide–ferrocyanide on nickel electrode, $3 \times 10^4 < Re < 1.3 \times 10^6$	Kishinevskii and Kornienko [108]
$Sh = 0.0791\,Re^{0.67}\,Sc^{0.356}$	Theoretical calculation with constant assumed to be 0.0791	Gabe and Robinson [109]
$Sh = 0.0791\,Re^{0.69}\,Sc^{0.41}$	Cathodic deposition of copper from copper sulfate $10^4 < Re \lesssim 5 \times 10^5$	Robinson and Gabe [110]
$Sh = 0.217\,Re^{0.6}\,Sc^{0.333}$	Copper-copper sulfate deposition on copper electrode $10^3 < Re < 1.9 \times 10^4$	Ariva, Carrozza, and Marchiano [111]
$Sh = 0.184\,Re^{0.62}\,Sc^{0.333}$	Compilation of 290 mass transfer data points $100 < Re < 10^5$	Singh and Mishra [112]
$Sh = 0.127\,Re^{0.64}\,Sc^{0.333}$	Oxygen reduction on monel under cathodic polarization $\sim 200 < Re < 2 \times 10^5$	Cornet and Kappesser [113]
$Sh = 0.165\,Re^{0.654}\,Sc^{0.333}$ $Sh = 0.144\,Re^{0.654}\,Sc^{0.356}$	Oxygen reduction in sodium chloride on monel and sodium sulfate on steel $2000 < Re < 1.2 \times 10^5$	Silverman [114] (exponent on Sc assumed)
$Sh = 0.0489\,Re^{0.748}\,Sc^{0.356}$	Derived from curve fit of Theodorsen and Regier [115] over range of $200 < Re < 4 \times 10^5$	Silverman [105] (exponent on Sc assumed to be 0.356)

rotating cylinder electrode coupled with electrochemical measurements is capable of demonstrating mass transfer–controlled corrosion and of providing reasonable predictions of field performance.

The deviation in slope in Figure 85.26 from that expected for a smooth cylinder was attributed at the time to an increase in surface roughness caused by corrosion [95]. That may not be the case. While Eq. (85.11) is reasonable, it is not the only equation that might be used to determine complete mass transfer control. A recent review [105] has revealed a number of mass transfer correlations that might be used. These are shown in Table 85.13. All are linearizations of what in reality is a curved relationship between the Sherwood and Reynolds numbers even when plotted on log–log coordinates. Figure 85.27 shows the relationship among the equations. The curved line was developed using the Chilton–Colburn

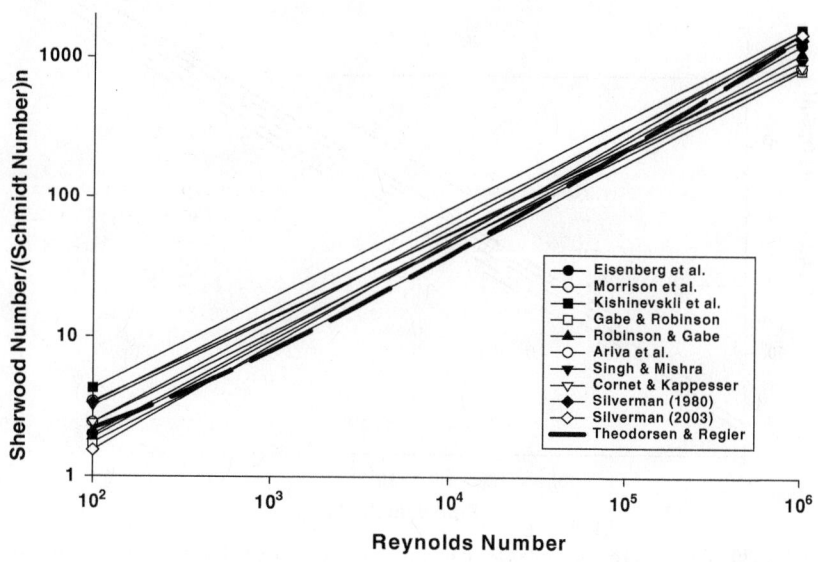

FIGURE 85.27. Comparison of Sherwood number vs. Reynolds number correlations for a hydraulically smooth rotating cylinder. See Table 85.13 to match references to correlation. (Reproduced from [119] with permission. © NACE International, 2008.)

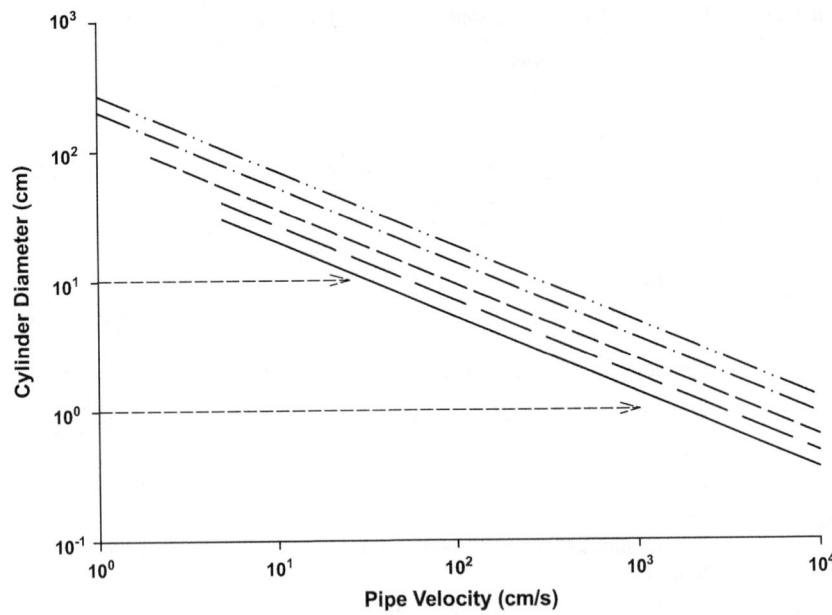

FIGURE 85.28. Cylinder diameter as function of pipe velocity to fulfill simultaneous equality of mass transfer and wall shear stress. The region of experimentally practical cylinder diameters lies between the two arrows. Pipe diameters: (——) 5 cm, (– –) 10 cm, (- - - - - -) 20 cm, (–·–·–) 50 cm, (–··–··–··) 100 cm. (Reproduced from [118] with permission. © NACE International, 2005.)

modification applied to the original friction factor equation developed by Theodorsen and Regier assuming that the exponent on the friction factor in the modification is 1 [105, 115]. The relative difference between the largest and

smallest estimates at a Reynolds number of 100 is about a factor of 3 and the difference at a Reynolds number of 10^6 is about a factor of 2. While the equation derived by Eisenberg et al. [Eq. (85.11)] does provide a reasonable representation

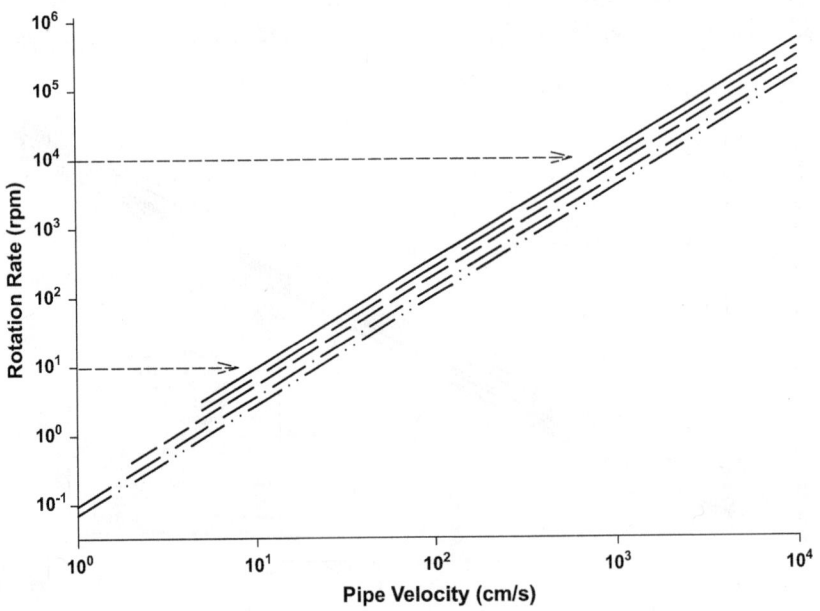

FIGURE 85.29. Cylinder rotation rates as function of pipe velocity to fulfill simultaneous equality of mass transfer coefficient and wall shear stress. The region of experimentally practical rotation rates lies between the two arrows. Pipe diameters: (——) 5 cm, (– –) 10 cm, (- - - - - -) 20 cm, (–·–·–) 50 cm, (–··–··–··) 100 cm. (Reproduced from [118] with permission. © NACE International, 2005.)

of the relationship and a reasonable estimate of the mass transfer coefficient, the estimate is just that, an estimate. At this point the "best" analytical equation of the form of Eq. (85.11) cannot be defined, because even if the best equation could be defined, it would still be a straight-line approximation (in log–log coordinates) to what is actually a curved-line relationship.

Requiring similarity in wall shear stress [116, 117] or similarity in mass transfer coefficients [98, 103] has been proposed as a way to establish flow conditions (rpm) within the rotating cylinder electrode that could enable predictions of velocity-sensitive corrosion mechanisms in other geometries. Which approach is most appropriate is a subject of debate and controversy. For fully developed turbulent flow, the two approaches are linked because of the relationship between the friction factor and the mass transfer coefficient. Recently, approximate equations were derived that could relate conditions in the rotating cylinder electrode to those in a pipe while maintaining simultaneous equality of both mass transfer coefficients and fluid wall shear stress [118]. Figures 85.28 and 85.29 show the relationship between diameters as a function of pipe velocity and the relationship between velocities as a function of pipe diameter. The lines in the figures have been drawn so that all pipe Reynolds numbers are turbulent. All rotating cylinder electrode Reynolds numbers are turbulent.

The results imply that some combinations of pipe diameter, pipe velocity, rotating cylinder diameter, and rotation rate exist so that the mass transfer coefficients and the wall shear stresses can be made similar simultaneously. Fulfilling both criteria where possible might circumvent the need to consider whether wall shear stress or mass transfer coefficient is the appropriate modeling parameter. This derivation is somewhat approximate [119]. The reader should consult [119] for details. Experimental confirmation is needed to show that conditions implied result in simultaneous similarity of both fluid shear stresses at the wall and mass transfer coefficients in the two geometries.

G. CONCLUSIONS

Electrochemical techniques provide a powerful assortment of tools that the corrosion practitioner can use to assess corrosion and make field predictions from that assessment The speed with which such information becomes available means that often the assessment can be made in real time or shortly after a question has been asked. All predictive techniques have shortcomings. The practitioner is cautioned that before using such techniques the pertinent literature be read and understood. The information and references provided here should be helpful in that regard Then, when using the techniques, the practitioner would be best served by making at least two independent measurements of corrosion, preferably one electrochemical and one nonelectrochemical (mass loss usually being the easiest to implement).

H. LIST OF SYMBOLS (by equation)

Equation (85.1)

C_{dl} = double-layer capacitance (μ_F/cm^2)

f_b = breakpoint frequency (Hz)

R_Ω = uncompensated solution resistance ($\Omega \cdot cm^2$)

R_p = polarization resistance ($\Omega \cdot cm^2$)

Equation (85.2)

f_b = breakpoint frequency (Hz)

V_{pp} = peak-to-peak voltage (V)

S_{max} = maximum scan rate (V/s)

Equation (85.3)

i = current density (A/cm^2)

i_{corr} = corrosion current density (A/cm^2)

V = applied potential (V)

V_{corr} = corrosion potential

b_a = anodic Tafel slope (V)

b_c = anodic Tafel slope (V)

Equation (85.4)

i_{corr} = corrosion current (A/cm^2)

b_a = anodic Tafel slope (V)

b_c = anodic Tafel slope (V)

R_P = polarization resistance ($\Omega \cdot cm^2$)

Equation (85.5)

W = Wagner number (dimensionless)

L = characteristic length dimension (cm)

κ = conductivity (mho/cm)

V = potential (V)

i = current density (A/cm^2)

Equation (85.6)

W = Wagner number (dimensionless)

R_Ω = uncompensated solution resistance ($\Omega \cdot cm^2$)

R_p = polarization resistance ($\Omega \cdot cm^2$)

Equations (85.7 and 85.8)

Z = impedance ($\Omega \cdot cm^2$)

R_Ω = uncompensated solution resistance ($\Omega \cdot cm^2$)

R_1 = resistance of one subcircuit ($\Omega \cdot cm^2$)

R_2 = resistance of one subcircuit ($\Omega \cdot cm^2$)

Q_1 = constant phase element ($\mu F/cm^2$)

Q_2 = constant phase element ($\mu F/cm^2$)

Equation (85.9)

Z = impedance ($\Omega \cdot cm^2$)

R_Ω = uncompensated solution resistance ($\Omega \cdot cm^2$)

R_t, = charge-transfer resistance ($\Omega \cdot cm^2$)

τ_1 = time constant (s)

τ_2 = time constant (s)

ω = frequency (rad/s)

ρ = resistance ($\Omega \cdot cm^2$)

$j = \sqrt{-1}$

Equation (85.10)

i_{corr} = corrosion current density (A/cm^2)

n = number of electrons transferred per atom (equiv/mol)

F = Faraday constant (9.649×10^4C/equiv)

k_{mt} = mass transfer coefficient (cm/s)

C_{bulk} = concentration in bulk fluid (mol/cm^3)

C_{wall} = concentration at surface (mol/cm^3)

Equation (85.11)

Re = Reynolds number (dimensionless), (Re = $\rho vd/\mu$)

Sc = Schmidt number (dimensionless), (Sc = $\mu/\rho d$)

Sh = Sherwood number (dimensionless), (Sh = $k_{mt}D/d$)

μ = absolute viscosity (g/cm-s)

ρ = density (g/cm^3)

d = characteristic length dimension (diameter) (cm)

k_{mt} = mass transfer coefficient (cm/s)

D = diffusion coefficient (cm^3/s)

v = velocity (cm/s)

REFERENCES

1. M. G. Fontana, Corrosion Engineering, 3rd ed., McGraw-Hill, New York, 1986.
2. G. S. Haynes and R. Baboian (Eds.), Laboratory Corrosion Tests and Standards, ASTM STP 866, American Society for Testing and Materials, Philadelphia, PA, 1985.
3. R. Baboian (Ed.), Electrochemical Techniques for Corrosion Engineering, NACE International, Houston, TX, 1987.
4. D. C. Silverman, Corros. Rev., **10**(1–2), 31 (1992).
5. M. Pourbaix, Atlas of Electrochemical Equilibria in Aqueous Solutions, NACE International, Houston, TX, 1974.
6. D. C. Silverman, "Derivation and Application of EMF–pH Diagrams," in Electrochemical Techniques for Corrosion Engineering, R. Baboian (Ed.), NACE International, Houston. TX, 1987.
7. R. W. Potter, and A. J. Clynne, Chem. Eng. Data, **25**, 50 (1980).
8. L. G. Sillen and A. E. Martell (Eds.), Stability Constants of Metal-Ion Complexes, Special Publication No. 17, The Chemical Society, Burlington House, London, 1964; Stability Constants of Metal-Ion Complexes, Special Publication No. 25, The Chemical Society, Burlington House, London, UK, 1971.
9. C. R. Kusik, and H. P. Meissner, AlChE Symp. Set, **173**, 14 (1978).
10. M. H. Fronig, M. E. Shanley, and E. D. Verink, Corros. Sci., **16**, 371 (1976).
11. HSC Chemistry for Windows, Version 3.0, Outakumpu Research Oy, Finland.
12. G. G. Naumov, B. N. Ryzenko, and I. L. Khodakovsky, Handbook of Thermodynamic Data, USGS Translation, USGS-WRD-74-001, 1974.
13. D. D. Wagman, W. H. Evans, V. B. Parker, I. Halow, S. M. Bailey, and R. H. Schumm,NBS Technical Notes 270-3, 270–4 (270-5, 270-6, 270-7, 270-8, NIST, Gaithersburg, MD, 1971–1978.
14. C. F. Baes, and R. E. Mesmer, Hydrolysis of Cations, Wiley, New York, 1976.
15. A. E. Martell, R. M. Smith, and R. J. Motekaitis, NIST Critical Stability Constants of Metal Complexes Database, NIST Standard Reference Database, No, 46, Gaithersburg, MD, 1993.
16. D. C. Silverman and A. L. Silverman,"Potential p-H (Pourbaix) Diagrams as Aids for Screening Corrosion Inhibitors and Sequestering Agents", Corrosion, **66**(5), 055003-1 (2010).
17. D. C. Silverman, Corrosion, **38**(10), 541 (1982).
18. ASTM, Standard G61 "Standard Test Method for Conducting Cyclic Potentiodynamic Polarization Measurements for Localized Coirosion Susceptibility of Iron-, Nickel-, or Cobalt-Based Alloys," Annual Book of ASTM Standards, Vol. 03.02, ASTM, West Conshohocken, PA, 2003.
19. M. Pourbaix, L. Klimzack-Mathieiu, Ch. Menens, J. Meunier, Cl. Vanleugenhaghe, L. de Munck, J. Laureys, L. Neelemans, and M. Warzee, Corros. Sci., **3**, 239 (1963); M. Pourbaix, Corrosion, **26**(10), 431 (1970).
20. E. M. Rosen, and D. C. Silverman, Corrosion, **48**(9), 734 (1992).
21. D. C. Silverman,"Tutorial on Cyclic Potentiodynamic Polarization Technique," presented at CORROSION/98, San Diego, CA. Match 23–28, 1998 (reprint of paper available from NACE International, Houston, TX).
22. B. E. Wilde and E. Williams, Electrochim. Acta, **16**, 1971 (1971).
23. N. G. Thompson and B. C. Syrett, Corrosioiu **48**(8), 649 (1992).
24. E. L. Liening, "Trouble Shooting Industrial Corrosion Problems with Electrochemical Testing Techniques," in Corrosion Monitoring in Industrial Plants Using Non-Destructive Testing and Electrochemical Methods, G. C Moran and P. Labine (Eds.), STP 908, ASTM, West Conshohocken, PA, 1986, p. 289.
25. R. L. Martin, Mater. Perform., **18**(3), 41 (1979).
26. S. D. Chyou and H. C. Shin, Bull. Hectrochera., **3**(1), 1 (1987).

27. R. E. McGuire and D. C. Silverman, Corrosion, **47**(11), 894 (1991).

28. ASTM, Standard G-5, "Standard Reference Test Method for Making Potentiostatic and Potentiodynamk Anodic Polarization Measurements," ASTM Annual Book of Standards, Vol. 03.02, ASTM, Philadelphia, PA, 2007.

29. F. Mansfeld, Corrosion, **38**(10), 556 (1982).

30. D. C. Silverman, Corrosion, **40**(5), 220 (1984).

31. F. Mansfeld, Corrosion, **37**(6), 301 (1981).

32. F. Mansfeld, and M. Kendig, Corrosion, **37**(9), 545 (1981).

33. O. L. Riggs, Jr. and C. E. Locke, Anodic Protection: Theory and Practice in the Prevention of Corrosion, Plenum, New York, 1981.

34. D. Bennion, Chem. Eng., CE419, Dec. 1970.

35. W. P. Banks and M. Hutchison, Mater. Perform., **7**(9), 31 (1968).

36. B. Under, L. Berthagen. B. Wallen, and G. Schou, MariChem. **4**, 159 (1983).

37. Z. A. Foroulis, I. & E. C. Process Design Development, **3**(1), 84 (1964).

38. Z. A. Foroulis, I. & E. C. Process Design Development, **4**(1), 20 (1965).

39. ASTM, Standard G-59, "Standard Practice for Conducting Potentiodynamic Polarization Resistance Measurements," ASTM Annual Book of Standards, Vol. 03.02, ASTM, Philadelphia, PA, 2007.

40. F. Mansfeld, "The Polarization Resistance Technique for Measuring Corrosion Currents," in Advances in Corrosion Engineering and Technology, M. G. Fontana and R. W. Staehle (Eds.), Vol. 6, Plenum, New York, 1976, Chapter 3.

41. F. Mansfeld, "Polarization Resistance Measurements-Today's Status," in Electrochemical Techniques for Corrosion Engineering, R. Baboian (Ed.), NACE International, Houston, TX, 1987.

42. G. L. Cooper, "Sensing Probes and Instruments for Electrochemical and Electrical Resistance Corrosion Monitoring," in Monitoring in Industrial Plants Using Non-Destructive Testing and Electrochemical Methods, G. C. Moran and P. Labine (Eds.), STP 908, ASTM, West Conshohocken, PA, 1986, p. 237.

43. F. Mansfeld and K. B. Oldham, Corros. Sd., **11**, 287 (1971).

44. J. R. Scully,"The Polarization Resistance Method fiw Determination of Instantaneous Corrosion Rates: A Review," Paper 304, presented at CORROSION/98, San Diego, Mar. 1998 (reprint of paper available from NACE International, Houston, TX). Additional information on experimentally induced artifacts in D. C. Silverman, "Measuring Corrosion Rates in Drinking Water by Linear Polarization—Assumptions and Watch-outs," Proceedings of the 1995 Water Quality Technology Conference, American Water Works Association, New Orleans, 1996.

45. G. Rocchini, Corros. Sci., **36**(6), 1063 (1994).

46. M. Jensen and D. Blitz, Corros. Sci., **32**(3), 285 (1991).

47. R. L. LeRoy, Corrosion, **31**(5), 173 (1975).

48. M. Stern and E. D. Weisart, Proc. Am. Soc. Testing Mater., **59**, 1280 (1959).

49. C. Wagner, J. Electrochem. Soc., **101**, 225 (1959).

50. L. L. Scribner and S. R. Taylor (Eds.), The Measurement and Correction of Electrolyte Resistance in Electrochemical Tests, ASTM STP 1056, American Society for Testing and Materials, West Conshohocken, PA, 1990.

51. D. C. Silverman and J. E. Carrico, Corrosion, **44**(5), 280 (1988).

52. G. Rocchini, Corros. Sd., **38**(6), 823 (1996).

53. E. Schmauch, and E. Buck, "Computerized Electrochemical Corrosion Monitoring," Paper No. 270, CORROSION/87, NACE International, Houston, TX, 1987.

54. ASTM, Standard G-102, "Standard Practice for Calculation of Corrosion Rates and Related Information from Electrochemical Measurements," ASTM Annual Book of Standards, Vol. 03.02, American Society for Testing and Materials, West Conshohocken, PA, 2007.

55. D. C. Silverman, "Corrosion Prediction from Circuit Models–Application to Evaluation of Corrosion Inhibitors," in Electrochemical Impedance-Analysis and Interpretation, J. R. Scully, D. C. Silverman, and M. W. Kendig (Eds,), ASTM STP 1188, American Society for Testing and Materials, West Conshohocken, PA, 1993.

56. C. Gabrielli, "Electrochemical Impedance Spectroscopy: Principles, Instrumentation, and Applications," in Physical Electrochemistry, I. Rubenstein (Ed.), Marcel Dekker, New York, 1995, p. 243.

57. C. Gabrielli, M. Keddam, and H. Takenouti, "The Use of AC Impedance Techniques in the Study of Corrosion and Passivity," in Treatise in Mater. Sci. Technol., H. Herman, (Ed.), Vol. 23, Academic, San Diego, CA, 1983, p. 395.

58. J. R. Scully, D. C. Silverman, and M. W. Kendig, Electrochemical Impedance: Analysis and Interpretation, ASTM STP 1188, American Society for Testing and Materials, West Conshohocken, PA, 1993.

59. J. R. Macdonald, Impedance Spectroscopy, Wiley, New York, 1987.

60. F. Mansfeld, "Evaluation of Corrosion Protection with Electrochemical Impedance Spectroscopy," Paper No. 481, CORROSION/87, San Francisco, CA, Mar. 1987 (preprint available from NACE International), Houston, TX.

61. M. E. Orazem,"A Tutorial on Impedance Spectroscopy," Paper No. 302, CORROSION/98, San Diego, CA, Mar., 1998 (preprint available from NACE International), Houston TX.

62. D. C. Silverman, "Primer on the AC Impedance Technique," in Electrochemical Techniques for Corrosion Engineering, R. Baboian (Ed.), NACE International, Houston, TX, 1986, p. 73.

63. B. Boukamp, E$_q$C Equivalent Circuit for Windows, Version 1.2, 2005. Further description is in B. A. Boukamp, Solid State Ionics, 20, 31 (1986).

64. J. R. Macdonald, CNLS Complex Non-Linear Least Squares Immittance Fining Program, LEVM, University of North Carolina, Chapel Hill, NC, 1990.

65. M. W. Kendig, E. M. Meyer, G. Lindbergh, and F. Mansfeld, Corros. Sci., **23**(9), 1007 (1983).

66. D, C. Silverman and J. E. Carrico, Corrosion, **44**(5), 280 (1988).

67. ASTM, Standard G106, "Standard Practice for Verification of Algorithm and Equipment for Electrochemical Impedance Measurements," Annual Book of ASTM Standards, Vol. 03.02, American Society for Testing and Materials, West Conshohocken, PA, 2007.

68. U. Rammelt and G. Reinhard, Prog. Org. Coatings, **21**, 205 (1992).

69. G. W. Walter, Corros. Sci., **26**(9), 681 (1986).

70. G. W. Walter, D. N. Nguyen, and M. A. D. Madurasinghe, Electrochim. Acta, **37**(2), 245 (1992).

71. H. P. Hack and J. R. Scully, J. Electrochem. Soc., **138**(1), 33 (1991).

72. J. H. W. de Wit, H. J. W. Lenderink, D. H. van der Weijde, and E. P. M. van Westing, Mater. Sci. Forum, **192–194**, 253 (1995).

73. J. N. Murray and H. P. Hack, "Electrochemical Impedance of Organic Coated Steel; Final Report—Correlation of Impedance Parameters with Long-Term Coating Performance," DTRC-SME-89/76, David Taylor Research Center, Bethesda, MD, Sept. 1990.

74. M. Urquidi-Macdonald, S. Real, and D. D. Maedonald, J. Electrocbem. Soc., **133**(10), 2018 (1986).

75. R. L. Van Meirhaeghe, E. C Dutoit, F. Cardon, and W. P. Gomes, Electrochim. Acta, **20**, 995 (1975).

76. H. Shin and F. Mansfeld, Corros. Sci., **28**(9), 933 (1988).

77. M. Duroha, M. E. Orazem, and L. H. Garcia-Rubio, J. Electrochem. Soc., **144**(1), 48 (1997).

78. S. Fletcher, J. Electrochem. Soc., **141**, 1823 (1994).

79. D. C. Silverman, Corrosion, **47**(2), 87 (1991).

80. I. Epelboin, M. Keddam, and H. Takenouti, J, Appl, Electrochem., **2**, 71 (1972).

81. W. J. Lorenz and E Mansfeld, Corros. Sci., **21**, 647 (1981).

82. M. W. Kendig, J. Electrochem, Soc., **131**(12), 2777 (1984).

83. D. C. Silverman, Corrosion, **45**(10), 824 (1989).

84. I. Epelboin and M. Keddam, J. Electrochem. Soc., **117**, 1052 (1970).

85. ASTM, Standard G-31, "Standard Practice for Laboratory Immersion Corrosion Testing of Metals," Annual Book of ASTM Standards, Vol. 03.02, American Society for Testing and Materials, West Conshohocken, PA, 2007.

86. R. A. Freeman and D. C. Silverman, Corrosion. **48**(6), 463 (1992).

87. D. C. Silverman, Corrosion, **46**(7), 589 (1990).

88. D. C. Silverman, in Proceedings of the 2nd International Symposium on Electrochemical Impedance Spectroscopy, D. D, Macdonald (Ed.), Hectrochim. Acta, **38**(14),2075 (1993). These proceedings of the 2nd International Symposium on Electrochemical Impedance Spectroscopy provide a rich source of additional applications of EIS.

89. W. J. Lorcnz and E Mansfeld, Electrochim. Acta, **31**, 467 (1986).

90. F. Mansfeld, M. W. Kendig, and W. J. Lorenz, J. Electrochem. Soc., **132**(2), 290 (1985).

91. E. Heitz, Corrosion. **47**(2), 135 (1991).

92. K. J. Kennelly, R. H. Hausler, and D. C. Silverman (Eds.), Flow-Induced Corrosion: Fundamental Studies and Industrial Experience, NACE International, Houston, TX, 1991.

93. B. T. Ellison and W. R. Schmeal, J. Electrochem. Soc., **125**, 524 (1978).

94. D. C. Silverman and M. E. Zen, Corrosion, **42**(11), 633 (1986).

95. B. K. Mabato, C. Y. Cha, and L. W. Shemilt, Corros. Sri., **20**, 421 (1980).

96. "State-of-the-Art Report on Conftollsd-Flow Laboratory Corrosion Tests," Technical Report prepared by NACE Task Group T5A-31 on "Fluid Enhanced Corrosion," NACE Publication 5A195, NACE International, Houston, TX, 1995.

97. B. Poulsou, Corros. Sci., **23**(4), 391 (1983).

98. D. C. Silverman, Corrosion, **44**(1), 42 (1988).

99. V. G. Levich, Physiochemical Hydrodynamics, Prentice-Hall, Englewood Cliffs, NJ, 1962.

100. R. C. Allrire, Chem. Eng. Commun., **38**, 401 (1985).

101. M. W. E. Coney, "Erosion–Corrosion: The Calculation of Mass-Transfer Coefficients," Laboratory Note No. RD/L/N 197/80, Central Electricity Research Laboratories, Leatherhead, Surrey, UK. 1980.

102. S. Nesic. G. T. Solvi, and J. Enerhaug, Corrosion, **51**(10), 773 (1995).

103. S. Nesic, G. T. Solvi, and S. Skjerve, Br. Corros. J., **32**(4), 269 (1997).

104. D. C. Silverman, Corrosion, **59**(3), 207 (2003).

105. D. C. Silverman, Corrosion, **60**(11), 1003 (2004).

106. M. Eisenberg, C. W. Tobias, and C. R. Wilke, J. Electrochem. Soc., **101**, 306 (1954).

107. B. Morrison, K. Striebel, and P. N. Ross, J. Electroanal. Chem., **215**, 151 (1986).

108. M. Kh. Kishinevskii and T. S. Kornienko, Elektrokhim., **25**(6), 821 (1989).

109. D. R. Gabe and D. J. Robinson, Electrochim. Acta, **17**, 1121 (1972).

110. D. J. Robinson and D. R. Gabe, Trans. Inst. Met Finish., **48**, 35 (1970).

111. A. J. Arvia, J. S. W. Carrozza, and S. L. Marchiano, Electrochim. Acta, **9**, 1483 (1964).

112. P. C. Singh and P. Mishra, Chem. Eng. Sci., **35**, 1657 (1980).

113. I. Cornet and R. Kappesser, Trans. Inst. Chem. Eng., **47**, T194 (1969).

114. D. C. Silverman, unpublished results, Monsanto, St. Louis, MO, 1980.

115. T. Theodorsen and A. Regier,"Experiments on Drag of Revolving Disks, Cylinders, and Streamline Rods at High Speeds," Nat. Advisory Comm. Aeronaut., Report No. 793, U.S. Government Printing Office, Washington, DC, 1945, p. 367.

116. D. C. Silverman, Corrosion, **40**, **5**, 221 (1984).

117. K. D. Efird, E. J. Wright, J. A. Boros, and T. G. Hailey, Corrosion, **49**, 2, 993 (1993).

118. D. C. Silverman, Corrosion, **61**, 6, 515 (2005).

119. D. C. Silverman, Corrosion, **64**, 8, 627 (2008).

86

ELECTROCHEMICAL NOISE

D. A. Eden*

Revised and updated by Q. J. Meng, M. Mendez, and M. Yunovich

Honeywell Corrosion Solutions, Houston, Texas

A. INTRODUCTION

Electrochemical noise (ECN) has become a powerful tool for both corrosion science and corrosion engineering. ECN consists of low-frequency, low-amplitude fluctuations of potential and current associated with electrochemical corrosion processes. It was observed many years ago that certain types of corrosion processes, in particular localized corrosion phenomena, had characteristic signatures. Changes in the free corrosion potential could be observed and correlated with localized attack. In itself, this proved to be a useful tool which could be used to identify changes in the behavior of materials in particular environments.

---

* Deceased

Uhlig's Corrosion Handbook, Third Edition, Edited by R. Winston Revie
Copyright © 2011 John Wiley & Sons, Inc.

The current noise is associated with discrete dissolution events on the corroding metal surface, while the potential noise is associated with the action of current noise on an interfacial impedance. The potential and current noise signals can be measured simultaneously or at different times. During the ECN measurements, no external potential or current signal is applied. As a result, the ECN measurements are passive monitoring of the corrosion system at *freely corroding* potential. Electrochemical noise resistance and corrosion mechanisms can be derived from the analysis of electrochemical noise.

The foundations of electrochemical noise technology for corrosion studies lie in the original work undertaken by Iverson in 1968 [1]. Initial work studied the fluctuations of electrochemical potential [2–5]. Subsequently the combination of electrochemical potential and current noise arising from the coupling current between two nominally identical electrodes was investigated in detail [6–9]. In 1986, Eden et al. introduced the concept of electrochemical noise resistance and described statistical methods for analysis of electrochemical noise [10, 11]. In the 1990s, important advancements in ECN analysis techniques were made. Mansfeld et al. [12, 13] introduced the concept of spectral noise resistance (impedance) and studied the spectral noise resistance in the frequency domain. Bertocci et al. [14–16] further used power spectral density to study the spectral noise resistance on a firmer theoretical basis. The spectral noise resistance at the zero-frequency limit and the polarization resistance were found to be well correlated [14, 15].

This chapter will overview the measurements, data analysis, and interpretation of ECN. In addition, applications of ECN for real-time corrosion monitoring in various environments will be briefly reviewed.

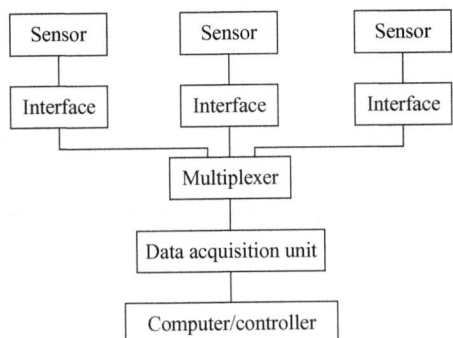

FIGURE 86.1. Electrochemical noise schematic.

B. NOISE MEASUREMENTS

Systems used for measuring electrochemical noise signals have several key components that depend on the type of noise measurements being undertaken. A typical schematic arrangement is shown in Figure 86.1.

It is common practice to use a three-electrode cell (sensor). The noise signals being measured may require some form of signal conditioning, and this is included in the interface. For the purposes of logging data from a series of sensors, a multiplexer is often incorporated. This may be configured in two particular ways, either before or after the signal conditioning units; the latter is shown in the diagram. The data acquisition unit and the computer controller may be either an integrated subsystem or separate devices.

B1. Electrode Assembly

The sensors may be configured so as to measure either potential or current noise in isolation or, as is more usual, for simultaneous measurements. Two typical three-electrode cell configurations are shown in Figures 86.2 and 86.3.

Figure 86.2 illustrates a system that could be used for *potentiostatic or galvanostatic* control (monitoring current or potential fluctuations, respectively). This type of

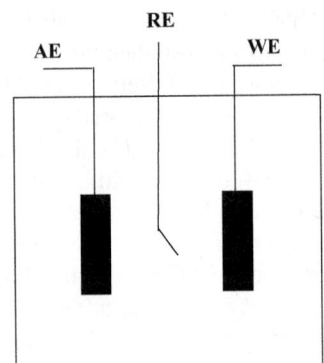

FIGURE 86.2. Three-electrode arrangement for potentiostatic or galvanostatic control: WE, working electrode; RE, reference electrode; AE, auxiliary electrode.

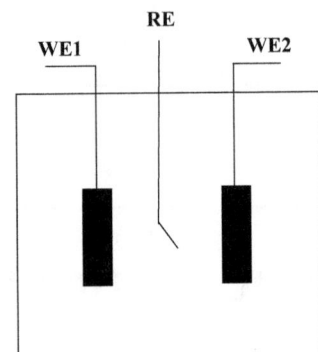

FIGURE 86.3. Electrode arrangement for electrochemical noise measurements.

arrangement is often used for laboratory studies, particularly for accelerated testing of material susceptibility to a variety of failure mechanisms (e.g., SCC and pitting attack) within defined potential regimes.

Figure 86.3 illustrates a typical configuration of electrodes used for noise measurements at the *corrosion potential* such that naturally occurring current and potential fluctuations may be monitored simultaneously. This type of arrangement is useful for studying the evolution of naturally occurring corrosion processes and is widely used in plant monitoring/surveillance.

Under certain circumstances it is possible to engineer a differential working electrode arrangement such that effects of stress, differential aeration, sensitization, and so on, can be simulated. One such case would be the study of susceptibility of materials to stress corrosion cracking; in this instance the sensor arrangement would be a sample of unstressed materials as one working electrode and a sample of stressed materials as the second working electrode.

B2. Interfaces

There are basically three different types of interfaces that can be used to follow the evolution of the electrochemical processes. Standard potentiostatic or galvanostatic interface arrangements are illustrated in Figures 86.4 and 86.5, respectively.

In the potentiostatic mode (Fig. 86.4), the potential of the working electrode with respect to the reference electrode

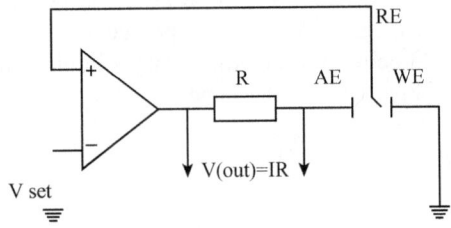

FIGURE 86.4. Potentiostatic arrangement for measurement of current noise at controlled potentials.

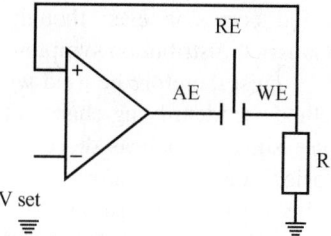

FIGURE 86.5. Galvanostatic arrangement for measurement of potential noise under constant current.

is controlled to the set potential. The current required to maintain the controlled potential flows through the sensing resistor R. It is advisable in these circumstances to monitor both the potential and current in the cell in order that the intrinsic noise arising from the potentiostat can be assessed. Evaluation of intrinsic instrumentation noise is easily achieved by substituting a resistor network for the electrode array.

In the galvanostatic mode (Fig. 86.5), the system acts to maintain the voltage across R at the set potential; thus, the current through the cell is kept at a constant value. The variation in potential of the working electrode is measured directly with respect to a reference electrode. The instrumentation component of the potential and current fluctuations can be assessed in a similar manner to that for the potentiostatic configuration by substituting a resistor network for the electrodes.

For measurements at the free corrosion potential, it is possible to measure potential and current fluctuations together or independently using an interface arrangement as illustrated in Figure 86.6. The Zero-resistance ammeter (ZRA) acts to maintain the potentials of the two working electrodes at the same potential (with modern devices the potential difference can be maintained within a microvolt). The current required to maintain the two working electrodes at the same potential flows through the feedback resistor R, and the voltage output of the ZRA is related to the current flowing through the system by $V = IR$. The potential buffer is

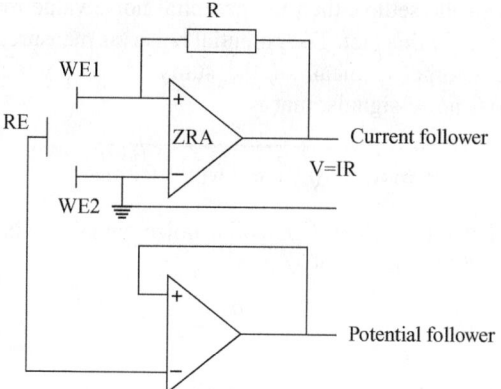

FIGURE 86.6. Interface for simultaneous monitoring of potential and current noise.

often used to ensure that there is no current drain on the reference electrode. Ideally, with no cells connected, the ZRA will give a signal of 0 V; however, intrinsic noise sources and offsets will inevitably be present. It is important to understand the limitations of such interfaces, and the baseline offset and intrinsic noise parameters should be measured as a matter of course.

Whereas it is possible to configure most potentiostats in a ZRA mode, this may not be desirable, particularly if low-current values are being measured. It is advisable to use potentiostats with low-noise and offset specification in all noise monitoring applications.

B3. Data Acquisition

There are many commercially available data acquisition systems (DASs), of which a few are suited to electrochemical noise measurements. Ideally the DAS unit should be able to resolve signals in the microvolt range and have sufficient resolution to accommodate direct-current (dc) offset voltages which will arise if reference electrodes are used for potential measurements. A basic minimum of a 16-bit analog-to-digital converter is recommended, although 20-bit converters (or better) may be required in certain circumstances. Ideally, the potential and current noise signals should be measured at the same time, but slight differences in timing are not considered to be problematic. In terms of the data acquisition rate, there does not appear to be any advantage in taking measurements at high rates, and a minimum logging period of 1 s has been found to be satisfactory for most circumstances. At higher rates of data acquisition, the amplitude of instrumentation noise approaches the electrochemical noise levels. In addition, electromagnetic interference at power supply frequencies becomes problematic. By measuring at low frequencies, it is possible to reject most spurious noise sources.

The advantage of using 16-bit converter technology is that the conversion itself is sufficiently fast to allow multiplexing of inputs from a number of sensors. The major disadvantage of this type of system is the amount of data which can be generated in a short period of time. Fortunately, data storage and manipulation are facilitated by the computing power available, and it is possible to process the data either on- or offline using, for example, digital filtering techniques, statistical analysis, and frequency domain transform techniques.

B4. Data Analysis

The analysis of electrochemical noise signals obtained as potential and current time record series may be undertaken by a variety of means:

1. Examination of the potential and current–time records
2. Statistical analysis
3. Frequency domain transforms

B4.1. Examination of Potential and Current–Time Records.

The potential and current–time records are typically used to identify, in a qualitative manner, short-term transient behavior in the data. These transients are usually indicative of spontaneous changes in corrosion mechanism, such as occurs, for example, during passive film breakdown due to pit initiation processes, cavitation attack, and certain types of SCC. The absolute magnitude of the current signal can be used qualitatively to assess the rate of corrosion.

B4.2. Statistical Analysis.

Statistical analysis is typically undertaken on batch data, although modifications can be made semicontinuous by data blocking or by discounting the past. The sample statistics may be derived, including mean, standard deviation, variance, root mean square(rms), third and fourth moments, skewness and kurtosis, and coefficient of variance. Another statistic routinely used in the analysis of current noise signals is the ratio of the standard deviation to the rms value of the current, referred to as the localization index (LI):

$$LI = \frac{\sigma_i}{i_{rms}} \qquad (86.1)$$

This statistic is used in preference to the coefficient of variance, since the current may exhibit a mean of zero having both positive and negative values.

The mean values of current or potential do not provide a great deal of information in typical plant monitoring situations. The mean current value may give a very broad appreciation of the rate of the process but generally may only be applied to provide a rough estimate of rate. The mean potential may have significance, but only if it is measured with respect to a reference electrode. If a pseudoreference is used, this is relatively meaningless.

The variance of the signal relates to the power in the data, and it is more usual to use the standard deviation of the signal. The standard deviation of the signal is a measure of the spread of the data around the mean value. In essence it relates to the broadband alternating-current (ac) component of the signal. The third central moment is a measure of the asymmetry of the data around the mean value, and the fourth moment is used to derive a value for kurtosis, which indicates the sharpness of the distribution.

The coefficient of variance, or more usually the LI, is a measure of the distribution of the data around the mean or rms value, respectively. As a general rule, if the LI has a value approaching 1, the corrosion process is unstable and therefore more likely to be stochastic. More uniform corrosion processes, on the other hand, have LI values that are typically of the order of 10^{-3}. The LI value relates to the basic statistics of a Poisson process, where the standard deviation is equal to the mean; that is, an LI value of 1. The LI is usually calculated from current data. Slow drift in the current through zero may give an artificially high value of LI and therefore indicates localized corrosion even though the corrosion process has a Gaussian distribution symptomatic of general corrosion. The LI must therefore be used with care.

Another method for identifying changes in the distribution of the noise signals [such as those occurring during localized corrosion, pit initiation/propagation, and stress corrosion cracking (SCC)] is to use the calculated kurtosis values. The value of kurtosis reflects the distribution of the signals. For data that exhibit spontaneous changes in amplitude distribution, the kurtosis value will typically be > 5. Kurtosis will also reflect sudden changes in corrosion rate that may occur due to changes in flow rate, pH, and so on.

In addition, the value of skewness may be used to identify whether the data have a nonuniform distribution, such as may occur, for example, during pitting attack, cavitation, and SCC.

The statistics can always be calculated for existing blocks of data, but this can be inefficient in terms of use of computer memory, since all the data must be available all the time. Running mean versions of these statistics for calculating their next or $(n + 1)$th values can be performed for online monitoring. In addition to the above methods, recursive techniques similar to those used for digital filtering may be used.

For simultaneous or correlated current and potential noise signals, certain derivatives are used to estimate the rate and mechanism of the processes occurring, in particular, the resistance noise value, which is defined as

$$R_n = \frac{\sigma_V}{\sigma_i} \qquad (86.2)$$

and the LI from Eq. 86.1.

For general corrosion, the resistance noise may be equated to a corrosion rate from knowledge of the Stern–Geary constant provided that the corrosion process is relatively stationary. The LI may be used to estimate the localized nature of the corrosion process.

If a pseudoreference electrode is used (i.e., three-identical-electrode setup), then the potential noise value must be corrected for this fact. The potential noise (as measured) will be the geometric mean of the sums of the uncorrelated potential noise signals, that is,

$$\sigma_{V(measured)} = \sqrt{\sigma_{V,WE1,WE2}^2 + \sigma_{V,REF}^2} \qquad (86.3)$$

such that the measured potential noise signal needs to be corrected by a factor of $\sqrt{2}$:

$$\sigma_{V,actual} = \frac{\sigma_{V,measured}}{\sqrt{2}} \qquad (86.4)$$

The current measurements must be normalized for area, and this is usually achieved by dividing, for example, the mean current, by the area of one electrode.

B4.3. Frequency Domain Transforms. Power spectra estimation can be conducted by transforming correlated ECN data to the frequency domain. Power spectral density (PSD) of the correlated potential and current noise (measured simultaneously) can be calculated and plotted as a function of frequency using the fast Fourier transform (FFT) or the maximum-entropy method (MEM). The PSD data are plotted over frequencies that range from $1/\tau$ (τ is the total sampling time) to one-half of data-sampling rate. The PSD value at a given frequency is proportional to the square of the amplitude at that frequency. The units for the PSD of the potential and current noise are V^2/Hz and A^2/Hz, respectively. The spectral noise resistance, $R_{sn}(f)$, is calculated by the square root of the ratio of the potential (or voltage) PSD to the current PSD:

$$R_{sn}(f) = \left(\frac{\psi_v(f)}{\psi_I(f)}\right)^{1/2} \tag{86.5}$$

where $\psi_v(f)$ and $\psi_I(f)$ are the PSD of potential and current noise, respectively.

The spectral noise resistance $R_{sn}(f)$ has been found to be proportional to the magnitude of the cell impedance $Z(f)$ in the two-electrode cell. For the ideally identical sized electrodes, the proportional factor was unity. Figure 86.7 shows that the spectral noise resistance $R_{sn}(f)$ coincided well with the cell impedance over frequencies for Fe electrodes in $1\,M$ Na_2SO_4 solution.

Under the certain conditions, noise resistance R_n can be related to the spectral noise resistance $R_{sn}(f)$ through the PSDs. The variance of a random signal in the time domain is the integral of its PSD in the frequency domain. As a result, the noise resistance R_n can be determined from the integrals of the potential and current PSDs:

$$R_n = \frac{\sigma_V}{\sigma_I} = \left(\frac{\int_{f_{min}}^{f_{max}} \psi_V(f)\,df}{\int_{f_{min}}^{f_{max}} \psi_I(f)\,df}\right)^{1/2} \tag{86.6}$$

where f_{min} is $1/T$ (T is the total sampling time) and f_{max} is one-half the sampling rate. The same symbols for PSDs are used as in Eq. 86.5.

Combining $R_{sn}(f)$ expressed in Eq. (86.5), the relationship between R_n and $R_{sn}(f)$ can be established through the potential and current PSDs as

$$R_n = \left(\frac{\int_{f_{min}}^{f_{max}} \psi_I(f) R_{sn}^2(f)\,df}{\int_{f_{min}}^{f_{max}} \psi_I(f)\,df}\right)^{1/2} \tag{86.7}$$

If $R_{sn}(f)$ is frequency independent in the range from f_{min} to f_{max}, then R_n will equal $R_{sn}(f)$. In most corrosion systems, $R_{sn}(f)$ is frequency dependent in the measured frequency range. As a result, R_n is usually lower than $R_{sn}(f)$ at the D C limit, $R_{sn}(f \to 0)$, or zero-frequency impedance, $R_p = |Z(f \to 0)|$.

In addition to the above theoretical aspects of frequency domain transforms, some practical aspects must be considered in applications such as online corrosion monitoring. It may prove possible to produce a limited spectral estimate on a continual basis using digital filtering techniques by assessing the bandpass characteristics at a number of discrete frequencies. For corrosion control purposes, a continuous update of corrosion-related information is required in order to provide operator feedback about corrosion rates and mechanisms. In order to achieve this, there must be continuous, online evaluation of data which will provide the operator with information as opposed to data. By conversion of the raw time record data into information, it should be possible to significantly reduce the amount of data stored provided that the data can be processed at a sufficiently fast rate. This, in essence, is a requirement for an online expert system, where the algorithms used to detect and extract the information should be derived from a working knowledge of a variety of corrosion processes.

The information most likely to be of use to a plant operator is that which is related to the rates and mechanisms of the corrosion processes, which can be correlated with plant

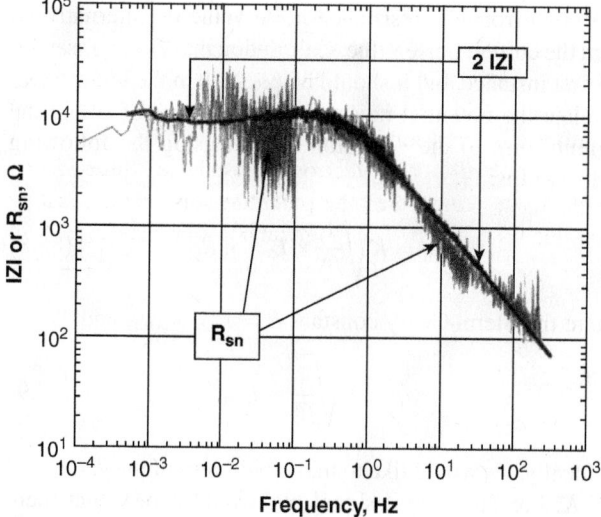

FIGURE 86.7. R_{sn} versus $|Z|$ for Fe electrodes in $1\,M$ Na_2SO_4 at pH 4. Impedance measurements were performed in a two-electrode arrangement using a noiseless reference (Fe) electrode. (From [15].)

variables. In essence, there are fundamentally two mechanisms that are of particular importance: general corrosion and processes involving stochastic events such as pitting and SCC. These two types of processes will be examined in order to identify methods by which information pertaining to corrosion rates and mechanisms may be extracted.

C. GENERAL CORROSION

General corrosion processes may be identified by several techniques, some of which will be described here. In the first instance, the time record data for these types of processes exhibit few signs of individual uncorrelated events, that is, there are few, if any, rapid transients. A statistical treatment of the data provides some indication of the rate and stability of the corrosion process in the short term. These types of processes have a typically Gaussian distribution of data, and examination for the types of distribution of both the current and potential data can be used to estimate if the process is indeed Gaussian.

This can be a time-consuming exercise but can usually be achieved simply by examining the statistical data, particularly of the current–time record. General corrosion processes can often be estimated from the moments, skewness, kurtosis, and coefficient of variance and its derivatives. Alternatively, the data may be transformed into the frequency domain. In this case the amplitudes of the signals at specific frequencies and the slope of the frequency spectrum can be used to distinguish between general and localized phenomena.

D. LOCALIZED PHENOMENA

Localized corrosion processes, pitting and SCC in particular, are characterized by transients that may be observed in the time record data. Pit initiation and SCC phenomena have distinguishing characteristics. The base process, however, is stochastic (i.e., random in nature) and therefore is essentially a Poisson process. The characteristics of Poisson processes are such that they can easily be distinguished from Gaussian processes by a variety of means, from both an appreciation of the statistics of the data and the frequency domain information. In addition, it should be possible to differentiate between the different operative mechanisms and to provide an indication of the severity of the corrosion problem. This can be achieved by an appreciation of how different processes contribute to the noise signals.

It should be possible therefore to provide an online data evaluation process which is either (a) tailored for a specific application or (b) application nonspecific.

The latter goal is desirable, but sufficient experience is required for this goal to be achieved. For the purposes of

extracting information from the data, we are concerned with (in order):

1. Identification of overall mechanism
2. Estimates of the general corrosion rate if the process is Gaussian
3. If the process is stochastic, then attribution of the process to a particular mechanism
4. Estimation of the severity of the stochastic process

The first step in identifying the overall mechanism is to decide if the data have a Poisson distribution and if possible over what frequency range. One of the simpler techniques (but by no means infallible) is to examine the ratio of the standard deviation divided by the rms of the signal. If this ratio approached 1, then the process is likely to have a Poisson distribution. If the ratio approached 10^{-3}, then the distribution is probably Gaussian and therefore indicative of relatively general corrosion. Another test for stochastic processes is to count the number of events in a time period as a function of different threshold intensities. This means that some arbitrary rate of change be applied as a filter to the data. The values of kurtosis for the potential and current signals are sensitive indicators to changes in corrosion rate and mechanisms. Alternatively, a series of bandpass-filtered signals at several frequencies can be compared in order to give an appreciation of the slope in the frequency domain. The expected slopes in the frequency domain (of the current signal) are zero from stochastic (Poisson) processes, -0.5 for diffusion-controlled processes, and -1 for processes having a Gaussian distribution. For corrosion processes, the current signals should be used for the above techniques.

If the process is Gaussian, the corrosion rate may be estimated from the resistance noise value or, alternatively, from the current noise value. Calibration may be necessary in the first instance, but it should be possible in the longer term to relate the potential noise signal to a value of B and the current noise to the corrosion current using the following relationships:

$$V_n = K \sqrt{\frac{1}{f^\alpha}} \times I_{corr} \times R_{ct} \qquad (86.8)$$

where the Stern–Geary constant $B = I_{corr} \times R_{ct}$, and

$$i_n = K \sqrt{\frac{1}{f^\alpha}} \times I_{corr} \qquad (86.9)$$

Ideally, we would like to make estimates of the value of α and K. For the current signal, we would expect that there would be three possible values for α, namely, 0, 1, and 2. In the frequency domain PSD, this would equate to slopes of 0, -1, and -2, respectively (or as amplitudes 0, -0.5, and -1, respectively). A slope of zero equates to a stochastic,

Poisson process, a slope of -1 to a diffusion-controlled process, and a slope of -2 to a Gaussian process. The value of the current noise may be estimated at a particular frequency, say 50 mHz, and compared with that of a lower frequency, say 5 mHz, to estimate the slope and hence the likely values of α. The factor K could be estimated from a knowledge of a general corrosion process, where $\alpha = 2$, and a knowledge of the measured potential noise and the B value. Initial estimates would suggest a value of $\sim 10^{-5}$ for K. In theory, provided that the relationships in Eqs. (86.8) and (86.9) are good for stochastic processes, then it should be possible to estimate a value of the corrosion current, I_{corr}, irrespective of the type of process occurring. In practice, this is unlikely to be the case, as inspection of Eq. (86.8) indicates that if stochastic processes are operative, then the term within the square-root sign reduced to a value of 1 irrespective of the frequency, and the corrosion current is directly related to the noise amplitude at any frequency through the factor K. However, for corrosion processes having a stochastic nature, it is difficult to estimate a localized pit penetration or crack propagation rate unless some assumptions are made regarding geometry and distribution of pits or cracks in the specimen.

For stochastic processes, the nature of the current and potential transients may provide an insight into the prevalent mechanism. Pitting systems often show bidirectional current transients associated with initiation and propagation events. The potential transient in these cases is predominantly negative going, associated with the depassivation of the metal and a shift of the potential into a more active corrosion regime. For pit initiation events, the current transients are short-lived pulses, whereas the potential transients exhibit a fast, negative-going spike, associated with the current followed by an exponential recovery to the passive value. It is possible to estimate the severity of the stochastic processes by several methods as discussed above. Stress corrosion cracking phenomena can, in certain cases, be very similar to pitting-type attack. For example, SCC of aluminum alloys exhibits transients very similar to pitting-type events during crack propagation as do certain intergranular SCC (IGSCC) systems. On the other hand, transgranular SCC (TGSCC) of carbon steel in CO/CO_2 environments exhibits positive potential transients associated with some cathodic process in the crack propagation step. The amplitude and frequency of the transients in both instances give an indication of crack propagation severity, and provided the data are correlated with fractographic examination after the exposure period, it may be possible to model effectively the crack propagation process.

The attribution of a signal type to a particular process may be relatively simple in practical situations. Pitting-type studies involve unstressed specimens, and it is expected that the pits will initiate and propagate with equal probability on both working electrodes, resulting in bidirectional current transients, associated with negative-going (less noble) potential transients. In the case of a stressed specimen, it is usual to make the current measurements between a stressed and an unstressed specimen. In this circumstance, the test is designed to give a differential view of the cracking process, where transients associated with crack initiation and propagation tend to dominate the noise signals. The current transients in this case can be related directly to anodic and cathodic processes associated with the cracking process. In favorable circumstances the skewness of the signals can be used to identify particular corrosion mechanism.

The severity of a particular stochastic process is directly related to the current associated with a particular event and the frequency of such events. This can be estimated by several methods. In the first instance the number of current excursions in some time period t and their mean amplitude may be used. In the past, the value of the localization index has been used to "weight" the apparent general corrosion rate of the system to take into account the localized nature of the process, particularly in pitting situations. For control purposes in plant surveillance situations, it may be simply necessary to identify the number of transients above some threshold amplitude in order to make the operator aware of the change in severity or mechanism of the corrosion process.

With regard to the potential noise spectra for freely corroding systems, it has been postulated that the potential signals arise from the interaction of the fluctuations in corrosion current and the interfacial impedance. It is possible to compare the estimate of the transfer function, or impedance $|Z|$, obtained from the current and potential signals with that obtained by more conventional methods, that is, electrochemical impedance spectroscopy. However, due to limitations in the high-frequency range over which it is possible to measure the spontaneous fluctuations of potential and current, this is more appropriate for slow corrosion processes. For fast corrosion processes, the low-frequency transfer function is usually almost purely resistive. By way of example, the power spectrum of a potential noise signal, arising from a current noise source having slope $\alpha = -2$, would depend on the interfacial time constant for the system. If the interfacial impedance were purely resistive, then the potential noise power spectrum would exhibit slope $\alpha = -2$. On the other hand, if the system behaves as a parallel RC network within the frequency range of the measurements, the potential noise power spectrum would exhibit slope $\alpha = -4$. In this latter case, the interfacial impedance is effectively acting as a low-pass filter to the current fluctuations.

E. APPLICATIONS FOR REAL-TIME CORROSION MONITORING

The electrochemical noise technology has been practically applied over the last 30 years to a wide variety of corrosion-related problems in major industrial sectors, such as oil and gas, fossil fuel power generation, petrochemicals, nuclear power, civil engineering, and aerospace industries [17–121].

As a nondestructive, in situ method, ECN has been extensively used to real-time monitor general and localized corrosion in a variety of environments. The following is by no means an exhaustive list of such applications:

- Drinking water [77]
- Concentrated chloride solutions [82]
- High-temperature aqueous solutions [78–81, 83–85]
- Sand-entrained solutions [4, 5, 86]
- Acids in vapor phase at dewpoint [52, 53]
- Concrete [47–51, 87, 88]
- Deposits of fouling and microbiology [57–61, 89–97]
- Water/oil mixtures in multiphase flow [98–102]
- Solutions containing inhibitors [54–56, 103–106]
- Coatings [67–75, 107–110]
- H$_2$S systems [111–113]
- Natural gas in transmission pipelines [114, 115]
- Process streams in chemical plants [116, 117]
- Nuclear waste environments [118–121]

Different ECN methods have been used to study and monitor corrosion processes and identify the corrosion mechanisms, with the most common being potential noise, current noise, and noise resistance monitoring.

PSD analysis in combination with time domain analysis is mainly used in laboratory studies. The following examples are given to show the use of different ECN methods in different environments.

In high-temperature aqueous systems (high subcritical and supercritical water), electrochemical current noise is the most used method. The standard deviation or rms of current noise has been found to be proportional to the corrosion rate determined by weight loss measurements [78–81, 83–85].

In gas–water or gas–water–oil multiphase flow, potential and current noise is used to monitor corrosion [98]. Noise fluctuations show trends, which can be related to the type of flow, such as full pipe flow and slug flow. The full pipe flow exhibits random fluctuations around the mean for both potential and current noise indicating uniform corrosion. In contrast to the full pipe flow, the slug flow exhibits transients in the potential noise and the frequency of transients corresponds to the observed frequency of slugs. The presence of these transients indicates pitting corrosion.

Recently, researchers have developed a new approach to signal collection and data analysis to study microbiologically induced corrosion [89]. The current noise level (CNL) and the potential noise level (PNL) as the trend of current or potential noise are collected by calculating the rms of a few hundred noise signal data points collected over a short period

(4–30 s). The PSD analysis of CNL or PNL is then conducted to obtain the linear slope in the low-frequency portion (less than 10^{-3} Hz) of the PSD. The degree of linear slope is correlated well with sustained localized pitting (SLP) and general corrosion processes. Based on the linear slope values, the SLP can be effectively differentiated from the general corrosion.

The recent trend is to integrate commercially developed corrosion monitoring tools based on electrochemical noise technology with process control systems. By continuously monitoring corrosion damage in plants, a variety of processes can be controlled and optimized in order to improve the safety and reliability.

REFERENCES

1. W. P. Iverson, J. Electrochem. Soc., **115**, 617 (1968).
2. K. Hladky, U.S. Patent No. 4,575,678, 1986.
3. G. Blanc, C. Gabrielle, and M. Keddham, C.R. Acd. Sci., **13**, 283C, 107 (1976).
4. K. Hladky and J. L. Dawson, Corr. Sci., **21**, 317 (1981).
5. K. Hladky and J. L. Dawson, Corr. Sci., **22**, 231 (1982).
6. J. L. Dawson, K. Hladky, and D. A. Eden,"Electrochemical Noise—Some New Developments in Corrosion Monitoring," UK Corrosion, 83, 1983, Proceedings of the Conference, The Institution of Corrosion Science and Technology, Exeter House, 48 Holloway Head, Birmingham, UK, 1983, pp. 99–108.
7. U. Bertocci, J. L. Mullen, and Y.-X. Ye, "Electrochemcial Noise Measurement for the Study of Localized Corrosion and Passivity Breakdown," in Passivity of Metals and Semiconductors, M. Froment, Ed., Elsevier Science, Amsterdam,, 1983.
8. K. Hladky, J. P. Lomas, D. G. John, D. A. Eden, and J. L. Dawson, "Corrosion Monitoring Using Electrochemical Noise: Theory and Practice," in Corrosion Monitoring and Inspection in the Oil, Petrochemical and Process Industries, Oyez Scientific and Technical Services, Bath House, London, 1984.
9. C. Gabrielli, F. Huet, M. Keddam, R. Oltra, and J. C. Colson, Mater. Sci. Forum, **8**, 491 (1986).
10. D. A. Eden, G. John, and J. L. Dawson, U.S. Patent No. 5,139,627, 1992.
11. D. A. Eden, R. N. Carr, and J. L. Dawson, U.S. Patent No. 5,425,867, 1995.
12. F. Mansfeld, C. Chen, C. Lee, and H. Xiao, Corros. Sci., **38**, 497 (1996).
13. F. Mansfeld, and H. Xiao, "Electrochemical Noise Measurements for Corrosion Applications," ASTM STP 1277, American Society for Testing and Materials, Philadelphia, PA, 1996.
14. U. Bertocci, C. Gabrielli, F. Huet, and M. Keddham, J. Electrochem. Soc., **144**, 31 (1997).

15. U. Bertocci, C. Gabrielli, F. Huet, M. Keddham, and P. Rousseau, J. Electrochem. Soc., **144**, 37 (1997).

16. U. Bertocci and F. Huet, J. Electrochem. Soc., **144**, 2786 (1997).

17. S. Smith and R. Francis, Br. Corros. J., **25**, 285 (1990).

18. C. Gabrielli, F. Huet, M. Keddham, and R. Oltra, "Localized Corrosion as a Stochastic Process: A Review," in Advances in Localized Corrosion, NACE, Houston, TX, 1990.

19. J. Gollner, I. Garz, and K. Meyer, Corrosion, **17**, 244 (1986).

20. C. Gabrielli, F. Huet, M. Keddam, and H. Takenouti, "Application of Electrochemcial Noise Measurement to the Study of Localized and Uniform Corrosion," 8th European Congress of Corrosion, Centre Français de la Corrosion, Société de Chimie Industrielle, Paris, 1985, Vol. 2, pp 37.1– 37.7.

21. I. A. Al-Zanki, J. S. Gill, and J. L. Dawson, Mater. Sci. Forum, **8**, 463 (1986).

22. T. Okada, J. Electrochem. Soc., **140**, 1261 (1993).

23. A. Legat and V. Delecek, J. Electrochem. Soc., **142**, 1851 (1995).

24. R. A. Cottis and T. Turgoose, Mater. Sci. Forum, **192**, 663 (1995).

25. U. Bertocci, J. Frydman, C. Gabrielli, F. Huet, and M. Keddham, J. Electrochem. Soc., **145**, 2780 (1998).

26. A. Bautista and F. Heut, J. Electrochem. Soc., **146**, 1736 (1999).

27. B. Bastos, F. Heut, R. P. Nogueira, and P. Rousseau, J. Electrochem. Soc., **147**, 671 (2001).

28. A. Bautista, U. Bertocci, and F. Huet, J. Electrochem. Soc., **148**, B412 (2001).

29. A. Aballe and F. Huet, J. Electrochem. Soc., **149**, B89 (2002).

30. U. Bertocci, F. Huet, B. Jaoul, and R. P. Nogueira, Corrosion, **56**, 675 (2000).

31. R. A. Cottis, Corrosion, **57**, 265 (2000).

32. A. Aballe, A. Bautista, U. Bertocci, and F. Huet, Corrosion, **57**, 35 (2001).

33. U. Bertocci, F. Huet, and R. P. Nogueira, Corrosion, **59**, 629 (2003).

34. M. Ferreira and J. L. Dawson, "Crevice Corrosion of an Austenitic Stainless Steel in 3% NaCl Solution," International Congress on Metallic Corrosion, National Research Coucil of Canada, Ottawa, Canada, 1984.

35. A. P. Simoes and M. Ferreira, Br. Corros. J., **22**, 21 (1987).

36. J. A. Wharto, B. G. Mellor, R. J. K. Wood, and C. J. E. Smith, J. Electrochem. Soc., **147**, 3294 (2000).

37. R. C. Newman and K. Sieradzki, Scripta Metall., **17**, 621 (1983).

38. R. A. Cottis and C. A. Loto, Mater. Sci. Forum, **8**, 201 (1986).

39. C. A. Loto and R. A. Cottis, Corrosion, **43**, 499 (1987).

40. C. A. Loto and R. A. Cottis, Corrosion, **45**, 136 (1989).

41. R. A. Cottis and C. A. Loto, Corrosion, **46**, 12 (1990).

42. J. Stewart, P. M. Scott, D. E. Williams, and D. B. Wells, Corros. Sci., **33**, 73 (1992).

43. E. Benzaid, F. Huet, C. Gabrielli, M. Jerome, F. Wenger, and J. Galland, "Investigation of Electrochemical Noise in the Study of Hydrogen Embrittlement of a 42CD4 Carbon Steel Electrode," in Progress in the Understanding and Prevention of Corrosion, Institute of Materials, London, 1993, Vol. II, pp. 1304–1311.

44. A. Benzaid, C. Gabrielli, F. Huet, M. Jerome, and F. Wenger, Mater. Sci. Forum, **111–112**, 167 (1992).

45. J. G. GonzalezRodriguez, V. M. Salina-Bravo, and J. M. Ramirez-Montano, Corros. Rev., **14**, 309 (1996).

46. J. L. Luo and L. J. Qiao, Corrosion, **55**, 870 (1999).

47. J. L. Dawson, Corrosion of Reinforcement in Concrete Construction, Ellis Horwood, Chichester, U.K., 1983, pp. 175–191.

48. S. G. McKenzie, Corros. Pre. Control, **34**, 1987 (1987).

49. D. G. John, D. A. Eden, J. L. Dawson, and P. E. Langford, "Corrosion Measurements on Reinforcing Steel and Monitoring of Concrete Structures," in Corrosion of Metals in Concrete, NACE, Houston, TX, 1987.

50. G. K. Glass, C. L. Page, and N. R. Short, "Monitoring the Corrosion of Steel Reinforcement," UK Corrosion'88– with Eurocorr, Brighton, U.K., October 3-5, 1988, Vol. II, pp. 243–247. 1988.

51. R. G. Hardon, P. Lambert, and C. L. Page, Br. Corros. J., **23**, 225 (1988).

52. J. L. Dawson, D. Gearey, and W. M. Cox, "Recent Experience of Monitoring Condensed Acid Corrosion in Boiler Flue Gas Ducts," UK National Corrosion Conference, The Institution of Corrosion Science and Technology, Exeter House, Birmingham, U.K., 1982, pp. 193–198.

53. W. M. Cox, D. Gearey, and J. L. Dawson, "Materials Evaluation and Corrosion Monitoring in Flue Gas Acid Dewpoint Environments," in Corrosion—Industrial Problems, Treatment, and Control Techniques, Kuwait, 1984, Pergamon, Headington Hill Hall, Oxford, 1987, pp. 85–89.

54. J. M. Bastida and J. M. Malo, Rev. Metal., **21**, 337 (1985).

55. C. Monticelli, G. Brunoro, A. Arignani, G. Trabanelli, J. Electrochem. Soc., **139**, 706 (1992).

56. W. Qafsaoui, F. Huet, and H. Takenouti, J. Electrochem. Soc., **156**, C67 (2009).

57. W. P. Iverson and L. F. Heverly, "Electrochemical Noise as an Indicator of Anaerobic Corrosion," in Corrosion Monitoring in Industrial Plants Using Non-Destructive Testing and Electrochemical Methods, G. C. Moran and P. Labine, eds., STP 908, American Society for Testing and Materials, Philadelphia, PA, 1986.

58. W. P. Iverson, G. J. Olson, and L. F. Heverly, "The Role of Phosphorus and Hydrogen Sulfide in the Anaerobic Corrosion of Iron and the Possible Detection of This Corrosion by an Electrochemical Noise Technique," in Biologically Induced Corrosion, S. C. Dexter, Ed., NACE, Houston, TX, 1986, pp. 154–161.

59. C. A. Sequeira, "Electrochemical Techniques for Studying Microbial Corrosion," in Microbial Corrosion, C. A. C.

Sequeira and A. K. Tiller, Eds., Elsevier, Applied Science, Barking, Essex, U.K., 1988, pp. 99–118.

60. A. M. Brennenstuhl, T. S. Gendron, and R. Cleland, Corros. Sci., **35**, 699 (1993).

61. A. M. Brennenstuhl and T. S. Gendron, "The Use of Field Tests and Electrochemical Noise to Define Conditions for Accelerated Microbiologically Induced Corrosion (MIC) Testing," in Microbiologically Influenced Corrosion Testing, ASTM STP 1232, American Society for Testing and Materials, Philadelphia, PA, 1994.

62. J. C. Uruchurtu and J. L. Dawson, Corrosion, **42**, 19 (1987).

63. N. Celati and C. Bataillon, "Pitting Corrosion and the Generation of Electrochemical Noise in Aluminium Alloys," in Durabilité de l'aluminium et ses alliages dans les industries electriques, Société des Electriciens et des Electroniciens, Paris, 1986, pp. 117–124.

64. J. Uruchurtu and J. L. Dawson, Mater. Sci. Forum, **8**, 113 (1985).

65. N. Celati, Report No. N90-27903/5/XAB, NASA, Washington, DC, Dec. 22, 1989.

66. K. Sasaki and H. S. Isaacs, J. Electrochem. Soc., **151**, B124 (2004).

67. D. A. Eden, M. Hoffman, and B. S. Skerry, "Application of Electrochemical Noise Measurements to Coated Systems," in Polymeric Materials for Corrosion Control, R. A. Dickie and F. L. Floyd, Eds., American Chemical Society, Washington, DC, 1986, pp. 36–47.

68. I. Pronto, A. M. P. Simoes, M. G. S. Ferreira, and M. C. Machado, Prot. Mater., **6**, 49 (1987).

69. B. S. Skerry and D. A. Eden, Prog. Organic Coat., **19**, 379 (1991).

70. H. Xiao, F. Mansfeld, J. Electrochem. Soc., **141**, 2332 (1994).

71. D. J. Mills, G. P. Bierwagen, B. S. Skerry, and D. Tallman, Mater. Perform., **34**, 33 (1995).

72. M. Moon and B. S. Skerry, J. Coat. Technol., **67**, 35 (1995).

73. F. Mansfeld, H. Xiao, L. T. Han, and C. C. Lee, Prog. Organic Coat., **30**, 89 (1997).

74. S. S. Wilks, Mathematical Statistics, Wiley, New York, 1962.

75. P. Z. Peebles, Random Variables and Random Signal Principles, McGraw-Hill International, New York, 1987.

76. Y. Watanabe, T. Shoji, and T. Kondo, "Electrochemical Noise Characteristics of IGSCC in Stainless Steels in Pressurized High-Temperature Water," Paper No. 129, Corrosion98, NACE, Houston, TX, 1998.

77. D. W. Townley, S. J. Duranceau, and D. D. Wilson, "On-Line Electrochemical Noise Corrosion Monitoring in Potable Water Distribution Systems," Paper No. 02328, Corrosion/2002, NACE, Houston, TX, 2002.

78. S. N. Lvov, "Advanced Techniques for High Temperature Electrochemical and Corrosion Studies," Paper No. 04497, Corrosion/2004, NACE, Houston, TX, 2004.

79. J. Macak, P. Sajdl, P. Kucera, and R. Novotny, "In-Situ Study of High Temperature Aqueous Corrosion by Electrochemical Techniques," Paper No. 05336, Corrosion/2005, NACE, Houston, TX, 2005.

80. X. Guan and D. D. Macdonald, Corrosion, **65**, 376 (2009).

81. D. D. Macdonald and X. Guan, Corrosion, **65**, 427 (2009).

82. F. Mansfeld and H. Xiao, J. Electrochem. Soc., **140**, 2205 (1993).

83. X. Y. Zhou, S. N. Lvov, X. J. Wei, L. G. Benning, and D. D. Macdonald, Corros. Sci., **44**, 841 (2002).

84. C. Liu, D. D. Macdonald, F. Medina, J. J. Villa, and J. M. Bueno, Corrosion, **50**, 687 (2002).

85. X. Guan, T. Zhu, and D. D. Macdonald, Paper No. 06449, Corrosion/2006, NACE, Houston, TX, 2006.

86. J. Chen, J. R. Shadley, and E. F. Rybicki, "Activation/ Repassivation Behavior of 13Cr in CO_2 and Sand Environments Using a Modified Electrochemical Noise Technique," Paper No. 02494, Corrosion/2002, NACE, Houston, TX, 2002.

87. M. Tullmin and C. M. Hansson, "Electrochemical Noise Measurements on Carbon and Stainless Steel Reinforcing Steels," Paper No. 372, Corrosion/1998, NACE, Houston, TX, 1998.

88. E. Garcia and J. Genesca, "Evaluation of a Polymeric Concrete by Electrochemical Noise," Paper No. 390, Corrosion/ 1998, NACE, Houston, TX, 1998.

89. Y. J. Lin, J. R. Frank, E. J. St. Martin, and D. H. Pope, "Electrochemical Noise Measurements of Sustained Microbially Influenced Pitting Corrosion in a Laboratory Flow Loop System," Paper No. 198, Corrosion/1999, NACE, Houston, TX, 1999.

90. G. J. Licina and G. Nekoksa, "Monitoring Biofilm Formation in a Brackish Water Cooled Power Plant Environment," Paper No. 222, Corrosion/1997, NACE, Houston, TX, 1997.

91. M. Enzien and B. Yang, "Effective Use of Biocide for MIC Control in Cooling Water Systems," Corrosion/2000, Paper No. 00384, NACE, Houston, TX, 2000.

92. F. Huet, N. M. Moros, R. P. Nogueira, B. Tribollet, and D. Festy, "Electrochemical Noise Analysis Applied to SRB-Induced Corrosion of Carbon Steel," Corrosion/2002, Paper No. 02449, NACE, Houston, TX, 2002.

93. M. Amaya, R. Perez, and L. Martinez, "Bacterial Influence on the Corrosion of API-XL70 Steel," Paper No. 02471, Corrosion/2002, NACE, Houston, TX, 2002.

94. M. Amaya, J. L. Villalobos, J. M. Romero, and E. Sosa, "Corrosion Behavior of the API X52 and API X65 Pipeline Steels in the Presence of Bacteria Consortia Using Electrochemical Noise," Paper No. 03414, Corrosion/2003, NACE, Houston, TX, 2003.

95. R. G. Martinez, G. G. Caloca, R. D. Romero, J. M. Flores, E. M. Nunez, and R. T. Sanchez, "Comparison of Electrochemical Techniques during the Corrosion of X52 Pipeline Steel in the Presence of Sulphate Reducing Bacteria (SRB)," Paper No. 03545, Corrosion/2003, NACE, Houston, TX, 2003.

96. A. P. Cviveros, E. G. Ochoa, and D. Alazard, "Electrochemical Behavior of Localized Corrosion in Steel by SRB under Oligotrophic Conditions," Paper No. 04583, Corrosion/2004, NACE, Houston, TX, 2004.

97. A. P. Viveros, E. G. Ochoa, R. G. Esquivel, D. Alazard, M. L. Fardeau, and B. Ollivier, "The Influence of the Culture Media on the Steel Corrosion Induced by *Desulfovibrio vulgaris* subsp. *oxamicus*: An electrochemical Noise Study," Paper No. 06522, Corrosion/2006, NACE, Houston, TX, 2006.

98. Y. P. Deva, M. Gopal, and W. P. Jepson, "Use of Electrochemical Noise to Monitor Multiphase Flow and Corrosion," Paper No. 337, Corrosion/1996, NACE, Houston, TX, 1996.

99. Y. Chen and M. Gopal, "Electrochemical Methods for Monitoring Performance of Corrosion Inhibitors under Multiphase Flow," Corrosion/1999, Paper No. 509, NACE, Houston, TX, 1999.

100. F. Huet and R. P. Nogueira, "Comparative Analysis of Potential, Current, and Electrolyte Resistance Fluctuations in Two-Phase Oil/Water Mixtures," Paper No. 03416, Corrosion/2003, NACE, Houston, TX, 2003.

101. H. Bouazaze, F. Huet, and R. P. Nogueira, "Monitoring Corrosion and Flow Characteristics in Oil/Brine Mixtures of Various Compositions," Paper No. 04466, Corrosion/2004, NACE, Houston, TX, 2004.

102. H. Wang, "Applications of Electrochemical Noise Technique in Multiphase Flow," Paper 05368, Corrosion/2005, NACE, Houston, TX, 2005.

103. S. L. Fu, A. M. Griffin, J. G. Garcia, and B. Yang, "A New Localized Corrosion Monitoring Technique for the Evaluation of Oilfield Inhibitors," Paper No. 346, Corrosion/1996, NACE, Houston, TX, 1996.

104. Y. J. Tan, B. Kinsella, and S. Bailey, "The Evaluation of Corrosion Inhibitor Film Persistency Using Electrochemical Impedance Spectroscopy and Electrochemical Noise Analysis," Paper No. 352, Corrosion/1996, NACE, Houston, TX, 1996.

105. E. E. Barr, A. H. Greenfield, and L. Pierrard, "Application of Electrochemical Noise Monitoring to Inhibitor Evaluation and Optimization in the Field: Results from the Kaybob South Sour Gas Field," Paper No. 01288, Corrosion/2001, NACE, Houston, TX, 2001.

106. B. F. Pots, E. L. Hendriksen, H. deReus, H. B. Pit, and S. J. Paterson, "Field Study of Corrosion Inhibition at Very High Flow Velocity," Paper No. 03321, Corrosion/2003, NACE, Houston, TX, 2003.

107. E. Garcia, J. Mojica, F. J. Rodriguez, J. Genesca, and J. J. Carpio, "Assessing of Three Industrial Paint Coatings by Electrochemical Noise," Paper No. 379, Corrosion/1998, NACE, Houston, TX, 1998.

108. G. P. Bierwagen, D. E. Tallman, S. Touzain, A. Smith, R. Twite, V. Balbyshev, and Y. Pae, "Electrochemical Noise Methods Applied to the Study of Organic Coatings and Pretreatments," Paper No. 380, Corroison/1998, NACE, Houston, TX, 1998.

109. R. L. Ruedisueli and B. D. Layer, "Practical Application of ECN/EIS for In-Service Coatings Assessment," Paper No. 02200, Corrosion/2002, NACE, Houston, TX, 2002.

110. R. L. Ruedisueli and J. N. Murray, "Evaluation of In-Service Navy Ship Hull Coating Utilizing Electrochemical Current and Impedance Measurement Techniques," Paper No. 04297, Corrosion/2004, NACE, Houston, TX, 2004.

111. P. J. Teevens, "Electrochemical Noise—A Potent Weapon in the Battle Against Sour Gas Plant Corrosion," Paper No. 388, Corrosion/1998, NACE, Houston, TX, 1998.

112. R. G. Maritnex, J. M. Flores, J. G. Llongueras, and R. D. Romero, "Electrochemical Noise Study on the Corrosion of X52 Pipeline Steel in Aqueous Solution Containing H_2S," Paper No. 04432, Corrosion/2004, NACE, Houston, TX, 2004.

113. S. A. Peralta, J. G. Llongueras, J. M. Flores, and R. D. Romero, "Corrosion Behavior of X70 Pipeline Steel in H_2S Containing Solutions," Paper No. 04627, Corrosion/2004, NACE, Houston, TX, 2004.

114. B. S. Covino, S. J. Bullard, S. D. Cramer, G. R. Holcomb, and M. Z. Moroz, "Evaluation of the Use of Electrochemical Noise Corrosion Sensors for Natural Gas Transmission Pipelines," Paper No. 04157, Corrosion/2004, NACE, Houston, TX, 2004.

115. S. J. Bullard, B. S. Covino, S. D. Cramer, G. R. Holcomb, M. Z. Moroz, B. Meidinger, R. D. Kane, and D. C. Eden, "Electrochemical Noise Monitoring of Corrosion in Natural Gas Production Plants," Paper No. 05359, Corrosion/2005, NACE, Houston, TX, 2005.

116. G. John and N. Rothwell, "Corrosion Monitoring of Process Plant Incorporating Electrochemical Noise," Paper No. 02337, Corrosion/2002, NACE, Houston, TX, 2002.

117. O. Ashiru, S. Ahmad, and A. A. Refaie, "Corrosion Monitoring by Electrochemical Noise Probes in a Purified Terephthalic Acid Production Plant," Paper No. 05363, Corrosion/2005, NACE, Houston, TX, 2005.

118. G. L. Edgemon, J. L. Nelson, and G. E. Bell, "Corrosion Data from Hanford High-Level Waste Tank 241-AN-107," Paper No. 469, Corrosion/1999, NACE, Houston, TX, 1999.

119. G. L. Edgemon, "Electrochemical Noise Based Corrosion Monitoring at the Hanford Site: Third Generation System Development, Design, and Data," Paper No. 01282, Corrosion/2001, NACE, Houston, TX, 2001.

120. E. E. Barr and G. Edgemon, "Managing Electrochemical Noise Data by Exception: Application of an On-Line En Data Analysis Technique to Data from a High Level Nuclear Waste Tank," Paper No. 04450, Corrosion/2004, NACE, Houston, TX, 2004.

121. G. L. Edgemon, "Electrochemical Noise Based Corrosion Monitoring: Hanford Site Program Status," Paper No. 05584, Corrosion/2005, NACE, Houston, TX, 2005.

PART VI

CORROSION MONITORING

87

CORROSION MONITORING

P. R. ROBERGE
Department of Chemistry and Chemical Engineering, Royal Military College of Canada, Kingston, Ontario, Canada

A. WHAT IS CORROSION MONITORING?

Corrosion monitoring generally refers to corrosion measurements performed under industrial or practical operating conditions. The assessment of corrosion in field conditions is complex due to the wide variety of process conditions and fluid phases that exist in industrial systems. Additionally, the expectations of a corrosion monitoring program vary greatly between organizations with well-established proactive corrosion management programs and other organizations where corrosion damage is simply a nuisance.

Advances in software and hardware tools have led to the evolution of corrosion monitoring tools toward real-time data acquisition in-line with other process control tools. However, corrosion monitoring is often more challenging than monitoring of other process parameters because:

- Different types of corrosion may be simultaneously present.
- Corrosion may be uniform over an area or concentrated in very small areas (pitting).
- General corrosion rates may vary substantially, even over relatively short distances.
- No single technique will detect all of these various conditions.

Before embarking on a corrosion monitoring program, it is therefore helpful to review historical data and consider the types of corrosion problems that need to be investigated. It is also advisable to use several complementary techniques rather than rely on a single monitoring method.

The dividing line between corrosion inspection and corrosion monitoring is not always clear. Usually inspection refers to short-term "once-off" measurements taken according to maintenance and inspection schedules while corrosion monitoring describes the measurement of corrosion damage over a longer time period and often involves an attempt to gain a deeper understanding of how and why the corrosion rate fluctuates over time. Corrosion inspection and monitoring are most beneficial and cost effective when they are utilized in an integrated manner since the associated techniques and methods are in reality complementary rather than substitutes for each other.

A considerable catalyst to the advancement of corrosion inspection and monitoring technology has been the

Uhlig's Corrosion Handbook, Third Edition, Edited by R. Winston Revie
Copyright © 2011 John Wiley & Sons, Inc.

exploitation of oil and gas resources in extreme environmental conditions. Work in these conditions has necessitated enhanced instrument reliability and the automation of many tasks, including inspection. In addition to the usual uncertainty of the onset or progression of corrosion of equipment, the oil industry has to face ever-changing corrosivity of processing streams. The corrosivity at a well head can oscillate many times during the life of an exploitation system, between being benign to becoming extremely corrosive [1].

Some modern corrosion monitoring technologies are particularly apt at revealing the highly time-dependent nature of corrosion processes. The integration of these corrosion monitoring technologies in existing systems can thus provide early warnings of costly corrosion damage. The latest technology allows operators to link corrosion processes to process conditions directly and in real time providing the ability to control and mitigate the rate of damage and reduce its impact on plant operations [2].

In an ideal corrosion control program, inspection and maintenance would be applied only where and when they are actually needed. In principle, the information obtained from corrosion monitoring systems can be of great assistance in reaching this goal. However, it is sometimes difficult for a corrosion engineer to get management's commitment to investing funds in such initiatives. The importance of corrosion monitoring in industrial plants and other engineering systems should be presented as an investment to achieve some of the following goals:

- Improved safety
- Reduced pollution and contamination risks
- Reduced downtime
- Production of early warnings before costly serious damage sets in
- Reduced maintenance costs
- Longer intervals between scheduled maintenance
- Reduced operating costs
- Life extension

Experience has shown that the potential cost savings resulting from the implementation of corrosion monitoring programs generally increase with the sophistication level (and cost) of the monitoring system. However, the emergence of online, real-time corrosion monitoring can improve the relevance of corrosion measurements and greatly reduce the manual effort and the high expenses required to obtain relevant information.

New integrated technologies provide the possibility of using existing data acquisition and automation systems already present in production facilities in order to monitor and control processes, trend key process information, and manage and optimize system productivity. By integrating the corrosion monitoring tools in such systems, data acquisition can be automated and viewed with other process variables. The main advantages of this approach over stand-alone systems include the following [2]:

- Improved cost effectiveness
- Less manual labor to accomplish key tasks
- Greater degree of integration with in-place systems to record, control, and optimize
- Efficient distribution of important information (corrosion and process data, related work instructions, and follow-up reports) among different groups required for increased work efficiency and ease of documentation.

B. CORROSION MONITORING TECHNIQUES

An extensive range of corrosion monitoring techniques and systems has evolved, particularly in the last two decades, for detecting, measuring, and predicting corrosion damage. The development of efficient corrosion monitoring techniques and user-friendly software have led to new field techniques that were until recently perceived to be mere laboratory curiosities. In several sectors, such as oil and gas production, sophisticated corrosion monitoring systems have gained successful track records and credibility, while in other sectors their application has made only limited progress.

Many of the possible corrosion monitoring and inspection techniques available have been recently organized by a group of experts and interested users in different categories, as shown in Tables 87.1 and 87.2 [3]. In the report produced by these experts, a direct technique is one that measures parameters directly affected by the corrosion processes while an indirect technique provides data on parameters that either affect or are affected by the corrosivity of the environment or by the products of the corrosion processes.

Additionally a technique can be described as being intrusive if it requires access through a pipe or vessel wall in order to make the measurements. Most commonly used intrusive techniques make use of some form of probe or test specimen, which include flush-mounted probe designs. Some indirect techniques can serve to monitor various parameters online in real time while others provide information offline after samples collected from process streams or other operational locations are further analyzed following an established method. A detailed description of these techniques is beyond the scope of the present chapter and the reader should consult the report itself for additional information or a recent book published on the subject [4].

TABLE 87.1. Direct Corrosion Measurement Techniques

Intrusive Techniques

Physical techniques
- Mass loss coupons
- Electrical resistance (ER)
- Visual inspection

Electrochemical DC techniques
- Linear polarization resistance (LPR)
- Zero-resistance ammeter (ZRA) between dissimilar alloy electrodes—galvanic
- Zero-resistance ammeter (ZRA) between the same alloy electrodes
- Potentiodynamic/galvanodynamic polarization
- Electrochemical noise (ECN)

Electrochemical AC techniques
- Electrochemical impedance spectroscopy (EIS)
- Harmonic distortion analysis

Nonintrusive Techniques

Physical techniques for metal loss
- Ultrasonics
- Magnetic flux leakage (MFL)
- Electromagnetic—eddy current
- Electromagnetic—remote field technique (RFT)
- Radiography
- Surface activation and gamma radiometry
- Electrical field mapping

Physical techniques for crack detection and propagation
- Acoustic emission
- Ultrasonics (flaw detection)
- Ultrasonics (flaw sizing)

TABLE 87.2. Indirect Corrosion Measurement Techniques

Online Techniques

Corrosion products
- Hydrogen monitoring

Electrochemical techniques
- Corrosion potential (E_{corr})

Water chemistry parameters.
- pH
- Conductivity
- Dissolved oxygen
- Oxidation reduction (Redox) potential

Fluid detection
- Flow regime
- Flow velocity

Process parameters
- Pressure
- Temperature
- Dewpoint

Deposition monitoring
- Fouling

External monitoring
- Thermography

Offline Techniques

Water chemistry parameters
- Alkalinity
- Metal ion analysis (iron, copper, nickel, zinc, manganese)
- Concentration of dissolved solids
- Gas analysis (hydrogen, H_2S, other dissolved gases)
- Residual oxidant (halogen, halides, and redox potential)
- Microbiological analysis (sulfide ion analysis)

Residual inhibitor
- Filming corrosion inhibitors
- Reactant corrosion inhibitors

Chemical analysis of process samples
- Total acid number
- Sulfur content
- Nitrogen content
- Salt content in crude oil

C. CORROSION MONITORING LOCATIONS

An important decision in setting up a corrosion monitoring system is the selection of the monitoring points where sensing elements will be located. As only a finite number of points can be considered for obvious economical reasons, it is usually desirable to monitor the "worst-case" conditions at points where corrosion damage is expected to be most severe. Often, such locations can be identified from basic corrosion principles, from analysis of in-service failure records, and in consultation with operational personnel. For example, the most corrosive conditions in water tanks are usually found at the water–air interface. Corrosion sensors could be attached to a floating platform to maintain these conditions independently of water-level changes in order to monitor corrosion under these conditions.

In any monitoring situation, the ideal probe placement is most often not possible. Invariably, the flush or protruding electrode probes should be placed in the most corrosive environment (e.g., at a 6 o'clock position in a pipeline, at the bottom of a vessel, or at a solution accumulation point in a separator tower). Almost equally invariably, the available

location is entirely at odds with these requirements. Although certain steps can be taken to provide the application with a modified design (e.g., a protruding electrode probe installed at the 12 o'clock position with a long body to extend to the aqueous phase), this is without question a compromise on the part of the technology provider [5].

It is obviously imperative that corrosion sensors be positioned to reflect the state of the actual component or system being monitored. If this requirement is not met, all subsequent signal processing or data analysis is negatively impacted and the value of information greatly diminished or even rendered worthless. For example, if turbulence is induced locally around a protruding corrosion sensor

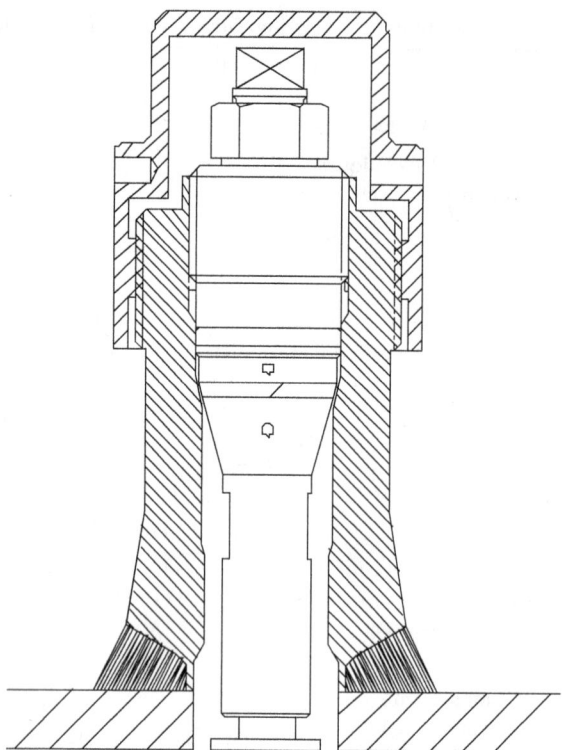

FIGURE 87.1. Flush-mounted corrosion sensor in an access fitting. (Courtesy of Metal Samples Company, www.metalsamples.com.)

mounted in a pipeline, the sensor will in all likelihood give a very poor indication of the risk of localized corrosion damage to the pipeline wall. In this particular case, a flush-mounted sensor should be used instead if the goal is to monitor localized corrosion (Fig. 87.1).

In practice, the choice of monitoring points is also dictated by the existence of suitable access points, especially in pressurized systems. It is usually preferable to use existing access points such as flanges for sensor installations. If it is difficult to install a suitable sensor in a given location, additional bypass lines with customized sensors and access fittings may represent a practical alternative. One advantage of a bypass is the opportunity of experimenting with local conditions of highly corrosive regimes in a controlled manner without affecting the actual operating plant.

The most important consideration when selecting corrosion monitoring locations within crude oil or wet gas production systems is to find locations near the end of the pipeline where the corrosion coupon or probe will be immersed in any produced water. This placement is typically at the 6 o'clock position on horizontal sections of pipeline because produced water is heavier than crude oil or gas condensates. In Figure 87.2(a), the monitoring probe installed at 6 o'clock is in an ideal position to sense corrosion processes.

Figure 87.2(b) illustrates a gas production line that contains a small quantity of condensate and produced water. The water is swept along the bottom of the pipe for horizontal runs. (Only at high fluid velocity is water swept along the circumference of the pipe.) Hence, monitoring locations on the side of the pipeline (3 or 9 o'clock) as shown in Figure 87.2(b) cannot accurately measure the corrosion rates associated with the aqueous phase at the bottom of the pipeline.

Unfortunately, many pipeline and facility designers have installed monitoring locations on the sides of the pipelines rather than the bottom. Although the side locations may

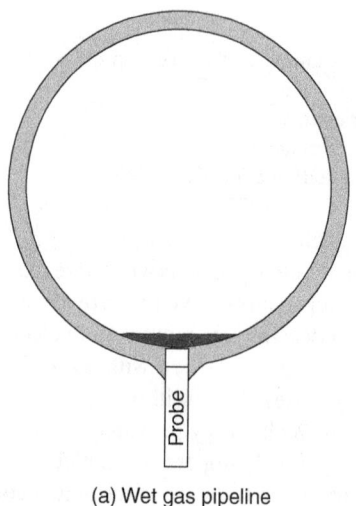

(a) Wet gas pipeline

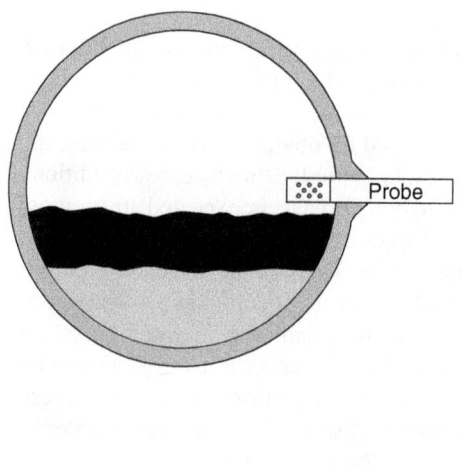

(b) Three-phase pipeline

FIGURE 87.2. (a) Probe at 6 o'clock. The flush-mounted probe is best positioned to monitor corrosion, even when there are small volumes of produced water, as in wet gas or some three-phase pipelines. (b) Probe at 3 o'clock. The intrusive probe is incorrectly positioned and cannot monitor corrosion associated with the water phase. This probe would yield invalid results because it is in the gas or oil phase.

provide easier access for coupon crews, these coupons or probes cannot provide accurate data unless the pipelines are essentially full of water. Hence, operator convenience must not come at the expense of obtaining valid results when selecting coupon or probe monitoring locations.

D. CORROSION MONITORING SYSTEMS

Corrosion monitoring systems vary significantly in complexity, from simple coupon exposures or hand-held data loggers (Fig. 87.3) to fully integrated plant process surveillance units with remote data access and data management capabilities. The first and most essential step to select a corrosion system is to define the general objective of the monitoring effort, a step often forgotten. If corrosion monitoring is done for corrosion control, the purpose is to assure that asset life is not jeopardized by too many high-corrosion-rate events. The main objective of corrosion monitoring is in this case to limit the "corrosion events" without using completely the corrosion allowance of a system before the end of its design life. The main factors that govern the design of a monitoring system in this case are [6]:

- Available corrosion allowance
- Uncontrolled corrosion rates
- Event rates
- Corrosion rate detection sensitivity and response rate
- Required service life

If corrosion monitoring is used for corrosion control, it is thus essential that the corrosion mechanism and the corrosion rates during times of noncompliance be relatively well known. Corrosion monitoring may also be used to optimize the corrosion control, for example, testing the efficiency of

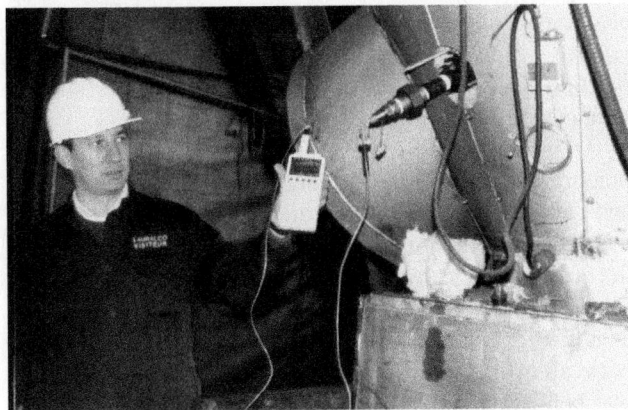

FIGURE 87.3. Field corrosion monitoring using electrochemical noise recorded with a hand-held data logger. (Courtesy of Kingston Technical Software.)

corrosion inhibitors, adjusting the corrosion inhibitor injection rate, or studying corrosion mechanisms. Such measurements can be carried out either on instrumented sections of the actual system or in a side stream.

Corrosion monitoring could also be needed in a broader context of integrity management to ensure that the operating envelope of a system is not exceeded. The time horizon of ongoing integrity activities can be much shorter than the plant lifetime. This can be satisfied by ensuring that integrity is maintained up to the next inspection date and reassessed for another period.

An effective corrosion monitoring system should exhibit the following characteristics [7]:

- *User Friendly.* The monitoring system must be simple to install, simple to use, and simple to interpret by system operators. At least some interpretation functions must be sufficiently developed so that the system can be interfaced to alarms and controllers for chemical treatment additions or online cleaning systems.
- *Rugged.* The monitoring system must be able to withstand the normal use and abuse if it is deployed in an industrial environment.
- *Sensitive.* The monitoring element or probe must be sensitive to the onset of a corrosion problem and provide a definitive indication in real time that may be used as a process control variable or to evaluate the effectiveness of a control measure.
- *Accurate.* False positives and negatives or any indications caused by interferences from effects such as flow, erosion, and fouling can be detrimental in many ways. Erroneous readings may seriously affect the credibility and straightforward usefulness of a corrosion monitoring program.
- *Maintainable.* Probes are expected to foul in service. A minimum time between servicing operations of several months to several years may be required for most applications. Periodic servicing and calibration should be simple and easy to perform.
- *Cost Effective.* The cost of the monitoring system must be significantly less than the cost of the downtime that is avoided or the treatment costs that are saved. The speed and accuracy of the technique are also factors in the cost effectiveness of the monitoring system.

E. INTEGRATION IN PROCESS CONTROL

In most field operations, corrosion is typically viewed as the difference between two measurements performed over a rather long interval of time. These corrosion measurements commonly come from measured changes in metal thickness obtained directly with ultrasonic inspection readings or by

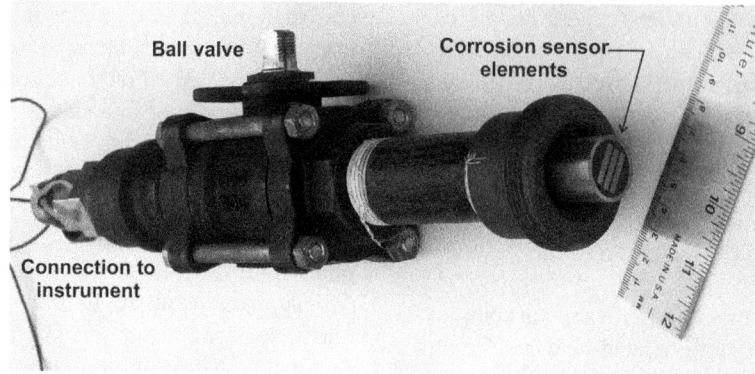

FIGURE 87.4. Corrosion sensor and access fitting used for thin-film corrosion monitoring. (Courtesy of Kingston Technical Software.)

monitoring the electrical resistance of probe elements or the mass loss of coupons. Such measurements are taken on the order of weeks, months, or sometimes years [2]. This traditional approach to corrosion monitoring can only provide a corrosion indication after the damage has accumulated and an average rate-of-metal loss during the measurement interval. With such a strategy, peak corrosion rates are not captured and documented in order to relate them to process conditions.

The following example illustrates how the corrosivity may vary at various points of an industrial gas scrubbing system where highly corrosive thin-film electrolytes are often prevalent [8]. These conditions arise when gas streams are cooled to a temperature below the dewpoint. The resulting thin electrolyte layer (moisture) is often highly concentrated in corrosive species.

The corrosion probe used in this example is illustrated in Figures 87.4 and 87.5. A retractable probe with flexible depth

FIGURE 87.5. Close-up of corrosion sensing elements used for thin-film corrosion monitoring. (Courtesy of Kingston Technical Software.)

was selected in order to mount the sensor surface flush with the internal scrubber wall surface. The close spacing of the carbon steel sensor elements was designed to work with a discontinuous thin surface electrolyte film. This corrosion sensor was connected to a hand-held multichannel data recorder by shielded multistrand cabling (Fig. 87.3). As the ducting of the gas scrubbing tower was heavily insulated, no special precautions were taken to cool the corrosion sensor surface.

Potential noise and current signals recorded during the first hour of exposure at the conical base of the gas scrubbing tower are presented in Figure 87.6. According to the operational history of the plant, condensate had a tendency to accumulate at this location where highly corrosive conditions had been noted. The high levels of potential noise and current noise in Figure 87.6 are indicative of a massive pitting attack which is consistent with the operational experience. It should be noted that the current noise is actually off scale for most of the monitoring period, in excess of 10 mA. The high corrosivity indicated by the electrochemical noise data from this sensor location was confirmed by direct evidence of severe pitting attack on the sensor elements, revealed by scanning electron microscopy (Fig. 87.7). In contrast, both current and voltage signals remained relatively constant and small at a position higher up in the tower, where the sensor surface remained mostly dry.

An improvement over this simple analysis is commonly practiced in industry by tracking the width of the potential and current signals as an indication of corrosion activity in the system being monitored. Figure 87.8 illustrates how the decrease in the current band obtained with a monitoring system was interpreted as a reduction in general corrosion activity in debutanizer overhead piping where the interaction between operational changes and the corrosion mechanism was being investigated.

In recent years, corrosion monitoring has developed from a manual, offline process to an online, real-time measurement (Fig. 87.9). The initial driving force for this migration was the

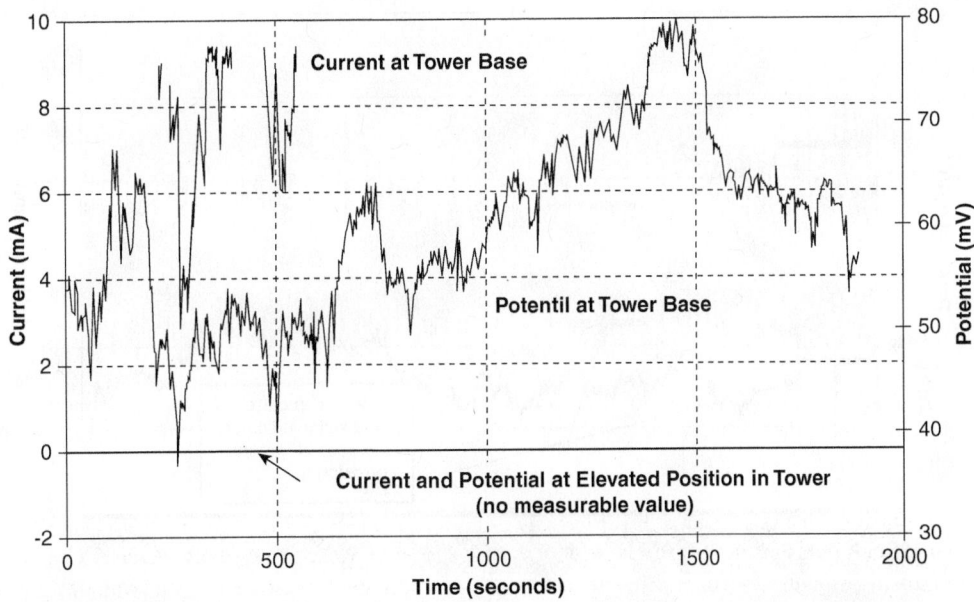

FIGURE 87.6. Potential and current noise records at two locations in a gas scrubbing tower.

benefit of automation permitting reduced time and effort to obtain corrosion data with a high level of data reliability. Corrosion monitoring takes on a new meaning when it can be viewed at a frequency that is consistent with the way process variables are measured and key performance indicators (KPIs) monitored.

In this regard, wireless technology has been an enabler for setting up much wider ranging networks of real-time corrosion data points in field assets and process plants

than was possible using conventional wired transmitters. Locations can now be dictated by critical need rather than by convenience of wire placement. Since corrosion can be a localized phenomenon, the ability to monitor more locations provides greater assurance that key locations have been included [2].

Management information is typically required on predicted costs of problems, the risks involved, the remaining life of the affected equipment, and what can be done to

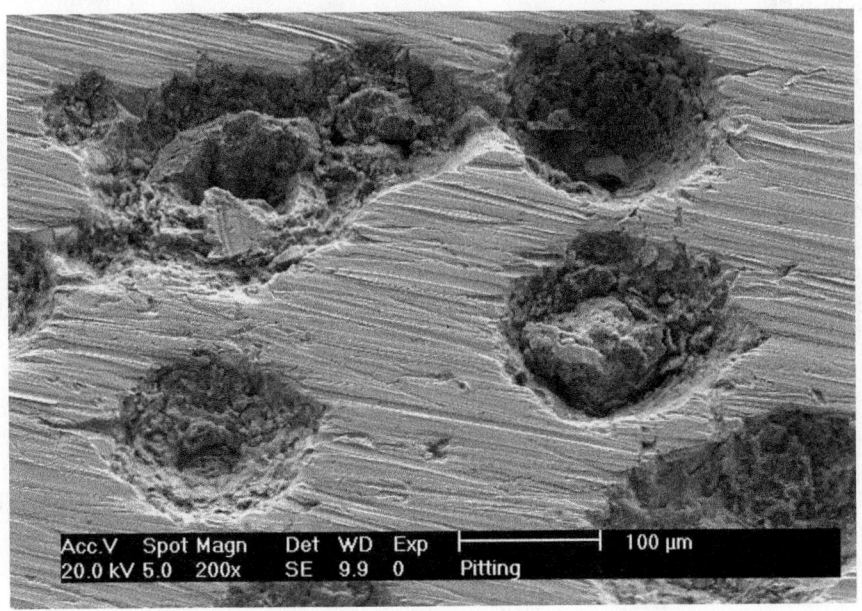

FIGURE 87.7. Scanning electron microscope image of a sensor element surface after exposure at the base of the scrubbing tower clearly showing corrosion pits.

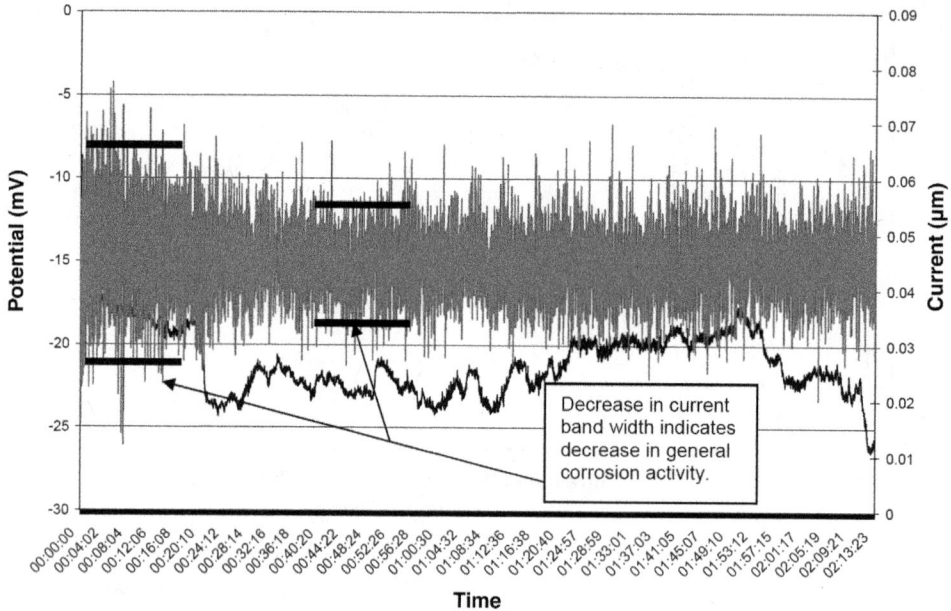

FIGURE 87.8. Electrochemical current noise (large band) and potential noise (lower signal) in debutanizer overhead piping obtained with the Concerto VT noise system. (Courtesy of CAPCIS Ltd.)

improve or eradicate these problems. The KPIs described in the following sections were developed specifically to measure the effect of corrosion on the technical and financial performance of assets involved in oil and gas production facilities and to address the performance of critical corrosion-related systems [9].

E1. Cost of Corrosion KPI

The cost of corrosion KPI allows converting the amount of corrosion damage sustained during a given period into a monetary figure to provide a clearer focus on corrosion management performance. Factors considered within this KPI are existing damage sustained prior to the period in question, the cost of repair or replacement, and the remaining service life of the plant. Performance can then be computed in terms of the cost of damage in the last period examined, the annual damage cost, and/or life-cycle costs. The cost of corrosion damage (C_{corr}) sustained in a given period can be derived with the equation

$$C_{corr} = \left(\frac{N_C R_{cost}}{FL}\right)\left(\frac{D_p}{365}\right) \qquad (87.1)$$

where C_{corr} = cost of corrosion damage in a specific time period
N_C = estimated number of replacement cycles to end of service life
R_{cost} = replacement cost (including lost product cost)

FL = required remaining field life (years)
D_p = days in monitoring period (days)

If however the calculated remaining life of a component (RL_C) as defined in Eq. (87.3) is greater than the field life (FL), C_{corr} can be assumed to be equal to zero. This is based on the assumption that the KPI is a performance indicator reflecting the effect on operating costs (Opex) and does not consider depreciation against the initial capital cost (Capex). The number of replacement cycles (N_C) can be estimated from the equation

$$N_C = \left(1 + \left(\frac{FL - RL_C}{RL_R}\right)\right) \qquad (87.2)$$

where RL_C = remaining life of current component (years)
RL_R = remaining life of replacement components (years)

RL_C and RL_R are respectively derived from Eqs. (87.3) and (87.4):

$$RL_C = \left(\frac{CA - DT}{CR}\right) \qquad (87.3)$$

$$RL_R = \left(\frac{CA}{CR}\right) \qquad (87.4)$$

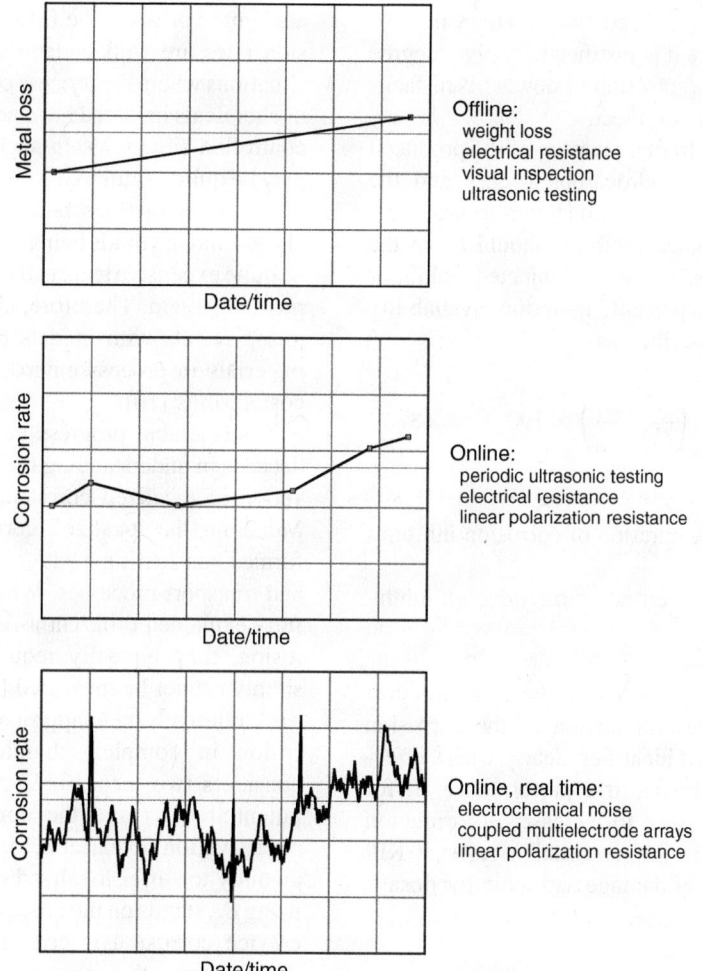

FIGURE 87.9. Corrosion monitoring has evolved from offline to online and online, real-time measurements. (Adapted from [2].)

where CA = corrosion allowance[1] (design or fitness for purpose) (mm)

DT = damage to date (mm)

CR = measured corrosion rate (mm/year)

The above formulas have been developed from an actual pipework monitoring and replacement program as a result of internal corrosion effects. Replacement cost (R_{cost}), component remaining life (RL_C), and field life (FL) are key factors in this method and reflect the need to understand accurate costing of installation and replacement activities. Required field life is not necessarily the difference between installation and design lives but more likely the time remaining until end-of-field or production life.

E2. Corrosion Inhibition Level KPI

This KPI is a measure of the availability of corrosion inhibitors to provide protection against corrosive processes. The inhibitor efficiency[2] itself should have been determined from a combination of previous laboratory and field testing to determine the optimum concentration that the chemical

---

[1] Corrosion allowance depends on the type of defect anticipated which needs to be identified by inspection. Once the defect geometry is known and the process parameters identified, the maximum allowable defect size may be calculated using fitness for purpose criteria to ensure that failure does not occur.

[2] The efficiency of an inhibitor is expressed as a measure of the improvement in lowering the corrosion rate of a system:

$$\text{Inhibitor efficiency } (\%) = \left(\frac{CR_{uninhibited} - CR_{inhibited}}{CR_{uninhibited}} \right) \times 100$$

where $CR_{uninhibited}$ = corrosion rate of the uninhibited system and $CR_{inhibited}$ = corrosion rate of the inhibited system.

inhibitor should be in the produced fluids. There may of course be applications where it is justified to apply a degree of overinjection to provide protection to downstream facilities where it is not practical to inject.

The KPI itself is derived from a measure of the produced fluids, including water and hydrocarbon phases and the inhibitor injected in the produced fluid stream to provide a correlation between how much inhibitor should be in the produced fluid stream versus actual injected inhibitor concentrations. The KPI percent inhibitor availability (Inhibitor$_{AV}$) function is described as

$$\text{Inhibitor}_{AV} = \left(\frac{C_{\text{actual}}}{C_{\text{required}}} \right) \times 100 \qquad (87.5)$$

where C_{actual} = actual concentration of corrosion inhibitor (ppm)

C_{required} = required concentration of corrosion inhibitor (ppm)

This KPI, when directly correlated with the corrosion cost KPI (C_{corr}), provides a clear indication of the corrosion performance of the asset and identifies clearly where effective action can be taken to improve performance if, for example, damage costs are seen to increase. A correlation between the cost-of-damage KPI and inhibition-level KPI may indicate how the cost of damage and inhibitor dosage level trend in actual performance.

E3. Completed Maintenance KPI

This KPI, which provides a measure of the reliability of the corrosion monitoring equipment, is determined from the asset's maintenance performance in repairing equipment faults reported during routine corrosion inspection visits. This measure reflects the recognized importance that equipment reliability is critical to the performance of corrosion inhibition systems and hence is also a key cost factor. The indicator is derived from a ratio of the number of maintenance actions raised versus the number of actions completed in a given monitoring period:

$$\% \text{ Completed maintenance} = \qquad (87.6)$$

$$\frac{\text{maintenance actions completed}}{\text{maintenance actions raised}} \times 100$$

F. MODELING CORROSION MONITORING RESPONSE

The interpretation of corrosion monitoring results may be relatively simple when the corrosion processes themselves are simple or when the factors causing the observed corrosion rates are well understood. However, there are many situations where the process conditions and associated fluids or chemicals involved have never been really investigated in controlled situations. In such cases, corrosion monitoring may be quite useful to develop a better understanding of the complex interactions between alloys being considered or chosen and the fluids being processed. Unfortunately, it may be quite expensive to test all the variations that occur in a real process stream. Therefore, comparing the corrosion monitoring results with models predicting the performance of materials in process environments can lead to significant cost savings [10].

Considerable progress has been made in the last three decades in understanding the initiation, growth, and repassivation of localized corrosion of many metallic materials. Modeling the localized corrosion process has been performed considering a battery of atomic/molecular processes and transport processes. While these models have successfully explained different aspects of pitting and crevice corrosion, they typically require too many parameters that simply cannot be measured [10].

A relatively new approach to predicting localized corrosion in complex chemical environments essentially considers two measurable components: the repassivation potential (E_{rp}) and the corrosion potential (E_{corr}). The repassivation potential E_{rp} is a measure of the tendency of an alloy to suffer localized corrosion. The justification for using E_{rp} stands on the observation that only stable pitting or crevice corrosion is critical to the life of a component whereas pits that nucleate without growing beyond an embryonic stage (metastable pits) do not adversely affect the performance of engineering structures. It has been shown that, for noble alloys, stable pitting or crevice corrosion does not occur below E_{rp} and that E_{rp} is relatively insensitive to prior pit depth and surface finish. The susceptibility of an alloy to localized corrosion in a given environment can be revealed by comparing the repassivation potential with E_{corr}.

The concept is illustrated schematically in Figure 87.10. For a given alloy, the repassivation potential E_{rp} decreases with an increase in chloride concentration. Three general types of E_{rp} behavior may be observed, but in some cases only a semilogarithmic decrease is observed. The corrosion potential itself is only slightly influenced by changes in chloride concentration unless significant localized corrosion occurs. The critical chloride level for localized corrosion to occur is when E_{rp} is lower than E_{corr} [Fig. 87.10(a)]. Similarly, for a given chloride concentration a critical temperature [Fig. 87.10(b)] and a critical inhibitor concentration exist [Fig. 87.10(c)]. In many processes, incidental contamination of the process fluid by redox species may increase E_{corr} values such that localized corrosion may occur beyond a critical concentration of redox species [(Fig. 87.10(d)]. The

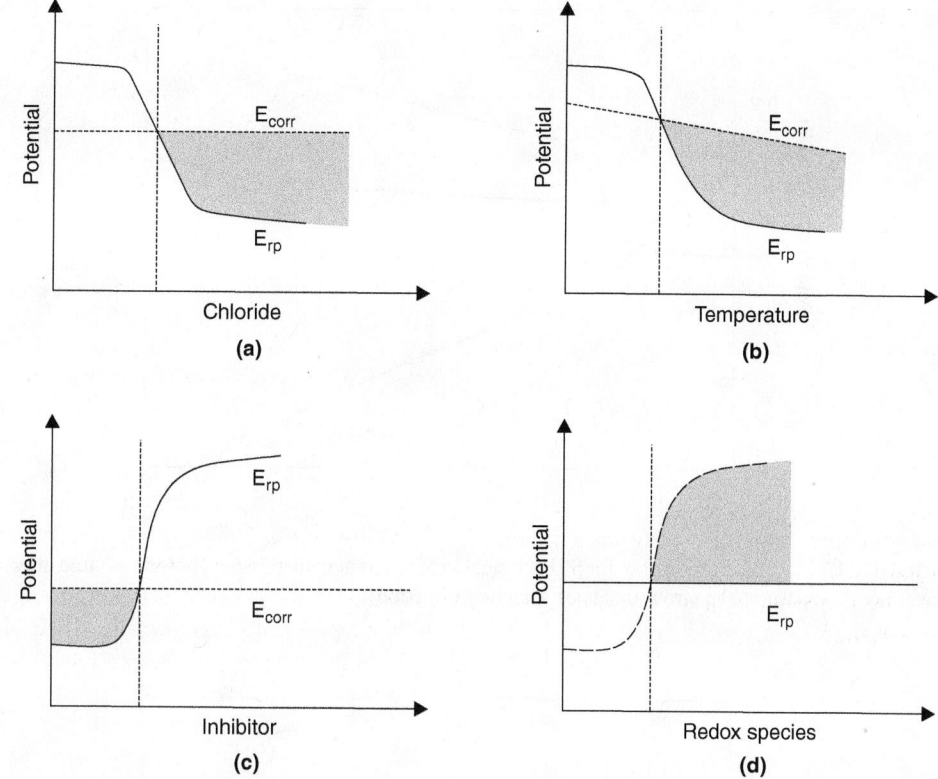

FIGURE 87.10. Comparison of repassivation (E_{rp}) and corrosion potentials (E_{corr}) for different environmental parameters. Shaded areas denote localized corrosion.

condition in a system will typically be a combination of the idealized cases shown in Figures 87.10(a)–(d).

Cyclic polarization tests are often used to reveal these characteristics in a practical manner and evaluate the pitting susceptibility of a material. The potential is swept in a single cycle or slightly less than one cycle usually starting the scan at the corrosion potential. The voltage is first increased in the anodic or noble direction (forward scan). The voltage scan direction is reversed at some chosen current or voltage toward the cathodic or active direction (backward or reverse scan) and terminated at another chosen voltage. The presence of the hysteresis between the currents measured in the forward and backward scans is believed to indicate pitting, while the size of the hysteresis loop itself has been related to the amount of pitting that has occurred during the scan.

This technique has been especially useful to assess localized corrosion for passivating alloys such as S31600 stainless steel, nickel-based alloys containing chromium, and other alloys such as titanium and zirconium. Though the generation of the polarization scan is simple, its interpretation can be difficult [11].

In the following example, the polarization scans were generated after one and four days of exposure to a chemical product maintained at 49°C. The goal of these tests was to examine if S31600 steel could be used for short-term

storage of a 50% commercial organic acid solution (aminotrimethylene phosphonic acid) in water. A small amount of chloride ion (1%) was also present in this acidic chemical.

In this example, the potential scan rate was 0.5 mV/s and the scan direction was reversed at 0.1 mA/cm². Coupon immersion tests were run in the same environment for 840 h. The S31600 steel specimens were exposed to the liquid, at the vapor–liquid interface, and in the vapor. The reason for the three exposures was that in most storage situations the containment vessel would be exposed to a vapor–liquid interface and a vapor phase at least part of the time. Corrosion in these regions can be very different from liquid exposures. The specimens were also fitted with artificial crevice formers.

Figure 87.11 shows the polarization scan generated after one day and Figure 87.12 shows the polarization scan generated after four days of exposure. The important parameters considered were the position of the "anodic-to-cathodic" transition relative to the corrosion potential, the existence of the repassivation potential and its value relative to the corrosion potential, the existence of the pitting potential and its value relative to the corrosion potential, and the hysteresis (positive or negative). The interpretation of the results is summarized in Table 87.3.

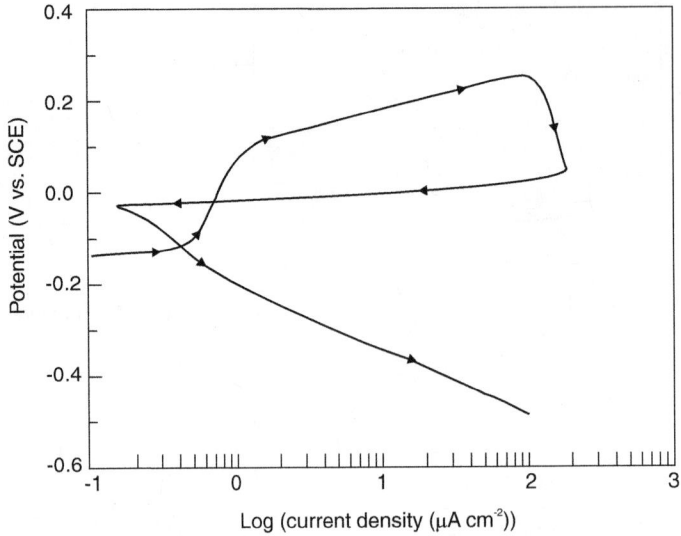

FIGURE 87.11. Polarization scan for S31600 steel in 50% aminotrimethylene phosphonic acid after one day of exposure (the arrow indicates scanning direction).

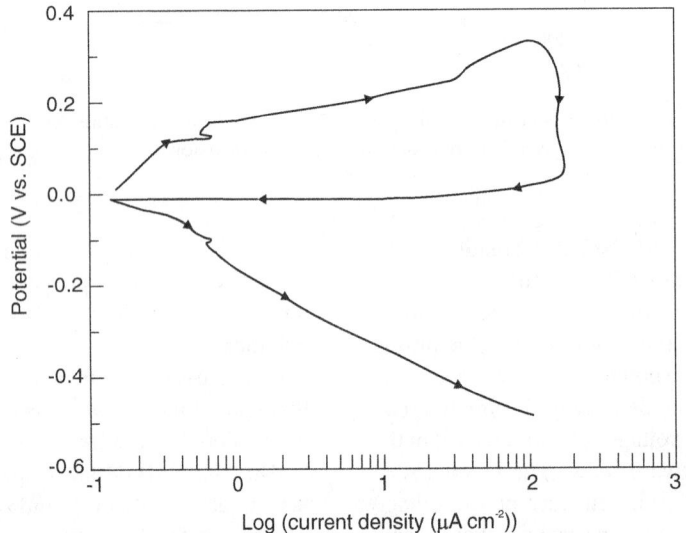

FIGURE 87.12. Polarization scan for S31600 steel in 50% aminotrimethylene phosphonic acid after four days of exposure (the arrow indicates scanning direction).

The presence of the negative hysteresis would typically suggest that localized corrosion is possible depending on the value of the corrosion potential relative to the characteristic potentials present in these polarization plots. After the first day of exposure, pitting was not expected to be a problem because the pitting potential was far away from the corrosion potential. The currents generated were much higher than those normally associated with S31600 steel in a passive state. These observations suggested that there was a risk of initiation of corrosion, particularly in localized areas where the pH can decrease drastically [11].

After four days, the risk of localized corrosion increased. At this time, the repassivation potential and the potential of the change from anodic to cathodic current were equal to the corrosion potential. The pitting potential was only about 0.1 V more noble than the corrosion potential and the hysteresis still negative. The risk of pitting had increased enough to become a concern.

Coupon immersion tests confirmed the long-term predictions. Slight attack was found under the artificial crevice formers in the complete liquid exposure. The practical conclusion of this in-service study was that, since localized

TABLE 87.3.   Features and Values Used to Interpret Figures 87.11 and 87.12

Feature	Value from Fig. 87.11	Value from Fig. 87.12
Repassivation potential—corrosion potential	0.12 V	0.0 V
Pitting potential—corrosion potential	0.22 V	0.12 V
Potential of anodic-to-cathodic transition—corrosion potential	0.12 V	0.0 V
Hysteresis	Negative	Negative
Active-to-passive transition	No	No

corrosion often takes time to develop, a few days of exposure to this chemical product could be acceptable. However, it was recommended to avoid long-term exposure since both pitting and crevice corrosion would be expected for longer exposure periods.

G.   PROBE DESIGN AND SELECTION

Corrosion sensors (probes) are an essential element of all corrosion monitoring systems. The nature of the sensors depends on the various individual techniques used for monitoring, but often a corrosion sensor can be viewed as an instrumented coupon. A single high-pressure access fitting for insertion of a retrievable corrosion probe can be used to accommodate most types of retrievable probes. With specialized tools sensor insertion and withdrawal can be possible under pressurized operating conditions.

All too often, the quality and relevance of the corrosion data measured can be severely compromised by inappropriate probe design. In this context, knowledge of the probe surface condition is particularly crucial during the initial design and obviously remains important for the duration of the exposure period.

Other factors to consider relate to surface roughness, residual stresses, corrosion products, surface deposits, preexisting corrosion damage, and temperature that can all have an important influence on corrosion damage and need to be taken into account for making representative probes. Considering these factors, it can be desirable to manufacture corrosion sensors from a precorroded material that has experienced actual operational conditions. Heating and cooling may also be applied to corrosion sensors, using special devices, for their surface conditions to reflect certain plant operating domains. Sensor designs such as spool pieces in pipes and heat exchanger tubes, flanged sections of candidate materials, or test paddles bolted to agitators also represent efforts to make these sensors representative of actual operational conditions.

The choice of a specific monitoring probe should also be based on the anticipated corrosion rates within the system as well as on the required sensitivity. When conducting a short-term corrosion test, probes with high sensitivity are desired. For long-term monitoring, however, a thicker probe element with a longer measurement lifetime may be desired.

G1.   Sensitivity and Response Time

The usefulness of a corrosion monitoring system strongly depends on how well it can deliver warnings of unwanted corrosion conditions. For measuring techniques this translates into two closely related properties:

- The sensitivity to detect a change in corrosion rate
- The time it requires to detect such a change, that is, the response time

By virtue of the measuring principle of many systems, sensitivity and response time have an inverse relation to each other. In order to compare corrosion monitoring systems on the basis of their sensitivity, it is important to distinguish between the accuracy of the measurement and the sensitivity to measure a change in the corrosion rate. The sensitivity to measure corrosion rate is the combined result of measurement accuracy and the elapsed time [6].

Sensitivity (S) and response time (R) of a certain technique are closely related and can be conveniently displayed in a single graph, as shown in Figure 87.13. It is most convenient to display such S–R curves on a log–log graph. The example in this figure has an application window which requires a response time of one to seven days while requiring corrosion rate measurement sensitivity in the range of 1–20 mm/year. The S–R curve for this example, based on a specific technique, lies below the application window and hence will satisfy each requirement for this application. In Table 87.4 typical applications are given with their characteristics. These applications are depicted in Figure 87.14 for their respective S–R windows.

A monitoring system can also be limited to measure accurately over long time scales because of inherent instability of the monitoring system. This might be the result of deterioration of the sensor or drift in the recording instrumentation. For systems that have to measure with high sensitivity and over a long interval it is recommended to perform routine verifications and calibrations of the monitoring equipment.

For rapid flow conditions or if there are concerns related to suspended solids, the sensing elements should be protected with a velocity shield. An ER probe can also be used to measure an "erosion rate" associated with production of sand or other solids. For this purpose, a noncorroding metal element should be selected [12].

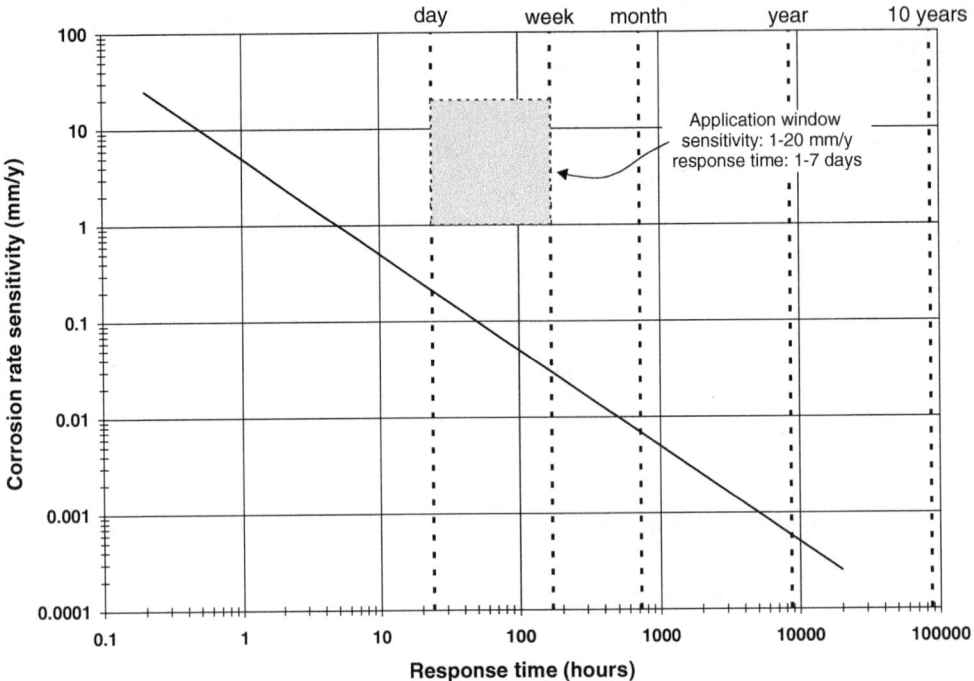

FIGURE 87.13. Corrosion monitoring system sensitivity (1–20 mm/year)/response time (1–7 days) application window for a given system performance threshold (solid line).

TABLE 87.4. Sensitivity and Response Time of Typical Corrosion Monitoring Applications

Application	Sensitivity Range (mm/year)	Response Time	System Characteristics
Corrosion tests	0.1–100	1 h–5 days	Continuous
Inhibition control	0.1–20	0.5–2 days	Continuous optimization
Corrosion control (upsets)	1–100	1 h–2 days	Continuous monitoring (upsets)
Corrosion control performance demonstration	1–10	1 week–1 month	Continuous/interval measurement
Inspection planning	0.2–10	1 month–0.5 year	Interval
Inspection	1–20	3 months–10 years	Interval

G2. Flush-Mounted Electrode Design

The flush-mounted electrode design is most appropriate for use in applications such as oil and gas flowlines where pigging operations are necessary. While the design is suited to this application operational need, it greatly limits the electrode exposed surface area and the accuracy of the measurements, particularly in low-conductivity environments or with low-sensitivity instrumentation. As with most measurement processes there is a trade-off between available area for measurement and the opportunity to actually measure corrosion events of low statistical probability [5].

Great care must also be taken in the manufacture of this type of probe as the opportunity exists to artificially create unwanted physical phenomena such as crevices. A crevice created between the outer circumference of an electrode and the surrounding insulating material could provide a focal point for localized corrosion activity and introduce a significant error in the measured data. This effect would be reduced by using larger surface area electrodes.

While this type of probe can be supplied with electrodes of sufficient material to last the lifetime of the component into which it is installed, the monitoring program undoubtedly will benefit from the ability to replace the probe on a regular basis for visual inspection and to confirm the measured data.

G3. Protruding-Electrode Design

The protruding-electrode design has more broad-ranging applications than the flush-mounted design. A major benefit

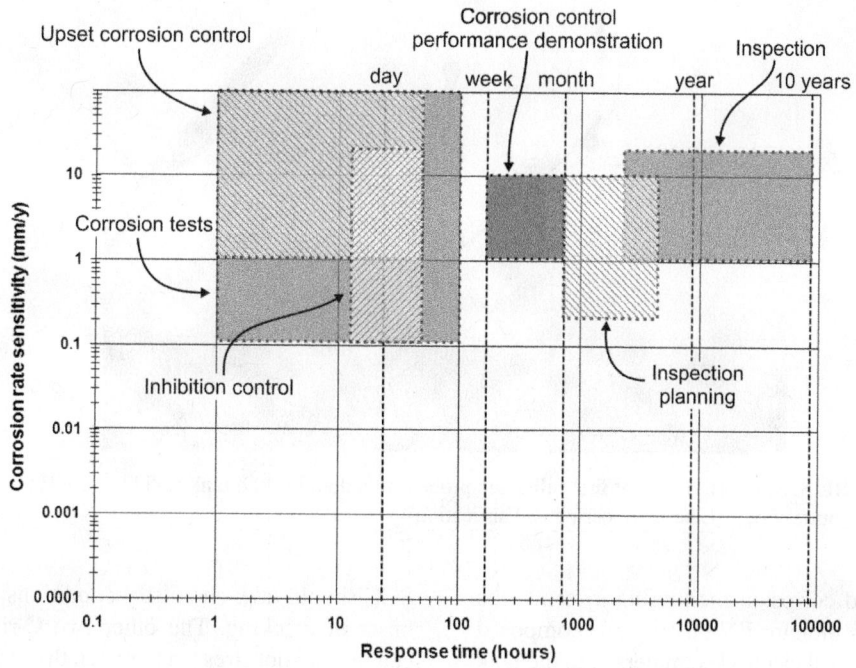

FIGURE 87.14. Application windows depicted in the *S–R* plot. The size range of the application windows is given in Table 86.4.

of this design is the possibility to use replaceable electrodes, as this provides a cost-effective solution. The possibility of crevice corrosion is also less important as the exposed length of an electrode increases the ratio of exposed surface area to the region where a crevice might occur, that is, the circumference of the electrode. However, this design relies on the complete exposure of the electrode surface to the corrosive environment, and so issues may arise in situations where the flow regime within the monitored system becomes turbulent or if the water cut is reduced significantly during operation [5].

G4. Probes to Suit the Application

Each corrosion monitoring application has its specific needs and requirements. The following sections describe a few probe designs that have been developed for specific degradation mechanisms and environments.

G4.1. Stress Corrosion Cracking Probe. Corrosion probes have been developed that enable the working electrode of a three-electrode arrangement to be prestressed in order to match the operating condition of a pipe or vessel. In the following example, corrosion monitoring and control of the double-shell tanks (DSTs) at Hanford had historically been provided through a waste chemistry sampling and analysis program. In this program, waste tank corrosion was inferred by comparing waste chemistry samples taken periodically from the DSTs with the results from a

series of laboratory tests done on tank steels immersed in a wide range of normal and off-normal waste chemistries [13].

This method has been effective but is expensive and time consuming and does not yield real-time data. The Hanford Site near Richland, Washington, has 177 underground waste tanks that store approximately 253 million liters of radioactive waste from 50 years of plutonium production. In 1996, the Department of Energy Tanks Focus Area launched an effort to improve Hanford's DST corrosion monitoring strategy and to help address questions concerning the remaining useful life of these tanks. Several new methods of online localized corrosion monitoring were evaluated. The electrochemical noise (EN) technique was selected for further study based on numerous reports that showed this technique to be the most appropriate for monitoring and identifying the onset of localized corrosion.

Based on a series of studies, a three-channel prototype field probe was designed constructed, and deployed in August 1996. Following the demonstration of the prototype for approximately a year, a longer, more advanced eight-channel system was designed and installed in September 1997. Figure 87.15 shows the installation of this system. Unlike the previous prototype, the in-tank probe on this system reached from tank top to tank bottom exposing two channels of EN electrodes in the sludge at the tank bottom, four channels in the tank supernate, and two channels in the tank vapor space. Four additional systems of similar design have been installed into other DSTs.

FIGURE 87.15. Installation of first full-scale probe into a double-shell tank (DST) at the Hanford site. (Courtesy of HiLine Engineering & Fabrication.)

Like most EN-based corrosion monitoring systems, the active Hanford systems monitor EN on channels composed of three nominally identical electrodes immersed in the tank waste. Each system is composed of an in-tank probe and ex-tank data collection hardware. The in-tank probe is fabricated from an ~17-m-long piece of 2.5-cm-diameter stainless steel tubing. Eight three-electrode channels are distributed along the probe body. Electrodes are fabricated from United Numbering System (UNS) K02400 steel that has been heat treated to match the tank wall heat treatment. Four channels on each probe are formed from sets of bullet-shaped electrodes ($25\,cm^2$/electrode). Four channels are formed from sets of thick-walled C-rings ($44\,cm^2$/electrode). Figure 87.16 shows two channels on the most recent probe. The unstressed bullet-shaped electrodes are used for pitting and uniform corrosion detection. The working electrode on each C-ring channel is notched, precracked, and stressed to yield prior to installation to facilitate the monitoring of

SCC should tank chemistry conditions change to allow the onset of cracking. The other two C-rings on each C-ring channel are not stressed to match the operating conditions of the vessel. Bullet and C-ring channels alternate up the length of the probe. Current DST waste levels in monitored tanks immerse three channels of bullet-shaped electrodes and three channels of C-ring electrodes.

In this way, the working electrode is allowed to behave in a manner most representative of the material in service, thus providing corrosion information reflecting the real-life situation of the plant equipment. The exposure of single or multiple corrosion probes can enable informed decisions to be made regarding the choice of a material or of a stress relief process.

G4.2. Corrosion in Hydrocarbon Environments. In hydrocarbon environments there must be an electrolytically conductive phase present which is generally provided by

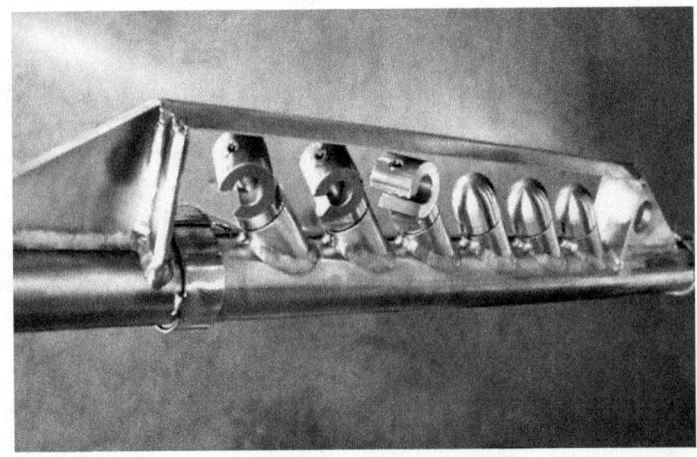

FIGURE 87.16. Detail of the Hanford site 25-cm^2 bullet and 47-cm^2 C-ring channel electrodes. (Courtesy of Glenn Hedgemon, HiLine Engineering & Fabrication.)

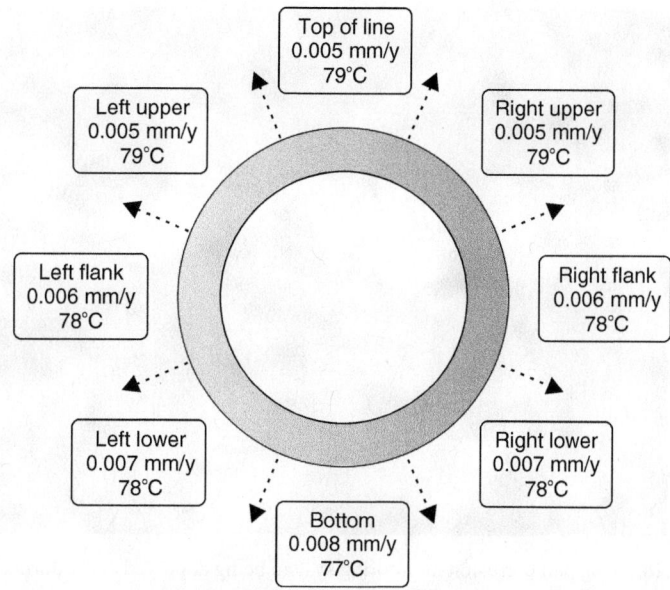

FIGURE 87.17. Principle of ring pair corrosion monitoring. (Courtesy of Cormon Ltd.)

an aqueous phase or by polar solvents for corrosion to occur. Examples may be flowlines in oil and gas applications or pipelines in chemical process environments where it may not be straightforward to introduce a complex probe system.

A circumferential spool probe has been developed especially for this application to allow a maximum contact of the electrodes with the process environment in "true" flowing conditions along flowlines or pipelines, especially in multiphase oil and gas applications.

The basic ring principle of these sensors allows elements to be made by "salami" slicing a pipe and reassembling pairs of the resulting rings, separated by insulation, to remake a pipe section capable of retaining line pressure. Each electrically isolated ring is measured using pick-up wires attached to the outer face. If wires are attached at equally spaced intervals around the ring, the instrumentation can be configured to measure the overall metal loss and the loss in the segment between each pick-up point Figure 87.17.

One of each pair of rings is kept from contact with the process stream by means of a high-integrity, thin-film, ceramic coating. The coated ring acts as the reference to the exposed sample ring.

A number of pairs of rings may be used enabling the study of different materials including weldment and heat-affected zone (HAZ) material if preferential weld corrosion is an issue. In addition to the temperature data from the elements, it is a simple matter to include pressure measurement in the device. Together these may add considerably to the understanding of the behavior of the fluid in the line.

Standard rings have the same wall thickness as the original line so no problem should arise with element life

in relation to the service life of the line. The thicker wall rings do have a lower speed of response that may not always meet the requirement, especially if real-time adjustment to chemical treatment is proposed. In this case a combination of two concentric rings may be used to provide a fast response from a thin element inside a thicker supporting ring.

The potentially limited life of the element is balanced by the ability to maintain low corrosion rates through active control, thereby extending the life of the asset and the sensor. The spool sensor is therefore a very versatile measurement tool for looking at the character and degree of corrosion of various materials in conditions that are a true representation of line flow. It is capable of being finely tuned to perform the required task with great precision.

The sensor housing uses a double-pressure barrier principle. The sensor rings, spacers, and isolators are held in compression by a clamp arrangement producing an inner pressure-tight cylinder. This cylinder is mounted in the outer housing using a pair of elastomer seal rings to complete the primary containment. The outer housing is a pressure-tight assembly in its own right, sealed using flanges, ring-type joints, and a spacer ring. Electronics, housings, power consumption, and telemetry are similar in most respects to those for intrusive probes. Figure 87.18 illustrates a monitoring system being deployed with a pipeline as it is submerged into the sea.

G4.3. Coupled Multielectrode Array Systems and Sensors.
The use of multielectrode array systems (CMASs) for corrosion monitoring is relatively new. The advantages of using multiple electrodes include the ability to obtain

FIGURE 87.18. Ring pair corrosion monitoring system being deployed at sea during the installation of a submerged pipeline. (Courtesy of Cormon Ltd.)

greater statistical sampling of current fluctuations, a greater ratio of cathode-to-anode areas that enhances the growth of localized corrosion once initiated, and, depending on the design, the ability to estimate the pit penetration rate and to obtain macroscopic spatial distribution of localized corrosion [14].

Figure 87.19 shows the principle of the CMAS in which a resistor is positioned between each electrode and the common coupling point. Electrons from a corroding or a relatively more corroding electrode flow through the resistor connected to the electrode and produce a small potential drop usually of the order of a few microvolts. This potential drop is measured by the high-resolution voltage-measuring instrument and used to derive the current of each electrode. The CMAS probes can be made in several configurations and sizes, depending on the applications. Figure 87.20 shows some of the typical probes that were reported for real-time corrosion monitoring.

The data from these CMAS probes are the large number of current values measured at a given time interval from all the electrodes. In a CMAS probe system, these data are reduced to a single parameter so that the probe can be conveniently used for real-time and online monitoring purposes. The most anodic current has been used as a

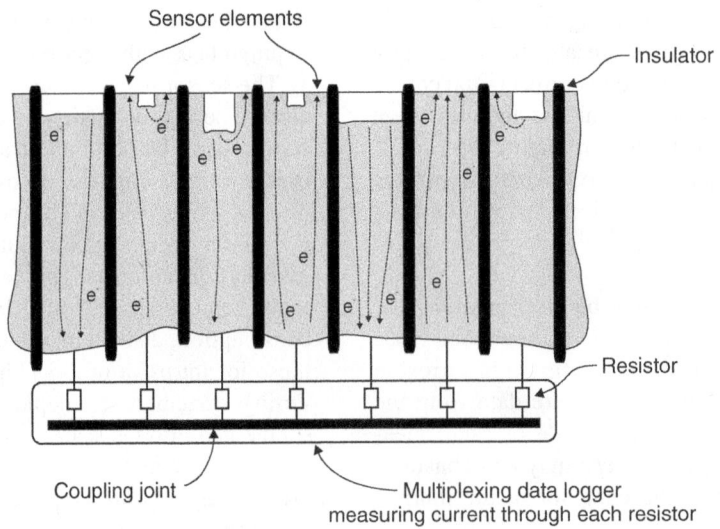

FIGURE 87.19. Multielectrode array multiplexed current and/or potential measurement. (Adapted from [17].)

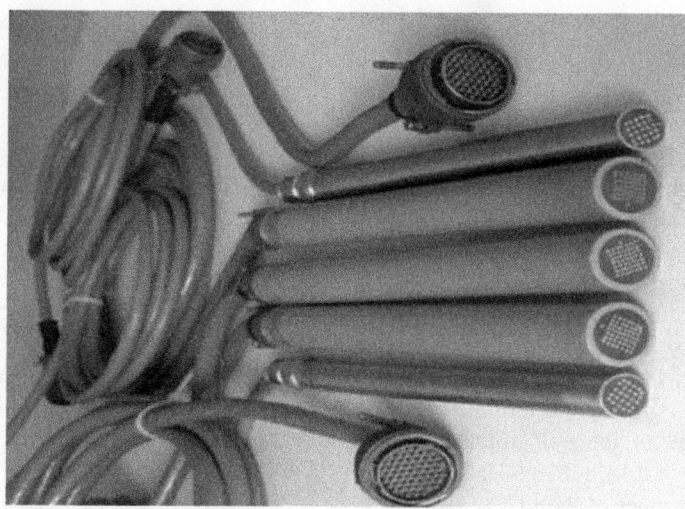

FIGURE 87.20. Typical CMAS probes used for real-time corrosion monitoring. (Courtesy of Corr Instruments.)

one-parameter signal for the CMAS probes. Because the anodic electrodes in a CMAS probe simulate the anodic sites on a metal surface, the most anodic current may be considered as the corrosion current from the most corroding site on the metal.

The value based on three times the standard deviation of currents is another way to represent the corrosion current from the most corroding site on the metal. Because the number of electrodes in a CMAS probe is always limited and usually far fewer than the number of corroding sites on the surface of a metal coupon, the value based on the statistical parameter, such as three times the standard deviation of current, was considered to be more appropriate than the single value of the most anodic current. The standard deviation value may be from the anodic currents or from both the anodic and cathodic currents.

In a less corrosive environment or with a more corrosion-resistant alloy, the most anodic electrode may not be fully covered by anodic sites until the electrode is fully corroded. Therefore, the most anodic electrode may still have cathodic sites available, and the electrons from the anodic sites may flow internally to the cathodic sites within the same electrode. The total anodic corrosion current, I_{corr}, and the measured anodic current, I_a^{ex}, may be related as

$$I_a^{ex} = \varepsilon I_{corr} \tag{87.7}$$

where ε is a current distribution factor that represents the fraction of electrons resulting from corrosion that flows through the external circuit. The value of ε may vary between zero and unity, depending on parameters such as surface heterogeneities on the metal, the environment, the electrode size, and the number of sensing electrodes. If an

electrode is severely corroded and significantly more anodic than the other electrodes in the probe, the ε value for this corroding electrode would be close to 1, and the measured external current would be equal to the localized corrosion current.

Because the electrode surface area is usually between 1 and 0.03 mm^2, which is approximately two to four orders of magnitude less than that of a typical LPR probe or a typical EN probe, the prediction of penetration rate or localized corrosion rate by assuming uniform corrosion on the small electrode is realistic in most applications. CMAS probes have been used for monitoring localized corrosion of a variety of metals and alloys in the following environments and conditions:

- Deposits of sulfate-reducing bacteria
- Deposits of salt in air
- High-pressure simulated natural gas systems
- H$_2$S systems
- Oil–water mixtures
- Cathodically protected systems
- Cooling water
- Simulated crevices in seawater
- Salt-saturated aqueous solutions
- Concentrated chloride solutions
- Concrete
- Soil
- Low-conductivity drinking water
- Process streams of chemical plants at elevated temperatures
- Coatings

H. DATA COMMUNICATION AND ANALYSIS REQUIREMENTS

The signal emanating from a corrosion sensor usually has to be processed and analyzed. Examples of signal processing include filtering, averaging, and unit conversions. Furthermore, in some corrosion sensing techniques, the sensor surface has to be perturbed by an input signal to generate a corrosion signal output. In older systems, electronic sensor leads were usually employed for these purposes and to relay the sensor signals to a signal processing unit. Advances in microelectronics have facilitated the sensor signal conditioning and processing by the introduction of microchips that have become an integral component of sensor units [15, 16]. Wireless data communication with such sensing units is also a product of the microelectronic revolution.

Irrespective of the sensor details, a data acquisition system is required for online and real-time corrosion monitoring. On several plants, the data acquisition system is housed in mobile laboratories, which can be made intrinsically safe. A computer system often performs a combined role of data acquisition, data processing, and information management. In data processing, a process is initiated to transform corrosion monitoring data (low intrinsic value) into process-relevant information (higher intrinsic value). Complementary data from other relevant sources such as process parameter logging and inspection reports can be acquired together with the data from corrosion sensors for use as input to a management information system.

It is important, at the onset of a corrosion monitoring program, to define the full data communication chain from signaling unacceptable corrosion to the implementation of a remedial action. The times for each step in the chain should be in balance; that is, it is obviously not very useful to invest in a system with a response time of one day if it requires weeks to process the information and/or months to implement follow-up remedial measures. The following individuals may be involved in the communication process [6]:

1. Process plant operator to collect data
2. Corrosion monitoring specialist (corrosion or inspection engineer) to process data
3. Corrosion engineer to assess the information and determine follow-up
4. Operations or maintenance engineer to plan and implement remedial action

The response time from sensor to desk for steps 1–3 determine the actual response time obtained from a corrosion monitoring system. For a highly critical monitoring task the data might go directly to the party responsible for remedial action (e.g., to the control room for action by an operator).

The perceived importance of the monitoring system and strategy has to be mirrored by commitment of all individuals involved in integrity management, that is, the asset holder, usually operations, but also maintenance and inspection staff, corrosion engineering, production chemists, and frequently the chemical treating contractor. It is essential that the approach is agreed upon and implemented by a team that includes these individuals, who together decide not only how corrosion should be controlled but also how the corrosion monitoring should be implemented.

REFERENCES

1. P. J. Moreland and J. G. Hines, Hydrocarbon Process. **57**, 543 (1978).

2. R. D. Kane, "A New Approach to Corrosion Monitoring," Chem. Eng. **114**, 34–41 (2007).

3. Techniques for Monitoring Corrosion and Related Parameters in Field Applications, NACE 3T199, NACE International, Houston, TX, 1999.

4. P. R. Roberge, Corrosion Inspection and Monitoring, Wiley, New York, 2007.

5. D. C. Eden, M. S. Cayard, J. D. Kintz, R. A. Schrecengost, B. P. Breen, and E. Kramer, "Making Credible Corrosion Measurements—Real Corrosion, Real Time," Paper No. 376, CORROSION 2003, NACE International, Houston, TX, 2003.

6. M. J. J. S. Thomas and S. Terpsta, "Corrosion Monitoring in Oil and Gas Production," Paper No. 431. CORROSION 2003, NACE International, Houston, TX, 2003.

7. T. P. Zintel, G. J. Licina, and T. R. Jack, "Techniques for MIC Monitoring," in A Practical Manual on Microbiologically Influenced Corrosion, J. G. Stoecker (Ed.), NACE international, Houston, TX, 2001.

8. P. R. Roberge, Handbook of Corrosion Engineering, McGraw-Hill, New York, 2000.

9. D. M. E. Queen, B. R. Ridd, and C. Packman, "Key Performance Indicators for Demonstrating Effective Corrosion Management in the Oil and Gas Industry," Paper No. 056, CORROSION 2001, NACE International, Houston, TX, 2001.

10. A. Anderko, N. Sridhar, L. T. Yang, S. L. Grise, B. J. Saldanha, and M. H. Dorsey, "Validation of Localised Corrosion Model Using Real Time Corrosion Monitoring in a Chemical Plant," Corros. Eng. Sci. Technol. **40**, 33–42 (2005).

11. D. C. Silverman, "Tutorial on Cyclic Potentiodynamic Polarization Technioue," Paper No. 299, CORROSION 98, NACE International, Houston, TX, 1998.

12. D. E. Powell, D. I. Ma'ruf, and I. Y. Rahman, "Practical Considerations in Establishing Corrosion Monitoring for

Upstream Oil and Gas Gathering Systems," Mater. Perform. **40**, 50–54 (2001).

13. G. L. Edgemon, "Electrochemical Noise Based Corrosion Monitoring: Hanford Site Program Status," Paper No. 584, CORROSION 2005, NACE International, Houston, TX, 2005.

14. L. Yang, N. Sridhar, O. Pensado, and D. S. Dunn, "An In-Situ Galvanically Coupled Multielectrode Array Sensor for Localized Corrosion," Corrosion, **58**, 1004–1014 (2002).

15. B. Zollars, N. Salazar, J. Gilbert, and M. Sanders, Remote Datalogger for Thin Film Sensors, NACE International, Houston, TX, 1997.

16. R. G. Kelly, J. Yuan, S. H. Jones, W. Blanke, J. H. Aylor, W. Wang, and A. P. Batson, "Embeddable Microinstruments for Corrosion Monitoring," Corrosion 97, NACE International, Houston, TX, 1997, pp. 1–12.

17. L. Yang and N. Sridhar, "Coupled Multielectrode Online Corrosion Sensor," Mater. Perform. **42**, 48–52 (2003).

88

DIAGNOSING, MEASURING, AND MONITORING MICROBIOLOGICALLY INFLUENCED CORROSION

B. J. Little, R. I. Ray, and J. S. Lee

Naval Research Laboratory, Stennis Space Center, Mississippi

A. INTRODUCTION

Many techniques have been described for diagnosing, measuring, and monitoring microbiologically influenced corrosion (MIC). However, none has been accepted as an industry standard or as a recommended practice by the American Society for Testing and Materials (ASTM) or NACE International. This chapter reviews the literature with an emphasis on the strengths and weaknesses of each technique. Diagnosing MIC after it has occurred requires a combination of microbiological, metallurgical, and chemical analyses. MIC investigations have typically attempted to (1) identify causative microorganisms in the bulk medium or associated with the corrosion products, (2) identify a pit morphology consistent with an MIC mechanism, and (3) identify a corrosion product chemistry that is consistent with the causative organisms. Electrochemical (EC) techniques can be used to measure and monitor MIC. The major limitation for MIC monitoring programs is the inability to relate microbiology to corrosion in real time. Some techniques can detect a specific modification in the system due to the presence and activities of microorganisms (e.g., heat transfer resistance, fluid friction resistance, galvanic current) and assume something about the corrosion. Others measure some electrochemical parameter (e.g., polarization resistance, electrochemical noise) and assume something about the microbiology. With experience and knowledge of a particular operating system either can be an effective monitoring tool, especially for evaluating a treatment regime (biocides or corrosion inhibitors).

B. DIAGNOSING

B1. Identification of Causative Organisms

The following are required for an accurate diagnosis of MIC:

1. A sample of the corrosion product or affected surface that has not been altered by collection or storage
2. Identification of a corrosion mechanism
3. Identification of microorganisms capable of growth and maintenance of the corrosion mechanism in the particular environment
4. Demonstration of an association of the microorganisms with the observed corrosion

Three types of evidence are used to diagnosis MIC: metallurgical, chemical, and biological. The objective is to have three independent types of measurements that are consistent with a mechanism for MIC.

The list of microorganisms involved in MIC and the resulting mechanisms are continuously growing. Causative microorganisms are from all three main branches of evolutionary descent, that is, bacteria, archaea, and eukaryotes. For many years the first step in identifying corrosion as MIC was to determine the presence of specific groups of bacteria in the bulk medium (planktonic cells) or associated with corrosion products (sessile cells). There are four approaches:

1. Culture the organisms on solid or in liquid media.
2. Extract and quantify a particular cell constituent.
3. Demonstrate/measure some activity.
4. Demonstrate a spatial relationship between microbial cells and corrosion products using microscopy.

B1.1. Culture Techniques. The method most often used for detecting and enumerating groups of bacteria is the serial dilution to extinction method using selective culture media. To culture microorganisms, a small amount of the sample of interest as a liquid or a suspension of a solid (the inoculum) is added to a solution or solid that contains nutrients (culture medium). There are three considerations when growing microorganisms: type of culture medium, incubation temperature, and length of incubation. The present trend in culture techniques is to attempt to culture several physiological groups, including aerobic, heterotrophic bacteria; facultative anaerobic bacteria; sulfate-reducing bacteria (SRB); and acid-producing bacteria (APB). Growth is detected as turbidity or a chemical reaction within the culture medium. Traditional SRB media contain sodium lactate as the carbon source [1]. When SRB are present in the sample, sulfate is reduced to sulfide, which reacts with iron (in either solution or solid) to produce black ferrous sulfide. Culture media are typically observed over several days (30 days may

be required for growth of SRB). There have been several attempts to improve culture media and to grow higher numbers of bacteria or to shorten the time required for some indication of growth. A complex SRB medium was developed containing multiple carbon sources that can be degraded to both acetate and lactate. In comparison tests, the complex medium produced higher counts of SRB from waters and surface deposits among five commercially available media [2]. Jhobalia et al. [3] developed an agar-based culture medium for accelerating the growth of SRB. The authors noted that over the range of 1.93–6.50 g/L, SRB grew best at the lowest sulfate concentration. Cowan [4] developed a rapid culture technique for SRB based on rehydration of dried nutrients with water from the system under investigation. The author claimed that using system water reduced the acclimation period for microorganisms by ensuring that the culture medium had the same salinity as the system water used to prepare the inoculum. The author reported quantification of SRB within one to seven days.

The distinct advantage of culturing techniques to detect specific microorganisms is that low numbers of cells grow to easily detectable higher numbers in the proper culture medium. However, there are numerous limitations for the detection and enumeration of cells by culturing techniques. Several investigators have followed the changes in microflora as a function of water storage. Zobell and Anderson [5] and Lloyd [6] demonstrated that when water is stored in glass bottles the bacterial numbers fall within the first few hours followed by an increase in the total bacterial population with a reduction in the number of species. If results from culturing techniques are to be related to the natural populations, culture media should be inoculated within hours of collection and the sample should be chilled during the interim. Under all circumstances culture techniques underestimate the organisms in a natural population [7, 8]. Kaeberlein et al. [9] suggest that 99% of microorganisms from the environment resist cultivation in the laboratory. One major problem in assessing microorganisms in natural environments is that viable microorganisms can enter into a nonculturable state [10]. Another problem is that culture media cannot approximate the complexity of a natural environment. Growth media tend to be strain specific. For example, lactate-based media sustain the growth of lactate oxidizers, but not acetate-oxidizing bacteria. Incubating at one temperature is further selective. The type of medium used to culture microorganisms determines to a large extent the numbers and types of microorganisms that grow. Zhu et al. [11] demonstrated dramatic changes in the microbial population from a gas pipeline after samples were introduced into liquid culture media. Similarly, Romero et al. [12] found that some bacteria present in small amounts in the original waters were enriched in the culture process. Archaea will not be detected using traditional culture techniques.

B1.2. Biochemical Assays.

Biochemical assays have been developed for the detection of specific microorganisms associated with MIC. Unlike culturing techniques, biochemical assays for detecting and quantifying bacteria do not require growth of the bacteria. Instead, biochemical assays measure constitutive properties including adenosine triphosphate (ATP) [13], phospholipid fatty acids (PLFAs) [14], cell-bound antibodies [15], and DNA [16]. Adenosine-5'-phosphosulfate (APS) reductase [17] and hydrogenase [18] have been used to estimate SRB populations.

B1.3. Physiological Activity.

Roszak and Colwell [10] reviewed techniques commonly used to detect microbial activities in natural environments, including transformations of radiolabeled metabolic precursors. Phelps et al. [19] and Mittelman et al. [20] used uptake or transformation of ^{14}C-labeled metabolic precursors to examine activities of sessile bacteria in natural environments and in laboratory models. Phelps et al. [19] used a variety of ^{14}C-labeled compounds to quantify catabolic and anabolic bacterial activities associated with corrosion tubercles in steel natural gas transmission pipelines. They demonstrated that organic acid was produced from hydrogen and carbon dioxide in natural gas by acetogenic bacteria and that acidification could lead to enhanced corrosion of the steel. Mittelman et al. [20] used measurement of lipid biosynthesis from ^{14}C-acetate, in conjunction with measurements of microbial biomass and extracellular polymer, to study effects of differential fluid shear on the physiology and metabolism of *Alteromonas* (formerly *Pseudomonas*) *atlantica*. Increasing shear force increased the rate of total lipid biosynthesis but decreased per cell biosynthesis. Increasing fluid shear also increased cellular biomass and greatly increased the ratio of extracellular polymer to cellular protein. Maxwell [21] developed a radiorespirometric technique for measuring SRB activity on metal surfaces that involved two distinct steps: incubation of the sample with ^{35}S sulfate and trapping the released sulfide.

B1.4. Molecular Techniques.

Molecular techniques have been used to identify and quantify microbial populations in natural environments [22–24]. These techniques involve amplification of 16S ribosomal RNA (rRNA) gene sequences by polymerase chain reaction (PCR) amplification of extracted and purified nucleic acids. The PCR products can be evaluated using community fingerprinting techniques such as denaturing gradient gel electrophoresis (DGGE). Each DGGE band is representative of a specific bacterial population and the number of distinctive bands is indicative of microbial diversity. The PCR products can also be sequenced, and the sequences are compared to the sequences in the Genbank database, which allows the identity of the species within an environmental sample. Horn et al. [25] identified the constituents of the microbial community within a proposed nuclear waste repository using two techniques: (1) isolation of DNA from growth culture and subsequent identification by 16S rRNA genes and (2) isolation of DNA directly from environmental samples followed by subsequent identification of the amplified 16S rRNA genes. Comparison of the data from the two techniques demonstrates that culture-dependent approaches underestimated the complexity of microbial communities. Zhu et al. [26, 27] used quantitative PCR and functional genes, that is, dissimilatory sulfite reductase for SRB, nitrite reductase gene for denitrifying bacteria, and methyl-coenzyme M reductase gene for methanogens, to characterize the types and abundance of bacterial species in gas pipeline samples. They found that methanogens were more abundant in most pipeline samples than denitrifying bacteria and that SRB were the least abundant bacteria.

Fluorescent in situ hybridization (FISH) uses specific fluorescent dye-labeled oligonucleotide probes to selectively identify and visualize SRB in both established and developing multispecies biofilms. Takai and Horikoshi [28] report that quantitative nucleic acid hybridization and FISH with archaeal rRNA-targeted nucleotide probes and competitive quantitative PCR could be used to detect and quantify archaea in a microbial community.

Restrictive fragment length polymorphism (RFLP) is a technique in which microorganisms can be differentiated by analysis of patterns derived from cleavage of their DNA. RFLP has been used to characterize bacterial communities in biofilms on copper pipes [29] and in concretions on the *USS Arizona* [30]. The conclusion in both studies was that the bacterial community played a role in corrosion.

B1.5. Microscopy.

Using light microscopy and proper staining, investigators [31, 32] have demonstrated a relationship between an unusual variety of copper pitting corrosion and gelatinous, polysaccharide-containing biofilms. Epifluorescence microscopy techniques have been developed for the identification of specific bacteria in biofilms [33, 34]. Epifluorescence cell surface antibody methods are based on the binding between cell-specific antibodies and subsequent detection with a secondary antibody. Antigenic structures of marine and terrestrial strains are distinctly different and therefore antibodies to either strain do not react with the other. Confocal laser scanning microscopy (CLSM) permits one to create three-dimensional images, determine surface contour in minute detail, and accurately measure critical dimensions by mechanically scanning the object with laser light. A sharply focused image of a single horizontal plane within a specimen is formed while light from out-of-focus areas is repressed from view. The process is repeated again and again at precise intervals on horizontal planes and the visual data from all images are compiled to create a single, multidimensional view of the subject. Geesey et al. [35] used CLSM to produce three-dimensional images of bacteria within scratches, milling lines, and grain boundaries. Atomic force microscopy (AFM) uses a microprobe mounted on a flexible cantilever to detect surface topography by scanning at a subnanometer scale. Repulsion by electrons overlapping at the tip of the microprobe causes deflections of the cantilever that can be detected by a laser beam. The signal is read by a feedback loop to maintain a constant tip displacement by varying voltage to a piezoelectric control. The variations in the voltage mimic the topography of the sample and together with the movement of the microprobe in the horizontal plane are converted to an image. Telegdi et al. [36] used AFM to image biofilm formation, extracellular polymer production, and subsequent corrosion.

Many of the conclusions about biofilm development, composition, distribution, and relationship to substratum/corrosion products have been derived from traditional scanning electron microscopy (SEM) and transmission electron microscopy (TEM). SEM has been used to image SRB from corrosion products on alloy 904L [37], microorganisms in corroding gas pipelines [26], and iron-oxidizing *Gallionella* in water distribution systems [38]. TEM has been used to demonstrate that bacteria are intimately associated with sulfide minerals and that on copper-containing surfaces the bacteria were found between alternate layers of corrosion products and attached to base metal [39].

Environmental electron microscopy includes both scanning (ESEM) and transmission (ETEM) techniques for the examination of biological materials with a minimum of manipulation, that is, fixation and dehydration. Little et al. [40] used ESEM to study marine biofilms on stainless steel surfaces. They observed a gelatinous layer in which bacteria and microalgae were embedded. Traditional SEM images of the same areas demonstrated a loss of cellular and extracellular material. Ray and Little [41] and Little et al. [42] used ESEM to demonstrate sulfide-encrusted SRB in corrosion layers on copper alloys and iron-depositing bacteria in tubercles on galvanized steel. Little et al. [43] used environmental TEM to image *Pseudomonas putida* on corroding iron filings and to demonstrate that the organisms were not directly in contact with the metal. Instead, the cells were attached to the substratum with extracellular material. Design and operation of the ESEM and ETEM have been described elsewhere [41].

There are fundamental problems in attempting to diagnose MIC by establishing a spatial relationship between numbers and types of microorganisms in the bulk medium or those associated with corrosion products using any of the techniques previously described. Zintel et al. [44] established that there were no relationships between the presence, type, or levels of planktonic or sessile bacteria and the occurrence of pits. Because microorganisms are ubiquitous, the presence of bacteria or other microorganisms does not necessarily indicate a causal relationship with corrosion.

B2. Pit Morphology

Pope [45] completed a study of gas pipelines to determine the relationship between extent of MIC and the levels/activities of SRB. He concluded that there was no relationship. Instead he found large numbers of APB and organic acids and identified the following metallurgical features in carbon steel:

- Large craters from 5 to 8 cm or greater in diameter surrounded by uncorroded metal
- Cup-type hemispherical pits on the pipe surface or in the craters
- Striations or contour lines in the pits or craters running parallel to longitudinal pipe axis (rolling direction)
- Tunnels at the ends of the craters also running parallel to the longitudinal axis of the pipe

Pope [45] reported that these metallurgical features were "fairly definitive for MIC." However, he did not advocate diagnosis of MIC based solely on pit morphology. Subsequent research has demonstrated that these features can be produced by abiotic reactions [46] and cannot be used to independently diagnose MIC.

Other investigators described ink-bottle-shaped pits in 300 series stainless steel that were supposed to be diagnostic of MIC. Borenstein and Lindsay [47, 48] reported that dendritic corrosion attack at welds was "characteristic of MIC." Hoffman [49] suggested that pit morphology was a "metallurgical fingerprint . . . definitive proof of the presence of MIC." Chung and Thomas [50, p. 2] compared MIC pit morphology with non-MIC chloride-induced pitting in 304/304L and E308 stainless steel base metals and welds. A faceted appearance was common to both types of pits in 304 and 304L base metal. Facets were located in the dendritic skeletons in MIC and non-MIC cavities of E308 weld metal. They concluded that there were no unique morphological characteristics for MIC pits in these materials. The problem that has resulted from the assumption that pits can be independently interpreted as MIC is that MIC is often misdiagnosed. For example, Welz and Tverberg [51] reported that leaks at welds in a 316L stainless steel hot water system in a brewery after six weeks in operation were due to MIC. The original diagnosis was based on the circumstantial evidence of attack at welds and the pitting morphology of scalloped pits within pits. However, a thorough investigation determined that no bacteria were associated with corrosion sites and deposits were too uniform to have been produced by bacteria. MIC was subsequently dismissed as the cause of localized corrosion. The hemispherical pits had been produced when CO_2 was liberated and low pH bubbles nucleated at surface discontinuities.

More recently several investigators have demonstrated that during the initial stages of pit formation due to certain types of bacteria, pits do have unique characteristics. Geiser et al. [52] found that pits in 316L stainless steel due to the manganese-oxidizing bacterium *Leptothrix discophora* had different morphologies than pits initiated by anodic polarization. The similarity between the dimensions of the bacterial cells attached to the surface and the dimensions of corrosion pits indicate a possibility that the pits were initiated at the sites where the microorganisms were attached. Eckert [53] used API 5L steel to demonstrate micromorphological characteristics that could be used to identify MIC initiation. Coupons were installed at various points in a pipeline system and were examined by SEM at 1000X and 2000X. They demonstrated that pit initiation and bacterial colonization were correlated and that pit locations physically matched the locations of cells. Telegdi et al. [36] demonstrated that pits produced by *Thiobacillus intermedius* had the same shape as the bacteria. None of these investigators claimed that these unique features can be detected with the unaided eye or that the features will be preserved as pits grow, propagate, and merge.

B3. Corrosion Product Composition

B3.1. Elemental Composition.
Elements in corrosion deposits can provide information about the cause of corrosion. Energy-dispersive X-ray (EDX) analysis coupled with SEM can be used to determine the elemental composition of corrosion deposits. Because all living organisms contain ATP, a phosphorus peak in an EDX analysis spectrum can be related to cells associated with the corrosion products. Other sources of phosphorus (e.g., phosphate water treatments) must be eliminated. The activities of SRB and manganese-oxidizing bacteria produce surface-bound sulfur and manganese, respectively. Chloride is typically found in crevices and pits and cannot be directly related to MIC. There are several limitations for EDX surface chemical analyses. Samples for EDX cannot be evaluated after heavy metal coating, so that EDX spectra must be collected prior to examination by SEM, making it difficult or impossible to match spectra with exact locations on images. This is not a problem with the ESEM because nonconducting samples can be imaged directly, meaning that EDX spectra can be collected of the area that is being imaged by ESEM. Little et al. [40] documented the changes in surface chemistry as a result of solvent extraction of water, a requirement for SEM. Other shortcomings of SEM/EDX include peak overlap. Peaks for sulfur overlap peaks for molybdenum and the characteristic peak for manganese coincides with the secondary peak for chromium. Wavelength-dispersive spectroscopy can be used to resolve overlapping EDX peaks. Peak heights cannot be used to determine the concentration of elements. It is also impossible to determine the valence state of an element with EDX.

B3.2. Mineralogical Fingerprints.
McNeil et al. [54] used mineralogical data determined by X-ray crystallography, thermodynamic stability diagrams (Pourbaix), and the simplexity principle for precipitation reactions to evaluate corrosion product mineralogy. They concluded that many sulfides under near-surface natural environmental conditions could only be produced by microbiological action on specific precursor metals. They reported that copper sulfides, djurleite, spinonkopite, and the high-temperature polymorph of chalcocite were mineralogical fingerprints for the SRB-induced corrosion of copper–nickel alloys. They also reported that the stability or tenacity of sulfide corrosion products determined their influence on corrosion. Jack et al. [55] reported that the mineralogy of corrosion products on pipelines could provide insight into the conditions under which the corrosion took place.

B3.3. Isotope Fractionation.
The stable isotopes of sulfur (^{32}S and ^{34}S), naturally present in any sulfate source, are selectively metabolized during sulfate reduction by SRB and the resulting sulfide is enriched in ^{32}S [56]. The ^{34}S is enriched in the starting sulfate as the ^{32}S is removed and becomes concentrated in the sulfide. Little et al. [57] demonstrated sulfur isotope fractionation in sulfide corrosion deposits resulting from activities of SRB within biofilms on copper surfaces. The ^{32}S accumulated in sulfide-rich corrosion products, and ^{34}S was concentrated in the residual sulfate in the culture medium. Accumulation of the lighter isotope was related to surface derivatization or corrosion as measured by weight loss. Use of this technique to identify SRB-related corrosion requires sophisticated laboratory procedures.

C. MEASURING AND MONITORING

EC techniques used to measure and monitor MIC include those in which no external signal is applied [e.g., measurement of redox potential ($E_{r/o}$) or corrosion potential (E_{corr}) and electrochemical noise analysis (ENA)], those in which only a small potential or current perturbation is applied [e.g., polarization resistance (R_p) and electrochemical impedance spectroscopy (EIS)], and those in which the potential is scanned over a wide range (e.g., anodic and cathodic polarization curves, pitting scans) [58]. The terms used to describe monitoring tools are *real time*, *online*, *in line*, and *side stream*. Real time refers to the measurements that are available at the actual time of collection and that are usually continuous or nearly continuous. On-line monitors are installed to provide real-time measurements and in line defines a measurement made in the bulk medium of a process stream. Side-stream devices are installed in parallel to the main system, taking a portion of the flow under identical operating conditions. The major limitation for MIC monitoring programs is the inability to relate microbiology to corrosion in real time.

C1. Techniques Requiring No External Signal

C1.1. Galvanic Couples.
Zero-resistance ammeter (ZRA) measurements of galvanic current have been used for many years to measure and monitor the electrochemical impact of microorganisms on metal surfaces in laboratory experiments and in field conditions using several designs, including concentric-ring electrodes, dual cells, and occluded cells mimicking small crevices.

Angell et al. [59] used a concentric-ring 304 stainless steel electrode to demonstrate that a consortium of SRB and a *Vibrio* sp. maintained a galvanic current between the anode and cathode. The anode ($0.031\ cm^2$) was concentric to and separated from the cathode ($4.87\ cm^2$) by a polytetrafluoroethylene spacer. Current was applied for 72 h either during or after microbial colonization. Once the applied current was removed, the resultant galvanic current flowing between the anode and the cathode was monitored by ZRA. They found that a current was maintained in the presence of a microbial consortium. No current was measured in a sterile control. The authors state that the concentric-ring electrode provides a technique by which MIC can be studied and is not intended to represent any natural situation.

In similar experiments, Campaignolle and Crolet [60] used a concentric-ring C1020 carbon steel (CS) electrode to examine stabilization of pitting corrosion by SRB. A current density of $1.5\ mA/cm^2$ was applied to the anode versus the cathode for 48 h to induce corrosion in deoxygenated nutrient-enriched synthetic seawater. During this preconditioning period, the test media was inoculated with *Desulfovibrio vulgaris* while a sterile condition was monitored as a control. SRB were shown to stabilize pitting corrosion of CS by maintaining a stable galvanic current between a local anode and a surrounding cathode. In the absence of SRB, the current was negligible. The authors noted that the sustained galvanic corrosion required both anode and cathode colonization by SRB. Corrosion rates were less than the observed rates in some field results.

Licina and Nekoksa [61] developed a probe consisting of ten 316 stainless steel concentric rings separated by epoxy. A potential (which varies according to experimental conditions) was imposed for 1 h each day between the electrodes so that the electrodes are alternately anodes and cathodes. The metal discs were polarized to produce an environment "conducive to biofilm formation." The applied current, required to achieve a preset potential between electrodes, remained stable until a biofilm formed. Once a biofilm was established, applied current increased. The emergence of generated current (galvanic current that continued to flow between the electrodes after the external polarization had been removed) was another indication of biofilm development. In the absence of a biofilm the generated current was zero. The probe is intended to provide information about an operating system (such as a flowing cooling water piping) that can be used to make decisions about cleaning or treatment, not to investigate localized corrosion mechanisms. The device has been used in fire protection systems, emergency service water stands, and equipment cooling water systems in nuclear power plants. The device provides a warning when the biofilm maintains a certain current so that the system can be cleaned or biocides can be added. This technique only indicates a corrosion risk. It does not measure any specific properties of biofilms or any parameter related to corrosion.

The dual-cell, split-cell, or biological battery can be used to monitor changes in corrosion rates due to the presence of a biofilm. A semipermeable membrane biologically separates two identical electrochemical cells. The two working electrodes are connected to a ZRA. Bacteria are added to one of the two cells and the sign and magnitude of the resulting galvanic current is monitored to determine details of the corrosive action of the bacteria [62].

Mollica and Ventura [63] used a galvanic couple between a stainless steel pipe and a copper/zinc pipe to monitor biofilm growth on surfaces exposed to natural seawater. Results of the field test showed a measurable increase in the galvanic current due to 10^7 cells/cm^2 and a current decrease after chlorination. They concluded that the device allowed optimization of antifouling treatments by controlling chlorine concentrations and frequency of injections to minimize biofilm recovery rate.

Uchida et al. [64] developed an electrochemical device for simulating a single pit consisting of an artificial anode and a carbon steel tube acting as the cathode coupled to a ZRA. The monitoring device was evaluated at a refinery plant-cooling tower. Pitting associated with biofilms had been observed on some heat exchangers during maintenance shutdowns. The goal was to improve the cooling water treatment program by reducing total maintenance costs. The authors reported that the galvanic current measurement was a sensitive indicator of biofilm formation and biocide effectiveness for real-time monitoring. Iimura et al. [65] reported that they had succeeded in predicting the penetration rate of carbon steel tubes of heat exchangers with a growth model of pitting corrosion and a similar device.

Galvanic current cannot be directly related to corrosion current [66]. The previously described techniques do not provide a means to calculate corrosion rates; rather they provide changes due to the presence of a biofilm.

C1.2. Open-Circuit or Corrosion Potential E_{corr}.
The E_{corr} measurements require a stable reference electrode—usually assumed to be unaffected by biofilm formation—and a high-impedance voltmeter. The E_{corr} values are difficult to interpret, especially when related to MIC [58]. Despite this

limitation, no other phenomenon has fascinated those studying MIC more than ennoblement, that is, the increase of E_{corr} due to formation of a biofilm on a metal surface. Although ennoblement has been observed for metals exposed to both freshwater (rivers and estuaries) and natural seawater, the mechanisms may be different. Little and Mansfeld [67] categorized the proposed mechanisms for ennoblement in marine environments into three categories: thermodynamic, kinetic, and alteration of the nature of the reduction reaction itself. It is not possible to determine the cause of ennoblement from E_{corr}–time curves.

C1.3. Electrochemical Noise Analysis. Electrochemical noise (EN) data can be obtained as fluctuations of potential and/or current. In laboratory studies, it is possible to measure potential and current fluctuations (EPN and ECN, respectively) simultaneously. In this approach two electrodes of the same material are coupled through a ZRA. Current fluctuations are measured with the ZRA, while the potential fluctuations are measured with a high-impedance voltmeter between the two coupled electrodes and a reference electrode which could be a stable reference electrode such as a saturated calomel electrode (SCE) or a third electrode of the same material as the two test electrodes. The main application of EN data has been in corrosion monitoring [68].

King et al. [69] interpreted noise measurements for steel pipes in environments containing SRB as being indicative of film formation and breakdown. Higher noise levels and greater fluctuations indicated localized corrosion. The magnitude of noise fluctuations depends on the total impedance of the system. A corroding metal undergoing uniform corrosion with fairly high corrosion rates might be less noisy than a passive metal showing occasional bursts of noise due to localized breakdown of the film followed by rapid repassivation.

Little et al. [70] used ENA for remote online monitoring of carbon steel electrodes in a test loop of a surge water tank at a gas storage field. Average corrosion rates (CR_{int}) for 19 noise electrodes were compared with the mean corrosion rates determined from weight loss data [Figs. 88.1(a)–(d)]. A least squares analysis was used to fit a straight line through the data. Results indicated that the slope of the line was 0.84 with the regression coefficient (R^2) equal to 0.86. The line with slope 1 is for the ideal case where results obtained using the electrochemical noise data and weight loss measurements are identical (Fig. 88.2).

C1.4. Microsensors. Lee and de Beer [71] described ideal microsensors as having the following qualities: small tip diameters to prevent distortion of the local environment, small sensor surfaces for optimal spatial resolution, low noise levels, stable signal, high selectivity, and strength to resist breakage. Specific microelectrode designs has been described elsewhere [72].

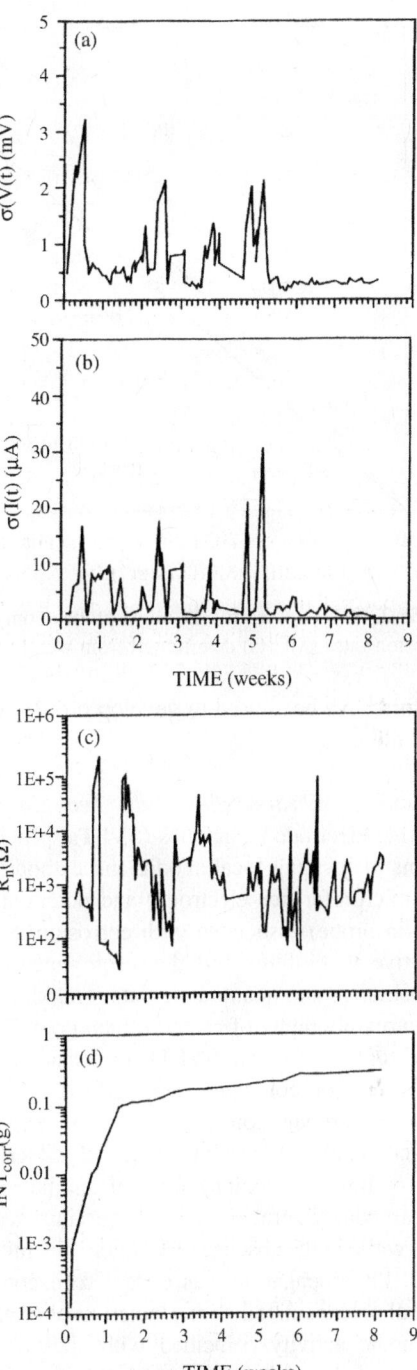

FIGURE 88.1. Representative noise data as monitored for steel electrodes exposed in a test loop at a gas storage field: (a) $\sigma[V(t)]$, (b) $\sigma[I(t)]$, (c) R_n, and (d) INT_{corr} [70].

Lewandowski et al. [73] and Lewandowski [74] used microelectrodes to determine the oxygen concentration around a microcolony. The microcolony was anoxic in the middle, but oxygen was detected at the bottom, demonstrating transport via channels and voids in addition to diffusion.

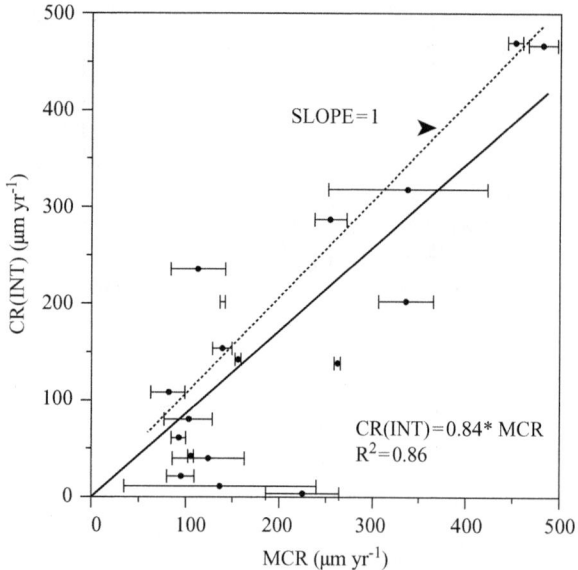

FIGURE 88.2. CR_{int} values for 19 noise probes compared with mean corrosion rates (MCRs) determined from weight loss [70].

Microsensors have been used to develop profiles in mixed-species biofilms.

C1.5. Scanning Vibrating-Electrode Techniques.
Scanning vibrating electrode techniques (SVETs) provide a sensitive means of locating local anodic and cathodic currents (vibrating microreference electrode) and potential distributions (Kelvin probe) associated with corrosion. SVETs are nondestructive to biofilms and their components and can provide qualitative and quantitative data. Franklin et al. [75, 76] used autoradiography of bacterial incorporated ^{14}C acetate to locate bacteria and SVETs to locate anodic and cathodic currents on colonized steel surface. They demonstrated that pit propagation in carbon steel exposed to a phosphate-containing electrolyte required either stagnant conditions or microbial colonization of anodic regions. In sterile, continuously aerated medium, pits initiated and repassivated, while in the absence of aeration, pits initiated and propagated. Pit propagation was observed in continuously aerated medium inoculated with a heterotrophic bacterium. Sites of anodic activity coincided with sites of bacterial activity (incorporation of ^{14}C acetate). Results suggest that bacteria may preferentially attach to the corrosion products formed over corrosion pits. Biofilms over anodic sites may create stagnant conditions within pits, resulting in pit propagation. SVETs are successfully used to monitor early phases of MIC but are usually complemented by other investigative methods for overview of corrosion processes.

C2. Techniques Requiring Small External Signal

C2.1. Polarization Resistance Technique.
Polarization resistance (R_p) techniques can be used to continuously monitor the instantaneous corrosion rate of a metal. Mansfeld [77] provided a thorough review of the use of the technique for the measurement of corrosion currents. Resistance R_p is defined as

$$R_p = \left(\frac{dE}{di}\right)i = 0 \qquad (88.1)$$

where R_p is the slope of a potential (E) – current density (i) curve at E_{corr}, $i = 0$. Corrosion current density (i_{corr}) is calculated from R_p by

$$i_{corr} = \frac{B}{R_p} \qquad (88.2)$$

where

$$B = \frac{b_a b_c}{2.303}(b_a + b_c) \qquad (88.3)$$

The exact calculation of i_{corr} for a given time requires simultaneous measurements of R_p and anodic and cathodic Tafel slopes (b_a and b_c) [78]. Computer programs have been developed for the determination of precise values of i_{corr} according to Eqs. (88.2) and (88.3). Experimental values of R_p (R'_p) contain a contribution from the uncompensated solution resistance (R_u) [78]:

$$R'_p = R_p + R_u \qquad (88.4)$$

Applications of R_p techniques have been reported by King et al. [79] in a study of the corrosion behavior of iron pipes in environments containing SRB. In a similar study, Kasahara and Kajiyama [80] used R_p measurements with compensation of the ohmic drop and reported results for active and inactive SRB. Nivens et al. [81] calculated the corrosion current density from experimental R_p data and Tafel slopes for 304 stainless steel exposed to a seawater medium containing the non-SRB *Vibrio natriegens*.

A simplification of the polarization resistance technique is the linear polarization resistance (LPR) technique in which it is assumed that the relationship between E and i is linear in a narrow range ($\pm 20\,mV$) around E_{corr}. This approach is used in field tests and forms the basis of commercial corrosion rate monitors. Mansfeld et al. [82] used the LPR technique to determine R_p for mild steel sensors embedded in concrete exposed to a sewer environment for about nine months. One sensor was periodically flushed with sewage in an attempt to remove the sulfuric acid produced by sulfur-oxidizing bacteria within a biofilm; another sensor was used as a control.

Significant errors in the calculation of corrosion rates can occur for electrolytes of low conductivity or systems with very high corrosion rates (low R_p) if a correction for R_u is not applied. Corrosion rates will be underestimated in these cases. Additional problems can arise from the effects of the sweep rate used to determine R_p. If the sweep rate is too high, the experimental value of R_p will be too low and the calculated corrosion rate will be too high. For localized corrosion, experimental R_p data should be used as a qualitative indication that rapid corrosion is occurring. Large fluctuations of R_p with time are often observed for systems undergoing pitting or crevice corrosion.

The R_p data are meaningful for general or uniform corrosion but less so for localized corrosion, including MIC. Additionally, the use of Stern–Geary theory where corrosion rate is inversely proportional to R_p at potentials close to E_{corr} is valid for conditions controlled by electron transfer but not for diffusion-controlled systems as frequently found in MIC.

C2.2. Electrochemical Impedance Spectroscopy (EIS).

EIS techniques record impedance data as a function of the frequency of an applied signal at a fixed potential. A large frequency range (1 mHz – 65 kHz) must be investigated to obtain a complete impedance spectrum. Dowling et al. [83] and Franklin et al. [84] demonstrated that the small signals required for EIS do not adversely affect the numbers, viability, and activity of microorganisms within a biofilm. EIS data may be used to determine R_p, the inverse of corrosion rate. EIS is commonly used for steady-state conditions (uniform corrosion); however, sophisticated models have been developed for localized corrosion [85, 86]. Several reports have been published in which EIS has been used to study the role of SRB in corrosion of buried pipes [69, 80, 87] and reinforced concrete [88]. Ferrante and Feron [89] used EIS data to conclude that the material composition of steels was more important for MIC resistance than bacterial population, incubation time, sulfide content, and other products of bacterial growth.

A disadvantage of EIS is the inability to quantify electrochemical parameters, such as R_p, from impedance spectra when MIC is the identified corrosion mechanism. Quantification requires a model electrical circuit for impedance analysis [90]. Mansfeld and Little [91, p. 52] state "this type of analysis is qualitative and no models for the impedance behavior have been presented for complicated systems encountered in MIC."

C3. Large-Signal Polarization

Recording polarization curves provides an overview of reactions for a given corrosion system—charge transfer or diffusion-controlled reactions, passivity, transpassivity, and localized corrosion phenomena. Large-signal polarization techniques require potential scans ranging from several hundred millivolts to several volts. Large-signal polarization is applied to obtain potentiostatic or potentiodynamic

polarization curves as well as pitting scans. Polarization curves can be used to determine i_{corr} by Tafel extrapolation, while mass-transport-related phenomena can be evaluated based on the limiting current density (i_{lim}). Mechanistic information can be obtained from experimental values of b_a and b_c. Pitting scans are used to determine pitting and protection potentials (E_{pit} and E_{prot}, respectively).

Numerous investigators have used polarization curves to determine the effects of microorganisms on the electrochemical properties of metal surfaces and the resulting corrosion behavior. In most of these studies comparisons have been made between polarization curves in sterile media with those obtained in the presence of bacteria and fungi [92, 93]. Disadvantages of large-signal polarizations are irreversible changes to surface properties of the metal and changes to biofilm structure and character.

C4. Multiple Device Monitors

Several investigators have used a combination of techniques to monitor MIC. Stokes et al. [94] described an online, real-time fouling and corrosion monitoring system that consisted of a miniature side-stream heat exchanger with an arrangement of corrosion sensors, flow and heat controllers, and a data collection device. Either heat transfer rate or wall temperature could be controlled. If the heat transfer rate was controlled, the wall temperature increased as the surface fouled to maintain the set heat rate. If the wall temperature was set, the heating rate decreased to maintain the set wall temperature as the resistance to heat transfer increased. Changes in heat transfer resistance from that of a clean surface were used as a measure of biofilm formation. Corrosion was monitored using four electrochemical techniques: ZRA, ECN, EPN, and LPR. The unit was used for one year on a cooling water system that used river water makeup to the cooling towers and had suffered severe underdeposit corrosion. Before cleaning the value of ECN and ZRA were always near full scale (no units). The trace of EPN showed transients typical of pit initiation. After cleaning, EPN was featureless and the ECN and ZRA values dropped by two and one orders of magnitude, respectively. By contrast, the LPR output indicated a larger active area of apparently increased corrosion. During the field trial, corrosion rates as high as 100 mpy due to underdeposit corrosion were indicated. Metallography confirmed extensive localized corrosion. The authors did not analyze fouling deposits, but based on the use of a natural water, they assumed biofilm formation.

Enzien and Yang [95] described a differential flow cell method for monitoring localized corrosion in an industrial water. In this technique, a combination of LPR and ZRA measurements were used to obtain the rate of localized corrosion for carbon steel in aqueous solutions. The measurements were carried out in an electrolytic flow cell with a large cathode placed in faster flow conditions and two small anodes placed in a slower flow condition. The anodes and cathode were electrically connected together via a ZRA. The technique was used in a pilot cooling water test focused on optimizing a scale and corrosion treatment program specifically for localized corrosion. They reported that monitoring planktonic bacteria was not effective at predicting microbial fouling or MIC. Additionally, general corrosion rates were low throughout experiments; therefore, linear polarization resistance measurements did not accurately predict a localized corrosion problem.

Smart et al. [96, 97] described an online internal corrosion and bacteria monitoring system used in a ship-unloading terminal that received crude oil and other hydrocarbons from tankers or barges. The continuously circulated by-pass unit took an oil–water suspension from a pipeline, separated the water for testing, and returned the oil to the pipeline. The instrumentation measured the corrosion rate and changes in corrosion characteristics of a pipeline in real time using multiple techniques including electrical resistance probes, LPR, galvanic probes, hydrogen probes, EIS, EN, pH, and conductivity measurements. Corrosion inhibitor residuals, dissolved iron, and bacteria were also measured. The system included coupons, which could be removed and examined. Analysis of the water chemistry and microbiology as well as examination of the coupons was offline and not in real time. The authors established that rapid MIC was occurring in the unloading pipeline and that the corrosion inhibitor was ineffective. They were able to control MIC by establishing an effective biocide treatment program.

Videla et al. [98] described a monitoring program in an oil field for assessing biodeterioration on mild steel and stainless steel in recirculating cooling water systems. The program was based on (1) water quality control, (2) corrosion monitoring in the field (weight loss and LPR), (3) laboratory corrosion tests (polarization techniques and E_{corr} vs. time measurements), and (4) use of a side-stream sampling device for monitoring sessile populations, biofilms, corrosion morphology and intensity, and biological and inorganic deposits analysis. Comparison of the corrosive attack on carbon steel coupons maintained with and without biocide indicated that there was little metal attack in the biocide-treated cooling system.

Kane and Campbell [99] described a technique for monitoring real-time MIC using LPR, harmonic distortion analysis for measurements of B value (Stern–Geary constant) for correction of LPR corrosion rates, and ECN for evaluation of pitting tendencies. Measurements were made in seawater using a three-electrode probe arrangement with three identical carbon steel (AISI 1018) electrodes placed in a test cell attached to a seawater loop designed to simulate water injection conditions to which SRB were added. They demonstrated that the ability to measure the actual B value in the

environment resulted in the determination of more accurate corrosion rates versus conventional corrosion monitoring techniques, which use a default B value that does not relate to actual values in the service environment. Their study indicated that corrosion rates did not correlate with H_2S production by SRB.

Brossia and Yang [100] developed a multielectrode array sensor system (MASS) to monitor corrosion in both laboratory tests and industrial processes. The probe was used to conduct a series of biotic and abiotic tests to determine if the probe could detect MIC. The MASS probe consists of multiple miniature electrodes made of metals to be studied. The miniature electrodes were coupled together by connecting each of them to a common joint through independent resistors, with each electrode simulating an area of corroding metal. The standard deviation of the currents from the different miniature electrodes was used as an indicator for localized corrosion. Using carbon steel (UNS G10100) electrodes, they were able to demonstrate that corrosion rates increased by an order of magnitude in the presence of SRB compared to sterile controls. For comparative purposes, several test cells were constructed using flat coupons and monitored using LPR to determine corrosion rates under similar conditions to those in the MASS probes. The corrosion rates obtained using the probe were much higher than those determined using LPR. The probe, however, cannot distinguish between biotic and abiotic pitting.

D. CONCLUSIONS

It is essential in diagnosing MIC to demonstrate a spatial relationship between the causative microorganisms and the corrosion phenomena. However, that relationship cannot be independently interpreted as MIC. Pitting due to MIC can initiate as small pits that have the same size and characteristics of the causative organisms. These features are not obvious to the unaided eye and are most often observed with an electron or atomic force microscope. MIC does not produce a macroscopic unique metallographic feature. Metallurgical features previously thought to be unique to MIC, for example, hemispherical pits in 300 series stainless steel localized at weld or tunneling in carbon steel, are consistent with some mechanisms for MIC but cannot be interpreted independently. Bacteria do produce corrosion products that could not be produced abiotically in near-surface environments, resulting in isotope fractionation and mineralogical fingerprints. The electrochemical techniques described in this chapter for measuring and monitoring MIC are useful for specific applications. All of the techniques are based on assumptions that can only be validated by a thorough understanding of the system that one is attempting to monitor.

ACKNOWLEDGMENTS

This work was supported by NRL 6.1 Program Element number 0601153N as publication number NRL/BC/7303/08/9047.

REFERENCES

1. J. R. Postgate, The Sulfate Reducing Bacteria, Cambridge University Press, New York, 1979.

2. P. J. B. Scott and M. Davies, "Survey of Field Kits for Sulfate-Reducing Bacteria," Mater. Perform., **31**(5), 64–68 (1992).

3. C. M. Jhobalia, A. Hu, T. Gu, and S. Nesic, "Biochemical Engineering Approaches to MIC," Paper No. 05500, CORROSION/2005, NACE International, Houston, TX, 2005.

4. J. K. Cowan, "Rapid Enumeration of Sulfate-Reducing Bacteria," Paper No. 05485, CORROSION/2005, NACE International, Houston, TX, 2005.

5. C. Zobell and D. Q. Anderson, "Effect of Volume on Bacterial Activity," Biolo. Bull., **71**, 324–342 (1936).

6. B. Lloyd, "Bacteria Stored in Seawater," J. Roy. Tech. Coll. Glasgow, **4**, 173 (1937).

7. S. J. Giovannoni, T. B. Britschgi, C. L. Moyer, and K. G. Field, "Genetic Diversity in Sargasso Sea Bacterioplankton," Nature, **344**, 60–63 (1990).

8. D. M. Ward, M. J. Ferris, S. C. Nold, and M. M. Bateson, "A Natural View of Microbial Diversity within Hot Spring Cyanobacterial Mat Communities," Microbiol. Mol. Biol. R **62**(4), 1353–1370 (1998).

9. T. Kaeberlein, K. Lewis, and S. S. Epstein, "Isolating Uncultivable' Microorganisms in Pure Culture in a Simulated Natural Environment," Science, **249**, 1127–1129 (2002).

10. D. B. Roszak and R. R. Cowell, "Survival Strategies of Bacteria in the Natural Environments," Microbiology, **51**(3), 365–379 (1987).

11. X. Zhu, A. Ayala, H. Modi, and J. J. Kilbane, "Application of Quantitative, Real-Time PCR in Monitoring Microbiologically Influenced Corrosion (MIC) in Gas Pipelines," Paper No. 05493, CORROSION/2005, NACE International, Houston, TX, 2005.

12. J. M. Romero, E. Velazquez, G.-V. J. L., M. Amaya, and S. Le Borgne, "Genetic Monitoring of Bacterial Populations in a Sewater Injection System, Identification of Biocide Resistant Bacteria and Study of Their Corrosive Effect," Paper No. 05483, CORROSION/2005, NACE International, Houston, TX, 2005.

13. E. S. Littmann, "Use of ATP Extraction in Oil Field Waters," in Oil Field Subsurface Injection of Water, Vol. STP 641, C. C. Wright, D. Cross, A. G. Ostroff, and J. R. Stanford (Eds.), American Society for Testing and Materials, Philadelphia, 1977, p. 79.

14. M. J. Franklin and D. C. White, "Biocorrosion," Curr. Opin. Biotechnol., **2**, 450–456 (1991).

15. D. H. Pope, "Discussion of Methods for the Detection of Microorganisms Involved in Microbiologically Influenced Corrosion," in Biologically Induced Corrosion, S. C. Dexter (Ed.), NACE International, Houston, TX, 1986, pp. 275–281.

16. J. J. Hogan, "A Rapid, Non-Radioactive DNA Probe for Detection of SRBs," in Institute of Gas Technology Symposium on Gas, Oil, Coal, and Environmental Biotechnology, Institute of Gas Technology, Chicago, IL, 1990.

17. R. E. Tatnall, K. M. Stanton, and R. C. Ebersole, "Methods of Testing for the Presence of Sulfate-Reducing Bacteria," Paper No. 88, CORROSION/88, NACE International, Houston, TX, 1988.

18. J. Boivin, E. J. Laishley, R. D. Bryant, and J. W. Costerton, "The Influence of Enzyme Systems on MIC," Paper No. 128, CORROSION/90, NACE International, Houston, TX, 1990.

19. T. J. Phelps, R. M. Schram, D. B. Ringelberg, N. J. E. Dowling, and D. C. White, "Anaerobic Microbial Activities Including Hydrogen-Mediated Acetogenesis within Natural Gas Transmission Lines," Biofouling, **3**, 265–276 (1991).

20. M. W. Mittleman, D. E. Nivens, C. Low, and D. C. White, "Differential Adhesion, Activity, and Carbohydrate: Protein Ratios of *Pseudomonas Atlantica* Monocultures Attaching to Stainless Steel in a Linear Shear Gradient," Microb. Ecol., **19**(3), 269–278 (1990).

21. S. Maxwell, "Assessment of Sulfide Corrosion Risks in Offshore Systems by Biological Monitoring," SPE Prod. Eng., **1**(5), 363–368 (1986).

22. R. I. Amann, J. Stromley, R. Devereux, R. Key, and D. A. Stahl, "Molecular and Microscopics Identification of Sulfate-Reducing Bacteria in Multispecies Biofilms," Appl. Environ. Microbiol., **58**(2), 614–623 (1992).

23. D. A. Stahl, B. Flesher, H. R. Mansfield, and L. Montgomery, "The Use of Phylogenetically Based Hybridization Probes for Studies of Ruminal Microbial Ecology," Appl. Environ. Microbiol, **54**(5)1079–1084 (1988).

24. D. A. Stahl, D. J. Lane, G. L. Olsen, and N. R. Pace, "Analysis of Hydrothermal Vent-Associated Symbionts by Ribosomal RNA Sequence," Science, **224**, 409–411 (1984).

25. J. Horn, C. Carrillo, and V. Dias, "Comparison of the Microbial Community Composition at Yucca Mountain and Laboratory Test Nuclear Repository Environments," Paper No. 03556, CORROSION/2003, NACE International, Houston, TX, 2003.

26. X. Zhu, J. Lubeck, and J. J. Kilbane, "Characterization of Microbial Communities in Gas Industry Pipelines," Appl. Environ. Microbiol., **69**(9), 5354–5363 (2003).

27. X. Zhu, J. Lubeck, K. Lowe, A. Daram, and J. J. Kilbane, "Improved Method for Monitoring Microbial Communities in Gas Pipelines," Paper No. 04592, CORROSION/2004, NACE International, Houston, TX, 2004.

28. K. Takai and K. Horikoshi, "Rapid Detection and Quantification of Members of the Archaeal Community by Quantitative PCR Using Fluorogenic Probes," Appl. Environ. Microbiol. **66**(11), 5066–5072 (2000).

29. A. Reyes, M. V. Letelier, R. De la Iglesia, B. Gonzalez, and G. Lagos, "Microbiologically Induced Corrosion of Copper Pipes in Low-pH Water," Int. Biodeter. Biodegr., **61**(2), 135–141 (2008).

30. C. J. McNamara, K. B. Lee, M. A. Russell, L. E. Murphy, and R. Mitchell, "Analysis of Bacterial Community Composition in Concretions Formed on the USS Arizona, Pearl Harbor, HI," J. Cult. Herit., **10**(2), 232–236 (2009).

31. A. H. L. Chamberlain, W. R. Fisher, U. Hinze, H. H. Paradies, C. A. C. Sequeira, H. Siedlarek, M. Thies, D. Wagner, and J. N. Wardell, "An Interdisciplinry Approach for Microbiologically Influenced Corrosion of Copper," in Microbial Corrosion, Proceedings of the 3rd International European Federation of Corrosion Workshop, Vol. No. 15, A. K. Tiller and C. A. C. Sequeira (Eds.), The Institute of Materials, London, 1995, p. 3.

32. A. H. L. Chamberlain, P. Angell, and H. S. Campbell, "Staining Procedures for Characterizing Biofilms in Corrosion Investigations," Br. Corros. J., **23**(3), 197–198 (1988).

33. A. R. Howgrave-Graham, and P. L. Steyn, "Application of the Fluorescent-Antibody Technique for the Detection of Sphaerotilus natans in Activated Sludge," Appl. Environ. Microbiol., **54**(3), 799–802 (1988).

34. J. J. Zambon, P. S. Huber, A. E. Meyer, J. Slots, M. S. Fornalik, and R. E. Baier, "In-Situ Identification of Bacterial Species in Marine Microfouling Films by Using an Immunofluorescence Technique," Appl. Environ. Microbiol., **48**(6), 1214–1220 (1984).

35. G. G. Geesey, Z. Lewandowski, and H. C. Flemming (Eds.), Biofouling and Biocorrosion in Industrial Water Systems, CRC Press, Boca Raton, FL, 1994.

36. J. Telegdi, Z. Keresztes, G. Paalinkas, E. Kalaman, and W. Sand, "Microbially Influenced Corrosion Visualized by Atomic Force Microscopy," Appl. Phys. A, **66**, S639–S649 (1998).

37. P. J. B. Scott, and M. Davies, "Microbiologically Influenced Corrosion of Alloy 904L," Mater. Perform., **28**(5), 57–63 (1989).

38. H. F. Ridgeway and B. H. Olson, "Scanning Electron Microscope Evidence for Bacterial Colonization of a Drinking-Water Distribution System," Appl. Environ. Microbiol., **41**(1), 274–287 (1981).

39. G. Blunn, "Biological Fouling of Copper and Copper Alloys," in Biodeterioration, Vol. 6, S. Barry, D. R. Houghton, G. C. Llewellyn, and C. E. O'Rear (Eds.), CAB International, Slough, UK, 1986, pp. 567–575.

40. B. J. Little, P. A. Wagner, R. I. Ray, R. K. Pope, and R. Scheetz, "Biofilms: an ESEM Evaluation of Artifacts Introduced during SEM Preparation," J. Ind. Microbiol., **8**, 213–222 (1991).

41. R. Ray and B. Little, "Environmental Electron Microscopy Applied to Biofilms," in Biofilms in Medicine, Industry and Environmental Biotechnology, P. Lens, A. P. Moran, T. Mahony, P. Stoodley, and V. O'FLaherty (Eds.), IWA Publishing, London, 2003, pp. 331–351.

42. B. J. Little, P. A. Wagner, and Z. Lewandowski, "The Role of Biomineralization in Microbiologically Influenced

Corrosion," Paper No. 294, CORROSION/98, NACE International, Houston, TX, 1998.

43. B. J. Little, R. K. Pope, T. L. Daulton, and R. I. Ray, "Application of Environmental Cell Transmission Electron Microscopy to Microbiologically Influenced Corrosion," Paper No. 1266, CORROSION/2001, NACE International, Houston, TX, 2001.

44. T. P. Zintel, D. A. Kostuck, and B. A. Cookingham, "Evaluation of Chemical Treatments in Natural Gas Systems versus MIC and Other Forms of Internal Corrosion Using Carbon Steel Coupons," Paper No. 03574, CORROSION/2003, NACE International, Houston, TX, 2003.

45. D. H. Pope, GRI-90/0299 Field Guide: Microbiologically Influenced Corrosion (MIC): Methods of Detection in the Field, Gas Research Institute, Chicago, IL, 1990.

46. R. B. Eckert, H. C. Aldrich, C. A. Edwards, and B. A. Cookingham, "Microscopic Differentiation of Internal Corrosion Initiation Mechanisms in Natural Gas Pipeline Systems," Paper No. 03544, CORROSION/2003, NACE International, Houston, TX, 2003.

47. J. T. Borenstein and P. B. Lindsay, "MIC Failure Analysis," Mater. Perform., **33**(4), 43–45 (1994).

48. S. W. Borenstein and P. B. Lindsay, "Microbiologically Influenced Corrosion Failure Analysis," Mater. Perform., **27**(3), 51–54 (1988).

49. R. A. Hoffman, "Case Histories of Microbiologically-Influenced Corrosion in Building and Power Generation Systems," Paper No. 317, CORROSION/93, NACE International, Houston, TX, 1993.

50. Y. Chung and L. K. Thomas, "Comparison of MIC Pit Morphology with Non-MIC Chloride Induced Pits in Types 304/304L/E308 Stainless Steel Base Metal/Welds," Paper No. 159, CORROSION/99, NACE International, Houston, TX, 1999.

51. J. C. Welz and J. T. Tverberg, "Case History: Corrosion of a Stainless Steel Hot Water System in a Brewery," Mater. Perform., **37**(5), 66–72 (1998).

52. M. Geiser, R. Avci, and Z. Lewandowski, "Pit Initiation on 316L Stainless Steel in the Presence of Bacteria Leptothrix discophora," Paper No. 01257, CORROSION/2001, NACE International, Houston, TX, 2001.

53. R. Eckert, FIeld Guide for Investigating Internal Corrosion of Pipelines, NACE International, Houston, TX, 2003.

54. M. B. McNeil, J. M. Jones, and B. J. Little, "Mineralogical Fingerprints for Corrosion Process Induced by Sulfate-Reducing Bacteria," Paper No. 580, CORROSION/91, NACE International, Houston, TX, 1991.

55. T. R. Jack, G. van Boven, M. Wilmot, and R. G. Worthingham, "Evaluating Performance of Coatings Exposed to Biologically Active Solis," Mater. Perform., **35**(3), 39–45 (1996).

56. A. Chambers and P. A. Trudinger, "Microbiological Fractionation of Stable Sulfur Isotopes: A Review and Critique," Geomicrobiol. J., **1**, 249–293 (1979).

57. B. J. Little, P. A. Wagner, R. I. Ray, M. B. McNeil, and J. Jones-Meehan, "Indicators of Microbiologically Influenced Corrosion in Copper Alloys," OEBALIA, **19**, 287–294 (1993).

58. B. J. Little and P. A. Wagner, "Application of Electrochemical Techniques to the Study of Microbiologically Influenced Corrosion," in Modern Aspects of Electrochemistry, Vol. 34, J. O. M. Bockris (Ed.), Kluwer Academic/Plenum, New York, 2001, pp. 205–246.

59. P. Angell, J.-S. Luo, and D. C. White, "Studies of the Reproducible Pitting of 304 Stainless Steel by a Consortium Containing Sulphate-Reducing Bacteria," International Conference on Microbially Influenced Corrosion, NACE International, Houston, TX, 1995, pp. 1/1–1/10.

60. X. Campaignolle and J.-L. Crolet, "Method for Studying Stabilization of Localized Corrosion on Carbon Steel by Sulfate-Reducing Bacteria," Corrosion, 53(6), 440–447 (1997).

61. G. J. Licina and G. Nekoksa, "On-Line Monitoring of Biofilm Formation for the Control and Prevention of Microbially Influenced Corrosion," 1995 International Conference on Microbially Influenced Corrosion, NACE International, Houston, TX, 1995, pp. 42/1–42/10.

62. B. J. Little, P. A. Wagner, K. R. Hart, R. I. Ray, D. M. Lavoie, K. Nealson, and C. Aguilar, "The Role of Metal-Reducing Bacteria in Microbiologically Influenced Corrosion," Paper No. 215, CORROSION/97, NACE International, Houston, TX, 1997.

63. A. Mollica and G. Ventura, "Use of a Biofilm Electrochemical Monitoring Device for an Automatic Application of Antifouling Procedures in Seawater," Proceedings of the 12th International Corrosion Congress, NACE International, Houston, TX, 1993, pp. 3807–3812.

64. T. Uchida, T. Umino, and T. Arai, "New Monitoring System for Microbiological Control Effectiveness on Pitting Corrosion of Carbon Steel," Paper No. 408, CORROSION/97, NACE International, Houston, TX, 1997.

65. A. Iimura, K. Takahashi, and T. Uchida, "Growth Model and Online Measurement of Pitting Corrosion on Carbon Steel," Mater. Perform., 35(12), 39–44 (1996).

66. C. Andrade, P. Garcésb, and I. Martínez, "Galvanic Currents and Corrosion Rates of Reinforcements Measured in Cells Simulating Different Pitting Areas Caused by Chloride Attack in Sodium Hydroxide," Corros. Sci., 50(10), 2959–2964 (2008).

67. B. J. Little and F. Mansfeld, "Passivity of Stainless Steels in Natural Seawater," in Proceedings of the H. H. Uhlig Memorial Symposium, Vol. 94–26, F. Mansfeld, A. Asphahani, H. Bohni, and R. M. Latanision (Eds.), The Electrochemical Society, Pennington, NJ, 1994, pp. 42–55.

68. D. A. Eden, "Electrochemical Noise—The First Two Octaves," Paper No. 386, CORROSION/98, NACE International, Houston, TX, 1998.

69. R. A. King, B. S. Skerry, D. C. A. Moore, J. F. D. Stott, and J. L. Dawson, "Corrosion Behaviour of Ductile and Grey Iron Pipelines in Environments Containing Sulphate-Reducing Bacteria," in Biologically Induced Corrosion, S. C. Dexter (Ed.), NACE International, Houston, TX, 1986, pp. 83–91.

70. B. J. Little, P. A. Wagner, and R. I. Ray, "New Experimental Techniques in the Study of MIC," Proceedings of CORROSION/97 Research Topical Symposium—Part I, Advanced Monitoring and Analytical Techniques, NACE International, Houston, TX, 1997, pp. 31–52.

71. W. C. Lee and D. de Beer, "Oxygen and pH Microprofiles above Corroding Mild Steel Covered with a Biofilm," Biofouling, 8, 273–280 (1995).

72. P. VanHoudt, Z. Lewandowski, and B. J. Little, "Construction and Application of Iridium Oxide pH Microelectrode," Bioeng. Biotechnol., 40(5), 601–608 (1992).

73. Z. Lewandowski, W. C. Lee, W. G. Characklis, and B. J. Little, "Dissolved Oxygen and pH Microelectrode Measurements at Water-Immersed Metal Surface," Corrosion, 45(2), 92–98 (1989).

74. Z. Lewandowski, "MIC and Biofilm Heterogeneity," Paper No. 00400, CORROSION/2000, NACE International, Houston, TX, 2000.

75. M. J. Franklin, D. C. White, and H. S. Isaacs, "The Use of Current Density Mapping in the Study of Microbial Influenced Corrosion," Paper No. 104, CORROSION/90, NACE International, Houston, TX, 1990.

76. M. J. Franklin, D. E. Nivens, J. B. Guckert, and D. C. White, "Effect of Electrochemical Impedance on Microbial Biofilm Cell Numbers, Viability, and Activity," Corrosion, 47(7), 519–522 (1991).

77. F. Mansfeld, "The Polarization Resistance Technique for Measuring Corrosion Currents," in Advances in Corrosion Science and Technology, Vol. 6, M. G. Fontana and R. W. Staehle (Eds.), Plenum, New York, 1976, pp. 163–262.

78. H. Shih and F. Mansfeld, "Software for Quantitative Analysis of Polarization Curves," in Computer Modeling and Corrosion, Vol. STP 1154, R. S. Munn (Ed.), American Society for Testing and Materials, Philadelphia, PA, 1992, pp. 174–185.

79. R. A. King, J. D. A. Miller, and J. F. D. Stott, "Subsea Pipelines: Internal and External Biological Corrosion," in Biologically Induced Corrosion, S. C. Dexter (Ed.), NACE International, Houston, TX, 1986, pp. 268–274.

80. K. Kasahara and F. Kajiyama, "Role of Sulfate-Reducing Bacteria in the Localized Corrosion of Buried Pipes," in Biologically Induced Corrosion, S. C. Dexter (Ed.), NACE International, Houston, TX, 1986, pp. 172–183.

81. D. E. Nivens, P. D. Nichols, J. M. Henson, G. G. Geesey, and D. C. White, "Reversible Acceleration of the Corrosion of AISI 304 Stainless Steel Exposed to Seawater Induced by Growth and Secretions of the Marine Bacterium *Vibrio natriegens*," Corrosion, 42(4), 204–210 (1986).

82. F. S. Mansfeld, H. A. Postyn, J. S. Devinny, R. L. Islander, and C. L. Chin, "Corrosion Monitoring and Control in Concrete Sewer Pipes," Corrosion, 47(5), 369–376 (1991).

83. N. J. E. Dowling, J. Guezennec, M. L. Lemoine, A. Tunlid, and D. C. White, "Corrosion Analysis of Carbon Steels Affected by Aerobic and Anaerobic Bacteria in Mono and Co-cultures Using AC Impedance and DC Techniques," Corrosion, 44(12), 869–874 (1988).

84. M. J. Franklin, D. C. White, and H. S. Isaacs, "Pitting Corrosion by Bacteria on Carbon Steel, Determined by the Scanning Vibrating Electrode Technique," Corros. Sci., **32**(9), 945–952 (1991).

85. M. Kendig, F. Mansfeld, and C. H. Tsai, "Determination of the Long-Term Corrosion Behavior of Coated Steel with AC Impedance Measurements," Corros. Sci., **23**(4), 317–329 (1983).

86. F. Mansfeld, M. Kendig, and C. H. Tsai, "Evaluation of Corrosion Behavior of Coated Metals with AC Impedance Measurements," Corrosion, **38**(9), 478–485 (1982).

87. K. Kasahara and F. Kajiyama, "Electrochemical Aspects of Microbiologically Influenced Corrosion on Buried Pipes," in Microbially Influenced Corrosion and Biodeterioration, N. J. Dowling, M. W. Mittleman, and J. C. Danko (Eds.), University of Tennessee, Knoxville, TN, 1991, pp. 2-33–2-39.

88. A. N. Moosavi, J. L. Dawson, C. J. Houghton, and R. A. King, "The Effect of Sulphate-Reducing Bacteria on the Corrosion of Reinforced Concrete," in Biologically Induced Corrosion, Vol. NACE 8, S. C. Dexter (Ed.), NACE International, Houston, TX, 1986, pp. 291–308.

89. V. Ferrante and D. Feron, "Microbially Influenced Corrosion of Steels Containing Molybdenum and Chromium: A Biological and Electrochemical Study," in Microbially Influenced Corrosion and Biodeterioration, N. J. Dowling, M. W. Mittleman, and J. C. Danko (Eds.), University of Tennessee, Knoxville, TN, 1991, pp. 3-55–3-63.

90. F. Mansfeld, "Don't Be Afraid of Electrochemical Techniques—But Use Them with Care!" Corrosion, **44**(12), 856–868 (1988).

91. F. Mansfeld and B. J. Little, "Electrochemical Techniques Applied to Studies of Microbiologically Influenced Corrosion (MIC)," Trends Electrochem., **1**, 47–61 (1992).

92. G. Schmitt, "Sophisticated Electrochemical Methods for MIC Investigation and Monitoring," Mater. Corros., **48**(9), 586–601 (1997).

93. M. B. Deshmukh, I. Akhtar, and C. P. De, "Influence of Sulphide Pollutants of Bacterial Origin on Corrosion Behaviour of Naval Brass," in 2nd International Symposium on Industrial & Basic Oriented Electrochemistry, IBH Publishing, Oxford, 1980, pp. 6.19.1–11.

94. P. S. N. Stokes, M. A. Winters, P. O. Zuniga, and D. J. Schlottenmier, "Developments in On-Line Fouling and Corrosion Surveillance," in Microbiologically Influenced Corrosion Testing, Vol. STP 1232, J. R. Kearns and B. J. Little (Eds.), American Society for Testing and Materials, Philadelphia, PA, 1994, pp. 99–107.

95. M. Enzien and B. Yang, "Effective Use of Monitoring Techniques for Use in Detecting and Controlling MIC in Cooling Water Systems," Biofouling, **17**(1), 47–57 (2001).

96. J. Smart, T. Pickthall, and T. G. Wright, "Field Experiences in On-Line Bacteria Monitoring," Paper No. 279, CORROSION/96, NACE International, Houston, TX, 1996.

97. J. Smart, T. Pickthall, and A. Carlile, "Using On-Line Monitoring to Solve Bacteria Corrosion Problems in the Field," Paper No. 212, CORROSION/1997, NACE International, Houston, TX, 1997.

98. H. A. Videla, F. Bianchi, M. M. S. Freitas, C. G. Canales, and J. F. Wilkes, "Monitoring Biocorrosion and Biofilms in Industrial Waters: A Practical Approach," in Microbiologically Influenced Corrosion Testing, Vol. STP 1232, J. R. Kearns and B. J. Little (Eds.), American Society for Testing and Materials, Philadelphia, PA, 1994, pp. 128–137.

99. R. D. Kane and S. Campbell, "Real-Time Corrosion Monitoring of Steel Influenced by Microbial Activity (SRB) in Simulated Seawater Injection Environments," Paper No. 04579, CORROSION/2004, NACE International, Houston, TX, 2004.

100. C. S. Brossia and L. Yang, "Studies of Microbiologically Influenced Corrosion Using a Coupled Multielectrode Array Sensor," Paper No. 03575, CORROSION/2003, NACE International, Houston, TX, 2003.

GLOSSARY OF SELECTED TERMS USED IN CORROSION SCIENCE AND ENGINEERING

For a more complete set of definitions of terms related to corrosion science and engineering, the reader may refer to:

M. S. Vukasovich (Ed.), *A Glossary of Corrosion-Related Terms Used in Science and Industry,* Society of Automotive Engineers, Warrendale, PA, 1995, 298 pp.

ASTM Standard G 15–97a, *Standard Terminology Relating to Corrosion and Corrosion Testing,* American Society for Testing and Materials, West Conshohocken, PA, 1997, 4 pp.

Abiotic: A reaction without the participation of enzymes or organisms.

Abrasive wear: Wear due to hard particles or hard protuberances forced against and moving along a solid surface. *Note:* In abrasive wear the force exerted by the particles is externally applied and approximately constant while in *solid-particle impingement erosion* the force exerted by the particles on the surface is due to their deceleration.

Acetogens: Bacteria capable of synthesizing acetate from molecules containing one carbon, for example, CO_2, CO, methanol, and a methoxy group.

Acidophile: An organism that grows best in acidic conditions.

Active–passive transition: Transition from an active to a passive condition on a metal surface—influenced by potential and environment.

Aerobic conditions: An environment or ecological niche in which molecular oxygen O_2 is present in the system.

Alkaliphile: A microorganism that requires alkalinity for optimal growth.

Anaerobic conditions: Contrary to aerobic conditions, no molecular oxygen (O_2) is available in the environment.

Anode: The electrode in an electrochemical cell at which oxidation occurs.

Anoxic: Absence of oxygen.

Archaea: A phylogenetic domain of prokaryotes consisting of the methanogens, mostly salt-loving and high-temperature microorganisms.

Assimilate: A process by which a chemical compound is altered and elements are incorporated in the synthesis of cellular components.

Autotroph: Microorganism capable of using light or inorganics as a sole source of energy.

Barophile: An organism that lives optimally under high hydrostatic pressure.

Biodegradable: Capable of being broken down to simple compounds by activity of biological species.

Biofilm: Intrinsic association of microorganisms and their exopolymeric materials on surfaces of material forming a solid–liquid interface.

Biofouling: A process referring to the adsorption of organics, micro- and macroorganisms, on a surface of a material in sequence.

Biological degradation/deterioration: Disintegration of a material caused by the activity of microorganisms; also referred to as biodegradation and biodeterioration.

Biopolymer: Complex biological long-chained material.

Bioremediation: A process involving the use of microorganisms to detoxify toxic chemicals in the environment.

Biosphere: The surface layer on Earth influenced by the activities of living organisms.

Boundary layer: The wall region in turbulent flow where the flow changes from fully turbulent to viscous, composed

Uhlig's Corrosion Handbook, Third Edition, Edited by R. Winston Revie
Copyright © 2011 John Wiley & Sons, Inc.

of the logarithmic region, the buffer region, and the viscous region.

Buffer region: In fluid flow, that portion of the wall region defined as $5 \leq y_+ \leq 30$.

Burst: A violent eruption or localized ejection of fluid away from the wall, caused by the passage of one or more vortices.

Cathode: The electrode in an electrochemical cell at which the reduction reaction occurs.

Cavitation: Repeated nucleation, growth, and violent collapse of cavities or bubbles containing vapor and/or gas within a liquid. Cavitation originates from a local decrease in hydrostatic pressure in the liquid produced by motion of the liquid (flow cavitation) or by pressure fluctuations within the liquid induced by vibration of a solid boundary (vibratory cavitation).

Cavitation cloud: A cloud of cavitation bubbles (< 1 mm diameter) that are considered to collapse in concert near or at a solid surface causing cavitation erosion. A cavitation cloud usually obscures a surface that is being eroded by cavitation.

Cavitation erosion: Progressive loss of material from a solid surface following *cavitation* when the cavities or bubbles collapse on or near the surface.

Cavitation erosion–corrosion: The conjoint action of cavitation erosion and corrosion.

Cavitation number σ: A measure of the intensity of cavitation:

$$\sigma = (p - p_v)/0.5\rho\, u^2$$

where p = static pressure (absolute)

p_v = the vapor pressure

ρ = density

u = free-stream velocity

When $\sigma = 0$, the pressure is reduced to the vapor pressure and cavitation will occur. *Note:* $\text{NPSH}_A = (\sigma + 1)\, u^2/2\,g$.

Chemoautotroph: An organism that obtains energy from oxidation of an inorganic compound.

Cloning: Insertion of a DNA fragment into a plasmid or phage for replication and expression.

Coherent structure: A morphologically invariant, three-dimensional region of the flow possessing an identifiable flow pattern. See also *turbulent structure*.

Cold climate regions: In the Northern Hemisphere—two regions where the average temperature of the coldest winter month is –18°C or less, or between 0 and –18°C. In the Southern Hemisphere—two global lines, the northernmost being an isotherm where the mean temperature of the warmest month is 10°C, and to the south, the polar front, previously known as the antarctic convergence.

Consortium: Several biological species coexisting to achieve survival.

Contact resistance: Partial interruption of electrical continuity between two surfaces caused by poorly conductive surface films or by excessive surface roughness.

Copolymer: A material consisting of more than one type of monomer as building blocks.

Copper-bearing steel: A low-carbon "mild" steel but containing ~ 0.20–0.25% copper—not a "weathering" steel—and with corrosion rate relatively insensitive to minor variations in composition.

Corrosion: The degradation of a material by its environment.

Corrosion fatigue: Cracking that results from the combined action of a corrosive environment and repeated or alternating stress.

Corrosion potential: The potential at which the rate of oxidation and the rate of reduction are equal; also known as the mixed potential and the rest potential.

Crevice corrosion: Corrosion that takes place inside a crevice.

Crystallographic texture: Orientation distribution of many crystals in a polycrystalline or nanocrystalline material.

Culturability: The ability of growing an organism under specific conditions.

Dealloying: A corrosion process whereby one constituent of an alloy is removed leaving an altered residual structure.

Depolymerase: Enzyme that cuts a polymer into smaller fragments.

Depolymerization: Breaking down of a polymer to release its monomers.

Deteriogens: Agents capable of causing the disintegration of materials.

Deterioration: A process by which the parent material is damaged by breaking of chemical bonds.

Differential aeration cell: An electrolytic cell consisting of zones with different levels of aeration (e.g., aerated and anoxic zones).

Diffuse double layer (DDL): An electrochemical envelope surrounding a particle in aqueous solution consisting of an inner tightly packed layer of adsorbed ions and an outer layer of loosely adsorbed ions.

Dimensionless groups:

Reynolds number: $\text{Re} = u_b \ell \rho/\mu$ is the ratio of inertial forces to viscous forces. Its size determines the nature of the flow. In single-phase pipe flow at $\text{Re} < 2000$ the flow is laminar. At $\text{Re} > 4000$ the flow is turbulent.

Schmidt number: $\text{Sc} = \mu/\rho D$ is the ratio of momentum diffusivity (kinematic viscosity, μ/D) to mass diffusivity.

Sherwood number: $\text{Sh} = kl/D$ is the ratio of convective mass transport to diffusive transport where l = characteristic dimension (e.g., pipe diameter) (m)

   μ = viscosity (Pa·s)

   u_b = bulk flow velocity (m/s)

   D = diffusion coefficient (m²/s)

   k = mass transfer coefficient (m/s)

   ρ = density (kg/m³)

Dissimilatory: An energy-generating property of certain organisms through processsing of a quantity of a chemical or element.

Ejection: The movement of fluid away from the wall at an angle to the nominal flow direction.

Electrodeposition: A process to deposit a metal or alloy from an aqueous plating bath containing metal ions in solution.

Environmental cracking: Brittle fracture of a normally ductile material in a corrosive environment. Environmental cracking is a general term that includes corrosion fatigue, hydrogen embrittlement, and stress corrosion cracking.

Erosion: Progressive loss of material from a solid surface due to the mechanical interaction between that surface and a flowing, single or multiphase fluid. Solid-particle impingement erosion, liquid-droplet impingement erosion, cavitation–erosion, and erosion of protective films in single-phase turbulent flow are included in this broad definition. Flow-enhanced dissolution and thinning of protective films can be considered a form of "chemical" erosion.

Erosion–corrosion: The conjoint action of *erosion* and *corrosion* in a flowing single or multiphase corrosive fluid leading to the accelerated loss of material. Encompasses a wide range of processes including solid-particle impingement erosion–corrosion, liquid-droplet impingement erosion–corrosion, cavitation erosion–corrosion, and single-phase erosion of protective films leading to accelerated corrosion. The relative contributions of erosion and corrosion to the total material loss vary with the type of erosion–corrosion. With metals, the mechanical or "chemical" erosion of protective films leading to accelerated corrosion is a major factor. Erosion predominates with cavitation and liquid impingement erosion–corrosion.

Eukaryote: A cell or organism having a unique membrane-enclosed nucleus.

Exodepolymerase: Extracellular enzymes capable of breaking chemical bonds of a polymer.

Exoenzyme: Enzyme produced externally by an organism.

Exopolymeric materials: Polymeric materials synthesized and excreted outside of the cell by the producing organisms.

Exopolysaccharides: Polysaccharides synthesized and excreted by the producing organisms.

Extracellular: Outside of a cell.

Fatigue life: The total number of cycles or time to failure; that is, to induce fatigue damage and to initiate a dominant fatigue flaw that propagates to failure.

Fatigue limit: The applied stress amplitude below which a material is expected to have an infinite fatigue life or can sustain a specified number of cycles without failure, usually 10^7 or 10^8 cycles; also called the endurance limit.

Fiber-reinforced polymeric composite: An engineering material consisting of several distinctive phases interacting to achieve maximum strength and mechanical properties.

Flow-accelerated corrosion (FAC): A term sometimes used to describe the accelerated corrosion caused by flow-enhanced dissolution and thinning of protective films in carbon steel pipes protected by magnetite carrying deaerated hot water or mixtures of water and steam. The FAC is more usually referred to as an erosion-corrosion process. (See *erosion* and *erosion–corrosion*).

Flow-dependent corrosion: Corrosion processes for which the corrosion rate varies with the velocity of flow of a single- or multiphase fluid. In erosion–corrosion processes the material loss increases with flow rate. In pitting corrosion of stainless alloys and corrosion processes involving transport of passivators and inhibitors to the solid surface, the effect of increased flow rate is beneficial.

Flow-induced corrosion: Corrosion resulting from increased fluid turbulence intensity and mass transfer as a result of the flow of a fluid over a surface. A breakaway velocity is often involved for a given system.

Fouling: Deposition of precipitates and organisms on surfaces.

Fungi: Nonphototrophic eukaryotic organisms possessing rigid cell walls.

Galvanic corrosion: When two dissimilar conducting materials in electrical contact with each other are exposed to an electrolyte, a current, called the galvanic current, flows from one to the other. Galvanic corrosion is that part of the corrosion that occurs at the anodic member of such a couple. The extent of galvanic corrosion is directly related to the galvanic current by Faraday's law.

Galvanostat: An electronic apparatus that controls the current between a sample under study (the working electrode) and a counter (or auxiliary) electrode.

Genetic probing: A technique using a unique piece of a genetic sequence to identify microorganisms in a specific environment or community.

Genetic recombination: A process by which genetic materials from two separate genomes are brought together in one unit.

Heterotroph: Organism obtaining energy from oxidation of organic compounds.

Homopolymer: A material composed of one type of repeating unit as its monomer.

Humification: A polymerization process occurring under natural conditions to form a natural polymer, humus.

Hydrogen embrittlement: A process resulting in a decrease in toughness or ductility of a metal due to the presence of atomic hydrogen.

Hydrogen-induced cracking (HIC): A type of cracking usually caused by hydrogen atoms that diffuse into steel and recombine at traps (such as elongated inclusions) causing microscopic blisters to form, which subsequently link and propagate; also known as stepwise cracking.

Hydrogenase: An enzyme system catalyzing the reaction from proton to molecular hydrogen.

Hydrophobicity: A condition in which water is repelled.

Impingement erosion (gas bubble): A term sometimes applied to the erosive role (removal of protective film) of impinging gas bubbles in the erosion–corrosion of heat exchanger tube inlets.

Impingement erosion (liquid droplet): The progressive loss of material from a solid surface due to continued exposure to repeated discrete impacts by liquid droplets which generate impulsive and destructive contact pressures on the solid target.

Impingement erosion (solid particle): Erosion by impinging solid particles suspended in a flowing fluid.

Impingement erosion–corrosion: A conjoint action of impingement erosion and corrosion.

Inhibitor: A chemical that reduces the rate of corrosion when added to the environment in small concentrations.

Inhibitor efficiency: The percentage reduction in corrosion rate caused by an inhibitor.

Intercrystalline defects: Includes grain boundaries separating crystals in different crystallographic orientations and triple junctions, the lines where three crystals are joined together.

Intergranular corrosion: Corrosion that takes place along grain boundaries of a metal. Unlike intergranular stress corrosion cracking, intergranular corrosion does not require applied or residual stress.

Lift-up: In fluid flow, movement of a wall streak away from the wall due to a local pressure gradient.

Liquid impingement erosion: See *impingement erosion.*

Logarithmic region: In fluid flow, that portion of the wall region defined as $30 \leq y_+ \leq 100$.

Mesophilic: Microorganisms growing optimally between 25 and 40°C.

Metabolites: Chemicals excreted during the normal life of organisms.

Methanogens: Bacteria capable of producing methane.

Microbial biofilm: A collection of microorganisms living on a solid surface.

Microbially influenced/induced corrosion (MIC): Corrosion caused by microbial activity.

Microflora: A community consisting of different groups of microorganisms in a specific niche.

Mineralization: Decomposition process resulting in the production of simple inorganic molecules (e.g., CO_2 and H_2O).

Morphological: Visually observable structure.

Mutation: An inheritable change in the base sequence of an organism's genome.

Nanocrystalline materials: Metals, alloys, composites, and other materials with grain sizes less than 100 nm.

Net positive suction head (available), $NPSH_A$: The difference between the total pressure (absolute) and vapor pressure at the pump suction, expressed in terms of equivalent height of fluid, or "head" by

$$NPSH_A = (p_0/\rho g) + (u^2/2g) - (p_v/\rho g)$$

where $p_0 =$ static pressure (absolute)

$p_v =$ vapor pressure of the flowing fluid

$\rho =$ density

$g =$ gravitational acceleration

$u =$ flow velocity

Net positive suction head (required), $NPSH_R$: The value of NPSH, given by pump manufacturer, at which the efficiency of the pump drops off. *Note:* Substantial cavitation may be occurring under this condition and a considerable safety margin ($NPSH_A - NPSH_R$) is normally used.

Nitrifying bacteria (nitric acid producing bacteria): Microorganisms converting ammonium to nitrate through a series of reactions.

Oligotrophic: Low nutrient environment.

Overpotential: Difference between the actual potential and the equilibrium, thermodynamic, or reversible potential.

Oxic: Aerated habitat or niche.

Oxidase: Enzyme capable of oxiding an organic compound.

Passivity: Definition 1: A metal active in the Emf Series, or an alloy composed of such metals, is considered passive when its electrochemical behavior approaches that of an appreciably less active or noble metal. Definition 2: A metal or alloy is passive if it substantially resists corrosion in an environment where thermodynamically there is a large free-energy decrease associated with its passage from the metallic state to appropriate corrosion products.

Peroxidase: An enzyme catalyzing a reaction to form H_2O_2.

pH: The negative logarithm (to the base 10) of hydrogen ion activity, or $pH = -\log_{10} a_{H^+}$.

Pitting: Localized corrosion of a metal surface at points or small areas.

Planktonic bacteria: Microbial cells living in a liquid, free from association with a surface.

Polarization: Change of potential caused by current flow.

Polymerase chain reaction (PCR): A technique used to amplify a specific DNA sequence in vitro.

Polymeric materials: Materials that are made up of many monomers condensed together through chemical linkages.

Potentiostat: An electronic device that controls the potential of an electrode (usually called the working electrode) with respect to a reference electrode by automatically applying the required current between the counter (or auxiliary) electrode and the working electrode.

Prokaryote: A cell or organism lacking a nucleus or other membrane-enclosed organelle.

Psychrophile: An organism with optimal growth below 15°C.

Recalcitrant: Difficult to assimilate or break down.

Recruiting: Attracting settlement selectively through chemical communication.

Residual stress: Stress remaining in a structure as a result of deformation during fabrication (e.g., forming and welding).

Respirometric biometry: Measurement of material balance through analysis of the compounds released.

Reynolds number: See *dimensionless groups*.

Ribosomal RNA: A type of RNA found in the ribosome, which may participate in protein synthesis.

Scale: An accumulation of corrosion products on a surface.

Schmidt number: See *dimensionless groups*.

Secondary metabolite: A product excreted by a microorganism near the end or at the stationary phase of its growth.

Sensitization: Precipitation of chromium carbides at grain boundaries in stainless steels, due to heating in certain temperature ranges and resulting in susceptibility to intergranular attack by some environments.

Sessile bacteria: Attached to surfaces.

Sherwood number: See *dimensionless groups*.

Sliding wear: Wear due to relative motion in the tangential plane of contact between two solids.

Slow-strain-rate testing: Tests in which a slow strain rate (typically $\sim 10^{-6}\, s^{-1}$) is applied to a metal while exposed to an environment.

Slurry: A suspension of solid particles in a flowing liquid.

Solid-particle impingement erosion: See *impingement erosion*.

Solid-particle impingement erosion–corrosion: The conjoint action of solid-particle impingement erosion and corrosion.

Stress corrosion cracking: Cracking produced by the combined actions of stress and an environment on a susceptible alloy.

Stress intensity factor: A measure of the intensification of a stress in the region of the tip of an existing crack.

Stress-oriented hydrogen-induced cracking (SOHIC): A form of hydrogen-induced cracking (HIC) in which the arrays of hydrogen blisters are oriented perpendicular to the orientation of the stress.

Sulfate-reducing bacteria: A group of microorganisms capable of reducing $SO_4{}^{2-}$ to H_2S.

Sulfide stress cracking (SSC): A form of hydrogen embrittlement that occurs in high-strength steels and in localized hard zones in weldments of susceptible materials, caused by the combined action of tensile stress and corrosion in the presence of water and hydrogen sulfide.

Sweep: In fluid flow, the movement of fluid toward the wall at an angle to the nominal flow direction.

Syntrophic: A nutritional situation that two or more organisms combine their metabolic capabilities to degrade a substance for carbon and energy sources.

Systematics: The science of classification and relationship between groups of organisms.

Taxonomy: The study of scientific classification and nomenclature.

Thermophile: Organism with an optimum growing temperature between 45 and 80°C.

Thiobacilli: A group of microorganisms capable of oxidizing reduced S and/or sulfur.

Time of wetness (TOW): The period during which a metallic surface is covered by adsorptive or liquid film of electrolyte enabling atmospheric corrosion to proceed.

Turbulent structure: A morphologically invariant, three-dimensional region of turbulent flow possessing an identifiable flow pattern. See also *coherent structure*.

Underdeposit corrosion: Corrosion occurring in a separated water phase beneath deposits of nonmetallic solids on a metal surface resulting from low flow turbulence.

Vibratory cavitation: See *cavitation*.

Viscous region: In fluid flow, that portion of the wall region defined as $0 \le y_+ \le 5$.

Vortex: An area of spiral or circular flow normal to a defined axis.

Wall region: In fluid flow, that portion of the boundary layer comprising the viscous sublayer, the buffer region, and the logarithmic region, defined as $y_+ \le 100$.

Wall streak: In fluid flow, a coherent segment of fluid near the wall moving parallel to the nominal flow direction but at a significantly different velocity.

INDEX

Uhlig's Corrosion Handbook, Third Edition, Edited by R. Winston Revie
Copyright © 2011 John Wiley & Sons, Inc.